AF757177

# A New General Catalogue of the Ants of the World

CENTRAL CONNECTICUT STATE UNIVERSITY
3 2530 05000 7879
WITHDRAWN

BARRY BOLTON

# A New General Catalogue of the Ants of the World

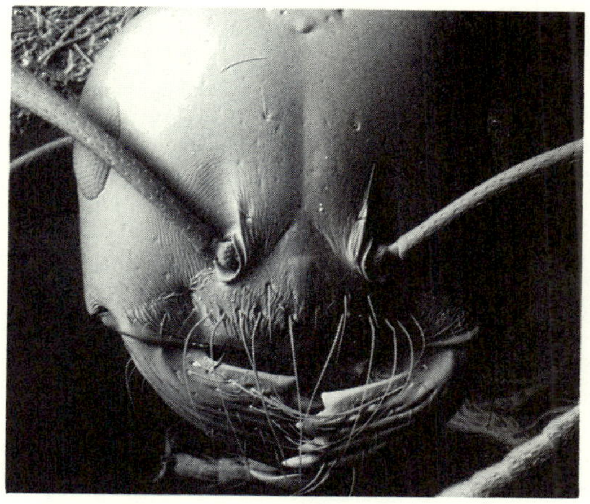

Harvard University Press
Cambridge, Massachusetts
London, England 1995

3/29
3722

QL
568
F7
B585
1995

Copyright © 1995 by the President and Fellows of Harvard College
All rights reserved
Printed in the United States of America

Scanning electron micrographs by Laraine Ficken.

Library of Congress Cataloging-in-Publication Data

Bolton, Barry.
A new general catalogue of the ants of the world / Barry Bolton.
p. cm.
Includes bibliographical references (p. ).
ISBN 0-674-61514-X (alk. paper)
1. Ants—Classification. 2. Ants, Fossil—Classification.
I. Title.
QL568.F7B585 1995
595.79'6'012—dc20
95-12908

# Contents

# A New General Catalogue of the Ants of the World

# Introduction

Taxonomy is a fascinating branch of biology, unique in that it must always be retrospective as well as prospective. This is because a taxonomic decision taken today projects forward in time and colours the decisions and opinions of taxonomists in the future. Similarly, any taxonomic decision taken since the inception of zoological nomenclature, in 1758, has relevance today, and on into the future. And it does not matter whether that decision was right or wrong; whatever its ultimate origin or fate, a taxonomic decision is a persistent fact that must be taken into account. I suppose that a quaint way of expressing this would be: a name is forever, or at least for as long as the science of taxonomy continues.

Thus any taxonomist working today must have available all the relevant decisions taken in the past, so that they can be assessed, so that errors and misinterpretations can be avoided, and so that correct decisions and opinions can be transmitted to the future. A useful way of collating all these past actions is to gather the references to all the taxonomic decisions taken in a given group and put them together in such a way that the relevant data can easily be found and extracted. This is a prime function of a catalogue, to show at a glance the condition and state of any previous taxonomic decision.

## History of ant catalogues

This volume is the first attempt to catalogue all the ants of the world, including the fossil forms, since 1893. Between that date and the present there have appeared a number of regional lists and catalogues dealing with larger or smaller areas, and one enormous work, the *Genera Insectorum,* that catalogued all the world's insect fauna except for the fossil taxa. Volumes in the *Genera Insectorum* series that deal with ants were published between 1910 and 1925. The series presented both a catalogue and a classification and in consequence was extremely valuable. A great deal has happened in myrmecology since 1925, however, and the series is outdated for modern needs. In particular, the classification offered there is now virtually useless, and taxonomic changes and additions have carried on apace since the appearance of the final volume.

The concept of a world catalogue of all names in the family commenced with Roger (1863b) and Mayr (1863), both of whom produced extremely valuable but now mostly forgotten works. Not only did they enumerate the taxa then known, but they also introduced many synonymies which accounted for numerous old and dubious names and dispersed lots of names from the old form-genera into the genus-rank taxa that we still recognise today. They were followed by Dalla Torre (1893), whose treatment of extinct as well as extant taxa added a dimension that unfortunately was not continued. Later world cataloguers tended to stick with either the fossil fauna (Handlirsch, 1906, 1907) or the extant fauna (Emery, 1910b, 1911d, 1912g, 1921f, 1922e, 1922f, 1925b; the *Genera Insectorum* series).

After these dates cataloguers specialised in regional surveys. Some of these dealt with entire zoogeographical regions, for instance Kempf (1972a) and Brando (1991) [Neotropical]; Smith, M.R. (1951a, 1958c) and Smith, D.R. (1979) [Nearctic]; Chapman & Capco (1951) [Oriental and Indo-Australian]; Wheeler, W.M. (1922a) [Afrotropical and Malagasy]; Taylor (1987a, 1987b) [Australasian]; Wheeler, W.M. (1935i) [Oceanian]. Mostly however they dealt only with single countries, for example Bruch (1914, 1915) [Argentina]; Borgmeier (1923, 1927c) [Brazil]; Taylor & Brown, D.R. (1985) [Australia]; Wheeler, W.M. (1930h) and Wu, C.F. (1941) [China]; Baltazar (1966) [Philippines]; Baroni Urbani (1971c) [Italy]; Ceballos (1956) [Spain]; Pisarski (1975) [Poland]; Onoyama (1980) [Japan]. More recently Shattuck (1994) has attempted to break away from this regional technique by producing a catalogue based on subfamily-rank taxa (Aneuretinae and Dolichoderinae) on a worldwide basis.

Among the fossils, lists of forms in amber have attracted some attention, with contributions by Burnham (1979), Keilbach (1982) and Spahr (1987). Fossils in rock have to a large extent been ignored, though Carpenter (1992) provides a review of fossil genera. Whole fossil faunas have of course been treateed in monographs by myrmecologists or palaeoentomologists, for example Emery (1891a), Wheeler, W.M. (1915h), Carpenter (1930), Théobald (1937b), Dlussky (1967b, 1975, 1981b, 1988), but these are mostly descriptive and are not cross-referred to other faunas. Considering that the ants have one of the best fossil records of any insect family, a modern worldwide review of the extinct ant taxa is long overdue.

The drift away from world to more limited regional catalogues was, I suspect, inspired mostly by strictures of time, availability of literature, and availability of financial support, though to a lesser extent a deficiency of imagination may also have played a part. Regional catalogues and lists have their advantages, but they also generate problems. Chief among the drawbacks is the tendency to produce and maintain homonymies and to overlook synonymies among those researchers who consult only the literature concerning the country or region in question and who ignore the extralimital possibilities.

In this new general catalogue I have attempted to examine every publication on ant taxonomy and related fields of study from 1758 to the present. The result is what you see here. This catalogue covers the entire family Formicidae, which currently contains the following numbers of valid extant taxa: 16 subfamilies, 296 genera, and 9,536 species (up to 31 December 1993). Among the extant forms there are also an uncounted number of subspecies, along with a very large number of junior synonyms, junior homonyms, and unavailable names. In addition to the living taxa, the family's impressive fossil record yields a further 408 species. These are distributed through 13 subfamilies, of which 9 are currently extant and 4 are wholly extinct, and 118 genera, of which 73 are totally extinct.

## History of ant classification and phylogeny

The compilation of a catalogue such as this relies heavily upon a reasonable classificatory framework for the group in question. As in many insect groups the classification of the Formicidae is in a state of flux, with the older classification being gradually updated and improved by the application of cladistic theories and methodologies. This means that parts of the classification within the family are more advanced than others, resulting in a variability which cannot be helped but which must be borne in mind by the reader.

The ant classification that we see today is the culmination of over a century of accumulated effort. Early authors such as Nylander (1846a, 1846b, 1849), Smith, F. (1858), Mayr (1861, 1863) and Roger (1863b) were concerned with the accumulation of named taxa at genus and species rank and did not show much interest in arranging these into a formal classification. The second generation of ant taxonomists, besides adding vast numbers of names to the rapidly growing list, began to attempt the arrangement of the names into a standard classification. The first moves toward what eventually developed into today's system were made by Emery (1877a, 1895j, 1896e, 1901a, 1914a, 1915i) and Forel (1878, 1893a). These were followed by the *Genera Insectorum* series of Emery (1910b, 1911d, 1912g, 1921f, 1922e, 1922f, 1925b), whose results were very closely followed by Wheeler, W.M. (1920, 1922a).

By 1925 a rough consensus had been reached, which has since been called the Emery-Wheeler classification (Brown, 1958g). This was extremely authoritative, as both authors also presented identification keys to the world's genera and higher taxa. So great was the influence of the Emery-Wheeler classification that study of the subject fell into abeyance for many years. Authors seemed content to cram any newly discovered taxon into the existing system, even when it was glaringly obvious that the classification had outlived its usefulness.

During the 1950s the flaws in the Emery-Wheeler system reached a critical point; the cracks could not be patched up or painted over any longer, and it began to fall apart. Two idiosyncratic classifications of ant subfamilies, proposed by Clark (1951) and Bernard (1951), were quickly followed by a new look at the topic by Brown (1954e). He proposed a radical change in the approach to ant classification and his paper forms the baseline for modern research into both the classification and the phylogeny of ants. In terms of classification these include the synopses by Brown (1973b), Snelling (1981), Wheeler, G.C. & Wheeler, J. (1985), Dlussky & Fedoseeva (1987), Hölldobler & Wilson (1990), and Bolton (1994).

The study of ant phylogeny, or at least an attempt to account for the origins of the various ant subfamilies, dates back to Wheeler, W.M. (1920), but the field did not advance much until Brown's (1954e) paper. A short history of modern works on ant phylogeny, which reflects the growing influence of cladistic methodology, is given by Bolton (1990c). Critical works that detail the progress of ant phylogenetic studies in recent years include Taylor (1978a), Dlussky & Fedoseeva (1987), Baroni Urbani (1989), Bolton (1990a, 1990b, 1990c), Ward (1990), Baroni Urbani, Bolton & Ward (1992), Shattuck (1992b), and Lattke (1994).

## General references and faunal surveys

The eusocial organisation of ants and their enormously diverse biologies have, in recent years, inspired a large number of books offering a synthesis of ant biology and behavior. Recommended books for anyone interested in ants in general, or in more specialised aspects of their biology, include the following and their included references: Beattie (1985), Blum (1984), Bolton (1994), Dlussky (1981a), Dumpert (1978), Gösswald (1985), Hermann (1979–1982, 1984), Hölldobler & Wilson (1990), Jolivet (1986, 1992), Passera (1984), Schneirla (1971), Sudd (1967), Sudd & Franks (1987), Wilson (1971).

Along with these general volumes, the period 1950–1994 has seen the appearance of a number of surveys of local ant faunas that include keys for the identification of species within the country or territory in question. Faunal surveys that antedate 1950 are not recommended, as their keys are clogged with unavailable names and infraspecific taxa and are of very limited use by modern standards. The following list indicates territories for which faunal surveys have been produced in the period 1950–1994, but references to earlier surveys and their contents, and to works dealing with individual taxa rather than geographical areas, are of course included within the corpus of the catalogue.

ARMENIA: Arakelian (1994).
BALKAN STATES: Agosti & Collingwood (1987a, 1987b).
BELGIUM: Boven (1977).
BULGARIA: Atanassov & Dlussky (1992).
CANADA: Francoeur (1977b, 1979) [Quebec].
CHILE: Snelling & Hunt (1976).
CORSICA: Casevitz-Weulersse (1990a, 1990b).
CUBA: Alayo (1974).
FENNOSCANDIA AND DENMARK: Collingwood (1979).
FRANCE: Bernard (1967).
GERMANY: Gösswald (1985).
GREAT BRITAIN: Bolton & Collingwood (1975).
IBERIAN PENINSULA: Collingwood (1978).
ITALY: Baroni Urbani (1964b, 1964c, 1969a, 1971c).
JAPAN: Morisita *et al.* (1989, 1991, 1992); Onoyama (1980); Onoyama & Terayama (1994) [list of references].

KIRGIZIA: Tarbinsky (1976).
MALTA: Baroni Urbani (1968b); Schembri & Collingwood (1981).
MELANESIA: Wilson (1958a, 1958d, 1959a, 1959b).
NEW ZEALAND: Brown (1958h).
NORTH AMERICA: Creighton (1950a); Gregg (1963) [Colorado]; Wheeler, G.C. & Wheeler, J. (1963) [North Dakota]; Wheeler, G.C. & Wheeler, J. (1986g) [Nevada].
POLYNESIA: Wilson & Taylor (1967).
PORTUGAL: see Iberian Peninsula.
SAUDI ARABIA: Collingwood (1985).
SPAIN: see Iberian Peninsula.
SWEDEN: Nilsson & Douwes (1987).
SWITZERLAND: Kutter (1977c).
TURKMENISTAN: Dlussky, Soyunov & Zabelin (1990).
USSR (former): Arnol'di & Dlussky (1978) [former European USSR]; Kupyanskaya (1990) [far eastern Russia]; Radchenko (1994b) [South Siberia].
WEST PALAEARCTIC: Bernard (1967).
WORLD GENERA: Hölldobler & Wilson (1990), Bolton (1994).

## Use of the catalogue

The catalogue is divided into three sections. The first deals with family-group (suprageneric) taxa: names applied to taxa with the rank of family, subfamily, tribe, or subtribe. The second deals with taxa in the genus-group: names applied to taxa with the rank of genus or subgenus. The third, and by far the largest, section deals with species-group taxa: names with the rank of species or subspecies.

Each section has its own introductory paragraphs, explaining layout, typography, special considerations, and so on. Within each section names are listed in alphabetical order, regardless of their current status, and are cross-referred to other names where necessary. Thus every subfamily within the catalogue of family-group names has a list of component tribes appended; all names of tribes that are currently valid include a list of component genera; all species with infraspecific taxa contain lists of their subspecies. Cross-referencing also works in the reverse direction, as the entry for any subspecies will provide an indication of the species to which it is appended, which in turn is included under a genus heading, which itself is referable back to the catalogue of genus-group names and to the catalogue of family-group names. Similarly, the provenance of junior and senior synonyms and homonyms can be traced to and fro within the catalogue.

Every entry strives to provide a taxonomic history of the name in question, and to provide additional references which supplement our understanding of the taxon. All changes of combination and status are recorded for each name, as are indications of synonymy and homonymy.

Two terms commonly encountered in the catalogue need, I think, a few words of explanation, as they have been widely ignored or misunderstood in the taxonomic literature dealing with ants.

1. *Combination.* The association of a genus-rank name and a species-rank name to form the name of a species is a *combination.* Similarly, the association of genus-rank name plus species-rank name plus subspecies-rank name forms a combination. A *new combination* occurs only when a previously published name in the species-group is first appended to a generic name other than the one it accompanied in its original description. A *revived combination* describes the state in which a species-group name is transferred back to a generic name from which, at some previous time, it had been excluded.

2. *Status.* A change of rank within a trinomen or a quadrinomen, (such as the elevation of a subgenus to genus rank, or of a subspecies to species rank; or *vice versa,* the demotion of a genus to subgenus rank or species to subspecies rank) is indicated by *new status.* A reversion in rank to one previously held, from one later designated, is a *revived status.*

The concepts and usages of combination and status are quite different and distinct. They should not be confused, although sadly they frequently are.

Three other concepts fundamental to an understanding of nomenclature, namely synonymy, homonymy and availability of names, have accrued a finely tuned series of regulations governing their use. The concepts and their applications are fully explained in the *International Code of Zoological Nomenclature* (Third Edition, 1985) and are applied throughout this catalogue. Their correct usage should be understood and adhered to by all taxonomists.

One common misuse of the term *synonymy* needs to be explained here. A secondary reference to a taxon, that directs material purportedly of that taxon to a different and senior taxon, is a misidentification, not a synonymy. For example, an entry in an imaginary taxonomic synopsis such as

*X-us y-us* Black, 1928: 4.
*X-us z-us* White; Green, 1930: 15, new synonymy.

is an incorrect and ignorant usage of the concept of synonymy. This entry should properly read:

*X-us y-us* Black, 1928: 4.
*X-us z-us* White, *sensu* (or "in the sense of") Green, 1930: 15, misidentification.

Synonymy in this instance applies only if the type-material of two (or more) taxa in the species-group has been compared and judged to be conspecific. Thus, in our imaginary example, if type-specimens of *y-us* and *z-us* had been compared and judged to be conspecific, the entry in the taxonomic synopsis would then indicate the material and nominate synonymy.

*X-us y-us* Black, 1928: 4. Holotype worker [examined].
*X-us z-us* White, 1930: 1. Holotype worker [examined], new synonymy.

# Checklist of Fossil Taxa

Throughout the catalogue the names of all fossil taxa are prefixed by the symbol *. Because the number of names applied to fossil forms is relatively small, their entries remain inconspicuous within the corpus of the catalogue. The following checklist has therefore been abstracted to allow those workers primarily interested in the fossil forms to gain rapid access to the taxa in question. The list here provides only the names of the taxa. Their individual entries and data relating to them are given in full in the catalogue proper.

---

## Suprageneric taxa known only from fossils

For further data see under the catalogue of family-group (suprageneric) taxa.

*ARCHIPONERINI (junior synonym of Ponerini).
***ARMANIINAE** (subfamily of Formicidae).
***FORMICIINAE** (subfamily of Formicidae).
*HYPOPOMYRMICITI (junior synonym of Pheidologetonini).
*MIOMYRMICINI (junior synonym of Dolichoderinae).
***PALEOSMINTHURINAE** (subfamily of Formicidae).
*PITYOMYRMECINI (junior synonym of Dolichoderinae).
***PRIONOMYRMECINI** (tribe of Myrmeciinae).
***SICILOMYRMECINI** (tribe of Formicinae).
***SPHECOMYRMINAE** (subfamily of Formicidae).
*ZHERICHINIINI (junior synonym of Dolichoderinae).

## Genus-rank and species-rank taxa known only from fossils

Names of species are noted under their current generic combinations only. For original combinations of species-rank taxa and for further data see under the catalogues of species-group names and genus-group names.

***ACANTHOGNATHUS*** (Myrmicinae: Dacetini). Included fossil species: **poinari.*

**ACROSTIGMA* (junior synonym of *Podomyrma*).
****AGROECOMYRMEX*** (Myrmicinae: Agroecomyrmecini). Included species: **duisburgi.*
****AMEGHINOIA*** (Myrmeciinae). Included species: **piatnitzkyi.*
****ANEURETELLUS*** (Aneuretinae). Included species: **deformis.*
***ANOCHETUS*** (Ponerinae: Ponerini). Included fossil species: **ambiguus, *brevidentatus, *conisquamis, *corayi, *dubius, *exstinctus, *intermedius, *lucidus.*
***ANONYCHOMYRMA*** (Dolichoderinae). Included fossil species: **constricta, *samlandica.*
***APHAENOGASTER*** (Myrmicinae: Pheidolini). Included fossil species: **avita, *donisthorpei, *livida, *longaeva, *maculata, *maculipes, *mayri, *mersa, *oligocenica, *pannonica, *sommerfeldti.*
****ARCHAEOPONE*** (*Armaniinae). Included species: **kzylzharica, *taylori.*
****ARCHIMYRMEX*** (Myrmicinae). Included species: **rostratus.*
****ARCHIPONERA*** (Ponerinae: Ponerini). Included species: **wheeleri.*
****ARMANIA*** (*Armaniinae). Included species: **robusta.*
****ARMANIELLA*** (*Armaniinae). Included species: **curiosa.*
****ASYMPHYLOMYRMEX*** (Dolichoderinae). Included species: **balticus.*
**ATTAICHNUS* (Myrmicinae: Attini (ichnotaxon)). Included ichnospecies: **kuenzelii.*
****ATTOPSIS*** (Myrmicinae). Included species: **longipennis, *longipes.*

***AZTECA*** (Dolichoderinae). Included fossil species: **alpha, *eumeces.*
****BAIKURIS*** (*Sphecomyrminae). Included species: **mandibularis, *mirabilis.*
****BRADOPONERA*** (Ponerinae: Proceratiini). Included species: **meieri.*
****CALYPTITES*** (dubiously Formicidae). Included species: **antediluvianum.*
****CAMPONOTITES*** (Formicinae: Camponotini). Included species: **macropterus.*
***CAMPONOTUS*** (Formicinae: Camponotini). Included fossil species: **brodiei, *compactus, *fuscipennis, *heracleus, *induratus, *lignitus, *longiventris, *mengei* (= **igneus*), **microcephalus, *novotnyi, *obesus, *oeningensis, *penninervis, *petrifactus, *pinguiculus, *pinguis, *saussurei* (unresolved homonym), **shanwangensis, *tokunagai, *ullrichi, *vehemens* (= **miserabilis*), **vetus.*
****CARIRIDRIS*** (Myrmeciinae). Included species: **bipetiolata.*
****CASALEIA*** (Ponerinae: Amblyoponini). Included species: **inversa.*
***CATAGLYPHIS*** (Formicinae: Formicini). Included fossil species: **constrictus.*
***CATAULACUS*** (Myrmicinae: Cataulacini). Included fossil species: **anthracinus* (= **nigra*), **planiceps, *silvestrii.*
****CEPHALOMYRMEX*** (Myrmicinae). Included species: **rotundatus.*
***CERAPACHYS*** (Cerapachyinae: Cerapachyini). Included fossil species: **annosus, *favosus.*

*CHIMAEROMYRMA* (Formicinae: Camponotini). Included species: **brachycephala*.

*CREMATOGASTER* (Myrmicinae: Crematogastrini). Included fossil species: **praecursor*, **pusilla*.

**CRETACOFORMICA* (dubiously Formicidae). Included species: **explicata*.

**CRETOMYRMA* (*Sphecomyrminae). Included species: **arnoldii*, **unicornis*.

**CRETOPONE* (*Armaniinae). Included species: **magna*.

**CTENOBETHYLUS* (junior synonym of *Liometopum*).

**DLUSSKYIDRIS* (*Sphecomyrminae). Included species: **zherichini*.

*DOLICHODERUS* (Dolichoderinae). Included fossil species: **affectus*, **anglicus*, **anthracinus*, **antiquus*, **balticus*, **britannicus*, **bruneti*, **caribbaeus*, **coquandi*, **cornutus*, **dibolius*, **elegans*, **explicans*, **gurnetensis*, **kutschlinicus*, **longipennis*, **mesosternalis*, **nitidus*, **obliteratus*, **oviformis*, **ovigerus*, **passalomma*, **primitivus*, **prolaminatus*, **rohweri*, **sculpturatus*, **tauricus*, **tertiarius*, **vectensis*, **vexillarius*.

**DOLICHOMYRMA* (*Armaniinae). Included species: **laticeps*, **longiceps*.

**DRYMOMYRMEX* (Formicinae: Camponotini). Included species: **claripennis*, **fuscipennis*.

*ECTATOMMA* (Ponerinae: Ectatommini). Included fossil species: **gracile*.

**ELAEOMYRMEX* (Dolichoderinae). Included species: **coloradensis*, **gracilis*.

**ELECTROMYRMEX* (Myrmicinae). Included species: **klebsi*.

**ELECTROPONERA* (Ponerinae: Ectatommini). Included species: **dubia*.

**EMPLASTUS* (Ponerinae: Ponerini). Included species: **emeryi*.

**ENNEAMERUS* (Myrmicinae: Myrmecinini (?)). Included species: **reticulatus*.

**EOCENIDRIS* (Myrmicinae).Included species: **crassa*.

**EOFORMICA* (Myrmicinae). Included species: **pinguis*.

**EOMYRMEX* (Myrmicinae). Included species: **guchengziensis*.

**EOPONERA* (junior synonym of **Formicium*).

**EOTAPINOMA* (Dolichoderinae). Included species: **compacta*, **gracilis*.

**EULITHOMYRMEX* (Myrmicinae: Agroecomyrmecini). Included species: **rugosus*, **striatus*.

*FORMICA* (Formicinae: Formicini). Included fossil species: **acuminata*, **aemula*, **alsatica*, **antiqua*, **arcana*, **atavina*, **auxillacensis*, **baltica*, **bauckhorni*, **buphthalma*, **cantalica*, **capito*, **cephalica*, **clymene*, **cockerelli*, **demersa*, **eoptera*, **flori*, **fragilis* (unresolved homonym), **gibbosa* (unresolved homonym), **globiventris*, **gracilis* (unresolved homonym), **grandis*, **gravida*, **heteroptera*, **horrida*, **immersa*, **kollari*, **latinodosa*, **lavateri*, **longicollis*, **longiventris*, **lucida*, **luteola*, **macrocephala* (unresolved homonym), **macrognatha*, **macrophthalma*, **maculipennis*, **martynovi*, **nigra* (unresolved homonym), **oblita*, **obtecta*, **obvoluta*, **ocella*, **oculata*, **ophthalmica*, **orbata*, **parvula*, **parvula* (unresolved homonym), **phaethusa*, **pitoni*, **primitiva*, **primordialis*, **procera*, **pulchella*, **pumila*, **quadrata*, **robusta*, **sepulta*, **serresi*, **seuberti*, **strangulata*, **surinamensis*, **trigona*, **tripartita*, **ungeri*.

**FORMICITES* (collective group name, without included named taxa).

**FORMICIUM* (*Formiciinae). Included species: **berryi*, **brodiei*, **giganteum*, **mirabile*, **simillimum*.

*GESOMYRMEX* (Formicinae: Gesomyrmecini). Included fossil species: **expectans*, **hoernesi* (= **annectens*, = **mayri*, = **theryi*), **miegi*.

**GLAPHYROMYRMEX* (Formicinae: Formicini). Included species: **oligocenicus*.

*GNAMPTOGENYS* (Ponerinae: Ectatommini). Included fossil species: **europaea*, **levinates*, **pristina*.

**HYPOPOMYRMEX* (Myrmicinae: Pheidologetonini). Included species: **bombiccii*.

**ILEMOMYRMEX* (Myrmicinae: Stenammini?). Included species: **caecus*.

**IMHOFFIA* (Formicinae). Included species: **nigra*, **pallida*.

*IRIDOMYRMEX* (Dolichoderinae). Included fossil species: **breviantennis*, **florissantius*, **geinitzi*, **haueri*, **mapesi*, **oblongiceps*, **obscurans*.

**KOTSHKORKIA* (Dolichoderinae). Included species: **laticeps*.

**LAMPROMYRMEX* (junior synonym of *Monomorium*).

*LASIUS* (Formicinae: Lasiini). Included fossil species: **chambonensis*, **crispus*, **epicentrus*, **globularis*, **longaevus*, **longipennis*, **minutulus*, **nemorivagus*, **obliteratus*, **oblongus*, **obscurus*, **occultatus*, **peritulus*, **pumilus* (= **punctulatus*), **redtenbacheri*, **schiefferdeckeri* (= **edentatus*), **terreus*, **tertiarius*, **vetulus*.

*LEPTOMYRMEX* (Dolichoderinae). Included fossil species: **neotropicus*.

**LEPTOMYRMULA* (Dolichoderinae). Included species: **maravignae*.

*LEPTOTHORAX* (Myrmicinae: Formicoxenini). Included fossil species: **glaesarius*, **gracilis*, **hystriculus*, **longaevus*, **placivus*, **praecreolus*.

**LEUCOTAPHUS* (Formicinae: Formicini). Included species: **cockerelli*, **gurnetensis*, **permancus*.

*LINEPITHEMA* (Dolichoderinae). Included fossil species: **hispaniolae*, **humiloides*.

*LIOMETOPUM* (Dolichoderinae). Included fossil species: **antiquum*, **goepperti* (= **succinalis*), **imhoffii* (= **schmidtii*), **miocenicum*, **oligocenicum*, **scudderi*.

**LITHOMYRMEX* (junior homonym, see **Eulithomyrmex*).

**LONCHOMYRMEX* (Myrmicinae: Pheidolini?). Included species: **freyeri*, **nigritus*.

**MEGAPTERITES* (junior synonym of **Formicium*).

*MESSOR* (Myrmicinae: Pheidolini). Included fossil species: **sculpturatus* (unresolved homonym).

**MIANEURETUS* (Aneuretinae). Included species: **mirabilis*.

**MIOMYRMEX* (Dolichoderinae). Included species: **impactus*, **striatus*.

*MONOMORIUM* (Myrmicinae: Solenopsidini). Included fossil species: **mayrianum* (= **gracillimus*), **pilipes*.

*MYRMICA* (Myrmicinae: Myrmicini). Included fossil species: **aemula*, **angusticollis*, **archaica*, **bremii*, **concinna*, **jurinei*, **longispinosa*, **macrocephala*, **molassica*, **nebulosa*, **obsoleta*, **tertiaria* (= **bicolor*), **venusta* (unresolved homonym).

**MYRMICITES* (collective group name, without included named taxa).

**MYRMICIUM* (Myrmicinae). Included species: **heerii*.

*NEIVAMYRMEX* (Ecitoninae: Ecitonini). Included fossil species: **ectopus*.

**NOTHOMYRMICA* (Myrmicinae: Myrmicini?). Included species: **intermedia*, **petiolata*, **rudis*, **rugosostriata*.

*ODONTOMACHUS* (Ponerinae: Ponerini). Included fossil species: **pseudobauri*, **spinifer*.

*OECOPHYLLA* (Formicinae: Oecophyllini). Included fossil species: **atavina*, **bartoniana*, **brischkei*, **crassinoda*, **leakeyi*, **megarche*, **obesa*, **perdita*, **praeclara*, **sicula*, **superba*.

*OLIGOMYRMEX* (Myrmicinae: Pheidologetonini). Included fossil species: **antiquus*, **bohemicus*, **sophiae*, **thorali*.

**OXYIDRIS* (Myrmicinae: Pheidologetonini). Included species: **antillana*.

*PACHYCONDYLA* (Ponerinae: Ponerini). Included fossil species: **calcarea*, **crawleyi*, **dubia*, **globiventris*, **gracilicornis*, **succinea*.

**PALAEOMYRMEX* (junior homonym, see **Dlusskyidris*).

**PALEOSMINTHURUS* (*Paleosminthurinae). Included species: **juliae*.

**PARAMERANOPLUS* (Myrmicinae: Meranoplini). Included species: **primaevus*.

**PARANEURETUS* (Aneuretinae). Included species: **longicornis*, **tornquisti*.

**PARAPHAENOGASTER* (Myrmicinae: Pheidolini). Included species: **microphthalma*, **shanwangensis*.

*PARAPONERA* (Ponerinae: Paraponerini). Included fossil species: **dieteri*.

**PETRAEOMYRMEX* (Dolichoderinae). Included species: **minimus*.

**PETROPONE* (*Armaniinae). Included species: **petiolata*.

*PHEIDOLE* (Myrmicinae: Pheidolini). Included fossil species: **cordata*, **tertiaria*, **tethepa*.

*PHEIDOLOGETON* (Myrmicinae: Pheidologetonini). Included fossil species: **rugiceps*, **schossnicensis*.

**PITYOMYRMEX* (Dolichoderinae). Included species: **tornquisti*.

*PLAGIOLEPIS* (Formicinae: Plagiolepidini). Included fossil species: **klinsmanni*, **kuenowi*, **labilis*, **singularis*, **solitaria*, **squamifera*, **succini*.

*PLATYTHYEA* (Ponerinae: Platythyreini). Included fossil species: **primaeva*.

*PODOMYRMA* (Myrmicinae: Formicoxenini). Included fossil species: **mayri*.

*POGONOMYRMEX* (Myrmicinae: Myrmicini). Included fossil species: **fossilis*.

*PONERA* (Ponerinae: Ponerini). Included fossil species: **atavia*, **elegantissima*, **leptocephala*, **minuta*.

**PONERITES* (Ponerinae: Ponerini). Included species: **karaganensis*, **stauropolitanus*, **umbra*.

**PONEROPSIS* (Ponerinae: Ponerini). Included species: **affinis*, **brunascens*, **crassinervis*, **croatica*, **elongata*, **elongatula*, **escheri*, **fuliginosa*, **globosa*, **hypolitha*, **longaeva*, **lugubris*, **morio*, **pallida*, **rhenana*, **stygia*, **tenuis*, **veneraria*, **ventrosa*.

**PONEROPTERUS* (*Armaniinae). Included species: **sphecoides*.

*PRENOLEPIS* (Formicinae: Lasiini). Included fossil species: **henschei*, **pygmaea*.

**PRIONOMYRMEX* (Myrmeciinae: Prionomyrmecini). Included species: **longiceps*.

*PROCERAPACHYS* (junior synonym of *Cerapachys*).

***PROCERATIUM*** (Ponerinae: Proceratiini). Included fossil species: **denticulatum.*

****PRODIMORPHOMYRMEX*** (Formicinae: Gesomyrmecini). Included species: **primigenius.*

****PROMYRMICIUM*** (Myrmicinae). Included species: **boreale.*

**PROPODOMYRMA* (junior synonym of *Vollenhovia*).

**PROTAMBLYOPONE* (junior homonym, see **Casaleia*).

****PROTANEURETUS*** (Aneuretinae). Included species: **succineus.*

****PROTAZTECA*** (Dolichoderinae). Included species: **capitata, *elongata, *hendersoni, *quadrata.*

****PROTOFORMICA*** (Formicinae: Formicini). Included species: **proformicoides.*

****PROTOPONE*** (Ponerinae: Ponerini). Included species: **primigena.*

****PROTRECHINA*** (Formicinae: Lasiini). Included species: **carpenteri.*

****PSEUDARMANIA*** (*Armaniinae). Included species: **aberrans, *rasnitsyni.*

****PSEUDOCAMPONOTUS*** (Formicinae: Camponotini). Included species: **elkoanus.*

***PSEUDOLASIUS*** (Formicinae: Lasiini). Included fossil species: **boreus.*

***PSEUDOMYRMEX*** (Pseudomyrmecinae). Included fossil species: **antiquus, *avitus, *baros, *coruscus, *extinctus, *macrops, *nexilis, *oryctus, *prioris, *succinus, *thecolor, *vicinus.*

****RHOPALOMYRMEX*** (Formicinae: Plagiolepidini). Included species: **pygmaeus.*

***RHYTIDOPONERA*** (Ponerinae: Ectatommini). Included fossil species: **kirghizorum.*

****SICILOMYRMEX*** (Formicinae: *Sicilomyrmecini). Included species: **corniger.*

***SMITHISTRUMA*** (Myrmicinae: Dacetini). Included fossil species: **electrina, *schleeorum.*

***SOLENOPSIS*** (Myrmicinae: Solenopsidini). Included fossil species: **blanda* (= **acuta*), **foersteri, *major* (unresolved homonym), **maxima, *moesta, *privata, *superba* (= **extensa*), **valida.*

****SPHECOMYRMA*** (*Sphecomyrminae). Included species: **canadensis, *freyi.*

***STENAMMA*** (Myrmicinae: Stenammini). Included fossil species: **berendti.*

****STIGMOMYRMEX*** (Myrmicinae: Formicoxenini?). Included species: **venustus.*

****STIPHROMYRMEX*** (Myrmicinae: Myrmecinini?). Included species: **robustus.*

****SYNTAPHUS*** (Ponerinae: Ectatommini). Included species: **wheeleri.*

***TAPINOMA*** (Dolichoderinae). Included fossil species: **minutissimum, *troche.*

***TECHNOMYRMEX*** (Dolichoderinae). Included fossil species: **deletus.*

***TETRAPONERA*** (Pseudomyrmecinae). Included fossil species: **angustata, *klebsi, *lacrimarum, *ocellata, *oligocenica, *simplex.*

***TRACHYMYRMEX*** (Myrmicinae: Attini). Included fossil species: **primaevus.*

***VOLLENHOVIA*** (Myrmicinae: Metaponini). Included fossil species: **beyrichi* (= **samlandica*), **prisca.*

****ZHERICHINIUS*** (Dolichoderinae). Included species: **horribilis, *rapax.*

# Catalogue of Family-Group (Suprageneric) Taxa

## Layout of the catalogue

In the alphabetically arranged list of family-group taxa below, each name is immediately followed by parentheses giving information on the current status of the name. Subsequent entries list references for the following:

i. The original description, designation, or nomination of the taxon, and its type-genus.
ii. In date order, any changes of spelling, rank-denoting suffix, stem, or assignment; as well as changes modifying the original content of the taxon.
iii. Revisions or reviews of the taxon.
iv. Junior or senior synonyms of the taxon.
v. List of valid lower-ranking taxa currently included within the higher-ranking taxon.
vi. Notes.

All family-group taxa are listed in alphabetical order regardless of their current status. Statements of synonymy are repeated under both junior and senior names, but junior synonyms appear in the form given at their inception. Taxonomically unavailable names are stated as such and are followed by a note explaining their provenance and dispersal. Names currently considered valid are in **BOLD CAPITALS**, all other names are in CAPITALS.

The symbol * preceding a name at any rank indicates that the taxon is known only from the fossil record.

No attempt has been made to record every occurrence of a name throughout the entire ant literature. Only references conveying new, modified, or synoptic information are given.

The appearance of simple misspellings caused by obvious typographical errors in the names has been ignored in the catalogue, except when it affects the nomenclature. For example, in Dlussky & Fedoseeva (1987: 78), Apomyrmini appears as Aromyrmini. As the only included genus is stated correctly as *Apomyrma*, such a minor error is unlikely to cause confusion. However, when a misspelling is so gross that it may tend to confuse, it is included in the catalogue; for example Myrmorhynchini for Myrmecorhynchini. The object of this survey is to present a concise and useful catalogue, not a long list of all the typographical errors ever perpetrated in the Formicidae.

## Application of the Code to family-group taxa

The provisions of the *International Code of Zoological Nomenclature* (Third Edition, 1985) have in the past been much neglected as regards their application to names in the family-group; they are implemented in this catalogue. It is not practical to repeat all the rules and provisions here, but some basic principles, particularly those which affect entries in this catalogue, should be clarified.

1. Names proposed in the family-group (suprageneric taxa) must be based upon the name-stem of an available genus rank taxon; otherwise they are taxonomically unavailable. For example, Formicidae, formed from the genus rank name *Formica;* Cerapachyinae, formed from the genus-rank name *Cerapachys;* and Brachymyrmecini, formed from the genus rank name *Brachymyrmex*, are all available. On the other hand, names such as Euformicinae, Rhagiomyrmicinae, and Taraxoponerinae are unavailable, as they are not based on a genus-rank name.

2. Standard suffixes indicating the rank of a suprageneric taxon are as follows:
*-idae* for taxa of family rank,
*-inae* for taxa of subfamily rank,
*-ini* for taxa of tribe rank.

A formal suffix to indicate subtribal rank has never been standardised, though *-ina* is much used elsewhere in the Hymenoptera. Brown (1957i: 193) has suggested *-iti*, but this does not appear to have gained wide recognition or support. As a personal, subjective, opinion I can see no obvious advantage to the use of subtribes. They merely interpose another formal taxonomic rank of very marginal value; their place should perhaps be taken by informal groups of genera which, however closely they are defined, do not add an additional stratum to the formal nomenclature.

Note that before about 1895 there was no general agreement on suffixes determining a particular rank. Thus, before that date *-idae*, for example, may be applied to suprageneric taxa below the rank of family.

3. A change of suffix in an available family-group name, indicating a change of rank, does not initiate any alteration of the name of the author, nor does it require a change of date, nor is the author's name placed in parentheses or in any other way modified. Irrespective of

the suffix, the name is always attributed to the author who first used it as an available name in the family-group, and with that date of publication.

4. The law of priority is applicable to suprageneric taxa. The valid name of a taxon is the oldest available name applied to it. For example, the name Leptothoracini Emery (1914a: 38) is antedated by Formicoxenini Forel (1893a: 164). Emery (1914a: 42) himself included the type-genus of Formicoxenini (*Formicoxenus*) in his tribe Leptothoracini. No subsequent author has questioned the fact that *Formicoxenus* and *Leptothorax* (the type-genus of Leptothoracini) belong in the same tribe; nor is it questioned here. Hence the better-known name Leptothoracini falls as a junior synonym of Formicoxenini.

5. The principle of homonymy is applicable to suprageneric taxa. For example Monomoriini Wheeler, W.M. 1915h: 45 is a junior homonym (and also a junior synonym) of Monomoriini Emery, 1914a: 41.

## History of synoptic classifications

The purpose of a synoptic classification is to present the reader with a systematic overview which shows at a glance the author's opinion of the worldwide classification of a particular group. Such synopses have a long history in the Formicidae and are divisible into two main categories: first, those with reasons given for the classification, either stated or implied by reference to earlier studies; and second, those taking the form of a simple list, without justification or argument for the system outlined.

In the main, later synopses were built upon, and relied heavily on, earlier studies. The first authors to come to grips with an overall classification of ants, in a recognisably modern system, were Forel (1893a) and Emery (1895j, 1896e), although these dated back to preliminary surveys by Emery (1877a) and Forel (1878). Further additions to the fledgling system thus produced were made by Emery (1901a, 1914a, 1915i). These culminated in his contributions to the remarkable *Genera Insectorum* volumes (Emery, 1910b, 1911d, 1912g, 1921f, 1922e, 1922f, 1925b), which offered both classification and catalogue. Other notable synopses made in the first quarter of the twentieth century included those of Forel (1917), Wheeler, W.M. (1922a), and the rather idiosyncratic classification of Ashmead (1905b). These publications account for the majority of subfamily and tribe names still in use in the Formicidae. Most of these contributions fall into the first category outlined above, but the works of Ashmead (1905b) and Forel (1917) fall into the second.

After 1925 the appeal of synoptic classifications appears to have waned, although a few regional synopses have been produced, for example by Kempf (1972a) at the end of his Neotropical catalogue. Fresh interest was kindled by Brown's (1973b) synoptic list of subfamilies and genera, and in recent years there has been a small deluge of synopses, all of which fall into the second category mentioned above. These include the works of Snelling (1981), Wheeler, G.C. & Wheeler, J. (1985), Dlussky & Fedoseeva (1987), Hölldobler & Wilson (1990), and Bolton (1994). For the most part these give a useful overview, but unfortunately the "simplified conspectus of Formicidae" produced by Wheeler, G.C. & Wheeler, J. (1985) is retrograde. Errors of omission and commision abound and the list is untrustworthy and not recommended.

## Note on tribe-rank classification

In subfamilies of Formicidae such as Cerapachyinae and Ecitoninae, and to a large extent Ponerinae, the tribe-rank classification is more or less accurate and reflects, to a reasonable extent, an acceptable classification at our current state of knowledge. The same could not be said of the subfamilies Formicinae and Myrmicinae. In these the tribe-rank classification is ancient, decrepit, largely speculative, and in dire need of serious modern analysis. The tribes of Formicinae listed here largely follow, with minor modifications, the classical systems of Wheeler, W.M. (1922a) and Emery (1925b), which, despite all their shortcomings, we must still use. The tribes of Myrmicinae, as presented here, reflect my current opinions, for what they are worth, but these should not be considered definite or final by any means.

---

## The catalogue of family-group (suprageneric) taxa

ACANTHOMYOPSINI (junior synonym of Lasiini)
Acanthomyopsini Donisthorpe, 1943f: 618 [as tribe of Formicinae]. Type-genus: *Acanthomyops*.
Junior synonym of Lasiini: Bolton, 1994: 50.

**ACANTHOSTICHINI** (tribe of Cerapachyinae)
Acanthostichii Emery, 1901a: 34 [as tribe of Dorylinae]. Type-genus: *Acanthostichus*.
Acanthostichinae: Ashmead, 1905b: 382 [as subfamily of Dorylidae].
Acanthostichini: Emery, 1911d: 12 [as tribe of Ponerinae; emended spelling of suffix].
Acanthostichini: Wheeler, W.M. 1922a: 640 [as tribe of Cerapachyinae].
Acanthostichini: Brown, 1975: 39 [as tribe of Ponerinae].
Acanthostichini: Bolton, 1990a: 67 [as tribe of Cerapachyinae].
Revision of tribe: Brown, 1975: 39.
Included genus: *Acanthostichus*.

**AENICTINAE** (subfamily of Formicidae)
Aenictii Emery, 1901a: 36 [as tribe of Dorylinae]. Type-genus: *Aenictus*.
Aenictini: Ashmead, 1905b: 381 [emended spelling of suffix].
Aenictii: Emery, 1910b: 28 [as junior synonym of Ecitoni]. (Wrongly retained as such by Wheeler, G.C. & Wheeler, J. 1985: 256).
Aenictini: Borgmeier, 1954c: 212 [revived from synonymy as tribe of Dorylinae].
Aenictinae: Bolton, 1990c: 1358 [as subfamily of Formicidae]; Baroni Urbani, Bolton & Ward, 1992: 315.
Included genus: *Aenictus*.

**AENICTOGITONINAE** (subfamily of Formicidae)
Aenictogitonini Ashmead, 1905b: 381 [as tribe of Dorylinae]. Type-genus: *Aenictogiton*.
Aenictogitini: Brown, 1975: 43 [as tribe of Ponerinae; unjustified emendation of spelling].
Aenictogitoninae: Baroni Urbani, Bolton & Ward, 1992: 315 [as subfamily of Formicidae].
Revision (as tribe of Ponerinae): Brown, 1975: 43.
Included genus: *Aenictogiton*.

**AGROECOMYRMECINI** (tribe of Myrmicinae)
Agroecomyrmicini Carpenter, 1930: 34 [as tribe of Myrmicinae]. Type-genus: **Agroecomyrmex*.
Agroecomyrmecini: Brown & Kempf, 1968: 184 [emended spelling].
Revision of tribe: Brown & Kempf, 1968: 184.
Included genera: **Agroecomyrmex*, **Eulithomyrmex*, *Tatuidris*.

ALLOFORMICINAE (unavailable name)
Alloformicinae Emery, 1925b: 9 [as "section" of Formicinae]. Section designated to include tribes Melophorini, Myrmelachistini and Plagiolepidini. Unavailable name; not based on genus rank taxon. Contained material referable to Formicinae: Bolton, 1994: 51.

**AMBLYOPONINI** (tribe of Ponerinae)
Amblyoponinae Forel, 1893a: 162 [as subfamily of Formicidae]. Type-genus: *Amblyopone*.
Amblyoponii: Emery, 1895c: 261; Emery, 1895j: 761 [as tribe of Ponerinae].
Amblyoponini: Ashmead, 1905b: 382 [emended spelling of suffix].
Amblyoponinae: Clark, 1951: 15 (in key) [as subfamily of Formicidae].
Amblyoponini: Brown, 1953c: 11 [as tribe of Ponerinae].
Revision of tribe: Brown, 1960a: 145; Ward, 1994: 167.
Senior synonym of
Ericapeltini (type-genus included in Amblyoponini by Brown, 1960a: 156, but not formally synonymised): Bolton, 1994: 164.
Examblyoponini: Brown, 1951: 102.
Onychomyrmicini (type-genus included in Amblyoponini by Brown, 1960a: 178, but not formally synonymised): Bolton, 1994: 164.
Reneini: Brown, 1953c: 11.
Included genera: *Adetomyrma, Amblyopone, *Casaleia, Concoctio, Myopopone, Mystrium, Onychomyrmex, Paraprionopelta, Prionopelta.*

ANERGATIDINI (junior synonym of Pheidolini)
Anergatidini Emery, 1922e: 116 [as subtribe of Pheidolini]. Type-genus: *Anergatides* (junior synonym of *Pheidole*).
Junior synonym of Pheidolini: Bolton, 1994: 106.

ANERGATINI (junior synonym of Tetramoriini)
Anergatini Emery, 1914a: 41 [as subtribe of Solenopsidini]. Type-genus: *Anergates*.
Junior synonym of Tetramoriini: Bolton, 1994: 106.

**ANEURETINAE** (subfamily of Formicidae)
Aneuretini Emery, 1912g: 6 [as tribe of Dolichoderinae]. Type-genus: *Aneuretus*.
Aneuretinae: Clark, 1951: 16 (footnote) [as subfamily of Formicidae].
Aneuretinae: Baroni Urbani, 1989: 147 [as junior synonym of Dolichoderinae].
Aneuretinae: Bolton, 1990c: 1361 [revived status as subfamily of Formicidae]; Shattuck, 1992b: 201; Baroni Urbani, Bolton & Ward, 1992: 315.
Review of subfamily: Wilson, Eisner, Wheeler & Wheeler, 1956: 81.
Catalogue and bibliography of subfamily: Shattuck, 1994: 1.
Included genera: **Aneuretellus, Aneuretus, *Mianeuretus, *Paraneuretus, *Protaneuretus.*
Note. Material originally included under the unavailable name Prodolichoderinae is referable in part to Aneuretinae: Bolton, 1994: 15.

**ANOMALOMYRMINI** (tribe of Leptanillinae)
Anomalomyrmini Taylor, in Bolton, 1990b: 278. Type-genus: *Anomalomyrma*.
Included genera: *Anomalomyrma, Protanilla.*

ANONYCHOMYRMINI (junior synonym of Dolichoderinae)
Anonychomyrmini Donisthorpe, 1947c: 588 [as tribe of Dolichoderinae]. Type-genus: *Anonychomyrma*.
Junior synonym of Dolichoderinae: Shattuck, 1992c: 5.

APHAENOGASTRINI (junior synonym of Pheidolini)
Aphaenogastrini Enzmann, J. 1947b: 147 [as tribe of Myrmicinae]. Type-genus: *Aphaenogaster.*
Junior synonym of Pheidolini: Brown, 1949a: 48.

**APOMYRMINAE** (subfamily of Formicidae)
Apomyrmini Dlussky & Fedoseeva, 1987: 78 [as tribe of Ponerinae]. Type-genus: *Apomyrma*.
Apomyrmini: Bolton, 1990b: 280 [as tribe of Leptanillinae].
Apomyrminae: Baroni Urbani, Bolton & Ward, 1992: 316 [as subfamily of Formicidae].
Included genus: *Apomyrma*.

ARCHAEOMYRMICINI (junior synonym of Myrmecinini)
Archaeomyrmicini Mann, 1921: 449 [as tribe of Myrmicinae]. Type-genus: *Archaeomyrmex* (junior synonym of *Myrmecina*).
Junior synonym of Myrmecinini: Brown, 1971a: 2

*ARCHIPONERINI (junior synonym of Ponerini)
*Archiponerini Dlussky & Fedoseeva, 1987: 78 [as tribe of Ponerinae]. Type-genus: **Archiponera*.
Junior synonym of Ponerini: Bolton, 1994: 164.

***ARMANIINAE** (subfamily of Formicidae)
*Armaniidae Dlussky, 1983: 66 [as family of Formicoidea]. Type-genus: **Armania*.
*Armaniidae: Wilson, 1987: 49 [as junior synonym of Formicidae].
*Armaniidae: Dlussky & Fedoseeva, 1987: 77 [revived status as family of Formicoidea].
*Armaniinae: Carpenter, 1992: 491; Bolton, 1994: 187 [as subfamily of Formicidae].
Included genera: **Archaeopone, *Armania, *Armaniella, *Cretopone, *Dolichomyrma, *Petropone, *Poneropterus, *Pseudarmania.*

**ATTINI** (tribe of Myrmicinae)
Attidae Smith, F. 1858b: 161 [as subfamily of Formicidae]. Type-genus: *Atta*.
Attidae: Smith, F. 1863: 21 [as family].
Attidae: Emery, 1877a: 72 [as "group" within Myrmicidae].
Attii: Forel, 1893a: 163 [as tribe of Myrmicinae].
Attinae: Ashmead, 1905b: 383 [as subfamily of Cryptoceridae].
Attini: Emery, 1914a: 39; Forel, 1917: 232 [emended spelling of suffix; as tribe of Myrmicinae].
Included genera: *Acromyrmex, Apterostigma, Atta, *Attaichnus* (ichnotaxon), *Cyphomyrmex, Mycetarotes, Mycetophylax, Mycetosoritis, Mycocepurus, Myrmicocrypta, Pseudoatta, Sericomyrmex, Trachymyrmex.*

AXINIDRINI (junior synonym of Dolichoderinae)
Axinidrini Weber, 1941a: 193 [as tribe of Dolichoderinae]. Type-genus: *Axinidris*.
Junior synonym of Dolichoderinae: Shattuck, 1992c: 5.

**BASICEROTINI** (tribe of Myrmicinae)
Basicerotini Brown, 1949f: 86 [as tribe of Myrmicinae]. Type-genus: *Basiceros*.
Revision of tribe: Brown & Kempf, 1960: 161.
Junior synonym of Dacetini: Baroni Urbani & De Andrade, 1994: 10.
Basicerotini revived from synonymy: **revived status**.
Included genera: *Basiceros, Creightonidris, Eurhopalothrix, Octostruma, Protalaridris, Rhopalothrix, Talaridris.*

**BLEPHARIDATTINI** (tribe of Myrmicinae)
Blepharidattini Wheeler, G.C. & Wheeler, J. 1991a: 133. Type-genus: *Blepharidatta*.
Included genera: *Blepharidatta, Wasmannia.*

**BRACHYMYRMECINI** (tribe of Formicinae)
Brachymyrmicini Emery, 1925b: 40 [as subtribe of Dimorphomyrmicini]. Type-genus: *Brachymyrmex*.
Brachymyrmicini: Wheeler, W.M. 1929a: 12 [as tribe of Formicinae].
Brachymyrmecini: Wheeler, G.C. & Wheeler, J. 1985: 258 [emended spelling].
Included genera: *Aphomomyrmex, Brachymyrmex, Cladomyrma, Petalomyrmex, Pseudaphomomyrmex.*

**BREGMATOMYRMINI** (tribe of Formicinae)
Bregmatomyrmii Wheeler, W.M. 1929b: 5 [as tribe of Formicinae]. Type-genus: *Bregmatomyrma*.
Bregmatomyrmini: Bolton, 1994: 50 [emended spelling of suffix].
Included genus: *Bregmatomyrma*.

CALYPTOMYRMECINI (junior synonym of Stenammini)
Calyptomyrmecini Dlussky & Fedoseeva, 1987: 80 [as tribe of Myrmicinae]. Type-genus: *Calyptomyrmex*.
Junior synonym of Stenammini: Bolton, 1994: 106.

**CAMPONOTINI** (tribe of Formicinae)
Camponotidae Forel, 1878: 364 [as subfamily of Formicidae]. Type-genus: *Camponotus*.
Camponotinae: Forel, 1893a: 165 [emended spelling of suffix].
Camponotii: Forel, 1893a: 165 [as tribe of Camponotinae].
Camponotini: Ashmead, 1905b: 384 [emended spelling of suffix].
Camponotini: Wheeler, W.M. 1922a: 693 [as tribe of Formicinae].
Senior synonym of Polyrhachidini: Bolton, 1994: 50.
Included genera: *Calomyrmex, *Camponotites, Camponotus, *Chimaeromyrma, Dendromyrmex, *Drymomyrmex, Echinopla, Forelophilus, Notostigma, Opisthopsis, Overbeckia, Phasmomyrmex, Polyrhachis, *Pseudocamponotus.*

CARDIOCONDYLINI (junior synonym of Formicoxenini)
Cardiocondylini Emery, 1914a: 36 [as tribe of Myrmicinae]. Type-genus: *Cardiocondyla*.
Junior synonym of Leptothoracini: Bolton, 1982: 311.
Junior synonym of Formicoxenini: Bolton, 1994: 105.

**CATAULACINI** (tribe of Myrmicinae)
Cataulacii Emery, 1895j: 771 [as tribe of Myrmicinae]. Type-genus: *Cataulacus.*
Cataulacinae: Ashmead, 1905b: 384 [as subfamily of Cryptoceridae].
Cataulacini: Emery, 1914a: 39 [emended spelling of suffix; as tribe of Myrmicinae].
Revision of tribe: Bolton, 1974a: 1.
Included genus: *Cataulacus.*

CENTROMYRMICINI (junior synonym of Ponerini)
Centromyrmicini Emery, 1911d: 57 [as subtribe of Ponerini]. Type-genus: *Centromyrmex.*
Junior synonym of Ponerini: Brown, 1953c: 9.

**CEPHALOTINI** (tribe of Myrmicinae)
Cephalotini Smith, M.R. 1949c: 18 [as tribe of Myrmicinae]. Type-genus: *Cephalotes.*
Revision of tribe: Kempf, 1951: 1; Kempf, 1952: 1; Kempf, 1958a: 1; Kempf, 1973c: 461.
Senior synonym of Cryptoceridae: Smith, M.R. 1949c: 18.
Included genera: *Cephalotes, Eucryptocerus, Procryptocerus, Zacryptocerus.*

**CERAPACHYINAE** (subfamily of Formicidae)
Cerapachysii Forel, 1893a: 162 [as tribe of Ponerinae]. Type-genus: *Cerapachys.*
Cerapachyinae: Wheeler, W.M. 1902d: 185 [as "group" of Ponerinae.]
Cerapachyinae: Wheeler, W.M. 1920: 51 [as subfamily of Formicidae].
Cerapachyidae: Bernard, 1951: 1046 [as family of Formicoidea].
Cerapachyinae: Clark, 1951: 15 (in key) [as subfamily of Formicidae].
Cerapachyinae: Brown, 1975: 14 [as junior synonym of Ponerinae].
Cerapachyinae: Bolton, 1990a: 66 [revived from synonymy as subfamily of Formicidae]; Bolton, 1990c: 1356; Baroni Urbani, Bolton & Ward, 1992: 316.
Revision of tribes: Brown, 1975: 11.
Included tribes: Acanthostichini, Cerapachyini, Cylindromyrmecini.
Note. Material originally included under the unavailable name Prodorylinae is referable to Cerapachyinae: Bolton, 1990a: 66.

**CERAPACHYINI** (tribe of Cerapachyinae)
Cerapachysii Forel, 1893a: 162 [as tribe of Ponerinae]. Type-genus: *Cerapachys.*
Cerapachyi: Emery, 1895j: 765 [as tribe of Dorylinae].
Cerapachyini: Ashmead, 1905b: 382 [emended spelling of suffix; as tribe of Ponerinae].
Cerapachyini: Wheeler, W.M. 1922a: 638 [as tribe of Cerapachyinae].
Cerapachyini: Brown, 1975: 14 [as tribe of Ponerinae].
Cerapachyini: Bolton, 1990a: 67 [as tribe of Cerapachyinae].
Revision of tribe: Brown, 1975: 14.
Senior synonym of
Eusphinctinae: Bolton, 1990a: 66.
Lioponerini (type-genus included in Cerapachyini by Brown, 1975: 19, but not formally synonymised): Bolton, 1994: 19.
Included genera: *Cerapachys, Simopone, Sphinctomyrmex.*

**CHELIOMYRMECINI** (tribe of Ecitoninae)
Cheliomyrmicini Wheeler, W.M. 1921d: 319 [as tribe of Dorylinae]. Type-genus: *Cheliomyrmex.*
Cheliomyrmecini: Wheeler, G.C. & Wheeler, J. 1985: 256 [emended spelling].
Cheliomyrmecini: Bolton, 1990c: 1357 [as tribe of Ecitoninae].
Revision of tribe: Borgmeier, 1955: 58.
Included genus: *Cheliomyrmex.*

**CREMATOGASTRINI** (tribe of Myrmicinae)
Cremastogastrii Forel, 1893a: 164 [as tribe of Myrmicinae]. Type-genus: *Crematogaster.*
Crematogastrii: Emery, 1895j: 769 [emended spelling].
Crematogastrini: Emery, 1914a: 37 [emended spelling of suffix].
Included genus: *Crematogaster.*

CRYPTOCERIDAE (junior synonym of Cephalotini)
Cryptoceridae Smith, F. 1854: 214 [as subfamily of Formicidae]. Type-genus: *Cryptocerus* (junior synonym of *Cephalotes*).
Cryptoceridae: Emery, 1877a: 72 [as "group" of Myrmicinae].
Cryptocerii: Forel, 1893a: 164; Emery, 1895j: 763 [as tribe of Myrmicinae].
Cryptoceridae: Ashmead, 1905b: 383 [as family].
Cryptocerini: Emery, 1914a: 39 [emended spelling of suffix; as tribe of Myrmicinae].
Junior synonym of Cephalotini: Smith, M.R. 1949c: 18.

**CYLINDROMYRMECINI** (tribe of Cerapachyinae)
Cylindromyrmii Emery, 1901a: 34 [as tribe of Dorylinae]. Type-genus: *Cylindromyrmex.*
Cylindromyrmicini: Ashmead, 1905b: 382 [emended spelling of suffix].
Cylindromyrmicini: Emery, 1911d: 14 [as tribe of Ponerinae].
Cylindromyrmecini: Brown, 1975: 36 [emended spelling].
Cylindromyrmecini: Bolton, 1990a: 67 [as tribe of Cerapachyinae].
Revision of tribe: Brown, 1975: 36.
Included genus: *Cylindromyrmex.*

DACETITI (junior synonym of Dacetini)
Dacetiti Brown, 1952g: 10 (footnote); Brown, 1954b: 465 (definition) [as subtribe of Dacetini]. Type-genus: *Daceton.*
Junior synonym of Dacetini: Baroni Urbani & De Andrade, 1994: 9.

**DACETINI** (tribe of Myrmicinae)
Dacetonini Forel, 1892d: 344 [as genus group]. Type-genus: *Daceton.*
Dacetonii: Forel, 1893a: 164 [as tribe of Myrmicinae].
Dacetii: Emery, 1895j: 770 [emended spelling].
Dacetini: Emery, 1914a: 13; Forel, 1917: 246 [emended spelling].
Revisionary studies: see publications of Brown, W.L.,Jr.
Senior synonym of Dacetiti, Epopostrumiti, Orectognathiti, Strumigeniti: Baroni Urbani & De Andrade, 1994: 9.
Included genera: *Acanthognathus, Asketogenys, Chelystruma, Cladarogenys, Codiomyrmex, Codioxenus, Colobostruma, Daceton, Dorisidris, Dysedrognathus, Epitritus, Epopostruma, Glamyromyrmex, Gymnomyrmex, Kyidris, Mesostruma, Microdaceton, Neostruma, Orectognathus, Pentastruma, Quadristruma, Serrastruma, Smithistruma, Strumigenys, Tingimyrmex, Trichoscapa.*

DIMORPHOMYRMII (junior synonym of Gesomymecini)
Dimorphomyrmii Emery, 1895j: 772 [as tribe of Camponotinae]. Type-genus: *Dimorphomyrmex* (junior synonym of *Gesomyrmex*).
Dimorphomyrmicini: Forel, 1917: 232 [emended spelling of suffix].
Dimorphomyrmicini: Wheeler, W.M. 1922a: 693 [as tribe of Formicinae].
Dimorphomyrmicini: Emery, 1925b: 39 [incorrectly as senior synonym of Gesomyrmecini].
Junior synonym of Gesomyrmecini: Wheeler, W.M. 1929a: 12.

DISCOTHYRINAE (junior synonym of Proceratiini)
Discothyrinae Clark, 1951: 15 (in key) [as subfamily of Formicinae]. Type-genus: *Discothyrea.*
Junior synonym of Ectatommini: Bolton, 1994: 164; of Proceratiini: Lattke, 1994: 112 (implied but not stated).

**DOLICHODERINAE** (subfamily of Formicidae)
Dolichoderidae Forel, 1878: 364 [as subfamily of Formicidae]. Type-genus: *Dolichoderus.*
Dolichoderinae: Forel, 1893a: 165 [emended spelling of suffix].
Dolichoderidae: Bernard, 1951: 1071 [as family of Formicoidea].
Dolichoderinae: Brown, 1954e: 29 [as subfamily of Formicidae]; Dlussky & Fedoseeva, 1987: 77; Shattuck, 1992b: 201; Baroni Urbani, Bolton & Ward, 1992: 316.
Revision of subfamily: Shattuck, 1992c: 1.
Catalogue and bibliography of subfamily: Shattuck, 1994: 3.
Senior synonym of
Anonychomyrmini, Axinidrini, Leptomyrmicini, Liometopini, *Pityomyrmecini, Tapinomini, *Zherichiniini: Shattuck, 1992c: 5.
*Miomyrmicini: Bolton, 1994: 26.
Included genera: *Anillidris, Anonychomyrma, *Asymphylomyrmex, Axinidris, Azteca, Bothriomyrmex, Doleromyrma, Dolichoderus, Dorymyrmex, Ecphorella, *Elaeomyrmex, *Eotapinoma, Forelius, Froggattella, Iridomyrmex, *Kotshkorkia, Leptomyrmex, *Leptomyrmula, Linepithema, Liometopum, Loweriella, *Miomyrmex, Ochetellus, Papyrius, *Petraeomyrmex, Philidris, *Pityomyrmex, *Protazteca, Tapinoma, Technomyrmex, Turneria, *Zherichinius.*
Note. Material originally included under the unavailable names Prodolichoderinae and Eudolichoderinae is referable to Dolichoderinae: Bolton, 1994: 27.

**DORYLINAE** (subfamily of Formicidae)
Dorylida Leach, 1815: 147 [as family-group name]. Type-genus: *Dorylus.*
Dorylidae: Haliday, 1836: 331 [emended spelling of suffix].
Dorylinae: Forel, 1893a: 163 [as subfamily of Formicidae].

Dorylii: Forel, 1893a: 163; Emery, 1895j: 758 [as tribe of Dorylinae].
Dorylini: Ashmead, 1905b: 381 [emended spelling of suffix].
Dorylidae: Bernard, 1951: 1046 [as family of Formicoidea].
Dorylinae: Brown, 1954e: 28 [as subfamily of Formicidae]; Dlussky & Fedoseeva, 1987: 79; Bolton, 1990c: 1358; Baroni Urbani, Bolton & Ward, 1992: 316.
Included genus: *Dorylus.*
Note. Material originally included under the unavailable name Eudorylinae is referable to Dorylinae: Bolton, 1990c: 1358.

DORYLOZELINI (junior synonym of Ponerini)
Dorylozelini Wheeler, W.M. 1922a: 646 [as tribe of Ponerinae]. Type-genus: *Dorylozelus* (junior synonym of *Leptogenys*).
Junior synonym of Ponerini: Taylor, 1969: 132.

DREPANOGNATHINI (junior synonym of Ponerini)
Drepanognathini Ashmead, 1905b: 382 [as tribe of Pachycondylinae]. Type-genus: *Drepanognathus* (junior synonym of *Harpegnathos*).
Junior synonym of Harpegnathini: Emery, 1911d: 58.
Junior synonym of Ponerini: Bolton, 1994: 164.

**ECITONINAE** (subfamily of Formicidae)
Ecitonii Forel, 1893a: 163 [as tribe of Dorylinae]. Type-genus: *Eciton.*
Ecitii: Emery, 1901a: 36 [emended spelling].
Ecitoninae: Ashmead, 1905b: 381 [as subfamily of Dorylidae].
Ecitoninae: Brown, 1973b: 166 [as subfamily of Formicidae]; Bolton, 1990c: 1357; Baroni Urbani, Bolton & Ward, 1992: 317.
Revision of subfamily: Borgmeier, 1955: 15; Watkins, 1976: 1.
Included tribes: Cheliomyrmecini, Ecitonini.

**ECITONINI** (tribe of Ecitoninae)
Ecitonii Forel, 1893a: 163 [as tribe of Dorylinae]. Type-genus: *Eciton.*
Ecitonini: Ashmead, 1905b: 381 [emended spelling of suffix].
Ecitini: Emery, 1910b: 15 [emended spelling, as tribe of Dorylinae].
Ecitonini: Wheeler, W.M. 1922a: 633 [as tribe of Dorylinae].
Ecitonini: Bolton, 1990c: 1357 [as tribe of Ecitoninae].
Revision of tribe: Borgmeier, 1955: 77.
Included genera: *Eciton, Labidus, Neivamyrmex, Nomamyrmex.*
Material originally included under the unavailable name Metadorylinae is referable in part to Ecitonini: Bolton, 1990c: 1357.

**ECTATOMMINI** (tribe of Ponerinae)
Ectatommii Emery, 1895j: 767 [as tribe of Ponerinae]. Type-genus: *Ectatomma.*
Ectatommini: Ashmead, 1905b: 382 [emended spelling of suffix].
Revision of tribe: Brown, 1958g: 175.
Phylogeny: Kugler, C. 1991: 153 (sting structure); Lattke, 1994: 105.
Senior synonym of
Stictoponerini (type-genus included in Ectatommini by Brown, 1958g: 211, but not formally synonymised): Bolton, 1994: 164.
Included genera: *Acanthoponera, Aulacopone, Ectatomma, *Electroponera, Gnamptogenys, Heteroponera, Rhytidoponera, *Syntaphus.*

EPOPOSTRUMITI (junior synonym of Dacetini)
Epopostrumiti Brown, 1952g: 10 (footnote); Brown, 1954b: 465 (definition) [as subtribe of Dacetini]. Type-genus: *Epopostruma.*
Junior synonym of Dacetini: Baroni Urbani & De Andrade, 1994: 9.

ERICAPELTINI (junior synonym of Amblyoponini)
Ericapeltini Kusnezov, 1955: 275 [as subtribe of Amblyoponini]. Type-genus: *Ericapelta* (junior synonym of *Amblyopone*).
Junior synonym of Amblyoponini: Bolton, 1994: 164.

EUCAMPONOTINAE (unavailable name)
Eucamponotinae Forel, 1912i: 88 [as "section" of Camponotinae]. Section designated to include tribes Camponotini, Formicini, Gesomyrmecini, Oecophyllini and Prenolepidini. Unavailable name, not based on genus rank taxon. Contained material referable to Formicinae: Bolton, 1994: 51.

EUDOLICHODERINAE (unavailable name)
Eudolichoderinae Forel, 1917: 247 [as "section" of Dolichoderinae]. Section designated to include tribe Tapinomini. Unavailable name, not based on genus rank taxon. Contained material referable to Dolichoderinae: Bolton, 1994: 27.

EUDORYLINAE (unavailable name)
Eudorylinae Forel, 1917: 239 [as "section" of Dorylinae]. Section designated to include tribe Dorylini. Unavailable name, not based on genus rank taxon. Contained material referable to Dorylinae: Bolton, 1990c: 1358.

EUFORMICINAE (unavailable name)
Euformicinae Emery, 1925b: 37 [as "section" of Formicinae]. Section designated to include tribes Brachymyrmecini, Camponotini, Dimorphomyrmicini, Formicini, Gigantiopini, Lasiini, Oecophyllini and Santschiellini. Unavailable name, not based on genus rank taxon. Contained material referable to Formicinae: Bolton, 1994: 51.

EUMYRMICINAE (unavailable name)
Eumyrmicinae Emery, 1914a: 35 [as "section" of Myrmicinae]. Section designated to include all Myrmicinae except Metaponini and Pseudomyrmecini. Unavailable name, not based on genus rank taxon. Contained material referable to Myrmicinae: Bolton, 1994: 106.

EUPONERINAE (junior synonym of Ponerini)
Euponerinae Emery, 1909c: 355 [as "group" of Ponerinae]. Type-genus: *Euponera* (junior synonym of *Pachycondyla*).
Junior synonym of Ponerini: Bolton, 1994: 164.

EUSPHINCTINAE (junior synonym of Cerapachyini)
Eusphinctinae Clark, 1951: 15 (in key) [as subfamily of Formicidae]. Type-genus: *Eusphinctus* (junior synonym of *Sphinctomyrmex*).
Junior synonym of Cerapachyini: Bolton, 1990a: 66.

EXAMBLYOPONINI (junior synonym of Amblyoponini)
Examblyoponini Donisthorpe, 1949g: 401 [as tribe of Ponerinae]. Type-genus: *Examblyopone* (junior synonym of *Prionopelta*).
Junior synonym of Amblyoponini: Brown, 1951: 102.

EXEUPONERINAE (unavailable name)
Exeuponerinae Donisthorpe, 1943d: 438 [as "section" of Ponerinae]. Section designated to include only *Pseudoneoponera* (junior synonym of *Pachycondyla*). Unavailable name, not based on genus rank taxon. Contained material referable to Ponerini: Bolton, 1994: 164.

**FORMICIDAE** (family of superfamily Vespoidea)
Formicariae Latreille, 1809: 124 [as family-group name]. Type-genus: *Formica.*
Formicidae: Stephens, 1829: 356 [emended spelling of suffix].
Included subfamilies: Aenictinae, Aenictogitoninae, Aneuretinae, Apomyrminae, *Armaniinae, Cerapachyinae, Dolichoderinae, Dorylinae, Ecitoninae, Formicinae, *Formiciinae, Leptanillinae, Leptanilloidinae, Myrmeciinae, Myrmicinae, Nothomyrmeciinae, *Paleosminthurinae, Ponerinae, Pseudomyrmecinae, *Sphecomyrminae.
Genera dubiously or perhaps wrongly placed in Formicidae: **Calyptites, Condylodon, *Cretacoformica, *Palaeomyrmex.*

***FORMICIINAE** (subfamily of Formicidae)
*Formiciinae Lutz, 1986: 181 [as subfamily of Formicidae]. Type-genus: **Formicium.*
Included genus: **Formicium.*

**FORMICINAE** (subfamily of Formicidae)
Formicariae Latreille, 1809: 124 [as family-group name]. Type-genus: *Formica.*
Formicidae: Mayr, 1865: 6 [as subfamily of Formicidae].
Formicinae: Wheeler, W.M. 1920: 53 [as subfamily of Formicidae].
Formicidae: Bernard, 1951: 1073 [as family of Formicoidea].
Formicinae: Brown, 1954e: 29 [as subfamily of Formicidae]; Dlussky & Fedoseeva, 1987: 77; Shattuck, 1992b: 201; Baroni Urbani, Bolton & Ward, 1992: 317; Agosti, 1994a: 95.
Contribution towards revision of subfamily: Agosti, 1991: 295.
Included tribes: Brachymyrmecini, Bregmatomyrmini, Camponotini, Formicini, Gesomyrmecini, Gigantiopini, Lasiini, Melophorini, Myrmelachistini, Myrmoteratini, Oecophyllini, Plagiolepidini, Santschiellini, *Sicilomyrmecini.
Formicinae genus unplaced to tribe: **Imhoffia.*
Formicinae collective group name: **Formicites.*
Note. Material originally included under the unavailable names Alloformicinae, Eucamponotinae, Euformicinae, Heteroformicinae and Mesocamponotinae is all referable to Formicinae: Bolton, 1994: 51.

**FORMICINI** (tribe of Formicinae)
Formicariae Latreille, 1809: 124 [as family-group name]. Type-genus: *Formica.*
Formicii: Forel, 1893a: 165 [as tribe of Camponotinae].
Formicini: Ashmead, 1905b: 384 [emended spelling of suffix].
Formicini: Wheeler, W.M. 1922a: 693 [as tribe of Formicinae].
Review of tribe: Agosti, 1990b: 295.
Revision and phylogeny of tribe: Agosti, 1994a: 95.
Included genera: *Alloformica, Bajcaridris, Cataglyphis, Formica, *Glaphyromyrmex, *Leucotaphus, Polyergus, Proformica, *Protoformica, Rossomyrmex.*

**FORMICOXENINI** (tribe of Myrmicinae)
Formicoxenii Forel, 1893a: 164 [as tribe of Myrmicinae]. Type-genus: *Formicoxenus.*
Formicoxenini: Bolton, 1994: 105 [emended spelling].
Senior synonym of
Cardiocondylini, Leptothoracini, Podomyrmini, Stereomyrmicini: Bolton, 1994: 105.
Included genera: *Ankylomyrma, Atopomyrmex, Cardiocondyla, Chalepoxenus, Dilobocondyla, Doronomyrmex, Epimyrma, Formicoxenus, Harpagoxenus, Ireneopone, Leptothorax, Paratopula, Peronomyrmex, Podomyrma, Poecilomyrma, Protomognathus, Romblonella, Rotastruma, Stereomyrmex, Terataner, Vombisidris, Willowsiella.*
Genera dubiously placed here: **Stigmomyrmex, Tricytarus.*

**GESOMYRMECINI** (tribe of Formicinae)
Gesomyrmicinae Ashmead, 1905b: 384 [as subfamily of Formicidae]. Type-genus: *Gesomyrmex.*
Gesomyrmicini: Ashmead, 1905b: 384 [as tribe of Gesomyrmicinae].
Gesomyrmecini: Wheeler, G.C. & Wheeler, J. 1985: 258 [emended spelling].
Senior synonym of
Gesomyrmini Forel: Bolton, 1994: 50.
Dimorphomyrmii: Wheeler, W.M. 1929a: 12.
Included genera: *Gesomyrmex, *Prodimorphomyrmex.*

GESOMYRMINI (junior synonym of the above; described as new independently of Ashmead, 1905b)
Gesomyrmini Forel, 1912i: 89 [as tribe of Camponotinae]. Type-genus: *Gesomyrmex.*
Gesomyrmicini: Wheeler, W.M. 1922a: 693 [emended spelling of suffix].
Junior synonym of Gesomyrmecini Ashmead: Bolton, 1994: 50.

**GIGANTIOPINI** (tribe of Formicinae)
Gigantiopini Ashmead, 1905b: 384 [as tribe of Gesomyrmicinae]. Type-genus: *Gigantiops.*
Gigantiopini: Wheeler, W.M. 1922a: 692 [as tribe of Formicinae].
Included genus: *Gigantiops.*

HARPEGNATHII (junior synonym of Ponerini)
Harpegnathii Forel, 1900c: 63 [as tribe of Ponerinae]. Type-genus: *Harpegnathos.*
Harpegnathini: Emery, 1911d: 58 [emended spelling of suffix; as subtribe of Ponerini].
Senior synonym of Drepanognathini: Emery, 1911d: 58.
Junior synonym of Ponerini: Bolton, 1994: 164.

HETEROFORMICINAE (unavailable name)
Heteroformicinae Emery, 1925b: 35 [as "section" of Formicinae]. Section designated to include tribes Myrmecorhynchini and Myrmoteratini. Unavailable name; not based on genus rank taxon. Contained material referable to Formicinae: Bolton, 1994: 51.

*HYPOPOMYRMICITI (junior synonym of Pheidologetonini)
*Hypopomyrmiciti Brown, 1952g: 10 (footnote) [as subtribe of Dacetini]. Type-species: **Hypopomyrmex.*
Junior synonym of Pheidologetonini: implied in Brown & Carpenter, 1979: 422 [synonymy not directly stated but **Hypopomyrmex* transferred to that tribe].

**LASIINI** (tribe of Formicinae)
Lasiini Ashmead, 1905b: 384 [as tribe of Formicinae]. Type-genus: *Lasius.*
Senior synonym of
Acanthomyopsini: Bolton, 1994: 50.
Prenolepidii: Emery, 1925b: 212.
Included genera: *Acanthomyops, Euprenolepis, Lasius, Myrmecocystus, Paratrechina, Prenolepis, *Protrechina, Pseudolasius, Teratomyrmex.*

LEPTALEINAE (junior synonym of Pseudomyrmecinae)
Leptaleinae Smith, M.R. 1951a: 788 [as subfamily of Formicidae]. Type-species: *Leptalea* (junior synonym of *Pseudomyrmex*).
Junior synonym of Pseudomyrmecinae: Smith, M.R. 1958c: 112.

**LEPTANILLINAE** (subfamily of Formicidae)
Leptanillini Emery, 1910b: 32 [as tribe of Dorylinae]. Type-genus: *Leptanilla.*
Leptanillinae: Wheeler, W.M. 1923f: 335 [as subfamily of Formicidae].
Leptanillidae: Bernard, 1951: 1053 [as family of Formicoidea].
Leptanillinae: Brown, 1954e: 28 [as subfamily of Formicidae]; Baroni Urbani, Bolton & Ward, 1992: 317; Kugler, C. 1992: 103.
Revision of tribes: Bolton, 1990b: 267.
Included tribes: Anomalomyrmini, Leptanillini.

**LEPTANILLINI** (tribe of Leptanillinae)
Leptanillini Emery, 1910b: 32 [as tribe of Dorylinae]. Type-genus: *Leptanilla.*
Leptanillini: Wheeler, W.M. 1923f: 335 [as tribe of Leptanillinae].
Leptanillini: Bolton, 1990b: 276 [as tribe of Leptanillinae].
Revision of tribe: Petersen, 1968: 577; Baroni Urbani, 1977c: 427.
Included genera: *Leptanilla, Noonilla, Phaulomyrma, Scyphodon, Yavnella.*
Material originally included under the unavailable name Metadorylinae is referable in part to Leptanillini: Bolton, 1994: 70.

**LEPTANILLOIDINAE** (subfamily of Formicidae)
Leptanilloidinae Bolton, in Baroni Urbani, Bolton & Ward, 1992: 317. Type-genus: *Leptanilloides.*
Included genus: *Leptanilloides.*

LEPTOGENYSII (junior synonym of Ponerini)
Leptogenysii Forel, 1893a: 162 [as tribe of Ponerini]. Type-genus: *Leptogenys.*
Leptogenyi: Emery, 1895j: 761; Emery, 1901a: 36 [emended spelling].
Leptogenyini: Ashmead, 1905b: 382 [emended spelling of suffix].
Junior synonym of Ponerini: Brown, 1963: 3.

LEPTOMYRMICINI (junior synonym of Dolichoderinae)
Leptomyrmicini Emery, 1912g: 15 [as tribe of Dolichoderinae]. Type-genus: *Leptomyrmex.*
Leptomyrmecini: Wheeler, G.C. & Wheeler, J. 1985: 258 [emended spelling].
Junior synonym of Dolichoderinae: Shattuck, 1992c: 5.

LEPTOTHORACINI (junior synonym of Formicoxenini)
Leptothoracini Emery, 1914a: 38 [as tribe of Myrmicinae]. Type-genus: *Leptothorax.*
Senior synonym of Cardiocondylini: Bolton, 1982: 311.
Junior synonym of Formicoxenini: Bolton, 1994: 105.

LIOMETOPINI (junior synonym of Dolichoderinae)
Liometopini Dlussky & Fedoseeva, 1987: 77 [as tribe of Dolichoderinae]. Type-genus: *Liometopum.*
Junior synonym of Dolichoderinae: Shattuck, 1992c: 5.

LIOPONERINI (junior synonym of Cerapachyini)
Lioponerini Ashmead, 1905b: 382 [as tribe of Pachycondylinae]. Type-genus: *Lioponera* (junior synonym of *Cerapachys*).
Junior synonym of Cerapachyini: Bolton, 1994: 19.

LOPHOMYRMICINI (junior synonym of Pheidolini)
Lophomyrmicini Emery, 1914a: 41 [as subtribe of Pheidologetonini]. Type-genus: *Lophomyrmex.*
Junior synonym of Pheidologetonini: Bolton, 1994: 106.
Junior synonym of Pheidolini (implied by Rigato, 1994: 51, but not stated): **new synonymy**.

MEGALOMYRMECINI (junior synonym of Solenopsidini)
Megalomyrmecini Dlussky & Fedoseeva: 1987: 80 [as tribe of Myrmicinae]. Type-genus: *Megalomyrmex.*
Junior synonym of Solenopsidini: Bolton, 1994: 106.

**MELISSOTARSINI** (tribe of Myrmicinae)
Melissotarsii Emery, 1901a: 36 [as tribe of Ponerinae]. Type-genus: *Melissotarsus.*
Melissotarsini: Ashmead, 1905b: 383 [emended spelling of suffix].
Melissotarsini: Forel, 1911h: 446 [as tribe of Myrmicinae].
Included genera: *Melissotarsus, Rhopalomastix.*

**MELOPHORINI** (tribe of Formicinae)
Melophorini Forel, 1912i: 88 [as tribe of Camponotinae]. Type-genus: *Melophorus.*

Melophorini: Wheeler, W.M. 1922a: 692 [as tribe of Formicinae].
Review of tribe: Brown, 1955d: 471.
Senior synonym of Myrmecorhynchini: Wheeler, W.M. 1935c: 69.
Included genera: *Lasiophanes, Melophorus, Myrmecorhynchus, Notoncus, Prolasius, Pseudonotoncus.*

**MERANOPLINI** (tribe of Myrmicinae)
Meranoplini Emery, 1914a: 37 [as tribe of Myrmicinae]. Type-genus: *Meranoplus.*
Review of tribe: Bolton, 1981a: 45.
Included genera: *Meranoplus, *Parameranoplus.*

MESOCAMPONOTINAE (unavailable name)
Mesocamponotinae Forel, 1912i: 88 [as "section" of Camponotinae]. Section designated to include tribes Melophorini and Plagiolepidini. Unavailable name; not based on genus rank taxon. Contained material referable to Formicinae: Bolton, 1994: 51.

METADORYLINAE (unavailable name)
Metadorylinae Forel, 1917: 240 [as "section" of Dorylinae]. Section designated to include tribes Ecitonini and Leptanillini. Unavailable name; not based on genus rank taxon. Contained material referable to Ecitoninae and Leptanillinae: Bolton, 1990c: 1357; Bolton, 1994: 70.

**METAPONINI** (tribe of Myrmicinae)
Metaponini Forel, 1911h: 447 [as tribe of Myrmicinae]. Type-genus: *Metapone.*
Included genera: *Liomyrmex, Metapone, Vollenhovia.*
Genus dubiously placed here: *Xenomyrmex.*
Note. Material originally referred to the unavailable name Promyrmicinae is referable to Metaponini: Bolton, 1994: 106.

*MIOMYRMICINI (junior synonym of Dolichoderinae)
*Miomyrmicini Carpenter, 1930: 51 [as tribe of Dolichoderinae]. Type-genus: **Miomyrmex.*
Junior synonym of Dolichoderinae: Bolton, 1994: 26.

MONOMORIINI (junior synonym of Solenopsidini)
Monomoriini Emery, 1914a: 41 [as subtribe of Solenopsidini]. Type-genus: *Monomorium.*
Monomoriini: Dlussky & Fedoseeva, 1987: 80 [as tribe of Myrmicinae].
Junior synonym of Solenopsidini: Bolton, 1994: 106.

MONOMORIINI (junior homonym and synonym of the above; junior synonym of Solenopsidini, described independently of Emery, 1914a)
Monomoriini Wheeler, W.M. 1915h: 45 [as tribe of Myrmicinae]. Type-genus: *Monomorium.*
Junior synonym of Solenopsidini: **new synonymy**.

MYCETOMYRMICINAE (unavailable name)
Mycetomyrmicinae Forel, 1917: 246 [as "section" of Myrmicinae]. Section designated to include tribes Attini and Proattini. Unavailable name; not based on genus rank taxon. Contained material referable to Myrmicinae: Bolton, 1994: 106.

**MYRMECIINAE** (subfamily of Formicidae)
Myrmeciidae Emery, 1877a: 71 [as "group" of Myrmicidae]. Type-genus: *Myrmecia.*
Myrmecii: Forel, 1893a: 162; Emery, 1895j: 766 [as tribe of Ponerinae].
Myrmeciinae: Ashmead, 1905b: 382 [as subfamily of Poneridae].
Myrmeciinae: Clark, 1951: 16 (in key) [as subfamily of Formicidae].
Included tribes: Myrmeciini, *Prionomyrmecini
Myrmeciinae genera unplaced to tribe: **Ameghinoia* and **Cariridris.*

**MYRMECIINI** (tribe of Myrmeciinae)
Myrmeciidae Emery, 1877a: 71 [as "group" of Myrmicidae]. Type-genus: *Myrmecia.*
Myrmecii: Forel, 1893a: 162 [as tribe of Ponerinae].
Myrmeciini: Emery, 1911d: 17 [emended spelling of suffix].
Myrmeciini: Brown, 1954e: 23 [as tribe of Myrmeciinae].
Revisions: Clark, 1951: 16; Brown, 1953j: 1; Ogata, 1991: 353; Ogata & Taylor, 1991: 1623.
Included genus: *Myrmecia.*
Note. Material originally included under the unavailable name Paleoponerinae is referable to Myrmeciini: Bolton, 1994: 73.

**MYRMECININI** (tribe of Myrmicinae)
Myrmecinini Ashmead, 1905b: 383 [as tribe of Myrmicinae]. Type-genus: *Myrmecina.*
Senior synonym of Archaeomyrmicini: Brown, 1971a: 2.
Included genera: *Acanthomyrmex, Myrmecina, Perissomyrmex, Pristomyrmex.*
Genera dubiously placed here: **Enneamerus, *Stiphromyrmex.*

MYRMECORHYNCHINI (junior synonym of Melophorini)
Myrmorhynchini (sic) Wheeler, W.M. 1917f: 19 [as tribe of Camponotinae]. Type-genus: *Myrmecorhynchus.*
Myrmecorhynchini: Emery, 1925b: 35 [emended spelling; as tribe of Formicinae].
Junior synonym of Melophorini: Wheeler, W.M. 1935c: 69.

**MYRMELACHISTINI** (tribe of Formicinae)
Myrmelachistini Forel, 1912i: 89 [as tribe of Camponotinae]. Type-genus: *Myrmelachista.*
Myrmelachistini: Wheeler, W.M. 1922a: 693 [as tribe of Formicinae].
Included genus: *Myrmelachista.*

**MYRMICARIINI** (tribe of Myrmicinae)
Myrmicarii Forel, 1893a: 165 [as tribe of Myrmicinae]. Type-genus: *Myrmicaria.*
Myrmicariini: Ashmead, 1905b: 383 [emended spelling of suffix].
Included genus: *Myrmicaria.*

**MYRMICINAE** (subfamily of Formicidae)
Myrmicites Lepeletier de Saint-Fargeau, 1835: 169 [family-group name]. Type-genus: *Myrmica.*
Myrmicidae: Smith, F. 1851: 4 [emended spelling].
Myrmicidae: Mayr, 1862: 738 [as subfamily of Formicidae].
Myrmicinae: Forel, 1893a: 163 [emended spelling of suffix].
Myrmicidae: Bernard, 1951: 1058 [as family of Formicoidea].
Myrmicinae: Brown, 1954e: 28 [as subfamily of Formicidae]; Dlussky & Fedoseeva, 1987: 79; Baroni Urbani, Bolton & Ward, 1992: 317.
Included tribes: Agroecomyrmecini, Attini, Basicerotini, Blepharidattini, Cataulacini, Cephalotini, Crematogastrini, Dacetini, Formicoxenini, Melissotarsini, Meranoplini, Metaponini, Myrmecinini, Myrmicariini, Myrmicini, Ochetomyrmecini, Phalacromyrmecini, Pheidolini, Pheidologetonini, Solenopsidini, Stegomyrmecini, Stenammini, Tetramoriini.
Genera of Myrmicinae unplaced to tribe: **Archimyrmex, *Attopsis, *Cephalomyrmex, *Electromyrmex, *Eocenidris, *Eoformica, *Eomyrmex, *Myrmicium, *Promyrmicium.*
Note. Material originally included under the unavailable names Eumyrmicinae, Mycetomyrmicinae and Rhagiomyrmicinae is all referable to Myrmicinae: Bolton, 1994: 106.

**MYRMICINI** (tribe of Myrmicinae)
Myrmicites Lepeletier de Saint-Fargeau, 1835: 169 [family-group name]. Type-genus: *Myrmica.*
Myrmicii: Forel, 1893a: 163 [as tribe of Myrmicinae].
Myrmicini: Ashmead, 1905b: 383 [emended spelling of suffix].
Included genera: *Eutetramorium, Hylomyrma, Manica, Myrmica, Pogonomyrmex.*
Genus dubiously placed here: **Nothomyrmica.*

**MYRMOTERATINI** (tribe of Formicinae)
Myrmoteratii Emery, 1895j: 772 [as tribe of Camponotinae]. Type-genus: *Myrmoteras.*
Myrmoteratini: Ashmead, 1905b: 384 [emended spelling of suffix; as tribe of Gesomyrmicinae].
Myrmoteratini: Wheeler, W.M. 1922a: 692 [as tribe of Formicinae].
Senior synonym of Myrmoteratini Forel: **new synonymy**.
Revision of tribe: Moffett, 1985b: 1.
Included genus: *Myrmoteras.*
Note. Material originally included under the unavailable name Procamponotinae is referable to Myrmoteratini: Bolton, 1994: 51.

MYRMOTERATINI (junior homonym and synonym of the above; described as new independently of Emery, 1895j)
Myrmoteratini Forel, 1912i: 88 [as tribe of Camponotinae]. Type-genus: *Myrmoteras.*
Junior synonym of Myrmoteratini Emery: **new synonymy**.

**NOTHOMYRMECIINAE** (subfamily of Formicidae)
Nothomyrmeciinae Clark, 1951: 16 (in key) [as subfamily of Formicidae]. Type-genus: *Nothomyrmecia.*
Nothomyrmeciini: Brown, 1953j: 1 [as tribe of Myrmeciinae].
Nothomyrmeciinae: Taylor, 1978a: 982 [as subfamily of Formicidae].
Nothomyrmeciini: Wheeler, G.C. & Wheeler, J. 1985: 256 [as tribe of Myrmeciinae].
Nothomyrmeciinae: Taylor & Brown, D.R. 1985: 5; Dlussky & Fedoseeva, 1987: 78; Bolton, 1990c: 1362; Hölldobler & Wilson, 1990: 11; Baroni Urbani, Bolton & Ward, 1992: 317 [as subfamily of Formicidae].

**NOTHOMYRMECIINAE** *(continued)*

Revision of subfamily: Taylor, 1978a: 979.
Included genus: *Nothomyrmecia.*

**OCHETOMYRMECINI** (tribe of Myrmicinae)
Ochetomyrmicini Emery, 1914a: 38 [as tribe of Myrmicinae]. Type-genus: *Ochetomyrmex.*
Ochetomyrmecini: Wheeler, G.C. & Wheeler, J. 1985: 257 [emended spelling].
Review of tribe: Bolton, 1987: 266.
Included genera: *Ochetomyrmex, Tranopelta.*

OCYMYRMICINI (junior synonym of Pheidolini)
Ocymyrmicini Emery, 1914a: 38 [as tribe of Myrmicinae]. Type-genus: *Ocymyrmex.*
Ocymyrmecini: Wheeler, G.C. & Wheeler, J. 1985: 257 [emended spelling].
Junior synonym of Pheidolini: Bolton & Marsh, 1989: 1281.

ODONTOMACHIDAE (junior synonym of Ponerini)
Odontomachidae Mayr, 1862: 708 [as subfamily of Formicidae]. Type-genus: *Odontomachus.*
Odontomachii: Forel, 1893a: 163; Emery, 1895j: 768 [as tribe of Ponerinae].
Odontomachidae: Ashmead, 1905b: 382 [as family].
Odontomachini: Emery, 1911d: 106 [emended spelling of suffix; as tribe of Ponerinae].
Odontomachinae: Clark, 1951: 15 (in key) [as subfamily of Formicidae].
Odontomachiti: Brown, 1976a: 71 [as subtribe of Ponerini], and revision of subtribe.
Odontomachini: Wheeler, G.C. & Wheeler, J. 1985: 256; Dlussky & Fedoseeva, 1987: 78; Hölldobler & Wilson, 1990: 11 [as tribe of Ponerinae].
Junior synonym of Ponerini: Bolton, 1994: 164.

**OECOPHYLLINI** (tribe of Formicinae)
Oecophyllii Emery, 1895j: 772 [as tribe of Camponotinae]. Type-genus: *Oecophylla.*
Oecophyllini: Ashmead, 1905b: 384 [emended spelling of suffix].
Oecophyllini: Wheeler, W.M. 1922a: 693 [as tribe of Formicinae].
Senior synonym of Oecophyllini Forel: **new synonymy**.
Included genus: *Oecophylla.*

OECOPHYLLINI (junior homonym and synonym of the above; described as new independently of Emery, 1895j)
Oecophyllini Forel, 1912i: 89 [as tribe of Camponotinae]. Type-genus: *Oecophylla.*
Junior synonym of Oecophyllini Emery: **new synonymy**.

ONYCHOMYRMICINI (junior synonym of Amblyoponini)
Onychomyrmicini Ashmead, 1905b: 382 [as tribe of Ponerinae]. Type-genus: *Onychomyrmex.*
Onychomyrmicini: Emery, 1911d: 96 [as subtribe of Ponerini].
Junior synonym of Amblyoponini: Bolton, 1994: 164.

ORECTOGNATHITI (junior synonym of Dacetini)
Orectognathiti Brown, 1952g: 10 (footnote); Brown, 1954b: 465 (definition) [as subtribe of Dacetini]. Type-genus: *Orectognathus.*
Junior synonym of Dacetini: Baroni Urbani & De Andrade, 1994: 9.

PACHYCONDYLINAE (junior synonym of Ponerini)
Pachycondylinae Ashmead, 1905b: 382 [as subfamily of Poneridae]. Type-genus: *Pachycondyla.*
Pachycondylini: Emery, 1911d: 59 [as subtribe of Ponerini].
Junior synonym of Ponerini: Bolton, 1994: 164.

PALEOPONERINAE (unavailable name)
Paleoponerinae Forel, 1917: 231 [as "section" of Ponerinae]. Section designated to include tribe Myrmeciini. Unavailable name; not based on genus rank taxon. Contained material referable to Myrmeciini: Bolton, 1994: 73.

***PALEOSMINTHURINAE** (subfamily of Formicidae)
*Paleosminthuridae Pierce & Gibron, 1962: 146 [as family of Collembola]. Type-genus: **Paleosminthurus.*
*Paleosminthuridae: Najt, 1987: 152 [transferred to Formicidae (*incertae sedis*)].
*Paleosminthurinae: Bolton, 1994: 187 [as subfamily of Formicidae].

**PARAPONERINI** (tribe of Ponerinae)
Paraponerii Emery, 1901a: 36 [as tribe of Ponerinae]. Type-genus: *Paraponera.*
Paraponerini: Emery, 1911d: 27 [emended spelling of suffix].
Junior synonym of Ectatommini: Brown, 1958g: 176; Bolton, 1994: 164.
Tribe of Ponerinae: Lattke, 1994: 111.
Included genus: *Paraponera.*

**PHALACROMYRMECINI** (tribe of Myrmicinae)
Phalacromyrmecini Wheeler, G.C. & Wheeler, J. 1985: 258 [as tribe of Myrmicinae]. Type-genus: *Phalacromyrmex.*
Revision of tribe: Bolton, 1984: 373.
Junior synonym of Dacetini: Baroni Urbani & De Andrade, 1994: 10.
Phalacromyrmecini revived from synonymy: **revived status**.
Included genera: *Ishakidris, Phalacromyrmex, Pilotrochus.*

**PHEIDOLINI** (tribe of Myrmicinae)
Pheidolidae Emery, 1877a: 72 [as "group" of Myrmicidae]. Type-genus: *Pheidole.*
Pheidolii: Forel, 1893a: 165 [as tribe of Myrmicinae].
Pheidolini: Emery, 1914a: 35 [emended spelling of suffix].
Senior synonym of
Anergatidini: Bolton, 1994: 106.
Aphaenogastrini: Brown, 1949a: 48.
Lophomyrmicini: implied by Rigato, 1994a: 51 but not stated; **new synonymy**.
Ocymyrmicini: Bolton & Marsh, 1989: 1281.
Included genera: *Aphaenogaster, Chimaeridris, Goniomma, Huberia, Kartidris, Lophomyrmex, Messor, Ocymyrmex, Oxyopomyrmex, *Paraphaenogaster, Pheidole.*
Genus dubiously placed here: **Lonchomyrmex.*

**PHEIDOLOGETONINI** (tribe of Myrmicinae)
Pheidologetini Emery, 1914a: 38 [as tribe of Myrmicinae]. Type-genus: *Pheidologeton.*
Pheidologetonini: Bolton, 1992: 38 [emended spelling.]
Review of tribe: Ettershank, 1966: 115.
Senior synonym of *Hypopomyrmiciti: Brown & Carpenter, 1979: 422 (by implication).
Included genera: *Adlerzia, Afroxyidris, Anisopheidole, Carebara, *Hypopomyrmex, Machomyrma, Oligomyrmex, *Oxyidris, Paedalgus, Pheidologeton.*
Genus dubiously placed here: *Recurvidris.*

*PITYOMYRMECINI (junior synonym of Dolichoderinae)
*Pityomyrmecini Wheeler, W.M. 1915h: 98 [as tribe of Dolichoderinae]. Type-genus: **Pityomyrmex.*
Junior synonym of Dolichoderinae: Shattuck, 1992c: 5.

**PLAGIOLEPIDINI** (tribe of Formicinae)
Plagiolepisii Forel, 1886f: 212 [as tribe of Camponotinae]. Type-genus: *Plagiolepis.*
Plagiolepidii: Emery, 1893i: 635 [emended spelling].
Plagiolepidini: Ashmead, 1905b: 384 [emended spelling of suffix].
Plagiolepidini: Wheeler, W.M. 1922a: 693 [as tribe of Formicinae].
Included genera: *Acropyga, Agraulomyrmex, Anoplolepis, Lepisiota, Plagiolepis, *Rhopalomyrmex, Stigmacros.*

**PLATYTHYREINI** (tribe of Ponerinae)
Platythyrei Emery, 1901a: 36 [as tribe of Ponerinae]. Type-genus: *Platythyrea.*
Platythyreini: Emery, 1911d: 28 [emended spelling of suffix].
Revision of tribe: Brown, 1975: 4.
Included genera: *Platythyrea, Probolomyrmex.*

PLECTROCTENINI (junior synonym of Ponerini)
Plectroctenini Emery, 1911d: 92 [as subtribe of Ponerini]. Type-genus: *Plectroctena.*
Junior synonym of Ponerini: Bolton, 1994: 164.

PODOMYRMINI (junior synonym of Formicoxenini)
Podomyrmini Emery, 1922f: 236 [as subtribe of Myrmecinini]. Type-genus: *Podomyrma.*
Podomyrmini: Dlussky & Fedoseeva, 1987: 79 [as tribe of Myrmicinae].
Junior synonym of Formicoxenini: Bolton, 1994: 105.

POLYRHACHIDINI (junior synonym of Camponotini)
Polyrhachidini Ashmead, 1905b: 384 [as tribe of Camponotinae]. Type-genus: *Polyrhachis.*
Junior synonym of Camponotini: Bolton, 1994: 50.

**PONERINAE** (subfamily of Formicidae)
Ponérites Lepeletier de Saint-Fargeau, 1835: 185 [family-group name]. Type-genus: *Ponera.*
Poneridae: Smith, F. 1851: 6 [emended spelling of suffix].
Poneridae: Mayr, 1862: 64 [as subfamily of Formicidae].
Ponerinae: Forel, 1893a: 162 [emended spelling of suffix].
Poneridae: Bernard, 1951: 1042 [as family of Formicoidea].
Ponerinae: Brown, 1954e: 24 [as subfamily of Formicidae]; Dlussky & Fedoseeva, 1987: 78; Baroni Urbani, Bolton & Ward, 1992: 317.

Included tribes: Amblyoponini, Ectatommini, Paraponerini, Platythyreini, Ponerini, Proceratiini, Thaumatomyrmecini, Typhlomyrmecini.
Note. Material originally included under the unavailable names Proponerinae and Taraxoponerinae is referable to Ponerinae: Bolton, 1994: 164.

**PONERINI** (tribe of Ponerinae)
Ponérites Lepeletier de Saint-Fargeau, 1835: 185 [family-group name]. Type-genus: *Ponera.*
Ponerii: Forel, 1893a: 163; Emery, 1895j: 767 [as tribe of Ponerinae].
Ponerini: Ashmead, 1905b: 382 [emended spelling of suffix].
Senior synonym of
Centromyrmicini: Brown, 1953c: 9.
Dorylozelini: Taylor, 1969: 132.
Leptogenysii: Brown, 1963: 3.
*Archiponerini, Drepanognathini, Euponerinae, Harpegnathii, Odontomachidae, Pachycondylinae, Plectroctenini: Bolton, 1994: 164.
Included genera: *Anochetus, *Archiponera, Asphinctopone, Belonopelta, Centromyrmex, Cryptopone, Diacamma, Dinoponera, Dolioponera, Emeryopone, *Emplastus, Harpegnathos, Hypoponera, Leptogenys, Myopias, Odontomachus, Odontoponera, Pachycondyla, Phrynoponera, Plectroctena, Ponera, *Ponerites, *Poneropsis, *Protopone, Psalidomyrmex, Simopelta, Streblognathus.*
Note. Material originally included under the unavailable name Exeuponerinae is referable to Ponerini: Bolton, 1994: 164.

PRENOLEPIDII (junior synonym of Lasiini)
Prenolepidii Forel, 1912i: 89 [as tribe of Camponotinae]. Type-genus: *Prenolepis.*
Prenolepidini: Emery, 1916b: 237 [emended spelling of suffix].
Prenolepidini: Wheeler, W.M. 1922a: 692 [as tribe of Formicinae].
Junior synonym of Lasiini: Emery, 1925b: 212.

***PRIONOMYRMECINI** (tribe of Myrmeciinae)
*Prionomyrmicini Wheeler, W.M. 1915h: 25 [as tribe of Ponerinae]. Type-genus: **Prionomyrmex.*
*Prionomyrmicini: Brown, 1954e: 22 [as tribe of Myrmeciinae].
*Prionomyrmecini: Bolton, 1994; 73 [emended spelling].
Included genus: **Prionomyrmex.*

PROATTINI (junior synonym of Stenammini)
Proattini Forel, 1917: 232 [as tribe of Myrmicinae]. Type-genus: *Proatta.*
Review of tribe: Moffett, 1987: 444.
Junior synonym of Stenammini: Bolton, 1994: 106.

PROCAMPONOTINAE (unavailable name)
Procamponotinae Forel, 1912i: 88 [as "section" of Camponotinae]. Section designated to include tribe Myrmoteratini. Unavailable name; not based on genus rank taxon. Contained material referable to Myrmoteratini: Bolton, 1994: 51.

**PROCERATIINI** (tribe of Ponerinae)
Proceratii Emery, 1895j: 765 [as tribe of Dorylinae]. Type-genus: *Proceratium.*
Proceratii: Emery, 1901a: 36 [as tribe of Ponerinae].
Proceratiini: Ashmead, 1905b: 382 [emended spelling of suffix].
Junior synonym of Ectatommini: Brown, 1958g: 176; Dlussky & Fedoseeva, 1987: 78; Hölldobler & Wilson, 1990: 10; Bolton, 1994: 164.
Proceratiini: Wheeler, G.C. & Wheeler, J. 1985: 256; Lattke, 1994: 112 [as tribe of Ponerinae].
Senior synonym of Discothyrinae: Lattke, 1994: 112 (implied but not stated).
Included genera: *Discothyea, *Bradoponera, Proceratium.*

PRODOLICHODERINAE (unavailable name)
Prodolichoderinae Forel, 1917: 247 [as "section" of Dolichoderinae]. Section designated to include tribes Aneuretini, Dolichoderini and Leptomyrmicini. Unavailable name; not based on genus rank taxon. Contained material referable to Aneuretinae and Dolichoderinae: Bolton, 1994: 15, 27.

PRODORYLINAE (unavailable name)
Prodorylinae Emery, 1909c: 355 [as "group" of Ponerinae]. Group designated to include tribe Cerapachyini. Unavailable name; not based on genus rank taxon. Contained material referable to Cerapachyinae: Bolton, 1990a: 66.

PROMYRMICINAE (unavailable name)
Promyrmicinae Forel, 1911h: 446 [as "section" of Ponerinae]. Section designated to include tribe Metaponini. Unavailable name; not based on genus rank taxon. Contained material referable to Metaponini: Bolton, 1994: 106.
Promyrmicinae: Forel, 1917: 240 [incorrect expansion of the above unavailable name to include tribes Metaponini and Pseudomyrmini]. Unavailable name.
Promyrmicidae: Bernard, 1951: 1053 [incorrectly treated as family, including only tribe Pseudomyrmini]. Unavailable name.

PROPONERINAE (unavailable name)
Proponerinae Emery, 1909c: 355 [as "group" of Ponerinae]. Group designated to include tribes Amblyoponini and Proceratiini. Unavailable name not based on genus rank taxon. Contained material referable to Ponerinae: Bolton, 1994: 164.

**PSEUDOMYRMECINAE** (subfamily of Formicidae)
Pseudomyrmecinae Smith, M.R. 1952a: 98 [as subfamily of Formicidae]. Type-genus: *Pseudomyrmex.*
Senior synonym of
Leptaleinae: Smith, M.R. 1958c: 112.
Pseudomyrmidae: Ward, 1990: 459.
Revision of subfamily: Ward, 1990: 449.
Included genera: *Myrcidris, Pseudomyrmex, Tetraponera.*

PSEUDOMYRMIDAE (junior synonym of Pseudomyrmecinae)
Pseudomyrmidae Forel, 1885a: 377 [as tribe of Myrmicinae]. Type-genus: *Pseudomyrma* (junior synonym of *Pseudomyrmex*).
Pseudomyrmii: Forel, 1893a: 164 [emended spelling].
Pseudomyrminae: Emery, 1899c: 8 [as subfamily of Formicidae].
Pseudomyrmini: Emery, 1914a: 34; Forel, 1917: 240 [as tribe of Myrmicinae].
Pseudomyrminae: Wheeler, W.M. 1922a: 632 [as subfamily of Formicidae].
Junior synonym of Pseudomyrmecinae: Ward, 1990: 459.

RENEINI (junior synonym of Amblyoponini)
Reneini Donisthorpe, 1947d: 183 [as tribe of Ponerinae]. Type-genus: *Renea* (junior synonym of *Prionopelta*).
Junior synonym of Amblyoponini: Brown, 1953c: 11.

RHAGIOMYRMICINAE (unavailable name)
Rhagiomyrmicinae Forel, 1917: 246 [as "section" of Myrmicinae]. Section designated to include tribes Cataulacini, Cephalotini [=Cryptocerini] and Dacetini. Unavailable name; not based on genus rank taxon. Contained material referable to Myrmicinae: Bolton, 1994: 106.

**SANTSCHIELLINI** (tribe of Formicinae)
Santschiellini Forel, 1917: 232 [as tribe of Camponotinae]. Type-genus: *Santschiella.*
Santschiellini: Wheeler, W.M. 1922a: 694 [as tribe of Formicinae].
Included genus: *Santschiella.*

***SICILOMYRMECINI** (tribe of Formicinae)
*Sicelomyrmicini Wheeler, W.M. 1929a: 12 [as tribe of Formicinae]. Type-genus: **Sicilomyrmex.*
*Sicilomyrmecini: Brown & Carpenter, 1979: 423 [emended spelling].
Included genus: **Sicilomyrmex.*

**SOLENOPSIDINI** (tribe of Myrmicinae)
Solenopsisii Forel, 1893a: 164 [as tribe of Myrmicinae]. Type-genus: *Solenopsis.*
Solenopsidii: Emery, 1895j: 769 [emended spelling].
Solenopsidini: Ashmead, 1905b: 383 [emended spelling of suffix].
Revisions of tribe: Ettershank, 1966: 73; Bolton, 1987: 263.
Senior synonym of
Megalomyrmecini, Monomoriini Emery: Bolton, 1994: 106.
Monomoriini Wheeler: **new synonymy**.
Included genera: *Allomerus, Anillomyrma, Antichthonidris, Bondroitia, Carebarella, Diplomorium, Epelysidris, Megalomyrmex, Monomorium, Nothidris, Oxyepoecus, Phacota, Solenopsis.*

***SPHECOMYRMINAE** (subfamily of Formicidae)
*Sphecomyrminae Wilson & Brown, in Wilson, Carpenter & Brown, 1967: 6 [as subfamily of Formicidae]. Type-genus: **Sphecomyrma.*
*Sphecomyrmidae: Dlussky, 1987: 132 [as family of Formicoidea].
*Sphecomyrminae: Hölldobler & Wilson, 1990: 9; Carpenter, 1992: 490; Bolton, 1994: 184 [as subfamily of Formicidae].
Included genera: **Baikuris, *Cretomyrma, *Dlusskyidris, *Sphecomyrma.*

**STEGOMYRMECINI** (tribe of Myrmicinae)
Stegomyrmicini Wheeler, W.M. 1922a: 668 [as tribe of Myrmicinae]. Type-genus: *Stegomyrmex.*
Stegomyrmecini: Wheeler, G.C. & Wheeler, J. 1985: 258 [emended spelling].
Revision of tribe: Diniz, 1990: 277.
Included genus: *Stegomyrmex.*

**STENAMMINI** (tribe of Myrmicinae)
Stenammini Ashmead, 1905b: 383 [as tribe of Myrmicinae]. Type-genus: *Stenamma.*
Stenammini: Emery, 1921f: 52 [as subtribe of Pheidolini].
Stenammini: Bolton, 1994: 106 [as tribe of Myrmicinae].
Senior synonym of Calyptomyrmecini, Proattini: Bolton, 1994: 106.
Included genera: *Ancyridris, Bariamyrma, Calyptomyrmex, Cyphoidris, Dacatria, Dacetinops, Dicroaspis, Indomyrma, Lachnomyrmex, Lordomyrma, Proatta, Rogeria, Rostromyrmex, Stenamma, Tetheamyrma.*
Genera dubiously included here: *Adelomyrmex, Baracidris, *Ilemomyrmex, Mayriella.*

STEREOMYRMICINI (junior synonym of Formicoxenini)
Stereomyrmicini Emery, 1914a: 36 [as tribe of Myrmicinae]. Type-genus: *Stereomyrmex.*
Stereomyrmecini: Wheeler, G.C. & Wheeler, J. 1985: 257 [emended spelling].
Junior synonym of Formicoxenini: Bolton, 1994: 105.

STICTOPONERINI (junior synonym of Ectatommini)
Stictoponerini Arnol'di, 1930d: 161 [as subtribe of Ectatommini]. Type-genus: *Stictoponera* (junior synonym of *Gnamptogenys*).
Junior synonym of Ectatommini: Bolton, 1994: 164.

STRUMIGENITI (junior synonym of Dacetini)
Strumigeniti Brown, 1952g: 10 (footnote); Brown, 1954b: 465 (definition) [as subtribe of Dacetini]. Type-genus: *Strumigenys.*
Junior synonym of Dacetini: Baroni Urbani & De Andrade, 1994: 9.

TAPINOMINI (junior synonym of Dolichoderinae)
Tapinomini Emery, 1912g: 17 [as tribe of Dolichoderinae]. Type-genus: *Tapinoma.*
Junior synonym of Dolichoderinae: Shattuck, 1992c: 5.

TARAXOPONERINAE (unavailable name)
Taraxoponerinae Forel, 1917: 231 [as "section" of Ponerinae]. Section designated to include tribes Ectatommini, Paraponerini, Platythyreini, Proceratiini and Thaumatomyrmecini. Unavailable name; not based on genus rank taxon. Contained material referable to Ponerinae: Bolton, 1994: 164.

TELEUTOMYRMINI (junior synonym of Tetramoriini)
Teleutomyrmini Kutter, 1950a: 81 [as subtribe of Tetramoriini]. Type-genus: *Teleutomyrmex.*
Junior synonym of Tetramoriini: Bolton, 1994: 106.

**TETRAMORIINI** (tribe of Myrmicinae)
Tetramorii Emery, 1895j: 770 [as tribe of Myrmicinae]. Type-genus: *Tetramorium.*
Tetramoriini: Ashmead, 1905b: 383 [emended spelling of suffix].
Revisions of tribe: Bolton, 1976: 281; Bolton, 1980: 193.
Senior synonym of Anergatini, Teleutomyrmini: Bolton, 1994: 106.
Included genera: *Anergates, Decamorium, Rhoptromyrmex, Secostruma, Strongylognathus, Teleutomyrmex, Tetramorium.*

**THAUMATOMYRMECINI** (tribe of Ponerinae)
Thaumatomyrmii Emery, 1901a: 36 [as tribe of Ponerinae]. Type-genus: *Thaumatomyrmex.*
Thaumatomyrmicini: Emery, 1911d: 48 [emended spelling of suffix].
Thaumatomyrmecini: Wheeler, G.C. & Wheeler, J. 1985: 256 [emended spelling].
Revisions of tribe: Kempf, 1975b: 95; Longino, 1988: 35.
Included genus: *Thaumatomyrmex.*

**TYPHLOMYRMECINI** (tribe of Ponerinae)
Typhlomyrmicini Emery, 1911d: 32 [as subtribe of Ectatommini]. Type-genus: *Typhlomyrmex.*
Typhlomyrmicini: Brown, 1953a: 104 [as tribe of Ponerinae].
Typhlomyrmecini: Brown, 1965c: 65 [emended spelling].
Revision of tribe: Brown, 1965c: 65.
Included genus: *Typhlomyrmex.*

*ZHERICHINIINI (junior synonym of Dolichoderinae)
*Zherichiniini Dlussky, 1988: 56 [as tribe of Dolichoderinae]. Type-genus: **Zherichinius.*
Junior synonym of Dolichoderinae: Shattuck, 1992c: 5.

# Catalogue of Genus-Group Taxa

## Layout of the catalogue

The provisions of the *International Code of Zoological Nomenclature* (Third Edition, 1985) are applied throughout this alphabetically arranged catalogue. Each name is immediately followed by information in parentheses on the current status of, or the subfamily and tribal placement of, the taxon in question. The notes on the current status, including junior synonyms, junior homonyms, subgeneric names, etc., provide a cross-reference within the catalogue of genus-group taxa. Inclusion of subfamily and tribal names, along with the currently valid genus-rank taxa, allows the name to be referred back to the catalogue of family-group taxa. Subsequent entries list references for the following, not all of which, after the second, may be applicable to any single name:

i. The original description, designation, or nomination of the taxon.
ii. The type-species of the taxon, together with synonymic or homonymic data concerning the type-species, where appropriate.
iii. Changes in rank of the taxon, in date order.
iv. Homonymy, or senior or junior synonymy, of the taxon.
v. References to any revision, review, or key concerning the taxon.
vi. List of lower-ranking taxa currently included within a higher-ranking taxon, in alphabetical order and with synonyms and homonyms indicated.

References to synonymies are repeated under both senior and junior names, as these may be widely separated in the catalogue. The earliest reference to a synonymy is always given. Where further references are included they present additional information, confirmation, or other data. Similarly, full data on junior homonyms are given, together with directions to the appropriate replacement names, where some of the information may be repeated. *Nomina nuda* and other taxonomically unavailable names are indicated as such and are followed by references or notes explaining their provenance and dispersal, where the latter is feasible.

Serious misspellings are entered in the catalogue in square brackets, and a note under the correctly spelled entry directs the reader's attention toward the misspelling. Simple typographical errors, which are unlikely to cause confusion, have usually been ignored, but some commonly perpetrated errors are included. Misspellings and deliberate emendations of spellings are differentiated. In general, any secondary spelling that is at variance with the original is treated as a misspelling unless its author has given a definite indication that an emendation in spelling is intended. Such emendations are regarded as justified or unjustified, according to the evidence of each case.

As in the catalogue of family-group names, no attempt has been made to record every occurrence of a name throughout the entire ant literature. Such a process would amass vast lists and enormous numbers of references, the majority of which would not convey any new information. The principal aim here is to provide a concise, useful, and accurate guide, rather than an exhaustive but mostly sterile list. Therefore, references initiating new taxa, synonymy, homonymy, etc., are all presented, but following these only references conveying new, additional, modified, or synoptic information are recorded.

References to reviews, revisions, and keys pertaining to whole genera or recognised subdivisions of genera throughout their known range, and studies of genus-group names applied to an entire zoogeographical region, are included. References to keys for small areas, for instance to a single genus in a single European country, are not normally given, but exceptions are made where such studies include radical new information or where a large genus has only been studied in a limited area. For a list of faunal studies the reader is referred to the introduction.

In general only relatively recent keys (post-1950) have been included as references. This is mainly because earlier keys laboured under the weird and worthless pentanomial system of ant nomenclature and are, by modern standards, clogged with unavailable infrasubspecific names. If early keys retain some value they have been included but are followed by the statement "out of date" where necessary.

The symbol * preceding a name indicates that the taxon is known only from the fossil record. At the initial entry for each genus-group name the following typography is used:

i. ***BOLD ITALIC CAPITALS*** indicate a currently valid name of genus or subgenus rank. To distinguish the two, all names currently accepted at the rank of subgenus are preceded by the symbol #.

ii. *ITALIC CAPITALS* indicate all other names in the genus-group, including junior synonyms, junior homonyms, *nomina nuda*, unavailable names, ichnotaxa, misspellings and unjustified emendations.

## Recent world lists of genera

Brown (1973b) published a list of world genera indicating all the names that he considered to be synonyms. He stressed, however, that these synonymies were provisional, termed them "projected synonymies," and accepted that some would be rejected when detailed analysis was carried out. In the present catalogue these proposals of Brown's are listed as "provisional junior synonym" where each occurs. When further work has either confirmed or rejected the provisional synonymy, the reference for that study is given. Where no additional information has accrued, the Brown entry is followed by the note "(unconfirmed)."

Later lists of genera, produced by Snelling (1981) and Hölldobler & Wilson (1990), used the Brown (1973b) paper as their template, with one important difference: the provisional synonyms of Brown are rendered as definite in both these publications, though none of the synonymies is claimed as new in either list. Both these lists are to a large extent ignored here, being noted only when they present information which is new or radically at variance with the Brown list.

The main result of this action is to deny the synonymy of subgenera in a number of speciose genera, such as *Camponotus* and *Polyrhachis*. Whilst feeling sure that Brown's provisional synonymies will in most cases prove to be justified, I consider that confirmation of synonymy should rest upon analysis or definitive statement, rather than merely upon inclusion in a later list.

The treatment of genus-group synonymies included in post-1973 catalogues (e.g. Smith, D.R. (1979); Taylor & Brown, D.R. (1985)), as opposed to simple lists, is at variance to the above. Synonymies recorded in catalogues are regarded here as confirming or denying the validity of Brown's (1973b) provisional synonyms. Such a decision may seem arbitrary but it is necessary, as catalogues present much more and more detailed information than lists do, which amounts to the making of a definitive statement. For example, Taylor & Brown, D.R. (1985) are considered to have confirmed the synonymy of *Dacryon* with *Podomyrma* by including all the species-group taxa originally assigned to the former under the latter name. Smith, D.R (1979) accepts Brown's synonymy of the subgenera of *Pheidole* by listing all taxa under one heading, but denies the synonymy of the subgenera of *Camponotus* and *Lasius* by cataloguing each subgenus and its contents separately.

Finally, a word must be said about the treatment of genera in the idiosyncratic and very unsatisfactory list given by Wheeler, G.C. & Wheeler, J. (1985). This appears to have been written without regard for, or perhaps in ignorance of, most preceding revisionary studies, although it is quoted as being "up to date according to our views." In the list large numbers of genera are resurrected from synonymy, apparently at whim. Others are returned to tribes from which they had expressly been excluded by earlier revisors, and still others are given as "*incertae sedis*" when their taxonomic position has previously been worked out. The number of such unjustified emendations, especially in the Myrmicinae, is regrettably large. In view of all these anomalies and anachronisms, I have chosen to omit this publication from reference in the generic catalogue which follows, whilst drawing attention to its existence here.

## The catalogue of genus-group taxa

*ACALAMA* (junior synonym of *Gauromyrmex*)

*Acalama* Smith, M.R. 1949a: 206. Type-species: *Acalama donisthorpei* (junior synonym of *Vollenhovia acanthinus*), by original designation. *Acalama* junior synonym of *Gauromyrmex:* Brown, 1953c: 10. [*Gauromyrmex* provisional junior synonym of *Vollenhovia:* Brown, 1973b: 178; confirmed by Bolton, 1994: 105].

*ACAMATUS* (junior homonym, see *Neivamyrmex*)

*Acamatus* Emery, 1894c: 181 [as subgenus of *Eciton*]. Type-species: *Eciton* (*Acamatus*) *schmitti* (junior synonym of *Neivamyrmex nigrescens*), by subsequent designation of Wheeler, W.M. 1911f: 157. [Junior homonym of *Acamatus* Schoenherr, 1833: 20 (Coleoptera).] *Neivamyrmex* Borgmeier, 1940: 606, replacement name for *Acamatus* Emery.

*ACANTHIDRIS* (junior synonym of *Rhopalothrix*)

*Acanthidris* Weber, 1941a: 188. Type-species: *Acanthidris isthmicus*, by original designation. *Acanthidris* junior synonym of *Rhopalothrix:* Brown & Kempf, 1960: 230.

*ACANTHOCLINEA* (junior synonym of *Dolichoderus*)

*Acanthoclinea* Wheeler, W.M. 1935c: 69 [as subgenus of *Dolichoderus*]. Type-species: *Dolichoderus doriae*, by original designation. *Acanthoclinea* raised to genus: Brown, 1950e: 249. *Acanthoclinea* provisional junior synonym of *Dolichoderus:* Brown, 1973b: 178. *Acanthoclinea* junior synonym of *Dolichoderus:* Taylor & Brown, D.R. 1985: 93; Shattuck, 1992c: 66.

***ACANTHOGNATHUS*** (Myrmicinae: Dacetini)

*Acanthognathus* Mayr, 1887: 578. Type-species: *Acanthognathus ocellatus*, by monotypy. Revision of genus: Brown & Kempf, 1969: 87.

*ACANTHOLEPIS* (junior homonym, see *Lepisiota*)

*Acantholepis* Mayr, 1861: 42. Type-species: *Hypoclinea frauenfeldi*, by monotypy. [Junior homonym of *Acantholepis* Kroyer, 1846: 98 (Pisces)]. *Lepisiota* Santschi, 1926a: 15; previously subgenus of *Acantholepis* Mayr, first available replacement name. [*Achantilepis:* see below.]

***ACANTHOMYOPS*** (Formicinae: Lasiini)

*Acanthomyops* Mayr, 1862: 699. Type-species: *Formica clavigera*, by monotypy. *Acanthomyops* junior synonym of *Lasius:* Mayr, 1866b: 888. *Acanthomyops* revived from synonymy as subgenus of *Lasius:* Emery, 1893i: 642. *Acanthomyops* incorrectly as senior synonym of *Lasius:* Forel, 1916: 460; Donisthorpe, 1916c: 276. *Acanthomyops* subgenus of *Formicina:* Forel, 1917: 249. *Acanthomyops* subgenus of *Lasius:* Wheeler, W.M. 1922a: 698; Emery, 1925b: 226. *Acanthomyops* revived status as genus: Creighton, 1950a: 426. Key to queens: Buren, 1950: 188. Revision of genus: Wing, 1968: 3.

***ACANTHOMYRMEX*** (Myrmicinae: Myrmecinini)

*Acanthomyrmex* Emery, 1893a: cclxxvi. Type-species: *Acanthomyrmex luciolae*, by subsequent designation of Bingham, 1903: 191. [*Acanthomyrmex* also described as new by Emery, 1893f: 244.] Revision of genus: Moffett, 1986c: 55.

***ACANTHOPONERA*** (Ponerinae: Ectatommini)

*Acanthoponera* Mayr, 1862: 732 [as subgenus of *Ectatomma*]. Type-species: *Ponera mucronata*, by subsequent designation of Wheeler, W.M. 1911f: 158. *Acanthoponera* raised to genus: Emery, 1895g: 346; Emery, 1911d: 35. Revision of genus: Brown, 1958g: 188. See also: Lattke, 1994: 110.

***ACANTHOSTICHUS*** (Cerapachyinae: Acanthostichini)

*Acanthostichus* Mayr, 1887: 549. Type-species: *Typhlopone serratula*, by monotypy. Key to species: Wheeler, W.M. 1934g: 162; see also Kusnezov, 1962a: 121; Kempf, 1964c: 263. Review of genus: Brown, 1975: 41. *Acanthostichus* senior synonym of *Ctenopyga:* MacKay, in preparation.

[*ACHANTILEPIS* (misspelling)
*Achantilepis* Santschi, 1935b: 274, incorrect subsequent spelling of *Acantholepis*.]

*ACIDOMYRMEX* (junior synonym of *Rhoptromyrmex*)
*Acidomyrmex* Emery, 1915i: 191 [as subgenus of *Rhoptromyrmex*]. Type-species: *Rhoptromyrmex wroughtonii*, by original designation. *Acidomyrmex* raised to genus: Wheeler, W.M. 1922a: 664. *Acidomyrmex* junior synonym of *Rhoptromyrmex*: Brown, 1964c: 11; see also Bolton, 1986a: 2.

*ACROCOELIA* (junior synonym of *Crematogaster*)
*Acrocoelia* Mayr, 1853a: 147. Type-species: *Acrocoelia ruficeps* (junior synonym of *Crematogaster scutellaris*), by subsequent designation of Wheeler, W.M. 1911f: 158. *Acrocoelia* junior synonym of *Crematogaster*: Mayr, 1863: 404; Wheeler, W.M. 1911f: 158; Wheeler, W.M. 1922a: 828. *Acrocoelia* revived from synonymy as subgenus of *Crematogaster*: Emery, 1922e: 140. *Acrocoelia* junior synonym of *Crematogaster*: Buren, 1959: 126. *Acrocoelia* revived from synonymy as genus: Soulié, 1964: 398. *Acrocoelia* junior synonym of *Crematogaster*: Brown, 1973b: 178. [The type-species of *Acrocoelia* and *Crematogaster* are synonymous, the generic synonymy is therefore absolute.]

***ACROMYRMEX*** (Myrmicinae: Attini)
*Acromyrmex* Mayr, 1865: 83 [as subgenus of *Atta*]. Type-species: *Formica hystrix*, by monotypy. *Acromyrmex* raised to genus: Emery, 1913: 251. Key to species: Santschi, 1925a: 388 (out of date). Revision of Brazilian fauna: Gonçalves, 1961: 113. Partial revision of genus: Fowler, 1988: 281. Bibliography: Cherrett & Cherrett, 1989: 1. Subgenera of *Acromyrmex*: nominal plus *Moellerius*.

***ACROPYGA*** (Formicinae: Plagiolepidini)
*Acropyga* Roger, 1862a: 242. Type-species: *Acropyga acutiventris*, by monotypy. Key to Neotropical species: Weber, 1944: 91. Subgenera of *Acropyga*: nominal plus *Atopodon*, *Malacomyrma*, *Rhizomyrma*.

**ACROSTIGMA* (junior synonym of *Podomyrma*)
**Acrostigma* Emery, 1891a: 149 [as subgenus of *Podomyrma*]. Type-species: **Podomyrma* (*Acrostigma*) *mayri*, by monotypy. **Acrostigma* junior synonym of *Podomyrma*: Dalla Torre, 1893: 159.

*ACROSTIGMA* (junior homonym, see *Stigmacros*)
*Acrostigma* Forel, 1902h: 477 [as subgenus of *Acantholepis*]. Type-species: *Acantholepis* (*Acrostigma*) *froggatti*, by subsequent designation of Wheeler, W.M. 1911f: 158. [Junior homonym of **Acrostigma* Emery, 1891a: 149, above.] *Stigmacros* Forel, 1905b: 179, replacement name for *Acrostigma* Forel.

***ADELOMYRMEX*** (Myrmicinae: Stenammini?)
*Adelomyrmex* Emery, 1897c: 590. Type-species: *Adelomyrmex biroi*, by monotypy. *Adelomyrmex* senior synonym of *Apsychomyrmex*: Kempf, 1972a: 18. *Adelomyrmex* provisional senior synonym of *Arctomyrmex*: Brown, 1973b: 178; confirmed Bolton, 1994: 106.

***ADETOMYRMA*** (Ponerinae: Amblyoponini)
*Adetomyrma* Ward, 1994: 160. Type-species: *Adetomyrma venatrix*, by original designation.

*ADFORMICA* (junior synonym of *Coptoformica*)
*Adformica* Lomnicki, 1925a: 164 [as subgenus of *Formica*]. Type-species: *Formica exsecta*, by subsequent designation of Donisthorpe, 1927b: 316. *Adformica* junior synonym of *Coptoformica*: Stitz, 1939: 306. [*Coptoformica* junior synonym of *Formica*: Smith, D.R. 1979: 1448.]

***ADLERZIA*** (Myrmicinae: Pheidologetonini)
*Adlerzia* Forel, 1902h: 445 [as subgenus of *Monomorium*]. Type-species: *Monomorium* (*Adlerzia*) *froggatti*, by monotypy. *Adlerzia* raised to genus: Brown, 1952j: 176. *Adlerzia* senior synonym of *Stenothorax*: Brown, 1952a: 110.

***AENICTOGITON*** (Aenictogitoninae)
*Aenictogiton* Emery, 1901d: 49. Type-species: *Aenictogiton fossiceps*, by monotypy. Key to species: Santschi, 1924b: 200. Review of genus: Brown, 1975: 43.

***AENICTUS*** (Aenictinae)
*Aenictus* Shuckard, 1840b: 266. Type-species: *Aenictus ambiguus*, by original designation. *Aenictus* senior synonym of *Typhlatta*: Forel, 1890b: ciii. *Aenictus* senior synonym of *Paraenictus*: Wilson, 1964a: 444. Key to Indo-Australian species: Wilson, 1964a: 436; see also Gotwald, 1982: 157. [*Enictus* Walker, 1860: 306; Smith, F. 1865: 79; incorrect subsequent spellings.]

*AEROMYRMA* (junior synonym of *Oligomyrmex*)
*Aeromyrma* Forel, 1891b: 198. Type-species: *Aeromyrma nosindambo*, by monotypy. [*Aeromyrma* also described as new by Forel, 1891a: cccvii; no species-rank taxa named.] *Aeromyrma* subgenus of *Oligomyrmex*: Emery, 1915c: 59 (footnote). *Aeromyrma* revived status as genus: Arnold, 1916: 256; maintained as genus: Wheeler, W.M. 1922a: 882. *Aeromyrma* subgenus of *Oligomyrmex*: Emery, 1922f: 215. *Aeromyrma* junior synonym of *Oligomyrmex*. Ettershank, 1966: 119. Key to Afrotropical species: Weber, 1952: 11.

*AETHIOPOPONE* (junior synonym of *Sphinctomyrmex*)
*Aethiopopone* Santschi, 1930a: 49. Type-species: *Sphinctomyrmex rufiventris*, by monotypy. *Aethiopopone* junior synonym of *Sphinctomyrmex*: Bolton, 1973a: 341; Brown, 1975: 31.

***AFROXYIDRIS*** (Myrmicinae: Pheidologetonini)
*Afroxyidris* Belshaw & Bolton, 1994: 631. Type-species: *Afroxyidris crigensis*, by original designation.

***AGRAULOMYRMEX*** (Formicinae: Plagiolepidini)
*Agraulomyrmex* Prins, 1983: 2. Type-species: *Agraulomyrmex meridionalis*, by original designation.

****AGROECOMYRMEX*** (Myrmicinae: Agroecomyrmecini)
**Agroecomyrmex* Wheeler, W.M. 1915h: 56. Type-species: **Myrmica duisburgi*, by original designation.

#***ALAOPONE*** (subgenus of *Dorylus*)
*Alaopone* Emery, 1881a: 274. Type-species: *Alaopone oberthueri* (junior synonym of *Dorylus orientalis*), by subsequent designation of Emery, 1910b: 15. *Alaopone* junior synonym of *Dorylus*: Dalla Torre, 1893: 9. *Alaopone* revived from synonymy as subgenus of *Dorylus*: Emery, 1895i: 702. *Alaopone* senior synonym of *Shuckardia*: Emery, 1910b: 14. *Alaopone* provisional junior synonym of *Dorylus*: Brown, 1973b: 178 (unconfirmed).

*ALFARIA* (junior synonym of *Gnamptogenys*)
*Alfaria* Emery, 1896e: 177 (diagnosis in key). Type-species: *Alfaria simulans*, by original designation. [*Alfaria* also described as new by Emery, 1896g: 41.] *Alfaria* senior synonym of *Opisthoscyphus*: Brown, in Borgmeier, 1957: 115. *Alfaria* junior synonym of *Gnamptogenys*: Brown, 1958g: 211.

*ALISTRUMA* (junior synonym of *Colobostruma*)
*Alistruma* Brown, 1948e: 117. Type-species: *Epopostruma foliacea*, by original designation. *Alistruma* junior synonym of *Colobostruma*: Brown, 1959c: 1.

***ALLOFORMICA*** (Formicinae: Formicini)
*Alloformica* Dlussky, 1969a: 219 [as subgenus of *Proformica*]. Type-species: *Formica aberrans*, by original designation. *Alloformica* provisional junior synonym of *Proformica*: Brown, 1973b: 178. *Alloformica* raised to genus: Dlussky & Fedoseeva, 1987: 77; maintained as genus: Agosti, 1991: 259; Agosti, 1994a: 99.

***ALLOMERUS*** (Myrmicinae: Solenopsidini)
*Allomerus* Mayr, 1878: 873. Type-species: *Allomerus decemarticulatus*, by subsequent designation of Wheeler, W.M. 1911f: 158. Reviews of genus: Ettershank, 1966: 111; Bolton, 1987: 282.

*ALLOPHEIDOLE* (junior synonym of *Pheidole*)
*Allopheidole* Forel, 1912f: 237 [as subgenus of *Pheidole*]. Type-species: *Pheidole kingi* (junior synonym of *Pheidole tepicana*), by subsequent designation of Wheeler, W.M. 1913a: 79. *Allopheidole* junior synonym of *Pheidole*: Emery, 1921f: 84.

*AMAUROMYRMEX* (junior synonym of *Pheidologeton*)
*Amauromyrmex* Wheeler, W.M. 1929b: 1. Type-species: *Amauromyrmex speculifrons* (junior synonym of *Pheidologeton silenus*), by original designation. *Amauromyrmex* junior synonym of *Pheidologeton*: Ettershank, 1966: 115.

***AMBLYOPONE*** (Ponerinae: Amblyoponini)
*Amblyopone* Erichson, 1842: 260. Type-species: *Amblyopone australis*, by monotypy. *Amblyopone* senior synonym of *Stigmatomma*: Emery & Forel, 1879: 455; Mayr, 1887: 546. *Amblyopone* senior synonym of *Arotropus*: Provancher, 1887: 240. *Amblyopone* senior synonym of *Amblyopopone*: Forel, 1893a: 166; Emery, 1911d: 25. *Amblyopone* senior synonym of *Neoamblyopone* and *Protamblyopone*: Brown, 1949c: 87. *Amblyopone* senior synonym of *Arotropus*, *Ericapelta*, *Lithomyrmex* Clark, *Neoamblyopone*, *Protamblyopone*, *Stigmatomma* (and its junior synonyms *Fulakora* and *Xymmer*): Brown, 1960a: 155. *Amblyopone* provisional senior synonym of *Paraprionopelta*: Brown, 1973b: 183 (unconfirmed). *Amblyopone* senior synonym of *Amblyopopona*: Bolton, 1994: 164. Revision of genus: Brown, 1960a: 155. Key to Mediterranean region: Baroni Urbani, 1978a: 41; to Neotropical: Lattke, 1991b: 6

*AMBLYOPOPONA* (junior synonym of *Amblyopone*)
*Amblyopopona* Schulz, 1906: 154, unjustified emendation of *Amblyopone*. *Amblyopopona* junior synonym of *Amblyopone*: Bolton, 1994: 164.

*AMBLYOPOPONE* (junior synonym of *Amblyopone*)
*Amblyopopone* Dalla Torre, 1893: 13, unjustified emendation of *Amblyopone*. *Amblyopopone* junior synonym of *Amblyopone*: Forel, 1893: 166.

****AMEGHINOIA*** (Myrmeciinae: unplaced to tribe)
**Ameghinoia* Viana & Haedo Rossi, 1957: 109. Type-species: **Ameghinoia piatnitzkyi*, by original designation.

*AMMOMYRMA* (junior synonym of *Araucomyrmex*)
*Ammomyrma* Santschi, 1922b: 365 [as subgenus of *Dorymyrmex*]. Type-species: *Dorymyrmex exsanguis*, by original designation. *Ammomyrma* junior synonym of *Araucomyrmex*: Snelling & Hunt, 1976: 93. [*Araucomyrmex* junior synonym of *Conomyrma*: Snelling, 1981: 402; *Conomyrma* junior synonym of *Dorymyrmex*: Shattuck, 1992c: 77.]
*AMYRMEX* (junior synonym of *Forelius*)
*Amyrmex* Kusnezov, 1953b: 333. Type-species: *Amyrmex golbachi*, by original designation. *Amyrmex* junior synonym of *Forelius*: Shattuck, 1992c: 87.
#***ANACANTHOLEPIS*** (subgenus of *Plagiolepis*)
*Anacantholepis* Santschi, 1914e: 36 [as subgenus of *Plagiolepis*]. Type-species: *Plagiolepis (Anacantholepis) decora*, by original designation. *Anacantholepis* provisional junior synonym of *Plagiolepis*: Brown, 1973b: 178 (unconfirmed).
*ANACANTHOPONERA* (junior synonym of *Heteroponera*)
*Anacanthoponera* Wheeler, W.M. 1923e: 176 [as subgenus of *Acanthoponera*]. Type-species: *Ponera dolo*, by original designation. *Anacanthoponera* junior synonym of *Heteroponera*: Brown, 1952h: 70.
*ANCYLOGNATHUS* (*nomen nudum*)
*Ancylognathus* Lund, 1831a: 121, 135. Type-species: *Ancylognathus lugubris*, *nomen nudum*. *Ancylognathus* material referred to *Eciton*: Smith, F. 1855c: 160.
***ANCYRIDRIS*** (Myrmicinae: Stenammini)
*Ancyridris* Wheeler, W.M. 1935a: 1. Type-species: *Ancyridris polyrhachioides*, by original designation. *Ancyridris* provisional junior synonym of *Lordomyrma*: Brown, 1973b: 178 (unconfirmed).
*ANDRAGNATHUS* (junior synonym of *Paratrechina*)
*Andragnathus* Emery, 1922d: 111. Type-species: *Andragnathus hubrechti*, by monotypy. *Andragnathus* junior synonym of *Paratrechina*: Agosti & Bolton, 1990a: 75.
*ANELEUS* (junior synonym of *Oligomyrmex*)
*Aneleus* Emery, 1900c: 327 [as subgenus of *Pheidologeton*]. Type-species: *Solenopsis similis*, by subsequent designation of Wheeler, W.M. 1911f: 158. [Type-species not *Pheidologeton pygmaeus*, unjustified subsequent designation by Wheeler, W.M. 1913a: 77, repeated in Emery, 1922f: 214.] *Aneleus* raised to genus: Emery, 1914a: 41; maintained as genus: Arnold, 1916: 254; Forel, 1917: 243; Wheeler, W.M. 1922a: 663; Emery, 1922f: 213. *Aneleus* senior synonym of *Sporocleptes*: Consani, 1951: 169; Arnold, 1952a: 460. *Aneleus* junior synonym of *Oligomyrmex*: Ettershank, 1966: 119.
***ANERGATES*** (Myrmicinae: Tetramoriini)
*Anergates* Forel, 1874: 67 (see also p. 93). Type-species: *Myrmica atratula*, by monotypy. Reviews of genus: Ettershank, 1966: 157; Bolton, 1976: 296.
*ANERGATIDES* (junior synonym of *Pheidole*)
*Anergatides* Wasmann, 1915b: 281. Type-species: *Anergatides kohli* (junior secondary homonym in *Pheidole*, replaced by *Pheidole neokohli*), by monotypy. *Anergatides* provisional junior synonym of *Pheidole*: Brown, 1973b: 178. *Anergatides* junior synonym of *Pheidole*: Smith, D.R. 1979: 1366; Wilson, 1984: 328.
****ANEURETELLUS*** (Aneuretinae)
**Aneuretellus* Dlussky, 1988: 54. Type-species: **Aneuretellus deformis*, by original designation.
***ANEURETUS*** (Aneuretinae)
*Aneuretus* Emery, 1893a: cclxxv. Type-species: *Aneuretus simoni*, by monotypy. [*Aneuretus* also described as new by Emery, 1893f: 241.] Review of genus: Wilson, Eisner, Wheeler & Wheeler, 1956: 81.
***ANILLIDRIS*** (Dolichoderinae)
*Anillidris* Santschi, 1936d: 414. Type-species: *Anillidris bruchi*, by original designation. *Anillidris* junior synonym of *Linepithema*: Kusnezov, 1958c: 273; Kusnezov, 1959: 50. *Anillidris* provisionally revived from synonymy: Brown, 1973b: 169. *Anillidris* revived from synonymy, and review of genus: Shattuck, 1992c: 33.
***ANILLOMYRMA*** (Myrmicinae: Solenopsidini)
*Anillomyrma* Emery, 1913: 261 [as subgenus of *Monomorium*]. Type-species: *Monomorium decamerum*, by monotypy. *Anillomyrma* raised to genus, and review of genus: Ettershank, 1966: 97. Revision of genus: Bolton, 1987: 274.
***ANISOPHEIDOLE*** (Myrmicinae: Pheidologetonini)
*Anisopheidole* Forel, 1914c: 616 (footnote) [as subgenus of *Pheidole*]. Type-species: *Pheidole froggatti* (junior synonym of *Anisopheidole antipodum*), by monotypy. [*Anisopheidole* also described as new by Forel, 1915b: 61.] *Anisopheidole* raised to genus, and review of genus: Ettershank, 1966: 129.
***ANKYLOMYRMA*** (Myrmicinae: Formicoxenini)
*Ankylomyrma* Bolton, 1973c: 235. Type-species: *Ankylomyrma coronacantha*, by original designation. Review of genus: Bolton, 1981b: 247.
***ANOCHETUS*** (Ponerinae: Ponerini)
*Anochetus* Mayr, 1861: 53. Type-species: *Odontomachus ghilianii*, by monotypy. *Anochetus* senior synonym of *Stenomyrmex*: Forel, 1887: 382. *Anochetus* senior synonym of *Myrmapatetes*: Brown, 1953h: 2. *Anochetus* provisional junior synonym of *Odontomachus*: Brown, 1973b: 178. *Anochetus* valid genus, senior synonym of *Stenomyrmex*, and revision of genus: Brown, 1978c: 549.
***ANOMALOMYRMA*** (Leptanillinae: Anomalomyrmini)
*Anomalomyrma* Taylor, in Bolton, 1990b: 278. Type-species: *Anomalomyrma taylori*, by original designation.
#***ANOMMA*** (subgenus of *Dorylus*)
*Anomma* Shuckard, 1840c: 326. Type-species: *Anomma burmeisteri*, by original designation. *Anomma* congeneric with *Dorylus*: Haldeman, 1849a: 201. *Anomma* senior synonym of *Sphegomyrmex*: Smith, F. 1858b: 112. *Anomma* senior synonym of *Sphecomyrmex*: **new synonymy**. *Anomma* subgenus of *Dorylus*: Emery, 1895j: 701. *Anomma* provisionally raised to genus: Brown, 1973b: 178 (unconfirmed); reverts here to subgenus of *Dorylus*.
***ANONYCHOMYRMA*** (Dolichoderinae)
*Anonychomyrma* Donisthorpe, 1947c: 588. Type-species: *Anonychomyrma myrmex*, by monotypy. *Anonychomyrma* provisional junior synonym of *Iridomyrmex*: Brown, 1973b: 178. *Anonychomyrma* valid genus: Shattuck, 1992a: 13. Review of genus: Shattuck, 1992c: 37. See also: Shattuck, 1994: 3.
***ANOPLOLEPIS*** (Formicinae: Plagiolepidini)
*Anoplolepis* Santschi, 1914b: 123 [as subgenus of *Plagiolepis*]. Type-species: *Formica longipes* (junior primary homonym in *Formica*, *Anoplolepis gracilipes* first available replacement name), by monotypy. *Anoplolepis* raised to genus: Emery, 1925b: 16. Partial revision of genus: Prins, 1982: 215. Subgenera of *Anoplolepis*: nominal plus *Mesanoplolepis*, *Tapinolepis*, *Zealleyella*.
#***ANOPLOMYRMA*** (subgenus of *Polyrhachis*)
*Anoplomyrma* Chapman, 1963: 258 [as subgenus of *Polyrhachis*]. Type-species: *Polyrhachis (Anoplomyrma) parabiotica*, by monotypy. *Anoplomyrma* provisional junior synonym of *Polyrhachis*: Brown, 1973b: 178 (unconfirmed).
***ANTICHTHONIDRIS*** (Myrmicinae: Solenopsidini)
*Antichthonidris* Snelling, 1975: 5. Type-species: *Monomorium denticulatum*, by original designation. Review of genus: Bolton, 1987: 283.
*ANTILLAEMYRMEX* (junior synonym of *Macromischa*)
*Antillaemyrmex* Mann, 1920: 408 [as subgenus of *Macromischa*]. Type-species: *Macromischa (Antillaemyrmex) terricola*, by original designation. *Antillaemyrmex* raised to genus: Wheeler, W.M. 1931b: 4 (in text). *Antillaemyrmex* junior synonym of *Macromischa*: Smith, M.R. 1937: 849; Smith, M.R. 1939e: 502. [*Macromischa* junior synonym of *Leptothorax*: Snelling, 1986: 154.]
*APEDUNCULATA* (junior synonym of *Pseudomyrmex*)
*Apedunculata* Enzmann, E.V. 1944: 62 [as "branch" of *Pseudomyrma*]. Type-species: *Pseudomyrma sericea*, by original designation. *Apedunculata* junior synonym of *Pseudomyrmex*: Ward, 1990: 469.
***APHAENOGASTER*** (Myrmicinae: Pheidolini)
*Aphaenogaster* Mayr, 1853b: 107. Type-species: *Aphaenogaster sardoa*, by subsequent designation of Bingham, 1903: 270. *Aphaenogaster* subgenus of *Stenamma*: Emery, 1895c: 298. *Aphaenogaster* revived status as genus: Emery, 1908c: 309. *Aphaenogaster* provisional senior synonym of *Attomyrma*, *Deromyrma*, *Nystalomyrma* and *Planimyrma*: Brown, 1973b: 178, 180, 183, 184 respectively. *Aphaenogaster* senior synonym of *Novomessor*: Brown, 1974b: 47. *Aphaenogaster* senior synonym of *Attomyrma*, *Deromyrma*, *Novomessor*, *Nystalomyrma*, *Planimyrma* and *Brunella*: Bolton, 1982: 364.
*APHANTOLEPIS* (junior synonym of *Technomyrmex*)
*Aphantolepis* Wheeler, W.M. 1930d: 44. Type-species: *Aphantolepis quadricolor*, by monotypy. *Aphantolepis* junior synonym of *Technomyrmex*: Brown, 1953h: 5.
***APHOMOMYRMEX*** (Formicinae: Brachymyrmecini)
*Aphomomyrmex* Emery, 1899e: 493. Type-species: *Aphomomyrmex afer*, by subsequent designation of Wheeler, W.M. 1911f: 159. Review of genus: Snelling, 1979b: 3. [*Aphomyrmex*: see below.]
[*APHOMYRMEX* (misspelling)
*Aphomyrmex* Ashmead, 1905a: 11, incorrect subsequent spelling of *Aphomomyrmex*.]
***APOMYRMA*** (Apomyrminae)
*Apomyrma* Brown, Gotwald & Lévieux, 1971: 259. Type-species: *Apomyrma stygia*, by original designation.
*APOROMYRMEX* (junior synonym of *Plagiolepis*)
*Aporomyrmex* Faber, 1969: 52. Type-species: *Aporomyrmex ampeloni*, by original designation. *Aporomyrmex* provisional junior synonym of

*Plagiolepis:* Brown, 1973b: 178; confirmed Bolton, 1994: 51.

*APSYCHOMYRMEX* (junior synonym of *Adelomyrmex*)

*Apsychomyrmex* Wheeler, W.M. 1910a: 261. Type-species: *Apsychomyrmex myops,* by monotypy. Review of genus: Smith, M.R. 1947b: 468. *Apsychomyrmex* junior synonym of *Adelomyrmex:* Kempf, 1972a: 18.

#**APTEROCREMA** (subgenus of *Crematogaster*)

*Apterocrema* Wheeler, W.M. 1936c: 45 [as subgenus of *Crematogaster*]. Type-species: *Crematogaster (Apterocrema) atitlanica,* by monotypy. *Apterocrema* provisional junior synonym of *Crematogaster:* Brown, 1973b: 178 (unconfirmed).

**APTEROSTIGMA** (Myrmicinae: Attini)

*Apterostigma* Mayr, 1865: 111. Type-species: *Apterostigma pilosum,* by monotypy.

*ARATROMYRMEX* (junior synonym of *Vollenhovia*)

*Aratromyrmex* Stitz, 1938: 105. Type-species: *Aratromyrmex luctuosus,* by subsequent designation of Donisthorpe, 1943f: 625. *Aratromyrmex* provisional junior synonym of *Liomyrmex:* Brown, 1973b: 178. *Aratromyrmex* junior synonym of *Vollenhovia:* Baroni Urbani, 1980a: 96.

*ARAUCOMYRMEX* (junior synonym of *Conomyrma*)

*Araucomyrmex* Gallardo, 1919b: 249. Type-species: *Dorymyrmex tener,* by original designation. *Araucomyrmex* subgenus of *Conomyrma:* Santschi, 1922b: 365. *Araucomyrmex* subgenus of *Dorymyrmex:* Kusnezov, 1952g: 428. *Araucomyrmex* revived status as genus: Kusnezov, 1959: 51; maintained as genus: Kempf, 1972a: 25. *Araucomyrmex* provisional junior synonym of *Dorymyrmex:* Brown, 1973b: 178. *Araucomyrmex* maintained as genus: Snelling, 1975: 9. *Araucomyrmex* senior synonym of *Ammomyrma:* Snelling & Hunt, 1976: 93. *Araucomyrmex* junior synonym of *Conomyrma:* Snelling, 1981: 402. [*Conomyrma* junior synonym of *Dorymyrmex:* Shattuck, 1992c: 77.]

*ARCHAEOMYRMEX* (junior synonym of *Myrmecina*)

*Archaeomyrmex* Mann, 1921: 448. Type-species: *Archaeomyrmex cacabau,* by original designation. *Archaeomyrmex* junior synonym of *Myrmecina:* Brown, 1971a: 1.

***ARCHAEOPONE** (*Armaniinae)

**Archaeopone* Dlussky, 1975: 120. Type-species: **Archaeopone kzylzharica,* by original designation. **Archaeopone* transferred from Ponerinae to *Armaniidae: Dlussky, 1983: 66. **Archaeopone* junior synonym of **Sphecomyrma:* Wilson, 1987: 49; Hölldobler & Wilson, 1990: 9. **Archaeopone* revived from synonymy: Dlussky & Fedoseeva, 1987: 77; Carpenter, 1992: 491.

*ARCHEATTA* (junior synonym of *Atta*)

*Archeatta* Gonçalves, 1942: 342 [as subgenus of *Atta*]. Type-species: *Oecodoma mexicana,* by original designation. *Archeatta* junior synonym of *Atta:* Smith, M.R. 1951a: 832; Weber, 1958a: 8. *Archeatta* revived from synonymy as subgenus of *Atta,* and revision of subgenus: Borgmeier, 1959b: 350. *Archeatta* provisional junior synonym of *Atta:* Brown, 1973b: 178; confirmed Bolton, 1994: 105.

***ARCHIMYRMEX** (Myrmicinae: unplaced to tribe)

**Archimyrmex* Cockerell, 1923a: 52. Type-species: **Archimyrmex rostratus,* by monotypy. **Archimyrmex* transferred from Ponerinae to Myrmicinae: Carpenter, 1930: 16.

***ARCHIPONERA** (Ponerinae: Ponerini)

**Archiponera* Carpenter, 1930: 27. Type-species: **Archiponera wheeleri,* by original designation.

*ARCTOMYRMEX* (junior synonym of *Adelomyrmex*)

*Arctomyrmex* Mann, 1921: 457 [as subgenus of *Adelomyrmex*]. Type-species: *Adelomyrmex (Arctomyrmex) hirsutus,* by original designation. *Arctomyrmex* provisional junior synonym of *Adelomyrmex:* Brown, 1973b: 178; confirmed Bolton, 1994: 106.

***ARMANIA** (*Armaniinae)

**Armania* Dlussky, 1983: 67. Type-species: **Armania robusta,* by original designation. **Armania* junior synonym of **Sphecomyrma:* Wilson, 1987: 49; Hölldobler & Wilson, 1990: 9. **Armania* revived from synonymy: Dlussky & Fedoseeva, 1987: 77; Carpenter, 1992: 491.

***ARMANIELLA** (*Armaniinae)

**Armaniella* Dlussky, 1983: 71. Type-species: **Armaniella curiosa,* by original designation. **Armaniella* junior synonym of **Sphecomyrma:* Wilson, 1987: 49; Hölldobler & Wilson, 1990: 9. **Armaniella* revived from synonymy: Dlussky & Fedoseeva, 1987: 77; Carpenter, 1992: 491.

*ARNOLDIDRIS* (junior synonym of *Orectognathus*)

*Arnoldidris* Brown, 1950c: 143. Type-species: *Orectognathus chyzeri,* by original designation. *Arnoldidris* provisional junior synonym of *Orectognathus:* Brown, 1973b: 178. *Arnoldidris* junior synonym of *Orectognathus:* Taylor, 1977: 581.

*AROTROPUS* (junior synonym of *Amblyopone*)

*Arotropus* Provancher, 1881a: 205. Type-species: *Arotropus binodosus* (junior synonym of *Amblyopone pallipes*), by monotypy. *Arotropus* junior synonym of *Amblyopone:* Provancher, 1887: 240.

*ASEMORHOPTRUM* (junior synonym of *Stenamma*)

*Asemorhoptrum* Mayr, 1861: 76. Type-species: *Myrmica lippula,* by monotypy. *Asemorhoptrum* junior synonym of *Stenamma:* André, 1883a: 310. [*Asemorhoptrum* incorrectly as junior synonym of *Proatta:* Snelling, 1981: 395, in error.]

**ASKETOGENYS** (Myrmicinae: Dacetini)

*Asketogenys* Brown, 1972: 23. Type-species: *Asketogenys acubecca,* by original designation. [See note under *Strumigenys.*]

**ASPHINCTOPONE** (Ponerinae: Ponerini)

*Asphinctopone* Santschi, 1914d: 318. Type-species: *Asphinctopone silvestrii,* by monotypy. *Asphinctopone* senior synonym of *Lepidopone:* Brown, 1953h: 2.

*ASPIDIDRIS* (junior synonym of *Basiceros*)

*Aspididris* Weber, 1950c: 3. Type-species: *Aspididris militaris,* by monotypy. Revision of genus: Brown & Kempf, 1960: 179. *Aspididris* junior synonym of *Basiceros:* Brown, 1974c: 132.

***ASYMPHYLOMYRMEX** (Dolichoderinae)

**Asymphylomyrmex* Wheeler, W.M. 1915h: 96. Type-species: **Asymphylomyrmex balticus,* by monotypy.

#**ATOPODON** (subgenus of *Acropyga*)

*Atopodon* Forel, 1912m: 771 [as subgenus of *Acropyga*]. Type-species: *Acropyga (Atopodon) inezae,* by subsequent designation of Wheeler, W.M. 1913a: 79. [*Atopodon* also described as new by Forel, 1913k: 100.] *Atopodon* subgenus of *Rhizomyrma:* Forel, 1917: 249. *Atopodon* raised to genus: Wheeler, W.M. 1922a: 695. *Atopodon* subgenus of *Acropyga:* Emery, 1925b: 30. *Atopodon* provisional junior synonym of *Acropyga:* Brown, 1973b: 178 (unconfirmed).

#**ATOPOGYNE** (subgenus of *Crematogaster*)

*Atopogyne* Forel, 1911d: 343 [as subgenus of *Crematogaster*]. Type-species: *Crematogaster (Atopogyne) hellenica,* by subsequent designation of Wheeler, W.M. 1911f: 159. [Type-species not *Crematogaster depressa,* unjustified subsequent designation by Emery, 1912d: 272; this error repeated in Emery, 1922e: 154 and in Soulié, 1965: 78.] *Atopogyne* raised to genus: Soulié, 1964: 398. *Atopogyne* provisional junior synonym of *Crematogaster:* Brown, 1973b: 178 (unconfirmed); reverts here to subgenus of *Crematogaster.*

**ATOPOMYRMEX** (Myrmicinae: Formicoxenini)

*Atopomyrmex* André, 1889: 226. Type-species: *Atopomyrmex mocquerysi,* by monotypy. Revision of genus: Bolton, 1981b: 249.

*ATOPULA* (junior synonym of *Tetramorium*)

*Atopula* Emery, 1912b: 104. Type-species: *Atopomyrmex nodifer,* by original designation. *Atopula* subgenus of *Vollenhovia:* Emery, 1912d: 273. *Atopula* revived status as genus: Emery, 1914a: 41. *Atopula* subgenus of *Terataner:* Arnold, 1952b: 129. *Atopula* junior synonym of *Tetramorium:* Bolton, 1976: 359; Bolton, 1980: 195.

**ATTA** (Myrmicinae: Attini)

*Atta* Fabricius, 1804: 421. Type-species: *Formica cephalotes,* by subsequent designation of Wheeler, W.M. 1911f: 159. *Atta* senior synonym of *Oecodoma:* Roger, 1863b: 35. *Atta* senior synonym of *Archeatta:* Smith, M.R. 1951a: 832. *Atta* senior synonym of *Archeatta, Epiatta, Neoatta, Palaeatta:* Weber, 1958a: 8. [Each of these names has subsequently been resurrected from synonymy, see under respective entries; they are all returned to the synonymy here.] *Myrmegis* (*nomen nudum*) referable to *Atta:* Brown, 1973b: 182; see under *Myrmegis.* Revision of genus: Borgmeier, 1959b: 321. Bibliography: Cherrett & Cherrett, 1989: 1; see also Lofgren & Vander Meer, 1986: 1.

**ATTAICHNUS* (Myrmicinae: Attini [ichnotaxon])

**Attaichnus* Laza, 1982: 112. Included ichnospecies: **Attaichnus kuenzelii.* [Ichnofossil, purportedly fossil traces of ant workings attributable to attines.]

*ATTOMYRMA* (junior synonym of *Aphaenogaster*)

*Attomyrma* Emery, 1915d: 70 [as subgenus of *Aphaenogaster*]. Type-species: *Formica subterranea,* by original designation. *Attomyrma* provisional junior synonym of *Aphaenogaster:* Brown, 1973b: 178. *Attomyrma* junior synonym of *Aphaenogaster:* Bolton, 1982: 364.

***ATTOPSIS** (Myrmicinae: unplaced to tribe)

**Attopsis* Heer, 1850: 155. Type-species: **Attopsis longipennis,* by subsequent designation of Wheeler, W.M. 1911f: 159.

#**AULACOMYRMA** (subgenus of *Polyrhachis*)

*Aulacomyrma* Emery, 1921e: 17 [as subgenus of *Polyrhachis*]. Type-species: *Polyrhachis porcata,* by original designation. *Aulacomyrma* senior synonym of *Johnia:* Hung, 1967b: 402. *Aulacomyrma* provisional junior synonym of *Polyrhachis:* Brown, 1973b: 178 (unconfirmed).

**AULACOPONE** (Ponerinae: Ectatommini)
*Aulacopone* Arnol'di, 1930a: 139. Type-species: *Aulacopone relicta,* by monotypy. [*Aulacopone* also described as new by Arnol'di, 1930d: 159.] Reviews of genus: Brown, 1958g: 206; Taylor, 1980c: 353. See also: Lattke, 1994: 113.

**#AUSTROLASIUS** (subgenus of *Lasius*)
*Austrolasius* Faber, 1967: 74 [as subgenus of *Lasius*]. Type-species: *Lasius carniolicus,* by original designation. *Austrolasius* provisional junior synonym of *Lasius:* Brown, 1973b: 179 (unconfirmed).

**AXINIDRIS** (Dolichoderinae)
*Axinidris* Weber, 1941a: 192. Type-species: *Axinidris acholli,* by original designation. Revision of genus: Shattuck, 1991: 105. Review of genus: Shattuck, 1992c: 42.

**AZTECA** (Dolichoderinae)
*Azteca* Forel, 1878: 384. Type-species: *Liometopum xanthochroum,* by monotypy. Revisionary studies: Longino, 1989: 1; Longino, 1991: 1571. Review of genus: Shattuck, 1992c: 49. [*Aztecum* Bertkau, 1879: 13, incorrect subsequent spelling.] See also: Shattuck, 1994: 10.

***BAIKURIS** (*Sphecomyrminae)
**Baikuris* Dlussky, 1987: 133. Type-species: **Baikuris mandibularis;* by original designation.

**BAJCARIDRIS** (Formicinae: Formicini)
*Bajcaridris* Agosti, 1994a: 99. Type-species: *Formica (Proformica) theryi,* by original designation. [*Bajcaridris* Agosti, 1991: 295, *nomen nudum.*]

**BARACIDRIS** (Myrmicinae: Stenammini?)
*Baracidris* Bolton, 1981b: 252. Type-species: *Baracidris meketra,* by original designation.

*BARBOURELLA* (junior synonym of *Gnamptogenys*)
*Barbourella* Wheeler, W.M. 1930f: 10 [as subgenus of *Emeryella*]. Type-species: *Emeryella (Barbourella) banksi,* by original designation. *Barbourella* junior synonym of *Gnamptogenys:* Brown, 1958g: 212.

**BARIAMYRMA** (Myrmicinae: Stenammini)
*Bariamyrma* Lattke, 1990a: 173. Type-species: *Bariamyrma hispidula,* by original designation.

*BARONIURBANIA* (junior synonym of *Lepisiota*)
*Baroniurbania* Pagliano & Scaramozzino, 1990: 4. Unnecessary replacement name for *Acantholepis* Mayr (junior homonym). [*Lepisiota* is first available replacement name, see there and under *Acantholepis.*] *Baroniurbania* junior synonym of *Lepisiota:* Bolton, 1994: 51.

**BASICEROS** (Myrmicinae: Basicerotini)
*Basiceros* Schulz, 1906: 156. Replacement name for *Ceratobasis* Smith, 1860c: 78. [Junior homonym of *Ceratobasis* Lacordaire, 1848: 362.] *Basiceros* senior synonym of *Aspididris:* Brown, 1974c: 132. Revisions of genus: Brown & Kempf, 1960: 171; Brown, 1974c: 139.

**BELONOPELTA** (Ponerinae: Ponerini)
*Belonopelta* Mayr, 1870a: 374. Type-species: *Belonopelta attenuata,* by monotypy. Key to species: Baroni Urbani, 1975b: 298. *Belonopelta* senior synonym of *Leiopelta:* Hölldobler & Wilson: 1990: 10.

*BICONOMYRMA* (junior synonym of *Conomyrma*)
*Biconomyrma* Kusnezov, 1952g: 429 [as subgenus of *Conomyrma*]. Type-species: *Dorymyrmex pyramicus* var. *brunneus* (now *Dorymyrmex brunneus*), by subsequent designation of Kusnezov, 1959: 51. *Biconomyrma* raised to genus: Kusnezov, 1959: 51. *Biconomyrma* junior synonym of *Conomyrma:* Smith, M.R. 1958c: 140; Kempf, 1972a: 78. [*Conomyrma* junior synonym of *Dorymyrmex:* Shattuck, 1992c: 78.]

*BISOLENOPSIS* (junior synonym of *Solenopsis*)
*Bisolenopsis* Kusnezov, 1953d: 9. Type-species: *Bisolenopsis sea,* by monotypy. *Bisolenopsis* junior synonym of *Solenopsis:* Ettershank, 1966: 134. [*Disolenopsis* Snelling, 1981: 397, incorrect subsequent spelling.]

**BLEPHARIDATTA** (Myrmicinae: Blepharidattini)
*Blepharidatta* Wheeler, W.M. 1915e: 484. Type-species: *Blepharidatta brasiliensis,* by monotypy. *Blepharidatta* provisional junior synonym of *Ochetomyrmex:* Brown, 1973b: 179. *Blepharidatta* valid genus: Kempf, 1975c: 369; Wheeler, G.C. & Wheeler, J. 1991a: 132.

**BONDROITIA** (Myrmicinae: Solenopsidini)
*Bondroitia* Forel, 1911c: 300 (footnote) [as subgenus of *Monomorium*]. Type-species: *Monomorium (Martia) coecum* (junior synonym of *Bondroitia lujae*), by monotypy. [*Bondroitia* also described as new, but as subgenus of *Diplomorium,* by Forel, 1911d: 398; see discussion in Bolton, 1987: 275.] *Bondroitia* raised to genus: Forel, 1915d: 38; maintained as genus by Wheeler, W.M. 1922a: 663. *Bondroitia* subgenus of *Diplomorium:* Emery, 1922e: 194. *Bondroitia* junior synonym of *Diplomorium:* Ettershank, 1966: 98. *Bondroitia* revived from synonymy, status as genus, and revision of genus: Bolton, 1987: 275.

*BORGMEIERITA* (junior synonym of *Glamyromyrmex*)
*Borgmeierita* Brown, 1953g: 23. Type-species: *Codiomyrmex excisus,* by original designation. *Borgmeierita* junior synonym of *Glamyromyrmex:* Brown, 1973a: 35.

**BOTHRIOMYRMEX** (Dolichoderinae)
*Bothriomyrmex* Emery, 1869c: 117. Type-species: *Bothriomyrmex costae,* by monotypy. Revision of genus: Emery, 1925e: 5 (out of date). *Bothriomyrmex* senior synonym of *Chronoxenus,* and review of genus: Shattuck, 1992c: 54. See also: Shattuck, 1994: 31.

*BOTHROPONERA* (junior synonym of *Pachycondyla*)
*Bothroponera* Mayr, 1862: 717. Type-species: *Ponera pumicosa,* by subsequent designation of Emery, 1901a: 42. *Bothroponera* subgenus of *Pachycondyla:* Emery, 1901a: 42. *Bothroponera* revived status as genus: Wheeler, W.M. 1918c: 299 (footnote). *Bothroponera* senior synonym of *Pseudoneoponera:* Wilson, 1958d: 361. *Bothroponera* provisional junior synonym of *Pachycondyla:* Brown, 1973b: 179. *Bothroponera* junior synonym of *Pachycondyla:* Brown, in preparation.

**BRACHYMYRMEX** (Formicinae: Brachymyrmecini)
*Brachymyrmex* Mayr, 1868b: 163. Type-species: *Brachymyrmex patagonicus,* by monotypy. Subgenera of *Brachymyrmex:* nominal plus *Bryscha.* Key to species: Santschi, 1923b: 652 (out of date).

*BRACHYPONERA* (junior synonym of *Pachycondyla*)
*Brachyponera* Emery, 1900c: 315 [as subgenus of *Euponera*]. Type-species: *Euponera (Brachyponera) croceicornis,* by monotypy. [*Brachyponera* also described as new by Emery, 1901a: 43. Type-species not *Ponera sennaarensis,* unjustified subsequent designation by Emery, 1901a: 43.] *Brachyponera* raised to genus: Bingham, 1903: 101. *Brachyponera* subgenus of *Euponera:* Emery, 1911d: 83; maintained as subgenus: Wheeler, W.M. 1922a: 777. *Brachyponera* raised to genus: Wilson, 1958d: 346. *Brachyponera* provisional junior synonym of *Pachycondyla:* Snelling, 1981: 389. *Brachyponera* junior synonym of *Pachycondyla:* Brown, in preparation.

***BRADOPONERA** (Ponerinae: Proceratiini)
**Bradoponera* Mayr, 1868c: 73. Type-species: **Bradoponera meieri,* by monotypy. See also: Lattke, 1994: 116.

*BRADYPONERA* (junior synonym of *Trapeziopelta*)
*Bradyponera* Mayr, 1886c: 362. Type-species: *Ponera nitida* (junior primary homonym in *Ponera,* replaced by *Myopias mayri*), by monotypy. [*Bradyponera* Smith, F. 1873: viii, *nomen nudum,* attributed to Mayr.] *Bradyponera* provisional junior synonym of *Trapeziopelta:* Emery, 1911d: 93. *Bradyponera* junior synonym of *Trapeziopelta:* Donisthorpe, 1932c: 468. [*Trapeziopelta* junior synonym of *Myopias:* Willey & Brown, 1983: 249.]

**BREGMATOMYRMA** (Formicinae: Bregmatomyrmini)
*Bregmatomyrma* Wheeler, W.M. 1929b: 3. Type-species: *Bregmatomyrma carnosa,* by original designation. [*Bregmatomyrmex* Wheeler, G.C. & Wheeler, J. 1985: 259, incorrect subsequent spelling.]

*BROWNIDRIS* (junior synonym of *Ochetomyrmex*)
*Brownidris* Kusnezov, 1957a: 275. Type-species: *Brownidris argentinus,* by original designation. Review of genus: Ettershank, 1966: 109. *Brownidris* junior synonym of *Ochetomyrmex:* Kempf, 1975c: 355.

*BRUCHOMYRMA* (junior synonym of *Pheidole*)
*Bruchomyrma* Santschi, 1922d: 248. Type-species: *Bruchomyrma acutidens,* by monotypy. *Bruchomyrma* provisional junior synonym of *Pheidole:* Brown, 1973b: 179. *Bruchomyrma* junior synonym of *Pheidole:* Smith, D.R. 1979: 1366; Wilson, 1984: 327.

*BRUNELLA* (junior synonym of *Aphaenogaster*)
*Brunella* Forel, 1917: 234. Type-species: *Aphaenogaster belti,* by monotypy. *Brunella* junior synonym of *Atopula:* Emery, 1922f: 242. *Brunella* junior synonym of *Aphaenogaster:* Bolton, 1982: 364.

**#BRYSCHA** (subgenus of *Brachymyrmex*)
*Bryscha* Santschi, 1923b: 652 [as subgenus of *Brachymyrmex*]. Type-species: *Brachymyrmex pilipes,* by original designation. *Bryscha* provisional junior synonym of *Brachymyrmex:* Brown, 1973b: 179 (unconfirmed).

*CACOPONE* (junior synonym of *Plectroctena*)
*Cacopone* Santschi, 1914d: 325. Type-species: *Cacopone hastifer,* by monotypy. *Cacopone* junior synonym of *Plectroctena:* Bolton, 1974c: 313.

**CALOMYRMEX** (Formicinae: Camponotini)
*Calomyrmex* Emery, 1895j: 772. Type-species: *Formica laevissima,* by monotypy.

***CALYPTITES** (dubiously Formicidae)
**Calyptites* Scudder, 1877b: 270 [as member of family Braconidae]. Type-species: **Calyptites antediluvianum,* by monotypy. **Calyptites* retained in Braconidae: Scudder, 1891: 629. **Calyptites* in Formicidae, position indeterminate: Carpenter, 1930: 21. **Calyptites* probably Braconidae: Donisthorpe, 1943f: 629. **Calyptites* in Formicidae:

Brown, 1973b: 179. Family assignment remains unresolved.

***CALYPTOMYRMEX*** (Myrmicinae: Stenammini)

*Calyptomyrmex* Emery, 1887b: 471. Type-species: *Calyptomyrmex beccarii*, by monotypy. *Calyptomyrmex* senior synonym of *Weberidris:* Donisthorpe, 1949f: 186; Brown, 1949f: 84. Revision of Oriental species: Baroni Urbani, 1975a: 410; of Afrotropical species: Bolton, 1981a: 61.

#***CAMPOMYRMA*** (subgenus of *Polyrhachis*)

*Campomyrma* Wheeler, W.M. 1911c: 860 [as subgenus of *Polyrhachis*]. Type-species: *Polyrhachis clypecta* (junior synonym of *Polyrhachis exercita*), by original designation. *Campomyrma* provisional junior synonym of *Polyrhachis:* Brown, 1973b: 179 (unconfirmed).

****CAMPONOTITES*** (Formicinae: Camponotini)

**Camponotites* Dlussky, 1981b: 76. Type-species: **Camponotites macropterus*, by monotypy. [It is unclear whether this was intended as a genus-group name or as a collective group name. However, as the presentation was originally "genus **Camponotites*", the former is accepted here.]

***CAMPONOTUS*** (Formicinae: Camponotini)

*Camponotus* Mayr, 1861: 35. Type-species: *Formica ligniperda*, by subsequent designation of Bingham, 1903: 347. Subgenera of *Camponotus* (alphabetical order with synonyms indicated): nominal plus *Colobopsis, Condylomyrma* (junior synonym of *Colobopsis*), *Dinomyrmex, Hypercolobopsis, Karavaievia, Manniella, Mayria, Myrmacrhaphe, Myrmamblys, Myrmaphaenus, Myrmentoma, Myrmepinotus, Myrmepomis, Myrmespera, Myrmeurynota, Myrmisolepis, Myrmobrachys, Myrmocamelus* (junior synonym of *Myrmophyma*), *Myrmocladoecus, Myrmodirachis, Myrmogigas* (junior synonym of *Dinomyrmex*), *Myrmogonia, Myrmolophus* (junior synonym of *Myrmepomis*), *Myrmomalis, Myrmonesites, Myrmopalpella, Myrmopelta, Myrmophyma, Myrmopiromis, Myrmoplatypus, Myrmoplatys, Myrmopsamma, Myrmopytia, Myrmosaga, Myrmosaulus, Myrmosericus, Myrmosphincta, Myrmostenus, Myrmotarsus, Myrmotemnus, Myrmothrix, Myrmotrema, Myrmoturba* (junior synonym of *Tanaemyrmex*), *Myrmoxygenys, Neocolobopsis* (junior synonym of *Hypercolobopsis*), *Neomyrmamblys* (junior synonym of *Myrmaphaenus*), *Orthonotomyrmex, Orthonotus* (junior homonym), *Paracolobopsis* (junior synonym of *Myrmaphaenus*), *Paramyrmamblys, Pseudocolobopsis, Rhinomyrmex, Tanaemyrmex, Thlipsepinotus*. Synopses of subgenera: Wheeler, W.M. 1922a: 703; Emery, 1925b: 62. All subgenera given as provisional junior synonyms of *Camponotus:* Brown, 1973b: 179–185 (unconfirmed); see under individual entries. [The entry of *Myrmophyma* and *Thlipsepinotus* under the synonymy of *Camponotus* by Taylor & Brown, D.R. 1985: 109, is not accepted as confirmation as not all taxa are included.] *Camponotus* (subgenus indeterminate) senior synonym of *Dolophra:* **new synonym**.

*CAMPOSTIGMACROS* (junior synonym of *Stigmacros*)

*Campostigmacros* McAreavey, 1957: 25 [as subgenus of *Stigmacros*]. Type-species: *Acantholepis aemula*, by original designation. *Campostigmacros* provisional junior synonym of *Stigmacros:* Brown, 1973b: 179. *Campostigmacros* junior synonym of *Stigmacros*. Taylor & Brown, D.R. 1985: 145 (accepted as confirmation).

*CAMPTOGNATHA* (junior synonym of *Eciton*)

*Camptognatha* Gray, G.R. 1832: 516. Type-species: *Camptognatha testacea* (junior synonym of *Eciton hamatum*), by monotypy. *Camptognatha* junior synonym of *Eciton:* Smith, F. 1855c: 160.

[*CAMPYLOMYRMA* (misspelling)

*Campylomyrma* Wheeler, W.M. 1934e: 421, incorrect subsequent spelling of *Condylomyrma*.]

***CARDIOCONDYLA*** (Myrmicinae: Formicoxenini)

*Cardiocondyla* Emery, 1869b: 20. Type-species: *Cardiocondyla elegans*, by monotypy. *Cardiocondyla* senior synonym of *Emeryia:* Forel, 1892g: 461; Forel, 1892i: 313. *Cardiocondyla* senior synonym of *Xenometra:* Baroni Urbani, 1973: 200. *Cardiocondyla* senior synonym of *Dyclona, Loncyda* and *Prosopidris:* Bolton, 1982: 309. Revision of Palaearctic species: Bernard, 1956: 299; of Afrotropical species: Bolton, 1982: 312.

*CARDIOPHEIDOLE* (junior synonym of *Pheidole*)

*Cardiopheidole* Wheeler, W.M. 1914b: 48 [as subgenus of *Pheidole*]. Type-species: *Pheidole vaslitii*, by original designation. *Cardiopheidole* junior synonym of *Pheidole:* Emery, 1921f: 84.

***CAREBARA*** (Myrmicinae: Pheidologetonini)

*Carebara* Westwood, 1840b: 86. Type-species: *Carebara lignata*, by monotypy. Key to Afrotropical species: Wheeler, W.M. 1922a: 172. Review of genus: Ettershank, 1966: 125.

***CAREBARELLA*** (Myrmicinae: Solenopsidini)

*Carebarella* Emery, 1906c: 137. Type-species: *Carebarella bicolor*, by monotypy. *Carebarella* senior synonym of *Carebarelloides:* Ettershank, 1966: 113. Reviews of genus: Ettershank, 1966: 113; Bolton, 1987: 286.

*CAREBARELLOIDES* (junior synonym of *Carebarella*)

*Carebarelloides* Borgmeier, 1937b: 236 [as subgenus of *Carebarella*]. Type-species: *Carebarella* (*Carebarelloides*) *condei*, by original designation. *Carebarelloides* junior synonym of *Carebarella:* Ettershank, 1966: 113.

****CARIRIDRIS*** (Myrmeciinae: unplaced to tribe)

**Cariridris* Brandão & Martins-Neto, 1990: 201. Type-species: **Cariridris bipetiolata*, by original designation.

****CASALEIA*** (Ponerinae: Amblyoponini)

**Casaleia* Pagliano & Scaramozzino, 1990: 5. Replacement name for **Protamblyopone* Dlussky, 1981b: 65. [Junior homonym of *Protamblyopone* Wheeler, W.M. 1927c: 1.]

***CATAGLYPHIS*** (Formicinae: Formicini)

*Cataglyphis* Foerster, 1850b: 493. Type-species: *Cataglyphis fairmairei* (junior synonym of *Formica bicolor*), by monotypy. *Cataglyphis* junior synonym of *Myrmecocystus:* Forel, 1878: 372; Forel, 1886f: 201; Mayr, 1886d: 424. *Cataglyphis* revived from synonymy: Wheeler, W.M. 1913a: 78. *Cataglyphis* senior synonym of *Monocombus:* Mayr, 1861: 44; synonymy confirmed by Agosti, 1990a: 1462. *Cataglyphis* senior synonym of *Machaeromyrma:* Baroni Urbani, 1969d: 218; synonymy confirmed by Agosti, 1990a: 1462. *Cataglyphis* senior synonym of *Eomonocombus* and *Paraformica:* Agosti, 1990a: 1462. Key to species: Santschi, 1929b: 56 (out of date). Review of genus: Agosti, 1990a: 1457; Agosti, 1994a: 103.

***CATAULACUS*** (Myrmicinae: Cataulacini)

*Cataulacus* Smith, F. 1854: 225. Type-species: *Cataulacus taprobanae*, by subsequent designation of Bingham, 1903: 120. *Cataulacus* senior synonym of *Otomyrmex*, and revision of genus: Bolton, 1974a: 1. Key to Afrotropical species: Bolton, 1982: 354.

*CAULOMYRMA* (junior synonym of *Nesomyrmex*)

*Caulomyrma* Forel, 1914d: 233 [as subgenus of *Leptothorax*]. Type-species: *Leptothorax echinatinodis*, by original designation. *Caulomyrma* provisional junior synonym of *Nesomyrmex:* Forel, 1915c: 364. *Caulomyrma* junior synonym of *Nesomyrmex:* Emery, 1915g: 24. [*Caulomyrma* and *Nesomyrmex* junior synonyms of *Goniothorax:* Emery, 1922f: 249 (incorrect procedure as *Goniothorax* is a junior homonym, of which *Nesomyrmex* is first available replacement name, see Smith, M.R. 1950: 30).] [*Nesomyrmex* junior synonym of *Leptothorax:* Bolton, 1982: 319.]

#***CAUTOLASIUS*** (subgenus of *Lasius*)

*Cautolasius* Wilson, 1955a: 13 [as subgenus of *Lasius*]. Type-species: *Formica flava*, by original designation. Revision of subgenus: Wilson, 1955a: 111. *Cautolasius* provisional synonym of *Lasius:* Brown, 1973b: 179 (unconfirmed).

***CENTROMYRMEX*** (Ponerinae: Ponerini)

*Centromyrmex* Mayr, 1866b: 894. Type-species: *Centromyrmex bohemanni* (junior synonym of *Centromyrmex brachycola*), by monotypy. *Centromyrmex* senior synonym of *Spalacomyrmex:* Emery, 1890b: 40. *Centromyrmex* senior synonym of *Typhloteras:* Brown, 1953c: 8. *Centromyrmex* senior synonym of *Glyphopone* and *Leptopone:* Brown, 1963: 9. *Centromyrmex* senior synonym of *Promyopias:* Bolton, 1994; 164. Revision of Neotropical species: Kempf, 1967b: 401.

*CEPHALOMORIUM* (junior synonym of *Hendecapheidole*)

*Cephalomorium* Forel, 1922: 91 [as subgenus of *Tetramorium*]. Type-species: *Tetramorium* (*Cephalomorium*) *bahai*, by monotypy. *Cephalomorium* junior synonym of *Hendecapheidole:* Santschi, 1925d: 228. [*Hendecapheidole* junior synonym of *Pheidole:* Smith, D.R. 1979: 1366.]

*CEPHALOMYRMA* (junior synonym of *Myrmhopla*)

*Cephalomyrma* Karavaiev, 1935a: 115 [as subgenus of *Polyrhachis*]. Type-species: *Polyrhachis* (*Cephalomyrma*) *stylifera*, by monotypy. *Cephalomyrma* junior synonym of *Myrmhopla:* Hung, 1967b: 402.

****CEPHALOMYRMEX*** (Myrmicinae: unplaced to tribe)

**Cephalomyrmex* Carpenter, 1930: 37. Type-species: **Cephalomyrmex rotundatus*, by original designation.

***CEPHALOTES*** (Myrmicinae: Cephalotini)

*Cephalotes* Latreille, 1802a: 357. Type-species: *Formica atrata*, by monotypy. *Cephalotes* junior synonym of *Cryptocerus:* Fabricius, 1804: 419. *Cephalotes* senior synonym of *Cryptocerus:* Wheeler, W.M. 1913a: 78; Smith, M.R. 1949c: 19; see also discussion in Kempf, 1951: 105. Revision of genus: Kempf, 1951: 107.

*CEPHALOXYS* (junior homonym, see *Smithistruma*)

*Cephaloxys* Smith, F. 1865: 76. Type-species: *Cephaloxys capitata*, by monotypy. [Junior homonym of

*Cephaloxys* Signoret, 1847: 294 (Hemiptera).] *Smithistruma* Brown, 1948e: 104, replacement name for *Cephaloxys* Smith.

*CEPOBROTICUS* (junior synonym of *Megalomyrmex*)

*Cepobroticus* Wheeler, W.M. 1925d: 168 [as subgenus of *Megalomyrmex*]. Type-species: *Megalomyrmex* (*Cepobroticus*) *symmetochus*, by monotypy. *Cepobroticus* junior synonym of *Megalomyrmex:* Ettershank, 1966: 101.

***CERAPACHYS*** (Cerapachyinae: Cerapachyini)

*Cerapachys* Smith, F. 1857a: 74. Type-species: *Cerapachys antennatus*, by subsequent designation of Bingham, 1903: 28. *Cerapachys* senior synonym of *Ceratopachys:* Wheeler, W.M. 1922a: 755. *Cerapachys* senior synonym of *Parasyscia* and *Syscia* (and its junior synonym *Cysias*): Kempf, 1972a: 76. *Cerapachys* senior synonym of *Chrysapace, Lioponera, Neophyracaces, Ooceraea, Phyracaces, *Procerapachys*, and revision of genus: Brown, 1975: 18.

*CERATOBASIS* (junior homonym, see *Basiceros*)

*Ceratobasis* Smith, F. 1860c: 78. Type-species: *Meranoplus singularis*, by monotypy. [Junior homonym of *Ceratobasis* Lacordaire, 1848: 362 (Coleoptera).] *Basiceros* Schulz,1906: 156, replacement name for *Ceratobasis* Smith.

*CERATOPACHYS* (junior synonym of *Cerapachys*)

*Ceratopachys* Schulz, 1906: 155, unjustified emendation of *Cerapachys*. *Ceratopachys* junior synonym of *Cerapachys:* Wheeler, W.M. 1922a: 755.

*CERATOPHEIDOLE* (junior synonym of *Pheidole*)

*Ceratopheidole* Pergande, 1896: 889 [as subgenus of *Pheidole*]. Type-species: *Pheidole* (*Ceratopheidole*) *granulata*, by monotypy. *Ceratopheidole* raised to genus: Emery, 1914a: 40; maintained as genus: Forel,1917: 241; Emery, 1922e: 112. *Ceratopheidole* provisional junior synonym of *Pheidole:* Brown, 1973b: 179. *Ceratopheidole* junior synonym of *Pheidole:* Smith, D.R. 1979: 1365.

*CHALCOPONERA* (junior synonym of *Rhytidoponera*)

*Chalcoponera* Emery, 1897d: 548. Type-species: *Ponera metallica*, by subsequent designation of Emery, 1911d: 39. *Chalcoponera* subgenus of *Rhytidoponera:* Forel, 1917: 236. *Chalcoponera* revived status as genus: Wheeler, W.M. 1922a: 643. *Chalcoponera* junior synonym of *Rhytidoponera:* Brown, 1953c: 2.

***CHALEPOXENUS*** (Myrmicinae: Formicoxenini)

*Chalepoxenus* Menozzi, 1923a: 257. Type-species: *Chalepoxenus gribodoi* (junior synonym of *Chalepoxenus muellerianus*), by monotypy. [Name of type-species misspelled *gridoboi*, p. 257; correct spelling as *gribodoi*, p. 258.] *Chalepoxenus* provisional junior synonym of *Leptothorax:* Brown, 1973b: 179. *Chalepoxenus* valid genus: Kutter, 1973d: 269. *Chalepoxenus* senior synonym of *Leonomyrma:* Buschinger, 1987: 117. Revision of genus: Kutter, 1973d: 269; Radchenko, 1989a: 37.

*CHAMPSOMYRMEX* (junior synonym of *Odontomachus*)

*Champsomyrmex* Emery, 1892d: 558 (footnote). Type-species: *Odontomachus coquereli*, by monotypy. *Champsomyrmex* subgenus of *Odontomachus:* Forel, 1917: 238. *Champsomyrmex* revived status as genus: Wheeler, W.M. 1922a: 653. *Champsomyrmex* provisional junior synonym of *Odontomachus:* Brown, 1973b: 179. *Champsomyrmex* junior synonym of *Odontomachus:* Brown, 1976a: 96.

*CHAPMANELLA* (junior synonym of *Euprenolepis*)

*Chapmanella* Wheeler, W.M. 1930d: 41. Type-species: *Chapmanella negrosensis*, by original designation. *Chapmanella* junior synonym of *Euprenolepis:* Brown, 1953h: 6.

#***CHARIOMYRMA*** (subgenus of *Polyrhachis*)

*Chariomyrma* Forel, 1915b: 107 [as subgenus of *Polyrhachis*]. Type-species: *Polyrhachis guerini*, by original designation. *Chariomyrma* provisional junior synonym of *Polyrhachis:* Brown, 1973b: 179 (unconfirmed).

*CHARIOSTIGMACROS* (junior synonym of *Stigmacros*)

*Chariostigmacros* McAreavey, 1957: 23 [as subgenus of *Stigmacros*]. Type-species: *Stigmacros* (*Chariostigmacros*) *hirsuta*, by original designation. *Chariostigmacros* provisional junior synonym of *Stigmacros:* Brown, 1973b: 179. *Chariostigmacros* junior synonym of *Stigmacros:* Taylor & Brown, D.R. 1985: 145 (accepted as confirmation).

*CHELANER* (junior synonym of *Monomorium*)

*Chelaner* Emery, 1914f: 410 [as subgenus of *Monomorium*]. Type-species: *Monomorium* (*Chelaner*) *forcipatum*, by subsequent designation of Emery, 1922e: 168. *Chelaner* raised to genus, and senior synonym of *Notomyrmex, Protholcomyrmex* and *Schizopelta:* Ettershank, 1966: 93. *Chelaner* junior synonym of *Monomorium:* Bolton, 1987: 300.

***CHELIOMYRMEX*** (Ecitoninae: Cheliomyrmecini)

*Cheliomyrmex* Mayr, 1870b: 968. Type-species: *Cheliomyrmex nortoni* (junior synonym of *Cheliomyrmex morosus*), by monotypy. Revision of genus: Borgmeier, 1955b: 62; see also Gotwald, 1971: 161. Key to species: Watkins, 1976: 6.

***CHELYSTRUMA*** (Myrmicinae: Dacetini)

*Chelystruma* Brown, 1950a: 33 [as subgenus of *Glamyromyrmex*]. Type-species: *Glamyromyrmex* (*Chelystruma*) *lilloana*, by monotypy. *Chelystruma* raised to genus: Kempf, 1959d: 338. *Chelystruma* provisional junior synonym of *Glamyromyrmex:* Brown, 1973b: 179 (unconfirmed). [See note under *Strumigenys*.]

***CHIMAERIDRIS*** (Myrmicinae: Pheidolini)

*Chimaeridris* Wilson, 1989: 63. Type-species: *Chimaeridris boltoni*, by original designation.

****CHIMAEROMYRMA*** (Formicinae: Camponotini)

**Chimaeromyrma* Dlussky, 1988: 58. Type-species: **Chimaeromyrma brachycephala*, by original designation.

*CHRONOXENUS* (junior synonym of *Bothriomyrmex*)

*Chronoxenus* Santschi, 1919i: 202 [as subgenus of *Bothriomyrmex*]. Type-species: *Bothriomyrmex myops*, by subsequent designation of Donisthorpe, 1944e: 102. *Chronoxenus* provisional junior synonym of *Bothriomyrmex:* Brown, 1973b: 179. *Chronoxenus* junior synonym of *Bothriomyrmex:* Shattuck, 1992c: 54.

*CHRYSAPACE* (junior synonym of *Cerapachys*)

*Chrysapace* Crawley, 1924: 380. Type-species: *Chrysapace jacobsoni* (junior secondary homonym in *Cerapachys*, replaced by *Cerapachys crawleyi*), by monotypy. *Chrysapace* junior synonym of *Cerapachys:* Brown, 1975: 19.

#***CHTHONOLASIUS*** (subgenus of *Lasius*)

*Chtonolasius* Ruzsky, 1912: 630 [as subgenus of *Lasius*]. Type-species: *Formica umbrata*, by subsequent designation of Emery, 1925b: 232. [*Chtonolasius* also described as new by Ruzsky, 1914a: 59.] *Chthonolasius*, justified emendation of spelling: Wheeler, W.M. 1916k: 170. [Donisthorpe's (1941f: 37) claim that Ruzsky nominated *Lasius flavus* as type-species is incorrect. Ruzsky (1912: 630) included those species "with the appearance of the yellow *Lasius* (*Las. flavus* D.G.) and its closely related species, and races (*Las. carniolicus, Las. umbratus, Las. myops, Las. mixtus*)". This is not a direct nomination of a type-species, nor is De Geer (D.G.) the author of *flavus*.] *Chthonolasius* subgenus of *Lasius:* Wheeler, W.M. 1922a: 698; Wilson, 1955a: 13. *Chthonolasius* provisional junior synonym of *Lasius:* Brown, 1973b: 179 (unconfirmed). Revisions of subgenus: Wilson, 1955a: 150. Seifert, 1988a: 143; Seifert, 1990: 1.

***CLADAROGENYS*** (Myrmicinae: Dacetini)

*Cladarogenys* Brown, 1976b: 33. Type-species: *Cladarogenys lasia*, by original designation. Review of genus: Bolton, 1983: 353. [See note under *Strumigenys*.]

***CLADOMYRMA*** (Formicinae: Brachymyrmecini)

*Cladomyrma* Wheeler, W.M. 1920: 53. Type-species: *Aphomomyrmex hewitti* (junior synonym of *Cladomyrma andrei*), by original designation. Revision of genus: Agosti, 1991: 293.

*CLARKISTRUMA* (junior synonym of *Colobostruma*)

*Clarkistruma* Brown, 1948e: 124. Type-species: *Epopostruma alinodis*, by original designation. *Clarkistruma* junior synonym of *Colobostruma:* Brown, 1959c: 1.

*CLAVANODA* (junior synonym of *Pseudomyrmex*)

*Clavanoda* Enzmann, E.V. 1944: 61 [as "branch" of *Pseudomyrma*]. Type-species: *Formica gracilis*, by original designation. *Clavanoda* junior synonym of *Pseudomyrmex:* Ward, 1990: 469.

***CODIOMYRMEX*** (Myrmicinae: Dacetini)

*Codiomyrmex* Wheeler, W.M. 1916d: 326. Type-species: *Codiomyrmex thaxteri*, by monotypy. Review of genus: Brown, 1948e: 115. *Codiomyrmex* provisional junior synonym of *Glamyromyrmex:* Brown, 1973b: 179 (unconfirmed). [See note under *Strumigenys*.]

***CODIOXENUS*** (Myrmicinae: Dacetini)

*Codioxenus* Santschi, 1931c: 278 [as subgenus of *Epitritus*]. Type-species: *Epitritus* (*Codioxenus*) *simulans*, by monotypy. *Codioxenus* raised to genus: Brown, 1948e: 123. [See note under *Strumigenys*.]

#***COLOBOCREMA*** (subgenus of *Crematogaster*)

*Colobocrema* Wheeler, W.M. 1927f: 31 [as subgenus of *Crematogaster*]. Type-species: *Crematogaster* (*Colobocrema*) *cylindriceps*, by monotypy. *Colobocrema* provisional junior synonym of *Crematogaster:* Brown, 1973b: 179 (unconfirmed).

#***COLOBOPSIS*** (subgenus of *Camponotus*)

*Colobopsis* Mayr, 1861: 38. Type-species: *Formica truncata*, by subsequent designation of Bingham, 1903: 342. *Colobopsis* subgenus of *Camponotus:* Emery, 1889b: 517. *Colobopsis* senior synonym of *Condylomyrma:* Wheeler, W.M. 1934e: 422 (in text). *Colobopsis* provisional junior synonym of *Camponotus:* Brown, 1973b: 179 (unconfirmed).

***COLOBOSTRUMA*** (Myrmicinae: Dacetini)

*Colobostruma* Wheeler, W.M. 1927f: 32 [as subgenus of *Epopostruma*]. Type-species: *Epopostruma* (*Col-*

*obostruma*) *leae*, by monotypy. *Colobostruma* raised to genus: Brown, 1948e: 118. *Colobostruma* senior synonym of *Alistruma* and *Clarkistruma:* Brown, 1959c: 1; Brown & Wilson, 1959b: 281.

*COMMATETA* (junior synonym of *Gnamptogenys*)

*Commateta* Santschi, 1929h: 476 [as subgenus of *Ectatomma*]. Type-species: *Ectatomma* (*Parectatomma*) *bruchi*, by original designation. *Commateta* junior synonym of *Gnamptogenys:* Brown, 1958g: 212.

***CONCOCTIO*** (Ponerinae: Amblyoponini)

*Concoctio* Brown, 1974d: 29. Type-species: *Concoctio concenta*, by original designation.

***CONDYLODON*** (dubiously Formicidae (Ponerinae?))

*Condylodon* Lund, 1831a: 131. Type-species: *Condylodon audouini*, by monotypy. *Condylodon* a member of family Mutillidae?: Swainson & Shuckard, 1840: 173. *Condylodon* junior synonym of *Pseudomyrma:* Dalla Torre, 1893: 55. *Condylodon* a member of Ponerinae?: Emery, 1921f: 28 (footnote). *Condylodon* not member of Pseudomyrmecinae, dubiously member of Ponerinae: Ward, 1990: 471.

*CONDYLOMYRMA* (junior synonym of *Colobopsis*)

*Condylomyrma* Santschi, 1928c: 72 [as subgenus of *Camponotus*]. Type-species: *Camponotus* (*Condylomyrma*) *bryani*, by monotypy. *Condylomyrma* junior synonym of *Colobopsis:* Wheeler, W.M. 1934e: 422 (in text). [*Campylomyrma:* see above.]

*CONOMYRMA* (junior synonym of *Dorymyrmex*)

*Conomyrma* Forel, 1913b: 350 (in text) [as subgenus of *Dorymyrmex*]. Type-species: *Prenolepis pyramica*, by subsequent designation of Santschi, 1922b: 365. *Conomyrma* raised to genus: Kusnezov, 1952g: 429. *Conomyrma* senior synonym of *Biconomyrma:* Smith, M.R. 1958c: 140; Kempf, 1972a: 78. *Conomyrma* provisional junior synonym of *Dorymyrmex:* Brown, 1973b: 179. *Conomyrma* senior synonym of *Araucomyrmex:* Snelling, 1981: 402. *Conomyrma* junior synonym of *Dorymyrmex:* Shattuck, 1992c: 77.

*CONOTHORACOIDES* (junior synonym of *Pheidole*)

*Conothoracoides* Strand, 1935: 176. Replacement name for *Conothorax* Karavaiev. [Junior homonym of *Conothorax* Jekel, 1854: 9 bis (foldout pages).] *Conothoracoides* provisional junior synonym of *Pheidole:* Brown, 1973b: 179. *Conothoracoides* junior synonym of *Pheidole:* Smith, D.R. 1979: 1366.

*CONOTHORAX* (junior homonym, see *Conothoracoides*)

*Conothorax* Karavaiev, 1935a: 75. Type-species: *Conothorax bilobum*, by monotypy. [Junior homonym of *Conothorax* Jekel, 1854: 9 bis (foldout pages) (Coleoptera).] *Conothoracoides* Strand, 1935: 176, replacement name for *Conothorax* Karavaiev.

*COPTOFORMICA* (junior synonym of *Formica*)

*Coptoformica* Müller, 1923: 146 [as subgenus of *Formica*]. Type-species: *Formica exsecta*, by subsequent designation of Donisthorpe, 1941f: 37. *Coptoformica* senior synonym of *Adformica:* Stitz, 1939: 306. *Coptoformica* junior synonym of *Formica:* Smith, M.R. 1951a: 860; Smith, D.R. 1979: 1448; Agosti, 1994a: 107.

*CORYNOMYRMEX* (junior synonym of *Monomorium*)

*Corynomyrmex* Viehmeyer, 1916a: 133 [as subgenus of *Monomorium*]. Type-species: *Monomorium* (*Corynomyrmex*) *hospitum*, by monotypy. *Corynomyrmex* provisional junior synonym of *Monomorium:* Ettershank, 1966:82. *Corynomyrmex* junior synonym of *Monomorium:* Bolton, 1987: 287.

*COSMAECETES* (junior synonym of *Typhlopone*)

*Cosmaecetes* Spinola, 1851b: 54. Type-species: *Cosmaecetes homalinus* (junior synonym of *Dorylus fulvus*), by monotypy. [Also described as new by Spinola, 1853: 70.] *Cosmaecetes* junior synonym of *Typhlopone:* Smith, F. 1858b: 110. [*Cosmaegetes* Dalla Torre, 1893: 9, incorrect subsequent spelling.]

*CRATEROPSIS* (junior synonym of *Oligomyrmex*)

*Crateropsis* Patrizi, 1948: 174 [as subgenus of *Solenopsis*]. Type-species: *Solenopsis* (*Crateropsis*) *elmenteitae*, by original designation. *Crateropsis* junior synonym of *Oligomyrmex:* Ettershank, 1966: 120.

*CRATOMYRMEX* (junior synonym of *Messor*)

*Cratomyrmex* Emery, 1892d: 572. Type-species: *Cratomyrmex regalis*, by monotypy. *Cratomyrmex* subgenus of *Messor:* Santschi, 1920d: 378. *Cratomyrmex* junior synonym of *Messor:* Emery, 1922f: 357. *Cratomyrmex* revived from synonymy with status as genus: Wheeler, W.M. 1922a: 802; maintained as genus: Bernard, 1971: 6. *Cratomyrmex* junior synonym of *Messor:* Bolton, 1982: 338.

***CREIGHTONIDRIS*** (Myrmicinae: Basicerotini)

*Creightonidris* Brown, 1949f: 89. Type-species: *Creightonidris scambognatha*, by original designation. Review of genus: Brown & Kempf, 1960: 178.

***CREMATOGASTER*** (Myrmicinae: Crematogastrini)

*Crematogaster* Lund, 1831a: 132. Type-species: *Formica scutellaris*, by subsequent designation of Bingham, 1903: 124. [Type-species not *Formica acuta*, unjustified subsequent designation by Emery, 1912d: 272; this error repeated in, for example, Emery, 1922e: 128 and Soulié, 1965: 78.] *Crematogaster* senior synonym of *Acrocoelia:* Mayr, 1863: 404 (see also Buren, 1959: 125; Kempf, 1972a: 81). *Crematogaster* senior synonym of *Tranopeltoides:* Kempf, 1960c: 173. [*Cremastogaster* Mayr, 1861: 74 (and many later authors), incorrect subsequent spelling. Discussion of spelling: Emery, 1912d: 272 (footnote); Wheeler, W.M. 1913a: 78; Donisthorpe, 1941f: 36 and Buren, 1959: 125.] Subgenera of *Crematogaster:* nominal plus *Apterocrema, Atopogyne, Colobocrema, Decacrema, Eucrema, Mesocrema, Nematocrema, Neocrema, Orthocrema, Oxygyne, Paracrema, Physocrema, Rhachiocrema, Sphaerocrema, Xiphocrema.*

****CRETACOFORMICA*** (dubiously Formicidae)

**Cretacoformica* Jell & Duncan, 1986: 190. Type-species: **Cretacoformica explicata*, by original designation. [This is almost certainly not an ant.]

****CRETOMYRMA*** (*Sphecomyrminae)

**Cretomyrma* Dlussky, 1975: 115. Type-species: **Cretomyrma arnoldii*, by original designation.

****CRETOPONE*** (*Armaniinae)

**Cretopone* Dlussky, 1975: 119. Type-species: **Cretopone magna*, by original designation. **Cretopone* transferred from Ponerinae to *Armaniidae: Dlussky & Fedoseeva, 1987: 77. **Cretopone* possible junior synonym of **Sphecomyrma:* Hölldobler & Wilson, 1990: 9.

*CROESOMYRMEX* (junior synonym of *Macromischa*)

*Croesomyrmex* Mann, 1920: 408 [as subgenus of *Macromischa*]. Type-species: *Macromischa* (*Croesomyrmex*) *wheeleri*, by original designation. *Croesomyrmex* raised to genus: Wheeler, W.M. 1931b: 4 (in text). *Croesomyrmex* junior synonym of *Macromischa:* Smith, M.R. 1937: 849; Smith, M.R. 1939e: 502. [*Macromischa* junior synonym of *Leptothorax:* Snelling, 1986: 154.]

[*CRYPTOCEPHALUS* (misspelling)

*Cryptocephalus* Lowne, 1865: 336, incorrect subsequent spelling of *Cryptocerus* in the combination *Cryptocephalus pubescens;* current combination *Meranoplus pubescens*.]

*CRYPTOCERUS* (junior synonym of *Cephalotes*)

*Cryptocerus* Latreille, 1803: 311. Type-species: *Formica atrata*, by subsequent designation of Latreille, 1810: 437. [Type-species not *Cryptocerus umbraculatus*, unjustified subsequent designation by Emery, 1914c: 38, repeated in Emery, 1922f: 305.] *Cryptocerus* junior synonym of *Cephalotes:* Wheeler, W.M. 1913a: 78; Smith, M.R. 1949c: 19; see also discussion in Kempf, 1951: 105.

***CRYPTOPONE*** (Ponerinae: Ponerini)

*Cryptopone* Emery, 1893a: cclxxv. Type-species: *Cryptopone testacea*, by monotypy. [Type-species not *Amblyopone testacea;* see discussion in Wilson, 1958d: 360.] [*Cryptopone* also described as new by Emery, 1893f: 240.] Key to Melanesian species: Wilson, 1958d: 357; see also notes in Brown, 1963: 6.

**CTENOBETHYLUS* (junior synonym of *Liometopum*)

**Ctenobethylus* Brues, 1939: 261 [as member of family Bethylidae]. Type-species: **Ctenobethylus succinalis*, by original designation. **Ctenobethylus* junior synonym of *Iridomyrmex:* Brown, 1977: 213. **Ctenobethylus* junior synonym of *Liometopum:* Shattuck, 1992c: 121.

*CTENOPYGA* (junior synonym of *Acanthostichus*)

*Ctenopyga* Ashmead, 1906: 29. Type-species: *Ctenopyga townsendi* (junior synonym of *Acanthostichus texanus*), by original designation. *Ctenopyga* subgenus of *Acanthostichus:* Emery, 1911d: 13. *Ctenopyga* provisional junior synonym of *Acanthostichus:* Brown, 1973b: 179. *Ctenopyga* revived status as genus, and review of genus: Brown, 1975: 42. *Ctenopyga* junior synonym of *Acanthostichus:* MacKay, in preparation.

*CYATHOCEPHALUS* (junior homonym, see *Cyathomyrmex*)

*Cyathocephalus* Emery, 1915i: 192 [as subgenus of *Cryptocerus*]. Type-species: *Cryptocerus pallens*, by original designation. [Junior homonym of *Cyathocephalus* Kessler, 1868: 135 (Cestoda).] *Cyathomyrmex* Creighton, 1933: 100, replacement name for *Cyathocephalus* Emery.

*CYATHOMYRMEX* (junior synonym of *Paracryptocerus*)

*Cyathomyrmex* Creighton, 1933: 100 [as subgenus of *Cryptocerus*]. Replacement name for *Cyathocephalus* Emery, above. [Junior homonym of *Cyathocephalus* Kessler, 1868: 135.] *Cyathomyrmex* subgenus of *Paracryptocerus:* Smith, M.R. 1949c: 21. *Cyathomyrmex* junior synonym of *Paracryptocerus:* Kempf, 1972a: 175. [*Paracryptocerus* junior synonym of *Zacryptocerus:* Kempf, 1973c: 460.]

***CYLINDROMYRMEX*** (Cerapachyinae: Cylindromyrmecini)

*Cylindromyrmex* Mayr, 1870b: 967. Type-species: *Cylindromyrmex striatus*, by monotypy. *Cylindromyrmex* subgenus of *Cerapachys:* Forel, 1892l: 243. *Cyl-*

ELIHU BURRITT LIBRARY
CENTRAL CONNECTICUT STATE UNIVERSITY
NEW BRITAIN, CONNECTICUT 06050

*indromyrmex* revived status as genus: Dalla Torre, 1983: 16. *Cylindromyrmex* senior synonym of *Holcoponera* Cameron (junior homonym): Forel, 1892f: 256. *Cylindromyrmex* senior synonym of *Hypocylindromyrmex* and *Metacylindromyrmex*, and revision of genus: Brown, 1975: 36.

***CYPHOIDRIS*** (Myrmicinae: Stenammini)

*Cyphoidris* Weber, 1952: 26. Type-species: *Cyphoidris spinosa*, by original designation. *Cyphoidris* provisional junior synonym of *Lordomyrma:* Brown, 1973b: 179. *Cyphoidris* valid genus, and revision of genus: Bolton, 1981b: 257.

*CYPHOMANNIA* (junior synonym of *Cyphomyrmex*)

*Cyphomannia* Weber, 1938b: 183 [as subgenus of *Cyphomyrmex*]. Type-species: *Cyphomyrmex* (*Cyphomannia*) *laevigatus*, by original designation. *Cyphomannia* junior synonym of *Cyphomyrmex:* Kempf, 1962b: 29. *Cyphomannia* revived from synonymy: Weber, 1966: 166. *Cyphomannia* junior synonym of *Cyphomyrmex:* Kempf, 1972a: 92.

***CYPHOMYRMEX*** (Myrmicinae: Attini)

*Cyphomyrmex* Mayr, 1862: 690. Type-species: *Cyphomyrmex minutus*, by monotypy. [Type-species not *Cyphomyrmex difformis*, unjustified subsequent designation by Emery, 1922f: 340.] *Cyphomyrmex* subgenus of *Atta:* Forel, 1912e: 188. *Cyphomyrmex* revived status as genus: Bruch, 1914: 217; Emery, 1914a: 42. *Cyphomyrmex* senior synonym of *Cyphomannia:* Kempf, 1962b: 29; Kempf, 1972a: 92. Revisions of genus: Kempf, 1964d: 1; Kempf, 1966: 161, Snelling & Longino, 1992: 479.

#CYRTOMYRMA (subgenus of *Polyrhachis*)

*Cyrtomyrma* Forel, 1915b: 107 [as subgenus of *Polyrhachis*]. Type-species: *Formica rastellata*, by original designation. *Cyrtomyrma* provisional junior synonym of *Polyrhachis:* Brown, 1973b: 179 (unconfirmed). Revision of subgenus: Donisthorpe, 1938c: 246 (out of date).

*CYRTOSTIGMACROS* (junior synonym of *Stigmacros*)

*Cyrtostigmacros* McAreavey, 1957: 35 [as subgenus of *Stigmacros*]. Type-species: *Acantholepis* (*Acrostigma*) *australis*, by original designation. *Cyrtostigmacros* provisional junior synonym of *Stigmacros:* Brown, 1973b: 179. *Cyrtostigmacros* junior synonym of *Stigmacros:* Taylor & Brown, D.R. 1985: 145 (accepted as confirmation).

*CYSIAS* (junior synonym of *Syscia*)

*Cysias* Emery, 1902c: 24 [as subgenus of *Cerapachys*]. Type-species: *Ooceraea papuana*, by original designation. *Cysias* raised to genus: Ashmead, 1905b: 382. *Cysias* junior synonym of *Syscia:* Emery, 1911d: 10. [*Syscia* junior synonym of *Cerapachys:* Kempf, 1972a: 76; see also Brown, 1975: 19.]

***DACATRIA*** (Myrmicinae: Stenammini)

*Dacatria* Rigato, 1994b: 155. Type-species: *Dacatria templaris*, by original designation.

***DACETINOPS*** (Myrmicinae: Stenammini)

*Dacetinops* Brown & Wilson, 1957a: 1. Type-species: *Dacetinops cibdelus*, by original designation. Revision of genus: Taylor, 1985: 41.

***DACETON*** (Myrmicinae: Dacetini)

*Daceton* Perty, 1833: 136. Type-species: *Formica armigera*, by monotypy. *Daceton* senior synonym of *Dacetum:* Brown, 1973b: 179.

*DACETUM* (junior synonym of *Daceton*)

*Dacetum* Agassiz, 1848: 332, unjustified emendation of *Daceton*. *Dacetum* junior synonym of *Daceton:* Brown, 1973b: 179.

*DACRYON* (junior synonym of *Podomyrma*)

*Dacryon* Forel, 1895f: 421. Type-species: *Dacryon omniparens*, by monotypy. *Dacryon* provisional junior synonym of *Podomyrma:* Brown, 1973b: 179. *Dacryon* junior synonym of *Podomyrma:* Taylor & Brown, D.R. 1985: 80 (accepted as confirmation).

#***DECACREMA*** (subgenus of *Crematogaster*)

*Decacrema* Forel, 1910a: 18 [as subgenus of *Crematogaster*]. Type-species: *Crematogaster schencki*, by subsequent designation of Wheeler, W.M. 1911f: 161. [*Decacrema* also described as new by Forel, 1910f: 9.] *Decacrema* raised to genus: Soulié, 1964: 398. *Decacrema* provisional junior synonym of *Crematogaster:* Brown, 1973b: 179 (unconfirmed); here reverts to subgenus of *Crematogaster*. [*Decracrema* Arnold, 1920a: 547, incorrect subsequent spelling.]

*DECAMERA* (junior homonym, see *Hincksidris*)

*Decamera* Roger, 1863a: 166. Type-species: *Decamera nigella*, by monotypy. [Junior homonym of *Decamera* Mulsant, 1842: 503 (Coleoptera).] *Hincksidris* Donisthorpe, 1944b: 59, replacement name for *Decamera* Roger. *Decamera* Roger junior synonym of *Myrmelachista:* Forel, 1878: 376; maintained as junior synonym: Mayr, 1887: 525; Dalla Torre, 1893: 174. *Decamera* Roger revived from synonymy as subgenus of *Myrmelachista:* Forel, 1886f: 214; maintained as subgenus: Forel, 1917: 249; Wheeler, W.M. 1922a: 696; Emery, 1925b: 32.

***DECAMORIUM*** (Myrmicinae: Tetramoriini)

*Decamorium* Forel, 1913a: 121 [as subgenus of *Tetramorium*]. Type-species: *Tetramorium* (*Decamorium*) *decem*, by monotypy. *Decamorium* raised to genus: Emery, 1914a: 42; maintained as genus: Wheeler, 1922a: 906. Revision of genus: Bolton, 1976: 297.

*DECAPHEIDOLE* (junior synonym of *Pheidole*)

*Decapheidole* Forel, 1912f: 237 [as subgenus of *Pheidole*]. Type-species: *Pheidole perpusilla*, by subsequent designation of Wheeler, W.M. 1913a: 80. *Decapheidole* raised to genus: Wheeler, W.M. 1922a: 661; maintained as genus: Borgmeier, 1927c: 78. *Decapheidole* provisional junior synonym of *Pheidole:* Brown, 1973b: 180. *Decapheidole* junior synonym of *Pheidole:* Smith, D.R. 1979: 1366.

#***DENDROLASIUS*** (subgenus of *Lasius*)

*Dendrolasius* Ruzsky, 1912: 630 [as subgenus of *Lasius*]. Type-species: *Formica fuliginosa*, by monotypy. [*Dendrolasius* also described as new by Ruzsky, 1914a: 59.] *Dendrolasius* subgenus of *Formicina:* Forel, 1917: 249. *Dendrolasius* subgenus of *Lasius:* Wheeler, W.M. 1922a: 698; maintained as subgenus: Wilson, 1955a: 14. *Dendrolasius* provisional junior synonym of *Lasius:* Brown, 1973b: 180 (unconfirmed). Revisions of subgenus: Wilson, 1955a: 138; Kupyanskaya, 1989: 779.

***DENDROMYRMEX*** (Formicinae: Camponotini)

*Dendromyrmex* Emery, 1895j: 772. Type-species: *Formica chartifex*, by subsequent designation of Wheeler, W.M. 1911f: 161.

*DEROMYRMA* (junior synonym of *Aphaenogaster*)

*Deromyrma* Forel, 1913b: 350 [as subgenus of *Aphaenogaster*]. Type-species: *Aphaenogaster* (*Ischnomyrmex*) *swammerdami*, by original designation. [*Deromyrma* also described as new by Forel, 1913k: 49.] *Deromyrma* provisional junior synonym of *Aphaenogaster:* Brown, 1973b: 180. *Deromyrma* junior synonym of *Aphaenogaster:* Bolton, 1982: 364.

*DESCOLEMYRMA* (junior synonym of *Mycocepurus*)

*Descolemyrma* Kusnezov, 1951d: 460. Type-species: *Descolemyrma ogloblini* (junior synonym of *Mycocepurus goeldii*), by monotypy. *Descolemyrma* junior synonym of *Mycocepurus:* Kempf, 1963b: 418.

*DIABOLUS* (junior homonym, see *Karawajewella*)

*Diabolus* Karavaiev, 1926d: 424 [as subgenus of *Dolichoderus*]. Type-species: *Dolichoderus* (*Diabolus*) *bifurcatus* (junior synonym of *Dolichoderus cuspidatus*), by original designation. [Junior homonym of *Diabolus* Gray, J.E. 1841: 400 (Mammalia).] *Karawajewella* Donisthorpe, 1944b: 59, replacement name for *Diabolus* Karavaiev.

***DIACAMMA*** (Ponerinae: Ponerini)

*Diacamma* Mayr, 1862: 718. Type-species: *Ponera rugosa*, by subsequent designation of Bingham, 1903: 75.

*DIAGYNE* (junior synonym of *Solenopsis*)

*Diagyne* Santschi, 1923c: 268 [as subgenus of *Solenopsis*]. Type-species: *Solenopsis succinea* by monotypy. *Diagyne* junior synonym of *Solenopsis:* Ettershank, 1966: 134.

*DICERATOCLINEA* (junior synonym of *Dolichoderus*)

*Diceratoclinea* Wheeler, W.M. 1935c: 69 [as subgenus of *Dolichoderus*]. Type-species: *Dolichoderus scabridus*, by original designation. *Diceratoclinea* raised to genus: Brown, 1950e: 249. *Diceratoclinea* provisional junior synonym of *Dolichoderus:* Brown, 1973b: 180. *Diceratoclinea* junior synonym of *Dolichoderus:* Shattuck, 1992c: 66.

*DICHOTHORAX* (junior synonym of *Leptothorax*)

*Dichothorax* Emery, 1895c: 323 [as subgenus of *Leptothorax*]. Type-species: *Leptothorax* (*Dichothorax*) *pergandei*, by subsequent designation of Wheeler, W.M. 1911f: 161. *Dichothorax* provisional junior synonym of *Leptothorax:* Brown,1973b: 180. *Dichothorax* junior synonym of *Leptothorax:* Bolton, 1982: 319.

#***DICHTHADIA*** (subgenus of *Dorylus*)

*Dichthadia* Gerstäcker, 1863: 93. Type-species: *Dichthadia glaberrima* (junior synonym of *Dorylus laevigatus*), by monotypy. *Dichthadia* junior synonym of *Dorylus:* Dalla Torre, 1893: 9. *Dichthadia* revived from synonymy as subgenus of *Dorylus:* Emery, 1895j: 702. *Dichthadia* provisional junior synonym of *Dorylus:* Brown, 1973b: 180 (unconfirmed).

***DICROASPIS*** (Myrmicinae: Stenammini)

*Dicroaspis* Emery, 1908b: 184. Type-species: *Dicroaspis cryptocera*, by monotypy. *Dicroaspis* subgenus of *Calyptomyrmex:* Emery, 1915g: 15. *Dicroaspis* revived status as genus: Wheeler, W.M. 1935a: 7 (in key). *Dicroaspis* provisional junior synonym of *Calyptomyrmex:* Brown, 1973b: 180. *Dicroaspis* valid genus, senior synonym of *Geognomicus*, and revision of genus: Bolton, 1981a: 56.

***DILOBOCONDYLA*** (Myrmicinae: Formicoxenini)

*Dilobocondyla:* Santschi, 1910h: 283. Type-species: *Atopomyrmex selebensis*, by subsequent designation of Wheeler, W.M. 1911f: 162. *Dilobocondyla* senior synonym of *Mesomyrma:* Emery, 1912b: 102. Key to species: Wheeler, W.M. 1924b: 248.

*DIMORPHOMYRMEX* (junior synonym of *Gesomyrmex*)

*Dimorphomyrmex* André, 1892b: 49. Type-species: *Dimorphomyrmex janeti* (junior synonym of *Gesomyrmex chaperi*), by monotypy. *Dimorphomyrmex* junior synonym of *Gesomyrmex:* Wheeler, W.M. 1929a: 1.

**#*DINOMYRMEX*** (subgenus of *Camponotus*)

*Dinomyrmex* Ashmead, 1905b: 384. Type-species: *Formica gigas,* by original designation. *Dinomyrmex* subgenus of *Camponotus:* Forel, 1913b: 350. *Dinomyrmex* senior synonym of *Myrmogigas:* Forel, 1913b: 350 (in text). *Dinomyrmex* provisional junior synonym of *Camponotus:* Brown, 1973b: 180 (unconfirmed).

***DINOPONERA*** (Ponerinae: Ponerini)

*Dinoponera* Roger, 1861a: 37. Type-species: *Ponera grandis* (junior synonym of *Dinoponera gigantea*), by monotypy. Revision of genus: Kempf, 1971: 369.

*DIODONTOLEPIS* (junior synonym of *Notoncus*)

*Diodontolepis* Wheeler, W.M. 1920: 53. Type-species: *Melophorus spinisquamis,* by original designation. *Diodontolepis* junior synonym of *Melophorus:* Emery, 1925b: 11. *Diodontolepis* revived from synonymy: Wheeler, W.M. 1935c: 71. *Diodontolepis* junior synonym of *Notoncus:* Brown, 1955d: 477.

***DIPLOMORIUM*** (Myrmicinae: Solenopsidini)

*Diplomorium* Mayr, 1901b: 16. Type-species: *Diplomorium longipenne,* by monotypy. Review of genus: Ettershank, 1966: 98. Revision of genus: Bolton, 1987: 278.

*DIPLORHOPTRUM* (junior synonym of *Solenopsis*)

*Diplorhoptrum* Mayr, 1855: 449. Type-species: *Formica fugax,* by monotypy. *Diplorhoptrum* junior synonym of *Solenopsis:* Mayr, 1862: 751; maintained as such: Forel, 1915d: 9; Emery, 1922e: 195; Wheeler, W.M. 1922a: 877. *Diplorhoptrum* revived from synonymy as subgenus of *Solenopsis:* Creighton, 1930b: 43; maintained as such: Kusnezov, 1962b: 160. *Diplorhoptrum* junior synonym of *Solenopsis:* Ettershank, 1966: 134. *Diplorhoptrum* revived from synonymy with status as genus: Baroni Urbani, 1968a: 68. *Diplorhoptrum* junior synonym of *Solenopsis:* Kempf, 1972a: 232; Bolton, 1987: 285.

***DISCOTHYREA*** (Ponerinae: Proceratiini)

*Discothyrea* Roger, 1863a: 176. Type-species: *Discothyrea testacea,* by monotypy. *Discothyrea* senior synonym of *Prodiscothyrea* and *Pseudosysphincta,* and revision of genus: Brown, 1958g: 248; see also Brown, 1958j: 115. Key to Neotropical species: Weber, 1939a: 99. See also: Lattke, 1994: 113.

***DLUSSKYIDRIS** (*Sphecomyrminae)

**Dlusskyidris* Bolton, 1994: 187. Replacement name for **Palaeomyrmex* Dlussky, 1975: 118. [Junior homonym of **Palaeomyrmex* Heer, 1865: 91.]

*DODECAMYRMICA* (junior synonym of *Myrmica*)

*Dodecamyrmica* Arnol'di, 1968: 1803 [as subgenus of *Myrmica*]. Type-species: *Myrmica arnoldii,* by original designation. *Dodecamyrmica* provisional junior synonym of *Myrmica:* Brown, 1973b: 180. *Dodecamyrmica* junior synonym of *Myrmica:* Francoeur, 1981: 759.

*DODOUS* (junior synonym of *Pristomyrmex*)

*Dodous* Donisthorpe, 1946f: 145. Type-species: *Dodous trispinosus,* by original designation. *Dodous* junior synonym of *Pristomyrmex:* Brown, 1971a: 3.

***DOLEROMYRMA*** (Dolichoderinae)

*Doleromyrma* Forel, 1907a: 28 [as subgenus of *Tapinoma*]. Type-species: *Tapinoma (Doleromyrma) darwinianum,* by monotypy. *Doleromyrma* subgenus of *Iridomyrmex:* Forel, 1912a: 81. *Doleromyrma* junior synonym of *Iridomyrmex:* Emery, 1912g: 22. *Doleromyrma* revived from synonymy with status as genus: Shattuck, 1992a: 14; review of genus: Shattuck, 1992c: 60.

***DOLICHODERUS*** (Dolichoderinae)

*Dolichoderus* Lund, 1831a: 130. Type-species: *Formica attelaboides,* by monotypy. *Dolichoderus* senior synonym of *Hypoclinea* and *Monacis:* Forel, 1878: 386. *Dolichoderus* senior synonym of *Acanthoclinea* and *Diceratoclinea:* Taylor & Brown, D.R. 1985: 93. *Dolichoderus* senior synonym of *Acanthoclinea, Diabolus* (junior homonym), *Diceratoclinea, Hypoclinea, Karawajewella, Monacis* and *Monoceratoclinea,* and review of genus: Shattuck, 1992c: 66. Key to Australian species: Clark, 1930b: 252. Keys to Neotropical species: Kempf, 1959b: 227; Kempf, 1969: 292; Kempf, 1972c: 253; Lattke, 1987b: 263; MacKay, 1993b: 17. See also: Shattuck, 1994: 39.

***DOLICHOMYRMA** (*Armaniinae)

**Dolichomyrma* Dlussky, 1975: 121. Type-species: **Dolichomyrma longiceps,* by original designation. **Dolichomyrma* placed in *Armaniidae: Dlussky, 1983: 66. **Dolichomyrma* junior synonym of **Sphecomyrma:* Wilson, 1987: 49; Hölldobler & Wilson, 1990: 9. **Dolichomyrma* revived from synonymy: Dlussky & Fedoseeva, 1987: 77; Carpenter, 1992: 491.

**#*DOLICHORHACHIS*** (subgenus of *Polyrhachis*)

*Dolichorhachis* Mann, 1919: 386 [as subgenus of *Polyrhachis*]. Type-species: *Polyrhachis (Dolichorhachis) malaensis,* by monotypy. *Dolichorhachis* provisional junior synonym of *Polyrhachis:* Brown, 1973b: 180 (unconfirmed).

***DOLIOPONERA*** (Ponerinae: Ponerini)

*Dolioponera* Brown, 1974e: 31. Type-species: *Dolioponera fustigera,* by original designation.

*DOLOPHRA* (junior synonym of *Camponotus*)

*Dolophra* Wu & Wang, 1994: 35. Type-species: *Dolophra politae,* by original designation. *Dolophra* junior synonym of *Camponotus* (subgenus indeterminate): **new synonym**.

*DONISTHORPEA* (junior synonym of *Lasius*)

*Donisthorpea* Morice & Durrant, 1915: 423. Type-species: *Formica nigra,* by original designation. [Unnecessary replacement name for *Lasius* Fabricius.] *Donisthorpea* subgenus of *Formicina:* Emery, 1916b: 239. *Donisthorpea* junior synonym of *Acanthomyops:* Forel, 1916: 460. *Donisthorpea* junior synonym of *Lasius:* Wheeler, W.M. 1916k: 172. [*Lasius* and *Donisthorpea* share the same type-species; synonymy is therefore absolute.]

***DORISIDRIS*** (Myrmicinae: Dacetini)

*Dorisidris* Brown, 1948e: 116. Type-species: *Strumigenys (Codiomyrmex) nitens,* by original designation. [See note under *Strumigenys.*]

***DORONOMYRMEX*** (Myrmicinae: Formicoxenini)

*Doronomyrmex* Kutter, 1945: 485. Type-species: *Doronomyrmex pacis,* by monotypy. *Doronomyrmex* provisional junior synonym of *Leptothorax:* Brown, 1973b: 180 (unconfirmed). Partial revision of genus: Buschinger, 1981: 211.

*DOROTHEA* (junior synonym of *Vollenhovia*)

*Dorothea* Donisthorpe, 1948c: 65. Type-species: *Dorothea novobritainae,* by monotypy. *Dorothea* junior synonym of *Vollenhovia:* Ettershank, 1966: 146.

*DORYLOZELUS* (junior synonym of *Leptogenys*)

*Dorylozelus* Forel, 1915b: 24. Type-species: *Dorylozelus mjobergi* (junior secondary homonym in *Leptogenys,* replaced by *Leptogenys tricosa*), by monotypy. *Dorylozelus* junior synonym of *Leptogenys:* Taylor, 1969: 132.

***DORYLUS*** (Dorylinae)

*Dorylus* Fabricius, 1793: 365. Type-species: *Vespa helvola,* by monotypy. Subgenera of *Dorylus* (alphabetical order with synonyms indicated): nominal plus *Alaopone, Anomma, Cosmaecetes* (junior synonym of *Typhlopone*), *Dichthadia, Rhogmus, Shuckardia* (junior synonym of *Alaopone*), *Sphegomyrmex* (junior synonym of *Anomma*), *Typhlopone.* Key to Indo-Australian species: Wilson, 1964a: 436; see also Gotwald, 1982: 157.

***DORYMYRMEX*** (Dolichoderinae)

*Dorymyrmex* Mayr, 1866a: 494. Type-species: *Dorymyrmex flavescens,* by monotypy. *Dorymyrmex* senior synonym of *Psammomyrma:* Forel, 1913b: 350 (in text). *Dorymyrmex* senior synonym of *Psammomyrma* and *Spinomyrma:* Snelling & Hunt, 1976: 93. *Dorymyrmex* senior synonymy of *Conomyrma* (and its junior synonyms *Biconomyrma* and *Araucomyrmex* [and its junior synonym *Ammomyrma*]), and review of genus: Shattuck, 1992c: 77. See also: Shattuck, 1994: 73.

*DREPANOGNATHUS* (junior synonym of *Harpegnathos*)

*Drepanognathus* Smith, F. 1858b: 81. Type-species: *Harpegnathos saltator,* by subsequent designation of Bingham, 1903: 49. [Unnecessary replacement name for *Harpegnathos.*] *Drepanognathus* junior synonym of *Harpegnathos:* Roger, 1863b: 22; Forel, 1900c: 63. [*Drepanognathus* incorrectly subsequently used as senior name: Bingham, 1903: 49.]

***DRYMOMYRMEX** (Formicinae: Camponotini)

**Drymomyrmex* Wheeler, W.M. 1915h: 135. Type-species: **Drymomyrmex fuscipennis,* by original designation. [Note: this name sometimes misspelled as **Dryomyrmex,* for example in Donisthorpe, 1920: 93.]

*DYCLONA* (junior synonym of *Cardiocondyla*)

*Dyclona* Santschi, 1930b: 70 (footnote) [as subgenus of *Cardiocondyla*]. Type-species: *Monomorium cristatum,* by original designation. *Dyclona* provisional junior synonym of *Cardiocondyla:* Brown, 1973b: 180. *Dyclona* junior synonym of *Cardiocondyla:* Bolton, 1982: 309.

*DYOMORIUM* (junior synonym of *Vollenhovia*)

*Dyomorium* Donisthorpe, 1947d: 191. Type-species: *Dyomorium ireneum,* by original designation. *Dyomorium* junior synonym of *Vollenhovia:* Ettershank, 1966: 146.

**DYSEDROGNATHUS** (Myrmicinae: Dacetini)

*Dysedrognathus* Taylor, 1968d: 132. Type-species: *Dysedrognathus extemenus*, by original designation. [See note under *Strumigenys*.]

**ECHINOPLA** (Formicinae: Camponotini)

*Echinopla* Smith, F. 1857a: 79. Type-species: *Echinopla melanarctos*, by subsequent designation of Wheeler, W.M. 1911f: 162. *Echinopla* senior synonym of *Mesoxena*: Bolton, 1994: 50.

**ECITON** (Ecitoninae: Ecitonini)

*Eciton* Latreille, 1804: 179. Type-species: *Formica hamata*, by subsequent designation of Shuckard, in Swainson & Shuckard, 1840: 173. *Eciton* senior synonym of *Camptognatha*, and material of *Ancylognathus* (*nomen nudum*) referred here: Smith, F. 1855c: 160. *Eciton* senior synonym of *Mayromyrmex*: Emery, 1906a: 718. *Eciton* senior synonym of *Holopone*: Borgmeier, 1936: 55. Revision of genus: Borgmeier, 1955b: 162. Key to species: Watkins, 1976: 9.

**ECPHORELLA** (Dolichoderinae)

*Ecphorella* Forel, 1909b: 65 [as subgenus of *Tapinoma*]. Type-species: *Tapinoma* (*Ecphorella*) *wellmani*, by monotypy. *Ecphorella* provisionally raised to genus: Brown, 1973b: 169. *Ecphorella* raised to genus, and review of genus: Shattuck, 1992c: 85.

**ECTATOMMA** (Ponerinae: Ectatommini)

*Ectatomma* Smith, F. 1858b: 102. Type-species: *Formica tuberculata*, by subsequent designation of Bingham, 1903: 82. Revisions of genus: Brown, 1958g: 206; Kugler, C. & Brown, 1982: 6. See also: Lattke, 1994: 109.

*ECTOMOMYRMEX* (junior synonym of *Pachycondyla*)

*Ectomomyrmex* Mayr, 1867a: 83. Type-species: *Ectomomyrmex javanus*, by subsequent designation of Bingham, 1903: 85. *Ectomomyrmex* subgenus of *Pachycondyla*: Emery, 1901a: 42. *Ectomomyrmex* revived status as genus: Wheeler, W.M. 1922a: 648. *Ectomomyrmex* provisional junior synonym of *Pachycondyla*: Brown, 1973b: 180. *Ectomomyrmex* junior synonym of *Pachycondyla*: Brown, in preparation. Key to Melanesian species: Wilson, 1958d: 362.

***ELAEOMYRMEX** (Dolichoderinae)

**Elaeomyrmex* Carpenter, 1930: 48. Type-species: **Elaeomyrmex gracilis*, by original designation.

*ELASMOPHEIDOLE* (junior synonym of *Pheidole*)

*Elasmopheidole* Forel, 1913k: 43 [as subgenus of *Pheidole*]. Type-species: *Pheidole aberrans*, by subsequent designation of Emery, 1922e: 110. *Elasmopheidole* provisional junior synonym of *Pheidole*: Brown, 1973b: 180. *Elasmopheidole* junior synonym of *Pheidole*: Smith, D.R. 1979: 1366. Key to species: Gallardo, 1932c: 179 (out of date).

***ELECTROMYRMEX** (Myrmicinae: unplaced to tribe)

**Electromyrmex* Wheeler, W.M. 1910g: 164 (by indication). Type-species: **Electromyrmex klebsi*, by monotypy. [**Electromyrmex* Wheeler, W.M. 1908g: 413, *nomen nudum*.] **Electromyrmex* description of genus: Wheeler, W.M. 1915h: 55.

*ELECTROPHEIDOLE* (junior synonym of *Pheidole*)

*Electropheidole* Mann, 1921: 438 [as subgenus of *Pheidole*]. Type-species: *Pheidole* (*Electropheidole*) *roosevelti*, by subsequent designation of Donisthorpe, 1943f: 642. *Electropheidole* provisional junior synonym of *Pheidole*: Brown, 1973b: 180. *Electropheidole* junior synonym of *Pheidole*: Smith, D.R. 1979: 1366.

***ELECTROPONERA** (Ponerinae: Ectatommini)

**Electroponera* Wheeler, W.M. 1915h: 34. Type-species: **Electroponera dubia*, by monotypy. See also: Lattke, 1994: 116.

*EMERYELLA* (junior synonym of *Gnamptogenys*)

*Emeryella* Forel, 1901e: 334. Type-species: *Emeryella schmitti*, by monotypy. *Emeryella* junior synonym of *Gnamptogenys*: Brown, 1958g: 211.

*EMERYIA* (junior synonym of *Cardiocondyla*)

*Emeryia* Forel, 1890b: cx. Type-species: *Emeryia wroughtonii*, by monotypy. *Emeryia* junior synonym of *Cardiocondyla*: Forel, 1892g: 461; Forel, 1892i: 313.

**EMERYOPONE** (Ponerinae: Ponerini)

*Emeryopone* Forel, 1912m: 761. Type-species: *Emeryopone buttelreepeni*, by monotypy. [*Emeryopone* also described as new by Forel, 1913k: 14.] *Emeryopone* junior synonym of *Belonopelta*: Baroni Urbani 1975b: 296 (key to species: 298). *Emeryopone* revived from synonymy: Brown, in preparation.

***EMPLASTUS** (Ponerinae: Ponerini)

**Emplastus* Donisthorpe, 1920: 86. Type-species: **Emplastus emeryi*, by original designation.

*ENDIODIOCTES* (junior synonym of *Myrmecocystus*)

*Endiodioctes* Snelling, 1976: 25 [as subgenus of *Myrmecocystus*]. Type-species: *Myrmecocystus melliger*, by original designation. Key to species: Snelling, 1976: 27. *Endiodioctes* junior synonym of *Myrmecocystus*: Snelling, 1981: 403.

*ENERIA* (junior synonym of *Strumigenys*)

*Eneria* Donisthorpe, 1948d: 598. Type-species: *Eneria excisa* (junior synonym of *Strumigenys loriae*), by original designation. *Eneria* junior synonym of *Strumigenys*: Brown, 1949d: 15.

*ENGRAMMA* (junior synonym of *Technomyrmex*)

*Engramma* Forel, 1905b: 180. Type-species: *Engramma lujae*, by monotypy. Key to species: Wheeler, W.M. 1922a: 202 (out of date). *Engramma* junior synonym of *Technomyrmex*: Shattuck, 1992c: 153.

***ENNEAMERUS** (Myrmicinae: Myrmecinini?)

**Enneamerus* Mayr, 1868c: 98. Type-species: **Enneamerus reticulatus*, by monotypy.

***EOCENIDRIS** (Myrmicinae: unplaced to tribe)

**Eocenidris* Wilson, 1985d: 209. Type-species: **Eocenidris crassa*, by original designation.

***EOFORMICA** (Myrmicinae: unplaced to tribe)

**Eoformica* Cockerell, 1921: 38. Type-species: **Eoformica eocenica* (junior synonym of **Eoformica pingue*), by monotypy. **Eoformica* transferred from Ponerinae to Myrmicinae: Carpenter, 1930: 17.

*EOMONOCOMBUS* (junior synonym of *Cataglyphis*)

*Eomonocombus* Arnol'di, 1968: 1815 [as subgenus of *Cataglyphis*]. Type-species: *Myrmecocystus cinnamomeus*, by original designation. *Eomonocombus* provisional junior synonym of *Cataglyphis*: Brown, 1973b: 180. *Eomonocombus* junior synonym of *Cataglyphis*: Agosti, 1990a: 1462; Agosti, 1994a: 103.

***EOMYRMEX** (Myrmicinae: unplaced to tribe)

**Eomyrmex* Hong, 1974: 138. Type-species: **Eomyrmex guchengziensis* (spelling here emended from **guchengzienis*), by monotypy.

**EOPONERA* (junior synonym of **Formicium*)

**Eoponera* Carpenter, 1929: 301. Type-species: **Eoponera berryi*, by original designation. **Eoponera* junior synonym of **Formicium*: Lutz, 1986: 189.

***EOTAPINOMA** (Dolichoderinae)

**Eotapinoma* Dlussky, 1988: 54. Type-species: **Eotapinoma gracilis*, by original designation.

**EPELYSIDRIS** (Myrmicinae: Solenopsidini)

*Epelysidris* Bolton, 1987: 279. Type-species: *Epelysidris brocha*, by original designation.

*EPHEBOMYRMEX* (junior synonym of *Pogonomyrmex*)

*Ephebomyrmex* Wheeler, W.M. 1902c: 390 [as subgenus of *Pogonomyrmex*]. Type-species: *Pogonomyrmex naegelii*, by subsequent designation of Wheeler, W.M. 1911f: 163. *Ephebomyrmex* raised to genus: Kusnezov, 1960b: 350. *Ephebomyrmex* subgenus of *Pogonomyrmex*: Cole, 1968: 22. *Ephebomyrmex* revived status as genus: Kempf, 1972a: 106; maintained as genus: Brown, 1973b: 180. Studies of North American species: Creighton, 1957a: 54; Cole, 1968: 22; Snelling, 1982a: 109. *Ephebomyrmex* junior synonym of *Pogonomyrmex*: Bolton, 1994: 106.

*EPIATTA* (junior synonym of *Neoatta*)

*Epiatta* Borgmeier, 1950d: 246 [as subgenus of *Atta*]. Type-species: *Oecodoma laevigata*, by original designation. *Epiatta* junior synonym of *Atta*: Weber, 1958a: 8. *Epiatta* junior synonym of *Neoatta*: Borgmeier, 1959b: 358. [*Neoatta* junior synonym of *Atta*: Bolton, 1994: 105.]

**EPIMYRMA** (Myrmicinae: Formicoxenini)

*Epimyrma* Emery, 1915a: 262. Type-species: *Epimyrma kraussei*, by original designation. *Epimyrma* provisional junior synonym of *Leptothorax*: Brown, 1973b: 180 (unconfirmed). *Epimyrma* provisional senior synonym of *Myrmoxenus* (and its junior synonym *Myrmetaerus*): Buschinger, 1990: 245; confirmed Bolton, 1994: 105. Revisions of genus: Menozzi, 1931c: 36; Kutter, 1973e: 281; Buschinger, 1989: 265. Subgenera of *Epimyrma*: nominal plus *Gonepimyrma*.

*EPIPHEIDOLE* (junior synonym of *Pheidole*)

*Epipheidole* Wheeler, W.M. 1903f: 664. Type-species: *Epipheidole inquilina*, by monotypy. [*Epipheidole* also described as new by Wheeler, W.M. 1904a: 14.] *Epipheidole* junior synonym of *Pheidole*: Cole, 1965: 174.

**EPITRITUS** (Myrmicinae: Dacetini)

*Epitritus* Emery, 1869a: 136. Type-species: *Epitritus argiolus*, by monotypy. Review of genus: Brown, 1948e: 122. Key to world species: Bolton, 1972: 208; to Afrotropical species: Bolton, 1983: 354. [See note under *Strumigenys*.]

*EPIXENUS* (junior synonym of *Monomorium*)

*Epixenus* Emery, 1908f: 556. Type-species: *Epixenus andrei* (junior secondary homonym in *Monomorium*, replaced by *Monomorium advena*), by subsequent designation of Wheeler, W.M. 1911f: 163. *Epixenus* junior synonym of *Monomorium*: Brown & Wilson, 1957b: 244; discussion and confirmation of synonymy: Bolton, 1987: 292; Tinaut & Ortiz, 1988: 167.

*EPOECUS* (junior synonym of *Monomorium*)

*Epoecus* Emery, 1893a: cclxxvi. Type-species: *Epoecus pergandei*, by monotypy. *Epoecus* junior synonym of *Monomorium*: Ettershank, 1966: 82; see also Bolton, 1987: 298.

**EPOPOSTRUMA** (Myrmicinae: Dacetini)
*Epopostruma* Forel, 1895f: 422 [as subgenus of *Strumigenys*]. Type-species: *Strumigenys (Epopostruma) quadrispinosa*, by subsequent designation of Wheeler, W.M. 1911f: 163. *Epopostruma* raised to genus: Emery, 1897c: 573. *Epopostruma* provisional senior synonym of *Hexadaceton:* Brown, 1973b: 181. *Epopostruma* senior synonym of *Hexadaceton:* Taylor & Brown, D.R. 1985: 63 (accepted as confirmation).

[*EQUESSIMESSOR* (misspelling)
*Equessimessor* Santschi, 1936a: 32, incorrect subsequent spelling of *Equestrimessor.*]

*EQUESTRIMESSOR* (junior synonym of *Monomorium*)
*Equestrimessor* Santschi, 1919d: 92 [as subgenus of *Monomorium*]. Type-species: *Holcomyrmex chobauti*, by subsequent designation of Donisthorpe, 1943f: 644. *Equestrimessor* junior synonym of *Monomorium:* Ettershank, 1966: 82; see also discussion in Bolton, 1987: 297. [*Equessimessor:* see above.]

*EREBOMYRMA* (junior synonym of *Oligomyrmex*)
*Erebomyrma* Wheeler, W.M. 1903a: 138. Type-species: *Erebomyrma longii*, by monotypy. *Erebomyrma* senior synonym of *Spelaeomyrmex:* Wilson, 1962a: 63. *Erebomyrma* junior synonym of *Oligomyrmex:* Ettershank, 1966: 119. *Erebomyrma* revived from synonymy: Wilson, 1986b: 61; returned to synonymy of *Oligomyrmex:* Bolton, 1994: 106.

*EREMNOCYSTUS* (junior synonym of *Myrmecocystus*)
*Eremnocystus* Snelling, 1976: 92 [as subgenus of *Myrmecocystus*]. Type-species: *Myrmecocystus creightoni*, by original designation. Key to species: Snelling, 1976: 93. *Eremnocystus* junior synonym of *Myrmecocystus:* Snelling, 1981: 403.

*ERICAPELTA* (junior synonym of *Amblyopone*)
*Ericapelta* Kusnezov, 1955: 273. Type-species: *Ericapelta egregia*, by monotypy. *Ericapelta* junior synonym of *Amblyopone:* Brown, 1960a: 156.

*ERIMELOPHORUS* (junior synonym of *Melophorus*)
*Erimelophorus* Wheeler, W.M. 1935c: 71 [as subgenus of *Melophorus*]. Type-species: *Melophorus wheeleri*, by original designation. *Erimelophorus* junior synonym of *Melophorus:* Brown, 1955d: 474.

*ERIOPHEIDOLE* (junior synonym of *Pheidole*)
*Eriopheidole* Kusnezov, 1952a: 10. Type-species: *Eriopheidole symbiotica* (unresolved junior secondary homonym in *Pheidole*), by monotypy. *Eriopheidole* provisional junior synonym of *Pheidole:* Brown,1973b: 180. *Eriopheidole* junior synonym of *Pheidole:* Smith, D.R. 1979: 1366.

*ESCHERICHIA* (junior synonym of *Probolomyrmex*)
*Escherichia* Forel, 1910c: 245. Type-species: *Escherichia brevirostris*, by monotypy. *Escherichia* junior synonym of *Probolomyrmex:* Taylor, 1965d: 346.

*EUBOTHROPONERA* (junior synonym of *Platythyrea*)
*Eubothroponera* Clark, 1930c: 8. Type-species: *Eubothroponera dentinodis*, by original designation. *Eubothroponera* junior synonym of *Platythyrea:* Brown, 1975: 6.

#**EUCREMA** (subgenus of *Crematogaster*)
*Eucrema* Santschi, 1918d: 182 [as subgenus of *Crematogaster*]. Type-species: *Formica acuta*, by original designation. *Eucrema* provisional junior synonym of *Crematogaster:* Brown, 1973b: 180 (unconfirmed). [Emery, 1919b: 62; Emery, 1922e: 137; and Soulié, 1965: 78 incorrectly treat *Eucrema* as a junior synonym of subgenus *Crematogaster sensu stricto*. This stems from Emery's (1912d: 272) unjustified subsequent designation of *Formica acuta* as type-species of *Crematogaster;* see above under that genus.]

**EUCRYPTOCERUS** (Myrmicinae: Cephalotini)
*Eucryptocerus* Kempf, 1951: 127. Type-species: *Cryptocerus oculatus*, by original designation. *Eucryptocerus* provisional junior synonym of *Cephalotes:* Brown, 1973b: 180. *Eucryptocerus* valid genus: Kempf, 1973c: 460. Key to species: Kempf, 1951: 128.

***EULITHOMYRMEX** (Myrmicinae: Agroecomyrmecini)
**Eulithomyrmex* Carpenter, 1935: 91. Replacement name for **Lithomyrmex* Carpenter, 1930: 34. [Junior homonym of *Lithomyrmex* Clark, 1928a: 30.]

*EUMECOPONE* (junior synonym of *Pachycondyla*)
*Eumecopone* Forel, 1901e: 335 [as subgenus of *Neoponera*]. Type-species: *Neoponera (Eumecopone) agilis*, by monotypy. *Eumecopone* provisional junior synonym of *Pachycondyla:* Brown, 1973b: 180. *Eumecopone* junior synonym of *Pachycondyla:* Brown, in preparation.

*EUOPHTHALMA* (junior synonym of *Solenopsis*)
*Euophthalma* Creighton, 1930b: 43 [as subgenus of *Solenopsis*]. Type-species: *Myrmica globularia*, by original designation. *Euophthalma* junior synonym of *Solenopsis:* Ettershank, 1966: 134.

*EUPONERA* (junior synonym of *Pachycondyla*)
*Euponera* Forel, 1891b: 126 [as subgenus of *Ponera*]. Type-species: *Ponera (Euponera) sikorae*, by monotypy. *Euponera* raised to genus: Emery, 1901a: 46. *Euponera* junior synonym of *Pachycondyla:* Brown, in preparation.

**EUPRENOLEPIS** (Formicinae: Lasiini)
*Euprenolepis* Emery, 1906b: 134 [as subgenus of *Prenolepis*]. Type-species: *Prenolepis procera*, by original designation. *Euprenolepis* raised to genus, and senior synonym of *Chapmanella:* Brown, 1953h: 6.

**EURHOPALOTHRIX** (Myrmicinae: Basicerotini)
*Eurhopalothrix* Brown & Kempf, 1961: 44. Type-species: *Rhopalothrix bolaui*, by original designation. [*Eurhopalothrix* Brown & Kempf, 1960: 202, unavailable name, proposed without designation of type-species.] Keys to world species: Brown & Kempf, 1960: 245; to Indo-Australian species: Taylor, 1968b: 334; Taylor, 1990b: 401.

*EUSPHINCTUS* (junior synonym of *Sphinctomyrmex*)
*Eusphinctus* Emery, 1893a: cclxxv. Type-species: *Eusphinctus furcatus*, by monotypy. *Eusphinctus* subgenus of *Sphinctomyrmex:* Emery, 1895k: 456; maintained as subgenus: Forel, 1917: 239. *Eusphinctus* revived status as genus: Wheeler, W.M. 1918a: 219. *Eusphinctus* provisional junior synonym of *Sphinctomyrmex:* Brown, 1973b: 180. *Eusphinctus* junior synonym of *Sphinctomyrmex:* Brown, 1975: 31.

**EUTETRAMORIUM** (Myrmicinae: Myrmicini)
*Eutetramorium* Emery, 1899f: 280. Type-species: *Eutetramorium mocquerysi*, by subsequent designation of Wheeler, W.M. 1911f: 163. Notes on genus: Bolton, 1976: 293.

*EVELYNA* (junior synonym of *Myrmothrinax*)
*Evelyna* Donisthorpe, 1937c: 273 [as subgenus of *Polyrhachis*]. Type-species: *Polyrhachis (Evelyna) cheesmanae*, by original designation. *Evelyna* junior synonym of *Myrmothrinax:* Hung, 1967b: 402.

*EXAMBLYOPONE* (junior synonym of *Prionopelta*)
*Examblyopone* Donisthorpe, 1949g: 401. Type-species: *Examblyopone churchilli* (junior synonym of *Prionopelta majuscula*), by original designation. *Examblyopone* junior synonym of *Prionopelta:* Brown, 1951: 102.

*FLORENCEA* (junior synonym of *Myrmhopla*)
*Florencea* Donisthorpe, 1937a: 624 [as subgenus of *Polyrhachis*]. Type-species: *Polyrhachis (Florencea) kirkae* (junior synonym of *Polyrhachis nigriceps*), by original designation. *Florencea* junior synonym of *Myrmhopla:* Hung, 1967b: 402.

*FORELIFIDIS* (junior synonym of *Oxyepoecus*)
*Forelifidis* Smith, M.R. 1954a: 17. Replacement name for *Martia* Forel, 1907a: 20. [Junior homonym of *Martia* Ragonot, 1887: 18.] *Forelifidis* junior synonym of *Oxyepoecus:* Brown, 1955b: 68.

**FORELIUS** (Dolichoderinae)
*Forelius* Emery, 1888d: 389. Type-species: *Iridomyrmex mccooki*, by monotypy. *Forelius* senior synonym of *Amyrmex* and *Neoforelius*, and review of genus: Shattuck, 1992c: 87. Key to Neotropical species: Kusnezov, 1957b: 16.

*FORELOMYRMEX* (junior synonym of *Pogonomyrmex*)
*Forelomyrmex* Wheeler, W.M. 1913a: 80 [as subgenus of *Pogonomyrmex*]. Replacement name for *Janetia* Forel, 1899c: 61 (footnote). [Junior homonym of *Janetia* Kieffer, 1896: 236.] *Forelomyrmex* provisional junior synonym of *Pogonomyrmex:* Brown, 1973b: 180; confirmed Bolton, 1994: 106.

**FORELOPHILUS** (Formicinae: Camponotini)
*Forelophilus* Kutter, 1931b: 193. Type-species: *Forelophilus overbecki*, by monotypy.

**FORMICA** (Formicinae: Formicini)
*Formica* Linnaeus, 1758: 579. Type-species: *Formica rufa*, by subsequent designation of Curtis, 1829: 752. *Formica* senior synonym of *Formicina:* Wheeler, W.M. 1911c: 860; Wheeler, W.M. 1911f: 164. *Formica* senior synonym of *Raptiformica* and *Serviformica:* Wheeler, W.M. 1922a: 699 (footnote). *Formica* senior synonym of *Coptoformica* (and its junior synonym *Adformica*), *Serviformica* and *Raptiformica:* Smith, M.R. 1951a: 860; Smith, D.R. 1979: 1448. *Formica* senior synonym of *Neoformica:* Buren, 1968a: 39. *Formica* senior synonym of *Serviformica:* Francoeur, 1973: 34. *Formica* senior synonym of *Coptoformica* (and its junior synonym *Adformica*), *Formicina, Hypochira, Iberoformica, Neoformica, Raptiformica* and *Serviformica:* Agosti, 1994a: 106. Keys to Nearctic fauna: Creighton, 1950a: 456; Francoeur, 1973: 272 (*fusca*-group); Buren, 1968a: 36 and Snelling & Buren, 1985: 71 (*sanguinea*-group); to Palaearctic fauna: Dlussky, 1967a: 39 (former U.S.S.R.); Dlussky, 1976b: 80 (*Baltic Amber); Dlussky, 1965a: 15 (Mongolia and Tibet); Arnol'di & Dlussky, 1978: 552 (former European U.S.S.R.); Bernard, 1967: 287 (France and Mediterranean basin); Collingwood, 1978: 93 (Iberian Peninsula); Collingwood, 1979: 112 (Fennoscandia and Denmark); Agosti & Collingwood, 1987b: 285 (Balkans); Kutter, 1977c: 236

(Switzerland); Gösswald, 1985: 274 (Germany). See also for *rufa*-group: Cotti, 1963: 1; Gösswald, 1989: 1; Gösswald, 1990: 1.

*FORMICINA* (junior synonym of *Formica*)

*Formicina* Shuckard, in Swainson & Shuckard, 1840: 172. Type-species: *Formica rufa*, by subsequent designation of Wheeler, W.M. 1911f: 164. *Formicina* junior synonym of *Formica:* Wheeler, W.M. 1911c: 860; Wheeler, W.M. 1911f: 164. *Formicina* unnecessarily and incorrectly used as replacement name for *Lasius* Fabricius: Emery, 1916b: 239; Bondroit, 1918: 19; see also Wheeler, W.M. 1916k: 172. *Formicina* incorrectly as junior synonym of *Lasius:* Emery, 1925b: 226. *Formicina* junior synonym of *Formica:* Donisthorpe, 1943f: 646; Agosti, 1994a: 106. *Formicina* incorrectly as provisional junior synonym of *Lasius:* Brown, 1973b: 180. [As *Formica* and *Formicina* share the same type-species the synonymy is absolute.]

**FORMICITES* (collective group name)

**Formicites* Dlussky, 1981b: 75 [collective group name, without included named taxa.]

****FORMICIUM*** (*Formiciinae)

**Formicium* Westwood, 1854: 393. Type-species: **Formicium brodiei*, by monotypy. **Formicium* junior synonym of **Pseudosirex* (family *Pseudosiricidae): Handlirsch, 1906: 577. **Formicium* revived from synonymy as genus of Formicidae: Cockerell, 1920: 279. **Formicium* senior synonym of **Eoponera*, and revision of genus: Lutz, 1986: 182. [**Formicium* incorrectly as senior synonym of **Pseudosirex:* Hölldobler & Wilson, 1990: 17; incorrectly retained in *Pseudosiricidae: Abe & Smith, 1991: 34.] **Formicium* senior synonym of **Megapterites:* Lutz, 1990: 61.

***FORMICOXENUS*** (Myrmicinae: Formicoxenini)

*Formicoxenus* Mayr, 1855: 413. Type-species: *Myrmica nitidula*, by monotypy. *Formicoxenus* provisional junior synonym of *Leptothorax:* Brown, 1973b: 180. *Formicoxenus* valid genus, senior synonym of *Symmyrmica*, and revision of genus: Francoeur, Loiselle & Buschinger, 1985: 343.

***FROGGATTELLA*** (Dolichoderinae)

*Froggattella* Forel, 1902h: 459. Type-species: *Acantholepis kirbii*, by original designation. Key to species: Wheeler, W.M. 1936a: 5. Review of genus: Shattuck, 1992c: 95.

*FULAKORA* (junior synonym of *Stigmatomma*)

*Fulakora* Mann, 1919: 279 [as subgenus of *Stigmatomma*]. Type-species: *Stigmatomma* (*Fulakora*) *celata*, by original designation. *Fulakora* junior synonym of *Stigmatomma:* Brown, 1949c: 88. [*Stigmatomma* junior synonym of *Amblyopone:* Brown, 1960a: 155.]

*GAESOMYRMEX* (junior synonym of *Gesomyrmex*)

*Gaesomyrmex* Dalla Torre, 1893: 175, unjustified emendation of *Gesomyrmex*. *Gaesomyrmex* junior synonym of *Gesomyrmex:* Forel, 1893a: 167.

*GALLARDOMYRMA* (junior synonym of *Pheidole*)

*Gallardomyrma* Bruch, 1932: 271. Type-species: *Gallardomyrma argentina*, by original designation. *Gallardomyrma* provisional junior synonym of *Pheidole:* Brown, 1973b: 180. *Gallardomyrma* junior synonym of *Pheidole:* Smith, D.R. 1979: 1366; Wilson, 1984: 327.

*GAUROMYRMEX* (junior synonym of *Vollenhovia*)

*Gauromyrmex* Menozzi, 1933c: 146. Type-species: *Gauromyrmex bengakalisi*, by monotypy. *Gauromyrmex* senior synonym of *Acalama* and *Solenomyrma:* Brown, 1953c: 10. *Gauromyrmex* provisional junior synonym of *Vollenhovia:* Brown, 1973b: 180; confirmed Bolton, 1994: 105.

*GEOGNOMICUS* (junior synonym of *Dicroaspis*)

*Geognomicus* Menozzi, 1924b: 220. Type-species: *Geognomicus wheeleri* (junior synonym of *Dicroaspis cryptocera*), by original designation. *Geognomicus* junior synonym of *Dicroaspis:* Bolton, 1981a: 56.

***GESOMYRMEX*** (Formicinae: Gesomyrmecini)

*Gesomyrmex* Mayr, 1868c: 50. Type-species: **Gesomyrmex hoernesi*, by monotypy. *Gesomyrmex* senior synonym of *Gaesomyrmex:* Forel, 1893a: 167; Emery, 1925b: 47. *Gesomyrmex* senior synonym of *Dimorphomyrmex:* Wheeler, W.M. 1929a: 1. Revision of genus: Cole, 1949: 71.

***GIGANTIOPS*** (Formicinae: Gigantiopini)

*Gigantiops* Roger, 1863b: 11. Type-species: *Formica destructor*, by monotypy.

***GLAMYROMYRMEX*** (Myrmicinae: Dacetini)

*Glamyromyrmex* Wheeler, W.M. 1915e: 487. Type-species: *Glamyromyrmex beebei*, by monotypy. *Glamyromyrmex* senior synonym of *Borgmeierita:* Brown, 1973a: 35. Key to Neotropical species: Kempf, 1960f: 447; to Afrotropical species: Bolton, 1983: 321. [See note under *Strumigenys*.]

****GLAPHYROMYRMEX*** (Formicinae: Formicini)

**Glaphyromyrmex* Wheeler, W.M. 1915h: 131. Type-species: **Glaphyromyrmex oligocenicus*, by monotypy.

*GLYPHOPONE* (junior synonym of *Centromyrmex*)

*Glyphopone* Forel, 1913b: 308. Type-species: *Glyphopone bequaerti*, by monotypy. *Glyphopone* junior synonym of *Centromyrmex:* Brown, 1963: 9.

*GLYPTOMYRMEX* (junior synonym of *Myrmicocrypta*)

*Glyptomyrmex* Forel, 1885a: 365. Type-species: *Glyptomyrmex dilaceratum*, by monotypy. *Glyptomyrmex* junior synonym of *Myrmicocrypta:* Emery, 1894c: 224.

***GNAMPTOGENYS*** (Ponerinae: Ectatommini)

*Gnamptogenys* Roger, 1863a: 174. Type-species: *Ponera tornata*, by subsequent designation of Emery, 1911d: 44. [Type-species not *Ectatomma concinna* as designated by Wheeler, W.M. 1911f: 164; see Wheeler, W.M. 1913a: 79.] *Gnamptogenys* subgenus of *Ectatomma:* Mayr, 1887: 540; maintained as subgenus: Dalla Torre, 1893: 23; Emery, 1911d: 44; Wheeler, W.M. 1922a: 644. *Gnamptogenys* revived status as genus, senior synonym of *Alfaria* (and its junior synonym *Opisthoscyphus*), *Barbourella*, *Commateta*, *Emeryella*, *Holcoponera*, *Parectatomma*, *Poneracantha*, *Rhopalopone* (and its junior synonym *Mictoponera*), *Spaniopone*, *Stictoponera*, *Tammoteca*, *Wheeleripone*, and revision of genus: Brown, 1958g: 211. Key to New World species: Brown, 1958g: 230; to Old World species: Brown, 1958g: 237.

**#*GONEPIMYRMA*** (subgenus of *Epimyrma*)

*Gonepimyrma* Bernard, 1948: 146 [as subgenus of *Epimyrma*]. Type-species: *Epimyrma* (*Gonepimyrma*) *africana*, by monotypy. *Gonepimyrma* provisional junior synonym of *Epimyrma:* Brown, 1973b: 181 (unconfirmed).

***GONIOMMA*** (Myrmicinae: Pheidolini)

*Goniomma* Emery, 1895c: 298 [as subgenus of *Stenamma*]. Type-species: *Aphaenogaster blanci*, by subsequent designation of Wheeler, W.M. 1911f: 164. *Goniomma* raised to genus: Emery, 1908e: 460. Key to species: Santschi, 1929e: 145 (out of date). Partial revision of genus: Acosta Salmerón, 1982: 7.

*GONIOTHORAX* (junior homonym, see *Nesomyrmex*)

*Goniothorax* Emery, 1896g: 58 (diagnosis in key) [as subgenus of *Leptothorax*]. Type-species: *Leptothorax vicinus*, by subsequent designation of Wheeler, W.M. 1911f: 164. [Junior homonym of *Goniothorax* Milne-Edwards, 1879: 103 (Crustacea).] *Goniothorax* Emery senior synonym of *Caulomyrma* and *Nesomyrmex:* Emery, 1922f: 249. *Nesomyrmex* first available replacement name for *Goniothorax* Emery: Smith, M.R. 1950: 30.

*GRANISOLENOPSIS* (junior synonym of *Solenopsis*)

*Granisolenopsis* Kusnezov, 1957a: 277 [as subgenus of *Solenopsis*]. Type-species: *Solenopsis* (*Granisolenopsis*) *granivora*, by original designation. *Granisolenopsis* junior synonym of *Solenopsis:* Ettershank, 1966: 134.

***GYMNOMYRMEX*** (Myrmicinae: Dacetini)

*Gymnomyrmex* Borgmeier, 1954b: 279. Type-species: *Gymnomyrmex splendens*, by original designation. Keys to species: Kempf, 1960f: 451; Perrault, 1986: 3. [See note under *Strumigenys*.]

*HAGENSIA* (junior synonym of *Pachycondyla*)

*Hagensia* Forel, 1901f: 333 [as subgenus of *Megaloponera* (sic)]. Type-species: *Megaloponera* (sic) (*Hagensia*) *havilandi*, by monotypy. *Hagensia* subgenus of *Euponera:* Forel, 1917: 237. *Hagensia* raised to genus: Arnold, 1926: 202. Revision of genus: Arnold, 1951: 53. *Hagensia* junior synonym of *Pachycondyla:* Brown, in preparation.

**#*HAGIOMYRMA*** (subgenus of *Polyrhachis*)

*Hagiomyrma* Wheeler, W.M. 1911c: 860 [as subgenus of *Polyrhachis*]. Type-species: *Formica ammon*, by original designation. *Hagiomyrma* provisional junior synonym of *Polyrhachis:* Brown, 1973b: 181 (unconfirmed).

*HAGIOSTIGMACROS* (junior synonym of *Stigmacros*)

*Hagiostigmacros* McAreavey, 1957: 19 [as subgenus of *Stigmacros*]. Type-species: *Stigmacros barretti*, by original designation. *Hagiostigmacros* provisional junior synonym of *Stigmacros:* Brown, 1973b: 181. *Hagiostigmacros* junior synonym of *Stigmacros:* Taylor & Brown, D.R. 1985: 145 (accepted as confirmation).

*HAGIOXENUS* (junior synonym of *Rhoptromyrmex*)

*Hagioxenus* Forel, 1910a: 8. Type-species: *Hagioxenus schmitzi*, by monotypy. *Hagioxenus* junior synonym of *Rhoptromyrmex:* Bolton, 1986a: 2.

*HALMAMYRMECIA* (junior synonym of *Promyrmecia*)

*Halmamyrmecia* Wheeler, W.M. 1922b: 195 [as subgenus of *Myrmecia*]. Type-species: *Myrmecia pilosula*, by original designation. *Halmamyrmecia* junior synonym of *Promyrmecia:* Clark, 1951: 119. [*Promyrmecia* junior synonym of *Myrmecia:* Brown, 1953j: 1.]

*HARNEDIA* (junior synonym of *Paracryptocerus*)

*Harnedia* Smith, M.R. 1949c: 20 [as subgenus of *Paracryptocerus*]. Type-species: *Cryptocerus umbraculatus*, by original designation. Revision of subgenus: Kempf, 1958a: 61. *Harnedia* junior synonym of *Paracryptocerus:* Kempf, 1972a: 175.

[*Paracryptocerus* junior synonym of *Zacryptocerus:* Kempf, 1973c: 460.]

**HARPAGOXENUS** (Myrmicinae: Formicoxenini)

*Harpagoxenus* Forel, 1893a: 167. Replacement name for *Tomognathus* Mayr, 1861: 56. [Junior homonym of *Tomognathus* Agassiz, 1850: 376.]

**HARPEGNATHOS** (Ponerinae: Ponerini)

*Harpegnathos* Jerdon, 1851: 116. Type-species: *Harpegnathos saltator,* by monotypy. *Harpegnathos* senior synonym of *Drepanognathus:* Roger, 1863b: 22. *Harpegnathos* incorrectly as junior synonym of *Drepanognathus:* Dalla Torre, 1893: 22. *Harpegnathos* senior synonym of *Drepanognathus:* Forel, 1893a: 166; Forel, 1900c: 63.

**#HEDOMYRMA** (subgenus of *Polyrhachis*)

*Hedomyrma* Forel, 1915b: 107 [as subgenus of *Polyrhachis*]. Type-species: *Polyrhachis ornata,* by original designation. *Hedomyrma* senior synonym of *Morleyidris:* Hung, 1967b: 402. *Hedomyrma* provisional junior synonym of *Polyrhachis:* Brown, 1973b: 181 (unconfirmed).

**#HEMIOPTICA** (subgenus of *Polyrhachis*)

*Hemioptica* Roger, 1862a: 238. Type-species: *Hemioptica scissa,* by monotypy. *Hemioptica* subgenus of *Polyrhachis:* Mayr, 1872: 138. *Hemioptica* revived status as genus: Dalla Torre, 1893: 271; maintained as genus: Emery, 1925b: 209. *Hemioptica* provisional junior synonym of *Polyrhachis:* Brown, 1973b: 181 (unconfirmed); here reverts to subgenus of *Polyrhachis.*

*HENDECAPHEIDOLE* (junior synonym of *Pheidole*)

*Hendecapheidole* Wheeler, W.M. 1922d: 3 [as subgenus of *Pheidole*]. Type-species: *Pheidole tachigaliae,* by original designation. *Hendecapheidole* senior synonym of *Cephalomorium:* Santschi, 1925d: 228. *Hendecapheidole* provisional junior synonym of *Pheidole:* Brown, 1973b: 181. *Hendecapheidole* junior synonym of *Pheidole:* Smith, D.R. 1979: 1366.

*HENDECATELLA* (junior synonym of *Oligomyrmex*)

*Hendecatella* Wheeler, W.M. 1927h: 93 [as subgenus of *Oligomyrmex*]. Type-species: *Oligomyrmex (Hendecatella) capreolus,* by monotypy. *Hendecatella* junior synonym of *Oligomyrmex:* Ettershank, 1966: 119.

*HEPTACONDYLUS* (junior synonym of *Myrmicaria*)

*Heptacondylus* Smith, F. 1857a: 71. Type-species: *Heptacondylus subcarinatus,* by subsequent designation of Wheeler, W.M. 1911f: 164. *Heptacondylus* senior synonym of *Physatta:* Mayr, 1862: 755 (in text); see also Roger, 1863b: 27. *Heptacondylus* junior synonym of *Myrmicaria:* Smith, F. 1865: 73.

*HEPTASTRUMA* (junior synonym of *Rhopalothrix*)

*Heptastruma* Weber, 1934a: 54. Type-species: *Heptastruma wheeleri* (junior secondary homonym in *Rhopalothrix,* replaced by *Rhopalothrix weberi*), by original designation. *Heptastruma* junior synonym of *Rhopalothrix:* Brown & Kempf, 1960: 230.

*HERCYNIA* (junior synonym of *Wasmannia*)

*Hercynia* Enzmann, J. 1947a: 43. Type-species: *Hercynia panamana* (junior synonym of *Wasmannia auropunctata*), by monotypy. *Hercynia* junior synonym of *Wasmannia:* Brown, 1948d: 102.

*HETEROMYRMEX* (junior synonym of *Vollenhovia*)

*Heteromyrmex* Wheeler, W.M. 1920: 53. Type-species: *Vollenhovia rufiventris,* by original designation. *Heteromyrmex* junior synonym of *Vollenhovia:* Ettershank, 1966: 146.

**HETEROPONERA** (Ponerinae: Ectatommini)

*Heteroponera* Mayr, 1887: 532. Type-species: *Heteroponera carinifrons,* by monotypy. *Heteroponera* junior synonym of *Acanthoponera:* Emery, 1911d: 35. *Heteroponera* revived from synonymy as subgenus of *Acanthoponera:* Forel, 1917: 236. *Heteroponera* revived status as genus: Brown, 1952h: 70. *Heteroponera* senior synonym of *Anacanthoponera:* Brown, 1952h: 70. *Heteroponera* senior synonym of *Paranomopone:* Brown, in Borgmeier, 1957: 112. Revision of genus: Brown, 1958g: 194. Key to Neotropical species: Kempf, 1962a: 31.

*HEXADACETON* (junior synonym of *Epopostruma*)

*Hexadaceton* Brown, 1948e: 120. Type-species: *Hexadaceton frosti,* by original designation. *Hexadaceton* provisional junior synonym of *Epopostruma:* Brown, 1973b: 181. *Hexadaceton* junior synonym of *Epopostruma:* Taylor & Brown, D.R. 1985: 63 (accepted as confirmation).

*HINCKSIDRIS* (junior synonym of *Myrmelachista*)

*Hincksidris* Donisthorpe, 1944b: 59 [as subgenus of *Myrmelachista*]. Replacement name for *Decamera* Roger, 1863a: 166. [Junior homonym of *Decamera* Mulsant, 1842: 503.] *Hincksidris* provisional junior synonym of *Myrmelachista:* Brown, 1973b: 181. *Hincksidris* junior synonym of *Myrmelachista:* Snelling & Hunt, 1976: 110.

[*HIPHOPELTA* (misspelling)

*Hiphopelta* Forel, 1913a: 108, incorrect subsequent spelling of *Xiphopelta.*]

*HOLCOMYRMEX* (junior synonym of *Monomorium*)

*Holcomyrmex* Mayr, 1879: 671. Type-species: *Holcomyrmex scabriceps,* by subsequent designation of Bingham, 1903: 280. [*Holcomyrmex* Smith, F. 1873: ix, *nomen nudum* attributed to Mayr.] *Holcomyrmex* junior synonym of *Monomorium:* Emery, 1908h: 667. *Holcomyrmex* revived from synonymy as subgenus of *Monomorium:* Emery, 1915i: 190. *Holcomyrmex* junior synonym of *Monomorium:* Ettershank, 1966: 82; see also discussion in Bolton, 1987: 295.

*HOLCOPONERA* (junior synonym of *Gnamptogenys*)

*Holcoponera* Mayr, 1887: 540 [as subgenus of *Ectatomma*]. Type-species: *Gnamptogenys striatula,* by subsequent designation of Emery, 1911d: 40. *Holcoponera* raised to genus: Emery, 1902b: 181; maintained as genus: Emery, 1911d: 40; Wheeler, W.M. 1922a: 643. Key to species: Santschi, 1929h: 471 (out of date). *Holcoponera* junior synonym of *Gnamptogenys:* Brown, 1958g: 211.

*HOLCOPONERA* (junior homonym, junior synonym of *Cylindromyrmex*)

*Holcoponera* Cameron, 1891: 92. Type-species: *Holcoponera whymperi* (junior synonym of *Cylindromyrmex striatus*), by monotypy. [Junior homonym of *Holcoponera* Mayr, above.] *Holcoponera* Cameron junior synonym of *Cylindromyrmex:* Forel, 1892f: 256.

*HOLOPONE* (junior synonym of *Eciton*)

*Holopone* Santschi, 1925b: 11 [as subgenus of *Eciton*]. Type-species: *Eciton rapax,* by original designation. *Holopone* junior synonym of *Eciton:* Borgmeier, 1936: 55.

*HOPLOMYRMUS* (junior synonym of *Myrma*)

*Hoplomyrmus* Gerstäcker, 1859: 262. Type-species: *Hoplomyrmus schistaceus,* by monotypy. [*Hoplomyrmus* also described as new by Gerstäcker, 1862: 508.] *Hoplomyrmus* junior synonym of *Polyrhachis:* Roger, 1861b: 174; Mayr, 1863: 446; Dalla Torre, 1893: 257. *Hoplomyrmus* junior synonym of *Myrma:* Wheeler, W.M. 1922a: 993; Emery, 1925b: 198; Bolton, 1973b: 288.

**HUBERIA** (Myrmicinae: Pheidolini)

*Huberia* Forel, 1890b: cv. Type-species: *Tetramorium striatum,* by monotypy. Review of genus: Ettershank, 1966: 155; see also notes in Brown, 1958h: 25.

*HYLIDRIS* (junior synonym of *Pristomyrmex*)

*Hylidris* Weber, 1941a: 190. Type-species: *Hylidris myersi* (junior synonym of *Pristomyrmex africanus*), by original designation. *Hylidris* junior synonym of *Pristomyrmex:* Brown, 1953c: 9.

**HYLOMYRMA** (Myrmicinae: Myrmicini)

*Hylomyrma* Forel, 1912g: 16 [as subgenus of *Pogonomyrmex*]. Type-species: *Pogonomyrmex (Hylomyrma) columbicus,* by original designation. *Hylomyrma* raised to genus: Wheeler, W.M. 1922a: 660 (in key); maintained as genus Brown, 1953h:3. *Hylomyrma* senior synonym of *Lundella:* Brown, 1953h: 3. Revision of genus: Kempf, 1973b: 225.

**#HYPERCOLOBOPSIS** (subgenus of *Camponotus*)

*Hypercolobopsis* Emery, 1920b: 250 [as subgenus of *Camponotus*]. Type-species: *Colobopsis paradoxa,* by original designation. *Hypercolobopsis* senior synonym of *Neocolobopsis:* Kempf, 1968b: 411. *Hypercolobopsis* provisional junior synonym of *Camponotus:* Brown, 1973b: 181 (unconfirmed).

*HYPOCHIRA* (not identifiable to subfamily)

*Hypochira* Buckley, 1866a: 169 [as subgenus of *Formica*]. Type-species: *Formica (Hypochira) subspinosa,* by monotypy. *Hypochira* junior synonym of *Formica:* Dalla Torre, 1893: 192; Agosti, 1994a: 106. *Hypochira* questionably junior synonym of *Dolichoderus:* Emery, 1895c: 338; Emery, 1925b: 271. *Hypochira* unplaced taxon of Dolichoderinae: Smith, D.R. 1979: 1422. Unrecognisable taxon.

*HYPOCLINEA* (junior synonym of *Dolichoderus*)

*Hypoclinea* Mayr, 1855: 377. Type-species: *Formica quadripunctata,* by subsequent designation of Wheeler, W.M. 1911f: 165. *Hypoclinea* junior synonym of *Dolichoderus:* Forel, 1878: 386. *Hypoclinea* revived from synonymy as subgenus of *Dolichoderus:* Emery, 1894c: 234; maintained as subgenus: Emery, 1912g: 10. *Hypoclinea* raised to genus: Kempf, 1972a: 118. *Hypoclinea* provisional junior synonym of *Dolichoderus:* Brown, 1973b: 181. *Hypoclinea* junior synonym of *Dolichoderus:* Shattuck, 1992c: 66.

*HYPOCRYPTOCERUS* (junior synonym of *Zacryptocerus*)

*Hypocryptocerus* Wheeler, W.M. 1920: 53 [as subgenus of *Cryptocerus*]. Type-species: *Formica haemorrhoidalis,* by original designation. *Hypocryptocerus* raised to genus: Wheeler, W.M. 1936b: 200. Revision of genus: Kempf, 1951: 143. *Hypocryptocerus* junior synonym of *Zacryptocerus:* Kempf, 1973c: 460.

*HYPOCYLINDROMYRMEX* (junior synonym of *Cylindromyrmex*)

*Hypocylindromyrmex* Wheeler, W.M. 1924a: 106 [as

subgenus of *Cylindromyrmex*]. Type-species: *Cylindromyrmex longiceps*, by original designation. *Hypocylindromyrmex* provisional junior synonym of *Cylindromyrmex:* Brown,1973b: 181. *Hypocylindromyrmex* junior synonym of *Cylindromyrmex:* Brown, 1975: 36.

HYPOPHEIDOLE (*nomen nudum*)

*Hypopheidole* Ashmead, 1905b: 383, *nomen nudum.*

***HYPOPOMYRMEX** (Myrmicinae: Pheidologetonini)

**Hypopomyrmex* Emery, 1891a: 148. Type-species: **Hypopomyrmex bombiccii*, by monotypy. **Hypopomyrmex* transferred from Dacetini to Pheidologetonini: Brown & Carpenter, 1979: 422.

**HYPOPONERA** (Ponerinae: Ponerini)

*Hypoponera* Santschi, 1938b: 79 [as subgenus of *Ponera*]. Type-species: *Ponera abeillei*, by original designation. *Hypoponera* raised to genus: Taylor, 1967a: 9.

IBEROFORMICA (junior synonym of *Formica*)

*Iberoformica* Tinaut, 1990: 282 [as subgenus of *Formica*]. Type-species: *Formica subrufa*, by original designation. *Iberoformica* junior synonym of *Formica:* Agosti, 1994a: 107.

ICOTHORAX (junior synonym of *Myrafant*)

*Icothorax* Hamann & Klemm, 1967: 415 [as subgenus of *Leptothorax*]. Type-species: *Leptothorax (Icothorax) megalops*, by monotypy. *Icothorax* junior synonym of *Myrafant:* Baroni Urbani, 1978b: 556. *Icothorax* junior synonym of *Chalepoxenus:* Hölldobler & Wilson, 1990: 13; reverts here to synonymy of *Myrafant.* [*Myrafant* junior synonym of *Leptothorax:* Bolton, 1982: 319.]

IDRISELLA (junior synonym of *Pheidologeton*)

*Idrisella* Santschi, 1937h: 372. Type-species: *Pheidologeton dentiviris*, by original designation. *Idrisella* junior synonym of *Pheidologeton:* Ettershank, 1966: 115.

***ILEMOMYRMEX** (Myrmicinae: Stenammini?)

**Ilemomyrmex* Wilson, 1985a: 2. Type-species: **Ilemomyrmex caecus*, by original designation.

***IMHOFFIA** (Formicinae: unplaced to tribe)

**Imhoffia* Heer, 1850: 153. Type-species: **Imhoffia nigra*, by monotypy.

**INDOMYRMA** (Myrmicinae: Stenammini)

*Indomyrma* Brown, 1986: 37. Type-species: *Indomyrma dasypyx*, by original designation.

IRENEA (junior synonym of *Myrmatopa*)

*Irenea* Donisthorpe, 1938e: 502 [as subgenus of *Dolichoderus*]. Type-species: *Dolichoderus (Irenea) omyrmex*, by original designation. *Irenea* raised to genus: Donisthorpe, 1941h: 59. *Irenea* provisional junior synonym of *Polyrhachis:* Brown, 1973b: 181. *Irenea* junior synonym of *Myrmatopa:* Bolton, 1994: 50.

IRENEELLA (junior synonym of *Rhoptromyrmex*)

*Ireneella* Donisthorpe, 1941i: 175. Type-species: *Ireneella papuensis* (junior synonym of *Rhoptromyrmex melleus*), by original designation. *Ireneella* junior synonym of *Rhoptromyrmex:* Bolton, 1976: 298.

IRENEIDRIS (junior synonym of *Monomorium*)

*Ireneidris* Donisthorpe, 1943c: 81. Type-species: *Ireneidris myops* (junior synonym of *Monomorium talpa*), by original designation. *Ireneidris* junior synonym of *Monomorium:* Ettershank, 1966: 82; see also notes in Bolton, 1987: 298.

**IRENEOPONE** (Myrmicinae: Formicoxenini)

*Ireneopone* Donisthorpe, 1946d: 242. Type-species: *Ireneopone gibber*, by original designation. See also notes in Bolton, 1976: 292.

**IRIDOMYRMEX** (Dolichoderinae)

*Iridomyrmex* Mayr, 1862: 702. Type-species: *Formica detecta* (junior synonym of *Iridomyrmex purpureus*), by subsequent designation of Bingham, 1903: 297. Review of genus: Shattuck, 1992a: 14; Shattuck, 1992c: 101. Keys to species: Shattuck, 1993a: 113 (*purpureus*-group); Shattuck, 1993b: 1303 (*calvus*-group). See also: Shattuck, 1994: 99.

IROGERA (junior synonym of *Rogeria*)

*Irogera* Emery, 1915i: 191 [as subgenus of *Rogeria*]. Type-species: *Rogeria procera*, by original designation. *Irogera* raised to genus: Brown, 1953h: 4 (in text); maintained as genus: Kempf, 1961d: 436. Key to species: Kempf, 1964e: 66. *Irogera* junior synonym of *Rogeria:* Kempf, 1965: 185.

ISCHNOMYRMEX (junior synonym of *Pheidole*)

*Ischnomyrmex* Mayr, 1862: 738. Type-species: *Myrmica longipes*, by monotypy. *Ischnomyrmex* subgenus of *Stenamma:* Emery, 1895c: 298. *Ischnomyrmex* revived status as genus: Stitz, 1911a: 367. *Ischnomyrmex* subgenus of *Pheidole:* Forel, 1913k: 49. *Ischnomyrmex* revived status as genus: Emery, 1914a: 40; maintained as genus: Wheeler, W.M. 1922a: 661; Emery, 1922e: 113. *Ischnomyrmex* senior synonym of *Isopheidole:* Forel, 1913k: 49. *Ischnomyrmex* provisional junior synonym of *Pheidole:* Brown, 1973b: 181. *Ischnomyrmex* junior synonym of *Pheidole:* Smith, D.R. 1979: 1365.

**ISHAKIDRIS** (Myrmicinae: Phalacromyrmecini)

*Ishakidris* Bolton, 1984: 374. Type-species: *Ishakidris ascitaspis*, by original designation.

ISOLCOMYRMEX (junior synonym of *Monomorium*)

*Isolcomyrmex* Santschi, 1917b: 296 [as subgenus of *Monomorium*]. Type-species: *Holcomyrmex santschii* (junior secondary homonym in *Monomorium*, replaced by *Monomorium santschianum*), by monotypy. *Isolcomyrmex* junior synonym of *Monomorium:* Ettershank, 1966: 82; see also notes in Bolton, 1987: 296. [*Isholcomyrmex* Santschi, 1936a: 32, incorrect subsequent spelling.]

ISOPHEIDOLE (junior synonym of *Ischnomyrmex*)

*Isopheidole* Forel, 1912m: 765 [as subgenus of *Pheidole*]. Type-species: *Myrmica longipes*, by monotypy. *Isopheidole* junior synonym of *Ischnomyrmex:* Forel, 1913k: 49. [*Ischnomyrmex* junior synonym of *Pheidole:* Smith, D.R. 1979: 1365.]

JANETIA (junior homonym, see *Forelomyrmex*)

*Janetia* Forel, 1899c: 61 (footnote) [as subgenus of *Pogonomyrmex*]. Type-species: *Pogonomyrmex (Janetia) mayri*, by monotypy. [Junior homonym of *Janetia* Kieffer, 1896: 236 (Diptera).] *Forelomyrmex* Wheeler, W.M. 1913a: 80, replacement name for *Janetia* Forel.

JOHNIA (junior synonym of *Aulacomyrma*)

*Johnia* Karavaiev, 1927e: 43 [as subgenus of *Polyrhachis*]. Type-species: *Polyrhachis (Johnia) schizospina*, by monotypy. *Johnia* junior synonym of *Aulacomyrma:* Hung, 1967b: 402.

**#KARAVAIEVIA** (subgenus of *Camponotus*)

*Karavaievia* Emery, 1925b: 115 [as subgenus of *Camponotus*]. Type-species: *Camponotus exsectus*, by original designation. *Karavaievia* provisional junior synonym of *Camponotus:* Brown, 1973b: 181 (unconfirmed). Key to species: Dumpert, 1986: 571.

KARAWAJEWELLA (junior synonym of *Dolichoderus*)

*Karawajewella* Donisthorpe, 1944b: 59. Replacement name for *Diabolus* Karavaiev, 1926d: 424. [Junior homonym of *Diabolus* Gray, J.E. 1841: 400.] *Karawajewella* provisional junior synonym of *Dolichoderus:* Brown, 1973b: 181. *Karawajewella* junior synonym of *Dolichoderus:* Shattuck, 1992c: 66.

**KARTIDRIS** (Myrmicinae: Pheidolini)

*Kartidris* Bolton, 1991: 10. Type-species: *Kartidris nyos*, by original designation. Key to species: Bolton, 1991: 12.

***KOTSHKORKIA** (Dolichoderinae)

**Kotshkorkia* Dlussky, 1981b: 71. Type-species: **Kotshkorkia laticeps*, by original designation.

**KYIDRIS** (Myrmicinae: Dacetini)

*Kyidris* Brown, 1949d: 3. Type-species: *Kyidris mutica*, by original designation. *Kyidris* senior synonym of *Polyhomoa:* Creighton, 1950b: 93; see also Brown & Yasumatsu, 1951: 93. Revision of genus: Wilson & Brown, 1956: 439. [See note under *Strumigenys.*]

LABAUCHENA (junior synonym of *Solenopsis*)

*Labauchena* Santschi, 1930e: 81. Type-species: *Labauchena daguerrei*, by monotypy. *Labauchena* junior synonym of *Solenopsis:* Ettershank, 1966: 134.

LABIDOGENYS (junior synonym of *Strumigenys*)

*Labidogenys* Roger, 1862a: 249. Type-species: *Labidogenys lyroessa*, by monotypy. *Labidogenys* junior synonym of *Strumigenys:* Roger, 1863b: 40; Dalla Torre, 1893: 145; Bingham, 1903: 147. *Labidogenys* revived from synonymy: Brown, 1948e: 102. *Labidogenys* junior synonym of *Strumigenys:* Brown & Wilson, 1959b: 281; Brown, 1960b: 38; Bolton, 1983: 358.

**LABIDUS** (Ecitoninae: Ecitonini)

*Labidus* Jurine, 1807: 282. Type-species: *Labidus latreillei* (junior synonym of *Labidus coecus*), by monotypy. *Labidus* junior synonym of *Eciton:* Dalla Torre, 1893: 1. *Labidus* revived from synonymy as subgenus of *Eciton:* Emery, 1910b: 21. *Labidus* revived status as genus: Borgmeier, 1953: 4. *Labidus* senior synonym of *Nycteresia* and *Pseudodichthadia:* Emery, 1910b: 21. Revision of genus: Borgmeier, 1955b: 80. Key to species: Watkins, 1976: 8.

**LACHNOMYRMEX** (Myrmicinae: Stenammini)

*Lachnomyrmex* Wheeler, W.M. 1910a: 263. Type-species: *Lachnomyrmex scrobiculatus*, by monotypy. Revision of genus: Smith, M.R. 1944e: 225. Key to species: Weber, 1950c: 3.

LAMPROMYRMEX (junior synonym of *Monomorium*)

*Lampromyrmex* Mayr, 1868c: 93. Type-species: **Lampromyrmex gracillimus* (junior secondary homonym in *Monomorium*, replaced by **Monomorium mayrianum*), by monotypy. *Lampromyrmex* junior synonym of *Monomorium:* Wheeler, W.M. 1915h: 45. *Lampromyrmex* revived from synonymy as subgenus of *Monomorium*, and senior synonym of *Mitara:* Emery, 1922e: 183. *Lampromyrmex* junior synonym of *Monomorium:* Ettershank, 1966: 82; see also notes in Bolton, 1987: 299.

*LAPAROMYRMEX* (junior synonym of *Liomyrmex*)
*Laparomyrmex* Emery, 1887b: 461. Type-species: *Laparomyrmex gestroi*, by monotypy. *Laparomyrmex* junior synonym of *Liomyrmex:* Dalla Torre, 1893: 63.
***LASIOPHANES*** (Formicinae: Melophorini)
*Lasiophanes* Emery, 1895i: 16 [as subgenus of *Melophorus*]. Type-species: *Formica nigriventris* (junior primary homonym in *Formica; Lasiophanes atriventris* first available replacement name), by subsequent designation of Wheeler, W.M. 1911f: 165. *Lasiophanes* raised to genus: Wheeler, W.M. 1922a: 695. Key to species: Snelling & Hunt, 1976: 104.
***LASIUS*** (Formicinae: Lasiini)
*Lasius* Fabricius, 1804: 415. Type-species: *Formica nigra*, by subsequent designation of Bingham, 1903: 338. [*Lasius* Jurine, 1801: 164 (Apidae): resurrected by Morice & Durrent, 1915: 421, suppressed by Act of Commission.] *Lasius* incorrectly as junior synonym of *Acanthomyops:* Forel, 1916: 460; Donisthorpe, 1916c: 276. *Lasius* incorrectly as junior synonym of *Formicina:* Emery, 1916a: 61; Bondroit, 1918: 19. *Lasius* valid genus and senior synonym of *Donisthorpea:* Wheeler, W.M. 1916k: 172; Emery, 1925b: 226. Revision of genus: Wilson, 1955a: 3. Subsequent revisionary studies with keys: Agosti & Collingwood, 1987b: 281; Arnol'di & Dlussky, 1978: 555; Bernard, 1967: 350; Collingwood, 1978: 89; Collingwood, 1979: 92; Collingwood, 1982: 283; Kupyanskaya, 1989: 779; Kutter, 1977c: 208; Seifert, 1988a: 143; Seifert, 1990: 1; Seifert, 1992b: 1; Tarbinsky, 1976: 133. See also Agosti & Bolton, 1990b: 149. Subgenera of *Lasius:* nominal plus *Austrolasius, Cautolasius, Chthonolasius, Dendrolasius.*
*LATINODA* (junior synonym of *Pseudomyrmex*)
*Latinoda* Enzmann, E.V. 1944: 62 [as "branch" of *Pseudomyrma*]. Type-species: *Pseudomyrma latinoda*, by original designation. *Latinoda* junior synonym of *Pseudomyrmex:* Ward, 1990: 469.
*LECANOMYRMA* (junior synonym of *Oligomyrmex*)
*Lecanomyrma* Forel, 1913k: 56 [as subgenus of *Pheidologeton*]. Type-species: *Pheidologeton (Lecanomyrma) butteli*, by monotypy. *Lecanomyrma* subgenus of *Aneleus:* Emery, 1922f: 215. *Lecanomyrma* junior synonym of *Oligomyrmex:* Ettershank, 1966: 119.
*LEIOPELTA* (junior synonym of *Belonopelta*)
*Leiopelta* Baroni Urbani, 1975b: 309. Type-species: *Belonopelta deletrix*, by original designation. *Leiopelta* junior synonym of *Belonopelta:* Hölldobler & Wilson, 1990: 10, here confirmed.
*LEONOMYRMA* (junior synonym of *Chalepoxenus*)
*Leonomyrma* Arnol'di, 1968: 1809. Type-species: *Leonomyrma spinosa*, by original designation. *Leonomyrma* provisional junior synonym of *Leptothorax:* Brown, 1973b: 181. *Leonomyrma* junior synonym of *Chalepoxenus:* Buschinger, 1987: 117.
*LEPIDOPONE* (junior synonym of *Asphinctopone*)
*Lepidopone* Bernard, 1953b: 207. Type-species: *Lepidopone lamottei* (junior synonym of *Asphinctopone silvestrii*), by monotypy. *Lepidopone* junior synonym of *Asphinctopone:* Brown, 1953h: 2.
***LEPISIOTA*** (Formicinae: Plagiolepidini)
*Lepisiota* Santschi, 1926a: 15 [as subgenus of *Acantholepis*]. Type-species: *Plagiolepis rothneyi*, by original designation. *Lepisiota* first available replacement name for *Acantholepis* Mayr (junior homonym, see *Acantholepis*, above), hence valid name of genus. *Lepisiota* senior synonym of *Baroniurbania* (unnecessary replacement name for *Acantholepis*): Bolton, 1994: 51. Taxa included in *Pseudacantholepis* (unavailable) referable to *Lepisiota.* [*Lepisiota* incorrectly as junior synonym of *Acanthomyrmex:* Brown, 1973b: 181; Snelling, 1981: 397, in error.]
*LEPTALEA* (junior synonym of *Pseudomyrma*)
*Leptalea* Erichson, 1839: 309. Type-species: *Formica gracilis*, by subsequent designation of Wheeler, W.M. 1911f: 166. *Leptalea* junior synonym of *Pseudomyrma:* Smith, F. 1858b: 153. [*Pseudomyrma* junior synonym of *Pseudomyrmex:* Smith, M.R. 1952a: 98.] [*Leptalaea* Spinola, 1851b: 52; Spinola, 1853: 68, incorrect subsequent spelling.]
***LEPTANILLA*** (Leptanillinae: Leptanillini)
*Leptanilla* Emery, 1870: 196. Type-species: *Leptanilla revelierii*, by monotypy. *Leptanilla* senior synonym of *Leptomesites:* Baroni Urbani, 1977c: 433. Revision of genus: Baroni Urbani, 1977c: 427.
***LEPTANILLOIDES*** (Leptanilloidinae)
*Leptanilloides* Mann, 1923: 13. Type-species: *Leptanilloides biconstricta*, by original designation.
***LEPTOGENYS*** (Ponerinae: Ponerini)
*Leptogenys* Roger, 1861a: 41. Type-species: *Leptogenys falcigera*, by subsequent designation of Bingham, 1903: 52. *Leptogenys* senior synonym of *Microbolbos:* Wilson, 1955c: 136. *Leptogenys* senior synonym of *Dorylozelus:* Taylor, 1969: 132. *Leptogenys* senior synonym of *Machaerogenys* and *Lobopelta:* Bolton, 1975a: 240. *Leptogenys* senior synonym of *Odontopelta:* Taylor & Brown, D.R. 1985: 32. *Leptogenys* senior synonym of *Prionogenys:* Taylor, 1988: 33. Keys to Melanesian species: Wilson, 1958a: 113; to New Caledonian species: Wilson, 1958a: 136; to Fijian species: Mann, 1921: 426; to Afrotropical and Malagasy species: Bolton, 1975a: 244, 293 respectively.
*LEPTOMESITES* (junior synonym of *Leptanilla*)
*Leptomesites* Kutter, 1948: 286. Type-species: *Leptomesites escheri*, by original designation. *Leptomesites* junior synonym of *Leptanilla:* Baroni Urbani, 1977c: 433.
*LEPTOMYRMA* (junior synonym of *Pheidole*)
*Leptomyrma* Motschoulsky, 1863: 17. Type-species: *Leptomyrma gracilipes*, by monotypy. *Leptomyrma* junior synonym of *Pheidole:* Emery, 1892b: 166.
***LEPTOMYRMEX*** (Dolichoderinae)
*Leptomyrmex* Mayr, 1862: 695. Type-species: *Formica erythrocephala*, by monotypy. Revision of genus: Wheeler, W.M. 1934c: 69. Review of genus: Shattuck, 1992c: 106.
*****LEPTOMYRMULA*** (Dolichoderinae)
**Leptomyrmula* Emery, 1912g: 16 (footnote). Type-species: **Leptomyrmex maravignae*, by monotypy.
*LEPTOPONE* (junior synonym of *Centromyrmex*)
*Leptopone* Arnold, 1916: 163 [as subgenus of *Glyphopone*]. Type-species: *Glyphopone (Leptopone) rufigaster* (junior synonym of *Centromyrmex bequaerti*), by original designation. *Leptopone* raised to genus: Wheeler, W.M. 1922a: 647. *Leptopone* junior synonym of *Centromyrmex:* Brown, 1963: 9.
***LEPTOTHORAX*** (Myrmicinae: Formicoxenini)
*Leptothorax* Mayr, 1855: 431. Type-species: *Formica acervorum*, by subsequent designation of Bingham, 1903: 214. [Type-species not *Myrmica clypeata*, unjustified subsequent designation by Emery, 1922f: 248.] *Leptothorax* senior synonym of *Temnothorax:* Forel, 1890a: lxxii. *Leptothorax* senior synonym of *Mychothorax:* Smith, M.R. 1950: 29. *Leptothorax* provisional senior synonym of *Antillaemyrmex, Caulomyrma, Croesomyrmex, Dichothorax, Goniothorax, Icothorax, Limnomyrmex, Macromischa, Mychothorax, Myrmammophilus, Myrafant, Nesomyrmex, Temnothorax:* Brown, 1973b: 178–185. *Leptothorax* senior synonym of *Dichothorax, Mychothorax, Myrafant* (and its junior synonym *Icothorax*), *Myrmammophilus, Nesomyrmex* (and its junior synonyms *Caulomyrma, Goniothorax* (junior homonym) and *Limnomyrmex*), *Temnothorax, Tetramyrma:* Bolton, 1982: 319. *Leptothorax* senior synonym of *Macromischa* (and its junior synonyms *Antillaemyrmex* and *Croesomyrmex*): Snelling, 1986: 154. *Leptothorax* senior synonym of *Nesomyrmex*, and hence of its junior synonym *Meia:* Bolton, 1994: 105. [One or more members of this tangle of current junior synonyms may be worthy of generic status, but until this group of names and the other related formicoxenine genera can be analysed together, on a world-wide basis, it is probably best to retain them as junior synonyms.] Keys to species formerly in *Nesomyrmex* (Neotropical): Kempf, 1959c: 394; to species formerly in *Macromischa* (Neotropical): Baroni Urbani, 1978b: 533. *Leptothorax*, keys to Nearctic species: Creighton, 1950a: 256; to Afrotropical species: Bolton, 1982: 323; to Palaearctic species: Agosti & Collingwood, 1987b: 273 (Balkans); Arnol'di & Dlussky, 1978: 540 (former European U.S.S.R., to the Urals); Bernard, 1967: 190 (France and Mediterranean basin); Collingwood, 1978: 83 (Spain); Collingwood, 1979: 69 (Fennoscandia and Denmark); Kutter, 1977c: 103 (Switzerland); Radchenko, 1994d: 146 (central & eastern Palaearctic).
*LEPTOXENUS (nomen nudum)*
*Leptoxenus* Forel, 1917: 245 (attributed to Santschi), *nomen nudum.*
****LEUCOTAPHUS*** (Formicinae: Formicini)
**Leucotaphus* Donisthorpe, 1920: 89. Type-species: **Leptothorax gurnetensis*, by original designation.
*LILIDRIS* (junior synonym of *Solenopsis*)
*Lilidris* Kusnezov, 1957a: 274. Type-species: *Lilidris metatarsalis*, by original designation. [*Lilidris* also described as new by Kusnezov, 1958b: 189.] *Lilidris* junior synonym of *Solenopsis:* Ettershank, 1966: 134.
*LIMNOMYRMEX* (junior synonym of *Nesomyrmex*)
*Limnomyrmex* Arnold, 1948: 222. Type-species: *Limnomyrmex stramineus*, by original designation. *Limnomyrmex* junior synonym of *Nesomyrmex:* Brown, 1971a: 4. [*Nesomyrmex* junior synonym of *Leptothorax:* Bolton, 1982: 319.]
***LINEPITHEMA*** (Dolichoderinae)
*Linepithema* Mayr, 1866a: 496. Type-species: *Linepithema fusca*, by monotypy. Review of genus: Shattuck, 1992a: 16; Shattuck, 1992c: 114.
***LIOMETOPUM*** (Dolichoderinae)
*Liometopum* Mayr, 1861: 38. Type-species: *Formica microcephala*, by monotypy. *Liometopum* senior syn-

onym of **Ctenobethylus*, and review of genus: Shattuck, 1992c: 121.

***LIOMYRMEX*** (Myrmicinae: Metaponini)

*Liomyrmex* Mayr, 1865: 23. Type-species: *Myrmica caeca* (junior primary homonym; replaced by *getroi*, first available replacement name). *Liomyrmex* senior synonym of *Laparomyrmex:* Dalla Torre, 1893: 63. *Liomyrmex* senior synonym of *Promyrma:* Forel, 1913k: 26. Review of genus: Ettershank, 1966: 152.

*LIOPONERA* (junior synonym of *Cerapachys*)

*Lioponera* Mayr, 1879: 666. Type-species: *Lioponera longitarsus*, by monotypy. *Lioponera* subgenus of *Cerapachys:* Forel, 1892l: 244. *Lioponera* revived status as genus: Dalla Torre, 1893: 17; maintained as genus: Bingham, 1903: 26; Emery, 1911d: 11. *Lioponera* junior synonym of *Cerapachys:* Brown, 1975: 19.

*LITHOMYRMEX* (junior synonym of *Amblyopone*)

*Lithomyrmex* Clark, 1928a: 30. Type-species: *Lithomyrmex glauerti*, by original designation. *Lithomyrmex* junior synonym of *Amblyopone:* Brown, 1960a: 156.

**LITHOMYRMEX* (junior homonym, see **Eulithomyrmex*)

**Lithomyrmex* Carpenter, 1930: 34. Type-species: **Lithomyrmex rugosus*, by original designation. [Junior homonym of *Lithomyrmex* Clark, above.] **Eulithomyrmex* Carpenter, 1935: 91, replacement name for **Lithomyrmex* Carpenter.

*LOBOGNATHUS* (junior synonym of *Veromessor*)

*Lobognathus* Enzmann, J. 1947b: 152 [as subgenus of *Veromessor*]. Erroneous entry for *Veromessor lobognathus* and hence junior synonym of *Veromessor:* Brown, 1949a: 49. [*Veromessor* junior synonym of *Messor:* Bolton, 1982: 338.]

*LOBOMYRMEX* (junior synonym of *Tetramorium*)

*Lobomyrmex* Kratochvil, 1941: 84 [as subgenus of *Tetramorium*]. Type-species: *Tetramorium* (*Lobomyrmex*) *ferox silhavyi* (junior synonym of *Tetramorium ferox*), by monotypy. *Lobomyrmex* provisional junior synonym of *Tetramorium:* Brown, 1973b: 181. *Lobomyrmex* junior synonym of *Tetramorium:* Bolton, 1976: 359.

*LOBOPELTA* (junior synonym of *Leptogenys*)

*Lobopelta* Mayr, 1862: 733. Type-species: *Ponera diminuta*, by subsequent designation of Bingham, 1903: 54. *Lobopelta* subgenus of *Leptogenys:* Forel, 1892k: 520. *Lobopelta* revived status as genus: Dalla Torre, 1893: 43; maintained as genus: Bingham, 1903: 54. *Lobopelta* junior synonym of *Leptogenys:* Emery, 1896e: 177 (footnote). *Lobopelta* revived from synonymy as subgenus of *Leptogenys:* Forel, 1900d: 304; maintained as subgenus: Emery, 1911d: 101; Forel, 1917: 238; Wheeler, W.M. 1922a: 653. *Lobopelta* provisional junior synonym of *Leptogenys:* Brown, 1973b: 181. *Lobopelta* junior synonym of *Leptogenys:* Bolton, 1975a: 240.

****LONCHOMYRMEX*** (Myrmicinae: Pheidolini?)

**Lonchomyrmex* Mayr, 1867b: 61. Type-species: **Formica freyeri*, by monotypy.

*LONCYDA* (junior synonym of *Cardiocondyla*)

*Loncyda* Santschi, 1930b: 70 [as subgenus of *Cardiocondyla*]. Type-species: *Cardiocondyla* (*Loncyda*) *monardi*, by monotypy. *Loncyda* provisional junior synonym of *Cardiocondyla:* Brown, 1973b: 181. *Loncyda* junior synonym of *Cardiocondyla:* Bolton, 1982: 309.

***LOPHOMYRMEX*** (Myrmicinae: Pheidolini)

*Lophomyrmex* Emery, 1892a: 114. Type-species: *Oecodoma quadrispinosa*, by monotypy. Review of genus: Ettershank, 1966: 132. Revision of genus: Rigato, 1994a: 47.

***LORDOMYRMA*** (Myrmicinae: Stenammini)

*Lordomyrma* Emery, 1897c: 591. Type-species: *Lordomyrma furcifera*, by subsequent designation of Wheeler, W.M. 1911f: 166. *Lordomyrma* senior synonym of *Prodicroaspis* and *Promeranoplus:* Bolton, 1994: 106.

***LOWERIELLA*** (Dolichoderinae)

*Loweriella* Shattuck, 1992c: 126. Type-species: *Loweriella boltoni*, by original designation.

*LUNDELLA* (junior synonym of *Hylomyrma*)

*Lundella* Emery, 1915i: 191. Type-species: *Tetramorium reitteri*, by original designation. *Lundella* junior synonym of *Hylomyrma:* Brown, 1953h: 3.

*MACHAEROGENYS* (junior synonym of *Leptogenys*)

*Machaerogenys* Emery, 1911d: 100 [as subgenus of *Leptogenys*]. Type-species: *Leptogenys truncatirostris*, by original designation. *Machaerogenys* provisional junior synonym of *Leptogenys:* Brown, 1973b: 181. *Machaerogenys* junior synonym of *Leptogenys:* Bolton, 1975a: 240.

*MACHAEROMYRMA* (junior synonym of *Cataglyphis*)

*Machaeromyrma* Forel, 1916: 441 [as subgenus of *Cataglyphis*]. Type-species: *Formica bombycina*, by original designation. Review of subgenus: Santschi, 1929b: 31 (out of date). *Machaeromyrma* junior synonym of *Cataglyphis:* Baroni Urbani, 1969d: 218; Agosti, 1994a: 103.

***MACHOMYRMA*** (Myrmicinae: Pheidologetonini)

*Machomyrma* Forel, 1895f: 425 [as subgenus of *Liomyrmex*]. Type-species: *Liomyrmex* (*Machomyrma*) *dispar*, by monotypy. *Machomyrma* raised to genus: Emery, 1896e: 184.

*MACROMISCHA* (junior synonym of *Leptothorax*)

*Macromischa* Roger, 1863a: 184. Type-species: *Macromischa purpurata*, by subsequent designation of Wheeler, W.M. 1911f: 166. *Macromischa* senior synonym of *Antillaemyrmex* and *Croesomyrmex:* Smith, M.R. 1937: 849; Smith, M.R. 1939e: 502; Baroni Urbani, 1978b: 398. *Macromischa* provisional junior synonym of *Leptothorax:* Brown, 1973b: 181. *Macromischa* subgenus of *Leptothorax*, and revision of subgenus: Baroni Urbani, 1978b: 398. *Macromischa* junior synonym of *Leptothorax:* Snelling, 1986: 154.

*MACROMISCHOIDES* (junior synonym of *Tetramorium*)

*Macromischoides* Wheeler, W.M. 1920: 53. Type-species: *Macromischa aculeata*, by original designation. Review of genus: Santschi, 1924b: 206. *Macromischoides* junior synonym of *Tetramorium:* Bolton, 1976: 359; Bolton, 1980: 196. [*Macromichoides* Santschi, 1924b: 206, incorrect subsequent spelling.]

*MACROPHEIDOLE* (junior synonym of *Pheidole*)

*Macropheidole* Emery, 1915i: 190 [as subgenus of *Pheidole*]. Type-species: *Pheidole fimbriata*, by monotypy. *Macropheidole* provisional junior synonym of *Pheidole:* Brown, 1973b: 181. *Macropheidole* junior synonym of *Pheidole:* Smith, D.R. 1979: 1366.

**#*MALACOMYRMA*** (subgenus of *Acropyga*)

*Malacomyrma* Emery, 1922d: 109 [as subgenus of *Acropyga*]. Type-species: *Acropyga silvestrii*, by monotypy. *Malacomyrma* provisional junior synonym of *Acropyga:* Brown, 1973b: 182 (unconfirmed).

***MANICA*** (Myrmicinae: Myrmicini)

*Manica* Jurine, 1807: 276. Type-species: *Formica rubida*, by subsequent designation of Wheeler, W.M. 1911f: 166. *Manica* junior synonym of *Myrmica:* Forel, 1915d: 9. *Manica* revived from synonymy as subgenus of *Myrmica*, and senior synonym of *Neomyrma* (and its junior synonym *Oreomyrma*): Emery, 1921f: 42; Wheeler, W.M. 1922a: 660. *Manica* revived status as genus: Weber, 1947: 440; Creighton, 1950a: 105.

**#*MANNIELLA*** (subgenus of *Camponotus*)

*Manniella* Wheeler, W.M. 1921a: 19 [as subgenus of *Camponotus*]. Type-species: *Camponotus sphaericus*, by original designation. *Manniella* provisional junior synonym of *Camponotus:* Brown, 1973b: 182 (unconfirmed).

*MARTIA* (junior homonym, see *Forelifidis*)

*Martia* Forel, 1907a: 20 [as subgenus of *Monomorium*]. Type-species: *Monomorium* (*Martia*) *vezenyii*, by monotypy. [Junior homonym of *Martia* Ragonot, 1887: 18 (Lepidoptera).] *Forelifidis* Smith, M.R. 1954a: 17, replacement name for *Martia* Forel.

**#*MAYRIA*** (subgenus of *Camponotus*)

*Mayria* Forel, 1878: 369. Type-species: *Mayria madagascarensis* (junior secondary homonym in *Camponotus*, replaced by *Camponotus repens*), by monotypy. *Mayria* subgenus of *Camponotus:* Forel, 1894e: 227. *Mayria* provisional junior synonym of *Camponotus:* Brown, 1973b: 182 (unconfirmed).

***MAYRIELLA*** (Myrmicinae: Stenammini?)

*Mayriella* Forel, 1902h: 452. Type-species: *Mayriella abstinens*, by monotypy. Key to species: Wheeler, W.M. 1935e: 154.

*MAYROMYRMEX* (junior synonym of *Eciton*)

*Mayromyrmex* Ashmead, 1905b: 381. Type-species: *Labidus fargeavii* (junior synonym of *Eciton quadriglume*), by original designation. *Mayromyrmex* junior synonym of *Eciton:* Emery, 1906a: 718.

***MEGALOMYRMEX*** (Myrmicinae: Solenopsidini)

*Megalomyrmex* Forel, 1885a: 371. Type-species: *Megalomyrmex leoninus*, by monotypy. *Megalomyrmex* senior synonym of *Cepobroticus* and *Wheelerimyrmex:* Ettershank, 1966: 101. Review of genus: Bolton, 1987: 285. Revision of genus: Brandão, 1990: 411.

[*MEGALOPONERA* (misspelling)

*Megaloponera* Emery, 1877b: 368, incorrect subsequent spelling of *Megaponera*.]

*MEGAPONERA* (junior synonym of *Pachycondyla*)

*Megaponera* Mayr, 1862: 734. Type-species: *Formica foetens* (junior primary homonym in *Formica*, replaced by *Formica analis*), by monotypy. *Megaponera* junior synonym of *Pachycondyla:* Brown, in preparation. [*Megaloponera:* see above.]

**MEGAPTERITES* (junior synonym of **Formicium*)

**Megapterites* Cockerell, 1920: 278. Type-species: **Megapterites mirabilis*, by monotypy (in family *Pseudosiricidae). **Megapterites* junior synonym of **Formicium:* Lutz, 1990: 61.

*MEIA* (junior synonym of *Leptothorax*)

*Meia* Pagliano & Scaramozzino, 1990: 5. Unnecessary replacement name for *Goniothorax* Emery, 1896g: 58. [Junior homonym of *Goniothorax* Milne-Edwards, 1879: 103.] [*Nesomyrmex* is first available replacement name for *Goniothorax* Emery, therefore *Meia* is automatically a junior synonym of *Nesomyrmex*, Bolton, 1994: 105. *Nesomyrmex* junior synonym of *Leptothorax:* Bolton, 1982: 319.]

***MELISSOTARSUS*** (Myrmicinae: Melissotarsini)

*Melissotarsus* Emery, 1877b: 378. Type-species: *Melissotarsus beccarii*, by monotypy. Review of Afrotropical species: Bolton, 1982: 335.

***MELOPHORUS*** (Formicinae: Melophorini)

*Melophorus* Lubbock, 1883: 51. Type-species: *Melophorus bagoti*, by monotypy. *Melophorus* senior synonym of *Erimelophorus* and *Trichomelophorus:* Brown, 1955d: 474.

***MERANOPLUS*** (Myrmicinae: Meranoplini)

*Meranoplus* Smith, F. 1854: 224. Type-species: *Cryptocerus bicolor*, by subsequent designation of Bingham, 1903: 166. Revision of Afrotropical species: Bolton, 1981a: 48. [*Cryptocephalus:* see above.]

#***MESANOPLOLEPIS*** (subgenus of *Anoplolepis*)

*Mesanoplolepis* Santschi, 1926a: 14 [as subgenus of *Anoplolepis*]. Type-species: *Plagiolepis simulans*, by original designation. *Mesanoplolepis* provisional junior synonym of *Anoplolepis:* Brown, 1973b: 182 (unconfirmed).

#***MESOCREMA*** (subgenus of *Crematogaster*)

*Mesocrema* Santschi, 1928b: 33 [as subgenus of *Crematogaster*]. Type-species: *Crematogaster rasoherinae*, by subsequent designation of Donisthorpe, 1943f: 661. *Mesocrema* provisional junior synonym of *Crematogaster:* Brown, 1973b: 182 (unconfirmed).

*MESOMYRMA* (junior synonym of *Dilobocondyla*)

*Mesomyrma* Stitz, 1911a: 363 [as subgenus of *Podomyrma*]. Type-species: *Podomyrma (Mesomyrma) cataulacoidea*, by monotypy. *Mesomyrma* junior synonym of *Dilobocondyla:* Emery, 1912b: 102.

*MESOPONERA* (junior synonym of *Pachycondyla*)

*Mesoponera* Emery, 1900d: 668 [as subgenus of *Euponera*]. Type-species: *Ponera melanaria*, by monotypy. [*Mesoponera* also described as new by Emery, 1901a: 43. Type-species not *Ponera caffraria*, unjustified subsequent designation by Emery, 1901a: 43.] *Mesoponera* raised to genus: Bingham, 1903: 99. *Mesoponera* subgenus of *Euponera:* Emery, 1911d: 80; maintained as subgenus: Wheeler, W.M. 1922a: 649. *Mesoponera* senior synonym of *Xiphopelta:* Wheeler, W.M. 1922a: 775. *Mesoponera* revived status as genus: Wilson, 1958d: 349. *Mesoponera* provisional junior synonym of *Pachycondyla:* Brown, 1973b: 182. *Mesoponera* junior synonym of *Pachycondyla:* Brown, in preparation.

***MESOSTRUMA*** (Myrmicinae: Dacetini)

*Mesostruma* Brown, 1948e: 118. Type-species: *Strumigenys (Epopostruma) turneri*, by original designation. Revisions of genus: Brown, 1952g: 9; Taylor, 1973: 24. *Mesostruma* junior synonym of *Colobostruma:* Baroni Urbani & De Andrade, 1994: 15. *Mesostruma* here: **revived from synonymy**.

*MESOXENA* (junior synonym of *Echinopla*)

*Mesoxena* Smith, F. 1860b: 106. Type-species: *Mesoxena mistura*, by monotypy. *Mesoxena* provisional junior synonym of *Echinopla:* Brown, 1973b: 182; confirmed Bolton, 1994: 50.

***MESSOR*** (Myrmicinae: Pheidolini)

*Messor* Forel, 1890a: lxviii [as subgenus of *Aphaenogaster*]. Type-species: *Formica barbara*, by subsequent designation of Bingham, 1903: 277. *Messor* subgenus of *Stenamma:* Emery, 1895c: 298. *Messor* raised to genus: Bingham, 1903: 277; maintained as genus: Emery, 1908e: 437. *Messor* senior synonym of *Cratomyrmex:* Emery, 1922f: 357. *Messor* senior synonym of *Cratomyrmex* and *Veromessor* (and its junior synonym *Lobognathus*): Bolton, 1982: 358. Species included in *Sphaeromessor* (unavailable) referable to *Messor.* Key to species of former U.S.S.R.: Arnol'di, 1977b: 1640; to Afrotropical species: Bolton, 1982: 342.

*METACYLINDROMYRMEX* (junior synonym of *Cylindromyrmex*)

*Metacylindromyrmex* Wheeler, W.M. 1924a: 106 [as subgenus of *Cylindromyrmex*]. Type-species: *Cylindromyrmex godmani*, by original designation. *Metacylindromyrmex* junior synonym of *Cylindromyrmex:* Kempf, 1972a: 91; Brown, 1975: 36.

***METAPONE*** (Myrmicinae: Metaponini)

*Metapone* Forel, 1911h: 447. Type-species: *Metapone greeni*, by monotypy. Review of genus: Wheeler, W.M. 1919h: 173; see also Kusnezov, 1960a: 119.

****MIANEURETUS*** (Aneuretinae)

**Mianeuretus* Carpenter, 1930: 38. Type-species: **Mianeuretus mirabilis*, by original designation.

*MICCOSTRUMA* (junior synonym of *Smithistruma*)

*Miccostruma* Brown, 1948e: 123. Type-species: *Epitritus mandibularis*, by original designation. *Miccostruma* junior synonym of *Smithistruma:* Bolton, 1983: 274.

*MICROBOLBOS* (junior synonym of *Leptogenys*)

*Microbolbos* Donisthorpe, 1948f: 170. Type-species: *Microbolbos testaceus*, by original designation. *Microbolbos* junior synonym of *Leptogenys:* Wilson, 1955c: 136.

***MICRODACETON*** (Myrmicinae: Dacetini)

*Microdaceton* Santschi, 1913g: 478. Type-species: *Microdaceton exornatum*, by monotypy. [*Microdaceton* also described as new by Santschi, 1914e: 33.] Revision of genus: Bolton, 1983: 401.

*MICROMYRMA* (junior synonym of *Tapinoma*)

*Micromyrma* Dufour, 1857: 60. Type-species: *Micromyrma pygmaea*, by monotypy. *Micromyrma* junior synonym of *Tapinoma:* Mayr, 1863: 455; Forel, 1878: 385; Emery, 1912g: 38; Smith, M.R. 1951a: 837; Shattuck, 1992c: 146.

*MICTOPONERA* (junior synonym of *Rhopalopone*)

*Mictoponera* Forel, 1901e: 372 [as subgenus of *Ectatomma*]. Type-species: *Ectatomma (Mictoponera) diehlii*, by monotypy. *Mictoponera* junior synonym of *Rhopalopone:* Emery, 1911d: 34. [*Rhopalopone* junior synonym of *Gnamptogenys:* Brown, 1958g: 211.]

****MIOMYRMEX*** (Dolichoderinae)

**Miomyrmex* Carpenter, 1930: 51. Type-species: **Formica impacta*, by original designation.

*MITARA* (junior synonym of *Monomorium*)

*Mitara* Emery, 1913: 261 [as subgenus of *Monomorium*]. Type-species: *Monomorium laeve*, by original designation. *Mitara* junior synonym of *Lampromyrmex:* Emery, 1922e: 183; Wheeler, W.M. 1922a: 162; see also notes in Bolton, 1987: 299.

#***MOELLERIUS*** (subgenus of *Acromyrmex*)

*Moellerius* Forel, 1893e: 589 [as subgenus of *Atta*]. Type-species: *Atta (Acromyrmex) landolti*, by subsequent designation of Wheeler, W.M. 1911f: 167. *Moellerius* subgenus of *Acromyrmex:* Emery, 1913: 251. *Moellerius* provisional junior synonym of *Acromyrmex:* Brown, 1973b: 182 (unconfirmed). Revisionary studies: see under *Acromyrmex*.

*MONACIS* (junior synonym of *Dolichoderus*)

*Monacis* Roger, 1862a: 233. Type-species: *Formica bispinosa*, by subsequent designation of Wheeler, W.M. 1911f: 167. *Monacis* junior synonym of *Hypoclinea:* Mayr, 1862: 704; of *Dolichoderus:* Forel, 1878: 386. *Monacis* revived from synonymy as subgenus of *Dolichoderus:* Emery, 1894c: 228; maintained as subgenus: Emery, 1912g: 9; Wheeler, W.M. 1922a: 688. *Monacis* revived status as genus: Brown, 1950e: 249; maintained as genus, and revision of genus: Kempf, 1959b: 227; emended key: Kempf, 1972c: 253. *Monacis* provisional junior synonym of *Dolichoderus:* Brown, 1973b: 182. *Monacis* junior synonym of *Dolichoderus:* Shattuck, 1992c: 66.

*MONOCERATOCLINEA* (junior synonym of *Dolichoderus*)

*Monoceratoclinea* Wheeler, W.M. 1935c: 68 [as subgenus of *Dolichoderus*]. Type-species: *Dolichoderus (Hypoclinea) monoceros*, by original designation. *Monoceratoclinea* raised to genus: Brown, 1950e: 249. *Monoceratoclinea* junior synonym of *Dolichoderus:* Shattuck, 1992c: 66.

*MONOCOMBUS* (junior synonym of *Cataglyphis*)

*Monocombus* Mayr, 1855: 381. Type-species: *Formica viatica*, by monotypy. *Monocombus* junior synonym of *Cataglyphis:* Mayr, 1861: 44. *Monocombus* junior synonym of *Myrmecocystus:* Mayr, 1886d: 424. *Monocombus* junior synonym of *Cataglyphis:* Dalla Torre, 1893: 216; Wheeler, W.M. 1922a: 944; Emery, 1925b: 261. *Monocombus* revived from synonymy as subgenus of *Cataglyphis:* Santschi, 1929b: 29. *Monocombus* junior synonym of *Cataglyphis:* Agosti, 1990a: 1462; Agosti, 1994a: 103.

***MONOMORIUM*** (Myrmicinae: Solenopsidini)

*Monomorium* Mayr, 1855: 452. Type-species: *Monomorium minutum* (junior secondary homonym in *Monomorium*, replaced by *Monomorium monomorium*), by monotypy. *Monomorium* senior synonym of *Lampromyrmex:* Wheeler, W.M. 1915h: 45. *Monomorium* senior synonym of *Paraphacota:* Santschi, 1927d: 243. *Monomorium* senior synonym of *Epixenus:* Brown & Wilson, 1957b: 244. *Monomorium* senior synonym of *Corynomyrmex, Epixenus, Epoecus, Equestrimessor, Holcomyrmex, Ireneidris, Isolcomyrmex, Lampromyrmex* (and its junior synonym *Mitara*), *Paraphacota, Parholcomyrmex, Trichomyrmex, Wheeleria* (homonym), *Wheeleriella, Xenhyboma*, and *Xeromyrmex:* Ettershank, 1966: 82. *Monomorium* senior synonym of *Chelaner* (and its junior synonyms *Notomyrmex, Protholcomyrmex, Schizopelta*), and *Syllophopsis:* Bolton, 1987: 287. Species included in *Pharaophanes* (unavailable) all referable to *Monomorium*. Reviews of genus: Ettershank, 1966: 82; Bolton, 1987: 287. Revision of Nearctic species: DuBois, 1986: 74; of Afrotropical species: Bolton, 1987: 307.

*MORLEYIDRIS* (junior synonym of *Hedomyrma*)

*Morleyidris* Donisthorpe, 1944c: 64 [as subgenus of *Polyrhachis*]. Type-species: *Polyrhachis* (*Morleyidris*) *trina*, by original designation. *Morleyidris* junior synonym of *Hedomyrma:* Hung, 1967b: 402.

#***MYAGROTERAS*** (subgenus of *Myrmoteras*)

*Myagroteras* Moffett, 1985b: 31 [as subgenus of *Myrmoteras*]. Type-species: *Myrmoteras donisthorpei,* by original designation.

***MYCETAROTES*** (Myrmicinae: Attini)

*Mycetarotes* Emery, 1913: 251 [as subgenus of *Cyphomyrmex*]. Type-species: *Cyphomyrmex parallelus,* by original designation. *Mycetarotes* raised to genus: Borgmeier, 1950b: 384. Review of genus: Kempf, 1960d: 277.

***MYCETOPHYLAX*** (Myrmicinae: Attini)

*Mycetophylax* Emery, 1913: 251 [as subgenus of *Cyphomyrmex*]. Type-species: *Myrmicocrypta brittoni* (junior synonym of *Mycetophylax conformis*), by original designation. *Mycetophylax* subgenus of *Myrmicocrypta:* Gallardo, 1916d: 320. *Mycetophylax* raised to genus: Santschi, 1923c: 268. *Mycetophylax* senior synonym of *Paramycetophylax:* Weber, 1958d: 262. Key to species: Santschi, 1922b: 357 (out of date).

[*MYCETOPURUS* (misspelling)

*Mycetopurus* Santschi, 1925b: 17, incorrect subsequent spelling of *Mycocepurus.* [Perhaps a combination in error of *Myceto*(phylax) + (Mycoce)*purus*.]

***MYCETOSORITIS*** (Myrmicinae: Attini)

*Mycetosoritis* Wheeler, W.M. 1907c: 714 [as subgenus of *Atta*]. Type-species: *Atta* (*Mycetosoritis*) *hartmanni,* by monotypy. *Mycetosoritis* subgenus of *Cyphomyrmex:* Emery, 1913: 251. *Mycetosoritis* subgenus of *Trachymyrmex:* Wheeler, W.M. 1922a: 669. *Mycetosoritis* raised to genus: Creighton, 1950a: 317.

*MYCHOTHORAX* (junior synonym of *Leptothorax*)

*Mychothorax* Ruzsky, 1904a: 288. Type-species: *Formica acervorum,* by original designation. *Mychothorax* subgenus of *Leptothorax:* Ruzsky, 1905b: 609; Emery, 1915g: 24. *Mychothorax* junior synonym of *Leptothorax:* Smith, M.R. 1950: 29. [*Leptothorax* and *Mychothorax* share the same type-species, synonymy is therefore absolute.]

***MYCOCEPURUS*** (Myrmicinae: Attini)

*Mycocepurus* Forel, 1893e: 602 [as subgenus of *Atta*]. Type-species: *Atta* (*Mycocepurus*) *smithii,* by subsequent designation of Wheeler, W.M. 1911f: 167. [*Mycocepurus* also described as new by Forel, 1893g: 369.] *Mycocepurus* raised to genus: Wheeler, W.M. 1911f: 167. *Mycocepurus* senior synonym of *Descolemyrma,* and revision of genus: Kempf, 1963b: 417. [*Mycetopurus:* see above.]

***MYOPIAS*** (Ponerinae: Ponerini)

*Myopias* Roger, 1861a: 39. Type-species: *Myopias amblyops,* by monotypy. *Myopias* senior synonym of *Trapeziopelta* (and its junior synonym *Bradyponera*), and key to Australasian species: Willey & Brown, 1983: 249.

***MYOPOPONE*** (Ponerinae: Amblyoponini)

*Myopopone* Roger, 1861a: 49. Type-species: *Myopopone maculata* (junior synonym of *Myopopone castanea*), by subsequent designation of Bingham, 1903: 33. Revision of genus: Brown, 1960a: 170.

*MYRAFANT* (junior synonym of *Leptothorax*)

*Myrafant* Smith, M.R. 1950: 30 [as subgenus of *Leptothorax*]. Type-species: *Leptothorax curvispinosus,* by original designation. *Myrafant* provisional junior synonym of *Leptothorax:* Brown, 1973b: 182. *Myrafant* senior synonym of *Icothorax:* Baroni Urbani, 1978b: 556. *Myrafant* junior synonym of *Leptothorax:* Bolton, 1982: 319.

***MYRCIDRIS*** (Pseudomyrmecinae)

*Myrcidris* Ward, 1990: 465. Type-species: *Myrcidris epicharis,* by original designation.

#***MYRMA*** (subgenus of *Polyrhachis*)

*Myrma* Billberg, 1820: 104. Type-species: *Formica militaris,* by subsequent designation of Wheeler, W.M. 1911c: 859; and Wheeler, W.M. 1911f: 168. *Myrma* senior synonym of *Hoplomyrmus:* Wheeler, W.M. 1922a: 993. *Myrma* senior synonym of *Pseudocyrtomyrma:* Bolton, 1973b: 288. [Although *Myrma* antedates *Polyrhachis,* it has been treated as a subgenus of the latter since Wheeler, W.M. 1922a: 993. See Hung, 1967b: 396 for a history of the name, and a plea for the conservation of *Polyrhachis.*]

#***MYRMACANTHA*** (subgenus of *Phasmomyrmex*)

*Myrmacantha* Emery, 1920b: 246 [as subgenus of *Camponotus*]. Type-species: *Camponotus aberrans,* by original designation. *Myrmacantha* junior synonym of *Myrmorhachis:* Wheeler, W.M. 1921a: 18. *Myrmacantha* revived status as subgenus of *Phasmomyrmex:* Emery, 1925b: 58.

#***MYRMACRHAPHE*** (subgenus of *Camponotus*)

*Myrmacrhaphe* Santschi, 1926c: 607 [as subgenus of *Camponotus*]. Type-species: *Camponotus conradti,* by original designation. *Myrmacrhaphe* provisional junior synonym of *Camponotus:* Brown, 1973b: 182 (unconfirmed).

#***MYRMAMBLYS*** (subgenus of *Camponotus*)

*Myrmamblys* Forel, 1912i: 90 [as subgenus of *Camponotus*]. Type-species: *Camponotus reticulatus,* by subsequent designation of Wheeler, W.M. 1913a: 80. [Type-species not *Camponotus fastigatus,* unjustified subsequent designation by Forel, 1914a: 263.] *Myrmamblys* provisional junior synonym of *Camponotus:* Brown, 1973b: 182 (unconfirmed).

*MYRMAMMOPHILUS* (junior synonym of *Leptothorax*)

*Myrmammophilus* Menozzi, 1925d: 29 [as subgenus of *Leptothorax*]. Type-species: *Leptothorax* (*Myrmammophilus*) *finzii,* by monotypy. *Myrmammophilus* provisional junior synonym of *Leptothorax:* Brown, 1973b: 182. *Myrmammophilus* junior synonym of *Leptothorax:* Bolton, 1982: 319.

*MYRMAPATETES* (junior synonym of *Anochetus*)

*Myrmapatetes* Wheeler, W.M. 1929b: 6. Type-species: *Myrmapatetes filicornis,* by original designation. *Myrmapatetes* junior synonym of *Anochetus:* Brown, 1953h: 2.

#***MYRMAPHAENUS*** (subgenus of *Camponotus*)

*Myrmaphaenus* Emery, 1920b: 237 [as subgenus of *Camponotus*]. Type-species: *Camponotus leydigi,* by original designation. *Myrmaphaenus* senior synonym of *Neomyrmamblys* and *Paracolobopsis:* Emery, 1925b: 152. *Myrmaphaenus* provisional junior synonym of *Camponotus:* Brown, 1973b: 182 (unconfirmed).

#***MYRMATOPA*** (subgenus of *Polyrhachis*)

*Myrmatopa* Forel, 1915b: 107 [as subgenus of *Polyrhachis*]. Type-species: *Polyrhachis schang,* by original designation. *Myrmatopa* provisional junior synonym of *Polyrhachis:* Brown, 1973b: 182 (unconfirmed). *Myrmatopa* senior synonym of *Irenea:* Bolton, 1994: 50.

***MYRMECIA*** (Myrmeciinae: Myrmeciini)

*Myrmecia* Fabricius, 1804: 423. Type-species: *Formica gulosa,* by subsequent designation of Shuckard, in Swainson & Shuckard, 1840: 173. *Myrmecia* senior synonym of *Promyrmecia* (and its junior synonyms *Halmamyrmecia, Pristomyrmecia*): Brown, 1953c: 1. Revisions of genus: Clark, 1943: 83; Clark, 1951: 1; Brown, 1953j: 1; Ogata, 1991: 353; Ogata & Taylor, 1991: 1623.

***MYRMECINA*** (Myrmicinae: Myrmecinini)

*Myrmecina* Curtis, 1829: 265. Type-species: *Myrmecina latreillii* (junior synonym of *Myrmecina graminicola*), by monotypy. *Myrmecina* senior synonym of *Archaeomyrmex:* Brown, 1971a: 1.

*MYRMECINELLA* (junior synonym of *Xenomyrmex*)

*Myrmecinella* Wheeler, W.M. 1922d: 1. Type-species: *Myrmecinella panamana,* by original designation. *Myrmecinella* junior synonym of *Xenomyrmex:* Wheeler, W.M. 1931a: 129.

***MYRMECOCYSTUS*** (Formicinae: Lasiini)

*Myrmecocystus* Wesmael, 1838: 770. Type-species: *Myrmecocystus mexicanus,* by monotypy. *Myrmecocystus* senior synonym of *Endiodioctes* and *Eremnocystus:* Snelling, 1981: 403. Revision of genus: Snelling, 1976: 1; Snelling, 1982b: 77.

*MYRMECOPSIS* (junior homonym, see *Opisthopsis*)

*Myrmecopsis* Smith, F. 1865: 68 [as subgenus of *Formica*]. Type-species: *Formica* (*Myrmecopsis*) *respiciens,* by monotypy. *Myrmecopsis* raised to genus: Mayr, 1876: 76. [Junior homonym of *Myrmecopsis* Newman, 1850: cxxii (Lepidoptera).] *Opisthopsis* Dalla Torre, 1893: 219, replacement name for *Myrmecopsis* Smith, F.

***MYRMECORHYNCHUS*** (Formicinae: Melophorini)

*Myrmecorhynchus* André, 1896b: 253. Type-species: *Myrmecorhynchus emeryi,* by monotypy.

*MYRMEGIS* (*nomen nudum*)

*Myrmegis* Rafinesque, 1815: 124, *nomen nudum.* [Brown, 1973b: 182, places *Myrmegis* as a junior synonym of *Atta,* because the entry in Rafinesque reads, "6.*Myrmegis* R. *Atta* Latr."]

***MYRMELACHISTA*** (Formicinae: Myrmelachistini)

*Myrmelachista* Roger, 1863a: 162. Type-species: *Myrmelachista kraatzii,* by monotypy. *Myrmelachista* senior synonym of *Decamyrma* Roger (junior homonym), *Hincksidris* and *Neaphomus:* Snelling & Hunt, 1976: 110. Key to species: Wheeler, W.M. 1934g: 188 (out of date).

#***MYRMENTOMA*** (subgenus of *Camponotus*)

*Myrmentoma* Forel, 1912i: 92 [as subgenus of *Camponotus*]. Type-species: *Formica lateralis,* by subsequent designation of Wheeler, W.M. 1913a: 80. *Myrmentoma* junior synonym of *Orthonotomyrmex:* Forel, 1913b: 350 (in text); maintained as junior synonym: Wheeler, W.M. 1922a: 970. *Myrmentoma* revived from synonymy as subgenus of *Camponotus:* Emery, 1920b: 257; maintained as subgenus: Emery, 1925a: 62; Emery, 1925b: 116. *Myrmentoma* provisional junior synonym of *Camponotus:* Brown, 1973b: 182 (unconfirmed). Revision of Nearctic species: Snelling, 1988: 55.

**#MYRMEPINOTUS** (subgenus of *Camponotus*)
*Myrmepinotus* Santschi, 1921f: 312 [as subgenus of *Camponotus*]. Type-species: *Camponotus echinoploides*, by original designation. *Myrmepinotus* provisional junior synonym of *Camponotus*: Brown, 1973b: 182 (unconfirmed).

**#MYRMEPOMIS** (subgenus of *Camponotus*)
*Myrmepomis* Forel, 1912i: 92 [as subgenus of *Camponotus*]. Type-species: *Formica sericeiventris*, by subsequent designation of Wheeler, W.M. 1913a: 81. [Type-species not *Formica fulvopilosa*, unjustified subsequent designation by Forel, 1914a: 265.] *Myrmepomis* senior synonym of *Myrmolophus*: Wheeler, W.M. 1921a: 17. *Myrmepomis* provisional junior synonym of *Camponotus*: Brown, 1973b: 182 (unconfirmed).

**#MYRMESPERA** (subgenus of *Camponotus*)
*Myrmespera* Santschi, 1926b: 247 [as subgenus of *Camponotus*]. Type-species: *Camponotus (Myrmespera) debellator*, by subsequent designation of Santschi, 1926c: 604. *Myrmespera* provisional junior synonym of *Camponotus*: Brown, 1973b: 182 (unconfirmed).

*MYRMETAERUS* (junior synonym of *Myrmoxenus*)
*Myrmetaerus* Soudek, 1925a: 33. Type-species: *Myrmetaerus microcellatus* (junior synonym of *Epimyrma gordiagini*), by monotypy. [*Myrmetaerus* also described as new by Soudek, 1925b: 14.] *Myrmetaerus* provisional junior synonym of *Leptothorax*: Brown, 1973b: 182 (unconfirmed). *Myrmetaerus* incorrectly as junior synonym of *Camponotus*: Snelling, 1981: 404. *Myrmetaerus* junior synonym of *Myrmoxenus*: Buschinger, Winter & Faber, 1984: 336. [*Myrmoxenus* junior synonym of *Epimyrma*: Bolton, 1994: 105.]

**#MYRMEURYNOTA** (subgenus of *Camponotus*)
*Myrmeurynota* Forel, 1912i: 92 [as subgenus of *Camponotus*]. Type-species: *Camponotus eurynotus*, by subsequent designation of Wheeler, W.M. 1913a: 81. [Type-species not *Camponotus gilviventris*, unjustified subsequent designation by Forel, 1914a: 266.] *Myrmeurynota* provisional junior synonym of *Camponotus*: Brown, 1973b: 182 (unconfirmed).

*MYRMEX* (junior homonym, junior synonym of *Pseudomyrma*)
*Myrmex* Guérin-Méneville, 1844a: 427. Type-species: *Myrmex perboscii*, by monotypy. [Junior homonym of *Myrmex* Sturm, 1826: 32 (Coleoptera).] *Myrmex* Guérin-Méneville junior synonym of *Pseudomyrma*: Smith, F. 1858b: 153. [*Pseudomyrma* junior synonym of *Pseudomyrmex*: Smith, M.R. 1952a: 98.]

**#MYRMHOPLA** (subgenus of *Polyrhachis*)
*Myrmhopla* Forel, 1915b: 107 [as subgenus of *Polyrhachis*]. Type-species: *Formica armata*, by original designation. *Myrmhopla* senior synonym of *Cephalomyrma* and *Florencea*: Hung, 1967b: 402. *Myrmhopla* provisional junior synonym of *Polyrhachis*: Brown, 1973b: 182 (unconfirmed).

***MYRMICA*** (Myrmicinae: Myrmicini)
*Myrmica* Latreille, 1804: 179. Type-species: *Formica rubra*, by subsequent designation of Latreille, 1810: 437. *Myrmica* provisional senior synonym of *Dodecamyrmica, Paramyrmica, Sifolinia, Sommimyrma, Symbiomyrma*: Brown, 1973b: 180–185. *Myrmica* senior synonym of *Dodecamyrmica*: Francoeur, 1981: 759. *Myrmica* senior synonym of *Paramyrmica, Sifolinia* (and its junior synonym *Symbiomyrma*), and *Sommimyrma*: Bolton, 1988a: 3. Revisionary studies: Arnol'di, 1970: 1832; Arnol'di, 1976a: 554; Finzi, 1926: 71; Menozzi, 1939a: 293; Seifert, 1988b: 1; Weber, 1947: 437; Weber, 1948a: 267; Weber, 1950b: 189; Radchenko, 1994a: 39; Radchenko, 1994c: 130. See also: Agosti & Collingwood, 1987b: 267; Arnol'di & Dlussky, 1978: 530; Collingwood, 1978: 78; Collingwood, 1979: 40; Creighton, 1950a: 87; Kutter, 1977c: 42.

***MYRMICARIA*** (Myrmicinae: Myrmicariini)
*Myrmicaria* Saunders, W.W. 1842: 57. Type-species: *Myrmicaria brunnea*, by monotypy. *Myrmicaria* senior synonym of *Heptacondylus* (and its junior synonym *Physatta*): Smith, F. 1865: 73. Revision of genus Santschi, 1925c: 133 (out of date).

**MYRMICITES* (collective group name)
**Myrmicites* Förster, 1891: 448 [collective group name, without included named taxa.]

****MYRMICIUM*** (Myrmicinae: unplaced to tribe)
**Myrmicium* Westwood, 1854: 396. Type-species: **Myrmicium heerii*, by monotypy. **Myrmicium* junior synonym of **Pseudosirex* (family *Pseudosiricidae): Handlirsch, 1906: 577; retained in *Pseudosiricidae: Abe & Smith, D.R. 1991: 53. **Myrmicium* junior synonym of *Myrmica*: Dalla Torre, 1893: 108; Forel, 1915d: 9. **Myrmicium*: provisionally revived from synonymy here and returned to Formicidae, subfamily Myrmicinae, pending detailed investigation. [**Myrmecium* Dalla Torre, 1893: 108, incorrect subsequent spelling.]

**MYRMICIUM* (junior homonym, see **Promyrmicium*)
**Myrmicium* Heer, 1870: 78. Type-species: **Myrmicium boreale*, by monotypy. [Junior homonym of **Myrmicium* Westwood, above.] **Promyrmicium* Baroni Urbani, 1971b: 362, replacement name for **Myrmicium* Westwood.

***MYRMICOCRYPTA*** (Myrmicinae: Attini)
*Myrmicocrypta* Smith, F. 1860c: 73. Type-species: *Myrmicocrypta squamosa*, by monotypy. *Myrmicocrypta* senior synonym of *Glyptomyrmex*: Emery, 1894c: 224.

**#MYRMISOLEPIS** (subgenus of *Camponotus*)
*Myrmisolepis* Santschi, 1921f: 310 [as subgenus of *Camponotus*]. Type-species: *Camponotus epinotalis*, by original designation. *Myrmisolepis* provisional junior synonym of *Camponotus*: Brown, 1973b: 182 (unconfirmed).

**#MYRMOBRACHYS** (subgenus of *Camponotus*)
*Myrmobrachys* Forel, 1912i: 91 [as subgenus of *Camponotus*]. Type-species: *Formica senex*, by subsequent designation of Wheeler, W.M. 1913a: 81. *Myrmobrachys* provisional junior synonym of *Camponotus*: Brown, 1973b: 182 (unconfirmed).

*MYRMOCAMELUS* (junior synonym of *Myrmophyma*)
*Myrmocamelus* Forel, 1914a: 261 [as subgenus of *Camponotus*]. Type-species: *Formica ephippium*, by original designation. [Type-species not *Camponotus gambeyi*, unjustified subsequent designation by Forel, 1922: 101. *Myrmocamelus* also described as new by Forel, 1915b: 102.] *Myrmocamelus* junior synonym of *Myrmophyma*: Emery, 1920b: 257; Wheeler, W.M. 1921a: 18.

**#MYRMOCLADOECUS** (subgenus of *Camponotus*)
*Myrmocladoecus* Wheeler, W.M. 1921a: 19 [as subgenus of *Camponotus*]. Type-species: *Camponotus latangulus*, by subsequent designation of Wheeler, W.M. 1922a: 709. *Myrmocladoecus* provisional junior synonym of *Camponotus*: Brown, 1973b: 182 (unconfirmed).

**#MYRMODIRACHIS** (subgenus of *Camponotus*)
*Myrmodirachis* Emery, 1925b: 168 [as subgenus of *Camponotus*]. Type-species: *Camponotus heathi*, by original designation. *Myrmodirachis* provisional junior synonym of *Camponotus*: Brown, 1973b: 182 (unconfirmed).

*MYRMOGIGAS* (junior synonym of *Dinomyrmex*)
*Myrmogigas* Forel, 1912i: 91 [as subgenus of *Camponotus*]. Type-species: *Formica gigas*, by subsequent designation of Wheeler, W.M. 1913a: 81. *Myrmogigas* junior synonym of *Dinomyrmex*: Forel, 1913c: 350 (in text); Forel, 1914a: 259.

**#MYRMOGONIA** (subgenus of *Camponotus*)
*Myrmogonia* Forel, 1912i: 92 [as subgenus of *Camponotus*]. Type-species: *Camponotus laminatus*, by subsequent designation of Wheeler, W.M. 1913a: 81. *Myrmogonia* provisional junior synonym of *Camponotus*: Brown, 1973b: 182 (unconfirmed).

*MYRMOLOPHUS* (junior synonym of *Myrmepomis*)
*Myrmolophus* Emery, 1920b: 237 [as subgenus of *Camponotus*]. Type-species: *Formica sericeiventris*, by original designation. *Myrmolophus* junior synonym of *Myrmepomis*: Wheeler, W.M. 1921a: 17.

**#MYRMOMALIS** (subgenus of *Camponotus*)
*Myrmomalis* Forel, 1914a: 263 [as subgenus of *Camponotus*]. Type-species: *Camponotus depressus*, by original designation. *Myrmomalis* provisional junior synonym of *Camponotus*: Brown, 1973b: 182 (unconfirmed).

**#MYRMONESITES** (subgenus of *Camponotus*)
*Myrmonesites* Emery, 1920b: 242 [as subgenus of *Camponotus*]. Type-species: *Camponotus putatus*, by original designation. *Myrmonesites* provisional junior synonym of *Camponotus*: Brown, 1973b: 182 (unconfirmed).

**#MYRMOPALPELLA** (subgenus of *Camponotus*)
*Myrmopalpella* Stärcke, 1934: 30 [as subgenus of *Camponotus*]. Type-species: *Camponotus megalonyx*, by monotypy. *Myrmopalpella* raised to genus: Wheeler, W.M. 1936e: 219 (in text). *Myrmopalpella* provisional junior synonym of *Camponotus*: Brown, 1973b: 182 (unconfirmed). *Myrmopalpella* here reverts to subgenus of *Camponotus*.

**#MYRMOPELTA** (subgenus of *Camponotus*)
*Myrmopelta* Santschi, 1921f: 310 [as subgenus of *Camponotus*]. Type-species: *Camponotus arminius*, by original designation. Review of subgenus: Santschi, 1926a: 16. *Myrmopelta* provisional junior synonym of *Camponotus*: Brown, 1973b: 182 (unconfirmed).

**#MYRMOPHYMA** (subgenus of *Camponotus*)
*Myrmophyma* Forel, 1912i: 91 [as subgenus of *Camponotus*]. Type-species: *Camponotus capito*, by subsequent designation of Wheeler, W.M. 1913a: 81. [Type-species not *Camponotus quadrisectus*, unjustified subsequent designation by Forel, 1914a: 261.] *Myrmophyma* senior synonym of *Myrmocamelus*: Emery, 1920b: 257; Wheeler, W.M. 1921a: 18. *Myrmophyma* provisional junior synonym of *Camponotus*: Brown, 1973b: 182 (unconfirmed).

**#*MYRMOPIROMIS*** (subgenus of *Camponotus*)
*Myrmopiromis* Wheeler, W.M. 1921a: 17 [as subgenus of *Camponotus*]. Type-species: *Formica fulvopilosa*, by subsequent designation of Wheeler, W.M. 1922a: 707. *Myrmopiromis* provisional junior synonym of *Camponotus:* Brown, 1973b: 182 (unconfirmed).

**#*MYRMOPLATYPUS*** (subgenus of *Camponotus*)
*Myrmoplatypus* Santschi, 1921f: 311 [as subgenus of *Camponotus*]. Type-species: *Camponotus platytarsus*, by original designation. *Myrmoplatypus* provisional junior synonym of *Camponotus:* Brown, 1973b: 182 (unconfirmed).

**#*MYRMOPLATYS*** (subgenus of *Camponotus*)
*Myrmoplatys* Forel, 1916: 460 [as subgenus of *Camponotus*]. Type-species: *Camponotus korthalsiae*, by original designation. *Myrmoplatys* raised to genus: Wheeler, W.M. 1936e: 219 (in text). *Myrmoplatys* provisional junior synonym of *Camponotus:* Brown, 1973b: 182 (unconfirmed). *Myrmoplatys* here reverts to subgenus of *Camponotus*.

**#*MYRMOPSAMMA*** (subgenus of *Camponotus*)
*Myrmopsamma* Forel, 1914a: 261 [as subgenus of *Camponotus*]. Type-species: *Camponotus mystaceus*, by original designation. *Myrmopsamma* provisional junior synonym of *Camponotus:* Brown, 1973b: 182 (unconfirmed).

**#*MYRMOPYTIA*** (subgenus of *Camponotus*)
*Myrmopytia* Emery, 1920b: 243 [as subgenus of *Camponotus*]. Type-species: *Camponotus imitator*, by original designation. *Myrmopytia* provisional junior synonym of *Camponotus:* Brown, 1973b: 182 (unconfirmed).

**#*MYRMORHACHIS*** (subgenus of *Phasmomyrmex*)
*Myrmorhachis* Forel, 1912i: 92 [as subgenus of *Camponotus*]. Type-species: *Camponotus polyrhachioides* (unnecessary replacement name for *Polyrhachis paradoxa*), by subsequent designation of Wheeler, W.M. 1913a: 81. [Type-species not *Camponotus latangulus*, unjustified subsequent designation by Forel, 1914a: 265.] *Myrmorhachis* subgenus of *Phasmomyrmex:* Emery, 1925b: 58. *Myrmorhachis* provisional junior synonym of *Phasmomyrmex:* Brown, 1973b: 182 (unconfirmed).

**#*MYRMOSAGA*** (subgenus of *Camponotus*)
*Myrmosaga* Forel, 1912i: 92 [as subgenus of *Camponotus*]. Type-species: *Camponotus kelleri*, by subsequent designation of Wheeler, W.M. 1913a: 81. [Type-species not *Camponotus quadrimaculatus*, unjustified subsequent designation by Forel, 1914a: 260.] *Myrmosaga* provisional junior synonym of *Camponotus:* Brown, 1973b: 182 (unconfirmed).

**#*MYRMOSAULUS*** (subgenus of *Camponotus*)
*Myrmosaulus* Wheeler, W.M. 1921a: 18 [as subgenus of *Camponotus*]. Type-species: *Formica cinerascens*, by subsequent designation of Wheeler, W.M. 1922a: 705. [Type-species not *Formica singularis*, unjustified subsequent designation by Emery, 1925b: 113.] *Myrmosaulus* provisional junior synonym of *Camponotus:* Brown, 1973b: 182 (unconfirmed). [Note: Donisthorpe, 1932c: 445, states that *cinerascens* belongs in subgenus *Myrmosericus*. If so *Myrmosaulus* falls as a junior synonym of *Myrmosericus*.]

**#*MYRMOSERICUS*** (subgenus of *Camponotus*)
*Myrmosericus* Forel, 1912i: 91 [as subgenus of *Camponotus*]. Type-species: *Formica rufoglauca*, by subsequent designation of Wheeler, W.M. 1913a: 81. *Myrmosericus* provisional junior synonym of *Camponotus:* Brown, 1973b: 182 (unconfirmed). [See note under *Myrmosaulus*, above.]

**#*MYRMOSPHINCTA*** (subgenus of *Camponotus*)
*Myrmosphincta* Forel, 1912i: 92 [as subgenus of *Camponotus*]. Type-species: *Formica sexguttata*, by subsequent designation of Wheeler, W.M. 1913a: 81. [Type-species not *Formica cinerascens*, unjustified subsequent designation by Forel, 1914a: 265.] *Myrmosphincta* provisional junior synonym of *Camponotus:* Brown, 1973b: 182 (unconfirmed).

**#*MYRMOSTENUS*** (subgenus of *Camponotus*)
*Myrmostenus* Emery, 1920b: 250 [as subgenus of *Camponotus*]. Type-species: *Camponotus mirabilis*, by original designation. *Myrmostenus* provisional junior synonym of *Camponotus:* Brown, 1973b: 182 (unconfirmed).

**#*MYRMOTARSUS*** (subgenus of *Camponotus*)
*Myrmotarsus* Forel, 1912i: 92 [as subgenus of *Camponotus*]. Type-species: *Formica mistura*, by subsequent designation of Wheeler, W.M. 1913a: 81. [Type-species not *Formica irritabilis*, unjustified subsequent designation by Forel, 1914a: 260.] *Myrmotarsus* provisional junior synonym of *Camponotus:* Brown, 1973b: 182 (unconfirmed).

**#*MYRMOTEMNUS*** (subgenus of *Camponotus*)
*Myrmotemnus* Emery, 1920b: 246 [as subgenus of *Camponotus*]. Type-species: *Camponotus moeschi*, by original designation. *Myrmotemnus* junior synonym of *Myrmamblys:* Wheeler, W.M. 1921a: 19; Emery, 1925b: 137. *Myrmotemnus* revived from synonymy: Santschi, 1926c: 601. *Myrmotemnus* provisional junior synonym of *Camponotus:* Brown, 1973b: 182 (unconfirmed).

***MYRMOTERAS*** (Formicinae: Myrmoteratini)
*Myrmoteras* Forel, 1893f: 607. Type-species: *Myrmoteras binghamii*, by monotypy. [*Myrmoteras* also described as new by Forel, 1894c: 418.] Revisions of genus: Moffett, 1985b: 1; Agosti, 1992: 405. Subgenera of *Myrmoteras:* nominal plus *Myagroteras*.

**#*MYRMOTHRINAX*** (subgenus of *Polyrhachis*)
*Myrmothrinax* Forel, 1915b: 107 [as subgenus of *Polyrhachis*]. Type-species: *Polyrhachis thrinax*, by original designation. *Myrmothrinax* senior synonym of *Evelyna:* Hung, 1967b: 402. *Myrmothrinax* provisional junior synonym of *Polyrhachis:* Brown, 1973b: 182 (unconfirmed).

**#*MYRMOTHRIX*** (subgenus of *Camponotus*)
*Myrmothrix* Forel, 1912i: 91 [as subgenus of *Camponotus*]. Type-species: *Formica abdominalis* (junior primary homonym; *Formica atriceps* first available replacement name), by subsequent designation of Wheeler, W.M. 1913a: 81. [Type-species not *Formica rufipes*, unjustified subsequent designation by Forel, 1914a: 260.] *Myrmothrix* provisional junior synonym of *Camponotus:* Brown, 1973b: 182. *Myrmothrix* subgenus of *Camponotus*, and revision of subgenus: Hashmi, 1973: 1.

**#*MYRMOTREMA*** (subgenus of *Camponotus*)
*Myrmotrema* Forel, 1912i: 91 [as subgenus of *Camponotus*]. Type-species: *Camponotus foraminosus*, by subsequent designation of Wheeler, W.M. 1913a: 81. *Myrmotrema* provisional junior synonym of *Camponotus:* Brown, 1973b: 182 (unconfirmed).

*MYRMOTURBA* (junior synonym of *Tanaemyrmex*)
*Myrmoturba* Forel, 1912i: 91 [as subgenus of *Camponotus*]. Type-species: *Formica maculata*, by subsequent designation of Wheeler, W.M. 1913a: 82. *Myrmoturba* junior synonym of *Tanaemyrmex:* Emery, 1925b: 75.

*MYRMOXENUS* (junior synonym of *Epimyrma*)
*Myrmoxenus* Ruzsky, 1902b: 474. Type-species: *Myrmoxenus gordiagini*, by monotypy. *Myrmoxenus* senior synonym of *Myrmetaerus:* Buschinger, Winter & Faber, 1984: 336. *Myrmoxenus* junior synonym of *Epimyrma:* Bolton, 1994: 105.

**#*MYRMOXYGENYS*** (subgenus of *Camponotus*)
*Myrmoxygenys* Emery, 1925b: 70 [as subgenus of *Camponotus*]. Type-species: *Camponotus caesar*, by monotypy. *Myrmoxygenys* provisional junior synonym of *Camponotus:* Brown, 1973b: 182 (unconfirmed).

*MYRMUS* (junior homonym, see *Strongylognathus*)
*Myrmus* Schenck, 1853: 188. Type-species: *Myrmus emarginatus* (junior synonym of *Strongylognathus testaceus*), by monotypy. [Junior homonym of *Myrmus* Hahn, 1832: 81 (Hemiptera).] *Strongylognathus* Mayr, 1853d: 389, replacement name for *Myrmus* Schenck.

*MYRTOTERAS* (junior synonym of *Odontomachus*)
*Myrtoteras* Matsumura, 1912: 191. Type-species: *Myrtoteras kuroiwae* (junior synonym of *Odontomachus monticola*), by monotypy. *Myrtoteras* provisional junior synonym of *Odontomachus:* Brown, 1973b: 182. *Myrtoteras* junior synonym of *Odontomachus:* Brown, 1976a: 96.

***MYSTRIUM*** (Ponerinae: Amblyoponini)
*Mystrium* Roger, 1862a: 245. Type-species: *Mystrium mysticum*, by monotypy. Revision of genus: Menozzi, 1929d: 518; Brown, 1960a: 169.

*NEAPHOMUS* (junior synonym of *Myrmelachista*)
*Neaphomus* Menozzi, 1935c: 324 [as subgenus of *Aphomomyrmex*]. Type-species: *Aphomomyrmex (Neaphomus) goetschi*, by monotypy. *Neaphomus* raised to genus: Kempf, 1972a: 152. *Neaphomus* provisional junior synonym of *Myrmelachista:* Brown, 1973b: 183. *Neaphomus* junior synonym of *Myrmelachista:* Snelling & Hunt, 1976: 110.

***NEIVAMYRMEX*** (Ecitoninae: Ecitonini)
*Neivamyrmex* Borgmeier, 1940: 606 [as subgenus of *Eciton*]. Replacement name for *Acamatus* Emery, 1894c: 181. [Junior homonym of *Acamatus* Schoenherr, 1833: 20.] *Neivamyrmex* raised to genus: Borgmeier, 1950c: 624. *Neivamyrmex* senior synonym of *Woitkowskia*, and revision of genus: Borgmeier, 1955b: 277. Key to species: Watkins, 1976: 11.

**#*NEMATOCREMA*** (subgenus of *Crematogaster*)
*Nematocrema* Santschi, 1918d: 182 [as subgenus of *Crematogaster*]. Type-species: *Crematogaster stadelmanni*, by original designation. *Nematocrema* raised to genus: Soulié, 1964: 398. *Nematocrema* provisional junior synonym of *Crematogaster:* Brown, 1973b: 183 (unconfirmed). *Nematocrema* reverts here to subgenus of *Crematogaster*.

*NEOAMBLYOPONE* (junior synonym of *Amblyopone*)
*Neoamblyopone* Wheeler, W.M. 1927c: 1 [as subgenus of *Amblyopone*]. Type-species: *Amblyopone (Neoamblyopone) clarki*, by monotypy. *Neoamblyopone* junior synonym of *Amblyopone*: Brown, 1949c: 87.
*NEOATTA* (junior synonym of *Atta*)
*Neoatta* Gonçalves, 1942: 346 [as subgenus of *Atta*]. Type-species: *Formica sexdens*, by original designation. *Neoatta* junior synonym of *Atta*: Weber, 1958a: 8. *Neoatta* revived from synonymy as subgenus of *Atta*, senior synonym of *Epiatta*, and revision of subgenus: Borgmeier, 1959b: 358. *Neoatta* provisional junior synonym of *Atta*: Brown, 1973b: 183; confirmed Bolton, 1994: 105.
*NEOCLYSTOPSENELLA* (junior synonym of *Tapinoma*)
*Neoclystopsenella* Kurian, 1955: 133 [in family Bethylidae]. Type-species: *Neoclystopsenella luffae*, by monotypy. *Neoclystopsenella* junior synonym of *Tapinoma*: Brown, 1988a: 337.
*NEOCOLOBOPSIS* (junior synonym of *Hypercolobopsis*)
*Neocolobopsis* Borgmeier, 1928b: 65 [as subgenus of *Camponotus*]. Type-species: *Camponotus (Neocolobopsis) scrobifer* (junior synonym of *Camponotus coriolanus*), by original designation. *Neocolobopsis* junior synonym of *Hypercolobopsis*: Kempf, 1968b: 411.
#***NEOCREMA*** (subgenus of *Crematogaster*)
*Neocrema* Santschi, 1918d: 182 [as subgenus of *Crematogaster*]. Type-species: *Crematogaster distans*, by original designation. *Neocrema* junior synonym of *Orthocrema*: Emery, 1922e: 130. *Neocrema* revived from synonymy and raised to genus: Soulié, 1964: 398. *Neocrema* subgenus of *Crematogaster*: Kempf, 1972a: 82. *Neocrema* provisional junior synonym of *Crematogaster*: Brown, 1973b: 183 (unconfirmed). *Neocrema* reverts here to subgenus of *Crematogaster*.
*NEOFORELIUS* (junior synonym of *Forelius*)
*Neoforelius* Kusnezov, 1953b: 327. Type-species: *Neoforelius tucumanus*, by monotypy. *Neoforelius* junior synonym of *Forelius*: Shattuck, 1992c: 87.
*NEOFORMICA* (junior synonym of *Formica*)
*Neoformica* Wheeler, W.M. 1913a: 82 [as subgenus of *Formica*]. Type-species: *Formica pallidefulva*, by original designation. [*Neoformica* also described as new by Wheeler, W.M. 1913f: 548.] *Neoformica* junior synonym of *Formica*: Buren, 1968a: 39; Agosti, 1994a: 107.
*NEOMYRMA* (junior synonym of *Manica*)
*Neomyrma* Forel, 1914a: 275 [as subgenus of *Aphaenogaster*]. Type-species: *Aphaenogaster (Neomyrma) calderoni* (junior synonym of *Manica bradleyi*), by monotypy. *Neomyrma* subgenus of *Myrmica*: Emery, 1915d: 69; Emery, 1916b: 120. *Neomyrma* senior synonym of *Oreomyrma*: Wheeler, W.M. 1915a: 50; Forel, 1915c: 364; Forel, 1915d: 9. *Neomyrma* raised to genus: Bondroit, 1918: 97. *Neomyrma* junior synonym of *Manica*: Emery, 1921f: 42.
*NEOMYRMAMBLYS* (junior synonym of *Myrmaphaenus*)
*Neomyrmamblys* Wheeler, W.M. 1921a: 19 [as subgenus of *Camponotus*]. Type-species: *Camponotus fastigatus*, by subsequent designation of Santschi, 1921f: 311. *Neomyrmamblys* junior synonym of *Myrmaphaenus*: Emery, 1925b: 152.
*NEOPHYRACACES* (junior synonym of *Cerapachys*)
*Neophyracaces* Clark, 1941: 76. Type-species: *Phyracaces clarus*, by original designation. *Neophyracaces* provisional junior synonym of *Cerapachys*: Brown, 1973b: 183. *Neophyracaces* junior synonym of *Cerapachys*: Brown, 1975: 19.
*NEOPONERA* (junior synonym of *Pachycondyla*)
*Neoponera* Emery, 1901a: 43. Type-species: *Formica villosa*, by original designation. *Neoponera* provisional junior synonym of *Pachycondyla*: Brown. 1973b: 183. *Neoponera* junior synonym of *Pachycondyla*: Brown, in preparation.
***NEOSTRUMA*** (Myrmicinae: Dacetini)
*Neostruma* Brown, 1948e: 111. Type-species: *Strumigenys crassicornis*, by original designation. Revision of genus: Brown, 1959b: 1. [See note under *Strumigenys*.]
*NESOLASIUS* (junior synonym of *Pseudolasius*)
*Nesolasius* Wheeler, W.M. 1935b: 50 [as subgenus of *Pseudolasius*]. Type-species: *Pseudolasius (Nesolasius) typhlops*, by original designation. *Nesolasius* provisional junior synonym of *Pseudolasius*: Brown, 1973b: 183; confirmed Bolton, 1994: 50.
*NESOMYRMEX* (junior synonym of *Leptothorax*)
*Nesomyrmex* Wheeler, W.M. 1910a: 259. Type-species: *Nesomyrmex clavipilis*, by monotypy. *Nesomyrmex* subgenus of *Leptothorax*: Emery, 1915g: 24; Smith, M.R. 1950: 30. *Nesomyrmex* senior synonym of *Caulomyrma*: Emery, 1915g: 24. *Nesomyrmex* junior synonym of *Goniothorax*: Emery, 1922f: 249 (incorrect procedure as *Goniothorax* a junior homonym, *Nesomyrmex* is first available replacement name, see Smith, M.R. 1950: 30). *Nesomyrmex* senior synonym of *Limnomyrmex*: Brown, 1971a: 4. *Nesomyrmex* senior synonym of *Meia* (unnecessary replacement name for *Goniothorax*): Bolton, 1994: 105. *Nesomyrmex* provisional junior synonym of *Leptothorax*: Brown, 1973b: 183. *Nesomyrmex* junior synonym of *Leptothorax*: Bolton, 1982: 319. [*Nesomyrmex* possibly valid genus: Francoeur & Loiselle, 1988b: 43. *Nesomyrmex* revived status as genus, but its junior synonyms left under *Leptothorax*: Hölldobler & Wilson, 1990: 14. *Nesomyrmex* reverts here to synonymy of *Leptothorax* until genus group taxonomy can be investigated.]
*NIMBAMYRMA* (junior synonym of *Oligomyrmex*)
*Nimbamyrma* Bernard, 1953b: 240. Type-species: *Nimbamyrma villiersi*, by monotypy. *Nimbamyrma* junior synonym of *Oligomyrmex*: Ettershank, 1966: 120.
***NOMAMYRMEX*** (Ecitoninae: Ecitonini)
*Nomamyrmex* Borgmeier, 1936: 55 [as subgenus of *Eciton*]. Type-species: *Eciton crassicornis*, by original designation. *Nomamyrmex* raised to genus: Borgmeier, 1953: 4. Revision of genus: Borgmeier, 1955b: 135. Keys to species: Watkins, 1976: 7; Watkins, 1977a: 203.
***NOONILLA*** (Leptanillinae: Leptanillini)
*Noonilla* Petersen, 1968: 582. Type-species: *Noonilla copiosa*, by original designation. Review of genus: Baroni Urbani, 1977c: 482.
***NOTHIDRIS*** (Myrmicinae: Solenopsidini)
*Nothidris* Ettershank, 1966: 105. Type-species: *Monomorium latastei*, by original designation. Key to species: Snelling & Hunt, 1976: 79. Review of genus: Bolton, 1987: 284.
***NOTHOMYRMECIA*** (Nothomyrmeciinae)
*Nothomyrmecia* Clark, 1934a: 17. Type-species: *Nothomyrmecia macrops*, by original designation. Review of genus: Taylor, 1978a: 979; see also Hölldobler & Taylor, 1984: 384.
****NOTHOMYRMICA*** (Myrmicinae: Myrmicini?)
**Nothomyrmica* Wheeler, W.M. 1915h: 60. Type-species: **Macromischa rudis*, by original designation. [**Nothomyrmica* Wheeler, W.M. 1908g: 413, and Wheeler, W.M. 1910g: 167, *nomina nuda*.]
*NOTHOSPHINCTUS* (junior synonym of *Sphinctomyrmex*)
*Nothosphinctus* Wheeler, W.M. 1918a: 219 [as subgenus of *Eusphinctus*]. Type-species: *Sphinctomyrmex froggatti*, by subsequent designation of Donisthorpe, 1943f: 675. *Nothosphinctus* provisional junior synonym of *Sphinctomyrmex*: Brown, 1973b: 183. *Nothosphinctus* junior synonym of *Sphinctomyrmex*: Brown, 1975: 31.
*NOTOMYRMEX* (junior synonym of *Chelaner*)
*Notomyrmex* Emery, 1915i: 190 [as subgenus of *Monomorium*]. Type-species: *Atta antarctica*, by original designation. *Notomyrmex* raised to genus: Kusnezov, 1957a: 269; Kusnezov, 1960b: 343. *Notomyrmex* junior synonym of *Chelaner*: Ettershank, 1966: 93. [*Chelaner* junior synonym of *Monomorium*: Bolton, 1987: 287.]
***NOTONCUS*** (Formicinae: Melophorini)
*Notoncus* Emery, 1895g: 352. Type-species: *Camponotus ectatommoides*, by original designation. *Notoncus* senior synonym of *Diodontolepis*, and revision of genus: Brown, 1955d: 471.
***NOTOSTIGMA*** (Formicinae: Camponotini)
*Notostigma* Emery, 1920b: 252. Type-species: *Camponotus carazzii*, by original designation.
*NOVOMESSOR* (junior synonym of *Aphaenogaster*)
*Novomessor* Emery, 1915d: 73. Type-species: *Aphaenogaster (Ischnomyrmex) cockerelli*, by original designation. Revision of genus: Wheeler, W.M. & Creighton, 1934: 343. *Novomessor* junior synonym of *Aphaenogaster*: Brown, 1974b: 47. *Novomessor* revived from synonymy: Hölldobler, Stanton & Engel, 1976: 32. *Novomessor* junior synonym of *Aphaenogaster*: Bolton, 1982: 364 (discussion pp. 339–341).
*NYCTERESIA* (junior synonym of *Labidus*)
*Nycteresia* Roger, 1861a: 21. Type-species: *Formica coeca*, by monotypy. *Nycteresia* junior synonym of *Eciton*: Mayr, 1865: 76. *Nycteresia* junior synonym of *Labidus*: Emery, 1910b: 21; Borgmeier, 1955b: 80.
*NYLANDERIA* (junior synonym of *Paratrechina*)
*Nylanderia* Emery, 1906b: 133 [as subgenus of *Prenolepis*]. Type-species: *Formica vividula*, by original designation. *Nylanderia* subgenus of *Paratrechina*: Emery, 1925b: 217. *Nylanderia* raised to genus: Wheeler, W.M. 1936b: 210; maintained as genus: Kempf, 1972a: 166. *Nylanderia* provisional junior synonym of *Paratrechina*: Brown, 1973b: 183. *Nylanderia* junior synonym of *Paratrechina*: Trager, 1984b: 51.
*NYSTALOMYRMA* (junior synonym of *Aphaenogaster*)
*Nystalomyrma* Wheeler, W.M. 1916j: 215 [as subgenus of *Aphaenogaster*]. Type-species: *Myrmica longiceps*, by original designation. *Nystalomyrma* provisional junior synonym of *Aphaenogaster*: Brown, 1973b:

183. *Nystalomyrma* junior synonym of *Aphaenogaster:* Bolton, 1982: 364.

***OCHETELLUS*** (Dolichoderinae)

*Ochetellus* Shattuck, 1992a: 16. Type-species: *Hypoclinea glabra*, by original designation. Review of genus: Shattuck, 1992c: 129.

***OCHETOMYRMEX*** (Myrmicinae: Ochetomyrmecini)

*Ochetomyrmex* Mayr, 1878: 871. Type-species: *Ochetomyrmex semipolitus*, by monotypy. *Ochetomyrmex* senior synonym of *Brownidris*, and review of genus: Kempf, 1975c: 355; see also notes in Bolton, 1987: 266.

*OCTELLA* (junior synonym of *Solenopsis*)

*Octella* Forel, 1915b: 70 [as subgenus of *Oligomyrmex*]. Type-species: *Oligomyrmex (Octella) pachycerus*, by original designation. *Octella* junior synonym of *Oligomyrmex:* Ettershank, 1966: 119. *Octella* junior synonym of *Solenopsis:* Taylor, 1991b: 611 (not directly stated but type-species transferred to *Solenopsis*).

***OCTOSTRUMA*** (Myrmicinae: Basicerotini)

*Octostruma* Forel, 1912e: 196 [as subgenus of *Rhopalothrix*]. Type-species: *Rhopalothrix simoni* (junior synonym of *Octostruma iheringi*), by subsequent designation of Wheeler, W.M. 1913a: 82. *Octostruma* raised to genus: Brown, 1948e: 102. Revision of genus: Brown & Kempf, 1960: 181.

***OCYMYRMEX*** (Myrmicinae: Pheidolini)

*Ocymyrmex* Emery, 1886: 364. Type-species: *Ocymyrmex barbiger*, by monotypy. Revision of genus: Bolton, 1981b: 260; Bolton & Marsh, 1989: 1267.

***ODONTOMACHUS*** (Ponerinae: Ponerini)

*Odontomachus* Latreille, 1804: 179. Type-species: *Formica haematoda*, by monotypy. *Odontomachus* senior synonym of *Pedetes:* Dalla Torre, 1893: 51. *Odontomachus* senior synonym of *Champsomyrmex* and *Myrtoteras*, and revision of genus: Brown, 1976a: 67.

*ODONTOMYRMEX* (junior synonym of *Pristomyrmex*)

*Odontomyrmex* André, 1905: 207. Type-species: *Odontomyrmex quadridentatus*, by monotypy. *Odontomyrmex* subgenus of *Pristomyrmex:* Forel, 1917: 244; maintained as subgenus: Wheeler, W.M. 1922a: 682. *Odontomyrmex* junior synonym of *Pristomyrmex:* Mann, 1919: 341; Emery, 1922f: 233; Taylor, 1965b: 35; Bolton, 1981b: 282.

*ODONTOPELTA* (junior synonym of *Leptogenys*)

*Odontopelta* Emery, 1911d: 101 [as subgenus of *Leptogenys*]. Type-species: *Leptogenys turneri*, by monotypy. *Odontopelta* provisional junior synonym of *Leptogenys:* Brown, 1973b: 183. *Odontopelta* junior synonym of *Leptogenys:* Taylor & Brown, D.R. 1985: 32 (accepted as confirmation).

***ODONTOPONERA*** (Ponerinae: Ponerini)

*Odontoponera* Mayr, 1862: 717. Type-species: *Ponera denticulata* (junior synonym of *Odontoponera transversa*), by monotypy.

*OECODOMA* (junior synonym of *Atta*)

*Oecodoma* Latreille, 1818: 222. Type-species: *Formica cephalotes*, by subsequent designation of Shuckard, in Swainson & Shuckard, 1840: 174. *Oecodoma* junior synonym of *Atta:* Roger, 1863b: 35.

*OECOPHTHORA* (junior synonym of *Pheidole*)

*Oecophthora* Heer, 1852: 15. Type-species: *Oecophthora pusilla* (junior synonym of *Pheidole megacephala*), by monotypy. *Oecophthora* junior synonym of *Pheidole:* Smith, F. 1858a: 282; Smith, F. 1858b: 172.

***OECOPHYLLA*** (Formicinae: Oecophyllini)

*Oecophylla* Smith, F. 1860b: 101. Type-species: *Formica virescens* (junior synonym of *Oecophylla smaragdina*), by subsequent designation of Bingham, 1903: 310.

*OEDALEOCERUS* (junior synonym of *Solenopsis*)

*Oedaleocerus* Creighton, 1930b: 43 [as subgenus of *Solenopsis*]. Type-species: *Solenopsis angulata*, by original designation. *Oedaleocerus* raised to genus: Kusnezov, 1957a: 272; Kusnezov, 1962b: 160. *Oedaleocerus* junior synonym of *Solenopsis:* Ettershank, 1966: 134.

***OLIGOMYRMEX*** (Myrmicinae: Pheidologetonini)

*Oligomyrmex* Mayr, 1867a: 110. Type-species: *Oligomyrmex concinnus*, by monotypy. *Oligomyrmex* senior synonym of *Aeromyrma, Aneleus* (and its junior synonym *Sporocleptes*), *Crateropsis, Hendecatella, Lecanomyrma, Nimbamyrma, Solenops*, and review of genus: Ettershank, 1966: 119. *Oligomyrmex* senior synonym of *Erebomyrma* (and its junior synonym *Spelaeomyrmex*): Bolton, 1994: 106.

***ONYCHOMYRMEX*** (Ponerinae: Amblyoponini)

*Onychomyrmex* Emery, 1895g: 349. Type-species: *Onychomyrmex hedleyi*, by monotypy. Revision of genus: Brown, 1960a: 178.

*OOCERAEA* (junior synonym of *Cerapachys*)

*Ooceraea* Roger, 1862a: 248. Type-species: *Ooceraea fragosa*, by monotypy. *Ooceraea* subgenus of *Cerapachys:* Emery, 1902d: 24; Wheeler, W.M. 1902d: 185. *Ooceraea* provisional junior synonym of *Cerapachys:* Brown, 1973b: 183. *Ooceraea* junior synonym of *Cerapachys:* Brown, 1975: 19.

*OPHTHALMOPONE* (junior synonym of *Pachycondyla*)

*Ophthalmopone* Forel, 1890b: cxi. Type-species: *Ophthalmopone berthoudi*, by monotypy. *Ophthalmopone* junior synonym of *Pachycondyla:* Brown, in preparation.

***OPISTHOPSIS*** (Formicinae: Camponotini)

*Opisthopsis* Dalla Torre, 1893: 219. Replacement name for *Myrmecopsis* Smith, F. 1865: 68. [Junior homonym of *Myrmecopsis* Newman, 1850: cxxii.] Revision of genus: Wheeler, W.M. 1918d: 341 (out of date).

*OPISTHOSCYPHUS* (junior synonym of *Alfaria*)

*Opisthoscyphus* Mann, 1922: 4. Type-species: *Opisthoscyphus scabrosus* (junior synonym of *Gnamptogenys minuta*), by original designation. *Opisthoscyphus* junior synonym of *Alfaria:* Brown, in Borgmeier, 1957: 115. [*Alfaria* junior synonym of *Gnamptogenys:* Brown, 1958g: 212.]

***ORECTOGNATHUS*** (Myrmicinae: Dacetini)

*Orectognathus* Smith, F. 1854: 227. Type-species: *Orectognathus antennatus*, by monotypy. *Orectognathus* senior synonym of *Arnoldidris:* Taylor, 1977: 581. Revisions of genus: Brown, 1953k: 84; Brown, 1958b: 17; Taylor, 1977: 585; Taylor, 1978b: 1.

*OREOMYRMA* (junior synonym of *Neomyrma*)

*Oreomyrma* Wheeler, W.M. 1914d: 118 [as subgenus of *Myrmica*]. Type-species: *Myrmica rubida*, by original designation. *Oreomyrma* junior synonym of *Neomyrma:* Wheeler, W.M. 1915a: 50. [*Neomyrma* junior synonym of *Manica:* Emery, 1921f: 42.]

*ORNATINODA* (junior synonym of *Pseudomyrmex*)

*Ornatinoda* Enzmann, E.V. 1944: 61 [as "branch" of *Pseudomyrma*]. Type-species: *Formica tenuis*, by original designation. *Ornatinoda* junior synonym of *Pseudomyrmex:* Ward, 1990: 469.

#***ORTHOCREMA*** (subgenus of *Crematogaster*)

*Orthocrema* Santschi, 1918d: 182 [as subgenus of *Crematogaster*]. Type-species: *Myrmica sordidula*, by original designation. *Orthocrema* raised to genus: Soulié, 1964: 398. *Orthocrema* provisional junior synonym of *Crematogaster:* Brown, 1973b: 183 (unconfirmed). *Orthocrema* reverts here to subgenus of *Crematogaster.* Key to Indo-Australian species: Menozzi, 1935a: 112.

#***ORTHONOTOMYRMEX*** (subgenus of *Camponotus*)

*Orthonotomyrmex* Ashmead, 1906: 31. Replacement name for *Orthonotus* Ashmead, 1905b: 384. [Junior homonym of *Orthonotus* Westwood, in Stephens, 1829: 344.] *Orthonotomyrmex* subgenus of *Camponotus:* Forel, 1913c: 350 (in text). *Orthonotomyrmex* provisional junior synonym of *Camponotus:* Brown, 1973b: 183 (unconfirmed). Notes on subgenus: Donisthorpe, 1942g: 248.

*ORTHONOTUS* (junior homonym, see *Orthonotomyrmex*)

*Orthonotus* Ashmead, 1905b: 384. Type-species: *Formica sericea*, by original designation. [Junior homonym of *Orthonotus* Westwood, in Stephens, 1829: 344 (Hemiptera).] *Orthonotomyrmex* Ashmead, 1906: 31, replacement name for *Orthonotus* Ashmead.

*OTOMYRMEX* (junior synonym of *Cataulacus*)

*Otomyrmex* Forel, 1891b: 147 [as subgenus of *Cataulacus*]. Type-species: *Cataulacus oberthueri*, by monotypy. *Otomyrmex* provisional junior synonym of *Cataulacus:* Brown, 1973b: 183. *Otomyrmex* junior synonym of *Cataulacus:* Bolton, 1974a: 7.

***OVERBECKIA*** (Formicinae: Camponotini)

*Overbeckia* Viehmeyer, 1916a: 151. Type-species: *Overbeckia subclavata*, by monotypy.

***OXYEPOECUS*** (Myrmicinae: Solenopsidini)

*Oxyepoecus* Santschi, 1926d: 6. Type-species: *Oxyepoecus bruchi*, by monotypy. *Oxyepoecus* senior synonym of *Forelifidis* (and the junior homonym *Martia* Forel): Brown, 1955b: 68. Review of genus: Ettershank, 1966: 144. Revision of genus: Kempf, 1974b: 471.

#***OXYGYNE*** (subgenus of *Crematogaster*)

*Oxygyne* Forel, 1901e: 375 [as subgenus of *Crematogaster*]. Type-species: *Crematogaster (Oxygyne) daisyi*, by subsequent designation of Wheeler, W.M. 1911f: 169. *Oxygyne* raised to genus: Soulié, 1964: 398. *Oxygyne* provisional junior synonym of *Crematogaster:* Brown, 1973b: 183 (unconfirmed). *Oxygyne* reverts here to subgenus of *Crematogaster.*

****OXYIDRIS*** (Myrmicinae: Pheidologetonini)

**Oxyidris* Wilson, 1985a: 5. Type-species: **Oxyidris antillana*, by original designation.

***OXYOPOMYRMEX*** (Myrmicinae: Pheidolini)

*Oxyopomyrmex* André, 1881b: 72. Type-species: *Oxyopomyrmex oculatus*, by monotypy.

***PACHYCONDYLA*** (Ponerinae: Ponerini)

*Pachycondyla* Smith, F. 1858b: 105. Type-species: *Formica crassinoda*, by subsequent designation of Emery, 1901a: 42. *Pachycondyla* provisional senior synonym of *Bothroponera* (and its junior synonym *Pseudoneoponera*), *Ectomomyrmex, Eumecopone,*

*Mesoponera* (and its junior synonym *Xiphopelta*), *Neoponera, Pseudoponera* (and its junior synonym *Trachymesopus*), *Termitopone* (and its junior synonym *Syntermitopone*), *Wadeura:* Brown, 1973b: 179–185. *Pachycondyla* provisional senior synonym of *Brachyponera:* Snelling, 1981: 389. Confirmation of these provisional synonyms, and *Pachycondyla* senior synonym of *Euponera, Hagensia, Megaponera, Ophthalmopone, Paltothyreus:* Brown, in preparation.

*PACHYSIMA* (junior synonym of *Tetraponera*)

*Pachysima* Emery, 1912b: 97 [as subgenus of *Sima*]. Type-species: *Tetraponera aethiops,* by monotypy. *Pachysima* raised to genus: Wheeler, W.M. 1918c: 308 (footnote). Revision of genus: Wheeler, W.M. 1922a: 112. *Pachysima* provisional junior synonym of *Tetraponera:* Brown, 1973b: 183. *Pachysima* junior synonym of *Tetraponera:* Ward, 1990: 470.

***PAEDALGUS*** (Myrmicinae: Pheidologetonini)

*Paedalgus* Forel, 1911i: 217. Type-species: *Paedalgus escherichi,* by monotypy. Review of genus: Ettershank, 1966: 128. Revision of genus: Bolton & Belshaw, 1993: 181.

*PALAEATTA* (junior synonym of *Atta*)

*Palaeatta* Borgmeier, 1950d: 244 [as subgenus of *Atta*]. Type-species: *Atta bisphaerica,* by original designation. *Palaeatta* junior synonym of *Atta:* Weber, 1958a: 8. *Palaeatta* revived from synonymy as subgenus of *Atta,* and revision of subgenus: Borgmeier, 1959b: 377. *Palaeatta* provisional junior synonym of *Atta:* Brown, 1973b: 183; confirmed Bolton, 1994: 105.

**PALAEOMYRMEX* (transferred to HOMOPTERA)

**Palaeomyrmex* Heer, 1865: 91. Type-species: **Palaeomyrmex prodromus,* by monotypy. **Palaeomyrmex* transferred from Formicidae to HOMOPTERA: Handlirsch, 1906: 507.

**PALAEOMYRMEX* (junior homonym, see **Dlusskyidris*)

**Palaeomyrmex* Dlussky, 1975: 118. Type-species: **Palaeomyrmex zherichini,* by original designation. [Junior homonym of **Palaeomyrmex* Heer, 1865: 91 (Homoptera).] **Dlusskyidris* Bolton, 1994: 187, replacement name for **Palaeomyrmex* Dlussky. **Palaeomyrmex* Dlussky transferred to *Sphecomyrmidae: Dlussky, 1983: 66. **Palaeomyrmex* Dlussky junior synonym of **Sphecomyrma:* Wilson, 1987: 49; Hölldobler & Wilson, 1990: 9. **Palaeomyrmex* Dlussky revived from synonymy: Dlussky & Fedoseeva, 1987: 77; Carpenter, 1992: 493.

****PALEOSMINTHURUS*** (*Paleosminthurinae)

**Paleosminthurus* Pierce & Gibron, 1962: 146. Type-species: **Paleosminthurus juliae,* by monotypy. **Paleosminthurus* transferred from *Paleosminthuridae to Formicidae: Najt, 1987: 152.

*PALTOTHYREUS* (junior synonym of *Pachycondyla*)

*Paltothyreus* Mayr, 1862: 735. Type-species: *Formica tarsata,* by monotypy. *Paltothyreus* junior synonym of *Pachycondyla:* Brown, in preparation.

***PAPYRIUS*** (Dolichoderinae)

*Papyrius* Shattuck, 1992a: 17. Type-species: *Iridomyrmex nitida,* by original designation. Review of genus: Shattuck, 1992c: 134.

*PARACOLOBOPSIS* (junior synonym of *Myrmaphaenus*)

*Paracolobopsis* Emery, 1920b: 249 [as subgenus of *Camponotus*]. Type-species: *Camponotus salvini,* by original designation. *Paracolobopsis* junior synonym of *Myrmaphaenus:* Emery, 1925b: 152.

#***PARACREMA*** (subgenus of *Crematogaster*)

*Paracrema* Santschi, 1918d: 182 [as subgenus of *Crematogaster*]. Type-species: *Crematogaster spengeli,* by original designation. *Paracrema* raised to genus: Soulié, 1964: 398. *Paracrema* provisional junior synonym of *Crematogaster:* Brown, 1973b: 183. *Paracrema* here reverts to subgenus of *Crematogaster.*

*PARACRYPTOCERUS* (junior synonym of *Zacryptocerus*)

*Paracryptocerus* Emery, 1915i: 192 [as subgenus of *Cryptocerus*]. Type-species: *Cryptocerus spinosus,* by original designation. *Paracryptocerus* raised to genus: Smith, M.R. 1949c: 20. *Paracryptocerus* senior synonym of *Cyathocephalus* (homonym), *Cyathomyrmex,* and *Harnedia:* Kempf, 1972a: 175. *Paracryptocerus* junior synonym of *Zacryptocerus:* Kempf, 1973c: 460. Revisionary studies: Kempf, 1951: 153; Kempf, 1952: 1; Kempf, 1958a: 61.

*PARAENICTUS* (junior synonym of *Aenictus*)

*Paraenictus* Wheeler, W.M. 1929g: 27 [as subgenus of *Aenictus*]. Type-species: *Aenictus (Paraenictus) silvestrii,* by monotypy. *Paraenictus* junior synonym of *Aenictus:* Wilson, 1964a: 444.

*PARAFORMICA* (junior synonym of *Cataglyphis*)

*Paraformica* Forel, 1915b: 95 (footnote) [as subgenus of *Formica*]. Type-species: *Formica (Paraformica) emmae,* by monotypy. *Paraformica* raised to genus: Emery, 1925b: 260; *Paraformica* subgenus of *Cataglyphis:* Santschi, 1925g: 353; Santschi, 1929b: 30. *Paraformica* subgenus of *Proformica:* Kuznetsov-Ugamsky, 1928a: 7. *Paraformica* provisional junior synonym of *Cataglyphis:* Brown, 1973b: 183. *Paraformica* junior synonym of *Cataglyphis:* Agosti, 1990a: 1462; Agosti, 1994a: 103.

****PARAMERANOPLUS*** (Myrmicinae: Meranoplini)

**Parameranoplus* Wheeler, W.M. 1915h: 69. Type-species: **Parameranoplus primaevus,* by monotypy.

*PARAMYCETOPHYLAX* (junior synonym of *Mycetophylax*)

*Paramycetophylax* Kusnezov, 1956: 23. Type-species: *Sericomyrmex bruchi,* by monotypy. *Paramycetophylax* junior synonym of *Mycetophylax:* Weber, 1958d: 262.

#***PARAMYRMAMBLYS*** (subgenus of *Camponotus*)

*Paramyrmamblys* Santschi, 1926c: 604 [as subgenus of *Camponotus*]. Type-species: *Camponotus (Myrmamblys) ostiarius,* by original designation. *Paramyrmamblys* provisional junior synonym of *Camponotus:* Brown, 1973b: 183 (unconfirmed).

*PARAMYRMICA* (junior synonym of *Myrmica*)

*Paramyrmica* Cole, 1957a: 37. Type-species: *Paramyrmica colax,* by original designation. *Paramyrmica* provisional junior synonym of *Myrmica:* Brown, 1973b: 183. *Paramyrmica* junior synonym of *Myrmica:* Bolton, 1988a: 3.

*PARANAMYRMA* (junior synonym of *Solenopsis*)

*Paranamyrma* Kusnezov, 1953d: 17. Type-species: *Paranamyrma solenopsidis,* by monotypy. [*Paranamyrma* also described as new by Kusnezov, 1954a: 9.] *Paranamyrma* junior synonym of *Solenopsis:* Ettershank, 1966: 134.

****PARANEURETUS*** (Aneuretinae)

**Paraneuretus* Wheeler, W.M. 1915h: 73. Type-species: **Paraneuretus tornquisti,* by original designation. [**Paraneuretus* Wheeler, W.M. 1908g: 413; Wheeler, W.M. 1910g: 148; Emery, 1912g: 2; *nomina nuda.*]

*PARANOMOPONE* (junior synonym of *Heteroponera*)

*Paranomopone* Wheeler, W.M. 1915c: 117. Type-species: *Paranomopone relicta,* by monotypy. *Paranomopone* junior synonym of *Heteroponera:* Brown, 1958g: 194.

*PARAPARATRECHINA* (junior synonym of *Paratrechina*)

*Paraparatrechina* Donisthorpe, 1947d: 192 [as subgenus of *Paratrechina*]. Type-species: *Paratrechina (Paraparatrechina) pallida,* by monotypy. *Paraparatrechina* provisional junior synonym of *Paratrechina:* Brown, 1973b: 183. *Paraparatrechina* junior synonym of *Paratrechina:* Trager, 1984b: 58.

*PARAPHACOTA* (junior synonym of *Monomorium*)

*Paraphacota* Santschi, 1919d: 90. Type-species: *Paraphacota surcoufi,* by monotypy. [Type-species not *Phacota noualhieri,* unjustified subsequent designation by Santschi, 1927d: 245; this error repeated in Ettershank, 1966: 82 and Bolton, 1973a: 352.] *Paraphacota* junior synonym of *Monomorium:* Santschi, 1927d: 243; Bolton, 1987: 293.

****PARAPHAENOGASTER*** (Myrmicinae: Pheidolini)

**Paraphaenogaster* Dlussky, 1981b: 68. Type-species: **Paraphaenogaster microphthalmus,* by original designation. [**Paraphaeogaster* Hong, 1984: 8, incorrect subsequent spelling.]

*PARAPHEIDOLE* (junior synonym of *Pheidole*)

*Parapheidole* Emery, 1915d: 68. Type-species: *Aphaenogaster oculata,* by monotypy. *Parapheidole* provisional junior synonym of *Pheidole:* Brown, 1973b: 183. *Parapheidole* junior synonym of *Pheidole:* Smith, D.R. 1979: 1366.

*PARAPLAGIOLEPIS* (junior synonym of *Plagiolepis*)

*Paraplagiolepis* Faber, 1969: 65 [as subgenus of *Plagiolepis*]. Type-species: *Plagiolepis xene,* by monotypy. *Paraplagiolepis* provisional junior synonym of *Plagiolepis:* Brown, 1973b: 183; confirmed Bolton, 1994: 51.

***PARAPONERA*** (Ponerinae: Paraponerini)

*Paraponera* Smith, F. 1858b: 100. Type-species: *Formica clavata,* by monotypy. Revision of genus: Brown, 1958g: 205. See also: Lattke, 1994: 111.

***PARAPRIONOPELTA*** (Ponerinae: Amblyoponini)

*Paraprionopelta* Kusnezov, 1955: 270. Type-species: *Paraprionopelta minima,* by monotypy. *Paraprionopelta* discussed: Brown, 1960a: 181. *Paraprionopelta* provisional junior synonym of *Amblyopone:* Brown, 1973b: 183 (unconfirmed).

*PARASIMA* (junior synonym of *Tetraponera*)

*Parasima* Donisthorpe, 1948d: 592 [as subgenus of *Tetraponera*]. Unnecessary replacement name for *Sima* in the sense of Emery, 1921f: 23. *Parasima* junior synonym of *Tetraponera:* Ward, 1990: 470.

*PARASYSCIA* (junior synonym of *Cerapachys*)

*Parasyscia* Emery, in André, 1882c: 235. Type-species: *Parasyscia piochardi,* by monotypy. *Parasyscia* subgenus of *Cerapachys:* Forel, 1892l: 243. *Parasyscia* junior synonym of *Cerapachys:* Kempf, 1972a: 76.

***PARATOPULA*** (Myrmicinae: Formicoxenini)

*Paratopula* Wheeler, W.M. 1919e: 144. Type-species: *Atopomyrmex ceylonicus,* by original designation. *Paratopula* provisional junior synonym of *Atopula:*

Brown, 1973b: 183. *Paratopula* valid genus, and revision of genus: Bolton, 1988b: 125.

***PARATRECHINA*** (Formicinae: Lasiini)

*Paratrechina* Motschoulsky, 1863: 13. Type-species: *Paratrechina currens* (junior synonym of *Paratrechina longicornis*), by subsequent designation of Wheeler, W.M. 1911f: 170. *Paratrechina* junior synonym of *Prenolepis:* Dalla Torre, 1893: 177; Wheeler, W.M. 1922a: 940. *Paratrechina* revived from synonymy: Emery, 1925b: 216. *Paratrechina* senior synonym of *Nylanderia* and *Paraparatrechina,* and revision of North American species: Trager, 1984b: 58. *Paratrechina* senior synonym of *Andragnathus:* Agosti & Bolton, 1990a: 75.

*PARECTATOMMA* (junior synonym of *Gnamptogenys*)

*Parectatomma* Emery, 1911d: 44 [as subgenus of *Ectatomma*]. Type-species: *Ectatomma triangulare,* by original designation. *Parectatomma* subgenus of *Gnamptogenys:* Forel, 1917: 236. *Parectatomma* junior synonym of *Gnamptogenys:* Brown, 1958g: 212.

*PARHOLCOMYRMEX* (junior synonym of *Monomorium*)

*Parholcomyrmex* Emery, 1915i: 190 [as subgenus of *Monomorium*]. Type-species: *Myrmica gracillima* (junior synonym of *Monomorium destructor*), by original designation. *Parholcomyrmex* junior synonym of *Monomorium:* Ettershank, 1966: 82; see also Bolton, 1987: 296. [*Paraholcomyrmex* Emery, 1915i: 191, incorrect subsequent spelling.]

*PEDETES* (junior synonym of *Odontomachus*)

*Pedetes* Bernstein, 1861: 7. Type-species: *Pedetes macrorhynchus,* by monotypy. *Pedetes* junior synonym of *Odontomachus:* Dalla Torre, 1893: 51.

***PENTASTRUMA*** (Myrmicinae: Dacetini)

*Pentastruma* Forel, 1912a: 50. Type-species: *Pentastruma sauteri,* by monotypy. Revision of genus: Brown & Boisvert, 1979: 201. [See note under *Strumigenys.*]

*PERGANDEA* (*nomen nudum*)

*Pergandea* Ashmead, 1905b: 382, *nomen nudum.*

***PERISSOMYRMEX*** (Myrmicinae: Myrmecinini)

*Perissomyrmex* Smith, M.R. 1947i: 281. Type-species: *Perissomyrmex snyderi,* by original designation.

***PERONOMYRMEX*** (Myrmicinae: Formicoxenini)

*Peronomyrmex* Viehmeyer, 1922: 212. Type-species: *Peronomyrmex overbecki,* by monotypy. Review of genus: Taylor, 1970b: 209.

***PETALOMYRMEX*** (Formicinae: Brachymyrmecini)

*Petalomyrmex* Snelling, 1979b: 5. Type-species: *Petalomyrmex phylax,* by original designation.

***PETRAEOMYRMEX** (Dolichoderinae)

**Petraeomyrmex* Carpenter, 1930: 55. Type-species: **Petraeomyrmex minimus,* by original designation.

***PETROPONE** (*Armaniinae)

**Petropone* Dlussky, 1975: 119. Type-species: **Petropone petiolata,* by original designation. **Petropone* placed in *Armaniidae: Dlussky & Fedoseeva, 1987: 77. **Petropone* possible junior synonym of **Sphecomyrma:* Hölldobler & Wilson, 1990: 9. **Petropone* status as genus: Carpenter, 1992: 493

***PHACOTA*** (Myrmicinae: Solenopsidini)

*Phacota* Roger, 1862b: 260. Type-species: *Phacota sichelii,* by monotypy. *Phacota* junior synonym of *Monomorium:* Ettershank, 1966: 82. *Phacota* revived from synonymy: Bolton, 1987: 281.

***PHALACROMYRMEX*** (Myrmicinae: Phalacromyrmecini)

*Phalacromyrmex* Kempf, 1960b: 89. Type-species: *Phalacromyrmex fugax,* by original designation. Review of genus: Bolton, 1984: 377.

*PHARAOPHANES* (unavailable name)

*Pharaophanes* Bernard, 1953b: 238 (attributed to Santschi) [as subgenus of *Monomorium*]. [Without designation of type-species; unavailable name.] Included species all referable to *Monomorium;* see also note in Baroni Urbani, 1971c: 90.

***PHASMOMYRMEX*** (Formicinae: Camponotini)

*Phasmomyrmex* Stitz, 1910: 146. Type-species: *Phasmomyrmex sericeus* (junior synonym of *Phasmomyrmex buchneri*), by monotypy. *Phasmomyrmex* subgenus of *Camponotus:* Forel, 1912i: 90. *Phasmomyrmex* revived status as genus: Wheeler, W.M. 1922a: 701. Subgenera of *Phasmomyrmex:* nominal plus *Myrmacantha, Myrmorhachis.*

***PHAULOMYRMA*** (Leptanillinae: Leptanillini)

*Phaulomyrma* Wheeler, G.C. & Wheeler, E.W. 1930: 193. Type-species: *Phaulomyrma javana,* by original designation. Review of genus: Petersen, 1968: 593; Baroni Urbani, 1977c: 480.

*PHEIDOLACANTHINUS* (junior synonym of *Pheidole*)

*Pheidolacanthinus* Smith, F. 1865: 75. Type-species: *Pheidolacanthinus armatus* (junior synonym of *Pheidole quadrispinosa*), by monotypy. *Pheidolacanthinus* subgenus of *Pheidole:* Forel, 1900a: 25. *Pheidolacanthinus* provisional junior synonym of *Pheidole:* Brown, 1973b: 183. *Pheidolacanthinus* junior synonym of *Pheidole:* Smith, D.R. 1979: 1365.

***PHEIDOLE*** (Myrmicinae: Pheidolini)

*Pheidole* Westwood, 1839: 219. Type-species: *Atta providens,* by monotypy. *Pheidole* senior synonym of *Oecophthora:* Smith, F. 1858a: 282; Smith, F. 1858b: 172. *Pheidole* senior synonym of *Leptomyrma:* Emery, 1892b: 166. *Pheidole* senior synonym of *Allopheidole* and *Cardiopheidole:* Emery, 1921f: 84. *Pheidole* senior synonym of *Phidole:* Wheeler, W.M. 1922a: 806. *Pheidole* senior synonym of *Macropheidole:* Smith, M.R. 1951a: 800. *Pheidole* senior synonym of *Epipheidole:* Cole, 1965: 174. *Pheidole* provisional senior synonym of *Anergatides, Bruchomyrma, Ceratopheidole, Conothoracoides, Conothorax* (homonym), *Decapheidole, Elasmopheidole, Electropheidole, Eriopheidole, Gallardomyrma, Hendecapheidole* (and its junior synonym *Cephalomorium*), *Ischnomyrmex* (and its junior synonym *Isopheidole*), *Parapheidole, Pheidolacanthinus, Scrobopheidole, Stegopheidole, Sympheidole,* and *Trachypheidole:* Brown, 1973b: 178–185. Confirmation of these synonyms: Smith, D.R. 1979: 1365. *Pheidole* senior synonym of *Anergatides* and *Bruchomyrma:* Wilson, 1984: 327. *Pheidole* senior synonym of *Xenoaphaenogaster:* Bolton, 1987: 291. *Pheidole* senior synonym of *Phidola:* Bolton, 1994: 106. Key to North American species: Gregg, 1959: 9.

***PHEIDOLOGETON*** (Myrmicinae: Pheidologetonini)

*Pheidologeton* Mayr, 1862: 750. Type-species: *Oecodoma diversa,* by subsequent designation of Bingham, 1903: 160. *Pheidologeton* senior synonym of *Phidologeton:* Wheeler, W.M. 1922a: 880. *Pheidologeton* senior synonym of *Amauromyrmex* and *Idrisella,* and review of genus: Ettershank, 1966: 115.

*PHIDOLA* (junior synonym of *Pheidole*)

*Phidola* Schulz, 1906: 155, unjustified emendation of *Pheidole. Phidola* junior synonym of *Pheidole:* Bolton, 1994: 106.

*PHIDOLE* (junior synonym of *Pheidole*)

*Phidole* Bingham, 1903: 220, unjustified emendation of *Pheidole. Phidole* junior synonym of *Pheidole:* Wheeler, 1922a: 806.

*PHIDOLOGETON* (junior synonym of *Pheidologeton*)

*Phidologeton* Bingham, 1903: 160, unjustified emendation of *Pheidologeton. Phidologeton* junior synonym of *Pheidologeton:* Wheeler, W.M. 1922a: 880.

***PHILIDRIS*** (Dolichoderinae)

*Philidris* Shattuck, 1992a: 17. Type-species: *Formica cordata,* by original designation. Review of genus: Shattuck, 1992c: 140.

***PHRYNOPONERA*** (Ponerinae: Ponerini)

*Phrynoponera* Wheeler, W.M. 1920: 53. Type-species: *Bothroponera gabonensis,* by original designation. Revision of genus: Wheeler, W.M. 1922a: 75 (out of date). *Phrynoponera* provisional junior synonym of *Pachycondyla:* Brown, 1973b: 183 (unconfirmed). *Phrynoponera* valid genus: here noted.

*PHYRACACES* (junior synonym of *Cerapachys*)

*Phyracaces* Emery, 1902c: 23. Type-species: *Cerapachys mayri,* by original designation. *Phyracaces* provisional junior synonym of *Cerapachys:* Brown, 1973b: 183. *Phyracaces* junior synonym of *Cerapachys:* Brown, 1975: 19.

*PHYSATTA* (junior synonym of *Heptacondylus*)

*Physatta* Smith, F. 1857a: 77. Type-species: *Physatta dromedarius* (junior synonym of *Myrmicaria carinata*), by monotypy. *Physatta* junior synonym of *Heptacondylus:* Mayr, 1862: 755 (in text). [*Heptacondylus* junior synonym of *Myrmicaria:* Smith, F. 1865: 73.]

**#PHYSOCREMA** (subgenus of *Crematogaster*)

*Physocrema* Forel, 1912f: 220 [as subgenus of *Crematogaster*]. Type-species: *Crematogaster inflata,* by subsequent designation of Wheeler, W.M. 1913a: 82. *Physocrema* raised to genus: Soulié, 1964: 398. *Physocrema* provisional junior synonym of *Crematogaster:* Brown, 1973b: 183. *Physocrema* here reverts to subgenus of *Crematogaster.*

***PILOTROCHUS*** (Myrmicinae: Phalacromyrmecini)

*Pilotrochus* Brown, 1978a: 218. Type-species: *Pilotrochus besmerus,* by original designation. Review of genus: Bolton, 1984: 380.

***PITYOMYRMEX** (Dolichoderinae)

**Pityomyrmex* Wheeler, W.M. 1915h: 98. Type-species: **Pityomyrmex tornquisti,* by monotypy.

***PLAGIOLEPIS*** (Formicinae: Plagiolepidini)

*Plagiolepis* Mayr, 1861: 42. Type-species: *Formica pygmaea,* by monotypy. *Plagiolepis* provisional senior synonym of *Anacantholepis, Aporomyrmex* and *Paraplagiolepis:* Brown, 1973b: 178, 183. Synonymy of *Aporomyrmex* and *Paraplagiolepis* confirmed Bolton, 1994: 51; synonymy of *Anacantholepis* remains unconfirmed.

*PLANIMYRMA* (junior synonym of *Aphaenogaster*)

*Planimyrma* Viehmeyer, 1914d: 604 [as subgenus of *Aphaenogaster*]. Type-species: *Stenamma* (*Ischnomyrmex*) *loriai,* by original designation.

*Planimyrma* provisional junior synonym of *Aphaenogaster:* Brown, 1973b: 184. *Planimyrma* junior synonym of *Aphaenogaster:* Bolton, 1982: 364.

*PLATYSTRUMA* (junior synonym of *Smithistruma*)

*Platystruma* Brown, 1953g: 112 [as subgenus of *Smithistruma*]. Type-species: *Strumigenys (Cephaloxys) depressiceps,* by original designation. *Platystruma* junior synonym of *Smithistruma:* Brown, 1973a: 35.

***PLATYTHYREA*** (Ponerinae: Platythyreini)

*Platythyrea* Roger, 1863a: 172. Type-species: *Pachycondyla punctata,* by subsequent designation of Bingham, 1903: 73. *Platythyrea* senior synonym of *Eubothroponera,* and revision of genus: Brown, 1975: 6.

***PLECTROCTENA*** (Ponerinae: Ponerini)

*Plectroctena* Smith, F. 1858b: 101. Type-species: *Plectroctena mandibularis,* by monotypy. *Plectroctena* senior synonym of *Cacopone:* Bolton, 1974c: 313. Revisions of genus: Santschi, 1924a: 155 (out of date); Bolton, 1974c: 309.

***PODOMYRMA*** (Myrmicinae: Formicoxenini)

*Podomyrma* Smith, F. 1859a: 145. Type-species: *Podomyrma femorata,* by subsequent designation of Wheeler, W.M. 1911f: 170. *Podomyrma* senior synonym of **Acrostigma:* Dalla Torre, 1893: 59. *Podomyrma* senior synonym of *Dacryon* and *Pseudopodomyrma:* Taylor & Brown, D.R. 1985: 80.

***POECILOMYRMA*** (Myrmicinae: Formicoxenini)

*Poecilomyrma* Mann, 1921: 445. Type-species: *Poecilomyrma senirewae,* by original designation.

***POGONOMYRMEX*** (Myrmicinae: Myrmicini)

*Pogonomyrmex* Mayr, 1868b: 169. Type-species: *Formica badia,* by subsequent designation of Wheeler, W.M. 1911f: 170. *Pogonomyrmex* senior synonym of *Ephebomyrmex* and *Forelomyrmex* (and the junior homonym *Janetia* Forel): Bolton, 1994: 106. Keys to species: Gallardo, 1932b: 95 and Kusnezov, 1951a: 254 (Argentina); Cole, 1968: 38 (North America); Snelling & Hunt, 1976: 72 (Chile); MacKay, W.P., MacKay, E.E., Dominguez, J.F.P., *et al.,* 1985: 39 (Mexico); Shattuck, 1987: 175 (*occidentalis*-complex); see also Hölldobler, 1976: 3; Taber, 1990: 307.

***POLYERGUS*** (Formicinae: Formicini)

*Polyergus* Latreille, 1804: 179. Type-species: *Formica rufescens,* by monotypy. Revision of North American species: Smith, M.R. 1947g: 150; Wheeler, J. 1968: 156; see also Creighton, 1950a: 552; Agosti, 1994a: 109.

*POLYHOMOA* (junior synonym of *Kyidris*)

*Polyhomoa* Azuma, 1950: 36. Type-species: *Polyhomoa itoi* (junior synonym of *Kyidris mutica*), by monotypy. *Polyhomoa* junior synonym of *Kyidris:* Creighton, 1950b: 93; see also Brown & Yasumatsu, 1951: 93.

***POLYRHACHIS*** (Formicinae: Camponotini)

*Polyrhachis* Smith, F. 1857a: 58. Type-species: *Formica bihamata,* by original designation. [*Polyrhachis* Shuckard, in Swainson & Shuckard, 1840: 172, *nomen nudum.*] *Polyrhachis* subgenus of *Myrma:* Wheeler, W.M. 1911f: 860. *Myrma* subgenus of *Polyrhachis:* Wheeler, W.M. 1922a: 993 (see under *Myrma*). Subgenera of *Polyrhachis* (alphabetical order with synonyms indicated): nominal plus *Anoplomyrma, Aulacomyrma, Campomyrma, Cephalomyrma* (junior synonym of *Myrmhopla*), *Chariomyrma, Cyrtomyrma, Dolichorhachis, Evelyna* (junior synonym of *Myrmothrinax*), *Florencea* (junior synonym of *Myrmhopla*), *Hagiomyrma, Hedomyrma, Hemioptica, Hoplomyrmus* (junior synonym of *Myrma*), *Irenea* (junior synonym of *Myrmatopa*), *Johnia* (junior synonym of *Aulacomyrma*), *Morleyidris* (junior synonym of *Hedomyrma*), *Myrma, Myrmatopa, Myrmhopla, Myrmothrinax, Pseudocyrtomyrma* (junior synonym of *Myrma*) Revision of subgenera: Hung, 1967b: 395. All subgenera given as provisional junior synonyms of *Polyrhachis:* Brown, 1973b: 178–184 (unconfirmed); see under individual entries. The entry of *Chariomyrma, Hagiomyrma,* and *Hedomyrma* under the synonymy of *Polyrhachis* by Taylor & Brown, D.R. 1985: 131, is not considered as confirmation as the three subgenera all range outside the Australasian Region. Key to Afrotropical species: Bolton, 1973b: 289. [*Polyrachis* Arnold, 1924: 741, incorrect subsequent spelling.]

***PONERA*** (Ponerinae: Ponerini)

*Ponera* Latreille, 1804: 179. Type-species: *Formica coarctata,* by subsequent designation of Westwood, 1840a: 83 (see discussion in Taylor, 1967a: 6). *Ponera* senior synonym of *Pseudocryptopone* and *Selenopone:* Wilson, 1957b: 356. Revision of genus: Taylor, 1967a: 1. *Ponera* senior synonym of *Pteroponera:* Brown, in preparation.

*PONERACANTHA* (junior synonym of *Gnamptogenys*)

*Poneracantha* Emery, 1897d: 547 [as subgenus of *Ectatomma*]. Type-specis: *Ectatomma (Holcoponera) bispinosum,* by original designation. *Poneracantha* junior synonym of *Gnamptogenys:* Brown, 1958g: 211.

**PONERITES* (Ponerinae: Ponerini)

**Ponerites* Dlussky, 1981b: 67 [collective group name].

****PONEROPSIS*** (Ponerinae: Ponerini)

**Poneropsis* Heer, 1867: 19. Type-species: **Ponera fuliginosa,* by subsequent designation of Wheeler, W.M. 1911f: 171.

****PONEROPTERUS*** (*Armaniinae)

**Poneropterus* Dlussky, 1983: 73. Type-species: **Poneropterus sphecoides,* by original designation. **Poneropterus* junior synonym of **Sphecomyrma:* Wilson, 1987: 49; Hölldobler & Wilson, 1990: 9. **Poneropterus* revived from synonymy: Dlussky & Fedoseeva, 1987: 77; Carpenter, 1992: 494.

***PRENOLEPIS*** (Formicinae: Lasiini)

*Prenolepis* Mayr, 1861: 52. Type-species: *Tapinoma nitens,* by subsequent designation of Bingham, 1903: 325.

*PRIONOGENYS* (junior synonym of *Leptogenys*)

*Prionogenys* Emery, 1895g: 348. Type-species: *Prionogenys podenzanai,* by monotypy. *Prionogenys* junior synonym of *Leptogenys:* Taylor, 1988: 33.

****PRIONOMYRMEX*** (Myrmeciinae: *Prionomyrmecini)

**Prionomyrmex* Mayr, 1868c: 77. Type-species: **Prionomyrmex longiceps,* by monotypy.

***PRIONOPELTA*** (Ponerinae: Amblyoponini)

*Prionopelta* Mayr, 1866a: 503. Type-species: *Prionopelta punctulata,* by monotypy. *Prionopelta* senior synonym of *Examblyopone:* Brown, 1951: 102. *Prionopelta* senior synonym of *Renea:* Brown, 1953c: 11. Revision of genus: Brown, 1960a: 173.

*PRISTOMYRMECIA* (junior synonym of *Promyrmecia*)

*Pristomyrmecia* Emery, 1911d: 21 [as subgenus of *Myrmecia*]. Type-species: *Myrmecia mandibularis,* by original designation. *Pristomyrmecia* junior synonym of *Promyrmecia:* Clark, 1951: 119. [*Promyrmecia* junior synonym of *Myrmecia:* Brown, 1953j: 1.]

***PRISTOMYRMEX*** (Myrmicinae: Myrmecinini)

*Pristomyrmex* Mayr, 1866b: 903. Type-species: *Pristomyrmex pungens,* by monotypy. *Pristomyrmex* senior synonym of *Odontomyrmex:* Mann, 1919: 341. *Pristomyrmex* senior synonym of *Hylidris:* Brown, 1953c: 9. *Pristomyrmex* senior synonym of *Dodous:* Brown, 1971a: 3. Revision of Australasian species: Taylor, 1965b: 35; Taylor, 1968c: 63; of Afrotropical species: Bolton, 1981b: 282.

***PROATTA*** (Myrmicinae: Stenammini)

*Proatta* Forel, 1912m: 768. Type-species: *Proatta butteli,* by monotypy. [*Proatta* also described as new by Forel, 1913k: 84.] [*Proatta* as senior synonym of *Asemorhoptrum* and *Theryella:* Snelling, 1981: 395, in error; these names properly attributable to synonymy of *Stenamma,* see there.]

***PROBOLOMYRMEX*** (Ponerinae: Platythyreini)

*Probolomyrmex* Mayr, 1901b: 2. Type-species: *Probolomyrmex filiformis,* by monotypy. *Probolomyrmex* senior synonym of *Escherichia,* and revision of genus: Taylor, 1965d: 346; see also Brown, 1975: 11.

**PROCERAPACHYS* (junior synonym of *Cerapachys*)

**Procerapachys* Wheeler, W.M. 1915h: 27. Type-species: **Procerapachys annosus,* by original designation. **Procerapachys* junior synonym of *Cerapachys:* Brown, 1975: 19.

***PROCERATIUM*** (Ponerinae: Proceratiini)

*Proceratium* Roger, 1863a: 171. Type-species: *Proceratium silaceum,* by monotypy. *Proceratium* senior synonym of *Sysphingta:* Brown, in Borgmeier, 1957: 118. Revision of genus: Brown, 1958g: 241. Keys to New World species: Snelling, 1967: 8; Brown, 1980b: 342; Ward, 1988: 116; to Afrotropical species: Terron, 1981: 102. See also: Lattke, 1994: 113.

***PROCRYPTOCERUS*** (Myrmicinae: Cephalotini)

*Procryptocerus* Emery, 1887b: 470 (footnote). Type-species: *Meranoplus striatus,* by subsequent designation of Wheeler, W.M. 1911f: 171. Revision of genus: Kempf, 1951: 15.

*PRODICROASPIS* (junior synonym of *Lordomyrma*)

*Prodicroaspis* Emery, 1914f: 414. Type-species: *Prodicroaspis sarasini,* by monotypy. *Prodicroaspis* junior synonym of *Lordomyrma:* Bolton, 1994: 106.

****PRODIMORPHOMYRMEX*** (Formicinae: Gesomyrmecini)

**Prodimorphomyrmex* Wheeler, W.M. 1915h: 111. Type-species: **Prodimorphomyrmex primigenius,* by monotypy.

*PRODISCOTHYREA* (junior synonym of *Discothyrea*)

*Prodiscothyrea* Wheeler, W.M. 1916i: 33. Type-species: *Prodiscothyrea velutina,* by monotypy. *Prodiscothyrea* junior synonym of *Discothyrea:* Brown, 1958g: 248.

***PROFORMICA*** (Formicinae: Formicini)

*Proformica* Ruzsky, 1902d: 13 [as subgenus of *Formica*]. Type-species: *Formica nasuta,* by monotypy. [*Proformica* also described as new by Ruzsky, 1903b: 303.] *Proformica* raised to genus: Bondroit, 1918: 40. Key to species of former U.S.S.R. and contiguous countries: Dlussky, 1969a: 231. See also: Agosti, 1994a: 112.

**PROLASIUS** (Formicinae: Melophorini)

*Prolasius* Forel, 1892c: 331 [as subgenus of *Lasius*]. Type-species: *Formica advena*, by monotypy. *Prolasius* subgenus of *Melophorus*: Forel, 1917: 248. *Prolasius* raised to genus: Wheeler, W.M. 1922a: 695. Key to species: McAreavey, 1947: 9.

*PROMERANOPLUS* (junior synonym of *Lordomyrma*)

*Promeranoplus* Emery, 1914f: 412. Type-species: *Promeranoplus rouxi*, by monotypy. *Promeranoplus* junior synonym of *Lordomyrma*: Bolton, 1994: 106.

*PROMYOPIAS* (junior synonym of *Centromyrmex*)

*Promyopias* Santschi, 1914d: 323 [as subgenus of *Myopias*]. Type-species: *Myopias* (*Promyopias*) *silvestrii*, by monotypy. *Promyopias* raised to genus: Emery, 1915g: 26. *Promyopias* subgenus of *Pseudoponera*: Wheeler, W.M. 1922a: 649. *Promyopias* provisional junior synonym of *Centromyrmex*: Brown, 1973b: 184; confirmed Bolton, 1994: 164.

*PROMYRMA* (junior synonym of *Liomyrmex*)

*Promyrma* Forel, 1912m: 764. Type-species: *Promyrma butteli*, by monotypy. *Promyrma* junior synonym of *Liomyrmex*: Forel, 1913k: 26; Emery, 1921f: 191.

*PROMYRMECIA* (junior synonym of *Myrmecia*)

*Promyrmecia* Emery, 1911d: 19 [as subgenus of *Myrmecia*]. Type-species: *Myrmecia aberrans*, by original designation. *Promyrmecia* senior synonym of *Halmamyrmecia* and *Pristomyrmecia*: Clark, 1927: 37; Clark, 1943: 84. *Promyrmecia* raised to genus, and revision of genus: Clark, 1943: 84. [*Promyrmecia* junior synonym of *Myrmecia*: Brown, 1953j: 1.]

***PROMYRMICIUM** (Myrmicinae: unplaced to tribe)

**Promyrmicium* Baroni Urbani, 1971b: 362. Replacement name for **Myrmicium* Heer, 1870: 78. [Junior homonym of **Myrmicium* Westwood, 1854: 396.]

**PROPODOMYRMA* (junior synonym of *Vollenhovia*)

**Propodomyrma* Wheeler, W.M. 1910g: 163. Type-species: **Propodomyrma samlandica* (junior synonym of **Vollenhovia beyrichi*), by monotypy. [**Propodomyrma* Wheeler, W.M. 1908g: 413, *nomen nudum*.] **Propodomyrma* junior synonym of *Vollenhovia*: Wheeler, W.M. 1915h: 51.

*PROSCOPOMYRMEX* (junior synonym of *Strumigenys*)

*Proscopomyrmex* Patrizi, 1946: 294. Type-species: *Proscopomyrmex londianensis*, by monotypy. *Proscopomyrmex* junior synonym of *Strumigenys*: Brown, 1949d: 15.

*PROSOPIDRIS* (junior synonym of *Cardiocondyla*)

*Prosopidris* Wheeler, W.M. 1935b: 40 [as subgenus of *Cardiocondyla*]. Type-species: *Cardiocondyla* (*Prosopidris*) *sima*, by original designation. *Prosopidris* raised to genus: Reiskind, 1965: 80. *Prosopidris* provisional junior synonym of *Cardiocondyla*: Brown, 1973b: 184. *Prosopidris* junior synonym of *Cardiocondyla*: Bolton, 1982: 309.

**PROTALARIDRIS** (Myrmicinae: Basicerotini)

*Protalaridris* Brown, 1980a: 36. Type-species: *Protalaridris armata*, by original designation.

*PROTAMBLYOPONE* (junior synonym of *Amblyopone*)

*Protamblyopone* Wheeler, W.M. 1927c: 1 [as subgenus of *Amblyopone*]. Type-species: *Amblyopone* (*Protamblyopone*) *aberrans*, by monotypy. *Protamblyopone* junior synonym of *Amblyopone*: Brown, 1949c: 87.

**PROTAMBLYOPONE* (junior homonym, see **Casaleia*)

**Protamblyopone* Dlussky, 1981b: 65. Type-species: **Protamblyopone inversa*, by original designation. [Junior homonym of *Protamblyopone* Wheeler, above.] **Casaleia* Pagliano & Scaramozzino, 1990: 5, replacement name for **Protamblyopone* Dlussky.

***PROTANEURETUS** (Aneuretinae)

**Protaneuretus* Wheeler, W.M. 1915h: 71. Type-species: **Protaneuretus succineus*, by monotypy. [**Protaneuretus* Wheeler, W.M. 1908g: 413; Wheeler, W.M. 1910g: 148; Emery, 1912g: 2, *nomina nuda*.]

**PROTANILLA** (Leptanillinae: Anomalomyrmini)

*Protanilla* Taylor, in Bolton, 1990b: 279. Type-species: *Protanilla rafflesi*, by original designation.

***PROTAZTECA** (Dolichoderinae)

**Protazteca* Carpenter, 1930: 41. Type-species: **Protazteca elongata*, by original designation.

*PROTHOLCOMYRMEX* (junior synonym of *Chelaner*)

*Protholcomyrmex* Wheeler, W.M. 1922a: 162 [as subgenus of *Monomorium*]. Type-species: *Monomorium rothsteini*, by original designation. *Protholcomyrmex* junior synonym of *Chelaner*: Ettershank, 1966: 93. [*Chelaner* junior synonym of *Monomorium*: Bolton, 1987: 287; see also discussion, p.300.]

***PROTOFORMICA** (Formicinae: Formicini)

**Protoformica* Dlussky, 1967b: 83 [as subgenus of *Formica*]. Type-species: **Formica* (*Protoformica*) *proformicoides*, by original designation. **Protoformica* raised to genus: Dlussky & Fedoseeva, 1987: 77.

**PROTOMOGNATHUS** (Myrmicinae: Formicoxenini)

*Protomognathus* Wheeler, W.M. 1905a: 3 [as subgenus of *Tomognathus*]. Type-species: *Tomognathus americanus*, by monotypy. *Protomognathus* junior synonym of *Harpagoxenus*: Emery, 1922f: 265. *Protomognathus* revived from synonymy and raised to genus: Hölldobler & Wilson, 1990: 65.

***PROTOPONE** (Ponerinae: Ponerini)

**Protopone* Dlussky, 1988: 52. Type-species: **Protopone primigena*, by original designation.

***PROTRECHINA** (Formicinae: Lasiini)

**Protrechina* Wilson, 1985d: 211. Type-species: **Protrechina carpenteri*, by original designation.

**PSALIDOMYRMEX** (Ponerinae: Ponerini)

*Psalidomyrmex* André, 1890: 313. Type-species: *Psalidomyrmex foveolatus*, by monotypy. Revision of genus: Bolton, 1975b: 1.

*PSAMMOMYRMA* (junior synonym of *Dorymyrmex*)

*Psammomyrma* Forel, 1912h: 43 [as subgenus of *Dorymyrmex*]. Type-species: *Dorymyrmex planidens*, by subsequent designation of Wheeler, W.M. 1913a: 82. *Psammomyrma* junior synonym of *Dorymyrmex*: Forel, 1913b: 350; Snelling & Hunt, 1976: 93.

*PSEUDACANTHOLEPIS* (unavailable name)

*Pseudacantholepis* Bernard, 1953b: 256 (attributed to Santschi) [as subgenus of *Acantholepis*]. [Without designation of type-species; unavailable name.] Included species referable to *Lepisiota*.

**PSEUDAPHOMOMYRMEX** (Formicinae: Brachymyrmecini)

*Pseudaphomomyrmex* Wheeler, W.M. 1920: 53. Type-species: *Aphomomyrmex emeryi*, by original designation. Diagnosis of genus: Wheeler, W.M. 1922a: 695 (in key). *Pseudaphomomyrmex* junior synonym of *Aphomomyrmex*: Emery, 1925b: 44. *Pseudaphomomyrmex* revived from synonymy: Chapman & Capco, 1951: 214.

***PSEUDARMANIA** (*Armaniinae)

**Pseudarmania* Dlussky, 1983: 69. Type-species: **Pseudarmania rasnitsyni*, by original designation. **Pseudarmania* junior synonym of **Sphecomyrma*: Wilson, 1987: 49; Hölldobler & Wilson, 1990: 9. **Pseudarmania* revived from synonymy: Dlussky & Fedoseeva, 1987: 77; Carpenter, 1992: 494 [misspelled as **Pseudoarmania*].

**PSEUDOATTA** (Myrmicinae: Attini)

*Pseudoatta* Gallardo, 1916d: 320. Type-species: *Pseudoatta argentina*, by monotypy. *Pseudoatta* provisional junior synonym of *Acromyrmex*: Brown, 1973b: 184 (unconfirmed).

***PSEUDOCAMPONOTUS** (Formicinae: Camponotini)

**Pseudocamponotus* Carpenter, 1930: 22. Type-species: **Pseudocamponotus elkoanus*, by original designation.

**#PSEUDOCOLOBOPSIS** (subgenus of *Camponotus*)

*Pseudocolobopsis* Emery, 1920b: 249 [as subgenus of *Camponotus*]. Type-species: *Camponotus macrocephalus*, by original designation. *Pseudocolobopsis* provisional junior synonym of *Camponotus*: Brown, 1973b: 184 (unconfirmed).

*PSEUDOCRYPTOPONE* (junior synonym of *Ponera*)

*Pseudocryptopone* Wheeler, W.M. 1933g: 12. Type-species: *Cryptopone tenuis*, by original designation. *Pseudocryptopone* junior synonym of *Ponera*: Wilson, 1957b: 356.

*PSEUDOCYRTOMYRMA* (junior synonym of *Myrma*)

*Pseudocyrtomyrma* Emery, 1921e: 18 [as subgenus of *Polyrhachis*]. Type-species: *Polyrhachis revoili*, by original designation. *Pseudocyrtomyrma* junior synonym of *Myrma*: Bolton, 1973b: 288.

*PSEUDODICHTHADIA* (junior synonym of *Labidus*)

*Pseudodichthadia* André, 1885: 838. Type-species: *Pseudodichthadia incerta* (junior synonym of *Labidus coecus*), by monotypy. *Pseudodichthadia* junior synonym of *Eciton*: André, in Forel, 1899c: 160 (footnote). *Pseudodichthadia* junior synonym of *Labidus*: Emery, 1910b: 21.

**PSEUDOLASIUS** (Formicinae: Lasiini)

*Pseudolasius* Emery, 1887a: 244. Type-species: *Formica familiaris*, by subsequent designation of Bingham, 1903: 337. *Pseudolasius* senior synonym of *Nesolasius*: Bolton, 1994: 50. Key to Afrotropical species: Menozzi, 1924b: 226 (out of date).

*PSEUDOMYRMA* (junior synonym of *Pseudomyrmex*)

*Pseudomyrma* Guérin-Méneville, 1844a: 427 [as subgenus of *Formica*]. Type-species: *Pseudomyrma bicolor*, by monotypy. *Pseudomyrma* senior synonym of *Leptalea* and *Myrmex*: Smith, F. 1858b: 153. *Pseudomyrma* junior synonym of *Pseudomyrmex*: Smith, M.R. 1952a: 98.

**PSEUDOMYRMEX** (Pseudomyrmecinae)

*Pseudomyrmex* Lund, 1831b: 106. Type-species: *Formica gracilis*, by subsequent designation of Smith, M.R. 1952a: 98. *Pseudomyrmex* senior synonym of *Pseudomyrma* (and its junior synonyms *Leptalea* and *Myrmex*): Smith, M.R. 1952a: 98. *Pseudomyrmex* senior synonym of *Apedunculata*, *Clavanoda*, *Latinoda*, *Ornatinoda*, and *Triangulinoda*: Ward, 1990: 469. Keys to species: Kempf, 1961a: 373 (*gracilis*-group); Ward, 1985b: 215 (Nearctic species); Ward, 1989: 407 (*oculatus*-group); Ward, 1989: 430 (*subtilissimus*-group); Ward, 1992: 76 (*Dominican Amber fauna); Ward, 1993: 117 (*Acacia*-associated taxa). Revision of genus: Ward, 1990: 469.

*PSEUDONEOPONERA* (junior synonym of *Bothroponera*)

*Pseudoneoponera* Donisthorpe, 1943d: 439. Type-species: *Pseudoneoponera verecundae*, by original designation. *Pseudoneoponera* junior synonym of *Bothroponera*: Wilson, 1958d: 361. [*Bothroponera* junior synonym of *Pachycondyla*: Brown, in preparation.]

***PSEUDONOTONCUS*** (Formicinae: Melophorini)

*Pseudonotoncus* Clark, 1934c: 64. Type-species: *Pseudonotoncus hirsutus*, by original designation.

*PSEUDOPODOMYRMA* (junior synonym of *Podomyrma*)

*Pseudopodomyrma* Crawley, 1925a: 40. Type-species: *Pseudopodomyrma clarki*, by monotypy. *Pseudopodomyrma* provisional junior synonym of *Podomyrma*: Brown, 1973b: 184. *Pseudopodomyrma* junior synonym of *Podomyrma*: Taylor & Brown, D.R. 1985: 80 (accepted as confirmation).

*PSEUDOPONERA* (junior synonym of *Pachycondyla*)

*Pseudoponera* Emery, 1900c: 314 [as subgenus of *Pachycondyla*]. Type-species: *Ponera quadridentata* (junior synonym of *Pachycondyla stigma*), by monotypy. [*Pseudoponera* also described as new by Emery, 1901a: 42. Type-species not *Ponera amblyops*, unjustified subsequent designation by Emery, 1901a: 42.] *Pseudoponera* subgenus of *Euponera*: Forel, 1901c: 141. *Pseudoponera* raised to genus: Bingham, 1903: 91; maintained as genus: Emery, 1911d: 86; Wheeler, W.M. 1922a: 649. *Pseudoponera* provisional junior synonym of *Pachycondyla*: Brown, 1973b: 184. *Pseudoponera* senior synonym of *Trachymesopus* (because of synonymous type-species): **new synonymy**. *Pseudoponera* junior synonym of *Pachycondyla*: Brown, in preparation.

*PSEUDOSTIGMACROS* (junior synonym of *Stigmacros*)

*Pseudostigmacros* McAreavey, 1957: 24 [as subgenus of *Stigmacros*]. Type-species: *Stigmacros* (*Pseudostigmacros*) *inermis*, by original designation. *Pseudostigmacros* provisional junior synonym of *Stigmacros*: Brown, 1973b: 184. *Pseudostigmacros* junior synonym of *Stigmacros*: Taylor & Brown, D.R. 1985: 145 (accepted as confirmation).

*PSEUDOSYSPHINCTA* (junior synonym of *Discothyrea*)

*Pseudosysphincta* Arnold, 1916: 161. Type-species: *Pseudosysphincta poweri*, by original designation. *Pseudosysphincta* junior synonym of *Discothyrea*: Brown, 1958g: 248. [*Pseudosphincta* Wheeler, W.M. 1922a: 645, incorrect subsequent spelling.]

*PTEROPONERA* (junior synonym of *Ponera*)

*Pteroponera* Bernard, 1950a: 3. Type-species: *Pteroponera sysphinctoides*, by monotypy. *Pteroponera* provisional junior synonym of *Ponera*: Brown, 1973b: 184. *Pteroponera* junior synonym of *Ponera*: Brown, in preparation.

*PYRAMICA* (junior synonym of *Strumigenys*)

*Pyramica* Roger, 1862a: 251. Type-species: *Pyramica gundlachi*, by monotypy. *Pyramica* junior synonym of *Strumigenys*: Roger, 1863b: 40; Dalla Torre, 1893: 145; Bingham, 1903: 147. *Pyramica* revived from synonymy as subgenus of *Strumigenys*: Brown, 1948e: 110. *Pyramica* junior synonym of *Strumigenys*: Brown & Wilson, 1959b: 281; Brown, 1960b: 37.

***QUADRISTRUMA*** (Myrmicinae: Dacetini)

*Quadristruma* Brown, 1949b: 47. Type-species: *Epitritus emmae*, by original designation. Review of genus: Bolton, 1983: 400. [See note under *Strumigenys*.]

*RAPTIFORMICA* (junior synonym of *Formica*)

*Raptiformica* Forel, 1913i: 361 [as subgenus of *Formica*]. Type-species: *Formica sanguinea*, by original designation. *Raptiformica* junior synonym of *Formica*: Wheeler, W.M. 1922a: 699 (footnote); Smith, D.R. 1979: 1448; Agosti, 1994a: 107.

***RECURVIDRIS*** (Myrmicinae: Pheidologetonini)

*Recurvidris* Bolton, 1992: 36. Replacement name for *Trigonogaster* Forel, 1890b: cviii. [Junior homonym of *Trigonogaster* Guérin-Méneville, 1844b: 1149.] Review of genus (as *Trigonogaster*): Ettershank, 1966: 158. Revision of genus: Bolton, 1992: 35.

*RENEA* (junior synonym of *Prionopelta*)

*Renea* Donisthorpe, 1947d: 183. Type-species: *Renea testacea* (junior synonym of *Prionopelta majuscula*), by original designation. *Renea* junior synonym of *Prionopelta*: Brown, 1953c: 11.

#***RHACHIOCREMA*** (subgenus of *Crematogaster*)

*Rhachiocrema* Mann, 1919: 318 [as subgenus of *Crematogaster*]. Type-species: *Crematogaster* (*Rhachiocrema*) *wheeleri*, by original designation. *Rhachiocrema* raised to genus: Soulié, 1964: 398. *Rhachiocrema* provisional junior synonym of *Crematogaster*: Brown, 1973b: 184 (unconfirmed). *Rhachiocrema* here reverts to subgenus of *Crematogaster*. Key to species: Creighton, 1945: 109.

#***RHINOMYRMEX*** (subgenus of *Camponotus*)

*Rhinomyrmex* Forel, 1886f: 192. Type-species: *Rhinomyrmex klaesii*, by monotypy. *Rhinomyrmex* subgenus of *Camponotus*: Emery, 1896d: 374. *Rhinomyrmex* provisional junior synonym of *Camponotus*: Brown, 1973b: 184 (unconfirmed).

#***RHIZOMYRMA*** (subgenus of *Acropyga*)

*Rhizomyrma* Forel, 1893g: 347 [as subgenus of *Acropyga*]. Type-species: *Acropyga* (*Rhizomyrma*) *goeldii*, by subsequent designation of Wheeler, W.M. 1911f: 172. *Rhizomyrma* raised to genus: Emery, 1906c: 182; maintained as genus: Forel, 1912i: 89; Forel, 1917: 249. *Rhizomyrma* subgenus of *Acropyga*: Emery, 1925b: 29. *Rhizomyrma* provisional junior synonym of *Acropyga*: Brown, 1973b: 184 (unconfirmed). Key to species: Weber, 1944: 91.

#***RHOGMUS*** (subgenus of *Dorylus*)

*Rhogmus* Shuckard, 1840c: 323. Type-species: *Rhogmus fimbriatus*, by original designation. *Rhogmus* subgenus of *Dorylus*: Emery, 1895j: 702. *Rhogmus* provisional junior synonym of *Dorylus*: Brown, 1973b: 184 (unconfirmed).

***RHOPALOMASTIX*** (Myrmicinae: Melissotarsini)

*Rhopalomastix* Forel, 1900a: 24. Type-species: *Rhopalomastix rothneyi*, by monotypy. Review of genus: Wheeler, W.M. 1929d: 95.

****RHOPALOMYRMEX*** (Formicinae: Plagiolepidini)

**Rhopalomyrmex* Mayr, 1868c: 41. Type-species: **Rhopalomyrmex pygmaeus*, by monotypy.

*RHOPALOPONE* (junior synonym of *Gnamptogenys*)

*Rhopalopone* Emery, 1897d: 549. Type-species: *Rhopalopone epinotalis*, by monotypy. *Rhopalopone* senior synonym of *Mictoponera*: Emery, 1911d: 34. *Rhopalopone* junior synonym of *Gnamptogenys*: Brown, 1958g: 211.

***RHOPALOTHRIX*** (Myrmicinae: Basicerotini)

*Rhopalothrix* Mayr, 1870a: 415. Type-species: *Rhopalothrix ciliata*, by subsequent designation of Wheeler, W.M. 1911f: 172. *Rhopalothrix* senior synonym of *Acanthidris* and *Heptastruma*, and revision of genus: Brown & Kempf, 1960: 230.

***RHOPTROMYRMEX*** (Myrmicinae: Tetramoriini)

*Rhoptromyrmex* Mayr, 1901b: 18. Type-species: *Rhoptromyrmex globinodis*, by subsequent designation of Wheeler, W.M. 1911f: 172. *Rhoptromyrmex* senior synonym of *Acidomyrmex*: Brown, 1964c: 11. *Rhoptromyrmex* senior synonym of *Ireneella*: Bolton, 1976: 298. *Rhoptromyrmex* senior synonym of *Hagioxenus*: Bolton, 1986a: 2. Revisions of genus: Brown, 1964c: 11; Bolton, 1976: 298; Bolton, 1986a: 1.

***RHYTIDOPONERA*** (Ponerinae: Ectatommini)

*Rhytidoponera* Mayr, 1862: 731 [as subgenus of *Ectatomma*]. Type-species: *Ponera araneoides*, by subsequent designation of Emery, 1911d: 37. [Type-species not *Ponera metallica* as designated by Wheeler, W.M. 1911f: 172; see Wheeler, W.M. 1913a: 79.] *Rhytidoponera* raised to genus: Emery, 1897d: 547; Emery, 1911d: 36. *Rhytidoponera* senior synonym of *Chalcoponera*: Brown, 1953c: 2. Revision of genus: Brown, 1958g: 288. Keys to species: Clark, 1936: 16 (Australia); Wilson, 1958d: 304 (Melanesia and Moluccas); Ward, 1980: 478 (*impressa*-group); Ward, 1984: 131 (New Caledonia).

***ROGERIA*** (Myrmicinae: Stenammini)

*Rogeria* Emery, 1894c: 188. Type-species: *Rogeria curvipubens*, by subsequent designation of Wheeler, W.M. 1911f: 172. *Rogeria* senior synonym of *Irogera*: Kempf, 1965: 185. Key to Neotropical species: Kempf, 1963a: 195. Revision of genus: Kugler, C., in press.

***ROMBLONELLA*** (Myrmicinae: Formicoxenini)

*Romblonella* Wheeler, W.M. 1935a: 5. Type-species: *Romblonella grandinodis* (junior synonym of *Romblonella opaca*), by original designation. Revision of genus: Smith, M.R. 1953a: 75.

***ROSSOMYRMEX*** (Formicinae: Formicini)

*Rossomyrmex* Arnol'di, 1928c: 299. Type-species: *Rossomyrmex proformicarum*, by original designation. See also: Agosti, 1994a: 112.

***ROSTROMYRMEX*** (Myrmicinae: Stenammini)

*Rostromyrmex* Rosciszewski, 1994: 160. Type-species: *Rostromyrmex pasohensis*, by original designation.

***ROTASTRUMA*** (Myrmicinae: Formicoxenini)

*Rotastruma* Bolton, 1991: 8. Type-species: *Rotastruma recava*, by original designation.

***SANTSCHIELLA*** (Formicinae: Santschiellini)

*Santschiella* Forel, 1916: 434. Type-species: *Santschiella kohli*, by monotypy.

*SCHIZOPELTA* (junior synonym of *Chelaner*)

*Schizopelta* McAreavey, 1949: 14. Type-species: *Schizopelta falcata*, by original designation. *Schizopelta* junior synonym of *Chelaner*: Ettershank, 1966: 93. [*Chelaner* junior synonym of *Monomorium*: Bolton, 1987: 287.

*SCROBOPHEIDOLE* (junior synonym of *Pheidole*)

*Scrobopheidole* Emery, 1915i: 190 [as subgenus of *Pheidole*]. Type-species: *Pheidole scrobifera*, by monotypy. *Scrobopheidole* provisional junior synonym of *Pheidole*: Brown, 1973b: 184. *Scrobopheidole* junior synonym of *Pheidole*: Smith, D.R. 1979: 1366.

***SCYPHODON*** (Leptanillinae: Leptanillini)
*Scyphodon* Brues, 1925: 93. Type-species: *Scyphodon anomalum,* by original designation. Reviews of genus: Petersen, 1968: 591; Baroni Urbani, 1977c: 482.
***SECOSTRUMA*** (Myrmicinae: Tetramoriini)
*Secostruma* Bolton, 1988c: 264. Type-species: *Secostruma lethifera,* by original designation.
*SELENOPONE* (junior synonym of *Ponera*)
*Selenopone* Wheeler, W.M. 1933g: 19. Type-species: *Ponera selenophora,* by original designation. *Selenopone* junior synonym of *Ponera:* Wilson, 1957b: 356.
*SEMONIUS* (junior synonym of *Tapinoma*)
*Semonius* Forel, 1910f: 21. Type-species: *Semonius schultzei,* by monotypy. *Semonius* junior synonym of *Tapinoma:* Shattuck, 1992c: 146.
***SERICOMYRMEX*** (Myrmicinae: Attini)
*Sericomyrmex* Mayr, 1865: 83. Type-species: *Sericomyrmex opacus,* by monotypy.
***SERRASTRUMA*** (Myrmicinae: Dacetini)
*Serrastruma* Brown, 1948e: 107 [as subgenus of *Smithistruma*]. Type-species: *Strumigenys simoni,* by original designation. *Serrastruma* raised to genus: Brown, 1949d: 6. Revisions of genus: Brown, 1952e: 67; Bolton, 1983: 335. [See note under *Strumigenys.*]
*SERVIFORMICA* (junior synonym of *Formica*)
*Serviformica* Forel, 1913i: 361. Type-species: *Formica fusca,* by original designation. *Serviformica* junior synonym of *Formica:* Wheeler, W.M. 1922a: 699 (footnote); Francoeur, 1973: 34; Smith, D.R. 1979: 1448; Agosti, 1994a: 106.
*SHUCKARDIA* (junior synonym of *Alaopone*)
*Shuckardia* Emery, 1895j: 703 [as subgenus of *Dorylus*]. Type-species: *Dorylus atriceps,* by original designation. *Shuckardia* junior synonym of *Alaopone:* Emery, 1910b: 15.
*****SICILOMYRMEX*** (Formicinae: *Sicilomyrmecini)
**Sicelomyrmex* Wheeler, W.M. 1915h: 111. Type-species: **Gesomyrmex corniger,* by original designation. **Sicilomyrmex:* Brown & Carpenter, 1979: 423, justified emendation of spelling.
*SIFOLINIA* (junior synonym of *Myrmica*)
*Sifolinia* Emery, 1907: 49. Type-species: *Sifolinia laurae,* by monotypy. *Sifolinia* senior synonym of *Symbiomyrma:* Samsinak, 1964: 156. *Sifolinia* junior synonym of *Myrmica:* Bolton, 1988a: 3.
*SIMA* (junior synonym of *Tetraponera*)
*Sima* Roger, 1863a: 178. Type-species: *Sima compressa* (junior synonym of *Tetraponera allaborans*), by monotypy. [Type-species not *Eciton rufonigrum,* unjustified subsequent designation by Emery, 1915b: 266 and Emery, 1921f: 22.] *Sima* junior synonym of *Tetraponera:* Smith, F. 1877b: 68; Donisthorpe, 1916b: 244; Ward, 1990: 470.
***SIMOPELTA*** (Ponerinae: Ponerini)
*Simopelta* Mann, 1922: 10 [as subgenus of *Belonopelta*]. Type-species: *Belonopelta* (*Simopelta*) *jeckylli,* by original designation. *Simopelta* raised to genus: Wheeler, W.M. 1935d: 8. *Simopelta* junior synonym of *Belonopelta:* Baroni Urbani, 1975b: 296. *Simopelta* revived from synonymy: Hölldobler & Wilson, 1990: 11. Revision of genus: Gotwald & Brown, 1967: 261.
***SIMOPONE*** (Cerapachyinae: Cerapachyini)
*Simopone* Forel, 1891b: 139. Type-species: *Simopone grandidieri,* by monotypy. *Simopone* subgenus of *Cerapachys:* Forel, 1892l: 243. *Simopone* revived status as genus: Dalla Torre, 1893: 17; Emery, 1911d: 15. Revision of genus: Brown, 1975: 35.
***SMITHISTRUMA*** (Myrmicinae: Dacetini)
*Smithistruma* Brown, 1948e: 104. Type-species: *Strumigenys pulchella,* by original designation. Replacement name for *Cephaloxys* Smith, F. 1865: 76. [Junior homonym of *Cephaloxys* Signoret, 1847: 294.] *Smithistruma* senior synonym of *Platystruma, Weberistruma,* and *Wessonistruma:* Brown, 1973a: 35. *Smithistruma* senior synonym of *Miccostruma:* Bolton, 1983: 274. Revision of genus: Brown, 1953g: 31; Brown, 1964a: 183. Keys to Neotropical species: Brown, 1953g: 112; to Nearctic species: Brown, 1953g: 96; Ward, 1988: 122; to Oriental species: Brown, 1953g: 125; to Afrotropical species: Brown, 1953g: 130; Bolton, 1983: 276. [See note under *Strumigenys.*]
*SOLENOMYRMA* (junior synonym of *Gauromyrmex*)
*Solenomyrma* Karavaiev, 1935a: 103. Type-species: *Solenomyrma acanthina,* by monotypy. *Solenomyrma* junior synonym of *Gauromyrmex:* Brown, 1953c: 10. [*Gauromyrmex* junior synonym of *Vollenhovia:* Bolton, 1994: 105.]
*SOLENOPS* (junior synonym of *Oligomyrmex*)
*Solenops* Karavaiev, 1930a: 207 [as subgenus of *Solenopsis*]. Type-species: *Solenopsis* (*Solenops*) *weyeri,* by monotypy. *Solenops* junior synonym of *Oligomyrmex:* Ettershank, 1966: 119.
***SOLENOPSIS*** (Myrmicinae: Solenopsidini)
*Solenopsis* Westwood, 1840b: 86. Type-species: *Solenopsis mandibularis* (junior synonym of *Solenopsis geminata*), by monotypy. *Solenopsis* senior synonym of *Diplorhoptrum:* Kempf, 1972a: 232; Bolton, 1987: 285. *Solenopsis* senior synonym of *Bisolenopsis, Diagyne, Euophthalma, Granisolenopsis, Labauchena, Lilidris, Oedaleocerus, Paranamyrma* and *Synsolenopsis:* Ettershank, 1966: 134. *Solenopsis* senior synonym of *Octella:* Taylor, 1991b: 611. Revision of New World species: Creighton, 1930b: 39 (out of date); of *geminata*-group: Trager, 1991: 141. Notes on synonymy: Bolton, 1987: 285.
*SOMMIMYRMA* (junior synonym of *Myrmica*)
*Sommimyrma* Menozzi, 1925d: 25. Type-species: *Sommimyrma symbiotica,* by original designation. *Sommimyrma* provisional junior synonym of *Myrmica:* Brown, 1973b: 184. *Sommimyrma* junior synonym of *Myrmica:* Bolton, 1988a: 3.
*SPALACOMYRMEX* (junior synonym of *Centromyrmex*)
*Spalacomyrmex* Emery, 1889b: 489. Type-species: *Spalacomyrmex feae,* by monotypy. *Spalacomyrmex* junior synonym of *Centromyrmex:* Emery, 1890b: 40.
*SPANIOPONE* (junior synonym of *Gnamptogenys*)
*Spaniopone* Wheeler, W.M. & Mann, 1914: 11. Type-species: *Spaniopone haytiana,* by monotypy. *Spaniopone* junior synonym of *Gnamptogenys:* Brown, 1958g: 212.
*SPELAEOMYRMEX* (junior synonym of *Erebomyrma*)
*Spelaeomyrmex* Wheeler, W.M. 1922c: 9. Type-species: *Spelaeomyrmex urichi,* by original designation. *Spelaeomyrmex* junior synonym of *Erebomyrma:* Wilson, 1962a: 63. [*Erebomyrma* junior synonym of *Oligomyrmex:* Bolton, 1994: 106.]
#***SPHAEROCREMA*** (subgenus of *Crematogaster*)
*Sphaerocrema* Santschi, 1918d: 182 [as subgenus of *Crematogaster*]. Type-species: *Crematogaster kneri,* by original designation. *Sphaerocrema* raised to genus: Soulié, 1964: 398. *Sphaerocrema* provisional junior synonym of *Crematogaster:* Brown, 1973b: 185 (unconfirmed). *Sphaerocrema* here reverts to subgenus of *Crematogaster.*
*SPHAEROMESSOR* (unavailable name)
*Sphaeromessor* Bernard, 1985: 48. [Without designation of type-species; unavailable name.] Included species referable to *Messor.*
****SPHECOMYRMA*** (*Sphecomyrminae)
**Sphecomyrma* Wilson & Brown, in Wilson, Carpenter & Brown, 1967: 8. Type-species: **Sphecomyrma freyi,* by original designation.
*SPHECOMYRMEX* (junior synonym of *Anomma*)
*Sphecomyrmex* Schulz, 1906: 154, unjustified emendation of *Sphegomyrmex. Sphecomyrmex* junior synonym of *Anomma:* **new synonymy**.
*SPHEGOMYRMEX* (junior synonym of *Anomma*)
*Sphegomyrmex* Imhoff, 1852: 176. Type-species: *Dorylus nigricans,* by subsequent designation of Donisthorpe, 1943g: 727. *Sphegomyrmex* junior synonym of *Anomma:* Smith, F. 1858b: 112.
***SPHINCTOMYRMEX*** (Cerapachyinae: Cerapachyini)
*Sphinctomyrmex* Mayr, 1866b: 895. Type-species: *Sphinctomyrmex stahli,* by monotypy. *Sphinctomyrmex* senior synonym of *Aethiopopone:* Bolton, 1973a: 341. *Sphinctomyrmex* senior synonym of *Eusphinctus, Nothosphinctus* and *Zasphinctus,* and revision of genus: Brown, 1975: 31.
*SPINOMYRMA* (junior synonym of *Dorymyrmex*)
*Spinomyrma* Kusnezov, 1952g: 429 (diagnosis in key) [as subgenus of *Dorymyrmex*]. Type-species: *Dorymyrmex alboniger,* by subsequent designation of Kusnezov, 1959: 51. *Spinomyrma* raised to genus: Kusnezov, 1959: 51. *Spinomyrma* subgenus of *Dorymyrmex:* Kempf, 1972a: 100. *Spinomyrma* junior synonym of *Dorymyrmex:* Snelling & Hunt, 1976: 93.
*SPOROCLEPTES* (junior synonym of *Aneleus*)
*Sporocleptes* Arnold, 1948: 219. Type-species: *Sporocleptes nicotianae,* by original designation. *Sporocleptes* junior synonym of *Aneleus:* Consani, 1951: 169; Arnold, 1952a: 460. [*Aneleus* junior synonym of *Oligomyrmex:* Ettershank, 1966: 119.]
***STEGOMYRMEX*** (Myrmicinae: Stegomyrmecini)
*Stegomyrmex* Emery, 1912b: 99. Type-species: *Stegomyrmex connectens,* by monotypy. Revision of genus: Diniz, 1990: 277.
*STEGOPHEIDOLE* (junior synonym of *Pheidole*)
*Stegopheidole* Emery, 1915i: 190 [as subgenus of *Pheidole*]. Type-species: *Pheidole* (*Elasmopheidole*) *upeneci,* by monotypy. *Stegopheidole* provisional junior synonym of *Pheidole:* Brown, 1973b: 185. *Stegopheidole* junior synonym of *Pheidole:* Smith, D.R. 1979: 1366.
***STENAMMA*** (Myrmicinae: Stenammini)
*Stenamma* Westwood, 1839: 219. Type-species: *Stenamma westwoodii,* by monotypy. *Stenamma* senior synonym of *Asemorhoptrum:* André, 1883a: 310. *Stenamma* senior synonym of *Theryella:* Santschi, 1923a: 136. Keys to species of New World: Smith,

M.R. 1957b: 133; Smith, M.R. 1962a: 38; Snelling, 1973c: 1; to species of former U.S.S.R.: Arnol'di, 1975: 1826.

*STENOMYRMEX* (junior synonym of *Anochetus*)

*Stenomyrmex* Mayr, 1862: 711. Type-species: *Myrmecia emarginata,* by subsequent designation of Wheeler, W.M. 1911f: 173. *Stenomyrmex* subgenus of *Anochetus:* Dalla Torre, 1893: 47; maintained as subgenus: Emery, 1911d: 110; Forel, 1917: 238; Kempf, 1972a: 21. Revision of subgenus: Kempf, 1964f: 237. *Stenomyrmex* junior synonym of *Anochetus:* Brown, 1978c: 552.

*STENOTHORAX* (junior synonym of *Adlerzia*)

*Stenothorax* McAreavey, 1949: 3. Type-species: *Stenothorax katerinae* (junior synonym of *Adlerzia froggatti*), by original designation. *Stenothorax* junior synonym of *Adlerzia:* Brown, 1952a: 110.

***STEREOMYRMEX*** (Myrmicinae: Formicoxenini)

*Stereomyrmex* Emery, 1901f: 115. Type-species: *Stereomyrmex horni,* by monotypy.

*STICTOPONERA* (junior synonym of *Gnamptogenys*)

*Stictoponera* Mayr, 1887: 539 [as subgenus of *Ectatomma*]. Type-species: *Ponera coxalis,* by subsequent designation of Bingham, 1903: 82. *Stictoponera* raised to genus: Emery, 1911d: 47. *Stictoponera* junior synonym of *Gnamptogenys:* Brown, 1958g: 211.

***STIGMACROS*** (Formicinae: Plagiolepidini)

*Stigmacros* Forel, 1905b: 179 [as subgenus of *Acantholepis*]. Replacement name for *Acrostigma* Forel, 1902h: 477. [Junior homonym of **Acrostigma* Emery, 1891a: 149.] *Stigmacros* raised to genus: Wheeler, W.M. 1922a: 696. *Stigmacros* senior synonym of *Campostigmacros, Chariostigmacros, Cyrtostigmacros, Hagiostigmacros,* and *Pseudostigmacros:* Taylor & Brown, D.R. 1985: 145. Revision of genus: McAreavey, 1957: 7.

*STIGMATOMMA* (junior synonym of *Amblyopone*)

*Stigmatomma* Roger, 1859: 250. Type-species: *Stigmatomma denticulatum,* by subsequent designation of Bingham, 1903: 36. *Stigmatomma* junior synonym of *Amblyopone:* Emery & Forel, 1879: 455; Mayr, 1887: 546. *Stigmatomma* revived from synonymy: Dalla Torre, 1893: 14; maintained as genus: Emery, 1911d: 23; Forel, 1917: 235; Wheeler, W.M. 1922a: 641. *Stigmatomma* subgenus of *Amblyopone:* Clark, 1934b: 27; Brown, 1949c: 87. *Stigmatomma* senior synonym of *Fulakora* and *Xymmer:* Brown, 1949c: 87. *Stigmatomma* junior synonym of *Amblyopone:* Brown, 1960a: 155.

****STIGMOMYRMEX*** (Myrmicinae: Formicoxenini?)

**Stigmomyrmex* Mayr, 1868c: 95. Type-species: *Stigmomyrmex venustus,* by subsequent designation of Wheeler, W.M. 1911f: 173.

****STIPHROMYRMEX*** (Myrmicinae: Myrmecinini?)

**Stiphromyrmex* Wheeler, W.M. 1915h: 67. Type-species: **Stigmomyrmex robustus,* by original designation.

***STREBLOGNATHUS*** (Ponerinae: Ponerini)

*Streblognathus* Mayr, 1862: 716. Type-species: *Ponera aethiopica,* by monotypy.

***STRONGYLOGNATHUS*** (Myrmicinae: Tetramoriini)

*Strongylognathus* Mayr, 1853d: 389. Type-species: *Eciton testaceum,* by monotypy. Replacement name for *Myrmus* Schenck, 1853: 188. [Junior homonym of *Myrmus* Hahn, 1832: 81.] Revision of *huberi*-group: Baroni Urbani, 1969c: 132. Review of genus: Bolton, 1976: 304. Key to fauna of former U.S S.R.: Radchenko, 1991: 89.

***STRUMIGENYS*** (Myrmicinae: Dacetini)

*Strumigenys* Smith, F. 1860c: 72. Type-species: *Strumigenys mandibularis,* by monotypy. *Strumigenys* senior synonym of *Labidogenys* and *Pyramica:* Roger, 1863b: 40; Brown & Wilson, 1959b: 281; Brown, 1960b: 37. *Strumigenys* senior synonym of *Enerxa* and *Proscopomyrmex:* Brown, 1949d: 15. Keys to Neotropical species: Brown, 1962c: 257; Kempf, 1976b: 43; to Afrotropical species: Brown, 1954k: 10; Bolton, 1983: 360; to the Fiji Is., Tonga and Samoa: Dlussky, 1993: 53. [Note. Baroni Urbani & De Andrade, 1994: 12–15, proposed that *Asketogenys, Chelystruma, Cladarogenys, Codiomyrmex, Codioxenus, Dorisidris, Dysedrognathus, Epitritus, Glamyromyrmex, Gymnomyrmex, Kyidris, Neostruma, Pentastruma, Quadristruma, Serrastruma, Smithistruma, Tingimyrmex* and *Trichoscapa,* along with all the previous junior synonyms of all these genera, be included as junior synonyms of *Strumigenys.* I consider this action too extreme by far as it is based on a grossly inadequate character set. In consequence their results are very insecure, and their proposed synonyms are all **revived from synonymy** here.]

*SULCOMYRMEX* (unavailable name)

*Sulcomyrmex* Kratochvíl, 1941: 84 [as subgenus of *Tetramorium*]. [Without designation of type-species; unavailable name.] Included species referable to *Tetramorium:* Bolton, 1976: 359.

*SYLLOPHOPSIS* (junior synonym of *Monomorium*)

*Syllophopsis* Santschi, 1915c: 259 [as subgenus of *Monomorium.*] Type-species: *Monomorium modestum,* by monotypy. *Syllophopsis* raised to genus: Santschi, 1921c: 119; maintained as genus: Ettershank, 1966: 100. *Syllophopsis* subgenus of *Monomorium:* Wheeler, W.M. 1922a: 676; Arnold, 1952a: 465. *Syllophopsis* junior synonym of *Monomorium:* Bolton, 1987: 287. [*Syllopsis* Santschi, 1921c: 120, incorrect subsequent spelling.]

*SYMBIOMYRMA* (junior synonym of *Sifolinia*)

*Symbiomyrma* Arnol'di, 1930c: 267. Type-species: *Symbiomyrma karavajevi,* by monotypy. [*Symbiomyrma* also described as new by Arnol'di, 1933a: 41.] *Symbiomyrma* junior synonym of *Sifolinia:* Samsinak, 1964: 156. [*Sifolinia* junior synonym of *Myrmica:* Bolton, 1988a: 3.]

*SYMMYRMICA* (junior synonym of *Formicoxenus*)

*Symmyrmica* Wheeler, W.M. 1904a: 3. Type-species: *Symmyrmica chamberlini,* by monotypy. *Symmyrmica* provisional junior synonym of *Leptothorax:* Brown, 1973b: 185. *Symmyrmica* junior synonym of *Formicoxenus:* Francoeur, Loiselle & Buschinger, 1985: 347.

*SYMPHEIDOLE* (junior synonym of *Pheidole*)

*Sympheidole* Wheeler, W.M. 1904a: 7. Type-species: *Sympheidole elecebra,* by monotypy. *Sympheidole* provisional junior synonym of *Pheidole:* Brown, 1973b: 185. *Sympheidole* junior synonym of *Pheidole:* Smith, D.R. 1979: 1365.

*SYNSOLENOPSIS* (junior synonym of *Solenopsis*)

*Synsolenopsis* Forel, 1918b: 155 [as subgenus of *Solenopsis*]. Type-species: *Solenopsis* (*Synsolenopsis*) *bruchi* (junior primary homonym in *Solenopsis,* replaced by *Solenopsis bruchiella*), by monotypy. *Synsolenopsis* raised to genus, and key to species: Kusnezov, 1953c: 341; maintained as genus: Kusnezov, 1962b: 160. *Synsolenopsis* junior synonym of *Solenopsis:* Ettershank, 1966: 134.

****SYNTAPHUS*** (Ponerinae: Ectatommini)

**Syntaphus* Donisthorpe, 1920: 84. Type-species: **Syntaphus wheeleri,* by original designation.

*SYNTERMITOPONE* (junior synonym of *Pachycondyla*)

*Syntermitopone* Wheeler, W.M. 1936d: 169 [as subgenus of *Termitopone.*] Type-species: *Ponera commutata,* by original designation. *Syntermitopone* junior synonym of *Termitopone:* Borgmeier, 1959a: 312. [*Termitopone* junior synonym of *Pachycondyla:* Brown, in preparation.]

*SYSCIA* (junior synonym of *Cerapachys*)

*Syscia* Roger, 1861a: 19. Type-species: *Syscia typhla,* by monotypy. *Syscia* subgenus of *Cerapachys:* Wheeler, W.M. 1902d: 185; Emery, 1902c: 24. *Syscia* senior synonym of *Cysias:* Emery, 1911d: 10. *Syscia* junior synonym of *Cerapachys:* Kempf, 1972a: 76.

*SYSPHINGTA* (junior synonym of *Proceratium*)

*Sysphingta* Roger, 1863a: 175. Type-species: *Sysphingta micrommata,* by monotypy. *Sysphingta* junior synonym of *Proceratium:* Mayr, 1886d: 437; Dalla Torre, 1893: 18. *Sysphingta* revived from synonymy: Emery, 1895c: 262. *Sysphingta* junior synonym of *Proceratium:* Brown, in Borgmeier, 1957: 118; Brown, 1958g: 241. [*Sysphincta* Mayr, 1865: 12 (and most subsequent authors), incorrect subsequent spelling.]

***TALARIDRIS*** (Myrmicinae: Basicerotini)

*Talaridris* Weber, 1941a: 184. Type-species: *Talaridris mandibularis,* by original designation. Revision of genus: Brown & Kempf, 1960: 241. *Talaridris* provisional junior synonym of *Rhopalothrix:* Brown, 1973b: 185 (unconfirmed).

*TAMMOTECA* (junior synonym of *Gnamptogenys*)

*Tammoteca* Santschi, 1929h: 476 [as subgenus of *Ectatomma*]. Type-species: *Ectatomma concinnum,* by original designation. *Tammoteca* junior synonym of *Gnamptogenys:* Brown, 1958g: 212.

#***TANAEMYRMEX*** (subgenus of *Camponotus*)

*Tanaemyrmex* Ashmead, 1905b: 384. Type-species: *Formica longipes* (junior primary homonym in *Formica,* replaced by *Camponotus etiolipes*), by original designation. *Tanaemyrmex* subgenus of *Camponotus,* and senior synonym of *Myrmoturba:* Emery, 1925b: 75. *Tanaemyrmex* provisional junior synonym of *Camponotus:* Brown, 1973b: 185 (unconfirmed).

#***TAPINOLEPIS*** (subgenus of *Anoplolepis*)

*Tapinolepis* Emery, 1925b: 18 [as subgenus of *Anoplolepis*]. Type-species: *Plagiolepis tumidula,* by monotypy. *Tapinolepis* provisional junior synonym of *Anoplolepis:* Brown, 1973b: 185 (unconfirmed).

***TAPINOMA*** (Dolichoderinae)

*Tapinoma* Foerster, 1850a: 43. Type-species: *Tapinoma collina* (junior synonym of *Tapinoma erraticum*), by monotypy. *Tapinoma* senior synonym of *Micromyrma:* Mayr, 1863: 455; Smith, M.R. 1951a: 837; Shattuck, 1992c: 146. *Tapinoma* senior synonym of *Neoclystopsenella:* Brown, 1988a: 337. *Tapinoma* senior synonym of *Semonius, Tapinoptera,*

*Zatapinoma*, and review of genus: Shattuck, 1992c: 146.

*TAPINOPTERA* (junior synonym of *Tapinoma*)

*Tapinoptera* Santschi, 1925g: 348 [as subgenus of *Tapinoma*]. Type-species: *Tapinoma vexatum*, by monotypy. *Tapinoptera* junior synonym of *Tapinoma:* Shattuck, 1992c: 146.

***TATUIDRIS*** (Myrmicinae: Agroecomyrmecini)

*Tatuidris* Brown & Kempf, 1968: 186. Type-species: *Tatuidris tatusia*, by original designation.

***TECHNOMYRMEX*** (Dolichoderinae)

*Technomyrmex* Mayr, 1872: 147. Type-species: *Technomyrmex strenuus*, by monotypy. *Technomyrmex* senior synonym of *Aphantolepis:* Brown, 1953h: 5. *Technomyrmex* senior synonym of *Engramma*, and review of genus: Shattuck, 1992c: 153.

***TELEUTOMYRMEX*** (Myrmicinae: Tetramoriini)

*Teleutomyrmex* Kutter, 1950a: 82. Type-species: *Teleutomyrmex schneideri*, by original designation.

*TEMNOTHORAX* (junior synonym of *Leptothorax*)

*Temnothorax* Mayr, 1861: 68. Type-species: *Myrmica recedens*, by monotypy. *Temnothorax* junior synonym of *Leptothorax:* Forel, 1890a: lxxii. *Temnothorax* revived from synonymy as subgenus of *Leptothorax:* Forel, 1892i: 315; maintained as subgenus: Emery, 1915g: 24; Forel, 1915d: 27; Bondroit, 1918: 117; Wheeler, W.M. 1922a: 679; Emery, 1922f: 259. *Temnothorax* raised to genus: Bernard, 1967: 185. *Temnothorax* junior synonym of *Leptothorax:* Baroni Urbani, 1971c: 96; Bolton, 1982: 319.

***TERATANER*** (Myrmicinae: Formicoxenini)

*Korataner* Emery, 1912b: 103. Type-species: *Atopomyrmex foreli*, by original designation. Review of genus: Arnold, 1952b: 129. *Terataner* senior synonym of *Tranetera*, and revision of genus: Bolton, 1981b: 288.

***TERATOMYRMEX*** (Formicinae: Lasiini)

*Teratomyrmex* McAreavey, 1957: 54. Type-species: *Teratomyrmex greavesi*, by original designation.

*TERMITOPONE* (junior synonym of *Pachycondyla*)

*Termitopone* Wheeler, W.M. 1936d: 159. Type-species: *Ponera laevigata*, by original designation. *Termitopone* senior synonym of *Syntermitopone:* Borgmeier, 1959a: 312. *Termitopone* provisional junior synonym of *Pachycondyla:* Brown, 1973b: 185. *Termitopone* junior synonym of *Pachycondyla:* Brown, in preparation.

***TETHEAMYRMA*** (Myrmicinae: Stenammini)

*Tetheamyrma* Bolton, 1991: 9. Type-species: *Tetheamyrma subspongia*, by original designation.

***TETRAMORIUM*** (Myrmicinae: Tetramoriini)

*Tetramorium* Mayr, 1855: 423. Type-species: *Formica caespitum*, by subsequent designation of Girard, 1879: 1016. *Tetramorium* senior synonym of *Tetrogmus:* Roger, 1862c: 297. *Tetramorium* senior synonym of *Atopula, Lobomyrmex, Macromischoides, Tetrogmus*, and *Xiphomyrmex:* Bolton, 1976: 359; Bolton, 1980: 195. Species included in *Sulcomyrmex* (unavailable) referable to *Tetramorium. Tetramorium* senior synonym of *Triglyphothrix:* Bolton, 1985: 247. Revision of genus: Bolton, 1976: 359. Keys to species: Bolton, 1976: 314, 341 (former *Triglyphothrix* species); Bolton, 1977: 67 (Oriental, Indo-Australian and Australasian); Bolton, 1979: 129 (Malagasy and New World); Bolton, 1980: 193 (Afrotropical); Radchenko, 1992a: 45 (former U.S.S.R.).

*TETRAMYRMA* (junior synonym of *Leptothorax*)

*Tetramyrma* Forel, 1912m: 766 [as subgenus of *Dilobocondyla*]. Type-species: *Dilobocondyla (Tetramyrma) braunsi*, by monotypy. *Tetramyrma* raised to genus: Forel, 1913a: 122. *Tetramyrma* junior synonym of *Leptothorax:* Bolton, 1982: 319.

***TETRAPONERA*** (Pseudomyrmecinae)

*Tetraponera* Smith, F. 1852: 44. Type-species: *Tetraponera atrata* (junior synonym of *Tetraponera nigra*), by subsequent designation of Wheeler, W.M. 1911f: 173. *Tetraponera* senior synonym of *Sima:* Smith, F. 1877b: 68. *Tetraponera* junior synonym of *Sima:* Dalla Torre, 1893: 53. *Tetraponera* subgenus of *Sima:* Emery, 1900d: 673; Emery, 1915b: 265. *Tetraponera* senior synonym of *Sima:* Donisthorpe, 1916b: 244; Wheeler, W.M. 1922a: 105. *Tetraponera* senior synonym of *Pachysima, Parasima, Sima, Viticicola*, and revision of genus: Ward, 1990: 470.

*TETROGMUS* (junior synonym of *Tetramorium*)

*Tetrogmus* Roger, 1857: 10. Type-species: *Tetrogmus caldarius*, by monotypy. *Tetrogmus* junior synonym of *Tetramorium:* Roger, 1862c: 297; Bolton, 1976: 359.

***THAUMATOMYRMEX*** (Ponerinae: Thaumatomyrmecini)

*Thaumatomyrmex* Mayr, 1887: 530. Type-species: *Thaumatomyrmex mutilatus*, by monotypy. Keys to species: Weber, 1942b: 65; Smith, M.R. 1944b: 97 (both out of date). Revisions of genus: Kempf, 1975b: 95; Longino, 1988: 35.

*THERYELLA* (junior synonym of *Stenamma*)

*Theryella* Santschi, 1921d: 68. Type-species: *Theryella myops* (provisional junior synonym of *Stenamma punctiventre*), by monotypy. *Theryella* junior synonym of *Stenamma:* Santschi, 1923a: 136. [*Theryella* incorrectly as junior synonym of *Proatta:* Snelling, 1981: 395, in error.]

#***THLIPSEPINOTUS*** (subgenus of *Camponotus*)

*Thlipsepinotus* Santschi, 1928e: 483 [as subgenus of *Camponotus*]. Type-species: *Camponotus claripes*, by original designation. *Thlipsepinotus* provisional junior synonym of *Camponotus:* Brown, 1973b: 185 (unconfirmed).

***TINGIMYRMEX*** (Myrmicinae: Dacetini)

*Tingimyrmex* Mann, 1926: 104 [as subgenus of *Strumigenys*]. Type-species: *Strumigenys (Tingimyrmex) mirabilis*, by monotypy. *Tingimyrmex* raised to genus: Brown, 1948e: 111. [See note under *Strumigenys*.]

*TITUSIA* (*nomen nudum*)

*Titusia* Ashmead, 1905b: 382, *nomen nudum*.

*TOMOGNATHUS* (junior homonym, see *Harpagoxenus*)

*Tomognathus* Mayr, 1861: 56. Type-species: *Myrmica sublaevis*, by monotypy. [Junior homonym of *Tomognathus* Agassiz, 1850: 376 (Pisces).] *Harpagoxenus* Forel, 1893a: 167, replacement name for *Tomognathus* Mayr.

*TRACHYMESOPUS* (junior synonym of *Pseudoponera*)

*Trachymesopus* Emery, 1911d: 84 [as subgenus of *Euponera*]. Type-species: *Formica stigma*, by original designation. *Trachymesopus* raised to genus: Wilson, 1958d: 352; see also Brown, 1963: 6. *Trachymesopus* provisional junior synonym of *Pachycondyla:* Brown, 1973b: 185. *Trachymesopus* junior synonym of *Pseudoponera:* **new synonymy**. [*Pseudoponera* junior synonym of *Pachycondyla:* Brown, in preparation.] [*Trachyponera:* see below.]

***TRACHYMYRMEX*** (Myrmicinae: Attini)

*Trachymyrmex* Forel, 1893e: 600 [as subgenus of *Atta*]. Type-species: *Atta septentrionalis*, by subsequent designation of Wheeler, W.M. 1911f: 174. *Trachymyrmex* subgenus of *Cyphomyrmex:* Emery, 1913: 251. *Trachymyrmex* raised to genus: Gallardo, 1916b: 242; Wheeler, W.M. 1916c: 11. *Trachymyrmex* subgenus of *Acromyrmex:* Forel, 1917: 247. *Trachymyrmex* subgenus of *Cyphomyrmex:* Emery, 1922f: 344. *Trachymyrmex* raised to genus: Wheeler, W.M. 1922a: 669; maintained as genus: Borgmeier, 1950b: 384; Kempf, 1972a: 252.

*TRACHYPHEIDOLE* (junior synonym of *Pheidole*)

*Trachypheidole* Emery, 1915i: 190 [as subgenus of *Pheidole*]. Type-species: *Pheidole bicornis*, by original designation. *Trachypheidole* provisional junior synonym of *Pheidole:* Brown, 1973b: 185. *Trachypheidole* junior synonym of *Pheidole:* Smith, D.R. 1979: 1366.

[*TRACHYPONERA* (misspelling)

*Trachyponera* Santschi, 1928a: 43, incorrect subsequent spelling of *Trachymesopus*.]

*TRANETERA* (junior synonym of *Terataner*)

*Tranetera* Arnold, 1952b: 130 [as subgenus of *Terataner*]. Type-species: *Terataner bottegoi*, by original designation. *Tranetera* provisional junior synonym of *Terataner:* Brown, 1973b: 185. *Tranetera* junior synonym of *Terataner:* Bolton, 1981b: 288.

***TRANOPELTA*** (Myrmicinae: Ochetomyrmecini)

*Tranopelta* Mayr, 1866a: 512. Type-species: *Tranopelta gilva*, by monotypy. Review of genus: Ettershank, 1966: 107.

*TRANOPELTOIDES* (junior synonym of *Crematogaster*)

*Tranopeltoides* Wheeler, W.M. 1922e: 10. Type-species: *Tranopelta huberi*, by original designation. *Tranopeltoides* junior synonym of *Crematogaster:* Kempf, 1960c: 173.

*TRAPEZIOPELTA* (junior synonym of *Myopias*)

*Trapeziopelta* Mayr, 1862: 715. Type-species: *Ponera maligna*, by monotypy. *Trapeziopelta* senior synonym of *Bradyponera:* Donisthorpe, 1932c: 468. *Trapeziopelta* provisional junior synonym of *Pachycondyla:* Brown, 1973b: 185. *Trapeziopelta* junior synonym of *Myopias:* Willey & Brown, 1983: 249.

*TRIANGULINODA* (junior synonym of *Pseudomyrmex*)

*Triangulinoda* Enzmann, E.V. 1944: 61 [as "branch" of *Pseudomyrma*]. Type-species: *Pseudomyrma spinicola*, by original designation. *Triangulinoda* junior synonym of *Pseudomyrmex:* Ward, 1990: 469.

*TRICHOMELOPHORUS* (junior synonym of *Melophorus*)

*Trichomelophorus* Wheeler, W.M. 1935c: 71 [as subgenus of *Melophorus*]. Type-species: *Melophorus hirsutus*, by original designation. *Trichomelophorus* junior synonym of *Melophorus:* Brown, 1955d: 474.

*TRICHOMYRMEX* (junior synonym of *Monomorium*)

*Trichomyrmex* Mayr, 1865: 19. Type-species: *Trichomyrmex rogeri*, by monotypy. *Trichomyrmex* junior synonym of *Monomorium:* Ettershank, 1966: 82; Bolton, 1987: 296.

***TRICHOSCAPA*** (Myrmicinae: Dacetini)
*Trichoscapa* Emery, 1869b: 24 [as subgenus of *Strumigenys*]. Type-species: *Strumigenys* (*Trichoscapa*) *membranifera*, by monotypy. *Trichoscapa* junior synonym of *Strumigenys:* Dalla Torre, 1893: 145. *Trichoscapa* junior synonym of *Cephaloxys:* Wheeler, W.M. 1922a: 668; Emery, 1922f: 323. *Trichoscapa* revived from synonymy: Brown, 1948e: 112. Review of genus: Bolton, 1983: 319. [See note under *Strumigenys*.]
***TRICYTARUS*** (Myrmicinae: Formicoxenini?)
*Tricytarus* Donisthorpe, 1947d: 187. Type-species: *Tricytarus parviumgulatus*, by original designation.
*TRIGLYPHOTHRIX* (junior synonym of *Tetramorium*)
*Triglyphothrix* Forel, 1890b: cvi. Type-species: *Triglyphothrix walshi*, by monotypy. *Triglyphothrix* subgenus of *Tetramorium:* Arnold, 1917: 333; maintained as subgenus: Arnold, 1926: 272. *Triglyphothrix* revived status as genus: Wheeler, W.M. 1922a: 909; Emery, 1922f: 272; maintained as genus, and revision of genus: Bolton, 1976: 310. *Triglyphothrix* junior synonym of *Tetramorium:* Bolton, 1985: 247.
*TRIGONOGASTER* (junior homonym; see *Recurvidris*)
*Trigonogaster* Forel, 1890b: cviii. Type-species: *Trigonogaster recurvispinosa*, by monotypy. [Junior homonym of *Trigonogaster* Guérin-Méneville, 1844b: 1149 (Hymenoptera, Pteromalidae).] *Recurvidris* Bolton, 1992: 36, replacement name for *Trigonogaster* Forel.
***TURNERIA*** (Dolichoderinae)
*Turneria* Forel, 1895f: 419. Type-species: *Turneria bidentata*, by monotypy. Revision of genus: Shattuck, 1990: 101; review of genus: Shattuck, 1992c: 161.
*TYPHLATTA* (junior synonym of *Aenictus*)
*Typhlatta* Smith, F. 1857a: 79. Type-species: *Typhlatta laeviceps*, by monotypy. *Typhlatta* junior synonym of *Aenictus:* Forel, 1890b: ciii. *Typhlatta* revived from synonymy as subgenus of *Aenictus:* Wheeler, W.M. 1930g: 198. *Typhlatta* junior synonym of *Aenictus:* Wilson, 1964a: 444.
***TYPHLOMYRMEX*** (Ponerinae: Typhlomyrmecini)
*Typhlomyrmex* Mayr, 1862: 736. Type-species: *Typhlomyrmex rogenhoferi*, by monotypy. [*Typhlomyrmex* Gistel, 1856: 447, *nomen nudum;* see Wheeler, W.M. 1911c: 858.] Revision of genus: Brown, 1965c: 65.
#***TYPHLOPONE*** (subgenus of *Dorylus*)
*Typhlopone* Westwood, 1839: 219. Type-species: *Typhlopone fulva*, by subsequent designation of Emery, 1895j: 701. *Typhlopone* senior synonym of *Cosmaecetes:* Smith, F. 1858b: 111. *Typhlopone* junior synonym of *Dorylus:* Dalla Torre, 1893: 9. *Typhlopone* revived from synonymy as subgenus of *Dorylus:* Emery, 1895j: 701. *Typhlopone* provisional junior synonym of *Dorylus:* Brown, 1973b: 185 (unconfirmed).
*TYPHLOTERAS* (junior synonym of *Centromyrmex*)
*Typhloteras* Karavaiev, 1925b: 128. Type-species: *Typhloteras hamulatum*, by monotypy. *Typhloteras* junior synonym of *Centromyrmex:* Brown, 1953c: 8.
*VEROMESSOR* (junior synonym of *Messor*)
*Veromessor* Forel, 1917: 235 [as subgenus of *Novomessor*]. Type-species: *Aphaenogaster andrei*, by subsequent designation of Emery, 1921f: 67. *Veromessor* raised to genus: Wheeler, W.M. 1922a: 661. Revision of genus: Wheeler, W.M. & Creighton, 1934: 354. *Veromessor* senior synonym of *Lobognathus:* Brown, 1949a: 49. Key to species: Smith, M.R. 1956a: 36. *Veromessor* junior synonym of *Messor:* Bolton, 1982: 338.
*VITICICOLA* (junior synonym of *Tetraponera*)
*Viticicola* Wheeler, W.M. 1919i: 130. Type-species: *Sima tessmanni*, by original designation. *Viticicola* provisional junior synonym of *Tetraponera:* Brown, 1973b: 185. *Viticicola* junior synonym of *Tetraponera:* Ward, 1990: 470.
*VOLLENHOVENIA* (junior synonym of *Vollenhovia*)
*Vollenhovenia* Dalla Torre, 1893: 61, unjustified subsequent emendation of *Vollenhovia*. *Vollenhovenia* junior synonym of *Vollenhovia:* Forel, 1893a: 166; Emery, 1922e: 163.
***VOLLENHOVIA*** (Myrmicinae: Metaponini)
*Vollenhovia* Mayr, 1865: 21. Type-species: *Vollenhovia punctatostriata*, by monotypy. *Vollenhovia* senior synonym of **Propodomyrma:* Wheeler, W.M. 1915h: 51. *Vollenhovia* senior synonym of *Vollenhovenia:* Forel, 1893a: 166. *Vollenhovia* senior synonym of *Dorothea*, *Dyomorium*, and *Heteromyrmex*, and review of genus: Ettershank, 1966: 146. *Vollenhovia* senior synonym of *Aratromyrmex:* Baroni Urbani, 1980a: 96. *Vollenhovia* senior synonym of *Gauromyrmex* (and its junior synonyms *Acalama* and *Solenomyrma*): Bolton, 1994: 105.
***VOMBISIDRIS*** (Myrmicinae: Formicoxenini)
*Vombisidris* Bolton, 1991: 1. Type-species: *Vombisidris philax*, by original designation. Key to species: Bolton, 1991: 2.
*WADEURA* (junior synonym of *Pachycondyla*)
*Wadeura* Weber, 1939a: 102. Type-species: *Wadeura guianensis*, by original designation. *Wadeura* provisional junior synonym of *Pachycondyla:* Brown, 1973b: 185. *Wadeura* junior synonym of *Pachycondyla:* Brown, in preparation.
***WASMANNIA*** (Myrmicinae: Blepharidattini)
*Wasmannia* Forel, 1893g: 383. Type-species: *Tetramorium auropunctatum*, by subsequent designation of Wheeler, W.M. 1911f: 174. *Wasmannia* senior synonym of *Hercynia:* Brown, 1948d: 102. *Wasmannia* provisional junior synonym of *Ochetomyrmex:* Brown, 1973b: 185. *Wasmannia* valid genus: Kempf, 1975c: 357; Snelling, 1981: 398.
*WEBERIDRIS* (junior synonym of *Calyptomyrmex*)
*Weberidris* Donisthorpe, 1949a: 281. Type-species: *Weberidris rufobrunnea* (junior synonym of *Calyptomyrmex beccarii*), by original designation. *Weberidris* junior synonym of *Calyptomyrmex:* Donisthorpe, 1949f: 186; Brown, 1949f: 84.
*WEBERISTRUMA* (junior synonym of *Smithistruma*)
*Weberistruma* Brown, 1948e: 106 [as subgenus of *Smithistruma*]. Type-species: *Strumigenys* (*Cephaloxys*) *leptothrix*, by original designation. *Weberistruma* raised to genus: Brown, 1949d: 7. *Weberistruma* junior synonym of *Smithistruma:* Brown, 1973a: 35.
*WESSONISTRUMA* (junior synonym of *Smithistruma*)
*Wessonistruma* Brown, 1948e: 106 [as subgenus of *Smithistruma*]. Type-species: *Strumigenys pergandei*, by original designation. Review of subgenus: Brown, 1953g: 50. *Wessonistruma* junior synonym of *Smithistruma:* Brown, 1973a: 35.
*WHEELERIA* (junior homonym; see *Wheeleriella*)
*Wheeleria* Forel, 1905b: 171. Type-species: *Wheeleria santschii*, by monotypy. [Junior homonym of *Wheeleria* Tutt, 1905: 37 (Lepidoptera).] *Wheeleriella* Forel, 1907f: 145, replacement name for *Wheeleria* Forel.
*WHEELERIELLA* (junior synonym of *Monomorium*)
*Wheeleriella* Forel, 1907f: 145. Replacement name for *Wheeleria* Forel, 1905b: 171. [Junior homonym of *Wheeleria* Tutt, 1905: 37.] *Wheeleriella* junior synonym of *Monomorium:* Ettershank, 1966: 82; Bolton, 1987: 291.
*WHEELERIMYRMEX* (junior synonym of *Megalomyrmex*)
*Wheelerimyrmex* Mann, 1922: 29 [as subgenus of *Megalomyrmex*]. Type-species: *Megalomyrmex silvestrii*, by original designation. *Wheelerimyrmex* junior synonym of *Megalomyrmex:* Ettershank, 1966: 101.
*WHEELERIPONE* (junior synonym of *Gnamptogenys*)
*Wheeleripone* Mann, 1919: 282. Type-species: *Wheeleripone albiclava*, by original designation. *Wheeleripone* junior synonym of *Gnamptogenys:* Brown, 1958g: 212.
***WILLOWSIELLA*** (Myrmicinae: Formicoxenini)
*Willowsiella* Wheeler, W.M. 1934a: 174. Type-species: *Willowsiella dispar*, by original designation.
*WOITKOWSKIA* (junior synonym of *Neivamyrmex*)
*Woitkowskia* Enzmann, E.V. 1952: 443. Type-species: *Woitkowskia connectens* (junior synonym of *Neivamyrmex walkeri*), by original designation. *Woitkowskia* junior synonym of *Neivamyrmex:* Borgmeier, 1955b: 277.
*XENHYBOMA* (junior synonym of *Monomorium*)
*Xenhyboma* Santschi, 1919g: 405. Type-species: *Xenhyboma mystes* (junior synonym of *Monomorium medinae*), by monotypy. *Xenhyboma* provisional junior synonym of *Monomorium:* Ettershank, 1966: 82. *Xenhyboma* junior synonym of *Monomorium:* Espadaler, 1982b: 112; Bolton, 1987: 293.
*XENOAPHAENOGASTER* (junior synonym of *Pheidole*)
*Xenoaphaenogaster* Baroni Urbani, 1964b: 50. Type-species: *Xenoaphaenogaster inquilina* (provisional junior synonym of *Pheidole pallidula*), by original designation. *Xenoaphaenogaster* provisional junior synonym of *Monomorium:* Brown, 1973b: 185. *Xenoaphaenogaster* junior synonym of *Pheidole:* Bolton, 1987: 291.
*XENOMETRA* (junior synonym of *Cardiocondyla*)
*Xenometra* Emery, 1917a: 96. Type-species: *Xenometra monilicornis* (junior synonym of *Cardiocondyla emeryi*), by monotypy. *Xenometra* junior synonym of *Cardiocondyla:* Baroni Urbani, 1973: 199.
***XENOMYRMEX*** (Myrmicinae: Metaponini?)
*Xenomyrmex* Forel, 1885a: 369. Type-species: *Xenomyrmex stollii*, by monotypy. *Xenomyrmex* senior synonym of *Myrmecinella*, and revision of genus: Wheeler, W.M. 1931a: 129. Key to species: Creighton, 1957c: 6. Review of genus: Ettershank, 1966: 150.
*XEROMYRMEX* (junior synonym of *Monomorium*)
*Xeromyrmex* Emery, 1915i: 190 [as subgenus of *Monomorium*]. Type-species: *Formica salomonis*, by original designation. *Xeromyrmex* junior synonym of *Monomorium:* Ettershank, 1966: 82; Bolton, 1987: 294.

**#XIPHOCREMA** (subgenus of *Crematogaster*)
*Xiphocrema* Forel, 1913k: 80 [as subgenus of *Crematogaster*]. Type-species: *Crematogaster tetracantha*, by subsequent designation of Emery, 1922e: 138. *Xiphocrema* raised to genus: Soulié, 1964: 398. *Xiphocrema* provisional junior synonym of *Crematogaster:* Brown, 1973b: 185 (unconfirmed). *Xiphocrema* here reverts to subgenus of *Crematogaster.*
*XIPHOMYRMEX* (junior synonym of *Tetramorium*)
*Xiphomyrmex* Forel, 1887: 385 [as subgenus of *Tetramorium*]. Type-species: *Tetramorium (Xiphomyrmex) kelleri*, by subsequent designation of Wheeler, W.M. 1911f: 175. *Xiphomyrmex* raised to genus: Emery, 1896e: 183 (in key). *Xiphomyrmex* junior synonym of *Tetramorium:* Bingham, 1903: 175. *Xiphomyrmex* revived from synonymy as subgenus of *Tetramorium:* Arnold, 1917: 345; maintained as subgenus: Arnold, 1926: 277. *Xiphomyrmex* raised to genus: Forel, 1917: 245; maintained as genus: Wheeler, W.M. 1922a: 193; Emery, 1922f: 286. *Xiphomyrmex* junior synonym of *Tetramorium:* Bolton, 1976: 359; Bolton, 1980: 195.
*XIPHOPELTA* (junior synonym of *Mesoponera*)
*Xiphopelta* Forel, 1913a: 108 [as subgenus of *Ponera*]. Type-species: *Ponera (Xiphopelta) arnoldi*, by monotypy. *Xiphopelta* subgenus of *Euponera:* Forel, 1917: 237. *Xiphopelta* junior synonym of *Mesoponera:* Wheeler, W.M. 1922a: 775. [*Mesoponera* junior synonym of *Pachycondyla:* Brown, in preparation.] [*Hiphopelta:* see above.]
*XYMMER* (junior synonym of *Stigmatomma*)
*Xymmer* Santschi, 1914d: 311 [as subgenus of *Stigmatomma*]. Type-species: *Stigmatomma (Xymmer) muticum*, by monotypy. *Xymmer* raised to genus: Wheeler, W.M. 1922a: 641. *Xymmer* subgenus of *Amblyopone:* Clark, 1934b: 27. *Xymmer* junior synonym of *Stigmatomma:* Brown, 1949c: 87. [*Stigmatomma* junior synonym of *Amblyopone:* Brown, 1960a: 155.]
**YAVNELLA** (Leptanillinae: Leptanillini)
*Yavnella* Kugler, J. 1987: 52. Type-species: *Yavnella argamani*, by original designation.
**ZACRYPTOCERUS** (Myrmicinae: Cephalotini)
*Zacryptocerus* Wheeler, W.M. 1911f: 175 (see also footnote). Type-species: *Cryptocerus clypeatus*, by original designation. [*Zacryptocerus* Ashmead, 1905b: 384, *nomen nudum.* (Based on a non-existent type-species: *Cryptocerus multistrigus, nomen nudum*, attributed to Smith, F.).] Revision of genus: Kempf, 1951: 134. *Zacryptocerus* senior synonym of *Hypocryptocerus* and *Paracryptocerus* (and its junior synonyms *Cyathomyrmex, Cyathocephalus* (homonym) and *Harnedia*): Kempf, 1973c: 460.
*ZASPHINCTUS* (junior synonym of *Sphinctomyrmex*)
*Zasphinctus* Wheeler, W.M. 1918a: 219 [as subgenus of *Eusphinctus*]. Type-species: *Sphinctomyrmex turneri*, by monotypy. *Zasphinctus* provisional junior synonym of *Sphinctomyrmex:* Brown, 1973b: 185. *Zasphinctus* junior synonym of *Sphinctomyrmex:* Brown, 1975: 31.
*ZATAPINOMA* (junior synonym of *Tapinoma*)
*Zatapinoma* Wheeler, W.M. 1928a: 20. Type-species: *Zatapinoma annandalei*, by original designation. *Zatapinoma* junior synonym of *Tapinoma:* Shattuck, 1992c: 146.
**#ZEALLEYELLA** (subgenus of *Anoplolepis*)
*Zealleyella* Arnold, 1922: 579 (diagnosis in key) [as subgenus of *Plagiolepis*]. Type-species: *Formica custodiens*, by subsequent designation of Santschi, 1926a: 14. *Zealleyella* subgenus of *Anoplolepis:* Santschi, 1926a: 14. *Zealleyella* provisional junior synonym of *Anoplolepis:* Brown, 1973b: 185 (unconfirmed).
***ZHERICHINIUS** (Dolichoderinae)
**Zherichinius* Dlussky, 1988: 57. Type-species: **Zherichinius horribilis*, by original designation.

# Catalogue of Species-Group Taxa

## Names in the genus-group

Names in the genus-group are listed in alphabetical order throughout, whatever their current status. For data on genus-group names, refer to the catalogue of genus-group taxa. Typography of genus-group names used here is as follows:

1. Genus-group names which are currently regarded as valid genera are listed in ***BOLD ITALIC CAPITALS*** and are followed by parentheses indicating the global distribution of each. Valid names of subgenus rank are prefixed by the symbol #. Species-group names contained within each genus follow immediately in alphabetical order; see notes below.

2. Genus-group names which are junior synonyms, junior homonyms, or any other category are listed in *ITALIC CAPITALS* and are followed by : "see under ——," plus the name in ***BOLD ITALIC CAPITALS*** of the currently valid genus where their included species-group names can be found. For example, the entry for *Triglyphothrix,* a junior synonym of *Tetramorium,* reads "*TRIGLYPHOTHRIX:* see under ***TETRAMORIUM,***" indicating that taxa originally described under the former name can be found under the latter.

## Names in the species-group

Every published name in the species-group, regardless of its rank and status, has its own entry and is cross-referred within the catalogue where necessary. As in previous sections serious misspellings of names, which may cause confusion, are listed, but simple and obvious typographical errors are omitted. The object of this catalogue is to be functional and useful, not to provide an inflated list of typographical errors for ant names. The following notes indicate the standardised system used throughout the catalogue.

1. Taxa within each currently recognised genus are listed alphabetically, regardless of their current status. Fossil forms are preceded by *. The incorrect appearance of diacritical marks, hyphens, capital letters, etc., in original descriptions of species-group names are corrected here in accordance with the rules of the *International Code of Zoological Nomenclature* (Third Edition, 1985), regardless of how the name first appeared in print.

2. Names with the current rank of species or subspecies are in ***bold italic lower case;*** the two are distinguished by the symbol #, which precedes all current subspecies-rank names. All other entries (junior synonyms, junior homonyms, unavailable names, etc.) are in *italic lower case;* see note 5, below.

3. Each name entered is immediately followed by its full original combination together with name(s) of author(s), bibliographical reference, sex or caste of material included in the original description (abbreviated in parentheses), and the name of the territory containing the type-locality (in CAPITALS).

Abbreviations denoting material mentioned in descriptions are as follows: (s.) = soldier caste or major worker, when appropriate; (w.) = worker; (q.) = queen; (m.) = male; (l.) = larva; (k.) = karyotype.

Details of type-locality are omitted, primarily because they can be obtained by reference to the original publication, and secondarily to economise on space.

4. Immediately following the basic data outlined above is the following supplementary information, when available:

- i. References to later descriptions of sexes, castes, etc., not covered by the original description.
- ii. References to changes in genus-group combination of the species-group name, introduced by "Combination in: ——."
- iii. References to changes in rank of the species-group name.
- iv. References to synonymy and/or homonymy of the species-group name, whether junior or senior.
- v. References to other literature for consultation, introduced by "See also: ——."
- vi. List of subspecies-rank names currently attached to a species-rank name, introduced by "Current subspecies: nominal plus ——," and followed by a list of their names which can be referred to under their individual headings.
- vii. A name originally described in one genus but subsequently transferred to another, either because the original combination was incorrect or because of genus-level synonymy, is listed under its original combination but is followed by "; see under ——." This is followed by the

currently recognised genus-level combination where full information may be found. For example in *Acromyrmex* the entry for *tucumana* reads: "*tucumana* Forel, 1914; see under ***TRACHYMYRMEX.***"

Where a name was originally described in a genus now synonymised but later transferred elsewhere, the genus of the original combination is quoted in square brackets. Thus in *Atta* the entry for *malabarica* reads: "*malabarica* Jerdon, 1851 [*Oecodoma*]; see under ***PHEIDOLE.***" This shows that the original combination was in *Oecodoma*, that the genus *Oecodoma* is now recognised as a junior synonym of *Atta*, but that the particular species *O. malabarica* should be sought under *Pheidole*.

In species whose names occur many times in the literature only critical references in categories ii–iv are recorded. The reason for this is that a complete list would be very repetitive and to a large extent sterile and uninformative. For example, a common name treated as a subspecies by one author throughout his life, but as a species by another, would show dozens of pointless entries in category iii. Similarly, frequent alternations of generic combination would produce a plethora of information-free entries in category ii.

viii. Some entries in categories i–iv above may show numerous references in certain taxa. In such cases the first reference is always the earliest to any particular event. Later references indicate confirmation or continuity of the decision indicated by the earliest reference.

5. All names other than those currently valid at species rank or subspecies rank are in *italic lower case*.

i. Old terms for ranks treated here as available names with the rank of subspecies (subsp.) include stirps (st.), race (r.), and variety (var.); the last only when applied directly to a name of species rank in an original description. The terms used in the original descriptions are reproduced in the catalogue, and their equivalence to subspecies must be borne in mind.

Although subspecies are listed in the catalogue, the reader should be aware that I am categorically opposed to the use of this taxonomic rank. Such names appear here only because time is not available to ascertain their real value. The uselessness of the subspecies rank was demonstrated by Wilson & Brown (1953: 97), and most serious myrmecologists immediately accepted the logic of their arguement. Unfortunately the taxonomic literature remains clogged with the infraspecific descriptive excesses of earlier authors, and even a few recent authors adhere to the sorry practice of creating subspecies.

ii. Variety names appended to any name of subspecies rank are infrasubspecific and hence taxonomically unavailable, unless made available by action of some later author. In such instances the name becomes available under the authorship and date of the later author, in accordance with the ICZN rules. Such names are entered in the catalogue under their first available usage; the reference to the unavailable original then follows the first available use in square brackets.

Permanantly unavailable names are stated as such. Where an earlier author has referred an unavailable name to the synonymy of an available name, an incorrect procedure as unavailable names have no status, the phrase "Unavailable name; material referred to ——," followed by available name plus reference(s), is utilised.

References to unavailable names, other than their original appearances, are ignored. Similarly, references to originally available names in an unavailable subsequent context, are ignored. For instance, the imaginary taxon *X-us x-us* subsp. *y-us* is available. A later reference in the unavailable form *X-us x-us* subsp. *z-us* var. *y-us* is ignored.

6. Newly detected junior homonyms in the species-group, primary or secondary, are treated in one of several ways.

i. In genera whose species-level taxonomy is comprehensively understood, or which have been recently revised, replacement names for newly detected junior homonyms at species rank are proposed.

ii. Junior homonyms at species rank in genera whose species-level taxonomy is not well understood are replaced only if there is minimal chance of later synonymy of the junior homonym and its replacement name, and only if the species in question has no complement of infraspecific taxa. If infraspecific taxa are present, revisionary work may show one of these to be available as a replacement name, which would make redundant any new name proposed here. Conversely, if subsequent taxonomic studies remove the infraspecific taxa from association with the species in question, proposal of a replacement name will become necessary at that time.

iii. Replacement names are not proposed for junior homonyms currently of infraspecific rank. Most infraspecific taxa turn out to be junior synonyms for which, if homonymous, replacement names are unnecessary (see iv, below). If on revision any homonymous infraspecific name is raised to the rank of species, then a replacement name can be proposed at that time.

iv. Replacement names are not proposed for junior homonyms that are currently also junior synonyms. If the name is later revived from synonymy, then a replacement name can be proposed at that time.

v. Replacement names are not proposed for junior homonyms in genera where I know that other workers are currently carrying out taxonomic studies.

vi. Replacement names are not proposed for junior homonyms that are unidentifiable to genus and appear in the text as *incertae sedis* in any given genus.

When a junior homonym is detected but not replaced, for any of the above reasons, its entry in the catalogue is followed by the statement "Unresolved junior primary (or secondary) homonym."

## The catalogue of species-group taxa

*ACALAMA:* see under ***VOLLENHOVIA.***

*ACAMATUS:* see under ***NEIVAMYRMEX.***

*ACANTHIDRIS:* see under ***RHOPALOTHRIX.***

*ACANTHOCLINEA:* see under ***DOLICHODERUS.***

***ACANTHOGNATHUS*** (Neotropical)

***brevicornis.*** *Acanthognathus brevicornis* Smith, M.R. 1944c: 151 (w.q.) PANAMA. See also: Brown & Kempf, 1969: 94.

***lentus.*** *Acanthognathus lentus* Mann, 1922: 34, fig. 16 (w.q.) HONDURAS. See also: Brown & Kempf, 1969: 95; Kempf, 1975a: 412.

***ocellatus.*** *Acanthognathus ocellatus* Mayr, 1887: 579 (w.) BRAZIL. Mann, 1916: 452 (q). See also: Brown & Kempf, 1969: 95.

****poinari.*** **Acanthognathus poinari* Baroni Urbani, in Baroni Urbani & De Andrade, 1994: 41, figs. 20, 21, 26, 27 (q.) DOMINICAN AMBER.

***rudis.*** *Acanthognathus rudis* Brown & Kempf, 1969: 100, figs. 1, 2, 4–10 (w.q.m.) BRAZIL. Wheeler, G.C. & Wheeler, J. 1969: 110 (l.). See also: Dietz & Brandão, 1993: 683.

***stipulosus.*** *Acanthognathus stipulosus* Brown & Kempf, 1969: 104, fig. 3 (w.) BRAZIL.

***teledectus.*** *Acanthognathus teledectus* Brown & Kempf, 1969: 105, fig. 11 (w.) COLOMBIA.

*ACANTHOLEPIS:* see under ***LEPISIOTA.***

***ACANTHOMYOPS*** (Nearctic)

***arizonicus.*** *Lasius* (*Acanthomyops*) *interjectus* subsp. *arizonicus* Wheeler, W.M. 1917a: 532 (w.) U.S.A. Wing, 1968: 90 (q.m.). Combination in *Acanthomyops:* Creighton, 1950a: 431. Raised to species: Buren, 1950: 184. See also: Wing, 1968: 90.

***bureni.*** *Acanthomyops bureni* Wing, 1968: 135, figs. 159–167 (w.q.m.) U.S.A.

***californicus.*** *Lasius* (*Acanthomyops*) *interjectus* subsp. *californicus* Wheeler, W.M. 1917a: 531 (w.q.) U.S.A. Wing, 1968: 85 (m.). Combination in *Acanthomyops,* and subspecies of *claviger:* Creighton, 1950a: 430. Raised to species: Buren, 1950: 185. See also: Wing, 1968: 85.

*capitatus* Kuznetsov-Ugamsky, 1927; see under ***LASIUS.***

***claviger.*** *Formica clavigera* Roger, 1862a: 241, pl. 1, fig. 13 (q.) U.S.A. Mayr, 1870b: 950 (w.m.); Wheeler G.C. & Wheeler, J. 1953c: 155 (l.). Combination in *Acanthomyops:* Mayr, 1862: 700; in *Lasius:* Mayr, 1870b: 950; in *Lasius* (*Acanthomyops*): Emery, 1893i: 642; in *Acanthomyops:* Donisthorpe, 1916c: 276. Senior synonym of *parvula:* Wing, 1968: 67. See also: Creighton, 1950a: 429; Regnier & Wilson, 1968: 955; Smith, D.R. 1979: 1440.

*clavigeroides. Lasius* (*Acanthomyops*) *clavigeroides* Buren, 1942: 406 (w.q.m.) U.S.A. Subspecies of *subglaber:* Buren, 1950: 185. Junior synonym of *subglaber:* Creighton, 1950a: 433.

***colei.*** *Acanthomyops colei* Wing, 1968: 88, figs. 64–67 (w.m.) U.S.A.

***coloradensis.*** *Lasius* (*Acanthomyops*) *interjectus* subsp. *coloradensis* Wheeler, W.M. 1917a: 532 (w.q.m.) U.S.A. Combination in *Acanthomyops,* and subspecies of *claviger:* Creighton, 1950a: 429. Raised to species: Buren, 1950: 185. See also: Wing, 1968: 78.

***creightoni.*** *Acanthomyops creightoni* Wing, 1968: 141, figs. 177–185 (w.q.m.) U.S.A.

*edwardsi* Donisthorpe, 1933; see under ***LASIOPHANES.***

*flavidus* Kuznetsov-Ugamsky, 1927; see under ***LASIUS.***

*hingstoni* Donisthorpe, 1929; see under ***LASIUS.***

***interjectus.*** *Lasius interjectus* Mayr, 1866b: 888, pl. 20, fig. 3 (q.) U.S.A. Emery, 1893i: 642 (w.); Wing, 1968: 93 (m.); Wheeler, G.C. & Wheeler, J. 1953c: 156 (l.). Combination in *Lasius* (*Acanthomyops*): Emery, 1893i: 642; Wheeler, W.M. 1917a: 531; in *Acanthomyops:* Creighton, 1950a: 430. See also: Wing, 1968: 92; Smith, D.R. 1979: 1441.

*kashmirensis* Donisthorpe, 1930; see under ***LASIUS.***

*kosswigi* Donisthorpe, 1950; see under ***PROFORMICA.***

***latipes.*** *Formica latipes* Walsh, 1863: 311 (w.q.m.) U.S.A. Combination in *Lasius:* Mayr, 1866b: 889; in *Lasius* (*Acanthomyops*): Emery, 1893i: 642; in *Acanthomyops:* Creighton, 1950a: 431. See also: Wheeler, W.M. & McClendon, 1903: 149; Wing, 1968: 98; Smith, D.R. 1979: 1441.

***mexicanus.*** *Lasius* (*Acanthomyops*) *interjectus* subsp. *mexicanus* Wheeler, W.M. 1914b: 55 (w.q.m.) MEXICO. Wheeler, G.C. & Wheeler, J. 1970: 648 (l.). Combination in *Acanthomyops:* Wing, 1968: 146. Raised to species: Buren, 1950: 185.

***murphyi.*** *Lasius* (*Acanthomyops*) *murphii* Forel, 1901e: 367 (w.q.m.) U.S.A. [Justified emendation of spelling to *murphyi:* Wheeler, W.M. 1905f: 398.] Combination in *Acanthomyops:* Creighton, 1950a: 431. See also: Wing, 1968: 113.

*negrensis* Donisthorpe, 1933; see under ***LASIOPHANES.***

*nigrobrunneus* Donisthorpe, 1926; see under ***LASIUS.***

*nitidus* Kuznetsov-Ugamsky, 1927; see under ***LASIUS.***

***occidentalis.*** *Lasius* (*Acanthomyops*) *occidentalis* Wheeler, W.M. 1909e: 83 (w.q.m.) U.S.A. Combination in *Acanthomyops:* Creighton, 1950a: 432. See also: Wing, 1968: 143.

*parvula. Lasius* (*Acanthomyops*) *parvula* Smith, M.R. 1934b: 213 (w.) U.S.A. Combination in *Acanthomyops:* Creighton, 1950a: 432. Junior synonym of *claviger:* Wing, 1968: 67.

*pilicornis* Kuznetsov-Ugamsky, 1927; see under ***LASIUS.***

***plumopilosus.*** *Lasius* (*Acanthomyops*) *plumopilosus* Buren, 1941: 231, fig. 1 (w.q.m.) U.S.A. Combination in *Acanthomyops:* Creighton, 1950a: 433. See also: Wing, 1968: 132.

***pogonogynus.*** *Lasius* (*Acanthomyops*) *pogonogynus* Buren, 1950: 186 (w.q.) U.S.A. Combination in *Acanthomyops:* Wing, 1968: 117. [According to Wing (1968: 117) *pogonogynus* is an *A. murphyi* X *A. latipes* hybrid; the name is not, however, formally synonymised.]

***pubescens.*** *Lasius* (*Acanthomyops*) *pubescens* Buren, 1942: 405 (w.q.) U.S.A. Combination in *Acanthomyops:* Creighton, 1950a: 433. See also: Wing, 1968: 138.

*rufoniger* Donisthorpe, 1933; see under ***LASIOPHANES.***

***subglaber.*** *Lasius* (*Acanthomyops*) *subglaber* Emery, 1893i: 642 (w.q.m.) U.S.A. Wheeler, G.C. & Wheeler, J. 1953c: 155 (l.). Combination in *Acanthomyops,* and senior synonym of *clavigeroides:* Creighton, 1950a: 433. See also: Buren, 1950: 185; Wing, 1968: 121.

*turkmenus* Kuznetsov-Ugamsky, 1927; see under ***LASIUS.***

***ACANTHOMYRMEX*** (Oriental, Indo-Australian)

***basispinosus.*** *Acanthomyrmex basispinosus* Moffett, 1986c: 67, figs. 8A, 9–14 (s.w.) SULAWESI.

***careoscrobis.*** *Acanthomyrmex careoscrobis* Moffett, 1986c: 78, figs. 39A, 40–43 (w.) BORNEO.

***concavus.*** *Acanthomyrmex concavus* Moffett, 1986c: 80, figs, 39B-C, 44–47 (w.) BORNEO.

***crassispinus.*** *Acanthomyrmex crassispina* Wheeler, W.M. 1930a: 101, fig. 2 (w.) TAIWAN. Moffett, 1986c: 69 (s.).

***dusun.*** *Acanthomyrmex dusun* Wheeler, W.M. 1919e: 89 (s.) BORNEO. See also: Moffett, 1986c: 70.

*dyak. Acanthomyrmex dyak* Wheeler, W.M. 1919e: 86 (s.w.) BORNEO. Junior synonym of *ferox:* Moffett, 1986c: 70.

***ferox.*** *Acanthomyrmex ferox* Emery, 1893f: 245 (footnote), pl. 6, fig. 11 (w.) WEST MALAYSIA. Moffett, 1986c: 72 (s.q.m.). Senior synonym of *dyak:* Moffett, 1986c: 70. See also: Moffett, 1985a: 165.

***foveolatus.*** *Acanthomyrmex foveolatus* Moffett, 1986c: 82, figs. 39D, 48–50 (s.) BORNEO.

*kochi* Emery, 1911; see under ***PHEIDOLE.***

***laevis.*** *Acanthomyrmex laevis* Moffett, 1986c: 76, figs. 8F, 29–32 (w.) WEST MALAYSIA.

***luciolae.*** *Acanthomyrmex luciolae* Emery, 1893a: cclxxvi (s.w.) SRI LANKA. [Also described as new by Emery, 1893f: 245.] See also: Moffett, 1986c: 76.

***mindanao.*** *Acanthomyrmex mindanao* Moffett, 1986c: 82, figs. 39E, 51–56 (s.w.) PHILIPPINES.

***notabilis.*** *Pheidole notabilis* Smith, F. 1860b: 111, pl. 1, figs. 3, 4 (s.w.) INDONESIA (Batjan I.). Wheeler, G.C. & Wheeler, J. 1983: 606 (l.). Combination in *Acanthomyrmex:* Emery, 1893a: cclxxvi. See also: Moffett, 1985a: 165; Moffett, 1986c: 86.

***ACANTHOPONERA*** (Neotropical)

*aurea* Forel, 1913; see under ***HETEROPONERA.***

*crassa. Acanthoponera crassa* Brown, 1958g: 255, fig. 10 (w.) ECUADOR. Junior synonym of *minor:* Kempf & Brown, 1968: 90.

***goeldii.*** *Acanthoponera goeldii* Forel, 1912c: 34 (w.) BRAZIL.

*goyana. Acanthoponera goeldii* subsp. *goyana* Borgmeier, 1939: 418 (w.) BRAZIL. Junior synonym of *mucronata:* Brown, 1958g: 194.

*imbellis* Emery, 1895; see under ***HETEROPONERA.***

*kirki* Wheeler, W.M. 1923; see under ***HETEROPONERA.***

*ACANTHOPONERA (continued)*

*leae* Wheeler, W.M. 1923; see under **HETEROPONERA.**

***minor.*** *Ectatomma (Acanthoponera) mucronatum* var. *minor* Forel, 1899c: 9 (w.) MEXICO. [*Acanthoponera minor* Emery, 1896g: 33. *Nomen nudum,* attributed to Forel.] Combination in *Acanthoponera:* Emery, 1911d: 36. Raised to species, and senior synonym of *schwarzi* and *spininodis:* Brown, 1958g: 194. Senior synonym of *crassa:* Kempf & Brown, 1968: 90.

***mucronata.*** *Ponera mucronata* Roger, 1860: 299 (q.) BRAZIL. Mayr, 1862: 733 (w.). Combination in *Ectatomma (Acanthoponera)*: Mayr, 1862: 733; in *Acanthoponera:* Emery, 1911d: 36. Senior synonym of *goyana, plaumanni* and *wagneri:* Brown, 1958g: 194.

*nigra* Clark, 1930; see under **HETEROPONERA.**

*occidentalis* Clark, 1926; see under **HETEROPONERA.**

***peruviana.*** *Acanthoponera peruviana* Brown, 1958g: 254, figs. 6,7 (w.) PERU.

*plaumanni. Acanthoponera plaumanni* Donisthorpe, 1938a: 140 (w.) BRAZIL. Junior synonym of *mucronata:* Brown, 1958g: 194.

*reichenspergeri* Santschi, 1929; see under **GNAMPTOGENYS.**

*scabra* Wheeler, W.M. 1923; see under **HETEROPONERA.**

*schwarzi. Acanthoponera goeldii* subsp. *schwarzi* Wheeler, W.M. 1923e: 188, fig. 5 (w.) GUATEMALA. Junior synonym of *minor:* Brown, 1958g: 194.

*schwebeli* Luederwaldt, 1918; see under **HETEROPONERA.**

*spininodis. Acanthoponera (Acanthoponera) spininodis* Weber, 1939a: 97, fig. 2 (w.) TRINIDAD. Junior synonym of *minor:* Brown, 1958g: 194.

*wagneri. Acanthoponera mucronata* var. *wagneri* Santschi, 1921g: 84 (w.) ARGENTINA. Junior synonym of *mucronata:* Brown, 1958g: 194.

***ACANTHOSTICHUS*** (Neotropical, southern Nearctic)

*afflictus* Gallardo, 1919; see under **NOMAMYRMEX.**

***brevicornis.*** *Acanthostichus brevicornis* Emery, 1894c: 142 (w.) FRENCH GUIANA. Senior synonym of *ramosmexiae:* Kusnezov, 1962a: 130.

***femoralis.*** *Acanthostichus femoralis* Kusnezov, 1962a: 126, figs. 2, 3 (w.) ARGENTINA.

***fuscipennis.*** *Acanthostichus fuscipennis* Emery, 1895j: 752, pl. 17, fig. 13 (m.) BRAZIL.

***kirbyi.*** *Acanthostichus kirbyi* Emery, 1895j: 751, fig. OO.B (w.) PARAGUAY.

***laticornis.*** *Acanthostichus laticornis* Forel, 1908c: 345 (w.) PARAGUAY. Santschi, 1933e: 106 (m.). Senior synonym of *obscuridens:* Kusnezov, 1962a: 131.

***niger.*** *Acanthostichus serratulus* var. *niger* Santschi, 1933e: 105 (q.) ARGENTINA. Junior synonym of *serratulus:* Kusnezov, 1962a: 132. Revived from synonymy and raised to species: Brown, 1975: 42.

*obscuridens. Acanthostichus laticornis* var. *obscuridens* Santschi, 1934c: 23 (w.) ARGENTINA. Bruch, 1934a: 4 (q.m.). Junior synonym of *laticornis:* Kusnezov, 1962a: 131.

***quadratus.*** *Acanthostichus quadratus* Emery, 1895j: 750, fig. OO.A; pl.16, figs. 5, 6; pl. 17, fig. 12 (w.q.) BOLIVIA.

*ramosmexiae. Acanthostichus ramosmexiae* Bruch, 1924b: 260, fig. (w.) ARGENTINA. Bruch, 1925a: 110 (m.l.). Junior synonym of *brevicornis:* Kusnezov, 1962a: 130.

***sanchezorum.*** *Acanthostichus sanchezorum* MacKay, 1985: 128, figs. 1–5 (w.) COLOMBIA.

***serratulus.*** *Typhlopone serratula* Smith, F. 1858b: 111 (w.) BRAZIL. Mayr, 1887: 551 (m.); Emery, 1899c: 6 (l.). Combination in *Acanthostichus:* Mayr, 1887: 549. See also: Gallardo, 1918b: 9.

***skwarrae.*** *Acanthostichus skwarrae* Wheeler, W.M. 1934g: 161 (w.) MEXICO.

***texanus.*** *Acanthostichus texanus* Forel, 1904d: 168 (q.) U.S.A. Combination in *Acanthostichus (Ctenopyga)*: Emery, 1911d: 14; in *Ctenopyga:* Brown, 1975: 42. Senior synonym of *townsendi:* Smith, M.R. 1955a: 49.

*townsendi. Ctenopyga townsendi* Ashmead, 1906: 29, fig. 4 (q.m.) MEXICO. Combination in *Acanthostichus (Ctenopyga)*: Emery, 1911d: 14. Junior synonym of *texanus:* Smith, M.R. 1955a: 49.

*ACHANTILEPIS:* see under **LEPISIOTA.**

*ACIDOMYRMEX:* see under **RHOPTROMYRMEX.**

*ACROCOELIA:* see under **CREMATOGASTER.**

***ACROMYRMEX*** (Neotropical, southern Nearctic)

*affinis. Acromyrmex aspersus* var. *affinis* Santschi, 1925a: 369 (w.) BRAZIL. Stirps of *aspersus:* Santschi, 1925d: 243. Junior synonym of *aspersus:* Gonçalves, 1961: 132.

#***ajax.*** *Atta (Acromyrmex) emilii* var. *ajax* Forel, 1909b: 58 (w.) "GUINEA" (in error; in text Forel states "probablement du Brésil"). Variety of *hystrix:* Santschi, 1925a: 358.

***ambiguus.*** *Atta (Acromyrmex) lundi* var. *ambigua* Emery, 1888c: 358 (w.) BRAZIL. Goni *et al.*, 1983: 365 (k.). Combination in *Acromyrmex:* Bruch, 1914: 216. Raised to species: Emery, 1905c: 50. Senior synonym of *erectus:* Gonçalves, 1961: 131.

#***andicola.*** *Acromyrmex moelleri* subsp. *andicola* Emery, 1922f: 349 (w.) ECUADOR. [First available use of *Atta (Acromyrmex) moelleri* st. *modesta* var. *andicola* Emery, 1905c: 45; unavailable name.] Subspecies of *coronatus:* Santschi, 1925a: 366; Santschi, 1925d: 239; Kempf, 1972a: 11.

*angustata. Atta (Acromyrmex) moelleri* subsp. *panamensis* var. *angustata* Forel, 1908b: 41 (w.q.) COSTA RICA. Unavailable name.

***aspersus.*** *Oecodoma aspersa* Smith, F. 1858b: 185, pl. 10, fig. 17 (q.) BRAZIL. Emery, 1905c: 49 (w.m.). Combination in *Atta:* Roger, 1863b: 35; in *Atta (Acromyrmex)*: Forel, 1893e: 590; in *Acromyrmex:* Gallardo, 1916d: 328. Junior synonym of *octospinosa:* Forel, 1893e: 590. Revived from synonymy: Emery, 1905c: 48. Senior synonym of *affinis, inquirens* and *mesonotalis,* and material of the unavailable name *clarus* referred here: Gonçalves, 1961: 132. Current subspecies: nominal plus *fuhrmanni.*

*atratus. Acromyrmex hispidus* st. *atratus* Santschi, 1925a: 376 (w.q.) ARGENTINA. Junior synonym of *crassispinus:* Gonçalves, 1961: 139.

***balzani.*** *Atta (Acromyrmex) balzani* Emery, 1890a: 67 (footnote) (w.) PARAGUAY. Combination in *Atta (Moellerius)*: Forel, 1893e: 589; in *Acromyrmex (Moellerius)*: Santschi, 1916e: 389; Gallardo, 1916d: 334. Subspecies of *landolti:* Gonçalves, 1961: 124; Kempf, 1972a: 15. Revived status as species: Fowler, 1988: 284. Senior synonym of *gallardoi:* Santschi, 1922b: 362; of *nivalis:* Fowler, 1988: 284; of *parens:* Gonçalves, 1961: 124. Current subspecies: nominal plus *multituber.*

*bigener. Acromyrmex rugosus* st. *bigener* Santschi, 1925d: 243 (w.q.m.) BRASIL. Junior synonym of *rugosus:* Gonçalves, 1961: 159.

*biscutatus. Formica biscutata* Fabricius, 1775: 394 (q.) FRENCH GUIANA. Combination in *Atta:* Fabricius, 1804: 422. *Incertae sedis* in *Acromyrmex:* Kempf, 1972a: 16.

#***boliviensis.*** *Atta (Acromyrmex) boliviensis* Emery, 1905c: 51, fig. 19 (w.m.) BOLIVIA. Combination in *Acromyrmex:* Santschi, 1916e: 386. Subspecies of *lundii:* Santschi, 1916e: 386; Kempf, 1972a: 13.

*bonariensis. Atta (Acromyrmex) pubescens* var. *bonariensis* Emery, 1905c: 52 (w.) ARGENTINA. Combination in *Acromyrmex:* Bruch, 1914: 216. Variety/stirps of *lundii:* Bruch, 1914: 216; Santschi, 1925a: 385. Junior synonym of *lundii:* Gallardo, 1916d: 331; Gonçalves, 1961: 150.

#***bruchi.*** *Atta (Moellerius) silvestrii* r. *bruchi* Forel, 1912e: 180 (w.) ARGENTINA. Forel, 1913l: 236 (q.m.). Combination in *Acromyrmex (Moellerius)*: Forel, 1913l: 236. Maintained as subspecies of *silvestrii:* Emery, 1922f: 351; Kempf, 1972a: 16.

#***brunneus.*** *Atta (Acromyrmex) subterranea* var. *brunnea* Forel, 1912e: 181 (w.q.m.) BRAZIL. [First available use of *Atta (Acromyrmex) coronata* subsp. *subterranea* var. *brunnea* Forel, 1911c: 291; unavailable name.] [Unresolved junior primary homonym of *Atta brunnea* Patton, 1894: 618 (now in *Odontomachus*).] Combination in *Acromyrmex:* Luederwaldt, 1918: 39. Subspecies of *subterraneus:* Gonçalves, 1961: 167; Kempf, 1972a: 15.

*bucki. Acromyrmex bucki* Wasmann, 1931: 106 (w.) BRAZIL. Junior synonym of *lobicornis:* Gonçalves, 1961: 148.

#***carli.*** *Acromyrmex lundi* subsp. *carli* Gonçalves, 1961: 152 (w.) MEXICO. [First available use of *Acromyrmex lundi* st. *pubescens* var. *carli* Santschi, 1925a: 385; unavailable name.]

*chacoensis. Acromyrmex lundi* st. *pubescens* var. *chacoensis* Santschi, 1916e: 387 (w.) ARGENTINA. Unavailable name.

#***chisosensis.*** *Atta (Moellerius) versicolor* subsp. *chisosensis* Wheeler, W.M. 1907c: 705 (w.) U.S.A. Combination in *Acromyrmex:* Emery, 1922f: 351.

*clarus. Acromyrmex aspersus* st. *mesonotalis* var. *clarus* Santschi, 1925d: 243 (w.) BRAZIL. Unavailable name; material referred to *aspersus* by Gonçalves, 1961: 132.

*cloosae. Atta (Moellerius) landolti* r. *cloosae* Forel, 1912e: 179 (w.) COLOMBIA. Combination in *Acromyrmex (Moellerius)*: Emery, 1922f: 351. Junior synonym of *landolti:* Fowler, 1988: 288.

#***cochlearis.*** *Acromyrmex lobicornis* var. *cochlearis* Santschi, 1933e: 120 (w.) ARGENTINA.

*corallina. Acromyrmex lundi* st. *decolor* var. *corallina* Santschi, 1919f: 51 (w.) ARGENTINA. Unavailable name.

*cordovana* Bruch, 1921; see under ***TRACHYMYRMEX***.

***coronatus.*** *Formica coronata* Fabricius, 1804: 413 (q.) SOUTH AMERICA. Forel, 1885a: 356 (w.). Combination in *Atta:* Roger, 1863b: 35; in *Atta (Acromyrmex)*: Forel, 1885a: 355; in *Acromyrmex:* Bruch, 1914: 216. Subspecies of *hystrix:* Forel, 1885a: 355. Revived status as species: Dalla Torre, 1893: 152. Senior synonym of: *meinerti, modesta, moelleri* and *ochraceolus,* and material of the unavailable names *flavescens, medianus, obscurior* and *ornatus* referred here: Gonçalves, 1961: 135. Current subspecies: nominal plus *andicola, globoculis, importunus, panamensis, rectispinus.*

***crassispinus.*** *Atta (Acromyrmex) mesonotalis* var. *crassispina* Forel, 1909a: 257 (w.) PARAGUAY. Santschi, 1925a: 374 (q.). Combination in *Acromyrmex:* Emery, 1922f: 349. Subspecies of *mesonotalis:* Forel, 1914e: 11. Raised to species: Santschi, 1925a: 374. Senior synonym of: *atratus, diabolica, insularis, mediocris* and *rusticus,* and material of the unavailable name *rufescens* referred here: Gonçalves, 1961: 139.

#***cubanus.*** *Acromyrmex octospinosus* subsp. *cubanus* Wheeler, W.M. 1937c: 74 (s.) CUBA.

#***decolor.*** *Atta (Acromyrmex) pubescens* subsp. *decolor* Emery, 1905c: 52 (w.) PARAGUAY. Combination in *Acromyrmex,* and subspecies of *lundii:* Bruch, 1914: 216.

*depressoculis. Acromyrmex subterranea* var. *depressoculis* Forel, 1913l: 237 (w.) BRAZIL. Stirps of *muticinoda:* Santschi, 1925a: 371. Junior synonym of *muticinoda:* Gonçalves, 1961: 153.

*diabolica. Acromyrmex nigrosetosa* var. *diabolica* Santschi, 1922b: 362 (w.) BRAZIL. Santschi, 1925d: 240 (m.). Stirps of *crassispinus:* Santschi, 1925a: 375. Raised to species: Santschi, 1925d: 240. Junior synonym of *crassispinus:* Gonçalves, 1961: 139.

***diasi.*** *Acromyrmex (Acromyrmex) diasi* Gonçalves, 1983a: 485, fig. 1 (w.q.m.) BRAZIL.

*dimidiata. Atta (Acromyrmex) aspersa* subsp. *dimidiata* Forel, 1911c: 292 (w.m.) BRAZIL. Combination in *Acromyrmex:* Emery, 1922f: 348. Stirps of *dimidiata:* Santschi, 1925a: 392. Junior synonym of *nigrosetosus:* Gonçalves, 1961: 146.

***disciger.*** *Atta discigera* Mayr, 1887: 551 (w.q.m.) BRAZIL. Combination in *Atta (Acromyrmex)*: Forel, 1893e: 590; in *Acromyrmex:* Luederwaldt, 1918: 38. See also: Gonçalves, 1961: 141.

*dubia. Atta (Acromyrmex) laticeps* var. *dubia* Forel, 1908c: 350 (w.) BRAZIL. Combination in *Acromyrmex:* Emery, 1922f: 349. Variety of *lundii:* Santschi, 1925a: 385. Junior synonym of *lundii:* Gonçalves, 1961: 150.

#***echinatior.*** *Atta (Acromyrmex) octospinosa* var. *echinatior* Forel, 1899c: 34 (w.q.) MEXICO. Wheeler, W.M. 1937c: 72 (m.). Combination in *Acromyrmex:* Emery, 1922f: 350. Subspecies of *octospinosus:* Wheeler, W.M. 1937c: 71.

*eidmanni. Acromyrmex (Acromyrmex) subterraneus* var. *eidmanni* Santschi, 1937f: 232, figs. 1, 6 (w.q.m.) BRAZIL. Junior synonym of *molestans:* Gonçalves, 1961: 168.

#***ekchuah.*** *Acromyrmex octospinosus* subsp. *ekchuah* Wheeler, W.M. 1937c: 73 (s.) MEXICO.

*emilii. Atta (Acromyrmex) emilii* Forel, 1904c: 32 (w.) BRAZIL. Wheeler, G.C. 1949: 675 (l.). Combination in *Acromyrmex:* Mann, 1916: 454. Junior synonym of *hystrix:* Santschi, 1925a: 357.

*erectus. Acromyrmex ambiguus* var. *erectus* Santschi, 1925a: 384, fig. 2 (w.) BRAZIL. Junior synonym of *ambiguus:* Gonçalves, 1961: 131.

***evenkul.*** *Acromyrmex evenkul* **new name**. Replacement name for *Acromyrmex gallardoi* Santschi, 1936d: 411. [Junior secondary homonym of *Sericomyrmex gallardoi* Santschi, 1920d: 379.]

#***fallax.*** *Acromyrmex hispidus* var. *fallax* Santschi, 1925a: 377 (w.) ARGENTINA. Subspecies of *hispidus:* Gonçalves, 1961: 142; Kempf, 1972a: 12.

#***ferrugineus.*** *Atta (Acromyrmex) lobicornis* var. *ferruginea* Emery, 1905c: 52 (w.) BRAZIL. Combination in *Acromyrmex:* Bruch, 1914: 216.

*flavescens. Acromyrmex coronatus* st. *andicola* var. *flavescens* Santschi, 1925d: 239 (w.) BRAZIL. Unavailable name; material referred to *coronatus* by Gonçalves, 1961: 135.

#***formosus.*** *Acromyrmex hispidus* st. *formosus* Santschi, 1925a: 377 (w.) ARGENTINA.

***fracticornis.*** *Atta (Moellerius) fracticornis* Forel, 1909a: 257 (w.) PARAGUAY. Fowler, 1988: 286 (m.). Combination in *Acromyrmex (Moellerius)*: Forel, 1913l: 236. Subspecies of *landolti,* and senior synonym of *jorgenseni:* Gonçalves, 1961: 126; Kempf, 1972a: 16. Revived status as species: Fowler, 1988: 286.

#***fuhrmanni.*** *Atta (Acromyrmex) mesonotalis* var. *fuhrmanni* Forel, 1914e: 10 (w.) COLOMBIA. Combination in *Acromyrmex:* Emery, 1922f: 349. Variety of *aspersus:* Santschi, 1925a: 369.

*gaigei* Forel, 1914; see under ***TRACHYMYRMEX***.

*gallardoi. Sericomyrmex gallardoi* Santschi, 1920d: 379 (w.) PARAGUAY. Combination in *Acromyrmex,* and junior synonym of *balzani:* Santschi, 1922b: 362.

*gallardoi. Acromyrmex gallardoi* Santschi, 1936d: 411 (w.q.m.) ARGENTINA. [Junior secondary homonym of *Sericomyrmex gallardoi* Santschi, above.] Replacement name: *evenkul* **new name.**

*garbei. Acromyrmex laticeps* st. *garbei* Santschi, 1925a: 381 (w.) BRAZIL. Stirps of *nigrosetosus:* Santschi, 1925b: 18 (also described as new in this publication). Junior synonym of *nigrosetosus:* Gonçalves, 1961: 146.

*gaudens. Acromyrmex (Moellerius) heyeri* var. *gaudens* Santschi, 1925e: 164 (w.) ARGENTINA. Junior synonym of *heyeri:* Fowler, 1988: 287.

#***globoculis.*** *Acromyrmex coronatus* subsp. *globoculis* Kempf, 1972a: 11 (w.) GUIANA. [First available use of *Acromyrmex moelleri* r. *meinerti* var. *globoculis* Forel. 1916: 429; unavailable name.] Weber, 1937: 408 (q.).

*guentheri. Atta (Acromyrmex) guentheri* Forel, 1893e: 594 (s.w.q.m.) TRINIDAD. Junior synonym of *octospinosus:* Emery, 1894c: 220.

***heyeri.*** *Atta (Moellerius) heyeri* Forel, 1899c: 31 (footnote) (s.w.) BRAZIL. Emery, 1905c: 43 (q.m.); Bruch, 1921: 199 (q.m.); Goni *et al.* 1983: 365 (k.). Combination in *Acromyrmex:* Bruch, 1914: 217. Senior synonym of *gaudens* and *lillensis:* Fowler, 1988: 287. See also: Gonçalves, 1961: 127.

***hispidus.*** *Acromyrmex hispidus* Santschi, 1925a: 376 (w.) BOLIVIA. Zolessi & Abenante, 1974: 154 (q.m.); Goni *et al.*, 1983: 365 (k.). Current subspecies: nominal plus *fallax, formosus.*

*homalops. Atta (Acromyrmex) muticinoda* var. *homalops* Emery, 1905c: 47, figs. 10, 11 (w.m.) BRAZIL. Combination in *Acromyrmex:* Luederwaldt, 1918: 38. Subspecies of *coronatus:* Forel, 1911c: 291; of *muticinoda:* Santschi, 1925a: 370. Junior synonym of *muticinoda:* Gonçalves, 1961: 153.

*hortulanus. Acromyrmex laticeps* var. *hortulanus* Santschi, 1925a: 381 (w.) URUGUAY. Junior synonym of *laticeps:* Gonçalves, 1961: 144.

***hystrix.*** *Formica hystrix* Latreille, 1802c: 230, pl. 10, fig. 61 (w.) FRENCH GUIANA. Emery, 1888c: 357 (m.). Combination in *Atta:* Haliday, 1836: 328; in *Atta (Acromyrmex)*: Mayr, 1865: 83; in *Acromyrmex:* Santschi, 1925a: 357. Subspecies of *octospinosus:* Forel, 1893e: 590; Emery, 1894c: 220. Junior synonym of *octospinosus:* Emery, 1922f: 350. Revived from synonymy, revived status as species, and senior synonym of *emilii:* Santschi, 1925a: 357. See also: Gonçalves, 1961: 143. Current subspecies: nominal plus *ajax.*

#***importunus.*** *Acromyrmex coronatus* st. *importunus* Santschi, 1925a: 362 (w.) COSTA RICA.

*inquirens. Atta (Acromyrmex) mesonotalis* var. *inquirens* Forel, 1914e: 11 (w.) BRAZIL. Combination in *Acromyrmex:* Emery, 1922f: 349. Variety of *aspersus:* Santschi, 1925a: 368. Junior synonym of *aspersus:* Gonçalves, 1961: 132.

*insularis. Acromyrmex aspersus* var. *insularis* Santschi, 1925d: 242 (w.) BRAZIL. Junior synonym of *crassispinus:* Gonçalves, 1961: 139.

#***inti.*** *Acromyrmex octospinosus* subsp. *inti* Wheeler, W.M. 1937c: 72 (s.) PERU.

*jorgenseni. Acromyrmex (Moellerius) fracticornis* var. *jorgenseni* Forel, 1913l: 236 (w.q.) ARGENTINA. Bruch, 1916: 325 (m.). Junior synonym of *fracticornis:* Gonçalves, 1961: 126.

*laeviventris. Acromyrmex (Moellerius) striatus* var. *laeviventris* Santschi, 1925a: 388 (w.) ARGENTINA. [First available use of *Acromyrmex (Moellerius) striatus* st. *silvestrii* var. *laeviventris* Santschi, 1920d: 380; unavailable name.] Junior synonym of *striatus:* Fowler, 1988: 290.

***landolti.*** *Atta (Acromyrmex) landolti* Forel, 1885a: 357 (w.) COLOMBIA. Forel, 1911c: 293 (q.). Combination in *Atta (Moellerius)*: Forel, 1893e: 589; in *Acromyrmex (Moellerius)*: Mann, 1916: 453. Senior synonym of *senex:* Gonçalves, 1961: 120; of *cloosae* and *planorum:* Fowler, 1988: 288. Current subspecies: nominal plus *myersi.*

***laticeps.*** *Atta (Acromyrmex) laticeps* Emery, 1905c: 50, figs. 16, 17 (w.q.m.) BRAZIL. Combination in *Acromyrmex:* Emery, 1922f: 349. Senior synonym of *hortulanus:* Gonçalves, 1961: 144. Current subspecies: nominal plus *nigrosetosus.*

*lillensis. Acromyrmex (Moellerius) heyeri* var. *lillensis* Santschi, 1925e: 165 (w.) ARGENTINA. Junior synonym of *heyeri:* Fowler, 1988: 287.

***lobicornis.*** *Atta (Acromyrmex) lobicornis* Emery, 1888c: 358 (w.) BRAZIL. Emery, 1905c: 52 (q.m.). Combination in *Acromyrmex:* Bruch, 1914: 216. [Incorrectly recorded from Ivory Coast: Santschi, 1928f:

209.] Senior synonym of *bucki* and *rufidens:* Gonçalves, 1961: 148. Current subspecies: nominal plus *cochlearis, ferrugineus, pencosensis, pruinosior.*

***lundii.*** *Myrmica lundii* Guérin-Méneville, 1838: 206 (q.m.) BRAZIL. Roger, 1863a: 201 (w.); Forel, 1885a: 356 (w.); Wheeler, G.C. 1949: 675 (l.). Combination in *Atta:* Roger, 1863a: 200; in *Atta* (*Acromyrmex*): Forel, 1885a: 356; in *Acromyrmex:* Forel, 1913l: 237. Senior synonym of *bonariensis:* Gallardo, 1916d: 331; of *risii:* Santschi, 1925a: 384; of *dubia:* Gonçalves, 1961: 150. See also: Bruch, 1921: 192. Current subspecies: nominal plus *boliviensis, carli, decolor, parallelus, pubescens.*

*medianus. Acromyrmex coronatus* st. *andicola* var. *medianus* Santschi, 1925a: 367 (w.) BRAZIL. Unavailable name; material referred to *coronatus* by Gonçalves, 1961: 135.

*mediocris. Acromyrmex diabolicus* var. *mediocris* Santschi, 1925d: 241 (w.) BRAZIL. Junior synonym of *crassispinus:* Gonçalves, 1961: 139.

*meinerti. Atta* (*Acromyrmex*) *moelleri* st. *meinerti* Forel, 1893e: 599 (q.) BRAZIL. Emery, 1905c: 45 (w.); Santschi, 1925a: 364 (w.). Combination in *Acromyrmex:* Forel, 1916: 429. Stirps of *coronatus:* Santschi, 1925a: 364. Junior synonym of *coronatus:* Gonçalves, 1961: 135.

*mesonotalis. Atta* (*Acromyrmex*) *mesonotalis* Emery, 1905c: 46, fig. 8 (w.) PERU. Combination in *Acromyrmex:* Luederwaldt, 1918: 38. Stirps of *aspersus:* Santschi, 1925d: 242. Junior synonym of *aspersus:* Gonçalves, 1961: 132.

*mesopotamicus. Acromyrmex* (*Moellerius*) *mesopotamicus* Gallardo, 1916d: 337, fig. 3 (w.) ARGENTINA. Junior synonym of *silvestrii:* Fowler, 1988: 290.

*mixtus. Acromyrmex subterraneus* var. *mixtus* Santschi, 1925a: 373 (w.) BRAZIL. Junior synonym of *subterraneus:* Gonçalves, 1961: 164.

*modesta. Atta* (*Acromyrmex*) *moelleri* r. *modesta* Forel, 1901h: 49 (s.) BRAZIL. Combination in *Acromyrmex:* Luederwaldt, 1918: 39. Junior synonym of *coronatus:* Gonçalves, 1961: 135.

*moelleri. Atta* (*Acromyrmex*) *moelleri* Forel, 1893e: 596 (s.w.q.m.) BRAZIL. Combination in *Acromyrmex:* Forel, 1916: 429. Stirps of *coronatus:* Santschi, 1925a: 363. Junior synonym of *coronatus:* Gonçalves, 1961: 135.

***#molestans.*** *Acromyrmex subterraneus* st. *molestans* Santschi, 1925a: 373 (w.) BRAZIL. Senior synonym of *eidmanni:* Gonçalves, 1961: 168.

***#multituber.*** *Acromyrmex* (*Moellerius*) *balzani* var. *multituber* Santschi, 1922b: 362 (w.) BOLIVIA. Synonym of *pampanus,* and subspecies of *balzani:* Fowler, 1988: 284. [Fowler incorrectly gives *pampanus* as senior synonym, but *multituber* has priority.]

*muticinoda. Atta* (*Acromyrmex*) *nigra* r. *muticinoda* Forel, 1901e: 336 (w.) BRAZIL. Combination in *Acromyrmex:* Luederwaldt, 1918: 39. Raised to species: Emery, 1905c: 47; Emery, 1922f: 350. Senior synonym of *depressoculis* and *homalops:* Gonçalves: 1961: 153. Junior synonym of *niger:* Gonçalves, 1967: 18.

***#myersi.*** *Acromyrmex* (*Moellerius*) *balzani* subsp. *myersi* Weber, 1937: 408 (w.) GUIANA. Subspecies of *landolti:* Kempf, 1972a: 16. See also: Fowler, 1988: 288.

*navarroi. Acromyrmex rugosus* var. *navarroi* Borgmeier, 1937b: 249 (w.) BRAZIL. Junior synonym of *rochai:* Gonçalves, 1961: 162.

***niger.*** *Oecodoma nigra* Smith, F. 1858b: 186 (q.) BRAZIL. Emery, 1905c: 48 (w.m.). Combination in *Atta:* Roger, 1863b: 35; in *Atta* (*Acromyrmex*): Forel, 1893e: 590; in *Acromyrmex:* Bruch, 1914: 216. Senior synonym of *muticinoda* (and its junior synonyms *depressoculis* and *homalops*): Gonçalves, 1967: 18.

*nigripes. Acromyrmex lundi* st. *boliviensis* var. *nigripes* Santschi, 1929d: 304 (w.q.) ARGENTINA. Unavailable name.

***#nigrosetosus.*** *Atta* (*Acromyrmex*) *nigrosetosa* Forel, 1908c: 350 (s.w.) BRAZIL. Santschi, 1922b: 361 (q.). Combination in *Acromyrmex:* Mann, 1916: 454. Stirps/subspecies of *laticeps:* Santschi, 1925a: 382; Gonçalves, 1961: 146; Kempf, 1972a: 12. Senior synonym of *dimidiata, garbei* and *pulchella:* Gonçalves, 1961: 146.

*nivalis. Acromyrmex* (*Moellerius*) *landolti* var. *nivalis* Santschi, 1922b: 362 (w.) BRAZIL. Junior synonym of *landolti:* Gonçalves, 1961: 120; of *balzani:* Fowler, 1988: 284.

***nobilis.*** *Acromyrmex* (*Acromyrmex*) *nobilis* Santschi, 1939e: 317 (w.) BRAZIL. [Also described as new by Santschi, 1939f: 164.] See also: Gonçalves, 1961: 156.

*obscurior. Acromyrmex coronatus* st. *moelleri* var. *obscurior* Santschi, 1925d: 240 (w.) BRAZIL. Unavailable name; material referred to *coronatus* by Gonçalves, 1961: 135.

*ochraceolus. Acromyrmex coronatus* st. *ochraceolus* Santschi, 1925a: 365 (w.) BRAZIL. [First available use of *Acromyrmex moelleri* r. *panamensis* var. *ochraceolus* Forel, 1922: 97; unavailable name.] Junior synonym of *coronatus:* Gonçalves, 1961: 135.

***octospinosus.*** *Formica octospinosa* Reich, 1793: 132 (w.) FRENCH GUIANA. Forel, 1893e: 590 (s.q.m.); Wheeler, G.C. 1949: 674 (l.); Wheeler, G.C. & Wheeler, J. 1986d: 496 (l.). Combination in *Atta:* Emery, 1892b: 163; in *Atta* (*Acromyrmex*): Forel, 1893e: 590; in *Acromyrmex:* Mann, 1916: 454. Senior synonym of *guentheri:* Emery, 1894c: 220; of *pallida* Crawley: Wheeler, W.M. 1937c: 70. See also: Gonçalves, 1961: 157. Current subspecies: nominal plus *cubanus, echinatior, ekchuah, inti, volcanus.*

***#ogloblini.*** *Acromyrmex subterraneus* var. *ogloblini* Santschi, 1933e: 121 (w.) ARGENTINA.

*ornatus. Acromyrmex coronatus* st. *ochraceolus* var. *ornatus* Santschi, 1925a: 365 (w.) BRAZIL. Unavailable name; material referred to *coronatus* by Gonçalves, 1961: 135.

*pallida. Oecodoma pallida* Smith, F. 1858b: 187 (s.) BRAZIL. Combination in *Atta:* Roger, 1863b: 35; in *Atta* (*Acromyrmex*): Forel, 1893e: 590. Junior synonym of *rugosus:* Forel, 1893e: 590.

*pallida. Acromyrmex octospinosa* var. *pallida* Crawley, 1921: 92 (s.w.) GUIANA. [Unresolved junior secondary homonym of *Oecodoma pallida* Smith, above.] Junior synonym of *octospinosus:* Wheeler, W.M. 1937c: 70.

*pampanus. Acromyrmex* (*Moellerius*) *balzani* subsp. *pampanus* Weber, 1938b: 200 (w.q.m.) BOLIVIA. Subspecies of *landolti:* Kempf, 1972a: 16. Synonym of *multituber:* Fowler, 1988: 284. [Fowler gives *pampanus* as senior name, but *multituber* correctly has priority.]

***#panamensis.*** *Atta* (*Acromyrmex*) *moelleri* var. *panamensis* Forel, 1899c: 35, pl. 1, figs. 16, 17 (w.q.) GUATEMALA. Combination in *Acromyrmex:* Emery, 1922f: 350. Subspecies of *moelleri:* Forel, 1908b: 40; of *coronatus:* Santschi, 1925a: 360; Kempf, 1972a: 11.

***#parallelus.*** *Acromyrmex lundi* var. *parallela* Santschi, 1916e: 387 (w.) ARGENTINA.

*parens. Acromyrmex* (*Moellerius*) *landolti* var. *parens* Santschi, 1925d: 239 (w.) BRAZIL. Junior synonym of *balzani:* Gonçalves, 1961: 124.

***#pencosensis.*** *Acromyrmex lobicornis* var. *pencosensis* Forel, 1914d: 282 (w.) ARGENTINA. Subspecies of *lobicornis:* Santschi, 1916e: 388.

***#peruvianus.*** *Acromyrmex* (*Acromyrmex*) *subterraneus* var. *peruvianus* Borgmeier, 1940: 606 (w.m.) PERU. Subspecies of *subterraneus:* Kempf, 1972a: 15.

*planorum. Acromyrmex* (*Moellerius*) *balzani* subsp. *planorum* Weber, 1937: 409 (w.) VENEZUELA. Subspecies of *landolti:* Kempf, 1972a: 16. Junior synonym of *landolti:* Fowler, 1988: 288.

***#pruinosior.*** *Acromyrmex lobicornis* var. *pruinosior* Santschi, 1916e: 388, fig. (w.) ARGENTINA.

***#pubescens.*** *Atta* (*Acromyrmex*) *pubescens* Emery, 1905c: 51, fig. 20 (w.) PARAGUAY. Combination in *Acromyrmex,* and variety of *lundii:* Bruch, 1914: 216. Subspecies of *lundii:* Gallardo, 1916d: 333; Santschi, 1916e: 386; Kempf, 1972a: 13.

*pulchella. Acromyrmex nigrosetosa* st. *pulchella* Santschi, 1925b: 18 (w.) BRAZIL. Junior synonym of *nigrosetosus:* Gonçalves, 1961: 146.

***pulvereus.*** *Acromyrmex* (*Moellerius*) *pulvereus* Santschi, 1919f: 52 (w.) ARGENTINA. See also: Fowler, 1988: 289 (who notes this as *nomen dubium*).

*purensis. Atta* (*Acromyrmex*) *subterranea* var. *purensis* Forel, 1912e: 181 (w.) BRAZIL. Combination in *Acromyrmex:* Emery, 1922f: 350. Junior synonym of *subterraneus:* Gonçalves, 1961: 164.

***#rectispinus.*** *Atta* (*Acromyrmex*) *moelleri* subsp. *rectispina* Forel, 1908b: 41 (w.) COSTA RICA. Combination in *Acromyrmex:* Emery, 1922f: 350. Stirps/subspecies of *coronatus:* Santschi, 1925a: 362; Kempf, 1972a: 12.

*risii. Atta* (*Acromyrmex*) *lundii* var. *risii* Forel, 1908c: 350. (w.) ARGENTINA. Combination in *Acromyrmex:* Forel, 1913l: 237. Junior synonym of *lundii:* Santschi, 1925a: 384.

***#rochai.*** *Atta* (*Acromyrmex*) *rugosus* var. *rochai* Forel, 1904c: 34 (w.) BRAZIL. Combination in *Acromyrmex:* Santschi, 1925a: 379. Junior synonym of *aspersus:* Emery, 1905c: 48. Revived from synonymy as variety of *rugosus:* Santschi, 1925a: 379. Subspecies of *rugosus,* and senior synonym of *navarroi:* Gonçalves, 1961: 162.

*rufescens. Acromyrmex hispidus* st. *formosus* var. *rufescens* Santschi, 1925a: 377 (w.) BRAZIL. Unavailable name; material referred to *crassispinus* by Gonçalves, 1961: 139.

*rufidens. Acromyrmex lobicornis* var. *rufidens* Santschi, 1933e: 120 (w.) BRAZIL. Junior synonym of *lobicornis:* Gonçalves, 1961: 148.

***rugosus.*** *Oecodoma rugosa* Smith, F. 1858b: 186 (q.) BRAZIL. Forel, 1904c: 33 (w.); Emery, 1905c: 109 (m.). Combination in *Atta:* Roger, 1863b: 35; in *Atta* (*Acromyrmex*): Forel, 1893e: 590; in *Acromyrmex:* Emery, 1922f: 348. Junior synonym of *coronatus:* Dalla Torre, 1893: 152; Forel, 1895b: 139. Revived from synonymy as variety of *coronatus:* Emery, 1905c: 49. Revived status as species: Santschi, 1925a: 378; Santschi, 1925d: 243. Senior synonym of *pallida* Smith: Forel, 1893e: 590; of *bigener* and *vestitus:* Gonçalves, 1961: 159. Current subspecies: nominal plus *rochai, santschii.*

*rusticus. Acromyrmex crassispinus* st. *rusticus* Santschi, 1925a: 375 (w.q.) BRAZIL. Junior synonym of *crassispinus:* Gonçalves, 1961: 139.

**#*santschii.*** *Atta* (*Acromyrmex*) *aspersa* var. *santschii* Forel, 1912e: 182 (w.) COLOMBIA. Combination in *Acromyrmex:* Emery, 1922f: 348. Variety of *rugosus:* Santschi, 1925a: 379.

*senex. Acromyrmex* (*Moellerius*) *balzani* var. *senex* Santschi, 1925b: 19 (w.) BRAZIL. Junior synonym of *landolti:* Gonçalves, 1961: 120.

***silvestrii.*** *Atta* (*Moellerius*) *silvestrii* Emery, 1905c: 42 (w.) ARGENTINA. Forel, 1911c: 292 (q.m.); Santschi 1912e: 530 (q.m.). Combination in *Acromyrmex* (*Moellerius*): Forel, 1913l: 236. Stirps of *striatus:* Santschi, 1920d: 380. Revived status as species: Emery, 1922f: 351. Senior synonym of *mesopotamicus:* Fowler, 1988: 290. Current subspecies: nominal plus *bruchi.*

***striatus.*** *Atta striata* Roger, 1863a: 202 (w.q.m.) URUGUAY. Combination in *Atta* (*Acromyrmex*): Emery, 1888c: 357; in *Atta* (*Moellerius*): Emery, 1905c: 42; in *Acromyrmex:* Bruch, 1914: 217. Senior synonym of *laeviventris:* Fowler, 1988: 290. See also: Gonçalves, 1961: 129.

***subterraneus.*** *Atta* (*Acromyrmex*) *subterranea* Forel, 1893e: 593 (s.w.q.) BRAZIL. Forel, 1901d: 301 (m.). Race/subspecies of *coronatus:* Forel, 1901d: 301; Forel, 1911c: 291; Bruch, 1914: 216. Status as species: Emery, 1905c: 48; Forel, 1912e: 181; Luederwaldt, 1918: 39; Santschi, 1925a: 372; Gonçalves, 1961: 164. Senior synonym of *mixtus* and *purensis:* Gonçalves, 1961: 164. Current subspecies: nominal plus *brunneus, molestans, oglobIini, peruvianus.*

*tucumana* Forel, 1914; see under ***TRACHYMYRMEX.***

***versicolor.*** *Atta versicolor* Pergande, 1894a: 31 (w.) MEXICO. Wheeler, W.M. 1907c: 704 (q.m.). Combination in *Atta* (*Acromyrmex*): Emery, 1895c: 330; in *Atta* (*Moellerius*): Emery, 1905c: 108; in *Acromyrmex* (*Moellerius*): Emery, 1922f: 351. See also: Fowler, 1988: 291. Current subspecies: nominal plus *chisosensis.*

*vestitus. Acromyrmex rugosus* var. *vestitus* Santschi, 1925a: 380 (w.) BRAZIL. Junior synonym of *rugosus:* Gonçalves, 1961: 159.

**#*volcanus.*** *Acromyrmex octospinosus* subsp. *volcanus* Wheeler, W.M. 1937c: 73 (s.) COSTA RICA.

***ACROPYGA*** (World tropics and subtropics)

***acutiventris.*** *Acropyga acutiventris* Roger, 1862a: 243 (w.q.) SRI LANKA. Imai, Brown *et al.* 1984: 68 (k.). Senior synonym of *flava:* Forel, 1894c: 418; of *australis:* Taylor, 1992a: 58. Current subspecies: nominal plus *bugnioni, carinata, javana, rubescens.*

***ambigua.*** *Acropyga* (*Atopodon*) *ambigua* Emery, 1922d: 107, fig. 1 (w.q.) NEW GUINEA.

***amblyops.*** *Acropyga* (*Atopodon*) *amblyops* Forel, 1915a: 34 (w.) SUMATRA.

***arnoldi.*** *Acropyga* (*Malacomyrma*) *arnoldi* Santschi, 1926b: 245 (w.q.) SOUTH AFRICA. Prins, 1982: 245 (m.). Senior synonym of *rhodesiana:* Prins, 1982: 238.

*australis. Acropyga moluccana* var. *australis* Forel, 1902h: 477 (w.) AUSTRALIA. Wheeler, G.C. & Wheeler, J. 1968: 209 (l.). Subspecies of *acutiventris:* Forel, 1911e: 286; of *moluccana:* Taylor & Brown, D.R. 1985: 108. Junior synonym of *acutiventris:* Taylor, 1992a: 58.

***baodaoensis.*** *Acropyga* (*Atopodon*) *baodaoensis* Terayama, 1985b: 284, figs. 1–13 (w.q.m.) TAIWAN.

***berwicki.*** *Acropyga* (*Rhizomyrma*) *berwicki* Wheeler, W.M. 1935f: 325, fig. 1 (w.) TRINIDAD.

***borgmeieri.*** *Acropyga* (*Rhizomyrma*) *borgmeieri* Donisthorpe, 1939c: 153 (m.) GUIANA.

***bruchi.*** *Acropyga* (*Rhizomyrma*) *bruchi* Santschi, 1925d: 308 (w.) ARGENTINA.

**#*bugnioni.*** *Acropyga acutiventris* subsp. *bugnioni* Forel, 1911i: 226 (w.q.) SRI LANKA.

***butteli.*** *Acropyga* (*Atopodon*) *butteli* Forel, 1912m: 772 (q.) WEST MALAYSIA. [Also described as new by Forel, 1913k: 101.]

**#*carinata.*** *Acropyga acutiventris* var. *carinata* Karavaiev, 1933a: 310 (w.) SUMATRA.

**#*columbica.*** *Rhizomyrma goeldii* var. *columbica* Forel, 1912i: 61 (w.) COLOMBIA. Combination in *Acropyga* (*Rhizomyrma*): Emery, 1925b: 29.

***crassicornis.*** *Acropyga crassicornis* Emery, 1900d: 698 (footnote), fig. 13 (w.) NEW GUINEA.

***decedens.*** *Brachymyrmex decedens* Mayr, 1887: 521 (w.q.m.) BRAZIL. Combination in *Rhizomyrma:* Emery, 1906c: 184; in *Acropyga* (*Rhizomyrma*): Emery, 1925b: 29.

***distinguenda.*** *Acropyga* (*Atopodon*) *distinguenda* Karavaiev, 1935a: 110, fig. 28 (w.q.m.) VIETNAM.

***dodo.*** *Pseudolasius dodo* Donisthorpe, 1946c: 782 (w.) MAURITIUS. Combination in *Acropyga:* **new combination**.

***donisthorpei.*** *Acropyga* (*Rhizomyrma*) *donisthorpei* Weber, 1944: 118, figs. 9, 20, 21 (w.) GUIANA.

***dubia.*** *Acropyga* (*Rhizomyrma*) *dubia* Karavaiev, 1933a: 311, fig. 3 (w.m.) SUMATRA.

***dubitata.*** *Rhizomyrma dubitata* Wheeler, W.M. & Mann, 1914: 47 (m.) HAITI. Combination in *Acropyga* (*Rhizomyrma*): Emery, 1925b: 29.

***emeryi.*** *Rhizomyrma emeryi* Forel, 1915c: 347 (w.) NEW GUINEA. Combination in *Acropyga* (*Rhizomyrma*): Emery, 1925b: 30.

***epedana.*** *Acropyga epedana* Snelling, 1973a: 7, fig. 1 (w.) U.S.A.

***exsanguis.*** *Rhizomyrma exsanguis* Wheeler, W.M. 1909b: 238 (w.) MEXICO. Combination in *Acropyga* (*Rhizomyrma*): Mann, 1922: 53.

*flava. Plagiolepis flava* Mayr, 1862: 699 (q.m.) NICOBAR IS. Combination in *Acropyga:* Mayr, 1862: 769. Junior synonym of *acutiventris:* Forel, 1894c: 418.

***fuhrmanni.*** *Rhizomyrma fuhrmanni* Forel, 1914e: 12 (w.q.) COLOMBIA. Weber, 1944: 97 (m.l.). Combination in *Acropyga* (*Rhizomyrma*): Emery, 1925b: 29.

***goeldii.*** *Acropyga* (*Rhizomyrma*) *goeldii* Forel, 1893g: 348 (footnote) (w.) BRAZIL. Combination in *Rhizomyrma:* Forel, 1912i: 60; in *Acropyga* (*Rhizomyrma*): Emery, 1925b: 29. Junior synonym of *decedens:* Costa Lima, 1931: 7. Revived from synonymy: Wheeler, W.M. 1935f: 322; Weber, 1944: 93. Current subspecies: nominal plus *columbica, tridentata.*

*gowdeyi. Acropyga gowdeyi* Gowdey, 1917: 187 UGANDA. *Nomen nudum.* [Note: these specimens were later described as *Pseudolasius gowdeyi;* see there.]

***guianensis.*** *Acropyga* (*Rhizomyrma*) *guianensis* Weber, 1944: 120, fig. 1 (w.) GUIANA.

***indistincta.*** *Acropyga indistincta* Crawley, 1923b: 178 (w.) AUSTRALIA.

***indosinensis.*** *Acropyga* (*Rhizomyrma*) *indosinensis* Wheeler, W.M. 1935c: 72. Replacement name for *Acropyga silvestrii* Wheeler, W.M. 1927h: 100. [Junior primary homonym of *Acropyga silvestrii* Emery, 1915g: 21.]

***inezae.*** *Acropyga* (*Atopodon*) *inezae* Forel, 1912m: 772 (w.) SUMATRA. Karavaiev, 1933a: 314 (q.m.). [Also described as new by Forel, 1913k: 102.]

**#*javana.*** *Acropyga* (*Acropyga*) *acutiventris* var. *javana* Karavaiev, 1933a: 308 (w.m.) JAVA.

***jiangxiensis.*** *Acropyga* (*Atopodon*) *jiangxiensis* Wang & Wu, 1992a: 226, figs. 1–4 (w.m.) CHINA.

***kathrynae.*** *Acropyga* (*Rhizomyrma*) *kathrynae* Weber, 1944: 114, fig. 12 (w.) TRINIDAD.

***lauta.*** *Acropyga* (*Rhizomyrma*) *lauta* Mann, 1919: 365 (w.q.) SOLOMON IS.

***major.*** *Acropyga* (*Rhizomyrma*) *major* Donisthorpe, 1949c: 755 (m.) NEW GUINEA.

***marshalli.*** *Rhizomyrma marshalli* Crawley, 1921: 93, fig. 2 (w.) BARBADOS. Combination in *Acropyga* (*Rhizomyrma*): Emery, 1925b: 30.

***meermohri.*** *Atopodon meermohri* Stärcke, 1930: 375, pl. 18, fig. oben (w.) INDONESIA (Palau I.). Combination in *Acropyga:* **new combination.**

***mesonotalis.*** *Acropyga* (*Rhizomyrma*) *mesonotalis* Weber, 1944: 111, fig. 13 (w.q.) HAITI.

***moluccana.*** *Acropyga moluccana* Mayr, 1879: 658 (w.) INDONESIA (Seram I.). Karavaiev, 1933a: 310 (q.m.); Wheeler, G.C. & Wheeler, J. 1953c: 137 (l.). Current subspecies: nominal plus *mysolensis, occipitalis, opaca, papuana.*

***myops.*** *Acropyga myops* Forel, 1910b: 59 (w.m.) AUSTRALIA. Combination in *Acropyga* (*Rhizomyrma*): Emery, 1925b: 30.

**#*mysolensis.*** *Acropyga moluccana* subsp. *mysolensis* Forel, 1911e: 285 (w.) INDONESIA (Mysol I.).

***nipponensis.*** *Acropyga* (*Atopodon*) *nipponensis* Terayama, 1985b: 287, figs. 14–18 (w.) JAPAN.

**#*occipitalis.*** *Acropyga moluccana* var. *occipitalis* Stitz, 1912: 510, fig. 12 (w.) NEW GUINEA.

***oceanica.*** *Acropyga* (*Rhizomyrma*) *oceanica* Emery, 1900c: 333, pl. 8, fig. 44 (w.q.m.) NEW GUINEA.

***oko.*** *Acropyga* (*Rhizomyrma*) *oko* Weber, 1944: 119, fig. 16 (w.) GUIANA.

**#*opaca.*** *Acropyga moluccana* var. *opaca* Stitz, 1911a: 370 (w.) NEW GUINEA. Donisthorpe, 1948g: 139 (m.).

***ACROPYGA** (continued)*

***pachycera.*** *Rhizomyrma pachycera* Emery, 1906c: 182 (w.q.) PARAGUAY. Combination in *Acropyga* (*Rhizomyrma*): Emery, 1925b: 30.

***paleartica.*** *Acropyga* (*Rhizomyrma*) *paleartica* Menozzi, 1936d: 298, fig. 17 (w.) GREECE.

***pallida.*** *Bothriomyrmex pallidus* Donisthorpe, 1938d: 598 (w.) NEW GUINEA. Combination in *Acropyga:* Shattuck, 1992c: 60.

***paludis.*** *Acropyga* (*Rhizomyrma*) *paludis* Weber, 1944: 121, fig. 15 (w.m.) GUIANA.

***panamensis.*** *Acropyga* (*Rhizomyrma*) *panamensis* Weber, 1944: 113, fig. 24 (w.) PANAMA.

#***papuana.*** *Acropyga* (*Acropyga*) *moluccana* subsp. *papuana* Mann, 1919: 364 (w.) SOLOMON IS. Wheeler, G.C. & Wheeler, J. 1953c: 136 (l.).

***paramaribensis.*** *Acropyga paramaribensis* Borgmeier, 1933c: 263 (w.q.) SURINAM. Borgmeier, 1934: 109 (m.); Bünzli, 1935: 519 (l.). Combination in *Acropyga* (*Rhizomyrma*): Kempf, 1972a: 17.

***parvidens.*** *Rhizomyrma parvidens* Wheeler, W.M. & Mann, 1914: 46 (w.) HAITI. Combination in *Acropyga* (*Rhizomyrma*): Emery, 1925b: 30.

***pickeli.*** *Acropyga* (*Rhizomyrma*) *pickeli* Borgmeier, 1927a: 287, figs. 1, 2 (w.q.) BRAZIL. Weber, 1944: 102 (m.). Junior synonym of *decedens:* Costa Lima, 1931: 7. Revived from synonymy: Borgmeier, 1932a: 238. See also: Wheeler, W.M. 1935f: 322; Weber, 1944: 102.

***quadriceps.*** *Acropyga* (*Rhizomyrma*) *quadriceps* Weber, 1944: 113, fig. 5 (w.) TRINIDAD.

*rhodesiana. Acropyga* (*Malacomyrma*) *rhodesiana* Santschi, 1928f: 211 (w.) ZIMBABWE. Junior synonym of *arnoldi:* Prins, 1982: 238.

***robae.*** *Acropyga* (*Rhizomyrma*) *robae* Donisthorpe, 1936b: 108, figs. 1–3 (w.q.m.) COLOMBIA.

*rubens. Acropyga acutiventris* subsp. *rubens* Forel, 1911e: 286 (w) no locality stated. [This name most probably in error for *rubescens.*] Junior synonym of *rubescens:* Emery, 1925b: 28.

#***rubescens.*** *Acropyga acutiventris* var. *rubescens* Forel, 1894c: 418 (w.q.m.) INDIA. Senior synonym of *rubens:* Emery, 1925b: 28.

***rutgersi.*** *Acropyga* (*Rhizomyrma*) *rutgersi* Bünzli, 1935: 464, figs. 8–11, 12, 14–16, 18, 20, 24 (w.q.m.) SURINAM.

***sauteri.*** *Acropyga* (*Rhizomyrma*) *sauteri* Forel, 1912a: 72 (w.) TAIWAN. Wheeler, W.M. 1928c: 31 (q.); Santschi, 1928b: 36 (q.m.).

***silvestrii.*** *Acropyga silvestrii* Emery, 1915g: 21, fig. 11 (w.) ETHIOPIA. Combination in *Acropyga* (*Malacomyrma*): Emery, 1922d: 109.

*silvestrii. Acropyga* (*Rhizomyrma*) *silvestrii* Wheeler, W.M. 1927h: 100, fig. 7 (w.) VIETNAM. [Junior primary homonym of *silvestrii* Emery, above.] Replacement name: *indosinensis* Wheeler, W.M. 1935c: 72.

***smithii.*** *Acropyga* (*Rhizomyrma*) *smithii* Forel, 1893g: 349 (q.) ANTILLES IS.

***termitobia.*** *Acropyga* (*Atopodon*) *termitobia* Forel, 1912m: 773 (q.) WEST MALAYSIA. [Also described as new by Forel, 1913k: 101.]

#***tridentata.*** *Rhizomyrma goeldii* var. *tridentata* Forel, 1912i: 61 (q.m.) PANAMA. Combination in *Acropyga* (*Rhizomyrma*): Emery, 1925b: 29.

***trinitatis.*** *Acropyga* (*Rhizomyrma*) *trinitatis* Weber, 1944: 117, figs. 7, 8, 17–19 (w.q.) TRINIDAD.

***undecema.*** *Pseudolasius undecemus* Donisthorpe, 1949c: 756 (w.) NEW GUINEA. Combination in *Acropyga:* **new combination**.

***urichi.*** *Acropyga* (*Rhizomyrma*) *urichi* Weber, 1944: 115, figs. 2,3 (w.q.m.) TRINIDAD.

***wheeleri.*** *Acropyga* (*Rhizomyrma*) *wheeleri* Mann, 1922: 52 (w.q.) HONDURAS.

---

**EXCURSUS**

**Species-rank taxa of *ACROPYGA*, grouped by subgenus.**

#***ACROPYGA:*** *acutiventris, crassicornis, indistincta, moluccana, undecema.*

#***ATOPODON:*** *ambigua, amblyops, baodaoensis, butteli, distinguenda, inezae, jiangxiensis, meermohri, nipponensis, termitobia.*

#***MALACOMYRMA:*** *arnoldi, dodo, silvestrii.*

#***RHIZOMYRMA:*** *berwicki, borgmeieri, bruchi, decedens, donisthorpei, dubia, dubitata, emeryi, epedana, exsanguis, fuhrmanni, goeldii, guianensis, indosinensis, kathrynae, lauta, major, marshalli, mesonotalis, myops, oceanica, oko, pachycera, paleartica, pallida, paludis, panamensis, paramaribensis, parvidens, pickeli, quadriceps, robae, rutgersi, sauteri, smithii, trinitatis, urichi, wheeleri.*

---

**ACROSTIGMA* Emery: see under ***PODOMYRMA.***

*ACROSTIGMA* Forel: see under ***STIGMACROS.***

***ADELOMYRMEX*** (Indo-Australian, Neotropical)

***biroi.*** *Adelomyrmex biroi* Emery, 1897c: 590, pl. 15, figs. 35–37 (w.q.) NEW GUINEA.

***hirsutus.*** *Adelomyrmex* (*Arctomyrmex*) *hirsutus* Mann, 1921: 458, fig. 21 (w.). FIJI IS.

***myops.*** *Apsychomyrmex myops* Wheeler, W.M. 1910a: 261, fig. 2 (w.) GUATEMALA. Wheeler, G.C. & Wheeler, J. 1955b: 29 (l.). Combination in *Adelomyrmex:* Kempf, 1972a: 18.

***samoanus.*** *Adelomyrmex* (*Arctomyrmex*) *samoanus* Wilson & Taylor, 1967: 77, fig. 62 (w.) SAMOA.

***silvestrii.*** *Apsychomyrmex silvestrii* Menozzi, 1931b: 270, fig.6 (w.) COSTA RICA. Combination in *Adelomyrmex:* Kempf, 1972a: 18.

***tristani.*** *Apsychomyrmex tristani* Menozzi, 1931b: 269, fig. 6 (w.) COSTA RICA. Combination in *Adelomyrmex:* Kempf, 1972a: 18.

***ADETOMYRMA*** (Madagascar)

***venatrix.*** *Adetomyrma venatrix* Ward, 1994: 161, figs. 1–7, 12, 13, 18, 24, 30, 36, 41 (w.) MADAGASCAR.

*ADFORMICA:* see under ***FORMICA.***

***ADLERZIA*** (Australia)

***froggatti.*** *Monomorium* (*Adlerzia*) *froggatti* Forel, 1902h: 445 (w) AUSTRALIA. Wheeler, G.C. & Wheeler, J. 1977: 583 (l.). Combination in *Adlerzia:* Brown, 1952j: 176. Senior synonym of *katerinae* and *silvestrii:* Brown, 1952a: 110.

*katerinae. Stenothorax katerinae* McAreavey, 1949: 3, figs. 7–16 (s.w.) AUSTRALIA. Junior synonym of *froggatti:* Brown, 1952a: 110.

*silvestrii. Machomyrma silvestrii* Emery, 1914b: 182, fig. 2 (s.) AUSTRALIA. Viehmeyer, 1924b: 311 (w.). Junior synonym of *froggatti:* Brown, 1952a: 110.

***AENICTOGITON*** (west and central Africa)

***attenuatus.*** *Aenictogiton elongatus* var. *attenuatus* Santschi, 1919c: 247, fig. 2 (m.) ZAIRE. Raised to species: Santschi, 1924b: 199.

***bequaerti.*** *Aenictogiton bequaerti* Forel, 1913b: 314 (m.) ZAIRE.

***elongatus.*** *Aenictogiton elongatus* Santschi, 1919c: 246, fig. 2 (m.) ZAIRE.

***emeryi.*** *Aenictogiton emeryi* Forel, 1913b: 315 (m.) ZAIRE.

***fossiceps.*** *Aenictogiton fossiceps* Emery, 1901d: 49, figs. 3,4 (m.) ZAIRE.

***schoutedeni.*** *Aenictogiton schoutedeni* Santschi, 1924b: 198, fig. 2 (m.) ZAIRE.

***sulcatus.*** *Aenictogiton sulcatus* Santschi, 1919c: 247, fig. 2 (m.) ZAIRE.

***AENICTUS*** (Old World tropics and temperates)

***abeillei.*** *Alaopone abeillei* André, 1886: 855 (q., not w.; see Emery, 1901d: 43) ALGERIA. Combination in *Dorylus* (*Shuckardia*): Emery, 1895j: 740; in *Aenictus:* Emery, 1901d: 43. Provisional junior synonym of *atriceps:* Emery, 1891b: 2. Revived from synonymy: Emery, 1895j: 740.

*aitkenii. Aenictus aitkenii* Forel, 1901a: 475 (w.) INDIA. Junior synonym of *aratus:* Wilson, 1964a: 446.

***alluaudi.*** *Aenictus bottegoi* var. *alluaudi* Santschi, 1910g: 754, fig. 10 (m.) KENYA. Raised to species: Santschi, 1924b: 203. Current subspecies: nominal plus *falcifer.*

***alticola.*** *Aenictus* (*Typhlatta*) *alticola* Wheeler, W.M. & Chapman, in Wheeler, W.M. 1930g: 205, fig. 5 (w.) PHILIPPINES. See also: Wilson, 1964a: 445.

***ambiguus.*** *Aenictus ambiguus* Shuckard, 1840b: 268 (m.) INDIA. See also: Wilson, 1964a: 444.

***anceps.*** *Aenictus bottegoi* subsp. *anceps* Forel, 1910c: 249 (m.) ETHIOPIA. Raised to species: Santschi, 1920d: 374.

#***andrieui.*** *Aenictus furculatus* st. *andrieui* Santschi, 1930a: 56, figs. 15–17 (m.) SUDAN.

***annae.*** *Aenictus annae* Forel, 1911b: 194 (m.) JAVA.

***aratus.*** *Aenictus aratus* Forel, 1900b: 74 (w.) AUSTRALIA. Wheeler, G.C. & Wheeler, J. 1984: 265 (l.). Senior synonym of *aitkenii, asiatica, levior* and *nesiotis,* and material of the unavailable name *fraterculus* referred here: Wilson, 1964a: 446.

***artipus.*** *Aenictus artipus* Wilson, 1964a: 449, fig. 60 (w.) THAILAND.

***arya.*** *Aenictus arya* Forel, 1901a: 472 (m.) INDIA.

***asantei.*** *Aenictus asantei* Campione, Novak & Gotwald, 1983: 873, 7 figs. (w.q.) GHANA.

*asiatica. Aenictus aratus* var. *asiatica* Forel, 1911h: 453 (w.) SRI LANKA. Junior synonym of *aratus:* Wilson, 1964a: 446.

#***ashaverus.*** *Aenictus gibbosus* r. *ashaverus* Forel, 1913k: 22 (m.) SUMATRA.

***asperivalvus.*** *Aenictus asperivalvus* Santschi, 1919b: 233 (m.) IVORY COAST.

#***atripennis.*** *Aenictus clavatus* r. *atripennis* Forel, 1913k: 24 (m.) SUMATRA.

***bakeri.*** *Aenictus bakeri* Menozzi, 1925c: 439, fig. 1 (m.) PHILIPPINES.

*batesi. Aenictus batesi* Forel, 1911e: 255 (w.) NIGERIA. Junior synonym of *decolor:* Gotwald & Leroux, 1980: 600.

**bayoni.** *Aenictus bayoni* Menozzi, 1933a: 94, fig. 1 (m.) UGANDA.

*bengalensis. Typhlatta bengalensis* Mayr, 1879: 669 (w.) INDIA. Combination in *Aenictus:* Dalla Torre, 1893: 7. Junior synonym of *pachycerus:* Bingham, 1903: 20.

*bidentatus. Aenictus bidentatus* Donisthorpe, 1942e: 701 (w.) GHANA. Junior synonym of *decolor:* Gotwald & Leroux, 1980: 600.

**binghami.** *Aenictus binghaniri* Forel, 1900b: 76 (w.) BURMA. *Aenictus binghami:* Forel, 1901e: 382, justified emendation of spelling. Senior synonym of *gatesi:* Wilson, 1964a: 450.

**biroi.** *Aenictus biroi* Forel, 1907a: 10 (w.) SRI LANKA. See also: Wilson, 1964a: 451.

*boelianensis. Aenictus martini* var. *boelianensis* Forel, 1913k: 20 (w.) SUMATRA. Junior synonym of *gracilis:* Wilson, 1964a: 463.

**bottegoi.** *Aenictus bottegoi* Emery, 1899a: 499 (m.) ETHIOPIA. Current subspecies: nominal plus *noctivagus.*

**brazzai.** *Aenictus eugenii* var. *brazzai* Santschi, 1910c: 355 (w.) CONGO. Raised to species: Santschi, 1924b: 204.

*breviceps. Aenictus fergusoni* var. *breviceps* Forel, 1912d: 105 (w.) JAVA. Junior synonym of *laeviceps:* Wilson, 1964a: 467.

**brevicornis.** *Typhlatta brevicornis* Mayr, 1879: 669 (w.) INDIA. Imai *et al.*, 1984: 8 (k.). [*Typhlatta brevicornis* Smith, F. 1873: ix. *Nomen nudum,* attributed to Mayr.] Combination in *Aenictus:* Dalla Torre, 1893: 7. See also: Wilson, 1964a: 451.

#**brunneus.** *Aenictus soudanicus* var. *brunnea* Forel, 1913b: 313 (m.) ZAIRE.

**buttelreepeni.** *Aenictus buttelreepeni* Forel, 1913k: 20, fig. F (m.) SUMATRA.

**buttgenbachi.** *Aenictus buttgenbachi* Forel, 1913b: 313 (m.) ZAIRE.

**camposi.** *Aenictus camposi* Wheeler, W.M. & Chapman, 1925: 48, pl. 1, figs. 3, 4 (w.) PHILIPPINES. See also: Wilson, 1964a: 452.

#**caroli.** *Aenictus eugenii* subsp. *caroli* Forel, 1910c: 248 (w.) ETHIOPIA.

**certus.** *Aenictus certus* Westwood, 1842: 79 (m.) no locality stated (may be India, see Bingham, 1903: 10).

**ceylonicus.** *Typhlatta ceylonica* Mayr, 1866a: 505 (w.) SRI LANKA. Wheeler, G.C. & Wheeler, J. 1984: 265 (l.). Combination in *Aenictus:* Dalla Torre, 1893: 7. Senior synonym of *formosensis, latro, orientalis, papuanus, similis* and *turneri* (and its junior synonyms *deuqueti* and *exiguus*): Wilson, 1964a: 452.

**chapmani.** *Aenictus chapmani* Wilson, 1964a: 456, figs. 45, 46 (w.) NEW GUINEA.

#**chevalieri.** *Aenictus humeralis* var. *chevalieri* Santschi, 1910g: 754 (m.) SENEGAL.

**clavatus.** *Aenictus clavatus* Forel, 1901a: 471 (m.) INDIA. Current subspecies: nominal plus *atripennis, kanariensis, sundaicus.*

**clavitibia.** *Aenictus clavitibia* Forel, 1901a: 472 (m.) INDIA. Current subspecies: nominal plus *facetus.*

**congolensis.** *Aenictus rixator* var. *congolensis* Santschi, 1911g: 207 (w.) CONGO. Santschi, 1917a: 277 (q.). Raised to species: Santschi, 1917a: 277.

*continuus. Aenictus bengalensis* var. *continuus* Forel, 1901a: 476 (w.) SRI LANKA. Junior synonym of *pachycerus:* Wilson, 1964a: 471.

**cornutus.** *Aenictus cornutus* Forel, 1900b: 75 (w.) BORNEO. See also: Wilson, 1964a: 457.

**crucifer.** *Aenictus crucifer* Santschi, 1914b: 68, fig 5 (m.) KENYA. Current subspecies: nominal plus *tuberculatus.*

**currax.** *Aenictus currax* Emery, 1900c: 310, pl. 8, fig. 1 (w.) NEW GUINEA. See also: Wilson, 1964a: 459.

**decolor.** *Typhlatta decolor* Mayr, 1879: 668 (diagnosis in key) (w.) EAST AFRICA. Gotwald & Leroux, 1980: 602 (q.). Combination in *Aenictus:* Dalla Torre, 1893: 7. Senior synonym of *batesi* and *bidentatus:* Gotwald & Leroux, 1980: 600.

**dentatus.** *Aenictus aitkeni* var. *dentata* Forel, 1911d: 383 (w.) WEST MALAYSIA. Raised to species: Wilson, 1964a: 460.

*deuqueti. Aenictus deuqueti* Crawley, 1923b: 177 (w.) AUSTRALIA. Junior synonym of *turneri:* Brown, 1952b: 123.

**dlusskyi.** *Aenictus dlusskyi* Arnol'di, 1968: 1800, fig. 1 (w.q.) ARMENIA.

**doryloides.** *Aenictus doryloides* Wilson, 1964a: 460, figs. 54, 55 (w.) INDIA.

*elongatus. Eciton (Aenictus) fergusoni* subsp. *elongatus* Karavaiev, 1926d: 424, fig. 3 (w.) SRI LANKA. Junior synonym of *gracilis:* Wilson, 1964a: 463.

**eugenii.** *Aenictus eugenii* Emery, 1895h: 17, pl. 2, figs. 1–4 (w.) SOUTH AFRICA. Gotwald & Cunningham-Van Someren, 1976: 183 (q.). Senior synonym of *kenyensis:* Gotwald & Cunningham-Van Someren, 1976: 183. Current subspecies: nominal plus *caroli, henrii.*

*exiguus. Aenictus exiguus* Clark, 1934b: 21, pl. 2, fig. 1 (w.) AUSTRALIA. Junior synonym of *ceylonicus:* Wilson, 1964a: 452.

**exilis.** *Aenictus exilis* Wilson, 1964a: 462, fig. 33 (w.) NEW GUINEA.

#**facetus.** *Aenictus clavitibia* subsp. *facetus* Forel, 1911b: 196 (m.) JAVA.

#**falcifer.** *Aenictus alluaudi* var. *falcifer* Santschi, 1924b: 203, fig. 6 (m.) ZAIRE.

**feae.** *Aenictus feae* Emery, 1889b: 486, pl. 10, figs. 4–9 (m.) BURMA.

**fergusoni.** *Aenictus fergusoni* Forel, 1901a: 473 (w.) INDIA. Senior synonym of *hodgsoni, karawaiewi, montanus* and *piltzi:* Wilson, 1964a: 462.

**foreli.** *Aenictus foreli* Santschi, 1919b: 234 (m.) IVORY COAST. [Also described as new by Santschi, 1920d: 369.]

*formosensis. Aenictus ceylonicus* var. *formosensis* Forel, 1913f: 188 (w.) TAIWAN. Junior synonym of *ceylonicus:* Wilson, 1964a: 452.

*fraterculus. Aenictus (Aenictus) aratus* subsp. *nesiotis* var. *fraterculus* Wheeler, W.M. & Chapman, in Wheeler, W.M. 1930g: 209 (w.) PHILIPPINES. Unavailable name; material referred to *aratus* by Wilson, 1964a: 446.

#**fumatus.** *Aenictus latiscapus* var. *fumatus* Wheeler, W.M. 1927d: 1 (m.) CHINA.

**furculatus.** *Aenictus forculatus* Santschi, 1919b: 233 (m.) SENEGAL. *Aenictus furculatus:* Santschi, 1919h: 91, justified emendation of spelling. [Also described as new by Santschi, 1920d: 372. Current subspecies: nominal plus *andrieui.*

**furibundus.** *Aenictus furibundus* Arnold, 1959: 335, fig. 20 (w.) ZIMBABWE.

**fuscipennis.** *Aenictus fuscipennis* Forel, 1913k: 24 (m.) SUMATRA.

**fuscovarius.** *Aenictus fuscovarius* Gerstäcker, 1859: 262 (m.) MOZAMBIQUE. [Also described as new by Gerstäcker, 1862: 501.] Current subspecies: nominal plus *laetior, magrettii, sagittarius.*

*gatesi. Aenictus binghami* var. *gatesi* Wheeler, W.M. 1927b: 42 (w.) BURMA. Junior synonym of *binghami:* Wilson, 1964a: 450.

**gibbosus.** *Aenictus gibbosus* Dalla Torre, 1893: 7 (m.) SUMATRA. Current subspecies: nominal plus *ashaverus.*

**gleadowii.** *Aenictus gleadowii* Forel, 1901a: 469 (m.) INDIA.

**gracilis.** *Aenictus gracilis* Emery, 1893e: 187, pl. 8, fig. 1 (w.) BORNEO. Wheeler, G.C. & Wheeler, J. 1990a: 460 (l.). Senior synonym of *boelianensis, elongatus* and *martini:* Wilson, 1964a: 463.

**grandis.** *Aenictus grandis* Bingham, 1903: 11, fig. 9 (m.) BURMA.

*greeni. Aenictus greeni* Bingham, 1903: 11 (m.) SRI LANKA. Junior synonym of *porizonoides:* Donisthorpe, 1932b: 576.

#**guineensis.** *Aenictus rotundatus* st. *guineensis* Santschi, 1924b: 204, fig. 7 (w.) GUINEA.

**hamifer.** *Aenictus hamifer* Emery, 1896i: 153, fig. (m.) KENYA. Current subspecies: nominal plus *spinosior.*

#**henrii.** *Aenictus eugenii* var. *henrii* Santschi, 1924b: 204 (w.) ZAIRE.

**hilli.** *Aenictus hilli* Clark, 1928a: 38, pl. 1, figs. 32–36 (m.) AUSTRALIA.

*hodgsoni. Aenictus fergusoni* var. *hodgsoni* Forel, 1901a: 474 (w.) BURMA. Junior synonym of *fergusoni:* Wilson, 1964a: 462.

**hottai.** *Aenictus hottai* Terayama & Yamane, 1989: 598, figs. 1, 2 (w.) SUMATRA.

**humeralis.** *Aenictus humeralis* Santschi, 1910g: 752, fig. 9 (m.) SUDAN. Current subspecies: nominal plus *chevalieri, viridans.*

**huonicus.** *Aenictus huonicus* Wilson, 1964a: 465, figs. 24, 25 (w.) NEW GUINEA.

**icarus.** *Aenictus icarus* Forel, 1911b: 197 (m.) JAVA. Current subspecies: nominal plus *incautus.*

**idoneus.** *Aenictus idoneus* Menozzi, 1928b: 1, figs. 1, 2 (m.) SUMATRA.

#**impressus.** *Aenictus bengalensis* r. *impressus* Forel, in Karavaiev, 1927e: 7 (w.) no locality given (probably India). Subspecies of *pachycerus:* Karavaiev, 1927e: 6.

#**incautus.** *Aenictus icarus* var. *incauta* Forel, 1911b: 198 (m.) JAVA.

**inconspicuus.** *Aenictus inconspicuus* Westwood, 1843: 85 (m.) SOUTH AFRICA. [Also described as new by Westwood, 1847a: 238.]

**jacobsoni.** *Aenictus jacobsoni* Forel, 1909d: 222 (m.) JAVA.

**javanus.** *Aenictus javanus* Emery, 1896f: 245, fig. (m.) (attributed to Ritsema) JAVA. Forel, 1909d: 222 (w.). See also: Wilson, 1964a: 467.

#**kanariensis.** *Aenictus clavatus* var. *kanariensis* Forel, 1901a: 467 (diagnosis in key) (m.) INDIA.

*karawaiewi. Aenictus fergusoni* var. *karawaiewi* Wheeler, W.M. & Chapman, in Wheeler, W.M. 1930g: 199 (diagnosis in key) (w.) no locality given. Junior synonym of *fergusoni:* Wilson, 1964a: 462.

*AENICTUS (continued)*

*kenyensis*. *Aenictus eugeniae* var. *kenyensis* Santschi, 1933b: 100 (w.) KENYA. Junior synonym of *eugenii:* Gotwald & Cunningham-Van Someren, 1976: 183.

**#*laetior*.** *Aenictus fuscovarius* var. *laetior* Forel, 1910c: 249 (m.) ETHIOPIA.

***laeviceps*.** *Typhlatta laeviceps* Smith, F. 1857a: 79 (w.) BORNEO. Wheeler, W.M. 1930g: 200 (q.); Wheeler, G.C. & Wheeler, J. 1984: 265 (l.). Combination in *Aenictus:* Forel, 1890b: ciii. Senior synonym of *breviceps, smythiesii* and *sundaica:* Wilson, 1964a: 467.

***latifemoratus*.** *Aenictus latifemoratus* Terayama & Yamane, 1989: 600, figs. 3–6 (w.) SUMATRA.

***latiscapus*.** *Aenictus latiscapus* Forel, 1901a: 471 (m.) INDIA. Current subspecies: nominal plus *fumatus, sauteri*.

*latro*. *Aenictus ceylonicus* var. *latro* Forel, 1901a: 477 (w.) INDIA. Junior synonym of *ceylonicus:* Wilson, 1964a: 452.

***leliepvrei*.** *Aenictus leliepvrei* Bernard, 1953a: 134, fig. 1 (m.) ALGERIA.

*levior*. *Eciton* (*Aenictus*) *impressus* var. *levior* Karavaiev, 1926d: 425 (w.) INDONESIA (Buru I.). Junior synonym of *aratus:* Wilson, 1964a: 446.

***lifuiae*.** *Aenictus lifuiae* Terayama, 1984: 13, figs. 1–13 (w.m.) TAIWAN.

***longi*.** *Aenictus longi* Forel, 1901a: 470 (m.) INDIA. Current subspecies: nominal plus *taivanae*.

***luteus*.** *Aenictus luteus* Emery, 1892d: 568, pl. 15, figs. 11, 12 (m.) SIERRA LEONE. Current subspecies: nominal plus *moestus*.

***luzoni*.** *Aenictus luzoni* Wheeler, W.M. & Chapman, 1925: 48, pl. 1, figs. 1, 2 (w.) PHILIPPINES. See also: Wilson, 1964a: 470.

**#*magrettii*.** *Aenictus magrettii* Emery, 1892d: 569, pl. 15, figs. 13, 14 (m.) SUDAN. Subspecies of *fuscovarius:* Emery, 1895j: 747.

***mariae*.** *Aenictus mariae* Emery, 1895h: 18, pl. 2, figs. 5–7 (w.) SOUTH AFRICA. Current subspecies: nominal plus *natalensis*.

***maroccanus*.** *Aenictus maroccanus* Santschi, 1926e: 232 (w.) MOROCCO.

*martini*. *Aenictus martini* Forel, 1901a: 473 (w.) WEST MALAYSIA. Wheeler, W.M. 1930g: 203 (q.); Chapman, 1963: 251 (m.); Wheeler, G.C. 1943: 326 (l.). Junior synonym of *gracilis:* Wilson, 1964a: 463.

***mauritanicus*.** *Aenictus mauritanicus* Santschi, 1910g: 756, fig. 11 (m.) "Côte atlantique du Sahara".

**#*maxillosus*.** *Aenictus sumatrensis* var. *maxillosa* Forel, 1913k: 24 (m.) SUMATRA.

***mentu*.** *Aenictus mentu* Weber, 1942a: 40, fig. 2 (w.) SUDAN.

**#*merwei*.** *Aenictus rotundatus* var. *merwei* Santschi, 1932a: 382 (w.) SOUTH AFRICA.

***minutulus*.** *Aenictus minutulus* Terayama & Yamane, 1989: 601, figs. 7–10 (w.) SUMATRA.

***mocsaryi*.** *Aenictus mocsaryi* Emery, 1901h: 152, fig. (m.) NEW GUINEA.

***moebii*.** *Aenictus moebii* Emery, 1895j: 747, fig. NN (m.) TOGO. Current subspecies: nominal plus *sankisianus*.

**#*moestus*.** *Aenictus luteus* var. *moestus* Santschi, 1930a: 56, figs. 18, 19 (m.) SUDAN.

*montanus*. *Aenictus fergusoni* var. *montanus* Forel, 1901a: 474 (w.) INDIA. Junior synonym of *fergusoni:* Wilson, 1964a: 462.

***mutatus*.** *Aenictus moebii* st. *mutatus* Santschi, 1913c: 304, fig. 2 (m.) IVORY COAST. Raised to species: Santschi, 1920d: 370. Current subspecies: nominal plus *pudicus*.

**#*natalensis*.** *Aenictus mariae* var. *natalensis* Forel, in Emery, 1901d: 49 (w.) SOUTH AFRICA.

*nesiotis*. *Aenictus* (*Aenictus*) *aratus* subsp. *nesiotis* Wheeler, W.M. & Chapman, in Wheeler, W.M. 1930g: 208, fig. 7 (w.) PHILIPPINES. Junior synonym of *aratus:* Wilson, 1964a: 446.

***nganduensis*.** *Aenictus nganduensis* Wilson, 1964a: 470, figs. 47, 48 (w.) NEW GUINEA.

**#*noctivagus*.** *Aenictus bottegoi* st. *noctivagus* Santschi, 1913c: 303, fig. 1 (m.) ETHIOPIA.

***obscurus*.** *Aenictus obscurus* Smith, F. 1865: 79 (m.) NEW GUINEA.

*orientalis*. *Eciton* (*Aenictus*) *ceylonicus* subsp. *orientalis* Karavaiev, 1926d: 423, fig. 2 (w.) INDONESIA (Aru I.). Junior synonym of *ceylonicus:* Wilson, 1964a: 452.

***pachycerus*.** *Eciton pachycerus* Smith, F. 1858b: 153 (w.) "South America ?" (in error; type-locality INDIA, see Bingham, 1903: 21). Combination in *Aenictus:* Dalla Torre, 1893: 8. Senior synonym of *bengalensis:* Bingham, 1903: 20; of *continuus:* Wilson, 1964a: 471. Current subspecies: nominal plus *impressus*.

*papuanus*. *Aenictus papuanus* Donisthorpe, 1941c: 129 (w.) NEW GUINEA. Junior synonym of *ceylonicus:* Wilson, 1964a: 452.

***peguensis*.** *Aenictus peguensis* Emery, 1895k: 452 (w.) BURMA. Subspecies of *ceylonicus:* Forel, 1901a: 477. Revived status as species: Bingham, 1903: 22. See also: Wilson, 1964a: 472.

***pharoa*.** *Aenictus pharao* Santschi, 1924b: 201, fig. 4 (m.) SUDAN.

***philiporum*.** *Aenictus philiporum* Wilson, 1964a: 473, fig. 74 (w.) AUSTRALIA.

***philippinensis*.** *Aenictus* (*Aenictus*) *philippinensis* Chapman, 1963: 247, fig. 1 (w.) PHILIPPINES. See also: Wilson, 1964a: 473.

***piercei*.** *Aenictus* (*Aenictus*) *piercei* Wheeler, W.M. & Chapman, in Wheeler, W.M. 1930g: 209, fig. 7 (w.) PHILIPPINES. See also: Wilson, 1964a: 474.

*piltzi*. *Aenictus fergusoni* var. *piltzi* Forel, 1901a: 474 (w.) INDIA. Junior synonym of *fergusoni:* Wilson, 1964a: 462.

***porizonoides*.** *Aenictus porizonoides* Walker, 1860: 306 (m.) SRI LANKA. Senior synonym of *greeni:* Donisthorpe, 1932b: 576.

***powersi*.** *Aenictus* (*Aenictus*) *powersi* Wheeler, W.M. & Chapman, in Wheeler, W.M. 1930g: 210, fig. 6 (w.) PHILIPPINES. See also: Wilson, 1964a: 475.

***pubescens*.** *Aenictus pubescens* Smith, F. 1859b: 10, pl. 2, fig. 17 (m.) INDIA.

**#*pudicus*.** *Aenictus mutatus* st. *pudicus* Santschi, 1919b: 234 (m.) IVORY COAST.

***punctiventris*.** *Aenictus punctiventris* Emery, 1901d: 47 (m.) BORNEO. Current subspecies: nominal plus *scutellaris*.

***punensis*.** *Aenictus punensis* Forel, 1901a: 476 (w.) INDIA. See also: Wilson, 1964a: 476.

***rabori*.** *Aenictus* (*Aenictus*) *rabori* Chapman, 1963: 249, fig. 2 (w.) PHILIPPINES. See also: Wilson, 1964a: 476.

***raptor*.** *Aenictus raptor* Forel, 1913b: 314 (m.) ZAIRE.

***reyesi*.** *Aenictus* (*Aenictus*) *reyesi* Chapman, 1963: 250, fig. 3 (w.) PHILIPPINES. See also: Wilson, 1964a: 477.

***rhodiensis*.** *Aenictus rhodiensis* Menozzi, 1936d: 266, fig. I,1–4 (w.) GREECE.

***rixator*.** *Aenictus rixator* Forel, in Emery, 1901d: 48 (w.) SOUTH AFRICA.

***rotundatus*.** *Aenictus rotundatus* Mayr, 1901b: 1 (w.) SOUTH AFRICA. Current subspecies: nominal plus *guineensis, merwei*.

***rougieri*.** *Aenictus rougieri* André, 1893a: cxci (m.) TUNISIA.

***sagei*.** *Aenictus wroughtonii* var. *sagei* Forel, 1901a: 469 (w.) INDIA. Raised to species: Wilson, 1964a: 477.

**#*sagittarius*.** *Aenictus fuscovarius* st. *sagittarius* Santschi, 1938a: 30, fig. 2 (m.) EGYPT.

**#*sankisianus*.** *Aenictus moebii* r. *sankisianus* Forel, 1913b: 312 (m.) ZAIRE.

*satoi*. *Aenictus latiscapus* st. *sauteri* var. *satoi* Santschi, 1937h: 367 (m.) TAIWAN. Unavailable name.

**#*sauteri*.** *Aenictus latiscapus* r. *sauteri* Forel, 1913f: 188 (m.) TAIWAN.

***schneirlai*.** *Aenictus schneirlai* Wilson, 1964a: 479, figs. 49, 50 (w.) NEW GUINEA.

**#*scutellaris*.** *Aenictus punctiventris* var. *scutellaris* Forel, 1912n: 53 (m.) SUMATRA.

***shuckardi*.** *Aenictus shuckardi* Forel, 1901a: 471 (m.) INDIA.

***silvestrii*.** *Aenictus* (*Paraenictus*) *silvestrii* Wheeler, W.M. 1929g: 28, fig. 1 (w.) WEST MALAYSIA. See also: Wilson, 1964a: 479.

*similis*. *Aenictus similis* Donisthorpe, 1948g: 131 (w.) NEW GUINEA. Junior synonym of *ceylonicus:* Wilson, 1964a: 452.

*smythiesii*. *Aenictus laeviceps* var. *smythiesii* Forel, 1901a: 465 (diagnosis in key) (w.) INDIA. Junior synonym of *laeviceps:* Wilson, 1964a: 467.

***soudanicus*.** *Aenictus soudanicus* Santschi, 1910g: 758, fig. 12 (m.) SUDAN. Current subspecies: nominal plus *brunneus*.

***spathifer*.** *Aenictus spathifer* Santschi, 1928h: 119, fig. 1 (m.) SUMATRA.

**#*spinosior*.** *Aenictus hamifer* r. *spinosior* Stitz, 1917: 335, fig. 1 (m.) ALGERIA.

***steindachneri*.** *Aenictus steindachneri* Mayr, 1901b: 2 (w.) SOUTH AFRICA.

***sumatrensis*.** *Aenictus sumatrensis* Forel, 1913k: 23 (m.) SUMATRA. Current subspecies: nominal plus *maxillosus*.

**#*sundaicus*.** *Aenictus clavatus* subsp. *sundaicus* Forel, 1909d: 223 (m.) JAVA.

*sundaica*. *Eciton* (*Aenictus*) *fergusoni* var. *sundaica* Karavaiev, 1927e: 7 (w.) JAVA. [Unresolved junior secondary homonym of *sundaicus* Forel, above.] Junior synonym of *laeviceps:* Wilson, 1964a: 467.

**#*taivanae*.** *Aenictus longi* r. *taivanae* Forel, 1913f: 189 (m.) TAIWAN.

***togoensis*.** *Aenictus togoensis* Santschi, 1915c: 248 (m.) TOGO.

*tricarinata*. *Typhlatta tricarinata* Smith, F. 1873: ix. *Nomen nudum*, attributed to Mayr.

***trigonus*.** *Aenictus trigonus* Forel, 1911b: 195 (m.) JAVA.

**#*tuberculatus*.** *Aenictus tuberculatus* Arnold, 1915: 142 (m.) ZIMBABWE. Subspecies of *crucifer:* Santschi, 1924b: 204.

*turneri*. *Aenictus turneri* Forel, 1900b: 75 (w.) AUSTRALIA. Wheeler, G.C. & Wheeler, J. 1964c: 131 (l.). Senior synonym of *deuqueti:* Brown, 1952b: 123; of *exiguus:* Brown, 1958a: 5. Junior synonym of *ceylonicus:* Wilson, 1964a: 452.

*unicolor*. *Aenictus unicolor* Radoszkovsky, 1881: 198. *Nomen nudum*, attributed to Smith, F.

**vagans.** *Aenictus vagans* Santschi, 1924b: 200, fig. 3 (m.) NIGER.

**vaucheri.** *Aenictus vaucheri* Emery, 1915a: 252, pl. 4, figs. 1, 2 (q.) MOROCCO. Santschi, 1936c: 198 (w.).

**villiersi.** *Aenictus villiersi* Bernard, 1953b: 221, fig. 7 (w.) GUINEA.

**#viridans.** *Aenictus humeralis* var. *viridans* Santschi, 1915c: 248 (m.) BENIN.

**weissi.** *Aenictus weissi* Santschi, 1910c: 354, fig. 2 (w.) CONGO. Santschi, 1920b: 9 (q.).

**westwoodi.** *Aenictus ambiguus* var. *westwoodi* Forel, 1901a: 469 (m.) INDIA. Raised to species: Bingham, 1903: 10.

**wroughtonii.** *Aenictus wroughtonii* Forel, 1890b: ciii (w.m.) INDIA. See also: Wilson, 1964a: 480.

*AEROMYRMA:* see under ***OLIGOMYRMEX.***

*AETHIOPOPONE:* see under ***SPHINCTOMYRMEX.***

***AFROXYIDRIS*** (west Africa)

**crigensis.** *Afroxyidris crigensis* Belshaw & Bolton, 1994: 632, figs. 1–3 (w.) GHANA.

***AGRAULOMYRMEX*** (southern Africa)

**meridionalis.** *Agraulomyrmex meridionalis* Prins, 1983: 2, fig. 1 (w.) SOUTH AFRICA.

**wilsoni.** *Agraulomyrmex wilsoni* Prins, 1983: 5, fig. 2 (w.q.) ZIMBABWE.

****AGROECOMYRMEX*** (Baltic Amber)

***duisburgi.** **Myrmica duisburgi* Mayr, 1868c: 87, pl. 5, figs. 87, 88 (w.) BALTIC AMBER (Oligocene). Wheeler, W.M. 1915h: 58 (q.). Combination in **Agroecomyrmex:* Wheeler, W.M. 1915h: 57.

***#ALAOPONE:*** see under ***DORYLUS.***

*ALFARIA:* see under ***GNAMPTOGENYS.***

*ALISTRUMA:* see under ***COLOBOSTRUMA.***

***ALLOFORMICA*** (Palaearctic)

**aberrans.** *Formica aberrans* Mayr, 1877: 7 (w.) TURKESTAN. [Redescribed in Latin: Mayr, 1880: 27.] Combination in *Formica* (*Proformica*): Ruzsky, 1905b: 421; in *Proformica:* Kuznetsov-Ugamsky, 1927b: 28; in *Proformica* (*Alloformica*): Dlussky, 1969a: 219; in *Alloformica:* Dlussky, Soyunov & Zabelin, 1990: 143. See also: Agosti, 1994a: 99.

**flavicornis.** *Cataglyphis cursor* subsp. *flavicornis* Kuznetsov-Ugamsky, 1926a: 72 (w.) UZBEKISTAN. Combination in *Alloformica:* Agosti, 1990a: 1495. Junior synonym of *Cataglyphis rockingeri:* Santschi, 1929b: 36; of *Cataglyphis aenescens:* Dlussky, Soyunov & Zabelin, 1990: 149. Revived from synonymy and raised to species: Agosti, 1990a: 1495. See also: Agosti, 1994a: 99.

**nitidior.** *Formica* (*Proformica*) *aberrans* var. *nitidior* Forel, 1904b: 383 (w.) TURKESTAN. Combination in *Proformica* (*Alloformica*): Dlussky, 1969a: 220; in *Alloformica:* Dlussky, Soyunov & Zabelin, 1990: 144. Raised to species: Dlussky, 1969a: 220. See also: Agosti, 1994a: 99.

**obscurior.** *Alloformica obscurior* Dlussky, in Dlussky, Soyunov & Zabelin, 1990: 145, fig. 15 (w.) TURKMENISTAN.

***ALLOMERUS*** (Neotropical)

*angulatus*. *Allomerus decemarticulatus* subsp. *octoarticulatus* var. *angulatus* Wheeler, W.M. & Mann, in Wheeler, W.M. 1942: 201 (w.) BOLIVIA. Unavailable name. [*Allomerus octoarticulatus* var. *angulatus:* Kempf, 1972a: 19; unavailable name.]

**decemarticulatus.** *Allomerus decemarticulatus* Mayr, 1878: 874 (w.) BRAZIL. Kempf, 1975c: 347 (q.). Current subspecies: nominal plus *novemarticulatus, septemarticulatus.*

**#demerarae.** *Allomerus octoarticulatus* var. *demerarae* Wheeler, W.M., in Wheeler, G.C. 1935: 92, pls. 7, 8 (l.) GUIANA. [*Allomerus octoarticulatus* var. *demerarae* Wheeler, W.M. 1929h: 343. *Nomen nudum.*] Wheeler, W.M. 1942: 200 (w.q.m.).

*exsanguis*. *Allomerus decemarticulatus* subsp. *octoarticulatus* var. *exsanguis* Wheeler, W.M. & Mann, in Wheeler, W.M. 1942: 200 (w.q.) BOLIVIA. Unavailable name. [*Allomerus octoarticulatus* var. *exsanguis:* Kempf, 1972a: 19; unavailable name.]

*melanoticus*. *Allomerus decemarticulatus* subsp. *octoarticulatus* var. *melanoticus* Wheeler, W.M. & Mann, in Wheeler, W.M. 1942: 202 (w.q.m.) BOLIVIA. Unavailable name. [*Allomerus octoarticulatus* var. *melanoticus:* Kempf, 1972a: 19; unavailable name.]

**#novemarticulatus.** *Allomerus decemarticulatus* subsp. *novemarticulatus* Wheeler, W.M. & Mann, in Wheeler, W.M. 1942: 199 (w.) BOLIVIA.

**octoarticulatus.** *Allomerus octoarticulatus* Mayr, 1878: 874 (w.) BRAZIL. Forel, 1904e: 679 (q.m.); Wheeler, G.C. & Wheeler, J. 1955c: 125 (l.). Subspecies of *decemarticulatus:* Wheeler, W.M. 1942: 199. Revived status as species: Kempf, 1972a: 19. Current subspecies: nominal plus *demerarae, tuberculatus.*

**#septemarticulatus.** *Allomerus septemarticulatus* Mayr, 1878: 874 (w.) BRAZIL. Forel, 1904e: 680 (q.). Variety of *octoarticulatus:* Forel, 1904e: 680. Subspecies of *decemarticulatus:* Wheeler, W.M. 1942: 203.

**#tuberculatus.** *Allomerus octoarticulatus* r. *tuberculatus* Forel, 1912g: 2 (w.m.) BRAZIL.

**vogeli.** *Allomerus vogeli* Kempf, 1975c: 348, figs. 1–4 (w.m.) BRAZIL.

*ALLOPHEIDOLE:* see under ***PHEIDOLE.***

*AMAUROMYRMEX:* see under ***PHEIDOLOGETON.***

***AMBLYOPONE*** (World tropics and temperate zones)

**aberrans.** *Amblyopone aberrans* Wheeler, W.M. 1927c: 26, fig. 8 (w.q.m.) AUSTRALIA.

**amblyops.** *Stigmatomma amblyops* Karavaiev, 1935a: 57, fig. 1 (w.) VIETNAM. Combination in *Amblyopone:* Brown, 1960a: 167.

**annae.** *Amblyopone* (*Stigmatomma*) *annae* Arnol'di, 1968: 1801, fig. 2 (w.) TURKMENIA. See also: Dlussky & Zabelin, 1985: 209.

*arizonense*. *Stigmatomma pallipes* subsp. *arizonense* Wheeler, W.M. 1915b: 389 (w.) U.S.A. Junior synonym of *pallipes:* Creighton, 1940b: 3.

**armigera.** *Amblyopone armigera* Mayr, 1887: 547 (w.) BRAZIL. Combination in *Stigmatomma:* Dalla Torre, 1893: 14; in *Stigmatomma* (*Fulakora*): Mann, 1919: 279; in *Amblyopone:* Borgmeier, 1957: 109.

**australis.** *Amblyopone australis* Erichson, 1842: 261, pl. 5, fig. 7 (w.) AUSTRALIA. Smith, F. 1858b: 109 (q.m.); Wheeler, G.C. & Wheeler, J. 1952a: 116 (l.); Imai, Crozier & Taylor, 1977: 347 (k.). Senior synonym of *laevidens* and *nana:* Wilson, 1958a: 142; of *cephalotes* (and its junior synonym *maculata*): Brown, 1958h: 14; of *fortis, foveolata, minor* and *obscura,* and material of the unavailable names *howensis, norfolkensis, pallens* and *queenslandica* referred here: Brown, 1960a: 167 (these previously provisional synonyms in Brown, 1958h: 13). See also: Taylor, 1979: 835.

*barretoi*. *Stigmatomma* (*Fulakora*) *elongata* st. *barretoi* Bruch, 1921: 184, fig. 6 (w.) ARGENTINA. Junior synonym of *elongata:* Borgmeier, 1957: 110.

**bellii.** *Amblyopone* (*Stigmatomma*) *bellii* Forel, 1900c: 55 (w.) INDIA. Combination in *Stigmatomma:* Bingham, 1903: 38; in *Amblyopone:* Brown, 1960a: 193.

**besucheti.** *Amblyopone besucheti* Baroni Urbani, 1978a: 49, figs. 15, 16 (w.) SEYCHELLES IS.

**bierigi.** *Stigmatomma* (*Fulakora*) *bierigi* Santschi, 1930c: 75 (w.) CUBA. Combination in *Amblyopone:* Brown, 1960a: 167.

*binodosus*. *Arotropus binodosus* Provancher, 1881a: 206, figs. 32, 33 (q.) CANADA. Provancher, 1887: 240 (w.). Combination in *Amblyopone:* Provancher, 1887: 240; in *Stigmatomma:* Dalla Torre, 1893: 14. Junior synonym of *pallipes:* Emery, 1895c: 261.

**bruni.** *Stigmatomma bruni* Forel, 1912a: 45 (w.) TAIWAN. Forel, 1913f: 183 (m.). Combination in *Amblyopone,* and senior synonym of *juergi:* Brown, 1960a: 168.

*castaneus* Smith, F. 1860; see under ***MYOPOPONE.***

**celata.** *Stigmatomma* (*Fulakora*) *celata* Mann, 1919: 279, fig. 2 (w.) SOLOMON IS. Combination in *Amblyopone:* Wilson, 1958a: 143. See also: Taylor, 1979: 830.

*cephalotes*. *Amblyopone cephalotes* Smith, F. 1876c: 490 (w.) NEW ZEALAND. Wheeler, G.C. & Wheeler, J. 1952a: 117 (l.). Subspecies of *australis* and senior synonym of *maculata:* Wheeler, W.M. 1927c: 13. Junior synonym of *australis:* Brown, 1958h: 14.

**chilensis.** *Amblyopone chilensis* Mayr, 1887: 547 (w.) CHILE. Combination in *Stigmatomma:* Dalla Torre, 1893: 14; in *Stigmatomma* (*Fulakora*): Mann, 1919: 279; in *Amblyopone:* Brown, 1960a: 183.

**clarki.** *Amblyopone clarki* Wheeler, W.M. 1927c: 24, fig. 7 (w.q.) AUSTRALIA.

**degenerata.** *Amblyopone degenerata* Borgmeier, 1957: 111, figs. 18–20 (w.) BRAZIL.

**denticulata.** *Stigmatomma denticulatum* Roger, 1859: 251, pl. 7, fig. 2 (w.) GREECE. Emery, 1875a: 74 (m.); Emery, 1909c: 357 (q.). Combination in *Amblyopone:* Emery & Forel, 1879: 455; in *Stigmatomma:* Dalla Torre, 1893: 14; in *Amblyopone:* Brown, 1960a: 168. Senior synonym of *gheorghieffi:* Emery, 1915a: 255, Emery, 1916b: 100; of *gracilicornis:* Brown, 1960a: 168. See also: Baroni Urbani, 1978a: 42; Atanassov & Dlussky, 1992: 61.

***AMBLYOPONE** (continued)*

***egregia.*** *Ericapelta egregia* Kusnezov, 1955: 274, figs. 5–7 (m.) ARGENTINA. Combination in *Amblyopone:* Brown, 1960a: 168.

***elongata.*** *Stigmatomma elongata* Santschi, 1912e: 519, fig. 1 (w.) URUGUAY. Borgmeier, 1957: 110 (q.). Combination in *Stigmatomma* (*Fulakora*): Santschi, 1922d: 241; in *Amblyopone:* Borgmeier, 1957: 110. Senior synonym of *barretoi, minor* and *paranense:* Borgmeier, 1957: 110.

***emeryi.*** *Stigmatomma emeryi* Saunders, 1890: 203, fig. (w.) MOROCCO. Emery, 1915a: 254 (m.). Combination in *Amblyopone:* Brown, 1960a: 168.

***exigua.*** *Amblyopone exigua* Clark, 1928a: 35, pl. 1, figs. 20–23 (q.) AUSTRALIA.

***falcata.*** *Amblyopone falcata* Lattke, 1991b: 1, fig. 1 (w.) PUERTO RICO.

***feae.*** *Stigmatomma feae* Emery, 1895k: 454 (w.) BURMA. Combination in *Amblyopone:* Forel, 1900c: 56.

***ferruginea.*** *Amblyopone ferruginea* Smith, F. 1858b: 110 (w.) AUSTRALIA. André, 1896b: 261 (q.). Senior synonym of *mandibularis:* Brown, 1952i: 265. See also: Brown, 1960a: 206.

*fortis. Amblyopone australis* var. *fortis* Forel, 1910b: 1 (w.) AUSTRALIA. Wheeler, W.M. 1927c: 9 (q.). Junior synonym of *australis:* Brown, 1960a: 167.

*foveolata. Amblyopone australis* subsp. *foveolata* Wheeler, W.M. 1927c: 9 (w.q.m.) AUSTRALIA. Junior synonym of *australis:* Brown, 1960a: 167.

***fulvida.*** *Amblyopone fulvida* Terayama, 1987: 31, figs. 1–5 (w.) JAPAN.

***gaetulica.*** *Amblyopone gaetulica* Baroni Urbani, 1978a: 47, figs. 6, 12 (w.) MOROCCO. See also: Tinaut, 1990d: 189.

*gheorghieffi. Amblyopone gheorghieffi* Forel, 1892i: 309 (m.) BULGARIA. Combination in *Stigmatomma:* Dalla Torre, 1893: 14. Junior synonym of *denticulata:* Emery, 1915a: 255.

***gingivalis.*** *Amblyopone gingivalis* Brown, 1960a: 205, figs. 30, 31 (w.) AUSTRALIA.

***glauerti.*** *Lithomyrmex glauerti* Clark, 1928a: 31, pl. 1, figs. 1–11 (w.q.m.) AUSTRALIA. Combination in *Amblyopone:* Brown, 1960a: 168.

***gnoma.*** *Amblyopone gnoma* Taylor, 1979: 829, figs. 13–16 (w.q.) SOLOMON IS.

*gracilicornis. Stigmatomma denticulatum* var. *gracilicornis* Menozzi, 1936d: 268 (w.) GREECE. Junior synonym of *denticulata:* Brown, 1960a: 168.

***gracilis.*** *Amblyopone* (*Fulakora*) *gracilis* Clark, 1934c: 52, pl. 4, figs. 3, 4 (w.q.) AUSTRALIA.

***hackeri.*** *Amblyopone hackeri* Wheeler, W.M. 1927c: 22, fig. 6 (w.) AUSTRALIA.

*howensis. Amblyopone australis* subsp. *cephalotes* var. *howensis* Wheeler, W.M. 1927c: 15 (w.q.m.) AUSTRALIA. Unavailable name (Taylor, 1986: 34); material referred to *australis* by Brown, 1960a: 167.

***impressifrons.*** *Stigmatomma impressifrons* Emery, 1869b: 13, pl. 1, fig. 8 (w.q.) ITALY. Emery, 1915a: 254 (m.). Combination in *Amblyopone:* Emery & Forel, 1879: 455. See also: Baroni Urbani, 1978a: 44; Atanassov & Dlussky, 1992: 62.

*juergi. Stigmatomma bruni* r. *juergi* Forel, 1922: 87 (w.) TAIWAN. Junior synonym of *bruni:* Brown, 1960a: 168.

*laevidens. Amblyopone laevidens* Emery, 1887b: 447 (w.) NEW GUINEA. Subspecies of *australis:* Wheeler, W.M. 1927c: 12. Junior synonym of *australis:* Wilson, 1958a: 142.

***leae.*** *Amblyopone leae* Wheeler, W.M. 1927c: 16, fig. 3 (w.) AUSTRALIA.

***longidens.*** *Amblyopone ferruginea* var. *longidens* Forel, 1910b: 1 (w.) AUSTRALIA. Wheeler, G.C. & Wheeler, J. 1964b: 446 (l.). Junior synonym of *ferruginea:* Wheeler, W.M. 1927c: 20. Revived from synonymy and raised to species: Brown, 1952i: 266. See also: Brown, 1960a: 207.

***lucida.*** *Amblyopone* (*Fulakora*) *lucida* Clark, 1934b: 27, pl. 2, fig. 8 (w.) AUSTRALIA.

***lurilabes.*** *Amblyopone lurilabes* Lattke, 1991b: 2, fig. 2 (w.q.) VENEZUELA.

***luzonica.*** *Stigmatomma* (*Fulakora*) *luzonicum* Wheeler, W.M. & Chapman, 1925: 56, pl. 1, fig. 18 (q.) PHILIPPINES. Combination in *Amblyopone,* and senior synonym of *williamsi:* Brown, 1960a: 168.

*maculata. Amblyopone cephalotes* var. *maculata* Stitz, 1911a: 351 (w.) NEW ZEALAND. Junior synonym of *cephalotes:* Wheeler, W.M. 1927c: 13.

*mandibularis. Amblyopone mandibularis* Clark, 1928a: 33, pl. 1, figs. 12–15 (w.) AUSTRALIA. Junior synonym of *ferruginea:* Brown, 1952i: 265.

***mercovichi.*** *Amblyopone mercovichi* Brown, 1960a: 201, figs. 29, 32 (w.) AUSTRALIA.

***michaelseni.*** *Amblyopone michaelseni* Forel, 1907h: 263 (w.) AUSTRALIA. Clark, 1928a: 33 (q.).

*minor. Amblyopone minor* Forel, 1915b: 1 (w.q.m.) AUSTRALIA. Junior synonym of *australis:* Brown, 1960a: 167.

*minor. Stigmatomma* (*Fulakora*) *elongata* var. *minor* Santschi, 1922d: 241 (w.) ARGENTINA. [Unresolved junior secondary homonym of *minor* Forel, above.] Junior synonym of *elongata:* Borgmeier, 1957: 110.

***minuta.*** *Stigmatomma minuta* Forel, 1913k: 4 (w.) SUMATRA. Combination in *Stigmatomma* (*Fulakora*): Mann, 1919: 279; in *Amblyopone:* Brown, 1960a: 168 (see also p. 196).

***monrosi.*** *Amblyopone monrosi* Brown, 1960a: 188, fig. 20 (w.) CHILE.

*montigena. Stigmatomma pallipes* subsp. *montigena* Creighton, 1940b: 7 (w.q.) U.S.A. Junior synonym of *pallipes:* Brown, 1949c: 84.

***mutica.*** *Stigmatomma* (*Xymmer*) *muticum* Santschi, 1914d: 311, fig. 2 (w.) NIGERIA. Combination in *Xymmer:* Wheeler, W.M. 1922a: 758; in *Amblyopone:* Brown, 1960a: 168.

***mystriops.*** *Amblyopone mystriops* Brown, 1960a: 185, fig. 19 (q.) GUATEMALA. Lattke, 1991b: 5 (w.).

*nana. Amblyopone nana* Emery, 1914f: 394 (w.q.) NEW CALEDONIA. Subspecies of *australis* Wheeler, W.M. 1927c: 16. Junior synonym of *australis:* Wilson, 1958a: 142.

***noonadan.*** *Amblyopone noonadan* Taylor, 1965a: 6, figs. 3, 4 (w.) NEW BRITAIN. Taylor, 1979: 826 (m.).

*norfolkensis. Amblyopone australis* subsp. *cephalotes* var. *norfolkensis* Wheeler, W.M. 1927c: 15 (w.) AUSTRALIA. Unavailable name (Taylor, 1986: 34); material referred to *australis* by Brown, 1960a: 167.

***normandi.*** *Stigmatomma normandi* Santschi, 1915a: 54, fig. 1 (w.) TUNISIA. Combination in *Amblyopone:* Brown, 1960a: 169. See also: Baroni Urbani, 1978a: 45.

*obscura. Amblyopone obscura* Smith, F. 1858b: 109 (w.q.) AUSTRALIA. Clark, 1925b: 138 (m.). Variety of *australis:* Froggatt, 1905: 8; Forel, 1910b: 2. Subspecies of *australis:* Wheeler, W.M. 1927c: 10. Junior synonym of *australis:* Brown, 1960a: 167.

***ophthalmica.*** *Amblyopone ophthalmica* Baroni Urbani, 1978a: 44, figs. 3, 9 (w.) ISRAEL.

***oregonensis.*** *Stigmatomma pallipes* subsp. *oregonense* Wheeler, W.M. 1915b: 389 (w.q.) U.S.A. Combination in *Amblyopone,* and raised to species: Brown, 1960a: 169. See also: Ward, 1988: 105.

***orizabana.*** *Amblyopone orizabana* Brown, 1960a: 190 (w.) MEXICO. Senior synonym of *tropicalis:* Lattke, 1991b: 5.

*pallens. Amblyopone australis* subsp. *obscura* var. *pallens* Wheeler, W.M. 1927c: 11 (w.) AUSTRALIA. Unavailable name (Taylor, 1986: 34); material referred to *australis* by Brown, 1960a: 167.

***pallipes.*** *Typhlopone pallipes* Haldeman, 1844: 54 (w.) U.S.A. Emery, 1895c: 261 (q.m.); Wheeler G.C. & Wheeler, J. 1952: 114 (l.). Combination in *Amblyopone:* Haldeman, 1849a: 201 (footnote); in *Stigmatomma:* Roger, 1863b: 20; in *Amblyopone:* Brown, 1960a: 169. [*Stigmatomma pallidipes:* Dalla Torre, 1893: 14, unjustified emendation.] Senior synonym of *serratum:* Mayr, 1886d: 439; of *binodosus:* Emery, 1895c: 261; of *arizonense* and *wheeleri:* Creighton, 1940b: 3; of *montigena:* Brown, 1949c: 84; of *subterranea:* Brown, 1960a: 169. See also: Whelden, 1958: 1; Traniello, 1982: 65.

***papuana.*** *Amblyopone papuana* Taylor, 1979: 828, figs. 5–8 (w.q.) NEW GUINEA.

*paranense. Stigmatomma* (*Fulakora*) *elongatum* var. *paranense* Santschi, 1925b: 6 (w.) BRAZIL. Junior synonym of *elongata:* Borgmeier, 1957: 110.

***pertinax.*** *Amblyopone pertinax* Baroni Urbani, 1978a: 48, figs. 13, 14 (w.) INDIA.

***pluto.*** *Amblyopone pluto* Gotwald & Levieux, 1972: 384, figs. 1–26 (w.q.m.) IVORY COAST.

***punctulata.*** *Amblyopone* (*Fulakora*) *punctulata* Clark, 1934b: 28, pl. 2, figs. 9, 10 (w.q.) AUSTRALIA.

***quadrata.*** *Stigmatomma quadratum* Karavaiev, 1935a: 58 (w.) VIETNAM. Combination in *Amblyopone:* Brown, 1960a: 169.

*queenslandica. Amblyopone australis* subsp. *laevidens* var. *queenslandica* Wheeler, W.M. 1927c: 12 (w.q.) AUSTRALIA. Unavailable name (Taylor, 1986: 34); material referred to *australis* by Brown, 1960a: 167.

***reclinata.*** *Amblyopone reclinata* Mayr, 1879: 667 (w.) JAVA. Stitz, 1925: 110 (m.); Imai, Brown *et al.* 1984: 67 (k.). Combination in *Stigmatomma:* Dalla Torre, 1893: 14; in *Amblyopone:* Brown, 1960a: 169 (see also p. 193).

***rothneyi.*** *Amblyopone* (*Stigmatomma*) *rothneyi* Forel, 1900c: 56 (w.) INDIA. Combination in *Stigmatomma:* Bingham, 1903: 37; in *Amblyopone:* Brown, 1960a: 169.

***rubiginoa.*** *Amblyopone rubiginous* Wu & Wang, 1992: 1303, fig. 4567 (w.) CHINA.

***sakaii.*** *Amblyopone sakaii* Terayama, 1989b: 344, figs. 1, 2 (w.) TAIWAN.

***santschii.*** *Stigmatomma santschii* Menozzi, 1922a: 347 (w.) SENEGAL. Combination in *Amblyopone:* Brown, 1960a: 169.

***saundersi.*** *Amblyopone saundersi* Forel, 1892c: 336 (q.) NEW ZEALAND. Clark, 1928a: 36 (w.). Combination in *Stigmatomma:* Dalla Torre, 1893: 15; in *Stigmatomma (Fulakora):* Mann, 1919: 279; in *Amblyopone:* Brown, 1958h: 15. See also: Brown, 1960a: 209.

*serratum. Stigmatomma serratum* Roger, 1859: 251 (w.) U.S.A. Junior synonym of *pallipes:* Mayr, 1886d: 439.

***silvestrii.*** *Stigmatomma silvestrii* Wheeler, W.M. 1928d: 97 (w.) JAPAN. Ogata, 1987: 103 (m.); Terayama, 1989b: 345 (q.); Masuko, 1990b: 221 (l.). Combination in *Amblyopone:* Brown, 1960a: 169 (see also p. 197). See also: Masuko, 1993: 35.

***smithi.*** *Amblyopone smithi* Brown, 1960a: 211, fig. 40 (w.) AUSTRALIA.

*subterranea. Stigmatomma pallipes* subsp. *subterranea* Creighton, 1940b: 8 (w.) U.S.A. Raised to species: Brown, 1949c: 85. Junior synonym of *pallipes:* Brown, 1960a: 169.

*testacea. ?Amblyopone testacea* Motschoulsky, 1863: 15 (q.) SRI LANKA. *Nomen dubium* (not congeneric with *Cryptopone testacea* Emery, 1893a: cclxxv), see Wilson, 1958d: 360.

***trigonignatha.*** *Amblyopone (Stigmatomma) trigonignatha* Brown, 1949c: 81, fig. 1 (w.) U.S.A.

*tropicalis. Amblyopone tropicalis* Brown, 1962a: 73 (w.q.) PANAMA. Junior synonym of *orizabana:* Lattke, 1991b: 5.

*wheeleri. Stigmatomma pallidipes* [sic] var. *wheeleri* Santschi, 1914a: 429 (w.q.m.) U.S.A. Junior synonym of *pallipes:* Creighton, 1940b: 3.

*williamsi. Stigmatomma (Fulakora) williamsi* Wheeler, W.M. & Chapman, 1925: 56, pl. 1, fig. 19 (w.q.) PHILIPPINES. Junior synonym of *luzonica:* Brown, 1960a: 168.

***wilsoni.*** *Amblyopone wilsoni* Clark, 1928a: 34, pl.1, figs. 16–19 (w.) AUSTRALIA.

***zwaluwenburgi.*** *Stigmatomma (Fulakora) zwaluwenburgi* Williams, 1946: 639 (w.) HAWAII. Combination in *Amblyopone:* Brown, 1960a: 169.

****AMEGHINOIA*** (Argentina)

****piatnitzkyi.*** **Ameghinoia piatnitzkyi* Viana & Haedo Rossi, 1957: 113, figs. 1–3, 3 pls. (q.) ARGENTINA (Upper Oligocene-Lower Miocene).

*AMMOMYRMA:* see under ***DORYMYRMEX.***

*AMYRMEX:* see under ***FORELIUS.***

*ANACANTHOLEPIS:* see under ***PLAGIOLEPIS.***

*ANACANTHOPONERA:* see under ***HETEROPONERA.***

*ANCYLOGNATHUS:* see under ***ECITON.***

***ANCYRIDRIS*** (New Guinea)

***polyrhachioides.*** *Ancyridris polyrhachioides* Wheeler, W.M. 1935a: 2, fig. 1 (w.) NEW GUINEA.

***rupicapra.*** *Pheidole (Pheidolacanthinus) rupicapra* Stitz, 1938: 99, fig. 1 (w.) NEW GUINEA. Combination in *Ancyridris:* **new combination**.

*ANDRAGNATHUS:* see under ***PARATRECHINA.***

*ANELEUS:* see under ***OLIGOMYRMEX.***

***ANERGATES*** (west Palaearctic, east Nearctic)

***atratulus.*** *Myrmica atratula* Schenck, 1852: 91 (q.m.) GERMANY. Wheeler, W.M. 1909g: 182 (l.). Combination in *Tetramorium:* Mayr, 1855: 429; in *Anergates:* Forel, 1874: 68 (see also p. 93). [Also described as new by Schenck, 1853: 186.] Senior synonym of *friedlandi:* Creighton, 1950a: 243. See also: Donisthorpe, 1915d: 89; Boven, 1977: 81; Wheeler, G.C. & Wheeler, J. 1955c: 128; Atanassov & Dlussky, 1992: 161.

*friedlandi. Anergates friedlandi* Creighton, 1934: 193 (q.) U.S.A. Junior synonym of *atratulus:* Creighton, 1950a: 243.

*ANERGATIDES:* see under ***PHEIDOLE.***

****ANEURETELLUS*** (Russia)

****deformis.*** **Aneuretellus deformis* Dlussky, 1988: 54, fig. 2 (w.q.) RUSSIA (Palaeocene?).

***ANEURETUS*** (Sri Lanka)

***simoni.*** *Aneuretus simoni* Emery, 1893a: cclxxvi (w.) SRI LANKA. Wilson, Eisner *et al.* 1956: 82 (q.m.l.). [Also described as new by Emery, 1893f: 242.] Senior synonym of *butteli:* Wilson, Eisner *et al.* 1956: 81. See also: Gotwald, 1970: 950; Jayasuriya & Traniello, 1986: 363; Traniello & Jayasuriya, 1986: 375; Shattuck, 1994: 1..

*butteli. Aneuretus butteli* Forel, 1912m: 770 (q.) SRI LANKA. [Also described as new by Forel, 1913k: 87.] Junior synonym of *simoni:* Wilson, Eisner *et al.* 1956: 81.

***ANILLIDRIS*** (Neotropical)

***bruchi.*** *Anillidris bruchi* Santschi, 1936d: 414, figs. 21–28 (w.) ARGENTINA. Santschi, 1937c: 69 (q.m.). Combination in *Linepithema:* Kempf, 1972a: 135; in *Anillidris:* Shattuck, 1994: 3.

*gallardoi. Anillidris gallardoi* Kusnezov, 1954b: 37. *Nomen nudum.*

***ANILLOMYRMA*** (Oriental, Indo-Australian)

#***continentis.*** *Anillomyrma decamera* subsp. *continentis* Wheeler, W.M. 1927h: 96 (w.) VIETNAM.

***decamera.*** *Monomorium decamerum* Emery, 1901e: 117 (w.) SRI LANKA. Combination in *Monomorium (Anillomyrma):* Emery, 1913: 261; in *Anillomyrma:* Ettershank, 1966: 98. See also: Bolton, 1987: 274. Current subspecies: nominal plus *continentis.*

***tridens.*** *Anillomyrma tridens* Bolton, 1987: 274, figs. 1, 4 (w.) BORNEO.

***ANISOPHEIDOLE*** (Australia)

***antipodum.*** *Atta antipodum* Smith, F. 1858b: 166 (q.) AUSTRALIA. Combination in *Pheidole:* Mayr, 1886c: 360; in *Aphaenogaster:* Wheeler, W.M. 1916j: 216; in *Pheidole:* Emery, 1922e: 97; in *Anisopheidole:* Ettershank, 1966: 132. Junior synonym of *Aphaenogaster longiceps:* Wheeler, W.M. 1916j: 216. Revived from synonymy, and senior synonym of *froggatti* (and its junior synonyms *lippulum* and *myops*): Ettershank, 1966: 132.

*froggatti. Pheidole froggatti* Forel, 1902h: 414 (q.m.) AUSTRALIA. Forel, 1907h: 282 (s.w.). Combination in *Pheidole (Allopheidole):* Forel, 1912f: 237; in *Pheidole (Anisopheidole):* Forel, 1914c: 616. Senior synonym of *lippulum:* Brown, 1953h: 7; of *myops:* Emery, 1921f: 83. Junior synonym of *antipodum:* Ettershank, 1966: 132.

*lippulum. Monomorium lippulum* Wheeler, W.M. 1927h: 89, fig. 3 (w.) AUSTRALIA. Junior synonym of *froggatti:* Brown, 1953h: 7.

*myops. Pheidole myops* Forel, 1902h: 421 (w.) AUSTRALIA. Junior synonym of *froggatti:* Emery, 1921f: 83.

***ANKYLOMYRMA*** (west Africa)

***coronacantha.*** *Ankylomyrma coronacantha* Bolton, 1973c: 235, figs. 1–3 (w.) GHANA. See also: Bolton, 1981b: 247.

***ANOCHETUS*** (World tropics and subtropics)

*abstracta. Anochetus bequaerti* st. *abstracta* Santschi, 1914e: 12 (w.) SOUTH AFRICA. Junior synonym of *bequaerti:* Brown, 1978c: 556.

***africanus.*** *Stenomyrmex africanus* Mayr, 1865: 11 (footnote) (w.) GHANA. André, 1892a: 47 (q.). Combination in *Anochetus:* Forel, 1887: 382. Senior synonym of *camerunensis:* Mayr, 1904b: 2; of *pasteuri:* Brown, 1978c: 556 (see also p. 602).

***agilis.*** *Anochetus agilis* Emery, 1901a: 52 (w.) BORNEO. Viehmeyer, 1916a: 116 (q.). See also: Brown, 1978c: 578.

***altisquamis.*** *Anochetus altisquamis* Mayr, 1887: 529 (w.) BRAZIL. Senior synonym of *fumata:* Kempf & Lenko, 1976: 58.

*amati. Anochetus amati* Karavaiev, 1925c: 285, fig. 8 (q.) INDONESIA (Aru I.). Junior synonym of *graeffei:* Wilson, 1959a: 507.

****ambiguus.*** **Anochetus ambiguus* De Andrade, 1994: 17, fig. 18 (w.) DOMINICAN AMBER.

***angolensis.*** *Anochetus angolensis* Brown, 1978c: 600, fig. 49 (w.) ANGOLA.

*angusticornis. Anochetus angusticornis* Arnold, 1946: 56 (w.q.) ZIMBABWE. Junior synonym of *traegaordhi:* Brown, 1978c: 559.

***armstrongi.*** *Anochetus armstrongi* McAreavey, 1949: 1, figs. 1–6 (w.q.) AUSTRALIA. See also: Brown, 1978c: 597.

*aurifrons. Anochetus pellucidus* var. *aurifrons* Santschi, 1910c: 351 (w.) CONGO. Junior synonym of *pellucidus:* Brown, 1978c: 558.

*australis. Anochetus australis* Emery, 1906c: 117, fig. 2 (w.) URUGUAY. Santschi, 1929d: 278 (q.). Subspecies of *mayri:* Emery, 1911d: 110; Bruch, 1915: 528. Junior synonym of *neglectus:* Brown, 1978c: 558.

*beccarii. Anochetus beccarii* Emery, 1884a: 379 (diagnosis in key) (w.) SUMATRA. Junior synonym of *rugosus:* Brown, 1964d: 214.

***bequaerti.*** *Anochetus bequaerti* Forel, 1913b: 309 (w.) ZAIRE. Arnold, 1915: 104 (q.m.). Senior synonym of *abstracta, estus* and *opaciventris:* Brown, 1978c: 556.

*bierigi. Anochetus bierigi* Santschi, 1931c: 268, figs. 4–6 (w.q.) PANAMA. Junior synonym of *diegensis:* Brown, 1964d: 215.

***bispinosus.*** *Odontomachus bispinosus* Smith, F. 1858b: 199 (w.) BRAZIL. Combination in *Stenomyrmex:* Mayr, 1886c: 361; in *Anochetus:* Dalla Torre, 1893: 47. See also: Brown, 1976a: 103.

****brevidentatus.*** **Anochetus brevidentatus* MacKay, 1991: 138, figs. 1–3, 8 (w.) DOMINICAN AMBER. See also: De Andrade, 1994: 19.

***ANOCHETUS*** *(continued)*

***brevis.*** *Anochetus brevis* Brown, 1978c: 581, fig. 2 (w.) PHILIPPINES.

*butteli. Anochetus longifossatus* r. *butteli* Forel, 1913k: 18 (w.) SRI LANKA. Junior synonym of *longifossatus:* Brown, 1978c: 557.

*cameroni. Anochetus cameroni* Forel, 1915c: 352 (w.) BRAZIL. Junior synonym of *ghilianii:* Brown, 1964d: 214.

*camerunensis. Anochetus africanus* var. *camerunensis* Mayr, 1896: 236 (w.m.) CAMEROUN. Emery, 1899e: 476 (q.). Junior synonym of *africanus:* Mayr, 1904b: 2.

***cato.*** *Anochetus cato* Forel, 1901b: 6 (w.) NEW BRITAIN. Senior synonym of *rossi* and *subfasciatus:* Wilson, 1959a: 504. See also Brown, 1978c: 585.

***chirichinii.*** *Anochetus chirichinii* Emery, 1897c: 597, pl. 15, figs. 46, 47 (w.) NEW GUINEA. See also: Brown, 1978c: 590.

*concinnus. Anochetus concinnus* Santschi, 1920b: 9 (q.) ZAIRE. Junior synonym of *katonae:* Brown, 1978c: 557.

****conisquamis.*** **Anochetus conisquamis* De Andrade, 1994: 22, fig. 20 (w.) DOMINICAN AMBER.

***consultans.*** *Formica consultans* Walker, 1859: 373 (m.) SRI LANKA. Combination in *Euponera:* Donisthorpe, 1932b: 576; in *Anochetus:* Brown, 1978c: 556 (see also p. 588).

****corayi.*** **Anochetus corayi* Baroni Urbani, 1980c: 2, figs. 1–4 (q.) DOMINICAN AMBER (30–40 M. years old).

*diabolus. Anochetus rectangularis* var. *diabolus* Forel, 1915b: 35 (w.) AUSTRALIA. Junior synonym of *rectangularis:* Brown, 1978c: 558.

***diegensis.*** *Anochetus diegensis* Forel, 1912c: 29 (w.) COLOMBIA. Senior synonym of *bierigi:* Brown, 1964d: 215. See also: Brown, 1978c: 614.

****dubius.*** **Anochetus dubius* De Andrade, 1994: 15, fig. 17 (w.) DOMINICAN AMBER.

*durbanensis. Anochetus punctatus* var. *durbanensis* Arnold, 1926: 215 (w.) SOUTH AFRICA. Junior synonym of *katonae:* Brown, 1978c: 557.

***elegans.*** *Anochetus elegans* Lattke, 1987a: 353, figs. 1, 2 (w.) COLOMBIA.

***emarginatus.*** *Myrmecia emarginata* Fabricius, 1804: 426 (w.) CENTRAL AMERICA. Wheeler, G.C. & Wheeler, J. 1952c: 644 (l.). Combination in *Odontomachus:* Illiger, 1807: 194; in *Stenomyrmex:* Mayr, 1862: 711; in *Anochetus:* Emery, 1890a: 63. Senior synonym of *quadrispinosus:* Roger, 1862c: 289; of *rugosus:* Kempf, 1964f: 238. See also: Brown, 1978c: 609.

*estus. Anochetus estus* Wheeler, W.M. 1922a: 98 (w.) ZAIRE. Junior synonym of *bequaerti:* Brown, 1978c: 556.

***evansi.*** *Anochetus evansi* Crawley, 1922a: 85 (w.) IRAN. See also: Brown, 1978c: 595.

****exstinctus.*** **Anochetus exstinctus* De Andrade, 1994: 11, fig. 15 (w.) DOMINICAN AMBER.

***faurei.*** *Anochetus faurei* Arnold, 1948: 215, fig. 4 (w.) SOUTH AFRICA. See also: Brown, 1978c: 609.

***filicornis.*** *Myrmapatetes filicornis* Wheeler, W.M. 1929b: 6, fig. 3 (m.) NEW GUINEA. Combination in *Anochetus:* Brown, 1953h: 2.

***fricatus.*** *Anochetus fricatus* Wilson, 1959a: 506 (w.) NEW GUINEA. See also: Brown, 1978c: 590.

*friederichsi. Anochetus africanus* var. *friederichsi* Forel, 1918b: 155 (w.) MADAGASCAR. Junior synonym of *madagascarensis:* Brown, 1978c: 557.

***fuliginosus.*** *Anochetus fuliginosus* Arnold, 1948: 214, fig. 3 (w.) SOUTH AFRICA. See also: Brown, 1978c: 605.

*fumata. Anochetus altisquamis* forma *fumata* Luederwaldt, 1918: 53 (w.) BRAZIL. Junior synonym of *altisquamis:* Kempf & Lenko, 1976: 58.

***ghilianii.*** *Odontomachus ghilianii* Spinola, 1851b: 55 (w.) MOROCCO. [Also described as new by Spinola, 1853: 71.] Emery, 1909c: 376 (q.m.). Combination in *Anochetus:* Mayr, 1861: 54. Senior synonym of *cameroni:* Brown, 1964d: 214. See also: Brown, 1978c: 598.

***gladiator.*** *Stenomyrmex gladiator* Mayr, 1862: 712 (w.) INDONESIA (Misool I.). Wheeler, G.C. & Wheeler, J. 1976a: 62 (l.). Combination in *Anochetus:* Dalla Torre, 1893: 48. Junior synonym of *Odontomachus tyrannicus* Smith, F. 1861b: 44 (Dalla Torre, 1893: 48), hence first available replacement name for *Odontomachus tyrannicus* Smith, F. 1861b: 44 [junior primary homonym of *Odontomachus tyrannicus* Smith, F. 1859a: 144], and senior synonym of *smithii* and *gladiator* Donisthorpe (unnecessary secondary replacement names): Brown, 1978c: 556. See also Brown, 1978c: 574).

*gladiator. Odontomachus gladiator* Donisthorpe, 1932c: 467. Unnecessary replacement name for *Odontomachus tyrannicus* Smith, F. 1861b: 44, and hence junior synonym of *gladiator* Mayr, 1862: 712, the first available replacement name: Brown, 1978c: 556.

*gnomulus. Anochetus gnomulus* Bernard, 1953b: 214, fig. 6 (w.) GUINEA. Junior synonym of *katonae:* Brown, 1978c: 557.

*gracilicornis. Anochetus gracilicornis* Viehmeyer, 1923: 87, fig. 1 (w.) SUDAN. Senior synonym of *sudanicus:* Brown, 1964d: 215. Junior synonym of *traegaordhi:* Brown, 1978c: 559.

*gracilis. Anochetus gracilis* Karavaiev, 1925c: 286, fig. 9 (w.q.) JAVA. Junior synonym of *risii:* Brown, 1978c: 558.

***graeffei.*** *Anochetus graeffei* Mayr, 1870b: 961 (w.) SAMOA. Mayr, 1876: 86 (q.); Wheeler, G.C. & Wheeler, J. 1971b: 1212 (l.); Imai *et al.* 1984: 5 (k.). Senior synonym of *amati, minutus, oceanicus* and *punctiventris:* Wilson, 1959a: 507; of *rudis, ruginotus* and *taylori:* Brown, 1978c: 557 (see also p. 586).

***grandidieri.*** *Anochetus grandidieri* Forel, 1891b: 108, pl. 3, fig. 9 (w.) MADAGASCAR. Senior synonym of *madecassus:* Brown, 1978c: 557.

***haytianus.*** *Anochetus* (*Stenomyrmex*) *haytianus* Wheeler, W.M. & Mann, 1914: 15, fig. 6 (w.) HAITI. See also: Brown, 1978c: 611.

***horridus.*** *Anochetus* (*Stenomyrmex*) *horridus* Kempf, 1964f: 239, figs. 4, 8, 12, 14 (gynecoid w.) BRAZIL. Wheeler, G.C. & Wheeler, J. 1971b: 1212 (l.). See also: Brown, 1978c: 612.

***inca.*** *Anochetus* (*Stenomyrmex*) *inca* Wheeler, W.M. 1925a: 8 (w.) PERU.

***incultus.*** *Anochetus incultus* Brown, 1978c: 578, fig. 37 (w.q.) PHILIPPINES.

*indicus. Anochetus sedilloti* var. *indicus* Forel, 1900c: 61 (w.q.m.) INDIA. Junior synonym of *sedilloti:* Brown, 1978c: 559.

*ineditus. Anochetus ineditus* Baroni Urbani, 1971b: 360. Unnecessary second replacement name for *Anochetus jacobsoni* Menozzi, 1939b: 178.

***inermis.*** *Anochetus* (*Stenomyrmex*) *inermis* André, 1889: 221 (w.) TRINIDAD. Forel, 1897b: 298 (m.); Wheeler, G.C. & Wheeler, J. 1976a: 60 (l.). Senior synonym of *meinerti:* Brown, 1978c: 557.

****intermedius.*** **Anochetus intermedius* De Andrade, 1994: 13, fig. 16 (w.) DOMINICAN AMBER.

***isolatus.*** *Anochetus cato* subsp. *isolatus* Mann, 1919: 302, fig. 11 (w.m.) SOLOMON IS. Raised to species: Wilson, 1959a: 508. Senior synonym of *rossi* and *splendens:* Brown, 1978c: 557 (see also p. 585).

*jacobsoni. Anochetus jacobsoni* Forel, 1911b: 193 (w.) JAVA. Forel, 1913k: 18 (q.). Junior synonym of *princeps:* Brown, 1964d: 213.

*jacobsoni. Anochetus jacobsoni* Menozzi, 1939b: 178, fig. 3 (w.) SUMATRA. [Junior primary homonym of *jacobsoni* Forel, above.] First replacement name: *menozzii* Donisthorpe, 1941k: 237. Second (unnecessary) replacement name: *ineditus* Baroni Urbani, 1971b: 360.

***jonesi.*** *Anochetus jonesi* Arnold, 1926: 216, fig. 61 (w.q.m.) SOUTH AFRICA.

***kanariensis.*** *Anochetus orientalis* r. *kanariensis* Forel, 1900c: 62 (w.) INDIA. Raised to species: Bingham, 1903: 44. See also: Brown, 1978c: 594.

***katonae.*** *Anochetus grandidieri* var. *katonae* Forel, 1907a: 1 (w.) TANZANIA. Raised to species, and senior synonym of *concinnus, durbanensis, gnomulus, lamottei, longiceps, occidentalis, parvus* and *punctatus:* Brown, 1978c: 557 (see also p. 607).

***kempfi.*** *Anochetus kempfi* Brown, 1978c: 611, fig. 52 (w.) PUERTO RICO.

*laeta. Anochetus princeps* var. *laeta* Forel, 1910a: 27 (w.) SUMATRA. Junior synonym of *princeps:* Brown, 1964d: 213.

*laeviusculus. Anochetus mayri* subsp. *laeviusculus* Wheeler, W.M. 1911a: 22 (w.q.) JAMAICA. Junior synonym of *mayri:* Brown, 1978c: 557.

*lamottei. Anochetus lamottei* Bernard, 1953b: 212, fig. 5 (w.) GUINEA. Junior synonym of *katonae:* Brown, 1978c: 557.

*latunei. Anochetus turneri* var. *latunei* Forel, 1915b: 35 (w.) AUSTRALIA. Junior synonym of *turneri:* Brown, 1978c: 559.

***levaillanti.*** *Anochetus levaillanti* Emery, 1895h: 21 (w.) SOUTH AFRICA. Arnold, 1915: 108 (q.). See also: Brown, 1978c: 594.

*longiceps. Anochetus parvus* var. *longiceps* Santschi, 1914d: 331 (q.) CAMEROUN. Junior synonym of *katonae:* Brown, 1978c: 557.

***longifossatus.*** *Anochetus longifossatus* Mayr, 1897: 425 (w.) SRI LANKA. Senior synonym of *butteli:* Brown, 1978c: 557.

***longispinus.*** *Anochetus* (*Stenomyrmex*) *haytianus* subsp. *longispina* Wheeler, W.M. 1936b: 196 (w.) HAITI. Raised to species: Brown, 1978c: 611.

****lucidus.*** **Anochetus lucidus* De Andrade, 1994: 19, fig. 19 (w.) DOMINICAN AMBER.

***madagascarensis.*** *Anochetus africanus* var. *madagascarensis* Forel, 1887: 382 (w.) MADAGASCAR. Raised to species, and senior synonym of *friederichsi:* Brown, 1978c: 557.

**madaraszi.** *Anochetus madaraszi* Mayr, 1897: 424 (w.) SRI LANKA. Forel, 1900c: 61 (q.m.); Imai *et al.* 1984: 5 (k.). See also: Brown, 1978c: 590.

*madecassus. Anochetus madecassus* Santschi, 1928d: 54 (q.) MADAGASCAR. Junior synonym of *grandidieri:* Brown, 1978c: 557.

**maynei.** *Anochetus pellucidus* var. *maynei* Forel, 1913h: 347 (q.) ZAIRE. Brown, 1978c: 601 (w.). Raised to species: Brown, 1978c: 557.

**mayri.** *Anochetus mayri* Emery, 1884a: 378 (diagnosis in key) (w.) ANTILLES IS. Emery, 1890a: 65 (q.); Wheeler, G.C. & Wheeler, J. 1964b: 455 (l.). Senior synonym of *laeviusculus:* Brown, 1978c: 557 (see also p. 617).

*meinerti. Anochetus inermis* var. *meinerti* Forel, 1905b: 156 (w.) VENEZUELA. Wheeler, W.M. 1916d: 323 (q.m.). Junior synonym of *inermis:* Brown, 1978c: 557.

*menozzii. Anochetus menozzii* Donisthorpe, 1941k: 237. Replacement name for *jacobsoni* Menozzi, 1939b: 178. [Junior primary homonym of *jacobsoni* Forel, 1911b: 193.] Junior synonym of *rugosus:* Brown, 1978c: 558.

**micans.** *Anochetus micans* Brown, 1978c: 557 (w.) COSTA RICA. [First available use of *Anochetus* (*Stenomyrmex*) *emarginatus* subsp. *testaceus* var. *micans* Forel, 1908b: 36; unavailable name.]

**minans.** *Anochetus minans* Mann, 1922: 17 (w.) HONDURAS.

*minutus. Anochetus minutus* Karavaiev, 1925c: 288, fig. 10 (w.q.) WEST MALAYSIA. Junior synonym of *graeffei:* Wilson, 1959a: 507.

**mixtus.** *Anochetus mixtus* Radchenko, 1993a: 77, figs. 3–5 (w.) VIETNAM.

**modicus.** *Anochetus modicus* Brown, 1978c: 582 (w.q.m.) BORNEO. Imai, Kubota *et al.* 1985: 47 (k.).

*mordax. Anochetus mordax* Donisthorpe, 1942d: 452 (w.) INDIA. Junior synonym of *rufus:* Brown, 1978c: 558.

**muzziolii.** *Anochetus muzziolii* Menozzi, 1932d: 4, fig. 2 (w.) SUMATRA. See also: Brown, 1978c: 576.

**myops.** *Anochetus myops* Emery, 1893e: 201, pl. 8, figs. 11, 12 (w.) WEST MALAYSIA.

**natalensis.** *Anochetus natalensis* Arnold, 1926: 215 (w.) SOUTH AFRICA. See also: Brown, 1978c: 603.

**neglectus.** *Anochetus mayri* subsp. *neglectus* Emery, 1894c: 188 (w.q.) BRAZIL. Raised to species, senior synonym of *australis,* and material of the unavailable name *nobilis* referred here: Brown, 1978c: 558.

**nietneri.** *Odontomachus nietneri* Roger, 1861a: 23 (w.) SRI LANKA. Combination in *Anochetus:* Mayr, 1865: 63 (footnote). See also: Brown, 1978c: 593.

*nobilis. Anochetus mayri* st. *neglectus* var. *nobilis* Santschi, 1925e: 155 (m.) ARGENTINA. Unavailable name; material referred to *neglectus* by Brown, 1978c: 558.

**obscuratus.** *Anochetus madagascarensis* var. *obscurata* Santschi, 1911c: 351 (w.q.) TANZANIA. Variety of *africanus:* Emery, 1911d: 108; Santschi, 1914b: 56. Raised to species: Santschi, 1923e: 265. Senior synonym of *schoutedeni* and *ustus:* Brown, 1978c: 558 (see also p. 602).

**obscurior.** *Anochetus obscurior* Brown, 1978c: 558 (w.) INDIA. [First available use of *Anochetus orientalis* r. *kanariensis* var. *obscurior* Forel, 1900c: 62; unavailable name.]

*occidentalis. Anochetus punctatus* var. *occidentalis* Santschi, 1914d: 330 (w.) CAMEROUN. Junior synonym of *katonae:* Brown, 1978c: 557.

*oceanicus. Anochetus punctiventris* subsp. *oceanicus* Emery, 1897c: 597 (w.) NEW GUINEA. Emery, 1914f: 400 (q.m.). Junior synonym of *graeffei:* Wilson, 1959a: 507.

*opaciventris. Anochetus opaciventris* Wheeler, W.M. 1922a: 98 (w.) ZAIRE. Junior synonym of *bequaerti:* Brown, 1978c: 556.

**orchidicola.** *Anochetus orchidicola* Brown, 1978c: 619 (w.) MEXICO.

**oriens.** *Anochetus* (*Stenomyrmex*) *oriens* Kempf, 1964f: 240, figs. 2, 6, 10, 13 (w.) BRAZIL.

**orientalis.** *Anochetus orientalis* André, 1887: 291 (w.) VIETNAM. See also: Brown, 1978c: 594.

**pangens.** *Formica pangens* Walker, 1859: 371 (m.) SRI LANKA. Combination in *Euponera* (*Brachyponera*): Donisthorpe, 1932b: 575; in *Anochetus:* Brown, 1978c: 558.

**paripungens.** *Anochetus paripungens* Brown, 1978c: 596, fig. 35 (w.) AUSTRALIA.

*parvus. Anochetus parvus* Santschi, 1914d: 330 (w.) NIGERIA. Junior synonym of *katonae:* Brown, 1978c: 557.

*pasteuri. Anochetus pasteuri* Santschi, 1923e: 265 (w.q.) ZAIRE. Junior synonym of *africanus:* Brown, 1978c: 556.

**pellucidus.** *Anochetus pellucidus* Emery, 1902d: 33 (w.) CAMEROUN. Senior synonym of *aurifrons:* Brown, 1978c: 558 (see also p. 605).

**peracer.** *Anochetus peracer* Brown, 1978c: 579, figs. 40, 53 (w.) NEW GUINEA.

**princeps.** *Anochetus princeps* Emery, 1884a: 379 (diagnosis in key) (w.) JAVA. Brown, 1978c: 577 (m.); Wheeler, G.C. & Wheeler, J. 1976a: 61 (l.). Senior synonym of *jacobsoni* Forel, *laeta, serratus* and *taipingensis:* Brown, 1964d: 213.

**pubescens.** *Anochetus pubescens* Brown, 1978c: 607, figs. 10, 16 (w.) ZIMBABWE.

**punctaticeps.** *Anochetus punctaticeps* Mayr, 1901b: 4 (w.) SOUTH AFRICA. See also: Brown, 1978c: 606.

*punctatus. Anochetus punctatus* Santschi, 1914e: 13 (w.) SOUTH AFRICA. Junior synonym of *katonae:* Brown, 1978c: 557.

*punctiventris. Anochetus punctiventris* Mayr, 1879: 659 (w.) INDIA. Forel, 1900c: 63 (q.). Junior synonym of *graeffei:* Wilson, 1959a: 507.

**pupulatus.** *Anochetus pupulatus* Brown, 1978c: 591, figs. 14, 21, 22 (w.) INDIA.

*quadrispinosus. Odontomachus quadrispinosus* Smith, F. 1858b: 78, pl. 5, figs. 15–17 (w.) BRAZIL. Junior synonym of *emarginatus:* Roger, 1862c: 289.

**rectangularis.** *Anochetus rectangularis* Mayr, 1876: 86 (w.) AUSTRALIA. Senior synonym of *diabolus:* Brown, 1978c: 558 (see also p. 597).

**risii.** *Anochetus risii* Forel, 1900c: 60 (w.) HONG KONG. Wheeler, W.M. 1927h: 84 (m.); Wheeler, W.M. 1928c: 8 (q.). Senior synonym of *gracilis:* Brown, 1978c: 558 (see also p. 577).

*rossi. Anochetus rossi* Donisthorpe, 1947d: 186 (w.q.) NEW GUINEA. Junior synonym of *cato:* Wilson, 1959a: 504.

*rossi. Anochetus rossi* Donisthorpe, 1949c: 747 (w.q.) NEW GUINEA. [Unresolved junior primary homonym of *rossi* Donisthorpe, above.] Junior synonym of *isolatus:* Brown, 1978c: 557.

**rothschildi.** *Anochetus rothschildi* Forel, 1907c: 129 (w.m.) SOMALIA.

*rudis. Anochetus rudis* Emery, 1889b: 499 (w.) BURMA. Subspecies of *punctiventris:* Forel, 1900c: 60; Emery, 1911d: 109. Retained as species: Bingham, 1903: 41. Junior synonym of *graeffei:* Brown, 1978c: 577.

**rufus.** *Odontomachus rufus* Jerdon, 1851: 116 (w.) INDIA. Combination in *Anochetus:* Emery, 1892d: 561. Senior synonym of *mordax:* Brown, 1978c: 558 (see also p. 593).

*ruginotus. Anochetus ruginotus* Stitz, 1925: 114 (w.) PHILIPPINES. Junior synonym of *graeffei:* Brown, 1978c: 557.

**rugosus.** *Odontomachus rugosus* Smith, F. 1857a: 65 (w.) SINGAPORE. Wheeler,G.C. & Wheeler, J. 1976: 61 (l.). Combination in *Anochetus:* Emery, 1911d: 109. Senior synonym of *beccarii:* Brown, 1964d: 214; of *ineditus* and *menozzii:* Brown, 1978c: 558. See also: Brown, 1978c: 575.

*rugosus. Anochetus* (*Stenomyrmex*) *emarginatus* r. *rugosus* Emery, 1890a: 63 (w.) BRAZIL. Mann, 1916: 418 (m.); Wheeler, G.C. & Wheeler, J. 1952c: 645 (l.). [Unresolved junior secondary homonym of *Odontomachus rugosus* Smith, above.] Junior synonym of *emarginatus:* Kempf, 1964f: 238.

*schoutedeni. Anochetus schoutedeni* Santschi, 1923e: 264 (w.) ZAIRE. Junior synonym of *obscuratus:* Brown, 1978c: 558.

**sedilloti.** *Anochetus sedilloti* Emery, 1884a: 377, fig. (w.) TUNISIA. Forel, 1907b: 202 (m.); Santschi, 1907: 325 (m.). Senior synonym of *indicus:* Brown, 1978c: 559 (see also p. 594).

**seminiger.** *Anochetus seminiger* Donisthorpe, 1943a: 170 (w.) NEW GUINEA.

*serratus. Anochetus serratus* Stitz, 1925: 113 (w.) JAVA. Junior synonym of *princeps:* Brown, 1964d: 213.

*silvaticus. Anochetus silvaticus* Bernard, 1953b: 212, fig. 5 (w.q.) GUINEA. Junior synonym of *traegaordhi:* Brown, 1964d: 215.

**simoni.** *Anochetus* (*Stenomyrmex*) *simoni* Emery, 1890a: 64 (w.) VENEZUELA. See also: Brown, 1978c: 613.

**siphneus.** *Anochetus siphneus* Brown, 1978c: 608, figs. 11, 19 (w.) IVORY COAST.

*smithii. Odontomachus smithii* Roger, 1863b: 21. Combination in *Anochetus:* Forel, 1911e: 250. [Unnecessary replacement name for *tyrannicus* Smith, F. 1861b: 44.] Junior synonym of *gladiator:* Brown, 1978c: 556.

*splendens. Anochetus splendens* Karavaiev, 1925c: 289, fig. 11 (q.) INDONESIA (Aru I.). Junior synonym of *isolatus:* Brown, 1978c: 557.

**splendidulus.** *Anochetus splendidulus* Yasumatsu, 1940b: 313, fig. 2 (w.q.) CAROLINA IS.

**striatulus.** *Anochetus* (*Stenomyrmex*) *emarginatus* r. *striatula* Emery, 1890a: 64 (w.) COSTA RICA. Raised to species: Brown, 1978c: 559 (see also p. 610).

**strigatellus.** *Anochetus strigatellus* Brown, 1978c: 582 (w.) WEST MALAYSIA.

**subcoecus.** *Anochetus subcoecus* Forel, 1912a: 46 (w.) TAIWAN. See also: Brown, 1978c: 590.

*subfasciatus. Anochetus cato* var. *subfasciatus* Mann, 1919: 301 (w.q.m.) SOLOMON IS. Junior synonym of *cato:* Wilson, 1959a: 504.

***ANOCHETUS** (continued)*

*sudanicus. Anochetus sudanicus* Weber, 1942a: 47, fig. 7 (w.) SUDAN. Junior synonym of *gracilicornis:* Brown, 1964d: 215.

*taipingensis. Anochetus jacobsoni* var. *taipingensis:* Forel, 1913k: 18 (w.m.) WEST MALAYSIA. Junior synonym of *princeps:* Brown, 1964d: 213.

***taiwaniensis.** Anochetus taiwaniensis* Terayama, 1989a: 26, figs. 1–6 (w.) TAIWAN.

***talpa.** Anochetus talpa* Forel, 1901f: 351 (w.) SOUTH AFRICA.

***targionii.** Anochetus targionii* Emery, 1894c: 187 (w.) BOLIVIA. See also: Brown, 1978c: 616.

*taylori. Anochetus punctiventris* r. *taylori* Forel, 1900c: 63 (w.) INDIA. Raised to species: Bingham, 1903: 43. Junior synonym of *graeffei:* Brown, 1978c: 557.

***testaceus.** Anochetus* (*Stenomyrmex*) *emarginatus* r. *testaceus* Forel, 1893g: 356 (w.m.) ANTILLES. Wheeler, G.C. & Wheeler, J. 1952c: 646 (l.). Raised to species: Brown, 1978c: 559 (see also p. 610).

***traegaordhi.** Anochetus traegaordhi* Mayr, 1904b: 2 (w.) SUDAN. Senior synonym of *silvaticus:* Brown, 1964d: 215; of *angusticornis* and *gracilicornis* (and its junior synonym *sudanicus*): Brown, 1978c: 559 (see also p. 598).

***tua.** Anochetus tua* Brown, 1978c: 580, figs. 1, 29 (w.) WEST MALAYSIA.

***turneri.** Anochetus turneri* Forel, 1900b: 55 (w.) AUSTRALIA. Senior synonym of *latunei:* Brown, 1978c: 559.

*tyrannicus. Odontomachus tyrannicus* Smith, F. 1861b: 44, pl. 1, fig. 4 (w.) SULAWESI. [Junior primary homonym of *Odontomachus tyrannicus* Smith, F. 1859a: 144.] Combination in *Anochetus,* and senior synonym of *Anochetus gladiator:* Dalla Torre, 1893: 48; hence *gladiator* first available replacement name. See also: Brown, 1978c: 574.

*ustus. Anochetus schoutedeni* var. *ustus* Santschi, 1923e: 265 (w.) ZAIRE. Junior synonym of *obscuratus:* Brown, 1978c: 558.

***vallensis.** Anochetus vallensis* Lattke, 1987a: 355, figs. 1, 2 (w.q.) COLOMBIA.

***variegatus.** Anochetus cato* var. *variegatus* Donisthorpe, 1938d: 597 (w.) NEW GUINEA. Raised to species: Wilson, 1959a: 509.

***vexator.** Anochetus* (*Stenomyrmex*) *vexator* Kempf, 1964f: 243, figs. 3, 7, 11, 15 (w.) BRAZIL.

***yerburyi.** Anochetus yerburyi* Forel, 1900c: 62 (w.) SRI LANKA. Imai *et al.* 1984: 5 (k.).

***yunnanensis.** Anochetus yunnanensis* Wang, M. 1993a: 226, fig. 7 (w.) CHINA.

***ANOMALOMYRMA*** (Borneo)

***taylori.** Anomalomyrma taylori* Bolton, 1990b: 279, fig. 7 (q.) BORNEO.

*#ANOMMA:* see under ***DORYLUS.***

***ANONYCHOMYRMA*** (Indo-Australian, Australia)

***anguliceps.** Iridomyrmex anguliceps* Forel, 1901b: 18 (q.m.) NEW GUINEA. Combination in *Anonychomyrma:* Shattuck, 1992a: 13.

***angusta.** Iridomyrmex angustus* Stitz, 1911a: 369, fig. 15 (w.) NEW GUINEA. Combination in *Anonychomyrma:* Shattuck, 1992a: 13.

***arcadia.** Iridomyrmex arcadius* Forel, 1915b: 82 (w.) AUSTRALIA. Combination in *Anonychomyrma:* Shattuck, 1992a: 13.

**#ballaratensis.** *Iridomyrmex itinerans* var. *ballaratensis* Forel, 1902h: 472 (w.m.) AUSTRALIA. Combination in *Anonychomyrma:* Shattuck, 1992a: 13. Subspecies of *itinerans:* Taylor & Brown, D.R. 1985: 100.

**#batesi.** *Iridomyrmex scrutator* subsp. *batesi* Forel, 1911e: 283 (w.) NEW GUINEA. Combination in *Anonychomyrma:* Shattuck, 1992a: 14.

***biconvexa.** Iridomyrmex biconvexus* Santschi, 1928e: 471 (w.) AUSTRALIA. Combination in *Anonychomyrma:* Shattuck, 1992a: 13. Senior synonym of *foetans:* Brown, 1954i: 67. [Misspelled as *triconvexus* by Baroni Urbani, 1977e: 76.]

***constricta.** **Hypoclinea constricta* Mayr, 1868c: 60, pl. 3, figs. 50, 51 (w.) BALTIC AMBER (Oligocene). Combination in *Bothriomyrmex:* Dalla Torre, 1893: 170; in *Iridomyrmex:* Wheeler, W.M. 1915h: 88; in *Anonychomyrma:* Shattuck, 1992a: 13.

**#contenta.** *Iridomyrmex dimorphus* subsp. *contenta* Viehmeyer, 1913: 146 (w., in copal) SULAWESI. Combination in *Anonychomyrma:* Shattuck, 1992a: 13.

**#depilis.** *Iridomyrmex itinerans* var. *depilis* Forel, 1902h: 471 (w.) AUSTRALIA. Subspecies of *itinerans:* Taylor & Brown, D.R. 1985: 100. Combination in *Anonychomyrma:* Shattuck, 1992a: 13.

***dimorpha.** Iridomyrmex dimorphus* Viehmeyer, 1912: 7, fig. 7 (w.) NEW GUINEA. Donisthorpe, 1941h: 60 (m.). Combination in *Anonychomyrma:* Shattuck, 1992a: 13. Redescription of w.: Mann, 1919: 363. Current subspecies: nominal plus *contenta.*

*foetans. Iridomyrmex foetans* Clark, 1929: 122, pl. 1, fig. 4 (w.) AUSTRALIA. Junior synonym of *biconvexa:* Brown, 1954i: 67.

***fornicata.** Iridomyrmex fornicatus* Emery, 1914b: 185, fig. 4 (w.) AUSTRALIA. Combination in *Anonychomyrma:* Shattuck, 1992a: 13.

***froggatti.** Iridomyrmex froggatti* Forel, 1902h: 470 (q.) AUSTRALIA. Combination in *Anonychomyrma:* Shattuck, 1992a: 13.

***gigantea.** Iridomyrmex giganteus* Donisthorpe, 1943d: 456 (q.) NEW GUINEA. Combination in *Anonychomyrma:* Shattuck, 1992a: 13.

***gilberti.** Iridomyrmex gilberti* Forel, 1902h: 470 (w.) AUSTRALIA. Combination in *Anonychomyrma:* Shattuck, 1992a: 13.

***glabrata.** Tapinoma glabrata* Smith, F. 1857a: 58 (w.) WEST MALAYSIA. Combination in *Iridomyrmex:* Mayr, 1886c: 356; in *Anonychomyrma:* Shattuck, 1992a: 13.

***incisa.** Iridomyrmex incisa* Stitz, 1932b: 734, fig. 1 (w.q.) SUMATRA. Combination in *Anonychomyrma:* Shattuck, 1992a: 13.

***itinerans.** Formica itinerans* Lowne, 1865a: 278 (w.) AUSTRALIA. Forel, 1913g: 189 (m.); Wheeler, G.C. & Wheeler, J. 1966: 729 (l.); Crozier, 1969: 249 (k.). Combination in *Hypoclinea:* Mayr, 1870b: 955; in *Iridomyrmex:* Emery, 1887a: 249; in *Anonychomyrma:* Shattuck, 1992a: 13. Current subspecies: nominal plus *ballaratensis, depilis, perthensis.*

***longicapitata.** Iridomyrmex longicapitatus* Donisthorpe, 1947c: 587 (q.) NEW GUINEA. Combination in *Anonychomyrma:* Shattuck, 1992a: 13.

***longiceps.** Iridomyrmex longiceps* Forel, 1907a: 27 (q.) AUSTRALIA. Combination in *Anonychomyrma:* Shattuck, 1992a: 13.

***malandana.** Iridomyrmex innocens* r. *malandanus* Forel, 1915b: 81 (w.) AUSTRALIA. Combination in *Anonychomyrma,* and raised to species: Shattuck, 1992a: 13.

***minuta.** Iridomyrmex minutus* Donisthorpe, 1943a: 171 (w.) NEW GUINEA. Combination in *Anonychomyrma:* Shattuck, 1992a: 14.

***murina.** Iridomyrmex murinus* Emery, 1911b: 536, fig. 4 (w.) NEW GUINEA. Combination in *Anonychomyrma:* Shattuck, 1992a: 14.

***myrmex.** Anonychomyrma myrmex* Donisthorpe, 1947c: 589 (m.) NEW GUINEA.

***nitidiceps.** Iridomyrmex nitidiceps* André, 1896b: 258 (w.) AUSTRALIA. Forel, 1907h: 292 (m.). Combination in *Anonychomyrma:* Shattuck, 1992a: 14. Subspecies of *itinerans:* Forel, 1902h: 472; Emery, 1911d: 25. Revived status as species: Taylor & Brown, D.R. 1985: 101. See also: Shattuck, 1994: 7.

**#perthensis.** *Iridomyrmex itinerans* r. *perthensis* Forel, 1902h: 472 (w.) AUSTRALIA. Combination in *Anonychomyrma:* Shattuck, 1992a: 13.

***polita.** Iridomyrmex politus* Stitz, 1912: 508, fig. 10 (w.) INDONESIA (Seram I.). Combination in *Anonychomyrma:* Shattuck, 1992a: 14.

***procidua.** Formica procidua* Erichson, 1842: 259 (w.) AUSTRALIA. Combination in *Liometopum:* Roger, 1863b: 14; in *Iridomyrmex:* Dalla Torre, 1893: 169; in *Anonychomyrma:* Shattuck & McArthur, 1995 (in press). See also: Shattuck, 1992a: 16.

***purpurescens.** Formica purpurescens* Lowne, 1865b: 331 (w.q.) AUSTRALIA. Combination in *Anonychomyrma:* Shattuck, 1992a: 14.

***samlandica.** **Iridomyrmex samlandicus* Wheeler, W.M. 1915h: 91, figs. 43, 44 (w.) BALTIC AMBER (Oligocene). Combination in *Anonychomyrma:* Shattuck, 1992a: 14.

***scrutator.** Formica scrutator* Smith, F. 1859a: 138 (w.) INDONESIA (Aru I.). Karavaiev, 1926d: 439 (q.m.). Combination in *Iridomyrmex:* Emery, 1887a: 250; in *Anonychomyrma:* Shattuck, 1992a: 14. Current subspecies: nominal plus *batesi.*

***sellata.** Iridomyrmex sellatus* Stitz, 1911a: 369, fig. 16 (w.) NEW GUINEA. Combination in *Anonychomyrma:* Shattuck, 1992a: 14.

***tigris.** Iridomyrmex tigris* Stitz, 1912: 507, fig. 9 (q.) NEW GUINEA. Combination in *Anonychomyrma:* Shattuck, 1992a: 14.

*triconvexus;* see under ***biconvexa.***

***ANOPLOLEPIS*** (Afrotropical, Malagasy; tramp-species)

**#angolensis.** *Anoplolepis* (*Mesanoplolepis*) *trimeni* var. *angolensis* Santschi, 1930b: 72 (w.) ANGOLA.

*berthoudi. Formica berthoudi* Forel, 1876: 33 (w.) LESOTHO. Junior synonym of *custodiens:* Forel, 1879a: 91.

**#biskrensis.** *Anoplolepis* (*Anoplolepis*) *simulans* var. *biskrensis:* Karavaiev, 1931a: 309 (w.) ALGERIA.

***bothae.** Plagiolepis bothae* Forel, 1907g: 86 (q.m.) KENYA. Combination in *Plagiolepis* (*Anacantholepis*): Wheeler, W.M. 1922: 930; in *Anoplolepis:* Emery, 1925b: 17.

*braunsi. Plagiolepis braunsi* Forel, 1913a: 141 (w.) SOUTH AFRICA. Combination in *Plagiolepis*

(*Anoplolepis*): Santschi, 1914b: 123; in *Plagiolepis* (*Zealleyella*): Arnold, 1922: 592; in *Anoplolepis:* Emery, 1925b: 17. Junior synonym of *steingroeveri:* Prins, 1982: 217.

***candida.*** *Anoplolepis* (*Tapinolepis*) *candida* Santschi, 1928f: 210 (w.) SOUTH AFRICA.

***carinata.*** *Plagiolepis carinata* Emery, 1899e: 491 (w.q.) CAMEROUN. Combination in *Plagiolepis* (*Anoplolepis*): Santschi, 1914b: 123; in *Anoplolepis:* Emery, 1925b: 17.

***custodiens.*** *Formica custodiens* Smith, F. 1858b: 33 (w.) SOUTH AFRICA. Mayr, 1895: 148 (m.); Emery, 1895h: 44 (q.); Wheeler, G.C. & Wheeler, J. 1953c: 135 (l.). Combination in *Plagiolepis:* Emery, 1892a: 117; in *Plagiolepis* (*Anoplolepis*): Santschi, 1914b: 123; in *Plagiolepis* (*Zealleyella*): Arnold, 1922: 586; in *Anoplolepis:* Emery, 1925b: 17. Senior synonym of *hendecarthrus:* Mayr, 1865: 54; of *berthoudi:* Forel, 1879a: 91. See also: Prins, 1982: 218. Current subspecies: nominal plus *detrita, hirsuta, pilipes.*

***deceptor.*** *Plagiolepis* (*Anoplolepis*) *deceptor* Arnold, 1922: 597 (w.) ZIMBABWE. Combination in *Anoplolepis:* Emery, 1925b: 17; in *Anoplolepis* (*Mesanoplolepis*): Santschi, 1926a: 14.

***decolor.*** *Plagiolepis decolor* Emery, 1895h: 44 (w.) SOUTH AFRICA. Forel, 1910f: 23 (q.m.). Combination in *Plagiolepis* (*Anoplolepis*): Arnold, 1922: 599; in *Anoplolepis:* Emery, 1925b: 17; in *Anoplolepis* (*Mesanoplolepis*): Santschi, 1926a: 14.

#***detrita.*** *Plagiolepis custodiens* var. *detrita* Emery, 1892a: 118 (w.) SOMALIA. Combination in *Anoplolepis:* Emery, 1925b: 17.

***fallax.*** *Plagiolepis fallax* Mayr, 1865: 54 (w.) SOUTH AFRICA. Forel, 1910f: 24 (q.). Combination in *Plagiolepis* (*Zealleyella*): Arnold, 1922: 589; in *Anoplolepis:* Emery, 1925b: 17. Race of *custodiens:* Mayr, 1893: 5; Arnold, 1922: 589. Revived status as species: Emery, 1925b: 17.

#***gertrudae.*** *Plagiolepis steingroeveri* r. *gertrudae* Forel, 1900b: 77 (w.) REUNION I. Combination in *Anoplolepis:* Emery, 1925b: 18.

***gracilipes.*** *Formica gracilipes* Smith, F. 1857a: 55 (w.) SINGAPORE. Mayr, 1867a: 73 (q.). Combination in *Prenolepis:* Mayr, 1862: 698; in *Plagiolepis,* and senior synonym of *trifasciata:* Mayr, 1867a: 73. Junior synonym of *longipes* Jerdon: Emery, 1887a: 247; hence first available replacement name for *Formica longipes* Jerdon, 1851: 122. [Junior primary homonym of *Formice longipes* Latreille, 1802c: 233 (now in *Pheidole*).]

*hendecarthrus. Camponotus hendecarthrus* Roger, 1863a: 132 (w.q.) SOUTH AFRICA. Junior synonym of *custodiens:* Mayr, 1865: 54.

#***hirsuta.*** *Plagiolepis custodiens* var. *hirsuta* Emery, 1892a: 118 (w.) SOMALIA. Combination in *Anoplolepis:* Emery, 1925b: 17.

#***karrooensis.*** *Plagiolepis* (*Anoplolepis*) *trimeni* var. *karrooensis* Arnold, 1922: 597 (w.q.) SOUTH AFRICA. Combination in *Anoplolepis:* Emery, 1925b: 18.

***litoralis.*** *Anoplolepis* (*Mesanoplolepis*) *litoralis* Arnold, 1958: 124 (w.) ZIMBABWE.

*longipes. Formica longipes* Jerdon, 1851: 122 (w.) INDIA. Wheeler, G.C. & Wheeler, J. 1953c: 132 (l.); Imai, Brown, *et al.* 1984: 68 (k.). Combination in *Plagiolepis:* Emery, 1887a: 247; in *Plagiolepis* (*Anoplolepis*): Santschi, 1914b: 123; in *Anoplolepis:* Emery, 1925b: 17. See also: Baker, 1976: 253; Haines & Haines, 1978: 109; Rao & Veeresh, 1991: 261. [Junior primary homonym of *Formica longipes* Latreille, 1802c: 233 (now in *Pheidole*).] Replacement name: *Formica gracilipes* Smith, F. 1857a: 55. [Note: this name is the oldest available junior synonym of *longipes* Jerdon (synonymy by Emery, 1887a: 247; confirmed by examination of syntypic workers), and hence is first available replacement name.]

***macgregori.*** *Plagiolepis* (*Anoplolepis*) *macgregori* Arnold, 1922: 598 (w.) ZIMBABWE. Combination in *Anoplolepis:* Emery, 1925b: 17.

***macrophthalma.*** *Anoplolepis macrophthalma* Arnold, 1962: 844, fig. 1 (w.q.m.) ZIMBABWE.

***mediterranea.*** *Plagiolepis mediterranea* Mayr, 1866a: 493 (w.) EGYPT. Combination in *Anoplolepis:* Emery, 1925b: 17.

***melanaria.*** *Plagiolepis* (*Anoplolepis*) *melanaria* Arnold, 1922: 599 (w.) ZIMBABWE. Combination in *Anoplolepis:* Emery, 1925b: 18; in *Anoplolepis* (*Mesanoplolepis*): Santschi, 1926a: 14. Current subspecies: nominal plus *ochraceotincta.*

***nuptialis.*** *Plagiolepis* (*Anoplolepis*) *nuptialis* Santschi, 1917b: 289 (m.) SOUTH AFRICA. Prins, 1982: 222 (q.). Combination in *Anoplolepis:* Emery, 1925b: 18; in *Anoplolepis* (*Zealleyella*): Santschi, 1926a: 14.

#***ochraceotincta.*** *Anoplolepis* (*Mesanoplolepis*) *melanaria* subsp. *ochraceotincta* Arnold, 1949: 268 (w.) ZIMBABWE.

***opaciventris.*** *Plagiolepis opaciventris* Emery, 1899e: 492 (q.) GABON. Combination in *Plagiolepis* (*Anoplolepis*): Wheeler, W.M. 1922a: 934; in *Anoplolepis:* Emery, 1925b: 18.

#***parsonsi.*** *Anoplolepis* (*Zealleyella*) *braunsi* st. *parsonsi* Santschi, 1937g: 83 (w.) SOUTH AFRICA. Subspecies of *steingroeveri:* current status.

***pernix.*** *Plagiolepis* (*Anoplolepis*) *pernix* Viehmeyer, 1923: 92 (w.) SUDAN. Combination in *Anoplolepis:* Emery, 1925b: 18.

#***pilipes.*** *Plagiolepis custodiens* var. *pilipes* Emery, 1892a: 118 (w.) SOMALIA. Combination in *Anoplolepis:* Emery, 1925b: 17.

***rufescens.*** *Plagiolepis* (*Anoplolepis*) *rufescens* Santschi, 1917b: 289 (s.w.) SOUTH AFRICA. Combination in *Anoplolepis:* Emery, 1925b: 18; in *Anoplolepis* (*Zealleyella*): Santschi, 1926a: 14.

***simulans.*** *Plagiolepis simulans* Santschi, 1908: 531, fig. 11 (w.q.m.) TUNISIA. Combination in *Plagiolepis* (*Anoplolepis*): Santschi, 1914b: 123; in *Anoplolepis:* Emery, 1925b: 18; in *Anoplolepis* (*Mesanoplolepis*): Santschi, 1926a: 14. Current subspecies: nominal plus *biskrensis.*

***steingroeveri.*** *Plagiolepis steingroeveri* Forel, 1894b: 72 (w.) NAMIBIA. Emery, 1895h: 44 (q.); Arnold, 1922: 591 (q.m.). Combination in *Plagiolepis* (*Anoplolepis*): Santschi, 1914b: 123; in *Plagiolepis* (*Zealleyella*): Arnold, 1922: 590; in *Anoplolepis:* Emery, 1925b: 18; in *Anoplolepis* (*Zealleyella*): Santschi, 1926a: 14. Senior synonym of *braunsi:* Prins, 1982: 217. Current subspecies: nominal plus *gertrudae, parsonsi.*

***tenella.*** *Plagiolepis tenella* Santschi, 1911c: 364 (w.) CONGO. [Also described as new by Santschi, 1911g: 210.] Wheeler, W.M. 1922a: 214 (q.m.). Combination in *Plagiolepis* (*Anoplolepis*): Santschi, 1914b: 123; in *Anoplolepis:* Emery, 1925b: 18.

*trifasciata. Formica trifasciata* Smith, F. 1858b: 27 (q.) JAVA. Junior synonym of *gracilipes:* Mayr, 1867a: 73.

***trimenii.*** *Plagiolepis trimenii* Forel, 1895g: 430 (w.) SOUTH AFRICA. Arnold, 1922: 596 (q.). Combination in *Plagiolepis* (*Anoplolepis*): Arnold, 1922: 595; in *Anoplolepis:* Emery, 1925b: 18; in *Anoplolepis* (*Mesanoplolepis*): Santschi, 1930b: 72. Current subspecies: nominal plus *angolensis, karrooensis.*

***tumidula.*** *Plagiolepis* (*Anoplolepis*) *tumidula* Emery, 1915g: 19, fig. 10 (w.m.) ETHIOPIA. Combination in *Anoplolepis* (*Tapinolepis*): Emery, 1925b: 18.

---

**EXCURSUS**

**Species-rank taxa of *ANOPLOLEPIS*, grouped by subgenus.**

#***ANOPLOLEPIS:*** *bothae, carinata, deceptor, gracilipes, macgregori, mediterranea, opaciventris, pernix, tenella.*

#***MESANOPLOLEPIS:*** *decolor, litoralis, macrophthalma, melanaria, simulans, trimenii.*

#***TAPINOLEPIS:*** *candida, tumidula.*

#***ZEALLEYELLA:*** *custodiens, fallax, nuptialis, rufescens, steingroeveri.*

---

#***ANOPLOMYRMA:*** see under ***POLYRHACHIS.***

***ANTICHTHONIDRIS*** (Chile, Argentina)

***bidentata.*** *Monomorium bidentatum* Mayr, 1887: 616 (w.q.) CHILE. Snelling, 1975: 5 (m.); Wheeler, G.C. & Wheeler, J. 1980: 533 (l.). Combination in *Monomorium* (*Notomyrmex*): Emery, 1922e: 169; in *Notomyrmex:* Kusnezov, 1960b: 345; in *Nothidris:* Ettershank, 1966: 107; in *Antichthonidris:* Snelling, 1975: 6. Senior synonym of *piceonigrum:* Kusnezov, 1960b: 345. See also: Kusnezov, 1949a: 431.

***denticulata.*** *Monomorium denticulatum* Mayr, 1887: 614 (w.q.) CHILE. Wheeler, G.C. & Wheeler, J. 1980: 532 (l.). Combination in *Monomorium* (*Notomyrmex*): Emery, 1922e: 169; in *Notomyrmex:* Kusnezov, 1960b: 347; in *Nothidris:* Ettershank, 1966: 107; in *Antichthonidris:* Snelling, 1975: 6. Senior synonym of *inerme:* Kusnezov, 1949a: 432; of *navarinensis* and *picea:* Kusnezov, 1960b: 347.

*inerme. Monomorium* (*Notomyrmex*) *denticulatum* subsp. *inerme* Borgmeier, 1948b: 469, figs. 18, 19 (w.) ARGENTINA. Junior synonym of *denticulata:* Kusnezov, 1949a: 432.

*navarinensis. Monomorium denticulatum* var. *navarinensis* Forel, 1904g: 7 (q.) CHILE. Combination in *Monomorium* (*Notomyrmex*): Emery, 1922e: 169. Junior synonym of *denticulata:* Kusnezov, 1960b: 347.

*picea. Monomorium denticulatum* var. *picea* Emery, 1906c: 120 (w.) CHILE. Combination in *Monomorium* (*Notomyrmex*): Emery, 1922e: 169. Junior synonym of *denticulata:* Kusnezov, 1960b: 347.

*piceonigrum. Monomorium* (*Notomyrmex*) *bidentatum* subsp. *piceonigrum* Borgmeier, 1948b: 468, figs. 16, 17 (w.) ARGENTINA. Junior synonym of *bidentata:* Kusnezov, 1960b: 345.

*ANTILLAEMYRMEX:* see under ***LEPTOTHORAX.***

*APEDUNCULATA:* see under ***PSEUDOMYRMEX.***

***APHAENOGASTER*** (Worldwide except Afrotropical)

*aciculata* Smith, F. 1874; see under ***MESSOR.***

*acoreensis. Aphaenogaster senilis* var. *acoreensis* Santschi, 1933a: 21 (q.) AZORES IS. [Also described as new by Santschi, 1933c: 397.] Junior synonym of *senilis:* Yarrow, 1967: 26.

*aegyptiaca* Emery, 1878; see under ***MESSOR.***

#***afra.*** *Aphaenogaster (Aphaenogaster) depilis* var. *afra* Santschi, 1933c: 395, fig. 32 (w.m.) ALGERIA. Subspecies of *depilis:* Cagniant, 1968a: 142.

*alabamensis. Aphaenogaster treatae* subsp. *pluteicornis* var. *alabamensis* Wheeler, G.C. & Wheeler, E.W. 1934: 10 (w.q.) U.S.A. Unavailable name; material referred to *treatae* by Creighton, 1950a: 153.

***albisetosa.*** *Aphaenogaster albisetosa* Mayr, 1886d: 446 (w.) U.S.A. Wheeler, W.M. & Creighton, 1934: 350 (q.m.); Wheeler, G.C. & Wheeler, J. 1953b: 70 (l.). Combination in *Stenamma (Aphaenogaster)*: Emery, 1895c: 306; in *Novomessor:* Emery, 1915d: 73; in *Aphaenogaster:* Brown, 1974b: 47. Senior synonym of *minor* Enzmann: Brown, 1949a: 49. See also: Hölldobler, Stanton & Engel, 1976: 32; Hölldobler, Stanton & Markl, 1978: 163; Smith, D.R. 1979: 1360; Bolton, 1982: 340.

*andrei* Mayr, 1886; see under ***MESSOR.***

#***angulata.*** *Aphaenogaster (Attomyrma) famelica* subsp. *angulata* Viehmeyer, 1922: 206 (w.) CHINA.

*anguliceps. Aphaenogaster (Deromyrma) longiceps* subsp. *ruginota* var. *anguliceps* Viehmeyer, 1924b: 311 (w.q.m.) AUSTRALIA. Unavailable name (Taylor, 1986: 34).

*angusta. Aphaenogaster angusta* Santschi, 1925g: 340, fig. 1 (w.) SPAIN. Junior synonym of *iberica:* Espadaler, 1983: 226.

***annandalei.*** *Aphaenogaster (Attomyrma) annandalei* Mukerjee, 1930: 151, fig. 1 (w.) INDIA.

#***anoemica.*** *Aphaenogaster sardoa* var. *anoemica* Santschi, 1910a: 43 (q.) MOROCCO.

*aquia. Myrmica (Monomorium) aquia* Buckley, 1867: 341 (w.q.) U.S.A. Combination in *Aphaenogaster,* and junior synonym of *fulva:* Mayr, 1886c: 365. Revived from synonymy as subspecies of *fulva:* Emery, 1895c: 304. Unrecognisable taxon in *Aphaenogaster:* Creighton, 1950a: 148; Smith, D.R. 1979: 1363.

*aralocaspia* Ruzsky, 1902; see under ***MESSOR.***

***araneoides.*** *Aphaenogaster araneoides* Emery, 1890b: 48, pl. 5, fig. f8 (w.) COSTA RICA. Borgmeier, 1949: 206 (m.). [Also described as new by Emery, 1894k: 54.] Combination in *Aphaenogaster (Ischnomyrmex)*: Forel, 1899c: 60; in *Stenamma:* Forel, 1907e: 4; in *Aphaenogaster (Deromyrma)*: Emery, 1915d: 71. Current subspecies: nominal plus *canalis, inermis, nitidiventris.*

#***armeniaca.*** *Aphaenogaster (Attomyrma) pallida* subsp. *armeniaca* Arnol'di, 1968: 1807 (w.) ARMENIA. Subspecies of *subterraneoides:* Arakelian, 1994: 33.

***ashmeadi.*** *Stenamma (Aphaenogaster) treatae* var. *ashmeadi* Emery, 1895c: 302 (w.) U.S.A. Combination in *Aphaenogaster:* Wheeler, W.M. 1913c: 114; in *Aphaenogaster (Attomyrma)*: Emery, 1921f: 60. Raised to species, and senior synonym of *harnedi:* Creighton, 1950a: 142.

***atlantis.*** *Aphaenogaster (Aphaenogaster) praedo* var. *atlantis* Santschi, 1929e: 139 (w.) MOROCCO. Cagniant, 1990b: 125 (q.m.). Raised to species: Cagniant, 1990b: 123.

*atra. Aphaenogaster (Planimyrma) loriai* var. *atra* Stitz, 1938: 101 (w.m.) NEW GUINEA. Junior synonym of *loriai:* Smith, M.R. 1961b: 223.

****avita.*** **Aphaenogaster (Deromyrma) avita* Fujiyama, 1970: 67, pl. 15, figs. 2, 3 (q.) JAPAN (Miocene).

#***azteca.*** *Aphaenogaster fulva* var. *azteca* Enzmann, J. 1947b: 150 (w.) MEXICO. [First available use of *Stenamma (Aphaenogaster) fulvum* subsp. *aquia* var. *aztecum* Emery, 1895c: 305; unavailable name.]

*bakeri. Stenamma (Aphaenogaster) patruele* subsp. *bakeri* Wheeler, W.M. 1904d: 270 (w.) U.S.A. Combination in *Aphaenogaster:* Wheeler, W.M. 1916h: 143; in *Aphaenogaster (Attomyrma)*: Emery, 1921f: 59. Junior synonym of *patruelis:* Smith, D.R. 1979: 1362.

***balcanica.*** *Stenamma (Aphaenogaster) testaceopilosa* var. *balcanica* Emery, 1898c: 136 (w.) ALBANIA. Menozzi, 1936d: 271 (q.m.l.). Combination in *Aphaenogaster:* Forel, 1911d: 346. Variety of *simonellii:* Santschi, 1933c: 390. Raised to species: Agosti & Collingwood, 1987a: 53.

*barbarastructor* Dalla Torre, 1893; see under ***MESSOR.***

***barbigula.*** *Aphaenogaster (Nystalomyrma) barbigula* Wheeler, W.M. 1916j: 221, pl. 21, figs. 7–9 (w.q.) AUSTRALIA. Forel, 1918a: 720 (m.).

*barcinensis. Aphaenogaster (Attomyrma) gibbosa* var. *barcinensis* Menozzi, 1922b: 325 (w.q.) SPAIN. Junior synonym of *gibbosa:* Collingwood, 1978: 67.

***baronii.*** *Aphaenogaster baronii* Cagniant, 1988c: 43, figs. 1–5 (w.q.m.) MOROCCO.

***beccarii.*** *Aphaenogaster (Ischnomyrmex) beccarii* Emery, 1887b: 456, pl. 1, fig. 12 (w.q.m.) SUMATRA. Imai, *et al.* 1984: 6 (k.). Combination in *Aphaenogaster (Deromyrma)*: Emery, 1915d: 71. See also: Bingham, 1903: 271.

***beesoni.*** *Aphaenogaster (Attomyrma) beesoni* Donisthorpe, 1933a: 24 (w.) INDIA.

***belti.*** *Aphaenogaster belti* Forel, 1895c: 248 (w.) MADAGASCAR. Combination in *Brunella:* Forel, 1917: 234; in *Atopula:* Emery, 1915d: 68; in *Aphaenogaster:* Bolton, 1982: 341 (see also p. 364).

**berendti* Mayr, 1868; see under ***STENAMMA.***

*blanci* André, 1881; see under ***GONIOMMA.***

*borealis. Aphaenogaster subterranea* subsp. *borealis* Wheeler, W.M. 1915b: 412 (w.) CANADA. Junior synonym of *occidentalis:* Creighton, 1950a: 150.

***boulderensis.*** *Aphaenogaster (Attomyrma) boulderensis* Smith, M.R. 1941: 120 (w.) U.S.A. Current subspecies: nominal plus *smithi.*

#***brevicollis.*** *Aphaenogaster (Ischnomyrmex) phalangium* var. *brevicollis* Forel, 1899c: 59 (w.m.). Combination in *Aphaenogaster (Deromyrma)*: Emery, 1921f: 65.

*brevicornis* Mayr, 1886; see under ***STENAMMA.***

*brevicornis. Aphaenogaster (Attomyrma) smythiesi* subsp. *brevicornis* Menozzi, 1941: 17 (w.m.) CHINA. *Nomen nudum.*

***burri.*** *Goniomma burri* Donisthorpe, 1950b: 60 (q.) TURKEY. Combination in *Aphaenogaster:* **new combination.**

#***cabylica.*** *Aphaenogaster testaceopilosa* r. *cabylica* Stitz, 1917: 341, fig. 10 (w.) ALGERIA. Santschi, 1929e: 141 (q.); Santschi, 1939d: 67 (m.).

***caeciliae.*** *Aphaenogaster (Attomyrma) caeciliae* Viehmeyer, 1922: 206 (w.) CHINA.

*calderoni* Forel, 1914; see under ***MANICA.***

***campana.*** *Aphaenogaster testaceopilosa* var. *campana* Emery, 1878b: 54, fig. (w.q.m.) ITALY. Retained as subspecies: Emery, 1908c: 322; Emery, 1916b: 131; Emery, 1921f: 62. Raised to species: Dalla Torre, 1893: 100; Santschi, 1933c: 400; Baroni Urbani, 1971c: 40.

#***canalis.*** *Aphaenogaster (Deromyrma) araneoides* var. *canalis* Enzmann, J. 1947b: 149, pl. 8 (w.) PANAMA.

#***canescens.*** *Stenamma (Aphaenogaster) testaceopilosa* subsp. *canescens* Emery, 1895b: 70 (w.) ALGERIA. Santschi, 1939d: 68 (m.). Combination in *Aphaenogaster:* Emery, 1908c: 322.

#***carbonaria.*** *Aphaenogaster carbonaria* Pergande, 1894b: 163 (w.) MEXICO. Combination in *Stenamma (Messor)*: Emery, 1895c: 308; in *Aphaenogaster:* Wheeler, W.M. 1917a: 516; in *Aphaenogaster (Attomyrma)*: Emery, 1921f: 59. Subspecies of *patruelis:* Forel, 1899c: 58; Emery, 1921f: 59.

***cardenai.*** *Aphaenogaster cardenai* Espadaler, 1981: 77, 2 figs. (w.) SPAIN.

#***carolinensis.*** *Aphaenogaster texana* var. *carolinensis* Wheeler, W.M. 1915b: 414 (w.q.) U.S.A. Combination in *Aphaenogaster (Attomyrma)*: Emery, 1921f: 60. Subspecies of *texana:* Creighton, 1950a: 153.

***cavernicola.*** *Aphaenogaster cavernicola* Donisthorpe, 1938e: 500 (w.) INDIA.

***cecconii.*** *Aphaenogaster (Ischnomyrmex) cecconii* Emery, 1894j: 7, fig. (w.) GREECE. Combination in *Aphaenogaster (Deromyrma)*: Emery, 1915d: 71.

*chorassanica. Aphaenogaster (Attomyrma) gibbosa* subsp. *chorassanica* Arnol'di, 1968: 1805 (w.q.m.) TURKMENIA. Junior synonym of *gibbosa:* Dlussky, Soyunov & Zabelin, 1990: 212.

#***clara.*** *Deromyrma swammerdami* var. *clara* Santschi, 1915c: 250 (w.) MADAGASCAR. Combination in *Aphaenogaster (Deromyrma)*: Emery, 1921f: 65; Wheeler, W.M. 1922a: 1017.

*clara. Aphaenogaster (Attomyrma) subterranea* var. *clara* Donisthorpe, 1926a: 6. *Nomen nudum,* attributed to Santschi.

*clavata* Emery, 1877; see under ***PHEIDOLE.***

***cockerelli.*** *Aphaenogaster (Ischnomyrmex) cockerelli* André, 1893b: 150 (w.) MEXICO. Wheeler, W.M. & Creighton, 1934: 350 (q.m.); Wheeler, G.C. & Wheeler, J. 1960b: 10 (l.). Combination in *Stenamma (Ischnomyrmex)*: Forel, 1901c: 128; in *Novomessor:* Emery, 1915d: 73; in *Aphaenogaster:* Brown, 1974b: 47. Senior synonym of *sonorae:* Wheeler, W.M. & Creighton, 1934: 352. See also: Wheeler, G.C. & Wheeler, J. 1972b: 239; Hölldobler, Stanton & Engel, 1976: 32; Hölldobler, Stanton & Markl, 1978: 163; Smith, D.R. 1979: 1360; Bolton, 1982: 340.

#***confinis.*** *Aphaenogaster (Aphaenogaster) praenoda* var. *confinis* Santschi, 1933c: 399 (m.) MOROCCO.

*conica*. *Aphaenogaster conica* Viehmeyer, 1914c: 40 (w.) SPAIN. Junior synonym of *iberica:* Viehmeyer, 1922: 205; Santschi, 1925g: 341.

*continentis* Forel, 1911; see under ***PHEIDOLE.***

*corsica*. *Aphaenogaster spinosa* var. *corsica* Santschi, 1933c: 390 (m.) CORSICA. *Nomen nudum.* [Material labelled as *corsica* has been attributed to *spinosa* by Baroni Urbani, 1971d: 1039.]

#***crinimera***. *Aphaenogaster* (*Attomyrma*) *huachucana* subsp. *crinimera* Cole, 1953b: 82, fig. 1 (w.q.m.) U.S.A.

***cristata***. *Stenamma* (*Aphaenogaster*) *cristatus* Forel, 1902c: 225 (w.) INDIA. Combination in *Aphaenogaster:* Bingham, 1903: 276; in *Aphaenogaster* (*Attomyrma*): Emery, 1921f: 57.

***crocea***. *Aphaenogaster crocea* André, 1881a: 61 and xlix (w.) ALGERIA. Emery, 1908c: 330 (q.); Cagniant, 1966c: 63 (m.). Combination in *Aphaenogaster* (*Attomyrma*): Emery, 1921f: 57. Race of *subterranea:* Emery, 1891b: 8. Revived status as species: Dalla Torre, 1893: 102. See also: Emery, 1916b: 130. Current subspecies: nominal plus *croceoides, lenis, splendidoides.*

#***croceoides***. *Aphaenogaster subterranea* var. *croceoides* Forel, 1890a: lxxi (w.) ALGERIA. Combination in *Aphaenogaster* (*Attomyrma*): Emery, 1921f: 57. Variety/subspecies of *crocea:* Dalla Torre, 1893: 102; Emery, 1908c: 331.

***curiosa***. *Aphaenogaster* (*Aphaenogaster*) *gemella* var. *curiosa* Santschi, 1933c: 403, figs. 7, 28 (m.) MOROCCO. Cagniant, 1990e: 461 (w.), 465 (q.). Raised to species: Cagniant, 1990e: 461.

#***curta***. *Aphaenogaster* (*Ischnomyrmex*) *swammerdami* var. *curta* Forel, 1891b: 169 (w.) MADAGASCAR. Combination in *Aphaenogaster* (*Deromyrma*): Emery, 1921f: 65.

*cypriotes*. *Aphaenogaster pallida* subsp. *cypriotes* Emery, 1908c: 329 (w.) CYPRUS. Junior synonym of *subterraneoides:* Emery, 1915a: 258.

*debilis* Walker, 1871; see under ***MONOMORIUM.***

***dejeani***. *Aphaenogaster dejeani* Cagniant, 1982: 281, figs. 1, 2 (w.m.) MOROCCO. Cagniant, 1992a: 67 (q.).

***depilis***. *Aphaenogaster testaceopilosa* var. *depilis* Santschi, 1911i: 283 (w.) ALGERIA. Santschi, 1933c: 394 (q.m.). Raised to species: Santschi, 1933c: 393. Current subspecies: nominal plus *afra, numida.*

***depressa***. *Aphaenogaster depressa* **new name**. Replacement name for *pergandei* Donisthorpe, 1950d: 639. [Junior secondary homonym of *pergandei* Mayr, 1886d: 448 (now in *Messor*).]

#***disjuncta***. *Aphaenogaster senilis* var. *disjuncta* Santschi, 1933c: 397, figs. 3, 36 (w.m.) MOROCCO.

***dlusskyi***. *Aphaenogaster dlusskyi* Radchenko & Arakelian, 1991: 72, figs. 1, 2 (w.) ARMENIA.

****donisthorpei***. **Aphaenogaster donisthorpei* Carpenter, 1930: 31, pl. 7, fig. 4 (q.) U.S.A. (Miocene).

***dromedaria***. *Ischnomyrmex dromedarius* Emery, 1900c: 322, pl. 8, figs. 23, 24 (w.) NEW GUINEA. Donisthorpe, 1938b: 32 (m.). Combination in *Aphaenogaster:* Viehmeyer, 1914d: 601; in *Aphaenogaster* (*Deromyrma*): Emery, 1915d: 71; in *Aphaenogaster* (*Planimyrma*): Emery, 1921f: 66. Senior synonym of *fusca* and *nigra:* Smith, M.R. 1961b: 219.

***dulciniae***. *Aphaenogaster* (*Attomyrma*) *dulciniae* Emery, 1924c: 164 (w.m.) SPAIN. [First available use of *Aphaenogaster* (*Attomyrma*) *pallida* st. *subterraneoides* var. *dulcineae* Santschi, 1919e: 245; unavailable name.] Santschi, 1932d: 344 (q.). Subspecies of *pallida:* Baroni Urbani, 1971c: 53. Raised to species: Collingwood & Yarrow, 1969: 59; Collingwood, 1978: 67.

*ecalcaratum*. *Stenamma* (*Aphaenogaster*) *tennesseense* var. *ecalcaratum* Emery, 1895c: 301 (w.) U.S.A. Combination in *Aphaenogaster:* Emery, 1921f: 60. Junior synonym of *tennesseensis:* Creighton, 1950a: 151.

#***ellipsoida***. *Aphaenogaster* (*Aphaenogaster*) *praedo* var. *ellipsoida* Santschi, 1933c: 400, fig. 13 (w.) MOROCCO.

***ensifera***. *Aphaenogaster* (*Ischnomyrmex*) *ensifera* Forel, 1899c: 59 (w.) MEXICO. Combination in *Aphaenogaster* (*Deromyrma*): Emery, 1915d: 71. Senior synonym of *manni:* Brown, 1974b: 46.

#***epinotalis***. *Aphaenogaster relicta* subsp. *epinotalis* Wheeler, W.M. & Mann, 1914: 27, figs. 8a,b (w.) HAITI. Combination in *Veromessor:* Wheeler, W.M. & Creighton, 1934: 384; in *Aphaenogaster:* **revived combination**.

***epirotes***. *Stenamma* (*Aphaenogaster*) *epirotes* Emery, 1895b: 72 (w.) ALBANIA. Müller, 1923: 50 (q.m.). Combination in *Aphaenogaster:* Emery, 1908c: 336; in *Aphaenogaster* (*Attomyrma*): Emery, 1921f: 58. Variety/subspecies of *obsidiana:* Emery, 1908c: 336; Müller, 1921: 49; Finzi, 1930d: 312; Baroni Urbani, 1971c: 50. Revived status as species: Agosti & Collingwood, 1987a: 53. See also: Atanassov & Dlussky, 1992: 108.

#***erabu***. *Aphaenogaster famelica* subsp. *erabu* Nishizono & Yamane, 1990: 29, figs. 1, 2, 5 (w.q.) JAPAN.

***espadaleri***. *Aphaenogaster espadaleri* Cagniant, 1984: 387, figs. 1–11 (w.q.m.) MOROCCO.

#***etrusca***. *Aphaenogaster spinosa* subsp. *etrusca* Baroni Urbani, 1969a: 328, fig. 1 (w.) ITALY.

***exasperata***. *Aphaenogaster exasperata* Wheeler, W.M. 1921c: 537 (w.) CHINA. Combination in *Aphaenogaster* (*Attomyrma*): Wheeler, W.M. 1930h: 62.

*expolitus*. *Aphaenogaster* (*Attomyrmex* [sic]) *expolitus* Azuma, 1950: 34, JAPAN. *Nomen nudum.* See Onoyama, 1980: 194.

***fabulosa***. *Aphaenogaster* (*Attomyrma*) *fabulosa* Arnol'di, 1968: 1803, fig. 5 (w.) TURKMENIA.

***fallax***. *Aphaenogaster fallax* Cagniant, 1992c: 199, figs. 2–4 (w.q.m.) MOROCCO.

***famelica***. *Ischnomyrmex famelicus* Smith, F. 1874: 405 (w.) JAPAN. Wheeler, W.M. 1928d: 102 (q.m.); Wheeler, G.C. & Wheeler, J. 1953b: 60 (l.); Imai, 1966: 131 (k.). Combination in *Aphaenogaster:* Mayr, 1879: 669; in *Stenamma* (*Ischnomyrmex*): Forel, 1901h: 61; in *Stenamma* (*Aphaenogaster*): Wheeler, W.M. 1906c: 315; in *Aphaenogaster:* Emery, 1908c: 324; in *Aphaenogaster* (*Attomyrma*): Emery, 1921f: 57. See also: Imai, 1971: 137. Current subspecies: nominal plus *angulata, erabu, frontosa.*

***faureli***. *Aphaenogaster crocea* subsp. *faureli* Cagniant, 1969a: 232, figs. (w.q.m.) ALGERIA. Raised to species: Cagniant, 1970a: 414.

***feae***. *Aphaenogaster* (*Ischnomyrmex*) *feae* Emery, 1889b: 503, pl. 11, fig. 26 (w.) BURMA. Wheeler, W.M. 1927b: 43 (q.m.). Combination in *Aphaenogaster* (*Deromyrma*): Emery, 1915d: 71. Current subspecies: nominal plus *imbellis, nicobarensis, simulans.*

***festae***. *Aphaenogaster splendida* subsp. *festae* Emery, 1915h: 2 (w.) GREECE. Menozzi, 1936d: 270 (q.). Combination in *Aphaenogaster* (*Attomyrma*): Emery, 1921f: 60. Raised to species: Arnol'di, 1976b: 1024 (in key); Agosti & Collingwood, 1987a: 53.

***finzii***. *Aphaenogaster pallida* subsp. *finzii* Müller, 1921: 47 (w.) YUGOSLAVIA. Müller, 1923: 53 (q.m.). Raised to species: Agosti & Collingwood, 1987a: 53. See also: Finzi, 1930a: 153.

#***fiorii***. *Aphaenogaster gibbosa* subsp. *fiorii* Emery, 1915a: 258 (w.) ITALY. Combination in *Aphaenogaster* (*Attomyrma*): Emery, 1921f: 58. See also: Baroni Urbani, 1971c: 49.

#***flava***. *Aphaenogaster* (*Nystalomyrma*) *longiceps* var. *flava* Emery, 1921f: 61 (m.) AUSTRALIA.

***flemingi***. *Aphaenogaster texana* subsp. *flemingi* Smith, M.R. 1928c: 275 (w.) U.S.A. Wheeler, G.C. & Wheeler, J. 1972b: 238 (l.). Combination in *Aphaenogaster* (*Attomyrma*), and raised to species: Creighton, 1950a: 143. Senior synonym of *macrospina:* Smith, M.R. 1958a: 113.

***floridana***. *Aphaenogaster* (*Attomyrma*) *floridana* Smith, M.R. 1941: 118 (w.) U.S.A. Wheeler, G.C. & Wheeler, J. 1960b: 7 (l.).

***friederichsi***. *Aphaenogaster friederichsi* Forel, 1918b: 151 (w.) MADAGASCAR. Combination in *Aphaenogaster* (*Attomyrma*): Emery, 1921f: 57.

#***frontosa***. *Aphaenogaster* (*Attomyrma*) *famelica* subsp. *frontosa* Wheeler, W.M. 1928d: 104 (w.) JAPAN.

*fuentei*. *Aphaenogaster testaceopilosa* var. *fuentei* Medina, 1893: 105 (w.) SPAIN. Junior synonym of *senilis:* Collingwood, 1978: 67.

***fulva***. *Aphaenogaster fulva* Roger, 1863a: 190 (w.) U.S.A. Mayr, 1886d: 445 (q.m.). Combination in *Stenamma* (*Aphaenogaster*): Emery, 1895c: 303; in *Aphaenogaster:* Wheeler, W.M. 1913c: 114; in *Aphaenogaster* (*Attomyrma*): Emery, 1921f: 57. Senior synonym of *rubida:* Brown, 1949a: 49. Current subspecies: nominal plus *azteca.* [See also entry under *pusilla.*]

*funkikoensis*. *Aphaenogaster funkikoensis* Creighton, 1950a: 152 (footnote). Unnecessary (third) replacement name for *silvestrii* Wheeler, W.M. 1929g: 37. [Junior primary homonym of *silvestrii* Menozzi, 1929b: 282.] Junior synonym of *lepida* (first replacement name): Brown, 1954d: 10.

*furvescens*. *Aphaenogaster texana* var. *furvescens* Wheeler, W.M. 1915b: 413 (w.q.) U.S.A. Combination in *Aphaenogaster* (*Attomyrma*): Emery, 1921f: 60. Junior synonym of *texana:* Creighton, 1950a: 152.

*fusca*. *Ischnomyrmex dromedarius* var. *fusca* Emery, 1900c: 322 (w.) NEW GUINEA. Combination in *Aphaenogaster* (*Planimyrma*): Emery, 1921f: 66. Junior synonym of *dromedaria:* Smith, M.R. 1961b: 219.

*gatesi* Wheeler, W.M. 1927; see under ***PHEIDOLE.***

***geei***. *Aphaenogaster geei* Wheeler, W.M. 1921c: 535 (w.q.m.) CHINA. Combination in *Aphaenogaster* (*Attomyrma*): Wheeler, W.M. 1929f: 2.

***gemella***. *Atta gemella* Roger, 1862b: 260 (w.m.) SPAIN. Santschi, 1929e: 140 (q.). Combination in *Aphaenogaster:* Roger, 1863b: 29. Subspe-

***APHAENOGASTER** (continued)*

cies/race/stirps of *testaceopilosa:* Emery & Forel, 1879: 463; Menozzi, 1926b: 181; Santschi, 1929e: 140. Revived status as species: Santschi, 1933c: 401; Cagniant, 1964: 86; Collingwood & Yarrow, 1969: 61. Current subspecies: nominal plus *marocana.*

**georgica.** *Aphaenogaster (Attomyrma) georgica* Arnol'di, 1968: 1804 (w.) GEORGIA.

**gibbosa.** *Formica gibbosa* Latreille, 1798: 50 (w.m.) FRANCE. Emery, 1908c: 333 (q.). Combination in *Aphaenogaster:* Dalla Torre, 1893: 107; in *Stenamma (Aphaenogaster)*: Emery, 1895c: 303; in *Aphaenogaster:* Emery, 1908c: 332; in *Aphaenogaster (Attomyrma)*: Menozzi, 1921: 26. Race of *testaceopilosa:* Emery & Forel, 1879: 463. Subspecies of *subterranea:* Dalla Torre, 1893: 107; Emery, 1895b: 71. Revived status as species: Emery, 1908c: 332. See also: Emery, 1916b: 130; Bondroit, 1918: 157; Santschi, 1923f: 328; Stitz, 1939: 116; Kutter, 1977c: 78; Attanasov & Dlussky, 1992: 109. Senior synonym of *striola:* Emery, 1895c: 303; of *barcinensis* and *laevior* Forel: Collingwood, 1978: 67; of *chorassanica* and *muschtaidica:* Dlussky, Soyunov & Zabelin, 1990: 212. Current subspecies: nominal plus *fiorii; homonyma, mauritanica.*

**gigantea.** *Aphaenogaster (Nystalomyrma) gigantea* Collingwood, 1962: 225, fig. 7 (w.q.) BURMA.

**gonacantha.** *Ischnomyrmex gonacantha* Emery, 1899f: 277, fig. (w.) MADAGASCAR. Combination in *Aphaenogaster (Deromyrma)*: Emery, 1915d: 71.

*gracilinodis* Emery, 1878; see under ***MESSOR.***

**#grata.** *Aphaenogaster senilis* var. *grata* Santschi, 1933c: 397, fig. 34 (m.) SPAIN.

**haarlovi.** *Aphaenogaster (Attomyrma) haarlovi* Collingwood, 1961a: 58, figs. 4, 5 (w.m.) AFGHANISTAN.

*harnedi. Aphaenogaster treatae* subsp. *harnedi* Wheeler, W.M. 1919b: 50 (w.) U.S.A. Junior synonym of *ashmeadi:* Creighton, 1950a: 142.

**hesperia.** *Aphaenogaster hesperia* Santschi, 1911i: 284 (w.) CANARY IS. Combination in *Aphaenogaster (Attomyrma)*: Emery, 1921f: 58.

*hispanica* André, 1883; see under ***GONIOMMA.***

**holtzi.** *Stenamma (Aphaenogaster) subterraneum* subsp. *holtzi* Emery, 1898c: 139 (w.) TURKEY. Combination in *Aphaenogaster,* and raised to species: Emery, 1908c: 328.

**#homonyma.** *Aphaenogaster (Attomyrma) gibbosa* var. *homonyma* Emery, 1921f: 58 (w.) TUNISIA.

**honduriana.** *Aphaenogaster (Deromyrma) honduriana* Mann, 1922: 23, fig. 11 (w.q.) HONDURAS.

**huachucana.** *Aphaenogaster (Attomyrma) huachucana* Creighton, 1934: 189 (w.) U.S.A. Creighton, 1952a: 94 (q.m.). Current subspecies: nominal plus *crinimera.*

**hunanensis.** *Aphaenogaster hunanensis* Wu & Wang, 1992: 1304, fig. 4575 (w.q.) CHINA.

**iberica.** *Aphaenogaster testaceopilosa* subsp. *iberica* Emery, 1908c: 321, fig. 10 (w.m.) SPAIN. Raised to species: Bondroit, 1918: 162. See also: Santschi, 1933c: 390. Senior synonym of *conica:* Viehmeyer, 1922: 205; Santschi, 1925g: 341; of *angusta* and *vieirai:* Espadaler, 1983: 226. See also: Collingwood & Yarrow, 1969: 61.

**#ichnusa.** *Aphaenogaster (Attomyrma) subterranea* var. *ichnusa* Santschi, 1925g: 341 (w.) SARDINIA.

**#imbellis.** *Ischnomyrmex feae* var. *imbellis* Emery, 1900d: 681 (w.) SUMATRA. Combination in *Aphaenogaster (Deromyrma)*: Emery, 1921f: 65.

**#inermis.** *Aphaenogaster (Ischnomyrmex) araneoides* var. *inermis* Forel, 1899c: 60 (w.m.) COSTA RICA. Wheeler, G.C. & Wheeler, J. 1953b: 64 (l.). Combination in *Aphaenogaster (Deromyrma)*: Kempf, 1972a: 23.

*inermis. Aphaenogaster semipolita* var. *inermis* Santschi, 1933c: 390 (w.) ITALY. [First available use of *Aphaenogaster testaceopilosa* subsp. *semipolita* var. *inermis* Emery, 1908c: 320; unavailable name.] Raised to species: Schembri & Collingwood, 1981: 423. [Junior primary homonym of *inermis* Forel, above.] Replacement name: *inermita* **new name**, below.

**inermita.** *Aphaenogaster inermita* **new name**. Replacement name for *inermis* Santschi, 1933c: 390. [Junior primary homonym of *inermis* Forel, 1899c: 60.]

**ionia.** *Aphaenogaster semipolita* var. *ionia* Santschi, 1933c: 390 (w.) ITALY. [First available use of *Aphaenogaster testaceopilosa* subsp. *semipolita* var. *ionia* Emery, 1915a: 257; unavailable name.] Raised to species: Schembri & Collingwood, 1981: 423. Senior synonym of *melitensis* and *ionia* Baroni Urbani, **new synonymy**.

*ionia. Aphaenogaster semipolita* subsp. *ionia* Baroni Urbani, 1968b: 422, figs. 10, 12 (w.) ITALY. Junior synonym of *ionia* Santschi: **new synonymy**.

**isekram.** *Aphaenogaster (Attomyrma) isekram* Bernard, 1977: 29, fig. 1 (w.) ALGERIA.

**italica.** *Aphaenogaster strioloides* var. *italica* Bondroit, 1918: 158 (w.) ITALY. [First available use of *Aphaenogaster (Attomyrma) gibbosa* subsp. *strioloides* var. *italica* Emery, 1916b: 134; unavailable name.] Stirps of *strioloides:* Santschi, 1932d: 339. Raised to species: Baroni Urbani, 1971c: 49. See also: Kutter, 1977c: 80.

**januschevi.** *Aphaenogaster januschevi* Arnol'di, 1976b: 1020, figs. 1–4 (w.q.m.) RUSSIA.

**#japonica.** *Aphaenogaster schmitzi* subsp. *japonica* Forel, 1911e: 267 (w.) JAPAN. Wheeler, W.M. 1928d: 102 (q.). Combination in *Aphaenogaster (Attomyrma)*: Emery, 1921f: 60. Subspecies of *syriaca:* Emery, 1921f: 60; of *smythiesii:* Wheeler, W.M. 1928d: 101; Nishizono & Yamane, 1990: 34.

*juliana* Pergande, 1894; see under ***MESSOR.***

**kervillei.** *Aphaenogaster kervillei* Forel, 1910a: 12 (w.) LEBANON. Combination in *Aphaenogaster (Attomyrma)*: Emery, 1921f: 58.

**kurdica.** *Aphaenogaster subterranea* var. *kurdica* Ruzsky, 1905b: 717 (w.) RUSSIA. Combination in *Aphaenogaster (Attomyrma)*: Emery, 1921f: 59. Subspecies of *smythiesii:* Emery, 1908c: 332. Raised to species: Arnol'di, 1948: 211.

**laevior.** *Aphaenogaster (Ischnomyrmex) longipes* var. *laevior* Emery, 1887b: 456 (w.) SULAWESI. Combination in *Aphaenogaster (Deromyrma)*: Emery, 1915d: 71. Raised to species: Emery, 1888a: 532.

*laevior. Aphaenogaster striola* var. *laevior* Forel, 1892g: 453 (w.) SPAIN. [Unresolved junior primary homonym of *laevior* Emery, above.] Variety of *gibbosa:* Emery, 1908c: 334. Junior synonym of *gibbosa:* Collingwood, 1978: 67.

*laevis. Atta laevis* Mayr, 1862: 743 (q.) U.S.A. Combination in *Aphaenogaster:* Roger, 1863b: 30. Junior synonym of *tennesseensis:* Mayr, 1886d: 446.

**lamellidens.** *Aphaenogaster lamellidens* Mayr, 1886d: 444 (w.q.m.) U.S.A. Taber & Cokendolpher, 1988: 95 (k.). Combination in *Stenamma (Aphaenogaster)*: Emery, 1895c: 302; in *Aphaenogaster:* Wheeler, W.M. 1913c: 114; in *Aphaenogaster (Attomyrma)*: Emery, 1921f: 58. Senior synonym of *nigripes:* Creighton, 1950a: 144.

**#laurenti.** *Aphaenogaster (Attomyrma) pallida* var. *laurenti* Santschi, 1939d: 69 (w.q.) ALGERIA.

**ledouxi.** *Aphaenogaster ledouxi* Tohmé, G. 1969: 3, fig. 1 (w.) LEBANON.

**#lenis.** *Aphaenogaster crocea* var. *lenis* Santschi, 1911i: 284 (w.) TUNISIA. Combination in *Aphaenogaster (Attomyrma)*: Emery, 1921f: 57.

**lepida.** *Aphaenogaster (Attomyrma) lepida* Wheeler, W.M. 1930a: 96. Replacement name for *silvestrii* Wheeler, W.M. 1929g: 37. [Junior primary homonym of *silvestrii* Menozzi, 1929b: 282.] Senior synonym of *phillipi* and *funkikoensis* (unnecessary second and third replacement names): Brown, 1954d: 10.

**lesbica.** *Aphaenogaster pallida* var. *lesbica* Forel, 1913d: 429 (w.) GREECE. Combination in *Aphaenogaster (Attomyrma)*: Emery, 1921f: 59. Raised to species: Agosti & Collingwood, 1987a: 53.

**#leveillei.** *Aphaenogaster pallida* r. *leveillei* Emery, 1881b: 534 (footnote) (w.) ALGERIA. Emery, 1884a: 382 (m.); Emery, 1908c: 329 (q.). Combination in *Aphaenogaster (Attomyrma)*: Emery, 1921f: 59. See also: Finzi, 1930a: 152.

***livida.** **Ponerops is livida* Heer, 1867: 25, pl. 2, fig. 5 (q.) YUGOSLAVIA. Combination in *Aphaenogaster:* Mayr, 1867b: 57.

***longaeva.** **Aphaenogaster longaeva* Scudder, 1877b: 267 (m.) CANADA (Miocene). Included as "(Myrmicinae) *longaeva*": Carpenter, 1930: 21.

**longiceps.** *Myrmica longiceps* Smith, F. 1858b: 128 (w.) AUSTRALIA. Mayr, 1876: 98 (q.m.); Wheeler, G.C. & Wheeler, J. 1972b: 238 (l.); Imai, Crozier & Taylor, 1977: 348 (k.). Combination in *Aphaenogaster:* Mayr, 1876: 98; in *Stenamma (Ischnomyrmex)*: Forel, 1899a: 118; in *Aphaenogaster (Deromyrma)*: Forel, 1913g: 187; in *Aphaenogaster (Nystalomyrma)*: Wheeler, W.M. 1916j: 216. Type-material redescribed: Crawley, 1922c: 19. Senior synonym of *ruginota:* Wheeler, W.M. 1916j: 217. Current subspecies: nominal plus *flava.*

*longispina. Aphaenogaster sardoa* var. *longispina* Wolf, 1915: 42, fig. 4 (w.) ALGERIA. Junior synonym of *sardoa:* Finzi, 1930a: 156.

**loriai.** *Stenamma (Ischnomyrmex) loriai* Emery, 1897d: 563, pl. 1, figs. 9, 10 (w.q.) NEW GUINEA. Viehmeyer, 1914d: 605 (m.). Combination in *Ischnomyrmex:* Stitz, 1911a: 367; in *Aphaenogaster (Planimyrma)*: Viehmeyer, 1914d: 604; in *Aphaenogaster (Deromyrma)*: Emery, 1915d: 71. Senior synonym of *atra:* Smith, M.R. 1961b: 223.

*lusitanica. Aphaenogaster senilis* var. *lusitanica* Schmitz, 1950: 15, PORTUGAL. *Nomen nudum.*

**lustrans.** *Aphaenogaster lustrans* Smith, M.R. 1961b: 234, pl. 16, fig. 3 (w.) NEW GUINEA.

*macrospina*. *Aphaenogaster texana* subsp. *macrospina* Smith, M.R. 1934a: 386, figs. 1, 2 (w.) U.S.A. Combination in *Aphaenogaster* (*Attomyrma*), and raised to species: Creighton, 1950a: 145. Junior synonym of *flemingi:* Smith, M.R. 1958a: 113.

***maculata.** **Aphaenogaster maculata* Théobald, 1937b: 313, pl. 25, fig. 5, pl. 26, fig. 14 (m.?) FRANCE (Oligocene).

***maculipes.** **Aphaenogaster maculipes* Théobald, 1937b: 204, pl. 14, fig. 7 (q.) FRANCE (Oligocene).

*manni*. *Aphaenogaster subterranea* subsp. *valida* var. *manni* Wheeler, W.M. 1917a: 516 (w.) U.S.A. Unavailable name; material referred to *occidentalis* by Creighton, 1950a: 150.

*manni*. *Novomessor manni* Wheeler, W.M. & Creighton, 1934: 353, pl. 1, fig. 2 (w.) MEXICO. Wheeler, G.C. & Wheeler, J. 1960b: 10 (l.). Junior synonym of *ensifera:* Brown, 1974b: 46.

**mariae.** *Aphaenogaster mariae* Forel, 1886b: xli (w.) U.S.A. Combination in *Stenamma* (*Aphaenogaster*): Emery, 1895c: 301; in *Aphaenogaster* (*Attomyrma*): Emery, 1921f: 58.

**#marocana.** *Stenamma* (*Aphaenogaster*) *testaceopilosum* var. *marocana* Forel, 1903c: 257 (w.m.) MOROCCO. Santschi, 1929e: 140 (q.). Combination in *Aphaenogaster:* Emery, 1908c: 319. Stirps of *gemella:* Santschi, 1933c: 404. Subspecies of *gemella*, and senior synonym of *tingitana:* Cagniant, 1990a: 47. See also: Cagniant, 1994: 25.

**#mauritanica.** *Aphaenogaster striola* var. *mauritanica* Dalla Torre, 1893: 105 (w.) TUNISIA. [First available use of *Aphaenogaster subterraneus* r. *striola* var. *mauritanicus* Emery, 1891b: 8; unavailable name.] Santschi, 1911i: 284 (m.); Santschi, 1932d: 342 (q.). Variety/stirps of *gibbosa:* Emery, 1908c: 334; Santschi, 1932d: 341.

***mayri.** **Aphaenogaster mayri* Carpenter, 1930: 30, pl. 5, fig. 5; pl. 8, fig. 4; pl. 9, fig. 5; pl. 11, fig. 4 (w.q.m.) U.S.A. (Miocene).

**megommata.** *Aphaenogaster* (*Attomyrma*) *megommatus* Smith, M.R. 1963: 244, fig. 1 (w.) U.S.A. Cole, 1966: 9 (q.m.); Wheeler, G.C. & Wheeler, J. 1972b: 238 (l.).

*melitensis*. *Aphaenogaster semipolita* var. *melitensis* Santschi, 1933c: 390 (w.) MALTA. [First available use of *Aphaenogaster testaceopilosa* subsp. *semipolita* var. *melitensis* Emery, 1924b: 12; unavailable name.] Junior synonym of *ionia:* **new synonymy**.

*meridionalis* André, 1883; see under ***MESSOR.***

***mersa.** **Aphaenogaster mersa* Wheeler, W.M. 1915h: 55, fig. 20 (w.) BALTIC AMBER (Oligocene).

**messoroides.** *Aphaenogaster messoroides* Dlussky, in Dlussky, Soyunov & Zabelin, 1990: 213, fig. 24 (w.m.) TURKMENISTAN.

**mexicana.** *Ischnomyrmex mexicanum* Pergande, 1896: 893 (w.) MEXICO. Combination in *Aphaenogaster* (*Ischnomyrmex*): Forel, 1899c: 60; in *Aphaenogaster* (*Deromyrma*): Emery, 1915d: 71.

**miamiana.** *Aphaenogaster* (*Attomyrma*) *texana* var. *miamiana* Wheeler, W.M. 1932a: 5 (w.q.m.) U.S.A. Raised to species: Creighton, 1950a: 145.

**miniata.** *Aphaenogaster miniata* Cagniant, 1990d: 144, figs. 1–4 (w.q.m.) MOROCCO.

*minor* André, 1883; see under ***MESSOR.***

*minor*. *Novomessor cockerelli* var. *minor* Enzmann, J. 1947b: 148, pl. 8 (w.) U.S.A. [Unresolved junior secondary homonym of *Aphaenogaster minor* André, 1883b: 355, above.] Junior synonym of *albisetosa:* Brown, 1949a: 49.

**muelleriana.** *Aphaenogaster ovaticeps* subsp. *muelleriana* Wolf, 1915: 44, figs. 5, 6 (w.) ITALY. Emery, 1916b: 140 (q.m.). Combination in *Aphaenogaster* (*Attomyrma*): Emery, 1921f: 58. [*Aphaenogaster ovaticeps* subsp. *muelleriana* Wolf, in Emery, 1914d: 156. *Nomen nudum.*] Raised to species: Agosti & Collingwood, 1987a: 54. See also: Finzi, 1927a: 8; Baroni Urbani, 1971c: 51.

*muschtaidica*. *Aphaenogaster gibbosa* subsp. *muschtaidica* Emery, 1908c: 334 (w.) RUSSIA. [First available use of *Aphaenogaster subterranea* subsp. *gibbosa* var. *muschtaidica* Ruzsky, 1905b: 719; unavailable name.] Combination in *Aphaenogaster* (*Attomyrma*): Emery, 1921f: 58. Raised to species: Collingwood, 1985: 246. Junior synonym of *gibbosa:* Dlussky, Soyunov & Zabelin, 1990: 212.

**mutica.** *Aphaenogaster mutica* Pergande, 1896: 891 (w.) MEXICO. Combination in *Aphaenogaster* (*Attomyrma*): Emery, 1921f: 58.

**nadigi.** *Aphaenogaster* (*Attomyrma*) *gibbosa* var. *nadigi* Santschi, 1923f: 328 (w.) MOROCCO. Cagniant, 1987a: 33 (q.m.). Raised to species: Cagniant, 1987a: 31.

*nana*. *Aphaenogaster* (*Attomyrma*) *texana* subsp. *nana* Wheeler, W.M. 1932a: 6 (w.) U.S.A. Unidentifiable: Creighton, 1950a: 151; unplaced taxon in *Aphaenogaster:* Smith, D.R. 1979: 1364.

**#nicobarensis.** *Stenamma* (*Ischnomyrmex*) *feae* var. *nicobarensis* Forel, 1903d: 405 (w.) NICOBAR IS. Combination in *Aphaenogaster* (*Deromyrma*): Emery, 1921f: 65.

*nigra* André, 1883; see under ***MESSOR.***

*nigra*. *Aphaenogaster* (*Planimyrma*) *dromedarius* var. *nigra* Donisthorpe, 1938b: 31 (w.) NEW GUINEA [Unresolved junior primary homonym of *Aphaenogaster nigra* André, 1883b: 355, above.] Junior synonym of *dromedaria:* Smith, M.R. 1961b: 219.

*nigripes*. *Aphaenogaster lamellidens* var. *nigripes* Smith, M.R. 1923b: 308 (w.) U.S.A. Junior synonym of *lamellidens:* Creighton, 1950a: 144.

*nitida*. *Aphaenogaster testaceopilosa* var. *nitida* Krausse, 1912b: 163 (w.) ITALY. [First available use of *Stenamma* (*Aphaenogaster*) *testaceopilosum* subsp. *spinosum* var. *nitidum* Emery, 1895b: 70; unavailable name.] Variety of *testaceopilosa:* Finzi, 1924a: 13; of *spinosa:* Bondroit, 1918: 162; Santschi, 1933c: 390. Junior synonym of *spinosa:* Baroni Urbani, 1971d: 1039; Casevitz-Weulersse, 1990a: 152.

**#nitidiventris.** *Aphaenogaster* (*Ischnomyrmex*) *araneoides* var. *nitidiventris* Forel, 1912g: 15 (w.m.) COSTA RICA. Combination in *Aphaenogaster* (*Deromyrma*): Kempf, 1972a: 23.

**#numida.** *Aphaenogaster* (*Aphaenogaster*) *depilis* var. *numida* Santschi, 1933c: 395, figs. 9, 22 (w.m.) TUNISIA.

**obsidiana.** *Atta obsidiana* Mayr, 1861: 67 (diagnosis in key) (w.) RUSSIA. Ruzsky, 1905b: 725 (q.); Karavaiev, 1926b: 98 (m.). Combination in *Aphaenogaster:* Roger, 1863b: 29; in *Stenamma* (*Aphaenogaster*): Ruzsky, 1902d: 26; in *Aphaenogaster:* Ruzsky, 1905b: 723; in *Aphaenogaster* (*Attomyrma*): Emery, 1921f: 58.

**occidentalis.** *Stenamma* (*Aphaenogaster*) *subterraneum* subsp. *occidentale* Emery, 1895c: 301 (w.) U.S.A. Wheeler, G.C. & Wheeler, J. 1953b: 61 (l.). Combination in *Aphaenogaster* (*Attomyrma*): Emery, 1921f: 60. Senior synonym of *borealis*, and material of the unavailable name *manni* referred here: Creighton, 1950a: 150. Raised to species: Hunt & Snelling, 1975: 21.

**#occidua.** *Aphaenogaster senilis* var. *occidua* Santschi, 1933c: 396, fig. 26 (w.m.) FRANCE.

*oculata* Emery, 1899; see under ***PHEIDOLE.***

*oklahomensis*. *Aphaenogaster treatae* subsp. *pluteicornis* var. *oklahomensis* Wheeler, G.C. & Wheeler, E.W. 1934: 10 (w.q.) U.S.A. Unavailable name; material referred to *pluteicornis* by Creighton, 1950a: 154.

***oligocenica.** **Aphaenogaster oligocenica* Wheeler, W.M. 1915h: 54, fig. 19 (w.) BALTIC AMBER (Oligocene).

*opposita*. *Myrmica opposita* Say, 1836: 292 (w.m.) Combination in *Aphaenogaster* (*Attomyrma*): Emery, 1921f: 58. Unrecognisable taxon: Smith, M.R. 1951a: 798; unplaced taxon in Myrmicinae: Smith, D.R. 1979: 1414.

**osimensis.** *Aphaenogaster famelica* var. *osimensis* Teranishi, 1940: 78 (w.q.) JAPAN. Raised to species: Imai, 1971: 139.

**ovaticeps.** *Stenamma* (*Aphaenogaster*) *ovaticeps* Emery, 1898c: 136, figs. (w.q.m.) ITALY. Combination in *Aphaenogaster:* Emery, 1908c: 324; in *Aphaenogaster* (*Attomyrma*): Emery, 1921f: 58.

**#pachei.** *Stenamma* (*Aphaenogaster*) *sagei* subsp. *pachei* Forel, 1906b: 82 (w.) NEPAL. Combination in *Aphaenogaster* (*Attomyrma*): Chapman & Capco, 1951: 132.

*pallescens*. *Aphaenogaster pallescens* Walker, 1871: 11 (q.) EGYPT. Unrecognisable taxon.

**pallida.** *Myrmica pallida* Nylander, 1849: 42 (w) ITALY. Roger, 1859: 254 (q.m.). Combination in *Aphaenogaster:* Roger, 1863b: 29; in *Stenamma* (*Aphaenogaster*): Emery, 1895c: 303; in *Aphaenogaster:* Emery, 1908c: 328; in *Aphaenogaster* (*Attomyrma*): Santschi, 1919e: 245. See also: Baroni Urbani, 1971c: 52. Current subspecies: nominal plus *laurenti, leveillei*.

***pannonica.** **Aphaenogaster pannonicus* Bachmayer, 1960: 12, pl. 1, fig. 1 (wing) AUSTRIA (Pliocene).

**patruelis.** *Aphaenogaster patruelis* Forel, 1886b: xli (w.) U.S.A. Wheeler, W.M. 1934f: 133 (m.). Combination in *Stenamma* (*Aphaenogaster*): Emery, 1895c: 302; in *Aphaenogaster* (*Attomyrma*): Emery, 1921f: 59. Subspecies of *subterranea:* Emery, 1895c: 302. Revived status as species: Wheeler, W.M. 1904d: 270; see also: Wheeler, W.M. 1934f: 132. Senior synonym of *willowsi:* Creighton, 1950a: 146; of *bakeri:* Smith, D.R. 1979: 1362. Current subspecies: nominal plus *carbonaria*.

*pergandei* Mayr, 1886; see under ***MESSOR.***

*pergandei*. *Ceratopheidole perganderi* Donisthorpe, 1950d: 639 (w.) TURKEY. [*Ceratopheidole pergandei:* Donisthorpe, 1950e: 1060, justified emendation of spelling.] Combination in *Aphaenogaster:* Brown, 1981: 529. [Junior secondary homonym of

*Aphaenogaster pergandei* Mayr, 1886d: 448, above.] Replacement name: *depressa:* **new name**.

***perplexa.*** *Aphaenogaster (Planimyrma) perplexus* Smith, M.R. 1961b: 228. Replacement name for *wheeleri* Donisthorpe, 1938b: 31. [Junior primary homonym of *wheeleri* Mann, 1915: 51.]

***phalangium.*** *Aphaenogaster phalangium* Emery, 1890b: 47, pl. 5, figs. 6, 7 (w.m.) COSTA RICA. Combination in *Aphaenogaster (Ischnomyrmex)*: Forel, 1899c: 59; in *Aphaenogaster (Deromyrma)*: Emery, 1915d: 71. [Also described as new by Emery, 1894k: 54.] Current subspecies: nominal plus *brevicollis*.

*phillipi. Aphaenogaster phillipi* Menozzi, 1932b: 311 (footnote). Unnecessary (second) replacement name for *silvestrii* Wheeler, W.M. 1929g: 37. [Junior primary homonym of *silvestrii* Menozzi, 1929b: 282.] Junior synonym of *lepida* (first replacement name): Brown, 1954d: 10.

***phillipsi.*** *Aphaenogaster (Deromyrma) phillipsi* Wheeler, W.M. & Mann, 1916: 168, fig. 1 (w.) JORDAN.

***picea.*** *Stenamma (Aphaenogaster) fulvum* var. *piceum* Wheeler, W.M. 1908f: 621 (w.q.m.) U.S.A. [First available use of *Stenamma (Aphaenogaster) fulvum* subsp. *aquia* var. *piceum* Emery, 1895c: 305; unavailable name.] Combination in *Aphaenogaster (Attomyrma)*: Emery, 1921f: 57. Subspecies of *fulva:* Buren, 1944a: 284; subspecies of *rudis:* Creighton, 1950a: 148 (incorrect as *picea* made available earlier than *rudis* and hence senior name). Raised to species: **new status**. Current subspecies: nominal plus *rudis*.

***picena.*** *Aphaenogaster picena* Baroni Urbani, 1971a: 32, figs. 1–9 (w.q.m.l.) ITALY. [*Aphaenogaster picena* Baroni Urbani, 1966: 69. *Nomen nudum.*] See also: Baroni Urbani, 1971c: 41.

***#pluteicornis.*** *Aphaenogaster treatae* subsp. *pluteicornis* Wheeler, G.C. & Wheeler, E.W. 1934: 7 (w.q.m.) U.S.A. Wheeler, G.C. & Wheeler, J. 1953b: 62 (l.). Combination in *Aphaenogaster (Attomyrma)*, and material of the unavailable name *oklahomensis* referred here: Creighton, 1950a: 154.

***poultoni.*** *Aphaenogaster poultoni* Crawley, 1922c: 17, figs. 11, 12 (w.) AUSTRALIA.

***praedo.*** *Aphaenogaster praedo* Emery, 1908c: 323, fig. 12 (w.) MOROCCO. Cagniant, 1969b: 105 (q.m.). See also: Cagniant, 1987d: 163. Current subspecies: nominal plus *ellipsoida*.

***praenoda.*** *Aphaenogaster (Aphaenogaster) praenoda* Santschi, 1933c: 398, figs. 1, 16, 30 (w.m.) MOROCCO. Current subspecies: nominal plus *confinis*.

***projectens.*** *Aphaenogaster (Deromyrma) projectens* Donisthorpe, 1947c: 581 (m.) NEW GUINEA.

***#prudens.*** *Stenamma (Aphaenogaster) smythiesii* var. *prudens* Forel, 1902c: 223 (w.) INDIA. Combination in *Aphaenogaster (Attomyrma)*: Emery, 1921f: 59.

*pseudoaegyptiaca* Emery, 1884; see under ***MESSOR.***

*punctata* Forel, 1886; see under ***MESSOR.***

***punctaticeps.*** *Aphaenogaster punctaticeps* MacKay, 1989: 47, 7 figs. (w.) U.S.A.

*punctithorax. Aphaenogaster texana* subsp. *punctithorax* Cole, 1938a: 239, fig. 2 (w.) U.S.A. Unidentifiable: Creighton, 1950a: 151; junior synonym of *picea:* Smith, M.R. 1951a: 796; unplaced taxon in *Aphaenogaster:* Smith, D.R. 1979: 1364.

*pusilla. Aphaenogaster fulva* var. *pusilla* Enzmann, J. 1947b: 150 (w.) U.S.A. [First available use of *Stenamma (Aphaenogaster) fulvum* subsp. *aquia* var. *pusillum* Emery, 1895c: 306; unavailable name.] Unidentifiable: Creighton, 1950a: 151; unplaced taxon in *Aphaenogaster:* Smith, D.R. 1979: 1364.

***pythia.*** *Aphaenogaster (Deromyrma) pythia* Forel, 1915b: 76 (w.) AUSTRALIA. Wheeler, W.M. 1916j: 220 (q.m.); Wheeler, G.C. & Wheeler, J. 1972b: 239 (l.). Combination in *Aphaenogaster (Nystalomyrma)*: Wheeler, W.M. 1916j: 219. See also: Smith, M.R. 1961b: 229.

***quadrispina.*** *Aphaenogaster (Ischnomyrmex) quadrispina* Emery, 1911c: 251 (w.m.) NEW GUINEA. Combination in *Aphaenogaster (Planimyrma)*: Viehmeyer, 1914b: 516; Emery, 1921f: 66; in *Aphaenogaster (Deromyrma)*: Emery, 1915d: 71.

***relicta.*** *Aphaenogaster relicta* Wheeler, W.M. & Mann, 1914: 25, fig. 8c (w.q.m.) HAITI. Combination in *Novomessor:* Emery, 1915d: 73; in *Veromessor:* Wheeler, W.M. & Creighton, 1934: 382; in *Aphaenogaster:* Bolton, 1982: 341. Current subspecies: nominal plus *epinotalis*.

***rhaphidiiceps.*** *Ischnomyrmex rhaphidiiceps* Mayr, 1877: 12 (w.) TURKESTAN. [Redescribed in Latin: Mayr, 1880: 31.] Arnol'di, 1976b: 1022 (q.). Combination in *Aphaenogaster:* André, 1883b: 349; in *Aphaenogaster (Deromyrma)*: Emery, 1915d: 71. See also: Pisarski, 1967: 379.

***rifensis.*** *Aphaenogaster rifensis* Cagniant, 1994: 23, figs. 3, 4 (w.q.m.) MOROCCO.

*romana. Aphaenogaster spinosa* var. *romana* Santschi, 1933c: 390 (w.) ITALY. [First available use of *Stenamma (Aphaenogaster) testaceopilosum* subsp. *spinosum* var. *romanum* Emery, 1895b: 70; unavailable name.] Junior synonym of *spinosa:* Baroni Urbani, 1971d: 1039.

***rothneyi.*** *Stenamma (Aphaenogaster) rothneyi* Forel, 1902c: 224 (w.) INDIA. Combination in *Aphaenogaster:* Bingham, 1903: 273; in *Aphaenogaster (Attomyrma)*: Emery, 1921f: 59.

*rubida. Aphaenogaster fulva* var. *rubida* Enzmann, J. 1947b: 147, pl. 8 (w.) U.S.A. Junior synonym of *fulva:* Brown, 1949a: 49.

***#rudis.*** *Aphaenogaster fulva* var. *rudis* Enzmann, J. 1947b: 150 (w.q.) U.S.A. [First available use of *Stenamma (Aphaenogaster) fulvum* subsp. *aquia* var. *rude* Emery, 1895c: 305; unavailable name.] Wheeler, G.C & Wheeler, J. 1953: 56 (l.); Crozier, 1970: 125 (k.). Raised to species: Creighton, 1950a: 147. Subspecies of *picea* as this name made available first (*picea* entered as subspecies of *rudis* in Creighton, 1950a: 147), and hence is senior name.

*rufotestacea. Aphaenogaster testaceopilosa* subsp. *semipolita* var. *rufotestacea* Donisthorpe, 1946a: 54 (w.) TURKEY. Unavailable name.

*ruginota. Stenamma (Ischnomyrmex) longiceps* r. *ruginota* Forel, 1902h: 440 (w.q.) AUSTRALIA. Emery, 1914b: 181 (m.). Combination in *Aphaenogaster:* Emery, 1914b: 181; in *Aphaenogaster (Deromyrma)*, and raised to species: Forel, 1915b: 75. Junior synonym of *longiceps:* Wheeler, W.M. 1916j: 217.

*rugosa* André, 1881; see under ***MESSOR.***

***#rugosoferruginea.*** *Aphaenogaster splendida* r. *rugosoferruginea* Forel, 1889: 260 (w.) GREECE. Combination in *Aphaenogaster (Attomyrma)*: Emery, 1921f: 60.

***ruida.*** *Aphaenogaster (Attomyrma) famelica* subsp. *ruida* Wheeler, W.M. 1928d: 104 (q.) JAPAN. Raised to species: Collingwood, 1976: 302. [Ogata, 1991b: 85, misspells the name as *rudia* and regards it as a subsp. of *famelica*. Collingwood's elevation to species is preferred here.]

***rupestris.*** *Aphaenogaster rupestris* Forel, 1909e: 379 (w.) ALGERIA. Forel, 1913d: 428 (m.); Cagniant, 1969b: 110 (q.).

***sagei.*** *Stenamma (Aphaenogaster) sagei* Forel, 1902c: 221 (w.q.m.) INDIA. Combination in *Aphaenogaster:* Bingham, 1903: 275; in *Aphaenogaster (Deromyrma)*: Forel, 1913k: 53; in *Aphaenogaster (Attomyrma)*: Emery, 1921f: 59. Current subspecies: nominal plus *pachei*.

***saharensis.*** *Aphaenogaster (Attomyrma) saharensis* Bernard, 1953a: 137, fig. 2 (m.) ALGERIA.

***sangiorgii.*** *Stenamma (Aphaenogaster) sangiorgii* Emery, 1901e: 57, fig. (q.) GREECE. Combination in *Aphaenogaster:* Emery, 1908c: 337.

***sardoa.*** *Aphaenogaster sardous* Mayr, 1853b: 107 (w) ITALY. André, 1883b: 369 (q.); Santschi, 1911i: 284 (m.). Senior synonym of *longispina:* Finzi, 1930a: 156. See also: Baroni Urbani,1971c: 41. Current subspecies: nominal plus *anoemica*.

*schaufussi* Forel, 1879; see under ***LEPTOTHORAX.***

***schmidti.*** *Aphaenogaster schmidti* Karavaiev, 1912b: 582, fig. 1 (w.) KOREA. Combination in *Aphaenogaster (Attomyrma)*: Emery, 1921f: 59.

***#schmitzi.*** *Aphaenogaster schmitzi* Forel, 1910a: 10 (w.) ISRAEL. Combination in *Aphaenogaster (Attomyrma)*:Emery, 1921f: 60. Subspecies of *syriaca:* Emery, 1921f: 60; Menozzi, 1933b: 51.

***schurri.*** *Stenamma (Aphaenogaster) schurri* Forel, 1902c: 223 (w.) INDIA. Combination in *Aphaenogaster:* Bingham, 1903: 274; in *Aphaenogaster (Attomyrma)*: Emery, 1921f: 59.

***semipolita.*** *Myrmica semipolita* Nylander, 1856b: 86 (w.) ITALY. Emery, 1878b: 53 (m.); André, 1883b: 370 (q.). Combination in *Aphaenogaster:* Roger, 1863b: 29. Variety/subspecies of *testaceopilosa:* Emery, 1878b: 53; Forel, 1910a: 24; Finzi, 1930d: 312. Revived status as species: Santschi, 1933c: 390.

*semirufa* André, 1883; see under ***MESSOR.***

***senilis.*** *Aphaenogaster senilis* Mayr, 1853b: 108 (w.q.) ITALY. Emery, 1916b: 140 (m.). Junior synonym of *testaceopilosa:* Nylander, 1856b: 86; Smith, F. 1858b: 165 (misspelled as *similis*); Roger, 1859: 255; Dalla Torre, 1893: 107. Variety/subspecies of *testaceopilosa:* Emery, 1878b: 53; Emery, 1916b: 131; Santschi, 1919e: 245. Revived status as species: Bondroit, 1918: 161; Santschi, 1933c: 396; Cagniant, 1964: 100; Bernard, 1967: 132. Baroni Urbani, 1971c: 44. Senior synonym of *acoreensis:* Yarrow, 1967: 26; of *fuentei:* Collingwood, 1978: 67. Current subspecies: nominal plus *disjuncta, grata, occidua*.

***sicardi.*** *Aphaenogaster sicardi* Cagniant, 1990c: 443, figs. 1, 2 (w.q.m.) MOROCCO.

***sicula.*** *Aphaenogaster crocea* subsp. *sicula* Emery, 1908c: 331 (w.q.) ITALY. Combination in *Aphaenogaster (Attomyrma)*: Emery, 1921f: 57. Raised to species:

Schembri & Collingwood, 1981: 423. Material of the unavailable name *subterraneosplendida* referred here by Emery, 1908c: 331.

*silvestrii. Aphaenogaster (Deromyrma) silvestrii* Menozzi, 1929b: 282, fig. 1 (w.q.) U.S.A. Combination in *Aphaenogaster (Attomyrma)*, and variety of *texana:* Wheeler, W.M. 1932a: 5. Junior synonym of *texana:* Creighton, 1950a: 152.

*silvestrii. Aphaenogaster (Attomyrma) silvestrii* Wheeler, W.M. 1929g: 37, fig. 3 (w.m.) TAIWAN. [Junior primary homonym of *silvestrii* Menozzi, above.] Replacement name: *lepida* Wheeler, W.M. 1930a: 96.

*similis* Smith, F. 1858; see under *senilis*, above.

***simonellii.*** *Aphaenogaster testaceopilosa* var. *simonellii* Emery, 1894j: 8 (w.) GREECE. Raised to species: Santschi, 1933c: 401. Current subspecies: nominal plus *sporadis.* See also: Schembri & Collingwood, 1987: 54.

#***simulans.*** *Aphaenogaster (Deromyrma) feae* r. *simulans* Forel, 1915a: 31 (w.q.m.) SUMATRA.

***sinensis.*** *Aphaenogaster smythiesi* subsp. *sinensis* Wheeler, W.M. 1928c: 9 (w.) CHINA. Wheeler, W.M. 1929f: 2 (q.m.). Raised to species: Arnol'di, 1976b: 1022. See also: Kupyanskaya, 1990: 119.

#***smithi.*** *Aphaenogaster boulderensis* subsp. *smithi* Gregg, 1949b: 171 (w.) U.S.A.

***smythiesii.*** *Stenamma (Aphaenogaster) smythiesii* Forel, 1902c: 222 (w.q.m.) INDIA. Imai, *et al.* 1984: 6 (k.). Combination in *Aphaenogaster:* Bingham, 1903: 276; in *Aphaenogaster (Attomyrma)*: Emery, 1921f: 59. See also: Imai, 1971: 137. Current subspecies: nominal plus *japonica, prudens.*

****sommerfeldti.*** **Aphaenogaster sommerfeldti* Mayr, 1868c: 81, pl. 4, figs. 76, 77 (w.) BALTIC AMBER (Oligocene).

*sonorae. Aphaenogaster sonorae* Pergande, 1894a: 34 (w.) MEXICO. Combination in *Novomessor:* Emery, 1915d: 73. Junior synonym of *cockerelli:* Wheeler, W.M. & Creighton, 1934: 352.

*sordida* Forel, 1892; see under ***MESSOR.***

#***spinipes.*** *Aphaenogaster (Ischnomyrmex) schwammerdami* [sic] var. *spinipes* Santschi, 1911e: 123 (w.) MADAGASCAR. Combination in *Aphaenogaster (Deromyrma)*: Emery, 1921f: 65.

***spinosa.*** *Aphaenogaster testaceopilosa* var. *spinosa* Emery, 1878b: 54, fig. (w.) ITALY. Emery, 1908c: 320 (q.m.). Raised to species: Dalla Torre, 1893: 105. Subspecies of *testaceopilosa:* Emery, 1908c: 320; Emery, 1916b: 131; Emery, 1921f: 63. Revived status as species: Bondroit, 1918: 162; Santschi, 1933c: 390; Bernard, 1967: 133; Baroni Urbani, 1969a: 326. Senior synonym of *nitida* and *romana*, and material of the unavailable name *corsica* referred here: Baroni Urbani, 1971d: 1039. See also: Casevitz-Weulersse, 1990a: 152. Current subspecies: nominal plus *etrusca.*

***splendida.*** *Atta splendida* Roger, 1859: 257 (w.) GREECE. Emery, 1869b: 17 (q.); Emery, 1908c: 325 (m.); Baroni Urbani, 1968b: 419 (l.). Combination in *Aphaenogaster:* Roger, 1863b: 29; in *Aphaenogaster (Attomyrma)*: Emery, 1921f: 59. Current subspecies: nominal plus *rugosoferruginea, transcaucasica.*

#***splendidoides.*** *Aphaenogaster subterranea* var. *splendidoides* Forel, 1890a: lxx (w.) ALGERIA. Variety of *crocea:* Dalla Torre, 1893: 102; Emery, 1908c: 331.

#***sporadis.*** *Aphaenogaster simonellii* var. *sporadis* Santschi, 1933c: 401, figs. 14, 37 (w.m.) GREECE.

*striaticeps* André, 1883; see under ***MESSOR.***

***striativentris.*** *Aphaenogaster striola* r. *striativentris* Forel, 1895d: 228 (w.) SPAIN. Tinaut & Jiminez-Rojas, 1991: 118 (q.m.). Combination in *Aphaenogaster (Attomyrma)*: Emery, 1921f: 58. Subspecies of *gibbosa:* Emery, 1908c: 334. Raised to species: Collingwod & Yarrow, 1969: 60.

*striatula* Emery, 1891; see under ***MESSOR.***

*striola. Atta striola* Roger, 1859: 252 (w.m.) SPAIN. Emery, 1869b: 17 (q.). Combination in *Aphaenogaster:* Roger, 1863b: 29. Junior synonym of *gibbosa:* Emery, 1895c: 303.

***strioloides.*** *Aphaenogaster subterranea* var. *strioloides* Forel, 1890a: lxxi (w.) ALGERIA. Emery, 1908c: 335 (q.); Santschi, 1932d: 341 (m.). Combination in *Aphaenogaster (Attomyrma)*: Menozzi, 1921: 26. Subspecies of *gibbosa:* Emery, 1908c: 335; Emery, 1916b: 130; Menozzi, 1921: 26; Emery, 1921f: 58. Raised to species: Bondroit, 1918: 158; revived status as species: Collingwood & Yarrow, 1969: 60.

***subcostata.*** *Aphaenogaster (Attomyrma) subcostata* Viehmeyer, 1922: 205 (w.) TURKEY.

*subrubra. Myrmica subrubra* Buckley, 1867: 336 (w.q.) U.S.A. Junior synonym of *tennesseensis:* Mayr, 1886c: 365.

***subterranea.*** *Formica subterranea* Latreille, 1798: 49 (w.q.m.) FRANCE. Latreille, 1802a: 220 (q.m.); Hauschteck, 1962: 219 (k.). Combination in *Atta:* Mayr, 1855: 460; in *Myrmica:* Nylander, 1856b: 85; in *Aphaenogaster:* Roger, 1863b: 29; in *Stenamma (Aphaenogaster)*: Emery, 1895c: 301; in *Aphaenogaster (Attomyrma)*: Emery, 1915d: 70. See also: Emery, 1908c: 326; Stitz, 1939: 113; Bernard, 1967: 134; Kutter, 1977c: 80; Atanassov & Dlussky, 1992: 111. Current subspecies: nominal plus *ichnusa, valida.*

***subterraneoides.*** *Aphaenogaster pallida* var. *subterraneoides* Emery, 1881b: 534 (w.) CYPRUS. Santschi, 1939c: 5 (q.). Combination in *Aphaenogaster (Attomyrma)*: Emery, 1921f: 59. Variety/subspecies of *striola:* Forel, 1890a: lxxi; of *gibbosa:* Karavaiev, 1912a: 5; of *pallida:* Emery, 1915a: 258; Finzi, 1930d: 312. Raised to species: Bondroit, 1918: 159. Junior synonym of *pallida:* Bernard, 1967: 134. Revived from synonymy and raised to species: Arnol'di, 1976b: 1023. Senior synonym of *cypriotes:* Emery, 1915a: 258. Current subspecies: nominal plus *armeniaca.*

*subterraneosplendida. Aphaenogaster subterranea* r. *striola* var. *subterraneosplendida* Emery & Forel, 1879: 462. *Nomen nudum.* [Later references to this name, eg. Emery, 1884a: 381; Emery, 1891b: 8, are *nomina nuda* or unavailable names.] Material referred to *sicula* by Emery, 1908c: 331.

*subterraneostriola. Aphaenogaster subterranea* r. *striola* var. *subterraneostriola* Emery & Forel, 1879: 462 (w.) ITALY. Unavailable name.

***swammerdami.*** *Aphaenogaster (Ischnomyrmex) swammerdami* Forel, 1886c: cvi (w.) MADAGASCAR. Forel, 1891b: 168 (q.m.); Wheeler, G.C. & Wheeler, J. 1977: 584 (l.). Combination in *Stenamma (Ischnomyrmex)*: Forel, 1907e: 14; in *Aphaenogaster (Deromyrma)*: Forel, 1913b: 350. Current subspecies: nominal plus *clara, curta, spinipes.*

*syria. Aphaenogaster schmitzi* var. *syria* Forel, 1910a: 12 (w.) SYRIA. Junior synonym of *syriaca:* Emery, 1921f: 60.

***syriaca.*** *Aphaenogaster gibbosa* subsp. *syriaca* Emery, 1908c: 335 (w.) LEBANON. Combination in *Aphaenogaster (Attomyrma)*: Emery, 1921f: 60. Raised to species: Emery, 1915a: 258. Senior synonym of *syria:* Emery, 1921f: 60. Current subspecies: nominal plus *schmitzi.*

***takahashii.*** *Aphaenogaster (Deromyrma) takahashii* Wheeler, W.M. 1930a: 96, fig. 1 (w.) TAIWAN.

***tennesseensis.*** *Atta tennesseensis* Mayr, 1862: 743 (w.) U.S.A. Wheeler, G.C. & Wheeler, J. 1953b: 61 (l.). Combination in *Aphaenogaster:* Roger, 1863b: 30; in *Stenamma (Aphaenogaster)*: Emery, 1895c: 301; in *Aphaenogaster (Attomyrma)*: Emery, 1921f: 60. Senior synonym of *subrubra:* Mayr, 1886c: 365; of *laevis:* Mayr, 1886d: 446; of *ecalcaratum:* Creighton, 1950a: 151.

***testaceopilosa.*** *Myrmica testaceopilosa* Lucas, 1849: 301, pl. 16, fig. 10 (w.) ALGERIA. Mayr, 1861: 67 (q.m.). Combination in *Aphaenogaster:* Roger, 1863b: 29; in *Stenamma (Aphaenogaster)*: Emery, 1895c: 303; in *Aphaenogaster:* Ruzsky, 1905b: 721. See also: Emery, 1908c: 314; Santschi, 1933c: 391. Current subspecies: nominal plus *cabylica, canescens.*

***texana.*** *Aphaenogaster texana* Wheeler, W.M. 1915b: 412 (w.q.m.) U.S.A. [First available use of *Stenamma (Aphaenogaster) fulvum* subsp. *aquia* var. *texanum* Emery, 1895c: 306; unavailable name.] Wheeler, G.C. & Wheeler, J. 1953b: 62 (l.). Senior synonym of *furvescens* and *silvestrii:* Creighton, 1950a: 152. Unidentifiable taxa originally attached to *texana:* see entries under *nana* and *punctithorax.* Current subspecies: nominal plus *carolinensis.*

***theryi.*** *Aphaenogaster (Attomyrma) theryi* Santschi, 1923f: 327, fig. 2 (w.) MOROCCO. Cagniant, 1986: 140 (q.m.).

***tibetana.*** *Aphaenogaster tibetana* Donisthorpe, 1929a: 447 (w.) TIBET.

***tinauti.*** *Aphaenogaster tinauti* Cagniant, 1992b: 365, figs. 1, 3, 5–11 (w.q.m.) MOROCCO.

*tingitana. Aphaenogaster gemella* var. *tingitana* Santschi, 1933c: 402, figs. 19, 27 (w.q.m.) MOROCCO. Junior synonym of *marocana:* Cagniant, 1990a: 47.

***tipuna.*** *Aphaenogaster rothneyi* r. *tipuna* Forel, 1913f: 195 (w.m.) TAIWAN. Hung, Imai & Kubota, 1972: 1024 (k.). Combination in *Aphaenogaster (Attomyrma)*: Emery, 1921f: 59. Raised to species: Hung, Imai & Kubota, 1972: 1024.

***torossiani.*** *Aphaenogaster torossiani* Cagniant, 1988a: 241, figs. 1–4 (w.q.m.) MOROCCO.

#***transcaucasica.*** *Aphaenogaster (Aphaenogaster) splendida* var. *transcaucasica* Karavaiev, 1926b: 97 (w.q.) former U.S.S.R. Subspecies of *splendida:* Arnol'di, 1948: 211.

***treatae.*** *Aphaenogaster treatae* Forel, 1886b: xl (w.q.m.) U.S.A. Combination in *Stenamma (Aphaenogaster)*: Emery, 1895c: 302; in *Aphaenogaster:* Wheeler, W.M. 1913c: 114; in *Aphaenogaster (Attomyrma)*: Emery, 1921f: 60. Senior synonym of *wheeleri*, and material of the unavailable name *alabamensis* referred here: Creighton, 1950a: 153.

***APHAENOGASTER*** *(continued)*

***turkestanica.*** *Aphaenogaster turkestanica* Arnol'di, 1976b: 1021 (q.) TURKESTAN.

***uinta.*** *Aphaenogaster uinta* Wheeler, W.M. 1917a: 517 (w.q.m.) U.S.A. Combination in *Aphaenogaster (Attomyrma)*: Emery, 1921f: 60.

***ujhelyii.*** *Aphaenogaster ujhelyii* Szabó, 1910a: 364, fig. 1 (w.) TUNISIA. Combination in *Aphaenogaster (Attomyrma)*: Emery, 1921f: 60. Possibly subspecies of *sardoa:* Santschi, 1933c: 390.

#***valida.*** *Aphaenogaster subterranea* subsp. *valida* Wheeler, W.M. 1915b: 411 (w.q.m.) U.S.A. Combination in *Aphaenogaster (Attomyrma)*: Emery, 1921f: 60.

***vapida.*** *Aphaenogaster (Attomyrma) vapida* Wheeler, W.M. 1928d: 105 (w.) JAPAN.

***verecunda.*** *Aphaenogaster (Attomyrma) verecunda* Wheeler, W.M. 1928d: 105 (w.) JAPAN.

*vieirai. Aphaenogaster iberica* var. *vieirai* Santschi, 1933c: 390 (w.m.) PORTUGAL. [First available use of *Aphaenogaster testaceopilosa* subsp. *iberica* var. *vieirai* Emery, 1908c: 321; unavailable name.] Junior synonym of *iberica:* Espadaler, 1983: 226.

***weigoldi.*** *Aphaenogaster (Attomyrma) weigoldi* Viehmeyer, 1922: 206 (w.) CHINA.

***weulersseae.*** *Aphaenogaster weulersseae* Cagniant, 1989: 113, figs. 1–4 (w.q.m.) MOROCCO.

*wheeleri. Aphaenogaster treatae* subsp. *wheeleri* Mann, 1915: 51, fig. 1b (w.q.) U.S.A. Junior synonym of *treatae:* Creighton, 1950a: 153.

*wheeleri. Aphaenogaster (Planimyrma) quadrispina* subsp. *wheeleri* Donisthorpe, 1938b: 31 (w.) NEW GUINEA. [Junior primary homonym of *wheeleri* Mann, above.] Replacement name: *perplexa* Smith, M.R. 1961b: 228.

*willowsi. Aphaenogaster patruelis* subsp. *willowsi* Wheeler, W.M. 1933a: 64 (w.) U.S.A. Junior synonym of *patruelis:* Creighton, 1950a: 146.

***wilsoni.*** *Aphaenogaster wilsoni* Cagniant, 1988b: 49, figs. 1, 2 (w.q.m.) MOROCCO.

*APHANTOLEPIS:* see under ***TECHNOMYRMEX.***

***APHOMOMYRMEX*** (Afrotropical)

***afer.*** *Aphomomyrmex afer* Emery, 1899e: 494, fig. (w.q.m.) CAMEROUN. Wheeler, G.C. & Wheeler, J. 1986d: 332 (l.). Senior synonym of *muralti:* Snelling, 1979b: 4 (in text).

*emeryi* Ashmead, 1905; see under ***PSEUDAPHOMOMYRMEX.***

*goetschi* Menozzi, 1935; see under ***MYRMELACHISTA.***

*hewitti* Wheeler, W.M. 1910; see under ***CLADOMYRMA.***

*muralti. Aphomomyrmex muralti* Forel, 1910e: 449 (w.) SOUTH AFRICA. Junior synonym of *afer:* Snelling, 1979b: 4.

***APOMYRMA*** (west Africa)

***stygia.*** *Apomyrma stygia* Brown, Gotwald & Lévieux, 1971: 264, figs. 1–9, 2 pls. (w.q.m.) IVORY COAST. Wheeler, G.C. & Wheeler, J. 1971a: 276 (l.).

*APOROMYRMEX:* see under ***PLAGIOLEPIS.***

*APSYCHOMYRMEX:* see under ***ADELOMYRMEX.***

#***APTEROCREMA:*** see under ***CREMATOGASTER.***

***APTEROSTIGMA*** (Neotropical)

*abdita. Apterostigma abdita* Mann, 1922: 50, fig. 22 (w.q.) HONDURAS. Junior synonym of *mayri:* Weber, 1958c: 248.

***affinis.*** *Apterostigma affinis* Santschi, 1922b: 354 (w.) PARAGUAY.

*amiae. Apterostigma amiae* Weber, 1937: 391 (w.) GUIANA. Junior synonym of *robustum:* Weber, 1958c: 249.

*angulatum. Apterostigma collare* subsp. *angulatum* Weber, 1938b: 169, figs. 4, 12 (w.q.) PANAMA. Wheeler, G.C. 1949: 667 (l.). Raised to species: Weber, 1941b: 111. Junior synonym of *dentigerum:* Weber, 1958c: 246.

***auriculatum.*** *Apterostigma auriculatum* Wheeler, W.M. 1925a: 49 (w.q.) TRINIDAD. Senior synonym of *demerarae* and *immobile:* Weber, 1958c: 243. Current subspecies: nominal plus *icta, petiolatum.*

***billi.*** *Apterostigma billi* Weber, 1938b: 165, figs. 5, 17 (w.) BOLIVIA.

***bolivianum.*** *Apterostigma bolivianum* Weber, 1938b: 181, fig. 15 (w.q.) BOLIVIA.

*branneri. Apterostigma branneri* Mann, 1916: 456, pl. 5, fig. 37 (w.q.m.) BRAZIL. Junior synonym of *robustum:* Weber, 1958c: 249.

***bruchi.*** *Apterostigma bruchi* Santschi, 1919f: 49, fig. 5 (w.) ARGENTINA. Bruch, 1921: 204 (q.m.).

***calverti.*** *Apterostigma calverti* Wheeler, W.M. 1911h: 206 (w.q.) COSTA RICA.

***collare.*** *Apterostigma collare* Emery, 1896g: 99, pl. 1, fig. 18 (w.q.) COSTA RICA. Forel, 1899c: 38 (m.).

*constrictum. Apterostigma robustum* subsp. *constrictum* Weber, 1938b: 173 (w.q.m.) BOLIVIA. Junior synonym of *robustum:* Weber, 1958c: 249.

*demerarae. Apterostigma auriculatum* var. *demerarae* Wheeler, W.M. 1925a: 51 (w.q.m.) GUIANA. Junior synonym of *auriculatum:* Weber, 1958c: 243.

***dentigerum.*** *Apterostigma dentigerum* Wheeler, W.M. 1925a: 51 (w.q.m.) COSTA RICA. Senior synonym of *angulatum* and *dubium:* Weber, 1958c: 246.

*discrepans. Apterostigma mayri* var. *discrepans* Forel, 1912e: 190 (w.q.m.) BRAZIL. Junior synonym of *mayri:* Weber, 1958c: 248.

***dorotheae.*** *Apterostigma dorotheae* Weber, 1937: 389, fig. 5 (w.) GUIANA.

*dubium. Apterostigma collare* subsp. *dubium* Weber, 1938b: 168, figs. 6, 14, 19 (w.q.) PANAMA. Raised to species: Weber, 1941b: 110. Junior synonym of *dentigerum:* Weber, 1958c: 246.

***epinotale.*** *Apterostigma epinotale* Weber, 1937: 390, fig. 6 (w.) GUIANA. Weber, 1946b: 135 (q.m.).

***fallax.*** *Apterostigma fallax* Borgmeier, 1934: 105, fig. 5 (w.q.) SURINAM.

*fitzgeraldi. Apterostigma fitzgeraldi* Weber, 1937: 393 (w.) TRINIDAD. Junior synonym of *ierense:* Weber, 1958c: 247.

***fusinodum.*** *Apterostigma fusinodum* Weber, 1938b: 177 (w.m.) BOLIVIA.

***gibbum.*** *Apterostigma gibbum* Weber, 1938b: 162, fig. 18 (w.q.m.) BOLIVIA.

*guianense. Apterostigma urichi* var. *guianense* Weber, 1937: 393 (q.) GUIANA. Junior synonym of *urichii:* Weber, 1958c: 250.

#***icta.*** *Apterostigma wasmanni* var. *icta* Weber, 1937: 393 (q.) TRINIDAD. Subspecies of *auriculatum:* Weber, 1958c: 244.

***ierense.*** *Apterostigma ierense* Weber, 1937: 387, fig. 4 (w.q.m.) TRINIDAD. Senior synonym of *fitzgeraldi:* Weber, 1958c: 247.

*immobile. Apterostigma immobile* Weber, 1940a: 418 (w.m.) PANAMA. Junior synonym of *auriculatum:* Weber, 1958c: 243.

*jubatum. Apterostigma jubatum* Wheeler, W.M. 1925a: 47 (w.q.) GUIANA. Junior synonym of *robustum:* Weber, 1958c: 249.

***luederwaldti.*** *Apterostigma luederwaldti* Santschi, 1923d: 66 (w.) BRAZIL.

***madidiense.*** *Apterostigma madidiense* Weber, 1938b: 179 (w.q.m.) BOLIVIA.

***manni.*** *Apterostigma manni* Weber, 1938b: 176, fig. 11 (w.) BOLIVIA.

***mayri.*** *Apterostigma mayri* Forel, 1893e: 604 (w.) TRINIDAD. Wheeler, G.C. 1949: 668 (l.). Senior synonym of *abdita, discrepans, pallidum* and *zip:* Weber, 1958c: 248.

***moelleri.*** *Apterostigma moelleri* Forel, 1892d: 348 (w.q.) BRAZIL.

*nitidum. Apterostigma urichi* var. *nitidum* Weber, 1937: 392 (w.m.) GUIANA. Junior synonym of *urichii:* Weber, 1958c: 250.

*pallidum. Apterostigma mayri* var. *pallidum* Weber, 1937: 388 (w.q.) TRINIDAD. Subspecies of *mayri:* Weber, 1938b: 168. Junior synonym of *mayri:* Weber, 1958c: 248.

***peruvianum.*** *Apterostigma peruvianum* Wheeler, W.M. 1925a: 46 (w.q.) PERU.

#***petiolatum.*** *Apterostigma wasmanni* subsp. *petiolatum* Weber, 1938b: 175, fig. 10 (w.m.) BOLIVIA. Subspecies of *auriculatum:* Weber, 1958c: 244.

***pilosum.*** *Apterostigma pilosum* Mayr, 1865: 113, pl. 4, fig. 35 (q.m.) BRAZIL. Mayr, 1887: 554 (w.).

***robustum.*** *Apterostigma robustum* Emery, 1896g: 98, pl. 1, fig. 17 (w.) COSTA RICA. Senior synonym of *amiae, branneri, constrictum, jubatum* and *tic:* Weber, 1958c: 249.

***scutellare.*** *Apterostigma pilosum* r. *scutellare* Forel, 1885a: 364 (m.) MEXICO. Raised to species: Dalla Torre, 1893: 149.

***steigeri.*** *Apterostigma steigeri* Santschi, 1911b: 52, figs. 1–3 (w.) ARGENTINA. Santschi, 1912e: 529 (q.).

*tic. Apterostigma robustum* subsp. *tic* Weber, 1938b: 170, fig. 7 (w.q.m.) COSTA RICA. Junior synonym of *robustum:* Weber, 1958c: 249.

***tramitis.*** *Apterostigma tramitis* Weber, 1940a: 417 (w.q.) PANAMA. Wheeler, G.C. 1949: 668 (l.).

*uncinatum* Mayr, 1887; see under ***MYRMICOCRYPTA.***

***urichii.*** *Apterostigma urichii* Forel, 1893e: 603 (w.) TRINIDAD. Senior synonym of *guianense* and *nitidum:* Weber, 1958c: 250.

***wasmannii.*** *Apterostigma wasmannii* Forel, 1892d: 345 (w.m.) BRAZIL. Forel, 1893e: 605 (q.).

*zip. Apterostigma mayri* subsp. *zip* Weber, 1938b: 166, figs. 13, 16 (w.q.m.) BOLIVIA. Junior synonym of *mayri:* Weber, 1958c: 248.

*ARATROMYRMEX:* see under ***VOLLENHOVIA.***

*ARAUCOMYRMEX:* see under ***DORYMYRMEX.***

*ARCHAEOMYRMEX:* see under ***MYRMECINA.***

*****ARCHAEOPONE*** (Russia)

****kzylzharica.*** **Archaeopone kzylzharica* Dlussky, 1975: 121, fig. 132 (m.) RUSSIA (Cretaceous).

****taylori.*** **Archaeopone taylori* Dlussky, 1983: 72, fig. 6 (m.) RUSSIA (Cretaceous).

*ARCHEATTA:* see under ***ATTA.***

*****ARCHIMYRMEX*** (U.S.A.)

****rostratus.*** **Archimyrmex rostratus* Cockerell, 1923a: 52, figs. a, b (w.) U.S.A. (Eocene). See also: Carpenter, 1930: 16.

*****ARCHIPONERA.*** (U.S.A.)

****wheeleri.*** **Archiponera wheeleri* Carpenter, 1930: 27, pl. 1, pl. 2, fig. 1 (w.) U.S.A. (Miocene).

*ARCTOMYRMEX:* see under ***ADELOMYRMEX.***

*****ARMANIA*** (Russia)

****robusta.*** **Armania robusta* Dlussky, 1983: 68, figs. 1, 2 (q.) RUSSIA (Cretaceous).

*****ARMANIELLA*** (Russia)

****curiosa.*** **Armaniella curiosa* Dlussky, 1983: 71, fig. 5 (q.) RUSSIA (Cretaceous).

*ARNOLDIDRIS:* see under ***ORECTOGNATHUS.***

*AROTROPUS:* see under ***AMBLYOPONE***

*ASEMORHOPTRUM:* see under ***STENAMMA.***

***ASKETOGENYS*** (Malaysia)

***acubecca.*** *Asketogenys acubecca* Brown, 1972: 23, figs. 1, 2 (w.) WEST MALAYSIA.

***ASPHINCTOPONE*** (Afrotropical)

***lamottei.*** *Lepidopone lamottei* Bernard, 1953b: 208, fig. 4 (w.) GUINEA. Combination in *Asphinctopone:* Brown, 1953h: 3.

***lucida.*** *Asphinctopone lucidus* Weber, 1949b: 7, figs. 5–7 (w.) CENTRAL AFRICAN REPUBLIC.

***silvestrii.*** *Asphinctopone silvestrii* Santschi, 1914d: 318, fig. 6 (w.) NIGERIA.

*ASPIDIDRIS:* see under ***BASICEROS.***

*****ASYMPHYLOMYRMEX*** (Baltic Amber)

****balticus.*** *Asymphylomyrmex balticus* Wheeler, W.M. 1915h: 97, fig. 47 (w.) BALTIC AMBER (Oligocene).

#***ATOPODON:*** see under ***ACROPYGA.***

#***ATOPOGYNE:*** see under ***CREMATOGASTER.***

***ATOPOMYRMEX*** (Afrotropical)

*alluaudi* Emery, 1895; see under ***TERATANER.***

*arnoldi. Atopomyrmex mocquerysi* var. *arnoldi* Santschi, 1923e: 283 (w.) ZAIRE. Junior synonym of *mocquerysi:* Bolton, 1981b: 251.

*australis. Atopomyrmex mocquerysi* var. *australis* Santschi, 1914e: 16 (w.) SOUTH AFRICA. Junior synonym of *mocquerysi:* Wheeler, W.M. 1922a: 885.

*bottegoi* Emery, 1896; see under ***TERATANER.***

***calpocalycola.*** *Atopomyrmex calpocalycolus* Snelling, 1992a: 96, figs. 1–5 (w.) CAMEROUN.

*ceylonicus* Emery, 1901; see under ***PARATOPULA.***

***cryptoceroides.*** *Atopomyrmex cryptoceroides* Emery, 1892d: 561, pl. 15, figs. 5, 6 (q.) IVORY COAST. Race of *mocquerysi:* Forel, 1913b: 336; Forel, 1915c: 343; Wheeler, W.M. 1922a: 886. Revived status as species: Emery, 1922f: 240; Santschi, 1925h: 163 (in key). Senior synonym of *deplanatus:* Emery, 1899e: 477; of *curvispina:* Emery, 1912d: 273; Emery, 1922f: 240; of *melanoticus:* **new synonymy**.

*curvispina. Atopomyrmex mocquerysi* var. *curvispina* Forel, 1911c: 311 (w.) ZAIRE. Arnold, 1916: 192 (q.). Junior synonym of *cryptoceroides:* Emery, 1912d: 273; Emery, 1922f: 240.

*deplanatus. Atopomyrmex deplanatus* Mayr, 1895: 113 (w.) SIERRA LEONE. Junior synonym of *cryptoceroides:* Emery, 1899e: 477.

*erigens. Atopomyrmex mocquerysi* var. *erigens* Santschi, 1924b: 205 (w.) ZAIRE. Junior synonym of *mocquerysi:* Bolton, 1981b: 251.

*escherichi* Forel, 1911; see under ***DILOBOCONDYLA.***

*foreli* Emery, 1899; see under ***TERATANER.***

*luteus* Emery, 1899; see under ***TERATANER.***

*melanoticus. Atopomyrmex cryptoceroides* var. *melanoticus* Santschi, 1925h: 153 (w.) ZAIRE. [First available use of *Atopomyrmex mocquerysi* subsp. *cryptoceroides* var. *melanoticus* Wheeler, W.M. 1922a: 182; unavailable name.] **New synonymy**. [Material of unavailable name formerly referred to *cryptoceroides* by Bolton, 1981b: 250.]

***mocquerysi.*** *Atopomyrmex mocquerysi* André, 1889: 227 (w.) SENEGAL. André, 1895a: 3 (q.); Forel, 1913b: 335 (m.). Senior synonym of *australis:* Wheeler, W.M. 1922a: 885; of *arnoldi, erigens, obscura* and *opaca,* and material of the unavailable name *nigellus* referred here: Bolton, 1981b: 251.

*nigellus. Atopomyrmex mocquerysi* st. *opacus* var. *nigellus* Santschi, 1930b: 72 (w.) ANGOLA. Unavailable name; material referred to *mocquerysi* by Bolton, 1981b: 251.

*nodifer* Emery, 1901; see under ***TETRAMORIUM.***

*obscura. Atopomyrmex mocquerysi* var. *obscura* Santschi, 1923e: 283 (w.) IVORY COAST. Junior synonym of *mocquerysi:* Bolton, 1981b: 251.

*opaca. Atopomyrmex mocquerysi* st. *opaca* Santschi, 1923e: 283 (w.) ANGOLA. [Also described as new by Santschi, 1925h: 163.] Junior synonym of *mocquerysi:* Bolton, 1981b: 251.

*scotti* Forel, 1912; see under ***TERATANER.***

*selebensis* Emery, 1898; see under ***DILOBOCONDYLA.***

*srilankensis* Cheng, Ye & Yang, 1992; see under ***PARATOPULA.***

*steinheili* Forel, 1895; see under ***TERATANER.***

*ATOPULA:* see under ***TETRAMORIUM.***

***ATTA*** (Neotropical, southern Nearctic)

*abdominalis. Oecodoma abdominalis* Smith, F. 1858b: 184, pl. 10, fig. 22 (q.) no locality given ("various parts of South America"). Combination in *Atta:* Roger, 1863b: 35. Junior synonym of *sexdens:* Mayr, 1865: 80.

*affinis* Jerdon, 1851 [*Oecodoma*]; see under ***PHEIDOLOGETON.***

*ajax* Forel, 1909; see under ***ACROMYRMEX.***

*ambigua* Emery, 1888; see under ***ACROMYRMEX.***

*andicola* Emery, 1905; see under ***ACROMYRMEX.***

*angustata* Forel, 1908; see under ***ACROMYRMEX.***

*antarctica* Smith, F. 1858; see under ***MONOMORIUM.***

*antipodum* Smith, F. 1858; see under ***ANISOPHEIDOLE.***

*arborea* Buckley, 1867 [*Oecodoma*]; see under ***CREMATOGASTER.***

*arizonensis* Wheeler, W.M. 1907; see under ***TRACHYMYRMEX.***

*aspersa* Smith, F. 1858 [*Oecodoma*]; see under ***ACROMYRMEX.***

*atrata* Forel, 1912; see under ***CYPHOMYRMEX.***

*autuorii. Atta* (*Neoatta*) *sexdens* subsp. *autuorii* Borgmeier, 1950d: 253, figs. 32–34 (w.q.m.) BRAZIL. Junior synonym of *sexdens:* Borgmeier, 1959b: 359.

*balzani* Emery, 1890; see under ***ACROMYRMEX.***

*bellicosa* Smith, F. 1858; see under ***PHEIDOLOGETON.***

*bicolor* Buckley, 1867 [*Oecodoma*]; see under ***CREMATOGASTER.***

***bisphaerica.*** *Atta sexdens* var. *bisphaerica* Forel, 1908c: 348 (w.q.m.) BRAZIL. Combination in *Atta* (*Neoatta*): Gonçalves, 1942: 347; in *Atta* (*Palaeatta*): Borgmeier, 1950d: 244. Raised to species: Borgmeier, 1939: 423. See also: Borgmeier, 1959b: 377.

*bispinosa* Spinola, 1851; see under ***POGONOMYRMEX.***

*bolchevista. Atta* (*Neoatta*) *sexdens* subsp. *bolchevista* Gonçalves, 1942: 350 (w.) PARAGUAY. [First available use of *Atta sexdens* r. *rubropilosa* var. *bolchevista* Forel, 1921a: 204; unavailable name.] Junior synonym of *laevigata:* Borgmeier, 1950d: 243.

*boliviensis* Emery, 1905; see under ***ACROMYRMEX.***

*bonariensis* Emery, 1905; see under ***ACROMYRMEX.***

*borinquenensis* Wheeler, W.M. 1907; see under ***MYCOCEPURUS.***

*brazoensis* Buckley, 1867; see under ***SOLENOPSIS.***

*bruchi* Forel, 1912; see under ***ACROMYRMEX.***

*brunnea* Patton, 1894; see under ***ODONTOMACHUS.***

*brunnea* Forel, 1912; see under ***ACROMYRMEX.***

*bugnioni* Forel, 1912; see under ***TRACHYMYRMEX.***

*capensis* Mayr, 1862; see under ***MESSOR.***

***capiguara.*** *Atta* (*Neoatta*) *capiguara* Gonçalves, 1944: 234, figs. 1–6 (s.w.q.m.) BRAZIL. Combination in *Atta* (*Epiatta*): Borgmeier, 1950d: 250. See also: Gonçalves, 1945: 209; Borgmeier, 1959b: 372.

*caroli* Wheeler, W.M. 1911; see under ***TRACHYMYRMEX.***

***cephalotes.*** *Formica cephalotes* Linnaeus, 1758: 581 (w.) SURINAM. Olivier, 1792: 500 (q.m.); Wheeler, G.C. 1949: 677 (l.). Combination in *Atta:* Fabricius, 1804: 421; in *Oecodoma:* Latreille, 1818: 224. Senior synonym of *migratoria:* Retzius, 1783: 76; of *grossa:* Latreille, 1802a: 224; of *fervens:* Smith, F. 1858b: 180; of *visitatrix:* Emery, 1892b: 162; of *integrior, isthmicola, lutea, oaxaquensis, opaca* and *polita:* Borgmeier, 1959b: 340.

*chisosensis* Wheeler, W.M. 1907; see under ***ACROMYRMEX.***

*cingulata* Smith, F. 1857; see under ***PHEIDOLE.***

*cloosae* Forel, 1912; see under ***ACROMYRMEX.***

*clypeata* Smith, F. 1858; see under ***SOLENOPSIS.***

*coloradensis* Buckley, 1867; see under ***SOLENOPSIS.***

***columbica.*** *Atta columbica* Guérin-Méneville, 1844a: 422 (w.) COLOMBIA. Forel, 1913l: 239 (m.); Borgmeier, 1959b: 348 (q.); Wheeler, G.C. 1949: 681 (l.). Variety of *lebasii:* Emery, 1890b: 55. Subspecies of *cephalotes:* Emery, 1913: 258. Revived status as species: Forel, 1913l: 239; Emery, 1922f: 353. See also: Gonçalves, 1942: 346; Borgmeier, 1959b: 345. Senior synonym of *lebasii:* Dalla Torre, 1893: 152; of *erecta* and *tonsipes:* Borgmeier, 1959b: 346.

***ATTA** (continued)*

*coptophylla*. *Atta coptophylla* Guérin-Méneville, 1844a: 422 (w.) BRAZIL. Junior synonym of *sexdens:* Mayr, 1865: 80.
*cornetzi* Forel, 1912; see under ***TRACHYMYRMEX.***
*crassispina* Forel, 1909; see under ***ACROMYRMEX.***
*crudelis* Smith, F. 1858; see under ***POGONOMYRMEX.***
*crystallina* Wheeler, W.M. 1911; see under ***TRACHYMYRMEX.***
*decolor* Emery, 1905; see under ***ACROMYRMEX.***
*dentigera* Roger, 1862; see under ***MONOMORIUM.***
*desertorum* Wheeler, W.M. 1911; see under ***TRACHYMYRMEX.***
*destructor* Jerdon, 1851; see under ***MONOMORIUM.***
*didita* Walker, 1859; see under ***DILOBOCONDYLA.***
*diffusa* Jerdon, 1851 [*Oecodoma*]; see under ***PHEIDOLE.***
*diligens* Smith, F. 1858; see under ***PHEIDOLE.***
*dimidiata* Forel, 1911: see under ***ACROMYRMEX.***
*discigera* Mayr, 1887; see under ***ACROMYRMEX.***
*dissimilis*. *Atta dissimilis* Jerdon, 1851: 107 (w.) INDIA. Unrecognisable taxon; *incertae sedis* in *Atta.*
*diversa* Jerdon, 1851 [*Oecodoma*]; see under ***PHEIDOLOGETON.***
*domicola*. *Atta domicola* Jerdon, 1851: 105 (w.) INDIA. Unrecognisable taxon; *incertae sedis* in *Atta.*
*dubia* Forel, 1908; see under ***ACROMYRMEX.***
*echinatior* Forel, 1899; see under ***ACROMYRMEX.***
*emilii* Forel, 1904; see under ***ACROMYRMEX.***
*erecta*. *Atta cephalotes* var. *erecta* Santschi, 1929f: 92 (diagnosis in key) (w.) COSTA RICA. Subspecies of *cephalotes:* Gonçalves, 1942: 345. Junior synonym of *columbica:* Borgmeier, 1959b: 346.
*fabricator* Smith, F. 1858; see under ***PHEIDOLE.***
*farinosa* Emery, 1894; see under ***TRACHYMYRMEX.***
*ferruginea* Emery, 1905; see under ***ACROMYRMEX.***
*fervens*. *Formica fervens* Drury, 1782: 58, pl. 42, fig. 3 (q.) MEXICO. Junior synonym of *cephalotes:* Smith, F. 1858b: 180.
*flavicornis*. *Formica flavicornis* Fabricius, 1789: 280 (q.) SURINAM. Junior synonym of *sexdens:* Forel, 1905b: 161.
*floricola* Jerdon, 1851; see under ***MONOMORIUM.***
*fracticornis* Forel, 1909; see under ***ACROMYRMEX.***
*fuhrmanni* Forel, 1912; see under ***ACROMYRMEX.***
*fumipennis* Smith, F. 1858; see under ***PHEIDOLE.***
*fusca* Emery, 1894; see under ***TRACHYMYRMEX.***
*fuscata*. *Atta sexdens* var. *fuscata* Santschi, 1922b: 362 (w.) BOLIVIA. Subspecies of *sexdens:* Gonçalves, 1942: 350. Junior synonym of *sexdens:* Borgmeier, 1959b: 359.
*gemella* Roger, 1862; see under ***APHAENOGASTER.***
*geminata* Fabricius, 1804; see under ***SOLENOPSIS.***
*goeldii* Forel, 1893; see under ***MYCOCEPURUS.***
**goiana.** *Atta* (*Neoatta*) *goiana* Gonçalves, 1942: 347 (m.) BRAZIL. Borgmeier, 1959b: 379 (w.). Combination in *Atta* (*Palaeatta*): Borgmeier, 1950d: 245.
*grossa*. *Formica grossa* Fabricius, 1787: 309 (q.) SURINAM. Junior synonym of *cephalotes:* Latreille, 1802a: 224.
*guentheri* Forel, 1893; see under ***ACROMYRMEX.***
*hartmanni* Wheeler, W.M. 1907; see under ***MYCETOSORITIS.***
*haytiana* Wheeler, W.M. & Mann, 1914; see under ***TRACHYMYRMEX.***
*heyeri* Forel, 1899; see under ***ACROMYRMEX.***
*homalops* Emery, 1905; see under ***ACROMYRMEX.***
*hostilis* Smith, F. 1858; see under ***PHEIDOLOGETON.***
*iheringi* Emery, 1888; see under ***TRACHYMYRMEX.***
*inquirens* Forel, 1912; see under ***ACROMYRMEX.***
*instabilis* Smith, F. 1858; see under ***MESSOR.***
**insularis.** *Atta insularis* Guérin-Méneville, 1844a: 422 (w.) CUBA. Mayr, 1865: 82 (q.m.). Combination in *Atta* (*Archeatta*): Gonçalves, 1942: 343. Subspecies of *fervens:* Borgmeier, 1939: 423. Revived status as species: Borgmeier, 1950d: 260. See also: Borgmeier, 1959b: 353.
*integrior*. *Atta cephalotes* var. *integrior* Forel, 1904c: 31 (w.) BRAZIL. Subspecies of *cephalotes:* Gonçalves, 1942: 346. Junior synonym of *cephalotes:* Borgmeier, 1959b: 340.
*intermedia* Forel, 1909; see under ***TRACHYMYRMEX.***
*irmgardae* Forel, 1912; see under ***TRACHYMYRMEX.***
*irrorata* Wheeler, W.M. 1911; see under ***TRACHYMYRMEX.***
*isthmicola*. *Atta cephalotes* subsp. *isthmicola* Weber, 1941b: 127 (w.q.) PANAMA. Junior synonym of *cephalotes:* Borgmeier, 1959b: 340.
*jamaicensis* André, 1893; see under ***TRACHYMYRMEX.***
**laevigata.** *Oecodoma laevigata* Smith, F. 1858b: 182, pl. 10, fig. 24 (s.w.) BRAZIL. Forel, 1913l: 239 (m.); Gonçalves, 1942: 336 (q.). Combination in *Atta:* Roger, 1863b: 35; in *Atta* (*Neoatta*): Gonçalves, 1942: 348; Borgmeier, 1959b: 365; in *Atta* (*Epiatta*): Borgmeier, 1950d: 247. Variety/subspecies of *sexdens:* Mayr, 1865: 80; Emery, 1913: 259; Wheeler, W.M. 1923a: 4. Revived status as species: Forel, 1913l: 239; Menozzi, 1935b: 197; Borgmeier, 1939: 423; Gonçalves, 1942: 348; Borgmeier, 1959b: 365. Senior synonym of *bolchevista:* Borgmeier, 1950d: 243; of *venezuelensis:* Borgmeier, 1959b: 365.
*laevis* Mayr, 1862; see under ***APHAENOGASTER.***
*landolti* Forel, 1895; see under ***ACROMYRMEX.***
*laticeps* Emery, 1905; see under ***ACROMYRMEX.***
*lebasii*. *Atta lebasii* Guérin-Méneville, 1844a: 422 (w.) COLOMBIA. Emery, 1890b: 54 (q.m.). Junior synonym of *columbica:* Dalla Torre, 1893: 152.
*lecta* Forel, 1911; see under ***CYPHOMYRMEX.***
*luederwaldti* Forel, 1911; see under ***MYCETAROTES.***
*lincecumii* Buckley, 1867; see under ***SOLENOPSIS.***
*lizeri*. *Atta polita* var. *lizeri* Santschi, 1920e: 123, fig. (s.w.) BOLIVIA. [Also described as new by Santschi, 1922b: 364.] Subspecies of *laevigata:* Gonçalves, 1942: 349; of *vollenweideri:* Borgmeier, 1950d: 243. Junior synonym of *saltensis:* Borgmeier, 1959b: 381.
*lobicornis* Emery, 1888; see under ***ACROMYRMEX.***
*lugens*. *Atta vollenweideri* var. *lugens* Borgmeier, 1939: 424, fig. 19 (w.) BRAZIL. Junior synonym of *piriventris:* Gonçalves, 1942: 351.
*lutea*. *Atta lutea* Forel, 1893e: 587 (s.w.) BARBADOS. Subspecies of *columbica:* Forel, 1908c: 349. Revived status as species: Borgmeier, 1950d: 259. Subspecies of *cephalotes:* Weber, 1958a: 10. Junior synonym of *cephalotes:* Borgmeier, 1959b: 340.
*malabarica* Jerdon, 1851 [*Oecodoma*]; see under ***PHEIDOLE.***
*maritima* Wheeler, W.M. 1905; see under ***TRACHYMYRMEX.***
*marthae* Forel, 1912; see under ***TRACHYMYRMEX.***
*meinerti* Forel, 1893; see under ***ACROMYRMEX.***
*mesonotalis* Emery, 1905; see under ***ACROMYRMEX.***
**mexicana.** *Oecodoma mexicana* Smith, F. 1858b: 185, pl. 10, fig. 20 (q.m.) MEXICO. Norton, 1868a: 66; Norton, 1868c: 9 (s.w.). Combination in *Atta:* Roger, 1863b: 35; in *Atta* (*Archeatta*): Gonçalves, 1942: 343. Junior synonym of *fervens:* Forel, 1885a: 362. Revived from synonymy as subspecies of *insularis:* Emery, 1913: 259. Revived status as species: Gonçalves, 1942: 335. See also: Borgmeier, 1959b: 351.
*migratoria*. *Formica migratoria* De Geer, 1773: 604, pl. 31, figs. 11–13 (w.) SURINAM. Junior synonym of *cephalotes:* Retzius, 1783: 76.
*minor* Jerdon, 1851 [*Oecodoma*]; see under ***PHEIDOLE.***
*minuta* Jerdon, 1851; see under ***MONOMORIUM.***
*modesta* Forel, 1901; see under ***ACROMYRMEX.***
*moelleri* Forel, 1893; see under ***ACROMYRMEX.***
*muticinoda* Forel, 1901; see under ***ACROMYRMEX.***
*naranjo* Forel, 1912; see under ***TRACHYMYRMEX.***
*nigra* Smith, F. 1858 [*Oecodoma*]; see under ***ACROMYRMEX.***
*nigriventris* Smith, F. 1858; see under ***PHEIDOLE.***
*nigrosetosa* Forel, 1908; see under ***ACROMYRMEX.***
*nodifer* Smith, F. 1858; see under ***PHEIDOLE.***
*oaxaquensis*. *Atta* (*Atta*) *cephalotes* subsp. *oaxaquensis* Gonçalves, 1942: 344 (w.) MEXICO. Junior synonym of *cephalotes:* Borgmeier, 1959b: 340.
*obscurata*. *Atta vollenweideri* var. *obscurata* Gallardo, 1916d: 344 (w.) ARGENTINA. Subspecies of *polita:* Santschi, 1929d: 305; of *vollenweideri:* Borgmeier, 1950d: 243. Junior synonym of *saltensis:* Borgmeier, 1959b: 381.
*obscurior* Wheeler, W.M. 1907; see under ***TRACHYMYRMEX.***
*obsidiana* Mayr, 1861; see under ***APHAENOGASTER.***
*oetkeri* Forel, 1908; see under ***TRACHYMYRMEX.***
*opaca*. *Atta cephalotes* var. *opaca* Forel, 1904c: 31 (w.) COLOMBIA. Subspecies of *cephalotes:* Gonçalves, 1942: 345. Junior synonym of *cephalotes:* Borgmeier, 1959b: 340.
**opaciceps.** *Atta bisphaerica* subsp. *opaciceps* Borgmeier, 1939: 427, fig.16 (w.) BRAZIL. Borgmeier, 1950d: 248 (m.); Borgmeier, 1959b: 368 (q.). Combination in *Atta* (*Neoatta*): Gonçalves, 1942: 340; Borgmeier, 1959b: 368; in *Atta* (*Epiatta*): Borgmeier, 1950d: 248. Raised to species: Gonçalves, 1945: 186; Borgmeier, 1959b: 369.
*pallida* Smith, F. 1858 [*Oecodoma*]; see under ***ACROMYRMEX.***
*panamensis* Forel, 1899; see under ***ACROMYRMEX.***
*penetrans* Smith, F. 1857; see under ***VOLLENHOVIA.***
*pennsylvanica* Buckley, 1867; see under ***PHEIDOLE.***
*picea* Buckley, 1867; see under ***PHEIDOLE.***
*piliventris* Smith, F. 1858; see under ***PHEIDOLE.***
*pilosa*. *Oecodoma pilosa* Buckley, 1867: 348 (w.) U.S.A. *Incertae sedis;* unrecognisable taxon. See notes in Wheeler, W.M. 1902f: 29. Unplaced myrmicine taxon: Smith, D.R. 1979: 1414.
*piriventris*. *Atta vollenweideri* var. *piriventris* Santschi, 1919f: 50 (w.) ARGENTINA. Subspecies of *sexdens,* and senior synonym of *lugens:* Gonçalves, 1942: 351. Junior synonym of *sexdens:* Borgmeier, 1959b: 359.

*polita. Atta cephalotes* subsp. *polita* Emery, 1905c: 54 (w.) BOLIVIA. Raised to species: Borgmeier, 1939: 423. Subspecies of *vollenweideri:* Borgmeier, 1950d: 243. Junior synonym of *cephalotes:* Borgmeier, 1959b: 340.
*providens* Sykes, 1835; see under ***PHEIDOLE.***
*pruinosa* Emery, 1906; see under ***TRACHYMYRMEX.***
*pubescens* Emery, 1905; see under ***ACROMYRMEX.***
*purensis* Forel, 1912; see under ***ACROMYRMEX.***
*quadriglumis* Haliday, 1836; see under ***ECITON.***
*quadrispinosa* Jerdon, 1851 [*Oecodoma*]; see under ***LOPHOMYRMEX.***
*rectispina* Forel, 1908; see under ***ACROMYRMEX.***
*risii* Forel, 1908; see under ***ACROMYRMEX.***
**robusta.** *Atta sexdens* subsp. *robusta* Borgmeier, 1939: 426, fig. 15 (w.q.m.) BRAZIL. Combination in *Atta* (*Neoatta*): Gonçalves, 1942: 348; Borgmeier, 1959b: 370; in *Atta* (*Epiatta*): Borgmeier, 1950d: 246. Raised to species: Gonçalves, 1942: 348. See also: Borgmeier, 1959b: 370.
*rochai* Forel, 1904; see under ***ACROMYRMEX.***
*rubra* Smith, F. 1858; see under ***PHEIDOLE.***
*rubropilosa. Atta sexdens* var. *rubropilosa* Forel, 1908c: 348 (w.q.m.) BRAZIL. Subspecies of *sexdens:* Santschi, 1929f: 93; Borgmeier, 1939: 422; Gonçalves, 1945: 187. Junior synonym of *sexdens:* Borgmeier, 1959b: 359.
*rufa* Jerdon, 1851; see under ***SOLENOPSIS.***
*rugosa* Smith, F. 1858 [*Oecodoma*]; see under ***ACROMYRMEX.***
**saltensis.** *Atta laevigata* var. *saltensis* Forel, 1913l: 237 (w.) ARGENTINA. Borgmeier, 1959b: 382 (q.m.). Combination in *Atta* (*Neoatta*): Gonçalves, 1942: 349; in *Atta* (*Epiatta*): Borgmeier, 1950d: 249; in *Atta* (*Palaeatta*): Borgmeier, 1959b: 381. Variety/subspecies of *vollenweideri:* Gallardo, 1916d: 344; Borgmeier, 1950d: 249; subspecies of *polita:* Santschi, 1929d: 305; Borgmeier, 1939: 423; subspecies of *laevigata:* Gonçalves, 1942: 349. Raised to species, and senior synonym of *lizeri* and *obscurata:* Borgmeier, 1959b: 381.
*santschii* Forel, 1912; see under ***ACROMYRMEX.***
*saussurei* Forel, 1885; see under ***TRACHYMYRMEX.***
*schuppi* Forel, 1901; see under ***MYCOCEPURUS.***
*seminole* Wheeler, W.M. 1911; see under ***TRACHYMYRMEX.***
*septentrionalis* McCook, 1881; see under ***TRACHYMYRMEX.***
**sexdens.** *Formica sexdens* Linnaeus, 1758: 581 (w) "America meridionali". Mayr, 1865: 82 (q.m.); Wheeler, G.C. 1949: 681 (l.). Combination in *Atta:* Fabricius, 1804: 422; in *Atta* (*Neoatta*): Gonçalves, 1942: 349. Senior synonym of *sexdentata:* Latreille, 1802a: 228; of *abdominalis* and *coptophylla:* Mayr, 1865: 80; of *flavicornis:* Forel, 1905b: 161; of *autuorii, fuscata, piriventris* (and its junior synonym *lugens*), and *rubropilosa:* Borgmeier, 1959b: 359.
*sexdentata. Formica sexdentata* Latreille, 1802a: 228, pl. 9, figs. 59, 60 (w.) FRENCH GUIANA. Junior synonym of *sexdens:* Latreille, 1802a: 228.
*sharpii* Forel, 1893; see under ***TRACHYMYRMEX.***
**silvae.** *Atta* (*Neoatta*) *silvae* Gonçalves, 1983b: 174, fig. (w.) BRAZIL.
*silvestrii* Emery, 1905; see under ***ACROMYRMEX.***
*smithii* Forel, 1893; see under ***MYCOCEPURUS.***
*spinoda* Smith, F. 1858; see under ***PHEIDOLE.***
*splendida* Roger, 1859; see under ***APHAENOGASTER.***
*squamulifera* Emery, 1896; see under ***TRACHYMYRMEX.***
*striata* Roger, 1863; see under ***ACROMYRMEX.***
*striola* Roger, 1859; see under ***APHAENOGASTER.***
*subterranea* Forel, 1893; see under ***ACROMYRMEX.***
*tardigrada. Oecodoma tardigrada* Buckley, 1867: 349 (w.q.m.) U.S.A. *Incertae sedis* in *Atta;* unrecognisable taxon. Included in *Acromyrmex:* Forel, 1885a: 358; in *Trachymyrmex:* Mayr, 1886d: 442; Forel, 1893e: 601; Wheeler, W.M. 1902f: 30. Unplaced myrmicine taxon: Smith, D.R. 1979: 1414.
*tennesseensis* Mayr, 1862; see under ***APHAENOGASTER.***
*testacea* Smith, F. 1858; see under ***PHEIDOLE.***
**texana.** *Myrmica* (*Atta*) *texana* Buckley, 1860a: 233 (w.q.m.) U.S.A. Wheeler, G.C. & Wheeler, J. 1974d: 79 (l.). Combination in *Atta:* Roger, 1863b: 35; in *Atta* (*Archeatta*): Gonçalves, 1942: 343. Junior synonym of *fervens:* Forel, 1885a: 362. Revived from synonymy: Wheeler, W.M. 1907c: 700. Subspecies of *insularis:* Emery, 1913: 259; Emery, 1922f: 354; of *fervens:* Borgmeier, 1939: 423. Revived status as species: Gonçalves, 1942: 335. See also: Borgmeier, 1959b: 356; Smith, D.R., 1979: 1413; Petralia & Vinson, 1980: 386.
*thoracica* Mayr, 1862; see under ***MESSOR.***
*tolteca* Wheeler, W.M. 1907; see under ***MYCOCEPURUS.***
*tonsipes. Atta columbica* var. *tonsipes* Santschi, 1929f: 92 (diagnosis in key) (w.) PANAMA. Subspecies of *columbica:* Gonçalves, 1942: 346. Junior synonym of *columbica:* Borgmeier, 1959b: 346.
*tristis. Atta vollenweideri* var. *tristis* Santschi, 1919f: 50 (w.) ARGENTINA. Subspecies of *sexdens:* Borgmeier, 1950d: 253. Junior synonym of *vollenweideri:* Borgmeier, 1959b: 374.
*turrifex* Wheeler, W.M. 1903; see under ***TRACHYMYRMEX.***
*urichii* Forel, 1893; see under ***TRACHYMYRMEX.***
*venezuelensis. Atta* (*Neoatta*) *laevigata* subsp. *venezuelensis* Gonçalves, 1942: 349 (w.q.) VENEZUELA. Combination in *Atta* (*Epiatta*): Borgmeier, 1950d: 243. Junior synonym of *laevigata:* Borgmeier, 1959b: 365.
*versicolor* Pergande, 1894; see under ***ACROMYRMEX.***
*vertebrata* Wheeler, W.M. 1911; see under ***TRACHYMYRMEX.***
*vigilans* Smith,F. 1858; see under ***PHEIDOLE.***
*virginiana* Buckley, 1867; see under ***TRACHYMYRMEX.***
*visitatrix. Formica visitatrix* Christ, 1791: 517 (w.) SURINAM. Junior synonym of *cephalotes:* Emery, 1892b: 162.
**vollenweideri.** *Atta sexdens* r. *vollenweideri* Forel, 1893e: 588 (s.w.q); Forel, 1912e: 179 (m.). Combination in *Atta* (*Neoatta*): Gonçalves, 1942: 348; Borgmeier, 1959b: 374; in *Atta* (*Epiatta*): Borgmeier, 1950d: 248. Raised to species: Forel, 1912e: 179. Subspecies of *laevigata:* Forel, 1913l: 237; Emery, 1922f: 354; of *sexdens:* Emery, 1913: 259. Revived status as species: Forel, 1915c: 358; Gallardo, 1916d: 341; Santschi, 1919f: 50; Santschi, 1922b: 363. See also: Borgmeier, 1939: 422; Gonçalves, 1942: 348; Borgmeier, 1959b: 374. Senior synonym of *tristis:* Borgmeier, 1959b: 374.

******ATTAICHNUS* (ichnotaxon)
**kuenzelii.* **Attaichnus kuenzelii* Laza, 1982: 112, figs. ARGENTINA (ichnospecies).

*ATTOMYRMA:* see under ***APHAENOGASTER.***

***ATTOPSIS** (Europe)
**acuta* Förster, 1891; see under ***SOLENOPSIS.***
**anthracina* Heer, 1850; see under ***CATAULACUS.***
**blanda* Förster, 1891; see under ***SOLENOPSIS.***
**extensa* Förster, 1891; see under ***SOLENOPSIS.***
****longipennis.*** *Attopsis longipennis* Heer, 1850: 155, pl. 12, fig. 11 (q.) YUGOSLAVIA (Miocene). See also: Förster, 1891: 437.
****longipes.*** *Attopsis longipes* Heer, 1867: 29, pl. 2, fig. 15 (m.?) YUGOSLAVIA (Miocene). See also: Förster, 1891: 442.
**maxima* Förster, 1891; see under ***SOLENOPSIS.***
**moesta* Förster, 1891; see under ***SOLENOPSIS.***
**nigra* Heer, 1850; see under ***CATAULACUS.***
**privata* Förster, 1891; see under ***SOLENOPSIS.***
**superba* Förster, 1891; see under ***SOLENOPSIS.***
**valida* Förster, 1891; see under ***SOLENOPSIS.***

**#*AULACOMYRMA:*** see under ***POLYRHACHIS.***

***AULACOPONE*** (Armenia)
**relicta.** *Aulacopone relicta* Arnol'di, 1930a: 140, figs. 1–5 (q.) ARMENIA. [Also described as new by Arnol'di, 1930d: 159.] See also: Brown, 1958g: 206; Taylor, 1980c: 353.

**#*AUSTROLASIUS:*** see under ***LASIUS.***

***AXINIDRIS*** (Afrotropical)
**acholli.** *Axinidris acholli* Weber, 1941a: 193, figs. 10–12 (w.) SUDAN. See also: Shattuck, 1991: 109.
**bidens.** *Axinidris bidens* Shattuck, 1991: 111, figs. 8–10, 35 (w.) GHANA.
**denticulata.** *Engramma denticulatum* Wheeler, W.M. 1922a: 205, fig. 51 (w.) ZAIRE. Combination in *Axinidris:* Shattuck, 1991: 112.
**ghanensis.** *Axinidris ghanensis* Shattuck, 1991: 114, fig. 35 (w.) GHANA.
**hylekoites.** *Axinidris hylekoites* Shattuck, 1991: 114, figs. 14–16, 35 (w.q.) GHANA.
**kakamegensis.** *Axinidris kakamegensis* Shattuck, 1991: 115, figs. 17–19, 35 (w.) KENYA.
**kinoin.** *Axinidris kinoin* Shattuck, 1991: 115, figs. 20–22 (w.) NIGERIA.
**murielae.** *Axinidris murielae* Shattuck, 1991: 116, figs. 23–25, 35 (w.q.) CAMEROUN.
**nigripes.** *Axinidris nigripes* Shattuck, 1991: 117, fig. 35 (w.) GHANA.
**occidentalis.** *Axinidris occidentalis* Shattuck, 1991: 118, figs. 26–28, 35 (w.) LIBERIA.
**palligastrion.** *Axinidris palligastrion* Shattuck, 1991: 118, figs. 29–31, 35 (w.) GHANA.
**parvus.** *Axinidris parvus* Shattuck, 1991: 118 (w.) LIBERIA.
**tridens.** *Engramma tridens* Arnold, 1946: 58, fig. 10 (q.) MALAWI. Combination in *Axinidris:* Shattuck, 1991: 119.

***AZTECA*** (Neotropical)

**adrepens.** *Azteca adrepens* Forel, 1911e: 284 (w.) PARAGUAY.

*aequalis. Azteca alfari* var. *aequalis* Forel, 1906d: 239 (w.) BRAZIL. Forel, 1908c: 387 (q.m.). Junior synonym of *ovaticeps:* Longino, 1989: 8.

*aequilata. Azteca alfari* var. *aequilata* Forel, 1904e: 691 (w.q.m.) BRAZIL. Junior synonym of *ovaticeps:* Longino, 1989: 8.

**aesopus.** ***Azteca aesopus*** Forel, 1908c: 392 (s.w.) BRAZIL.

**alfari.** *Azteca alfari* Emery, 1893b: 138, pl. 2, figs. 48, 49 (s.w.) COSTA RICA. Emery, 1896c: 3 (q.); Wheeler, G.C. & Wheeler, J. 1951: 192 (l.). Senior synonym of *breviscapa, cecropiae, curtiscapa, fumaticeps, langi, lucidula, lynchi* (and its junior synonym *argentina*), *mixta, virens,* and material of the unavailable name *zonalis* referred here: Longino, 1989: 5; senior synonym of *bicolor:* Longino, 1992: 1575. See also: Wheeler, W.M. 1942: 215. [Note: in the older literature the name of this taxon is frequently misspelled as *alfaroi,* for example Emery, 1896c: 4; Forel, 1899c: 113.] Current subspecies: nominal plus *beltii.*

**alpha.* **Azteca alpha* Wilson, 1985c: 58, figs. 5–8 (w.q.m.) DOMINICAN AMBER (Oligocene).

**angusticeps.** *Azteca angusticeps* Emery, 1893b: 144, pl. 2, figs. 66–71 (s.w.q.) BRAZIL.

**#antillana.** *Azteca delpini* r. *antillana* Forel, 1899c: 111 (w.) ANTILLES.

**aragua.** *Azteca aragua* Longino, 1992: 1575, figs. 11, 31, 45 (q.) VENEZUELA.

*argentina. Azteca alfari* var. *argentina* Forel, 1914d: 287 (w.) ARGENTINA. Gallardo, 1916a: 115 (q.). Junior synonym of *lynchi:* Bruch, 1915: 534; Gallardo, 1915: 20.

**aurita.** *Azteca aurita* Emery, 1893b: 146, pl. 2, figs. 85, 86 (q.) BRAZIL. Senior synonym of *lacrymosa:* Forel, 1906d: 237; of *pilosula:* Forel, 1912h: 54. Current subspecies: nominal plus *silvae.*

**australis.** *Azteca xanthochroa* subsp. *australis* Wheeler, W.M. 1942: 249 (q.) PERU. Longino, 1992: 1582 (w.). Raised to species: Longino, 1992: 1578.

**#bahiana.** *Azteca polymorpha* var. *bahiana* Forel, 1912h: 56 (w.) BRAZIL.

**barbifex.** *Azteca barbifex* Forel, 1906d: 242 (s.w.q.) BRAZIL.

**#beltii.** *Azteca bicolor* r. *beltii* Emery, 1893b: 142, pl. 2, figs. 61–63 (s.w.) COSTA RICA. Subspecies of *alfari:* implied by Longino, 1992: 1575 (as *bicolor* is given as a junior synonym of *alfari*).

**bequaerti.** *Azteca bequaerti* Wheeler, W.M. & Bequaert, 1929: 35 (w.q.) BRAZIL.

*bicolor. Azteca bicolor* Emery, 1893b: 141, pl. 2, figs. 54–60 (s.w.q.) BRAZIL. Junior synonym of *alfari:* Longino, 1992: 1575.

**#boliviana.** *Azteca brevicornis* var. *boliviana* Wheeler, W.M. & Mann, in Wheeler, W.M. 1942: 225 (w.) BOLIVIA.

**#bondari.** *Azteca paraensis* var. *bondari* Borgmeier, 1937b: 250 (w.q.) BRAZIL.

**brevicornis.** *Liometopum brevicorne* Mayr, 1878: 870 (w.) BRAZIL. Combination in *Azteca:* Emery, 1893b: 145. Current subspecies: nominal plus *boliviana.*

**brevis.** *Azteca brevis* Forel, 1899c: 121, pl. 4, fig. 17 (s.w.) COSTA RICA.

*breviscapa. Azteca foreli* r. *breviscapa* Forel, 1912h: 51 (w.) COSTA RICA. [First available use of *Azteca foreli* r. *championi* var. *breviscapa* Forel, 1899c: 112; unavailable name.] Junior synonym of *alfari:* Longino, 1989: 5.

*brunni. Azteca muelleri* var. *brunni* Forel, 1909a: 251 (w.) EL SALVADOR. Junior synonym of *muelleri:* Longino, 1992: 1591.

**#cayennensis.** *Azteca gnava* r. *cayennensis* Forel, 1912h: 49 (w.) FRENCH GUIANA.

**#cearensis.** *Azteca chartifex* r. *cearensis* Forel, 1903c: 259 (w.) BRAZIL.

*cecropiae. Azteca alfari* r. *cecropiae* Forel, 1906d: 240 (w.) BRAZIL. Forel, 1908c: 387 (q.m.). See also: Forel, 1912h: 51; Stitz, 1913: 212. Junior synonym of *alfari:* Longino, 1989: 5.

**#championi.** *Azteca foreli* r. *championi* Forel, 1899c: 112 (w.) COLOMBIA. See also: Forel, 1912h: 50.

**chartifex.** *Azteca chartifex* Forel, in Emery, 1896c: 4 (footnote), fig. 5 (w.) TRINIDAD. Current subspecies: nominal plus *cearensis, decipiens, lanians, laticeps, multinida, spiriti, stalactitica.*

**christopherseni.** *Azteca christopherseni* Forel, 1912h: 52 (w.) PANAMA.

**#clariceps.** *Azteca lanuginosa* var. *clariceps* Santschi, 1933e: 121 (w.) ARGENTINA.

**coeruleipennis.** *Azteca coeruleipennis* Emery, 1893b: 130, pl. 1, figs. 1–11 (w.q.m.) COSTA RICA. See also: Longino, 1992: 1583.

**#columbica.** *Azteca tonduzi* var. *columbica* Forel, 1912h: 51 (w.) COLOMBIA.

**constructor.** *Azteca constructor* Emery, 1896c: 2, figs. 2, 3 (w.q.m.) COSTA RICA. Senior synonym of *emmae* and *guianae:* Longino, 1992: 1585.

**#cordiae.** *Azteca ulei* var. *cordiae* Forel, 1904e: 695 (s.w.) BRAZIL. Wheeler, W.M. 1942: 246 (q.).

**#cordincola.** *Azteca longiceps* r. *cordincola* Forel, 1921a: 203 (w.) BOLIVIA. Wheeler, W.M. 1942: 232 (s.q.).

*costaricensis. Azteca xanthochroa* var. *costaricensis* Wheeler, W.M. 1942: 248 (w.) COSTA RICA. Junior synonym of *xanthochroa:* Longino, 1992: 1594.

**coussapoae.** *Azteca coussapoae* Forel, 1904e: 700 (w.) BRAZIL.

**crassicornis.** *Azteca crassicornis* Emery, 1893b: 147 (q.) BRAZIL.

*croceiscapa. Azteca goeldii* st. *croceiscapa* Forel, 1915c: 361. *Nomen nudum,* attributed to Emery.

*curtiscapa. Azteca alfari* var. *curtiscapa* Forel, 1912h: 51 (w.q.) COSTA RICA. Junior synonym of *alfari:* Longino, 1989: 5.

**#decipiens.** *Azteca chartifex* r. *decipiens* Forel, 1906d: 236 (w.) BRAZIL. [First available use of *Azteca chartifex* r. *laticeps* var. *decipiens* Forel, 1904c: 44; unavailable name.]

**delpini.** *Azteca delpini* Emery, 1893b: 133, pl. 1, figs. 30–32 (s.w.) BRAZIL. Current subspecies: nominal plus *antillana, trinidadensis.*

**depilis.** *Azteca depilis* Emery, 1893b: 139, pl. 2, figs. 37–42 (s.w.q.m.) BRAZIL. [*Azteca depilis* Schumann, 1881: 417. *Nomen nudum.*]

*diegensis. Azteca delpini* r. *antillana* var. *diegensis* Forel, 1912h: 50 (w.) COLOMBIA. Unavailable name (Shattuck, 1994: 30).

*dominicensis. Azteca delphini* subsp. *antillana* var. *dominicensis* Wheeler, W.M. 1913d: 243 (w.) DOMINICA. Unavailable name (Shattuck, 1994: 30).

**#dubia.** *Azteca schumanni* subsp. *dubia* Mann, 1916: 469, pl. 2, fig. 15 (s.w.m.) BRAZIL.

**duckei.** *Azteca duckei* Forel, 1906d: 243, fig. 2 (s.w.q.) BRAZIL.

**duroiae.** *Azteca duroiae* Forel, 1904e: 697 (s.w.) BRAZIL. Forel, 1906d: 241 (m.).

**#eiseni.** *Azteca eiseni* Pergande, 1896: 868 (s.w.) MEXICO. Variety of *forelii:* Forel, 1899c: 111.

*elatior. Azteca traili* r. *tococae* var. *elatior* Forel, 1906d: 239 (s.w.m.) BRAZIL. Wheeler, W.M. & Bequaert, 1929: 34 (q.). Unavailable name (Shattuck, 1994: 30).

**#emarginatisquamis.** *Azteca pittieri* var. *emarginatisquamis* Forel, 1921a: 204 (w.) COSTA RICA.

**emeryi.** *Azteca emeryi* Forel, 1904e: 698 (s.w.q.) BRAZIL.

*emmae. Azteca emmae* Forel, 1904e: 702 (s.w.q.m.) COSTA RICA. Junior synonym of *constructor:* Longino, 1992: 1585.

**eumeces.* **Azteca eumeces* Wilson, 1985c: 31, fig. 9 (w.) DOMINICAN AMBER (Oligocene).

**fasciata.** *Azteca fasciata* Emery, 1893b: 143, pl. 2, figs. 83, 84 (q.) BOLIVIA. Emery, 1896c: 6 (w.). Current subspecies: nominal plus *laeta, nigricans, similis.*

*fasciata. Azteca coeruleipennis* var. *fasciata* Pergande, 1896: 867 (s.w.) MEXICO. [Junior primary homonym of *fasciata* Emery, above.] Replacement name: *velox* Forel, 1899c: 108.

*festai. Azteca festai* Emery, 1896b: 4, fig. 3 (w.) PANAMA. Forel, 1899c: 118 (s.). Junior synonym of *trigona:* Forel, 1906d: 237.

**fiebrigi.** *Azteca fiebrigi* Forel, 1909a: 261 (s.w.q.m.) PARAGUAY.

**#filicis.** *Azteca traili* var. *filicis* Forel, 1904e: 692 (w.) PERU.

**forelii.** *Azteca forelii* Emery, 1893b: 137, pl. 2, figs. 50, 51 (s.w.) COSTA RICA. Current subspecies: nominal plus *championi, eiseni, raptrix, ursina, xysticola.*

**foveiceps.** *Azteca foveiceps* Wheeler, W.M. 1921f: 163, fig. 16 (w.q.m.) GUIANA.

*fumaticeps. Azteca alfaroi* var. *fumaticeps* Forel, 1909a: 250 (w.) MEXICO. Wheeler, W.M. 1942: 218 (q.). Junior synonym of *alfari:* Longino, 1989: 5.

**#gagatina.** *Azteca ulei* var. *gagatina* Wheeler, W.M. & Mann, in Wheeler, W.M. 1942: 246 (w.) BOLIVIA.

**#gaigei.** *Azteca trigona* r. *gaigei* Forel, 1914c: 618 (s.w.) COLOMBIA.

*gibbifera. Azteca ulei* var. *gibbifera* Forel, 1908c: 392 (w.) BRAZIL. Junior synonym of *muelleri:* Longino, 1992: 1591.

**gnava.** *Azteca paraensis* r. *gnava* Forel, 1906d: 241 (s.w.) COSTA RICA. Raised to species: Forel, 1912h: 49. Current subspecies: nominal plus *cayennensis, surubrensis.*

**godmani.** *Azteca godmani* Forel, 1899c: 114 (q.) PANAMA.

**goeldii.** *Azteca goeldii* Forel, 1906d: 245 (s.w.) BRAZIL.

*guadeloupensis. Azteca delpini* r. *antillana* var. *guadeloupensis* Forel, 1899c: 111 (w.) ANTILLES. Unavailable name (Shattuck, 1994: 30).

*guianae*. *Azteca constructor* var *guianae* Wheeler, W.M. 1942: 226 (w.q.) GUIANA. Junior synonym of *constructor:* Longino, 1992: 1585.

*heinii*. *Azteca alfaroi* var. *heinii* Menozzi, 1935b: 199. *Nomen nudum*, attributed to Forel.

**huberi.** *Azteca huberi* Forel, 1906d: 244 (s.w.) BRAZIL.

**hypophylla.** *Azteca hypophylla* Forel, 1899c: 120 (footnote) (s.w.) COLOMBIA.

**iheringi.** *Azteca iheringi* Forel, 1915c: 359 (s.w.) BRAZIL.

**instabilis.** *Tapinoma instabilis* Smith, F. 1862b: 31 (w.) PANAMA. Emery, 1893b: 135 (s.); Emery, 1896b: 2 (q.); Wheeler, G.C. & Wheeler, J. 1951: 193 (l.). Combination in *Liometopum:* Mayr, 1878: 870; in *Azteca:* Emery, 1893b: 135. See also: Wheeler, W.M. 1907a: 276. Current subspecies: nominal plus *major, mexicana.*

**isthmica.** *Azteca xanthochroa* subsp. *isthmica* Wheeler, W.M. 1942: 249 (w.q.) PANAMA. Raised to species: Longino, 1992: 1588.

*janeirensis*. *Azteca muelleri* var. *janeirensis* Forel, 1912h: 48 (w.) BRAZIL. Junior synonym of *muelleri:* Longino, 1992: 1591.

**jelskii.** *Azteca jelskii* Emery, 1893b: 138, pl. 2, figs. 52, 53 (s.w.) FRENCH GUIANA.

**#juruensis.** *Azteca longiceps* var. *juruensis* Forel, 1904e: 699 (s.w.q.m.) BRAZIL. See also: Wheeler, W.M. 1942: 233.

*lacrymosa*. *Azteca lacrymosa* Forel, 1899c: 116 (s.w.) COSTA RICA. Junior synonym of *aurita:* Forel, 1906d: 237.

**#laeta.** *Azteca fasciata* subsp. *laeta* Wheeler, W.M. 1942: 227 (q.) PANAMA.

**lallemandi.** *Azteca lallemandi* Forel, 1899c: 119 (footnote) (s.w.) COLOMBIA.

*langi*. *Azteca alfari* var. *langi* Wheeler, W.M. 1942: 218 (w.q.m.) GUIANA. Junior synonym of *alfari:* Longino, 1989: 5.

**#lanians.** *Azteca chartifex* subsp. *lanians* Emery, 1912g: 32 (s.w.) BRAZIL. [First available use of *Azteca chartifex* r. *decipiens* var. *laniens* Forel, 1906d: 237; unavailable name.]

**lanuginosa.** *Azteca lanuginosa* Emery, 1893b: 141, pl. 2, fig. 47 (w.) BRAZIL. Forel, 1908c: 389 (q.m.). Current subspecies: nominal plus *clariceps, pruinosa.*

**#laticeps.** *Azteca chartifex* r. *laticeps* Forel, 1899c: 117 (w.) COSTA RICA.

**lattkei.** *Azteca lattke* Longino, 1992: 1588, figs. 17, 32, 45 (w.) VENEZUELA.

**longiceps.** *Azteca longiceps* Emery, 1893b: 144, pl. 2, figs. 64, 65 (q.) COSTA RICA. Wheeler, W.M. 1942: 230 (s.w.m.); Wheeler, G.C. & Wheeler, J. 1951: 193 (l.). Current subspecies: nominal plus *cordincola, juruensis, patruelis, sapii.*

**lucida.** *Azteca alfaroi* r. *lucida* Forel, 1899c: 113 (s.w.) GUATEMALA. Raised to species: Forel, 1912h: 51. Subspecies of *alfari:* Wheeler, W.M. 1942: 223. Revived status as species: Longino, 1989: 13.

*lucidula*. *Azteca alfaroi* r. *lucidula* Forel, 1899c: 113 (w.q.m.) GUATEMALA. Raised to species: Forel, 1912h: 51. Subspecies of *alfari:* Wheeler, W.M. 1942: 222; Kempf, 1972a; 29. Junior synonym of *alfari:* Longino, 1989: 5.

**luederwaldti.** *Azteca luederwaldti* Forel, 1909a: 261 (s.w.) PARAGUAY.

*lynchi*. *Azteca lynchi* Brèthes, 1914a: 93 (w.) BOLIVIA. Variety of *alfari*, and senior synonym of *argentina:* Bruch, 1915: 534; Gallardo, 1915: 20. Junior synonym of *alfari:* Longino, 1989: 5.

**#major.** *Azteca instabilis* var. *major* Forel, 1899c: 107 (w.) PANAMA.

**#mathildae.** *Azteca trigona* r. *mathildae* Forel, 1906d: 238 (s.w.) BRAZIL. Subspecies of *tonduzi:* Wheeler, W.M. 1942: 245; of *trigona:* Kempf, 1972a: 35; Shattuck, 1994: 27.

**mayrii.** *Azteca mayrii* Emery, 1893b: 143, pl. 2, figs. 81, 82 (q.) BRAZIL.

**#mediops.** *Azteca festai* r. *mediops* Forel, 1904c: 43 (w.) BRAZIL. Race of *trigona:* Forel, 1906d: 237; Wheeler, W.M. 1916d: 330. Subspecies of *tonduzi:* Wheeler, W.M. 1942: 245; of *trigona:* Kempf, 1972a: 35.

**#menceps.** *Azteca theresiae* var. *menceps* Forel, 1912h: 54 (w.) PANAMA. Wheeler, W.M. 1942: 240 (s.q.m.).

**merida.** *Azteca merida* Longino, 1992: 1589, figs. 13, 14, 30, 45 (w.q.) VENEZUELA.

**#mexicana.** *Azteca instabilis* var. *mexicana* Emery, 1896b: 3 (w.) MEXICO.

**minor.** *Azteca minor* Forel, 1904e: 696 (w.q.) BRAZIL.

*mixta*. *Azteca alfari* var. *mixta* Forel, 1908c: 386 (w.q.m.) PARAGUAY. Junior synonym of *alfari:* Longino, 1989: 5.

**muelleri.** *Azteca muelleri* Emery, 1893b: 131, pl. 1, figs. 12–15 (s.w.q.) BRAZIL. Senior synonym of *nigella:* Wheeler, W.M. 1942: 236; of *brunni, gibbifera, janeirensis, nigella, nigridens,* and *wacketi:* Longino, 1992: 1591. Current subspecies: nominal plus *pallida, terminalis.*

**#multinida.** *Azteca chartifex* r. *multinida* Forel, 1899c: 117 (footnote) (w.) COLOMBIA.

*nigella*. *Azteca nigella* Emery, 1893b: 132, pl. 1, figs. 23, 24 (w.) BRAZIL. Variety of *muelleri:* Emery, 1912g: 34. Junior synonym of *muelleri:* Wheeler, W.M. 1942: 236; Longino, 1992: 1591.

**#nigra.** *Azteca velox* r. *nigra* Forel, 1912h: 48 (w.q.m.) PANAMA.

**#nigricans.** *Azteca fasciata* var. *nigricans* Forel, 1899c: 122 (q.) PANAMA.

**#nigricornis.** *Azteca ulei* r. *nigricornis* Forel, 1904e: 695 (s.w.) BRAZIL.

*nigridens*. *Azteca muelleri* var. *nigridens* Forel, 1908c: 392 (w.) BRAZIL. Junior synonym of *muelleri:* Longino, 1992: 1591.

**#nigriventris.** *Azteca velox* var. *nigriventris* Forel, 1899c: 109 (s.w.q.) COLOMBIA. Race of *velox:* Forel, 1906d: 241; Forel, 1912h: 48.

**olitrix.** *Azteca olitrix* Forel, 1904e: 693 (w.q.) BRAZIL.

**ovaticeps.** *Azteca alfaroi* var. *ovaticeps* Forel, 1904c: 44 (w.q.) BRAZIL. Raised to species, and senior synonym of *aequalis, aequilata,* and *tuberosa:* Longino, 1989: 8.

**#pallida.** *Azteca muelleri* var. *pallida* Stitz, 1937: 135 (w.) MEXICO.

**paraensis.** *Azteca velox* r. *paraensis* Forel, 1904c: 45 (w.) BRAZIL. Raised to species: Forel, 1906d: 240. Current subspecies: nominal plus *bondari.*

**#patruelis.** *Azteca longiceps* subsp. *patruelis* Forel, 1908c: 392 (footnote) (s.w.q.) MEXICO.

**petalocephala.** *Azteca petalocephala* Longino, 1992: 1592, figs. 12, 27, 45 (w.q.) BOLIVIA.

*pilosula*. *Azteca aurita* r. *pilosula* Forel, 1899c: 115, pl. 4, fig. 15 (q.) COSTA RICA. Junior synonym of *aurita:* Forel, 1912h: 54.

**pittieri.** *Azteca pittieri* Forel, 1899c: 120, pl. 4, fig. 16 (w.) COSTA RICA. Menozzi, 1927d: 337 (q.m.). See also: Wheeler, W.M. 1942: 237. Current subspecies: nominal plus *emarginatisquamis.*

**polymorpha.** *Azteca polymorpha* Forel, 1899c: 121 (footnote) (s.w.) COLOMBIA. Forel, 1912h: 56 (q.). Current subspecies: nominal plus *bahiana.*

**prorsa.** *Azteca prorsa* Wheeler, W.M. 1942: 238, pl. 56, fig. a (s.w.) GUATEMALA.

**#pruinosa.** *Azteca lanuginosa* subsp. *pruinosa* Mann, 1916: 472 (w.) BRAZIL.

**#raptrix.** *Azteca foreli* var. *raptrix* Forel, 1912h: 50 (w.) COLOMBIA.

**#rectinota.** *Azteca velox* var. *rectinota* Forel, 1908b: 61 (w.) COSTA RICA.

**#rochai.** *Azteca velox* r. *rochai* Forel, 1912h: 49 (w.) BRAZIL.

**#rossi.** *Azteca ulei* subsp. *rossi* Forel, 1909a: 251 (w.) MEXICO.

**salti.** *Azteca xanthochroa* subsp. *salti* Wheeler, W.M., in Wheeler, W.M. & Darlington, 1930: 114 (w.q.) COLOMBIA. [Also described as new by Wheeler, W.M. 1942: 250.] Raised to species: Longino, 1992: 1593.

**#sapii.** *Azteca longiceps* r. *sapii* Forel, 1912h: 56 (w.q.) BRAZIL.

**schimperi.** *Azteca schimperi* Emery, 1893b: 140, pl. 2, figs. 43–46 (s.w.) COSTA RICA. Forel, 1908c: 387 (q.m.).

**schumannii.** *Azteca schumannii* Emery, 1893b: 145, pl. 2, figs. 72–75 (w.q.) VENEZUELA. Current subspecies: nominal plus *dubia, taediosa.*

**sericea.** *Iridomyrmex sericeus* Mayr, 1866a: 498, pl. fig. 8 (w.) MEXICO. Wheeler, W.M. 1942: 239 (q.). Combination in *Azteca:* Forel, 1878: 384. Junior synonym of *instabilis:* Mayr, 1878: 870. Revived from synonymy: Emery, 1893b: 134.

**severini.** *Azteca severini* Emery, 1896c: 5, fig. 7 (w.) BRAZIL.

**#silvae.** *Azteca lacrymosa* r. *silvae* Forel, 1899c: 116 (s.w.) BRAZIL. Race of *aurita:* Forel, 1906d: 237.

**#similis.** *Azteca fasciata* var. *similis* Mann, 1916: 472 (s.w.) BRAZIL.

**#spiriti.** *Azteca chartifex* var. *spiriti* Forel, 1912h: 52 (w.) BRAZIL.

*spuria*. *Azteca trigona* r. *mathildae* var. *spuria* Forel, 1906d: 238 (s.w.) BRAZIL. Unavailable name (Shattuck, 1994: 31).

**#stalactitica.** *Azteca chartifex* subsp. *stalactitica* Emery, 1896c: 4, fig. 6 (w.) BRAZIL.

**stanleyuli.** *Azteca stanleyuli* Forel, 1921a: 202 (w.) BRAZIL.

**stigmatica.** *Azteca stigmatica* Emery, 1896c: 6, fig. 8 (w.) BRAZIL.

**stolli.** *Azteca stolli* Forel, 1912h: 54 (w.) GUATEMALA.

**#subdentata.** *Azteca festai* r. *subdentata* Forel, 1904c: 43 (w.) BRAZIL. Menozzi, 1935b: 200 (q.). Subspecies of *trigona:* Forel, 1906d: 237.

**subopaca.** *Azteca nigella* var. *subopaca* Forel, 1899c: 110 (q.) PANAMA. Raised to species: Emery, 1912g: 34.

**#surubrensis.** *Azteca gnava* var. *surubrensis* Forel, 1912h: 49 (w.) PANAMA. [First available use of *Azteca*

**AZTECA** *(continued)*

*paraensis* subsp. *gnava* var. *surubrensis* Forel, 1908b: 62; unavailable name.]

***tachigaliae.*** *Azteca tachigaliae* Forel, 1904e: 701 (w.) PERU.

#***taediosa.*** *Azteca schumanni* var. *taediosa* Forel, 1904c: 42 (w.) BRAZIL.

#***terminalis.*** *Azteca muelleri* subsp. *terminalis* Mann, 1916: 470, pl. 2, fig. 16 (s.) BRAZIL.

***theresiae.*** *Azteca theresiae* Forel, 1899b: 303 (w.) COLOMBIA. [Also described as new by Forel, 1899d: 276.] Current subspecies: nominal plus *menceps.*

#***tillandsiarum.*** *Azteca traili* var. *tillandsiarum* Wheeler, W.M. 1942: 243 (w.q.) GUIANA.

#***tococae.*** *Azteca traili* var. *tococae* Forel, 1904e: 693 (w.m.) BRAZIL. Subspecies of *traili:* Forel, 1906d: 239.

***tonduzi.*** *Azteca tonduzi* Forel, 1899c: 114 (w.m.) COSTA RICA. Current subspecies: nominal plus *columbica.*

***trailii.*** *Azteca trailii* Emery, 1893b: 133, pl. 1, figs. 33–36 (s.w.m.) BRAZIL. Forel, 1904e: 692 (q.). See also: Wheeler, W.M. 1942: 243. Current subspecies: nominal plus *filicis, tillandsiarum, tococae.*

***trianguliceps.*** *Azteca trianguliceps* Forel, 1912h: 53 (w.q.) BRAZIL.

***trigona.*** *Azteca trigona* Emery, 1893b: 146, pl. 2, figs. 79, 80 (q.) BRAZIL. Emery, 1896b: 3 (w.); Forel, 1907e: 8 (m.). Senior synonym of *festai:* Forel, 1906d: 237. Current subspecies: nominal plus *gaigei, mathildae, mediops, subdentata.*

#***trinidadensis.*** *Azteca velox* var. *trinidadensis* Forel, 1899c: 109 (s.w.q.) TRINIDAD. Race of *velox:* Forel, 1912h: 49; subspecies of *delpini:* Forel, 1908c: 391; Emery, 1912g: 32; Wheeler, W.M. 1922c: 14. See also: Wheeler, W.M. 1942: 227.

*tuberosa. Azteca alfari* r. *tuberosa* Forel, 1906d: 240 (s.w.q.) BRAZIL. [*Azteca alfari* r. *tuberosa* Forel, 1905b: 159. *Nomen nudum.*] Junior synonym of *ovaticeps:* Longino, 1989: 8.

***ulei.*** *Azteca ulei* Forel, 1904e: 694 (s.w.) BRAZIL. Current subspecies: nominal plus *cordiae, gagatina, nigricornis, rossi.*

#***ursina.*** *Azteca foreli* r. *ursina* Forel, 1899c: 112 (s.w.m.) TRINIDAD.

***velox.*** *Azteca velox* Forel, 1899c: 108, pl. 4, fig. 13 (s.w.q.m.) MEXICO. Replacement name for *fasciata* Pergande, 1896: 867. [Junior primary homonym of *fasciata* Emery, 1893b: 143.] Current subspecies: nominal plus *nigra, nigriventris, rectinota, rochai.*

*virens. Azteca virens* Forel, 1899c: 115 (footnote) (w.) BRAZIL. Junior synonym of *alfari:* Longino, 1989: 5.

*wacketi. Azteca muelleri* var. *wacketi* Emery, in Forel, 1908c: 391 (footnote) (w.m.) BRAZIL. Junior synonym of *muelleri:* Longino, 1992: 1591.

***xanthochroa.*** *Liometopum xanthochroum* Roger, 1863a: 167 (q.) MEXICO. Mayr, 1866a: 497 (w.); Emery, 1896c: 2 (w.); Wheeler, W.M. 1942: 247 (l.); Wheeler, G.C. & Wheeler, J. 1951: 193 (l.). Combination in *Azteca:* Forel, 1878: 384. Junior synonym of *instabilis:* Mayr, 1878: 870. Revived from synonymy as variety of *instabilis:* Emery, 1893b: 137. Revived status as species: Emery, 1896c: 2. Senior synonym of *costaricensis:* Longino, 1992: 1594.

#***xysticola.*** *Azteca foreli* var. *xysticola* Forel, 1899c: 111 (w.) COLOMBIA.

*zonalis. Azteca alfari* subsp. *lucidula* var. *zonalis* Wheeler, W.M. 1942: 222 (w.q.) PANAMA. Unavailable name; material referred to *alfari* by Longino, 1989: 7.

***BAIKURIS** (Russia)

****mandibularis.*** **Baikuris mandibularis* Dlussky, 1987: 134, fig. 2 (m.) RUSSIA (Cretaceous).

****mirabilis.*** **Baikuris mirabilis* Dlussky, 1987: 134, fig. 3 (m.) RUSSIA (Cretaceous).

**BAJCARIDRIS** (North Africa)

***kraussii.*** *Formica kraussii* Forel, 1895d: 231 (w.) ALGERIA. Emery, 1899b: 18 (m.); Forel, 1902a: 155 (m.). Combination in *Formica* (*Proformica*): Emery, 1909b: 204; in *Bajcaridris:* Agosti, 1994a: 99.

***menozzii.*** *Proformica menozzii* Santschi, 1923a: 137 (w.) ALGERIA. Combination in *Bajcaridris:* Agosti, 1994a: 99.

***theryi.*** *Formica* (*Proformica*) *theryi* Santschi, 1936c: 209, fig. 6 (w.) MOROCCO. Espadaler & Cagniant, 1987: 133 (m.). Combination in *Proformica:* Cagniant, 1964: 110; in *Bajcaridris:* Agosti, 1994a: 99.

**BARACIDRIS** (Afrotropical)

***meketra.*** *Baracidris meketra* Bolton, 1981b: 255, figs. 12, 13 (w.q.) NIGERIA.

***sitra.*** *Baracidris sitra* Bolton, 1981b: 256, fig. 14 (w.) GABON.

**BARIAMYRMA** (Venezuela)

***hispidula.*** *Bariamyrma hispidula* Lattke, 1990a: 174, figs. 1–4 (q.) VENEZUELA.

*BARBOURELLA:* see under ***GNAMPTOGENYS.***

*BARONIURBANIA:* see under ***LEPISIOTA.***

**BASICEROS** (Neotropical)

***conjugans.*** *Basiceros conjugans* Brown, 1974c: 134, figs. 1, 2 (w.q.m.) EQUADOR.

***convexiceps.*** *Ceratobasis convexiceps* Mayr, 1887: 581 (w.) BRAZIL. Emery, 1922f: 328 (m.). Combination in *Basiceros:* Emery, 1922f: 328. Senior synonym of *squamifer:* Brown & Kempf, 1960: 172.

***disciger.*** *Ceratobasis disciger* Mayr, 1887: 581 (w.) BRAZIL. Brown & Kempf, 1960: 180 (q.); Brown, 1974c: 140 (m.). Combination in *Basiceros:* Emery, 1922f: 328; in *Aspididris:* Brown & Kempf, 1960: 179; in *Basiceros:* Brown, 1974c: 140.

***manni.*** *Basiceros manni* Brown & Kempf, 1960: 177, figs. 3, 11 (w.q.) COSTA RICA. Wheeler, G.C. & Wheeler, J. 1983: 607 (l.). See also: Hölldobler & Wilson, 1986: 12; Wilson & Hölldobler, 1986: 70.

***militaris.*** *Aspididris militaris* Weber, 1950c: 3, fig. 1 (w.q.) TRINIDAD. Wheeler, G.C. & Wheeler, J. 1955a: 114 (l.). Combination in *Basiceros:* Brown, 1974c: 134. See also: Brown & Kempf, 1960: 181.

***redux.*** *Rhopalothrix redux* Donisthorpe, 1939c: 152 (m.) GUIANA. Combination in *Basiceros:* **new combination**. See also: Brown & Kempf, 1960: 164.

***singularis.*** *Meranoplus singularis* Smith, F. 1858b: 195, pl. 13, figs. 6–10 (w., not q.; see Brown & Kempf, 1960: 175) BRAZIL. Wheeler, W.M. 1916c: 9 (q.): Brown, 1974c: 140 (m.); Wheeler, G.C. & Wheeler, J. 1980: 537 (l.). Combination in *Ceratobasis:* Smith, F. 1860c: 78; in *Basiceros:* Wheeler, W.M. 1916c: 9. See also: Brown, 1949f: 88.

*squamifer. Basiceros squamifer* Borgmeier, 1937b: 245, figs. 30–33 (w.q.) BRAZIL. Junior synonym of *convexiceps:* Brown & Kempf, 1960: 172.

**BELONOPELTA** (Neotropical)

***attenuata.*** *Belonopelta attenuata* Mayr, 1870a: 395, pl. 1, fig. 11 (w.) COLOMBIA. See also: Baroni Urbani, 1975b: 300.

*crassicornis* Emery, 1897; see under ***CRYPTOPONE.***

*curvata* Mayr, 1887; see under ***SIMOPELTA.***

*darwinii* Forel, 1893; see under ***PACHYCONDYLA.***

***deletrix.*** *Belonopelta* (*Belonopelta*) *deletrix* Mann, 1922: 9, fig. 4 (w.) HONDURAS. Wilson, 1955b: 82 (q.); Wheeler, G.C. & Wheeler, J. 1964b: 452 (l.). Combination in *Leiopelta:* Baroni Urbani, 1975b: 309; in *Belonopelta:* **revived combination.**

*franzi* Baroni Urbani, 1975; see under ***EMERYOPONE.***

*indica* Emery, 1899; see under ***PACHYCONDYLA.***

*indica* Forel, 1900; see under ***PACHYCONDYLA.***

*jeckylli* Mann, 1916; see under ***SIMOPELTA.***

*loebli* Baroni Urbani, 1975; see under ***EMERYOPONE.***

*madecassa* Emery, 1899; see under ***PACHYCONDYLA.***

*minima* Brandão, 1989; see under ***SIMOPELTA.***

*pergandei* Forel, 1909; see under ***SIMOPELTA.***

*BICONOMYRMA:* see under ***DORYMYRMEX.***

*BISOLENOPSIS:* see under ***SOLENOPSIS.***

**BLEPHARIDATTA** (Neotropical)

***brasiliensis.*** *Blepharidatta brasiliensis* Wheeler, W.M. 1915e: 484, fig. 1 (w.) BRAZIL. Kempf, 1975c: 369 (m.); Wheeler, G.C. & Wheeler, J. 1991a: 134 (l.).

***conops.*** *Blepharidatta conops* Kempf, 1967d: 355, figs. 4, 5 (w.) BRAZIL.

**BONDROITIA** (Afrotropical)

*coecum. Monomorium* (*Martia*) *coecum* Forel, 1911c: 299 (footnote) (w.) SWITZERLAND (locality in error, see Bolton, 1987: 277). Combination in *Monomorium* (*Bondroitia*): Forel, 1911c: 300; in *Diplomorium* (*Bondroitia*): Forel, 1911d: 398; in *Bondroitia:* Forel, 1915d: 38; in *Diplomorium:* Emery, 1922e: 195. Junior synonym of *lujae:* Bolton, 1987: 277.

***lujae.*** *Diplomorium lujae* Forel, 1909b: 72 (w.q.m.) ZAIRE. Combination in *Diplomorium* (*Bondroitia*): Forel, 1911d: 398; in *Bondroitia:* Wheeler, W.M. 1922a: 877; in *Diplomorium:* Ettershank, 1966: 100; in *Bondroitia,* and senior synonym of *coecum:* Bolton, 1987: 277.

***saharensis.*** *Diplomorium saharensis* Santschi, 1923e: 278 (q.m.) NIGER. Combination in *Bondroitia:* Bolton, 1987: 277.

*BORGMEIERITA:* see under ***GLAMYROMYRMEX.***

**BOTHRIOMYRMEX** (Palaearctic, Oriental, Australia)

***adriacus.*** *Bothriomyrmex meridionalis* var. *adriaca* Santschi, 1922a: 66 (w., not q., see Emery, 1925e; 15) ITALY. Emery, 1925e: 16 (q.m.). Raised to species: Emery, 1925e: 15. Senior synonym of *ionius:* Agosti & Collingwood, 1987a: 57. Current subspecies: nominal plus *anatolicus.*

#***aequalis.*** *Bothriomyrmex pusillus* var. *aequalis* Forel, 1902h: 476 (w.q.m.) AUSTRALIA. Subspecies of *pusillus:* Forel, 1907h: 287.

#***anatolicus.*** *Bothriomyrmex adriacus* subsp. *anatolicus* Emery, 1925e: 16, fig. A11 (w.) TURKEY.

***atlantis.*** *Bothriomyrmex meridionalis* r. *atlantis* Forel, 1894d: 14 (w.q.) ALGERIA. Raised to species: Santschi, 1911f: 79. See also: Santschi, 1919i: 210. Current subspecies: nominal plus *inquilinus, perfidus.*

*bernardi. Bothriomyrmex meridionalis bernardi* Bernard, 1950a: 20. *Nomen nudum.* [Material apparently attributed to *Bothriomyrmex corsicus* by Bernard, 1967: 266.]

***breviceps.*** *Bothriomyrmex breviceps* Santschi, 1919i: 207, pl. 2, figs. 3, 4 (w.q.m.) TUNISIA.

***communistus.*** *Bothriomyrmex meridionalis* var. *communista* Santschi, 1919i: 206, pl. 2, fig. 2 (w.) UKRAINE. Emery, 1925e: 11 (m.); Karavaiev, 1927d: 341 (q.). Raised to species: Emery, 1925e: 10. See also: Dlussky, 1962: 180; Atanassov & Dlussky, 1992: 195.

***corsicus.*** *Bothriomyrmex meridionalis* var. *corsica* Santschi, 1923a: 136 (q.) CORSICA. Emery, 1925e: 12 (w.); Bernard, 1967: 266 (m.). Raised to species: Emery, 1925e: 12. Current subspecies: nominal plus *liguricus, mohelensis.*

***costae.*** *Bothriomyrmex costae* Emery, 1869c: 118 (q.m.) ITALY. Junior synonym of *meridionalis:* Mayr, 1870b: 954; Emery, 1912g: 28. Revived from synonymy as variety of *meridionalis:* Santschi, 1919i: 205. Revived status as species: Baroni Urbani, 1971c: 164.

***crosi.*** *Bothriomyrmex crosi* Santschi, 1919i: 212, pl. 3, figs. 32–37 (w.q.m.) ALGERIA.

***cuculus.*** *Bothriomyrmex cuculus* Santschi, 1919i: 208, pl. 2, figs. 17, 18 (w.q.) TUNISIA.

#***dalyi.*** *Bothriomyrmex wroughtoni* r. *dalyi* Forel, 1895e: 469 (diagnosis in key) (w.) INDIA. Raised to species: Bingham, 1903: 307. Subspecies of *wroughtonii:* Emery, 1912g: 29; Santschi, 1919i: 202.

***decapitans.*** *Bothriomyrmex decapitans* Santschi, 1911f: 78, fig. 10c (w.q.m.) TUNISIA.

*dimmocki* Wheeler, W.M. 1915; see under ***TAPINOMA.***

***emarginatus.*** *Bothriomyrmex emarginatus* Santschi, 1919i: 209, pl. 2, fig. 11 (q.) TUNISIA.

***flavus.*** *Bothriomyrmex flavus* Crawley, 1922c: 27, fig. 15 (w.q.m.) AUSTRALIA.

#***formosensis.*** *Bothriomyrmex wroughtoni* r. *formosensis* Forel, 1913f: 197 (w.) TAIWAN. [Misspelled as *formosae* by Santschi, 1919i: 202.]

***gallicus.*** *Bothriomyrmex corsicus* subsp. *gallicus* Emery, 1925e: 13, figs. A8, B5, C5 (w.q.m.) SWITZERLAND. Junior synonym of *corsicus:* Bernard, 1967: 266. Revived from synonymy and raised to species: Kutter, 1971: 269. See also: Kutter, 1977c: 176.

***gibbus.*** *Bothriomyrmex meridionalis* subsp. *gibbus* Soudek, 1925c: 216, figs. 1–10 (w.q.m.) CZECHOSLOVAKIA. Raised to species: Emery, 1925e: 14. See also: Stitz, 1939: 226; Agosti & Collingwood, 1987a: 57; Shattuck, 1994: 34.

***hispanicus.*** *Bothriomyrmex meridionalis* var. *hispanicus* Santschi, 1922a: 66 (w.q.) SPAIN. Menozzi, 1922b: 330 (m.). Raised to species: Emery, 1925e: 10.

#***hungaricus.*** *Bothriomyrmex meridionalis* var. *hungarica* Röszler, 1942b: 39 (w.) HUNGARY.

#***inquilinus.*** *Bothriomyrmex inquilinus* Santschi, 1919i: 209, pl. 2, figs. 14–16 (w.q.) TUNISIA. Wheeler, G.C. & Wheeler, J. 1951: 190 (l.). Variety of *atlantis:* Santschi, 1922a: 66.

*ionius. Bothriomyrmex adriacus* subsp. *ionius* Emery, 1925e: 16 (w.) GREECE. Junior synonym of *adriacus:* Agosti & Collingwood, 1987a: 57.

*iranicus. Bothriomyrmex iranicus* Arnol'di, 1948: 212. *Nomen nudum.*

***jannonei.*** *Bothriomyrmex jannonei* Menozzi, 1936d: 295, fig. 15, 1–4 (w.m.) GREECE.

#***javanus.*** *Bothriomyrmex wroughtoni* var. *javana* Forel, 1909d: 229 (q.) JAVA.

***kusnezovi.*** *Bothriomyrmex kusnezovi* Emery, 1925e: 12, fig. A5 (w.) TURKESTAN. Tarbinsky, 1976: 125 (q.m.).

***laticeps.*** *Bothriomyrmex corsicus* subsp. *laticeps* Emery, 1925e: 14, fig. A9 (w.) FRANCE. Santschi, 1927b: 127 (q.m.). Junior synonym of *corsicus:* Bernard, 1967: 266. Revived from synonymy and raised to species: Collingwood & Yarrow, 1969: 76.

#***liguricus.*** *Bothriomyrmex corsicus* var. *ligurica* Emery, 1925e: 13, fig. A7 (w.) ITALY. Menozzi, 1932a: 9 (q.m.l.); Wheeler, G.C. & Wheeler, J. 1951: 190 (l.).

#***marocanus.*** *Bothriomyrmex meridionalis* var. *marocanus* Santschi, 1923f: 335 (q.m.) MOROCCO.

***menozzii.*** *Bothriomyrmex menozzii* Emery, 1925e: 17, figs. B8, C8 (q.m.) ITALY. Kutter, 1971: 270 (w.). See also: Kutter, 1977c: 176.

***meridionalis.*** *Tapinoma meridionale* Roger, 1863a: 165 (w.) FRANCE. André, 1882b: 218 (q.m.); Wheeler, G.C. & Wheeler, J. 1951: 190 (l.). Combination in *Bothriomyrmex:* Mayr, 1870b: 954. See also: Santschi, 1919i: 204; Emery, 1925e: 7. Current subspecies: nominal plus *hungaricus, marocanus.*

***modestus.*** *Bothriomyrmex modestus* Radchenko, 1985b: 64, figs. 1–13 (w.q.m.) UKRAINE.

#***mohelensis.*** *Bothriomyrmex corsicus* subsp. *mohelensis* Novak, in Novak & Sadil, 1941: 97 (w.q.m.) CZECHOSLOVAKIA. [Also described as new by Novak, in Kratochvil, Novak & Snoflak, 1944: 106.]

***myops.*** *Bothriomyrmex myops* Forel, 1895e: 471 (w.q.) INDIA. Combination in *Bothriomyrmex* (*Chronoxenus*): Donisthorpe, 1944e: 102. See also: Bingham, 1903: 306.

*nigrescens. Bothriomyrmex nigrescens* Kutter, 1971: 268. *Nomen nudum;* attributed to Santschi.

*pallidus* Donisthorpe, 1938; see under ***ACROPYGA.***

#***perfidus.*** *Bothriomyrmex atlantis* var. *perfidus* Santschi, 1926e: 235 (w.q.) TUNISIA.

***pubens.*** *Bothriomyrmex pubens* Santschi, 1919i: 210, pl. 3, figs. 19–25 (w.q.m.) TUNISIA.

*pumicatus. Bothriomyrmex pumicatus* Wasmann, 1894: 100. *Nomen nudum.* [The name appears to be a gross error for *Bothroponera pumicata* Ray, in Raffray, 1887: 21, itself a misinterpretation of *Bothroponera pumicosa;* see footnote in Wheeler, W.M. 1922a: 921.]

***pusillus.*** *Tapinoma pusillum* Mayr, 1876: 83 (w.q.m.) AUSTRALIA. Wheeler, G.C. & Wheeler, J. 1951: 189 (l.); Imai, Crozier & Taylor, 1977: 345 (k.). Combination in *Bothriomyrmex:* Forel, 1879a: 127. Current subspecies: nominal plus *aequalis.*

***regicidus.*** *Bothriomyrmex regicidus* Santschi, 1919i: 206, pl. 2, figs. 5–10 (w.q.m.) TUNISIA.

*rogeri. Bothriomyrmex rogeri* Emery, 1924c: 168, fig. B, 1–7 (w.m.) SPAIN. Junior synonym of *saundersi:* Emery, 1925e: 9.

***rossi.*** *Bothriomyrmex rossi* Donisthorpe, 1950a: 340 (w.) NEW GUINEA.

***salsurae.*** *Bothriomyrmex salsurae* Donisthorpe, 1944e: 100 (w.q.m.) ALGERIA.

***saundersi.*** *Bothriomyrmex regicidus* var. *saundersi* Santschi, 1922a: 67 (q.m.) SPAIN. Raised to species, and senior synonym of *rogeri:* Emery, 1925e: 9.

***scissor.*** *Bothriomyrmex scissor* Crawley, 1922c: 29, fig. 16 (q.) AUSTRALIA.

*sicula. Bothriomyrmex adriacus* subsp. *ionius* var. *sicula* Emery, 1925e: 17 (w.) ITALY. Unavailable name (Shattuck, 1994: 38).

***syrius.*** *Bothriomyrmex meridionalis* var. *syria* Forel, 1910a: 13 (w.) LEBANON. Tohmé, H. & Tohmé, G. 1981: 97 (q.m.l.). Raised to species: Emery, 1925e: 11. See also: Menozzi, 1933b: 78; Shattuck, 1994: 37.

***turcomenicus.*** *Bothriomyrmex syrius* subsp. *turcomenicus* Emery, 1925e: 11, fig. A4 (w.) TURKESTAN. Raised to species: Dlussky, Soyunov & Zabelin, 1990: 171.

#***victoriae.*** *Bothriomyrmex wroughtonii* var. *victoriae* Forel, 1895e: 469 (diagnosis in key) (w.) HONG KONG.

***walshi.*** *Bothriomyrmex walshi* Forel, 1895e: 471 (w.q.) INDIA. See also: Bingham, 1903: 306.

***wilsoni.*** *Bothriomyrmex wilsoni* Clark, 1934b: 39, pl. 3, fig. 5 (w.) AUSTRALIA.

***wroughtonii.*** *Bothriomyrmex wroughtonii* Forel, 1895e: 470 (w.q.m.) INDIA. See also: Bingham, 1903: 307. Current subspecies: nominal plus *dalyi, formosensis, javanus, victoriae.*

*BOTHROPONERA:* see under ***PACHYCONDYLA.***

***BRACHYMYRMEX*** (Neotropical, Nearctic; tramp species)

***admotus.*** *Brachymyrmex admotus* Mayr, 1887: 523 (w.q.) BRAZIL. Wheeler, G.C. & Wheeler, J. 1982: 178 (l.).

#***andinus.*** *Brachymyrmex levis* var. *andina* Santschi, 1923b: 659 (w.) ARGENTINA.

***antennatus.*** *Brachymyrmex* (*Bryscha*) *antennatus* Santschi, 1929d: 312, fig. 35 (w.q.) BRAZIL.

#***aphidicola.*** *Brachymyrmex heeri* var. *aphidicola* Forel, 1909a: 263 (w.) PARAGUAY. Forel, 1912i: 62 (q.).

#***atratulus.*** *Brachymyrmex patagonicus* var. *atratula* Santschi, 1923b: 657, fig. 3 (w.) ARGENTINA.

#***attenuatus.*** *Brachymyrmex luederwaldti* st. *attenuatus* Santschi, 1929d: 310 (w.) BRAZIL.

***australis.*** *Brachymyrmex minutus* r. *australis* Forel, 1901d: 302 (w.) BRAZIL. Raised to species: Santschi, 1922d: 260. See also: Santschi, 1923b: 662. Current subspecies: nominal plus *curtus.*

#***balboae.*** *Brachymyrmex pictus* subsp. *balboae* Wheeler, W.M. 1942: 253 (w.q.m.) PANAMA.

#***basalis.*** *Brachymyrmex heeri* var. *basalis* Wheeler, W.M. 1921f: 166 (w.) GUIANA.

#***bonariensis.*** *Brachymyrmex constrictus* st. *bonariensis* Santschi, 1933e: 122, figs. 20, 21 (w.) ARGENTINA.

***brevicornoeides.*** *Brachymyrmex patagonicus* var. *brevicornoeides* Forel, 1914d: 287 (w.q.m.) ARGENTINA. Treated as junior synonym of *nigricans:* Santschi, 1923b: 657. [Incorrect procedure as

***BRACHYMYRMEX** (continued)*

*brevicornoeides* is senior name and *nigricans* its junior synonym.]

***brevicornis.*** *Brachymyrmex brevicornis* Emery, 1906c: 180, figs. 38, 40, 41 (w.q.m.) ARGENTINA.

***bruchi.*** *Brachymyrmex bruchi* Forel, 1912i: 64 (w.m.) ARGENTINA. Santschi, 1929d: 309 (q.). Current subspecies: nominal plus *rufipes.*

***cavernicola.*** *Brachymyrmex cavernicola* Wheeler, W.M. 1938: 252 (w.m.) MEXICO.

***coactus.*** *Brachymyrmex coactus* Mayr, 1887: 523 (w.q.m.) BRAZIL. Current subspecies: nominal plus *nictitans, robustus.*

***constrictus.*** *Brachymyrmex constrictus* Santschi, 1923b: 671, figs. 5, 38, 61 (w.) BOLIVIA. Current subspecies: nominal plus *bonariensis.*

***cordemoyi.*** *Brachymyrmex patagonicus* var. *cordemoyi* Forel, 1895a: 49 (w.) REUNION I. Emery, 1906c: 180 (q.m.). Raised to species: Emery, 1906c: 179. Subspecies of *patagonicus:* Forel, 1908c: 399; Forel, 1912k: 165; Santschi, 1912e: 533. Raised to species: Wheeler, W.M. 1922a: 1036; Emery, 1925b: 41. Current subspecies: nominal plus *distinctus.*

***#cordobensis.*** *Brachymyrmex giardi* var. *cordobensis* Santschi, 1929d: 309 (w.) ARGENTINA.

***#curtus.*** *Brachymyrmex australis* var. *curta* Santschi, 1922d: 260 (q.m.) ARGENTINA. [Also described as new by Santschi, 1923b: 663.]

*decedens* Mayr, 1887; see under ***ACROPYGA.***

***degener.*** *Brachymyrmex coactus* subsp. *degener* Emery, 1906c: 177 (w.) BRAZIL. Raised to species: Santschi, 1923b: 670. Current subspecies: nominal plus *niger.*

***depilis.*** *Brachymyrmex heeri* subsp. *depilis* Emery, 1893i: 635 (w.q.) U.S.A. Wheeler, G.C. & Wheeler, J. 1968: 210 (l.). Raised to species: Santschi, 1923b: 663. Senior synonym of *nanellus:* Creighton, 1950a: 359; of *flavescens:* Cole, 1953g: 266.

***#distinctus.*** *Brachymyrmex cordemoyi* var. *distincta* Santschi, 1923b: 658, figs. 6, 50, 59 (w.q.) ARGENTINA.

***donisthorpei.*** *Brachymyrmex donisthorpei* Santschi, 1939e: 320, figs. 4, 5 (w.) COLOMBIA.

***#fallax.*** *Brachymyrmex heeri* var. *fallax* Santschi, 1923b: 665, fig. 29 (w.) PARAGUAY. Santschi, 1925e: 168 (q.).

***fiebrigi.*** *Brachymyrmex fiebrigi* Forel, 1908c: 400 (w.) PARAGUAY. Santschi, 1922d: 260 (q.m.). Current subspecies: nominal plus *fumidus, funicularis.*

*flavescens. Brachymyrmex depilis* subsp. *flavescens* Grundmann, 1952: 117 (w.) U.S.A. Junior synonym of *depilis:* Cole, 1953g: 266.

***flavidulus.*** *Plagiolepis flavidula* Roger, 1863a: 162 (w.) CUBA. Combination in *Brachymyrmex:* Smith, M.R. 1955b: 99.

***#fumidus.*** *Brachymyrmex fiebrigi* var. *fumida* Santschi, 1923b: 661 (w.) ARGENTINA.

***#funicularis.*** *Brachymyrmex fiebrigi* var. *funicularis* Santschi, 1922d: 260 (w.) ARGENTINA.

***#fusculus.*** *Brachymyrmex levis* var. *fuscula* Emery, 1906c: 178 (w.q.) ARGENTINA.

***gagates.*** *Brachymyrmex gagates* Wheeler, W.M. 1934g: 206 (w.) MEXICO.

***gaucho.*** *Brachymyrmex gaucho* Santschi, 1917f: 283 (w.) ARGENTINA. Combination in *Brachymyrmex (Bryscha)*: Santschi, 1923b: 674.

***giardi.*** *Brachymyrmex giardi* Emery, 1895d: 215 (w.q.) CHILE. Emery, 1906c: 178 (m.). Senior synonym of *nitida:* Snelling & Hunt, 1976: 114. Current subspecies: nominal plus *cordobensis.*

***goeldii.*** *Brachymyrmex goeldii* Forel, 1912i: 65 (w.) BRAZIL.

***heeri.*** *Brachymyrmex heeri* Forel, 1874: 91, figs. 16, 20 (w.) SWITZERLAND. Forel, 1876: 52 (q.m.). See also: Santschi, 1923b: 664. Current subspecies: nominal plus *aphidicola, basalis, fallax, termitophilus.*

***#hemiops.*** *Brachymyrmex longicornis* var. *hemiops* Santschi, 1923b: 668, fig. 20 (w.) BRAZIL.

***#immunis.*** *Brachymyrmex longicornis* var. *immunis* Forel, 1908c: 400 (w.q.m.) BRAZIL. See also: Forel, 1911c: 308; Santschi, 1923b: 668.

***incisus.*** *Brachymyrmex incisus* Forel, 1912i: 63 (w.m.) COLOMBIA.

***laevis.*** *Brachymyrmex laevis* Emery, 1895d: 216 (w.) CHILE. Emery, 1906c: 178 (q.m.). [Note: in the literature sometimes misspelled as *levis,* for example Emery, 1906c: 178; Santschi, 1923b: 659.] See also: Forel, 1908c: 400; Forel, 1912i: 62. Current subspecies: nominal plus *andinus, fusculus.*

***longicornis.*** *Brachymyrmex longicornis* Forel, 1907e: 9 (w.) BRAZIL. Forel, 1912i: 62 (q.). Current subspecies: nominal plus *hemiops, immunis, pullus.*

***luederwaldti.*** *Brachymyrmex luederwaldti* Santschi, 1923b: 672, figs. 36, 66 (w.) BRAZIL. Current subspecies: nominal plus *attenuatus.*

***melensis.*** *Brachymyrmex melensis* Zolessi, Abenante & Gonzalez, 1978: 26, figs. 3–22 (w.q.l.) URUGUAY.

***micromegas.*** *Brachymyrmex (Bryscha) micromegas* Emery, in Santschi, 1923b: 675, figs. 30, 32 (w) BRAZIL.

***minutus.*** *Brachymyrmex minutus* Forel, 1893g: 346 (w.q.) ANTILLES.

***modestus.*** *Brachymyrmex modestus* Santschi, 1923c: 271 (w.) BRAZIL.

***musculus.*** *Brachymyrmex tristis* r. *musculus* Forel, 1899c: 124 (w.) COSTA RICA. Forel, 1901c: 135 (q.). Raised to species: Forel, 1901c: 135. See also: Santschi, 1923b: 673.

***myops.*** *Brachymyrmex myops* Emery, 1906c: 181 (footnote), fig. 42 (w.m.) BRAZIL.

*nanellus. Brachymyrmex nanellus* Wheeler, W.M. 1903b: 102, fig. 7b (w.m.) U.S.A. Junior synonym of *depilis:* Creighton, 1950a: 359.

***#nictitans.*** *Brachymyrmex coactus* var. *nictitans* Emery, 1906c: 178 (w.) COSTA RICA.

***#niger.*** *Brachymyrmex admotus* r. *niger* Forel, 1912i: 62 (w.) BRAZIL. Subspecies of *degener:* Santschi, 1923b: 671.

*nigricans. Brachymyrmex cordemoyi* var. *nigricans* Santschi, 1916e: 395 (w.) ARGENTINA. [First available use of *Brachymyrmex patagonicus* st. *cordemoyi* var. *nigricans* Santschi, 1912e: 533; unavailable name.] Raised to species: Santschi, 1923b: 657. Junior synonym of *brevicornoeides:* Santschi, 1923b: 657 [*nigricans* incorrectly given as senior name].

*nitida. Brachymyrmex giardi* var. *nitida* Santschi, 1922d: 261 (w.) CHILE. Junior synonym of *giardi:* Snelling & Hunt, 1976: 114.

***obscurior.*** *Brachymyrmex heeri* var. *obscurior* Forel, 1893g: 345 (w.q.m.) ANTILLES. Race of *heeri:* Forel, 1897b: 298; Forel, 1912i: 62. Raised to species: Wilson & Taylor, 1967: 92.

***oculatus.*** *Brachymyrmex oculatus* Santschi, 1919f: 55 (w.) ARGENTINA.

***patagonicus.*** *Brachymyrmex patagonicus* Mayr, 1868b: 164 (w.m.) ARGENTINA. Emery, 1906c: 179 (q.). Current subspecies: nominal plus *atratulus.*

***physogaster.*** *Brachymyrmex physogaster* Kusnezov, 1960c: 382, figs. 1–4 (w.) ARGENTINA.

***pictus.*** *Brachymyrmex pictus* Mayr, 1887: 522 (w.q.) BRAZIL. Current subspecies: nominal plus *balboae.*

***pilipes.*** *Brachymyrmex pilipes* Mayr, 1887: 524 (q.m.) BRAZIL. Santschi, 1929d: 310 (w.). Combination in *Brachymyrmex (Bryscha)*: Santschi, 1923b: 674.

***#pullus.*** *Brachymyrmex longicornis* var. *pullus* Santschi, 1933e: 122 (w.) ARGENTINA.

***#robustus.*** *Brachymyrmex coactus* var. *robustus* Santschi, 1923c: 272 (w.) BRAZIL.

***#rufipes.*** *Brachymyrmex bruchi* var. *rufipes* Forel, 1912i: 65 (w.) ARGENTINA.

***santschii.*** *Brachymyrmex santschii* Menozzi, 1927d: 338, fig. 5 (w.) COSTA RICA.

***#termitophilus.*** *Brachymyrmex heeri* var. *termitophilus* Forel, 1896: 179 (w.) BRAZIL.

***tristis.*** *Brachymyrmex tristis* Mayr, 1870a: 389 (w.) COLOMBIA.

*BRACHYPONERA:* see under ***PACHYCONDYLA.***

****BRADOPONERA.*** (Baltic Amber)

****meieri.*** **Bradoponera meieri* Mayr, 1868c: 74, pl. 4, figs. 70, 71 (w.) BALTIC AMBER. Wheeler, W.M. 1915h: 32 (q.).

*BRADYPONERA:* see under **MYOPIAS.**

***BREGMATOMYRMA*** (Borneo)

***carnosa.*** *Bregmatomyrma carnosa* Wheeler, W.M. 1929b: 4, fig. 2 (q.) BORNEO.

*BROWNIDRIS:* see under ***OCHETOMYRMEX.***

*BRUCHOMYRMA:* see under ***PHEIDOLE.***

*BRUNELLA:* see under ***APHAENOGASTER.***

*BRYSCHA:* see under ***BRACHYMYRMEX.***

*CACOPONE:* see under ***PLECTROCTENA.***

***CALOMYRMEX*** (Indo-Australian, Australia)

***albertisi.*** *Camponotus albertisi* Emery, 1887a: 221 (w.) NEW GUINEA. Combination in *Calomyrmex:* Emery, 1896d: 378.

***albopilosus.*** *Camponotus albopilosus* Mayr, 1876: 61 (w.q.m.) AUSTRALIA. Wheeler, G.C. & Wheeler, J. 1953e: 200 (l.). Combination in *Calomyrmex:* Emery, 1896d: 378. Current subspecies: nominal plus *wienandsi.*

*cyanea. Camponotus splendidus* subsp. *purpureus* var. *cyanea* Forel, 1910b: 83 (w.) AUSTRALIA. Unavailable name (Taylor, 1986: 34).

*eremophilus. Calomyrmex splendidus* subsp. *purpureus* var. *eremophilus* Wheeler, W.M. 1915g: 820 (w.) AUSTRALIA. Clark, 1930c: 17 (q.m.). Unavailable name (Taylor, 1986: 34).

***glauerti.*** *Calomyrmex glauerti* Clark, 1930a: 125, fig. 1, no. 13 (w.) AUSTRALIA.

**impavidus.** *Camponotus impavidus* Forel, 1893d: 455 (w.) AUSTRALIA. Wheeler, G.C. & Wheeler, J. 1968: 220 (l.). Combination in *Calomyrmex:* Emery, 1896d: 378.

**laevissimus.** *Formica laevissima* Smith, F. 1859a: 138 (w.) INDONESIA (Aru I.). Mayr, 1897: 435 (q.). Combination in *Camponotus:* Mayr, 1867a: 39; in *Calomyrmex:* Emery, 1895j: 772. Current subspecies: nominal plus *trochanteralis.*

#**mutans.** *Camponotus (Calomyrmex) splendidus* var. *mutans* Forel, 1910b: 83 (w.q.) AUSTRALIA. Combination in *Calomyrmex:* Emery, 1925b: 175. Subspecies of *splendidus:* Taylor & Brown, D.R. 1985: 109.

**purpureus.** *Camponotus purpureus* Mayr, 1876: 62 (w.) AUSTRALIA. Combination in *Calomyrmex:* Emery, 1896d: 378. Subspecies of *splendidus:* Forel, 1910b: 83; Wheeler, W.M. 1915g: 820; Emery, 1925b: 175. Revived status as species: Taylor & Brown, D.R. 1985: 109. Current subspecies: nominal plus *smaragdinus.*

**similis.** *Camponotus similis* Mayr, 1876: 61 (w.) AUSTRALIA. Combination in *Calomyrmex:* Emery, 1896d: 378.

#**smaragdinus.** *Calomyrmex purpureus* var. *smaragdina* Emery, 1898b: 238 (w.) AUSTRALIA. Subspecies of *purpureus:* Taylor & Brown, D.R. 1985: 109.

**splendidus.** *Camponotus splendidus* Mayr, 1876: 61 (w.) AUSTRALIA. Forel, 1910b: 82 (q.m.). Combination in *Calomyrmex:* Emery, 1896d: 378. Current subspecies: nominal plus *mutans, viridiventris.*

#**trochanteralis.** *Calomyrmex laevissimus* var. *trochanteralis* Emery, 1911b: 536 (w.) NEW GUINEA.

**tropicus.** *Formica tropica* Smith, F. 1861b: 37 (w.) INDONESIA (Halmahera I.). Combination in *Calomyrmex:* Donisthorpe, 1932c: 465.

#**viridiventris.** *Calomyrmex splendidus* var. *viridiventris* Forel, 1915b: 106 (w.) AUSTRALIA. Subspecies of *splendidus:* Taylor & Brown, D.R. 1985: 109.

#**wienandsi.** *Camponotus (Calomyrmex) albopilosus* var. *wienandsi* Forel, 1910b: 82 (w.q.) AUSTRALIA. Combination in *Calomyrmex:* Emery, 1925b: 175. Subspecies of *albopilosus:* Taylor & Brown, D.R. 1985: 108.

*__CALYPTITES__ (Canada)

*__antediluvianum.__ **Calyptites antediluvianum* Scudder, 1877b: 270 (forewing) CANADA (Miocene). [Note: may not belong in Formicidae; see **Calyptites* in catalogue of genus-group taxa.]

***CALYPTOMYRMEX*** (Old World Tropics)

**arnoldi.** *Dicroaspis arnoldi* Forel, 1913a: 115 (w.) ZIMBABWE. Combination in *Calyptomyrmex:* Arnold, 1917: 360. See also: Bolton, 1981a: 66.

**barak.** *Calyptomyrmex barak* Bolton, 1981a: 62, figs. 28, 33 (w.q.) GHANA.

**beccarii.** *Calyptomyrmex beccarii* Emery, 1887b: 472, pl. 2, fig. 23 (w.) NEW GUINEA. Szabó, 1910a: 365 (q.). Senior synonym of *rufobrunnea:* Brown, 1951: 101; of *schraderi:* Taylor, 1991b: 600.

**brevis.** *Calyptomyrmex (Calyptomyrmex) brevis* Weber, 1943c: 366, pl. 15, fig. 1 (w.) SUDAN. See also: Bolton, 1981a: 63.

**brunneus.** *Calyptomyrmex brunneus* Arnold, 1948: 221 (w.) SOUTH AFRICA. See also: Bolton, 1981a: 70.

*cataractae. Calyptomyrmex cataractae* Arnold, 1926: 283 (w.) ZIMBABWE. Wheeler, G.C. & Wheeler, J. 1973a: 35 (l.). Junior synonym of *piripilis:* Bolton, 1981a: 65.

**clavatus.** *Calyptomyrmex (Calyptomyrmex) clavatus* Weber, 1952: 23, figs. 20, 22, 24 (w.) KENYA. Bolton, 1981a: 65 (q.).

**claviseta.** *Dicroaspis claviseta* Santschi, 1914e: 27 (w.) SOUTH AFRICA. Combination in *Calyptomyrmex (Dicroaspis):* Emery, 1915g: 15; in *Calyptomyrmex:* Bolton, 1981a: 71.

**duhun.** *Calyptomyrmex duhun* Bolton, 1981a: 72, fig. 37, 43 (w.) UGANDA.

**emeryi.** *Calyptomyrmex emeryi* Forel, 1901h: 51 (w.m.) BORNEO. Wheeler, W.M. 1919e: 90 (q.).

*emeryi. Dicroaspis emeryi* Forel, 1910c: 262 (w.) ETHIOPIA. Santschi, 1913c: 311 (q.). [Junior secondary homonym of *emeryi* Forel, above.] Replacement name: *foreli* Emery, 1915g: 15.

**foreli.** *Calyptomyrmex (Dicroaspis) foreli* Emery, 1915g: 15. Replacement name for *Dicroaspis emeryi* Forel, 1910c: 262. [Junior secondary homonym of *Calyptomyrmex emeryi* Forel, 1901h: 51.] Senior synonym of *hartwigi* and *pusillus:* Bolton, 1981a: 67.

**friederikae.** *Calyptomyrmex friederikae* Kutter, 1976a: 269, figs. 1–7 (w.) BHUTAN.

**glabratus.** *Calyptomyrmex beccarii* var. *glabratus* Viehmeyer, 1916a: 128 (w.) SINGAPORE. Raised to species: Baroni Urbani, 1975a: 410 (in key).

*hartwigi. Calyptomyrmex arnoldi* subsp. *hartwigi* Arnold, 1948: 220 (w.m.) SOUTH AFRICA. Junior synonym of *foreli:* Bolton, 1981a: 67.

**kaurus.** *Calyptomyrmex kaurus* Bolton, 1981a: 64 (w.) GHANA.

*laevidens* Santschi, 1919; see under ***DICROASPIS.***

*litoralis. Calyptomyrmex cataractae* subsp. *litoralis* Arnold, 1948: 221 (w.) SOUTH AFRICA. Junior synonym of *piripilis:* Bolton, 1981a: 65.

**nedjem.** *Calyptomyrmex nedjem* Bolton, 1981a: 68 (w.) ZIMBABWE.

**nummuliticus.** *Calyptomyrmex nummulitica* Santschi, 1914d: 352 (w.) CAMEROUN. Bolton, 1981a: 65 (q.); Wheeler, G.C. & Wheeler, J. 1983: 606 (l.). Senior synonym of *reticulatus:* Bolton, 1981a: 65.

**piripilis.** *Calyptomyrmex piripilis* Santschi, 1923e: 282 (w.) ZAIRE. Bolton, 1981a: 65 (q.). Senior synonym of *cataractae, litoralis* and *punctatus:* Bolton, 1981a: 65.

*punctatus. Calyptomyrmex (Calyptomyrmex) punctatus* Weber, 1952: 24, figs. 19, 21, 23 (w.) KENYA. Junior synonym of *piripilis:* Bolton, 1981a: 65.

*pusillus. Calyptomyrmex (Dicroaspis) pusillus* Santschi, 1915c: 256 (w.) GABON. Junior synonym of *foreli:* Bolton, 1981a: 67.

**rennefer.** *Calyptomyrmex rennefer* Bolton, 1981a: 72, fig. 35 (w.q.) ZAIRE.

*reticulatus. Calyptomyrmex (Calyptomyrmex) reticulatus* Weber, 1952: 25, fig. 17 (w.) ZAIRE. Junior synonym of *nummuliticus:* Bolton, 1981a: 65.

*rufobrunnea. Weberidris rufobrunnea* Donisthorpe, 1949a: 281, fig. 1 (w.) NEW GUINEA. Combination in *Calyptomyrmex:* Donisthorpe, 1949f: 186; Brown, 1949f: 84. Junior synonym of *beccarii:* Brown, 1951: 101.

*schraderi. Calyptomyrmex schraderi* Forel, 1901h: 50 (q.m.) AUSTRALIA. Junior synonym of *beccarii:* Taylor, 1991b: 600.

**shasu.** *Calyptomyrmex shasu* Bolton, 1981a: 73, fig. 38 (w.q.) BURUNDI.

**singalensis.** *Calyptomyrmex singalensis* Baroni Urbani, 1975a: 402, figs. 3, 7 (w.) SRI LANKA.

**stellatus.** *Calyptomyrmex stellatus* Santschi, 1915c: 255, fig. 5 (w.) GABON. See also: Bolton, 1981a: 69.

**tamil.** *Calyptomyrmex tamil* Baroni Urbani, 1975a: 398, figs. 2, 6 (w.) SRI LANKA.

**tensus.** *Calyptomyrmex tensus* Bolton, 1981a: 74, figs. 31, 40 (w.) UGANDA.

**vedda.** *Calyptomyrmex vedda* Baroni Urbani, 1975a: 404, figs. 4, 8 (w.) SRI LANKA.

**wittmeri.** *Calyptomyrmex wittmeri* Baroni Urbani, 1975a: 396, figs. 1, 5 (w.) BHUTAN.

#***CAMPOMYRMA:*** see under ***POLYRHACHIS.***

*__CAMPONOTITES__ (Russia)

*__macropterus.__ **Camponotites macropterus* Dlussky, 1981b: 76, fig. 53 (q.) RUSSIA.

***CAMPONOTUS*** (Worldwide)

**abditus.** *Camponotus (Colobopsis) abditus* Forel, 1899c: 158 (q.) GUATEMALA. Wheeler, W.M. 1934g: 216 (s.w.m.); Wheeler, G.C. & Wheeler, J. 1991b: 96 (l.).

*abdominalis. Formica abdominalis* Fabricius, 1804: 409 (w.) CENTRAL AMERICA. Wheeler, G.C. & Wheeler, J. 1953e: 196 (l.). Combination in *Camponotus:* Roger, 1862c: 285; in *C. (Myrmothrix):* Forel, 1912i: 91. [Junior primary homonym of *Formica abdominalis* Latreille, 1802c: 175.] Replacement name: *Formica atriceps* Smith, F. 1858b: 44. [Note: *atriceps* was given as a junior synonym of *abdominalis* Fabricius by Roger, 1862c: 285, and the synonymy was confirmed by Hashmi, 1973: 83; *atriceps* is first available replacement name.] See under *atriceps* for synonyms formerly applied to *abdominalis* Fabricius.

*aberrans* Mayr, 1895; see under ***PHASMOMYRMEX.***

**abjectus.** *Camponotus maculatus* st. *abjectus* Santschi, 1937d: 242, figs. 47, 48 (s.w.) ANGOLA. Current subgeneric combination: *Camponotus (Tanaemyrmex).* Raised to species: Baroni Urbani, 1972: 131.

**abrahami.** *Camponotus (Orthonotomyrmex) libanicus* var. *abrahami* Forel, 1913d: 435 (s.w.q.) LEBANON. Tohmé, G. 1969: 11 (m.). Combination in *C. (Myrmentoma):* Emery, 1925b: 121. Raised to species: Tohmé, G. 1969: 11.

#**absalon.** *Camponotus (Myrmotrema) auropubens* var. *absalon* Santschi, 1915c: 274 (s.w.q.) SOUTH AFRICA.

**abscisus.** *Camponotus abscisus* Roger, 1863a: 151 (q.) MEXICO. Forel, 1885a: 341 (s.w.m.). Combination in *C. (Myrmamblys):* Forel, 1914a: 271; in *C. (Myrmobrachys):* Emery, 1920b: 260.

**abunanus.** *Camponotus (Myrmoturba) maculatus* subsp. *abunanus* Mann, 1916: 475, pl. 6, fig. 44 (s.) BRAZIL. Santschi, 1925d: 245 (w.). Combination in *C. (Tanaemyrmex):* Emery, 1925b: 84. Subspecies of *picipes:* Emery, 1920b: 233 (footnote). Raised to species: Santschi, 1925d: 245; Emery, 1925b: 84.

*CAMPONOTUS (continued)*

**#*abyssinicus*.** *Camponotus* (*Myrmotrema*) *troglodytes* var *abyssinica* Santschi, 1913c: 314 (q.) ETHIOPIA. Subspecies of *robecchii:* current status.

**#*acaciae*.** *Camponotus* (*Myrmobrachys*) *planatus* var. *acaciae* Emery, 1920c: 39 (w.) COSTA RICA.

*achaeus*. *Camponotus* (*Myrmoturba*) *maculatus* r. *novaehollandae* var. *achaeus* Forel, 1915b: 98 (w.) AUSTRALIA. Unavailable name (Taylor, 1986: 34).

*acoma* Forel, 1908; see under ***DENDROMYRMEX.***

***acutirostris*.** *Camponotus acutirostris* Wheeler, W.M. 1910d: 317 (s.w.q.m.) U.S.A. Combination in *C.* (*Myrmoturba*): Forel, 1914a: 266; in *C.* (*Camponotus*): Emery, 1925b: 74; in *C.* (*Tanaemyrmex*), and senior synonym of *clarigaster:* Creighton, 1950a: 375.

**#*acutisquamis*.** *Camponotus acutisquamis* Mayr, 1902: 296 (s.w.) CAMEROUN. Forel, 1913e: 670 (q.); Forel, 1916: 444 (m.). Combination in *C.* (*Orthonotomyrmex*): Forel, 1913h: 357; Wheeler, W.M. 1922a: 970; in *C.* (*Myrmopiromis*): Emery, 1925b: 129; in *C.* (*Myrmopelta*): Santschi, 1921f: 311; Santschi, 1926a: 20. Subspecies of *chrysurus:* Forel, 1910c: 269; Forel, 1911f: 281; Forel, 1913h: 357. Raised to species: Emery, 1920b: 257; Wheeler, W.M. 1922a: 970. Subspecies of *chrysurus:* Santschi, 1925h: 167; Emery, 1925b: 129; Santschi, 1926a: 20.

***acvapimensis*.** *Camponotus acvapimensis* Mayr, 1862: 664 (w.) GHANA. Forel, 1915c: 348 (m.); Arnold, 1922: 656 (s.q.). Combination in *C.* (*Myrmoturba*): Forel, 1913b: 340; in *C.* (*Tanaemyrmex*): Emery, 1925b: 91. Senior synonym of *mombassae:* Wheeler, W.M. 1922a: 948; of *flavosetosus* and *poultoni:* Brown, 1956a: 39. See also: Menozzi, 1930b: 118.

***adami*.** *Camponotus adami* Forel, 1910b: 70 (s.w.) AUSTRALIA. Wheeler, G.C. & Wheeler, J. 1968: 217 (l.). Combination in *C.* (*Myrmogonia*): Forel, 1914a: 269; in *C.* (*Myrmophyma*): Emery, 1925b: 110; in *C.* (*Thlipsepinotus*): Santschi, 1928e: 483.

***adenensis*.** *Camponotus maculatus* subsp. *adenensis* Emery, 1893f: 257 (s.w.q.) YEMEN. Combination in *C.* (*Tanaemyrmex*): Emery, 1925b: 92. Junior synonym of *thraso:* Emery, 1901f: 121. Revived from synonymy as variety of *thraso:* Emery, 1925b: 92. Raised to species: Collingwood, 1985: 277.

**#*adi*.** *Camponotus* (*Tanaemyrmex*) *fedtschenkoi* var. *adi* Stitz, 1930: 239 (s.w.) TURKESTAN.

*adlerzii*. *Prenolepis adlerzii* Forel, 1886f: 209 (w.) AUSTRALIA. Junior synonym of *vitreus:* Forel, 1895e: 458.

*adpressisetosus*. *Camponotus adpressisetosus* Forel, 1879a: 101 (s.) BRAZIL. Combination in *C.* (*Myrmobrachys*): Forel, 1914a: 270. See also: Mann, 1916: 480. Junior synonym of *arboreus:* Kempf, 1965: 186.

**#*adultus*.** *Camponotus* (*Myrmomalis*) *hospes* subsp. *adultus* Viehmeyer, 1916a: 157 (w.) EAST MALAYSIA. Combination in *C.* (*Myrmoplatys*): Emery, 1925b: 136.

**#*adustus*.** *Camponotus* (*Myrmamblys*) *bellus* subsp. *adustus* Viehmeyer, 1916a: 159 (w.) SINGAPORE.

**#*aegaeus*.** *Camponotus* (*Orthonotomyrmex*) *libanicus* var. *aegaea* Emery, 1915h: 4, figs. 1, 2 (s.w.q.m.) GREECE. Combination in *C.* (*Myrmentoma*): Emery, 1925b: 121.

***aegyptiacus*.** *Camponotus maculatus* subsp. *aegyptiacus* Emery, 1915f: 79 (w.) EGYPT. Menozzi, 1932e: 453 (q.m.). [Also described as new by Emery, 1915g: 22.] Combination in *C.* (*Tanaemyrmex*): Emery, 1925b: 87. Variety/subspecies of *maculatus:* Forel, 1916: 459; Emery, 1920c: 5; Santschi, 1935b: 275. Raised to species, senior synonym of *infrasquameus*, and material of the unavailable names *claveaui* and *sudanicus* referred here: Baroni Urbani, 1972: 130.

***aeneopilosus*.** *Camponotus aeneopilosus* Mayr, 1862: 665 (w.) AUSTRALIA. Mayr, 1865: 34 (q.); Wheeler, G.C. & Wheeler, J. 1953e: 195 (l.). Combination in *C.* (*Myrmosericus*): Forel, 1914a: 268; in *C.* (*Myrmophyma*): Wheeler, W.M. 1922a: 706. Current subspecies: nominal plus *flavidopubescens*.

***aequatorialis*.** *Camponotus aequatorialis* Roger, 1863a: 135 (w.) GABON. Combination in *C.* (*Dinomyrmex*): Forel, 1914a: 268; in *C.* (*Tanaemyrmex*): Emery, 1925b: 84. Subspecies of *maculatus:* Emery, 1896d: 371. Revived status as species: Santschi, 1911g: 211. Current subspecies: nominal plus *kohli*.

**#*aequinotus*.** *Camponotus pellitus* var. *aequinotus* Forel, 1902b: 173 (w.) VENEZUELA. Combination in *C.* (*Myrmaphaenus*), and variety of *blandus:* Emery, 1925b: 153.

***aequitas*.** *Camponotus* (*Myrmamblys*) *aequitas* Santschi, 1920d: 387, fig. 16 (s.w.q.m.) ZIMBABWE. Combination in *C.* (*Myrmosphincta*): Santschi, 1921f: 311; in *C.* (*Myrmotrema*): Santschi, 1926c: 599.

**#*aeschylus*.** *Camponotus* (*Dinomyrmex*) *perroti* var. *aeschylus* Forel, 1913j: 224 (q.) MADAGASCAR. Combination in *C.* (*Tanaemyrmex*): Emery, 1925b: 89.

**#*aesopus*.** *Camponotus arcuatus* var. *aesopus* Forel, 1907h: 302 (w.) AUSTRALIA. Combination in *C.* (*Myrmophyma*): Emery, 1925b: 110. Subspecies of *arcuatus:* Taylor & Brown, D.R. 1985: 110.

***aethiops*.** *Formica aethiops* Latreille, 1798: 35 (w.q.m.) FRANCE. Combination in *Camponotus:* Mayr, 1861: 36; in *C.* (*Myrmoturba*): Forel, 1914a: 266; in *C.* (*Tanaemyrmex*): Emery, 1925b: 97. Race/subspecies of *sylvaticus:* Forel, 1874: 38 (in key); Nasonov, 1889: 11; of *rubripes:* Forel, 1886f: 143; of *maculatus:* Forel, 1892i: 306; Ruzsky, 1902d: 6; Emery, 1908a: 199; Forel, 1915d: 70; Finzi, 1924a: 14; Menozzi, 1918: 87; Karavaiev, 1927c: 276; Kuznetsov-Ugamsky, 1929b: 36. Revived status as species: Emery, 1898c: 125; Bondroit, 1918: 72; Emery, 1920b: 255; Emery, 1920c: 7; Santschi, 1921a: 116; Menozzi, 1922b: 331; Finzi, 1930d: 317; Santschi, 1931a: 11; Menozzi, 1936d: 300; Karavaiev, 1936: 185; Stitz, 1939: 248; Novak & Sadil, 1941: 110; Bernard, 1967: 336; Kutter, 1977c: 206; Arnol'di & Dlussky, 1978: 551; Atanassov & Dlussky, 1992: 217. Senior synonym of *nigrata:* Mayr, 1855: 313; of *angustata:* Emery, 1908a: 199; of *hyalinipennis:* Emery, 1914e: 2; Emery, 1915a: 250; of *marginata* (and its junior synonym *sylvaticoaethiops*): Pisarski, 1975: 31; Agosti & Collingwood, 1987a: 58; of *concava* and *sylvaticoides:* Agosti & Collingwood, 1987a: 58; of *glaber:* Dlussky, Soyunov & Zabelin, 1990: 128. Current subspecies: nominal plus *cachmiriensis*, *escherichi*, *rubicolor*, *vitiosus*.

**#*afer*.** *Camponotus* (*Camponotus*) *ligniperda* ab. *afer* Stärcke, 1942c: xxviii (q.) CZECHOSLOVAKIA.

***afflatus*.** *Camponotus* (*Myrmosaga*) *afflatus* Viehmeyer, 1925b: 140 (w.) AUSTRALIA.

*africanus*. *Camponotus* (*Myrmentoma*) *piceus* st. *spissinodis* var. *africanus* Santschi, 1929e: 164 (w.) MOROCCO. Unavailable name.

***ager*.** *Formica agra* Smith, F. 1858b: 47 (w.) BRAZIL. Combination in *Camponotus:* Roger, 1863b: 5; in *C.* (*Myromgigas*): Forel, 1912i: 91; in *C.* (*Dinomyrmex*): Forel, 1914a: 268; in *C.* (*Tanaemyrmex*): Emery, 1925b: 83. [Forel, 1886f: 167 suggests that *ager* may be minor worker of *egregius*. See also: Emery, 1903: 77.]

*agilis*. *Formica agilis* Smith, F. 1858b: 37 (w.) AUSTRALIA. Combination in *Camponotus*, and junior synonym of *intrepidus:* Roger, 1863b: 4.

*agnatus*. *Camponotus agnatus* Roger, 1863a: 137 (s.w.) SRI LANKA. Combination in *C.* (*Myrmoturba*): Forel, 1914a: 266. Junior synonym of *irritans:* Bingham, 1903: 353.

***agonius*.** *Camponotus* (*Myrmoturba*) *roubaudi* var. *agonia* Santschi, 1915c: 282 (w.) ZAIRE. Combination in *C.* (*Tanaemyrmex*): Emery, 1925b: 103. Raised to species: Santschi, 1926b: 252. See also: Bernard, 1953b: 262.

**#*agricola*.** *Camponotus maculatus* subsp. *agricola* Forel, 1910e: 454 (s.w.) SOUTH AFRICA. Combination in *C.* (*Tanaemyrmex*): Emery, 1925b: 92. Race of *maculatus:* Arnold, 1922: 647. Subspecies of *thraso:* Emery, 1920c: 6; Emery, 1925b: 92.

***aguilerai*.** *Camponotus aguilerai* Kusnezov, 1952d: 198, fig. 12 (w.) ARGENTINA. Combination in *C.* (*Tanaemyrmex*): Kempf, 1972a: 65.

**#*ajjer*.** *Camponotus foleyi* subsp. *ajjer* Bernard, 1953a: 192 (s.w.q.m.) ALGERIA.

**#*akka*.** *Camponotus* (*Myrmamblys*) *ferreri* r. *akka* Forel, 1916: 446 (w.q.) ZAIRE.

***alacer*.** *Camponotus alacer* Forel, 1912i: 76 (s.w.) BRAZIL. Combination in *C.* (*Myrmothrix*): Forel, 1914a: 268; in *C.* (*Tanaemyrmex*): Kempf, 1972a: 65.

**#*albemarlensis*.** *Camponotus* (*Myrmamblys*) *macilentus* var. *albemarlensis* Wheeler, W.M. 1919c: 284 (s.) GALAPAGOS IS. Combination in *C.* (*Pseudocolobopsis*): Emery, 1925b: 158.

*albertisi* Emery, 1887; see under ***CALOMYRMEX.***

***albicoxis*.** *Camponotus landolti* r. *albicoxis* Forel, 1899c: 136 (w.) COSTA RICA. Combination in *C.* (*Tanaemyrmex*), and raised to species: Emery, 1925b: 83.

***albipes*.** *Camponotus albipes* Emery, 1893f: 253, pl. 6, fig. 15 (w.) SRI LANKA. Combination in *C.* (*Myrmosaga*): Forel, 1914a: 269; in *C.* (*Myrmotemnus*): Emery, 1920b: 258; in *C.* (*Myrmamblys*): Emery, 1925b: 138.

*albisectus*. *Camponotus micans* r. *albisectus* Emery, 1892a: 120 (w.) SOMALIA. Combination in *C.* (*Myrmosericus*): Forel, 1914a: 268. Raised to species: Dalla Torre, 1893: 221; Forel, 1914a: 268. Junior synonym of *flavomarginatus:* Emery, 1895h: 48; Wheeler, W.M. 1922a: 968; Menozzi, 1930b: 118.

***albistramineus*.** *Camponotus* (*Myrmeurynota*) *albistramineus* Wheeler, W.M. 1936b: 208 (w.) HAITI.

*albiventris*. *Camponotus mutilatus* var. *albiventris* Baroni Urbani, 1971b: 361. Incorrectly proposed replacement name for *luteiventris* Stitz, 1938: 116. [Junior primary homonym of *luteiventris* Emery, 1897d: 574.] Unavailable name.

**alboannulatus.** *Camponotus alboannulatus* Mayr, 1887: 511 (w.q.) BRAZIL. Wheeler, G.C. & Wheeler, J. 1968: 219 (l.). Combination in *C.* (*Myrmamblys*): Forel, 1912i: 90; in *C.* (*Pseudocolobopsis*): Santschi, 1921f: 311. Current subspecies: nominal plus *montanus, nessus*.

**albocinctus.** *Colobopsis albocincta* Ashmead, 1905a: 110 (w.q.) PHILLIPINES. Combination in *Camponotus* (*Colobopsis*): Wheeler, W.M. 1919a: 5; in *C.* (*Myrmamblys*): Emery, 1925b: 138.

*albofasciata. Formica albofasciata* Smith, F. 1862b: 29 (w.) PANAMA. Combination in *Camponotus*, and junior synonym of *ruficeps:* Mayr, 1886c: 362.

*albopilosus* Mayr, 1876; see under ***CALOMYRMEX.***

**albosparsus.** *Camponotus taylori* var. *albosparsus* Bingham, 1903: 354 (s.w.) INDIA. [First available use of *Camponotus maculatus* r. *taylori* var. *albosparsus* Forel, 1894c: 397, unavailable name.] Combination in *C.* (*Tanaemyrmex*): Emery, 1925b: 93. Race/subspecies of *taylori:* Wheeler, W.M. 1913e: 237; of *barbatus:* Forel, 1906b: 84; Forel, 1912b: 75; Wheeler, W.M. 1921c: 544; Santschi, 1937h: 379. Raised to species: Wang, C., Xiao & Wu, 1989b: 327.

**#albotaeniolatus.** *Camponotus* (*Myrmosphincta*) *sexguttatus* var. *albotaeniolata* Forel, 1921b: 135 (w.) ECUADOR.

**#aldabrensis.** *Camponotus foraminosus* r. *aldabrensis* Forel, 1897c: 203 (s.w.) ALDABRA I. Combination in *C.* (*Myrmotrema*): Santschi, 1915c: 267. Subspecies of *auropubens:* Santschi, 1915c: 267 (in key); of *foraminosus:* Wheeler, W.M. 1922a: 1050.

*alfaroi. Camponotus alfaroi* Emery, 1890b: 57 (s.w.) COSTA RICA. Race/variety of *striatus:* Forel, 1899c: 147; Wheeler, W.M. 1942: 259. Junior synonym of *striatus:* Emery, 1892b: 167; Kempf, 1972a: 54.

*algirica. Camponotus sylvaticus* st. *barbarus* var. *algirica* Santschi, 1921d: 73 (s.w.) ALGERIA. Unavailable name.

**alii.** *Camponotus rubripes* r. *alii* Forel, 1890a: lxi (s.w.) ALGERIA. Combination in *C.* (*Tanaemyrmex*): Emery, 1925b: 91. Race of *maculatus:* Emery, 1891b: 19. Raised to species: Dalla Torre, 1893: 221; Emery, 1920c: 7. Stirps of *nylanderi:* Santschi, 1931a: 11. Revived status as species: Santschi, 1936c: 207; Santschi, 1939c: 5. Senior synonym of *seurati:* Bernard, 1953a: 187. Current subspecies: nominal plus *auresi, concolor* (unresolved junior homonym), *hoggarensis, laurenti*.

*aloysii. Camponotus* (*Myrmaphaenus*) *personatus* subsp. *aloysii* Emery, 1925b: 156 (w.) PARAGUAY. Junior synonym of *personatus:* Santschi, 1939e: 330.

*alpina. Camponotus rufipes* var. *alpina* Emery, 1906c: 187 (w.) ARGENTINA. Combination in *C.* (*Myrmothrix*): Bruch, 1914: 230. Junior synonym of *rufipes:* Hashmi, 1973: 52.

*altaica. Camponotus herculeanus* var. *altaica* Ruzsky, 1915b: 6 (w.) RUSSIA. Subspecies of *herculeanus:* Ruzsky, 1926: 108; of *saxatilis:* Ruzsky, 1936: 90. Junior synonym of *sachalinensis:* Arnol'di, 1967: 1821.

**#altinotus.** *Camponotus* (*Pseudocolobopsis*) *macilentus* var. *altinota* Stitz, 1932a: 370, fig. 3 (w.) GALAPAGOS IS.

*altior. Camponotus* (*Myrmoturba*) *radamae* st. *becki* var. *altior* Santschi, 1923e: 292 (w.) MADAGASCAR. Unavailable name.

**altivagans.** *Camponotus* (*Myrmeurynota*) *altivagans* Wheeler, W.M. 1936b: 207 (s.w.) HAITI.

**amamianus.** *Camponotus* (*Paramyrmamblys*) *amamianus* Terayama, 1991: 165, fig. 1 (s.w.) JAPAN.

**#amazonicus.** *Camponotus* (*Myrmobrachys*) *crassus* st. *amazonicus* Santschi, 1922c: 122, fig. 3 (s.w.) BRAZIL. [Misspelled *amazonensis* by Emery, 1925b: 163.]

**#ambonensis.** *Camponotus* (*Tanaemyrmex*) *variegatus* var. *ambonensis* Karavaiev, 1930a: 211 (w.q.m.) INDONESIA (Ambon I.).

**#ambustus.** *Camponotus christi* var. *ambustus* Forel, 1892l: 233 (s.w.q.) MADAGASCAR. Combination in *C.* (*Mayria*): Emery, 1925b: 122.

**americanus.** *Camponotus americanus* Mayr, 1862: 661 (w.q.) U.S.A. Emery, 1893i: 674 (m.); Wheeler, G.C. & Wheeler, J. 1968: 216 (l.). Combination in *C.* (*Camponotus*): Emery, 1925b: 74. Junior synonym of *castaneus:* Mayr, 1886d: 420. Revived from synonymy as subspecies of *castaneus:* Emery, 1893i: 674; Wheeler, W.M. 1913c: 117; Wheeler, W.M. 1917i: 465; Wheeler, W.M. 1932a: 13. Revived status as species, and senior synonym of *rufinasis:* Creighton, 1950a: 365. See also: Eisner & Wilson, 1952: 47.

*americana. Formica americana* Buckley, 1866: 154 (w.q.) U.S.A. Combination in *Camponotus:* Emery, 1895c: 337. [Unresolved junior secondary homonym of *Camponotus americanus* Mayr, above.] Unrecognisable taxon; *incertae sedis* in *Camponotus*. [Note: Emery, 1895c: 337, and Wheeler, W.M. 1902f: 19, suspected this name was cognate with *marginatus*. Emery, 1925b: 117 provisionally synonymised it with *minutus*. Smith, M.R. 1951a: 845, and Smith, D.R. 1979: 1432, dubiously place it near *nearcticus*.]

**#amius.** *Camponotus friedae* var. *amia* Forel, 1912a: 78 (s.w.) TAIWAN. Combination in *C.* (*Tanaemyrmex*): Emery, 1925b: 102.

**amoris.** *Camponotus amoris* Forel, 1904c: 51 (s.) BRAZIL. Combination in *C.* (*Myrmamblys*): Forel, 1912i: 90; in *C.* (*Dinomyrmex*): Forel, 1914a: 268; in *C.* (*Tanaemyrmex*): Emery, 1925b: 83.

**#amperei.** *Camponotus* (*Myrmocamelus*) *cinereus* var. *amperei* Forel, 1913g: 192 (w.) AUSTRALIA. Combination in *C.* (*Myrmophyma*): Emery, 1925b: 110. Subspecies of *cinereus:* Taylor & Brown, D.R. 1985: 111.

**amphidus.** *Camponotus* (*Myrmisolepis*) *amphidus* Santschi, 1926b: 260, fig. 6 (s.w.) ZIMBABWE.

**#amplior.** *Camponotus* (*Myrmosericus*) *eugeniae* r. *amplior* Forel, 1913b: 341 (w.) ZAIRE.

**#andigenus.** *Camponotus punctulatus* subsp. *andigenus* Emery, 1903: 71 (w.) PERU. Combination in *C.* (*Myrmoturba*): Forel, 1913l: 248; in *C.* (*Tanaemyrmex*): Emery, 1925b: 78.

**andrei.** *Camponotus andrei* Forel, 1885a: 343 (s.w.) MEXICO. Combination in *C.* (*Myrmamblys*): Forel, 1914a: 271; in *C.* (*Myrmaphaenus*): Emery, 1925b: 154. Current subspecies: nominal plus *cholericus*.

**andrewsi.** *Camponotus* (*Myrmamblys*) *andrewsi* Donisthorpe, 1936e: 527 (s.m.) CHRISTMAS I.

**#andrius.** *Camponotus oertzeni* var. *andrius* Dalla Torre, 1893: 246 (s.w.) GREECE. [First available use of *Camponotus rubripes* r. *oertzeni* var. *andrius* Forel, 1889: 265; unavailable name (Agosti & Collingwood, 1987a: 61).] Menozzi, 1936d: 302 (q.l.). Combination in *C.* (*Tanaemyrmex*): Emery, 1925b: 97.

*angolensis. Camponotus angolensis* Radoszkovsky, 1881: 197 ANGOLA. *Nomen nudum*, attributed to Smith, F.

*angulata. Formica angulata* Smith, F. 1859a: 139 (w.) INDONESIA (Aru I.). Junior synonym of *quadriceps:* Emery, 1896d: 375.

*angulatus. Camponotus angulatus* Mayr, 1870a: 382 (w.) COLOMBIA. Combination in *C.* (*Myrmeurynota*): Forel, 1912i: 92. [Junior secondary homonym of *angulata* Smith, F., above.] Replacement name: *linnaei* Forel, 1886f: 171.

**anguliceps.** *Camponotus* (*Tanaemyrmex*) *anguliceps* Stitz, 1938: 113 (q.) BORNEO.

*angulicollis. Camponotus angulicollis* Emery, 1887a: 218. Unnecessary (second) replacement name for *angulatus* Mayr, 1870a: 382. [Junior secondary homonym of *angulata* Smith, F. 1859a: 139.] Junior synonym of *linnaei* (first replacement name): Dalla Torre, 1893: 240.

*angustata. Formica angustata* Latreille, 1798: 34 (w.m.) FRANCE. Junior synonym of *aethiops:* Emery, 1908a: 199.

*angustata. Colobopsis angustata* Mayr, 1870b: 942 (q.) SINGAPORE. Forel, 1893b: 436 (w.); Viehmeyer, 1916a: 160 (s.m.). Combination in *Camponotus:* Dalla Torre, 1893: 221; in *C.* (*Colobopsis*): Forel, 1893b: 435. Variety/subspecies of *vitreus:* Emery, 1900d: 708; Forel, 1909d: 229. [Junior secondary homonym of *angustata* Latreille, above.] Replacement name: *angustulus* Emery, 1925b: 148.

*angustatus. Camponotus kiesenwetteri* r. *angustatus* Forel, 1889: 261 (w.) GREECE. [Junior secondary homonym of *angustata* Latreille, above.] Replacement name: *stenoticus* Emery, 1915h: 7.

**angusticeps.** *Camponotus angusticeps* Emery, 1886: 355, pl. 17, fig. 1 (w.) SOUTH AFRICA. Arnold, 1924: 688 (s.). Combination in *C.* (*Myrmosericus*): Forel, 1914a: 268.

**angusticollis.** *Formica angusticollis* Jerdon, 1851: 120 (w.q.) INDIA. Combination in *Camponotus:* Roger, 1863b: 3; in *C.* (*Dinomyrmex*): Forel, 1913d: 438; in *C.* (*Tanaemyrmex*): Emery, 1925b: 89. Junior synonym of *gigas:* Smith, F. 1858b: 14. Revived from synonymy: Roger, 1863b: 7. Senior synonym of *ardens* (and its junior synonym *impetuosa*) and *prismaticus:* Forel, 1892j: 235. See also: Bingham, 1903: 366. Current subspecies: nominal plus *sanguinolentus*.

**#angustinodus.** *Camponotus* (*Tanaemyrmex*) *discors* var. *angustinoda* Emery, 1925b: 102 (w.q.) AUSTRALIA. [First available use of *Camponotus maculatus* subsp. *discors* var. *angustinota* Forel, 1907h: 300; unavailable name (Taylor, 1986: 34).]

**#angustulus.** *Camponotus* (*Colobopsis*) *vitreus* var. *angustula* Emery, 1925b: 148. Replacement name for *angustata* Mayr, 1870b: 942. [Junior secondary homonym of *angustata* Latreille, 1798: 34.] Material of the unavailable name *continentalis* referred here by Viehmeyer, 1925b: 145.

**anningensis.** *Camponotus anningensis* Wu & Wang; in Wang, C., Xiao & Wu, 1989b: 323, figs. 10, 11 (s.w.) CHINA.

*annobonensis. Camponotus* (*Myrmotrema*) *foraminosus* var. *annobonensis* Santschi, 1920i: 4 (w.) ANNOBON

I. [First available use of *Camponotus* (*Myrmotrema*) *foraminosus* st. *chrysogaster* var. *annobonensis* Santschi, 1915c: 277; unavailable name.] Junior synonym of *insularis:* Santschi, 1920i: 4.

***annulatus.*** *Camponotus* (*Myrmamblys*) *annulatus* Karavaiev, 1929c: 238, fig. 4 (s.w.) INDONESIA (Aru I.).

#***antennatus.*** *Camponotus* (*Myrmobrachys*) *canescens* var. *antennata* Santschi, 1920e: 126, fig. 3 (w.) BOLIVIA.

#***antennatus.*** *Camponotus* (*Myrmobrachys*) *brasiliensis* var. *antennatus* Santschi, 1925d: 246 (w.) COSTA RICA. [Unresolved junior primary homonym of *antennatus* Santschi, above.]

*antespectans* Forel, 1913; see under ***EUPRENOLEPIS.***

#***anthracinus.*** *Camponotus* (*Myrmosericus*) *vestitus* var. *anthracinus* Santschi, 1930a: 74 (s.w.) ANGOLA. Variety of *cosmicus:* Santschi, 1935a: 280; of *vestitus:* Santschi, 1937d: 244.

***anthrax.*** *Camponotus anthrax* Wheeler, W.M. 1911d: 96 (s.w.q.m.) U.S.A. Wheeler, G.C. & Wheeler, J. 1970: 650 (l.). Combination in *C.* (*Camponotus*): Forel, 1914a: 266; in *C.* (*Myrmentoma*): Emery, 1920b: 257. See also: Snelling, 1988: 59.

#***antiguanus.*** *Camponotus sexguttatus* var. *antiguanus* Wheeler, W.M. 1923c: 7 (s.w.m.) ANTIGUA. Combination in *C.* (*Myrmosphincta*): Kempf, 1972a: 58.

#***apellis.*** *Camponotus* (*Orthonotomyrmex*) *chrysurus* var. *apellis* Wheeler, W.M. 1922a: 972 (s.w.) ANGOLA. [First available use of *Camponotus chrysurus* subsp. *acutisquamis* var. *apellis* Forel, 1910c: 269; unavailable name.] Combination in *C.* (*Myrmopiromis*): Emery, 1925b: 129; in *C.* (*Myrmopelta*): Santschi, 1930b: 79. Subspecies of *chrysurus:* Santschi, 1926c: 603; Santschi, 1930b: 79.

***apostolus.*** *Camponotus apostolus* Forel, 1901e: 369 (s.w.q.) COSTA RICA. Combination in *C.* (*Myrmamblys*): Forel, 1912i: 90; in *C.* (*Myrmoturba*): Emery, 1920b: 255; in *C.* (*Tanaemyrmex*): Emery, 1925b: 79.

***arabicus.*** *Camponotus arabicus* Collingwood, 1985: 278, figs. 67, 72, 74 (w.q.) SAUDI ARABIA. Current subgenus: *C.* (*Tanaemyrmex*).

***arboreus.*** *Formica arborea* Smith, F. 1858b: 44 (w.) BRAZIL. Combination in *Camponotus:* Mayr, 1862: 666; in *C.* (*Myrmobrachys*), and senior synonym of *adpressisetosus:* Kempf, 1965: 186.

***arcuatus.*** *Camponotus arcuatus* Mayr, 1876: 63 (s.w.) AUSTRALIA. Wheeler, G.C. & Wheeler, J. 1968: 217 (l.). Combination in *C.* (*Myrmoturba*): Forel, 1914a: 266; in *C.* (*Myrmophyma*): Emery, 1920b: 257. Current subspecies: nominal plus *aesopus.*

*ardens. Formica ardens* Smith, F. 1858b: 17 (q.) INDIA. Combination in *Camponotus:* Roger, 1863b: 3. Senior synonym of *impetuosa:* Mayr, 1886c: 353. Junior synonym of *angusticollis:* Forel, 1892j: 235.

*argentatus* Radoszkovsky, 1876; see under ***CATAGLYPHIS.***

#***argentopubens.*** *Camponotus* (*Myrmotrema*) *auropubens* st. *argentopubens* Santschi, 1915c: 268 (q.) ANGOLA. [First available use of *Camponotus foraminosus* subsp. *auropubens* var. *argentopubens* Forel, 1909b: 66; unavailable name.]

***argus.*** *Camponotus* (*Myrmotrema*) *argus* Santschi, 1935a: 282 (w.) ZAIRE.

#***arhuacus.*** *Camponotus ustus* var. *arhuacus* Forel, 1902b: 173 (w.) COLOMBIA. Combination in *C.* (*Pseudocolobopsis*): Emery, 1925b: 158; in *C.* (*Tanaemyrmex*): Kempf, 1972a: 73.

#***armeniacus.*** *Camponotus micans* subsp. *armeniacus* Arnol'di, 1967: 1828, fig. 3 (s.w.q.m.) ARMENIA.

***arminius.*** *Camponotus barbarossa* subsp. *arminius* Forel, 1910e: 457 (s.w.) MADAGASCAR. Combination in *C.* (*Orthonotomyrmex*): Forel, 1914d: 256; in *C.* (*Myrmamblys*): Emery, 1925b: 142; in *C.* (*Myrmopelta*): Santschi, 1921f: 311; Santschi, 1926c: 603. Raised to species: Santschi, 1921f: 311. Current subspecies: nominal plus *bicontractus.*

*armouri. Camponotus* (*Myrmentoma*) *lateralis* subsp. *armouri* Wheeler, W.M. 1926: 5 (w.q.m.) BALEARIC IS. Senior synonym of *balearis:* Santschi, 1936c: 207 (given as junior synonym but *armouri* has priority over *balearis*). Junior synonym of *lateralis:* Collingwood, 1978: 73.

***armstrongi.*** *Camponotus* (*Myrmogonia*) *armstrongi* McAreavey, 1949: 19, figs. 53–56 (s.w.) AUSTRALIA.

#***arnoldi.*** *Camponotus truebi* var. *arnoldi* Forel, 1912a: 75 (w.) TAIWAN. Combination in *C.* (*Myrmentoma*): Emery, 1925b: 119.

#***arnoldi.*** *Camponotus transvaalensis* subsp. *arnoldi* Prins. 1965a: 156, figs. 6–10 (s.w.) ZIMBABWE. [Unresolved junior primary homonym of *arnoldi* Forel, above.]

*arnoldi. Camponotus* (*Tanaemyrmex*) *arnoldi* Pisarski, 1971a: 671, figs. 1–4 (w.) IRAQ. [Junior primary homonym of *arnoldi* Forel, above.] Replacement name: *babau* **new name**.

***arnoldinus.*** *Camponotus* (*Myrmoturba*) *arnoldinus* Forel, 1914d: 251 (s.w.) SOUTH AFRICA. Combination in *C.* (*Tanaemyrmex*): Emery, 1925b: 91. Stirps of *maculatus:* Santschi, 1923e: 290. Revived status as species: Emery, 1925b: 91. See also: Arnold, 1922: 639.

***arrogans.*** *Formica arrogans* Smith, F. 1858b: 23 (w.) SINGAPORE. Combination in *Camponotus:* Roger, 1863b: 3; in *C.* (*Myrmoturba*): Forel, 1914a: 266; in *C.* (*Tanaemyrmex*): Emery, 1925b: 101. Subspecies of *sylvaticus:* Mayr, 1886c: 354; of *maculatus:* Emery, 1896d: 371. Revived status as species: Dalla Torre, 1893: 222; Bingham, 1903: 357; Emery, 1925b: 101. Senior synonym of *junctus:* Bingham, 1903: 357; Rothney, 1903: 99.

***aruensis.*** *Camponotus* (*Colobopsis*) *aruensis* Karavaiev, 1933a: 317, fig. 5 (w.) INDONESIA (Aru I.).

***asli.*** *Camponotus* (*Karavaievia*) *asli* Dumpert, in Dumpert, Maschwitz, *et al.* 1989: 218, figs. 1–4 (w.q.) WEST MALAYSIA.

#***asniensis.*** *Camponotus* (*Myrmosericus*) *micans* var. *asniensis* Santschi, 1939c: 5 (w.) MOROCCO.

#***asper.*** *Camponotus* (*Myrmosericus*) *cruentatus* var. *aspera* Menozzi, 1925b: 371 (w.) AUSTRALIA.

#***assabensis.*** *Camponotus* (*Tanaemyrmex*) *thraso* var. *assabensis* Emery, 1925b: 92 (w.) ETHIOPIA. [First available use of *Camponotus maculatus* subsp. *adenensis* var. *assabensis* Emery, 1893f: 258; unavailable name.]

#***assumptionis.*** *Camponotus* (*Myrmotrema*) *troglodytes* r. *assumptionis* Crawley, 1921: 95 (s.w.) ALDABRA I. Subspecies of *robecchii:* current status.

#***astilbus.*** *Camponotus* (*Tanaemyrmex*) *coruscus* var. *astilbus* Stitz, 1938: 115 (w.) SURINAM.

#***aterrimus.*** *Camponotus pennsylvanicus* var. *aterrimus* Emery, 1895k: 478 (footnote) (w.q.) RUSSIA. Ruzsky, 1905b: 238 (m.); Karavaiev, 1912b: 594 (s.). Combination in *C.* (*Camponotus*): Emery, 1925b: 73. Subspecies of *herculeanus:* Ruzsky, 1915a: 419. Raised to species: Ruzsky, 1926: 108. Junior synonym of *japonicus:* Kuznetsov-Ugamsky, 1929a: 17; Yasumatsu & Brown, 1951: 36. Variety/subspecies of *japonicus:* Santschi, 1925f: 88; Wheeler, W.M. 1929f: 9; Arnol'di, 1967: 1827; Kupyanskaya, 1990: 170. Senior synonym of *manczshuricus:* Arnol'di, 1967: 1827.

***atlantis.*** *Camponotus rubripes* r. *atlantis* Forel, 1890a: lxiii (s.w.q.) TUNISIA. Combination in *C.* (*Myrmoturba*): Forel, 1914a: 266; in *C.* (*Tanaemyrmex*): Emery, 1925b: 91. Race of *maculatus:* Emery, 1891b: 20. Raised to species: Emery, 1920b: 255; Emery, 1925b: 91; Santschi, 1931a: 11. Current subspecies: nominal plus *atrioris, hesperius, marocanus, nigrovarius, planitae.*

*atra. Formica atra* Zetterstedt, 1838: 450 (m.) SWEDEN. Junior synonym of *herculeanus:* Nylander, 1846a: 894.

*atra. Formica atra* Buckley, 1866: 160 (w.) U.S.A. [Unresolved junior primary homonym of *atra* Zetterstedt, above.] Unrecognisable taxon; *incertae sedis* in *Camponotus:* Emery, 1895c: 337; Wheeler, W.M. 1902f: 20. [Note: Wheeler, W.M. 1917c: 28 suggests synonymy with *caryae,* but Smith, D.R. 1979: 1467 retains *atra* among the unplaced taxa.]

#***atrabilis.*** *Camponotus* (*Myrmotrema*) *grandidieri* var. *atrabilis* Santschi, 1915c: 273 (w.) MADAGASCAR.

*atramentarius. Camponotus maculatus* var. *atramentarius* Forel, 1904b: 379 (w.) ETHIOPIA. Forel, 1910f: 26 (q.). Combination in *C.* (*Myrmoturba*): Forel, 1913b: 340; in *C.* (*Tanaemyrmex*): Emery, 1925b: 88. Subspecies of *maculatus:* Forel, 1905b: 180; Emery, 1920c: 5. Junior synonym of *maculatus:* Baroni Urbani, 1972: 125.

#***atratior.*** *Camponotus* (*Myrmoturba*) *simillimum* st. *atratior* Santschi, 1922c: 98, fig. 1 (s.w.) BRAZIL. Combination in *C.* (*Tanaemyrmex*): Emery, 1925b: 82.

***atriceps.*** *Formica atriceps* Smith, F. 1858b: 44 (s.w.) BRAZIL. Mayr, 1862: 660 (q.); Forel, 1878: 76 (m.). Combination in *Camponotus:* Roger, 1862c: 285. Junior synonym of *abdominalis:* Roger, 1862c: 285; Santschi, 1913h: 42; Hashmi, 1973: 83. Variety/subspecies of *abdominalis:* Emery, 1906c: 187; Forel, 1911c: 310. Senior synonym of *taeniatus:* Mayr, 1870a: 378. Oldest synonym of, and hence first available replacement name for, *Formica abdominalis* Fabricius, 1804: 409. [Junior primary homonym of *Formica abdominalis* Latreille, 1802c: 175.] Junior synonyms attached to *abdominalis* Fabricius by Hashmi, 1973: 82 are here transferred to *atriceps,* which hence is senior synonym of: *atricipitoesuriens, costaricensis, cupiens, depressidens, epistomatus, esuriens* (and its junior synonym *vulpinus*), *fuchsae, fulvacea, fumata, laevilata, mediopallidus, rabidus, romani, stercorarius, taeniatus, transvectus* and *ustulatus.* Material of the unavailable names *defrictus, flavobasalis, luisae, misionensis* and *panamanus*, referred to *abdominalis* Fabricius by Hashmi, 1973: 82, is here referred to *atriceps*. Current subspecies: nominal plus *nocens* (status uncertain, see Hashmi, 1973: 120).

*atricipitoesuriens. Camponotus atriceps* var. *atricipitoesuriens* Forel, 1879a: 77 (w.) MEXICO. Junior synonym of *atriceps:* Hashmi, 1973: 82.

*atricolor. Formica atricolor* Nylander, 1849: 36 (w.) RUSSIA. Combination in *Camponotus:* Roger, 1863b: 1; in *C.* (*Myrmentoma*): Emery, 1925a: 67; Emery, 1925b: 120. Junior synonym of *lateralis:* Mayr, 1855: 322; Roger, 1863b: 1. Revived from synonymy as subspecies of *lateralis:* Forel, 1892i: 306; Ruzsky, 1905b: 254; Kuznetsov-Ugamsky, 1929b: 36. Junior synonym of *piceus:* Dalla Torre, 1893: 238. Revived from synonymy as subspecies of *piceus:* Emery, 1925a: 63; Karavaiev, 1927c: 277. Variety of *merulus:* Novak & Sadil, 1941: 109. Junior synonym of *merulus:* Bernard, 1967: 344. Revived from synonymy, and revived status as species: Arnol'di & Dlussky, 1978: 552; Agosti & Collingwood, 1987a: 58. Senior synonym of *rectus:* Emery, 1896d: 373; Karavaiev, 1936: 191. Junior synonym of *piceus:* Atanassov & Dlussky, 1992: 222.

*atrigenatus. Camponotus* (*Myrmamblys*) *tokioensis* var. *atrigenatus* Santschi, 1937h: 384, fig. 13 (s.w.) JAPAN. Junior synonym of *tokioensis:* Terayama & Satoh, 1990a: 412.

#***atrioris.*** *Camponotus atlantis* var. *atrioris* Santschi, 1922a: 68 (w.) ALGERIA. Combination in *C.* (*Tanaemyrmex*): Emery, 1925b: 91.

***atriscapus.*** *Camponotus* (*Myrmotrema*) *atriscapa* Santschi, 1926c: 614 (w.m.) ZAIRE.

***atrox.*** *Camponotus* (*Camponotus*) *japonicus* var. *atrox* Emery, 1925b: 73 (s.w.q.) KOREA. Replacement name for *cruentata* Emery, 1925b: 73. [First available use of *Camponotus herculeanus* subsp. *japonicus* var. *cruentata* Karavaiev, 1912b: 595; unavailable name.] [Junior secondary homonym of *cruentata* Latreille, 1802c: 116.] Junior synonym of *japonicus:* Yasumatsu & Brown, 1951: 37. Revived from synonymy and raised to species: Yasumatsu & Brown, 1957: 49.

***attila.*** *Camponotus* (*Tanaemyrmex*) *attila* Stärcke, 1935: 264, fig. 3 (s.) TURKMENIA.

***augustei.*** *Camponotus christophei* subsp. *augustei* Wheeler, W.M. & Mann, 1914: 59 (s.w.) HAITI. Combination in *C.* (*Myrmeurynota*): Emery, 1925b: 168. Raised to species: Wheeler, W.M. 1936b: 207.

#***augustulus.*** *Camponotus* (*Colobopsis*) *fictor* subsp. *augustulus* Viehmeyer, 1925b: 145 (s.w.q.) AUSTRALIA. Material of the unavailable name *continentalis* referred here: Viehmeyer, 1925b: 145. Subspecies of *macrocephalus:* current status.

#***aulicus.*** *Camponotus* (*Myrmocladoecus*) *rectangularis* var. *aulicus* Wheeler, W.M. 1933a: 64 (s.w.) GUATEMALA.

***auratus.*** *Camponotus* (*Colobopsis*) *auratus* Karavaiev, 1935a: 113 (w.) VIETNAM.

***aurelianus.*** *Camponotus* (*Colobopsis*) *aurelianus* Forel, 1912n: 68 (w.) SUMATRA. Combination in *C.* (*Myrmotemnus*): Emery, 1920b: 258; in *C.* (*Colobopsis*): Emery, 1925b: 148.

***aureopilus.*** *Camponotus* (*Myrmogonia*) *aureopilus* Viehmeyer, 1914b: 531 (w.) NEW GUINEA. Stitz, 1938: 119 (q.). Combination in *C.* (*Myrmophyma*): Emery, 1925b: 110; in *C.* (*Thlipsepinotus*): Santschi, 1928e: 483. Current subspecies: nominal plus *velutinus*.

#***auresi.*** *Camponotus* (*Myrmoturba*) *alii* var. *auresi* Emery, 1920c: 7 (w.) ALGERIA. [First available use of *Camponotus maculatus* subsp. *alii* var. *auresi* Forel, 1909e: 375; unavailable name.]

***auricomus.*** *Camponotus auricomus* Roger, 1862c: 283 (footnote) (w.q.) MEXICO. Forel, 1879a: 98 (m.). Combination in *C.* (*Myrmobrachys*): Forel, 1914a: 270. Race of *senex:* Forel, 1879a: 97. Revived status as species: Forel, 1893g: 334; Dalla Torre, 1893: 222. Current subspecies: nominal plus *lucianus, vincentensis*.

***auriculatus.*** *Camponotus auriculatus* Mayr, 1897: 432 (s.w.) SRI LANKA. Combination in *C.* (*Myrmamblys*): Forel, 1914a: 271.

***auriventris.*** *Camponotus auriventris* Emery, 1889b: 514 (s.w.q.) BURMA. Combination in *C.* (*Myrmosphincta*): Forel, 1912i: 92; in *C.* (*Myrmosaulus*): Emery, 1925b: 113.

***aurocinctus.*** *Formica aurocincta* Smith, F. 1858b: 39 (w.) AUSTRALIA. Viehmeyer, 1914c: 45 (s.). Combination in *Camponotus:* Mayr, 1886c: 355; in *C.* (*Myrmosphincta*): Forel, 1914a: 273; in *C.* (*Myrmosaulus*): Wheeler, W.M. 1922a: 705.

***aurofasciatus.*** *Camponotus* (*Myrmotrema*) *aurofasciatus* Santschi, 1915c: 276 (w.) SAO THOME I. Emery, 1920c: 29 (s.).

*aurofasciatus. Camponotus* (*Myrmamblys*) *aurofasciatus* Wheeler, W.M. 1915g: 817, pl. 66, fig. 7 (w.) AUSTRALIA. [Unresolved junior primary homonym of *aurofasciatus* Santschi, above.] Junior synonym of *inflatus:* Wheeler, W.M. 1916e: 40.

***auropubens.*** *Camponotus foraminosus* r. *auropubens* Forel, 1894b: 67 (w.) MOZAMBIQUE. Combination in *C.* (*Myrmotrema*): Santschi, 1914e: 41. Subspecies of *grandidieri:* Mayr, 1895: 150; of *foraminosus:* Emery, 1896d: 376; Wheeler, W.M. 1922a: 250. Raised to species: Santschi, 1915c: 267 (in key); Arnold, 1924: 728; Emery, 1925b: 131. Current subspecies: nominal plus *absalon, argentopubens, jacob*.

***aurosus.*** *Camponotus aurosus* Roger, 1863a: 134 (w.) MAURITIUS. Combination in *C.* (*Myrmepomis*): Emery, 1920b: 258; in *C.* (*Myrmoserícus*): Forel, 1914a: 268; Wheeler, W.M. 1922a: 1044.

*aurulenta. Formica aurulenta* Latreille, 1802c: 114, pl. 3, fig. 9 (q.) SENEGAL. Junior synonym of *sericeus:* Roger, 1863b: 2.

***autrani.*** *Camponotus autrani* Forel, 1886f: 165 (s.w.) SUMATRA. Combination in *C.* (*Myrmoturba*): Forel, 1913k: 129; in *C.* (*Tanaemyrmex*): Emery, 1925b: 89. Subspecies of *festinus:* Emery, 1888a: 528. Revived status as species: Dalla Torre, 1893: 222; Emery, 1896d: 370; Forel, 1914a: 266.

*avarus. Camponotus* (*Myrmoturba*) *simillimus* st. *indianus* var. *avarus* Santschi, 1936b: 206 (w.) PANAMA. Unavailable name.

***avius.*** *Camponotus* (*Myrmotrema*) *avius* Santschi, 1926c: 615 (s.w.) ZIMBABWE. Current subspecies: nominal plus *hertigi*.

*axillaris. Formica axillaris* Spinola, 1808: 243 (w.) ITALY. Junior synonym of *lateralis:* Mayr, 1855: 322.

***babau.*** *Camponotus babau* **new name**. Replacement name for *arnoldi* Pisarski, 1971a: 671. [Junior primary homonym of *arnoldi* Forel, 1912a: 75.]

#***babiensis.*** *Camponotus* (*Tanaemyrmex*) *carin* var. *babiensis* Emery, 1925b: 90 (s.w.) SUMATRA. [First available use of *Camponotus* (*Dinomyrmex*) *dorycus* r. *carin* var. *babiensis* Forel, 1915a: 37; unavailable name.]

#***bacchus.*** *Formica bacchus* Smith, F. 1858b: 21 (w.) SRI LANKA. Smith, F. 1878b: 10 (q.m.). Combination in *Camponotus:* Roger, 1863b: 3; in *C.* (*Tanaemyrmex*): Emery, 1925b: 95. Senior synonym of *ventralis:* Mayr, 1865: 27. Junior synonym of *mitis:* Mayr, 1879: 646; Dalla Torre, 1893: 243. Race of *sylvaticus:* Forel, 1879a: 67. Revived from synonymy as variety of *mitis:* Bingham, 1903: 356. Variety of *variegatus:* Emery, 1925b: 95.

***bactrianus.*** *Camponotus bactrianus* Pisarski, 1967: 411, figs. 77–81 (s.w.q.) AFGHANISTAN. Combination in *C.* (*Tanaemyrmex*): Pisarski, 1970: 322.

***badius.*** *Formica badia* Smith, F. 1857a: 54 (w.) SINGAPORE. Viehmeyer, 1916a: 161 (m.). Combination in *Camponotus:* Roger, 1863b: 3; in *C.* (*Colobopsis*): Forel, 1913k: 133. [Unresolved junior primary homonym of *Formica badia* Latreille, 1802c: 238 (now in *Pogonomyrmex*).] Current subspecies: nominal plus *saginatus*.

*baeticus. Camponotus maculatus* subsp. *barbaricus* var. *baeticus* Emery, 1908a: 197 (s.w.q.) SPAIN. Unavailable name; material referred to *barbaricus* by Collingwood, 1978: 72.

#***bajulus.*** *Camponotus* (*Myrmobrachys*) *iheringi* var. *bajulus* Emery, 1925b: 165. Replacement name for *latinota* Forel, 1909a: 264. [Junior primary homonym of *latinotus* Forel, 1907c: 144.]

***bakeri.*** *Camponotus hyatti* var. *bakeri* Wheeler, W.M. 1904d: 271 (w.q.) U.S.A. Combination in *C.* (*Myrmentoma*): Emery, 1925b: 119. Subspecies of *hyatti:* Creighton, 1950a: 387; Smith, M.R. 1951a: 844. Raised to species: Snelling, 1985: 24; Snelling, 1988: 60.

***baldaccii.*** *Camponotus maculatus* subsp. *baldaccii* Emery, 1908a: 198 (s.w.) GREECE. [First available use of *Camponotus maculatus* subsp. *dichrous* var. *baldaccii* Emery, 1894j: 9; unavailable name.] Combination in *C.* (*Tanaemyrmex*): Emery, 1925b: 100. Subspecies of *sylvaticus:* Emery, 1925b: 100; Menozzi, 1936d: 300. Raised to species: Collingwood, 1985: 279.

*balearis. Camponotus* (*Myrmentoma*) *lateralis* var. *balearis* Santschi, 1929e: 164 (w.) BALEARIC IS. [First available use of *Camponotus lateralis* st. *spissinodis* var. *balearis* Santschi, 1925g: 360; unavailable name.] Junior synonym of *armouri:* Santschi, 1936c: 207 (given as senior synonym but *armouri* has priority).

*ballioni. Camponotus maculatus* r. *ballioni* Forel, 1904d: 176 (w.q.) SOUTH AFRICA. Combination in *C.* (*Tanaemyrmex*): Emery, 1925b: 87. Raised to species: Santschi, 1925h: 166. Junior synonym of *maculatus:* Baroni Urbani, 1972: 125.

***balzani.*** *Camponotus balzani* Emery, 1894c: 166 (s.w.) BOLIVIA. Forel, 1918a: 726 (q.). Combination in *C.* (*Myrmothrix*): Forel, 1914a: 268; in *C.* (*Tanaemyrmex*): Emery, 1925b: 79.

***CAMPONOTUS** (continued)*

*bambusarum. Camponotus cingulatus* var. *bambusarum* Forel, 1902b: 176 (w.) BRAZIL. Santschi, 1922c: 109 (s.). Combination in *C.* (*Myrmothrix*): Emery, 1925b: 108. Junior synonym of *cingulatus:* Hashmi, 1973: 31.

***banghaasi.*** *Camponotus banghaasi* Emery, 1903: 79 (s.w.) PERU. Combination in *C.* (*Myrmothrix*): Forel, 1914a: 268; in *C.* (*Myrmaphaenus*): Emery, 1920b: 256; in *C.* (*Myrmoplatypus*): Santschi, 1921f: 311.

***barbaricus.*** *Camponotus maculatus* subsp. *barbaricus* Emery, 1905d: 31 (s.w.) GIBRALTAR. Emery, 1908a: 197 (q.); Baroni Urbani, 1976: 217 (m.); Baroni Urbani, 1968b: 483 (l.). Combination in *C.* (*Myrmoturba*): Santschi, 1923f: 335; in *C.* (*Tanaemyrmex*): Emery, 1925b: 100. Subspecies of *sylvaticus:* Emery, 1920c: 7; Santschi, 1921d: 72 (misspelled as *barbarus*); Santschi, 1931a: 11; Cagniant, 1964: 109. Raised to species: Baroni Urbani, 1964b: 61; Baroni Urbani, 1968b: 482; Cagniant, 1970c: 32; Schembri & Collingwood, 1981: 437. Material of the unavailable name *baeticus* referred here by Collingwood, 1978: 72. See also: Emery, 1916b: 229. Current subspecies: nominal plus *eubarbaricus, xanthomelas.*

***barbarossa.*** *Camponotus* (*Orthonotomyrmex*) *arminius* subsp. *barbarossa* Emery, 1920c: 26 (s.w.q.) CAMEROUN. [*Camponotus barbarossa* Forel, 1910e: 457, and Forel, 1912i: 92, *nomina nuda.*] Combination in *C.* (*Myrmamblys*): Emery, 1925b: 142; in *C.* (*Myrmopelta*), and raised to species: Santschi, 1926a: 21; Santschi, 1926c: 603. Current subspecies: nominal plus *micipsa, sulcatinasis.*

***barbatus.*** *Camponotus barbatus* Roger, 1863a: 138 (w.q.) SRI LANKA. Emery, 1901f: 121 (s.). Combination in *C.* (*Myrmoturba*): Forel, 1914a: 267; in *C.* (*Tanaemyrmex*): Emery, 1925b: 92. Current subspecies: nominal plus *infuscoides, samarus, taylori.*

*barbatus. Camponotus* (*Myrmotarsus*) *barbatus* Menozzi, 1926e: 94 (q.) PHILIPPINES. [Junior primary homonym of *barbatus* Roger, above.] Replacement name: *barbosus* Baroni Urbani, 1972: 362.

*barbatus. Camponotus fallax* var. *barbatus* Vashkevich, 1926: 118 (s.w.q.m.) RUSSIA. [Junior primary homonym of *barbatus* Roger, above.] Replacement name: *barbiger* Donisthorpe, 1941f: 38.

#***barbiculus.*** *Camponotus* (*Neomyrmamblys*) *fastigatus* st. *barbiculus* Santschi, 1922c: 113 (s.w.) BOLIVIA. Combination in *C.* (*Myrmaphaenus*): Emery, 1925b: 155.

***barbiger.*** *Camponotus barbiger* Donisthorpe, 1941f: 38. Replacement name for (and elevation to species of) *Camponotus fallax* var. *barbatus* Vashkevich, 1926: 118. [Junior primary homonym of *barbatus* Roger, 1863a: 138.] Current subgenus: *C.* (*Myrmentoma*).

***barbosus.*** *Camponotus barbosus* Baroni Urbani, 1972: 362. Replacement name for *barbatus* Menozzi, 1926e: 94. [Junior primary homonym of *barbatus* Roger, 1863a: 138.]

#***bardus.*** *Camponotus walkeri* subsp. *bardus* Forel, 1910b: 73 (s.) AUSTRALIA. Combination in *C.* (*Myrmophyma*): Emery, 1925b: 112.

#***baretoi.*** *Camponotus* (*Myrmobrachys*) *propinquus* var. *baretoi* Santschi, 1922c: 123 (w.) ARGENTINA.

***baronii.*** *Camponotus baronii* Alayo & Zayas Montero, 1977: 1, fig. 1 (w.) CUBA. Subgenus indeterminate.

#***barringtonensis.*** *Camponotus* (*Myrmamblys*) *macilentus* var. *barringtonensis* Wheeler, W.M. 1919c: 282 (w.q.) GALAPAGOS IS. Combination in *C.* (*Pseudocolobopsis*): Emery, 1925b: 158.

#***basalis.*** *Camponotus basalis* Smith, F. 1878b: 9, pl., fig. 1 (q.) INDIA. Bingham, 1903: 352 (s.w.). Combination in *C.* (*Tanaemyrmex*): Emery, 1925b: 101. Senior synonym of *lobinieri:* Bingham, 1903: 352. Race/subspecies of *maculatus:* Forel, 1904c: 29; Forel, 1906b: 84; of *sylvaticus:* Emery, 1920c: 7; Eidmann, 1942: 251.

#***basirectus.*** *Camponotus sexguttatus* subsp. *basirectus* Wheeler, W.M. 1923c: 7 (w.) BRAZIL. Combination in *C.* (*Myrmosphincta*): Kempf, 1972a: 59.

***basuto.*** *Camponotus* (*Tanaemyrmex*) *basuto* Arnold, 1958: 125, fig. 6 (s.w.) LESOTHO.

***batesii.*** *Camponotus batesii* Forel, 1895c: 243 (w.) MADAGASCAR. Combination in *C.* (*Dinomyrmex*): Forel, 1914a: 268; in *C.* (*Myrmosphincta*): Emery, 1920b: 259; in *C.* (*Myrmosaulus*): Wheeler, W.M. 1922a: 705.

#***batta.*** *Camponotus* (*Tanaemyrmex*) *variegatus* subsp. *batta* Menozzi, 1930a: 3, fig. 2 (s.w.) SUMATRA.

***bayeri.*** *Camponotus* (*Myrmotrema*) *bayeri* Forel, 1913h: 356 (s.) ZAIRE.

***baynei.*** *Camponotus maculatus* r. *baynei* Arnold, 1922: 632 (s.w.) SOUTH AFRICA. Combination in *C.* (*Tanaemyrmex*), and raised to species: Emery, 1925b: 92.

***beccarii.*** *Camponotus beccarii* Emery, 1887a: 218, pl. 3, fig. 7 (s.w.) SUMATRA. Combination in *C.* (*Myrmosphincta*): Forel, 1914a: 273; in *C.* (*Myrmoplatys*): Emery, 1920b: 256.

#***becki.*** *Camponotus* (*Myrmoturba*) *radamae* st. *becki* Santschi, 1923e: 292 (w.) MADAGASCAR. [First available use of *Camponotus* (*Myrmoturba*) *maculatus* r. *radamae* var. *becki* Forel, 1914d: 251; unavailable name.] Combination in *C.* (*Tanaemyrmex*): Emery, 1925b: 86. Subspecies of *hova:* current status.

***bedoti.*** *Camponotus bedoti* Emery, 1893e: 196, pl. 8. fig. 2 (w.) INDONESIA (Batjan I.). Viehmeyer, 1916a: 157 (s.); Karavaiev, 1929c: 240 (q.). Combination in *C.* (*Myrmotemnus*): Emery, 1920b: 258; in *C.* (*Myrmamblys*): Forel, 1914a: 271; Wheeler, W.M. 1922a: 988. Subspecies of *reticulatus:* Forel, 1901b: 27; Forel, 1909d: 229. Revived status as species: Viehmeyer, 1916a: 157; Wheeler, W.M. 1922a: 988. Subspecies of *reticulatus:* Emery, 1925b: 139; Stitz, 1934: 5. Revived status as species: Wang, C., Xiao & Wu, 1989a: 222 (in key).

***beebei.*** *Camponotus* (*Myrmobrachys*) *beebei* Wheeler, W.M. 1918b: 27 (w.) GUIANA.

#***belli.*** *Camponotus* (*Colobopsis*) *loa* var. *belli* Mann, 1919: 375 (s.w.) SOLOMON IS.

#***belliceps.*** *Camponotus* (*Myrmosericus*) *cinctellus* var. *belliceps* Santschi, 1939c: 10 (w.) ZAIRE.

#***bellicosus.*** *Camponotus intrepidus* r. *bellicosus* Forel, 1902h: 493 (s.w.) AUSTRALIA. Combination in *C.* (*Myrmosaulus*): Emery, 1925b: 114.

***belligerus.*** *Camponotus belligerum* Santschi, 1920d: 384, fig. 15 (s.w.) NAMIBIA. Combination in *C.* (*Myrmoturba*): Wheeler, W.M. 1922a: 949; in *C.* (*Myrmamblys*): Emery, 1925b: 140; in *C.* (*Myrmespera*): Santschi, 1926b: 248.

***bellus.*** *Camponotus bellus* Forel, 1908c: 404. Replacement name for *ornatus* Forel, 1907a: 36. [Junior primary homonym of *ornatus* Emery, 1894c: 173.] Forel, 1912n: 65 (q.); Viehmeyer, 1916a: 158 (w.m.). Combination in *C.* (*Myrmotemnus*): Emery, 1920b: 258; in *C.* (*Myrmosphincta*): Santschi, 1921f: 311; in *C.* (*Myrmamblys*): Forel, 1912a: 90; Emery, 1925b: 138. Current subspecies: nominal plus *adustus, leucodiscus.*

#***bendingensis.*** *Camponotus suffusus* var. *bendingensis* Forel, 1902h: 493 (w.) AUSTRALIA. Crawley, 1921: 96 (m.). Combination in *C.* (*Myrmosaulus*): Emery, 1925b: 114.

***benguelensis.*** *Camponotus foraminosus* st. *benguelensis* Santschi, 1911g: 213 (s.w.q.m.) ANGOLA. Combination in *C.* (*Myrmotrema*): Santschi, 1915c: 269. Subspecies of *grandidieri:* Emery, 1925b: 132. Raised to species: Santschi, 1937d: 244.

#***bequaerti.*** *Camponotus* (*Myrmamblys*) *confluens* var. *bequaerti* Forel, 1913b: 343 (s.w.q.) ZAIRE.

*beritchi. Camponotus* (*Myrmoturba*) *maculatus* r. *negus* var. *beritchi* Forel, 1915c: 348 (w.) SOUTH AFRICA. Unavailable name.

*berkeleyensis. Camponotus maculatus* r. *maccooki* var. *berkeleyensis* Forel, 1914c: 619 (w.) U.S.A. Unavailable name; material referred to *vicinus:* Snelling, 1970: 396.

***bermudezi.*** *Camponotus* (*Tanaemyrmex*) *bermudezi* Aguayo, 1932: 225 (w.) CUBA.

***berthoudi.*** *Camponotus berthoudi* Forel, 1879a: 108, pl. 1, fig. 6 (w.) SOUTH AFRICA. Arnold, 1924: 700. Combination in *C.* (*Myrmamblys*): Forel, 1914a: 271; in *C.* (*Myrmopiromis*): Emery, 1925b: 128.

***bertolonii.*** *Camponotus bertolonii* Emery, 1895h: 51 (s.w.q.) SOUTH AFRICA. Mayr, 1901b: 28 (m.). Combination in *C.* (*Myrmoturba*): Forel, 1914a: 267; Arnold, 1922: 651; in *C.* (*Myrmotemnus*): Emery, 1920b: 258; in *C.* (*Myrmamblys*): Santschi, 1914b: 134; Wheeler, W.M. 1922a: 989; Emery, 1925b: 141; in *C.* (*Paramyrmamblys*): Santschi, 1926c: 604.

*beta. Camponotus compressus* subsp. *thoracica* var. *beta* Santschi, 1907: 333 (q.) ALGERIA. Unavailable name.

***bianconii.*** *Camponotus bianconii* Emery, 1895h: 52 (s.) SOUTH AFRICA. Combination in *C.* (*Tanaemyrmex*): Emery, 1925b: 92.

*bicolor. Formica bicolor* Latreille, 1798: 43 (w.) FRANCE. [Unresolved junior primary homonym of *Formica bicolor* Fabricius, 1793: 351 (now in *Cataglyphis*).] Junior synonym of *lateralis:* Mayr, 1853b: 102.

*bicolor. Camponotus sayi* subsp. *bicolor* Pergande, 1894b: 161 (s.w.q.m.) MEXICO. [Junior secondary homonym of *Formica bicolor* Latreille, 1798, above.] Replacement name: *californicus* Emery, 1925b: 118.

#***bicontractus.*** *Camponotus* (*Orthonotomyrmex*) *arminius* var. *bicontractus* Wheeler, W.M. 1922a: 971 (w.) SOUTH AFRICA. [First available use of *Camponotus* (*Orthonotomyrmex*) *barbarossa* r. *arminius* var. *bicontracta* Forel, 1914d: 256; unavailable name.] Arnold, 1924: 721 (s.). Combination in *C.* (*Myrmamblys*): Emery, 1925b: 142; in *C.* (*Myrmopelta*): Santschi, 1926c: 603 (misspelled as *biconstrictus*).

***bidens.*** *Camponotus bidens* Mayr, 1870a: 387 (w.) COLOMBIA. Forel, 1899c: 145 (footnote) (s.);

Wheeler, W.M. 1934g: 225 (q.). Combination in *C. (Myrmeurynota)*: Forel, 1912i: 92; in *C. (Myrmorhachis)*: Forel, 1914a: 274; in *C. (Myrmocladoecus)*: Emery, 1925b: 166; Wheeler, W.M. 1934e: 424. Current subspecies: nominal plus *repressus*.

**bifossus.** *Camponotus (Colobopsis) bifossus* Santschi, 1917b: 293 (s.) SOUTH AFRICA.

**bigenus.** *Camponotus (Myrmocamelus) bigenus* Santschi, 1919a: 333 (s.w.m.) AUSTRALIA. Combination in *C. (Myrmophyma)*: Emery, 1925b: 110; Santschi, 1928e: 481.

#**biguttatus.** *Camponotus biguttatus* Emery, 1894c: 174 (s.w.) BOLIVIA. Combination in *C. (Myrmosphincta)*: Emery, 1925b: 152. Senior synonym of *grenadensis*: Emery, 1898a: 226; Emery, 1925b: 152. Variety of *sexguttatus*: Wheeler, W.M. 1923c: 8.

*bimaculata. Formica bimaculata* Smith, F. 1858b: 50 (w.) ANTILLES. Junior synonym of *ruficeps*: Roger, 1862c: 285. Revived from synonymy as variety of *sexguttatus*: Forel, 1907a: 34; Forel, 1912i: 87. Junior synonym of *sexguttatus*: Emery, 1925b: 152; Kempf, 1972a: 58.

*binaluana. Camponotus quadrisectus* subsp. *hians* var. *binaluana* Baroni Urbani, 1971b: 361. Unavailable name. [Unnecessary replacement name for *orthocephalus* Stitz; unavailable name.]

#**bindloensis.** *Camponotus (Myrmamblys) macilentus* var. *bindloensis* Wheeler, W.M. 1919c: 286 (s.w.) GALAPAGOS IS. Combination in *C. (Pseudocolobopsis)*: Emery, 1925b: 158.

**binghamii.** *Camponotus binghamii* Forel, 1894c: 398 (w.) BURMA. Forel, 1902d: 285 (s.). Combination in *C. (Myrmotemnus)*: Emery, 1920b: 258; in *C. (Myrmamblys)*: Forel, 1914a: 271; Emery, 1925b: 138.

#**binominatus.** *Camponotus (Myrmoturba) oblongus* var. *binominata* Forel, 1916: 458. Replacement name for *opaca* Forel, 1907a: 32. [Junior primary homonym of *opaca* Emery, 1899f: 290.] Combination in *C. (Tanaemyrmex)*: Emery, 1925b: 95.

**biolleyi.** *Camponotus biolleyi* Forel, 1902b: 177 (s.w.m.) COSTA RICA. Combination in *C. (Myrmobrachys)*: Forel, 1914a: 270.

*birmana. Cataglyphis birmana* Collingwood, 1962: 227, figs. 8, 9 (w.) BURMA. Junior synonym of *singularis*: Brown, 1967a: 277.

**bispinosus.** *Camponotus bispinosus* Mayr, 1870a: 388, pl., fig. 6 (w.) COLOMBIA. Combination in *C. (Myrmeurynota)*: Forel, 1912i: 92; in *C. (Myrmorhachis)*: Forel, 1914a: 274; in *C. (Myrmocladoecus)*: Emery, 1925b: 166; Wheeler, W.M. 1934e: 424.

**bituberculatus.** *Camponotus bituberculatus* André, 1889: 217 (w.) SENEGAL. André, 1890: 311 (q.). Combination in *C. (Myrmotrema)*: Forel, 1914a: 270.

**blandus.** *Formica blanda* Smith, F. 1858b: 43 (w.) BRAZIL. Combination in *Camponotus*: Roger, 1863b: 5; in *C. (Myrmocamelus)*: Forel, 1913l: 249; in *C. (Myrmaphaenus)*: Emery, 1920b: 256. Current subspecies: nominal plus *aequinotus, crispulus, denudatus, pellitus, pronotalis, rosariensis, scintillans*.

**bocki.** *Camponotus bocki* Forel, 1907e: 12 (s.w.q.m.) BOLIVIA. Combination in *C. (Camponotus)*: Forel, 1914a: 266; in *C. (Tanaemyrmex)*: Emery, 1925b: 78.

*boera. Camponotus (Myrmoturba) ballioni* var. *boera* Santschi, 1925h: 166 (s.w.q.) SOUTH AFRICA. [First available use of *Camponotus maculatus* subsp. *cognatus* var. *boera* Forel, 1910f: 27; unavailable name.] Junior synonym of *maculatus*: Baroni Urbani, 1972: 125.

**boghossiani.** *Camponotus boghossiani* Forel, 1911d: 357 (s.w.) GREECE. Combination in *C. (Orthonotomyrmex)*: Forel, 1914a: 273; in *C. (Myrmentoma)*: Emery, 1925b: 121. Current subspecies: nominal plus *stenoticus*.

#**boivini.** *Camponotus maculatus* r. *boivini* Forel, 1891b: 34 (w.q.) MADAGASCAR. Forel, 1891b: 213 (m.). Combination in *C. (Tanaemyrmex)*: Emery, 1925b: 86. Raised to species: Dalla Torre, 1893: 223. Subspecies of *maculatus*: Emery, 1895f: 337; Wheeler, W.M. 1922a: 1040. Subspecies of *hova*: Emery, 1920c: 6; Emery, 1925b: 86.

*boliviana. Camponotus (Tanaemyrmex) distinguendus* r. *boliviana* Stitz, 1933: 71, fig. 3 (s.) BOLIVIA. Junior synonym of *bruchi*: Kusnezov, 1953e: 225.

#**bombycinus.** *Camponotus (Myrmosericus) vestitus* st. *bombycinus* Santschi, 1930b: 77 (w.) ANGOLA.

**bonariensis.** *Camponotus bonariensis* Mayr, 1868b: 162 (w.) ARGENTINA. Forel, 1913l: 248 (q.m.); Goni *et al.* 1983: 366 (k.). Combination in *C. (Myrmoturba)*: Forel, 1913l: 248; in *C. (Tanaemyrmex)*: Emery, 1925b: 79. Race of *rubripes*: Emery, 1888c: 365; of *maculatus*: Forel, 1895b: 101; Forel, 1912i: 70. Revived status as species: Emery, 1894g: 373; Forel, 1914a: 267; Emery, 1920b: 233; Santschi, 1922d: 261. See also: Bruch, 1916: 352. Current subspecies: nominal plus *curtulus, garbei, luteolus, parvulus, tucumanus, weiseri*.

**borellii.** *Camponotus maculatus* subsp. *borellii* Emery, 1894e: 1 (s.w.q.) ARGENTINA. Combination in *C. (Myrmoturba)*: Forel, 1913l: 248; in *C. (Tanaemyrmex)*: Emery, 1925b: 79. Raised to species: Emery, 1920b: 233. Current subspecies: nominal plus *saltensis*.

#**borneensis.** *Camponotus gigas* var. *borneensis* Emery, 1887a: 214 (w.q.m.) BORNEO. Combination in *C. (Dinomyrmex)*: Emery, 1925b: 70. Race of *gigas*: Emery, 1893e: 195.

*bosphoricus. Camponotus maculatus* r. *samius* var. *bosphoricus* Forel, 1906c: 187 (w.) TURKEY. Unavailable name; material referred to *samius* by Atanassov & Dlussky, 1992: 219.

**bottegoi.** *Camponotus bottegoi* Emery, 1895a: 181 (s.w.q.) SOMALIA. Combination in *C. (Myrmoturba)*: Forel, 1914a: 267; in *C. (Myrmotrema)*: Santschi, 1914b: 136.

**brachycephalus.** *Camponotus (Colobopsis) brachycephalus* Santschi, 1920g: 182 (q.) VIETNAM.

#**bradleyi.** *Camponotus (Myrmocladoecus) callistus* subsp. *bradleyi* Wheeler, W.M. 1934g: 234 (w.) PERU.

**branneri.** *Dendromyrmex branneri* Mann, 1916: 488, pl. 6, fig. 47 (w.) BRAZIL. Kempf, 1976a: 107 (s.). Combination in *Camponotus*, subgenus indeterminate: Kempf, 1976a: 107.

**brasiliensis.** *Camponotus brasiliensis* Mayr, 1862: 671 (w.) BRAZIL. Kutter, 1932: 210 (q.). Combination in *C. (Myrmobrachys)*: Emery, 1925b: 163. Variety/race of *crassus*: Forel, 1885a: 346; Forel, 1912i: 78. Revived status as species: Santschi, 1919f: 56; Santschi, 1925d: 246. Current subspecies: nominal plus *antennatus* (unresolved junior homonym), *chacoensis, clivius*.

**braunsi.** *Camponotus braunsi* Mayr, 1895: 151 (w.) MOZAMBIQUE. Mayr, 1901b: 28 (footnote) (q.); Arnold, 1924: 715 (s.). Combination in *C. (Myrmentoma)*: Forel, 1913a: 146; in *C. (Orthonotomyrmex)*: Forel, 1914a: 273; Wheeler, W.M. 1922a: 971; in *C. (Myrmisolepis)*: Santschi, 1921f: 310; Emery, 1925b: 134. Current subspecies: nominal plus *candidus, epinotalis, erythromelus*.

**brettesi.** *Camponotus brettesi* Forel, 1899c: 141 (w.) GUATEMALA. Combination in *C. (Myrmobrachys)*: Forel, 1914a: 270. Current subspecies: nominal plus *canalis*.

#**brevibarbis.** *Camponotus (Myrmoturba) punctulatus* var. *brevibarbis* Santschi, 1939e: 325 (w.) BOLIVIA. Combination in *C. (Tanaemyrmex)*: Kempf, 1972a: 70.

**brevicollis.** *Camponotus (Dinomyrmex) brevicollis* Stitz, 1916: 401, fig. 13 (w.) CONGO. Combination in *C. (Tanaemyrmex)*: Emery, 1925b: 84.

**brevis.** *Camponotus brevis* Forel, 1899c: 140, pl. 4, fig. 20 (s.w.q.m.) PANAMA. Combination in *C. (Myrmobrachys)*: Forel, 1914a: 270. Current subspecies: nominal plus *obscurifrons, pauli*.

**brevisetosus.** *Camponotus fulvopilosus* subsp. *brevisetosus* Forel, 1910f: 25 (w.q.) SOUTH AFRICA. Combination in *C. (Myrmopiromis)*: Wheeler, W.M. 1922a: 986. Raised to species: Robertson, 1990: 327.

*brevitibialis. Camponotus (Myrmoturba) maculatus* st. *radamoides* var. *brevitibialis* Santschi, 1920g: 181 (s.w.) SOUTH AFRICA. Unavailable name; material referred to *guttatus* by Baroni Urbani, 1972: 132.

**britteni.** *Camponotus (Tanaemyrmex) britteni* Donisthorpe, 1931a: 129 (s.w.) GREAT BRITAIN.

***brodiei.** **Camponotus (Colobopsis) brodiei* Donisthorpe, 1920: 93, pl. 5, fig. 13 (s.) GREAT BRITAIN (Oligocene).

**brookei.** *Camponotus (Myrmamblys) brookei* Forel, 1914d: 259 (s.w.) SOUTH AFRICA. Arnold, 1924: 706 (s.q.). Combination in *C. (Paramyrmamblys)*: Santschi, 1926c: 604.

**bruchi.** *Camponotus bruchi* Forel, 1912i: 69 (s.w.q.m.) ARGENTINA. Combination in *C. (Myrmoturba)*: Forel, 1914a: 267; in *C. (Tanaemyrmex)*: Emery, 1925b: 78. Senior synonym of *boliviana*: Kusnezov, 1953e: 225. Current subspecies: nominal plus *lysistrata, titicacensis*.

*bruesi. Camponotus bruesi* Wheeler, W.M. 1910d: 349 (s.w.) U.S.A. Combination in *C. (Myrmamblys)*: Forel, 1914a: 271; in *C. (Myrmaphaenus)*: Emery, 1925b: 154. Junior synonym of *ulcerosus*: Creighton, 1951: 53.

#**brullei.** *Formica brullei* Smith, F. 1858b: 35. Replacement name for *Formica carinata* Brullé, 1840: 84. [Junior primary homonym of *Formica carinata* Fabricius, 1804: 413.] Combination in *Camponotus*: Roger, 1863b: 2. Subspecies of *compressus*: Santschi, 1939d: 82 (as *carinata*).

**bruneiensis.** *Camponotus (Myrmamblys) bruneiensis* Viehmeyer, 1922: 217 (w.) BORNEO.

*brunneigaster. Camponotus (Myrmoturba) punctulatus* st. *minutior* var. *brunneigaster* Santschi, 1929d: 313 (s.w.q.) BRAZIL. Unavailable name.

***CAMPONOTUS** (continued)*

*brunneiventris. Camponotus (Myrmothrix) cingulatus* var. *brunneiventris* Santschi, 1920e: 125 (w.) BOLIVIA. [Also described as new by Santschi, 1922c: 109.] Stirps of *cingulatus:* Santschi, 1925d: 245. Junior synonym of *cingulatus:* Hashmi, 1973: 31.

***brunni.** Camponotus marginatus* r. *brunni* Forel, 1901h: 70 (w.) JAPAN. Combination in *C. (Myrmentoma)*: Emery, 1925b: 118. Subspecies/variety of *fallax:* Forel, 1911e: 294; of *caryae:* Wheeler, W.M. 1917c: 29; of *marginatus:* Ruzsky, 1925b: 42. Raised to species: Ruzsky, 1926: 109.

***brutus.** Camponotus rubripes* r. *brutus* Forel, 1886f: 155 (s.w.q.) ANGOLA. Mayr, 1895: 150 (m.). Combination in *C. (Myrmoturba)*: Forel, 1913b: 340; in *C. (Tanaemyrmex)*: Emery, 1925b: 100. Subspecies of *maculatus:* Emery, 1896d: 371; Wheeler, W.M. 1922a: 236; of *solon:* Emery, 1925b: 100; Menozzi, 1942: 180. Raised to species: **new status**.

***bryani.** Camponotus (Condylomyrma) bryani* Santschi, 1928c: 72, fig. 2 (w.) FIJI IS. Combination in *C. (Colobopsis)*: Wheeler, W.M. 1934e: 423.

*bucephala. Formica bucephala* Roger, 1859: 228. Unavailable name; material referred to *cruentatus* by Roger, 1859: 228.

***buchholzi.** Camponotus buchholzi* Mayr, 1902: 297 (s.w.q.) CAMEROUN. Combination in *C. (Myrmamblys)*: Forel, 1914a: 271.

*buchneri* Forel, 1886; see under ***PHASMOMYRMEX.***

***buddhae.** Camponotus buddhae* Forel, 1892j: 238 (s.w.) INDIA. Ruzsky, 1905b: 191 (q.); Viehmeyer, 1916b: 290 (m.). Combination in *C.(Myrmoturba)*: Forel, 1914a: 267; in *C. (Tanaemyrmex)*: Emery, 1925b: 93.

***bugnioni.** Camponotus bugnioni* Forel, 1899c: 131 (footnote) (s.w.) COLOMBIA. Kutter, 1931a: 62 (q.); Santschi, 1936b: 196 (m.). Combination in *C. (Myrmothrix)*: Forel, 1912i: 91. See also: Hashmi, 1973: 24.

*bulimosus. Camponotus maculatus* subsp. *bulimosus* Wheeler, W.M. 1910d: 308 (s.w.q.m.) U.S.A. Combination in *C. (Myrmoturba)*: Forel, 1914a: 267; in *C. (Camponotus)*: Emery, 1925b: 75; in *C. (Tanaemyrmex)*: Creighton, 1950a: 379. Subspecies of *sansabeanus:* Emery, 1920b: 232 (footnote); Creighton, 1950a: 379. Junior synonym of *sansabeanus:* Wheeler, G.C. & Wheeler, J. 1986g: 14.

***burgeoni.** Camponotus (Myrmacrhaphe) burgeoni* Santschi, 1926c: 609, fig. 3 (w.) ZAIRE.

***burtoni.** Camponotus (Myrmamblys) burtoni* Mann, 1916: 479, pl. 6, figs. 45, 46 (s.) BRAZIL. Combination in *C. (Hypercolobopsis)*: Emery, 1920b: 260; in *C. (Myrmobrachys)*: Emery, 1925b: 166.

***butteli.** Camponotus butteli* Forel, 1905b: 164 (w.) MADAGASCAR. Combination in *C. (Myrmocamelus)*: Forel, 1914a: 270; in *C. (Tanaemyrmex)*: Emery, 1925b: 104.

#***buttesi.** Camponotus contractus* var. *buttesi* Forel, 1902g: 463 (s.w.) WEST MALAYSIA. Combination in *C. (Myrmoplatys)*: Emery, 1925b: 136.

***buttikeri.** Camponotus (Tanaemyrmex) buttikeri* Arnold, 1958: 126, fig. 7 (s.w.) SOUTH AFRICA.

*buxtoni. Camponotus (Colobopsis) buxtoni* Santschi, 1928a: 56, fig. 9 (w.) SAMOA. Junior synonym of *flavolimbatus:* Wilson & Taylor, 1967: 97.

*cabrerai. Camponotus alii* st. *hesperius* var. *cabrerai* Santschi, 1939d: 84 (w.) CANARY IS. Unavailable name.

#***cachmiriensis.** Camponotus (Tanaemyrmex) aethiops* var. *cachmiriensis* Emery, 1925b: 97 (s.w.m.) INDIA. [First available use of *Camponotus maculatus* r. *aethiops* var. *cachmiriensis* Forel, 1904c: 29; unavailable name.]

***cacicus.** Camponotus cacicus* Emery, 1903: 76 (s.w.) PERU. Combination in *C. (Myrmogigas)*: Forel, 1912i: 91; in *C. (Dinomyrmex)*: Forel, 1914a: 268; in *C (Tanaemyrmex)*: Emery, 1925b: 84.

*caeciliae. Camponotus maculatus* r. *caeciliae* Forel, 1906c: 188 (s.w.) TURKEY. Junior synonym of *festai:* Emery, 1908a: 187.

***caesar.** Camponotus caesar* Forel, 1886f: 162 (w.) ANGOLA. Emery, 1899e: 503 (s.q.). Combination in *C. (Myrmogigas)*: Forel, 1912i: 91; in *C. (Dinomyrmex)*: Forel, 1914a: 268; in *C. (Myrmoxygenys)*: Emery, 1925b: 70. Current subspecies: nominal plus *imperator.*

***caffer.** Camponotus caffer* Emery, 1895h: 47, pl. 2, figs. 25, 26 (s.w.) SOUTH AFRICA. Combination in *C. (Myrmoturba)*: Forel, 1914a: 267; in *C. (Dinomyrmex)*: Emery, 1920b: 256; in *C. (Tanaemyrmex)*: Emery, 1925b: 84. Race of *maculatus:* Forel, 1901h: 69. Revived status as species: Forel, 1914a: 267; Emery, 1920b: 256; Arnold, 1922: 650; Wheeler, W.M. 1922a: 963.

*cajurensis. Camponotus (Myrmothrix) rufipes* forma *cajurensis* Luederwaldt, 1918: 53 (w.) BRAZIL. Junior synonym of *rufipes:* Hashmi, 1973: 52.

*calceatus. Camponotus (Myrmoturba) maculatus* st. *radamoides* var. *calceatus* Santschi, 1930b: 76 (s.w.) ANGOLA. Unavailable name; material referred to *maculatus* by Baroni Urbani, 1972: 125.

*californica. Camponotus (Myrmentoma) sayi* subsp. *californica* Emery, 1925b: 118. Replacement name for *Camponotus sayi* subsp. *bicolor* Pergande, 1894b: 161. [Junior secondary homonym of *Formica bicolor* Latreille, 1798: 43.] Junior synonym of *sayi:* Snelling, 1968b: 355; of *hyatti:* Snelling, 1988: 69.

*callida. Formica callida* Smith, F. 1858b: 18 (w.) INDIA. Combination in *Camponotus:* Roger, 1863b: 2. Junior synonym of *ardens:* Mayr, 1886c: 353; of *angusticollis:* Dalla Torre, 1893: 221; Emery, 1896d: 370; Bingham, 1903: 366; Wheeler, W.M. 1919e: 106; of *compressus:* Roger, 1863b: 2; Emery, 1925b: 98.

***callistus.** Camponotus callistus* Emery, 1911a: 223 (s.w.m.) BOLIVIA. Combination in *C. (Myrmorhachis)*: Forel, 1914a: 274; in *C. (Myrmocladoecus)*: Emery, 1925b: 166; Wheeler, W.M. 1934e: 424. Current subspecies: nominal plus *bradleyi.*

***callmorphus.** Camponotus (Myrmosaga) callmorphus* Stitz, 1923: 166 (w.) NAMIBIA.

#***caloratus.** Camponotus (Colobopsis) gasseri* subsp. *caloratus* Wheeler, W.M. 1934d: 162 (s.w.q.m.) AUSTRALIA.

***calvus.** Camponotus (Colobopsis) calvus* Emery, 1920c: 32 (q.) WEST MALAYSIA.

***cambouei.** Camponotus cambouei* Forel, 1891b: 68, pl. 1, fig. 13; pl. 2, fig. 1 (s.w.) MADAGASCAR. Combination in *C. (Camponotus)*: Forel, 1914a: 266; in *C. (Myrmosaga)*: Emery, 1920b: 257; in *C. (Mayria)*: Emery, 1925b: 122.

***camelinus.** Formica camelina* Smith, F. 1857a: 57 (w.) SINGAPORE. Bingham, 1903: 373 (s.q.). Combination in *Camponotus:* Roger, 1863b: 3; in *C. (Myrmosphincta)*: Forel, 1912i: 92; Emery, 1920b: 256; in *C. (Myrmocamelus)*: Forel, 1914a: 270; in *C. (Myrmosaulus)*: Wheeler, W.M. 1922a: 705. Variety of *singularis:* Emery, 1887a: 210; of *cinerascens:* Dalla Torre, 1893: 224. Revived status as species: Emery, 1895f: 355; Bingham, 1903: 373; Forel, 1912i: 92; Wheeler, 1919e: 112. Senior synonym of *senilis:* Forel, 1892j: 232.

***camelus.** Camponotus camelus* Emery, 1883: 146, fig. (w.) NEW CALEDONIA. Combination in *C. (Myrmosphincta)*: Forel, 1914a: 273; in *C. (Myrmotemnus)*: Emery, 1920b: 258; in *C. (Colobopsis)*: Emery, 1925b: 149

***cameranoi.** Camponotus senex* subsp. *cameranoi* Emery, 1894e: 3 (s.w.) ARGENTINA. Forel, 1909a: 264 (q.). Combination in *C. (Myrmobrachys)*: Forel, 1914a: 270. Raised to species: Forel, 1914a: 270; Bruch, 1914: 228. Current subspecies: nominal plus *eugaster, inlex.*

***cameratus.** Camponotus (Myrmogonia) cameratus* Viehmeyer, 1925b: 146 (s.w.) AUSTRALIA.

***cameroni.** Camponotus cameroni* Forel, 1892f: 256. Replacement name for *mayri* Cameron, 1891: 89. [Junior primary homonym of *mayri* Forel, 1879a: 92.] Combination in *C. (Myrmobrachys)*: Forel, 1914a: 270. Subspecies of *senex:* Emery, 1896d: 377. Revived status as species: Forel, 1914a: 270.

#***canalis.** Camponotus brettesi* var. *canalis* Forel, 1912i: 79 (w.) PANAMA. Combination in *C. (Myrmobrachys)*: Emery, 1925b: 163.

#***candidus.** Camponotus (Myrmisolepis) braunsi* var. *candidus* Santschi, 1926c: 618 (w.) TANZANIA.

***candiotes.** Camponotus lateralis* var. *candiotes* Emery, 1894j: 10 (w.) GREECE. Combination in *C. (Myrmentoma)*: Emery, 1925a: 68; Emery, 1925b: 120. Variety of *piceus:* Emery, 1925a: 68; Menozzi, 1936d: 304. Raised to species: Agosti & Collingwood, 1987a: 58.

***canescens.** Camponotus canescens* Mayr, 1870a: 386 (w.) COLOMBIA. Menozzi, 1935b: 203 (q.); Wheeler, G.C. & Wheeler, J. 1953e: 194 (l.). Combination in *C. (Myrmobrachys)*: Forel, 1914a: 270. Current subspecies: nominal plus *antennatus, stomatus.*

***capito.** Camponotus capito* Mayr, 1876: 64 (s.w.) AUSTRALIA. Forel, 1915b: 99 (w.); Wheeler, G.C. & Wheeler, J. 1970: 650 (l.). Combination in *C. (Myrmophyma)*: Forel, 1912i: 91. Provisional junior synonym of *cinereus:* Santschi, 1919a: 330. Retained as species: Taylor & Brown, D.R. 1985: 110. Current subspecies: nominal plus *ebeninithorax.*

***capperi.** Camponotus capperi* Forel, 1899c: 138 (s.w.) JAMAICA. Combination in *C. (Myrmothrix)*: Forel, 1914a: 268; in *C. (Myrmobrachys)*: Wheeler, W.M. 1917g: 468; in *C. (Myrmamblys)*: Emery, 1920b: 259; in *C. (Myrmaphaenus)*: Emery, 1925b: 156. Current subspecies: nominal plus *corticalis, formosulus, subdepilis, unctulus.*

***caracalla.** Camponotus caracalla* Forel, 1912i: 80 (w.) BRAZIL. Combination in *C. (Myrmamblys)*: Forel,

1914a: 271; in *C.* (*Myrmobrachys*): Emery, 1925b: 165.

*carazzii* Emery, 1895; see under ***NOTOSTIGMA.***

***carbo.*** *Camponotus carbo* Emery, 1877b: 364, fig. (w.) ETHIOPIA. Combination in *C.* (*Myrmotrema*): Forel, 1913b: 341. Current subspecies: nominal plus *occidentalis.*

***carbonarius.*** *Formica carbonaria* Latreille, 1802c: 114, pl. 3, fig. 8 (w.) "GRAND-INDES". Combination in *Camponotus:* Roger, 1863b: 2. Subgenus indeterminate: Emery, 1925b: 171.

#***cardini.*** *Camponotus* (*Colobopsis*) *sphaericus* var. *cardini* Mann, 1920: 438 (w.) CUBA. Combination in *C.* (*Manniella*): Emery, 1925b: 160.

#***carensis.*** *Camponotus* (*Myrmoturba*) *irritans* subsp. *carensis* Emery, 1920c: 16, fig. 3 (w.) NICOBAR IS. Combination in *C.* (*Tanaemyrmex*): Emery, 1925b: 94.

***carin.*** *Camponotus dorycus* r. *carin* Emery, 1889b: 512 (s.w.) BURMA. Combination in *C.* (*Dinomyrmex*): Forel, 1915a: 37; in *C.* (*Tanaemyrmex*): Emery, 1925b: 90. Raised to species: Dalla Torre, 1893: 223; Bingham, 1903: 365. Subspecies of *dorycus:* Forel, 1915a: 37. Raised to species: Emery, 1925b: 90; Wheeler, W.M. 1927d: 9; Wheeler, W.M. 1928c: 12; Wheeler, W.M. 1930a: 105; Wang, C., Xiao & Wu, 1989a: 223. Current subspecies: nominal plus *babiensis, tenuisquamis, tipunus.*

*carinata. Formica carinata* Brullé, 1840: 84 (w.) CANARY IS. [Junior primary homonym of *Formica carinata* Fabricius, 1804: 413 (now in *Polyrhachis*).] Replacement name: *Formica brullei* Smith, F. 1858b: 35. [Note: the replacement name, although proposed long ago, appears to have been mostly ignored by later authors.] Combination in *Camponotus:* Roger, 1863b: 2. Junior synonym of *maculatus:* Roger, 1863b: 2; Mayr, 1865: 27. Variety of *cognatus:* Dalla Torre, 1893: 225. Subspecies of *maculatus:* Emery, 1893c: 85; Emery, 1895h: 46. Stirps of *compressus:* Santschi, 1939d: 82.

*carinata. Colobopsis carinata* Mayr, 1870b: 943 (s.) FIJI IS. Combination in *Camponotus* (*Colobopsis*): Emery, 1889b: 517; in *C.* (*Myrmamblys*): Forel, 1914a: 271; in *C.* (*Myrmotemnus*): Emery, 1920b: 258. [Junior secondary homonym of *Formica carinata* Brullé, above.] Replacement name: *polynesicus* Emery, 1896d: 374.

#***carinatus.*** *Camponotus* (*Colobopsis*) *vitreus* r. *carinata* Stitz, 1938: 116 (q.) NEW GUINEA. [Unresolved junior secondary homonym of *Formica carinata* Brullé, above. Note that the proposed replacement name, *neocarinata* Baroni Urbani, 1971b: 361, is unavailable.]

#***carinifer.*** *Camponotus* (*Myrmoturba*) *tinctus* var. *carinifer* Viehmeyer, 1916a: 156 (s.q.) SINGAPORE. Subspecies of *irritans:* current status.

*carolinus. Camponotus* (*Myrmoturba*) *maculatus* r. *carolinus* Forel, 1916: 458. Subspecies of *bonariensis:* Emery, 1925b: 79; Kempf, 1972a: 65. Unnecessarily proposed replacement name for *luteolus* Emery, 1906c: 188, hence junior synonym of the latter: **new synonymy**.

***caryae.*** *Formica caryae* Fitch, 1855: 855 (w.q.m.) U.S.A. Combination in *Camponotus:* Roger, 1863b: 6; in *C.* (*Myrmentoma*): Emery, 1920b: 257. Junior synonym of *pennsylvanicus:* Forel, 1879a: 56; Dalla Torre, 1893: 247. Revived from synonymy: Wheeler, W.M. 1917c: 28. Material of the unavailable name *cnemidatus* referred here: Smith, M.R. 1940b: 139. See also: Creighton, 1950a: 385; Snelling, 1988: 60. Current subspecies: nominal plus *himalayanus, kamensis.*

***casicus.*** *Camponotus* (*Dinomyrmex*) *casica* Santschi, 1920g: 180 (attributed to Emery) (q.) FRENCH GUIANA. Subgeneric placement: probably *Tanaemyrmex.*

#***cassius.*** *Camponotus* (*Dinomyrmex*) *pompeius* subsp. *cassius* Wheeler, W.M. 1922a: 239, figs. 60, 61 (s.w.) ZAIRE. Combination in *C.* (*Tanaemyrmex*): Emery, 1925b: 89.

***castaneus.*** *Formica castanea* Latreille, 1802c: 118, pl. 3. figs. 11, 12 (w.q.m.) U.S.A. Wheeler, W.M. 1910d: 321 (s.). Combination in *Camponotus:* Roger, 1863b: 6; in *C.* (*Camponotus*): Forel, 1914a: 266; in *C.* (*Tanaemyrmex*): Creighton, 1950a: 375. Senior synonym of *mellea* (and its junior synonym *clarus*): Mayr, 1886d: 420; Forel, 1886f: 141.

*castanea. Formica castanea* Lepeletier de Saint-Fargeau, 1835: 215 (w.q.m.) U.S.A. [Unresolved junior primary homonym of *castanea* Latreille, above.] Junior synonym of *herculeanus:* Mayr, 1863: 399.

*castanea. Camponotus* (*Myrmacantha*) *castanea* Stitz, 1925: 128, fig. 7 (w.) PHILIPPINES. [Junior secondary homonym of *Formica castanea* Latreille, above.] Replacement name: *castanicola* Donisthorpe, 1943e: 116.

***castanicola.*** *Camponotus* (*Myrmacantha*) *castanicolus* Donisthorpe, 1943e: 116. Replacement name for *castanea* Stitz, 1925: 128. [Junior secondary homonym of *castanea* Latreille, 1802c: 118.] Subgenus indeterminate.

*castanipes. Formica castanipes* Leach, 1825: 290 (w.q.m.) FRANCE. Junior synonym of *sylvaticus:* Roger, 1863b: 1.

#***castellanus.*** *Camponotus* (*Pseudocolobopsis*) *macilentus* var. *castanellus* Wheeler, W.M. 1924a: 116, fig. 25 (s.w.q.m.) GALAPAGOS IS.

***catalanus.*** *Camponotus* (*Myrmoturba*) *sylvaticus* var. *catalana* Emery, 1924c: 170 (w.) SPAIN. Raised to species: Collingwood & Yarrow, 1969: 83.

*cataractae. Camponotus* (*Myrmoturba*) *maculatus* st. *radamoides* var. *cataractae* Santschi, 1919b: 239 (s.w.) ZIMBABWE. Unavailable name; material referred to *maculatus* by Baroni Urbani, 1972: 125.

#***catharinae.*** *Camponotus melanoticus* var. *catharinae* Santschi, 1939e: 323 (s.w.) BRAZIL. Combination in *C.* (*Tanaemyrmex*): Kempf, 1972a: 69.

#***cato.*** *Camponotus* (*Orthonotomyrmex*) *meinerti* r. *cato* Forel, 1913b: 346 (s.w.) ZAIRE. Combination in *C.* (*Myrmamblys*): Emery, 1925b: 142; in *C.* (*Myrmopelta*): Santschi, 1926a: 18. Subspecies of *vividus:* Wheeler, W.M. 1922a: 248; Santschi, 1926a: 18. Senior synonym of *laevithorax:* Menozzi, 1933a: 113.

*caucasicus. Camponotus herculeanus* subsp. *caucasicus* Arnol'di, 1967: 1822 (s.w.q.m.) CAUCASUS. Junior synonym of *herculeanus:* Arakelian, 1994: 85.

*cavallus. Camponotus maculatus* st. *cavallus* Santschi, 1911g: 211 (s.w.q.) ANGOLA. Combination in *C.* (*Tanaemyrmex*): Emery, 1925b: 88. Junior synonym of *maculatus:* Baroni Urbani, 1972: 125.

#***cavisquamis.*** *Camponotus* (*Paramyrmamblys*) *ferreri* st. *cavisquamis* Santschi, 1926c: 604 (q.) KENYA.

***cecconii.*** *Camponotus maculatus* subsp. *cecconii* Emery, 1908a: 198 (s.w.) CYPRUS. Combination in *C.* (*Tanaemyrmex*): Emery, 1925b: 101. Subspecies of *sylvaticus:* Emery, 1920c: 7. Raised to species: Agosti & Collingwood, 1987a: 58.

***cerberulus.*** *Camponotus* (*Colobopsis*) *cerberulus* Emery, 1920c: 34 (q.) MEXICO. Wheeler, W.M. 1934g: 214 (s.m.).

***ceriseipes.*** *Camponotus* (*Myrmophyma*) *ceriseipes* Clark, 1938: 378, figs. 17–19 (s.w.) AUSTRALIA. Wheeler, G.C. & Wheeler, J. 1974a: 60 (l.).

***cervicalis.*** *Camponotus cervicalis* Roger, 1863a: 134 (w.) MADAGASCAR. Forel, 1891b: 19 (s.w.q.m.). Combination in *C.* (*Dinomyrmex*): Forel, 1914a: 268; in *C.* (*Tanaemyrmex*): Emery, 1925b: 84. Current subspecies: nominal plus *gaullei.*

#***cetegus.*** *Camponotus festinus* subsp. *cetegus* Forel, 1911a: 50 (s.) SUMATRA. Combination in *C.* (*Tanaemyrmex*): Emery, 1925b: 90.

***ceylonicus.*** *Camponotus* (*Colobopsis*) *ceylonicus* Emery, 1925b: 145. Replacement name for *testacea* Bingham, 1903: 347. [Junior secondary homonym of *testaceus* Emery, 1894f: 3.]

*chaboti. Camponotus* (*Myrmosericus*) *rufoglaucus* st. *controversus* var. *chaboti* Santschi, 1925h: 167 (s.w.q.) ANGOLA. Unavailable name.

*chacoanus. Camponotus* (*Myrmaphaenus*) *blandus* st. *rosariensis* var. *chacoanus* Santschi, 1929d: 314 (w.) ARGENTINA. Unavailable name.

#***chacoensis.*** *Camponotus* (*Myrmobrachys*) *brasiliensis* var. *chacoensis* Santschi, 1919f: 56 (s.w.) ARGENTINA.

*chalaconotus. Camponotus* (*Orthonotomyrmex*) *chalaconotus* Menozzi, 1929c: 430, figs. 1, 2 (s.) INDIA. Junior synonym of *megalonyx:* Wheeler, W.M. 1936e: 217.

***chalceoides.*** *Camponotus* (*Myrmophyma*) *chalceoides* Clark, 1938: 376, figs. 14–16 (s.w.) AUSTRALIA.

***chalceus.*** *Camponotus* (*Myrmosaga*) *chalceus* Crawley, 1915b: 236 (s.w.q.) AUSTRALIA. Combination in *C.* (*Myrmophyma*): Emery, 1920b: 257; Santschi, 1928e: 482.

***championi.*** *Camponotus championi* Forel, 1899c: 157, pl. 4, figs. 21, 22 (s.w.) MEXICO. Combination in *C.* (*Myrmamblys*): Forel, 1914a: 271; in *C.* (*Myrmobrachys*): Emery, 1925b: 165; in *C.* (*Manniella*): Wheeler, W.M. 1934g: 210; in *C.* (*Myrmaphaenus*): Creighton, 1951: 57.

***chapini.*** *Camponotus* (*Myrmamblys*) *chapini* Wheeler, W.M. 1922a: 254, fig. 68 (s.w.) ZAIRE. Combination in *C.* (*Myrmoturba*): Santschi, 1926c: 600; in *C.* (*Myrmamblys*): Weber, 1943c: 386. Variety of *agonius:* Santschi, 1926b: 252; Santschi, 1926c: 600. Revived status as species: Weber, 1943c: 386. Current subspecies: nominal plus *ganzii.*

***chazaliei.*** *Camponotus novogranadensis* r. *chazaliei* Forel, 1899c: 149 (s.w.) ANTILLES. Combination in *C.* (*Myrmamblys*): Forel, 1912i: 90; in *C.* (*Myrmaphaenus*): Emery, 1925b: 155. Raised to species: Forel, 1902b: 172; Emery, 1903: 74.

***cheesmanae.*** *Camponotus* (*Myrmobrachys*) *cheesmanae* Donisthorpe, 1932a: 88 (s.w.) COLOMBIA.

***chilensis.*** *Formica chilensis* Spinola, 1851a: 237 (w.q.) CHILE. Combination in *Camponotus:* Roger, 1863b:

***CAMPONOTUS** (continued)*

5; in *C.* (*Myrmosericus*): Forel, 1914a: 268; in *C.* (*Myrmamblys*): Bruch, 1914: 228; in *C.* (*Myrmoturba*): Emery, 1920b: 255; in *C.* (*Tanaemyrmex*): Emery, 1925b: 78. See also: Snelling & Hunt, 1976: 117.

#***chimporensis.*** *Camponotus* (*Orthonotomyrmex*) *mayri* var. *chimporensis* Santschi, 1930b: 80 (w.) ANGOLA.

#***chiton.*** *Camponotus* (*Tanaemyrmex*) *solon* var. *chiton* Emery, 1925b: 100 (s.w.q.) SENEGAL. [First available use of *Camponotus maculatus* r. *solon* var. *chilon* Forel, 1901h: 68; unavailable name.]

*chlorogaster. Camponotus* (*Myrmoturba*) *irritans* subsp. *chlorogaster* Emery, 1920c: 7 (s.w.) VANUATU. [First available use of *Camponotus* (*Myrmoturba*) *maculatus* subsp. *chlorotica* var. *chlorogaster* Emery, 1914f: 429; unavailable name.] Combination in *C.* (*Tanaemyrmex*): Emery, 1925b: 94. Junior synonym of *chloroticus:* Wilson & Taylor, 1967: 93.

***chloroticus.*** *Camponotus maculatus* subsp. *chloroticus* Emery, 1897d: 574 (w.) NEW GUINEA. Emery, 1914f: 424 (m.). Combination in *C.* (*Myrmoturba*): Emery, 1914f: 424; in *C.* (*Tanaemyrmex*): Emery, 1925b: 94. Subspecies of *irritans:* Emery, 1920c: 7; Karavaiev, 1933a: 316. Raised to species, senior synonym of *chlorogaster* and *sanctaecrucis,* and material of the unavailable name *samoaensis* referred here: Wilson & Taylor, 1967: 93.

#***cholericus.*** *Camponotus andrei* var. *cholericus* Wheeler, W.M. 1914b: 59 (s.w.m.) MEXICO. Combination in *C.* (*Myrmaphaenus*): Emery, 1925b: 154.

***chongqingensis.*** *Camponotus chongqingensis* Wu & Wang; in Wang, C., Xaio & Wu, 1989a: 226, figs. 15, 16 (s.w.) CHINA.

***christi.*** *Camponotus christi* Forel, 1886f: 184 (w.) MADAGASCAR. Combination in *C.* (*Myrmocamelus*): Forel, 1914a: 270; in *C.* (*Myrmosaga*): Emery, 1920b: 257; in *C.* (*Mayria*): Emery, 1925b: 122. Current subspecies: nominal plus *ambustus, ferrugineus* (unresolved homonym), *foersteri, maculiventris.*

#***christoides.*** *Camponotus lubbocki* r. *christoides* Forel, 1891b: 61 (w.) MADAGASCAR. Combination in *C.* (*Mayria*): Emery, 1925b: 123. Raised to species: Dalla Torre, 1893: 224. Subspecies of *lubbocki:* Emery, 1896d: 772; Wheeler, W.M. 1922a: 1046.

***christophei.*** *Camponotus christophei* Wheeler, W.M. & Mann, 1914: 57, figs. 25, 26 (s.) HAITI. Combination in *C.* (*Myrmeurynota*): Forel, 1914a: 274.

***christopherseni.*** *Camponotus* (*Colobopsis*) *christopherseni* Forel, 1912i: 85 (s.w.q.) PANAMA. Combination in *C.* (*Hypercolobopsis*): Emery, 1925b: 160; in *C.* (*Myrmeurynota*), and possible junior synonym of *linnaei:* Kempf, 1960f: 464.

***chromaiodes.*** *Camponotus chromaiodes* **new name**. Replacement name for *Formica ferruginea* Fabricius, 1798: 279. [Junior primary homonym of *Formica ferruginea* Christ, 1791: 512.] [For earlier history of the taxon see under *ferruginea* Fabricius.]

#***chrysogaster.*** *Camponotus foraminosus* subsp. *chrysogaster* Emery, 1895a: 182 (s.) ETHIOPIA. Forel, 1907g: 90 (q.m.). Combination in *C.* (*Myrmotrema*): Santschi, 1915c: 277. Raised to species: Emery, 1899a: 501. Subspecies of *foraminosus:* Forel, 1907g: 90; Wheeler, W.M. 1922a: 979.

*chrysothrix. Camponotus* (*Myrmobrachys*) *crassus* st. *amazonicus* var. *chrysothrix* Santschi, 1922c: 123 (w.) BRAZIL. Unavailable name.

***chrysurus.*** *Camponotus chrysurus* Gerstäcker, 1871: 355 (w.) KENYA. Forel, 1899b: 308 (s.); Santschi, 1914b: 135 (q.); Forel, 1915c: 350 (q.). Combination in *C.* (*Myrmentoma*): Forel, 1912i: 92; in *C.* (*Orthonotomyrmex*): Forel, 1913b: 346; Wheeler, W.M. 1922a: 971; in *C.* (*Myrmopiromis*): Emery, 1925b: 129; in *C.* (*Myrmopelta*): Santschi, 1921f: 311; Santschi, 1926a: 20; Santschi, 1926c: 603. Current subspecies: nominal plus *acutisquamis, apellis, securifer, yvonnae.*

#***chubutensis.*** *Camponotus* (*Myrmoturba*) *punctulatus* var. *chubutensis* Forel, 1913l: 248 (s.). ARGENTINA. Combination in *C.* (*Tanaemyrmex*): Emery, 1925b: 78.

***cilicicus.*** *Camponotus cilicicus* Emery, 1908a: 186 (w.) TURKEY. Combination in *C.* (*Camponotus*): Forel, 1914a: 266; Emery, 1925b: 72.

***cillae.*** *Camponotus cillae* Forel, 1912i: 74 (s.w.) BRAZIL. Combination in *C.* (*Myrmoturba*): Forel, 1914a: 267; in *C.* (*Tanaemyrmex*): Emery, 1925b: 79.

***cinctellus.*** *Formica cinctella* Gerstäcker, 1859: 262 (w.q.m.) MOZAMBIQUE. [Also described as new by Gerstäcker, 1862: 510.] Arnold, 1924: 689 (s.). Combination in *Camponotus:* Roger, 1863b: 2; in *C.* (*Myrmosericus*): Forel, 1913b: 340. Subspecies/race of *rufoglaucus:* Mayr, 1893: 195; Emery, 1895a: 180; Forel, 1909b: 67; Forel, 1910c: 271; Forel, 1913h: 356; Wheeler, W.M. 1922a: 966; Arnold, 1924: 689; Emery, 1925b: 105; Menozzi, 1939c: 107. Revived status as species: Santschi, 1939c: 10. Senior synonym of *venustus:* Emery, 1895h: 48; Wheeler, W.M. 1922a: 967; Emery, 1925b: 106. Current subspecies: nominal plus *belliceps.*

***cinerascens.*** *Formica cinerascens* Fabricius, 1787: 308 (q.) INDIA. Combination in *Camponotus:* Roger, 1963b: 3; in *C.* (*Myrmosphincta*): Forel, 1912i: 92; in *C.* (*Myrmosaulus*): Wheeler, W.M. 1922a: 705; in *C.* (*Myrmosericus*): Emery, 1925b: 105; Donisthorpe, 1932c: 445 (see note under *Myrmosaulus* in catalogue of genus-group taxa). Junior synonym of *rufoglaucus:* Roger, 1863b: 3; Forel, 1892j: 237; Emery, 1925b: 105 (questionable synonym). Status as species: Dalla Torre, 1893: 224; Emery, 1895k: 479; Emery, 1896d: 373; Wheeler, W.M. 1922a: 705; Donisthorpe, 1932c: 444.

***cinereus.*** *Camponotus cinereus* Mayr, 1876: 62 (w.) AUSTRALIA. Combination in *C.* (*Myrmocamelus*): Forel, 1913g: 192; Forel, 1914a: 270; in *C.* (*Myrmophyma*): Emery, 1920b: 192; Santschi, 1928e: 481. Current subspecies: nominal plus *amperei, notterae.*

***cingulatus.*** *Camponotus cingulatus* Mayr, 1862: 661 (w.) BRAZIL. Forel, 1879a: 73 (s.); Hashmi, 1973: 41 (q.). Combination in *C.* (*Myrmothrix*): Forel, 1914a: 268. Subspecies of *maculatus:* Bruch, 1914: 229 (in error). Revived status as species: Santschi, 1922c: 109. Senior synonym of *bambusarum, brunneiventris, damocles, fissa* and *myster,* and material of the unavailable name *postniger* referred here: Hashmi, 1973: 31.

***circularis.*** *Camponotus circularis* Mayr, 1870a: 380, pl., fig. 2 (w.) COLOMBIA. Combination in *C.* (*Myrmorhachis*): Emery, 1920b: 260; in *C.* (*Myrmobrachys*): Forel, 1914a: 271; Emery, 1920c: 44. Current subspecies: nominal plus *rufitibia.*

***circumspectus.*** *Formica circumspecta* Smith, F. 1861b: 37 (w.q.) SULAWESI. Combination in *Camponotus:* Roger, 1863b: 4; in *C.* (*Tanaemyrmex*): Donisthorpe, 1932c: 465.

*citinus. Camponotus* (*Myrmoturba*) *maculatus* st. *radamoides* var. *citinus* Santschi, 1930b: 77 (footnote) (s.w.) ZAIRE. Unavailable name; material referred to *maculatus* by Baroni Urbani, 1972: 125.

*clarigaster. Camponotus acutirostris* var. *clarigaster* Wheeler, W.M. 1915b: 420 (s.) U.S.A. Junior synonym of *acutirostris:* Creighton, 1950a: 375.

#***clarior.*** *Camponotus nigriceps* var. *clarior* Forel, 1902h: 506 (s.w.) AUSTRALIA. Combination in *C.* (*Tanaemyrmex*): Emery, 1925b: 103. Junior synonym of *pallidiceps:* Wheeler, W.M. 1933b: 23. Revived from synonymy as subspecies of *nigriceps:* Clark, 1934c: 71; Taylor & Brown, D.R. 1985: 117.

***claripes.*** *Camponotus claripes* Mayr, 1876: 64 (s.w.) AUSTRALIA. Forel, 1902h: 496 (q.). Combination in *C.* (*Myrmophyma*): Forel, 1914a: 269; in *C.* (*Thlipsepinotus*): Santschi, 1928e: 483. Current subspecies: nominal plus *elegans, inverallensis, marcens, minimus, nudimalis, orbiculatopunctatus, piperatus.*

***clarithorax.*** *Camponotus caryae* subsp. *clarithorax* Creighton, 1950a: 386 (s.w.q.m.) U.S.A. [First available use of *Camponotus marginatus* subsp. *discolor* var. *clarithorax* Emery, 1893i: 678; unavailable name.] Raised to species: Smith, M.R. 1951a: 844. See also: Snelling, 1988: 61.

*clarus. Camponotus clarus* Mayr, 1862: 661 (w.) U.S.A. Junior synonym of *mellea:* Mayr, 1866a: 485 (in text).

*clara. Camponotus* (*Tanaemyrmex*) *maculatus* subsp. *aethiops* var. *clara* Karavaiev, 1926e: 190 (s.w.q.) RUSSIA. Unavailable name.

#***clarus.*** *Camponotus* (*Tanaemyrmex*) *substitutus* var. *clara* Stitz, 1938: 116 (q.) ARGENTINA. [Unresolved junior primary homonym of *clarus* Mayr, 1862: 661. Note that proposed replacement name, *lacteus* Baroni Urbani, 1971b: 361, is unavailable.]

*claveaui. Camponotus* (*Myrmoturba*) *maculatus* st. *aegyptiacus* var. *claveaui* Santschi, 1926b: 250 (w.) SENEGAL. Unavailable name; material referred to *aegyptiacus* by Baroni Urbani, 1972: 130.

***claviscapus.*** *Camponotus claviscapus* Forel, 1899c: 152 (s.w.q.) TRINIDAD. [*Camponotus claviscapus* Emery, 1896g: 36. *Nomen nudum,* attributed to Forel.] Wheeler, G.C. & Wheeler, J. 1968: 219 (l.). Combination in *C.* (*Myrmamblys*): Forel, 1912i: 90; in *C.* (*Pseudocolobopsis*): Emery, 1920b: 259. Current subspecies: nominal plus *lehmanni, occultus, subcarinatus.*

#***cleliae.*** *Camponotus* (*Myrmotemnus*) *nutans* var. *cleliae* Santschi, 1926c: 601 (w.) SUMATRA.

***cleobulus.*** *Camponotus* (*Myrmoturba*) *cleobulus* Santschi, 1919b: 238 (s.w.) SOUTH AFRICA. Combination in *C.* (*Tanaemyrmex*): Emery, 1925b: 92.

#***cleon.*** *Camponotus* (*Myrmoturba*) *maculatus* r. *cleon* Forel, 1913k: 126 (s.w.q.) SUMATRA. Combination in *C.* (*Tanaemyrmex*): Emery, 1925b: 96. Subspecies of *variegatus:* Emery, 1920c: 7.

***clerodendri.*** *Colobopsis clerodendri* Emery, 1887a: 241, fig. (s.w.) BORNEO. Combination in *Camponotus* (*Colobopsis*): Emery, 1889b: 517; Forel, 1912i: 90.

**#cliens.** *Camponotus maculatus* subsp. *cliens* Forel, 1911b: 210 (s.w.) JAVA. Combination in *C.* (*Tanaemyrmex*): Emery, 1925b: 94. Subspecies of *irritans:* Emery, 1920c: 7.

**#clivius.** *Camponotus* (*Myrmobrachys*) *crassus* var. *clivia* Emery, 1925b: 163 (w.) BRAZIL. [First available use of *Camponotus crassus* r. *brasiliensis* var. *clivia* Forel, 1912i: 78; unavailable name.] Subspecies of *brasiliensis:* Kempf, 1972a: 50.

*cluis.* *Camponotus maculatus* subsp. *atramentarius* var. *cluis* Forel, 1909b: 67 (w.) ANGOLA. Unavailable name; material referred to *maculatus* by Baroni Urbani, 1972: 125.

*cluisoides.* *Camponotus* (*Myrmoturba*) *maculatus* var. *cluisoides* Forel, 1913h: 354 (s.w.) KENYA. Combination in *C.* (*Tanaemyrmex*): Emery, 1925b: 87. Junior synonym of *maculatus:* Baroni Urbani, 1972: 125.

**clypeatus.** *Camponotus clypeatus* Mayr, 1866a: 487 (w.) BRAZIL. Combination in *C.* (*Camponotus*): Forel, 1914a: 266; in *C.* (*Myrmopelta*): Santschi, 1921f: 311; in *C.* (*Myrmaphaenus*): Emery, 1925b: 155.

*cnemidatus.* *Camponotus marginatus* subsp. *discolor* var. *cnemidatus* Emery, 1893i: 678 (w.) U.S.A. Unavailable name; material referred to *caryae* by Smith, M.R. 1940b: 139.

**cocosensis.** *Camponotus* (*Myrmamblys*) *cocosensis* Wheeler, W.M. 1919d: 305 (s.w.) COCOS IS. Wheeler, W.M. 1933a: 61 (q.m.). Combination in *C.* (*Myrmaphaenus*): Emery, 1925b: 155.

*cognata.* *Formica cognata* Smith, F. 1858b: 35 (s.w.) SOUTH AFRICA. Mayr, 1862: 655 (q.m.). Combination in *Camponotus:* Mayr, 1862: 655; in *C.* (*Tanaemyrmex*): Emery, 1925b: 88. Subspecies of *sylvaticus:* Forel, 1879a: 65; of *rubripes:* Forel, 1886f: 143; of *maculatus:* Mayr, 1862: 655; Emery, 1891b: 19; Emery, 1920c: 5; Wheeler, W.M. 1922a: 953; Arnold, 1922: 631. Junior synonym of *maculatus:* Roger, 1863b: 2; Baroni Urbani, 1972: 125.

**cognatocompressus.** *Camponotus maculatus* r. *cognatocompressus* Forel, 1904b: 379 (w.q.m.) "AFRICA" (no locality). [Date of publication (31).iii.1904.] [First available use of *Camponotus rubripes* r. *cognatus* var. *cognatocompressus* Forel, 1886f: 151; unavailable name.] Senior synonym of *xerxes:* Karavaiev, 1910b: 9 (*xerxes* given as senior, but *cognatocompressus* has priority). Raised to species (under the name of *xerxes*): Pisarski, 1967: 414.

*cognatodichrous.* *Camponotus maculatus* var. *cognatodichrous* Forel, 1904c: 13. *Nomen nudum.*

*cognatomaculatodichrous.* *Camponotus rubripes* r. *cognatus* var. *cognatomaculatodichrous* Forel, 1889: 255. Unavailable name.

*cognatomaculatus.* *Camponotus rubripes* r. *cognatus* var. *cognatomaculatus* Forel, 1889: 255. Unavailable name.

*cognatooasium.* *Camponotus maculatus* r. *oasium* var. *cognatooasium* Forel, 1902a: 158 (w.) ALGERIA. Unavailable name.

*cognatopilicornis.* *Camponotus rubripes* r. *cognatus* var. *cognatopilicornis* Forel, 1889: 262. Unavailable name.

*collegianus.* *Camponotus* (*Myrmocladoecus*) *tripartitus* var. *collegianus* Santschi, 1929d: 315 (w.) BRAZIL. Provisional junior synonym of *tripartitus:* Kempf, 1968b: 404.

**#colombicus.** *Camponotus planatus* var. *colombicus* Forel, 1899c: 141 (w.q.) COLOMBIA. Combination in *C.* (*Myrmobrachys*): Emery, 1925b: 164. Race of *planatus:* Forel, 1902b: 171.

**#coloratus.** *Camponotus melanoticus* var. *colorata* Forel, 1904c: 49 (s.w.q.) COLOMBIA. Combination in *C.* (*Myrmoturba*): Bruch, 1914: 229; in *C.* (*Tanaemyrmex*), and variety of *substitutus:* Emery, 1925b: 83.

**#comoedus.** *Camponotus* (*Manniella*) *linnaei* subsp. *comoedus* Wheeler, W.M. 1934g: 213 (s.w.) MEXICO. Combination in *C.* (*Myrmeurynota*): Kempf, 1972a: 49.

**#comorensis.** *Camponotus* (*Myrmotrema*) *grandidieri* var. *comorensis* Santschi, 1915c: 274 (s.w.) COMORO IS.

**#comottoi.** *Camponotus sylvaticus* r. *comottoi* Emery, 1887a: 213 (w.q.) BURMA. Combination in *C.* (*Tanaemyrmex*): Emery, 1925b: 96. Raised to species: Dalla Torre, 1893: 225. Subspecies of *mitis:* Bingham, 1903: 356; of *variegatus:* Emery, 1925b: 96.

***compactus.** **Camponotus compactus* Förster, 1891: 431, pl. 13, fig. 5 (q.) GERMANY (Oligocene).

**#completus.** *Camponotus* (*Myrmotrema*) *rhamses* var. *completus* Santschi, 1930a: 76 (w.) ETHIOPIA.

**compositor.** *Camponotus* (*Neomyrmamblys*) *compositor* Santschi, 1922c: 111 (s.w.) BRAZIL. Combination in *C.* (*Myrmaphaenus*): Emery, 1925b: 155.

**compressiscapus.** *Camponotus compressiscapus* André, 1889: 218 (w.) SIERRA LEONE. Combination in *C.* (*Myrmoserícus*): Forel, 1914a: 268; in *C.* (*Myrmotrema*): Santschi, 1915c: 271.

**compressus.** *Formica compressa* Fabricius, 1787: 307 (w.) INDIA. Smith, F. 1858b: 14 (q.m.); Kumbkarni, 1965: 305 (k.). Combination in *Camponotus:* Roger, 1863b: 2; in *C.* (*Tanaemyrmex*): Emery, 1925b: 98; in *C.* (*Myrmoturba*): Santschi, 1929g: 165. Race of *rubripes:* Forel, 1886f: 143. Subspecies of *maculatus:* Emery, 1895k: 473; Emery, 1896d: 371; Wheeler, W.M. 1919e: 108. Status as species: Dalla Torre, 1893: 225; Bingham, 1903: 351; Emery, 1920c: 6; Emery, 1925b: 98; Santschi, 1929g: 165; Santschi, 1936c: 207. Senior synonym of *indefessa:* Jerdon, 1851: 119; of *quadrilaterus:* Forel, 1892j: 240; of *callida:* Emery, 1925b: 98. Current subspecies: nominal plus *brullei, cosensis, irakensis, martensi, nitens* (unresolved junior homonym), *occipitalis, probativus, pupillus, rectinotus, symiensis.*

**#comptus.** *Camponotus* (*Myrmosericus*) *vestitus* var. *comptus* Santschi, 1926b: 259 (q.) IVORY COAST. Santschi, 1930a: 76 (m.).

*conakryensis.* *Camponotus* (*Myrmoturba*) *maculatus* var. *conakryensis* Emery, 1920c: 13 (s.w.) GUINEA. Junior synonym of *maculatus:* Baroni Urbani, 1972: 125.

*concava.* *Camponotus aethiops* var. *concava* Dalla Torre, 1893: 221 (s.w.) GREECE. [First available use of *Camponotus rubripes* r. *aethiops* var. *concavus* Forel, 1889: 263; unavailable name.] Finzi, 1930d: 317 (q.); Menozzi, 1936d: 301 (l.). Combination in *C.* (*Tanaemyrmex*): Emery, 1925b: 97. Subspecies of *aethiops:* Arnol'di, 1967: 1829; Pisarski, 1970: 322. Junior synonym of *aethiops:* Agosti & Collingwood, 1987a: 58.

**#concilians.** *Camponotus* (*Myrmomalis*) *korthalsiae* st. *concilians* Forel, 1915a: 37 (w.q.) SUMATRA. Combination in *C.* (*Myrmoplatys*), and senior synonym of *jacobsoni:* Emery, 1925b: 136.

**concolor.** *Camponotus concolor* Forel, 1891b: 214, pl. 6, fig. 12 (s.w.) MADAGASCAR. Senior synonym of *gallienii* (unnecessary replacement name); **new synonymy**.

**#concolor.** *Camponotus alii* var. *concolor* Dalla Torre, 1893: 221 (s.w.) ALGERIA. [First available use of *Camponotus rubripes* r. *alii* var. *concolor* Forel, 1890a: lxiii; unavailable name.] Emery, 1891b: 19 (q.). Combination in *C.* (*Tanaemyrmex*): Emery, 1925b: 91. Variety of *alii:* Cagniant, 1964: 90. [Unresolved junior primary homonym of *concolor* Forel, above.]

**#concordius.** *Camponotus* (*Myrmotrema*) *olivieri* st. *concordia* Santschi, 1915c: 272 (s.w.) ANGOLA.

*confellah.* *Camponotus* (*Myrmoturba*) *compressus* st. *sanctus* var. *confellah* Santschi, 1939d: 83 (s.w.q.) SYRIA. Unavailable name.

**confluens.** *Camponotus* (*Myrmamblys*) *confluens* Forel, 1913b: 342 (s.w.) ZAIRE. Combination in *C.* (*Myrmamblys*): Emery, 1925b: 141; in *C.* (*Myrmotrema*): Emery, 1920b: 258; Santschi, 1926c: 599. Current subspecies: nominal plus *bequaerti, trematogaster.*

**confucii.** *Camponotus confucii* Forel, 1894c: 396 (s.w.) INDIA. Combination in *C.* (*Myrmosphincta*): Forel, 1914a: 273; in *C.* (*Orthonotomyrmex*): Emery, 1920b: 257; in *C.* (*Myrmentoma*): Emery, 1925b: 119.

**#confusus.** *Camponotus dorycus* r. *confusus* Emery, 1887a: 215 (w.q.) AUSTRALIA. Combination in *C.* (*Tanaemyrmex*): Emery, 1925b: 90. Raised to species: Dalla Torre, 1893: 226. Subspecies of *dorycus:* Emery, 1925b: 90; Taylor & Brown, D.R. 1985: 113.

**congolensis.** *Camponotus maculatus* subsp. *congolensis* Emery, 1899e: 499, fig. (s.w.q.) ZAIRE. Santschi, 1910c: 393 (q.); Forel, 1913b: 340 (m.). Combination in *C.* (*Myrmoturba*): Forel, 1913b: 340; in *C.* (*Tanaemyrmex*): Emery, 1925b: 99. Subspecies of *maculatus:* Wheeler, W.M. 1922a: 646; Arnold, 1922: 235. Raised to species: Emery, 1920c: 6; Emery, 1925b: 99. Current subspecies: nominal plus *weissi.*

**coniceps.** *Camponotus* (*Tanaemyrmex*) *coniceps* Santschi, 1926b: 258 (s.) CONGO.

**conicus.** *Colobopsis conica* Mayr, 1876: 67 (w.) TONGA IS. Combination in *Camponotus:* Dalla Torre, 1893: 226; in *C.* (*Myrmotemnus*): Emery, 1920b: 258; in *C.* (*Colobopsis*): Forel, 1914a: 272; Emery, 1925b: 148.

**conithorax.** *Camponotus* (*Colobopsis*) *conithorax* Emery, 1914f: 430, pl. 13, fig. 18 (w.q.m.) VANUATU. Combination in *C.* (*Myrmotemnus*): Emery, 1920b: 258; in *C.* (*Colobopsis*): Emery, 1925b: 148.

**conradti.** *Camponotus conradti* Forel, 1914d: 256 (s.w.) CAMEROUN. Combination in *C.* (*Myrmepomis*): Emery, 1920b: 258; in *C.* (*Myrmopiromis*): Wheeler, W.M. 1922a: 985; in *C.* (*Myrmamblys*): Emery, 1925b: 142; in *C.* (*Myrmacrhaphe*): Santschi, 1926c: 607. Current subspecies: nominal plus *fimbriatipes.*

**consanguineus.** *Formica consanguinea* Smith, F. 1861b: 36 (w.) SULAWESI. Combination in *Camponotus:* Roger, 1863b: 4; in *C.* (*Tanaemyrmex*): Donisthorpe, 1932c: 465.

**CAMPONOTUS** *(continued)*

***consectator.*** *Formica consectator* Smith, F. 1858b: 38 (q.) AUSTRALIA. Combination in *Camponotus:* Roger, 1863b: 4; current subgeneric placement in *C.* (*Tanaemyrmex*).

*consensi. Camponotus compressus* subsp. *sancta* var. *consensi* Finzi, 1928c: 791. Unavailable name.

***consobrinus.*** *Formica consobrina* Erichson, 1842: 258 (q.) AUSTRALIA. Smith, F. 1858b: 41 (w.); Imai, Crozier & Taylor, 1977: 346 (k.). Combination in *Camponotus:* Roger, 1863b: 4; in *C.* (*Tanaemyrmex*): Wheeler, W.M. 1933b: 22. Senior synonym of *dimidiatus:* Wheeler, W.M. 1933b: 23; Clark, 1934c: 70. Current subspecies: nominal plus *perthianus.*

***conspicuus.*** *Formica conspicua* Smith, F. 1858b: 48 (q.) JAMAICA. Wheeler, W.M. 1917g: 462 (s.w.m.). Combination in *Camponotus:* Roger, 1863b: 5; in *C.* (*Myrmoturba*): Forel, 1914a: 267; in *C.* (*Tanaemyrmex*): Emery, 1925b: 80. Current subspecies: nominal plus *inaequalis, sharpi, williamsi, zonatus.*

**constrictus* Mayr, 1868; see under ***CATAGLYPHIS.***

***constructor.*** *Camponotus constructor* Forel, 1899c: 155, pl. 4, fig. 25 (w.q.) PANAMA. Combination in *C.* (*Myrmotemnus*): Emery, 1920b: 259; in *C.* (*Myrmosphincta*): Forel, 1912i: 92; Emery, 1925b: 151.

*consul. Camponotus* (*Dinomyrmex*) *caesar* st. *imperator* var. *consul* Santschi, 1920g: 180 (w.) CAMEROUN. Unavailable name.

*contaminatus. Camponotus* (*Myrmoturba*) *maculatus* st. *manzer* var. *contaminatus* Santschi, 1917b: 291 (s.w.) ZIMBABWE. Unavailable name; material referred to *maculatus* by Baroni Urbani, 1972: 125.

*continentalis. Camponotus* (*Colobopsis*) *sommeri* subsp. *continentalis* Viehmeyer, 1925b: 145. Unavailable name (published in synonymy); material referred to *augustulus* by Viehmeyer, 1925b: 145.

#***continentis.*** *Camponotus planatus* var. *continentis* Forel, 1901e: 371 (in text) (w.) MEXICO. Combination in *C.* (*Myrmobrachys*): Emery, 1925b: 164.

***contractus.*** *Camponotus contractus* Mayr, 1872: 137 (w.) BORNEO. Combination in *C.* (*Myrmomalis*): Forel, 1914a: 271; in *C.* (*Myrmoplatys*): Emery, 1920b: 256. Current subspecies: nominal plus *buttesi, scortechinii.*

*contraria. Camponotus* (*Myrmosericus*) *rufoglaucus* st. *flavomarginatus* var. *contraria* Santschi, 1916b: 509 (w.) ZIMBABWE. Unavailable name.

#***controversus.*** *Camponotus* (*Myrmosericus*) *rufoglaucus* st. *controversus* Santschi, 1916b: 509. Replacement name for *flavopilosus* Viehmeyer, 1913: 47. [Junior primary homonym of *flavopilosus* Emery, 1895h: 54.]

***conulus.*** *Camponotus conulus* Mayr, 1870a: 382, pl. fig. 3 (w.) COLOMBIA. Combination in *C.* (*Myrmamblys*): Emery, 1920b: 259; in *C.* (*Myrmobrachys*): Forel, 1914a: 270; Emery, 1925b: 163.

#***convexinodis.*** *Camponotus* (*Myrmocladoecus*) *sanctaefidei* subsp. *convexinodis* Wheeler, W.M. 1934g: 233 (w.) BRAZIL.

#***convexior.*** *Camponotus* (*Colobopsis*) *quadriceps* var. *convexior* Viehmeyer, 1914a: 115, fig. (s.w.) INDONESIA (Seram I.).

*cooperi* Gregg, 1951; see under ***MYRMELACHISTA.***

***coptobregma.*** *Camponotus* (*Hypercolobopsis*) *coptobregma* Kempf, 1968b: 411, figs. 23–26 (s.w.) BRAZIL.

***corallinus.*** *Colobopsis corallina* Roger, 1863a: 159 (w.) PHILIPPINES. Combination in *Camponotus:* Dalla Torre, 1893: 226; in *C.* (*Myrmamblys*): Forel, 1914a: 271; in *C.* (*Myrmotemnus*): Emery, 1920b: 258; in *C.* (*Colobopsis*): Emery, 1925b: 150.

***cordiceps.*** *Camponotus* (*Myrmothrix*) *cordiceps* Santschi, 1939e: 327, figs. 6, 7 (s.w.) ARGENTINA. See also: Hashmi, 1973: 121.

***cordincola.*** *Camponotus* (*Colobopsis*) *cordincola* Wheeler, W.M. 1934g: 218 (s.) PANAMA.

***coriolanus.*** *Camponotus* (*Colobopsis?*) *coriolanus* Forel, 1912i: 84 (w.) BRAZIL. Combination in *C.* (*Hypercolobopsis*): Emery, 1925b: 160. Senior synonym of *scrobifer:* Kempf, 1968b: 410.

***corniculatus.*** *Camponotus* (*Myrmocladoecus*) *corniculatus* Wheeler, W.M. 1934g: 236, fig. 6 (s.w.q.) GUIANA. [*Camponotus corniculatus* Wheeler, W.M. 1934e: 424, *nomen nudum.*]

***cornis.*** *Camponotus cornis* Wang & Wu, 1994: 25, figs. 1, 14 (w.) CHINA.

#***cornutus.*** *Camponotus* (*Myrmamblys*) *lilianae* r. *cornutus* Forel, 1913b: 345 (s.) ZAIRE.

#***coronatus.*** *Camponotus* (*Myrmocladoecus*) *quadrilaterus* st. *coronata* Santschi, 1922c: 123 (w.) BRAZIL. Variety of *sanctaefidei:* Emery, 1925b: 167; Wheeler, W.M. 1934g: 232.

#***corticalis.*** *Camponotus capperi* r. *corticalis* Forel, 1899c: 138 (w.) JAMAICA. Combination in *C.* (*Myrmobrachys*): Wheeler, W.M. 1917g: 470; in *C.* (*Myrmaphaenus*): Emery, 1925b: 156.

***coruscus.*** *Formica corusca* Smith, F. 1862b: 30 (w.) PANAMA. Combination in *Camponotus:* Mayr, 1870a: 373; in *C.* (*Myrmothrix*): Forel, 1914a: 268; in *C.* (*Myrmaphaenus*): Emery, 1920b: 256; in *C.* (*Tanaemyrmex*): Emery, 1925b: 80. Current subspecies: nominal plus *astilbus, fulgens.*

#***corvus.*** *Camponotus natalensis* r. *corvus* Forel, 1879a: 79 (s.w.q.m.) SOUTH AFRICA. Combination in *C.* (*Tanaemyrmex*): Emery, 1925b: 103. Raised to species: Dalla Torre, 1893: 227. Subspecies of *natalensis:* Emery, 1895h: 51; Arnold, 1922: 654; Wheeler, W.M. 1922a: 960.

#***cosensis.*** *Camponotus* (*Tanaemyrmex*) *compressus* var. *cosensis* Finzi, 1939c: 159 (s.w.) GREECE. [First available use of *Camponotus maculatus* r. *sanctus* var. *cosensis* Forel, 1904c: 20; unavailable name (Agosti & Collingwood, 1987a: 61).] Finzi, 1932: 25 (q.m.).

***cosmicus.*** *Formica cosmica* Smith, F. 1858b: 34 (s.w.) SOUTH AFRICA. Combination in *Camponotus:* Roger, 1863b: 2; in *C.* (*Myrmosericus*): Emery, 1925b: 106. Subspecies of *rufoglaucus:* Emery, 1895h: 49 (in key); Stitz, 1916: 401; Arnold, 1924: 693. Revived status as species: Santschi, 1935a: 280.

*costaricensis. Camponotus abdominalis* var. *costaricensis* Forel, 1899c: 132 (w.) COSTA RICA. Menozzi, 1927d: 339 (q.). Combination in *C.* (*Myrmothrix*): Emery, 1925b: 107. Junior synonym of *atriceps:* Hashmi, 1973: 82.

***cotesii.*** *Camponotus* (*Colobopsis*) *cotesii* Forel, 1893b: 438 (w.) INDIA. Bingham, 1903: 345 (s.).

***cowlei.*** *Camponotus cowlei* Froggatt, 1896: 387, pl. 27, figs. 1–5 (w.q.m.) AUSTRALIA. Combination in *C.* (*Myrmophyma*): Emery, 1925b: 110. Junior synonym of *Melophorus bagoti:* Clark, 1930c: 22. Combination in *Camponotus,* and revived from synonymy: Taylor & Brown, D.R. 1985: 112.

***coxalis.*** *Formica coxalis* Smith, F. 1859a: 136 (s.) INDONESIA (Aru I.). Combination in *Camponotus:* Roger, 1863b: 4; in *C.* (*Tanaemyrmex*): Emery, 1925b: 90. Race of *dorycus:* Emery, 1887a: 216. Revived status as species: Emery, 1925b: 90.

***crassicornis.*** *Camponotus* (*Myrmobrachys*) *crassicornis* Emery, 1920c: 41 (s.w.q.) BOLIVIA.

#***crassinodis.*** *Camponotus mitis* var. *crassinodis* Forel, 1892j: 230 (diagnosis in key) (w.) BURMA. Combination in *C.* (*Tanaemyrmex*), and variety of *variegatus:* Emery, 1925b: 96.

*crassinodis. Camponotus lateralis* var. *crassinodis* Forel, 1894d: 3 (w.) ALGERIA. [Junior primary homonym of *crassinodis,* above.] Replacement name: *spissinodis* Forel, 1909e: 374.

***crassisquamis.*** *Camponotus crassisquamis* Forel, 1902d: 286 (s.w.) INDIA. Imai, *et al.* 1984: 9 (k.). Combination in *C.* (*Myrmoturba*): Forel, 1914a: 267; in *C.* (*Tanaemyrmex*): Emery, 1925b: 93.

***crassus.*** *Camponotus crassus* Mayr, 1862: 670 (w.) BRAZIL. Forel, 1912i: 78 (q.). Combination in *C.* (*Myrmobrachys*): Forel, 1914a: 270. Race of *senex:* Forel, 1879a: 99. Revived status as species: Dalla Torre, 1893: 227; Emery, 1906c: 192. Senior synonym of *flexus:* Mayr, 1870a: 387. Current subspecies: nominal plus *amazonicus, delabiatus, picticornis.*

***crawleyi.*** *Camponotus* (*Dinomyrmex*) *crawleyi* Emery, 1920c: 24, fig. 6 (w.) CAMEROUN. Combination in *C.* (*Tanaemyrmex*): Emery, 1925b: 85.

***crenatus.*** *Camponotus crenatus* Mayr, 1876: 64 (s.) AUSTRALIA. Subgenus indeterminate.

***crepusculi.*** *Camponotus* (*Myrmophyma*) *crepusculi* Arnold, 1922: 665, pl. 9, fig. 155 (s.w.q.) ZIMBABWE. Combination in *C.* (*Myrmamblys*): Emery, 1925b: 140; in *C.* (*Myrmespera*): Santschi, 1926c: 604.

***cressoni.*** *Camponotus cressoni* André, 1887: 282 (s.w.) MEXICO. Combination in *C.* (*Myrmamblys*): Forel, 1912i: 90; in *C.* (*Paracolobopsis*): Emery, 1920b: 259; in *C.* (*Myrmaphaenus*): Emery, 1925b: 153; Wheeler, W.M. 1934g: 207. Current subspecies: nominal plus *purensis.*

*creticus. Camponotus gestroi* st. *creticus* Forel, 1886e: clix, fig. 1 (w.) GREECE. Combination in *C.* (*Orthonotomyrmex*): Forel, 1913d: 435; in *C.* (*Myrmentoma*): Emery, 1925b: 119. Junior synonym of *gestroi:* Agosti & Collingwood, 1987a: 58.

***criniticeps.*** *Camponotus* (*Myrmamblys*) *criniticeps* Menozzi, 1939c: 109, fig. 1 (w.) ETHIOPIA.

#***crispulus.*** *Camponotus* (*Myrmaphaenus*) *blandus* var. *crispula* Santschi, 1922c: 110 (w.) ARGENTINA. Subspecies of *blandus:* Santschi, 1929d: 315.

***cristatus.*** *Camponotus cristatus* Mayr, 1866a: 489, fig. 3 (w.) FIJI IS. Mann, 1921: 479 (s.q.m.). Combination in *C.* (*Myrmogonia*): Forel, 1912i: 92. Current subspecies: nominal plus *nagasau, sadinus.*

#***criton.*** *Camponotus* (*Tanaemyrmex*) *zenon* var. *criton* Emery, 1925b: 84 (s.) BRAZIL. [First available use of *Camponotus maculatus* r. *zenon* var. *criton* Forel, 1912i: 73; unavailable name.]

#***croceomaculatus.*** *Camponotus* (*Myrmoturba*) *croceomaculatus* Emery, 1914f: 452 (q.) NEW CALE-

DONIA. Combination in *C.* (*Tanaemyrmex*): Emery, 1925b: 94. Subspecies of *irritans:* Emery, 1920c: 7.

***crucheti.*** *Camponotus perrisii* st. *crucheti* Santschi, 1911g: 215 (s.w.q.m.) ANGOLA. Combination in *C.* (*Myrmotrema*), and raised to species: Santschi, 1915c: 271 (in key).

***cruentatus.*** *Formica cruentata* Latreille, 1802c: 116, pl. 3, fig. 10 (w.) "L'habitation est inconnue. Je soupçonne qu'elle est d'Afrique." [Distribution is Mediterranean.] Roger, 1862b: 255 (q.m.); Emery, 1916b: 230 (s.). Combination in *Camponotus:* Mayr, 1861: 36; in *C.* (*Myrmosericus*): Forel, 1914a: 268. Senior synonym of *opaca,* and material of the unavailable name *bucephala* referred here: Roger, 1859: 228; Mayr, 1861: 36. Senior synonym of *gigas* Leach: Dalla Torre, 1893: 227; of *cruentior* and *lindbergi:* Cagniant, 1991b: 51. See also: André, 1882b: 155; Cagniant, 1985b: 77; Cagniant, 1991b: 42. Current subspecies: nominal plus *asper.*

*cruentata. Camponotus japonicus* var. *cruentata* Emery, 1925b: 73 (s.w.q.) KOREA. [First available use of *Camponotus herculeanus* subsp. *japonicus* var. *cruentata* Karavaiev, 1912b: 595; unavailable name.] [Junior secondary homonym of *Formica cruentata* Latreille, 1802c: 116, above.] Replacement name: *atrox* Emery, 1925b: 73.

*cruentior. Camponotus* (*Myrmosericus*) *cruentatus* var. *cruentior* Santschi, 1931a: 12 (s.w.) MOROCCO. Junior synonym of *cruentatus:* Cagniant, 1991b: 51.

#***cruentus.*** *Camponotus* (*Myrmoturba*) *punctulatus* st. *cruenta* Santschi, 1922c: 104 (w.) ARGENTINA. [First available use of *Camponotus punctulatus* subsp. *imberbis* var. *cruenta* Emery, 1906c: 190; unavailable name.] Material of the unavailable name *mediorufa* referred here: Santschi, 1922c: 104.

*cualatensis. Camponotus* (*Myrmepomis*) *sericeiventris* subsp. *rex* var. *cualatensis* Wheeler, W.M. 1931c: 96 (s.w.) MEXICO. Unavailable name.

***cuauhtemoc.*** *Camponotus* (*Myrmentoma*) *cuauhtemoc* Snelling, 1988: 62, figs. 10–12, 22, 23 (s.w.q.m.) MEXICO.

***cubangensis.*** *Camponotus mayri* var. *cubangensis* Forel, 1901d: 309 (w.q.) ANGOLA. [Also described as new, but spelled *kubangensis,* by Forel, 1903e: 563.] Combination in *C.* (*Orthonotomyrmex*): Emery, 1925b: 125. Raised to species: Santschi, 1930b: 80. Current subspecies: nominal plus *dofleini.*

#***cuitensis.*** *Camponotus foraminosus* r. *cuitensis* Forel, 1901d: 309 (w.) ANGOLA. [Also described as new by Forel, 1903e: 563.] Combination in *C.* (*Myrmotrema*): Santschi, 1915c: 271.

***culmicola.*** *Camponotus* (*Colobopsis*) *culmicola* Wheeler, W.M. 1905b: 117, figs. V, W; pl. 7, figs. 10, 11 (s.w.q.m.) BAHAMAS. Wheeler, G.C. & Wheeler, J. 1953e: 190 (l.). Current subspecies: nominal plus *haweisi.*

*cuneata. Formica cuneata* Perty, 1833: 134, pl. 27, fig. 1 (w.) BRAZIL. Junior synonym of *sericeiventris:* Smith, F. 1858b: 46.

***cuneidorsus.*** *Camponotus* (*Myrmobrachys*) *cuneidorsus* Emery, 1920c: 40 (s.w.) PERU.

***cuneiscapus.*** *Camponotus cuneiscapus* Forel, 1910f: 29, pl. 1, fig. 2 (s.w.) SOUTH AFRICA. Arnold, 1924: 679 (q.). Combination in *C.* (*Myrmopsamma*): Forel, 1914a: 270; Wheeler, W.M. 1922a: 988; Arnold, 1924: 679; in *C.* (*Camponotus*): Santschi, 1923e: 292; in *C.* (*Myrmamblys*): Emery, 1925b: 142; in *C.* (*Myrmespera*): Santschi, 1926b: 248.

*cupiens. Camponotus abdominalis* subsp. *cupiens* Forel. 1908c: 410 (s.w.q.) BRAZIL. Junior synonym of *atriceps:* Hashmi, 1973: 82.

#***curtior.*** *Camponotus somalinus* var. *curtior* Forel, 1894b: 65 (s.w.) ETHIOPIA. Combination in *C.* (*Tanaemyrmex*): Emery, 1925b: 100.

#***curtulus.*** *Camponotus* (*Myrmoturba*) *bonariensis* st. *curtulus* Santschi, 1936b: 205 (s.w.) BRAZIL.

#***curtus.*** *Camponotus maculatus* subsp. *curtus* Emery, 1900d: 704 (s.w.) INDONESIA (Mentawei I.). Combination in *C.* (*Tanaemyrmex*): Emery, 1925b: 94. Subspecies of *irritans:* Emery, 1920c: 7.

#***curvatus.*** *Camponotus quadriceps* var. *curvatus* Stitz, 1911a: 372, fig. 18 (w.) NEW GUINEA. Combination in *C.* (*Colobopsis*): Emery, 1925b: 150.

***curviscapus.*** *Camponotus curviscapus* Emery, 1896g: 103, pl. 1, fig. 20 (s.w.q.) COSTA RICA. Combination in *C.* (*Colobopsis*): Forel, 1914a: 272; in *C.* (*Pseudocolobopsis*): Emery, 1920b: 259.

***custodulus.*** *Camponotus* (*Colobopsis*) *custodulus* Emery, 1911b: 537 (q.) NEW GUINEA.

*cyaneus* Forel, 1910; see under ***CALOMYRMEX.***

***cylindricus.*** *Formica cylindrica* Fabricius, 1798: 280 (q.) BORNEO. Combination in *Colobopsis:* Mayr, 1862: 691; in *Camponotus* (*Colobopsis*): Emery, 1889b: 517 Senior synonym of *doriae:* Emery, 1917a: 97.

*cypriacus. Camponotus maculatus* r. *sanctus* var. *cypriacus* Forel, 1904d: 176 (w.q.m.) CYPRUS. Unavailable name (Agosti & Collingwood, 1987a: 61).

#***cypridis.*** *Camponotus* (*Myrmentoma*) *lateralis* var. *cypridis* Santschi, 1939c: 6 (s.w.q.m.) CYPRUS.

#***cyprius.*** *Camponotus* (*Orthonotomyrmex*) *kiesenwetteri* var. *cypria* Emery, 1920c: 26 (w.) CYPRUS. Combination in *C.* (*Myrmentoma*): Emery, 1925b: 121.

***cyrtomyrmodes.*** *Camponotus* (*Myrmophyma*) *cyrtomyrmodes* Donisthorpe, 1941c: 139 (w.m.) NEW GUINEA.

*daliensis. Camponotus abdominalis* var. *daliensis* Forel, 1901h: 70 (w.q.) COSTA RICA. *Nomen nudum.*

***dalmasi.*** *Camponotus dalmasi* Forel, 1899c: 145 (footnote) (w.) COLOMBIA. Combination in *C.* (*Myrmorhachis*): Forel, 1914a: 274; in *C.* (*Myrmocladoecus*): Emery, 1925b: 166; Wheeler, W.M. 1934e: 424.

***dalmaticus.*** *Formica dalmatica* Nylander, 1849: 37 (w.) YUGOSLAVIA. Forel, 1913d: 436 (q.m.). Combination in *Camponotus:* Mayr, 1863: 399; in *C.* (*Orthonotomyrmex*): Müller, 1923: 164; in *C.* (*Myrmentoma*): Menozzi, 1921: 32; Emery, 1925b: 120. Junior synonym of *lateralis:* Mayr, 1855: 322. Revived from synonymy as subspecies of *lateralis:* Forel, 1874: 40; Forel, 1892i: 306; Müller, 1923: 164; Emery, 1925a: 64. Revived status as species: Finzi, 1927b: 52; Finzi, 1930d: 318; Santschi, 1934d: 281; Kutter, 1977c: 207; Agosti & Collingwood, 1987a: 58; Atanassov & Dlussky, 1992: 224.

*damocles. Camponotus cingulatus* var. *damocles* Forel, 1909a: 264 (s.w.m.) PARAGUAY. Combination in *C.* (*Myrmothrix*): Emery, 1925b: 108. Junior synonym of *cingulatus:* Hashmi, 1973: 31.

*darlingtoni. Camponotus* (*Myrmophyma*) *darlingtoni* Wheeler, W.M. 1934d: 160 (s.w.q.) AUSTRALIA. [Date of publication: 5.x.1934.] Senior synonym of the unnecessary replacement name *rottnesti,* and junior synonym of *testaceipes:* Brown, 1956a: 39.

*darlingtoni. Camponotus* (*Myrmocladoecus*) *sanctaefidei* subsp. *darlingtoni* Wheeler, W.M. 1934g: 232 (w.) TRINIDAD. [Date of publication: (30).xi.1934.] [Junior primary homonym of *darlingtoni,* above.] Replacement name: *weberi* Wheeler, W.M. 1935c: 72.

***darwinii.*** *Camponotus darwinii* Forel, 1886f: 179 (s.w.) MADAGASCAR. Combination in *C.* (*Myrmobrachys*): Forel, 1914a: 270; in *C.* (*Myrmepomis*): Emery, 1920b: 258; in *C.* (*Myrmopiromis*): Wheeler, W.M. 1922a: 1051. See also: Forel, 1891b: 43. Current subspecies: nominal plus *rubropilosus, themistocles.*

***debellator.*** *Camponotus* (*Myrmespera*) *debellator* Santschi, 1926b: 247 (s.w.q.) ZIMBABWE.

#***decessor.*** *Camponotus emeryodicatus* subsp. *decessor* Forel, 1908c: 413 (w.q.) BRAZIL. Combination in *C.* (*Myrmomalis*): Emery, 1925b: 169.

***decipiens.*** *Camponotus marginatus* var. *decipiens* Emery, 1893i: 676 (w.q.) U.S.A. Wheeler, W.M. 1910f: 227 (s.w.q.m.). Combination in *C.* (*Myrmentoma*): Emery, 1925b: 117. Variety of *fallax:* Wheeler, W.M. 1910f: 227; Wheeler, W.M. 1917a: 558; of *caryae:* Wheeler, W.M. 1917c: 28; Emery, 1925b: 117. Junior synonym of *nearcticus:* Creighton, 1950a: 388. Revived from synonymy, raised to species, and senior synonym of *rasilis:* Snelling, 1988: 66.

***declivus.*** *Camponotus* (*Paracolobopsis*) *declivus* Santschi, 1922c: 120 (w.) MEXICO. Combination in *C.* (*Myrmaphaenus*): Emery, 1925b: 154.

#***decorus.*** *Formica decora* Smith, F. 1858b: 43 (q.) BRAZIL. Combination in *Camponotus:* Roger, 1863b: 4; in *C.* (*Myrmosphincta*): Emery, 1925b: 152. Junior synonym of *ruficeps:* Mayr, 1865: 38; Dalla Torre, 1893: 250. Revived from synonymy as variety of *sexguttatus:* Emery, 1894f: 2; Emery, 1896d: 375; Kempf, 1972a: 58.

***dedalus.*** *Camponotus dedalus* Forel, 1911b: 208 (q.) JAVA. Combination in *C.* (*Myrmamblys*): Forel, 1914a: 271. Current subspecies: nominal plus *moeschiellus.*

*deductrix. Camponotus* (*Myrmoturba*) *maculatus* r. *parvulus* var. *deductrix* Forel, 1913l: 248 (s.) ARGENTINA. Unavailable name.

#***deductus.*** *Camponotus* (*Myrmotrema*) *foraminosus* var. *deducta* Santschi, 1915c: 268 (in key) (w.) NIGERIA.

*defrictus. Camponotus* (*Myrmothrix*) *abdominalis* st. *floridanus* var. *defrictus* Santschi, 1936b: 214 (s.w.) MEXICO. Unavailable name; material referred to *atriceps* by Hashmi, 1973: 82.

#***delabiatus.*** *Camponotus* (*Myrmobrachys*) *crassus* var. *delabiatus* Santschi, 1925d: 247 (w.) BRAZIL.

#***delagoensis.*** *Camponotus foraminosus* r. *delagoensis* Forel, 1894b: 68 (s.w.) MOZAMBIQUE. Combination in *C.* (*Myrmotrema*): Santschi, 1915c: 270. Subspecies of *foraminosus:* Wheeler, W.M. 1922a: 980; of *olivieri:* Santschi, 1915c: 270; Arnold, 1924: 733; Emery, 1925b: 132.

***deletangi.*** *Camponotus* (*Myrmoturba*) *deletangi* Santschi, 1920e: 124, fig. 2 (w.) BOLIVIA. [Also described as new by Santschi, 1922c: 105.] Combination in *C.* (*Tanaemyrmex*): Emery, 1925b: 78.

*delta. Camponotus (Myrmoturba) compressus* st. *thoracicus* var. *delta* Santschi, 1938a: 41 (s.m.) EGYPT. Unavailable name.

#***densipunctatus.*** *Camponotus (Myrmotrema) perrisii* var. *densipunctatus* Stitz, 1916: 399 (s.w.m.) ZAIRE.

***dentatus.*** *Colobopsis dentata* Mayr, 1866a: 492, fig. 5 (w.) FIJI IS. Mann, 1921: 491 (s.q.m.). Combination in *Camponotus:* Dalla Torre, 1893: 228; in *C. (Myrmamblys)*: Forel, 1914a: 271; in *C. (Myrmotemnus)*: Emery, 1920b: 258; in *C. (Colobopsis)*: Mann, 1921: 491. Current subspecies: nominal plus *humeralis* (unresolved junior homonym).

***denticulatus.*** *Camponotus denticulatus* Kirby, W.F. 1896: 204 (s.w.) AUSTRALIA. Combination in *C. (Tanaemyrmex)*: Clark, 1930c: 19.

#***denudatus.*** *Camponotus blandus* subsp. *denudatus* Emery, 1903: 67 (w.q.) BOLIVIA. Combination in *C. (Myrmocamelus)*: Forel, 1913l: 249; Bruch, 1914: 228; in *C. (Myrmaphaenus)*: Emery, 1925b: 154.

*denudata. Camponotus distinguendus* var. *denudata* Emery, 1906c: 191 (w.) CHILE. [Junior primary homonym of *denudatus*, above.] Replacement name: *tenuipubens* Santschi, 1916d: 242.

#***depolitus.*** *Camponotus (Tanaemyrmex) ustus* var. *depolitus* Wheeler, W.M. 1936b: 205 (s.w.m.) DOMINICA.

***depressiceps.*** *Camponotus depressiceps* Forel, 1879a: 106, pl. 1, fig. 2 (q.) BRAZIL. Combination in *C. (Myrmomalis)*: Forel, 1914a: 271.

*depressidens. Camponotus (Myrmothrix) abdominalis* st. *depressidens* Santschi, 1922c: 107 (s.w.) FRENCH GUIANA. Junior synonym of *atriceps:* Hashmi, 1973: 82.

***depressus.*** *Camponotus depressus* Mayr, 1866a: 487, pl., fig. 1 (w.) BRAZIL. Menozzi, 1925b: 371 (q.). Combination in *C. (Myrmomalis)*: Forel, 1914a: 263.

***desantii.*** *Camponotus (Myrmoturba) maculatus* st. *desantii* Santschi, 1915c: 280, fig. 14 (s.w.q.m.) BENIN. Combination in *C. (Tanaemyrmex)*: Emery, 1925b: 87. Raised to species: Baroni Urbani, 1972: 131.

***descarpentriesi.*** *Camponotus (Myrmopiromis) descarpentriesi* Santschi, 1926b: 263 (s.w.m.) MADAGASCAR.

***desectus.*** *Formica desecta* Smith, F. 1860b: 97 (q.) NEW GUINEA. Combination in *Colobopsis:* Mayr, 1862: 691; in *Camponotus:* Dalla Torre, 1893: 228; in *C. (Colobopsis)*: Forel, 1914a: 272.

*detritoides. Camponotus fulvopilosus* var. *detritoides* Forel, 1910f: 25 (w.) SOUTH AFRICA. Combination in *C. (Myrmopiromis)*: Wheeler, W.M. 1922a: 986. Junior synonym of *fulvopilosus:* Robertson, 1990: 327.

***detritus.*** *Camponotus detritus* Emery, 1886: 357 (w.) SOUTH AFRICA. Forel, 1910f: 26 (s.). Combination in *C. (Myrmepomis)*: Forel, 1914a: 273; in *C. (Myrmopiromis)*: Wheeler, W.M. 1922a: 986. Subspecies of *fulvopilosus:* Forel, 1910f: 26; Wheeler, W.M. 1922a: 986; Stitz, 1923: 166; Arnold, 1924: 739. Revived status as species: Forel, 1914a: 273; Emery, 1920b: 258; Emery, 1925b: 128; Robertson, 1990: 327.

*devalata. Camponotus (Myrmepomis) sericeiventris* subsp. *rex* ab. *devalata* Stärcke, 1944b: xxii (w.) NETHERLANDS (possible origin in Jamaica). Unavailable name.

***devestitus.*** *Camponotus (Tanaemyrmex) devestitus* Wheeler, W.M. 1928d: 117 (s.w.) JAPAN.

***dewitzii.*** *Camponotus dewitzii* Forel, 1886f: 187 (w.) ZAIRE. Combination in *C. (Myrmosaga)*: Forel, 1914a: 269; in *C. (Mayria)*: Emery, 1925b: 123.

#***diabolus.*** *Camponotus natalensis* r. *diabolus* Forel, 1879a: 81 (s.w.) SOUTH AFRICA. Combination in *C. (Myrmoturba)*: Forel, 1914a: 267; in *C. (Tanaemyrmex)*: Emery, 1925b: 103. Raised to species: Dalla Torre, 1893: 228; Forel, 1914a: 267. Subspecies of *natalensis:* Emery, 1896d: 373; Wheeler, W.M. 1922a: 961; Emery, 1925b: 103. See also: Arnold, 1922: 646.

#***dichromothrix.*** *Camponotus (Dinomyrmex) hildebrandti* subsp. *dichromothrix* Emery, 1920c: 25 (s.) MADAGASCAR. Combination in *C. (Tanaemyrmex)*: Emery, 1925b: 85.

*dichrous. Camponotus sylvaticus* var. *dichrous* André, 1882a: 145 (w.) ALGERIA. [First available use of *Camponotus sylvaticus* subsp. *maculatus* var. *dichrous* Forel, 1879a: 65; unavailable name.] Forel, 1886e: clxi (m.); Forel, 1890a: lxv (q.); Bingham, 1903: 356 (s.). Variety of *rubripes:* Forel, 1890a: clxi. Subspecies of *maculatus:* Emery, 1891b: 19; Forel, 1894d: 41; Ruzsky, 1905b: 199. Raised to species: Dalla Torre, 1893: 228; revived status as species: Bingham, 1903: 356. Junior synonym of *thoracicus:* Emery, 1908a: 193.

***dicksoni.*** *Camponotus (Tanaemyrmex) dicksoni* Arnold, 1948: 227, fig. 13 (s.w.) SOUTH AFRICA.

***difformis.*** *Camponotus (Myrmocamelus) difformis* Stitz, 1938: 120, fig. 13 (w.) NEW GUINEA. Current subgeneric placement: *C. (Myrmophyma)*.

*diffusus. Camponotus (Myrmoturba) maculatus* st. *radamoides* var. *diffusus* Santschi, 1917b: 292 (s.w.) ZIMBABWE. Unavailable name; material referred to *maculatus* by Baroni Urbani, 1972: 125.

#***diligens.*** *Formica diligens* Smith, F. 1857a: 55 (q.) WEST MALAYSIA. Combination in *Camponotus:* Roger, 1863b: 3; in *C. (Tanaemyrmex)*: Emery, 1925b: 90. Variety of *festinus:* Forel, 1892j: 240; Emery, 1896d: 370. Revived status as species: Bingham, 1903: 262. Variety of *festinus:* Emery, 1925b: 90.

*dimidiatus. Camponotus dimidiatus* Roger, 1863b: 4 (w.) AUSTRALIA. Combination in *C. (Myrmoturba)*: Forel, 1913g: 181; in *C. (Tanaemyrmex)*: Emery, 1925b: 103. Junior synonym of *nigriceps:* Mayr, 1876: 63. Revived from synonymy as variety/race of *nigriceps:* Emery, 1887a: 211; Stitz, 1911a: 372; Forel, 1915b: 97; Crawley, 1922c: 35. Junior synonym of *consobrinus:* Wheeler, W.M. 1933b: 23; Clark, 1934c: 70.

***dimorphus.*** *Camponotus dimorphus* Emery, 1894c: 170 (s.w.q.) BRAZIL. Combination in *C. (Myrmamblys)*: Forel, 1914a: 271; in *C. (Myrmorhachis)*: Emery, 1920b: 260; in *C. (Myrmobrachys)*: Emery, 1925b: 165.

#***diogenes.*** *Camponotus maculatus* subsp. *diogenes* Forel, 1909e: 396 (s.w.q.m.) SRI LANKA. Combination in *C. (Tanaemyrmex)*: Emery, 1925b: 92. Subspecies of *thraso:* Emery, 1920c: 6.

***diplopunctatus.*** *Camponotus (Myrmotrema) diplopunctatus* Emery, 1915g: 22, fig. 12 (s.q.) ETHIOPIA. Current subspecies: nominal plus *subconvexus.*

***discolor.*** *Formica discolor* Buckley, 1866: 166 (w.q.) U.S.A. Emery, 1893i: 678 (m.); Wheeler, W.M. 1910f: 230 (s.). Combination in *Camponotus:* Mayr, 1886c: 365; in *C. (Myrmentoma)*: Emery, 1925b: 117. Junior synonym of *marginatus:* Mayr, 1886d: 423. Revived from synonymy as subspecies of *marginatus:* Emery, 1893i: 677. Subspecies of *fallax:* Wheeler, W.M. 1910f: 230; Wheeler, W.M. 1917a: 558; of *caryae:* Wheeler, W.M. 1917c: 28; Creighton, 1950a: 386. Revived status as species: Snelling, 1988: 68.

***discors.*** *Camponotus maculatus* st. *discors* Forel, 1902h: 497 (s.w.) AUSTRALIA. Combination in *C. (Myrmoturba)*: Forel, 1915b: 99; Wheeler, W.M. 1915g: 814; in *C. (Tanaemyrmex)*: Emery, 1925b: 102. Raised to species: Emery, 1920c: 8. Current subspecies: nominal plus *angustinodus, laetus, yarrabahensis.*

#***dispar.*** *Camponotus (Myrmotrema) troglodytes* var. *dispar* Stitz, 1923: 165 (w.) NAMIBIA. Subspecies of *robecchii:* current status.

***distinguendus.*** *Formica distinguenda* Spinola, 1851a: 235 (w.) CHILE. Combination in *Camponotus:* Roger, 1863b: 5; in *C. (Myrmosericus)*: Forel, 1914a: 268; in *C. (Myrmamblys)*: Bruch, 1914: 228; in *C. (Myrmoturba)*: Emery, 1920b: 255; in *C. (Tanaemyrmex)*: Emery, 1925b: 78. Senior synonym of *santacruziana:* Kusnezov, 1960b: 397; of *tenuipubens:* Snelling & Hunt, 1976: 120.

*disturbans. Camponotus barbatus* subsp. *taylori* var. *disturbans* Forel, 1911i: 228 (s.) SRI LANKA. Unavailable name.

***divergens.*** *Camponotus divergens* Mayr, 1887: 516 (s.) BRAZIL. Combination in *C. (Myrmamblys)*: Forel, 1914a: 271; in *C. (Pseudocolobopsis)*: Emery, 1925b: 158; in *C. (Hypercolobopsis)*: Kempf, 1960f: 458.

***diversipalpus.*** *Camponotus (Dinomyrmex) diversipalpus* Santschi, 1922c: 106 (w.) BRAZIL. Combination in *C. (Tanaemyrmex)*: Emery, 1925b: 83.

#***divus.*** *Camponotus nigroaeneus* subsp. *divus* Forel, 1907a: 34 (w.) AUSTRALIA. Combination in *C. (Myrmophyma)*: Emery, 1925b: 111.

#***dofleini.*** *Camponotus dofleini* Forel, 1911e: 291 (w.) SOUTH AFRICA. Combination in *C. (Myrmosphincta)*: Forel, 1914a: 273; in *C. (Orthonotomyrmex)*: Emery, 1920b: 257. Arnold, 1924: 714 (s.). Senior synonym of *ledieui:* Emery, 1920c: 29. Variety of *cubangensis:* Santschi, 1930b: 81 (in text).

***dolabratus.*** *Camponotus (Myrmobrachys) dolabratus* Menozzi, 1927d: 341, fig. 6 (s.w.) COSTA RICA.

***dolendus.*** *Camponotus rufoglaucus* r. *dolendus* Forel, 1892j: 238 (w.q.m.) INDIA. Imai, *et al.* 1984: 9 (k.). Combination in *C. (Myrmosericus)*: Forel, 1914a: 268. Raised to species: Bingham, 1903: 364; Forel, 1906b: 84; Santschi, 1920h: 174. Subspecies of *rufoglaucus:* Emery, 1925b: 106. Revived status as species: Wang, C., Xiao & Wu, 1989a: 222 (in key).

#***dolentulus.*** *Camponotus (Paracolobopsis) patimae* var. *dolentulus* Wheeler, W.M. 1942: 262 (s.w.) GUIANA. Combination in *C. (Myrmaphaenus)*: Kempf, 1972a: 47.

***dolichoderoides.*** *Camponotus dolichoderoides* Forel, 1911a: 51 (w.) BORNEO. Combination in *C.* (*Myrmosphincta*): Forel, 1914a: 273; in *C.* (*Myrmotemnus*): Emery, 1920b: 258; in *C.* (*Karavaievia*): Emery, 1925b: 116. See also: Dumpert, 1985: 568.

#***dominicensis.*** *Camponotus maculatus* subsp. *dominicensis* Wheeler, W.M. 1913d: 244 (s.w.) DOMINICA. Combination in *C.* (*Tanaemyrmex*): Emery, 1925b: 81. Race of *fumidus:* Santschi, 1922c: 101.

***donisthorpei.*** *Camponotus* (*Dinomyrmex*) *donisthorpei* Emery, 1920c: 23, fig. 6 (w.) CAMEROUN. Combination in *C.* (*Tanaemyrmex*): Emery, 1925b: 85.

*doriae. Colobopsis doriae* Mayr, 1872: 137 (w.) BORNEO. Emery, 1900d: 707 (s.q.). Combination in *Camponotus:* Dalla Torre, 1893: 228; in *C.* (*Colobopsis*): Forel, 1913k: 133. Junior synonym of *cylindricus:* Emery, 1917a: 97.

#***dorsalis.*** *Camponotus* (*Myrmotrema*) *foraminosus* st. *dorsalis* Santschi, 1926b: 265 (s.w.q.) TANZANIA.

***dorycus.*** *Formica dorycus* Smith, F. 1860b: 96 (w.) NEW GUINEA. Combination in *Camponotus:* Roger, 1863b: 4; in *C.* (*Myrmogigas*): Forel, 1912i: 91; in *C.* (*Dinomyrmex*): Forel, 1914a: 268; in *C.* (*Tanaemyrmex*): Emery, 1925b: 90. Current subspecies: nominal plus *confusus, recticeps.*

*dotalis Camponotus* (*Myrmopelta*) *chrysurus* st. *acutisquamis* var. *dotalis* Santschi, 1926a: 20 (w.) CONGO. Unavailable name.

***dracocephalus.*** *Camponotus* (*Tanaemyrmex*) *dracocephalus* Stitz, 1938: 112, fig. 10 (q.) BORNEO.

***dromas.*** *Camponotus* (*Myrmocamelus*) *dromas* Santschi, 1919a: 332 (s.w.m.) AUSTRALIA. Combination in *C.* (*Myrmophyma*): Emery, 1925b: 111; Santschi, 1928e: 481.

***dromedarius.*** *Camponotus dromedarius* Forel, 1891b: 65, pl. 2, fig. 5 (w.m.) MADAGASCAR. Combination in *C.* (*Myrmocamelus*): Forel, 1914a: 270; in *C.* (*Myrmosaga*): Emery, 1920b: 257; in *C.* (*Mayria*): Emery, 1925b: 122. Current subspecies: nominal plus *pulcher.*

***druryi.*** *Camponotus druryi* Forel, 1886f: 160 (w.) TANZANIA. Combination in *C.* (*Myrmosericus*): Forel, 1914a: 268.

*dubitatus. Camponotus rufoglaucus* r. *feai* var. *dubitatus* Emery, 1893c: 87 (w.) CANARY IS. Unavailable name.

***dufouri.*** *Camponotus dufouri* Forel, 1891b: 16, pl. 1, fig. 2 (s.w.) MADAGASCAR. Forel, 1895c: 244 (q.); Forel, 1895h: 488 (m.). Combination in *C.* (*Dinomyrmex*): Forel, 1914a: 268; in *C.* (*Tanaemyrmex*): Emery, 1925b: 85. Current subspecies: nominal plus *imerinensis.*

#***dulcis.*** *Camponotus mitis* var. *dulcis* Dalla Torre, 1893: 243 (w.) BURMA. [First available use of *Camponotus rubripes* r. *mitis* var. *dulcis* Emery, 1889b: 511; unavailable name.] Variety of *mitis:* Bingham, 1903: 356; of *variegatus:* Emery, 1925b: 96, Wheeler, W.M. 1927d: 10; Wheeler, W.M. 1928c: 32; Wheeler, W.M. 1929f: 9.

***dumetorum.*** *Camponotus maculatus* subsp. *dumetorum* Wheeler, W.M. 1910d: 354 (s.w.m.) U.S.A. Combination in *C.* (*Myrmoturba*): Wheeler, W.M. 1917a: 560; in *C.* (*Camponotus*): Emery, 1925b: 75; in *C.* (*Tanaemyrmex*): Creighton, 1950a: 377. Subspecies of *sansabeanus:* Emery, 1925b: 75. Junior synonym of *maccooki:* Creighton, 1950a: 377. Revived from synonymy and raised to species: Snelling, 1970: 396.

#***duncanensis.*** *Camponotus* (*Myrmamblys*) *macilentus* var. *duncanensis* Wheeler, W.M. 1919c: 283 (s.w.q.m.) GALAPAGOS IS. Combination in *C.* (*Pseudocolobopsis*): Emery, 1925b: 158.

*dusmeti. Camponotus* (*Myrmentoma*) *piceus* st. *spissinodis* var. *dusmeti* Santschi, 1932c: 71 (w.) SPAIN. Unavailable name.

#***ebeninithorax.*** *Camponotus* (*Myrmophyma*) *capito* var. *ebeninithorax* Forel, 1915b: 100 (w.) AUSTRALIA.

*ebeninus. Camponotus ebeninus* Emery, 1869b: 2, pl. 1, fig. 2 (w.m.) ITALY. Junior synonym of *foveolata:* Forel, 1874: 40.

#***ebneri.*** *Camponotus* (*Myrmentoma*) *lateralis* var. *ebneri* Finzi, 1930c: 24 (w.) LEBANON.

*eburneus. Camponotus landolti* subsp. *zonatus* var. *eburneus* Wheeler, W.M. 1905b: 117 (s.w.q.) BAHAMAS. Unavailable name; material referred to *inaequalis* by Wheeler, W.M. 1913b: 503 (in text).

***echinoploides.*** *Camponotus echinoploides* Forel, 1891b: 51, pl. 1, fig. 8 (w.) MADAGASCAR. Combination in *C.* (*Myrmobrachys*): Forel, 1914a: 270; in *C.* (*Orthonotomyrmex*): Emery, 1920b: 258; in *C.* (*Myrmepinotus*): Santschi, 1921f: 312.

*ectatommoides* Forel, 1892; see under ***NOTONCUS.***

***edmondi.*** *Camponotus edmondi* André, 1887: 281 (w.) MADAGASCAR. Forel, 1891b: 49 (s.). Combination in *C.* (*Myrmobrachys*): Forel, 1914a: 270; in *C.* (*Orthonotomyrmex*): Emery, 1920b: 258; in *C.* (*Myrmisolepis*): Santschi, 1921f: 310; in *C.* (*Myrmepinotus*) Emery, 1925b: 127. Current subspecies: nominal plus *ernesti.*

***egregius.*** *Formica egregia* Smith, F. 1858b: 45 (w.) BRAZIL. Forel, 1879a: 98 (s.); Emery, 1903: 77 (q.). Combination in *Camponotus:* Roger, 1863b: 3; in *C.* (*Myrmogigas*): Forel, 1912i: 91; in *C.* (*Dinomyrmex*): Forel, 1914a: 268; in *C.* (*Tanaemyrmex*): Emery, 1925b: 83. Race of *compressus:* Forel, 1879a: 89. Subspecies of *angusticollis:* Emery, 1896d: 370. Revived status as species: Forel, 1886f: 166; Emery, 1903: 77; Emery, 1920b: 256.

#***elegans.*** *Camponotus claripes* r. *elegans* Forel, 1902h: 496 (s.w.) AUSTRALIA. Combination in *C.* (*Myrmophyma*): Emery, 1925b: 110.

***elevatus.*** *Camponotus elevatus* Forel, 1899c: 144 (w.) COSTA RICA. Combination in *C.* (*Myrmamblys*): Forel, 1914a: 271; in *C.* (*Myrmobrachys*): Emery, 1920b: 260.

***ellioti.*** *Camponotus ellioti* Forel, 1891b: 37 (w.) MADAGASCAR. Forel, 1904b: 377 (s.). Combination in *C.* (*Myrmepomis*): Forel, 1912i: 92; in *C.* (*Myrmopiromis*): Wheeler, W.M. 1922a: 707. Current subspecies: nominal plus *relucens.*

*elongata. Camponotus* (*Myrmophyma*) *walkeri* r. *bardus* var. *elongata* Crawley, 1923b: 178 (s.w.) AUSTRALIA. Unavailable name (Taylor, 1986: 34).

***elysii.*** *Camponotus* (*Colobopsis*) *elysii* Mann, 1919: 372 (s.w.) SOLOMON IS.

***emarginatus.*** *Camponotus emarginatus* Emery, 1886: 358, pl. 17, fig. 2 (s.w.q.m.) SOUTH AFRICA. Combination in *C.* (*Myrmophyma*): Forel, 1914a: 269; Arnold, 1922: 663; in *C.* (*Myrmotemnus*): Emery, 1920b: 258; in *C.* (*Myrmamblys*): Wheeler, W.M. 1922a: 989; in *C.* (*Myrmespera*): Santschi, 1926c: 604.

*emeryi* Forel, 1879; see under ***POLYRHACHIS.***

*emeryi. Camponotus emeryi* Forel, 1901h: 72 (w.) PARAGUAY. [Junior primary homonym of *emeryi* Forel, 1879a: 113, above.] Replacement name: *emeryodicatus* Forel, 1901e: 371.

***emeryodicatus.*** *Camponotus emeryodicatus* Forel, 1901e: 371. Replacement name for *emeryi* Forel, 1901h: 72. [Junior primary homonym of *emeryi* Forel, 1879a: 113.] Combination in *C.* (*Myrmomalis*): Forel, 1914a: 271. Current subspecies: nominal plus *decessor.*

***empedocles.*** *Camponotus* (*Myrmoturba*) *thales* var. *empedocles* Emery, 1920c: 6 (s.w.) ZIMBABWE. [First available use of *Camponotus* (*Myrmoturba*) *maculatus* r. *thales* var. *empedocles* Forel, 1913j: 224; unavailable name.] Subspecies of *maculatus:* Arnold, 1922: 645. Raised to species: Collingwood, 1985: 279.

#***enganensis.*** *Camponotus* (*Myrmotarsus*) *nigricans* r. *enganensis* Forel, 1916: 443 (q.) SUMATRA. Senior synonym of *nitidiceps:* Viehmeyer, 1916b: 291.

***ephippiatus.*** *Camponotus* (*Myrmamblys*) *ephippiatus* Viehmeyer, 1916a: 160, fig. 13 (w.) SINGAPORE.

***ephippium.*** *Formica ephippium* Smith, F. 1858b: 39 (w.) AUSTRALIA. Wheeler, G.C. & Wheeler, J. 1968: 218 (l.). Combination in *Camponotus*): Roger, 1863b: 4; in *C.* (*Myrmocamelus*): Forel, 1914a: 270; in *C.* (*Myrmophyma*): Emery, 1920b: 257; Santschi, 1928e: 481. Current subspecies: nominal plus *narses.*

#***epinotalis.*** *Camponotus* (*Orthonotomyrmex*) *epinotalis* Santschi, 1916b: 508 (s.w.) ZIMBABWE. Combination in *C.* (*Myrmisolepis*): Santschi, 1921f: 310; Emery, 1925b: 134. Variety of *braunsi:* Emery, 1925b: 134. See also: Arnold, 1924: 716.

*epistomatus. Camponotus* (*Myrmothrix*) *abdominalis* st. *epistomatus* Santschi, 1936b: 208 (w.) COLOMBIA. Junior synonym of *atriceps:* Hashmi, 1973: 82.

***equus.*** *Camponotus* (*Colobopsis*) *equus* Stitz, 1932b: 734, fig. 2 (q.) SUMATRA.

***eremicus.*** *Camponotus* (*Myrmogonia*) *eremicus* Wheeler, W.M. 1915g: 815, pl. 66, figs. 5, 6 (s.w.) AUSTRALIA. Combination in *C.* (*Myrmophyma*): Emery, 1925b: 111; in *C.* (*Thlipsepinotus*): Santschi, 1928e: 483.

***erigens.*** *Camponotus maculatus* r. *erigens* Forel, 1894d: 6 (s.w.q.) ALGERIA. Combination in *C.* (*Tanaemyrmex*): Emery, 1925b: 99. Raised to species: Emery, 1920c: 6; Santschi, 1929g: 166. Current subspecies: nominal plus *subconcolor.*

***erinaceus.*** *Camponotus erinaceus* Gerstäcker, 1871: 355 (w.) TANZANIA. Emery, 1895a: 183 (q.); Forel, 1899b: 306 (s.). Combination in *C.* (*Myrmotrema*): Forel, 1912i: 91; in *C.* (*Orthonotomyrmex*): Forel, 1914a: 273; in *C.* (*Myrmisolepis*): Santschi, 1921f: 310.

*ermanii. Camponotus* (*Tanaemyrmex*) *picipes* subsp. *ermanii* Menozzi, 1936b: 94 (w.q.m.) BRAZIL. *Nomen nudum.*

#***ernesti.*** *Camponotus edmondi* var. *ernesti* Forel, 1891b: 50 (s.w.) MADAGASCAR. Combination in *C.* (*Myrmepinotus*): Emery, 1925b: 127.

***errabundus.*** *Camponotus* (*Tanaemyrmex*) *errabundus* Arnold, 1949: 270, fig. 11 (s.w.) ZIMBABWE.

***CAMPONOTUS** (continued)*

#***errans.*** *Camponotus (Myrmotarsus) pressipes* var. *errans* Forel, 1913k: 129 (q.) SUMATRA.

*erythraea. Camponotus (Myrmoturba) maculatus* var. *erythraea* Emery, 1920c: 14 (s.) ETHIOPIA. Junior synonym of *maculatus:* Baroni Urbani, 1972: 125.

#***erythromelus.*** *Camponotus erythromelus* Emery, 1896i: 158, fig. (w.) SOMALIA. Menozzi, 1927f: 361 (q.). Combination in *C. (Orthonotomyrmex)*: Santschi, 1914b: 135; in *C. (Myrmisolepis)*: Emery, 1925b: 134; Santschi, 1926b: 262. Subspecies of *braunsi:* Emery, 1897e: 605; Santschi, 1914b: 135; Wheeler, W.M. 1922a: 971. [Misspelled as *erythroderus* by Menozzi, 1939c: 108.]

*erythropus. Camponotus erythropus* Pergande, 1894a: 28 (s.w.) MEXICO. Junior synonym of *mina:* Emery, 1895c: 336.

*erythropus. Camponotus (Myrmosericus) feae* var. *erythropus* Stitz, 1917: 350 (w.) CANARY IS. [First available use of *Camponotus rufoglaucus* r. *feai* var. *erythropus* Emery, 1893c: 87; unavailable name.] [Unresolved junior primary homonym of *erythropus* Pergande, above.] [*Camponotus feae* var. *erythropes* Forel, 1893d: 466; Forel, 1894a: 162. *Nomina nuda.*] Junior synonym of *feae:* Barquin, 1981: 411.

*erythropus. Camponotus (Myrmosaga) erythropus* Viehmeyer, 1925b: 141 (w.q.) AUSTRALIA. [Junior primary homonym of *erythropus* Pergande, above.] Replacement name: *ezotus* **new name**.

***erythrostoma.*** *Camponotus (Myrmamblys) erythrostoma* Emery, 1920c: 34 (s.w.) BRAZIL. Combination in *C. (Pseudocolobopsis)*: Emery, 1925b: 158.

***esau.*** *Camponotus (Myrmocamelus) esau* Forel, 1915b: 103, fig. 6 (w.) AUSTRALIA. Dubiously in *C. (Myrmophyma)*: Emery, 1925b: 112.

*esau. Camponotus (Myrmotrema) auropubens* var. *esau* Santschi, 1915c: 275. *Nomen nudum.*

#***escherichi.*** *Camponotus (Tanaemyrmex) aethiops* var. *escherichi* Emery, 1925b: 97 (s.w.) TURKEY. [First available use of *Camponotus maculatus* subsp. *oertzeni* var. *escherichi* Emery, 1897f: 239; unavailable name.]

*escherichi. Camponotus (Myrmaphaenus) escherichi* Menozzi, 1926c: 71, fig. 2 (w.) BRAZIL. [Junior primary homonym of *escherichi* Emery, above.] Replacement name: *fallatus* **new name**.

#***esdras.*** *Camponotus (Myrmobrachys) planatus* var. *esdras* Forel, 1916: 458 (w.) MEXICO.

***essigi.*** *Camponotus caryae* var. *essigi* Smith, M.R. 1923b: 306 (s.w.) U.S.A. Combination in *C. (Myrmentoma)*, and raised to species: Creighton, 1950a: 387. Senior synonym of *nevadensis:* Wheeler, G.C. & Wheeler, J. 1986g: 61. See also: Smith, M.R. 1951a: 844; Snelling, 1988: 68.

*esuriens. Formica esuriens* Smith, F. 1858b: 54 (s.w.) MEXICO. Norton, 1868c: 1 (q.m.). Combination in *Camponotus:* Roger, 1863b: 5; in *C. (Myrmothrix)*: Emery, 1925b: 107. Race of *atriceps:* Forel, 1879a: 76; Forel, 1885a: 340. Revived status as species: Pergande, 1896: 862. Subspecies of *abdominalis:* Emery, 1896d: 372; Forel, 1899c: 133; Kempf, 1972a: 60. Senior synonym of *vulpinus:* Roger, 1863b: 5; of *fulvacea:* Forel, 1885a: 340. Junior synonym of *atriceps:* Hashmi, 1973: 82.

***ethicus.*** *Camponotus ethicus* Forel, 1897c: 200 (w.m.) MADAGASCAR. Combination in *C. (Myrmentoma)*: Forel, 1912i: 92; in *C. (Orthonotomyrmex)*: Forel, 1914a: 273; in *C. (Myrmisolepis)*: Santschi, 1921f: 310; in *C. (Myrmepinotus)*: Emery, 1925b: 127.

*etiennei. Camponotus maculatus* subsp. *negus* var. *etiennei* Forel, 1910e: 453 (s.w.) ZAIRE. Unavailable name.

***etiolatus.*** *Camponotus (Colobopsis) abditus* var. *etiolatus* Wheeler, W.M. 1904b: 150, fig. 5 (s.w.q.m.) U.S.A. Wheeler, G.C. & Wheeler, J. 1970: 650 (l.). Raised to species: Wheeler, W.M. 1927f: 31; Wheeler, W.M. 1934g: 216.

***etiolipes.*** *Camponotus etiolipes* **new name**. Replacement name for *Formica longipes* Gerstäcker, 1859: 262. [Junior primary homonym of *Formica longipes* Latreille, 1802c: 233 (now in *Pheidole*).]

#***eubarbaricus.*** *Camponotus barbaricus* subsp. *eubarbaricus* Cagniant, 1970c: 32 (w.q.m.) ALGERIA.

#***euchrous.*** *Camponotus (Orthonotomyrmex) sericeus* var. *euchrous* Santschi, 1926b: 267 (q.) SUDAN.

#***eudokiae.*** *Camponotus herculeanus* var. *eudokiae* Ruzsky, 1926: 108 (w.) RUSSIA.

#***eugaster.*** *Camponotus (Myrmobrachys) cameranoi* var. *eugaster* Santschi, 1922c: 121 (s.) BRAZIL.

***eugeniae.*** *Camponotus micans* r. *eugeniae* Forel, 1879a: 78 (s.w.) SOUTH AFRICA. Forel, 1886f: 174 (q.); Arnold, 1924: 687 (m.). Combination in *C. (Myrmosericus)*: Forel, 1912i: 91. Race of *rufoglaucus:* Mayr, 1893: 195. Raised to species: Forel, 1886f: 174; Emery, 1895h: 50. Current subspecies: nominal plus *amplior.*

***eurynotus.*** *Camponotus eurynotus* Forel, 1907a: 35 (w.) BRAZIL. Combination in *C. (Myrmeurynota)*: Forel, 1912i: 92.

***evae.*** *Camponotus evae* Forel, 1910b: 74 (w.) AUSTRALIA. Forel, 1915b: 101 (s.); Wheeler, G.C. & Wheeler, J. 1968: 218 (l.). Combination in *C. (Myrmogonia)*: Forel, 1914a: 269; in *C. (Myrmophyma)*: Emery, 1925b: 111; in *C. (Thlipsepinotus)*: Santschi, 1928e: 483. Current subspecies: nominal plus *zeuxis.*

***evansi.*** *Camponotus (Myrmoturba) maculatus* subsp. *evansi* Crawley, 1920b: 178 (s.w.) IRAQ. Combination in *C. (Tanaemyrmex)*: Emery, 1925b: 89. Raised to species: Pisarski, 1971b: 727.

*exasperata. Formica exasperata* Smith, F. 1857a: 56 (w.) BORNEO. Mayr, 1867a: 37 (q.). Combination in *Camponotus:* Mayr, 1862: 659. Junior synonym of *misturus:* Mayr, 1886c: 354.

***excavatus.*** *Camponotus (Colobopsis) excavatus* Donisthorpe, 1948e: 121, figs. a, b (q.) NEW GUINEA.

*excelsus. Camponotus (Myrmosericus) feae* var. *excelsus* Stitz, 1917: 350 (w.) CANARY IS. [First available use of *Camponotus rufoglaucus* r. *feai* var. *excelsus* Emery, 1893c: 88; unavailable name.] Junior synonym of *feae:* Barquin, 1981: 411.

***excisus.*** *Camponotus excisus* Mayr, 1870a: 384, pl. fig. 4 (w.) COLOMBIA. Combination in *C. (Myrmobrachys)*: Forel, 1914a: 270.

***exiguoguttatus.*** *Camponotus sexguttatus* subsp. *exiguoguttatus* Forel, 1886d: 239 (s.w.q.) INDIA. Combination in *C. (Tanaemyrmex)*: Emery, 1925b: 95. Raised to species: Dalla Torre, 1893: 230. Subspecies of *nicobarensis:* Emery, 1889b: 513; Emery, 1895k: 478; Emery, 1896d: 371; Bingham, 1903: 364; Viehmeyer, 1916a: 156; Forel, 1922: 100. Raised to species: Wang, C., Xiao & Wu, 1989b: 327.

#***eximius.*** *Camponotus festinus* subsp. *eximius* Emery, 1900d: 701 (s.w.q.) SUMATRA. Forel, 1912n: 67 (m.). Combination in *C. (Myrmoturba)*: Forel, 1913k: 128; in *C. (Tanaemyrmex)*: Emery, 1925b: 90.

#***exsanguis.*** *Camponotus mystaceus* var. *exsanguis* Forel, 1910f: 28 (s.w.q.m.) NAMIBIA. Combination in *C. (Myrmopsamma)*: Wheeler, W.M. 1922a: 988.

***exsectus.*** *Camponotus exsectus* Emery, 1900d: 709, fig. 9 (w.q.) INDONESIA (Mentawei I.). Combination in *C. (Camponotus)*: Forel, 1914a: 266; in *C. (Myrmophyma)*: Wheeler, W.M. 1919e: 111; in *C. (Myrmotemnus)*: Emery, 1920b: 258; in *C. (Karavaievia)*: Emery, 1925b: 116. See also: Dumpert, 1985: 569.

***extensus.*** *Camponotus extensus* Mayr, 1876: 65 (s.w.) AUSTRALIA. Combination in *C. (Myrmoturba)*: Forel, 1914a: 267; in *C. (Tanaemyrmex)*: Emery, 1925b: 90. Race of *sexguttatus:* Forel, 1879a: 73. Revived status as species: Dalla Torre, 1893: 230; Emery, 1894f: 3; Emery, 1896d: 371; Emery, 1920b: 255.

***ezotus.*** *Camponotus ezotus* **new name**. Replacement name for *erythropus* Viehmeyer, 1925b: 141. [Junior primary homonym of *erythropus* Pergande, 1894a: 28.]

***fabricator.*** *Formica fabricator* Smith, F. 1858b: 35 (s.w.q.) ST HELENA I. Combination in *Camponotus:* Roger, 1863b: 2; in *C. (Myrmoturba)*: Forel, 1914a: 267; in *C. (Tanaemyrmex)*: Emery, 1925b: 104. Variety of *sylvaticus:* Mayr, 1886c: 355. Subspecies of *maculatus:* Emery, 1896d: 371. Revived status as species: Dalla Torre, 1893: 230; Forel, 1914a: 267.

*fabricii* Roger, 1862; see under ***DENDROMYRMEX.***

#***fairmairei.*** *Camponotus maculatus* st. *fairmairei* Santschi, 1911e: 130, fig. 3 (s.w.) MADAGASCAR. Combination in *C. (Tanaemyrmex)*: Emery, 1925b: 86. Subspecies of *maculatus:* Wheeler, W.M. 1922a: 1040; of *hova:* Emery, 1920c: 6; Emery, 1925b: 86.

***falco.*** *Camponotus falco* Forel, 1902b: 175 (w.) COLOMBIA. Combination in *C. (Myrmocamelus)*: Forel, 1914a: 270; in *C. (Myrmamblys)*: Emery, 1920b: 259; in *C. (Myrmaphaenus)*: Emery, 1925b: 155.

***fallatus.*** *Camponotus fallatus* **new name**. Replacement name for *escherichi* Menozzi, 1926c: 71. [Junior primary homonym of *escherichi* Emery, 1925b: 97.]

***fallax.*** *Formica fallax* Nylander, 1856b: 57 (s.w.) FRANCE. Roger, 1859: 230 (q.m.). Combination in *Camponotus:* Mayr, 1861: 36; in *C. (Camponotus)*: Forel, 1914a: 266; in *C. (Myrmentoma)*: Emery, 1920b: 257. Junior synonym of *marginatus:* Roger, 1863b: 1; Mayr, 1886d: 423; Nasonov, 1889: 10. Revived from synonymy: Wheeler, W.M. 1913c: 117; Forel, 1915d: 67; Emery, 1916b: 226. Variety of *caryae:* Wheeler, W.M. 1917c: 27. Revived status as species: Wheeler, W.M. 1917a: 558; Bondroit, 1918: 71; Finzi, 1924a: 14; Karavaiev, 1927c: 277. See also: Kutter, 1977c: 206; Collingwood, 1979: 88; Atanassov & Dlussky, 1992: 220. Current subspecies: nominal plus *pageti.*

#***famelicus.*** *Camponotus subnitidus* st. *famelicus* Emery, 1887a: 214 (footnote), pl. 3, fig. 5 (w.) AUSTRALIA.

Wheeler, G.C. & Wheeler, J. 1968: 217 (l.). Combination in *C.* (*Tanaemyrmex*): Emery, 1925b: 91.

***fasciatellus.*** *Camponotus fasciatellus* Dalla Torre, 1892: 91. Replacement name for *fasciatus* Mayr, 1870a: 379. [Junior secondary homonym of *fasciata* Mayr, 1867a: 69, below.] Combination in *C.* (*Tanaemyrmex*): Emery, 1925b: 80.

***fasciatus.*** *Colobopsis fasciata* Mayr, 1867a: 69 (q.) JAVA. Combination in *Camponotus* (*Colobopsis*): Emery, 1889b: 517; in *C.* (*Myrmamblys*): Forel, 1914a: 271.

*fasciatus. Camponotus fasciatus* Mayr, 1870a: 379 (s.w.) COLOMBIA. [Junior secondary homonym of *Colobopsis fasciata* Mayr, above.] Replacement name: *fasciatellus* Dalla Torre, 1892: 91.

***fastigatus.*** *Camponotus fastigatus* Roger, 1863b: 5 (w.) BRAZIL. Emery, 1903: 73 (q.); Forel, 1908c: 403 (m.). Combination in *C.* (*Myrmamblys*): Forel, 1912i: 90; in *C.* (*Neomyrmamblys*): Wheeler, W.M. 1922a: 708; Santschi, 1922c: 113; in *C.* (*Myrmaphaenus*): Emery, 1925b: 155. Current subspecies: nominal plus *barbiculus, schmalzi, vagulus, verae.*

**#*fatuus.*** *Camponotus kubaryi* r. *fatuus* Forel, 1886f: 163 (w.) SUMATRA. Viehmeyer, 1916a: 156 (s.q.m.). Race of *irritans:* Emery, 1888a: 528. Raised to species: Dalla Torre, 1893: 230. Subspecies of *maculatus:* Emery, 1896d: 370; Viehmeyer, 1916a: 156; of *irritans:* Emery, 1920c: 7.

***favorabilis.*** *Camponotus* (*Myrmosaga*) *favorabilis* Santschi, 1919a: 349 (w.) ZIMBABWE. Combination in *C.* (*Myrmamblys*): Emery, 1925b: 141; in *C.* (*Myrmosaga*): Santschi, 1926c: 600.

***fayfaensis.*** *Camponotus fayfaensis* Collingwood, 1985: 279, fig. 70 (w.) SAUDI ARABIA. Current subgeneric placement: *C.* (*Myrmotrema*).

**#*feae.*** *Camponotus feae* Emery, 1882: 449, fig. (s.w.) CANARY IS. Junior synonym of, and first available replacement name for, *pubescens* Brullé, 1840: 84. [Junior primary homonym of *pubescens* Fabricius, 1775: 392.] Combination in *C.* (*Myrmoserícus*): Forel, 1914a: 268. Race of *rufoglaucus:* Emery, 1893c: 87; Crawley, 1923a: 31; Emery, 1925b: 106. Senior synonym of *erythropus* and *excelsus:* Barquin, 1981: 411.

***fedtschenkoi.*** *Camponotus fedtschenkoi* Mayr, 1877: 3 (s.w.q.m.) TURKESTAN. [Redescribed in Latin: Mayr, 1880: 23.] Combination in *C.* (*Tanaemyrmex*): Emery, 1925b: 93. Subspecies of *sylvaticus:* Forel, 1879a: 67; of *rubripes:* Forel, 1886f: 143; of *maculatus:* Emery, 1896d: 370; Emery, 1898c: 127; Ruzsky, 1902c: 6; Forel, 1904b: 378; Ruzsky, 1905b: 194; Emery, 1908a: 203. Revived status as species: Emery, 1920c: 7; Emery, 1925b: 93; Stitz, 1930: 239; Tarbinsky, 1976: 153; Arnol'di & Dlussky, 1978: 551. Current subspecies: nominal plus *adi.*

***fellah.*** *Camponotus oasium* var. *fellah* Dalla Torre, 1893: 245 (w.) EGYPT. [First available use of *Camponotus maculatus* r. *oasium* var. *fellah* Emery, 1891b: 18; unavailable name.] Emery, 1908a: 194 (s.q.m.). Combination in *C.* (*Tanaemyrmex*): Emery, 1925b: 99. Subspecies of *maculatus:* Forel, 1910a: 14; Forel, 1910c: 267; of *compressus:* Santschi, 1939d: 80. Raised to species: Pisarski, 1971b: 728.

***femoratus.*** *Formica femorata* Fabricius, 1804: 397 (q.) BRAZIL. Emery, 1894c: 174 (w.); Forel, 1904c: 49 (s.). Combination in *Camponotus:* Roger, 1862c: 284; in *C.* (*Myrmothrix*): Forel, 1914a: 269. Junior synonym of *rufipes:* Roger, 1861b: 164. Revived from synonymy: Mayr, 1863: 14. See also: Hashmi, 1973: 44.

**#*fenestralis.*** *Camponotus* (*Myrmespera*) *nasutus* var. *fenestralis* Santschi, 1937a: 65 (w.q.) NAMIBIA.

*ferganensis. Camponotus ferganensis* Pisarski, 1967: 414 (s.w.) TURKESTAN. [First available use of *Camponotus maculatus* subsp. *dichrous* var. *ferganensis* Ruzsky, 1905b: 200; unavailable name.] Current subgeneric placement: *C.* (*Tanaemyrmex*). Junior synonym of *turkestanus:* Dlussky, Soyunov & Zabelin, 1990: 132. See also: Tarbinsky, 1976: 164.

**#*fernandinensis.*** *Camponotus* (*Myrmorhachis*) *planus* var. *fernandinensis* Wheeler, W.M. 1919c: 296 (s.) GALAPAGOS IS. Combination in *C.* (*Myrmocladoecus*): Emery, 1925b: 167.

***ferreri.*** *Camponotus* (*Myrmamblys*) *ferreri* Forel, 1913e: 671 (q.) CAMEROUN. Combination in *C.* (*Paramyrmamblys*): Santschi, 1926c: 604. Current subspecies: nominal plus *akka, cavisquamis.*

*ferruginea. Formica ferruginea* Fabricius, 1798: 279 (w.q.) U.S.A. [Junior primary homonym of *Formica ferruginea* Christ, 1791: 512.] Replacement name: *Camponotus chromaiodes* **new name**. Wheeler, W.M. 1910d: 338 (s.m.). Combination in *Camponotus:* Roger, 1863b: 6; in *C.* (*Camponotus*): Emery, 1925b: 72. Junior synonym of *pennsylvanicus:* Roger, 1863b: 6. Revived from synonymy as variety of *pennsylvanicus:* Forel, 1879a: 56. Variety of *herculeanus:* Emery, 1925b: 72. Subspecies of *pennsylvanicus:* Creighton, 1950a: 368. Revived status as species: Brown, 1950d: 158.

**#*ferrugineus.*** *Camponotus christi* var. *ferruginea* Emery, 1899f: 290 (w.) MADAGASCAR. [Unresolved junior secondary homonym of *Formica ferruginea* Fabricius, above.] Combination in *C.* (*Mayria*): Emery, 1925b: 122.

*ferruginipes. Camponotus* (*Myrmosaga*) *ferruginipes* Crawley, in Poulton & Crawley, 1922: 125 (s.) AUSTRALIA. Combination in *C.*(*Myrmophyma*): Emery, 1925b: 111. Junior synonym of *hartogi:* Brown, 1956a: 40.

*fervens. Formica fervens* Smith, F. 1857a: 55 (w.) BORNEO. [Junior primary homonym of *Formica fervens* Drury, 1782: 58 (now in *Atta*).] Replacement name: *fervidus* Donisthorpe, 1943e: 116. Combination in *Camponotus:* Roger, 1863b: 3; in *C.* (*Myrmoturba*): Forel, 1914a: 267.

***fervidus.*** *Camponotus* (*Tanaemyrmex*) *fervidus* Donisthorpe, 1943e: 116. Replacement name for *fervens* Smith, above. [Junior primary homonym of *fervens* Drury, 1782: 58.]

***festai.*** *Camponotus maculatus* subsp. *festai* Emery, 1894d: 1 (s.w.) LEBANON. Emery, 1898c: 149 (q.). Combination in *C.* (*Myrmoturba*): Forel, 1913d: 434; in *C.* (*Tanaemyrmex*): Emery, 1925b: 99. Raised to species: Emery, 1905d: 31 (in text); Emery, 1908a: 187; Emery, 1920c: 6. Senior synonym of *caeciliae:* Emery, 1908a: 187. Current subspecies: nominal plus *riedeli.*

***festinatus.*** *Formica festinata* Buckley, 1866: 164 (w.q.) U.S.A. Wheeler, W.M. 1910d: 312 (s.m.); Wheeler, G.C. & Wheeler, J. 1970: 651 (l.). Combination in *Camponotus:* Dalla Torre, 1893: 231; in *C.* (*Myrmoturba*): Wheeler, W.M. 1917a: 561; in *C.* (*Tanaemyrmex*): Emery, 1925b: 80. Variety/subspecies of *fumidus:* Wheeler, W.M. 1902f: 22; Creighton, 1950a: 376. Revived status as species: Snelling, 1968b: 350. Senior synonym of *pubicornis:* Wheeler, W.M. 1910d: 312; of *spurcus:* Creighton, 1950a: 376; of *fragilis:* Snelling, 1968b: 351.

***festinus.*** *Formica festina* Smith, F. 1857a: 53 (q.) BORNEO. Emery, 1887a: 216 (w.); Emery, 1900d: 701 (s.); Imai, Brown, *et al.* 1984: 68 (k.). Combination in *Camponotus:* Roger, 1863b: 3; in *C.* (*Myrmoturba*): Forel, 1913k: 128; in *C.* (*Dinomyrmex*): Santschi, 1921f: 312; in *C.* (*Tanaemyrmex*): Emery, 1925b: 90. Current subspecies: nominal plus *cetegus, diligens, eximius, inezae, simaluranus.*

**#*fezzanensis.*** *Camponotus* (*Tanaemyrmex*) *foleyi* subsp. *fezzanensis* Bernard, 1953a: 190 (s.w.q.m.) ALGERIA. [First available use of *Camponotus* (*Tanaemyrmex*) *compressus* subsp. *foleyi* var. *fezzanensis* Bernard, 1948: 159; unavailable name.]

*fictor. Camponotus* (*Colobopsis*) *fictor* Forel, 1902h: 508 (s.w.) AUSTRALIA. Wheeler, G.C. & Wheeler, J. 1968: 217 (l.). Junior synonym of *macrocephalus* Erichson: Shattuck & McArthur, 1995 (in press).

**#*fidelis.*** *Camponotus* (*Myrmorhachis*) *planus* var. *fidelis* Wheeler, W.M. 1919c: 295 (s.w.q.) GALAPAGOS IS. Combination in *C.* (*Myrmocladoecus*): Emery, 1925b: 167.

***fiebrigi.*** *Camponotus fiebrigi* Forel, 1906d: 247 (s.w.) PARAGUAY. Combination in *C.* (*Myrmamblys*): Forel, 1914a: 271; in *C.* (*Myrmaphaenus*): Emery, 1925b: 155. Race of *alboannulatus:* Forel, 1912i: 83. Revived status as species: Emery, 1925b: 155.

***fieldeae.*** *Camponotus fieldeae* Forel, 1902h: 495 (s.w.) AUSTRALIA. Combination in *C.* (*Myrmoturba*): Forel, 1914a: 267; in *C.* (*Tanaemyrmex*): Emery, 1925b: 90.

***fieldellus.*** *Camponotus fieldellus* Forel, 1910b: 79 (s.w.q.) AUSTRALIA. Forel, 1915: 100 (m.). Combination in *C.* (*Myrmophyma*): Forel, 1912i: 91.

***figaro.*** *Camponotus* (*Myrmentoma*) *figaro* Collingwood & Yarrow, 1969: 84 (w.) SPAIN. [First available use of *Camponotus* (*Myrmoturba*) *lateralis* subsp. *piceus* var. *figaro* Emery, 1924c: 170; unavailable name.]

**#*fimbriatipes.*** *Camponotus* (*Myrmotrema*) *fimbriatipes* Santschi, 1920g: 181 (w.) CAMEROUN. Junior synonym of *conradti:* Emery, 1920c: 31. Revived from synonymy as variety of *conradti:* Santschi, 1926c: 612.

*fissa. Camponotus* (*Myrmothrix*) *cingulatus* var. *fissa* Stitz, 1933: 72, fig. 4 (w.) BRAZIL. Junior synonym of *cingulatus:* Hashmi, 1973: 31.

***flavescens.*** *Formica flavescens* Fabricius, 1793: 353 (w.) FRENCH GUIANA. Combination in *Camponotus:* Roger, 1863b: 4; in *C.* (*Tanaemyrmex*): Emery, 1925b: 80.

**#*flavidopubescens.*** *Camponotus aeneopilosus* var. *flavidopubescens* Forel, 1902h: 504 (w.) AUSTRALIA. Combination in *C.* (*Myrmophyma*): Emery, 1925b: 110. Raised to species: Santschi, 1928e: 482. Subspecies of *aeneopilosus:* Taylor & Brown, D.R. 1985: 110.

*flavifemur. Camponotus* (*Myrmoturba*) *maculatus* var. *flavifemur* Santschi, 1937g: 84 (w.q.) TANZANIA. Junior synonym of *maculatus:* Baroni Urbani, 1972: 125.

*CAMPONOTUS (continued)*

*flavobasalis. Camponotus* (*Myrmothrix*) *abdominalis* st. *mediopallidus* var. *flavobasalis* Santschi, 1936b: 209 (w.q.) BRAZIL. Unavailable name; material referred to *atriceps* by Hashmi, 1973: 82.

***flavocassis.*** *Camponotus* (*Myrmamblys*) *flavocassis* Donisthorpe, 1941c: 137 (s.w.) NEW GUINEA.

***flavocrines.*** *Camponotus* (*Myrmophyma*) *flavocrines* Donisthorpe, 1941j: 207 (w.) NEW GUINEA.

***flavolimbatus.*** *Camponotus* (*Colobopsis*) *flavolimbatus* Viehmeyer, 1922: 218 (q.) SAMOA. Senior synonym of *buxtoni* and *nautarum:* Wilson & Taylor, 1967: 97.

***flavomarginatus.*** *Camponotus flavomarginatus* Mayr, 1862: 664 (w.) GHANA. Forel, 1907c: 144 (q.m.); Menozzi, 1926a: 38 (q.m.). Combination in *C.* (*Myrmosericus*): Santschi, 1916b: 509. Variety of *micans:* Emery, 1881b: 526. Junior synonym of *micans:* Dalla Torre, 1893: 243. Subspecies of *rufoglaucus:* Forel, 1892j: 237; Emery, 1895h: 48; Santschi, 1911g: 212; Arnold, 1924: 692; Santschi, 1937d: 244. Senior synonym of *albisectus:* Emery, 1895h: 48; Wheeler, W.M. 1922a: 968; Menozzi, 1930b: 118. Revived status as species: Collingwood, 1985: 281.

*flavominor. Camponotus* (*Tanaemyrmex*) *maculatus* var. *flavominor* Emery, 1925b: 87 (s.w.) BENIN. [First available use of *Camponotus* (*Myrmoturba*) *maculatus* st. *melanocnemis* var. *flavominor* Santschi, 1920i: 4; unavailable name.] Junior synonym of *maculatus:* Baroni Urbani, 1972: 125.

*flavonitidus. Camponotus* (*Tanaemyrmex*) *turkestanus* var. *flavonitidus* Emery, 1925b: 101 (w.) TURKESTAN. [First available use of *Camponotus maculatus* subsp. *dichrous* var. *flavonitidus* Ruzsky, 1905b: 201; unavailable name.] Junior synonym of *turkestanus:* Dlussky, Soyunov & Zabelin, 1990: 132.

*flavopilosus. Camponotus fulvopilosus* var. *flavopilosus* Emery, 1895h: 54 (w.) NAMIBIA. Forel, 1910f: 25 (m.); Santschi, 1930b: 78 (q.). Combination in *C.* (*Myrmopiromis*): Wheeler, W.M. 1922a: 986. See also: Arnold, 1922: 738. Junior synonym of *fulvopilosus:* Robertson, 1990: 327.

*flavopilosus. Camponotus rufoglaucus* subsp. *flavopilosus* Viehmeyer, 1913: 47 (w.) TANZANIA. [Junior primary homonym of *flavopilosus* Emery, above.] Replacement name: *controversus* Santschi, 1916b: 509.

#***flavopubens.*** *Camponotus* (*Tanaemyrmex*) *melanoticus* subsp. *flavopubens* Emery, 1925b: 81 (w.) MEXICO. [First available use of *Camponotus maculatus* r. *simillimus* var. *flavopubens* Forel, 1901c: 135; unavailable name.]

*flavosetosus. Camponotus* (*Myrmopiromis*) *flavosetosus* Donisthorpe, 1945d: 271 (s.w.) TOGO. Junior synonym of *acvapimensis:* Brown, 1956a: 39.

#***flavotestaceus.*** *Camponotus* (*Tanaemyrmex*) *variegatus* var. *flavotestaceus* Donisthorpe, 1948b: 311 (s.w.q.) NEW GUINEA.

*flavotibialis. Camponotus* (*Myrmophyma*) *claripes* subsp. *orbiculatopunctatus* var. *flavotibialis* Viehmeyer, 1925b: 143 (s.) AUSTRALIA. Unavailable name.

#***flavus.*** *Camponotus foraminosus* r. *flavus* Stitz, 1916: 397, fig. 11 (s.w.) CONGO. Combination in *C.* (*Myrmotrema*): Wheeler, W.M. 1922a: 980.

***fletcheri.*** *Camponotus* (*Orthonotomyrmex*) *fletcheri* Donisthorpe, 1942g: 250 (w.) SRI LANKA.

*flexus. Camponotus flexus* Mayr, 1862: 671 (w.) BRAZIL. Junior synonym of *crassus:* Mayr, 1870a: 387.

***floridanus.*** *Formica floridana* Buckley, 1866: 161 (w.) U.S.A. Mayr, 1886d: 423 (q.m.). Combination in *Camponotus:* Mayr, 1886d: 423; in *C.* (*Myrmothrix*): Emery, 1925b: 108. Variety of *atriceps:* Mayr, 1886d: 423. Subspecies of *abdominalis:* Emery, 1893i: 670. Revived status as species: Wheeler, W.M. 1910d: 325. Subspecies of *abdominalis:* Wheeler, W.M. 1913c: 117. Senior synonym of *yankee:* Mayr, 1886d: 423. Junior synonym of *atriceps:* Hashmi, 1973: 82. Revived from synonymy as subspecies of *abdominalis:* Deyrup & Trager, 1986: 219. Revived status as species: Deyrup, Trager, *et al.* 1989: 100.

***florius.*** *Camponotus* (*Myrmacrhaphe*) *florius* Santschi, 1926c: 608 (w.) CONGO.

#***foersteri.*** *Camponotus christi* r. *foersteri* Forel, 1886f: 185 (w.) MADAGASCAR. Forel, 1891b: 64 (s.). Combination in *C.* (*Mayria*): Emery, 1925b: 122. Raised to species: Dalla Torre, 1893: 231. Variety/subspecies of *christi:* Emery, 1896d: 771; Wheeler, W.M. 1922a: 1045.

***foleyi.*** *Camponotus compressus* st. *foleyi* Santschi, 1939d: 82, figs. 15–17 (s.) ALGERIA. Bernard, 1953a: 191 (m.). Raised to species: Bernard, 1953a: 189. Current subspecies: nominal plus *ajjer, fezzanensis, grasi, hoggarensis* (unresolved junior homonym), *lelubrei, rufescens.*

#***folicola.*** *Camponotus urichii* r. *folicola* Forel, 1904c: 47 (s.w.m.) BRAZIL. Combination in *C.* (*Myrmosphincta*): Emery, 1925b: 152.

***foraminosus.*** *Camponotus foraminosus* Forel, 1879a: 87 (s.w.) SENEGAMBIA. Forel, 1913e: 671 and Forel, 1916: 444 (q.); Wheeler, G.C. & Wheeler, J. 1953e: 196 (l.). Combination in *C.* (*Myrmotrema*): Forel, 1912i: 91. See also: Forel, 1910c: 272. Current subspecies: nominal plus *aldabrensis, chrysogaster, cuitensis, deductus, dorsalis, flavus, honorus.*

***foreli.*** *Camponotus foreli* Emery, 1881b: 526 (footnote) fig. (w.) ALGERIA. Combination in *C.* (*Myrmosphincta*): Forel, 1914a: 273; in *C.* (*Orthonotomyrmex*): Emery, 1920b: 258; in *C.* (*Tanaemyrmex*): Emery, 1925b: 97. Current subspecies: nominal plus *tingitanus.*

#***formaster.*** *Camponotus* (*Myrmocladoecus*) *mucronatus* subsp. *formaster* Wheeler, W.M. 1934g: 227 (s.w.) GUATEMALA.

***formiciformis.*** *Camponotus senex* r. *formiciformis* Forel, 1885a: 345 (s.w.) GUATEMALA. Combination in *C.* (*Myrmobrachys*): Emery, 1925b: 164. Raised to species: Dalla Torre, 1893: 232; Forel, 1899c: 141.

*formosae. Camponotus maculatus* subsp. *taylori* var. *formosae* Wheeler, W.M. 1909d: 336 (s.w.) TAIWAN. Unavailable name.

***formosensis.*** *Camponotus punctatissimus* subsp. *formosensis* Wheeler, W.M. 1927e: 3 (s.) TAIWAN. Raised to species: Yasumatsu & Brown, 1951: 42. See also: Brown, 1952d: 19.

#***formosulus.*** *Camponotus* (*Myrmobrachys*) *capperi* var. *formosulus* Wheeler, W.M. 1917g: 468, fig. 3d (w.m.) JAMAICA. Combination in *C.* (*Myrmaphaenus*): Emery, 1925b: 156.

#***fornaronis.*** *Camponotus mistura* subsp. *fornaronis* Forel, 1892j: 232 (s.) INDIA. Combination in *C.* (*Myrmotarsus*): Emery, 1925b: 135. Raised to species: Bingham, 1903: 360. Subspecies of *misturus:* Forel, 1911e: 290.

***fornasinii.*** *Camponotus maculatus* subsp. *fornasinii* Emery, 1895h: 46 (w.) MOZAMBIQUE. Combination in *C.* (*Myrmoturba*): Forel, 1914a: 267; in *C.* (*Tanaemyrmex*): Emery, 1925b: 99. Subspecies of *maculatus:* Arnold, 1922: 641; Wheeler, W.M. 1922a: 954. Raised to species: Forel, 1914a: 267; Emery, 1920c: 6; Emery, 1925b: 99.

*foveolata. Formica foveolata* Mayr, 1853c: 277 (w.q.m.) HUNGARY. Combination in *Camponotus:* Roger, 1863b: 1; in *C.* (*Orthonotomyrmex*): Forel, 1915d: 72. Junior synonym of *lateralis:* Mayr, 1855: 322; Nylander, 1856b: 58. Revived from synonymy as variety of *lateralis:* Forel, 1874: 40; Forel, 1892i: 306. Senior synonym of *ebeninus:* Forel, 1874: 40; Forel, 1879a: 94. Junior synonym of *merulus:* Emery, 1896d: 373; Emery, 1916b: 226. Junior synonym of *piceus:* Dalla Torre, 1893: 238; Ruzsky, 1905b: 255; Kutter, 1977c: 207; Atanassov & Dlussky, 1992: 222.

#***foveolatus.*** *Camponotus* (*Myrmoturba*) *maculatus* r. *foveolatus* Stitz, 1925: 125 (s.w.) PHILIPPINES. [Unresolved junior secondary homonym of *Formica foveolata* Mayr, above.]

*fragilis. Camponotus fragilis* Pergande, 1894a: 26 (s.w.) MEXICO. Combination in *C.* (*Myrmoturba*): Forel, 1914a: 267. Variety of *fumidus:* Emery, 1895c: 336. Junior synonym of *festinatus:* Snelling, 1968b: 351.

*frater. Camponotus maculatus* subsp. *fatuus* var. *frater* Emery, 1911c: 258 (w.) NEW GUINEA. Unavailable name.

#***fraterculus.*** *Camponotus maculatus* subsp. *fraterculus* Wheeler, W.M. & Mann, 1914: 50, fig. 21 (s.w.) HAITI. Combination in *C.* (*Tanaemyrmex*), and subspecies of *fumidus:* Emery, 1925b: 81.

#***fraxinicola.*** *Camponotus* (*Colobopsis*) *pylartes* subsp. *fraxinicola* Smith, M.R. 1923a: 86 (s.w.) U.S.A.

#***freyeri.*** *Camponotus* (*Myrmotrema*) *olivieri* st. *freyeri* Santschi, 1915c: 270 (w.) MADAGASCAR. [First available use of *Camponotus foraminosus* subsp. *aldabrensis* var. *freyeri* Forel, 1912k: 166; unavailable name.]

*fricatus. Camponotus* (*Myrmotrema*) *tauricolis* st. *osiris* var. *fricatus* Santschi, 1926c: 616 (w.) ZAIRE. Unavailable name.

***friedae.*** *Camponotus friedae* Forel, 1912a: 77 (s.w.q.m.) TAIWAN. Combination in *C.* (*Myrmoturba*): Forel, 1914a: 267; in *C.* (*Tanaemyrmex*): Emery, 1925b: 102. Current subspecies: nominal plus *amius.*

***froggatti.*** *Camponotus froggatti* Forel, 1902h: 504 (s.w.) AUSTRALIA. Wheeler, G.C. & Wheeler, J. 1968: 218 (l.). Combination in *C.* (*Myrmosaga*): Forel, 1912i: 92; in *C.* (*Myrmophyma*): Emery, 1920b: 257. See also: Santschi, 1928e: 482.

***frontalis.*** *Camponotus frontalis* Pergande, 1896: 862 (s.w.) MEXICO. Combination in *C.* (*Myrmamblys*): Forel, 1914a: 271; in *C.* (*Myrmaphaenus*): Emery, 1925b: 155.

***fryi.*** *Camponotus* (*Myrmoturba*) *maculatus* subsp. *fryi* Mann, 1916: 474, pl. 6, fig. 52 (s.) BRAZIL. Combination in *C.* (*Tanaemyrmex*): Emery, 1925b: 84. Subspecies of *picipes:* Emery, 1920b: 233 (footnote). Raised to species: Emery, 1925b: 84.

*fuchsae. Camponotus abdominalis* subsp. *fuchsae* Forel, 1908c: 409 (s.w.q.m.) BRAZIL. Combination in *C. (Myrmothrix)*: Bruch, 1914: 228. Junior synonym of *atriceps:* Hashmi, 1973: 82.

***fugax. Camponotus fugax*** Forel, 1902b: 173 (s.w.) JAMAICA. Combination in *C. (Myrmamblys)*: Forel, 1914a: 271; in *C. (Pseudocolobopsis)*: Emery, 1920b: 259.

#***fuhrmanni.*** *Camponotus (Myrmoturba) nitens* var. *furmanni* Forel, 1921b: 135 (s.w.) COLOMBIA. Combination in *C. (Tanaemyrmex)*, and justified emendation of spelling to *fuhrmanni:* Emery, 1925b: 81.

#***fulgens.*** *Camponotus coruscus* var. *fulgens* Forel, 1885a: 345 (w.) GUATEMALA. Combination in *C. (Tanaemyrmex)*: Emery, 1925b: 80. Subspecies of *coruscus:* Emery, 1896d: 375.

#***fullawayi.*** *Camponotus reticulatus* subsp. *fullawayi* Wheeler, W.M. 1912a: 47 (s.w.) GUAM.

*fulvacea. Formica fulvacea* Norton, 1868a: 60, pl. 2, fig. 1 (s.w.) MEXICO. Combination in *Camponotus:* Norton, 1868c: 2. Junior synonym of *atriceps:* Mayr, 1870a: 377. Junior synonym of *esuriens:* Forel, 1885a: 340. Revived from synonymy as variety of *esuriens:* Dalla Torre, 1893: 230. Junior synonym of *atriceps:* Hashmi, 1973: 82.

#***fulvicornis.*** *Camponotus westermanni* var. *fulvicornis* Emery, 1903: 74, fig. 11 (s.w.) BRAZIL. Combination in *C. (Myrmaphaenus)*: Emery, 1925b: 156.

#***fulvipes.*** *Camponotus (Myrmoturba) fulvipes* Forel, 1914a: 267 (w.) SOUTH AFRICA. [First available use of *Camponotus natalensis* subsp. *corvus* var. *fulvipes* Emery, 1895h: 51; unavailable name.] Santschi, 1914e: 38 (q.m.). Variety of *natalensis:* Santschi, 1914e: 38; Arnold, 1922: 654; Emery, 1925b: 103.

*fulvooctonotata. Camponotus maculatus* subsp. *fulva* var. *fulvooctonotata* Forel, 1907d: 93 SEYCHELLES. Unavailable name.

***fulvopilosus.*** *Formica fulvopilosa* De Geer, 1778: 613, pl. 45, figs. 13, 14 (w.) SOUTH AFRICA. Forel, 1879a: 108 (s.); Forel, 1910f: 25 (m.); Arnold, 1924: 737 (q.). Combination in *Camponotus:* Mayr, 1862: 668; in *C. (Myrmepomis)*: Forel, 1912i: 92; in *C. (Myrmopiromis)*: Wheeler, W.M. 1922a: 985. Senior synonym of *pilosa* and *rufiventris:* Lepeletier, 1835: 213; Smith, F. 1858b: 32; Wheeler, W.M. 1922a: 985; of *detritoides* and *flavopilosus:* Robertson, 1990: 327.

#***fulvulus.*** *Camponotus (Myrmoturba) medea* subsp. *fulvula* Emery, 1920c: 21 (q.) PERU. Combination in *C. (Pseudocolobopsis)*: Emery, 1925b: 159.

#***fulvus.*** *Camponotus maculatus* subsp. *fulvus* Emery, 1894a: 72 (s.w.q.) SEYCHELLE IS. Combination in *C. (Tanaemyrmex)*: Emery, 1925b: 86. Subspecies of *maculatus:* Wheeler, W.M. 1922a: 1040; of *hova:* Emery, 1920c: 6; Emery, 1925b: 86.

*fumata. Camponotus (Tanaemyrmex) abdominalis* var. *fumata* Stitz, 1933: 71 (w.) VENEZUELA. Junior synonym of *atriceps:* Hashmi, 1973: 82.

#***fumeus.*** *Camponotus (Myrmoturba) traegaordhi* st. *fumeus* Santschi, 1926b: 252 (s.w.) ZIMBABWE.

***fumidus.*** *Camponotus fumidus* Roger, 1863a: 151 (s.) VENEZUELA. Combination in *C. (Myrmoturba)*: Forel, 1914a: 267; in *C. (Tanaemyrmex)*: Emery, 1925b: 80. Current subspecies: nominal plus *dominicensis, fraterculus, haytianus, illitus, imbecillus, soulouquei, toltecus.*

*furmanni;* see *fuhrmanni,* above.

#***furnissi.*** *Camponotus ustus* var. *furnissi* Wheeler, W.M. & Mann, 1914: 55 (s.w.q.) HAITI. Combination in *C. (Pseudocolobopsis)*: Emery, 1925b: 158; in *C. (Tanaemyrmex)*: Kempf, 1972a: 73.

***furvus.*** *Camponotus buchholzi* var. *furva* Santschi, 1911g: 213 (w.) CONGO. Combination in *C. (Myrmamblys)*: Emery, 1925b: 142; in *C. (Myrmacrhaphe)*, and raised to species: Santschi, 1926c: 613.

*fusca. Camponotus (Tanaemyrmex) irritans* r. *tinctus* var. *fusca* Stitz, 1932b: 736 (w.) SUMATRA. Unavailable name.

#***fusciceps.*** *Camponotus sexguttatus* var. *fusciceps* Emery, 1906c: 192 (s.w.) BRAZIL. Combination in *C. (Myrmosphincta)*: Wheeler, W.M. 1916c: 14.

*****fuscipennis.*** **Camponotus fuscipennis* Carpenter, 1930: 59, pl. 5, fig. 4; pl. 11, fig. 6 (q.) U.S.A. (Miocene).

*fuscipes. Formica fuscipes* Mayr, 1853c: 280 (w.) AUSTRIA. Combination in *Colobopsis:* Mayr, 1861: 38. Junior synonym of *truncatus:* Emery & Forel, 1879: 449.

#***fuscithorax.*** *Camponotus mitis* var. *fuscithorax* Dalla Torre, 1893: 243 (w.) INDIA. [First available use of *Camponotus rubripes* r. *mitis* var. *fuscithorax* Forel, 1886f: 154; unavailable name.] Viehmeyer, 1916a: 156 (s.q.m.). Combination in *C. (Tanaemyrmex)*, and variety of *variegatus:* Emery, 1925b: 96.

***fuscivillosus.*** *Camponotus fuscivillosus* Xiao & Wang; in Wang, C., Xiao, & Wu, 1989a: 225, figs. 27, 28 (s.w.) CHINA.

***fuscocinctus.*** *Camponotus rubripes* r. *fuscocinctus* Emery, 1888c: 364 (w.) BRAZIL. Combination in *C. (Tanaemyrmex)*: Emery, 1925b: 80. Raised to species: Dalla Torre, 1893: 232; Emery, 1894g: 373; Forel, 1895b: 603. Subspecies of *maculatus:* Emery, 1896c: 371; Forel, 1908c: 406; Forel, 1911c: 311; of *picipes:* Emery, 1920b: 233 (footnote); of *conspicuus:* Santschi, 1922c: 101. Raised to species: Kusnezov, 1952d: 210.

#***fuscogaster.*** *Camponotus (Myrmobrachys) pittieri* var. *fuscogaster* Emery, 1920c: 39 (s.) PERU.

*fuscoptera. Formica fuscoptera* Geoffroy, in Fourcroy, 1785: 452 (q.) FRANCE. Junior synonym of *vagus:* Latreille, 1802c: 96.

***gabonensis.*** *Camponotus (Myrmacrhaphe) gabonensis* Santschi, 1926c: 611, fig. 3 (w.q.) GABON.

*gaetulus. Camponotus (Myrmoturba) sylvaticus* st. *barbaricus* var. *gaetulus* Santschi, 1923f: 335 (w.) ALGERIA Unavailable name.

***galla.*** *Camponotus galla* Forel, 1894b: 68 (w.) ETHIOPIA. Combination in *C. (Orthonotomyrmex)*: Forel, 1914a: 273; in *C. (Myrmotrema)*: Forel, 1912i: 91; Santschi, 1915c: 267 (in key). Subspecies of *erinaceus:* Forel, 1910c: 271. Revived status as species: Forel, 1911d: 372. Senior synonym of *latinotus:* Forel, 1910c: 271.

*gallienii. Camponotus (Myrmoturba) gallienii* Forel, 1916: 457. [Unnecessary replacement name for *concolor* Forel, 1891b: 214.] Junior synonym of *concolor:* **new synonymy**.

***gambeyi.*** *Camponotus gambeyi* Emery, 1883: 145 (w.) NEW CALEDONIA. Combination in *C. (Myrmocamelus)*: Forel, 1914a: 270; in *C. (Myrmophyma)*: Emery, 1920b: 257; in *C. (Tanaemyrmex)*: Emery, 1925b: 104. Current subspecies: nominal plus *marthae.*

#***gamma.*** *Camponotus (Tanaemyrmex) wellmani* var. *gamma* Santschi, 1926b: 257 (s.w.) ZAIRE.

#***ganzii.*** *Camponotus (Myrmamblys) chapini* subsp. *ganzii* Weber, 1943c: 386 (s.w.) SUDAN.

#***garbei.*** *Camponotus (Myrmoturba) bonariensis* st. *garbei* Santschi, 1922c: 102 (s.w.) BRAZIL. Combination in *C. (Tanaemyrmex)*: Emery, 1925b: 79.

***gasseri.*** *Colobopsis gasseri* Forel, 1894e: 233 (s.w.) AUSTRALIA. Forel, 1902h: 507 (q.); Wheeler, G.C. & Wheeler, J. 1953e: 190 (l.). Combination in *Camponotus (Colobopsis)*: Forel, 1902h: 507; Forel, 1912i: 90. Current subspecies: nominal plus *caloratus, lysias, obtusitruncatus.*

#***gaullei.*** *Camponotus gaullei* Santschi, 1911e: 128 (w.) MADAGASCAR. Combination in *C. (Dinomyrmex)*: Wheeler, W.M. 1922a: 1043; in *C. (Tanaemyrmex)*, and variety of *cervicalis:* Emery, 1925b: 84.

***geayi.*** *Camponotus (Pseudocolobopsis) geayi* Santschi, 1922c: 117, fig. 2 (s.w.) FRENCH GUIANA.

#***genaiai.*** *Camponotus (Myrmamblys) itoi* var. *genaiai* Santschi, 1928b: 41 (w.) TAIWAN. Variety of *truebi:* Santschi, 1937h: 380.

***genatus.*** *Camponotus (Neomyrmamblys) genatus* Santschi, 1922c: 114, fig. 2 (s.w.q.m.) BRAZIL. Combination in *C. (Myrmaphaenus)*: Emery, 1925b: 155.

*georgei. Camponotus (Myrmoturba) maculatus* st. *arnoldius* [sic] var. *georgei* Santschi, 1923e: 290, fig. 5 (s.w.) ZIMBABWE. Unavailable name.

*geralensis. Camponotus (Pseudocolobopsis) macrocephalus* subsp. *geralensis* Emery, 1920c: 36 (s.w.q.) BRAZIL. Junior synonym of *macrocephalus:* Kempf, 1968b: 408.

***gerberti.*** *Camponotus (Tanaemyrmex) gerberti* Donisthorpe, 1949e: 271 (s.w.) MADAGASCAR.

***germaini.*** *Camponotus germaini* Emery, 1903: 71, fig. 6 (s.w.) BRAZIL. Combination in *C. (Myrmamblys)*: Forel, 1914a: 271; in *C. (Neomyrmamblys)*: Santschi, 1921f: 311; in *C. (Myrmaphaenus)*: Emery, 1925b: 155. Current subspecies: nominal plus *tacuruensis.*

#***gestiens.*** *Camponotus (Myrmamblys) reticulatus* r. *gestiens* Forel, 1915a: 39 (w.) SUMATRA.

***gestroi.*** *Camponotus gestroi* Emery, 1878b: 44, fig. (s.w.) SARDINIA. Forel, 1894d: 4 (q.m.); Emery, 1916b: 160 (q.). Combination in *C. (Orthonotomyrmex)*: Forel, 1913d: 435; in *C. (Myrmentoma)*: Emery, 1920b: 257. Senior synonym of *creticus:* Agosti & Collingwood, 1987a: 58. Current subspecies: nominal plus *ponariensis.*

***gibber.*** *Camponotus quadrimaculatus* var. *gibber* Forel, 1891b: 59 (s.) MADAGASCAR. Forel, 1891b: 215 (w.); Forel, 1892l: 232 (q.m.). Combination in *C. (Myrmosphincta)*: Forel, 1914a: 273; in *C. (Mayria)*: Emery, 1925b: 122. Raised to species: Forel, 1891b: 215.

***gibbinotus.*** *Camponotus gibbinotus* Forel, 1902h: 498 (w.) AUSTRALIA. Forel, 1915b: 102 (s.). Combination in *C. (Myrmogonia)*: Forel, 1914a: 209; in *C.*

(*Myrmophyma*): Emery, 1925b: 111; in *C.* (*Thlipsepinotus*): Santschi, 1928e: 483.

***gibbosus.*** *Camponotus* (*Myrmamblys*) *gibbosus* Karavaiev, 1929c: 240, fig. 5 (s.w.) INDONESIA (Aru I.).

***gigas.*** *Formica gigas* Latreille, 1802c: 105, pl. 2, fig. 6 (s.) "Grandes-Indes". Smith, F. 1858b: 14 (w.q.m.). Combination in *Camponotus:* Mayr, 1862: 669; in *Dinomyrmex:* Ashmead, 1905b: 384; in *C.* (*Myrmogigas*): Forel, 1912i: 91; in *C.* (*Dinomyrmex*): Forel, 1914a: 268. See also: Bingham, 1903: 369. Current subspecies: nominal plus *borneensis.*

*gigas. Formica gigas* Leach, 1825: 292 (w.q.m.) FRANCE. [Unresolved junior primary homonym of *gigas* Latreille, above.] Junior synonym of *herculeanus:* Roger, 1859: 229; of *cruentatus:* Dalla Torre, 1893: 227.

***gilviceps.*** *Camponotus gilviceps* Roger, 1863b: 3. Replacement name for *Formica ruficeps* Smith, F. 1857a: 54. [Junior primary homonym of *Formica ruficeps* Fabricius, 1804: 404.] Forel, 1886f: 171 (q.). Combination in *C.* (*Myrmotarsus*), and junior synonym of *mistura:* Emery, 1925b: 135. Revived from synonymy: Donisthorpe, 1932c: 443. Senior synonym of *smithi* (unnecessary second replacement name for *ruficeps* Smith, F.): **new synonymy**.

#***gilvigaster.*** *Camponotus* (*Myrmeurynota*) *heathi* var. *gilvigaster* Wheeler, W.M. 1923a: 5 (s.w.) BRAZIL. Combination in *C.* (*Myrmodirachis*): Emery, 1925b: 168.

*gilvinotata. Camponotus* (*Myrmoturba*) *maculatus* r. *subnudus* var. *gilvinotata* Forel, 1913k: 126 (q.) SUMATRA. Unavailable name.

***gilviventris.*** *Camponotus gilviventris* Roger, 1863a: 145 (w.) CUBA. Mann, 1920: 435 (s.). Combination in *C.* (*Myrmeurynota*): Forel, 1914a: 274. Current subspecies: nominal plus *refectus, renormatus.*

*glaber. Camponotus* (*Tanaemyrmex*) *aethiops* var. *glaber* Emery, 1925b: 97 (w.) RUSSIA. [First available use of *Camponotus maculatus* subsp. *aethiops* var. *glaber* Ruzsky, 1905b: 213; unavailable name.] Junior synonym of *aethiops:* Dlussky, Soyunov & Zabelin, 1990: 128.

***glabrisquamis.*** *Camponotus glabrisquamis* Emery, 1911c: 255 (s.w.q.) NEW GUINEA. Combination in *C.* (*Myrmoturba*): Forel, 1914a: 267; in *C.* (*Tanaemyrmex*): Emery, 1925b: 93.

***godmani.*** *Camponotus godmani* Forel, 1899c: 143 (w.q.) MEXICO. Combination in *C.* (*Myrmobrachys*): Forel, 1914a: 270. Current subspecies: nominal plus *palliolatus.*

***goeldii.*** *Camponotus goeldii* Forel, 1894d: 43, pl. 2, fig. 5 (s.w.) BRAZIL. Combination in *C.* (*Myrmoturba*): Forel, 1914a: 267; in *C.* (*Tanaemyrmex*): Emery, 1925b: 81.

***gombaki.*** *Camponotus* (*Karavaievia*) *gombaki* Dumpert, 1986: 563, figs. 3, 4 (w.q.) WEST MALAYSIA.

***gouldi.*** *Camponotus egregius* r. *gouldi* Forel, 1886c: civ (s.) MADAGASCAR. [Also described as new by Forel, 1886f: 167. *Camponotus gouldi* Forel, 1879a: 115, *nomen nudum.*] Forel, 1897c: 210 (w.q.); Forel, 1907g: 90 (m.). Combination in *C.* (*Myrmogigas*): Forel, 1912i: 91; in *C.* (*Dinomyrmex*): Forel, 1914a: 268; in *C.* (*Tanaemyrmex*): Emery, 1925b: 85. Raised to species: Dalla Torre, 1893: 233; Forel, 1897c: 201; Forel, 1904b: 377.

***gouldianus.*** *Camponotus* (*Myrmophyma*) *gouldianus* Forel, 1922: 100 (w.) AUSTRALIA.

#***gracilentus.*** *Camponotus* (*Colobopsis*) *pilosus* subsp. *gracilentus* Viehmeyer, 1916a: 161 (q.) SINGAPORE. Subspecies of *leonardi:* Emery, 1925b: 149.

***grandidieri.*** *Camponotus grandidieri* Forel, 1886c: ciii (s.w.) MADAGASCAR. [*Camponotus grandidieri* Forel, 1879a: 115, *nomen nudum.*] Forel, 1891b: 39 (s.); Forel, 1892l: 232 (m.); Forel, 1912k: 166 (q.). Combination in *C.* (*Myrmotrema*): Forel, 1913a: 145. Variety/subspecies of *foraminosus:* Mayr, 1893: 195; Forel, 1907e: 14. Revived status as species: Forel, 1914a: 270. Current subspecies: nominal plus *atrabilis, comorensis, mendax* (unresolved junior homonym), *ruspolii.*

*grandior. Camponotus* (*Myrmotrema*) *foraminosus* r. *perrisii* var. *grandior* Forel, 1913b: 342 (s.w.) ZAIRE. Unavailable name.

*granulatus. Dolichoderus granulatus* Pergande, 1896: 866 (w.) MEXICO. Junior synonym of *striatus:* Smith, M.R. 1953b: 211.

#***grasi.*** *Camponotus* (*Tanaemyrmex*) *foleyi* subsp. *grasi* Bernard, 1953a: 191 (s.w.q.m.) ALGERIA. [First available use of *Camponotus* (*Tanaemyrmex*) *compressus* subsp. *foleyi* var. *grasi* Bernard, 1948: 160; unavailable name.]

***greeni.*** *Camponotus greeni* Forel, 1911a: 54 (s.w.) SRI LANKA. Combination in *C.* (*Myrmotemnus*): Emery, 1920b: 258; in *C.* (*Myrmamblys*): Forel, 1914a: 271; Emery, 1925b: 139.

*grenadensis. Camponotus sexguttatus* var. *grenadensis* Forel, 1897b: 297 (s.w.q.) GRENADA, W.I. Combination in *C.* (*Myrmosphincta*): Emery, 1925b: 152. Junior synonym of *biguttatus:* Emery, 1898a: 226; Emery, 1925b: 152.

***gretae.*** *Camponotus gretae* Forel, 1902d: 288 (q.) INDIA. Combination in *C.* (*Myrmamblys*): Forel, 1914a: 271.

#***griqua.*** *Camponotus* (*Myrmespera*) *transvaalensis* subsp. *griqua* Arnold, 1952a: 465 (s.w.q.m.) SOUTH AFRICA.

#***griseus.*** *Camponotus* (*Colobopsis*) *leonardi* var. *grisea* Karavaiev, 1929c: 243 (s.w.) JAVA.

***guanchus.*** *Camponotus sicheli* var. *guancha* Santschi, 1908: 534 (w.q.) "North Africa". Cagniant & Espadaler, 1993: 427 (m.). Combination in *C.* (*Myrmentoma*), and variety of *lateralis:* Emery, 1925a: 64; Emery, 1925b: 121. Raised to species: Cagniant & Espadaler, 1993: 426.

#***guatemalensis.*** *Camponotus sylvaticus* r. *guatemalensis* Forel, 1885a: 364 (w.) GUATEMALA. Combination in *C.* (*Tanaemyrmex*): Emery, 1925b: 81. Race of *rubripes:* Forel, 1886f: 143. Revived status as species: Dalla Torre, 1893: 233. Subspecies of *maculatus:* Emery, 1896d: 371; Forel, 1899c: 135; of *picipes:* Emery, 1920b: 233; Kempf, 1972a: 69.

***guayapa.*** *Camponotus guayapa* Kusnezov, 1952d: 211 (w.) fig. 13 ARGENTINA. Combination in *C.* (*Myrmobrachys*): Kempf, 1972a: 52.

***gundlachi.*** *Camponotus* (*Colobopsis*) *gundlachi* Mann, 1920: 438 (w.) CUBA.

***guppyi.*** *Camponotus* (*Colobopsis*) *guppyi* Mann, 1920: 370, figs. 43, 44 (s.w.) SOLOMON IS.

***guttatus.*** *Camponotus maculatus* subsp. *guttatus* Emery, 1899e: 498 (s.w.q.) CAMEROUN. Forel, 1913e: 670 (m.). Combination in *C.* (*Tanaemyrmex*): Emery, 1925b: 85. Subspecies of *maculatus:* Wheeler, W.M. 1922a: 954. Raised to species: Emery, 1920c: 6; Emery, 1925b: 85. Material of the unavailable name *brevitibialis* referred here: Baroni Urbani, 1972: 132. Current subspecies: nominal plus *minusculus.*

***habereri.*** *Camponotus habereri* Forel, 1911e: 293 (w.) JAPAN. Forel, 1912a: 76 (s.); Forel, 1913f: 200 (q.). Combination in *C.* (*Myrmoturba*): Forel, 1913f: 200; Emery, 1920b: 255; in *C.* (*Myrmothrix*): Forel, 1914a: 269; in *C.* (*Tanaemyrmex*): Emery, 1925b: 93.

***haematocephalus.*** *Camponotus haematocephalus* Emery, 1903: 78 (s.) PERU. Combination in *C.* (*Myrmoturba*): Forel, 1914a: 267; in *C.* (*Tanaemyrmex*): Emery, 1925b: 83.

***haereticus.*** *Camponotus* (*Myrmotrema*) *foraminosus* st. *haereticus* Santschi, 1914d: 382 (s.w.) NIGERIA. Subspecies of *foraminosus* Wheeler, W.M. 1922a: 250. Raised to species: Santschi, 1915c: 269 (in key); Emery, 1925b: 132.

***hagensii.*** *Camponotus rubripes* r. *hagensii* Forel, 1886f: 158 (s.w.) MADAGASCAR. Forel, 1895c: 244 (q.m.). Combination in *C.* (*Myrmoturba*): Forel, 1914a: 267; in *C.* (*Tanaemyrmex*): Emery, 1925b: 92. Race of *maculatus:* Forel, 1891b: 27; of *fumidus:* Santschi, 1922c: 101. Raised to species: Dalla Torre, 1893: 233; Forel, 1914a: 267; Emery, 1920c: 6; Wheeler, W.M. 1922a: 1039; Emery, 1925b: 92.

*hagmanni. Camponotus hagmanni* Santschi, 1922d: 262. *Nomen nudum,* attributed to Forel.

*hannae. Camponotus* (*Myrmoturba*) *maculatus* st. *hannae* Santschi, 1919a: 349, fig. 4 (s.w.) ZIMBABWE. Combination in *C.* (*Tanaemyrmex*): Emery, 1925b: 88. Junior synonym of *maculatus:* Baroni Urbani, 1972: 125.

***hannani.*** *Camponotus hannani* Forel, 1899c: 131 (footnote) (w.) JAMAICA. Forel, 1907e: 11 (m.); Wheeler, W.M. 1917g: 468 (q.). Combination in *C.* (*Myrmothrix*): Forel, 1912i: 91. Senior synonym of *willardi:* Hashmi, 1973: 12.

*hansingi. Camponotus maculatus* subsp. *liengmei* var. *hansingi* Forel, 1910e: 452 (s.) MOZAMBIQUE. Unavailable name; material referred to *maculatus* by Baroni Urbani, 1972: 125.

***hapi.*** *Camponotus* (*Myrmamblys*) *hapi* Weber, 1943c: 386, pl. 16, fig. 21 (s.w.q.) SUDAN.

***hartogi.*** *Camponotus hartogi* Forel, 1902h: 500 (w.) AUSTRALIA. Wheeler, G.C. & Wheeler, J. 1974a: 61 (l.). Combination in *C.* (*Myrmepomis*): Forel, 1914a: 273; in *C.* (*Myrmophyma*): Emery, 1925b: 111. Senior synonym of *ferruginipes:* Brown, 1956a: 40.

***hastifer.*** *Camponotus hastifer* Emery, 1911c: 254 (s.w.) NEW GUINEA. Combination in *C.* (*Dinomyrmex*): Forel, 1914a: 268; in *C.* (*Tanaemyrmex*): Emery, 1925b: 90.

***havilandi.*** *Camponotus* (*Myrmophyma*) *havilandi* Arnold, 1922: 666 (s.w.) SOUTH AFRICA. Combination in *C.* (*Myrmamblys*): Emery, 1925b: 140; in *C.* (*Myrmespera*): Santschi, 1926c: 604.

*hawaiensis*. *Camponotus* (*Myrmoturba*) *variegatus* subsp. *hawaiensis* Emery, 1920c: 7 (w.q.m.) HAWAII. [First available use of *Camponotus maculatus* r. *mitis* var. *hawaiensis* Forel, 1899a: 122; unavailable name.] Combination in *C.* (*Tanaemyrmex*): Emery, 1925b: 96. Junior synonym of *variegatus*: Wilson & Taylor, 1967: 95.

#***haweisi***. *Camponotus* (*Colobopsis*) *culmicola* var. *haweisi* Mann, 1920: 439 (s.q.) BAHAMAS.

#***haytianus***. *Camponotus maculatus* subsp. *haytianus* Wheeler, W.M. & Mann, 1914: 48, fig. 20 (s.w.) HAITI. Combination in *C.* (*Tanaemyrmex*): Emery, 1925b: 81. Subspecies of *fumidus*: Santschi, 1922c: 101.

***heathi***. *Camponotus* (*Myrmeurynota*) *heathi* Mann, 1916: 481, pl. 5, figs. 40, 41 (s.w.) BRAZIL. Combination in *C.* (*Myrmodirachis*): Emery, 1925b: 168. See also: Kempf, 1975c: 377. Current subspecies: nominal plus *gilvigaster*.

***hedwigae***. *Camponotus hedwigae* Forel, 1912i: 79 (s.w.) BRAZIL. Combination in *C.* (*Myrmorhachis*): Forel, 1914a: 274; in *C.* (*Myrmocladoecus*): Emery, 1925b: 166. See also: Wheeler, W.M. 1934e: 423; Kempf, 1968b: 406.

*heliades*. *Camponotus* (*Myrmoturba*) *punctulatus* st. *termitaria* var. *heliades* Santschi, 1922b: 377 (w.) ARGENTINA. Unavailable name.

***helleri***. *Camponotus helleri* Emery, 1903: 66 (w.) PERU. Combination in *C.* (*Myrmocamelus*): Forel, 1914a: 270; in *C.* (*Paracolobopsis*): Emery, 1920b: 259; in *C.* (*Myrmaphaenus*): Emery, 1925b: 154.

***hellmichi***. *Camponotus* (*Tanaemyrmex*) *morosus* var. *hellmichi* Menozzi, 1935c: 331 (diagnosis in key) (w.) CHILE. Raised to species: Snelling & Hunt, 1976: 120.

***helvus***. *Camponotus helvus* Xiao & Wang; in Wang, C., Xiao & Wu, 1989b: 322, figs. 6–8 (s.w.) CHINA.

***hemichlaena***. *Camponotus obscuripes* subsp. *hemichlaena* Yasumatsu & Brown, 1951: 40 (w.) JAPAN. Raised to species: Yasumatsu & Brown, 1957: 50.

*hemipsila*. *Formica hemipsila* Foerster, 1850b: 492 (q.) ALGERIA. Junior synonym of *lateralis*: Forel, 1894d: 5.

*hendecarthrus* Roger, 1863; see under ***ANOPLOLEPIS***.

#***hephaestus***. *Camponotus* (*Myrmocladoecus*) *planus* var. *hephaestus* Wheeler, W.M. 1933a: 59 (s.w.) GALAPAGOS IS.

****heracleus***. **Formica heraclea* Heer, 1850: 116, pl. 11, fig. 3 (m.) SWITZERLAND (Miocene). Combination in *Camponotus*: Mayr, 1867b: 52.

*herculeanoligniperdus*. *Camponotus herculeanus* var. *herculeanoligniperdus* Forel, 1874: 39 (w.q.m.) SWITZERLAND. Variety of *ligniperdus*: Menozzi, 1922c: 144; Emery, 1925b: 74. Raised to species: Stitz, 1939: 242. Junior synonym of *ligniperdus*: **new synonymy**.

*herculeanopennsylvanicus*. *Camponotus herculeanus* var. *herculeanopennsylvanicus* Forel, 1879a: 57 (w.) U.S.A. Senior synonym of *mahican* (unnecessary replacement name): Emery, 1925b: 72. Junior synonym of *pennsylvanicus*: Creighton, 1950a: 367.

***herculeanus***. *Formica herculeana* Linnaeus, 1758: 579 (q.) EUROPE. Lepeletier, 1835: 209 (w.q.m.); Mayr, 1855: 308 (w.q.m.); Wheeler, G.C. & Wheeler, J. 1953e: 185 (l.). Combination in *Camponotus*: Mayr, 1861: 36; in *C.* (*Camponotus*): Forel, 1914a: 259. Senior synonym of *atra* and *intermedia*: Nylander, 1846a: 894; of *castanea*: Mayr, 1863: 399; of *whymperi*: Creighton, 1950a: 366; of *nadigi*: Yasumatsu & Brown, 1951: 30; of *montanus* and *shitkowi*: Emery, 1925b: 72; Karavaiev, 1936: 178; Arnol'di, 1967: 1819; of *caucasicus*: Arakelian, 1994: 85. See also: Tarbinsky, 1976: 148; Kutter, 1977c: 204; Collingwood, 1979: 90; Smith, D.R. 1979: 1426; Atanassov & Dlussky, 1992: 210. Current subspecies: nominal plus *eudokiae*.

***hermanni***. *Camponotus hermanni* Emery, 1911a: 221 (s.w.) BRAZIL. Combination in *C.* (*Myrmamblys*): Forel, 1914a: 271; in *C.* (*Myrmaphaenus*): Emery, 1925b: 155.

*herodes*. *Camponotus* (*Myrmoturba*) *compressus* st. *thoracicus* var. *herodes* Santschi, 1938a: 42 (s.w.m.) EGYPT. Unavailable name.

***heros***. *Camponotus* (*Tanaemyrmex*) *heros* Santschi, 1926b: 253 (s.) TANZANIA.

*herrichi*. *Formica herrichi* Mayr, 1853b: 113 (w.) BRAZIL. Junior synonym of *rufipes*: Roger, 1861b: 164.

#***hertigi***. *Camponotus* (*Myrmotrema*) *avius* var. *hertigi* Santschi, 1937d: 244 (w.) ANGOLA.

#***hesperius***. *Camponotus maculatus* r. *hesperius* Emery, 1893c: 85 (s.w.q.) CANARY IS. Combination in *C.* (*Myrmoturba*): Forel, 1914a: 267; in *C.* (*Tanaemyrmex*): Emery, 1925b: 91. [Misspelled as *hesperinus* by Forel, 1893d: 464.] Raised to species: Emery, 1920b: 255. Subspecies of *pallens*: Emery, 1920c: 7; of *atlantis*: Emery, 1925b: 91; Wheeler, W.M. 1927g: 115; of *alii*: Santschi, 1939d: 84; of *atlantis*: Barquin, 1981: 397; Hohmann, La Roche, *et al.*, 1993: 160.

***heteroclitus***. *Camponotus heteroclitus* Forel, 1895c: 243 (s.) MADAGASCAR. Combination in *Calomyrmex*: Forel, 1914a: 274; in *Camponotus* (*Myrmonesites*): Emery, 1920b: 257.

#***hians***. *Camponotus quadrisectus* subsp. *hians* Forel, 1907a: 32 (s.w.) SINGAPORE. Combination in *C.* (*Myrmotarsus*): Emery, 1925b: 135.

*hieroglyphicus*. *Camponotus maculatus* st. *hieroglyphicus* Santschi, 1917b: 290 (s.w.q.) ANGOLA. Combination in *C.* (*Tanaemyrmex*): Emery, 1925b: 88. Junior synonym of *maculatus*: Baroni Urbani, 1972: 125.

***hildebrandti***. *Camponotus hildebrandti* Forel, 1886f: 161 (w.q.) MADAGASCAR. Combination in *C.* (*Dinomyrmex*): Forel, 1914a: 268; in *C.* (*Tanaemyrmex*): Emery, 1925b: 85. Current subspecies: nominal plus *dichromothrix*.

#***himalayanus***. *Camponotus marginatus* var. *himalayanus* Forel, 1893b: 431 (w.) INDIA. Combination in *C.* (*Myrmentoma*): Emery, 1925b: 118. Variety of *fallax*: Forel, 1906b: 84; of *caryae*: Wheeler, 1917c: 29; Emery, 1925b: 118.

***hippocrepis***. *Camponotus* (*Myrmorhachis*) *hippocrepis* Emery, 1920c: 48, fig. 8 (w.) BOLIVIA. Combination in *C.* (*Myrmocladoecus*): Emery, 1925b: 167; Wheeler, W.M. 1934e: 424.

#***hirsutinasus***. *Camponotus* (*Myrmocladoecus*) *mucronatus* subsp. *hirsutinasus* Wheeler, W.M. 1934g: 226 (s.w.) GUATEMALA.

#***hispidus***. *Camponotus punctulatus* var. *hispida* Emery, 1906c: 190 (s.w.) PARAGUAY. Combination in *C.* (*Tanaemyrmex*): Emery, 1925b: 78.

***hoelldobleri***. *Camponotus hoelldobleri* Cagniant, 1991a: 37, figs. 2, 3 (s.w.q.m.) MOROCCO. [Given as related to *cruentatus*, *obscuriventris*, *pexus*; hence subgenus *C.* (*Myrmosericus*).]

#***hoggarensis***. *Camponotus* (*Myrmoturba*) *seurati* var. *hoggarensis* Santschi, 1929c: 104, figs. 6, 7 (s.w.) ALGERIA. [Also described as new by Santschi, 1934f: 174.] Subspecies of *alii*: Bernard, 1953a: 187.

#***hoggarensis***. *Camponotus* (*Tanaemyrmex*) *foleyi* subsp. *hoggarensis* Bernard, 1953a: 191 (s.m.) ALGERIA. [First available use of *Camponotus* (*Tanaemyrmex*) *compressus* subsp. *foleyi* var. *hoggarensis* Bernard, 1948: 158; unavailable name.] [Unresolved junior primary homonym of *hoggarensis* Santschi, above.]

#***holmgreni***. *Camponotus* (*Myrmepomis*) *sericeiventris* var. *holmgreni* Wheeler, W.M. 1931c: 93 (s.w.) BOLIVIA.

***holosericeus***. *Camponotus holosericeus* Emery, 1889b: 515 (s.w.) BURMA. Combination in *C.* (*Myrmosphincta*): Forel, 1914a: 273; in *C.* (*Myrmosaulus*): Emery, 1925b: 113.

***holzi***. *Camponotus holzi* Forel, 1921b: 134 (s.w.) ECUADOR. Combination in *C.* (*Tanaemyrmex*): Emery, 1925b: 81.

#***hondurianus***. *Camponotus* (*Myrmorhachis*) *quadrilaterus* var. *hondurianus* Mann, 1922: 53 (w.) HONDURAS. Combination in *C.* (*Myrmocladoecus*): Emery, 1925b: 167. Variety of *sanctaefidei*: Emery, 1925b: 167; Wheeler, W.M. 1934g: 231.

#***hongkongensis***. *Camponotus maculatus* subsp. *hongkongensis* Forel, 1912n: 62 (s.w.) HONG KONG. Combination in *C.* (*Tanaemyrmex*): Emery, 1925b: 94. Subspecies of *irritans*: Emery, 1920c: 7; Wheeler, W.M. 1930h: 76.

#***honorus***. *Camponotus carbo* subsp. *honorus* Forel, 1910c: 268 (w.) ETHIOPIA. Combination in *C.* (*Myrmotrema*), and stirps of *foraminosus*: Santschi, 1915c: 268 (in key).

#***hoodensis***. *Camponotus* (*Myrmamblys*) *macilentus* var. *hoodensis* Wheeler, W.M. 1919c: 285 (s.w.) GALAPAGOS IS. Combination in *C.* (*Pseudocolobopsis*): Emery, 1925b: 158.

***hoplites***. *Camponotus* (*Myrmocamelus*) *hoplites* Emery, 1914f: 426, pl. 13, fig. 16 (s.w.) NEW CALEDONIA. Combination in *C.* (*Myrmophyma*): Emery, 1920b: 257; in *C.* (*Tanaemyrmex*): Emery, 1925b: 104.

*horni* Kirby, W.F. 1896; see under ***IRIDOMYRMEX***.

***horni***. *Camponotus* (*Tanaemyrmex*) *horni* Clark, 1930c: 20 (q.) AUSTRALIA. [Unresolved junior primary homonym of *horni* Kirby, W.F. 1896: 205, above.] See also: Shattuck, 1993: 128.

***horrens***. *Camponotus horrens* Forel, 1910d: 127 (w.) PHILIPPINES. Combination in *C.* (*Myrmosphincta*): Forel, 1912i: 92; in *C.* (*Colobopsis*): Emery, 1925b: 149.

***horripilus***. *Camponotus horripilus* Emery, 1900d: 708 (w.) SUMATRA. Combination in *C.* (*Colobopsis*): Forel, 1914a: 272.

***horseshoetus***. *Camponotus horseshoetus* Datta & Raychaudhuri, 1985: 271, fig. 1, pl. 1 (w.) INDIA. Subgenus indeterminate.

***hosei***. *Camponotus hosei* Forel, 1911a: 55 (w.) BORNEO. Combination in *C.* (*Colobopsis*): Forel, 1914a: 272. Current subspecies: nominal plus *mimus*.

***hospes***. *Iridomyrmex hospes* Emery, 1884b: 64, fig. 9 (w.) SUMATRA. Combination in *Camponotus*:

Emery, 1887a: 219; in *C.* (*Myrmomalis*): Forel, 1914a: 271; in *C.* (*Myrmoplatys*): Emery, 1920b: 256. Current subspecies: nominal plus *adultus*.

***hova.** Camponotus maculatus* r. *hova* Forel, 1891b: 35 (s.w.q.) MADAGASCAR. [First available use of *Camponotus rubripes* r. *maculatus* var. *hova* Forel, 1886f: 150; unavailable name.] Combination in *C.* (*Tanaemyrmex*): Emery, 1925b: 85. Race/subspecies of *maculatus:* Forel, 1897c: 200; Forel, 1907e: 14; Wheeler, W.M. 1922a: 1040. Raised to species: Dalla Torre, 1893: 235; Emery, 1920c: 5; Emery, 1925b: 85. Current subspecies: nominal plus *becki, boivini, fairmairei, fulvus, hovahovoides, hovoides, mixtellus, obscuratus, pictiventris, radamae*.

#***hovahovoides.** Camponotus hovahovoides* Forel, 1892l: 232 (w.q.m.) MADAGASCAR. Variety of *hova:* Dalla Torre, 1893: 235.

#***hovoides.** Camponotus radamae* var. *hovoides* Dalla Torre, 1893: 249 (s.w.q.m.) MADAGASCAR. [First available use of *Camponotus maculatus* r. *radamae* var. *hovoides* Forel, 1891b: 33; unavailable name.] Combination in *C.* (*Tanaemyrmex*): Emery, 1925b: 85. Variety of *hova:* Santschi, 1911e: 132; Emery, 1925b: 85.

***howensis.** Camponotus* (*Colobopsis*) *howensis* Wheeler, W.M. 1927i: 152, fig. 12 (w.) AUSTRALIA.

***humeralis.** Camponotus* (*Myrmeurynota*) *humeralis* Emery, 1920c: 47 (w.) PERU.

#***humeralis.** Camponotus* (*Colobopsis*) *dentatus* subsp. *humeralis* Wheeler, W.M. 1934e: 416 (w.) FIJI IS. [Unresolved junior primary homonym of *humeralis* Emery, above.]

***humerus.** Camponotus humerus* Wang & Wu, 1994: 25, figs. 5, 12 (w.) CHINA.

#***humilior.** Camponotus maculatus* var. *humilior* Forel, 1902h: 497 (s.w.) AUSTRALIA. Wheeler, G.C. & Wheeler, J. 1968: 220 (l.). Subspecies of *maculatus:* Taylor & Brown, D.R. 1985: 116.

***hunteri.** Camponotus* (*Colobopsis*) *pylartes* var. *hunteri* Wheeler, W.M. 1910d: 353 (s.w.) U.S.A. Raised to species: Creighton, 1950a: 393.

*hyalinipennis. Camponotus marginatus* var. *hyalinipennis* Costa, 1884: 56 (q.) SARDINIA. Junior synonym of *aethiops:* Emery, 1914e: 2; Emery, 1915a: 250.

***hyatti.** Camponotus hyatti* Emery, 1893i: 680, pl. 22, figs. 25, 26 (s.w.) U.S.A. Combination in *C.* (*Camponotus*): Forel, 1914a: 266; in *C.* (*Myrmentoma*): Emery, 1920b: 257. Senior synonym of *californica:* Snelling, 1988: 69.

#***hybridus.** Camponotus tenuiscapus* r. *hybridus* Forel, 1901h: 69 (s.w.q.) BRAZIL. Combination in *C.* (*Myrmoturba*): Forel, 1915c: 363; Bruch, 1915: 535; in *C.* (*Tanaemyrmex*): Emery, 1925b: 78. Race/subspecies of *punctulatus:* Forel, 1915c: 363; Bruch, 1915: 535.

*hyksos. Camponotus* (*Myrmoturba*) *maculatus* st. *aegyptiacus* var. *hyksos* Santschi, 1938a: 41 (w.) EGYPT. Unavailable name.

***hypoclineoides.** Camponotus* (*Myrmosphincta*) *hypoclineoides* Wheeler, W.M. 1919e: 112 (w.) BORNEO. Combination in *C.* (*Myrmamblys*): Emery, 1925b: 140; in *C.* (*Myrmotemnus*): Santschi, 1926c: 601.

***icarus.** Camponotus icarus* Forel, 1912n: 66 (q.) SUMATRA. Combination in *C.* (*Myrmamblys*): Forel, 1914a: 271.

#***ifranensis.** Camponotus vagus* subsp. *ifranensis* Cagniant, 1987b: 117, figs. A-D (w.q.m.) MOROCCO.

***ignestii.** Camponotus* (*Tanaemyrmex*) *ignestii* Menozzi, 1935b: 200, fig. 4 (s.) GUIANA.

**igneus. *Camponotus igneus* Mayr, 1868c: 28, pl. 1, figs. 9, 10 (w.) BALTIC AMBER (Oligocene). Junior synonym of **mengei:* Wheeler, W.M. 1915h: 138.

*igniceps. Camponotus sericeus* r. *peguensis* var. *igniceps* Forel, 1913e: 664 (w.) SRI LANKA. [Also described as new by Forel, 1913k: 129.] Unavailable name.

***iheringi.** Camponotus iheringi* Forel, 1908c: 412 (s.w.) BRAZIL. Combination in *C.* (*Myrmamblys*): Forel, 1914a: 271; in *C.* (*Myrmobrachys*): Emery, 1920b: 260. [Frequently misspelled as *jheringi*, for example by Emery, 1925b: 165.] Current subspecies: nominal plus *bajulus*.

***ilgii.** Camponotus ilgii* Forel, 1894b: 64 (w.) ETHIOPIA. Menozzi, 1930b: 120 (q.). Combination in *C.* (*Myrmotrema*): Santschi, 1913c: 314.

#***illitus.** Camponotus fumidus* var. *illitus* Wheeler, W.M. & Mann, 1914: 51, fig. 22 (s.w.) HAITI. Combination in *C.* (*Tanaemyrmex*): Emery, 1925b: 80.

#***imbecillus.** Camponotus fumidus* var. *imbecillus* Wheeler, W.M. & Mann, 1914: 52 (s.w.) HAITI. Combination in *C.* (*Tanaemyrmex*): Emery, 1925b: 80.

#***imberbis.** Camponotus punctulatus* subsp. *imberbis* Emery, 1906c: 190 (s.w.q.) ARGENTINA.Combination in *C.* (*Myrmoturba*): Bruch, 1914: 229; in *C.* (*Tanaemyrmex*): Emery, 1925b: 78.

#***imerinensis.** Camponotus dufouri* var. *imerinensis* Forel, 1891b: 18 (q.) MADAGASCAR. Combination in *C.* (*Tanaemyrmex*): Emery, 1925b: 85.

***imitator.** Camponotus imitator* Forel, 1891b: 209, pl. 4, fig. 15; pl. 5, fig. 8 (w.) MADAGASCAR. Combination in *C.* (*Myrmosphincta*): Forel, 1912i: 92; in *C.* (*Myrmopytia*): Emery, 1920b: 257. Current subspecies: nominal plus *resinicola*.

#***immaculatus.** Camponotus quadrimaculatus* var. *immaculatus* Forel, 1892l: 233 (w.q.) MADAGASCAR. Combination in *C.* (*Mayria*): Emery, 1925b: 123.

*immaculifrons. Camponotus* (*Myrmoturba*) *maculatus* st. *pictiventris* var. *immaculifrons* Santschi, 1919a: 349 (w.) SOUTH AFRICA. Unavailable name.

***immigrans.** Camponotus* (*Myrmothrix*) *immigrans* Santschi, 1914a: 436 (s.w.q.m.) CAMEROUN. Combination in *C.* (*Dinomyrmex*): Santschi, 1921f: 312; in *C.* (*Tanaemyrmex*): Emery, 1925b: 86.

#***imparilis.** Camponotus* (*Myrmamblys*) *reticulatus* r. *imparilis* Forel, 1915a: 40 (w.) SUMATRA.

#***impatibilis.** Camponotus maculatus* r. *impatibilis* Forel, 1899c: 135 (w.) COSTA RICA. Combination in *C.* (*Tanaemyrmex*): Emery, 1925b: 82. Subspecies of *picipes:* Emery, 1920b: 233; of *simillimus:* Santschi, 1922c: 101.

*impavidus* Forel, 1893; see under ***CALOMYRMEX.***

#***imperator.** Camponotus caesar* subsp. *imperator* Emery, 1899e: 503, fig. (w.) CAMEROUN. Combination in *C.* (*Myrmoxygenys*): Emery, 1925b: 170.

*imperator. Camponotus* (*Myrmepomis*) *sericeiventris* subsp. *imperator* Wheeler, W.M. 1934f: 140 (w.) MEXICO. [Junior primary homonym of *imperator* Emery, above.] Replacement name: *pontifex* Santschi, 1936b: 216.

*impetuosa. Formica impetuosa* Smith, F. 1858b: 18 (w.) INDIA. Combination in *Camponotus:* Roger, 1863b: 3. Junior synonym of *ardens:* Mayr, 1886c: 353.

*impolitus. Camponotus foreli* r. *impolitus* Forel, 1902a: 158 (w.) ALGERIA. Junior synonym of *mozabensis:* Forel, 1902g: 463; Emery, 1925b: 105.

*importunoides. Camponotus* (*Myrmoturba*) *maculatus* r. *liengmei* var. *importunoides* Forel, 1914d: 249 (s.w.) SOUTH AFRICA. Unavailable name; material referred to *maculatus* by Baroni Urbani, 1972: 125.

***importunus.** Camponotus maculatus* subsp. *importunus* Forel, 1911e: 288 (s.w.q.m.) SOUTH AFRICA. [First available use of *Camponotus maculatus* subsp. *sexpunctatus* var. *importuna* Forel, 1910f: 27; unavailable name.] Combination in *C.* (*Myrmoturba*): Emery, 1920c: 5; in *C.* (*Tanaemyrmex*): Emery, 1925b: 87. Subspecies of *maculatus:* Emery, 1920c: 5; Arnold, 1922: 635. Raised to species: Baroni Urbani, 1972: 131.

***impressilabris.** Camponotus* (*Myrmotemnus*) *impressilabris* Stitz, 1938: 117, fig. 12 (s.q.) WEST MALAYSIA.

***impressus.** Colobopsis impressa* Roger, 1863a: 160 (w.) U.S.A. Mayr, 1886d: 423 (s.q.); Emery, 1893i: 681 (m.). Combination in *Camponotus* (*Colobopsis*): Emery, 1889b: 517.

***improprius.** Colobopsis impropria* Forel, 1879a: 125 (s.) COLOMBIA. Combination in *Camponotus:* Dalla Torre, 1893: 235; in *C.* (*Myrmamblys*): Forel, 1912i: 90; in *C.* (*Pseudocolobopsis*): Emery, 1920b: 259.

#***inaequalis.** Camponotus inaequalis* Roger, 1863a: 147 (w.q.) CUBA. Combination in *C.* (*Myrmoturba*): Forel, 1914a: 267; in *C.* (*Tanaemyrmex*): Emery, 1925b: 80. Material of the unavailable name *eburneus* referred here by Wheeler, W.M. 1913b: 503 (in text). Subspecies of *conspicuus:* Forel, 1914a: 267.

***inca.** Camponotus inca* Emery, 1903: 75, fig. 13 (w.) PERU. Combination in *C.* (*Myrmoturba*): Forel, 1914a: 267; in *C.* (*Myrmamblys*): Emery, 1920b: 259; in *C.* (*Neomyrmamblys*): Santschi, 1923d: 68; in *C.* (*Myrmaphaenus*): Emery, 1925b: 155. Current subspecies: nominal plus *rector*.

***incensus.** Camponotus* (*Tanaemyrmex*) *incensus* Wheeler, W.M. 1932a: 14 (s.w.) U.S.A.

*incommoda. Camponotus* (*Myrmoturba*) *maculatus* r. *thoracicus* var. *incommoda* Forel, 1914d: 250 (w.) SOUTH AFRICA. Unavailable name.

***inconspicuus.** Camponotus inconspicuus* Mayr, 1872: 136 (w.) BORNEO. Combination in *C.* (*Tanaemyrmex*): Emery, 1925b: 93. Junior synonym of *irritans:* Mayr, 1879: 646. Revived from synonymy: Donisthorpe, 1932c: 444; Donisthorpe, 1943b: 204.

*inconstans. Camponotus* (*Myrmamblys*) *tokioensis* var. *inconstans* Santschi, 1937h: 385 (w.) JAPAN. Junior synonym of *tokioensis:* Terayama & Satoh, 1990a: 412.

*incursor. Formica incursor* Smith, F. 1860b: 95 (w.) INDONESIA (Batjan I.). Junior synonym of *vitreus:* Donisthorpe, 1932c: 459.

*indefessa. Formica indefessa* Sykes, 1835: 104, pl. 13, fig. 6 (w.q.) INDIA. Junior synonym of *compressus:* Jerdon, 1851: 119.

**#indefessus.** *Camponotus* (*Myrmorhachis*) *planus* var. *indefessus* Wheeler, W.M. 1919c: 294 (s.w.) GALAPAGOS IS. Combination in *C.* (*Myrmocladoecus*): Wheeler, W.M. 1924a: 119. [Unresolved junior secondary homonym of *indefessa* Sykes, above.]

**indefinitus.** *Camponotus* (*Tanaemyrmex*) *indefinitus* Karavaiev, 1929c: 235, fig. 2 (s.w.) INDONESIA (Ambon I.).

**indeflexus.** *Formica indeflexa* Walker, 1859: 373 (w., not q.) SRI LANKA. Combination in *Camponotus* (*Myrmamblys*): Donisthorpe, 1932b: 576.

**#indianus.** *Camponotus sylvaticus* r. *indianus* Forel, 1879a: 68 (s.w.q.) COLOMBIA. Wheeler, G.C. & Wheeler, J. 1953e: 197 (l.). Combination in *C.* (*Tanaemyrmex*): Emery, 1925b: 82. Race of *rubripes:* Forel, 1886f: 143. Raised to species: Dalla Torre, 1893: 263. Race/subspecies of *maculatus:* Emery, 1896d: 371; Forel, 1899c: 134 (footnote); Forel, 1904d: 177. Subspecies of *simillimus:* Santschi, 1922c: 101.

**indicatus.** *Camponotus* (*Myrmoturba*) *fumidus* st. *indicatus* Santschi, 1922c: 100 (w.q.) FRENCH GUIANA. Combination in *C.* (*Tanaemyrmex*), and raised to species: Emery, 1925b: 83.

***induratus.** **Formica indurata* Heer, 1850: 116, pl. 11, fig. 2 (m.) YUGOSLAVIA. Combination in *Camponotus:* Mayr, 1867b: 52.

**#inezae.** *Camponotus* (*Dinomyrmex*) *inezae* Forel, 1913k: 122, fig. L1 (s.w.) WEST MALAYSIA. Combination in *C.* (*Tanaemyrmex*), and subspecies of *festinus:* Emery, 1925b: 90; Santschi, 1932b: 18.

**#infelix.** *Camponotus* (*Myrmotrema*) *olivieri* st. *infelix* Santschi, 1915c: 270 (s.w.) SOUTH AFRICA. [First available use of *Camponotus* (*Myrmotrema*) *foraminosus* r. *lemma* var. *infelix* Forel, 1914d: 254; unavailable name.]

**#inferior.** *Camponotus* (*Tanaemyrmex*) *irritans* var. *inferior* Emery, 1925b: 93 (w.q.) BORNEO. [First available use of *Camponotus maculatus* subsp. *irritans* var. *inferior* Forel, 1911a: 48; unavailable name.] See also: Wheeler, W.M. 1937a: 24.

*infernalis. Camponotus maculatus* subsp. *vicinus* var. *infernalis* Wheeler, W.M. 1910d: 305 (w.) U.S.A. [Material of *subrostrata* referable to *infernalis:* Wheeler, W.M. 1917a: 560.] Unavailable name; material referred to *vicinus* by Creighton, 1950a: 381.

**inflatus.** *Camponotus inflatus* Lubbock, 1880: 186, pl. 8 (s.) AUSTRALIA. Froggatt, 1896: 389 (q.m.); Forel, 1910b: 73 (w.m.). Combination in *C.* (*Myrmophyma*): Forel, 1914a: 269. Senior synonym of *aurofasciatus:* Wheeler, W.M. 1916e: 40; Wheeler, W.M. 1916i: 37.

*infrasquameus. Camponotus* (*Myrmoturba*) *maculatus* st. *infrasquameus* Santschi, 1926b: 249 (s.w.q.) BENIN. Junior synonym of *aegyptiacus:* Baroni Urbani, 1972: 130.

**#infuscoides.** *Camponotus taylori* var. *infuscoides* Bingham, 1903: 354 (w.) SRI LANKA. [First available use of *Camponotus maculatus* r. *taylori* var. *infuscoides* Forel, 1893b: 433; unavailable name.] Subspecies of *barbatus: current status.*

**#infuscus.** *Camponotus maculatus* r. *infuscus* Forel, 1892j: 242 (w.) SRI LANKA. Bingham, 1903: 354 (s.); Forel, 1908a: 6 (q.). Combination in *C.* (*Myrmoturba*): Forel, 1913k: 126; in *C.* (*Tanaemyrmex*): Emery, 1925b: 96. Raised to species: Bingham, 1903: 354. Race of *maculatus:* Forel, 1913k: 126. Subspecies of *variegatus:* Emery, 1920c: 7; Emery, 1925b: 96.

**#inlex.** *Camponotus cameranoi* var. *inlex* Forel, 1901d: 302 (s.) BRAZIL. Combination in *C.* (*Myrmobrachys*): Emery, 1925b: 163.

**innexus.** *Camponotus innexus* Forel, 1902h: 499 (s.w.q.m.) AUSTRALIA. Wheeler, G.C. & Wheeler, J. 1968: 218 (l.). Combination in *C.* (*Myrmepomis*): Forel, 1914a: 273; in *C.* (*Myrmophyma*): Emery, 1920b: 257; Santschi, 1928e: 482.

**#innocens.** *Camponotus trapeziceps* var. *innocens* Forel, 1909a: 264 (w.) PARAGUAY. Combination in *C.* (*Myrmaphaenus*): Emery, 1925b: 156.

**#insignis.** *Camponotus* (*Myrmamblys*) *scipio* var. *insignis* Santschi, 1922c: 120 (w.) PARAGUAY.

**insipidus.** *Camponotus insipidus* Forel, 1893d: 454 (s.w.) AUSTRALIA. Wheeler, G.C. & Wheeler, J. 1974a: 61 (l.). Combination in *C.* (*Myrmophyma*): Forel, 1914a: 269.

**#insularis.** *Camponotus* (*Myrmotrema*) *perrisii* var. *insularis* Stitz, 1916: 399, fig. 12 (s.w.q.m.) ANNOBON I. Senior synonym of *annobonensis:* Santschi, 1920i: 4.

**integellus.** *Camponotus integellus* Forel, 1899c: 150 (w.) COSTA RICA. Combination in *C.* (*Myrmeurynota*): Forel, 1912i: 92; in *C.* (*Myrmobrachys*): Forel, 1914a: 270; in *C.* (*Paracolobopsis*): Emery, 1920b: 259; in *C.* (*Myrmaphaenus*): Emery, 1925b: 154. Senior synonym of *rudigenis:* Forel, 1904c: 49.

**#integer.** *Camponotus sericeus* var. *integer* Forel, 1895e: 454 (w.) SRI LANKA. Forel, 1909e: 397 (s.q.). Combination in *C.* (*Orthonotomyrmex*): Forel, 1913k: 129. Variety of *mendax:* Bingham, 1903: 370; of *sericeus:* Forel, 1908a: 6; of *pseudolus:* Emery, 1925b: 125. Currently subspecies of *mendax* as *pseudolus* is an unnecessary replacement name; see there.

**interjectus.** *Camponotus interjectus* Mayr, 1877: 4 (w.q.) TURKESTAN. [Redescribed in Latin: Mayr, 1880: 24.] Karavaiev, 1910b: 12 (s.). Combination in *C.* (*Orthonotomyrmex*): Forel, 1914a: 273; in *C.* (*Myrmentoma*): Emery, 1920b. 257. Subspecies of *lateralis:* Karavaiev, 1910b: 12. Revived status as species: Forel, 1914a: 273; Emery, 1925a: 64. See also: Ruzsky, 1902d: 8; Tarbinsky, 1976: 161.

*intermedia. Formica intermedia* Zetterstedt, 1838: 448 (w.) SWEDEN. Junior synonym of *herculeanus:* Nylander, 1846a: 894.

*intermedius. Camponotus herculeanus* subsp. *shitkowi* var. *intermedius* Ruzsky, 1914b: 100 (w.q.) RUSSIA. Unavailable name.

*intonsus. Camponotus maculatus* subsp. *intonsus* Emery, 1905d: 29 (footnote) (s.w.) SOUTH AFRICA. Junior synonym of *maculatus:* Baroni Urbani, 1972: 125.

**#intrans.** *Camponotus maculatus* subsp. *intrans* Forel, 1911a: 49 (s.w.) SRI LANKA. Combination in *C.* (*Tanaemyrmex*): Emery, 1925b: 96. Subspecies of *variegatus:* Emery, 1920c: 7.

**intrepidus.** *Formica intrepida* Kirby, W. 1819: 477 (w.) AUSTRALIA. Lowne, 1865a: 277 (s.); Mayr, 1862: 659 (q.); Wheeler, G.C. & Wheeler, J. 1968: 218 (l.). Combination in *Camponotus:* Mayr, 1862: 659; in *C.* (*Myrmosphincta*): Forel, 1912i: 92; in *C.* (*Myrmosaulus*): Emery, 1925b: 114. Senior synonym of *agilis:* Roger, 1863b: 4; Forel, 1902h: 493; of *magnus:* Roger, 1863b: 4. Current subspecies: nominal plus *bellicosus.*

**#intuens.** *Camponotus* (*Myrmosericus*) *vestitus* var. *intuens* Santschi, 1926b: 258 (w.) ANGOLA. [First available use of *Camponotus rufoglaucus* subsp. *cinctellus* var. *intuens* Forel, 1909b: 67; unavailable name.] Variety of *cosmicus:* Santschi, 1935a: 280; of *vestitus:* Santschi, 1937d: 244.

**#invalidus.** *Camponotus kelleri* var. *invalidus* Forel, 1897c: 200 (w.) MADAGASCAR. Combination in *C.* (*Mayria*): Emery, 1925b: 123.

**#inverallensis.** *Camponotus claripes* var. *inverallensis* Forel, 1910b: 72 (s.w.) AUSTRALIA. Combination in *C.* (*Myrmophyma*): Emery, 1925b: 110. Subspecies of *claripes:* Taylor & Brown, D.R. 1985: 111.

*inversa. Camponotus maculatus* subsp. *barbaricus* var. *inversa* Forel, 1910a: 26 (w.) ALGERIA. Unavailable name.

**invidus.** *Camponotus invidus* Forel, 1892j: 234 (w.) INDIA. Combination in *C.* (*Myrmoturba*): Forel, 1914a: 267; in *C.* (*Tanaemyrmex*): Emery, 1925b: 93. See also: Bingham, 1903: 367.

**ionius.** *Camponotus* (*Myrmoturba*) *samius* var. *ionia* Emery, 1920c: 6 (w.) GREECE. [First available use of *Camponotus maculatus* subsp. *samius* var. *ionia* Emery, 1901e: 59; unavailable name.] Santschi, 1921a: 116 (q.). Combination in *C.* (*Tanaemyrmex*): Emery, 1925b: 100. Variety of *samius:* Menozzi, 1936d: 300. Raised to species: Aktaç, 1976: 125; Agosti & Collingwood, 1987a: 58.

**#iota.** *Camponotus* (*Tanaemyrmex*) *pompeius* var. *iota* Santschi, 1926b: 257 (w.) CONGO.

**#irakensis.** *Camponotus* (*Tanaemyrmex*) *compressus* subsp. *irakensis* Menozzi, 1927e: 117, fig. 1 (s.q.m.) IRAQ.

**iridis.** *Camponotus* (*Neomyrmamblys*) *iridis* Santschi, 1922c: 115 (s.w.) BRAZIL. Combination in *C.* (*Myrmaphaenus*): Emery, 1925b: 155.

**#irredux.** *Camponotus irredux* Forel, 1910f: 30 (s.w.) SOUTH AFRICA. Combination in *C.* (*Myrmopiromis*): Wheeler, W.M. 1922a: 987. Subspecies of *niveosetosus:* Forel, 1910e: 451; Arnold, 1922: 659.

**irritabilis.** *Formica irritabilis* Smith, F. 1857a: 56 (w.) BORNEO. Combination in *Camponotus:* Roger, 1863b: 3; in *C.* (*Myrmotarsus*): Forel, 1912i: 92. Current subspecies: nominal plus *winkleri.*

**irritans.** *Formica irritans* Smith, F. 1857a: 55 (w.) WEST MALAYSIA. [Described a second time on p. 58 of same publication.] Wheeler, W.M. 1924b: 253 (q.m.). Combination in *Camponotus:* Roger, 1863b: 3; in *C.* (*Myrmoturba*): Forel, 1913k: 125; in *C.* (*Tanaemyrmex*): Emery, 1925b: 93. Subspecies of *maculatus:* Forel, 1892j: 242; Emery, 1896d: 370; Forel, 1901b: 27; Forel, 1909d: 229; Wheeler, W.M. 1924b: 253. Status as species: Dalla Torre, 1893: 236; Bingham, 1903: 353; Emery, 1920b: 255; Emery, 1920c: 7; Santschi, 1924c: 114; Emery, 1925b: 93; Wheeler, W.M. 1929g: 57; Stärcke, 1930: 376; Karavaiev, 1933a: 316. Senior synonym of *agnatus:* Bingham, 1903: 353. [Donisthorpe, 1932c: 444, suggests that *irritans* may be the worker of *diligens.*] Current subspecies: nominal plus *carensis, carinifer, cliens, croceomaculatus, curtus, fatuus, hongkongensis, inferior, kubaryi, melanogaster, pallidus,* *procax, puberulus* (unresolved junior homonym), *tinctus*.

**#*isabelensis.*** *Camponotus* (*Myrmorhachis*) *planus* var. *isabelensis* Wheeler, W.M. 1919c: 293 (s.w.q.m.) GALAPAGOS IS. Combination in *C.* (*Myrmocladoecus*): Wheeler, W.M. 1924a: 119.

***isabellae.*** *Camponotus isabellae* Forel, 1909e: 399 (q.m.) SRI LANKA. Combination in *C.* (*Myrmamblys*): Forel, 1914a: 271.

***itoi.*** *Camponotus* (*Myrmamblys*) *itoi* Forel, 1912l: 340 (s.w.) JAPAN. Combination in *C.* (*Myrmotemnus*): Emery, 1920b: 258; in *C.* (*Myrmamblys*): Emery, 1925b: 138; Santschi, 1925f: 89. Current subspecies: nominal plus *kwansienensis*.

**#*jacob.*** *Camponotus* (*Myrmotrema*) *auropubens* var. *jacob* Santschi, 1915c: 275 (w.) MOZAMBIQUE.

**#*jacobensis.*** *Camponotus* (*Myrmamblys*) *macilentus* var. *jacobensis* Wheeler, W.M. 1919c: 280 (s.w.q.m.) GALAPAGOS IS. Combination in *C.* (*Pseudocolobopsis*): Emery, 1925b: 158.

*jacobsoni. Camponotus* (*Myrmomalis*) *jacobsoni* Forel, 1915a: 39 (q.) SUMATRA. Junior synonym of *concilians:* Emery, 1925b: 136.

*jacuticus. Camponotus* (*Camponotus*) *herculeanus* var. *jacuticus* Karavaiev, 1929b: 210 (s.w.q.m.) RUSSIA. Junior synonym of *herculeanus:* Yasumatsu & Brown, 1951: 30; of *sachalinensis:* Arnol'di, 1967: 1821.

**#*jagori.*** *Camponotus* (*Myrmamblys*) *reticulatus* r. *jagori* Stitz, 1925: 126 (w.) PHILIPPINES.

***jaliensis.*** *Camponotus oertzeni* var. *jaliensis* Dalla Torre, 1893: 246 (s.w.) GREECE. [First available use of *Camponotus rubripes* r. *oertzeni* var. *jaliensis* Forel, 1889: 264; unavailable name.] Combination in *C.* (*Tanaemyrmex*): Emery, 1925b: 97. Subspecies of *maculatus:* Emery, 1908a: 201; of *aethiops:* Emery, 1920c: 7. Raised to species: Pisarski, 1971a: 672; Agosti & Collingwood, 1987a: 58. Current subspecies: nominal plus *shaqualavensis*.

*jalpiguri. Camponotus jalpiguri* Forel, 1902d: 287 (q.). *Nomen nudum*.

**#*jamaicensis.*** *Camponotus* (*Myrmoturba*) *maculatus* subsp. *jamaicensis* Wheeler, W.M. 1917g: 465, fig. 2 (s.w.m.) JAMAICA. Combination in *C.* (*Tanaemyrmex*): Emery, 1925b: 82. Subspecies of *picipes:* Emery, 1920b: 233.

***janeti.*** *Camponotus janeti* Forel, 1895f: 417 (s.w.) AUSTRALIA. Combination in *C.* (*Myrmamblys*): Forel, 1914a: 271.

*janitor. Camponotus* (*Colobopsis*) *paradoxus* subsp. *janitor* Forel, 1908c: 415 (s.w.q.m.) BRAZIL. Combination in *C.* (*Hypercolobopsis*): Emery, 1925b: 160. Junior synonym of *paradoxus:* Kempf, 1960f: 460.

**#*janus.*** *Camponotus petersi* var. *janus* Forel, 1911e: 228 (w.) SOUTH AFRICA. Combination in *C.* (*Myrmosericus*): Emery, 1925b: 105.

*janus. Camponotus* (*Colobopsis*) *janus* Mann, 1921: 498 (s.w.) FIJI IS. [Junior primary homonym of *janus* Forel, above.] Replacement name: *janussus* **new name**.

***janussus.*** *Camponotus janussus* **new name**. Replacement name for *janus* Mann, 1921: 498. [Junior primary homonym of *janus* Forel, 1911e: 228.]

***japonicus.*** *Camponotus japonicus* Mayr, 1866b: 885 (w.) JAPAN. Forel, 1904b: 380 (m.); Wheeler, W.M. 1906c: 324 (s.q.); Imai & Kubota, 1972: 195 (k.). Combination in *C.* (*Camponotus*): Forel, 1914a: 266. Variety of *pennsylvanicus:* Forel, 1879a: 56; Forel, 1904b: 380; Ruzsky, 1905b: 239. Subspecies of *herculeanus:* Emery, 1908a: 186; Wheeler, W.M. 1909d: 343; Wheeler, W.M. 1921c: 544; Ruzsky, 1925b: 42; Kuznetsov-Ugamsky, 1929a: 17; Yasumatsu & Brown, 1951: 36. Status as species: Bingham, 1903: 390; Santschi, 1920h: 255; Emery, 1925b: 73; Santschi, 1925f: 88; Ruzsky, 1926: 108; Wheeler, W.M. 1927e: 3; Wheeler, W.M. 1928d: 116; Karavaiev, 1929b: 212; Stitz, 1934: 4; Yasumatsu & Brown, 1957: 49; Arnol'di, 1967: 1826; Kupyanskaya, 1990: 169. Senior synonym of *jeholensis, koreanus, miltotus, sanguinea* and *wui:* Yasumatsu & Brown, 1951: 36. Current subspecies: nominal plus *aterrimus*.

**#*jason.*** *Camponotus maculatus* r. *jason* Forel, 1912i: 70 (s.w.) BOLIVIA. Combination in *C.* (*Tanaemyrmex*): Emery, 1925b: 82. Subspecies of *picipes:* Emery, 1920b: 233.

***jeanneli.*** *Camponotus* (*Myrmosaga*) *jeanneli* Santschi, 1914b: 137, fig. 30 (s.w.) KENYA. Combination in *C.* (*Mayria*): Emery, 1925b: 123.

*jeholensis. Camponotus herculeanus* var. *jeholensis* Teranishi, 1936: 4 (w.) KOREA. Junior synonym of *japonicus:* Yasumatsu & Brown, 1951: 36.

***jejuensis.*** *Camponotus jejuensis* Kim & Kim, 1986: 140, figs. 1–16 (w.) KOREA. Current subgeneric placement: *C.* (*Myrmentoma*).

**#*jejunus.*** *Camponotus* (*Tanaemyrmex*) *langi* var. *jejunus* Santschi, 1926b: 254 (w.) ZAIRE.

***jheringi:*** see under *iheringi*.

***jianghuaensis.*** *Camponotus jianghuaensis* Xiao & Wang; in Wang, C., Xiao & Wu, 1989b: 321, figs. 22, 23 (s.w.) CHINA.

***jizani.*** *Camponotus jizani* Collingwood, 1985: 281, figs. 71, 73 (w.) SAUDI ARABIA. Subgenus indeterminate.

***johnclarki.*** *Camponotus johnclarki* Taylor, 1992a: 59. Replacement name for *Notostigma sanguinea* Clark, 1930a: 116. [Junior secondary homonym of *Camponotus sanguinea* Karavaiev, 1929b: 212.] Subgenus indeterminate.

**#*jucundus.*** *Camponotus perrisii* st. *jucundus* Santschi, 1911g: 216 (s.w.) CONGO. Wheeler, W.M. 1922a: 251 (q.). Combination in *C.* (*Myrmotrema*): Santschi, 1913c: 314.

**#*jugurtha.*** *Camponotus* (*Tanaemyrmex*) *solon* var. *jugurtha* Emery, 1925b: 100 (s.) ZAIRE. [First available use of *Camponotus* (*Myrmoturba*) *maculatus* subsp. *solon* var. *jugurtha* Wheeler, W.M. 1922a: 236; unavailable name.]

***juliae.*** *Camponotus juliae* Emery, 1903: 68, fig. 3 (s.w.) COSTA RICA. Combination in *C.* (*Myrmothrix*): Forel, 1914a: 269; in *C.* (*Myrmamblys*): Emery, 1920b: 259; in *C.* (*Myrmaphaenus*): Emery, 1925b: 155.

*junctus. Camponotus maculatus* r. *junctus* Forel, 1892j: 243 (w.) INDIA. Junior synonym of *arrogans:* Bingham, 1903: 357.

**#*kadi.*** *Camponotus* (*Myrmogonia*) *schmeltzii* var. *kadi* Mann, 1921: 485 (s.w.m.) FIJI IS.

*kairouana. Camponotus* (*Myrmoturba*) *sylvaticus* st. *barbarus* var. *kairouana* Santschi, 1921d: 72 (s.w.q.) TUNISIA. Unavailable name.

**#*kamae.*** *Camponotus mystaceus* var. *kamae* Forel, 1910f: 27 (w.) NAMIBIA. Arnold, 1924: 677 (s.). Combination in *C.* (*Myrmopsamma*): Wheeler, W.M. 1922a: 677.

**#*kamensis.*** *Camponotus marginatus* var. *kamensis* Ruzsky, 1903b: 302 (w.) RUSSIA. Combination in *C.* (*Myrmentoma*): Emery, 1925b: 118. Variety of *caryae:* Wheeler, W.M. 1917c: 29.

**#*kappariensis.*** *Camponotus oertzeni* var. *kappariensis* Dalla Torre, 1893: 246 (s.w.) GREECE. [First available use of *Camponotus rubripes* r. *oertzeni* var. *kappariensis* Forel, 1889: 264; unavailable name (Agosti & Collingwood, 1987a: 61).]

***karawaiewi.*** *Camponotus* (*Colobopsis*) *karawaiewi* Menozzi, 1926e: 96 (q.) PHILIPPINES.

*kaschgariensis. Camponotus maculatus* r. *kaschgariensis* Forel, 1901h: 67 (s.w.q.) CHINA. Junior synonym of *turkestanicus:* Forel, 1904b: 378.

**#*kattensis.*** *Camponotus dichrous* var. *kattensis* Bingham, 1903: 356 (w.) INDIA. [First available use of *Camponotus maculatus* r. *dichrous* var. *kattensis* Forel, 1892j: 231; unavailable name.] Subspecies of *maculatus:* Forel, 1906b: 84; of *variegatus:* Emery, 1920c: 7.

***keiferi.*** *Camponotus* (*Camponotus*) *keiferi* Wheeler, W.M. 1934f: 137 (s.w.) MEXICO.

***keihitoi.*** *Camponotus fallax* var. *keihitoi* Forel, 1913e: 663 (w.q.) JAPAN. Raised to species, and senior synonym of *teranishii:* Terayama & Satoh, 1990b: 532.

***kelleri.*** *Camponotus kelleri* Forel, 1886f: 186 (s.w.) MADAGASCAR. Combination in *C.* (*Myrmosaga*): Forel, 1912i: 92; in *C.* (*Mayria*): Emery, 1925b: 123. Current subspecies: nominal plus *invalidus*.

***kersteni.*** *Camponotus kersteni* Gerstäcker, 1871: 355 (w.) KENYA. Santschi, 1914b: 131 (s.q.m.). Combination in *C.* (*Myrmoturba*): Forel, 1914a: 267; in *C.* (*Tanaemyrmex*): Emery, 1925b: 92. Subspecies of *maculatus:* Forel, 1899b: 309; Wheeler, W.M. 1922a: 955. Status as species: Dalla Torre, 1893: 237; Santschi, 1914b: 131; Emery, 1925b: 92.

***kiesenwetteri.*** *Formica* (*Hypoclinea*) *kiesenwetteri* Roger, 1859: 241 (w., not m., see Mayr, 1861: 39) GREECE. Forel, 1886e: clxi (q.m.). Combination in *Camponotus:* Mayr, 1861: 36; in *C.* (*Orthonotomyrmex*): Forel, 1914a: 273; in *C.* (*Myrmentoma*): Forel, 1912i: 92; Emery, 1925b: 121. Current subspecies: nominal plus *cyprius, nitidescens*.

***kiusiuensis.*** *Camponotus* (*Paramyrmamblys*) *kiusiuensis* Santschi, 1937h: 379, figs. 11, 14 (w.q.) JAPAN. Imai, 1966: 119 (k.). See also: Terayama, 1991: 167.

***klaesii.*** *Rhinomyrmex klaesii* Forel, 1886f: 193 (w.) SUMATRA. Combination in *Camponotus* (*Rhinomyrmex*): Emery, 1896d: 374.

***klugii.*** *Camponotus bedoti* var. *klugii* Emery, 1895h: 51 (w.) SOUTH AFRICA. Combination in *C.* (*Myrmamblys*): Wheeler, W.M. 1922a: 989. Raised to species: Emery, 1925b: 141.

***knysnae.*** *Camponotus* (*Myrmoturba*) *maculatus* r. *knysnae* Arnold, 1922: 642 (s.w.q.) SOUTH AFRICA. Combination in *C.* (*Tanaemyrmex*), and raised to species: Emery, 1925b: 100 (misspelled here as *kaysnae*).

#***kodoricus.*** *Camponotus vagus* var. *kodorica* Forel, 1913a: 145 (w.) RUSSIA.

#***kohli.*** *Camponotus* (*Dinomyrmex*) *aequatorialis* r. *kohli* Forel, 1915c: 349 (s.w.) ZAIRE. Combination in *C.* (*Tanaemyrmex*): Emery, 1925b: 84.

***kollbrunneri.*** *Camponotus chrysurus* subsp. *kollbrunneri* Forel, 1910c: 268 (s.) ETHIOPIA. Combination in *C.* (*Orthonotomyrmex*): Wheeler, W.M. 1922a: 972; in *C.* (*Myrmopiromis*): Emery, 1925b: 129; in *C.* (*Myrmopelta*), and raised to species: Santschi, 1926a: 20; Santschi, 1926c: 603.

***kolthoffi.*** *Camponotus* (*Myrmentoma*) *kolthoffi* Stitz, 1934: 4, fig. 2 (w.) CHINA. See also: Wang, C., Xiao & Wu, 1989a: 222.

***kopetdaghensis.*** *Camponotus* (*Myrmentoma*) *kopetdaghensis* Dlussky & Zabelin, 1985: 236, fig. 8 (s.w.) RUSSIA.

*koreanus. Camponotus herculeanus* var. *koreanus* Teranishi, 1940: 71 (s.w.) KOREA. Junior synonym of *japonicus:* Yasumatsu & Brown, 1951: 36.

***korthalsiae.*** *Camponotus korthalsiae* Emery, 1887a: 220, pl. 3, fig. 9 (w.) SUMATRA. Combination in *C.* (*Myrmomalis*): Forel, 1914a: 271; in *C.* (*Myrmoplatys*): Forel, 1916: 460. Current subspecies: nominal plus *concilians.*

***koseritzi.*** *Camponotus tenuiscapus* r. *koseritzi* Emery, 1888c: 366 (s.w.) BRAZIL. Combination in *C.* (*Myrmoturba*): Bruch, 1914: 229; in *C.* (*Tanaemyrmex*): Emery, 1925b: 78. Race of *tenuiscapus:* Forel, 1901h: 70. Status as species: Dalla Torre, 1893: 237; Emery, 1894g: 375; Forel, 1895b: 103; Emery, 1903: 68; Kempf, 1972a: 68.

***kosswigi.*** *Camponotus* (*Orthonotomyrmex*) *kosswigi* Donisthorpe, 1950c: 68 (w.) TURKEY.

***kraepelini.*** *Camponotus kraepelini* Forel, 1901h: 73 (w.) BOLIVIA. Combination in *C.* (*Myrmeurynota*): Forel, 1914a: 274.

#***krafti.*** *Camponotus* (*Colobopsis*) *rothneyi* r. *krafti* Forel, 1901h: 66 (q.) SINGAPORE.

#***krama.*** *Camponotus* (*Colobopsis*) *badius* var. *krama* Forel, 1912d: 110 (s.) JAVA. Variety of *saundersi:* Emery, 1925b: 150.

#***kubaryi.*** *Camponotus kubaryi* Mayr, 1876: 66 (s.w.) PALAU IS. Combination in *C.* (*Myrmoturba*): Forel, 1914a: 267; in *C.* (*Tanaemyrmex*): Emery, 1925b: 94. Subspecies of *maculatus:* Emery, 1896d: 370; of *irritans:* Emery, 1920c: 7; Karavaiev, 1929c: 236.

***kurdistanicus.*** *Camponotus gestroi* subsp. *kurdistanicus* Emery, 1898c: 151 (w.q.) RUSSIA. Combination in *C.* (*Myrmentoma*), and raised to species: Emery, 1920b: 257; Emery, 1920c: 25. See also: Emery, 1925a: 70.

***kutteri.*** *Camponotus* (*Myrmamblys*) *kutteri* Forel, 1915a: 40 (s.w.q.m.) SUMATRA.

***kutterianus.*** *Camponotus kutterianus* Baroni Urbani, 1972: 132, figs. 1–3 (w.) CUBA. Subgenus indeterminate.

#***kwansienensis.*** *Camponotus* (*Myrmamblys*) *itoi* subsp. *kwansienensis* Viehmeyer, 1922: 216 (w.) CHINA.

*laboriosa. Formica laboriosa* Smith, F. 1858b: 32 (q.) SIERRA LEONE. Combination in *Camponotus:* Roger, 1863b: 2. Junior synonym of *vividus:* Mayr, 1863: 416; Wheeler, W.M. 1922a: 975.

***laconicus.*** *Camponotus* (*Myrmoturba*) *samius* var. *laconicus* Emery, 1920c: 6 (s.w.) GREECE. [First available use of *Camponotus maculatus* subsp. *samius* var. *laconicus* Emery, 1908a: 192; unavailable name.] Combination in *C.* (*Tanaemyrmex*): Emery, 1925b: 100. Raised to species: Agosti & Collingwood, 1987a: 58.

*lacteipennis. Formica lacteipennis* Smith, F. 1858b: 34 (s.w.q.m.) SOUTH AFRICA. Combination in *C.* (*Myrmoturba*): Forel, 1914a: 267; in *C.* (*Tanaemyrmex*): Emery, 1925b: 90. Subspecies of *sylvaticus:* Mayr, 1886c: 355; of *maculatus:* Emery, 1895h: 46; Emery, 1920c: 5. Junior synonym of *maculatus:* Baroni Urbani, 1972: 125. See also: Arnold, 1922: 626.

*lacteus. Camponotus substitutus* var. *lacteus* Baroni Urbani, 1971b: 361. [Incorrectly proposed as replacement name for *clarus* Stitz, 1938: 116.] Unavailable name.

#***laetus.*** *Camponotus* (*Tanaemyrmex*) *discors* var. *laeta* Emery, 1925b: 102 (s.) AUSTRALIA. [First available use of *Camponotus maculatus* subsp. *discors* var. *laeta* Forel, 1910b: 70; unavailable name (Taylor, 1986: 34).]

***laevigatus.*** *Formica laevigata* Smith, F. 1858b: 55 (w q.) U.S.A. Wheeler, W.M. 1910d: 327 (s.m.); Wheeler, G.C. & Wheeler, J. 1968: 216 (l.). Combination in *Camponotus:* Roger, 1863b: 5; in *C.* (*Camponotus*): Forel, 1914a: 266.

*laevilata. Camponotus* (*Myrmothrix*) *abdominalis* var. *laevilata* Santschi, 1922c: 108 (s.q.) BRAZIL. Junior synonym of *atriceps:* Hashmi, 1973: 82.

*laevithorax. Camponotus vividus* var. *laevithorax* Menozzi, 1924b: 227 (s.) UGANDA. Junior synonym of *cato:* Menozzi, 1933a: 113.

***lamarckii.*** *Camponotus lamarckii* Forel, 1892j: 236 (s.w.) INDIA. Combination in *C.* (*Myrmoturba*): Forel, 1914a: 267; in *C.* (*Tanaemyrmex*): Emery, 1925b: 94.

***lamborni.*** *Camponotus* (*Myrmotrema*) *lamborni* Donisthorpe, 1933b: 196 (s.w.m.) MALAWI.

***lameerei.*** *Camponotus marginatus* var. *lameerei* Emery, 1898c: 150 (s.w.q.m.) RUSSIA. Combination in *C.* (*Myrmentoma*): Emery, 1925b: 118. Variety/subspecies of *marginatus:* Ruzsky, 1905b: 249; Kuznetsov-Ugamsky, 1923: 242; of *lateralis:* Forel, 1904b: 380; of *caryae:* Wheeler, W.M. 1917c: 29. Raised to species: Tarbinsky, 1976: 146; Dlussky & Zabelin, 1985: 237.

***laminatus.*** *Camponotus laminatus* Mayr, 1866a: 489, fig. 4 (w.) FIJI IS. Mann, 1921: 477 (s.q.m.). Combination in *C.* (*Myrmogonia*): Forel, 1912i: 92. Current subspecies: nominal plus *levuanus.*

***lancifer.*** *Camponotus lancifer* Emery, 1894c: 172, pl. 1, fig. 4 (w.) BRAZIL. Combination in *C.* (*Myrmotemnus*): Emery, 1920b: 259; in *C.* (*Myrmosphincta*): Forel, 1914a: 273; Emery, 1925b: 151.

***landolti.*** *Camponotus sexguttatus* r. *landolti* Forel, 1879a: 71 (s.w.) COLOMBIA. Combination in *C.* (*Myrmoturba*): Forel, 1914a: 267; in *C.* (*Tanaemyrmex*): Emery, 1925b: 83. Subspecies of *extensus:* Emery, 1894f: 3; Emery, 1896d: 371. Raised to species: Forel, 1886f: 168; Forel, 1899c: 135.

***langi.*** *Camponotus* (*Dinomyrmex*) *langi* Wheeler, W.M. 1922a: 241, fig. 62 (s.w.q.m.) ZAIRE. Combination in *C.* (*Tanaemyrmex*): Emery, 1925b: 86. Current subspecies: nominal plus *jejunus.*

***laotzei.*** *Camponotus* (*Colobopsis*) *laotzei* Wheeler, W.M. 1921c: 545 (s.w.) CHINA.

*lapidicola. Camponotus punctulatus* st. *imberbis* var. *lapidicola* Santschi, 1939e: 326 (w.) ARGENTINA. Unavailable name.

***largiceps.*** *Camponotus largiceps* Wu & Wang; in Wang, C., Xiao & Wu 1989a: 227, figs. 20, 21 (s.w.) CHINA.

***larvigerus.*** *Camponotus larvigerus* Wheeler, W.M. & Mann, 1914: 52, fig. 23 (s.w.q.) HAITI. Combination in *C.* (*Myrmoturba*): Forel, 1914a: 267; in *C.* (*Tanaemyrmex*): Emery, 1925b: 81. Current subspecies: nominal plus *maculifrons.*

***lasiselene.*** *Camponotus lasiselene* Wang & Wu, 1994: 24, fig. 3 (w.) CHINA.

***latangulus.*** *Camponotus latangulus* Roger, 1863a: 142 (w.) SURINAM. Emery, 1894c: 172 (s.); Wheeler, W.M. 1934g: 224 (q.); Wheeler, G.C. & Wheeler, J. 1953e: 195 (l.). Combination in *C.* (*Myrmorhachis*): Forel, 1914a: 274; in *C.* (*Myrmocladoecus*): Wheeler, W.M. 1922a: 709; Wheeler, W.M. 1934e: 424.

***latebrosus.*** *Formica latebrosa* Walker, 1859: 371 (m.) SRI LANKA. Combination in *C.* (*Tanaemyrmex*): Donisthorpe, 1932b: 575.

***lateralis.*** *Formica lateralis* Olivier, 1792: 497 (q.) FRANCE. Mayr, 1853b: 103 (s.w.m.); Hauschteck, 1962: 219 (k.). Combination in *Camponotus:* Mayr, 1861: 36; in *C.* (*Orthonotomyrmex*): Forel, 1913d: 436; in *C.* (*Myrmentoma*): Forel, 1912i: 92; Emery, 1925b: 120. Senior synonym of *melanogastes* and *bicolor:* Mayr, 1853b: 102; of *axillaris:* Mayr, 1855: 322; of *pallidinervis:* Mayr, 1863: 399; of *hemipsila:* Forel, 1894d: 5; of *armouri* (and its junior synonyn *balearis*): Collingwood, 1978: 73. See also: Bondroit, 1918: 78; Emery, 1924a: 68; Finzi, 1930d: 318; Stitz, 1939: 254; Bernard, 1967: 344; Baroni Urbani, 1971c: 191; Kutter, 1977c: 207; Atanassov & Dlussky, 1992: 225. Current subspecies: nominal plus *cypridis, ebneri, purius, rebeccae, rhodius.*

#***latericius.*** *Camponotus* (*Myrmosericus*) *rufoglaucus* r. *latericius* Stitz, 1923: 165 (w.) NAMIBIA.

*latinotus. Camponotus foraminosus* subsp. *latinotus* Forel, 1907c: 144 (w.) KENYA. Junior synonym of *galla:* Forel, 1910c: 271.

*latinota. Camponotus iheringi* var. *latinota* Forel, 1909a: 264 (w.) PARAGUAY. [Junior primary homonym of *latinotus* Forel, above.] Replacement name: *bajulus* Emery, 1925b: 165.

#***latinotus.*** *Camponotus* (*Colobopsis*) *vitreus* var. *latinota* Stitz, 1925: 127 (q.) BORNEO. [Unresolved junior primary homonym of *latinotus* Forel, 1907c: 144.]

#***latitans.*** *Camponotus reticulatus* var. *latitans* Forel, 1893b: 431 (w.) INDIA. Forel, 1915a: 40 (footnote) (q.). Combination in *C.* (*Myrmamblys*): Forel, 1915a: 40 (footnote).

***latrunculus.*** *Camponotus* (*Myrmoturba*) *latrunculus* Wheeler, W.M. 1915g: 814, pl. 66, figs. 3, 4 (s.) AUSTRALIA. Combination in *C.* (*Tanaemyrmex*): Emery, 1925b: 102. Current subspecies: nominal plus *victoriensis.*

***lauensis.*** *Camponotus* (*Myrmogonia*) *lauensis* Mann, 1921: 488, fig. 35 (w.) FIJI IS.

#***laurenti.*** *Camponotus* (*Myrmoturba*) *alii* st. *laurenti* Santschi, 1939d: 84, figs. 18, 19, 22 (s.w.q.) ALGERIA.

*lautus*. *Formica lauta* Say, 1836: 286 (q.m.) U.S.A. Combination in *Camponotus:* Dalla Torre, 1893: 238. Unrecognisable taxon: Smith, D.R. 1979: 1467.

***leae***. *Camponotus* (*Myrmosphincta*) *leae* Wheeler, W.M. 1915g: 819, pl. 66, fig. 9 (w.) AUSTRALIA. Combination in *C.* (*Myrmosaulus*): Emery, 1925b: 114.

*ledieui*. *Camponotus* (*Orthonotomyrmex*) *mayri* r. *ledieui* Forel, 1916: 444 (w.) SOUTH AFRICA. Junior synonym of *dofleini:* Emery, 1920c: 29.

***legionarium***. *Camponotus maculatus* st. *legionarium* Santschi, 1911i: 283 (s.) MADAGASCAR. Combination in *C.* (*Tanaemyrmex*): Emery, 1925b: 89. Raised to species: Baroni Urbani, 1972: 132.

#***lehmanni***. *Camponotus claviscapus* var. *lehmanni* Forel, 1911e: 294 (q.) NICARAGUA. Combination in *C.* (*Pseudocolobopsis*): Emery, 1925b: 157.

#***lelubrei***. *Camponotus foleyi* subsp. *lelubrei* Bernard, 1953a: 193 (s.w.) ALGERIA.

#***lemma***. *Camponotus foraminosus* r. *lemma* Forel, 1886f: 176 (w.q.) ANGOLA. Combination in *C.* (*Myrmotrema*): Santschi, 1915c: 270. Raised to species: Dalla Torre, 1893: 238. Subspecies of *foraminosus:* Emery, 1896d: 376; Wheeler, W.M. 1922a: 980; of *olivieri:* Santschi, 1915c: 270; Emery, 1925b: 132.

***lenkoi***. *Camponotus* (*Myrmothrix*) *lenkoi* Kempf, 1960e: 398, figs. 10–12 (s.w.) BRAZIL. See also: Hashmi, 1973: 116.

***leonardi***. *Camponotus* (*Colobopsis*) *leonardi* Emery, 1889b: 515, pl. 11, figs. 22, 23 (s.w.) BURMA. Karavaiev, 1929c: 243 (m.). [Junior synonyn of *pilosa* Smith, F. 1857a: 54 (and its junior synonym *pubescens* Mayr, 1862: 691). However, as both these earlier names are junior homonyms within *Camponotus, leonardi* becomes the first available name for this species. See: Forel, 1893b: 437; Dalla Torre, 1893: 248; Emery, 1900d: 706; Wheeler, W.M. 1919e: 115; Emery, 1921a: 25; Emery, 1925b: 149.] Current subspecies: nominal plus *gracilentus, griseus*.

#***leonhardi***. *Camponotus quadrilaterus* r. *leonhardi* Forel, 1901h: 73 (w.) BOLIVIA. Combination in *C.* (*Myrmocladoecus*), and subspecies of *sanctaefidei:* Emery, 1925b: 167; Wheeler, W.M. 1934g: 232.

***leptocephalus***. *Camponotus* (*Pseudocolobopsis*) *leptocephalus* Emery, 1923: 62, fig. 3 (q.) BRAZIL.

***lespesii***. *Camponotus sexguttatus* r. *lespesii* Forel, 1886f: 169 (w.) COLOMBIA. Mayr, 1887: 513 (q.m.). Combination in *C.* (*Dinomyrmex*): Forel, 1914a: 268; in *C.* (*Tanaemyrmex*): Emery, 1925b: 83. Raised to species: Mayr, 1887: 513. Current subspecies: nominal plus *melancholicus*.

***lessonai***. *Camponotus rufipes* subsp. *lessonai* Emery, 1894e: 3 (s.w.) PARAGUAY. Combination in *C.* (*Myrmothrix*): Emery, 1925b: 109. Referred to as "species" and excluded from *Myrmothrix* (hence subgenus indeterminate): Hashmi, 1973: 129.

#***leucodiscus***. *Camponotus* (*Colobopsis*) *leucodiscus* Wheeler, W.M. 1919e: 117 (q.) BORNEO. Wheeler, W.M. 1919e: 118 (footnote) (s.w.). Combination in *C.* (*Myrmamblys*), and subspecies of *bellus:* Emery, 1925b: 138.

***leucophaeus***. *Formica leucophaea* Smith, F. 1861b: 37 (w.) SULAWESI. Combination in *Camponotus:* Dalla Torre, 1893: 238; in *C.* (*Myrmosericus*): Donisthorpe, 1932c: 465.

#***leucopus***. *Camponotus* (*Colobopsis*) *rufifrons* var. *leucopus* Emery, 1914f: 427, pl. 13, fig. 17 (s.w.) NEW CALEDONIA. Santschi, 1928a: 58 (q.).

*leucozona*. *Camponotus* (*Myrmoturba*) *punctulatus* st. *termitarius* var. *leucozona* Santschi, 1923c: 273 (w.) BRAZIL. Unavailable name.

***leveillei***. *Camponotus leveillei* Emery, 1895f: 344 (w.) MADAGASCAR. Combination in *C.* (*Myrmobrachys*): Forel, 1914a: 270; in *C.* (*Myrmonesites*): Emery, 1920b: 257.

#***levuanus***. *Camponotus* (*Myrmogonia*) *laminatus* var. *levuanus* Mann, 1921: 479 (w.) FIJI IS.

***leydigi***. *Camponotus leydigi* Forel, 1886f: 169 (w.) BRAZIL. Combination in *C.* (*Myrmothrix*): Forel, 1914a: 269; in *C.* (*Myrmaphaenus*): Emery, 1920b: 256. Senior synonym of *nobilis:* Emery, 1925b: 153.

***libanicus***. *Camponotus libanicus* André, 1881b: 54, pl. 3, figs. 14, 15 (w.) LEBANON. Forel, 1911d: 355 (q.); Tohmé, G. 1969: 11 (m.). Combination in *C.* (*Orthonotomyrmex*): Forel, 1913d: 435; in *C.* (*Myrmentoma*): Emery, 1925b: 121; Menozzi, 1936d: 304. Current subspecies: nominal plus *aegaeus, sahlbergi*.

*lichtensteini*. *Camponotus lichtensteini* Bondroit, 1918: 74 (s.w.) FRANCE. Junior synonym of *sylvaticus:* Bernard, 1967: 337.

*liengmei*. *Camponotus sexpunctatus* var. *liengmei* Forel, 1894b: 67 (s.w.) MOZAMBIQUE. Forel, 1907g: 88 (m.). Combination in *C.* (*Myrmoturba*): Forel, 1914a: 267; in *C.* (*Dinomyrmex*): Santschi, 1921f: 312; in *C.* (*Tanaemyrmex*): Emery, 1925b: 87. Subspecies/race of *maculatus:* Forel, 1901h: 69; Forel, 1911e: 288; Arnold, 1922: 636; Emery, 1920c: 5; Emery, 1925b: 87. Raised to species: Santschi, 1923e: 292. Junior synonym of *maculatus* Baroni Urbani, 1972: 125.

*ligatus*. *Camponotus* (*Myrmocladoecus*) *rectangularis* subsp. *setipes* var. *ligatus* Wheeler, 1934g: 223 (w.) GUIANA. Unavailable name.

***ligeus***. *Camponotus* (*Tanaemyrmex*) *ligea* Donisthorpe, 1931b: 500 (s.w.) ETHIOPIA.

***lighti***. *Camponotus* (*Myrmamblys*) *itoi* subsp. *lighti* Wheeler, W.M. 1927d: 10 (s.w.) CHINA. Subspecies of *tokioensis:* Santschi, 1937h: 385. Raised to species: Wu & Wang, 1992: 1315.

***ligniperdus***. *Formica ligniperda* Latreille, 1802c: 88, pl. 1, figs. A-N (s.w.q.m.) FRANCE. Wheeler, G.C. & Wheeler, J. 1953e: 187 (l.); Hauschteck, 1961: 221 (k.). Combination in *Camponotus:* Mayr, 1861: 36; in *C.* (*Camponotus*): Forel, 1914a: 266. Subspecies/race of *herculeanus:* Forel, 1874: 39; Mayr, 1886d: 420; Emery, 1893i: 674; Forel, 1915d: 68; Karavaiev, 1927c: 275. Status as species: André, 1882a: 142; Forel, 1889: 255; Nasonov, 1889: 10; Forel, 1899c: 130; Bondroit, 1912: 352; Bondroit, 1918: 69; Emery, 1920b: 255; Menozzi, 1922c: 143; Emery, 1925b: 73; Karavaiev, 1936: 181; Stitz, 1939: 237; Röszler, 1942a: 54; Holgersen, 1942: 11; Kutter, 1977c: 205; Collingwood, 1979: 91; Atanassov & Dlussky, 1992: 212. Junior synonym of *herculeanus:* Bernard, 1967: 340; revived from synonymy: Collingwood & Yarrow, 1969: 81. Senior synonym of *obsoleta:* Emery, 1892b: 161; of *herculeanoligniperdus:* **new synonymy**. Current subspecies: nominal plus *afer, nigrescens*.

****lignitus***. **Formica lignitum* Germar, 1837: pl. 19, fig. 19 (q.) GERMANY (Miocene?). Combination in *Camponotus:* Mayr, 1867b: 51. See also: Heer, 1850: 112.

***lilianae***. *Camponotus* (*Myrmamblys*) *lilianae* Forel, 1913b: 344 (s.w.) ZAIRE. Combination in *C.* (*Paramyrmamblys*): Santschi, 1926c: 604. Current subspecies: nominal plus *cornutus*.

#***lillii***. *Camponotus punctulatus* subsp. *lillii* Forel, 1908c: 402 (s.w.) BRAZIL. Combination in *C.* (*Tanaemyrmex*): Emery, 1925b: 78.

***limbiventris***. *Camponotus limbiventris* Santschi, 1911c: 366, fig. (w.) TANZANIA. Combination in *C.* (*Camponotus*): Forel, 1914a: 266; in *C.* (*Myrmamblys*): Santschi, 1914b: 134; Wheeler, W.M. 1922a: 989; Emery, 1925b: 141; in *C.* (*Myrmotemnus*): Emery, 1920b: 258; in *C.* (*Paramyrmamblys*): Santschi, 1926c: 604.

*lindbergi*. *Camponotus* (*Myrmosericus*) *cruentatus* var. *lindbergi* Santschi, 1931a: 12 (s.w.) MOROCCO. Junior synonym of *cruentatus:* Cagniant, 1991b: 51.

***lindigi***. *Camponotus lindigi* Mayr, 1870a: 385, pl. fig. 5 (s.w.q.) COLOMBIA. Santschi, 1936b: 217 (m.). Combination in *C.* (*Myrmobrachys*): Forel, 1914a: 270.

***linnaei***. *Camponotus linnaei* Forel, 1886f: 171. Replacement name for *Camponotus angulatus* Mayr, 1870a: 382. [Junior secondary homonym of *Formica angulata* Smith, F. 1859a: 139.] Wheeler, W.M. 1934g: 210 (s.m.); Wheeler, G.C. & Wheeler, J. 1953e: 193 (l.). Combination in *C.* (*Manniella*): Wheeler, W.M. 1934g: 210; in *C.* (*Myrmaphaenus*): Creighton, 1951: 57; in *C.* (*Myrmeurynota*): Forel, 1914a: 274; Kempf, 1972a: 49. Senior synonym of *angulicollis* (unnecessary second replacement name): Dalla Torre, 1893: 240; Emery, 1896d: 377; Forel, 1899c: 147. Current subspecies: nominal plus *comoedus, maccus, muticus*.

*liocnemis*. *Camponotus maculatus* subsp. *liocnemis* Emery, 1905d: 30 (footnote) (s.w.) SOUTH AFRICA. Santschi, 1914e: 38 (q.m.). Junior synonym of *maculatus:* Baroni Urbani, 1972: 125.

***liogaster***. *Camponotus* (*Orthonotomyrmex*) *liogaster* Santschi, 1932a: 390 (w.) SOUTH AFRICA.

***lividicoxis***. *Camponotus* (*Myrmophyma*) *lividicoxis* Viehmeyer, 1925b: 142 (w.) AUSTRALIA.

*lividior*. *Camponotus maculatus* var. *lividior* Santschi, 1911e: 128 (s.w.q.m.) COMORO IS. Junior synonym of *maculatus:* Baroni Urbani, 1972: 125.

#***lividipes***. *Camponotus nigriceps* st. *lividipes* Emery, 1887a: 211 (footnote) (s.w.) AUSTRALIA. Combination in *C.* (*Myrmoturba*): Forel, 1914a: 267; in *C.* (*Tanaemyrmex*): Emery, 1925b: 103. Raised to species: Dalla Torre, 1893: 240; Forel, 1914a: 267. Subspecies of *consobrinus:* Wheeler, W.M. 1933b: 23; Clark, 1934c: 71; of *nigriceps:* Taylor & Brown, D.R. 1985: 117. Senior synonym of *prostans:* Wheeler, W.M. 1933b: 23; Clark, 1934c: 71.

#***lizeri***. *Camponotus* (*Myrmoturba*) *punctulatus* st. *lizeri* Santschi, 1922c: 104 (s.w.) BOLIVIA. Combination in *C.* (*Tanaemyrmex*): Emery, 1925b: 79.

***loa.*** *Camponotus (Colobopsis) loa* Mann, 1919: 373, figs. 45, 46 (s.w.) SOLOMON IS. Current subspecies: nominal plus *belli.*

*lobinieri. Camponotus maculatus* r. *lobinieri* Forel, 1902d: 287 (w.) INDIA. Junior synonym of *basalis:* Bingham, 1903: 352.

*lohieri. Camponotus (Myrmoturba) maculatus* var. *lohieri* Emery, 1915g: 22 (w.) IVORY COAST. [First available use of *Camponotus (Myrmoturba) maculatus* st. *melanocnemis* var. *lohieri* Santschi, 1913c: 313; unavailable name.] Junior synonym of *maculatus:* Baroni Urbani, 1972: 125.

#***loloma.*** *Camponotus (Myrmogonia) schmeltzi* var. *loloma* Mann, 1921: 486, fig. 33 (w.) FIJI IS.

***longi.*** *Camponotus (Colobopsis) longi* Forel, 1902d: 284 (s.w.) INDIA.

***longiceps.*** *Formica longiceps* Smith, F. 1863: 13 (w.) INDONESIA (Waigeo I.). Combination in *Camponotus:* Roger, 1863b: 43; in *C. (Myrmamblys):* Donisthorpe, 1932c: 469.

#***longinodis.*** *Camponotus (Dinomyrmex) subnitidus* var. *longinodis* Forel, 1915b: 96 (w.) AUSTRALIA. Clark, 1930a: 119 (s.q.). Combination in *C. (Tanaemyrmex):* Emery, 1925b: 91. Raised to species: Clark, 1930a: 119. Subspecies of *subnitidus:* Taylor & Brown, D.R. 1985: 120.

***longipalpis.*** *Camponotus (Myrmacrhaphe) longipalpis* Santschi, 1926c: 612, fig. 3 (w.) ZAIRE.

*longipes. Formica longipes* Gerstäcker, 1859: 262 (w.) MOZAMBIQUE. [Junior primary homonym of *Formica longipes* Latreille, 1802c: 233 (now in *Pheidole*).] [Also described as new by Gerstäcker, 1862: 509.] Replacement name: *Camponotus etiolipes* **new name**. Forel, 1886f: 164 (s.); Forel, 1907g: 91 (m.). Combination in *Camponotus:* Roger, 1863b: 2; in *Tanaemyrmex:* Ashmead, 1905b: 384; in *C. (Myrmogigas):* Forel, 1912i: 91; in *C. (Dinomyrmex):* Forel, 1913b: 340; in *C. (Tanaemyrmex):* Emery, 1925b: 86.

***longipilis.*** *Camponotus longipilis* Emery, 1911a: 224 (q.) PERU. Combination in *C. (Myrmamblys):* Forel, 1914a: 271; in *C. (Myrmostenus):* Emery, 1920b: 260. Current subspecies: nominal plus *postangulatus.*

****longiventris.*** **Camponotus longiventris* Théobald, 1937b: 315, pl. 25, figs. 3, 19 (q.m.) FRANCE (Oligocene).

***lownei.*** *Camponotus lownei* Forel, 1895a: 43. Replacement name for *Formica nitida* Lowne, 1865a: 277. [Junior primary homonym of *Formica nitida* Razoumowsky, 1789: 427.] Crawley, 1922c: 33 (q.m.). Combination in *C. (Myrmophyma):* Forel, 1914a: 269; in *C. (Myrmogonia):* Emery, 1920b: 257; in *C. (Thlipsepinotus):* Santschi, 1928e: 483.

***lubbocki.*** *Camponotus lubbocki* Forel, 1886f: 186 (w.) MADAGASCAR. Forel, 1891b: 60 (q.); Forel, 1891b: 216 (s.). Combination in *C. (Myrmosaga):* Forel, 1912i: 92; in *C. (Mayria):* Emery, 1925b: 123. Current subspecies: nominal plus *christoides, rectus.*

***lucayanus.*** *Camponotus maculatus* subsp. *lucayanus* Wheeler, W.M. 1905b: 112, figs. P-R (s.w.q.) BAHAMAS. Combination in *C. (Tanaemyrmex):* Emery, 1925b: 81. Race of *fumidus:* Santschi, 1922c: 101. Raised to species: Smith, M.R. 1954c: 12. See also: Snelling, 1968b: 351.

#***lucianus.*** *Camponotus auricomus* var. *lucianus* Forel, 1899c: 139 (w.q.) ANTILLES. Wheeler, W.M. 1913d: 243 (m.). Combination in *C. (Myrmobrachys):* Emery, 1925b: 163.

#***lucidior.*** *Camponotus wiederkehri* var. *lucidior* Forel, 1910b: 81 (s.w.m.) AUSTRALIA. Crawley, 1915a: 136 (q.). Combination in *C. (Tanaemyrmex):* Emery, 1925b: 102. See also: Forel, 1915b: 100.

***luctuosus.*** *Formica luctuosa* Smith, F. 1858b: 27 (q.) SUMATRA. Combination in *Camponotus:* Roger, 1863b: 3; in *C. (Myrmamblys):* Forel, 1914a: 271. Subgenus indeterminate: Emery, 1925b: 172.

*luederwaldti. Camponotus (Pseudocolobopsis) luederwaldti* Santschi, 1922c: 118 (s.w.) BRAZIL. Junior synonym of *macrocephalus:* Kempf, 1968b: 408.

*luisae. Camponotus (Myrmothrix) abdominalis* st. *cupiens* var. *luisae* Santschi, 1936b: 210 (s.) BRAZIL. Unavailable name; material referred to *atriceps* by Hashmi, 1973: 82.

#***lujai.*** *Camponotus (Myrmosericus) vestitus* var. *lujai* Santschi, 1930a: 75 (s.w.m.) ZAIRE.

*luteangulus. Camponotus maculatus* subsp. *vicinus* var. *luteangulus* Wheeler, W.M. 1910d: 304 (w.m.) U.S.A. Unavailable name; material referred to *vicinus* by Creighton, 1950a: 381.

***luteiventris.*** *Camponotus luteiventris* Emery, 1897d: 574 (w.) NEW GUINEA. Combination in *C. (Myrmocamelus):* Forel, 1914a: 270; in *C. (Myrmophyma):* Emery, 1920b: 257; in *C. (Tanaemyrmex):* Emery, 1925b: 104.

#***luteiventris.*** *Camponotus (Colobopsis) mutilatus* var. *luteiventris* Stitz, 1938: 116 (q.) NEW GUINEA. [Unresolved junior primary homonym of *luteiventris* Emery, above. Proposed replacement name, *albiventris* Baroni Urbani, 1971b: 361, is unavailable.]

#***luteolus.*** *Camponotus maculatus* subsp. *luteolus* Emery, 1906c: 188 (s.w.) ARGENTINA. Combination in *C. (Tanaemyrmex):* Emery, 1925b: 79. Subspecies of *bonariensis:* Emery, 1920b: 233. Senior synonym of *carolinus* (unnecessary replacement name): **new synonymy**.

#***luteolus.*** *Camponotus (Tanaemyrmex) hova* var. *luteolus* Emery, 1925b: 85 (w.) MADAGASCAR. [First available use of *Camponotus maculatus* r. *hova* var. *luteolus* Forel, 1897c: 202; unavailable name.] [Unresolved junior primary homonym of *luteolus* Emery, above.]

***luteus.*** *Formica lutea* Smith, F. 1858b: 19 (q.) INDIA. Combination in *Camponotus:* Roger, 1863b: 3; in *C. (Tanaemyrmex):* Emery, 1925b: 95.

***lutzi.*** *Camponotus lutzi* Forel, 1905b: 169 (s.w.q.) BRAZIL. Combination in *C. (Myrmothrix):* Forel, 1914a: 269; in *C. (Tanaemyrmex):* Emery, 1925b: 81.

*lycurgus. Camponotus maculatus* subsp. *brutus* var. *lycurgus* Emery, 1899e: 501 (w.) CAMEROUN. Unavailable name.

#***lygaeus.*** *Camponotus (Myrmamblys) moeschi* var. *lygaea* Viehmeyer, 1916a: 157 (w.) SINGAPORE.

#***lysias.*** *Camponotus (Colobopsis) gasseri* var. *lysias* Forel, 1913g: 193 (w.) AUSTRALIA.

#***lysistrata.*** *Camponotus (Myrmoturba) bruchi* st. *lysistrata* Santschi, 1919f: 56 (w.q.) ARGENTINA. Combination in *C. (Tanaemyrmex):* Emery, 1925b: 78.

***maafui.*** *Camponotus (Myrmogonia) maafui* Mann, 1921: 482, fig. 31 (s.w.) FIJI IS.

***macareaveyi.*** *Camponotus macareaveyi* Taylor, 1992a: 60. Replacement name for *sanguinea* McAreavey, 1949: 18. [Junior primary homonym of *sanguinea* Karavaiev, 1929b: 212.]

***maccooki.*** *Camponotus sylvaticus* r. *maccooki* Forel, 1879a: 69 (w.q.m.) MEXICO. Wheeler, W.M. 1910d: 306 (s.). Combination in *C. (Myrmoturba):* Wheeler, 1917a: 560; in *C. (Camponotus):* Emery, 1925b: 75; in *C. (Tanaemyrmex):* Creighton, 1950a: 377. Race of *rubripes:* Forel, 1886f: 143. Raised to species: Dalla Torre, 1893: 241. Subspecies of *maculatus:* Emery, 1893i: 672; of *sansabeanus:* Emery, 1920b: 232 (footnote). Revived status as species: Creighton, 1950a: 377; Snelling, 1970: 396.

#***maccus.*** *Camponotus (Manniella) linnaei* subsp. *maccus* Wheeler, W.M. 1934g: 213 (s.w.) GUIANA. Combination in *C. (Myrmeurynota):* Kempf, 1972a: 49.

***macilentus.*** *Camponotus macilentus* Smith, F. 1877a: 83 (w.) GALAPAGOS IS. Combination in *C. (Myrmamblys):* Wheeler, W.M. 1919c: 279; in *C. (Pseudocolobopsis):* Emery, 1920b: 259. Current subspecies: nominal plus *albemarlensis, altinotus, barringtonensis, bindloensis, castellanus, duncanensis, hoodensis, jacobensis, narboroensis, pervicus, sapphirinus, vulcanalis, wollebaeki.*

#***mackayensis.*** *Camponotus reticulatus* r. *mackayensis* Forel, 1902h: 506 (s.w.) AUSTRALIA. Combination in *C. (Myrmamblys):* Emery, 1925b: 139.

***macrocephalus.*** *Formica macrocephala* Erichson, 1842: 259 (q.) AUSTRALIA. Combination in *Iridomyrmex:* Taylor & Brown, D.R. 1985: 100; in *Camponotus*, and senior synonym of *fictor:* Shattuck & McArthur, 1995 (in press). Current subspecies: nominal plus *augustulus.* See also: Shattuck, 1992a: 16.

***macrocephalus.*** *Camponotus macrocephalus* Emery, 1894c: 169 (s.) BRAZIL. [Unresolved junior secondary homonym of *macrocephala* Erichson, above.] Emery, 1920c: 36 (q.). Combination in *C. (Myrmamblys):* Forel, 1912i: 90; in *C. (Pseudocolobopsis):* Emery, 1920b: 259. Senior synonym of *geralensis* and *luederwaldti:* Kempf, 1968b: 408.

***macrochaeta.*** *Camponotus macrochaeta* Emery, 1903: 78 (w.) BRAZIL. Combination in *C. (Dinomyrmex):* Forel, 1914a: 268; in *C. (Tanaemyrmex):* Emery, 1925b: 84.

*maculatocognatus. Camponotus rubripes* r. *maculatus* var. *maculatocognatus* Forel, 1886f: 143. Unavailable name.

*maculatodichrous. Camponotus rubripes* r. *cognatus* var. *maculatodichrous* Forel, 1889: 262. Unavailable name.

*maculatoides. Camponotus (Myrmoturba) hova* subsp. *maculatoides* Emery, 1920c: 6 (w.) MADAGASCAR. [First available use of *Camponotus maculatus* r. *hova* var. *maculatoides* Forel, 1897c: 200; unavailable name.] Combination in *C. (Dinomyrmex):* Santschi, 1921f: 312. Junior synonym of *strangulatus:* Emery, 1920c: 6.

*maculatosylvaticus. Camponotus rubripes* r. *maculatus* var. *maculatosylvaticus* Forel, 1886f: 143. Unavailable name.

***maculatus.*** *Formica maculata* Fabricius, 1782: 491 (w.) WEST AFRICA. Mayr, 1862: 654 (q.m.); Forel, 1891b: 29 (s.w.q.m.); Donisthorpe, 1915a: 221 (redescription of holotype). Combination in *Camponotus:* Mayr, 1862: 654; in *C. (Myrmoturba):* Forel, 1912i: 91; in *C. (Tanaemyrmex):* Emery,

***CAMPONOTUS** (continued)*

1925b: 86. Race of *sylvaticus:* Forel, 1879a: 64; of *rubripes:* Forel, 1886f: 143. Revived status as species: Emery, 1891b: 17. Senior synonym of *atramentarius, ballioni, boera, cavallus, cluisoides, cognata, conakryensis, erythraea, flavifemur, flavominor, hannae, hieroglyphicus, intonsus, lacteipennis, liengmei, liocnemis, lividior, lohieri, manzer, mathildae, melanocnemis* (and its junior synonym *schultzei*), *nubis, proletaria, radamoides, sarmentus, schereri, semispicatus, sudanicus, thomensis, tuckeri* and *zumpti*, and material of the unavailable names *calceatus, cataractae, citinus, cluis, contaminatus, diffusus, hansingi, importunoides, madecassa* and *pessimus* referred here: Baroni Urbani, 1972: 125–127. Current subspecies: nominal plus *foveolatus* (unresolved junior homonym), *humilior, miserabilis* (unresolved junior homonym), *obfuscatus, ruzskyi* (unresolved junior homonym), *strangulatus, subnudus, sylvaticomaculatus, ugandensis.*

#***maculifrons.*** *Camponotus* (*Tanaemyrmex*) *larvigerus* var. *maculifrons* Menozzi & Russo, 1930: 167 (s.w.) DOMINICA.

#***maculiventris.*** *Camponotus christi* var. *maculiventris* Emery, 1895f: 344 (w.q.) MADAGASCAR. Combination in *C.* (*Mayria*): Emery, 1925b: 122.

#***madagascarensis.*** *Camponotus niveosetosus* r. *madagascarensis* Forel, 1886c: civ (w.) MADAGASCAR. [*Camponotus niveosetosus* var. *madagascarensis* Forel, 1879a: 115; *nomen nudum.*] Santschi, 1926b: 262 (q.). Combination in *C.* (*Myrmopiromis*): Wheeler, W.M. 1922a: 1052.

*madagascarensis. Mayria madagascarensis* Forel, 1886c: civ (w.) MADAGASCAR. Forel, 1894e: 227 (s.). Combination in *Camponotus* (*Mayria*): Forel, 1894e: 227. [Junior secondary homonym of *madagascarensis* Forel, above.] Replacement name: *repens* Forel, 1897c: 187.

*madecassa. Camponotus maculatus* subsp. *liocnemis* var. *madecassa* Emery, 1905d: 30 (footnote) (w.) MADAGASCAR. Unavailable name.

***magister.*** *Camponotus* (*Myrmoturba*) *erigens* st. *magister* Santschi, 1925g: 357 (s.w.) MOROCCO. Raised to species: Santschi, 1932f: 519. Current subspecies: nominal plus *tibestiensis.*

*magnifica. Camponotus* (*Myrmothrix*) *rufipes* var. *magnifica* Forel, 1913l: 249 (s.w.) ARGENTINA. Santschi, 1916e: 397 (q.). Junior synonym of *rufipes:* Hashmi, 1973: 52.

*magnus. Camponotus magnus* Mayr, 1862: 673 (w.q.) AUSTRALIA. Junior synonym of *intrepidus:* Roger, 1863b: 4.

***maguassa.*** *Camponotus* (*Myrmoturba*) *maguassa* Wheeler, W.M. 1922a: 237 (s.w.) ZAIRE. Combination in *C.* (*Tanaemyrmex*): Emery, 1925b: 92.

*mahican. Camponotus herculeanus* var. *mahican* Emery, 1925b: 72 (w.) U.S.A. [First available use of *Camponotus herculeanus* subsp. *pennsylvanicus* var. *mahican* Wheeler, W.M. 1910d: 338; unavailable name.] Unnecessary replacement name for *herculeanopennsylvanicus,* and hence junior synonym of the latter. See also: Creighton, 1950a: 367; Smith, M.R. 1951a: 842.

#***makilingi.*** *Camponotus* (*Colobopsis*) *rothneyi* var. *makilingi* Viehmeyer, 1916b: 286 (q.) PHILIPPINES.

*manczshuricus. Camponotus* (*Camponotus*) *japonicus* var. *manczshuricus* Emery, 1925b: 73 (w.) RUSSIA. [First available use of *Camponotus herculeanus* subsp. *japonicus* var. *manczshuricus* Ruzsky, 1915a: 420; unavailable name.] Subspecies of *japonicus:* Ruzsky, 1926: 109. Junior synonym of *herculeanus:* Yasumatsu & Brown, 1957: 47; of *aterrimus:* Arnol'di, 1967: 1827.

#***manidis.*** *Camponotus manidis* Forel, 1909b: 62 (s.w.) ZAIRE. Combination in *C.* (*Myrmamblys*): Wheeler, W.M. 1922a: 990. Subspecies of *simus:* Forel, 1911f: 282.

***manni.*** *Camponotus* (*Colobopsis*) *manni* Wheeler, W.M. 1934e: 418, fig. 4 (w.) FIJI IS. Current subspecies: nominal plus *umbratilis.*

*manzer. Camponotus maculatus* subsp. *manzer* Forel, 1910e: 452 (s.) ZIMBABWE. Forel, 1911f: 280 (w.); Santschi, 1916b: 509 (m.). Combination in *C.* (*Myrmoturba*): Santschi, 1916b: 509; in *C.* (*Tanaemyrmex*): Emery, 1925b: 88. Junior synonym of *maculatus:* Baroni Urbani, 1972: 125.

#***marcens.*** *Camponotus claripes* subsp. *marcens* Forel, 1907h: 300 (s.w.) AUSTRALIA. Combination in *C.* (*Myrmophyma*): Emery, 1925b: 110.

#***marcidus.*** *Camponotus inaequalis* var. *marcidus* Wheeler, W.M. 1905b: 116 (s.w.q.m.) BAHAMAS. Combination in *C.* (*Tanaemyrmex*), and subspecies of *ramulorum:* Emery, 1925b: 82.

#***margaritae.*** *Camponotus quadrisectus* subsp. *margaritae* Forel, 1907a: 33 (w.) SINGAPORE. Combination in *C.* (*Myrmotarsus*): Emery, 1925c: 135.

*marginata. Formica marginata* Latreille, 1798: 35 (w.q.) FRANCE. Nylander, 1856b: 54 (m.). Combination in *Camponotus:* Mayr, 1861: 36; in *C.* (*Myrmoturba*): Forel, 1915d: 70; in *C.* (*Tanaemyrmex*): Emery, 1925b: 97. Junior synonym of *sylvaticus:* Mayr, 1865: 27. Revived from synonymy: André, 1874: 176; Forel, 1874: 40; Ruzsky, 1895: 9; Baroni Urbani, 1971c: 184. Variety/subspecies of *aethiops:* Bondroit, 1918: 73; Menozzi, 1925d: 35; Finzi, 1928c: 791; Stitz, 1939: 251. Senior synonym of *sylvaticoaethiops:* Forel, 1915d: 70. Junior synonym of *fallax:* Bernard, 1967: 342; of *aethiops:* Pisarski, 1975: 31; Agosti & Collingwood, 1987a: 58.

*maritimus. Camponotus maculatus* subsp. *vicinus* var. *maritimus* Wheeler, W.M. 1910d: 305 (w.q.m.) U.S.A. Unavailable name; material referred to *vicinus* by Creighton, 1950a: 381.

#***marius.*** *Camponotus pompeius* subsp. *marius* Emery, 1899e: 502, fig. (w.) CAMEROUN. Santschi, 1926b: 257 (s.). Combination in *C.* (*Tanaemyrmex*): Emery, 1925b: 89.

#***marocanus.*** *Camponotus* (*Tanaemyrmex*) *atlantis* var. *marocanus* Emery, 1925b: 91 (s.w.) MOROCCO. [First available use of *Camponotus maculatus* subsp. *atlantis* var. *maroccana* Emery, 1908a: 202; unavailable name.] Santschi, 1929e: 159 (q.).

*marshalli. Camponotus* (*Myrmisolepis*) *marshalli* Santschi, 1926b: 262. *Nomen nudum,* attributed to Crawley.

#***martensi.*** *Camponotus maculatus* subsp. *martensi* Forel, 1907e: 16 (s.w.) ALGERIA. Santschi, 1929g: 165 (q.). Combination in *C.* (*Tanaemyrmex*): Emery, 1925b: 98. Subspecies of *compressus:* Emery, 1920c: 6; Santschi, 1936c: 207.

#***marthae.*** *Camponotus gambeyi* r. *marthae* Forel, 1894e: 231 (w.) NEW CALEDONIA. Emery, 1914f: 426 (q.m.). Combination in *C.* (*Myrmocamelus*): Emery, 1914f: 426; in *C.* (*Tanaemyrmex*): Emery, 1925b: 104.

***massiliensis.*** *Camponotus pilicornis* var. *massiliensis* Schmitz, 1950: 13 (w.) FRANCE. [First available use of *Camponotus maculatus* r. *dichrous* var. *massiliensis* Forel, 1894d: 41; unavailable name.] Combination in *C.* (*Tanaemyrmex*), and raised to species: Collingwood & Yarrow, 1969: 83.

***massinissa.*** *Camponotus* (*Dinomyrmex*) *massinissa* Wheeler, W.M. 1922a: 244 (q.) ZAIRE. Combination in *C.* (*Tanaemyrmex*): Emery, 1925b: 89.

*mathildae. Camponotus maculatus* subsp. *mathildae* Forel, 1910c: 266 (s.w.q.) SOUTH AFRICA. Combination in *C.* (*Tanaemyrmex*): Emery, 1925b: 88. Junior synonym of *maculatus:* Baroni Urbani, 1972: 125.

*mathildeae. Camponotus* (*Colobopsis*) *mathildeae* Smith, M.R. 1949d: 178, figs. 1–3 (s.w.) MEXICO.

***maudella.*** *Camponotus* (*Colobopsis*) *maudella* Mann, 1921: 496, fig. 38 (s.w.q.m.) FIJI IS. Current subspecies: nominal plus *seemanni.*

***maxwellensis.*** *Camponotus* (*Myrmoturba*) *maculatus* r. *maxwellensis* Forel, 1913k: 127 (s.w.) WEST MALAYSIA. Combination in *C.* (*Tanaemyrmex*): Emery, 1925b: 102. Raised to species: Emery, 1920c: 7.

***maynei.*** *Camponotus* (*Orthonotomyrmex*) *maynei* Forel, 1916: 445 (s.w.q.) ZAIRE. Combination in *C.* (*Myrmepomis*): Emery, 1920b: 258; in *C.* (*Myrmopiromis*): Wheeler, W.M. 1922a: 986; in *C.* (*Myrmisolepis*): Emery, 1925b: 134; Santschi, 1926b: 262.

***mayri.*** *Camponotus mayri* Forel, 1879a: 92, pl. 1, fig. 1 (w.) SOUTH AFRICA. Combination in *C.* (*Myrmentoma*): Forel, 1913a: 146; in *C.* (*Orthonotomyrmex*): Forel, 1913b: 347; Emery, 1925b: 125. Current subspecies: nominal plus *chimporensis.*

*mayri. Camponotus mayri* Cameron, 1891: 89, fig. (w.) ECUADOR. [Junior primary homonym of *mayri* Forel, above.] Replacement name: *cameroni* Forel, 1892f: 256.

*mayriella. Camponotus* (*Colobopsis*) *mayriella* Mann, 1921: 494. Unnecessary (second) replacement name for *carinata* Mayr, 1870b: 943. Junior synonym of *polynesicus* Emery, 1896d: 374, the first replacement name.

***medeus.*** *Camponotus* (*Myrmoturba*) *medea* Emery, 1920c: 19, fig. 5 (s.w.q.) PERU. Combination in *C.* (*Pseudocolobopsis*): Emery, 1925b: 159. Current subspecies: nominal plus *fulvulus.*

*mediopallidus. Camponotus abdominalis* r. *mediopallidus* Forel, 1901h: 70 (w.q.) COLOMBIA. [First available use of *Camponotus abdominalis* r. *ustulatus* var. *mediopallidus* Forel, 1899c: 132; unavailable name.] Junior synonym of *atriceps:* Hashmi, 1973: 82.

*mediorufa. Camponotus* (*Myrmoturba*) *punctulatus* r. *minutior* var. *mediorufa* Forel, 1913l: 249 (s.w.) ARGENTINA. Unavailable name; material referred to *cruentus* by Santschi, 1922c: 104.

***megalonyx.*** *Camponotus* (*Myrmosphincta*) *megalonyx* Wheeler, W.M. 1919e: 114 (s.w.) BORNEO. Stärcke, 1934: 24 (q.m.). Combination in *C.* (*Myrmopalpella*): Stärcke, 1934: 30. Senior synonym of *chalaconotus:* Stärcke, 1934: 29 (provisional); Wheeler, W.M. 1936e: 217.

**#meinerti.** *Camponotus meinerti* Forel, 1886f: 189 (s.w.q.) ANGOLA. Combination in *C.* (*Myrmentoma*): Forel, 1912i: 92; in *C.* (*Orthonotomyrmex*): Forel, 1913b: 346; in *C.* (*Myrmamblys*): Emery, 1925b: 142; in *C.* (*Myrmopelta*): Santschi, 1926a: 17. Junior synonym of *vividus:* Santschi, 1915c: 264; Wheeler, W.M. 1922a: 975. Revived from synonymy as subspecies of *vividus:* Santschi, 1926a: 17; Santschi, 1935a: 283.

**#melancholicus.** *Camponotus lespesi* subsp. *melancholicus* Emery, 1894e: 2 (s.w.) PARAGUAY. Combination in *C.* (*Tanaemyrmex*): Emery, 1925b: 83.

**melanocephalus.** *Camponotus melanocephalus* Roger, 1863a: 144 (s.) VENEZUELA. Combination in *C.* (*Myrmothrix*): Forel, 1914a: 269; in *C.* (*Tanaemyrmex*): Emery, 1925b: 81.

*melanocnemis. Camponotus maculatus* var. *melanocnemis* Santschi, 1911c: 368 (s.) ZAIRE. Santschi, 1915c: 278 (w.). Race/stirps of *maculatus:* Forel, 1913h: 355; Santschi, 1913c: 313. See also: Arnold, 1922: 637. Senior synonym of *schultzei:* Forel, 1914d: 250. Junior synonym of *maculatus:* Baroni Urbani, 1972: 125.

**#melanogaster.** *Camponotus* (*Tanaemyrmex*) *irritans* var. *melanogaster* Stitz, 1938: 114 (q.) CAROLINE IS.

*melanogastes. Formica melanogastes* Latreille, 1802c: 171, pl. 7, fig. 39 (w.) FRANCE. Junior synonym of *lateralis:* Mayr, 1853b: 102.

**melanoticus.** *Camponotus sexguttatus* var. *melanoticus* Emery, 1894c: 167 (w.) BOLIVIA. Forel, 1911c: 310 (m.). Combination in *C.* (*Myrmoturba*): Forel, 1914a: 267; in *C.* (*Tanaemyrmex*): Emery, 1925b: 81. Subspecies of *extensus:* Emery, 1894f: 3; Emery, 1896d: 371. Raised to species: Forel, 1899c: 136; Emery, 1920b: 255; Santschi, 1922c: 99. Current subspecies: nominal plus *catharinae, flavopubens, nigrescens* (unresolved junior homonym), *paranaensis, publicola, valerius.*

**melichloros.** *Camponotus melichloros* Kirby, W.F. 1888: 551 (w.) CHRISTMAS I. Combination in *C.* (*Tanaemyrmex*): Donisthorpe, 1935: 635.

*mellea. Formica mellea* Say, 1836: 286 (m.) U.S.A. Mayr, 1866a: 485 (w.q.). Combination in *Camponotus:* Roger, 1863b: 5. Senior synonym of *clarus:* Mayr, 1866a: 485 (in text). Junior synonym of *castaneus:* Mayr, 1886d: 420; Forel, 1886f: 141.

**mendax.** *Camponotus sericeus* var. *mendax* Forel, 1895e: 454 (s.) INDIA. [Date of publication 20.vi.1895.] Bingham, 1903: 370 (q.). Combination in *C.* (*Orthonotomyrmex*): Emery, 1925b: 125. Senior synonym of *pseudolus* (unnecessary replacement name as this *mendax* is senior to that noted below). Raised to species: Bingham, 1903: 370. Current subspecies: nominal plus *integer.*

**#mendax.** *Camponotus foraminosus* subsp. *mendax* Emery, 1895h: 54 (w.) SOUTH AFRICA. [Date of publication 16.viii.1895.] [Unresolved junior primary homonym of *mendax* Forel, above.] Forel, 1910f: 30 (q.). Combination in *C.* (*Myrmotrema*): Santschi, 1915c: 274. Stirps/race of *grandidieri:* Santschi, 1915c: 274; Stitz, 1923: 165.

**#mendozanus.** *Camponotus* (*Myrmobrachys*) *mus* var. *mendozana* Santschi, 1922b: 377 (s.w.) ARGENTINA.

*mendozensis. Camponotus* (*Myrmocamelus*) *blandus* r. *rosariensis* var. *mendozensis* Forel, 1913l: 249 (s.w.) ARGENTINA. Unavailable name.

***mengei.** **Camponotus mengei* Mayr, 1868c: 27, pl. 1, figs. 1, 8 (w.) BALTIC AMBER (Oligocene). Senior synonym of **igneus:* Wheeler, W.M. 1915h: 138.

*meningensis. Camponotus* (*Myrmoturba*) *compressus* st. *rectinota* var. *meningensis* Santschi, 1921d: 75 (s.w.m.) ALGERIA (?). Unavailable name.

*merdicola. Formica merdicola* Lund, 1831a: 129 (w.) BRAZIL. Junior synonym of *rufipes:* Dalla Torre, 1893: 250; Forel, 1895b: 104.

*merula. Formica merula* Losana, 1834: 313, pl.36, fig. 2 (w.) ITALY. Bondroit, 1918: 77 (s.q.m.). Combination in *Camponotus:* Dalla Torre, 1893: 238; in *C.* (*Orthonotomyrmex*): Müller, 1923: 162; in *C.* (*Myrmentoma*): Emery, 1925b: 120. Junior synonym of *piceus:* Dalla Torre, 1893: 238; Ruzsky, 1905b: 255; Baroni Urbani, 1971c: 194; Kutter, 1977c: 207. Variety/subspecies of *lateralis:* Emery, 1896d: 373; Emery, 1898c: 125; Ruzsky, 1902d: 7; Emery, 1916b: 226; Ruzsky, 1925b: 42. Status as species: Bondroit, 1918: 77; Müller, 1923: 162; Novak & Sadil, 1941: 109; Bernard, 1967: 344; Arnol'di & Dlussky, 1978: 552. Junior synonym of *piceus:* Atanassov & Dlussky, 1992: 222.

**#mestrei.** *Camponotus ramulorum* var. *mestrei* Wheeler, W.M. 1913b: 503 (s.w.) CUBA. Combination in *C.* (*Tanaemyrmex*): Emery, 1925b: 82.

**micans.** *Formica micans* Nylander, 1856b: 55 (w.) ALGERIA. Roger, 1859: 229 (q.); Emery, 1916b: 236 (q.). Combination in *Camponotus:* Mayr, 1861: 36; in *C.* (*Myrmosericus*): Menozzi, 1921: 32. Subspecies/race of *rufoglaucus:* Forel, 1892j: 237; Forel, 1894d: 3; Emery, 1895h: 49; Forel, 1895d: 227; Santschi, 1919e: 247. Revived status as species: Santschi, 1931a: 11; Santschi, 1939c: 5. Current subspecies: nominal plus *armeniacus, asniensis.*

**michaelseni.** *Camponotus michaelseni* Forel, 1907h: 303 (w.) AUSTRALIA. Combination in *C.* (*Myrmogonia*): Forel, 1914a: 269; in *C.* (*Myrmophyma*): Emery, 1925b: 111; in *C.* (*Thlipsepinotus*): Santschi, 1928e: 483.

**#micipsa.** *Camponotus* (*Myrmotrema*) *micipsa* Wheeler, W.M. 1922a: 252, fig. 67 (s.w.) ZAIRE. Combination in *C.* (*Myrmisolepis*): Emery, 1925b: 134; in *C.* (*Myrmopelta*), and variety of *barbarossa:* Santschi, 1926a: 21.

***microcephalus.** **Camponotus microcephalus* Carpenter, 1930: 60, pl. 6, fig. 4; pl. 11, fig. 8 (q.) U.S.A. (Miocene).

**micrositus.** *Camponotus* (*Manniella*) *micrositus* Wheeler, W.M. 1937b: 462 (w.) CUBA.

**midas.** *Camponotus midas* Froggatt, 1896: 390, pl. 27, figs. 6–9 (s.w.q.) AUSTRALIA. Combination in *C.* (*Myrmophyma*): Emery, 1925b: 111; in *C.* (*Myrmosaulus*): Clark, 1930c: 22.

*miltotus. Camponotus japonicus* var. *miltotus* Wheeler, W.M. 1929f: 9 (s.) CHINA. Junior synonym of *japonicus:* Yasumatsu & Brown, 1951: 36.

**#mimus.** *Camponotus hosei* subsp. *mimus* Forel, 1911a: 56 (q.) BORNEO. Combination in *C.* (*Colobopsis*): Emery, 1925b: 148.

**mina.** *Camponotus mina* Forel, 1879a: 83 (s.) MEXICO. Creighton, 1965b: 6 (w.q.m.). Combination in *C.* (*Myrmobrachys*): Forel, 1914a: 270. Subspecies of *senex:* Emery, 1896d: 377. Revived status as species: Wheeler, W.M. 1910d: 346. Senior synonym of *erythropus:* Emery, 1895c: 336; of *zuni:* Creighton, 1965b: 3.

*miniata. Camponotus rufoglaucus* r. *flavomarginata* var. *miniata* Arnold, 1947: 136 (w.) ZAMBIA. Unavailable name.

**#minimus.** *Camponotus* (*Myrmophyma*) *claripes* r. *minima* Crawley, 1922c: 31 (s.w.q.m.) AUSTRALIA. Combination in *C.* (*Thlipsepinotus*): Santschi, 1928e: 483.

*minor. Camponotus maculatus* subsp. *thoracicus* var. *minor* Forel, 1909e: 389 (w.q.m.) ALGERIA. Unavailable name.

**minozzii.** *Camponotus* (*Myrmoturba*) *minozzii* Emery, 1920c: 21 (w.) PERU. Combination in *C.* (*Tanaemyrmex*): Emery, 1925b: 81.

**minus.** *Camponotus minus* Wang & Wu, 1994: 26, figs. 2, 7, 11 (w.) CHINA.

*minusculus. Camponotus mutilatus* var. *minusculus* Forel, 1901b: 27 (q.). *Nomen nudum,* attributed to Emery.

**#minusculus.** *Camponotus maculatus* subsp. *minusculus* Viehmeyer, 1914c: 44 (s.w.q.) TANZANIA. Combination in *C.* (*Tanaemyrmex*): Emery, 1925b: 85. Variety of *guttatus:* Emery, 1920c: 6.

**#minutior.** *Camponotus punctulatus* r. *minutior* Forel, 1886f: 173 (s.w.) ARGENTINA. Combination in *C.* (*Myrmoturba*): Forel, 1913l: 249; in *C.* (*Tanaemyrmex*): Emery, 1925b: 79. Race of *tenuiscapus:* Emery, 1888c: 366; of *punctulatus:* Emery, 1894g: 376; Santschi, 1922c: 104. Raised to species: Dalla Torre, 1893: 243; Emery, 1903: 71; Emery, 1906c: 190. Provisional synonym of *punctulatus:* Wheeler, W.M. 1929e: 104 (footnote). Subspecies of *punctulatus:* Kempf, 1972a: 71.

*minutus. Camponotus marginatus* var. *minutus* Emery, 1893i: 676 (w.q.m.) U.S.A. Wheeler, W.M. 1910f: 224 (s.w.q.m.). Combination in *C.* (*Myrmentoma*): Emery, 1925b: 117. Variety of *fallax:* Wheeler, W.M. 1910f: 224; Wheeler, W.M. 1917a: 558; of *caryae:* Wheeler, W.M. 1917c: 28. Junior synonym of *nearcticus:* Creighton, 1950a: 388.

**mirabilis.** *Camponotus mirabilis* Emery, 1903: 80, fig. 15 (q.) PERU. Combination in *C.* (*Myrmomalis*): Forel, 1914a: 271; in *C.* (*Myrmostenus*): Emery, 1920b: 260.

**miserabilis.* **Camponotus miserabilis* Förster, 1891: 430, pl. 13, fig. 4 (q.) GERMANY (Oligocene). Junior synonym of **vehemens:* Théobald, 1937b: 218.

*miserabilis. Camponotus* (*Myrmoturba*) *maculatus* st. *miserabilis* Santschi, 1914d: 379 (s.w.) GUINEA. [Junior primary homonym of **miserabilis* Förster, above.] Replacement name: *proletaria* Baroni Urbani, 1971b: 362.

*misionensis. Camponotus abdominalis* st. *cupiens* var. *misionensis* Santschi, 1939e: 328 (s.) ARGENTINA. Unavailable name; material referred to *atriceps* by Hashmi, 1973: 82.

**mississippiensis.** *Camponotus* (*Colobopsis*) *mississippiensis* Smith, M.R. 1923a: 83 (s.w.) U.S.A. Wheeler, G.C. & Wheeler, J. 1953e: 188 (l.).

**misturus.** *Formica mistura* Smith, F. 1857a: 53 (q.) BORNEO. Combination in *Camponotus:* Roger, 1863b: 3; in *C.* (*Myrmotarsus*): Forel, 1912i: 92.

**CAMPONOTUS** *(continued)*

Senior synonym of *exasperata:* Mayr, 1886c: 354. Current subspecies: nominal plus *fornaronis.*

***mitis.*** *Formica mitis* Smith, F. 1858b: 20 (w.) SRI LANKA. Karavaiev, 1929c: 237 (q.m.); Imai *et al.* 1984: 9 (k.). Combination in *Camponotus:* Roger, 1863b: 3; in *C.* (*Myrmoturba*): Forel, 1913k: 125; in *C.* (*Tanaemyrmex*): Emery, 1925b: 96. Race/variety of *rubripes:* Forel, 1886f: 143; of *sylvaticus:* Emery, 1887a: 213; of *maculatus:* Forel, 1892j: 230; Emery, 1901f: 121; Forel, 1907e: 19; of *variegatus:* Emery, 1920c: 7; Santschi, 1920h: 173; Emery, 1925b: 96; Wheeler, W.M. 1927d: 10; Karavaiev, 1933a: 316. Status as species: Dalla Torre, 1893: 243; Bingham, 1903: 355; Emery, 1920b: 255; Mukerjee, 1934: 10; Wang, C., Xiao & Wu, 1989a: 224.

#***mixtellus.*** *Camponotus radamae* var. *mixtellus* Dalla Torre, 1893: 249 (s.w.) MADAGASCAR. [First available use of *Camponotus maculatus* r. *radamae* var. *mixtellus* Forel, 1891b: 433; unavailable name.] Forel, 1912k: 165 (q.m.). Combination in *C.* (*Tanaemyrmex*), and variety of *hova:* Emery, 1925b: 85.

***mocquerysi.*** *Camponotus mocquerysi* Emery, 1899f: 289, fig. (w.) MADAGASCAR. Combination in *C.* (*Orthonotomyrmex*): Forel, 1914a: 273; in *C.* (*Myrmonesites*): Emery, 1920b: 257.

***mocsaryi.*** *Camponotus mocsaryi* Forel, 1902b: 178 (w.) BRAZIL. Emery, 1903: 79 (q.). Combination in *C.* (*Dinomyrmex*): Forel, 1914a: 268; in *C.* (*Myrmaphaenus*): Emery, 1920b: 256; in *C.* (*Myrmoplatypus*): Santschi, 1921f: 311.

***moderatus.*** *Camponotus* (*Paramyrmamblys*) *moderatus* Santschi, 1930a: 76 (w.) ETHIOPIA.

#***modestior.*** *Camponotus novogranadensis* var. *modestior* Forel, 1904c: 51 (w.q.) COSTA RICA. Combination in *C.* (*Myrmaphaenus*): Emery, 1925b: 156.

***modoc.*** *Camponotus herculeanus* var. *modoc* Wheeler, W.M. 1910d: 333 (s.w.q.m.) U.S.A. Combination in *C.* (*Camponotus*): Emery, 1925b: 72. Subspecies of *pennsylvanicus:* Creighton, 1950a: 369; of *herculeanus:* Brown, 1950d: 158. Raised to species: Hunt & Snelling, 1975: 22; Smith, D.R. 1979: 1426; MacKay, Lowrie, *et al.* 1988: 104 (in key).

***moelleri.*** *Camponotus moelleri* Forel, 1912i: 83 (w.m.) BRAZIL. Combination in *C.* (*Myrmamblys*): Forel, 1914a: 271; in *C.* (*Pseudocolobopsis*): Emery, 1925b: 158.

***moeschi.*** *Camponotus moeschi* Forel, 1910d: 129 (w.) SUMATRA. Forel, 1912n: 64 (s.). Combination in *C.* (*Myrmamblys*): Forel, 1914a: 272; Emery, 1925b: 140; Emery, 1926: 8; in *C.* (*Myrmosphinctus*): Santschi, 1921f: 311; in *C.* (*Myrmotemnus*): Emery, 1920b: 258; Santschi, 1926c: 601. Current subspecies: nominal plus *lygaeus.*

#***moeschiellus.*** *Camponotus dedalus* var. *moeschiella* Forel, 1912n: 65 (q.) SUMATRA. Combination in *C.* (*Myrmamblys*): Emery, 1925b: 139.

***molossus.*** *Camponotus molossus* Forel, 1907h: 306 (s.w.) AUSTRALIA. Wheeler, G.C. & Wheeler, J. 1968: 218 (l.). Combination in *C.* (*Myrmosphincta*): Forel, 1914a: 273; in *C.* (*Myrmosaulus*): Emery, 1925b: 114.

*mombassae. Camponotus mombassae* Forel, 1886f: 180 (s.) KENYA. Subspecies of *kersteni:* Emery, 1896d: 372. Junior synonym of *acvapimensis:* Wheeler, W.M. 1922a: 948; Menozzi, 1930b: 118.

***monardi.*** *Camponotus* (*Myrmotrema*) *monardi* Santschi, 1930b: 78 (s.) ANGOLA.

#***montanus.*** *Camponotus alboannulatus* subsp. *montanus* Emery, 1894c: 168 (w.) BOLIVIA. Combination in *C.* (*Pseudocolobopsis*): Emery, 1925b: 157.

*montanus. Camponotus herculeanus* var. *montanus* Ruzsky, 1904a: 293 (w.) RUSSIA. Karavaiev, 1926e: 192 (s.w.q.). [Unresolved junior primary homonym of *montanus* Emery, above.] Subspecies of *herculeanus:* Ruzsky, 1926: 108. Junior synonym of *herculeanus:* Emery, 1925b: 72; Karavaiev, 1936: 178.

*montanus. Camponotus* (*Karavaievia*) *montanus* Dumpert, 1989: 221, figs. 4–9 (w.q.m.) WEST MALAYSIA. [Junior primary homonym of *montanus* Emery, 1894c: 168.] Replacement name: *orinus* Dumpert, **new name**.

#***monticola.*** *Camponotus nicobarensis* var. *monticola* Bingham, 1903: 365 (w.) BURMA. [First available use of *Camponotus nicobarensis* subsp. *exiguoguttatus* var. *monticola* Emery, 1895k: 479; unavailable name.] Combination in *C.* (*Tanaemyrmex*): Emery, 1925b: 95.

#***montinanus.*** *Camponotus* (*Myrmoturba*) *thraso* st. *montinanus* Santschi, 1926b: 251 (s.w.) ZIMBABWE.

***montivagus.*** *Camponotus nitidus* r. *montivagus* Forel, 1885a: 347 (w.s.) GUATEMALA. Combination in *C.* (*Myrmamblys*): Forel, 1914a: 272; in *C.* (*Myrmentoma*): Emery, 1925b: 118; in *C.* (*Pseudocolobopsis*): Snelling, 1988: 57. Raised to species: Dalla Torre, 1893: 243. Current subspecies: nominal plus *nuperus, rectithorax.*

#***montserratensis.*** *Camponotus sexguttatus* var. *montserratensis* Wheeler, W.M. 1923c: 7 (w.q.m.) ANTILLES. Combination in *C.* (*Myrmosphincta*): Kempf, 1972a: 58.

***morosus.*** *Formica morosa* Smith, F. 1858b: 50 (w.) CHILE. Mayr, 1865: 32 (s.). Combination in *Camponotus:* Mayr, 1862: 665; in *C.* (*Tanaemyrmex*): Emery, 1925b: 78. Race of *rubripes:* Forel, 1886f: 144. Junior synonym of *distinguendus:* Dalla Torre, 1893: 228. Revived from synonymy as variety/subspecies of *distinguendus:* Emery, 1895i: 18; Emery, 1906c: 191; Santschi, 1916e: 396. Revived status as species: Menozzi, 1935c: 329; Kempf, 1972a: 69; Snelling & Hunt, 1976: 120.

*mortis. Camponotus maculatus* subsp. *thoracicus* var. *mortis* Forel, 1910a: 13 (s.w.) ISRAEL. Unavailable name.

#***moshianus.*** *Camponotus* (*Myrmotrema*) *olivieri* var. *moshiana* Santschi, 1915c: 270 (w.) TANZANIA. [First available use of *Camponotus foraminosus* subsp. *lemma* var. *moshiana* Forel, 1907a: 31; unavailable name.]

*motschulskii. Camponotus bedoti* r. *motschulskii* Emery, 1893e: 197 (w.) SRI LANKA. [*motschulskii* Emery, 1893f: 253, *nomen nudum.*] Junior synonym of *yerburyi:* Emery, 1896d: 374.

***mozabensis.*** *Camponotus mozabensis* Emery, 1899b: 17, fig. (w.) ALGERIA. Combination in *C.* (*Myrmosphincta*): Forel, 1914a: 273; in *C.* (*Myrmosericus*): Emery, 1920b: 256. Senior synonym of *impolitus:* Forel, 1902g: 463; Emery, 1925b: 105.

***mucronatus.*** *Camponotus mucronatus* Emery, 1890b: 58 (s.w.) COSTA RICA. [Also described as new by Emery, 1894k: 63.] Combination in *C.* (*Myrmorhachis*): Forel, 1914a: 274; in *C.* (*Myrmocladoecus*): Emery, 1925b: 167; Wheeler, W.M. 1934e: 424. Current subspecies: nominal plus *formaster, hirsutinasus, santschii.*

*muger. Camponotus* (*Myrmoturba*) *maculatus* r. *traegaordhi* var. *muger* Forel, 1914d: 250 (s.w.) SOUTH AFRICA. Unavailable name.

#***multipilis.*** *Camponotus melanoticus* var. *multipilis* Forel, 1907e: 10 (w.q.m.) PARAGUAY. Combination in *C.* (*Myrmoturba*): Bruch, 1914: 229; in *C.* (*Tanaemyrmex*), and variety of *substitutus:* Emery, 1925b: 83.

***mus.*** *Camponotus mus* Roger, 1863a: 143 (w.q.) URUGUAY. Forel, 1879a: 98 (s.); Emery, 1894g: 374 (q.m.); Goni *et al.* 1983: 366 (k.). Combination in *C.* (*Myrmobrachys*): Forel, 1914a: 270. Race of *senex:* Forel, 1879a: 98; Forel, 1886f: 180. Revived status as species: Dalla Torre, 1893: 243; Forel, 1914a: 270; Emery, 1920b: 260. See also: Bruch, 1916: 353. Senior synonym of *ogloblini:* Kempf, 1978: 37. Current subspecies: nominal plus *mendozanus, mustelus.*

***mussolinii.*** *Camponotus* (*Myrmophyma*) *mussolinii* Donisthorpe, 1936e: 528 (w.) NEW GUINEA.

#***mustelus.*** *Camponotus* (*Myrmobrachys*) *mus* subsp. *mustela* Emery, 1920c: 39 (w.q.) BOLIVIA.

*mutans* Forel, 1910; see under ***CALOMYRMEX.***

#***muticus.*** *Camponotus angulatus* r. *muticus* Forel, 1899c: 147 (w.) COSTA RICA. Combination in *C.* (*Myrmeurynota*), and variety of *linnaei:* Emery, 1925b: 168.

***mutilatus.*** *Formica mutilata* Smith, F. 1859a: 137 (w.) INDONESIA (Aru I.). Mayr, 1876: 68 (q.). Combination in *Colobopsis:* Mayr, 1862: 691; in *Camponotus* (*Colobopsis*): Dalla Torre, 1893: 244; Emery, 1896d: 376. Current subspecies: nominal plus *luteiventris* (unresolved junior homonym).

#***mutilarius.*** *Camponotus wasmanni* var. *mutilarius* Emery, 1893e: 224 (w.q.) BURMA. Combination in *C.* (*Orthonotomyrmex*): Emery, 1925b: 126. [Misspelled as *mutilloides* by Bingham, 1903: 376.]

***myoporus.*** *Camponotus* (*Tanaemyrmex*) *myoporus* Clark, 1938: 379, figs. 20–22 (s.w.) AUSTRALIA.

***mystaceus.*** *Camponotus mystaceus* Emery, 1886: 356, pl. 17, fig. 3 (s.w.m.) SOUTH AFRICA. Arnold, 1924: 675 (q.). Combination in *C.* (*Myrmopsamma*): Forel, 1914a: 261. Current subspecies: nominal plus *exsanguis, kamae.*

*myster. Camponotus* (*Myrmothrix*) *cingulatus* var. *myster* Santschi, 1922c: 109 (s.w.) BRAZIL. Junior synonym of *cingulatus:* Hashmi, 1973: 31.

***nacerdus.*** *Camponotus* (*Formica*) *nacerdus* Norton, 1868c: 2 (w.) MEXICO. [*Formica nacerda* Norton, 1868a: 60, *nomen nudum.*] Subgenus indeterminate: Emery, 1925b: 172.

*nadigi. Camponotus herculeanus* var. *nadigi* Menozzi, 1922c: 142 (s.w.q.m.) ITALY. Material of the *nomen nudum vagusherculeanus* referred here by Emery, 1925b: 72; Kutter, 1977c: 205. Junior synonym of *herculeanus:* Yasumatsu & Brown, 1951: 30; Baroni Urbani, 1971c: 176.

***nadimi.*** *Camponotus* (*Myrmentoma*) *nadimi* Tohmé, G. 1969: 6, figs. 3, 4 (s.w.) LEBANON.

**naegelii.** *Camponotus naegelii* Forel, 1879a: 84 (s.w.q.) BRAZIL. Combination in *C.* (*Myrmamblys*): Forel, 1912i: 90; in *C.* (*Myrmaphaenus*): Emery, 1925b: 155. Race of *fastigatus*: Forel, 1886f: 172; Emery, 1894g: 376. Revived status as species: Emery, 1903: 65.

*naevius*. *Camponotus maculatus* r. *parvulus* var. *naevius* Forel, 1912i: 73 (s.w.) BRAZIL. Unavailable name.

#**nagasau.** *Camponotus* (*Myrmogonia*) *cristatus* var. *nagasau* Mann, 1921: 482 (s.w.) FIJI IS.

**namacola.** *Camponotus* (*Mayria*) *namacolus* Prins, 1973: 20, figs. 23, 24, 33 (w.) NAMIBIA.

*nana*. *Formica nana* Smith, F. 1858b: 41 (w.) BRAZIL. [Junior primary homonym of *Formica nana* Latreille, 1802c: 263.] Replacement name: *nywet* **new name**. Combination in *Camponotus*: Roger, 1863b: 5; in *C.* (*Myrmocamelus*): Forel, 1914a: 270; in *C.* (*Myrmaphaenus*): Emery, 1925b: 154.

*nana*. *Camponotus quadriceps* var. *nana* Emery, 1911b: 537 (s.w.) NEW GUINEA. [Junior secondary homonym of *nana* Smith, F., above.] Replacement name: *nanulus* Emery, below.

#**nanulus.** *Camponotus* (*Colobopsis*) *quadriceps* var. *nanula* Emery, 1925b: 150. Replacement name for *nana* Emery, above. [Junior secondary homonym of *Formica nana* Smith, F. 1858b: 41.]

#**narboroensis.** *Camponotus* (*Myrmamblys*) *macilentus* var. *narboroensis* Wheeler, W.M. 1919c: 286 (q.) GALAPAGOS IS. Combination in *C.* (*Pseudocolobopsis*): Emery, 1925b: 158.

#**narses.** *Camponotus* (*Myrmocamelus*) *ephippium* r. *narses* Forel, 1915b: 103 (w.) AUSTRALIA. Crawley, 1925b: 596 (s.). Combination in *C.* (*Myrmophyma*): Emery, 1925b: 111.

**nasicus.** *Camponotus nasica* Forel, 1891b: 67 (w.) MADAGASCAR. Combination in *C.* (*Myrmocamelus*): Forel, 1914a: 270; in *C.* (*Tanaemyrmex*): Emery, 1925b: 104.

**nasutus.** *Camponotus nasutus* Emery, 1895h: 53, pl. 2, fig. 27 (w.) SOUTH AFRICA. Combination in *C.* (*Myrmophyma*): Forel, 1914a: 269; in *C.* (*Myrmotemnus*): Emery, 1920b: 258; in *C.* (*Myrmamblys*): Santschi, 1921f: 311; in *C.* (*Myrmespera*): Santschi, 1926c: 604. Current subspecies: nominal plus *fenestralis, pretiosus, quadridentatus, quinquedentatus, subnasutus.*

**natalensis.** *Formica natalensis* Smith, F. 1858b: 33 (s.w.q.m.) SOUTH AFRICA. Combination in *Camponotus*: Roger, 1863b: 2; in *C.* (*Myrmoturba*): Forel, 1914a: 267; in *C.* (*Tanaemyrmex*): Emery, 1925b: 103. See also: Arnold, 1922: 652. Current subspecies: nominal plus *corvus, diabolus, fulvipes, politiceps.*

*nautarum*. *Camponotus* (*Colobopsis*) *conithorax* var. *nautarum* Santschi, 1919a: 327 (w.) SAMOA. Santschi, 1928a: 57 (q.). Junior synonym of *flavolimbatus*: Wilson & Taylor, 1967: 97.

**navigator.** *Camponotus* (*Myrmoturba*) *navigator* Wilson & Taylor, 1967: 94, fig. 77 (s.w.q.) SAMOA.

**nawai.** *Camponotus fallax* var. *nawai* Ito, 1914: 44 (s.) JAPAN. Combination in *C.* (*Myrmamblys*): Emery, 1925b: 138. Variety of *caryae*: Wheeler, W.M. 1917c: 29. Raised to species: Terayama & Satoh, 1990a: 405.

**nearcticus.** *Camponotus marginatus* var. *nearcticus* Emery, 1893i: 675 (w.q.) U.S.A. Wheeler, W.M. 1910f: 223 (s.m.); Wheeler, G.C. & Wheeler, J. 1953e: 193 (l.). Combination in *C.* (*Myrmentoma*): Emery, 1925b: 117. Variety of *fallax*: Wheeler, W.M. 1908g: 406; Wheeler, W.M. 1917a: 558. Junior synonym of *caryae*: Wheeler, W.M. 1917c: 27. Revived from synonymy as subspecies of *caryae*: Buren, 1944a: 294. Raised to species: Creighton, 1950a: 388; Smith, M.R. 1951a: 845. See also: Snelling, 1988: 70. Senior synonym of *minutus, pardus* and *tanquaryi* Creighton, 1950a: 388.

#**nefasitensis.** *Camponotus thraso* var. *nefasitensis* Menozzi, 1931a: 155 (s.w.) ETHIOPIA. [First available use of *Camponotus maculatus* subsp. *negus* var. *nefassitensis* Forel, 1910c: 266; unavailable name.]

#**negus.** *Camponotus maculatus* subsp. *negus* Forel, 1907c: 142 (s.w.q.m.) ETHIOPIA. Combination in *C.* (*Myrmoturba*): Forel, 1915c: 348; in *C.* (*Tanaemyrmex*): Emery, 1925b: 92. Subspecies of *thraso*: Emery, 1920c: 6.

*neocarinata*. *Camponotus vitreus* var. *neocarinata* Baroni Urbani, 1971b: 361. [Incorrectly proposed replacement name for *carinata* Stitz, 1938: 116.] Unavailable name.

**nepos.** *Camponotus nepos* Forel, 1912i: 75 (w.) BRAZIL. Combination in *C.* (*Myrmoturba*): Forel, 1914a: 267; in *C.* (*Tanaemyrmex*): Emery, 1925b: 84.

#**nessus.** *Camponotus alboannulatus* r. *nessus* Forel, 1912i: 81 (s.w.q.m.) BRAZIL. Combination in *C.* (*Pseudocolobopsis*): Emery, 1925b: 157.

*nevadensis*. *Camponotus* (*Myrmentoma*) *nevadensis* Gregg, 1973: 39, figs. 1, 2 (s.w.) U.S.A. Junior synonym of *essigi*: Wheeler, G.C. & Wheeler, J. 1986g: 61.

**newzealandicus.** *Camponotus* (*Colobopsis*) *newzealandicus* Donisthorpe, 1940a: 47 (q.) NEW ZEALAND. See also: Brown, 1958h: 45.

**nicobarensis.** *Camponotus nicobarensis* Mayr, 1865: 31, pl. 1, fig. 1 (w.) NICOBAR IS. Combination in *C.* (*Myrmothrix*): Forel, 1914a: 269; in *C.* (*Myrmoturba*): Emery, 1920b: 255; in *C.* (*Tanaemyrmex*): Emery, 1925b: 95. Current subspecies: nominal plus *monticola, rabbani.*

#**niger.** *Camponotus* (*Myrmentoma*) *sicheli* var. *nigra* Emery, 1925b: 121 (w.) ALGERIA (?). [First available use of *Camponotus lateralis* subsp. *sicheli* var. *nigra* Karavaiev, 1912a: 21; unavailable name.]

#**nigeriensis.** *Camponotus* (*Myrmotrema*) *bayeri* st. *nigeriensis* Santschi, 1914d: 383 (w.) NIGERIA. Santschi, 1915c: 277 (s.w.q.m.). Stirps of *perrisii*: Santschi, 1915c: 277.

*nigrata*. *Formica nigrata* Nylander, 1849: 35 (w.q.m.) SICILY. Junior synonym of *aethiops*: Mayr, 1855: 313.

#**nigrescens.** *Camponotus ligniperdus* var. *nigrescens* Gösswald, 1932: 147 (w.) GERMANY.

#**nigrescens.** *Camponotus melanoticus* var. *nigrescens* Santschi, 1939e: 324 (w.) BRAZIL. [Unresolved junior primary homonym of *nigrescens* Gösswald, above.] Combination in *C.* (*Tanaemyrmex*): Kempf, 1972a: 69.

**nigricans.** *Camponotus platypus* var. *nigricans* Roger, 1863a: 140 (q.) PHILIPPINES. Emery, 1893g: 268 (w.). Combination in *C.* (*Myrmotarsus*): Forel, 1912i: 92. Raised to species: Emery, 1893g: 268; Wheeler, W.M. 1929g: 57. Current subspecies: nominal plus *enganensis, pantiensis.*

**nigriceps.** *Formica nigriceps* Smith, F. 1858b: 38 (w.) AUSTRALIA. Mayr, 1876: 63 (m.); Forel, 1879a: 75 (q.m.); Wheeler, G.C. & Wheeler, J. 1974a: 61 (l.). Combination in *Camponotus*: Roger, 1863b: 4; in *C.* (*Myrmoturba*): Forel, 1913g: 191; in *C.* (*Tanaemyrmex*): Emery, 1925b: 103. Subspecies of *consobrinus*: Wheeler, W.M. 1933b: 23. Revived status as species: Clark, 1934c: 71. Current subspecies: nominal plus *clarior, lividipes, obniger, pallidiceps.*

*nigricolor*. *Camponotus quadrinotatus* var. *nigricolor* Ruzsky, 1926: 109 (w.) RUSSIA. [First available use of *Camponotus marginatus* subsp. *quadrinotatus* var. *nigricolor* Ruzsky, 1925b: 42; unavailable name.] Junior synonym of *quadrinotatus*: Kuznetsov-Ugamsky, 1929a: 18.

**nigrifrons.** *Colobopsis nigrifrons* Mayr, 1870b: 943 (q.) TONGA. Combination in *Camponotus* (*Colobopsis*): Dalla Torre, 1893: 244; in *C.* (*Myrmamblys*): Forel, 1914a: 272; in *C.* (*Colobopsis*): Emery, 1925b: 146.

*nigriscapa*. *Camponotus* (*Myrmoturba*) *punctulatus* st. *andigena* var. *nigriscapa* Santschi, 1922c: 104 (s.) ARGENTINA. Unavailable name.

*nigrithorax*. *Camponotus compressus* st. *thoracicus* var. *nigrithorax* Santschi, 1921d: 74 (s.w.q.m.) ALGERIA. Unavailable name.

#**nigrithorax.** *Camponotus* (*Myrmobrachys*) *rubrithorax* var. *nigrithorax* Stitz, 1937: 135 (w.) MEXICO.

**nigroaeneus.** *Formica nigroaenea* Smith, F. 1858b: 40 (s.) AUSTRALIA. Lowne, 1865a: 277 (w.); Wheeler, G.C. & Wheeler, J. 1970: 650 (l.). Combination in *Camponotus*: Mayr, 1862: 663; in *C.* (*Myrmoturba*): Forel, 1914a: 267; in *C.* (*Myrmophyma*): Emery, 1925b: 111. Current subspecies: nominal plus *divus, xuthus.*

**nigronitidus.** *Camponotus* (*Myrmamblys*) *nigronitidus* Azuma, 1951: 89 (footnote) (w.) JAPAN.

#**nigrovarius.** *Camponotus atlantis* var. *nigrovarius* Forel, 1894d: 5 (w.) ALGERIA. Combination in *C.* (*Tanaemyrmex*): Emery, 1925b: 91.

#**ninivae.** *Camponotus* (*Tanaemyrmex*) *thoracicus* subsp. *ninivae* Pisarski, 1971a: 674, figs. 10–14 (s.w.) IRAQ.

**nipponensis.** *Camponotus* (*Myrmamblys*) *nipponensis* Santschi, 1937h: 381, fig. 12 (s.w.q.m.) JAPAN. [Grossly erroneous combination in *Lasius*: Kim, 1988: 119.]

**nipponicus.** *Camponotus* (*Colobopsis*) *nipponicus* Wheeler, W.M. 1928d: 118 (s.q.) JAPAN.

**nirvanae.** *Camponotus nirvanae* Forel, 1893b: 433 (s.w.q.) INDIA. Combination in *C.* (*Myrmamblys*): Forel, 1914a: 272. See also: Bingham, 1903: 377.

**nitens.** *Camponotus nitens* Mayr, 1870a: 378 (w.q.) COLOMBIA. Combination in *C.* (*Camponotus*): Forel, 1914a: 266; in *C.* (*Myrmoturba*): Emery, 1920b: 255; in *C.* (*Tanaemyrmex*): Emery, 1925b: 81. Current subspecies: nominal plus *fuhrmanni.*

#**nitens.** *Camponotus compressus* subsp. *nitens* Bernard, 1953a: 194 (q.) ALGERIA. [Unresolved junior primary homonym of *nitens* Mayr, above.]

*nitida*. *Formica nitida* Smith, F. 1859a: 138 (w.) INDONESIA (Aru I.). [Unresolved junior primary homonym of *Formica nitida* Razoumowsky, 1789: 427.] Junior synonym of *quadriceps*: Mayr, 1863: 403.

*nitida*. *Tapinoma nitida* Smith, F. 1860a: 69 (q.) SULAWESI. [Junior secondary homonym of *Formica nitida* Smith, F. 1859a: 138, above.] Replacement name: *perneser* **new name**. Combination in *Camponotus* (*Colobopsis*): Donisthorpe, 1932c: 457.

*nitida. Formica nitida* Lowne, 1865a: 277 (s.) AUSTRALIA. [Junior primary homonym of *Formica nitida* Razoumowsky, 1789: 427.] Replacement name: *lownei* Forel, 1895a: 43.

*nitida. Camponotus* (*Formica*) *nitida* Norton, 1868c: 2 (w.) MEXICO. [*Formica nitida* Norton, 1868a: 60; *nomen nudum*.] [Junior secondary homonym of *Formica nitida* Smith, F. 1859a: 138.] Replacement name: *rectithorax* Forel, 1895a: 44.

#***nitidescens.*** *Camponotus kiesenwetteri* r. *nitidescens* Forel, 1889: 260 (w.) GREECE. Combination in *C.* (*Myrmentoma*): Emery, 1925b: 121.

*nitidiceps. Camponotus* (*Myrmotarsus*) *nigricans* subsp. *nitidiceps* Viehmeyer, 1916a: 156 (q.) SINGAPORE. Junior synonym of *enganensis:* Viehmeyer, 1916b: 291.

*nitidiceps. Camponotus* (*Myrmophyma*) *nitidiceps* Viehmeyer, 1925b: 141 (s.w.q.) AUSTRALIA. [Junior primary homonym of *nitidiceps* Viehmeyer, above.] Replacement name: *peseshus* **new name**.

#***nitidior.*** *Camponotus* (*Myrmotrema*) *olivieri* var. *nitidior* Santschi, 1926c: 617 (w.) ZAIRE.

*nitidiventris. Camponotus maculatus* subsp. *vicinus* var. *nitidiventris* Emery, 1893i: 672 (w.) U.S.A. Unavailable name; material referred to *vicinus* by Creighton, 1950a: 381.

***nitidus:*** see under *nitida*, above.

***niveosetosus.*** *Camponotus niveosetosus* Mayr, 1862: 665 (w.) SOUTH AFRICA. Mayr, 1866b: 885 (q.); Forel, 1891b: 41 (s.); Arnold, 1922: 659 (m.). Combination in *C.* (*Myrmoturba*): Forel, 1914a: 267; in *C.* (*Myrmepomis*): Emery, 1920b: 259; in *C.* (*Myrmopiromis*): Wheeler, W.M. 1922a: 986. Current subspecies: nominal plus *irredux, madagascarensis.*

*nobilis. Camponotus* (*Myrmothrix*) *nobilis* Santschi, 1916e: 396 (w.) PARAGUAY. Junior synonym of *leydigi:* Emery, 1925b: 153.

#***nocens.*** *Camponotus abdominalis* subsp. *nocens* Wheeler, W.M. 1911b: 171 (s.w.q.) GRANADA I. Combination in *C.* (*Myrmothrix*): Emery, 1925b: 108. Status uncertain: Hashmi, 1973: 120; currently subspecies of *atriceps*, see under *abdominalis* and *atriceps.*

***normatus.*** *Camponotus normatus* Forel, 1899c: 146 (w.) COSTA RICA. Combination in *C.* (*Myrmobrachys*): Forel, 1914a: 270.

***nossibeensis.*** *Camponotus nossibeensis* André, 1887: 281 (w.) MADAGASCAR. Combination in *C.* (*Myrmobrachys*): Forel, 1912i: 91; in *C.* (*Myrmepomis*): Emery, 1920b: 258; in *C.* (*Myrmopiromis*): Wheeler, W.M. 1922a: 1052.

#***notterae.*** *Camponotus cinereus* subsp. *notterae* Forel, 1907h: 303 (w.) AUSTRALIA. Combination in *C.* (*Myrmophyma*): Emery, 1925b: 110.

***novaehollandiae.*** *Camponotus novaehollandiae* Mayr, 1870b: 939 (w.) AUSTRALIA. Mayr, 1876: 66 (q.m.). Combination in *C.* (*Myrmoturba*): Forel, 1915b: 98; Wheeler, W.M. 1915g: 814; in *C.* (*Tanaemyrmex*): Emery, 1925b: 96. Race of *sylvaticus:* Forel, 1879a: 67; of *rubripes:* Forel, 1886f: 143; of *maculatus:* Forel, 1915b: 98; of *variegatus:* Emery, 1920c: 7; Emery, 1925b: 96. Revived status as species: Dalla Torre, 1893: 245; Clark, 1930c: 19; Taylor & Brown, D.R. 1985: 118.

***noveboracensis.*** *Formica noveboracensis* Fitch, 1854: 52 (w.) U.S.A. Wheeler, W.M. 1910d: 340 (s.w.q.m.); Wheeler, G.C. & Wheeler, J. 1953e: 182 (l.). Combination in *Camponotus:* Roger, 1863b: 6; in *C.* (*Camponotus*): Emery, 1925b: 72. Junior synonym of *pennsylvanicus:* Dalla Torre, 1893: 247; Emery, 1896d: 372. Revived from synonymy as variety of *ligniperdus:* Forel, 1899c: 130. Variety/subspecies of *herculeanus:* Emery, 1925b: 72; Buren, 1944a: 293. Revived status as species: Creighton, 1950a: 369. Senior synonym of *pictus:* Forel, 1899c: 130; of *rubens:* Creighton, 1950a: 370.

***novogranadensis.*** *Camponotus novogranadensis* Mayr, 1870a: 380 (s.w.) COLOMBIA. Emery, 1906c: 191 (q.); Wheeler, G.C. & Wheeler, J. 1953e: 192 (l.). Combination in *C.* (*Myrmamblys*): Forel, 1912i: 90; in *C.* (*Neomyrmamblys*): Santschi, 1921f: 311; in *C.* (*Myrmaphaenus*): Emery, 1925b: 156. Current subspecies: nominal plus *modestior.*

*****novotnyi.*** **Camponotus novotnyi* Samsinak, 1967: 365, pls, 1, 2 (q.) CZECHOSLOVAKIA (Miocene).

*nubis. Camponotus* (*Myrmoturba*) *maculatus* subsp. *nubis* Weber, 1943c: 385, pl. 16, fig. 30 (s.w.) SUDAN. Junior synonym of *maculatus:* Baroni Urbani, 1972: 125.

*nuda. Camponotus* (*Myrmoturba*) *maculatus* subsp. *fedtshenkoi* var. *nuda* Ruzsky, 1923: 2 (w.) RUSSIA. Unavailable name.

#***nudimalis.*** *Camponotus claripes* var. *nudimalis* Forel, 1913g: 191 (s.) AUSTRALIA. Combination in *C.* (*Myrmophyma*): Emery, 1925b: 110. Subspecies of *claripes:* Taylor & Brown, D.R. 1985: 112.

#***nuntius.*** *Camponotus scratius* var. *nuntius* Forel, 1907h: 306 (w.) AUSTRALIA. Combination in *C.* (*Myrmophyma*): Emery, 1925b: 112. Subspecies of *scratius:* Taylor & Brown, D.R. 1985: 119.

#***nuperus.*** *Camponotus nitidus* var. *nuperus* Wheeler, W.M. 1914b: 58 (s.w.q.) MEXICO. Combination in *C.* (*Myrmentoma*), and subspecies of *montivagus:* Emery, 1925b: 118.

***nutans.*** *Camponotus nutans* Mayr, 1867c: 440 (w.) "am Schiffe gefunden" (type-locality SUMATRA, see Emery, 1896d: 374). Combination in *C.* (*Myrmosphincta*): Forel, 1914a: 273; in *C.* (*Myrmamblys*): Emery, 1925b: 140; Emery, 1926: 8; in *C.* (*Myrmotemnus*): Santschi, 1926c: 601. Current subspecies: nominal plus *cleliae.*

***nylanderi.*** *Camponotus nylanderi* Emery, 1921a: 25. Replacement name for *Formica pallens* Nylander, 1849: 36. [Junior primary homonym of *Formica pallens* Le Guillou, 1842: 316.] Combination in *C.* (*Tanaemyrmex*), and subspecies of *atlantis:* Emery, 1925b: 91. Subspecies of *alii:* Santschi, 1939d: 86. Revived status as species: Santschi, 1931a: 11; Baroni Urbani, 1964b: 59; Baroni Urbani, 1971c: 185.

***nywet.*** *Camponotus nywet* **new name**. Replacement name for *Formica nana* Smith, F. 1858b: 41. [Junior primary homonym of *Formica nana* Latreille, 1802c: 263.]

***oasium.*** *Camponotus rubripes* r. *oasium* Forel, 1890a: lxv (s.w.q.) ALGERIA. Forel, 1902a: 158 (m.). Race of *maculatus:* Emery, 1891b: 18; Forel, 1902a: 158. Raised to species: Dalla Torre, 1893: 245. Variety of *compressus:* Menozzi, 1934: 164. Revived status as species: Collingwood, 1961a: 75.

*****obesus.*** **Camponotus obesus* Piton, in Piton & Théobald, 1935: 68, pl. 2, fig. 5 (q.) FRANCE (Mio-pliocene).

#***obfuscatus.*** *Camponotus* (*Myrmoturba*) *subnudus* var. *obfuscata* Viehmeyer, 1916a: 154 (s.w.) SINGAPORE. Subspecies of *maculatus:* current status.

***obliquipilosus.*** *Camponotus fedtschenkoi* var. *obliquipilosa* Emery, 1920c: 7 (q.) RUSSIA. [First available use of *Camponotus maculatus* r. *fedtschenkoi* var. *obliquipilosa* Forel, 1904b: 378; unavailable name.] Raised to species: Pisarski, 1967: 411. See also: Dlussky & Zabelin, 1985: 235.

***obliquus.*** *Camponotus* (*Colobopsis*) *obliquus* Smith, M.R. 1930b: 567 (s.) U.S.A.

***oblongus.*** *Formica oblonga* Smith, F. 1858b: 21 (q.) BURMA. Forel, 1892j: 227 (w.). Combination in *Camponotus:* Roger, 1863b: 3; in *C.* (*Myrmoturba*): Forel, 1914a: 267; in *C.* (*Tanaemyrmex*): Emery, 1925b: 95. See also: Bingham, 1903: 358. Current subspecies: nominal plus *binominatus.*

#***obniger.*** *Camponotus nigriceps* r. *obniger* Forel, 1902h: 506 (s.w.) AUSTRALIA. Forel, 1910b: 72 (q.). Combination in *C.* (*Tanaemyrmex*): Emery, 1925b: 103. Subspecies of *consobrinus:* Wheeler, W.M. 1933b: 23; Clark, 1934c: 71; of *nigriceps:* Taylor & Brown, D.R. 1985: 117.

***obreptivus.*** *Camponotus obreptivus* Forel, 1899c: 137 (s.w.) COSTA RICA. Combination in *C.* (*Dinomyrmex*): Forel, 1914a: 268; in *C.* (*Tanaemyrmex*): Emery, 1925b: 84.

#***obscuratus.*** *Camponotus* (*Tanaemyrmex*) *hova* var. *obscurata* Emery, 1925b: 85 (w.m.) MADAGASCAR. [First available use of *Camponotus maculatus* subsp. *radamae* var. *obscurata* Forel, 1907g: 89; unavailable name.] Material of the unavailable name *obscurior* referred here by Emery, 1925b: 85.

#***obscurifrons.*** *Camponotus* (*Myrmobrachys*) *brevis* var. *obscurifrons* Santschi, 1936b: 216 (q.) PANAMA. Wheeler, G.C. & Wheeler, J. 1953e: 194 (l.).

*obscurior. Camponotus maculatus* subsp. *hova* var. *obscurior* Santschi, 1911e: 132. Unavailable name; material referred to *obscuratus* by Emery, 1925b: 85.

***obscuripes.*** *Camponotus ligniperdus* var. *obscuripes* Mayr, 1879: 645 (s.w.) JAPAN. André, 1903: 128 (q.m.). Combination in *C.* (*Camponotus*): Emery, 1925b: 73. Subspecies of *ligniperdus:* Forel, 1907e: 19; of *herculeanus:* Emery, 1908a: 185; Emery, 1925b: 73. Raised to species: Yasumatsu & Brown, 1951: 38; Arnol'di, 1967: 1823; Kupyanskaya, 1990: 171.

*obscuripes. Camponotus* (*Myrmobrachys*) *senex* subsp. *textor* var. *obscuripes* Borgmeier, 1929: 214 (w.) BRAZIL. Unavailable name.

***obscuriventris.*** *Camponotus obscuriventris* Cagniant, 1991b: 47, figs. 4, 6 (w.q.m.) MOROCCO. [Described as *cruentatus*-group, hence subgenus *Myrmoserics.*]

#***obscurus.*** *Camponotus* (*Myrmotarsus*) *rufifemur* var. *obscura* Stitz, 1925: 126 (w.) PHILIPPINES.

*obsoleta. Formica obsoleta* Christ, 1791: 509, pl. 60, fig. 5 (q.). [Unresolved junior primary homonym of *Formica obsoleta* Linnaeus, 1758: 580.] Junior synonym of *ligniperdus:* Emery, 1892b: 161.

**obtritus.** *Camponotus obtritus* Emery, 1911a: 222 (s.w.) PERU. Combination in *C.* (*Myrmomalis*): Forel, 1914a: 271.

*obtusa. Formica obtusa* Smith, F. 1858b: 30 (w.) EGYPT. Junior synonym of *sericeus:* Roger, 1863b: 2.

#**obtusatus.** *Polyrhachis selene* subsp. *obtusata* Emery, 1895k: 480 (w.) BURMA. Combination in *Camponotus:* Emery, 1896d: 376; in *C.* (*Orthonotomyrmex*): Emery, 1925b: 125.

#**obtusitruncatus.** *Camponotus* (*Colobopsis*) *gasseri* r. *obtusitrumcatus* Forel, 1902h: 508 (s.w.q.m.) AUSTRALIA. [*C. gasseri* subsp. *obtusitruncatus:* Emery, 1925b: 148; justified emendation of spelling.]

**occasus.** *Camponotus* (*Myrmoturba*) *occasus* Emery, 1920c: 17, fig. 4 (s.w.) CAPE VERDE IS.

#**occidentalis.** *Camponotus carbo* var. *occidentalis* Mayr, 1902: 300 (s.w.) GHANA. Combination in *C.* (*Myrmotrema*): Wheeler, W.M. 1922a: 977.

#**occipitalis.** *Camponotus* (*Myrmoturba*) *maculatus* r. *occipitalis* Stitz, 1917: 352, fig. 14 (w.) ALGERIA. Combination in *C.* (*Tanaemyrmex*): Emery, 1925b: 98. Subspecies of *compressus:* Emery, 1925b: 98; Santschi, 1939d: 80; Bernard, 1953a: 189.

#**occultus.** *Camponotus claviscapus* subsp. *occultus* Wheeler, W.M. & Mann, 1914: 56, fig. 24 (s.w.q.m.) HAITI. Combination in *C.* (*Pseudocolobopsis*): Emery, 1925b: 157.

**oceanicus.** *Colobopsis oceanica* Mayr, 1870b: 943 (q.) FIJI IS. Combination in *Camponotus:* Dalla Torre, 1893: 245; in *C.* (*Myrmamblys*): Forel, 1914a: 272; in *C.*(*Myrmotemnus*): Emery, 1920b: 258; in *C.* (*Colobopsis*): Emery, 1925b: 146.

**ocreatus.** *Camponotus maculatus* subsp. *ocreatus* Emery, 1893i: 673 (s.w.) U.S.A. Wheeler, W.M. 1910d: 310 (q.); Wheeler, G.C. & Wheeler, J. 1970: 651 (l.). Combination in *C.* (*Myrmoturba*): Wheeler, W.M. 1917a: 561; in *C.* (*Camponotus*): Emery, 1920b: 255; in *C.* (*Tanaemyrmex*): Creighton, 1950a: 378. Raised to species: Wheeler, W.M. 1917a: 561. Senior synonym of *primipilaris:* Creighton, 1950a: 378.

*octomaculata. Camponotus maculatus* subsp. *pictiventris* var. *octomaculata* Forel, 1910e: 452 (w.) SOUTH AFRICA. Unavailable name.

*octonotatus. Camponotus maculatus* r. *fulvus* var. *octonotatus* Forel, 1897c: 202 (w.q.) SEYCHELLE IS. Unavailable name.

**oculatior.** *Camponotus* (*Myrmachraphe*) *oculatior* Santschi, 1935a: 281, figs. 33, 34 (w.) ZAIRE.

**odiosus.** *Camponotus rubripes* r. *odiosus* Forel, 1886f: 154 (s.w.) SUMATRA. Combination in *C.* (*Myrmoturba*): Forel, 1913k: 126; in *C.* (*Tanaemyrmex*): Emery, 1925b: 102. Raised to species: Dalla Torre, 1893: 245. Subspecies of *maculatus:* Emery, 1896d: 371; Emery, 1900d: 701; Forel, 1911e: 290. Raised to species: Emery, 1920c: 7.

#**oebalis.** *Camponotus* (*Colobopsis*) *vitreus* var. *oebalis* Forel, 1911b: 208 (w.) JAVA.

*****oeningensis.** **Formica obesa oeningensis* Heer, 1850: 110, pl. 8, fig. 2 (q.) SWITZERLAND. Combination in *Oecophylla:* Mayr, 1867b: 51; in *Camponotus,* and raised to species: Cockerell, 1915: 486.

**oertzeni.** *Camponotus rubripes* r. *oertzeni* Forel, 1889: 263 (s.w.) GREECE. Finzi, 1939c: 158 (q.m.). Combination in *C.* (*Tanaemyrmex*): Emery, 1925b: 97. Raised to species: Dalla Torre, 1893: 246. Subspecies of *maculatus:* Emery, 1896d: 370; Emery, 1898c: 148; Forel, 1911e: 290. Subspecies of *aethiops:* Emery, 1920c: 7. Raised to species: Agosti & Collingwood, 1987a: 59. Current subspecies: nominal plus *andrius, kappariensis.*

**oetkeri.** *Camponotus oetkeri* Forel, 1910b: 75 (s.w.) AUSTRALIA. Combination in *C.* (*Myrmogonia*): Forel, 1913g: 191; in *C.* (*Myrmophyma*): Emery, 1925b: 112; in *C.* (*Thlipsepinotus*): Santschi, 1928e: 483. Current subspecies: nominal plus *voltai.*

**ogasawarensis.** *Camponotus* (*Myrmamblys*) *ogasawarensis* Terayama & Satoh, 1990c: 117, figs. 1–26 (s.w.q.m.) JAPAN.

*ogloblini. Camponotus* (*Myrmobrachys*) *ogloblini* Santschi, 1934c: 32, fig. 20 (w.) ARGENTINA. Junior synonym of *mus:* Kempf, 1978: 37.

*okamotoi. Camponotus* (*Myrmamblys*) *itoi* subsp. *lighti* var. *okamotoi* Wheeler, W.M. 1927d: 10 (w.) CHINA. Unavailable name.

**olivieri.** *Camponotus foraminosus* r. *olivieri* Forel, 1886f: 175 (s.w.q.) ANGOLA. Combination in *C.* (*Myrmotrema*): Santschi, 1915c: 269. Raised to species: Dalla Torre, 1893: 246. Subspecies of *foraminosus:* Emery, 1895h: 54; Santschi, 1910c: 393. Raised to species: Santschi, 1915c: 269 (in key); Santschi, 1935a: 283. Current subspecies: nominal plus *concordius, delagoensis, freyeri, infelix, lemma, moshianus, nitidior, osiris, patersoni, pax, sorptus, tenuipilis.*

*ombilicata. Camponotus reticulatus* subsp. *bedoti* var. *ombilicata* Baroni Urbani, 1971b: 362. [Unnecessary replacement name for *punctulata* Karavaiev, 1929c: 240, unavailable name.] Unavailable name.

**ominosus.** *Camponotus ominosus* Forel, 1911a: 52 (s.w.) SRI LANKA. Combination in *C.* (*Myrmosaga*): Forel, 1914a: 269; in *C.* (*Myrmotemnus*): Emery, 1920b: 258; in *C.* (*Myrmamblys*): Emery, 1925b: 138.

*opaca. Formica opaca* Nylander, 1856b: 55, pl. 3, fig. 16 (w.) FRANCE. Junior synonym of *cruentatus:* Roger, 1859: 228; Mayr, 1861: 36.

*opaca. Camponotus quadrimaculatus* var. *opaca* Emery, 1899f: 290 (w.) MADAGASCAR. [Junior secondary homonym of *Formica opaca* Nylander, above.] Replacement name: *opacatus* Emery, 1925b: 123.

*opaca. Camponotus oblongus* var. *opaca* Forel, 1907a: 32 (w.) INDIA. [Junior secondary homonym of *Formica opaca* Nylander, above.] Replacement name: *binominatus* Forel, 1916: 458.

#**opacatus.** *Camponotus* (*Mayria*) *quadrimaculatus* var. *opacata* Emery, 1925b: 123. Replacement name for *opaca* Emery, 1899f: 290. [Junior secondary homonym of *Formica opaca* Nylander, 1856b: 55.]

**opaciceps.** *Camponotus opaciceps* Roger, 1863a: 141 (q.) BRAZIL. Combination in *C.* (*Myrmothrix*): Emery, 1925b: 108. Junior synonym of *esuriens:* Emery, 1896d: 372. Revived from synonymy as race of *abdominalis:* Forel, 1897b: 297. Revived status as species: Emery, 1925b: 108. Status uncertain: Hashmi, 1973: 122.

#**opaciventris.** *Camponotus opaciventris* Mayr, 1879: 648 (w.) INDIA. [*Camponotus opaciventris* Smith, F. 1873: viii, *nomen nudum,* attributed to Mayr.] Combination in *C.* (*Orthonotomyrmex*): Emery, 1925b: 126. Variety/subspecies of *sericeus:* Forel, 1892j: 223; Emery, 1896d: 376. Junior synonym of *sericeus:* Bingham, 1903: 376. Revived from synonymy as variety of *sericeus:* Forel, 1908a: 6; Emery, 1925b: 126.

*opica. Camponotus maculatus* subsp. *parvulus* var. *opica* Forel, 1911c: 311 (w.) BRAZIL. Unavailable name.

*opitrix. Camponotus emeryodicatus* subsp. *decessor* var. *opitrix* Forel, 1908c: 414 (w.) BRAZIL. Unavailable name.

#**orbiculatopunctatus.** *Camponotus* (*Myrmophyma*) *claripes* subsp. *orbiculatopunctatus* Viehmeyer, 1925b: 143 (s.w.) AUSTRALIA.

**orinobates.** *Camponotus* (*Myrmamblys*) *orinobates* Santschi, 1919a: 348, fig. 4 (s.w.) KENYA. Combination in *C.* (*Paramyrmamblys*): Santschi, 1926c: 604.

*orinodromus. Camponotus* (*Myrmamblys*) *orinodromus* Santschi, 1919a: 347, fig. 4 (w.) SOUTH AFRICA. Combination in *C.* (*Paramyrmamblys*): Santschi, 1926c: 604. Junior synonym of *orites:* Arnold, 1926: 291.

**orinus.** *Camponotus* (*Karavaievia*) *orinus* Dumpert, **new name** for *montanus* Dumpert, 1989: 221. [Junior primary homonym of *montanus* Emery, 1894c: 168.]

**orites.** *Camponotus* (*Myrmamblys*) *orites* Santschi, 1919a: 346, fig. 4 (w.) SOUTH AFRICA. Arnold, 1926: 290 (s.). Combination in *C.* (*Paramyrmamblys*): Santschi, 1926c: 604. Senior synonym of *orinodromus:* Arnold, 1926: 291.

#**ornatus.** *Camponotus ruficeps* var. *ornatus* Emery, 1894c: 173 (s.w.) BOLIVIA. Combination in *C.* (*Myrmosphincta*): Crawley, 1916b: 376. Variety of *sexguttatus:* Emery, 1894f: 2; Emery, 1896d: 375; Wheeler, W.M. 1923c: 8; Crawley, 1916b: 376.

*ornatus. Camponotus ornatus* Forel, 1907a: 36 (s.) INDONESIA (Ambon I.). [Junior primary homonym of *ornatus* Emery, above.] Replacement name: *bellus* Forel, 1908c: 404.

**orthocephalus.** *Camponotus orthocephalus* Emery, 1894c: 169 (s.) BRAZIL. Emery, 1920c: 37 (q.); Menozzi, 1927d: 339 (w.). Combination in *C.* (*Myrmamblys*): Forel, 1914a: 90; in *C.* (*Pseudocolobopsis*): Emery, 1920b: 259. See also: Kempf, 1968b: 410.

*orthocephalus. Camponotus* (*Myrmophyma*) *quadrisectus* r. *hians* var. *orthocephalus* Stitz, 1925: 127 (w.) PHILIPPINES. Unavailable name.

**orthodoxus.** *Camponotus* (*Myrmotrema*) *foraminosus* st. *orthodoxus* Santschi, 1914d: 381 (s.w.) GUINEA. Subspecies of *foraminosus:* Wheeler, W.M. 1922a: 981. Raised to species: Santschi, 1915c: 267 (in key); Emery, 1925b: 133.

*osceola. Camponotus* (*Tanaemyrmex*) *socius* var. *osceola* Wheeler, W.M. 1932a: 15 (w.) U.S.A. Junior synonym of *socius:* Creighton, 1950a: 379.

#**osiris.** *Camponotus carbo* subsp. *osiris* Forel, 1911f: 281 (w.) ZAIRE. Combination in *C.* (*Myrmotrema*): Santschi, 1915c: 270. Subspecies of *tauricollis:* Santschi, 1926c: 616; of *olivieri:* Santschi, 1915c: 270; Santschi, 1935a: 283.

**ostiarius.** *Camponotus* (*Myrmamblys*) *ostiarius* Forel, 1914d: 258 (s.w.) SOUTH AFRICA. Arnold, 1924: 702 (q.m.). Combination in *C.* (*Paramyrmamblys*): Santschi, 1926c: 604.

#**otoquensis.** *Camponotus* (*Myrmepomis*) *sericeiventris* var. *otoquensis* Wheeler, W.M. 1931c: 94 (s.w.) PANAMA.

*CAMPONOTUS (continued)*

***ovaticeps.*** *Formica ovaticeps* Spinola, 1851a: 238 (w.) CHILE. Combination in *Camponotus:* Roger, 1863b: 5; in *C.* (*Myrmoturba*): Bruch, 1914: 230; in *C.* (*Tanaemyrmex*): Emery, 1925b: 78. Variety of *chilensis:* Emery, 1895d: 214; Kempf, 1972a: 66. Revived status as species: Snelling & Hunt, 1976: 122.

***overbecki.*** *Camponotus* (*Myrmosphincta*) *dolichoderoides* subsp. *overbecki* Viehmeyer, 1916a: 162, fig. 14 (w.q.m.) SINGAPORE. Combination in *C.* (*Karavaievia*): Emery, 1925b: 116. Raised to species: Dumpert, 1986: 567.

***oxleyi.*** *Camponotus oxleyi* Forel, 1902h: 501 (s.w.q.) AUSTRALIA. Combination in *C.* (*Myrmocamelus*): Forel, 1914a: 270; in *C.* (*Myrmophyma*): Emery, 1920b: 257.

***pachylepis.*** *Camponotus* (*Myrmobrachys*) *pachylepis* Emery, 1920c: 45, fig. 7 (w.) PERU.

#***pageti.*** *Camponotus* (*Myrmentoma*) *fallax* subsp. *pageti* Hamann & Klemm, 1976: 674, fig. 1 (q.) GREECE.

***pallens.*** *Formica pallens* Le Guillou, 1842: 316 (q.) VAVAO I. Combination in *Camponotus:* Emery, 1921a: 25; in *C.* (*Tanaemyrmex*): Emery, 1925b: 95.

*pallens. Formica pallens* Nylander, 1849: 36 (w.) SICILY. [Junior primary homonym of *pallens* Guillou, above.] Replacement name: *nylanderi* Emery, 1921a: 25. Forel, 1886f: 149 (s.); Emery, 1908a: 201 (q.). Combination in *Camponotus:* Roger, 1863b: 1; in *C.* (*Myrmoturba*): Emery, 1920b: 255. Junior synonym of *sylvaticus:* Roger, 1863b: 1; Mayr, 1865: 27; Nasonov, 1889: 11. Race of *rubripes:* Forel, 1886f: 143; of *maculatus:* Emery, 1896d: 370; Emery, 1914d: 159. Status as species: Dalla Torre, 1893: 246; Emery, 1920b: 255; Emery, 1920c: 7. [For later history of name see under *nylanderi.*]

***pallescens.*** *Camponotus pallescens* Mayr, 1887: 512 (s.w.) BRAZIL. Wheeler, G.C. & Wheeler, J. 1968: 219 (l.). Combination in *C.* (*Camponotus*): Forel, 1914a: 266; in *C.* (*Pseudocolobopsis*): Emery, 1920b: 259; in *C.* (*Tanaemyrmex*): Kempf, 1960f: 453.

#***pallidiceps.*** *Camponotus nigriceps* var. *pallidiceps* Emery, 1887a: 211 (w.q.) AUSTRALIA. Combination in *C.* (*Tanaemyrmex*): Emery, 1925b: 103. See also: Clark, 1934c: 71.

*pallidinervis. Formica pallidinervis* Brullé, 1832: 326 (m.) GREECE. Combination in *Camponotus,* and junior synonym of *aethiops:* Roger, 1863b: 1. Junior synonym of *lateralis:* Mayr, 1863: 399.

#***pallidus.*** *Formica pallida* Smith, F. 1857a: 57 (s.) BORNEO. [Unresolved junior primary homonym of *Formica pallida* Latreille, 1798: 41 (now in *Lasius*).] Smith, F. 1860a: 68 (w.q.). Combination in *Camponotus:* Mayr, 1862: 656; in *C.* (*Myrmoturba*): Forel, 1913k: 125; in *C.* (*Tanaemyrmex*): Emery, 1925b: 94. Race/subspecies of *maculatus:* Forel, 1903d: 408; Santschi, 1919a: 326; Mann, 1921: 477. Revived status as species: Wheeler, W.M. 1919e: 108. Subspecies of *irritans:* Emery, 1920c: 7; Emery, 1925b: 94.

#***palliolatus.*** *Camponotus* (*Myrmobrachys*) *godmani* var. *palliolatus* Wheeler, W.M. 1922c: 16 (w.) TRINIDAD.

***palpatus.*** *Camponotus palpatus* Emery, 1897d: 575 (s.w.q.) NEW GUINEA. Combination in *C.* (*Camponotus*): Forel, 1914a: 266; in *C.* (*Myrmoturba*): Emery, 1920b: 255; in *C.* (*Tanaemyrmex*): Emery, 1925b: 102.

*panamanus. Camponotus* (*Myrmothrix*) *abdominalis* st. *floridanus* var. *panamanus* Santschi, 1936b: 214 (w.) PANAMA. Unavailable name; material referred to *atriceps* by Hashmi, 1973: 82.

#***pansus.*** *Camponotus* (*Myrmosericus*) *zimmermanni* var. *pansus* Santschi, 1926c: 613 (w.) KENYA.

#***pantiensis.*** *Camponotus* (*Myrmotarsus*) *nigricans* var. *pantiensis* Santschi, 1932b: 18 (q.) SUMATRA.

***papago.*** *Camponotus* (*Colobopsis*) *papago* Creighton, 1953c: 153, figs. 1, 2; pl. 13, figs. 1–3 (s.w.q.m.) U.S.A.

*papua. Camponotus maculatus* subsp. *novaehollandiae* var. *papua* Emery, 1911c: 256 (s.w.) NEW GUINEA. Unavailable name.

#***paradichrous.*** *Camponotus* (*Tanaemyrmex*) *sylvaticus* var. *paradichroa* Emery, 1925b: 101 (w.) INDIA. Menozzi, 1939a: 316 (q.m.).

***paradoxus.*** *Colobopsis paradoxa* Mayr, 1866b: 887, pl. 20, fig. 2 (q., not w. as originally stated, see Kempf, 1960f: 460) BRAZIL. Combination in *Camponotus:* Dalla Torre, 1893: 246; in *C.* (*Colobopsis*): Forel, 1912i: 90; in *C.* (*Hypercolobopsis*): Emery, 1920b: 260. Senior synonym of *janitor:* Kempf, 1960f: 460.

#***paranaensis.*** *Camponotus* (*Myrmoturba*) *melanoticus* var. *paranaensis* Santschi, 1939e: 322 (s.w.) BRAZIL.

*pardus. Camponotus fallax* var. *pardus* Wheeler, W.M. 1910f: 225 (s.w.q.m.) U.S.A. Variety of *caryae:* Wheeler, W.M. 1917c: 28. Junior synonym of *nearcticus:* Creighton, 1950a: 388.

*paris. Camponotus paris* Forel, 1912i: 80 (s.) BRAZIL. Combination in *C.* (*Myrmamblys*): Forel, 1914a: 272; in *C.* (*Myrmobrachys*): Emery, 1925b: 165. Junior synonym of *propinquus:* Kempf, 1960f: 461.

***parius.*** *Camponotus micans* r. *paria* Emery, 1889b: 513 (w.) BURMA. Bingham, 1903: 364 (q.); Imai, *et al.* 1984: 9 (k.). Combination in *C.* (*Myrmosericus*): Forel, 1913k: 129. Raised to species: Bingham, 1903: 364. Subspecies/race of *rufoglaucus:* Forel, 1892j: 238; Emery, 1895h: 48 (in key); Forel, 1907e: 19; Santschi, 1920h: 174; Wheeler, W.M. 1921c: 545; Wheeler, W.M. 1923b: 6; Wheeler, W.M. 1929f: 9. Raised to species: Donisthorpe, 1929a: 448.

#***parvulus.*** *Camponotus maculatus* subsp. *parvulus* Emery, 1894c: 167 (s.w.) BRAZIL. Combination in *C.* (*Myrmoturba*): Forel, 1913l: 248; in *C.* (*Tanaemyrmex*): Emery, 1925b: 79. Subspecies of *bonariensis:* Emery, 1920b: 233.

#***patersoni.*** *Camponotus* (*Myrmotrema*) *olivieri* subsp. *patersoni* Arnold, 1959: 337, fig. 23 (s.w.) BOTSWANA.

***patimae.*** *Camponotus* (*Paracolobopsis*) *patimae* Wheeler, W.M. 1942: 260, pl. 57, fig. b (s.w.m.) GUIANA. Combination in *C.* (*Myrmaphaenus*): Kempf, 1972a: 46. Current subspecies: nominal plus *dolentulus.*

#***pattensis.*** *Camponotus somalinus* var. *pattensis* Forel, 1907g: 91 (w.) KENYA. Combination in *C.* (*Tanaemyrmex*): Emery, 1925b: 100.

*paucipilis. Camponotus marginatus* subsp. *subbarbatus* var. *paucipilis* Emery, 1893i: 677 (w.m.) U.S.A. Unavailable name; material referred to *subbarbatus* by Creighton, 1950a: 390.

*paucipubens. Camponotus rufoglaucus* st. *flavomarginatus* var *paucipubens* Santschi, 1911g: 212 (w.) ANGOLA. Unavailable name.

#***pauli.*** *Camponotus brevis* var. *pauli* Forel, 1908b: 71 (w.m.) COSTA RICA. Combination in *C.* (*Myrmobrachys*): Emery, 1925b: 163.

***pavidus.*** *Formica pavida* Smith, F. 1860b: 96 (q.) INDONESIA (Batjan I.). Combination in *Camponotus* (*Myrmamblys*): Donisthorpe, 1932c: 459.

*pavidus. Camponotus* (*Myrmentoma*) *pavidus* Snelling, 1988: 71 (s.w.q.) U.S.A. [First available use of *Camponotus fallax* subsp. *rasilis* var. *pavidus* Wheeler, W.M. 1910f: 228; unavailable name.] [Junior secondary homonym of *Formica pavida* Smith, F., above.] Replacement name: *snellingi* **new name**.

#***pax.*** *Camponotus* (*Myrmotrema*) *olivieri* var. *pax* Menozzi, 1942: 180 (s.w.) CAMEROUN. [First available use of *Camponotus* (*Myrmotrema*) *olivieri* st. *tauricollis* var. *pax* Santschi, 1915c: 273; unavailable name.]

#***pazosi.*** *Camponotus* (*Myrmoturba*) *santosi* var. *pazosi* Santschi, 1922c: 101 (w.) CUBA. Combination in *C.* (*Tanaemyrmex*): Emery, 1925b: 82.

#***pectitus.*** *Camponotus* (*Myrmosericus*) *vestitus* var. *pectitus* Santschi, 1930a: 74 (s.w.q.) ANGOLA. [First available use of *Camponotus rufoglaucus* st. *vestitus* var. *pectita* Santschi, 1911g: 212; unavailable name.] See also: Arnold, 1924: 695; Santschi, 1926b: 259 (indicates that material of *pectitus* and *intuens* may be identical).

#***peguensis.*** *Camponotus sericeus* subsp. *peguensis* Emery, 1895k: 479 (w.) BURMA. Combination in *C.* (*Orthonotomyrmex*): Forel, 1913k: 129. Material of the *nomen nudum sericeoopaciventris* referred here: Forel, 1895e: 455.

***pellarius.*** *Camponotus pellarius* Wheeler, W.M. 1914b: 60 (s.w.) MEXICO. Combination in *C.* (*Myrmobrachys*): Emery, 1920b: 260; in *C.* (*Myrmaphaenus*): Emery, 1925b: 156.

***pellax.*** *Camponotus* (*Myrmocamelus*) *pellax* Santschi, 1919a: 330 (s.w.) AUSTRALIA. [*Camponotus pellax* Santschi, 1918e: 63; *nomen nudum.*] Combination in *C.* (*Myrmophyma*): Emery, 1925b: 112. See also: Santschi, 1928e: 482.

#***pellitus.*** *Camponotus pellitus* Mayr, 1862: 668 (w.) BRAZIL. Mayr, 1866a: 486 (s.w.q.). Combination in *C.* (*Myrmamblys*): Bruch, 1914: 229; in *C.* (*Myrmaphaenus*): Emery, 1925b: 153. Junior synonym of *blandus:* Emery, 1903: 67. Revived from synonymy as subspecies of *blandus:* Santschi, 1922c: 111.

****penninervis.*** **Camponotus penninervis* Théobald, 1937b: 317, pl. 26, fig. 15 (m.) FRANCE (Oligocene).

*pennsylvanicopubescens. Camponotus pennsylvanicus* var. *pennsylvanicopubescens* Ruzsky, 1896: 68, *nomen nudum.*

***pennsylvanicus.*** *Formica pensylvanica* De Geer, 1773: 603, pl. 31, figs. 9, 10 (s.w.q.m.) U.S.A. Wheeler, G.C. & Wheeler, J. 1953e: 187 (l.). Combination in *Camponotus:* Mayr, 1862: 666; in *C.* (*Camponotus*): Forel, 1914a: 266. Subspecies/race of *herculeanus:* Forel, 1879a: 57; Mayr, 1886d: 420; Wheeler, W.M. 1913c: 117; Wheeler, W.M. 1917i: 465; Wheeler, W.M. 1932a: 13; Buren, 1944a: 293. Status as species: Ruzsky, 1896: 67; Forel, 1901h: 70; Forel,

1907e: 10; Wheeler, W.M. 1910d: 335; Emery, 1920b: 255; Creighton, 1950a: 367; Smith, D.R. 1979: 1427. Senior synonym of *semipunctata:* Mayr, 1886d: 420; of *herculeanopennsylvanicus* (and its junior synonym *mahican*): Creighton, 1950a: 367.

***peperi.*** *Camponotus (Myrmobrachys) peperi* Forel, 1913l: 246 (s.w.) GUATEMALA.

#***peregrinus.*** *Camponotus peregrinus* Emery, 1893d: 91 (w.) GALAPAGOS IS. Wheeler, W.M. 1924a: 118 (q.m.). Combination in *C. (Myrmorhachis)*: Forel, 1914a: 274; in *C. (Myrmocladoecus)*: Wheeler, W.M. 1924a: 118. Subspecies of *planus:* Wheeler, W.M. 1919c: 291.

#***pergandei.*** *Camponotus (Tanaemyrmex) punctulatus* var. *pergandei* Emery, 1925b: 78. Replacement name for *ruficornis* Pergande, 1896: 864. [Junior primary homonym of *ruficornis* Emery, 1895d: 214.]

*pericles. Camponotus maculatus* subsp. *oertzeni* var. *pericles* Forel, 1911e: 290 (w.) GREECE. Unavailable name (Agosti & Collingwood, 1987a: 61).

***perneser.*** *Camponotus perneser* **new name**. Replacement name for *Tapinoma nitida* Smith, F. 1860a: 69. [Junior secondary homonym of *Formica nitida* Smith, F. 1859a: 138.]

#***perpectitus.*** *Camponotus (Myrmosericus) vestitus* var. *perpectitus* Santschi, 1926b: 259 (q.) BENIN.

***perrisii.*** *Camponotus foraminosus* r. *perrisii* Forel, 1886f: 177 (s.w.q.) ANGOLA. Combination in *C. (Myrmotrema)*: Forel, 1913b: 342. Subspecies of *foraminosus:* Santschi, 1910c: 393. Status as species: Dalla Torre, 1893: 247; Santschi, 1911g: 216; Forel, 1914a: 270; Santschi, 1915c: 277; Stitz, 1916: 399; Wheeler, W.M. 1922a: 983. Current subspecies: nominal plus *densipunctatus, insularis, jucundus, nigeriensis.*

***perroti.*** *Camponotus perroti* Forel, 1897c: 202 (s.) MADAGASCAR. [*Camponotus perroti* Wasmann, 1897: 250, *nomen nudum,* attributed to Forel.] Combination in *C. (Dinomyrmex)*: Forel, 1913j: 224; in *C. (Myrmoturba)*: Forel, 1914a: 267; in *C. (Tanaemyrmex)*: Emery, 1925b: 89. Current subspecies: nominal plus *aeschylus.*

***personatus.*** *Camponotus personatus* Emery, 1894g: 373 (footnote) (s.w.) BRAZIL. Combination in *C. (Myrmoturba)*: Forel, 1914a: 267; in *C. (Myrmamblys)*: Bruch, 1914: 229; in *C. (Neomyrmamblys)*: Wheeler, W.M. 1922a: 708; Santschi, 1939e: 330; in *C. (Myrmaphaenus)*: Emery, 1925b: 156; Kempf, 1972a: 47. Senior synonym of *aloysii:* Santschi, 1939e: 330.

#***perthianus.*** *Camponotus (Tanaemyrmex) consobrinus* var. *perthianus* Wheeler, W.M. 1933b: 23 (s.w.) AUSTRALIA. [First available use of *Camponotus (Myrmoturba) nigriceps* r. *dimidiatus* var. *perthiana* Forel, 1915b: 97; unavailable name (Taylor, 1986: 34).] Subspecies of *consobrinus:* Clark, 1934c: 71.

#***perturbans.*** *Camponotus (Myrmosphincta) sexguttatus* var. *perturbans* Kutter, 1931a: 62 (s.w.q.m.) COLOMBIA.

#***pervicus.*** *Camponotus (Paracolobopsis) macilentus* var. *pervicus* Wheeler, W.M. 1924a: 115 (s.w.q.m.) GALAPAGOS IS.

***peseshus.*** *Camponotus peseshus* **new name**. Replacement name for *nitidiceps* Viehmeyer, 1925b: 141. [Junior primary homonym of *nitidiceps* Viehmeyer, 1916a: 156.]

*pessimus. Camponotus (Myrmoturba) maculatus* subsp *miserabilis* var. *pessimus* Wheeler, W.M. 1922a: 235 (s.w.) ZAIRE. Unavilable name; material referred to *maculatus* by Baroni Urbani, 1972: 125.

***petersii.*** *Camponotus petersii* Emery, 1895h: 50, pl. 2, fig. 28 (w.) SOUTH AFRICA. Forel, 1914d: 254 (s.); Arnold, 1924: 684 (q.m.). Combination in *C. (Myrmosericus)*: Forel, 1912i: 91. Current subspecies: nominal plus *janus.*

****petrifactus.*** **Camponotus petrifactus* Carpenter, 1930: 60, pl. 6, fig. 5 (w.) U.S.A. (Miocene).

***pexus.*** *Camponotus (Myrmosericus) pexus* Santschi, 1929e: 161 (s.w.q.) MOROCCO. See also: Cagniant, 1985b: 77; Cagniant, 1991b: 52.

*phaeogaster* Walker, 1871; see under ***CATAGLYPHIS.***

***phragmaticola.*** *Camponotus (Colobopsis) phragmaticola* Donisthorpe, 1943b: 205 (q.) INDIA.

***phytophilus.*** *Camponotus (Myrmobrachys) phytophilus* Wheeler, W.M. 1934g: 207 (s,w.q.m.) MEXICO.

***piceatus.*** *Camponotus (Tapinoma) piceatus* Norton, 1868c: 3 (s.) MEXICO. [*Tapinoma piceata* Norton, 1868a: 60, *nomen nudum.*] Combination in *C. (Myrmobrachys)*: Forel, 1914a: 270. Subspecies of *senex:* Emery, 1896d: 377. Revived status as species: Forel, 1914a: 270.

***piceus.*** *Formica picea* Leach, 1825: 292 (w.q.m.) FRANCE. Combination in *Camponotus:* Roger, 1863b: 1; in *C. (Myrmentoma)*: Menozzi, 1921: 32. Race/subspecies of *lateralis:* Emery, 1891b: 21; Dalla Torre, 1893: 238; Finzi, 1924a: 14. Status as species: Emery, 1925a: 67; Karavaiev, 1927c: 277; Finzi, 1930d: 318; Kutter, 1977c: 207. Senior synonym of *foveolata* (and its junior synonym *ebeninus*): Dalla Torre, 1893: 238; Ruzsky, 1905b: 255; Kutter, 1977c: 207; of *atricolor* (and its junior synonym *rectus*), *foveolata,* and *merula:* Atanassov & Dlussky, 1992: 222. Current subspecies: nominal plus *spissinodis.*

***picipes.*** *Formica picipes* Olivier, 1792: 501 (w.) FRENCH GUIANA. Forel, 1879a: 67 (s.q.m.). Combination in *Camponotus:* Mayr, 1862: 657; in *C. (Myrmoturba)*: Emery, 1920b: 255; in *C. (Tanaemyrmex)*: Emery, 1925b: 81. Race/subspecies of *sylvaticus:* Forel, 1879a: 67; of *rubripes:* Forel, 1886f: 143; of *maculatus:* Emery, 1896d: 371. Status as species: Dalla Torre, 1893: 247; Santschi, 1913h: 42; Emery, 1920b: 233 (footnote); Santschi, 1922c: 101. Current subspecies: nominal plus *guatemalensis, jamaicensis, jason, pilosulus, plombyi, pudorosus, spengleri.*

#***picticornis.*** *Camponotus (Myrmobrachys) crassus* var. *picticornis* Santschi, 1936d: 421 (w.) ARGENTINA.

***pictipes.*** *Camponotus pictipes* Forel, 1891b: 217 (w.) MADAGASCAR. Combination in *C. (Myrmocamelus)*: Forel, 1914a: 270; in *C. (Myrmosaga)*: Emery, 1920b: 257; in *C. (Mayria)*: Emery, 1925b: 123.

#***pictiventris.*** *Camponotus maculatus* subsp. *pictiventris* Mayr, 1901b: 28 (s.w.) SOUTH AFRICA. Combination in *C. (Tanaemyrmex)*: Emery, 1925b: 86. Subspecies of *maculatus:* Arnold, 1922: 629; Wheeler, W.M. 1922a: 958; of *hova:* Emery, 1920c: 6; Emery, 1925b: 86.

***pictostriatus.*** *Camponotus (Myrmamblys) pictostriatus* Karavaiev, 1933b: 103, fig. 6 (w.) JAVA.

*pictus. Camponotus ligniperdus* var. *pictus* Forel, 1886f: 141 (w.q.m.) U.S.A. [First available use of *Camponotus herculeanus* r. *ligniperdus* var. *pictus* Forel, 1879a: 59; unavailable name.] Variety of *herculeanus:* Mayr, 1886d: 420. Junior synonym of *noveboracensis:* Forel, 1899c: 130.

***pilicornis.*** *Formica marginata* var. *pilicornis* Roger, 1859: 228 (s.w.q.) SPAIN. Emery, 1908a: 197 (s.q.m.). Combination in *Camponotus:* Mayr, 1863: 419; in *C. (Tanaemyrmex)*: Emery, 1925b: 101. Variety/race/subspecies of *sylvaticus:* André, 1882a: 146; Emery, 1920c: 7; Santschi, 1925g: 356; of *rubripes:* Forel, 1886f: 143; of *maculatus:* Emery, 1905d: 31; Emery, 1916b: 150. Raised to species: Bondroit, 1918: 76; Schmitz, 1950: 13; Collingwood & Yarrow, 1969: 82, Baroni Urbani, 1971c: 186. Current subspecies: nominal plus *siculus.*

*piliventris. Formica piliventris* Smith, F. 1858b: 39 (w.) AUSTRALIA. Combination in *Camponotus:* Roger, 1863b: 4. Junior synonym of *suffusus:* Mayr, 1886c: 355.

*pilosa. Formica pilosa* Olivier, 1792: 498 (w.) SOUTH AFRICA. Junior synonym of *fulvopilosus:* Lepeletier, 1835: 213; Smith, F. 1858b: 32; Emery, 1896d: 376; Wheeler, W.M. 1922a: 985.

*pilosa. Formica pilosa* Smith, F. 1857a: 54 (w.) BORNEO. [Junior primary homonym of *pilosa* Olivier, above.] Replacement name (first available): *leonardi* Emery, 1889b: 515 (see there). Combination in *Colobopsis:* Mayr, 1862: 691; in *Camponotus:* Dalla Torre, 1893: 247; in *C. (Colobopsis)*: Forel, 1912i: 90. Senior synonym of *pubescens:* Emery, 1900d: 706.

#***pilosulus.*** *Camponotus (Tanaemyrmex) picipes* var. *pilosula* Emery, 1925b: 81 (w.) MEXICO. [First available use of *Camponotus maculatus* r. *picipes* var. *pilosula* Forel, 1901c: 135; unavailable name.]

****pinguiculus.*** **Formica pinguicula* Heer, 1850: 118, pl. 9, fig. 9; pl. 11, fig. 4 (m.) SWITZERLAND (Miocene). Combination in *Camponotus:* Mayr, 1867b: 52.

****pinguis.*** **Formica pinguis* Heer, 1850: 110, pl. 8, figs. 3, 4 (q.) YUGOSLAVIA (Miocene). Combination in *Camponotus:* Mayr, 1867b: 51.

#***pinzonensis.*** *Camponotus (Myrmorhachis) planus* var. *pinzonensis* Wheeler, W.M. 1919c: 297 (s.w.q.) GALAPAGOS IS. Combination in *C. (Myrmocladoecus)*: Emery, 1925b: 167.

#***piperatus.*** *Camponotus (Myrmophyma) claripes* subsp. *piperatus* Wheeler, W.M. 1933b: 26, fig. 2 (s.w.q.) AUSTRALIA.

***pittieri.*** *Camponotus pittieri* Forel, 1899c: 143 (s.w.) COSTA RICA. Combination in *C. (Myrmobrachys)*: Forel, 1914a: 271. Current subspecies: nominal plus *fuscogaster, poenalis.*

***placidus.*** *Formica placida* Smith, F. 1858b: 27 (q.) JAVA. Combination in *Camponotus:* Roger, 1863b: 3. Subgenus indeterminate.

***planatus.*** *Camponotus planatus* Roger, 1863a: 148 (s.w.q.m.) CUBA. Wheeler, G.C. & Wheeler, J. 1953e: 194 (l.). Combination in *C. (Myrmobrachys)*: Forel, 1914a: 271. Race of *senex:* Forel, 1879a: 97; Emery, 1890b: 56. Revived status as species: Dalla Torre, 1893: 248; Forel, 1899c: 141; Forel, 1901e:

371. Current subspecies: nominal plus *acaciae, colombicus, continentis, esdras.*

#***planitae.*** *Camponotus (Myrmoturba) atlantis* var. *planitae* Santschi, 1929e: 159 (w.) TUNISIA.

***planus.*** *Camponotus planus* Smith, F. 1877a: 83 (w.) GALAPAGOS IS. Wheeler, W.M. 1919c: 287 (s.w.q.m.). Combination in *C. (Myrmorhachis)*: Emery, 1920b: 260; in *C. (Myrmocladoecus)*: Wheeler, W.M. 1924a: 118. Current subspecies: nominal plus *fernandinensis, fidelis, hephaestus, indefessus* (unresolved junior homonym), *isabelensis, peregrinus, pinzonensis, sansalvadorensis, santacruzensis.*

***platypus.*** *Camponotus platypus* Roger, 1863a: 140 (q.) PHILIPPINES. Combination in *C. (Myrmoplatypus)*: Santschi, 1921f: 311; in *C. (Myrmotarsus)*: Forel, 1914a: 269; Emery, 1925b: 135.

***platytarsus.*** *Camponotus platytarsus* Roger, 1863a: 150 (q.) MEXICO. Combination in *C. (Myrmaphaenus)*: Emery, 1920b: 256; in *C. (Myrmoplatypus)*: Santschi, 1921f: 311.

#***plombyi.*** *Camponotus maculatus* subsp. *plombyi* Wheeler, W.M. & Mann, 1914: 47, fig. 19 (s.w.) HAITI. Combination in *C. (Tanaemyrmex)*: Emery, 1925b: 82. Subspecies of *picipes:* Emery, 1920b: 233.

*plorabilis. Camponotus maculatus* subsp. *vicinus* var. *plorabilis* Wheeler, W.M. 1910d: 303 (w.q.m.) U.S.A. Unavailable name; material referred to *vicinus* by Creighton, 1950a: 381.

***plutus.*** *Camponotus (Myrmoplatypus) plutus* Santschi, 1922c: 109 (s.) FRENCH GUIANA. [*Camponotus plutus* Santschi, 1921f: 311, *nomen nudum.*]

*podenzanai* Emery, 1895; see under ***NOTOSTIGMA.***

***poecilus.*** *Camponotus poecilus* Emery, 1893e: 225, pl. 8, fig. 3 (w.) INDONESIA (Buru I.). Combination in *C. (Myrmosaga)*: Forel, 1914a: 269; in *C. (Myrmamblys)*: Emery, 1925b: 139.

#***poenalis.*** *Camponotus (Myrmobrachys) pittieri* var. *poenalis* Wheeler, W.M. 1921f: 167 (w.q.) GUIANA.

***politae.*** *Dolophra politae* Wu & Wang, 1994: 37, figs. 1–3 (w.) CHINA. Combination in *Camponotus:* **new combination**.

#***politiceps.*** *Camponotus (Myrmoturba) natalensis* var. *politiceps* Santschi, 1914e: 38 (w.m.) SOUTH AFRICA. Combination in *C. (Tanaemyrmex)*: Emery, 1925b: 103.

***polynesicus.*** *Camponotus polynesicus* Emery, 1896d: 374. Replacement name for *Colobopsis carinata* Mayr, 1870b: 943. [Junior secondary homonym of *Formica carinata* Brullé, 1840: 84.] Combination in *C. (Colobopsis)*: Emery, 1925b: 146 (as *mayriella*). Senior synonym of *mayriella* (unnecessary second replacement name for *carinata* Mayr): **new synonymy**.

*polyrhachioides* Emery, 1898; see under ***PHASMOMYRMEX.***

***pompeius.*** *Camponotus rubripes* r. *pompeius* Forel, 1886f: 157 (s.w.) GABON. Emery, 1899e: 501 (q.); Menozzi, 1933a: 110 (m.). Combination in *C. (Myrmoturba)*: Forel, 1914a: 267; in *C. (Dinomyrmex)*: Emery, 1920b: 256; in *C. (Tanaemyrmex)*: Emery, 1925b: 89. Subspecies of *maculatus:* Emery, 1896d: 371. Raised to species: Emery, 1899e: 501; Stitz, 1910: 149; Santschi, 1910c: 393; Wheeler, W.M. 1922a: 239. Current subspecies: nominal plus *cassius, iota, marius.*

#***ponariensis.*** *Camponotus (Myrmentoma) gestroi* var. *ponariensis* Santschi, 1921b: 435 (w.) ALGERIA.

#***pontifex.*** *Camponotus (Myrmepomis) sericeiventris* st. *pontifex* Santschi, 1936b: 216. Replacement name for *imperator* Wheeler, W.M. 1934f: 140. [Junior primary homonym of *imperator* Emery, 1899e: 503.]

#***postangulatus.*** *Camponotus longipilis* var. *postangulata* Emery, 1911a: 225 (q.) PERU. Combination in *C. (Myrmostenus)*: Emery, 1925b: 161.

***postcornutus.*** *Camponotus (Tanaemyrmex) postcornutus* Clark, 1930a: 121, fig. 1 (s.w.) AUSTRALIA. Wheeler, G.C. & Wheeler, J. 1974a: 62 (l.).

***posticus.*** *Camponotus (Tanaemyrmex) posticus* Santschi, 1926b: 254, fig. 5 (s.w.) ZAIRE.

*postniger. Camponotus (Myrmothrix) cingulatus* st. *brunneiventris* var. *postniger* Santschi, 1925d: 246 (w.) BRAZIL. Unavailable name; material referred to *cingulatus* by Hashmi, 1973: 31.

***postoculatus.*** *Camponotus (Myrmotrema) postoculatus* Forel, 1914d: 254 (s.w.) SOUTH AFRICA.

*poultoni. Camponotus acwapimensis* var. *poultoni* Forel, 1913c: 353 (w.) NIGERIA. Forel, 1915c: 348 (m.). Junior synonym of *acvapimensis:* Brown, 1956a: 39.

#***praeluteus.*** *Camponotus (Colobopsis) vitreus* var. *praelutea* Karavaiev, 1929c: 242 (s.) SUMATRA.

#***praerufus.*** *Camponotus vitreus* var. *praerufa* Emery, 1900d: 707 (s.w.) INDONESIA (Mentawei I.). Combination in *C. (Colobopsis)*: Emery, 1925b: 149.

***pressipes.*** *Camponotus pressipes* Emery, 1893g: 268 (footnote) (q.) BORNEO. Combination in *C. (Myrmotarsus)*: Forel, 1912i: 92. Current subspecies: nominal plus *errans.*

#***pretiosus.*** *Camponotus (Myrmophyma) nasutus* var. *pretiosus* Arnold, 1922: 669 (s.w.) ZIMBABWE. Santschi, 1937a: 65 (q.). Combination in *C. (Myrmamblys)*: Emery, 1925b: 140.

*primipilaris. Camponotus acutirostris* subsp. *primipilaris* Wheeler, W.M. 1910d: 319 (s.w.q.) U.S.A. Combination in *C. (Myrmoturba)*, and subspecies of *ocreatus:* Wheeler, W.M. 1917a: 562. Junior synonym of *ocreatus:* Creighton, 1950a: 378.

*prismaticus. Camponotus prismaticus* Mayr, 1862: 669 (s.w.) BORNEO. Junior synonym of *egregius:* Roger, 1863b: 3; of *angusticollis:* Forel, 1892j: 235.

#***probativus.*** *Camponotus (Myrmoturba) compressus* st. *probativus* Santschi, 1921d: 73 (footnote) (s.w.) MOZAMBIQUE. Combination in *C. (Tanaemyrmex)*: Emery, 1925b: 98.

#***procax.*** *Camponotus (Myrmoturba) maculatus* st. *procax* Santschi, 1919a: 335 (s.w.) VIETNAM. Combination in *C. (Tanaemyrmex)*, and misspelled as *praecox:* Emery, 1925b: 94. Subspecies of *irritans:* Santschi, 1920h: 174; Emery, 1920c: 7.

#***proles.*** *Camponotus (Tanaemyrmex) variegatus* var. *proles* Emery, 1925b: 96 (s.w.q.) LAOS. [First available use of *Camponotus (Myrmoturba) variegatus* st. *mitis* var. *proles* Santschi, 1920h: 173; unavailable name. Also described as new, and also unavailable, in Santschi, 1924c: 114.] See also: Wheeler, W.M. 1928c: 32.

*proletaria. Camponotus maculatus* subsp. *proletaria* Baroni Urbani, 1971b: 362. Replacement name for *miserabilis* Santschi, 1914d: 379. [Junior primary homonym of **miserabilis* Förster, 1891: 430.] Junior synonym of *maculatus:* Baroni Urbani, 1972: 125.

#***pronotalis.*** *Camponotus (Myrmaphaenus) blandus* var. *pronotalis* Santschi, 1936b: 215 (w.) PANAMA.

***propinquellus.*** *Camponotus (Myrmobrachys) propinquellus* Emery, 1920c: 46, fig. 7 (w.) PERU.

***propinquus.*** *Camponotus propinquus* Mayr, 1887: 517 (w.) BRAZIL. Kusnezov, 1952d: 194 (s.q.). Combination in *C. (Myrmamblys)*: Forel, 1914a: 272; in *C. (Myrmobrachys)*: Emery, 1920c: 46. Senior synonym of *paris:* Kempf, 1960f: 461. Current subspecies: nominal plus *baretoi.*

#***prosaicus.*** *Camponotus (Neomyrmamblys) trapeziceps* var. *prosaica* Santschi, 1922c: 112 (s.w.) BRAZIL. Combination in *C. (Myrmaphaenus)*: Emery, 1925b: 156.

*prostans. Camponotus nigriceps* subsp. *prostans* Forel, 1910b: 72 (w.q.) AUSTRALIA. [First available use of *Camponotus nigriceps* subsp. *obnigra* var. *prostans* Forel, 1907h: 301; unavailable name.] Junior synonym of *lividipes:* Wheeler, W.M. 1933b: 23; Clark, 1934c: 71.

***prosulcatus.*** *Camponotus (Paramyrmamblys) prosulcatus* Santschi, 1935a: 284, figs. 38–40 (w.) ZAIRE.

***pseudoirritans.*** *Camponotus pseudoirritans* Wu & Wang; in Wang, C., Xiao & Wu, 1989b: 321, figs. 24, 25 (s.w.) CHINA.

***pseudolendus.*** *Camponotus pseudolendus* Wu & Wang; in Wang, C., Xiao & Wu, 1989a: 225, figs. 13, 14 (s.w.) CHINA.

*pseudolus. Camponotus pseudolus* Forel, 1907a: 31. Unnecessary replacement name for *mendax* Forel, 1895e: 454, and hence its junior synonym.

***puberulus.*** *Camponotus carbo* subsp. *puberulus* Emery, 1897e: 604 (w.) SOMALIA. Santschi, 1915c: 278 (s.). Combination in *C. (Myrmotrema)*: Forel, 1913b: 341. Subspecies of *carbo:* Wheeler, W.M. 1922a: 977. Raised to species: Santschi, 1915c: 278; Emery, 1925b: 131.

#***puberulus.*** *Camponotus irritans* subsp. *puberula* Emery, 1920c: 7 (s.w.) SUMATRA. [First available use of *Camponotus maculatus* subsp. *subnudus* var. *puberula* Emery, 1900d: 703; unavailable name.] [Unresolved junior primary homonym of *puberulus* Emery, above.] Karavaiev, 1929c: 236 (q.m.). Combination in *C. (Tanaemyrmex)*: Emery, 1925b: 94.

*pubescens. Formica pubescens* Fabricius, 1775: 392 (w.) HUNGARY. Lepeletier, 1835: 211 (w.q.m.); Schenck, 1852: (q.m.). Combination in *Camponotus:* Mayr, 1861: 36. Race of *herculeanus:* Forel, 1879a: 59. Status as species: André, 1874: 176; André, 1882a: 142; Nasonov, 1889: 12; Ruzsky, 1895: 8. Junior synonym of *vagus:* Olivier, 1792: 492; Forel, 1892i: 306; Emery, 1916b: 225; Emery, 1925b: 74; Bernard, 1967: 341; Kutter, 1977c: 205.

*pubescens. Formica pubescens* Brullé, 1840: 84 (w.) CANARY IS. [Junior primary homonym of *pubescens* Fabricius, above.] Replacement name: *feae* Emery, 1882: 449, junior synonym of *pubescens* Brullé (synonymy by Emery, 1893c: 88) and hence first available replacement name.

*pubescens. Colobopsis pubescens* Mayr, 1862: 691 (q.) SULAWESI. Mayr, 1867a: 68 (w.). [Unresolved junior secondary homonym of *Formica pubescens* Fabricius, above.] Junior synonym of *pilosa:* Emery, 1900d: 706.

*pubicornis. Camponotus fumidus* var. *pubicornis* Emery, 1893i: 670 (s.w.) U.S.A. Junior synonym of *festinatus:* Wheeler, W.M. 1910d: 312.

**#publicola.** *Camponotus maculatus* r. *publicola* Forel, 1912i: 71 (s.) BRAZIL. Combination in *C. (Tanaemyrmex)*: Emery, 1925b: 81. Subspecies of *picipes:* Emery, 1920b: 233; of *melanoticus:* Santschi, 1922c: 101; Emery, 1925b: 81; Kempf, 1972a: 69.

**#pudorosus.** *Camponotus (Tanaemyrmex) picipes* var. *pudorosus* Emery, 1925b: 81 (s.w.q.m.) MEXICO. [First available use of *Camponotus maculatus* subsp. *picipes* var. *pudorosus* Wheeler, W.M. 1914b: 57; unavailable name.]

**pulchellus.** *Camponotus pulchellus* Forel, 1894e: 232 (w.) NEW CALEDONIA. Combination in *C. (Myrmamblys)*: Forel, 1914a: 272.

**#pulcher.** *Camponotus dromedarius* var. *pulcher* Forel, 1892l: 234 (w.) MADAGASCAR. Combination in *C. (Mayria)*: Emery, 1925b: 122.

**pullatus.** *Camponotus pullatus* Mayr, 1866a: 484 (w.) MEXICO. Combination in *C. (Myrmothrix)*: Forel, 1914a: 269; in *C. (Tanaemyrmex)*: Emery, 1925b: 84.

**#pullulus.** *Camponotus (Tanaemyrmex) substitutus* var. *pullula* Emery, 1925b: 83 (w.) ARGENTINA. [First available use of *Camponotus melanoticus* st. *substitutus* var. *pullula* Santschi, 1922d; 262, unavailable name.]

**pulvinatus.** *Camponotus maculatus* subsp. *pulvinatus* Mayr, 1907b: 19 (w.) TANZANIA. Combination in *C. (Tanaemyrmex)*: Emery, 1925b: 88. Raised to species: Baroni Urbani, 1972: 131.

**punctaticeps.** *Colobopsis punctaticeps* Mayr, 1867a: 69 (q.) JAVA. Combination in *Camponotus:* Dalla Torre, 1893: 248; in *C. (Colobopsis)*: Emery, 1896d: 376; in *C. (Myrmamblys)*: Forel, 1914a: 272.

**punctatissimus.** *Camponotus pennsylvanicus* subsp. *punctatissimus* Forel, 1907e: 19 (s.) BURMA. [First available use of *Camponotus herculeanus* subsp. *pennsylvanicus* var. *punctatissimus* Emery, 1895k: 477; unavailable name.] Combination in *C. (Camponotus)*, and raised to species: Emery, 1925b: 74; Wheeler, W.M. 1927e: 3.

**punctatus.** *Camponotus punctatus* Forel, 1912i: 76 (w.) BRAZIL. Combination in *C. (Myrmothrix)*: Forel, 1914a: 269. See also: Hashmi, 1973: 119.

**punctiventris.** *Camponotus (Myrmogonia) punctiventris* Emery, 1920c: 31 (w.) AUSTRALIA. Combination in *C. (Myrmophyma)*: Emery, 1925b: 112.

**punctulatus.** *Camponotus punctulatus* Mayr, 1868b: 161 (w.q.) ARGENTINA. Goni *et al.*, 1983: 366 (k.). Combination in *C. (Myrmoturba)*: Forel, 1913l: 248; in *C. (Neomyrmamblys)*: Wheeler, W.M. 1922a: 708; in *C. (Tanaemyrmex)*: Emery, 1925b: 78. Race of *tenuiscapus:* Emery, 1888c: 365. Status as species: Dalla Torre, 1893: 248; Emery, 1894g: 375; Pergande, 1896: 864; Emery, 1903: 65; Santschi, 1922c: 104. Current subspecies: nominal plus *andigenus, brevibarbis, chubutensis, cruentus, hispidus, hybridus, imberbis, lillii, lizeri, minutior, pergandei, tenuibarbis, termitarius.*

*punctulata. Camponotus reticulatus* subsp. *bedoti* var. *punctulata* Karavaiev, 1929c: 240 (w.). Unavailable name.

**puniceps.** *Camponotus (Orthonotomyrmex) puniceps* Donisthorpe, 1942d: 458 (w.) INDIA.

**#pupillus.** *Camponotus (Myrmoturba) compressus* st. *pupillus* Santschi, 1939d: 83 (w.) MOROCCO. [First available use of *Camponotus (Myrmoturba) erigens* st. *magister* var. *pupillus* Santschi, 1925g: 358; unavailable name.] See also: Bernard, 1953a: 189

**#purensis.** *Camponotus cressoni* var. *purensis* Forel, 1912i: 84 (w.) BRAZIL. Combination in *C. (Myrmaphaenus)*: Emery, 1925b: 154.

**#purius.** *Camponotus (Myrmentoma) lateralis* var. *purius* Santschi, 1929e: 163 (w.q.m.) ALGERIA.

*purpureus* Mayr, 1876; see under ***CALOMYRMEX.***

**putatus.** *Camponotus putatus* Forel, 1892l: 234 (w.) MADAGASCAR. Combination in *C. (Myrmentoma)*: Forel, 1912i: 92; in *Calomyrmex:* Forel, 1914a: 275; in *Camponotus (Myrmonesites)*: Emery, 1920b: 257.

**pylartes.** *Camponotus (Colobopsis) pylartes* Wheeler, W.M. 1904b: 147, fig. 4 (s.w.q.) U.S.A. Petralia & Vinson, 1980: 386 (l.). Current subspecies: nominal plus *fraxinicola.*

**pylorus.** *Camponotus (Colobopsis) pylorus* Santschi, 1920g: 184 (q.) VIETNAM.

*pyrrhocephala. Formica pyrrhocephala* Motschoulsky, 1863: 11 (w.) SRI LANKA. Junior synonym of *sericeus:* Emery, 1893f: 254.

**quadriceps.** *Formica quadriceps* Smith, F. 1859a: 137 (s.w.) INDONESIA (Aru I.). Mayr, 1862: 692 (q.); Karavaiev, 1929c: 243 (m.). Combination in *Colobopsis:* Mayr, 1862: 692; in *Camponotus:* Dalla Torre, 1893: 249; in *C. (Myrmamblys)*: Forel, 1914a: 272; in *C. (Colobopsis)*: Emery, 1920b: 259. Senior synonym of *nitida:* Mayr, 1863: 403; of *angulata:* Emery, 1896d: 375; Donisthorpe, 1932c: 453. Current subspecies: nominal plus *convexior, curvatus, nanulus.*

**#quadridentatus.** *Camponotus (Myrmespera) nasutus* var. *quadridentatus* Santschi, 1937a: 65 (w.) NAMIBIA.

*quadrilaterus. Camponotus quadrilaterus* Roger, 1863a: 136 (w.) INDIA. Junior synonym of *compressus:* Forel, 1892j: 240.

*quadrilaterus. Camponotus quadrilaterus* Mayr, 1887: 519 (w.) COLOMBIA. Emery, 1894c: 171 (s.q.). [Junior primary homonym of *quadrilaterus* Roger, above.] Replacement name: *sanctaefidei* Dalla Torre, 1892: 92.

**quadrimaculatus.** *Camponotus quadrimaculatus* Forel, 1886c: cii (s.w.) MADAGASCAR. Forel, 1891b: 58 (s.q.m.). [*Camponotus quadrimaculatus* Forel, 1879a: 115, *nomen nudum.*] Combination in *C. (Myrmosaga)*: Forel, 1912i: 92; in *C. (Mayria)*: Emery, 1925b: 123. Current subspecies: nominal plus *immaculatus, opacatus, sellaris.*

**quadrinotatus.** *Camponotus marginatus* var. *quadrinotatus* Forel, 1886f: 142 (s.w.q.) JAPAN. Santschi, 1925f: 89 (m.). Combination in *C. (Myrmentoma)*: Emery, 1925b: 118. Subspecies of *fallax:* Forel, 1907e: 19; of *caryae:* Wheeler, W.M. 1917c: 29; Santschi, 1925f: 89; of *marginatus:* Ruzsky, 1925b: 42; Kuznetsov-Ugamsky, 1928b: 5. Raised to species: Ruzsky, 1926: 109. Status as species: Collingwood, 1976: 307; Wang, C., Xiao & Wu, 1989b: 327; Kupyanskaya, 1990: 173. Senior synonym of *nigricolor:* Kuznetsov-Ugamsky, 1929a: 18.

**quadrisectus.** *Formica quadrisecta* Smith, F. 1858b: 28 (q.) PHILIPPINES. Combination in *Camponotus:* Roger, 1863b: 3; in *C. (Myrmophyma)*: Forel, 1912i: 91; in *C. (Myrmotarsus)*: Emery, 1920b: 256. Current subspecies: nominal plus *hians, margaritae.*

**quercicola.** *Camponotus (Camponotus) quercicola* Smith, M.R. 1954b: 211 (w.) U.S.A.

**#quinquedentatus.** *Camponotus nasutus* var. *quinquedentata* Forel, 1910e: 451 (w.) SOUTH AFRICA. Combination in *C. (Myrmamblys)*: Wheeler, W.M. 1922a: 990.

*quiriguensis. Camponotus (Myrmepomis) sericeiventris* subsp. *rex* var. *quiriguensis* Wheeler, W.M. 1931c: 97 (s.w.) GUATEMALA. Unavailable name.

**#rabbani.** *Camponotus (Tanaemyrmex) nicobarensis* var. *rabbani* Emery, 1925b: 95 (w.) BURMA. [First available use of *Camponotus (Myrmothrix) nicobarensis* r. *exiguoguttatus* var. *rabbani* Forel, 1922: 100; unavailable name.]

*rabidus. Camponotus (Myrmothrix) abdominalis* st. *rabidus* Santschi, 1922c: 107 (s.w.q.) PERU. Junior synonym of *atriceps:* Hashmi, 1973: 82.

**#radamae.** *Camponotus maculatus* r. *radamae* Forel, 1891b: 31, pl. 1, fig. 4 (s.w.) MADAGASCAR. Combination in *C. (Myrmoturba)*: Forel, 1914a: 267; in *C. (Tanaemyrmex)*: Emery, 1925b: 86. Subspecies of *maculatus:* Emery, 1895f: 337; Arnold, 1922: 628; Wheeler, W.M. 1922a: 1041. Status as species: Dalla Torre, 1893: 249; Emery, 1920b: 255; Santschi, 1923e: 292. Subspecies of *hova:* Emery, 1920c: 6; Emery, 1925b: 86

*radamoides Camponotus maculatus* var. *radamoides* Forel, 1891b: 213 (s.w.) MADAGASCAR. Combination in *C. (Tanaemyrmex)*: Emery, 1925b: 88. Stirps of *maculatus:* Santschi, 1917b: 292; Santschi, 1935a: 279. Junior synonym of *maculatus:* Baroni Urbani, 1972: 125.

**radiatus.** *Camponotus radiatus* Forel, 1892j: 233 (w.) INDIA. Combination in *C. (Camponotus)*: Forel, 1914a: 266; in *C. (Myrmotrema)*: Emery, 1920b: 258.

**radovae.** *Camponotus radovae* Forel, 1886c: civ (s.) MADAGASCAR. [*Camponotus radovae* Forel, 1879a: 115, *nomen nudum.*] Forel, 1886f: 178 (w.). Combination in *C. (Myrmobrachys)*: Forel, 1912i: 91; in *C. (Myrmepomis)*: Emery, 1920b: 238; in *C. (Myrmopiromis)*: Wheeler, W.M. 1922a: 1052. Current subspecies: nominal plus *radovaedarwinii.*

**#radovaedarwinii.** *Camponotus radovae* var. *radovaedarwinii* Forel, 1891b: 46 (s.w.) MADAGASCAR.

**ramulorum.** *Camponotus inaequalis* var. *ramulorum* Wheeler, W.M. 1905b: 114, figs. S-U (s.w.q.m.) BAHAMAS. Combination in *C. (Myrmoturba)*: Forel, 1914a: 267; in *C. (Tanaemyrmex)*: Emery, 1925b: 82. Raised to species: Wheeler, W.M. 1913b: 503 (in text). Current subspecies: nominal plus *marcidus, mestrei, vernulus.*

**rapax.** *Formica rapax* Fabricius, 1804: 398 (w.) BRAZIL. Mann, 1916: 476 (s.). Combination in *Camponotus:* Roger, 1863b: 5; in *C. (Myrmothrix)*: Forel, 1914a: 269; in *C. (Myrmaphaenus)*: Emery, 1920b: 256; in *C. (Tanaemyrmex)*: Emery, 1925b: 82.

**raphaelis.** *Camponotus raphaelis* Forel, 1899c: 146 (w.) COSTA RICA. Combination in *C. (Myrmorhachis)*: Forel, 1914a: 274; in *C. (Myrmocladoecus)*: Emery, 1925b: 167; Wheeler, W.M. 1934e: 424.

*rasilis. Camponotus fallax* subsp. *rasilis* Wheeler, W.M. 1910f: 227 (s.w.q.m.) U.S.A. Petralia & Vinson,

1980: 386 (l.). Combination in *C.* (*Myrmentoma*): Emery, 1925b: 117. Subspecies of *fallax:* Wheeler, W.M. 1917a: 558; of *caryae:* Wheeler, W.M. 1917d: 28. Raised to species: Creighton, 1950a: 389; Smith, M.R. 1951a: 845. Junior synonym of *sayi:* Snelling, 1968b: 355; of *decipiens:* Snelling, 1988: 66.

***reaumuri.*** *Camponotus reaumuri* Forel, 1892l: 236 (w.) MADAGASCAR. Combination in *C.* (*Myrmosphincta*): Forel, 1912i: 92; in *C.* (*Myrmonesites*): Emery, 1920b: 257.

#***rebeccae.*** *Camponotus* (*Orthonotomyrmex*) *lateralis* var. *rebeccae* Forel, 1913d: 436 (s.) SYRIA. Menozzi, 1933b: 81 (q.m.). Combination in *C.* (*Myrmentoma*): Emery, 1925a: 69; Emery, 1925b: 120.

***rectangularis.*** *Camponotus rectangularis* Emery, 1890b: 57 (w.) COSTA RICA. [Also described as new by Emery, 1894k: 62.] Combination in *C.* (*Myrmorhachis*): Forel, 1914a: 274; in *C.* (*Myrmocladoecus*): Emery, 1925b: 167; Wheeler, W.M. 1933a: 63. Current subspecies: nominal plus *aulicus, rubroniger, setipes, sordidatus, willowsi.*

#***recticeps.*** *Camponotus* (*Tanaemyrmex*) *dorycus* var. *recticeps* Stitz, 1938: 115, fig. 11 (w.) NEW GUINEA.

#***rectinotus.*** *Camponotus compressus* st. *rectinota* Santschi, 1921d: 75 (w.) LIBYA. [First available use of *Camponotus* (*Myrmoturba*) *maculatus* r. *samius* var. *rectinotus* Stitz, 1917: 351; unavailable name.] Combination in *C.* (*Tanaemyrmex*): Emery, 1925b: 98.

#***rectithorax.*** *Camponotus montivagus* r. *rectithorax* Forel, 1895a: 44 (w.) MEXICO. Replacement name for *Camponotus nitida* Norton, 1868c: 2. [Junior secondary homonym of *Formica nitida* Smith, F. 1859a: 138.] Combination in *C.* (*Myrmentoma*): Emery, 1925b: 118.

#***rector.*** *Camponotus* (*Neomyrmamblys*) *inca* st. *rector* Santschi, 1923d: 68 (s.w.q.) BRAZIL.

#***rectus.*** *Camponotus lubbocki* var. *rectus* Forel, 1891b: 217 (w.) MADAGASCAR. Combination in *C.* (*Mayria*): Emery, 1925b: 123.

*rectus. Camponotus lateralis* var. *rectus* Forel, 1892i: 306 (w.) BULGARIA. [Unresolved junior primary homonym of *rectus* Forel, above.] Variety of *piceus:* Dalla Torre, 1893: 238. Junior synonym of *atricolor:* Emery, 1896d: 373; Karavaiev, 1936: 191.

*redtenbacheri. Camponotus redtenbacheri* Mayr, 1862: 667 (w.) SRI LANKA. Combination in *C.* (*Myrmosericus*): Forel, 1913k: 129. Variety of *rufoglaucus:* Forel, 1913k: 129. Junior synonym of *rufoglaucus:* Forel, 1892j: 237; Bingham, 1903: 363; Emery, 1925b: 105 (dubious synonym).

***reepeni.*** *Camponotus* (*Colobopsis*) *reepeni* Forel, 1913k: 131 (q.) SUMATRA.

***reevei.*** *Camponotus* (*Myrmophyma*) *reevei* Arnold, 1922: 671, fig. 58 (s.w.) SOUTH AFRICA. Combination in *C.* (*Myrmamblys*): Emery, 1925b: 141; in *C.* (*Myrmespera*): Santschi, 1926c: 604.

#***refectus.*** *Camponotus* (*Myrmeurynota*) *gilviventris* var. *refectus* Wheeler, W.M. 1937b: 460 (w.) CUBA.

#***reginae.*** *Camponotus reginae* Forel, 1901d: 307 (s.w.) ANGOLA. [Also described as new by Forel, 1903e: 562.] Arnold, 1924: 719 (q.). Combination in *C.* (*Orthonotomyrmex*): Forel, 1913b: 346; in *C.* (*Myrmamblys*): Emery, 1925b: 142; in *C.* (*Myrmopelta*): Santschi, 1926a: 18. Variety/race/subspecies of *meinerti:* Forel, 1911e: 291; Forel, 1911f: 282; Forel, 1913e: 670; of *vividus:* Wheeler, W.M. 1922a: 976; Arnold, 1924: 718; Santschi, 1926a: 18.

***reichardti.*** *Camponotus reichardti* Arnol'di, 1967: 1824, fig. 1 (s.) UZBEKISTAN. See also: Tarbinsky, 1976: 151.

***reichenspergeri.*** *Camponotus* (*Myrmotemnus*) *reichenspergeri* Santschi, 1926c: 602, fig. 1 (w.) SUMATRA.

***reinaldi.*** *Camponotus* (*Myrmosphincta*) *reinaldi* Kempf, 1960f: 456, figs. 36, 37, 39 (s.w.) BRAZIL.

#***relucens.*** *Camponotus ellioti* var. *relucens* Santschi, 1911e: 133 (w.) MADAGASCAR. Combination in *C.*(*Myrmopiromis*): Wheeler, W.M. 1922a: 1052.

***renggeri.*** *Camponotus rufipes* subsp. *renggeri* Emery, 1894e: 3 (w.q.) PARAGUAY. Forel, 1909a: 265 (m.). Combination in *C.* (*Myrmothrix*), and raised to species: Forel, 1912i: 91. Subspecies of *rufipes:* Bruch, 1914: 230; Santschi, 1916e: 397; Wheeler, W.M. 1923a: 5. Status as species: Luederwaldt, 1918: 53; Emery, 1920b: 256; Emery, 1925b: 109; Kusnezov, 1952d: 219; Hashmi, 1973: 67.

#***renormatus.*** *Camponotus* (*Myrmeurynota*) *gilviventris* var. *renormatus* Wheeler, W.M. 1937b: 461 (w.) CUBA.

***repens.*** *Camponotus* (*Mayria*) *repens* Forel, 1897c: 187. Replacement name for *Mayria madagascarensis* Forel, 1886c: civ. [Junior secondary homonym of *Camponotus niveosetosus* r. *madagascarensis* Forel, 1886c: civ.]

#***repressus.*** *Camponotus bidens* r. *repressus* Forel, 1912i: 79 (w.) BRAZIL. Combination in *C.* (*Myrmocladoecus*): Emery, 1925b: 166.

#***resinicola.*** *Camponotus imitator* var. *resinicola* Santschi, 1911e: 133 (w.) MADAGASCAR. Combination in *C.* (*Myrmopytia*): Emery, 1925b: 115.

***reticulatus.*** *Camponotus reticulatus* Roger, 1863a: 139 (w.) SRI LANKA. Combination in *C.* (*Myrmotemnus*): Emery, 1920b: 258; in *C.* (*Myrmamblys*): Forel, 1912i: 90; Emery, 1925b: 139. Junior synonym of *indeflexus:* Donisthorpe, 1932b: 576. Revived from synonymy: Wheeler, W.M. 1934a: 179. Current subspecies: nominal plus *fullawayi, gestiens, imparilis, jagori, latitans, mackayensis, sericellus, yerburyi.*

*reticulatus. Camponotus reticulatus* Kirby, W.F. 1896: 204 (s.w.) AUSTRALIA. [Junior primary homonym of *reticulatus* Roger, above.] Replacement name: *spenceri* Clark, 1930c: 18.

#***rex.*** *Camponotus sericeiventris* subsp. *rex* Forel, 1907e: 13 (w.) COSTA RICA. Wheeler, W.M. 1931c: 95 (q.). Combination in *C.* (*Myrmepomis*): Emery, 1925b: 171.

***rhamses.*** *Camponotus* (*Myrmotrema*) *rhamses* Santschi, 1915c: 275 (s.w.) EGYPT. Current subspecies: nominal plus *completus.*

#***rhodesianus.*** *Camponotus* (*Myrmotrema*) *troglodytes* var. *rhodesiana* Forel, 1913a: 145 (s.w.) ZIMBABWE. Arnold, 1924: 726 (q.). Variety of *troglodytes:* Wheeler, W.M. 1922a: 985; variety/race/subspecies of *robecchii:* Santschi, 1915c: 271; Arnold, 1924: 725; Emery, 1925b: 133.

#***rhodius.*** *Camponotus* (*Myrmentoma*) *lateralis* var. *rhodia* Santschi, 1934d: 280 (w.) GREECE. [First available use of *Camponotus* (*Myrmentoma*) *lateralis* subsp. *dalmaticus* var. *rhodia* Emery, 1925a: 69; unavailable name.] Menozzi, 1936d: 304 (q.).

#***riedeli.*** *Camponotus* (*Tanaemyrmex*) *festai* subsp. *riedeli* Pisarski, 1971b: 728, figs. 1–5 (s.w.) IRAQ.

***riehlii.*** *Colobopsis riehlii* Roger, 1863a: 159 (q.) CUBA. Combination in *Camponotus* (*Colobopsis*): Emery, 1889b: 517.

#***riograndensis.*** *Camponotus rubripes* r. *riograndensis* Emery, 1888c: 364 (s.w.q.m.) BRAZIL. Combination in *C.* (*Myrmoturba*): Santschi, 1922c: 97; in *C.* (*Tanaemyrmex*): Emery, 1925b: 83. Raised to species: Dalla Torre, 1893: 250; Emery, 1894g: 373; Forel, 1895b: 104. Subspecies of *picipes:* Emery, 1920b: 233; of *simillimus:* Santschi, 1922c: 97.

***robecchii.*** *Camponotus robecchii* Emery, 1892a: 120, fig. (s.w.) SOMALIA. Emery, 1901e: 63 (q.). Combination in *C.* (*Myrmotrema*): Forel, 1914a: 270. Subspecies of *foraminosus:* Forel, 1907g: 90. Revived status as species: Santschi, 1915c: 271; Wheeler, W.M. 1922a: 984; Arnold, 1924: 752; Emery, 1925b: 133. Current subspecies: nominal plus *abyssinicus, assumptionis, dispar, rhodesianus, troglodytes.*

***robertae.*** *Camponotus* (*Paramyrmamblys*) *robertae* Santschi, 1926c: 605 (q.) IVORY COAST.

*robustior. Camponotus darwinii* r. *rubropilosus* var. *robustior* Forel, 1905b: 165 (w.) MADAGASCAR. Unavailable name.

***robustus.*** *Camponotus robustus* Roger, 1863a: 135 (w.) MADAGASCAR. Forel, 1891b: 54 (s.). Combination in *C.* (*Myrmentoma*): Forel, 1912i: 92; in *C.* (*Orthonotomyrmex*): Forel, 1914a: 273; in *C.* (*Myrmisolepis*): Santschi, 1921f: 310; in *C.* (*Myrmepinotus*): Emery, 1925b: 127.

***roeseli.*** *Camponotus roeseli* Forel, 1910a: 20 (s.w.) MADAGASCAR. Combination in *C.* (*Myrmoturba*): Forel, 1914a: 267; in *C.* (*Tanaemyrmex*): Emery, 1925b: 89.

*rollei. Camponotus foraminosus* subsp. *ruspolii* var. *rollei* Forel, 1910c: 270 (s.) SOUTH AFRICA. Unavailable name.

*romani. Camponotus* (*Myrmothrix*) *abdominalis* subsp. *romani* Wheeler, W.M. 1923a: 5 (s.w.) BRAZIL. Junior synonym of *atriceps:* Hashmi, 1973: 82.

#***rosariensis.*** *Camponotus blandus* r. *rosariensis* Forel, 1912i: 75 (s.w.) ARGENTINA. Combination in *C.* (*Myrmocamelus*): Forel, 1913l: 249; in *C.* (*Myrmaphaenus*): Emery, 1925b: 154.

***rothneyi.*** *Camponotus* (*Colobopsis*) *rothneyi* Forel, 1893b: 435 (s.w.q.) INDIA. See also: Forel, 1913k: 132 (footnote). Current subspecies: nominal plus *krafti, makilingi, taivanae.*

*rottnesti. Camponotus* (*Myrmophyma*) *rottnesti* Donisthorpe, 1941k: 239. Unnecessary replacement name for *darlingtoni* Wheeler, W.M. 1934d: 160, hence its junior synonym: Brown, 1956a: 39.

***rotumanus.*** *Camponotus* (*Myrmamblys*) *rotumanus* Wilson & Taylor, 1967: 98, fig. 82 (w.) ROTUMA I.

*rotundata. Camponotus foraminosus* subsp. *ruspolii* var. *rotundata* Forel, 1907g: 90 (w.) MADAGASCAR. Unavailable name.

***rotundinodis.*** *Camponotus* (*Myrmachraphe*) *rotundinodis* Santschi, 1935a: 281, fig. 35 (w.) ZAIRE.

***roubaudi.*** *Camponotus roubaudi* Santschi, 1911c: 365, fig. (s.w.) CONGO. Combination in *C.* (*Myrmoturba*): Forel, 1913b: 340; in *C.* (*Tanaemyrmex*): Emery,

1925b: 103. [Misspelled as *ronbaudi* by Emery, 1920b: 255.]

*rubens. Camponotus (Camponotus) herculeanus* var. *rubens* Emery, 1925b: 73 (w.m.) U.S.A. [First available use of *Camponotus herculeanus* subsp. *ligniperdus* var. *rubens* Wheeler, W.M. 1906a: 41; unavailable name.] Junior synonym of *noveboracensis:* Creighton, 1950a: 370.

#***rubicolor.*** *Camponotus marginatus* var. *rubicolor* Ruzsky, 1925b: 42 (w.) RUSSIA. [Also described as new by Ruzsky, 1926: 109.] Subspecies of *aethiops:* current status.

***rubidus.*** *Camponotus rubidus* Xiao & Wang; in Wang, C., Xiao & Wu, 1989a: 227, figs. 18, 19 (s.w.) CHINA.

***rubiginosus.*** *Camponotus rubiginosus* Mayr, 1876: 66 (s.w.) AUSTRALIA. Combination in *C. (Myrmogonia)*: Forel, 1914a: 269; in *C. (Myrmophyma)*: Emery, 1920b: 257; in *C. (Thlipsepinotus)*: Santschi, 1928e: 483.

*rubra. Camponotus (Myrmotrema) sicheli* var. *rubra* Emery, 1925b: 121 (w.q.) ALGERIA. [First available use of *Camponotus lateralis* subsp. *sicheli* var. *rubra* Karavaiev, 1912a: 20; unavailable name.] See also: Santschi, 1925g: 359; Baroni Urbani, 1971c: 195. Junior synonym of *sicheli:* Collingwood, 1978: 72.

***rubripes.*** *Formica rubripes* Latreille, 1802c: 112 (w.) SIERRA LEONE. Combination in *Camponotus:* Roger, 1863b: 2; in *C. (Tanaemyrmex)*: Emery, 1925b: 104. Junior synonym of *maculatus:* Mayr, 1865: 27. Revived from synonymy: Forel, 1886f: 143. See also: Emery, 1891b: 17.

***rubrithorax.*** *Camponotus rubrithorax* Forel, 1899c: 142 (w.) MEXICO. Combination in *C. (Myrmobrachys)*: Forel, 1914a: 271. Current subspecies: nominal plus *nigrithorax.*

*rubrolabiatus. Camponotus (Myrmotrema) foraminosus* r. *ruspolii* var. *rubrolabiatus* Stitz, 1923: 165 (w.q.) NAMIBIA. Unavailable name.

#***rubroniger.*** *Camponotus rectangularis* var. *rubroniger* Forel, 1899c: 144 (w.) EL SALVADOR. Wheeler, W.M. 1934g: 221 (q.m.). Combination in *C. (Myrmocladoecus)*: Emery, 1925b: 167.

#***rubropilosus.*** *Camponotus darwinii* var. *rubropilosus* Forel, 1891b: 44 (s.w.) MADAGASCAR. Combination in *C. (Myrmopiromis)*: Wheeler, W.M. 1922a: 1051. See also: Forel, 1905b: 165.

*rudigenis. Camponotus rudigenis* Emery, 1903: 66, fig. 1 (s.w.) VENEZUELA. Junior synonym of *integellus:* Forel, 1904c: 49.

#***rufescens.*** *Camponotus (Tanaemyrmex) foleyi* subsp. *rufescens* Bernard, 1953a: 190 (in key) (s.) ALGERIA. [First available use of *Camponotus (Tanaemyrmex) compressus* subsp. *foleyi* var. *rufescens* Bernard, 1948: 160; unavailable name.]

*ruficeps. Formica ruficeps* Fabricius, 1804: 404 (s.) SOUTH AMERICA. Combination in *Camponotus:* Roger, 1862c: 285. Senior synonym of *albofasciata:* Mayr, 1886c: 362. Junior synonym of *sexguttatus:* Emery, 1894f: 2; Emery, 1896d: 375.

*ruficeps. Formica ruficeps* Smith, F. 1857a: 54 (s.w.) BORNEO. [Junior primary homonym of *ruficeps* Fabricius, above.] Replacement name: *gilviceps* Roger, 1863b: 3.

*ruficlypeus. Camponotus (Myrmobrachys) senex* subsp. *textor* var. *ruficlypeus* Emery, 1920c: 38 (w.m.) BRAZIL. Unavailable name.

***ruficornis.*** *Camponotus chilensis* var. *ruficornis* Emery, 1895d: 214 (s.w.) CHILE. Menozzi, 1935c: 329 (q.m.). Combination in *C. (Tanaemyrmex)*: Emery, 1925b: 79. Subspecies of *chilensis:* Forel, 1907e: 10. Raised to species: Emery, 1903: 69; Menozzi, 1935c: 329. See also: Snelling & Hunt, 1976: 117.

*ruficornis. Camponotus punctulatus* var. *ruficornis* Pergande, 1896: 864 (w.) MEXICO. [Junior primary homonym of *ruficornis* Emery, above.] Replacement name: *pergandei* Emery, 1925b: 78.

***rufifemur.*** *Camponotus rufifemur* Emery, 1900d: 705 (s.q.) INDONESIA (Mentawei I.). Combination in *C. (Myrmotarsus)*: Forel, 1912i: 92. Current subspecies: nominal plus *obscurus.*

***rufifrons.*** *Formica rufifrons* Smith, F. 1860b: 95 (w.q.) INDONESIA (Batjan I.). Mayr, 1870b: 945 (w.). Combination in *Colobopsis:* Mayr, 1862: 691; in *Camponotus:* Dalla Torre, 1893: 250; in *C. (Colobopsis)*: Forel, 1914a: 272. Current subspecies: nominal plus *leucopus.*

***rufigaster.*** *Camponotus (Myrmotarsus) rufigaster* Menozzi, 1928b: 2, fig. 4 (w.) SUMATRA.

*rufigenis. Camponotus (Myrmosericus) rufoglaucus* r. *cinctellus* var. *rufigenis* Forel, 1913b: 341 (w.) ZAIRE. Unavailable name.

*rufinasis. Camponotus (Camponotus) castaneus* st. *rufinasis* Santschi, 1936b: 204 (s.w.) U.S.A. Raised to species: Santschi, 1937h: 380. Junior synonym of *americanus:* Creighton, 1950a: 365.

#***rufipartis.*** *Camponotus (Dinomyrmex) wellmani* var. *rufipartis* Forel, 1916: 441 (w.q.m.) ZAIRE. Combination in *C. (Tanaemyrmex)*: Emery, 1925b: 89.

***rufipes.*** *Formica rufipes* Fabricius, 1775: 391 (s.) BRAZIL. Smith, F. 1858b: 44 (w.); Mayr, 1862: 663 (q.m.); Goni *et al.*, 1983: 366 (k.). Combination in *Camponotus:* Mayr, 1862: 663; in *C. (Myrmothrix)*: Forel, 1912i: 91. Senior synonym of *herrichi:* Roger, 1861b: 164; of *merdicola:* Dalla Torre, 1893: 250; Forel, 1895b: 104; of *alpina, cajurensis, magnifica, subrufescens:* Hashmi, 1973: 52.

#***rufipilis.*** *Camponotus (Manniella) sphaericus* subsp. *rufipilis* Aguayo, 1932: 226 (w.) CUBA.

#***rufitibia.*** *Camponotus (Myrmobrachys) circularis* subsp. *rutifibia* Emery, 1920c: 44 (w.) PERU. [Emended to *rufitibia:* Emery, 1925b: 165, justified emendation of spelling.]

*rufiventris. Formica rufiventris* Fabricius, 1804: 409 (w.) SOUTH AFRICA. Junior synonym of *fulvopilosus:* Lepeletier, 1835: 213.

***rufoglaucus.*** *Formica rufoglauca* Jerdon, 1851: 124 (w.) INDIA. Combination in *Camponotus:* Roger, 1863b: 3; in *C. (Myrmosericus)*: Forel, 1912i: 91. Senior synonym of *redtenbacheri:* Forel, 1892j: 237. Current subspecies: nominal plus *controversus, feae, latericus, syphax, tenuis, zanzibaricus, zulu.*

#***rufomaculatus.*** *Camponotus (Myrmosaulus) singularis* var. *rufomaculatus* Donisthorpe, 1941j: 208 (s.w.) BORNEO.

***rufus.*** *Camponotus (Dinomyrmex) rufus* Crawley, 1925b: 596 (s.w.q.) AUSTRALIA. Current subgeneric combination: *C. (Tanaemyrmex)*.

*rugosior. Camponotus (Myrmamblys) reticulatus* r. *sericellus* var. *rugosior* Crawley, 1923a: 30 (w.q.m.) SUMATRA. Unavailable name.

#***ruspolii.*** *Camponotus foraminosus* r. *ruspolii* Forel, 1892d: 349 (w.) SOMALIA. Menozzi, 1926a: 39 (q.). Combination in *C. (Myrmotrema)*: Forel, 1913b: 341. Subspecies of *foraminosus:* Emery, 1896d: 376; Wheeler, W.M. 1922a: 981. Status as species: Dalla Torre, 1893: 250; Emery, 1920b: 258. Subspecies of *grandidieri:* Santschi, 1915c: 269; Emery, 1925b: 132; Menozzi, 1926a: 39.

***rusticus.*** *Camponotus (Myrmamblys) rusticus* Santschi, 1916e: 398, fig. (w.) ARGENTINA. Combination in *C. (Myrmaphaenus)*: Emery, 1925b: 156.

***ruzskyellus.*** *Camponotus (Myrmoturba) maculatus* r. *ruzskyellus* Forel, 1922: 98 (w.) TURKESTAN. Combination in *C. (Tanaemyrmex)*, and raised to species: Emery, 1925b: 104.

***ruzskyi.*** *Camponotus marginatus* var. *ruzskyi* Emery, 1898c: 150 (w.) RUSSIA. Combination in *C. (Myrmentoma)*: Emery, 1925b: 118. Variety of *caryae:* Wheeler, W.M. 1917d: 29. Subspecies of *marginatus:* Kuznetsov-Ugamsky, 1929b: 36. Raised to species: Arakelian, 1994: 88.

*ruzskyi. Camponotus maculatus* subsp. *ruzskyi* Vashkevich, 1924a: 144 (w.) RUSSIA. [Unresolved junior primary homonym of *ruzskyi* Emery, above.] Junior synonym of *turkestanus:* Dlussky, Soyunov & Zabelin, 1990: 132.

***sacchii.*** *Camponotus sacchii* Emery, 1899a: 501 (s.) SOMALIA. Combination in *C. (Myrmoturba)*: Forel, 1914a: 267; in *C. (Tanaemyrmex)*: Emery, 1925b: 103.

***sachalinensis.*** *Camponotus herculeanus* var. *sachalinensis* Forel, 1904b: 38 (q.) RUSSIA. Karavaiev, 1912b: 592 (m.). Combination in *C. (Camponotus)*: Emery, 1925b: 72. Subspecies of *herculeanus:* Ruzsky, 1926: 108. Provisional junior synonym of *saxatilis:* Kuznetsov-Ugamsky, 1928b: 4; Kuznetsov-Ugamsky, 1929a: 18. Junior synonym of *herculeanus:* Yasumatsu & Brown, 1957: 49. Revived from synonymy as subspecies of *herculeanus:* Arnol'di, 1967: 1821; subspecies of *herculeanus:* Collingwood, 1976: 306; Kupyanskaya, 1990: 166. Raised to species: Collingwood, 1981: 29, retained as such here. Senior synonym of *altaica, jacuticus:* Arnol'di, 1967: 1821.

#***sadinus.*** *Camponotus (Myrmogonia) cristatus* subsp. *sadina* Mann, 1921: 482 (s.w.) FIJI IS.

#***saginatus.*** *Camponotus (Colobopsis) badius* var. *saginata* Stitz, 1925: 127 (q.) BORNEO.

*saharensis. Camponotus (Myrmoturba) compressus* st. *thoracicus* var. *saharensis* Santschi, 1921d: 73 (s.w.q.) TUNISIA. Unavailable name.

#***sahlbergi.*** *Camponotus (Orthonotomyrmex) libanicus* r. *sahlbergi* Forel, 1913d: 435 (s.w.) TURKEY. Combination in *C. (Myrmentoma)*: Emery, 1925b: 121.

#***saltensis.*** *Camponotus (Tanaemyrmex) borellii* var. *saltensis* Emery, 1925b: 79 (s.) ARGENTINA. [First available use of *Camponotus (Myrmoturba) maculatus* r. *borelli* var. *saltensis* Forel, 1913l: 248; unavailable name.]

***salvini.*** *Camponotus salvini* Forel, 1899c: 153, pl. 4, fig. 26 (s.w.) PANAMA. [*Camponotus salvini* Emery, 1896g: 36, *nomen nudum*, attributed to Forel.] Combination in *C. (Myrmamblys)*: Forel, 1914a: 272; in *C.*

(*Paracolobopsis*): Emery, 1920b: 259; in *C.* (*Myrmaphaenus*): Emery, 1925b: 154.

**#*samarus.*** *Camponotus* (*Myrmotarsus*) *barbatus* var. *samarus* Santschi, 1932b: 19, fig. 4 (q.) BORNEO.

***samius.*** *Camponotus rubripes* r. *samius* Forel, 1889: 262 (s.w.) GREECE. Combination in *C.* (*Tanaemyrmex*): Emery, 1925b: 100. Raised to species: Dalla Torre, 1893: 251. Subspecies of *maculatus:* Emery, 1896d: 370; Emery, 1905d: 35; Forel, 1906c: 187. Revived status as species: Emery, 1920c: 6; Emery, 1925b: 100; Finzi, 1930d: 318. Senior synonym of *spagnolinii,* and material of the unavailable name *bosphoricus* referred here: Atanassov & Dlussky, 1992: 219..

*samoaensis.* *Camponotus* (*Myrmoturba*) *maculatus* st. *pallidus* var. *samoaensis* Santschi, 1919a: 326 (w.) SAMOA. Unavailable name; material referred to *chloroticus* by Wilson & Taylor, 1967: 93.

*sanctaecrucis.* *Camponotus* (*Myrmoturba*) *maculatus* subsp. *sanctaecrucis* Mann, 1919: 369, fig. 42 (s.) SOLOMON IS. Combination in *C.* (*Tanaemyrmex*): Emery, 1925b: 94. Subspecies of *irritans:* Emery, 1920c: 7. Junior synonym of *chloroticus:* Wilson & Taylor, 1967: 93.

***sanctaefidei.*** *Camponotus sanctaefidei* Dalla Torre, 1892: 92. Replacement name for *quadrilaterus* Mayr, 1887: 519. [Junior primary homonym of *quadrilaterus* Roger, 1863a: 136.] Wheeler, W.M. 1934e: 228 (s.w.q.). Combination in *C.* (*Myrmocladoecus*): Emery, 1925b: 167. Current subspecies: nominal plus *convexinodus, coronatus, hondurianus, leonhardi, weberi.*

*sanctoides.* *Camponotus maculatus* subsp. *thoracicus* var. *sanctoides* Forel, 1910a: 13 (w.q.) ISRAEL. Unavailable name.

***sanctus.*** *Camponotus maculatus* r. *sanctus* Forel, 1904c: 18 (s.w.q.m.) ISRAEL. Combination in *C.* (*Tanaemyrmex*): Emery, 1925b: 98. Subspecies of *compressus:* Emery, 1920c: 6; Santschi, 1921a: 116. Raised to species: Collingwood, 1961a: 76; Agosti & Collingwood, 1987a: 59.

*sanguinea.* *Camponotus* (*Camponotus*) *japonicus* var. *sanguinea* Karavaiev, 1929b: 212 (w.q.) RUSSIA. Junior synonym of *japonicus:* Yasumatsu & Brown, 1951: 36.

*sanguinea.* *Notostigma sanguinea* Clark, 1930a: 116, fig. 1 (s.w.) AUSTRALIA. [Junior secondary homonym of *sanguinea* Karavaiev, above.] Replacement name: *johnclarki* Taylor, 1992a: 59.

*sanguinea.* *Camponotus* (*Myrmogonia*) *sanguinea* McAreavey, 1949: 18, figs. 50–52 (s.w.) AUSTRALIA. [Junior primary homonym of *sanguinea* Karavaiev, above.] Replacement name: *macareaveyi* Taylor, 1992a: 60.

**#*sanguiniceps.*** *Camponotus* (*Orthonotomyrmex*) *sericeus* var. *sanguiniceps* Donisthorpe, 1942d: 458 (w.) BURMA [Publication date (30).vi.1942.] [Also described as new by Donisthorpe, 1942g: 250; publication date 2.xi.1942.]

***sanguinifrons.*** *Camponotus* (*Colobopsis*) *sanguinifrons* Viehmeyer, 1925b: 143 (s.w.) AUSTRALIA.

**#*sanguinolentus.*** *Camponotus angusticollis* var. *sanguinolentus* Forel, 1895e: 454 (w.) INDIA. Combination in *C.* (*Tanaemyrmex*): Emery, 1925b: 89.

***sankisianus.*** *Camponotus* (*Orthonotomyrmex*) *mayri* r. *sankisianus* Forel, 1913b: 347 (w.) ZAIRE. Arnold, 1946: 65 (s.). Raised to species: Arnold, 1946: 65.

***sansabeanus.*** *Formica sansabeana* Buckley, 1866: 167 (s.w.q.m.) U.S.A. Wheeler, G.C. & Wheeler, J. 1953e: 196 (l.). Combination in *Camponotus:* Mayr, 1886c: 365; in *C.* (*Myrmoturba*): Forel, 1914a: 267; in *C.* (*Camponotus*): Emery, 1920b: 255; in *C.* (*Tanaemyrmex*): Creighton, 1950a: 378. Junior synonym of *marginatus:* Mayr, 1886c: 423. Revived from synonymy as variety of *marginatus:* Dalla Torre, 1893: 242. Subspecies of *maculatus:* Emery, 1896d: 371; Wheeler, W.M. 1910d: 307. Revived status as species: Emery, 1920b: 232 (footnote); Cole, 1942: 387; Creighton, 1950a: 378. Senior synonym of *bulimosus* and *torrefactus:* Wheeler, G.C. & Wheeler, J. 1986g: 14.

**#*sansalvadorensis.*** *Camponotus* (*Myrmocladoecus*) *planus* var. *sansalvadorensis* Wheeler, W.M. 1924a: 119, fig. 27 (w.m.) GALAPAGOS IS.

**#*santacruzensis.*** *Camponotus* (*Myrmorhachis*) *planus* var. *santacruzensis* Wheeler, W.M. 1919c: 294 (s.w.q.m.) GALAPAGOS IS. Combination in *C.* (*Myrmocladoecus*): Wheeler, W.M. 1924a: 119.

*santacruziana.* *Camponotus distinguendus* var. *santacruziana* Emery, 1906c: 191 (s.w.) ARGENTINA. Combination in *C.* (*Myrmamblys*): Bruch, 1914: 228; in *C.* (*Myrmoserícus*): Santschi, 1920d: 386; in *C.* (*Tanaemyrmex*): Emery, 1925b: 78. Junior synonym of *distinguendus:* Kusnezov, 1960b: 397.

***santosi.*** *Camponotus santosi* Forel, 1908c: 408 (s.w.) CUBA. Wheeler, W.M. 1913b: 502 (q.); Wheeler, G.C. & Wheeler, J. 1968: 220 (l.). Combination in *C.* (*Myrmoturba*): Forel, 1914a: 267; in *C.* (*Tanaemyrmex*): Emery, 1925b: 82. Current subspecies: nominal plus *pazosi.*

**#*santschii.*** *Camponotus mucronatus* r. *santschi* Forel, 1899c: 145 (w.) COLOMBIA. Combination in *C.* (*Myrmocladoecus*): Emery, 1925b: 167.

**#*sapphirinus.*** *Camponotus* (*Pseudocolobopsis*) *macilentus* var. *sapphirinus* Wheeler, W.M. 1924a: 114, fig. 24 (s.w.q.m.) GALAPAGOS IS.

*sarmentus.* *Camponotus* (*Myrmoturba*) *maculatus* subsp. *sarmentus* Emery, 1920c: 14, fig. 2 (s.w.) TANZANIA. Combination in *C.* (*Tanaemyrmex*): Emery, 1925b: 88. Junior synonym of *maculatus:* Baroni Urbani, 1972: 125.

***satan.*** *Camponotus* (*Myrmotarsus*) *satan* Wheeler, W.M. 1919e: 146 (q.) BORNEO.

**#*satrapus.*** *Camponotus* (*Myrmepomis*) *sericeiventris* var. *satrapa* Wheeler, W.M. 1931c: 92 (s.w.q.) PANAMA.

***saundersi.*** *Camponotus* (*Colobopsis*) *saundersi* Emery, 1889b: 516 (s.w.) BURMA. Forel, 1893b: 438 (q.). Junior synonym of *badius:* Emery, 1896d: 375. Revived from synonymy: Bingham, 1903: 346. Current subspecies: nominal plus *krama.*

***saussurei.*** *Camponotus saussurei* Forel, 1879a: 103, pl. 1, fig. 3 (w.) ANTILLES. Combination in *C.* (*Myrmeurynota*): Forel, 1912i: 92.

***saussurei.** **Camponotus saussurei* Théobald, 1937b: 316, pl. 25, fig. 15 (m.) FRANCE (Oligocene). [Unresolved junior primary homonym of *saussurei* Forel, above.]

*saxatileaterrimmus.* *Camponotus herculeanus* subsp. *saxatilis* var. *saxatileaterrimmus* Ruzsky, 1914b: 101 (w.) RUSSIA. Unavailable name.

***saxatilis.*** *Camponotus herculeanus* subsp. *saxatilis* Ruzsky, 1895: 7 (w.q.m.) RUSSIA. Arnol'di, 1967: 1825 (s.). Combination in *C.* (*Camponotus*): Emery, 1925b: 73. Subspecies of *pennsylvanicus:* Ruzsky, 1905b: 228; of *herculeanus:* Ruzsky, 1925b: 42. Junior synonym of *japonicus:* Yasumatsu & Brown, 1951: 36. Revived from synonymy: Arnol'di, 1967: 1825; Kupyanskaya, 1990: 167.

***sayi.*** *Camponotus sayi* Emery, 1893i: 679, pl. 22, figs. 27, 28 (s.w.) U.S.A. Wheeler, W.M. 1910d: 343 (s.w.). Combination in *C.* (*Camponotus*): Forel, 1914a: 266; in *C.* (*Myrmentoma*): Emery, 1920b: 257. See also: Snelling, 1988: 72.

***scabrinodis.*** *Camponotus* (*Orthonotomyrmex*) *scabrinodis* Arnold, 1924: 717, fig. 59 (w.) ZIMBABWE.

***scalaris.*** *Camponotus scalaris* Forel, 1901d: 308 (s.w.q.) SOUTH AFRICA. Combination in *C.* (*Myrmosphincta*): Forel, 1912i: 92; in *C.* (*Orthonotomyrmex*): Emery, 1920b: 258; in *C.* (*Myrmisolepis*): Santschi, 1921f: 310; in *C.* (*Myrmopiromis*): Emery, 1925b: 129.

***schaefferi.*** *Camponotus schaefferi* Wheeler, W.M. 1909e: 88 (w.q.) U.S.A. Wheeler, W.M. 1910d: 344 (s.). Combination in *C.* (*Myrmentoma*): Emery, 1920b: 257; in *C.* (*Camponotus*): Forel, 1914a: 266; Creighton, 1950a: 371.

*scheffleri.* *Camponotus maculatus* subsp. *guatemalensis* var. *scheffleri* Forel, 1908c: 407 (w.m.) BRAZIL. Unavailable name.

*schencki.* *Camponotus schencki* Mayr, 1862: 674 (w.) AUSTRALIA. Junior synonym of *suffusus:* Dalla Torre, 1893: 254; Emery, 1896d: 373.

*schereri.* *Camponotus maculatus* subsp. *schereri* Forel, 1911e: 289 (w.q.) LIBERIA. Combination in *C.* (*Myrmoturba*): Santschi, 1915c: 280; in *C.* (*Tanaemyrmex*): Emery, 1925b: 88. Junior synonym of *maculatus:* Baroni Urbani, 1972: 125.

**#*schmalzi.*** *Camponotus fastigatus* var. *schmalzi* Emery, 1894g: 376 (footnote) (w.) BRAZIL. Combination in *C.* (*Myrmaphaenus*): Emery, 1925b: 155. Raised to species: Emery, 1903: 73. Subspecies of *fastigatus:* Forel, 1908c: 404; Emery, 1925b: 155.

***schmeltzi.*** *Camponotus schmeltzi* Mayr, 1866a: 490 (w.) FIJI IS. Mann, 1921: 484 (s.). Combination in *C.* (*Myrmogonia*): Forel, 1912i: 92. Current subspecies: nominal plus *kadi, loloma, trotteri.*

***schmitzi.*** *Camponotus* (*Colobopsis*) *schmitzi* Stärcke, 1933a: 29, figs. 1–6 (s.w.q.l.) BORNEO.

***schneei.*** *Camponotus schneei* Mayr, 1903: 401 (s.w.q.) MARSHALL IS. Combiation in *C.* (*Myrmamblys*): Forel, 1914a: 272.

***schoutedeni.*** *Camponotus schoutedeni* Forel, 1911f: 281 (w.) ZAIRE. Forel, 1913a: 146 (s.). Combination in *C.* (*Myrmosaga*): Forel, 1912i: 92; in *C.* (*Myrmamblys*): Wheeler, W.M. 1922a: 990; in *C.* (*Mayria*): Emery, 1925b: 123.

*schultzei.* *Camponotus maculatus* var. *schultzei* Forel, 1912j: 179 (s.w.m.) UGANDA. Junior synonym of *melanocnemis:* Forel, 1914d: 250.

*schweinfurti*. *Camponotus maculatus* subsp. *negus* var. *schweinfurti* Forel, 1910e: 453 (s.w.) SAUDI ARABIA. Unavailable name.

#***scintillans.*** *Camponotus pellitus* r. *scintillans* Forel, 1901h: 72 (w.) PARAGUAY. Combination in *C.* (*Myrmaphaenus*): Emery, 1925b: 154. Stirps of *blandus:* Santschi, 1922c: 111.

***scipio.*** *Camponotus scipio* Forel, 1908c: 410 (s.w.q.m.) PARAGUAY. Combination in *C.* (*Myrmobrachys*): Forel, 1914a: 271. Current subspecies: nominal plus *insignis.*

***scissus.*** *Camponotus scissus* Mayr, 1887: 518 (w.) BRAZIL. Combination in *C.* (*Colobopsis*): Forel, 1914a: 272; in *C.* (*Myrmorhachis*): Emery, 1920b: 260; in *C.* (*Myrmobrachys*): Emery, 1925b: 165.

#***scortechinii.*** *Camponotus contractus* var. *scortechinii* Emery, 1888a: 529 (footnote) (w.) WEST MALAYSIA. Combination in *C.* (*Myrmoplatys*): Emery, 1925b: 136.

***scratius.*** *Camponotus scratius* Forel, 1907h: 304 (s.w.q.) AUSTRALIA. Combination in *C.* (*Myrmophyma*): Forel, 1914a: 269; in *C.* (*Thlipsepinotus*): Santschi, 1928e: 483. Current subspecies: nominal plus *nuntius.*

*scrobifer*. *Camponotus* (*Neocolobopsis*) *scrobifer* Borgmeier, 1928b: 66, figs. 2–4 (s.w.) BRAZIL. Junior synonym of *coriolanus:* Kempf, 1968b: 410.

#***sculnus.*** *Camponotus urichi* var. *sculna* Forel, 1904c: 47 (s.w.m.) BRAZIL. Combination in *C.* (*Myrmosphincta*): Emery, 1925b: 152.

***sculptor.*** *Camponotus* (*Myrmamblys*) *sculptor* Santschi, 1920d: 386 (s.w.) ARGENTINA. Combination in *C.* (*Neomyrmamblys*): Santschi, 1921f: 311; in *C.* (*Myrmaphaenus*): Emery, 1925b: 156.

*scutellus*. *Camponotus scutellus* Clark, 1930a: 123, fig. 1 (s.w.) AUSTRALIA. Junior synonym of *whitei:* Brown, 1956a: 40.

#***securifer.*** *Camponotus* (*Orthonotomyrmex*) *yvonnae* subsp. *securifer* Emery, 1920c: 28 (s.w.q.) CAMEROUN. Combination in *C.* (*Myrmamblys*): Emery, 1925b: 142. Subspecies of *chrysurus:* current status.

***sedulus.*** *Formica sedula* Smith, F. 1857a: 56 (w.) BORNEO. Mayr, 1872: 137 (q.). Combination in *Camponotus:* Roger, 1863b: 3; in *C.* (*Myrmotarsus*): Wheeler, W.M. 1919e: 110. Junior synonym of *irritabilis:* Mayr, 1879: 648; Dalla Torre, 1893: 236. Revived from synonymy as variety of *irritabilis:* Forel, 1911a: 51; Wheeler, W.M. 1919e: 110; Emery, 1925b: 135. Revived status as species: Donisthorpe, 1932c: 444.

#***seemanni.*** *Camponotus* (*Colobopsis*) *maudella* var. *seemanni* Mann, 1921: 498 (s.w.q.) FIJI IS.

***selene.*** *Polyrhachis selene* Emery, 1889b: 518, pl. 11, figs. 18, 19 (w.) BURMA. Combination in *Camponotus:* Emery, 1896d: 376; in *C.* (*Myrmorhachis*): Forel, 1912i: 92; in *C.* (*Myrmacantha*): Emery, 1920b: 258; in *C.* (*Orthonotomyrmex*): Santschi, 1921f: 310. Current subspecies: nominal plus *obtusatus.*

#***sellaris.*** *Camponotus quadrimaculatus* subsp. *sellaris* Emery, 1895f: 344, fig. 3 (w.) MADAGASCAR. Combination in *C.* (*Mayria*): Emery, 1925b: 123.

***sellidorsatus.*** *Camponotus* (*Mayria*) *sellidorsatus* Prins, 1973: 21, figs. 25, 26, 34 (w.) SOUTH AFRICA.

***semicarinatus.*** *Colobopsis rufifrons* r. *semicarinata* Forel, 1895f: 418 (s.) AUSTRALIA. Combination in *Camponotus* (*Colobopsis*), and raised to species: Forel, 1902h: 508.

*semiceps*. *Camponotus* (*Myrmaphaenus*) *blandus* st. *pellitus* var. *semiceps* Santschi, 1922c: 111 (w.) COSTA RICA. Unavailable name.

#***semidepilis.*** *Camponotus* (*Orthonotomyrmex*) *vividus* var. *semidepilis* Wheeler, W.M. 1922a: 248 (s.w.) ZAIRE. Combination in *C.* (*Myrmamblys*): Emery, 1925b: 142; in *C.* (*Myrmopelta*): Santschi, 1926a: 20.

*semipunctata*. *Formica semipunctata* Kirby, W. 1837: 262 (q.) U.S.A. Combination in *Camponotus:* Roger, 1863b: 43. Junior synonym of *pennsylvanicus:* Mayr, 1886d: 420. See also: Creighton, 1950a: 367, who associates this name with *herculeanus.*

*semirex*. *Camponotus* (*Myrmepomis*) *sericeiventris* subsp. *rex* var. *semirex* Wheeler, W.M. 1931c: 96 (w.) HONDURAS. Unavailable name.

***semirufus.*** *Camponotus* (*Myrmentoma*) *interjectus* subsp. *semirufus* Emery, 1925a: 65, fig. 2 (w.) TURKESTAN. [First available use of *Camponotus lateralis* subsp. *interjectus* var. *semirufus* Kuznetsov-Ugamsky, 1923: 243; unavailable name.] Raised to species: Tarbinsky, 1976: 163; Dlussky & Zabelin, 1985: 237.

*semispicatus*. *Camponotus* (*Myrmoturba*) *maculatus* var. *semispicatus* Emery, 1920c: 5 (w.) KENYA. [First available use of *Camponotus* (*Myrmoturba*) *maculatus* st. *mathildae* var. *semispicata* Santschi, 1914b: 130; unavailable name.] Menozzi, 1930b: 116 (q.). Combination in *C.* (*Tanaemyrmex*): Emery, 1925b: 87. Variety of *maculatus:* Wheeler, W.M. 1922a: 957. Junior synonym of *maculatus:* Baroni Urbani, 1972: 125.

***semitestaceus.*** *Camponotus* (*Tanaemyrmex*) *semitestaceus* Snelling, 1970: 396 (w.) U.S.A. [First available use of *Camponotus maculatus* subsp. *vicinus* var. *semitestaceus* Emery, 1893i: 672; unavailable name.]

***semoni.*** *Camponotus semoni* Forel, 1905b: 160 (s.w.) VENEZUELA. Combination in *C.* (*Myrmoturba*): Forel, 1914a: 267; in *C.* (*Pseudocolobopsis*): Emery, 1925b: 158.

***senex.*** *Formica senex* Smith, F. 1858b: 47 (w.q.) BRAZIL. Mayr, 1878: 868 (m.); Forel, 1879a: 96 (s.); Wheeler, G.C. & Wheeler, J. 1953e: 194 (l.). Combination in *Camponotus:* Mayr, 1862: 676; in *C.* (*Myrmobrachys*): Forel, 1912i: 91. Senior synonym of *tomentosa:* Emery, 1892b: 167. Current subspecies: nominal plus *textor.*

*senilis*. *Camponotus senilis* Mayr, 1862: 675, pl. 19, fig. 3 (w.) BORNEO. Junior synonym of *singularis:* Mayr, 1867a: 39; of *camelinus:* Forel, 1892j: 232.

*septentrionalis*. *Formica septentrionale* Buckley, 1866: 161 (w.q.) U.S.A. Combination in *Camponotus:* Emery, 1895c: 338; Wheeler, W.M. 1902f: 21. Unrecognisable taxon; *incertae sedis* in *Camponotus.*

***sericatus.*** *Camponotus sericatus* Mayr, 1887: 515 (w.) BRAZIL. Combination in *C.* (*Myrmothrix*): Forel, 1914a: 269. See also: Kempf, 1960e: 385; Hashmi, 1973: 79.

***sericeiventris.*** *Formica sericeiventris* Guérin-Méneville, 1838: 205 (w.) BRAZIL. Mayr, 1862: 667 (q.). Combination in *Camponotus:* Mayr, 1862: 667; in *C.* (*Myrmolophus*): Emery, 1920b: 256; in *C.* (*Myrmepomis*): Forel, 1912i: 92; Emery, 1925b: 171. Senior synonym of *cuneata:* Smith, F. 1858b: 46. Current subspecies: nominal plus *holmgreni, otoquensis, pontifex, rex, satrapus.*

#***sericellus.*** *Camponotus* (*Myrmamblys*) *bedoti* subsp. *sericellus* Viehmeyer, 1916a: 158 (s.w.q.) SINGAPORE. Race of *reticulatus:* Crawley, 1923a: 30.

*sericeoopaciventris*. *Camponotus sericeus* var. *sericeoopaciventris* Forel, 1895e: 455. *Nomen nudum.*

***sericeus.*** *Formica sericea* Fabricius, 1798: 279 (w.) SENEGAL. Mayr, 1866b: 886 (q.); Forel, 1886f: 192 (m.); Forel, 1891b: 56 (s.w.q.m.); Imai, *et al.* 1984: 9 (k.). Combination in *Camponotus:* Mayr, 1862: 675; in *Orthonotomyrmex:* Ashmead, 1906: 31; in *C.* (*Myrmentoma*): Forel, 1912i: 92; in *C.* (*Orthonotomyrmex*): Forel, 1913k: 129. Senior synonym of *aurulenta* and *obtusa:* Roger, 1863b: 2; of *pyrrhocephala:* Emery, 1893f: 254. Current subspecies: nominal plus *euchrous, opaciventris, peguensis, sanguiniceps, sulgeri.*

*serotina*. *Camponotus sylvaticus* subsp. *barbarica* var. *serotina* Menozzi, 1922a: 355 (s.q.) MOROCCO. Unavailable name.

***sesquipedalis.*** *Camponotus sesquipedalis* Roger, 1863a: 137 (w.) SRI LANKA. Combination in *C.* (*Dinomyrmex*): Forel, 1914a: 268; in *C.* (*Tanaemyrmex*): Emery, 1925b: 90.

#***setipes.*** *Camponotus rectangularis* var. *setipes* Forel, 1904c: 51 (w.) TRINIDAD. Combination in *C.* (*Myrmocladoecus*): Emery, 1925b: 167.

***setitibia.*** *Camponotus maculatus* r. *setitibia* Forel, 1901h: 68 (s.w.) BORNEO. Combination in *C.* (*Tanaemyrmex*): Emery, 1925b: 95. Raised to species: Emery, 1920c: 8.

*seurati*. *Camponotus alii* var. *seurati* Emery, 1920c: 7 (s.) ALGERIA. [First available use of *Camponotus* (*Myrmoturba*) *maculatus* st. *alii* var. *seurati* Santschi, 1915a: 62; unavailable name.] Combination in *C.* (*Tanaemyrmex*): Emery, 1925b: 91. Raised to species: Santschi, 1929c: 104. Junior synonym of *alii:* Bernard, 1953a: 187.

***severini.*** *Camponotus* (*Colobopsis*) *severini* Forel, 1909b: 55 (s.w.q.m.) BORNEO.

***sexguttatus.*** *Formica sexguttata* Fabricius, 1793: 354 (q.) ANTILLES. Forel, 1879a: 71 (s.w.); Wheeler, G.C. & Wheeler, J. 1953e: 195 (l.). Combination in *Camponotus:* Mayr, 1862: 656; in *C.* (*Myrmosphincta*): Forel, 1912i: 92; in *C.* (*Myrmothrix*): Bruch, 1914: 230; in *C.* (*Myrmotemnus*): Emery, 1920b: 259; in *C.* (*Myrmosphincta*): Emery, 1925b: 151. Variety of *sylvaticus:* Mayr, 1880: 23. Revived status as species: Forel, 1885a: 346; Forel, 1895b: 104; Emery, 1896d: 375. Senior synonym of *ruficeps* (and its junior synonym *albofasciata*): Emery, 1894f: 2; Emery, 1896d: 375; of *bimaculata:* Emery, 1925b: 152; Kempf, 1972a: 58. Current subspecies: nominal plus *albotaeniolatus, antiguanus, basirectus, biguttatus, decorus, fusciceps, montserratensis, ornatus, perturbans, unitaeniatus.*

***sexpuctatus.*** *Camponotus sexpunctatus* Forel, 1894b: 66 (s.w.q.) MOZAMBIQUE. Combination in *C.* (*Dinomyrmex*): Forel, 1914a: 268; in *C.* (*Tanaemyrmex*): Emery, 1925b: 89. Subspecies of *maculatus:* Emery, 1896d: 371; Arnold, 1922: 633.

*CAMPONOTUS (continued)*

Revived status as species: Emery, 1920b: 256; Emery, 1925b: 89.

***shanwangensis.** **Camponotus shanwangense* Hong, 1984: 9, pl. 1, fig. 5; pl. 3, fig. 4 (q.) CHINA (Miocene).

**#shaqualavensis.** *Camponotus (Tanaemyrmex) jaliensis* subsp. *shaqualavensis* Pisarski, 1971a: 672, figs. 5–9 (s.w.) IRAQ.

**#sharpi.** *Camponotus sharpi* Forel, 1893g: 335 (s.w.q.m.) ANTILLES. Combination in *C. (Tanaemyrmex)*: Emery, 1925b: 80. Race of *abdominalis:* Forel, 1899c: 133; of *conspicuus:* Santschi, 1922c: 101.

*shitkowi. Camponotus herculeanus* var. *shitkowi* Ruzsky, 1904a: 292 (w.q.) RUSSIA. Junior synonym of *herculeanus:* Emery, 1925b: 72; Karavaiev, 1936: 178; Arnol'di, 1967: 1819.

**sibreei.** *Camponotus sibreei* Forel, 1891b: 53, pl. 1, fig. 9 (w.) MADAGASCAR. Combination in *C. (Myrmobrachys)*: Forel, 1914a: 271; in *C. (Orthonotomyrmex)*: Emery, 1920b: 258; in *C. (Myrmepinotus)*: Emery, 1925b: 127.

**sicheli.** *Camponotus sicheli* Mayr, 1866a: 488, pl., fig. 2 (w.) ALGERIA. Forel, 1890a: lxvi (s.q.m.). Combination in *C. (Orthonotomyrmex)*: Forel, 1914a: 273; in *C. (Myrmentoma)*: Emery, 1920b: 257. Subspecies of *lateralis:* Karavaiev, 1912a: 21; of *piceus:* Emery, 1925a: 63. Revived status as species: Santschi, 1919e: 248; Emery, 1925b: 121; Santschi, 1925g: 359; Santschi, 1931a: 13. Senior synonym of *rubra:* Collingwood, 1978: 72. Current subspecies: nominal plus *niger.*

**#siculus.** *Camponotus sylvaticus* var. *sicula* Grandi, 1935: 102 (s.w.q.) SICILY. [First available use of *Camponotus maculatus* subsp. *pilicornis* var. *sicula* Emery, 1908a: 198; unavailable name.] Subspecies of *pilicornis:* Baroni Urbani, 1971c: 186.

**siemsseni.** *Camponotus siemsseni* Forel, 1901h: 70 (s.q.) SUMATRA. Forel, 1912a: 76 (w.); Santschi, 1937h: 378 (m.). Combination in *C. (Myrmoturba)*: Forel, 1913k: 129; in *C. (Tanaemyrmex)*: Emery, 1925b: 102. See also: Wang, C., Xiao & Wu, 1989a: 223.

*siggii. Camponotus (Colobopsis) angustata* var. *siggii* Forel, 1893b: 436 (s.w.) THAILAND. Junior synonym of *vitreus:* Forel, 1895e: 455.

**sikorai.** *Camponotus (Myrmonesites) sikorae* Emery, 1920b: 242, fig. (w.) MADAGASCAR.

**silvestrii.** *Camponotus silvestrii* Emery, 1906c: 193, fig. 47 (s.w.q.) BRAZIL. Combination in *C. (Myrmamblys)*: Forel, 1914a: 272; in *C. (Paracolobopsis)*: Emery, 1920b: 259; in *C. (Myrmaphaenus)*: Emery, 1925b: 154.

**silvicola.** *Camponotus silvicola* Forel, 1902b: 174 (s.w.) COLOMBIA. Combination in *C. (Myrmoturba)*: Forel, 1914a: 267; in *C. (Tanaemyrmex)*: Emery, 1925b: 84. Race of *fumidus:* Santschi, 1922c: 101. Revived status as species: Emery, 1925b: 84.

**#simaluranus.** *Camponotus (Dinomyrmex) inezae* var. *simalurana* Forel, 1915a: 37 (w.q.) SUMATRA. Subspecies of *festinus:* current status.

*similis* Mayr, 1876; see under ***CALOMYRMEX.***

*similis. Camponotus (Tanaemyrmex) similis* Donisthorpe, 1933b: 198 (s.w.) BURMA. [Junior primary homonym of *similis* Mayr, 1876: 61, above.] Replacement name: *sklarus* **new name**.

**simillimus.** *Formica simillima* Smith, F. 1862b: 30 (q.) PANAMA. Forel, 1886f: 152 (s.w.). Combination in *Camponotus:* Mayr, 1870a: 374; in *C. (Myrmoturba)*: Santschi, 1922c: 97; in *C. (Tanaemyrmex)*: Emery, 1925b: 82. Variety of *sylvaticus:* Mayr, 1880: 23. Race of *rubripes:* Forel, 1886f: 143; of *maculatus:* Emery, 1890b: 56; Forel, 1912i: 70; of *picipes:* Emery, 1920b: 233. Status as species: Dalla Torre, 1893: 253; Emery, 1894g: 372; Santschi, 1922c: 97; Emery, 1925b: 82. Current subspecies: nominal plus *atratior, impatibilis, indianus, riograndensis.*

**simoni.** *Camponotus simoni* Emery, 1893f: 250 (s.w.) SRI LANKA. Combination in *C. (Tanaemyrmex)*: Emery, 1925b: 90. Race of *dorycus:* Forel, 1902d: 288; Forel, 1908a: 6. Revived status as species: Emery, 1925b: 90.

**simulans.** *Camponotus simulans* Forel, 1910f: 28, pl. 1, fig. 4 (w.) NAMIBIA. Arnold, 1944: 17 (s.). Combination in *C. (Myrmopsamma)*: Forel, 1914a: 270.

**simulator.** *Camponotus (Dinomyrmex) simulator* Forel, 1915b: 96 (s.w.) AUSTRALIA. Combination in *C. (Tanaemyrmex)*: Emery, 1925b: 90.

**simus.** *Camponotus simus* Emery, 1908b: 188, fig. 2 (s.w.q.) ZAIRE. Combination in *C. (Myrmamblys)*: Forel, 1914a: 272; Wheeler, W.M. 1922a: 990; Emery, 1925b: 141; in *C. (Myrmotemnus)*: Emery, 1920b: 258; in *C. (Paramyrmamblys)*: Santschi, 1926c: 604. Current subspecies: nominal plus *manidis.*

**singularis.** *Formica singularis* Smith, F. 1858b: 27 (w.) JAVA. Combination in *Camponotus:* Roger, 1863b: 3; in *C. (Myrmosphincta)*: Wheeler, W.M. 1919e: 112; in *C. (Myrmosaulus)*: Emery, 1925b: 113. Junior synonym of *cinerascens:* Dalla Torre, 1893: 224; Emery, 1896d: 373; of *camelinus:* Bingham, 1903: 373. Revived from synonymy as race of *camelinus:* Forel, 1905c: 26; Forel, 1907e: 19; Forel, 1909d: 229; Wheeler, W.M. 1919e: 112; Santschi, 1920h: 174. Revived status as species: Emery, 1925b: 113. Senior synonym of *birmana:* Brown, 1967a: 277. Current subspecies: nominal plus *rufomaculatus.*

**#skaifei.** *Camponotus (Tanaemyrmex) werthi* var. *skaifei* Arnold, 1959: 336, fig. 22 (s.w.) SOUTH AFRICA.

**sklarus.** *Camponotus sklarus* **new name**. Replacement name for *similis* Donisthorpe, 1933b: 198. [Junior primary homonym of *similis* Mayr, 1876: 61.]

*smithi. Camponotus (Myrmotarsus) smithi* Donisthorpe, 1932c: 443. Unnecessary (second) replacement name for *ruficeps* Smith, F. 1857a: 54 (junior homonym), and hence junior synonym of *gilviceps* Roger, 1863b: 3, the first replacement name.

**smithianus.** *Camponotus (Colobopsis) smithianus* Wheeler, W.M. 1919e: 119 (s.w.) BORNEO.

**snellingi.** *Camponotus snellingi* **new name**. Replacement name for *pavidus* Snelling, 1988: 71. [Junior secondary homonym of *pavida* Smith, F. 1860b: 96.]

**socius.** *Camponotus socius* Roger, 1863a: 140 (w.) BRAZIL. Forel, 1879a: 74 (s.); Mayr, 1886d: 422 (q.); Wheeler, W.M. 1910d: 319 (m.). Combination in *C. (Myrmothrix)*: Forel, 1914a: 269; in *C. (Tanaemyrmex)*: Emery, 1925b: 82. Senior synonym of *osceola:* Creighton, 1950a: 379.

**socorroensis.** *Camponotus (Myrmobrachys) socorroensis* Wheeler, W.M. 1934f: 142 (s.w.m.) MEXICO.

**socrates.** *Camponotus socrates* Forel, 1904c: 27 (s.w.) INDIA. Combination in *C. (Myrmophyma)*: Forel, 1912i: 91; in *C. (Myrmoturba)*: Forel, 1914a: 267; in *C. (Tanaemyrmex)*: Emery, 1925b: 100.

**solenobius.** *Camponotus (Colobopsis) solenobius* Menozzi, 1926e: 97 (q.) PHILIPPINES.

**solon.** *Camponotus rubripes* r. *solon* Forel, 1886f: 152 (s.w.) ANGOLA. Emery, 1899e: 500 (q.). Combination in *C. (Myrmoturba)*: Forel, 1913h: 355; in *C. (Tanaemyrmex)*: Emery, 1925b: 100. Raised to species: Dalla Torre, 1893: 253. Subspecies of *maculatus:* Emery, 1896d: 371; Wheeler, W.M. 1922a: 235. Status as species: Emery, 1920b: 255; Emery, 1920c: 6; Emery, 1925b: 100; Santschi, 1935a: 279. Current subspecies: nominal plus *chiton, jugurtha.*

**somalinus.** *Camponotus somalinus* André, 1887: 280 (w.) SOMALIA. Menozzi, 1930b: 117 (m.). Combination in *C. (Myrmoturba)*: Forel, 1914a: 267; in *C. (Tanaemyrmex)*: Emery, 1925b: 100. Current subspecies: nominal plus *curtior, pattensis.*

**sommeri.** *Colobopsis sommeri* Forel, 1894e: 230 (s.w.q.m.) NEW CALEDONIA. Combination in *Camponotus (Colobopsis)*: Forel, 1914a: 272.

**#somnificus.** *Camponotus maculatus* r. *somnificus* Forel, 1902d: 287 (w.) INDIA. Forel, 1908a: 6 (q.m.). Combination in *C. (Myrmoturba)*: Forel, 1913k: 128; in *C. (Tanaemyrmex)*, and subspecies of *variegatus:* Emery, 1925b: 96.

**#sordidatus.** *Camponotus (Myrmocladoecus) rectangularis* var. *sordidatus* Wheeler, W.M. 1934g: 221 (w.) PERU.

**#sorptus.** *Camponotus (Myrmotrema) olivieri* var. *sorpta* Santschi, 1915c: 270 (s.w.q.m.) ZAIRE. [First available use of *Camponotus foraminosus* subsp. *delagoensis* var. *sorpta* Forel, 1909b: 62; unavailable name.]

**#soulouquei.** *Camponotus maculatus* r. *soulouquei* Forel, 1901h: 68 (s.w.q.) HAITI. Combination in *C. (Tanaemyrmex)*, and subspecies of *fumidus:* Emery, 1925b: 81.

*spagnolinii. Camponotus samius* var. *spagnolinii* Emery, 1920c: 6 (w.) TURKEY. [First available use of *Camponotus maculatus* subsp. *samius* var. *spagnolinii* Emery, 1905d: 36; unavailable name.] Junior synonym of *samius:* Atanassov & Dlussky, 1992: 219.

*spahis. Camponotus (Myrmoturba) compressus* st. *thoracicus* var. *spahis* Santschi, 1925g: 357 (s.q.m.) ALGERIA. Unavailable name.

**spanis.** *Camponotus spanis* Xiao & Wang; in Wang, C., Xiao & Wu 1989a: 224, fig. 26 (w.) CHINA.

**spenceri.** *Camponotus (Tanaemyrmex) spenceri* Clark, 1930c: 18. Replacement name for *reticulatus* Kirby, W.F. 1896: 204. [Junior primary homonym of *reticulatus* Roger, 1863a: 139.]

**#spengleri.** *Camponotus maculatus* subsp. *spengleri* Forel, 1908c: 406 (s.w.) BRAZIL. Combination in *C. (Tanaemyrmex)*: Emery, 1925b: 82. Subspecies of *picipes:* Emery, 1920b: 233.

**#sphaeralis.** *Camponotus sphaeralis* Roger, 1863a: 147 (w.) CUBA. Mann, 1920: 438 (s.). Combination in *C. (Myrmeurynota)*: Forel, 1914a: 274; in *C. (Colobopsis)*: Mann, 1920: 438; in *C. (Manniella)*: Emery, 1925b: 160. Subspecies of *sphaericus:* Mann, 1920: 438; Wheeler, W.M. 1921a: 19.

***sphaericus.*** *Camponotus sphaericus* Roger, 1863a: 146 (w.) CUBA. Mann, 1920: 436 (s.). Combination in *C.* (*Myrmeurynota*): Forel, 1914a: 274; in *C.* (*Colobopsis*): Mann, 1920: 436; in *C.* (*Manniella*): Wheeler, W.M. 1921a: 19. Current subspecies: nominal plus *cardini, rufipilis, sphaeralis.*

***sphenocephalus.*** *Camponotus sphenocephalus* Emery, 1911a: 225 (q.) PERU. Combination in *C.* (*Myrmamblys*): Forel, 1914a: 272; in *C.* (*Myrmostenus*): Emery, 1920b: 260.

***sphenoidalis.*** *Camponotus sphenoidalis* Mayr, 1870a: 383 (w.) COLOMBIA. Combination in *C.* (*Myrmamblys*): Emery, 1920b: 259; in *C.* (*Myrmobrachys*): Forel, 1914a: 271; Emery, 1925b: 164.

***spinitarsus.*** *Camponotus* (*Dinomyrmex*) *spinitarsus* Emery, 1920c: 22 (s.) AUSTRALIA. Combination in *C.* (*Tanaemyrmex*): Emery, 1925b: 91.

***spinolae.*** *Camponotus spinolae* Roger, 1863a: 144 (w.) CHILE. Combination in *C.* (*Myrmosericus*): Forel, 1914a: 268; in *C.* (*Tanaemyrmex*): Emery, 1925b: 79. Senior synonym of *ruficornis:* Snelling & Hunt, 1976: 122.

#***spissinodis.*** *Camponotus lateralis* r. *spissinodis* Forel, 1909e: 374. Replacement name for *crassinodis* Forel, 1894d: 3. [Junior primary homonym of *crassinodis* Forel, 1892j: 230.] Karavaiev, 1912a: 20 (m.); Santschi, 1929e: 164 (q.). Combination in *C.* (*Orthonotomyrmex*): Forel, 1916: 458; in *C.* (*Myrmentoma*): Emery, 1925a: 63. Subspecies of *lateralis:* Cagniant, 1964: 109; of *piceus:* Emery, 1925a: 63; Santschi, 1929e: 164; Baroni Urbani, 1971c: 195.

*splendidus* Mayr, 1876; see under ***CALOMYRMEX.***

***sponsorum.*** *Camponotus sponsorum* Forel, 1910b: 76 (s.w.m.) AUSTRALIA. Combination in *C.* (*Myrmophyma*): Forel, 1912i: 91.

*spurcus. Camponotus fumidus* var. *spurcus* Wheeler, W.M. 1910d: 315 (s.w.q.) U.S.A. Combination in *C.* (*Tanaemyrmex*): Emery, 1925b: 80. Junior synonym of *festinatus:* Creighton, 1950a: 376.

***staryi.*** *Camponotus* (*Myrmentoma*) *staryi* Pisarski, 1971b: 730, figs. 6–9 (w.) IRAQ.

#***stenonotus.*** *Camponotus* (*Tanaemyrmex*) *variegatus* r. *stenonota* Stitz, 1938: 114 (q.) BORNEO.

#***stenoticus.*** *Camponotus boghossiani* var. *stenotica* Emery, 1915h: 7. Replacement name for *angustatus* Forel, 1889: 261. [Junior secondary homonym of *angustata* Mayr, 1870b: 942.]

*stercorarius. Camponotus atriceps* r. *stercorarius* Forel, 1885a: 340 (w.q.m.) GUATEMALA. Wheeler, G.C. & Wheeler, J. 1953e: 196 (l.). Combination in *C.* (*Myrmothrix*): Crawley, 1916b: 376. Raised to species: Dalla Torre, 1893: 253. Subspecies/race of *abdominalis:* Emery, 1896d: 372; Forel, 1899c: 133; Forel, 1922: 99. Junior synonym of *atriceps:* Hashmi, 1973: 82.

#***stomatus.*** *Camponotus* (*Myrmobrachys*) *canescens* st. *stomatus* Santschi, 1925d: 246 (w.) PANAMA.

***storeatus.*** *Camponotus fulvopilosus* subsp. *storeatus* Forel, 1910f: 26 (w.) SOUTH AFRICA. Arnold, 1924: 739 (s.q.). Combination in *C.* (*Myrmopiromis*): Wheeler, W.M. 1922a: 986. Raised to species: Robertson, 1990: 327.

#***strangulatus.*** *Camponotus maculatus* st. *strangulatus* Santschi, 1911e: 129, fig. 3 (s.w.q.) MADAGASCAR. Senior synonym of *maculatoides:* Emery, 1920c: 6. [Emery gives *strangulatus* as the junior synonym of *maculatoides* but the former name has priority.]

***striatus.*** *Formica striata* Smith, F. 1862b: 30 (w.) PANAMA. Menozzi, 1927d: 340 (q.m.). Combination in *Camponotus:* Emery, 1892b: 167; in *C.* (*Myrmamblys*): Forel, 1914a: 272; in *C.* (*Myrmorhachis*): Emery, 1920b: 260; in *C.* (*Myrmobrachys*): Emery, 1925b: 166. Senior synonym of *alfaroi:* Emery, 1892b: 167; Kempf, 1972a: 54; of *granulatus:* Smith, M.R. 1953b: 211.

***strictus.*** *Formica stricta* Jerdon, 1851: 123 (w.) INDIA. Bingham, 1903: 344 (s.q.). Combination in *Colobopsis:* Mayr, 1863: 403; in *Camponotus:* Dalla Torre, 1893: 253; in *C.* (*Colobopsis*): Forel, 1893b: 438.

#***strophiatus.*** *Camponotus* (*Myrmosericus*) *vestitus* var. *strophiatus* Santschi, 1926b: 259 (w.) SOUTH AFRICA.

***subbarbatus.*** *Camponotus marginatus* subsp. *subbarbatus* Emery, 1893i: 676 (w.q.m.) U.S.A. Wheeler, W.M. 1910f: 229 (s.). Combination in *C.* (*Myrmentoma*): Emery, 1925b: 117. Subspecies of *fallax:* Wheeler, W.M. 1910f: 229; of *caryae:* Wheeler, W.M. 1917c: 28. Raised to species, and material of the unavailable name *paucipilis* referred here: Creighton, 1950a: 390. See also: Snelling, 1988: 73.

#***subcarinatus.*** *Camponotus claviscapus* var. *subcarinatus* Forel, 1899c: 153, pl. 4, fig. 24 (s.) GUATEMALA. Combination in *C.* (*Myrmamblys*): Mann, 1922: 53; in *C.* (*Pseudocolobopsis*): Emery, 1925b: 157.

***subcircularis.*** *Camponotus* (*Myrmobrachys*) *subcircularis* Emery, 1920c: 45, fig. 6 (w.) BOLIVIA.

#***subconcolor.*** *Camponotus* (*Myrmoturba*) *erigens* var. *subconcolor* Emery, 1920c: 6 (s.w.q.) ALGERIA. [First available use of *Camponotus* (*Myrmoturba*) *maculatus* subsp. *erigens* var. *subconcolor* Viehmeyer, 1916b: 289; unavailable name.] Combination in *C.* (*Tanaemyrmex*): Emery, 1925b: 99.

#***subconvexus.*** *Camponotus* (*Myrmotrema*) *diplopunctatus* subsp. *subconvexus* Viehmeyer, 1923: 93, fig. 4 (s.w.) SUDAN.

#***subdepilis.*** *Camponotus* (*Myrmobrachys*) *capperi* subsp. *subdepilis* Wheeler, W.M. 1917g: 470 (s.w.) JAMAICA. [First available use of *Camponotus capperi* subsp. *corticalis* var. *subdepilis* Wheeler, W.M. 1908b: 163; unavailable name.] Combination in *C.* (*Myrmaphaenus*): Emery, 1925b: 156.

#***sublautus.*** *Camponotus ustus* var. *sublautus* Wheeler, W.M. & Mann, 1914: 55 (s.w.) HAITI. Combination in *C.* (*Pseudocolobopsis*): Emery, 1925b: 158; in *C.* (*Tanaemyrmex*): Kempf, 1972a: 73.

#***subnasutus.*** *Camponotus* (*Myrmophyma*) *nasutus* var. *subnasutus* Arnold, 1922: 670 (s.w.) SOUTH AFRICA. Combination in *C.* (*Myrmamblys*): Emery, 1925b: 141.

***subnitidus.*** *Camponotus subnitidus* Mayr, 1876: 65 (s.w.) AUSTRALIA. Combination in *C.* (*Dinomyrmex*): Forel, 1914a: 268; in *C.* (*Tanaemyrmex*): Emery, 1925b: 91. Current subspecies: nominal plus *famelicus, longinodis.*

#***subnudus.*** *Camponotus rubripes* r. *subnudus* Emery, 1889b: 511 (s.w.) BURMA. Forel, 1913k: 125 (q.). Combination in *C.* (*Myrmoturba*): Forel, 1913k: 125; in *C.* (*Tanaemyrmex*): Emery, 1925b: 94. Raised to species: Dalla Torre, 1893: 254. Variety/subspecies of *pallidus:* Emery, 1893g: 267; Emery, 1900d: 703; of *maculatus:* Forel, 1903d: 408. Junior synonym of *pallidus:* Bingham, 1903: 357. Revived from synonymy as subspecies of *maculatus:* Forel, 1907e: 19; Forel, 1910d: 127. Subspecies of *pallidus:* Forel, 1907a: 32; Wheeler, W.M. 1909d: 342. Revived status as species: Viehmeyer, 1916a: 154. Subspecies of *maculatus:* Stitz, 1925: 124.

*subrostrata. Camponotus maculatus* r. *vicinus* var. *subrostrata* Forel, 1914c: 620 (w.) U.S.A. Unavailable name; material referred to *vicinus* by Creighton, 1950a: 381.

*subrufescens. Camponotus* (*Myrmothrix*) *rufipes* var. *subrufescens* Santschi, 1929d: 314 (w.m.) BRAZIL. Junior synonym of *rufipes:* Hashmi, 1973: 52.

***substitutus.*** *Camponotus melanoticus* var. *substitutus* Emery, 1894f: 3 (s.w.) BRAZIL. Combination in *C.* (*Myrmoturba*): Bruch, 1914: 229; in *C.* (*Tanaemyrmex*): Emery, 1925b: 82. Race/subspecies of *landolti:* Forel, 1899c: 136; of *melanoticus:* Emery, 1906c: 188; of *maculatus:* Santschi, 1916e: 396; of *fumidus:* Santschi, 1922c: 101. Raised to species: Emery, 1925b: 82; Santschi, 1929d: 313. Current subspecies: nominal plus *clarus* (unresolved junior homonym), *coloratus, multipilis, pullulus.*

***subtilis.*** *Formica subtilis* Smith, F. 1860b: 94 (s.w.) INDONESIA (Batjan I.). Combination in *Camponotus:* Roger, 1863b: 4; in *C.* (*Myrmamblys*): Forel, 1914a: 272; in *C.* (*Tanaemyrmex*): Donisthorpe, 1932c: 459.

***subtruncatus.*** *Camponotus* (*Pseudocolobopsis*) *subtruncatus* Borgmeier, 1929: 212 (s.w.) BRAZIL.

*succinipes. Camponotus bonariensis* st. *parvulus* var. *succinipes* Santschi, 1939e: 321 (s.w.) ARGENTINA. Unavailable name.

***sucki.*** *Camponotus sucki* Forel, 1901h: 71 (s.) BORNEO. Combination in *C.* (*Myrmoturba*): Forel, 1914a: 267; in *C.* (*Tanaemyrmex*): Emery, 1925b: 102.

*sudanicus. Camponotus* (*Myrmoturba*) *maculatus* st. *aegyptiacus* var. *sudanicus* Santschi, 1926b: 250 (s.w.) CONGO. Unavailable name; material referred to *aegyptiacus* by Baroni Urbani, 1972: 130.

*sudanicus. Camponotus* (*Myrmoturba*) *maculatus* subsp. *sudanicus* Weber, 1943c: 385 (s.w.) SUDAN. Junior synonym of *maculatus:* Baroni Urbani, 1972: 125.

***suffusus.*** *Formica suffusa* Smith, F. 1858b: 38 (q.) AUSTRALIA. Forel, 1915b: 105 (m.); Wheeler, G.C. & Wheeler, J. 1968: 219 (l.). Combination in *Camponotus:* Roger, 1863b: 4; in *C.* (*Myrmosphincta*): Forel, 1912i: 92; in *C.* (*Myrmosaulus*): Emery, 1925b: 114. Senior synonym of *piliventris:* Mayr, 1886c: 355; of *schencki:* Dalla Torre, 1893: 254; Emery, 1896d: 373. Current subspecies: nominal plus *bendigensis.*

#***sulcatinasis.*** *Camponotus* (*Myrmopelta*) *barbarossa* var. *sulcatinasis* Santschi, 1926a: 21 (w.) ZAIRE.

#***sulgeri.*** *Camponotus* (*Orthonotomyrmex*) *sericeus* var. *sulgeri* Santschi, 1913c: 313 (w.) CONGO. Race of *sericeus:* Stitz, 1916: 397.

***CAMPONOTUS** (continued)*

*sylvaticoaethiops. Camponotus sylvaticus* var. *sylvaticoaethiops* Forel, 1874: 39 (w.q.m.) SWITZERLAND. Variety of *aethiops:* Dalla Torre, 1893: 221. Junior synonym of *marginata:* Forel, 1915d: 70.

*sylvaticocognatus. Camponotus maculatus* r. *cognatus* var. *sylvaticocognatus* Forel, 1894d: 5 (w.q.) ALGERIA. Unavailable name.

*sylvaticoides. Camponotus aethiops* var. *sylvaticoides* Dalla Torre, 1893: 221 (s.w.) BULGARIA. [First available use of *Camponotus maculatus* r. *aethiops* var. *sylvaticoides* Forel, 1892i: 306; unavailable name.] Variety/subspecies of *aethiops:* Müller, 1923: 156; Emery, 1925b: 97. Junior synonym of *aethiops:* Agosti & Collingwood, 1987a: 58.

**#*sylvaticomaculatus.*** *Camponotus maculatus* var. *sylvaticomaculatus* Dalla Torre, 1893: 241 (w.) GREECE. [First available use of *Camponotus rubripes* r. *maculatus* var. *sylvaticomaculatus* Forel, 1886f: 150; unavailable name.]

***sylvaticus.*** *Formica sylvatica* Olivier, 1792: 491 (q.) FRANCE. Forel, 1874: 38 (s.w.m.); Emery, 1916b: 229 (s.w., in key). Combination in *Camponotus:* Roger, 1862c: 291; in *C.* (*Myrmoturba*): Menozzi, 1922b: 330; in *C.* (*Tanaemyrmex*): Emery, 1925b: 100. Race of *rubripes:* Forel, 1886f: 143; of *maculatus:* Forel, 1890c: 219; Ruzsky, 1902d: 7; Emery, 1916b: 150. Revived status as species: Bondroit, 1918: 75; Emery, 1920c: 6; Menozzi, 1922b: 330; Santschi, 1931a: 11; Santschi, 1936c: 207; Menozzi, 1939a: 315. Senior synonym of *castanipes:* Roger, 1863b: 1; of *lichtensteini:* Bernard, 1967: 337. Current subspecies: nominal plus *basalis, paradichrous.*

**#*symiensis.*** *Camponotus maculatus* subsp. *symiensis* Forel, 1910a: 26 (w.) GREECE. Combination in *C.* (*Tanaemyrmex*): Emery, 1925b: 98. Subspecies of *compressus:* Emery, 1920c: 6.

*sympatrica. Camponotus rubrithorax* var. *sympatrica* Baroni Urbani, 1971b: 362. [Incorrectly proposed and unnecessary replacement name for *nigrithorax* Stitz, 1937: 135.] Unavailable name.

**#*syphax.*** *Camponotus* (*Myrmosericus*) *rufoglaucus* subsp. *syphax* Wheeler, W.M. 1922a: 246 (s.w.) ZAIRE.

*tacitus. Camponotus* (*Myrmoturba*) *maculatus* st. *agricola* var. *tacitus* Santschi, 1919b: 240 (s.w.) SOUTH AFRICA. Unavailable name.

**#*tacuruensis.*** *Camponotus germaini* subsp. *tacuruensis* Emery, 1911a: 222 (s.w.) PARAGUAY. Combination in *C.* (*Myrmaphaenus*): Emery, 1925b: 155.

*taeniatus. Camponotus taeniatus* Roger, 1863a: 149 (q.) MEXICO. Junior synonym of *atriceps:* Mayr, 1870a: 378; Hashmi, 1973: 82.

***tahatensis.*** *Camponotus* (*Myrmoturba*) *erigens* var. *tahaensis* Santschi, 1929g: 166 (s.w.) ALGERIA. [First available use of *Camponotus* (*Myrmoturba*) *erigens* st. *magister* var. *tahatensis* Santschi, 1929c: 103; unavailable name.] [Also described as new, but *Camponotus compressus* st. *thoracicus* var. *tahatensis* Santschi, 1934f: 171, based on same material; unavailable name.] Subspecies of *compressus,* and justified emendation of spelling to *tahatensis:* Santschi, 1939d: 82. Raised to species: Bernard, 1953a: 189.

***taipingensis.*** *Camponotus* (*Dinomyrmex*) *taipingensis* Forel, 1913k: 124 (w.) WEST MALAYSIA. Combination in *C.* (*Tanaemyrmex*): Emery, 1925b: 91.

**#*taivanae.*** *Camponotus* (*Colobopsis*) *rothneyi* var. *taivanae* Forel, 1913f: 201 (q.) TAIWAN.

*tambourinensis. Camponotus* (*Myrmoturba*) *maculatus* r. *novaehollandiae* var. *tambourinensis* Forel, 1915b: 98 (s.w.) AUSTRALIA. Unavailable name (Taylor, 1986: 34).

***tameri.*** *Camponotus tameri* **new name**. Replacement name for *tricolor* Weber, 1943c: 387. [Junior secondary homonym of *tricolor* Stitz, 1912: 511.]

*tanquaryi. Camponotus fallax* var. *tanquaryi* Wheeler, W.M. 1910f: 226 (s.w.q.m.) U.S.A. Combination in *C.* (*Myrmentoma*): Emery, 1925b: 117. Variety of *caryae:* Wheeler, W.M. 1917c: 28. Junior synonym of *nearcticus:* Creighton, 1950a: 388.

***tashcumiri.*** *Camponotus tashcumiri* Tarbinsky, 1976: 158, figs. 280, 281 (s.w.) KIRGIZIA. Subgenus indeterminate.

***tasmani.*** *Camponotus tasmani* Forel, 1902h: 503 (s.w.) AUSTRALIA. Combination in *C.* (*Myrmocamelus*): Forel, 1914a: 270; in *C.* (*Myrmophyma*): Emery, 1920b: 257; Santschi, 1928e: 482.

***tauricollis.*** *Camponotus foraminosus* r. *tauricollis* Forel, 1894b: 68 (s.w.) MOZAMBIQUE. Combination in *C.* (*Myrmotrema*), and subspecies of *olivieri:* Santschi, 1915c: 273. Raised to species: Santschi, 1926c: 616.

**#*taylori.*** *Camponotus maculatus* r. *taylori* Forel, 1892j: 241 (s.w.) INDIA. Imai *et al.*, 1984: 9 (k.). Combination in *C.* (*Tanaemyrmex*): Emery, 1925b: 93. Raised to species: Bingham, 1903: 353; Wheeler, W.M. 1913e: 237. Subspecies/race of *barbatus:* Forel, 1902d: 288; Forel, 1904c: 29; Forel, 1907e: 19; Emery, 1925b: 93.

*tejonia. Formica tejonia* Buckley, 1866: 161 (m.) U.S.A. Combination in *Camponotus:* Emery, 1895c: 338; Wheeler, W.M. 1902f: 21. Unrecognisable taxon; *incertae sedis* in *Camponotus.*

**#*tenuibarbis.*** *Camponotus punctulatus* var. *tenuibarbis* Santschi, 1939e: 325 (w.) ARGENTINA.

***tenuipes.*** *Formica tenuipes* Smith, F. 1857a: 57 (q.) BORNEO. Combination in *Camponotus:* Roger, 1863b: 3; in *C.* (*Tanaemyrmex*): Donisthorpe, 1932c: 444.

**#*tenuipilis.*** *Camponotus* (*Myrmotrema*) *olivieri* var. *tenuipilis* Santschi, 1915c: 270 (in key) (w.) SOUTH AFRICA. [First available use of *Camponotus foraminosus* subsp. *olivieri* var. *tenuipilis* Emery, 1895h: 54; unavailable name.]

*tenuipubens. Camponotus* (*Myrmoturba*) *distinguendus* var. *tenuipubens* Santschi, 1916d: 242. Replacement name for *denudata* Emery, 1906c: 191. [Junior primary homonym of *denudata* Emery, 1903: 67.] Combination in *C.* (*Tanaemyrmex*), and name misspelled as *tenuipubescens:* Emery, 1925b: 78. Junior synonym of *distinguendus:* Snelling & Hunt, 1976: 120.

**#*tenuis.*** *Camponotus rufoglaucus* var. *tenuis* Forel, 1907a: 32 (w.) INDIA. Combination in *C.* (*Myrmosericus*): Emery, 1925b: 105.

***tenuiscapus.*** *Camponotus tenuiscapus* Roger, 1863a: 143 (w.) URUGUAY. Emery, 1888c: 79 (s.). Combination in *C.* (*Myrmamblys*): Bruch, 1914: 230; in *C.* (*Myrmoturba*): Forel, 1914a: 267; Emery, 1920b: 255; in *C.* (*Tanaemyrmex*): Emery, 1925b: 79.

**#*tenuisquamis.*** *Camponotus carin* var. *tenuisquamis* Forel, 1901h: 67 (w.q.) SUMATRA. Combination in *C.* (*Tanaemyrmex*): Emery, 1925b: 90.

*tenutina. Camponotus sylvaticus* var. *tenutina* Menozzi, 1934: 165. *Nomen nudum.*

*tephronotus. Camponotus maculatus* subsp. *lucayanus* var. *tephronotus* Wheeler, W.M. 1905b: 114 (s.w.q.m.) BAHAMAS. Unavailable name.

***tepicanus.*** *Camponotus tepicanus* Pergnade, 1896: 865 (s.w.) MEXICO. Combination in *C.* (*Myrmoturba*): Forel, 1914a: 267; in *C.* (*Tanaemyrmex*): Emery, 1925b: 82.

*teranishii. Camponotus* (*Myrmentoma*) *caryae* var. *teranishii* Wheeler, W.M. 1928d: 118. Replacement name for *tokyoensis* Teranishi, 1915: 137. [Junior primary homonym of *tokioensis* Ito, 1912: 341.] Junior synonym of *keihitoi:* Terayama & Satoh, 1990b: 532.

***terbimaculatus.*** *Camponotus* (*Pseudocolobopsis*) *terbimaculatus* Emery, 1920c: 37 (q.) BRAZIL.

***terebrans.*** *Formica terebrans* Lowne, 1865a: 278 (s.w.q.) AUSTRALIA. Junior synonym of *testaceipes* Smith, F.: Mayr, 1876: 65. [*Formica testaceipes* Smith, F. 1858b: 39 is junior primary homonym of *Formica testaceipes* Leach, 1825: 290.] Hence *terebrans* first available replacement name. See also under *testaceipes.*

***tergestinus.*** *Camponotus tergestinus* Müller, 1921: 46 (w.q.) ITALY. Combination in *C.* (*Myrmentoma*): Emery, 1925b: 119. See also: Baroni Urbani, 1971c: 196.

**#*termitarius.*** *Camponotus punctulatus* subsp. *termitarius* Emery, 1902b: 297 (footnote) (w.q.) BRAZIL. Combination in *C.* (*Myrmoturba*): Forel, 1913l: 249; in *C.* (*Tanaemyrmex*): Emery, 1925b: 79. See also: Emery, 1903: 70.

***terricola.*** *Camponotus* (*Tanaemyrmex*) *terricola* Karavaiev, 1929c: 234, fig. 1 (s.w.) JAVA.

***territus.*** *Camponotus* (*Paracolobopsis*) *territus* Santschi, 1939e: 329 (w.) ARGENTINA. Combination in *C.* (*Myrmaphaenus*): Kempf, 1972a: 47.

*testaceipes. Formica testaceipes* Smith, F. 1858b: 39 (w.) AUSTRALIA. [Junior primary homonym of *Formica testaceipes* Leach, 1825: 290.] Mayr, 1862: 662 (q.); Forel, 1902h: 494 (s.m.). Combination in *Camponotus:* Mayr, 1862: 662; in *C.* (*Myrmophyma*): Forel, 1914a: 269; in *C.* (*Tanaemyrmex*): Emery, 1925b: 102. Senior synonym of *terebrans:* Mayr, 1876: 65 (hence *terebrans* first available replacement name and valid name of taxon). Senior synonym of *darlingtoni* and *rottnesti:* Brown, 1956a: 39.

***testaceus.*** *Camponotus extensus* subsp. *testaceus* Emery, 1894f: 3 (s.w.q.) BRAZIL. Combination in *C.* (*Tanaemyrmex*), and raised to species: Emery, 1925b: 84.

*testacea. Colobopsis testacea* Bingham, 1903: 347 (q.) SRI LANKA. [Junior secondary homonym of *testaceus* Emery, above.] Replacement name: *ceylonicus* Emery, 1925b: 145.

***texanus.*** *Camponotus texanus* Wheeler, W.M. 1903b: 108, fig. 10 (s.w.q.m.) U.S.A. Combination in *C.* (*Myrmentoma*): Emery, 1920b: 257; in *C.* (*Camponotus*): Forel, 1914a: 266; Creighton, 1950a: 371.

***texens.*** *Camponotus* (*Karavaievia*) *texens* Dumpert, 1986: 559, figs. 1, 2, 3 (w.q.m.) WEST MALAYSIA.

#***textor.*** *Camponotus senex* r. *textor* Forel, 1899c: 138 (w.q.) COSTA RICA. Combination in *C.* (*Myrmobrachys*): Emery, 1925b: 164.

***thales.*** *Camponotus maculatus* subsp. *thales* Forel, 1910e: 453 (s.w.q.m.) LESOTHO. Combination in *C.* (*Myrmoturba*): Forel, 1913a: 145; in *C.* (*Tanaemyrmex*): Emery, 1925b: 92. Race of *maculatus:* Arnold, 1922: 644. Raised to species: Emery, 1920c: 6; Emery, 1925b: 92.

#***themistocles.*** *Camponotus themistocles* Forel, 1910e: 456 (w.) MADAGASCAR. Santschi, 1911e: 133 (s.). Combination in *C.* (*Myrmobrachys*): Forel, 1914a: 271; in *C.* (*Myrmepomis*): Emery, 1920b: 258; in *C.* (*Myrmopiromis*): Wheeler, W.M. 1922a: 707. Subspecies of *darwinii:* Santschi, 1911e: 133.

***thomasseti.*** *Camponotus thomasseti* Forel, 1912k: 166 (w.m.) SEYCHELLE IS. Combination in *C.* (*Myrmamblys*): Forel, 1914a: 272.

*thomensis. Camponotus* (*Myrmoturba*) *maculatus* var. *thomensis* Santschi, 1920i: 3 (s.w.m.) SAO THOME I. Junior synonym of *maculatus:* Baroni Urbani, 1972: 125.

***thoracicus.*** *Formica thoracica* Fabricius, 1804: 397 (q.) ALGERIA. [Unresolved junior primary homonym of *Formica thoracica* Olivier, 1792: 503.] Emery, 1908a: 193 (s.w.m.). Combination in *Camponotus:* Roger, 1863b: 2; in *C.* (*Myrmoturba*): Forel, 1913d: 434; in *C.* (*Tanaemyrmex*): Emery, 1925b: 99. Junior synonym of *maculatus:* Roger, 1863b: 2; Mayr, 1865: 27. Revived from synonymy as subspecies of *maculatus:* Forel, 1906c: 189. Revived status as species: Emery, 1920b: 256. Subspecies of *maculatus:* Arnold, 1922: 630; Karavaiev, 1926e: 189; of *compressus:* Emery, 1920c: 6; Santschi, 1921d: 73; Wheeler, W.M. 1927g: 114; Santschi, 1929g: 165. Revived status as species: Pisarski, 1971a: 674; Dlussky & Zabelin, 1985: 236. Senior synonym of *dichrous:* Emery, 1908a: 193. Current subspecies: nominal plus *ninivae.*

***thraso.*** *Camponotus maculatus* r. *thraso* Forel, 1893b: 432 (w.) SRI LANKA. Imai *et al.*, 1984: 9 (k.). Combination in *C.* (*Tanaemyrmex*): Emery, 1925b: 92. Raised to species: Bingham, 1903: 356; Menozzi, 1931a: 155. Current subspecies: nominal plus *agricola, assabensis, diogenes, montinanus, nefasitensis, negus.*

***thysanopus.*** *Camponotus* (*Myrmeurynota*) *thysanopus* Wheeler, W.M. 1937b: 461 (w.) CUBA.

#***tibestiensis.*** *Camponotus* (*Myrmoturba*) *magister* var. *tibestiensis* Santschi, 1932f: 519 (w.) LIBYA. [Also described as new by Santschi, 1936e: 81.]

***tichomirovi.*** *Camponotus tichomirovi* Ruzsky, 1905b: 188 (w.) RUSSIA. Combination in *C.* (*Myrmoturba*): Forel, 1914a: 267; in *C.* (*Tanaemyrmex*): Emery, 1925b: 95. See also: Tarbinsky, 1976: 160.

***tilhoi.*** *Camponotus* (*Myrmotrema*) *tilhoi* Santschi, 1926b: 264 (s.w.) CHAD.

***timidus.*** *Formica timida* Jerdon, 1851: 122 (w.q.) INDIA. [Unresolved junior primary homonym of *Formica timida* Foerster, 1850a: 35 (now in *Lasius*).] Combination in *Camponotus:* Roger, 1863b: 3. Subgenus indeterminate: Emery, 1925b: 172.

#***tinctus.*** *Formica tincta* Smith, F. 1858b: 21 (q.) BURMA. Viehmeyer, 1916a: 155 (s.w.); Karavaiev, 1929c: 237 (m.). Combination in *Camponotus:* Mayr, 1862: 676; in *C.* (*Myrmoturba*): Forel, 1914a: 267; in *C.* (*Tanaemyrmex*): Emery, 1925b: 94. Junior synonym of *luteus:* Bingham, 1903: 379. Subspecies of *maculatus:* Emery, 1900d: 704; Stitz, 1925: 124. Revived status as species: Viehmeyer, 1916a: 154. Subspecies of *irritans:* Emery, 1920c: 7; Karavaiev, 1929c: 237.

#***tingitanus.*** *Camponotus* (*Myrmentoma*) *gestroi* st. *tingitana* Santschi, 1921b: 435 (w.q.) MOROCCO. Combination in *C.* (*Tanaemyrmex*), and subspecies of *foreli:* Emery, 1925b: 98; Santschi, 1929e: 161.

#***tipunus.*** *Camponotus* (*Dinomyrmex*) *dorycus* var. *tipuna* Forel, 1913f: 201 (s.w.m.) TAIWAN. Combination in *C.* (*Tanaemyrmex*): Emery, 1925b: 90. Variety of *carin:* Wheeler, W.M. 1930a: 105.

#***titicacensis.*** *Camponotus* (*Myrmoturba*) *bruchi* subsp. *titicacensis* Emery, 1920c: 19 (s.) PERU. Combination in *C.* (*Tanaemyrmex*): Emery, 1925b: 78.

***tokioensis.*** *Camponotus* (*Myrmamblys*) *itoi* r. *tokioensis* Ito, 1912: 341 (s.w.) JAPAN. Forel, 1913f: 201 (q., misspelled as *toxioensis*); Santschi, 1928b: 40 (q.m.); Imai & Kubota, 1972: 195 (k.). Raised to species: Santschi, 1937h: 384. Senior synonym of *atrigenatus* and *inconstans:* Terayama & Satoh, 1990a: 412.

****tokunagai.*** **Camponotus tokunagai* Naora, 1933: 208, pl. 19 (caste?) CHINA.

*tokyoensis. Camponotus tokyoensis* Teranishi, 1915: 137, fig. (w.) JAPAN. [Junior primary homonym of *tokioensis* Ito, above.] Replacement name: *teranishii* Wheeler, W.M. 1928d: 118.

#***toltecus.*** *Camponotus maculatus* r. *toltecus* Forel, 1899c: 135 (s.w.q.) MEXICO. Combination in *C.* (*Tanaemyrmex*): Emery, 1925b: 81. Subspecies of *picipes:* Emery, 1920b: 233; of *fumidus:* Emery, 1925b: 81.

*tomentosa. Tapinoma tomentosa* Norton, 1868a: 60, pl. 2, fig. 2 (w.) MEXICO. Combination in *Camponotus:* Norton, 1868c: 3. Junior synonym of *senex:* Emery, 1892b: 167.

***tonduzi.*** *Camponotus tonduzi* Forel, 1899c: 151 (s.w.q.) COSTA RICA. Combination in *C.* (*Myrmamblys*): Forel, 1912i: 90; in *C.* (*Hypercolobopsis*): Emery, 1920b: 260; in *C.* (*Pseudocolobopsis*): Emery, 1925b: 158.

***tonkinus.*** *Camponotus* (*Camponotus*) *japonicus* var. *tonkinus* Santschi, 1925f: 88 (w.) CHINA. Raised to species: Wang, C., Xiao & Wu, 1989b: 327.

*torrefactus. Camponotus* (*Camponotus*) *sansabeanus* var. *torrefactus* Emery, 1925b: 75 (s.w.m.) U.S.A. [First available use of *Camponotus maculatus* subsp. *sansabeanus* var. *torrefactus* Wheeler, W.M. 1910d: 308; unavailable name.] Combination in *C.* (*Tanaemyrmex*): Creighton, 1950a: 379. Variety/subspecies of *sansabeanus:* Cole, 1942: 388; Creighton, 1950a: 379. Junior synonym of *sansabeanus:* Wheeler, G.C. & Wheeler, J. 1986g: 14.

***torrei.*** *Camponotus torrei* Aguayo, 1932: 227 (w.) CUBA. Combination in *C.* (*Manniella*): Kempf, 1972a: 43.

***tortuganus.*** *Camponotus maculatus* subsp. *tortuganus* Emery, 1895c: 336 (w.) U.S.A. Wheeler, W.M. 1910d: 310 (s.q.m.). Combination in *C.* (*Tanaemyrmex*): Emery, 1925b: 80. Subspecies of *picipes:* Emery, 1920b: 233. Raised to species: Wheeler, W.M. 1932a: 13; Creighton, 1950a: 380; Wilson, 1964b: 12.

***toussainti.*** *Camponotus toussainti* Wheeler, W.M. & Mann, 1914: 60, fig. 27 (w.) HAITI. Combination in *C.* (*Myrmeurynota*): Forel, 1914a: 274.

***traegaordhi.*** *Camponotus* (*Myrmoturba*) *maculatus* st. *traegaordhi* Santschi, 1914e: 38 (s.w.q.m.) SOUTH AFRICA. Combination in *C.* (*Tanaemyrmex*): Emery, 1925b: 85. Subspecies of *guttatus:* Emery, 1920c: 6; of *maculatus:* Arnold, 1922: 627. Raised to species: Santschi, 1926b: 252. Current subspecies: nominal plus *fumeus.*

*traili* Mayr, 1878; see under ***DENDROMYRMEX.***

***trajanus.*** *Camponotus* (*Colobopsis*) *trajanus* Forel, 1912n: 68 (w.) SUMATRA.

*transcaucasicus. Camponotus* (*Myrmentoma*) *interjectus* subsp. *transcaucasicus* Arnol'di, 1948: 212. *Nomen nudum.*

*transilis. Camponotus* (*Myrmocladoecus*) *sanctaefidei* subsp. *convexinodis* var. *transilis* Wheeler, W.M. 1934g: 234 (w.) TRINIDAD. Unavailable name.

*transitoria. Camponotus* (*Orthonotomyrmex*) *braunsi* st. *erythromelus* var. *transitoria* Santschi, 1914b: 135 (w.) KENYA. Unavailable name.

***transvaalensis.*** *Camponotus* (*Myrmespera*) *transvaalensis* Arnold, 1948: 228, fig. 14 (s.w.) SOUTH AFRICA. Current subspecies: nominal plus *arnoldi* (unresolved junior homonym), *griqua.*

*transvectus. Camponotus abdominalis* subsp. *transvectus* Wheeler, W.M. 1910d: 326 (s.w.q.m.) U.S.A. Combination in *C.* (*Myrmothrix*): Emery, 1925b: 108. Junior synonym of *atriceps:* Hashmi, 1973: 82.

***trapeziceps.*** *Camponotus trapeziceps* Forel, 1908c: 405 (s.w.q.) BRAZIL. Forel, 1911c: 310 (m.). Combination in *C.* (*Myrmamblys*): Forel, 1914a: 272; in *C.* (*Neomyrmamblys*): Santschi, 1921f: 311; in *C.* (*Myrmaphaenus*): Emery, 1925b: 156. Current subspecies: nominal plus *innocens, prosaicus.*

***trapezoideus.*** *Camponotus trapezoideus* Mayr, 1870a: 385 (w.) COLOMBIA. Mayr, 1887: 517 (q.m.). Combination in *C.* (*Myrmobrachys*): Forel, 1914a: 271.

#***trematogaster.*** *Camponotus* (*Myrmamblys*) *confluens* var. *trematogaster* Santschi, 1915c: 266 (w.) ZAIRE.

***trepidulus.*** *Camponotus* (*Myrmobrachys*) *trepidulus* Creighton, 1965b: 7, figs. 5–8 (s.w.q.m.) U.S.A.

***tricolor.*** *Colobopsis tricolor* Stitz, 1912: 511, fig. 13 (q.) NEW GUINEA. Combination in *Camponotus* (*Colobopsis*): Forel, 1914a: 272.

*tricolor. Camponotus* (*Myrmopiromis*) *tricolor* Weber, 1943c: 387, pl. 16, fig. 33 (w.) SUDAN. [Junior secondary homonym of *tricolor* Stitz, above.] Replacement name: *tameri* **new name**.

***tricoloratus.*** *Camponotus* (*Tanaemyrmex*) *tricoloratus* Clark, 1941: 90, pl. 13, figs. 22, 23 (s.w.) AUSTRALIA.

***trietericus.*** *Camponotus* (*Myrmamblys*) *trietericus* Menozzi, 1926e: 95 (w.) PHILIPPINES.

***trifasciatus.*** *Camponotus* (*Myrmespera*) *debellator* var. *trifasciatus* Santschi, 1926b: 248 (q.) SOUTH AFRICA. Raised to species: Santschi, 1926c: 604.

***tripartitus.*** *Camponotus tripartitus* Mayr, 1887: 519 (s.w.q.) BRAZIL. Combination in *C.* (*Myrmeurynota*): Forel, 1912i: 92; in *C.* (*Myrmorhachis*): Forel, 1914a: 274; in *C.* (*Myrmocladoecus*): Emery, 1925b: 167;

Wheeler, W.M. 1934e: 423. Provisional senior synonym of *collegianus*, and discussion: Kempf, 1968b: 404.

***tristis.*** *Camponotus* (*Myrmophyma*) *tristis* Clark, 1930a: 124, fig. 1 (s.w.) AUSTRALIA. Wheeler, G.C. & Wheeler, J. 1974a: 60 (l.).

***triton.*** *Camponotus* (*Colobopsis*) *triton* Wheeler, W.M. 1934g: 217 (q.) BAHAMAS.

***tritschleri.*** *Camponotus tritschleri* Forel, 1912n: 63 (w.) SUMATRA. Combination in *C.* (*Colobopsis*): Forel, 1914a: 272.

#***troglodytes.*** *Camponotus troglodites* Forel, 1894b: 69 (w.) MOZAMBIQUE. Forel, 1904b: 377 (q.); Santschi, 1914e: 40 (s.); Arnold, 1924: 727 (m.). Combination in *C.* (*Myrmotrema*): Forel, 1913a: 145. [*Camponotus troglodytes:* Forel, 1904b: 377, justified emendation of spelling.] Status as species: Emery, 1920b: 258; Crawley, 1921: 95; Wheeler, W.M. 1922a: 984. Subspecies/race of *robecchii:* Santschi, 1915c: 271; Arnold, 1924: 727; Emery, 1925b: 133.

#***trotteri.*** *Camponotus* (*Myrmogonia*) *schmeltzii* subsp. *trotteri* Mann, 1921: 486, fig. 34 (w.) FIJI IS.

***truebi.*** *Camponotus truebi* Forel, 1910a: 31 (w.) TAIWAN. Forel, 1912: 75 (q.). Combination in *C.* (*Camponotus*): Forel, 1914a: 266; *C.* (*Myrmentoma*): Emery, 1925b: 119. Current subspecies: nominal plus *arnoldi, genaiai.*

***truncatus.*** *Formica truncata* Spinola, 1808: 244 (q.) ITALY. Forel, 1874: 45 (s.w.); Mayr, 1861: 38 (w.q.); Emery, 1916b: 235 (m., in key). Combination in *Colobopsis:* Mayr, 1861: 38; in *Camponotus* (*Colobopsis*): Forel, 1893b: 435. Senior synonym of *fuscipes:* Emery & Forel, 1879: 449; Forel, 1915d: 73. See also: Kutter, 1977c: 207.

*tuckeri. Camponotus* (*Myrmoturba*) *maculatus* var. *tuckeri* Santschi, 1932a: 391 (s.w.m.) NAMIBIA. Junior synonym of *maculatus:* Baroni Urbani, 1972: 125.

#***tucumanus.*** *Camponotus* (*Myrmoturba*) *bonariensis* var. *tucumana* Santschi, 1922c: 102 (w.) ARGENTINA. Santschi, 1922d: 261 (q.m.). Combination in *C.* (*Tanaemyrmex*): Emery, 1925b: 79.

***tumidus.*** *Camponotus* (*Myrmogonia*) *tumidus* Crawley, 1922c: 34, fig. 17 (s.w.) AUSTRALIA. Crawley, 1925b: 595 (q.m.). Combination in *C.* (*Myrmophyma*): Emery, 1925b: 112; in *C.* (*Thlipsepinotus*): Santschi, 1928e: 483.

*tunetica. Camponotus sylvaticus* st. *barbarus* var. *tunetica* Santschi, 1921d: 73 (w.q.) TUNISIA. Unavailable name.

*turkestanicocognatus. Camponotus maculatus* subsp. *turkestanicus* var. *turkestanicocognatus* Ruzsky, 1902c: 5 (w.) TURKESTAN. Unavailable name.

***turkestanicus.*** *Camponotus sylvaticus* st. *turkestanicus* Emery, 1887a: 212 (w.) TURKESTAN. Ruzsky, 1905b: 203 (q.m.); Emery, 1908a: 203 (s.w.q.m.). Combination in *C.* (*Tanaemyrmex*): Emery, 1925b: 95. Subspecies of *maculatus:* Emery, 1896d: 370; Ruzsky, 1902c: 3; Forel, 1911e: 290. Raised to species: Dalla Torre, 1893: 255; Emery, 1920c: 8. See also: Tarbinsky, 1976: 158; Arnol'di & Dlussky, 1978: 551. Senior synonym of *kaschgariensis:* Forel, 1904b: 378.

***turkestanus.*** *Camponotus sylvaticus* var. *turkestanus* André, 1882a: 145 (w.) TURKESTAN. Forel, 1904b: 380 (q.); Collingwood, 1961a: 74 (m.). Combination in *C.* (*Tanaemyrmex*): Emery, 1925b: 101. Subspecies of *maculatus:* Emery, 1896d: 371; Forel, 1904b: 380. Raised to species: Emery, 1920c: 6; Tarbinsky, 1976: 156. Senior synonym of *ferganensis, flavonitidus* and *ruzskyi:* Dlussky, Soyunov & Zabelin, 1990: 132.

#***ugandensis.*** *Camponotus* (*Dinomyrmex*) *liengmei* var. *ugandensis* Santschi, 1923e: 292 (w.) UGANDA. Subspecies of *maculatus:* current status.

***ulcerosus.*** *Camponotus ulcerosus* Wheeler, W.M. 1910d: 351 (s.) U.S.A. Creighton, 1951: 58 (w.q.). Combination in *C.* (*Colobopsis*): Forel, 1914a: 272; in *C.* (*Myrmamblys*): Emery, 1920b: 259; in *C.* (*Manniella*): Emery, 1925b: 160; in *C.* (*Myrmaphaenus*): Creighton, 1951: 57. Senior synonym of *bruesi:* Creighton, 1951: 53. See also: Creighton, 1953d: 82.

***ulei.*** *Camponotus ulei* Forel, 1904e: 706 (s.w.q.) PERU. Combination in *C.* (*Myrmamblys*): Forel, 1912i: 90; in *C.* (*Myrmoturba*): Emery, 1920b: 255; in *C.* (*Pseudocolobopsis*): Santschi, 1921f: 311.

*****ullrichi.*** **Camponotus ullrichi* Bachmayer, 1960: 13, pl. 1, fig. 2 (wing) AUSTRIA (Pliocene).

***ulvarum.*** *Camponotus ulvarum* Forel, 1899c: 153 (s.) COSTA RICA. Combination in *C.* (*Myrmamblys*): Forel, 1914a: 272; in *C.* (*Pseudocolobopsis*): Emery, 1920b: 259.

#***ulysses.*** *Camponotus ustus* var. *ulysses* Forel, 1907e: 11 (s.w.q.m.). HAITI. Combination in *C.*(*Pseudocolobopsis*): Emery, 1925b: 158; in *C.* (*Tanaemyrmex*): Kempf, 1972a: 73.

#***umbratilis.*** *Camponotus* (*Colobopsis*) *manni* subsp. *umbratilis* Wheeler, W.M. 1934e: 420, fig. 4 (w.m.) FIJI IS.

#***unctulus.*** *Camponotus* (*Myrmobrachys*) *capperi* subsp. *unctulus* Wheeler, W.M. 1917g: 470, fig. 3 (s.w.q.) JAMAICA. Combination in *C.* (*Myrmaphaenus*): Emery, 1925b: 156.

#***unitaeniatus.*** *Camponotus sexguttatus* var. *unitaeniatus* Wheeler, W.M. 1923c: 7 (w.) PERU. Combination in *C.* (*Myrmosphincta*): Kempf, 1972a: 59.

***universitatis.*** *Camponotus universitatis* Forel, 1890c: 218 (w.) FRANCE. Tinaut, Espadaler & Jimenez, 1992: 234 (q.m.). Combination in *C.* (*Myrmoturba*): Forel, 1914a: 267; in *C.* (*Tanaemyrmex*): Emery, 1925b: 98. See also: Baroni Urbani, 1971c: 187; Kutter, 1977c: 206.

***urichi.*** *Camponotus urichi* Forel, 1899c: 155 (footnote) (s.w.q.) TRINIDAD. Combination in *C.* (*Myrmotemnus*): Emery, 1920b: 259; in *C.* (*Myrmosphincta*): Forel, 1912i: 92; Emery, 1925b: 152. Current subspecies: nominal plus *folicola, sculnus.*

***ursus.*** *Camponotus ursus* Forel, 1886c: ci (s.w.) MADAGASCAR. [*Camponotus ursus* Forel, 1879a: 115, *nomen nudum.*] Combination in *C.* (*Myrmobrachys*): Forel, 1912i: 91; in *C.* (*Myrmepomis*): Emery, 1920b: 258; in *C.* (*Myrmopiromis*): Wheeler, W.M. 1922a: 1053. See also: Forel, 1891b: 46.

*ustithorax. Camponotus rufoglaucus* subsp. *cinctellus* var. *ustithorax* Forel, 1910c: 271 (w.) ETHIOPIA. Unavailable name.

*ustulatus. Camponotus atriceps* r. *ustulatus* Forel, 1885a: 339 (s.w.) GUATEMALA. Santschi, 1936b: 211 (q.m.). Combination in *C.* (*Myrmothrix*): Emery, 1925b: 108. Raised to species: Dalla Torre, 1893: 255. Subspecies of *abdominalis:* Emery, 1896d: 372; Forel, 1899c: 132. Junior synonym of *atriceps:* Hashmi, 1973: 82.

***ustus.*** *Camponotus ustus* Forel, 1879a: 75 (s.w.q.m.) ANTILLES. Wheeler, G.C. & Wheeler, J. 1974a: 61 (l.). Combination in *C.* (*Myrmoturba*): Forel, 1914a: 267; in *C.* (*Pseudocolobopsis*): Emery, 1920b: 259; in *C.* (*Tanaemyrmex*): Wheeler, W.M. 1936b: 205. Current subspecies: nominal plus *arhuacus, depolitus, furnissi, sublautus, ulysses.*

***vafer.*** *Camponotus vafer* Wheeler, W.M. 1910d: 315 (s.w.q.) U.S.A. Combination in *C.* (*Myrmoturba*): Forel, 1914a: 268; in *C.* (*Camponotus*): Emery, 1925b: 75; in *C.* (*Tanaemyrmex*): Creighton, 1950a: 381.

#***vagulus.*** *Camponotus fastigatus* subsp. *vagulus* Forel, 1908c: 403 (s.w.q.) BRAZIL. Combination in *C.* (*Myrmaphaenus*): Emery, 1925b: 155.

***vagus.*** *Formica vaga* Scopoli, 1763: 312 (w.) AUSTRIA. Latreille, 1802c: 96 (w.q.m.). Hauschteck, 1961: 221 (k.). Combination in *Camponotus:* Roger, 1863b: 1; in *C.* (*Camponotus*): Forel, 1914a: 266. [Misspelled as *vagans:* Emery, 1891b: 20.] Subspecies of *herculeanus:* Emery, 1896d: 372; Emery, 1908a: 185; Bondroit, 1910: 488. Revived status as species: Ruzsky, 1905b: 241; Forel, 1915d: 68; Emery, 1916b: 225; Emery, 1920b: 255; Finzi, 1924a: 14; Karavaiev, 1927c: 275; Finzi, 1930d: 317; Santschi, 1931a: 11; Stitz, 1939: 246. See also: Yasumatsu & Brown, 1951: 31; Arnol'di & Dlussky, 1978: 551; Atanassov & Dlussky, 1992: 214. Senior synonym of *pubescens:* Olivier, 1792: 492; Emery, 1891b: 20; Forel, 1892i: 306; of *fuscoptera:* Latreille, 1802c: 96. Current subspecies: nominal plus *ifranensis, kodoricus.*

*vagusherculeanus. Camponotus vagusherculeanus* Nadig, 1918: 340. *Nomen nudum;* material referred to *nadigi* by Emery, 1925b: 72.

***valdeziae.*** *Camponotus valdeziae* Forel, 1879a: 90 (w.q.) SOUTH AFRICA. Arnold, 1924: 696 (s.). Combination in *C.* (*Myrmosericus*): Forel, 1914a: 268.

#***valerius.*** *Camponotus* (*Myrmoturba*) *melanoticus* st. *valerius* Santschi, 1922c: 99, fig. 1 (s.w.) BRAZIL. Combination in *C.* (*Tanaemyrmex*): Emery, 1925b: 81.

***varians.*** *Camponotus varians* Roger, 1863a: 138 (w.) SRI LANKA. Emery, 1893f: 252 (s.q.). Combination in *C.* (*Myrmotemnus*): Emery, 1920b: 258; in *C.* (*Myrmamblys*): Forel, 1914a: 272; Emery, 1925b: 139.

***variegatus.*** *Formica variegata* Smith, F. 1858b: 19 (w.q.) SRI LANKA. Imai *et al.*, 1984: 9 (k.). Combination in *Camponotus:* Mayr, 1862: 656; in *C.* (*Myrmoturba*): Forel, 1913k: 125; in *C.* (*Tanaemyrmex*): Emery, 1925b: 95. Variety of *sylvaticus:* Mayr, 1880: 21; of *maculatus:* Forel, 1886f: 149. Revived status as species: Dalla Torre, 1893: 256; Santschi, 1920h: 173; Emery, 1920c: 7. Senior synonym of *hawaiensis:* Wilson & Taylor, 1967: 95. Current subspecies: nominal plus *ambonensis, bacchus, batta, cleon, comottoi, crassinodis, dulcis, flavotestaceus, fuscithorax, infuscus, intrans, kattensis, proles, somnificus, stenonotus.*

**varius.** *Camponotus* (*Myrmentoma*) *varius* Donisthorpe, 1943b: 204 (w.) INDIA.

**varus.** *Camponotus varus* Forel, 1910e: 455 (w.) ANGOLA. Combination in *C.* (*Dinomyrmex*): Forel, 1914a: 268; in *C.* (*Tanaemyrmex*): Emery, 1925b: 89.

***vehemens.** **Camponotus vehemens* Förster, 1891: 428, pl. 13, fig. 3 (m.) GERMANY (Oligocene). Théobald, 1937b: 218 (w.q.m.). Senior synonym of **miserabilis:* Théobald, 1937b: 218.

**velox.** *Formica velox* Jerdon, 1851: 124 (w.) INDIA. Combination in *Camponotus:* Roger, 1863b: 3. Subgenus indeterminate: Emery, 1925b: 172.

**#velutinus.** *Camponotus* (*Myrmophyma*) *aureopilus* var. *velutina* Stitz, 1938: 120 (w.) NEW GUINEA.

*ventralis. Formica ventralis* Smith, F. 1858b: 20 (q.) SRI LANKA. Combination in *Camponotus:* Roger, 1863b: 3. Junior synonym of *bacchus:* Mayr, 1865: 27.

*venustus. Camponotus venustus* Mayr, 1867c: 441 (w.) [no locality given]. Junior synonym of *cinctellus:* Emery, 1895h: 48.

**#verae.** *Camponotus fastigatus* subsp. *verae* Forel, 1908c: 403 (w.) BRAZIL. Forel, 1911c: 311 (m.); Wheeler, G.C. & Wheeler, J. 1968: 217 (l.). Combination in *C.* (*Myrmaphaenus*): Emery, 1925b: 155.

**#vernulus.** *Camponotus* (*Tanaemyrmex*) *ramulorum* var. *vernula* Wheeler, W.M. 1936b: 206 (s.w.) HAITI.

**versicolor.** *Camponotus* (*Myrmosaulus*) *versicolor* Clark, 1930a: 122 (s.w.) AUSTRALIA.

*vertex. Camponotus* (*Myrmosericus*) *rufoglaucus* st. *cinctellus* var. *vertex* Santschi, 1935a: 280 (w.q.) ZAIRE. Unavailable name.

**vespertinus.** *Camponotus vespertinus* Arnold, 1960b: 456, fig. 6 (w.) ZIMBABWE.

**vestitus.** *Formica vestita* Smith, F. 1858b: 32 (q.) SOUTH AFRICA. Arnold, 1924: 694 (s.w.). Combination in *Camponotus:* Roger, 1863b: 2; in *C.* (*Myrmosericus*): Emery, 1925b: 106. Subspecies of *rufoglaucus:* Emery, 1894i: 47; Emery, 1895h: 48; Forel, 1907e: 16; Santschi, 1914d: 381; Arnold, 1924: 694. Revived status as species: Santschi, 1926b: 258; Santschi, 1930a: 74; Bernard, 1953b: 262. Current subspecies: nominal plus *anthracinus, bombycinus, comptus, intuens, lujai, pectitus, perpectitus, strophiatus.*

***vetus.** **Camponotus vetus* Scudder, 1877a: 742 (q.?) U.S.A. (Eocene).

**vezenyii.** *Camponotus vezenyii* Forel, 1907a: 34 (w.) PARAGUAY. Combination in *C.* (*Myrmobrachys*): Forel, 1914a: 271. Subspecies of *crassus:* Forel, 1911e: 291. Revived status as species: Emery, 1920b: 260; Kempf, 1972a: 54.

**vicinus.** *Camponotus vicinus* Mayr, 1870b: 940 (w.) U.S.A. Emery, 1893i: 671 (s.q.); Wheeler, W.M. 1910d: 301 (m.); Wheeler, G.C. & Wheeler, J. 1968: 220 (l.). Combination in *C.* (*Myrmoturba*): Forel, 1914a: 268; in *C.* (*Camponotus*): Emery, 1925b: 75; in *C.* (*Tanaemyrmex*): Creighton, 1950a: 381. Variety/subspecies of *sylvaticus:* Mayr, 1886d: 422; of *maculatus:* Emery, 1893i: 671; of *sansabeanus:* Emery, 1920b: 232 (footnote); Cole, 1942: 387. Revived status as species: Creighton, 1950a: 381; Snelling, 1970: 396. Material of the unavailable names *infernalis, luteangulus, maritimus, nitidiventris, plorabilis* and *subrostrata* referred here by Creighton, 1950a: 381; of the unavailable name *berkeleyensis* referred here by Snelling, 1970: 396.

**victoriae.** *Camponotus* (*Myrmopiromis*) *victoriae* Arnold, 1959: 338, fig. 24 (s.w.) SOUTH AFRICA.

**#victoriensis.** *Camponotus latrunculus* var. *victoriensis* Santschi, 1928e: 479 (s.w.m.) AUSTRALIA.

**viehmeyeri.** *Camponotus viehmeyeri* Forel, 1911e: 292 (w.) NEW GUINEA. [Also as new by Forel, in Viehmeyer, 1912: 7.] Combination in *C.* (*Myrmogonia*): Forel, 1914a: 269; in *C.* (*Myrmophyma*): Emery, 1925c: 112; in *C.* (*Thlipsepinotus*): Santschi, 1928e: 483.

**vigilans.** *Formica vigilans* Smith, F. 1858b: 26 (w.) BORNEO. Combination in *Camponotus:* Roger, 1863b: 3, in *C.* (*Colobopsis*): Mayr, 1886c: 354; in *C.* (*Myrmamblys*): Forel, 1914a: 272; subgenus indeterminate: Emery, 1925b: 172.

**villosus.** *Camponotus* (*Myrmoturba*) *villosa* Crawley, 1915a: 136 (s.w.) AUSTRALIA. Combination in *C.* (*Tanaemyrmex*): Emery, 1925b: 102.

**#vincentensis.** *Camponotus auricomus* var. *vincentensis* Forel, 1899c: 139 (w.) ANTILLES. Combination in *C.* (*Myrmobrachys*): Emery, 1925b: 163.

**vinosus.** *Formica vinosa* Smith, F. 1858b: 42 (w.) BRAZIL. Combination in *Camponotus:* Roger, 1863b: 5; subgenus indeterminate: Emery, 1925b: 172.

**viri.** *Camponotus* (*Myrmamblys*) *viri* Santschi, 1915c: 265, fig. 13 (q.) KENYA. Combination in *C.* (*Myrmotemnus*): Emery, 1920b: 258; in *C.* (*Myrmamblys*): Wheeler, W.M. 1922a: 990.

**virulens.** *Formica virulens* Smith, F. 1861b: 38 (w.) SULAWESI. Combination in *Camponotus:* Roger, 1863b: 4; in *C.* (*Tanaemyrmex*): Donisthorpe, 1932c: 465.

**vitiensis.** *Camponotus* (*Colobopsis*) *vitiensis* Mann, 1921: 490, fig. 36 (w.) FIJI IS. Donisthorpe, 1946b: 69 (s.q.m.).

**#vitiosus.** *Camponotus vitiosus* Smith, F. 1874: 403 (w.) JAPAN. Wheeler, W.M. 1906c: 326 (s.). Junior synonym of *marginata:* Mayr, 1879: 645. Revived from synonymy as subspecies of *marginata:* Emery, 1893i: 675; of *fallax:* Forel, 1907e: 19; of *caryae:* Wheeler, W.M. 1917c: 29; Emery, 1925b: 118; of *marginata:* Onoyama, 1980: 201. Subspecies of *aethiops:* current status.

**vitreus.** *Formica vitrea* Smith, F. 1860b: 94 (w.) INDONESIA (Batjan I.). Emery, 1899c: 7 (l.); Viehmeyer, 1916a: 160 (s.q.m.); Karavaiev, 1933a: 319 (m.). Combination in *Camponotus:* Dalla Torre, 1893: 257; in *C.* (*Colobopsis*): Emery, 1893e: 225. Senior synonym of *siggii:* Forel, 1895e: 455; of *adlerzii:* Forel, 1895e: 458; of *incursor:* Donisthorpe, 1932c: 459. Current subspecies: nominal plus *angustulus, carinatus* (unresolved junior homonym), *latinotus* (unresolved junior homonym), *oebalis, praeluteus, praerufus, vittatulus.*

**vittatus.** *Camponotus melanoticus* var. *vittata* Forel, 1904c: 49 (w.) BRAZIL. Emery, 1906c: 188 (q.). Combination in *C.* (*Myrmoturba*): Bruch, 1915: 535; in *C.* (*Tanaemyrmex*): Emery, 1925b: 81. Race of *fumidus:* Santschi, 1922c: 101. Raised to species: Snelling, 1968b: 351.

**#vittatulus.** *Camponotus* (*Colobopsis*) *vitreus* var. *vittatula* Forel, 1913k: 133 (q.) SUMATRA.

**vividus.** *Formica vivida* Smith, F. 1858b: 31 (w.) SIERRA LEONE. Wheeler, G.C. & Wheeler, J. 1953e: 192 (l.). Combination in *Camponotus:* Roger, 1863b: 2; in *C.* (*Orthonotomyrmex*): Santschi, 1915c: 264; in *C.* (*Myrmamblys*): Emery, 1925b: 142; in *C.* (*Myrmopelta*): Santschi, 1921f: 311; Santschi, 1926a: 17; Santschi, 1926c: 603. Senior synonym of *laboriosa:* Mayr, 1863: 416; Wheeler, W.M. 1922a: 975. Current subspecies: nominal plus *cato, meinerti, reginae, semidepilis.*

**voeltzkowii.** *Camponotus voeltzkowii* Forel, 1894e: 226 (w.) MADAGASCAR. Combination in *C.*(*Myrmobrachys*): Forel, 1914a: 271; in *C.* (*Myrmepomis*): Emery, 1920b: 258; in *C.* (*Myrmopiromis*): Wheeler, W.M. 1922a: 1053.

**vogti.** *Camponotus vogti* Forel, 1906c: 187 (w.) TURKEY. Combination in *C.* (*Orthonotomyrmex*): Forel, 1914a: 273; in *C.* (*Myrmentoma*), as junior synonym of *kurdistanicus:* Emery, 1920b: 257. Revived from synonymy as variety of *kurdistanicus:* Emery, 1920c: 26; Emery, 1925a: 70. Revived status as species: Agosti & Collingwood, 1987a: 59.

*volkensi. Camponotus* (*Myrmoturba*) *maculatus* r. *tinctus* var. *volkensi* Stitz, 1925: 124, fig. 6 (w.) CAROLINE IS. Unavailable name.

**#voltai.** *Camponotus* (*Myrmogonia*) *oetkeri* r. *voltai* Forel, 1913g: 191 (w.) AUSTRALIA. Combination in *C.* (*Myrmophyma*): Emery, 1925b: 112.

**#vulcanalis.** *Camponotus* (*Myrmamblys*) *macilentus* var. *vulcanalis* Wheeler, W.M. 1919c: 284 (s.w.q.m.) GALAPAGOS IS. Combination in *C.* (*Pseudocolobopsis*): Emery, 1925b: 158.

*vulpinus. Camponotus vulpinus* Mayr, 1862: 658 (w.q.) VENEZUELA. Junior synonym of *esuriens:* Roger, 1863b: 5.

**vulpus.** *Camponotus* (*Paramyrmamblys*) *vulpus* Santschi, 1926c: 606, fig. 2 (w.) ZAIRE.

*wagneri. Camponotus* (*Myrmothrix*) *abdominalis* r. *stercorarius* var. *wagneri* Forel, 1922: 99 (w.) PERU. Unavailable name.

**walkeri.** *Camponotus walkeri* Forel, 1893d: 454 (s.) AUSTRALIA. Forel, 1910b: 72 (w.). Combination in *C.* (*Myrmophyma*): Forel, 1914a: 269. Current subspecies: nominal plus *bardus.*

**wasmanni.** *Camponotus wasmanni* Emery, 1893e: 224 (s.w.q.) INDIA. Forel, 1895e: 453 (m.). Combination in *C.* (*Myrmepomis*): Forel, 1912i: 92; in *C.* (*Orthonotomyrmex*): Emery, 1920b: 258. Current subspecies: nominal plus *mutilarius.*

**#weberi.** *Camponotus* (*Myrmocladoecus*) *sanctaefidei* subsp. *weberi* Wheeler, W.M. 1935c: 72. Replacement name for *darlingtoni* Wheeler, W.M. 1934g: 232. [Junior primary homonym of *darlingtoni* Wheeler, W.M. 1934d: 160.]

**wedda.** *Camponotus wedda* Forel, 1908a: 6 (s.) SRI LANKA. Forel, 1909e: 397 (w.q.); Forel, 1912n: 67 (w.q.). Combination in *C.* (*Myrmotemnus*): Emery, 1920b: 258; in *C.* (*Myrmosphincta*): Santschi, 1921f: 311; in *C.* (*Myrmamblys*): Forel, 1914a: 272; Emery, 1925b: 139.

**#weiseri.** *Camponotus* (*Myrmoturba*) *bonariensis* st. *weiseri* Santschi, 1922c: 102 (s.w.q.m.) ARGENTINA. Combination in *C.* (*Tanaemyrmex*): Emery, 1925b: 79.

**weismanni.** *Camponotus weismanni* Forel, 1901b: 28 (w.) NEW GUINEA. Combination in *C.* (*Myrmamblys*): Forel, 1914a: 272.

***CAMPONOTUS*** *(continued)*

**#*weissi.*** *Camponotus maculatus* st. *weissi* Santschi, 1911c: 368 (s.w.) CONGO. Combination in *C.* (*Tanaemyrmex*): Emery, 1925b: 99. Subspecies of *congolensis:* Emery, 1920c: 6.

***wellmani.*** *Camponotus wellmani* Forel, 1909b: 67 (s.w.q.) ANGOLA. Combination in *C.* (*Myrmogigas*): Forel, 1912i: 90; in *C.* (*Dinomyrmex*): Forel, 1913b: 340; in *C.* (*Tanaemyrmex*): Emery, 1925b: 89. Current subspecies: nominal plus *gamma, rufipartis.*

***werthi.*** *Camponotus werthi* Forel, 1908a: 18 (w.) KERGUELEN I. Forel, 1914d: 253 (s.); Arnold, 1922: 661 (q.). Combination in *C.* (*Myrmoturba*): Forel, 1914a: 268; in *C.* (*Tanaemyrmex*): Emery, 1925b: 104. Current subspecies: nominal plus *skaifei.*

***westermanni.*** *Camponotus westermanni* Mayr, 1862: 665 (w.) BRAZIL. Combination in *C.* (*Myrmamblys*): Forel, 1914a: 272; in *C.* (*Neomyrmamblys*): Santschi, 1922c: 115; in *C.* (*Myrmaphaenus*): Emery, 1925b: 156. Current subspecies: nominal plus *fulvicornis.*

***wheeleri.*** *Camponotus* (*Myrmothrix*) *wheeleri* Mann, 1916: 477, pl. 6, figs. 49–51 (s.w.) BRAZIL. Combination in *C.* (*Myrmoplatypus*): Santschi, 1921f: 311.

***whitei.*** *Camponotus* (*Myrmosphincta*) *whitei* Wheeler, W.M. 1915g: 818, pl. 66, fig. 8 (w.) AUSTRALIA. Combination in *C.* (*Myrmosaulus*): Emery, 1925b: 114. Senior synonym of *scutellus:* Brown, 1956a: 40.

*whymperi. Camponotus herculeanus* var. *whymperi* Forel, 1902i: 699 (w.q.) CANADA. Wheeler, W.M. 1910d: 330 (s.m.). Raised to species: Ruzsky, 1926: 108. Subspecies of *herculeanus:* Ruzsky, 1936: 90. Junior synonym of *herculeanus:* Creighton, 1950a: 366.

***wiederkehri.*** *Camponotus weiderkehri* Forel, 1894e: 232 (s.w.) AUSTRALIA. Combination in *C.* (*Myrmosphincta*): Forel, 1912i: 92; in *C.* (*Myrmophyma*): Forel, 1914a: 269; in *C.* (*Tanaemyrmex*): Emery, 1925b: 102. Current subspecies: nominal plus *lucidior.*

*wienandsi* Forel, 1910; see under ***CALOMYRMEX.***

***wildae.*** *Camponotus* (*Colobopsis*) *wildae* Donisthorpe, 1948b: 312 (q.) NEW GUINEA.

*willardi. Camponotus hannani* r. *willardi* Forel, 1899c: 132 (footnote) (w.) JAMAICA. Combination in *C.* (*Myrmothrix*): Emery, 1925b: 108. Junior synonym of *hannani:* Hashmi, 1973: 12.

**#*williamsi.*** *Camponotus* (*Myrmoturba*) *conspicuus* r. *williamsi* Forel, 1916: 442 (s.) ANTILLES. Combination in *C.* (*Tanaemyrmex*): Emery, 1925b: 80.

**#*willowsi.*** *Camponotus* (*Myrmocladoecus*) *rectangularis* var. *willowsi* Wheeler, W.M. 1933a: 63 (w.) MEXICO.

**#*winkleri.*** *Camponotus irritabilis* var. *winkleri* Forel, 1911a: 51 (w.) BORNEO. Combination in *C.* (*Myrmotarsus*): Wheeler, W.M. 1919e: 110.

*wittei. Camponotus* (*Myrmoturba*) *punctulatus* r. *hybridus* var. *wittei* Forel, 1915c: 363 (w.) ARGENTINA. Unavailable name.

*wolfi* Emery, 1920; see under ***PHASMOMYRMEX.***

**#*wollebaeki.*** *Camponotus* (*Pseusocolobopsis*) *macilentus* var. *wollebaeki* Stitz, 1932a: 371, fig. 4 (w.q.) GALAPAGOS IS.

***wroughtonii.*** *Camponotus wroughtonii* Forel, 1893b: 430 (w.m.) INDIA. Bingham, 1903: 372 (s.q.). Combination in *C.* (*Camponotus*): Forel, 1914a: 266; in *C.* (*Myrmentoma*): Emery, 1920b: 257.

*wui. Camponotus japonicus* subsp. *wui* Wheeler, W.M. 1929f: 9 (w.) CHINA. Junior synonym of *japonicus:* Yasumatsu & Brown, 1951: 36.

***wytsmani.*** *Camponotus* (*Myrmobrachys*) *wytsmani* Emery, 1920c: 42, fig. 7 (w.) BOLIVIA.

***xanthogaster.*** *Camponotus* (*Myrmoturba*) *xanthogaster* Santschi, 1925b: 19 (s.w.q.m.) BRAZIL. Combination in *C.* (*Tanaemyrmex*): Kempf, 1972a: 73.

**#*xanthomelas.*** *Camponotus maculatus* subsp. *xanthomelas* Emery, 1905d: 34 (s.w.) ALGERIA. Cagniant, 1970c: 32 (q.m.). Combination in *C.* (*Tanaemyrmex*): Emery, 1925b: 101. Subspecies of *sylvaticus:* Emery, 1920c: 7; of *barbaricus:* Cagniant, 1970c: 32.

*xerxes. Camponotus maculatus* r. *xerxes* Forel, 1904f: 424 (w.q.) IRAN. [Date of publication 23.xii.1904.] Karavaiev, 1910b: 9 (s.); Forel, 1913d: 434 (m.). Combination in *C.* (*Tanaemyrmex*): Emery, 1925b: 99. Junior synonym of *cognatocompressus:* Karavaiev, 1910b: 9 (*xerxes* given as senior, but *cognatocompressus* has priority).

**#*xuthus.*** *Camponotus* (*Myrmophyma*) *nigroaeneus* subsp. *xuthus* Emery, 1925b: 111 (s.w.) AUSTRALIA. [First available use of *Camponotus* (*Myrmoturba*) *nigroaeneus* r. *divus* var. *xuthus* Forel, 1915b: 97; unavailable name.]

***yala.*** *Camponotus yala* Kusnezov, 1952d: 226, fig. 12 (s.w.) ARGENTINA. Combination in *C.* (*Myrmobrachys*): Kempf, 1972a: 54.

***yamaokai.*** *Camponotus* (*Myrmamblys*) *yamaokai* Terayama & Satoh, 1990a: 406, figs. 1–13, 15, 17, 19, 21, 22 (s.w.q.m.) JAPAN.

*yankee. Camponotus atriceps* r. *yankee* Forel, 1885a: 340 (w.) U.S.A. Junior synonym of *floridanus:* Mayr, 1886d: 423.

**#*yarrabahensis.*** *Camponotus* (*Myrmoturba*) *maculatus* r. *yarrabahensis* Forel, 1915b: 98 (s.w.) AUSTRALIA. Combination in *C.* (*Tanaemyrmex*): Emery, 1925b: 102. Subspecies of *discors:* Emery, 1920c: 8.

**#*yerburyi.*** *Camponotus reticulatus* r. *yerburyi* Forel, 1893b: 431 (s.w.) SRI LANKA. Combination in *C.* (*Myrmamblys*): Emery, 1925b: 139. Raised to species: Bingham, 1903: 372. Subspecies of *reticulatus:* Emery, 1925b: 139. Senior synonym of *motschulskii:* Emery, 1896d: 374.

***yessensis.*** *Camponotus yessensis* Yasumatsu & Brown, 1951: 41 (s.w.) JAPAN. [First available use of *Camponotus herculeanus* subsp. *vagus* var. *yessensis* Teranishi, 1940: 72; unavailable name.] See also: Kupyanskaya, 1990: 172.

***yiningensis.*** *Camponotus yiningensis* Wang & Wu, 1994: 27, figs. 6, 13 (w.) CHINA.

***yogi.*** *Camponotus yogi* Wheeler, W.M. 1915b: 420 (s.w.) U.S.A. Creighton & Snelling, 1967: 195 (q.m.); Wheeler, G.C. & Wheeler, J. 1970: 650 (l.). Combination in *C.* (*Colobopsis*): Wheeler, W.M. 1917a: 562; in *C.* (*Myrmaphaenus*): Emery, 1925b: 156.

**#*yvonnae.*** *Camponotus* (*Myrmamblys*) *yvonnae* Forel, 1920: 480 (q.) ZAIRE. Combination in *C.* (*Myrmamblys*): Emery, 1925b: 142; in *C.* (*Myrmopelta*): Santschi, 1921f: 311; Santschi, 1926a: 20. Variety of *chrysurus:* Santschi, 1926a: 20.

*zacapensis. Camponotus* (*Myrmepomis*) *sericeiventris* subsp. *rex* var. *zacapensis* Wheeler, W.M. 1931c: 96 (s.w.) GUATEMALA. Unavailable name.

**#*zanzibaricus.*** *Camponotus rufoglaucus* subsp. *zanzibaricus* Forel, 1911e: 287 (w.) TANZANIA. Combination in *C.* (*Myrmosericus*): Emery, 1925b: 107.

***zenon.*** *Camponotus maculatus* r. *zenon* Forel, 1912i: 72 (s.w.) BRAZIL. Combination in *C.* (*Tanaemyrmex*): Emery, 1925b: 84. Subspecies of *picipes:* Emery, 1920b: 233. Raised to species: Emery, 1925b: 84. Current subspecies: nominal plus *criton.*

**#*zeuxis.*** *Camponotus* (*Myrmogonia*) *evae* r. *zeuxis* Forel, 1915b: 101 (w.) AUSTRALIA. Combination in *C.* (*Myrmophyma*): Emery, 1925b: 111.

***zimmermanni.*** *Camponotus zimmermanni* Forel, 1894b: 66 (w.) ETHIOPIA. Combination in *C.* (*Myrmosericus*): Forel, 1914a: 268. Subspecies of *rufoglaucus:* Emery, 1895h: 49 (in key); Wheeler, W.M. 1922a: 969; Emery, 1925b: 107. Revived status as species: Santschi, 1926c: 613. Current subspecies: nominal plus *pansus.*

***zoc.*** *Camponotus senex* st. *zoc* Forel, 1879a: 100 (s.w.) VENEZUELA. Wheeler, G.C. & Wheeler, J. 1953e: 195 (l.). Combination in *C.* (*Myrmobrachys*): Forel, 1914a: 271. Raised to species: Dalla Torre, 1893: 257; Forel, 1899c: 142.

**#*zonatus.*** *Camponotus extensus* subsp. *zonatus* Emery, 1894f: 3 (s.w.) COSTA RICA. Forel, 1908a: 72 (m.). Combination in *C.* (*Myrmoturba*): Forel, 1914a: 268; in *C.* (*Tanaemyrmex*): Emery, 1925b: 80. Race of *landolti:* Forel, 1899c: 136; Wheeler, W.M. 1905b: 117. Raised to species: Forel, 1901e: 371; Emery, 1920b: 255. Junior synonym of *inaequalis:* Wheeler, W.M. 1913b: 503 (in text). Race/subspecies of *conspicuus:* Forel, 1902b: 171; Forel, 1908a: 72; Forel, 1914a: 268; Santschi, 1922c: 101; Emery, 1925b: 80; Kempf, 1972a: 66.

**#*zulu.*** *Camponotus rufoglaucus* subsp. *zulu* Emery, 1895h: 50 (s.w.) SOUTH AFRICA. Combination in *C.* (*Myrmosericus*): Emery, 1925b: 107.

*zumpti. Camponotus* (*Myrmoturba*) *maculatus* var. *zumpti* Santschi, 1937b: 103 (w.) CAMEROUN. Junior synonym of *maculatus:* Baroni Urbani, 1972: 125.

*zuni. Camponotus mina* subsp. *zuni* Wheeler, W.M. 1910d: 346 (s.w.) U.S.A. Combination in *C.* (*Myrmobrachys*): Wheeler, W.M. 1917a: 562. Junior synonym of *mina:* Creighton, 1965b: 5.

---

**EXCURSUS**

**Species-rank taxa of *CAMPONOTUS*, grouped by subgenus.**

The subgenera of *Camponotus* are arranged alphabetically below in ***BOLD ITALIC CAPITALS.*** The species currently assigned to each subgenus follow the subgeneric name in alphabetical order. It should be stressed that many of the subgenera are weak or only poorly defined, and are untrustworthy. A comprehensive review of *Camponotus,* to determine genuine species groups and their relationships, is long overdue and very necessary.

**#*CAMPONOTUS:*** *americanus, atrox, chromaiodes, cilicicus, formosensis, fuscovillosus, hemichlaena, herculeanus, japonicus, keiferi, laevigatus, ligniperdus, modoc, noveboracensis, obscuripes, pennsylvanicus, punctatissimus, quercicola, reichardti, sachalinensis, saxatilis, schaefferi, spanis, texanus, tonkinus, vagus, yessensis.*

**#COLOBOPSIS:** *abditus, aruensis, auratus, aurelianus, badius, bifossus, brachycephalus, bryani, calvus, camelus, cerberulus, ceylonicus, clerodendri, conicus, conithorax, corallinus, cordincola, cotesii, culmicola, custodulus, cylindricus, dentatus, desectus, elysii, equus, etiolatus, excavatus, flavolimbatus, gasseri, gundlachi, guppyi, horrens, horripilus, hosei, howensis, hunteri, impressus, janussus, karawaiewi, laotzei, leonardi, loa, longi, macrocephalus, manni, mathildeae, maudella, mississippiensis, mutilatus, newzealandicus, nigrifrons, nipponicus, obliquus, oceanicus, papago, perneser, phragmaticola, polynesicus, pylartes, pylorus, quadriceps, reepeni, riehlii, rothneyi, rufifrons, sanguinifrons, saundersi, schmitzi, semicarinatus, severini, smithianus, solenobius, sommeri, strictus, trajanus, tricolor, triton, tritschleri, truncatus, vitiensis, vitreus, wildae.*

**#DINOMYRMEX:** *gigas.*

**#HYPERCOLOBOPSIS:** *coptobregma, coriolanus, divergens, paradoxus.*

**#KARAVAIEVIA:** *asli, dolichoderoides, exsectus, gombaki, orinus, overbecki, texens.*

**#MANNIELLA:** *micrositus, sphaericus, torrei.*

**#MAYRIA:** *cambouei, christi, dewitzii, dromedarius, gibber, kelleri, lubbocki, namacola, pictipes, quadrimaculatus, repens, schoutedeni, sellidorsatus.*

**#MYRMACRHAPHE:** *burgeoni, conradti, florius, furvus, gabonensis, longipalpalis, oculatior, rotundinodis.*

**#MYRMAMBLYS:** *albipes, albocinctus, andrewsi, anningensis, annulatus, auriculatus, bedoti, bellus, binghamii, bruneiensis, buchholzi, chapini, criniticeps, dedalus, ephippiatus, fasciatus, flavocassis, gibbosus, greeni, gretae, hapi, humerus, icarus, indeflexus, isabellae, itoi, janeti, klugii, kutteri, lighti, longiceps, nawai, nigronitidus, nipponensis, nirvanae, ogasawarensis, ominosus, pavidus, pictostriatus, poecilus, pulchellus, punctaticeps, reticulatus, rotumanus, schneei, thomasseti, tokioensis, trietericus, varians, viri, wedda, weismanni, yamaokai.*

**#MYRMAPHAENUS:** *andrei, blandus, capperi, championi, chazaliei, clypeatus, cocosensis, compositor, cressoni, declivus, falco, fallatus, fastigatus, fiebrigi, frontalis, genatus, germaini, helleri, hermanni, inca, integellus, iridis, juliae, leydigi, naegelii, novogranadensis, nywet, patimae, pellarius, personatus, rusticus, salvini, sculptor, silvestrii, territus, trapeziceps, ulcerosus, westermanni, yogi.*

**#MYRMENTOMA**: *abrahami, anthrax, atricolor, bakeri, barbiger, boghossiani, brunni, candiotes, caryae, clarithorax, confucii, cornis, cuauhtemoc, dalmaticus, decipiens, discolor, essigi, fallax, figaro, gestroi, hyatti, interjectus, jejuensis, keihitoi, kiesenwetteri, kolthoffi, kopetdaghensis, kurdistanicus, lameerei, lateralis, libanicus, merulus, minus, nadimi, nearcticus, piceus, quadrinotatus, sayi, semirufus, sicheli, snellingi, staryi, subbarbatus, tergestinus, truebi, varius, vogti, wroughtonii.*

**#MYRMEPINOTUS:** *echinoploides, edmondi, ethicus, robustus, sibreei.*

**#MYRMEPOMIS:** *sericeiventris.*

**#MYRMESPERA:** *belligerus, crepusculi, cuneiscapus, debellator, emarginatus, havilandi, nasutus, reevei, transvaalensis, trifasciatus.*

**#MYRMEURYNOTA:** *albistramineus, altivagans, augustei, christophei, christopherseni, eurynotus, gilviventris, humeralis, kraepelini, linnaei, saussurei, thysanopus, toussainti.*

**#MYRMISOLEPIS:** *amphidus, braunsi, erinaceus, maynei.*

**#MYRMOBRACHYS:** *abscissus, arboreus, auricomus, beebei, biolleyi, brasiliensis, brettesi, brevis, burtoni, cameranoi, cameroni, canescens, caracalla, cheesmanae, circularis, conulus, crassicornis, crassus, cuneidorsus, dimorphus, dolabratus, elevatus, excisus, formiciformis, godmani, guayapa, iheringi, lindigi, mina, mus, normatus, pachylepis, peperi, phytophilus, piceatus, pittieri, planatus, propinquellus, propinquus, rubrithorax, scipio, scissus, senex, socorroensis, sphenoidalis, striatus, subcircularis, trapezoideus, trepidulus, vezenyii, wytsmani, yala, zoc.*

**#MYRMOCLADOECUS:** *bidens, bispinosus, callistus, corniculatus, dalmasi, hedwigae, hippocrepis, latangulus, mucronatus, planus, raphaelis, rectangularis, sanctaefidei, tripartitus.*

**#MYRMODIRACHIS:** *heathi.*

**#MYRMOGONIA:** *armstrongi, cameratus, cristatus, laminatus, lauensis, maafui, macareaveyi, schmeltzi.*

**#MYRMOMALIS:** *depressiceps, depressus, emeryodicatus, obtritus.*

**#MYRMONESITES:** *heteroclitus, leveillei, mocquerysi, putatus, reaumuri, sikorae.*

**#MYRMOPALPELLA:** *megalonyx.*

**#MYRMOPELTA:** *arminius, barbarossa, chrysurus, kollbrunneri, vividus.*

**#MYRMOPHYMA:** *aeneopilosus, arcuatus, bigenus, capito, ceriseipes, chalceoides, chalceus, cinereus, cowlei, cyrtomyrmodes, difformis, dromas, ephippium, esau, fieldellus, flavocrines, froggatti, gouldianus, hartogi, inflatus, innexus, insipidus, lividicoxis, mussolinii, nigroaeneus, oxleyi, pellax, peseshus, punctiventris, sponsorum, tasmani, tristis, walkeri.*

**#MYRMOPIROMIS:** *berthoudi, brevisetosus, darwinii, descarpentriesi, detritus, ellioti, fulvopilosus, niveosetosus, nossibeensis, radovae, scalaris, storeatus, tameri, ursus, victoriae, voeltzkowii.*

**#MYRMOPLATYPUS:** *banghaasi, mocsaryi, platytarsus, plutus, wheeleri.*

**#MYRMOPLATYS:** *beccarii, contractus, hospes, korthalsiae.*

**#MYRMOPSAMMA:** *mystaceus, simulans.*

**#MYRMOPYTIA:** *imitator.*

**#MYRMOSAGA:** *afflatus, callmorphus, ezotus, favorabilis, jeanneli.*

**#MYRMOSAULUS:** *auriventris, aurocinctus, batesii, camelinus, holosericeus, intrepidus, leae, midas, molossus, singularis, suffusus, versicolor, whitei.*

**#MYRMOSERICUS:** *angusticeps, aurosus, cinctellus, cinerascens, compressiscapus, cosmicus, cruentatus, dolendus, druryi, eugeniae, flavomarginatus, hoelldobleri, leucophaeus, micans, mozabensis, obscuriventris, parius, petersii, pexus, pseudolendus, rufoglaucus, valdeziae, vestitus, zimmermanni.*

**#MYRMOSPHINCTA:** *constructor, lancifer, reinaldi, sexguttatus, urichi.*

**#MYRMOSTENUS:** *longipilis, mirabilis, sphenocephalus.*

**#MYRMOTARSUS:** *barbosus, gilviceps, irritabilis, misturus, nigricans, platypus, pressipes, quadrisectus, rufifemur, rufigaster, satan, sedulus.*

**#MYRMOTEMNUS:** *hypoclinoides, impressilabris, moeschi, nutans, reichenspergeri.*

**#MYRMOTHRIX:** *atriceps, bugnioni, cingulatus, cordiceps, femoratus, floridanus, hannani, lenkoi, opaciceps, punctatus, renggeri, rufipes, sericatus.*

**#MYRMOTREMA:** *aequitas, argus, atriscapus, aurofasciatus, auropubens, avius, bayeri, benguelensis, bituberculatus, bottegoi, carbo, confluens, crucheti, diplopunctatus, fayfaensis, foraminosus, galla, grandidieri, haereticus, ilgii, lamborni, monardi, olivieri, orthodoxus, perrisii, postoculatus, puberulus, radiatus, rhamses, robecchii, tauricollis, tilhoi.*

**#MYRMOXYGENYS:** *caesar.*

**#ORTHONOTOMYRMEX:** *cubangensis, fletcheri, kosswigi, lasiselene, liogaster, mayri, mendax, puniceps, sankisianus, scabrinodis, selene, sericeus, wasmanni, yiningensis.*

**#PARAMYRMAMBLYS:** *amamianus, bertolonii, brookei, ferreri, kiusiuensis, lilianae, limbiventris, moderatus, orinobates, orites, ostiarius, prosulcatus, robertae, simus, vulpus.*

**#PSEUDOCOLOBOPSIS:** *alboannulatus, claviscapus, curviscapus, erythrostoma, fugax, geayi, improprius, leptocephalus, macilentus, macrocephalus* (junior homonym), *medeus, moelleri, montivagus, orthocephalus, semoni, subtruncatus, terbimaculatus, tonduzi, ulei, ulvarum.*

**#RHINOMYRMEX:** *klaesii.*

**#TANAEMYRMEX:** *abjectus, abunanus, acutirostris, acvapimensis, adenensis, aegyptiacus, aequatorialis, aethiops, ager, agonius, aguilerai, alacer, albicoxis, albosparsus, alii, amoris, anguliceps, angusticollis, apostolus, arabicus, arnoldinus, arrogans, atlantis, attila, autrani, babau, bactrianus, baldaccii, balzani, barbaricus, barbatus, basuto, baynei, bermudezi, bianconii, bocki, bonariensis, borellii, brevicollis, britteni, bruchi, brutus, buddhae, butteli, buttikeri, cacicus, caffer, carin, cacicus, castaneus, catalanus, cecconii, cervicalis, chilensis, chloroticus, chongqingensis, cillae, circumspectus, cleobulus, cognatocompressus, compressus, concolor, congolensis, coniceps, consanguineus, consectator, consobrinus, conspicuus, coruscus, coxalis, crassisquamis, crawleyi, deletangi, denticulatus, desantii, devestitus, dicksoni, discors, distinguendus, diversipalpus, donisthorpei, dorycus, dracocephalus, dufouri, dumetorum, egregius, empedocles, erigens, errabundus, etiolipes, evansi, exiguoguttatus, extensus, fabricator, fasciatellus, fedtschenkoi, fellah, ferganensis, fervidus, festai, festinatus, festinus, fieldeae, flavescens, foleyi, foreli, fornasinii, friedae, fryi, fumidus, fuscocinctus, gambeyi, gerberti, glabrisquamis, goeldii, gouldi, guttatus, habereri, haematocephalus, hagensii, hastifer, hellmichi, heros, hildebrandti, holzi, hoplites, horni, hova, ignestii, immigrans, importunus, incensus, inconspicuus, indefinitus, indicatus, invidus, ionius, irritans, jaliensis, jianghuaensis, kersteni, knysnae, koseritzi, laconicus, lamarckii, landolti, langi, largiceps, larvigerus, latebrosus, latrunculus, legionarum, lespesii, ligeus, lucayanus, luteiventris, luteus, lutzi, maccooki, macrochaeta, maculatus, magister, maguassa, massiliensis, massinissa, maxwellensis, melanocephalus, melanoticus, melichloros, minozzii, mitis, morosus, myoporus, nasicus, natalensis, navigator, nepos, nicobarensis, nigriceps, nitens, novaehollandiae, nylanderi, oasium, obliquipilosus, oblongus, obreptivus, occasus, ocreatus, odiosus,*

***CAMPONOTUS** (continued)*

*oertzeni, ovaticeps, pallens, pallescens, palpatus, perroti, picipes, pilicornis, pompeius, postcornutus, posticus, pseudoirritans, pullatus, pulvinatus, punctulatus, ramulorum, rapax, roeseli, roubaudi, rubidus, rubripes, ruficornis, rufus, ruzskyi, ruzskyellus, sacchii, samius, sanctus, sansabeanus, santosi, semitestaceus, sesquipedalis, setitibia, sexpunctatus, siemsseni, silvicola, simillimus, simoni, simulator, skalus, socius, socrates, solon, somalinus, spenceri, spinitarsus, spinolae, subnitidus, substitutus, subtilis, sucki, sylvaticus, tahatensis, taipingensis, tenuipes, tenuiscapus, tepicanus, terebrans, terricola, testaceus, thales, thoracicus, thraso, tichomirovi, tortuganus, traegaordhi, tricoloratus, turkestanicus, turkestanus, universitatis, ustus, vafer, variegatus, varus, vicinus, villosus, virulens, vittatus, wellmani, werthi, wiederkehri, xanthogaster, zenon.*

***#THLIPSEPINOTUS:** adami, aureopilus, claripes, eremicus, evae, gibbinotus, lownei, michaelseni, oetkeri, rubiginosus, scratius, tumidus, viehmeyeri.*

**UNPLACED TO SUBGENUS OR SUBGENUS UNKNOWN**: *baronii, branneri, carbonarius, castanicola, crenatus, helvus, horseshoetus, jizani, johnclarki, kutterianus, lessonai, luctuosus, nacerdus, placidus, politae, tashcumiri, timidus, velox, vespertinus, vigilans, vinosus.*

**UNRECOGNISABLE TAXA, *INCERTAE SEDIS* IN *CAMPONOTUS*:** *americana, atra, septentrionalis, tejonia.*

---

*CAMPOSTIGMACROS:* see under ***STIGMACROS.***

*CAMPTOGNATHA:* see under ***ECITON.***

*CAMPYLOMYRMA:* see under *CONDYLOMYRMA.*

***CARDIOCONDYLA*** (Worldwide)

*atalanta. Cardiocondyla nuda* var. *atalanta* Forel, 1915b: 75 (w.) AUSTRALIA. Junior synonym of *nuda:* Taylor, 1991b: 601.

*badonei. Cardiocondyla badonei* Arnold, 1926: 225, fig. 64 (w.) MOZAMBIQUE. Stirps of *shuckardi:* Santschi, 1932a: 387. Junior synonym of *shuckardi:* Bolton, 1982: 316.

***batesii.*** *Cardiocondyla batesii* Forel, 1894d: 17 (w.q.) ALGERIA.

*bicolor. Cardiocondyla bicolor* Donisthorpe, 1930b: 366 (w.) ISRAEL. Junior synonym of *wroughtonii:* Kugler, J. 1984: 6.

*bimaculata. Cardiocondyla wroughtoni* var. *bimaculata* Wheeler, W.M. 1929g: 43 (w.q.) TAIWAN. Subspecies of *wroughtonii:* Smith, M.R. 1944a: 40; Creighton, 1950a: 199. Junior synonym of *wroughtonii:* Smith, D.R. 1979: 1376.

***bogdanovi.*** *Cardiocondyla bogdanovi* Ruzsky, 1905b: 630, figs. 155, 157, 158 (w.q.) CAUCASUS.

*brevispinosa. Pheidole* (*Pheidolacanthinus*) *brevispinosa* Donisthorpe, 1948d: 593 (w.) NEW GUINEA. Junior synonym of *paradoxa:* Smith, M.R. 1955d: 305.

*brevispinosa. Cardiocondyla brevispinosa* Weber, 1952: 6 (w.) ZAIRE. [Unresolved junior secondary homonym of *Pheidole brevispinosa* Donisthorpe, above.] Junior synonym of *shuckardi:* Bolton, 1982: 316.

***britteni.*** *Cardiocondyla britteni* Crawley, 1920c: 180, fig. (w.) GREAT BRITAIN.

***bulgarica.*** *Cardiocondyla elegans* var. *bulgarica* Forel, 1892i: 312, pl. 5, fig. 4 (w.q.) BULGARIA. Raised to species: Agosti & Collingwood, 1987a: 56; Atanassov & Dlussky, 1992: 174.

***carbonaria.*** *Cardiocondyla carbonaria* Forel, 1907a: 17 (w.) INDIA.

*caucasica. Cardiocondyla elegans* subsp. *caucasica* Arnol'di, 1948: 211. *Nomen nudum.*

*chlorotica. Cardiocondyla emeryi* subsp. *chlorotica* Menozzi, 1930b: 84 (w.q.) SOMALIA. Junior synonym of *wroughtonii:* Bolton, 1982: 317.

*clarior. Cardiocondyla elegans* subsp. *ulianini* var. *clarior* Kuznetsov-Ugamsky, 1927d: 38 (w.) RUSSIA. Unavailable name.

***cristata.*** *Monomorium cristatum* Santschi, 1912b: 163, fig. D (w.) MADAGASCAR. Combination in *Cardiocondyla:* Santschi, 1913c: 306; in *C.* (*Dyclona*): Santschi, 1930b: 70.

***#dalmatica.*** *Cardiocondyla elegans* var. *dalmatica* Soudek, 1925a: 34 (w.) YUGOSLAVIA. [Also described as new by Soudek, 1925b: 14.]

***ectopia.*** *Cardiocondyla ectopia* Snelling, 1974: 76, figs. 1–5 (w.q. ergatoid m.) U.S.A.

***elegans.*** *Cardiocondyla elegans* Emery, 1869b: 21, pl. 1, fig. 10 (w.q.) ITALY. Menozzi, 1918: 83 (ergatoid m.); Kugler, J. 1984: 14 (ergatoid m.). Senior synonym of *santschii:* Bondroit, 1918: 147; Emery, 1922e: 125; of *gallica:* Baroni Urbani, 1973: 200; of *ulianini:* Dlussky, Soyunov & Zabelin, 1990: 194. Current subspecies: nominal plus *dalmatica, eleonorae, schkaffi, torretassoi.* See also: Atanassov & Dlussky, 1992: 172.

***#eleonorae.*** *Cardiocondyla elegans* var. *eleonorae* Forel, 1911d: 337 (w.) TURKEY.

***emeryi.*** *Cardiocondyla emeryi* Forel, 1881: 5 (w.) VIRGIN IS. André, 1881b: 69 (m.); Forel, 1904f: 422 (q.); Emery, 1909a: 26 (m. ergatoid m., not q.); Arnold, 1916: 201 (q.). Senior synonym of *nereis:* Wilson & Taylor, 1967: 53; of *monilicornis:* Baroni Urbani, 1973: 200; of *mahdii, mauritia* and *rasalamae:* Bolton, 1982: 312. See also: Smith, M.R. 1944a: 33; Kugler, J. 1984: 3. Current subspecies: nominal plus *fezzanensis, schatzmayri.*

***#fajumensis.*** *Cardiocondyla nuda* var. *fajumensis* Forel, 1913d: 429 (q.) EGYPT.

***#fezzanensis.*** *Cardiocondyla emeryi* subsp. *fezzanensis* Bernard, 1948: 142 (q.) ALGERIA.

*fusca. Cardiocondyla fusca* Weber, 1952: 7 (w.) UGANDA. Junior synonym of *shuckardi:* Bolton, 1982: 316.

*gallica. Xenometra gallica* Bernard, 1957b: 101, fig. 1 (ergatoid m., not q.) FRANCE. Junior synonym of *elegans:* Baroni Urbani, 1973: 200.

*gibbosa. Cardiocondyla elegans* subsp. *gibbosa* Kuznetsov-Ugamsky, 1927d: 37 (w.) TURKESTAN. Junior synonym of *koshewnikovi:* Dlussky, Soyunov & Zabelin, 1990: 195.

*globinodis. Cardiocondyla globinodis* Stitz, 1923: 154 (w.) NAMIBIA. Junior synonym of *shuckardi:* Bolton, 1982: 316.

*hawaiensis. Cardiocondyla wroughtonii* var. *hawaiensis* Forel, 1899a: 119 (w.) HAWAII. Forel, 1902h: 440 (q.); Santschi, 1919a: 328 (m.); Smith, M.R. 1944a: 30 (ergatoid m.). Junior synonym of *wroughtonii:* Wilson & Taylor, 1967: 56.

***jacquemini.*** *Cardiocondyla jacquemini* Bernard, 1953a: 149, fig. 3 (w.) ALGERIA.

***koshewnikovi.*** *Cardiocondyla koshewnikovi* Ruzsky, 1902b: 480 (w.q.) RUSSIA. [Also described as new by Ruzsky, 1902c: 16.] Variety/subspecies of *stambuloffii:* Forel, 1902h: 440; Kuznetsov-Ugamsky, 1927d: 37. Junior synonym of *stambuloffii:* Pisarski, 1970: 308. Revived from synonymy as subspecies of *stambuloffii:* Tarbinsky, 1976: 72; Dlussky & Zabelin, 1985: 213. Revived status as species, and senior synonym of *gibbosa:* Dlussky, Soyunov & Zabelin, 1990: 195.

***kushanica.*** *Cardiocondyla kushanica* Pisarski, 1967: 386, figs. 16–22 (w.q.) AFGHANISTAN.

***longispina.*** *Cardiocondyla longispina* Karavaiev, 1935a: 88, fig. 14 (w.) JAVA.

*mahdii. Cardiocondyla emeryi* subsp. *mahdii* Karavaiev, 1911: 8 (w.) SUDAN. Finzi, 1936: 169 (q.m.). Junior synonym of *emeryi:* Bolton, 1982: 312.

***mauritanica.*** *Cardiocondyla nuda* var. *mauritanica* Forel, 1890a: lxxv (w.) TUNISIA. Forel, 1901e: 378 (q.); Forel, 1904c: 6 (ergatoid m.). Race of *nuda:* Forel, 1901e: 378. Raised to species: Ortiz & Tinaut, 1987: 32. Senior synonym of *nitida:* Bernard, 1956c: 305.

*mauritia. Cardiocondyla mauritia* Donisthorpe, 1946c: 776 (w.) MAURITIUS. Junior synonym of *emeryi:* Bolton, 1982: 312.

*minutior. Cardiocondyla nuda* var. *minutior* Forel, 1899a: 120 (w.) HAWAII. Wheeler, W.M. 1922f: 317 (q.). Subspecies of *nuda:* Smith, M.R. 1944a: 38. Creighton, 1950a: 198. Junior synonym of *nuda:* Wilson & Taylor, 1967: 55.

***monardi.*** *Cardiocondyla* (*Loncyda*) *monardi* Santschi, 1930b: 70, fig. 5 (w.) ANGOLA. See also: Bolton, 1982: 314.

*monilicornis. Xenometra monilicornis* Emery, 1917a: 96 (ergatoid m., not q.) VIRGIN IS. Junior synonym of *emeryi:* Baroni Urbani, 1973: 200.

*montandoni. Cardiocondyla montandoni* Santschi, 1912g: 657, figs. 1, 3 (w.) ROMANIA. Junior synonym of *stambuloffii:* Pisarski, 1962a: 331.

***neferka.*** *Cardiocondyla neferka* Bolton, 1982: 314, fig. 7 (w.) GHANA.

*nereis. Cardiocondyla nuda* subsp. *nereis* Wheeler, W.M. 1927i: 140 (w.q.) AUSTRALIA. Junior synonym of *emeryi:* Wilson & Taylor, 1967: 53.

***nigra.*** *Cardiocondyla batesii* var. *nigra* Forel, 1905b: 174 (w.q.) TUNISIA. Santschi, 1907: 324 (m. ergatoid m. and gynandromorph). Raised to species: Schembri & Collingwood, 1981: 428.

***nigrocerea.*** *Cardiocondyla nigrocerea* Karavaiev, 1935a: 89, fig. 15 (w.) INDONESIA (Ambon I.).

***nilotica.*** *Cardiocondyla nilotica* Weber, 1952: 8, fig. 13 (w.) SUDAN. See also: Bolton, 1982: 315.

*nitida. Cardiocondyla emeryi* subsp. *nitida* Bernard, 1948: 142 (w.) ALGERIA. Junior synonym of *mauritanica:* Bernard, 1956c: 305.

***nivalis.*** *Cardiocondyla nivalis* Mann, 1919: 317 (w.) SOLOMON IS.

**nuda.** *Leptothorax nudus* Mayr, 1866a: 508 (w.) FIJI IS. Emery, 1897c: 588 (q. ergatoid m.); Wheeler, G.C. & Wheeler, J. 1973a: 27 (l.); Imai, *et al.* 1984: 6 (k.). Combination in *Cardiocondyla:* Forel, 1881: 6. Senior synonym of *minutior:* Wilson & Taylor, 1967: 55; of *atalanta:* Taylor, 1991b: 601. Current subspecies: nominal plus *fajumensis, sculptinodis, shuckardoides, strigifrons.*

**#obscurior.** *Cardiocondyla wroughtoni* var. *obscurior* Wheeler, W.M. 1929g: 44 (w.q.) TAIWAN.

**papuana.** *Prosopidris papuana* Reiskind, 1965: 81, figs. 1–8 (w.q. ergatoid m.) NEW GUINEA. Combination in *Cardiocondyla:* Bolton, 1982: 311 (in text).

**paradoxa.** *Cardiocondyla paradoxa* Emery, 1897c: 589, pl. 15, figs. 22, 23 (w.) NEW GUINEA. Senior synonym of *brevispinosa:* Smith, M.R. 1955d: 305.

**parvinoda.** *Cardiocondyla parvinoda* Forel, 1902c: 213 (w.) INDIA.

**provincialis.** *Cardiocondyla provincialis* Bernard, 1956c: 303, fig. 6 (w.) FRANCE.

**#quadraticeps.** *Cardiocondyla wroughtonii* subsp. *quadraticeps* Forel, 1912n: 57 (w.) SINGAPORE. Viehmeyer, 1916a: 122 (q.).

*rasalamae. Cardiocondyla emeryi* var. *rasalamae* Forel, 1891b: 161 (w.) MADAGASCAR. Forel, 1912k: 163 (q.). Junior synonym of *emeryi:* Bolton, 1982: 312.

*sabulosa. Cardiocondyla stambulovi* subsp. *taurica* morpha *sabulosa* Arnol'di, 1928a: 131. Unavailable name.

**sahlbergi.** *Cardiocondyla elegans* var. *sahlbergi* Forel, 1913d: 429 (q.) ISRAEL. Finzi, 1936: 166 (w.). Raised to species: Agosti & Collingwood, 1987a: 56.

*salina. Cardiocondyla stambulovi* subsp. *taurica* morpha *salina* Arnol'di, 1928a: 131. Unavailable name.

*santschii. Cardiocondyla elegans* r. *santschii* Forel, 1905b: 174 (w.) FRANCE. Forel, 1910a: 23 (q.). Junior synonym of *elegans:* Bondroit, 1918: 147.

**#schatzmayri.** *Cardiocondyla emeryi* subsp. *schatzmayri* Finzi, 1936: 170, fig. 5 (w.q.) EGYPT.

**schkaffi.** *Cardiocondyla elegans* subsp. *schkaffi* Arnol'di, 1933b: 599 (w.q.) RUSSIA.

**#sculptinodis.** *Cardiocondyla shuckardi* st. *sculptinodis* Santschi, 1913c: 305 (q.) MADAGASCAR. Subspecies of *shuckardi:* Wheeler, W.M. 1922a: 1021; of *nuda:* Emery, 1922e: 126.

*sculptior. Cardiocondyla wasmanni* var. *sculptior* Santschi, 1926b: 241 (w.) GABON. Junior synonym of *shuckardi:* Bolton, 1982: 316.

**sekhemka.** *Cardiocondyla sekhemka* Bolton, 1982: 315 (w.) GHANA.

*semenowi. Cardiocondyla elegans* var. *semenowi* Forel, 1911d: 337 (w.). *Nomen nudum,* attributed to Ruzsky.

**shuckardi.** *Cardiocondyla shuckardi* Forel, 1891b: 161 (w.) MADAGASCAR. Race of *nuda:* Forel, 1895c: 251. Revived status as species: Wheeler, W.M. 1922a: 1021. Senior synonym of *badonei, brevispinosa, fusca, globinodis, sculptior* and *wassmanni:* Bolton, 1982: 316.

**#shuckardoides.** *Cardiocondyla nuda* var. *shuckardoides* Forel, 1895c: 250 (w.) MADAGASCAR.

**sima.** *Cardiocondyla (Prosopidris) sima* Wheeler, W.M. 1935b: 41, fig. 1 (w.q.) PHILIPPINES. Combination in *Prosopidris:* Reiskind, 1965: 84; in *Cardiocondyla:* Bolton, 1982: 309.

**stambuloffii.** *Cardiocondyla stambuloffii* Forel, 1892i: 310, pl. 5, fig. 1 (w.q. ergatoid m.) BULGARIA. Kugler, J. 1984: 8 (ergatoid m.). Senior synonym of *montandoni:* Pisarski, 1962a: 331; of *taurica:* Arnol'di & Dlussky, 1978: 538. See also: Atanassov & Dlussky, 1992: 174.

**#strigifrons.** *Cardiocondyla nuda* subp. *strigifrons* Viehmeyer, 1922: 211 (w.) JAVA.

*taurica. Cardiocondyla stambuloffi* subsp. *taurica* Karavaiev, 1927a: 288, fig. 2 (w.) UKRAINE. Junior synonym of *stambuloffii:* Arnol'di & Dlussky, 1978: 538.

**thoracica.** *Myrmica thoracica* Smith, F. 1859a: 148 (w.) INDONESIA (Aru I.). Combination in *Cardiocondyla:* Donisthorpe, 1932c: 455.

**tjibodana.** *Cardiocondyla tjibodana* Karavaiev, 1935a: 87, fig. 13 (w.) JAVA.

**#torretassoi.** *Cardiocondyla elegans* var. *torretassoi* Finzi, 1936: 167 (w.) EGYPT.

*ulianini. Cardiocondyla elegans* var. *ulianini* Emery, 1889a: 441 (w.) RUSSIA. Karavaiev, 1910b: 57 (q.); Wheeler, G.C. & Wheeler, J. 1953d: 188 (l.). Subspecies of *elegans:* Kuznetsov-Ugamsky, 1927d: 38; Pisarski, 1967: 388; Dlussky & Zabelin, 1985: 213. Raised to species: Agosti & Collingwood, 1987a: 56. Junior synonym of *elegans:* Dlussky, Soyunov & Zabelin, 1990: 194.

**venustula.** *Cardiocondyla venustula* Wheeler, W.M. 1908a: 128, pl. 11, fig. 5 (w.q.) PUERTO RICO. See also: Smith, M.R. 1944a: 37.

*wassmanni. Cardiocondyla wassmanni* Santschi, 1926b: 241 (w.) CAMEROUN. Junior synonym of *shuckardi:* Bolton, 1982: 316.

**weserka.** *Cardiocondyla weserka* Bolton, 1982: 317, fig. 6 (w.) CAMEROUN.

**wheeleri.** *Cardiocondyla wheeleri* Viehmeyer, 1914b: 518, fig. 2 (w.) NEW GUINEA.

**wroughtonii.** *Emeryia wroughtonii* Forel, 1890b: cxi (ergatoid m., not w.) INDIA. Forel, 1903a: 689 (w.q.); Borgmeier, 1937a: 129 (ergatoid m.); Kugler, J. 1984: 7 (m. ergatoid m.). Combination in *Cardiocondyla:* Forel, 1892h: 461; Forel, 1892i: 313. Senior synonym of *hawaiensis:* Wilson & Taylor, 1967: 56; of *bimaculata:* Smith, D.R. 1979: 1376; of *chlorotica:* Bolton, 1982: 317; of *bicolor:* Kugler, J. 1984: 6. Current subspecies: nominal plus *obscurior, quadraticeps.*

**zoserka.** *Cardiocondyla zoserka* Bolton, 1982: 318 (q.) NIGERIA.

*CARDIOPHEIDOLE:* see under ***PHEIDOLE.***

**CAREBARA** (Old and New World Tropics)

*abdominalis. Carebara vidua* var. *abdominalis* Santschi, 1912d: 285 (q.) KENYA. Stirps of *vidua:* Santschi, 1914b: 84. Junior synonym of *vidua:* Forel, 1913b: 337; Arnold, 1916: 252; Wheeler, W.M. 1922a: 884.

**ampla.** *Carebara ampla* Santschi, 1912d: 284 (q.m.) ZAIRE. Current subspecies: nominal plus *cincta, obscurithorax, rugosa.*

**anophthalma.** *Oligomyrmex anophthalmus* Emery, 1906c: 138 (footnote) (w.) BRAZIL. Combination in *Carebara:* Wheeler, W.M. 1922a: 170; Emery, 1922f: 221.

**arnoldi.** *Oligomyrmex arnoldi* Forel, 1913a: 123 (w.) ZIMBABWE. Combination in *Carebara:* Santschi, 1914d: 363. See also: Arnold, 1916: 253.

**#augustata.** *Carebara castanea* var. *augustata* Santschi, 1920h: 162 (q.m.) LAOS. [Also described as new by Santschi, 1924c: 100.]

**bartrumi.** *Carebara bartrumi* Weber, 1943c: 364, pl. 15, figs. 17, 20 (q.m.) SUDAN.

**bicarinata.** *Carebara bicarinata* Santschi, 1912a: 139, figs. 1, 2 (q.m.) GUIANA.

**castanea.** *Carebara castanea* Smith, F. 1858b: 178 (q.) HONG KONG. Current subspecies: nominal plus *augustata.*

**#cincta.** *Carebara ampla* var. *cincta* Santschi, 1926b: 230 (q.) ZAIRE.

*colossus. Carebara colossus* Gerstäcker, 1859: 263 (q.) MOZAMBIQUE. [Also as new in Gerstäcker, 1862: 512.] Junior synonym of *vidua:* Mayr, 1866b: 904.

*dux. Carebara dux* Smith, F. 1858b: 179 (q.) SOUTH AFRICA. Variety of *vidua:* Forel, 1913b: 337. Junior synonym of *vidua:* Mayr, 1866b: 904; Santschi, 1913c: 307.

**#fur.** *Carebara vidua* var. *fur* Santschi, 1928f: 198 (q.) CONGO.

**incerta.** *Carebara incerta* Santschi, 1923d: 66 (m.) BRAZIL.

**junodi.** *Carebara junodi* Forel, 1904a: 154 (q.) SOUTH AFRICA. Forel, 1909b: 53 (m.); Forel, 1913b: 336 (w.). Subspecies/race of *vidua:* Forel, 1909b: 53; Forel, 1913b: 336; Arnold, 1916: 252. Revived status as species: Santschi, 1914d: 364 (in key); Wheeler, W.M. 1922a: 883; Emery, 1922f: 220; Santschi, 1935a: 263.

**langi.** *Carebara langi* Wheeler, W.M. 1922a: 173 (q.) ZAIRE.

**lignata.** *Carebara lignata* Westwood, 1840b: 86, pl. 2, fig. 6 (q.) JAVA. Emery, 1889b: 505 (w.m.); Forel, 1903a: 691 (w.m.); Wheeler, G.C. & Wheeler, J. 1954a: 142 (l.). See also: Bingham, 1903: 151.

**mayri.** *Tranopelta mayri* Forel, 1901h: 61 (m.) PARAGUAY. Combination in *Carebara:* Santschi, 1912a: 140.

**#obscurithorax.** *Carebara ampla* var. *obscurithorax* Santschi, 1926b: 230 (q.) ZAIRE. Santschi, 1928f: 197 (m.).

**osborni.** *Carebara osborni* Wheeler, W.M. 1922a: 174, fig. 41 (w.q.m.) ZAIRE.

**patrizii.** *Carebara patrizii* Menozzi, 1927f: 357, figs. 1, 2 (m.) SOMALIA.

**#rugosa.** *Carebara ampla* var. *rugosa* Santschi, 1928f: 197 (q.) ZAIRE.

**sicheli.** *Carebara sicheli* Mayr, 1862: 754 (q.) SENEGAL.

**silvestrii.** *Carebara silvestrii* Santschi, 1914d: 362, fig. 23 (w.) GUINEA.

**sudanica.** *Carebara sudanica* Santschi, 1933b: 103 (q.m.) SUDAN.

**vidua.** *Carebara vidua* Smith, F. 1858b: 179 (m.) SOUTH AFRICA. Mayr, 1866b: 904 (q.); Forel, 1901g: 392 (w.). Senior synonym of *colossus* and *dux:* Mayr, 1866b: 904; Santschi, 1913c: 307; of *abdominalis:* Forel, 1913b: 337; Arnold, 1916: 252; Wheeler, W.M. 1922a: 884. See also: Arnold, 1916: 249; Robertson & Villet, 1989: 121. Current subspecies: nominal plus *fur.*

**CAREBARA** *(continued)*

**winifredae.** *Carebara winifredae* Wheeler, W.M. 1922e: 2, fig. 1 (w.q.m.) GUIANA. Wheeler, G.C. & Wheeler, J. 1954a: 142 (l.).

**CAREBARELLA** (Neotropical)

**alvarengai.** *Carebarella alvarengai* Kempf, 1975c: 352, figs. 6–8 (q.) BRAZIL.

**bicolor.** *Carebarella bicolor* Emery, 1906c: 138 (q.) ARGENTINA. Borgmeier, 1948b: 465 (w.m.). Senior synonym of *punctatorugosa:* Kempf, 1969: 281.

**condei.** *Carebarella* (*Carebarelloides*) *condei* Borgmeier, 1937b: 236, figs. 17–20, 22–24, pl. 2, figs. 2, 3 (q.m.) BRAZIL.

*punctatorugosa. Carebarella bicolor* var. *punctatorugosa* Emery, 1906c: 139 (footnote), fig. 18 (q.) ARGENTINA. Eidmann, 1936: 43 (w.l.). Raised to species: Ettershank, 1966: 115. Junior synonym of *bicolor:* Kempf, 1969: 281.

*CAREBARELLOIDES:* see under **CAREBARELLA.**

***CARIRIDRIS** (Brazil)

***bipetiolata.** **Cariridris bipetiolata* Brandão & Martins-Neto, 1990: 201, figs. 1–3 (q.) BRAZIL (Cretaceous).

***CASALEIA** (Russia)

***inversa.** **Protamblyopone inversa* Dlussky, 1981b: 65, fig. 50 (w.) RUSSIA (Miocene). Combination in *Casaleia:* **new combination**. [Note. Pagliano & Scaramozzino, 1990: 5, did not indicate the combination of this species in *Casaleia.*]

**CATAGLYPHIS** (Palaearctic, Afrotropical, Oriental)

**#abdominalis.** *Cataglyphis* (*Monocombus*) *emeryi* var. *abdominalis* Santschi, 1929b: 37 (w.) TURKMENIA.

**abyssinicus.** *Myrmecocystus viaticus* r. *abyssinicus* Forel, 1904b: 382 (w.q.) ETHIOPIA. Combination in *Cataglyphis:* Wheeler, W.M. 1922a: 944. Variety\stirps of *bicolor:* Emery, 1908g: 217; Santschi, 1929b: 51. Raised to species: Collingwood, 1985: 286.

**adenensis.** *Myrmecocystus viaticus* r. *adenensis* Forel, 1904b: 382 (footnote) (w.) YEMEN. Combination in *Cataglyphis:* Wheeler, W.M. 1922a: 944. Subspecies of *bicolor:* Emery, 1906d: 58; Emery, 1908g: 217. Raised to species: Santschi, 1929b: 41. Current subspecies: nominal plus *bugnioni.*

**#adustus.** *Cataglyphis bicolor* var. *adusta* Santschi, 1929b: 43 (w.q.m.) TUNISIA.

**aenescens.** *Formica aenescens* Nylander, 1849: 37 (w.q.m.) RUSSIA. Combination in *Myrmecocystus:* Emery & Forel, 1879: 449; in *Cataglyphis:* Mayr, 1861: 45; Müller, 1923: 149; in *C.* (*Monocombus*): Santschi, 1929b: 35. Junior synonym of *cursor:* Mayr, 1861: 45. Revived from synonymy: Emery & Forel, 1879: 449. Subspecies of *cursor:* Emery, 1906d: 47; Wheeler, W.M. & Mann, 1916: 173; Santschi, 1921a: 116; Karavaiev, 1924: 301; Santschi, 1929b: 35; Karavaiev, 1936: 263. Revived status as species: Pisarski, 1967: 416; Tarbinsky, 1976: 203; Arnol'di & Dlussky, 1978: 554; Dlussky & Zabelin, 1985: 239; Agosti, 1990a: 1480; Atanassov & Dlussky, 1992: 294. Senior synonym of *caspius:* Karavaiev, 1910b: 18; Lomnicki, 1925a: 160; of *jakobsoni,* and material of the unavailable name *flavigastra* referred here: Dlussky, Soyunov & Zabelin, 1990: 149. Current subspecies: nominal plus *chatkalensis, georgicus, tancrei.*

*agilis. Cataglyphis* (*Cataglyphis*) *albicans* st. *rubra* var. *agilis* Santschi, 1929b: 39 (w.) TUNISIA. Unavailable name (Agosti, 1990a: 1498).

*agnata. Cataglyphis* (*Cataglyphis*) *albicans* st. *livida* var. *agnata* Santschi, 1929b: 62 (w.) ALGERIA. [Also described as new by Santschi, 1929c: 107; and by Santschi, 1934f: 176.] Unavailable names (Agosti, 1990a: 1498).

**albicans.** *Formica albicans* Roger, 1859: 235 (w.) NORTH AFRICA. André, 1882b: 174 (m.); Forel, 1894d: 11 (q.); Santschi, 1921d: 76 (gynandromorph). Combination in *Cataglyphis:* Mayr, 1862: 701; Santschi, 1912c: 149; Menozzi, 1922b: 332; Santschi, 1929b: 37; in *Myrmecocystus:* Emery, 1891b: 17; in *Myrmecocystus* (*Cataglyphis*): Karavaiev, 1924: 305. Race of *viaticus:* Emery & Forel, 1879: 449. Revived status as species: André, 1882b: 168; Emery, 1891b: 17; Emery, 1906d: 51. Current subspecies: nominal plus *arenarius, armenus, auratus, cubicus, fezzanensis, franchettii, kairuanus, mixtus, opacus, targuius, vaucheri.*

**alibabae.** *Cataglyphis alibabae* Pisarski, 1965: 419, figs. 11–14 (w.) IRAQ.

**altisquamis.** *Myrmecocystus altisquamis* André, 1881b: 56, pl. 3, figs. 6, 7 (w.) SYRIA. André, 1882b: 171 (q.). Combination in *Cataglyphis:* Emery, 1925b: 264.

*ambigua. Cataglyphis* (*Cataglyphis*) *albicans* st. *livida* var. *ambigua* Santschi, 1929b: 40 (w.) EGYPT. Unavailable name (Agosti, 1990a: 1498).

*arabica. Myrmecocystus albicans* subsp. *livida* var. *arabica* Emery, 1906d: 54, fig. 16 (w.) YEMEN. Unavailable name (Agosti, 1990a: 1498).

**#arenarius.** *Cataglyphis albicans* var. *arenaria* Finzi, 1940: 164 (w.) ALGERIA. [First available use of *Myrmecocystus albicans* subsp. *lividus* var. *arenaria* Forel, 1909e: 384; unavailable name (Agosti, 1990a: 1498).]

**argentatus.** *Camponotus argentata* Radoszkowsky, 1876: 140 (w.) EGYPT. Junior synonym of *bombycinus:* Dalla Torre, 1893: 217. Revived from synonymy: Agosti, 1990a: 1492.

**#armenus.** *Cataglyphis albicans* subsp. *armena* Arnol'di, 1964: 1808, figs. 7, 8 (w.q.m.) ARMENIA.

**asiriensis.** *Cataglyphis asiriensis* Collingwood, 1985: 286, fig. 76 (w.) SAUDI ARABIA.

*assyria. Cataglyphis bicolor* st. *nodus* var. *assyria* Santschi, 1929b: 44, fig. 7 (w.) IRAN. Unavailable name (Agosti, 1990a: 1498).

**#aterrimus.** *Cataglyphis aenescens* subsp. *aterrima* Pisarski, 1967: 417 (w.) RUSSIA. [First available use of *Myrmecocystus* (*Cataglyphis*) *cursor* subsp. *aenescens* var. *aterrima* Karavaiev, 1916: 507; unavailable name.] Subspecies of *cursor:* Agosti, 1990a: 1482.

**#auratus.** *Cataglyphis albicans* var. *aurata* Menozzi, 1932c: 95 (w.) SUDAN. [First available use of *Myrmecocystus albicans* subsp. *lividus* var. *auratus* Karavaiev, 1911: 10; unavailable name.]

**#basalis.** *Cataglyphis bicolor* var. *basalis* Santschi, 1929b: 42 (w.q.m.) ALGERIA.

**#bellicosus.** *Myrmecocystus* (*Cataglyphis*) *bicolor* subsp. *bellicosus* Karavaiev, 1924: 307 (w.) IRAN. Combination in *Cataglyphis:* Agosti, 1990a: 1487.

**#bergianus.** *Cataglyphis setipes* subsp. *bergiana* Arnol'di, 1964: 1805 (w.) KAZAKHSTAN.

**bicolor.** *Formica bicolor* Fabricius, 1793: 351 (m.) NORTH AFRICA. Wheeler, G.C. & Wheeler, J. 1980: 538 (l.). Combination in *Cataglyphis:* Roger, 1863b: 12; Mayr, 1863: 402; Emery, 1912f: 98; Emery, 1915g: 22; in *Myrmecocystus:* Emery & Forel, 1879: 449. Junior synonym of *viaticus:* Smith, F. 1861a: 32; Nasonov, 1889: 14. Revived from synonymy: Emery & Forel, 1879: 449; Emery, 1906d: 58; Forel, 1907e: 15. Subspecies of *viaticus:* Forel, 1907e: 15. Revived status as species: Karavaiev, 1912b: 590; Emery, 1925b: 264; Santschi, 1929b: 41. Senior synonym of *fairmairei:* Santschi, 1929b: 41; of *megalocola:* Mayr, 1855: 382; Santschi, 1929b: 41; of *rotundinodis:* Santschi, 1929b: 41. Current subspecies: nominal plus *adustus, basalis, bellicosus, congolensis, oasium, protuberatus, pubens, rufiventris, seticornis, sudanicus.*

**bicoloripes.** *Cataglyphis bicoloripes* Walker, 1871: 10 (q.) EGYPT.

*birmana* Collingwood, 1962; see under **CAMPONOTUS.**

**bombycinus.** *Formica bombycina* Roger, 1859: 232, pl. 7, fig. 1 (s.w.q.m.) LIBYA. Combination in *Cataglyphis:* Mayr, 1862: 701; in *Myrmecocystus:* Emery, 1891b: 17; in *Cataglyphis:* Emery, 1912f: 98; in *C.* (*Machaeromyrma*): Santschi, 1929b: 30. Senior synonym of *phaeogaster:* Dalla Torre, 1893: 217; Agosti, 1990a: 1484. Current subspecies: nominal plus *bruneipes, sinaiticus.*

**#bruneipes.** *Myrmecocystus* (*Cataglyphis*) *bombycinus* var. *bruneipes* Santschi, 1911i: 287 (w.) TUNISIA. Combination in *Cataglyphis:* Emery, 1925b: 267.

*bucculenta. Cataglyphis bicolor* st. *saharae* var. *bucculenta* Santschi, 1929b: 48 (w) MOROCCO. Unavailable name (Agosti, 1990a: 1498).

**#bucharicus.** *Cataglyphis altisquamis* var. *bucharica* Emery, 1925b: 264 (w.) UZBEKISTAN. Arnol'di, 1964: 1811 (q.). [First available use of *Myrmecocystus altisquamis* r. *foreli* var. *bucharica* Forel, 1904b: 393; unavailable name.] Subspecies of *altisquamis:* Arnol'di, 1964: 1811; of *foreli:* Pisarski, 1967: 418.

**#bugnioni.** *Cataglyphis adenensis* var. *bugnioni* Santschi, 1929b: 41 (w.) EGYPT. [First available use of *Myrmecocystus viaticus* r. *adenensis* var. *bugnioni* Forel, 1908a: 16; unavailable name.]

**#bulgaricus.** *Cataglyphis* (*Cataglyphis*) *livida* subsp. *bulgarica* Atanassov, 1982: 213, fig. 4 (w.q.) BULGARIA. See also: Atanassov & Dlussky, 1992: 293.

*caerulescens. Cataglyphis bicolor* st. *nigra* var. *caerulescens* Santschi, 1929b: 50, fig. 10 (w.) SYRIA. Unavailable name (Agosti, 1990a: 1498).

**cana.** *Cataglyphis albicans* var. *cana* Santschi, 1925g: 356 (w.) MOROCCO. [First available use of *Myrmecocystus albicans* subsp. *vaucheri* var. *cana* Emery, 1906d: 54; unavailable name.] Variety of *albicans:* Santschi, 1929b: 38. Raised to species: Collingwood, 1978: 73.

*caspius. Myrmecocystus cursor* var. *caspius* Ruzsky, 1902b: 470 (q.m.) RUSSIA. Ruzsky, 1905b: 447 (w.).

Junior synonym of *aenescens:* Karavaiev, 1910b: 18; Lomnicki, 1925a: 160.

*caucasea* Santschi, 1925; see under ***PROFORMICA***.

*caucasicola. Cataglyphis nodus* subsp. *caucasicola* Arnol'di, 1964: 1803, figs. 1, 2 (w.m.) GEORGIA. Junior synonym of *nodus:* Dlussky, Soyunov & Zabelin, 1990: 153.

**#*chatkalensis.*** *Cataglyphis aenescens* subsp. *chatkalensis* Tarbinsky, 1976: 202 (w.) KIRGIZIA.

***cinnamomeus.*** *Myrmecocystus albicans* subsp. *cinnamomeus* Karavaiev, 1910a: 269 (w.) RUSSIA. [Also described as new, as *Myrmecocystus albicans* subsp. *viaticoides* var. *cinnamomea* Karavaiev, 1910b: 37; unavailable name.] Combination in *Cataglyphis:* Emery, 1925b: 263. Raised to species: Pisarski, 1967: 418; Arnol'di, 1968: 1815.

**#*congolensis.*** *Myrmecocystus bicolor* var. *congolensis* Stitz, 1916: 396, fig. 10 (w.) CHAD. Combination in *Cataglyphis:* Wheeler, W.M. 1922a: 945.

***constrictus.*** **Camponotus constrictus* Mayr, 1868c: 29, pl. 1, fig. 11 (w.) BALTIC AMBER (Oligocene). Combination in *Formica:* Wheeler, W.M. 1915h: 128; in *Cataglyphis:* Dlussky, 1967b: 85.

**#*creticus.*** *Myrmecocystus cursor* var. *cretica* Forel, 1910a: 23 (w.m.) GREECE. [First available use of *Myrmecocystus cursor* subsp. *aenescens* var. *cretica* Emery, 1906d: 48; unavailable name (Agosti & Collingwood, 1987a: 61).] Combination in *Cataglyphis:* Emery, 1925b: 264. See also: Agosti, 1990a: 1482.

**#*cubicus.*** *Myrmecocystus albicans* r. *cubicus* Forel, 1903c: 267 (w.q.) MOROCCO. Combination in *Cataglyphis:* Emery, 1925b: 263.

***cugiai.*** *Cataglyphis* (*Monocombus*) *cugiai* Menozzi, 1939a: 323, figs. 13–15 (w.q.) INDIA.

**#*cuneinodis.*** *Cataglyphis viaticoides* subsp. *cuneinodis* Arnol'di, 1964: 1810 (w.) TRANSCAUCASIA. [First available use of *Myrmecocystus* (*Cataglyphis*) *albicans* subsp. *ruber* var. *cuneinodis* Karavaiev, 1924: 305; unavailable name.]

***cursor.*** *Formica cursor* Fonscolombe, 1846: 41 (w.m.) FRANCE. Mayr, 1861: 45 (w.q.m.). Combination in *Cataglyphis:* Mayr, 1861: 45; Emery, 1912f: 99; Emery, 1916b: 267; Emery, 1925b: 263; in *Myrmecocystus:* Emery & Forel, 1879: 449; in *Myrmecocystus* (*Cataglyphis*): Karavaiev, 1924: 301; in *Cataglyphis* (*Monocombus*): Santschi, 1929b: 29. Current subspecies: nominal plus *aterrimus, creticus, rockingeri.*

*deserta. Proformica deserta* Kuznetsov-Ugamsky, 1926a: 73, figs. 12–15 (w.) TURKESTAN. Collingwood, 1961a: 67 (m.); Collingwood, 1962: 222 (q.). Junior synonym of *emeryi:* Dlussky, Soyunov & Zabelin, 1990: 151.

*desertorum. Myrmecocystus viaticus* var. *desertorum* Forel, 1894c: 402 (w.) SPAIN. Karavaiev, 1912a: 17 (m.); Santschi, 1929b: 46 (m.); Karavaiev, 1924: 306 (q.). Combination in *Myrmecocystus* (*Cataglyphis*): Karavaiev, 1924: 306; in *Cataglyphis:* Wheeler, W.M. 1922a: 944; Emery, 1925b: 265. Race/subspecies of *viaticus:* Emery, 1898c: 126; Forel, 1902a: 156. Junior synonym of *bicolor:* Forel, 1907e: 15; Santschi, 1910f: 234; Emery, 1925b: 265. Variety/subspecies of *bicolor:* Karavaiev, 1911: 10; Karavaiev, 1912a: 17; Karavaiev, 1924: 306. Raised to species: Collingwood, 1985: 286. Junior synonym of *viaticus:* Agosti, 1990a: 1491.

***diehlii.*** *Myrmecocystus viaticus* r. *diehlii* Forel, 1902g: 462 (w.q.m.) ALGERIA. [First available use of *Myrmecocystus viaticus* r. *desertorum* var. *diehlii* Forel, 1902a: 156; unavailable name.] Combination in *Myrmecocystus* (*Cataglyphis*): Forel, 1913d: 434; in *Cataglyphis:* Emery, 1912f: 98; Emery, 1925b: 265. Subspecies of *bicolor:* Emery, 1906d: 58; Emery, 1925b: 265; Santschi, 1929b: 51. Raised to species: Forel, 1913d: 434; Collingwood, 1985: 287; Wehner, 1986: 108.

*dorica. Cataglyphis* (*Monocombus*) *cursor* st. *helenica* var. *dorica* Santschi, 1929b: 35 (w.) GREECE. Unavailable name (Agosti & Collingwood, 1987a: 61).

*drusa. Cataglyphis bicolor* st. *nodus* var. *drusa* Santschi, 1929b: 44 (w.) SYRIA. Unavailable name (Agosti, 1990a: 1498).

**#*dschambulicus.*** *Cataglyphis setipes* subsp. *dschambulica* Tarbinsky, 1976: 200 (w.) KIRGIZIA.

***elegantissimus.*** *Cataglyphis* (*Eomonocombus*) *elegantissima* Arnol'di, 1968: 1815, figs. 13, 14 (w.) TURKMENIA. [Originally misspelled *eiegantissima*, on same page.]

***emeryi.*** *Myrmecocystus emeryi* Karavaiev, 1910b: 34, figs. 11–13 (w.q.m.) TURKESTAN. Combination in *Cataglyphis:* Emery, 1912f: 103; in *C.* (*Monocombus*): Santschi, 1929b: 30. Senior synonym of *deserta* and *pilosulus:* Dlussky, Soyunov & Zabelin, 1990: 151. Current subspecies: nominal plus *abdominalis, karavaiewi.*

***emmae.*** *Formica* (*Proformica*) *emmae* Forel, 1909e: 381 (w.) ALGERIA. Delye, 1965: 52 (q.m.). Combination in *Proformica:* Emery, 1912f: 103; in *Formica* (*Paraformica*): Forel, 1915b: 95; in *Paraformica:* Emery, 1925b: 261; in *Cataglyphis* (*Paraformica*): Santschi, 1929b: 30. Current subspecies: nominal plus *hoggarensis.*

*europaea. Formica cephalotes* subsp. *europaea* Christ, 1791: 511, pl. 60, fig. 9 (w.) EUROPE. Junior synonym of *viaticus:* Emery, 1892b: 161.

*fairmairei. Cataglyphis fairmairei* Foerster, 1850b: 494 (m.) ALGERIA. Junior synonym of *megalocola:* Mayr, 1863: 402.

**#*fezzanensis.*** *Cataglyphis albicans* var. *fezzanensis* Bernard, 1948: 163 (w.q.m.) ALGERIA.

*flavicornis* Kuznetsov-Ugamsky, 1926; see under ***ALLOFORMICA.***

*flavigastra. Myrmecocystus* (*Cataglyphis*) *cursor* subsp. *aenescens* var. *flavigastra* Karavaiev, 1924: 303 (q.m.) TRANSCAUCASIA. Unavailable name (Agosti, 1990a: 1498); material referred to *aenescens:* Dlussky, Soyunov & Zabelin, 1990: 149.

***floricola.*** *Cataglyphis floricola* Tinaut, 1993: 125, figs. 1–4 (w.q.m.) SPAIN. [*Cataglyphis floricola* Cerda, Retana, Carpintero, & Cros, 1992: 315. *Nomen nudum.*]

***foreli.*** *Myrmecocystus foreli* Ruzsky, 1903a: 36 (w.) RUSSIA. Arnol'di, 1964: 1803 (q.m.). Combination in *Cataglyphis:* Emery, 1912f: 104; in *C.* (*Monocombus*): Santschi, 1929b: 29. Race of *altisquamis:* Forel, 1904b: 382. Revived status as species: Emery, 1912f: 104; Arnol'di, 1964: 1812; Pisarski, 1967: 418. Current subspecies: nominal plus *bucharicus, murgabicus.*

***fortis.*** *Myrmecocystus albicans* var. *fortis* Forel, 1902a: 156 (w.) ALGERIA. Wehner, 1983: 95 (q.m.). [Also described as new by Forel, 1903b: 455.] Combination in *Cataglyphis:* Emery, 1925b: 262. Stirps of *albicans:* Santschi, 1929b: 61. Raised to species, and senior synonym of *halophila:* Wehner, 1983: 95.

**#*franchettii.*** *Cataglyphis albicans* subsp. *franchettii* Menozzi, 1931a: 155, fig. (w.) ETHIOPIA.

***frigidus.*** *Myrmecocystus cursor* var. *frigidus* André, 1881b: 59 (in text) (w.) SYRIA. Tohmé, H. & Tohmé, G. 1985: 84 (l.). Combination in *Cataglyphis:* Emery, 1912f: 103; in *C.* (*Monocombus*): Santschi, 1929b: 29. Raised to species: Emery, 1906d: 49. Current subspecies: nominal plus *persicus.*

***gaetulus.*** *Cataglyphis* (*Monocombus*) *gaetula* Santschi, 1929b: 33, figs. 20, 22, 29 (w.q.m.) MOROCCO. Current subspecies: nominal plus *pilisquamis.*

**#*georgicus.*** *Cataglyphis* (*Monocombus*) *aenescens* subsp. *georgica* Arnol'di, 1968: 1816, figs. 15, 16, 17 (q.m.) GEORGIA.

***gracilens.*** *Cataglyphis gracilens* Santschi, 1929b: 65 (w.) RUSSIA. [First available use of *Myrmecocystus altisquamis* subsp. *foreli* var. *gracilens* Ruzsky, 1905b: 439; unavailable name.]

*halophila. Cataglyphis halophila* Bernard, 1954: 48, figs. c, f, h (w.) ALGERIA. Junior synonym of *fortis:* Wehner, 1983: 95.

***hannae.*** *Cataglyphis hannae* Agosti, 1994b: 914, fig. 1 (q.m.) TUNISIA.

*helladica. Cataglyphis* (*Cataglyphis*) *bicolor* st. *nodus* var. *helladica* Santschi, 1934d: 281 (w.) GREECE. Unavailable name (Agosti & Collingwood, 1987a: 61).

***hellenicus.*** *Myrmecocystus cursor* var. *hellenicus* Forel, 1886f: 204 (w.) GREECE. Combination in *Cataglyphis:* Müller, 1923: 150; in *C.* (*Monocombus*): Santschi, 1929b: 35. Raised to species: Agosti & Collingwood, 1987a: 59.

***hispanicus.*** *Myrmecocystus viaticus* subsp. *hispanica* Emery, 1906d: 57 (w.) SPAIN. Tinaut, 1991: 218 (q.m.). [First available use of *Myrmecocystus viaticus* r. *niger* var. *hispanica* Forel, 1903c: 267; unavailable name.] Combination in *Cataglyphis:* Santschi, 1919e: 246; in *C.* (*Monocombus*): Santschi, 1929b: 29. Senior synonym of *nigroides:* Tinaut, 1991: 220.

*hispidus. Cataglyphis hispidus* Menozzi, in Eidmann, 1942: 255 (w.) INDIA. *Nomen nudum.*

**#*hoggarensis.*** *Cataglyphis* (*Paraformica*) *emmae* var. *hoggarensis* Santschi, 1929b: 53 (footnote) (w.) ALGERIA. [Also described as new by Santschi, 1929c: 106, and Santschi, 1934f: 174.]

***humeya.*** *Cataglyphis humeya* Tinaut, 1991: 220, figs. 4, 5, 7 (w.q.m.) SPAIN.

***ibericus.*** *Myrmecocystus albicans* subsp. *ibericus* Emery, 1906d: 52, fig. 12 (w.m.) SPAIN. Combination in *Cataglyphis:* Emery, 1912f: 99. Raised to species: Collingwood & Yarrow, 1969: 85; Tinaut & Plaza, 1990: 196.

***indicus.*** *Cataglyphis indica* Pisarski, 1961: 515, figs. 1–3 (w.) INDIA.

***isis.*** *Myrmecocystus* (*Cataglyphis*) *diehlii* var. *isis* Forel, 1913d: 434 (w.) EGYPT. [First available use of *Myrmecocystus viaticus* subsp. *diehlii* var. *isis* Forel, 1909e: 385; unavailable name.] Combination in *Cataglyphis:* Emery, 1925b: 265. Subspecies of *diehlii:* Pisarski, 1965: 421. Raised to species: Collingwood, 1985: 288.

***CATAGLYPHIS** (continued)*

**italicus.** *Myrmecocystus cursor* subsp. *italicus* Emery, 1906d: 48, fig. 5 (w.) ITALY. Emery, 1916b: 268 (q.m.). Combination in *Cataglyphis:* Emery, 1914e: 2. Junior synonym of *cursor:* Bernard, 1967: 281. Revived from synonymy and raised to species: Agosti & Collingwood, 1987b: 262.

*jakobsoni. Myrmecocystus altisquamis* var. *jakobsoni* Ruzsky, 1905b: 436 (w.) TURKESTAN. Karavaiev, 1910b: 22 (m.). Combination in *Cataglyphis:* Emery, 1925b: 264. Subspecies of *cursor:* Karavaiev, 1910b: 21; Santschi, 1929b: 58; of *aenescens:* Pisarski, 1967: 417. Raised to species: Pisarski, 1970: 323. Junior synonym of *aenescens:* Dlussky, Soyunov & Zabelin, 1990: 149.

**#kairuanus.** *Cataglyphis albicans* var. *kairuana* Santschi, 1912c: 149 (w.) TUNISIA.

**karakalensis.** *Cataglyphis setipes* subsp. *karakalensis* Arnol'di, 1964: 1805, fig. 5 (w.m.) TURKMENIA. Raised to species: Dlussky, Soyunov & Zabelin, 1990: 152. See also: Pisarski, 1970: 325.

**#karavaiewi.** *Cataglyphis karavaiewi* Santschi, 1925g: 354 (w.) RUSSIA. Stirps of *emeryi:* Santschi, 1929b: 37.

*kuenlunensis. Cataglyphis (Monocombus) cursor* subsp. *aenescens* var. *kuenlunensis* Stärcke, 1935: 269 (w.) KARAKORUM. Unavailable name (Agosti, 1990a: 1498).

**kurdistanicus.** *Cataglyphis kurdistanicus* Pisarski, 1965: 417, figs. 1–10 (s.w.) IRAQ.

**laevior.** *Cataglyphis bicolor* st. *laevior* Santschi, 1929b: 47 (w.) ALGERIA. [First available use of *Myrmecocystus viaticus* r. *bicolor* var. *laevior* Stitz, 1917: 348; unavailable name.] Raised to species: Collingwood, 1985: 288.

*lameerei. Myrmecocystus lameerei* Forel, 1902a: 156 (w.q.) ALGERIA. Junior synonym of *lucasi:* Emery, in Forel, 1902g: 463.

**lividus.** *Myrmecocystus albicans* var. *lividus* André, 1881b: 58 (w.) ISRAEL. Menozzi, 1933b: 84 (q.m.). Combination in *Cataglyphis:* Wheeler, W.M. & Mann, 1916: 173. Race/stirps of *albicans:* Forel, 1902a: 156; Santschi, 1929b: 62. Raised to species: Arnol'di, 1964: 1810; Pisarski, 1965: 420. Current subspecies: nominal plus *bulgaricus, luteus.*

**longipedem.** *Formica longipedem* Eichwald, 1841: 224 (w.) RUSSIA. Senior synonym of *turcomanica:* Kolossov, 1932: 118; Agosti, 1990a: 1490 [Agosti repeats the synonymy but incorrectly gives *turcomanica* seniority.]

**lucasi.** *Myrmecocystus lucasi* Emery, 1898c: 147 (w.) TUNISIA. Bernard, 1953a: 203 (m.). Combination in *Cataglyphis:* Emery, 1912f: 104; in *C. (Machaeromyrma)*: Santschi, 1929b: 30. Senior synonym of *lameerei:* Emery, in Forel, 1902g: 463.

**lunaticus.** *Cataglyphis lunatica* Baroni Urbani, 1969d: 213, figs. 1, 2, 3 (w.) TURKEY.

**#luteus.** *Cataglyphis livida* subsp. *lutea* Pisarski, 1967: 418 (w.) IRAN. [First available use of *Myrmecocystus albicans* subsp. *viaticoides* var. *lutea* Emery, 1906d: 53; unavailable name.]

**machmal.** *Cataglyphis machmal* Radchenko & Arakelian, 1991: 73, figs. 3–5 (w.) ARMENIA.

*maeotica. Cataglyphis (Monocombus) cursor* subsp. *aenescens* var. *maeotica* Karavaiev, 1935b: 110 (m.) UKRAINE. Unavailable name (Agosti, 1990a: 1498).

**mauritanicus.** *Myrmecocystus viaticus* subsp. *mauritanicus* Emery, 1906d: 57, figs. 23, 26 (w.m.) TUNISIA. Santschi, 1929b: 31 (q.). Combination in *Cataglyphis:* Emery, 1912f: 98; in *C. (Monocombus)*: Santschi, 1929b: 29. Raised to species: Agosti, 1990a: 1483.

*megalocola. Formica megalocola* Foerster, 1850b: 490 (w.) ALGERIA. Emery, 1891b: 16 (q.m.). Combination in *Cataglyphis:* Roger, 1863b: 12; Mayr, 1863: 402; in *Myrmecocystus:* Emery & Forel, 1879: 449; Forel, 1890a: lxvii; in *Cataglyphis:* Emery, 1912f: 98; Emery, 1925b: 265. Variety/race of *viaticus:* Emery & Forel, 1879: 449; Forel, 1890a: lxvii; Forel, 1892i: 306. Subspecies of *bicolor:* Emery, 1906d: 58; Emery, 1908g: 217. Senior synonym of *fairmairei:* Mayr, 1863: 402. Junior synonym of *bicolor:* Mayr, 1855: 382; Santschi, 1929b: 41; Agosti, 1990a: 1486.

*mesasiaticus. Cataglyphis nodus* subsp. *mesasiatica* Arnol'di, 1964: 1803, figs. 1, 3 (w.m.) UZBEKISTAN. Junior synonym of *nodus:* Dlussky, Soyunov & Zabelin, 1990: 153.

**minimus.** *Cataglyphis minima* Collingwood, 1985: 289, fig. 75 (w.q.) SAUDI ARABIA.

**#mixtus.** *Myrmecocystus albicans* var. *mixtus* Forel, 1895d: 229 (w.) TURKEY. Combination in *Cataglyphis:* Emery, 1925b: 263.

**#murgabicus.** *Cataglyphis foreli* subsp. *murgabica* Arnol'di, 1964: 1814 (w.) TURKMENIA.

**niger.** *Myrmecocystus viaticus* var. *niger* André, 1881b: 56 (w.) ISRAEL. Forel, 1913d: 434 (q.). Combination in *Cataglyphis:* Emery, 1925b: 265. Subspecies of *viaticus:* Emery, 1898c: 126; of *bicolor:* Forel, 1908a: 15; Menozzi, 1927f: 380; Santschi, 1929b: 50. Raised to species: Collingwood, 1985: 290.

**nigripes.** *Cataglyphis setipes* subsp. *nigripes* Arnol'di, 1964: 1806 (w.m.) GEORGIA. [First available use of *Cataglyphis bicolor* st. *setipes* var. *nigripes* Santschi, 1929b: 50; unavailable name.] Raised to species: Agosti, 1990a: 1486.

*nigroides. Cataglyphis (Monocombus) hispanica* var. *nigroides* Santschi, 1929b: 59 (diagnosis in key) (w.) PORTUGAL. [First available use of *Cataglyphis viaticus* st. *hispanicus* var. *nigroides* Santschi, 1925g: 346; unavailable name.] Junior synonym of *hispanicus:* Tinaut: 1991: 220.

**nodus.** *Formica nodus* Brullé, 1832: 326, pl. 48, fig. 1 (w.) GREECE. Combination in *Cataglyphis:* Mayr, 1861: 44. Junior synonym of *viaticus:* Roger, 1861b: 163; Dalla Torre, 1893: 218. Revived from synonymy as variety of *bicolor:* Emery, 1925b: 265. Variety/subspecies of *bicolor:* Menozzi, 1927e: 119; Santschi, 1929b: 47; Menozzi, 1936d: 305; Bernard, 1948: 162. Raised to species: Arnol'di, 1964: 1802; Pisarski, 1965: 421. Junior synonym of *bicolor:* Wehner, Harkness & Schmid-Hempel, 1983: 7. Revived from synonymy: Collingwood, 1985: 290; Agosti, 1990a: 1489; Atanassov & Dlussky, 1992: 291. Senior synonym of *orientalis:* Emery, 1925b: 265; Finzi, 1928c: 791; of *caucasicola* and *mesasiaticus:* Dlussky, Soyunov & Zabelin, 1990: 153. Misspelled as *nuda:* Santschi, 1933b: 108.

**#oasium.** *Cataglyphis bicolor* var. *oasium* Menozzi, 1932c: 95 (w.q.m.) TUNISIA. [First available use of *Cataglyphis bicolor* st. *nodus* var. *oasium* Santschi, 1929b: 46; unavailable name.]

*occidentalis. Cataglyphis (Monocombus) viaticus* st. *mauritanica* var. *occidentalis* Santschi, 1929b: 32, figs. 19, 25 (w.q.) MOROCCO. Unavailable name (Agosti, 1990a: 1498).

**#opacus.** *Cataglyphis albicans* var. *opaca* Santschi, 1912c: 149 (w.) TUNISIA.

*opaciventris. Cataglyphis (Monocombus) viaticus* st. *mauritanica* var. *opaciventris* Santschi, 1929b: 31, figs. 24, 28 (w.q.m.) TUNISIA. Unavailable name (Agosti, 1990a: 1498).

*orientalis. Myrmecocystus viaticus* var. *orientalis* Forel, 1895d: 228 (w.) TURKEY. Combination in *Cataglyphis:* Emery, 1912f: 99. Junior synonym of *setipes:* Ruzsky, 1905b: 430. Revived from synonymy as subspecies of *viaticus:* Forel, 1910a: 24. Stirps of *bicolor:* Santschi, 1921a: 116. Junior synonym of *nodus:* Emery, 1925b: 265; Finzi, 1928c: 791.

**otini.** *Cataglyphis (Cataglyphis) albicans* st. *otini* Santschi, 1929b: 38 (w.) MOROCCO. [Misspelled as *odini:* Santschi, 1929b: 60.] Raised to species: Collingwood, 1978: 73.

**oxianus.** *Cataglyphis foreli* subsp. *oxiana* Arnol'di, 1964: 1813, figs. 13, 14 (w.) TURKMENIA. Raised to species: Dlussky, Soyunov & Zabelin, 1990: 153.

**pallidus.** *Cataglyphis pallida* Mayr, 1877: 9 (w.m.) TURKESTAN. Karavaiev, 1910b: 25 (q.). [Redescribed in Latin: Mayr, 1880: 28.] Combination in *Myrmecocystus:* Emery, 1898c: 126; in *Cataglyphis:* Emery, 1912f: 103; in *C. (Monocombus)*: Santschi, 1929b: 29.

**#persicus.** *Myrmecocystus frigidus* var. *persica* Emery, 1906d: 50, fig. 7 (w.) IRAN. Combination in *Cataglyphis:* Emery, 1925b: 264.

*phaeogaster. Camponotus phaeogaster* Walker, 1871: 10 (w.) SAUDI ARABIA. Junior synonym of *bombycinus:* Dalla Torre, 1893: 217.

*pharao. Cataglyphis bicolor* st. *nigra* var. *pharao* Santschi, 1929b: 51 (w.) EGYPT. Unavailable name (Agosti, 1990a: 1498).

**piliger.** *Cataglyphis foreli* subsp. *piligera* Arnol'di, 1964: 1813 (w.) UZBEKISTAN. Raised to species: Dlussky, Soyunov & Zabelin, 1990: 154.

**piliscapus.** *Myrmecocystus cursor* var. *piliscapa* Forel, 1901h: 66 (w.) FRANCE. Santschi, 1929b: 36 (m.). Combination in *Cataglyphis (Monocombus)*: Santschi, 1929b: 36. [Misspelled as *pilicornis:* Santschi, 1932c: 73.] Raised to species: Agosti & Collingwood, 1987b: 285 (in key). Senior synonym of *tibialis:* Agosti, 1990a: 1482.

**#pilisquamis.** *Cataglyphis (Monocombus) gaetula* var. *pilisquamis* Santschi, 1929b: 34 (w.) MOROCCO.

*pilosulus. Cataglyphis pilosulus* Kuznetsov-Ugamsky, 1926a: 71, figs. 1–5 (w.) TURKESTAN. Junior synonym of *emeryi:* Dlussky, Soyunov & Zabelin, 1990: 151.

**#protuberatus.** *Cataglyphis bicolor* var. *protuberata* Crawley, 1920b: 177 (w.) IRAQ. Stirps of *bicolor:* Santschi, 1929b: 53.

**#pubens.** *Cataglyphis bicolor* var. *pubens* Santschi, 1929b: 43 (w.m.) MOROCCO.

**#rockingeri.** *Myrmecocystus* (*Cataglyphis*) *cursor* subsp. *rockingeri* Forel, 1911e: 287 (w.) RUSSIA. Combination in *Cataglyphis:* Emery, 1925b: 264. Subspecies of *aenescens:* Pisarski, 1969b: 315; Dlussky & Pisarski, 1970: 89; of *cursor:* Agosti, 1990a: 1482.

**rosenhaueri.** *Cataglyphis albicans* st. *rosenhaueri* Santschi, 1925g: 356 (w.) SPAIN. [First available use of *Myrmecocystus albicans* subsp. *ruber* var. *rosenhaueri* Emery, 1906d: 53; unavailable name.] Raised to species: Collingwood & Yarrow, 1969: 85; Tinaut & Plaza, 1990: 192.

*rotundinodis. Myrmecocystus albicans* subsp. *rotundinodis* Karavaiev, 1912a: 16 (w.) ALGERIA. Combination in *Cataglyphis:* Emery, 1925b: 262. See also: Karavaiev, 1924: 305. Junior synonym of *bicolor:* Santschi, 1929b: 41.

**ruber.** *Myrmecocystus albicans* r. *ruber* Forel, 1903c: 268 (w.) ALGERIA. Santschi, 1929b: 39 (q.m.). Combination in *Cataglyphis:* Emery, 1912f: 99. Raised to species: Collingwood, 1985: 291; Wehner, 1986: 109.

*rufidens. Cataglyphis* (*Cataglyphis*) *bicolor* st. *laevior* var. *rufidens* Santschi, 1932f: 518 (w.q.) ALGERIA. Unavailable name (Agosti, 1990a: 1498).

**#rufiventris.** *Cataglyphis bicolor* var. *rufiventris* Emery, 1925b: 265 (w.) GREECE. [First available use of *Myrmecocystus* (*Cataglyphis*) *viaticus* subsp. *orientalis* var. *rufiventris* Forel, 1911d: 353; unavailable name.]

**sabulosus.** *Cataglyphis sabulosa* Kugler, J. 1981: 84, figs. 1–13 (w.q.m.) ISRAEL.

**saharae.** *Cataglyphis* (*Cataglyphis*) *bicolor* st. *saharae* Santschi, 1929b: 48, fig. 3 (w.) ALGERIA. Raised to species: Collingwood, 1985: 291.

**savignyi.** *Formica savignyi* Dufour, 1862: 141 (w.m.) EGYPT. Combination in *Myrmecocystus:* Emery & Forel, 1879: 449; in *Cataglyphis:* Roger, 1863b: 12; Emery, 1925b: 265. Junior synonym of *viaticus:* Emery & Forel, 1879: 449; of *bicolor:* Emery, 1925b: 265. Revived from synonymy: Billen, 1989: 301; Agosti, 1990a: 1490.

**semitonsus.** *Cataglyphis albicans* var. *semitonsa* Santschi, 1929b: 61 (w.) ALGERIA. [First available use of *Cataglyphis albicans* st. *vaucheri* var. *semitonsa* Santschi, 1926e: 236; unavailable name.] Raised to species: Collingwood, 1985: 291.

**#seticornis.** *Myrmecocystus bicolor* subsp. *seticornis* Emery, 1906d: 58 (w.) GHANA. [First available use of *Myrmecocystus viaticus* subsp. *setipes* var. *seticornis* Emery, 1898c: 147 (footnote); unavailable name.] Combination in *Cataglyphis:* Wheeler, W.M. 1922a: 945. Subspecies of *bicolor:* Emery, 1908g: 217; Santschi, 1929b: 47.

*setipedesertorum. Myrmecocystus viaticus* subsp. *setipes* var. *setipedesertorum* Ruzsky, 1905b: 433 (w.) RUSSIA. Unavailable name (Agosti, 1990a: 1469).

**setipes.** *Myrmecocystus viaticus* r. *setipes* Forel, 1894c: 401 (w.) INDIA. Ruzsky, 1902d: 9 (q.m.); Imai, *et al.* 1984: 9 (k.). Combination in *Cataglyphis:* Wheeler, W.M. 1922a: 945. Subspecies of *bicolor:* Emery, 1906d: 58; Emery, 1908g: 217; Santschi, 1929b: 49. Raised to species: Bingham, 1903: 312; Collingwood, 1961: 289; Collingwood, 1961a: 65; Arnol'di, 1964: 1804; Arnol'di & Dlussky, 1978: 554. Current subspecies: nominal plus *bergianus, dschambulicus.*

**#sinaiticus.** *Cataglyphis bombycina* var. *sinaitica* Wheeler, W.M. & Mann, 1916: 173 (s.w.) EGYPT.

*subareolata. Cataglyphis* (*Monocombus*) *cursor* subsp. *subareolata* Arnol'di, 1948: 212. *Nomen nudum.*

**#sudanicus.** *Myrmecocystus bicolor* var. *sudanica* Karavaiev, 1912b: 590, fig. 2 (w.m.) SUDAN. Santschi, 1933b: 108 (q.). Combination in *Cataglyphis:* Emery, 1925b: 265.

**takyrica.** *Cataglyphis takyrica* Dlussky, in Dlussky, Soyunov & Zabelin, 1990: 155, fig. 45 (w.q.) TURKMENISTAN.

**#tancrei.** *Myrmecocystus cursor* r. *tancrei* Forel, 1901h: 66 (w.) CHINA. Forel, 1904b: 383 (m.); Ruzsky, 1905b: 446 (q.). Combination in *Cataglyphis:* Emery, 1925b: 263. Variety of *aenescens:* Santschi, 1929b: 58.

**#targuius.** *Cataglyphis* (*Cataglyphis*) *albicans* var. *targuia* Santschi, 1929b: 55 (footnote) (w.) ALGERIA. [Also described as new by Santschi, 1929c: 106; Santschi, 1934f: 175.]

**theryi.** *Cataglyphis theryi* Santschi, 1921d: 76 (m.) MOROCCO.

*tibialis. Cataglyphis tibialis* Bondroit, 1918: 39 (w.) FRANCE. Variety of *cursor:* Santschi, 1925g: 355; Emery, 1925b: 263. Junior synonym of *cursor:* Bernard, 1967: 281; Collingwood, 1978: 73; of *piliscapus:* Santschi, 1929b: 36; Agosti, 1990a: 1482.

**#tonsilis.** *Cataglyphis* (*Monocombus*) *viaticus* var. *tonsilis* Santschi, 1936a: 209 (w.q.m.) MOROCCO. [First available use of *Cataglyphis* (*Monocombus*) *viatica* st. *mauritanica* var. *tonsilis* Santschi, 1929b: 32; unavailable name.]

*turcomanica. Cataglyphis bicolor* var. *turcomanica* Crawley, 1920b: 177 (w.) TURKMENIA. [First available use of *Myrmecocystus viaticus* subsp. *desertorum* var. *turcomanica* Emery, 1898c: 147; unavailable name.] Combination in *Cataglyphis:* Emery, 1925b: 266. Subspecies of *setipes:* Arnol'di, 1964: 1805; Pisarski, 1967: 418. Junior synonym of *setipes:* Dlussky, Soyunov & Zabelin, 1990: 155; of *longipedem:* Kolossov, 1932: 118; Agosti, 1990a: 1490.

**urens.** *Cataglyphis urens* Collingwood, 1985: 290 (w.m.) OMAN. [First available use of *Cataglyphis bicolor* st. *abyssinica* var. *urens* Santschi, 1929b: 52; unavailable name.]

**#vaucheri.** *Myrmecocystus albicans* subsp. *vaucheri* Emery, 1906d: 54, fig. 17 (w.) MOROCCO. Combination in *Cataglyphis:* Emery, 1925b: 263.

**velox.** *Cataglyphis* (*Monocombus*) *viatica* var. *velox* Santschi, 1929b: 30, figs. 6, 34 (w.) SPAIN. Tinaut, 1990c: 54 (q.m.). Junior synonym of *viaticus:* Collingwood, 1979: 73. Revived from synonymy and raised to species: Tinaut, 1990c: 52.

**viaticus.** *Formica viatica* Fabricius, 1787: 308 (w.) SPAIN. Mayr, 1861: 45 (q.m.). Combination in *Monocombus:* Mayr, 1855: 382; in *Myrmecocystus:* Emery & Forel, 1879: 449; in *Cataglyphis:* Mayr, 1861: 45; Emery, 1912f: 98; Santschi, 1919e: 246; in *C.* (*Monocombus*): Santschi, 1929b: 29. Senior synonym of *europaea:* Emery, 1892b: 161; of *desertorum:* Agosti, 1990a: 1491. See also: Tinaut, 1990c: 50. Current subspecies: nominal plus *tonsilis.*

**viaticoides.** *Myrmecocystus albicans* var. *viaticoides* André, 1881b: 57 (w.) LEBANON. Arnol'di, 1964: 1809 (m.). Combination in *Cataglyphis:* Emery, 1925b: 263. Race/stirps of *albicans:* Forel, 1902a: 156; Santschi, 1929b: 61. Raised to species: Arnol'di, 1964: 1809; Collingwood & Yarrow, 1969: 85. See also: Tinaut & Plaza, 1990: 192. Current subspecies: nominal plus *cuneinodis.*

***CATAULACUS*** (Old World Tropics)

*adlerzi* Mayr, 1887; see under ***PROCRYPTOCERUS.***

**adpressus.** *Cataulacus adpressus* Bolton, 1974a: 30, figs. 15, 18 (w.) GHANA.

*aethiops. Cataulacus jeanneli* var. *aethiops* Santschi, 1924b: 220 (w.) ZAIRE. Junior synonym of *weissi:* Bolton, 1982: 358.

*alenensis. Cataulacus sulcatus* var. *alenensis* Stitz, 1910: 137 (w.) EQUATORIAL GUINEA. Variety of *guineensis:* Wheeler, W.M. 1922a: 913. Junior synonym of *guineensis:* Bolton, 1974a: 55.

*andamanensis. Cataulacus granulatus* r. *andamanensis* Forel, 1903d: 406 (w.) ANDAMAN IS. Junior synonym of *simoni:* Bolton, 1974a: 72.

***anthracinus.** **Attopsis anthracina* Heer, 1850: 156, pl. 12, fig. 12 (m.) YUGOSLAVIA. Combination in *Cataulacus,* and senior synonym of *nigra:* Mayr, 1867b: 58. See also: Bolton, 1974a: 87.

*bakusuensis. Cataulacus pygmaeus* var. *bakusuensis* Forel, 1913h: 350 (w.m.) ZAIRE. Junior synonym of *pygmaeus:* Bolton, 1974a: 48.

*batonga. Cataulacus baumi* r. *batonga* Forel, 1913a: 114 (w.) ZIMBABWE. Junior synonym of *intrudens:* Bolton, 1974a: 43.

*baumi. Cataulacus baumi* Forel, 1901d: 304 (w.q.m.) ANGOLA. [Also described as new by Forel, 1903e: 560.] Junior synonym of *intrudens:* Bolton, 1974a: 43.

**bequaerti.** *Cataulacus bequaerti* Forel, 1913b: 316 (w.) ZAIRE. See also: Bolton, 1974a: 40.

**boltoni.** *Cataulacus boltoni* Snelling, 1979a: 3, figs. 1–8 (w.) NIGERIA.

*brazzavillensis. Cataulacus kohli* st. *brazzavillensis* Santschi, 1910c: 389, fig. 15 (w.m.) CONGO. Junior synonym of *kohli:* Bolton, 1974a: 22.

**brevisetosus.** *Cataulacus brevisetosus* Forel, 1901d: 305 (w.) ANGOLA. [Also described as new by Forel, 1903e: 560.] Subspecies of *pygmaeus:* Santschi, 1916b: 508; Wheeler, W.M. 1922a: 298. Revived status as species: Santschi, 1939b: 246; Bolton, 1974a: 31. Senior synonym of *meduseus,* and material of the unavailable name *plebeja* referred here: Bolton, 1974a: 31.

*brookei. Cataulacus brookei* Forel, 1901e: 378 (w.q.m.) BORNEO. Junior synonym of *hispidulus:* Bolton, 1974a: 66.

*bulawayensis. Cataulacus baumi* var. *bulawayensis* Arnold, 1917: 391 (w.q.) ZIMBABWE. [First available use of *Cataulacus baumi* r. *batonga* var. *bulawayensis* Forel, 1914d: 218; unavailable name (Bolton, 1974a: 43).] Junior synonym of *intrudens:* **new synonymy.**

*carbonarius* Mayr, 1870; see under ***PROCRYPTOCERUS.***

**catuvolcus.** *Cataulacus catuvolcus* Bolton, 1974a: 74, figs. 4, 37 (w.q.m.) PHILIPPINES.

**centrurus.** *Cataulacus centrurus* Bolton, 1982: 359, fig. 33 (w.) CAMEROUN.

**cestus.** *Cataulacus cestus* Bolton, 1982: 360 (w.) ZAIRE.

***CATAULACUS** (continued)*

***chapmani.*** *Cataulacus chapmani* Bolton, 1974a: 75, fig. 30 (w.m.) PHILIPPINES.

*chariensis. Cataulacus pygmaeus* var. *chariensis* Santschi, 1911c: 358 (w.) CHAD. Junior synonym of *pygmaeus:* Bolton, 1974a: 48.

*convergens* Mayr, 1887; see under ***PROCRYPTOCERUS.***

*coriaceus. Cataulacus coriaceus* Stitz, 1910: 138, fig. 7 (w.) CAMEROUN. Junior synonym of *pullus:* Bolton, 1974a: 26.

*coriarius* Mayr, 1870; see under ***PROCRYPTOCERUS.***

*crassispina. Cataulacus erinaceus* var. *crassispina* Santschi, 1917b: 287 (w.) CONGO. Junior synonym of *erinaceus:* Bolton, 1974a: 52.

*deformis* Roger, 1863; see under ***CYPHOMYRMEX.***

*degener. Cataulacus pygmaeus* st. *degener* Santschi, 1916b: 507 (w.) ANGOLA. Junior synonym of *jeanneli:* Bolton, 1982: 358.

*densipunctatus. Cataulacus johannae* r. *densipunctatus* Stitz, 1923: 163 (w.) NAMIBIA. Junior synonym of *intrudens:* Bolton, 1974a: 43.

***difficilis.*** *Cataulacus pygmaeus* st. *difficilis* Santschi, 1916b: 506 (w.) BENIN. Raised to species: Bolton, 1974a: 33.

*donisthorpei. Cataulacus donisthorpei* Santschi, 1937a: 61 (w.) KENYA. Junior synonym of *striativentris:* Bolton, 1974a: 36.

*durbanensis. Cataulacus micans* r. *durbanensis* Forel, 1914d: 219 (w.) SOUTH AFRICA. Arnold, 1917: 395 (q.m.). Junior synonym of *wissmannii:* Bolton, 1974a: 51.

***ebrardi.*** *Cataulacus ebrardi* Forel, 1886c: cv (w.) MADAGASCAR. Forel, 1910a: 20 (q.). See also: Bolton, 1974a: 41.

***egenus.*** *Cataulacus egenus* Santschi, 1911c: 359, fig (w.) CONGO. Wheeler, G.C. & Wheeler, J. 1954b: 151 (l.); Bolton, 1974a: 18 (q.m.). Senior synonym of *simplex:* Bolton, 1974a: 18.

***elongatus.*** *Cataulacus elongatus* Santschi, 1924b: 221 (w.) ANGOLA. Bolton, 1974a: 34 (q.).

***erinaceus.*** *Cataulacus erinaceus* Stitz, 1910: 134, fig. 3 (w.) EQUATORIAL GUINEA. Forel, 1913h: 350 (q.). Material of the nomen nudum *princeps* referred here: Wheeler, W.M. 1922a: 913. Senior synonym of *crassispina:* Bolton, 1974a: 52.

*fernandensis. Cataulacus sulcatus* var. *fernandensis* Stitz, 1910: 137 (w.) FERNANDO PO I. Variety of *guineensis:* Wheeler, W.M. 1922a: 913. Junior synonym of *guineensis:* Bolton, 1974a: 55.

***flagitiosus.*** *Cataulacus flagitiosus* Smith, F. 1862a: 49 (w.) SULAWESI. See also: Bolton, 1974a: 76.

*foveolatus. Cataulacus foveolatus* Stitz, 1910: 140 (w.) EQUATORIAL GUINEA. Junior synonym of *kohli:* Bolton, 1974a: 22.

*foveosquamosus. Cataulacus foveosquamosus* Santschi, 1937a: 58, figs. 8, 9 (q.) SOUTH AFRICA. Junior synonym of *intrudens:* Bolton, 1974a: 43.

***fricatidorsus.*** *Cataulacus otii* st. *fricatidorsus* Santschi, 1914e: 26 (w.) SOUTH AFRICA. Raised to species: Bolton, 1974a: 42.

*gazanus. Cataulacus baumi* var. *gazanus* Santschi, 1928f: 208 (w.) MOZAMBIQUE. Junior synonym of *intrudens:* Bolton, 1974a: 43.

*gilviventris. Cataulacus lujae* var. *gilviventris* Forel, 1913b: 316 (q.) ZAIRE. Junior synonym of *lujae:* Bolton, 1982: 358.

***granulatus.*** *Formica granulata* Latreille, 1802c: 275, pl. 12, fig. 75 (w.) "GRAND INDES". Emery, 1893e: 215 (q.m.); Bingham, 1903: 123 (q.); Bolton, 1974a: 65 (m.). Combination in *Cataulacus:* Smith, F. 1853: 226. Senior synonym of *hispidus:* Bolton, 1974a: 64.

***greggi.*** *Cataulacus greggi* Bolton, 1974a: 54, fig. 22 (w.) ZAIRE.

*guilelmi. Cataulacus huberi* subsp. *guilelmi* Wheeler, W.M. 1925b: 1 (w.) ZAIRE. Junior synonym of *huberi:* Bolton, 1974a: 19.

***guineensis.*** *Cataulacus guineensis* Smith, F. 1853: 225, pl. 20, fig. 5 (w.) "TROPICAL WEST AFRICA". Santschi, 1910c: 390 (q.); Wheeler, W.M. 1922a: 199 (m.). Senior synonym of *sulcatus:* Forel, 1910e: 421; of *alenensis, fernandensis, parallelus, sulcatus,* and *sulcinodis:* Bolton, 1974a: 55.

*hararicus. Cataulacus hararicus* Forel, 1894b: 79 (w.) ETHIOPIA. Junior synonym of *intrudens:* Bolton, 1974a: 43.

*herteri. Cataulacus huberi* r. *herteri* Forel, 1913b: 315 (w.) ZAIRE. Junior synonym of *huberi:* Bolton, 1974a: 19.

***hispidulus.*** *Cataulacus hispidulus* Smith, F. 1865: 76, pl. 4, fig. 7 (w.) SUMATRA. Bolton, 1974a: 66 (q.m.). Variety of *granulatus:* Dalla Torre, 1893: 138; Emery, 1922f: 298. Revived status as species: Donisthorpe, 1932c: 474. Senior synonym of *brookei:* Bolton, 1974a: 66.

*hispidus. Cataulacus hispidus* Smith, F. 1876d: 611, pl. 11, fig. 11 (w.) SINGAPORE. Subspecies of *granulatus:* Forel, 1911a: 24; Viehmeyer, 1916a: 140. Junior synonym of *granulatus:* Bolton, 1974a: 64.

***horridus.*** *Cataulacus horridus* Smith, F. 1857a: 81, pl. 2, fig. 3 (w.) BORNEO. Crawley, 1924: 401 (q.); Wheeler, G.C. & Wheeler, J. 1954b: 151 (l.). Senior synonym of *insularis:* Wheeler, W.M. 1919e: 93; Bolton, 1974a: 84. [Bolton gives priority to *insularis* but priority should rightly be given to *horridus,* on first revisor principle based on Wheeler's attribution.]

***huberi.*** *Cataulacus huberi* André, 1890: 326 (w.) SIERRA LEONE. André, 1895a: 4 (q.); Bolton, 1974a: 20 (m.). Senior synonym of *guilelmi, herteri, longispinus,* and material of the unavailable name *luebensis* referred here: Bolton, 1974a: 19.

***impressus.*** *Cataulacus impressus* Bolton, 1974a: 35 (w.) UGANDA.

***inermis.*** *Cataulacus inermis* Santschi, 1924b: 218, fig. 10 (w.) ZAIRE. See also: Bolton, 1974a: 21.

*insularis. Cataulacus insularis* Smith, F. 1857a: 80, pl. 2, fig. 4 (m.) BORNEO. Junior synonym of *horridus:* Wheeler, W.M. 1919e: 93; Bolton, 1974a: 84 [see note under *horridus*].

*intermedius. Cataulacus intrudens* st. *intermedius* Santschi, 1917b: 287 (w.) ZIMBABWE. Junior synonym of *intrudens:* Bolton, 1974a: 43.

***intrudens.*** *Meranoplus intrudens* Smith, F. 1876d: 609, pl. 11, fig. 7 (w.q.m.) SOUTH AFRICA. Combination in *Cataulacus:* Mayr, 1886c: 364. Senior synonym of *batonga, baumi, densipunctatus, foveosquamosus, gazanus, hararicus, intermedius, johannae, pseudotrema, rugosus, subrugosus, umbilicatus;* and material of the unavailable names *krugeri* and *tangana* referred here: Bolton, 1974a: 43; of *bulawayensis:* **new synonymy**.

***jacksoni.*** *Cataulacus jacksoni* Bolton, 1982: 360 (w.) CAMEROUN.

***jeanneli.*** *Cataulacus jeanneli* Santschi, 1914b: 108, fig. 16 (w.) KENYA. Stirps of *pygmaeus:* Santschi, 1916b: 506. Revived status as species: Santschi, 1924b: 220. Misspelled as *janneli:* Santschi, 1926b: 244. Junior synonym of *brevisetosus:* Bolton, 1974a: 31. Revived from synonymy, and senior synonym of *degener* and *loveridgei:* Bolton, 1982: 358.

*johannae. Cataulacus johannae* Forel, 1895c: 250 (w.q.) MADAGASCAR. Junior synonym of *intrudens:* Bolton, 1974a: 43.

***kenyensis.*** *Cataulacus jeanneli* st. *kenyensis* Santschi, 1935b: 272, fig. 6 (w.) KENYA. Junior synonym of *weissi:* Bolton, 1974a: 39. Revived from synonymy and raised to species: Bolton, 1982: 358.

***kohli.*** *Cataulacus kohli* Mayr, 1895: 127, fig. 2 (w.) SIERRA LEONE. Senior synonym of *brazzavillensis, foveolatus* and *latipes:* Bolton, 1974a: 22.

*krugeri. Cataulacus rugosus* var. *krugeri* Prins, 1965d: 104 (w.q.m.) SOUTH AFRICA. Unavailable name; material referred to *intrudens* by Bolton, 1974a: 43.

*latipes. Cataulacus latipes* Menozzi, 1933a: 106, fig. 4 (w.) UGANDA. Junior synonym of *kohli:* Bolton, 1974a: 22.

***latissimus.*** *Cataulacus latissimus* Emery, 1893e: 215, pl. 8, fig. 10 (w.) WEST MALAYSIA. Viehmeyer, 1922: 212 (q.); Bolton, 1974a: 78 (m.). Senior synonym of *mimula:* Bolton, 1974a: 77.

***latus.*** *Cataulacus latus* Forel, 1891b: 144 (m.) INDIA. Forel, 1903a: 706 (w., in key); Bingham, 1903: 122 (q.). See also: Bolton, 1974a: 78.

*linearis. Cataulacus wissmanni* st. *linearis* Santschi, 1914b: 109, fig. 17 (w.) KENYA. Junior synonym of *wissmannii:* Bolton, 1974a: 51.

***lobatus.*** *Cataulacus lobatus* Mayr, 1895: 126, fig. 1 (w.) CAMEROUN. See also: Bolton, 1974a: 23.

***longinodus.*** *Cataulacus granulatus* var. *longinoda* Forel, 1912n: 60 (q.) SUMATRA. Raised to species: Bolton, 1974a: 67.

*longispinus. Cataulacus huberi* var. *longispinus* Stitz, 1910: 139, fig. 8 (w.) CAMEROUN. Subspecies of *huberi:* Wheeler, W.M. 1925b: 2. Junior synonym of *huberi:* Bolton, 1974a: 19.

*loveridgei. Cataulacus jeanneli* var. *loveridgei* Santschi, 1926b: 244 (w.) TANZANIA. Junior synonym of *jeanneli:* Bolton, 1982: 358.

*luebensis. Cataulacus huberi* st. *herteri* var. *luebensis* Santschi, 1937a: 57 (w.) ZAIRE. Unavailable name; material referred to *huberi* by Bolton, 1974a: 19.

***lujae.*** *Cataulacus lujae* Forel, 1911c: 311 (w.) ZAIRE. Forel, 1914d: 220 (q.m.); Forel, 1916: 427 (q.m.). Race of *brevisetosus:* Forel, 1914d: 220. Stirps of *pygmaeus:* Santschi, 1916b: 506. Revived status as species: Santschi, 1924b: 219. Junior synonym of *brevisetosus:* Bolton, 1974a: 31. Revived from synonymy, and senior synonym of *gilviventris:* Bolton, 1982: 358.

***marginatus.*** *Cataulacus marginatus* Bolton, 1974a: 68, fig. 38 (w.) CHINA.

*marleyi*. *Cataulacus marleyi* Forel, 1914d: 219 (w.) SOUTH AFRICA. Arnold, 1917: 399 (q.). Junior synonym of *traegaordhi:* Bolton, 1982: 359.

***mckeyi***. *Cataulacus mckeyi* Snelling, 1979a: 5, figs. 9–18 (w.q.m.) CAMEROUN. See also: McKey, 1984: 81.

*meduseus*. *Cataulacus meduseus* Santschi, 1939b: 245 (w.) SOUTH AFRICA. Junior synonym of *brevisetosus:* Bolton, 1974a: 31.

***micans***. *Cataulacus rugosus* subsp. *micans* Mayr, 1901b: 27 (w.q.m.) SOUTH AFRICA. Raised to species: Forel, 1914d: 219. Senior synonym of *tristiculus:* Bolton, 1974a: 46.

*mimula*. *Cataulacus latissimus* var. *mimula* Menozzi, 1923b: 210 (w.) BORNEO. Junior synonym of *latissimus:* Bolton, 1974a: 77.

*minor*. *Cataulacus reticulatus* var. *minor* Smith, F. 1857a: 81 (w.) BORNEO. Junior synonym of *reticulatus:* Bolton, 1974a: 82.

***mocquerysi***. *Cataulacus mocquerysi* André, 1889: 229 (w.) SIERRA LEONE. Bolton, 1974a: 48 (q.). Senior synonym of *nainei:* Bolton, 1974a: 47.

***moloch***. *Cataulacus moloch* Bolton, 1982: 361, fig. 34 (w.) GHANA.

***muticus***. *Cataulacus muticus* Emery, 1889b: 507, pl. 10, fig. 17 (w.) BURMA.

*nainei*. *Cataulacus mocquerysi* var. *nainei* Forel, 1918a: 724 (w.) ZAIRE. Junior synonym of *mocquerysi:* Bolton, 1974a: 47.

***nenassus***. *Cataulacus nenassus* Bolton, 1974a: 70, fig. 41 (w.q.m.) JAVA.

**nigra*. **Attopsis nigra* Heer, 1850: 157, pl. 12, fig. 13 (q.) YUGOSLAVIA. Combination in *Cataulacus,* and junior synonym of *anthracinus:* Mayr, 1867b: 58.

***oberthueri***. *Cataulacus* (*Otomyrmex*) *oberthueri* Emery, in Forel, 1891b: 146, pl. 4, fig. 9 (w.) MADAGASCAR. See also: Bolton, 1974a: 24.

*orientalis*. *Cataulacus pullus* var. *orientalis* Santschi, 1914b: 108 (w.) KENYA. Junior synonym of *pullus:* Bolton, 1974a: 26.

*otii*. *Cataulacus wissmanni* r. *otii* Forel, 1901d: 304 (w.q.) SOUTH AFRICA. Raised to species: Santschi, 1914e: 26; Arnold, 1917: 401. Junior synonym of *wissmannii:* Bolton, 1974a: 51.

*parallelus*. *Cataulacus parallelus* Smith, F. 1853: 226, pl. 19, fig. 6 (q.) SOUTH AFRICA (locality in error). Junior synonym of *guineensis:* Bolton, 1974a: 55.

***pilosus***. *Cataulacus pilosus* Santschi, 1920c: 118 (w.q.) ZAIRE. See also: Bolton, 1974a: 35.

****planiceps***. **Cataulacus planiceps* Emery, 1891a: 148, pl. 1, figs. 8, 9 (w.) SICILIAN AMBER. See also: Bolton, 1974a: 87.

*plebeja*. *Cataulacus pygmaeus* st. *lujae* var. *plebeja* Santschi, 1916b: 508 (w.) ZIMBABWE. Unavailable name; material referred to *brevisetosus* by Bolton, 1974a: 31.

*plectroniae*. *Cataulacus traegaordhi* var. *plectroniae* Wheeler, W.M. 1922a: 199 (w.) ZAIRE. Junior synonym of *weissi:* Bolton, 1974a: 39.

***pompom***. *Cataulacus pompom* Snelling, 1979a: 8, figs. 19–26 (w.) WEST MALAYSIA.

***porcatus***. *Cataulacus porcatus* Emery, 1899f: 286 (w.q.) MADAGASCAR. See also: Bolton, 1974a: 25.

***praetextus***. *Cataulacus praetextus* Smith, F. 1867: 528, fig. 5 (w.) BORNEO. Bolton, 1974a: 81 (q.m.). Senior synonym of *sumatrensis:* Bolton, 1974a: 80.

*princeps*. *Cataulacus princeps* Forel, 1909b: 71. *Nomen nudum,* attributed to Emery; material referred to *erinaceus* by Wheeler, W.M. 1922a: 913.

*pseudotrema*. *Cataulacus baumi* st. *pseudotrema* Santschi, 1926b: 244 (w.) TANZANIA. Junior synonym of *intrudens:* Bolton, 1974a: 43.

***pullus***. *Cataulacus pullus* Santschi, 1910c: 387, fig. 13 (w.) CONGO. Senior synonym of *coriaceus* and *orientalis:* Bolton, 1974a: 26.

***pygmaeus***. *Cataulacus pygmaeus* André, 1890: 325 (w.) SIERRA LEONE. Senior synonym of *bakusuensis* and *chariensis:* Bolton, 1974a: 48.

***regularis***. *Cataulacus regularis* Forel, 1892l: 252 (w.) MADAGASCAR. See also: Bolton, 1974a: 27.

***resinosus***. *Cataulacus taprobanae* var. *resinosa* Viehmeyer, 1913: 145 (w.) SULAWESI (in copal). Raised to species: Bolton, 1974a: 86.

***reticulatus***. *Cataulacus reticulatus* Smith, F. 1857a: 81, pl. 2, fig. 8 (w.) BORNEO. Junior synonym of *granulatus:* Dalla Torre, 1893: 138; Wheeler, W.M. 1919e: 92. Revived status as species: Donisthorpe, 1932c: 451. Senior synonym of *minor:* Bolton, 1974a: 82.

*rudis* Mayr, 1870; see under ***PROCRYPTOCERUS***.

*rugosus*. *Cataulacus intrudens* var. *rugosus* Forel, 1894b: 78 (w.) MOZAMBIQUE. Mayr, 1895: 129 (q.). Raised to species: Mayr, 1895: 129; Prins, 1963: 104. Junior synonym of *intrudens:* Bolton, 1974a: 43.

***satrap***. *Cataulacus satrap* Bolton, 1982: 363 (w.) CAMEROUN.

*schoutedeni*. *Cataulacus schoutedeni* Santschi, 1919c: 248 (w.q.) ZAIRE. Junior synonym of *tardus:* Bolton, 1974a: 27.

***setosus***. *Cataulacus setosus* Smith, F. 1860b: 114, pl. 1, fig. 7 (w.) INDONESIA (Batjan I.). Smith, F. 1863: 24 (q.). See also: Bolton, 1974a: 71.

****silvestrii***. **Cataulacus silvestrii* Emery, 1891a: 147, pl. 1, figs. 5–7 (w.) SICILIAN AMBER. See also: Bolton, 1974a: 86.

***simoni***. *Cataulacus simoni* Emery, 1893f: 248 (w.) SRI LANKA. Forel, 1909e: 393 (q.). Senior synonym of *andamanensis:* Bolton, 1974a: 72.

*simplex*. *Cataulacus egenus* st. *simplex* Santschi, 1914b: 111, fig 18 (w.) UGANDA. Junior synonym of *egenus:* Bolton, 1974a: 18.

***striativentris***. *Cataulacus wissmanni* var. *striativentris* Santschi, 1924b: 219 (w.) ZAIRE. Raised to species, and senior synonym of *donisthorpei:* Bolton, 1974a: 36.

*subrugosus*. *Cataulacus rugosus* var. *subrugosus* Santschi, 1914e: 26 (w.) SOUTH AFRICA. Santschi, 1937d: 236 (q.). Junior synonym of *intrudens:* Bolton, 1974a: 43.

*suddensis*. *Cataulacus pygmaeus* subsp. *suddensis* Weber, 1943c: 378 (w.m.) SUDAN. Junior synonym of *traegaordhi:* Bolton, 1982: 359.

*sulcatus*. *Cataulacus sulcatus* Stitz, 1910: 136, figs. 4–6 (w.q.m.) CAMEROUN. Variety of *guineensis:* Wheeler, W.M. 1922a: 913. Junior synonym of *guineensis:* Forel, 1910e: 421; Bolton, 1974a: 55.

*sulcinodis*. *Cataulacus guineensis* r. *sulcinodis* Emery, 1892d: 563, pl. 15, fig. 8 (w.) IVORY COAST. Raised to species: Dalla Torre, 1893: 139. Subspecies of *guineensis:* Wheeler, W.M. 1922a: 913. Junior synonym of *guineensis:* Bolton, 1974a: 55.

*sumatrensis*. *Cataulacus praetextus* var. *sumatrensis* Forel, 1912n: 60 (w.) SUMATRA. Viehmeyer, 1916a: 140 (q.). Junior synonym of *praetextus:* Bolton, 1974a: 80.

*tangana*. *Cataulacus baumi* st. *pseudotrema* var. *tangana* Santschi, 1928f: 209 (w.q.) TANZANIA. Unavailable name; material referred to *intrudens* by Bolton, 1974a: 43.

***taprobanae***. *Cataulacus taprobanae* Smith, F. 1853: 225, pl. 20, fig. 10 (w.) SRI LANKA. Forel, 1909e: 393 (q.); Wheeler, G.C. & Wheeler, J. 1954b: 150 (l.); Bolton, 1974a: 83 (m.).

***tardus***. *Cataulacus tardus* Santschi, 1914d: 372, fig. 33 (w.) GUINEA. Senior synonym of *schoutedeni:* Bolton, 1974a: 27.

***taylori***. *Cataulacus taylori* Bolton, 1982: 364 (w.) NIGERIA.

***tenuis***. *Cataulacus tenuis* Emery, 1899f: 288 (q.) MADAGASCAR. Santschi, 1913c: 310 (w.). See also: Bolton, 1974a: 37.

***theobromicola***. *Cataulacus theobromicola* Santschi, 1939c: 8, fig. 2 (w.) ZAIRE. See also: Bolton, 1974a: 28.

***traegaordhi***. *Cataulacus traegaordhi* Santschi, 1914e: 24, fig. 3 (w.q.m.) SOUTH AFRICA. Junior synonym of *pygmaeus:* Bolton, 1974a: 48. Revived from synonymy, and senior synonym of *marleyi, suddensis* and *ugandensis:* Bolton, 1982: 359.

*tristiculus*. *Cataulacus intrudens* st. *tristiculus* Santschi, 1919b: 237 (w.q.m.) SOUTH AFRICA. Junior synonym of *micans:* Bolton, 1974a: 46.

*ugandensis*. *Cataulacus traegaordhi* var. *ugandensis* Santschi, 1914b: 110 (w.) UGANDA. [Misspelled in original as *trägardhi*]. Junior synonym of *traegaordhi:* Bolton, 1982: 359.

*umbilicatus*. *Cataulacus umbilicatus* Santschi, 1937a: 59, figs. 10–12 (q.) MOZAMBIQUE. Junior synonym of *intrudens:* Bolton, 1974a: 43.

***voeltzkowi***. *Cataulacus voeltzkowi* Forel, 1907g: 84 (w.) MADAGASCAR. See also: Bolton, 1974a: 50.

***vorticus***. *Cataulacus vorticus* Bolton, 1974a: 38, figs. 16, 19 (w.q.m.) NIGERIA.

***wasmanni***. *Cataulacus* (*Otomyrmex*) *wasmanni* Forel, 1897c: 193 (w.) MADAGASCAR. [*Otomyrmex wasmanni* Wasmann, 1897: 250; *nomen nudum,* attributed to Forel.] See also: Bolton, 1974a: 29.

***weissi***. *Cataulacus weissi* Santschi, 1913c: 310 (w.) CONGO. Forel, 1916: 427 (q.m.). Stirps of *pygmaeus:* Santschi, 1916b: 506. Variety of *lujae:* Santschi, 1924b: 219. Revived status as species, and senior synonym of *plectroniae:* Bolton, 1974a: 39. Senior synonym of *aethiops:* Bolton, 1982: 358.

***wissmannii***. *Cataulacus wissmannii* Forel, 1894b: 78 (w.) MOZAMBIQUE. Santschi, 1937a: 61 (q.). Senior synonym of *durbanensis, linearis* and *otii:* Bolton, 1974a: 51.

*CAULOMYRMA:* see under ***LEPTOTHORAX***.

#***CAUTOLASIUS:*** see under ***LASIUS***.

***CENTROMYRMEX*** (Old and New World Tropics)

***alfaroi***. *Centromyrmex alfaroi* Emery, 1890b: 40 (w.) COSTA RICA. Emery, 1906c: 115 (q.). [Also

***CENTROMYRMEX** (continued)*

described as new by Emery, 1894k: 46.] See also: Kempf, 1967b: 404.

**#angolensis.** *Centromyrmex constanciae* var. *angolensis* Santschi, 1937d: 216 (w.) ANGOLA.

*arnoldi. Centromyrmex arnoldi* Santschi, 1919b: 229, figs. a-d (w.m.) MOZAMBIQUE. Variety of *constanciae:* Santschi, 1920b: 8. Junior synonym of *constanciae:* Arnold, 1926: 199 (in text).

*asili. Promyopias asili* Crawley, 1916a: 30, fig. (q.) MALAWI. Combination in *Pseudoponera* (*Promyopias*): Wheeler, W.M. 1922a: 779. Junior synonym of *silvestrii:* Brown, 1963: 10.

**bequaerti.** *Glyphopone bequaerti* Forel, 1913b: 308, fig. 1 (q.) ZAIRE. Wheeler, G.C. & Wheeler, J. 1976a: 48 (l.). Combination in *Centromyrmex,* and senior synonym of *rufigaster:* Brown, 1963: 10.

*bohemanni. Centromyrmex bohemanni* Mayr, 1866b: 895, pl. 20, fig. 7 (w.) BRAZIL. Junior synonym of *brachycola:* Kempf, 1967b: 405.

**brachycola.** *Ponera brachycola* Roger, 1861a: 5 (q.) BRAZIL. Emery, 1906c: 114 (w.). Combination in *Centromyrmex:* Emery, 1890a: 74; Emery, 1890b: 40 (footnote). Senior synonym of *bohemanni* and *paulina:* Kempf, 1967b: 405.

**#ceylonicus.** *Centromyrmex feae* var. *ceylonicus* Forel, 1900d: 303 (w.m.) SRI LANKA. [Misspelled as *ceylonensis:* Forel, 1911i: 215.]

**congolensis.** *Centromyrmex congolensis* Weber, 1949b: 5, figs. 3, 4 (w.) ZAIRE.

**constanciae.** *Centromyrmex constanciae* Arnold, 1915: 38, pl. 2, fig. 14 (w.q.) ZIMBABWE. Arnold, 1926: 199 (m.). Senior synonym of *arnoldi:* Arnold, 1926: 199 (in text). Current subspecies: nominal plus *angolensis, guineensis.*

*donisthorpei. Centromyrmex donisthorpei* Menozzi, 1925c: 443, fig. 4 (q.) PHILIPPINES. Junior synonym of *feae:* Brown, 1955f: 103.

**feae.** *Spalacomyrmex feae* Emery, 1889b: 491, pl. 10, figs. 11–15 (w.) BURMA. Forel, 1900d: 303 (q.m.); Wheeler, W.M. 1936d: 207 (m.); Wheeler, G.C. & Wheeler, J. 1952c: 604 (l.); Imai, *et al.* 1984: 5 (k.). Combination in *Centromyrmex:* Emery, 1890b: 40 (footnote). Senior synonym of *donisthorpei:* Brown, 1955f: 103. Current subspecies: nominal plus *ceylonicus, greeni.*

**gigas.** *Centromyrmex gigas* Forel, 1911c: 287 (w.) BRAZIL. Borgmeier, 1937b: 223 (q.). See also: Kempf, 1967b: 407.

**#greeni.** *Centromyrmex greeni* Forel, 1901f: 325 (w.) SINGAPORE. Variety of *feae:* Emery, 1911d: 58.

**#guineensis.** *Centromyrmex arnoldi* r. *guineensis* Bernard, 1953b: 186, fig. 1 (w.) GUINEA. Subspecies of *constanciae:* current status.

**hamulatus.** *Typhloteras hamulatum* Karavaiev, 1925b: 129, fig. 5 (w.) SUMATRA. Karavaiev, 1926d: 419 (q.m.). Combination in *Centromyrmex:* Brown, 1953c: 8.

**#longiventris.** *Centromyrmex sellaris* var. *longiventris* Santschi, 1919b: 229 (w.) CAMEROUN.

*paulina. Centromyrmex brachycola* var. *paulina* Forel, 1911c: 287 (w.) BRAZIL. Junior synonym of *brachycola:* Kempf, 1967b: 405.

*rufigaster. Glyphopone* (*Leptopone*) *rufigaster* Arnold, 1916: 163, fig. 10 (q.) ZIMBABWE. Combination in *Leptopone:* Wheeler, W.M. 1922a: 766. Junior synonym of *bequaerti:* Brown, 1963: 10.

*sculpturatus* Santschi, 1931; see under ***TYPHLOMYRMEX.***

**sellaris.** *Centromyrmex sellaris* Mayr, 1896: 230 (w.) CAMEROUN. Current subspecies: nominal plus *longiventris.*

**silvestrii.** *Myopias* (*Promyopias*) *silvestrii* Santschi, 1914d: 324, fig. 10 (w.) GUINEA. Combination in *Promyopias:* Emery, 1915g: 26; in *Pseudoponera* (*Promyopias*): Wheeler, W.M. 1922a: 779; in *Centromyrmex:* Brown, 1963: 10. Senior synonym of *asili:* Brown, 1963: 10.

*CEPHALOMORIUM:* see under ***PHEIDOLE.***

*CEPHALOMYRMA:* see under ***POLYRHACHIS.***

***CEPHALOMYRMEX** (U.S.A.)

***rotundatus.** **Cephalomyrmex rotundatus* Carpenter, 1930: 37, pl. 7, fig. 5; pl. 10, fig. 10 (q.) U.S.A. (Miocene).

***CEPHALOTES*** (Neotropical)

*abdominalis* Santschi, 1929; see under ***EUCRYPTOCERUS.***

*adolphi* Emery, 1906 [*Cryptocerus*]; see under ***ZACRYPTOCERUS.***

*aethiops* Smith, F. 1853 [*Cryptocerus*]; see under ***EUCRYPTOCERUS.***

*alfaroi. Cryptocerus alfaroi* Emery, 1890b: 76 (s.w.) COSTA RICA. [Also described as new by Emery, 1894k: 59.] Combination in *Cephalotes:* Emery, 1914c: 39. See also: Kempf, 1951: 112. Junior synonym of *serraticeps:* Kempf, 1963c: 437.

*amazonensis* Forel, 1911 [*Cryptocerus*]; see under ***ZACRYPTOCERUS.***

*angulatus* Smith, F. 1858 [*Cryptocerus*]; see under ***ZACRYPTOCERUS.***

*angulosus* Smith, F. 1857 [*Cryptocerus*]; see under ***ZACRYPTOCERUS.***

*angustus* Mayr, 1862 [*Cryptocerus*]; see under ***ZACRYPTOCERUS.***

*araneolus* Smith, F. 1853 [*Cryptocerus*]; see under ***ZACRYPTOCERUS.***

*argentatus* Smith, F. 1853 [*Cryptocerus*]; see under ***ZACRYPTOCERUS.***

**atratus.** *Formica atrata* Linnaeus, 1758: 581 (w.) "AMERICA MERIDIONALI". Latreille, 1802c: 274 (q.); Spinola, 1851b: 47; Spinola, 1853: 63 (m.); Wheeler, G.C. & Wheeler, J. 1954b: 154 (l.). Combination in *Cephalotes:* Latreille, 1802a: 358; in *Cryptocerus:* Latreille, 1803: 280; in *Cryptocerus* (*Cephalotes*): Wheeler, W.M. 1916c: 12; in *Cephalotes:* Smith, M.R. 1949c: 19. Senior synonym of *quadridens:* Retzius, 1783: 76; Latreille, 1802c: 272; Kempf, 1951: 115; of *marginatus:* Klug, 1824: 200; Roger, 1862c: 291; of *dubitatus:* Smith, F. 1858b: 188; of *crassispina, nitidiventris, rufiventris,* and material of the unavailable name *dehnowi* referred here: Kempf, 1951: 115. Current subspecies: nominal plus *erectus.*

*aztecus* Forel, 1901 [*Cryptocerus*]; see under ***ZACRYPTOCERUS.***

*basalis* Smith, F. 1876 [*Cryptocerus*]; see under ***ZACRYPTOCERUS.***

*bicolor* Guérin-Méneville, 1844 [*Cryptocerus*]; see under ***MERANOPLUS.***

*biguttatus* Emery, 1890 [*Cryptocerus*]; see under ***ZACRYPTOCERUS.***

*bimaculatus* Smith, F. 1860 [*Cryptocerus*]; see under ***ZACRYPTOCERUS.***

*bivestitus* Santschi, 1922 [*Cryptocerus*]; see under ***ZACRYPTOCERUS.***

*bohlsi* Emery, 1896 [*Cryptocerus*]; see under ***ZACRYPTOCERUS.***

*boliviensis* Santschi, 1921 [*Cryptocerus*]; see under ***ZACRYPTOCERUS.***

*brevispinosa* Santschi, 1921 [*Cryptocerus*]; see under ***ZACRYPTOCERUS.***

*bruchi* Forel, 1912 [*Cryptocerus*]; see under ***ZACRYPTOCERUS.***

*cearensis* Forel, 1908 [*Cryptocerus*]; see under ***ZACRYPTOCERUS.***

*christopherseni* Forel, 1912 [*Cryptocerus*]; see under ***ZACRYPTOCERUS.***

*clarior* Forel, 1912 [*Cryptocerus*]; see under ***ZACRYPTOCERUS.***

*clypeatus* Fabricius, 1804 [*Cryptocerus*]; see under ***ZACRYPTOCERUS.***

*cognatus* Smith, F. 1862 [*Cryptocerus*]; see under ***ZACRYPTOCERUS.***

*columbica* Forel, 1912 [*Cryptocerus*]; see under ***ZACRYPTOCERUS.***

*complanatus* Guérin-Méneville, 1844 [*Cryptocerus*]; see under ***ZACRYPTOCERUS.***

*conspersus* Smith, F. 1867 [*Cryptocerus*]; see under ***ZACRYPTOCERUS.***

*convexus* Santschi, 1916 [*Cryptocerus*]; see under ***ZACRYPTOCERUS.***

*cordatus* Smith, F. 1853 [*Cryptocerus*]; see under ***ZACRYPTOCERUS.***

*cordiae* Stitz, 1913 [*Cryptocerus*]; see under ***ZACRYPTOCERUS.***

*cordiventris* Santschi, 1931 [*Cryptocerus*]; see under ***ZACRYPTOCERUS.***

*crassispina. Cephalotes atratus* st. *crassispina* Santschi, 1920f: 148 (w.) BRAZIL. Junior synonym of *atratus:* Kempf, 1951: 115.

*crenaticeps* Mayr, 1866 [*Cryptocerus*]; see under ***ZACRYPTOCERUS.***

*cristatus* Emery, 1890 [*Cryptocerus*]; see under ***ZACRYPTOCERUS.***

*curvispinosus. Cryptocerus curvispinosus* Mayr, 1863: 459. *Nomen nudum;* attributed to Roger.

*curvistriatus* Forel, 1899 [*Cryptocerus*]; see under ***ZACRYPTOCERUS.***

**decemspinosus.** *Cephalotes decemspinosus* Santschi, 1920f: 148 (w.) FRENCH GUIANA. Senior synonym of *spissus:* Kempf, 1951: 125.

*dehnowi. Cephalotes atratus* r. *quadridens* var. *dehnowi* Forel, 1922: 97 (w.) "AMERIQUE". Unavailable name; material referred to *atratus* by Kempf, 1951: 115.

*denticulatus* Emery, 1894 [*Cryptocerus*]; see under ***ZACRYPTOCERUS.***

*depressus* Klug, 1824 [*Cryptocerus*]; see under ***ZACRYPTOCERUS.***

*discocephalus* Smith, F. 1853 [*Cryptocerus*]; see under ***ZACRYPTOCERUS.***
*dorbignyanus* Smith, F. 1853 [*Cryptocerus*]; see under ***ZACRYPTOCERUS.***
*dubitatus. Cryptocerus dubitatus* Smith, F. 1853: 216, pl. 20, fig. 1 (m.) BRAZIL. Junior synonym of *atratus:* Smith, F. 1858b: 188.
*duckei* Forel, 1906 [*Cryptocerus*]; see under ***ZACRYPTOCERUS.***
*eduarduli* Forel, 1920 [*Cryptocerus*]; see under ***ZACRYPTOCERUS.***
*elegans* Smith, F. 1853 [*Cryptocerus*]; see under ***ZACRYPTOCERUS.***
*ellenriederi* Forel, 1911 [*Cryptocerus*]; see under ***ZACRYPTOCERUS.***
*elongatus* Klug, 1824 [*Cryptocerus*]; see under ***ZACRYPTOCERUS.***
*emeryi* Forel, 1912 [*Cryptocerus*]; see under ***ZACRYPTOCERUS.***
**#*erectus.*** *Cephalotes atratus* subsp. *erectus* Kempf, 1951: 124 (w.q.) PANAMA. [First available use of *Cephalotes atratus* st. *quadridens* var. *erecta* Santschi, 1920f: 149; unavailable name.]
*exiguus* Smith, F. 1867 [*Cryptocerus*]; see under ***ZACRYPTOCERUS.***
*femoralis* Smith, F. 1853 [*Cryptocerus*]; see under ***ZACRYPTOCERUS.***
*fenestralis* Smith, F. 1876 [*Cryptocerus*]; see under ***EUCRYPTOCERUS.***
*fervidus* Smith, F. 1876 [*Cryptocerus*]; see under ***ZACRYPTOCERUS.***
*fides* Santschi, 1925 [*Cryptocerus*]; see under ***ZACRYPTOCERUS.***
*fiebrigi* Forel, 1906 [*Cryptocerus*]; see under ***ZACRYPTOCERUS***
*flavomaculatus* Mayr, 1862 [*Cryptocerus*]; see under ***ZACRYPTOCERUS.***
*foliaceus* Emery, 1906 [*Cryptocerus*]; see under ***ZACRYPTOCERUS.***
*fossithorax* Santschi, 1921 [*Cryptocerus*]; see under ***ZACRYPTOCERUS.***
*gaudens* Santschi, 1922 [*Cryptocerus*]; see under ***ZACRYPTOCERUS.***
*gibbosus* Smith, F. 1876 [*Cryptocerus*]; see under ***ZACRYPTOCERUS.***
*goeldii* Forel, 1912 [*Cryptocerus*]; see under ***ZACRYPTOCERUS.***
*grandinosus* Smith, F. 1860 [*Cryptocerus*]; see under ***ZACRYPTOCERUS.***
*guttifer* Santschi, 1919 [*Cryptocerus*]; see under ***ZACRYPTOCERUS.***
*hamulus* Roger, 1863 [*Cryptocerus*]; see under ***ZACRYPTOCERUS.***
*haytianus* Forel, 1901 [*Cryptocerus*]; see under ***ZACRYPTOCERUS.***
*inaequalis* Mann, 1916 [*Cryptocerus*]; see under ***ZACRYPTOCERUS.***
*inca* Santschi, 1911 [*Cryptocerus*]; see under ***ZACRYPTOCERUS.***
*incertus* Emery, 1906 [*Cryptocerus*]; see under ***ZACRYPTOCERUS.***
*insularis* Wheeler, W.M. 1934 [*Cryptocerus*]; see under ***ZACRYPTOCERUS***
*jamaicensis* Forel, 1922 [*Cryptocerus*]; see under ***ZACRYPTOCERUS.***
*jheringi* Emery, 1894 [*Cryptocerus*]; see under ***ZACRYPTOCERUS.***
*jocans* Forel, 1913 [*Cryptocerus*]; see under ***ZACRYPTOCERUS.***
*jucundus* Smith, F. 1876 [*Cryptocerus*]; see under ***ZACRYPTOCERUS.***
*klugi* Emery, 1894 [*Cryptocerus*]; see under ***ZACRYPTOCERUS.***
*laminatus* Smith, F. 1860 [*Cryptocerus*]; see under ***ZACRYPTOCERUS.***
*lanuginosus* Santschi, 1919 [*Cryptocerus*]; see under ***ZACRYPTOCERUS.***
*liogaster* Santschi, 1916 [*Cryptocerus*]; see under ***ZACRYPTOCERUS.***
*lobigaster* Santschi, 1919 [*Cryptocerus*]; see under ***ZACRYPTOCERUS.***
*maculatus* Smith, F. 1876 [*Cryptocerus*]; see under ***ZACRYPTOCERUS.***
*magdalenensis* Forel, 1899 [*Cryptocerus*]; see under ***ZACRYPTOCERUS.***
*marginatus. Cryptocerus marginatus* Fabricius, 1804: 419 (q.) "CENTRAL AMERICA". Junior synonym of *atratus:* Klug, 1824: 200; Roger, 1862c: 291.
*marginata* Wheeler, W.M. & Mann, 1914 [*Cryptocerus*]; see under ***ZACRYPTOCERUS.***
*medusa* Santschi, 1919 [*Cryptocerus*]; see under ***ZACRYPTOCERUS.***
*membranaceus* Klug, 1824 [*Cryptocerus*]; see under ***ZACRYPTOCERUS.***
*minutus* Fabricius, 1804 [*Cryptocerus*]; see under ***ZACRYPTOCERUS.***
*multispinosus* Norton, 1868 [*Cryptocerus*]; see under ***ZACRYPTOCERUS.***
*multispinus* Emery, 1890 [*Cryptocerus*]; see under ***ZACRYPTOCERUS.***
*multistrigus. Cryptocerus multistrigus* Ashmead, 1905b: 384. *Nomen nudum,* attributed to Smith, F.
*nanus* Forel, 1912 [*Cryptocerus*]; see under ***ZACRYPTOCERUS.***
*nevadensis* Forel, 1912 [*Cryptocerus*]; see under ***ZACRYPTOCERUS.***
*nitidiventris. Cephalotes atratus* var. *nitidiventris* Santschi, 1920f: 148 (w.) FRENCH GUIANA. Junior synonym of *atratus:* Kempf, 1951: 115.
*notatus* Mayr, 1866 [*Cryptocerus*]; see under ***ZACRYPTOCERUS.***
*obtusus* Smith, F. 1858 [*Cryptocerus*]; see under ***ZACRYPTOCERUS.***
*oculatus* Spinola, 1851 [*Cryptoceros*]; see under ***EUCRYPTOCERUS.***
*opacus* Santschi, 1920; see under ***EUCRYPTOCERUS.***
*orbis* Forel, 1906 [*Cryptocerus*]; see under ***ZACRYPTOCERUS.***
*pallens* Klug, 1824 [*Cryptocerus*]; see under ***ZACRYPTOCERUS.***
*pallidicephalus* Smith, F. 1876 [*Cryptocerus*]; see under ***ZACRYPTOCERUS.***
*pampaensis* Santschi, 1931 [*Cryptocerus*]; see under ***ZACRYPTOCERUS.***
*patellaris* Mayr, 1866 [*Cryptocerus*]; see under ***ZACRYPTOCERUS.***
*pavonii* Latreille, 1809 [*Cryptocerus*]; see under ***ZACRYPTOCERUS.***
*peltatus* Emery, 1896 [*Cryptocerus*]; see under ***ZACRYPTOCERUS.***
*peruvianus* Forel, 1911 [*Cryptocerus*]; see under ***ZACRYPTOCERUS.***
*pilosus* Emery, 1896 [*Cryptocerus*]; see under ***ZACRYPTOCERUS.***
*pinelii* Guérin-Méneville, 1844 [*Cryptocerus*]; see under ***ZACRYPTOCERUS.***
*placidus* Smith, F. 1860 [*Cryptocerus*]; see under ***EUCRYPTOCERUS.***
*porrasi* Wheeler, W.M. 1942 [*Cryptocerus*]; see under ***ZACRYPTOCERUS.***
*prodigiosus* Santschi, 1921 [*Cryptocerus*]; see under ***ZACRYPTOCERUS.***
*pubescens* Smith, F. 1853 [*Cryptocerus*]; see under ***MERANOPLUS.***
*punctatus* Mayr, 1862 [*Cryptocerus*]; see under ***ZACRYPTOCERUS.***
*pusillus* Klug, 1824 [*Cryptocerus*]; see under ***ZACRYPTOCERUS.***
*quadratus* Mayr, 1868 [*Cryptocerus*]; see under ***ZACRYPTOCERUS.***
*quadridens. Formica quadridens* De Geer, 1773: 609, pl. 31, figs.17–20 (w.) SURINAM. Combination in *Cephalotes:* Santschi, 1919f: 44. Junior synonym of *atratus:* Retzius, 1783: 76; Latreille, 1802c: 272. Revived from synonymy: Santschi, 1919f: 44. Race/stirps of *atratus:* Santschi, 1920f: 148; Forel, 1922: 97. Junior synonym of *atratus:* Kempf, 1951: 115.
*quadriguttatus* Guérin-Méneville, 1844 [*Cryptocerus*]; see under ***ZACRYPTOCERUS.***
*quadrimaculatus* Klug, 1824 [*Cryptocerus*]; see under ***ZACRYPTOCERUS.***
*ramiphilus* Forel, 1904 [*Cryptocerus*]; see under ***ZACRYPTOCERUS.***
*ridiculus* Santschi, 1915 [*Cryptocerus*]; see under ***ZACRYPTOCERUS.***
*rimosus* Spinola, 1851 [*Cryptocerus*]; see under ***CYPHOMYRMEX.***
*rohweri* Wheeler, W.M. 1916 [*Cryptocerus*]; see under ***ZACRYPTOCERUS.***
*rufescens* Santschi, 1921 [*Cryptocerus*]; see under ***ZACRYPTOCERUS.***
*ruficeps* Santschi, 1929 [*Cryptocerus*]; see under ***ZACRYPTOCERUS.***
*rufiventris. Cryptocerus atratus* ab. *rufiventris* Emery, 1894e: 4 (w.) ARGENTINA. Junior synonym of *atratus:* Kempf, 1951: 115.
*scutulatus* Smith, F. 1867 [*Cryptocerus*]; see under ***ZACRYPTOCERUS.***
***serraticeps.*** *Cryptocerus serraticeps* Smith, F. 1858b: 188, pl. 11, fig. 7 (q.) BRAZIL. Wheeler, G.C. & Wheeler, J. 1983: 607 (l.). Combination in *Eucryptocerus:* Kempf, 1951: 132; in *Cephalotes,* and senior synonym of *alfaroi:* Kempf, 1963c: 437.
*setulifer* Emery, 1894 [*Cryptocerus*]; see under ***ZACRYPTOCERUS.***
*silvae* Forel, 1906 [*Cryptocerus*]; see under ***ZACRYPTOCERUS.***
*sorocabensis* Forel, 1912 [*Cryptocerus*]; see under ***ZACRYPTOCERUS.***
*spinosus* Mayr, 1862 [*Cryptocerus*]; see under ***ZACRYPTOCERUS.***
*spissus. Cephalotes decemspinosus* var. *spissus* Santschi, 1929d: 302 (w.) SURINAM. Junior synonym of *decemspinosus:* Kempf, 1951: 125.

***CEPHALOTES** (continued)*

*steinheili* Forel, 1908 [*Cryptocerus*]; see under **ZACRYPTOCERUS**

*striativentris* Emery, 1894 [*Cryptocerus*]; see under **ZACRYPTOCERUS.**

*targionii* Emery, 1894 [*Cryptocerus*]; see under **ZACRYPTOCERUS.**

*texanus* Santschi, 1915 [*Cryptocerus*]; see under **ZACRYPTOCERUS.**

*trivialis* Santschi, 1929 [*Cryptocerus*]; see under **ZACRYPTOCERUS.**

*tucumana* Forel, 1914 [*Cryptocerus*]; see under **ZACRYPTOCERUS.**

*umbraculatus* Fabricius, 1804 [*Cryptocerus*]; see under **ZACRYPTOCERUS.**

*unimaculatus* Smith, F. 1853 [*Cryptocerus*]; see under **ZACRYPTOCERUS.**

*varians* Smith, F. 1876 [*Cryptocerus*]; see under **ZACRYPTOCERUS.**

*variegata* Forel, 1911 [*Cryptocerus*]; see under **ZACRYPTOCERUS.**

*volxemi* Emery, 1878 [*Cryptocerus*]; see under **ZACRYPTOCERUS.**

*wheeleri* Forel, 1901 [*Cryptocerus*]; see under **ZACRYPTOCERUS.**

*CEPHALOXYS:* see under ***SMITHISTRUMA.***

*CEPOBROTICUS:* see under ***MEGALOMYRMEX.***

**CERAPACHYS** (World Tropics and Subtropics)

***aberrans.*** *Phyracaces aberrans* Clark, 1934b: 25, pl. 2, fig. 6 (w.) AUSTRALIA. Combination in *Cerapachys:* Brown, 1975: 22.

***adamus.*** *Cerapachys* (*Phyracaces*) *adamus* Forel, 1910b: 18 (w.) AUSTRALIA. Combination in *Phyracaces:* Emery, 1911d: 11; in *Cerapachys:* Brown, 1975: 67.

*aegyptiacus. Cerapachys aegyptiacus* Brown, 1975: 22. Replacement name for *Lioponera cooperi* Donisthorpe, 1939a: 255. [Junior secondary homonym of *Cerapachys cooperi* Arnold, 1915: 14.] Junior synonym of *longitarsus:* **new synonymy**.

***afer.*** *Cerapachys afer* Forel, 1907a: 9 (w.q.) TANZANIA. Senior synonym of *cooperi:* Brown, 1975: 22.

***aitkenii.*** *Cerapachys aitkenii* Forel, 1900d: 332 (w.m.) INDIA.

*alfierii. Lioponera alfierii* Donisthorpe, 1939a: 256 (m.) EGYPT. Combination in *Cerapachys:* Brown, 1975: 22. Junior synonym of *longitarsus:* **new synonymy**.

***angustatus.*** *Phyracaces angustatus* Clark, 1924: 76, pl. 6, fig. 7 (q.) AUSTRALIA. Combination in *Cerapachys:* Brown, 1975: 22.

*****annosus.*** **Procerapachys annosus* Wheeler, W.M. 1915h: 28, fig. 2 (w.m.) BALTIC AMBER. Combination in *Cerapachys:* Brown, 1975: 22.

***antennatus.*** *Cerapachys antennatus* Smith, F. 1857a: 74 (w.) BORNEO. Emery, 1887b: 445 (q.); Wheeler, W.M. 1919e: 140 (m.). Type-material redescribed: Crawley, 1926: 387. See also: Donisthorpe, 1941j: 199. Senior synonym of *wheeleri:* Brown, 1975: 22.

***aranus.*** *Cerapachys aranus* **new name**. Replacement name for *emeryi* Viehmeyer, 1914c: 26. [Junior secondary homonym of *emeryi* Forel, 1893d: 461.]

***arnoldi.*** *Cerapachys arnoldi* Forel, 1914d: 211 (w.) SOUTH AFRICA. Senior synonym of *hewitti:* Brown, 1975: 22. See also: Arnold, 1926: 191.

***augustae.*** *Cerapachys* (*Parasyscia*) *augustae* Wheeler, W.M. 1902d: 182, figs. 1, 2 (w.q.) U.S.A. Wheeler, W.M. 1903g: 206 (l.); Wheeler, G.C. 1950: 106 (l.); Smith, M.R. 1942b: 63 (m.).

*australis. Lioponera longitarsus* var. *australis* Forel, 1895f: 422 (w.) AUSTRALIA. Raised to species: Wheeler, W.M. 1918a: 264. Junior synonym of *longitarsus:* Brown, 1975: 23.

*australis. Syscia australis* Forel, 1900b: 68 (w.) AUSTRALIA. Wheeler, G.C & Wheeler, J. 1973d: 205 (l.). Combination in *Cerapachys* (*Syscia*): Emery, 1902c: 24. [Junior secondary homonym of *australis* Forel, above.] Replacement name: *edentatus* Forel, 1900b: 69 [junior synonym of *australis* Forel, 1900b: 68, hence first available replacement name.]

***bakeri.*** *Lioponera bakeri* Wheeler, W.M. & Chapman, 1925: 55, pl. 1, fig. 17 (w.) PHILIPPINES. Combination in *Cerapachys:* Brown, 1975: 22.

***besucheti.*** *Cerapachys besucheti* Brown, 1975: 73, figs. 79–83, 86 (w.q.) INDIA.

***bicolor.*** *Phyracaces bicolor* Clark, 1924: 77, pl. 6, figs. 9, 10 (w.q.) AUSTRALIA. Combination in *Cerapachys:* Brown, 1975: 22.

*bicolor. Lioponera bicolor* Wheeler, W.M. & Chapman, 1925: 54, pl. 1, figs. 15, 16 (w.q.m.) PHILIPPINES. [Unresolved junior secondary homonym of *bicolor* Clark, above.] Junior synonym of *longitarsus:* Brown, 1975: 23.

***binodis.*** *Cerapachys* (*Phyracaces*) *binodis* Forel, 1910b: 20 (w.) AUSTRALIA. Combination in *Phyracaces:* Emery, 1911d: 11; in *Cerapachys:* Brown, 1975: 22. See also: Wheeler, W.M. 1918a: 261.

***biroi.*** *Cerapachys* (*Syscia*) *biroi* Forel, 1907a: 7 (w.) SINGAPORE. Imai *et al.* 1984: 5 (k.). Senior synonym of *ierensis, silvestrii* (and its junior synonym *seini*) and *sinensis:* Brown, 1975: 22. See also: Ogata, 1983: 136.

*brachynodus. Cerapachys mayri* var. *brachynodus* Forel, 1892k: 520 (w.) MADAGASCAR. Junior synonym of *mayri:* Brown, 1975: 23.

***braunsi.*** *Phyracaces braunsi* Emery, 1902c: 27 (w.) SOUTH AFRICA. Combination in *Cerapachys:* Brown, 1975: 22.

***braytoni.*** *Phyracaces braytoni* Weber, 1949a: 3, fig. 4 (w.) KENYA. Combination in *Cerapachys:* Brown, 1975: 22.

***brevicollis.*** *Phyracaces brevicollis* Clark, 1923: 78, fig. 4 (w.) AUSTRALIA. Combination in *Cerapachys:* Brown, 1975: 22.

***brevis.*** *Phyracaces brevis* Clark, 1924: 78, pl. 6, fig. 5 (w.) AUSTRALIA. Imai, Crozier & Taylor, 1977: 348 (k.). Combination in *Cerapachys:* Brown, 1975: 22.

***bryanti.*** *Cerapachys bryanti* Wheeler, W.M. 1919e: 47 (w.) BORNEO.

*butteli. Cerapachys butteli* Forel, 1913k: 3 (w.q.) SUMATRA. Junior synonym of *sulcinodis:* Brown, 1975: 23.

*castaneus. Phyracaces castaneus* Clark, 1924: 79, pl. 7, figs. 7–14 (w.q.m.) AUSTRALIA. Junior synonym of *clarki:* Brown, 1975: 22.

***centurio.*** *Cerapachys centurio* Brown, 1975: 62, figs. 66–71 (w.) ZAIRE.

*ceylonica* Motschoulsky, 1863; see under ***TETRAPONERA.***

***clarki.*** *Phyracaces clarki* Crawley, 1922b: 433 (w.) AUSTRALIA. Combination in *Cerapachys*, and senior synonym of *castaneus:* Brown, 1975: 22.

*clarus* Forel, 1893; see under ***SPHINCTOMYRMEX.***

*clarus. Phyracaces clarus* Clark, 1930c: 3, fig. 1 (w.q.m.) AUSTRALIA. Combination in *Neophyracaces:* Clark, 1941: 76. [Junior secondary homonym of *Cerapachys clarus* Forel, 1893d: 462, above.] Replacement name: *princeps* [junior synonym of *clarus* Clark, hence first available replacement name.]

***coecus.*** *Ooceraea coeca* Mayr, 1897: 420 (w.) SRI LANKA. Forel, 1900d: 329 (q.). Combination in *Cerapachys:* Emery, 1902c: 24.

***cohici.*** *Phyracaces cohici* Wilson, 1957a: 1, fig. (w.m.) NEW CALEDONIA. Combination in *Cerapachys:* Brown, 1975: 22. See also: Wilson, 1958c: 131.

*congolensis. Phyracaces cooperi* r. *congolensis* Forel, 1916: 401 (w.) ZAIRE. Junior synonym of *nkomoensis:* Brown, 1975: 23.

***conservatus.*** *Cerapachys conservatus* Viehmeyer, 1913: 142 (w.) SULAWESI (in copal). See also: Brown, 1975: 68.

***constrictus.*** *Phyracaces constricta* Clark, 1923: 79, fig. 5 (q.) AUSTRALIA. Combination in *Cerapachys:* Brown, 1975: 22.

*cooperi. Cerapachys cooperi* Arnold, 1915: 14 (w.q.) SOUTH AFRICA. Junior synonym of *afer:* Brown, 1975: 22.

*cooperi. Phyracaces cooperi* Arnold, 1915: 18 (w.) SOUTH AFRICA. [Unresolved junior secondary homonym of *cooperi* Arnold, above.] Junior synonym of *vespula:* Brown, 1975: 24.

*cooperi. Lioponera cooperi* Donisthorpe, 1939a: 255 (m.) EGYPT. [Junior secondary homonym of *cooperi* Arnold, 1915: 14.] Replacement name: *aegyptiacus* Brown, 1975: 22.

***coxalis.*** *Phyracaces coxalis* Arnold, 1926: 193, fig. 60 (w.) ZIMBABWE. Combination in *Cerapachys:* Brown, 1975: 22.

***crassus.*** *Phyracaces crassus* Clark, 1941: 74, pl. 13, fig. 1 (w.) AUSTRALIA. Combination in *Cerapachys:* Brown, 1975: 22.

***crawleyi.*** *Cerapachys* (*Chrysapace*) *crawleyi* Wheeler, W.M. 1924d: 225. Replacement name for *Chrysapace jacobsoni* Crawley, 1924: 381. [Junior secondary homonym of *Cerapachys jacobsoni* Forel, 1912d: 103.] Senior synonym of *mirandus:* Brown, 1975: 22.

***cribrinodis.*** *Parasyscia cribrinodis* Emery, 1899e: 463, fig. (w.) CAMEROUN. Combination in *Cerapachys:* Emery, 1902c: 24. See also: Brown, 1975: 62.

***cryptus.*** *Cerapachys* (*Syscia*) *cryptus* Mann, 1921: 408, fig. 1 (w.q.) FIJI IS. Wheeler, G.C. 1950: 105 (l.).

***davisi.*** *Cerapachys* (*Parasyscia*) *davisi* Smith, M.R. 1942b: 64 (m.) U.S.A.

***decorsei.*** *Lioponera decorsei* Santschi, 1912b: 150 (m.) CHAD. Combination in *Cerapachys:* Brown, 1975: 22.

***desertorum.*** *Cerapachys desertorum* Dlussky, in Dlussky, Soyunov & Zabelin, 1990: 177, fig. 43 (w.m.) TURKMENISTAN.

***desposyne.*** *Cerapachys* (*Cerapachys*) *desposyne* Wilson, 1959b: 45 (w.) NEW GUINEA.

**dohertyi.** *Cerapachys dohertyi* Emery, 1902c: 25 (w.) BORNEO. Senior synonym of *parvula:* Wilson, 1959b: 57.

**dominulus.** *Cerapachys (Cerapachys) dominula* Wilson, 1959b: 46 (w.) NEW GUINEA.

*dromus. Phyracaces dromus* Clark, 1941: 75, pl. 13, figs. 2, 3 (w.q.) AUSTRALIA. Junior synonym of *fervidus:* Brown, 1975: 22.

**dumbletoni.** *Phyracaces dumbletoni* Wilson, 1957a: 5, fig. (w.q.) NEW CALEDONIA. Combination in *Cerapachys:* Brown, 1975: 22. See also: Wilson, 1958c: 133.

**edentatus.** *Syscia australis* var. *edentata* Forel, 1900b: 69 (w.) AUSTRALIA. Combination in *Cerapachys (Syscia)*: Emery, 1911d: 10. Junior synonym of *australis* Forel, 1900b: 68 [junior secondary homonym of *australis* Forel, 1895f: 422] and hence first available replacement name: Brown, 1975: 22.

*eidmanni. Phyracaces eidmanni* Menozzi, 1942: 165 (w.) FERNANDO PO I. Junior synonym of *nkomoensis:* Brown, 1975: 23.

**elegans.** *Phyracaces elegans* Wheeler, W.M. 1918a: 254, fig. 13 (w.q.) AUSTRALIA. Combination in *Cerapachys:* Brown, 1975: 22.

*emeryi* Forel, 1893; see under ***SPHINCTOMYRMEX.***

*emeryi. Phyracaces emeryi* Viehmeyer, 1914c: 26 (w.) AUSTRALIA. Combination in *Cerapachys:* Brown, 1975: 22. [Junior secondary homonym of *emeryi* Forel, 1893d: 461, above.] Replacement name: *aranus* **new name**.

**faurei.** *Cerapachys faurei* Arnold, 1949: 262, fig. 2 (w.) SOUTH AFRICA.

***favosus.** **Procerapachys favosus* Wheeler, W.M. 1915h: 31, fig. 3 (w.) BALTIC AMBER (Oligocene). Combination in *Cerapachys:* Brown, 1975: 22.

*femoralis* Motschoulsky, 1863; see under ***TETRAPONERA.***

**fervidus.** *Phyracaces fervidus* Wheeler, W.M. 1918a: 245, fig. 9 (w.) AUSTRALIA. Combination in *Cerapachys,* and senior synonym of *dromus, fici, flavescens, leae, newmani* and *scrutator:* Brown, 1975: 22.

*fici. Phyracaces fici* Viehmeyer, 1924a: 222 (w.q.) AUSTRALIA. Junior synonym of *fervidus:* Brown, 1975: 22.

**ficosus.** *Phyracaces ficosus* Wheeler, W.M. 1918a: 252, fig. 12 (w.) AUSTRALIA. Wheeler, G.C. & Wheeler, J. 1973d: 205 (l.). Combination in *Cerapachys:* Brown, 1975: 22.

**flammeus.** *Phyracaces flammeus* Clark, 1930c: 4, fig. 1 (w.q.) AUSTRALIA. Combination in *Cerapachys:* Brown, 1975: 22.

**flavaclavatus.** *Cerapachys (Cerapachys) flavaclavata* Donisthorpe, 1938e: 499 (w.) NEW GUINEA. See also: Wilson, 1958c: 129.

*flavescens. Phyracaces flavescens* Clark, 1930c: 5, fig. 1 (w.q.) AUSTRALIA. Junior synonym of *fervidus:* Brown, 1975: 22.

**foreli.** *Phyracaces foreli* Santschi, 1914d: 309 (w.) GHANA. Santschi, 1915c: 245 (q.). Combination in *Cerapachys,* and senior synonym of *langi, occipitalis* and *santschii:* Brown, 1975: 22.

**fossulatus.** *Cerapachys fossulatus* Forel, 1895a: 48 (w.) SRI LANKA.

**foveolatus.** *Cerapachys foveolatus* Radchenko, 1993b: 43, figs. 1–3 (w.) VIETNAM.

**fragosus.** *Ooceraea fragosa* Roger, 1862a: 249, pl. 1, fig. 16 (ergatoid q., not w.) SRI LANKA. Combination in *Cerapachys (Ooceraea)*: Emery, 1902c: 24. See also: Brown, 1975: 73.

**fuscior.** *Cerapachys (Syscia) cryptus* subsp. *fuscior* Mann, 1921: 410 (w.) FIJI IS. Raised to species: Wilson, 1959b: 55.

**gilesi.** *Phyracaces gilesi* Clark, 1923: 81, fig. 6 (w.q.l.) AUSTRALIA. Combination in *Cerapachys:* Brown, 1975: 22.

**grandis.** *Phyracaces grandis* Clark, 1934b: 22, pl. 2, fig. 2 (w.) AUSTRALIA. Combination in *Cerapachys:* Brown, 1975: 22.

**greavesi.** *Phyracaces greavesi* Clark, 1934b: 25, pl. 2, fig. 5 (w.) AUSTRALIA. Combination in *Cerapachys:* Brown, 1975: 22.

**gwynethae.** *Neophyracaces gwynethae* Clark, 1941: 77, pl. 13, figs. 6, 7 (w.m.) AUSTRALIA. Combination in *Cerapachys:* Brown, 1975: 22.

**heros.** *Phyracaces heros* Wheeler, W.M. 1918a: 240, fig. 7 (w.) AUSTRALIA. Combination in *Cerapachys:* Brown, 1975: 22.

**hewitti.** *Phyracaces hewitti* Wheeler, W.M. 1919e: 48 (w.q.) BORNEO. Combination in *Cerapachys:* Brown, 1975: 22.

*hewitti. Cerapachys arnoldi* var. *hewitti* Arnold, 1926: 192 (w.) SOUTH AFRICA. [Unresolved junior secondary homonym of *hewitti* Wheeler, above.] Junior synonym of *arnoldi:* Brown, 1975: 22.

*hewitti. Phyracaces hewitti* Donisthorpe, 1931b: 494 (w.m.) BORNEO. [Unresolved junior primary homonym of *hewitti* Wheeler, above.] Junior synonym of *hewitti* Wheeler: Baroni Urbani, 1971b: 361; of *suscitatus:* Brown, 1975: 23.

**hondurianus.** *Cerapachys (Parasyscia) hondurianus* Mann, 1922: 1 (w.q.) HONDURAS.

**humicola.** *Cerapachys humicola* Ogata, 1983: 132, figs. 1–13 (w.q.m.) JAPAN.

*ierensis. Cerapachys (Syscia) ierensis* Weber, 1939a: 94 (w.) TRINIDAD. Junior synonym of *biroi:* Brown, 1975: 22.

**imerinensis.** *Parasyscia imerinensis* Forel, 1891b: 138, pl. 3, fig. 12 (q.) MADAGASCAR. Combination in *Cerapachys:* Emery, 1911d: 9.

**inconspicuus.** *Cerapachys inconspicua* Emery, 1901h: 153 (w.) NEW GUINEA.

*inconspicuus. Phyracaces inconspicuus* Clark, 1924: 82, pl. 6, figs. 1, 2 (w.q.) AUSTRALIA. [Junior secondary homonym of *inconspicuus* Emery, above.] Replacement name: *incontentus* Brown, 1975: 23.

**incontentus.** *Cerapachys incontentus* Brown, 1975: 23. Replacement name for *inconspicuus* Clark, 1924: 82. [Junior secondary homonym of *inconspicuus* Emery, 1901h: 153.]

**indicus.** *Cerapachys indicus* Brown, 1975: 69, figs. 58–65, 72 (w.q.) INDIA.

**jacobsoni.** *Cerapachys jacobsoni* Forel, 1912d: 103 (w.) JAVA. Senior synonym of *sumatrensis* and *vandermeermohri:* Brown, 1975: 23.

*jacobsoni. Chrysapace jacobsoni* Crawley, 1924: 381, fig. 1 (q.) SUMATRA. [Junior secondary homonym of *jacobsoni* Forel, above.] Replacement name: *crawleyi* Wheeler, W.M. 1924d: 225.

**jovis.** *Cerapachys (Phyracaces) jovis* Forel, 1915b: 20, pl. 1, fig. 1 (w.) AUSTRALIA. Combination in *Phyracaces:* Wheeler, W.M. 1918a: 243; in *Cerapachys:* Brown, 1975: 23.

**kenyensis.** *Cerapachys kenyensis* Consani, 1951: 169, fig. 2 (w.) KENYA.

**kodecorum.** *Cerapachys kodecorum* Brown, 1975: 70, figs. 74–78 (w.) BORNEO.

**kraepelinii.** *Cerapachys kraepelinii* Forel, 1895c: 246 (w.) MADAGASCAR.

**krombeini.** *Phyracaces krombeini* Donisthorpe, 1947c: 580 (w.) NEW GUINEA. Combination in *Cerapachys:* Brown, 1975: 23.

**lamborni.** *Cerapachys lamborni* Crawley, 1923a: 29 (w.) MALAWI. Senior synonym of *pigra:* Brown, 1975: 23.

*langi. Phyracaces langi* Wheeler, W.M. 1922a: 54, fig. 4 (w.q.) ZAIRE. Junior synonym of *foreli:* Brown, 1975: 22.

**larvatus.** *Phyracaces larvatus* Wheeler, W.M. 1918a: 257, fig. 15 (w.) AUSTRALIA. Clark, 1924: 83 (q.m.); Wheeler, G.C. & Wheeler, J. 1964a: 69 (l.). Combination in *Cerapachys:* Brown, 1975: 23.

*latiuscula. Parasyscia peringueyi* var. *latiuscula* Emery, 1895h: 19 (w.) SOUTH AFRICA. Combination in *Cerapachys:* Emery, 1911d: 9. Junior synonym of *peringueyi:* Brown, 1975: 23.

**latus.** *Cerapachys latus* Brown, 1975: 23. Replacement name for *Phyracaces reticulatus* Clark, 1926: 45. [Junior secondary homonym of *Cerapachys reticulatus* Emery, 1923: 60.]

*leae. Phyracaces leae* Wheeler, W.M. 1918a: 243, fig. 8 (w.) AUSTRALIA. Junior synonym of *fervidus:* Brown, 1975: 22.

**lindrothi.** *Cerapachys (Cerapachys) lindrothi* Wilson, 1959b: 52 (w.q.) FIJI IS.

**lividus.** *Cerapachys lividus* Brown, 1975: 64, figs. 46, 47, 50 (w.) MADAGASCAR.

*longicornis. Cerapachys opacus* subsp. *longicornis* Viehmeyer, 1914c: 27 (w.) NEW GUINEA. Junior synonym of *opacus:* Wilson, 1959b: 48.

**longitarsus.** *Lioponera longitarsus* Mayr, 1879: 667 (w.q.) INDIA. Forel, 1900d: 329 (m.). Combination in *Cerapachys:* Brown, 1975: 23. Senior synonym of *australis, bicolor, parva* and *pygmaeus:* Brown, 1975: 23; of *aegyptiacus* and *alfierii:* **new synonymy**.

**luteoviger.** *Cerapachys luteoviger* Brown, 1975: 70, figs. 54–57, 90, 91, 93 (w.q.) SRI LANKA.

**luzuriagae.** *Lioponera luzuriagae* Wheeler, W.M. & Chapman, 1925: 53, pl. 1, figs. 13, 14 (w.q.) PHILIPPINES. Wheeler, G.C. 1950: 108 (l.). Combination in *Cerapachys:* Brown, 1975: 23.

**macrops.** *Neophyracaces macrops* Clark, 1941: 79, pl. 13, fig. 8 (w.) AUSTRALIA. Combination in *Cerapachys:* Brown, 1975: 23.

**majusculus.** *Cerapachys (Cerapachys) majusculus* Mann, 1921: 408 (w.) FIJI IS.

**manni.** *Cerapachys manni* Crawley, 1926: 390 (q.) SUMATRA.

**marginatus.** *Cerapachys marginata* Emery, 1897c: 594, pl. 15, fig. 41 (w.) NEW GUINEA. Combination in *Phyracaces:* Emery, 1902c: 24; in *Cerapachys:* Brown, 1975: 23.

**mayri.** *Cerapachys mayri* Forel, 1892l: 244 (w.) MADAGASCAR. Emery, 1899f: 264 (m.). Combination in

***CERAPACHYS** (continued)*

*Phyracaces:* Emery, 1902c: 23; in *Cerapachys,* and senior synonym of *brachynodus:* Brown, 1975: 23.

*mayri* Emery, 1900; see under ***SIMOPONE.***

*mirandus. Cerapachys mirandus* Wheeler, W.M. 1924d: 224 (q.) SUMATRA. Junior synonym of *crawleyi:* Brown, 1975: 22.

***mjobergi.*** *Cerapachys (Phyracaces) mjobergi* Forel, 1915b: 18, pl. 1, fig. 9 (w.) AUSTRALIA. Combination in *Phyracaces:* Wheeler, W.M. 1918a: 243; in *Cerapachys:* Brown, 1975: 23.

***muiri.*** *Cerapachys muiri* Wheeler, W.M. & Chapman, 1925: 52, pl. 1, figs. 11, 12 (w.) PHILIPPINES.

***mullewanus.*** *Phyracaces mullewanus* Wheeler, W.M. 1918a: 251 (m.) AUSTRALIA. Combination in *Cerapachys:* Brown, 1975: 23.

***natalensis.*** *Cerapachys cribrinodis* var. *natalensis* Forel, 1901f: 335 (w.) SOUTH AFRICA. [Also described as new by Forel, 1910e: 423.] Raised to species: Brown, 1975: 23.

*negrosensis. Cerapachys negrosensis* Wheeler, W.M. & Chapman, 1925: 51, pl. 1, fig. 8 (w.) PHILIPPINES. Junior synonym of *rufithorax:* Brown, 1955e: 52.

***neotropicus.*** *Cerapachys (Cerapachys) neotropicus* Weber, 1939a: 93, fig. 1 (w.) TRINIDAD.

*newmani. Phyracaces newmani* Clark, 1923: 82, fig. 7 (w.) AUSTRALIA. Clark, 1924: 84 (q.). Junior synonym of *fervidus:* Brown, 1975: 22.

***niger.*** *Lioponera nigra* Santschi, 1914b: 46 (m.) KENYA. Combination in *Cerapachys:* Brown, 1975: 23.

***nigriventris.*** *Phyracaces nigriventris* Clark, 1924: 84, pl. 6, figs. 3, 4 (w.q.) AUSTRALIA. Combination in *Cerapachys:* Brown, 1975: 23.

***nitens.*** *Cerapachys (Cerapachys) nitens* Donisthorpe, 1949b: 487 (m.) NEW GUINEA.

*nitida. Cerapachys nitida* Wheeler, W.M. & Chapman, 1925: 52, pl. 1, figs. 9, 10 (w.) PHILIPPINES. Junior synonym of *rufithorax:* Brown, 1955e: 52.

*nitida. Cerapachys (Parasyscia) nitida* Weber, 1949a: 2, figs. 1–3 (w.q.) ZAIRE. [Junior primary homonym of *nitida* Wheeler & Chapman, above.] Replacement name: *nitidulus* Brown, 1975: 23.

***nitidulus.*** *Cerapachys nitidulus* Brown, 1975: 23. Replacement name for *nitida* Weber, 1949a: 2. [Junior primary homonym of *nitida* Wheeler, W.M. & Chapman, 1925: 52.]

***nkomoensis.*** *Phyracaces nkomoensis* Forel, 1916: 400 (w.) ZAIRE. Combination in *Cerapachys,* and senior synonym of *congolensis* and *eidmanni:* Brown, 1975: 23.

***noctambulus.*** *Liponera noctambula* Santschi, 1910d: 70 (m.) TUNISIA. Combination in *Cerapachys:* Brown, 1975: 23.

*occipitalis. Phyracaces occipitalis* Bernard, 1953b: 216, fig. 6 (w.) GUINEA. Junior synonym of *foreli:* Brown, 1975: 22.

*oculatus* Smith, F. 1857; see under ***PARATOPULA.***

***opacus.*** *Cerapachys opaca* Emery, 1901h: 153 (w.) NEW GUINEA. Wheeler, G.C. & Wheeler, J. 1964a: 67 (l.). Senior synonym of *longicornis:* Wilson, 1959b: 48. See also: Wilson, 1958c: 130.

***papuanus.*** *Ooceraea papuana* Emery, 1897c: 594, pl. 15, figs. 42–43 (w.) NEW GUINEA. Combination in *Cerapachys (Cysias):* Emery, 1902c: 24; in *C. (Syscia):* Emery, 1911d: 10. See also: Brown, 1975: 72.

*parva. Lioponera longitarsus* r. *parva* Forel, 1900d: 330 (w.m.) INDIA. Raised to species: Bingham, 1903: 27. Junior synonym of *longitarsus:* Brown, 1975: 23.

*parvula. Cerapachys dohertyi* var. *parvula* Emery, 1902c: 25 (w.) BORNEO. Raised to species: Wheeler, W.M. 1919e: 47. Junior synonym of *dohertyi:* Wilson, 1959b: 57.

***pawa.*** *Cerapachys (Syscia) pusilla* subsp. *pawa* Mann, 1919: 277 (w.) SOLOMON IS. Raised to species: Wilson, 1959b: 51.

***peringueyi.*** *Parasyscia peringueyi* Emery, 1886: 360 (w.) SOUTH AFRICA. Emery, 1895j: 757 (q.). Combination in *Cerapachys:* Emery, 1902c: 24. Senior synonym of *latiuscula:* Brown, 1975: 23. See also: Arnold, 1915: 13.

***picipes.*** *Phyracaces picipes* Clark, 1924: 86, pl. 6, fig. 8 (w.) AUSTRALIA. Combination in *Cerapachys:* Brown, 1975: 23.

***pictus.*** *Phyracaces pictus* Clark, 1934b: 23, pl. 2, fig. 3 (w.) AUSTRALIA. Combination in *Cerapachys:* Brown, 1975: 23.

*pigra. Cerapachys (Cerapachys) pigra* Weber, 1942a: 41, fig. 11 (w.) SUDAN. Junior synonym of *lamborni:* Brown, 1975: 23.

***piliventris.*** *Neophyracaces piliventris* Clark, 1941: 80, pl. 13, fig. 9 (w.) AUSTRALIA. Combination in *Cerapachys:* Brown, 1975: 23.

***piochardi.*** *Parasyscia piochardi* Emery, in André, 1882c: 236, pl. 14, figs. 6, 7 (w.) SYRIA. Combination in *Cerapachys (Parasyscia):* Emery, 1902c: 24.

***polynikes.*** *Cerapachys (Cerapachys) polynikes* Wilson, 1959b: 50 (w.) NEW GUINEA.

***potteri.*** *Neophyracaces potteri* Clark, 1941: 76, pl. 13, figs. 4, 5 (w.m.) AUSTRALIA. Combination in *Cerapachys:* Brown, 1975: 23.

***princeps.*** *Phyracaces princeps* Clark, 1934b: 24, pl. 2, fig. 4 (w.) AUSTRALIA. Combination in *Cerapachys,* junior synonym of *clarus* Clark, and hence first available replacement name: Brown, 1975: 23.

***pruinosus.*** *Cerapachys pruinosus* Brown, 1975: 71, figs. 51–53 (w.) PHILIPPINES.

***pubescens.*** *Phyracaces pubescens* Emery, 1902c: 26 (q.) BORNEO. Combination in *Cerapachys:* Brown, 1975: 23.

***punctatissimus.*** *Phyracaces punctatissima* Clark, 1923: 84, fig. 8 (w.q.) AUSTRALIA. Combination in *Cerapachys:* Brown, 1975: 23.

***pusillus.*** *Ooceraea pusilla* Emery, 1897c: 595 (w.) NEW GUINEA. Combination in *Cerapachys (Cysias):* Emery, 1902c: 24; in *C. (Syscia):* Emery, 1911d: 10; Mann, 1919: 277.

*pygmaeus. Phyracaces pygmaeus* Clark, 1934b: 26, pl. 2, fig. 7 (w.) AUSTRALIA. Junior synonym of *longitarsus:* Brown, 1975: 23.

***reticulatus.*** *Cerapachys reticulatus* Emery, 1923: 60, fig. 1 (q.) TAIWAN.

*reticulatus. Phyracaces reticulatus* Clark, 1926: 45, pl. 6, fig. 2 (w.) AUSTRALIA. [Junior secondary homonym of *reticulatus* Emery, above.] Replacement name: *latus* Brown, 1975: 23.

*rhodesiana. Cerapachys wroughtoni* var. *rhodesiana* Forel, 1913j: 212 (w.) ZIMBABWE. Junior synonym of *wroughtoni:* Brown, 1975: 24.

*risii. Cerapachys risii* Forel, 1892l: 244 (footnote) (w.) HONG KONG. Junior synonym of *sulcinodis:* Brown, 1975: 23.

*roberti. Cerapachys roberti* Forel, 1914d: 212 (w.) SOUTH AFRICA. Junior synonym of *wroughtoni:* Brown, 1975: 24. See also: Arnold, 1926: 192.

*rotula. Cerapachys (Phyracaces) singularis* subsp. *rotula* Forel, 1910b: 21 (w.) AUSTRALIA. Combination in *Phyracaces:* Emery, 1911d: 11. Junior synonym of *singularis:* Brown, 1975: 23.

***ruficornis.*** *Phyracaces ruficornis* Clark, 1923: 86, fig. 9 (w.l.) AUSTRALIA. Combination in *Cerapachys:* Brown, 1975: 23.

***rufithorax.*** *Cerapachys rufithorax* Wheeler, W.M. & Chapman, 1925: 50, pl. 1, figs. 5–7 (w.) PHILIPPINES. Senior synonym of *negrosensis* and *nitida:* Brown, 1955e: 52.

***rugulinodis.*** *Phyracaces rugulinodis* Wheeler, W.M. 1918a: 249, fig. 11 (m.) AUSTRALIA. Combination in *Cerapachys:* Brown, 1975: 23.

***salimani.*** *Cerapachys (Cerapachys) salimani* Karavaiev, 1925a: 72 (w.) JAVA.

*santschii. Phyracaces santschii* Wheeler, W.M. 1922a: 56 (q.) GABON. Junior synonym of *foreli:* Brown, 1975: 22.

***sauteri.*** *Cerapachys sauteri* Forel, 1913f: 187 (w.) TAIWAN. Radchenko, 1993a: 76 (q.). See also: Terayama, Kubota, Sakai & Kawazoe, 1988: 35.

*scrutator. Phyracaces scrutator* Wheeler, W.M. 1918a: 247, fig. 10 (w.) AUSTRALIA. Junior synonym of *fervidus:* Brown, 1975: 22.

***sculpturatus.*** *Cerapachys (Cerapachys) vitiensis* subsp. *sculpturatus* Mann, 1921: 407 (w.) FIJI IS. Raised to species: Wilson, 1959b: 53.

*seini. Cerapachys (Syscia) seini* Mann, 1931: 440, fig. 1 (w.) PUERTO RICO. Junior synonym of *silvestrii:* Wilson & Taylor, 1967: 33.

***senescens.*** *Phyracaces senescens* Wheeler, W.M. 1918a: 259, fig. 16 (w.) AUSTRALIA. Wheeler, G.C. & Wheeler, J. 1964a: 70 (l.). Combination in *Cerapachys:* Brown, 1975: 23.

*silvestrii. Cerapachys (Syscia) silvestrii* Wheeler, W.M. 1909c: 269 (w.) HAWAII. Senior synonym of *seini:* Wilson & Taylor, 1967: 33. Junior synonym of *biroi:* Brown, 1975: 22.

***similis.*** *Lioponera similis* Santschi, 1930a: 51, figs. 8–10 (m.) IVORY COAST. Combination in *Cerapachys:* Brown, 1975: 23.

***simmonsae.*** *Phyracaces simmonsae* Clark, 1923: 87, fig. 10 (w.q.l.) AUSTRALIA. Combination in *Cerapachys:* Brown, 1975: 23.

*sinensis. Cerapachys (Syscia) sinensis* Wheeler, W.M. 1928c: 3 (w.) CHINA. Junior synonym of *biroi:* Brown, 1975: 22.

***singaporensis.*** *Phyracaces singaporensis* Viehmeyer, 1916a: 110 (w., ergatandromorph) SINGAPORE. Combination in *Cerapachys:* Brown, 1975: 23.

***singularis.*** *Cerapachys singularis* Forel, 1900b: 69 (w.) AUSTRALIA. Combination in *Phyracaces:* Emery, 1902c: 24; in *Cerapachys,* and senior synonym of *rotula:* Brown, 1975: 23.

***sjostedti.*** *Cerapachys (Phyracaces) sjostedti* Forel, 1915b: 19, pl. 1, fig. 6 (w.) AUSTRALIA. Combination in *Phyracaces:* Wheeler, W.M. 1918a: 243; in *Cerapachys:* Brown, 1975: 23.

***splendens.*** *Cerapachys splendens* Borgmeier, 1957: 107, figs. 14–17 (w.) BRAZIL.
***sudanensis.*** *Cerapachys sudanensis* Weber, 1942a: 42, figs. 1, 12 (w.) SUDAN. Senior synonym of *variolosus:* Brown, 1975: 23.
***sulcinodis.*** *Cerapachys sulcinodis* Emery, 1889b: 493 (w.) BURMA. Senior synonym of *butteli* and *risii:* Brown, 1975: 23.
*sumatrensis. Cerapachys jacobsoni* subsp. *sumatrensis* Crawley, 1926: 391 (w.) SUMATRA. Junior synonym of *jacobsoni:* Brown, 1975: 23.
***superatus.*** *Cerapachys (Cerapachys) superata* Wilson, 1959b: 50 (w.) NEW GUINEA.
***suscitatus.*** *Phyracaces suscitatus* Viehmeyer, 1913: 143 (w.) SULAWESI (in copal). Combination in *Cerapachys,* and senior synonym of *hewitti:* Brown, 1975: 23.
***sylvicola.*** *Cerapachys sylvicola* Arnold, 1955: 761, fig. 26 (w.) ZIMBABWE.
***terricola.*** *Cerapachys (Cerapachys) terricola* Mann, 1919: 277, fig. 1 (w.m.) SOLOMON IS. Junior synonym of *inconspicuus:* Wilson, 1959b: 47. Revived from synonymy, and senior synonym of *tulagi:* Brown, 1975: 23.
***toltecus.*** *Cerapachys toltecus* Forel, 1909a: 247 (w.) GUATEMALA.
*tulagi. Cerapachys (Cerapachys) terricola* subsp. *tulagi* Mann, 1919: 279 (w.m.) SOLOMON IS. Junior synonym of *terricola:* Brown, 1975: 23.
***turneri.*** *Cerapachys (Phyracaces) turneri* Forel, 1902h: 405 (w.q.) AUSTRALIA. Combination in *Phyracaces:* Emery, 1911d: 11; in *Cerapachys:* Brown, 1975: 23. See also: Wheeler, W.M. 1918a: 256.
***typhlus.*** *Syscia typhla* Roger, 1861a: 20 (w.) SRI LANKA. Forel, 1900d: 329 (q.). Combination in *Cerapachys (Syscia):* Emery, 1902c: 24. See also: Brown, 1975: 74.
***validus.*** *Cerapachys validus* Arnold, 1960a: 79, fig. 1 (w.) SOUTH AFRICA.
*vandermeermohri. Phyracaces vandermeermohri* Menozzi, 1932d: 2, fig. 1 (w.) SUMATRA. Junior synonym of *jacobsoni:* Brown, 1975: 23.
***varians.*** *Phyracaces varians* Clark, 1924: 87, pl. 6, fig. 6 (w.) AUSTRALIA. Combination in *Cerapachys:* Brown, 1975: 24.
*variolosus. Cerapachys variolosus* Arnold, 1949: 261, fig. 1 (w.q.) ZIMBABWE. Junior synonym of *sudanensis:* Brown, 1975: 23.
***versicolor.*** *Lioponera versicolor* Donisthorpe, 1948d: 589 (m.) NEW GUINEA. Combination in *Cerapachys:* Brown, 1975: 24.
***vespula.*** *Phyracaces vespula* Weber, 1949a: 5, fig. 5 (w.) KENYA. Combination in *Cerapachys,* and senior synonym of *cooperi:* Brown, 1975: 24.
***villiersi.*** *Cerapachys villiersi* Bernard, 1953b: 215, fig. 6 (w.) GUINEA.
***vitiensis.*** *Cerapachys (Cerapachys) vitiensis* Mann, 1921: 406 (w.) FIJI IS.
*wheeleri. Cerapachys wheeleri* Crawley, 1926: 389 (w.) BORNEO. Junior synonym of *antennatus:* Brown, 1975: 22.
***wittmeri.*** *Cerapachys wittmeri* Collingwood, 1985: 237, fig. 8 (w.) SAUDI ARABIA.
***wroughtoni.*** *Cerapachys wroughtoni* Forel, 1910e: 422 (w.) SOUTH AFRICA. Senior synonym of *rhodesiana* and *roberti:* Brown, 1975: 24.
***xizangensis.*** *Cerapachys xizangensis* Tang & Li, 1982: 372 (w.) TIBET.
***zimmermani.*** *Cerapachys (Cerapachys) zimmermani* Wilson, 1959b: 54 (w.) FIJI IS.

*CERATOBASIS:* see under ***BASICEROS.***

*CERATOPHEIDOLE:* see under ***PHEIDOLE.***

*CHALCOPONERA:* see under ***RHYTIDOPONERA.***

***CHALEPOXENUS*** (Palaearctic)
***brunneus.*** *Chalepoxenus brunneus* Cagniant, 1985a: 141, figs. 1–5 (q.m.) MOROCCO. See also: Buschinger, Cagniant, Erhardt & Heinze, 1989: 253.
*gribodoi. Chalepoxenus gribodoi* Menozzi, 1923a: 257, figs (w.q.) ITALY. [Spelled *gridoboi* p. 257, corrected to *gribodoi* p. 258.] Menozzi, 1925a: 21 (m.). Junior synonym of *muellerianus:* Kutter, 1973d: 276.
*insubricus. Chalepoxenus insubricus* Kutter, 1950b: 337. figs. (m.) SWITZERLAND. Kutter, 1973d: 279 (w.q.). See also: Kutter, 1977c: 136. Provisional junior synonym of *muellerianus:* Buschinger, Erhardt, Fischer & Ofer, 1988: 398.
***kutteri.*** *Chalepoxenus kutteri* Cagniant, 1973: 148, figs. 1–4 (w.q.m.) FRANCE.
*longipilosus. Leptothorax longipilosus* Tarbinsky, 1976: 88, figs. 130–133 (q.) KIRGIZIA. [Junior primary homonym of *Leptothorax longipilosus* Santschi, 1912f: 174.] Replacement name: *Leonomyrma tarbinskii* Arnol'di, in Tarbinsky, 1976: 88.
***muellerianus.*** *Leptothorax (Temnothorax) muellerianus* Finzi, 1922: 118, fig. 1 (w.) ITALY. Combination in *Chalepoxenus:* Finzi, 1924b: 123. Senior synonym of *gribodoi:* Kutter, 1973d: 276; provisional senior synonym of *insubricus* and *siciliensis:* Buschinger, Erhardt, Fischer & Ofer, 1988: 398.
*siciliensis. Chalepoxenus siciliensis* Kutter, 1973d: 278, figs. 1, 3, 4, 6 (w.q.) SICILY. Provisional junior synonym of *muellerianus:* Buschinger, Erhardt, Fischer & Ofer, 1988: 398.
***spinosus.*** *Leonomyrma spinosa* Arnol'di, 1968: 1809, figs. 9, 10 (q.m.) KAZAKHSTAN. Combination in *Chalepoxenus:* Buschinger, 1987: 121.
***tarbinskii.*** *Leonomyrma tarbinskii* Arnol'di, in Tarbinsky, 1976: 88 (footnote). Replacement name for *Leptothorax longipilosus* Tarbinsky, 1976: 88. [Junior primary homonym of *Leptothorax longipilosus* Santschi, 1912f: 174.] Combination in *Chalepoxenus:* Radchenko, 1989a: 40.
***tauricus.*** *Chalepoxenus tauricus* Radchenko, 1989a: 37, fig. 1 (w.q.) UKRAINE.
***tramieri.*** *Chalepoxenus tramieri* Cagniant, 1983: 319, figs. (w.q.) MOROCCO.
***zabelini.*** *Chalepoxenus zabelini* Radchenko, 1989a: 39, fig. 1 (q.m.) TURKMENIA.

*CHAMPSOMYRMEX:* see under ***ODONTOMACHUS.***

*CHAPMANELLA:* see under ***EUPRENOLEPIS.***

#***CHARIOMYRMA:*** see under ***POLYRHACHIS.***

*CHARIOSTIGMACROS:* see under ***STIGMACROS.***

*CHELANER:* see under ***MONOMORIUM.***

***CHELIOMYRMEX*** (Neotropical)
***andicola.*** *Cheliomyrmex nortoni* subp. *andicola* Emery, 1894c: 185 (s.) PERU. Forel, 1914e: 10 (w.). Raised to species: Forel, 1914e: 10. Senior synonym of *ursinum* Kempf, 1975c: 372. See also: Borgmeier, 1955b: 71.
***audax.*** *Cheliomyrmex morosus* st. *audax* Santschi, 1921g: 92 (m.) ECUADOR. Raised to species: Borgmeier, 1955b: 74.
***megalonyx.*** *Cheliomyrmex megalonyx* Wheeler, W.M. 1921d: 322, figs. 5, 9, 10 (s.w.m.) GUIANA. Wheeler, G.C. 1943: 320 (l.); Wheeler, G.C. & Wheeler, J. 1984: 267 (l.). See also: Borgmeier, 1955b: 67.
***morosus.*** *Labidus morosus* Smith, F. 1859b: 6 (m.) MEXICO. Combination in *Eciton (Labidus):* Mayr, 1886d: 441; in *Cheliomyrmex:* Wheeler, W.M. 1921d: 318. Senior synonym of *nortoni:* Wheeler, W.M. 1921d: 324; of *payarum:* Borgmeier, 1955b: 63. See also: Gotwald & Kupiec, 1975: 961.
*nortoni. Cheliomyrmex nortoni* Mayr, 1870b: 969 (w.) MEXICO. André, 1887: 294 (s.w.). Junior synonym of *morosus:* Wheeler, W.M. 1921d: 324.
*payarum. Eciton morosum* var. *payarum* Forel, 1899c: 27 (m.) HONDURAS. Combination in *Cheliomyrmex:* Wheeler, W.M. 1921d: 324. Junior synonym of *morosus:* Borgmeier, 1955b: 63.
*ursinum. Eciton morosum* subsp. *ursinum* Emery, 1901d: 52 (m.) BRAZIL. Combination in *Cheliomyrmex,* and raised to species: Wheeler, W.M. 1921d: 324. Junior synonym of *andicola:* Kempf, 1975c: 372.

***CHELYSTRUMA*** (Argentina)
***lilloana.*** *Glamyromyrmex (Chelystruma) lilloana* Brown, 1950a: 23, fig. 1 (w.) ARGENTINA. Combination in *Chelystruma:* Kempf, 1959d: 338.

***CHIMAERIDRIS*** (Indo-Australian)
***boltoni.*** *Chimaeridris boltoni* Wilson, 1989: 63, fig. 1 (w.) SULAWESI.
***burckhardti.*** *Chimaeridris burckhardti* Wilson, 1989: 66, fig. 2 (w.) BORNEO.

****CHIMAEROMYRMA*** (Russia)
****brachycephala.*** **Chimaeromyrma brachycephala* Dlussky, 1988: 60, fig. 5 (w.) RUSSIA (Paleocene).

*CHRONOXENUS:* see under ***BOTHRIOMYRMEX.***

*CHRYSAPACE:* see under ***CERAPACHYS.***

#***CHTHONOLASIUS:*** see under ***LASIUS.***

***CLADAROGENYS*** (Gabon)
***lasia.*** *Cladarogenys lasia* Brown, 1976b: 34, figs. 1–5, A-D (w.) GABON.

***CLADOMYRMA*** (Indo-Australian)
***andrei.*** *Dimorphomyrmex andrei* Emery, 1894b: 73, fig. (q.) BORNEO. Combination in *Aphomomyrmex:* Emery, 1899e: 494; in *Cladomyrma:* Emery, 1925b: 45. Senior synonym of *hewitti* and *hobbyi:* Agosti, 1991: 303.
***cryptata.*** *Cladomyrma cryptata* Agosti, 1991: 304, figs. 2, 18, 23 (s.w.q.m.) BORNEO.
*hewitti. Aphomomyrmex hewitti* Wheeler, W.M. 1910c: 132, fig. 1 (s.q.) BORNEO. Combination in *Cladomyrma:* Wheeler, W.M. 1920: 53. Junior synonym of *andrei:* Agosti, 1991: 303.

***CLADOMYRMA** (continued)*

*hobbyi. Cladomyrma hobbyi* Donisthorpe, 1937a: 620, figs. 3, 4 (q.) BORNEO. Junior synonym of *andrei:* Agosti, 1991: 303.

***maschwitzi.*** *Cladomyrma maschwitzi* Agosti, 1991: 306, figs. 5, 8–12, 15, 19, 21, 24 (s.w.q.m.) WEST MALAYSIA.

***mossyna.*** *Cladomyrma mossyna* Agosti, 1991: 307, figs. 14, 17, 22 (s.w.q.) BORNEO.

***petalae.*** *Cladomyrma petalae* Agosti, 1991: 308, figs. 13, 16, 20, 25, 26 (s.w.q.m.) WEST MALAYSIA.

*CLARKISTRUMA:* see under ***COLOBOSTRUMA.***

*CLAVANODA:* see under ***PSEUDOMYRMEX.***

***CODIOMYRMEX*** (Trinidad)

*excisus* Weber, 1934; see under ***GLAMYROMYRMEX.***

*flagellatus* Taylor, 1962; see under ***GLAMYROMYRMEX.***

*loveridgei* Brown, 1953; see under ***GLAMYROMYRMEX.***

*reticeps* Kempf, 1969; see under ***GLAMYROMYRMEX.***

*semicomptus* Brown, 1959; see under ***GLAMYROMYRMEX.***

*tetragnathus* Taylor, 1965; see under ***GLAMYROMYRMEX.***

***thaxteri.*** *Codiomyrmex thaxteri* Wheeler, W.M. 1916d: 327, fig. 1 (w.) TRINIDAD. See also: Brown, 1953g: 21.

***CODIOXENUS*** (Cuba)

***simulans.*** *Epitritus* (*Codioxenus*) *simulans* Santschi, 1931c: 278, figs. 11, 12 (w.) CUBA. Combination in *Codioxenus:* Brown, 1948e: 123.

#*COLOBOCREMA:* see under ***CREMATOGASTER.***

#*COLOBOPSIS:* see under ***CAMPONOTUS.***

***COLOBOSTRUMA*** (Australia, New Guinea).

***alinodis.*** *Epopostruma alinodis* Forel, 1913g: 179, pl. 2, figs. 1, 2 (w.q.) AUSTRALIA. Crozier, 1968b: 88 (k.); Wheeler, G.C. & Wheeler, J. 1973c: 211 (l., misspelled as *alata*). Combination in *Clarkistruma:* Brown, 1948e: 124; in *Colobostruma:* Brown & Wilson, 1959b: 281.

***australis.*** *Colobostruma australis* Brown, 1959c: 4, fig. 3 (w.q.m.) AUSTRALIA.

***cerornata.*** *Colobostruma cerornata* Brown, 1959c: 1, figs. 1, 2 (w.) AUSTRALIA.

***elliotti.*** *Epitritus elliotti* Clark, 1928c: 42, figs. 1, 2 (w.q.) AUSTRALIA. Combination in *Clarkistruma:* Brown, 1948e: 124; in *Colobostruma:* Taylor & Brown, D.R. 1985: 60.

***foliacea.*** *Epopostruma foliacea* Emery, 1897c: 573, pl. 15, figs. 20, 21 (w.) NEW GUINEA. Combination in *Alistruma:* Brown, 1948e: 117; in *Colobostruma:* **new combination**.

***froggatti.*** *Epopostruma froggatti* Forel, 1913g: 177, pl. 2, figs. 3, 4 (w.) AUSTRALIA. Combination in *Alistruma:* Brown, 1948e: 117; in *Colobostruma:* Taylor & Brown, D.R. 1985: 60.

***leae.*** *Epopostruma* (*Colobostruma*) *leae* Wheeler, W.M. 1927f: 32, fig. 40 (q.) AUSTRALIA. Combination in *Colobostruma:* Brown, 1948e: 118.

***nancyae.*** *Colobostruma nancyae* Brown, 1965b: 22, figs. 1–3 (w.) AUSTRALIA.

***papulata.*** *Colobostruma papulata* Brown, 1965a: 21, fig. 1 (w.) AUSTRALIA.

*COMMATETA:* see under ***GNAMPTOGENYS.***

***CONCOCTIO*** (Afrotropical)

***concenta.*** *Concoctio concenta* Brown, 1974d: 30, figs. 1–5 (w.q.) GABON.

***CONDYLODON*** (Brazil)

***audouini.*** *Condylodon audouini* Lund, 1831a: 132 (w.) BRAZIL. Member of family Mutillidae?: Swainson & Shuckard, 1840: 173; combination in *Pseudomyrma:* Dalla Torre, 1893: 56; member of subfamily Ponerinae?: Emery, 1921f: 28 (footnote); *incertae sedis* in Ponerinae: Ward, 1990: 471.

*CONDYLOMYRMA:* see under ***CAMPONOTUS.***

*CONOMYRMA:* see under ***DORYMYRMEX.***

*CONOTHORACOIDES:* see under ***PHEIDOLE.***

*CONOTHORAX:* see under ***PHEIDOLE.***

*COPTOFORMICA:* see under ***FORMICA.***

*CORYNOMYRMEX:* see under ***MONOMORIUM.***

*COSMAECETES:* see under ***DORYLUS.***

*CRATEROPSIS:* see under ***OLIGOMYRMEX.***

*CRATOMYRMEX:* see under ***MESSOR.***

***CREIGHTONIDRIS*** (Brazil)

***scambognatha.*** *Creightonidris scambognatha* Brown, 1949f: 89 (q.) BRAZIL. See also: Brown & Kempf, 1960: 178.

***CREMATOGASTER*** (Worldwide)

[Notes. (i) The original spelling *Crematogaster* is used throughout, the incorrect subsequent spelling *Cremastogaster* is ignored; see catalogue of genus-group names. (ii) The spurious paper by Soulié & Dicko, 1965: 85, is ignored. This publication merely repeats, inaccurately, the Wheeler, W.M. 1922a: 828 catalogue of Afrotopical *Crematogaster,* but treats the subgenera as genera; see catalogue of genus-group names.]

***abdominalis.*** *Crematogaster abdominalis* Motschoulsky, 1863: 20 (w.) INDIA. Combination in *C.* (*Acrocoelia*): Emery, 1922e: 149.

***aberrans.*** *Crematogaster aberrans* Forel, 1892k: 532 (q.),: 533 (w.),: 534 (m.) (diagnoses in keys) INDIA. Combination in *C.* (*Oxygyne*): Forel, 1901e: 375. Current subspecies: nominal plus *assmuthi, inglebyi.* [See note under *diffusa.*]

***abrupta.*** *Crematogaster abrupta* Mann, 1919: 320 (w.) SOLOMON IS.

***abstinens.*** *Crematogaster abstinens* Forel, 1899c: 85 (w.) PANAMA. Combination in *C.* (*Orthocrema*): Emery, 1922e: 133. Current subspecies: nominal plus *chacoana, suturalis.*

***acaciae.*** *Crematogaster acaciae* Forel, 1892b: 141 (w.) ETHIOPIA. Combination in *C.* (*Acrocoelia*): Emery, 1922e: 144. Subspecies of *brunneipennis:* Wheeler, W.M. 1922a: 152. Revived status as species: Emery, 1922e: 144; Collingwood, 1985: 260. Current subspecies: nominal plus *generosa, gloriosa, victoriosa.*

*acanthobia. Crematogaster excisa* r. *andrei* var. *acanthobia* Forel, 1914d: 236 (w.) SOUTH AFRICA. Unavailable name.

#***accola.*** *Crematogaster* (*Orthocrema*) *sculpturata* subsp. *accola* Wheeler, W.M. 1934g: 175 (w.) MEXICO.

#***acis.*** *Crematogaster sewellii* r. *acis* Forel, 1913h: 353 (w.) ZAIRE. Combination in *C.* (*Acrocoelia*): Emery, 1922e: 148.

***aculeata.*** *Crematogaster* (*Rhachiocrema*) *aculeata* Donisthorpe, 1941j: 203 (w.) NEW GUINEA.

***acuta.*** *Formica acuta* Fabricius, 1804: 411 (w.) CENTRAL AMERICA. Wheeler, G.C. & Wheeler, J. 1952b: 258 (l.). Combination in *Crematogaster:* Roger, 1862c: 291; in *C.* (*Eucrema*): Santschi, 1918d: 182. Senior synonym of *quadriceps:* Roger, 1862c: 291. Current subspecies: nominal plus *centralis.*

#***acutidens.*** *Crematogaster delagoense* var. *acutidens* Arnold, 1944: 7, fig. 7 (w.) SOUTH AFRICA.

*admota. Crematogaster admota* Sjöstedt, 1908: 116 (w.) KENYA. *Nomen nudum* [attributed to Mayr].

***adrepens.*** *Crematogaster adrepens* Forel, 1897c: 197 (w.) MADAGASCAR. Combination in *C.* (*Crematogaster*): Wheeler, W.M. 1922a: 1022; in *C.* (*Acrocoelia*): Emery, 1922e: 144.

#***adultera.*** *Crematogaster depressa* var. *adultera* Santschi, 1915c: 252 (w.) CONGO. Combination in *C.* (*Atopogyne*): Wheeler, W.M. 1922a: 856.

#***adusta.*** *Crematogaster castanea* st. *adusta* Santschi, 1935a: 257 (w.) NAMIBIA. [First available use of *Crematogaster tricolor* r. *ferruginea* var. *adusta* Stitz, 1923: 155; unavailable name.]

***aegyptiaca.*** *Crematogaster aegyptiaca* Mayr, 1862: 765 (w.) EGYPT. Combination in *C.* (*Crematogaster*): Wheeler, W.M. 1922a: 828; in *C.* (*Acrocoelia*): Emery, 1922e: 144. Current subspecies: nominal plus *pharaonis, robusta, turkanensis.*

#***aeolia.*** *Crematogaster sordidula* var. *aeolia* Forel, 1911d: 343 (w.) TURKEY. Combination in *C.* (*Orthocrema*): Emery, 1922e: 131. Subspecies of *sordidula:* Emery, 1912e: 671.

#***aethiops.*** *Crematogaster chiarinii* var. *aethiops* Forel, 1907c: 142 (footnote) (w.) SOMALIA. Menozzi, 1930b: 90 (m.). Combination in *C.* (*Acrocoelia*): Emery, 1922e: 146.

***affabilis.*** *Crematogaster chiarinii* var. *affabilis* Forel, 1907c: 142 (w.) SOMALIA. Combination in *C.* (*Acrocoelia*): Emery, 1922e: 146. Raised to species: Collingwood, 1985: 260.

***afghanica.*** *Crematogaster afghanica* Pisarski, 1967: 389, figs. 23–25 (w.) AFGHANISTAN.

***africana.*** *Crematogaster africana* Mayr, 1895: 142 (w.) CAMEROUN. Santschi, 1937g: 78 (q.). Combination in *C.* (*Atopogyne*): Santschi, 1916b: 501. Race of *buchneri:* Forel, 1913a: 125. Revived status as species: Santschi, 1918d: 183. Current subspecies: nominal plus *alligatrix, biemarginata, camena, polymorphica, schumanni, stanleyi, stolonis, thoracica, tibialis, variegata.*

*aglaea. Crematogaster excisa* subsp. *impressa* var. *aglaea* Santschi, in Wheeler, W.M. 1922a: 153 (w.) IVORY COAST. Unavailable name.

***agnetis.*** *Crematogaster agnetis* Forel, 1892k: 531 (q.), : 533 (w.), : 534 (m.) (diagnoses in keys) MADAGASCAR. Combination in *C.* (*Oxygyne*): Wheeler, W.M. 1922a: 1024.

***agniae.*** *Crematogaster* (*Acrocoelia*) *agniae* Karavaiev, 1935a: 90, fig. 16 (w.) VIETNAM.

***agnita.*** *Crematogaster* (*Orthocrema*) *agnita* Wheeler, W.M. 1934g: 175 (w.q.) GUATEMALA.

#***airensis.*** *Crematogaster laestrygon* st. *airensis* Santschi, 1937e: 309 (w.m.) NIGER. [First available use of *Crematogaster* (*Acrocoelia*) *auberti* st. *oasium* var. *airensis* Santschi, 1932f: 517; unavailable name.]

***aitkenii.*** *Crematogaster biroi* var. *aitkenii* Forel, 1902c: 203 (w.) INDIA. Combination in *C.* (*Orthocrema*): Emery, 1922e: 132. Raised to species: Wu & Wang, 1992: 1319.

***algirica.*** *Myrmica algirica* Lucas, 1849: 300, pl. 16, fig. 9 (w.q.) ALGERIA. Combination in *Crematogaster:* Emery, 1891b: 14; in *C.* (*Acrocoelia*): Emery, 1922e: 143. Junior synonym of *scutellaris:* Mayr, 1856: xxi; Nylander, 1856b: 100; Smith, F. 1858b: 135; Mayr, 1861: 74. Revived from synonymy as variety of *scutellaris:* Emery, 1891b: 14; Karavaiev, 1912a: 11; Santschi, 1929e: 148. Revived status as species: Collingwood, 1978: 68.

#***alii.*** *Crematogaster scutellaris* var. *alii* Forel, 1907b: 207 (w.) ALGERIA. Combination in *C.* (*Acrocoelia*): Emery, 1922e: 144.

#***allegrensis.*** *Crematogaster victima* r. *allegrensis* Forel, 1901d: 300 (w.q.m.) BRAZIL. Combination in *C.* (*Orthocrema*): Emery, 1922e: 136.

#***alligatrix.*** *Crematogaster buchneri* subsp. *alligatrix* Forel, 1911e: 271 (w.) NIGERIA. Combination in *C.* (*Atopogyne*), and subspecies of *africana:* Wheeler, W.M. 1922a: 852.

***alluaudi.*** *Crematogaster alluaudi* Emery, 1893c: 83 (w.) CANARY IS. Santschi, 1921e: 168 (q.); Santschi, 1937e: 299 (m.). Combination in *C.* (*Atopogyne*): Santschi, 1916b: 501; in *C.* (*Acrocoelia*): Emery, 1922e: 142. Current subspecies: nominal plus *noualhieri.*

***aloysiisabaudiae.*** *Crematogaster* (*Acrocoelia*) *aloysiisabaudiae* Menozzi, 1930b: 86, fig. 1 (w.) SOMALIA.

***alulai.*** *Crematogaster menileki* var. *alulai* Emery, 1901e: 61 (w.) ETHIOPIA. Combination in *C.* (*Crematogaster*): Wheeler, W.M. 1922a: 829; in *C.* (*Acrocoelia*): Emery, 1922e: 144. Race of *buchneri:* Forel, 1916: 411. Raised to species: Emery, 1915g: 11; Emery, 1922e: 144; Menozzi, 1922a: 354. Current subspecies: nominal plus *scrutans.*

***amabilis.*** *Crematogaster amabilis* Santschi, 1911c: 355 (w.) TANZANIA. Santschi, 1914b: 91 (q.m.). Combination in *C.* (*Crematogaster*): Wheeler, W.M. 1922a: 829; in *C.* (*Acrocoelia*): Emery, 1922e: 145. Current subspecies: nominal plus *retiaria.*

***amapaensis.*** *Crematogaster* (*Orthocrema*) *amapaensis* Kempf, 1960e: 394, figs. 7–9 (w.) BRAZIL.

#***amazonensis.*** *Crematogaster stollii* var. *amazonensis* Forel, 1904e: 682 (w.) BRAZIL. Combination in *C.* (*Orthocrema*): Emery, 1922e: 136.

***ambigua.*** *Crematogaster* (*Atopogyne*) *ambigua* Santschi, 1926b: 228 (w.) ZAIRE.

#***amia.*** *Crematogaster popohana* r. *amia* Forel, 1913f: 194 (q.m.) TAIWAN. Combination in *C.* (*Acrocoelia*): Emery, 1922e: 151.

***amita.*** *Crematogaster kneri* var. *amita* Forel, 1913a: 125 (w.) ZIMBABWE. Combination in *C.* (*Sphaerocrema*): Wheeler, W.M. 1922a: 847. Race of *kneri:* Forel, 1922: 94. Raised to species: Santschi, 1925h: 161; Santschi, 1926b: 226. Current subspecies: nominal plus *bushimana, caffra, makololo, matabele.*

#***ampla.*** *Crematogaster brevispinosa* var. *ampla* Forel, 1912f: 211 (w.) COLOMBIA. Combination in *C.* (*Orthocrema*): Emery, 1922e: 134.

***ampullaris.*** *Crematogaster ampullaris* Smith, F. 1861b: 47 (w.) SULAWESI. Combination in *C.* (*Physocrema*): Donisthorpe, 1932c: 469. Junior synonym of *difformis:* Emery, 1901g: 574; Wheeler, W.M. 1919e: 74. Revived from synonymy: Donisthorpe, 1932c: 469.

#***analis.*** *Crematogaster flaviventris* var. *analis* Santschi, 1910c: 372 (w.) CONGO. Forel, 1910e: 433 (q.m.). Subspecies of *castanea:* current status.

#***anamita.*** *Crematogaster* (*Paracrema*) *modiglianii* var. *anamita* Santschi, 1925f: 90 (w.) BURMA [attributed to Forel]. [*Crematogaster modiglianii* var. *annamita* Forel, 1903a: 680. *Nomen nudum,* attributed to Emery.]

***ancipitula.*** *Crematogaster ancipitula* Forel, 1917: 252 (q.) ANGOLA. Combination in *C.* (*Acrocoelia*): Emery, 1922e: 145.

#***andelis.*** *Crematogaster biroi* var. *andelis* Santschi, 1928h: 129 (w.) SUMATRA. Stärcke, 1930: 373 (m.).

*andrei. Crematogaster excisa* subsp. *andrei* Forel, 1911f: 277 (w.q.) ZAIRE. Combination in *C.* (*Acrocoelia*): Emery, 1922e: 147. Race/stirps of *excisa:* Forel, 1913a: 126; Wheeler, W.M. 1922a: 836; Santschi, 1928d: 63; of *impressa:* Santschi, 1913a: 412 (in key); Santschi, 1935a: 257. Junior synonym of *impressa:* Santschi, 1937b: 98.

#***anguliceps.*** *Crematogaster* (*Oxygyne*) *stadelmanni* var. *anguliceps* Stitz, 1916: 384, fig. 4 (w.) CAMEROUN. Combination in *C.* (*Nematocrema*): Wheeler, W.M. 1922a: 859.

***angulosa.*** *Crematogaster angulosa* André, 1896b: 264 (w.) BORNEO. Combination in *C.* (*Decacrema*): Forel, 1911a: 32.

#***angustata.*** *Crematogaster stadelmanni* var. *angustata* Mayr, 1895: 142 (w.) CAMEROUN. Combination in *C.* (*Nematocrema*): Wheeler, W.M. 1922a: 859.

***angusticeps.*** *Crematogaster angusticeps* Santschi, 1911c: 354, fig. (q.) SUDAN. Combination in *C.* (*Atopogyne*): Wheeler, W.M. 1922a: 853; in *C.* (*Acrocoelia*): Emery, 1922e: 145; in *C.* (*Sphaerocrema*): Santschi, 1926b: 227.

#***angustior.*** *Crematogaster peringueyi* var. *angustior* Arnold, 1920a: 510, fig. 40 (w.q.m.) SOUTH AFRICA.

#***anoemica.*** *Crematogaster* (*Paracrema*) *modiglianii* var. *anoemica* Santschi, 1925f: 91 (w.) BURMA.

***antaris.*** *Crematogaster auberti* r. *antaris* Forel, 1894d: 26 (w.q.) ALGERIA. Combination in *C.* (*Acrocoelia*): Emery, 1922e: 142. Subspecies of *inermis:* Emery, 1926: 2; Menozzi, 1927g: 379; of *auberti:* Emery, 1924a: 8; Finzi, 1930b: 15; Santschi, 1938a: 38; Cagniant, 1964: 103. Raised to species: Santschi, 1921d: 71; Collingwood, 1985: 260. Current subspecies: nominal plus *sordida.*

***anthracina.*** *Crematogaster anthracinus* Smith, F. 1857a: 75 (w.) SINGAPORE. Combination in *C.* (*Acrocoelia*): Donisthorpe, 1932c: 450.

#***antillana.*** *Crematogaster curvispinosa* var. *antillana* Forel, 1893g: 399 (w.) ANTILLES IS. Forel, 1912f: 216 (q.). Combination in *C.* (*Orthocrema*): Emery, 1922e: 134.

#***aphrodite.*** *Crematogaster* (*Acrocoelia*) *inermis* var. *aphrodite* Santschi, 1937e: 298, figs. 2, 17 (w.q.m.) CYPRUS.

***apicalis.*** *Crematogaster apicalis* Motschoulsky, 1863: 20 (w.) SRI LANKA. Combination in *C.* (*Acrocoelia*): Emery, 1922e: 149.

*apicalis. Crematogaster apicalis* Smith, F. 1878b: 12 (q.) INDIA. [Junior primary homonym of *apicalis* Motschoulsky, above.] Replacement name: *betapicalis:* **new name**. Combination in *C.* (*Acrocoelia*), and possible synonym of *subdentata:* Emery, 1922e: 144.

#***apilis.*** *Crematogaster treubi* r. *apilis* Forel, 1913f: 195 (w.) TAIWAN. Combination in *C.* (*Orthocrema*): Emery, 1922e: 132.

#***aquila.*** *Crematogaster ferruginea* var. *aquila* Forel, 1907c: 139 (w.) KENYA. Santschi, 1914b: 87 (m.). Combination in *C.* (*Acrocoelia*): Emery, 1922e: 145. Subspecies of *castanea:* current status.

***arata.*** *Crematogaster arata* Emery, 1906c: 140 (w.) ARGENTINA. Combination in *C.* (*Orthocrema*): Emery, 1922e: 133; in *C.* (*Neocrema*): Santschi, 1918d: 182; Kempf, 1972a: 83.

#***arawak.*** *Crematogaster brasiliensis* subsp. *arawak* Weber, 1938a: 208 (w.) GUIANA.

#***arborea.*** *Crematogaster arboreus* Smith, F. 1858b: 138, pl. 14 (nest) (w.) SOUTH AFRICA. Combination in *C.* (*Acrocoelia*): Emery, 1922e: 145. Junior synonym of *tricolor:* Mayr, 1895: 136 (footnote); Arnold, 1920a: 506 (note that *arborea* is really the senior name); of *castanea:* Wheeler, W.M. 1922a: 830. Revived from synonymy as subspecies of *castanea:* Emery, 1922e: 145. Variety of *castanea,* and senior synonym of *decolorata:* Santschi, 1930b: 69.

*arborea. Oecodoma* (*Atta*) *arborea* Buckley, 1867: 349 (w.q.) U.S.A. [Unresolved junior secondary homonym of *arboreus* Smith, F. above.] Junior synonym of *lineolata:* McCook, 1879: 187; Dalla Torre, 1893: 83; of *laeviuscula:* Smith, M.R. 1951a: 809; Buren, in Smith, M.R. 1958c: 126.

***arcuata.*** *Crematogaster sulcata* r. *arcuata* Forel, 1899c: 84, pl. 4, fig. 3 (w.) PANAMA. Combination in *C.* (*Physocrema*): Forel, 1913l: 234; in *C.* (*Orthocrema*): Emery, 1922e: 134; in *C.* (*Neocrema*): Santschi, 1922d: 244; Kempf, 1972a: 83. Raised to species: Forel, 1913l: 234. Current subspecies: nominal plus *aruga.*

#***argentina.*** *Crematogaster steinheili* var. *argentina* Forel, 1915c: 356 (w.q.) ARGENTINA. Combination in *C.* (*Orthocrema*): Emery, 1922e: 136.

***arizonensis.*** *Crematogaster arizonensis* Wheeler, W.M. 1908e: 482, pl. 27, fig. 40 (w.) U.S.A. Wheeler, W.M. 1912b: 132 (q.m.). Combination in *C.* (*Orthocrema*): Emery, 1922e: 134.

***armandi.*** *Crematogaster armandi* Forel, 1921a: 208 (w.q.) BRAZIL. Combination in *C.* (*Orthocrema*): Kempf, 1972a: 84.

#***armatula.*** *Crematogaster inermis* var. *armatula* Emery, 1926: 3 (w.) SYRIA. Menozzi, 1933b: 59 (q.).

***arnoldi.*** *Crematogaster arnoldi* Forel, 1914d: 238 (w.) SOUTH AFRICA. Combination in *C.* (*Acrocoelia*):

***CREMATOGASTER*** *(continued)*

Emery, 1922e: 145. Variety of *godfreyi:* Arnold, 1920a: 504. Revived status as species: Santschi, 1926b: 214. Current subspecies: nominal plus *loveridgei.*

***aroensis.*** *Crematogaster (Acrocoelia) aroensis* Menozzi, 1935a: 114, fig. 7 (w.) NEW GUINEA.

***arthurimuelleri.*** *Crematogaster gallicola* r. *arthurimuelleri* Forel, 1894b: 96 (w.) MOZAMBIQUE. Combination in *C. (Decacrema):* Arnold, 1920a: 547; Wheeler, W.M. 1922a: 860; in *C. (Acrocoelia):* Emery, 1922e: 147. Subspecies of *gallicola:* Emery, 1922e: 147. Raised to species: Mayr, 1895: 144; Arnold, 1920a: 547; Wheeler, W.M. 1922a: 860; current status.

#***artifex.*** *Crematogaster artifex* Mayr, 1879: 684 (w.m.) THAILAND. Wheeler, W.M. 1927d: 4 (q.). Combination in *C. (Acrocoelia):* Emery, 1922e: 150. Stirps of *dohrni:* Forel, 1902c: 203. Revived status as species: Bingham, 1903: 139. Subspecies of *dohrni:* Emery, 1922e: 150; Wheeler, W.M. 1924b: 246; Wheeler, W.M. 1928c: 16; Wheeler, W.M. 1930h: 65.

#***aruga.*** *Crematogaster (Physocrema) arcuata* var. *aruga* Forel, 1913l: 234 (w.) BRAZIL. Combination in *C. (Orthocrema):* Emery, 1922e: 134; in *C. (Neocrema):* Kempf, 1972a: 83.

#***ascendens.*** *Crematogaster (Eucrema) inca* subsp. *ascendens* Wheeler, W.M. 1925a: 28 (w.) PERU.

***ashmeadi.*** *Crematogaster ashmeadi* Mayr, 1886d: 463 (w.m.) U.S.A. Combination in *C. (Acrocoelia):* Emery, 1922e: 141; in *C. (Crematogaster):* Buren, 1968b: 92. Senior synonym of *matura:* Creighton, 1950a: 206. See also: Johnson, 1988: 315.

#***assmuthi.*** *Crematogaster (Oxygyne) aberrans* var. *assmuthi* Forel, 1913e: 662 (w.q.) INDIA.

#***atavista.*** *Crematogaster sanguinea* var. *atavista* Mann, 1920: 425 (w.) CUBA.

#***aterrima.*** *Crematogaster formosa* var. *aterrima* Wheeler, W.M. 1909b: 234 (w.m.) MEXICO. Combination in *C. (Orthocrema):* Emery, 1922e: 135.

***atitlanica.*** *Crematogaster (Apterocrema) atitlanica* Wheeler, W.M. 1936c: 47, fig. 1 (q.m.) GUATEMALA. Wheeler, G.C. & Wheeler, J. 1960b: 13 (l.).

***atkinsoni.*** *Crematogaster atkinsoni* Wheeler, W.M. 1919g: 108, fig. 16 (w.) U.S.A. Wheeler, W.M. 1932a: 9 (q.m.). Combination in *C. (Acrocoelia):* Creighton, 1950a: 207; in *C. (Crematogaster):* Buren, 1968b: 92. Senior synonym of *helveola:* Creighton, 1950a: 207. See also: Johnson, 1988: 316.

#***atlantis.*** *Crematogaster (Acrocoelia) laestrygon* var. *atlantis* Santschi, 1937e: 306 (w.q.m.) ALGERIA. [First available use of *Crematogaster auberti* subsp. *laestrygon* var. *atlantis* Forel, 1911d: 341; unavailable name.]

***atra.*** *Crematogaster atra* Mayr, 1870b: 994 (w.) MEXICO. Combination in *C. (Orthocrema):* Emery, 1922e: 134. Current subspecies: nominal plus *uruguayensis.*

#***atraplex.*** *Crematogaster (Sphaerocrema) bequaerti* var. *atraplex* Santschi, in Wheeler, W.M. 1922a: 155 (w.) ZAIRE.

*atratula. Crematogaster scutellaris* subsp. *schmidti* var. *atratula* Zimmermann, 1935: 21 (w.) YUGOSLAVIA. Unavailable name.

*atrigaster. Crematogaster meneliki* r. *occidentalis* var. *atrigaster* Forel, 1913h: 353 (w.) ZAIRE. Unavailable name.

#***atriscapis.*** *Crematogaster ruspolii* var. *atriscapis* Forel, 1915c: 338 (w.) ZAIRE. Combination in *C. (Acrocoelia):* Emery, 1922e: 148.

***auberti.*** *Crematogaster auberti* Emery, 1869b: 23 (footnote) (w.) FRANCE. André, 1883b: 395 (q.); Emery, 1891b: 14 (m.). Combination in *C. (Acrocoelia):* Emery, 1922e: 142. Variety of *scutellaris:* Emery & Forel, 1879: 464; of *schmidti:* Emery, 1891b: 14; of *inermis:* Emery, 1926: 3. Status as species: Forel, 1894d: 24; Forel, 1902a: 152; Bondroit, 1918: 115; Karavaiev, 1927b: 104; Finzi, 1930b: 15; Santschi, 1937e: 300. Senior synonym of *iberica:* Collingwood, 1978: 69. Current subspecies: nominal plus *karawaewi, levithorax, nigripes, regilla, savinae, vogti.*

***augusti.*** *Crematogaster augusti* Emery, 1895h: 31 (footnote), pl. 2, figs. 19, 20 (q.) SUMATRA. Combination in *C. (Oxygyne):* Emery, 1922e: 157.

***aurita.*** *Crematogaster (Physocrema) aurita* Karavaiev, 1935a: 92, fig. 18 (w.) CAMBODIA.

***australis.*** *Crematogaster australis* Mayr, 1876: 108 (w.q.m.) AUSTRALIA. Wheeler, G.C. & Wheeler, J. 1973a: 28 (l.). Combination in *C. (Acrocoelia):* Emery, 1922e: 152. Current subspecies: nominal plus *chillagoensis, sycites.*

#***autruni.*** *Crematogaster (Crematogaster) stolli* subsp. *autruni* Mann, 1916: 441 (w.) BRAZIL. Combination in *C. (Orthocrema):* Emery, 1922e: 136.

*azumai. Crematogaster (Acrocoelia) brunnea* st. *ruginota* var. *azumai* Santschi, 1941: 275 (w.) JAPAN. Unavailable name.

***baduvi.*** *Crematogaster baduvi* Forel, 1912d: 106 (w.) JAVA. Forel, 1913k: 77 (q.); Menozzi, 1935a: 104 (m.). Combination in *C. (Orthocrema):* Emery, 1922e: 131. [Misspelled as *batuvi:* Crawley, 1924: 394.]

*baetica. Crematogaster laestrygon* subsp. *vaucheri* var. *baetica* Emery, 1926: 3 (w.) SPAIN. Unavailable name.

***bakeri.*** *Crematogaster (Physocrema) bakeri* Menozzi, 1925c: 447, fig. 6 (w.) PHILIPPINES.

#***bandarensis.*** *Crematogaster biroi* var. *bandarensis* Forel, 1913k: 76 (w.q.) SUMATRA. Combination in *C. (Orthocrema):* Emery, 1922e: 132.

*bantuana. Crematogaster amita* st. *matabele* var. *bantuana* Santschi, 1925h: 161 (w.) ANGOLA. Unavailable name.

***barbouri.*** *Crematogaster barbouri* Weber, 1934a: 24, fig. 1 (w.q.m.) CUBA. Combination in *C. (Orthocrema):* Kempf, 1972a: 84.

#***bassuto.*** *Crematogaster (Sphaerocrema) hottentota* var. *bassuto* Santschi, 1925h: 161 (footnote) (w.) LESOTHO. [Also described as new by Santschi, 1926b: 224.]

***batesi.*** *Crematogaster batesi* Forel, 1911e: 270 (w.) NIGERIA. Combination in *C. (Atopogyne):* Wheeler, W.M. 1922a: 853.

#***bayeri.*** *Crematogaster (Acrocoelia) chiarinii* st. *bayeri* Santschi, 1926b: 215 (w.) KENYA.

#***behanzini.*** *Crematogaster pronotalis* var. *behanzini* Santschi, 1915c: 251 (w.) BENIN. Combination in *C. (Sphaerocrema):* Wheeler, W.M. 1922a: 848. Variety of *kneri:* Santschi, 1926b: 228.

#***benitensis.*** *Crematogaster striatula* var. *benitensis* Santschi, 1910c: 374 (w.) CONGO. Combination in *C. (Sphaerocrema):* Wheeler, W.M. 1922a: 849.

***bequaerti.*** *Crematogaster bequaerti* Forel, 1913b: 319 (w.q.) ZAIRE. Combination in *C. (Sphaerocrema):* Santschi, 1918d: 182. Current subspecies: nominal plus *atraplex, gerardi, ludia, modica, mutabilis, semiclara.*

***betapicalis.*** *Crematogaster betapicalis* **new name**. Replacement name for *apicalis* Smith, F. 1878b: 12. [Junior primary homonym of *apicalis* Motschoulsky, 1863: 20.]

***bicolor.*** *Crematogaster bicolor* Smith, F. 1860b: 109 (w.) INDONESIA (Batjan I.). Mayr, 1867a: 104 (q.m.). Combination in *C. (Orthocrema):* Emery, 1922e: 131; in *C. (Acrocoelia):* Donisthorpe, 1932c: 463. Current subspecies: nominal plus *imbellis.*

*bicolor. Oecodoma (Atta) bicolor* Buckley, 1867: 350 (w.) U.S.A. [Junior secondary homonym of *bicolor* Smith, above.] Senior synonym of *clara:* McCook, 1879: 188; Mayr, 1886d: 463 [*clara* oldest synonym and hence first available replacement name.]

#***biemarginata.*** *Crematogaster africana* var. *biemarginata* Forel, 1910e: 433 (w.) CAMEROUN. Combination in *C. (Atopogyne):* Wheeler, 1922a: 852.

***biformis.*** *Crematogaster biformis* André, 1892b: 53 (w.) BORNEO. Combination in *C. (Decacrema):* Forel, 1911a: 32.

#***biimpressa.*** *Crematogaster buchneri* subsp. *biimpressa* Mayr, 1895: 139 (w.) CONGO. Combination in *C. (Atopogyne):* Wheeler, W.M. 1922a: 854. Stirps of *clariventris:* Santschi, 1930b: 69.

***binghamii.*** *Crematogaster binghamii* Forel, 1904c: 24 (w.) INDIA. Combination in *C. (Orthocrema):* Santschi, 1918d: 182.

***bingo.*** *Crematogaster bingo* Forel, 1908c: 368 (w.) BRAZIL. Combination in *C. (Eucrema):* Santschi, 1922d: 244.

#***bipartita.*** *Crematogaster (Orthocrema) dispar* var. *bipartita* Emery, 1922e: 132 (w.) AUSTRALIA. [First available use of *Crematogaster sordidula* r. *dispar* var. *bipartita* Forel, 1915b: 54; unavailable name (Taylor, 1986: 34).]

***biroi.*** *Crematogaster biroi* Mayr, 1897: 428 (w.) SRI LANKA. Bingham, 1903: 138 (q.); Imai, *et al.* 1984: 6 (k.). Combination in *C. (Orthocrema):* Emery, 1922e: 131. Current subspecies: nominal plus *andelis, bandarensis, quadriruga, smythiesii.*

*biroi. Crematogaster biroi* Emery, 1900c: 332, pl. 8, figs. 39–41 (w.) NEW GUINEA. [Junior primary homonym of *biroi* Mayr, above.] Replacement name: *emeryi* Forel, 1907a: 25.

***bison.*** *Crematogaster bison* Forel, 1913f: 195 (w.) TAIWAN. Combination in *C. (Acrocoelia):* Emery, 1922e: 149.

***boera.*** *Crematogaster (Orthocrema) boera* Santschi, 1926b: 211 (w.) ZIMBABWE.

***bogojawlenskii.*** *Crematogaster sordidula* subsp. *bogojawlenskii* Ruzsky, 1905b: 506 (w.m.) UZBEKISTAN. Raised to species: Dlussky, Soyunov & Zabelin, 1990: 243. Junior synonym of *sordidula:* Atanassov & Dlussky, 1992: 182. Revived from synonymy, revived status as species, and senior synonym of *kosti:* Arakelian, 1994: 44.

***boliviana.*** *Tranopeltoides bolivianus* Wheeler, W.M. 1922e: 13, fig. 3 (m.) BOLIVIA. Combination in *Crematogaster:* Kempf, 1960c: 174; in *C.* (*Orthocrema*): Kempf, 1972a: 84.

***#boliviana.*** *Crematogaster* (*Orthocrema*) *longispina* var. *boliviana* Wheeler, W.M. 1925a: 25 (w.q.) BOLIVIA. [Unresolved junior secondary homonym of *boliviana* Wheeler, above.]

***#bomaella.*** *Crematogaster excisa* st. *bomaella* Santschi, 1935a: 258 (w.) ZAIRE.

***borneensis.*** *Crematogaster borneensis* André, 1896b: 263 (w.) BORNEO. Forel, 1911d: 386 (m.). Combination in *C.* (*Decacrema*): Forel, 1910a: 18. Current subspecies: nominal plus *capax, harpyia, hosei, insulsa, macarangae, novem, sembilana, symbia.*

***#bosqi.*** *Crematogaster* (*Orthocrema*) *iheringi* st. *bosqi* Santschi, 1934c: 29 (w.q.) ARGENTINA.

***#bouvardi.*** *Crematogaster walshi* st. *bouvardi* Santschi, 1920h: 160 (w.q.) VIETNAM. [Also described as new by Santschi, 1924c: 98.]

*boxi. Crematogaster* (*Sphaerocrema*) *boxi* Donisthorpe, 1945a: 10 (w.) GHANA. Junior synonym of *wellmani:* **new synonymy**.

***brasiliensis.*** *Crematogaster brasiliensis* Mayr, 1878: 875 (w.) BRAZIL. Forel, 1912f: 217 (q.). Combination in *C.* (*Orthocrema*): Emery, 1922e: 136. Current subspecies: nominal plus *arawak, cocciphila, ludio.*

***#braunsi.*** *Crematogaster* (*Orthocrema*) *natalensis* var. *braunsi* Emery, 1922e: 131 (w.) SOUTH AFRICA. [First available use of *Crematogaster sordidula* subsp. *natalensis* var. *braunsi* Forel, 1911f: 277; unavailable name.]

***#brazzai.*** *Crematogaster impressa* st. *brazzai* Santschi, 1910c: 373 (w.q.) CONGO. Combination in *C.* (*Acrocoelia*): Emery, 1922e: 147.

***#brevarmata.*** *Crematogaster santschii* var. *brevarmata* Forel, 1915c: 337 (w.m.) ZAIRE. Santschi, 1930a: 65 (q.). Combination in *C.* (*Oxygyne*): Wheeler, W.M. 1922a: 858; in *C.* (*Acrocoelia*): Emery, 1922e: 148. Variety of *margaritae:* Santschi, 1930a: 65.

***#brevidentata.*** *Crematogaster brevispinosa* var. *brevidentata* Forel, 1912f: 212 (w.) BARBADOS. Combination in *C.* (*Orthocrema*): Emery, 1922e: 134.

***brevimandibularis.*** *Crematogaster* (*Orthocrema*) *brevimandibularis* Donisthorpe, 1943d: 451 (m.) NEW GUINEA.

***brevis.*** *Crematogaster brevis* Emery, 1887b: 467 (footnote), pl. 2, fig. 19 (w.q.) JAVA. Combination in *C.* (*Orthocrema*): Emery, 1922e: 132; in *C.* (*Mesocrema*): Santschi, 1928b: 33.

***brevispinosa.*** *Crematogaster brevispinosa* Mayr, 1870a: 403 (w.) COLOMBIA. Wheeler, G.C. & Wheeler, J. 1952b: 260 (l.). Combination in *C.* (*Orthocrema*): Santschi, 1918d: 182. Current subspecies: nominal plus *ampla, brevidentata, chathamensis, crucis, malevolens, mancocapaci, minutior, moelleri, montana, projecta, recurvispina, rochai, russata, sampaioi, schuppi, sericea, striatinota, subtonsa, thalia, townsendi, tumulifera, vicina.*

***breviventris.*** *Crematogaster* (*Nematocrema*) *breviventris* Santschi, 1920d: 379 (q.) CAMEROUN.

*brieyi. Crematogaster africana* r. *winkleri* var. *brieyi* Forel, 1913h: 352 (w.) ZAIRE. Unavailable name.

*brookei. Crematogaster brookei* Menozzi, in Eidmann, 1941: 18 (w.) CHINA. *Nomen nudum.*

***#broomensis.*** *Crematogaster leviceps* var. *broomensis* Forel, 1915b: 56 (w.) AUSTRALIA. Subspecies of *laeviceps:* Taylor & Brown, D.R. 1985: 61.

***browni.*** *Crematogaster* (*Crematogaster*) *browni* Buren, 1968b: 100 (w.q.) U.S.A.

***bruchi.*** *Crematogaster bruchi* Forel, 1912f: 219 (w.) ARGENTINA. Combination in *C.* (*Orthocrema*): Emery, 1922e: 134.

***#bruesi.*** *Crematogaster laevis* var. *bruesi* Wheeler, W.M. 1911b: 169 (w.q.) GRENADA I. Combination in *C.* (*Orthocrema*): Emery, 1922e: 135.

***brunnea.*** *Crematogaster brunneus* Smith, F. 1857a: 75 (w.) BORNEO. Combination in *C.* (*Acrocoelia*): Emery, 1922e: 149. Current subspecies: nominal plus *contemta, latipetiolata, nicevillei, nilgirica, rabula, ruginota, sundaica, teranishii.*

*brunnea. Crematogaster rasoherinae* var. *brunnea* Forel, 1907g: 79 (w.q.) MADAGASCAR. [Junior primary homonym of *brunnea* Smith, above.] Replacement name: *brunneola* Emery, 1922e: 131.

***brunneipennis.*** *Crematogaster brunneipennis* André, 1890: 323 (w.q.) SIERRA LEONE. Santschi, 1933b: 101 (m.). Combination in *C.* (*Crematogaster*): Wheeler, W.M. 1922a: 829; in *C.* (*Acrocoelia*): Emery, 1922e: 145. Current subspecies: nominal plus *omniparens, yorubosa.*

***#brunneola.*** *Crematogaster* (*Orthocrema*) *rasoherinae* var. *brunneola* Emery, 1922e: 131. Replacement name for *brunnea* Forel, 1907g: 79. [Junior primary homonym of *brunnea* Smith, F. 1857a: 75.] Combination in *C.* (*Mesocrema*): Santschi, 1928b: 33.

***brunnescens.*** *Crematogaster brunnescens* Motschoulsky, 1863: 20 (w.) SRI LANKA. Combination in *C.* (*Acrocoelia*): Emery, 1922e: 150.

***#bruta.*** *Crematogaster castanea* st. *bruta* Santschi, 1926b: 213 (w.) SOUTH AFRICA. [First available use of *Crematogaster tricolor* st. *ferruginea* var. *bruta* Santschi, 1913a: 413; unavailable name.]

***buchneri.*** *Crematogaster buchneri* Forel, 1894b: 94 (w.) ANGOLA. Emery, 1899e: 479 (q.). Combination in *C.* (*Atopogyne*): Santschi, 1916b: 501. Senior synonym of *halli:* **new synonymy**. Current subspecies: nominal plus *composita, graeteri, uasina.*

***buddhae.*** *Crematogaster buddhae* Forel, 1902c: 206 (w.) INDIA. Combination in *C.* (*Acrocoelia*): Emery, 1922e: 150.

*bulawayensis. Crematogaster bulawayensis* Santschi, 1916b: 498 (w.) ZIMBABWE. [First available use of *Crematogaster buchneri* r. *africana* var. *bulawayensis* Forel, 1913a: 125; unavailable name.] Combination in *C.* (*Atopogyne*): Wheeler, W.M. 1922a: 854; in *C.* (*Acrocoelia*): Emery, 1922e: 148. Raised to species: Arnold, 1920a: 521; Wheeler, W.M. 1922a: 854. Subspecies of *sjostedti:* Santschi, 1918d: 185; Emery, 1922e: 148; Santschi, 1926b: 214; of *gerstaeckeri:* Santschi, 1930b: 69. Senior synonym of *foraminicipoides:* Santschi, 1926b: 214. Synonym of *oraclum:* Arnold, 1920a: 521 [Arnold gives *bulawayensis* as senior name but *oraclum* has seniority and hence is first available name for this taxon].

***#bushimana.*** *Crematogaster* (*Sphaerocrema*) *amita* var. *bushimana* Santschi, 1926b: 226 (w.q.) SOUTH AFRICA.

***#busschodtsi.*** *Crematogaster tricolor* var. *busschodtsi* Emery, 1899e: 487 (w.) ZAIRE. Santschi, 1930a: 64 (q.m.). Combination in *C.* (*Acrocoelia*): Emery, 1922e: 146. Stirps of *castanea:* Santschi, 1928d: 61.

***butteli.*** *Crematogaster* (*Oxygyne*) *butteli* Forel, 1913k: 78 (w.q.m.) SUMATRA.

***#cacochyma.*** *Crematogaster peringueyi* r. *cacochyma* Forel, 1914d: 235 (w.q.) SOUTH AFRICA. Combination in *C.* (*Acrocoelia*): Emery, 1922e: 512.

***#cacodaemon.*** *Crematogaster peringueyi* r. *cacodaemon* Forel, 1914d: 234 (w.q.m.) SOUTH AFRICA. Combination in *C.* (*Acrocoelia*): Emery, 1922e: 148. Stirps of *opaciceps:* Santschi, 1926b: 216.

***#cacozela.*** *Crematogaster cacozela* Santschi, 1914d: 342 (w.) GUINEA. Combination in *C.* (*Acrocoelia*): Emery, 1922e: 149. Variety of *werneri:* Santschi, 1915c: 253.

***#caculata.*** *Crematogaster peringueyi* r. *caculata* Forel, 1915c: 341 (w.) SOUTH AFRICA. Combination in *C.* (*Decacrema*): Arnold, 1920a: 550; Wheeler, W.M. 1922a: 861; in *C.* (*Acrocoelia*): Emery, 1922e: 148. Race of *liengmei:* Arnold, 1920a: 550; Wheeler, W.M. 1922a: 861.

***#caffra.*** *Crematogaster* (*Sphaerocrema*) *amita* var. *caffra* Santschi, 1926b: 226 (w.q.m.) SOUTH AFRICA. [First available use of *Crematogaster* (*Sphaerocrema*) *kneri* st. *amita* var. *caffra* Santschi, in Forel, 1922: 94 (footnote); unavailable name.]

***#calens.*** *Crematogaster capensis* var. *calens* Forel, 1910e: 435 (w.) MOZAMBIQUE. Combination in *C.* (*Acrocoelia*): Emery, 1922e: 145.

***californica.*** *Crematogaster laeviuscula* var. *californica* Wheeler, W.M. 1919g: 111 (w.) U.S.A. [First available use of *Crematogaster lineolata* subsp. *laeviuscula* var. *californica* Emery, 1895c: 285; unavailable name.] Combination in *C.* (*Crematogaster*): Buren, 1968b: 94. Subspecies of *lineolata:* Wheeler, W.M. 1934f: 135. Junior synonym of *coarctata:* Creighton, 1950a: 207. Revived from synonymy and raised to species: Smith, M.R. 1951a: 808; Buren, 1968b: 94 (in key).

*calychroa. Crematogaster* (*Acrocoelia*) *castanea* st. *rufimembrum* var. *calychroa* Santschi, 1926b: 214 (w.) TANZANIA. Unavailable name.

***#camena.*** *Crematogaster* (*Atopogyne*) *africana* var. *camena* Wheeler, W.M. 1922a: 852 (w.) ZAIRE. [First available use of *Crematogaster buchneri* r. *africana* var. *camena* Forel, 1916: 410; unavailable name.]

***#canariensis.*** *Crematogaster laestrygon* subsp. *canariensis* Barquin, 1981: 146 (w.) CANARY IS. [First available use of *Crematogaster laestrygon* subsp. *maura* var. *canariensis* Emery, 1926: 3; unavailable name.]

***#capax.*** *Crematogaster* (*Decacrema*) *borneensis* subsp. *capax* Forel, 1911a: 37 (w.q.) BORNEO.

***capensis.*** *Crematogaster capensis* Mayr, 1862: 764 (w.q.) SOUTH AFRICA. Combination in *C.* (*Crematogaster*): Wheeler, W.M. 1922a: 830; in *C.* (*Acrocoelia*): Emery, 1922e: 145. Current subspecies: nominal plus *calens, tropicorum.*

***captiosa.*** *Crematogaster* (*Decacrema*) *captiosa* Forel, 1911a: 37 (w.) BORNEO.

***#carbonescens.*** *Crematogaster evallans* var. *carbonescens* Forel, 1913l: 233 (w.) BRAZIL. Combination in *C.* (*Eucrema*): Emery, 1922e: 137.

***carinata.*** *Crematogaster carinata* Mayr, 1862: 768 (w.) BRAZIL. Combination in *C.* (*Orthocrema*): Emery, 1922e: 136.

*carininotum. Crematogaster neuvillei* st. *carininotum* Santschi, 1917b: 279 (w.) ZIMBABWE. Junior synonym of *godefreyi:* Santschi, 1918d: 184.

*carminis. Crematogaster brevispinosa* r. *crucis* var. *carminis* Forel, 1914d: 270 (w.) ARGENTINA. Unavailable name.

*caspica. Crematogaster sordidula* var. *caspica* Emery, 1912e: 670, figs. 7, 8 (w.) Caspian Sea coast of former U.S.S.R. Junior synonym of *sordidula:* Atanassov & Dlussky, 1992: 182.

***castanea.*** *Crematogaster castaneus* Smith, F. 1858b: 136, pl. 9, fig. 2 (w.q.) SOUTH AFRICA (not Albania). Combination in *C.* (*Crematogaster*): Wheeler, W.M. 1922a: 830; in *C.* (*Acrocoelia*): Emery, 1922e: 145. Variety of *tricolor:* Forel, 1911f: 279; Santschi, 1913a: 412 (in key). Revived status as species: Emery, 1915g: 12; Arnold, 1920a: 488; Santschi, 1920d: 378. Senior synonym of *decolor:* Forel, 1907g: 80. Current subspecies: nominal plus *adusta, analis, aquila, arborea, bruta, busschodtsi, durbanensis, ferruginea, flaviventris, hararica, insidiosa, inversa, mediorufa, museisapientiae, rufimembrum, rufonigra, simia, tricolor, ulugurensis, yambatensis.*

#***cavinota.*** *Crematogaster excisa* r. *cavinota* Stitz, 1916: 385, fig. 5 (w.) CONGO. Combination in *C.* (*Acrocoelia*): Emery, 1922e: 147.

#***cedrosensis.*** *Crematogaster lineolata* subsp. *cedrosensis* Wheeler, W.M. 1934f: 136 (w.) MEXICO. Subspecies of *opaca:* Enzmann, J. 1946c: 94.

***censor.*** *Crematogaster senegalensis* subsp. *censor* Forel, 1910e: 434 (w.) SENEGAL. Combination in *C.* (*Crematogaster*): Wheeler, W.M. 1922a: 834; in *C.* (*Acrocoelia*): Emery, 1922e: 146. Raised to species: Forel, 1916: 406; Wheeler, W.M. 1922a: 834. Current subspecies: nominal plus *junodi.*

#***centralis.*** *Crematogaster* (*Eucrema*) *acuta* var. *centralis* Santschi, 1932e: 412 (w.q.) PANAMA.

***cephalotes.*** *Crematogaster cephalotes* Smith, F. 1857a: 75 (w.) BORNEO. Combination in *C.* (*Decacrema*): Donisthorpe, 1932c: 450.

*cephalotes. Crematogaster cephalotes* Gerstäcker, 1871: 356 (w.q.) KENYA. [Junior primary homonym of *cephalotes* Smith, F. above.] Replacement name: *gerstaeckeri* Dalla Torre, 1892: 90.

***cerasi.*** *Myrmica cerasi* Fitch, 1854: 835 (w.) U.S.A. Wheeler, G.C. & Wheeler, J. 1973a: 28 (l.). Combination in *Crematogaster:* Roger, 1863b: 37; in *C.* (*Acrocoelia*): Emery, 1922e: 141; in *C.* (*Crematogaster*): Buren, 1968b: 92. Junior synonym of *lineolata:* Dalla Torre, 1893: 83; Creighton, 1950a: 213. Revived from synonymy as variety/subspecies of *lineolata:* Emery, 1895c: 282; Smith, M.R. 1951a: 809. Revived status as species, senior synonym of *kennedyi,* and material of the unavailable name *punctinodis* referred here: Buren, in Smith, M.R. 1958c: 125. See also: Buren, 1968b: 92; Johnson, 1988: 318.

#***chacoana.*** *Crematogaster* (*Orthocrema*) *abstinens* st. *chacoana* Santschi, 1933e: 111, fig. 6 (w.) ARGENTINA.

*charon. Crematogaster* (*Acrocoelia*) *charon* Santschi, 1935a: 257 (w.) ZAIRE. *Nomen nudum.*

#***chasei.*** *Crematogaster laeviceps* var. *chasei* Forel, 1902h: 413 (w.) AUSTRALIA. Wheeler, W.M. 1934d: 144 (q.m.). Combination in *C.* (*Acrocoelia*): Emery, 1922e: 152. Subspecies of *laeviceps:* Taylor & Brown, D.R. 1985: 61.

#***chathamensis.*** *Crematogaster* (*Orthocrema*) *brevispinosa* subsp. *chathamensis* Wheeler, W.M. 1933a: 58 (w.) GALAPAGOS IS.

***chiarinii.*** *Crematogaster chiarinii* Emery, 1881a: 271, fig. (w.) ETHIOPIA. Forel, 1892e: 353 (q.m.). Combination in *C.* (*Crematogaster*): Wheeler, W.M. 1922a: 835; in *C.* (*Acrocoelia*): Emery, 1922e: 146. Current subspecies: nominal plus *aethiops, bayeri, cincta, nigra, sellula, subsulcata, taediosa, v-nigra.*

#***chillagoensis.*** *Crematogaster australis* var. *chillagoensis* Forel, 1915b: 57 (w.) AUSTRALIA. Combination in *C.* (*Acrocoelia*): Emery, 1922e: 152. Subspecies of *australis:* Taylor & Brown, D.R. 1985: 60.

***chlorotica.*** *Crematogaster chlorotica* Emery, 1899e: 483 (w.q.) CAMEROUN. Combination in *C.* (*Sphaerocrema*): Santschi, 1918d: 182.

#***chodati.*** *Crematogaster goeldii* var. *chodati* Forel, 1921a: 207 (w.) PARAGUAY. Combination in *C.* (*Orthocrema*): Kempf, 1972a: 89. Subspecies of *torosa:* current status.

***chopardi.*** *Crematogaster chopardi* Bernard, 1950b: 288, fig. 1 (w.) NIGER.

*christowitchii. Crematogaster scutellaris* var. *christowitchii* Forel, 1892i: 316 (w.) BULGARIA. Junior synonym of *schmidti:* Emery, 1922e: 143.

***chungi.*** *Crematogaster* (*Acrocoelia*) *chungi* Brown, 1949e: 38 (w.) CHINA. [First available use of *Crematogaster* (*Acrocoelia*) *brunnea* subsp. *ruginota* var. *chungi* Wheeler, W.M. 1929f: 6; unavailable name.]

***cicatriculosa.*** *Crematogaster cicatriculosa* Roger, 1863a: 207 (w.) SOUTH AFRICA. Mayr, 1866b: 901 (q.). Combination in *C.* (*Crematogaster*): Wheeler, W.M. 1922a: 836; in *C.* (*Acrocoelia*): Emery, 1922e: 146.

#***cincta.*** *Crematogaster chiarinii* var. *cincta* Emery, 1896i: 157 (w.) SOMALIA. Combination in *C.* (*Acrocoelia*): Emery, 1922e: 146.

#***cisplatinalis.*** *Crematogaster victima* var. *cisplatinalis* Mayr, 1887: 624 (w.) URUGUAY. Combination in *C.* (*Orthocrema*): Emery, 1922e: 136. Raised to species: Dalla Torre, 1893: 80. Subspecies of *victima:* Emery, 1922e: 136; Kempf, 1972a: 90.

#***civa.*** *Crematogaster rothneyi* var. *civa* Forel, 1902c: 203 (w.) INDIA. Combination in *C.* (*Acrocoelia*): Emery, 1922e: 151.

*clara. Crematogaster clara* Mayr, 1870b: 993 (w.) U.S.A. Wheeler, W.M. 1908e: 481 (q.m.). Combination in *C.* (*Acrocoelia*): Emery, 1922e: 141; in *C.* (*Crematogaster*): Buren, 1968b: 92. Junior synonym of *bicolor* Buckley (junior homonym): McCook, 1879: 188; Mayr, 1886d: 463, and hence first available replacement name. Variety of *laeviuscula:* Dalla Torre, 1893: 83; Wheeler, W.M. 1919g: 111; Enzmann, J. 1946c: 93. Junior synonym of *laeviuscula:* Creighton, 1950a: 210. Revived from synonymy and revived status as species: Buren, in Smith, M.R. 1958c: 126; Buren, 1968b: 92 (in key); Smith, D.R. 1979: 1378. Junior synonym of *laeviuscula:* Johnson, 1988: 322.

#***clarior.*** *Crematogaster laeviceps* var. *clarior* Forel, 1902h: 414 (w.) AUSTRALIA. Combination in *C.* (*Acrocoelia*): Emery, 1922e: 152. Subspecies of *laeviceps:* Taylor & Brown, D.R. 1985: 61.

***clariventris.*** *Crematogaster buchneri* subsp. *clariventris* Mayr, 1895: 139 (w.) CONGO. Forel, 1911e: 271 (q.). Combination in *C.* (*Atopogyne*): Wheeler, W.M. 1922a: 854. Raised to species: Emery, 1922e: 155; Santschi, 1930a: 65. Current subspecies: nominal plus *biimpressa.*

#***clemensae.*** *Crematogaster modiglianii* var. *clemensae* Forel, 1910d: 125 (w.) PHILIPPINES. Combination in *C.* (*Paracrema*): Emery, 1922e: 156.

#***clepens.*** *Crematogaster opaciceps* var. *clepens* Forel, 1913h: 353 (w.) ZAIRE. Combination in *C.* (*Acrocoelia*): Emery, 1922e: 148.

***clydia.*** *Crematogaster clydia* Forel, 1912f: 219 (w.) BRAZIL. Combination in *C.* (*Acrocoelia*): Emery, 1922e: 142.

#***clymene.*** *Crematogaster santschii* var. *clymene* Forel, 1915c: 337 (w.) SOUTH AFRICA. Combination in *C.* (*Oxygyne*): Wheeler, W.M. 1922a: 858; in *C.* (*Acrocoelia*): Emery, 1922e: 148.

*coachellai. Crematogaster* (*Acrocoelia*) *lineolata* subsp. *coachellai* Enzmann, J. 1946c: 93, U.S.A. *Nomen nudum;* material referred to *vermiculata* by Brown, 1949a: 48.

***coarctata.*** *Crematogaster coarctata* Mayr, 1870b: 992 (w.) U.S.A. Combination in *C.* (*Acrocoelia*): Emery, 1922e: 141; in *C.* (*Crematogaster*): Buren, 1968b: 94. Junior synonym of *lineolata:* Mayr, 1886d: 462. Revived from synonymy as subspecies of *lineolata:* Emery, 1895c: 283; Wheeler, W.M. 1904d: 270. Revived status as species: Wheeler, W.M. 1919g: 111. Subspecies of *lineolata:* Emery, 1922e: 141; of *opaca:* Enzmann, J. 1946c: 94. Revived status as species: Creighton, 1950a: 207; Buren, 1968b: 94 (in key).

#***cocciphila.*** *Crematogaster* (*Orthocrema*) *brasiliensis* var. *cocciphila* Borgmeier, 1934: 102 (w.q.) SURINAM.

***coelestis.*** *Crematogaster coelestis* Santschi, 1911c: 353 (w.) GUINEA. Combination in *C.* (*Acrocoelia*): Emery, 1922e: 147. Subspecies/race of *excisa:* Forel, 1915c: 342; Wheeler, W.M. 1922a: 837; of *gallicola:* Emery, 1922e: 147. Revived status as species: Santschi, 1939f: 162. Current subspecies: nominal plus *mercatori.*

***colei.*** *Crematogaster* (*Crematogaster*) *colei* Buren, 1968b: 108 (w.q.) U.S.A.

*columbiana. Myrmica* (*Monomorium*) *columbiana* Buckley, 1867: 340 (w.q.) U.S.A. Junior synonym of *lineolata:* Mayr, 1886d: 462.

#***compacta.*** *Crematogaster* (*Paracrema*) *spengeli* subsp. *compacta* Crawley, 1924: 398 (w.) SUMATRA.

#***completa.*** *Crematogaster* (*Acrocoelia*) *menileki* st. *completa* Santschi, 1928d: 63 (w.) ZIMBABWE.

#***composita.*** *Crematogaster* (*Atopogyne*) *buchneri* var. *composita* Santschi, 1933b: 101 (w.) ZAIRE.

***concava.*** *Crematogaster concava* Emery, 1899e: 480 (w.) CAMEROUN. Santschi, 1910c: 375 (q.m.); Forel, 1916: 404 (m.). Combination in *C.* (*Sphaerocrema*): Santschi, 1918d: 182.

*confucii. Crematogaster* (*Acrocoelia*) *laboriosa* st. *politula* var. *confucii* Santschi, 1928b: 32 (w.q.m.) CHINA. Unavailable name.

#***confusa.*** *Crematogaster wilwerthi* var. *confusa* Santschi, 1911g: 208 (w.) CONGO. Combination in *C.* (*Sphaerocrema*): Wheeler, W.M. 1922a: 849.

***constructor.*** *Crematogaster constructor* Emery, 1895h: 29, pl. 2, figs. 12, 13 (w.) SOUTH AFRICA. [Also described as new by Emery, in Mayr, 1895: 135 (footnote).] Combination in *C.* (*Crematogaster*): Wheeler, W.M. 1922a: 836; in *C.* (*Acrocoelia*): Emery, 1922e: 146. Current subspecies: nominal plus *kirbyella*.

#***contemta.*** *Crematogaster contemta* Mayr, 1879: 685 (w.) INDIA. Bingham, 1903: 130 (q.). Combination in *C.* (*Acrocoelia*): Emery, 1922e: 150. Subspecies/stirps of *subnuda:* Forel, 1902c: 208; Forel, 1907e: 17; of *laboriosa:* Santschi, 1928b: 33; of *brunnea:* Emery, 1922e: 150; Wheeler, W.M. 1929f: 5; Wheeler, W.M. 1930h: 64.

*contemtior. Crematogaster* (*Acrocoelia*) *brunnea* subsp. *contemta* var. *contemtior* Wheeler, W.M. 1929f: 5 (w.) CHINA. Unavailable name.

*convexiclypea. Crematogaster* (*Decacrema*) *solenopsisoides* [misspelling of *solenopsides*] r. *flavida* var. *convexiclypea* Forel, 1916: 411 (w.q.m.) ZAIRE. Unavailable name.

*convicta. Crematogaster brevispinosa* r. *tumulifera* var. *convicta* Forel, 1912f: 215 (w.m.) ARGENTINA. Unavailable name.

#***cooperi.*** *Crematogaster neuvillei* r. *cooperi* Forel, 1914d: 233 (w.) SOUTH AFRICA. Arnold, 1920a: 498 (q.). Combination in *C.* (*Acrocoelia*): Emery, 1922e: 148.

#***corax.*** *Crematogaster ebenina* var. *corax* Forel, 1902c: 200 (w.) BURMA. Combination in *C.* (*Oxygyne*): Forel, 1903a: 683.

*cordinoda. Crematogaster distans* var. *cordinoda* Forel, 1914e: 12 (w.) COLOMBIA. Combination in *C.* (*Orthocrema*): Emery, 1922e: 135. Junior synonym of *distans:* Kempf, 1968b: 390.

***coriaria.*** *Crematogaster coriaria* Mayr, 1872: 154 (w.) BORNEO. Combination in *C.* (*Paracrema*): Emery, 1922e: 155. Current subspecies: nominal plus *fraudatrix*.

***cornigera.*** *Crematogaster cornigera* Forel, 1902h: 407 (w.q.) AUSTRALIA. Combination in *C.* (*Acrocoelia*): Emery, 1922e: 152.

***cornuta.*** *Crematogaster* (*Acrocoelia*) *cornuta* Crawley, 1924: 395 (w.q.m.) SUMATRA.

***corporaali.*** *Crematogaster* (*Acrocoelia*) *corporaali* Santschi, 1928h: 126 (w.q.) SUMATRA.

*corsica. Crematogaster scutellaris* var. *corsica* Santschi, 1921b: 435 (w.) CORSICA. Junior synonym of *scutellaris:* Casevitz-Weulersse, 1990a: 137.

***corticicola.*** *Crematogaster distans* var. *corticicola* Mayr, 1887: 625 (w.q.m.) BRAZIL. Combination in *C.* (*Orthocrema*): Emery, 1922e: 135; in *C.* (*Neocrema*): Kempf, 1972a: 83. Variety/subspecies of *curvispinosa:* Forel, 1895b: 131; of *distans:* Emery, 1906c: 139. Raised to species, senior synonym of *rugiceps*, and material of the unavailable name *pullipes* referred here: Kempf, 1968b: 394.

***corvina.*** *Crematogaster corvina* Mayr, 1870b: 994 (w.) MEXICO. Combination in *C.* (*Orthocrema*): Emery, 1922e: 134; in *C.* (*Neocrema*): Santschi, 1933e: 114.

#***costeboriensis.*** *Crematogaster* (*Decacrema*) *solenopsoides* st. *costeboriensis* Santschi, 1919h: 86 (w.) IVORY COAST.

#***costulata.*** *Crematogaster rogenhoferi* var. *costulata* Emery, 1895k: 467 (w.) BURMA.

***crassicornis.*** *Crematogaster crassicornis* Emery, 1893g: 265, pl. 2, figs. 4, 5 (w.) PHILIPPINES. Combination in *C.* (*Orthocrema*): Emery, 1922e: 132.

#***crawleyi.*** *Crematogaster jehovae* subsp. *crawleyi* Emery, 1926: 4 (w.) IRAQ.

*creightoni. Crematogaster* (*Acrocoelia*) *creightoni* Wheeler, W.M. 1933d: 86 (q.) U.S.A. Junior synonym of *pilosa:* Buren, in Smith, M.R. 1958c: 127.

*cretica. Crematogaster auberti* subsp. *laestrigon* var. *cretica* Karavaiev, 1927b: 106, fig. 2 (w.) GREECE. Unavailable name.

***crinosa.*** *Crematogaster crinosa* Mayr, 1862: 767 (w.) BRAZIL. Mayr, 1887: 627 (q.m.). Combination in *C.* (*Orthocrema*): Emery, 1922e: 134.

***cristata.*** *Crematogaster* (*Orthocrema*) *cristata* Santschi, 1929d: 294 (w.) BRAZIL. Senior synonym of *terricola:* Kempf, 1968b: 384.

*cristulata. Crematogaster* (*Neocrema*) *montezumia* st. *cristulata* Santschi, 1925d: 233 (w.q.) BRAZIL. Junior synonym of *montezumia:* Kempf, 1968b: 386.

#***crucis.*** *Crematogaster brevispinosa* r. *crucis* Forel, 1912f: 212 (w.) COLOMBIA. Combination in *C.* (*Orthocrema*): Emery, 1922e: 134.

#***cubaensis.*** *Crematogaster victima* subsp. *cubaensis* Mann, 1920: 425 (w.q.) CUBA. Subspecies of *steinheili:* Menozzi, 1929a: 2; of *laeviuscula:* Enzmann, J. 1946c: 93; of *victima:* Kempf, 1972a: 90.

#***cupida.*** *Crematogaster* (*Oxygyne*) *margaritae* st. *cupida* Santschi, 1935a: 261, fig. 3 (w.) ZAIRE.

#***curticeps.*** *Crematogaster longiceps* var. *curticeps* Wheeler, W.M. 1915g: 809 (w.) AUSTRALIA. Combination in *C.* (*Acrocoelia*): Emery, 1922e: 152. Subspecies of *longiceps:* Taylor & Brown, D.R. 1985: 62.

***curvispinosa.*** *Crematogaster curvispinosa* Mayr, 1862: 768 (w.) BRAZIL. Forel, 1911c: 301 (q.); Forel, 1912f: 216 (m.). Combination in *C.* (*Orthocrema*): Emery, 1922e: 134. Senior synonym of *fuliginea:* Santschi, 1929f: 88. Current subspecies: nominal plus *antillana, kemali, obscura* (unresolved junior homonym), *panamana*.

***cuvierae.*** *Crematogaster* (*Atopogyne*) *cuvierae* Donisthorpe, 1945a: 10 (w.) GHANA.

***cylindriceps.*** *Crematogaster* (*Colobocrema*) *cylindriceps* Wheeler, W.M. 1927f: 31, fig. 39 (q.) PHILIPPINES.

#***cypria.*** *Crematogaster* (*Acrocoelia*) *jehovae* var. *cypria* Santschi, 1930d: 266 (w.) CYPRUS.

***dahlii.*** *Crematogaster dahlii* Forel, 1901b: 15 (w.) BISMARCK ARCHIPELAGO. Combination in *C.* (*Xiphocrema*): Forel, 1913k: 80.

***daisyi.*** *Crematogaster* (*Oxygyne*) *daisyi* Forel, 1901e: 376 (w.q.) BORNEO.

#***dakarensis.*** *Crematogaster pronotalis* var. *dakarensis* Santschi, 1914d: 349 (w.) SENEGAL. Combination in *C.* (*Sphaerocrema*): Wheeler, W.M. 1922a: 848. Variety of *kneri:* Santschi, 1926b: 228.

***dalyi.*** *Crematogaster* (*Oxygyne*) *dalyi* Forel, 1902c: 201 (w.) INDIA. Current subspecies: nominal plus *sikkimensis*.

***decamera.*** *Crematogaster* (*Decacrema*) *decamera* Forel, 1910a: 18 (footnote) (w.q.m.) BORNEO.

*decolor. Crematogaster tricolor* var. *decolor* Forel, 1891b: 188 (w.) MADAGASCAR. Junior synonym of *castanea:* Forel, 1907g: 80.

*decolorata. Crematogaster* (*Acrocoelia*) *castanea* var. *decolorata* Santschi, 1926b: 213 (w.) SOUTH AFRICA. Junior synonym of *arborea:* Santschi, 1930b: 69.

#***defleta.*** *Crematogaster opaciceps* subsp. *defletus* Forel, 1910e: 434 (w.) SOUTH AFRICA. Combination in *C.* (*Acrocoelia*): Emery, 1922e: 148.

*deformis*, Roger, 1863; see under ***difformis.***

***degeeri.*** *Crematogaster degeeri* Forel, 1886c: cvii (w.) MADAGASCAR. Forel, 1891b: 190 (q.m.). Combination in *C.* (*Crematogaster*): Wheeler, W.M. 1922a: 1022; in *C.* (*Acrocoelia*): Emery, 1922e: 146. [*Crematogaster geeri* Schulz, 1906: 155; unjustified subsequent emendation.] Current subspecies: nominal plus *lunaris*.

*degener. Crematogaster scutellaris* var. *degener* Santschi, 1937e: 310 (w.) TUNISIA. Junior synonym of *scutellaris:* Casevitz-Weulersse, 1990a: 137.

***delagoensis.*** *Crematogaster inermis* r. *delagoensis* Forel, 1894b: 99 (w.) MOZAMBIQUE. Combination in *C.* (*Acrocoelia*), and raised to species: Emery, 1922e: 146. Current subspecies: nominal plus *acutidens, merwei, rhodesiana*.

***delitescens.*** *Crematogaster* (*Orthocrema*) *delitescens* Wheeler, W.M. 1921f: 152 (w.q.) GUIANA.

#***dentata.*** *Crematogaster sewellei* var. *dentata* Dalla Torre, 1893: 86 (w.) MADAGASCAR. [First available use of *Crematogaster inermis* r. *sewellii* var. *dentatus* Forel, 1891b: 196; unavailable name.] Combination in *C.* (*Acrocoelia*): Emery, 1922e: 148. Variety/subspecies of *sewellei:* Forel, 1907g: 79; Donisthorpe, 1946e: 30.

#***dentinasis.*** *Crematogaster* (*Orthocrema*) *rufotestacea* st. *dentinasis* Santschi, 1929f: 89 (w.q.m.) AUSTRALIA.

***dentinodis.*** *Crematogaster opaca* var. *dentinodis* Forel, 1901c: 130 (w.) MEXICO. Combination in *C.* (*Crematogaster*): Buren, 1968b: 93. Raised to species: Smith, M.R. 1958c: 126; Buren, 1968b: 93 (in key).

#***dentulata.*** *Crematogaster peringueyi* r. *dentulata* Stitz, 1923: 156 (w.) SOUTH AFRICA.

***depilis.*** *Crematogaster opaca* var. *depilis* Wheeler, W.M. 1919g: 111 (w.) U.S.A. [First available use of *Crematogaster lineolata* subsp. *opaca* var. *depilis* Wheeler, W.M. 1908e: 478; unavailable name.] Wheeler, G.C. & Wheeler, J. 1973a: 28 (l.). Combination in *C.* (*Acrocoelia*): Emery, 1922e: 141; in *C.* (*Crematogaster*): Buren, 1968b: 93. Variety of *punctulata:* Enzmann, J. 1946c: 93. Raised to species: Creighton, 1950a: 209; Buren, in Smith, M.R. 1958c: 126.

*deponens* Walker, 1859; see under ***OLIGOMYRMEX.***

***depressa.*** *Formica depressa* Latreille, 1802c: 268, pl. 11, fig. 73 (q.) GUINEA. Forel, 1910f: 5 (m.). Combination in *Crematogaster:* Emery, 1892b: 164; in *C.* (*Oxygyne*): Forel, 1901e: 375; in *C.* (*Atopogyne*): Forel, 1911d: 343; Santschi, 1918d: 183. Senior synonym of *platygnatha* (and its junior synonym *mandibularis*): Emery, 1892b: 164; of *foreli:* **new synonymy**. Current subspecies: nominal plus *adultera, fuscipennis*.

***descarpentriesi.*** *Crematogaster* (*Oxygyne*) *descarpentriesi* Santschi, 1928d: 68 (w.) MADAGASCAR.

***descolei.*** *Crematogaster* (*Neocrema*) *descolei* Kusnezov, 1949e: 587, figs. 1–4 (q.) ARGENTINA.

***desecta.*** *Crematogaster desecta* Forel, 1911a: 27 (w.q.) SRI LANKA. Combination in *C.* (*Acrocoelia*): Emery, 1922e: 150.

***CREMATOGASTER** (continued)*

***desperans.*** *Crematogaster gabonensis* var. *desperans* Forel, 1914d: 233 (w.) SOUTH AFRICA. Combination in *C.* (*Atopogyne*): Wheeler, W.M. 1922a: 854; in *C.* (*Acrocoelia*): Emery, 1922e: 146. Variety of *bulawayensis:* Arnold, 1920a: 523; Wheeler, W.M. 1922a: 854. Raised to species: Santschi, 1918d: 185; Emery, 1922e: 146.

**#*detritinodis.*** *Crematogaster* (*Physocrema*) *stethogompha* var. *detritinodis* Wheeler, W.M. 1919e: 77 (w.) BORNEO.

*devincta. Crematogaster aegyptiaca* st. *senegalensis* var. *devincta* Santschi, 1914d: 343 (w.) GUINEA. Unavailable name.

**#*dextella.*** *Crematogaster* (*Orthocrema*) *limata* st. *dextella* Santschi, 1929f: 88, fig. 2 (w.) PANAMA. Wheeler, G.C. & Wheeler, J. 1952b: 258 (l.).

*dewasi. Sphaerocrema dewasi* Miré, 1966: 240, fig. (w.) CAMEROUN. Junior synonym of *wellmani:* **new synonymy**.

***difformis.*** *Crematogaster difformis* Smith, F. 1857a: 76 (w.) SINGAPORE. Emery, 1901g: 574 (q.); Forel, 1909d: 225 (m.); Wheeler, G.C. & Wheeler, J. 1952b: 261 (l.). Combination in *C.* (*Physocrema*): Forel, 1912f: 220. Senior synonym of *edentata:* Emery, 1901g: 574. [Name misspelled as *deformis* by Roger, 1863b: 37, and many subsequent authors.] Current subspecies: nominal plus *physothorax, sewardi, vacca.*

***diffusa.*** *Myrmica diffusa* Jerdon, 1851: 113 (w.q.m.) INDIA. Combination in *Crematogaster:* Roger, 1863b: 37; in *C.* (*Acrocoelia*): Emery, 1922e: 150. [Bingham, 1903: 136 suggests that *diffusa* may be a senior synonym of *aberrans.*]

**#*diminuta.*** *Crematogaster laestrygon* var. *diminuta* Santschi, 1910a: 43 (w.) TUNISIA. Karavaiev, 1927b: 105 (q.). Combination in *C.* (*Acrocoelia*): Emery, 1922e: 142.

*dirce. Crematogaster excisa* r. *coelestis* var. *dirce* Forel, 1915c: 342 (w.) ZAIRE. Unavailable name.

**#*discinodis.*** *Crematogaster discinodis* Emery, 1893g: 261, pl. 6, fig. 9 (w.) SINGAPORE. Combination in *C.* (*Acrocoelia*): Emery, 1922e: 150. Variety of *subnuda:* Emery, 1900d: 688.

***dispar.*** *Crematogaster sordidula* r. *dispar* Forel, 1902h: 412 (w.q.m.) AUSTRALIA. Combination in *C.* (*Orthocrema*), and raised to species: Emery, 1922e: 132. Current subspecies: nominal plus *bipartita.*

***distans.*** *Crematogaster distans* Mayr, 1870a: 402 (w.) COLOMBIA. Kempf, 1968b: 392 (q.). Combination in *C.* (*Orthocrema*): Emery, 1922e: 134; in *C.* (*Neocrema*): Santschi, 1918d: 182; Kempf, 1972a: 83. Senior synonym of *cordinoda, paraensis, parviceps, pevsnerae* and *vanda:* Kempf, 1968b: 390.

*diversa. Crematogaster* (*Acrocoelia*) *nigriceps* st. *prelli* var. *diversa* Santschi, 1928d: 65 (w.) TANZANIA. Unavailable name.

***dohrni.*** *Crematogaster dohrni* Mayr, 1879: 682 (w.) SRI LANKA. Combination in *C.*(*Acrocoelia*): Emery, 1922e: 150. Current subspecies: nominal plus *artifex, fabricans, gigas, kerri, kiangsiensis.*

***dolens.*** *Crematogaster dolens* Forel, 1910e: 433 (w.) TANZANIA. Combination in *C.* (*Orthocrema*), and subspecies of *pauciseta:* Wheeler, W.M. 1922a: 850. Revived status as species: Emery, 1922e: 131.

**#*dolichocephala.*** *Crematogaster* (*Oxygyne*) *stadelmanni* var. *dolichocephala* Santschi, 1911g: 208 (w.q.m.) CONGO. [*C. stadelmanni* st. *dolichocephala* Santschi, 1910c: 378; *nomen nudum*, attributed to Emery.] Combination in *C.* (*Nematocrema*): Wheeler, W.M. 1922a: 859.

***donisthorpei.*** *Crematogaster* (*Oxygyne*) *donisthorpei* Santschi, 1934a: 188 (w.q.m.) ZIMBABWE.

***dorsidens.*** *Crematogaster* (*Orthocrema*) *dorsidens* Santschi, 1925d: 233 (w.) BRAZIL.

***dubia.*** *Crematogaster* (*Paracrema*) *dubia* Karavaiev, 1935a: 93, fig. 19 (w.) CAMBODIA.

**#*dulcis.*** *Crematogaster* (*Orthocrema*) *natalensis* var. *dulcis* Santschi, 1926b: 212 (w.) SOUTH AFRICA.

**#*durbanensis.*** *Crematogaster ferruginea* var. *durbanensis* Forel, 1914d: 234 (w.) SOUTH AFRICA. Arnold, 1920a: 491 (q.). Combination in *C.* (*Acrocoelia*): Emery, 1922e: 145. Subspecies of *castanea:* current status.

*dusmeti. Crematogaster* (*Acrocoelia*) *auberti* st. *fuenti* var. *dusmeti* Santschi, 1925g: 348 (w.) SPAIN. Unavailable name.

***ebenina.*** *Crematogaster* (*Oxygyne*) *ebenina* Forel, 1902c: 199 (w.q.) INDIA. Current subspecies: nominal plus *corax.*

*edentata. Crematogaster edentata* Mayr, 1867a: 104, pl. 2, fig. 10 (q.) JAVA. Junior synonym of *difformis:* Emery, 1901g: 574.

***edentula.*** *Crematogaster* (*Decacrema*) *edentula* Santschi, 1914d: 349, fig. 18 (w.m.) GUINEA. Santschi, 1915c: 253 (q.).

**#*edmeeae.*** *Crematogaster* (*Sphaerocrema*) *zavathari* [misspelling of *zavattarii*] st. *edmeeae* Santschi, 1937g: 77 (w.) CONGO.

***egidyi.*** *Crematogaster egidyi* Forel, 1903c: 254 (w.) CHINA. Combination in *C.* (*Acrocoelia*): Emery, 1922e: 151. Current subspecies: nominal plus *ingvei, spinozai, szaboi.*

***egregior.*** *Crematogaster longispina* subsp. *egregior* Forel, 1912f: 218 (w.) BRAZIL. Combination in *C.* (*Orthocrema*): Emery, 1922e: 136. Raised to species: Kempf, 1970a: 332.

***elegans.*** *Crematogaster elegans* Smith, F. 1859a: 149 (w.) INDONESIA (Aru I.). Combination in *C.* (*Xiphocrema*): Emery, 1922e: 139; Donisthorpe, 1932c: 456.

*elgona. Crematogaster* (*Acrocoelia*) *castanea* st. *inversa* var. *elgona* Santschi, 1935b: 269 (w.) KENYA. Unavailable name.

***elysii.*** *Crematogaster elysii* Mann, 1919: 319, fig. 22 (w.) SOLOMON IS.

*emacerata. Crematogaster* (*Acrocoelia*) *castanea* st. *busschodtsi* var. *emacerata* Santschi, 1928d: 62 (w.) SOUTH AFRICA. Unavailable name.

***emeryana.*** *Crematogaster* (*Acrocoelia*) *lineolata* subsp. *emeryana* Creighton, 1950a: 213 (w.) U.S.A. Combination in *C.* (*Crematogaster*), and raised to species: Buren, 1968b: 94 (in key).

***emeryi.*** *Crematogaster emeryi* Forel, 1907a: 25. Replacement name for *biroi* Emery, 1900c: 332. [Junior primary homonym of *biroi* Mayr, 1897: 428.] Combination in *C.* (*Orthocrema*): Emery, 1922e: 133.

***emmae.*** *Crematogaster emmae* Forel, 1891b: 227, pl. 6, fig. 11 (q.) MADAGASCAR. Santschi, 1928d: 69 (w.). Combination in *C.* (*Oxygyne*): Forel, 1901e: 375. Current subspecies: nominal plus *laticeps.*

***enneamera.*** *Crematogaster enneamera* Emery, 1900c: 330, pl. 8, figs. 33–35 (w.m.) NEW GUINEA. Combination in *C.* (*Decacrema*): Forel, 1910a: 18; Forel, 1910f: 9.

***ensifera.*** *Crematogaster* (*Decacrema*) *ensifera* Forel, 1910a: 18 (w.) MADAGASCAR.

***erecta.*** *Crematogaster erecta* Mayr, 1866b: 902, pl. 20, fig. 12 (w.) COSTA RICA. Combination in *C.* (*Orthocrema*): Emery, 1922e: 135. Senior synonym of *virgula:* Kempf, 1968b: 381.

**#*erici.*** *Crematogaster* (*Orthocrema*) *quadriformis* st. *erici* Santschi, 1929d: 291 (w.) ARGENTINA.

***esterelana.*** *Orthocrema esterelana* Bernard, 1978b: 43, fig. 1 (w.q.m.) FRANCE. Combination in *Crematogaster:* **new combination**.

**#*etiolata.*** *Crematogaster lucayana* subsp. *etiolata* Wheeler, W.M. 1905b: 95 (w.q.) BAHAMAS. Combination in *C.* (*Acrocoelia*): Emery, 1922e: 141.

*euphrosyne. Crematogaster excisa* subsp. *impressa* var. *euphrosyne* Santschi, in Wheeler, W.M. 1922a: 153 (w.) ZAIRE. Unavailable name.

***eurydice.*** *Crematogaster* (*Atopogyne*) *eurydice* Forel, 1915b: 56 (q.) AUSTRALIA. Combination in *C.* (*Acrocoelia*): Emery, 1922e: 152.

***euterpe.*** *Crematogaster* (*Neocrema*) *euterpe* Santschi, 1922d: 246 (w.) ARGENTINA.

***evallans.*** *Crematogaster evallans* Forel, 1907e: 6 (footnote) (q.) BRAZIL. Forel, 1912f: 218 (w.). Combination in *C.* (*Eucrema*): Emery, 1922e: 137. Current subspecies: nominal plus *carbonescens.*

***excisa.*** *Crematogaster excisa* Mayr, 1895: 139 (w.) CONGO. Combination in *C.* (*Crematogaster*): Wheeler, W.M. 1922a: 836; in *C.* (*Acrocoelia*): Emery, 1922e: 146. Race of *buchneri:* Forel, 1916: 411. Revived status as species: Stitz, 1916: 385; Wheeler, W.M. 1922a: 153; Santschi, 1928d: 63. Current subspecies: nominal plus *bomaella, cavinota, lacustris.*

**#*exigua.*** *Crematogaster xerophila* var. *exigua* Wheeler, W.M. 1915g: 811 (w.) AUSTRALIA. Combination in *C.* (*Orthocrema*): Emery, 1922e: 135. Subspecies of *xerophila:* Taylor & Brown, D.R. 1985: 63.

**#*fabricans.*** *Crematogaster rogenhoferi* var. *fabricans* Forel, 1911b: 201 (w.) JAVA. Wheeler, W.M. 1924b: 246 (q.m.). Combination in *C.* (*Acrocoelia*): Emery, 1922e: 150. Subspecies/race of *rogenhoferi:* Forel, 1913f: 194; of *dohrni:* Emery, 1922e: 150; Wheeler, W.M. 1927d: 4; Wheeler, W.M. 1930h: 66.

**#*fauconneti.*** *Crematogaster wilwerthi* var. *fauconneti* Forel, 1910e: 434 (w.) ZAIRE. Combination in *C.* (*Sphaerocrema*): Wheeler, W.M. 1922a: 849.

*fautrix. Crematogaster brevispinosa* subsp. *tumulifera* var. *fautrix* Forel, 1908c: 369 (w.) BRAZIL. Unavailable name.

***ferrarii.*** *Crematogaster ferrarii* Emery, 1888a: 533 (w.) SUMATRA. Combination in *C.* (*Acrocoelia*): Emery, 1922e: 151. Current subspecies: nominal plus *lumpurensis.*

**#*ferruginea.*** *Crematogaster ferruginea* Forel, 1892e: 353 (w.) SOMALIA. Combination in *C.* (*Acrocoelia*): Emery, 1922e: 145. Stirps/subspecies of *tricolor:* Santschi, 1913a: 413 (in key); Santschi, 1914b: 86;

of *castanea:* Emery, 1915g: 12; Wheeler, W.M. 1922a: 831; Santschi, 1935b: 269.

*fickendeyi. Crematogaster buchneri* r. *winkleri* var. *fickendeyi* Forel, 1913a: 125 (w.) CAMEROUN. Unavailable name.

**#fictrix.** *Crematogaster rogenhoferi* var. *fictrix* Forel, 1911a: 27 (w.) BORNEO.

*flaccida. Crematogaster (Acrocoelia) dohrni* subsp. *fabricans* var. *flaccida* Wheeler, W.M. 1927d: 5 (w.) CHINA. Unavailable name.

*flachii. Crematogaster sordidula* var. *flachii* Forel, 1895d: 229 (w.) YUGOSLAVIA. Combination in *C. (Orthocrema)*: Emery, 1922e: 130. Junior synonym of *mayri:* Müller, 1923: 74. Revived from synonymy as variety of *sordidula:* Karavaiev, 1927b: 106. Junior synonym of *sordidula:* Agosti & Collingwood, 1987a: 54.

**#flava.** *Crematogaster flava* Forel, 1886d: 248 (w.) INDIA. Variety of *rogenhoferi:* Forel, 1903a: 682.

**#flavicapilla.** *Crematogaster subdentata* var. *flavicapilla* Ruzsky, 1905b: 502 (w.) TRANSCASPIA. Combination in *C. (Acrocoelia)*: Emery, 1922e: 144.

**flavicornis.** *Crematogaster flavicornis* Emery, 1897d: 567 (w.) NEW GUINEA. Combination in *C. (Orthocrema)*: Emery, 1922e: 132; in *C. (Mesocrema)*: Santschi, 1937e: 310.

**#flavida.** *Crematogaster solenopsides* subsp. *flavida* Mayr, 1907b: 15 (w.) TANZANIA. Combination in *C. (Decacrema)*: Wheeler, W.M. 1922a: 862.

**#flavior.** *Crematogaster (Orthocrema) major* subsp. *flavior* Donisthorpe, 1941b: 40 (w.) NEW GUINEA.

*flaviscapis. Crematogaster (Decacrema) solenopsides* st. *flavida* var. *flaviscapis* Santschi, 1919h: 85 (w.) ANGOLA. Unavailable name.

**flavitarsis.** *Crematogaster flavitarsis* Emery, 1900c: 332, pl. 8, fig. 42 (w.) NEW GUINEA. Combination in *C. (Xiphocrema)*: Emery, 1922e: 139.

**#flaviventris.** *Crematogaster flaviventris* Santschi, 1910c: 370 (w.q.m.) ZAIRE. Variety/stirps of *inversa:* Santschi, 1914b: 86; of *castanea:* Santschi, 1935a: 257.

**#florida.** *Crematogaster lutzi* var. *florida* Forel, 1908c: 369 (w.) BRAZIL. Combination in *C. (Orthocrema)*: Emery, 1922e: 136.

**foraminiceps.** *Crematogaster foraminiceps* Santschi, 1913a: 412 (w.) (diagnosis in key) KENYA. Santschi, 1914b: 89 (w.q.m.). Combination in *C. (Crematogaster)*: Wheeler, W.M. 1922a: 838; in *C. (Acrocoelia)*: Emery, 1922e: 147. Current subspecies: nominal plus *mirmillo, staitchi.*

*foraminicipoides. Crematogaster godfreyi* var. *foraminicipoides* Arnold, 1920a: 505 (w.) ZIMBABWE. [First available use of *Crematogaster vulcania* r. *godfreyi* var. *foraminicipoides* Forel, 1916: 406; unavailable name.] Junior synonym of *bulawayensis:* Santschi, 1926b: 214.

*foreli. Crematogaster buchneri* subsp. *foreli* Mayr, 1895: 138 (w.) GUINEA. Combination in *C. (Atopogyne)*: Santschi, 1916b: 501. Junior synonym of *depressa:* Emery, 1899e: 478. Revived from synonymy as subspecies of *buchneri:* Forel, 1909b: 53. Raised to species: Santschi, 1918d: 183. Junior synonym of *depressa:* Wheeler, W.M. 1922a: 855; Emery, 1922e: 155. Variety of *buchneri:* Santschi, 1928d: 68; of *depressa:* Santschi, 1933b: 100. Junior synonym of *depressa:* **new synonymy**.

**formosa.** *Crematogaster formosa* Mayr, 1870b: 994 (w.) MEXICO. Combination in *C. (Orthocrema)*: Emery, 1922e: 135. Current subspecies: nominal plus *aterrima.*

**#formosae.** *Crematogaster subnuda* var. *formosae* Wheeler, W.M. 1909d: 336 (w.) TAIWAN.

*forticulus* Walker, 1859; see under ***TECHNOMYRMEX.***

**foxi.** *Crematogaster foxi* Mann, 1919: 321, fig. 23 (w.q.) SOLOMON IS.

**#fraudatrix.** *Crematogaster coriaria* subsp. *fraudatrix* Forel, 1911d: 376 (w.) SUMATRA. Combination in *C. (Paracrema)*: Emery, 1922e: 155.

**fraxatrix.** *Crematogaster fraxatrix* Forel, 1911a: 28 (w.) BORNEO. Combination in *C. (Acrocoelia)*: Emery, 1922e: 151.

**fritzi.** *Crematogaster fritzi* Emery, 1901g: 576, fig. B (w.m.) SULAWESI. Combination in *C. (Orthocrema)*: Emery, 1922e: 132. Race of *sordidula:* Forel, 1902h: 410. Revived status as species: Emery, 1912e: 668.

**frivola.** *Crematogaster frivolus* Forel, 1902h: 412 (w.) AUSTRALIA. Combination in *C. (Orthocrema)*: Emery, 1922e: 132; in *C. (Acrocoelia)*: Santschi, 1929f: 90. Current subspecies: nominal plus *sculpticeps.*

**#froggatti.** *Crematogaster sordidula* r. *froggatti* Forel, 1902h: 410 (w.q.) AUSTRALIA. Combination in *C. (Orthocrema)*: Emery, 1922e: 133; in *C. (Mesocrema)*, and raised to species: Santschi, 1929f: 90. Subspecies of *queenslandica:* Emery, 1914b: 185; Taylor & Brown, D.R. 1985: 62.

**#frontalis.** *Crematogaster impressiceps* var. *frontalis* Santschi, in Wheeler, W.M. 1922a: 154 (w.) ZAIRE.

**fruhstorferi.** *Crematogaster fruhstorferi* Emery, 1901g: 577, fig. C (w.) SULAWESI. Combination in *C. (Acrocoelia)*: Emery, 1922e: 151.

**fuentei.** *Crematogaster (Acrocoelia) auberti* subsp. *fuentei* Menozzi, 1922b: 327 (w.) SPAIN. Subspecies of *inermis:* Emery, 1926: 2; of *auberti:* Santschi, 1931a: 5. Raised to species: Collingwood, 1978: 69.

*fuliginea. Crematogaster (Orthocrema) fuliginea* Santschi, 1925d: 231 (w.) BRAZIL. Junior synonym of *curvispinosa:* Santschi, 1929f: 88.

*fuliginosa. Crematogaster meneliki* st. *completa* var. *fuliginosa* Arnold, 1947: 135, fig. 4 (w.) ZAMBIA. Unavailable name.

**fulmeki.** *Crematogaster (Physocrema) fulmeki* Forel, 1922: 95 (w.) SUMATRA. See also: Santschi, 1928h: 127.

*fulva. Crematogaster (Sphaerocrema) fulva* Donisthorpe, 1945d: 265 (w.) GHANA. Junior synonym of *gambiensis:* **new synonymy**.

*functa. Crematogaster montezumia* var. *functa* Forel, 1911c: 300 (w.) BRAZIL. Combination in *C. (Orthocrema)*: Emery, 1922e: 135. Junior synonym of *montezumia:* Kempf, 1968b: 386.

**#funerea.** *Crematogaster pronotalis* var. *funerea* Santschi, 1915c: 251 (w.) CAMEROUN. Combination in *C. (Sphaerocrema)*: Wheeler, W.M. 1922a: 848. Variety of *kneri:* Santschi, 1926b: 228.

**fusca.** *Crematogaster fusca* Mayr, 1876: 107 (w.) AUSTRALIA. Combination in *C. (Acrocoelia)*: Emery, 1922e: 152.

**#fuscipennis.** *Crematogaster depressa* var. *fuscipennis* Emery, 1899e: 479 (w.q.) CAMEROUN. Forel, 1909b: 53 (m.). Combination in *C. (Atopogyne)*: Wheeler, W.M. 1922a: 856.

**#fuscitatis.** *Crematogaster gabonensis* var. *fuscitatis* Forel, 1913h: 351 (w.) ZAIRE. Combination in *C. (Atopogyne)*: Wheeler, W.M. 1922a: 856; in *C. (Sphaerocrema)*: Emery, 1922e: 153.

**gabonensis.** *Crematogaster gabonensis* Emery, 1899e: 484 (w.) GABON. Santschi, 1911i: 281 (m.). Combination in *C. (Atopogyne)*: Wheeler, W.M. 1922a: 856; in *C. (Sphaerocrema)*: Emery, 1922e: 153; Santschi, 1926b: 224. Current subspecies: nominal plus *fuscitatis.*

*gallarum. Crematogaster solenopsides* st. *flavida* var. *gallarum* Santschi, 1910c: 370 (w.) CONGO. Unavailable name.

**gallicola.** *Crematogaster gallicola* Forel, 1894b: 95 (w.) MOZAMBIQUE. Arnold, 1920a: 532 (q.). Combination in *C. (Crematogaster)*: Wheeler, W.M. 1922a: 838; in *C. (Acrocoelia)*: Emery, 1922e: 147. Current subspecies: nominal plus *latro, rauana.*

**gambiensis.** *Crematogaster gambiensis* André, 1889: 228 (w.) GAMBIA. Forel, 1913b: 324 (q.). Combination in *C. (Sphaerocrema)*: Wheeler, W.M. 1922a: 846; Santschi, 1926b: 227. Senior synonym of *longiruga:* Santschi, 1914d: 346; of *fulva:* **new synonymy**. Current subspecies: nominal plus *krantziana, sejuncta, transversiruga.*

**gavapiga.** *Crematogaster (Orthocrema) gavapiga* Menozzi, 1935a: 109, fig. 5 (w.) JAVA.

**#gedeon.** *Crematogaster peringueyi* var. *gedeon* Arnold, 1920a: 512 (w.) SOUTH AFRICA. [First available use of *Crematogaster peringueyi* r. *cacodaemon* var. *gedeon* Forel, 1915c: 341; unavailable name.] Combination in *C. (Acrocoelia)*: Emery, 1922e: 148.

*geeri* Schulz, 1906; see under ***degeeri.***

**#generosa.** *Crematogaster acaciae* var. *generosa* Santschi, 1919a: 343 (w.q.) SOUTH AFRICA.

**#gerardi.** *Crematogaster bequaerti* var. *gerardi* Santschi, 1915c: 252 (w.) ZAIRE. Combination in *C. (Sphaerocrema)*: Wheeler, W.M. 1922a: 845.

**gerstaeckeri.** *Crematogaster gerstaeckeri* Dalla Torre, 1892: 90. Replacement name for *cephalotes* Gerstäcker, 1871: 356. [Junior primary homonym of *cephalotes* Smith, F. 1857a: 75.] Menozzi, 1930b: 92 (w.q.m.). Combination in *C. (Crematogaster)*: Wheeler, 1922a: 838; in *C. (Acrocoelia)*: Emery, 1922e: 147. Current subspecies: nominal plus *godefreyi, infaceta, inquieta, kohliella, maledicta, oraclum, pudica, pulla, rufescens, sjostedti, zulu.*

**gibba.** *Crematogaster gibba* Emery, 1894a: 70, fig. (w.) SEYCHELLE IS. Combination in *C. (Crematogaster)*: Wheeler, W.M. 1922a: 1022; in *C. (Acrocoelia)*: Emery, 1922e: 147.

*gigantea. Crematogaster sewellii* subsp. *lobata* var. *gigantea* Forel, 1907g: 80 (footnote) (w.) MADAGASCAR. Unavailable name.

**#gigas.** *Crematogaster dohrni* var. *gigas* Forel, 1913k: 74 (w.) SRI LANKA. Combination in *C. (Acrocoelia)*: Emery, 1922e: 150.

**#gilberti.** *Crematogaster (Orthocrema) queenslandica* var. *gilberti* Emery, 1922e: 133 (w.) AUSTRALIA. [First available use of *Crematogaster sordidula* subsp. *queenslandica* var. *gilberti* Forel, 1910b: 32; unavailable name (Taylor, 1986: 34).] Subspecies of *queenslandica:* Taylor & Brown, D.R. 1985: 62.

**#*gloriosa.*** *Crematogaster acaciae* var. *gloriosa* Santschi, 1914d: 346 (w.) ZAIRE. Combination in *C.* (*Acrocoelia*): Emery, 1922e: 144.

**#*godefreyi.*** *Crematogaster foraminiceps* r. *godefreyi* Forel, 1914d: 237 (w.) SOUTH AFRICA. Combination in *C.* (*Crematogaster*): Wheeler, W.M. 1922a: 839; in *C.* (*Acrocoelia*): Emery, 1922e: 148. Race/strips of *vulcania:* Forel, 1916: 406; of *sjostedti:* Santschi, 1918d: 184; Emery, 1922e: 148. Raised to species: Arnold, 1920a: 503; Wheeler, W.M. 1922a: 839. Stirps of *gerstaeckeri:* Santschi, 1930b: 69. Senior synonym of *carininotum:* Santschi, 1918d: 184.

**#*goeldii.*** *Crematogaster goeldii* Forel, 1903c: 255 (w.) BRAZIL. Gallardo, 1931b: 300 (q.). Combination in *C.* (*Orthocrema*): Emery, 1922e: 135. Variety of *torosa:* Santschi, 1925d: 229; Santschi, 1929d: 294; Gallardo, 1931b: 300.

**#*goliathula.*** *Crematogaster senegalensis* r. *goliathula* Forel, 1922: 93 (w.) ANGOLA.

*gordonensis. Crematogaster excisa* r. *andrei* var. *gordonensis* Forel, 1914d: 236 (w.) SOUTH AFRICA. Unavailable name.

**#*gracilenta.*** *Crematogaster* (*Oxygyne*) *stadelmanni* var. *gracilenta* Viehmeyer, 1922: 210 (w.q.) CAMEROUN.

**#*gracilior.*** *Crematogaster quadriformis* var. *gracilior* Forel, 1901g: 391 (w.) BRAZIL. Combination in *C.* (*Orthocrema*): Emery, 1922e: 135. Race of *quadriformis:* Forel, 1912f: 218.

**#*graeteri.*** *Crematogaster buchneri* var. *graeteri* Forel, 1916: 409 (w.) ZAIRE. Combination in *C.* (*Atopogyne*): Wheeler, W.M. 1922a: 854.

**#*granulata.*** *Crematogaster laestrygon* var. *granulata* Santschi, 1910a: 43 (w.) TUNISIA.

***gratiosa.*** *Crematogaster* (*Orthocrema*) *gratiosa* Santschi, 1926b: 210 (w.) ZAIRE.

***grevei.*** *Crematogaster grevei* Forel, 1891b: 183, pl. 6, fig. 8 (w.) MADAGASCAR. Combination in *C.* (*Decacrema*): Forel, 1910a: 18; Forel, 1910f: 9.

**#*grossulior.*** *Crematogaster pauciseta* r. *grossulior* Forel, 1916: 404 (w.) ZAIRE. Combination in *C.* (*Orthocrema*): Wheeler, W.M. 1922a: 850.

*grouvellei. Crematogaster scutellaris* var. *grouvellei* Bondroit, 1918: 115 (w.q.) FRANCE. Junior synonym of *scutellaris:* Bernard, 1967: 164.

**#*guianensis.*** *Crematogaster stolli* var. *guianensis* Crawley, 1916b: 369 (w.) GUIANA. Combination in *C.* (*Orthocrema*): Emery, 1922e: 136.

***gutenbergi.*** *Crematogaster gutenbergi* Santschi, 1914a: 433 (w.) NAMIBIA. Combination in *C.* (*Crematogaster*): Wheeler, W.M. 1922a: 839; in *C.* (*Acrocoelia*): Emery, 1922e: 147.

*haematocephala. Formica haematocephala* Leach, 1825: 291 (w.q.m.) FRANCE. Junior synonym of *scutellaris:* Roger, 1863b: 36.

*halli. Crematogaster* (*Atopogyne*) *halli* Donisthorpe, 1945d: 268 (w.) GHANA. Junior synonym of *buchneri:* **new synonymy**.

**#*hammi.*** *Crematogaster transvaalensis* var. *hammi* Arnold, 1920a: 534, fig. 48 (w.) ZIMBABWE. Combination in *C.* (*Orthocrema*): Wheeler, W.M. 1922a: 851.

**#*haputalensis.*** *Crematogaster rothneyi* var. *haputalensis* Forel, 1913k: 75 (q.) SRI LANKA. Combination in *C.* (*Acrocoelia*): Emery, 1922e: 152.

**#*hararica.*** *Crematogaster ferruginea* var. *hararica* Forel, 1894b: 97 (w.) ETHIOPIA. Combination in *C.* (*Acrocoelia*): Emery, 1922e: 145. Subspecies of *castanea:* current status.

**#*harpyia.*** *Crematogaster* (*Decacrema*) *borneensis* var. *harpyia* Forel, 1911a: 33 (w.q.) BORNEO.

*hayvepana. Crematogaster egidyi* subsp. *spinozae* var. *hayvepana* Forel, 1911a: 26 (w.q.) BORNEO. Unavailable name.

***heathi.*** *Crematogaster* (*Crematogaster*) *heathi* Mann, 1916: 442, pl. 3, fig. 27 (w.) BRAZIL. Combination in *C.* (*Orthocrema*): Emery, 1922e: 135.

**#*hellenica.*** *Crematogaster* (*Atopogyne*) *hellenica* Forel, 1911d: 342 (q.) GREECE. Combination in *C.* (*Acrocoelia*): Emery, 1922e: 143. Variety of *lorteti:* Forel, 1911g: 284.

*helveola. Crematogaster atkinsoni* var. *helveola* Wheeler, W.M. 1919g: 109 (w.q.m.) U.S.A. Junior synonym of *atkinsoni:* Creighton, 1950a: 207.

***hemiceros.*** *Crematogaster* (*Sphaerocrema*) *hemiceros* Santschi, 1926b: 221, fig. 2F (w.) ZIMBABWE.

***hespera.*** *Crematogaster* (*Crematogaster*) *hespera* Buren, 1968b: 98 (w.q.) U.S.A.

***hezaradjatica.*** *Crematogaster hezaradjaticus* Pisarski, 1967: 391, figs. 26–36 (w.m.) AFGHANISTAN.

***himalayana.*** *Crematogaster himalayanus* Forel, 1902c: 202 (w.) INDIA. Combination in *C.*(*Acrocoelia*): Emery, 1922e: 151.

***hogsoni.*** *Crematogaster hogsoni* Forel, 1902c: 204 (w.) BURMA. Combination in *C.* (*Acrocoelia*): Emery, 1922e: 151.

***homeri.*** *Crematogaster* (*Atopogyne*) *homeri* Forel, 1913e: 668 (q.) CAMEROUN. Combination in *C.* (*Acrocoelia*): Emery, 1922e: 147.

**#*homonyma.*** *Crematogaster* (*Acrocoelia*) *melanogaster* var. *homonyma* Emery, 1922e: 147 (w.q.) SOUTH AFRICA.

**#*horatii.*** *Crematogaster* (*Sphaerocrema*) *striatula* var. *horatii* Santschi, 1937a: 55 (w.) KENYA.

**#*hosei.*** *Crematogaster* (*Decacrema*) *borneensis* subsp. *hosei* Forel, 1911a: 35 (w.q.m.) BORNEO.

***hottentota.*** *Crematogaster kneri* subsp. *hottentota* Emery, 1899e: 482 (w.) SOUTH AFRICA. Combination in *C.* (*Sphaerocrema*): Wheeler, W.M. 1922a: 847. Raised to species: Santschi, 1925h: 161 (footnote); Santschi, 1926b: 224. Current subspecies: nominal plus *bassuto*.

***hova.*** *Crematogaster hova* Forel, 1887: 387 (w.) MADAGASCAR. Combination in *C.* (*Decacrema*): Forel, 1910f: 9. Current subspecies: nominal plus *latinoda, nossibeensis*.

***huberi.*** *Tranopelta huberi* Forel, 1907e: 5 (q.) SURINAM. Combination in *Tranopeltoides:* Wheeler, W.M. 1922e: 11; in *Crematogaster:* Kempf, 1960c: 173; in *C.* (*Orthocrema*): Kempf, 1972a: 87.

*hybrida. Crematogaster scutellaris* subsp. *schmidti* var. *hybrida* Emery, 1916b: 157 (w.) ITALY. Unavailable name (Agosti & Collingwood, 1987a: 61); material referred to *scutellaris* by Baroni Urbani, 1964a: 4.

*hyperphyes. Crematogaster hyperphyes* Kusnezov, 1953e: 215 (m.) BOLIVIA. Junior synonym of *stollii:* Kempf, 1968b: 384.

*iberica. Crematogaster auberti* var. *iberica* Forel, 1909c: 103 (w.) SPAIN. Menozzi, 1922b: 327 (m.). Junior synonym of *auberti:* Collingwood, 1978: 69.

***iheringi.*** *Crematogaster iheringi* Forel, 1908c: 366 (w.q.) BRAZIL. Combination in *C.* (*Orthocrema*): Emery, 1922e: 135. Current subspecies: nominal plus *bosqi, nemesis*.

***ilgii.*** *Crematogaster scrutans* var. *ilgii* Forel, 1910c: 259 (w.) ETHIOPIA. Combination in *C.* (*Crematogaster*): Wheeler, W.M. 1922a: 839. Raised to species: Santschi, 1916b: 502; Wheeler, W.M. 1922a: 839.

**#*imbellis.*** *Crematogaster bicolor* subsp. *imbellis* Emery, 1893g: 266 (w.q.) PHILIPPINES. Combination in *C.* (*Orthocrema*): Emery, 1922e: 131.

***impressa.*** *Crematogaster impressa* Emery, 1899e: 486 (w.q.) CAMEROUN. Forel, 1916: 408 (m.). Combination in *C.* (*Acrocoelia*): Emery, 1922e: 147. Race/subspecies of *buchneri:* Forel, 1916: 411; of *excisa:* Wheeler, W.M. 1922a: 153. Revived status as species: Emery, 1922e: 147; Santschi, 1935a: 257. Senior synonym of *andrei:* Santschi, 1937b: 98. Current subspecies: nominal plus *brazzai, maynei, sapora*.

***impressiceps.*** *Crematogaster impressiceps* Mayr, 1902: 294 (w.) EQUATORIAL GUINEA. Combination in *C.* (*Crematogaster*): Wheeler, W.M. 1922a: 839; in *C.* (*Acrocoelia*): Emery, 1922e: 147. Race of *buchneri:* Forel, 1916: 411. Revived status as species: Stitz, 1916: 384; Wheeler, W.M. 1922a: 839; Santschi, 1930a: 65. Current subspecies: nominal plus *frontalis, longiscapa, lujana*.

**#*improba.*** *Crematogaster sewelli* var. *improba* Forel, 1907g: 80 (footnote) (w.) MADAGASCAR. Combination in *C.* (*Acrocoelia*): Emery, 1922e: 148.

***inca.*** *Crematogaster* (*Eucrema*) *inca* Wheeler, W.M. 1925a: 27 (w.) PERU. Current subspecies: nominal plus *ascendens*.

***inconspicua.*** *Crematogaster inconspicua* Mayr, 1896: 246 (w.q.) CAMEROUN. Combination in *C.* (*Crematogaster*): Wheeler, W.M. 1922a: 840; in *C.* (*Acrocoelia*): Emery, 1922e: 147.

***incorrecta.*** *Crematogaster inconspicua* st. *incorrecta* Santschi, 1917b: 280 (w.) GABON. Combination in *C.* (*Acrocoelia*): Emery, 1922e: 147. Raised to species: Santschi, 1937g: 77.

***indefensa.*** *Crematogaster indefensa* Kempf, 1968b: 388, fig. 7 (w.q.) BOLIVIA. Combination in *C.* (*Neocrema*): Kempf, 1972a: 83.

***inermis.*** *Crematogaster inermis* Mayr, 1862: 766 (w.) EGYPT. André, 1883b: 395 (q.); Santschi, 1937e: 296 (q.); Santschi, 1938a: 38 (m.). Combination in *C.* (*Crematogaster*): Wheeler, W.M. 1922a: 840; in *C.*(*Acrocoelia*): Emery, 1922e: 143. Current subspecies: nominal plus *aphrodite, armatula, lucida*.

**#*infaceta.*** *Crematogaster bulawayensis* st. *infaceta* Santschi, 1916b: 498 (w.) ZIMBABWE. Combination in *C.* (*Atopogyne*): Wheeler, W.M. 1922a: 855; in *C.* (*Acrocoelia*): Emery, 1922e: 149. Subspecies of *bulawayensis:* Wheeler, W.M. 1922a: 855; of *sjostedti:* Santschi, 1918d: 184; Emery, 1922e: 149; of *gerstaeckeri:* current status.

***inflata.*** *Crematogaster inflatus* Smith, F. 1857a: 76, pl. 2, fig. 2 (w.) SINGAPORE. Santschi, 1928h: 128 (q.). Combination in *C.* (*Physocrema*): Forel, 1912f: 220.

**#*inglebyi.*** *Crematogaster* (*Oxygyne*) *aberrans* var. *inglebyi* Forel, 1902c: 201 (w.) INDIA.

*ingravis*. *Crematogaster neuvillei* r. *cooperi* var. *ingravis* Forel, 1915c: 341 (w.) SOUTH AFRICA. Unavailable name.

**#*ingvei*.** *Crematogaster egidyi* var. *ingvei* Forel, 1911g: 284 (w.) JAVA. Combination in *C.* (*Acrocoelia*): Emery, 1922e: 151.

***innocens*.** *Crematogaster innocens* Forel, 1911a: 30 (w.q.) BORNEO. Combination in *C.* (*Acrocoelia*): Emery, 1922e: 151.

***inops*.** *Crematogaster inops* Forel, 1892l: 254 (w.) MADAGASCAR. Combination in *C.* (*Oxygyne*): Wheeler, W.M. 1922a: 1024.

**#*inquieta*.** *Crematogaster* (*Acrocoelia*) *sjostedti* var. *inquieta* Santschi, 1928d: 65 (w.) ZIMBABWE. Subspecies of *gerstaeckeri:* current status.

**#*insidiosa*.** *Crematogaster castanea* st. *insidiosa* Santschi, 1920d: 378 (w.) NAMIBIA.

***insularis*.** *Crematogaster insularis* Smith, F. 1859a: 149 (w.) NEW GUINEA. Combination in *C.* (*Orthocrema*): Donisthorpe, 1932c: 456.

**#*insulsa*.** *Crematogaster* (*Decacrema*) *borneensis* var. *insulsa* Forel, 1911a: 33 (w.q.m.) BORNEO.

**#*intermedia*.** *Crematogaster stadelmanni* var. *intermedia* Mayr, 1896: 242 (w.q.) CAMEROUN. Combination in *C.* (*Nematocrema*): Wheeler, W.M. 1922a: 860.

**#*inversa*.** *Crematogaster tricolor* var. *inversa* Forel, 1907g: 81 (w.) KENYA. Emery, 1912c: 2 (q.m.); Santschi, 1930b: 69 (q.m.). Combination in *C.* (*Acrocoelia*): Emery, 1922e: 146. Stirps/subspecies of *tricolor:* Santschi, 1913a: 413 (in key); Santschi, 1914b: 86; of *castanea:* Wheeler, W.M. 1922a: 152; Santschi, 1930b: 69.

***ionia*.** *Crematogaster scutellaris* var. *ionia* Forel, 1911d: 340 (w.q.) TURKEY. Combination in *C.* (*Acrocoelia*): Emery, 1922e: 144. Raised to species: Agosti & Collingwood, 1987a: 54.

***iridipennis*.** *Crematogaster iridipennis* Smith, F. 1865: 74 (q.) NEW GUINEA. Combination in *C.* (*Orthocrema*): Donisthorpe, 1932c: 473.

***irritabilis*.** *Crematogaster irritabilis* Smith, F. 1860b: 109 (w.) NEW GUINEA. Combination in *C.* (*Orthocrema*): Emery, 1922e: 133; Donisthorpe, 1932c: 463. Current subspecies: nominal plus *leguilloui, subtilis.*

***isolata*.** *Crematogaster* (*Crematogaster*) *isolata* Buren, 1968b: 106 (w.) U.S.A.

**#*iwatensis*.** *Crematogaster* (*Acrocoelia*) *matsumurai* var. *iwatensis* Santschi, 1937h: 371 (w.) JAPAN. [First available use of *Crematogaster* (*Acrocoelia*) *brunnea* st. *matsumurai* var. *iwatensis* Santschi, 1930d: 264; unavailable name.]

***jacobsoni*.** *Crematogaster jacobsoni* Forel, 1911b: 201 (w.) JAVA. Combination in *C.* (*Orthocrema*): Emery, 1922e: 132; in *C.* (*Mesocrema*): Santschi, 1937e: 310.

*japonica*. *Crematogaster sordidula* var. *japonica* Forel, 1912l: 339 (w.) JAPAN. Junior synonym of *osakensis:* Brown, 1949e: 37.

***javanica*.** *Crematogaster* (*Orthocrema*) *javanica* Menozzi, 1935a: 108, fig. 4 (w.) JAVA.

***jeanneli*.** *Crematogaster jeanneli* Santschi, 1914b: 100 (w.) KENYA. Combination in *C.* (*Orthocrema*): Wheeler, W.M. 1922a: 849.

***jehovae*.** *Crematogaster auberti* subsp. *jehovae* Forel, 1907b: 207 (w.) ISRAEL. Menozzi, 1933b: 58 (q.). Combination in *C.* (*Acrocoelia*): Emery, 1922e: 142; Santschi, 1930d: 266. Raised to species: Emery, 1926: 3; Menozzi, 1933b: 58. Current subspecies: nominal plus *crawleyi, cypria, mosis.*

***jullieni*.** *Crematogaster jullieni* Santschi, 1910c: 378, fig. 8 (w.q.) GABON. Combination in *C.* (*Atopogyne*): Forel, 1911d: 343; Santschi, 1916b: 502.

**#*junodi*.** *Crematogaster censor* r. *junodi* Forel, 1916: 406 (w.) SOUTH AFRICA. Combination in *C.* (*Acrocoelia*): Emery, 1922e: 146.

***juventa*.** *Crematogaster* (*Sphaerocrema*) *juventa* Santschi, 1926b: 219 (w.) ZAIRE.

***kachelibae*.** *Crematogaster kachelibae* Arnold, 1954: 294, fig. 5 (w.) KENYA.

*karawaewi*. *Crematogaster auberti* subsp. *karawaewi* Ruzsky, 1905b: 497 (w.) UKRAINE. Combination in *C.* (*Acrocoelia*): Emery, 1922e: 142. Subspecies of *inermis:* Emery, 1926: 3; of *auberti:* Arnol'di & Dlussky, 1978: 538. Junior synonym of *schmidti:* Atanassov & Dlussky, 1992: 184.

***karawaiewi*.** *Crematogaster* (*Orthocrema*) *karawaiewi* Menozzi, 1935a: 106, fig. 3 (w.) NEW GUINEA. Combination in *C.* (*Mesocrema*): Santschi, 1937e: 310. Senior synonym of *wladimiri:* unnecessary replacement name.

***kasaiensis*.** *Crematogaster kasaiensis* Forel, 1913b: 321 (w.q.) ZAIRE. Combination in *C.* (*Atopogyne*): Wheeler, W.M. 1922a: 856. Race of *buchneri:* Forel, 1916: 411. Revived status as species: Wheeler, W.M. 1922a: 856.

**#*kaschgariensis*.** *Crematogaster inermis* r. *kaschgariensis* Forel, 1901h: 63 (w.) TURKESTAN. Combination in *C.* (*Acrocoelia*): Emery, 1922e: 144. Variety of *subdentata:* Forel, 1904b: 372.

***kelleri*.** *Crematogaster kelleri* Forel, 1891b: 197, pl. 6, fig. 10 (w.) MADAGASCAR. Combination in *C.* (*Crematogaster*): Wheeler, W.M. 1922a: 1022; in *C.* (*Acrocoelia*): Emery, 1922e: 147.

**#*kemali*.** *Crematogaster* (*Orthocrema*) *curvispinosa* var. *kemali* Santschi, 1923c: 250 (w.) BRAZIL.

*kennedyi*. *Crematogaster* (*Acrocoelia*) *kennedyi* Wheeler, W.M. 1930e: 58 (q.m.) U.S.A. Junior synonym of *cerasi:* Buren, in Smith, M.R. 1958c: 126.

**#*kerri*.** *Crematogaster rogenhoferi* subsp. *kerri* Forel, 1911g: 284 (w.) THAILAND. Combination in *C.* (*Acrocoelia*), and subspecies of *dohrni:* Emery, 1922e: 150.

**#*kiangsiensis*.** *Crematogaster dohrni* r. *kiangsiensis* Forel, 1903c: 254 (w.) CHINA.

***kirbii*.** *Myrmica kirbii* Sykes, 1835: 99, pl. 13, figs. 1–4 (w.q.m.) INDIA. Combination in *Crematogaster:* Smith, F. 1858b: 136; in *C.* (*Acrocoelia*): Emery, 1922e: 151.

**#*kirbyella*.** *Crematogaster* (*Acrocoelia*) *constructor* var. *kirbyella* Emery, 1922e: 146. Replacement name for *kirbyi* Mayr, below. [Junior secondary homonym of *Myrmica kirbii* Sykes, above.]

*kirbyi*. *Crematogaster constructor* var. *kirbyi* Mayr, 1895: 137, fig. 3 (w.) SOUTH AFRICA. [Junior secondary homonym of *kirbii* Sykes, 1835: 99.] Replacement name: *kirbyella* Emery, above.

*kloofensis*. *Crematogaster coelestis* var. *kloofensis* Forel, 1914d: 237 (w.) SOUTH AFRICA. Junior synonym of *weitzeckeri:* Arnold, 1920a: 549.

***kneri*.** *Crematogaster kneri* Mayr, 1862: 764 (w.) GHANA. Forel, 1913a: 125 (q.). Combination in *C.* (*Sphaerocrema*): Santschi, 1918d: 182. See also: Arnold, 1920a: 527. Current subspecies: nominal plus *behanzini, dakarensis, funerea, pronotalis.*

***kohli*.** *Crematogaster kohli* Forel, 1909b: 70 (w.) ZAIRE. Forel, 1915c: 340 (q.). Combination in *C.* (*Acrocoelia*): Emery, 1922e: 147; in *C.* (*Atopogyne*): Wheeler, W.M. 1922a: 856; Santschi, 1937g: 80. Race of *buchneri:* Forel, 1916: 411. Revived status as species: Wheeler, W.M. 1922a: 856; Santschi, 1937g: 79. Current subspecies: nominal plus *winkleri.*

**#*kohliella*.** *Crematogaster sjostedti* st. *kohliella* Santschi, 1918d: 185 (w.) ZAIRE. [First available use of *Crematogaster excisa* r. *andrei* var. *kohliella* Forel, 1916: 409; unavailable name.] Combination in *C.* (*Acrocoelia*): Emery, 1922e: 149. Subspecies of *gerstaeckeri:* current status.

*koratii*. *Crematogaster striatula* var. *koratii* Menozzi, 1942: 169. *Nomen nudum;* attributed to Santschi.

*kosti*. *Crematogaster sordidula* var. *kosti* Ruzsky, 1905b: 506 CAUCASUS. Combination in *C.* (*Orthocrema*): Emery, 1922e: 131. Junior synonym of *sordidula:* Atanassov & Dlussky, 1992: 182; of *bogojawlenskii:* Arakelian, 1994: 44.

**#*krantziana*.** *Crematogaster gambiensis* var. *krantziana* Forel, 1914d: 234 (w.m.) SOUTH AFRICA. Arnold, 1920a: 530 (q.). Combination in *C.* (*Sphaerocrema*): Wheeler, W.M. 1922a: 846.

*kusnezovi*. *Crematogaster sordidula* subsp. *mayri* var. *kusnezovi* Emery, 1926: 8 (w.m.) TURKESTAN. Unavailable name.

***kutteri*.** *Crematogaster kutteri* Viehmeyer, 1924b: 314 (w.) AUSTRALIA.

**#*kuznetzovi*.** *Crematogaster subdentata* subsp. *kuznetzovi* Stärcke, 1935: 263 (footnote), fig. 2 (w.) TURKESTAN.

*laboriosus* Smith, F. 1860; see under ***SOLENOPSIS*.**

***laboriosa*.** *Crematogaster laboriosa* Smith, F. 1874: 407 (w.) JAPAN. [Unresolved junior primary homonym of *laboriosus* Smith, F. 1860b: 109, above.] Wheeler, W.M. 1928d: 109 (m.); Imai, 1966: 119 (k.). Combination in *C.* (*Acrocoelia*): Emery, 1922e: 143. Current subspecies: nominal plus *nawai, zoceensis.*

**#*lacustris*.** *Crematogaster excisa* st. *lacustris* Santschi, 1914b: 99 (w.) KENYA. Combination in *C.* (*Acrocoelia*): Emery, 1922e: 147.

***laestrygon*.** *Crematogaster laestrygon* Emery, 1869a: 135 (w.) SICILY. Baroni Urbani, 1964b: 43 (q.). Combination in *C.* (*Acrocoelia*): Emery, 1922e: 142. Variety/subspecies of *scutellaris:* Emery & Forel, 1879: 464; André, 1883b: 393; of *schmidti:* Emery, 1891b: 15; of *auberti:* Forel, 1902a: 154; Ruzsky, 1902d: 31; Stitz, 1917: 341; Santschi, 1921d: 70; Emery, 1922e: 142; Santschi, 1934f: 167. Status as species: Forel, 1909c: 104; Santschi, 1915a: 59; Emery, 1926: 2; Wheeler, W.M. 1927g: 106; Santschi, 1936c: 201. Senior synonym of *vaucheri:* Collingwood, 1978: 69. Current subspecies: nominal plus *airensis, atlantis, canariensis, diminuta, granulata, maura, striaticeps, submaura, theryi.*

***laeviceps*.** *Crematogaster laeviceps* Smith, F. 1858b: 138 (w.) AUSTRALIA. Combination in *C.* (*Acrocoelia*): Emery, 1922e: 152. [Misspelled as *leviceps* by Dalla Torre, 1893: 83, and others.] Current subspecies: nominal plus *broomensis, chasei, clarior.*

***CREMATOGASTER** (continued)*

#***laevinota.*** *Crematogaster sagei* var. *laevinota* Forel, 1902c: 205 (w.) INDIA. Combination in *C. (Acrocoelia)*: Emery, 1922e: 152.

*laevior. Crematogaster africana* subsp. *laevior* Mayr, 1895: 136. *Nomen nudum.*

***laevis.*** *Crematogaster laevis* Mayr, 1878: 876 (w.) BRAZIL. Combination in *C. (Orthocrema)*: Emery, 1922e: 135. Current subspecies: nominal plus *bruesi.*

***laevissima.*** *Crematogaster laevissimus* Smith, F. 1860a: 74 (q.) SULAWESI. Combination in *C. (Acrocoelia)*: Donisthorpe, 1932c: 458.

***laeviuscula.*** *Crematogaster laeviuscula* Mayr, 1870b: 993 (w.) U.S.A. Wheeler, W.M. 1908e: 480 (q.); Petralia & Vinson, 1980: 383 (l.). Combination in *C. (Acrocoelia)*: Emery, 1922e: 141; in *C. (Crematogaster)*: Buren, 1968b: 92. Subspecies of *lineolata:* Emery, 1895c: 284; Wheeler, W.M. 1904e: 301; Emery, 1922e: 141. Status as species: Wheeler, W.M. 1919g: 111; Creighton, 1950a: 210. Senior synonym of *clara:* Creighton, 1950a: 210; Johnson, 1988: 322; of *arborea:* Smith, M.R. 1951a: 809 (provisional); Buren, in Smith, M.R. 1958c: 126.

***lamottei.*** *Crematogaster (Sphaerocrema) lamottei* Bernard, 1953b: 232, fig. 9 (w.q.m.) GUINEA.

#***langi.*** *Crematogaster (Sphaerocrema) kneri* st. *langi* Santschi, 1926b: 225 (w.) ZAIRE. Subspecies of *striatula:* Menozzi, 1942: 169.

***lango.*** *Crematogaster (Decacrema) lango* Weber, 1943c: 356, pl. 15, figs. 6, 7 (w.) SUDAN.

***larreae.*** *Crematogaster (Crematogaster) larreae* Buren, 1968b: 117 (w.q.m.) U.S.A.

#***laticeps.*** *Crematogaster emmae* var. *laticeps* Forel, 1892k: 529 (q.), :534 (w.), :535 (m.) (diagnoses in keys) MADAGASCAR. Combination in *C. (Oxygyne)*: Emery, 1922e: 156.

#***latinoda.*** *Crematogaster hova* var. *latinoda* Forel, 1892k: 535 (w.) MADAGASCAR. Combination in *C. (Decacrema)*: Wheeler, W.M. 1922a: 1026.

#***latipetiolata.*** *Crematogaster (Acrocoelia) brunnea* subsp. *latipetiolata* Karavaiev, 1935a: 91, fig. 17 (w.) VIETNAM.

#***latro.*** *Crematogaster gallicola* subsp. *latro* Forel, 1910f: 6 (w.) BOTSWANA. Combination in *C. (Acrocoelia)*: Emery, 1922e: 147.

***latuka.*** *Crematogaster (Crematogaster) latuka* Weber, 1943c: 355, pl. 15, fig. 11 (w.) SUDAN.

***laurenti.*** *Crematogaster africana* subsp. *laurenti* Forel, 1909b: 69 (w.) ZAIRE. Combination in *C. (Atopogyne)*: Wheeler, W.M. 1922a: 852. Race of *buchneri:* Forel, 1913a: 125; Forel, 1913c: 352; Forel, 1916: 410; of *africana:* Wheeler, W.M. 1922a: 852. Raised to species: Emery, 1922e: 155; Santschi, 1935a: 262. Current subspecies: nominal plus *zeta.*

***ledouxi.*** *Crematogaster (Acrocoelia) ledouxi* Soulié, in Fromantin & Soulié, 1961: 102, figs. 5, 6, 8 (w.) CAMBODIA.

*lefevrei. Crematogaster (Acrocoelia) brunnea* subsp. *ruginota* var. *lefevrei* Wheeler, W.M. 1929f: 6 (w.) CHINA. Unavailable name.

#***leguilloui.*** *Crematogaster leguilloui* Emery, 1900c: 331 (footnote) (w.) NEW GUINEA. Combination in *C. (Orthocrema)*: Emery, 1922e: 133. Variety of *irritabilis:* Forel, 1911e: 270.

*levior. Crematogaster limata* subsp. *parabiotica* var. *levior* Forel, 1911e: 274 (w.) BRAZIL. Unavailable name.

#***levithorax.*** *Crematogaster auberti* var. *levithorax* Forel, 1902g: 462 (w.q.) ALGERIA. Santschi, 1937e: 301 (m.). Combination in *C. (Acrocoelia)*: Emery, 1922e: 142. Subspecies of *inermis:* Emery, 1926: 2; of *auberti:* Finzi, 1930b: 15; Santschi, 1937e: 301.

***libengensis.*** *Crematogaster libengensis* Stitz, 1916: 389, fig. 7 (w.) ZAIRE. Combination in *C. (Sphaerocrema)*: Wheeler, W.M. 1922a: 847; Santschi, 1926b: 227. Current subspecies: nominal plus *rufula.*

*lichtensteini. Crtematogaster scutellaris* var. *lichtensteini* Bondroit, 1918: 114 (w.) FRANCE. Junior synonym of *scutellaris:* Bernard, 1967: 164.

#***liebknechti.*** *Crematogaster pronotalis* var. *liebknechti* Forel, 1915c: 340 (w.) ZAIRE. Combination in *C. (Sphaerocrema)*: Wheeler, W.M. 1922a: 848. Variety of *rugosior:* Santschi, 1926b: 223; of *luctans:* Menozzi, 1942: 169.

***liengmei.*** *Crematogaster gallicola* var. *liengmei* Forel, 1894b: 96 (w.) MOZAMBIQUE. Combination in *C. (Decacrema)*: Forel, 1910a: 18; Forel, 1910f: 9. Raised to species: Mayr, 1895: 135; Arnold, 1920a: 548; Wheeler, W.M. 1922a: 860. Current subspecies: nominal plus *caculata, weitzeckeri.*

***limata.*** *Crematogaster limatus* Smith, F. 1858b: 139 (w.q.m.) BRAZIL. Combination in *C. (Orthocrema)*: Wheeler, W.M. 1921f: 151. Current subspecies: nominal plus *dextella, palans, parabiotica.*

***lineolata.*** *Myrmica lineolata* Say, 1836: 290 (w.q.m.) U.S.A. Wheeler, G.C. & Wheeler, J. 1952b: 250 (l.). Combination in *Crematogaster:* Roger, 1863b: 37; in *C. (Acrocoelia)*: Emery, 1922e: 141; in *C. (Crematogaster)*: Buren, 1968b: 93. Senior synonym of *columbiana, marylandica* and *novaeboracensis:* Mayr, 1886d: 462; of *lutescens:* Creighton, 1950a: 213; of *subopaca:* Johnson, 1988: 320. Material of the unavailable name *wheldeni* referred here: Brown, 1949a: 48.

***litoralis.*** *Crematogaster (Orthocrema) litoralis* Arnold, 1955: 733, fig. 1 (w.) KENYA.

#***livingstonei.*** *Crematogaster (Orthocrema) muralti* st. *livingstonei* Santschi, 1919a: 344 (w.) ZAMBIA.

***lobata.*** *Crematogaster sevellei* var. *lobata* Emery, 1895f: 342 (w.) MADAGASCAR. Combination in *C. (Acrocoelia)*: Emery, 1922e: 148. Subspecies of *sewellii:* Forel, 1907g: 80 (footnote). Raised to species: Santschi, 1930a: 63. Current subspecies: nominal plus *pacifica.*

***longiceps.*** *Crematogaster longiceps* Santschi, 1910c: 376 (q.) SUDAN.

***longiceps.*** *Crematogaster longiceps* Forel, 1910b: 32 (w.q.) AUSTRALIA. [Unresolved junior primary homonym of *longiceps* Santschi, above.] Combination in *C. (Acrocoelia)*: Emery, 1922e: 152. Current subspecies: nominal plus *curticeps.*

***longiclava.*** *Crematogaster longiclava* Emery, 1893g: 264, pl. 6, figs. 2, 3 (w.) PHILIPPINES. Combination in *C. (Acrocoelia)*: Emery, 1922e: 151. Current subspecies: nominal plus *placens.*

***longipilosa.*** *Crematogaster longipilosa* Forel, 1907a: 24 (w.) WEST MALAYSIA. Forel, 1911d: 383 (q.m.); Viehmeyer, 1916a: 124 (q.). Combination in *C. (Orthocrema)*: Emery, 1922e: 132.

*longiruga. Crematogaster gambiensis* var. *longiruga* Forel, 1907c: 140 (w.q.m.) KENYA. Junior synonym of *gambiensis:* Santschi, 1914d: 346.

#***longiscapa.*** *Crematogaster impressiceps* var. *longiscapa* Stitz, 1916: 384 (w.) ZAIRE. Combination in *C. (Acrocoelia)*: Emery, 1922e: 147.

***longispina.*** *Crematogaster longispina* Emery, 1890b: 53 (w.) COSTA RICA. [Also described as new by Emery, 1894k: 57.] Combination in *C. (Eucrema)*: Santschi, 1918d: 182 [misspelled here as *longissima*]; in *C. (Orthocrema)*: Emery, 1922e: 136; Santschi, 1923c: 250. Current subspecies: nominal plus *boliviana, naumannae, tenuicula.*

***lorteti.*** *Crematogaster lorteti* Forel, 1910e: 435 (w.q.) SYRIA. Combination in *C. (Acrocoelia)*: Emery, 1922e: 143. Current subspecies: *hellenica.*

***lotti.*** *Crematogaster (Sphaerocrema) lotti* Weber, 1943c: 358, pl. 15, fig. 2 (w.) SUDAN.

#***loveridgei.*** *Crematogaster (Acrocoelia) arnoldi* var. *loveridgei* Santschi, 1926b: 214 (w.) TANZANIA.

***lucayana.*** *Crematogaster lucayana* Wheeler, W.M. 1905b: 94, figs. G, H (w.q.) BAHAMAS. Combination in *C. (Acrocoelia)*: Emery, 1922e: 141. Subspecies of *sanguinea:* Wheeler, W.M. 1913b: 490 (in text). Revived status as species: Emery, 1922e: 141. Current subspecies: nominal plus *etiolata.*

#***luciae.*** *Crematogaster (Atopogyne) luciae* Forel, 1913b: 325 (q.) CAMEROUN. Combination in *C. (Sphaerocrema)*: Emery, 1922e: 153. Variety of *wellmani:* Santschi, 1916b: 500; Santschi, 1926b: 218.

#***lucida.*** *Crematogaster inermis* var. *lucidus* Forel, 1890c: 220 (w.) TUNISIA. Forel, 1904c: 6 (m.); Santschi, 1937e: 297 (q.). Combination in *C. (Acrocoelia)*: Emery, 1922e: 143. Variety of *wellmani:* Santschi, 1937b: 100; of *inermis:* Santschi, 1937e: 297; Santschi, 1938a: 38; Bernard, 1948: 143.

***luctans.*** *Crematogaster luctans* Forel, 1907a: 22 (w.q.) KENYA. Combination in *C. (Sphaerocrema)*: Santschi, 1918d: 182. Current subspecies: nominal plus *liebknechti, sordidoides.*

#***luctuosa.*** *Crematogaster (Acrocoelia) rivai* var. *luctuosa* Menozzi, 1930b: 96, figs. 2, 3 (w.l.) SOMALIA.

#***ludia.*** *Crematogaster bequaerti* r. *ludia* Forel, 1913b: 321 (w.) ZAIRE. Combination in *C. (Sphaerocrema)*: Wheeler, W.M. 1922a: 845.

#***ludio.*** *Crematogaster limata* r. *ludio* Forel, 1912f: 217 (w.) BRAZIL. Wheeler, W.M. 1921f: 151 (q.). Combination in *C. (Orthocrema)*: Emery, 1922e: 136. Variety of *brasiliensis:* Forel, 1913l: 233.

*luisae. Crematogaster (Acrocoelia) scutellaris* subsp. *schmidti* var. *luisae* Kutter, 1928: 66 (w.) LIBYA. Unavailable name.

#***lujae.*** *Crematogaster (Oxygyne) margaritae* r. *lujae* Forel, 1913b: 325 (q.) ZAIRE.

#***lujana.*** *Crematogaster impressiceps* var. *lujana* Forel, 1915c: 340 (w.) ZAIRE. Combination in *C. (Acrocoelia)*: Emery, 1922e: 147.

#***lumpurensis.*** *Crematogaster ferrarii* subsp. *lumpurensis* Forel, 1907a: 24 (w.) WEST MALAYSIA. Combination in *C. (Acrocoelia)*: Emery, 1922e: 151.

#***lunaris.*** *Crematogaster (Acrocoelia) degeeri* var. *lunaris* Santschi, 1928d: 64 (w.) MADAGASCAR.

*lusitanica*. *Crematogaster inermis* subsp. *auberti* var. *lusitanica* Emery, 1926: 4 (w.) PORTUGAL. Unavailable name.

**#lutea.** *Crematogaster rogenhoferi* var. *lutea* Emery, 1893e: 193 (w.) SUMATRA. Subspecies of *rogenhoferi:* Emery, 1900d: 688; of *artifex:* Forel, 1907a: 23; of *rogenhoferi:* current status.

*lutescens*. *Crematogaster lineolata* var. *lutescens* Emery, 1895c: 282 (w.) U.S.A. Junior synonym of *lineolata:* Creighton, 1950a: 213.

**lutzi.** *Crematogaster lutzi* Forel, 1905b: 167 (w.q.) BRAZIL. Forel, 1911c: 301 (m.). Combination in *C.* (*Orthocrema*): Emery, 1922e: 136. Current subspecies: nominal plus *florida*.

*macaoensis*. *Crematogaster* (*Acrocoelia*) *dohrni* subsp. *artifex* var. *macaoensis* Wheeler, W.M. 1928c: 16 (w.) CHINA. Unavailable name.

**#macarangae.** *Crematogaster* (*Decacrema*) *borneensis* subsp. *macarangae* Viehmeyer, 1916a: 126 (w.) SINGAPORE.

**macracantha.** *Crematogaster* (*Rhachiocrema*) *macracantha* Creighton, 1945: 114, pl. 12, figs. 4, 5 (w.) BORNEO.

**madagascariensis.** *Crematogaster madagascariensis* André, 1887: 297 (w.q.m.) MADAGASCAR. Combination in *C.* (*Crematogaster*): Wheeler, W.M. 1922a: 1023; in *C.* (*Acrocoelia*): Emery, 1922e: 147.

**madecassa.** *Crematogaster sordidula* var. *madecassa* Emery, 1895f: 342 (w.q.) MADAGASCAR. Combination in *C.* (*Orthocrema*): Wheeler, W.M. 1922a: 1024. Variety of *sordidula:* Wheeler, W.M. 1922a: 1024. Raised to species: Emery, 1912e: 668; Emery, 1922e: 131.

**magitae.** *Crematogaster* (*Oxygyne*) *magitae* Forel, 1910f: 9 (q.) WEST AFRICA. Combination in *C.* (*Nematocrema*): Emery, 1922e: 157.

**magnifica.** *Crematogaster* (*Neocrema*) *magnifica* Santschi, 1925d: 234 (w.m.) BRAZIL. Senior synonym of *nociva:* Kempf, 1968b: 385.

**major.** *Crematogaster* (*Orthocrema*) *major* Donisthorpe, 1941b: 39 (w.) NEW GUINEA. Current subspecies: nominal plus *flavior.*

**#makololo.** *Crematogaster* (*Sphaerocrema*) *amita* var. *makololo* Santschi, 1926b: 226 (w.) ZIMBABWE.

**#maledicta.** *Crematogaster excisa* var. *maledicta* Forel, 1914d: 236 (w.q.) ZIMBABWE. Combination in *C.* (*Acrocoelia*): Emery, 1922e: 147. Variety of *sjostedti:* Santschi, 1920b: 15. Subspecies of *gerstaeckeri:* current status.

**#malevolens.** *Crematogaster* (*Orthocrema*) *brevispinosa* st. *malevolens* Santschi, 1919f: 41 (w.) ARGENTINA. Gallardo, 1931b: 298 (q.m.).

**#mancocapaci.** *Crematogaster brevispinosa* [misspelled here as *brevispina*] st. *mancocapaci* Santschi, 1911i: 280 (w.q.m.) ECUADOR. [Also described as new by Santschi, 1913h: 38.] Combination in *C.* (*Orthocrema*): Emery, 1922e: 134.

*mandibularis*. *Crematogaster mandibularis* André, 1889: 229 (q.) SIERRA LEONE. Junior synonym of *platygnatha:* André, 1890: 323.

**#mandonbii.** *Crematogaster* (*Decacrema*) *solenopsides* st. *mandonbii* Santschi, 1926b: 229 (w.) UGANDA.

**manni.** *Crematogaster* (*Crematogaster*) *manni* Buren, 1968b: 95 (w.) CUBA.

**margaritae.** *Crematogaster margaritae* Emery, 1895h: 31 (q.) ZAIRE. Mayr, 1896: 243 (w.m.). Combination in *C.* (*Oxygyne*): Santschi, 1910c: 381. Current subspecies: nominal plus *brevarmata, cupida, lujae.*

**marioni.** *Crematogaster* (*Crematogaster*) *marioni* Buren, 1968b: 105 (w.) U.S.A.

**#marnoi.** *Crematogaster sewellei* var. *marnoi* Mayr, 1895: 138 (w.) SUDAN. Combination in *C.* (*Acrocoelia*): Emery, 1922e: 148. Subspecies of *sewellii:* Forel, 1907g: 79; Santschi, 1910c: 376.

**#marocana.** *Crematogaster* (*Orthocrema*) *sordidula* st. *marocana* Santschi, 1921d: 70 (w.) MOROCCO.

**marthae.** *Crematogaster marthae* Forel, 1892k: 529 (q.), :534 (w.) (diagnoses in keys) MADAGASCAR. Combination in *C.* (*Oxygyne*): Wheeler, W.M. 1922a: 1025.

*marylandica*. *Myrmica* (*Monomorium*) *marylandica* Buckley, 1867: 339 (w.) U.S.A. Junior synonym of *lineolata:* Mayr, 1886d: 462.

**#matabele.** *Crematogaster kneri* var. *matabele* Arnold, 1920a: 528, fig. 46 (w.) ZIMBABWE. Combination in *C.* (*Sphaerocrema*): Wheeler, W.M. 1922a: 847. Stirps of *amita:* Santschi, 1925h: 161.

**matsumurai.** *Crematogaster laboriosa* var. *matsumurai* Forel, 1901e: 372 (w.) JAPAN. Wheeler, W.M. 1906c: 312 (q.). Combination in *C.* (*Acrocoelia*): Emery, 1922e: 143. Subspecies of *brunnea:* Wheeler, W.M. 1929f: 5; Wheeler, W.M. 1930h: 65. Raised to species: Wheeler, W.M. 1928d: 110; Santschi, 1937h: 371; Collingwood, 1976: 303. Current subspecies: nominal plus *iwatensis, vagula.*

*matura*. *Crematogaster* (*Acrocoelia*) *ashmeadi* var. *matura* Wheeler, W.M. 1932a: 8 (w.) U.S.A. Junior synonym of *ashmeadi:* Creighton, 1950a: 206.

**#maura.** *Crematogaster laestrygon* var. *maura* Forel, 1909c: 104 (w.) ALGERIA. [First available use of *Crematogaster auberti* r. *laestrygon* var. *maura* Forel, 1894d: 25; unavailable name.] Subspecies of *laestrygon:* Emery, 1926: 2; Cagniant, 1964: 103.

**#mauritiana.** *Crematogaster sewelli* subsp. *mauritiana* Forel, 1907g: 79 (w.) MAURITIUS. Combination in *C.* (*Acrocoelia*): Emery, 1922e: 148.

**#maya.** *Crematogaster* (*Orthocrema*) *sumichrasti* subsp. *maya* Wheeler, W.M. 1936c: 48, fig. 1 (w.) GUATEMALA.

**#maynei.** *Crematogaster impressa* r. *maynei* Forel, 1913h: 352 (w.) ZAIRE. Combination in *C.* (*Acrocoelia*): Emery, 1922e: 147. Subspecies of *excisa:* Wheeler, W.M. 1922a: 838; of *impressa:* Emery, 1922e: 147.

*mayri*. *Acrocoelia mayri* Mayr, 1853b: 114 (w.) (attributed to Schmidt) YUGOSLAVIA. Müller, 1923: 76 (q.). Combination in *Crematogaster:* Mayr, 1855: 472. Junior synonym of *sordidula:* Mayr, 1855: 472; Dalla Torre, 1893: 86; Emery, 1922e: 130. Revived from synonymy as subspecies of *sordidula:* Müller, 1923: 74; Finzi, 1927a: 10. Junior synonym of *sordidula:* Agosti & Collingwood, 1987a: 54.

**#mediorufa.** *Crematogaster tricolor* var. *mediorufa* Forel, 1907g: 81 (footnote) (w.m.) SOUTH AFRICA. Forel, 1913a: 124 (q.); Arnold, 1920a: 493 (w.q.m.). Subspecies of *castanea:* current status.

*medispina*. *Crematogaster scutellaris* var. *medispina* Forel, 1905b: 178 (w.q.) ITALY. Junior synonym of *schmidti:* Emery, 1922e: 143; Müller, 1923: 72.

**meijerei.** *Crematogaster meijerei* Emery, 1911b: 534, fig. 2 (w.) NEW GUINEA. Combination in *C.* (*Acrocoelia*): Emery, 1922e: 152.

*melancholica*. *Crematogaster* (*Acrocoelia*) *auberti* st. *laevithorax* var. *melancholica* Santschi, 1921e: 168 (w.) MOROCCO. Unavailable name.

**melanogaster.** *Crematogaster arborea* subsp. *melanogaster* Emery, 1895h: 29 (w.q.) SOUTH AFRICA. Combination in *C.* (*Crematogaster*): Wheeler, W.M. 1922a: 840; in *C.* (*Acrocoelia*): Emery, 1922e: 147. Raised to species: Wheeler, W.M. 1922a: 840. Current subspecies: nominal plus *homonyma.*

**menilekii.** *Crematogaster menilekii* Forel, 1894b: 97 (w.) ETHIOPIA. Combination in *C.* (*Crematogaster*): Wheeler, W.M. 1922a: 840; in *C.* (*Acrocoelia*): Emery, 1922e: 147. Race of *buchneri:* Forel, 1916: 411. Revived status as species: Santschi, 1919h: 87; Wheeler, W.M. 1922a: 154; Santschi, 1928d: 63. Current subspecies: nominal plus *completa, occidentalis, proserpina, satan, spuria, suddensis, viehmeyeri.*

**#mercatori.** *Crematogaster* (*Acrocoelia*) *coelestis* st. *mercatori* Santschi, 1939f: 162 (w.) GUINEA.

**#merwei.** *Crematogaster* (*Acrocoelia*) *delagoensis* st. *merwei* Santschi, 1932a: 385 (w.) SOUTH AFRICA.

**mesonotalis.** *Crematogaster mesonotalis* Emery, 1911b: 535, fig. 3 (w.) NEW GUINEA. Combination in *C.* (*Orthocrema*): Emery, 1922e: 133.

**microspina.** *Crematogaster* (*Sphaerocrema*) *microspina* Menozzi, 1942: 167 (w.) EQUATORIAL GUINEA.

**millardi.** *Crematogaster millardi* Forel, 1902c: 205 (w.) BURMA. Viehmeyer, 1916a: 125 (q.). Combination in *C.* (*Orthocrema*): Emery, 1922e: 132; in *C.* (*Mesocrema*): Santschi, 1928b: 33.

**mimicans.** *Crematogaster* (*Orthocrema*) *mimicans* Donisthorpe, 1932c: 475 (w.) NEW GUINEA.

**mimosae.** *Crematogaster mimosae* Santschi, 1914b: 87, fig. 11 (w.) KENYA. Menozzi, 1939c: 105 (q.). Combination in *C.* (*Crematogaster*): Wheeler, W.M. 1922a: 841; in *C.* (*Acrocoelia*): Emery, 1922e: 148. Current subspecies: nominal plus *tenuipilis.*

**#minutior.** *Crematogaster brevispinosa* var. *minutior* Forel, 1893g: 399 (w.q.) ANTILLES IS. Combination in *C.* (*Orthocrema*): Emery, 1922e: 134. Race of *brevispinosa:* Forel, 1897b: 300.

**minutissima.** *Crematogaster minutissima* Mayr, 1870b: 995 (w.q.) U.S.A. Wheeler, W.M. 1908e: 484 (m.); Wheeler, G.C. & Wheeler, J. 1960b: 14 (l.). Combination in *C.* (*Orthocrema*): Emery, 1922e: 135. Current subspecies: nominal plus *missuriensis, smithi.*

**#mirmillo.** *Crematogaster* (*Acrocoelia*) *foraminiceps* var. *mirmillo* Santschi, 1935b: 269, fig. 5 (w.) KENYA.

**misella.** *Crematogaster misella* Arnold, 1920a: 513, fig. 40 (w.) ZIMBABWE. Combination in *C.* (*Crematogaster*): Wheeler, W.M. 1922a: 841.

**#missuriensis.** *Crematogaster victima* subsp. *missuriensis* Emery, 1895c: 287 (attributed to Pergande) (w.) U.S.A. Wheeler, G.C. & Wheeler, J. 1952b: 260 (l.). Combination in *C.* (*Orthocrema*): Emery, 1922e: 136. Subspecies of *minutissima:* Creighton, 1950a: 138.

**#mixta.** *Crematogaster scita* r. *mixta* Forel, 1902h: 409 (w.) AUSTRALIA. Combination in *C.* (*Orthocrema*): Emery, 1922e: 133.

***CREMATOGASTER*** *(continued)*

***mjobergi.*** *Crematogaster mjobergi* Forel, 1915b: 54 (w.) AUSTRALIA. Combination in *C.* (*Acrocoelia*): Emery, 1922e: 152.

**#*modica.*** *Crematogaster* (*Sphaerocrema*) *modica* Santschi, 1926b: 222, fig. 2D (w.) UGANDA. Variety of *bequaerti*, and senior synonym of *saga:* Santschi, 1937g: 77.

***modiglianii.*** *Crematogaster modiglianii* Emery, 1900d: 688 (w.) INDONESIA (Mentawei I.). Combination in *C.* (*Paracrema*): Emery, 1922e: 156; Santschi, 1925f: 90. Current subspecies: nominal plus *anamita, anoemica, clemensae, sarawakana, surbeki.*

**#*moelleri.*** *Crematogaster brevispinosa* r. *moelleri* Forel, 1912f: 214 (w.m.) BRAZIL. Combination in *C.* (*Orthocrema*): Emery, 1922e: 134; Santschi, 1923c: 250.

**#*molongori.*** *Crematogaster* (*Orthocrema*) *sordidula* subsp. *molognori* Weber, 1943c: 357 (w.) SUDAN.

*mongolica. Crematogaster* (*Acrocoelia*) *laboriosa* st. *politula* var. *mongolica* Santschi, 1928b: 33 (w.) MONGOLIA. Unavailable name.

**#*montana.*** *Crematogaster* (*Orthocrema*) *brevispinosa* subsp. *montana* Borgmeier, 1939: 421 (w.) BRAZIL.

***montezumia.*** *Crematogaster montezumia* Smith, F. 1858b: 139, pl. 1, fig. 1 (nest) (w.q.m.) MEXICO. Combination in *C.* (*Physocrema*): Forel, 1912f: 220; in *C.* (*Orthocrema*): Emery, 1922e: 135; in *C.* (*Neocrema*): Santschi, 1922d: 244; Kempf, 1972a: 83. Senior synonym of *sulcata:* Forel, 1901h: 65; Kempf, 1968b: 386; of *cristulata, functa, proletaria* and *ramulinida:* Kempf, 1968b: 386.

***monticola.*** *Crematogaster monticola* Arnold, 1920a: 517, fig. 40 (w.) SOUTH AFRICA. Combination in *C.* (*Crematogaster*): Wheeler, W.M. 1922a: 841.

***moorei.*** *Crematogaster* (*Physocrema*) *moorei* Donisthorpe, 1941l: 225 (w.) PHILIPPINES.

***moqorensis.*** *Crematogaster moqorensis* Pisarski, 1967: 394, figs. 37–49 (w.m.) AFGHANISTAN.

***mormonum.*** *Crematogaster coarctata* var. *mormonum* Wheeler, W.M. 1919g: 111 (w.) U.S.A. [First available use of *Crematogaster lineolata* subsp. *coarctata* var. *mormonum* Emery, 1895c: 284; unavailable name.] Combination in *C.* (*Acrocoelia*): Emery, 1922e: 141; in *C.* (*Crematogaster*): Buren, 1968b: 94. Raised to species: Creighton, 1950a: 215.

**#*mosis.*** *Crematogaster auberti* var. *mosis* Forel, 1909c: 104 (w.q.) ISRAEL. [First available use of *Crematogaster auberti* subsp. *jehovae* var. *mosis* Forel, 1907b: 207; unavailable name.] Variety of *jehovae:* Menozzi, 1933b: 61 (in key).

*mossamedana. Crematogaster* (*Acrocoelia*) *castanea* st. *inversa* var. *mossamedana* Santschi, 1930b: 69 (w.q.) ANGOLA. Unavailable name.

***mottazi.*** *Crematogaster* (*Atopogyne*) *mottazi* Santschi, 1928d: 67 (w.q.) IVORY COAST.

***mucronata.*** *Crematogaster deformis* var. *mucronata* Emery, 1900d: 690 (w.) SUMATRA. Combination in *C.* (*Physocrema*), and raised to species: Emery, 1922e: 140; Donisthorpe, 1941l: 226.

**#*muellerianus.*** *Crematogaster* (*Acrocoelia*) *muellerianus* Finzi, 1939a: 157 (q.) ETHIOPIA. Variety of *pseudinermis:* Menozzi, 1942: 168.

***muralti.*** *Crematogaster muralti* Forel, 1910e: 432 (w.) CAMEROUN. Combination in *C.* (*Orthocrema*): Wheeler, W.M. 1922a: 850. Current subspecies: nominal plus *livingstonei, ugandensis.*

**#*museisapientiae.*** *Crematogaster ferruginea* subsp. *museisapientiae* Forel, 1911e: 272 (w.q.m.) "Tropical East Africa". Combination in *C.* (*Acrocoelia*): Emery, 1922e: 146. Stirps/subspecies of *tricolor:* Santschi, 1913a: 413 (in key); of *castanea:* Wheeler, W.M. 1922a: 833.

**#*mutabilis.*** *Crematogaster bequaerti* var. *mutabilis* Santschi, 1914b: 93, fig. 12 (w.) KENYA. Combination in *C.* (*Sphaerocrema*): Wheeler, W.M. 1922a: 845.

***mutans.*** *Crematogaster* (*Crematogaster*) *mutans* Buren, 1968b: 115 (w.q.) U.S.A.

***myops.*** *Crematogaster myops* Forel, 1911a: 31 (w.q.) BORNEO. Combination in *C.* (*Orthocrema*): Santschi, 1918d: 182.

*nana. Crematogaster* (*Orthocrema*) *victima* st. *nitidiceps* var. *nana* Santschi, 1934c: 29 (w.) ARGENTINA. Unavailable name.

***natalensis.*** *Crematogaster sordidula* subsp. *natalensis* Forel, 1910e: 431 (w.q.m.) SOUTH AFRICA. Combination in *C.* (*Orthocrema*): Wheeler, W.M. 1922a: 850. Raised to species: Emery, 1912e: 668. Race/subspecies of *sordidula:* Arnold, 1920a: 539; Wheeler, W.M. 1922a: 850. Revived status as species: Emery, 1922e: 131; Santschi, 1926b: 212; Menozzi, 1930b: 85. Current subspecies: nominal plus *braunsi, dulcis.*

**#*naumannae.*** *Crematogaster longispina* r. *naumannae* Forel, 1921b: 133 (w.) ECUADOR. Combination in *C.*(*Orthocrema*): Kempf, 1972a: 88.

***navajoa.*** *Crematogaster* (*Crematogaster*) *navajoa* Buren, 1968b: 102 (w.q.) U.S.A.

**#*nawai.*** *Crematogaster auberti* var. *nawai* Ito, 1914: 41 (w.) JAPAN. Combination in *C.* (*Acrocoelia*), and raised to species: Emery, 1922e: 143. Subspecies of *laboriosa:* Santschi, 1930d: 263.

**#*nemesis.*** *Crematogaster* (*Orthocrema*) *iheringi* var. *nemesis* Santschi, 1934c: 28 (w.q.) ARGENTINA.

***nesiotis.*** *Crematogaster nesiotis* Mann, 1919: 322, fig. 24 (w.) SOLOMON IS.

***neuvillei.*** *Crematogaster neuvillei* Forel, 1907c: 140 (w.) ETHIOPIA. Combination in *C.* (*Crematogaster*): Wheeler, W.M. 1922a: 841; in *C.* (*Acrocoelia*): Emery, 1922e: 148. Current subspecies: nominal plus *cooperi.*

**#*nicevillei.*** *Crematogaster* (*Acrocoelia*) *brunnea* var. *nicevillei* Emery, 1922e: 149 (w.) INDIA. [First available use of *Crematogaster subnuda* r. *rabula* var. *nicevillei* Forel, 1902c: 208; unavailable name.] Variety of *brunnea:* Wheeler, W.M. 1927d: 4; Santschi, 1928b: 33.

***nigeriensis.*** *Crematogaster nigeriensis* Santschi, 1914d: 346, fig. 16 (w.) NIGERIA. Combination in *C.* (*Sphaerocrema*): Santschi, 1918d: 182. Current subspecies: nominal plus *wilnigra.*

**#*nigra.*** *Crematogaster chiarinii* var. *nigrum* Forel, 1910e: 434 (w.) CONGO.

**#*nigra.*** *Crematogaster scutellaris* var. *nigra* Krausse, 1912b: 164 (w.) SARDINIA. [Unresolved junior primary homonym of *nigra* Forel, above.] [Also described as new by Emery, 1912e: 654, based on the same material.]

***nigrans.*** *Crematogaster foraminiceps* r. *nigrans* Forel, 1915c: 338 (w.q.) ZAIRE. Combination in *C.* (*Acrocoelia*): Emery, 1922e: 147. Raised to species: Santschi, 1921c: 117 [misspelled as *nigricans*]; Santschi, 1926b: 227.

***nigriceps.*** *Crematogaster nigriceps* Emery, 1897e: 600, fig. (w.) SOMALIA. Viehmeyer, 1923: 89 (q.m.). Combination in *C.* (*Crematogaster*): Wheeler, W.M. 1922a: 842; in *C.* (*Acrocoelia*): Emery, 1922e: 148. Current subspecies: nominal plus *prelli, saganensis.*

**#*nigripes.*** *Crematogaster auberti* var. *nigripes* Menozzi, 1940: 268 (w.) LIBYA. [First available use of *Crematogaster auberti* subsp. *antaris* var. *nigripes* Emery, 1924a: 9; unavailable name.]

**#*nigriventris.*** *Crematogaster* (*Sphaerocrema*) *rugosa* var. *nigriventris* Santschi, 1919h: 87 (w.) IVORY COAST.

*nigromedia. Crematogaster* (*Orthocrema*) *major* subsp. *flavior* var. *nigromedia* Donisthorpe, 1941b: 40 (w.) NEW GUINEA. Unavailable name.

***nigronitens.*** *Crematogaster nigronitens* Santschi, 1917b: 281 (w.) ZIMBABWE. Combination in *C.* (*Crematogaster*): Wheeler, W.M. 1922a: 842; in *C.* (*Acrocoelia*): Emery, 1922e: 148.

***nigropilosa.*** *Crematogaster nigropilosa* Mayr, 1870a: 405 (w.) COLOMBIA. Mayr, 1887: 624 (q.). Combination in *C.* (*Orthocrema*): Emery, 1922e: 136.

**#*nilgirica.*** *Crematogaster* (*Acrocoelia*) *brunnea* var. *nilgirica* Emery, 1922e: 149 (w.) INDIA. [First available use of *Crematogaster subnuda* r. *rabula* var. *nilgirica* Forel, 1902c: 208; unavailable name.]

*nitidus. Crematogaster nitidus* Radoszkovsky, 1881: 197. *Nomen nudum*, attributed to Smith, F.

*nitida. Crematogaster* (*Acrocoelia*) *dohrni* subsp. *rogenhoferi* var. *nitida* Crawley, 1924: 395 (w.) SUMATRA. Unavailable name.

**#*nitidiceps.*** *Crematogaster victima* var. *nitidiceps* Emery, 1895c: 288 (footnote) (w.) BRAZIL. Combination in *C.* (*Orthocrema*): Emery, 1922e: 136. Subspecies of *victima:* Forel, 1911e: 274.

*nociva. Crematogaster* (*Neocrema*) *magnifica* var. *nociva* Borgmeier, 1937b: 234 (w.q.) BRAZIL. Junior synonym of *magnifica:* Kempf, 1968b: 385.

***nocturna.*** *Crematogaster* (*Crematogaster*) *nocturna* Buren, 1968b: 112 (w.m.) U.S.A.

*normandi. Crematogaster auberti* st. *laestrygon* var. *normandi* Santschi, 1921d: 71 (w.) TUNISIA. Unavailable name.

**#*nossibeensis.*** *Crematogaster hova* r. *nossibeensis* Forel, 1891b: 181 (q.) MADAGASCAR. Combination in *C.* (*Decacrema*): Wheeler, W.M. 1922a: 1026. Raised to species: Dalla Torre, 1893: 84. Subspecies of *hova:* Wheeler, W.M. 1922a: 1026; Emery, 1922e: 138.

*notabilis. Crematogaster subnuda* r. *contemta* var. *notabilis* Forel, 1902c: 208 (w.q.) INDIA. Unavailable name.

**#*noualhieri.*** *Crematogaster alluaudi* st. *noualhieri* Emery, 1893c: 83 (w.) CANARY IS. Santschi, 1921e: 168 (q.); Santschi, 1937e: 300 (m.). Combination in *C.* (*Acrocoelia*): Emery, 1922e: 142. Raised to species: Santschi, 1937e: 300. Subspecies of *alluaudi:* Barquin, 1981: 142.

*novaeboracensis. Crematogaster novaeboracensis* Buckley, 1867: 337 (q.) U.S.A. Junior synonym of *lineolata:* Mayr, 1886d: 462.

**#novem.** *Crematogaster* (*Decacrema*) *borneensis* subsp. *novem* Forel, 1911a: 35 (w.q.) BORNEO.

**oasium.** *Crematogaster auberti* st. *oasium* Santschi, 1911f: 84 (w.) TUNISIA. Santschi, 1937e: 303 (q.); Bernard, 1953a: 156 (m.). Combination in *C.* (*Acrocoelia*): Emery, 1922e: 142. Variety/subspecies of *antaris:* Santschi, 1921d: 71; of *auberti:* Santschi, 1929c: 99. Raised to species: Santschi, 1937e: 303; Collingwood, 1985: 261. Current subspecies: nominal plus *saharensis.*

**obnigra.** *Crematogaster obnigra* Mann, 1919: 323 (w.) SOLOMON IS.

**obscura.** *Crematogaster obscurus* Smith, F. 1857a: 76 (w.) BORNEO. Combination in *C.* (*Acrocoelia*): Emery, 1922e: 151; Donisthorpe, 1932c: 450.

*obscurus. Crematogaster obscurus* Smith, F. 1860b: 109 (w.) INDONESIA (Batjan I.). [Junior primary homonym of *obscura* Smith, F. above.] Replacement name: *obscurior* Dalla Torre, 1892: 90.

**#obscura.** *Crematogaster* (*Orthocrema*) *curvispinosa* var. *obscura* Santschi, 1929d: 293 (w.) ARGENTINA. [Unresolved junior primary homonym of *obscura* Smith, F. 1857a: 76.] Gallardo, 1934: 34 (q.).

**#obscurata.** *Crematogaster victima* var. *obscurata* Emery, 1895c: 287 (footnote) (w.) VENEZUELA. Combination in *C.* (*Orthocrema*): Emery, 1922e: 136.

**obscurior.** *Crematogaster obscurior* Dalla Torre, 1892: 90. Replacement name for *obscurus* Smith, F. 1860b: 109. [Junior primary homonym of *obscurus* Smith, F. 1857a: 76.]

**#obstinata.** *Crematogaster striatula* var. *obstinata* Santschi, 1911g: 207 (w.) CONGO. Combination in *C.* (*Sphaerocrema*): Wheeler, W.M. 1922a: 849. Stirps of *striatula:* Santschi, 1926b: 223.

**#occidentalis.** *Crematogaster menileki* subsp. *occidentalis* Mayr, 1902: 293 (w.) EQUATORIAL GUINEA. Combination in *C.* (*Acrocoelia*): Emery, 1922e: 148.

**ochracea.** *Crematogaster ochracea* Mayr, 1862: 766 (w.) PHILIPPINES. Combination in *C.* (*Acrocoelia*): Emery, 1922e: 151.

**ochraceiventris.** *Crematogaster ochraceiventris* Stitz, 1916: 391, fig. 8 (w.) CONGO. Combination in *C.* (*Crematogaster*): Wheeler, W.M. 1922a: 842; in *C.* (*Acrocoelia*): Emery, 1922e: 148.

**#oeagria.** *Crematogaster* (*Neocrema*) *polymnia* var. *oeagria* Santschi, 1933e: 113 (w.) ARGENTINA.

**#omega.** *Crematogaster* (*Sphaerocrema*) *striatula* var. *omega* Santschi, 1935a: 259 (w.) ZAIRE. [First available use of *Crematogaster* (*Sphaerocrema*) *striatula* st. *obstinata* var. *omega* Santschi, 1926b: 224; unavailable name.]

**#omniparens.** *Crematogaster brunneipennis* r. *omniparens* Forel, 1914d: 237 (w.) SOUTH AFRICA. Combination in *C.* (*Acrocoelia*): Emery, 1922e: 145.

**onusta.** *Crematogaster* (*Physocrema*) *onusta* Stitz, 1925: 118 (w.) PHILIPPINES. Combination in *C.* (*Paracrema*): Donisthorpe, 1941l: 226.

**opaca.** *Crematogaster opaca* Mayr, 1870b: 992 (w.) MEXICO. Combination in *C.* (*Acrocoelia*): Emery, 1922e: 141. Subspecies of *lineolata:* Emery, 1895c: 283; Wheeler, W.M. 1908e: 478; Wheeler, W.M. 1914b: 44; Emery, 1922e: 141. Revived status as species: Wheeler, W.M. 1919g: 111; Wheeler, W.M. 1934f: 139. Current subspecies: nominal plus *cedrosensis.*

**opaciceps.** *Crematogaster opaciceps* Mayr, 1901b: 12 (w.) SOUTH AFRICA. Combination in *C.* (*Crematogaster*): Wheeler, W.M. 1922a: 842; in *C.* (*Acrocoelia*): Emery, 1922e: 148. Current subspecies: nominal plus *cacodaemon, clepens, defleta.*

**opuntiae.** *Crematogaster* (*Crematogaster*) *opuntiae* Buren, 1968b: 120 (w.) U.S.A.

**#oraclum.** *Crematogaster gallicola* var. *oraclum* Forel, 1913b: 323 (w.) ZAIRE. Combination in *C.* (*Acrocoelia*): Emery, 1922e: 149. Stirps of *sjostedti:* Santschi, 1918d: 185; of *gerstaeckeri:* Santschi, 1930b: 69. Senior synonym of *bulawayensis* (and its junior synonym *foraminicipoides*): Arnold, 1920a: 521. [Arnold gives *bulawayensis* as the senior name, but *oraclum* has priority.]

**ornatipilis.** *Crematogaster ornatipilis* Wheeler, W.M. 1918b: 25 (w.) GUIANA. Combination in *C.* (*Orthocrema*): Emery, 1922e: 135.

**orobia.** *Crematogaster orobia* Santschi, 1919a: 343 (w.) SOUTH AFRICA. Combination in *C.* (*Crematogaster*): Wheeler, W.M. 1922a: 842.

**osakensis.** *Crematogaster sordidula* var. *osakensis* Forel, 1900e: 269 (w.) JAPAN. Wheeler, W.M. 1928d: 111 (q.m.). Combination in *C.* (*Orthocrema*): Emery, 1922e: 131. Subspecies of *sordidula:* Emery, 1912e: 671; Wheeler, W.M. 1928d: 111. Raised to species: Collingwood, 1976: 303; Kupyanskaya, 1990: 129. Senior synonym of *japonica:* Brown, 1949e: 37.

**oscaris.** *Crematogaster* (*Oxygyne*) *oscaris* Forel, 1910f: 7, pl. 1, figs. 5, 6 (w.q.m.) SOUTH AFRICA.

**#osmanica.** *Crematogaster* (*Orthocrema*) *sordidula* var. *osmanica* Santschi, 1921a: 114 (w.) TURKEY.

**overbecki.** *Crematogaster overbecki* Viehmeyer, 1916a: 123, fig. 3 (w.q.m.) SINGAPORE. Combination in *C.* (*Orthocrema*): Emery, 1922e: 132; in *C.* (*Mesocrema*): Santschi, 1928b: 33.

**#ovinodis.** *Crematogaster* (*Oxygyne*) *stadelmanni* var. *ovinodis* Stitz, 1916: 383, fig. 4 (w.) ZAIRE. Combination in *C.* (*Nematocrema*): Wheeler, W.M. 1922a: 860.

**oxygynoides.** *Crematogaster* (*Neocrema*) *oxygynoides* Santschi, 1934a: 189 (w.q.) ARGENTINA.

**#pacifica.** *Crematogaster pacifica* Santschi, 1919b: 236 (w.) MADAGASCAR. Combination in *C.* (*Crematogaster*): Wheeler, W.M. 1922a: 1023. Variety of *lobata:* Santschi, 1930a: 63.

**painei.** *Crematogaster* (*Acrocoelia*) *painei* Donisthorpe, 1945d: 267 (w.) GHANA.

**#palans.** *Crematogaster limata* var. *palans* Forel, 1912f: 216 (w.) PANAMA. Wheeler, W.M. 1921f: 151 (q.); Wheeler, W.M. 1942: 197 (m.). Combination in *C.* (*Orthocrema*): Emery, 1922e: 136.

**pallida.** *Crematogaster pallidus* Lowne, 1865: 335 (w.q.) AUSTRALIA. Combination in *C.* (*Orthocrema*): Emery, 1922e: 132.

**pallipes.** *Crematogaster pallipes* Mayr, 1862: 768 (w.) AUSTRALIA. Mayr, 1876: 108 (q.m.). Combination in *C.* (*Orthocrema*): Emery, 1922e: 132; in *C.* (*Mesocrema*): Santschi, 1929f: 90. [Misspelled as *pallidipes* by Dalla Torre, 1893: 84.] Senior synonym of *piceus:* Mayr, 1876: 108.

**#panamana.** *Crematogaster* (*Orthocrema*) *curvispinosa* var. *panamana* Wheeler, W.M. 1942: 195 (w.q.) PANAMA.

**paolii.** *Crematogaster* (*Sphaerocrema*) *paolii* Menozzi, 1930b: 102, fig. 4 (w.) SOMALIA.

**#parabiotica.** *Crematogaster limata* r. *parabiotica* Forel, 1904e: 683 (w.) COLOMBIA. Forel, 1912f: 211 (q.). Combination in *C.* (*Orthocrema*): Emery, 1922e: 136. Raised to species: Wheeler, W.M. 1921b: 99 (in text); Wheeler, W.M. & Darlington, 1930: 109. Subspecies of *limata:* Wheeler, W.M. 1921f: 152; Wheeler, W.M. 1923a: 3; Kutter, 1931a: 61; Santschi, 1939f: 161; Kempf, 1972a: 88.

**paradoxa.** *Crematogaster paradoxa* Emery, 1894b: 72, fig. (w.) NEW GUINEA. Combination in *C.* (*Rhachiocrema*): Mann, 1919: 318.

*paraensis. Crematogaster distans* r. *paraensis* Forel, 1904c: 37 (w.) BRAZIL. Junior synonym of *distans:* Kempf, 1968b: 390.

**parallela.** *Crematogaster* (*Orthocrema*) *parallela* Santschi, 1925b: 14 (w.) BRAZIL.

*parviceps. Crematogaster distans* subsp. *parviceps* Forel, 1908c: 369 (w.) BRAZIL. Combination in *C.* (*Orthocrema*): Emery, 1922e: 135. Junior synonym of *distans:* Kempf, 1968b: 390.

*parvispina. Tranopeltoides parvispina* Wheeler, W.M. 1922e: 12, fig. 3 (q.) GUIANA. Junior synonym of *stollii:* Kempf, 1960c: 174.

**#pasithea.** *Crematogaster werneri* var. *pasithea* Santschi, 1915c: 253 (w.) BENIN. Combination in *C.* (*Acrocoelia*): Emery, 1922e: 149.

**patei.** *Crematogaster* (*Crematogaster*) *patei* Buren, 1968b: 107 (w.) MEXICO.

**pauciseta.** *Crematogaster pauciseta* Emery, 1899e: 485, fig. (w.q.) CAMEROUN. Forel, 1913e: 669 (m.). Combination in *C.* (*Orthocrema*): Santschi, 1918d: 182. Current subspecies: nominal plus *grossulior.*

**pauli.** *Crematogaster pauli* Emery, 1901g: 575, fig. A (w.) SULAWESI. Combination in *C.* (*Orthocrema*): Emery, 1922e: 132.

**#paulinae.** *Crematogaster paulinae* Forel, 1892k: 530 (q.) (diagnosis in key) MADAGASCAR. Forel, 1903c: 254 (w.). Combination in *C.* (*Oxygyne*): Wheeler, W.M. 1922a: 1025. Race/subspecies of *ranavalonae:* Emery, 1897a: 14; Forel, 1903c: 254; Wheeler, W.M. 1922a: 1025.

*paulinaeranavalonae. Crematogaster ranavalonae* var. *paulinaeranavalonae* Wasmann, 1897: 249. *Nomen nudum.* [Also noted by Forel, 1897c: 198. *Nomen nudum.*]

**pellens.** *Crematogaster pellens* Walker, 1859: 374 (q.) SRI LANKA. Combination in *C.* (*Acrocoelia*): Donisthorpe, 1932b: 576.

*pensitata. Crematogaster excisa* r. *andrei* var. *pensitata* Forel, 1915c: 341 (w.) SOUTH AFRICA. Unavailable name.

**#pepo.** *Crematogaster* (*Oxygyne*) *ranavalonae* var. *pepo* Forel, 1922: 96 (w.q.m.) MADAGASCAR.

**perelegans.** *Crematogaster perlegans* Forel, 1902c: 202 (w.q.) INDIA. [Justified emendation of spelling to *perelegans:* Forel, 1903a: 683.] Bingham, 1903: 143 (m.). Combination in *C.* (*Acrocoelia*): Emery, 1922e: 151.

*pergens. Crematogaster victima* subsp. *nitidiceps* var. *pergens* Forel, 1911e: 274 (w.) PARAGUAY. Unavailable name.

**peringueyi.** *Crematogaster peringueyi* Emery, 1895h: 27, figs. 16, 17 (w.q.m.) SOUTH AFRICA. Combination in *C.* (*Crematogaster*): Wheeler, W.M. 1922a: 842; in *C.* (*Acrocoelia*): Emery, 1922e: 148. Current subspecies: nominal plus *angustior, cacochyma, dentulata, gedeon*.

***peristerica***. *Crematogaster (Orthocrema) peristericus* Menozzi, 1925b: 369, fig. 2 (w.q.) BRAZIL.

***perthensis***. *Crematogaster perthensis* Crawley, 1922c: 21 (w.m.) AUSTRALIA.

***peruviana***. *Tranopeltoides peruvianus* Wheeler, W.M. 1922e: 14 (m.) PERU. Combination in *Crematogaster*: Kempf, 1960c: 175; in *C. (Orthocrema)*: Kempf, 1972a: 88.

***petiolidens***. *Crematogaster (Decacrema) petiolidens* Forel, 1916: 412 (w.) ZAIRE.

*pevsnerae*. *Crematogaster distans* r. *pevsnerae* Forel, 1912f: 218 (w.) VENEZUELA. Combination in *C. (Orthocrema)*: Emery, 1922e: 135. Junior synonym of *distans*: Kempf, 1968b: 390.

#***pharaonis***. *Crematogaster aegyptiaca* st. *pharaonis* Santschi, 1914a: 432 (w.) EGYPT. Combination in *C.(Acrocoelia)*: Emery, 1922e: 144.

***phoenica***. *Crematogaster laestrygon* st. *phoenica* Santschi, 1915a: 59 (w.) SYRIA. Combination in *C.(Acrocoelia)*: Emery, 1922e: 142. Subspecies of *jehovae*: Emery, 1926: 3. Raised to species: Santschi, 1929e: 149; Santschi, 1934d: 276. Current subspecies: nominal plus *pygmalion*.

***phoenix***. *Crematogaster (Sphaerocrema) phoenix* Santschi, 1921c: 116 (w.) TANZANIA.

#***physothorax***. *Crematogaster deformis* r. *physothorax* Emery, 1889b: 506 (w.) BURMA. Combination in *C. (Physocrema)*: Emery, 1922e: 140. Raised to species: Dalla Torre, 1893: 84; Bingham, 1903: 146. Subspecies of *difformis*: Wheeler, W.M. 1919e: 75; Emery, 1922e: 140.

#***phytoeca***. *Crematogaster sculpturata* subsp. *phytoeca* Wheeler, W.M. 1934g: 173 (w.q.m.) MEXICO.

***pia***. *Crematogaster tumidula* subsp. *pia* Forel, 1911d: 384 (w.) WEST MALAYSIA. Combination in *C. (Acrocoelia)*, and raised to species: Emery, 1922e: 151. Current subspecies: nominal plus *soengeiensis, taivanae*.

#***picata***. *Crematogaster (Paracrema) spengeli* var. *picata* Emery, 1922e: 156 (w.) WEST MALAYSIA. [First available use of *Crematogaster spengeli* subp. *taipingensis* var. *picata* Viehmeyer, 1916a: 125; unavailable name.]

*piceus*. *Crematogaster piceus* Lowne, 1865: 335 (w.) AUSTRALIA. Junior synonym of *pallipes*: Mayr, 1876: 108.

***pilosa***. *Crematogaster lineolata* subsp. *pilosa* Emery, 1895c: 285 (w.) (attributed to Pergande) U.S.A. Combination in *C. (Acrocoelia)*: Emery, 1922e: 141; in *C. (Crematogaster)*: Buren, 1968b: 92. Raised to species: Wheeler, W.M. 1919g: 111; Creighton, 1950a: 216; Johnson, 1988: 321. Senior synonym of *creightoni*, and material of the *nomen nudum subpilosa* referred here: Buren, in Smith, M.R. 1958c: 127.

#***placens***. *Crematogaster longiclava* var. *placens* Forel, 1911d: 384 (w.) WEST MALAYSIA. Combination in *C. (Acrocoelia)*: Emery, 1922e: 151.

*platygnatha*. *Formica platygnatha* Roger, 1863a: 168 (q.) SIERRA LEONE. Combination in *Crematogaster*, and senior synonym of *mandibularis*: André, 1890: 323. Junior synonym of *depressa*: Emery, 1892b: 164.

*pluton*. *Crematogaster (Crematogaster) meneliki* st. *proserpina* var. *pluton* Santschi, 1919h: 87 (w.) ZAIRE. Unavailable name.

***polita***. *Crematogaster politus* Smith, F. 1865: 74 (w.) NEW GUINEA. Combination in *C. (Orthocrema)*: Emery, 1922e: 133; Donisthorpe, 1932c: 473.

***politula***. *Crematogaster subnuda* r. *politula* Forel, 1902c: 207 (w.) INDIA. Combination in *C. (Acrocoelia)*: Emery, 1922e: 150. Subspecies/stirps of *brunnea*: Emery, 1922e: 150; Wheeler, W.M. 1930h: 65; of *laboriosa*: Santschi, 1928b: 33. Raised to species: Bingham, 1903: 131; Mukerjee, 1930: 153.

***polymnia***. *Crematogaster (Neocrema) polymnia* Santschi, 1922d: 247 (w.) ARGENTINA. Current subspecies: nominal plus *oeagria*.

#***polymorphica***. *Crematogaster (Atopogyne) africana* subsp. *polymorphica* Weber, 1943c: 357 (w.) SUDAN.

***popohana***. *Crematogaster popohana* Forel, 1912a: 69 (w.) TAIWAN. Forel, 1913f: 193 (q.m.). Combination in *C. (Acrocoelia)*: Emery, 1922e: 151. Current subspecies: nominal plus *amia*.

*poststriata*. *Crematogaster (Decacrema) solenopsides* st. *flavida* var. *poststriata* Santschi, 1921c: 118 (w.) TANZANIA. Unavailable name.

****praecursor***. **Crematogaster praecursor* Emery, 1891a: 152, pl. 2, figs. 19–21 (m.) SICILIAN AMBER (Miocene).

#***prelli***. *Crematogaster nigriceps* subsp. *prelli* Forel, 1911d: 368 (w.) TANZANIA. Menozzi, 1939c: 103 (q.). Combination in *C.(Acrocoelia)*: Emery, 1922e: 148.

#***projecta***. *Crematogaster (Orthocrema) brevispinosa* st. *projecta* Santschi, 1925d: 230 (w.) SURINAM.

*proletaria*. *Crematogaster (Neocrema) montezumia* var. *proletaria* Santschi, 1933e: 113 (w.q.) ARGENTINA. Junior synonym of *montezumia*: Kempf, 1968b: 386.

#***pronotalis***. *Crematogaster pronotalis* Santschi, 1914d: 347, fig. 17 (w.) NIGERIA. Combination in *C. (Sphaerocrema)*: Santschi, 1918d: 182. Variety of *kneri*: Santschi, 1926b: 228.

#***proserpina***. *Crematogaster (Crematogaster) meneliki* st. *proserpina* Santschi, 1919h: 87 (w.) ZAIRE. [Also described as new by Santschi, in Wheeler, W.M. 1922a: 154.] Wheeler, G.C. & Wheeler, J. 1952b: 256 (l.).

***pseudinermis***. *Crematogaster pseudinermis* Viehmeyer, 1923: 88, fig. 2 (w.q.) SUDAN. Combination in *C. (Sphaerocrema)*, and variety of *wellmani*: Santschi, 1926b: 219. Revived status as species: Santschi, 1935a: 260. Current subspecies: nominal plus *muellerianus*.

#***pudica***. *Crematogaster (Acrocoelia) gerstaeckeri* var. *pudica* Santschi, 1930b: 69 (w.) ZIMBABWE. [First available use of *Crematogaster bulawayensis* st. *infaceta* var. *pudica* Santschi, 1916b: 499; unavailable name.]

***pulchella***. *Crematogaster (Orthocrema) pulchella* Bernard, 1953b: 230, fig. 9 (w.q.) GUINEA.

#***pulla***. *Crematogaster (Crematogaster) sjostedti* var. *pulla* Wheeler, W.M. 1922a: 844 (w.) KENYA. [First available use of *Crematogaster gallicola* st. *sjostedti* var. *pulla* Santschi, 1914b: 94; unavailable name.] Combination in *C. (Acrocoelia)*: Emery, 1922e: 149. Variety of *gerstaeckeri*: Santschi, 1930b: 69.

*pullipes*. *Crematogaster (Neocrema) distans* st. *rugiceps* var. *pullipes* Santschi, 1920d: 379 (w.) ARGENTINA. Unavailable name; material referred to *corticicola* by Kempf, 1968b: 394.

*punctinodis*. *Crematogaster (Acrocoelia) lineolata* subsp. *cerasi* var. *punctinodis* Enzmann, J. 1946c: 91 (w.q.m.) U.S.A. Unavailable name; material referred to *lineolata* by Brown, 1949a: 48 and Creighton, 1950a: 213; to *cerasi* by Buren, in Smith, M.R. 1958c: 126.

***punctulata***. *Crematogaster punctulata* Emery, 1895c: 287 (w.) U.S.A. Combination in *C. (Acrocoelia)*: Emery, 1922e: 141; in *C. (Crematogaster)*: Buren, 1968b: 92. Variety/subspecies of *opaca*: Wheeler, W.M. 1919g: 111; Creighton, 1950a: 214. Revived status as species: Enzmann, J. 1946c: 93; Buren, in Smith, M.R. 1958c: 127; Johnson, 1988: 319. Senior synonym of *texana*: Creighton, 1950a: 214.

****pusilla***. **Myrmica pusilla* Heer, 1850: 164, pl. 13, fig. 9 (q.) YUGOSLAVIA (Miocene). Combination in *Crematogaster*: Handlirsch, 1907: 873. See also: Mayr, 1867b: 59.

***pygmaea***. *Crematogaster pygmaea* Forel, 1904c: 37 (w.) BRAZIL. Combination in *C. (Orthocrema)*: Emery, 1922e: 135. Variety of *brevispinosa*: Forel, 1909a: 259. Revived status as species: Forel, 1912f: 235; Emery, 1922e: 135.

#***pygmalion***. *Crematogaster (Acrocoelia) phoenica* var. *pygmalion* Santschi, 1934d: 276 (w.) SYRIA.

***pythia***. *Crematogaster pythia* Forel, 1915b: 53 (w.) AUSTRALIA. Combination in *C. (Acrocoelia)*: Emery, 1922e: 152.

*quadriceps*. *Crematogaster quadriceps* Smith, F. 1858b: 140 (w.) BRAZIL. Junior synonym of *acuta*: Roger, 1862c: 291.

***quadriformis***. *Crematogaster quadriformis* Roger, 1863a: 207 (w.) BRAZIL. Combination in *C. (Orthocrema)*: Santschi, 1918d: 182. Current subspecies: nominal plus *erici, gracilior, roveretoi, vezenyii*.

#***quadriruga***. *Crematogaster biroi* var. *quadriruga* Forel, 1911h: 455 (w.) THAILAND. Combination in *C. (Orthocrema)*: Emery, 1922e: 132.

***quadrispinosa***. *Crematogaster quadrispinosa* Roger, 1863a: 208 (q.) MEXICO. Combination in *C. (Eucrema)*: Emery, 1922e: 137.

***queenslandica***. *Crematogaster sordidula* var. *queenslandica* Forel, 1902h: 410 (w.q.) AUSTRALIA. Combination in *C. (Orthocrema)*: Emery, 1922e: 133; in *C. (Mesocrema)*: Santschi, 1928b: 33. Subspecies of *sordidula*: Forel, 1910b: 32. Raised to species: Emery, 1912e: 668. Current subspecies: nominal plus *froggatti, gilberti, rogans, scabrula*.

*quisquilia*. *Crematogaster castanea* st. *rufonigra* var. *quisquilia* Santschi, 1916b: 502 (w.) ZIMBABWE. Unavailable name.

#***rabula***. *Crematogaster subnuda* r. *rabula* Forel, 1902c: 207 (w.) INDIA. Imai, *et al.* 1984: 6 (k.). Combination in *C. (Acrocoelia)*: Emery, 1922e: 150. Variety of *brunnea*: Emery, 1922e: 150; Wheeler, W.M. 1927e: 1; Santschi, 1928b: 33.

***ralumensis***. *Crematogaster ralumensis* Forel, 1901b: 16 (w.) NEW GUINEA. Combination in *C. (Orthocrema)*: Emery, 1922e: 133.

*ramulinida*. *Crematogaster sulcata* var. *ramulinida* Forel, 1899c: 84 (footnote) (w.) COLOMBIA. Combination in *C.* (*Orthocrema*): Emery, 1922e: 135. Variety of *montezumia:* Forel, 1904c: 36; Forel, 1912f: 220. Junior synonym of *montezumia:* Kempf, 1968b: 386.

***ranavalonae.*** *Crematogaster ranavalonis* Forel, 1887: 388 (w.) MADAGASCAR. [Justified emendation of spelling to *ranavalonae:* Forel, 1891b: 184.] Forel, 1891b: 185 (m.); Emery, 1897a: 13 (q.). Combination in *C.* (*Oxygyne*): Forel, 1901e: 375. Current subspecies: nominal plus *paulinae, pepo.*

***ransonneti.*** *Crematogaster ransonneti* Mayr, 1868d: 287 (w.) SRI LANKA. Combination in *C.* (*Acrocoelia*): Emery, 1922e: 151.

***rasoherinae.*** *Crematogaster rasoherinae* Forel, 1891b: 194 (w.) MADAGASCAR. Forel, 1912k: 164 (q.m.). Combination in *C.* (*Neocrema*): Santschi, 1918d: 182; in *C.* (*Crematogaster*): Wheeler, W.M. 1922a: 1023; in *C.* (*Orthocrema*): Emery, 1922e: 131; in *C.* (*Mesocrema*): Santschi, 1928b: 33; Donisthorpe, 1943f: 661. Current subspecies: nominal plus *brunneola.*

#***rauana.*** *Crematogaster gallicola* var. *rauana* Forel, 1907a: 22 (w.) TANZANIA. Combination in *C.* (*Acrocoelia*): Emery, 1922e: 147.

*rectangulata. Crematogaster scutellaris* subsp. *schmidti* var. *rectangulata* Viehmeyer, 1923: 89 (w.) SUDAN. Unavailable name.

***rectinota.*** *Crematogaster sordidula* r. *rectinota* Forel, 1913a: 126 (w.) ZIMBABWE. Arnold, 1920a: 538 (q.). Combination in *C.* (*Orthocrema*): Wheeler, W.M. 1922a: 851. Raised to species: Emery, 1922e: 131.

***recurva.*** *Crematogaster recurva* Emery, 1897d: 556, pl. 1, figs. 11, 12 (w.) NEW GUINEA. Combination in *C.* (*Xiphocrema*): Emery, 1922e: 139.

#***recurvispina.*** *Crematogaster brevispinosa* r. *recurvispina* Forel, 1912f: 212 (w.) BRAZIL. Combination in *C.* (*Orthocrema*): Emery, 1922e: 134; Santschi, 1925d: 230.

*rediana. Myrmica rediana* Gené, 1841: 399 (w.) ITALY. Junior synonym of *scutellaris:* Mayr, 1855: 469.

#***regilla.*** *Crematogaster* (*Acrocoelia*) *auberti* st. *regilla* Santschi, 1937e: 302 (w.) MOROCCO. [*C. auberti* st. *regilla* Santschi, 1936c: 201; *nomen nudum.*]

***resulcata.*** *Crematogaster resulcata* **new name**. Replacement name for *subsulcata* Santschi, 1932a: 386. [Junior primary homonym of *subsulcata* Santschi, 1914a: 432.]

#***retiaria.*** *Crematogaster* (*Acrocoelia*) *amabilis* var. *retiaria* Santschi, 1933b: 101 (w.) KENYA.

***retifera.*** *Crematogaster* (*Acrocoelia*) *retifera* Santschi, 1926b: 216 (w.q.) ZIMBABWE.

*retusa. Crematogaster* (*Atopogyne*) *wellmani* st. *retusa* Santschi, 1916b: 500 (w.q.m.) ZIMBABWE. Combination in *C.* (*Acrocoelia*): Emery, 1922e: 149; in *C.*(*Sphaerocrema*): Santschi, 1926b: 219. Junior synonym of *wellmani:* **new synonymy**.

#***rhodesiana.*** *Crematogaster bulawayensis* var. *rhodesiana* Arnold, 1920a: 532 (w.) ZIMBABWE. [First available use of *Crematogaster inermis* r. *delagoensis* var. *rhodesiana* Forel, 1913a: 126; unavailable name.] Combination in *C.* (*Atopogyne*): Wheeler, W.M. 1922a: 854; in *C.* (*Acrocoelia*): Emery, 1922e: 146. Variety of *bulawayensis:* Wheeler, W.M. 1922a: 854; of *delagoensis:* Emery, 1922e: 146.

***rifelna.*** *Crematogaster* (*Crematogaster*) *rifelna* Buren, 1968b: 96 (w.q.m.) U.S.A.

***rivai.*** *Crematogaster rivai* Emery, 1897e: 598, fig. (w.) ETHIOPIA. Combination in *C.* (*Crematogaster*): Wheeler, W.M. 1922a: 843; in *C.* (*Acrocoelia*): Emery, 1922e: 148. Current subspecies: nominal plus *luctuosa.*

#***robusta.*** *Crematogaster robusta* Emery, 1877b: 379 (q.m.) ETHIOPIA. Emery, 1915g: 10 (w.). Combination in *C.* (*Acrocoelia*): Emery, 1922e: 144. Variety of *aegyptiaca:* Emery, 1915g: 10.

#***rochai.*** *Crematogaster rochai* Forel, 1903c: 255 (w.q.m.) BRAZIL. Combination in *C.* (*Orthocrema*): Emery, 1922e: 134. Race/subspecies of *brevispinosa:* Forel, 1912f: 213; Gallardo, 1934: 21.

#***rogans.*** *Crematogaster sordidula* r. *rogans* Forel, 1902h: 411 (w.q.) AUSTRALIA. Combination in *C.* (*Orthocrema*): Emery, 1922e: 133; in *C.* (*Mesocrema*): Santschi, 1929f: 90. Subspecies of *froggatti:* Santschi, 1929f: 90; of *queenslandica:* Taylor & Brown, D.R. 1985: 63.

***rogenhoferi.*** *Crematogaster rogenhoferi* Mayr, 1879: 683 (w.) BURMA. Forel, 1913f: 194 (q.m.). Combination in *C.* (*Acrocoelia*): Emery, 1922e: 150. Subspecies of *dohrni:* Emery, 1922e: 150; Crawley, 1924: 395; Wheeler, W.M. 1924b: 246. Revived status as species: Wheeler, W.M. 1930h: 66. Current subspecies: nominal plus *costulata, fictrix, flava, lutea.*

***rogeri.*** *Crematogaster* (*Acrocoelia*) *rogeri* Emery, 1922e: 151 (w.) SRI LANKA.

***rossi.*** *Crematogaster* (*Crematogaster*) *rossi* Buren, 1968b: 111 (w.q.) MEXICO.

***rothneyi.*** *Crematogaster rothneyi* Mayr, 1879: 685 (w.) INDIA. [*Crematogaster rothneyi* Smith, F. 1873: viii; *nomen nudum,* attributed to Mayr.] Imai, *et al.* 1984: 6 (k.). Combination in *C.* (*Acrocoelia*): Emery, 1922e: 151. Current subspecies: nominal plus *civa, haputalensis.*

#***roveretoi.*** *Crematogaster quadriformis* r. *roveretoi* Forel, 1913l: 234 (w.) ARGENTINA. Santschi, 1929d: 292 (q.m.). Combination in *C.* (*Orthocrema*): Emery, 1922e: 135.

*rubriceps. Myrmica rubriceps* Nylander, 1849: 44 (w.) SICILY. Junior synonym of *scutellaris:* Mayr, 1855: 469.

***rudis.*** *Crematogaster rudis* Emery, 1894g: 395 (footnote) (w.) BRAZIL. Combination in *C.* (*Orthocrema*): Emery, 1922e: 135. Race/subspecies of *quadriformis:* Forel, 1913l: 234; Bruch, 1914: 219. Revived status as species: Emery, 1922e: 135; Kempf, 1972a: 89.

***rufa.*** *Myrmica rufa* Jerdon, 1851: 114 (w.) INDIA. Combination in *Crematogaster:* Roger, 1863b: 37; in *C.* (*Acrocoelia*): Emery, 1922e: 152.

#***rufescens.*** *Crematogaster* (*Crematogaster*) *sjostedti* var. *rufescens* Wheeler, W.M. 1922a: 844 (w.) SOUTH AFRICA. [First available use of *Crematogaster sjostedti* st. *bulawayensis* var. *rufescens* Santschi, 1919b: 236; unavailable name.] Variety of *bulawayensis:* Arnold, 1926: 289. Subspecies of *gerstaeckeri:* current status.

*ruficeps. Acrocoelia ruficeps* Mayr, 1853a: 147 (w.) AUSTRIA. Junior synonym of *scutellaris:* Mayr, 1855: 469.

***rufigena.*** *Crematogaster rufigena* Arnold, 1958: 123, fig. 4 (w.) ZIMBABWE.

#***rufimembrum.*** *Crematogaster tricolor* st. *rufimembrum* Santschi, 1914a: 431 (w.) TANZANIA. Combination in *C.* (*Acrocoelia*): Emery, 1922e: 146. Subspecies of *castanea:* Wheeler, W.M. 1922a: 833; Santschi, 1926b: 214.

#***rufonigra.*** *Crematogaster tricolor* subsp. *rufonigra* Emery, 1895h: 27 (w.) SOUTH AFRICA. Arnold, 1920a: 492 (q.m.). Combination in *C.* (*Acrocoelia*): Emery, 1922e: 146. Stirps of *castanea:* Santschi, 1916b: 502.

***rufotestacea.*** *Crematogaster rufotestacea* Mayr, 1876: 109 (w.) AUSTRALIA. Emery, 1914b: 184 (q.); Crawley, 1922c: 23 (m.). Combination in *C.* (*Orthocrema*): Emery, 1922e: 133. Current subspecies: nominal plus *dentinasis.*

#***rufula.*** *Crematogaster* (*Sphaerocrema*) *libengensis* var. *rufula* Santschi, 1926b: 227 (w.) ZAIRE.

#***rugaticeps.*** *Crematogaster* (*Sphaerocrema*) *rugosa* var. *rugaticeps* Santschi, 1919h: 87 (w.) CAMEROUN.

*rugiceps. Crematogaster distans* var. *rugiceps* Forel, 1911c: 301 (attributed to Mayr) (w.q.) BRAZIL. [Misspelled as *rugifrons* by Forel, 1914e: 11.] Gallardo, 1931b: 304 (m.). Combination in *C.* (*Neocrema*): Santschi, 1920d: 379; in *C.* (*Orthocrema*): Emery, 1922e: 135. Race of *distans:* Forel, 1912f: 218. Junior synonym of *corticicola:* Kempf, 1968b: 394.

#***ruginota.*** *Crematogaster laboriosa* st. *ruginota* Santschi, 1928b: 33 (w.) INDIA. [First available use of *Crematogaster subnuda* r. *politula* var. *ruginota* Forel, 1902c: 207; unavailable name.] Subspecies of *brunnea:* Wheeler, W.M. 1929f: 5 (in text).

***rugosa.*** *Crematogaster rugosa* André, 1895a: 3 (w.) CONGO. Emery, 1899e: 484 (q.). Combination in *C.* (*Sphaerocrema*): Wheeler, W.M. 1922a: 848; Santschi, 1926b: 228. Current subspecies: nominal plus *nigriventris, rugaticeps.*

***rugosior.*** *Crematogaster luctans* st. *rugosior* Santschi, 1910c: 375 (w.) CONGO. Wheeler, W.M. 1922a: 156 (q.). Combination in *C.* (*Sphaerocrema*), and raised to species: Wheeler, W.M. 1922a: 156; Santschi, 1926b: 223.

***ruspolii.*** *Crematogaster ruspolii* Forel, 1892b: 142 (w.m.) ETHIOPIA. Combination in *C.* (*Crematogaster*): Wheeler, W.M. 1922a: 843; in *C.* (*Acrocoelia*): Emery, 1922e: 148. Current subspecies: nominal plus *atriscapis.*

#***russata.*** *Crematogaster* (*Orthocrema*) *brevispinosa* subsp. *russata* Wheeler, W.M. 1925a: 24 (w.) BOLIVIA.

***russoi.*** *Crematogaster* (*Orthocrema*) *russoi* Menozzi, in Menozzi & Russo, 1930: 155, fig. 2 (w.) DOMINICA.

***rustica.*** *Crematogaster* (*Sphaerocrema*) *rustica* Santschi, 1935a: 259 (w.) ZAIRE.

*saga. Crematogaster* (*Sphaerocrema*) *bequaerti* var. *saga* Santschi, 1937a: 56 (w.) KENYA. Junior synonym of *modica:* Santschi, 1937g: 77.

#***saganensis.*** *Crematogaster* (*Acrocoelia*) *nigriceps* subsp. *saganensis* Consani, in Menozzi & Consani, 1952: 65, fig. 1 (w.) ETHIOPIA.

***sagei.*** *Crematogaster sagei* Forel, 1902c: 204 (w.q.) INDIA. Combination in *C.* (*Acrocoelia*): Emery, 1922e: 152. Current subspecies: nominal plus *laevinota.*

#***saharensis.*** *Crematogaster aegyptiaca* var. *saharensis* Santschi, 1921d: 72 (w.) ALGERIA. Santschi, 1937e: 303 (q.m.). Stirps of *oasium* Santschi, 1937e: 303.

*salvazai. Crematogaster walshi* st. *bouvardi* var. *salvazai* Santschi, 1920h: 161 (w.) LAOS. Unavailable name.

[Also described as new by Santschi, 1924c: 99; unavailable name.]

**#*sampaioi.*** *Crematogaster brevispinosa* r. *sampaioi* Forel, 1912f: 213 (w.q.m.) BRAZIL. Combination in *C.* (*Orthocrema*): Emery, 1922e: 134.

***sanguinea.*** *Crematogaster sanguinea* Roger, 1863a: 208 (w.) CUBA. Combination in *C.* (*Acrocoelia*): Emery, 1922e: 141. Current subspecies: nominal plus *atavista, sotolongoi, torrei.*

***santschii.*** *Crematogaster santschii* Forel, 1913b: 322 (w.) ZAIRE. Combination in *C.* (*Oxygyne*): Wheeler, W.M. 1922a: 858; in *C.* (*Acrocoelia*): Emery, 1922e: 148. Current subspecies: nominal plus *clymene.*

**#*sapora.*** *Crematogaster impressa* var. *sapora* Forel, 1916: 408 (w.q.) ZAIRE. Combination in *C.* (*Acrocoelia*): Emery, 1922e: 147.

**#*sarawakana.*** *Crematogaster modiglianii* var. *sarawakana* Forel, 1911a: 25 (w.) BORNEO. Combination in *C.* (*Paracrema*): Emery, 1922e: 156.

**#*satan.*** *Crematogaster menileki* r. *satan* Forel, 1916: 407 (w.) ZAIRE. Combination in *C.* (*Acrocoelia*): Emery, 1922e: 148.

*satanula. Crematogaster menileki* r. *satan* var. *satanula* Forel, 1916: 407 (w.) ZAIRE. Unavailable name.

***saussurei.*** *Crematogaster saussurei* Forel, 1899c: 86 (w.) MEXICO. Combination in *C.* (*Acrocoelia*): Emery, 1922e: 141.

**#*savinae.*** *Crematogaster auberti* subsp. *savinae* Zimmermann, 1935: 25 (w.) YUGOSLAVIA.

*scabrida. Crematogaster subdentata* subsp. *kaschgariensis* var. *scabrida* Ruzsky, 1905b: 501 (w.) CAUCASUS. Unavailable name; material referred to *subdentata* by Arakelian, 1994: 45.

**#*scabrula.*** *Crematogaster froggatti* var. *scabrula* Emery, 1914b: 184 (w.) AUSTRALIA. Subspecies of *queenslandica:* Taylor & Brown, D.R. 1985: 63.

***scapamaris.*** *Crematogaster* (*Neocrema*) *scapamaris* Santschi, 1922d: 244 (w.q.m.) ARGENTINA.

***scelerata.*** *Crematogaster* (*Oxygyne*) *scelerata* Santschi, 1917f: 279 (w.q.m.) ARGENTINA. Combination in *C.* (*Neocrema*): Santschi, 1922d: 244. Current subspecies: nominal plus *taperensis.*

***schencki.*** *Crematogaster schencki* Forel, 1891b: 182, pl. 6, fig. 2 (w.q.) MADAGASCAR. Combination in *C.* (*Decacrema*): Forel, 1910a: 18; Forel, 1910f: 9.

**#*schereri.*** *Crematogaster stadelmanni* var. *schereri* Forel, 1911e: 273 (w.) LIBERIA. Combination in *C.* (*Nematocrema*): Wheeler, W.M. 1922a: 860.

***schimmeri.*** *Crematogaster schimmeri* Forel, 1912a: 69 (w.q.) TAIWAN. Combination in *C.* (*Decacrema*): Emery, 1922e: 132.

***schmidti.*** *Acrocoelia schmidti* Mayr, 1853a: 149 (w.q.) AUSTRIA. Ruzsky, 1905b: 491 (m.). Combination in *Crematogaster:* Mayr, 1855: 469; in *C.* (*Acrocoelia*): Emery, 1922e: 143. Junior synonym of *scutellaris:* Mayr, 1855: 469; Mayr, 1856: xxi; Mayr, 1863: 405. Revived from synonymy as species: Emery, 1891b: 14; Dalla Torre, 1893: 85. Variety/subspecies of *scutellaris:* Ruzsky, 1905b: 491; Forel, 1911d: 340; Emery, 1912e: 654; Emery, 1914d: 157; Santschi, 1929e: 148. Junior synonym of *scutellaris:* Baroni Urbani, 1964a: 4. Revived from synonymy, and revived status as species: Agosti & Collingwood, 1987a: 54. Senior synonym of *christowitchii* and *medispina:* Emery, 1922e: 143, the latter confirmed in Müller, 1923: 72; of *karawaewi:* Atanassov & Dlussky, 1992: 184.

***schultzei.*** *Crematogaster schultzei* Forel, 1910f: 6 (w.q.) SOUTH AFRICA. Combination in *C.* (*Crematogaster*): Wheeler, W.M. 1922a: 843; in *C.* (*Acrocoelia*): Emery, 1922e: 148.

**#*schumanni.*** *Crematogaster africana* subsp. *schumanni* Mayr, 1895: 144 (w.) CAMEROUN. Combination in *C.* (*Atopogyne*): Wheeler, W.M. 1922a: 852.

**#*schuppi.*** *Crematogaster brevispinosa* var. *schuppi* Forel, 1901d: 299 (w.q.) BRAZIL. Combination in *C.* (*Orthocrema*): Emery, 1922e: 134.

***scita.*** *Crematogaster scita* Forel, 1902h: 409 (w.) AUSTRALIA. Combination in *C.* (*Orthocrema*): Emery, 1922e: 133; in *C.* (*Mesocrema*): Santschi, 1928b: 33. Current subspecies: nominal plus *mixta.*

**#*scrutans.*** *Crematogaster scrutans* Forel, 1910c: 258 (w.) ETHIOPIA. Menozzi, 1922a: 354 (q.). Combination in *C.* (*Acrocoelia*): Emery, 1922e: 144. Subspecies of *alulai:* Emery, 1915g: 11; Wheeler, W.M. 1922a: 829.

**#*sculpticeps.*** *Crematogaster frivola* var. *sculpticeps* Forel, 1907h: 279 (w.) AUSTRALIA. Combination in *C.* (*Orthocrema*): Emery, 1922e: 132. Subspecies of *frivola:* Taylor & Brown, D.R. 1985: 61.

***sculpturata.*** *Crematogaster sculpturata* Pergande, 1896: 876 (w.) MEXICO. Wheeler, W.M. 1934g: 171 (w. redescribed). Combination in *C.* (*Orthocrema*): Emery, 1922e: 136. Current subspecies: nominal plus *accola, phytoeca.*

*scurra. Crematogaster brevispinosa* subsp. *tumulifera* var. *scurra* Forel, 1914c: 615 (w.) COLOMBIA. Unavailable name.

***scutellaris.*** *Formica scutellaris* Olivier, 1792: 497 (w.) FRANCE. Mayr, 1855: 469 (q.m.). Combination in *Crematogaster:* Mayr, 1855: 469; in *C.* (*Acrocoelia*): Emery, 1922e: 143. Senior synonym of *rediana, rubriceps* and *ruficeps:* Mayr, 1855: 469; of *haematocephala:* Roger, 1863b: 36; of *grouvellei* and *lichtensteini:* Bernard, 1967: 164; of *corsica* and *degener:* Casevitz-Weulersse, 1990a: 137. Material of the unavailable name *hybrida* referred here: Baroni Urbani, 1964a: 4. See also: Emery, 1912e: 652; Stitz, 1939: 128; Baroni Urbani, 1971c: 79; Kutter, 1977c: 88. Current subspecies: nominal plus *alii, nigra* (unresolved junior homonym), *tenuispina.*

**#*sejuncta.*** *Crematogaster sejuncta* Stitz, 1916: 387, fig. 6 (w.) ZAIRE. Combination in *C.* (*Sphaerocrema*): Wheeler, W.M. 1922a: 848. Stirps of *gambiensis:* Santschi, 1926b: 227.

**#*sellula.*** *Crematogaster chiarinii* st. *sellula* Santschi, 1914a: 431 (w.) TANZANIA. Combination in *C.* (*Acrocoelia*): Emery, 1922e: 146.

**#*sembilana.*** *Crematogaster* (*Decacrema*) *borneensis* subsp. *sembilana* Forel, 1911d: 385 (q.m.) WEST MALAYSIA.

**#*semiclara.*** *Crematogaster* (*Sphaerocrema*) *bequaerti* var. *semiclara* Santschi, 1928d: 66 (w.) ZIMBABWE.

*semisericea. Crematogaster* (*Orthocrema*) *brevispinosa* st. *sericea* var. *semisericea* Santschi, 1923c: 249 (w.) ARGENTINA. Unavailable name.

***semperi.*** *Crematogaster semperi* Emery, 1893g: 263, pl. 2, fig. 7 (w.) PHILIPPINES. Combination in *C.* (*Acrocoelia*): Emery, 1922e: 152.

***senegalensis.*** *Crematogaster senegalensis* Roger, 1863a: 206 (w.q.) SENEGAL. Wheeler, G.C. & Wheeler, J. 1952b: 254 (l.). Combination in *C.*(*Acrocoelia*): Emery, 1922e: 144. Subspecies/race/stirps of *aegyptiaca:* Santschi, 1914d: 344; Emery, 1915g: 10; Wheeler, W.M. 1922a: 828; Emery, 1922e: 144; Santschi, 1929e: 148. Revived status as species: Forel, 1922: 93; Bernard, 1950b: 288. Current subspecies: nominal plus *goliathula.*

**#*sericea.*** *Crematogaster brevispinosa* r. *sericea* Forel, 1912f: 214 (w.) BRAZIL. Combination in *C.* (*Orthocrema*): Emery, 1922e: 134; Santschi, 1923c: 249.

**#*sewardi.*** *Crematogaster deformis* r. *sewardi* Forel, 1901h: 64 (w.q.) BORNEO. Combination in *C.* (*Physocrema*): Emery, 1922e: 140.

***sewellii.*** *Crematogaster inermis* r. *sewellii* Forel, 1891b: 196, pl. 6, fig. 9 (w.) MADAGASCAR. Combination in *C.* (*Crematogaster*): Wheeler, W.M. 1922a: 844; in *C.* (*Acrocoelia*): Emery, 1922e: 148. Raised to species: Dalla Torre, 1893: 86; Mayr, 1895: 138; Emery, 1895f: 342. Current subspecies: nominal plus *acis, dentata, improba, marnoi, mauritiana.*

*sic. Crematogaster buddhae* r. *sic* Forel, 1907a: 24 (w.) WEST MALAYSIA. Junior synonym of *treubi:* Emery, 1922e: 132.

**#*sikkimensis.*** *Crematogaster* (*Oxygyne*) *dalyi* var. *sikkimensis* Forel, 1904c: 24 (w.) INDIA.

**#*simia.*** *Crematogaster tricolor* var. *simia* Forel, 1910f: 6 (w.q.m.) BOTSWANA. Combination in *C.* (*Acrocoelia*): Emery, 1922e: 145. Variety of *castanea:* Wheeler, W.M. 1922a: 831; Santschi, 1937d: 229.

***similis.*** *Crematogaster similis* Stitz, 1911b: 388, fig. 7 (w.) ZAIRE. Combination in *C.* (*Sphaerocrema*): Santschi, 1918d: 182.

***simoni.*** *Crematogaster simoni* Emery, 1893g: 263, pl. 6, fig. 6 (w.) PHILIPPINES. Combination in *C.* (*Acrocoelia*): Emery, 1922e: 152.

**#*sjostedti.*** *Crematogaster sjostedti* Mayr, 1907b: 17 (w.) TANZANIA. Combination in *C.* (*Crematogaster*): Wheeler, W.M. 1922a: 844; in *C.* (*Acrocoelia*): Emery, 1922e: 148. Race of *gallicola:* Forel, 1913a: 124 (in text). Revived status as species: Santschi, 1919b: 236; Wheeler, W.M. 1922a: 844; Santschi, 1928d: 65. Stirps of *gerstaeckeri:* Santschi, 1930b: 69; Santschi, 1937a: 55.

***skounensis.*** *Crematogaster skounensis* Soulié, in Fromantin & Soulié, 1961: 99, figs. 3, 6, 7, 8 (w.q.m.) CAMBODIA. [*Crematogaster skounensis* Soulié, 1960: 284; *nomen nudum.*]

**#*smithi.*** *Crematogaster* (*Orthocrema*) *minutissima* subsp. *smithi* Creighton, 1950a: 205. Replacement name for *thoracica* Creighton, 1939b: 138. [Junior primary homonym of *thoracica* Santschi, 1921c: 118.]

**#*smythiesii.*** *Crematogaster biroi* var. *smythiesii* Forel, 1902c: 203 (w.q.) INDIA. Combination in *C.* (*Orthocrema*): Emery, 1922e: 132.

**#*soengeiensis.*** *Crematogaster tumidula* var. *soengeiensis* Forel, 1913k: 77 (w.) SUMATRA. Combination in *C.* (*Acrocoelia*), and variety of *pia:* Emery, 1922e: 151.

***solenopsides.*** *Crematogaster solenopsides* Emery, 1899e: 487 (w.) TANZANIA. Menozzi, 1930b: 105 (m.). Combination in *C.* (*Decacrema*): Forel, 1916: 411. Current subspecies: nominal plus *costeboriensis, flavida, mandonbii.*

**solers.** *Crematogaster solers* Forel, 1910f: 7 (w.) NAMIBIA. Combination in *C.* (*Crematogaster*): Wheeler, W.M. 1922a: 844; in *C.* (*Acrocoelia*): Emery, 1922e: 149.

**#sordida.** *Crematogaster auberti* var. *sordida* Forel, 1909c: 104 (w.) ALGERIA. [First available use of *Crematogaster auberti* r. *laestrygon* var. *sordida* Forel, 1894d: 26; unavailable name.] Wheeler, G.C. & Wheeler, J. 1952b: 255 (l.). Combination in *C.* (*Acrocoelia*): Emery, 1922e: 142. Variety of *antaris:* Santschi, 1921d: 71.

**#sordidoides.** *Crematogaster* (*Sphaerocrema*) *luctans* var. *sordidoides* Santschi, 1930d: 269 (w.) SIERRA LEONE.

**sordidula.** *Myrmica sordidula* Nylander, 1849: 44 (w.) SICILY. Mayr, 1855: 472 (q.); Forel, 1870: 308 (m.). Combination in *Crematogaster:* Mayr, 1855: 472; in *C.* (*Orthocrema*): Santschi, 1918d: 182. Senior synonym of *mayri:* Mayr, 1855: 472; of *flachii:* Finzi, 1930d: 313; both synonyms confirmed: Agosti & Collingwood, 1987a: 54; of *caspica:* Atanassov & Dlussky, 1992: 182. See also: Emery, 1912e: 667; Kutter, 1977c: 89. Current subspecies: nominal plus *aeolia, marocana, molongori, osmanica.*

**sorokini.** *Crematogaster auberti* var. *sorokini* Ruzsky, 1905b: 495 (w.) TURKESTAN. Combination in *C.* (*Acrocoelia*): Emery, 1922e: 142. Subspecies of *inermis:* Emery, 1926: 3. Raised to species: Dlussky & Zabelin, 1985: 215.

**soror.** *Crematogaster* (*Oxygyne*) *soror* Forel, 1902c: 200 (w.q.) INDIA.

**#sotolongoi.** *Crematogaster sanguinea* subsp. *sotolongoi* Mann, 1920: 425 (w.) CUBA.

**spengeli.** *Crematogaster spengeli* Forel, 1912n: 56 (w.) SUMATRA. Combination in *C.* (*Paracrema*): Santschi, 1918d: 182. Current subspecies: nominal plus *compacta, picata, taipingensis.*

**#spinozai.** *Crematogaster egidyi* subsp. *spinozai* Forel, 1911a: 25 (q.w.) BORNEO. Combination in *C.* (*Acrocoelia*): Emery, 1922e: 151.

**#spissata.** *Crematogaster* (*Nematocrema*) *stadelmanni* var. *spissata* Santschi, 1937g: 80, fig. 5 (w.) ZAIRE.

**#spuria.** *Crematogaster gallicola* r. *spuria* Forel, 1913b: 323 (w.) ZAIRE. Combination in *C.* (*Acrocoelia*): Emery, 1922e: 147. Variety of *menilekii:* Santschi, 1939c: 7.

**stadelmanni.** *Crematogaster stadelmanni* Mayr, 1895: 140 (w.q.) LIBERIA. Delage-Darchen, 1972b: 259 (l.). Combination in *C.* (*Oxygyne*): Forel, 1915c: 337; in *C.* (*Nematocrema*): Santschi, 1918d: 182. Current subspecies: nominal plus *anguliceps, angustata, dolichocephala, gracilenta, intermedia, ovinodis, schereri, spissata.*

**#staitchi.** *Crematogaster foraminiceps* r. *staitchi* Forel, 1915c: 339 (w.q.) ZAIRE. Combination in *C.* (*Acrocoelia*): Emery, 1922e: 147.

**#stanleyi.** *Crematogaster* (*Atopogyne*) *africana* var. *stanleyi* Wheeler, W.M. 1922a: 852 (w.) CONGO. [First available use of *Crematogaster buchneri* st. *africana* var. *stanleyi* Santschi, 1915c: 253; unavailable name.] [*Crematogaster africana* var. *stanleyi* Forel, 1911f: 278; *nomen nudum.*] Santschi, 1937g: 79 (q.).

**steinheili.** *Crematogaster steinheili* Forel, 1881: 15 (w.) ANTILLES. Forel, 1912f: 216 (q.m.); Menozzi & Russo, 1930: 154 (q.m.). Combination in *C.* (*Orthocrema*): Emery, 1922e: 136. Junior synonym of *victima:* Mayr, 1887: 624. Revived from synonymy as variety of *victima:* Dalla Torre, 1893: 87; Emery, 1895c: 288 (footnote). Revived status as species: Forel, 1912f: 216; Wheeler, W.M. 1917g: 460; Gallardo, 1934: 45. Current subspecies: nominal plus *argentina.*

**stenocephala.** *Crematogaster* (*Acrocoelia*) *stenocephala* Emery, 1922e: 145 (footnote) (q.) SENEGAL. Combination in *C.* (*Sphaerocrema*): Santschi, 1926b: 227; Santschi, 1930a: 65.

**stethogompha.** *Crematogaster* (*Physocrema*) *stethogompha* Wheeler, W.M. 1919e: 75 (w.) BORNEO. Current subspecies: nominal plus *detritinodis.*

**stigmata.** *Crematogaster stigmata* Santschi, 1914d: 344, fig. 15 (w.m.) NIGERIA. Combination in *C.* (*Crematogaster*): Wheeler, W.M. 1922a: 845; in *C.* (*Acrocoelia*): Emery, 1922e: 149.

**#stigmatica.** *Crematogaster goeldii* var. *stigmatica* Forel, 1911c: 301 (w.m.) BRAZIL. Combination in *C.* (*Orthocrema*): Emery, 1922e: 135. Subspecies of *torosa:* current status.

**stollii.** *Crematogaster stollii* Forel, 1885a: 373 (w.) GUATEMALA. Forel, 1899c: 83 (q.). Combination in *C.* (*Neocrema*): Santschi, 1918d: 182; in *C.* (*Oxygyne*): Santschi, 1922d: 244; in *C.* (*Orthocrema*): Emery, 1922e: 136; Kempf, 1972a: 89. Senior synonym on *parvispina:* Kempf, 1960c: 174; of *hyperphyes:* Kempf, 1968b: 384. Current subspecies: nominal plus *amazonensis, autruni, guianensis.*

**#stolonis.** *Crematogaster* (*Atopogyne*) *africana* st. *stolonis* Santschi, 1937g: 78, fig. 4 (w.) ZAIRE.

**#striaticeps.** *Crematogaster laestrygon* var. *striaticeps* Forel, 1909c: 104 (w.) ALGERIA. [First available use of *Crematogaster auberti* r. *laestrygon* var. *striaticeps* Forel, 1902g: 462; unavailable name.] Santschi, 1937e: 308 (q.m.). Combination in *C.* (*Acrocoelia*): Emery, 1922e: 142.

**#striatinota.** *Crematogaster brevispinosa* var. *striatinota* Forel, 1912f: 211 (w.) COLOMBIA.

**striatula.** *Crematogaster striatulus* Emery, 1892c: liii (w.) IVORY COAST. Emery, 1899e: 482 (q.); Delage-Darchen, 1978: 293 (l.). Combination in *C.* (*Sphaerocrema*): Santschi, 1918d: 182. Current subspecies: nominal plus *benitensis, horatii, langi, obstinata, omega.*

**subcircularis.** *Crematogaster subcircularis* Mayr, 1879: 685 (w.) BORNEO. Combination in *C.* (*Acrocoelia*): Emery, 1922e: 152.

**subdentata.** *Crematogaster subdentata* Mayr, 1877: 19 (w.q.) TURKESTAN. Forel, 1904b: 372 (m.). [Redescribed in Latin: Mayr, 1880: 39.] Combination in *C.* (*Acrocoelia*): Emery, 1922e: 144. Material of the unavailable name *scabrida* referred here by Arakelian, 1994: 45. See also: Arnol'di & Dlussky, 1978: 538. Current subspecies: nominal plus *flavicapilla, kaschgariensis, kuznetzovi.*

**#submaura.** *Crematogaster laestrygon* var. *submaura* Lomnicki, 1925b: 1 (w.) BALEARIC IS. Variety/subspecies of *auberti:* Menozzi, 1926b: 182; of *laestrygon:* Santschi, 1937e: 310; Collingwood, 1978: 69.

**subnuda.** *Crematogaster subnuda* Mayr, 1879: 682 (w.) INDIA. Imai, *et al.* 1984: 6 (k.). Combination in *C.* (*Acrocoelia*): Emery, 1922e: 150. Subspecies/stirps of *laboriosa:* Santschi, 1928b: 33; Wheeler, W.M. 1929f: 5 (in text); of *brunnea:* Emery, 1922e: 150; Santschi, 1925f: 85; Wheeler, W.M. 1930a: 98. Revived status as species: Mukerjee, 1930: 153. Current subspecies: nominal plus *discinodis, formosae.*

*subopaca. Crematogaster lineolata* var. *subopaca* Emery, 1895c: 283 (w.q.m.) U.S.A. Wheeler, G.C. & Wheeler, J. 1952b: 252 (l.). Combination in *C.* (*Acrocoelia*): Emery, 1922e: 141. Subspecies of *lineolata:* Creighton, 1950a: 214. Junior synonym of *punctulata:* Buren, in Smith, M.R. 1958c: 127; Smith, D.R. 1979: 1380; of *lineolata:* Johnson, 1988: 320.

*subpilosa. Crematogaster lineolata* subsp. *subpilosa* Wheeler, W.M. 1913c: 115 U.S.A. *Nomen nudum;* material referred to *pilosa:* Buren, in Smith, M.R. 1958c: 127.

**#subsulcata.** *Crematogaster chiarinii* st. *subsulcata* Santschi, 1914a: 432 (w.) SUDAN. Combination in *C.* (*Acrocoelia*): Emery, 1922e: 146.

*subsulcata. Crematogaster* (*Atopogyne*) *subsulcata* Santschi, 1932a: 386 (w.) SOUTH AFRICA. [Junior primary homonym of *subsulcata* Santschi, above.] Replacement name: *resulcata* **new name**.

**#subtilis.** *Crematogaster irritabilis* var. *subtilis* Viehmeyer, 1914c: 32 (w.) NEW GUINEA. Combination in *C.* (*Orthocrema*): Emery, 1922e: 133.

**#subtonsa.** *Crematogaster* (*Orthocrema*) *brevispinosa* var. *subtonsa* Santschi, 1925d: 229 (w.) BRAZIL.

**#suddensis.** *Crematogaster* (*Crematogaster*) *menileki* subsp. *suddensis* Weber, 1943c: 356 (w.q.) SUDAN.

*sulcata. Crematogaster sulcata* Mayr, 1870a: 403 (w.) COLOMBIA. Forel, 1899c: 84 (q.). Combination in *C.* (*Neocrema*): Santschi, 1922d: 244; in *C.* (*Orthocrema*): Emery, 1922e: 135. Junior synonym of *montezumia:* Forel, 1901h: 65; Santschi, 1922d: 244. Revived from synonymy as variety of *montezumia:* Forel, 1904c: 36; Emery, 1922e: 135. Junior synonym of *montezumia:* Kempf, 1968b: 386.

**sumichrasti.** *Crematogaster sumichrasti* Mayr, 1870b: 993 (w.) MEXICO. Forel, 1908b: 46 (q.m.). Combination in *C.* (*Orthocrema*): Emery, 1922e: 136. Current subspecies: nominal plus *maya, surdior.*

**#sundaica.** *Crematogaster subnuda* var. *sundaica* Forel, 1909d: 225 (w.) JAVA. Variety of *brunnea:* Santschi, 1928b: 33.

**#surbeki.** *Crematogaster* (*Paracrema*) *modiglianii* var. *surbeki* Santschi, 1925f: 90 (w.) SUMATRA.

*surcoufi. Crematogaster auberti* st. *laestrygon* var. *surcoufi* Santschi, 1921d: 71 (w.) ALGERIA. Unavailable name.

**#surdior.** *Crematogaster sumichrasti* st. *surdior* Forel, 1885a: 375 (w.q.) GUATEMALA. Combination in *C.* (*Orthocrema*): Emery, 1922e: 136. Raised to species: Dalla Torre, 1893: 87. Race/subspecies of *sumichrasti:* Forel, 1899c: 82; Emery, 1922e: 136; Kempf, 1972a: 89.

**#suturalis.** *Crematogaster abstinens* var. *suturalis* Forel, 1912f: 216 (w.) VENEZUELA. Combination in *C.* (*Orthocrema*): Emery, 1922e: 133.

**#sycites.** *Crematogaster australis* var. *sycites* Forel, 1916: 406 (w.) AUSTRALIA. Combination in *C.* (*Acrocoelia*): Emery, 1922e: 152. Subspecies of *australis:* Taylor & Brown, D.R. 1985: 60.

**#symbia.** *Crematogaster* (*Decacrema*) *borneensis* subsp. *symbia* Forel, 1911a: 34 (w.q.m.) BORNEO.

**#*szaboi*.** *Crematogaster egidyi* var. *szaboi* Forel, 1913d: 437 (w.) SINGAPORE. Combination in *C.* (*Acrocoelia*): Emery, 1922e: 151.

**#*taediosa*.** *Crematogaster taediosa* Forel, 1894b: 98 (w.) ETHIOPIA. Combination in *C.* (*Acrocoelia*): Emery, 1922e: 146. Subspecies of *chiarinii:* Emery, 1915g: 12.

*tagala*. *Crematogaster subnuda* subsp. *politula* var. *tagala* Forel, 1910d: 124 (w.q.m.) PHILIPPINES. Unavailable name.

**#*taipingensis*.** *Crematogaster spengeli* r. *taipingensis* Forel, 1913k: 75, fig. V (w.q.m.) WEST MALAYSIA. Combination in *C.* (*Paracrema*): Emery, 1922e: 156.

**#*taivanae*.** *Crematogaster tumidula* var. *taivanae* Forel, 1913f: 194 (w.q.) TAIWAN. Combination in *C.* (*Acrocoelia*), and variety of *pia:* Emery, 1922e: 151.

*tanganikana*. *Crematogaster* (*Acrocoelia*) *castanea* st. *bruta* var. *tanganikana* Santschi, 1926b: 214 (w.) MALAWI. Unavailable name.

**#*taperensis*.** *Crematogaster* (*Neocrema*) *scelerata* var. *taperensis* Borgmeier, 1937b: 234 (w.q.) BRAZIL.

*targuia*. *Crematogaster* (*Acrocoelia*) *aegyptiaca* st. *senegalensis* var. *targuia* Santschi, 1929e: 148 (w.) ALGERIA. Unavailable name.

***tarsata*.** *Crematogaster tarsata* Smith, F. 1865: 74 (w.) INDONESIA (Morotai I.). Combination in *C.* (*Xiphocrema*): Emery, 1922e: 139; Donisthorpe, 1932c: 473.

**#*tenuicula*.** *Crematogaster longispina* r. *tenuicula* Forel, 1904c: 36 (w.) BRAZIL. Combination in *C.* (*Orthocrema*): Emery, 1922e: 136.

**#*tenuipilis*.** *Crematogaster* (*Acrocoelia*) *mimosae* var. *tenuipilis* Santschi, 1937a: 55 (w.) KENYA.

**#*tenuispina*.** *Crematogaster scutellaris* r. *tenuispina* Forel, 1902a: 154 (w.) ALGERIA.

**#*teranishii*.** *Crematogaster* (*Acrocoelia*) *brunnea* st. *teranishii* Santschi, 1930d: 265 (w.) JAPAN.

***terminalis*.** *Myrmica terminalis* Shuckard, 1838: 627 (w.) GREAT BRITAIN. Combination in *Crematogaster:* Dalla Torre, 1893: 87. See also: Donisthorpe, 1915d: 338.

*terricola*. *Crematogaster* (*Orthocrema*) *terricola* Borgmeier, 1929: 210, pl. 6, fig. 1; pl. 7, fig. 7 (w.) BRAZIL. Junior synonym of *cristata:* Kempf, 1968b: 384.

***tetracantha*.** *Crematogaster tetracantha* Emery, 1887b: 467, pl. 2, fig. 18 (w.) NEW GUINEA. Combination in *C.* (*Xiphocrema*): Forel, 1913k: 80.

*texana*. *Crematogaster* (*Acrocoelia*) *opaca* var. *texana* Santschi, 1929f: 91 (w.) U.S.A. Junior synonym of *punctulata:* Creighton, 1950a: 214.

*thais*. *Crematogaster excisa* r. *andrei* var. *thais* Forel, 1913a: 126 (w.) SOUTH AFRICA. Unavailable name.

**#*thalia*.** *Crematogaster brevispinosa* var. *thalia* Forel, 1911e: 273 (w.q.) PARAGUAY. Combination in *C.* (*Orhtocrema*): Emery, 1922e: 134.

*themis*. *Crematogaster* (*Orthocrema*) *brevispinosa* st. *minutior* var. *themis* Santschi, 1925d: 230 (w.) BRAZIL. Unavailable name.

**#*theryi*.** *Crematogaster auberti* st. *theryi* Santschi, 1921d: 71 (w.) MOROCCO. Santschi, 1937e: 309 (q.). Stirps/subspecies of *auberti:* Santschi, 1931a: 5; of *laestrygon:* Emery, 1926: 2; Santschi, 1936c: 200; Cagniant, 1964: 89.

***theta*.** *Crematogaster africana* var. *theta* Forel, 1911f: 278 (w.) ZAIRE. Forel, 1916: 410 (q.). Combination in *C.* (*Atopogyne*): Wheeler, W.M. 1922a: 857. Raised to species: Wheeler, W.M. 1922a: 857; Emery, 1922e: 155; Santschi, 1935a: 262.

**#*thoracica*.** *Crematogaster* (*Atopogyne*) *africana* st. *thoracica* Santschi, 1921c: 118 (w.) ZAIRE.

*thoracica*. *Crematogaster* (*Orthocrema*) *minutissima* subsp. *thoracica* Creighton, 1939b: 138 (w.) U.S.A. [Junior primary homonym of *thoracica* Santschi, above.] Replacement name: *smithi* Creighton, 1950a: 205.

**#*tibialis*.** *Crematogaster* (*Atopogyne*) *africana* subsp. *tibialis* Santschi, in Wheeler, W.M. 1922a: 157 (w.) ZAIRE.

***togoensis*.** *Crematogaster* (*Atopogyne*) *togoensis* Donisthorpe, 1945d: 270 (w.) TOGO.

***torosa*.** *Crematogaster torosa* Mayr, 1870a: 404 (w.) COLOMBIA. Combination in *C.* (*Orthocrema*): Emery, 1922e: 136; Santschi, 1925d: 229. Current subspecies: nominal plus *chodati, goeldii, stigmatica.*

**#*torrei*.** *Crematogaster sanguinea* var. *torrei* Wheeler, W.M. 1913b: 490 (w.q.) CUBA. Combination in *C.* (*Acrocoelia*): Emery, 1922e: 141.

**#*townsendi*.** *Crematogaster* (*Orthocrema*) *brevispinosa* subsp. *townsendi* Wheeler, W.M. 1925a: 25 (w.) PERU.

***transiens*.** *Crematogaster buchneri* r. *transiens* Forel, 1913b: 324 (w.) ZAIRE. Combination in *C.* (*Atopogyne*): Santschi, 1916b: 501. Raised to species: Wheeler, W.M. 1922a: 160; Emery, 1922e: 155.

***transvaalensis*.** *Crematogaster transvaalensis* Forel, 1894b: 96 (w.) SOUTH AFRICA. Combination in *C.* (*Orthocrema*): Wheeler, W.M. 1922a: 851. Current subspecies: nominal plus *hammi.*

**#*transversiruga*.** *Crematogaster gambiensis* var. *transversiruga* Santschi, 1914d: 346 (w.) SUDAN. Combination in *C.* (*Sphaerocrema*): Wheeler, W.M. 1922a: 846.

*transversiruginota*. *Crematogaster buchneri* r. *winkleri* var. *transversiruginota* Forel, 1916: 410 (w.) ZAIRE. Unavailable name.

***trautweini*.** *Crematogaster* (*Oxygyne*) *trautweini* Viehmeyer, 1914c: 33 (w.) CAMEROUN. Combination in *C.* (*Acrocoelia*): Emery, 1922e: 148.

***travancorensis*.** *Crematogaster* (*Oxygyne*) *travancorensis* Forel, 1902c: 200 (w.) INDIA.

***treubi*.** *Crematogaster treubi* Emery, 1896f: 246 (w.) JAVA. Combination in *C.* (*Orthocrema*): Emery, 1922e: 132; in *C.* (*Mesocrema*): Santschi, 1937e: 310. Senior synonym of *sic:* Emery, 1922e: 132. Current subspecies: nominal plus *apilis, vastatrix.*

**#*tricolor*.** *Crematogaster tricolor* Gerstäcker, 1859: 263 (w.) MOZAMBIQUE. [Also described as new by Gerstäcker, 1862: 515.] Forel, 1887: 388 (q.m.). Combination in *C.* (*Acrocoelia*): Emery, 1922e: 145. Junior synonym of *castanea:* Mayr, 1907b: 16; Emery, 1915g: 12; Wheeler, W.M. 1922a: 830. Revived from synonymy as variety/subspecies of *castanea:* Arnold, 1920a: 489; Emery, 1922e: 145; Santschi, 1926b: 213.

*tricoloroides*. *Crematogaster* (*Acrocoelia*) *gerstaeckeri* st. *sjostedti* var. *tricoloroides* Santschi, 1937a: 55 (w.) KENYA. Unavailable name.

**#*tropicorum*.** *Crematogaster capensis* subsp. *tropicorum* Forel, 1910e: 435 (w.) MOZAMBIQUE. Combination in *C.* (*Acrocoelia*): Emery, 1922e: 145.

*tubuli*. *Crematogaster* (*Decacrema*) *borneensis* subsp. *hosei* var. *tubuli* Viehmeyer, 1916a: 126 (w.) SINGAPORE. Unavailable name.

*tucumanensis*. *Crematogaster brevispinosa* r. *moelleri* var. *tucumanensis* Forel, 1914d: 271 (w.q.) ARGENTINA. Unavailable name.

***tumidula*.** *Crematogaster tumidula* Emery, 1900d: 689, fig. 8 (w.) SUMATRA. Combination in *C.* (*Physocrema*): Emery, 1922e: 140; Donisthorpe, 1941l: 226.

*tumulicula*. *Crematogaster brevispinosa* subsp. *tumulifera* var. *tumulicula* Forel, 1909a: 258 (w.) PARAGUAY. Unavailable name.

**#*tumulifera*.** *Crematogaster brevispinosa* var. *tumulifera* Forel, 1899c: 84 (w.) NICARAGUA. Forel, 1908b: 47 (q.m.); Wheeler, G.C. & Wheeler, J. 1952b: 260 (l.). Combination in *C.* (*Orthocrema*): Emery, 1922e: 134. Subspecies of *brevispinosa:* Forel, 1907a: 25; Forel, 1908c: 369; Forel, 1909a: 258; Forel, 1912f: 215.

*turbida*. *Crematogaster* (*Orthocrema*) *victima* st. *nitidiceps* var. *turbida* Santschi, 1925d: 231 (w.) BRAZIL. Unavailable name.

**#*turkanensis*.** *Crematogaster* (*Acrocoelia*) *aegyptiaca* var. *turkanensis* Santschi, 1935b: 271 (w.) KENYA.

**#*uasina*.** *Crematogaster* (*Atopogyne*) *buchneri* var. *uasina* Santschi, 1935b: 271 (w.) KENYA.

***udo*.** *Crematogaster sordidula* var. *udo* Forel, 1905c: 20 (w.) JAVA. Combination in *C.* (*Orthocrema*), and raised to species: Emery, 1922e: 132.

**#*ugandensis*.** *Crematogaster muralti* st. *ugandensis* Santschi, 1914b: 99 (w.) UGANDA. Combination in *C.* (*Orthocrema*): Wheeler, W.M. 1922a: 850.

**#*ulugurensis*.** *Crematogaster ferruginea* var. *ulugurensis* Forel, 1911d: 369 (w.) TANZANIA. Santschi, 1914b: 86 (q.m.). Combination in *C.* (*Acrocoelia*): Emery, 1922e: 145. Subspecies of *castanea:* current status.

***unciata*.** *Crematogaster* (*Orthocrema*) *unciata* Santschi, 1925d: 228 (w.) BRAZIL.

**#*uruguayensis*.** *Crematogaster atra* st. *uruguayensis* Santschi, 1912e: 527 (w.) URUGUAY.

***ustiventris*.** *Crematogaster* (*Orthocrema*) *ustiventris* Menozzi, 1935a: 110, fig. 6 (w.) NEW GUINEA.

**#*vacca*.** *Crematogaster deformis* subsp. *vacca* Forel, 1911d: 384 (w.) WEST MALAYSIA.

**#*vagula*.** *Crematogaster* (*Acrocoelia*) *matsumurai* subsp. *vagula* Wheeler, W.M. 1928d: 110 (w.q.) JAPAN.

*vanda*. *Crematogaster distans* subsp. *vanda* Borgmeier, 1929: 209 (w.) BRAZIL. Junior synonym of *distans:* Kempf, 1968b: 390.

***vandeli*.** *Crematogaster* (*Acrocoelia*) *vandeli* Soulié, in Soulié & Fromantin, 1961: 100, figs. 4, 6, 7, 8 (w.q.m.) CAMBODIA. [*Crematogaster vandeli* Soulié, 1961: 284. *Nomen nudum.*]

***vandermeermohri*.** *Crematogaster* (*Paracrema*) *vandermeermohri* Menozzi, 1930a: 1, fig. 1 (w.) SUMATRA.

*vanderpyli*. *Crematogaster vanderpyli* Stärcke, 1944b: xix NETHERLANDS. *Nomen nudum.*

**#*variegata*.** *Crematogaster africana* var. *variegata* Mayr, 1902: 294 (w.) CAMEROUN. Combination in *C.* (*Atopogyne*): Wheeler, W.M. 1922a: 852.

**#vastatrix.** *Crematogaster treubi* var. *vastatrix* Forel, 1911a: 24 (w.q.) JAVA. Combination in *C.* (*Orthocrema*): Emery, 1922e: 132.

*vaucheri. Crematogaster laestrygon* subsp. *vaucheri* Emery, 1926: 2 (w.) MOROCCO. [First available use of *Crematogaster auberti* st. *laestrygon* var. *vaucheri* Santschi, 1921d: 71; unavailable name.] Junior synonym of *laestrygon:* Collingwood, 1978: 69.

**vermiculata.** *Crematogaster vermiculata* Emery, 1895c: 286 (w.) U.S.A. Wheeler, G.C. & Wheeler, J. 1960b: 13 (l.). Combination in *C.* (*Acrocoelia*): Emery, 1922e: 141; in *C.* (*Crematogaster*): Buren, 1968b: 91. Subspecies of *coarctata:* Creighton, 1950a: 209. Revived status as species: Buren, 1968b: 91 (in key); Johnson, 1988: 323. Material of the *nomen nudum coachellai* referred here: Brown, 1949a: 48.

**#vezenyii.** *Crematogaster quadriformis* subsp. *vezenyii* Forel, 1907a: 25 (w.) PARAGUAY. Combination in *C.* (*Orthocrema*): Emery, 1922e: 135.

**#vicina.** *Crematogaster vicina* André, 1893b: 151 (w.) JAMAICA. Wheeler, W.M. 1917g: 461 (q.m.). Combination in *C.* (*Orthocrema*): Emery, 1922e: 134. Race of *brevispinosa:* Forel, 1912f: 212. Senior synonym of *wighti:* Wheeler, W.M. 1917g: 461 (in text).

**victima.** *Crematogaster victima* Smith, F. 1858b: 140 (w.) BRAZIL. Mayr, 1870b: 995 (q.); Wheeler, G.C. & Wheeler, J. 1973a: 30 (l.). Combination in *C.* (*Orthocrema*): Emery, 1922e: 136; Santschi, 1925d: 231. Current subspecies: nominal plus *allegrensis, cisplatinalis, cubaensis, nitidiceps, obscurata.*

**#victoriosa.** *Crematogaster acaciae* st. *victoriosa* Santschi, 1916b: 497 (w.) ZIMBABWE. Combination in *C.* (*Acrocoelia*): Emery, 1922e: 144.

**vidua.** *Crematogaster* (*Acrocoelia*) *vidua* Santschi, 1928d: 62 (w.) IVORY COAST.

**#viehmeyeri.** *Crematogaster* (*Acrocoelia*) *viehmeyeri* Santschi, 1921c: 117 (w.) TANZANIA. Stirps of *menilekii:* Santschi, 1926b: 217.

*virgula. Crematogaster virgula* Forel, 1899c: 85, pl. 4, fig. 4 (w.) COSTA RICA. Forel, 1908b: 48 (q.). Combination in *C.* (*Orthocrema*): Emery, 1922e: 136. Junior synonym of *erecta:* Kempf, 1968b: 381.

**vitalisi.** *Crematogaster* (*Atopogyne*) *vitalisi* Menozzi, 1925c: 448 (q.) PHILIPPINES.

*vivax. Crematogaster* (*Acrocoelia*) *auberti* st. *laestrygon* var. *vivax* Santschi, 1929e: 149 (w.) MOROCCO. Unavailable name.

**#v-nigra.** *Crematogaster chiarinii* var. *v-nigrum* Forel, 1910e: 434: (w.) ZAIRE. Combination in *C.* (*Acrocoelia*): Emery, 1922e: 146.

**voeltzkowi.** *Crematogaster voeltzkowi* Forel, 1907g: 78 (w.) COMORO IS. Combination in *C.* (*Neocrema*): Santschi, 1918d: 182 (misspelled as *woelzkowi*); in *C.* (*Crematogaster*): Wheeler, W.M. 1922a: 1024; in *C.* (*Orthocrema*): Emery, 1922e: 131.

**#vogti.** *Crematogaster auberti* var. *vogti* Forel, 1909c: 103 (w.) SPAIN. Combination in *C.* (*Acrocoelia*): Emery, 1922e: 142.

*vogtiiberica. Crematogaster inermis* subsp. *auberti* var. *vogtiiberica* Emery, 1926: 4 (attributed to Forel) (w.) FRANCE. Unavailable name.

**vulcania.** *Crematogaster vulcania* Santschi, 1913a: 413 (diagnosis in key) (w.) KENYA. Santschi, 1914b: 94 (w.q.m.). Combination in *C.* (*Crematogaster*): Wheeler, W.M. 1922a: 845; in *C.* (*Acrocoelia*): Emery, 1922e: 149. Race of *buchneri:* Forel, 1916: 411. Revived status as species: Wheeler, W.M. 1922a: 845.

**walshi.** *Crematogaster walshi* Forel, 1902c: 205 (w.) INDIA. Combination in *C.* (*Orthocrema*): Emery, 1922e: 132; in *C.* (*Mesocrema*): Santschi, 1928b: 33. Current subspecies: nominal plus *bouvardi.*

**warburgi.** *Crematogaster* (*Acrocoelia*) *warburgi* Menozzi, 1933b: 59, fig. 1 (w.) ISRAEL.

**wasmanni.** *Crematogaster wasmanni* Santschi, 1910c: 376, fig. 7 (w.) ZAIRE. Combination in *C.* (*Atopogyne*): Santschi, 1918d: 183.

**weberi.** *Crematogaster weberi* Emery, 1911c: 254, fig. 3 (w.) NEW GUINEA. Combination in *C.* (*Xiphocrema*): Emery, 1922e: 139.

**#weissi.** *Crematogaster wellmani* var. *weissi* Santschi, 1910c: 376 (w.) CONGO. Santschi, 1916b: 500 (q.). Combination in *C.* (*Atopogyne*): Santschi, 1916b: 500; in *C.* (*Acrocoelia*): Emery, 1922e: 149; in *C.* (*Sphaerocrema*): Santschi, 1926b: 219.

**#weitzeckeri.** *Crematogaster weitzeckeri* Emery, 1895h: 28, pl. 2, figs. 14, 15 (w.) SOUTH AFRICA. Combination in *C.* (*Decacrema*): Forel, 1910a: 18; Forel, 1910f: 9. Variety/subspecies of *liengmei:* Forel, 1910a: 18; Santschi, 1914e: 21; Arnold, 1920a: 548. Senior synonym of *kloofensis:* Arnold, 1920a: 549.

**wellmani.** *Crematogaster wellmani* Forel, 1909b: 64 (w.) ANGOLA. Santschi, 1916b: 500 (q.). Combination in *C.* (*Atopogyne*): Santschi, 1916b: 500; in *C.* (*Acrocoelia*): Emery, 1922e: 149; in *C.* (*Sphaerocrema*): Santschi, 1926b: 218. Senior synonym of *boxi, dewasi* and *retusa:* **new synonymy**. Current subspecies: nominal plus *luciae, weissi.*

**werneri.** *Crematogaster werneri* Mayr, 1907a: 388 (w.) UGANDA. Combination in *C.* (*Crematogaster*): Wheeler, W.M. 1922a: 845; in *C.* (*Acrocoelia*): Emery, 1922e: 149. Subspecies of *alulai:* Emery, 1915g: 11. Revived status as species: Wheeler, W.M. 1922a: 845. Current subspecies: nominal plus *cacozela, pasithea.*

**wheeleri.** *Crematogaster* (*Rhachiocrema*) *wheeleri* Mann, 1919: 318, fig. 21 (w.) SOLOMON IS.

*wheldeni. Crematogaster* (*Acrocoelia*) *lineolata* subsp. *cerasi* var. *wheldeni* Enzmann, J. 1946c: 92 (w.) U.S.A. Unavailable name; material referred to *lineolata:* Brown, 1949a: 47.

**whitei.** *Crematogaster whitei* Wheeler, W.M. 1915g: 808 (w.) AUSTRALIA. Combination in *C.* (*Acrocoelia*): Emery, 1922e: 152.

*wighti. Crematogaster vicina* var. *wighti* Wheeler, W.M. 1908b: 161 (w.) JAMAICA. Junior synonym of *vicina:* Wheeler, W.M. 1917g: 461 (in text).

**#wilnigra.** *Crematogaster nigeriensis* var. *wilniger* Forel, 1916: 405 (w.q.) ZAIRE. Combination in *C.* (*Sphaerocrema*): Wheeler, W.M. 1922a: 847.

**wilwerthi.** *Crematogaster wilwerthi* Santschi, 1910c: 372, fig. 6 (w.) ZAIRE. Combination in *C.* (*Sphaerocrema*): Wheeler, W.M. 1922a: 849. Current subspecies: nominal plus *confusa, fauconneti.*

**#winkleri.** *Crematogaster africana* var. *winkleri* Forel, 1909b: 69 (w.) CAMEROUN. Forel, 1913a: 124 (q.). Combination in *C.* (*Atopogyne*): Wheeler, W.M. 1922a: 158; Santschi, 1937g: 80. Race of *buchneri:* Forel, 1913a: 124; Forel, 1913b: 324; Forel, 1913e: 670; of *africana:* Wheeler, W.M. 1922a: 158; of *kohli:* Santschi, 1937g: 80.

*wladimiri. Crematogaster wladimiri* Consani, 1953: 100. [Unnecessary replacement name for *karawaiewi* Menozzi, 1935a: 106.]

**wroughtonii.** *Crematogaster wroughtonii* Forel, 1902c: 206 (w.q.) INDIA. Combination in *C.* (*Paracrema*): Emery, 1922e: 156.

**xerophila.** *Cematogaster xerophila* Wheeler, W.M. 1915g: 810 (w.) AUSTRALIA. Combination in *C.* (*Orthocrema*): Emery, 1922e: 133. Current subspecies: nominal plus *exigua.*

**#yambatensis.** *Crematogaster ferruginea* var. *yambatensis* Forel, 1913h: 353 (w.) ZAIRE. Subspecies of *castanea:* current status.

**yappi.** *Crematogaster yappi* Forel, 1901e: 374 (w.) WEST MALAYSIA. Combination in *C.* (*Acrocoelia*): Emery, 1922e: 152.

**#yorubosa.** *Crematogaster* (*Acrocoelia*) *brunneipennis* var. *yorubosa* Santschi, 1933b: 102 (w.) NIGERIA.

**zavattarii.** *Crematogaster* (*Sphaerocrema*) *zavattarii* Menozzi, 1926a: 39 (w.) NIGERIA. Current subspecies: nominal plus *edmeeae.*

**#zeta.** *Crematogaster* (*Atopogyne*) *laurenti* var. *zeta* Emery, 1922e: 155 (w.) ZAIRE. [First available use of *Crematogaster africana* subsp. *laurenti* var. *zeta* Forel, 1909b: 70; unavailable name.] Variety of *laurenti:* Santschi, 1937g: 79.

**#zoceensis.** *Crematogaster* (*Acrocoelia*) *laboriosa* var. *zoceensis* Santschi, 1925f: 85 (w.q.m.) CHINA.

**zonacaciae.** *Crematogaster* (*Sphaerocrema*) *zonacaciae* Weber, 1943c: 359, pl. 15, fig. 3 (w.) SUDAN.

**#zulu.** *Crematogaster* (*Atopogyne*) *bulawayensis* var. *zulu* Wheeler, W.M. 1922a: 855 (w.q.) SOUTH AFRICA. [First available use of *Crematogaster sjostedti* st. *bulawayensis* var. *zulu* Santschi, 1920b: 15; unavailable name.] Variety/subspecies of *bulawayensis:* Arnold, 1926: 289; of *gerstaeckeri:* current status.

*zumpti. Crematogaster* (*Acrocoelia*) *impressa* st. *brazzai* var. *zumpti* Santschi, 1937b: 99 (w.) CAMEROUN. Unavailable name.

---

**EXCURSUS**

**Species-rank taxa of *CREMATOGASTER*, grouped by subgenus.**

The subgenera of *Crematogaster* are arranged alphabetically below in ***BOLD ITALIC CAPITALS.*** The species currently assigned to each subgenus follow the subgeneric name in alphabetical order.

Most, probably all, of the subgenera are untrustworthy and may be best abandoned. While a few of them may define reasonable species-groups others are certainly artificial, being aggregations grouped by variable, convergent, plesiomorphic or otherwise unreliable characters. It is certain that many species are misplaced in the arrangement bequeathed to us and which we are currently constrained to employ.

***#APTEROCREMA:*** *atitlanica.*

***#ATOPOGYNE:*** *africana, ambigua, batesi, buchneri, clariventris, cuvierae, depressa, jullieni, kasaiensis, kohli, laurenti, lorteti, mottazi, resulcata, theta, togoensis, transiens, vitalisi, wasmanni.*

***#COLOBOCREMA:*** *cylindriceps.*

*CREMATOGASTER* (continued)

#*CREMATOGASTER: abdominalis, abrupta, acaciae, adrepens, aegyptiaca, affabilis, afghanica, agniae, algirica, alluaudi, aloysiisabaudiae, alulai, amabilis, ancipitula, antaris, anthracina, apicalis, arnoldi, aroensis, arthurimuelleri* (?subgen. *Decacrema*), *ashmeadi, atkinsoni, auberti, australis, betapicalis, bicolor, bison, browni, brunnea, brunneipennis, brunnescens, buddhae, californica, capensis, castanea, censor, cerasi, chiarinii, chopardi, chungi, cicatriculosa, civa, clydia, coarctata, coelestis, colei, constructor, cornigera, cornuta, corporaali, degeeri, delagoensis, dentinodis, depilis, desecta, desperans, diffusa, dohrni, egidyi, elysii, emeryana, eurydice, excisa, ferrarii, foraminiceps, foxi, fraxatrix, frivola, fruhstorferi, fuentei, fusca, gallicola, gerstaeckeri, gibba, gutenbergi, hespera, hezaradjatica, himalayana, hogsoni, homeri, ilgii, impressa, impressiceps, inconspicua, incorrecta, inermis, innocens, ionia, isolata, jehovae, kachelibae, kelleri, kirbii, kutteri, laboriosa* (junior homonym), *laestrygon, laeviceps, laevissima, laeviuscula, larreae, latuka, ledouxi, lineolata, lobata, longiceps* (junior homonym), *longiclava, lucayana, madagascariensis, manni, marioni, matsumurai, meijerei, melanogaster, menilekii, mimosae, misella, mjobergi, monticola, moqorensis, mormonum, mutans, navahoa, nesiotis, neuvillei, nigrans, nigriceps, nigronitens, nocturna, oasium, obnigra, obscura, ochracea, ochraceiventris, opaca, opaciceps, opuntiae, orobia, painei, patei, pellens, perelegans, peringueyi, perthensis, phoenica, pia, pilosa, politula, popohana, punctulata, pythia, ransonneti, retifera, rifelna, rivai, rogenhoferi, rogeri, rossi, rothneyi, rufa, rufigena, ruspolii, sagei, sanguinea, santschii, saussurei, schmidti, schultzei, scutellaris, semperi, senegalensis, sewellii, simoni, skounensis, solers, sorokini, stigmata, subcircularis, subdentata, subnuda, trautweini, vandeli, vermiculata, vidua, vulcania, warburgi, werneri, whitei, yappi.*

#*DECACREMA: angulosa, arthurimuelleri* (?subgen. *Crematogaster*), *biformis, borneensis, caculata* (?subgen. *Crematogaster*), *captiosa, cephalotes, decamera, edentula, enneamera, ensifera, grevei, hova, lango, liengmei, petiolidens, schencki, schimmeri, solenopsides.*

#*EUCREMA: acuta, bingo, evallans, inca, quadrispinosa.*

#*MESOCREMA: brevis, flavicornis, jacobsoni, karawaiewi, millardi, overbecki, pallipes, queenslandica, rasoherinae, scita, treubi, walshi.*

#*NEMATOCREMA: breviventris, magitae, stadelmanni.*

#*NEOCREMA: arata, arcuata, corticicola, corvina, descolei, distans, euterpe, indefensa, magnifica, montezumia, oxygynoides, polymnia, scapamaris, scelerata.*

#*ORTHOCREMA: abstinens, agnita, aitkenii, amapaensis, arizonensis, armandi, atra, baduvi, barbouri, binghamii, biroi, boera, bogokawlenskii, boliviana, brasiliensis, brevimandibularis, brevispinosa, bruchi, carinata, crassicornis, crinosa, cristata, curvispinosa, delitescens, dispar, dolens, dorsidens, egregior, emeryi, erecta, esterelana, formosa, fritzi, gavapiga, gratiosa, heathi, huberi, iheringi, insularis, iridipennis, irritabilis, javanica, jeanneli, laevis, limata, litoralis, longipilosa, longispina, lutzi, madecassa, major, mesonotalis, mimicans, minutissima, muralti, myops, natalensis, nigropilosa, ornatipilis, osakensis, pallida, parallela, pauciseta, pauli, peristerica, peruviana, polita, pulchella, pygmaea, quadriformis, ralumensis, rectinota, rudis, rufotestacea, russoi, sculpturata, sordidula, steinheili, stollii, sumichrasti, torosa, transvaalensis, udo, unciata, ustiventris, victima, voeltzkowi, xerophila.*

#*OXYGYNE: aberrans, agnetis, augusti, butteli, daisyi, dalyi, descarpentriesi, donisthorpei, ebenina, emmae, inops, margaritae, marthae, oscaris, ranavalonae, soror, travancorensis.*

#*PARACREMA: coriaria, dubia, modiglianii, onusta, spengeli, vandermeermohri, wroughtonii.*

#*PHYSOCREMA: ampullaris, aurita, bakeri, difformis, fulmeki, inflata, moorei, mucronata, stethogompha, tumidula.*

#*RHACHIOCREMA: aculeata, macracantha, paradoxa, wheeleri.*

#*SPHAEROCREMA: amita, angusticeps, bequaerti, chlorotica, concava, gabonensis, gambiensis, hemiceros, hottentota, juventa, kneri, lamottei, libengensis, lotti, luctans, microspina, nigeriensis, paolii, phoenix, pseudinermis, rugosa, rugosior, rustica, similis, stenocephala, striatula, wellmani, wilwerthi, zavattarii, zonacaciae.*

#*XIPHOCREMA: dahlii, elegans, flavitarsis, recurva, tarsata, tetracantha, weberi.*

**UNPLACED TO SUBGENUS OR SUBGENUS UNKNOWN**: *longiceps, obscurior, *praecursor, *pusilla, terminalis.*

---

**CRETACOFORMICA* (Australia)

*****explicata.*** **Cretacoformica explicata* Jell & Duncan, 1986: 191, fig. 69 (m.) AUSTRALIA (Lower Cretaceous). Family uncertain but probably not Formicidae: Naumann, 1993: 355. [This taxon does not belong in Formicidae.]

**CRETOMYRMA* (Russia)

*****arnoldii.*** **Cretomyrma arnoldii* Dlussky, 1975: 116, fig. 128 (w.) RUSSIA (Cretaceous).

*****unicornis.*** **Cretomyrma unicornis* Dlussky, 1975: 116, fig. 129 (w.) RUSSIA (Cretaceous).

**CRETOPONE* (Russia)

*****magna.*** **Cretopone magna* Dlussky, 1975: 119, fig. 131; pl. 7, fig. 30 (w.) RUSSIA (Cretaceous).

*CROESOMYRMEX:* see under ***LEPTOTHORAX.***

*CRYPTOCEPHALUS:* see under ***MERANOPLUS.***

*CRYPTOCERUS:* see under ***CEPHALOTES.***

**CRYPTOPONE** (Worldwide)

*angustata* Santschi, 1914; see under ***HYPOPONERA.***

*anommata. Ponera anommata* Donisthorpe, 1948b: 301 (w.m.) NEW GUINEA. Junior synonym of *testacea:* Wilson, 1958d: 360.

***butteli.*** *Cryptopone butteli* Forel, 1913k: 9, fig. C (w.) SUMATRA. Wheeler, W.M. 1933g: 10 (q.m.). See also: Wilson, 1958d: 358.

***crassicornis.*** *Belonopelta crassicornis* Emery, 1897d: 553 (w.) NEW GUINEA. Combination in *Pachycondyla* (*Pseudoponera*): Emery, 1901a: 46; in *Euponera* (*Trachymesopus*): Emery, 1911d: 86; in *Trachymesopus:* Wilson, 1958d: 353; in *Cryptopone:* Brown, 1963: 6.

*emeryi. Cryptopone emeryi* Donisthorpe, 1943a: 168 (footnote). Unnecessary replacement name for, and junior synonym of, *testacea:* Wilson, 1958d: 360.

***fusciceps.*** *Cryptopone fusciceps* Emery, 1900c: 321, pl. 8, figs. 19, 20 (w.q.) NEW GUINEA. [Misspelled as *fuscipes:* Emery, 1911d: 88.] Senior synonym of *fuscior* and *minor:* Wilson, 1958d: 358.

*fuscior. Cryptopone mayri* var. *fuscior* Mann, 1919: 294 (w.) SOLOMON IS. Raised to species: Wheeler, W.M. 1933g: 8. Junior synonym of *fusciceps:* Wilson, 1958d: 358.

***gilva.*** *Ponera gilva* Roger, 1863a: 170 (w.) U.S.A. Creighton & Tulloch, 1930: 74 (q.m.); Wheeler, G.C. & Wheeler, J. 1952c: 625 (l.). Combination in *Pachycondyla* (*Pseudoponera*): Emery, 1901a: 46; in *Euponera* (*Trachymesopus*): Emery, 1911d: 86; in *Trachymesopus:* Kempf, 1960f: 424; in *Cryptopone:* Brown, 1963: 3. Senior synonym of *harnedi:* Creighton & Tulloch, 1930: 74.

***guatemalensis.*** *Ponera ochracea* r. *guatemalensis* Forel, 1899c: 16 (w.) GUATEMALA. Combination in *Pachycondyla* (*Pseudoponera*): Emery, 1901a: 46; in *Euponera* (*Trachymesopus*): Emery, 1911d: 86; in *Trachymesopus:* Kempf, 1960f: 424; in *Cryptopone,* and raised to species: Brown, 1963: 6.

*harnedi. Euponera* (*Trachymesopus*) *gilva* subsp. *harnedi* Smith, M.R. 1929: 543 (w.) U.S.A. Junior synonym of *gilva:* Creighton & Tulloch, 1930: 74.

***hartwigi.*** *Cryptopone hartwigi* Arnold, 1948: 213, fig. 2 (w.) SOUTH AFRICA. Combination in *Hypoponera:* Taylor, 1967a: 12; in *Cryptopone:* **revived combination.**

*mayri. Cryptopone mayri* Mann, 1919: 293 (w.q.) SOLOMON IS. Wheeler, G.C. & Wheeler, J. 1952c: 630 (l.). Junior synonym of *testacea:* Wilson, 1958d: 360.

*minor. Cryptopone mayri* var. *minor* Mann, 1919: 294 (diagnosis in key) (w.) SOLOMON IS. Junior synonym of *fusciceps:* Wilson, 1958d: 358.

*mjobergi. Ponera mjobergi* Forel, 1915b: 22, pl. 1, fig. 2 (w.q.) AUSTRALIA. Junior synonym of *rotundiceps:* Brown, 1963: 6.

*mocsaryi* Szabó, 1910; see under ***PONERA.***

***motschulskyi.*** *Cryptopone motschulskyi* Donisthorpe, 1943a: 167 (q.) NEW GUINEA.

***nicobarensis.*** *Cryptopone nicobarensis* Forel, 1905b: 166 (q.) NICOBAR IS.

***ochracea.*** *Ponera ochracea* Mayr, 1855: 390 (footnote) (q.) ITALY. Emery, 1869b: 12 (w.); André, 1882c: 243 (m.). Combination in *Pachycondyla* (*Pseudoponera*): Emery, 1901a: 46; in *Euponera* (*Pseudoponera*): Emery, 1909c: 364; in *Euponera* (*Trachymesopus*): Emery, 1911d: 86; in *Trachymesopus:* Kempf, 1960f: 424; in *Cryptopone:* Brown, 1963: 6. See also: Emery, 1916b: 106; Kutter, 1977c: 28; Atanassov & Dlussky, 1992: 67. Current subspecies: nominal plus *sicula.*

***rotundiceps.*** *Euponera* (*Trachymesopus*) *rotundiceps* Emery, 1914f: 397 (q.) NEW CALEDONIA. Wheeler, G.C. & Wheeler, J. 1971b: 1208 (l.); Imai, Crozier & Taylor, 1977: 348 (k.). Combination in *Trachymesopus:* Wilson, 1958d: 354; in *Cryptopone,* and senior synonym of *mjobergi:* Brown, 1963: 6.

*rufotestaceus* Donisthorpe, 1943; see under ***PACHYCONDYLA.***

*sarawakana. Cryptopone sarawakana* Wheeler, W.M. 1933g: 11, fig. 4 (w.m.) EAST MALAYSIA. Junior synonym of *testacea:* Wilson, 1958d: 360.
***sauteri.*** *Pachycondyla (Pseudoponera) sauteri* Wheeler, W.M. 1906c: 304, pl. 41, fig. 6 (w.q.) JAPAN. Ogata, 1987: 118 (m.); Imai & Kubota, 1972: 194 (k.). Combination in *Euponera (Pseudoponera):* Emery, 1909c: 365; in *Euponera (Trachymesopus):* Teranishi, 1940: 8; in *Cryptopone:* Brown, 1963: 6.
**#*sicula.*** *Euponera (Pseudoponera) ochracea* subsp. *sicula* Emery, 1909c: 365 (w.) SICILY. Emery, 1916b: 105 (q.m.). Combination in *Euponera (Trachymesopus):* Emery, 1911d: 86; in *Cryptopone:* Baroni Urbani, 1971c: 14.
***taivanae.*** *Euponera (Trachymesopus) taivanae* Forel, 1913f: 184 (q.) TAIWAN. Combination in *Cryptopone:* Brown, 1963: 6.
***takahashii.*** *Euponera (Trachymesopus) takahashii* Wheeler, W.M. 1930a: 94 (q.) TAIWAN. Combination in *Cryptopone:* Brown, 1963: 6.
*tenuis* Emery, 1900; see under ***PONERA.***
***testacea.*** *Cryptopone testacea* Emery, 1893a: cclxxv (w.) SRI LANKA. [Also described as new by Emery, 1893f: 240.] Donisthorpe, 1942d: 451 (m.); Wheeler, G.C. & Wheeler, J. 1986c: 88 (l.); Imai, Brown *et al.* 1984: 67 (k.); Tjan *et al.* 1986: 57 (k.). Senior synonym of *anommata, emeryi, mayri* and *sarawakana:* Wilson, 1958d: 360.
***typhlos.*** *Ponera typhlos* Karavaiev, 1935a: 73, fig. 6 (w.) VIETNAM. Combination in *Cryptopone:* Brown, 1963: 6.

**CTENOBETHYLUS:* see under ***LIOMETOPUM.***

*CTENOPYGA:* see under ***ACANTHOSTICHUS.***

*CYATHOCEPHALUS:* see under ***ZACRYPTOCERUS.***

*CYATHOMYRMEX:* see under ***ZACRYPTOCERUS.***

***CYLINDROMYRMEX*** (Neotropical)
***boliviae.*** *Cylindromyrmex (Metacylindromyrmex) boliviae* Wheeler, W.M. 1924a: 104, fig. 20 (q.) BOLIVIA. See also: Brown, 1975: 38.
***brasiliensis.*** *Cylindromyrmex brasiliensis* Emery, 1901a: 53 (w.m.) BRAZIL. Borgmeier, 1937b: 218 (q.).
***brevitarsus.*** *Cylindromyrmex brevitarsus* Santschi, 1925b: 5 (w.) BRAZIL. Combination in *C. (Hypocylindromyrmex):* Kempf, 1972a: 91. See also: Brown, 1975: 38.
***darlingtoni.*** *Cylindromyrmex (Hypocylindromyrmex) darlingtoni* Wheeler, W.M. 1937b: 441 (w.q.) CUBA.
***godmani.*** *Cylindromyrmex godmani* Forel, 1899c: 4, pl. 1, fig. 2 (q.) PANAMA. Combination in *C. (Metacylindromyrmex):* Wheeler, 1924a: 106. See also: Brown, 1975: 38.
***longiceps.*** *Cylindromyrmex longiceps* André, 1892a: 47 (w.) BRAZIL. Kempf, 1968b: 372 (q.). Combination in *C. (Hypocylindromyrmex):* Wheeler, W.M. 1924a: 106. See also: Brown, 1975: 38.
***meinerti.*** *Cylindromyrmex meinerti* Forel, 1905b: 155 (w.) VENEZUELA. Combination in *C. (Hypocylindromyrmex):* Kempf, 1972a: 91. See also: Brown, 1975: 38.
***parallelus.*** *Cylindromyrmex parallelus* Santschi, 1932e: 410, fig. 1 (q.) PANAMA. Combination in *C. (Hypocylindromyrmex):* Kempf, 1972a: 91. See also: Brown, 1975: 38.
***schmidti.*** *Cylindromyrmex (Metacylindromyrmex) schmidti* Menozzi, 1931d: 191, figs. 3, 4 (w.q.) COSTA RICA. See also: Brown, 1975: 38.
***striatus.*** *Cylindromyrmex striatus* Mayr, 1870b: 967 (q.) SURINAM. Mayr, 1887: 545 (w.m.). Senior synonym of *whymperi:* Emery, 1901a: 54; of *tibialis* and *williamsi:* Brown, 1975: 38.
*tibialis. Cylindromyrmex (Hypocylindromyrmex) striatus* var. *tibialis* Stitz, 1932a: 367 (q.) GALAPAGOS IS. Junior synonym of *striatus:* Brown, 1975: 38.
*whymperi. Holcoponera whymperi* Cameron, 1891: 92, fig. (w.) ECUADOR. Combination in *Cylindromyrmex:* Forel, 1892f: 256. Junior synonym of *striatus:* Emery, 1901a: 54.
*williamsi. Cylindromyrmex williamsi* Wheeler, W.M. 1924a: 101, fig. 19 (w.q.) GALAPAGOS IS. Junior synonym of *striatus:* Brown, 1975: 38.

***CYPHOIDRIS*** (Afrotropical)
***exalta.*** *Cyphoidris exalta* Bolton, 1981b: 258, fig. 17 (w.) CAMEROUN.
***parissa.*** *Cyphoidris parissa* Bolton, 1981b: 258 (w.) LIBERIA.
***spinosa.*** *Cyphoidris spinosus* Weber, 1952: 26, figs. 7, 8 (w.) ZAIRE. See also: Bolton, 1981b: 257.
***werneri.*** *Cyphoidris werneri* Bolton, 1981b: 259 (w.) RWANDA.

*CYPHOMANNIA:* see under ***CYPHOMYRMEX.***

***CYPHOMYRMEX*** (Neotropical, Nearctic)
*acutus. Cyphomyrmex acutus* Weber, 1940a: 409 (diagnosis in key) (w.) PANAMA. Subspecies of *salvini:* Weber, 1958d: 261. Junior synonym of *salvini:* Kempf, 1966: 190.
*arnoldi. Cyphomyrmex rimosus* var. *arnoldi* Aguayo, 1932: 223 (w.) JAMAICA. Junior synonym of *minutus:* Snelling & Longino, 1992: 490.
*asper* Mayr, 1887; see under ***MYCETOSORITIS.***
*atrata. Atta (Cyphomyrmex) rimosa* r. *atrata* Forel, 1912e: 188 (w.q.m.) COLOMBIA. Combination in *Cyphomyrmex:* Emery, 1922f: 342. Junior synonym of *minutus:* Snelling & Longino, 1992: 490.
***auritus.*** *Cyphomyrmex auritus* Mayr, 1887: 559 (w.q.m.) BRAZIL. See also: Kempf, 1964d: 9.
***bicarinatus.*** *Cyphomyrmex bicarinatus* Snelling & Longino, 1992: 492, figs. 1, 10, 19 (w.q.) COLOMBIA.
***bicornis.*** *Cyphomyrmex bicornis* Forel, 1896: 179 (w.) BRAZIL. See also: Kempf, 1966: 177.
***bigibbosus.*** *Cyphomyrmex bigibbosus* Emery, 1894c: 226 (w.) BRAZIL. Wheeler, G.C. 1949: 670 (l.). Senior synonym of *tumulus:* Kempf, 1972a: 92. See also: Kempf, 1964d: 20.
*breviscapus. Cyphomyrmex rimosus* subsp. *breviscapus* Weber, 1940a: 412 (diagnosis in key) (w.) PANAMA. Junior synonym of *minutus:* Snelling & Longino, 1992: 490.
***bruchi.*** *Cyphomyrmex bruchi* Santschi, 1917f: 282 (w.) ARGENTINA. See also: Kempf, 1964d: 34.
***castagnei.*** *Cyphomyrmex castagnei* MacKay & Baena, 1993: 32, figs. 1, 2, 4–7, 9–12 (w.q.m.) COLOMBIA.
*championi. Cyphomyrmex championi* Forel, 1899c: 41, pl. 3, fig. 3 (m.) PANAMA. See also: Kempf, 1966: 192. Junior synonym of *salvini:* Snelling & Longino, 1992: 492.
*clorindae* Kusnezov, 1949; see under ***MYCETOSORITIS.***
*cochunae. Cyphomyrmex rimosus* subsp. *cochunae* Kusnezov, 1949d: 439 (w.) ARGENTINA. Raised to species: Kusnezov, 1957b: 10 (in key). Subspecies of *rimosus:* Kempf, 1966: 162. Junior synonym of *rimosus:* Snelling & Longino, 1992: 491.
*colombianus. Cyphomyrmex colombianus* Weber, 1940a: 413 (w.q.) COLOMBIA. Junior synonym of *costatus:* Kempf, 1966: 172.
*comalensis. Cyphomyrmex rimosus* var. *comalensis* Wheeler, W.M. 1907c: 719, pl. 49, fig. 1 (w.q.m.) U.S.A. Junior synonym of *minutus:* Creighton, 1950a: 315.
*conformis* Mayr, 1884; see under ***MYCETOPHYLAX.***
***cornutus.*** *Cyphomyrmex cornutus* Kempf, 1968a: 35, figs. 1–4 (w.q.) COLOMBIA.
***costatus.*** *Cyphomyrmex costatus* Mann, 1922: 44 (w.) HONDURAS. Kempf, 1966: 175 (q.). Senior synonym of *colombianus:* Kempf, 1966: 172.
*curiapensis. Cyphomyrmex rimosus* subsp. *curiapensis* Weber, 1938b: 190 (w.q.m.) VENEZUELA. Junior synonym of *fuscus:* Weber, 1958d: 260.
***daguerrei.*** *Cyphomyrmex (Cyphomyrmex) daguerrei* Santschi, 1933e: 118 (w.) ARGENTINA. See also: Kempf, 1964d: 28.
*dentatus. Cyphomyrmex rimosus* r. *dentatus* Forel, 1901c: 124 (w.) MEXICO. Wheeler, W.M. 1907c: 723 (q.). Raised to species: Kempf, 1966: 184. Junior synonym of *flavidus:* Snelling & Longino, 1992: 485.
*difformis. Meranoplus difformis* Smith, F. 1858b: 195 (w.) BRAZIL. Roger, 1863a: 210 (q.m.). Combination in *Cataulacus:* Roger, 1863a: 210; in *Cyphomyrmex:* Mayr, 1887: 558. Junior synonym of *minutus:* Roger, 1863b: 40; of *rimosus:* Forel, 1893e: 607; Emery, 1894c: 224; Snelling & Longino, 1992: 491. [Note. The name was misspelled as *deformis* by Roger, 1863a: 210, and many subsequent authors.]
***dixus.*** *Cyphomyrmex dixus* Snelling & Longino, 1992: 489, figs. 3, 11, 18 (w.) COSTA RICA.
***faunulus.*** *Cyphomyrmex bigibbosus* subsp. *faunulus* Wheeler, W.M. 1925a: 44 (w.q.) GUIANA. Raised to species: Kempf, 1964d: 23. Senior synonym of *petiolatus:* Kempf, 1972a: 92.
*fiebrigi* Santschi, 1916; see under ***TRACHYMYRMEX.***
*flavescens. Cyphomyrmex rimosus* subsp. *flavescens* Weber, 1940a: 411 (in key and footnote) (w.) HAITI. [First available name for *Cyphomyrmex rimosus* subp. *minutus* var. *flavidus* Wheeler, W.M. 1936b: 204; unavailable name.] Junior synonym of *minutus:* Snelling & Longino, 1992: 490.
***flavidus.*** *Cyphomyrmex flavidus* Pergande, 1896: 895 (w.) MEXICO. See also: Wheeler, W.M. 1907c: 726; Kempf, 1966: 172. Senior synonym of *dentatus:* Snelling & Longino, 1992: 485.
*flavidus. Cyphomyrmex rimosus* subsp. *minutus* var. *flavidus* Wheeler, W.M. 1936b: 204 (w.) HAITI. Unavailable name. [See *flavescens,* above.]
***foxi.*** *Cyphomyrmex foxi* André, 1892a: 55 (w.) JAMAICA. Wheeler, W.M. 1917g: 461 (q.). See also: Kempf, 1966: 186; Snelling & Longino, 1992: 487.

***CYPHOMYRMEX** (continued)*

*fuscula. Cyphomyrmex rimosus* var. *fuscula* Emery, 1922f: 342. [Unnecessary replacement name for *fuscus* Emery, 1894c: 225.] Subspecies of *rimosus:* Santschi, 1931e: 279; Weber, 1940a: 411 (in key). Junior synonym of *fuscus:* Weber, 1958d: 260.

*fuscus. Cyphomyrmex rimosus* var. *fuscus* Emery, 1894c: 225 (w.q.m.) BRAZIL. Senior synonym of *curiapensis* and *fuscula:* Weber, 1958d: 260. Junior synonym of *rimosus:* Snelling & Longino, 1992: 491.

***hamulatus.*** *Cyphomyrmex rimosus* subsp. *hamulatus* Weber, 1938b: 190 (w.) BOLIVIA. Wheeler, G.C. & Wheeler, J. 1983: 608 (l.). Raised to species: Kempf, 1966: 197. Senior synonym of *trinitatis:* Snelling & Longino, 1992: 490.

***kirbyi.*** *Cyphomyrmex kirbyi* Mayr, 1887: 557 (w.) COLOMBIA. Santschi, 1921g: 97 (q.). See also: Kempf, 1966: 179; Snelling & Longino, 1992: 485.

***laevigatus.*** *Cyphomyrmex* (*Cyphomannia*) *laevigatus* Weber, 1938b: 184, figs. 20, 21 (w.) BOLIVIA. See also: Kempf, 1966: 178; Snelling & Longino, 1992: 493.

***lectus.*** *Atta* (*Cyphomyrmex*) *olitrix* subsp. *lecta* Forel, 1911c: 295 (w.) BRAZIL. Combination in *Cyphomyrmex:* Luederwaldt, 1918: 39. Raised to species: Kempf, 1964d: 36.

***lilloanus.*** *Cyphomyrmex lilloanus* Kusnezov, 1949d: 442, figs. 1–3 (w.) ARGENTINA. See also: Kempf, 1964d: 40.

***longiscapus.*** *Cyphomyrmex longiscapus* Weber, 1940a: 410 (diagnosis in key) (w.) COLOMBIA. See also: Kempf, 1966: 165.

***major.*** *Cyphomyrmex rimosus* var. *major* Forel, 1901c: 125 (w.) GUATEMALA. Raised to species: Snelling & Longino, 1992: 490.

***minutus.*** *Cyphomyrmex minutus* Mayr, 1862: 691 (w.) CUBA. Mayr, 1887: 558 (q.m.). Junior synonym of *rimosus:* Dalla Torre, 1893: 150; Forel, 1895b: 137. Revived from synonymy as subspecies of *rimosus:* Emery, 1894c: 225; Wheeler, W.M. 1905b: 106. Subspecies of *rimosus:* Emery, 1922f: 342; Creighton, 1950a: 316; Kempf, 1966: 162. Revived status as species: Snelling & Longino, 1992: 490. Senior synonym of *steinheili:* Mayr, 1887: 558; of *comalensis:* Creighton, 1950a: 315; of *arnoldi, atrata, breviscapus, flavescens* and *venezuelensis:* Snelling & Longino, 1992: 490.

***morschi.*** *Cyphomyrmex morschi* Emery, 1888c: 360 (w.) BRAZIL. Senior synonym of *personatus:* Kempf, 1964d: 25.

***nemei.*** *Cyphomyrmex nemei* Kusnezov, 1957b: 7 (w.q.m.) ARGENTINA. See also: Kempf, 1964d: 39.

***nesiotus.*** *Cyphomyrmex nesiotus* Snelling & Longino, 1992: 485, figs. 7, 15, 27, 28 (w.q.m.) ECUADOR.

***occultus.*** *Cyphomyrmex occultus* Kempf, 1964d: 41, figs. 37–39 (q.m.) BRAZIL.

*olindanus. Cyphomyrmex rimosus* r. *olindanus* Forel, 1901e: 337 (w.) BRAZIL. Junior synonym of *transversus:* Wheeler, W.M. 1907c: 723.

***olitor.*** *Cyphomyrmex olitor* Forel, 1893e: 605 (w.q.) BRAZIL. Wheeler, G.C. 1949: 670 (l.). Combination in *Atta* (*Cyphomyrmex*), with unjustified emendation of spelling to *olitrix:* Forel, 1911c: 295; in *Cyphomyrmex:* Luederwaldt, 1918: 39. Senior synonym of *quebradae:* Kempf, 1964d: 30.

***paniscus.*** *Cyphomyrmex paniscus* Wheeler, W.M. 1925a: 42 (w.q.m.) BRAZIL. See also: Kempf, 1964d: 18.

*parallelus* Emery, 1906; see under ***MYCETAROTES.***

***peltatus.*** *Cyphomyrmex peltatus* Kempf, 1966: 181, figs. 13, 20, 35, 50 (w.q.) BRAZIL.

*pencosensis. Cyphomyrmex rimosus* var. *pencosensis* Forel, 1914d: 281 (w.) ARGENTINA. Santschi, 1931e: 278 (q.m.). Subspecies of *rimosus:* Weber, 1940a: 411 (in key). Junior synonym of *transversus:* Kempf, 1966: 193.

*personatus. Cyphomyrmex* (*Mycetoritis* [sic]) *personatus* Santschi, 1923c: 268 (q.) ARGENTINA. See also: Santschi, 1925e: 164. Junior synonym of *morschi:* Kempf, 1964d: 25.

*petiolatus. Cyphomyrmex bigibbosus* subsp. *petiolatus* Weber, 1938b: 187 (w.q.) BOLIVIA. Junior synonym of *faunulus:* Kempf, 1964d: 23. Revived from synonymy as subspecies of *bigibbosus:* Weber, 1966: 167. Junior synonym of *faunulus:* Kempf, 1972a: 92.

***plaumanni.*** *Cyphomyrmex plaumanni* Kempf, 1962b: 31, figs. 29–31 (w.) BRAZIL. See also: Kempf, 1964d: 16.

***podargus.*** *Cyphomyrmex podargus* Snelling & Longino, 1992: 487, figs. 8, 16, 20 (w.) JAMAICA.

*quebradae. Cyphomyrmex quebradae* Kusnezov, 1949d: 445, figs. 6–8, 11 (w.q.m.) ARGENTINA. Junior synonym of *olitor:* Kempf, 1964d: 30.

***rimosus.*** *Cryptocerus rimosus* Spinola, 1851b: 49 (w.m.) BRAZIL. [Also described as new by Spinola, 1853: 65.] Emery, 1894c: 224 (q.); Wheeler, G.C. 1949: 668 (l.). Combination in *Cyphomyrmex:* Emery, 1893h: 2; in *Atta* (*Cyphomyrmex*): Forel, 1912e: 188; in *Cyphomyrmex:* Bruch, 1914: 217. Senior synonym of *difformis:* Forel, 1893e: 607; Emery, 1894c: 224; of *cochunae* and *fuscus* (and its junior synonyms *curiapensis* and *fuscula*): Snelling & Longino, 1992: 491. See also: Smith, D.R. 1979: 1410.

***salvini.*** *Cyphomyrmex rimosus* r. *salvini* Forel, 1899c: 40, pl. 3, fig. 2 (q.) PANAMA. Wheeler, W.M. 1907c: 724 (w.m.). Raised to species: Weber, 1958d: 261. Senior synonym of *acutus:* Kempf, 1966: 190; of *championi:* Snelling & Longino, 1992: 492.

*simplex* Emery, 1888; see under ***MYCETOPHYLAX.***

*steinheili. Cyphomyrmex steinheili* Forel, 1885a: 368 (w.) MEXICO. Junior synonym of *minutus:* Mayr, 1887: 558.

***strigatus.*** *Cyphomyrmex strigatus* Mayr, 1887: 558 (w.) BRAZIL. Forel, 1893e: 606 (q.m.); Wheeler, G.C. 1949: 669 (l.). See also: Kempf, 1964d: 14.

***transversus.*** *Cyphomyrmex rimosus* subsp. *transversus* Emery, 1894c: 226 (w.q.m.) BRAZIL. Raised to species: Kempf, 1966: 193. Senior synonym of *olindanus:* Wheeler, W.M. 1907c: 723; Wheeler, W.M. 1925a: 45; of *pencosensis:* Kempf, 1966: 193.

*trinitatis. Cyphomyrmex rimosus* subsp. *trinitatis* Weber, 1938b: 189 (w.q.) TRINIDAD. Junior synonym of *hamulatus:* Snelling & Longino, 1992: 490.

*tumulus. Cyphomyrmex bigibbosus* subsp. *tumulus* Weber, 1938b: 185 (w.q.m.) GUIANA. Junior synonym of *bigibbosus:* Kempf, 1964d: 20. Revived from synonymy and raised to species: Weber, 1966: 167. Junior synonym of *bigibbosus:* Kempf, 1972a: 92.

***vallensis.*** *Cyphomyrmex daguerrei* subsp. *vallense* Kusnezov, 1949d: 450, figs. 12–15 (w.) ARGENTINA. Raised to species: Kusnezov, 1957b: 11 (in key). See also: Kempf, 1964d: 34.

*venezuelensis. Cyphomyrmex rimosus* subsp. *venezuelensis* Weber, 1938b: 188 (w.) VENEZUELA. Junior synonym of *transversus:* Weber, 1958d: 261. Revived from synonymy as subspecies of *rimosus:* Kempf, 1966: 162. Junior synonym of *minutus:* Snelling & Longino, 1992: 490.

***vorticis.*** *Cyphomyrmex vorticis* Weber, 1940a: 410 (diagnosis in key) (w.) ST HELENA. See also: Kempf, 1966: 188.

***wheeleri.*** *Cyphomyrmex wheeleri* Forel, 1900e: 282 (w.q.) U.S.A. Wheeler, W.M. 1907c: 726 (m.). See also: Kempf, 1966: 167.

***#CYRTOMYRMA:*** see under ***POLYRHACHIS.***

*CYRTOSTIGMACROS:* see under ***STIGMACROS.***

*CYSIAS:* see under ***CERAPACHYS.***

***DACATRIA*** (Korea)

*templaris. Dacatria templaris* Rigato, 1994b: 157, figs. 1–3 (w.) REPUBLIC OF KOREA.

***DACETINOPS*** (Indo-Australian)

***cibdelus.*** *Dacetinops cibdela* Brown & Wilson, 1957: 4, figs. 1–3 (w.q.) NEW GUINEA. Wheeler, G.C. & Wheeler, J. 1957a: 1 (l.). See also: Taylor, 1985: 53.

***cirrosus.*** *Dacetinops cirrosus* Taylor, 1985: 56, figs. 1, 16–18 (w.q.m.) BORNEO.

***concinnus.*** *Dacetinops concinna* Taylor, 1965: 1, figs. 1–2 (w.) BORNEO. Taylor, 1985: 60 (q.m.); Imai, Brown, *et al.* 1984: 68 (k.).

***darlingtoni.*** *Dacetinops darlingtoni* Taylor, 1985: 61, figs. 22–24 (w.) NEW GUINEA.

***ignotus.*** *Dacetinops ignotus* Taylor, 1985: 62, figs. 25–27 (w.q.) NEW GUINEA.

***solivagus.*** *Dacetinops solivagus* Taylor, 1985: 63, figs. 28–30 (w.q.) BORNEO.

***wilsoni.*** *Dacetinops wilsoni* Taylor, 1985: 64, figs. 7–9, 31–33 (w.q.) BORNEO.

***DACETON*** (Neotropical)

***armigerum.*** *Formica armigera* Latreille, 1802c: 244, pl. 9, fig. 58 (w.) BRAZIL. Smith, F. 1853: 226 (q.m.); Wheeler, G.C. & Wheeler, J. 1955a: 122 (l.). Combination in *Daceton:* Perty, 1833: 136. Senior synonym of *cordata:* Roger, 1862c: 290. See also: Wilson, 1962b: 401.

*cordata. Myrmecia cordata* Fabricius, 1804: 425 (w.) CENTRAL AMERICA. Junior synonym of *armigerum:* Roger, 1862c: 290.

*DACRYON:* see under ***PODOMYRMA.***

***#DECACREMA:*** see under ***CREMATOGASTER.***

*DECAMERA:* see under ***MYRMELACHISTA.***

***DECAMORIUM*** (Afrotropical)

***decem.*** *Tetramorium* (*Decamorium*) *decem* Forel, 1913a: 121 (w.) ZIMBABWE. Arnold, 1917: 350 (q.m.). Combination in *Decamorium:* Wheeler, W.M. 1922a: 906. Senior synonym of *ultor:* Bolton, 1976: 298.

*nimba. Decamorium decem* r. *ultor* Bernard, 1953b: 250 (w.) GUINEA. Junior synonym of *uelense:* Bolton, 1976: 298.

***uelense.*** *Tetramorium* (*Decamorium*) *decem* st. *uelense* Santschi, 1923e: 258 (w.q.) ZAIRE. Combination in *Decamorium:* Bolton, 1976: 298. Raised to species, and senior synonym of *nimba:* Bolton, 1976: 298. See also: Longhurst, Johnson & Wood, 1979: 83.

*ultor. Tetramorium* (*Decamorium*) *decem* var. *ultor* Forel, 1913j: 217 (w.) ZIMBABWE. Combination in *Decamorium:* Wheeler, W.M. 1922a: 906. Junior synonym of *decem:* Bolton, 1976: 298.

*DECAPHEIDOLE:* see under ***PHEIDOLE.***

***#DENDROLASIUS:*** see under ***LASIUS.***

***DENDROMYRMEX*** (Neotropical)

***#acomus.*** *Camponotus* (*Dendromyrmex*) *fabricii* var. *acoma* Forel, 1908c: 418 (w.) BRAZIL. Combination in *Dendromyrmex:* Emery, 1925b: 173.

***apicalis.*** *Dendromyrmex apicalis* Mann, 1916: 487, pl. 6, fig. 43 (w.) BRAZIL. Current subspecies: nominal plus *filiae, guyanensis, opaciceps.*

*branneri* Mann, 1916; see under ***CAMPONOTUS.***

***chartifex.*** *Formica chartifex* Smith, F. 1860c: 68 (w.m.) BRAZIL. Forel, 1899c: 158 (q.); Wheeler, G.C. & Wheeler, J. 1986d: 337 (l.). Combination in *Camponotus:* Mayr, 1862: 673; in *Dendromyrmex:* Emery, 1895j: 772. Current subspecies: nominal plus *felis, mamoreensis, vestitus.*

***fabricii.*** *Camponotus fabricii* Roger, 1862c: 285 (footnote) (w.) SURINAM. Combination in *Dendromyrmex:* Emery, 1895j: 772. Current subspecies: nominal plus *acomus, io, isthmicus, niger, nitidior, pictus, rufescens.*

***#felis.*** *Dendromyrmex chartifex* var. *felis* Wheeler, W.M. 1916c: 13 (w.q.) GUIANA. [Also described as new by Mann, 1916: 484.]

***#filiae.*** *Dendromyrmex apicalis* subsp. *filiae* Weber, 1943b: 74 (w.q.) GUIANA.

*graciliceps. Dendromyrmex graciliceps* Viehmeyer, 1916b: 290 (w.) BOLIVIA. Junior synonym of *rufogaster:* Santschi, 1921g: 102.

***#guyanensis.*** *Dendromyrmex apicalis* var. *guyanensis* Santschi, 1921g: 101 (w.q.) FRENCH GUIANA.

***#io.*** *Dendromyrmex fabricii* subsp. *io* Weber, 1943b: 77 (w.) GUIANA.

***#isthmicus.*** *Dendromyrmex fabricii* subsp. *isthmicus* Weber, 1943b: 76 (q.) PANAMA.

***madeirensis.*** *Dendromyrmex madeirensis* Mann, 1916: 486 (w.) BRAZIL. Current subspecies: nominal plus *melinoni, romani.*

***#mamoreensis.*** *Dendromyrmex chartifex* subsp. *mamoreensis* Mann, 1916: 485 (w.) BRAZIL. Raised to species: Santschi, 1921g: 102. Subspecies of *chartifex:* Emery, 1925b: 173; Kempf, 1972a: 95.

***#melinoni.*** *Dendromyrmex madeirensis* var. *melinoni* Santschi, 1921g: 102 (w.) FRENCH GUIANA.

***nidulans.*** *Formica nidulans* Smith, F. 1860c: 69 (w.q.m.) BRAZIL. Combination in *Camponotus:* Mayr, 1862: 672; in *Dendromyrmex:* Emery, 1896d: 378. Current subspecies: nominal plus *nigripes.*

***#niger.*** *Dendromyrmex fabricii* subsp. *niger* Weber, 1943b: 76 (w.) SURINAM.

***#nigripes.*** *Dendromyrmex nidulans* var. *nigripes* Wheeler, W.M. 1916c: 13 (w.q.) GUIANA.

***#nitidior.*** *Dendromyrmex fabricii* subsp. *nitidior* Santschi, 1921g: 102 (w.) PANAMA.

***#opaciceps.*** *Dendromyrmex apicalis* var. *opaciceps* Wheeler, W.M. 1923a: 6 (q.) BRAZIL.

***#pictus.*** *Dendromyrmex fabricii* var. *pictus* Wheeler, W.M. 1923a: 6 (w.) BRAZIL. Wheeler, G.C. & Wheeler, J. 1953e: 202 (l.).

***#romani.*** *Dendromyrmex madeirensis* var. *romani* Wheeler, W.M. 1923a: 6 (w.) BRAZIL.

***#rufescens.*** *Dendromyrmex fabricii* var. *rufescens* Wheeler, W.M. 1916c: 13 (w.) GUIANA.

***#rufogaster.*** *Dendromyrmex traili* subsp. *rufogaster* Mann, 1916: 484 (w.) BOLIVIA. Senior synonym of *graciliceps:* Santschi, 1921g: 102.

***traili.*** *Camponotus traili* Mayr, 1878: 868 (w.) BRAZIL. Combination in *Dendromyrmex:* Emery, 1896d: 378. Current subspecies: nominal plus *rufogaster.*

***#vestitus.*** *Dendromyrmex mamoreensis* st. *vestita* Santschi, 1921g: 102 (w.) FRENCH GUIANA. Subspecies of *chartifex:* Emery, 1925b: 173.

***wheeleri.*** *Dendromyrmex wheeleri* Donisthorpe, 1937a: 623 (w.) TRINIDAD.

*DEROMYRMA:* see under ***APHAENOGASTER.***

*DESCOLEMYRMA:* see under ***MYCOCEPURUS.***

*DIABOLUS:* see under ***DOLICHODERUS.***

***DIACAMMA*** (Oriental, Indo-Australian, Australia)

***#anceps.*** *Diacamma rugosum* var. *anceps* Matsumura & Uchida, 1926: 51 (w.) HONG KONG. [First available use of *Diacamma rugosum* subsp. *geometricum* var. *anceps* Emery, 1897b: 155; unavailable name.]

*andamana. Diacamma rugosum* r. *vagans* var. *andamana* Forel, 1922: 90 (w.m.) ANDAMAN IS. Unavailable name.

***#arcuatum.*** *Diacamma rugosum* subsp. *arcuata* Karavaiev, 1925b: 118 (w.) SUMATRA.

***assamense.*** *Diacamma tritschleri* r. *assamensis* Forel, in Emery, 1897b: 151 (footnote) (w.) INDIA. Raised to species: Forel, 1900d: 319; Bingham, 1903: 79; Emery, 1911d: 65.

***australe.*** *Formica australis* Fabricius, 1775: 393 (w.) AUSTRALIA. Forel, 1915b: 26 (m.); Wheeler, G.C. & Wheeler, J. 1952c: 609 (l.). Combination in *Diacamma:* Mayr, 1862: 718. Senior synonym of *colosseensis* and *levis:* Taylor & Brown, D.R. 1985: 29.

***baguiense.*** *Diacamma baguiensis* Wheeler, W.M. & Chapman, 1925: 66, pl. 2, figs. 11, 12 (w.) PHILIPPINES.

***#balinense.*** *Diacamma rugosum* subsp. *balinensis* Karavaiev, 1925b: 116 (w.) INDONESIA (Bali I.).

***#birmanum.*** *Diacamma vagans* var. *birmanum* Emery, 1887b: 441 (w.) BURMA. Subspecies of *rugosum:* current status.

***bispinosum.*** *Ponera bispinosa* Le Guillou, 1842: 317 (w.) INDONESIA (Ternate I.). Combination in *Diacamma:* Mayr, 1862: 718. Senior synonym of *buruensis* and *subsulcata:* Wilson, 1958d: 367.

*brevispinum. Diacamma rugosum* r. *geometricum* var. *brevispinum* Forel, 1900d: 319 (w.) BURMA. Unavailable name.

*buruensis. Diacamma rugosum* subsp. *buruensis* Karavaiev, 1925b: 117 (w.) INDONESIA (Buru I.). Junior synonym of *bispinosum:* Wilson, 1958d: 367.

***#celebense.*** *Diacamma celebense* Emery, 1887b: 441 (w.) SULAWESI. Subspecies of *rugosum:* Emery, 1897b: 156; Forel, 1907e: 17.

***ceylonense.*** *Diacamma ceylonense* Emery, 1897b: 159 (w.) SRI LANKA. Subspecies of *rugosum:* Forel, 1900d: 318; Emery, 1911d: 66. Revived status as species: Bingham, 1903: 79; Santschi, 1932b: 14. Current subspecies: nominal plus *orbiculatum.*

*coincidens. Diacamma rugosum* subsp. *geometrica* var. *coincidens* Baroni Urbani, 1971b: 360. [Unnecessary replacement name for *concentrica* Karavaiev, 1935a: 65.] Unavailable name.

*colosseensis. Diacamma australe* var. *colosseensis* Forel, 1915b: 26 (w.) AUSTRALIA. Junior synonym of *australe:* Taylor & Brown, D.R. 1985: 29.

*compressum. Diacamma compressum* Mayr, 1879: 660 (w.) INDIA. Junior synonym of *scalpratum:* Dalla Torre, 1893: 29.

***#concentricum.*** *Diacamma palawanicum* var. *concentricum* Wheeler, W.M. & Chapman, 1925: 64, pl. 2, figs. 7, 8 (w.) PHILIPPINES.

*concentrica. Diacamma rugosum* subsp. *geometricum* var. *concentrica* Karavaiev, 1935a: 65 (w.) VIETNAM. Unavailable name.

*convexa. Diacamma rugosum* st. *sculpturata* var. *convexa* Santschi, 1920g: 179 (w.) VIETNAM. Unavailable name.

***cupreum.*** *Ponera cuprea* Smith, F. 1860b: 104 (w.) NEW GUINEA. Combination in *Diacamma:* Mayr, 1886c: 361. Race of *geometricum:* Emery, 1887b: 438. Revived status as species: Donisthorpe, 1932c: 462.

*curtula. Diacamma rugosum* r. *geometricum* var. *curtula* Forel, 1912d: 100 (w.) JAVA. Unavailable name.

***cyaneiventre.*** *Diacamma cyaneiventre* André, 1887: 293 (w.) INDIA. Race of *rugosum:* Forel, 1900d: 318. Revived status as species: Bingham, 1903: 78.

*debilior. Diacamma rugosum* subsp. *hortensis* var. *debilior* Karavaiev, 1927e: 5 (w.) INDONESIA (Prinsen I.). Unavailable name.

*divergens. Diacamma rugosum* subsp. *sculpturatum* var. *divergens* Emery, 1897b: 160 (w.) NEW GUINEA. Unavailable name; material referred to *rugosum* by Wilson, 1958d: 368.

***#doveri.*** *Diacamma vagans* var. *doveri* Mukerjee, 1934: 3 (w.) INDIA. Subspecies of *rugosum:* current status.

*frontalis. Diacamma vagans* var. *frontalis* Stitz, 1911a: 355 (w.) NEW GUINEA. Junior synonym of *rugosum:* Wilson, 1958d: 368.

*geminatum. Diacamma rugosum* subsp. *geminatum* Emery, 1897b: 161, figs. 6, 12 (w.) INDONESIA (Halmahera I.). Junior synonym of *rugosum:* Wilson, 1958d: 368.

*geometrica. Ponera geometrica* Smith, F. 1857a: 67 (w.) SINGAPORE. Wheeler, W.M. 1919e: 54 (m.); Emery, 1899c: 5 (l.). Combination in *Diacamma:* Mayr, 1862: 718. Status as species: Santschi, 1932b: 14; Donisthorpe, 1932c: 448. Subspecies of *rugosum:* Emery, 1897b: 154; Wheeler, W.M. 1928c: 5; Karaviev, 1935: 62; Yasumatsu, 1940a: 67. Junior synonym of *rugosum:* Wilson, 1958d: 368.

***#gibbosum.*** *Diacamma rugosum* subsp. *gibbosum* Karavaiev, 1935a: 68, fig. 4 (w.) VIETNAM.

***holosericeum.*** *Ponera holoserica* Roger, 1860: 302 (w.) JAVA. Combination in *Diacamma,* and spelling emended to *holosericea:* Mayr, 1862: 718 [spelling *D. holosericeum* in Mayr, 1863: 407]. See also: Karavaiev, 1925b: 115.

***DIACAMMA** (continued)*

*homonyma. Diacamma rugosum* subsp. *sculpturata* var. *homonyma* Baroni Urbani, 1971b: 360. [Unnecessary replacement name for *papuana* Karavaiev, 1935a: 67.] Unavailable name.

*horizontalis. Diacamma rugosum* subsp. *geometricum* var. *horizontalis* Karavaiev, 1935a: 63 (w.) VIETNAM. Unavailable name.

#***hortense.*** *Diacamma rugosum* subsp. *hortensis* Karavaiev, 1925b: 119 (w.) JAVA. [First available use of *Diacamma rugosum* r. *vagans* var. *hortensis* Forel, 1913k: 7; unavailable name.]

*indica. Diacamma rugosum* var. *indica* Santschi, 1920g: 179 (w.) INDIA. [First available use of *Diacamma rugosum* r. *vagans* var. *indicum* Forel, 1903d: 400; unavailable name.] Junior synonym of *vagans:* Mukerjee & Ribeiro, 1925: 205.

***intricatum.*** *Ponera intricata* Smith, F. 1857a: 67 (w.) BORNEO. Combination in *Diacamma:* Mayr, 1862: 718. Current subspecies: nominal plus *kershawi.*

***jacobsoni.*** *Diacamma jacobsoni* Forel, 1912d: 100 (w.) JAVA.

*japensis. Diacamma rugosum* subsp.*japensis* Donisthorpe, 1941h: 52 (w.) NEW GUINEA. Junior synonym of *rugosum:* Wilson, 1958d: 368.

#***javanum.*** *Diacamma javanum* Emery, 1887b: 439, pl. 1, fig. 6 (w.) JAVA. Junior synonym of *geometrica:* Emery, 1897b: 154; Forel, 1900d: 320. Revived from synonymy as variety of *rugosum:* Forel, 1905c: 6.

#***jerdoni.*** *Diacamma rugosum* var. *jerdoni* Forel, 1903d: 401 (w.) INDIA. Provisional synonym of *sculptum:* Emery, 1911d: 66. [Emery gives priority to *jerdoni.* If the synonymy correct then *sculptum* has priority.]

#***kershawi.*** *Diacamma intricatum* subsp. *kershawi* Wheeler, W.M. 1919e: 53 (w.) BORNEO.

#***latispinum.*** *Diacamma rugosum* subsp. *latispina* Karavaiev, 1925b: 119 (w.) JAVA. See also: Karavaiev, 1935a: 67.

*levis. Diacamma australe* var. *levis* Crawley, 1915a: 134 (w.) AUSTRALIA. Junior synonym of *australe:* Taylor & Brown, D.R. 1985: 29.

#***lombokense.*** *Diacamma rugosum* var. *lombokensis* Emery, 1897b: 154 (w.) INDONESIA (Lombok I.).

#***longiceps.*** *Diacamma geometricum* var. *longiceps* Santschi, 1932b: 14 (w.) VIETNAM. Subspecies of *rugosum:* current status.

***longitudinale.*** *Diacamma longitudinale* Emery, 1889b: 496 (footnote) (w.) VIETNAM. Status as species: Emery, 1897b: 163; Forel, 1911i: 216 (footnote); Santschi, 1920h: 158; Santschi, 1924c: 95.

*macreta. Diacamma rugosum* subsp. *sculpturatum* var. *macreta* Viehmeyer, 1912: 6 (w.) NEW GUINEA. Unavailable name; material referred to *rugosum* by Wilson, 1958d: 368.

*moica. Diacamma rugosum* subsp. *ceylonense* var. *moica* Forel, 1911d: 377 (w.) SOUTH EAST ASIA ("pays de Moïs"). Unavailable name.

*nitidiventris. Diacamma rugosum* subsp. *timorense* var. *nitidiventris* Emery, 1897b: 158 (w.) TIMOR. Unavailable name.

*oblica. Diacamma rugosum* st. *sculpturata* var. *oblica* Santschi, 1920g: 178 (w.) VIETNAM. Unavailable name.

#***orbiculatum.*** *Diacamma ceylonensis* var. *orbiculatum* Santschi, 1932b: 14 (w.) LAOS.

#***ovale.*** *Diacamma rugosum* var. *ovalis* Karavaiev, 1935a: 62 (w.) VIETNAM.

***palawanicum.*** *Diacamma palawanicum* Emery, 1900d: 666 (footnote) (w.) PHILIPPINES. Current subspecies: nominal plus *concentricum.*

***panayense.*** *Diacamma panayensis* Wheeler, W.M. & Chapman, 1925: 65, pl. 2, figs. 9, 10 (w.) PHILIPPINES.

*papuanum. Diacamma vagans* var. *papuanum* Stitz, 1911a: 354, fig. 5 (w.) NEW GUINEA. Junior synonym of *rugosum:* Wilson, 1958d: 368.

*papuana. Diacamma rugosum* subsp. *sculpturatum* var. *papuana* Karavaiev, 1935a: 67 (w.) NEW GUINEA. Unavailable name.

*paralleliruga. Diacamma rugosum* subsp. *geometricum* var. *paralleliruga* Karavaiev, 1935a: 64 (w.) VIETNAM. Unavailable name.

*pubescens. Diacamma rugosum* subsp. *sculpturatum* var. *pubescens* Emery, 1897b: 160 (w.) INDONESIA (Mentawei I.). Unavailable name.

*pulchellum. Diacamma rugosum* st. *sculpturatum* var. *pulchellum* Santschi, 1932b: 12, fig. 1 (w.) NEW GUINEA. Unavailable name; material referred to *rugosum* by Wilson, 1958d: 368.

***purpureum.*** *Ponera purpurea* Smith, F. 1863: 18 (w.) INDONESIA (Halmahera I.). Combination in *Diacamma:* Emery, 1887b: 439. Variety of *cupreum:* Emery, 1887b: 439. Revived status as species: Emery, 1897b: 152; Emery, 1911d: 66. Senior synonym of *waigeuensis:* **new synonymy**.

#***rothneyi.*** *Diacamma rugosum* var. *rothneyi* Forel, 1900d: 318 (diagnosis in key) (w.) INDIA.

***rugivertex.*** *Diacamma rugivertex* Emery, 1902c: 28 (w.) TIMOR.

***rugosum.*** *Ponera rugosa* Le Guillou, 1842: 318 (w.) BORNEO. Ogata, 1987: 111 (m.). Combination in *Diacamma:* Mayr, 1862: 718. Senior synonym of *versicolor:* Roger, 1863b: 16; Emery, 1897b: 153; of *frontalis, geminatum, geometrica, japensis, papuanum, saussurei, sculpturata, smithi, striata, vagans* (and its junior synonym *indica*), and material of the unavailable names *divergens, macreta* and *pulchellum* referred here: Wilson, 1958d: 368. Current subspecies: nominal plus *anceps, arcuatum, balinense, birmanum, celebense, doveri, gibbosum, hortense, javanum, jerdoni, latispinum, lombokense, longiceps, ovale, rothneyi, sculptum, sikkimense, timorense, viridipurpureum.*

*saussurei. Diacamma bispinosum* var. *saussurei* Forel, 1922: 89 (w.) INDONESIA (Halmahera I.). Junior synonym of *rugosum:* Wilson, 1958d: 368.

***scalpratum.*** *Ponera scalprata* Smith, F. 1858b: 84, pl. 6, figs. 21, 22 (w.) INDIA. Emery, 1889b: 496 (m.); Wheeler, G.C. & Wheeler, J. 1952c: 610 (l.). Combination in *Diacamma:* Mayr, 1862: 718. Senior synonym of *compressum:* Dalla Torre, 1893: 29; Emery, 1911d: 65. Current subspecies: nominal plus *violaceum.*

#***sculptum.*** *Ponera sculpta* Jerdon, 1851: 117 (w.) INDIA. Combination in *Diacamma:* Mayr, 1863: 407. Junior synonym of *rugosum:* Roger, 1863b: 16; Mayr, 1863: 407; Mayr, 1865: 65; of *vagans:* Emery, 1887b: 440. Revived from synonymy: Dalla Torre, 1893: 29. Subspecies of *rugosum:* Emery, 1897b: 156. Revived status as species Bingham, 1903: 80. Subspecies of *rugosum:* Wheeler, W.M. 1909d: 334. Provisional synonym of *jerdoni:* Emery, 1911d: 66 [Emery gives *jerdoni* as the senior name. If the synonymy correct then *sculptum* has priority.] Subspecies of *rugosum:* current status.

*sculpturata. Ponera sculpturata* Smith, F. 1859a: 142 (w.) INDONESIA (Aru I.). Karavaiev, 1926d: 417 (m.). Combination in *Diacamma:* Mayr, 1862: 718. Variety of *geometricum:* Dalla Torre, 1893: 28. Subspecies of *rugosum:* Emery, 1897b: 159; Forel, 1901f: 338; Santschi, 1920g: 178; Crawley, 1924: 384; Karavaiev, 1925b: 120; Santschi, 1932b: 12; Karavaiev, 1935a: 65. Junior synonym of *rugosum:* Wilson, 1958d: 368.

***sericeiventre.*** *Diacamma sericeiventre* Stitz, 1925: 111, fig. 1 (w.) PHILIPPINES.

#***sikkimense.*** *Diacamma rugosum* var. *sikkimensis* Forel, 1903d: 401 (w.) INDIA.

*smithi. Diacamma rugosum* subsp. *smithi* Donisthorpe, 1943e: 115. Replacement name for *Ponera tortuolosa* Smith, F. 1863: 18. [Junior primary homonym of *Ponera tortuolosa* Smith, F. 1858b: 99.] Junior synonym of *rugosum:* Wilson, 1958d: 368.

*striata. Ponera striata* Smith, F. 1860b: 104 (w.) INDONESIA (Batjan I.). Combination in *Diacamma:* Mayr, 1886c: 361. Variety of *versicolor:* Emery, 1887b: 436 (footnote). Subspecies of *rugosum:* Emery, 1897b: 161. Junior synonym of *rugosum:* Wilson, 1958d: 368.

*subsulcata. Diacamma bispinosum* var. *subsulcata* Emery, 1897b: 152 (w.) INDONESIA (Halmahera I.). Junior synonym of *bispinosum:* Wilson, 1958d: 367.

#***timorense.*** *Diacamma javanum* r. *timorense* Emery, 1887b: 439 (w.) TIMOR. Subspecies of *rugosum:* Emery, 1897b: 158.

*tonkina. Diacamma rugosum* st. *vagans* var. *tonkina* Santschi, 1920g: 179 (w.) VIETNAM. Unavailable name.

*tortuolosa. Ponera tortuolosa* Smith, F. 1863: 18 (w.) INDONESIA (Seram I.). [Junior primary homonym of *Ponera tortuolosa* Smith, F. 1858b: 99 (now in *Gnamptogenys*).] Replacement name: *smithi:* Donisthorpe, 1943e: 115. Race/subspecies of *geometricum:* Emery, 1887b: 438; of *rugosum:* Emery, 1897b: 550. Revived status as species: Donisthorpe, 1932c: 471.

***tritschleri.*** *Diacamma tritschleri* Forel, in Emery, 1897b: 151 (footnote) (w.) SUMATRA.

*vagans. Ponera vagans* Smith, F. 1860b: 103 (w.) INDONESIA (Batjan I.). Bingham, 1903: 81 (m.); Wheeler, G.C. & Wheeler, J. 1952c: 611 (l.); Imai *et al.* 1984: 5 (k.). Combination in *Diacamma:* Mayr, 1862: 718. Subspecies/race of *rugosum:* Forel, 1908a: 1; Forel, 1912a: 49; Forel, 1922: 90. Revived status as species: Mukerjee, 1934: 3. Senior synonym of *indica:* Mukerjee & Ribeiro, 1925: 205. Junior synonym of *rugosum:* Wilson, 1958d: 368.

*versicolor. Ponera versicolor* Smith, F. 1857a: 65 (w.) BORNEO. Combination in *Diacamma:* Emery, 1887b: 438. Race of *geometricum:* Emery, 1887b: 438. Junior synonym of *rugosum:* Roger, 1863b: 16; Emery, 1897b: 153; Emery, 1911d: 66.

**#violaceum.** *Diacamma scalpratum* var. *violaceum* Forel, 1900d: 317 (diagnosis in key) (w.) BURMA.
**#viridipurpureum.** *Diacamma geometricum* var. *viridipurpureum* Emery, 1893g: 261 (w.) PHILIPPINES. Wheeler, G.C. & Wheeler, J. 1952c: 610 (l.). Subspecies of *rugosum:* current status.
*waigeuensis. Diacamma rugosum* subsp. *waigeuensis* Donisthorpe, 1942e: 703 (w.) NEW GUINEA. Junior synonym of *purpureum:* **new synonymy**.

*DIAGYNE:* see under ***SOLENOPSIS.***

*DICERATOCLINEA:* see under ***DOLICHODERUS.***

*DICHOTHORAX:* see under ***LEPTOTHORAX.***

***#DICHTHADIA:*** see under ***DORYLUS.***

***DICROASPIS*** (Afrotropical)
*arnoldi* Forel, 1913; see under ***CALYPTOMYRMEX.***
*claviseta* Santschi, 1914; see under ***CALYPTOMYRMEX.***
**cryptocera.** *Dicroaspis cryptocera* Emery, 1908b: 185, fig. 1 (w.q.) ZAIRE. Combination in *Calyptomyrmex* (*Dicroaspis*): Wheeler, W.M. 1922a: 887; in *Dicroaspis:* Bolton, 1981a: 58. Senior synonym of *wheeleri:* Bolton, 1981a: 58.
*emeryi* Forel, 1910; see under ***CALYPTOMYRMEX.***
**laevidens.** *Calyptomyrmex* (*Dicroaspis*) *cryptocerus* var. *laevidens* Santschi, 1919h: 88 (w.) ZAIRE. Combination in *Dicroaspis,* and raised to species: Bolton, 1981a: 59.
*wheeleri. Geognomicus wheeleri* Menozzi, 1924b: 220, figs. 1–3 (w.) PRINCIPE I. Junior synonym of *cryptocera:* Bolton, 1981a: 58.

***DILOBOCONDYLA*** (Oriental, Indo-Australian)
**borneensis.** *Dilobocondyla borneensis* Wheeler, W.M. 1916b: 12, fig. 2 (w.) BORNEO.
*braunsi* Forel, 1912; see under ***LEPTOTHORAX.***
**cataulacoidea.** *Podomyrma* (*Mesomyrma*) *cataulacoidea* Stitz, 1911a: 364, figs. 10–12 (w.) NEW GUINEA. Combination in *Dilobocondyla:* Emery, 1912b: 102. Senior synonym of *concolor:* Taylor, 1991b: 601.
**chapmani.** *Dilobocondyla chapmani* Wheeler, W.M. 1924b: 250 (w.) PHILIPPINES. Wheeler, G.C. & Wheeler, J. 1954c: 128 (l.). Current subspecies: nominal plus *rufobrunnea.*
*concolor. Dilobocondyla cataulacoidea* var. *concolor* Viehmeyer, 1914c: 40 (w.) NEW GUINEA. Junior synonym of *cataulacoidea:* Taylor, 1991b: 601.
**didita.** *Atta didita* Walker, 1859: 375 (w.) SRI LANKA. Combination in *Aphaenogaster:* Dalla Torre, 1893: 102; in *Dilobocondyla,* and senior synonym of *escherichi:* Donisthorpe: 1932b: 576.
*escherichi. Atopomyrmex escherichi* Forel, 1911i: 223 (w.) SRI LANKA. Combination in *Dilobocondyla:* Forel, 1913k: 64. Junior synonym of *didita:* Donisthorpe, 1932b: 576.
**fouqueti.** *Dilobocondyla fouqueti* Santschi, 1910h: 283 (w.) VIETNAM. Senior synonym of *lighti:* Wang & Wu, 1992b: 562.
**fulva.** *Dilobocondyla cataulacoidea* var. *fulva* Viehmeyer, 1916a: 127 (w.q.) SINGAPORE. Raised to species: Taylor, 1991b: 601.
**karnyi.** *Dilobocondyla karnyi* Wheeler, W.M. 1924b: 249 (w.) JAVA.
*lighti. Dilobocondyla fouqueti* var. *lighti* Wheeler, W.M. 1927d: 6 (w.) CHINA. Junior synonym of *fouqueti:* Wang & Wu 1992b: 562.
**#rufobrunnea.** *Dilobocondyla chapmani* subsp. *rufobrunnea* Wheeler, W.M. 1935b: 43 (w.q.m.) PHILIPPINES.
**sebesiana.** *Dilobocondyla sebesiana* Wheeler, W.M. 1924b: 248 (q.) INDONESIA (Sebesi I.).
**selebensis.** *Atopomyrmex selebensis* Emery, 1898b: 234, figs. 4, 5 (q.) SULAWESI. Combination in *Dilobocondyla:* Santschi, 1910h: 283. Current subspecies: nominal plus *simalurana.*
**#simalurana.** *Dilobocondyla selebensis* st. *simalurana* Forel, 1915a: 25 (w.) SUMATRA.

*DIMORPHOMYRMEX:* see under ***GESOMYRMEX.***

***#DINOMYRMEX:*** see under ***CAMPONOTUS.***

***DINOPONERA*** (Neotropical)
**australis.** *Dinoponera grandis* subsp. *australis* Emery, 1901a: 48 (w.) PARAGUAY. Santschi, 1921g: 85 (m.). Raised to species: Borgmeier, 1937b: 227. Senior synonym of *brevis:* Kempf, 1971: 382. Current subspecies: nominal plus *bucki, nigricolor.*
*brevis. Dinoponera australis* var. *brevis* Borgmeier, 1937b: 227 (w.) PARAGUAY. [First available use of *Dinoponera grandis* subsp. *australis* var. *brevis* Santschi, 1929a: 416; unavailable name.] Junior synonym of *australis:* Kempf, 1971: 382.
**#bucki.** *Dinoponera australis* subsp. *bucki* Borgmeier, 1937b: 228, figs. 7, 8 (w.m.) BRAZIL. See also: Kempf, 1971: 386.
**gigantea.** *Ponera gigantea* Perty, 1833: 135, pl. 27, fig. 3 (w.) BRAZIL. Kempf, 1971: 372 (m.). Combination in *Dinoponera:* Roger, 1861a: 38. Senior synonym of *grandis:* Roger, 1861a: 38; Kempf, 1971: 371.
*grandis. Ponera grandis* Guérin-Méneville, 1838: 206 (w.) BRAZIL. Emery, 1911a: 219 (m.); Wheeler, G.C. & Wheeler, J. 1986a: 387 (l.). Combination in *Dinoponera:* Roger, 1861a: 38. Junior synonym of *gigantea:* Roger, 1861a: 38; Kempf, 1971: 371. See also: Haskins & Zahl, 1971: 1.
**longipes.** *Dinoponera grandis* subsp. *longipes* Emery, 1901a: 48 (w.) PERU. Raised to species: Kempf, 1971: 375.
**lucida.** *Dinoponera grandis* subsp. *lucida* Emery, 1901a: 48 (w.) BRAZIL. Raised to species: Kempf, 1971: 376.
**mutica.** *Dinoponera grandis* subsp. *mutica* Emery, 1901a: 48 (w.) BRAZIL. Wheeler, G.C. & Wheeler, J. 1952c: 607 (l.). Raised to species: Kempf, 1971: 378.
**#nigricolor.** *Dinoponera australis* subsp. *nigricolor* Borgmeier, 1937b: 228, figs. 5, 6, 8 (w.q.m.) BRAZIL. See also: Kempf, 1971: 387.
*opaca. Dinoponera opaca* Kempf, 1971: 379 (w.) BRAZIL. [First available use of *Dinoponera grandis* st. *mutica* var. *opaca* Santschi, 1921g: 84; unavailable name.] Junior synonym of *quadriceps:* Kempf, 1975c: 344.
**quadriceps.** *Dinoponera quadriceps* Kempf, 1971: 380 (w.) BRAZIL. [First available use of *Dinoponera grandis* st. *mutica* var. *quadriceps* Santschi, 1921g: 84; unavailable name.] Senior synonym of *opaca:* Kempf, 1975c: 344.

*DIODONTOLEPIS:* see under ***NOTONCUS.***

***DIPLOMORIUM*** (Afrotropical)
*cotterelli* Donisthorpe, 1942; see under ***MONOMORIUM.***
**longipenne.** *Diplomorium longipenne* Mayr, 1901b: 18 (w.q.) SOUTH AFRICA. See also: Arnold, 1916: 240; Bolton, 1987: 278.
*lujae* Forel, 1909; see under ***BONDROITIA.***
*saharensis* Santschi, 1923; see under ***BONDROITIA.***

*DIPLORHOPTRUM:* see under ***SOLENOPSIS.***

***DISCOTHYREA*** (World tropics and subtropics)
**antarctica.** *Discothyrea antarctica* Emery, 1895c: 266 (footnote) (w.) NEW ZEALAND. [Also described as new by Emery, 1895e: 636.] Brown, 1958h: 19 (q.); Wheeler, G.C. & Wheeler, J. 1971b: 1202 (l.).
**bidens.** *Discothyrea bidens* Clark, 1928a: 38, pl. 1 figs. 30, 31 (w.) AUSTRALIA. See also: Brown, 1958g: 339.
**bryanti.** *Prodiscothyrea bryanti* Wheeler, W.M. 1917d: 29 (w.) WEST MALAYSIA. Combination in *Discothyrea:* Brown, 1958g: 253.
**clavicornis.** *Discothyrea clavicornis* Emery, 1897c: 593, pl. 15, figs. 39, 40 (w.) NEW GUINEA. Mann, 1919: 288 (q.).
**crassicornis.** *Discothyrea crassicornis* Clark, 1926: 46, pl. 6, fig. 4 (w.) AUSTRALIA. See also: Brown, 1958g: 340.
**denticulata.** *Discothyrea denticulata* Weber, 1939a: 100, fig. 4 (w.) GUIANA.
**globa.** *Discothyrea globus* Forel, 1905c: 4 (w.) JAVA.
**hewitti.** *Discothyrea hewitti* Arnold, 1916: 160 (q.) SOUTH AFRICA.
**horni.** *Discothyrea horni* Menozzi, 1927c: 270, fig. 1 (q.) COSTA RICA. Menozzi, 1931b: 261 (w.).
**humilis.** *Discothyrea humilis* Weber, 1939a: 100 (q.) PANAMA.
**icta.** *Discothyrea icta* Weber, 1939a: 101 (q.) TRINIDAD.
**isthmica.** *Discothyrea isthmica* Weber, 1940b: 78 (q.) PANAMA.
**leae.** *Discothyrea leae* Clark, 1934b: 29, pl. 2, fig. 11 (q.) AUSTRALIA.
**mixta.** *Discothyrea mixta* Brown, 1958g: 343, figs. 47, 48 (w.) LIBERIA.
**neotropica.** *Discothyrea neotropica* Bruch, 1919: 400, figs. (q.) ARGENTINA.
**oculata.** *Discothyrea oculata* Emery, 1901a: 52 (w.q.m.) CAMEROUN. See also: Brown, 1958g: 340.
**patrizii.** *Discothyrea patrizzii* Weber, 1949b: 2, fig. (w.q.) KENYA. Justified emendation of spelling to *patrizii:* Brown, 1958g: 253.
**poweri.** *Pseudosysphincta poweri* Arnold, 1916: 162, fig. 9 (w.) ZIMBABWE. Combination in *Discothyrea:* Brown, 1958g: 253.
**remingtoni.** *Discothyrea remingtoni* Brown, 1948c: 38, fig. 1 (w.) NEW CALEDONIA. See also: Brown, 1958g: 342.
**sauteri.** *Discothyrea globus* var. *sauteri* Forel, 1912a: 47 (w.q.) TAIWAN. Ogata, 1987: 109 (m.). Raised to species: Brown, 1958g: 253.
**sculptior.** *Discothyrea oculata* var. *sculptior* Santschi, 1913c: 302 (w.) CONGO. Raised to species: Brown, 1958g: 253.
**sexarticulata.** *Discothyrea sexarticulata* Borgmeier, 1954a: 191, figs. 1–9 (w.q.) BRAZIL.

***DISCOTHYREA** (continued)*

**stumperi.** *Discothyrea stumperi* Baroni Urbani, 1977f: 97, figs. 1, 2 (w.) BHUTAN.

**testacea.** *Discothyrea testacea* Roger, 1863a: 177 (w.q.) U.S.A. See also: Smith, M.R. & Wing, 1955: 105.

**traegaordhi.** *Discothyrea traegaordhi* Santschi, 1914e: 3 (w.) SOUTH AFRICA. See also: Brown, 1958g: 340.

**turtoni.** *Discothyrea turtoni* Clark, 1934c: 53, pl. 4, figs. 5, 6 (w.q.) AUSTRALIA. See also: Brown, 1958g: 339.

**velutina.** *Prodiscothyrea velutina* Wheeler, W.M. 1916i: 34, pl. 4, figs. 1–9 (w.q.) AUSTRALIA. Combination in *Discothyrea:* Brown, 1958g: 253.

***DLUSSKYIDRIS** (Russia)

***zherichini.** **Palaeomyrmex zherichini* Dlussky, 1975: 118, fig. 130, pl. 7, fig. 29 (m.) RUSSIA (Cretaceous). Combination in **Dlusskyidris:* **new combination**.

*DODECAMYRMICA:* see under ***MYRMICA.***

*DODOUS:* see under ***PRISTOMYRMEX.***

**DOLEROMYRMA** (Australia)

**darwiniana.** *Tapinoma (Doleromyrma) darwinianus* Forel, 1907a: 28 (q.m.) AUSTRALIA. Imai, Crozier & Taylor, 1977: 345 (k.). Combination in *Iridomyrmex:* Emery, 1912g: 24; in *Doleromyrma:* Shattuck, 1992a: 14. Current subspecies: nominal plus *fida, leae.*

#**fida.** *Tapinoma (Doleromyrma) darwinianum* var. *fida* Forel, 1907h: 286 (w.q.) AUSTRALIA. Combination in *Iridomyrmex:* Emery, 1912g: 24; in *Doleromyrma:* Shattuck, 1992a: 14. Subspecies of *darwiniana:* Taylor & Brown, D.R. 1985: 98.

#**leae.** *Iridomyrmex darwinianus* r. *leae* Forel, 1913g: 189 (w.) AUSTRALIA. Combination in *Doleromyrma:* Shattuck, 1992a: 14.

**DOLICHODERUS** (Worldwide except Afrotropical and Malagasy)

*abietis. Dolichoderus (Hypoclinea) abietis* Kono & Sugihara, 1939: 12 (w.) JAPAN. Junior synonym of *sibiricus:* Yasumatsu, 1941: 182.

**abruptus.** *Formica abrupta* Smith, F. 1858b: 45 (w.) BRAZIL. MacKay, 1993b: 32 (q.m.). Combination in *Hypoclinea:* Mayr, 1863: 423; Kempf, 1972a: 118; in *Dolichoderus:* Dalla Torre, 1893: 156; in *Dolichoderus (Hypoclinea):* Emery, 1894c: 228; in *Dolichoderus:* Shattuck, 1992c: 77. Junior synonym of *bidens:* Roger, 1862b: 285. Revived from synonymy: Mayr, 1863: 423. Senior synonym of *ursus:* MacKay, 1993b: 33.

***affectus.** **Dolichoderus affectus* Théobald, 1937b: 207, pl. 4, fig. 4; pl. 14, figs. 15, 16 (m.) FRANCE (Oligocene).

**affinis.** *Dolichoderus affinis* Emery, 1889b: 508, pl. 11, fig. 20 (w.) BURMA. Viehmeyer, 1916a: 141 (q.). Combination in *D. (Hypoclinea):* Emery, 1894c: 229. Current subspecies: nominal plus *glabripes, mus, nigricans.*

*albatus. Dolichoderus (Hypoclinea) bidens* subsp. *albatus* Viehmeyer, 1922: 213 (w.) PERU. Combination in *Hypoclinea:* Kempf, 1972a: 119; in *Dolichoderus:* Shattuck, 1992c: 77. Junior synonym of *spurius:* MacKay, 1993b: 97.

*analis. Dolichoderus (Hypoclinea) analis* Emery, 1894c: 236 (w.q.) BRAZIL. Combination in *Hypoclinea:* Kempf, 1972a: 118; in *Dolichoderus:* Shattuck, 1992c: 77. Race of *gibbosus:* Forel, 1912h: 34. Revived status as species: Mann, 1916: 465. Junior synonym of *quadridenticulatus:* MacKay, 1993b: 86.

**andinus.** *Monacis andina* Kempf, 1962b: 36, figs. 38–40 (w.) PERU. Combination in *Dolichoderus:* Shattuck, 1992c: 77. See also: MacKay, 1993b: 34.

***anglicus.** **Dolichoderus anglicus* Cockerell, 1915: 484, pl. 65, fig. 8 (q. wing) GREAT BRITAIN (Oligocene).

**angusticornis.** *Dolichoderus (Hypoclinea) angusticornis* Clark, 1930b: 260, fig. 10 (w.) AUSTRALIA. Combination in *D. (Diceratoclinea):* Wheeler, W.M. 1935c: 69.

***anthracinus.** **Poneropsis anthracina* Heer, 1867: 22, pl. 2, fig. 8 (q.) YUGOSLAVIA (Miocene). Combination in *Hypoclinea:* Mayr, 1867b: 57; in *Dolichoderus:* Handlirsch, 1907: 870. Given as "Formicidae *incertae generis*" by Taylor, 1964: 138; retained in *Dolichoderus* by Shattuck, 1994: 42.

***antiquus.** **Dolichoderus (Hypoclinea) antiquus* Carpenter, 1930: 39, pl. 4, fig. 6; pl. 9, fig. 1; pl. 10, fig. 2 (w.q.) U.S.A. (Miocene).

*arboricola. Polyrhachis arboricola* Norton, 1868a: 60, fig. 3 (w.) MEXICO (fig. only). [Description by Norton, 1868c: 4.] Junior synonym of *bispinosus:* Emery, 1892b: 167.

**armstrongi.** *Dolichoderus (Hypoclinea) armstrongi* McAreavey, 1949: 17, figs. 48, 49 (w.) AUSTRALIA.

#**ater.** *Dolichoderus cuspidatus* var *atra* Menozzi, 1930a: 3 (w.) SUMATRA. Variety of *cuspidatus:* Shattuck, 1994: 47; of *furcifer:* Menozzi, 1932d: 6; and current status.

*aterrimus. Dolichoderus (Hypoclinea) taschenbergi* var. *aterrimus* Wheeler, W.M. 1915b: 417. Replacement name for *gagates* Wheeler, W.M. 1905d: 310. [Junior primary homonym of *gagates* Emery, 1890a: 69.] Junior synonym of *taschenbergi:* Creighton, 1950a: 336.

*atriceps. Dolichoderus (Hypoclinea) sinensis* var. *atriceps* Wheeler, W.M. 1928c: 29 (q.) CHINA. Junior synonym of *sibiricus:* Yasumatsu, 1962: 96.

**attelaboides.** *Formica attelaboides* Fabricius, 1775: 394 (w.) BRAZIL. Smith, F. 1858b: 75 (q.m.); Mayr, 1865: 59 (m.); Wheeler, G.C. & Wheeler, J. 1951: 178 (l.). Combination in *Dolichoderus:* Lund, 1831a: 130. Senior synonym of *pulla:* Kempf, 1969: 289; of *heterogaster* and *imbecillus:* MacKay, 1993b: 35. See also: Shattuck, 1994: 42.

*attenuatus. Dolichoderus bidens* r. *attenuatus* Forel, 1903c: 258 (w.q.) BRAZIL. Combination in *Hypoclinea:* Kempf, 1972a: 119; in *Dolichoderus:* Shattuck, 1992c: 77. Junior synonym of *bidens:* MacKay, 1993b: 40.

*auromaculatus. Dolichoderus auromaculatus* Forel, 1885a: 350 (q.) BRAZIL. Junior synonym of *bidens:* Emery, 1894c: 234.

**australis.** *Dolichoderus australis* André, 1896b: 257 (w.) AUSTRALIA. Wheeler, G.C. & Wheeler, J. 1974b: 396 (l.). Combination in *D. (Hypoclinea):* Emery, 1912g: 12. See also: Clark, 1930b: 263.

**baenae.** *Dolichoderus baenae* MacKay, 1993b: 38, fig. 36 (w.) COLOMBIA.

*bahiana. Dolichoderus (Hypoclinea) bidens* var. *bahiana* Santschi, 1921g: 101 (w.) BRAZIL. Combination in *Hypoclinea:* Kempf, 1972a: 119; in *Dolichoderus:* Shattuck, 1992c: 77. Junior synonym of *bidens:* MacKay, 1993b: 40.

***balticus.** *Hypoclinea baltica* Mayr, 1868c: 64, pl. 3, figs. 61–63; pl. 4, fig. 64 (w.q.m.) BALTIC AMBER (Oligocene). Combination in *Dolichoderus:* Dalla Torre, 1893: 157.

**beccarii.** *Dolichoderus beccarii* Emery, 1887a: 253 (q.) SUMATRA. Combination in *D. (Hypoclinea):* Emery, 1894c: 229; in *D. (Diceratoclinea):* Wheeler, W.M. 1935c: 69.

*beutenmuelleri. Dolichoderus plagiatus* var. *beutenmuelleri* Wheeler, W.M. 1904e: 304 (w.) U.S.A. Variety of *pustulatus:* Wheeler, W.M. 1913c: 116. Junior synonym of *pustulatus:* Creighton, 1950a: 335.

#**bicolor.** *Dolichoderus (Hypoclinea) moggridgei* var. *bicolor* Santschi, 1920h: 169 (w.) INDIA. [Also described as new by Santschi, 1924c: 109.]

**bidens.** *Formica bidens* Linnaeus, 1758: 581 (w.) "America meridionali". Wheeler, G.C. & Wheeler, J. 1951: 173 (l.); MacKay, 1993b: 40 (q.m.). Combination in *Hypoclinea:* Mayr, 1862: 707; Kempf, 1972a: 119; in *Dolichoderus (Hypoclinea):* Emery, 1894c: 228; in *Dolichoderus:* Dalla Torre, 1893: 157; Shattuck, 1992c: 77. Senior synonym of *perditor:* Roger, 1862b: 285; of *auromaculatus:* Emery, 1894c: 234; of *attenuatus* and *bahiana:* MacKay, 1993b: 40.

*bifurcatus. Dolichoderus (Diabolus) bifurcatus* Karavaiev, 1926d: 425, fig. 3A (w.) WEST MALAYSIA. Junior synonym of *surbecki:* Wheeler, W.M. 1935c: 68.

#**bilikanus.** *Dolichoderus (Hypoclinea) bituberculatus* var. *bilikana* Santschi, 1925f: 92 (w.) SUMATRA. Subspecies of *thoracicus:* Shattuck, 1994: 70.

*biolleyi. Dolichoderus biolleyi* Forel, 1908b: 61 (w.) COSTA RICA. Combination in *D. (Monacis):* Emery, 1912g: 9. Race of *schulzi:* Forel, 1912h: 35. Junior synonym of *schulzi:* Kempf, 1959b: 254.

**bispinosus.** *Formica bispinosa* Olivier, 1792: 502 (w.) FRENCH GUIANA. Roger, 1862a: 236 (q.); Forel, 1908b: 60 (m.); Wheeler, G.C. & Wheeler, J. 1966: 728 (l.). Combination in *Hypoclinea:* Mayr, 1862: 708; in *Monacis:* Roger, 1862a: 235; Kempf, 1959b: 240; in *Dolichoderus:* Emery, 1890a: 69; Dalla Torre, 1893: 158; in *D. (Monacis):* Emery, 1894c: 228. Senior synonym of *arboricola:* Emery, 1892b: 167; of *fungosa:* Latreille, 1802c: 133; of *vestita:* Emery, 1894c: 232. See also: Kempf, 1959b: 240; MacKay, 1993b: 42; Shattuck, 1994: 44.

*bituberculatus. Hypoclinea bituberculata* Mayr, 1862: 705 (w.) PHILIPPINES. Emery, 1887a: 254 (q.); Karavaiev, 1926d: 426 (q.); Wheeler, G.C. & Wheeler, J. 1951: 174 (l.); Imai, Brown, *et al.* 1984: 68 (k.). Combination in *Dolichoderus:* Emery, 1887a: 254; in *D. (Hypoclinea):* Emery, 1894c: 229. Senior synonym of *sellaris:* Mayr, 1867a: 76. Junior synonym of *thoracicus:* Donisthorpe, 1932c: 457. See also: Bingham, 1903: 295.

*blatchleyi. Dolichoderus (Hypoclinea) mariae* var. *blatchleyi* Wheeler, W.M. 1917i: 462 (w.) U.S.A. Junior synonym of *mariae:* Creighton, 1950a: 334.

*borealis*. *Dolichoderus borealis* Provancher, 1888: 408 (w.) CANADA. Junior synonym of *plagiatus*: Emery, 1895c: 330.

**#borneensis.** *Dolichoderus taprobanae* var. *borneensis* Forel, 1911a: 46 (w.) BORNEO. Variety\subspecies of *bituberculatus*: Santschi, 1928h: 131; of *taprobanae*: Shattuck, 1994: 68.

**#borneonensis.** *Hypoclinea sellaris* var. *borneonensis* Roger, 1863a: 214 (w.) BORNEO. Combination in *Dolichoderus*: Shattuck, 1992c: 77. Variety of *bituberculatus*: Wheeler, W.M. 1919e: 97. Subspecies of *thoracicus*: Shattuck, 1992c: 77.

**brevis.** *Dolichoderus* (*Hypoclinea*) *brevis* Santschi, 1920h: 167 (w.) LAOS. [Also described as new by Santschi, 1924c: 107.]

**#brevithorax.** *Dolichoderus furcifer* subsp. *brevithorax*: Menozzi, 1928b: 2, fig. 3 (w.) SUMATRA.

***britannicus.** **Dolichoderus britannicus* Cockerell, 1915: 483, pl. 65, figs. 6, 7 (q.) GREAT BRITAIN (Oligocene).

***bruneti.** **Dolichoderus bruneti* Théobald, 1937b: 206, pl. 14, fig. 12 (q.) FRANCE (Oligocene).

**burmanicus.** *Dolichoderus burmanicus* Bingham, 1903: 293, fig. 88 (w.) BURMA. Combination in *D.* (*Hypoclinea*): Emery, 1912g: 13.

**butteli.** *Dolichoderus* (*Hypoclinea*) *butteli* Forel, 1913k: 89 (w.) SUMATRA.

**#caligatus.** *Dolichoderus* (*Hypoclinea*) *feae* var. *caligatus* Wheeler, W.M. 1927h: 99 (w.) THAILAND.

*capitatus*. *Dolichoderus capitatus* Santschi, 1921g: 99 (w.) FRENCH GUIANA. Junior synonym of *decollatus*: MacKay, 1993b: 48.

**carbonarius.** *Dolichoderus carbonarius* Emery, 1895k: 474 (footnote) (w.) WEST MALAYSIA. Combination in *D.* (*Hypoclinea*): Emery, 1912g: 13. Current subspecies: nominal plus *latisquamis*.

***caribbaeus.** **Monacis caribbaea* Wilson, 1985c: 19, fig. 2 (w.) DOMINICAN AMBER (Oligocene). Combination in *Dolichoderus*: Shattuck, 1992c: 77. See also: MacKay, 1993b: 44.

**#ceramensis.** *Dolichoderus taprobanae* var. *ceramensis* Stitz, 1912: 509, fig. 11 (w.) INDONESIA (Seram I.).

*championi*. *Dolichoderus championi* Forel, 1899c: 100 (w.q.m.) MEXICO. Wheeler, G.C. & Wheeler, J. 1951: 175 (l.). Combination in *D.* (*Hypoclinea*): Emery, 1912g: 12; in *Hypoclinea*: Kempf, 1972a: 119; in *Dolichoderus*: Shattuck, 1992c: 77. Junior synonym of *diversus*: MacKay, 1993b: 50.

*cingulata*. *Hypoclinea cingulata* Mayr, 1862: 705 (w.) BRAZIL. Junior synonym of *lutosus*: Mayr, 1886c: 356.

**clarki.** *Dolichoderus* (*Acanthoclinea*) *clarki* Wheeler, W.M. 1935c: 69. Replacement name for *tristis* Clark, 1930b: 254. [Junior primary homonym of *tristis* Mann, 1916: 463.] See also: Shattuck, 1994: 45.

**clusor.** *Dolichoderus clusor* Forel, 1907h: 285 (w.) AUSTRALIA. Combination in *D.* (*Hypoclinea*): Emery, 1912g: 12. See also: Clark, 1930b: 262.

**cogitans.** *Dolichoderus bidens* r. *cogitans* Forel, 1912h: 34 (w.) BRAZIL. Combination in *Hypoclinea*: Kempf, 1972a: 119; in *Dolichoderus*: Shattuck, 1992c: 77. Raised to species: MacKay, 1993b: 44.

*columbica*. *Dolichoderus schulzi* var. *columbica* Forel, 1914e: 12 (w.) COLOMBIA. [First available use of *Dolichoderus schulzi* r. *biolleyi* var. *columbica* Forel, 1912h: 35; unavailable name.] Junior synonym of *schulzi*: Kempf, 1959b: 254.

**coniger.** *Hypoclinea conigera* Mayr, 1870b: 956 (diagnosis in key) (w.) BORNEO. Combination in *Dolichoderus*: Emery, 1887a: 256; in *D.* (*Hypoclinea*): Emery, 1894c: 229; in *D.* (*Diabolus*): Wheeler, W.M. 1935c: 68.

**constricta* Mayr, 1868 [*Hypoclinea*]; see under ***ANONYCHOMYRMA.***

***coquandi.** **Dolichoderus coquandi* Théobald, 1937b: 315, pl. 4, figs. 9–11; pl. 14, fig. 11; pl. 26, fig. 18 (m.) FRANCE (Oligocene).

***cornutus.** **Hypoclinea cornuta* Mayr, 1868c: 61, pl. 3, fig. 52 (w.) BALTIC AMBER (Oligocene). Combination in *Dolichoderus*: Dalla Torre, 1893: 158; in *D.* (*Diceratoclinea*): Wheeler, W.M. 1935c: 69.

**crawleyi.** *Dolichoderus* (*Hypoclinea*) *crawleyi* Donisthorpe, 1917: 201 (w.q.) SINGAPORE.

**curvilobus.** *Hypoclinea curviloba* Lattke, 1987b: 259, fig. 1 (w.) COLOMBIA. MacKay, 1993b: 45 (q.). Combination in *Dolichoderus*: Shattuck, 1992c: 77.

**cuspidatus.** *Polyrhachis cuspidatus* Smith, F. 1857a: 63 (w.) BORNEO. Forel, 1913k: 88 (m.). Combination in *Hypoclinea*: Mayr, 1870b: 955 (in key); in *Dolichoderus*: Forel, 1878: 382; in *D.* (*Hypoclinea*): Emery, 1894c: 229; in *D.* (*Diabolus*): Wheeler, W.M. 1935c: 68. Senior synonym of *surbecki* (and its junior synonym *bifurcatus*): Chapman & Capco, 1951: 181. Current subspecies: nominal plus *tenuispinus*. See also: Shattuck, 1994: 47.

*davisi*. *Dolichoderus mariae* subsp. *davisi* Wheeler, W.M. 1905d: 306 (w.) U.S.A. Junior synonym of *mariae*: Creighton, 1950a: 334.

**debilis.** *Dolichoderus debilis* Emery, 1890a: 69 (w.) VENEZUELA. Wheeler, G.C. & Wheeler, J. 1951: 176 (l.). Combination in *D.* (*Monacis*): Emery, 1894c: 228; in *Monacis*: Kempf, 1959b: 247; in *Hypoclinea*: Lattke, 1987b: 264; in *Dolichoderus*: Shattuck, 1992c: 77. Senior synonym of *parabiotica* and *sieversi*: Kempf, 1959b: 247. See also: MacKay, 1993b: 47.

**decollatus.** *Dolichoderus decollatus* Smith, F. 1858b: 75 (w.q.) GUIANA. Wheeler, G.C. & Wheeler, J. 1951: 178 (l.). Senior synonym of *capitatus* and *neglectus*: MacKay, 1993b: 48.

**dentatus.** *Dolichoderus doriae* var. *dentata* Forel, 1902h: 461 (w.) AUSTRALIA. Combination in *D.* (*Hypoclinea*): Emery, 1912g: 12; in *D.* (*Acanthoclinea*), and misspelled as *edentata*: Wheeler, W.M. 1935c: 69. Raised to species: Clark, 1930b: 255. See also: Shattuck, 1994: 48.

***dibolius.** *Dolichoderus dibolia* Wilson, 1985c: 18, fig. 1 (w.) DOMINICAN AMBER (Oligocene). See also: MacKay, 1993b: 49.

**diversus.** *Dolichoderus* (*Hypoclinea*) *diversus* Emery, 1894c: 237 (w.) COLOMBIA. MacKay, 1993b: 52 (q.m.) Combination in *Hypoclinea*: Kempf, 1972a: 119; in *Dolichoderus*: Shattuck, 1992c: 77. Senior synonym of *championi, garbei, leviusculus, mazaruni, ornatus, taeniatus* and *trinidadensis*: MacKay, 1993b: 50. See also: Shattuck, 1994: 48.

**doloniger.** *Monacis dolonigera* Roger, 1862a: 237 (w.) VENEZUELA. Lattke, 1987b: 262 (q.). Combination in *Hypoclinea*: Mayr, 1862: 707; in *Dolichoderus*: Dalla Torre, 1893: 158; in *D.* (*Monacis*): Emery, 1894c: 228; in *Monacis*: Kempf, 1959b: 246; in *Dolichoderus*: Shattuck, 1992c: 77. See also: Kempf, 1959b: 246; MacKay, 1993b: 53.

**doriae.** *Dolichoderus doriae* Emery, 1887a: 252 (w.) AUSTRALIA. Forel, 1902h: 461 (m.); Clark, 1934b: 40 (q.); Wheeler, G.C. & Wheeler, J. 1966: 726 (l.). Combination in *D.* (*Hypoclinea*): Emery, 1894c: 228; in *D.* (*Acanthoclinea*): Wheeler, W.M. 1935c: 69. See also: Clark, 1930b: 254; Shattuck, 1994: 50.

***elegans.** **Dolichoderus* (*Hypoclinea*) *elegans* Wheeler, W.M. 1915h: 80, fig. 35 (w.) BALTIC AMBER (Oligocene).

**#emarginatus.** *Dolichoderus* (*Hypoclinea*) *bituberculatus* var. *emarginata* Santschi, 1920h: 168 (w.) LAOS. [Also described as new by Santschi, 1924c: 108.] Subspecies of *thoracicus*: Shattuck, 1994: 70.

*ensiger*. *Dolichoderus spinicollis* subsp. *ensiger* Forel, 1910a: 31 (w.) SURINAM. Junior synonym of *mucronifer*: Forel, 1911c: 306.

**epetreius.** *Hypoclinea epetreia* Lattke, 1987b: 262, fig. 4 (w.) VENEZUELA. Combination in *Dolichoderus*: Shattuck, 1992c: 77. See also: MacKay, 1993b: 54.

**erectilobus.** *Dolichoderus* (*Hypoclinea*) *erectilobus* Santschi, 1920h: 171, fig. 2 (w.) VIETNAM. [Also described as new by Santschi, 1924c: 111.]

***explicans.** **Hypoclinea explicans* Förster, B. 1891: 434, pl. 13, fig. 7 (q.) GERMANY (Oligocene). Combination in *Dolichoderus*: Handlirsch, 1907: 870. See also: Théobald, 1937b: 207.

**extensispinus.** *Dolichoderus doriae* r. *extensispina* Forel, 1915b: 76 (w.) AUSTRALIA. Combination in *D.* (*Acanthoclinea*): Wheeler, W.M. 1935c: 69. Raised to species: Clark, 1930b: 256. [Both Wheeler and Clark misspell the name as *extensispinosa*.]

**feae.** *Dolichoderus feae* Emery, 1889b: 509, pl. 11, fig. 21 (w.) BURMA. Combination in *D.* (*Hypoclinea*): Emery, 1894c: 229. Current subspecies: *caligatus, fuscus*.

**fernandezi.** *Dolichoderus fernandezi* MacKay, 1993b: 54, figs. 5, 59 (w.) COLOMBIA.

**ferrugineus.** *Dolichoderus bidens* r. *ferrugineus* Forel, 1903c: 258 (w.) BRAZIL. MacKay, 1993b: 56 (q.). Combination in *Hypoclinea*: Kempf, 1972a: 119; in *Dolichoderus*: Shattuck, 1992c: 77. Raised to species, and senior synonym of *inferior*: MacKay, 1993b: 56.

*flavipes* Kirby, W.F. 1896 [*Hypoclinea*]; see under ***OCHETELLUS.***

**formosus.** *Dolichoderus* (*Hypoclinea*) *formosus* Clark, 1930b: 265, fig. 15 (w.q.) AUSTRALIA.

*foveolatus*. *Polyrhachis foveolatus* Lowne, 1865b: 334 (w.) AUSTRALIA. Junior synonym of *scabridus*: Mayr, 1870b: 953.

*frauenfeldi* Mayr, 1855 [*Hypoclinea*]; see under ***LEPISIOTA.***

**#friedrichsi.** *Dolichoderus taprobanae* var. *friedrichsi* Forel, 1914c: 618 (w.m.) VIETNAM.

*fungosa*. *Formica fungosa* Fabricius, 1798: 281 (w.) SURINAM. Junior synonym of *bispinosus*: Latreille, 1802c: 133.

**furcifer.** *Dolichoderus furcifer* Emery, 1887a: 256 (w.) SUMATRA. Combination in *D.* (*Hypoclinea*): Emery, 1894c: 229; in *D.* (*Diabolus*): Wheeler, W.M. 1935c: 68. Current subspecies: nominal plus *ater, brevithorax*.

**#fuscus.** *Dolichoderus feae* r. *fusca* Emery, 1889b: 509 (w.) BURMA. Raised to species: Dalla Torre, 1893: 159;

Bingham, 1903: 294. Subspecies of *feae:* Emery, 1912g: 14.

***gagates.*** *Dolichoderus gagates* Emery, 1890a: 69 (footnote) (w.) BRAZIL. Combination in *D.* (*Monacis*): Emery, 1894c: 228; in *Monacis:* Kempf, 1959b: 251; in *Dolichoderus:* Shattuck, 1992c: 77. See also: Kempf, 1959b: 251; MacKay, 1993b: 57.

*gagates. Dolichoderus taschenbergi* var. *gagates* Wheeler, W.M. 1905d: 310 (w.) U.S.A. [Junior primary homonym of *gagates* Emery, above.] Replacement name (first): *aterrimus* Wheeler, W.M. 1915b: 417. Replacement name (second, unnecessary): *wheeleriella* Forel, 1916: 458.

*garbei. Dolichoderus germaini* subsp. *garbei* Forel, 1911c: 305 (w.) BRAZIL. Combination in *Hypoclinea:* Kempf, 1972a: 119; in *Dolichoderus:* Shattuck, 1992c: 77. Junior synonym of *diversus:* MacKay, 1993b: 50.

**geinitzi* Mayr, 1868 [*Hypoclinea*]; see under ***IRIDOMYRMEX.***

***germaini.*** *Dolichoderus* (*Hypoclinea*) *germaini* Emery, 1894c: 237 (w.) BRAZIL. Wheeler, G.C. & Wheeler, J. 1974b: 397 (l.). Combination in *Hypoclinea:* Kempf, 1972a: 119; in *Dolichoderus:* Shattuck, 1992c: 77. See also: MacKay, 1993b: 57.

***ghilianii.*** *Dolichoderus* (*Hypoclinea*) *ghilianii* Emery, 1894c: 238, pl. 1, fig. 15 (w.) BRAZIL. MacKay, 1993b: 59 (q.m.). Combination in *Hypoclinea:* Kempf, 1969: 293; in *Dolichoderus:* Shattuck, 1992c: 77. See also: Mann, 1916: 465; MacKay, 1993b: 58.

***gibbifer.*** *Dolichoderus gibbifer* Emery, 1887a: 254, pl. 4, fig. 28 (w.) JAVA. Combination in *D.* (*Hypoclinea*): Emery, 1894c: 229. Current subspecies: nominal plus *gibbosior.*

#***gibbosior.*** *Dolichoderus gibbifer* var. *gibbosior* Forel, 1912d: 108 (w.) JAVA.

*gibbosoanalis. Dolichoderus* (*Hypoclinea*) *gibbosus* var. *gibbosoanalis* Forel, 1922: 98 (w.) VENEZUELA. Subspecies of *quadridenticulatus:* Shattuck, 1992c: 77. Junior synonym of *quadridenticulatus:* MacKay, 1993b: 86.

*gibbosa. Formica gibbosa* Smith, F. 1858b: 19 (w.) BRAZIL (not India). [Junior primary homonym of *Formica gibbosa* Latreille, 1798: 50 (now in *Aphaenogaster*).] Combination in *Dolichoderus:* Mayr, 1886c: 354; in *Hypoclinea:* Kempf, 1972a: 119. Senior synonym of *quadridenticulatus:* Mayr, 1886c: 354; Emery, 1894c: 235, and hence *quadridenticulatus* first available replacement name for *gibbosa* Smith, F.

***gibbus.*** *Formica* (*Tapinoma*) *gibba* Smith, F. 1861b: 38 (w.) SULAWESI. Combination in *Dolichoderus:* Dalla Torre, 1893: 159.

*glabra* Mayr, 1862 [*Hypoclinea*]; see under ***OCHETELLUS.***

#***glabripes*** *Dolichoderus affinis* var. *glabripes* Forel, 1895e: 466 (w.) INDIA.

***glauerti.*** *Dolichoderus* (*Hypoclinea*) *glauerti* Wheeler, W.M. 1934d: 147 (w.m.) AUSTRALIA.

**goepperti* Mayr, 1868 [*Hypoclinea*]; see under ***LIOMETOPUM.***

***goudiei.*** *Dolichoderus* (*Hypoclinea*) *goudiei* Clark, 1930b: 264, fig. 17 (w.) AUSTRALIA.

#***gracilipes.*** *Hypoclinea gracilipes* Mayr, 1879: 658 (w.) INDIA. Combination in *Dolichoderus:* Dalla Torre, 1893: 159; in *D.* (*Hypoclinea*): Emery, 1894c: 229. Junior synonym of *taprobanae:* Forel, 1895e: 466. Revived from synonymy as variety of *taprobanae:* Bingham, 1903: 296; Emery, 1912g: 14; Forel, 1912n: 61.

*gracilis. Hypoclinea gracilis* Motschoulsky, 1863: 14 (w.) SRI LANKA. Junior synonym of *taprobanae:* Emery, 1912g: 14.

*grandii. Dolichoderus* (*Hypoclinea*) *grandii* Menozzi, 1924c: 119 (w.) PERU. Combination in *Hypoclinea:* Kempf, 1972a: 120; in *Dolichoderus:* Shattuck, 1992c: 77. Junior synonym of *mesonotalis:* MacKay, 1993b: 77.

*granulatus* Pergande, 1896; see under ***CAMPONOTUS.***

****gurnetensis.*** *Dolichoderus gurnetensis* Donisthorpe, 1920: 88, pl. 5, fig. 7 (q.) GREAT BRITAIN (Oligocene).

***haradae.*** *Dolichoderus haradae* MacKay, 1993b: 60, fig. 158 (w.) BRAZIL.

*haueri* Mayr, 1867 [*Hypoclinea*]; see under ***IRIDOMYRMEX.***

*heterogaster. Dolichoderus imbecillus* var. *heterogaster* Santschi, 1923c: 269 (w.) BRAZIL. Junior synonym of *attelaboides:* MacKay, 1993b: 35.

*humilis* Mayr, 1868 [*Hypoclinea*]; see under ***LINEPITHEMA.***

*imbecillus. Dolichoderus* (*Dolichoderus*) *imbecillus* Mann, 1916: 459, pl. 2, fig. 18 (w.) BRAZIL. Junior synonym of *attelaboides:* MacKay, 1993b: 35.

***imitator.*** *Dolichoderus* (*Dolichoderus*) *imitator* Emery, 1894c: 230, pl. 1, fig. 12 (w.q.m.) BRAZIL. See also: MacKay, 1993b: 61.

***indrapurensis.*** *Dolichoderus indrapurensis* Forel, 1912n: 61 (w.) SUMATRA. Combination in *D.* (*Hypoclinea*): Forel, 1913k: 90; in *D.* (*Diceratoclinea*): Wheeler, W.M. 1935c: 69. Current subspecies: nominal plus *nigrogaster.*

***inermis.*** *Dolichoderus inermis* MacKay, 1993b: 63, figs. 9, 10 (w.) COSTA RICA.

*inferior. Dolichoderus bidens* var. *inferior* Mann, 1916: 465 (w.) BRAZIL. Combination in *Hypoclinea:* Kempf, 1972a: 119; in *Dolichoderus:* Shattuck, 1992c: 77. Junior synonym of *ferrugineus:* MacKay, 1993b: 56.

*ingruens. Formica ingruens* Walker, 1859: 372 (q.) SRI LANKA. Junior synonym of *taprobanae:* Donisthorpe, 1932b: 575.

*iniqua* Mayr, 1870 [*Hypoclinea*]; see under ***LINEPITHEMA.***

*inornatus. Dolichoderus plagiatus* var. *inornatus* Wheeler, W.M. 1905d: 313 (w.) U.S.A. Junior synonym of *plagiatus:* Creighton, 1950a: 335.

***inpai.*** *Monacis inpai* Harada, 1987: 600, figs. 1, 2 (w.) BRAZIL. Combination in *Dolichoderus:* Shattuck, 1992c: 77. See also: MacKay, 1993b: 64.

*integra. Dolichoderus gibbosus* var. *integra* Forel, 1911c: 306 (w.) BRAZIL. Combination in *Hypoclinea:* Kempf, 1972a: 120; in *Dolichoderus,* and subspecies of *quadridenticulatus:* Shattuck, 1992c: 77. Junior synonym of *quadridenticulatus:* MacKay, 1993b: 86.

***intermedius.*** *Dolichoderus intermedius* MacKay, 1993b: 64, fig. 6 (w.) DOMINICA.

***jacobsoni.*** *Dolichoderus jacobsoni* Forel, 1909d: 226 (q.) JAVA. Combination in *D.* (*Hypoclinea*): Emery, 1912g: 14.

*japonicus. Dolichoderus quadripunctatus* subsp. *japonicus* Yoshioka, 1939: 70 (w.) JAPAN. Junior synonym of *sibiricus:* Yasumatsu, 1941: 182.

*karawaiewi. Dolichoderus patens* r. *pubiventris* var. *karawaiewi* Forel, 1901d: 303 (w.q.m.) JAVA. Unavailable name (Shattuck, 1994: 73).

#***kratochvili.*** *Dolichoderus* (*Hypoclinea*) *quadripunctatus* var. *kratochvili* Novak, V. 1941: 47, figs. 2–4 (w.) CZECHOSLOVAKIA.

****kutschlinicus.*** **Hypoclinea kutschlinicus* Deichmüller, 1881: 322, pl. 21, fig. 13 (q.) CZECHOSLOVAKIA (Tertiary). Combination in *Dolichoderus:* Dalla Torre, 1893: 159.

#***lacciperdus.*** *Dolichoderus* (*Hypoclinea*) *bituberculatus* var. *lacciperda* Santschi, 1925f: 91 (w.) CAMBODIA. Subspecies of *thoracicus:* Shattuck, 1994: 70.

***lactarius.*** *Formica lactaria* Smith, F. 1860b: 95 (w.) INDONESIA (Batjan I.). Combination in *Dolichoderus* (*Hypoclinea*): Donisthorpe, 1932c: 459 [and suggests that this name may be synonymous with *thoracicus*].

***lamellosus.*** *Hypoclinea lamellosa* Mayr, 1870a: 390 (q.) COLOMBIA. Emery, 1894c: 233 (w.); Kempf, 1959b: 264 (m.). Combination in *Dolichoderus:* Emery, 1890a: 70; in *D.* (*Monacis*): Emery, 1894c: 228; in *Monacis:* Kempf, 1959b: 263; in *Dolichoderus:* Shattuck, 1992c: 77. Senior synonym of *missionensis:* Kempf, 1959b: 263. See also: MacKay, 1993b: 65.

***laminatus.*** *Hypoclinea laminata* Mayr, 1870a: 389, fig. 8 (w.) COLOMBIA. [Also described as new by Mayr, 1870b: 956 (in key).] Forel, 1899c: 99 (q.); Wheeler, G.C. & Wheeler, J. 1951: 177 (l.). Combination in *Dolichoderus:* Emery, 1890a: 70; in *D.* (*Monacis*): Emery, 1894c: 228; in *Monacis:* Kempf, 1959b: 259; in *Dolichoderus:* Shattuck, 1992c: 77. Senior synonym of *luteiventris:* Kempf, 1959b: 259. See also: MacKay, 1993b: 66.

***laotius.*** *Dolichoderus* (*Hypoclinea*) *laotius* Santschi, 1920h: 170 (w.) LAOS. [Also described as new by Santschi, 1924c: 110.]

#***latisquamis.*** *Dolichoderus carbonarius* subsp. *latisquamis* Emery, 1900d: 694 (w.) SUMATRA.

***laurae.*** *Dolichoderus laurae* MacKay, 1993b: 67, figs. 150–157 (w.q.m.) COLOMBIA.

#***levior.*** *Dolichoderus* (*Hypoclinea*) *bituberculatus* var. *levior* Karavaiev, 1926d: 427 (w.) INDONESIA. Subspecies of *thoracicus:* Shattuck, 1994: 71.

*leviusculus. Dolichoderus germaini* subsp. *leviusculus* Emery, 1906c: 173 (w.) BRAZIL. Forel, 1909a: 260 (q.). Combination in *Hypoclinea:* Kempf, 1972a: 119; in *Dolichoderus:* Shattuck, 1992c: 77. Junior synonym of *diversus:* MacKay, 1993b: 50.

***lobicornis.*** *Monacis lobicornis* Kempf, 1959b: 267, pl. 1, fig. 10; pl. 2, fig. 13 (w.) BRAZIL. Combination in *Dolichoderus:* Shattuck, 1992c: 77. See also: MacKay, 1993b: 69.

***longicollis.*** *Dolichoderus longicollis* MacKay, 1993b: 69, figs. 17, 57 (w.) COLOMBIA.

****longipennis.*** **Hypoclinea longipennis* Mayr, 1868c: 67, pl. 4, fig. 65 (m.) BALTIC AMBER (Oligocene). Combination in *Dolichoderus:* Dalla Torre, 1893: 159.

***luederwaldti.*** *Dolichoderus (Hypoclinea) luederwaldti* Santschi, 1921g: 100 (w.) BRAZIL. Combination in *Hypoclinea:* Kempf, 1969: 293; in *Dolichoderus:* Shattuck, 1992c: 77. See also: MacKay, 1993b: 71.

***lugens.*** *Dolichoderus (Hypoclinea) lugens* Emery, 1894c: 239, pl. 1, fig. 17 (w.) BOLIVIA. MacKay, 1993b: 72 (q.m.). Combination in *Hypoclinea:* Kempf, 1972a: 120; in *Dolichoderus:* Shattuck, 1992c: 77.

#***lugubris.*** *Dolichoderus (Hypoclinea) moggridgei* var. *lugubris* Santschi, 1920h: 169 (w.) INDIA.

***lujae.*** *Dolichoderus (Hypoclinea) luederwaldti* st. *lujae* Santschi, 1923c: 269 (w.) BRAZIL. Combination in *Hypoclinea:* Kempf, 1972a: 120; in *Dolichoderus:* Shattuck, 1992c: 77. Raised to species: MacKay, 1993b: 73.

*luteiventris. Dolichoderus (Monacis) laminatus* subsp. *luteiventris* Emery, 1894c: 232 (w.) BRAZIL. Junior synonym of *laminatus:* Kempf, 1959b: 259.

***lutosus.*** *Formica lutosa* Smith, F. 1858b: 42 (w.) BRAZIL. Mann, 1916: 468 (q.m.). Combination in *Dolichoderus:* Mayr, 1886c: 356; in *D. (Hypoclinea)*: Emery, 1894c: 228; in *Hypoclinea:* Kempf, 1972a: 120; in *Dolichoderus:* Shattuck, 1992c: 77. Senior synonym of *cingulata:* Mayr, 1886c: 356. Senior synonym of *nigriventris,* and material of the *nomen nudum ruficauda* referred here: MacKay, 1993b: 74.

***mariae.*** *Dolichoderus mariae* Forel, 1885a: 349 (w.) U.S.A. Mayr, 1886d: 436 (q.); Wheeler, W.M. 1905d: 306 (m.); Wheeler, G.C. & Wheeler, J. 1966: 726 (l.). Combination in *D. (Hypoclinea)*: Emery, 1894c: 229. Senior synonym of *blatchleyi* and *davisi:* Creighton, 1950a: 334. See also: Smith, D.R. 1979: 1415; Johnson, 1989a: 2; MacKay, 1993b: 75; Shattuck, 1994: 58.

*mazaruni. Iridomyrmex mazaruni* Donisthorpe, 1939c: 152 (w.q.) GUIANA. Combination in *Dolichoderus:* Shattuck, 1992c: 77. Junior synonym of *diversus:* MacKay, 1993b: 50.

***mesonotalis.*** *Dolichoderus mesonotalis* Forel, 1907a: 26 (w.) BRAZIL. MacKay, 1993b: 77 (m.). Combination in *D. (Hypoclinea)*: Emery, 1912g: 12; in *Hypoclinea:* Kempf, 1972a: 120; in *Dolichoderus:* Shattuck, 1992c: 77. Senior synonym of *grandii* and *simplex:* MacKay, 1993b: 77.

****mesosternalis.*** **Dolichoderus (Hypoclinea) mesosternalis* Wheeler, W.M. 1915h: 81, fig. 36 (w.) BALTIC AMBER (Oligocene).

*missionensis. Dolichoderus (Monacis) lamellosa* var. *missionensis* Santschi, 1916e: 390, fig. (w.) ARGENTINA. Variety of *schulzi:* Santschi, 1923c: 269. Junior synonym of *lamellosus:* Kempf, 1959b: 263.

***modiglianii.*** *Dolichoderus modiglianii* Emery, 1900d: 693, fig. 9 (w.) INDONESIA (Mentawei I.). Combination in *D. (Hypoclinea)*: Emery, 1912g: 14. Current subspecies: nominal plus *rubescens.*

***moggridgei.*** *Dolichoderus moggridgei* Forel, 1886d: 243 (w.) INDIA. Combination in *D. (Hypoclinea)*: Emery, 1894c: 229. Current subspecies: nominal plus *bicolor, lugubris.*

***monoceros.*** *Dolichoderus monoceros* Emery, 1897d: 570, pl. 1, fig. 16 (w.) NEW GUINEA. Combination in *D. (Hypoclinea)*: Emery, 1912g: 13; in *D. (Monoceratoclinea)*: Wheeler, W.M. 1935c: 68.

***mucronifer.*** *Monacis mucronifera* Roger, 1862a: 236 (w.) FRENCH GUIANA. Santschi, 1921g: 100 (q.); Kempf, 1959b: 236 (m.). Combination in *Hypoclinea:* Mayr, 1863: 423; in *Dolichoderus:* Dalla Torre, 1893: 160; in *D. (Monacis)*: Emery, 1894c: 228; in *Monacis:* Kempf, 1959b: 235; in *Dolichoderus:* Shattuck, 1992c: 77. Senior synonym of *ensiger:* Forel, 1911c: 306. See also: MacKay, 1993b: 79.

#***mus.*** *Dolichoderus (Hypoclinea) affinis* var. *mus* Santschi, 1920h: 169 (w.) VIETNAM. [Also described as new by Santschi, 1924c: 108.]

#***nasutus.*** *Dolichoderus (Hypoclinea) bituberculatus* var. *nasuta* Karavaiev, 1935a: 108, fig. 26 (w.) CAMBODIA. Subspecies of *thoracicus:* Shattuck, 1994: 71.

*neglectus. Dolichoderus decollatus* subsp. *neglectus* Menozzi, 1935b: 197, fig. 3B (w.q.) GUIANA. Raised to species: Kempf, 1969: 291. Junior synonym of *decollatus:* MacKay, 1993b: 48.

#***niger.*** *Dolichoderus (Hypoclinea) ypsilon* var. *nigra* Crawley, 1922c: 25 (w.) AUSTRALIA.

#***nigricans.*** *Dolichoderus affinis* var. *nigricans* Emery, 1895k: 474 (w.) BURMA.

***nigricornis.*** *Dolichoderus (Hypoclinea) nigricornis* Clark, 1930b: 264, fig. 14 (w.) AUSTRALIA.

*nigriventris. Dolichoderus lutosus* var. *nigriventris* Forel, 1893g: 351 (w.) ANTILLES IS. Combination in *Hypoclinea:* Kempf, 1972a: 120; in *Dolichoderus:* Shattuck, 1992c: 77. Junior synonym of *lutosus:* MacKay, 1993b: 74.

#***nigriventris.*** *Dolichoderus (Hypoclinea) sulcaticeps* var. *nigriventris* Santschi, 1920h: 171 (w.) VIETNAM. [Unresolved junior primary homonym of *nigriventris* Forel, above.] [Also described as new by Santschi, 1924c: 111.]

#***nigrogaster.*** *Dolichoderus (Hypoclinea) indrapurensis* var. *nigrogaster* Viehmeyer, 1922: 214 (w.) BORNEO.

*nitidior. Dolichoderus (Hypoclinea) gibbosus* var. *nitidior* Emery, 1894c: 235 (w.) BRAZIL. Combination in *Hypoclinea:* Kempf, 1972a: 120; in *Dolichoderus,* and subspecies of *quadridenticulatus:* Shattuck, 1992c: 77. Junior synonym of *quadridenticulatus:* MacKay, 1993b: 86.

****nitidus.*** **Ponera nitida* Heer, 1850: 149, pl. 12, fig. 4 (q.) YUGOSLAVIA (Miocene). Combination in *Poneropsis:* Heer, 1867: 20; Taylor, 1964: 138; in *Hypoclinea:* Mayr, 1867b: 57; in *Dolichoderus:* Handlirsch, 1907: 870; Shattuck, 1994: 59.

****obliteratus.*** **Hypoclinea obliterata* Scudder, 1877b: 267 (wing) CANADA (Miocene). Combination in *Dolichoderus:* Dalla Torre, 1893: 160. Included as "(Dolichoderinae) *obliterata*" by Carpenter, 1930: 21. See also: MacKay, 1993b: 79; Shattuck, 1994: 171.

*obscura. Formica obscura* Smith, F. 1858b: 42 (q.) BRAZIL. [Junior primary homonym of **Formica obscura* Heer, 1850: 119.] Replacement name: *satanus* **new name**. Combination in *Dolichoderus:* Mayr, 1886c: 356; in *D. (Monacis)*: Emery, 1894c: 228; in *Monacis:* Kempf, 1959b: 268; in *Dolichoderus:* Shattuck, 1992c: 77. See also: MacKay, 1993b: 80; Shattuck, 1994: 59.

#***obscuripes.*** *Dolichoderus (Hypoclinea) taprobanae* var. *obscuripes* Santschi, 1920h: 169 (w.q.) LAOS. [Also described as new by Santschi, 1924c: 109.]

***occidentalis.*** *Dolichoderus (Hypoclinea) occidentalis* Clark, 1930b: 268, fig. 12 (w.) AUSTRALIA.

***omacanthus.*** *Monacis omacantha* Kempf, 1972c: 251, figs. 1–3 (w.) BRAZIL. Combination in *Dolichoderus:* Shattuck, 1992c: 77. See also: MacKay, 1993b: 80.

*omyrmex* Donisthorpe, 1938; see under ***POLYRHACHIS.***

*ornatus. Dolichoderus (Hypoclinea) championi* var. *ornatus* Mann, 1916: 466 (w.q.) BRAZIL. Combination in *Hypoclinea:* Kempf, 1972a: 119; in *Dolichoderus:* Shattuck, 1992c: 77. Junior synonym of *diversus:* MacKay, 1993b: 50.

****oviformis.*** **Dolichoderus oviformis* Théobald, 1937b: 128, pl. 9, figs. 11, 12; pl. 10, figs. 14, 15 (q.m.) FRANCE (Oligocene).

****ovigerus.*** **Dolichoderus ovigerus* Cockerell, 1915: 484, pl. 65, fig. 9 (q.) GREAT BRITAIN (Oligocene).

*parabiotica. Dolichoderus debilis* var. *parabiotica* Forel, 1912h: 33 (w.q.m.) COLOMBIA. Junior synonym of *debilis:* Kempf, 1959b: 247.

***parvus.*** *Dolichoderus (Hypoclinea) parvus* Clark, 1930b: 263, fig. 16 (w.) AUSTRALIA.

****passalomma.*** **Dolichoderus (Hypoclinea) passalomma* Wheeler, W.M. 1915h: 79, fig. 34 (w.) BALTIC AMBER (Oligocene).

***patens.*** *Hypoclinea patens* Mayr, 1870b: 957 (diagnosis in key) (w.) BORNEO. Combination in *Dolichoderus (Hypoclinea)*: Emery, 1887a: 254. Current subspecies: nominal plus *pubiventris.*

*perditor. Formica perditor* Fabricius, 1804: 402 (w.) SOUTH AMERICA. Junior synonym of *bidens:* Roger, 1862c: 285.

***piceus.*** *Dolichoderus piceus* MacKay, 1993b: 81, figs. 38, 40, 75 (w.) COLOMBIA.

*pilifera* Mayr, 1870 [*Hypoclinea*]; see under ***LINEPITHEMA.***

***plagiatus.*** *Hypoclinea plagiata* Mayr, 1870b: 960 (w.) U.S.A. Wheeler, W.M. 1905d: 311 (q.m.); Wheeler, G.C. & Wheeler, J. 1951: 175 (l.). Combination in *Dolichoderus:* Mayr, 1886d: 436; in *D. (Hypoclinea)*: Emery, 1894c: 229. Senior synonym of *borealis:* Emery, 1895c: 330; of *inornatus:* Creighton, 1950a: 335. See also: Wheeler, G.C. & Wheeler, J. 1966: 727; Smith, D.R. 1979: 1416; Johnson, 1989a: 3; MacKay, 1993b: 81; Shattuck, 1994: 61.

****primitivus.*** **Hypoclinea primitiva* Wilson, 1985c: 22, fig. 4 (w.) DOMINICAN AMBER. Combination in *Dolichoderus:* Shattuck, 1992c: 77. See also: MacKay, 1993b: 83.

****prolaminatus.*** **Monacis prolaminata* Wilson, 1985c: 20, fig. 3 (w.) DOMINICAN AMBER. Combination in *Dolichoderus:* Shattuck, 1992c: 77.

#***pubiventris.*** *Dolichoderus patens* subsp. *pubiventris* Emery, 1900d: 694 (w.) INDONESIA (Mentawei I.). Combination in *D. (Hypoclinea)*: Emery, 1912g: 14.

*pulla. Dolichoderus attelaboides* var. *pulla* Santschi, 1923c: 269 (w.) BRAZIL. Junior synonym of *attelaboides:* Kempf, 1969: 289.

***pustulatus.*** *Dolichoderus pustulatus* Mayr, 1886d: 436 (w.q.) U.S.A. Wheeler, G.C. & Wheeler, J. 1966: 727 (l.); MacKay, 1993b: 85 (m.). Combination in *D. (Hypoclinea)*: Emery, 1894c: 229. Subspecies of *plagiatus:* Wheeler, W.M. 1905d: 313. Revived status as species, and senior synonym of *beutenmuelleri:* Creighton, 1950a: 335. See also: Smith, D.R. 1979: 1416; Johnson, 1989a: 3; MacKay, 1993b: 84.

***quadridenticulatus.*** *Formica quadridenticulata* Roger, 1862c: 287 (q.) "Südamerika". Mayr, 1862: 706 (w.). MacKay, 1993b: 87 (q.m.). Combination in *Hypoclinea:* Mayr, 1862: 706; in *Dolichoderus:* Dalla Torre, 1893: 159. Junior synonym of *gibbosa:* Mayr, 1886c: 354; Emery, 1894c: 235, hence first available replacement name for *Formica gibbosa* Smith, F. 1858b: 19. [Junior primary homonym of *Formica gibbosa* Latreille, 1798: 50.] Senior synonym of *analis, gibbosoanalis, integra* and *nitidior:* MacKay, 1993b: 86.

***quadripunctatus.*** *Formica quadripunctata* Linnaeus, 1771: 541 (w.) GERMANY. Latreille, 1802c: 179 (q.); Mayr, 1855: 379 (w.q.m.). Combination in *Hypoclinea:* Mayr, 1855: 379; in *Dolichoderus:* Emery & Forel, 1879: 455; in *D.* (*Hypoclinea*): Emery, 1894c: 229. Senior synonym of *unicolor:* Arakelian, 1994: 72. Current subspecies: nominal plus *kratochvili.* See also: Bernard, 1967: 249; Kutter, 1977c: 172; Atanassov & Dlussky, 1992: 186; Shattuck, 1994: 63.

***reflexus.*** *Dolichoderus* (*Hypoclinea*) *reflexus* Clark, 1930b: 261, fig. 13 (w.) AUSTRALIA.

****rohweri.*** **Dolichoderus* (*Hypoclinea*) *rohweri* Carpenter, 1930: 40, pl. 4, fig. 1; pl. 9, fig. 7 (q.) U.S.A. (Miocene). See also: MacKay, 1993b: 88.

***rosenbergi.*** *Dolichoderus rosenbergi* Forel, 1911e: 282 (w.) ECUADOR. See also: MacKay, 1993b: 89.

#***rubescens.*** *Dolichoderus modiglianii* var. *rubescens* Emery, 1900d: 693 (w.) INDONESIA (Mentawei I.). Combination in *D.* (*Hypoclinea*): Emery, 1912g: 14.

***rufescens.*** *Dolichoderus* (*Monacis*) *debilis* var. *rufescens* Mann, 1912: 40 (w.q.m.) BRAZIL. Combination in *Monacis:* Kempf, 1959b: 252; in *Dolichoderus:* Shattuck, 1992c: 77. Raised to species: Kempf, 1959b: 252. See also: MacKay, 1993b: 90.

#***rufescens.*** *Dolichoderus* (*Hypoclinea*) *bituberculatus* var. *rufescens* Stitz, 1925: 123 (w.) BORNEO. [Unresolved junior primary homonym of *rufescens* Mann, above.] Subspecies of *thoracicus:* Shattuck, 1994: 71.

*ruficauda. Dolichoderus* (*Hypoclinea*) *lutosus* var. *ruficauda* Wheeler, W.M. 1936d: 229. *Nomen nudum;* material referred to *lutosus* by MacKay, 1993b: 74.

*ruficornis. Ponera ruficornis* Spinola, 1851b: 53 (w.) BRAZIL. [Also described as new by Spinola, 1853: 96.] [Emery, 1893h: 2 notes that the "specimen is head and prothorax of *Dolichoderus* sp.; rest of body is *Platythyrea* sp.] See also: Brown, 1975: 9, who refers the name to *Platythyrea.*]

#***ruficornis.*** *Dolichoderus* (*Hypoclinea*) *scabridus* var. *ruficornis* Santschi, 1916c: 175 (w.) AUSTRALIA. Clark, 1930b: 258 (m.).

#***rufotibialis.*** *Dolichoderus* (*Hypoclinea*) *ypsilon* var. *rufotibialis* Clark, 1930b: 259, fig. 9 (w.) AUSTRALIA.

***rugosus.*** *Polyrhachis rugosus* Smith, F. 1858b: 74 (w.) BRAZIL. MacKay, 1993b: 91 (q.m.). Combination in *Dolichoderus:* Mayr, 1886c: 357.

***satanus.*** *Dolichoderus satanus* **new name**. Replacement name for *Formica obscura* Smith, F. 1858b: 42. [Junior primary homonym of **Formica obscura* Heer, 1850: 119.] [For earlier data see under *obscura* Smith.]

***scabridus.*** *Dolichoderus scabridus* Roger, 1862a: 244 (w.) AUSTRALIA. Clark, 1930b: 257 (m.); Wheeler, G.C. & Wheeler, J. 1951: 176 (l.); Crozier, 1970: 119 (k.). Combination in *D.* (*Hypoclinea*): Emery, 1894c: 228; in *D.* (*Diceratoclinea*): Wheeler, W.M. 1935c: 69. Senior synonym of *foveolatus:* Mayr, 1870b: 953. Current subspecies: nominal plus *ruficornis.* See also: Shattuck, 1994: 64.

***schulzi.*** *Dolichoderus* (*Monacis*) *schulzi* Emery, 1894c: 233 (w.) BRAZIL. Kempf, 1959b: 255 (q.); MacKay, 1993b: 92 (m.). Combination in *Monacis:* Kempf, 1959b: 254; in *Dolichoderus:* Shattuck, 1992c: 77. Senior synonym of *biolleyi* and *columbica:* Kempf, 1959b: 254.

***scrobiculatus.*** *Hypoclinea scrobiculata* Mayr, 1876: 80 (w.) AUSTRALIA. Combination in *Dolichoderus:* Dalla Torre, 1893: 161; in *D.* (*Hypoclinea*): Emery, 1894c: 228. See also: Clark, 1930b: 260.

****sculpturatus.*** **Hypoclinea sculpturata* Mayr, 1868c: 62, pl. 3, figs. 53–55 (w.) BALTIC AMBER (Oligocene). Combination in *Dolichoderus:* Dalla Torre, 1893: 161.

*sellaris. Hypoclinea sellaris* Roger, 1863a: 213 (w.) SINGAPORE. Junior synonym of *bituberculatus:* Mayr, 1867a: 76.

*semirufus. Dolichoderus semirufus* André, 1887: 288 (w.) VIETNAM. Junior synonym of *gracilipes:* Emery, 1889b: 509 (footnote); of *taprobanae:* Forel, 1895e: 466; Bingham, 1903: 296.

***semirugosus.*** *Hypoclinea semirugosa* Mayr, 1870b: 956 (diagnosis in key) (w.) BORNEO. Combination in *Dolichoderus:* Emery, 1887a: 254; Shattuck, 1992c: 77.

***septemspinosus.*** *Dolichoderus* (*Monacis*) *septemspinosus* Emery, 1894c: 231 (w.) BRAZIL. Combination in *Monacis:* Kempf, 1959b: 237; in *Dolichoderus:* Shattuck, 1992c: 77. See also: MacKay, 1993b: 93.

***setosus.*** *Monacis setosa* Kempf, 1959b: 267, pl. 2, fig. 8; pl. 4, fig. 3 (w.) BRAZIL. Combination in *Dolichoderus:* Shattuck, 1992c: 77.

***shattucki.*** *Dolichoderus shattucki* MacKay, 1993b: 94, figs. 1, 60, 90, 91 (w.q.) PANAMA.

#***siamensis.*** *Dolichoderus taprobanae* var. *siamensis* Forel, 1911a: 46 (w.q.) THAILAND. [Also described as new by Forel, 1911g: 285.]

***sibiricus.*** *Dolichoderus quadripunctatus* subsp. *sibiricus* Emery, 1889a: 442 (w.) RUSSIA. Kupyanskaya, 1990: 155 (q.m.). Combination in *D.* (*Hypoclinea*): Emery, 1894c: 229. Raised to species: Yasumatsu, 1962: 96. Senior synonym of *abietis, japonicus* and *yoshiokae:* Yasumatsu, 1941: 182; of *atriceps* and *sinensis:* Yasumatsu, 1962: 96.

*sieversi. Dolichoderus debilis* var. *sieversi* Forel, 1901h: 65 (w.) VENEZUELA. Combination in *D.* (*Monacis*): Emery, 1912g: 9. Junior synonym of *debilis:* Kempf, 1959b: 247.

***siggii.*** *Dolichoderus siggii* Forel, 1895e: 465 (w.) THAILAND. Combination in *D.* (*Hypoclinea*): Emery, 1912g: 14.

*simplex. Dolichoderus simplex* Forel, 1912h: 35 (w.) PERU. Combination in *Hypoclinea:* Kempf, 1972a: 120; in *Dolichoderus:* Shattuck, 1992c: 77. Junior synonym of *mesonotalis:* MacKay, 1993b: 77.

*sinensis. Dolichoderus* (*Hypoclinea*) *sinensis* Wheeler, W.M. 1921e: 111, fig. 1 (w.) CHINA. Junior synonym of *sibiricus:* Yasumatsu, 1962: 96.

***smithi.*** *Dolichoderus smithi* MacKay, 1993b: 96, fig. 37 (w.) BOLIVIA.

***spinicollis.*** *Formica spinicollis* Latreille, 1817: 99, pl. 38, figs. 6–12 (w.) BRAZIL. Combination in *Monacis:* Roger, 1862a: 234; Kempf, 1959b: 232; in *Hypoclinea:* Mayr, 1862: 708; in *Dolichoderus* (*Monacis*): Emery, 1894c: 228; in *Dolichoderus:* Dalla Torre, 1893: 161; Shattuck, 1992c: 77. See also: Kempf, 1959b: 232; MacKay, 1993b: 97.

***spurius.*** *Dolichoderus bidens* var. *spurius* Forel, 1903c: 258 (w.q.) TRINIDAD. Combination in *Hypoclinea:* Kempf, 1972a: 119; in *Dolichoderus:* Shattuck, 1992c: 77. Race of *bidens:* Forel, 1912h: 34. Raised to species, and senior synonym of *albatus:* MacKay, 1993b: 97.

[*subspinosa. Formica* (*Hypochira*) *subspinosa* Buckley, 1866: 169 (w.) U.S.A. Combination in *Dolichoderus?:* Emery, 1895c: 338. Unrecognisable and unplaceable taxon; may possibly be a dolichoderine. See: Emery, 1925b: 271; Smith, D.R. 1979: 1422.]

***sulcaticeps.*** *Hypoclinea sulcaticeps* Mayr, 1870b: 957 (diagnosis in key) (w.) BORNEO. Viehmeyer, 1916a: 141 (q.m.); Karavaiev, 1926d: 428 (q.m.). Combination in *Dolichoderus:* Emery, 1887a: 254; in *D.* (*Hypoclinea*): Emery, 1894c: 229. See also: Bingham, 1903: 293. Current subspecies: nominal plus *nigriventris* (unresolved junior homonym).

***superaculus.*** *Hypoclinea superacula* Lattke, 1987b: 261, fig. 2 (w.) COLOMBIA. Combination in *Dolichoderus:* Shattuck, 1992c: 77. See also: MacKay, 1993b: 99.

*surbecki. Dolichoderus* (*Hypoclinea*) *cuspidatus* st. *surbecki* Santschi, 1925f: 92 (w.) SUMATRA. Combination in *D.* (*Diabolus*), and senior synonym of *bifurcatus:* Wheeler, W.M. 1935c: 68. Junior synonym of *cuspidatus:* Chapman & Capco, 1951: 181.

*taeniatus. Dolichoderus* (*Hypoclinea*) *championi* var. *taeniatus* Wheeler, W.M. 1916d: 329 (w.q.) COLOMBIA. [First available use of *Dolichoderus championi* r. *trinidadensis* var. *taeniatus* Forel, 1899c: 101; unavailable name.] Junior synonym of *diversus:* MacKay, 1993b: 50.

***taprobanae.*** *Formica taprobane* Smith, F. 1858b: 13 (q.) SRI LANKA. Bingham, 1903: 296 (w.). Combination in *Dolichoderus:* Mayr, 1886c: 353; in *D.* (*Hypoclinea*): Emery, 1912g: 14. Senior synonym of *gracilis:* Emery, 1912g: 14; Wheeler, W.M. 1919e: 99; of *ingruens:* Donisthorpe, 1932b: 575; of *semirufus:* Forel, 1895e: 466; Bingham, 1903: 296. Current subspecies: nominal plus *borneensis, ceramensis, friedrichsi, gracilipes, obscuripes, siamensis, tonkinus.*

***taschenbergi.*** *Hypoclinea taschenbergi* Mayr, 1866a: 498 (w.) U.S.A. Wheeler, G.C. & Wheeler, J. 1951: 173 (l.); MacKay, 1993b: 100 (q.m.). Combination in *Dolichoderus:* Mayr, 1886d: 437; in *D.* (*Hypoclinea*): Emery, 1894c: 229. Senior synonym of *aterrimus:* Creighton, 1950a: 336; of the unnecessary replacement name *wheeleriella:* Smith, M.R. 1951a: 834. See also: Wheeler, G.C & Wheeler, J. 1966: 728; Smith, D.R. 1979: 1416; Francoeur & Elias, 1985: 303; Johnson, 1989a: 4; MacKay, 1993b: 99; Shattuck, 1994: 69.

****tauricus.*** **Dolichoderus tauricus* Dlussky, 1981b: 70, fig. 52 (q.) RUSSIA (Miocene).

**#tenuispinus.** *Dolichoderus (Hypoclinea) cuspidatus* subsp. *tenuispina* Viehmeyer, 1922: 214 (w.) BORNEO.
***tertiarius.** **Hypoclinea tertiaria* Mayr, 1868c: 62, pl. 3, figs. 56–60 (w.q.m.) BALTIC AMBER (Oligocene). Combination in *Dolichoderus:* Dalla Torre, 1893: 162.
**thoracicus.** *Tapinoma thoracica* Smith, F. 1860a: 69 (w.) SULAWESI. Combination in *Dolichoderus:* Dalla Torre, 1893: 162; in *D. (Hypoclinea)*: Emery, 1912g: 14. Senior synonym of *bituberculatus* (and its junior synonym *sellaris*): Donisthorpe, 1932c: 457. Current subspecies: nominal plus *bilikanus, borneonensis, emarginatus, lacciperdus, levior, nasutus, rufescens* (unresolved junior homonym). See also: Shattuck, 1994: 70.
**#tonkinus.** *Dolichoderus (Hypoclinea) taprobanae* var. *tonkina* Santschi, 1920h: 170 (w.) VIETNAM. [Also described as new by Santschi, 1924c: 109.]
**tricolor.** *Dolichoderus tricolor* Emery, 1914f: 417, pl.13, fig. 11 (w.q.m.) NEW CALEDONIA.
**tricornis.** *Dolichoderus tricornis* Emery, 1897d: 571 (w.) NEW GUINEA. Combination in *D. (Hypoclinea)*: Emery, 1912g: 13; in *D. (Monoceratoclinea)*: Wheeler, W.M. 1935c: 68.
*trinidadensis. Dolichoderus championi* r. *trinidadensis* Forel, 1899c: 100 (w.) TRINIDAD. Combination in *D. (Hypoclinea)*: Emery, 1912g: 12; in *Hypoclinea:* Kempf, 1972a: 119; in *Dolichoderus:* Shattuck, 1992c: 77. Junior synonym of *diversus:* MacKay, 1993b: 50.
**tristis.** *Dolichoderus (Monacis) tristis* Mann, 1916: 463, pl. 2, fig. 17 (w.) BRAZIL. Combination in *Monacis:* Kempf, 1959b: 256; in *Dolichoderus:* Shattuck, 1992c: 77. See also: MacKay, 1993b: 103.
*tristis. Dolichoderus (Hypoclinea) tristis* Clark, 1930b: 254, fig. 2 (w.) AUSTRALIA. [Junior primary homonym of *tristis* Mann, above.] Replacement name: *clarki* Wheeler, W.M. 1935c: 69.
**tuberifer.** *Dolichoderus tuberifer* Emery, 1887a: 255, pl. 4, fig. 29 (w.) SUMATRA. Combination in *D. (Hypoclinea)*: Emery, 1894c: 229.
**turneri.** *Dolichoderus turneri* Forel, 1902h: 462 (w.) AUSTRALIA. Combination in *D. (Hypoclinea)*: Emery, 1912g: 13. See also: Clark, 1930b: 261.
*unicolor. Dolichoderus quadripunctatus* var. *unicolor* Ruzsky, 1905b: 471 (w.q.) CAUCASUS. Junior synonym of *quadripunctatus:* Arakelian, 1994: 72.
*ursus. Hypoclinea ursus* Mayr, 1866a: 499 (q.) ECUADOR. Combination in *Dolichoderus:* Dalla Torre, 1893: 162; in *D. (Hypoclinea)*: Emery, 1894c: 228; in *Hypoclinea:* Kempf, 1972a: 120; in *Dolichoderus:* Shattuck, 1992c: 77. Junior synonym of *abruptus:* MacKay, 1993b: 32.
**validus.** *Monacis validus* Kempf, 1959b: 244, pl. 1, fig. 5; pl. 2, fig. 9; pl. 3, fig. 7; pl. 5, fig. 4 (w.q.) COSTA RICA. Combination in *Dolichoderus:* Shattuck, 1992c: 77. See also: MacKay, 1993b: 104.
**varians.** *Dolichoderus (Monacis) varians* Mann, 1916: 462 (w.) BRAZIL. Combination in *Monacis:* Kempf, 1959b: 261; in *Dolichoderus:* Shattuck, 1992c: 77. See also: MacKay, 1993b: 105.
***vectensis.** **Dolichoderus vectensis* Donisthorpe, 1920: 88, pl. 5, fig. 6 (q.) GREAT BRITAIN (Oligocene).
*vestita. Hypoclinea vestita* Mayr, 1862: 707 (q.) ANTILLES IS. Junior synonym of *bispinosus:* Emery, 1894c: 232.
***vexillarius.** **Dolichoderus (Hypoclinea) vexillarius* Wheeler, W.M. 1915h: 83, fig. 37 (w.) BALTIC AMBER (Oligocene).
**voraginosus.** *Dolichoderus voraginosus* MacKay, 1993b: 106, figs, 49, 51, 72 (w.) BRAZIL.
*wheeleriella. Dolichoderus taschenbergi* var. *wheeleriella* Forel, 1916: 458. Unnecessary (second) replacement name for *gagates* Wheeler, W.M. 1905d: 310. Junior synonym of *taschenbergi:* Smith, M.R. 1951a: 834.
*yoshiokae. Dolichoderus (Hypoclinea) quadripuctatus* subsp. *yoshiokae* Wheeler, W.M. 1933c: 67 (w.) JAPAN. Junior synonym of *sibiricus:* Yasumatsu, 1941: 182.
**ypsilon.** *Dolichoderus scabridus* r. *ypsilon* Forel, 1902h: 461 (w.) AUSTRALIA. Forel, 1907h: 284 (m.). Combination in *D. (Hypoclinea)*: Emery, 1912g: 13; in *D. (Diceratoclinea)*: Wheeler, W.M. 1935c: 69. Race of *scabridus:* Forel, 1915b: 76. Raised to species: Forel, 1907h: 284; Clark, 1930b: 258. Current subspecies: nominal plus *niger, rufotibialis.*

***DOLICHOMYRMA** (Russia)
***latipes.** **Dolichomyrma latipes* Dlussky, 1975: 122, fig. 134 (w.) RUSSIA (Cretaceous).
***longiceps.** **Dolichomyrma longiceps* Dlussky, 1975: 121, fig. 133 (w.) RUSSIA (Cretaceous).
**#DOLICHORHACHIS:** see under **POLYRHACHIS.**

**DOLIOPONERA** (Afrotropical)
**fustigera.** *Dolioponera fustigera* Brown, 1974e: 32, figs. 1–4 (w.) GABON.

*DOLOPHRA:* see under **CAMPONOTUS.**

*DONISTHORPEA:* see under **LASIUS.**

**DORISIDRIS** (Cuba)
**nitens.** *Strumigenys (Codiomyrmex) nitens* Santschi, 1932e: 413, fig. 2 (w.) CUBA. Combination in *Dorisidris:* Brown, 1948e: 116.

**DORONOMYRMEX** (Holarctic)
**goesswaldi.** *Leptothorax goesswaldi* Kutter, 1967b: 82, figs. 3–8, 10 (q.) GERMANY. Kutter, 1977c: 129 (m.). Combination in *Doronomyrmex:* Buschinger, 1981: 215.
**kutteri.** *Leptothorax (Mychothorax) kutteri* Buschinger, 1966a: 327, figs. 1, 2, 3 (q.m.) GERMANY. Combination in *Doronomyrmex:* Buschinger, 1981: 215. See also: Kutter, 1967b: 85; Kutter, 1977c: 130; Buschinger, 1972: 169.
**pacis.** *Doronomyrmex pacis* Kutter, 1945: 486, 3 figs. (q.) SWITZERLAND. Kutter, 1950c: 348 (m.). See also: Buschinger, 1972: 169; Kutter, 1977c: 146.
**pocahontas.** *Doronomyrmex pocahontas* Buschinger, 1979b: 216, figs. 1–4, 7 (q.) CANADA. Buschinger & Heinze, 1993: 428 (w.m.).

*DOROTHEA:* see under **VOLLENHOVIA.**

*DORYLOZELUS:* see under **LEPTOGENYS.**

**DORYLUS** (Old World Tropics except Madagascar and Australia)
*abeillei* André, 1886; see under **AENICTUS.**
*abyssinicus. Dorylus (Dorylus) brevinodosus* var. *abyssinicus* Emery, 1895j: 717 (w.) ETHIOPIA. Variety of *affinis:* Emery, 1910b: 10. Junior synonym of *aegyptiacus:* Forel, 1910c: 248.
**#acherontus.** *Dorylus (Anomma) funereus* var. *acherontus* Santschi, 1937b: 97 (m.) CONGO.
**#acuminatus.** *Dorylus (Alaopone) attenuatus* var. *acuminata* Emery, 1899e: 462 (m.) SOUTH AFRICA.
**acutus.** *Dorylus (Alaopone) diadema* st. *acutus* Santschi, 1937a: 49, figs. 4–7 (m.) ZAIRE. Raised to species: Santschi, 1939a: 150.
**#aegyptiacus.** *Dorylus aegyptiacus* Mayr, 1865: 76 (footnote) (m.) EGYPT. Emery, 1915g: 3 (w.). Junior synonym of *affinis:* Emery, 1892a: 110. Revived from synonymy as variety/subspecies of *affinis:* Emery, 1892c: lv; Emery, 1895j: 720; Emery, 1915g: 5; Santschi, 1917c: 19. Senior synonym of *abyssinicus:* Forel, 1910c: 248.
**aethiopicus.** *Dorylus (Shuckardia) atriceps* subsp. *aethiopicus* Emery, 1895j: 741, figs. JJ, KK (m.) SUDAN. Forel, 1907b: 201 (w.). Combination in *D. (Alaopone)*: Forel, 1907b: 201. Raised to species: Santschi, 1923e: 277.
**affinis.** *Dorylus affinis* Shuckard, 1840c: 316 (m.) GAMBIA. Emery, 1901c: 189 (w.l.). Senior synonym of *brevinodosus:* Emery, 1901c: 189; of *planiceps:* Emery, 1895j: 719. Current subspecies: nominal plus *aegyptiacus, denudatus, exilis, hirsutus, loewyi, parapsidalis, pulliceps, sudanicus, ugandensis.* See also: Arnold, 1915: 120; Santschi, 1917c: 18.
**agressor.** *Dorylus agressor* Santschi, 1923e: 268 (s.w.) ZAIRE.
**alluaudi.** *Dorylus alluaudi* Santschi, 1914b: 60, fig. 4 (m.) UGANDA. Current subspecies: nominal plus *lobatus.*
**#anceps.** *Dorylus braunsi* r. *anceps* Forel, 1914d: 215 (w.) ZIMBABWE.
*antinorii. Alaopone antinorii* Emery, 1881a: 275, fig. (w.) ETHIOPIA. Combination in *Dorylus:* Dalla Torre, 1893: 10; in *D. (Alaopone)*: Emery, 1895j: 736. Junior synonym of *molestus:* Emery, 1910b: 12.
**#arcens.** *Anomma arcens* Westwood, 1847b: 16, pl. 1, fig. 3 (w.) LIBERIA. Combination in *Dorylus (Anomma)*: Emery, 1895j: 710. Subspecies of *burmeisteri:* Emery, 1892d: 554; of *nigricans:* Emery, 1895j: 710; Wheeler, W.M. 1922a: 738. Senior synonym of *pubescens:* Emery, 1895j: 710.
**#arnoldi.** *Dorylus (Alaopone) diadema* r. *arnoldi* Forel, 1914d: 216 (m.) ZIMBABWE. Arnold, 1948: 217 (w.).
**atratus.** *Dorylus atratus* Smith, F. 1859b: 4, pl. 1, fig. 15 (m.) NIGERIA. Combination in *Anomma:* Menozzi, 1942: 165. See also: Emery, 1895j: 722.
**atriceps.** *Dorylus atriceps* Shuckard, 1840c: 323 (m.) GAMBIA. Combination in *D. (Shuckardia)*: Emery, 1895j: 740; in *D. (Alaopone)*: Emery, 1910b: 15. Senior synonym of *ritsemae:* Emery, 1895j: 740.
**attenuatus.** *Dorylus attenuatus* Shuckard, 1840c: 322 (m.) GAMBIA. Combination in *D. (Alaopone)*: Emery, 1895j: 733. Senior synonym of *umbratipennis:* Santschi, 1939a: 148. Current subspecies: nominal plus *acuminatus, australis, bondroiti, latinodis.*
**#australis.** *Dorylus (Alaopone) montanum* st. *australis* Santschi, 1919b: 231 (w.) SOUTH AFRICA. [Also described as new, but as species, by Santschi, 1920d: 362; based on same specimens.] Subspecies of

*montanus:* Wheeler, W.M. 1922a: 750; of *attenuatus:* Santschi, 1939a: 147.

*badior. Dorylus (Dorylus) affinis* st. *exilis* var. *badior* Santschi, 1939b: 237 (m.) ZAIRE. Unavailable name.

**#*badius.*** *Dorylus badius* Gerstäcker, 1859: 261 (w.) MOZAMBIQUE. [Also described as new by Gerstäcker, 1862: 499.] Gerstäcker, 1862: 499 (m.); Santschi, 1914b: 63 (s.). Combination in *D. (Typhlopone)*: Emery, 1895j: 724. Variety/subspecies of *juvenculus:* Emery, 1894i: 46; of *fulvus:* Emery, 1895j: 724; Emery, 1910b: 13; Stitz, 1911b: 375. Senior synonym of *rhodesiae:* Arnold, 1944: 2. See also: Arnold, 1915: 125.

***bequaerti.*** *Dorylus bequaerti* Forel, 1913b: 311 (w.) ZAIRE.

**#*berlandi.*** *Dorylus (Alaopone) conradti* var. *berlandi* Santschi, 1926b: 209 (w.) IVORY COAST.

***bishyiganus.*** *Anomma bishyiganum* Boven, 1972: 140, figs. 3, 4, 5 (q.) RWANDA. Combination in *Dorylus:* **new combination**.

**#*bondroiti.*** *Dorylus (Alaopone) montanus* var. *bondroiti* Santschi, 1912b: 162 (m.) SOUTH AFRICA. Stirps of *montanus:* Santschi, 1915c: 248; of *attenuatus:* Santschi, 1939a: 148.

***braunsi.*** *Dorylus (Dorylus) braunsi* Emery, 1895j: 718, fig. G (w.) LIBERIA. Senior synonym of *impressus:* Santschi, 1924b: 198. Current subspecies: nominal plus *anceps.*

*breviceps. Dorylus (Typhlopone) breviceps* Emery, 1889b: 488 (w.) BURMA. Combination in *D. (Dichthadia)*: Emery, 1895j: 729. Junior synonym of *laevigatus:* Emery, 1895j: 729; Bingham, 1903: 3. Revived from synonymy as variety of *laevigatus:* Emery, 1910b: 9. Junior synonym of *laevigatus:* Wilson, 1964a: 440.

*brevinodosus. Typhlopone oraniensis* var. *brevinodosus* Mayr, 1862: 737 (w.) GHANA. Combination in *Dorylus:* Dalla Torre, 1893: 10. Raised to species: Emery, 1877b: 381. Junior synonym of *affinis:* Emery, 1901c: 189.

***brevipennis.*** *Dorylus (Dorylus) brevipennis* Emery, 1895j: 721, figs. K, L (m.) TANZANIA. Emery, 1901c: 192 (w.); Boven, 1975: 193 (q.). [Misspelled as *brevispinus* by Santschi, 1933b: 97.] Current subspecies: nominal plus *marshalli, zimmermanni.*

***brevis.*** *Dorylus (Alaopone) montanus* st. *brevis* Santschi, 1919c: 245, fig. 1 (m.) ZAIRE. Raised to species: Santschi, 1939a: 147.

*bulawayensis. Dorylus affinis* r. *zimmermanni* var. *bulawayensis* Forel, 1914d: 215 (w.) ZIMBABWE. Unavailable name.

**#*burmeisteri.*** *Anomma burmeisteri* Shuckard, 1840c: 326 (w.) SIERRA LEONE. Emery, 1895j: 710 (m.); Raignier & Boven, 1955: 93 (q.). Combination in *Dorylus (Anomma)*: Emery, 1895j: 710. Junior synonym of *nigricans:* Emery, 1895j: 710; Emery, 1910b: 12. Revived from synonymy as subspecies of *nigricans:* Wasmann, 1904b: 671; Mayr, 1907b: 7; Santschi, 1910c: 352; Santschi, 1914b: 62. Raised to species: Stitz, 1916: 373; Santschi, 1921c: 115. Subspecies of *nigricans:* Wheeler, W.M. 1922a: 738; Santschi, 1937b: 96.

***buyssoni.*** *Dorylus (Alaopone) buyssoni* Santschi, 1910g: 748, fig. 7 (m.) KENYA. Combination in *Alaopone:* Menozzi, 1942: 165. Current subspecies: nominal plus *conjugens.*

**#*chapini.*** *Dorylus (Anomma) kohli* var. *chapini* Wheeler, W.M. 1922a: 45 (w.) ZAIRE.

**#*clarior.*** *Dorylus affinis* st. *clarior* Santschi, 1917c: 19 (m.) ZAIRE. [First available use of *Dorylus affinis* st. *depilis* var. *clarior* Santschi, 1915c: 246; unavailable name.] Variety of *depilis:* Menozzi, 1942: 165.

**#*claripennis.*** *Dorylus moestus* var. *claripennis* Santschi, 1919c: 243 (m.) ZAIRE. Santschi, 1930b: 60 (w.).

*clausii. Typhlopone clausii* Joseph, 1882: 47 (w.) NORTH AFRICA. Combination in *Dorylus:* André, 1882c: 252. Junior synonym of *juvenculus:* Emery, 1891b: 1. Revived from synonymy as variety of *juvenculus:* Dalla Torre, 1893: 12. Junior synonym of *fulvus:* Emery, 1895j: 723.

**#*confusus.*** *Dorylus gribodoi* var. *confusa* Santschi, 1915c: 246 (m.) IVORY COAST.

***congolensis.*** *Dorylus (Anomma) kohli* var. *congolensis* Santschi, 1910c: 352, fig. 1 (s.) CONGO. Raised to species: Boven & Lévieux, 1970: 356.

**#*conjugens.*** *Dorylus (Alaopone) buyssoni* var. *conjugens* Santschi, 1910g: 750, fig. 7 (m.) KENYA.

***conradti.*** *Dorylus (Alaopone) conradti* Emery, 1895j: 734, pl. 16, figs. 1–4; pl. 17, figs. 7–10 (w.q.) TOGO. Current subspecies: nominal plus *berlandi.*

**#*crampeli.*** *Dorylus (Rhogmus) fimbriatus* var. *crampeli* Santschi, 1919b: 232 (m.) CONGO.

**#*crosi.*** *Dorylus (Typhlopone) fulvus* var. *crosi* Santschi, 1926e: 230 (w.m.) ALGERIA.

*curtisii. Labidus (Typhlopone) curtisii* Shuckard, 1840b: 265 (w.) SRI LANKA. Combination in *Dorylus:* Dalla Torre, 1893: 10; in *D. (Alaopone)*, and senior synonym of *oberthueri:* Emery, 1895j: 730. Junior synonym of *orientalis:* Forel, 1901a: 463; Bingham, 1903: 4.

*dahlbomii. Typhlopone dahlbomii* Westwood, 1840b: 88, pl. 2, fig. 3 (w.) no locality given. Combination in *Dorylus:* Dalla Torre, 1893: 10. Junior synonym of *fulvus:* Emery, 1895j: 723.

**#*dentifrons.*** *Dorylus fulvus* subsp. *dentifrons* Wasmann, 1904b: 673 (w.) ZAIRE. Combination in *D. (Typhlopone)*: Emery, 1910b: 13. Provisional junior synonym of *glabratus:* Emery, 1915g: 6. Subspecies of *fulvus,* and senior synonym of *stramineus:* Emery, 1919c: 102; Wheeler, W.M. 1922a: 745.

**#*denudatus.*** *Dorylus denudatus* Santschi, 1910g: 742, figs. 1, 4 (m.) NIGER. Subspecies of *affinis:* Santschi, 1917c: 19; Santschi, 1937a: 49.

***depilis.*** *Dorylus (Dorylus) affinis* var. *depilis* Emery, 1895j: 720 (m.) CAMEROUN. Forel, 1911e: 254 (w.). Raised to species: Emery, 1910b: 10. Subspecies of *affinis:* Forel, 1909b: 53; Santschi, 1915c: 246; Santschi, 1917c: 19 (in key). Status as species: Santschi, 1910g: 737; Forel, 1911e: 254; Wheeler, W.M. 1922a: 731; Santschi, 1923e: 271; Menozzi, 1927f: 356; Menozzi, 1942: 165. Current subspecies: nominal plus *clarior.*

***diadema.*** *Dorylus diadema* Gerstäcker, 1859: 261 (m.) MOZAMBIQUE. [Also described as new by Gerstäcker, 1862: 500.] Combination in *D. (Shuckardia)*: Emery, 1895j: 742; in *D. (Alaopone)*: Emery, 1910b: 15. Current subspecies: nominal plus *arnoldi, fusciceps.*

***distinctus.*** *Dorylus (Alaopone) distinctus* Santschi, 1910g: 747, fig. 6 (m.) GUINEA.

*dorylus. Mutilla dorylus* Lamarck, 1817: 101 (m.) "Africa". Junior synonym of *helvolus:* Emery, 1888b: 350.

***ductor.*** *Dorylus (Alaopone) ductor* Santschi, 1939a: 149, figs. 8, 9, 17 (m.) CONGO.

***emeryi.*** *Dorylus emeryi* Mayr, 1896: 225 (w.) CAMEROUN. Combination in *D. (Anomma)*: Emery, 1910b: 11; in *Anomma:* Menozzi, 1942: 165. Current subspecies: nominal plus *opacus, pulsi.* See also: Raignier & Boven, 1955: 83.

***erraticus.*** *Anomma erratica* Smith, F. 1865: 71 (w.) AFRICA (not New Guinea as stated; see: Wilson, 1964a: 443). Combination in *Dorylus:* Emery, 1910b: 15; in *D. (Anomma)*: Donisthorpe, 1932c: 473.

**#*eurous.*** *Dorylus (Typhlopone) fulvus* subsp. *euroa* Emery, 1915g: 5, fig. 2 (w.m.) ETHIOPIA.

*europaea. Typhlopone europaea* Roger, 1859: 248, pl. 7, fig. 3 (w.) ITALY. Junior synonym of *helvolus:* Emery, 1888b: 350.

**#*exilis.*** *Dorylus affinis* var. *exilis* Santschi, 1914b: 59 (m.) TANZANIA. [First available use of *Dorylus affinis* st. *loewyi* var. *exilis* Santschi, 1910g: 744; unavailable name.]

***faurei.*** *Dorylus faurei* Arnold, 1946: 60, fig. 12 (q.) SOUTH AFRICA.

***fimbriatus.*** *Rhogmus fimbriatus* Shuckard, 1840c: 325 (m.) GAMBIA. Emery, 1895j: 736 (w.); Brauns, 1903: 294 (q.). Combination in *Dorylus (Rhogmus)*: Emery, 1895j: 736. Current subspecies: nominal plus *crampeli, laevipodex, poweri.* See also: Emery, 1901c: 187; Forel, 1901h: 47; Arnold, 1915: 129; Boven, 1967: 55.

**#*frenisyi.*** *Dorylus (Anomma) kohli* var. *frenisyi* Forel, 1916: 402 (w.) ZAIRE.

***fulvus.*** *Typhlopone fulva* Westwood, 1839: 219, fig. 86 (w.) NORTH AFRICA. Emery, 1895j: 723 (m.). Combination in *Dorylus:* Dalla Torre, 1893: 10; in *D. (Typhlopone)*: Emery, 1895j: 723. Senior synonym of *kirbii:* Smith, F. 1858b: 110; Mayr, 1863: 457; of *clausii, dahlbomii, shuckardi, spinolae* and *thwaitsii:* Emery, 1895j: 723. Current subspecies: nominal plus *badius, crosi, dentifrons, eurous, glabratus, juvenculus, mordax, obscurior, punicus, ruficeps, saharensis.*

***funereus.*** *Dorylus (Anomma) nigricans* var. *funereus* Emery, 1895j: 712 (m.) GHANA. Raised to species: Emery, 1910b: 11; Forel, 1916: 402; Santschi, 1935a: 256. Current subspecies: nominal plus *acherontus, pardus, stygis, zumpti.*

***furcatus.*** *Dichthadia furcata* Gerstäcker, 1872: 267 (q.) SOUTH AFRICA. Combination in *Dorylus:* Emery, 1888b: 345.

**#*fusciceps.*** *Dorylus (Shuckardia) diadema* var. *fusciceps* Emery, 1899e: 463 (m.) MALAWI. Combination in *D. (Alaopone)*: Emery, 1910b: 15.

***fuscipennis.*** *Rhogmus fuscipennis* Emery, 1892d: 570 (m.) GHANA. Forel, 1909d: 224 (footnote) (w.). Combination in *Dorylus (Rhogmus)*: Emery, 1895j: 737. Current subspecies: nominal plus *lugubris, marginiventris.* See also: Forel, 1913h: 349.

*fuscus*. *Dorylus fuscus* Emery, 1889b: 487 (m.) BURMA. Combination in *D.* (*Alaopone*): Emery, 1895j: 731. Variety/subspecies of *orientalis:* Dalla Torre, 1893: 13; Emery, 1895k: 453; Bingham, 1903: 5; Emery, 1910b: 15. Junior synonym of *orientalis:* Wilson, 1964a: 442.

***gaudens***. *Dorylus gaudens* Santschi, 1919c: 244 (m.) ZAIRE.

***gerstaeckeri***. *Dorylus* (*Anomma*) *gerstaeckeri* Emery, 1895j: 713, pl. 17, fig. 11 (w.) GHANA. Current subspecies: nominal plus *quadratus*. See also: Raignier & Boven, 1955: 85.

***ghanensis***. *Dorylus ghanensis* Boven, 1975: 191, figs. 6–8 (q.) GHANA.

*glaberrima*. *Dichthadia glaberrima* Gerstäcker, 1863: 93, pl. 1, fig. 2 (q.) JAVA. Combination in *Dorylus* (*Dichthadia*): Emery, 1888b: 345. Junior synonym of *laevigatus:* Forel, 1901a: 464.

#***glabratus***. *Dorylus glabratus* Shuckard, 1840c: 317 (m.) GAMBIA. Junior synonym of *labiatus:* Emery, 1895j: 724. Revived from synonymy as subspecies of *fulvus:* Emery, 1910b: 13; Santschi, 1931d: 407.

***gribodoi***. *Dorylus gribodoi* Emery, 1892d: 570, pl. 15, fig. 15 (m.) TOGO. Current subspecies: nominal plus *confusus, insularis*.

***helvolus***. *Vespa helvola* Linnaeus, 1764: 412 (m.) SOUTH AFRICA. Linnaeus, 1767: 968 (w.); Emery, 1888b: 344 (q.). Combination in *Dorylus:* Fabricius, 1793: 365. Senior synonym of *dorylus, europaea, planifrons* and *punctata:* Emery, 1888b: 350; Emery, 1895j: 714. Current subspecies: nominal plus *pretoriae*. See also: Arnold, 1915: 116.

*hindostanus*. *Dorylus hindostanus* Smith, F. 1859b: 3, pl. 1, fig. 1, 16 (m.) INDIA. Possible junior synonym of *juvenculus:* Dalla Torre, 1893: 12. Junior synonym of *labiatus:* Emery, 1895j: 724; Bingham, 1903: 2.

#***hirsutus***. *Dorylus affinis* var. *hirsutus* Wheeler, W.M. 1922a: 729 (m.) ETHIOPIA. [First available use of *Dorylus affinis* subsp. *aegyptiaca* var. *hirsuta* Santschi, 1917c: 20; unavailable name.]

*homalinus*. *Cosmaecetes homalinus* Spinola, 1851b: 55 (w.) MOROCCO. [Also described as new by Spinola, 1853: 71.] Combination in *Typhlopone:* Smith, F. 1858b: 111; in *Dorylus:* Dalla Torre, 1893: 11. Junior synonym of *juvenculus:* Emery, 1893h: 1.

*hybrida*. *Dorylus* (*Anomma*) *nigricans* st. *burmeisteri* var. *hybrida* Santschi, 1912b: 161, fig. 14 (s.w.) SENEGAL. Unavailable name.

*impressus*. *Dorylus* (*Dorylus*) *helvolus* var. *impressus* Stitz, 1910: 127 (s.w.) EQUATORIAL GUINEA. Junior synonym of *braunsi:* Santschi, 1924b: 198.

*impressus*. *Dorylus* (*Typhlopone*) *fulvus* var. *impressus* Santschi, 1928d: 56 (w.) IVORY COAST. [Junior primary homonym of *impressus* Stitz, above.] Replacement name: *mordax* Santschi, 1931d: 407.

#***indocilis***. *Dorylus* (*Anomma*) *kohli* var. *indocilis* Santschi, 1933b: 98 (s.w.) ZAIRE.

#***insularis***. *Dorylus* (*Anomma*) *gribodoi* var. *insularis* Santschi, 1937b: 98 (m.) FERNANDO PO I.

#***juvenculus***. *Dorylus juvenculus* Shuckard, 1840c: 318 (m.) MOROCCO. Senior synonym of *oraniensis:* Emery, 1888b: 350; Forel, 1890a: lxviii; of *homalinus:* Emery, 1893h: 1. Junior synonym of *fulvus:* Emery, 1895j: 723. Revived from synonymy as subspecies of *fulvus:* Santschi, 1931d: 401.

***katanensis***. *Dorylus* (*Alaopone*) *atriceps* var. *katanensis* Stitz, 1911b: 375 (m.) ZAIRE. Raised to species: Santschi, 1939a: 147.

*kirbii*. *Labidus* (*Typhlopone*) *kirbii* Shuckard, 1840b: 265 (w.) no locality given. Junior synonym of *fulvus:* Smith, F. 1858b: 110; Mayr, 1863: 457.

*klugi*. *Dorylus klugi* Emery, 1887b: 448, pl. 1, fig. 10 (m.) SUMATRA (attributed to Hagenbach). Junior synonym of *laevigatus:* Forel, 1901a: 464.

***kohli***. *Dorylus* (*Anomma*) *kohli* Wasmann, 1904b: 669 (s.w.) ZAIRE. Santschi, 1921c: 114 (s.); Boven, 1968: 732 (q.). Current subspecies: nominal plus *chapini, frenisyi, indocilis, langi, militaris, minor, victoriae*. See also: Raignier & Boven, 1955: 78.

***labiatus***. *Dorylus labiatus* Shuckard, 1840c: 319 (m.) INDIA. Forel, 1901a: 464 (w.). Combination in *D.* (*Typhlopone*): Emery, 1895j: 724. Possible junior synonym of *juvenculus:* Dalla Torre, 1893: 11. Variety of *fulvus:* Emery, 1895j: 724. Revived status as species: Forel, 1901a: 464; Bingham, 1903: 2. Senior synonym of *hindostanus:* Emery, 1895j: 724; of *laeviceps:* Forel, 1901a: 464. See also: Bingham, 1903: 2; Wilson, 1964a: 440.

*laeviceps*. *Dorylus* (*Typhlopone*) *laeviceps* Smith, F. 1878b: 13, fig. 2 (w.) INDIA. Junior synonym of *fulvus:* Emery, 1895j: 723; of *labiatus:* Forel, 1901a: 464; Wilson, 1964a: 440.

***laevigatus***. *Typhlopone laevigata* Smith, F. 1857a: 70 (w.) BORNEO. Forel, 1901a: 464 (q.m.). Combination in *Dorylus* (*Dichthadia*): Emery, 1895j: 729. Senior synonym of *breviceps:* Dalla Torre, 1893: 12; Emery, 1895j: 729; of *klugi* and *glaberrima:* Forel, 1901a: 464. See also: Emery, 1895k: 453; Bingham, 1903: 3; Wilson, 1964a: 440.

#***laevipodex***. *Dorylus* (*Rhogmus*) *fimbriatus* var. *laevipodex* Santschi, 1919b: 232 (m.) KENYA.

***lamottei***. *Dorylus* (*Anomma*) *lamottei* Bernard, 1953b: 219, fig. 7 (w.) GUINEA.

#***langi***. *Dorylus* (*Anomma*) *kohli* var. *langi* Wheeler, W.M. 1922a: 45 (w.) ZAIRE.

#***latinodis***. *Dorylus* (*Alaopone*) *attenuatus* r. *latinodis* Forel, 1920: 479 (m.) ZAIRE.

***leo***. *Dorylus* (*Rhogmus*) *leo* Santschi, 1919b: 232 (m.) IVORY COAST.

#***lobatus***. *Dorylus lobatus* Santschi, 1919c: 244 (m.) ZAIRE. Stirps/subspecies of *alluaudi:* Santschi, 1930d: 268.

#***longiceps***. *Dorylus* (*Dorylus*) *spininodis* subsp. *longiceps* Viehmeyer, 1914c: 26, fig. 1 (w.) TANZANIA.

*longicornis*. *Dorylus longicornis* Shuckard, 1840c: 321 (m.) INDIA. Combination in *D.* (*Alaopone*): Emery, 1895j: 731. Variety of *orientalis:* Dalla Torre, 1893: 13; Emery, 1895j: 731. Junior synonym of *orientalis:* Bingham, 1903: 4. Revived from synonymy as subspecies of *orientalis:* Forel, 1907e: 17; Santschi, 1924c: 97. Junior synonym of *orientalis:* Wilson, 1964a: 442.

#***loewyi***. *Dorylus affinis* subsp. *loewyi* Forel, 1907b: 202 (m.) TANZANIA. [Misspelled as *lowgi* by Santschi, 1917c: 19 (in key).]

#***lugubris***. *Dorylus* (*Rhogmus*) *fimbriatus* var. *lugubris* Santschi, 1919b: 232 (m.) IVORY COAST. Variety of *fuscipennis:* Santschi, 1919b: 231 (in key); Santschi, 1919h: 91.

***mandibularis***. *Dorylus mandibularis* Mayr, 1896: 228 (m.) CAMEROUN. Current subspecies: nominal plus *pulchellus*.

#***marginiventris***. *Dorylus* (*Rhogmus*) *fimbriatus* var. *marginiventris* Santschi, 1919b: 233 (m.) IVORY COAST. Variety of *fuscipennis:* Santschi, 1919b: 231 (in key); Santschi, 1919h: 91.

#***marshalli***. *Dorylus brevipennis* var. *marshalli* Emery, 1901c: 193, pl. 1, fig. 13 (w.m.) ZIMBABWE.

***mayri***. *Dorylus* (*Anomma*) *mayri* Santschi, 1912b: 159, fig. 13 (w.) CAMEROUN.

*mediatus*. *Dorylus mediatus* Fabricius, 1804: 428 (m.) CENTRAL AMERICA. Combination in *Labidus:* Shuckard, 1840b: 261. Transferred to family MUTILLIDAE: Westwood, 1842: 78. [Incorrectly continued listed under *Dorylus* by Mayr, 1863: 408, and Roger, 1863b: 41.]

#***militaris***. *Dorylus* (*Anomma*) *kohli* var. *militaris* Santschi, 1923e: 274 (s.w.) ZAIRE.

#***minor***. *Dorylus* (*Anomma*) *kohli* var. *minor* Santschi, 1911g: 206 (w.) ANGOLA.

***moestus***. *Dorylus* (*Dorylus*) *affinis* var. *moestus* Emery, 1895j: 720 (m.) ZAIRE. Raised to species: Forel, 1909b: 71; Santschi, 1910g: 741; Emery, 1910b: 10; Stitz, 1916: 374. Current subspecies: nominal plus *claripennis, morio, schereri*.

#***molestus***. *Anomma molesta* Gerstäcker, 1859: 262 (w.) MOZAMBIQUE. [Also described as new by Gerstäcker, 1862: 502.] Menozzi, 1927a: 264 (q.); Boven, 1972: 135 (q.); Wheeler, G.C. 1943: 322 (l.). Combination in *Dorylus* (*Anomma*): Emery, 1895j: 710. Variety/subspecies of *burmeisteri:* Emery, 1892d: 554; of *nigricans:* Emery, 1895j: 710; Stitz, 1911b: 375; Santschi, 1935b: 264. Senior synonym of *antinorii:* Emery, 1910b: 12. See also: Raignier & Boven, 1955: 103; Wheeler, G.C. & Wheeler, J. 1984: 268.

***montanus***. *Dorylus* (*Alaopone*) *montanus* Santschi, 1910g: 750, fig. 8 (m.) TANZANIA. Santschi, 1914b: 66 (w.).

#***mordax***. *Dorylus fulvus* var. *mordax* Santschi, 1931d: 407. Replacement name for *impressus* Santschi, 1928d: 56. [Junior primary homonym of *impressus* Stitz, 1910: 127.]

#***morio***. *Dorylus moestus* var. *morio* Santschi, 1919c: 243 (m.) CONGO.

#***mucronatus***. *Dorylus* (*Rhogmus*) *savagei* var. *mucronata* Emery, 1899e: 462 (m.) NIGERIA.

***niarembensis***. *Anomma niarembense* Boven, 1972: 143, fig. 5 (diagnosis in key) (q.) ZAIRE. Combination in *Dorylus:* **new combination**.

***nigricans***. *Dorylus nigricans* Illiger, 1802: 188 (m.) SIERRA LEONE. Mayr, 1896: 225 (w.); Forel, 1912j: 177 (q.); Wheeler, G.C. & Wheeler, J. 1984: 268 (l.). Combination in *D.* (*Anomma*): Emery, 1895j: 710; in *Anomma:* Menozzi, 1942: 165. Current subspecies: nominal plus *arcens, burmeisteri, molestus, pallidus, rubellus, sjoestedti, sjostedtiwilverthi, terrificus*. See also: Santschi, 1930a: 55; Raignier & Boven, 1955: 70.

*nigritarsis*. *Dorylus nigritarsis* Strand, 1911: 118 (q.) CAMEROUN. Combination in *D.* (*Anomma*): Santschi, 1915c: 248. Junior synonym of *nigricans:* Wheeler, W.M. 1922a: 737; of *wilverthi:* Boven, 1972: 144.

***DORYLUS** (continued)*

*nomadas. Dorylus (Anomma) nomadas* Santschi, 1935a: 254, fig. 1 (q.) ZAIRE. Junior synonym of *wilverthi:* Raignier & Boven, 1955: 93.

*oberthueri. Alaopone oberthueri* Emery, 1881a: 274, fig. (w.) INDIA. Combination in *Dorylus:* Dalla Torre, 1893: 12. Junior synonym of *curtisii:* Emery, 1895j: 730.

#***obscuriceps.*** *Dorylus (Alaopone) orientalis* var. *obscuriceps* Santschi, 1920h: 160 (m.) INDIA. [Also described as new by Santschi, 1924c: 97.]

#***obscurior.*** *Dorylus fulvus* var. *obscurior* Wheeler, W.M. 1925b: 1 (s.w.) GUINEA. [First available use of *Dorylus (Typhlopone) fulvus* st. *badius* var. *obscurior* Santschi, 1914d: 333; unavailable name.]

***ocellatus.*** *Dichthadia ocellata* Stitz, 1910: 128 (q.) CAMEROUN. [Specimen described but not named: Stitz, 1909: 231.] Combination in *Dorylus (Rhogmus)*: Wheeler, W.M. 1922a: 747.

#***opacus.*** *Dorylus (Anomma) emeryi* subsp. *opaca* Forel, 1909b: 58 (w.) ZAIRE. Forel, 1911h: 452 (m.). Raised to species: Santschi, 1912b: 154 (in key); Forel, 1916: 402. Subspecies of *emeryi:* Wheeler, W.M. 1922a: 43. See also: Stitz, 1910: 127; Forel, 1913h: 348.

*oraniensis. Typhlopone oraniensis* Lucas, 1849: 302, pl. 16, fig. 11 (w.) ALGERIA. Combination in *Dorylus:* André, 1882c: 252. Junior synonym of *juvenculus:* Emery, 1888b: 350; Forel, 1890a: lxviii.

***orientalis.*** *Dorylus orientalis* Westwood, 1835: 72 (m.) INDIA. Forel, 1901a: 463 (w.). Combination in *D. (Alaopone)*: Emery, 1895j: 731. Senior synonym of *curtisii* (and its junior synonym *oberthueri*): Forel, 1901a: 463; of *longicornis:* Bingham, 1903: 4; Wilson, 1964a: 442; of *fuscus:* Wilson, 1964a: 442. Current subspecies: nominal plus *obscuriceps.*

*ornata. Dorylus (Anomma) nigricans* st. *burmeisteri* var. *ornata* Santschi, 1912b: 161 (in text) (w.) CAMEROUN. Unavailable name.

#***pallidus.*** *Dorylus (Anomma) burmeisteri* var. *pallida* Santschi, 1921c: 115 (s.w.) CAMEROUN. [First available use of *Dorylus (Anomma) nigricans* st. *burmeisteri* var. *pallida* Santschi, 1912b: 160; unavailable name.] Subspecies of *nigricans:* current status.

*pallipes* Haldeman, 1844 [*Typhlopone*]; see under ***AMBLYOPONE.***

#***parapsidalis.*** *Dorylus affinis* var. *parapsidalis* Santschi, 1917c: 20 (m.) MALAWI.

#***pardus.*** *Dorylus (Anomma) funereus* var. *pardus* Santschi, 1937b: 97 (m.) ZAIRE.

*planiceps. Dorylus planiceps* Haldeman, 1849b: 204 (m.) "West Africa". Junior synonym of *affinis:* Emery, 1895j: 719.

*planifrons. Dorylus planifrons* Mayr, 1865: 74, pl. 3, fig. 21 (m.) SOUTH AFRICA. Junior synonym of *helvolus:* Emery, 1888b: 350.

***politus.*** *Dorylus (Dorylus) politus* Emery, 1901c: 194, pl. 1, figs. 14–17 (w.) CAMEROUN.

#***poweri.*** *Dorylus (Rhogmus) fimbriatus* var. *poweri* Forel, 1914d: 217 (w.) SOUTH AFRICA.

#***pretoriae.*** *Dorylus helvolus* subsp. *pretoriae* Arnold, 1946: 59, fig. 11 (w.q.m.) SOUTH AFRICA.

*pubescens. Anomma pubescens* Roger, 1861a: 47 (w.) LIBERIA. Junior synonym of *arcens:* Emery, 1895j: 710.

#***pulchellus.*** *Dorylus mandibularis* var. *pulchella* Santschi, 1920d: 361 (m.) IVORY COAST.

#***pulliceps.*** *Dorylus affinis* var. *pulliceps* Santschi, 1917c: 19 (m.) IVORY COAST.

#***pulsi.*** *Anomma emeryi* var. *pulsi* Forel, 1904d: 170 (w.) "West Africa". Santschi, 1923e: 273 (s.). Combination in *Dorylus (Anomma)*: Emery, 1910b: 11.

*punctata. Typhlopone punctata* Smith, F. 1858b: 112 (w.) SOUTH AFRICA. Combination in *Dorylus:* André, 1882c: 251. Junior synonym of *helvolus:* Emery, 1888b: 350.

*punctus. Dorylus (Dorylus) spininodis* st. *longiceps* var. *punctus* Santschi, 1933b: 97 (w.) ZAIRE. Unavailable name.

#***punicus.*** *Dorylus (Typhlopone) fulvus* var. *punicus* Santschi, 1926e: 231, fig. 1A (w.m.) TUNISIA. Santschi, 1931d: 401 (q.).

#***quadratus.*** *Dorylus (Anomma) gerstaeckeri* st. *quadratus* Santschi, 1914a: 430 (s.) TANZANIA.

*rhodesiae. Dorylus fulvus* r. *rhodesiae* Forel, 1913a: 111 (w.) ZIMBABWE. Junior synonym of *badius:* Arnold, 1944: 2.

*ritsemae. Dorylus ritsemae* Dalla Torre, 1892: 89. Replacement name for *shuckardi* Ritsema, 1874: 182. [Junior secondary homonym of *shuckardi* Westwood, 1840b: 88.] Junior synonym of *atriceps:* Emery, 1895j: 740.

#***rubellus.*** *Anomma rubella* Savage, 1849: 196 (w.) "West Africa". Combination in *Dorylus (Anomma)*: Emery, 1895j: 710. Variety of *burmeisteri:* Emery, 1892d: 554; of *nigricans:* Emery, 1895j: 701; Emery, 1910b: 12; Forel, 1913h: 348.

***rufescens.*** *Dorylus nigricans* st. *rufescens* Santschi, 1915c: 247 (w.) CAMEROUN. [First available use of *Dorylus (Anomma) nigricans* subsp. *sjostedti* var. *rufescens* Wasmann, 1904b: 673; unavailable name.] Santschi, 1915c: 247 (q.). Raised to species: Santschi, 1931d: 407.

#***ruficeps.*** *Dorylus (Typhlopone) fulvus* var. *ruficeps* Santschi, 1926e: 232 (m.) LEBANON. Junior synonym of *punicus:* Menozzi, 1933b: 50. Revived from synonymy as race of *fulvus:* Santschi, 1938a: 29.

#***saharensis.*** *Dorylus (Typhlopone) fulvus* var. *saharensis* Santschi, 1926e: 231, fig. 1 (w.m.) "Sahara".

***savagei.*** *Dorylus (Rhogmus) savagei* Emery, 1895j: 738, figs. EE, FF (m.) GABON. Current subspecies: nominal plus *mucronatus.*

#***schereri.*** *Dorylus moestus* subsp. *schereri* Forel, 1911e: 254 (m.) LIBERIA.

***schoutedeni.*** *Dorylus schoutedeni* Santschi, 1923e: 267, fig. 2 (m.) ZAIRE.

*serratula* Smith, F. 1858 [*Typhlopone*]; see under ***ACANTHOSTICHUS.***

*shuckardi. Typhlopone shuckardi* Westwood, 1840b: 88, pl.2, fig. 2 (w.) no locality given. Junior synonym of *fulvus:* Emery, 1895j: 723.

*shuckardi. Dorylus shuckardi* Ritsema, 1874: 182 (m.) GUINEA. [Junior secondary homonym of *Typhlopone shuckardi* Westwood, above.] Replacement name: *ritsemae* Dalla Torre, 1892: 89.

#***sjoestedti.*** *Dorylus nigricans* subsp. *sjoestedti* Emery, 1899e: 461 (w.) CAMEROUN. Santschi, 1912b: 160 (s.); Raignier & Boven, 1955: 98 (q.). Raised to species: Wasmann, 1916: 136. Subspecies of *nigricans:* Wheeler, W.M. 1922a: 741.

#***sjostedtiwilverthi.*** *Anomma sjostedti* var. *sjostedtiwilverthi* Wasmann, 1917: 305 (w.) CAMEROUN. [*A. sjostedti* var. *sjostedtiwilverthi* Wasmann, 1916: 136; *nomen nudum.*] Subspecies of *nigricans:* current status.

***spininodis.*** *Dorylus (Dorylus) spininodis* Emery, 1901c: 194, pl. 1, figs. 18–20 (w.) CAMEROUN. Current subspecies: nominal plus *longiceps.*

*spinolae. Labidus (Typhlopone) spinolae* Shuckard, 1840c: 327 (w.) no locality given. Combination in *Dorylus:* Dalla Torre, 1893: 13. Junior synonym of *fulvus:* Emery, 1895j: 723.

***stadelmanni.*** *Dorylus (Dorylus) stadelmanni* Emery, 1895j: 722 (m.) ZAIRE.

***stanleyi.*** *Dorylus (Anomma) stanleyi* Forel, 1909b: 52 (m.) ZAIRE. Junior synonym (provisional) of *rubellus:* Santschi, 1930a: 56. Revived from synonymy as species: Santschi, 1935b: 264; Bernard, 1953b: 218.

***staudingeri.*** *Dorylus (Dorylus) staudingeri* Emery, 1895j: 721 (m.) ZAIRE.

*stramineus. Dorylus (Typhlopone) fulvus* var. *stramineus* Stitz, 1910: 128 (s.w.) CAMEROUN. Junior synonym of *dentifrons:* Emery, 1919c: 102.

***striatidens.*** *Dorylus striatidens* Santschi, 1910g: 745, fig. 5 (w.) SENEGAL.

#***stygis.*** *Dorylus (Anomma) funereus* var. *stygis* Santschi, 1937b: 96 (m.) ZAIRE.

#***sudanicus.*** *Dorylus affinis* var. *sudanica* Santschi, 1917c: 19 (m.) CONGO.

***termitarius.*** *Dorylus (Rhogmus) termitarius* Wasmann, 1911: 111, pl. 1, fig. 4 (w.q.). See also: Boven, 1967: 57.

#***terrificus.*** *Dorylus (Anomma) nigricans* var. *terrificus* Santschi, 1923e: 275 (s.w.) ZAIRE.

*thwaitsii. Labidus (Typhlopone) thwaitsii* Shuckard, 1840c: 326 (w.) no locality given. Combination in *Dorylus:* Dalla Torre, 1893: 13. Junior synonym of *fulvus:* Emery, 1895j: 723.

***titan.*** *Dorylus (Anomma) titan* Santschi, 1923e: 271 (s.w.) ZAIRE. Current subspecies: nominal plus *vinalli.* See also: Raignier & Boven, 1955: 85.

#***ugandensis.*** *Dorylus affinis* var. *ugandensis* Santschi, 1914b: 59 (m.) UGANDA.

*umbratipennis. Dorylus (Alaopone) attenuatus* var. *umbratipennis* Forel, 1909b: 52 (m.) "Nini", probably ZAIRE (see Wheeler, W.M. 1922a: 749). Junior synonym of *attenuatus:* Santschi, 1939a: 148.

#***victoriae.*** *Dorylus (Anomma) kohli* var. *victoriae* Santschi, 1921c: 115 (w.) UGANDA. Santschi, 1933b: 98 (s.).

#***vinalli.*** *Dorylus (Anomma) titan* var. *vinalli* Santschi, 1933b: 97 (w.) ZAIRE. [Also described as new by Santschi, in Poulton, 1933: 58 (footnote).]

***vishnui.*** *Dorylus (Alaopone) vishnui* Wheeler, W.M. 1913e: 233 (w.) BURMA. See also: Wilson, 1964a: 443.

***westwoodii.*** *Labidus (Typhlopone) westwoodii* Shuckard, 1840b: 266 (w.) "South America" (locality incorrect). Combination in *Dorylus:* Dalla Torre, 1893: 13. See also: Emery, 1895j: 743; Emery, 1910b: 15.

***wilverthi.*** *Dorylus* (*Anomma*) *wilverthi* Emery, 1899e: 459, fig. (w.) ZAIRE. Forel, 1909b: 51 (m.); Raignier & Boven, 1955: 86 (q.); Wheeler, G.C. 1943: 321 (l.). Senior synonym of *nomadas:* Raignier & Boven, 1955: 93; of *nigritarsis:* Boven, 1972: 144. See also: Wheeler, G.C. & Wheeler, J. 1984: 268; Raignier & Boven, 1955: 115. [This name frequently misspelled as *wilwerthi;* for example by Santschi, 1910c: 352.]

#***zimmermanni.*** *Dorylus* (*Dorylus*) *brevipennis* st. *zimmermanni* Santschi, 1910g: 738, figs. 1, 2 (s.w.m.) CONGO. Race/subspecies of *affinis:* Forel, 1914d: 215; of *brevipennis:* Arnold, 1915: 123; Wheeler, W.M. 1922a: 731.

#***zumpti.*** *Dorylus* (*Anomma*) *funereus* var. *zumpti* Santschi, 1937b: 96, fig. 2 (m.) CAMEROUN.

---

**EXCURSUS**

**Species-rank taxa of *DORYLUS*, grouped by subgenus.**

The subgenera of *Dorylus* are arranged alphabetically below in ***BOLD ITALIC CAPITALS.*** The species currently assigned to each subgenus follow the subgeneric name in alphabetical order. The status of the subgenera is dubious. Gotwald & Barr, 1980: 231, imply that some may be distinct and perhaps worthy of elevation to the rank of genus; but others would appear to have little or no taxonomic value. My current opinion is that the subgenera should all be synonymised, as *Dorylus* forms a single monophyletic group with a unique diagnosis.

#***ALAOPONE:*** *acutus, aethiopicus, atriceps, attenuatus, brevis, buyssoni, conradti, diadema, distinctus, ductor, katanensis, montanus, orientalis, vishnui.*

#***ANOMMA:*** *atratus, bishyiganus, congolensis, emeryi, erraticus, funereus, gerstaeckeri, kohli, lamottei, mayri, niarembensis, nigricans, rufescens, stanleyi, titan, wilverthi.*

#***DICHTHADIA:*** *laevigatus.*

#***DORYLUS:*** *affinis, agressor, alluaudi, bequaerti, braunsi, brevipennis, depilis, faurei, furcatus, gaudens, ghanensis, gribodoi, helvolus, mandibularis, moestus, politus, schoutedeni, spininodis, stadelmanni, staudingeri, striatidens.*

#***RHOGMUS:*** *fimbriatus, fuscipennis, leo, ocellatus, savagei, termitarius.*

#***TYPHLOPONE:*** *fulvus, labiatus.*

**SUBGENUS UNKNOWN**: *westwoodii.*

---

***DORYMYRMEX*** (Neotropical, Nearctic)

***agallardoi.*** *Dorymyrmex agallardoi* Snelling, 1975: 6, figs. 12, 13 (w.) CHILE.

#***albemarlensis.*** *Dorymyrmex pyramicus* var. *albemarlensis* Wheeler, W.M. 1919c: 275 (w.) GALAPAGOS IS. Combination in *D.* (*Conomyrma*): Wheeler, W.M. 1924a: 110; in *Conomyrma:* Kempf, 1972a: 80.

*albipes. Dorymyrmex flavescens* st. *joergenseni* var. *albipes* Santschi, 1925e: 165 (w.) ARGENTINA. Unavailable name (Shattuck, 1994: 90).

***alboniger.*** *Dorymyrmex alboniger* Forel, 1914d: 283 (q.) ARGENTINA. Combination in *D.* (*Spinomyrma*): Kusnezov, 1952g: 429. See also: Gallardo, 1916a: 42.

#***alticonis.*** *Dorymyrmex pyramicus* var. *alticonis* Forel, 1912h: 36 (w.) BRAZIL. Combination in *Conomyrma:* Kempf, 1972a: 79.

#***alvarezi.*** *Dorymyrmex* (*Conomyrma*) *breviscapis* var. *alvarezi* Santschi, 1922b: 371 (w.) ARGENTINA. Combination in *Conomyrma:* Kempf, 1972a: 78.

#***anaemicus.*** *Dorymyrmex* (*Ammomyrma*) *exsanguis* var. *anaemica* Santschi, 1922b: 368 (w.q.) ARGENTINA. Combination in *Araucomyrmex:* Brandão, 1991: 327.

***antarcticus.*** *Dorymyrmex antarcticus* Forel, 1904g: 6 (w.q.) CHILE. Combination in *D.* (*Conomyrma*): Gallardo, 1916a: 48; in *D.* (*Ammomyrma*): Kusnezov, 1952g: 429; in *Araucomyrmex:* Kusnezov, 1960b: 369; in *Dorymyrmex:* Shattuck, 1992c: 85. Senior synonym of *pallidipes:* Gallardo, 1916a: 51; material of the unavailable name *depilitibia* referred here: Kusnezov, 1960b: 369. See also: Snelling & Hunt, 1976: 96.

*antillana. Dorymyrmex pyramicus* subsp. *brunneus* var. *antillana* Forel, 1911c: 306 (w.) ANTILLES IS. Unavailable name; material referred to *insanus* by Snelling, 1973b: 5.

*azulensis. Dorymyrmex flavescens* st. *joergenseni* var. *azulensis* Santschi, 1925e: 165 (w.) ARGENTINA. Unavailable name (Shattuck, 1994: 90).

***baeri.*** *Dorymyrmex baeri* André, 1903: 364 (w.) ARGENTINA. Santschi, 1925e: 166 (q.m.). Combination in *D.* (*Conomyrma*): Gallardo, 1916a: 47; in *D.* (*Ammomyrma*): Santschi, 1925e: 152; in *Araucomyrmex:* Brandão, 1991: 327; in *Dorymyrmex:* Shattuck, 1992c: 85.

***bicolor.*** *Dorymyrmex pyramicus* var. *bicolor* Wheeler, W.M. 1906d: 342 (w.) U.S.A. Gallardo, 1916a: 63 (q.); Crozier, 1970: 114 (k.). Combination in *D.* (*Conomyrma*): Gallardo, 1916a: 63; in *D.* (*Biconomyrma*): Kusnezov, 1952g: 430; in *Conomyrma:* Snelling, 1973b: 4; in *Dorymyrmex:* Shattuck, 1992c: 85. Subspecies of *pyramicus:* Creighton, 1950a: 349. Raised to species: Kusnezov, 1952g: 430; Cole, 1958b: 130. See also: Wilson, 1958b: 76; Johnson, 1989b: 192; Shattuck, 1994: 75.

*bicolor. Iridomyrmex bicolor* Donisthorpe, 1933c: 534 (w.) ARGENTINA. [Junior primary homonym of *Iridomyrmex iniquus* var. *bicolor* Forel, 1912h: 47.] Replacement name: *Iridomyrmex donisthorpei* Santschi, 1936d: 417.

***biconis.*** *Dorymyrmex biconis* Forel, 1912h: 37 (w.) COLOMBIA. Combination in *Conomyrma* (*Biconomyrma*): Kusnezov, 1952g: 430; in *Dorymyrmex:* Shattuck, 1992c: 85.

***bituber.*** *Dorymyrmex* (*Conomyrma*) *bituber* Santschi, 1916e: 394, fig. (w.) ARGENTINA. Combination in *Conomyrma* (*Biconomyrma*): Kusnezov, 1952g: 430; in *Dorymyrmex:* Shattuck, 1992c: 85. Current subspecies: nominal plus *laticeps.*

***bossutus.*** *Conomyrma bossuta* Trager, 1988: 15, figs. 1, 8 (w.q.) U.S.A. Combination in *Dorymyrmex:* Shattuck, 1992c: 85. See also: Johnson, 1989b: 192.

***breviscapis.*** *Dorymyrmex breviscapis* Forel, 1912h: 40 (w.) ARGENTINA. Combination in *D.* (*Conomyrma*): Forel, 1913l: 244; in *Conomyrma:* Kusnezov, 1952g: 430; in *Dorymyrmex:* Shattuck, 1992c: 84. Current subspecies: nominal plus *alvarezi, elongatulus, speculiceps.*

***bruchi.*** *Dorymyrmex bruchi* Forel, 1912h: 42 (w.) ARGENTINA. Combination in *D.* (*Psammomyrma*): Forel, 1912h: 43; in *D.* (*Spinomyrma*): Kusnezov, 1952g: 429. See also: Gallardo, 1916a: 40. Current subspecies: nominal plus *ebeninus.*

***brunneus.*** *Dorymyrmex pyramicus* var. *brunnea* Forel, 1908c: 385 (w.) ARGENTINA. Gallardo, 1916a: 60 (q.). Combination in *D.* (*Conomyrma*): Forel, 1913l: 244; in *Conomyrma* (*Biconomyrma*): Kusnezov, 1952g: 430; in *Biconomyrma:* Kusnezov, 1959: 51; in *Dorymyrmex:* Shattuck, 1992c: 85. Subspecies of *pyramicus:* Forel, 1911c: 306; Santschi, 1912e: 531. Raised to species: Santschi, 1929d: 305; Kusnezov, 1952g: 430; Kempf, 1975c: 375.

*bureni. Conomyrma bureni* Trager, 1988: 19, figs. 2, 9, 19, 20 (w.q.) U.S.A. Material of the *nomen nudum edeni* referable here: Trager, 1988: 19. Junior synonym of *flavus:* Johnson, 1989b: 187.

#***carbonarius.*** *Dorymyrmex* (*Conomyrma*) *exsanguis* var. *carbonaria* Forel, 1913l: 243 (w.m.) ARGENTINA. Gallardo, 1918a: 29 (q.). Combination in *D.* (*Ammomyrma*): Santschi, 1922b: 369; in *Araucomyrmex:* Brandão, 1991: 327.

***carettei.*** *Dorymyrmex* (*Conomyrma*) *carettei* Forel, 1913l: 244 (w.) ARGENTINA. Combination in *Conomyrma:* Kusnezov, 1952g: 430; in *Dorymyrmex:* Shattuck, 1992c: 85. See also: Gallardo, 1916a: 67.

***caretteoides.*** *Dorymyrmex* (*Conomyrma*) *breviscapis* var. *caretteoides* Forel, 1914d: 284 (w.) ARGENTINA. Combination in *Conomyrma:* Kusnezov, 1952g: 430; in *Dorymyrmex:* Shattuck, 1992c: 85. Raised to species: Kusnezov, 1952g: 430.

***chilensis.*** *Dorymyrmex tener* var. *chilensis* Forel, 1911c: 307 (w.q.) CHILE. Combination in *Araucomyrmex:* Kempf, 1972a: 25; in *Dorymyrmex:* Shattuck, 1992c: 85. Raised to species: Snelling & Hunt, 1976: 96.

***confusus.*** *Conomyrma confusa* Kusnezov, 1952g: 430. Replacement name for *gallardoi* Santschi, 1919f: 55. [Junior primary homonym of *gallardoi* Brèthes, 1914a: 95.] Combination in *Dorymyrmex:* Shattuck, 1992c: 85.

***coniculus.*** *Dorymyrmex* (*Ammomyrma*) *coniculus* Santschi, 1922b: 370 (w.) ARGENTINA. Combination in *Araucomyrmex:* Brandão, 1991: 327; in *Dorymyrmex:* Shattuck, 1992c: 85.

*connexa. Dorymyrmex* (*Conomyrma*) *pyramicus* st. *brunneus* var. *connexa* Santschi, 1922b: 373 (w.) BOLIVIA. Unavailable name (Shattuck, 1994: 90).

*depilitibia. Dorymyrmex* (*Conomyrma*) *tener* r. *richteri* var. *depilitibia* Forel, 1915c: 359 (w.) ARGENTINA. Unavailable name; material referred to *antarcticus* by Kusnezov, 1960b: 368.

#***donisthorpei.*** *Iridomyrmex donisthorpei* Santschi, 1936d: 417. Replacement name for *Iridomyrmex bicolor* Donisthorpe, 1933c: 534. [Junior primary homonym of *Iridomyrmex iniquus* var. *bicolor* Forel, 1912h: 47.] Combination in *Araucomyrmex:* Kusnezov, 1960b: 372; in *Dorymyrmex:* Shattuck, 1992c: 85. Senior synonym of *obscurior:* Kusnezov, 1960b: 372. Subspecies of *tener:* Shattuck, 1992c: 85.

#***dubius.*** *Dorymyrmex goeldii* var. *dubius* Forel, 1912h: 38 (w.) BRAZIL. Combination in *Conomyrma:* Kempf, 1972a: 79.

**#*ebeninus*.** *Dorymyrmex bruchi* var. *ebenina* Forel, 1914d: 282 (w.) ARGENTINA. [Possibly a junior synonym of *alboniger:* Kusnezov, 1952g: 429.]

*edeni. Conomyrma edeni* Tryon, 1986: 340 (attributed to Buren), U.S.A. *Nomen nudum;* material referred to *bureni* by Trager, 1988:19.

***elegans*.** *Conomyrma elegans* Trager, 1988: 21, figs. 3, 10 (w.q.) U.S.A. Combination in *Dorymyrmex:* Shattuck, 1992c: 85. See also: Johnson, 1989b: 191.

**#*elongatulus*.** *Dorymyrmex* (*Conomyrma*) *breviscapis* var. *elongatula* Santschi, 1919f: 53 (w.) ARGENTINA. Combination in *Conomyrma:* Kempf, 1972a: 78.

***emmaericaellus*.** *Dorymyrmex emmaericaellus* Kusnezov, 1951c: 59 (w.) BOLIVIA. Combination in *D.* (*Ammomyrma*): Kempf, 1972a: 100; in *Araucomyrmex:* Brandão, 1991: 327.

***ensifer*.** *Dorymyrmex ensifer* Forel, 1912h: 40 (w.) ARGENTINA. Combination in *D.* (*Psammomyrma*): Forel, 1912h: 43. Stirps of *mucronatus:* Santschi, 1916e: 392. Revived status as species: Gallardo, 1916a: 37; Santschi, 1922b: 367. Current subspecies: nominal plus *laevigatus, weiseri.*

***exsanguis*.** *Dorymyrmex exsanguis* Forel, 1912h: 38 (w.) ARGENTINA. Forel, 1913l: 242 (q.m.). Combination in *D.* (*Conomyrma*): Forel, 1913l: 242; in *D.* (*Ammomyrma*): Santschi, 1922b: 368; in *Araucomyrmex:* Brandão, 1991: 327; in *Dorymyrmex:* Shattuck, 1992c: 85. Current subspecies: nominal plus *anaemicus, carbonarius, sordidus.*

***flavescens*.** *Dorymyrmex flavescens* Mayr, 1866a: 495 (q.m.) ARGENTINA. Santschi, 1922b: 366 (w.). Senior synonym of *mucronatus:* Gallardo, 1930: 144. Current subspecies: nominal plus *jactans, mandibularis.* See also: Gallardo, 1916a: 28.

***flavopectus*.** *Dorymyrmex pyramicus* subsp. *flavopectus* Smith, M.R. 1944d: 15 (w.) U.S.A. Combination in *Conomyrma:* Kusnezov, 1952g: 430; in *Dorymyrmex:* Shattuck, 1992c: 85. Raised to species: Kusnezov, 1952g: 430. See also: Snelling, 1973b: 4; Buren, Nickerson & Thompson, 1976: 306; Trager, 1988: 22; Johnson, 1989b: 192 (in key).

***flavus*.** *Dorymyrmex flavus* McCook, 1879: 188, figs. a-c (w.) U.S.A. Johnson, 1989b: 190 (q.m.); Cokendolpher & Francke, 1985: 349 (k.). Combination in *Conomyrma:* Kempf, 1972a: 80; in *Dorymyrmex:* Shattuck, 1992c: 85. Variety/subspecies of *pyramicus:* Mayr, 1886d: 433; Forel, 1911e: 285. Junior synonym of *pyramicus:* Creighton, 1950a: 348; of *insanus:* Snelling, 1973b: 5. Revived from synonymy, revived status as species, and senior synonym of *bureni* and *nigra:* Johnson, 1989b: 187.

**#*fumigatus*.** *Dorymyrmex goeldii* subsp. *fumigatus* Forel, 1908c: 385 (w.m.) BRAZIL. Combination in *Conomyrma:* Kempf, 1972a: 79.

***fusculus*.** *Dorymyrmex* (*Ammomyrma*) *fusculus* Santschi, 1922b: 369 (w.) ARGENTINA. Combination in *Araucomyrmex:* Brandão, 1991: 328; in *Dorymyrmex:* Shattuck, 1992c: 85.

*gallardoi* Brèthes, 1914; see under ***LINEPITHEMA*.**

*gallardoi. Dorymyrmex* (*Conomyrma*) *gallardoi* Santschi, 1919f: 55 (w.) ARGENTINA. [Junior primary homonym of *gallardoi* Brèthes, 1914a: 95, above.] Replacement name: *confusus* Kusnezov, 1952g: 430.

**#*garbei*.** *Dorymyrmex pyramicus* subsp. *garbei* Forel, 1911c: 306 (w.q.) BRAZIL. Gallardo, 1916a: 61 (m.). Combination in *D.* (*Conomyrma*): Gallardo, 1916a: 60; in *Conomyrma:* Kempf, 1972a: 79. Junior synonym of *pyramicus:* Santschi, 1922b: 372. Revived from synonymy as variety of *pyramicus:* Santschi, 1925d: 245.

***goeldii*.** *Dorymyrmex goeldii* Forel, 1904c: 41 (w.) BRAZIL. Combination in *D.* (*Ammomyrma*): Kusnezov, 1952g: 429; in *Conomyrma:* Kempf, 1972a: 79; in *Dorymyrmex:* Shattuck, 1992c: 85. Current subspecies: nominal plus *dubius, fumigatus.*

***goetschi*.** *Dorymyrmex goetschi* Goetsch, 1933: 27, fig. 1 (w.) CHILE (indication only; attributed to Menozzi). Menozzi, 1935c: 322 (w., full description). Combination in *D.* (*Araucomyrmex*): Menozzi, 1935c: 322; in *Araucomyrmex:* Snelling & Hunt, 1976: 97; in *Dorymyrmex:* Shattuck, 1992c: 85. Subspecies of *tener:* Menozzi, 1935c: 322. Junior synonym of *tener:* Kusnezov, 1959: 370. Revived from synonymy and revived status as species: Snelling & Hunt, 1976: 97.

**#*gracilis*.** *Dorymyrmex* (*Conomyrma*) *thoracicus* var. *gracilis* Santschi, 1920d: 381 (w.) ARGENTINA. Combination in *Conomyrma:* Kempf, 1972a: 80.

***grandulus*.** *Prenolepis* (*Nylanderia*) *parvula* var. *grandula* Forel, 1922: 98 (w.) U.S.A. Trager, 1988: 24 (q.). Combination in *Paratrechina* (*Nylanderia*): Emery, 1925b: 222; Creighton, 1950a: 409; in *Conomyrma:* Trager, 1984: 64; in *Dorymyrmex:* Shattuck, 1992c: 85. Junior synonym of *parvula:* Creighton, 1950a: 409. Revived from synonymy and raised to species: Trager, 1984: 64. See also: Johnson, 1989b: 192 (in key).

**#*guyanensis*.** *Dorymyrmex* (*Conomyrma*) *pyramicus* var. *guyanensis* Santschi, 1922b: 372 (w.) FRENCH GUIANA. Combination in *Conomyrma:* Kempf, 1972a: 79.

***hunti*.** *Araucomyrmex hunti* Snelling, 1975: 10, figs. 14, 15 (w.) CHILE. Combination in *Dorymyrmex:* Shattuck, 1992c: 85.

***hypocritus*.** *Araucomyrmex hypocritus* Snelling, 1975: 12, figs. 16, 17 (w.) CHILE. Combination in *Dorymyrmex:* Shattuck, 1992c: 85.

***incomptus*.** *Araucomyrmex incomptus* Snelling, 1975: 13, figs. 18, 19 (w.) CHILE. Combination in *Dorymyrmex:* Shattuck, 1992c: 85.

***insanus*.** *Formica insana* Buckley, 1866: 165 (w.q.) U.S.A. Combination in *Dorymyrmex:* McCook, 1879: 185; in *D.* (*Conomyrma*): Smith, M.R. 1951a: 837; in *Conomyrma:* Snelling, 1973b: 4; in *Dorymyrmex:* Shattuck, 1992c: 85. Junior synonym of *pyramicus:* Mayr, 1886d: 433. Revived from synonymy as variety of *pyramicus:* Santschi, 1920d: 381. Junior synonym of *pyramicus:* Creighton, 1950a: 348. Revived from synonymy as species: Snelling, 1973b: 5. Senior synonym of *smithi* and *wheeleri,* and material of the unavailable name *antillana* referred here: Snelling, 1973b: 5. Senior synonym of *medeis* and *reginicula:* Johnson, 1989b: 185. See also: Buren, Nickerson & Thompson, 1976: 306; Shattuck, 1994: 84.

**#*jactans*.** *Dorymyrmex mucronatus* var. *jactans* Santschi, 1916e: 392, fig. (w.) ARGENTINA. Variety of *flavescens:* Santschi, 1922b: 367.

***jheringi*.** *Dorymyrmex jheringi* Forel, 1912h: 39 (w.) BRAZIL. Combination in *D.* (*Conomyrma*): Forel, 1913l: 243 (in text); in *D.* (*Ammomyrma*): Kusnezov, 1952g: 429; in *Conomyrma:* Kempf, 1972a: 79; in *Dorymyrmex:* Shattuck, 1992c: 85.

***joergenseni*.** *Dorymyrmex mucronatus* var. *joergenseni* Bruch, 1917d: 431, figs. 1, 2 (w.) ARGENTINA. Stirps of *flavescens:* Santschi, 1925e: 165. Raised to species: Kusnezov, 1952g: 429.

**#*laevigatus*.** *Dorymyrmex ensifer* var. *laevigata* Gallardo, 1916a: 39 (w.) ARGENTINA. Material of the unavailable name *taeniata* referred here by Santschi, 1919f: 54.

**#*laticeps*.** *Dorymyrmex* (*Conomyrma*) *bituber* var. *laticeps* Santschi, 1919f: 54 (w.) ARGENTINA. Combination in *Conomyrma:* Kempf, 1972a: 78.

**#*mandibularis*.** *Dorymyrmex flavescens* var. *mandibularis* Santschi, 1925e: 166 (w.) ARGENTINA.

*medeis. Conomyrma medeis* Trager, 1988: 25, figs. 6, 13, 15, 16 (w.q.) U.S.A. Junior synonym of *insanus:* Johnson, 1989b: 185.

**#*mesonotalis*.** *Dorymyrmex pyramicus* var. *mesonotalis* Forel, 1912h: 37 (w.) PANAMA. Combination in *Conomyrma:* Kempf, 1972a: 79.

***minutus*.** *Dorymyrmex minutus* Emery, 1895i: 15, fig. (w.) CHILE. Combination in *D.* (*Conomyrma*): Gallardo, 1916a: 43; in *D.* (*Ammomyrma*): Kusnezov, 1952g: 429; in *Araucomyrmex:* Snelling & Hunt, 1976: 100; in *Dorymyrmex:* Shattuck, 1992c: 85.

***morenoi*.** *Dorymyrmex morenoi* Bruch, 1921: 187, fig. 7 (w.q.m.) ARGENTINA. Current subspecies: nominal plus *patagon.*

*mucronatus. Dorymyrmex mucronatus* Emery, 1906c: 174 (w.) ARGENTINA. Gallardo, 1916a: 35 (m.). Combination in *D.* (*Psammomyrma*): Forel, 1912h: 43. Variety of *flavescens:* Santschi, 1922b: 367. Junior synonym of *flavescens:* Gallardo, 1930: 144.

*nigra. Dorymyrmex pyramicus* var. *nigra* Pergande, 1896: 871 (w.) MEXICO. Forel, 1912h: 36 (q.). Combination in *D.* (*Conomyrma*): Creighton, 1950a: 348; in *Conomyrma:* Kempf, 1972a: 79. Junior synonym of *pyramicus:* Creighton, 1950a: 348; of *insanus:* Snelling, 1973b: 5; of *flavus:* Johnson, 1989b: 187.

**#*nigriventris*.** *Dorymyrmex* (*Conomyrma*) *pyramicus* var. *nigriventris* Santschi, 1922b: 373 (w.) ARGENTINA. [First available use of *Dorymyrmex* (*Conomyrma*) *pyramicus* st. *garbei* var. *nigriventris* Santschi, 1916e: 393; unavailable name.] Combination in *Conomyrma:* Kempf, 1972a: 80; in *Dorymyrmex:* Shattuck, 1992c: 85.

*obscurior. Araucomyrmex tener* subsp. *obscurior* Kusnezov, 1953a: 111 (diagnosis in key) (w.) ARGENTINA. Junior synonym of *donisthorpei:* Kusnezov, 1960b: 372.

*pallidipes. Dorymyrmex tener* var. *pallidipes* Brèthes, 1914a: 96 (w.) ARGENTINA. Junior synonym of *antarcticus:* Gallardo, 1916a: 51.

***pappodes*.** *Araucomyrmex pappodes* Snelling, 1975: 14, figs. 20, 21 (w.) CHILE. Combination in *Dorymyrmex:* Shattuck, 1992c: 85.

**#*paranensis*.** *Dorymyrmex* (*Conomyrma*) *pyramicus* var. *paranensis* Santschi, 1922b: 372 (w.) PARAGUAY.

[First available use of *Dorymyrmex pyramicus* subsp. *flava* var. *paranensis* Forel, 1911e: 285; unavailable name.] Combination in *Conomyrma:* Kempf, 1972a: 79.

**#patagon.** *Dorymyrmex morenoi* st. *patagon* Santschi, 1922b: 367 (w.) ARGENTINA.

**#peruvianus.** *Dorymyrmex pyramicus* var. *peruvianus* Wheeler, W.M. 1919c: 275 (w.) PERU. Combination in *Conomyrma:* Kempf, 1972a: 79.

**planidens.** *Dorymyrmex planidens* Mayr, 1868b: 165 (w.) ARGENTINA. Brèthes, 1914b: 231 (q.m.). Combination in *D.* (*Psammomyrma*): Forel, 1912h: 43. See also: Gallardo, 1916a: 28.

**#platensis.** *Dorymyrmex pyramicus* var. *platensis* Gallardo, 1916a: 55 (w.) ARGENTINA. Santschi, 1920d: 381 (q.m.). Combination in *Conomyrma:* Kempf, 1972a: 80; in *Dorymyrmex:* Shattuck, 1992c: 85. Subspecies of *steigeri:* Kempf, 1972a: 80.

**pogonius.** *Araucomyrmex pogonius* Snelling, 1975: 15, figs. 22–28 (w.m.) CHILE. Combination in *Dorymyrmex:* Shattuck, 1992c: 85.

**pulchellus.** *Dorymyrmex* (*Conomyrma*) *pulchellus* Santschi, 1922b: 371 (w.) ARGENTINA. Crozier, 1970: 114 (k.). Combination in *Conomyrma* (*Biconomyrma*): Kusnezov, 1952g: 430; in *Dorymyrmex:* Shattuck, 1992c: 85.

**pyramicus.** *Prenolepis pyramica* Roger, 1863a: 160 (w.) BRAZIL. Emery, 1888c: 362 (m.); Gallardo, 1916a: 56 (q.); Wheeler, G.C. & Wheeler, J. 1951: 183 (l.); Goni, *et al.* 1983: 366 (k.). Combination in *Dorymyrmex:* Mayr, 1870b: 947; in *D.* (*Conomyrma*): Forel, 1913l: 244; in *Conomyrma:* Kusnezov, 1952g: 430; in *Dorymyrmex:* Shattuck, 1992c: 85. Current subspecies: nominal plus *albemarlensis, alticonis, garbei, guyanensis, mesonotalis, nigriventris, paranensis, peruvianus, rubriceps.* See also: Emery, 1895c: 331; Snelling, 1973b: 4.

*reginicula. Conomyrma reginicula* Trager, 1988: 27, figs. 7, 14, 17, 18 (w.q.) U.S.A. Junior synonym of *insanus:* Johnson, 1989b: 185.

**richteri.** *Dorymyrmex tener* subsp. *richteri* Forel, 1911c: 308 (w.) ARGENTINA. Gallardo, 1918a: 29 (m.). Combination in *D.* (*Conomyrma*): Forel, 1915c: 359; in *D.* (*Araucomyrmex*): Kusnezov, 1952g: 429; in *Araucomyrmex:* Kusnezov, 1953a: 110; in *Dorymyrmex:* Shattuck, 1992c: 85. Raised to species: Kusnezov, 1952g: 429. Junior synonym of *antarcticus:* Kusnezov, 1960b: 369; Kempf, 1972a: 25. Revived from synonymy and revived status as species: Snelling & Hunt, 1976: 96 (in text).

**#rubriceps.** *Dorymyrmex pyramicus* var. *rubriceps* Forel, 1912h: 36 (w.) BRAZIL. Combination in *Conomyrma:* Kempf, 1972a: 80.

**santschii.** *Dorymyrmex santschii* Gallardo, 1917: 50, figs. (w.) ARGENTINA. Combination in *D.* (*Conomyrma*): Santschi, 1922d: 257; in *Conomyrma:* Kusnezov, 1952g: 431; in *Dorymyrmex:* Shattuck, 1992c: 85. Variety of *wolffhuegeli:* Santschi, 1922d: 257. Revived status as species: Kusnezov, 1952g: 431.

*schilleri. Dorymyrmex schilleri* Kusnezov, 1952g: 429. *Nomen nudum,* attributed to Santschi.

**silvestrii.** *Dorymyrmex silvestrii* Gallardo, 1916c: 259, figs. 3, 4 (w.q.) ARGENTINA. Combination in *Conomyrma:* Kusnezov, 1952g: 431; in *Dorymyrmex:* Shattuck, 1992c: 85.

*smithi. Dorymyrmex pyramicus* var. *smithi* Cole, 1936b: 120 (w.) U.S.A. Combination in *D.* (*Conomyrma*): Creighton, 1950a: 349; in *Conomyrma:* Snelling, 1973b: 4. Junior synonym of *pyramicus:* Creighton, 1950a: 249; of *insanus:* Snelling, 1973b: 5.

**#sordidus.** *Dorymyrmex* (*Conomyrma*) *exsanguis* st. *sordida* Santschi, 1919f: 54 (w.) ARGENTINA. Santschi, 1922b: 369 (q.); Santschi, 1925e: 167 (m.). Combination in *D.* (*Ammomyrma*): Santschi, 1922b: 369; in *Araucomyrmex:* Brandão, 1991: 327.

**#speculiceps.** *Dorymyrmex* (*Conomyrma*) *breviscapis* st. *speculiceps* Santschi, 1922b: 371 (w.) ARGENTINA Combination in *Conomyrma:* Kempf, 1972a: 78.

**spurius.** *Dorymyrmex* (*Conomyrma*) *brunneus* var. *spurius* Santschi, 1929d: 305 (w.) PARAGUAY. [First available use of *Dorymyrmex pyramicus* subsp. *brunneus* var. *spuria* Forel, 1911e: 285; unavailable name.] Combination in *Conomyrma:* Kempf, 1972a: 78; in *Dorymyrmex:* Shattuck, 1992c: 85. Raised to species: Kempf, 1975c: 376.

**steigeri.** *Dorymyrmex steigeri* Santschi, 1912e: 532, fig. 4 (w.) ARGENTINA. Santschi, 1916e: 393 (m.). Combination in *D.* (*Conomyrma*): Santschi, 1916e: 393; Gallardo, 1916a: 66; in *Conomyrma:* Kusnezov, 1952g: 430; in *Dorymyrmex:* Shattuck, 1992c: 85. Stirps of *wolffhuegeli:* Santschi, 1920d: 381. Revived status as species: Kusnezov, 1952g: 430. Current subspecies: nominal plus *platensis.*

*taeniata. Dorymyrmex mucronatus* st. *ensifer* var. *taeniata* Santschi, 1916e: 392, fig. (w.) ARGENTINA. Unavailable name; material referred to *laevigatus* by Santschi, 1919f: 54.

**tener.** *Dorymyrmex tener* Mayr, 1868b: 166 (w.) CHILE. Emery, 1906c: 175 (q.m.); Wheeler, G.C. & Wheeler, J. 1951: 184 (l.). Combination in *D.* (*Conomyrma*): Forel, 1915c: 359; in *Araucomyrmex:* Gallardo, 1919b: 250; in *Dorymyrmex:* Shattuck, 1992c: 85. Current subspecies: nominal plus *donisthorpei.*

**thoracicus.** *Dorymyrmex* (*Conomyrma*) *thoracicus* Gallardo, 1916a: 52 (w.) ARGENTINA. [First available use of *Dorymyrmex pyramicus* st. *brunneus* var. *thoracica* Santschi, 1912e: 531; unavailable name.] Crozier, 1970: 114 (k.). Combination in *Conomyrma* (*Biconomyrma*): Kusnezov, 1952g: 430; in *Dorymyrmex:* Shattuck, 1992c: 85. Current subspecies: nominal plus *gracilis, tigris.*

**#tigris.** *Dorymyrmex* (*Conomyrma*) *thoracicus* var. *tigris* Santschi, 1925e: 166 (w.) ARGENTINA. Combination in *Conomyrma:* Kempf, 1972a: 80.

**#weiseri.** *Dorymyrmex ensifer* var. *weiseri* Santschi, 1922b: 367 (w.) ARGENTINA.

*wheeleri. Conomyrma* (*Biconomyrma*) *wheeleri* Kusnezov, 1952g: 438 (w.) U.S.A. Junior synonym of *insanus:* Snelling, 1973b: 5.

**wolffhuegeli.** *Dorymyrmex wolffhuegeli* Forel, 1911c: 307 (w.) ARGENTINA. Forel, 1914d: 284 (q.). Combination in *D.* (*Conomyrma*): Forel, 1913l: 244; in *Conomyrma:* Kusnezov, 1952g: 430; in *Dorymyrmex:* Shattuck, 1992c: 85.

*DREPANOGNATHUS:* see under ***HARPEGNATHOS.***

**DRYMOMYRMEX** (Baltic Amber)

***claripennis.** **Drymomyrmex claripennis* Wheeler, W.M. 1915h: 138 (q.) BALTIC AMBER (Oligocene).

***fuscipennis.** **Drymomyrmex fuscipennis* Wheeler, W.M. 1915h: 136, fig. 56 (q.) BALTIC AMBER (Oligocene). Théobald, 1937b: 216 (w.m.).

*DYCLONA:* see under ***CARDIOCONDYLA.***

*DYOMORIUM:* see under ***VOLLENHOVIA.***

***DYSEDROGNATHUS*** (Indo-Australian)

**extemenus.** *Dysedrognathus extemenus* Taylor, 1968d: 133, figs. 3, 4 (w.) SINGAPORE.

***ECHINOPLA*** (Indo-Australian, Australasian)

**#aciculata.** *Polyrhachis aciculatus* Smith, F. 1858b: 70, pl. 4, figs. 17, 18 (w.) PHILIPPINES. Combination in *Echinopla:* Emery, 1900d: 721 (footnote). Variety of *striata:* Emery, 1925b: 211.

**arfaki.** *Echinopla arfaki* Donisthorpe, 1943a: 176. Replacement name for *rugosa* Stitz, 1938: 110. [Junior primary homonym of *rugosa* André, 1892b: 47.] Senior synonym of *corrugata* Donisthorpe, 1943e: 116 (unnecessary (second) replacement name for *rugosa* Stitz): **new synonymy**.

**australis.** *Echinopla australis* Forel, 1901h: 75 (w.) AUSTRALIA. [*Echinopla australis* Emery, 1897d: 577. *Nomen nudum.*]

*corrugata. Echinopla corrugata* Donisthorpe, 1943e: 116. Unnecessary (second) replacement name for *rugosa* Stitz, 1938: 110. Junior synonym of *arfaki* (first available replacement name): **new synonymy**.

**crenulata.** *Echinopla crenulata* Donisthorpe, 1941g: 115 (w.) NEW GUINEA.

**deceptor.** *Echinopla deceptor* Smith, F. 1863: 23 (q.) INDONESIA (Buru I.).

**densistriata.** *Echinopla densistriata* Stitz, 1938: 110, fig. 7 (w.) BORNEO.

**dubitata.** *Echinopla dubitata* Smith, F. 1862a: 50 (w.) SULAWESI.

**#gibbosa.** *Echinopla striata* subsp. *gibbosa* Karavaiev, 1927e: 51 (w.) INDONESIA (Ambon I.).

**#goramensis.** *Echinopla striata* var. *goramensis* Emery, 1887a: 241 (w.) INDONESIA (Goram I.).

**lineata.** *Echinopla lineata* Mayr, 1862: 689 (w.) JAVA. Mayr, 1872: 143 (q.). Senior synonym of *sucki:* Emery, 1925b: 211. Current subspecies: nominal plus *senilis.* See also: Mayr, 1865: 48.

**maeandrina.** *Echinopla maeandrina* Stitz, 1938: 111, fig. 9 (q.) NEW GUINEA.

**melanarctos.** *Echinopla melanarctos* Smith, F. 1857a: 79, pl. 1, figs. 25–29 (w.) SINGAPORE.

**mistura.** *Mesoxena mistura* Smith, F. 1860b: 107, pl. 1, fig. 10 (w.) INDONESIA (Batjan I.). Combination in *Echinopla:* **new combination**. Redescription of (w.): Donisthorpe, 1936d: 119.

**nitida.** *Echinopla nitida* Smith, F. 1863: 23 (w.) INDONESIA (Seram I.).

**octodentata.** *Echinopla australis* var. *octodentata* Stitz, 1911a: 381 (w.) NEW GUINEA. Raised to species: Stitz, 1938: 109.

**pallipes.** *Echinopla pallipes* Smith, F. 1857a: 80 (w.) BORNEO. Forel, 1913k: 133 (q.). [Misspelled as *pallidipes* by Dalla Torre, 1893: 272; Emery, 1925b: 211.]

*pictipes. Echinopla turneri* var. *pictipes* Forel, 1901h: 76 (w.q.) AUSTRALIA. Junior synonym of *turneri:* Taylor, 1992a: 61.

***ECHINOPLA** (continued)*

***praetexta.*** *Echinopla praetexta* Smith, F. 1860b: 113, pl. 1, fig. 5 (w.) no locality given. [Emery, 1896d: 382 gives INDONESIA (Batjan I.) as type-locality, but I am unaware of how he obtained this information.]

***pseudostriata.*** *Echinopla pseudostriata* Donisthorpe, 1943a: 174 (w.) NEW GUINEA.

***rugosa.*** *Echinopla rugosa* André, 1892b: 47 (w.) BORNEO.

*rugosa. Echinopla rugosa* Stitz, 1938: 110, fig. 8 (w.) NEW GUINEA. [Junior primary homonym of *rugosa* André, above.] Replacement name: *arfaki* Donisthorpe, 1943a: 176.

#***senilis.*** *Echinopla senilis* Mayr, 1862: 690 (w.) NICOBAR IS. Variety of *lineata:* Emery, 1896d: 382.

***serrata.*** *Polyrhachis serratus* Smith, F. 1859a: 140 (w.) INDONESIA (Aru I.). Combination in *Polyrhachis* (*Aulacomyrma*): Emery, 1925b: 198; in *Echinopla:* Donisthorpe, 1932c: 453.

***silvestrii.*** *Echinopla silvestrii* Donisthorpe, 1936e: 529 (q.) NEW GUINEA.

***striata.*** *Echinopla striata* Smith, F. 1857a: 80 (w.) WEST MALAYSIA. Donisthorpe, 1942e: 709 (q.). Current subspecies: nominal plus *aciculata, gibbosa, goramensis.*

*sucki. Echinopla sucki* Forel, 1901h: 75 (w.) BORNEO. Junior synonym of *lineata:* Emery, 1925b: 211.

***tritschleri.*** *Echinopla tritschleri* Forel, 1901h: 74 (w.) SUMATRA. Forel, 1913: 134 (q.).

***turneri.*** *Echinopla turneri* Forel, 1901h: 76 (w.q.) AUSTRALIA. Senior synonym of *pictipes:* Taylor, 1992a: 61.

***vermiculata.*** *Echinopla striata* var. *vermiculata* Emery, 1898b: 244 (q.) SULAWESI. Emery, 1900d: 721 (w.). Raised to species: Emery, 1900d: 721.

***ECITON*** (Neotropical)

*abstinens* Ihering, 1912; see under ***NEIVAMYRMEX.***

*adnepos* Wheeler, W.M. 1922; see under ***NEIVAMYRMEX.***

*alfaroi* Emery, 1890; see under ***NEIVAMYRMEX.***

#***allognathum.*** *Eciton vagans* subsp. *allognathum* Borgmeier, 1955b: 239, pl. 17, fig. 7; pl. 19, fig. 13; pl. 64, fig. 2 (s.m.) VENEZUELA.

*amazona. Eciton amazona* Santschi, 1923d: 68 (s.w.) BRAZIL. Variety of *mattogrosense:* Santschi, 1925d: 221. Junior synonym of *hamatum:* Borgmeier, 1953: 14.

*andrei* Emery, 1901; see under ***NEIVAMYRMEX***

#***angustatum.*** *Eciton angustata* Roger, 1863a: 204 (w.) MEXICO. Wheeler, W.M. 1912e: 207 (m.); Menozzi, 1931d: 188 (q.). Combination in *E.* (*Holopone*): Santschi, 1925b: 11. Variety of *burchellii:* Emery, 1900a: 185. Revived status as species: Emery, 1910b: 20. Junior synonym of *vagans:* Borgmeier, 1939: 405. Revived from synonymy as subspecies of *vagans:* Borgmeier, 1955b: 229. Senior synonym of *brunnea:* Borgmeier, 1955b: 229.

*angustinode* Emery, 1888; see under ***NEIVAMYRMEX.***

*angustius* Forel, 1909; see under ***NEIVAMYRMEX.***

*antillanum* Forel, 1897; see under ***NEIVAMYRMEX.***

*apicifer* Santschi, 1916; see under ***NOMAMYRMEX.***

#***argentinum.*** *Eciton mexicanum* subsp. *argentinum* Borgmeier, 1955b: 259, pl. 55, figs. 7, 11; pl. 84, fig. 2 (s.q.) ARGENTINA.

*arizonense* Wheeler, W.M. 1908; see under ***NEIVAMYRMEX.***

*auropubens* Santschi, 1920; see under ***LABIDUS.***

*aztecum* Forel, 1901; see under ***NEIVAMYRMEX.***

*balzani* Emery, 1894; see under ***NEIVAMYRMEX.***

*beebei* Wheeler, W.M. 1921; see under ***NEIVAMYRMEX.***

*bellicosum. Eciton bellicosum* Santschi, 1921g: 90 (s.w.) FRENCH GUIANA. Junior synonym of *drepanophorum:* Borgmeier, 1936: 52.

*biloba* Emery, 1901; see under ***LABIDUS.***

*bohlsi* Emery, 1896; see under ***NEIVAMYRMEX.***

*bruchi* Forel, 1912; see under ***NEIVAMYRMEX.***

*brunnea. Eciton brunnea* Norton, 1868c: 6 (w.) MEXICO. [*Eciton brunnea* Norton, 1868a: 62. *Nomen nudum.*] Junior synonym of *rapax:* Norton, 1868b: 45; of *vagans:* Emery, 1910b: 21; of *angustatum:* Borgmeier, 1955b: 229.

***burchellii.*** *Labidus burchellii* Westwood, 1842: 74, pl. 20, fig. 1 (m.) BRAZIL. Emery, 1896g: 39 (s.w.); Wheeler, W.M. 1921d: 304 (q.); Emery, 1899c: 6 (l.). Combination in *Eciton:* Emery, 1890b: 39. Current subspecies: nominal plus *cupiens, foreli, parvispinum, urichi.* See also: Borgmeier, 1939: 404; Borgmeier, 1955b: 178; Wheeler, G.C. 1943: 328; Wheeler, G.C. & Wheeler, J. 1974c: 169; Wheeler, G.C. & Wheeler, J. 1984: 270.

*californicum* Mayr, 1870; see under ***NEIVAMYRMEX.***

*caretti* Forel, 1913; see under ***NEIVAMYRMEX.***

*carolinensis* Emery, 1894; see under ***NEIVAMYRMEX.***

*castaneum* Borgmeier, 1939; see under ***NEIVAMYRMEX.***

*clavicornis* Norton, 1868; see under ***NEIVAMYRMEX.***

*cloosae* Forel, 1912; see under ***NEIVAMYRMEX.***

*cocula* Mann, 1925; see under ***NEIVAMYRMEX.***

*colombi* Santschi, 1921; see under ***NEIVAMYRMEX.***

*columbicum* Forel, 1912; see under ***NEIVAMYRMEX.***

*commutatum* Emery, 1900; see under ***NEIVAMYRMEX.***

#***conquistador.*** *Eciton* (*Eciton*) *conquistador* Weber, 1949c: 3, figs. 1–7 (s.w.l.) PANAMA. Reichensperger, 1934: 240 (q.). Subspecies of *lucanoides:* Borgmeier, 1955b: 264.

*crassicornis* Smith, F. 1855; see under ***NOMAMYRMEX.***

#***crassinode.*** *Eciton dulcius* subsp. *crassinode* Borgmeier, 1955b: 209 (s.q.) PANAMA.

*crenulatum* Mann, 1916; see under ***NEIVAMYRMEX.***

*cristatum* André, 1889; see under ***NEIVAMYRMEX.***

*crudele* Santschi, 1929; see under ***NOMAMYRMEX.***

#***cupiens.*** *Eciton burchelli* st. *cupiens* Santschi, 1923c: 249 (m.) FRENCH GUIANA. Wheeler, W.M. 1921d: 293 (s.w.q.) [as *burchellii;* see Borgmeier, 1955b: 178]. Senior synonym of *jeanae:* Borgmeier, 1955b: 178.

*curvidentata. Formica curvidentata* Latreille, 1802c: 269, pl. 8, fig. 55 (w.) FRENCH GUIANA. Combination in *Eciton:* Lepeletier, 1835: 179. Junior synonym of *hamatum:* Smith, F. 1858b: 148.

*curvipes* Emery, 1900; see under ***LABIDUS.***

*densepunctatum* Borgmeier, 1933; see under ***NEIVAMYRMEX.***

*desartsi* Forel, 1907; see under ***NEIVAMYRMEX.***

*diabolus* Forel, 1912; see under ***NEIVAMYRMEX.***

*diana* Forel, 1912; see under ***NEIVAMYRMEX.***

#***dispar.*** *Eciton vagans* subsp. *dispar* Borgmeier, 1955b: 244, pl. 10, fig. 4; pl. 12, fig. 9; pl. 17, fig. 5; pl. 19, fig. 17; pl. 21, figs. 4, 6; pl. 22, figs. 8, 10; pl. 64, fig. 1 (s.w.q.m.) BRAZIL.

*diversinode* Borgmeier, 1933; see under ***NEIVAMYRMEX.***

***drepanophorum.*** *Eciton drepanophorum* Smith, F. 1858b: 149 (s.w.) BRAZIL. Junior synonym of *hamatum:* Mayr, 1886b: 117; Forel, 1899c: 23; Emery, 1910b: 21. Revived from synonymy as variety of *hamatum:* Santschi, 1921g: 89. Revived status as species: Borgmeier, 1936: 52. Senior synonym of *bellicosum:* Borgmeier, 1936: 52; of *ferox* and *quitensis:* Borgmeier, 1953: 13 and :9 respectively. See also: Borgmeier, 1955b: 223.

*dubia* Creighton, 1932; see under ***NEIVAMYRMEX.***

#***dubitatum.*** *Eciton dubitatum* Emery, 1896g: 40, pl. 1, fig. 4 (m.) PARAGUAY. Subspecies of *vagans:* Borgmeier, 1953: 6. Senior synonym of *francanum* and *reichenspergeri:* Borgmeier, 1955b: 229.

***dulcium.*** *Eciton quadriglume* r. *dulcius* Forel, 1912c: 42 (w.) BRAZIL. Borgmeier, 1939: 406 (m.); Borgmeier, 1955b: 203 (s.q.). Combination in *E.* (*Holopone*): Santschi, 1925b: 11. Raised to species: Santschi, 1921g: 90; Borgmeier, 1939: 406. Senior synonym of *hirsutum* (and its junior synonym *jujuyensis*): Borgmeier, 1948b: 459. Current subspecies: nominal plus *crassinode.* See also: Borgmeier, 1953: 9; Borgmeier, 1955b: 203.

*elongatus* Karavaiev, 1926; see under ***AENICTUS.***

*elsbethae* Forel, 1922; see under ***LABIDUS.***

*emersoni* Wheeler, W.M. 1921, see under ***NEIVAMYRMEX.***

*emeryi* Santschi, 1921; see under ***NEIVAMYRMEX.***

*emiliae* Mann, 1916; see under ***LABIDUS.***

*erratica* Smith, F. 1860; see under ***LABIDUS.***

*exagona* Santschi, 1923; see under ***NEIVAMYRMEX.***

*falciferum* Emery, 1900; see under ***NEIVAMYRMEX.***

*fargeavii. Labidus fargeavii* Shuckard, 1840a: 198 (m.) no locality given. Combination in *Eciton:* Dalla Torre, 1893: 2; in *Mayromyrmex:* Ashmead, 1905b: 381; in *Eciton:* Emery, 1906a: 718. Junior synonym of *quadriglume:* Emery, 1896g: 38; Borgmeier, 1953: 15.

*ferox. Eciton ferox* Santschi, 1920e: 122 (s.w.) BOLIVIA. [Also described as new by Santschi, 1921g: 89.] Junior synonym of *drepanophorum:* Borgmeier, 1953: 13.

*ferruginea* Norton, 1868; see under ***LABIDUS.***

*fiebrigi* Forel, 1909; see under ***NEIVAMYRMEX.***

*filicorne* Santschi, 1916; see under ***NEIVAMYRMEX.***

*fimbriatum* Santschi, 1920; see under ***NOMAMYRMEX.***

#***foreli.*** *Eciton foreli* Mayr, 1886b: 116 (diagnosis in key) (s.w.) MEXICO. Borgmeier, 1936: 154 (q.). Junior synonym of *burchellii:* Emery, 1896g: 39. Revived from synonymy as variety/subspecies of *burchellii:* Santschi, 1925d: 222; Borgmeier, 1936: 54. Senior synonym of *viator:* Borgmeier, 1936: 54. See also: Borgmeier, 1955b: 188.

*francanum. Eciton vagans* var. *francanum* Santschi, 1923g: 1260 (w.) BRAZIL. [*Eciton francanum* Ihering, 1912: 299, and *Eciton vagans* subsp. *francanum* Forel, 1913l: 213. *Nomina nuda.*] Junior synonym of *vagans:* Borgmeier, 1939: 405; of *dubitatum:* Borgmeier, 1955b: 229.

*frontalis* Menozzi, 1924; see under ***NEIVAMYRMEX.***

*fumosa* Forel, 1913; see under ***NEIVAMYRMEX.***
*funesta. Eciton hamatum* var. *funesta* Santschi, 1921g: 89 (s.w.) COSTA RICA. Junior synonym of *hamatum:* Borgmeier, 1953: 13.
#***fur.*** *Eciton vagans* subsp. *fur* Borgmeier, 1955b: 245, pl. 19, fig. 20; pl. 20, fig. 2; pl. 21, fig. 10; pl. 22, figs. 7, 15; pl. 63, fig. 4 (s.w.m.) BRAZIL.
*furiosum* Santschi, 1929; see under ***NOMAMYRMEX.***
*fuscipennis* Smith, M.R. 1942; see under ***NEIVAMYRMEX.***
*garbei* Forel, 1912; see under ***NEIVAMYRMEX.***
*goeldii* Forel, 1901; see under ***NEIVAMYRMEX.***
#***goianum.*** *Eciton mexicanum* subsp. *goianum* Borgmeier, 1955b: 258, pl. 10, fig. 5; pl. 11, fig. 10; pl. 12, fig. 8; pl. 55, figs. 6, 10 (s.w.q.m.) BRAZIL.
*gracellae* Mann, 1926; see under ***NEIVAMYRMEX.***
*grandipseudops* Forel, 1915; see under ***NEIVAMYRMEX.***
*grassator* Forel, 1911; see under ***LABIDUS.***
*guianense* Wheeler, W.M. 1921; see under ***LABIDUS.***
*guyanensis* Santschi, 1916; see under ***NEIVAMYRMEX.***
***hamatum.*** *Formica hamata* Fabricius, 1782: 494 (s.) FRENCH GUIANA. Mayr, 1886b: 121 (w.); Emery, 1896g: 39 (m.); Wheeler, W.M. 1925c: 145 (q.); Wheeler, G.C. 1943: 327 (l.). Combination in *Eciton:* Latreille, 1804: 179; in *Camptognatha:* Gray, G.R. 1832: 16; in *Eciton:* Smith, F. 1855c: 161. Senior synonym of *curvidentata:* Smith, F. 1858b: 148; of *testacea:* Kempf, 1972a: 102; of *amazona, funesta, mattogrossensis* and *pittieri:* Borgmeier, 1953: 14, 13, 10, 9 repectively. See also: Borgmeier, 1955b: 214; Tafuri, 1957: 21; Wheeler, G.C. & Wheeler, J. 1964c: 134; Wheeler, G.C. & Wheeler, J. 1984: 269.
*hansi* Forel, 1912; see under ***NOMAMYRMEX.***
*haskinsi* Weber, 1943; see under ***NEIVAMYRMEX.***
*hetschkoi* Mayr, 1886; see under ***NEIVAMYRMEX.***
*hirsutipodex* Santschi, 1920; see under ***NOMAMYRMEX.***
*hirsutum. Eciton hirsutum* Santschi, 1912e: 523, fig. 2 (m.) ARGENTINA. Variety of *dulcium,* and senior synonym of *jujuyensis:* Santschi, 1936d: 402. Junior synonym of *dulcium:* Borgmeier, 1948b: 459.
*hospes* Emery, 1906; see under ***NEIVAMYRMEX.***
*hostilis* Santschi, 1920; see under ***LABIDUS.***
*humile* Borgmeier, 1939; see under ***NEIVAMYRMEX.***
*ierense* Weber, 1938; see under ***NEIVAMYRMEX.***
*iheringi* Forel, 1908; see under ***NEIVAMYRMEX.***
*imbellis* Emery, 1900; see under ***NEIVAMYRMEX.***
*impudens* Mann, 1922; see under ***NEIVAMYRMEX.***
*inca* Santschi, 1921; see under ***NEIVAMYRMEX.***
*indomabile* Santschi, 1930; see under ***NOMAMYRMEX.***
*infumatum. Eciton burchelli* var. *infumatum* Wheeler, W.M. 1909b: 231 (s.w.) MEXICO. [Misspelled as *infuscatum* by Santschi, 1925d: 222.] Junior synonym of *parvispinum:* Borgmeier, 1955b: 178.
*integrum. Eciton integrum* Santschi, 1921g: 91 (m.) FRENCH GUIANA. Junior synonym of *vagans:* Borgmeier, 1953: 13.
*interrupta* Santschi, 1923; see under ***NOMAMYRMEX.***
*intrepidum* Santschi, 1929; see under ***NOMAMYRMEX.***
***jansoni.*** *Eciton jansoni* Forel, 1912c: 42 (m.) NICARAGUA. Senior synonym of *velutatus:* Borgmeier, 1953: 11. See also: Borgmeier, 1955b: 270.
*jeanae. Eciton burchelli* subsp. *jeanae* Weber, 1941c: 329 (s.w.q.) GUIANA. Weber, 1943a: 89 (q.). Junior synonym of *cupiens:* Borgmeier, 1955b: 178.
*jerrmanni* Forel, 1901; see under ***NEIVAMYRMEX.***
*jujuyensis. Eciton dulcius* var. *jujuyensis* Santschi, 1921g: 91 (w.) ARGENTINA. [First available use of *Eciton (Eciton) quadriglume* r. *dulcius* var. *jujuyensis* Forel, 1913l: 208; unavailable name.] Bruch, 1923: 177 (q.); Bruch, 1934c: 264 (m.). Junior synonym of *hirsutum:* Santschi, 1936d: 402.
*kulowi* Forel, 1901; see under ***LABIDUS.***
*laevigatum* Borgmeier, 1948; see under ***NEIVAMYRMEX.***
#***latidens.*** *Eciton latidens* Santschi, 1911h: 113, fig. b (m.) FRENCH GUIANA. Subspecies of *mexicanum:* Borgmeier, 1955b: 250.
*latiscapum* Emery, 1901; see under ***NEIVAMYRMEX.***
*legionis* Smith, F. 1855; see under ***NEIVAMYRMEX.***
*lemoulti* Santschi, 1912; see under ***NEIVAMYRMEX.***
*leonardi* Wheeler, W.M. 1915; see under ***NEIVAMYRMEX.***
*leptognathum* Emery, 1900; see under ***NEIVAMYRMEX.***
*levior* Karavaiev, 1926; see under ***AENICTUS.***
*lieselae* Forel, 1913; see under ***NEIVAMYRMEX.***
***lucanoides.*** *Eciton lucanoides* Emery, 1894c: 176, pl.2, fig. 1 (s.) PERU. Current subspecies: nominal plus *conquistador.* See also: Borgmeier, 1955b: 262.
*luederwaldti* Emery, 1911; see under ***NEIVAMYRMEX.***
*lugubris. Ancylognathus lugubris* Lund, 1831a: 121 (w.) BRAZIL. *Nomen nudum.* Combination in *Eciton:* Smith, F. 1855c: 160.
*lugubris. Eciton lugubris* Roger, 1863a: 203 (w.) BRAZIL. Forel, 1886a: 217 (s.). Junior synonym of *quadriglume:* Mayr, 1886b: 118.
*macrodentatus* Menozzi, 1931; see under ***NEIVAMYRMEX.***
*mandibularis* Smith, M.R. 1942; see under ***NEIVAMYRMEX.***
*manni* Wheeler, W.M. 1914; see under ***NEIVAMYRMEX.***
*marginatum* Borgmeier, 1939; see under ***NEIVAMYRMEX.***
*mars* Forel, 1912; see under ***LABIDUS.***
*mattogrossensis. Eciton hamatum* var. *mattogrossensis* Luederwaldt, 1920: 4 (w.) BRAZIL. Santschi, 1925d: 221 (s.m.); Reichensperger, 1926: 401 (q.). Raised to species: Santschi, 1925d: 221. Junior synonym of *hamatum:* Borgmeier, 1953: 10.
*maxillosum* Emery, 1900; see under ***NEIVAMYRMEX.***
*melanocephalum* Emery, 1895; see under ***NEIVAMYRMEX.***
*mesognathum* Santschi, 1920; see under ***NEIVAMYRMEX.***
***mexicanum.*** *Eciton mexicana* Roger, 1863a: 205 (w.) MEXICO. Emery, 1894c: 178 (s.); Borgmeier, 1933a: 92 (q., as *rogeri*); Borgmeier, 1955b: 252 (m.). Senior synonym of *rogeri:* Borgmeier, 1955b: 250. Current subspecies: nominal plus *argentinum, goianum, latidens, moralum, panamense.*
*militarium* Santschi, 1929; see under ***NEIVAMYRMEX.***
*minensis* Borgmeier, 1928; see under ***NEIVAMYRMEX.***
*minutum* Jerdon, 1851; see under ***TETRAPONERA.***
*modestum* Borgmeier, 1933; see under ***NEIVAMYRMEX.***
*modificata* Forel, 1913; see under ***NEIVAMYRMEX.***
*mojave* Smith, M.R. 1943; see under ***NEIVAMYRMEX.***
#***moralum.*** *Eciton (Eciton) latidens* var. *moralus* Santschi, 1923g: 1259 (m.) FRENCH GUIANA. Raised to species: Borgmeier, 1953: 13. Subspecies of *mexicanum:* Borgmeier, 1955b: 250 [misspelled as *morulum*].
*mordax* Santschi, 1929; see under ***NOMAMYRMEX.***
#***mutatum.*** *Eciton vagans* subsp. *mutatum* Borgmeier, 1955b: 242, pl. 11, fig. 7; pl. 14, fig. 8; pl. 19, fig. 12 (s.w.q.m.) COSTA RICA.
*nero* Santschi, 1930; see under ***LABIDUS.***
*nigrita* Emery, 1901; see under ***LABIDUS.***
*nigrum* Jerdon, 1851; see under ***TETRAPONERA.***
*nitens* Mayr, 1868; see under ***NEIVAMYRMEX.***
*nocua* Santschi, 1921; see under ***NEIVAMYRMEX.***
*obscura* Forel, 1914; see under ***NEIVAMYRMEX.***
*obscurum* Forel, 1911; see under ***NEIVAMYRMEX.***
*ogloblini* Santschi, 1934; see under ***NEIVAMYRMEX.***
*opacifrons* Wheeler, W.M. 1921; see under ***LABIDUS.***
*opacithorax* Emery, 1894; see under ***NEIVAMYRMEX.***
*orientalis* Karavaiev, 1926; see under ***AENICTUS.***
*orthonotum* Borgmeier, 1933; see under ***NEIVAMYRMEX.***
*oslari* Wheeler, W.M. 1908; see under ***NEIVAMYRMEX.***
*pachycerus* Smith, F. 1858; see under ***AENICTUS.***
#***panamense.*** *Eciton mexicanum* subsp. *panamense* Borgmeier, 1955b: 257, pl. 55, figs. 5, 9 (s.w.) PANAMA.
*paraense* Forel, 1912; see under ***NEIVAMYRMEX.***
#***parvispinum.*** *Eciton burchelli* var. *parvispina* Forel, 1899c: 23 (w.) GUATEMALA. Subspecies of *burchellii:* Borgmeier, 1953: 8. Senior synonym of *infumatum:* Borgmeier, 1955b: 178.
*pauxillum* Wheeler, W.M. 1903; see under ***NEIVAMYRMEX.***
*payarum* Forel, 1899; see under ***CHELIOMYRMEX.***
*peninsulare* Mann, 1926; see under ***NEIVAMYRMEX.***
*physognatha* Emery, 1900; see under ***NEIVAMYRMEX.***
*pilosa* Smith, F. 1858; see under ***NEIVAMYRMEX.***
*pittieri. Eciton pittieri* Forel, 1899c: 24, pl. 1, fig. 15 (w.) COSTA RICA. Combination in *E. (Labidus):* Emery, 1910b: 23; in *E. (Holopone):* Santschi, 1925b: 11. Junior synonym of *hamatum:* Borgmeier, 1953: 9.
*planidorsum* Emery, 1906; see under ***NEIVAMYRMEX.***
*porrectognathum* Borgmeier, 1933; see under ***NEIVAMYRMEX.***
*postangulatum* Borgmeier, 1934; see under ***NEIVAMYRMEX.***
*praedator* Smith, F. 1858; see under ***LABIDUS.***
*pseudops* Forel, 1909; see under ***NEIVAMYRMEX.***
*punctaticeps* Emery, 1894; see under ***NEIVAMYRMEX.***
***quadriglume.*** *Atta quadriglumis* Haliday, 1836: 328 (w.) BRAZIL. Mayr, 1886b: 122 (s.); Emery, 1896g: 38 (m.); Reichensperger, 1924: 203 (q.). Combination in *Eciton:* Smith, F. 1858b: 151. Junior synonym of *rapax:* Smith, F. 1858b: 151. Revived from synonymy: Roger, 1863b: 36. Senior synonym of *lugubris:* Mayr, 1886b: 118; of *fargeavii:* Emery, 1896g: 38. See also: Borgmeier, 1953: 15; Borgmeier, 1955b: 196.
*quitensis. Eciton hamatum* var. *quitensis* Forel, 1921b: 135 (s.w.) ECUADOR. Junior synonym of *drepanophorum:* Borgmeier, 1953: 9.

**ECITON** *(continued)*

*radoszkowskii* Emery, 1900; see under ***NEIVAMYRMEX.***
***rapax.*** *Eciton rapax* Smith, F. 1855c: 163 (s.) BRAZIL. [w. excluded from type-series, see: Borgmeier, 1955b: 211.] [In the figure legends of Smith, F. 1855c: 169 the name *rapax* is misspelled as *raptor;* incorrect subsequent spelling.] Mayr, 1886b: 118 (w.); Wheeler, G.C. & Wheeler, J. 1974c: 169 (l.); Rettenmeyer, 1974: 297 (q.m.). See also: Borgmeier, 1955b: 211; Wheeler, G.C. & Wheeler, J. 1984: 270.
*raptans* Forel, 1911; see under ***NEIVAMYRMEX.***
*raptor* Smith, F. 1855; see under ***rapax,*** above.
*raptor* Forel, 1911; see under ***NEIVAMYRMEX.***
*rebellatum* Santschi, 1934; see under ***NEIVAMYRMEX.***
*reichenspergeri. Eciton* (*Holopone*) *angustatum* st. *reichenspergeri* Santschi, 1925b: 12 (w.) BRAZIL. Junior synonym of *vagans:* Borgmeier, 1939: 405; Borgmeier, 1953: 13; of *dubitatum:* Borgmeier, 1955b: 229.
*richteri* Forel, 1913; see under ***NEIVAMYRMEX.***
*rogeri. Eciton rogeri* Dalla Torre, 1892: 89. Unnecessary replacement name for *mexicanum* Roger, 1863b: 205. Junior synonym of *mexicanum:* Borgmeier, 1955b: 250.
*rosenbergi* Forel, 1911; see under ***NEIVAMYRMEX.***
*rufipes* Jerdon, 1851; see under ***TETRAPONERA.***
*rufonigrum* Jerdon, 1851; see under ***TETRAPONERA.***
*schlechtendali* Mayr, 1887; see under ***NOMAMYRMEX.***
*schmitti* Emery, 1894; see under ***NEIVAMYRMEX.***
*sedulus* Menozzi, 1926; see under ***LABIDUS.***
*selysi* Forel, 1904; see under ***LABIDUS.***
*serpentis* Weber, 1938; see under ***LABIDUS.***
***setigaster.*** *Eciton setigaster* Borgmeier, 1953: 22 (m.) BRAZIL. See also: Borgmeier, 1955b: 273.
*shuckardi* Emery, 1900; see under ***NEIVAMYRMEX.***
*silvestrii* Emery, 1906; see under ***NEIVAMYRMEX.***
*simillima. Eciton simillima* Smith, F. 1855c: 164 (w.) BRAZIL. Junior synonym of *vagans:* Smith, F. 1858b: 151.
*smithii* Dalla Torre, 1892; see under ***LABIDUS.***
*spatulatum* Borgmeier, 1939; see under ***NEIVAMYRMEX.***
*spegazzinii* Emery, 1888; see under ***NEIVAMYRMEX.***
*spininode* Emery, 1890; see under ***LABIDUS.***
*spoliator* Forel, 1899; see under ***NEIVAMYRMEX.***
*subsulcatum* Mayr, 1886; see under ***NEIVAMYRMEX.***
*sumichrasti* Norton, 1868; see under ***NEIVAMYRMEX.***
*sundaica* Karavaiev, 1926; see under ***AENICTUS.***
*taunayi* Borgmeier, 1928; see under ***NEIVAMYRMEX.***
*tepeguas* Norton, 1868; see under ***LABIDUS.***
*terrificum* Santschi, 1930; see under ***NOMAMYRMEX.***
*testacea. Camptognatha testacea* Gray, G.R. 1832: 516, pl 76, fig. 4 (w.) no locality given. Junior synonym of *hamatum:* Kempf, 1972a: 102.
*testaceum* Schenck, 1852; see under ***STRONGYLOGNATHUS.***
*tristani* Borgmeier, 1939; see under ***NEIVAMYRMEX.***
*triste* Forel, 1901; see under ***NEIVAMYRMEX.***
*truncatidens* Santschi, 1920; see under ***LABIDUS.***
***uncinatum.*** *Eciton uncinatum* Borgmeier, 1953: 21, fig. 1 (m.) ECUADOR. See also: Borgmeier, 1955b: 275.
#***urichi.*** *Eciton burchelli* r. *urichi* Forel, 1899c: 23 (w.) TRINIDAD. Weber, 1941c: 329 (q.). Junior synonym of *foreli:* Borgmeier, 1939: 404. Revived from synonymy as subspecies of *burchellii:* Weber, 1941c: 329. See also: Borgmeier, 1955b: 190.
*ursinum* Emery, 1901; see under ***CHELIOMYRMEX.***
***vagans.*** *Formica vagans* Olivier, 1792: 501 (w.) FRENCH GUIANA. Smith, F. 1855c: 162 (s.); Wheeler, W.M. 1912e: 206 (m.); Menozzi, 1931d: 188 (q.); Wheeler, G.C. & Wheeler, J. 1974c: 170 (l.). Combination in *Eciton:* Smith, 1855c: 162. Senior synonym of *simillima:* Smith, F. 1858b: 151; of *integrum:* Borgmeier, 1953: 13. Current subspecies: nominal plus *allognathum, angustatum, dispar, dubitatum, fur, mutatum.* See also: Borgmeier, 1955b: 229; Wheeler, G.C. & Wheeler, J. 1984: 270.
*vastator* Smith, F. 1860; see under ***LABIDUS.***
*velutatus. Eciton* (*Eciton*) *velutatus* Menozzi, 1926d: 30, fig. 1 (m.) COSTA RICA. Junior synonym of *jansoni:* Borgmeier, 1953: 11.
*viator. Eciton burchelli* var. *viator* Santschi, 1925d: 221 (s.w.) COLOMBIA. Junior synonym of *foreli:* Borgmeier, 1936: 54.
*westwoodi* Emery, 1900; see under ***LABIDUS.***
*wheeleri* Emery, 1901; see under ***NEIVAMYRMEX.***
*wilsoni* Santschi, 1920; see under ***NOMAMYRMEX.***
*xipe* Wheeler, W.M. 1914; see under ***NEIVAMYRMEX.***

**ECPHORELLA** (Afrotropical)

***wellmani.*** *Tapinoma* (*Ecphorella*) *wellmani* Forel, 1909b: 66 (w.) ANGOLA. Combination in *Ecphorella:* Shattuck, 1992c: 85.

**ECTATOMMA** (Neotropical)

*aciculata* Smith, F. 1858; see under ***RHYTIDOPONERA.***
*acrista. Ectatomma tuberculatum* subsp. *acrista* Forel, 1909a: 254 (w.q.) PARAGUAY. Junior synonym of *tuberculatum:* Weber, 1946a: 2. Revived from synonymy, and raised to species: Brown, 1958g: 208. Junior synonym of *tuberculatum:* Kempf, 1962b: 2.
*aculeaticoxae* Santschi, 1921; see under ***GNAMPTOGENYS.***
*acuminatum* Emery, 1896; see under ***GNAMPTOGENYS.***
*aerea. Ectatomma strigosum* var. *aerea* Forel, 1912c: 32 (w.) BRAZIL. Junior synonym of *permagnum:* Brown, 1958g: 209.
*alfaroi* Emery, 1894; see under ***GNAMPTOGENYS.***
*angustiloba* Forel, 1908; see under ***GNAMPTOGENYS.***
*angustipleura* Forel, 1908; see under ***GNAMPTOGENYS.***
*annulata* Mayr, 1887; see under ***GNAMPTOGENYS.***
*aztecum. Ectatomma aztecum* Emery, 1901a: 50 (w.) MEXICO. Junior synonym of *ruidum:* Kugler, C. & Brown, 1982: 2. See also: Brown, 1958g: 295.
*bicolor* Emery, 1889; see under ***GNAMPTOGENYS.***
*binghamii* Forel, 1900; see under ***GNAMPTOGENYS.***
*bispinosum* Emery, 1890; see under ***GNAMPTOGENYS.***
*brouni* Forel, 1892; see under ***HETEROPONERA.***
*bruchi* Santschi, 1922; see under ***GNAMPTOGENYS.***
***brunneum.*** *Ectatomma brunnea* Smith, F. 1858b: 103 (w.q.) BRAZIL. Junior synonym of *quadridens:* Roger, 1860: 307, and hence first available replacement name for *Formica quadridens* Fabricius, 1793: 362. [Junior primary homonym of *Formica quadridens* De Geer, 1773: 609 (now in *Cephalotes*).]
*caro* Forel, 1910; see under ***RHYTIDOPONERA.***
*cearensis* Forel, 1912; see under ***GNAMPTOGENYS.***
*concinna* Smith, F. 1858; see under ***GNAMPTOGENYS.***
*concolor. Ectatomma opaciventre* var. *concolor* Santschi, 1919f: 37, fig. 1 (w.) ARGENTINA. Santschi, 1921g: 83 (q.). Junior synonym of *opaciventre:* Brown, 1958g: 208.
***confine.*** *Ectatomma confine* Mayr, 1870a: 397 (w.) COLOMBIA. See also: Brown, 1958g: 295; Kugler, C. & Brown, 1982: 2.
*confusa. Ectatomma strigosum* var. *confusa* Forel, 1909a: 266 (w.) PARAGUAY. Junior synonym of *permagnum:* Brown, 1958g: 209.
*continuum* Mayr, 1887; see under ***GNAMPTOGENYS.***
*convexum* Mayr, 1876; see under ***RHYTIDOPONERA.***
*cornutum* Emery, 1895; see under ***RHYTIDOPONERA.***
*costatum* Emery, 1889; see under ***GNAMPTOGENYS.***
*crassicorne* Forel, 1912; see under ***GNAMPTOGENYS.***
*crassinode* Forel, 1907; see under ***RHYTIDOPONERA.***
*cristatum* Mayr, 1876; see under ***RHYTIDOPONERA.***
*cristulatum* Forel, 1900; see under ***RHYTIDOPONERA.***
*curtulum* Emery, 1896; see under ***GNAMPTOGENYS.***
*cyrus* Forel, 1910; see under ***RHYTIDOPONERA.***
*densestriata. Ectatomma edentatum* r. *muticum* var. *densestriata* Forel, 1912c: 31 (w.) BRAZIL. Unavailable name; material referred to *edentatum* by Brown, 1958g: 208.
*dentinode* Mayr, 1887; see under ***HETEROPONERA.***
*diehlii* Forel, 1901; see under ***GNAMPTOGENYS.***
***edentatum.*** *Ectatomma edentatum* Roger, 1863a: 173 (w.) ARGENTINA. Emery, 1901c: 199 (l.); Almeida, 1987: 175 (q.). Senior synonym of *inversa* and *iris,* and material of the unavailable name *densestriata* referred here: Brown, 1958g: 208; senior synonym of *macdonaghi* and *morgani:* Kugler, C. & Brown, 1982: 4.
*ericae* Forel, 1912; see under ***GNAMPTOGENYS.***
**europaeum* Mayr, 1868; see under ***GNAMPTOGENYS.***
*exaratum* Emery, 1901; see under ***GNAMPTOGENYS.***
*ferrugineus. Ectatomma ferrugineus* Norton, 1868a: 61, pl. 2, fig. 4 (w.m.) MEXICO. Norton, 1968c: 5 (w.m., descriptions). Junior synonym of *tuberculatum:* Emery, 1892b: 167.
*fiebrigi* Forel, 1909; see under ***GNAMPTOGENYS.***
*froggatti* Forel, 1910; see under ***RHYTIDOPONERA.***
*fulgens* Emery, 1883; see under ***RHYTIDOPONERA.***
*gemma* Forel, 1907; see under ***RHYTIDOPONERA.***
***gibbum.*** *Ectatomma gibbum* Kugler, C. & Brown, 1982: 4, fig. 1 (w.) COSTA RICA.
*glabrius* Forel, 1907; see under ***RHYTIDOPONERA.***
***goninion.*** *Ectatomma goninion* Kugler, C. & Brown, 1982: 5, fig. 1 (w.) COLOMBIA.
****gracile.*** **Ectatomma gracile* Emery, 1891a: 145, pl. 1, figs. 1, 2 (m.) SICILIAN AMBER. [Probably not correct generic combination, hence *incertae sedis* in *Ectatomma.*]
*haeckeli* Forel, 1910; see under ***RHYTIDOPONERA.***
*haenschi* Emery, 1902; see under ***GNAMPTOGENYS.***
*hartmani* Wheeler, W.M. 1915; see under ***GNAMPTOGENYS.***
*hilare* Forel, 1895; see under ***HETEROPONERA.***
*horni* Santschi, 1929; see under ***GNAMPTOGENYS.***
*impressum* Mayr, 1876; see under ***RHYTIDOPONERA.***

*inerme* Emery, 1894; see under ***HETEROPONERA.***
*interruptum* Mayr, 1887; see under ***GNAMPTOGENYS.***
*inversa. Ectatomma edentatum* st. *inversa* Santschi, 1912e: 521 (w.) URUGUAY. Junior synonym of *edentatum:* Brown, 1958g: 208.
*iris. Ectatomma edentatum* var. *iris* Forel, 1909a: 253 (w.m.) PARAGUAY. Junior synonym of *edentatum:* Brown, 1958g: 208.
*irregularis. Ectatomma tuberculatum* var. *irregularis* Santschi, 1921g: 83 (w.) COLOMBIA. Junior synonym of *tuberculatum:* Weber, 1946a: 2.
*laevius* Forel, 1905; see under ***GNAMPTOGENYS.***
*lobulifera. Ectatomma muticum* var. *lobulifera* Forel, 1909a: 254 (w.) BRAZIL. Race of *edentatum:* Forel, 1912c: 31. Junior synonym of *muticum:* Brown, 1958g: 208.
***lugens.*** *Ectatomma opaciventre* var. *lugens* Emery, 1894c: 144 (w.) BRAZIL. Almeida, 1987: 177 (q.). Variety of *strigosum:* Forel, 1909a: 265. Raised to species: Brown, 1958g: 208.
*macdonaghi. Ectatomma edentatum* r. *macdonaghi* Forel, 1915c: 351 (w.) ARGENTINA. Raised to species: Brown, 1958g: 208. Junior synonym of *edentatum:* Kugler, C. & Brown, 1982: 4.
*maniae* Forel, 1900; see under ***RHYTIDOPONERA.***
*mayri* Emery, 1883; see under ***RHYTIDOPONERA.***
*menadensis* Mayr, 1887; see under ***GNAMPTOGENYS.***
*minor* Forel, 1899; see under ***ACANTHOPONERA.***
*minor* Forel, 1900; see under ***GNAMPTOGENYS.***
*modestum* Emery, 1895; see under ***RHYTIDOPONERA.***
*morgani. Ectatomma morgani* Forel, 1912c: 31 (w.) PERU. Junior synonym of *edentatum:* Kugler, C. & Brown, 1982: 4. See also: Brown, 1958g: 297.
***muticum.*** *Ectatomma muticum* Mayr, 1870b: 962 (w.) BRAZIL. Almeida, 1986: 26 (m.); Almeida, 1987: 176 (q.). Race of *edentatum:* Forel, 1912c: 31. Revived status as species, and senior synonym of *lobulifera:* Brown, 1958g: 208.
*nigrifrons* Borgmeier, 1948; see under ***GNAMPTOGENYS.***
*nitens* Mann, 1916; see under ***GNAMPTOGENYS.***
*nodiferum* Emery, 1895; see under ***RHYTIDOPONERA.***
*nudatum* Mayr, 1876; see under ***RHYTIDOPONERA.***
*numeensis* André, 1889; see under ***RHYTIDOPONERA.***
*obscurum* Emery, 1896; see under ***GNAMPTOGENYS.***
*obscurum* Forel, 1900; see under ***RHYTIDOPONERA.***
***opaciventre.*** *Ponera* (*Ectatomma*) *opaciventris* Roger, 1861b: 169 (w.) BRAZIL. Mayr, 1866a: 503 (q.); Almeida, 1986: 27 (m.). Combination in *Ectatomma:* Emery, 1894c: 144. Senior synonym of *concolor:* Brown, 1958g: 208.
*panamense* Forel, 1899; see under ***HETEROPONERA.***
*panamensis* Santschi, 1931; see under ***GNAMPTOGENYS.***
*paulina* Forel, 1908; see under ***GNAMPTOGENYS.***
***permagnum.*** *Ectatomma opaciventre* var. *permagna* Forel, 1908c: 342 (w.) BRAZIL. Santschi, 1921g: 83 (q.); Almeida, 1986: 28 (m.). Variety of *strigosum:* Forel, 1911c: 286. Raised to species, first available replacement name for *strigosum* Emery, 1894c: 144 [Junior primary homonym of *strigosum* Emery, 1887b: 444.], and senior synonym of *aerea* and *confusa:* Brown, 1958g: 209. Senior synonym of *regis:* Kempf, 1962b: 4; Kempf, 1975c: 343. See also: Almeida, 1987: 178.
***planidens.*** *Ectatomma planidens* Borgmeier, 1939: 418 (w.) BRAZIL. Kempf, 1960f: 420 (q.). See also: Brown, 1958g: 298; Kempf, 1960f: 418.
*pleurodon* Emery, 1896; see under ***GNAMPTOGENYS.***
*porcatum* Emery, 1896; see under ***GNAMPTOGENYS.***
*pulchellum* Emery, 1883; see under ***RHYTIDOPONERA.***
*punctata* Smith, F. 1858; see under ***RHYTIDOPONERA.***
*punctigerum. Ectatomma tuberculatum* var. *punctigerum* Emery, 1890a: 56 (w.) VENEZUELA. Lever, 1930: 214 (q.). Junior synonym of *tuberculatum:* Weber, 1946a: 2.
*punctiventris* Forel, 1900; see under ***RHYTIDOPONERA.***
*purensis* Forel, 1912; see under ***GNAMPTOGENYS.***
*purpureum* Emery, 1887; see under ***RHYTIDOPONERA.***
*quadridens. Formica quadridens* Fabricius, 1793: 362 (w.) FRENCH GUIANA. [Junior primary homonym of *Formica quadridens* De Geer, 1773: 609 (now in *Cephalotes*).] Junior synonym and hence first available replacement name: *Ectatomma brunnea* Smith, F. 1858b: 103. Mann, 1916: 405 (m.); Santschi, 1921g: 83 (q.); Wheeler, G.C. & Wheeler, J. 1952a: 131 (l.). Combination in *Ectatomma:* Mayr, 1862: 732. Senior synonym of *brunnea:* Roger, 1860: 307.
*rastratum* Mayr, 1866; see under ***GNAMPTOGENYS.***
*regis. Ectatomma regis* Kusnezov, 1957b: 12 (w.) ARGENTINA. Junior synonym of *permagnum:* Kempf, 1975c: 343.
*reticulatum* Forel, 1893; see under ***RHYTIDOPONERA.***
*richteri* Forel, 1913; see under ***GNAMPTOGENYS.***
*romani* Wheeler, W.M. 1923; see under ***GNAMPTOGENYS.***
*rothneyi* Forel, 1900; see under ***RHYTIDOPONERA.***
*rufescens* Forel, 1900; see under ***RHYTIDOPONERA.***
*rugosa* Smith, F. 1859; see under ***RHYTIDOPONERA.***
***ruidum.*** *Ponera* (*Ectatomma*) *ruida* Roger, 1860: 306 (w.q.) BRAZIL. Borgmeier, 1934: 95 (m.); Wheeler, G.C. & Wheeler, J. 1952a: 131 (l.). Combination in *Ectatomma:* Mayr, 1862: 732. Senior synonym of *scabrosa:* Roger, 1862c: 292; of *aztecum:* Kugler, C. & Brown, 1982: 2.
*satzgeri* Forel, 1908; see under ***GNAMPTOGENYS.***
*scaberrimum* Emery, 1895; see under ***RHYTIDOPONERA.***
*scabrosa. Ectatomma scabrosa* Smith, F. 1862b: 31 (w.) PANAMA. Junior synonym of *ruidum:* Roger, 1862c: 292.
*scabrum* Mayr, 1876; see under ***RHYTIDOPONERA.***
*schubarti* Borgmeier, 1948; see under ***GNAMPTOGENYS.***
*scrobiculatum* Forel, 1900; see under ***RHYTIDOPONERA.***
*sebastiani* Borgmeier, 1937; see under ***GNAMPTOGENYS.***
*simplex* Emery, 1896; see under ***GNAMPTOGENYS.***
*simplicoides* Forel, 1908; see under ***GNAMPTOGENYS.***
*socialis* MacLeay, 1873; see under ***PACHYCONDYLA.***
*socrus* Forel, 1894; see under ***RHYTIDOPONERA.***
*spatiatum* Forel, 1900; see under ***RHYTIDOPONERA.***
*splendidum* Pergande, 1895; see under ***GNAMPTOGENYS.***
*splendidum* Forel, 1910; see under ***RHYTIDOPONERA.***
*spoliatum* Emery, 1895; see under ***RHYTIDOPONERA.***
*spurium* Forel, 1908; see under ***GNAMPTOGENYS.***
*stolli* Forel, 1899; see under ***GNAMPTOGENYS.***
*strigosum* Emery, 1887; see under ***RHYTIDOPONERA.***
*strigosum. Ectatomma opaciventre* var. *strigosum* Emery, 1894c: 144 (w.) BOLIVIA. Raised to species: Forel, 1909a: 265. [Junior primary homonym of *strigosum* Emery, 1887b: 444, above.] First available replacement name: *permagnum* Forel, 1908c: 342.
***suzanae.*** *Ectatomma suzanae* Almeida, 1986: 24 (m.) BRAZIL. Almeida, 1987: 179 (q.).
*taurus* Forel, 1910; see under ***RHYTIDOPONERA.***
*tenue* Forel, 1900; see under ***RHYTIDOPONERA.***
*triangulare* Mayr, 1887; see under ***GNAMPTOGENYS.***
*tridentata. Formica tridentata* Fabricius, 1804: 412 (w.) "America meridionali". Junior synonym of *tuberculatum:* Smith, F. 1858b: 102.
*trigonum* Emery, 1906; see under ***GNAMPTOGENYS.***
***tuberculatum.*** *Formica tuberculata* Olivier, 1792: 498 (w.q.) TRINIDAD. Smith, F. 1858b: 102 (m.); Wheeler, G.C. & Wheeler, J. 1952a: 129 (l.). Combination in *Ectatomma:* Smith, F. 1858b: 102. Senior synonym of *tridentata:* Smith, F. 1858b: 102; of *ferrugineus:* Emery, 1892b: 167; of *irregularis* and *punctigerum:* Weber, 1946a: 2; of *acrista:* Kempf, 1962b: 2.
*turneri* Forel, 1910; see under ***RHYTIDOPONERA.***
*victoriae* André, 1896; see under ***RHYTIDOPONERA.***
*violaceum* Forel, 1907; see under ***RHYTIDOPONERA.***
***vizottoi.*** *Ectatomma vizottoi* Almeida, 1987: 180 (q.) no locality given.

*ECTOMOMYRMEX:* see under ***PACHYCONDYLA.***

****ELAEOMYRMEX*** (U.S.A.)
***coloradensis.** **Elaeomyrmex coloradensis* Carpenter, 1930: 49, pl. 3, fig. 2; pl. 9, fig. 2 (w.q.) U.S.A. (Miocene).
***gracilis.** **Elaeomyrmex gracilis* Carpenter, 1930: 48, pl. 3, fig. 7; pl. 6, fig. 2; pl. 11, fig. 1 (w.q.) U.S.A. (Miocene).

*ELASMOPHEIDOLE:* see under ***PHEIDOLE.***

****ELECTROMYRMEX*** (Baltic Amber)
***klebsi.** **Electromyrmex klebsi* Wheeler, W.M. 1910g: 164, fig. 94 (by indication) (w.) BALTIC AMBER (Oligocene). See also: Wheeler, W.M. 1915h: 56 (full description).

*ELECTROPHEIDOLE:* see under ***PHEIDOLE.***

****ELECTROPONERA*** (Baltic Amber)
***dubia.** **Electroponera dubia* Wheeler, W.M. 1915h: 35, fig. 6 (w.) BALTIC AMBER (Oligocene).

*EMERYELLA:* see under ***GNAMPTOGENYS.***

*EMERYIA:* see under ***CARDIOCONDYLA.***

***EMERYOPONE*** (East Mediterranean to Indonesia)
***buttelreepeni.*** *Emeryopone buttelreepeni* Forel, 1912m: 762 (w.) SUMATRA. [Also described as new by Forel, 1913k: 15.] Combination in *Belonopelta:* Baroni Urbani, 1975b: 303; in *Emeryopone:* **revived combination**.
***franzi.*** *Belonopelta franzi* Baroni Urbani, 1975b: 305, figs. 4, 6 (w.) NEPAL. Combination in *Emeryopone:* **new combination**.

***EMERYOPONE*** *(continued)*

***loebli.*** *Belonopelta loebli* Baroni Urbani, 1975b: 307, figs. 4, 7 (w.) ISRAEL. Combination in *Emeryopone:* **new combination**.

***EMPLASTUS*** (Great Britain)
****emeryi.*** **Emplastus emeryi* Donisthorpe, 1920: 86, pl. 5, fig. 5 (q.) GREAT BRITAIN (Oligocene).

*ENDIODIOCTES:* see under ***MYRMECOCYSTUS.***

*ENERIA:* see under ***STRUMIGENYS.***

*ENGRAMMA:* see under ***TECHNOMYRMEX.***

****ENNEAMERUS*** (Baltic Amber)
****reticulatus.*** **Enneamerus reticulatus* Mayr, 1868c: 100, pl. 5, figs. 102, 103 (w.) BALTIC AMBER (Oligocene).

****EOCENIDRIS*** (U.S.A.)
****crassa.*** **Eocenidris crassa* Wilson, 1985d: 209, fig. 3 (w.) U.S.A. (Eocene).

****EOFORMICA*** (U.S.A.)
**eocenica. *Eoformica eocenica* Cockerell, 1921: 38, fig. 9, pl. 8, fig. 11 (m.) U.S.A. (Eocene). Junior synonym of *pinguis:* Carpenter, 1930: 17.
****pinguis.*** **Liometopum pingue* Scudder, 1877a: 742 (m.) U.S.A. (Eocene). Combination in *Eoformica,* and senior synonym of *eocenica:* Carpenter, 1930: 17.

*EOMONOCOMBUS:* see under ***CATAGLYPHIS.***

****EOMYRMEX*** (China)
****guchengziensis.*** **Eomyrmex guchengziensis* Hong, 1974: 138 [English translation p. 147] (q.) CHINA (Eocene). Spelling here emended from *guchengzienis.*

**EOPONERA:* see under ****FORMICIUM.***

****EOTAPINOMA*** (Russia)
****compacta.*** **Eotapinoma compacta* Dlussky, 1988: 55, fig. 3 (w.) RUSSIA (Palaeocene?).
****gracilis.*** **Eotapinoma gracilis* Dlussky, 1988: 55, fig. 3 (w.) RUSSIA (Palaeocene?).

***EPELYSIDRIS*** (Borneo)
***brocha.*** *Epelysidris brocha* Bolton, 1987: 280, figs. 16, 17 (w.) BORNEO.

*EPHEBOMYRMEX:* see under ***POGONOMYRMEX.***

*EPIATTA:* see under ***ATTA.***

***EPIMYRMA*** (Palaearctic)
***adlerzi.*** *Epimyrma adlerzi* Douwes, Jessen & Buschinger, 1988: 240, figs. 2–5 (q.m.) GREECE.
***africana.*** *Epimyrma* (*Gonepimyrma*) *africana* Bernard, 1948: 148, fig. 7 (w.) ALGERIA.
***algeriana.*** *Epimyrma algeriana* Cagniant, 1968c: 157, figs. 1–4 (w.q.m.) ALGERIA. See also: Kutter, 1973e: 281; Buschinger, Jessen & Cagniant, 1990: 23.
***bernardi.*** *Epimyrma bernardi* Espadaler, 1982a: 1, figs. 1–11 (w.q.) SPAIN.
***corsica.*** *Formicoxenus corsicus* Emery, 1895b: 68, fig. 2 (q.) CORSICA. Buschinger & Winter, 1985: 70 (m.). Combination in *Epimyrma:* Emery, 1915a: 262. See also: Menozzi, 1931c: 48; Kutter, 1973e: 281.
*foreli. Epimyrma foreli* Menozzi, 1921: 29, fig. 2 (w.q.) ITALY. Junior synonym of *kraussei:* Buschinger, Fischer, *et al.* 1987: 272. See also: Menozzi, 1931c: 45; Kutter, 1973e: 281.
*goesswaldi. Epimyrma goesswaldi* Menozzi, in Gösswald, 1930: 464, figs. 1–3 (w.q.m.) GERMANY. [Also described as new by Menozzi, 1931c: 41.] Junior synonym of *ravouxi:* Buschinger, 1982: 352. See also: Stitz, 1939: 186; Kutter, 1973e: 281.
***gordiagini.*** *Myrmoxenus gordiagini* Ruzsky, 1902b: 475, figs. A-H (1–8) (w.q.m.) RUSSIA. Senior synonym of *menozzii,* and provisional senior synonym of *microcellatus:* Buschinger, Winter & Faber, 1984: 336. See also: Kupyanskaya, 1990: 145.
***kraussei.*** *Epimyrma kraussei* Emery, 1915a: 262, pl. 4, figs. 14–16 (w.q.) SARDINIA. Senior synonym of *foreli* and *vandeli:* Buschinger, Fischer, *et al.* 1987: 272. See also: Menozzi, 1931c: 39; Kutter, 1973e: 281.
*menozzii. Myrmoxenus gordiagini* subsp. *menozzii* Finzi, 1924b: 120 (q.m.) ITALY. Junior synonym of *gordiagini:* Buschinger, Winter & Faber, 1984: 336.
*microcellatus. Myrmetaerus microcellatus* Soudek, 1925a: 33, figs. 1–6 (w.q.) YUGOSLAVIA. [Also described as new by Soudek, 1925b: 14.] Provisional junior synonym of *gordiagini:* Buschinger, Winter & Faber, 1984: 336.
***ravouxi.*** *Formicoxenus ravouxi* André, 1896a: 367 (q.) FRANCE. Combination in *Epimyrma:* Emery, 1915a: 262. Senior synonym of *goesswaldi:* Buschinger, 1982: 352. See also: Menozzi, 1931c: 44; Kutter, 1973e: 281.
***stumperi.*** *Epimyrma stumperi* Kutter, 1950b: 340, figs. 4–7 (m.) SWITZERLAND. Kutter, 1951: 153 (w.q.). See also: Stumper & Kutter, 1951: 983; Kutter, 1973e: 281.
***tamarae.*** *Epimyrma tamarae* Arnol'di, 1968: 1811, fig. 11 (w.) GEORGIA.
*vandeli. Epimyrma vandeli* Santschi, 1927b: 126 (q.m.) FRANCE. [m. described but not named: Vandel, 1926: 198.] See also: Menozzi, 1931c: 47; Kutter, 1973e: 281. Junior synonym of *kraussei:* Buschinger, Fischer, *et al.* 1987: 272.
***zaleskyi.*** *Epimyrma zaleskyi* Sadil, 1953a: 188, figs. 1, 2 (w.q.) CZECHOSLOVAKIA. See also: Kutter, 1973e: 281.

*EPIPHEIDOLE:* see under ***PHEIDOLE.***

***EPITRITUS*** (Old World tropics and temperates)
***argiolus.*** *Epitritus argiolus* Emery, 1869a: 136, fig. 1 (q.) ITALY. Emery, 1875a: 75 (w.); Emery, 1875b: 473 (w.); André, 1883b: 404 (m.). Senior synonym of *barbara:* Brown, 1949b: 44. See also: Kutter, 1977c: 168.
*barbara. Epitritus argiolus* var. *barbara* Santschi, 1923a: 136 (w.) TUNISIA. Junior synonym of *argiolus:* Brown, 1949b: 44.
*baudueri* Emery, 1875; see under ***SMITHISTRUMA.***
*clypeatus* Szabó, 1909; see under ***QUADRISTRUMA.***
*elliotti* Clark, 1928; see under ***COLOBOSTRUMA.***
*emmae* Emery, 1890; see under ***QUADRISTRUMA.***
*eurycerus* Emery, 1897; see under ***QUADRISTRUMA.***
***hexamerus.*** *Epitritus hexamerus* Brown, 1958d: 70, figs. 1–3 (w.q.) JAPAN.
***hirashimai.*** *Epitritus hirashimai* Ogata, 1990: 197, figs. 1–3 (w.) JAPAN.
***laticeps.*** *Epitritus laticeps* Brown, 1962b: 77, figs. 1–4 (w.) NIGERIA. See also: Bolton, 1983: 355.
*malesiana* Forel, 1913; see under ***QUADRISTRUMA.***
*mandibularis* Szabó, 1909; see under ***SMITHISTRUMA.***
*marginatus* Santschi, 1914; see under ***SMITHISTRUMA.***
***minimus.*** *Epitritus minimus* Bolton, 1972: 205, figs. 1, 2 (w.) GHANA. See also: Bolton, 1983: 355.
***murphyi.*** *Epitritus murphyi* Taylor, 1968d: 130, figs. 1, 2 (w.) WEST MALAYSIA.
***roomi.*** *Epitritus roomi* Bolton, 1972: 206, figs. 3, 4 (w.q.) GHANA. See also: Bolton, 1983: 356.
*simulans* Santschi, 1931; see under ***CODIOXENUS.***
***tiglath.*** *Epitritus tiglath* Bolton, 1983: 357 (w.) IVORY COAST.
*wheeleri* Donisthorpe, 1916; see under ***QUADRISTRUMA.***

*EPIXENUS:* see under ***MONOMORIUM.***

*EPOECUS:* see under ***MONOMORIUM.***

***EPOPOSTRUMA*** (Australia)
*alata* Wheeler, G.C. & Wheeler, J. 1973; see under ***COLOBOSTRUMA.***
*alinodis* Forel, 1913; see under ***COLOBOSTRUMA.***
*ferruginea. Epopostruma quadrispinosa* subsp. *ferruginea* Forel, 1910b: 51 (q.) AUSTRALIA. Raised to species: Brown, 1948e: 120. Junior synonym of *quadrispinosa:* Taylor, 1991b: 602.
*foliacea* Emery, 1897; see under ***COLOBOSTRUMA.***
*froggatti* Forel, 1913; see under ***COLOBOSTRUMA.***
***frosti.*** *Hexadaceton frosti* Brown, 1948e: 120, fig. 2 (w.) AUSTRALIA. Wheeler, G.C. & Wheeler, J. 1973c: 211 (l.). Combination in *Epopostruma:* Taylor, 1973: 26.
*leae* Wheeler, W.M. 1927; see under ***COLOBOSTRUMA.***
***monstrosa.*** *Epopostruma monstrosa* Viehmeyer, 1925a: 30 (q.) AUSTRALIA. Combination in *Mesostruma:* Brown, 1952g: 13; in *Epopostruma:* Taylor, 1973: 25. See also: Taylor, 1991b: 602.
***quadrispinosa.*** *Strumigenys* (*Epopostruma*) *quadrispinosa* Forel, 1895f: 422 (w.) AUSTRALIA. Wheeler, G.C. & Wheeler, J. 1973c: 212 (l.). Combination in *Epopostruma:* Forel, 1910b: 51. Senior synonym of *ferruginea:* Taylor, 1991b: 602. See also: Brown, 1948e: 120.

*EQUESTRIMESSOR:* see under ***MONOMORIUM.***

*EREBOMYRMA:* see under ***OLIGOMYRMEX.***

*EREMNOCYSTUS:* see under ***MYRMECOCYSTUS.***

*ERICAPELTA:* see under ***AMBLYOPONE.***

*ERIMELOPHORUS:* see under ***MELOPHORUS.***

*ERIOPHEIDOLE:* see under ***PHEIDOLE.***

*ESCHERICHIA:* see under ***PROBOLOMYRMEX.***

*EUBOTHROPONERA:* see under ***PLATYTHYREA.***

***#EUCREMA:*** see under ***CREMATOGASTER.***

***EUCRYPTOCERUS*** (Neotropical)

***abdominalis.*** *Cephalotes abdominalis* Santschi, 1929d: 302 (w.) FRENCH GUIANA. Combination in *Eucryptocerus:* Kempf, 1951: 129.

*aethiops. Cryptocerus aethiops* Smith, F. 1853: 216, pl. 20, fig. 9 (w.) BRAZIL. Junior synonym of *oculatus:* Smith, F. 1858b: 188.

*fenestralis. Cryptocerus fenestralis* Smith, F. 1876d: 607 (q.) BRAZIL. Combination in *Paracryptocerus:* Kempf, 1951: 232. Junior synonym of *placidus:* Kempf, 1963c: 437.

***oculatus.*** *Cryptocerus oculatus* Spinola, 1851b: 49 (w.) BRAZIL. [Also described as new by Spinola, 1853: 65.] Combination in *Cephalotes:* Emery, 1914c: 39; in *Eucryptocerus:* Kempf, 1951: 130. Senior synonym of *aethiops:* Smith, F. 1858b: 188.

*opacus. Cephalotes opacus* Santschi, 1920f: 147 (w.) GUIANA. Combination in *Eucryptocerus:* Kempf, 1951: 131. Junior synonym of *placidus:* Kempf, 1959a: 92.

***placidus.*** *Cryptocerus placidus* Smith, F. 1860c: 76 (m.) BRAZIL. Snelling, 1966: 38 (w.q.). Combination in *Cephalotes:* Emery, 1922f: 304; in *Eucryptocerus:* Kempf, 1959a: 92. Senior synonym of *opacus:* Kempf, 1959a: 92; of *fenestralis:* Kempf, 1963c: 437.

*****EULITHOMYRMEX*** (U.S.A.)

****rugosus.*** **Lithomyrmex rugosus* Carpenter, 1930: 35, pl. 5, figs. 1, 3; pl. 8, fig. 2; pl. 11, fig. 3 (q.m.) U.S.A. (Miocene). Combination in *Eulithomyrmex:* Carpenter, 1935: 91.

****striatus.*** **Lithomyrmex striatus* Carpenter, 1930: 36, pl. 6, fig. 1 (w.q.) U.S.A. (Miocene). Combination in *Eulithomyrmex:* Carpenter, 1935: 91.

*EUMECOPONE:* see under ***PACHYCONDYLA.***

*EUOPHTHALMA:* see under **SOLENOPSIS.**

*EUPONERA:* see under ***PACHYCONDYLA.***

***EUPRENOLEPIS*** (Indo-Australian)

***antespectans.*** *Camponotus (Myrmosphincta) antespectans* Forel, 1913k: 130, fig. MI (w.) SUMATRA. Combination in *Paratrechina (Euprenolepis)*: Emery, 1925b: 223; in *Euprenolepis:* **new combination**.

*geei* Brown, 1953; see under ***PRENOLEPIS*** (*P. emmae*).

***helleri.*** *Prenolepis (Euprenolepis) helleri* Viehmeyer, 1914c: 41, fig. 6 (w.q.m.) NEW GUINEA. Combination in *Paratrechina (Euprenolepis)*: Emery, 1925b: 224; in *Euprenolepis:* **new combination**.

***negrosensis.*** *Chapmanella negrosensis* Wheeler, W.M. 1930d: 42, fig. 1 (w.) PHILIPPINES. Combination in *Euprenolepis:* Brown, 1953h: 6.

***procera.*** *Prenolepis procera* Emery, 1900d: 699, fig. 14 (w.) SUMATRA. Combination in *Prenolepis (Euprenolepis)*: Emery, 1906b: 134; in *Paratrechina (Euprenolepis)*: Emery, 1925b: 224; in *Euprenolepis:* **new combination**.

***steeli.*** *Prenolepis (Nylanderia) steeli* Forel, 1910b: 69 (w.) NEW GUINEA. Combination in *Paratrechina (Euprenolepis)*: Emery, 1925b: 224; in *Euprenolepis:* **new combination**.

***stigmatica.*** *Prenolepis (Nylanderia) stigmaticus* Mann, 1919: 367, fig. 41 (w.) SOLOMON IS. Combination in *Paratrechina (Nylanderia)*: Emery, 1925b: 221; in *Euprenolepis:* Donisthorpe, 1941c: 42.

***EURHOPALOTHRIX*** (U.S.A., Neotropical, Australasian, Indo-Australian)

***alopeciosa.*** *Eurhopalothrix alopeciosa* Brown & Kempf, 1960: 206 (w.) TRINIDAD.

*amoena. Rhopalothrix (Rhopalothrix) amoena* Mann, 1922: 39 (w.q.) HONDURAS. Wheeler, G.C. & Wheeler, J. 1960b: 25 (l.). Junior synonym of *bolaui:* Brown & Kempf, 1960: 210.

*angulinodis. Rhopalothrix angulinodis* Stitz, 1925: 122 (w.) PALAWAN I. Junior synonym of *procera:* Brown & Kempf, 1960: 225.

***apharogonia.*** *Eurhopalothrix apharogonia* Snelling, 1968c: 1 (w.) EL SALVADOR.

***australis.*** *Eurhopalothrix australis* Brown & Kempf, 1960: 218, figs. 46, 51 (w.q.) AUSTRALIA. Wheeler, G.C. & Wheeler, J. 1973c: 207 (l.).

*ballionii. Rhopalothrix procera* var. *ballionii* Forel, 1904d: 175 (w.) INDONESIA (Morotai I.). Junior synonym of *procera:* Brown & Kempf, 1960: 225.

***biroi.*** *Rhopalothrix biroi* Szabó, 1910a: 365, fig. 2 (w.) NEW GUINEA. Brown & Kempf, 1960: 223 (m.). Combination in *Eurhopalothrix:* Brown & Kempf, 1960: 222.

***bolaui.*** *Rhopalothrix bolaui* Mayr, 1870a: 415 (footnote) (q.) SURINAM. Combination in *Eurhopalothrix,* and senior synonym of *amoena:* Brown & Kempf, 1960: 210.

*borneensis. Rhopalothrix borneensis* Wheeler, W.M. 1919e: 96 (q.) BORNEO. Junior synonym of *procera:* Brown & Kempf, 1960: 225.

***brevicornis.*** *Rhopalothrix brevicornis* Emery, 1897c: 572, fig. 19 (w.) NEW GUINEA. Szabó, 1910a: 367 (q.). Combination in *Eurhopalothrix,* and senior synonym of *mixta:* Brown & Kempf, 1960: 215.

***browni.*** *Eurhopalothrix browni* Taylor, 1990b: 404, figs. 2–4, 44 (w.) BORNEO.

***bruchi.*** *Rhopalothrix bruchi* Santschi, 1922d: 256 (w.) ARGENTINA. Combination in *Eurhopalothrix:* Brown & Kempf, 1960: 214.

***caledonica.*** *Eurhopalothrix caledonica* Brown & Kempf, 1960: 220, figs. 44, 51 (w.) NEW CALEDONIA.

***chapmani.*** *Eurhopalothrix chapmani* Taylor, 1990b: 406, figs. 5–7, 23, 45 (w.) PHILIPPINES.

***cinnamea.*** *Eurhopalothrix cinnamea* Taylor, 1970a: 50, figs. 1–4 (w.) BISMARCK ARCHIPELAGO.

***clypeata.*** *Eurhopalothrix clypeata* Brown & Kempf, 1960: 205, fig. 37 (w.) GUIANA.

***coronata.*** *Eurhopalothrix coronata* Taylor, 1990b: 407, figs. 8–12, 46 (w.) BORNEO.

***dubia.*** *Eurhopalothrix dubia* Taylor, 1990b: 409, figs. 13–15, 47 (w.q.) BORNEO.

*elegans. Rhopalothrix (Rhopalothrix) elegans* Mann, 1921: 467, fig. 25 (w.) FIJI IS. Junior synonym of *emeryi:* Brown & Kempf, 1960: 230.

***emeryi.*** *Rhopalothrix emeryi* Forel, 1912n: 58 (q.) AUSTRALIA. Combination in *Eurhopalothrix,* and senior synonym of *elegans:* Brown & Kempf, 1960: 230.

***floridana.*** *Eurhopalothrix floridana* Brown & Kempf, 1960: 207, fig. 34 (w.) U.S.A.

***gravis.*** *Rhopalothrix (Rhopalothrix) gravis* Mann, 1922: 40, fig. 19 (w.q.m.) HONDURAS. Wheeler, G.C. & Wheeler, J. 1955a: 117 (l.). Combination in *Eurhopalothrix,* and senior synonym of *reichenspergeri* and *schmidti:* Brown & Kempf, 1960: 211.

***greensladei.*** *Eurhopalothrix greensladei* Taylor, 1968b: 342, figs. 6, 7 (w.) SOLOMON IS.

***heliscata.*** *Eurhopalothrix heliscata* Wilson & Brown, 1985: 410, figs. 1–3 (w.q.) SINGAPORE.

***hoplites.*** *Eurhopalothrix hoplites* Taylor, 1980b: 231, figs. 1–3 (w.) NEW GUINEA.

***insidiatrix.*** *Eurhopalothix insidiatrix* Taylor, 1980b: 238, figs. 11–14 (w.q.) FIJI IS.

***isabellae.*** *Rhopalothrix isabellae* Mann, 1919: 357, fig. 35 (w.) SOLOMON IS. Combination in *Eurhopalothrix:* Brown & Kempf, 1960: 225. See also: Taylor, 1968b: 344.

***jennya.*** *Eurhopalothrix jennya* Taylor, 1990b: 413, figs. 16–18, 48 (w.q.) BORNEO.

*kokodensis. Rhopalothrix (Rhopalothrix) kokodensis* Donisthorpe, 1936e: 524 (q.m.) NEW GUINEA. Junior synonym of *procera:* Brown & Kempf, 1960: 225.

***lenkoi.*** *Eurhopalothrix lenkoi* Kempf, 1967d: 358, figs. 6, 7 (w.q.) BRAZIL.

*malua. Rhopalothrix procera* subsp. *malua* Mann, 1919: 358, figs. 36–38 (w.q.m.) SOLOMON IS. Junior synonym of *procera:* Brown & Kempf, 1960: 225.

*manni. Rhopalothrix manni* Menozzi, 1923b: 210, fig. 2 (w.) PHILIPPINES. Junior synonym of *procera:* Brown & Kempf, 1960: 225.

*melanotica. Rhopalothrix procera* subsp. *malua* var. *melanotica* Mann, 1919: 360 (w.) SOLOMON IS. Unavailable name; material referred to *procera* by Brown & Kempf, 1960: 225.

*mixta. Rhopalothrix mixta* Szabó, 1910a: 366, fig. 4 (w.) NEW GUINEA. Junior synonym of *brevicornis:* Brown & Kempf, 1960: 215.

***omnivaga.*** *Eurhopalothrix omnivaga* Taylor, 1990b: 413, figs. 24–26, 50 (w.q.) WEST MALAYSIA.

***philippina.*** *Eurhopalothrix philippina* Brown & Kempf, 1960: 224, figs. 47, 50 (w.q.) PHILIPPINES.

***pilulifera.*** *Eurhopalothrix pilulifera* Brown & Kempf, 1960: 208, fig. 38 (w.q.) MEXICO.

***platisquama.*** *Eurhopalothrix platisquama* Taylor, 1990b: 417, figs. 30–33, 52 (w.) WEST MALAYSIA.

***procera.*** *Rhopalothrix procera* Emery, 1897c: 572, pl. 14, fig. 18 (w.q.) NEW GUINEA. Combination in *Eurhopalothrix,* senior synonym of *angulinodis, ballionii, borneensis, kokodensis, malua, manni, samoana, subdentatus,* and material of the unavailable name *melanotica* referred here: Brown & Kempf, 1960: 225.

***punctata.*** *Rhopalothrix punctata* Szabó, 1910a: 366, fig. 3 (w.) NEW GUINEA. Combination in *Eurhopalothrix:* Brown & Kempf, 1960: 221.

*reichenspergeri. Rhopalothrix reichenspergeri* Santschi, 1923g: 1263 (w.) BRAZIL. Combination in *Octostruma:* Brown, 1949f: 92. Junior synonym of *gravis:* Brown & Kempf, 1960: 211.

***rothschildi.*** *Eurhopalothrix rothschildi* Taylor, 1990b: 418, figs. 38–40, 54 (w.) BORNEO.

*samoana. Rhopalothrix procera* st. *samoana* Santschi, 1928a: 51, fig. 5 (w.m.) SAMOA. Junior synonym of *procera:* Brown & Kempf, 1960: 225.

*schmidti. Rhopalothrix (Rhopalothrix) schmidti* Menozzi, 1936c: 82, fig. 1 (w.q.l.) COSTA RICA. Junior synonym of *gravis:* Brown & Kempf, 1960: 211.

***seguensis.*** *Eurhopalothrix seguensis* Taylor, 1990b: 421, figs. 34, 41–43, 55 (w.) BORNEO.

***speciosa.*** *Eurhopalothrix speciosa* Brown & Kempf, 1960: 203, figs. 31–33 (w.) BRAZIL.
***spectabilis.*** *Eurhopalothrix spectabilis* Kempf, 1962b: 27, figs. 25–27 (q.) BRAZIL. Kempf, 1968b: 398 (w.).
*subdentatus. Rhopalothrix* (*Rhopalothrix*) *subdentatus* Donisthorpe, 1942b: 67 (q.) PHILIPPINES. Junior synonym of *procera:* Brown & Kempf, 1960: 225.
***szentivanyi.*** *Eurhopalothrix szentivanyi* Taylor, 1968b: 346, figs. 12–15 (w.q.) NEW GUINEA.

*EUSPHINCTUS:* see under ***SPHINCTOMYRMEX.***

***EUTETRAMORIUM*** (Malagasy)
***mocquerysi.*** *Eutetramorium mocquerysi* Emery, 1899f: 281, figs. (w.) MADAGASCAR. Wheeler, G.C. & Wheeler, J. 1980: 537 (l.).
***monticellii.*** *Eutetramorium monticellii* Emery, 1899f: 283 (w.q.) MADAGASCAR.

*EVELYNA:* see under ***POLYRHACHIS.***

*EXAMBLYOPONE:* see under ***PRIONOPELTA.***

*FLORENCEA:* see under ***POLYRHACHIS.***

*FORELIFIDIS:* see under ***OXYEPOECUS.***

***FORELIUS*** (New World)
***albiventris.*** *Forelius chalybaeus* r. *albiventris* Forel, 1912h: 45 (w.) ARGENTINA. Raised to species: Kusnezov, 1954b: 37; Kusnezov, 1957b: 16 (in key). See also: Gallardo, 1916a: 89.
*analis. Tapinoma anale* André, 1893b: 148 (w.) MEXICO. Combination in *Iridomyrmex:* Forel, 1908b: 62; in *Forelius:* Wheeler, G.C. & Wheeler, J. 1986g: 13. Variety of *pruinosus:* Emery, 1895c: 333. Revived status as species: Forel, 1908b: 62; Emery, 1912g: 26. Variety/subspecies of *pruinosus:* Wheeler, W.M. 1913b: 497; Creighton, 1950a: 343. Senior synonym of *testaceus:* Creighton, 1950a: 343. Junior synonym of *pruinosus:* Wheeler, G.C. & Wheeler, J. 1986g: 13. [Shattuck, 1994: 96, treats *analis* as a subspecies of *pruinosus,* but the Wheeler synonymy is preferred here.]
***andinus.*** *Forelius andinus* Kusnezov, 1957b: 15 (w.q.) ARGENTINA.
*andrei. Forelius maccooki* r. *andrei* Forel, 1912h: 44 (w.) U.S.A. Subspecies of *foetidus:* Smith, M.R. 1951a: 836. Junior synonym of *foetidus:* Wheeler, W.M. 1901b: 520; of *mccooki:* Smith, D.R. 1979: 1419.
*basalis. Forelius chalybaeus* st. *grandis* var. *basalis* Santschi, 1922b: 375 (w.q.m.) ARGENTINA. Unavailable name (Shattuck, 1994: 97).
***brasiliensis.*** *Iridomyrmex* (*Forelius*) *maccooki* var. *brasiliensis* Forel, 1908c: 396 (w.) BRAZIL. Wheeler, G.C. & Wheeler, J. 1966: 730 (l.). Combination in *Forelius:* Emery, 1912g: 36. Race of *mccooki:* Forel, 1912h: 43. Raised to species: Santschi, 1922b: 373; Kusnezov, 1957b: 17. Current subspecies: nominal plus *pilipes.*
***breviscapus.*** *Forelius maccooki* r. *breviscapus* Forel, 1914d: 286 (w.) ARGENTINA. [First available use of *Forelius maccooki* r. *fiebrigi* var. *breviscapa* Forel, 1913l: 241; unavailable name.] Raised to species: Santschi, 1922b: 374; Kusnezov, 1957b: 16. Current subspecies: nominal plus *obscuratus, pusillus.*
*carmelitana. Forelius maccooki* subsp. *brasiliensis* var. *carmelitana* Gallardo, 1916a: 79, fig. 35 (w.) ARGENTINA. Unavailable name (Shattuck, 1994: 97).
***chalybaeus.*** *Forelius chalybaeus* Emery, 1906c: 176 (w.q.m.) ARGENTINA. Current subspecies: nominal plus *paucistrictus, personatus.* See also: Gallardo, 1916a: 84.
***#debenedettii.*** *Forelius chalybaeus* st. *debenedettii* Santschi, 1919f: 53 (w.) ARGENTINA. Subspecies of *grandis:* Kempf, 1972a: 109.
*eidmanni* Goetsch, 1933; see under ***TAPINOMA.***
***fiebrigi.*** *Forelius maccooki* r. *fiebrigi* Forel, 1912h: 44 (w.) PARAGUAY. Raised to species: Kempf, 1972a: 109.
*foetida. Formica foetida* Buckley, 1866: 167 (w.q.) U.S.A. Crozier, 1970: 113 (k.); Wheeler, G.C. & Wheeler, J. 1974b: 401 (l.). [Junior primary homonym of *Formica foetida* Linnaeus, 1758: 582 (now in *Pachycondyla*).] Replacement name: *mccooki* McCook, 1879: 187.
***golbachi.*** *Amyrmex golbachi* Kusnezov, 1953b: 335, 6 figs. (w.m.) ARGENTINA. Combination in *Forelius:* Shattuck, 1992c: 95.
***grandis.*** *Forelius chalybaeus* r. *grandis* Forel, 1912h: 46 (w.) ARGENTINA. Forel, 1914d: 285 (q.m.). Raised to species: Kusnezov, 1954b: 37; Kusnezov, 1957b: 18 (in key). [Santschi, 1916e: 393 erroneously refers to this taxon as *major.*] Current subspecies: nominal plus *debenedettii.*
***instabilis.*** *Forelius maccooki* st. *instabilis* Santschi, 1920d: 382 (w.q.m.) ARGENTINA. Junior synonym of *obscuratus:* Santschi, 1922b: 374. Revived from synonymy, and raised to species: Kusnezov, 1957b: 16 (in key).
***#keiferi.*** *Forelius foetidus* subsp. *keiferi* Wheeler, W.M. 1934f: 141 (w.q.m.) MEXICO. Subspecies of *mccooki:* Shattuck, 1994: 94.
***macrops.*** *Forelius macrops* Kusnezov, 1957b: 14 (w.) ARGENTINA.
*major* Santschi, 1916; see under ***grandis,*** above.
***mccooki.*** *Iridomyrmex mccooki* McCook, 1879: 187 (w.) U.S.A. Junior synonym of *foetida* Buckley: Wheeler, W.M. 1901b: 520, and hence first available replacement name for *Formica foetida* Buckley, 1866: 167. [Junior primary homonym of *Formica foetida* Linnaeus, 1758: 582.] [*Iridomyrmex maccooki* Forel, 1878: 382. *Nomen nudum.*] Mayr, 1886d: 432 (q.m.); Forel, 1886b: xxxix (w.). Combination in *Forelius:* Emery, 1888d: 389. Senior synonym of *andrei:* Smith, D.R. 1979: 1419. Current subspecies: nominal plus *keiferi.*
***minor.*** *Forelius chalybaeus* var. *minor* Forel, 1913l: 241 (w.) ARGENTINA. Raised to species: Kusnezov, 1957b: 16 (in key).
***#modestus.*** *Forelius nigriventris* var. *modesta* Santschi, 1922b: 374 (w.) ARGENTINA.
***nigriventris.*** *Forelius nigriventris* Forel, 1912h: 44 (w.) ARGENTINA. Brèthes, 1914b: 233 (q.). Current subspecies: nominal plus *modestus, transiens.* See also: Gallardo, 1916a: 81.
***#obscuratus.*** *Forelius breviscapus* var. *obscurata* Santschi, 1922b: 374 (w.q.m.) ARGENTINA. [First available use of *Forelius maccooki* r. *breviscapus* var. *obscurata* Forel, 1914d: 286; unavailable name.]
***#paucistrictus.*** *Forelius chalybaeus* var. *paucistricta* Santschi, 1916e: 391 (w.) ARGENTINA. [Misspelled as *paucistriatus* by Kempf, 1972a: 109.]
***#personatus.*** *Forelius chalybaeus* var. *personata* Santschi, 1922b: 375 (w.) ARGENTINA.
***#pilipes.*** *Forelius rufus* var. *pilipes* Santschi, 1920d: 382 (w.) ARGENTINA. [*Forelius rufus* var. *pilipes* Santschi, 1919f: 53. *Nomen nudum.*] Variety of *brasiliensis:* Santschi, 1922b: 374.
***pruinosus.*** *Tapinoma pruinosum* Roger, 1863a: 165 (w.) CUBA. Wheeler, G.C. & Wheeler, J. 1951: 185 (l.). Combination in *Iridomyrmex:* Wheeler, W.M. 1913b: 497; in *Forelius:* Wheeler, G.C. & Wheeler, J. 1986g: 13; Shattuck, 1992c: 95. Senior synonym of *analis* (and its junior synonym *testaceus*): Wheeler, G.C. & Wheeler, J. 1986g: 13. See also: Creighton, 1950a: 342; Petralia & Vinson, 1980: 386.
***#pusillus.*** *Forelius breviscapus* var. *pusillus* Santschi, 1922b: 374 (w.) ARGENTINA.
***rubriceps.*** *Forelius chalybaeus* var. *rubriceps* Gallardo, 1916a: 85 (w.q.m.) ARGENTINA. [First available use of *Forelius chalybaeus* r. *grandis* var. *rubriceps* Forel, 1914d: 286; unavailable name.] Raised to species: Kusnezov, 1957b: 17 (in key).
***rufus.*** *Forelius rufus* Gallardo, 1916a: 77 (w.) ARGENTINA. Santschi, 1922b: 373 (q.m.).
*symbiotica. Forelius chalybaeus* st. *grandis* var. *symbiotica* Santschi, 1916e: 391 (w.q.) ARGENTINA. Unavailable name (Shattuck, 1994: 97).
*testaceus. Iridomyrmex pruinosus* var. *testaceus* Cole, 1936b: 121 (w.q.) U.S.A. Junior synonym of *analis:* Creighton, 1950a: 343.
***#transiens.*** *Forelius nigriventris* var. *transiens* Santschi, 1922b: 374 (w.) ARGENTINA.
***tucumanus.*** *Neoforelius tucumanus* Kusnezov, 1953b: 330, figs. 1–12 (w.q.m.) ARGENTINA. Combination in *Forelius:* Shattuck, 1992c: 95.

*FORELOMYRMEX:* see under ***POGONOMYRMEX.***

***FORELOPHILUS*** (Indo-Australian)
***overbecki.*** *Forelophilus overbecki* Kutter, 1931b: 193, figs. (w.q.) JAVA.

***FORMICA*** (Holarctic)
*abdominalis. Formica abdominalis* Latreille, 1802c: 175, pl. 3, fig. 13 (q.) "Grandes Indes". Unidentifiable to genus; *incertae sedis* in *Formica.*
*abdominalis* Fabricius, 1804; see under ***CAMPONOTUS.***
*aberrans* Mayr, 1877; see under ***ALLOFORMICA.***
*abrupta* Smith, F. 1858; see under ***DOLICHODERUS.***
***accreta.*** *Formica accreta* Francoeur, 1973: 182, figs. 308–323 (w.q.m.) CANADA. Junior synonym of *fusca:* Wheeler, G.C. & Wheeler, J. 1986g: 16. Revived from synonymy: **revived status.**
*acervorum* Fabricius, 1793; see under ***LEPTOTHORAX.***
*aculeata. Formica rubra aculeata* Retzius, 1783: 76 (w.) no locality given. Unavailable name (published as junior synonym).
*aculeata. Formica fusca aculeata* Retzius, 1783: 76 (w.) no locality given. Unavailable name (published as junior synonym).
*aculeata* Olivier, 1792; see under ***PARAPONERA.***
**acuminata.* **Formica acuminata* Heer, 1850: 142, pl. 11, figs. 13, 14 (m.) YUGOSLAVIA (Miocene).

*acuta* Fabricius, 1804; see under ***CREMATOGASTER.***
***adamsi.*** *Formica adamsi* Wheeler, W.M. 1909e: 84 (w.) U.S.A. [*Formica adamsi* Wheeler, W.M. 1908g: 408. *Nomen nudum.*] Variety/subspecies of *whymperi:* Wheeler, W.M. 1917a: 544; Creighton, 1950a: 509. [First available name is *adamsi* as this has priority over *whymperi* (see there).] Current subspecies: nominal plus *alpina, whymperi.*
***adelungi.*** *Formica adelungi* Forel, 1904b: 385 (m.) MONGOLIA. See also: Wheeler, W.M. 1913f: 450.
*advena* Smith, F. 1862; see under ***PROLASIUS.***
*aedificator* Schilling, 1839; see under ***MESSOR.***
*aegyptiaca. Formica aegyptiaca* Fabricius, 1775: 393 (w.) EGYPT. Unidentifiable to genus; *incertae sedis* in *Formica.*
*****aemula.*** **Formica aemula* Heer, 1867: 18, pl. 1, fig. 19 (caste?) YUGOSLAVIA (Miocene).
*aeneovirens* Lowne, 1865; see under ***MELOPHORUS.***
*aenescens* Nylander, 1849; see under ***CATAGLYPHIS.***
*aequalis. Formica aequalis* Walker, 1871: 9 (q.) SAUDI ARABIA. Unidentifiable to genus; *incertae sedis* in *Formica.*
***aerata.*** *Formica aerata* Francoeur, 1973: 116, figs. 183–189 (w.q.) U.S.A.
*aerea* Roger, 1859; see under ***PROFORMICA.***
*aethiops* Latreille, 1798; see under ***CAMPONOTUS.***
*affinis. Formica affinis* Leach, 1825: 290 (w.q.) FRANCE. Unidentifiable to genus; *incertae sedis* in *Formica.*
*affinis* Le Guillou, 1842; see under ***POLYRHACHIS.***
*affinis* Schenck, 1852; see under ***LASIUS.***
*aggerans. Formica rufa* subsp. *aggerans* Wheeler, W.M. 1912c: 90 (w.) U.S.A. Wheeler, W.M. 1913f: 430 (q.m.). Junior synonym of *obscuripes:* Forel, 1914c: 619; Creighton, 1940a: 1.
*agilis* Smith, F. 1858; see under ***CAMPONOTUS.***
*agra* Smith, F. 1858; see under ***CAMPONOTUS.***
*albicans* Roger, 1859; see under ***CATAGLYPHIS.***
*albipennis. Formica albipennis* Fabricius, 1793: 354 (m.) ANTILLES IS. Unidentifiable to genus; *incertae sedis* in *Formica.*
*albipes* Smith, F. 1861; see under ***TECHNOMYRMEX.***
*albofasciata* Smith, F. 1862; see under ***CAMPONOTUS.***
*algida. Formica fusca* var. *algida* Wheeler, W.M. 1915f: 205 (w.q.) U.S.A. Combination in *F.* (*Serviformica*): Emery, 1925b: 248. Subspecies of *neorufibarbis:* Creighton, 1950a: 537. Junior synonym of *neorufibarbis:* Francoeur, 1973: 215.
*aliena* Foerster, 1850; see under ***LASIUS.***
#***alpicola.*** *Formica fusca* var. *alpicola* Gredler, 1858: 10 (w.) AUSTRIA.
#***alpina.*** *Formica adamsi* var. *alpina* Wheeler, W.M. 1909e: 85 (w.) U.S.A. Variety/subspecies of *whymperi:* Wheeler, W.M. 1917a: 544; Creighton, 1950a: 509; Smith, D.R. 1979: 1463. Relationship with *densiventris:* Cole, 1954a: 89. Subspecies of *adamsi:* current status.
*alpina. Formica rufa* var. *alpina* Santschi, 1911j: 349 (w.) ITALY. [Junior primary homonym of *alpina,* above.] Replacement name: *santschii* Wheeler, 1913f: 428. Raised to species: Bondroit, 1918: 59.
****alsatica.*** **Formica alsatica* Théobald, 1937b: 215, pl. 15, figs. 9–13; pl. 4, fig. 3 (q.) FRANCE (Oligocene).
*alticola. Formica munda* var. *alticola* Wheeler, W.M. 1917a: 534 (w.) U.S.A. Combination in *F.* (*Raptiformica*), and subspecies of *obtusopilosa:* Creighton, 1950a: 466. Junior synonym of *obtusopilosa:* Wilson & Brown, 1955: 128.
***altipetens.*** *Formica cinerea* var. *altipetens* Wheeler, W.M. 1913f: 399 (diagnosis in key) (w.) U.S.A. [*Formica cinerea* subsp. *cinerea* var. *altipetens* Wheeler, W.M. 1913f: 523 (w.q.m.); unavailable name.] Wheeler, G.C. & Wheeler, J. 1953c: 160 (l.). Variety/subspecies of *cinerea:* Wheeler, W.M. 1917a: 550; Wheeler, G.C. & Wheeler, E.W. 1944: 259. Raised to species: Creighton, 1950a: 531. See also: Francoeur, 1973: 52.
*americana* Buckley, 1866; see under ***CAMPONOTUS.***
*ammon* Fabricius, 1775; see under ***POLYRHACHIS.***
*amyoti. Formica amyoti* Le Guillou, 1842: 315 (w.) AUSTRALIA. Unidentifiable to genus; *incertae sedis* in *Formica.*
*analis* Latreille, 1802; see under ***PACHYCONDYLA.***
*anceps* Roger, 1863; see under ***IRIDOMYRMEX.***
*angulata* Smith, F. 1859; see under ***CAMPONOTUS.***
*angustata* Latreille, 1798; see under ***CAMPONOTUS.***
#***angusticeps.*** *Formica rufa* var. *angusticeps* Stärcke, 1947: 146 (w.) NETHERLANDS.
*angusticollis* Jerdon, 1851; see under ***CAMPONOTUS.***
*animosa. Formica animosa* Forskål, 1775: 85 (w.) EGYPT ?. Unidentifiable to genus; *incertae sedis* in *Formica.*
*antiguensis* Fabricius, 1793; see under ***MONOMORIUM.***
****antiqua.*** **Formica* (*Serviformica*) *antiqua* Dlussky, 1967b: 82, fig. 1 (w.) BALTIC AMBER.
*aphidicola* Walsh, 1863; see under ***LASIUS.***
*apicalis* Latreille, 1802; see under ***PACHYCONDYLA.***
*apicalis. Formica apicalis* Smith, F. 1858b: 49 (q.) "South America"; locality in error. [Unresolved junior primary homonym of *apicalis* Latreille, 1802c: 204, above.] Junior synonym of *rufa:* Roger, 1862c: 287.
***approximans.*** *Formica truncorum* var. *approximans* Wheeler, W.M. 1933c: 65 (w.) CHINA. Raised to species: Wu, 1990: 4.
***aquilonia.*** *Formica aquilonia* Yarrow, 1955a: 31 (diagnosis in key), figs. 21, 25, 29, 33, 37, 41, 45, 49, 53, 57, 61, 65 (w.q.m.) GREAT BRITAIN. Synonymous with *schmidti* Ruzsky, 1920: 78 [Junior primary homonym of **schmidtii* Heer, 1850: 138.]: Dlussky, 1967a: 90, and hence first available replacement name. See also: Kutter, 1977c: 271; Kupyanskaya, 1986a: 98; Gösswald, 1989: 19; Kupyanskaya, 1990: 197; Atanassov & Dlussky, 1992: 276.
*arborea* Smith, F. 1858; see under ***CAMPONOTUS.***
****arcana.*** **Formica arcana* Scudder, 1877b: 266 (wing) CANADA. Carpenter, 1930: 21 treats this taxon as (Formicidae) *arcana* (Scudder).
***archboldi.*** *Formica pallidefulva* subsp. *archboldi* Smith, M.R. 1944d: 16 (w.) U.S.A. Combination in *F.* (*Neoformica*), and raised to species: Creighton, 1950a: 549.
*arcuata* Le Guillou, 1842; see under ***POLYRHACHIS.***
*ardens* Smith, F. 1858; see under ***CAMPONOTUS.***
*arenaria* Fabricius, 1787; see under ***MESSOR.***
*arenicola. Formica arenicola* Buckley, 1866: 160 (w.) U.S.A. Unidentifiable to genus; *incertae sedis* in *Formica.*
*arenicola. Formica sanguinea* subsp. *arenicola* Kuznetsov-Ugamsky, 1928b: 15 (w.) RUSSIA. [Unresolved junior primary homonym of *arenicola* Buckley, above.] Junior synonym of *sanguinea:* Dlussky, 1965a: 16.
*argentata* Fabricius, 1804; see under ***POLYRHACHIS.***
*argentata. Formica fusca* var. *argentata* Wheeler, W.M. 1902e: 952 (footnote) (w.) U.S.A. [Junior primary homonym of *argentata* Fabricius, above.] Replacement name: *argentea* Wheeler, W.M. 1912c: 90.
***argentea.*** *Formica fusca* var. *argentea* Wheeler, W.M. 1912c: 90. Replacement name for *argentata* Wheeler, W.M., above. [Junior primary homonym of *argentata* Fabricius, 1804: 413.] Wheeler, W.M. 1913f: 502 (q.m.). Combination in *F.* (*Serviformica*): Emery, 1925b: 248. Subspecies of *fusca:* Buren, 1944a: 301. Junior synonym of *fusca:* Creighton, 1950a: 532. Revived from synonymy, raised to species, and senior synonym of *blanda:* Francoeur, 1973: 141.
*armata* Le Guillou, 1842; see under ***POLYRHACHIS.***
#***armenica.*** *Formica cinerea* var. *armenica* Ruzsky, 1905b: 406 (w.q.) ARMENIA. Combination in *F.* (*Serviformica*): Emery, 1925b: 246. Subspecies of *cinerea:* Dlussky, 1967a: 66; Arakelian, 1994: 92.
*armigera* Latreille, 1802; see under ***DACETON.***
*arrogans* Smith, F. 1858; see under ***CAMPONOTUS.***
***aserva.*** *Formica sanguinea* r. *aserva* Forel, 1901g: 395 (w.q.) CANADA. Wheeler, W.M. 1913f: 405 (m.). Combination in *F.* (*Raptiformica*): Emery, 1925b: 260. Synonym of *subnuda:* Creighton, 1950a: 469; Wilson & Brown, 1955: 118; Buren, 1968a: 33. [All these authors give *subnuda* as the senior name but the date of availability of *subnuda* makes it clear that *aserva* has seniority; see under *subnuda.*]
*assimilis* Jerdon, 1851; see under ***PARATRECHINA.***
****atavina.*** **Formica atavina* Heer, 1850: 143, pl. 11, fig. 10 (m.) YUGOSLAVIA (Miocene).
***aterrima.*** *Formica aterrima* Cresson, 1865: 426 (m.) U.S.A. See also: Brown, 1947: 8.
*atomus* Latreille, 1798; see under ***TAPINOMA.***
*atra* Zetterstedt, 1838; see under ***CAMPONOTUS.***
*atra. Formica atra* Schilling, 1839: 54 (w.) POLAND. [Unresolved junior primary homonym of *atra* Zetterstedt, 1838: 450, above.] Unidentifiable to genus; *incertae sedis* in *Formica.*
*atra* Buckley, 1866; see under ***CAMPONOTUS.***
*atrata* Linnaeus, 1758; see under ***CEPHALOTES.***
*atriceps* Smith, F. 1858; see under ***CAMPONOTUS.***
*atricolor* Nylander, 1849; see under ***CAMPONOTUS.***
*atriventris* Smith, F. 1858; see under ***LASIOPHANES.***
*attelaboides* Fabricius, 1775; see under ***DOLICHODERUS.***
*attenuata. Formica attenuata* Schilling, 1830: 55. *Nomen nudum.*
*aurocincta* Smith, F. 1858; see under ***CAMPONOTUS.***
*aurulenta* Latreille, 1802; see under ***CAMPONOTUS.***
*australis* Fabricius, 1775; see under ***DIACAMMA.***
*austriaca* Mayr, 1853; see under ***LIOMETOPUM.***
****auxillacensis.*** **Formica auxillacensis* Piton, in Piton & Théobald, 1935: 91, pl. 3, fig. 8 (wing) FRANCE (Miocene).
*axillaris* Spinola, 1808; see under ***CAMPONOTUS.***
*bacchus* Smith, F. 1858; see under ***CAMPONOTUS.***
*badia* Latreille, 1802; see under ***POGONOMYRMEX.***
*badia* Smith, F. 1857; see under ***CAMPONOTUS.***
***balcanica.*** *Formica balcanica* Petrov & Collingwood, 1993: 349, figs. 1–5 (w.q.m.) YUGOSLAVIA.
****baltica.*** **Formica* (*Serviformica*) *baltica* Dlussky, 1967b: 81, fig. 1 (w.) BALTIC AMBER.

***FORMICA** (continued)*

*barbara* Linnaeus, 1767; see under ***MESSOR.***
*barbata. Formica barbata* Razoumowski, 1789: 225, fig. 12 (w.) SWITZERLAND. Junior synonym of *fusca:* Emery & Forel, 1879: 451.
***bauckhorni.** *Formica bauckhorni* Meunier, 1915: 214, pl. 23, fig. 3 (m.) GERMANY. See also: Meunier, 1917: 8.
***beijingensis.*** *Formica beijingensis* Wu, 1990: 2, figs. 1, 15 (w.) CHINA.
*belgarum* Bondroit, 1918 [*Formicina*]; see under ***LASIUS.***
*berthoudi* Forel, 1876; see under ***ANOPLOLEPIS.***
*bicolor* Fabricius, 1793; see under ***CATAGLYPHIS.***
*bicolor* Latreille, 1798; see under ***CAMPONOTUS.***
*bicolor. Formica bicolor* Leach, 1825: 289 (q.m.) FRANCE. [Unresolved junior primary homonym of *bicolor* Fabricius, 1793: 351, above.] Unidentifiable to genus; *incertae sedis* in *Formica.*
*bicolor. Formica bicolor* Schilling, 1830: 55. *Nomen nudum.*
*bicornis* Foerster, 1850; see under ***LASIUS.***
*bidens* Linnaeus, 1758; see under ***DOLICHODERUS.***
*bihamata* Drury, 1773; see under ***POLYRHACHIS.***
*bimaculata* Smith, F. 1858; see under ***CAMPONOTUS.***
*binodis* Linnaeus, 1763; see under ***TETRAMORIUM.***
*binodis* Fabricius, 1775; see under ***MESSOR.***
*bipartita* Smith, F. 1861; see under ***LEPISIOTA.***
*bipilosa. Formica* (*Serviformica*) *cinerea* var. *bipilosa* Karavaiev, 1926e: 199 (w.) RUSSIA. Raised to species: Collingwood, 1961a: 70; Collingwood, 1961b: 289. Junior synonym of *subpilosa:* Dlussky, 1965a: 28.
*biscutata* Fabricius, 1775; see under ***ACROMYRMEX.***
*bispinosa* Olivier, 1792; see under ***DOLICHODERUS.***
*bituberculata* Fabricius, 1798; see under ***MEGALOMYRMEX.***
*blanda* Smith, F. 1858; see under ***CAMPONOTUS.***
*blanda. Formica fusca* var. *blanda* Wheeler, W.M. 1913f: 398 (in key) (w.) U.S.A. [Unresolved junior primary homonym of *blanda* Smith, F. 1858b: 43, above.] Variety of *fusca:* Wheeler, W.M. 1917a: 548. Junior synonym of *argentea:* Francoeur, 1973: 141.
*bombycina* Roger, 1859; see under ***CATAGLYPHIS.***
*bondroiti. Formica rufa* subsp. *polyctena* ab. *bondroiti* Stärcke, 1942b: 43 (w.) SPAIN. Unavailable name; material referred to *rufa* by Yarrow, 1955a: 4.
*borea. Formica* (*Raptiformica*) *sanguinea* var. *borea* Santschi, 1925g: 351 (w.m.) FINLAND. Junior synonym of *sanguinea:* Dlussky, 1965a: 16.
*borealis. Formica fusca* subsp. *borealis* Vashkevich, 1924b: 147 (w.) RUSSIA. Junior synonym of *lemani:* Dlussky, 1965a: 35.
***bradleyi.*** *Formica bradleyi* Wheeler, W.M. 1913f: 423 (w.m.) U.S.A. Wheeler, W.M. 1917a: 535 (q.); Wheeler, G.C. & Wheeler, J. 1953c: 168 (l.). Combination in *F.* (*Raptiformica*): Emery, 1925b: 259. Senior synonym of *morbida:* Wilson & Brown, 1955: 126. See also: Smith, D.R. 1979: 1449.
#**brevisetosa.** *Formica rufa* subsp. *brevisetosa* Ruzsky, 1926: 110 (w.) RUSSIA.
*brevisetosa. Formica* (*Serviformica*) *cinerea* var. *brevisetosa* Karavaiev, 1927a: 302 (w.) UKRAINE. [Unresolved junior primary homonym of *brevisetosa* Ruzsky, above.] Junior synonym of *cinerea:* Finzi, 1928a: 68.
***browni.*** *Formica browni* Francoeur, 1973: 229, figs. 398–404 (w.q.) MEXICO.
*brullei* Smith, F. 1858; see under ***CAMPONOTUS.***
***bruni.*** *Formica* (*Coptoformica*) *bruni* Kutter, 1967a: 229, figs. 1–14 (w.q.m.) SWITZERLAND. See also: Kutter, 1977c: 284.
*brunnea* Latreille, 1798; see under ***LASIUS.***
***brunneonitida.*** *Formica* (*Coptoformica*) *brunneonitida* Dlussky, 1964: 1034, figs. (w.q.) MONGOLIA. See also: Dlussky, 1965a: 23.
*bucephala* Roger, 1859; see under ***CAMPONOTUS.***
*buddhaensis* Ruzsky, 1915; see under ***PROFORMICA.***
***buphthalma.** *Formica buphthalma* Novak, O. 1878: 91, pl. 3, fig. 2 (m.?) GERMANY (Oligocene).
*caduca* Victor, 1839; see under ***MESSOR.***
*caerulescens* Losana, 1834; see under ***TAPINOMA.***
*caespitum* Linnaeus, 1758; see under ***TETRAMORIUM.***
*californica. Formica microgyna* subsp. *californica* Wheeler, W.M. 1917a: 543 (w.) U.S.A. Subspecies of *whymperi,* and material of the unavailable name *hybrida* referred here: Creighton, 1950a: 510. Junior synonym of *whymperi:* Wheeler, G.C. & Wheeler, J. 1986g: 15.
***californica.*** *Myrmecocystus melliger* subsp. *californicus* Creighton, 1950a: 445 (w.) U.S.A. [First available use of *Myrmecocystus melliger* subsp. *mimicus* var. *californicus* Cole, 1936b: 118; unavailable name.] [Unresolved junior secondary homonym of *californica* Wheeler, W.M., above.] Combination in *Formica:* Snelling, 1969a: 2.
*callida* Smith, F. 1858; see under ***CAMPONOTUS.***
***calviceps.*** *Formica calviceps* Cole, 1954c: 164 (w.) U.S.A.
*camelina* Smith, F. 1857; see under ***CAMPONOTUS.***
*camponoticeps. Formica subpolita* var. *camponoticeps* Wheeler, W.M. 1913f: 535 (w.) U.S.A. Subspecies of *subpolita:* Creighton, 1950a: 542. Junior synonym of *subpolita:* Francoeur, 1973: 107.
***canadensis.*** *Formica cinerea* var. *canadensis* Santschi, 1914a: 435 (w.q.) CANADA. Wheeler, W.M. 1917a: 550 (q.); Francoeur, 1973: 63 (m.). Combination in *F.* (*Serviformica*): Emery, 1925b: 246. Junior synonym of *lepida:* Smith, M.R. 1961a: 252. Revived from synonymy, and raised to species: Francoeur, 1973: 60.
***candida.*** *Formica candida* Smith, F. 1878b: 11 (q.) CHINA. [Smith's description is repeated by Bingham, 1903: 335 (footnote).] Probable synonym of *picea* Nylander: Emery, 1925b: 249. Junior synonym of *picea* Nylander: Dlussky, 1967a: 61. Hence *candida* is first available replacement name for *Formica picea* Nylander, 1846a: 917. [Junior primary homonym of *Formica picea* Leach, 1825: 292 (now in *Camponotus*).] Current synonymy of *candida:* see under *picea* Nylander, to which all older synonymy was referred. Current subspecies: nominal plus *formosae.*
***cantalica.** *Formica cantalica* Piton, in Piton & Théobald, 1935: 68, pl. 1, fig. 10 (m.?) FRANCE (Mio-Pliocene).
*capitata* Latreille, 1798; see under ***MESSOR.***
***capito.** *Formica capito* Heer, 1867: 14, pl. 1, fig. 13 (q.) YUGOSLAVIA (Miocene).
*capsincola. Formica capsincola* Schilling, 1839: 54 (w.q.) POLAND. Junior synonym of *gagates:* Mayr, 1855: 248; Nylander, 1856b: 65.
*carbonaria* Latreille, 1802; see under ***CAMPONOTUS.***
*carinata* Fabricius, 1804; see under ***POLYRHACHIS.***
*carinata* Brullé, 1840; see under ***CAMPONOTUS.***
*caryae* Fitch, 1855; see under ***CAMPONOTUS.***
*castanea* Latreille, 1802; see under ***CAMPONOTUS.***
*castanea* Lepeletier, 1835; see under ***CAMPONOTUS.***
*castaneipes* Leach, 1825; see under ***CAMPONOTUS.***
*caucasica. Formica rufibarbis* var. *caucasica* Wheeler, W.M. 1913f: 517 (w.) CAUCASUS. [First available use of *Formica rufibarbis* subsp. *clara* var. *caucasica* Ruzsky, 1905b: 401; unavailable name.] Subspecies of *rufibarbis:* Stitz, 1930: 238. Junior synonym of *glauca:* Dlussky, 1967a: 74.
*caustica* Kollar, 1832; see under ***ZACRYPTOCERUS.***
**cephalica. *Formica cephalica* Scudder, 1891: 699, no caste mentioned, BALTIC AMBER. *Nomen nudum;* attributed to Burmeister. [Name may be based on a misinterpretation. Burmeister, 1831: 1100, does not name a species but mentions that he has a number of ant heads (*formicae cephalicae*) in amber.]*
*cephalotes* Linnaeus, 1758; see under ***ATTA.***
*chartifex* Smith, F. 1860; see under ***DENDROMYRMEX.***
*chelifera* Latreille, 1802; see under ***ODONTOMACHUS.***
*chevrolatii. Formica chevrolatii* Romand, 1846: xxxii, pl. 2, fig. 5 (w.) FRANCE. Junior synonym of *fusca:* Dalla Torre, 1893: 193.
*chilensis* Spinola, 1851; see under ***CAMPONOTUS.***
*chufejif. Formica chufejif* Forskål, 1775: xxiii (w.) YEMEN. Unidentifiable to genus; *incertae sedis* in *Formica.* See: Emery, 1892b: 160.
***ciliata.*** *Formica ciliata* Mayr, 1886d: 428 (q.) U.S.A. Wheeler, W.M. 1903e: 640 (w.m.). See also: Wheeler, W.M. 1913f: 452.
*ciliata. Formica pratensis* var. *ciliata* Ruzsky, 1926: 110 (w.) RUSSIA. [First available use of *Formica rufa* subsp. *pratensis* var. *ciliata* Ruzsky, 1915b: (pagination unknown, reference not seen; data from Dlussky & Pisarski, 1971: 177).] [Unresolved junior primary homonym of *ciliata* Mayr, above.] Junior synonym of *pratensis:* Dlussky & Pisarski, 1971: 177.
*cinctella* Gerstäcker, 1859; see under ***CAMPONOTUS.***
*cinerascens* Fabricius, 1787; see under ***CAMPONOTUS.***
***cinerea.*** *Formica cinerea* Mayr, 1853c: 281 (w.q.) AUSTRIA. Mayr, 1855: 344 (m.). Combination in *F.* (*Serviformica*): Forel, 1915d: 64. Variety/race/subspecies of *fusca:* Forel, 1874: 54; Emery & Forel, 1879: 451; Mayr, 1886d: 427; Forel, 1892i: 307; Emery, 1909b: 199; Bondroit, 1910: 483; Emery, 1914d: 159. Status as species: Mayr, 1861: 48; Emery, 1898c: 126; Ruzsky, 1902d: 12; Forel, 1904b: 385; Wheeler, W.M. 1913f: 521; Forel, 1915d: 64; Emery, 1916b: 255; Wheeler, W.M. 1917a: 550; Menozzi, 1918: 88; Bondroit, 1918: 53; Müller, 1923: 141; Emery, 1925b: 246; Karavaiev, 1929b: 214; Karavaiev, 1936: 224; Dlussky, 1967a: 65; Bernard, 1967: 300; Dlussky & Pisarski, 1971: 157; Pisarski, 1975: 42; Kutter, 1977c: 252; Collingwood, 1979: 124; Atanassov & Dlussky, 1992: 264. Senior synonym of *brevisetosa:* Finzi, 1928a: 68; of *subrufoides:* Agosti & Collingwood, 1987a: 59. Material of the *nomen nudum cinereoglebaria* referred here: Dlussky & Pisarski, 1971: 157. Current subspecies:

nominal plus *armenica, cinereoimitans, fuscocinerea, iberica, italica, novaki.*

***cinereofusca.*** *Formica* (*Serviformica*) *cinerea* var. *cinereofusca* Karavaiev, 1929b: 214 (w.) CAUCASUS. Raised to species: Dlussky, 1967a: 60; Arnol'di & Dlussky, 1978: 553; Agosti & Collingwood, 1987a: 59.

*cinereofuscoides. Formica fusca* r. *gagates* var. *cinereofuscoides* Forel, 1892i: 307. Unavailable name.

*cinereoglebaria. Formica cinerea* var. *cinereoglebaria* Kulmatycki, 1922: 85, no caste given, POLAND. *Nomen nudum;* material referred to *cinerea* by Dlussky & Pisarski, 1971: 157.

*cinereoides. Formica fusca* subsp. *rufibarbis* var. *cinereoides* Forel, 1874: 312, no caste given, SWITZERLAND. Unavailable name.

#***cinereoimitans.*** *Formica cinerea* var. *cinereoimitans* Ruzsky, 1905b: 405 (w.) RUSSIA.

*cinereorufibarbis. Formica fusca* var. *cinereorufibarbis* Forel, 1874: 55 (w.q.m.) SWITZERLAND. Combination in *F.* (*Serviformica*): Forel, 1915d: 64. Variety of *rufibarbis:* Dalla Torre, 1893: 210; of *cinerea:* Forel, 1915d: 64; Emery, 1916b: 255; Emery, 1925b: 246. Junior synonym of *rufibarbis:* Bernard, 1967: 297; of *fuscocinerea:* Dlussky & Pisarski, 1971: 161. Revived from synonymy, and raised to species: Kutter, 1977c: 253. Junior synonym of *rufibarbis:* Collingwood, 1978: 73.

*circumspecta* Smith, F. 1861; see under ***CAMPONOTUS.***

***clara.*** *Formica rufibarbis* var. *clara* Forel, 1886f: 206 (w.) RUSSIA. Ruzsky, 1902b: 471 (q.m.). Combination in *F.* (*Serviformica*): Emery, 1925b: 250. Race of *rufibarbis:* Forel, 1901h: 65. Raised to species: Collingwood, 1961a: 68; Dlussky, 1967a: 77. Material of the unavailable names *transversa* and *clara* referred here: Dlussky, 1967a: 77. Senior synonym of *rizadaghi:* Dlussky, Soyunov & Zabelin, 1990: 139.

*clara. Formica rufibarbis* subsp. *montana* var. *clara* Kuznetsov-Ugamsky, 1926b: 98 (w.) TURKESTAN. Unavailable name; material referred to *clara* Forel (above) by Dlussky, 1967a: 77.

*clara. Formica rufibarbis* subsp. *subpilosa* var. *clara* Kuznetsov-Ugamsky, 1926b: 98 (w.) TURKESTAN. Unavailable name; material referred to *subpilosa* by Dlussky, 1965a: 28.

*clara. Formica* (*Raptiformica*) *sanguinea* var. *clara* Karavaiev, 1927d: 345 (w.) UKRAINE. [Junior primary homonym of *clara* Forel, 1886f: 206.] Replacement name: *leninei* Santschi, 1928b: 46.

*clarior. Formica* (*Raptiformica*) *sanguinea* var. *clarior* Ruzsky, 1905b: 420 (w.) CAUCASUS. Junior synonym of *sanguinea:* Dlussky, 1965a: 16.

*clarior. Formica rufibarbis* subsp. *subpilosa* var. *clarior* Ruzsky, 1915a: 421 (w.) RUSSIA. Unavailable name; material referred to *clarissima* Emery, 1925b: 250.

#***clarissima.*** *Formica* (*Serviformica*) *rufibarbis* var. *clarissima* Emery, 1925b: 250 (w.) RUSSIA. [First available name for material formerly referred to *clarior* Ruzsky, 1915a: 421, above; unavailable name.] Senior synonym of *ruzskyi* (unnecessary replacement name): **new synonymy**. Subspecies of *subpilosa:* Dlussky, 1965a: 30 (under the name *ruzskyi*).

#***clarorufibarbis.*** *Formica rufibarbis* var. *clarorufibarbis* Wheeler, W.M. & Mann, 1916: 173 (w.) RUSSIA. [First available use of *Formica rufibarbis* subsp. *clara* var. *clarorufibarbis* Ruzsky, 1905b: 400; unavailable name.]

*clavata* Fabricius, 1775; see under ***PARAPONERA.***

*clavigera* Roger, 1862; see under ***ACANTHOMYOPS.***

#***clivia.*** *Formica rufa* subsp. *clivia* Creighton, 1940a: 8, fig. 1 (w.q.m.) U.S.A. Subspecies of *obscuriventris:* Creighton, 1950a: 494.

****clymene.*** **Formica clymene* Wheeler, W.M. 1915h: 127 (w.) BALTIC AMBER (Oligocene). [**Formica clymene* Wheeler, W.M. 1913f: 383. *Nomen nudum.*] See also: Dlussky, 1967b: 80.

*coarctata* Latreille, 1802; see under ***PONERA.***

****cockerelli.*** **Formica cockerelli* Carpenter, 1930: 57, pl. 4, fig. 3 (q.m.) U.S.A. (Miocene).

*coeca* Latreille, 1802; see under ***LABIDUS.***

*cognata. Formica cognata* Stephens, 1829: 357. *Nomen nudum.*

*cognata* Smith, F. 1858; see under ***CAMPONOTUS.***

***coloradensis.*** *Formica rufa* subsp. *coloradensis* Creighton, 1940a: 1 (w.q.) U.S.A. [First available use of *Formica truncicola* subsp. *integroides* var. *coloradensis* Wheeler, W.M. 1913f: 440; unavailable name.] Subspecies of *integroides:* Creighton, 1950a: 489. Raised to species: Wheeler, G.C. & Wheeler, J. 1986g: 81; MacKay, Lowrie, *et al.* 1988: 111 (in key).

***comata.*** *Formica comata* Wheeler, W.M. 1909e: 85 (w.q.m.) U.S.A. See also: Wheeler, W.M. 1913f: 454.

*compressa* Fabricius, 1787; see under ***CAMPONOTUS.***

#***comptula.*** *Formica oreas* var. *comptula* Wheeler, W.M. 1913f: 460 (w.q.) U.S.A. Subspecies of *oreas:* Creighton, 1950a: 494.

*congerens. Formica congerens* Nylander, 1846a: 906 (w.) FINLAND. Nylander, 1849: 30 (m.); Foerster, 1850a: 17 (q.). Junior synonym of *pratensis:* Emery & Forel, 1879: 450; Nasonov, 1889: 17; Wheeler, W.M. 1913f: 428; Forel, 1915d: 57; Emery, 1916b: 256; Müller, 1923: 142. Revived from synonymy: Betrem, 1953: 324. Junior synonym of *lugubris:* Yarrow, 1955a: 5; Dlussky, 1967a: 91.

*conica. Formica conica* Fabricius, 1798: 279 (m.) INDIA. Unidentifiable to genus; *incertae sedis* in *Formica.* [Dalla Torre, 1893: 183 places this in *Lasius,* which is certainly incorrect. Forel, 1905b: 161 suggests that it may be the male of *Camponotus compressus,* but there is little in the original description to confirm this.]

*connecticutensis. Formica connecticutensis* Buckley, 1866: 154 (w.q.) U.S.A. Unidentifiable but possibly a *Formica;* see Wheeler, W.M. 1902f: 19. *Incertae sedis* in *Formica.*

*consanguinea* Smith, F. 1861; see under ***CAMPONOTUS.***

*consectator* Smith, F. 1858; see under ***CAMPONOTUS.***

*consobrina* Erichson, 1842; see under ***CAMPONOTUS.***

*consocians. Formica difficilis* var. *consocians* Wheeler, W.M. 1904f: 371 (w.q.m.) U.S.A. Junior synonym of *difficilis:* Cole, 1943a: 389.

*conspicua* Smith, F. 1858; see under ***CAMPONOTUS.***

#***constricta.*** *Formica* (*Formica*) *rufa* var. *constricta* Karavaiev, 1929b: 215, fig. 6 (w.) RUSSIA.

*consultans* Walker, 1859; see under ***ANOCHETUS.***

*contracta* Latreille, 1802; see under ***PONERA.***

**cordata* Holl, 1829; see under ***PHEIDOLE.***

*cordata* Smith, F. 1859; see under ***PHILIDRIS.***

*cordieri. Formica pratensis* var. *cordieri* Bondroit, 1917a: 174 (diagnosis in key) (q.) FRANCE. Raised to species: Betrem, 1960b: 77. Junior synonym of *nigricans:* Betrem, 1962: 38; Kutter, 1965: 140.

*coronata* Fabricius, 1804; see under ***ACROMYRMEX.***

*corusca* Smith, F. 1862; see under ***CAMPONOTUS.***

*cosmica* Smith, F. 1858; see under ***CAMPONOTUS.***

*coxalis* Smith, F. 1859; see under ***CAMPONOTUS.***

*crassinoda* Latreille, 1802; see under ***PACHYCONDYLA.***

***creightoni.*** *Formica* (*Raptiformica*) *creightoni* Buren, 1968a: 35, fig. 3 (w.) U.S.A.

*crepusculascens* Roger, 1859; see under ***PRENOLEPIS.***

*crinita* Smith, F. 1858; see under ***LASIUS.***

*crinita. Formica crinita* Wheeler, W.M. 1909e: 87 (w.q.) U.S.A. [Junior primary homonym of *crinita* Smith, F. 1858b: 13, above.] Replacement name: *criniventris* Wheeler, W.M. 1912c: 90. [*Formica crinita* Wheeler, W.M. 1908g: 408. *Nomen nudum.*]

***criniventris.*** *Formica criniventris* Wheeler, W.M. 1912c: 90. Replacement name for *crinita* Wheeler, W.M., above. [Junior primary homonym of *crinita* Smith, F. 1858b: 13.]

*cruda* Smith, F. 1860; see under ***PHILIDRIS.***

*cruentata* Latreille, 1802; see under ***CAMPONOTUS.***

*cuneata* Perty, 1833; see under ***CAMPONOTUS.***

***cunicularia.*** *Formica cunicularia* Latreille, 1798: 40 (w.q.m.) FRANCE. Combination in *F.* (*Serviformica*): Forel, 1915d: 64. Junior synonym of *rufibarbis:* Walckenaer, 1802: 161; Dalla Torre, 1893: 209; Ruzsky, 1905b: 385; Forel, 1915d: 64; Emery, 1916b: 255; Emery, 1925b: 250. Revived from synonymy, and status as species: Yarrow, 1954a: 231. Senior synonym of *rubescens:* Yarrow, 1954a: 231; Dlussky, 1967a: 73; Bernard, 1967: 296; of *fuscorufibarbis:* Dlussky, 1967a: 73; Dlussky & Pisarski, 1971: 166; of *glebaria:* Bernard, 1967: 296; Boven, 1977: 164; Agosti & Collingwood, 1987a: 59; of *glabridorsis:* Wu, 1990: 4; of *glauca* (and its junior synonyms *caucasica, katuniensis, montivaga, montaniformis* and *volgensis*): Atanassov & Dlussky, 1992: 267; of *fuscoides:* Arakelian, 1994: 94.

***curiosa.*** *Formica curiosa* Creighton, 1935: 5, fig. 2 (w.q.) U.S.A. Senior synonym of *parcipappa:* Snelling, 1969b: 194.

*cursor* Fonscolombe, 1846; see under ***CATAGLYPHIS.***

*curvidentata* Latreille, 1802; see under ***ECITON.***

*custodiens* Smith, F. 1858; see under ***ANOPLOLEPIS.***

*cylindrica* Fabricius, 1798; see under ***CAMPONOTUS.***

***dakotensis.*** *Formica dakotensis* Emery, 1893i: 652, pl. 22, fig. 5 (w.) U.S.A. Wasmann, 1902a: 4 (q.m.); Wheeler, W.M. 1913f: 462 (q.). Senior synonym of *saturata* and *wasmanni* (and its junior synonym *specularis*): Creighton, 1950a: 484; of *montigena:* Brown, 1957f: 167.

*dalailamae. Formica* (*Neoformica*) *dalailamae* Ruzsky, 1915a: 431, fig. 9 (w.) TIBET. Combination in *F.* (*Serviformica*): Emery, 1925b: 246. Junior synonym of *sentschuensis:* Dlussky, 1965a: 17.

*dalcqi. Formica dalcqi* Bondroit, 1918: 63 (w.) FRANCE. Subspecies of *exsecta:* Emery, 1925b: 257. Junior synonym of *exsecta:* Bernard, 1967: 323.

*dalmatica* Nylander, 1849; see under ***CAMPONOTUS.***

*davisi. Formica exsectoides* var. *davisi* Wheeler, W.M. 1913f: 396 (in key) (w.q.) CANADA. [*Formica exsectoides* subsp. *exsectoides* var. *davisi* Wheeler, W.M. 1913f: 484; unavailable name.] Junior synonym of *exsectoides:* Creighton, 1950a: 513.

***decipiens.*** *Formica decipiens* Bondroit, 1918: 52 (w.q.) FRANCE. Espadaler, 1986b: 46 (m.). Combination in *F.* (*Serviformica*): Emery, 1925b: 247. Variety of *fusca:* Menozzi, 1922b: 331. Revived status as species: Bernard, 1967: 298. See also: Kutter, 1977c: 255.

*decora* Smith, F. 1858; see under ***CAMPONOTUS.***

*defensor. Formica defensor* Smith, F. 1878b: 11 (w.) CHINA. Junior synonym of *rufibarbis:* Forel, 1894c: 403.

*delicata. Formica pallidefulva* subsp. *delicata* Cole, 1938c: 369 (w.q.) U.S.A. Junior synonym of *nitidiventris:* Creighton, 1950a: 551.

****demersa.*** **Formica demersa* Heer, 1850: 140, pl.11, fig. 7 (m.) SWITZERLAND (Miocene).

***densiventris.*** *Formica fusca* var. *densiventris* Viereck, 1903: 74 (w.) U.S.A. Cole, 1954a: 90 (m.). Junior synonym of *subaenescens:* Wheeler, W.M. 1913f: 504. Revived from synonymy as subspecies of *fusca:* Brown, 1947: 7. Raised to species: Creighton, 1950a: 499. Senior synonym of *spicata:* Cole, 1954a: 89; Cole, 1955b: 50; of *rasilis:* MacKay, Lowrie, *et al.* 1988: 114.

*depilis* Santschi, 1925; see under ***PROFORMICA.***

*depressa* Latreille, 1802; see under ***CREMATOGASTER.***

*desecta* Smith, F. 1860; see under ***CAMPONOTUS.***

*destructor* Fabricius, 1804; see under ***GIGANTIOPS.***

*detecta* Smith, F. 1858; see under ***IRIDOMYRMEX.***

*detorquens* Walker, 1859; see under ***TECH-NOMYRMEX.***

*devastator. Formica devastator* Christ, 1791: 517 (w.) INDIA. *Nomen nudum.*

*didyma. Formica didyma* Fabricius, 1782: 489 (w.) ITALY. Unidentifiable to genus; *incertae sedis* in *Formica.*

***difficilis.*** *Formica rufa* subsp. *difficilis* Emery, 1893i: 651, pl. 22, figs. 9, 14 (w.q.m.) U.S.A. Wheeler, G.C. & Wheeler, J. 1953c: 164 (l.). Raised to species: Wheeler, W.M. 1904e: 305; Wheeler, W.M. 1904f: 348. Senior synonym of *consocians* and *habrogyna:* Cole, 1943a: 389.

*diffidens* Walker, 1859; see under ***PHEIDOLE.***

*diligens* Smith, F. 1857; see under ***CAMPONOTUS.***

***dirksi.*** *Formica dirksi* Wing, 1949: 13 (q.) U.S.A.

*discolor* Buckley, 1866; see under ***CAMPONOTUS.***

*dislocata. Formica dislocata* Say, 1836: 288 (q.m.) U.S.A. Unidentifiable to genus; *incertae sedis* in *Formica.*

*distinguenda* Spinola, 1851; see under ***CAMPONOTUS.***

*distinguenda* Emery, 1916 [*Formicina*]; see under ***LASIUS.***

***dlusskyi.*** *Formica dlusskyi* **new name**. Replacement name for *longiceps* Dlussky, 1964: 1036. [Junior primary homonym of *longiceps* Smith, F. 1863: 13.]

**#dolosa.*** *Formica* (*Neoformica*) *pallidefulva* subsp. *dolosa* Buren, 1944a: 309 (w.) U.S.A. [First available use of *Formica pallidefulva* subsp. *schaufussi* var. *dolosa* Wheeler, W.M. 1912c: 90; unavailable name and unnecessary replacement name for *meridionalis* Wheeler, W.M. 1904f: 370, unavailable name; see there.] Subspecies of *schaufussi:* Creighton, 1950a: 551.

*dominula. Formica dominula* Nylander, 1846a: 905, pl. 18, fig. 15 (w.q.m.) FINLAND. Junior synonym of *sanguinea:* Smith, F. 1851: 115.

*dorsata. Formica dorsata* Panzer, 1798: no. 1651 (q.) AUSTRIA. Junior synonym of *rufa:* Stephens, 1829: 357; Nylander, 1846a: 902.

*dorycus* Smith, F. 1860; see under ***CAMPONOTUS.***

*dryas. Formica dryas* Wheeler, W.M. 1905c: 268 (w.q.) U.S.A. Junior synonym of *obscuriventris:* Wheeler, W.M. 1913f: 446.

***dusmeti.*** *Formica rufa* subsp. *dusmeti* Emery, 1909b: 188 (w.) SPAIN. Bondroit, 1920a: 145 (q.); Santschi, 1932g: 5 (q.m.). Subspecies of *truncicola:* Wheeler, W.M. 1913f: 438; Stitz, 1939: 346; of *rufa:* Emery, 1925b: 254. Raised to species: Bondroit, 1918: 60; Santschi, 1931a: 13. Junior synonym of *nigricans:* Bernard, 1967: 314. Revived from synonymy and status as species: Collingwood & Yarrow, 1969: 94; Collingwood, 1978: 74.

*edax* Forskål, 1775; see under ***PHEIDOLE.***

*egregia* Smith, F. 1858; see under ***CAMPONOTUS.***

*elata. Formica elata* Lund, 1831a: 130 (w.) BRAZIL. *Nomen nudum.*

*elevata. Formica elevata* Fabricius, 1782: 492 (w.) "East Indies". Unidentifiable to genus; *incertae sedis* in *Formica.*

*elongata. Formica elongata* Fabricius, 1787: 308 (w.) INDIA. Unidentifiable to genus; *incertae sedis* in *Formica.*

*emarginata* Olivier, 1792; see under ***LASIUS.***

***emeryi.*** *Formica emeryi* Wheeler, 1913f: 419 (w.q.) U.S.A. Combination in *F.* (*Raptiformica*): Emery, 1925b: 259. Junior synonym of *subnuda:* Wilson & Brown, 1955: 118. Revived from synonymy: Buren, 1968a: 34.

*emeryi. Formica rufa* ab. *emeryi* Krausse, 1926a: 108 (w.) GERMANY. See also: Stitz, 1939: 337. Junior synonym of *rufa:* Yarrow, 1955a: 4.

*emmae* Forel, 1909; see under ***CATAGLYPHIS.***

****eoptera.*** **Formica eoptera* Cockerell, 1923b: 399, fig. 1 (wing) U.S.A. (Eocene).

*ephippium* Smith, F. 1858; see under ***CAMPONOTUS.***

*erratica* Latreille, 1798; see under ***TAPINOMA.***

*erythrocephala* Fabricius, 1775; see under ***LEPTOMYRMEX.***

*esuriens* Smith, F. 1858; see under ***CAMPONOTUS.***

*etrusca. Formica exsecta* var. *etrusca* Emery, 1909b: 191, fig. 5 (w.) ITALY. Raised to species: Bondroit, 1918: 64. Variety of *exsecta:* Emery, 1925b: 257; Karavaiev, 1926e: 196. Junior synonym of *exsecta:* Bernard, 1967: 323.

*europaea* Christ, 1791; see under ***CATAGLYPHIS.***

*exasperata* Smith, 1857; see under ***CAMPONOTUS.***

*exercita* Walker, 1859; see under ***POLYRHACHIS.***

***exsecta.*** *Formica exsecta* Nylander, 1846a: 909, pl. 18, fig. 20 (w.q.m.) FINLAND. Combination in *F.* (*Coptoformica*): Müller, 1923: 146. Senior synonym of *dalcqi* and *etrusca:* Bernard, 1967: 323; of *exsectopressilabris:* Bernard, 1967: 323; Dlussky & Pisarski, 1971: 194; of *kontuniemii, sudetica* and *wheeleri:* Dlussky, 1967a: 100; Dlussky & Pisarski, 1971: 194; of *rubens:* Dlussky, 1964: 1027. Current subspecies: nominal plus *manchu.* See also: Donisthorpe, 1915d: 273; Parapura, 1972: 763; Kutter, 1977c: 283; Agosti & Hauschteck-Jungen, 1988: 280; Kupyanskaya, 1990: 200; Atanassov & Dlussky, 1992: 282.

***exsectoides.*** *Formica exsectoides* Forel, 1886b: xxxviii (w.q.) U.S.A. Emery, 1893i: 653 (m.). Senior synonym of *davisi* and *hesperia:* Creighton, 1950a: 513. See also: Wheeler, W.M. 1913f: 481; Smith, D.R. 1979: 1456.

*exsectopressilabris. Formica exsecta* var. *exsectopressilabris* Forel, 1874: 52 (w.q.m.) SWITZERLAND. Variety of *pressilabris:* Dalla Torre, 1893: 205. Raised to species: Bondroit, 1912: 352; Bondroit, 1918: 63. Variety of *exsecta:* Karavaiev, 1927c: 284; Karavaiev, 1936: 255; Boven, 1947: 189. Junior synonym of *exsecta:* Bernard, 1967: 323; Dlussky & Pisarski, 1971: 194.

*exudans* Walker, 1859; see under ***LEPTOGENYS.***

*fabricator* Smith, F. 1858; see under ***CAMPONOTUS.***

*fairchildi. Formica fusca* var. *fairchildi* Wheeler, W.M. 1927a: 40 (w.) SUMATRA (locality in error). Junior synonym of *glacialis:* Francoeur, 1977a: 11.

*fallax* Nylander, 1856; see under ***CAMPONOTUS.***

*fallax. Formica* (*Serviformica*) *rufibarbis* st. *subpilosa* var. *fallax* Santschi, 1928b: 45. Unnecessary and unavailable replacement name for *clara* Kuznetsov-Ugamsky; unavailable name. Material referred to *subpilosa* by Dlussky, 1967a: 70.

*familiaris* Smith, F. 1860: 68; see under ***PSEUDOLASIUS.***

*familiaris* Smith, F. 1860: 96; see under ***TAPINOMA.***

*fatale. Formica fatale* Christ, 1791: 515 (w.) INDIA. Unidentifiable to genus; *incertae sedis* in *Formica.* [Emery, 1892b: 162, suspected that this may be a termite.]

*femorata* Fabricius, 1804; see under ***CAMPONOTUS.***

***ferocula.*** *Formica ferocula* Wheeler, W.M. 1913f: 461 (w.) U.S.A.

*ferruginea. Formica ferruginea* Christ, 1791: 512, pl. 60, fig. 10 (w.) no locality. Junior synonym of *rufa:* Emery, 1892b: 161.

*ferruginea* Fabricius, 1798; see under ***CAMPONOTUS.***

*fervens* Drury, 1782; see under ***ATTA.***

*fervens* Smith, F. 1857; see under ***CAMPONOTUS.***

*festina* Smith, F. 1857; see under ***CAMPONOTUS.***

*festinata* Buckley, 1866; see under ***CAMPONOTUS.***

*ficticia. Formica subpolita* var. *ficticia* Wheeler, W.M. 1913f: 561 (w.q.m.) U.S.A. Subspecies of *subpolita:* Creighton, 1950a: 542. Junior synonym of *subpolita:* Francoeur, 1973: 107.

*filchneri. Formica fusca* subsp. *gagates* var. *filchneri* Forel, 1907b: 208 (w.) CHINA. [Also described as new by Forel, 1908d: 105.] Unavailable name; material referred to *picea* Nylander: by Emery, 1909b: 159.

*filiformis* Fabricius, 1804; see under ***PSEUDOMYRMEX.***

**#finzii.*** *Formica truncorum* var. *finzii* Stitz, 1939: 347 (w.q.) GERMANY. [First available use of *Formica rufa* subsp. *truncicola* var. *finzii* Krausse, 1926b: 114; unavailable name.]

*flammiventris. Formica flammiventris* Wheeler, W.M. 1912c: 90. Replacement name for *rufiventris* Emery, 1893i: 665. [Junior primary homonym of *rufiventris* Fabricius, 1804: 409.] Junior synonym of *subpolita:* Wheeler, W.M. 1913f: 532.

*flava* Fabricius, 1782; see under ***LASIUS.***

*flavescens* Fabricius, 1793; see under ***CAMPONOTUS.***
*flavicornis* Fabricius, 1789; see under ***ATTA.***
*flavipes. Formica flavipes* Geoffroy, in Fourcroy, 1785: 452 (m.) FRANCE. Junior synonym of *fusca:* Latreille, 1802c: 161.
*flavitarsa* Smith, F. 1858; see under ***LEPTOMYRMEX.***
*flavorubra. Formica sanguinea* var. *flavorubra* Forel, 1909c: 105 (w.) SPAIN. Junior synonym of *sanguinea:* Dlussky, 1967a: 97; Collingwood, 1978: 74.
*flavosetosa* Viehmeyer, 1922; see under ***PROFORMICA.***
*****flori.*** **Formica flori* Mayr, 1868c: 48, pl. 2, figs. 35–37 (w.q.m.) BALTIC AMBER (Oligocene). See also: Dlussky, 1967b: 81; Baroni Urbani & Graeser, 1987: 1.
*floridana* Buckley, 1866; see under ***CAMPONOTUS.***
*foetens. Formica foetens* Olivier, 1792: 503. Replacement name for *foetida* De Geer, 1773: 611. [Junior primary homonym of *foetida* Linnaeus, 1758: 582.] Unidentifiable to genus; *incertae sedis* in *Formica.* [It is apparent that this taxon is properly referable to subfamily Myrmicinae.]
*foetens* Fabricius, 1793; see under ***PACHYCONDYLA.***
*foetida* Linnaeus, 1758; see under ***PACHYCONDYLA.***
*foetida. Formica foetida* De Geer, 1773: 611, pl. 31, figs. 21, 22 (q.) SURINAM. [Junior primary homonym of *foetida* Linnaeus, 1758: 582, above.] Replacement name: *foetens* Olivier, 1792: 503.
*foetida* Buckley, 1866; see under ***FORELIUS.***
*foreli. Formica foreli* Bondroit, 1918: 65 (w.q.m.) SWITZERLAND. [First available use of *Formica exsecta* subsp. *pressilabris* var. *foreli* Emery, 1909b: 192; unavailable name.] Combination in *F.* (*Coptoformica*), and subspecies of *pressilabris:* Müller, 1923: 146. Revived status as species: Dlussky, 1964: 1033; Bernard, 1967: 324; Kutter, 1977c: 284. Junior synonym of *pressilabris:* Arakelian, 1994: 97.
*foreli. Formica rufa* subsp. *pratensis* var. *foreli* Krausse, 1926a: 108 (w.) GERMANY. Unavailable name; material referred to *nigropratensis* by Betrem, 1962: 38.
***foreliana.*** *Formica foreliana* Wheeler, W.M. 1913f: 451 (w.) U.S.A. Francoeur, 1973: 235 (q.m.). Junior synonym of *gnava:* Creighton, 1950a: 539. Revived from synonymy: Francoeur, 1973: 235.
*forficata* Fabricius, 1787; see under ***MYRMECIA.***
#***formosae.*** *Formica* (*Serviformica*) *picea* var. *formosae* Emery, 1925b: 249 (w.) TAIWAN. [First available use of *Formica fusca* r. *picea* var. *formosae* Forel, 1913f: 200.] Subspecies of *candida:* current status.
***forsslundi.*** *Formica* (*Coptoformica*) *forsslundi* Lohmander, 1949: 163 (w.q.m.) SWEDEN. Senior synonym of *nemoralis:* Dlussky, 1967a: 105; of *strawinskii:* Dlussky & Pisarski, 1971: 199. See also: Kutter, 1977c: 285.
***fossaceps.*** *Formica fossaceps* Buren, 1942: 402 (w.q.m.) U.S.A.
***fossilabris.*** *Formica* (*Coptoformica*) *fossilabris* Dlussky, 1965a: 25, fig. 16 (w.) TIBET.
*foveolata* Mayr, 1853; see under ***CAMPONOTUS.***
*fragilis* Smith, F. 1859; see under ***LEPTOMYRMEX.***
*****fragilis.*** **Formica fragilis* Heer, 1867: 8, pl. 1, figs. 4, 5 (q.m.) YUGOSLAVIA (Miocene). [Unresolved junior primary homonym of *fragilis* Smith, F. 1859a: 136, above.] Combination in *Lasius* (part) and *Plagiolepis* (part): Mayr, 1867b: 56; in *Lasius:* Scudder, 1891: 713; in *Plagiolepis:* Handlirsch, 1907: 859. Unidentifiable to genus; *incertae sedis* in *Formica.*
***francoeuri.*** *Formica francoeuri* **new name.** Replacement name for *pilicornis* Emery, 1893i: 664. [Junior primary homonym of *pilicornis* Roger, 1859: 228.] For previous history see under *pilicornis* Emery.
*fraterna. Formica fraterna* Smith, F. 1878b: 11 (w.) CHINA. Junior synonym of *rufibarbis:* Forel, 1894c: 403; Bingham, 1903: 335.
**freyeri* Heer, 1867; see under ****LONCHOMYRMEX.***
#***frontalis.*** *Formica truncorum* var. *frontalis* Santschi, 1919e: 246 (w.) SPAIN.
*fugax* Latreille, 1798; see under ***SOLENOPSIS.***
***fukaii.*** *Formica exsecta* var. *fukaii* Wheeler, W.M. 1914a: 26 (w.) JAPAN. Kupyanskaya, 1990: 201 (q.m.). Raised to species, and senior synonym of *mesasiatica:* Sonobe & Dlussky, 1977: 23. [Misspelled as *fukali* by Emery, 1925b: 257; and as *fukalii* by Wu, 1990: 5.]
*fuliginosa* Latreille, 1798; see under ***LASIUS.***
***fuliginothorax.*** *Formica fuliginothorax* Blacker, 1992: 11 (w.) CANADA.
*fulvacea* Norton, 1868; see under ***CAMPONOTUS.***
*fulvopilosa* De Geer, 1778; see under ***CAMPONOTUS.***
*fungosa* Fabricius, 1798; see under ***DOLICHODERUS.***
***fusca.*** *Formica fusca* Linnaeus, 1758: 580 (w.) EUROPE. Latreille, 1802c: 159 (q.m.); Wheeler, G.C. & Wheeler, J. 1953c: 164 (l.). Combination in *F.* (*Serviformica*): Forel, 1913i: 361. Senior synonym of *barbata:* Emery & Forel, 1879: 451; of *chevrolatii:* Dalla Torre, 1893: 193; of *flavipes* and *libera:* Latreille, 1802c: 159; Smith, F. 1851: 3; Roger, 1863b: 13; of *marcida:* Francoeur, 1973: 189; of *pallipes* and *rufipes:* Dlussky, 1967a: 58; of *tristis:* Emery, 1892b: 162. Current subspecies: nominal plus *alpicola, fuscolemani, hyrcana, maura, tombeuri.* See also: Wheeler, W.M. 1913f: 494; Donisthorpe, 1915d: 304; Stitz, 1939: 348; Dlussky, 1967a: 58; Dlussky & Pisarski, 1971: 148; Francoeur, 1973: 189; Tarbinsky, 1976: 180; Collingwood, 1979: 120; Kupyanskaya, 1990: 183; Atanassov & Dlussky, 1992: 260.
*fusca* Leach, 1825; see under ***TETRAMORIUM.***
*fuscata. Formica pallidefulva* subsp. *fuscata* Emery, 1893i: 656 (w.q.) U.S.A. Combination in *F.* (*Neoformica*): Wheeler, W.M. 1913f: 557. Junior synonym of *nitidiventris:* Creighton, 1950a: 551.
*fuscescens Formica fuscescens* Gmelin, in Linnaeus, 1790: 2804 (w.) EUROPE. Unidentifiable to genus; *incertae sedis* in *Formica.*
*fuscicauda. Formica fuscicauda* Motschoulsky, 1863: 12 (w.) SRI LANKA. Unidentifiable to genus; *incertae sedis* in *Formica.*
*fusciceps. Formica sanguinea* var. *fusciceps* Emery, 1895c: 335 (footnote) (w.) JAPAN. Junior synonym of *sanguinea:* Dlussky, 1965a: 16.
*fuscipes* Mayr, 1853; see under ***CAMPONOTUS.***
#***fuscocinerea.*** *Formica fusca* var. *fuscocinerea* Forel, 1874: 55 (w.q.m.) SWITZERLAND. Variety/subspecies of *cinerea:* Wheeler, W.M. 1913f: 522; Dlussky, 1967a: 65; Dlussky & Pisarski, 1971: 161; Baroni Urbani, 1971c: 234; Pisarski, 1975: 43.
*fuscogagates. Formica fusca* var. *fuscogagates* Forel, 1874: 54 (w.q.) SWITZERLAND. Combination in *F.* (*Serviformica*): Forel, 1915d: 63. Raised to species: Bondroit, 1918: 47. Variety of *gagates:* Emery, 1909b: 195; Emery, 1925b: 249. Junior synonym of *gagates:* Kutter, 1977c: 254.
*fuscoides. Formica* (*Serviformica*) *cunicularia* subsp. *fuscoides* Dlussky, 1967a: 74 (w.q.m.) ARMENIA. Junior synonym of *cunicularia:* Arakelian, 1994: 94.
#***fuscolemani.*** *Formica fusca* form. *fuscolemani* Samsinak, 1951: 126 (m.) CZECHOSLOVAKIA.
*fuscoptera* Geoffroy, in Fourcroy, 1785; see under ***CAMPONOTUS.***
*fuscorufibarbis. Formica fusca* var. *fuscorufibarbis* Forel, 1874: 54 (w.q.) SWITZERLAND. Combination in *F.* (*Serviformica*): Forel, 1915d: 63. Raised to species: Forel, 1906c: 189. Variety of *glebaria:* Bondroit, 1918: 50; of *rufibarbis:* Dalla Torre, 1893: 210; Stitz, 1939: 357; Novak & Sadil, 1941: 107. Junior synonym of *rufibarbis:* Bernard, 1967: 297; of *cunicularia:* Dlussky, 1967a: 73; Dlussky & Pisarski, 1971: 166.
***gagates.*** *Formica gagates* Latreille, 1798: 36 (w.q.) FRANCE. Mayr, 1855: 349 (m.). Combination in *F.* (*Serviformica*): Forel, 1915d: 63. Variety/race of *fusca:* Forel, 1874: 53; Emery & Forel, 1879: 451; Mayr, 1886d: 426; Forel, 1892i: 307; Forel, 1907b: 208; Bondroit, 1910: 483. Status as species: André, 1882b: 182; Nasonov, 1889: 19; Ruzsky, 1902d: 13; Forel, 1904b: 384; Wheeler, W.M. 1913f: 513; Forel, 1915d: 63; Emery, 1916b: 254; Bondroit, 1918: 46; Karavaiev, 1927c: 285; Karavaiev, 1936: 229; Stitz, 1939: 362; Bernard, 1967: 298; Dlussky, 1967a: 79; Kutter, 1977c: 250; Atanassov & Dlussky, 1992: 269. Senior synonym of *capsincola:* Mayr, 1855: 348; Nylander, 1856b: 65; of *morio:* Emery, 1925b: 248; Karavaiev, 1936: 230; Dlussky & Pisarski, 1971: 169; of *fuscogagates:* Kutter, 1977c: 254. Material of the unavailable name *muralewiczi* referred here: Emery, 1909b: 194; Karavaiev, 1927a: 301. Current subspecies: nominal plus *piceogagates.*
***gagatoides.*** *Formica fusca* var. *gagatoides* Ruzsky, 1904a: 289 (w.q.) RUSSIA. Ruzsky, 1915a: 423 (m.); Stärcke, 1935: 267 (footnote) (m.). Combination in *F.* (*Serviformica*): Emery, 1925b: 249. Variety of *picea* Nylander: Emery, 1925b: 249; Stitz, 1939: 368. Raised to species: Holgersen, 1942: 15; Holgersen, 1943: 9. Junior synonym of *picea* Nylander: Bernard, 1967: 299. Revived from synonymy and status as species: Kutter, 1977c: 255; Collingwood, 1979: 121; Kupyanskaya, 1990: 188.
*gaullei. Formica gaullei* Bondroit, 1917a: 176, fig. 1 (w.) FRANCE. Variety of *rufa:* Emery, 1925b: 253; Stitz, 1939: 339; Röszler, 1950: 213. Junior synonym of *rufa:* Yarrow, 1955a: 4.
*gelida. Formica fusca* var. *gelida* Wheeler, W.M. 1913f: 399 (in key) (w.q.m.) U.S.A. [*Formica fusca* subsp. *fusca* var. *gelida* Wheeler, W.M. 1913f: 505; unavailable name.] Variety of *fusca:* Wheeler, W.M. 1917a: 546; of *neorufibarbis:* Creighton, 1950a: 537. Junior synonym of *neorufibarbis:* Francoeur, 1973: 215.
***gerardi.*** *Formica gerardi* Bondroit, 1917b: 186 (w.) FRANCE. Provisional junior synonym of *subrufa:* Bernard, 1967: 301. Revived status as species: Collingwood & Yarrow, 1969: 89; Collingwood, 1978: 73.
*gibba* Smith, F. 1861; see under ***DOLICHODERUS.***
*gibbosa* Latreille, 1798; see under ***APHAENOGASTER.***

***gibbosa.** *Formica gibbosa* Presl, 1822: 197 (no caste) BALTIC AMBER (Oligocene). [Unresolved junior primary homonym of *gibbosa* Latreille, 1798: 50, above.]

*gibbosa* Smith, F. 1858; see under ***DOLICHODERUS.***

*gigas* Latreille, 1802; see under ***CAMPONOTUS.***

*gigas* Leach, 1825; see under ***CAMPONOTUS.***

*gilvescens. Formica sanguinea* subsp. *subintegra* var. *gilvescens* Wheeler, W.M. 1913f: 412 (w.) U.S.A. Unavailable name; material referred to *subintegra* by Creighton, 1950a: 470.

*glabra. Formica glabra* Gmelin, in Linnaeus, 1790: 2804 (w.) EUROPE. Unidentifiable to genus; *incertae sedis* in *Formica.*

*glabra. Formica glabra* White, W.F. 1884: 253 (w.) GREAT BRITAIN. [Unresolved junior primary homonym of *glabra* Gmelin, above.] Junior synonym of *gagates:* Dalla Torre, 1893: 188. Junior synonym of *picea* Nylander: Donisthorpe, 1918a: 9; Dlussky, 1967a: 61.

*glabrella* Nylander, 1849; see under ***TAPINOMA.***

*glabridorsis. Formica (Serviformica) rufibarbis* var. *glabridorsis* Santschi, 1925f: 95 (w.) CHINA. Junior synonym of *glauca:* Dlussky, 1967a: 74; of *cunicularia:* Wu, 1990: 4.

**glacialis.** *Formica fusca* var. *glacialis* Wheeler, W.M. 1908f: 624 (w.q.m.) U.S.A. Combination in *F. (Serviformica)*: Emery, 1925b: 247. Junior synonym of *fusca:* Wheeler, W.M. 1913f: 494; Creighton, 1950a: 532. Revived from synonymy and raised to species: Francoeur, 1973: 152. Senior synonym of *fairchildi:* Francoeur, 1977a: 11. Junior synonym of *subsericea:* Wheeler, G.C. & Wheeler, J. 1986g: 92. Revived from synonymy: **revived status.**

*glauca. Formica rufibarbis* var. *glauca* Ruzsky, 1896: 70 (w.) RUSSIA. [First available use of *Formica fusca* subsp. *rufibarbis* var. *glauca* Ruzsky, 1895: 20; unavailable name.] Emery, 1909b: 198 (q.m.). Combination in *F. (Serviformica)*: Emery, 1925b: 250. Variety/subspecies of *rufibarbis:* Emery, 1925b: 250; of *cunicularia:* Pisarski, 1967: 416; Dlussky, 1967a: 74; Tarbinsky, 1976: 184. Raised to species: Agosti & Collingwood, 1987a: 59. Senior synonym of *caucasica, katuniensis, montaniformis, montivaga,* and *volgensis,* and material of the unavailable name *minor* referred here: Dlussky, 1967a: 74 (note that the name *montaniformis* is misspelled as *montanoides* here). Junior synonym of *cunicularia:* Atanassov & Dlussky, 1992: 267.

*glebaria. Formica glebaria* Nylander, 1846a: 917, pl. 18, fig. 23 (w.q.) FINLAND. Foerster, 1850a: 31 (m.). Combination in *F. (Serviformica)*: Forel, 1915d: 63. Variety/subspecies of *fusca:* Emery & Forel, 1879: 451; Wheeler, W.M. 1908g: 408; Emery, 1909b: 196; Ruzsky, 1914b: 104; Forel, 1915d: 63; Emery, 1916b: 254; Santschi, 1925g: 353; Karavaiev, 1927c: 286; Stitz, 1939: 353; Novak & Sadil, 1941: 108; of *rufibarbis:* Karavaiev, 1936: 237. Status as species: Bondroit, 1918: 49; Müller, 1923: 139; Boven, 1947: 188. Junior synonym of *fusca:* Mayr, 1855: 346; Yarrow, 1954a: 230; Dlussky, 1967a: 58; of *cunicularia:* Bernard, 1967: 296; Boven, 1977: 164; Agosti & Collingwood, 1987a: 59.

***globiventris.** *Formica globiventris* Heer, 1850: 131, pl. 9, fig. 13 (q.) SWITZERLAND (Miocene).

**globularis* Heer, 1850; see under ***LASIUS.***

**gnava.** *Formica gnava* Buckley, 1866: 156 (w.q.m.) U.S.A. Combination in *F. (Serviformica)*: Emery, 1925b: 250. Variety/subspecies of *fusca:* Wheeler, W.M. 1902f: 20; of *rufibarbis:* Wheeler, W.M. 1913f: 518; Creighton, 1950a: 539. Revived status as species: Francoeur, 1973: 238. Material of the *nomen nudum subsericeoneorufibarbis* (misspelled as *subsericeorufibarbis* by Emery, 1925b: 250) referred here: Wheeler, W.M. 1913f: 518.

**goesswaldi.** *Formica (Coptoformica) goesswaldi* Kutter, 1967a: 234, figs. 15–22 (w.q.m.) GERMANY. See also: Kutter, 1977c: 284.

*gracilescens* Nylander, 1856; see under ***PARATRECHINA.***

*gracilipes* Smith, F. 1857; see under ***ANOPLOLEPIS.***

*gracilis* Fabricius, 1804; see under ***PSEUDOMYRMEX.***

*gracilis* Lowne, 1865; see under ***IRIDOMYRMEX.***

*gracilis* Buckley, 1866; see under ***TAPINOMA.***

***gracilis.** *Formica gracilis* Heer, 1867: 7, pl. 1, fig. 3 (q.) YUGOSLAVIA (Miocene). [Unresolved junior primary homonym of *gracilis* Fabricius, 1804: 405, above.]

*graminicola* Latreille, 1802; see under ***MYRMECINA.***

***grandis.** *Formica grandis* Carpenter, 1930: 57, pl. 6, fig. 3 (q.) U.S.A. (Miocene).

*granulata* Latreille, 1802; see under ***CATAULACUS.***

**gravelyi.** *Formica gravelyi* Mukerjee, 1930: 162, fig. 6 (w.) INDIA.

***gravida.** *Formica gravida* Heer, 1850: 114, pl. 9, fig. 1 (q.) SWITZERLAND (Miocene).

*grisea* Le Guillou, 1842; see under ***POLYRHACHIS.***

*griseopubescens. Formica sanguinea* var. *griseopubescens* Kuznetsov-Ugamsky, 1926b: 95 (w.) TURKESTAN. Junior synonym of *sanguinea:* Dlussky, 1965a: 16.

*grossa* Fabricius, 1787; see under ***ATTA.***

**#grouvellei.** *Formica rufa* var. *grouvellei* Bondroit, 1918: 60 (q.) FRANCE.

*grundmanni. Formica moki* subsp. *grundmanni* Cole, 1943b: 184 (w.) U.S.A. Junior synonym of *xerophila:* Francoeur, 1973: 259.

*guineense* Fabricius, 1793; see under ***PHEIDOLE.***

*gulosa* Fabricius, 1775; see under ***MYRMECIA.***

*gymnomma. Formica dryas* var. *gymnomma* Wheeler, W.M. 1905c: 269 (w.) U.S.A. Subspecies of *rufa:* Creighton, 1940a: 3. Junior synonym of *obscuriventris:* Creighton, 1950a: 493.

**gynocrates.** *Formica gynocrates* Snelling & Buren, 1985: 72, figs. 7–15 (w.q.m.) U.S.A.

*habrogyna. Formica habrogyna* Cole, 1939: 413, figs. A-D (w.q.) U.S.A. Junior synonym of *difficilis:* Cole, 1943a: 389.

*haematocephala* Leach, 1825; see under ***CREMATOGASTER.***

*haematoda* Linnaeus, 1758; see under ***ODONTOMACHUS.***

*haemorrhoidalis* Latreille, 1802; see under ***ZACRYPTOCERUS.***

*haemorrhoidalis. Formica rufa* subsp. *haemorrhoidalis* Creighton, 1940a: 1 (w.) U.S.A. [First available use of *Formica rufa* subsp. *integra* var. *haemorrhoidalis* Emery, 1893i: 652 (w.); Wheeler, W.M. 1913f: 441 (q.m.); unavailable name.] [Junior primary homonym of *Formica haemorrhoidalis* Latreille, 1802c: 276, above.] Replacement name: *ravida* Creighton, 1940a: 1 (here designated first available replacement name as *haemorrhoidalis* Creighton was raised to species and made senior synonym of *ravida* and *tahoensis* by Brown, 1965d: 181).

*hamata* Fabricius, 1782; see under ***ECITON.***

*harpax* Fabricius, 1804; see under ***PACHYCONDYLA.***

*hastata* Latreille, 1802; see under ***POLYRHACHIS.***

*hebescens. Formica ulkei* var. *hebescens* Wheeler, W.M. 1913f: 487 (w.) U.S.A. Junior synonym of *ulkei:* Creighton, 1950a: 515.

*hemipsila* Foerster, 1850; see under ***CAMPONOTUS.***

**heraclea* Heer, 1850; see under ***CAMPONOTUS.***

*herculeana* Linnaeus, 1758; see under ***CAMPONOTUS.***

*herrichi* Mayr, 1853; see under ***CAMPONOTUS.***

*hesperia. Formica exsectoides* var. *hesperia* Wheeler, W.M. 1913f: 396 (in key) (w.) U.S.A. [*Formica exsectoides* subsp. *exsectoides* var. *hesperia* Wheeler, W.M. 1913f: 484; unavailable name.] Variety of *exsectoides:* Wheeler, W.M. 1917a: 544. Junior synonym of *exsectoides:* Creighton, 1950a: 513.

***heteroptera.** *Formica heteroptera* Cockerell, 1920: 278, pl. 16, fig. 11 (wing) GREAT BRITAIN (Eocene).

**hewitti.** *Formica hewitti* Wheeler, W.M. 1917a: 552 (w.q.m.) CANADA. See also: Francoeur, 1973: 208.

*hexacantha* Erichson, 1842; see under ***POLYRHACHIS.***

*hirta. Formica hirta* Gravenhorst, 1807: 286 (w.) no locality given. Unidentifiable to genus; *incertae sedis* in *Formica.*

***horrida.** *Formica horrida* Wheeler, W.M. 1915h: 125, fig. 59 (w.) BALTIC AMBER (Oligocene). See also: Dlussky, 1967b: 80.

*huberiana* Leach, 1825; see under ***MESSOR.***

*hybrida. Formica microgyna* subsp. *californica* var. *hybrida* Wheeler, W.M. 1917a: 543 (w.) U.S.A. Unavailable name; material referred to *californica* by Creighton, 1950a: 510.

*hybrida* Emery, 1916 [*Formicina*]; see under ***LASIUS.***

**#hyrcana.** *Formica (Serviformica) fusca* subsp. *hyrcana* Arnol'di, 1968: 1821 (w.q.) RUSSIA. [*Formica fusca* subsp. *hyrcana* Arnol'di, 1948: 212; *nomen nudum.*]

*hystrix* Latreille, 1802; see under ***ACROMYRMEX.***

**#iberica.** *Formica cinerea* var. *iberica* Finzi, 1928a: 71 (w.) SPAIN.

*ignivora. Formica ignivora* Rafin, 1885: 403. *Nomen nudum.*

**imhoffii* Heer, 1850; see under ***LIOMETOPUM.***

**imitans.** *Formica cinerea* var. *imitans* Ruzsky, 1902b: 472 (w.q.) RUSSIA. [Also described as new by Ruzsky, 1902c: 10 (footnote).] Karavaiev, 1936: 226 (m.). Junior synonym of *rufibarbis:* Kuznetsov-Ugamsky, 1926b: 97. Revived from synonymy as variety of *cinerea:* Karavaiev, 1927a: 302; Kuznetsov-Ugamsky, 1929b: 39. Raised to species: Agosti & Collingwood, 1987a: 59. Senior synonym of *ochracea* and *sabulosa:* Dlussky, 1967a: 65.

***immersa.** *Formica immersa* Heer, 1850: 122, pl. 9, fig. 5 (q.) SWITZERLAND (Miocene).

**impacta* Cockerell, 1927; see under ****MIOMYRMEX.***

*imparis* Say, 1836; see under ***PRENOLEPIS.***

*impetuosa* Smith, F. 1858; see under ***CAMPONOTUS.***

**impexa.** *Formica impexa* Wheeler, W.M. 1905c: 273 (w.) U.S.A. Wheeler, W.M. 1906a: 40 (q.). See also: Wheeler, W.M. 1913f: 472.

*incerta*. *Formica* (*Neoformica*) *pallidefulva* subsp. *incerta* Buren, 1944a: 309 (w.q.m.) U.S.A. [First available use of *Formica pallidefulva* subsp. *schaufussi* var. *incerta* Emery, 1893i: 655; unavailable name.] Junior synonym of *nitidiventris:* Creighton, 1950a: 551.

*incisa* Schenck, 1852; see under ***LASIUS.***

*incisa*. *Formica incisa* Smith, F. 1858b: 52 (q.) MEXICO. [Unresolved junior primary homonym of *incisa* Schenck, 1852: 63, above.] Unidentifiable to genus; *incertae sedis* in *Formica*.

*incisa*. *Formica rufa* subsp. *pratensis* var. *incisa* Krausse, 1922: 155 (w.) GERMANY. Unavailable name; material referred to *nigropratensis* by Betrem, 1962: 38.

*incursor* Smith, F. 1860; see under ***CAMPONOTUS.***

*indefessa* Sykes, 1835; see under ***CAMPONOTUS.***

*indeflexa* Walker, 1859; see under ***CAMPONOTUS.***

***indianensis.*** *Formica indianensis* Cole, 1940: 224 (w.m.) U.S.A.

*indificans* Jerdon, 1851; see under ***POLYRHACHIS.***

**indurata* Heer, 1850; see under ***CAMPONOTUS.***

*inequalis*. *Formica inequalis* Lowne, 1865b: 331 (q.m.) AUSTRALIA. Unidentifiable to genus; *incertae sedis* in *Formica*.

*ingruens* Walker, 1859; see under ***DOLICHODERUS.***

*inplana*. *Formica* (*Serviformica*) *picea* var. *inplana* Emery, 1925b: 249 (w.) TIBET. [First available use of *Formica fusca* subsp. *picea* var. *inplana* Ruzsky, 1915a: 426; unavailable name.] Subspecies of *fusca:* Ruzsky, 1936: 92. Junior synonym of *picea* Nylander: Dlussky, 1965a: 31.

*insana* Buckley, 1866; see under ***DORYMYRMEX.***

*insultans*. *Formica insultans* Forskål, 1775: 85 (w.) EGYPT. Unidentifiable to genus; *incertae sedis* in *Formica*. See also: Emery, 1892b: 160.

***integra.*** *Formica integra* Nylander, 1856b: 62 (footnote) (w.) U.S.A. Mayr, 1886d: 425 (q.m.). Subspecies of *rufa:* Emery, 1893i: 652; Wheeler, W.M. 1908g: 406; Creighton, 1940a: 1; of *truncicola:* Wheeler, W.M. 1913c: 116; Wheeler, W.M. 1913f: 444; Wheeler, W.M. 1917a: 464. Revived status as species: Creighton, 1950a: 487. Senior synonym of *similis:* Emery, 1893i: 652; Wheeler, W.M. 1906b: 16.

***integroides.*** *Formica truncicola* subsp. *integroides* Wheeler, W.M. 1913f: 438 (w.q.) U.S.A. [First available use of *Formica rufa* subsp. *obscuriventris* var. *integroides* Emery, 1893i: 649; unavailable name.] Subspecies of *rufa:* Creighton, 1940a: 3. Raised to species, and senior synonym of *subcaviceps:* Creighton, 1950a: 489. Senior synonym of *subfasciata:* Brown, 1965d: 185.

*intermedia* Zetterstedt, 1838; see under ***CAMPONOTUS.***

*intrepida* Kirby, W. 1819; see under ***CAMPONOTUS.***

*irritabilis* Smith, F. 1857; see under ***CAMPONOTUS.***

*irritans* Smith, F. 1857; see under ***CAMPONOTUS.***

#***italica.*** *Formica cinerea* var. *italica* Finzi, 1928a: 70 (w.) ITALY.

*itinerans* Lowne, 1865; see under ***ANONYCHOMYRMA.***

*jacoti* Wheeler, W.M. 1923; see under ***PROFORMICA.***

***japonica.*** *Formica japonica* Motschoulsky, 1866: 183 (w.) JAPAN. Imai, 1966: 119 (k.). Combination in *F.* (*Serviformica*): Emery, 1925b: 247. Variety/subspecies of *fusca:* Emery, 1909b: 197; Emery, 1923: 62; Wheeler, W.M. 1929f: 10; Kuznetsov-Ugamsky, 1929a: 19; Stitz, 1934: 9. Revived status as species: Dlussky, 1967a: 78; Collingwood, 1976: 305; Wu, 1990: 5; Kupyanskaya, 1990: 185. Senior synonym of *nipponensis:* Emery, 1909b: 197; Karavaiev, 1927b: 111; Kuznetsov-Ugamsky, 1929a: 19.

*juvenilis* Fabricius, 1804; see under ***MESSOR.***

#***kashmirica.*** *Formica* (*Serviformica*) *rufibarbis* var. *kashmirica* Stärcke, 1935: 268 (w.q.m.) CHINA/KASHMIR?

*katuniensis*. *Formica katuniensis* Ruzsky, 1915b: ? [pagination unknown, reference not seen; this reference from Dlussky, 1967a: 74.] Junior synonym of *glauca:* Dlussky, 1967a: 74.

*kiesenwetteri* Roger, 1859; see under ***CAMPONOTUS.***

***knighti.*** *Formica* (*Formica*) *knighti* Buren, 1944a: 303 (w.) U.S.A.

****kollari.*** **Formica kollari* Heer, 1867: 15, pl. 1, fig. 14 (q.) YUGOSLAVIA (Miocene).

*kontuniemii*. *Formica kontuniemii* Betrem, 1954: 230 (w.) FINLAND. Junior synonym of *exsecta:* Dlussky, 1967a: 100; Dlussky & Pisarski, 1971: 194.

*korbi* Emery, 1909; see under ***PROFORMICA.***

***kozlovi.*** *Formica* (*Serviformica*) *kozlovi* Dlussky, 1965a: 32 (w.q.) MONGOLIA.

*kraussei* Forel, 1895; see under ***BAJCARIDRIS.***

***kupyanskayae.*** *Formica kupyanskayae* **new name.** Replacement name for *opacus* Kupyanskaya, 1980: 106. [Junior primary homonym of *opaca* Nylander 1856b: 55.]

*kusnezowi* Santschi, 1928; see under ***PROFORMICA.***

*laboriosa* Smith, F. 1858; see under ***CAMPONOTUS.***

*lactaria* Smith, F. 1860; see under ***DOLICHODERUS.***

*lacteipennis* Smith, F. 1858; see under ***CAMPONOTUS.***

***laeviceps.*** *Formica rufa* subsp. *laeviceps* Creighton, 1940a: 9 (w.q.) U.S.A. Raised to species: Creighton, 1950a: 491.

*laevigata* Smith, F. 1858; see under ***CAMPONOTUS.***

*laevissima* Smith, F. 1859; see under ***CALOMYRMEX.***

*lapidum* Fabricius, 1804; see under ***MESSOR.***

***lasioides.*** *Formica lasioides* Emery, 1893i: 664 (w.) U.S.A. Combination in *Proformica:* Emery, 1912f: 100; in *Formica* (*Proformica*): Wheeler, W.M. 1913f: 539; in *Formica:* Francoeur, 1973: 114. Subspecies of *fusca:* Wheeler, W.M. 1908g: 409; of *neogagates:* Wheeler, W.M. 1913f: 539. Revived status as species, and senior synonym of *vetula:* Creighton, 1950a: 457.

*latebrosa* Walker, 1859; see under ***CAMPONOTUS.***

*lateralis* Olivier, 1792; see under ***CAMPONOTUS.***

****latinodosa.*** **Formica latinodosa* Théobald, 1937b: 318, pl. 25, fig. 13 (m.) FRANCE (Oligocene).

*latipes* Walsh, 1863; see under ***ACANTHOMYOPS.***

*latreillei* Guérin-Méneville, 1838; see under ***POLYRHACHIS.***

*lauta* Say, 1836; see under ***CAMPONOTUS.***

****lavateri.*** **Formica lavateri* Heer, 1850: 127, pl. 9, fig. 11 (q.) SWITZERLAND (Miocene). Current subspecies: nominal plus **major* (unresolved junior homonym).

*lecontei*. *Formica lecontei* Kennedy & Dennis, 1937: 542 (w.q.m.) U.S.A. Junior synonym of *fusca:* Creighton, 1950a: 532; of *subsericea:* Francoeur, 1973: 172.

*lefevrei* Wheeler, W.M. 1929; see under ***PROFORMICA.***

***lefrancoisi.*** *Formica lefrancoisi* Bondroit, 1918: 54 (w.m.) FRANCE. Combination in *F.* (*Serviformica*), and variety of *cinerea:* Emery, 1925b: 246. Revived status as species: Kutter, 1977c: 253.

***lemani.*** *Formica lemani* Bondroit, 1917b: 186 (w.q.) FRANCE. Bondroit, 1920a: 145 (m.). Combination in *F.* (*Serviformica*): Emery, 1925b: 247. Variety/subspecies of *fusca:* Müller, 1923: 139; Santschi, 1925g: 352; Novak & Sadil, 1941: 107. Revived status as species: Yarrow, 1954a: 231. Senior synonym of *borealis:* Dlussky, 1965a: 35. See also: Bernard, 1967: 295; Dlussky, 1967a: 60; Dlussky & Pisarski, 1971: 150; Kutter, 1977c: 251; Kupyanskaya, 1990: 184; Atanassov & Dlussky, 1992: 261.

*leninei*. *Formica* (*Raptiformica*) *sanguinea* var. *leninei* Santschi, 1928b: 46. Replacement name for *clara* Karavaiev, 1927d: 345. [Junior primary homonym of *clara* Forel, 1886f: 206.] Junior synonym of *sanguinea:* Dlussky, 1965a: 16.

***lepida.*** *Formica cinerea* var. *lepida* Wheeler, W.M. 1913f: 399 (diagnosis in key) (w.) U.S.A. [*Formica cinerea* subsp. *cinerea* var. *lepida* Wheeler, W.M. 1913f: 526; unavailable name.] Wheeler, W.M. 1917a: 551 (q.). Variety/subspecies of *cinerea:* Wheeler, W.M. 1917a: 551; Creighton, 1950a: 531. Raised to species: Francoeur, 1973: 122.

*leucophaea* Smith, F. 1861; see under ***CAMPONOTUS.***

*libera*. *Formica libera* Scopoli, 1763: 313 (w.) AUSTRIA. Junior synonym of *fusca:* Latreille, 1802: 159.

*ligniperda* Latreille, 1802; see under ***CAMPONOTUS.***

**lignitum* Germar, 1837; see under ***CAMPONOTUS.***

***limata.*** *Formica* (*Proformica*) *limata* Wheeler, W.M. 1913f: 541 (w.) U.S.A.

*lincecumii*. *Formica lincecumii* Buckley, 1866: 163 (w.q.m.) U.S.A. Unidentifiable to genus; *incertae sedis* in *Formica*. See also: Wheeler, W.M. 1902f: 22.

#***litoralis.*** *Formica rufibarbis* subsp. *litoralis* Kuznetsov-Ugamsky, 1926b: 100, figs. 13, 14 (w.) TURKESTAN. Dlussky, 1965a: 29 (q.m.). Subspecies of *subpilosa:* Dlussky, 1965a: 29; Tarbinsky, 1976: 190.

*lobata* De Geer, 1773; see under ***PACHYCONDYLA.***

*lochmatteri*. *Formica* (*Serviformica*) *picea* var. *lochmatteri* Stärcke, 1935: 265, fig. 4 (w.q.) CHINA/KASHMIR? Junior synonym of *picea* Nylander: Dlussky, 1965a: 31.

**longaeva* Heer, 1850; see under ***LASIUS.***

*longiceps* Smith, F. 1863; see under ***CAMPONOTUS.***

*longiceps*. *Formica* (*Coptoformica*) *longiceps* Dlussky, 1964: 1036 (w.q.) MONGOLIA. [Junior primary homonym of *longiceps* Smith, F. 1863: 13, above.] Replacement name: *dlusskyi* **new name.** See also: Dlussky, 1965a: 19; Kupyanskaya, 1990: 202.

****longicollis.*** **Formica longicollis* Heer, 1850: 115, pl. 11, fig. 1 (m.) SWITZERLAND (Miocene).

*longicornis* Latreille, 1802; see under ***PARATRECHINA.***

*longinoda* Latreille, 1802; see under ***OECOPHYLLA.***

*longipedem* Eichwald, 1841; see under ***CATAGLYPHIS.***

**longipennis* Heer, 1850; see under ***LASIUS.***

*longipes* Latreille, 1802; see under ***PHEIDOLE.***

*longipes* Jerdon, 1851; see under ***ANOPLOLEPIS.***

*longipes* Gerstäcker, 1859; see under ***CAMPONOTUS.***

***longipilosa.*** *Formica longipilosa* Francoeur, 1973: 125, figs. 208–214 (w.) U.S.A. Francoeur & Snelling, 1979: 4 (m.).

****longiventris.*** **Formica longiventris* Heer, 1850: 123, pl. 9, fig. 6 (q.) YUGOSLAVIA (Miocene).

***FORMICA (continued)***

****lucida.*** **Formica lucida* Giebel, 1856: 163 (q.) BALTIC AMBER (Oligocene). [Mayr, 1868c: 19, suggests that this taxon should be transferred to family BRACONIDAE.]

*lucidula. Formica sanguinea* subsp. *rubicunda* var. *lucidula* Wheeler, W.M. 1913f: 390 (in key) (w.) U.S.A. Unavailable name.

*luctuosa* Smith, F. 1858; see under ***CAMPONOTUS.***

***lugubris.*** *Formica lugubris* Zetterstedt, 1838: 449 (m.) NORWAY. Junior synonym of *rufa:* Nylander, 1856b: 60; Emery & Forel, 1879: 450; Wheeler, W.M. 1913f: 425; Emery, 1925b: 253; Stitz, 1939: 328. Revived from synonymy and status as species: Yarrow, 1955a: 5; Betrem, 1960b: 77; Dlussky, 1967a: 91; Dlussky & Pisarski, 1971: 180; Baroni Urbani, 1971c: 218; Kutter, 1977c: 271; Gösswald, 1989: 19; Kupyanskaya, 1990: 198; Atanassov & Dlussky, 1992: 274. Senior synonym of *congerens:* Yarrow, 1955a: 5; Dlussky, 1967a: 91; of *nylanderi* and *santschii:* Yarrow, 1955a: 5; of *montana* Sadil: Samsinak, 1964: 157; of *unicolor:* Dlussky, 1967a: 91; Dlussky & Pisarski, 1971: 180. Material of the unavailable name *tir* referred here by Yarrow, 1955a: 5.

*lusitanica. Formica fusca* subsp. *pyrenaea* var. *lusitanica* Schmitz, 1950: 13 (w.) PORTUGAL. Unavailable name.

*lutaria. Formica lutaria* Stephens, 1829: 357. *Nomen nudum.*

*lutea* Smith, F. 1858; see under ***CAMPONOTUS.***

*lutea. Formica lutea* Duméril, 1860: 930 (w.q.m.) FRANCE. [Unresolved junior primary homonym of *lutea* Smith, F. 1858b: 19, above.] Unidentifiable to genus; *incertae sedis* in *Formica.*

****luteola.*** **Formica luteola* Presl, 1822: 197 (no sex or caste indicated) BALTIC AMBER (Oligocene).

*lutescens. Formica fusca* subsp. *pruinosa* var. *lutescens* Wheeler, W.M. 1917a: 549 (w.) U.S.A. Unavailable name; material referred to *neoclara* by Creighton, 1950a: 535.

*lutosa* Smith, F. 1858; see under ***DOLICHODERUS.***

*macra* Guérin-Méneville, 1831; see under ***OECOPHYLLA.***

*macrocephala* Erichson, 1842; see under ***CAMPONOTUS.***

****macrocephala.*** **Formica macrocephala* Heer, 1850: 125, pl. 9, fig. 10 (q.) SWITZERLAND (Miocene). [Unresolved junior primary homonym of *macrocephala* Erichson, 1842: 259, above.]

****macrognatha.*** **Formica macrognatha* Presl, 1822: 198 (no caste or sex given) BALTIC AMBER (Oligocene).

****macrophthalma.*** **Formica macrophthalma* Heer, 1850: 124, pl. 9, fig. 8 (q.) SWITZERLAND (Miocene).

*maculata* Fabricius, 1782; see under ***CAMPONOTUS.***

*maculata. Formica maculata* Geoffroy, in Fourcroy, 1785: 452 (w.?) FRANCE. [Unresolved junior primary homonym of *maculata* Fabricius, 1782: 491, above.] Unidentifiable to genus; *incertae sedis* in *Formica.*

****maculipennis.*** **Formica maculipennis* Théobald, in Théobald & Piton, 1935: 91, pl. 5, fig. 1 (wing) FRANCE (Miocene).

*major. Formica nigra major* Retzius, 1783: 75 (w.) no locality given. Unavailable name (published as junior synoynm).

*major. Formica major* Nylander, 1849: 29 (w.) FINLAND. Junior synonym of *rufa:* Emery & Forel, 1879: 450. Revived from synonymy: Betrem, 1926: 213. Senior synonym of *piniphila:* Betrem, 1953: 325. Junior synonym of *rufa:* Yarrow, 1955a: 3; Betrem, 1960b: 76; Kutter, 1977c: 273.

*#***major.*** **Formica lavateri* var. *major* Heer, 1867: 11, pl. 1, fig. 106 (q.) YUGOSLAVIA (Miocene). [Unresolved junior primary homonym of *major* Nylander, 1849: 29, above.]

*major. Formica rufa* subsp. *pratensis* var. *major* Gösswald, 1942: 78 (w.q.m.) GERMANY. Unavailable name; material referred to *rufa* by Dlussky, 1967a: 93.

*malabarica. Formica malabarica* Schrank, 1837: 16 (w.) INDIA. Unidentifiable to genus; *incertae sedis* in *Formica.*

*maligna. Formica maligna* Forskål, 1775: 84 (w.) EGYPT? Unidentifiable to genus; *incertae sedis* in *Formica.* See also: Emery, 1892b: 160.

#***manchu.*** *Formica* (*Coptoformica*) *exsecta* subsp. *manchu* Wheeler, W.M. 1929f: 10 (w.) CHINA.

***manni.*** *Formica manni* Wheeler, W.M. 1913f: 420 (w.q.) U.S.A. Combination in *F.* (*Raptiformica*): Emery, 1925b: 259.

*marcida. Formica fusca* var. *marcida* Wheeler, 1913f: 398 (in key) (w.q.) CANADA. [*Formica fusca* subsp. *fusca* var. *marcida* Wheeler, W.M. 1913f: 503; unavailable name.] Variety of *fusca:* Wheeler, W.M. 1917a: 546. Raised to species: Creighton, 1950a: 534. Junior synonym of *fusca:* Francoeur, 1973: 189.

*marginata* Latreille, 1798; see under ***CAMPONOTUS.***

****martynovi.*** **Formica martynovi* Popov, 1933: 19, figs. 2, 3 (q.) RUSSIA. See also: Dlussky, 1981b: 75.

*masonia* Buckley, 1866; see under ***LIOMETOPUM.***

#***maura.*** *Formica glebaria* var. *maura* Santschi, 1929e: 164 (w.q.) MOROCCO. Variety of *fusca:* Santschi, 1931a: 13.

*maxillosa* De Geer, 1773; see under ***ODONTOMACHUS.***

*maxillosa. Formica maxillosa* Fabricius, 1775: 396 (w.) INDIA. [Unresolved junior primary homonym of *maxillosa* De Geer, 1773: 601, above.] Unidentifiable to genus; *incertae sedis* in *Formica.*

*maxima* Moore, 1842; see under ***MYRMECIA.***

*megacephala* Fabricius, 1793; see under ***PHEIDOLE.***

*megacephala* Leach, 1825; see under ***MESSOR.***

*megacephala* Losana, 1834; see under ***PHEIDOLE.***

*megalocola* Foerster, 1850; see under ***CATAGLYPHIS.***

*melanocephala* Fabricius, 1793; see under ***TAPINOMA.***

*melanogastes* Latreille, 1802; see under ***CAMPONOTUS.***

*melanophthalma. Formica melanophthalma* Reich, 1793: 132 (no sex or caste given) FRENCH GUIANA. Unidentifiable to genus; *incertae sedis* in *Formica.*

*melanopis. Formica melanopis* Gmelin, in Linnaeus, 1790: 2804 (w.) EUROPE. Unidentifiable to genus; *incertae sedis* in *Formica.*

*melanotica. Formica rufa* subsp. *melanotica* Creighton, 1940a: 1 (w.q.m.) U.S.A. [First available use of *Formica rufa* subsp. *obscuriventris* var. *melanotica* Emery, 1893i: 650; unavailable name.] Junior synonym of *obscuripes:* Creighton, 1950a: 492.

*mellea* Say, 1836; see under ***CAMPONOTUS.***

*mellea* Provancher, 1881; see under ***LASIUS.***

*melligera* Llave, 1832; see under ***MYRMECOCYSTUS.***

*menozzii. Formica truncorum* var. *menozzii* Stitz, 1939: 347 (w.) GERMANY. [First available use of *Formica rufa* subsp. *truncicola* var. *menozzii* Krausse, 1926c: 336; unavailable name. Note: *truncicola* is misspelled *aruncicola* in this paper.] Junior synonym of *truncorum:* Dlussky, 1967a: 81.

*merdicola* Lund, 1831; see under ***CAMPONOTUS.***

*meridionalis. Formica rufa* var. *meridionalis* Nasonov, 1889: 17 (w.) RUSSIA. Variety/subspecies of *rufa:* Ruzsky, 1905b: 330; Emery, 1909b: 186; Karavaiev, 1936: 244. Junior synonym of *rufa:* Yarrow, 1955a: 4; Dlussky & Pisarski, 1971: 184.

*meridionalis. Formica pallidefulva* subsp. *schaufussi* var. *meridionalis* Wheeler, W.M. 1904f: 370 (diagnosis in key) (w.) U.S.A. Unavailable name.

*meridionalis* Bondroit, 1920 [*Formicina*]; see under ***LASIUS.***

*meritans* Walker, 1859; see under ***LEPTOGENYS.***

*merula* Losana, 1834; see under ***CAMPONOTUS.***

*mesasiatica. Formica* (*Coptoformica*) *mesasiatica* Dlussky, 1964: 1030, figs. (w.q.m.) RUSSIA. Junior synonym of *fukaii:* Sonobe & Dlussky, 1977: 23. See also: Tarbinsky, 1976: 179.

*micans* Nylander, 1856; see under ***CAMPONOTUS.***

*microcephala* Panzer, 1798; see under ***LIOMETOPUM.***

***microgyna.*** *Formica microgyna* Wheeler, W.M. 1903e: 645, fig. 3 (w.q.m.) U.S.A. Senior synonym of *recidiva:* Creighton, 1950a: 504. See also: Wheeler, W.M. 1903f: 656; Wheeler, W.M. 1913f: 465; Letendre & Huot, 1972: 117.

*microgyna* Bondroit, 1918 [*Formicina*]; see under ***LASIUS.***

***microphthalma.*** *Formica microphthalma* Francoeur, 1973: 201, figs. 344–359 (w.m.) U.S.A.

*migratoria* De Geer, 1773; see under ***ATTA.***

*militaris* Fabricius, 1782; see under ***POLYRHACHIS.***

*minor. Formica nigra minor* Retzius, 1783: 75 (w.) no locality given). Unavailable name (published as junior synonym).

*minor. Formica rufibarbis* subsp. *montana* var. *minor* Kuznetsov-Ugamsky, 1926b: 100, fig. 12 (w.) TURKESTAN. Unavailable name; material referred to *glauca* by Dlussky, 1967a: 74.

*minor. Formica minor* Gösswald, 1951: 436 (w.q.) GERMANY. [First available use of *Formica rufa* subsp. *pratensis* var. *minor* Gösswald, 1942: 78; unavailable name.] Junior synonym of *polyctena:* Betrem, 1960b: 64; Dlussky, 1967a: 93.

*minuta. Formica minuta* Lowne, 1865b: 331 (w.) AUSTRALIA. Unidentifiable to genus; *incertae sedis* in *Formica.*

*minuta. Formica sanguinea* subsp. *monticola* var. *minuta* Kuznetsov-Ugamsky, 1926b: 95, figs. 5, 6 (w.) TURKESTAN. Unavailable name; material referred to *sanguinea* by Dlussky, 1965a: 16.

**minutula* Heer, 1850; see under ***LASIUS.***

*mistura* Smith, F. 1857; see under ***CAMPONOTUS.***

*mitis* Smith, F. 1858; see under ***CAMPONOTUS.***

*mixta* Nylander, 1846; see under ***LASIUS.***

***moki.*** *Formica moki* Wheeler, W.M. 1906d: 343 (w.) U.S.A. Cole, 1966: 25 (m.). Combination in *F.* (*Neoformica*): Wheeler, W.M. 1913f: 558. Junior synonym of *occidua:* Francoeur, 1973: 254. [Note that *moki* has seniority over *occidua* and hence is the valid name; see Smith, D.R. 1979: 1453.] See also: Smith, M.R. 1939f: 582.

*molestans. Formica molestans* Latreille, 1802c: 241 (w.) FRENCH GUIANA. Unidentifiable to genus; *incertae sedis* in *Formica.*

*mollesonae. Formica sanguinea* var. *mollesonae* Ruzsky, 1903c: 206 (w.) RUSSIA. Junior synonym of *sanguinea:* Dlussky, 1965a: 16.

*mongolica* Emery, 1901; see under ***PROFORMICA.***

***montana.*** *Formica subpolita* var. *montana* Wheeler, W.M. 1910g: 571 (w.) U.S.A. [First available use of *Formica fusca* subsp. *subpolita* var. *montana* Emery, 1893i: 663; unavailable name.] Francoeur, 1973: 70 (q.m.); Hung, 1969: 455 (k.). Raised to species: Wheeler, W.M. 1913f: 529. Senior synonym of *neocinerea* and *rutilans:* Creighton, 1950a: 534; Francoeur, 1973: 67.

*montana. Formica rufibarbis* natio *montana* Kuznetsov-Ugamsky, 1923: 245 (w.) TURKESTAN. [Junior primary homonym of *montana* Wheeler, W.M., above.] Replacement name: *montivaga* Santschi, 1928b: 45.

*montana. Formica rufa* var. *montana* Sadil, 1953b: 198, fig. 1 (q.) CZECHOSLOVAKIA. [Unresolved junior primary homonym of *montana* Wheeler, W.M., above.] Junior synonym of *lugubris:* Samsinak, 1964: 157.

*montaniformis. Formica rufibarbis* var. *montaniformis* Kuznetsov-Ugamsky, 1929b: 39 (w.) DAGHESTAN. Junior synonym of *glauca:* Dlussky: 1967a: 74 [misspelled as *montanoides.*]

*monticola* Buckley, 1866; see under ***LASIUS.***

*monticola. Formica sanguinea* subsp. *monticola* Kuznetsov-Ugamsky, 1926b: 95, figs. 3, 4 (w.) TURKESTAN. [Unresolved junior primary homonym of *monticola* Buckley, 1866: 157, above.] Junior synonym of *sanguinea:* Dlussky, 1965a: 16.

*montigena. Formica montigena* Wheeler, W.M. 1904f: 374 (w.q.m.) U.S.A. Variety/subspecies of *dakotensis:* Wheeler, W.M. 1913f: 463; Wheeler, G.C. & Wheeler, J. 1944: 260. Junior synonym of *dakotensis:* Brown, 1957f: 167.

*montivaga. Formica (Serviformica) rufibarbis* st. *montivaga* Santschi, 1928b: 45. Replacement name for *montana* Kuznetsov-Ugamsky, 1923: 245. [Junior primary homonym of *montana* Wheeler, W.M. 1910g: 571.] Junior synonym of *glauca:* Dlussky, 1967a: 74.

*morbida. Formica (Proformica) neogagates* var. *morbida* Wheeler, W.M. 1913f: 400 (in key) (w.q.) U.S.A. [*Formica (Proformica) neogagates* subsp. *neogagates* var. *morbida* Wheeler, W.M. 1913f: 538; unavailable name.] Subspecies of *neogagates:* Buren, 1944a: 309. Junior synonym of *neogagates:* Creighton, 1950a: 459; of *bradleyi:* Wilson & Brown, 1955: 126.

*morbosa* Bondroit, 1918 [*Formicina*]; see under ***LASIUS.***

*morio. Formica morio* Latreille, 1798: 36 (w.) FRANCE. Junior synonym of *gagates:* Emery, 1925b: 248; Dlussky & Pisarski, 1971: 169.

*morosa* Smith, F. 1858; see under ***CAMPONOTUS.***

***morsei.*** *Formica morsei* Wheeler, W.M. 1906a: 39, pl. 4, fig. 1 (w.) U.S.A. See also: Wheeler, W.M. 1913f: 480.

***mucescens.*** *Formica truncicola* subsp. *mucescens* Wheeler, W.M. 1913f: 442 (w.q.m.) U.S.A. Subspecies of *rufa:* Creighton, 1940a: 3. Raised to species: Creighton, 1950a: 491.

*munda. Formica munda* Wheeler, W.M. 1905c: 267 (w.q.) U.S.A. Junior synonym of *obtusopilosa:* Emery, 1925b: 259; Creighton, 1950a: 465. See also: Wheeler, W.M. 1913f: 416.

*muralewiczi. Formica fusca* subsp. *gagates* var. *muralewiczi* Ruzsky, 1905b: 384 (w.) RUSSIA. Unavailable name; material referred to *gagates* by Emery, 1909b: 194; Karavaiev, 1927a: 301.

*mutilata* Smith, F. 1859; see under ***CAMPONOTUS.***

*nacerda* Norton, 1868; see under ***CAMPONOTUS.***

***naefi.*** *Formica (Coptoformica) naefi* Kutter, 1957: 4, figs. 1–6 (w.q.m.) SWITZERLAND. See also: Kutter, 1977c: 285.

*nahua. Formica microgyna* subsp. *rasilis* var. *nahua* Wheeler, W.M. 1913f: 562 (w.q.m.) MEXICO. Unavailable name.

*nana. Formica nana* Latreille, 1802c: 263 (w.) FRENCH GUIANA. Unidentifiable to genus; *incertae sedis* in *Formica.* [See note under *pusilla.*]

*nana* Jerdon, 1851; see under ***TAPINOMA.***

*nana* Smith, F. 1858; see under ***CAMPONOTUS.***

*nasuta* Nylander, 1856; see under ***PROFORMICA.***

*natalensis* Smith, F. 1858; see under ***CAMPONOTUS.***

*nemoralis. Formica nemoralis* Dlussky, 1964: 1037 (w.m.) UKRAINE. Junior synonym of *forsslundi:* Dlussky, 1967a: 105.

*neocinerea. Formica cinerea* var. *neocinerea* Wheeler, W.M. 1913f: 399 (in key) (w.q.m.) U.S.A. [*Formica cinerea* var. *neocinerea* Wheeler, W.M. 1910g: 571. *Nomen nudum.*] Subspecies of *cinerea:* Buren, 1944a 301. Junior synonym of *montana:* Creighton, 1950a: 534.

***neoclara.*** *Formica fusca* var. *neoclara* Emery, 1893i: 661 (w.) U.S.A. Wheeler, W.M. 1913f: 509 (q.m.); Wheeler, G.C. & Wheeler, J. 1953c: 165 (l.). Combination in *F. (Serviformica)*: Emery, 1925b: 248. Subspecies of *fusca:* Buren, 1944a: 301. Raised to species, and material of the unavailable name *lutescens* referred here: Creighton, 1950a: 535. Senior synonym of *pruinosa:* Francoeur, 1973: 84.

***neogagates.*** *Formica fusca* var. *neogagates* Viereck, 1903: 74 (w.q.m.) U.S.A. [First available use of *Formica fusca* subsp. *subpolita* var. *neogagates* Emery, 1893i: 661; unavailable name.] Wheeler, G.C. & Wheeler, J. 1953c: 158 (l.). Combination in *F. (Proformica)*: Wheeler, W.M. 1913f: 536. Variety of *subpolita:* Wheeler, W.M. 1908f: 625. Raised to species: Wheeler, W.M. 1913f: 536.

***neorufibarbis.*** *Formica fusca* var. *neorufibarbis* Emery, 1893i: 660 (w.) U.S.A. Forel, 1902i: 699 (q.); Wheeler, W.M. 1917a: 547 (m.); Wheeler, G.C. & Wheeler, J. 1953c: 165 (l.). Combination in *F. (Serviformica)*: Emery, 1925b: 248. Subspecies of *fusca:* Wheeler, G.C. & Wheeler, E.W. 1944: 261. Raised to species: Creighton, 1950a: 537. Senior synonym of *algida* and *gelida:* Francoeur, 1973: 215.

***nepticula.*** *Formica nepticula* Wheeler, W.M. 1905c: 270 (w.q.m.) U.S.A. See also: Wheeler, W.M. 1913f: 475.

**neutra* Heer, 1850; see under ***LASIUS.***

***nevadensis.*** *Formica microgyna* var. *nevadensis* Wheeler, W.M. 1904f: 373 (q.) U.S.A. Cole, 1956f: 256 (w.). Raised to species: Wheeler, W.M. 1905c: 272; Cole, 1956f: 257.

*nicaeensis. Formica nicaeensis* Leach, 1825: 291 (w.q.m.) FRANCE. Junior synonym of *rufibarbis:* Roger, 1863b: 13.

*nidulans* Smith, F. 1860; see under ***DENDROMYRMEX.***

*nigerrima* Christ, 1791; see under ***LASIUS.***

*nigerrima* Nylander, 1856; see under ***TAPINOMA.***

*nigra* Linnaeus, 1758; see under ***LASIUS.***

****nigra.*** **Formica nigra* Presl, 1822: 196, no caste or sex given, BALTIC AMBER (Oligocene). [Unresolved junior primary homonym of *nigra* Linnaeus, 1758: 580, above.]

*nigrata* Nylander, 1849; see under ***CAMPONOTUS.***

*nigricans. Formica pratensis* var. *nigricans* Bondroit, 1912: 352 (w.) SPAIN. [First available use of *Formica rufa* subsp. *pratensis* var. *nigricans* Emery, 1909b: 187; unavailable name.] Variety of *pratensis:* Novak & Sadil, 1941: 105; Boven, 1947: 189. Raised to species: Yarrow, 1955a: 4. Status as species: Betrem, 1960b: 77; Bernard, 1967: 314; Baroni Urbani, 1971c: 220; Paraschivescu, 1972: 534; Collingwood, 1979: 153. Junior synonym of *pratensis:* Dlussky, 1967a: 84; Dlussky & Pisarski, 1971: 177; Kutter, 1977c: 272; Seifert, 1992a: 225. Senior synonym of *pratensoides* and *thyssei:* Yarrow, 1955a: 4; of *cordieri:* Betrem, 1962: 38; Kutter, 1965: 140.

*nigriceps* Smith, F. 1858; see under ***CAMPONOTUS.***

*nigriventris* Guérin-Méneville, 1831; see under ***LEPTOMYRMEX.***

*nigriventris* Spinola, 1851; see under ***LASIOPHANES.***

*nigroaenea* Smith, F. 1858; see under ***CAMPONOTUS.***

***nigropratensis.*** *Formica nigropratensis* Betrem, 1962: 38 (w.) EUROPE. Material of the unavailable names *foreli* and *incisa* referred here: Betrem, 1962: 38.

*nipponensis. Formica fusca* var. *nipponensis* Forel, 1900e: 270 (w.) JAPAN. Forel, 1901h: 66 (q.). Subspecies of *fusca:* Ruzsky, 1925b: 44. Junior synonym of *japonica:* Emery, 1909b: 197; Karavaiev, 1927b: 111; Kuznetsov-Ugamsky, 1929a: 19.

*nitida. Formica nitida* Razoumowsky, 1789: 427 (no caste or sex given) SWITZERLAND. Unidentifiable to genus; *incertae sedis* in *Formica.*

*nitida* Smith, F, 1859; see under ***CAMPONOTUS.***

*nitida* Lowne, 1865; see under ***CAMPONOTUS.***

*nitida* Norton, 1868; see under ***CAMPONOTUS.***

*nitidior* Forel, 1904; see under ***ALLOFORMICA.***

*nitidissima* André, 1896; see under ***PROLASIUS.***

***nitidiventris.*** *Formica pallidefulva* subsp. *nitidiventris* Emery, 1893i: 656, pl. 22, figs. 13, 19 (w.q.m.) U.S.A. Wheeler, G.C. & Wheeler, J. 1953c: 159 (l.). Combination in *F. (Neoformica)*: Wheeler, W.M. 1913f: 555. Raised to species: Francoeur, 1977b: 208. Senior synonym of *delicata, fuscata* and *incerta:* Creighton, 1950a: 551.

*nodosa* Latreille, 1802; see under ***GNAMPTOGENYS.***

*nodus* Brullé, 1832; see under ***CATAGLYPHIS.***

*nortonii. Formica nortonii* Buckley, 1866: 153 (w.q.) U.S.A. Unidentifiable to genus; *incertae sedis* in *Formica.* See also: Wheeler, W.M. 1902f: 19.

*novaeanglae. Formica novaeanglae* Buckley, 1866: 153 (w.q.) U.S.A. Unidentifiable to genus; *incertae sedis* in *Formica.*

***FORMICA** (continued)*

**#*novaki*.** *Formica cinerea* var. *novaki* Kratochvil, in Novak & Sadil, 1941: 106 (w.q.) CZECHOSLOVAKIA.

*noveboracensis* Fitch, 1854; see under ***CAMPONOTUS.***

*nuda* Bondroit, 1918 [*Formicina*]; see under ***LASIUS.***

**#*nuda*.** *Formica pratensis* subsp. *nuda* Ruzsky, 1926: 110 (w.) RUSSIA.

*nuda*. *Formica* (*Formica*) *rufa* var. *nuda* Karavaiev, 1930b: 148 (w.) SWEDEN. [Unresolved junior primary homonym of *nuda* Ruzsky, above.] Junior synonym of *rufa:* Karavaiev, 1936: 240; Yarrow, 1955a: 4; of *polyctena:* Dlussky, 1967a: 93; Dlussky & Pisarski, 1971: 187.

*nylanderi*. *Formica rufa* var. *nylanderi* Bondroit, 1920a: 145 (q.) FRANCE. [Also described as new by Bondroit, 1920b: 300.] Junior synonym of *lugubris:* Yarrow, 1955a: 5.

*obcordata*. *Formica obcordata* Schilling, 1830: 55 (no caste given) no locality given. *Nomen nudum.*

**obesa* Heer, 1850; see under ***OECOPHYLLA.***

***oblita*.** **Formica oblita* Heer, 1867: 13, pl. 1, fig. 12 (q.) YUGOSLAVIA.

**obliterata* Heer, 1850; see under ***LASIUS.***

*oblonga* Smith, F. 1858; see under ***CAMPONOTUS.***

**obscura* Heer, 1850; see under ***LASIUS.***

*obscura* Smith, F. 1858; see under ***DOLICHODERUS.***

*obscurans* Walker, 1859; see under ***PACHYCONDYLA.***

**#*obscurata*.** *Formica rufa* var. *obscurata* Santschi, 1925g: 351 (w.m.) SPAIN.

***obscuripes*.** *Formica rufa* r. *obscuripes* Forel, 1886b: xxxix (w.) U.S.A. Emery, 1893i: 650 (q.m.); Wheeler, G.C. & Wheeler, J. 1953c: 165 (l.); Hung, 1969: 455 (k.). Raised to species: Creighton, 1950a: 492. Senior synonym of *aggerans:* Forel, 1914c: 619; Creighton, 1940a: 1; of *melanotica:* Creighton, 1950a: 492. Material of the unavailable name *rubiginosa* referred here: Creighton, 1940a: 1. See also: Weber, 1935: 165.

***obscuriventris*.** *Formica truncicola* var. *obscuriventris* Mayr, 1870b: 951 (w.) U.S.A. Wheeler, W.M. 1913f: 446 (q.m.). Variety/subspecies of *rufa:* Mayr, 1886d: 426; Emery, 1893i: 649; Wheeler, W.M. 1908g: 408; Creighton, 1940a: 1; of *truncicola:* Wheeler, W.M. 1913c: 116; Wheeler, W.M. 1913f: 445; Wheeler, W.M. 1917i: 464. Raised to species: Creighton, 1950a: 493. Senior synonym of *dryas:* Wheeler, W.M. 1913f: 446; of *gymnomma:* Creighton, 1950a: 493. Current subspecies: nominal plus *clivia.*

***obsidiana*.** *Formica* (*Serviformica*) *obsidiana* Emery, 1923: 61 (q.) TAIWAN.

*obsoleta*. *Formica obsoleta* Linnaeus, 1758: 580 (w.) EUROPE. Unidentifiable to genus; *incertae sedis* in *Formica.*

*obsoleta* Christ, 1791; see under ***CAMPONOTUS.***

***obtecta*.** **Formica obtecta* Heer, 1850: 123, pl. 9, fig. 7 (q.) YUGOSLAVIA (Miocene).

*obtusa* Smith, F. 1858; see under ***CAMPONOTUS.***

***obtusopilosa*.** *Formica obtusopilosa* Emery, 1893i: 648 (w.) U.S.A. Combination in *F.* (*Raptiformica*): Emery, 1925b: 259. Subspecies of *sanguinea:* Wheeler, W.M. 1901c: 713; Wheeler, W.M. 1913f: 414. Revived status as species: Emery, 1925b: 259; Cole, 1942: 377. Senior synonym of *munda:* Emery, 1925b: 259; Creighton, 1950a: 465; of *alticola:* Wilson & Brown, 1955: 128.

***obvoluta*.** **Formica obvoluta* Heer, 1850: 141, pl. 10, fig. 9 (m.) YUGOSLAVIA (Miocene).

*occidentalis*. *Formica occidentalis* Buckley, 1866: 157 (w.q.m.) U.S.A. Unidentifiable to genus; *incertae sedis* in *Formica.* See also: Emery, 1895c: 337; Wheeler, W.M. 1902f: 20.

*occidentalis*. *Formica fusca* var. *occidentalis* Wheeler, W.M. 1908g: 409 (w.) U.S.A. *Nomen nudum. F. rufibarbis* var. *occidentalis* Wheeler, W.M. 1910g: 570; *nomen nudum. F. rufibarbis* var. *occidua* Wheeler, W.M. 1912c: 90; unnecessary replacement name; *nomen nudum.*

*occidua*. *Formica rufibarbis* var. *occidua* Wheeler, W.M. 1913f: 517 (w.q.) U.S.A. [*F. rufibarbis* var. *occidua* Wheeler, W.M. 1912c: 90; *nomen nudum.*] Subspecies of *rufibarbis:* Creighton, 1950a: 539. Raised to species, and stated as senior synonym of *moki:* Francoeur, 1973: 254 [the latter has priority and hence is the valid name: Smith, D.R. 1979: 1453].

***occulta*.** *Formica occulta* Francoeur, 1973: 94, figs. 132–147 (w.q.m.) U.S.A.

**occultata* Heer, 1850; see under ***LASIUS.***

***ocella*.** **Formica ocella* Heer, 1850: 133, pl. 10, fig. 5 (q.m.) YUGOSLAVIA (Miocene).

*ochracea*. *Formica* (*Serviformica*) *cinerea* var. *ochracea* Karavaiev, 1937: 177 (w.) UKRAINE. Junior synonym of *imitans:* Dlussky, 1967a: 65.

*octospinosa* Reich, 1793; see under ***ACROMYRMEX.***

***oculata*.** **Formica oculata* Heer, 1850: 143, pl. 10, fig. 9 (m.) YUGOSLAVIA (Miocene). See also: Théobald, 1937b: 318.

*oculata* Smith, F. 1859; see under ***OPISTHOPSIS.***

*oculatissima* Forel, 1886; see under ***PROFORMICA.***

**oeningensis* Heer, 1850; see under ***CAMPONOTUS.***

*omnivora*. *Formica omnivora* Linnaeus, 1758: 581 (w.) U.S.A. Unidentifiable to genus; *incertae sedis* in *Formica.* [Wheeler, W.M. 1908b: 160, gives a combination in *Monomorium* but there is no evidence that this is correct. See also note under *pusilla.*]

*opaca* Nylander, 1856; see under ***CAMPONOTUS***

*opacus*. *Formica opacus* Kupyanskaya, 1980: 106, figs. 21–25 (w.q.m.) RUSSIA. [Junior primary homonym of *opaca* Nylander, 1856b: 55, above.] Replacement name: *kupyanskayae* **new name**. See also: Kupyanskaya, 1990: 199.

***opaciventris*.** *Formica exsectoides* var. *opaciventris* Emery, 1893i: 653 (w.m.) U.S.A. Gregg, 1952b: 14 (q.). Subspecies of *exsectoides:* Wheeler, W.M. 1913f: 484. Raised to species: Creighton, 1950a: 514.

***ophthalmica*.** **Formica ophthalmica* Heer, 1850: 125, pl. 9, fig. 9 (q.) YUGOSLAVIA (Miocene). See also: Mayr, 1867b: 52.

***orbata*.** **Formica orbata* Heer, 1850: 141, pl. 11, fig. 8 (m.) SWITZERLAND (Miocene).

***oreas*.** *Formica oreas* Wheeler, W.M. 1903e: 643, fig. 2 (w.q.m.) U.S.A. Current subspecies: nominal plus *comptula.* See also: Wheeler, W.M. 1913f: 458.

***oregonensis*.** *Formica oregonensis* Cole, 1938c: 368 (w.) U.S.A. Combination in *F.* (*Raptiformica*): Creighton, 1950a: 465. Junior synonym of *manni:* Wilson & Brown, 1955: 127. Revived from synonymy: Cole, 1956d: 212.

*orientalis*. *Formica fusca* subsp. *orientalis* Ruzsky, 1915a: 427, fig. 5 (w.) TIBET. Junior synonym of *picea* Nylander: Dlussky, 1965a: 31.

*orientalis*. *Formica rufibarbis* var. *orientalis* Wheeler, W.M. 1923b: 4 (w.) CHINA. [Junior primary homonym of *orientalis* Ruzsky, above.] Replacement name: *sinae* Emery, 1925b: 250.

*orites* Santschi, 1928; see under ***PROFORMICA.***

*ovata*. *Formica ovata* Reich, 1793: 132 (w.) FRENCH GUIANA. Unidentifiable to genus; *incertae sedis* in *Formica.*

*ovaticeps* Spinola, 1851; see under ***CAMPONOTUS.***

***pachucana*.** *Formica pachucana* Francoeur, 1973: 128, figs. 216–219 (w.q.) MEXICO.

***pacifica*.** *Formica pacifica* Francoeur, 1973: 131, figs. 190–197 (w.q.) U.S.A.

*pallens* Le Guillou, 1842; see under ***CAMPONOTUS.***

*pallens* Nylander, 1849; see under ***CAMPONOTUS.***

*pallescens* Schenck, 1852; see under ***LASIUS.***

*pallida* Latreille, 1798; see under ***LASIUS.***

*pallida* Smith, F. 1857; see under ***CAMPONOTUS.***

***pallidefulva*.** *Formica pallidefulva* Latreille, 1802c: 174 (w.) U.S.A. Mayr, 1866b: 889 (q.); Emery, 1893i: 657 (m.). Combination in *F.* (*Neoformica*): Wheeler, W.M. 1913a: 82. Senior synonym of *succinea:* Creighton, 1950a: 550. See also: Wheeler, W.M. 1913f: 548.

*pallidelutea*. *Formica pallidelutea* Latreille, 1802c: 241 (w.) SENEGAL. Possible combination in *Pheidole:* Roger, 1863b: 30. Unidentifiable to genus; *incertae sedis* in *Formica.*

*pallidinervis* Brullé, 1832; see under ***CAMPONOTUS.***

*pallipes*. *Formica pallipes* Latreille, 1787: 309 (m.) FRENCH GUIANA. [Misspelled as *pallidipes* by Dalla Torre, 1893: 40.] Unidentifiable to genus; *incertae sedis* in *Formica.*

*pallipes*. *Formica fusca* var. *pallipes* Kuznetsov-Ugamsky, 1926b: 97 (w.) TURKESTAN. [Unresolved junior primary homonym of *pallipes* Latreille, above.] Junior synonym of *fusca:* Dlussky, 1967a: 58; Tarbinsky, 1976: 180.

*pallitarsis* Provancher, 1881; see under ***LASIUS.***

**#*pamirica*.** *Formica subpilosa* subsp. *pamirica* Dlussky, 1965a: 30 (w.q.) TURKMENIA. See also: Tarbinsky, 1976: 190.

*pangens* Walker, 1859; see under ***ANOCHETUS.***

*parcipappa*. *Formica parcipappa* Cole, 1947: 616 (w.) U.S.A. Junior synonym of *curiosa:* Snelling, 1969b: 194. See also: Buren, 1968a: 33.

**parschlugiana* Heer, 1850; see under ***LASIUS.***

*parva* Buckley, 1866; see under ***TAPINOMA.***

***parvula*.** **Formica parvula* Presl, 1822: 196 (no caste or sex given) BALTIC AMBER (Oligocene).

***parvula*.** **Formica parvula* Dlussky, 1967b: 83, fig. 1 (m.) BALTIC AMBER (Oligocene). [Unresolved junior primary homonym of **parvula* Presl, above.]

*pavida* Smith, F. 1860; see under ***CAMPONOTUS.***

*pediculus*. *Formica pediculus* Christ, 1791: 518 (w.) no locality given. [According to Emery, 1892b: 162, this is a termite, Order ISOPTERA].

*pennsylvanica* De Geer, 1773; see under ***CAMPONOTUS.***

*perditor* Fabricius, 1804; see under ***DOLICHODERUS.***

***pergandei*.** *Formica pergandei* Emery, 1893i: 646, pl. 22, fig. 1 (w.) U.S.A. Combination in *F.* (*Raptiformica*):

Emery, 1925b: 259. Junior synonym of *subintegra:* Wilson & Brown, 1955: 120. Revived from synonymy, and senior synonym of *sublucida:* Buren, 1968a: 28. See also: Snelling & Buren, 1985: 72.

*perminuta* Buckley, 1866; see under ***PARATRECHINA.***

*perniciosa. Formica perniciosa* Christ, 1791: 517 (no caste given) GUINEA. *Nomen nudum.*

***perpilosa.*** *Formica perpilosa* Wheeler, W.M. 1913f: 421 (w.q.m.) U.S.A. [First available use of *Formica fusca* subsp. *subpolita* var. *perpilosa* Wheeler, W.M. 1902b: 141; unavailable name.]

****phaethusa.*** **Formica phaethusa* Wheeler, W.M. 1915h: 126, fig. 60 (w.) BALTIC AMBER (Oligocene). See also: Dlussky, 1967b: 80.

*pharaonis* Linnaeus, 1758; see under ***MONOMORIUM.***

*phyllophila. Formica phyllophila* Jerdon, 1851: 125 (w.) INDIA. Unidentifiable to genus; *incertae sedis* in *Formica.*

*picea* Leach, 1825; see under ***CAMPONOTUS.***

*picea. Formica picea* Nylander, 1846a: 917 (w.q.) FINLAND. [Junior primary homonym of *picea* Leach, 1825: 292, above.] First available replacement name: *candida* Smith, F. 1878b: 11. Forel, 1914b: 1 (m.). Junior synonym of *gagates:* Nylander, 1856b: 65; Smith, F. 1858b: 6; Dalla Torre, 1893: 198. Revived from synonymy as variety of *fusca:* Emery, 1909b: 195. Revived status as species: Bondroit, 1912: 352. Subspecies of *fusca:* Kuznetsov-Ugamsky, 1929a: 19. Status as species: Emery, 1916b: 254; Emery, 1925b: 249; Karavaiev, 1929b: 213; Karavaiev, 1936: 231; Dlussky, 1965a: 31; Dlussky, 1967a: 61; Dlussky & Pisarski, 1971: 153; Pisarski, 1975: 42; Tarbinsky, 1976: 183; Kutter, 1977c: 250; Kupyanskaya, 1990: 186. Senior synonym of *candida:* Emery, 1925b: 249; Karavaiev, 1936: 231; Dlussky, 1967a: 61; of *glabra* White: Donisthorpe, 1918a: 9; Dlussky, 1967a: 61; of *transkaucasica:* Emery, 1909b: 195; Donisthorpe, 1918a: 9; Karavaiev, 1936: 231; Dlussky, 1965a: 31; Dlussky, 1967a: 61; of *inplana, lochmatteri* and *orientalis:* Dlussky, 1965: 31; of *piceoinplana:* Dlussky, 1967a: 61. Material of the following unavailable names referred here: *filchneri:* Emery, 1909b: 195; *piceoorientalis:* Dlussky, 1965a: 31; *transcaucasicogagates:* Dlussky, 1967a: 61.

*picea* Buckley, 1866; see under ***PARATRECHINA.***

*picea. Formica lasioides* var. *picea* Emery, 1895c: 335 (w.) U.S.A. [Junior primary homonym of *picea* Leach, 1825: 292, above.] Replacement name: *vetula* Wheeler, W.M. 1912c: 90.

*piceogagates. Formica fusca* subsp. *picea* var. *piceogagates* Ruzsky, 1915a: 427 (w.) RUSSIA. Unavailable name.

*piceogagates. Formica* (*Serviformica*) *gagates* var. *piceogagates* Karavaiev, 1926e: 197 (w.q.) RUSSIA. Junior synonym of *transkaucasica:* Arakelian, 1994: 102.

*piceoinplana. Formica* (*Serviformica*) *picea* var. *piceoinplana* Emery, 1925b: 249 (w.) CHINA. [First available use of *Formica fusca* subsp. *picea* var. *piceoinplana* Ruzsky, 1915a: 426; unavailable name.] Junior synonym of *picea* Nylander: Dlussky, 1967a: 61.

*piceoorientalis. Formica fusca* subsp. *orientalis* var. *piceoorientalis* Ruzsky, 1915a: 427, fig. 6 (w.) TIBET. Unavailable name; material referred to *picea* Nylander by Dlussky, 1965a: 31.

*picipes* Olivier, 1792; see under ***CAMPONOTUS.***

*picipes. Formica picipes* Reich, 1793: 132 (no caste given) FRENCH GUIANA. [Unresolved junior primary homonym of *picipes* Olivier, 1792: 501, above.] Unidentifiable to genus; *incertae sedis* in *Formica.*

*pilicornis* Roger, 1859; see under ***CAMPONOTUS.***

*pilicornis. Formica pilicornis* Emery, 1893i: 664 (w.q.m.) U.S.A. [Junior primary homonym of *pilicornis* Roger, 1859: 228, above.] Replacement name: *francoeuri* **new name**. Subspecies of *cinerea:* Wheeler, W.M. 1913f: 527. Revived status as species: Creighton, 1950a: 538. See also: Francoeur, 1973: 74.

*piligera. Formica rufibarbis* var. *piligera* Lomnicki, 1925a: 175 (w.q.) POLAND. Junior synonym of *rufibarbis:* Lomnicki, 1928: 9; Dlussky & Pisarski, 1971: 163.

*piliventris* Smith, F. 1858; see under ***CAMPONOTUS.***

*pilosa* Olivier, 1792; see under ***CAMPONOTUS.***

*pilosa* Smith, F. 1857; see under ***CAMPONOTUS.***

*pinetorum. Formica microgyna* subsp. *rasilis* var. *pinetorum* Wheeler, W.M. 1917a: 542 (w.) U.S.A. Unavailable name; material referred to *spicata* by Creighton, 1950a: 507.

**pinguicula* Heer, 1850; see under ***CAMPONOTUS.***

**pinguis* Heer, 1850; see under ***CAMPONOTUS.***

*piniphila. Formica piniphila* Schenck, 1852: 28 (w.q.m.) GERMANY. Junior synonym of *rufa:* Mayr, 1855: 328; Emery & Forel, 1879: 450; Dalla Torre, 1893: 208. Revived from synonymy as variety of *rufa:* Forel, 1915d: 57. Revived status as species: Bondroit, 1918: 57. Variety of *rufa:* Emery, 1925b: 253; Stitz, 1939: 338. Junior synonym of *major:* Betrem, 1953: 325; of *rufa:* Yarrow, 1955a: 4; Betrem, 1960b: 76; Dlussky, 1967a: 93; Bernard, 1967: 308; Kutter, 1977c: 273.

***pisarskii.*** *Formica* (*Coptoformica*) *pisarskii* Dlussky, 1964: 1034, figs. (w.q.) MONGOLIA. See also: Kupyanskaya, 1990: 204.

****pitoni.*** **Formica pitoni* Théobald, in Piton & Théobald, 1935: 82, pl. 5, fig. 11 (q.?) FRANCE.

*placida* Smith, F. 1858; see under ***CAMPONOTUS.***

***planipilis.*** *Formica rufa* subsp. *planipilis* Creighton, 1940a: 9 (w.) U.S.A. Cole, 1956g: 259 (q.). Subspecies of *integroides:* Creighton, 1950a: 490. Raised to species: Cole, 1956g: 259.

*platygnatha* Roger, 1863; see under ***CREMATOGASTER.***

***podzolica.*** *Formica podzolica* Francoeur, 1973: 162, figs. 272–287 (w.q.m.) CANADA. Junior synonym of *subsericea:* Wheeler, G.C. & Wheeler, J. 1986g: 92. Revived from synonymy: **revived status**.

*politurata. Formica politurata* Buckley, 1866: 160 (w.) U.S.A. Unidentifiable to genus; *incertae sedis* in *Formica.*

***polyctena.*** *Formica polyctena* Foerster, 1850a: 15 (w.q.m.) GERMANY. Junior synonym of *rufa:* Nylander, 1856b: 60; Emery & Forel, 1879: 450; Dalla Torre, 1893: 208; Yarrow, 1955a: 3. Variety/subspecies of *rufa:* Forel, 1915d: 58; Emery, 1925b: 253; Stitz, 1939: 339; Boven, 1947: 189. Status as species: Bondroit, 1917a: 174; Müller, 1923: 144; Betrem, 1926: 212; Betrem, 1960b: 64; Dlussky, 1967a: 93; Dlussky & Pisarski, 1971: 187; Kutter, 1977c: 272; Collingwood, 1979: 144. Senior synonym of *minor:* Betrem, 1960b: 64; Dlussky, 1967a: 93; of *nuda:* Dlussky, 1967a: 93; Dlussky & Pisarski, 1971: 187. See also: Mabelis, 1979: 451; Gösswald, 1989: 18; Atanassov & Dlussky, 1992: 279.

***postoculata.*** *Formica postoculata* Kennedy & Dennis, 1937: 540, figs. 16–18 (w.) U.S.A.

***pratensis.*** *Formica pratensis* Retzius, 1783: 75 (w.) no locality given. Forel, 1874: 52 (q.m.). Subspecies of *rufa:* Forel, 1874: 52; Forel, 1892i: 307; Ruzsky, 1905b: 337; Emery, 1909b: 186; Wheeler, W.M. 1913f: 428; Forel, 1915d: 57; Emery, 1916b: 256; Emery, 1925b: 254; Ruzsky, 1925b: 43; Karavaiev, 1929b: 217; Stitz, 1939: 341; Gösswald, 1942: 81; Holgersen, 1942: 13. Status as species: André, 1882b: 184; Nasonov, 1889: 17; Ruzsky, 1902d: 10; Bondroit, 1912: 352; Donisthorpe, 1915d: 267; Bondroit, 1917a: 174; Müller, 1923: 142; Ruzsky, 1926: 110; Novak & Sadil, 1941: 105; Boven, 1947: 188; Yarrow, 1955a: 4; Kutter, 1965: 140; Dlussky, 1967a: 84; Dlussky & Pisarski, 1971: 177; Tarbinsky, 1976: 194; Kutter, 1977c: 272; Collingwood, 1979: 152; Gösswald, 1989: 19; Atanassov & Dlussky, 1992: 272; Seifert, 1992a: 225. Senior synonym of *nigricans* (and its junior synonyms *cordieri, pratensoides* and *thyssei*): Dlussky, 1967a: 84; Kutter, 1977c: 272; Seifert, 1992a: 225; of *ciliata:* Dlussky & Pisarski, 1971: 177; Seifert, 1992a: 225. Current subspecies: nominal plus *nuda, staerckei.*

*pratensoides. Formica minor* subsp. *pratensoides* Gösswald, 1951: 436 (w.q.m.) GERMANY. Junior synonym of *nigricans:* Yarrow, 1955a: 4. [Gösswald, 1989: 20, revives *pratensoides* from synonymy, to which it is here returned.]

***pressilabris.*** *Formica pressilabris* Nylander, 1846a: 911, pl. 18, fig. 21 (w.q.m.) FINLAND. Combination in *F.* (*Coptoformica*): Müller, 1923: 146. Subspecies of *exsecta:* Forel, 1874: 51; Emery & Forel, 1879: 450; Ruzsky, 1904b: 5; Emery, 1909b: 191; Wheeler, W.M. 1913f: 491; Forel, 1915d: 59; Emery, 1916b: 257; Ruzsky, 1925b: 43; Emery, 1925b: 257; Karavaiev, 1929b: 217; Karavaiev, 1936: 255; Stitz, 1939: 312. Status as species: André, 1882b: 179; Ruzsky, 1895: 12; Bondroit, 1912: 352; Bondroit, 1918: 64; Kutter, 1957: 11; Bernard, 1967: 324; Dlussky, 1964: 1032; Dlussky, 1967a: 103; Dlussky & Pisarski, 1971: 201; Kutter, 1977c: 285; Collingwood, 1979: 132; Atanassov & Dlussky, 1992: 285. Senior synonym of *foreli* and *tamarae:* Arakelian, 1994: 97.

****primitiva.*** **Formica primitiva* Heer, 1850: 139, pl. 11, fig. 6 (m.) SWITZERLAND (Miocene).

****primordialis.*** **Formica primordialis* Heer, 1850: 120, pl. 9, fig. 3 (q.) SWITZERLAND (Miocene).

****procera.*** **Formica procera* Heer, 1850: 111, pl. 8, fig. 5 (q.) SWITZERLAND (Miocene).

*procidua* Erichson, 1842; see under ***ANONYCHOMYRMA.***

***prociliata.*** *Formica prociliata* Kennedy & Dennis, 1937: 531, figs. 1–9 (w.q.m.) U.S.A.

**proformicoides* Dlussky, 1967; see under ****PROTOFORMICA.***

***propatula.*** *Formica propatula* Francoeur, 1973: 101, figs. 150–165 (w.q.m.) MEXICO.

***propinqua.*** *Formica rufa* subsp. *propinqua* Creighton, 1940a: 1 (w.) U.S.A. [First available use of *Formica truncicola* subsp. *integroides* var. *propinqua* Wheeler, W.M. 1917a: 538; unavailable name.] Cole, 1956g: 257 (q.m.). Subspecies of *integroides:* Creighton, 1950a: 490. Raised to species: Wheeler, G.C. & Wheeler, J. 1986g: 85; MacKay, Lowrie, *et al.* 1988: 112 (in key).

*pruinosa. Formica fusca* subsp. *pruinosa* Wheeler, W.M. 1917a: 548 (w.q.m.) U.S.A. Raised to species: Creighton, 1950a: 538. Junior synonym of *neoclara:* Francoeur, 1973: 84.

***puberula.*** *Formica sanguinea* subsp. *puberula* Emery, 1893i: 648 (w.) U.S.A. Wheeler, W.M. 1913f: 413 (q.m.). Combination in *F.* (*Raptiformica*): Emery, 1925b: 260. Raised to species: Creighton, 1950a: 468. Junior synonym of *subintegra:* Wilson & Brown, 1955: 120. Revived from synonymy: Buren, 1968a: 30.

*pubescens* Fabricius, 1775; see under ***CAMPONOTUS.***

*pubescens* Brullé, 1840; see under ***CAMPONOTUS.***

****pulchella.*** **Formica pulchella* Heer, 1850: 142, pl. 11, fig. 9 (m.) SWITZERLAND (Miocene).

***pulla.*** *Formica pulla* Francoeur, 1973: 80, figs. 107–113 (w.q.) MEXICO.

*pullula. Formica microgyna* subsp. *rasilis* var. *pullula* Wheeler, W.M. 1913f: 562 (w.q.) U.S.A. Unavailable name; material referred to *whymperi* by Creighton, 1950a: 509.

****pumila.*** **Formica pumila* Heer, 1850: 137, pl. 11, fig. 5 (q.m.) YUGOSLAVIA (Miocene).

*purpurea* Smith, F. 1858; see under ***IRIDOMYRMEX.***

*purpurescens* Lowne, 1865; see under ***ANONYCHOMYRMA.***

*pusilla. Formica pusilla* De Geer, 1773: 612, pl. 31, figs. 23, 24 (w.) SURINAM. Unidentifiable to genus; *incertae sedis* in *Formica.* [Latreille, 1802c: 263, gives *pusilla* as a senior synonym of *nana* Latreille, but the latter is also unrecognisable. Similarly Retzius, 1783: 76, gives *pusilla* as a junior synonym of *omnivora,* another unidentifiable name.]

*pygmaea* Latreille, 1798; see under ***PLAGIOLEPIS.***

***pyrenaea.*** *Formica pyrenaea* Bondroit, 1918: 52 (w.) FRANCE. Subspecies of *fusca:* Schmitz, 1950: 13. Possible junior synonym of *subrufa:* Bernard, 1967: 301. Revived status as species: Collingwood & Yarrow, 1969: 88; Collingwood, 1978: 73.

*pyrrhocephala* Motschoulsky, 1863; see under ***CAMPONOTUS.***

****quadrata.*** **Formica quadrata* Holl, 1829: 140 (no caste given) BALTIC AMBER (Oligocene).

*quadriceps* Smith, F. 1859; see under ***CAMPONOTUS.***

*quadridens* De Geer, 1773; see under ***CEPHALOTES.***

*quadridens* Fabricius, 1793; see under ***ECTATOMMA.***

*quadridenticulata* Roger, 1862; see under ***DOLICHODERUS.***

*quadrinotata. Formica quadrinotata* Losana, 1834: 320, pl. 36, fig. 4 (w.) ITALY. Unidentifiable to genus; *incertae sedis* in *Formica.* [Mayr, 1855: 389 (footnote) suggests a combination in *Ponera.* Emery, 1916b: 105, suspected that this was synonymous with *Cryptopone ochracea,* but the description is not convincing.]

*quadripunctata* Linnaeus, 1771; see under ***DOLICHODERUS.***

*quadrisecta* Smith, F. 1858; see under ***CAMPONOTUS.***

***querquetulana.*** *Formica querquetulana* Kennedy & Dennis, 1937: 536, figs. 10–15 (w.q.) U.S.A.

*rabaudi* Bondroit, 1917 [*Formicina*]; see under ***LASIUS.***

*rapax* Fabricius, 1804; see under ***CAMPONOTUS.***

*rasilis. Formica microgyna* var. *rasilis* Wheeler, W.M. 1903e: 648 (w.q.m.) U.S.A. Subspecies of *microgyna:* Wheeler, W.M. 1913f: 468. Raised to species: Creighton, 1950a: 506. Junior synonym of *densiventris:* MacKay, Lowrie, *et al.* 1988: 114.

*rastellata* Latreille, 1802; see under ***POLYRHACHIS.***

***ravida.*** *Formica rufa* subsp. *ravida* Creighton, 1940a: 1 (w.q.) U.S.A. [First available use of *Formica truncicola* subsp. *integroides* var. *ravida* Wheeler, W.M. 1913f: 560; unavailable name.] Subspecies of *obscuripes:* Creighton, 1950a: 493. Junior synonym of *haemorrhoidalis* Creighton: Brown, 1965d: 181. [*F. haemorrhoidalis* Creighton, 1940a: 1, junior primary homonym of *F. haemorrhoidalis* Latreille, 1802c: 276.] *F. ravida* here designated as first available replacement name. Senior synonym of *tahoensis:* **new synonymy**.

*recidiva. Formica microgyna* var. *recidiva* Wheeler, W.M. 1913f: 395 (in key) (w.) U.S.A. [*Formica microgyna* subsp. *microgyna* var. *recidiva* Wheeler, W.M. 1913f: 467 (w.m.); unavailable name.] Variety of *microgyna:* Wheeler, W.M. 1917a: 542. Junior synonym of *microgyna:* Creighton, 1950a: 504.

*rediana. Formica rediana* Leach, 1825: 291 (w.q.m.) FRANCE. Unidentifiable to genus; *incertae sedis* in *Formica.*

**redtenbacheri* Heer, 1850; see under ***LASIUS.***

***reflexa.*** *Formica reflexa* Buren, 1942: 399 (w.q.m.) U.S.A.

*relucens* Latreille, 1802; see under ***POLYRHACHIS.***

*respiciens* Smith, F. 1865; see under ***OPISTHOPSIS.***

***retecta.*** *Formica retecta* Francoeur, 1973: 245, figs. 405–408 (w.q.) MEXICO.

*rizadaghi. Formica* (*Serviformica*) *rizadaghi* Arnol'di, 1968: 1819, figs. 19, 20 (w.q.) TURKMENIA. Junior synonym of *clara:* Dlussky, Soyunov & Zabelin, 1990: 139.

****robusta.*** **Formica robusta* Carpenter, 1930: 56, pl. 6, fig. 6; pl. 7, fig. 6 (q.m.) U.S.A. (Miocene).

*rostrata. Formica rostrata* Fabricius, 1787: 310 (w.) FRENCH GUIANA. Unidentifiable to genus; *incertae sedis* in *Formica.*

*rotundata. Formica sanguinea* var. *rotundata* Kuznetsov-Ugamsky, 1926b: 95, fig. 2 (w.) TURKESTAN. Raised to species: Dlussky, 1962: 182. Junior synonym of *sanguinea:* Dlussky, 1965a: 16.

*rubens. Formica exsecta* var. *rubens* Forel, 1874: 51 (w.) SWITZERLAND. Raised to species: Bondroit, 1918: 62. Variety of *exsecta:* Ruzsky, 1925b: 43; Emery, 1925b: 257. Junior synonym of *exsecta:* Dlussky, 1964: 1027.

*rubescens. Formica rubescens* Leach, 1825: 289 (w.q.m.) FRANCE. Unidentifiable to genus; *incertae sedis* in *Formica.*

*rubescens. Formica fusca* var. *rubescens* Forel, 1904f: 423 (w.) SWITZERLAND. [Unresolved junior primary homonym of *rubescens* Leach, above.] Emery, 1909b: 196 (q.); Wheeler, W.M. 1913f: 498 (m.). Variety of *glebaria:* Bondroit, 1918: 50; Boven, 1947: 188. Junior synonym of *cunicularia:* Yarrow, 1954a: 231; Dlussky, 1967a: 73; Bernard, 1967: 296.

***rubicunda.*** *Formica sanguinea* subsp. *rubicunda* Emery, 1893i: 647, pl. 22, fig. 2 (w.q.) U.S.A. Wheeler, W.M. 1913f: 407 (m.). Combination in *F.* (*Raptiformica*): Emery, 1925b: 260. Raised to species: Creighton, 1950a: 468. Junior synonym of *subintegra:* Wilson & Brown, 1955: 120. Revived from synonymy: Buren, 1968a: 30.

*rubida* Latreille, 1802; see under ***MANICA.***

*rubiginosa* Latreille, 1802; see under ***LASIUS.***

*rubiginosa* Le Guillou, 1842; see under ***POLYRHACHIS.***

*rubiginosa. Formica rufa* subsp. *obscuriventris* var. *rubiginosa* Emery, 1893i: 650 (w.q.) U.S.A. Unavailable name; material referred to *obscuripes* by Creighton, 1940a: 1.

*rubra* Linnaeus, 1758; see under ***MYRMICA.***

*rubripes* Latreille, 1802; see under ***CAMPONOTUS.***

***rufa.*** *Formica rufa* Linnaeus, 1761: 426 (q.m.) EUROPE. [Not Linnaeus, 1758: 580 (w.); see Yarrow, 1954b: 313; Yarrow, 1955a: 5.] Latreille, 1802c: 143 (w.q.m.); Wheeler, G.C. & Wheeler, J. 1953c: 166 (l.). Senior synonym of *apicalis:* Roger, 1862c: 287; of *dorsata:* Stephens, 1829: 357; Nylander, 1846a: 902; of *emeryi:* Yarrow, 1955a: 4; of *ferruginea:* Emery, 1892b: 161; of *gaullei:* Yarrow, 1955a: 4; Bernard, 1967: 308; of *major:* Emery & Forel, 1879: 450; Yarrow, 1955a: 3; Dlussky, 1967a: 93; of *meridionalis:* Yarrow, 1955a: 4; Dlussky & Pisarski, 1971: 184; of *piniphila:* Mayr, 1855: 328; Yarrow, 1955a: 4; Betrem, 1960b: 76; of *rufopratensis:* Karavaiev, 1936: 240; Dlussky, 1967a: 93. Material of the unavailable name *bonroiti* referred here by Yarrow, 1955a: 4; of *major* Gösswald by Dlussky, 1967a: 93. Current subspecies: nominal plus *angusticeps, brevisetosa, constricta, grouvellei, obscurata, rufopratensoides, rufobrevisetosa, tshugunovi.* See also: Wheeler, W.M. 1913f: 425; Donisthorpe, 1915d: 245; Yarrow, 1955a: 3; Dlussky, 1967a: 93; Kutter, 1977c: 273; Collingwood, 1979: 141; Douwes, 1979: 187; Douwes, 1981: 213; Pavan, 1981: 1; Gösswald, 1989: 16; Atanassov & Dlussky, 1992: 277.

*rufescens* Latreille, 1798; see under ***POLYERGUS.***

***rufibarbis.*** *Formica rufibarbis* Fabricius, 1793: 355 (w.) FRANCE. Jurine, 1807: 273 (q.m.); Emery, 1909b: 197 (q.m.). Combination in *F.* (*Serviformica*): Forel, 1915d: 64. Variety/race/subspecies of *fusca:* Forel, 1874: 54; Mayr, 1886d: 427; Forel, 1892i: 307; Ruzsky, 1904b: 4; Wheeler, W.M. 1908g: 406; Emery, 1909b: 197; Forel, 1915d: 64; Emery, 1916b: 255; Santschi, 1919e: 247. Status as species: André, 1882b: 182; Nasonov, 1889: 19; Dalla Torre, 1893: 209; Emery, 1898c: 126; Ruzsky, 1902d: 11; Bingham, 1903: 335; Wheeler, W.M. 1913f: 514; Donisthorpe, 1915d: 320; Wheeler, W.M. 1917a: 550; Bondroit, 1918: 51; Emery, 1925b: 249; Karavaiev, 1927c: 286; Stitz, 1930: 238; Karavaiev, 1936: 234; Stitz, 1939: 355; Novak & Sadil, 1941: 107; Yarrow, 1954a: 231; Dlussky, 1967a: 73; Bernard, 1967: 297; Francoeur, 1973: 228; Collingwood, 1979: 128. Senior synonym of *cinereorufibarbis:* Bernard, 1967: 297; Collingwood, 1978: 73; of *defensor* and *fraterna:* Forel, 1894c: 403; Bingham, 1903: 335; of *nicaeensis:* Roger, 1863b: 13; of *piligera:* Lomnicki, 1928: 9; Dlussky & Pisarski, 1971: 163; of *stenoptera:* Dalla Torre, 1893: 210; Yarrow, 1954a: 231; Dlussky, 1967a: 73. Current subspecies: nominal plus *clarorufibarbis, kashmirica, subpilosorufibarbis.*

*ruficeps* Fabricius, 1804; see under ***CAMPONOTUS.***

*ruficeps* Smith, F. 1857; see under ***CAMPONOTUS.***

*ruficeps. Formica ruficeps* Motschoulsky, 1863: 12 (w.) JAVA. [Unresolved junior primary homonym of *ruficeps* Fabricius, 1804: 404, above.] Unidentifiable to genus; *incertae sedis* in *Formica.*

*ruficornis. Formica ruficornis* Gmelin, in Linnaeus, 1790: 2805 (q.) EUROPE. Unidentifiable to genus; *incertae sedis* in *Formica.*

*ruficornis* Fabricius, 1804; see under ***LASIUS.***

*rufifrons* Smith, F. 1860; see under ***CAMPONOTUS.***

*rufipes* Fabricius, 1775; see under ***CAMPONOTUS.***

*rufipes. Formica fusca* var. *rufipes* Stitz, 1930: 238 (w.q.) RUSSIA. [Unresolved junior primary homonym of *rufipes* Fabricius, 1775: 391, above.] Junior synonym of *fusca:* Dlussky, 1967a: 58.

*rufitarsis* Fabricius, 1804; see under ***MESSOR.***

*rufiventris* Fabricius, 1804; see under ***CAMPONOTUS.***

*rufiventris. Formica rufiventris* Emery, 1893i: 665, pl. 22, fig. 11 (m.) U.S.A. [Junior primary homonym of *rufiventris* Fabricius, 1804: 409, above.] Replacement name: *flammiventris* Wheeler, W.M. 1912c: 90.

*rufobarboides. Formica fusca* var. *rufobarboides* Emery, 1897f: 239, TURKEY. *Nomen nudum* (attributed to Forel).

#***rufobrevisetosa.*** *Formica rufa* var. *rufobrevisetosa* Ruzsky, 1926: 110 (w.) RUSSIA.

*rufoglauca* Jerdon, 1851; see under ***CAMPONOTUS.***

*rufoides. Formica rufa* r. *pratensis* var. *rufoides* Forel, 1874: 368 (w.) SWITZERLAND. Unavailable name.

***rufolucida.*** *Formica* (*Neoformica*) *rufolucida* Collingwood, 1962: 227, figs. 10, 11 (w.) BURMA. [Incorrect generic combination?]

***rufomaculata.*** *Formica exsecta* var. *rufomaculata* Ruzsky, 1895: 13 (w.) RUSSIA. [Also described as new by Ruzsky, 1896: 68.] Variety of *pressilabris:* Ruzsky, 1902e: 16. Raised to species: Dlussky, 1964: 1035. See also: Dlussky, 1967a: 108; Arnol'di & Dlussky, 1978: 552; Kupyanskaya, 1990: 203.

*rufonigra* Lowne, 1865; see under ***IRIDOMYRMEX.***

*rufopratensis. Formica rufa* var. *rufopratensis* Forel, 1874: 53 (w.q.) SWITZERLAND. Stitz, 1939: 338 (m.). Variety of *pratensis:* Dalla Torre, 1893: 205; of *rufa:* Emery, 1925b: 253; Karavaiev, 1927c: 282. Junior synonym of *rufa:* Karavaiev, 1936: 240; Dlussky, 1967a: 93.

#***rufopratensoides.*** *Formica rufa* var. *rufopratensoides* Forel, 1874: 368 (w.) SWITZERLAND.

*rufotruncicola. Formica rufa* var. *rufotruncicola* Ruzsky, 1896: 68 (w.) RUSSIA. [First available use of *Formica rufa* subsp. *pratensis* form. *rufotruncicola* Ruzsky, 1895: 11; unavailable name.] Junior synonym of *truncorum:* Dlussky, 1967a: 81.

*rupestris. Formica rupestris* Leach, 1825: 291 (w.q.m.) FRANCE. Unidentifiable to genus; *incertae sedis* in *Formica.*

*rutilans. Formica cinerea* var. *rutilans* Wheeler, W.M. 1913f: 399 (diagnosis in key) (w.) U.S.A. [*Formica cinerea* subsp. *cinerea* var. *rutilans* Wheeler, W.M. 1913f: 525; unavailable name.] Junior synonym of *montana:* Creighton, 1950a: 534.

*ruzskyi. Formica subpilosa* subsp. *ruzskyi* Dlussky, 1965a: 30. Unnecessary replacement name for *clarior* Ruzsky, 1915a: 421; unavailable name. Junior synonym of *clarissima:* **new synonymy**. See also: Kupyanskaya, 1990: 188.

*sabularum* Bondroit, 1918 [*Formicina*]; see under ***LASIUS.***

*sabulosa. Formica* (*Serviformica*) *cinerea* var. *sabulosa* Karavaiev, 1931a: 315 (w.) UKRAINE. Junior synonym of *imitans:* Dlussky, 1967a: 65.

*saccharivora. Formica saccharivora* Linnaeus, 1758: 580 (w.) U.S.A. Unidentifiable to genus; *incertae sedis* in *Formica.*

*salomonis* Linnaeus, 1758; see under ***MONOMORIUM.***

***sanguinea.*** *Formica sanguinea* Latreille, 1798: 37 (w.) FRANCE. Jurine, 1807: 272 (q.m.); Wheeler, G.C. & Wheeler, J. 1953c: 169 (l.); Imai, 1966: 119 (k.). Combination in *F.* (*Raptiformica*): Forel, 1913i: 361. Senior synonym of *dominula:* Smith, F. 1851: 115; Schenck, 1852: 36; Mayr, 1855: 337; of *arenicola, borea, clarior, fusciceps, griseopubescens, leninei, mollesonae, monticola, rotundata, strennua* and *tristis:* Dlussky, 1965a: 16; of *flavorubra:* Dlussky, 1967a: 97; Collingwood, 1978: 74. Material of the unavailable name *minuta* referred here by Dlussky, 1965a 16. See also: Wheeler, W.M. 1913f: 401; Donisthorpe, 1915d: 280; Stitz, 1939: 316; Dlussky, 1967a: 97; Tarbinsky, 1976: 186; Kutter, 1977c: 289; Collingwood, 1979: 137; Kupyanskaya, 1990: 191; Atanassov & Dlussky, 1992: 280.

*sansabeana* Buckley, 1866; see under ***CAMPONOTUS.***

*santschii. Formica rufa* var. *santschii* Wheeler, W.M. 1913f: 390 (in key). Replacement name for *alpina* Santschi, 1911j: 349. [Junior primary homonym of *alpina* Wheeler, W.M. 1909e: 85.] Junior synonym of *lugubris:* Yarrow, 1955a: 5.

*santschii. Formica rufa* subsp. *pratensis* var. *santschii* Krausse, 1926b: 115 (w.q.) GERMANY. Unavailable name.

*saturata. Formica dakotensis* var. *saturata* Wheeler, W.M 1917a: 542 (w.) U.S.A. Junior synonym of *dakotensis:* Creighton, 1950a: 484.

*savignyi* Dufour, 1862; see under ***CATAGLYPHIS.***

*saxicola. Formica saxicola* Buckley, 1866: 166 (w.q.) U.S.A. Unidentifiable to genus; *incertae sedis* in *Formica.* See also: Emery, 1895c: 338; Wheeler, W.M. 1902f: 23.

*schardj. Formica schardj* Forskål, 1775: xxiii (w.) ISRAEL? Unidentifiable to genus; *incertae sedis* in *Formica.* See also: Emery, 1892b: 160.

***schaufussi.*** *Formica schaufussi* Mayr, 1866a: 493, fig. 6 (w.) U.S.A. Mayr, 1886d: 427 (q.m.). Combination in *F.* (*Neoformica*): Wheeler, W.M. 1913f: 552. Subspecies of *pallidefulva:* Emery, 1893i: 654; Wheeler, W.M. 1908g: 408. Revived status as species: Creighton, 1950a: 551. Current subspecies: nominal plus *dolosa.*

**schmidtii* Heer, 1850; see under ***LIOMETOPUM.***

*schmidti. Formica rufa* var. *schmidti* Ruzsky, 1920: 77 (w.) RUSSIA. Subspecies of *rufa:* Ruzsky, 1926: 109. [Junior primary homonym of **schmidtii* Heer, 1850: 138, above.] Synonymous with *aquilonia:* Dlussky, 1967a: 90, hence *aquilonia* the first available replacement name.

***scitula.*** *Formica microgyna* subsp. *scitula* Wheeler, W.M. 1913f: 470 (q.) U.S.A. Raised to species: Creighton, 1950a: 507.

*scorpiomajor. Formica scorpiomajor* Christ, 1791: 518 (no caste given) U.S.A.? *Nomen nudum.*

*scorpiominor. Formica scorpiominor* Christ, 1791: 518 (no caste given) U.S.A.? *Nomen nudum.*

*scrutator* Smith, F. 1859; see under ***ANONYCHOMYRMA.***

*scutellaris* Olivier, 1792; see under ***CREMATOGASTER.***

*sedula* Smith, F. 1857; see under ***CAMPONOTUS.***

***selysi.*** *Formica cinerea* var. *selysi* Bondroit, 1918: 54 (w.) FRANCE. Combination in *F.* (*Serviformica*): Emery, 1925b: 246. Raised to species: Kutter, 1977c: 253.

*semipunctata* Kirby, W. 1837; see under ***CAMPONOTUS.***

*senex* Smith, F. 1858; see under ***CAMPONOTUS.***

***sentschuensis.*** *Formica* (*Raptiformica*) *sentschuensis* Ruzsky, 1915a: 428, figs. 7, 8 (w.q.) TIBET. Senior synonym of *dalailamae:* Dlussky, 1965a: 17.

*septentrionale* Buckley, 1866; see under ***CAMPONOTUS.***

****sepulta.*** **Formica sepulta* Théobald, 1937b: 319, pl. 15, fig. 14; pl. 25, fig. 16; pl. 26, fig. 16 (q.m.) FRANCE (Oligocene).

*sericata* Guérin-Méneville, 1831; see under ***POLYRHACHIS.***

*sericea* Fabricius, 1798; see under ***CAMPONOTUS.***

*sericeiventris* Guérin-Méneville, 1838; see under ***CAMPONOTUS.***

****serresi.*** **Formica serresi* Théobald, 1937b: 317, pl. 26, fig. 19 (m.) FRANCE (Oligocene).

*sessilis* Say, 1836; see under ***TAPINOMA.***

****seuberti.*** **Formica seuberti* Heer, 1850: 127, pl. 9, fig. 12 (q.) SWITZERLAND (Miocene).

*sexdens* Linnaeus, 1758; see under ***ATTA.***

*sexdentata* Latreille, 1802; see under ***ATTA.***

*sexguttata* Fabricius, 1793; see under ***CAMPONOTUS.***

*sexspinosa* Latreille, 1802; see under ***POLYRHACHIS.***

*siamicaalba. Formica siamicaalba* Christ, 1791: 518 (no caste given) SENEGAL. *Nomen nudum.* See also: Emery, 1892b: 162.

*siamicarubra. Formica siamicarubra* Christ, 1791: 518 (w.) "Barbary". Unidentifiable to genus; *incertae sedis* in *Formica.* See also: Emery, 1892b: 162.

***sibylla.*** *Formica sibylla* Wheeler, W.M. 1913f: 530 (w.m.) U.S.A. See also: Wheeler, W.M. 1917a: 551; Francoeur, 1973: 263.

*similis. Formica integra* var. *similis* Mayr, 1886d: 425 (w.q.m.) U.S.A. Junior synonym of *integra:* Emery, 1893i: 652; Wheeler, W.M. 1906b: 16; Creighton, 1950a: 487.

*simillima* Smith, F. 1862; see under ***CAMPONOTUS.***

*simulata. Formica simulata* Smith, F. 1878b: 10 (w.) CHINA. Junior synonym of *truncicola:* Forel, 1894c: 403.

***sinae.*** *Formica* (*Serviformica*) *rufibarbis* var. *sinae* Emery, 1925b: 250. Replacement name for *orientalis* Wheeler, W.M. 1923b: 4. [Junior primary homonym of *orientalis* Ruzsky, 1915a: 427.] Santschi, 1928b: 42 (q.). Junior synonym of *glauca:* Dlussky,

1967a: 74. Revived from synonymy, and raised to species: Wu, 1990: 4.

**sinensis.** *Formica truncicola* var. *sinensis* Wheeler, W.M. 1913f: 391 (diagnosis in key) (w.) CHINA. [*Formica truncicola* subsp. *truncicola* var. *sinensis* Wheeler, W.M. 1913f: 437; unavailable name.] Raised to species: Wu, 1990: 6 (in key).

*singularis* Smith, F. 1858; see under ***CAMPONOTUS.***

*smaragdina* Fabricius, 1775; see under ***OECOPHYLLA.***

*smithii* Lowne, 1865; see under ***IRIDOMYRMEX.***

*solitaria* Smith, 1858; see under ***GIGANTIOPS.***

**spatulata.** *Formica microgyna* subsp. *spatulata* Buren, 1944a: 305 (w.q.m.) U.S.A. Raised to species: Creighton, 1950a: 508.

*specularis. Formica specularis* Wheeler, W.M. 1908g: 408 (q.) U.S.A. [First available use of *Formica fusca* subsp. *subpolita* var. *specularis* Emery, 1893i: 633; unavailable name.] Junior synonym of *wasmanni:* Wheeler, W.M. 1913f: 464 (Wheeler incorrectly gives seniority to *specularis*). Junior synonym of *dakotensis:* Creighton, 1950a: 484.

*spicata. Formica rasilis* subsp. *spicata* Creighton, 1950a: 507 (w.q.m.) U.S.A. [First available use of *Formica microgyna* subsp. *rasilis* var. *spicata* Wheeler, W.M. 1913f: 469; unavailable name.] Material of the unavailable name *pinetorum* referred here by Creighton, 1950a: 507. Junior synonym of *densiventris:* Cole, 1954a: 89; Cole, 1955b: 50.

*spinicollis* Latreille, 1817; see under ***DOLICHODERUS.***

*spininoda* Latreille, 1802; see under ***PARAPONERA.***

*staegeri. Formica truncorum* var. *staegeri* Stitz, 1939: 347 (w.) GERMANY. [First available use of *Formica rufa* subsp. *truncicola* ab. *staegeri* Krausse, 1926d: 264; unavailable name.] Variety of *truncorum:* Betrem, 1960b: 76. Junior synonym of *truncorum:* Dlussky, 1967a: 81; Dlussky & Pisarski, 1971: 174.

**#staerckei.** *Formica nigricans* var. *staerckei* Betrem, 1960a: 132 (q.) NETHERLANDS. [Also described as new by Betrem, 1960b: 70.] Subspecies of *pratensis:* current status.

*stenoptera. Formica stenoptera* Foerster, 1850a: 26 (w.q.) GERMANY. Junior synonym of *cunicularia:* Smith, F. 1855a: 103; Mayr, 1855: 343; of *rufibarbis:* Dalla Torre, 1893: 210; Stitz, 1939: 355; Dlussky, 1967a: 73.

*stigma* Fabricius, 1804; see under ***PACHYCONDYLA.***

*stitzi. Formica truncorum* ab. *stitzi* Stitz, 1939: 347 (w.) GERMANY. [First available use of *Formica rufa* subsp. *truncicola* ab. *stitzi* Krausse, 1926d: 264; unavailable name.] Junior synonym of *truncorum:* Dlussky, 1967a: 81.

***strangulata.** **Formica strangulata* Wheeler, W.M. 1915h: 130, fig. 62 (w.) BALTIC AMBER (Oligocene). [*Formica strangulata* Wheeler, W.M. 1913f: 383 (footnote); *nomen nudum.*] See also: Dlussky, 1967b: 80.

*strawinskii. Formica forsslundi* subsp. *strawinskii* Petal, 1963: 195, figs. 1a, 2a, 3, 4a, 5a, 6a, 7, 8a, 9a, 10a, 11 (w.q.m.) POLAND. Junior synonym of *forsslundi:* Dlussky & Pisarski, 1971: 199.

*strenua. Formica strenua* Haliday, 1836: 329 (w.) CHILE. Unidentifiable to genus; *incertae sedis* in *Formica.*

*strennua. Formica* (*Raptiformica*) *sanguinea* var. *strennua* Santschi, 1925g: 352 (w.) SPAIN. [Unresolved junior primary homonym of *strenua* Haliday, above.] Junior synonym of *sanguinea:* Dlussky, 1965a: 16.

*striata* Smith, F. 1862; see under ***CAMPONOTUS.***

*striaticeps* Forel, 1911; see under ***PROFORMICA.***

*stricta* Jerdon, 1851; see under ***CAMPONOTUS.***

*structor* Latreille, 1798; see under ***MESSOR.***

**subaenescens.** *Formica fusca* var. *subaenescens* Emery, 1893i: 659 (w.) U.S.A. Wheeler, W.M. 1913f: 504 (q.m.). Subspecies of *fusca:* Buren, 1944a: 300. Junior synonym of *fusca:* Creighton, 1950a: 532; Francoeur, 1973: 189; Wheeler, G.C. & Wheeler, J. 1986g: 16. Revived from synonymy, and raised to species: Francoeur, 1977b: 208; and current status.

*subcaviceps. Formica rufa* subsp. *subcaviceps* Creighton, 1940a: 1 (w.m.) U.S.A. [First available use of *Formica truncicola* subsp. *integra* var. *subcaviceps* Wheeler, W.M. 1917a: 540; unavailable name.] Junior synonym of *integroides:* Creighton, 1950a: 489.

**subcyanea.** *Formica subcyanea* Wheeler, W.M. 1913f: 563 (w.q.m.) MEXICO. See also: Francoeur, 1973: 249.

**subelongata.** *Formica subelongata* Francoeur, 1973: 268, figs. 482–485 (w.) U.S.A.

*subfasciata. Formica rufa* subsp. *subfasciata* Creighton, 1940a: 1 (w.) U.S.A. [First available use of *Formica truncicola* subsp. *integroides* var. *subfasciata* Wheeler, W.M. 1917a: 539; unavailable name.] Subspecies of *integroides:* Creighton, 1950a: 490. Junior synonym of *integroides:* Brown, 1965d: 185.

**subintegra.** *Formica sanguinea* subsp. *subintegra* Wheeler, W.M. 1908f: 627 (w.q.) U.S.A. [First available use of *Formica sanguinea* subsp. *rubicunda* var. *subintegra* Emery, 1893i: 648, pl. 22, fig. 3; unavailable name.] Wheeler, W.M. 1913f: 411 (m.); Hung, 1969: 455 (k.). Combination in *F.* (*Raptiformica*): Emery, 1925b: 260. Subspecies of *sanguinea:* Wheeler, W.M. 1913f: 410. Raised to species, and material of the unavailable name *gilvescens* referred here: Creighton, 1950a: 470. See also: Talbot & Kennedy, 1940: 560.

*sublucida. Formica sublucida* Creighton, 1950a: 472 (w.q.) U.S.A. [First available use of *Formica sanguinea* subsp. *rubicunda* var. *sublucida* Wheeler, W.M. 1913f: 408; unavailable name.] Junior synonym of *subintegra:* Wilson & Brown, 1955: 120; of *pergandei:* Buren, 1968a: 28.

**subnitens.** *Formica rufa* subsp. *subnitens* Creighton, 1940a: 10, fig. 1 (w.) U.S.A. Miller, 1957: 255 (q.m.). Subspecies of *integroides:* Creighton, 1950a: 490. Raised to species: Cole, 1955b: 50; Miller, 1957: 253; MacKay, Lowrie, *et al.* 1988: 111 (in key).

*subnuda. Formica sanguinea* subsp. *subnuda* Wheeler, W.M. 1910g: 570 (w.) CANADA. [First available use of *Formica sanguinea* subsp. *rubicunda* var. *subnuda* Emery, 1895c: 335; unavailable name.] Wheeler, W.M. 1913f: 409 (q.m.); Wheeler, G.C. & Wheeler, J. 1953c: 170 (l). Combination in *F.* (*Raptiformica*): Emery, 1925b: 260. Subspecies of *sanguinea:* Wheeler, W.M. 1913f: 469; Wheeler, W.M. 1917a: 533. Raised to species: Creighton, 1950a: 469. Synonym of *aserva:* Creighton, 1950a: 469; Wilson & Brown, 1955: 118; Buren, 1968a: 33. [Creighton, Wilson & Brown, and Buren incorrectly make *subnuda* the senior name; *aserva* has seniority and is therefore the valid name of this taxon.]

*subpicea. Formica subpicea* Motschoulsky, 1863: 12 (w.) SRI LANKA. Unidentifiable to genus; *incertae sedis* in *Formica.*

**subpilosa.** *Formica rufibarbis* subsp. *subpilosa* Ruzsky, 1902b: 472 (w.) RUSSIA. Karavaiev, 1910b: 17 (q.m.). Combination in *F.* (*Serviformica*): Emery, 1925b: 250. Raised to species: Bondroit, 1918: 51; Collingwood, 1961a: 70. Junior synonym of *rufibarbis:* Bernard, 1967: 297. Revived from synonymy, with status as species: Dlussky, 1965a: 28; Dlussky, 1967a: 70; Tarbinsky, 1976: 189. Senior synonym of *bipilosa* and material of the unavailable names *clara* and *fallax* referred here: Dlussky, 1965a: 28. Material of the unavailable name *subpilosoclara* referred here: Dlussky, 1967a: 70. Current subspecies: nominal plus *clarissima, litoralis, pamirica.*

*subpilosoclara. Formica rufibarbis* subsp. *clara* var. *subpilosoclara* Ruzsky, 1905b: 400 (w.) RUSSIA. Unavailable name; material referred to *subpilosa* by Dlussky, 1967a: 70.

**#subpilosorufibarbis.** *Formica rufibarbis* var. *subpilosorufibarbis* Ruzsky, 1905b: 398 (w.) RUSSIA.

**subpolita.** *Formica fusca* var. *subpolita* Mayr, 1886d: 426 (w.q.) U.S.A. Wheeler, W.M. 1913f: 534 (m.). Combination in *F.* (*Serviformica*): Emery, 1925b: 251. Variety of *gagates:* Dalla Torre, 1893: 199. Subspecies of *fusca:* Emery, 1893i: 661. Raised to species: Wheeler, W.M. 1908g: 625; Wheeler, W.M. 1913f: 532. Senior synonym of *flammiventris:* Wheeler, W.M. 1913f: 532; Creighton, 1950a: 541; of *camponoticeps* and *ficticia:* Francoeur, 1973: 107.

**subrufa.** *Formica subrufa* Roger, 1859: 236 (w.) SPAIN. Emery, 1909b: 200 (q.m.); Tinaut, 1990a: 283 (m.). Combination in *F.* (*Serviformica*): Santschi, 1919e: 247; in *F.* (*Iberoformica*): Tinaut, 1990a: 282. Race/subspecies of *fusca:* Emery & Forel, 1879: 451; Wheeler, W.M. 1908g: 409. Status as species: André, 1882b: 181; Nasonov, 1889: 20; Ruzsky, 1902d: 13; Emery, 1909b: 199; Wheeler, W.M. 1913f: 531; Santschi, 1919e: 247; Bernard, 1967: 301; Collingwood, 1978: 93; Tinaut, 1990a: 282.

*subrufoides. Formica cinerea* var. *subrufoides* Forel, 1913i: 360 (w.) AUSTRIA. Röszler, 1950: 219 (q.). Raised to species: Bondroit, 1918: 55; Kutter, 1977c: 255. Junior synonym of *cinerea:* Agosti & Collingwood, 1987a: 59.

**subsericea.** *Formica subsericea* Say, 1836: 289 (w.m.) U.S.A. Junior synonym of *fusca:* Mayr, 1886d: 426; Creighton, 1950a: 532. Revived from synonymy as variety/subspecies of *fusca:* Emery, 1893i: 659; Buren, 1944a: 300. Revived from synonymy, raised to species, and senior synonym of *lecontei:* Francoeur, 1973: 172.

*subsericeoneorufibarbis. Formica subsericeoneorufibarbis* Wheeler, W.M. 1901b: 519. *Nomen nudum,* attributed to Emery. Material referred to *gnava:* Wheeler, W.M. 1913f: 518. [Name misspelled *subsericeorufibarbis* by Emery, 1925b: 250.]

*subspinosa. Formica* (*Hypochira*) *subspinosa* Buckley, 1866: 169 (w.) U.S.A. Unidentifiable to genus; *incertae sedis* in *Formica.* [May possibly be a dolichoderine, see: Emery, 1895c: 338; Emery, 1925b: 271; Smith, D.R. 1979: 1422.]

*subterranea* Latreille, 1798; see under ***APHAENOGASTER.***

*subtilis* Smith, F. 1860; see under ***ANOPLOLEPIS*** and ***CAMPONOTUS.***

*succinea. Formica pallidefulva* var. *succinea* Wheeler, W.M. 1904f: 369 (diagnosis in key) (w.) U.S.A. Wheeler, W.M. 1913f: 551 (q.m.). Junior synonym of *pallidefulva:* Creighton, 1950a: 550.

*sudetica. Formica exsecta* var. *sudetica* Scholz, 1924: 48 (w.) POLAND. Junior synonym of *exsecta:* Dlussky & Pisarski, 1971: 194.

**suecica.** *Formica suecica* Adlerz, 1902: 263 (w.q.m.) SWEDEN. Subspecies of *exsecta:* Wheeler, W.M. 1908g: 408; Emery, 1909b: 193. Status as species: Wheeler, W.M. 1913f: 492; Emery, 1925b: 258; Stitz, 1939: 314; Dlussky, 1964: 1035; Bernard, 1967: 326; Dlussky, 1967a: 107; Collingwood, 1979: 133.

*suffusa* Smith, F. 1858; see under ***CAMPONOTUS.***

*sulci. Formica fusca* r. *lemani* var. *sulci* Samsinak, 1951: 126 (m.) CZECHOSLOVAKIA. Unavailable name.

*superba. Formica pratensis* var. *superba* Wheeler, W.M. 1933c: 65 (w.) CHINA. Raised to species: Santschi, 1941: 278. Junior synonym of *uralensis:* Wu, 1990: 4.

***surinamensis.** *Formica surinamensis* Berendt, 1830: 37 (no caste given) BALTIC AMBER.

*suzukii. Formica rufa* subsp. *dusmeti* var. *suzukii* Teranishi, 1940: 23 (w.) KOREA. Unavailable name.

*sylvicola* Jerdon, 1851 [original spelling *syloicola*]; see under ***POLYRHACHIS.***

*sylvatica* Olivier, 1792; see under ***CAMPONOTUS.***

*syrdariana* Santschi, 1928; see under ***PROFORMICA.***

*tahoensis. Formica rufa* subsp. *tahoensis* Creighton, 1940a: 1 (w.q.) U.S.A. [First available use of *Formica truncicola* subsp. *integroides* var. *tahoensis* Wheeler, W.M. 1917a: 538; unavailable name.] Cole, 1956g: 260 (q.m.). Subspecies of *integra:* Creighton, 1950a: 488. Junior synonym of *haemorrhoidalis:* Brown, 1965d: 181; of *ravida:* **new synonymy**.

**talbotae.** *Formica talbotae* Wilson, 1977: 277, fig. 1 (q.m.) U.S.A. See also: Talbot, 1977: 282.

*tamarae. Formica* (*Coptoformica*) *tamarae* Dlussky, 1964: 1033, figs. (w.q.m.) CAUCASUS. Junior synonym of *pressilabris:* Arakelian, 1994: 97.

*taprobanae* Smith, F. 1858; see under ***DOLICHODERUS.***

*tarsata* Fabricius, 1798; see under ***PACHYCONDYLA.***

*tejonia* Buckley, 1866; see under ***CAMPONOTUS.***

*tenuipes* Smith, F. 1857; see under ***CAMPONOTUS.***

*tenuis* Fabricius, 1804; see under ***PSEUDOMYRMEX.***

*tenuissima. Formica tenuissima* Buckley, 1866: 162 (w.) U.S.A. Unidentifiable to genus; *incertae sedis* in *Formica.* See also: Mayr, 1886d: 432; Emery, 1895c: 338; Wheeler, W.M. 1902f: 21.

*terebrans* Lowne, 1865; see under ***CAMPONOTUS.***

*terricola* Buckley, 1866; see under ***PARATRECHINA.***

*testacea. Formica testacea* Gmelin, in Linnaeus, 1790: 2804 (w.) EUROPE. Unidentifiable to genus; *incertae sedis* in *Formica.*

*testacea* Fabricius, 1804; see under ***POLYERGUS.***

*testaceipes. Formica testaceipes* Leach, 1825: 290 (w.) FRANCE. Unidentifiable to genus; *incertae sedis* in *Formica.*

*testaceipes* Smith, F. 1858; see under ***CAMPONOTUS.***

*theryi* Santschi, 1936; see under ***BAJCARIDRIS.***

*thoracica. Formica thoracica* Olivier, 1792: 503 (q.m.) FRANCE. Unidentifiable to genus; *incertae sedis* in *Formica.*

*thoracica* Fabricius, 1804; see under ***CAMPONOTUS.***

*thyssei. Formica pratensis* ab. *thyssei* Stärcke, 1942b: 43 (footnote) (q.) BASSES ALPES. Raised to species (and misspelled as *thijssei*): Betrem, 1955: 392. Junior synonym of *nigricans:* Yarrow, 1955a: 4.

*timida* Foerster, 1850; see under ***LASIUS.***

*timida* Jerdon, 1851; see under ***CAMPONOTUS.***

*tincta* Smith, F. 1858; see under ***CAMPONOTUS.***

*tir. Formica rufa* subsp. *rufa* ab. *tir* Stärcke, 1942a: 22 (q.) NORWAY. Unavailable name; material referred to *lugubris* by Yarrow, 1955a: 5.

#**tombeuri.** *Formica tombeuri* Bondroit, 1917b: 187 (w.) PORTUGAL. Variety of *fusca:* Santschi, 1925g: 352.

*tomentosa. Formica tomentosa* Reich, 1793: 132 (no caste given) FRENCH GUIANA. Unidentifiable to genus; *incertae sedis* in *Formica.*

**torrentium.** *Formica torrentium* Bernard, 1967: 300 (w.q.m.) FRANCE. [Bernard, 1967: 300, indicates that Bernard, 1960: 108 contains the original description of *torrentium.* The name is not mentioned in the latter publication.]

*transkaucasica. Formica transkaucasica* Nasonov, 1889: 21 (w.) RUSSIA. Race/subspecies of *gagates:* Ruzsky, 1902d: 13; of *fusca:* Ruzsky, 1905b: 384. Junior synonym of *picea* Nylander: Emery, 1909b: 195; Donisthorpe, 1918a: 9; Emery, 1925b: 249; Dlussky, 1965a: 31; Dlussky, 1967a: 61. First available replacement name for *picea* Nylander: Donisthorpe, 1918a: 9; Yarrow, 1954a: 233; Atanassov & Dlussky, 1992: 263. Senior synonym of *piceogagates:* Arakelian, 1994: 102. Junior synonym of *candida:* current status.

*transcaucasicogagates. Formica fusca* subsp. *gagates* var. *transcaucasicogagates* Ruzsky, 1905b: 379 (w.) RUSSIA. Unavailable name; material referred to *picea* Nylander by Dlussky, 1965a: 31.

**transmontanis.** *Formica transmontanis* Francoeur, 1973: 135, figs. 220–232 (w.q.m.) CANADA.

*transversa. Formica* (*Serviformica*) *rufibarbis* st. *montivaga* var. *transversa* Santschi, 1928b: 45. Unavailable name; unnecessary replacement name for *clara* Kuznetsov-Ugamsky; unavailable name. Material referred to *clara* Forel by Dlussky, 1967a: 77.

*triangularis. Formica triangularis* Say, 1836: 288 (q.m.) U.S.A. Unidentifiable to genus; *incertae sedis* in *Formica.*

*tridentata* Fabricius, 1804; see under ***ECTATOMMA.***

*trifasciata* Smith, F. 1858; see under ***ANOPLOLEPIS.***

***trigona.** *Formica trigona* Presl, 1822: 198 (no caste given) BALTIC AMBER (Oligocene).

***tripartita.** *Formica tripartita* Théobald, 1937b: 214, pl. 15, figs. 4–8 (q.m.) FRANCE (Oligocene).

*tristis. Formica tristis* Christ, 1791: 513, pl. 60, fig. 13 (m.) no locality given. Junior synonym of *fusca:* Emery, 1892b: 162.

*tristis. Formica* (*Raptiformica*) *sanguinea* var. *tristis* Karavaiev, 1929b: 217 (w.) RUSSIA. [Unresolved junior primary homonym of *tristis* Christ, above.] Junior synonym of *sanguinea:* Dlussky, 1965a: 16.

*tropica* Smith, F. 1861; see under ***CALOMYRMEX.***

*truncata* Spinola, 1808; see under ***CAMPONOTUS.***

*truncicola. Formica truncicola* Nylander, 1846a: 907 (w.q.) FINLAND. Nylander, 1849: 29 (m.). Subspecies/race of *rufa:* Forel, 1874: 52; André, 1903: 128; Ruzsky, 1905b: 330; Wheeler, W.M. 1908g: 406; Emery, 1909b: 187; Forel, 1915d: 57; Emery, 1916b: 256; Santschi, 1925f: 96. Status as species: Bingham, 1903: 334; Wheeler, W.M. 1913f: 434. Senior synonym of *simulata:* Forel, 1894c: 403. Junior synonym of *truncorum:* Roger, 1863b: 13; Bondroit, 1918: 60; Emery, 1925b: 255; Bernard, 1967: 307; Dlussky, 1967a: 81; Dlussky & Pisarski, 1971: 174; Tarbinsky, 1976: 192; Kutter, 1977c: 274.

*truncicolopratensis. Formica rufa* var. *truncicolopratensis* Forel, 1874: 53 (w.q.m.) SWITZERLAND. Variety/subspecies of *pratensis:* Dalla Torre, 1893: 205; of *rufa:* Karavaiev, 1912b: 589; of *truncorum:* Stitz, 1939: 346; Holgersen, 1942: 14. Raised to species: Bondroit, 1918: 61. Junior synonym of *truncorum:* Bernard, 1967: 307; Dlussky, 1967a: 81; Dlussky & Pisarski, 1971: 174.

**truncorum.** *Formica truncorum* Fabricius, 1804: 403 (q.) CZECHOSLOVAKIA. Subspecies/race of *rufa:* Emery & Forel, 1879: 450; Santschi, 1925g: 351; Karavaiev, 1936: 247. Status as species: Bondroit, 1917a: 174; Stitz, 1939: 344; Novak & Sadil, 1941: 105; Holgersen, 1942: 13; Dlussky, 1967a: 81; Tarbinsky, 1976: 192; Kutter, 1977c: 274; Gösswald, 1989: 21; Kupyanskaya, 1990: 192; Atanassov & Dlussky, 1992: 270. Senior synonym of *truncicola:* Roger, 1863b: 13; Bondroit, 1918: 60; Dlussky, 1967a: 81; of *truncicolopratensis:* Dlussky, 1967a: 81; Bernard, 1967: 307; of *menozzii, rufotruncicola, staegeri, stitzi:* Dlussky, 1967a: 81. Current subspecies: nominal plus *finzii, frontalis.*

#**tshugunovi.** *Formica rufa* var. *tshugunovi* Ruzsky, 1914b: 102 (w.) RUSSIA. Subspecies of *rufa:* Ruzsky, 1926: 110.

*tuberculata* Olivier, 1792; see under ***ECTATOMMA.***

*tuberosa* Latreille, 1802; see under ***LEPTOTHORAX.***

*tuberum* Fabricius, 1775; see under ***LEPTOTHORAX.***

*typhlos. Formica typhlos* Lund, 1831a: 135 (w.) BRAZIL. *Nomen nudum.*

**ulkei.** *Formica ulkei* Emery, 1893i: 653, pl. 22, fig. 7 (w.) U.S.A. Wheeler, W.M. 1913f: 486 (q.m.); Wheeler, G.C. & Wheeler, J. 1953c: 168 (l.); Hung, 1969: 455 (k.). Senior synonym of *hebescens:* Creighton, 1950a: 515. See also: Smith, D.R. 1979: 1456.

*umbrata* Nylander, 1846; see under ***LASIUS.***

***ungeri.** *Formica ungeri* Heer, 1850: 128, pl. 10, fig. 1 (q.) YUGOSLAVIA (Miocene). See also: Mayr, 1867b: 53.

*unicolor. Formica pratensis* subsp. *unicolor* Ruzsky, 1926: 110 (w.) RUSSIA. [First available use of *Formica rufa* subsp. *pratensis* var. *unicolor* Ruzsky, 1914b: 102; unavailable name.] Junior synonym of *lugubris:* Dlussky, 1967a: 91.

*unifasciata* Latreille, 1798; see under ***LEPTOTHORAX.***

*unispinosa* Fabricius, 1793; see under ***ODONTOMACHUS.***

**uralensis.** *Formica uralensis* Ruzsky, 1895: 13 (w.q.m.) RUSSIA. [Also described as new by Ruzsky, 1896: 69.] Senior synonym of *superba:* Wu, 1990: 4. See also: Emery, 1909b: 189; Wheeler, W.M. 1913f: 448;

Karavaiev, 1936: 249; Stitz, 1939: 325; Dlussky, 1967a: 79; Kutter, 1977c: 274; Collingwood, 1979: 135; Kupyanskaya, 1990: 189.

*vaga* Scopoli, 1763; see under ***CAMPONOTUS.***

*vagans* Olivier, 1792; see under ***ECITON.***

*vagans. Formica vagans* Fabricius, 1793: 358 (w.) GERMANY. [Unresolved junior primary homonym of *vagans* Olivier, 1792: 501.] Unidentifiable to genus; *incertae sedis* in *Formica.* See also: Curtis, 1854: 213 [combination in *Myrmica;* misidentification]; Smith, F. 1855b: 117; Mayr, 1855: 407.

*vagans* Jerdon, 1851; see under ***PARATRECHINA.***

*variegata* Smith, F. 1858; see under ***CAMPONOTUS.***

*velox* Jerdon, 1851; see under ***CAMPONOTUS.***

*venosa. Formica venosa* Gmelin, in Linnaeus, 1790: 2804 (q.) no locality given. Unidentifiable to genus; *incertae sedis* in *Formica.*

*ventralis* Smith, F. 1858; see under ***CAMPONOTUS.***

*vestita* Smith, F. 1858; see under ***CAMPONOTUS.***

*vetula. Formica lasioides* var. *vetula* Wheeler, W.M. 1912c: 90. Replacement name for *picea* Emery, 1895c: 335. [Junior primary homonym of *picea* Leach, 1825: 292.] Wheeler, W.M. 1917a: 554 (q.). Combination in *F.* (*Proformica*): Wheeler, W.M. 1913f: 540. Subspecies of *neogagates:* Buren, 1944a: 308. Junior synonym of *lasioides:* Creighton, 1950a: 458.

*viatica* Fabricius, 1787; see under ***CATAGLYPHIS.***

*vigilans* Smith, F. 1858; see under ***CAMPONOTUS.***

*villosa* Fabricius, 1804; see under ***PACHYCONDYLA.***

***vinculans.*** *Formica* (*Proformica*) *neogagates* var. *vinculans* Wheeler, W.M. 1913f: 400 (in key) (w.q.) U.S.A. [*Formica* (*Proformica*) *neogagates* subsp. *neogagates* var. *vinculans* Wheeler, W.M. 1913f: 539; unavailable name.] Subspecies of *neogagates:* Buren, 1944a: 309. Junior synonym of *neogagates:* Creighton, 1950a: 459. Revived from synonymy, and raised to species: Snelling & Buren, 1985: 76.

*vinosa* Smith, 1858; see under ***CAMPONOTUS.***

*vinsonnella* Dufour, 1864; see under ***LEPTOGENYS.***

*virescens* Fabricius, 1775; see under ***OECOPHYLLA.***

*virginiana. Formica virginiana* Buckley, 1866: 159 (w.) U.S.A. Unidentifiable to genus; *incertae sedis* in *Formica.* See also: Emery, 1895c: 337; Wheeler, W.M. 1902f: 20.

*viridis* Kirby, W. 1819; see under ***OECOPHYLLA.***

*virulens* Smith, F. 1861; see under ***CAMPONOTUS.***

*visitatrix* Christ, 1791; see under ***ATTA.***

*vitrea* Smith, F. 1860; see under ***CAMPONOTUS.***

*vivida* Smith, F. 1858; see under ***CAMPONOTUS.***

*vividula* Nylander, 1846; see under ***PARATRECHINA.***

*volgensis. Formica rufibarbis* subsp. *volgensis* Ruzsky, 1914c: 323, fig. (w.) RUSSIA. Junior synonym of *glauca:* Dlussky, 1967a: 74.

*volitans. Formica volitans* Christ, 1791: 518 (no caste given) INDIA. *Nomen nudum.*

*vorax* Fabricius, 1804; see under ***PHEIDOLE.***

*wasmanni. Formica dakotensis* var. *wasmanni* Forel, 1904a: 153 (footnote) (w.q.m.) U.S.A. Synonym of *specularis:* Wheeler, W.M. 1913f: 464 (Wheeler gives *specularis* as senior name, but *wasmanni* has seniority). Junior synonym of *dakotensis:* Creighton, 1950a: 484.

*wheeleri. Formica wheeleri* Creighton, 1935: 1, fig. 1 (w.q.) U.S.A. Combination in *F.* (*Raptiformica*): Creighton, 1950a: 472.

*wheeleri. Formica exsecta* var. *wheeleri* Stitz, 1939: 311 (w.) GERMANY. [First available use of *Formica exsecta* subsp. *exsecta* var. *wheeleri* Krausse, 1926d: 264; unavailable name.] Junior synonym of *exsecta:* Dlussky, 1967a: 100.

#***whymperi.*** *Formica whymperi* Wheeler, W.M. 1917a: 544 (w.) CANADA. [First available use of *Formica rufa* r. *obscuripes* var. *whymperi* Forel, 1904a: 152; unavailable name.] Material of the unavailable name *pullula* referred here: Creighton, 1950a: 509. Senior synonym of *californica:* Wheeler, G.C. & Wheeler, J. 1986g: 15. Subspecies of *adamsi:* current status (as the latter name has priority over *whymperi*).

*wichita* Buckley, 1866; see under ***PRENOLEPIS.***

***wongi.*** *Formica wongi* Wu, 1990: 2, figs. 10, 18 (w.) CHINA.

***xerophila.*** *Formica* (*Neoformica*) *moki* subsp. *xerophila* Smith, M.R. 1939f: 583 (w.) U.S.A. Raised to species, and senior synonym of *grundmanni:* Francoeur, 1973: 259.

*yatsuensis. Formica fusca* subsp. *picea* var. *yatsuensis* Teranishi, 1940: 80 (w.) JAPAN. Unavailable name.

***yessensis.*** *Formica truncicola* var. *yessensis* Wheeler, W.M. 1913f: 391 (in key) (w.) JAPAN. [First available use of *Formica rufa* r. *truncicola* var. *yessensis* Forel, 1901h: 66; unavailable name.] Kupyanskaya, 1984: 103 (q.m.); Imai, 1966: 119 (k.). Subspecies of *truncicola:* Ruzsky, 1926: 110. Variety of *truncorum:* Wheeler, W.M. 1933c: 66. Junior synonym of *lugubris:* Dlussky, 1967a: 91. Revived from synonymy, and raised to species: Collingwood, 1976: 304. See also: Kupyanskaya, 1990: 194.

***yoshiokae.*** *Formica yoshiokae* Wheeler, W.M. 1933c: 66 (w.) JAPAN.

*zealandica* Smith, 1878; see under ***PROLASIUS.***

*zonata* Guérin-Méneville, 1838; see under ***OECOPHYLLA.***

---

**EXCURSUS**

**Species-rank taxa of *FORMICA*, grouped by species group.**

The species groups noted here are to a large extent traditional. Most are not well defined and are untrustworthy; for instance the limits of the large *rufa* and *fusca* groups have never been accurately defined, and other groups tend to grade together.

***F. exsecta* group**: *beijingensis, bruni, brunneonitida, dlusskyi, exsecta, exsectoides, forsslundi, fossilabris, fukaii, goesswaldi, naefi, opaciventris, pisarskii, pressilabris, rufomaculata, suecica, ulkei.*

***F. fusca* group**: *accreta, aerata, altipetens, *antiqua, argentea, *baltica, browni, canadensis, candida, cinerea, cinereofusca, clara, cunicularia, decipiens, *flori, foreliana, francoeuri, fuliginothorax, fusca, gagates, gagatoides, gerardi, glacialis, gnava, gravelyi, hewitti, *horrida, imitans, japonica, kozlovi, lefrancoisi, lemani, lepida, longipilosa, *martynovi, microphthalma, moki, montana, neoclara, neorufibarbis, obsidiana, occulta, pachucana, pacifica, *parvula* (junior homonym), *podzolica, propatula, pulla, pyrenaea, retecta, rizadaghi, rufibarbis, selysi, sibylla, sinae, *strangulata, subaenescens, subcyanea, subelongata, subpilosa, subpolita, subrufa, subsericea, torrentium, transmontanis, xerophila, yoshiokae.*

***F. microgyna* group**: *adamsi, densiventris, difficilis, dirksi, impexa, indianensis, knighti, microgyna, morsei, nepticula, nevadensis, postoculata, querquetulana, scitula, spatulata, talbotae.*

***F. neogagates* group**: *bradleyi, lasioides, limata, manni, neogagates, oregonensis, perpilosa, vinculans.*

***F. pallidefulva* group**: *archboldi, nitidiventris, pallidefulva, rufolucida, schaufussi.*

***F. rufa* group**: *adelungi, approximans, aquilonia, aterrima, calviceps, ciliata, *clymene, coloradensis, comata, criniventris, dakotensis, dusmeti, ferocula, fossaceps, integra, integroides, kupyanskayae, laeviceps, lugubris, mucescens, nigropratensis, obscuripes, obscuriventris, oreas, *phaethusa, planipilis, polyctena, pratensis, prociliata, propinqua, ravida, reflexa, rufa, sinensis, subnitens, truncorum, uralensis, wongi, yessensis.*

***F. sanguinea* group**: *aserva, creightoni, curiosa, emeryi, gynocrates, obtusopilosa, pergandei, puberula, rubicunda, sanguinea, sentschuensis, subintegra, wheeleri.*

**Species unplaced to group or group unknown**

(i) extinct forms: **acuminata, *aemula, *alsatica, *arcana, *atavina, *auxillacensis, *bauckhorni, *buphthalma, *cantalica, *capito, *cockerelli, *demersa, *eoptera, *fragilis* (junior homonym), **gibbosa* (junior homonym), **globiventris, *gracilis* (junior homonym), **grandis, *gravida, *heteroptera, *immersa, *kollari, *latinodosa, *lavateri, *longicollis, *longiventris, *lucida, *luteola, *macrocephala, *macrognatha, *macrophthalma, *maculipennis, *nigra* (junior homonym), **oblita, *obtecta, *obvoluta, *ocella, *oculata, *ophthalmica, *orbata, *parvula, *pitoni, *primitiva, *primordialis, *procera, *pulchella, *pumila, *quadrata, *robusta, *sepulta, *serresi, *seuberti, *surinamensis, *trigona, *tripartita, *ungeri.*

(ii) extant forms: *californica* (junior homonym),

---

*FORMICINA:* see under ***FORMICA.***

*FORMICITES Nomen nudum.*

***FORMICIUM*** (Holarctic)

****berryi.*** **Eoponera berryi* Carpenter, 1929: 301, fig. 1 (wing) U.S.A. (Eocene). Combination in **Formicium:* Lutz, 1986: 187.

****brodiei.*** **Formicium brodiei* Westwood, 1854: 393, pl. 14, fig. 8 (wing) GREAT BRITAIN (Eocene). See also: Lutz, 1986: 184.

****giganteum.*** **Formicium giganteum* Lutz, 1986: 190, figs. 3–8 (q.m.) GERMANY (Eocene).

****mirabile.*** **Megapterites mirabilis* Cockerell, 1920: 278, fig. 3 (wing) GREAT BRITAIN (Eocene). Combination in **Formicium:* Lutz, 1990: 61.

****simillimum.*** **Formicium simillimum* Lutz, 1986: 198, figs. 8–10 (q.m.) GERMANY (Eocene).

***FORMICOXENUS*** (Holarctic)

***chamberlini.*** *Symmyrmica chamberlini* Wheeler, W.M. 1904a: 5, pl. 1, figs. 1–7 (w.q.m.) U.S.A. Combination in *Formicoxenus:* Francoeur, Loiselle & Buschinger, 1985: 379. See also: Buschinger & Francoeur, 1983: 297.

*corsicus* Emery, 1895; see under ***EPIMYRMA.***

***diversipilosus.*** *Leptothorax* (*Mychothorax*) *diversipilosus* Smith, M.R. 1939c: 179 (w.q.) U.S.A. Smith, M.R. 1956c: 274 (q.m.); Wheeler, G.C. & Wheeler, J. 1989a: 321 (l.). Combination in *Formicoxenus:* Buschinger, 1979a: 61. See also: Francoeur, Loiselle & Buschinger, 1985: 381.

*emersoni. Leptothorax emersoni* Wheeler, W.M. 1901a: 433, figs. 2, 3 (w.q.m.) U.S.A. Combination in *L.* (*Mychothorax*): Emery, 1922f: 261. Junior synonym of *provancheri:* Creighton, 1950a: 279. See also: Wheeler, W.M. 1903c: 230.

*formidolosus. Leptothorax* (*Mychothorax*) *hirticornis* var. *formidolosus* Wheeler, W.M. 1915b: 415 (w.q.) U.S.A. Junior synonym of *hirticornis:* Smith, M.R. 1939c: 177.

*glacialis. Leptothorax emersoni* subsp. *glacialis* Wheeler, W.M. 1907b: 71 (w.q.m.) U.S.A. Subspecies of *provancheri:* Creighton, 1950a: 280. Junior synonym of *provancheri:* Cole, 1954d: 241.

***hirticornis.*** *Leptothorax hirticornis* Emery, 1895c: 319 (w.) U.S.A. Smith, M.R. 1939c: 176 (q. ergatoid); Snelling, 1965a: 18 (q.m. ergatoid). Combination in *L.* (*Mychothorax*): Emery, 1922f: 262; in *Mychothorax:* Ruzsky, 1936: 96; in *Formicoxenus:* Buschinger, 1979a: 61. Senior synonym of *formidolosus:* Smith, M.R. 1939c: 177. See also: Wheeler, W.M. 1903c: 224; Francoeur, Loiselle & Buschinger, 1985: 382.

*hirtipilis. Leptothorax emersoni* subsp. *hirtipilis* Wheeler, W.M. 1917a: 515 (w.) CANADA. Junior synonym of *provancheri:* Creighton, 1950a: 279.

*laeviuscula. Myrmica laeviuscula* Foerster, 1850a: 54 (q.) GERMANY. Junior synonym of *nitidulus:* Mayr, 1855: 418.

*lucidula. Myrmica lucidula* Smith, F. 1857b: 39 (w.) GREAT BRITAIN. *Nomen nudum.* Material referred to *nitidulus* by Donisthorpe, 1915d: 83.

***nitidulus.*** *Myrmica nitidula* Nylander, 1846b: 1058 (w.) FINLAND. Nylander, 1849: 34 (q.); Mayr, 1855: 419 (m.). Combination in *Formicoxenus:* Mayr, 1855: 418. Senior synonym of *laeviuscula:* Mayr, 1855: 418; of *picea:* Francoeur, Loiselle & Buschinger, 1985: 380. Material of the *nomen nudum lucidula* referred here by Donisthorpe, 1915d: 83. See also: Kutter, 1977c: 144; Francoeur, Loiselle & Buschinger, 1985: 380.

*orientalis. Formicoxenus orientalis* Dlussky, 1963: 193, figs. 6, 8 (w.) RUSSIA. Junior synonym of *sibiricus:* Francoeur, Loiselle & Buschinger, 1985: 382. [Kupyanskaya, 1990: 149, uses the name *orientalis,* apparently unaware of the revision by Francoeur, *et al;* their synonymy is affirmed by Radchenko, 1994b: 111.]

*picea. Leptothorax nitidulus* var. *picea* Wasmann, 1906: 120 (w.) LUXEMBOURG. Junior synonym of *nitidulus:* Francoeur, Loiselle & Buschinger, 1985: 380.

***provancheri.*** *Leptothorax provancheri* Emery, 1895c: 320 (w.) CANADA. Francoeur, Loiselle & Buschinger, 1985: 377 (q.m.); Wheeler, G.C. & Wheeler, J. 1973b: 73 (l.); Buschinger, Francoeur & Fischer, 1981: 8 (k.). Combination in *L.* (*Mychothorax*): Emery, 1922f: 262; in *Formicoxenus:* Francoeur, Loiselle & Buschinger, 1985: 377. Senior synonym of *emersoni* and *hirtipilis:* Creighton, 1950a: 279; of *glacialis:* Cole, 1954d: 241. See also: Wheeler, W.M. 1903c: 229. [MacKay, Lowrie, *et al.* 1988: 89 (in key), and Wheeler, G.C. & Wheeler, J. 1989a: 323, recombine *provancheri* in *Leptothorax,* apparently unaware of the revision by Francoeur *et al,* whose combination is affirmed here.]

***quebecensis.*** *Formicoxenus quebecensis* Francoeur, in Francoeur, Loiselle & Buschinger, 1985: 378, figs. (q.m.) CANADA. Wheeler, G.C. & Wheeler, J. 1989a: 323 (l.).

*ravouxi* André, 1896; see under ***EPIMYRMA.***

***sibiricus.*** *Leptothorax sibiricus* Forel, 1899b: 306 (w.) RUSSIA. Combination in *L.* (*Mychothorax*): Emery, 1922f: 262; in *Formicoxenus:* Francoeur, Loiselle & Buschinger, 1985: 382. Junior synonym of *hirticornis:* Ruzsky, 1905b: 622. Revived from synonymy as variety of *hirticornis:* Emery, 1922f: 262. Revived status as species, and senior synonym of *orientalis:* Francoeur, Loiselle & Buschinger, 1985: 382.

***FROGGATTELLA*** (Australia)

***#bispinosa.*** *Froggattella kirbyi* var. *bispinosa* Forel, 1902h: 460 (w.) AUSTRALIA.

***#ianthina.*** *Froggattella kirbyi* subsp. *ianthina* Wheeler, W.M. 1936a: 8 (w.) AUSTRALIA.

***kirbii.*** *Acantholepis kirbii* Lowne, 1865b: 333 (w.) AUSTRALIA. Wheeler, W.M. 1936a: 6 (q.m.); Wheeler, G.C. & Wheeler, J. 1951: 182 (l.). Combination in *Hypoclinea:* Mayr, 1870b: 954; in *Dolichoderus:* Dalla Torre, 1893: 159; Emery, 1894c: 228; in *Froggattella:* Forel, 1902h: 459. [Unjustified emendation of spelling to *kirbyi:* Mayr, 1876: 80; Forel, 1902h: 459.] See also: Wheeler, G.C. & Wheeler, J. 1966: 728. Current subspecies: nominal plus *bispinosa, ianthina, laticeps, lutescens, nigripes.*

***#laticeps.*** *Froggattella kirbyi* subsp. *laticeps* Wheeler, W.M. 1936a: 10 (w.) AUSTRALIA.

***latispina.*** *Froggattella latispina* Wheeler, W.M. 1936a: 10 (w.) AUSTRALIA.

***#lutescens.*** *Froggattella kirbyi* subsp. *lutescens* Wheeler, W.M. 1936a: 9 (w.) AUSTRALIA.

***#nigripes.*** *Froggattella kirbyi* subsp. *nigripes* Wheeler, W.M. 1936a: 8 (w.) AUSTRALIA.

*FULAKORA:* see under ***AMBLYOPONE.***

*GALLARDOMYRMA:* see under ***PHEIDOLE.***

*GAUROMYRMEX:* see under ***VOLLENHOVIA.***

*GEOGNOMICUS:* see under ***DICROASPIS.***

***GESOMYRMEX*** (Oriental, Indo-Australian)

*andrei* Emery, 1894 [*Dimorphomyrmex*]; see under ***CLADOMYRMA.***

**annectens. *Gesomyrmex annectens* Wheeler, W.M. 1915h: 107, fig. 52 (w.) BALTIC AMBER (Oligocene). Junior synonym of **hoernesi:* Wheeler, W.M. 1929a: 11.

***chaperi.*** *Gesomyrmex chaperi* André, 1892b: 47, fig. 1–3 (w.) BORNEO. Senior synonym of *janeti:* Wheeler, W.M. 1929a: 10.

*chapmani. Gesomyrmex luzonensis* var. *chapmani* Wheeler, W.M. 1930c: 35, fig. 1 (s.w.q.m.) PHILIPPINES. Junior synonym of *luzonensis:* Brown, 1950e: 249.

**corniger* Emery, 1891; see under ***SICILOMYRMEX.***

****expectans.*** **Gesomyrmex expectans* Théobald, 1937b: 210, pl. 4, fig. 13; pl. 14, fig. 21 (m.) FRANCE (Oligocene).

***#gracilis.*** *Gesomyrmex kalshoveni* subsp. *gracilis* Hamann, 1957: 3, figs. 1–5 (w.) JAVA.

****hoernesi.*** **Gesomyrmex hoernesi* Mayr, 1868c: 52, pl. 2, figs. 38–41 (w.m.) BALTIC AMBER (Oligocene). Senior synonym of **annectens, *mayri* and **theryi:* Wheeler, W.M. 1929a: 11.

***howardi.*** *Gesomyrmex howardi* Wheeler, W.M. 1921e: 112, fig. 2 (s.w.) CHINA.

*janeti. Dimorphomyrmex janeti* André, 1892b: 51, figs. 4, 5 (s.w.) BORNEO. Junior synonym of *chaperi:* Wheeler, W.M. 1929a: 10.

***kalshoveni.*** *Gesomyrmex kalshoveni* Wheeler, W.M. 1929a: 2, fig. 1 (s.w.) JAVA. Wheeler, G.C. & Wheeler, J. 1953c: 140 (l.). Current subspecies: nominal plus *gracilis.* See also: Wheeler, W.M. 1929c: 91.

***luzonensis.*** *Dimorphomyrmex luzonensis* Wheeler, W.M. 1916b: 16, fig. 4 (q.) PHILIPPINES. Wheeler, G.C. & Wheeler, J. 1968: 210 (l.). Combination in *Gesomyrmex:* Wheeler, W.M. 1929a: 10. Senior synonym of *chapmani:* Brown, 1950e: 249.

**mayri. *Dimorphomyrmex mayri* Wheeler, W.M. 1915h: 106, fig. 51 (w.) BALTIC AMBER (Oligocene). Junior synonym of **hoernesi:* Wheeler, W.M. 1929a: 11.

****miegi.*** **Gesomyrmex miegi* Théobald, 1937b: 211, pl. 14, figs. 22, 23 (q.m.) FRANCE (Oligocene).

***spatulatus.*** *Gesomyrmex spatulatus* Cole, 1949: 72, fig. 1 (s.w.) INDIA.

**theryi. *Dimorphomyrmex theryi* Emery, 1905b: 188, fig. 1 (w.) BALTIC AMBER (Oligocene). Junior synonym of **hoernesi:* Wheeler, W.M. 1929a: 11.

***GIGANTIOPS*** (Neotropical)

***destructor.*** *Formica destructor* Fabricius, 1804: 402 (w.) SOUTH AMERICA. Wheeler, G.C. & Wheeler, J. 1953c: 170 (l.). Combination in *Gigantiops:* Roger, 1863b: 11. Senior synonym of *solitaria:* Roger, 1862c: 287. See also: Wheeler, W.M. 1922b: 185; Wheeler, G.C. & Wheeler, J. 1968: 215.

*solitaria. Formica solitaria* Smith, F. 1858b: 45, pl.13, figs. 4, 5 (w.q.) BRAZIL. Junior synonym of *destructor:* Roger, 1862c: 287.

***GLAMYROMYRMEX*** (Neotropical, Afrotropical, Australasian)

***africanus.*** *Glamyromyrmex africanus* Bolton, 1983: 322, fig. 27 (w.) GABON.

***appretiatus.*** *Glamyromyrmex appretiatus* Borgmeier, 1954b: 282, figs. 7–9 (w.) BRAZIL.

***aztecus.*** *Glamyromyrmex aztecus* Kempf, 1960f: 444, figs. 22, 27, 28 (w.) MEXICO.

***beebei.*** *Glamyromyrmex beebei* Wheeler, W.M. 1915e: 488, fig. 2 (w.q.m.) BRAZIL. See also: Brown, 1950a: 32.

***convexiceps.*** *Strumigenys* (*Codiomyrmex*) *convexiceps* Santschi, 1931c: 277, figs. 9, 10 (w.) CUBA. Combination in *Codiomyrmex:* Weber, 1934a: 52; in *Glamyromyrmex:* Brown, 1948e: 116.

***crypturus.*** *Glamyromyrmex crypturus* Bolton, 1983: 327, fig. 23 (w.) GHANA.

***dagon.*** *Glamyromyrmex dagon* Bolton, 1983: 325, figs. 25, 32 (w.) RWANDA.

***excisus.*** *Codiomyrmex excisus* Weber, 1934a: 51, fig. 9 (w.) NICARAGUA. Combination in *Borgmeierita:* Brown, 1953g: 23; in *Glamyromyrmex:* Brandão, 1991: 344.

***flagellatus.*** *Codiomyrmex flagellatus* Taylor, 1962: 7, figs. 6–9 (w.) AUSTRALIA. Combination in *Glamyromyrmex:* Taylor & Brown, D.R. 1985: 64.

*lilloana* Brown, 1950; see under ***CHELYSTRUMA.***

***loveridgei.*** *Codiomyrmex loveridgei* Brown, 1953g: 21 (w.) MALAWI. Combination in *Glamyromyrmex:* Bolton, 1983: 328.

***ravidurus.*** *Glamyromyrmex ravidurus* Bolton, 1983: 331, fig. 28 (w.) CAMEROUN.

***reticeps.*** *Codiomyrmex reticeps* Kempf, 1969: 286, figs. 7–9 (w.) BRAZIL. Combination in *Glamyromyrmex:* **new combination**.

***sahurus.*** *Glamyromyrmex sahurus* Bolton, 1983: 326, figs. 26, 33 (w.) RWANDA.

***semicomptus.*** *Codiomyrmex semicomptus* Brown, 1959c: 9, fig. 5 (w.) AUSTRALIA. Combination in *Glamyromyrmex:* Taylor & Brown, D.R. 1985: 64.

***sistrurus.*** *Glamyromyrmex sistrurus* Bolton, 1983: 329, fig. 24 (w.) CAMEROUN.

***substrictus.*** *Glamyromyrmex substrictus* Kempf, 1964e: 68, figs. 21–23 (w.) BRAZIL.

***tetragnathus.*** *Codiomyrmex tetragnathus* Taylor, 1966a: 225, figs. 1, 2 (w.) ANGOLA. Combination in *Glamyromyrmex:* Bolton, 1983: 323.

***thuvidus.*** *Glamyromyrmex thuvidus* Bolton, 1983: 332, fig. 31 (w.) KENYA.

***trymalus.*** *Glamyromyrmex trymalus* Bolton, 1983: 333 (w.) CAMEROUN.

***tukultus.*** *Glamyromyrmex tukultus* Bolton, 1983: 334, fig. 29 (w.) IVORY COAST.

***wheeleri.*** *Glamyromyrmex wheeleri* Smith, M.R. 1944f: 254, figs. 1, 2 (w.) PANAMA. See also: Brown, 1950a: 32.

*****GLAPHYROMYRMEX*** (Baltic Amber)

*****oligocenicus.*** *Glaphyromyrmex oligocenicus* Wheeler, W.M. 1915h: 132, fig. 63 (w.) BALTIC AMBER (Oligocene).

*GLYPHOPONE:* see under ***CENTROMYRMEX.***

*GLYPTOMYRMEX:* see under ***MYRMICOCRYPTA.***

***GNAMPTOGENYS*** (Neotropical, s. Nearctic, Oriental, Indo-Australian, Australasian)

***aculeaticoxae.*** *Ectatomma* (*Parectatomma*) *aculeaticoxae* Santschi, 1921g: 82 (w.m.) FRENCH GUIANA. Wheeler, G.C. & Wheeler, J. 1971b: 1201 (l.). Combination in *E.* (*Poneracantha*): Santschi, 1929h: 476; in *Gnamptogenys:* Brown, 1958g: 227. See also: Lattke, 1990b: 5.

***acuminata.*** *Ectatomma* (*Gnamptogenys*) *acuminatum* Emery, 1896g: 50 (w.q.) BRAZIL. Combination in *Gnamptogenys:* Brown, 1958g: 227.

***acuta.*** *Holcoponera acuta* Brown, 1956b: 491, fig. 1 (w.q.m.) BOLIVIA. Combination in *Gnamptogenys:* Brown, 1958g: 227.

***albiclava.*** *Wheeleripone albiclava* Mann, 1919: 283, fig. 3 (w.) SOLOMON IS. Combination in *Gnamptogenys:* Brown, 1958g: 227.

***alfaroi.*** *Ectatomma* (*Gnamptogenys*) *alfaroi* Emery, 1894c: 145 (q.) COSTA RICA. Combination in *Gnamptogenys:* Brown, 1958g: 227. See also: Brown, 1958g: 301.

***ammophila.*** *Gnamptogenys ammophila* Lattke, 1990b: 6, figs. 6, 9 (w.) VENEZUELA.

*angustiloba. Ectatomma* (*Holcoponera*) *striatulum* var. *angustiloba* Forel, 1908c: 341 (w.) BRAZIL. Forel, 1911e: 254 (q.); Gallardo, 1918b: 23 (m.). Combination in *Holcoponera:* Emery, 1911d: 41. Stirps of *brasiliensis,* and senior synonym of *vollenweideri:* Santschi, 1929d: 275; Santschi, 1929h: 453. Junior synonym of *striatula:* Brown, 1958g: 229.

*angustipleura. Ectatomma* (*Holcoponera*) *striatulum* var. *angustipleura* Forel, 1908c: 342 (w.) BRAZIL. Combination in *Holcoponera:* Emery, 1911d: 41. Raised to species: Santschi, 1929h: 456. Junior synonym of *striatula:* Brown, 1958g: 229.

***annulata.*** *Ectatomma* (*Gnamptogenys*) *rimulosa* var. *annulata* Mayr, 1887: 543 (w.q.m.) BRAZIL. Combination in *Gnamptogenys:* Mann, 1922: 3. Raised to species: Emery, 1890b: 41.

*antillana. Holcoponera striatula* var. *antillana* Santschi, 1929h: 444, figs. 16, 21 (w.m.) MARTINIQUE I. Junior synonym of *striatula:* Brown, 1956b: 490.

***arcuata.*** *Holcoponera regularis* var. *arcuata* Santschi, 1929h: 457 (w.) BRAZIL. Combination in *Gnamptogenys:* Brown, 1958g: 227. Junior synonym of, and hence first available replacement name for *Holcoponera regularis* Santschi, 1929d: 275 [Junior secondary homonym of *Gnamptogenys regularis* Mayr, 1870b: 965.], and raised to species: Brown, 1958g: 227.

***aterrima.*** *Wheeleripone aterrima* Mann, 1921: 411, fig. 2 (w.q.m.) FIJI IS. Combination in *Gnamptogenys:* Brown, 1958g: 227.

*avia. Stictoponera laevior* var. *avia* Forel, 1912d: 98 (w.q.m.) JAVA. Junior synonym of *laevior:* Brown, 1954h: 2.

***banksi.*** *Emeryella* (*Barbourella*) *banksi* Wheeler, W.M. 1930f: 10, fig. 1 (q.) PANAMA. Combination in *Gnamptogenys:* Brown, 1958g: 227.

***bicolor.*** *Ectatomma* (*Stictoponera*) *bicolor* Emery, 1889b: 493 (w.) BURMA. Combination in *Stictoponera:* Emery, 1911d: 48; in *Gnamptogenys:* Brown, 1958g: 227. Variety/subspecies of *coxalis:* Emery, 1895k: 458; of *menadensis:* Emery, 1911d: 48; Wheeler, W.M. 1921c: 529. Status as species: Forel, 1900d: 316; Bingham, 1903: 83; Brown, 1950e: 245. Senior synonym of *minor:* Brown, 1950e: 245. See also: Brown, 1954h: 4.

***binghamii.*** *Ectatomma* (*Stictoponera*) *binghamii* Forel, 1900d: 317 (w.) BURMA. Viehmeyer, 1916a: 112 (q.m.); Wheeler, G.C. & Wheeler, J. 1976a: 43 (l.); Imai, Brown, *et al.* 1984: 67 (k.). Combination in *Stictoponera:* Emery, 1911d: 47; in *Gnamptogenys:* Brown, 1958g: 227. Senior synonym of *borneensis:* Brown, 1954h: 5.

***biroi.*** *Stictoponera biroi* Emery, 1901h: 154 (w.) NEW GUINEA. Combination in *Gnamptogenys:* Brown, 1958g: 227. See also: Brown, 1954h: 6.

***bispinosa.*** *Ectatomma* (*Holcoponera*) *bispinosum* Emery, 1890b: 40 (w.) COSTA RICA. [Also described as new by Emery, 1894k: 47.] Wheeler, G.C. & Wheeler, J. 1952a: 133 (l.). Combination in *E.* (*Ectatomma*): Emery, 1896g: 37; in *E.* (*Poneracantha*): Emery, 1897d: 547; in *Gnamptogenys:* Brown, 1958g: 227. See also: Brown, 1958g: 301.

***bisulca.*** *Gnamptogenys bisulca* Kempf & Brown, 1968: 92 (w.) COLOMBIA. Wheeler, G.C. & Wheeler, J. 1971b: 1202 (l.).

*borneensis. Stictoponera borneensis* Emery, 1900d: 662 (footnote) (w.) BORNEO. Junior synonym of *binghamii:* Brown, 1954h: 5.

*brasiliensis. Holcoponera brasiliensis* Emery, 1902a: 181 (w.q.m.) BRAZIL. Wheeler, G.C. & Wheeler, J. 1952a: 124 (l.). Junior synonym of *striatula:* Brown, 1958g: 229.

***bruchi.*** *Ectatomma* (*Parectatomma*) *bruchi* Santschi, 1922d: 241 (w.) ARGENTINA. Combination in *E.* (*Commateta*): Santschi, 1929h: 476; in *Gnamptogenys:* Brown, 1958g: 227.

***bufonis.*** *Alfaria bufonis* Mann, 1926: 101 (w.) MEXICO. Combination in *Gnamptogenys:* Brown, 1958g: 227.

*bufonum. Gnamptogenys sulcatum* subsp. *bufonum* Weber, 1938a: 208 (w.) GUIANA. Junior synonym of *sulcata:* Brown, 1958g: 229.

***caelata.*** *Gnamptogenys caelata* Kempf, 1967a: 121, figs. 1–3 (w.) BRAZIL. Senior synonym of *soror:* Lattke, 1992a: 126.

*calcarata. Holcoponera brasiliensis* var. *calcarata* Santschi, 1929h: 452, fig. 28 (w.) BRAZIL. Junior synonym of *striatula:* Brown, 1958g: 229.

*carinata. Alfaria carinata* Weber, 1940b: 82, fig. 3 (q.) GUIANA. Junior synonym of *minuta:* Brown, in Borgmeier, 1957: 115.

*cearensis. Ectatomma* (*Gnamptogenys*) *sulcatum* var. *cearensis* Forel, 1912c: 33 (w.) BRAZIL. Junior synonym of *sulcata:* Brown, 1958g: 229.

***chapmani.*** *Gnamptogenys chapmani* Brown, 1958g: 305, figs. 19, 44 (w.q.) PHILIPPINES.

*concentrica. Gnamptogenys concentrica* Mayr, 1870b: 964 (w.) MEXICO. Combination in *Ectatomma* (*Holcoponera*): Mayr, 1887: 540. Junior synonym of *strigata:* Dalla Torre, 1893: 26; Emery, 1896g: 44 (in key).

***concinna.*** *Ectatomma concinna* Smith, F. 1858b: 103 (q.) BRAZIL. Mayr, 1866b: 892 (w.). Combination in *Gnamptogenys:* Mayr, 1866b: 892; in *Ectatomma* (*Gnamptogenys*): Mayr, 1887: 541; in *E.* (*Tammoteca*): Santschi, 1929h: 476; in *Gnamptogenys,* and senior synonym of *conica, romani* and *semicircularis:* Brown, 1958g: 227.

*concinna. Holcoponera teffensis* var. *concinna* Santschi, 1929h: 450, fig. 23 (w.) GUIANA. [Unresolved junior secondary homonym of *concinna* Smith, F. above.] Junior synonym of *teffensis:* Brown, 1958g: 229.

*conica. Gnamptogenys concinna* var. *conica* Borgmeier, 1929: 196 (w.) BRAZIL. Junior synonym of *concinna:* Brown, 1958g: 227.

***continua.*** *Ectatomma* (*Gnamptogenys*) *continuum* Mayr, 1887: 544 (w.) BRAZIL. Combination in *Gnamptogenys,* and senior synonym of *panamensis:* Brown, 1958g: 227.

***costata.*** *Ectatomma* (*Stictoponera*) *costatum* Emery, 1889b: 494 (w.) BURMA. Wheeler, G.C. & Wheeler, J. 1976a: 44 (l.). Combination in *Stictoponera:* Emery, 1900d: 662; in *Gnamptogenys:* Brown, 1958g:

227. Senior synonym of *parva, pinealis, simalurensis, unicolor* and *wallacei:* Brown, 1954h: 7.

***coxalis.*** *Ponera coxalis* Roger, 1860: 308 (w.) SRI LANKA. Combination in *Ectatomma* (*Rhytidoponera*): Mayr, 1862: 732; in *E.* (*Stictoponera*): Mayr, 1887: 539; in *Stictoponera:* Emery, 1900d: 662; in *Gnamptogenys:* Brown, 1958g: 228. See also: Brown, 1954h: 9.

***crassicornis.*** *Ectatomma* (*Stictoponera*) *binghami* subsp. *crassicorne* Forel, 1912n: 51 (w.) SUMATRA. Combination in *Gnamptogenys:* Brown, 1958g: 228. Raised to species: Brown, 1954h: 6.

***crenaticeps.*** *Wheeleripone crenaticeps* Mann, 1919: 285, fig. 4 (w.) SOLOMON IS. Combination in *Gnamptogenys:* Brown, 1958g: 228.

***cribrata.*** *Rhopalopone cribrata* Emery, 1900c: 311 (w.) NEW GUINEA. Combination in *Gnamptogenys:* Brown, 1958g: 228.

***curtula.*** *Ectatomma* (*Holcoponera*) *curtulum* Emery, 1896g: 47, pl. 1, fig. 8 (w.) COSTA RICA. Forel, 1899c: 7 (q.). Combination in *Holcoponera:* Emery, 1911d: 40; in *Gnamptogenys,* and senior synonym of *stolli:* Brown, 1958g: 228.

***curvoclypeata.*** *Gnamptogenys curvoclypeata* Lattke, 1990b: 9, figs. 7, 13 (w.) VENEZUELA.

***dammermani.*** *Rhopalopone dammermani* Wheeler, W.M. 1924b: 240 (w.) INDONESIA (Sebesi I.). Combination in *Gnamptogenys:* Brown, 1958g: 228.

***diehlii.*** *Ectatomma* (*Mictoponera*) *diehlii* Forel, 1901e: 372 (w.) BORNEO. Combination in *Rhopalopone:* Emery, 1911d: 35; in *Gnamptogenys:* Brown, 1958g: 228.

*dina. Parectatomma dina* Kusnezov, 1969: 35 (diagnosis in key) (w.) ARGENTINA. Junior synonym of *nana:* Kempf, 1978: 35.

*emeryi. Alfaria emeryi* Forel, 1901f: 326 (w.) COLOMBIA. Junior synonym of *minuta:* Brown, in Borgmeier, 1957: 115.

*emeryi. Holcoponera emeryi* Santschi, 1929h: 463, fig. 45 (w.) BOLIVIA. [Unresolved junior secondary homonym of *emeryi* Forel, above.] Junior synonym of *pleurodon:* Brown, 1958g: 229.

***epinotalis.*** *Rhopalopone epinotalis* Emery, 1897d: 550 (w.q.) NEW GUINEA. [Also described as new by Emery, 1900c: 311.] Combination in *Gnamptogenys:* Brown, 1958g: 228.

*ericae. Ectatomma* (*Gnamptogenys*) *tornatum* var. *ericae* Forel, 1912c: 33 (w.) COLOMBIA. Junior synonym of *tornata:* Brown, 1958g: 229.

****europaea.*** **Ectatomma europaeum* Mayr, 1868c: 76, pl. 4, figs. 72, 73 (q.) BALTIC AMBER (Oligocene). Wheeler, W.M. 1915h: 33 (m.). Combination in *Gnamptogenys:* Brown, 1958g: 181.

***exarata.*** *Ectatomma* (*Gnamptogenys*) *exaratum* Emery, 1901a: 50 (q.) PERU. See also: Brown, 1958g: 303.

***falcifera.*** *Gnamptogenys falcifera* Kempf, 1967d: 353, figs. 1–3 (q.) PERU. Perrault, 1986b: 158 (w.).

***fernandezi.*** *Gnamptogenys fernandezi* Lattke, 1990b: 11, fig. 10 (w.q.) VENEZUELA.

*fiebrigi. Ectatomma* (*Gnamptogenys*) *fiebrigi* Forel, 1909a: 253 (w.) PARAGUAY. Junior synonym of *regularis:* Brown, 1958g: 229.

***fieldi.*** *Gnamptogenys fieldi* Lattke, 1990b: 12, fig. 14 (w.) VENEZUELA.

*foreli. Holcoponera simplex* st. *foreli* Santschi, 1929h: 460, fig. 34 (w.m.) COSTA RICA. Junior synonym of *simplex:* Brown, 1956b: 490.

***gracilis.*** *Holcoponera gracilis* Santschi, 1929h: 468, figs. 14, 41 (w.) GUIANA. Combination in *Gnamptogenys:* Brown, 1958g: 228.

***grammodes.*** *Gnamptogenys grammodes* Brown, 1958g: 310, fig. 41 (w.) NEW GUINEA.

***haenschei.*** *Ectatomma* (*Gnamptogenys*) *haenschei* Emery, 1902c: 27 (q.) ECUADOR. Brown, 1958g: 302 (w.).

***hartmani.*** *Ectatomma* (*Parectatomma*) *hartmani* Wheeler, W.M. 1915b: 390 (w.) U.S.A. Brown, 1961b: 69 (q.); Wheeler, G.C. & Wheeler, J. 1964b: 450 (l.). Combination in *Gnamptogenys:* Brown, 1958g: 228.

***haytiana.*** *Spaniopone haytiana* Wheeler, W.M. & Mann, 1914: 11, fig. 4 (w.) HAITI. Combination in *Gnamptogenys:* Brown, 1958g: 228.

***horni.*** *Ectatomma* (*Gnamptogenys*) *regulare* st. *horni* Santschi, 1929h: 475, fig. 47 (w.) GUIANA. Combination in *Gnamptogenys,* and raised to species: Brown, 1958g: 228.

*hybrida. Holcoponera brasiliensis* st. *simplicoides* var. *hybrida* Santschi, 1929h: 455 (w.) BRAZIL. Unavailable name; material referred to *striatula* by Brown, 1958g: 229.

***ingeborgae.*** *Gnamptogenys ingeborgae* Brown, 1993: 279, figs. 3–7 (w.q.) COLOMBIA.

***interrupta.*** *Ectatomma* (*Gnamptogenys*) *interruptum* Mayr, 1887: 543 (w.) no locality given ("ohne Vaterlandsangabe. . . . .wohl jedenfalls aus Südamerika"). Wheeler, W.M. 1909b: 228 (q., MEXICO). Combination in *Gnamptogenys:* Mann, 1922: 3. See also: Brown, 1958g: 303.

*isthmica. Holcoponera wasmanni* var. *isthmica* Santschi, 1929h: 467 (w.) PANAMA. Junior synonym of *wasmanni:* Brown, 1958g: 230.

***kalabit.*** *Gnamptogenys kalabit* Brown, 1958g: 308, fig. 43 (w.q.m.) BORNEO.

***kempfi.*** *Gnamptogenys kempfi* Lenko, 1964: 257, figs. 1, 2 (w.) BRAZIL.

***laevior.*** *Ectatomma* (*Stictoponera*) *laevius* Forel, 1905c: 7 (w.) JAVA. Combination in *Stictoponera,* and spelling emended to *laevior:* Emery, 1911d: 48; in *Gnamptogenys:* Brown, 1958g: 228. Senior synonym of *avia:* Brown, 1954h: 2.

***lanei.*** *Gnamptogenys lanei* Kempf, 1960e: 388, figs. 1–3 (w.) BRAZIL.

****levinates.*** **Gnamptogenys levinates* Baroni Urbani, 1980e: 5, figs. 5–8 (w.) DOMINICAN AMBER (Miocene).

*lineata. Gnamptogenys lineata* Mayr, 1870b: 965 (w.) BRAZIL. Combination in *Ectatomma* (*Gnamptogenys*): Mayr, 1887: 541. Variety of *sulcata:* Emery, 1896g: 51; Forel, 1912c: 32. Junior synonym of *sulcata:* Brown, 1958g: 229.

***lineolata.*** *Gnamptogenys lineolata* Brown, 1993: 275, figs. 1, 2 (w.) DOMINICA.

***lucaris.*** *Gnamptogenys lucaris* Kempf, 1968b: 379, fig. 5 (w.) BRAZIL.

***lucida.*** *Wheeleripone lucida* Mann, 1919: 285 (w.) SOLOMON IS. Combination in *Gnamptogenys:* Brown, 1958g: 228.

***luzonensis.*** *Rhopalopone luzonensis* Wheeler, W.M. 1929g: 30 (w.q.) PHILIPPINES. Combination in *Gnamptogenys:* Brown, 1958g: 228.

***macretes.*** *Gnamptogenys macretes* Brown, 1958g: 313 (w.q.m.) NEW GUINEA.

***magnifica.*** *Holcoponera magnifica* Santschi, 1921g: 81 (w.) BRAZIL. Combination in *Gnamptogenys:* Brown, 1958g: 228.

***major.*** *Rhopalopone major* Emery, 1901h: 154 (w.) NEW GUINEA. Combination in *Gnamptogenys:* Brown, 1958g: 228.

***malaensis.*** *Rhopalopone malaensis* Mann, 1919: 281 (w.) SOLOMON IS. Combination in *Gnamptogenys:* Brown, 1958g: 228.

*mayri. Holcoponera brasiliensis* var. *mayri* Santschi, 1929h: 453 (w.) COSTA RICA. Junior synonym of *wheeleri:* Brown, 1958g: 230.

***mecotyle.*** *Gnamptogenys mecotyle* Brown, 1958g: 318, fig. 42 (w.) BOLIVIA.

***mediatrix.*** *Gnamptogenys mediatrix* Brown, 1958g: 326, fig. 15 (w.q.) BRAZIL.

***menadensis.*** *Ectatomma* (*Stictoponera*) *menadensis* Mayr, 1887: 539 (footnote) (w.) SULAWESI. Wheeler, G.C. & Wheeler, J. 1964b: 450 (l.); Imai, Brown, *et al.* 1984: 67 (k.). Combination in *Stictoponera:* Emery, 1900d: 663; in *Gnamptogenys:* Brown, 1958g: 228. Senior synonym of *obscura* and *stylata:* Brown, 1954h: 2.

***menozzii.*** *Holcoponera menozzii* Borgmeier, 1928a: 32 (w.q.) BRAZIL. Combination in *Ectatomma* (*Parectatomma*): Santschi, 1929h: 476; in *Gnamptogenys:* Brown, 1958g: 228.

***mina.*** *Holcoponera mina* Brown, 1956b: 494, fig. 2 (w.q.) BOLIVIA. Combination in *Gnamptogenys:* Brown, 1958g: 228.

*minor. Ectatomma* (*Stictoponera*) *bicolor* var. *minor* Forel, 1900d: 317 (w.) BURMA. [Unresolved junior primary homonym of *Ectatomma mucronatum* var. *minor* Forel, 1899c: 9 (now in *Acanthoponera.*] Junior synonym of *bicolor:* Brown, 1950e: 245.

*minor. Emeryella schmitti* subsp. *minor* Wheeler, W.M. 1936b: 195 (w.) HAITI. [Unresolved junior secondary homonym of *minor* Forel, 1900d: 317, above.] Junior synonym of *schmitti:* Brown, 1958g: 229.

***minuta.*** *Alfaria minuta* Emery, 1896g: 106 (q.) BOLIVIA. Combination in *Gnamptogenys:* Brown, 1958g: 228. Senior synonym of *carinata, emeryi, mus, panamensis* and *scabrosus:* Brown, in Borgmeier, 1957: 115; of *pneodonax:* Lattke, 1992a: 127.

***moelleri.*** *Holcoponera moelleri* Forel, 1912c: 34 (w.q.) BRAZIL. Combination in *Gnamptogenys,* and senior synonym of *splendens:* Brown, 1958g: 228.

***mordax.*** *Ponera mordax* Smith, F. 1858b: 98 (w.q.) COSTA RICA. Borgmeier, 1937b: 220 (m.); Wheeler, G.C. & Wheeler, J. 1952a: 134 (l.). Combination in *Ectatomma* (*Gnamptogenys*): Emery, 1894c: 145; in *Gnamptogenys:* Mann, 1922: 3. Senior synonym of *nodosa, purensis* and *sebastiani:* Brown, 1958g: 228.

*mus. Alfaria mus* Santschi, 1931c: 265 (w.) PANAMA. Junior synonym of *minuta:* Brown, in Borgmeier, 1957: 115.

***nana.*** *Gnamptogenys nana* Kempf, 1960f: 422, figs. 7–9 (w.) BRAZIL. Senior synonym of *dina:* Kempf, 1978: 35.

**nigrifrons.** *Ectatomma (Gnamptogenys) nigrifrons* Borgmeier, 1948a: 199, figs. 24, 25 (w.q.) PERU. Combination in *Gnamptogenys:* Brown, 1958g: 228.
*nitens. Ectatomma (Gnamptogenys) sulcatum* var. *nitens* Mann, 1916: 407 (w.) BRAZIL. Junior synonym of *sulcata:* Brown, 1958g: 229.
*nodosa. Formica nodosa* Latreille, 1802c: 217, pl. 8, fig. 48 (w.) FRENCH GUIANA. Combination in *Gnamptogenys:* Mayr, 1866b: 892 (in text); in *Ectatomma (Gnamptogenys):* Dalla Torre, 1893: 25. Variety of *mordax:* Emery, 1896g: 50. Junior synonym of *mordax:* Brown, 1958g: 228.
*obscurum. Ectatomma (Holcoponera) obscurum* Emery, 1896g: 48 (w.) BRAZIL. Combination in *Holcoponera:* Emery, 1911d: 41. Variety of *striatula:* Forel, 1908c: 341. Junior synonym of *striatula:* Santschi, 1929h: 442.
*obscura. Stictoponera menadensis* var. *obscura* Santschi, 1932b: 11 (w.) SULAWESI. [Unresolved junior secondary homonym of *obscurum* Emery, above.] Junior synonym of *menadensis:* Brown, 1954h: 2.
*panamensis. Ectatomma (Gnamptogenys) continuum* var. *panamensis* Santschi, 1931c: 265, fig. 1 (w.) PANAMA. [Unresolved junior primary homonym of *Ectatomma dentinode* var. *panamense* Forel, 1899c: 9 (now in *Heteroponera*).] Junior synonym of *continua:* Brown, 1958g: 227.
*panamensis. Alfaria panamensis* Weber, 1940b: 80 (w.q.) PANAMA. [Unresolved junior secondary homonym of *panamensis* Santschi, above.] Junior synonym of *minuta:* Brown, in Borgmeier, 1957: 115.
**panda.** *Stictoponera panda* Brown, 1948a: 263 (w.) CHINA. Combination in *Gnamptogenys:* Brown, 1958g: 228.
*parva. Stictoponera rugosa* r. *parva* Forel, 1913k: 6 (w.) SUMATRA. Junior synonym of *costata:* Brown, 1954h: 7.
*paulina. Ectatomma (Holcomyrmex) curtula* var. *paulina* Forel, 1908c: 342 (*Holcomyrmex* here is an error for *Holcoponera*) (w.) BRAZIL. Combination in *Holcoponera:* Emery, 1911d: 40. Stirps of *brasiliensis:* Santschi, 1929h: 456. Junior synonym of *striatula:* Brown, 1958g: 229.
*pernambucana. Holcoponera brasiliensis* var. *pernambucana* Santschi, 1929h: 452, fig. 27 (w.) BRAZIL (attributed to Borgmeier). Junior synonym of *striatula:* Brown, 1958g: 229.
**perspicax.** *Gnamptogenys perspicax* Kempf & Brown, 1970: 316, figs. 3–5 (w.q.) COLOMBIA.
**petiscapa.** *Gnamptogenys petiscapa* Lattke, 1990b: 16, fig. 15 (w.) VENEZUELA.
*pinealis. Stictoponera costata* var. *pinealis* Wheeler, W.M. 1929g: 31 (q.) WEST MALAYSIA. Junior synonym of *costata:* Brown, 1954h: 7.
**pittieri.** *Gnamptogenys pittieri* Lattke, 1990b: 18, fig. 11 (w.q.) VENEZUELA.
**pleurodon.** *Ectatomma (Holcoponera) pleurodon* Emery, 1896g: 47 (w.q.) BRAZIL. Combination in *Holcoponera:* Emery, 1911d: 41; in *Gnamptogenys:* Brown, 1958g: 229. Variety of *striatula:* Forel, 1908c: 340. Revived status as species, and senior synonym of *emeryi, recta* and *vidua:* Brown, 1958g: 229.
*pneodonax. Gnamptogenys pneodonax* Kempf, 1968b: 375, figs. 2, 3 (w.) BRAZIL. Junior synonym of *minuta:* Lattke, 1992a: 127.
**porcata.** *Ectatomma (Holcoponera) porcatum* Emery, 1896g: 48 (w.) COSTA RICA. Forel, 1908b: 38 (q.m.). Combination in *Holcoponera:* Emery, 1911d: 40; in *Gnamptogenys:* Brown, 1958g: 229.
**posteropsis.** *Stictoponera posteropsis* Gregg, 1951: 77, figs. 1, 3 (q.) SUMATRA. Combination in *Gnamptogenys:* Brown, 1958g: 229.
***pristina.** **Gnamptogenys pristina* Baroni Urbani, 1980e: 2, figs. 1–4 (w.) DOMINICAN AMBER (Miocene).
*purensis. Ectatomma (Gnamptogenys) mordax* var. *purensis* Forel, 1912c: 33 (w.) BRAZIL. Junior synonym of *mordax:* Brown, 1958g: 228.
*quitensis. Gnamptogenys tortuolosa* var. *quitensis* Forel, 1921b: 133 (w.) ECUADOR. Junior synonym of *tortuolosa:* Brown, 1958g: 230.
**rastrata.** *Ectatomma rastratum* Mayr, 1866b: 890 (q.) BRAZIL. Combination in *E. (Gnamptogenys):* Emery, 1890b: 41; in *E. (Parectatomma):* Emery, 1911d: 44; in *Gnamptogenys:* Brown, 1958g: 229.
*recta. Holcoponera emeryi* var. *recta* Santschi, 1929h: 465 (w.) BOLIVIA. Junior synonym of *pleurodon:* Brown, 1958g: 229.
**regularis.** *Gnamptogenys regularis* Mayr, 1870b: 965 (w.q.) MEXICO. Combination in *Ectatomma (Gnamptogenys):* Mayr, 1887: 541; in *Gnamptogenys:* Mann, 1922: 3. Senior synonym of *fiebrigi* and *splendidum:* Brown, 1958g: 229.
*regularis. Holcoponera regularis* Santschi, 1929d: 275 (w.) BRAZIL. [Also described as new by Santschi, 1929h: 457.] [Junior secondary homonym of *regularis* Mayr, above.] Senior synonym of *arcuata* and hence the latter the first available replacement name for *regularis* Santschi: Brown, 1958g: 227.
**reichenspergeri.** *Acanthoponera (Anacanthoponera) reichenspergeri* Santschi, 1929d: 274 (w.) BRAZIL. Combination in *Gnamptogenys:* Brown, 1958g: 229.
**relicta.** *Rhopalopone relicta* Mann, 1916: 403, pl. 1, figs. 4, 5 (w.) BRAZIL. Wheeler, G.C. & Wheeler, J. 1964b: 446 (l.). Combination in *Holcoponera:* Brown, 1956b: 491; in *Gnamptogenys:* Brown, 1958g: 229.
*richteri. Ectatomma (Parectatomma) triangulare* r. *richteri* Forel, 1913l: 203 (w.) ARGENTINA. Forel, 1914d: 265 (q.). Junior synonym of *triangularis:* Brown, 1958g: 230.
**rimulosa.** *Ponera rimulosa* Roger, 1861a: 18 (w.) BRAZIL. Combination in *Ectatomma (Gnamptogenys):* Mayr, 1887: 541; Emery, 1896g: 52; in *Gnamptogenys:* Roger, 1863a: 174; Brown, 1958g: 229.
*romani. Ectatomma (Gnamptogenys) concinnum* var. *romani* Wheeler, W.M. 1923: 2 (w.) BRAZIL. Junior synonym of *concinna:* Brown, 1958g: 227.
*rugosa. Ponera rugosa* Smith, F. 1857a: 66 (q.) BORNEO. [Junior primary homonym of *Ponera rugosa* Le Guillou, 1842: 318 (now in *Diacamma*).] Replacement name: *wallacei:* Donisthorpe, 1932c: 447. Combination in *Stictoponera:* Emery, 1911d: 48.
**rustica.** *Holcoponera rustica* Santschi, 1929h: 446, figs. 11, 18 (w.) PARAGUAY. Combination in *Gnamptogenys:* Brown, 1958g: 229.
*satzgeri. Ectatomma (Holcoponera) satzgeri* Forel, 1908b: 39 (w.q.) COSTA RICA. Combination in *Holcoponera:* Emery, 1911d: 40. Stirps of *simplex:* Santschi, 1929h: 461. Junior synonym of *simplex:* Brown, 1956b: 490.
*sauteri. Stictoponera sauteri* Chapman & Capco, 1951: 30. [Combination proposed in error for *Pachycondyla sauteri* Forel, 1912a: 49; see Brown, 1954h: 10.]
*scabrosus. Opisthoscyphus scabrosus* Mann, 1922: 5, figs. 2, 3 (w.) HONDURAS. Junior synonym of *minuta:* Brown, in Borgmeier, 1957: 115. See also: Brown, 1958g: 317.
**schmitti.** *Emeryella schmitti* Forel, 1901e: 334 (w.) HAITI. Wheeler, W.M. & Mann, 1914: 9 (m.l.); Wheeler, G.C. & Wheeler, J. 1952a: 127 (l.). Combination in *Gnamptogenys*, and senior synonym of *minor:* Brown, 1958g: 229.
**schubarti.** *Ectatomma (Parectatomma) schubarti* Borgmeier, 1948a: 198, fig. 22 (w.) BRAZIL. Combination in *Gnamptogenys:* Brown, 1958g: 229.
*sebastiani. Ectatomma (Gnamptogenys) mordax* subsp. *sebastiani* Borgmeier, 1937b: 220 (w.) BRAZIL. Junior synonym of *mordax:* Brown, 1958g: 228.
*semicircularis. Gnamptogenys concinna* var. *semicircularis* Borgmeier, 1929: 195 (w.) BRAZIL. Junior synonym of *concinna:* Brown, 1958g: 227.
**semiferox.** *Gnamptogenys semiferox* Brown, 1958g: 324, fig. 14 (w.) DOMINICA.
*simalurensis. Stictoponera costata* st. *simalurensis* Forel, 1915a: 23 (w.) SUMATRA. Junior synonym of *costata:* Brown, 1954h: 7.
*simplex. Ectatomma (Holcoponera) simplex* Emery, 1896g: 46, pl. 1, fig. 7 (w.q.) COSTA RICA. Combination in *Holcoponera:* Emery, 1911d: 40; in *Gnamptogenys:* Brown, 1958g: 229. Senior synonym of *foreli, satzgeri* and *spurium:* Brown, 1956b: 490. Junior synonym of *strigata:* Kempf & Brown, 1968: 90.
**simplicoides.** *Ectatomma (Holcoponera) striatulum* var. *simplicoides* Forel, 1908c: 341 (w.) BRAZIL. Santschi, 1929h: 455 (m.). Combination in *Holcoponera:* Emery, 1911d: 41; in *Gnamptogenys:* Brown, 1958g: 229. Stirps of *brasiliensis:* Santschi, 1929h: 454. Raised to species: Brown, 1958g: 229.
**simulans.** *Alfaria simulans* Emery, 1896e: 177 (footnote) (w.) COSTA RICA. Emery, 1896g: 42 (q.m., species also described as new here). Combination in *Gnamptogenys:* Brown, 1958g: 229 (see also p. 316).
**sinensis.** *Gnamptogenys sinensis* Wu & Xiao, 1987: 303, figs. 1–3 (w.) CHINA.
*soror. Gnamptogenys soror* Kempf & Brown, 1968: 90 (w.) BRAZIL. Junior synonym of *caelata:* Lattke, 1992a: 126.
**spiralis.** *Stictoponera spiralis* Karavaiev, 1925a: 79 (w.) JAVA. Combination in *Gnamptogenys:* Brown, 1958g: 229.
*splendens. Holcoponera moelleri* var. *splendens* Santschi, 1929h: 450, fig. 15 (w.m.) BRAZIL. Junior synonym of *moelleri:* Brown, 1958g: 228.
*splendidum. Ectatomma (Gnamptogenys) rimulosum* var. *splendidum* Pergande, 1896: 871 (w.) MEXICO. Raised to species: Forel, 1899c: 8. Variety of *regularis:* Emery, 1911d: 45. Junior synonym of *regularis:* Brown, 1958g: 229.
*spurium. Ectatomma (Holcoponera) simplex* subsp. *spurium* Forel, 1908b: 39 (w.) GUATEMALA. Combina-

tion in *Holcoponera:* Emery, 1911d: 40. Junior synonym of *simplex:* Brown, 1956b: 490.

*stolli. Ectatomma (Holcoponera) curtulum* var. *stolli* Forel, 1899c: 7 (w.m.) GUATEMALA. Combination in *Holcoponera:* Emery, 1911d: 40. Stirps of *curtula:* Santschi, 1929h: 470. Junior synonym of *curtula:* Brown, 1958g: 228.

***striatula.*** *Gnamptogenys striatula* Mayr, 1884: 32 (w.) FRENCH GUIANA. Mayr, 1887: 542 (q.m.); Gallardo, 1918b: 26 (q.m.); Wheeler, G.C. & Wheeler, J. 1952a: 123 (l.). Combination in *Ectatomma (Holcoponera):* Mayr, 1887: 541; in *Holcoponera:* Emery, 1911d: 40; in *Gnamptogenys:* Brown, 1958g: 229. Senior synonym of *obscurum:* Santschi, 1929h: 442; of *antillana:* Brown, 1956b: 490; of *angustiloba* (and its junior synonym *vollenweideri*), *angustipleura, brasiliensis, calcarata, paulina, pernambucana,* and material of the unavailable name *hybrida* referred here: Brown, 1958g: 229.

***strigata.*** *Polyrhachis strigata* Norton, 1868c: 4 (w.) MEXICO. Wheeler, G.C. & Wheeler, J. 1964b: 450 (l.). Combination in *Ectatomma (Holcoponera):* Emery, 1892b: 167; in *Holcoponera:* Emery, 1911d: 41; in *Gnamptogenys:* Brown, 1958g: 229. Senior synonym of *concentrica:* Dalla Torre, 1893: 26; Emery, 1896g: 44 (in key); of *simplex* (and its junior synonyms *foreli, satzgeri* and *spurium*): Kempf & Brown, 1968: 90.

***striolata.*** *Alfaria striolata* Borgmeier, 1957: 116, figs. 27–29 (w.) BRAZIL. Combination in *Gnamptogenys:* Brown, 1958g: 229.

*stylata. Stictoponera stylata* Menozzi, 1925c: 440, fig. 2 (w.) PHILIPPINES. Junior synonym of *menadensis:* Brown, 1954h: 2.

***sulcata.*** *Ponera sulcata* Smith, F. 1858b: 99 (w.) BRAZIL. Forel, 1899c: 8 (q.m.). Combination in *Ectatomma (Gnamptogenys):* Dalla Torre, 1893: 26; Emery, 1896g: 51; in *Gnamptogenys:* Mayr, 1886c: 358; Brown, 1958g: 229. Senior synonym of *bufonum, cearensis, lineata, nitens* and *ypirangensis:* Brown, 1958g: 229.

***taivanensis.*** *Stictoponera taivanensis* Wheeler, W.M. 1929g: 32 (w.) TAIWAN. Combination in *Gnamptogenys:* Brown, 1958g: 229.

***teffensis.*** *Holcoponera teffensis* Santschi, 1929h: 449, figs. 17, 22 (w.) BRAZIL (attributed to Borgmeier). Combination in *Gnamptogenys,* and senior synonym of *concinna:* Brown, 1958g: 229.

***tornata.*** *Ponera tornata* Roger, 1861a: 15 (w.q.m.) MEXICO. Combination in *Ectatomma (Gnamptogenys):* Mayr, 1887: 541; Emery, 1896g: 46; in *Gnamptogenys:* Roger, 1863a: 174; Mann, 1922: 3; Brown, 1958g: 229. Senior synonym of *ericae:* Brown, 1958g: 229. See also: Kempf, 1968b: 378.

***tortuolosa.*** *Ponera tortuolosa* Smith, F. 1858b: 99 (w.) BRAZIL. Wheeler, G.C. & Wheeler, J. 1952a: 134 (l.). Combination in *Ectatomma (Gnamptogenys):* Dalla Torre, 1893: 26; Emery, 1896g: 51; in *Gnamptogenys:* Mayr, 1886c: 358; Brown, 1958g: 230. Senior synonym of *quitensis:* Brown, 1958g: 230.

***triangularis.*** *Ectatomma (Gnamptogenys) triangulare* Mayr, 1887: 544 (q.) URUGUAY. Emery, 1906c: 113 (w.). Combination in *E. (Parectatomma):* Emery, 1911d: 44; in *Gnamptogenys:* Brown, 1958g: 230. Senior synonym of *richteri:* Brown, 1958g: 230.

***trigona.*** *Ectatomma (Gnamptogenys) trigonum* Emery, 1906c: 114 (footnote) (q.) BRAZIL. Combination in *E. (Parectatomma):* Emery, 1911d: 44; in *Gnamptogenys:* Brown, 1958g: 230.

***turmalis.*** *Gnamptogenys turmalis* Kempf & Brown, 1968: 93 (w.) PANAMA.

*unicolor. Stictoponera costata* var. *unicolor* Forel, 1901f: 335 (w.) BORNEO. Junior synonym of *costata:* Brown, 1954h: 7.

*vidua. Holcoponera vidua* Santschi, 1929h: 467, fig. 3 (q.) BRAZIL. Junior synonym of *pleurodon:* Brown, 1958g: 229.

*vollenweideri. Holcoponera curtula* var. *vollenweideri* Forel, 1912c: 33 (w.) ARGENTINA. Junior synonym of *angustiloba:* Santschi, 1929d: 275; Santschi, 1929h: 453.

***vriesi.*** *Gnamptogenys vriesi* Brandão & Lattke, 1990: 490, figs. 1–7 (w.q.) ECUADOR.

*wallacei. Stictoponera wallacei* Donisthorpe, 1932c: 447. Replacement name for *Ponera rugosa* Smith, F. 1857a: 66. [Junior primary homonym of *Ponera rugosa* Le Guillou, 1842: 318.] Junior synonym of *costata:* Brown, 1954h: 7.

***wasmanni.*** *Holcoponera wasmanni* Santschi, 1929h: 466, fig. 42 (w.) BRAZIL. Combination in *Gnamptogenys,* and senior synonym of *isthmica:* Brown, 1958g: 230.

***wheeleri.*** *Holcoponera rustica* st. *wheeleri* Santschi, 1929h: 448, figs. 9, 19 (w.) COSTA RICA. Wheeler, G.C. & Wheeler, J. 1952a: 124 (l.). Combination in *Gnamptogenys,* raised to species, and senior synonym of *mayri:* Brown, 1958g: 230.

*ypirangensis. Gnamptogenys ypirangensis* Borgmeier, 1928b: 60 (w.) BRAZIL. Junior synonym of *sulcata:* Brown, 1958g: 229.

#*GONEPIMYRMA:* see under ***EPIMYRMA.***

***GONIOMMA*** (Palaearctic)

***baeticum.*** *Goniomma baeticum* Reyes, Espadaler & Rodriguez, 1987: 269, figs. 1–4 (w.q.m.) SPAIN.

*barbaricum. Goniomma hispanicum* st. *tuneticum* var. *barbaricum* Santschi, 1929e: 145 (w.) TUNISIA. Unavailable name; material referred to *hispanicum* by Acosta Salmerón, 1981: 13.

***blanci.*** *Aphaenogaster blanci* André, 1881a: 60 and xlviii (w.) FRANCE. Espadaler & Batet, 1979: 11 (q.); Espadaler, 1984: 138 (m.). Combination in *Stenamma (Goniomma):* Emery, 1895c: 298; in *Goniomma:* Emery, 1908e: 460.

*burri* Donisthorpe, 1950; see under ***APHAENOGASTER.***

*gallica. Goniomma hispanicum* var. *gallica* Emery, 1908e: 461 (w.) FRANCE. Stirps of *hispanicum:* Santschi, 1929e: 145. Junior synonym of *hispanicum:* Acosta Salmerón, 1981: 13.

***hispanicum.*** *Aphaenogaster hispanica* André, 1883b: 365, pl. 23, figs. 11, 13, 21 (q.m.) SPAIN. Forel, 1892g: 453 (w.); Wheeler, G.C. & Wheeler, J. 1983: 602 (l.). Combination in *Stenamma (Goniomma):* Emery, 1895c: 298; in *Goniomma:* Emery, 1908e: 461. Senior synonym of *gallica, maurum, otini* and *tuneticus,* and material of the unavailable names *barbaricum* and *nitidifrons* referred here: Acosta Salmerón, 1981: 13.

***kugleri.*** *Goniomma kugleri* Espadaler, 1986a: 61, figs. 1–15 (w.q.m.) SPAIN.

*maurum. Goniomma maurum* Santschi, 1926e: 235 (w.) MOROCCO. Junior synonym of *hispanicum:* Acosta Salmerón, 1981: 13.

*nitidifrons. Goniomma hispanicum* st. *tuneticum* var. *nitidifrons* Santschi, 1910a: 45 (w.q.m.) TUNISIA. Unavailable name; material referred to *hispanicum* by Acosta Salmerón, 1981: 13.

*otini. Goniomma otini* Santschi, 1929e: 144 (w.) MOROCCO. Junior synonym of *hispanicum:* Acosta Salmerón, 1981: 13.

***punicum.*** *Oxyopomyrmex (Goniomma) punicus* Forel, 1907b: 205 (w.) TUNISIA. Santschi, 1910a: 46 (q.). Combination in *Goniomma:* Emery, 1908e: 462.

*thoracica. Oxyopomyrmex (Goniomma) hispanicus* r. *tuneticus* var. *thoracica* Santschi, 1907: 330 (w.) TUNISIA. Unavailable name.

*tuneticus. Oxyopomyrmex (Goniomma) blanci* r. *tuneticus* Forel, 1905b: 175 (w.) TUNISIA. Combination in *Goniomma:* Emery, 1908e: 462. Subspecies of *hispanicum:* Emery, 1908e: 462; Santschi, 1929e: 145. Raised to species: Collingwood & Yarrow, 1969: 64. Junior synonym of *hispanicum:* Acosta Salmerón, 1981: 13.

*GONIOTHORAX:* see under ***LEPTOTHORAX.***

*GRANISOLENOPSIS:* see under ***SOLENOPSIS.***

***GYMNOMYRMEX*** (Neotropical)

***abditus.*** *Gymnomyrmex abditus* Kempf, 1960f: 448, figs. 23, 25, 29, 31 (w.) BRAZIL.

***comis.*** *Gymnomyrmex comis* Kempf, 1959d: 342, figs. 3, 5, 6, 9 (w.) BRAZIL.

***dentinasis.*** *Gymnomyrmex dentinasis* Kempf, 1960f: 450, figs. 24, 26, 30 (w.) BRAZIL.

***minusculus.*** *Gymnomyrmex minusculus* Kempf, 1962b: 24, figs. 22–24 (w.) BRAZIL.

***rugithorax.*** *Gymnomyrmex rugithorax* Kempf, 1959d: 339, figs. 2, 4, 7 (w.) BRAZIL.

***splendens.*** *Gymnomyrmex splendens* Borgmeier, 1954b: 279, figs. 1–6 (w.) BRAZIL. Kempf, 1959d: 339 (q.).

***villiersi.*** *Gymnomyrmex villiersi* Perrault, 1986a: 1, figs. 1–3 (w.) FRENCH GUIANA.

*HAGENSIA:* see under ***PACHYCONDYLA.***

#***HAGIOMYRMA:*** see under ***POLYRHACHIS.***

*HAGIOSTIGMACROS:* see under ***STIGMACROS.***

*HAGIOXENUS:* see under ***RHOPTROMYRMEX.***

*HALMAMYRMECIA:* see under ***MYRMECIA.***

*HARNEDIA:* see under ***ZACRYPTOCERUS.***

***HARPAGOXENUS*** (Holarctic)

*americanus* Emery, 1895 [*Tomognathus*]; see under ***PROTOMOGNATHUS.***

***canadensis.*** *Harpagoxenus canadensis* Smith, M.R. 1939b: 168, figs. 1, 2 (q., not w.) CANADA. Gregg, 1945a: 74 (w.).

#***caucasicus.*** *Harpagoxenus sublaevis* subsp. *caucasicus* Arnol'di, 1968: 1812, fig. 11 (w.) GEORGIA.

#***hirtulus.*** *Myrmica hirtula* Nylander, 1849: 45 (w.) FINLAND. Combination in *Harpagoxenus:* Emery, 1922f:

***HARPAGOXENUS** (continued)*

266. Junior synonym of *sublaevis:* Mayr, 1861: 56; Dalla Torre, 1893: 64. Revived from synonymy as variety of *sublaevis:* Emery, 1922f: 266; Stitz, 1939: 154.

***sublaevis.*** *Myrmica sublaevis* Nylander, 1849: 33 (w.) FINLAND. Adlerz, 1896: 62 (q.m.); Viehmeyer, 1906: 58 (q.m.); Wheeler, G.C. & Wheeler, J. 1955b: 27 (l.). Combination in *Tomognathus:* Mayr, 1861: 56; in *Harpagoxenus:* Forel, 1893a: 167. See also: Stitz, 1939: 151; Kutter, 1977c: 146; Buschinger, 1981: 211. Current subspecies: nominal plus *caucasicus, hirtulus.*

***zaisanicus.*** *Harpagoxenus zaisanicus* Pisarski, 1963: 39, figs. 1–4 (w.) MONGOLIA.

***HARPEGNATHOS*** (Oriental, Indo-Australian)

#***chapmani.*** *Harpegnathos venator* var. *chapmani* Donisthorpe, 1937b: 199 (w.) PHILIPPINES.

#***cruentatus.*** *Drepanognathus cruentatus* Smith, F. 1858b: 82 (w.) HONG KONG. Forel, 1900c: 65 (q.m.). Combination in *Harpegnathos:* Forel, 1900c: 65. Junior synonym of *saltator:* Bingham, 1903: 50; Emery, 1911d: 59. Revived from synonymy as variety of *saltator:* Donisthorpe, 1937b: 198.

***empesoi.*** *Harpegnathos empesoi* Chapman, 1963: 254, fig. 5 (w.) PHILIPPINES.

***hobbyi.*** *Harpegnathos hobbyi* Donisthorpe, 1937b: 196 (q.) BORNEO.

***macgregori.*** *Harpegnathos macgregori* Wheeler, W.M. & Chapman, 1925: 61 (w.) PHILIPPINES.

***medioniger.*** *Harpegnathos medioniger* Donisthorpe, 1942b: 64 (m.) PHILIPPINES.

***pallipes.*** *Ponera pallipes* Smith, F. 1858b: 87 (m.) JAVA. Combination in *Harpegnathos:* Donisthorpe, 1932c: 474. [Misspelled as *pallidipes* by Dalla Torre, 1893: 41.]

#***rugosus.*** *Drepanognathus rugosus* Mayr, 1862: 723 (q.) HONG KONG. Combination in *Harpegnathos:* Forel, 1900c: 65. Variety of *venator:* Forel, 1900c: 65; Wheeler, W.M. 1928c: 4.

***saltator.*** *Harpegnathos saltator* Jerdon, 1851: 117 (w.) INDIA. Combination in *Drepanognathus:* Smith, F. 1858b: 82; in *Harpegnathos:* Forel, 1900c: 64; Emery, 1911d: 59. Forel, 1913e: 660 (q.m.). Current subspecies: nominal plus *cruentatus, taprobanae.* See also: Bingham, 1903: 50.

#***taprobanae.*** *Harpegnathos cruentatus* var. *taprobanae* Forel, 1909e: 392 (w.q.) SRI LANKA. Variety of *saltator:* Emery, 1911d: 59; Donisthorpe, 1937b: 198.

***venator.*** *Drepanognathus venator* Smith, F. 1858b: 82 (w.) INDIA. Forel, 1900c: 64 (q.m.). Combination in *Harpegnathos:* Emery, 1889b: 494. Current subspecies: nominal plus *chapmani, rugosus.* See also: Bingham, 1903: 51.

#***HEDOMYRMA:*** see under ***POLYRHACHIS.***

#***HEMIOPTICA:*** see under ***POLYRHACHIS.***

*HENDECAPHEIDOLE:* see under ***PHEIDOLE.***

*HENDECATELLA:* see under ***OLIGOMYRMEX.***

*HEPTACONDYLUS:* see under ***MYRMICARIA.***

*HEPTASTRUMA:* see under ***RHOPALOTHRIX.***

*HERCYNIA:* see under ***WASMANNIA.***

*HETEROMYRMEX:* see under ***VOLLENHOVIA.***

***HETEROPONERA*** (Neotropical, Australasian)

***angulata.*** *Heteroponera angulata* Borgmeier, 1959a: 310, figs. 12, 13 (w.) BRAZIL.

*aurea. Acanthoponera dolo* var. *aurea* Forel, 1913l: 203 (w.m.) ARGENTINA. Combination in *A.* (*Anacanthoponera*): Wheeler, W.M. 1923e: 187. Junior synonym of *dolo:* Brown, 1958g: 195.

***brounii.*** *Ectatomma* (*Acanthoponera*) *brownii* Forel, 1892c: 335 (w.) NEW ZEALAND. Combination in *Acanthoponera,* and spelling emended to *brounii:* Forel, 1905a: 353; in *Heteroponera:* Brown, in Borgmeier, 1957: 112. Senior synonym of *kirki:* Brown, 1958h: 18. See also: Wheeler, W.M. 1923e: 183; Brown, 1958g: 195.

***carinifrons.*** *Heteroponera carinifrons* Mayr, 1887: 533 (w.) CHILE. Emery, 1906c: 112 (q.). Combination in *Acanthoponera:* Emery, 1895g: 347; in *A.* (*Anacanthoponera*): Wheeler, W.M. 1923e: 186; in *Heteroponera:* Brown, 1952h: 70. See also: Brown, 1958g: 257.

***dentinodis.*** *Ectatomma* (*Acanthoponera*) *dentinode* Mayr, 1887: 541 (w.q.m.) BRAZIL. Combination in *Acanthoponera:* Emery, 1911d: 36; in *A.* (*Anacanthoponera*): Wheeler, W.M. 1923e: 186; in *Heteroponera:* Brown, 1958g: 195. See also: Brown, 1958g: 257; Kempf, 1962a: 40.

***dolo.*** *Ponera dolo* Roger, 1860: 293 (w.q.) BRAZIL. Wheeler, G.C. & Wheeler, J. 1962a: 33 (l.). Combination in *Ectatomma* (*Acanthoponera*): Mayr, 1862: 733; in *Acanthoponera:* Emery, 1906c: 112; in *A.* (*Anacanthoponera*): Wheeler, W.M. 1923e: 187; in *Heteroponera:* Brown, 1952h: 70. Senior synonym of *aurea:* Brown, 1958g: 195. See also: Kempf, 1962a: 33.

***flava.*** *Heteroponera flava* Kempf, 1962a: 38, figs. 2, 9 (w.q.) BRAZIL.

*hilare. Ectatomma* (*Acanthoponera*) *imbellis* var. *hilare* Forel, 1895f: 421 (w.) AUSTRALIA. Combination in *Acanthoponera:* Emery, 1911d: 36; in *A.* (*Anacanthoponera*): Wheeler, W.M. 1923e: 180. Junior synonym of *imbellis:* Brown, 1958g: 196.

***imbellis.*** *Acanthoponera imbellis* Emery, 1895g: 346 (w.) AUSTRALIA. Wheeler, G.C. & Wheeler, J. 1964b: 447 (l.). Combination in *A.* (*Anacanthoponera*): Wheeler, W.M. 1923e: 179; in *Heteroponera:* Brown, in Borgmeier, 1957: 112. Senior synonym of *hilare, nigra, occidentalis* and *scabra:* Brown, 1958g: 196. See also: Brown, 1958g: 261; Wheeler, G.C. & Wheeler, J. 1971b: 1200.

***inca.*** *Heteroponera inca* Brown, 1958g: 259, figs. 12, 13 (w.) COLOMBIA. Wheeler, G.C. & Wheeler, J. 1971b: 1201 (l.).

***inermis.*** *Ectatomma* (*Acanthoponera*) *dentinode* var. *inerme* Emery, 1894c: 143 (q.) BRAZIL. Combination in *Acanthoponera:* Emery, 1911d: 36; in *A.* (*Anacanthoponera*): Wheeler, W.M. 1923e: 186; in *Heteroponera,* and raised to species: Brown, 1958g: 196. Senior synonym of *schwebeli:* Kempf, 1962a: 36.

*kirki. Acanthoponera* (*Anacanthoponera*) *brouni* subsp. *kirki* Wheeler, W.M. 1923e: 184, fig. 4 (w.q.) NEW ZEALAND. Junior synonym of *brounii:* Brown, 1958h: 18.

***leae.*** *Acanthoponera* (*Anacanthoponera*) *leae* Wheeler, W.M. 1923e: 181, fig. 3 (w.) AUSTRALIA. Combination in *Heteroponera:* Brown, 1958g: 196.

***mayri.*** *Heteroponera mayri* Kempf, 1962a: 43, figs. 5, 12 (w.q.) BRAZIL.

***microps.*** *Heteroponera microps* Borgmeier, 1957: 112, figs. 23–26 (w.) BRAZIL.

*minuta. Heteroponera minuta* Kusnezov, 1954b: 34. *Nomen nudum.*

***monticola.*** *Heteroponera monticola* Kempf & Brown, 1970: 312, figs. 1, 2 (w.) COLOMBIA.

*nigra. Acanthoponera nigra* Clark, 1930c: 6 (w.) AUSTRALIA. Junior synonym of *imbellis:* Brown, 1958g: 196.

*occidentalis. Acanthoponera occidentalis* Clark, 1926: 47, pl. 6, fig. 3 (w.) AUSTRALIA. Junior synonym of *imbellis:* Brown, 1958g: 196.

***panamensis.*** *Ectatomma* (*Acanthoponera*) *dentinode* var. *panamense* Forel, 1899c: 9 (w.) PANAMA. Combination in *Acanthoponera:* Emery, 1911d: 36; in *A.* (*Anacanthoponera*): Wheeler, W.M. 1923e: 187; in *Heteroponera,* and raised to species: Brown, 1958g: 196. See also: Brown, 1958g: 259; Kempf, 1962a: 40.

***relicta.*** *Paranomopone relicta* Wheeler, W.M. 1915c: 188, pl. 8 (w.q.l.) AUSTRALIA. Imai, Crozier & Taylor, 1977: 347 (k.). Combination in *Heteroponera:* Brown, 1958g: 196.

***robusta.*** *Heteroponera robusta* Kempf, 1962a: 36 (w.q.) BRAZIL.

*scabra. Acanthoponera* (*Anacanthoponera*) *imbellis* var. *scabra* Wheeler, W.M. 1923e: 181 (w.) AUSTRALIA. Junior synonym of *imbellis:* Brown, 1958g: 196.

*schwebeli. Acanthoponera dolo* var. *schwebeli* Luederwaldt, 1918: 54 (w.) BRAZIL. Combination in *A.* (*Anacanthoponera*): Wheeler, W.M. 1923e: 188; in *Heteroponera:* Brown, 1958g: 196. Junior synonym of *inermis:* Kempf, 1962a: 36.

*HEXADACETON:* see under ***EPOPOSTRUMA.***

*HINCKSIDRIS:* see under ***MYRMELACHISTA.***

*HOLCOMYRMEX:* see under ***MONOMORIUM.***

*HOLCOPONERA* Mayr: see under ***GNAMPTOGENYS.***

*HOLCOPONERA* Cameron: see under ***CYLINDROMYRMEX.***

*HOLOPONE:* see under ***ECITON.***

*HOPLOMYRMUS:* see under ***POLYRHACHIS.***

***HUBERIA*** (New Zealand)

***brounii.*** *Huberia brounii* Forel, 1895a: 41 (w.) NEW ZEALAND. [Name misspelled *braumi* by Forel, 1905a: 354.] See also: Brown, 1958h: 27.

*rufescens. Huberia striata* var. *rufescens* Forel, 1892c: 339 (w.) NEW ZEALAND. Forel, 1902h: 447 (q.). Junior synonym of *striata:* Brown, 1958h: 26.

***striata.*** *Tetramorium striatum* Smith, F. 1876b: 481 (w.q.m.) NEW ZEALAND. Forel, 1894e: 229 (m.); Emery, 1899c: 8 (l.); Wheeler, G.C. & Wheeler, J. 1960b: 15 (l.). Combination in *Huberia:* Forel, 1890b: cvi. Senior synonym of *rufescens:* Brown, 1958h: 26.

*HYLIDRIS:* see under **PRISTOMYRMEX.**

***HYLOMYRMA*** (Neotropical)

***balzani.*** *Tetramorium balzani* Emery, 1894c: 165 (w.) PARAGUAY. Kempf, 1960f: 434 (q.). Combination in *Lundella:* Emery, 1915i: 191; in *Hylomyrma,* and senior synonym of *speciosa:* Kempf, 1960f: 434. See also: Kempf, 1973b: 230.

***blandiens.*** *Hylomyrma blandiens* Kempf, 1961b: 500, fig. 8 (w.q.) SURINAM. See also: Kempf, 1973b: 231.

***columbica.*** *Pogonomyrmex* (*Hylomyrma*) *columbicus* Forel, 1912g: 16 (w.) COLOMBIA. Wheeler, G.C. & Wheeler, J. 1960b: 4 (l.). Combination in *Hylomyrma:* Kempf, 1973b: 234.

***dentiloba.*** *Lundella dentiloba* Santschi, 1931c: 271 (w.) PANAMA. Combination in *Hylomyrma:* Kempf, 1960f: 430. See also: Kempf, 1964e: 52; Kempf, 1973b: 238.

***dolichops.*** *Hylomyrma dolichops* Kempf, 1973b: 239, figs. 4, 12 (w.q.) COLOMBIA.

*goeldii. Pogonomyrmex* (*Hylomyrma*) *goeldii* Forel, 1912g: 17 (q.) BRAZIL. Junior synonym of *reitteri:* Kempf, 1960f: 430.

***immanis.*** *Hylomyrma immanis* Kempf, 1973b: 241, fig. 2 (w.q.) COLOMBIA.

***longiscapa.*** *Hylomyrma longiscapa* Kempf, 1961b: 498, fig. 7 (w.) SURINAM. See also: Kempf, 1973b: 244.

***praepotens.*** *Hylomyrma praepotens* Kempf, 1973b: 245, figs. 1, 15 (w.) COLOMBIA.

***reginae.*** *Hylomyrma reginae* Kutter, 1977a: 85, figs. 1–4 (w.) GUIANA.

***reitteri.*** *Tetramorium reitteri* Mayr, 1887: 621 (w.) BRAZIL. Kempf, 1960f: 430 (q.); Wheeler, G.C. & Wheeler, J. 1977: 581 (l.). Combination in *Lundella:* Emery, 1915i: 191; in *Hylomyrma,* and senior synonym of *goeldii:* Kempf, 1960f: 430. See also: Kempf, 1973b: 247.

***sagax.*** *Hylomyrma sagax* Kempf, 1973b: 249, fig. 3 (w.q.) COLOMBIA.

*speciosa. Lundella speciosa* Borgmeier, 1937b: 241, figs. 25–29 (w.) BRAZIL. Junior synonym of *balzani:* Kempf, 1960f: 434.

***transversa.*** *Hylomyrma transversa* Kempf, 1973b: 250, fig. 9 (w.q.) PERU.

***versuta.*** *Hylomyrma versuta* Kempf, 1973b: 253, fig. 6 (w.q.) BELIZE.

#***HYPERCOLOBOPSIS:*** see under **CAMPONOTUS.**

*HYPOCHIRA:* see under **FORMICA.**

*HYPOCLINEA:* see under **DOLICHODERUS.**

*HYPOCRYPTOCERUS:* see under **ZACRYPTOCERUS.**

*HYPOCYLINDROMYRMEX:* see under **CYLINDROMYRMEX.**

*HYPOPHEIDOLE: Nomen nudum.*

****HYPOPOMYRMEX*** (Sicilian Amber)

****bombiccii.*** **Hypopomyrmex bombiccii* Emery, 1891a: 149, pl. 1, figs. 10, 11 (q.) SICILIAN AMBER (Miocene). See also: Brown & Carpenter, 1979: 417.

***HYPOPONERA*** (World wide)

***abeillei.*** *Ponera abeillei* André, 1881a: 61 and xlviii (w.) CORSICA. Santschi, 1921e: 167 (m.). Combination in *P.* (*Hypoponera*): Santschi, 1938b: 79; in *Hypoponera:* Taylor, 1967a: 12. Current subspecies: nominal plus *assmuthi.* See also: Baroni Urbani, 1971c: 15.

***abyssinica.*** *Ponera* (*Hypoponera*) *jeanneli* st. *abyssinica* Santschi, 1938b: 80, fig. 2 (w.q.) ETHIOPIA. [Unresolved junior primary homonym of *Ponera abyssinica* Guérin-Méneville, 1849: 352 (now in *Pachycondyla*).] Combination in *Hypoponera:* **new combination.** Raised to species: Bernard, 1953b: 205.

#***aemula.*** *Ponera dulcis* var. *aemula* Santschi, 1911c: 351 (w.) TANZANIA. Combination in *Hypoponera:* **new combination.**

***aethiopica.*** *Ponera gleadowi* subsp. *aethiopica* Forel, 1907c: 132 (q.) ETHIOPIA. [Unresolved junior primary homonym of *Ponera aethiopica* Smith, F. 1858b: 91 (now in *Streblognathus*).] Combination in *Hypoponera:* **new combination.** Raised to species: Bernard, 1953b: 199 (in key).

***agilis.*** *Ponera agilis* Borgmeier, 1934: 96 (w.) SURINAM. Combination in *Hypoponera:* Kempf, 1972a: 121.

#***aitkenii.*** *Ponera confinis* var. *aitkenii* Forel, 1900d: 325 (diagnosis in key) (w.) INDIA. Forel, 1900d: 327 (q.). Combination in *Hypoponera:* **new combination.**

***albopubescens.*** *Ponera albopubescens* Menozzi, 1939b: 177, fig. 2 (w.) SUMATRA. Combination in *Hypoponera:* **new combination.**

***aliena.*** *Ponera aliena* Smith, F. 1858b: 99 (w.) BRAZIL. Combination in *Hypoponera:* Kempf, 1972a: 121.

*andrei. Ponera andrei* Emery, 1900c: 318 (footnote), pl. 8, fig. 47 (w.) NEW CALEDONIA. Junior synonym of *perkinsi:* Wilson, 1958d: 334.

*androgyna. Ponera androgyna* Roger, 1859: 246 (footnote) (m.) GERMANY. Junior synonym of *punctatissima:* Emery & Forel, 1879: 455.

***angustata.*** *Cryptopone angustata* Santschi, 1914d: 319, fig. 7 (w.) GUINEA. Combination in *Pseudocryptopone:* Santschi, 1937h: 366; in *Ponera:* Brown, 1963: 6; in *Hypoponera:* Taylor, 1967a: 12.

*antipodum. Ponera antipodum* Forel, 1895a: 43 (q., not w.; see Brown, 1958h: 23) NEW ZEALAND. Junior synonym of *eduardi:* Brown, 1958h: 23.

#***antoniensis.*** *Ponera fiebrigi* var. *antoniensis* Forel, 1912c: 41 (w.) COLOMBIA. Combination in *Hypoponera:* Kempf, 1972a: 122.

***argentina.*** *Ponera distinguenda* var. *argentina* Santschi, 1922d: 242 (w.) ARGENTINA. Combination in *Hypoponera:* Kempf, 1972a: 121. Raised to species: Santschi, 1925b: 8; Santschi, 1925e: 154.

***argonautorum.*** *Ponera argonautorum* Arnol'di, 1932b: 66, figs. 6, 7 (q.) RUSSIA. Combination in *Hypoponera:* **new combination.**

#***assmuthi.*** *Ponera abeillei* r. *assmuthi* Forel, 1905b: 182 (w.) INDIA. Combination in *Hypoponera:* **new combination.**

***beebei.*** *Ponera beebei* Wheeler, W.M. 1924a: 107, fig. 21 (q.) GALAPAGOS IS. Combination in *Hypoponera:* Kempf, 1972a: 121.

#***bellicosa.*** *Ponera indigens* var. *bellicosa* Forel, 1895h: 487 (w.q.m.) MADAGASCAR. Combination in *Hypoponera:* **new combination.**

***biroi.*** *Ponera biroi* Emery, 1900c: 318, pl. 8, fig. 9 (w.q.) NEW GUINEA. Combination in *P.* (*Hypoponera*): Santschi, 1938b: 79; in *Hypoponera:* Taylor, 1967a: 12. See also: Wilson, 1958d: 327.

***boerorum.*** *Ponera coarctata* r. *boerorum* Forel, 1901f: 339 (w.) SOUTH AFRICA. Combination in *Hypoponera:* Taylor, 1967a: 12. Raised to species: Santschi, 1938b: 78.

***bondroiti.*** *Ponera ergatandria* subsp. *bondroiti* Forel, 1911c: 285 (w.q.m.) BELGIUM. Combination in *Hypoponera:* Onoyama, 1989b: 2. Raised to species: Santschi, 1937h: 364.

***breviceps.*** *Ponera breviceps* Bernard, 1953b: 202, fig. 3 (q.m.) GUINEA. Combination in *Hypoponera:* **new combination.**

***brevis.*** *Ponera brevis* Santschi, 1921c: 113, fig. 1 (w.) BENIN. Combination in *Hypoponera:* **new combination.**

***bugnioni.*** *Ponera bugnioni* Forel, 1912n: 52 (q.) SUMATRA. Combination in *Hypoponera:* **new combination.**

#***bulawayensis.*** *Ponera ragusai* st. *bulawayensis* Forel, 1913j: 203 (w.) ZIMBABWE. Combination in *Hypoponera:* **new combination.** See also: Arnold, 1915: 80.

***butteli.*** *Ponera butteli* Forel, 1913k: 12 (w.) SUMATRA. Combination in *Hypoponera:* **new combination.**

***cammerunensis.*** *Ponera abeillei* var. *cammerunensis* Santschi, 1914d: 321 (w.) CAMEROUN. Combination in *P.* (*Hypoponera*): Santschi, 1938b: 79; in *Hypoponera,* and raised to species: Taylor, 1967a: 12.

#***cauta.*** *Ponera trigona* var. *cauta* Forel, 1912c: 40 (w.) BRAZIL. Combination in *Hypoponera:* Kempf, 1972a: 124.

***ceylonensis.*** *Ponera ceylonensis* Mayr, 1897: 422 (w.) SRI LANKA. Combination in *Hypoponera:* Taylor, 1967a: 12. See also: Bingham, 1903: 90.

*chilensis. Ponera opaciceps* r. *chilensis* Forel, 1914d: 264 (w.) CHILE. Combination in *Hypoponera:* Kempf, 1972a: 123. Junior synonym of *opacior:* Snelling & Hunt, 1976: 66.

***clavatula.*** *Ponera clavatula* Emery, 1906c: 116, fig. 1 (w.) ARGENTINA. Santschi, 1929d: 276 (q.). Combination in *Hypoponera:* Kempf, 1972a: 121.

***coeca.*** *Ponera coeca* Santschi, 1914d: 322, fig. 9 (w.q.) CAMEROUN. Combination in *P.* (*Hypoponera*): Santschi, 1938b: 79; in *Hypoponera:* **new combination.**

***cognata.*** *Ponera ergatandria* st. *cognata* Santschi, 1912b: 153 (w.) ANGOLA. [Unresolved junior primary homonym of *Ponera cognata* Emery, 1896g: 56 (now in *Pachycondyla*).] Bernard, 1953b: 202 (q.). Combination in *Hypoponera:* **new combination.** Raised to species: Bernard, 1953b: 202. See also: Arnold, 1915: 81.

***collegiana.*** *Ponera collegiana* Santschi, 1925b: 9 (w.) BRAZIL. Combination in *P.* (*Hypoponera*): Santschi, 1938b: 79; in *Hypoponera:* Kempf, 1972a: 121. Current subspecies: nominal plus *paranensis*

***confinis.*** *Ponera confinis* Roger, 1860: 284 (w.) SRI LANKA. Forel, 1900d: 327 (q.); Imai, Kubota, *et al.* 1985: 47 (k.). Combination in *Hypoponera,* and senior synonym of *nautarum:* Wilson & Taylor, 1967: 26. Current subspecies: nominal plus *aitkenii,* *epinotalis, javana, singaporensis, wroughtonii.* See also: Bingham, 1903: 91.

***congrua.*** *Ponera congrua* Wheeler, W.M. 1934d: 142 (w.q.) AUSTRALIA. Combination in *Hypoponera:* Taylor & Brown, D.R. 1985: 31.

***convexiuscula.*** *Ponera trigona* r. *convexiuscula* Forel, 1900b: 60 (w.) AUSTRALIA. Combination in *Hypoponera:* Taylor & Brown, D.R. 1985: 31. Raised to species: Santschi, 1938b: 79.

***creola.*** *Ponera creola* Menozzi, 1931b: 263, fig. 3 (w.) COSTA RICA. Combination in *Hypoponera:* Kempf, 1972a: 121.

#***cubana.*** *Ponera opaciceps* var. *cubana* Santschi, 1930c: 76 (w.) CUBA. Combination in *Hypoponera:* Kempf, 1972a: 123.

*decipiens. Ponera gleadowii* r. *decipiens* Forel, 1899a: 118 (w.) HAWAII. Junior synonym of *gleadowi:* Wilson, 1958d: 328.

***decora.*** *Ponera decora* Clark, 1934c: 56, pl. 4, figs. 12, 13 (w.q.) AUSTRALIA. Combination in *Hypoponera:* Taylor & Brown, D.R. 1985: 31.

#***devota.*** *Ponera spei* st. *devota* Santschi, 1914e: 5 (w.) SOUTH AFRICA. Combination in *Hypoponera:* **new combination**.

***dideroti.*** *Ponera dideroti* Forel, 1913j: 203 (w.q.m.) SOUTH AFRICA. Combination in *Hypoponera:* **new combination**. See also: Arnold, 1915: 78.

#***dispar.*** *Ponera distinguenda* var. *dispar* Santschi, 1925b: 8 (w.) BRAZIL. Combination in *Hypoponera:* Kempf, 1972a: 122.

***distinguenda.*** *Ponera distinguenda* Emery, 1890a: 61 (w.) VENEZUELA. Santschi, 1929d: 277 (q.). Combination in *Hypoponera:* Kempf, 1972a: 121. Current subspecies: nominal plus *dispar, histrio, inexpedita, vana.*

***dulcis.*** *Ponera dulcis* Forel, 1907a: 2 (w.) TANZANIA. Arnold, 1915: 82 (q.). Combination in *Hypoponera:* **new combination**. Current subspecies: nominal plus *aemula, uncta.*

#***durbanensis.*** *Ponera sulcatinasis* r. *durbanensis* Forel, 1914d: 213 (w.) SOUTH AFRICA. Arnold, 1926: 206 (q.). Combination in *Hypoponera:* **new combination**.

***eduardi.*** *Ponera eduardi* Forel, 1894d: 15 (w.) ALGERIA. Emery, 1895b: 61 (q.m.); Forel, 1904f: 421 (m. ergatoid); Emery, 1909: 373 (q.m.); Wheeler, G.C. & Wheeler, J. 1990a: 458 (l.). Combination in *Hypoponera:* Taylor, 1967a: 12. Subspecies of *confinis:* Emery, 1895b: 64. Revived status as species: Forel, 1904f: 421. Senior synonym of *antipodum:* Brown, 1958h: 23. See also: Emery, 1916b: 109; Wheeler, W.M. 1937c: 59; Brown, 1958h: 24; Baroni Urbani, 1971c: 15; Kutter, 1977c: 27; Atanassov & Dlussky, 1992: 70.

***elliptica.*** *Ponera truncata* var. *elliptica* Forel, 1900b: 62 (w.) AUSTRALIA. Combination in *Hypoponera:* Taylor & Brown, D.R. 1985: 31. Subspecies of *truncata:* Emery, 1914f: 397. Raised to species: Wilson, 1958d: 345.

***emeryi.*** *Ponera emergi* Donisthorpe, 1943d: 443 (m.) NEW GUINEA. Justified emendation of spelling to *emeryi:* Wilson, 1958d: 328. Combination in *Hypoponera:* **new combination**.

#***epinotalis.*** *Ponera confinis* subsp. *epinotalis* Viehmeyer, 1916a: 115 (w.) SINGAPORE. Combination in *Hypoponera:* **new combination**.

***ergatandria.*** *Ponera ergatandria* Forel, 1893g: 365 (w.q.m.) WEST INDIES. Combination in *Hypoponera:* Kempf, 1972a: 122. See also: Wheeler, W.M. 1937c: 59.

***eutrepta.*** *Ponera eutrepta* Wilson, 1958d: 344. Replacement name for *Ponera biroi* subsp. *rugosa* Mann, 1921: 415. [Junior primary homonym of *Ponera rugosa* Le Guillou, 1842: 318 (now in *Diacamma*).] Combination in *Hypoponera:* **new combination**.

*exacta. Ponera punctatissima* var. *exacta* Santschi, 1923a: 134 (w.) MOROCCO. Junior synonym of *punctatissima:* Atanassov & Dlussky, 1992: 71.

#***excelsior.*** *Ponera sakalava* r. *excelsior* Forel, 1892l: 248 (w.) MADAGASCAR. Combination in *Hypoponera:* **new combination**. Raised to species: Dalla Torre, 1893: 39. Subspecies of *sakalava:* Wheeler, W.M. 1922a: 1010.

***exoecata.*** *Ponera exoecata* Wheeler, W.M. 1928c: 7 (w.) CHINA. Combination in *Hypoponera:* Ogata, 1987: 123. See also: Onoyama, 1989b: 4.

***faceta.*** *Ponera faceta* Menozzi, 1931b: 264, fig. 4 (w.) COSTA RICA. Combination in *Hypoponera:* Kempf, 1972a: 122.

#***fallax.*** *Ponera inexorata* subsp. *fallax* Forel, 1909a: 245 (w.) GUATEMALA. Combination in *Hypoponera:* Kempf, 1972a: 122.

#***famini.*** *Ponera fiebrigi* var. *famini* Forel, 1912c: 41 (w.) COLOMBIA. Combination in *Hypoponera:* Kempf, 1972a: 122.

***fenestralis.*** *Ponera fenestralis* Gallardo, 1918b: 82, fig. 20 (q.) ARGENTINA. Combination in *Hypoponera:* Kempf, 1972a: 122.

#***fidelis.*** *Ponera spei* var. *fidelis* Santschi, 1926b: 207, fig. 1 (w.q.) SOUTH AFRICA. Combination in *Hypoponera:* **new combination**.

***fiebrigi.*** *Ponera fiebrigi* Forel, 1908c: 343 (w.q.m.) PARAGUAY. Combination in *Hypoponera:* Kempf, 1972a: 122. Current subspecies: nominal plus *antoniensis, famini, transiens.*

***foeda.*** *Ponera foeda* Forel, 1893: 364 (w.q.) WEST INDIES. Combination in *Hypoponera:* Kempf, 1972a: 122. Current subspecies: nominal plus *saroltae.*

***foreli.*** *Ponera foreli* Mayr, 1887: 534 (w.q.) BRAZIL. Combination in *Hypoponera:* Kempf, 1972a: 122. See also: Kempf, 1962b: 12.

*formosae. Ponera japonica* r. *formosae* Forel, 1913f: 186 (w.) TAIWAN. Junior synonym of *gleadowi:* Taylor, 1967a: 76.

#***fugitans.*** *Ponera schmalzi* var. *fugitans* Forel, 1912c: 40 (w.) BRAZIL. Combination in *Hypoponera:* Kempf, 1972a: 123.

*fuscula. Ponera pallidula* var. *fuscula* Emery, 1901h: 158 (w.q.) NEW GUINEA. Junior synonym of *pallidula:* Wilson, 1958d: 332.

#***gaigei.*** *Ponera opaciceps* var. *gaigei* Forel, 1914: 615 (w.) COLOMBIA. Combination in *Hypoponera:* Kempf, 1972a: 123.

#***gibbinota.*** *Ponera opaciceps* var. *gibbinota* Forel, 1912n: 52 (w.) GREAT BRITAIN. Combination in *Hypoponera:* **new combination**.

*glabrata. Ponera punctatissima* r. *jugata* var. *glabrata* Forel, 1895c: 245 (w.) MADAGASCAR. Unavailable name.

***gleadowi.*** *Ponera gleadowi* Forel, in Emery, 1895b: 60 (footnote and in key) (w.) INDIA. Forel, 1900d: 327 (q.m.). [*Ponera gleadowi* Emery, 1893f: 242; *nomen nudum*, attributed to Forel.] Combination in *Hypoponera:* Taylor, 1967a: 12. Senior synonym of *decipiens, kalakauae* and *mumfordi:* Wilson, 1958d: 328; of *formosae:* Taylor, 1967a: 76; of *oblongiceps:* Taylor, 1968a: 65. See also: Bingham, 1903: 91; Onoyama, 1989b: 5.

***gracilicornis.*** *Ponera gracilicornis* Menozzi, 1931d: 197 (w.q.) COSTA RICA. [Unresolved junior primary homonym of **Ponera gracilicornis* Mayr, 1868c: 72.] Combination in *Hypoponera:* Kempf, 1972a: 122.

***grandidieri.*** *Ponera grandidieri* Santschi, 1921c: 114, fig. 1 (w.) MADAGASCAR. Combination in *Hypoponera:* **new combination**.

*gyptis. Ponera gyptis* Santschi, 1921b: 435. Unnecessary (second) replacement name for *parva* Bondroit, 1918: 85. [Junior primary homonym of *parva* Forel, 1909a: 244.] Junior synonym of *massiliensis:* **new synonymy**.

***herbertonensis.*** *Ponera pruinosa* r. *herbertonensis* Forel, 1915b: 24 (w.q.m.) AUSTRALIA. Combination in *Hypoponera,* and raised to species: Taylor & Brown, D.R. 1985: 31.

#***histrio.*** *Ponera distinguenda* var. *histrio* Forel, 1912c: 40 (w.) BRAZIL. Combination in *Hypoponera:* Kempf, 1972a: 122.

***idelettae.*** *Ponera idelettae* Santschi, 1923g: 1258 (w.) BRAZIL. Combination in *Hypoponera:* Kempf, 1972a: 122. See also: Kempf, 1962b: 13.

***ignigera.*** *Ponera ignigera* Menozzi, 1927c: 271 (w.q.) COSTA RICA. Combination in *Hypoponera:* Kempf, 1972a: 122.

***iheringi.*** *Ponera iheringi* Forel, 1908c: 344 (w.) BRAZIL. Forel, 1913: 204 (q.m.); Wheeler, G.C. & Wheeler, J. 1971b: 1211 (l.). Combination in *Hypoponera:* Kempf, 1972a: 122. [This name frequently misspelled as *jheringi,* for example by Emery, 1911d: 92; Kempf, 1972a: 122.]

***imatongica.*** *Ponera coarctata* subsp. *imatongica* Weber, 1942a: 44, fig. 3 (w.) SUDAN. Combination in *Hypoponera,* and raised to species: Taylor, 1967a: 12.

***inaudax.*** *Ponera inaudax* Santschi, 1919h: 83 (w.) ZAIRE. Combination in *P.* (*Hypoponera*): Santschi, 1938b: 79; in *Hypoponera:* **new combination**.

***incisa.*** *Ponera incisa* Santschi, 1914d: 320, fig. 8 (w.) NIGERIA. Combination in *Hypoponera:* **new combination**.

#***indifferens.*** *Ponera punctatissima* r. *indifferens* Forel, 1895c: 245 (w.q.) MADAGASCAR. Combination in *Hypoponera:* **new combination**.

***indigens.*** *Ponera indigens* Forel, 1895c: 246 (w., not q., see Wheeler, W.M. 1922a: 1009) MADAGASCAR. Combination in *Hypoponera:* **new combination**. Current subspecies: nominal plus *bellicosa.*

***inexorata.*** *Ponera inexorata* Wheeler, W.M. 1903b: 94, fig. 2 (w.q.) U.S.A. Combination in *Hypoponera:* Taylor, 1968a: 65. Current subspecies: nominal plus *fallax.*

#***inexpedita.*** *Ponera inexorata* var. *inexpedita* Forel, 1911c: 285 (w.) BRAZIL. Combination in

*Hypoponera:* Kempf, 1972a: 122. Stirps of *distinguenda:* Santschi, 1923c: 247; Santschi, 1929d: 277.

***intermedia.*** *Ponera (Hypoponera) intermedia* Bernard, 1953b: 206, fig. 3 (w.) GUINEA. [Unresolved junior primary homonym of *Ponera piliventris* var. *intermedia* Forel, 1900b: 63 (now in *Pachycondyla*).] Combination in *Hypoponera:* **new combination**.

***#jamaicensis.*** *Ponera opaciceps* var. *jamaicensis* Aguayo, 1932: 216 (w.) JAMAICA. Combination in *Hypoponera:* Kempf, 1972a: 123. Subspecies of *opaciceps:* Wheeler, W.M. 1937b: 445.

***#javana.*** *Ponera confinis* var. *javana* Forel, 1905c: 6 (w.q.) JAVA. Combination in *Hypoponera:* **new combination**.

***jeanneli.*** *Ponera jeanneli* Santschi, 1935b: 262, fig. 3 (w.) KENYA. Combination in *P. (Hypoponera):* Santschi, 1938b: 79; in *Hypoponera:* **new combination**.

*jheringi;* see under ***iheringi,*** above.

***johannae.*** *Ponera johannae* Forel, 1891b: 220, pl. 5, fig. 11 (w.) MADAGASCAR. Emery, 1894a: 68 (q.). Combination in *Hypoponera:* **new combination**.

***#jugata.*** *Ponera punctatissima* r. *jugata* Forel, 1892l: 251 (q.) MADAGASCAR. Combination in *Hypoponera:* **new combination**. Raised to species: Dalla Torre, 1893: 39. Subspecies of *punctatissima:* Emery, 1899f: 268.

*kalakauae. Ponera kalakauae* Forel, 1899a: 115 (w.q.m.) HAWAII. Junior synonym of *gleadowi:* Wilson, 1958d: 328.

***lamellosa.*** *Ponera lamellosa* Forel, 1907a: 4 (w.) WEST MALAYSIA. [Unresolved junior primary homonym of *Ponera lamellosa* Roger, 1860: 295 (now in *Platythyrea*).] Combination in *Hypoponera:* **new combination**.

***lamottei.*** *Ponera lamottei* Bernard, 1953b: 204, fig. 3 (w.) GUINEA. Combination in *Hypoponera:* **new combination**.

***lea.*** *Ponera lea* Santschi, 1937g: 73, fig. 2 (w.) SIERRA LEONE. Combination in *Hypoponera:* **new combination**.

***leninei.*** *Ponera leninei* Santschi, 1925b: 7 (w.) BRAZIL. Santschi, 1929d: 276 (q.). Combination in *Hypoponera:* Kempf, 1972a: 123.

*lesnei. Ponera lesnei* Bondroit, 1916: 212, fig. (w.) FRANCE. Junior synonym of *ragusai:* Bernard, 1967: 88.

***longiceps.*** *Ponera longiceps* Forel, 1913k: 13 (w.) JAVA. Combination in *Hypoponera:* **new combination**.

*longiceps. Ponera longiceps* Menozzi, 1931b: 262, fig. 2 (w.) COSTA RICA. [Junior primary homonym of *longiceps* Forel, above.] Replacement name: *menozzii* Santschi, 1932e: 412.

***lotti.*** *Ponera lotti* Weber, 1942a: 45, fig. 5 (w.) SUDAN. Combination in *Hypoponera:* **new combination**.

***ludovicae.*** *Ponera ludovicae* Forel, 1892l: 249 (w.) MADAGASCAR. Combination in *Hypoponera:* **new combination**.

***lumpurensis.*** *Ponera lumpurensis* Forel, 1907a: 3 (w.) WEST MALAYSIA. Combination in *Hypoponera:* **new combination**. Current subspecies: nominal plus *slamatana*.

***mackayensis.*** *Ponera coarctata* var. *mackayensis* Forel, 1900b: 61 (w.) AUSTRALIA. Combination in *Hypoponera,* and raised to species: Taylor, 1967a: 12.

***macradelphe.*** *Ponera macradelphe* Wilson, 1958d: 330 (w.) NEW GUINEA. Combination in *Hypoponera:* **new combination**.

***madecassa.*** *Ponera (Hypoponera) madecassa* Santschi, 1938b: 79 (w.) MADAGASCAR. Combination in *Hypoponera:* **new combination**.

***malayana.*** *Ponera malayana* Wheeler, W.M. 1929g: 35 (w.m.) WEST MALAYSIA. Combination in *Hypoponera:* **new combination**.

***mandibularis.*** *Ponera mandibularis* Bernard, 1953b: 205, fig. 3 (w.q.) GUINEA. Combination in *Hypoponera:* **new combination**.

*massiliensis. Ponera massiliensis* Bondroit, 1920a: 158. Replacement name for *parva* Bondroit, 1918: 85. [Junior primary homonym of *parva* Forel, 1909a: 244.] Senior synonym of *gyptis* (unnecessary second replacement name): **new synonymy**. Junior synonym of *ragusai:* Bernard, 1967: 88.

***menozzii.*** *Ponera menozzii* Santschi, 1932e: 412. Replacement name for *longiceps* Menozzi, 1931b: 262. [Junior primary homonym of *longiceps* Forel, 1913k: 13.] Combination in *Hypoponera:* Kempf, 1972a: 123.

***mesoepinotalis.*** *Ponera mesoepinotalis* Weber, 1942a: 44, fig. 4 (w.) SUDAN. Combination in *Hypoponera:* **new combination**.

***mina.*** *Ponera mina* Wheeler, W.M. 1927i: 131, fig. 2 (w.q.m.) AUSTRALIA. Combination in *Hypoponera:* Taylor & Brown, D.R. 1985: 31. Junior synonym of *gleadowi:* Wilson, 1958d: 329. Revived from synonymy: Taylor & Brown, D.R. 1985: 31. See also: Wheeler, W.M. 1937c: 59.

*mocsaryi. Ponera mocsaryi* Emery, 1900c: 320, pl. 8, figs. 15, 16 (w.) NEW GUINEA. Junior synonym of *pruinosa:* Wilson, 1958d: 335.

***monticola.*** *Ponera monticola* Mann, 1921: 418, fig. 5 (w.) FIJI IS. Santschi, 1928c: 67 (q.). Combination in *Hypoponera:* Taylor, 1967a: 12.

*mumfordi. Ponera mumfordi* Wheeler, W.M. 1933f: 141 (w.) MARQUESAS IS. Wheeler, W.M. 1936f: 4 (q.). Junior synonym of *gleadowi:* Wilson, 1958d: 328.

***muscicola.*** *Ponera muscicola* Weber, 1942a: 45, fig. 13 (q.) SUDAN. Combination in *Hypoponera:* **new combination**.

***myrmicariae.*** *Ponera myrmicariae* Wasmann, 1918b: 144 (w.) CAMEROUN. Combination in *P. (Hypoponera):* Santschi, 1938b: 79; in *Hypoponera:* **new combination**.

***natalensis.*** *Ponera coarctata* st. *natalensis* Santschi, 1914e: 7 (w.) SOUTH AFRICA. Combination in *P. (Hypoponera):* Santschi, 1938b: 79; in *Hypoponera:* Taylor, 1967a: 12. Raised to species: Santschi, 1938b: 79; Taylor, 1967a: 12.

*nautarum. Ponera convexiuscula* var. *nautarum* Santschi, 1938b: 79 (w.q.) SAMOA. [First available use of *Ponera trigona* st. *convexiuscula* var. *nautarum* Santschi, 1920g: 180; unavailable name.] Junior synonym of *confinis:* Wilson & Taylor, 1967: 26.

***neglecta.*** *Ponera neglecta* Santschi, 1923g: 1257 (w.) BRAZIL. Combination in *Hypoponera:* Kempf, 1972a: 123.

***nippona.*** *Ponera nippona* Santschi, 1937h: 364, figs. 3, 4 (w.) JAPAN. Onoyama, 1989b: 6 (q.m.). Combination in *Hypoponera:* Ogata, 1987: 123.

***nitidula.*** *Ponera nitidula* Emery, 1890b: 43 (w.) COSTA RICA. [Also described as new by Emery, 1894k: 49.] Wheeler, G.C. & Wheeler, J. 1964b: 453 (l.). Combination in *Hypoponera:* Kempf, 1972a: 123.

***nivariana.*** *Ponera nivariana* Santschi, 1908: 518 (w.) CANARY IS. Combination in *Hypoponera:* Hohmann, La Roche, *et al.,* 1993: 145.

*oblongiceps. Ponera oblongiceps* Smith, M.R. 1939a: 76, figs. 1–3 (w.q.m.) U.S.A. Junior synonym of *gleadowi:* Taylor, 1968a: 65.

***occidentalis.*** *Ponera (Hypoponera) occidentalis* Bernard, 1953b: 205, fig. 3 (w.) GUINEA. Combination in *Hypoponera:* **new combination**.

***opaciceps.*** *Ponera opaciceps* Mayr, 1887: 536 (w.q.) BRAZIL. Smith, M.R. 1929: 545 (m.); Wheeler, G.C. & Wheeler, J. 1964b: 453 (l.). Combination in *Hypoponera:* Taylor, 1967a: 11. Senior synonym of *perkinsi* (and its junior synonym *andrei*): Wilson & Taylor, 1967: 28. Current subspecies: nominal plus *cubana, gaigei, gibbinota, jamaicensis, pampana, postangustata*. See also: Smith, M.R. 1936: 428; Wheeler, W.M. 1937c: 59; Kempf, 1962b: 7.

***opacior.*** *Ponera trigona* var. *opacior* Forel, 1893g: 363 (w.q.) WEST INDIES. Emery, 1895c: 268 (m.); Wheeler, G.C. & Wheeler, J. 1964b: 454 (l.). Combination in *Hypoponera:* Taylor, 1968a: 65. Subspecies of *trigona:* Emery, 1906c: 116. Raised to species: Kempf, 1962b: 10. Senior synonym of *chilensis:* Snelling & Hunt, 1976: 66.

***orba.*** *Ponera orba* Emery, 1915g: 7, fig. 4 (w.) ETHIOPIA. Combination in *P. (Hypoponera):* Santschi, 1938b: 79; in *Hypoponera:* **new combination**.

***pallidula.*** *Ponera pallidula* Emery, 1900c: 320, pl. 8, figs. 17, 18 (w.) NEW GUINEA. Mann, 1919: 295 (q.). Combination in *Hypoponera:* **new combination**. Senior synonym of *fuscula:* Wilson, 1958d: 332.

***#pampana.*** *Ponera opaciceps* var. *pampana* Santschi, 1925e: 153 (w.q.m.) ARGENTINA. Combination in *Hypoponera:* Kempf, 1972a: 123.

***papuana.*** *Ponera papuana* Emery, 1900c: 319, pl. 8, figs. 10, 11 (w.q.) NEW GUINEA. Combination in *P. (Hypoponera):* Santschi, 1938b: 79; in *Hypoponera:* **new combination**. See also: Wilson, 1958d: 332.

***#paranensis.*** *Ponera collegiana* var. *paranensis* Santschi, 1925b: 9 (w.) BRAZIL. Combination in *Hypoponera:* Kempf, 1972a: 121.

***parva.*** *Ponera parva* Forel, 1909a: 244 (w.) GUATEMALA. Combination in *Hypoponera:* Kempf, 1972a: 123.

*parva. Ponera parva* Bondroit, 1918: 85 (w.) FRANCE. [Junior primary homonym of *parva* Forel, above.] Replacement name: *massiliensis:* Bondroit, 1920a: 158.

***#paulina.*** *Ponera schmaltzi* var. *paulina* Forel, 1913l: 205 (q.m.) BRAZIL. Combination in *Hypoponera:* Kempf, 1972a: 123.

*perkinsi. Ponera perkinsi* Forel, 1899a: 117 (w.q.m.) HAWAII. Senior synonym of *andrei:* Wilson, 1958d: 334. Junior synonym of *opaciceps:* Wilson & Taylor, 1967: 28. See also: Wheeler, W.M. 1909c: 271.

*HYPOPONERA (continued)*

***perplexa.*** *Ponera perplexa* Mann, 1922: 8 (w.) HONDURAS. Combination in *Hypoponera:* Kempf, 1972a: 123.

***petiolata.*** *Ponera petiolata* Bernard, 1953b: 203, fig. 3 (w.) GUINEA. Combination in *Hypoponera:* **new combination**.

***petri.*** *Ponera ergatandria* r. *petri* Forel, 1916: 397 (w.) ZAIRE. Combination in *Hypoponera:* **new combination**. Raised to species: Santschi, 1938b: 78.

***pia.*** *Ponera pia* Forel, 1901b: 9 (q.) NEW GUINEA. Combination in *Hypoponera:* **new combination**.

#***postangustata.*** *Ponera opaciceps* var. *postangustata* Forel, 1908c: 343 (w.) PARAGUAY. Combination in *Hypoponera:* Kempf, 1972a: 123.

***pruinosa.*** *Ponera pruinosa* Emery, 1900c: 319, pl. 8, figs. 13, 14 (w.) NEW GUINEA. Viehmeyer, 1916b: 284 (q.m.); Imai, Brown, *et al.* 1984: 67 (k.). Combination in *Hypoponera:* Imai, Brown *et al.* 1984: 67. Senior synonym of *mocsaryi:* Wilson, 1958d: 335. See also: Mann, 1919: 294.

***punctatissima.*** *Ponera punctatissima* Roger, 1859: 246, pl. 7, fig. 7 (w.q.) GERMANY. Forel, 1874: 92 (m.). Combination in *Hypoponera:* Taylor, 1967a: 12. Senior synonym of *androgyna:* Emery & Forel, 1879: 455; of *tarda:* Dalla Torre, 1893: 41; of *schauinslandi:* Wilson & Taylor, 1967: 28; of *exacta:* Atanassov & Dlussky, 1992: 71. Current subspecies: nominal plus *indifferens, jugata.* See also: Emery, 1916b: 110; Wheeler, W.M. 1937c: 59; Collingwood, 1979: 30.

***punctiventris.*** *Ponera punctiventris* Emery, 1901h: 157, fig. (w.) NEW GUINEA. Combination in *Hypoponera:* **new combination**.

***pygmaea.*** *Ponera pygmaea* Forel, 1907a: 4 (w.) WEST MALAYSIA. Combination in *Hypoponera:* **new combination**.

***queenslandensis.*** *Ponera queenslandensis* Forel, 1900b: 61 (w.q.) AUSTRALIA. Combination in *Hypoponera:* Taylor, 1967a: 12.

***ragusai.*** *Ponera ragusai* Emery, 1894h: 28 (w.) SICILY. [Also described as new by Emery, 1895b: 65.] Santschi, 1907: 313 (q.m.); Santschi, 1908: 519 (m.); Emery, 1916b: 109 (q.). Combination in *Hypoponera:* Baroni Urbani, 1971c: 18. Senior synonym of *lesnei* and *massiliensis* (and its junior synonym *gyptis*): Bernard, 1967: 88. Current subspecies: nominal plus *bulawayensis, santschii, sordida.* See also: Wheeler, W.M. 1937c: 59.

***rectidens.*** *Ponera rectidens* Clark, 1934c: 57, pl. 4, fig. 14 (w.) AUSTRALIA. Combination in *Hypoponera:* Taylor & Brown, D.R. 1985: 31.

***reichenspergeri.*** *Ponera reichenspergeri* Santschi, 1923g: 1255 (w.) BRAZIL. Combination in *P.* (*Hypoponera*): Santschi, 1938b: 79; in *Hypoponera:* Kempf, 1972a: 123.

***rothkirchi.*** *Ponera rothkirchi* Wasmann, 1918b: 145 (w.) CAMEROUN. Menozzi, 1944: 434 (q.). Combination in *Hypoponera:* **new combination**.

*rugosa. Ponera biroi* subsp. *rugosa* Mann, 1921: 415 (w.) FIJI IS. [Junior primary homonym of *Ponera rugosa* Le Guillou, 1842: 318 (now in *Diacamma*).] Replacement name: *eutrepta* Wilson, 1958d: 344.

***sabronae.*** *Ponera sabronae* Donisthorpe, 1941c: 130 (w.m.) NEW GUINEA. Combination in *Hypoponera:* **new combination**. See also: Wilson, 1958d: 338.

***sakalava.*** *Ponera sakalava* Forel, 1891b: 124, pl. 4, fig. 3 (q.) MADAGASCAR. Forel, 1891b: 222 (w.). Combination in *Hypoponera:* **new combination**. Current subspecies: nominal plus *excelsior.*

*sancta. Ponera spei* r. *devota* var. *sancta* Forel, 1914d: 213 (w.) SOUTH AFRICA. Unavailable name.

#***santschii.*** *Ponera ragusai* var. *santschii* Emery, 1909c: 371, fig. 9 (w.q.m.) ALGERIA. Combination in *Hypoponera:* **new combination**.

#***saroltae.*** *Ponera foeda* r. *saroltae* Forel, 1912c: 41 (w.) BRAZIL. Combination in *Hypoponera:* Kempf, 1972a: 122.

***sauteri.*** *Hypoponera sauteri* Onoyama, 1989b: 7 (w.q.) TAIWAN. [First available use of *Ponera gleadowi* subsp. *decipiens* var. *sauteri* Forel, 1912a: 48; unavailable name.]

*schauinslandi. Ponera punctatissima* subsp. *schauinslandi* Emery, 1899d: 439 (q.) HAWAII. Junior synonym of *punctatissima:* Wilson & Taylor, 1967: 28.

***schmalzi.*** *Ponera schmalzi* Emery, 1896g: 57 (w.) BRAZIL. Combination in *Hypoponera:* Kempf, 1972a: 123. Current subspecies: nominal plus *fugitans, paulina.*

***schwebeli.*** *Ponera parva* var. *schwebeli* Forel, 1913l: 205 (q.) BRAZIL. Santschi, 1923g: 1256 (w.). Combination in *Hypoponera:* Kempf, 1972a: 124. Raised to species: Santschi, 1923g: 1256.

***scitula.*** *Ponera scitula* Clark, 1934c: 55, pl. 4, figs. 9–11 (w.q.) AUSTRALIA. Combination in *Hypoponera:* Taylor, 1964: 140.

***silvestrii.*** *Ponera silvestrii* Donisthorpe, 1947a: 284 (w.q.) CHINA. Combination in *Hypoponera:* **new combination**.

#***singaporensis.*** *Ponera confinis* var. *singaporensis* Viehmeyer, 1916a: 115 (w.q.) SINGAPORE. Combination in *Hypoponera:* **new combination**.

***sinuosa.*** *Ponera sinuosa* Bernard, 1953b: 204, fig. 3 (w.q.) GUINEA. Combination in *Hypoponera:* **new combination**.

***siremps.*** *Ponera siremps* Forel, 1901b: 8 (w.) BISMARCK ARCHIPELAGO. Combination in *Hypoponera:* **new combination**.

#***slamatana.*** *Ponera lumpurensis* r. *slamatana* Forel, 1913k: 10 (q.) SUMATRA. Combination in *Hypoponera:* **new combination**.

#***sordida.*** *Ponera ragusai* var. *sordida* Santschi, 1914b: 54 (w.) KENYA. Combination in *Hypoponera:* **new combination**.

***sororcula.*** *Ponera sororcula* Wilson, 1958d: 338 (w.) NEW GUINEA. Combination in *Hypoponera:* **new combination**.

***spei.*** *Ponera spei* Forel, 1910e: 423 (w.) SOUTH AFRICA. Combination in *Hypoponera:* **new combination**. Current subspecies: nominal plus *devota, fidelis.* See also: Arnold, 1915: 77.

***stoica.*** *Ponera stoica* Santschi, 1912e: 522 (w.) URUGUAY. Combination in *Hypoponera:* Kempf, 1972a: 124.

***sulcatinasis.*** *Ponera sulcatinasis* Santschi, 1914e: 4 (w.) SOUTH AFRICA. Combination in *Hypoponera:* **new combination**. Current subspecies: nominal plus *durbanensis.*

***sulciceps.*** *Ponera sulciceps* Clark, 1928c: 40, fig. (w.) AUSTRALIA. Combination in *Hypoponera:* Taylor & Brown, D.R. 1985: 32.

***sulcitana.*** *Ponera sulcitana* Stefani, 1970: 1, figs. 1–12 (w.q.m.l.) ITALY. Combination in *Hypoponera:* **new combination**.

***taprobanae.*** *Ponera taprobanae* Forel, 1913k: 11 (q.) SRI LANKA. Combination in *Hypoponera:* **new combination**.

*tarda. Ponera tarda* Charsley, 1877: 162 (w.q.) GREAT BRITAIN. Junior synonym of *punctatissima:* Dalla Torre, 1893: 41.

***tenella.*** *Ponera tenella* Emery, 1901h: 158, fig. (w.q.) NEW GUINEA. Combination in *Hypoponera:* **new combination**. See also: Wilson, 1958d: 340.

***traegaordhi.*** *Ponera traegaordhi* Santschi, 1914e: 6 (w.) SOUTH AFRICA. Combination in *P.* (*Hypoponera*): Santschi, 1938b: 79; in *Hypoponera:* **new combination**.

#***transiens.*** *Ponera fiebrigi* var. *transiens* Santschi, 1925e: 155 (w.) ARGENTINA. Combination in *Hypoponera:* Kempf, 1972a: 122.

***transvaalensis.*** *Ponera transvaalensis* Arnold, 1947: 132, fig. 2 (w.) SOUTH AFRICA. Combination in *Hypoponera:* **new combination**.

***trigona.*** *Ponera punctatissima* var. *trigona* Mayr, 1887: 537 (w.q.) BRAZIL. Combination in *P.* (*Hypoponera*): Santschi, 1938b: 79; in *Hypoponera:* Taylor, 1968a: 65. Subspecies of *confinis:* Emery, 1895b: 64. Raised to species: Forel, 1893g: 364; Emery, 1911d: 92; Gallardo, 1918b: 80. Current subspecies: nominal plus *cauta.* See also: Kempf, 1962b: 10.

***truncata.*** *Ponera truncata* Smith, F. 1860a: 72 (q.) SULAWESI. Forel, 1905c: 6 (w.). Combination in *Hypoponera:* **new combination**. See also: Bingham, 1903: 90.

***turaga.*** *Ponera turaga* Mann, 1921: 416 (w.q.) FIJI IS. Combination in *Hypoponera:* **new combination**.

#***uncta.*** *Ponera dulcis* var. *uncta* Santschi, 1914e: 7 (w.) SOUTH AFRICA. Combination in *Hypoponera:* **new combination**.

***ursa.*** *Ponera ursa* Santschi, 1924b: 197 (w.q.) ZAIRE. Combination in *P.* (*Hypoponera*): Santschi, 1938b: 79; in *Hypoponera:* **new combination**.

***ursoidea.*** *Ponera ursoidea* Bernard, 1953b: 203, fig. 3 (w.q.) GUINEA. Combination in *Hypoponera:* **new combination**.

#***vana.*** *Ponera distinguenda* subsp. *vana* Forel, 1909a: 245 (w.) GUATEMALA. Combination in *Hypoponera:* Kempf, 1972a: 122.

***vanreesi.*** *Ponera vanreesi* Forel, 1912d: 97 (w.) JAVA. Combination in *Hypoponera:* **new combination**.

***vernacula.*** *Ponera vernacula* Kempf, 1962b: 13, figs. 13, 16 (w.) BRAZIL. Combination in *Hypoponera:* Kempf, 1972a: 124.

***villiersi.*** *Ponera* (*Hypoponera*) *villiersi* Bernard, 1953b: 206, fig. 3 (w.) GUINEA. Combination in *Hypoponera:* **new combination**.

***viri.*** *Ponera viri* Santschi, 1923c: 247 (w.) BRAZIL. Combination in *Hypoponera:* Kempf, 1972a: 124.

***vitiensis.*** *Ponera vitiensis* Mann, 1921: 414, fig. 4 (w.q.) FIJI IS. Combination in *Hypoponera:* **new combination**.

***wilsoni.*** *Ponera wilsoni* Santschi, 1925b: 10 (w.) BRAZIL. Combination in *Hypoponera:* Kempf, 1972a: 124.

**#wroughtonii.** *Ponera confinis* var. *wroughtonii* Forel, 1900d: 325 (diagnosis in key) (w.) INDIA. Combination in *Hypoponera:* **new combination.**

**zwaluwenburgi.** *Pseudocryptopone zwaluwenburgi* Wheeler, W.M. 1933g: 14, fig. 5 (w.) HAWAII. Combination in *Ponera:* Wilson, 1957b: 375; in *Hypoponera:* Taylor, 1967a: 12.

*IBEROFORMICA:* see under ***FORMICA.***

*ICOTHORAX:* see under ***LEPTOTHORAX.***

*IDRISELLA:* see under ***PHEIDOLOGETON.***

***ILEMOMYRMEX** (Dominican Amber)

***caecus.** **Ilemomyrmex caecus* Wilson, 1985a: 3, figs. 1, 2 (w.q.) DOMINICAN AMBER.

***IMHOFFIA** (Europe)

***nigra.** **Imhoffia nigra* Heer, 1850: 153, pl. 12, fig. 10 (q.) SWITZERLAND (Miocene).

***pallida.** **Imhoffia pallida* Heer, 1867: 28, pl. 2, fig. 9 (q.?) SWITZERLAND (Miocene)

**INDOMYRMA** (Oriental)

**dasypyx.** *Indomyrma dasypyx* Brown, 1986: 46, figs. 1–21 (w.q.m.) INDIA.

*IRENEA:* see under ***POLYRHACHIS.***

*IRENEELLA:* see under ***RHOPTROMYRMEX.***

*IRENEIDRIS:* see under ***MONOMORIUM.***

**IRENEOPONE** (Mauritius)

**gibber.** *Ireneopone gibber* Donisthorpe, 1946d: 242, figs. 1, 2 (w.) MAURITIUS.

**IRIDOMYRMEX** (Oriental, Indo-Australian, Australasian)

*abbotti* Wheeler, W.M. 1906; see under ***OCHETELLUS.***

**#aeneogaster.** *Iridomyrmex discors* var. *aeneogaster* Wheeler, W.M. 1915g: 811 (w.) AUSTRALIA.

*aeneum. Liometopum aeneum* Mayr, 1862: 704 (q.) AUSTRALIA. Junior synonym of *purpureus:* Mayr, 1876: 81.

**agilis.** *Iridomyrmex agilis* Forel, 1907h: 295 (w.) AUSTRALIA.

*agnatus* Mann, 1921; see under ***PHILIDRIS.***

**albitarsus.** *Iridomyrmex albitarsus* Wheeler, W.M. 1927i: 147, fig. 9 (w.q.m.) AUSTRALIA. See also: Shattuck, 1993b: 1307.

*alticola* Mann, 1921; see under ***PHILIDRIS.***

**anceps.** *Formica anceps* Roger, 1863a: 164 (w.) WEST MALAYSIA. Karavaiev, 1926d: 433 (m.); Imai, Baroni Urbani, *et al.* 1984: 8 (k.). Combination in *Iridomyrmex:* Dalla Torre, 1893: 168. Senior synonym of *excisus:* Dalla Torre, 1893: 168; Forel, 1895e: 469; Emery, 1895k: 475; of *papuana* (and its junior synonym *discoidalis*): Wilson & Taylor, 1967: 78. Material of the unavailable name *neocaledonica* referred here by Taylor, 1987a: 33. Current subspecies: nominal plus *ignobilis, sikkimensis, watsonii.*

*andamanensis* Forel, 1903; see under ***PHILIDRIS.***

**anderseni.** *Iridomyrmex anderseni* Shattuck, 1993b: 1310, figs. 22, 23, 38 (w.) AUSTRALIA.

*angulatus* Emery, 1894; see under ***LINEPITHEMA.***

*anguliceps* Forel, 1901; see under ***ANONYCHOMYRMA.***

**angusticeps.** *Iridomyrmex angusticeps* Forel, 1901b: 19 (w.) BISMARCK ARCHIPELAGO.

*angustus* Stitz, 1911; see under ***ANONYCHOMYRMA.***

**anteroinclinus.** *Iridomyrmex anteroinclinus* Shattuck, 1993b: 1311, figs. 28, 29, 37 (w.) AUSTRALIA.

*arcadius* Forel, 1915; see under ***ANONYCHOMYRMA.***

**argutus.** *Iridomyrmex argutus* Shattuck, 1993b: 1312, figs. 8, 9, 36 (w.) AUSTRALIA.

*arrogans* Santschi, 1929; see under ***LINEPITHEMA.***

*aspidocoptus* Kempf, 1969; see under ***LINEPITHEMA.***

**#azureus.** *Iridomyrmex bicknelli* var. *azureus* Viehmeyer, 1914c: 41 (w.) AUSTRALIA.

*ballaratensis* Forel, 1902; see under ***ANONYCHOMYRMA.***

*batesi* Forel, 1911; see under ***ANONYCHOMYRMA.***

**bicknelli.** *Iridomyrmex bicknelli* Emery, 1898b: 236, figs. 6, 7 (w.) AUSTRALIA. Current subspecies: nominal plus *azureus, brunneus, formosae, luteus, splendidus.*

*bicolor* Forel, 1912; see under ***LINEPITHEMA.***

*bicolor* Donisthorpe, 1933; see under ***DORYMYRMEX.***

*biconvexus* Santschi, 1928; see under ***ANONYCHOMYRMA.***

**bigi.** *Iridomyrmex bigi* Shattuck, 1993a: 119, fig. 9 (w.) AUSTRALIA.

*blackalensis. Iridomyrmex rufoniger* r. *suchieri* var. *blackalensis* Forel, 1915b: 78 (w.q.m.) AUSTRALIA. Unavailable name (Taylor, 1986: 34).

*brasiliensis* Forel, 1908; see under ***FORELIUS.***

***breviantennis.** **Iridomyrmex breviantennis* Théobald, 1937b: 209, pl. 4, figs. 5, 7; pl. 14, fig. 20 (q.) FRANCE (Oligocene). Provisionally retained in *Iridomyrmex:* Shattuck, 1992a: 15.

*breviscapa* Forel, 1914; see under ***LINEPITHEMA.***

**#brunneus.** *Iridomyrmex bicknelli* r. *brunneus* Forel, 1902h: 469 (w.) AUSTRALIA.

*butteli* Forel, 1913; see under ***PHILIDRIS.***

**butteli.** *Turneria butteli* Forel, 1913k: 92 (q.) SUMATRA. Combination in *Iridomyrmex:* Shattuck, 1990: 105. [Not a junior homonym as the name *butteli* Forel, 1913 above, is unavailable.]

**calvus.** *Iridomyrmex calvus* Emery, 1914f: 419, pl. 13, fig. 12 (w.q.) NEW CALEDONIA. See also: Shattuck, 1993b: 1313.

**cappoinclinus.** *Iridomyrmex cappoinclinus* Shattuck, 1993b: 1315, figs. 30, 31, 37 (w.) AUSTRALIA.

*castrae. Iridomyrmex detectus* var. *castrae* Viehmeyer, 1925a: 31 (w.) AUSTRALIA. Subspecies of *purpureus:* Taylor & Brown, D.R. 1985: 102. Junior synonym of *purpureus:* Shattuck, 1993a: 128.

*cellarum. Iridomyrmex cellarum* Stärcke, 1933b: xxvi. *Nomen nudum.*

*centralis. Iridomyrmex rufoniger* subsp. *suchieri* var. *centralis* Forel, 1910b: 52 (w.) AUSTRALIA. Unavailable name (Taylor, 1986: 34).

**cephaloinclinus.** *Iridomyrmex cephaloinclinus* Shattuck, 1993b: 1315, figs. 32, 33, 37 (w.) AUSTRALIA.

**chasei.** *Iridomyrmex chasei* Forel, 1902h: 467 (w.) AUSTRALIA. Forel, 1907h: 288 (m.); Forel, 1910b: 52 (q.). Current subspecies: nominal plus *concolor, yalgooensis.*

*cinerea. Iridomyrmex gracilis* subsp. *papuanus* var. *cinerea* Emery, 1897d: 572 (w.) NEW GUINEA. Unavailable name (Shattuck, 1994: 113).

*clarithorax* Forel, 1902; see under ***OCHETELLUS.***

*clitellarius* Viehmeyer, 1925; see under ***PAPYRIUS.***

**#concolor.** *Iridomyrmex chasei* var. *concolor* Forel, 1902h: 468 (w.) AUSTRALIA. Forel, 1907h: 289 (q.); Forel, 1910b: 53 (m.). Subspecies of *chasei:* Forel, 1907h: 289.

**conifer.** *Iridomyrmex conifer* Forel, 1902h: 463 (w.) AUSTRALIA. Forel, 1907h: 297 (q.); Crawley, 1922c: 26 (m.); Wheeler, G.C. & Wheeler, J. 1974b: 398 (l.). See also: Shattuck, 1994: 103.

*consimilis* Viehmeyer, 1914; see under ***OCHETELLUS.***

*contenta* Viehmeyer, 1913; see under ***ANONYCHOMYRMA.***

**#continentis.** *Iridomyrmex mattiroloi* var. *continentis* Forel, 1907h: 290 (w.q.m.) AUSTRALIA.

**cyaneus.** *Iridomyrmex cyaneus* Wheeler, W.M. 1915g: 812 (w.) AUSTRALIA.

*decipiens* Forel, 1901; see under ***PHILIDRIS.***

*depilis* Forel, 1902; see under ***ANONYCHOMYRMA.***

*detecta. Formica detecta* Smith, F. 1858b: 36 (q.) AUSTRALIA. Wheeler, G.C. & Wheeler, J. 1974b: 398 (l.); Crozier, 1968a: 25 (k.). Combination in *Iridomyrmex:* Dalla Torre, 1893: 168. Junior synonym of *purpureus:* Lowne, 1865a: 275; Mayr, 1876: 81; Taylor & Brown, D.R. 1985: 102.

*dimorphus* Viehmeyer, 1912; see under ***ANONYCHOMYRMA.***

*discoidalis. Prenolepis discoidalis* Donisthorpe, 1947c: 594 (m.) NEW GUINEA. Junior synonym of *papuana:* Smith, M.R. 1957a: 347.

**discors.** *Iridomyrmex discors* Forel, 1902h: 464 (w.) AUSTRALIA. Current subspecies: nominal plus *aeneogaster, obscurior, occipitalis.*

*dispertitus* Forel, 1895; see under ***LINEPITHEMA.***

**#domesticus.** *Iridomyrmex rufoniger* var. *domestica* Forel, 1907h: 291 (w.) AUSTRALIA. Forel, 1910b: 51 (q.m.) (the taxon also described as new in this publication).

*dominicensis* Wheeler, W.M. 1913; see under ***LINEPITHEMA.***

*donisthorpei* Santschi, 1936; see under ***DORYMYRMEX.***

**dromus.** *Iridomyrmex dromus* Clark, 1938: 374, fig. 12 (w.) AUSTRALIA.

**emeryi.** *Iridomyrmex emeryi* Crawley, 1918: 90, fig. (w.q.m.) AUSTRALIA.

*epinotalis* Viehmeyer, 1914; see under ***OCHETELLUS.***

*eteocles. Iridomyrmex gracilis* r. *mayri* var. *eteocles* Forel, 1915b: 80 (w.) AUSTRALIA. Unavailable name (Taylor, 1986: 34).

*excisus. Iridomyrmex excisus* Mayr, 1867a: 77 (w.q.) INDONESIA. Junior synonym of *anceps:* Dalla Torre, 1893: 168; Forel, 1895e: 469; Emery, 1895k: 475.

*exilior. Iridomyrmex discors* subsp. *occipitalis* var. *exilior* Forel, 1907h: 294 (w.q.m.) AUSTRALIA. Unavailable name (Taylor, 1986: 34).

**exsanguis.** *Iridomyrmex exsanguis* Forel, 1907h: 296 (w.q.) AUSTRALIA. Crawley, 1922c: 26 (m.).

**extensus.** *Iridomyrmex extensus* Emery, 1887a: 251, pl. 4, fig. 26 (w.) NEW GUINEA.

*flava. Iridomyrmex rufoniger* subsp. *pallidus* var. *flava* Karavaiev, 1926d: 438 (w.) INDONESIA (Ambon I.). Unavailable name (Shattuck, 1994: 114).

*flavescens* Wheeler, W.M. & Mann, 1914; see under ***LINEPITHEMA.***

*flavus* Mayr, 1865; see under ***PAPYRIUS.***

***IRIDOMYRMEX** (continued)*

***florissantius.** **Iridomyrmex florissantius* Carpenter, 1930: 50, pl. 2, fig. 4; pl. 10, fig. 7 (q.) U.S.A. Provisionally retained in *Iridomyrmex:* Shattuck, 1992a: 15.
*foetans* Clark, 1929; see under ***ANONYCHOMYRMA.***
**#formosae.** *Iridomyrmex bicknelli* r. *formosae* Forel, 1912a: 70 (w.q.) TAIWAN.
*fornicatus* Emery, 1914; see under ***ANONYCHOMYRMA.***
*froggatti* Forel, 1902; see under ***ANONYCHOMYRMA.***
*fuscecens* Wheeler, W.M. 1908; see under ***LINEPITHEMA.***
**#fusciventris.** *Iridomyrmex gracilis* var. *fusciventris* Forel, 1913g: 188 (w.) AUSTRALIA.
*fuscus* Forel, 1901; see under ***PHILIDRIS.***
**galbanus.** *Iridomyrmex galbanus* Shattuck, 1993a: 121, fig. 10 (w.) AUSTRALIA.
***geinitzi.** **Hypoclinea geinitzi* Mayr, 1868c: 58, pl. 3, figs. 47–49 (w.q.m.) BALTIC AMBER (Oligocene). Combination in *Bothriomyrmex:* Dalla Torre, 1893: 170; in *Iridomyrmex:* Wheeler, W.M. 1915h: 86; provisionally retained in *Iridomyrmex:* Shattuck, 1992a: 15.
*giganteus* Donisthorpe, 1943; see under ***ANONYCHOMYRMA.***
*gilberti* Forel, 1902; see under ***ANONYCHOMYRMA.***
**gracilis.** *Formica gracilis* Lowne, 1865a: 280 (w.) AUSTRALIA. [Unresolved junior primary homonym of *Formica gracilis* Fabricius, 1804: 405 (now in *Pseudomyrmex*).] Mayr, 1876: 82 (q.m.); Wheeler, G.C. & Wheeler, J. 1966: 729; Crozier, 1969: 248 (k.). Combination in *Hypoclinea:* Mayr, 1870b: 955; in *Iridomyrmex:* Dalla Torre, 1893: 169. Current subspecies: nominal plus *fusciventris, mayri, minor, rubriceps, spurcus.*
**greensladei.** *Iridomyrmex greensladei* Shattuck, 1993a: 122, fig. 11 (w.q.m.) AUSTRALIA.
**hartmeyeri.** *Iridomyrmex hartmeyeri* Forel, 1907h: 296 (w.) AUSTRALIA.
***haueri.** **Hypoclinea haueri* Mayr, 1867b: 60, pl. 1, fig. 11 (q.) YUGOSLAVIA (Miocene). Combination in *Iridomyrmex:* Dalla Torre, 1893: 169; provisionally retained in *Iridomyrmex:* Shattuck, 1992a: 15.
**hesperus.** *Iridomyrmex hesperus* Shattuck, 1993b: 1316, figs. 24, 25, 38 (w.) AUSTRALIA.
**hispaniolae* Wilson, 1985; see under ***LINEPITHEMA.***
*horni. Camponotus horni* Kirby, W.F. 1896: 205 (w., not q.) AUSTRALIA. Junior synonym of *purpureus:* Clark, 1930c: 20. See also: Shattuck, 1993a: 128.
*hospes* Emery, 1884; see under ***CAMPONOTUS.***
**humiloides* Wilson, 1985; see under ***LINEPITHEMA.***
**#ignobilis.** *Iridomyrmex anceps* subsp. *ignobilis* Mann, 1921: 472 (w.) FIJI IS.
*impotens* Santschi, 1923; see under ***LINEPITHEMA.***
**#incertus.** *Iridomyrmex rufoniger* var. *incertus* Forel, 1902h: 466 (w.) BISMARCK ARCHIPELAGO. [First available use of *Iridomyrmex rufoniger* r. *pallidus* var. *incertus* Forel, 1901b: 22; unavailable name.]
*incisa* Stitz, 1932; see under ***ANONYCHOMYRMA.***
**innocens.** *Iridomyrmex innocens* Forel, 1907h: 292 (w.q.m.) AUSTRALIA.
*itoi* Forel, 1900; see under ***OCHETELLUS.***
*jactans* Forel, 1911; see under ***PHILIDRIS.***
*janthina* Emery, 1914; see under ***OCHETELLUS.***
*keiteli* Forel, 1907; see under ***LINEPITHEMA.***
**krakatauae.** *Iridomyrmex krakatauae* Wheeler, W.M. 1924b: 252 (w.) INDONESIA (Krakatau I.). See also: Shattuck, 1992a: 15.
*laevigatus* Emery, 1895; see under ***PHILIDRIS.***
**latifrons.** *Iridomyrmex latifrons* Karavaiev, 1933a: 305, fig. 1 (w.) JAVA. See also: Shattuck, 1992a: 15.
*leae* Forel, 1913; see under ***DOLEROMYRMA.***
*leucomelas* Emery, 1894; see under ***LINEPITHEMA.***
*linae. Iridomyrmex gracilis* subsp. *rubriceps* var. *linae* Santschi, 1928e: 472 (w.) AUSTRALIA. Unavailable name (Taylor, 1986: 34).
**lividus.** *Iridomyrmex lividus* Shattuck, 1993a: 125, fig. 12 (w.) AUSTRALIA.
*longicapitatus* Donisthorpe, 1947; see under ***ANONYCHOMYRMA.***
*longiceps* Forel, 1907; see under ***ANONYCHOMYRMA.***
**#luteus.** *Iridomyrmex bicknelli* var. *lutea* Forel, 1915b: 77 (w.) AUSTRALIA.
*malandanus* Forel, 1915; see under ***ANONYCHOMYRMA.***
*mamillatus. Acantholepis mamillatus* Lowne, 1865b: 333 (w.) AUSTRALIA. Combination in *Hypoclinea:* Mayr, 1870b: 955. Junior synonym of *rufoniger:* Mayr, 1876: 82.
*mandibularis* Santschi, 1928; see under ***PHILIDRIS.***
***mapesi.** **Iridomyrmex mapesi* Wilson, 1985d: 211, figs. 4, 5 (w.) U.S.A. Provisionally retained in *Iridomyrmex:* Shattuck, 1992a: 16.
**mattiroloi.** *Iridomyrmex mattiroloi* Emery, 1898b: 236, fig. 8 (w.) AUSTRALIA. Crozier, 1969: 249 (k.). Current subspecies: nominal plus *continentis, parcens, splendens.*
**#mayri.** *Iridomyrmex gracilis* r. *mayri* Forel, 1915b: 80 (w.) AUSTRALIA.
*mazaruni* Donisthorpe, 1939; see under ***DOLICHODERUS.***
*mccooki* McCook, 1879; see under ***FORELIUS.***
**meinerti.** *Iridomyrmex meinerti* Forel, 1901b: 22 (q.) BISMARCK ARCHIPELAGO.
*melleus* Wheeler, W.M. 1908; see under ***LINEPITHEMA.***
**#metallescens.** *Iridomyrmex rufoniger* var. *metallescens* Emery, 1893e: 194 (w.) INDONESIA (Ambon I.). Donisthorpe, 1948b: 309 (q.m.).
*micans* Forel, 1908; see under ***LINEPITHEMA.***
**mimulus.** *Iridomyrmex mimulus* Shattuck, 1993b: 1317, figs. 14, 15, 36 (w.) AUSTRALIA.
**#minor.** *Iridomyrmex gracilis* var. *minor* Forel, 1915b: 80 (w.q.) AUSTRALIA.
*minutus* Donisthorpe, 1943; see under ***ANONYCHOMYRMA.***
**mjobergi.** *Iridomyrmex mjobergi* Forel, 1915b: 77 (w.) AUSTRALIA.
*murinus* Emery, 1911; see under ***ANONYCHOMYRMA.***
*myrmecodiae* Emery, 1887; see under ***PHILIDRIS.***
*nagasau* Mann, 1921; see under ***PHILIDRIS.***
*neocaledonica. Iridomyrmex anceps* subsp. *papuana* var. *neocaledonica* Emery, 1914f: 419 (w.) NEW CALEDONIA. Unavailable name; material referred to *anceps* by Taylor, 1987a: 33.
*nigellus* Emery, 1890; see under ***LINEPITHEMA.***
*nigriventris* Donisthorpe, 1941; see under ***PHILIDRIS.***
*nitida* Mayr, 1862; see under ***PAPYRIUS.***
*nitidiceps* André, 1896; see under ***ANONYCHOMYRMA.***
**notialis.** *Iridomyrmex notialis* Shattuck, 1993b: 1318, figs. 16, 17, 36 (w.) AUSTRALIA.
***oblongiceps.** **Iridomyrmex oblongiceps* Wheeler, W.M. 1915h: 93, fig. 45 (w.) BALTIC AMBER (Oligocene). Provisionally retained in *Iridomyrmex:* Shattuck, 1992a: 16.
*oblongus* Santschi, 1929; see under ***LINEPITHEMA.***
***obscurans.** **Iridomyrmex obscurans* Carpenter, 1930: 51 (q.) U.S.A. (Miocene). Provisionally retained in *Iridomyrmex:* Shattuck, 1992a: 16.
**#obscurior.** *Iridomyrmex discors* var. *obscurior* Forel, 1902h: 465 (w.) AUSTRALIA.
**obscurus.** *Iridomyrmex obscurus* Crawley, 1921: 92, fig. 1 (w.) AUSTRALIA.
**obsidianus.** *Iridomyrmex obsidianus* Emery, 1914f: 419, pl. 13, fig. 13 (w.) NEW CALEDONIA. See also: Shattuck, 1993b: 1319.
**occiduus.** *Iridomyrmex occiduus* Shattuck, 1993b: 1320, figs. 18, 19, 36 (w.q.m.) AUSTRALIA.
**#occipitalis.** *Iridomyrmex discors* subsp. *occipitalis* Forel, 1907h: 294 (w.) AUSTRALIA.
*oceanicus* Forel, 1901; see under ***PAPYRIUS.***
*pallidior. Iridomyrmex rufoniger* r. *pallidus* var. *pallidior* Forel, 1915b: 78 (w.q.m.) AUSTRALIA. Unavailable name (Taylor, 1986: 34).
**#pallidus.** *Iridomyrmex rufoniger* r. *pallidus* Forel, 1901b: 22 (footnote) (w.) AUSTRALIA. Forel, 1902h: 466 (q.).
*papuana. Iridomyrmex gracilis* subsp. *papuana* Emery, 1897d: 572 (w.) NEW GUINEA. Forel, 1901b: 20 (q.m.). Race of *anceps:* Forel, 1901b: 19. Senior synonym of *discoidalis:* Smith, M.R. 1957a: 347. Junior synonym of *anceps:* Wilson & Taylor, 1967: 78.
*paranaensis* Santschi, 1929; see under ***LINEPITHEMA.***
**#parcens.** *Iridomyrmex mattiroloi* subsp. *parcens* Forel, 1907a: 27 (w.) AUSTRALIA.
*pertaesta* Santschi, 1916; see under ***LINEPITHEMA.***
*perthensis* Forel, 1902; see under ***ANONYCHOMYRMA.***
*platensis* Forel, 1912; see under ***LINEPITHEMA.***
*politus* Stitz, 1912; see under ***ANONYCHOMYRMA.***
*pordescens* Wheeler, W.M. 1942; see under ***LINEPITHEMA.***
*praenotandum* Schmitz, 1950; see under ***LINEPITHEMA.***
**prismatis.** *Iridomyrmex prismatis* Shattuck, 1993b: 1321, figs. 20, 21, 36 (w.) AUSTRALIA.
*protensus* Forel, 1911; see under ***PHILIDRIS.***
*pubescens* Donisthorpe, 1949; see under ***PHILIDRIS.***
*pulex* Santschi, 1923; see under ***LINEPITHEMA.***
*punctatissimus* Emery, 1887; see under ***OCHETELLUS.***
**purpureus.** *Formica purpurea* Smith, F. 1858b: 40 (w.) AUSTRALIA. Lowne, 1865a: 275 (m.); Imai, Crozier & Taylor, 1977: 345 (k.). Combination in *Iridomyrmex:* Mayr, 1862: 702. Senior synonym of *detecta:* Lowne, 1865a: 275; of *smithii:* Mayr, 1870b: 955; of *aeneum:* Mayr, 1876: 81; of *horni:* Clark, 1930c: 20; Shattuck, 1993a: 128; of *castrae:* Shattuck, 1993a: 128. See also: Greaves, 1971: 15; Shattuck, 1993a: 128; Shattuck, 1994: 109.
*queenslandensis* Forel, 1901; see under ***PAPYRIUS.***
**reburrus.** *Iridomyrmex reburrus* Shattuck, 1993a: 132, figs. 8, 14 (w.) AUSTRALIA.

*riograndensis* Borgmeier, 1928; see under ***LINEPITHEMA.***
*rostrinotus* Forel, 1910; see under ***OCHETELLUS.***
**#rubriceps.** *Iridomyrmex gracilis* r. *rubriceps* Forel, 1902h: 468 (w.) AUSTRALIA.
**rufoinclinus.** *Iridomyrmex rufoinclinus* Shattuck, 1993b: 1322, figs. 34, 35, 37 (w.) AUSTRALIA.
**rufoniger.** *Formica rufonigra* Lowne, 1865a: 279 (w.) AUSTRALIA. Mayr, 1876: 82 (q.m.). Combination in *Hypoclinea:* Mayr, 1870b: 955; in *Iridomyrmex:* Emery, 1887a: 251. Senior synonym of *mamillatus:* Mayr, 1876: 82. Current subspecies: nominal plus *domesticus, incertus, metallescens, pallidus, septentrionalis, suchieri, victorianus.* See also: Shattuck, 1994: 110.
**samlandicus* Wheeler, W.M. 1915; see under ***ANONYCHOMYRMA.***
**sanguineus.** *Iridomyrmex detectus* var. *sanguinea* Forel, 1910b: 53 (w.) AUSTRALIA. Forel, 1915b: 79 (q.). Subspecies of *purpureus:* Greenslade, 1974: 247; Taylor & Brown, D.R. 1985: 102. Raised to species: Shattuck, 1993a: 134.
*scotti* Santschi, 1919; see under ***LINEPITHEMA.***
*sellatus* Stitz, 1911; see under ***ANONYCHOMYRMA.***
**#septentrionalis.** *Iridomyrmex rufoniger* var. *septentrionalis* Forel, 1902h: 465 (w.) AUSTRALIA. [First available use of *Iridomyrmex rufoniger* r. *pallidus* var. *septentrionalis* Forel, 1901b: 23; unavailable name.]
*sericeus* Mayr, 1866; see under ***AZTECA.***
**#sikkimensis.** *Iridomyrmex anceps* var. *sikkimensis* Forel, 1904c: 27 (w.) INDIA. Forel, 1907e: 18 (q.).
*simalurana* Forel, 1915; see under ***PHILIDRIS.***
*smithii. Formica smithii* Lowne, 1865a: 276 (w.) AUSTRALIA. Junior synonym of *purpureus:* Mayr, 1870b: 955; Taylor & Brown, D.R. 1985: 102.
*smithi* Wheeler, W.M. 1909; see under ***TECHNOMYRMEX.***
*sommeri* Forel, 1902; see under ***OCHETELLUS.***
*sororis* Mann, 1921; see under ***OCHETELLUS.***
**spadius.** *Iridomyrmex spadius* Shattuck, 1993a: 138, figs. 7, 17 (w.) AUSTRALIA.
**#splendens.** *Iridomyrmex mattiroloi* var. *splendens* Forel, 1907h: 290 (w.) AUSTRALIA.
**#splendidus.** *Iridomyrmex bicknelli* var. *splendidus* Forel, 1902h: 468 (w.) AUSTRALIA. Forel, 1907h: 290 (q.).
**spodipilus.** *Iridomyrmex spodipilus* Shattuck, 1993a: 139, fig. 18 (w.) AUSTRALIA.
**#spurcus.** *Iridomyrmex gracilis* subsp. *spurcus* Wheeler, W.M. 1915g: 813 (w.) AUSTRALIA.
*stewartii* Forel, 1893; see under ***PHILIDRIS.***
*subfasciatus* Wheeler, W.M. & Mann, 1914; see under ***LINEPITHEMA.***
*succineus* Forel, 1908; see under ***LINEPITHEMA.***
**#suchieri.** *Iridomyrmex rufoniger* subsp. *suchieri* Forel, 1907h: 291 (w.q.m.) AUSTRALIA.
*testaceus* Cole, 1936; see under ***FORELIUS.***
*tigris* Stitz, 1912; see under ***ANONYCHOMYRMA.***
*transiens* Forel, 1913; see under ***LINEPITHEMA.***
*triconvexus* (misspelling); see under ***ANONYCHOMYRMA.***
**variscapus.** *Iridomyrmex variscapus* Shattuck, 1993a: 141, fig. 9 (w.) AUSTRALIA.
**vicinus.** *Iridomyrmex vicina* Clark, 1934c: 62, pl. 4, fig. 22 (w.q.) AUSTRALIA.
**#victorianus.** *Iridomyrmex rufoniger* var. *victorianus* Forel, 1902h: 466 (w.q.) AUSTRALIA. See also: Forel, 1907a: 27.
*victorianus* Stitz, 1911; see under ***PAPYRIUS.***
*vinsoni* Donisthorpe, 1946; see under ***OCHETELLUS.***
**viridiaeneus.** *Iridomyrmex detectus* var. *viridiaeneus* Viehmeyer, 1914c: 41 (w.) AUSTRALIA. Wheeler, W.M. 1915g: 811 (q.). Wheeler, G.C. & Wheeler, J. 1974b: 400 (l.). Raised to species: Wheeler, G.C. & Wheeler, J. 1974b: 400. Subspecies of *purpureus:* Greenslade, 1974: 247; Taylor & Brown, D.R. 1985: 102. Raised to species: Shattuck, 1993a: 142. See also: Shattuck, 1994: 113.
**viridigaster.** *Iridomyrmex viridigaster* Clark, 1941: 87, pl. 13, fig. 18 (w.) AUSTRALIA. See also: Shattuck, 1993b: 1323.
*waldoi* Forel, 1911; see under ***PHILIDRIS.***
**#watsonii.** *Iridomyrmex anceps* var. *watsonii* Forel, 1895e: 468 (diagnosis in key) (w.) BURMA.
*wheeleri* Emery, 1912; see under ***TECHNOMYRMEX.***
**wingi.** *Iridomyrmex wingi* Donisthorpe, 1949c: 754 (w.) NEW GUINEA.
**#yalgooensis.** *Iridomyrmex chasei* var. *yalgooensis* Forel, 1907h: 288 (w.q.) AUSTRALIA. See also: Wheeler, W.M. 1934d: 149.
*yarrabahana. Iridomyrmex innocens* r. *malandanus* var. *yarrabahana* Forel, 1915b: 81 (w.) AUSTRALIA. Unavailable name (Taylor, 1986: 34).

*IROGERA:* see under ***ROGERIA.***

*ISCHNOMYRMEX:* see under ***PHEIDOLE.***

**ISHAKIDRIS** (Borneo)
**ascitaspis.** *Ishakidris ascitaspis* Bolton, 1984: 375, figs. 1–4 (w.) BORNEO.

*ISOLCOMYRMEX:* see under ***MONOMORIUM.***

*ISOPHEIDOLE:* see under ***PHEIDOLE.***

*JANETIA:* see under ***POGONOMYRMEX.***

*JOHNIA:* see under ***POLYRHACHIS.***

***#KARAVAIEVIA:*** see under ***CAMPONOTUS.***

*KARAWAJEWELLA:* see under ***DOLICHODERUS.***

**KARTIDRIS** (Oriental)
**fujianensis.** *Kartidris fujianensis* Wang, M. 1993b: 434, fig. 2 (w.) CHINA.
**galos.** *Kartidris galos* Bolton, 1991: 13, fig. 20 (w.) CHINA.
**matertera.** *Kartidris matertera* Bolton, 1991: 13, fig. 21 (w.) THAILAND.
**nyos.** *Kartidris nyos* Bolton, 1991: 12, figs. 18, 19 (w.) INDIA.

***KOTSHKORKIA** (Russia)
***laticeps.** **Kotshkorkia laticeps* Dlussky, 1981b: 71, fig. 52 (q.) RUSSIA (Miocene).

**KYIDRIS** (Palaearctic, Oriental, Indo-Australian)
*itoi. Polyhomoa itoi* Azuma, 1950: 36, figs. (w.) JAPAN. Junior synonym of *mutica:* Brown & Yasumatsu, 1951: 94.
**media.** *Kyidris media* Wilson & Brown, 1956: 445, figs. 2, 5 (w.q.m.) NEW GUINEA.
**mutica.** *Kyidris mutica* Brown, 1949d: 3, fig. 1 (w.) JAPAN. Wilson & Brown, 1956: 442 (q.m.); Imai, Kubota, *et al.* 1985: 47 (k.). Senior synonym of *itoi:* Brown & Yasumatsu, 1951: 94; of *nuda:* Brown, 1952c: 124.
*nuda. Kyidris nuda* Brown, 1949d: 23 (q.) TAIWAN. Junior synonym of *mutica:* Brown, 1952c: 124.
**yaleogyna.** *Kyidris yaleogyna* Wilson & Brown, 1956: 443, fig. 3 (w.q.m.) NEW GUINEA.

*LABAUCHENA:* see under ***SOLENOPSIS.***

*LABIDOGENYS:* see under ***STRUMIGENYS.***

**LABIDUS** (Neotropical, s. Nearctic)
*amplipennis* Smith, F. 1859; see under ***NEIVAMYRMEX.***
*atriceps. Labidus atriceps* Smith, F. 1859b: 5 (m.) BRAZIL. Junior synonym of *coecus:* Emery, 1900a: 186; Borgmeier, 1955b: 86.
**auropubens.** *Eciton* (*Labidus*) *praedator* st. *auropubens* Santschi, 1920d: 366, fig. 3 (m.) FRENCH GUIANA. Combination in *Labidus,* and raised to species: Borgmeier, 1953: 14. See also: Borgmeier, 1955b: 131.
*biloba. Eciton coecum* var. *biloba* Emery, 1901d: 51 (m.) ECUADOR. Junior synonym of *coecus:* Borgmeier, 1953: 16.
*burchellii* Westwood, 1842; see under ***ECITON.***
**coecus.** *Formica coeca* Latreille, 1802c: 270, pl. 9, fig. 56 (s.) CENTRAL AMERICA. Roger, 1861a: 22 (w.); Emery, 1895c: 258 (m.); Bruch, 1934b: 116 (q.); Weber, 1941c: 325 (q.); Wheeler, G.C. 1943: 332 (l.). Combination in *Labidus:* Jurine, 1807: 282; in *Nycteresia:* Roger, 1861a: 22; in *Eciton:* Mayr, 1886b: 119; in *Eciton* (*Labidus*): Santschi, 1924b: 11; in *Labidus:* Borgmeier, 1955b: 86. Senior synonym of *vastator:* Mayr, 1865: 78; Borgmeier, 1955b: 86; of *jurinii:* Mayr, 1886a: 33; Borgmeier, 1953: 15; of *erratica:* Mayr, 1886b: 119; Borgmeier, 1955b: 86; of *latreillii, rubra* and *saji:* Emery, 1895c: 258; Borgmeier, 1955b: 86; of *atriceps:* Emery, 1900a: 186; Borgmeier, 1955b: 86; of *incerta:* André, in Forel, 1899c: 160; Emery, 1900a: 186; Borgmeier, 1955b: 86; of *fulvescens:* Emery, 1910b: 22; Borgmeier, 1955b: 86; of *pilosus* and *smithii:* Santschi, 1913h: 35; Borgmeier, 1955b: 86; of *grassator:* Borgmeier, 1923: 41; of *panzeri, biloba, elsbethae, servillei* and *opacifrons:* Borgmeier, 1953: 16, 16, 16, 17, 19, repectively; of *nigrita, kulowi, selysi* and *serpentis:* Borgmeier, 1955b: 86; material of the unavailable name *hostilis* referred here: Borgmeier, 1953: 14. See also: Gallardo, 1920: 326; Wheeler, G.C. & Wheeler, J. 1984: 272; Watkins, 1976: 8.
*curtisii* Shuckard, 1840; see under ***DORYLUS.***
**curvipes.** *Eciton curvipes* Emery, 1900a: 180 (m.) COSTA RICA. Combination in *Labidus:* Borgmeier, 1953: 6. See also: Borgmeier, 1955b: 133.
**#denticulatus.** *Labidus nero* subsp. *denticulatus* Borgmeier, 1955b: 129 (m.) BRAZIL.
*dorbignii* Shuckard, 1840; see under ***NEIVAMYRMEX.***
*elsbethae. Eciton* (*Labidus*) *coecum* var. *elsbethae* Forel, 1922: 91 (m.) COLOMBIA. Junior synonym of *coecus:* Borgmeier, 1953: 16.
*emiliae. Eciton* (*Labidus*) *praedator* subsp. *emiliae* Mann, 1916: 421 (s.w.) BRAZIL. Junior synonym of *praedator:* Borgmeier, 1953: 10.

***LABIDUS** (continued)*

*erichsonii* Westwood, 1842; see under **NEIVAMYRMEX.**
*erratica. Eciton erratica* Smith, F. 1860c: 71 (w.) BRAZIL. Junior synonym of *coecus:* Mayr, 1886b: 119; Borgmeier, 1955b: 86.
*esenbeckii* Westwood, 1842; see under **NOMAMYRMEX.**
*fargeavii* Shuckard, 1840; see under **ECITON.**
*ferruginea. Eciton tepeguas* var. *ferruginea* Norton, 1868b: 46 (s.) MEXICO. Variety of *praedator:* Dalla Torre, 1893: 5; Forel, 1899c: 26. Junior synonym of *praedator:* Borgmeier, 1955b: 86.
*fonscolombii* Westwood, 1842; see under **NEIVAMYRMEX.**
*fulvescens. Mutilla (Labidus) fulvescens* Blanchard, in Cuvier, 1846, pl. 118, fig. 2 (m.) BRAZIL. Combination in *Eciton:* Dalla Torre, 1893: 2. Junior synonym of *coecus:* Emery, 1910b: 22; Borgmeier, 1955b: 86.
*grassator. Eciton grassator* Forel, 1911c: 288 (w.) BRAZIL. Variety of *coecus:* Forel, 1911d: 396 (footnote); Santschi, 1923g: 1260; Wheeler, W.M. 1925b: 2. Junior synonym of *coecus:* Borgmeier, 1923: 41.
*gravenhorstii* Westwood, 1842; see under **NEIVAMYRMEX.**
*guerinii* Shuckard, 1840; see under **NEIVAMYRMEX.**
*guianense. Eciton (Labidus) praedator* var. *guianense* Wheeler, W.M. 1921d: 311 (s.w.) GUIANA. Junior synonym of *praedator:* Borgmeier, 1953: 19.
*halidaii* Shuckard, 1840; see under **NEIVAMYRMEX.**
*harrisii* Haldeman, 1852; see under **NEIVAMYRMEX.**
*hartigii* Westwood, 1842; see under **NOMAMYRMEX.**
*hopei* Shuckard, 1840; see under **NEIVAMYRMEX.**
*hostilis. Eciton (Labidus) caecum* st. *servillei* var. *hostilis* Santschi, 1920d: 368 (m.) FRENCH GUIANA. Unavailable name; material referred to *coecus* by Borgmeier, 1953: 14.
*illigeri* Shuckard, 1840; see under **NEIVAMYRMEX.**
*incerta. Pseudodichthadia incerta* André, 1885: 840, figs. 1–5 (q.) MEXICO. Junior synonym of *coecus:* André, in Forel, 1899c: 160; Emery, 1900a: 186; Borgmeier, 1955b: 86.
*jurinii. Labidus jurinii* Shuckard, 1840a: 198 (m.) GUIANA. Combination in *Eciton:* Dalla Torre, 1893: 4. Variety of *coecus:* Emery, 1910b: 22; Forel, 1912c: 43; Gallardo, 1920: 332. Junior synonym of *coecus:* Mayr, 1886a: 33; Borgmeier, 1953: 15; Borgmeier, 1955b: 86.
*kirbii* Shuckard, 1840; see under **DORYLUS.**
*klugii* Shuckard, 1840; see under **NEIVAMYRMEX.**
*kulowi. Eciton coecum* r. *kulowi* Forel, 1901h: 47 (m.) MEXICO. Junior synonym of *coecus:* Borgmeier, 1955b: 86.
*latreillii. Labidus latreillii* Jurine, 1807: 283 (m.) SURINAM. Combination in *Eciton:* Dalla Torre, 1893: 4. Junior synonym of *coecus:* Emery, 1895c: 258; Borgmeier, 1955b: 86.
*latreillei* Perty, 1833; see under **NEIVAMYRMEX.**
***mars.*** *Eciton (Labidus) mars* Forel, 1912c: 44 (s.w.) BRAZIL. Combination in *Labidus:* Borgmeier, 1953: 8. See also: Borgmeier, 1955b: 121; Watkins, 1976: 8.
*melshaemeri* Haldeman, 1852; see under **NEIVAMYRMEX.**
*mexicanus* Smith, F. 1859; see under **NEIVAMYRMEX.**
*minor* Cresson, 1872; see under **NEIVAMYRMEX.**
*morosus* Smith, F. 1859; see under **CHELIOMYRMEX.**
***nero.*** *Eciton (Labidus) coecum* st. *nero* Santschi, 1930c: 82 (m.) BRAZIL. Combination in *Labidus,* and raised to species: Borgmeier, 1953: 14. Current subspecies: nominal plus *denticulatus.* See also: Borgmeier, 1955b: 127.
*nigrescens* Cresson, 1872; see under **NEIVAMYRMEX.**
*nigrita. Eciton nigrita* Emery, 1901d: 52, fig. 5 (m.) GUATEMALA. Combination in *E. (Labidus):* Emery, 1910b: 23; in *Labidus:* Borgmeier, 1953: 6. Stirps/subspecies of *praedator:* Santschi, 1925b: 11; of *coecus:* Borgmeier, 1953: 6. Junior synonym of *coecus:* Borgmeier, 1955b: 86.
*opacifrons. Eciton (Labidus) coecum* var. *opacifrons* Wheeler, W.M. 1921d: 310 (w.) BRAZIL. Junior synonym of *coecus:* Borgmeier, 1953: 19.
*panzeri. Labidus panzeri* Smith, F, 1859b: 71, pl. 2, fig. 2 (m.) COLOMBIA. Junior synonym of *coecus:* Borgmeier, 1953: 16.
*pertii* Shuckard, 1840; see under **NEIVAMYRMEX.**
*pilosus. Labidus pilosus* Smith, F. 1859b: 7 (m.) BRAZIL. Junior synonym of *coecus:* Santschi, 1913h: 35; Borgmeier, 1955b: 86.
***praedator.*** *Eciton praedator* Smith, F. 1858b: 152 (s.w.) BRAZIL. Forel, 1906d: 246 (m.); Luederwaldt, 1918: 54 (q.). Combination in *E. (Labidus):* Bruch, 1914: 215; in *Labidus:* Borgmeier, 1953: 16. Senior synonym of *tepeguas:* Mayr, 1886b: 118; Borgmeier, 1955b: 103; of *westwoodi:* Emery, 1910b: 23; Borgmeier, 1955b: 103; of *emiliae* and *guianense:* Borgmeier, 1953: 10, 19, respectively; of *ferruginea:* Borgmeier, 1955b: 103. Current subspecies: nominal plus *sedulus.* See also: Gallardo, 1920: 333; Bruch, 1921: 181; Watkins, 1976: 8.
*romandii* Shuckard, 1840; see under **NEIVAMYRMEX.**
*rubra. Myrmica rubra* Buckley, 1867: 335 (w.) U.S.A. Junior synonym of *coecus:* Emery, 1895c: 258; Borgmeier, 1955b: 86.
*saji. Labidus saji* Haldeman, 1852: 367, pl. 9, figs. 1–3 (m.) U.S.A. [Spelling emended to *sayi* by Roger, 1863b: 42, repeated by all later authors.] Combination in *Eciton (Labidus):* Mayr, 1886d: 440. Variety of *servillei:* Emery, 1890b: 39; of *latreillii:* Dalla Torre, 1893: 4. Junior synonym of *coecus:* Emery, 1895c: 258. Revived from synonymy as variety of *coecus:* Santschi, 1923g: 1261. Junior synonym of *coecus:* Borgmeier, 1955b: 86.
#***sedulus.*** *Eciton (Labidus) sedulus* Menozzi, 1926d: 30, fig. 2 (m.) COLOMBIA. Borgmeier, 1955b: 112 (s.w.); Borgmeier, 1958: 198 (q.). Combination in *Labidus:* Borgmeier, 1953: 11. Subspecies of *praedator:* Borgmeier, 1953: 11.
*selysi. Eciton selysi* Forel, 1904d: 169 (w.) BRAZIL. Combination in *Labidus:* Borgmeier, 1953: 8. Subspecies of *coecus:* Forel, 1905b: 157; Santschi, 1923g: 1260. Junior synonym of *coecus:* Borgmeier, 1955b: 86.
*serpentis. Eciton (Labidus) serpentis* Weber, 1938a: 209 (w.) GUIANA. Junior synonym of *coecus:* Borgmeier, 1955b: 86.
*servillei. Labidus servillei* Westwood, 1842: 75, pl. 20, fig. 2 (m.) BRAZIL. Combination in *Eciton:* Dalla Torre, 1893: 4. Variety of *latreillii:* Dalla Torre, 1893: 4; of *coecus:* Forel, 1912c: 43; Santschi, 1920d: 368. Junior synonym of *coecus:* Borgmeier, 1953: 17.
*smithii. Eciton smithii* Dalla Torre, 1892: 89. [Unnecessary replacement name for *Labidus pilosus* Smith, F. 1859b: 7.] Junior synonym of *coecus:* Santschi, 1913h: 35.
***spininodis.*** *Eciton spininode* Emery, 1890b: 38 (w.) COSTA RICA. [Also described as new by Emery, 1894k: 45.] Borgmeier, 1955b: 118 (s.). Combination in *Eciton (Labidus):* Emery, 1910b: 23; in *Labidus:* Borgmeier, 1953: 6.
*spinolae* Shuckard, 1840; see under **DORYLUS.**
*spinolae* Westwood, 1842; see under **NEIVAMYRMEX.**
*strobeli* Mayr, 1868; see under **NEIVAMYRMEX.**
*sulcatus* Mayr, 1868; see under **NEIVAMYRMEX.**
*swainsonii* Shuckard, 1840; see under **NEIVAMYRMEX.**
*tepeguas. Eciton tepeguas* Norton, 1868b: 46 (s.) MEXICO. Junior synonym of *praedator:* Mayr, 1886b: 118; Borgmeier, 1955b: 103.
*thwaitsii* Shuckard, 1840; see under **DORYLUS.**
***truncatidens.*** *Eciton (Labidus) truncatidens* Santschi, 1920d: 367, fig. 4 (m.) FRENCH GUIANA. Combination in *Labidus:* Borgmeier, 1953: 13. See also: Borgmeier, 1955b: 124.
*vastator. Eciton vastator* Smith, F. 1860c: 71 (w.) BRAZIL. Junior synonym of *coecus:* Mayr, 1865: 78; Borgmeier, 1955b: 86.
*walkerii* Westwood, 1842; see under **NEIVAMYRMEX.**
*westwoodii* Shuckard, 1840; see under **DORYLUS.**
*westwoodi. Eciton westwoodi* Emery, 1900a: 180 (m.) MEXICO. Junior synonym of *praedator:* Emery, 1910b: 23; Borgmeier, 1955b: 103.

***LACHNOMYRMEX*** (Neotropical)
***haskinsi.*** *Lachnomyrmex haskinsi* Smith, M.R. 1944e: 227, figs. 1, 2 (w.) PANAMA.
***pilosus.*** *Lachnomyrmex pilosus* Weber, 1950c: 1 (w.) TRINIDAD.
***plaumanni.*** *Lachnomyrmex plaumanni* Borgmeier, 1957: 125, figs. 47–52 (w.q.) BRAZIL.
***scrobiculatus.*** *Lachnomyrmex scrobiculatus* Wheeler, W.M. 1910a: 263, fig. 3 (w.q.) GUATEMALA. Wheeler, G.C & Wheeler, J. 1989a: 321 (l.).

*LAMPROMYRMEX:* see under **MONOMORIUM.**

*LAPAROMYRMEX:* see under **LIOMYRMEX.**

***LASIOPHANES*** (Neotropical)
***atriventris.*** *Formica atriventris* Smith, F. 1858b: 51 (w.) CHILE. Earliest junior synonym of *Formica nigriventris* Spinola, 1851a: 239 [Junior primary homonym of *Formica nigriventris* Guérin-Méneville, 1831, pl. 8, fig. 4 (now in *Leptomyrmex*).], hence first available replacement name for *nigriventris* Spinola (see there for other synonymy, etc.).
*bidens. Melophorus (Lasiophanes) picinus* var. *bidens* Emery, 1895i: 17 (w.q.m.) CHILE. Junior synonym of *picinus:* Menozzi, 1935c: 323.
*bolivari. Prenolepis (Nylanderia) bolivari* Santschi, 1916b: 512 (w.) ARGENTINA. Combination in *Melophorus (Lasiophanes):* Santschi, 1922d: 259; in *Lasiophanes:* Wheeler, W.M. 1935c: 70. Junior synonym of *picinus:* Kusnezov, 1952c: 92; of *valdiviensis:* Snelling & Hunt, 1976: 108.
*bruchi. Prenolepis bruchi* Forel, 1915c: 361 (w.) ARGENTINA. Combination in *Melophorus (Lasiophanes):* Santschi, 1920d: 383; in *Lasiophanes:* Wheeler, W.M.

1935c: 70. Junior synonym of *picinus:* Snelling & Hunt, 1976: 107.

*dichrous. Lasius dichrous* Roger, 1863a: 164 (w.) CHILE. Forel, 1886f: 208 (q.). Junior synonym of *nigriventris:* Emery, 1893h: 2; Dalla Torre, 1893: 190.

*edwardsi. Acanthomyops (Donisthorpea) edwardsi* Donisthorpe, 1933c: 535 (w.q.m.) ARGENTINA. Combination in *Lasiophanes:* Wheeler, W.M. 1935c: 70; Donisthorpe, 1937a: 628. Junior synonym of *picinus:* Kusnezov, 1952c: 92.

***hoffmanni.*** *Melophorus hoffmanni* Forel, 1903c: 266 (w.) CHILE. [*Lasius hoffmanni* Emery, 1895i: 18; *nomen nudum.*] Combination in *Melophorus (Lasiophanes)*: Emery, 1906c: 185; in *Lasiophanes:* Wheeler, W.M. 1935c: 70. See also: Snelling & Hunt, 1976: 104.

*negrensis. Acanthomyops (Chthonolasius) negrensis* Donisthorpe, 1933c: 537 (q.) ARGENTINA. Combination in *Lasiophanes:* Wheeler, W.M. 1935c: 70; Donisthorpe, 1937a: 628. Junior synonym of *nigriventris:* Kusnezov, 1952c: 94.

*nigriventris. Formica nigriventris* Spinola, 1851a: 239 (w.) CHILE. [Junior primary homonym of *Formica nigriventris* Guérin-Méneville, 1831, pl. 8, fig. 4 (now in *Leptomyrmex*).] First available replacement name: *atriventris* Smith, F. 1858b: 51. Combination in *Melophorus (Lasiophanes)*: Emery, 1895i: 16; in *Lasiophanes:* Wheeler, W.M. 1935c: 70. Senior synonym of *atriventris:* Mayr, 1863: 412; Snelling & Hunt, 1976: 107; of *dichrous:* Emery, 1893h: 2; Dalla Torre, 1893: 190; of *negrensis* and *rufoniger:* Kusnezov, 1952c: 94.

***perplexus.*** *Melophorus (Lasiophanes) perplexus* Santschi, 1920d: 384 (q.) CHILE. Combination in *Lasiophanes:* Wheeler, W.M. 1935c: 70. Junior synonym of *nigriventris:* Kusnezov, 1952c: 92. Revived from synonymy: Snelling & Hunt, 1976: 107.

***picinus.*** *Lasius picinus* Roger, 1863a: 163 (w.) CHILE. Combination in *Melophorus (Lasiophanes)*: Emery, 1895i: 17; in *Lasiophanes:* Wheeler, W.M. 1935c: 70. Senior synonym of *bidens:* Menozzi, 1935c: 323; of *bruchi, edwardsi* and *sauberi:* Kusnezov, 1952c: 92; of *bruchi:* Snelling & Hunt, 1976: 107.

*pilosulus. Melophorus (Lasiophanes) pilosulus* Emery, 1922a: 93 (w.) CHILE. [Misspelled as *pilosa* by Santschi, 1922d: 259.] Junior synonym of *picinus:* Kusnezov, 1952c: 92; of *valdiviensis:* Snelling & Hunt, 1976: 108.

*rufoniger. Acanthomyops (Chthonolasius) rufoniger* Donisthorpe, 1933c: 537 (w.) ARGENTINA. Combination in *Lasiophanes:* Wheeler, W.M. 1935c: 70; Donisthorpe, 1937a: 628. Junior synonym of *nigriventris:* Kusnezov, 1952c: 94.

*sauberi. Melophorus sauberi* Forel, 1903c: 266 (w.) CHILE. Forel, 1904g: 4 (q.m.). [Also described as new by Forel, 1904g: 4.] Combination in *M. (Lasiophanes)*: Bruch, 1914: 230; in *Lasiophanes:* Wheeler, W.M. 1935c: 70. Junior synonym of *picinus:* Kusnezov, 1952c: 92.

*uxorius. Melophorus (Lasiophanes) uxorius* Emery, 1922a: 93 (q.) CHILE. Combination in *Lasiophanes:* Wheeler, W.M. 1935c: 70. Junior synonym of *valdiviensis:* Snelling & Hunt, 1976: 108.

***valdiviensis.*** *Melophorus valdiviensis* Forel, 1904g: 6 (w.) CHILE. Combination in *Lasiophanes:* Wheeler, W.M. 1935c: 70. Senior synonym of *bolivari, pilosulus* and *uxorius:* Snelling & Hunt, 1976: 108.

***LASIUS*** (Holarctic, Oriental)

*affinis. Formica affinis* Schenck, 1852: 62 (w.q.m.) GERMANY. [Junior primary homonym of *Formica affinis* Leach, 1825: 290.] Replacement name: *citrinus* Emery, 1922b: 12, oldest synonym and hence first available replacement name: Seifert, 1990: 7. Combination in *Lasius:* Mayr, 1861: 50; in *Formicina:* Emery, 1916b: 242; in *Acanthomyops:* Ruzsky, 1925b: 45; in *Lasius (Chthonolasius)*: Ruzsky, 1914a: 61; Emery, 1925b: 233. Junior synonym of *bicornis:* Smith, F. 1858b: 8. Revived from synonymy: Mayr, 1863; 425. Variety/subspecies of *umbratus:* Forel, 1874: 47; Mayr, 1886d: 430; Emery, 1925b: 233; Karavaiev, 1936: 214; of *bicornis:* André, 1882b: 196; Nasonov, 1889: 24; Forel, 1915d: 52; Emery, 1916b: 242. Status as species: Ruzsky, 1902d: 15; Bondroit, 1912: 352; Emery, 1914d: 159; Bondroit, 1918: 31; Stärcke, 1937: 40; Stitz, 1939: 299; Novak & Sadil, 1941: 103. Junior synonym of *umbratus:* Wilson, 1955a: 150. Revived from synonymy: Collingwood, 1963: 145. Status as species: Collingwood & Yarrow, 1969: 80; Baroni Urbani, 1971c: 209; Kutter, 1977c: 232; Collingwood, 1979: 102; Seifert, 1988: 163; Atanassov & Dlussky, 1992: 248.

*affinoumbratus. Lasius umbratus* var. *affinoumbratus* Donisthorpe, 1914: 40 (w.) GREAT BRITAIN. Junior synonym of *umbratus:* Wilson, 1955a: 151.

*alienoamericanus. Lasius niger* subsp. *alienus* var. *alienoamericanus* Wheeler, W.M. 1917a: 526 (q.) U.S.A. Unavailable name; material referred to *alienus* by Wilson, 1955a: 78.

*alienobrunneus. Lasius niger* var. *alienobrunneus* Forel, 1874: 47 (w.) SWITZERLAND. Emery, 1916b: 249 (q.). Junior synonym of *alienus:* Dalla Torre, 1893: 182. Revived from synonymy as variety of *brunneus:* Ruzsky, 1902d: 17. Raised to species: Bingham, 1903: 340. Variety of *brunneus:* Forel, 1915d: 53; Karavaiev, 1927c: 279. Junior synonym of *brunneus:* Stärcke, 1944a: 157; Wilson, 1955a: 47; Seifert, 1992b: 6.

***alienoflavus.*** *Lasius alienoflavus* Bingham, 1903: 341 (w.q.) INDIA. Combination in *L. (Cautolasius)*: Wilson, 1955a: 111. See also: Collingwood, 1982: 290.

*alienoides. Lasius niger* subsp. *alienoides* Forel, 1890a: lxvii (no caste given) TUNISIA. *Nomen nudum.* Emery, 1891b: 19; *nomen nudum,* attributed to Forel.

*alienoniger. Lasius niger* var. *alienoniger* Forel, 1874: 47 (w.q.) SWITZERLAND. Variety of *niger:* Finzi, 1924a: 14; Stärcke, 1944a: 154. Raised to species: Röszler, 1943: 44. Junior synonym of *niger:* Wilson, 1955a: 59. Unrecognisable taxon: Seifert, 1992b: 48.

***alienus.*** *Formica aliena* Foerster, 1850a: 36 (w.m.) GERMANY. Foerster, 1850a: 71 (q.); Hauschteck, 1962: 219 (k.). Combination in *Lasius:* Mayr, 1861: 49; in *Donisthorpea:* Donisthorpe, 1915d: 212; in *Formicina:* Emery, 1916b: 240; in *Acanthomyops:* Ruzsky, 1925b: 44; in *Lasius:* Menozzi, 1921: 32; Müller, 1923: 125; Emery, 1925b: 230; Kuznetsov-Ugamsky, 1929a: 27. Variety/race/subspecies of *niger:* Forel, 1874: 46; Mayr, 1886d: 429; Forel, 1892i: 307; Forel, 1904b: 386; Wheeler, W.M. 1906c: 322; Forel, 1913d: 438; Forel, 1915d: 53; Emery, 1916b: 240; Santschi, 1925g: 349; Karavaiev, 1927c: 280; Menozzi, 1936d: 305; Menozzi, 1939a: 312. Status as species: Saunders, E. 1880: 209; André, 1882b: 192; Nasonov, 1889: 22; Emery, 1897f: 238; Ruzsky, 1902d: 16; Emery, 1908d: 24; Bondroit, 1911: 11; Donisthorpe, 1915d: 212; Bondroit, 1918: 25; Stitz, 1939: 279; Novak & Sadil, 1941: 101; Röszler, 1942a: 53; Stärcke, 1944a: 153; Wilson, 1955a: 77; Baroni Urbani, 1971c: 200; Kutter, 1977c: 227; Collingwood, 1979: 97; Collingwood, 1982: 285; Kupyanskaya, 1990: 218; Atanassov & Dlussky, 1992: 237; Seifert, 1992b: 13. Senior synonym of *americanus* and *pannonica,* and material of the unavailable names *alienoamericanus, flavidus* and *turkmenus* referred here: Wilson, 1955a: 77.

*americanus. Lasius niger* var. *americanus* Emery, 1893i: 639 (w.q.m.) U.S.A. Wheeler, G.C. & Wheeler, J. 1953c: 147 (l.). Variety/subspecies of *niger:* Wheeler, W.M. 1904e: 305; Forel, 1900e: 285; Buren, 1944a: 296. Raised to species: Gregg, E.V. 1945: 530. Subspecies of *alienus:* Creighton, 1950a: 419. Junior synonym of *alienus:* Wilson, 1955a: 77.

*apennina. Lasius (Chthonolasius) umbratus* var. *apennina* Menozzi, 1925d: 34 (w.) ITALY. Menozzi, 1932a: 8 (q.m.). Variety of *flavus:* Menozzi, 1932a: 8. Junior synonym of *flavus:* Wilson, 1955a: 112.

*aphidicola. Formica aphidicola* Walsh, 1863: 310 (w.q.m.) U.S.A. Wheeler, G.C. & Wheeler, J. 1953c: 150 (l.). Combination in *Lasius:* Mayr, 1886d: 429; in *L. (Chthonolasius)*: Ruzsky, 1914a: 61. Junior synonym of *flavus:* Mayr, 1886d: 429. Revived from synonymy: Dalla Torre, 1893: 182. Variety/subspecies of *umbratus:* Wheeler, W.M. 1908g: 406; Buren, 1944a: 297. Junior synonym of *umbratus:* Wilson, 1955a: 150.

*arizonicus* Wheeler, W.M. 1917; see under ***ACANTHOMYOPS.***

***atopus.*** *Lasius (Chthonolasius) atopus* Cole, 1958a: 75, fig. 1 (w.) U.S.A.

***balcanicus.*** *Lasius (Chthonolasius) balcanicus* Seifert, 1988: 152, figs. 9, 19, 28, 37, 42, 46–52 (w.q.m.) BULGARIA.

*barbarus. Lasius alienus* var. *barbarus* Santschi, 1931a: 11 (w.) MOROCCO. [First available use of *Lasius niger* st. *lasioides* var. *barbara* Santschi, 1921e: 170; unavailable name.] Variety of *brunneus:* Santschi, 1936c: 208. Junior synonym of *alienus:* Wilson, 1955a: 78; of *lasioides:* Seifert, 1992b: 8.

*belgarum. Formicina belgarum* Bondroit, 1918: 31, fig. 17 (w.q.) BELGIUM. Combination in *Lasius (Chthonolasius)*: Emery, 1925b: 234. Variety of *umbratus:* Stärcke, 1937: 57; of *mixtus:* Stitz, 1939: 299. Junior synonym of *umbratus:* Wilson, 1955a: 151; Seifert, 1988: 150.

***bicornis.*** *Formica bicornis* Foerster, 1850a: 41 (q.) GERMANY. Forel, 1874: 47 (w.); Forel, 1915d: 56 (m.). Combination in *Lasius:* Mayr, 1861: 51; in *Formicina:* Emery, 1916b: 242; Menozzi, 1918: 87; in *Acanthomyops:* Ruzsky, 1925b: 45; in *Lasius (Chthonolasius)*: Ruzsky, 1914a: 61; Emery, 1925b: 232; Wilson, 1955a: 183. Race/subspecies of *umbratus:* Forel, 1874: 47; Emery & Forel, 1879: 453; Mayr, 1886d: 430; Forel, 1886f: 208; Forel, 1904b: 387; Bondroit, 1910: 485. Status as species:

André, 1882b: 196; Ruzsky, 1902d: 15; Bondroit, 1912: 352; Emery, 1916b: 242; Bondroit, 1918: 34; Emery, 1922b: 13; Karavaiev, 1927c: 279; Stärcke, 1937: 55; Stitz, 1939: 302; Wilson, 1955a: 183; Bernard, 1967: 365; Kutter, 1977c: 231; Collingwood, 1982: 293; Seifert, 1988: 161; Atanassov & Dlussky, 1992: 251. Senior synonym of *incisa:* Smith, F. 1858b: 8; Emery, 1922b: 13; Seifert, 1988: 161; of *microgyna:* Vandel, 1926: 197; Wilson, 1955a: 183; of *kashmirensis, neapolitana* and *oertzeni:* Wilson, 1955a: 183.

*brevicornis. Lasius brevicornis* Emery, 1893i: 639, pl. 22, fig. 22 (w.q.m.) U.S.A. Junior synonym of *flavus:* Wilson, 1955a: 112.

***breviscapus.*** *Lasius (Lasius) breviscapus* Seifert, 1992b: 24, fig. 17 (w.) INDIA.

*brunneoalienus. Lasius niger* r. *alienus* var. *brunneoalienus* Kulmatycki, 1922: 81 (no caste given) POLAND. Unavailable name.

*brunneoemarginatus. Lasius emarginatus* var. *brunneoemarginatus* Forel, 1874: 47 (w.) SWITZERLAND. Junior synonym of *emarginatus:* Wilson, 1955a: 89.

*brunneoides. Lasius niger* var. *brunneoides* Forel, 1874: 47 (w.) SWITZERLAND. Junior synonym of *emarginatus:* Wilson, 1955a: 89.

***brunneus.*** *Formica brunnea* Latreille, 1798: 41 (w.q.) FRANCE. Mayr, 1855: 358 (m.). Combination in *Lasius:* Mayr, 1861: 50; in *Donisthorpea:* Donisthorpe, 1915d: 347; in *Formicina:* Emery, 1916b: 241; in *Lasius:* Müller, 1923: 127; in *Acanthomyops:* Kuznetsov-Ugamsky, 1927e: 188; in *Lasius (Lasius)*: Wilson, 1955a: 47. Race/subspecies of *niger:* Forel, 1874: 47; Forel, 1892i: 307; Bondroit, 1910: 486; Kuznetsov-Ugamsky, 1927e: 188. Status as species: André, 1881b: 60; Dalla Torre, 1893: 182; Bondroit, 1912: 352; Forel, 1915d: 52; Bondroit, 1918: 26; Müller, 1923: 127; Karavaiev, 1927c: 279; Finzi, 1930d: 316; Menozzi, 1939a: 313; Wilson, 1955a: 47; Bernard, 1967: 358; Kutter, 1977c: 228; Collingwood, 1982: 285; Atanassov & Dlussky, 1992: 238; Seifert, 1992b: 6. Senior synonym of *pallida:* Latreille, 1802c: 169; Wilson, 1955a: 47; of *timida:* Smith, F. 1858b: 7; Seifert, 1992b: 6; of *alienobrunneus:* Stärcke, 1944a: 157; Wilson, 1955a: 47; Seifert, 1992b: 6; of *nigrobrunneus:* Wilson, 1955a: 47; Seifert, 1992b: 6. Current subspecies: nominal plus *emarginatobrunneus.*

***buccatus.*** *Lasius (Dendrolasius) buccatus* Stärcke, 1942c: xxvii, figs. 6, 7 (q.m.) YUGOSLAVIA.

*californicus* Wheeler, W.M. 1917; see under ***ACANTHOMYOPS.***

***capitatus.*** *Acanthomyops fuliginosus* subsp. *capitatus* Kuznetsov-Ugamsky, 1927e: 187 (w.) TURKESTAN. [Also described as new by Kuznetsov-Ugamsky, 1928b: 18.] Combination in *Lasius:* Wilson, 1955a: 138. Junior synonym of *fuliginosus:* Wilson, 1955a: 138. Revived from synonymy, raised to species and senior synonym of *crispus* Wilson: Kupyanskaya, 1989: 785.

***carniolicus.*** *Lasius carniolicus* Mayr, 1861: 51 (diagnosis in key) (q.) YUGOSLAVIA. Forel, 1886f: 206 (w.m.). Combination in *Formicina:* Emery, 1916b: 242; in *Acanthomyops:* Ruzsky, 1925b: 45; in *Lasius (Chthonolasius)*: Ruzsky, 1914a: 59; Müller, 1923: 132; Wilson, 1955a: 189; in *L. (Austrolasius)*: Faber, 1967: 74. Senior synonym of *kusnezovi:* Wilson, 1955a: 189. See also: Bernard, 1967: 365; Kutter, 1977c: 230; Collingwood, 1982: 292; Kupyanskaya, 1990: 221; Atanassov & Dlussky, 1992: 252.

*cereomicans. Lasius umbratus* subsp. *distinguendus* var. *cereomicans* Stärcke, 1937: 49 (w.q.m.) ITALY. Unavailable name; material referred to *umbratus* by Wilson, 1955a: 152, and later to *distinguendus* by Seifert, 1990: 10.

**chambonensis.* **Lasius chambonensis* Théobald, in Piton & Théobald, 1935: 82, pl. 5, fig. 5 (wing) FRANCE (Mio-Pliocene).

***cinereus.*** *Lasius (Lasius) cinereus* Seifert, 1992b: 33, fig. 25 (w.) SPAIN.

***citrinus.*** *Lasius bicornis* var. *citrina* Emery, 1922b: 12 (w.) ITALY. Combination in *L. (Chthonolasius)*: Emery, 1925b: 232. Junior synonym of *umbratus:* Wilson, 1955a: 151; of *affinis* Schenck, and hence first available replacement name for *Formica affinis* Schenck, 1852: 62 [Junior primary homonym of *Formica affinis* Leach, 1825: 290.], and raised to species: Seifert, 1990: 7.

*claripennis. Lasius (Formicina) flavus* subsp. *claripennis* Wheeler, W.M. 1917a: 527 (w.q.m.) U.S.A. Junior synonym of *microps:* Creighton, 1950a: 422.

*clavigera* Roger, 1862; see under ***ACANTHOMYOPS.***

*clavigeroides* Buren, 1942; see under ***ACANTHOMYOPS.***

*coloradensis* Wheeler, W.M. 1917; see under ***ACANTHOMYOPS.***

***coloratus.*** *Lasius niger* st. *coloratus* Santschi, 1937h: 387 (w.q.) TAIWAN. Combination in *L. (Lasius)*, and junior synonym of *niger:* Wilson, 1955a: 60. Revived from synonymy, and raised to species: Seifert, 1992b: 44.

***crinitus.*** *Formica crinita* Smith, F. 1858b: 13 (q.) INDIA. Wilson, 1955a: 188 (w.). Combination in *Lasius:* Mayr, 1862: 700; in *L. (Chthonolasius)*: Emery, 1925b: 233. Senior synonym of *hingstoni:* Wilson, 1955a: 187. See also: Bingham, 1903: 339; Collingwood, 1982: 293.

**crispus.* **Lasius crispus* Théobald, in Piton & Théobald, 1935: 68, pl. 5, fig. 3 (wing) FRANCE (Mio-Pliocene).

*crispus. Lasius (Dendrolasius) crispus* Wilson, 1955a: 144 (w.q.m.) JAPAN. [Unresolved junior primary homonym of *crispus* Théobald, above.] Junior synonym of *capitatus:* Kupyanskaya, 1989: 785.

***crypticus.*** *Lasius (Lasius) crypticus* Wilson, 1955a: 104 (w.q.m.) U.S.A.

*dalmatica. Lasius viehmeyeri* var. *dalmatica* Stärcke, 1937: 53 (q.m.) YUGOSLAVIA. Junior synonym of *umbratus:* Wilson, 1955a: 152; of *viehmeyeri:* Seifert, 1990: 11.

*dichrous* Roger, 1863; see under ***LASIOPHANES.***

***distinguendus.*** *Formicina umbrata* subsp. *distinguenda* Emery, 1916a: 64, figs. VI,4; VII,6 (w.q.) ITALY. Karavaiev, 1936: 214 (m.). [Also described as new, but *bicornis* susbp. *distinguenda,* by Emery, 1916b: 242.] Combination in *Lasius:* Müller, 1923: 13; in *L. (Chthonolasius)*: Emery, 1925b: 234. Raised to species: Bondroit, 1918: 32. Subspecies of *umbratus:* Emery, 1922b: 13; Stärcke, 1937: 47. Status as species: Müller, 1923: 130. Junior synonym of *umbratus:* Wilson, 1955a: 151. Revived from synonymy as subspecies of *umbratus:* Pisarski, 1975: 39. Revived status as species: Collingwood, 1978: 72. See also: Collingwood, 1979: 102; Agosti & Collingwood, 1987a: 58; Seifert, 1988: 149; Seifert, 1990: 10. Senior synonym of *hybrida:* Stärcke, 1937: 57; Seifert, 1988: 149; of *nuda,* and material of the unavailable name *cereomicans* referred here: Seifert, 1990: 10.

***draco.*** *Lasius draco* Collingwood, 1982: 294, figs. 9, 22 (w.) BHUTAN.

**edentatus.* **Lasius edentatus* Mayr, 1868c: 46 (m.) BALTIC AMBER (Oligocene). Junior synonym of **schiefferdeckeri:* Wilson, 1955a: 52.

*elegantulus. Lasius lasioides* subsp. *elegantulus* Arnol'di, 1948: 212. *Nomen nudum.*

***#emarginatobrunneus.*** *Lasius brunneus* var. *emarginatobrunneus* Ruzsky, 1902d: 17 (w.) RUSSIA (attributed to Forel).

***emarginatus.*** *Formica emarginata* Olivier, 1792: 494 (w.q.m.) FRANCE. Hauschteck, 1962: 219 (k.). Combination in *Lasius:* Fabricius, 1804: 416; in *Formicina (Donisthorpea)*: Emery, 1916b: 240; in *Lasius:* Menozzi, 1921: 32; Müller, 1923: 123; in *L. (Lasius)*: Wilson, 1955a: 89. Subspecies/race of *niger:* Forel, 1874: 46; Emery & Forel, 1879: 452; Ruzsky, 1905b: 302; Bondroit, 1910: 486; Stitz, 1914: 85. Status as species: André, 1882b: 193; Nasonov, 1889: 23; Ruzsky, 1902d: 16; Forel, 1915d: 53; Emery, 1916b: 240; Bondroit, 1918: 24; Müller, 1923: 123; Finzi, 1924a: 14; Karavaiev, 1927c: 279; Stitz, 1939: 283; Stärcke, 1944a: 155; Wilson, 1955a: 89; Bernard, 1967: 357; Kutter, 1977c: 228; Atanassov & Dlussky, 1992: 239; Seifert, 1992b: 34. Senior synonym of *brunneoemarginatus, brunneoides* and *nigroemarginatus:* Wilson, 1955a: 89; of *illyricus* and *pontica:* Seifert, 1992b: 34.

*emeryi. Lasius niger* subsp. *emeryi* Ruzsky, 1905b: 313 (w.) RUSSIA. Junior synonym of *niger:* Wilson, 1955a: 59. Unrecognisable taxon: Seifert, 1992b: 48.

****epicentrus.*** **Lasius epicentrus* Théobald, 1937b: 317, pl. 25, fig. 20 (m.) FRANCE (Oligocene).

*epinotalis. Lasius (Chthonolasius) umbratus* subsp. *epinotalis* Buren, 1944a: 297 (w.) U.S.A. Junior synonym of *umbratus:* Wilson, 1955a: 152.

*escamole. Lasius escamole* Reza, 1925: 12, figs. 1–14 (w.q.m.) MEXICO. Unidentifiable to genus; *incertae sedis* in *Lasius.* [Wilson, 1955a: 192, comments that this is not a *Lasius* but its correct generic combination cannot be ascertained.]

*exacutus. Lasius umbratus* var. *exacutus* Ruzsky, 1902d: 15 (w.) RUSSIA. Combination in *Acanthomyops:* Ruzsky, 1925b: 44; in *Lasius (Chthonolasius)*: Ruzsky, 1914a: 61; Emery, 1925b: 234. Subspecies of *umbratus:* Ruzsky, 1905b: 287. Raised to species: Ruzsky, 1925b: 44. Subspecies of *bicornis:* Menozzi, 1939a: 314 (in key). Junior synonym of *umbratus:* Wilson, 1955a: 150.

*exulans. Lasius exulans* Fabricius, 1804: 416 (q.) CENTRAL AMERICA. Unidentifiable to genus; *incertae sedis* in *Lasius.*

**fallax.** *Lasius* (*Cautolasius*) *fallax* Wilson, 1955a: 130 (w.q.m.) U.S.A.

**flavescens.** *Lasius niger* r. *flavescens* Forel, 1904b: 386 (w.) UZBEKISTAN. Ruzsky, 1905b: 304 (q.). Junior synonym of *niger:* Wilson, 1955a: 59. Revived from synonymy, and raised to species: Collingwood, 1961a: 72. See also: Dlussky, 1962: 181; Tarbinsky, 1976: 136; Seifert, 1992b: 26.

*flavidus. Acanthomyops niger* subsp. *alienus* var. *flavidus* Kuznetsov-Ugamsky, 1927e: 189 (w.) TURKESTAN. Unavailable name; material referred to *alienus* by Wilson, 1955a: 78.

*flavoides. Lasius flavus* var. *flavoides* Emery, 1925b: 231 (w.) SWITZERLAND. [First available use of *Lasius flavus* subsp. *myops* var. *flavoides* Forel, 1894d: 12; unavailable name.] Röszler, 1943: 47 (q.m.). Variety of *myops:* Ruzsky, 1925a: 288; of *flavus:* Karavaiev, 1936: 209. Junior synonym of *flavus:* Wilson, 1955a: 112; of *myops:* Kutter, 1977c: 229.

*flavomyops. Formicina flava* var. *flavomyops* Emery, 1916b: 241 (w.) SWITZERLAND. [First available use of *Lasius flavus* r. *myops* var. *flavomyops* Forel, 1915d: 52 (in key); unavailable name.] Combination in *Lasius:* Emery, 1925b: 231. Junior synonym of *flavoides:* Emery, 1925b: 231. Revived from synonymy, and raised to species: Röszler, 1943: 47. Junior synonym of *flavus:* Wilson, 1955a: 112; of *myops:* Kutter, 1977c: 229.

**flavoniger.** *Lasius* (*Lasius*) *flavoniger* Seifert, 1992b: 27, fig. 20 (w.) TURKEY.

**flavus.** *Formica flava* Fabricius, 1782: 491 (w.) EUROPE. Latreille, 1798: 42 (q.m.); Wheeler, G.C. & Wheeler, J. 1953c: 152 (l.); Hauschteck, 1962: 219 (k.); Imai, 1966: 120 (k.). Combination in *Lasius:* Mayr, 1861: 50; Emery, 1925b: 231; Kuznetsov-Ugamsky, 1929b: 36; in *Donisthorpea:* Donisthorpe, 1915d: 216; in *Formicina:* Emery, 1916b: 241; in *Acanthomyops:* Forel, 1916: 460; Kuznetsov-Ugamsky, 1927e: 187; in *Lasius* (*Chthonolasius*): Ruzsky, 1914a: 59; in *Chthonolasius:* Ruzsky, 1925a: 288; Ruzsky, 1936: 90; in *Lasius* (*Cautolasius*): Wilson, 1955a: 112. Senior synonym of *ruficornis:* Roger, 1862c: 285; of *ibericus,* and material of the unavailable name *sancho* referred here: Wilson, 1955a: 112; Seifert, 1990: 12; of *apennina, brevicornis, fuscoides, helvus, microps* (and its junior synonym *claripennis*), *morbosa, odoratus* and *olivacea:* Wilson, 1955a: 112. See also: Bernard, 1967: 359; Kutter, 1977c: 229; Collingwood, 1979: 96; Kupyanskaya, 1990: 222; Atanassov & Dlussky, 1992: 241.

**fuliginosus.** *Formica fuliginosa* Latreille, 1798: 36 (w.q.m.) FRANCE. Hauschteck, 1962: 219 (k.). Combination in *Lasius:* Mayr, 1861: 49; in *Donisthorpea:* Donisthorpe, 1915d: 188; in *Formicina:* Emery, 1916b: 242; in *Acanthomyops:* Forel, 1916: 460; in *Lasius* (*Dendrolasius*): Ruzsky, 1912: 630; Müller, 1923: 132; Emery, 1925b: 236; Wilson, 1955a: 138. Senior synonym of *nipponensis* and *orientalis:* Wilson, 1955a: 138. See also: Collingwood, 1982: 292; Kupyanskaya, 1989: 783; Atanassov & Dlussky, 1992: 243.

*fumatus. Lasius fumatus* Emery, 1869b: 26, fig. 3 (w.q.) ITALY. Junior synonym of *lasioides:* Emery, 1870: 194. [The species referred to as *Prenolepis lasioides* in the text of Emery, 1869b: 6, is referred to in the list of figures for that paper (p. 26) as *Lasius fumatus.* Both names refer to the same specimens and hence are automatically synonymous.]

*fuscoides. Lasius flavus* var. *fuscoides* Ruzsky, 1902e: 16 (w.) RUSSIA. Combination in *L.* (*Chthonolasius*): Ruzsky, 1914a: 61; in *Chthonolasius:* Ruzsky, 1925a: 288; in *Acanthomyops:* Kuznetsov-Ugamsky, 1927e: 187. Junior synonym of *flavus:* Wilson, 1955a: 112.

*fuscula. Prenolepis fusculus* Emery, 1869b: 8, pl. 1, fig. 4 (w.) ITALY. [Also referred to as *Lasius fusculus* by Emery, 1869b: 26.] Junior synonym of *lasioides:* Dalla Torre, 1893: 182.

**gebaueri.** *Lasius* (*Lasius*) *gebaueri* Seifert, 1992b: 22, fig. 15 (w.) TIBET.

*__globularis.__ *Formica globularis* Heer, 1850: 131, pl. 10, fig. 3 (q.) YUGOSLAVIA (Miocene). Combination in *Lasius:* Mayr, 1867b: 54.

**grandis.** *Lasius niger* var. *grandis* Forel, 1909c: 104 (w.) SPAIN. Seifert, 1992b: 32 (q.). Junior synonym of *alienus:* Wilson, 1955a: 77; of *niger:* Yarrow, 1967: 30. Revived from synonymy, and raised to species: Seifert, 1992b: 32.

**hayashi.** *Lasius hayashi* Yamauchi & Hayashida, 1970: 510, figs. 1, 5, 7, 8, 11 (w.q.m.) JAPAN. See also: Collingwood, 1976: 305; Kupyanskaya, 1990: 220; Seifert, 1992b: 45.

*helvus. Lasius helvus* Cook, 1953: 326, figs. (w.) U.S.A. [Also spelled *helveolus,* on p. 327.] Junior synonym of *flavus:* Wilson, 1955a: 113.

**hikosanus.** *Lasius* (*Chthonolasius*) *hikosanus* Yamauchi, 1979: 169 (w.) JAPAN.

**himalayanus.** *Lasius himalayanus* Bingham, 1903: 340 (w.) INDIA. [First available use of *Lasius niger* r. *brunneus* var. *himalayanus* Forel, 1894c: 404; unavailable name.] Variety/race of *brunneus:* Forel, 1906b: 85; Emery, 1925b: 229; Menozzi, 1939a: 313. Junior synonym of *brunneus:* Wilson, 1955a: 47. Revived from synonymy: Seifert, 1992b: 8 [ostensibly as a subspecies of *brunneus,* but Seifert states, on the same page, that *himalayanus* may be a valid species; so treated here].

*hingstoni. Acanthomyops* (*Chthonolasius*) *hingstoni* Donisthorpe, 1929a: 448 (w.) INDIA. Junior synonym of *crinitus:* Wilson, 1955a: 187.

**hirsutus.** *Lasius* (*Lasius*) *hirsutus* Seifert, 1992b: 41, fig. 31 (w.) PAKISTAN.

*hirtiscapus. Lasius umbratus* var. *hirtiscapus* Stärcke, 1937: 43 (q.) CZECHOSLOVAKIA. Variety of *mixtus:* Stitz, 1939: 299. Junior synonym of *umbratus:* Wilson, 1955a: 152.

*hoffmanni* Emery, 1895; see under ***LASIOPHANES.***

**humilis.** *Lasius* (*Formicina*) *humilis* Wheeler, W.M. 1917a: 528 (w.q.) U.S.A. Combination in *L.* (*Chthonolasius*): Emery, 1925b: 233. See also: Wilson, 1955a: 185.

*hybrida. Formicina distinguenda* var. *hybrida* Bondroit, 1918: 33 (w.q.) ITALY. [First available use of *Formicina umbrata* subsp. *distinguenda* var. *hybrida* Emery, 1916a: 66; unavailable name.] Combination in *Lasius* (*Chthonolasius*): Emery, 1925b: 234. Variety of *umbratus:* Emery, 1922b: 13. Junior synonym of *distinguendus:* Stärcke, 1937: 57. Revived from synonymy, and raised to species: Arnol'di & Dlussky, 1978: 555 (in key). Junior synonym of *distinguendus:* Seifert, 1988: 149.

*ibericus. Lasius* (*Chthonolasius*) *umbratus* st. *ibericus* Santschi, 1925g: 349, fig. 2 (w.) SPAIN. Junior synonym of *flavus:* Wilson, 1955a: 112.

*illyricus. Lasius alienus* subsp. *illyricus* Zimmermann, 1935: 50 (w.q.m.) YUGOSLAVIA. Junior synonym of *alienus:* Wilson, 1955a: 78; of *emarginatus:* Seifert, 1992b: 34.

*incisa. Formica incisa* Schenck, 1852: 63 (w.) GERMANY. Junior synonym of *bicornis:* Smith, F. 1858b: 8; Emery, 1922b: 13; Seifert, 1988: 161.

*interjectus* Mayr, 1866; see under ***ACANTHOMYOPS.***

**japonicus.** *Lasius* (*Lasius*) *emarginatus* var. *japonicus* Santschi, 1941: 277 (w.q.) JAPAN. Junior synonym of *niger:* Wilson, 1955a: 60. Revived from synonymy, and raised to species: Seifert, 1992b: 30.

**jensi.** *Lasius* (*Chthonolasius*) *jensi* Seifert, 1982: 85, 21 figs. (w.q.m.) GERMANY. Current subspecies: nominal plus *longiceps.* See also: Seifert, 1988: 155.

**karpinisi.** *Lasius* (*Lasius*) *karpinisi* Seifert, 1992b: 23, fig. 16 (w.) GREECE.

*kashmirensis. Acanthomyops* (*Chthonolasius*) *bicornis* subsp. *kashmirensis* Donisthorpe, 1930a: 225 (q.m.) INDIA. Junior synonym of *bicornis:* Wilson, 1955a: 183.

**koreanus.** *Lasius* (*Lasius*) *koreanus* Seifert, 1992b: 47, fig. 39 (w.q.) KOREA.

*kusnezovi. Lasius* (*Chthonolasius*) *carniolicus* var. *kusnezovi* Karavaiev, 1929b: 212, fig. 5 (w.) RUSSIA. Junior synonym of *carniolicus:* Wilson, 1955a: 189.

**lasioides.** *Prenolepis lasioides* Emery, 1869b: 6, pl. 1, fig. 3 (w.q.m.) ITALY. Combination in *Lasius:* Emery, 1870: 194; in *Formicina* (*Donisthorpea*): Emery, 1916b: 241; in *Acanthomyops:* Kuznetsov-Ugamsky, 1927e: 189; in *Lasius:* Müller, 1923: 126; Kuznetsov-Ugamsky, 1929b: 38. Variety/subspecies of *alienus:* Dalla Torre, 1893: 182; of *niger:* Forel, 1913d: 434; Emery, 1916b: 241; Menozzi, 1925d: 34. Revived status as species: Stärcke, 1944a: 155. Junior synonym of *alienus:* Wilson, 1955a: 77; Agosti & Collingwood, 1987a: 58 [Repeat of this synonymy in Arakelian, 1994: 117, is ignored]. Revived from synonymy: Seifert, 1992b: 8. Senior synonym of *fumatus:* Emery, 1870: 194; of *fuscula:* Dalla Torre, 1893: 182; of *barbarus:* Seifert, 1992b: 8.

**lawarai.** *Lasius* (*Lasius*) *lawarai* Seifert, 1992b:38, fig. 28 (w.) PAKISTAN.

*__longaevus.__ *Formica longaeva* Heer, 1850: 132, pl. 10, fig. 4 (q.) YUGOSLAVIA (Miocene). Combination in *Lasius:* Mayr, 1867b: 54.

#**longiceps.** *Lasius* (*Chthonolasius*) *jensi* subsp. *longiceps* Seifert, 1988: 158, figs. 12, 22, 31 (w.q.) RUSSIA.

*__longipennis.__ *Formica longipennis* Heer, 1850: 136, pl. 10, fig. 7 (q.) YUGOSLAVIA (Miocene). Combination in *Lasius:* Mayr, 1867b: 55.

**magnus.** *Lasius* (*Lasius*) *magnus* Seifert, 1992b: 43, figs. 33–35, 40 (w.q.) BHUTAN.

**martynovi* Popov, 1933; see under ***FORMICA.***

*mellea. Formica mellea* Provancher, 1881b: 356 (w.) CANADA. [Unresolved junior primary homonym of *Formica mellea* Say, 1836: 286 (now in *Camponotus*).] [Also described as new by Provancher, 1883: 599.] Junior synonym of *flavus:*

Dalla Torre, 1893: 185; Provancher, 1887: 236; of *nearcticus:* Francoeur & Béique, 1966: 144.

***meridionalis.*** *Formicina meridionalis* Bondroit, 1920a: 143 (q.) FRANCE. Stärcke, 1937: 52 (w.m.). Combination in *Lasius:* Emery, 1922b: 13; in *L.* (*Chthonolasius*): Emery, 1925b: 234. Variety of *umbratus:* Emery, 1922b: 13; Menozzi, 1925d: 34; Stitz, 1939: 295; Novak & Sadil, 1941: 103. Status as species: Stärcke, 1937: 50. Junior synonym of *rabaudi:* Wilson, 1955a: 168; Bernard, 1967: 364. Revived from synonymy, status as species: Pisarski, 1975: 37; Collingwood, 1979: 102; Seifert, 1988: 154; Kupyanskaya, 1990: 226; Atanassov & Dlussky, 1992: 247.

*mexicanus* Wheeler, W.M. 1914; see under ***ACANTHOMYOPS.***

*microgyna. Formicina microgyna* Bondroit, 1918: 33, figs. 16, 17 (q.m.) FRANCE. Combination in *Lasius* (*Chthonolasius*): Emery, 1925b: 233. Variety of *bicornis:* Emery, 1925b: 233; Stärcke, 1937: 56. Junior synonym of *bicornis:* Vandel, 1926: 197; Wilson, 1955a: 183.

*microps. Lasius* (*Formicina*) *brevicornis* var. *microps* Wheeler, W.M. 1917a: 526 (w.) U.S.A. Wheeler, G.C. & Wheeler, J. 1953c: 152 (l.). Combination in *L.* (*Chthonolasius*): Creighton, 1950a: 422. Subspecies of *brevicornis:* Wheeler, G.C. & Wheeler, E.W. 1944: 253; of *flavus:* Creighton, 1950a: 422. Senior synonym of *claripennis:* Creighton, 1950a: 422. Junior synonym of *flavus:* Wilson, 1955a: 112.

***mikir.*** *Lasius mikir* Collingwood, 1982: 295, figs. 10, 23 (w.) INDIA.

*minimus. Acanthomyops niger* var. *minimus* Kuznetsov-Ugamsky, 1928b: 20 (w.) RUSSIA. Combination in *Lasius:* Kuznetsov-Ugamsky, 1929a: 26. Junior synonym of *niger:* Wilson, 1955a: 59. Unrecognisable taxon: Seifert, 1992b: 48.

****minutulus.*** **Formica minutula* Heer, 1850: 136, pl. 10, fig. 8 (q.) YUGOSLAVIA (Miocene). Combination in *Lasius:* Mayr, 1867b: 55. See also: Théobald, 1937b: 319.

***minutus.*** *Lasius umbratus* subsp. *minutus* Emery, 1893i: 641 (w.q.m.) U.S.A. Combination in *L.* (*Chthonolasius*): Ruzsky, 1914a: 61. Subspecies of *bicornis:* Emery, 1925b: 233. Raised to species: Wilson, 1955a: 180. See also: Wheeler, W.M. 1910e: 241; Smith, D.R. 1979: 1483.

*mixtoaffinis. Lasius umbratus* var. *mixtoaffinis* Ruzsky, 1902d: 15 (w.) RUSSIA. *Nomen nudum;* material referred to *umbratus* by Wilson, 1955a: 151.

*mixtobicornis. Lasius umbratus* var. *mixtobicornis* Ruzsky, 1905b: 292, RUSSIA. *Nomen nudum;* material referred to *umbratus* by Wilson, 1955a: 151.

*mixtoumbratus. Lasius umbratus* var. *mixtoumbratus* Forel, 1874: 48 (w.q.) SWITZERLAND. Combination in *Donisthorpea:* Donisthorpe, 1915d: 233; in *Formicina:* Emery, 1916b: 242; in *Lasius* (*Chthonolasius*): Emery, 1925b: 234. Raised to species: Bondroit, 1918: 30. Variety of *mixtus:* Stitz, 1939: 298; of *umbratus:* Novak & Sadil, 1941: 103. Junior synonym of *umbratus:* Wilson, 1955a: 151; of *mixtus:* Pisarski, 1975: 38. Unrecognisable taxon: Seifert, 1990: 6.

***mixtus.*** *Formica mixta* Nylander, 1846b: 1050 (w.q.m.) FINLAND. Combination in *Lasius:* Mayr, 1861: 50; in *Donisthorpea:* Donisthorpe, 1915d: 236; in *Formicina:* Emery, 1916b: 242; in *Acanthomyops:* Ruzsky, 1925b: 45; in *Lasius* (*Chthonolasius*): Ruzsky, 1914a: 59; Emery, 1925b: 234; Wilson, 1955a: 150. Variety/subspecies of *umbratus:* Forel, 1874: 47; Mayr, 1886d: 430; Wheeler, W.M. 1904e: 305; Bondroit, 1910: 485; Emery, 1914d: 159; Wheeler, W.M. 1917a: 528; Wheeler, W.M. 1917h: 167; Emery, 1922b: 13; Emery, 1925b: 234; Karavaiev, 1927c: 279. Status as species: André, 1882b: 196; Nasonov, 1889: 24; Ruzsky, 1902d: 15; Bondroit, 1912: 352; Donisthorpe, 1915d: 236; Emery, 1916b: 242; Bondroit, 1918: 32; Finzi, 1930d: 317; Stärcke, 1937: 52; Stitz, 1939: 295; Holgersen, 1940: 185; Novak & Sadil, 1941: 103. Junior synonym of *umbratus:* Wilson, 1955a: 150. Revived from synonymy, and status as species: Collingwood, 1962: 219; Collingwood, 1963: 149; Collingwood & Yarrow, 1969: 80; Kutter, 1977c: 233; Collingwood, 1979: 106; Seifert, 1988: 145; Atanassov & Dlussky, 1992: 249. Senior synonym of *mixtoumbratus:* Pisarski, 1975: 38.

*monticola. Formica monticola* Buckley, 1866: 157 (w.q.m.) U.S.A. Combination in *Lasius:* Emery, 1895c: 337; Wheeler, W.M. 1902f: 20; Emery, 1925b: 237. Unrecognisable taxon; *incertae sedis* in *Lasius.*

*morbosa. Formicina flava* var. *morbosa* Bondroit, 1918: 28 (w.q.) FRANCE. Junior synonym of *flavoides:* Emery, 1925b: 231; of *flavus:* Wilson, 1955a: 112.

***morisitai.*** *Lasius* (*Dendrolasius*) *morisitai* Yamauchi, 1979: 176 (w.q.) JAPAN. See also: Kupyanskaya, 1989: 788.

*murphyi* Forel, 1901; see under ***ACANTHOMYOPS.***

***myops.*** *Lasius flavus* r. *myops* Forel, 1894d: 12 (w.) ALGERIA. Forel, 1915d: 54 (in key) (q.); Seifert, 1983: 13 (m.). Combination in *Formicina:* Emery, 1916b: 241; in *Chthonolasius:* Ruzsky, 1925a: 288; in *Acanthomyops:* Ruzsky, 1925b: 44; in *Lasius:* Müller, 1923: 128; in *L.* (*Chthonolasius*): Ruzsky, 1914a: 59; in *L.* (*Cautolasius*): Wilson, 1955a: 112. Variety/subspecies of *flavus:* Forel, 1915d: 51; Emery, 1916b: 241; Müller, 1923: 128; Karavaiev, 1927c: 278. Status as species: Wheeler, W.M. 1905f: 397; Bondroit, 1918: 29; Ruzsky, 1925a: 288. Junior synonym of *flavus:* Wilson, 1955a: 112; Atanassov & Dlussky, 1992: 241; Arakelian, 1994: 120. Revived from synonymy, raised to species, and senior synonym of *flavoides* and *flavomyops:* Kutter, 1977c: 229; Seifert, 1983: 13; and current status.

*neapolitana. Lasius bicornis* var. *neapolitana* Emery, 1922b: 13 (q.) ITALY. Junior synonym of *bicornis:* Wilson, 1955a: 183.

***nearcticus.*** *Lasius flavus* subsp. *nearcticus* Wheeler, W.M. 1906a: 38 (w.q.m.) U.S.A. Wheeler, G.C. & Wheeler, J. 1953c: 151 (l.); Crozier, 1970: 114 (k.). Combination in *L.* (*Chthonolasius*): Creighton, 1950a: 422; in *L.* (*Cautolasius*), and raised to species: Wilson, 1955a: 133. Senior synonym of *mellea:* Francoeur & Béique, 1966: 144.

*neglectus. Lasius* (*Lasius*) *neglectus* Van Loon, Boomsma & Andrasfalvy, 1990: 350, fig. 5 (w.q.m.) HUNGARY. Junior synonym of *turcicus:* Seifert, 1992b: 10.

****nemorivagus.*** **Lasius nemorivagus* Wheeler, W.M. 1915h: 123 (q.) BALTIC AMBER (Oligocene). See also: Wilson, 1955a: 150.

***neoniger.*** *Lasius niger* var. *neoniger* Emery, 1893i: 639 (w.q.m.) U.S.A. Variety/subspecies of *niger:* Wheeler, W.M. 1917a: 463; Buren, 1944a: 296; Creighton, 1950a: 420. Status as species: Gregg, E.V. 1945: 530; Wilson, 1955a: 97; Smith, D.R. 1979: 1436.

#****neuter.*** **Formica redtenbacheri* subsp. *neutra* Heer, 1850: 130 (q.?). Combination in *Lasius:* **new combination** [as *redtenbacheri* referred to *Lasius* by Mayr, 1867b: 54].

***nevadensis.*** *Lasius* (*Chthonolasius*) *nevadensis* Cole, 1956a: 26 (w.q.m.) U.S.A.

***niger.*** *Formica nigra* Linnaeus, 1758: 580 (w.) EUROPE. Latreille, 1798: 43 (q.m.); Wheeler, G.C. & Wheeler, J. 1953c: 148 (l.); Hauschteck, 1962: 219 (k.); Imai & Kubota, 1972: 196 (k.). Combination in *Lasius:* Fabricius, 1804: 415; in *Donisthorpea:* Donisthorpe, 1915d: 200; in *Formicina* (*Donisthorpea*): Emery, 1916b: 240; in *Acanthomyops:* Forel, 1916: 460; in *Lasius:* Menozzi, 1921: 32; Müller, 1923: 124; Kuznetsov-Ugamsky, 1929a: 26; Wilson, 1955a: 59. Senior synonym of *pallescens:* Mayr, 1861: 49; of *nigerrima:* Emery, 1892b: 162; of *alienoniger, emeryi, nitidus, minimus* and *transylvanica,* and material of the unavailable name *pilicornis* referred here: Wilson, 1955a: 59. Current subspecies: nominal plus *pinetorum.* See also: Imai, 1966: 119; Tarbinsky, 1976: 134; Kutter, 1977c: 227; Collingwood, 1982: 287; Kupyanskaya, 1990: 219; Seifert, 1992b: 27.

*nigerrima. Formica nigerrima* Christ, 1791: 513 (w.) no locality given. Junior synonym of *niger:* Emery, 1892b: 162.

***nigrescens.*** *Lasius emarginatus* var. *nigrescens* Stitz, 1930: 240 (q.) RUSSIA. Junior synonym of *niger:* Wilson, 1955a: 60. Revived from synonymy, and raised to species: Seifert, 1992b: 38.

*nigrobrunneus. Acanthomyops* (*Dendrolasius*) *brunneus* var. *nigrobrunneus* Donisthorpe, 1926b: 18 (w.) ITALY. Junior synonym of *brunneus:* Wilson, 1955a: 47.

*nigroemarginatus. Lasius niger* var. *nigroemarginatus* Forel, 1874: 47 (w.q.) SWITZERLAND. Variety of *emarginatus:* Dalla Torre, 1893: 184; Forel, 1915d: 53 (in key); Karavaiev, 1927c: 279. Junior synonym of *emarginatus:* Wilson, 1955a: 89.

*nipponensis. Lasius fuliginosus* var. *nipponensis* Forel, 1912l: 339 (w.q.) JAPAN. Junior synonym of *fuliginosus:* Wilson, 1955a: 138.

*nitidus. Acanthomyops niger* subsp. *nitidus* Kuznetsov-Ugamsky, 1927e: 188 (w.) TURKESTAN. Junior synonym of *niger:* Wilson, 1955a: 59. Unrecognisable taxon: Seifert, 1992b: 48.

*nuda. Formicina umbrata* var. *nuda* Bondroit, 1917a: 176, fig. 2 (q.) no locality given. Junior synonym of *affinis:* Emery, 1925b: 234; of *umbratus:* Stärcke, 1937: 57; Agosti & Collingwood, 1987a: 58; of *distinguendus:* Seifert, 1990: 10.

*nyaradi. Chthonolasius affinis* var. *nyaradi* Röszler, 1943: 47 (w.q.) ROMANIA. Junior synonym of *umbratus:* Wilson, 1955a: 152.

***obliteratus.** *Formica obliterata* Heer, 1850: 144, pl. 11, figs. 11, 12 (m.) YUGOSLAVIA (Miocene). Combination in *Lasius:* Mayr, 1867b: 56.

***oblongus.** **Lasius oblongus* Assmann, 1870: 38, pl. 1, fig. 1 (wing) GERMANY (Miocene).

***obscurus.** **Formica obscura* Heer, 1850: 119, pl. 9, fig. 2 (q.) YUGOSLAVIA (Miocene). Combination in *Lasius:* Mayr, 1867b: 52.

**obscuratus.** *Lasius brunneus* var. *obscuratus* Stitz, 1930: 239 (w.q.) CAUCASUS. Junior synonym of *alienus:* Wilson, 1955a: 78. Revived from synonymy, and raised to species: Seifert, 1992b: 21.

*occidentalis* Wheeler, W.M. 1909; see under ***ACANTHOMYOPS.***

***occultatus.** **Formica occultata* Heer, 1850: 134, pl. 10, fig. 6; pl. 11, fig. 11 (q.m.) YUGOSLAVIA (Miocene). Combination in *Lasius:* Mayr, 1867b: 55. Current subspecies: nominal plus **parschlugianus.*

*odoratus. Lasius flavus* var. *odoratus* Ruzsky, 1905b: 282 (w.) RUSSIA. Junior synonym of *flavus:* Wilson, 1955a: 112.

*oertzeni. Lasius bicornis* subsp. *oertzeni* Forel, 1910a: 26 (w.q.m.) GREECE. Combination in *L.* (*Chthonolasius*): Emery, 1925b: 233. Raised to species: Stärcke, 1937: 56. Junior synonym of *bicornis:* Wilson, 1955a: 183.

*olivacea. Lasius* (*Lasius*) *flavus* var. *olivacea* Karavaiev, 1926e: 194 (w.) CAUCASUS. Junior synonym of *flavus:* Wilson, 1955a: 113.

*orientalis. Lasius fuliginosus* var. *orientalis* Karavaiev, 1912b: 586 (w.) KOREA. Junior synonym of *fuliginosus:* Wilson, 1955a: 138.

*osakana. Lasius* (*Chthonolasius*) *silvestrii* var. *osakana* Santschi, 1941: 278 (q.) JAPAN. Junior synonym of *umbratus:* Wilson, 1955a: 152.

*ouchii. Lasius* (*Chthonolasius*) *ouchii* Teranishi, 1940: 76 (q.) JAPAN. Junior synonym of *teranishii:* Morisita, 1945: 23.

*pallescens. Formica pallescens* Schenck, 1852: 55 (q.m.) GERMANY. Junior synonym of *niger:* Mayr, 1861: 49.

*pallida. Formica pallida* Latreille, 1798: 41 (w.q.) FRANCE. Junior synonym of *brunneus:* Latreille, 1802c: 169. Unrecognisable taxon: Seifert, 1992b: 48. [As the taxa *brunneus* and *pallida* were both described by Latreille, and the synonymy was also established by Latreille, it seems reasonable to let it stand.]

**pallitarsis.** *Formica pallitarsis* Provancher, 1881b: 355 (w.m.) CANADA. [Also described as new by Provancher, 1883: 598.] Hung, 1969: 456 (k.). Combination in *Lasius:* André, 1887: 288. Junior synonym of *alienus:* André, 1887: 288; Wilson, 1955a: 77. Revived from synonymy, and senior synonym of *sitkaensis:* Francoeur & Béique, 1966: 144. See also: Smith, D.R. 1979: 1437.

*pannonica. Lasius alienus* var. *pannonica* Röszler, 1942b: 40 (w.) HUNGARY. Junior synonym of *alienus:* Wilson, 1955a: 78. Unrecognisable taxon: Seifert, 1992b: 48.

**paralienus.** *Lasius* (*Lasius*) *paralienus* Seifert, 1992b: 16, figs. 5, 6, 12 (w.q.) GERMANY.

#***parschlugianus.** **Formica occultata* subsp. *parschlugiana* Heer, 1850: 135 (caste?) AUSTRIA ? Combination in *Lasius:* Handlirsch, 1907: 862.

*parvula* Smith, M.R. 1934; see under ***ACANTHOMYOPS.***

***peritulus.** **Tetramorium peritulum* Cockerell, 1927: 165 (m.) U.S.A. (Miocene). Combination in *Lasius:* Carpenter, 1930: 58. See also: Wilson, 1955a: 58.

*picinus* Roger, 1863; see under ***LASIOPHANES.***

*pilicornis. Acanthomyops niger* subsp. *alienus* var. *pilicornis* Kuznetsov-Ugamsky, 1927e: 189 (w.) TURKESTAN. Unavailable name; material referred to *niger* by Wilson, 1955a: 59.

**piliferus.** *Lasius* (*Lasius*) *piliferus* Seifert, 1992b: 20, fig. 13 (w.) SPAIN.

*pilosus. Lasius* (*Chthonolasius*) *pilosus* Smith, M.R. 1934a: 384 (w.) U.S.A. Junior synonym of *vestitus:* Wilson, 1955a: 173.

#**pinetorum.** *Lasius niger* var. *pinetorum* Ruzsky, 1907: 68 (footnote) (w.) RUSSIA.

**platythorax.** *Lasius* (*Lasius*) *platythorax* Seifert, 1991: 73, figs. 3, 6, 7 (w.q.) GERMANY. See also: Seifert, 1992b: 29.

*plumopilosus* Buren, 1941; see under ***ACANTHOMYOPS.***

*pogonogynus* Buren, 1950; see under ***ACANTHOMYOPS.***

*pontica. Lasius alienus* var. *pontica* Stärcke, 1944a: 157 (w.) CAUCASUS. Junior synonym of *alienus:* Wilson, 1955a: 78 [repeat of this synonymy in Arakelian, 1994: 117, is ignored]; of *emarginatus:* Seifert, 1992b: 34.

**productus.** *Lasius* (*Lasius*) *productus* Wilson, 1955a: 95 (w.q.m.) JAPAN. See also: Seifert, 1992b: 46.

**przewalskii.** *Lasius* (*Chthonolasius*) *umbratus* subsp. *przewalskii* Ruzsky, 1915a: 434 (w.) TIBET. Junior synonym of *umbratus:* Wilson, 1955a: 151. Revived from synonymy, and raised to species: Pisarski & Krzysztofiak, 1981: 161.

**psammophilus.** *Lasius* (*Lasius*) *psammophilus* Seifert, 1992b: 15, figs. 4, 6, 11 (w.q.) GERMANY.

*pubescens* Buren, 1942; see under ***ACANTHOMYOPS.***

***pumilus.** **Lasius pumilus* Mayr, 1868c: 46, pl. 2, fig. 33 (w.) BALTIC AMBER (Oligocene). Senior synonym of **punctulatus:* Wilson, 1955a: 57. [The name **pumilus* misspelled as **pusillus* by Wheeler, W.M. 1915h: 142.]

**punctulatus. *Lasius punctulatus* Mayr, 1868c: 46, pl. 2, fig. 34 (q.) BALTIC AMBER (Oligocene). Junior synonym of **pumilus:* Wilson, 1955a: 57.

**pusillus* Wheeler, W.M. 1915h: 142; see under ***pumilus,** above.

**rabaudi.** *Formicina rabaudi* Bondroit, 1917a: 177, fig. 2 (q.) FRANCE. Emery, 1924c: 170 (w.); Wilson, 1955a: 168 (m.). Combination in *Lasius* (*Chthonolasius*): Emery, 1925b: 233. Junior synonym of *umbratus:* Bourne, 1973: 25. Revived from synonymy: Kutter, 1977c: 234. Senior synonym of *tibialis:* Wilson, 1955a: 168. See also: Bernard, 1967: 364; Arnol'di & Dlussky, 1978: 555; Seifert, 1988: 159.

***redtenbacheri.** **Formica redtenbacheri* Heer, 1850: 129, pl. 10, fig. 2 (w.q.) YUGOSLAVIA (Miocene). Combination in *Lasius:* Mayr, 1867b: 54. Current subspecies: nominal plus **neuter.*

**reginae.** *Lasius* (*Austrolasius*) *reginae* Faber, 1967: 75, figs. 2–6, 8–13, 16–19, 21–23 (w.q.m.) AUSTRIA.

*rubiginosa. Formica rubiginosa* Latreille, 1802c: 170 (q.) FRANCE. Combination in *Lasius:* Emery, 1925b: 230. [Bondroit, 1918: 25, suggests a possible correlation of this name with *L. alienus,* but *rubiginosa* is best regarded as an unrecognisable taxon, *incertae sedis* in *Lasius.*]

*ruficornis. Formica ruficornis* Fabricius, 1804: 397 (q.) INDIA. Junior synonym of *flavus:* Roger, 1862c: 285. Unrecognisable taxon; *incertae sedis* in *Lasius.* [Identity of this taxon is unknowable (material lost, description very vague), so Roger's synonymy may stand. Emery's, 1925b: 231, provisional synonymy of *ruficornis* with *nearcticus* is certainly incorrect.]

**sabularum.** *Formicina umbrata* var. *sabularum* Bondroit, 1918: 30 (q.) FRANCE. Seifert, 1988: 147 (w.m.). Combination in *Lasius* (*Chthonolasius*): Emery, 1925b: 234. Variety of *mixtus:* Bondroit, 1920a: 143; Stitz, 1939: 299. Junior synonym of *umbratus:* Stärcke, 1937: 56; Wilson, 1955a: 151. Revived from synonymy, and raised to species: Seifert, 1988: 147.

**sakagamii.** *Lasius sakagamii* Yamauchi & Hayashida, 1970: 504, figs. 3, 4, 6, 9, 10 (w.q.m.) JAPAN. See also: Seifert, 1992b: 42.

*sancho. Lasius* (*Chthonolasius*) *umbratus* st. *ibericus* var. *sancho* Santschi, 1925g: 350 (w.) SPAIN. Unavailable name; material referred to *flavus* by Wilson, 1955a: 113.

**schaeferi.** *Lasius* (*Lasius*) *schaeferi* Seifert, 1992b: 40, fig. 30 (w.q.m.) TIBET.

***schiefferdeckeri.** **Lasius schiefferdeckeri* Mayr, 1868c: 44, pl. 1, fig. 2; pl. 2, figs. 27–32 (w.q.m.) BALTIC AMBER (Oligocene). Senior synonym of **edentatus:* Wilson, 1955a: 52.

**schulzi.** *Lasius* (*Lasius*) *schulzi* Seifert, 1992b: 24, fig. 18 (w.q.) TURKEY.

*silvestrii. Lasius* (*Chthonolasius*) *silvestrii* Wheeler, W.M. 1928d: 120 (q.) JAPAN. Junior synonym of *umbratus:* Wilson, 1955a: 152.

**sitiens.** *Lasius* (*Lasius*) *sitiens* Wilson, 1955a: 108 (w.q.m.) U.S.A.

*sitkaensis. Lasius niger* subsp. *sitkaensis* Pergande, 1900: 519 (w.) U.S.A. Wheeler, W.M. 1917a: 525 (q.m.). Junior synonym of *neoniger:* Creighton, 1950a: 420. Revived from synonymy, and raised to species: Wilson, 1955a: 36. Junior synonym of *pallitarsis:* Francoeur & Béique, 1966: 144.

**sonobei.** *Lasius* (*Cautolasius*) *sonobei* Yamauchi, 1979: 163 (w.q.) JAPAN.

**spathepus.** *Lasius spathepus* Wheeler, W.M. 1910b: 130, fig. 1 (q.) JAPAN. Wheeler, W.M. 1928d: 122 (w.); Wilson, 1955a: 149 (m.). Combination in *L.* (*Dendrolasius*): Emery, 1925b: 236. Variety of *fuliginosus:* Teranishi, 1927: 90. Revived status as species: Wheeler, W.M. 1928d: 122. See also: Teranishi, 1940: 76; Wilson, 1955a: 149; Kupyanskaya, 1989: 787; Kupyanskaya, 1990: 231.

**speculiventris.** *Lasius speculiventris* Emery, 1893i: 641 (w.m.) U.S.A. Wilson, 1955a: 172 (q.). Combination in *L.* (*Chthonolasius*): Emery, 1925b: 235. Subspecies of *umbratus:* Wheeler, W.M. 1910e: 237; Wheeler, W.M. 1917h: 167. Revived status as species: Creighton, 1950a: 423.

*subglaber* Emery, 1893; see under ***ACANTHOMYOPS.***

***LASIUS** (continued)*

***subumbratus.*** *Lasius umbratus* subsp. *subumbratus* Viereck, 1903: 73 (q.) U.S.A. Wheeler, W.M. 1910e: 238 (w.m.). Combination in *L.* (*Chthonolasius*): Emery, 1925b: 235. Raised to species: Creighton, 1950a: 424. See also: Wheeler, W.M. 1917h: 167; Wilson, 1955a: 175.

*subvillosus. Lasius* (*Chthonolasius*) *affinis* subsp. *subvillosus* Arnol'di, 1948: 212. *Nomen nudum.*

***talpa.*** *Lasius* (*Cautolasius*) *talpa* Wilson, 1955a: 136 (w.q.m.) JAPAN. Imai, 1966: 119 (k.). See also: Collingwood, 1982: 291.

***tebessae.*** *Lasius* (*Lasius*) *tebessae* Seifert, 1992b: 37, fig. 27 (w.q.) ALGERIA.

***teranishii.*** *Lasius* (*Chthonolasius*) *teranishii* Wheeler, W.M. 1928d: 120 (q.) JAPAN. Yamauchi & Hayashida, 1968: 399 (w.m.). Combination in *L.* (*Dendrolasius*), and senior synonym of *ouchii:* Wilson, 1955a: 146. See also: Kupyanskaya, 1989: 786.

****terreus.*** **Lasius terreus* Scudder, 1878: 747 (w.) U.S.A. (Eocene). Carpenter, 1930: 19 has this as *incertae sedis* in Formicidae.

****tertiarius.*** **Lasius tertiarius* Zalessky, 1949: 51 (caste?) RUSSIA.

*tibialis. Lasius* (*Chthonolasius*) *tibialis* Santschi, 1936c: 208 (q.) MOROCCO. Junior synonym of *rabaudi:* Wilson, 1955a: 168.

*timida. Formica timida* Foerster, 1850a: 35 (w.) GERMANY. Schenck, 1852: 54 (q.m.). Junior synonym of *brunneus:* Smith, F. 1858b: 7; Seifert, 1992b: 6.

*transylvanica. Lasius transylvanica* Röszler, 1943: 44 (w.q.m.) ROMANIA. Junior synonym of *niger:* Wilson, 1955a: 60. Unrecognisable taxon: Seifert, 1992b: 48.

***turcicus.*** *Lasius niger* st. *turcica* Santschi, 1921a: 115 (w.q.) TURKEY. Subspecies of *alienus:* Emery, 1925b: 230; Stärcke, 1944a: 157 (in key). Junior synonym of *alienus:* Wilson, 1955a: 78 [Repeat of this synonymy in Arakelian, 1994: 117, is ignored.]. Revived from synonymy, raised to species, and senior synonym of *neglectus:* Seifert, 1992b: 10.

*turkmenus. Acanthomyops niger* subsp. *alienus* var. *turkmenus* Kuznetsov-Ugamsky, 1927e: 189 (w.) TURKESTAN. Unavailable name; material referred to *alienus* by Wilson, 1955a: 78.

*umbratoaffinis. Lasius umbratus* subsp. *affinis* f.interm. *umbratoaffinis* Ruzsky, 1895: 24 (w.) RUSSIA. Unavailable name.

***umbratus.*** *Formica umbrata* Nylander, 1846b: 1048 (q.m.) FINLAND. Schenck, 1852: 59 (w.); Hauschteck, 1962: 219 (k.); Hung, 1969: 456 (k.). Combination in *Lasius:* Mayr, 1861: 50; in *Donisthorpea:* Donisthorpe, 1915d: 223; in *Formicina:* Emery, 1916b: 241; in *Acanthomyops:* Ruzsky, 1925b: 44; in *Chthonolasius:* Ruzsky, 1936: 91; in *Lasius* (*Chthonolasius*): Ruzsky, 1914a: 59; Müller, 1923: 129; Emery, 1925b: 233; Wilson, 1955a: 150. Senior synonym of *belgarum:* Wilson, 1955a: 150; Seifert, 1988: 150; of *hirtiscapus:* Wilson, 1955a: 150; Seifert, 1990: 11; of *affinoumbratus, aphidicola, epinotalis, exacutus, nyaradi, osakana* and *silvestrii,* and material of the unavailable names *mixtoaffinis* and *mixtobicornis* referred here: Wilson, 1955a: 150. See also: Stitz, 1939: 291; Bernard, 1967: 363; Kutter, 1977c: 232; Seifert, 1988: 150; Atanassov & Dlussky, 1992: 245.

***uzbeki.*** *Lasius* (*Lasius*) *uzbeki* Seifert, 1992b: 25, fig. 41 (w.) UZBEKISTAN.

***vestitus.*** *Lasius umbratus* subsp. *vestitus* Wheeler, W.M. 1910e: 242 (q.) U.S.A. Wilson, 1955a: 174 (w.). Combination in *L.* (*Chthonolasius*): Emery, 1925b: 235. Raised to species: Creighton, 1950a: 425. Senior synonym of *pilosus:* Wilson, 1955a: 173.

****vetulus.*** **Lasius vetulus* Dlussky, 1981b: 72, figs. 53, 54 (q.) RUSSIA (Miocene).

***viehmeyeri.*** *Lasius umbratus* var. *viehmeyeri* Emery, 1922b: 13, fig. 2 (w.q.) GREECE. Combination in *L.* (*Chthonolasius*): Emery, 1925b: 234. Raised to species: Stärcke, 1937: 53. Junior synonym of *umbratus:* Wilson, 1955a: 152. Revived from synonymy, and senior synonym of *dalmatica:* Seifert, 1990: 3, 11.

***wittmeri.*** *Lasius* (*Lasius*) *wittmeri* Seifert, 1992b: 39, fig. 29 (w.) PAKISTAN.

***xerophilus.*** *Lasius xerophilus* MacKay, W.P. & MacKay, E.E., 1994: 37, figs. 1–7, 9 (w.q.m.) U.S.A.

---

**EXCURSUS**

**Species-rank taxa of *LASIUS,* grouped by subgenus.**

#***AUSTROLASIUS:*** *carniolicus, reginae.*

#***CAUTOLASIUS:*** *alienoflavus, fallax, flavus, myops, nearcticus, sonobei, talpa.*

#***CHTHONOLASIUS:*** *atopus, balcanicus, bicornis, citrinus, crinitus, distinguendus, draco, hikosanus, humilis, jensi, meridionalis, mikir, minutus, mixtus, *nemorivagus, nevadensis, przewalskii, rabaudi, sabularum, speculiventris, subumbratus, umbratus, vestitus, viehmeyeri.*

#***DENDROLASIUS:*** *buccatus, capitatus, fuliginosus, morisitai, spathepus, teranishii.*

#***LASIUS:*** *alienus, breviscapus, brunneus, cinereus, coloratus, crypticus, emarginatus, flavescens, flavoniger, gebaueri, grandis, hayashi, himalayanus, hirsutus, japonicus, karpinisi, koreanus, lasioides, lawarai, magnus, neoniger, niger, nigrescens, obscuratus, pallitarsis, paralienus, *peritulus, piliferus, platythorax, productus, psammophilus, *pumilus, sakagamii, schaeferi, *schiefferdeckeri, schulzi, sitiens, tebessae, turcicus, uzbeki, wittmeri, xerophilus.*

Unplaced to subgenus or subgenus unknown: **chambonensis, *crispus, *epicentrus, *globularis, *longaevus, *longipennis, *minutulus, *obliteratus, *oblongus, *obscurus, *occultatus, *redtenbacheri, *terreus, *tertiarius, *vetulus.*

---

*LATINODA:* see under ***PSEUDOMYRMEX.***

*LECANOMYRMA:* see under ***OLIGOMYRMEX.***

*LEIOPELTA:* see under ***BELONOPELTA.***

*LEONOMYRMA:* see under ***CHALEPOXENUS.***

*LEPIDOPONE:* see under ***ASPHINCTOPONE.***

***LEPISIOTA*** (Palaearctic, Afrotropical, Oriental, Indo-Australian)

[Note: except where otherwise stated all names retained in this genus which formerly were called *Acantholepis* are here designated as **new combinations** in *Lepisiota.*]

#***abdominalis.*** *Acantholepis gracilicornis* r. *abdominalis* Forel, 1894b: 73 (w.) ETHIOPIA.

#***acholli.*** *Acantholepis capensis* subsp. *acholli* Weber, 1943c: 382, pl. 16, fig. 26 (w.q.) SUDAN.

#***aegyptiaca.*** *Acantholepis frauenfeldi* var. *aegyptiaca* Finzi, 1936: 184, fig. 9 (w.) EGYPT.

*aemula* Forel, 1907 [*Acantholepis*]; see under ***STIGMACROS.***

***affinis.*** *Acantholepis affinis* Santschi, 1937a: 63, figs. 13–15 (w.) KENYA.

***ajjer.*** *Acantholepis ajjer* Bernard, 1953a: 181, fig. 7 (w.m.) ALGERIA. See also: Bernard & Cagniant, 1963: 162.

***albata.*** *Acantholepis albata* Santschi, 1935a: 269, figs. 11–13 (w.) ZAIRE.

***alexis.*** *Acantholepis alexis* Santschi, 1937d: 239 (w.) ZIMBABWE. [First available use of *Acantholepis capensis* r. *laevis* var. *alexis* Arnold, 1920a: 573; unavailable name.] Current subspecies: nominal plus *dulcis.*

***ambigua.*** *Acantholepis ambigua* Santschi, 1935a: 270, figs. 15–18 (w.) ZAIRE [original entry is spelled *ambiga* but figure legends (same page) are all *ambigua;* the latter is accepted as the valid spelling].

#***anceps.*** *Acantholepis capensis* var. *anceps* Forel, 1916: 438 (w.) ZAIRE.

***angolensis.*** *Acantholepis angolensis* Santschi, 1937d: 238, fig. 43 (w.) ANGOLA.

***annandalei.*** *Acantholepis annandalei* Mukerjee, 1930: 156, fig. 4 (w.) INDIA.

*anodon. Acantholepis capensis* st. *canescens* var. *anodon* Santschi, 1933b: 107 (w.) KENYA. Unavailable name.

***arabica.*** *Acantholepis arabica* Collingwood, 1985: 293, fig. 89 (w.) SAUDI ARABIA.

***arenaria.*** *Acantholepis arenaria* Arnold, 1920a: 560, fig. 56 (w.) ZIMBABWE.

***arnoldi.*** *Acantholepis arnoldi* Forel, 1913a: 142 (w.) ZIMBABWE. Current subspecies: nominal plus *mota.* See also: Arnold, 1920a: 563.

*arnoldovi. Acantholepis frauenfeldi* var. *arnoldovi* Ruzsky, 1905b: 462 (w.) TURKESTAN. Junior synonym of *semenovi:* Dlussky, Soyunov & Zabelin, 1990: 163.

#***aspera.*** *Acantholepis submetallica* var. *aspera* Arnold, 1920a: 568 (w.) MOZAMBIQUE.

#***atlantis.*** *Acantholepis frauenfeldi* var. *atlantis* Santschi, 1917d: 42 (w.q.) ALGERIA.

***aurea.*** *Acantholepis aurea* Karavaiev, 1933a: 306, fig. 2 (w.) JAVA. Current subspecies: nominal plus *punctaticeps.*

*australis* Forel, 1902 [*Acantholepis*]; see under ***STIGMACROS.***

#***australis.*** *Acantholepis monardi* var. *australis* Santschi, 1930b: 75, fig. 7 (w.) ANGOLA. [Unresolved junior primary homonym of *australis* Forel, 1902h: 479, above.]

*avunculus. Acantholepis rubrovaria* r. *pilosa* var. *avunculus* Arnold, 1920a: 563 (w.) ZIMBABWE. Unavailable name.

*azerbeidzhanica. Acantholepis frauenfeldi* var. *azerbeidzhanica* Karavaiev, 1932: 250, fig. 1 (w.) AZERBAIJAN. Junior synonym of *frauenfeldi:* Arakelian, 1994: 81.

#**ballaensis.** *Acantholepis spinosior* var. *ballaensis* Arnold, 1920a: 558 (w.) ZIMBABWE.

#**barbara.** *Acantholepis frauenfeldi* var. *barbara* Santschi, 1917d: 44, fig. 5 (w.) TUNISIA.

#**baumi.** *Acantholepis carbonaria* subsp. *baumi* Forel, 1910e: 449 (w.) ANGOLA.

#**bipartita.** *Formica bipartita* Smith, F. 1861a: 33 (w.) LEBANON. André, 1882b: 211 (q.); Karavaiev, 1910b: 44 (m.). Combination in *Acantholepis:* Roger, 1863b: 11; in *Lepisiota:* Baroni Urbani, Bolton & Ward, 1992: 303. Junior synonym of *frauenfeldi:* Mayr, 1863: 394. Revived from synonymy as variety of *frauenfeldi:* André, 1882b: 211. Variety/subspecies of *frauenfeldi:* Emery, 1891b: 16; Emery, 1898c: 126; Forel, 1902a: 155; Ruzsky, 1905b: 463; Santschi, 1917d: 47 (in key).

*bossii* Forel, 1902 [*Acantholepis*]; see under ***STIGMACROS.***

**cacozela.** *Acantholepis capensis* var. *cacozela* Stitz, 1916: 395 (w.) CONGO. [First available use of *Acantholepis capensis* st. *canescens* var. *cacozela* Santschi, 1914b: 126; unavailable name.] Raised to species: Santschi, 1935a: 270.

**canescens.** *Acantholepis capensis* var. *canescens* Emery, 1897e: 601 (w.) SOMALIA. Stirps of *capensis:* Santschi, 1914b: 124; Santschi, 1914d: 378. Raised to species: Santschi, 1935a: 268; Bernard & Cagniant, 1963: 163. Current subspecies: nominal plus *latior.*

**capensis.** *Acantholepis capensis* Mayr, 1862: 699 (w.) SOUTH AFRICA. Mayr, 1865: 57 (m.); Emery, 1877b: 366 (q.); Wheeler, G.C. & Wheeler, J. 1980: 542 (l.). Current subspecies: nominal plus *acholli, anceps, guineensis, issore, junodi, laevis, lunaris, minuta, simplex, simplicoides, specularis, subopaciceps, thoth, validiuscula.* See also: Bingham, 1903: 316; Arnold, 1920a: 568.

**capitata.** *Acantholepis capitata* Forel, 1913b: 338 (w.) ZAIRE.

**carbonaria.** *Acantholepis carbonaria* Emery, 1892a: 119 (w.q.) SOMALIA. Current subspecies: nominal plus *baumi.*

*caucasica. Acantholepis frauenfeldi* var. *caucasica* Santschi, 1917d: 43 (w.) CAUCASUS. Raised to species: Agosti & Collingwood, 1987a: 57. Junior synonym of *frauenfeldi:* Arakelian, 1994: 81.

**chapmani.** *Acantholepis chapmani* Wheeler, W.M. 1935b: 47, fig. 2 (w.q.) PHILIPPINES.

*clivispina* Forel, 1902 [*Acantholepis*]; see under ***STIGMACROS.***

#**congolensis.** *Acantholepis imperfecta* var. *congolensis* Santschi, 1935a: 278 (w.) ZAIRE.

#**convexa.** *Acantholepis foreli* var. *convexa* Arnold, 1920a: 565 (w.) ZIMBABWE.

*coriacea. Acantholepis frauenfeldi* subsp. *coriacea* Kuznetsov-Ugamsky, 1929c: 487, fig. 10 (w.) TURKESTAN. Junior synonym of *semenovi:* Dlussky, Soyunov & Zabelin, 1990: 165.

**crinita.** *Acantholepis crinita* Mayr, 1895: 149 (w.) SOUTH AFRICA.

**curta.** *Acantholepis curta* Emery, 1897e: 603, fig. (w.) SOMALIA.

**dendrophila.** *Acantholepis dendrophila* Arnold, 1949: 269, fig. 10 (w.) ZIMBABWE.

**depilis.** *Acantholepis capensis* subsp. *depilis* Emery, 1897e: 602 (w.q.) SOMALIA. Raised to species: Collingwood, 1985: 294.

**deplanata.** *Acantholepis deplanatus* Stitz, 1911b: 390, fig. 8 (w.) TANZANIA.

**depressa.** *Acantholepis depressa* Santschi, 1914b: 126, fig. 28 (w.) KENYA.

*deserticola. Acantholepis frauenfeldi* subsp. *deserticola* Kuznetsov-Ugamsky, 1929c: 485, fig. 5 (w.) TURKESTAN. Junior synonym of *semenovi:* Dlussky, Soyunov & Zabelin, 1990: 165.

**dolabellae.** *Acantholepis frauenfeldi* var. *dolabellae* Forel, 1911d: 351 (w.) TURKEY. Forel, 1913d: 433 (q.m.). Raised to species: Collingwood, 1985: 294.

#**dulcis.** *Acantholepis alexis* st. *dulcis* Santschi, 1937d: 239, figs. 37–40 (w.) ANGOLA.

**egregia.** *Acantholepis egregia* Forel, 1913a: 143 (w.) ZIMBABWE. Current subspecies: nominal plus *santschii.* See also: Arnold, 1920a: 574.

*elevata. Acantholepis capensis* r. *hirsuta* var. *elevata* Forel, 1914d: 249 (w.) ZIMBABWE. Unavailable name.

**emmelii.** *Acantholepis emmelii* Kutter, 1932: 208, fig. (w.) JAVA.

**erythraea.** *Acantholepis carbonaria* var. *erythraea* Forel, 1910c: 265 (w.q.) ETHIOPIA. Raised to species: Collingwood, 1985: 294.

#**ferganica.** *Acantholepis frauenfeldi* subsp. *ferganica* Kuznetsov-Ugamsky, 1929c: 490, fig. 14 (w.) TURKESTAN.

**fergusoni.** *Acantholepis fergusoni* Forel, 1895e: 459 (w.) INDIA. See also: Bingham, 1903: 319.

**foreli.** *Acantholepis foreli* Arnold, 1920a: 564, pl. 18, fig. 121 (w.) ZIMBABWE. Current subspecies: nominal plus *convexa, impressa.*

*foreli* Viehmeyer, 1925 [*Acantholepis*]; see under ***STIGMACROS.***

*fossulata* Viehmeyer, 1925 [*Acantholepis*]; see under ***STIGMACROS.***

**frauenfeldi.** *Hypoclinea frauenfeldi* Mayr, 1855: 378 (w.) YUGOSLAVIA. Roger, 1859: 243 (q.); Emery, 1878b: 46 (m.); Wheeler, G.C. & Wheeler, J. 1968: 209 (l.). Combination in *Acantholepis:* Mayr, 1861: 42; in *Lepisiota:* Baroni Urbani, Bolton & Ward, 1992: 303. Senior synonym of *azerbeidzhanica* and *caucasica:* Arakelian, 1994: 81. Current subspecies: nominal plus *aegyptiaca, atlantis, barbara, bipartita, ferganica, integra, kantarensis, kassansai, libanica, marocana, opaciventris, pubescens, saharensis, surchanica, truncata, variabilis, velox.* See also: Bingham 1903: 316; Santschi, 1917d: 42; Tohmé, H. 1981: 1. Atanassov & Dulssky, 1992: 205.

*froggatti* Forel, 1902 [*Acantholepis*]; see under ***STIGMACROS.***

**gerardi.** *Acantholepis gerardi* Santschi, 1915c: 262 (w.) ZAIRE.

**gracilicornis.** *Acantholepis gracilicornis* Forel, 1892a: 42 (diagnosis in key) (w.) YEMEN. Emery, 1893f: 257 (q.). Current subspecies: nominal plus *abdominalis.*

#**guineensis.** *Acantholepis capensis* var. *guineensis* Mayr, 1902: 296 (w.) GHANA.

*halinae. Acantholepis frauenfeldi* subsp. *halinae* Kuznetsov-Ugamsky, 1929c: 489, fig. 13 (w.) TURKESTAN. Junior synonym of *sericea:* Pisarski, 1967: 408; of *semenovi:* Dlussky, Soyunov & Zabelin, 1990: 165.

**hirsuta.** *Acantholepis capensis* st. *hirsuta* Santschi, 1914b: 124, fig. 26 (w.) KENYA. Raised to species: Santschi, 1935a: 273. Current subspecies: nominal plus *setosella.*

**imperfecta.** *Acantholepis imperfecta* Santschi, 1926a: 27 (w.) ZIMBABWE. Current subspecies: nominal plus *congolensis.*

#**impressa.** *Acantholepis foreli* var. *impressa* Arnold, 1920a: 565 (w.) ZIMBABWE.

**incisa.** *Acantholepis capensis* r. *incisa* Forel, 1913b: 338 (w.) ZAIRE. Raised to species: Collingwood, 1985: 295.

*infumata. Acantholepis frauenfeldi* subsp. *turkmena* var. *infumata* Kuznetsov-Ugamsky, 1929c: 489 (w.) TURKESTAN. Unavailable name; material referred to *semenovi* by Dlussky, Soyunov & Zabelin, 1990: 165.

*intacta* Viehmeyer, 1925 [*Acantholepis*]; see under ***STIGMACROS.***

#**integra.** *Acantholepis frauenfeldi* var. *integra* Forel, 1894c: 411 (diagnosis in key) (w.) INDIA. Race/stirps of *frauenfeldi:* Forel, 1906b: 86; Santschi, 1917d: 44.

*integrisquama. Acantholepis frauenfeldi* subsp. *integrisquama* Kuznetsov-Ugamsky, 1929c: 486, fig. 8 (w.) TURKESTAN. Junior synonym of *semenovi:* Dlussky, Soyunov & Zabelin, 1990: 165.

#**issore.** *Acantholepis capensis* subsp. *issore* Weber, 1943c: 383, pl. 16, fig. 27 (w.) SUDAN.

#**junodi.** *Acantholepis capensis* r. *junodi* Forel, 1916: 438 (w.) SOUTH AFRICA.

**kabulica.** *Acantholepis kabulica* Pisarski, 1967: 409, figs. 73–76 (w.) AFGHANISTAN.

#**kantarensis.** *Acantholepis frauenfeldi* var. *kantarensis* Forel, 1911d: 351 (w.) ALGERIA.

*karawaiewi. Acantholepis frauenfeldi* subsp. *karawaiewi* Kuznetsov-Ugamsky, 1929c: 483 (w.q.m.) TURKESTAN. [First available use of *Acantholepis frauenfeldi* st. *melas* var. *karawaiewi* Santschi, 1917d: 44; unavailable name.] Raised to species: Pisarski, 1967: 410. Junior synonym of *semenovi:* Dlussky, Soyunov & Zabelin, 1990: 163.

#**kassansai.** *Acantholepis frauenfeldi* subsp. *kassansai* Tarbinsky, 1976: 132, figs. 235–237 (w.) KIRGIZIA.

*kirbii* Lowne, 1865 [*Acantholepis*]; see under ***FROGGATTELLA.***

#**laevis.** *Acantholepis laevis* Santschi, 1913c: 312, fig. 5 (w.) SENEGAL. Stirps of *simplex:* Santschi, 1914d: 378. Status as species: Santschi, 1935a: 278. Subspecies of *capensis:* Arnold, 1920a: 573; Wheeler, W.M. 1922a: 937; Bernard & Cagniant, 1963: 163.

#**latior.** *Acantholepis canescens* var. *latior* Santschi, 1935a: 268 (w.) ZAIRE.

*latisquama. Acantholepis frauenfeldi* subsp. *latisquama* Kuznetsov-Ugamsky, 1929c: 487, fig. 9 (w.) TURKESTAN. Junior synonym of *semenovi:* Dlussky, Soyunov & Zabelin, 1990: 165.

#**libanica.** *Acantholepis frauenfeldi* var. *libanica* Santschi, 1921a: 115 (w.) SYRIA.

*litoralis. Acantholepis frauenfeldi* subsp. *litoralis* Kuznetsov-Ugamsky, 1929c: 488, fig. 11 (w.) TURKESTAN. Junior synonym of *semenovi:* Dlussky, Soyunov & Zabelin, 1990: 165.

**longinoda.** *Acantholepis longinoda* Arnold, 1920a: 559, pl. 8, fig. 120 (w.q.) ZIMBABWE.

**#longipilosa.** *Acantholepis piliscapa* var. *longipilosa* Santschi, 1935a: 274 (w.) ZAIRE.

**#lunaris.** *Acantholepis lunaris* Emery, 1893f: 250, pl. 6, fig. 12 (w.) SRI LANKA. Imai, Baroni Urbani, *et al.* 1984: 9 (k.). Race/subspecies of *capensis:* Forel, 1895e: 458; Forel, 1906b: 86; Forel, 1909e: 395.

*mamillatus* Lowne, 1865 [*Acantholepis*]; see under ***IRIDOMYRMEX.***

**#marocana.** *Acantholepis frauenfeldi* var. *marocana* Santschi, 1936c: 206, fig. 5 (w.) MOROCCO.

*medioreticulata* Viehmeyer, 1925 [*Acantholepis*]; see under ***STIGMACROS.***

*mediorubra. Acantholepis frauenfeldi* subsp. *mediorubra* Kuznetsov-Ugamsky, 1929c: 486, fig. 7 (w.) TURKESTAN. Junior synonym of *semenovi:* Dlussky, Soyunov & Zabelin, 1990: 165.

**megacephala.** *Acantholepis megacephala* Weber, 1943c: 384, pl. 16, fig. 23 (w.) SUDAN.

**melanogaster.** *Acantholepis frauenfeldi* var. *melanogaster* Emery, 1915h: 4 (w.) TURKESTAN. Karavaiev, 1926e: 189 (q.m.). Stirps of *frauenfeldi:* Santschi, 1917d: 47 (in key). Raised to species: Pisarski, 1967: 410; Arnol'di & Dlussky, 1978: 551 (in key).

**melas.** *Acantholepis frauenfeldi* var. *melas* Emery, 1915h: 3 (w.) GREECE. Finzi, 1932: 25 (q.); Finzi, 1939c: 158 (m.). Stirps of *frauenfeldi:* Santschi, 1917d: 44. Raised to species: Agosti & Collingwood, 1987a: 57.

*minima. Acantholepis frauenfeldi* st. *integra* var. *minima* Santschi, 1917d: 45, fig. 4 (w.) TUNISIA. Unavailable name.

**#minuta.** *Acantholepis simplex* var. *minuta* Forel, 1916: 438 (w.) SOUTH AFRICA. Variety of *capensis:* Arnold, 1920a: 572.

*minuta. Acantholepis frauenfeldi* subsp. *karawajewi* var. *minuta* Kuznetsov-Ugamsky, 1929c: 484 (w.) TURKESTAN. Unavailable name; material referred to *semenovi* by Dlussky, Soyunov & Zabelin, 1990: 163.

**mlanjiensis.** *Acantholepis mlanjiensis* Arnold, 1946: 64, fig. 16 (w.) MALAWI.

**modesta.** *Acantholepis modesta* Forel, 1894c: 412 (diagnosis in key) (w.) INDIA.

**monardi.** *Acantholepis monardi* Santschi, 1930b: 74, figs. 7, 8 (w.) ANGOLA. Current subspecies: nominal plus *australis* (unresolved junior homonym).

**#mota.** *Acantholepis arnoldi* st. *mota* Santschi, 1937d: 241, figs. 30, 31 (w.) ANGOLA.

**#natalensis.** *Acantholepis spinosior* var. *natalensis* Arnold, 1920a: 559 (w.) SOUTH AFRICA.

**ngangela.** *Acantholepis ngangela* Santschi, 1937d: 239, figs. 32–36 (w.) ANGOLA.

**nigra.** *Acantholepis frauenfeldi* var. *nigra* Dalla Torre, 1893: 171 (w.q.m.) ITALY. Emery, 1916b: 222 (w.q.m.). Raised to species: Zimmermann, 1935: 45; Baroni Urbani, 1971c: 175. Junior synonym of *frauenfeldi:* Collingwood, 1978: 71. Revived from synonymy: Agosti & Collingwood, 1987a: 57.

**nigrescens.** *Acantholepis frauenfeldi* var. *nigrescens* Karavaiev, 1912a: 15, fig. 6 (w.). Menozzi, 1927g: 380 (q.) TUNISIA. Raised to species: Collingwood, 1985: 295. [This name misspelled as *nigricans* by Santschi, 1917d: 43; see Karavaiev, 1926e: 189.]

**nigrisetosa.** *Acantholepis nigrisetosa* Santschi, 1935a: 271, figs. 21, 22, 24, 26 (w.) ZAIRE.

**nigriventris.** *Acantholepis nigriventris* Emery, 1899e: 493 (w.) CAMEROUN.

**obtusa.** *Acantholepis carbonaria* var. *obtusa* Emery, 1901e: 63 (w.) ETHIOPIA. Raised to species: Collingwood, 1985: 296.

*occidentalis* Crawley, 1922 [*Acantholepis*]; see under ***STIGMACROS.***

**oculata.** *Acantholepis oculata* Santschi, 1935a: 277, figs. 31–33 (w.) ZAIRE.

**opaca.** *Acantholepis opaca* Forel, 1892a: 43 (diagnosis in key) (w.) INDIA. Current subspecies: nominal plus *pulchella.* See also: Bingham, 1903: 318.

**#opaciventris.** *Acantholepis frauenfeldi* var. *opaciventris* Finzi, 1936: 187, fig. 11 (w.) EGYPT.

**palpalis.** *Acantholepis palpalis* Santschi, 1935a: 276, figs. 27–30 (w.) ZAIRE.

**piliscapa.** *Acantholepis piliscapa* Santschi, 1935a: 273 (w.) ZAIRE. Current subspecies: nominal plus *longipilosa, punctifrons.*

**#pilosa.** *Acantholepis rubrovaria* r. *pilosa* Forel, 1913a: 141 (w.) ZIMBABWE.

*pilosella* Viehmeyer, 1925 [*Acantholepis*]; see under ***STIGMACROS.***

**#pubescens.** *Acantholepis frauenfeldi* var. *pubescens* Forel, 1892a: 41 (diagnosis in key) (w.) TUNISIA. Stirps of *frauenfeldi:* Santschi, 1917d: 43.

**#pulchella.** *Acantholepis opaca* r. *pulchella* Forel, 1892a: 43 (diagnosis in key) (w.) INDIA. Raised to species: Bingham, 1903: 318. Subspecies of *opaca:* Emery, 1925b: 27.

**#punctaticeps.** *Acantholepis aurea* subsp. *punctaticeps* Wheeler, W.M. 1935b: 46 (w.) PHILIPPINES.

**#punctifrons.** *Acantholepis piliscapa* var. *punctifrons* Santschi, 1935a: 274 (w.) ZAIRE.

**quadraticeps.** *Acantholepis quadraticeps* Arnold, 1944: 15, fig. 20 (w.m.) ZIMBABWE.

**rothneyi.** *Plagiolepis rothneyi* Forel, 1894c: 415 (w.) INDIA. Combination in *Acantholepis* (*Lepisiota*): Santschi, 1926a: 15. Current subspecies: nominal plus *splendida, sundaica, taivanae, watsonii, wroughtonii.*

**rubrovaria.** *Acantholepis simplex* var. *rubrovaria* Forel, 1910e: 449 (w.m.) LESOTHO. Raised to species: Forel, 1913a: 141. Current subspecies: nominal plus *pilosa.* See also: Arnold, 1920a: 562.

**rugithorax.** *Acantholepis rugithorax* Santschi, 1930b: 73, fig. 6 (w.) ANGOLA.

**#saharensis.** *Acantholepis frauenfeldi* st. *saharensis* Santschi, 1917d: 44, fig. 2 (w.) ALGERIA.

**#santschii.** *Acantholepis egregia* r. *santschii* Arnold, 1920a: 575 (w.) ZIMBABWE.

**schoutedeni.** *Acantholepis schoutedeni* Santschi, 1935a: 271, fig. 20 (w.) ZAIRE.

**semenovi.** *Acantholepis frauenfeldi* var. *semenovi* Ruzsky, 1905b: 461 (w.) RUSSIA. Subspecies of *frauenfeldi:* Kuznetsov-Ugamsky, 1929c: 485. Raised to species: Pisarski, 1967: 410; Dlussky & Zabelin, 1985: 235. Senior synonym of *arnoldovi, coriacea, deserticola, halinae, integrisquama, karawaiewi, latisquama, litoralis, mediorubra, spinisquama, transcaspiensis, turkmena,* and material of the unavailable names *infumata* and *minuta* referred here: Dlussky, Soyunov & Zabelin, 1990: 163–165.

**sericea.** *Acantholepis frauenfeldi* var. *sericea* Forel, 1892a: 41 (diagnosis in key) (w.) INDIA. Raised to species: Pisarski, 1967: 408.

**#setosella.** *Acantholepis hirsuta* var. *setosella* Santschi, 1935a: 273 (w.) ZAIRE.

**silvicola.** *Acantholepis silvicola* Arnold, 1920a: 576, fig. 57 (w.q.m.) MOZAMBIQUE.

**#simplex.** *Acantholepis simplex* Forel, 1892a: 43 (diagnosis in key) (w.) SOMALIA. Forel, 1892e: 350 (q.m.). Status as species: Bingham, 1903: 317; Forel, 1910e: 449; Forel, 1916: 438; Santschi, 1914d: 378. Subspecies/race of *capensis:* Forel, 1907g: 86 (footnote); Arnold, 1920a: 572; Wheeler, W.M. 1922a: 937.

**#simplicoides.** *Acantholepis capensis* var. *simplicoides* Forel, 1907g: 86 (footnote) (w.) SOUTH AFRICA.

**somalica.** *Acantholepis somalica* Menozzi, 1927f: 359, fig. 3 (w.q.) SOMALIA.

**#specularis.** *Acantholepis laevior* st. *specularis* Santschi, 1935a: 278 (w.) ZAIRE [*laevior* is a misspelling of *laevis.*] Subspecies of *capensis:* current status.

*spinisquama. Acantholepis frauenfeldi* subsp. *spinisquama* Kuznetsov-Ugamsky, 1929c: 483, fig. 3 (w.) TURKESTAN. Raised to species: Pisarski, 1967: 410. Junior synonym of *semenovi:* Dlussky, Soyunov & Zabelin, 1990: 163.

**spinosior.** *Acantholepis spinosior* Forel, 1913a: 144 (w.) ZIMBABWE. Arnold, 1920a: 558 (q.). Current subspecies: nominal plus *ballaensis, natalensis.*

**splendens.** *Acantholepis frauenfeldi* var. *splendens* Karavaiev, 1912b: 586 (w.q.) GREECE. Raised to species: Agosti & Collingwood, 1987a: 57.

**#splendida.** *Acantholepis rothneyi* subsp. *splendida* Viehmeyer, 1916a: 146 (w.) SINGAPORE.

**submetallica.** *Acantholepis submetallica* Arnold, 1920a: 566 (w.q.m.) ZIMBABWE. Current subspecies: nominal plus *aspera.*

**#subopaciceps.** *Acantholepis capensis* var. *subopaciceps* Santschi, 1937d: 237 (w.) ANGOLA.

**#sundaica.** *Plagiolepis* (*Anacantholepis*) *rothneyi* subsp. *sundaica* Emery, 1921d: 318 (w.) BORNEO.

**#surchanica.** *Acantholepis frauenfeldi* subsp. *surchanica* Kuznetsov-Ugamsky, 1929c: 491, fig. 15 (w.) TURKESTAN.

**syriaca.** *Acantholepis frauenfeldi* var. *syriaca* André, 1881b: 61, pl. 3, fig. 8 (w.q.m.) SYRIA. Tohmé, H. & Tohmé, G. 1980c: 521 (l.). Raised to species: Tohmé, H. & Tohmé, G. 1980c: 518. See also: Viehmeyer, 1922: 215.

**#taivanae.** *Plagiolepis rothneyi* r. *taivanae* Forel, 1913f: 197 (w.) TAIWAN.

**tenuipilis.** *Acantholepis tenuipilis* Santschi, 1935a: 275 (w.) ZAIRE.

**#thoth.** *Acantholepis capensis* subsp. *thoth* Weber, 1943c: 383, pl. 16, figs. 28, 39 (w.q.) SUDAN.

*transcaspiensis. Acantholepis frauenfeldi* var. *transcaspiensis* Santschi, 1917d: 43 (w.) RUSSIA. Karavaiev, 1926e: 188 (q.m.). Subspecies of *frauenfeldi:* Kuznetsov-Ugamsky, 1929c: 484. Junior synonym of *semenovi:* Dlussky, Soyunov & Zabelin, 1990: 163.

**#truncata.** *Acantholepis frauenfeldi* var. *truncata* Finzi, 1940: 162 (w.q.) TUNISIA. [First available use of *Acantholepis frauenfeldi* st. *integra* var. *truncata* Santschi, 1917d: 45; unavailable name.]

*tuberculatus* Lowne, 1865 [*Acantholepis*]; see under ***PAPYRIUS.***

*turkmena*. *Acantholepis frauenfeldi* subsp. *turkmena* Kuznetsov-Ugamsky, 1929c: 488, fig. 12 (w.) TURKESTAN. Junior synonym of *semenovi*: Dlussky, Soyunov & Zabelin, 1990: 165.

#***validiuscula***. *Acantholepis capensis* var. *validiuscula* Emery, 1897e: 602 (w.) SOMALIA. Arnold, 1920a: 570 (q.m.).

#***variabilis***. *Acantholepis frauenfeldi* var. *variabilis* Santschi, 1917d: 43 (w.) TUNISIA.

#***velox***. *Acantholepis frauenfeldi* subsp. *velox* Baroni Urbani, 1968b: 480 (w.q.m.) TUNISIA. [First available use of *Acantholepis frauenfeldi* st. *integra* var. *velox* Santschi, 1917d: 44; unavailable name.]

*vestita*. *Acantholepis vestita* Dixey & Longstaff, 1907: 333. *Nomen nudum;* attributed to Smith, F.

#***watsonii***. *Plagiolepis rothneyi* r. *watsonii* Forel, 1894c: 416 (w.) BURMA. Raised to species: Bingham, 1903: 322. Subspecies of *rothneyi*: Wheeler, W.M. 1921c: 542.

#***wroughtonii***. *Plagiolepis rothneyi* r. *wroughtonii* Forel, 1902d: 292 (w.) INDIA. Raised to species: Bingham, 1903: 321. Subspecies of *rothneyi*: Emery, 1925b: 23.

*LEPTALEA*: see under ***PSEUDOMYRMEX***.

***LEPTANILLA*** (Old World tropics and subtropics)

***africana***. *Leptanilla africana* Baroni Urbani, 1977c: 468, figs. 34–37, 43 (m.) NIGERIA.

***alexandri***. *Leptanilla alexandri* Dlussky, 1969b: 1667, fig. 1 (m.) UZBEKHISTAN.

***astylina***. *Leptanilla astylina* Petersen, 1968: 578, figs. 1–5 (m.) PHILIPPINES.

***australis***. *Leptanilla australis* Baroni Urbani, 1977c: 471, figs. 38, 40–42, 44 (m.) SOUTH AFRICA.

***besucheti***. *Leptanilla besucheti* Baroni Urbani, 1977c: 448, figs. 10, 17 (w.) SRI LANKA.

***bifurcata***. *Leptanilla bifurcata* Kugler, J. 1987: 48, figs. 5–9 (m.) ISRAEL.

*bimaculata*. *Leptanilla revelierei* var. *bimaculata* Emery, 1899b: 20, fig. (w.) CORSICA. Junior synonym of *revelierii*: Baroni Urbani, 1977c: 437.

***boltoni***. *Leptanilla boltoni* Baroni Urbani, 1977c: 465, figs. 27, 30 (w.) GHANA.

***buddhista***. *Leptanilla buddhista* Baroni Urbani, 1977c: 446, figs. 7, 9 (w.) NEPAL.

***butteli***. *Leptanilla butteli* Forel, 1913k: 25 (w.) MALAYSIA. See also: Baroni Urbani, 1977c: 456.

***charonea***. *Leptanilla charonea* López, Martínez & Barandica, 1994: 187, figs. 5–7, 11 (w.q.) SPAIN.

*chobauti*. *Laptanilla revelierei* subsp. *chobauti* Emery, 1899b: 20 (w.) MOROCCO. Junior synonym of *revelierii*: Baroni Urbani, 1977c: 437.

***doderoi***. *Leptanilla doderoi* Emery, 1915a: 253, pl. 4, figs. 3, 4 (w.) SARDINIA. Junior synonym of *revelierii*: Baroni Urbani, 1977c: 437. Revived from synonymy: Leo & Fancello, 1990: 130.

***escheri***. *Leptomesites escheri* Kutter, 1948: 287, figs. 1–4 (w.q.l.) INDIA. Wheeler, G.C. & Wheeler, J. 1965: 31 (l.). Combination in *Leptanilla*: Baroni Urbani, 1977c: 462.

***exigua***. *Leptanilla exigua* Santschi, 1908: 519, figs. 1, 3 (m.) TUNISIA.

***havilandi***. *Leptanilla havilandi* Forel, 1901e: 373 (w.) SINGAPORE. See also: Baroni Urbani, 1977c: 452.

***islamica***. *Leptanilla islamica* Baroni Urbani, 1977c: 474, figs. 39, 45, 46 (m.) YEMEN.

***israelis***. *Leptanilla israelis* Kugler, J. 1987: 50, figs. 10–13 (m.) ISRAEL.

***japonica***. *Leptanilla japonica* Baroni Urbani, 1977c: 460, figs. 13, 312 (w.) JAPAN. Wheeler, G.C. & Wheeler, J. 1989c: 186 (l.). See also: Masuko, 1987: 597; Masuko, 1990: 31.

***judaica***. *Leptanilla judaica* Kugler, J. 1987: 46, figs. 1–4 (w.l.) ISRAEL.

***kubotai***. *Leptanilla kubotai* Baroni Urbani, 1977c: 444 figs. 6, 8 (w.) JAPAN.

***minuscula***. *Leptanilla minuscula* Santschi, 1907: 309, fig. 2 (m.) TUNISIA.

***morimotoi***. *Leptanilla morimotoi* Yasumatsu, 1960: 17, pl. 5, fig. 1 (w.) JAPAN. See also: Baroni Urbani, 1977c: 445.

***nana***. *Leptanilla nana* Santschi, 1915a: 55, fig. 2 (w.) TUNISIA. See also: Baroni Urbani, 1977c: 466.

*nordenskioldii* Holmgren, 1908; see under ***NEIVAMYRMEX***.

***oceanica***. *Leptanilla oceanica* Baroni Urbani, 1977c: 450, figs. 21, 22 (w.) JAPAN.

***ortunoi***. *Leptanilla ortunoi* López, Martínez, & Barandica, 1994: 197, figs. 9, 11 (w.) SPAIN.

***palauensis***. *Probolomyrmex palauensis* Smith, M.R. 1953c: 128, figs. 1, 2 (m.) PALAU IS. Combination in *Leptanilla*: Taylor, 1965d: 363. See also: Petersen, 1968: 596.

***plutonia***. *Leptanilla plutonia* López, Martínez & Barandica, 1994: 194, figs. 8, 11 (w.) SPAIN.

***revelierii***. *Leptanilla revelierii* Emery, 1870: 196, pl. 2, figs. 2, 7 (w.q.) CORSICA. Senior synonym of *bimaculata*, *chobauti* and *sardoa*: Baroni Urbani, 1977c: 437.

***santschii***. *Leptanilla santschii* Wheeler, G.C. & Wheeler, E.W. 1930: 197, fig. 2 (m.) JAVA. See also: Petersen, 1968: 595.

*sardoa*. *Leptanilla revelierei* subsp. *sardoa* Emery, 1916b: 97 (diagnosis in key), fig. 8 (w.) SARDINIA. Wheeler, G.C. & Wheeler, J. 1965: 27 (l.). Junior synonym of *revelierii*: Baroni Urbani, 1977c: 437.

***swani***. *Leptanilla swani* Wheeler, W.M. 1932b: 54, fig. 1 (w.q.) AUSTRALIA. Wheeler, G.C. & Wheeler, J. 1965: 30 (l.). See also: Baroni Urbani, 1977c: 457.

***tanakai***. *Leptanilla tanakai* Baroni Urbani, 1977c: 458, figs. 11, 33 (w.) JAPAN.

***tanit***. *Leptanilla tanit* Santschi, 1907: 310, fig. 2 (m.) TUNISIA. Combination in *Phaulomyrma*: Wheeler, G.C. & Wheeler, E.W. 1930: 201; in *Leptanilla*: Petersen, 1968: 587. See also: Baroni Urbani, 1977c: 477.

***tenuis***. *Leptanilla tenuis* Santschi, 1907: 307, fig. 1 (m.) TUNISIA.

***thai***. *Leptanilla thai* Baroni Urbani, 1977c: 454, figs. 21, 23 (w.) THAILAND.

***theryi***. *Leptanilla theryi* Forel, 1903c: 252 (w.) ALGERIA. Santschi, 1915a: 57 (q.). See also: Baroni Urbani, 1977c: 442.

***vaucheri***. *Leptanilla vaucheri* Emery, 1899b: 19, fig. (w.) MOROCCO. See also: Baroni Urbani, 1977c: 464.

***zaballosi***. *Leptanilla zaballosi* López, Martínez & Barandica, 1994: 183, figs. 1–4, 11 (w.) SPAIN.

***LEPTANILLOIDES*** (Neotropical)

***biconstricta***. *Leptanilloides biconstricta* Mann, 1923: 14, fig. 1 (w.) BOLIVIA. See also: Borgmeier, 1955b: 653.

***LEPTOGENYS*** (World tropics)

***acutangula***. *Leptogenys* (*Lobopelta*) *acutangula* Emery, 1914f: 398, pl. 13, fig. 2 (w.) NEW CALEDONIA. Senior synonym of *brevinoda*: Wilson, 1958a: 136.

***acutirostris***. *Leptogenys saussurei* st. *acutirostris* Santschi, 1912b: 150 (w.) MADAGASCAR. Raised to species: Bolton, 1975a: 297.

***adlerzi***. *Leptogenys* (*Lobopelta*) *conigera* r. *adlerzi* Forel, 1900b: 65 (w.) AUSTRALIA. Raised to species: Taylor, 1988: 36.

*adpressa*. *Leptogenys* (*Lobopelta*) *nitida* var. *adpressa* Forel, 1914d: 214 (w.) SOUTH AFRICA. Junior synonym of *nitida*: Bolton, 1975a: 289.

*aena*. *Leptogenys* (*Lobopelta*) *nitida* var. *aena* Forel, 1914d: 215 (w.) SOUTH AFRICA. Junior synonym of *nitida*: Bolton, 1975a: 289.

*africanus*. *Leptogenys* (*Leptogenys*) *africanus* Weber, 1942a: 47, fig. 8 (w.) SUDAN. Junior synonym of *crustosa*: Bolton, 1975a: 251.

***alluaudi***. *Leptogenys alluaudi* Emery, 1895f: 338 (w.) MADAGASCAR. See also: Bolton, 1975a: 297.

#***altisquamis***. *Leptogenys* (*Lobopelta*) *kitteli* r. *altisquamis* Forel, 1900d: 306 (diagnosis in key) (w.) BURMA. Combination in *Lobopelta*: Bingham, 1903: 61; in *Leptogenys*: Emery, 1911d: 104. Raised to species: Bingham, 1903: 61. Subspecies of *kitteli*: Emery, 1911d: 104; Wheeler, W.M. 1921c: 530.

***amazonica***. *Leptogenys* (*Leptogenys*) *amazonica* Borgmeier, 1930: 25, pl. 3, fig. 10; pl. 4, figs. 11, 12 (w.) BRAZIL.

***ambigua***. *Leptogenys* (*Lobopelta*) *ambigua* Santschi, 1931c: 267, figs. 2, 3 (w.) PANAMA.

*amboinensis*. *Leptogenys* (*Lobopelta*) *diminuta* subsp. *fruhstorferi* var. *amboinensis* Karavaiev, 1925c: 277 (w.) INDONESIA (Ambon I.). Unavailable name; material referred to *diminuta* by Wilson, 1958a: 118.

***amon***. *Leptogenys amon* Bolton, 1975a: 283, fig. 18 (w.) CAMEROUN.

***anacleti***. *Leptogenys* (*Lobopelta*) *anacleti* Borgmeier, 1930: 30, pl. 3, fig. 9; pl. 4, figs. 18, 19 (w.) BRAZIL.

*andrei*. *Lobopelta distinguenda* var. *andrei* Emery, 1887b: 431 (w.) VIETNAM. Junior synonym of *ocellifera*: Emery, 1895k: 461 (footnote).

***angusta***. *Lobopelta angusta* Forel, 1892k: 519 (w.) MADAGASCAR. Combination in *Leptogenys* (*Lobopelta*): Emery, 1911d: 101. See also: Bolton, 1975a: 296.

*angusticeps*. *Leptogenys stuhlmanni* r. *camerunensis* var. *angusticeps* Forel, 1916: 399 (w.) CAMEROUN. Unavailable name; material referred to *camerunensis* by Bolton, 1975a: 256.

***angustinoda***. *Leptogenys* (*Lobopelta*) *angustinoda* Clark, 1934b: 34, pl. 2, figs. 18, 19 (w.q.) AUSTRALIA.

***anitae***. *Leptogenys* (*Lobopelta*) *anitae* Forel, 1915b: 29 (w.) AUSTRALIA.

***ankhesa***. *Leptogenys ankhesa* Bolton, 1975a: 284, fig. 25 (w.) ZAIRE.

***antillana***. *Leptogenys* (*Lobopelta*) *antillana* Wheeler, W.M. & Mann, 1914: 14, fig. 5 (w.) HAITI.

***LEPTOGENYS*** *(continued)*

***antongilensis.*** *Leptogenys incisa* var. *antongilensis* Emery, 1899f: 272 (w.) MADAGASCAR. Combination in *L.* (*Machaerogenys*), and raised to species: Emery, 1911d: 100.

***arcirostris.*** *Leptogenys* (*Machaerogenys*) *arcirostris* Santschi, 1926a: 25 (w.) MADAGASCAR. See also: Bolton, 1975a: 295.

***arcuata.*** *Leptogenys arcuata* Roger, 1861a: 44 (w.) SURINAM. Forel, 1893g: 359 (m.). Current subspecies: nominal plus *deletangi.*

***arnoldi.*** *Leptogenys* (*Lobopelta*) *arnoldi* Forel, 1913a: 110 (w.m.) ZIMBABWE. Arnold, 1915: 99 (q.). See also: Bolton, 1975a: 281.

***aspera.*** *Lobopelta aspera* André, 1889: 222 (w.) VIETNAM. Wheeler, G.C. & Wheeler, J. 1976a: 52 (l.). Combination in *Leptogenys:* Forel, 1900d: 310. See also: Bingham, 1903: 60.

***assamensis.*** *Leptogenys* (*Lobopelta*) *assamensis* Forel, 1900d: 313 (w.) INDIA. Combination in *Lobopelta:* Bingham, 1903: 70; in *Leptogenys* (*Lobopelta.*): Emery, 1911d: 102.

***attenuata.*** *Ponera attenuata* Smith, F. 1858b: 91 (w.) SOUTH AFRICA. Arnold, 1915: 98 (m.). Combination in *Lobopelta:* Roger, 1863b: 19; in *Leptogenys:* Mayr, 1895: 124. Senior synonym of *jaegerskjoeldi:* Arnold, 1926: 210. See also: Bolton, 1975a: 274.

***australis.*** *Lobopelta australis* Emery, 1888e: 693 (w.) ARGENTINA. Combination in *Leptogenys* (*Lobopelta*): Emery, 1911d: 105.

#***baccha.*** *Leptogenys* (*Lobopelta*) *kraepelini* st. *baccha* Santschi, 1919a: 336 (w.) VIETNAM.

#***bahiana.*** *Leptogenys unistimulosa* var. *bahiana* Santschi, 1929a: 416 (w.) BRAZIL.

***bellii.*** *Leptogenys bellii* Emery, 1901e: 59 (w.) ETHIOPIA. See also: Bolton, 1975a: 265.

*bellua. Leptogenys* (*Lobopelta*) *parva* var. *bellua* Forel, 1914d: 214 (w.) SOUTH AFRICA. Arnold, 1926: 213 (q.). Junior synonym of *castanea:* Bolton, 1975a: 285.

*bernardi. Leptogenys ferrarii* subsp. *bernardi* Baroni Urbani, 1971b: 360. Replacement name for *sulcinodis* Bernard, 1953b: 210. [Junior secondary homonym of *sulcinoda* André, 1892a: 48.] Junior synonym of *occidentalis:* Bolton, 1975a: 270.

***bidentata.*** *Leptogenys* (*Lobopelta*) *bidentata* Forel, 1900b: 66 (w.) AUSTRALIA.

***binghamii.*** *Leptogenys* (*Lobopelta*) *binghamii* Forel, 1900d: 310 (w.) BURMA. Combination in *Lobopelta:* Bingham, 1903: 58; in *Leptogenys* (*Lobopelta*): Emery, 1911d: 102.

***birmana.*** *Leptogenys* (*Lobopelta*) *birmana* Forel, 1900d: 310 (w.) BURMA. Bingham, 1903: 58 (m.).

*bismarckensis. Leptogenys* (*Lobopelta*) *diminuta* var. *bismarckensis* Forel, 1901b: 7 (w.) BISMARCK ARCHIPELAGO. Junior synonym of *diminuta:* Wilson, 1958a: 118.

***bituberculata.*** *Leptogenys bituberculata* Emery, 1901h: 160 (w.) NEW GUINEA. See also: Wilson, 1958a: 116.

***bohlsi.*** *Leptogenys bohlsi* Emery, 1896h: 627 (w.) PARAGUAY. Wheeler, G.C. & Wheeler, J. 1986c: 89 (l.). Current subspecies: nominal plus *weiseri.*

***borneensis.*** *Leptogenys* (*Lobopelta*) *borneensis* Wheeler, W.M. 1919e: 59 (w.) BORNEO. Imai, Brown, *et al.* 1984: 67 (k.).

***breviceps.*** *Leptogenys* (*Lobopelta*) *breviceps* Viehmeyer, 1914c: 30, fig. 4 (w.) NEW GUINEA. Senior synomnym of *lubbocki* and *niger:* Wilson, 1958a: 116.

*brevinoda. Leptogenys* (*Lobopelta*) *acutangula* var. *brevinoda* Emery, 1914f: 399, pl. 13, fig. 26 (ergatoid q.) NEW CALEDONIA. Junior synonym of *acutangula:* Wilson, 1958a: 136.

*brevinodis. Leptogenys* (*Lobopelta*) *nitida* r. *brevinodis* Forel, 1915c: 335 (w.) SOUTH AFRICA. Raised to species: Santschi, 1926b: 209. [Junior primary homonym of *brevinoda* Emery, above.] Replacement name: *capensis* Baroni Urbani, 1971b: 360.

***bubastis.*** *Leptogenys bubastis* Bolton, 1975a: 275 (w.) CAMEROUN.

*bucki* Borgmeier, 1927; see under ***PACHYCONDYLA.***

***buyssoni.*** *Leptogenys* (*Lobopelta*) *buyssoni* Forel, 1907c: 131 (w.) ETHIOPIA. See also: Bolton, 1975a: 285.

***caeciliae.*** *Leptogenys* (*Lobopelta*) *caeciliae* Viehmeyer, 1912: 6, fig. 5 (w.) NEW GUINEA.

***camerunensis.*** *Leptogenys camerunensis* Stitz, 1910: 130 (w.) CAMEROUN. Race/subspecies of *stuhlmanni:* Forel, 1916: 399; Wheeler, W.M. 1922a: 787. Revived status as species, and material of the unavailable names *angusticeps* and *opalescens* referred here: Bolton, 1975a: 256.

*capensis. Leptogenys nitida* subsp. *capensis* Baroni Urbani, 1971b: 360. Replacement name for *brevinodis* Forel, 1915c: 335. [Junior primary homonym of *brevinoda* Emery, 1914f: 399.] Junior synonym of *nitida:* Bolton, 1975a: 289.

***carinata.*** *Leptogenys* (*Lobopelta*) *carinata* Donisthorpe, 1943b: 200 (m.) INDIA.

***castanea.*** *Lobopelta castanea* Mayr, 1862: 734 (w.) SOUTH AFRICA. Wheeler, G.C. & Wheeler, J. 1986c: 89 (l.). Combination in *Leptogenys:* Emery, 1895h: 20. Senior synonym of *bellua, dispar, hewitti* and *parva:* Bolton, 1975a: 285.

***centralis.*** *Leptogenys* (*Lobopelta*) *centralis* Wheeler, W.M. 1915g: 805 (w.m.) AUSTRALIA. Raised to species: Taylor, 1988: 36.

***chalybaea.*** *Lobopelta chalybaea* Emery, 1887b: 432 (w.) BORNEO. Combination in *Leptogenys* (*Lobopelta*): Emery, 1911d: 102.

***chelifera.*** *Pseudoponera chelifer* Santschi, 1928e: 466 (w.) AUSTRALIA. Combination in *Leptogenys:* Taylor & Brown, D.R. 1985: 32.

***chinensis.*** *Lobopelta chinensis* Mayr, 1870b: 965 (w.) CHINA. Forel, 1900d: 313 (m.). Combination in *Leptogenys:* Forel, 1900d: 313; Emery, 1911d: 103. See also: Bingham, 1903: 69.

***clarki.*** *Leptogenys clarki* Wheeler, W.M. 1933i: 82, fig. 26 (w.) AUSTRALIA.

***coerulescens.*** *Leptogenys coerulescens* Emery, 1895f: 339 (w.) MADAGASCAR. See also: Bolton, 1975a: 296.

#***columbica.*** *Leptogenys mucronata* var. *columbica* Forel, 1901f: 328 (w.) COLOMBIA.

*comorensis. Leptogenys comorensis* Forel, 1907g: 76 (w.) COMORO IS. Junior synonym of *stuhlmanni:* Bolton, 1975a: 262.

***confucii.*** *Leptogenys* (*Lobopelta*) *confucii* Forel, 1912a: 48 (w.) TAIWAN. Ogata, 1987: 123 (m.).

***conigera.*** *Lobopelta conigera* Mayr, 1876: 89 (w.) AUSTRALIA. Combination in *Leptogenys:* Forel, 1900b: 65. Senior synonym of *mutans:* Taylor, 1988: 34.

***conradti.*** *Leptogenys conradti* Forel, 1913e: 666 (w.) CAMEROUN. See also: Bolton, 1975a: 250.

***consanguinea.*** *Leptogenys* (*Lobopelta*) *consanguinea* Wheeler, W.M. 1909b: 229 (w.m.) MEXICO. Wheeler, G.C. & Wheeler, J. 1964b: 455 (l.).

#***coonoorensis.*** *Leptogenys* (*Lobopelta*) *roberti* r. *coonoorensis* Forel, 1900d: 306 (diagnosis in key) (w.) INDIA. Raised to species: Bingham, 1903: 68. Subspecies of *roberti:* Emery, 1911d: 105.

***crassicornis.*** *Leptogenys crassicornis* Emery, 1895k: 462 (w.) BURMA. See also: Bingham, 1903: 59.

***crassinoda.*** *Leptogenys attenuata* var. *crassinoda* Arnold, 1926: 210 (w.) SOUTH AFRICA. Raised to species: Bolton, 1975a: 276.

*cribrata. Leptogenys cribrata* Emery, 1895h: 20 (w.) SOUTH AFRICA. Junior synonym of *maxillosa:* Bolton, 1975a: 254.

***crudelis.*** *Ponera crudelis* Smith, F. 1858b: 97 (w.) BRAZIL. Combination in *Lobopelta:* Mayr, 1886c: 358; in *Leptogenys:* Emery, 1911d: 105. Senior synonym of *rubicunda:* Borgmeier, 1932b: 485.

***crustosa.*** *Leptogenys crustosa* Santschi, 1914d: 326, fig. 12 (w.) GUINEA. Senior synonym of *africanus* and *rufipes:* Bolton, 1975a: 251.

***cryptica.*** *Leptogenys cryptica* Bolton, 1975a: 286 (w.) GHANA.

#***cubaensis.*** *Leptogenys pubiceps* st. *cubaensis* Santschi, 1930c: 76 (w.m.) CUBA.

#***currens.*** *Leptogenys* (*Lobopelta*) *iridescens* st. *currens* Forel, 1901f: 329 (w.) BORNEO.

*cursor. Leptogenys* (*Lobopelta*) *cursor* Arnold, 1954: 293, fig. 4 (w.) UGANDA. Junior synonym of *ergatogyna:* Bolton, 1975a: 276.

***dalyi.*** *Leptogenys* (*Lobopelta*) *dalyi* Forel, 1900d: 311 (w.) INDIA. See also: Bingham, 1903: 67.

***darlingtoni.*** *Leptogenys* (*Lobopelta*) *darlingtoni* Wheeler, W.M. 1933i: 90, fig. 28 (w.q.) AUSTRALIA.

***dasygyna.*** *Leptogenys* (*Lobopelta*) *dasygyna* Wheeler, W.M. 1923d: 8, fig. 3 (w.q.) GUIANA.

***davydovi.*** *Leptogenys* (*Lobopelta*) *davydovi* Karavaiev, 1935a: 74, fig. 7 (w.) VIETNAM.

#***deceptrix.*** *Leptogenys* (*Lobopelta*) *diminuta* r. *deceptrix* Forel, 1901h: 46 (w.) INDIA.

*deflocata. Leptogenys* (*Lobopelta*) *brevinodis* var. *deflocata* Santschi, 1926b: 209 (w.) MOZAMBIQUE. Junior synonym of *nitida:* Bolton, 1975a: 289.

#***defuga.*** *Leptogenys* (*Lobopelta*) *gaigei* var. *defuga* Wheeler, W.M. 1923d: 8 (w.) GUIANA.

#***deletangi.*** *Leptogenys arcuata* st. *deletangi* Santschi, 1921g: 87 (w.) BOLIVIA.

*dentatula. Leptogenys ferrari* st. *dentatula* Santschi, 1915c: 246 (w.) ZAIRE. [Name misspelled as *dentulata* by Bernard, 1953b: 210.] Junior synonym of *ferrarii:* Bolton, 1975a: 267.

***dentilobis.*** *Leptogenys* (*Lobopelta*) *dentilobis* Forel, 1900d: 305 (diagnosis in key) (w.) INDIA. See also: Bingham, 1903: 58.

***diatra.*** *Leptogenys diatra* Bolton, 1975a: 287 (w.) CAMEROUN.

***diminuta.*** *Ponera diminuta* Smith, F. 1857a: 69 (w.) BORNEO. Forel, 1900d: 312 (q.m.); Wheeler, G.C. & Wheeler, J. 1976a: 49 (l.); Imai, Brown, *et al.*

1984: 67 (k.); Imai, Kubota, *et al.* 1985: 46 (k.). Combination in *Lobopelta:* Mayr, 1862: 734; in *Leptogenys:* Emery, 1895k: 461. Senior synonym of *simillima:* Mayr, 1863: 428; of *bismarckensis, ferox, stitzi* and *santschii,* and material of the unavailable names *amboinensis* and *huruensis* referred here: Wilson, 1958a: 118; of *yarrabahna:* Taylor, 1988: 34. Current subspecies: nominal plus *deceptrix, diminutolaeviceps, fruhstorferi, laeviceps, nongnongi, opacinodis, palliseri, sarasinorum, striatula, tjibodana, woodmasoni.* See also: Bingham, 1903: 61.

#***diminutolaeviceps.*** *Leptogenys* (*Lobopelta*) *diminuta* var. *diminutolaeviceps* Forel, 1900d: 312 (w.) INDIA.

*dispar. Leptogenys* (*Lobopelta*) *parva* var. *dispar* Santschi, 1914e: 10 (w.) SOUTH AFRICA. Junior synonym of *castanea:* Bolton, 1975a: 285.

#***distinguenda.*** *Lobopelta distinguenda* Emery, 1887b: 430, pl. 1, fig. 4 (w.) BORNEO. Combination in *Leptogenys:* Emery, 1895k: 461. Subspecies of *ocellifera:* Emery, 1893f: 243; Forel, 1901h: 46; of *processionalis:* Emery, 1911d: 105.

***donisthorpei.*** *Leptogenys* (*Leptogenys*) *donisthorpei* Mann, 1922: 11, fig. 6 (w.) HONDURAS.

***drepanon.*** *Leptogenys drepanon* Wilson, 1958a: 123, fig. 5 (w.) NEW GUINEA.

***ebenina.*** *Leptogenys* (*Lobopelta*) *ebenina* Forel, 1915b: 30 (w.m.) AUSTRALIA.

***elegans.*** *Leptogenys elegans* Bolton, 1975a: 265, fig. 9 (w.) NIGERIA.

***elongata.*** *Ponera elongata* Buckley, 1866: 172 (w.) U.S.A. Wheeler, W.M. 1900b: 7 (q.m.); Wheeler, W.M. 1904c: 258 (w.q.m.); Wheeler, G.C. & Wheeler, J. 1952c: 641 (l.); Petralia & Vinson, 1980: 378 (l.). Combination in *Leptogenys:* Emery, 1895c: 338; Wheeler, W.M. 1900b: 2. Senior synonym of *septentrionalis* and *texana:* Wheeler, W.M. 1902f: 25; Wheeler, W.M. 1904c: 258.

***emeryi.*** *Leptogenys* (*Leptogenys*) *emeryi* Forel, 1901b: 7 (w.) NEW BRITAIN.

***emiliae.*** *Leptogenys* (*Lobopelta*) *emiliae* Forel, 1902d: 294 (w.) INDIA. See also: Bingham, 1903: 66.

***ergatogyna.*** *Leptogenys* (*Lobopelta*) *ergatogyna* Wheeler, W.M. 1922a: 95, fig. 20 (q.) ZAIRE. Senior synonym of *cursor:* Bolton, 1975a: 276.

***erythraea.*** *Leptogenys stuhlmanni* subsp. *erylrhaea* Emery, 1902c: 33 (w.) ETHIOPIA. Forel, 1907c: 131 (m.). [Justified emendation of spelling to *erythraea:* Forel, 1907c: 131.] Raised to species: Bolton, 1975a: 257.

#***esae.*** *Leptogenys* (*Lobopelta*) *kraepelini* r. *esae* Forel, 1912d: 102 (w.m.) JAVA.

***excellens.*** *Leptogenys excellens* Bolton, 1975a: 266 (w.) ZAIRE.

***excisa.*** *Lobopelta excisa* Mayr, 1876: 89 (w.) AUSTRALIA. Combination in *Leptogenys:* Forel, 1910b: 18. Senior synonym of *major:* Taylor, 1988: 36.

***exigua.*** *Leptogenys* (*Lobopelta*) *conigera* r. *exigua* Crawley, 1921: 89 (w.) AUSTRALIA. Raised to species: Taylor, 1988: 37.

***exudans.*** *Formica exudans* Walker, 1859: 371 (m.) SRI LANKA. Combination in *Leptogenys:* Donisthorpe, 1932b: 575.

*falcata. Leptogenys falcata* Roger, 1861a: 42 (w.) CUBA. Roger, 1862a: 244 (m.). Junior synonym of *maxillosa:* Forel, 1901h: 46. Revived from synonymy as variety of *maxillosa:* Emery, 1911d: 99; Wheeler, W.M. 1913b: 482. Junior synonym of *maxillosa:* Bolton, 1975a: 254.

***falcigera.*** *Leptogenys falcigera* Roger, 1861a: 42 (w.) SRI LANKA. Forel, 1891b: 111 (m.). Senior synonym of *insularis:* Wilson & Taylor, 1967: 30. See also: Bolton, 1975a: 252.

***fallax.*** *Lobopelta fallax* Mayr, 1876: 88 (w.m.) AUSTRALIA. Combination in *Leptogenys:* Forel, 1900b: 64

***famelica.*** *Leptogenys famelica* Emery, 1896g: 91, pl. 1, fig. 6 (w.) COSTA RICA. See also: Forel, 1899c: 13.

*ferox. Ponera ferox* Smith, F. 1865: 70 (w.) INDONESIA (Salawaty I.). Combination in *Leptogenys:* Donisthorpe, 1932c: 472. Junior synonym of *diminuta:* Wilson, 1958a: 118.

***ferrarii.*** *Leptogenys ferrarii* Forel, 1913j: 209 (w.) ZIMBABWE. Senior synonym of *dentatula:* Bolton, 1975a: 267.

***foreli.*** *Leptogenys* (*Leptogenys*) *foreli* Mann, 1919: 297, fig. 9 (not fig. 10 as given in the original description) (w.m.) SOLOMON IS. Senior synonym of *walkeri:* Wilson, 1958a: 126.

***fortior.*** *Leptogenys* (*Lobopelta*) *fallax* r. *fortior* Forel, 1900b: 64 (w.m.) AUSTRALIA. Wheeler, W.M. 1916a: 48 (q.; name misspelled as *fortis*); Wheeler, G.C. & Wheeler, J. 1974g: 281. Raised to species: Taylor, 1988: 37.

*foveolata. Leptogenys* (*Lobopelta*) *kitteli* subsp. *levis* var. *foveolata* Emery, 1900d: 670 (w.) SUMATRA. Unavailable name.

***foveopunctata.*** *Leptogenys* (*Lobopelta*) *foveopunctata* Mann, 1921: 421 (w.) FIJI IS.

#***fruhstorferi.*** *Leptogenys diminuta* subsp. *fruhstorferi* Emery, 1896f: 248 (w.) JAVA. [Also described as new by Emery, 1898b: 233.] Forel, 1913k: 17 (m.); Crawley, 1924: 388 (q.). Race/subspecies of *diminuta:* Forel, 1913k: 17; Karavaiev, 1925c: 276.

***fugax.*** *Leptogenys* (*Lobopelta*) *fugax* Mann, 1921: 422, fig. 7 (w.) FIJI IS.

***furtiva.*** *Leptogenys havilandi* r. *furtiva* Arnold, 1926: 209 (w.) ZIMBABWE. Raised to species: Bolton, 1975a: 281.

***gagates.*** *Leptogenys* (*Leptogenys*) *gagates* Mann, 1922: 11, fig. 5 (w.) HONDURAS.

***gaigei.*** *Leptogenys* (*Lobopelta*) *gaigei* Wheeler, W.M. 1923d: 7, fig. 2 (w.) GUIANA. Current subspecies: nominal plus *defuga.*

***gracilis.*** *Leptogenys gracilis* Emery, 1899f: 271 (w.) MADAGASCAR.

*gracilis. Leptogenys* (*Lobopelta*) *nitida* var. *gracilis* Santschi, 1914e: 11 (w.) ZAIRE. [Unresolved junior primary homonym of *gracilis* Emery, above.] Junior synonym of *nitida:* Bolton, 1975a: 289.

***grandidieri.*** *Leptogenys* (*Lobopelta*) *grandidieri* Forel, 1910a: 17 (w.) MADAGASCAR. See also: Bolton, 1975a: 296.

*grandior. Leptogenys* (*Lobopelta*) *nitida* var. *grandior* Forel, 1915c: 335 (w.) SOUTH AFRICA. Junior synonym of *nitida:* Bolton, 1975a: 289.

***guianensis.*** *Leptogenys* (*Lobopelta*) *guianensis* Wheeler, W.M. 1923d: 12, fig. 5 (w.) GUIANA.

***guineensis.*** *Leptogenys* (*Lobopelta*) *guineensis* Santschi, 1914d: 329, fig. 14 (w.) GUINEA. See also: Bolton, 1975a: 278.

***hackeri.*** *Leptogenys* (*Lobopelta*) *hackeri* Clark, 1934b: 35, pl. 2, fig. 20 (w.) AUSTRALIA.

***hanseni.*** *Leptogenys* (*Lobopelta*) *hanseni* Borgmeier, 1930: 31, pl. 4, figs. 16, 23, 24, 27, 28 (w.q.) BRAZIL.

***harmsi.*** *Leptogenys* (*Lobopelta*) *harmsi* Donisthorpe, 1935: 631 (w.m.) CHRISTMAS I.

***havilandi.*** *Leptogenys* (*Lobopelta*) *havilandi* Forel, 1901f: 332 (w.) SOUTH AFRICA. See also: Bolton, 1975a: 282.

***hebrideana.*** *Leptogenys hebrideana* Wilson, 1958a: 127 (w.) VANUATU.

***hemioptica.*** *Leptogenys* (*Lobopelta*) *hemioptica* Forel, 1901f: 331 (w.) SUMATRA.

*hewitti. Leptogenys hewitti* Santschi, 1932a: 381, fig. 1 (w.) SOUTH AFRICA. Junior synonym of *castanea:* Bolton, 1975a: 285.

***hodgsoni.*** *Leptogenys* (*Lobopelta*) *diminuta* r. *hodgsoni* Forel, 1900d: 308 (diagnosis in key) (w.) BURMA. Raised to species: Bingham, 1903: 62.

***honduriana.*** *Leptogenys* (*Lobopelta*) *honduriana* Mann, 1922: 13, fig. 7 (w.) HONDURAS.

***honoria.*** *Leptogenys honoria* Bolton, 1975a: 287 (w.) CAMEROUN.

***humiliata.*** *Leptogenys* (*Lobopelta*) *humiliata* Mann, 1921: 421 (w.) FIJI IS.

*huruensis. Leptogenys* (*Lobopelta*) *diminuta* subsp. *fruhstorferi* var. *huruensis* Karavaiev, 1925c: 278 (w.) INDONESIA (Buru I.). Unavailable name; material referred to *diminuta* by Wilson, 1958a: 118.

***hysterica.*** *Leptogenys* (*Lobopelta*) *hysterica* Forel, 1900d: 311 (w.) SRI LANKA. Imai, Baroni Urbani, *et al.* 1984: 5 (k.). See also: Bingham, 1903: 64.

***iheringi.*** *Leptogenys* (*Lobopelta*) *iheringi* Forel, 1911c: 286 (w.q.) BRAZIL. Wheeler, G.C. & Wheeler, J. 1971b: 1212 (l.).

*imerinensis. Leptogenys incisa* var. *imerinensis* Forel, 1892l: 242 (w.m.) MADAGASCAR. Junior synonym of *incisa:* Bolton, 1975a: 298.

***imperatrix.*** *Leptogenys* (*Lobopelta*) *imperatrix* Mann, 1922: 15, fig. 8 (w.q.) HONDURAS.

***incisa.*** *Leptogenys incisa* Forel, 1891b: 113, pl. 4, fig. 1 (w.) MADAGASCAR. Senior synonym of *imerinensis* and *suarensis:* Bolton, 1975a: 298.

***indigatrix.*** *Leptogenys indigatrix* Wilson, 1958a: 128 (w.) NEW GUINEA.

***ingens.*** *Leptogenys ingens* Mayr, 1866a: 503 (w.) COLOMBIA.

*insinuata. Leptogenys* (*Lobopelta*) *nitida* st. *insinuata* Santschi, 1914e: 11 (w.) SOUTH AFRICA. Junior synonym of *nitida:* Bolton, 1975a: 289.

*insularis. Leptogenys insularis* Smith, F. 1879: 675 (w.) HAWAII. Forel, 1899a: 118 (m.). Variety of *falcigera:* Forel, 1899a: 118. Junior synonym of *falcigera:* Wilson & Taylor, 1967: 30.

***intermedia.*** *Leptogenys intermedia* Emery, 1902c: 32 (w.) SOUTH AFRICA. See also: Forel, 1914d: 214. Junior synonym of *nitida* Smith: Bolton, 1975a: 289. First available replacement name for *nitida* Smith, F. 1858b: 92. [Junior primary homonym of **nitida* Heer, 1850: 149.] For earlier history of the name and current synonymy see under *nitida.*

***intricata.*** *Leptogenys* (*Lobopelta*) *intricata* Viehmeyer, 1924a: 228 (w.) AUSTRALIA. Viehmeyer, 1924b: 310 (m.).

***iridescens.*** *Ponera iridescens* Smith, F. 1857a: 66 (w.) BORNEO. Wheeler, G.C. & Wheeler, J. 1976a: 51 (l.); Imai, Brown, *et al.* 1984: 67 (k.). Combination in *Lobopelta:* Roger, 1863b: 19; in *Leptogenys:* Forel, 1901f: 329. Current subspecies: nominal plus *currens.*

***iridipennis.*** *Ponera iridipennis* Smith, F. 1858b: 85 (m.) INDIA. Combination in *Leptogenys:* Donisthorpe, 1932c: 474.

*jaegerskjoeldi. Leptogenys (Lobopelta) jaegerskjoeldi* Santschi, 1914e: 9 (w.) SOUTH AFRICA. Variety of *attenuata:* Santschi, 1917b: 279. Junior synonym of *attenuata:* Arnold, 1926: 210; Bolton, 1975a: 274.

***jeanneli.*** *Leptogenys jeanneli* Santschi, 1914b: 55, fig. 3 (w.) TANZANIA. See also: Bolton, 1975a: 253.

*jonesii* Forel, 1891; see under ***PACHYCONDYLA.***

***karawaiewi.*** *Leptogenys (Lobopelta) karawaiewi* Santschi, 1928h: 124 (m.) JAVA.

***keysseri.*** *Leptogenys (Lobopelta) keysseri* Viehmeyer, 1914c: 29, fig. 3 (w.) NEW GUINEA.

***khaura.*** *Leptogenys khaura* Bolton, 1975a: 288 (w.) NIGERIA.

***kitteli.*** *Lobopelta kitteli* Mayr, 1870b: 966 (w.) INDIA. Wheeler, G.C. & Wheeler, J. 1976a: 52 (l.). Combination in *Leptogenys:* Emery, 1895k: 461. See also: Bingham, 1903: 60. Current subspecies: nominal plus *altisquamis, laevis, minor, siemsseni, transiens.*

***kraepelini.*** *Leptogenys (Lobopelta) kraepelini* Forel, 1905c: 5 (w.) JAVA. Karavaiev, 1925c: 280 (m.); Imai, Kubota, *et al.* 1985: 46 (k.). Current subspecies: nominal plus *baccha, esae, nero.*

#***laeviceps.*** *Ponera laeviceps* Smith, F. 1857a: 69 (w.) BORNEO. Forel, 1900d: 312 (m.); Imai, Baroni Urbani, *et al.* 1984: 5 (k.). Combination in *Lobopelta:* Mayr, 1862: 734; in *Leptogenys:* Emery, 1895k: 461. [Name misspelled as *leviceps* by Dalla Torre, 1893: 44.] Variety/subspecies of *diminuta:* Emery, 1887b: 433; Bingham, 1903: 62; Forel, 1912n: 52.

#***laevis.*** *Lobopelta kitteli* var. *laevis* Mayr, 1879: 664 (in key) (w.) JAVA. Combination in *Leptogenys:* Emery, 1900d: 670. Raised to species: Dalla Torre, 1893: 45 (misspelled as *levis*). Variety/subspecies of *kitteli:* Emery, 1900d: 670; Forel, 1901f: 329; Forel, 1911e: 253; Karavaiev, 1925c: 279.

***langi.*** *Leptogenys (Lobopelta) langi* Wheeler, W.M. 1923d: 5, fig. 1 (w.q.) GUIANA.

***leiothorax.*** *Leptogenys stuhlmanni* subsp. *leiothorax* Prins, 1965a: 154, fig. 1 (w.) MOZAMBIQUE. Raised to species: Bolton, 1975a: 258.

***letilae.*** *Leptogenys (Lobopelta) letilae* Mann, 1921: 419, fig. 6 (w.q.m.) FIJI IS.

***linearis.*** *Ponera linearis* Smith, F. 1858b: 96 (w.m.) BRAZIL. Combination in *Leptogenys:* Borgmeier, 1923: 75.

***longensis.*** *Leptogenys (Lobopelta) turneri* var. *longensis* Forel, 1915b: 33 (w.) AUSTRALIA. Raised to species: Taylor, 1988: 37.

***longiceps.*** *Leptogenys longiceps* Santschi, 1914d: 328, fig. 13 (w.) SENEGAL. See also: Bolton, 1975a: 268.

*longinoda. Leptogenys (Lobopelta) diminuta* subsp. *fruhstorferi* var. *longinoda* Karavaiev, 1925c: 278 (w.) MALAYSIA. Unavailable name.

***longiscapa.*** *Leptogenys (Lobopelta) longiscapus* Donisthorpe, 1943b: 199 (m.) INDIA.

*longitudinalis. Leptogenys (Lobopelta) diminuta* r. *palliseri* var. *longitudinalis* Forel, 1913k: 17 (w.) JAVA. Unavailable name.

*lubbocki. Pseudoponera lubbocki* Donisthorpe, 1938d: 593, fig. (w.) NEW GUINEA. Junior synonym of *breviceps:* Wilson, 1958a: 116.

***lucidula.*** *Leptogenys lucidula* Emery, 1895k: 462 (w.) BURMA. See also: Bingham, 1903: 66.

***luederwaldti.*** *Leptogenys (Lobopelta) luederwaldti* Forel, 1913l: 206 (w.) BRAZIL.

***mactans.*** *Leptogenys mactans* Bolton, 1975a: 288, fig. 17 (w.) SOUTH AFRICA.

***magna.*** *Leptogenys (Lobopelta) magna* Forel, 1900b: 65 (w.m.) AUSTRALIA.

*major. Leptogenys (Lobopelta) excisa* var. *major* Forel, 1910b: 18 (w.) AUSTRALIA. Junior synonym of *excisa:* Taylor, 1988: 36.

*malandensis. Prionogenys podenzanai* var. *malandensis* Forel, 1915b: 27 (w.) AUSTRALIA. Junior synonym of *podenzanai:* Taylor, 1988: 33.

***manni.*** *Leptogenys (Lobopelta) elongata* subsp. *manni* Wheeler, W.M. 1923d: 14 (w.) U.S.A. Trager & Johnson, 1988: 31 (q.m.). Raised to species: Trager & Johnson, 1988: 31.

***mastax.*** *Leptogenys mastax* Bolton, 1975a: 268, figs. 12, 27 (w.q.) GHANA.

***maxillosa.*** *Ponera maxillosa* Smith, F. 1858b: 93 (w.m.) MAURITIUS. Combination in *Leptogenys:* Roger, 1861a: 43. Senior synonym of *falcata:* Forel, 1901h: 46; Bolton, 1975a: 254; of *cribrata* and *vinsonnella:* Bolton, 1975a: 254.

***melzeri.*** *Leptogenys (Lobopelta) melzeri* Borgmeier, 1930: 27, pl. 4, figs. 15, 20, 22 (w.) BRAZIL.

***meritans.*** *Formica meritans* Walker, 1859: 371 (m.) SRI LANKA. Combination in *Leptogenys:* Donisthorpe, 1932b: 575.

***mexicana.*** *Lobopelta mexicana* Mayr, 1870b: 966 (w.) MEXICO. Combination in *Leptogenys:* Forel, 1899c: 18.

***microps.*** *Leptogenys microps* Bolton, 1975a: 278 (w.) IVORY COAST.

***minchinii.*** *Leptogenys (Lobopelta) minchinii* Forel, 1900d: 308 (diagnosis in key) (w.) INDIA. Forel, 1900d: 314 (m.); Imai, Baroni Urbani, *et al.* 1984: 5 (k.). See also: Bingham, 1903: 70; Rothney, 1903: 97.

#***minor.*** *Leptogenys (Lobopelta) kitteli* var. *minor* Forel, 1900d: 307 (diagnosis in key) (w.) INDIA.

***mjobergi.*** *Leptogenys (Lobopelta) mjobergi* Forel, 1915b: 32, pl. 2, fig. 5 (w.) AUSTRALIA.

*mjobergi. Dorylozelus mjobergi* Forel, 1915b: 25, fig. 4 (w.) AUSTRALIA. [Junior secondary homonym of *mjobergi* Forel, above.] Replacement name: *tricosa* Taylor, 1969: 132.

***modiglianii.*** *Leptogenys modiglianii* Emery, 1900d: 669, fig. 3 (w.) INDONESIA (Engano I.).

***moelleri.*** *Lobopelta moelleri* Bingham, 1903: 67 (w.) INDIA. Combination in *Leptogenys:* Emery, 1911d: 104.

***mucronata.*** *Leptogenys mucronata* Forel, 1893g: 360 (w.) ANTILLES IS. Current subspecies: nominal plus *columbica.*

***mutabilis.*** *Ponera mutabilis* Smith, F. 1861b: 45 (w.) SULAWESI. Combination in *Lobopelta:* Mayr, 1862: 734; in *Leptogenys:* Emery, 1900d: 671.

*mutans. Leptogenys (Lobopelta) conigera* r. *mutans* Forel, 1900b: 65 (w.) AUSTRALIA. Junior synonym of *conigera:* Taylor, 1988: 34.

***myops.*** *Lobopelta myops* Emery, 1887b: 432 (w.) JAVA. Karavaiev, 1925c: 283 (q.); Karavaiev, 1926d: 421 (m.); Imai, Kubota, *et al.* 1985: 46 (k.); Tjan, Imai, *et al.* 1986: 57 (k.). Combination in *Leptogenys:* Emery, 1900d: 671.

***navua.*** *Leptogenys (Lobopelta) navua* Mann, 1921: 423, figs. 8, 9 (w.m.) FIJI IS.

***nebra.*** *Leptogenys nebra* Bolton, 1975a: 258 (w.) CAMEROUN.

#***nero.*** *Leptogenys (Lobopelta) kraepelini* var. *nero* Forel, 1913k: 16 (w.) JAVA.

***neutralis.*** *Leptogenys (Lobopelta) neutralis* Forel, 1907h: 271 (w.) AUSTRALIA. Crawley, 1922b: 434 (q.). See also: Wheeler, W.M. 1933i: 86.

*niger. Euponera (Brachyponera) niger* Donisthorpe, 1949g: 405 (m.) NEW GUINEA. Junior synonym of *breviceps:* Wilson, 1958a: 116.

***nitens.*** *Leptogenys (Lobopelta) nitens* Donisthorpe, 1943a: 169 (w.) NEW GUINEA. Donisthorpe, 1943d: 444 (m.).

*nitida. Ponera nitida* Smith, F. 1858b: 92 (w.) SOUTH AFRICA. [Junior primary homonym of **Ponera nitida* Heer, 1850: 149.] First available replacement name: *intermedia* Emery, 1902c: 32 (synonymy with *nitida* Smith by Bolton, 1975a: 289). Arnold, 1926: 211 (q.). Combination in *Lobopelta:* Mayr, 1886c: 358; in *Leptogenys:* Forel, 1901f: 329. Senior synonym of *tenuis:* Forel, 1912n: 52; Bolton, 1975a: 289; of *adpressa, aena, capensis, deflocata, gracilis, grandior, insinuata, intermedia* and *speculans:* Bolton, 1975a: 289.

*nitida. Leptogenys (Lobopelta) diminuta* subsp. *fruhstorferi* var. *nitida* Karavaiev, 1927e: 6 (w.) INDONESIA (Prinsen I.). Unavailable name.

*nitidiceps. Leptogenys (Lobopelta) diminuta* subsp. *sarasinorum* var. *nitidiceps* Emery, 1911d: 103 (w.) INDIA. Unavailable name.

*nitidula. Leptogenys (Lobopelta) diminuta* st. *fruhstorferi* var. *nitidula* Santschi, 1928h: 139. Unnecessary replacement name for *nitida* Karavaiev, above. Unavailable name.

#***nongnongi.*** *Leptogenys (Lobopelta) diminuta* subsp. *nongnongi* Karavaiev, 1925c: 278 (w.) JAVA.

***nuserra.*** *Leptogenys nuserra* Bolton, 1975a: 269 (w.q.) GHANA.

***occidentalis.*** *Leptogenys (Lobopelta) nitida* r. *occidentalis* Bernard, 1953b: 211, fig. 5 (w.) GUINEA. Raised to species, and senior synonym of *bernardi:* Bolton, 1975a: 270.

*ocellifera. Ponera ocellifera* Roger, 1861a: 13 (w.) SRI LANKA. Forel, 1900d: 309 (m.). Combination in *Lobopelta:* Roger, 1863b: 19; in *Leptogenys:* Emery, 1895k: 461. Junior synonym of *mutabilis:* Mayr, 1879: 665. Revived from synonymy: Dalla Torre, 1893: 45. Senior synonym of *andrei:* Emery, 1895k: 461 (footnote). Junior synonym of *processionalis:* Emery, 1911d: 105. See also: Bingham, 1903: 57.

**#opacinodis.** *Lobopelta diminuta* var. *opacinodis* Emery, 1887b: 433 (w.) SUMATRA. Combination in *Leptogenys:* Emery, 1895k: 461.

*opalescens. Leptogenys stuhlmanni* subsp. *camerunensis* var. *opalescens* Wheeler, W.M. 1922a: 94 (w.) ZAIRE. Unavailable name; material referred to *camerunensis* by Bolton, 1975a: 256.

**optica.** *Leptogenys* (*Lobopelta*) *caeciliae* var. *optica* Viehmeyer, 1914d: 609 (w.) NEW GUINEA. Raised to species: Wilson, 1958a: 130.

**oresbia.** *Leptogenys oresbia* Wilson, 1958a: 131 (w.) SOLOMON IS.

**oswaldi.** *Lobopelta oswaldi* Forel, 1891b: 119, pl. 4, fig. 2 (w.) MADAGASCAR. Combination in *Leptogenys:* Emery, 1911d: 102. See also: Bolton, 1975a: 297.

**#palliseri.** *Leptogenys* (*Lobopelta*) *diminuta* r. *palliseri* Forel, 1900d: 307 (diagnosis in key) (w.) INDIA. Forel, 1900d: 312 (m.). Raised to species: Bingham, 1903: 63. Subspecies/race of *diminuta:* Emery, 1911d: 103; Forel, 1913k: 17; Karavaiev, 1925c: 279.

**#panamana.** *Leptogenys* (*Lobopelta*) *pusilla* var. *panamana* Wheeler, W.M. 1923d: 14 (w.) PANAMA.

**papuana.** *Leptogenys papuana* Emery, 1897d: 556, pl. 1, figs. 5, 6 (w.) NEW GUINEA.

*papuana. Leptogenys* (*Lobopelta*) *diminuta* var. *papuana* Stitz, 1912: 498, fig. 1 (w.) NEW GUINEA. [Junior primary homonym of *papuana* Emery, above.] Replacement name: *stitzi* Viehmeyer, 1924b: 310.

*parva. Leptogenys* (*Lobopelta*) *parva* Forel, 1901f: 330 (w.m.) SOUTH AFRICA. Junior synonym of *castanea:* Bolton, 1975a: 285.

**parvula.** *Leptogenys* (*Lobopelta*) *parvula* Emery, 1900d: 670 (w.) SUMATRA.

**pavesii.** *Leptogenys pavesii* Emery, 1892a: 111 (w.) SOMALIA. Senior synonym of *sericeus:* Bolton, 1975a: 255.

**peninsularis.** *Leptogenys* (*Lobopelta*) *peninsularis* Mann, 1926: 102 (w.) MEXICO.

**peringueyi.** *Leptogenys* (*Lobopelta*) *havilandi* st. *peringueyi* Forel, 1913j: 210 (w.) SOUTH AFRICA. Raised to species: Arnold, 1915: 96. See also: Bolton, 1975a: 282.

**peuqueti.** *Lobopelta peuqueti* André, 1887: 292 (w.) VIETNAM. Imai, Kubota, *et al.* 1985: 46 (k.). Combination in *Leptogenys:* Emery, 1895k: 461. See also: Bingham, 1903: 71.

**piroskae.** *Leptogenys* (*Lobopelta*) *piroskae* Forel, 1910c: 247 (w.) ETHIOPIA. See also: Bolton, 1975a: 290.

**podenzanai.** *Prionogenys podenzanai* Emery, 1895g: 349, fig. 1 (w.) AUSTRALIA. Combination in *Leptogenys,* and senior synonym of *malandensis:* Taylor, 1988: 33.

**pompiloides.** *Ponera pompiloides* Smith, F. 1857a: 69 (m.) BORNEO. Combination in *Leptogenys:* Donisthorpe, 1932c: 448.

**princeps.** *Leptogenys princeps* Bolton, 1975a: 271, fig. 26 (w.) GHANA.

**processionalis.** *Ponera processionalis* Jerdon, 1851: 118 (w.) INDIA. Imai, Baroni Urbani, *et al.* 1984: 5 (k.). Combination in *Leptogenys,* and senior synonym of *ocellifera* (and its junior synonym *andrei*): Emery, 1911d: 105. Current subspecies: nominal plus *distinguenda.*

**pruinosa.** *Leptogenys* (*Lobopelta*) *pruinosa* Forel, 1900d: 304 (diagnosis in key) (w.) SRI LANKA. See also: Bingham, 1903: 53.

**pubiceps.** *Leptogenys pubiceps* Emery, 1890a: 62 (w.) VENEZUELA. Forel, 1893g: 362 (q.m.). Current subspecies: nominal plus *cubaensis, vincentensis.*

**punctata.** *Leptogenys* (*Lobopelta*) *punctata* Emery, 1914f: 398 (w.) NEW CALEDONIA.

**punctaticeps.** *Leptogenys punctaticeps* Emery, 1890a: 62 (footnote) (w.) COSTA RICA. [Also described as new by Emery, 1894k: 49.] Wheeler, W.M. 1911b: 168 (m.); Menozzi, 1927c: 272 (m.); Wheeler, G.C. & Wheeler, J. 1964b: 455 (l.). [Name misspelled as *puncticeps* by Emery, 1911d: 100; Wheeler, W.M. & Mann, 1914: 14; see Kempf, 1972a: 130.]

**punctiventris.** *Lobopelta punctiventris* Mayr, 1879: 666 (w.) INDIA. [*Lobopelta punctiventris* Smith, F. 1873: viii. *Nomen nudum;* attributed to Mayr.] Combination in *Leptogenys:* Forel, 1900d: 311. See also: Bingham, 1903: 64.

**purpurea.** *Lobopelta kitteli* var. *purpurea* Emery, 1887b: 433 (w.) NEW GUINEA. Combination in *Leptogenys:* Emery, 1911d: 104. Raised to species: Wilson, 1958a: 132.

**pusilla.** *Lobopelta pusilla* Emery, 1890b: 43 (w.) COSTA RICA. [Also described as new by Emery, 1894k: 49.] Combination in *Leptogenys:* Forel, 1899c: 18. Current subspecies: nominal plus *panamana.*

**quiriguana.** *Leptogenys* (*Lobopelta*) *quiriguana* Wheeler, W.M. 1923d: 10, fig. 4 (w.) GUATEMALA.

**ravida.** *Leptogenys ravida* Bolton, 1975a: 272, fig. 10 (w.) ZAIRE.

**regis.** *Leptogenys regis* Bolton, 1975a: 259, fig. 8 (w.) KENYA.

**ridens.** *Leptogenys ridens* Forel, 1910a: 16 (w.) MADAGASCAR. Combination in *L.* (*Machaerogenys*): Emery, 1911d: 100. See also: Bolton, 1975a: 295.

**ritae.** *Leptogenys ritae* Forel, 1899c: 17 (w.) PANAMA.

**roberti.** *Leptogenys* (*Lobopelta*) *roberti* Forel, 1900d: 311 (w.) INDIA. Forel, 1913k: 17 (m.). Current subspecies: nominal plus *coonoorensis.* See also: Bingham, 1903: 68.

**rouxi.** *Prionogenys rouxi* Emery, 1914f: 399, pl. 13, fig. 3 (w.) NEW CALEDONIA. Combination in *Leptogenys:* Taylor, 1988: 34.

*rubicunda. Leptogenys* (*Lobopelta*) *rubicunda* Borgmeier, 1930: 28, pl. 4, figs. 13, 14, 17, 21, 25, 26 (w.q.) BRAZIL. Junior synonym of *crudelis:* Borgmeier, 1932b: 485.

**rufa.** *Leptogenys* (*Lobopelta*) *rufa* Mann, 1922: 14 (w.) HONDURAS.

*rufipes. Leptogenys* (*Leptogenys*) *conradti* var. *rufipes* Santschi, 1937g: 75 (w.) GABON. Junior synonym of *crustosa:* Bolton, 1975a: 251.

**rugosopunctata.** *Leptogenys* (*Lobopelta*) *rugosopunctata* Karavaiev, 1925c: 283 (w.q.) JAVA.

**sagaris.** *Leptogenys sagaris* Wilson, 1958a: 137, fig. 5 (w.) NEW CALEDONIA.

*santschii. Leptogenys* (*Lobopelta*) *diminuta* subsp. *santschii* Mann, 1919: 299 (w.q.m.) SOLOMON IS. Junior synonym of *diminuta:* Wilson, 1958a: 118.

**#sarasinorum.** *Leptogenys* (*Lobopelta*) *diminuta* var. *sarasinorum* Forel, 1900d: 307 (diagnosis in key) (w.) SRI LANKA. Subspecies/race of *diminuta:* Bingham, 1903: 62; Emery, 1911d: 103; Forel, 1913e: 661.

**saussurei.** *Lobopelta saussurei* Forel, 1891b: 121 (w.) MADAGASCAR. Forel, 1891b: 218 (m.). Combination in *Leptogenys:* Emery, 1911d: 102. See also: Bolton, 1975a: 297.

**schwabi.** *Leptogenys schwabi* Forel, 1913j: 208 (w.) ZIMBABWE. Subspecies of *stuhlmanni:* Prins, 1965a: 153. Revived status as species: Bolton, 1975a: 259.

*septentrionalis. Lobopelta septentrionalis* Mayr, 1886d: 438 (w.) U.S.A. Combination in *Leptogenys:* Emery, 1895c: 268. Junior synonym of *elongata:* Wheeler, W.M. 1902f: 25.

*sericeus. Leptogenys* (*Leptogenys*) *maxillosa* subsp. *sericeus* Weber, 1942a: 46 (w.) SUDAN. Junior synonym of *pavesii:* Bolton, 1975a: 255.

**#siemsseni.** *Leptogenys* (*Lobopelta*) *kitteli* subsp. *siemsseni* Viehmeyer, 1922: 203, fig. 1 (w.) CHINA.

*simillima. Ponera simillima* Smith, F. 1860b: 104 (w.) INDONESIA (Batjan I.). Combination in *Lobopelta:* Mayr, 1862: 734. Junior synonym of *diminuta:* Mayr, 1863: 428.

**sjostedti.** *Leptogenys sjostedti* Forel, 1915b: 27, fig. 5 (w.) AUSTRALIA.

**spandax.** *Leptogenys spandax* Bolton, 1975a: 279, fig. 28 (w.) GHANA.

*speculans. Leptogenys* (*Lobopelta*) *nitida* st. *speculans* Santschi, 1926b: 209 (w.) ZAIRE. Junior synonym of *nitida:* Bolton, 1975a: 289.

*stadelmanni. Leptogenys* (*Leptogenys*) *stadelmanni* Santschi, 1937a: 47. *Nomen nudum;* attributed to Mayr. [Name perhaps in error for *stuhlmanni.*]

**stenocheilos.** *Ponera stenocheilos* Jerdon, 1851: 118 (w.) INDIA. Probable combination in *Leptogenys:* Emery, 1911d: 106. [From the descriptions in Jerdon, 1851: 118 and Jerdon, 1854: 101, I suspect that this combination is correct.]

**sterops.** *Leptogenys sterops* Bolton, 1975a: 261, fig. 20 (w.) IVORY COAST.

*stitzi. Leptogenys* (*Lobopelta*) *diminuta* var. *stitzi* Viehmeyer, 1924b: 310. Replacement name for *papuana* Stitz, 1912: 498. [Junior primary homonym of *papuana* Emery, 1897d: 556.] Junior synonym of *diminuta:* Wilson, 1958a: 118.

**strator.** *Leptogenys strator* Bolton, 1975a: 291 (w.) ZAIRE.

**striatidens.** *Leptogenys striatidens* Bolton, 1975a: 292 (w.) CAMEROUN.

**#striatula.** *Leptogenys striatula* Emery, 1895k: 461 (w.) BURMA. Subspecies/race of *diminuta:* Forel, 1900d: 307; Emery, 1911d: 103. See also: Bingham, 1903: 63.

**stuhlmanni.** *Leptogenys* (*Lobopelta*) *stuhlmanni* Mayr, 1893: 198 (w.) MOZAMBIQUE. Santschi, 1914b: 55 (m.). Senior synonym of *comorensis:* Bolton, 1975a: 262.

**stygia.** *Leptogenys stygia* Bolton, 1975a: 292 (w.) NIGERIA.

*suarensis. Leptogenys incisa* subsp. *suarensis* Emery, 1895f: 338 (w.) MOZAMBIQUE. Junior synonym of *incisa:* Bolton, 1975a: 298.

**sulcinoda.** *Lobopelta sulcinoda* André, 1892a: 48 (w.) GABON. Combination in *Leptogenys:* Emery, 1911d: 102. See also: Bolton, 1975a: 272.

***LEPTOGENYS** (continued)*

*sulcinodis. Leptogenys ferrarii* r. *sulcinodis* Bernard, 1953b: 210, fig. 5 (w.) GUINEA. [Junior secondary homonym of *sulcinoda* André, above.] Replacement name: *bernardi* Baroni Urbani, 1971b: 360.

*tenuis. Leptogenys tenuis* Stitz, 1911b: 376, fig. 1 (w.) ZAIRE. Junior synonym of *nitida:* Forel, 1912n: 52; Bolton, 1975a: 289.

***terroni.*** *Leptogenys terroni* Bolton, 1975a: 273 (w.) CAMEROUN.

***testacea.*** *Microbolbos testacea* Donisthorpe, 1948f: 170 (w.) GHANA. Combination in *Leptogenys:* Wilson, 1955c: 136. See also: Bolton, 1975a: 280.

*texana. Ponera texana* Buckley, 1866: 170 (w.) U.S.A. Combination in *Leptogenys:* Emery, 1895c: 338. Junior synonym of *elongata:* Wheeler, W.M. 1902f: 25; Wheeler, W.M. 1904c: 258.

***titan.*** *Leptogenys titan* Bolton, 1975a: 263 (w.) CAMEROUN.

#***tjibodana.*** *Leptogenys* (*Lobopelta*) *diminuta* var. *tjibodana* Karavaiev, 1926d: 421 (w.) JAVA.

#***transiens.*** *Leptogenys* (*Lobopelta*) *kitteli* var. *transiens* Forel, 1911e: 253 (w.) BORNEO.

***tricosa.*** *Leptogenys tricosa* Taylor, 1969: 132. Replacement name for *Dorylozelus mjobergi* Forel, 1915b: 25. [Junior secondary homonym of *Leptogenys mjobergi* Forel, 1915b: 32.]

***triloba.*** *Leptogenys triloba* Emery, 1901h: 159 (w.) NEW GUINEA.

***trilobata.*** *Leptogenys trilobatus* Santschi, 1924b: 196, fig. 1 (w.) ZAIRE. See also: Bolton, 1975a: 273.

#***trinidadensis.*** *Leptogenys unistimulosa* var. *trinidenysadensis* Forel, 1901f: 328 (w.) TRINIDAD. [Justified emendation of spelling to *trinidadensis:* Forel, 1905b: 156.]

***truncatirostris.*** *Leptogenys truncatirostris* Forel, 1897c: 195, fig. 2 (w.) MADAGASCAR. Forel, 1907g: 76 (m.). Combination in *L.* (*Machaerogenys*): Emery, 1911d: 101. See also: Bolton, 1975a: 295.

***truncata.*** *Leptogenys* (*Leptogenys*) *truncatus* Mann, 1919: 296, fig. 10 (not fig. 9 as indicated in the original description) (w.) SOLOMON IS.

***turneri.*** *Leptogenys turneri* Forel, 1900b: 67 (w.) AUSTRALIA. Wheeler, G.C. & Wheeler, J. 1952c: 642 (l.). Combination in *L.* (*Odontopelta*): Emery, 1911d: 101.

***unistimulosa.*** *Leptogenys unistimulosa* Roger, 1863a: 175 (w.) BRAZIL. Current subspecies: nominal plus *bahiana, trinidadensis.*

***varicosa.*** *Leptogenys* (*Lobopelta*) *varicosa* Stitz, 1925: 112 (w.) PHILIPPINES.

***venatrix.*** *Leptogenys venatrix* Forel, 1899c: 17 (w.) PANAMA.

#***vincentensis.*** *Leptogenys pubiceps* var. *vincentensis* Forel, 1901f: 328 (w.) ANTILLES IS.

***vindicis.*** *Leptogenys vindicis* Bolton, 1975a: 263 (w.) CAMEROUN.

*vinsonnella. Formica vinsonnella* Dufour, 1864: 210 (w.) REUNION I. Combination in *Leptogenys:* Emery, 1894a: 68. Variety of *maxillosa:* Emery, 1894a: 68; Forel, 1895a: 49. Junior synonym of *maxillosa:* Bolton, 1975a: 254.

***violacea.*** *Leptogenys* (*Lobopelta*) *violacea* Donisthorpe, 1942e: 705 (w.) NEW GUINEA.

***vitiensis.*** *Leptogenys* (*Lobopelta*) *vitiensis* Mann, 1921: 424, fig. 10 (w.) FIJI IS.

***voeltzkowi.*** *Leptogenys voeltzkowi* Forel, 1897c: 194 (w.) MADAGASCAR.

***vogeli.*** *Leptogenys* (*Lobopelta*) *vogeli* Borgmeier, 1933b: 226 (w.) BRAZIL.

*walkeri. Leptogenys* (*Leptogenys*) *walkeri* Donisthorpe, 1942e: 704 (w.) VANUATU. Junior synonym of *foreli:* Wilson, 1958a: 126.

***watsoni.*** *Leptogenys peuqueti* r. *watsoni* Forel, 1900d: 309 (diagnosis in key) (w.) BURMA. Raised to species: Bingham, 1903: 72.

#***weiseri.*** *Leptogenys bohlsi* st. *weiseri* Santschi, 1925e: 153, fig. 1 (w.) ARGENTINA.

***wheeleri.*** *Leptogenys wheeleri* Forel, 1901c: 123 (w.) MEXICO.

#***woodmasoni.*** *Lobopelta woodmasoni* Forel, 1886d: 246 (w.) INDIA. Combination in *Leptogenys:* Forel, 1900d: 313. Subspecies of *diminuta:* Forel, 1900d: 313; Emery, 1911d: 103. See also: Bingham, 1903: 62.

*yarrabahna. Leptogenys* (*Lobopelta*) *diminuta* var. *yarrabahna* Forel, 1915b: 29 (w.m.) AUSTRALIA. Junior synonym of *diminuta:* Taylor, 1988: 34.

***yerburyi.*** *Leptogenys* (*Lobopelta*) *yerburyi* Forel, 1900d: 311 (w.) SRI LANKA. See also: Bingham, 1903: 69.

***zapyxis.*** *Leptogenys zapyxis* Bolton, 1975a: 264, figs. 4, 19, 21 (w.) GHANA.

*LEPTOMESITES:* see under ***LEPTANILLA.***

*LEPTOMYRMA:* see under ***PHEIDOLE.***

***LEPTOMYRMEX*** (Sulawesi, New Guinea, Australasian)

#***angusticeps.*** *Leptomyrmex varians* st. *angusticeps* Santschi, 1929f: 93, fig. 3 (m.) AUSTRALIA.

#***basirufus.*** *Leptomyrmex erythrocephalus* subsp. *basirufus* Wheeler, W.M. 1934c: 90 (w.) AUSTRALIA.

#***brunneiceps.*** *Leptomyrmex erythrocephalus* subsp. *brunneiceps* Wheeler, W.M. 1934c: 88 (w.) AUSTRALIA.

#***clarki.*** *Leptomyrmex erythrocephalus* subsp. *clarki* Wheeler, W.M. 1934c: 117 (w.) AUSTRALIA.

#***cnemidatus.*** *Leptomyrmex erythrocephalus* var. *cnemidatus* Wheeler, W.M. 1915d: 268 (w.) AUSTRALIA. Wheeler, W.M. 1934c: 92 (m.). Subspecies of *erythrocephalus:* Wheeler, W.M. 1934c: 92.

***contractus.*** *Leptomyrmex contractus* Donisthorpe, 1947c: 586 (w.) NEW GUINEA.

***darlingtoni.*** *Leptomyrmex darlingtoni* Wheeler, W.M. 1934c: 104, fig. 13 (w.m.) AUSTRALIA. Current subspecies: nominal plus *fascigaster, jucundus.*

#***decipiens.*** *Leptomyrmex erythrocephalus* var. *decipiens* Wheeler, W.M. 1915d: 268 (w.) AUSTRALIA. Wheeler, W.M. 1934c: 90 (m.). Subspecies of *erythrocephalus:* Wheeler, W.M. 1934c: 90.

***erythrocephalus.*** *Formica erythrocephala* Fabricius, 1775: 391 (w.) AUSTRALIA. Emery, 1883: 147 (q.); Emery, 1891a: 152 (m.); Wheeler, G.C. & Wheeler, J. 1951: 179 (l.); Imai, Crozier & Taylor, 1977: 345 (k.). Combination in *Leptomyrmex:* Mayr, 1862: 696. Current subspecies: nominal plus *basirufus, brunneiceps, clarki, cnemidatus, decipiens, mandibularis, rufithorax, unctus, venustus.* See also: Wheeler, W.M. 1915d: 265; Wheeler, W.M. 1934c: 85.

#***fascigaster.*** *Leptomyrmex darlingtoni* subsp. *fascigaster* Wheeler, W.M. 1934c: 107 (w.) AUSTRALIA.

#***femoratus.*** *Leptomyrmex fragilis* var. *femorata* Santschi, 1932b: 17, fig. 3 (w.) NEW GUINEA. Subspecies of *fragilis:* Wheeler, W.M. 1934c: 113.

***flavitarsus.*** *Formica flavitarsus* Smith, F. 1859a: 136 (w.) INDONESIA (Aru I.). Combination in *Camponotus:* Roger, 1863b: 4; in *Camponotus* (*Tanaemyrmex*): Emery, 1925b: 90; in *Leptomyrmex:* Shattuck, 1992c: 114.

***fragilis.*** *Formica fragilis* Smith, F. 1859a: 136 (w.) INDONESIA (Aru I.). Emery, 1897d: 571 (m.); Stitz, 1912: 507 (m.). Combination in *Leptomyrmex:* Emery, 1897d: 571. Senior synonym of *gracillimus:* Baroni Urbani & Wilson, 1987: 2. See also: Wheeler, W.M. 1934c: 113. Current subspecies: nominal plus *femoratus, maculatus, melanoticus.*

***froggatti.*** *Leptomyrmex froggatti* Forel, 1910b: 57 (w.m.) AUSTRALIA. See also: Wheeler, W.M. 1934c: 101.

#***geniculatus.*** *Leptomyrmex pallens* var. *geniculata* Emery, 1914f: 418 (w.) NEW CALEDONIA. Subspecies of *pallens:* Wheeler, W.M. 1934c: 109.

*gracillimus. Leptomyrmex gracillimus* Wheeler, W.M. 1934c: 115, fig. 16 (w.) NEW GUINEA. Junior synonym of *fragilis:* Baroni Urbani & Wilson, 1987: 2.

#***hackeri.*** *Leptomyrmex nigriventris* subsp. *hackeri* Wheeler, W.M. 1934c: 99 (w.) AUSTRALIA.

#***jucundus.*** *Leptomyrmex darlingtoni* subsp. *jucundus* Wheeler, W.M. 1934c: 107 (w.) AUSTRALIA.

***lugubris.*** *Leptomyrmex lugubris* Wheeler, W.M. 1934c: 110, fig. 14 (w.) NEW GUINEA.

#***maculatus.*** *Leptomyrmex fragilis* var. *maculata* Stitz, 1938: 108 (w.) NEW GUINEA.

#***mandibularis.*** *Leptomyrmex erythrocephalus* var. *mandibularis* Wheeler, W.M. 1915d: 268 (w.) AUSTRALIA. Subspecies of *erythrocephalus:* Wheeler, W.M. 1934c: 87.

**maravignae* Emery, 1891; see under ****LEPTOMYRMULA.***

#***melanoticus.*** *Leptomyrmex fragilis* subsp. *melanoticus* Wheeler, W.M. 1934c: 113, fig. 15 (w.m.) NEW GUINEA.

***mjobergi.*** *Leptomyrmex mjobergi* Forel, 1915b: 84 (w.) AUSTRALIA.

****neotropicus.*** **Leptomyrmex neotropicus* Baroni Urbani, 1980d: 4, figs. 1–5 (w.) DOMINICAN AMBER (Miocene). Combination in *Camponotus:* Wilson, 1985c: 34; in *Leptomyrmex:* Baroni Urbani & Wilson, 1987: 2.

***niger.*** *Leptomyrmex niger* Emery, 1900c: 333, pl. 8, figs. 3, 4 (w.) NEW GUINEA. Karavaiev, 1926d: 430 (m.). See also: Wheeler, W.M. 1934c: 109.

#***nigriceps.*** *Leptomyrmex pallens* var. *nigriceps* Emery, 1914f: 418 (w.) NEW CALEDONIA. Subspecies of *pallens:* Wheeler, W.M. 1934c: 109.

***nigriventris.*** *Formica nigriventris* Guérin-Méneville, 1831, pl. 8, fig. 4 (w.) AUSTRALIA. Wheeler, W.M. 1915d: 261 (l.); Wheeler, W.M. 1934c: 95 (m.). Combination in *Leptomyrmex:* Mayr, 1862: 696. Junior synonym of *erythrocephalus:* Mayr, 1876: 77. Revived from synonymy as variety of *erythrocephalus:* Emery, 1887a: 252. Revived status as species: Emery, 1895g: 351. Current subspecies: nominal plus *hackeri, tibialis.*

**pallens.** *Leptomyrmex pallens* Emery, 1883: 147, fig. (w.) NEW CALEDONIA. André, 1887: 290 (m.). Current subspecies: nominal plus *geniculatus, nigriceps.* See also: Wheeler, W.M. 1915d: 276; Wheeler, W.M. 1934c: 108.

#**pictus.** *Leptomyrmex wiburdi* var. *pictus* Wheeler, W.M. 1915d: 273 (w.) AUSTRALIA. Wheeler, G.C. & Wheeler, J. 1966: 728 (l.). Subspecies of *wiburdi:* Wheeler, W.M. 1934c: 101.

**puberulus.** *Leptomyrmex puberulus* Wheeler, W.M. 1934c: 112, fig. 14 (w.) NEW GUINEA.

#**quadricolor.** *Leptomyrmex varians* subsp. *quadricolor* Wheeler, W.M. 1934c: 104 (w.) AUSTRALIA.

#**rothneyi.** *Leptomyrmex varians* var. *rothneyi* Forel, 1902h: 473 (w.) AUSTRALIA. Subspecies of *varians:* Wheeler, W.M. 1934c: 102.

#**ruficeps.** *Leptomyrmex varians* var. *ruficeps* Emery, 1895g: 352 (w.) AUSTRALIA. Forel, 1915b: 84 (m.); Wheeler, W.M. 1915d: 261 (l.). Subspecies of *varians:* Wheeler, W.M. 1934c: 102.

#**rufipes.** *Leptomyrmex varians* var. *rufipes* Emery, 1895g: 352 (w.) AUSTRALIA. Subspecies of *varians:* Wheeler, W.M. 1934c: 104.

#**rufithorax.** *Leptomyrmex erythrocephalus* var. *rufithorax* Forel, 1915b: 83 (w.) AUSTRALIA. Wheeler, W.M. 1934c: 91 (m.). Subspecies of *erythrocephalus:* Wheeler, W.M. 1934c: 91.

#**tibialis.** *Leptomyrmex nigriventris* var. *tibialis* Emery, 1895g: 351 (w.) AUSTRALIA. Wheeler, W.M. 1934c: 96 (q.m.). Subspecies of *nigriventris:* Wheeler, W.M. 1934c: 96.

#**unctus.** *Leptomyrmex erythrocephalus* subsp. *unctus* Wheeler, W.M. 1934c: 87 (w.) AUSTRALIA.

**unicolor.** *Leptomyrmex unicolor* Emery, 1895g: 352, figs. 3, 4 (w.) AUSTRALIA. Wheeler, G.C. & Wheeler, J. 1951: 180 (l.). See also: Wheeler, W.M. 1915d: 261; Wheeler, W.M. 1934c: 108.

**varians.** *Leptomyrmex varians* Emery, 1895g: 352, fig. 3 (w.) AUSTRALIA. Current subspecies: nominal plus *angusticeps, quadricolor, rothneyi, ruficeps, rufipes.* See also: Wheeler, W.M. 1934c: 101.

#**venustus.** *Leptomyrmex erythrocephalus* subsp. *venustus* Wheeler, W.M. 1934c: 87, fig. 5 (w.q.) AUSTRALIA.

**wheeleri.** *Leptomyrmex wheeleri* Donisthorpe, 1948d: 600 (w.m.) NEW GUINEA.

**wiburdi.** *Leptomyrmex wiburdi* Wheeler, W.M. 1915d: 272, fig. 7 (w.) AUSTRALIA. Wheeler, W.M. 1934c: 99 (m.). Current subspecies: nominal plus *pictus.*

**LEPTOMYRMULA** (Sicilian Amber)

***maravignae.** **Leptomyrmex maravignae* Emery, 1891a: 152, pl. 2, fig. 22 (m.) SICILIAN AMBER (Miocene). Combination in **Leptomyrmula:* Emery, 1912g: 16 (footnote).

*LEPTOPONE:* see under ***CENTROMYRMEX.***

***LEPTOTHORAX*** (Neotropical, Holarctic, Afrotropical, Malagasy, Oriental)

**acervorum.** *Formica acervorum* Fabricius, 1793: 358 (w.) DENMARK. Latreille, 1798: 49 (q.m.); Wheeler, G.C. & Wheeler, J. 1955b: 21 (l.). Combination in *Leptothorax:* Mayr, 1855: 436; in *L.* (*Mychothorax*): Ruzsky, 1904a: 288. Senior synonym of *lacteipennis:* Nylander, 1846a: 936; of *kamtschaticum:* Kupyanskaya, 1986b: 96; of *orientalis:* Kupyanskaya, 1990: 137. Material of the *nomen nudum melanocephala* referred here by Mayr, 1855: 411. Current subspecies: nominal plus *nigrescens, superus, vandeli.* See also: Donisthorpe, 1915d: 147; Stitz, 1939: 158; Tarbinsky, 1976: 86; Kutter, 1977c: 128; Atanassov & Dlussky, 1992: 130.

*aculeata* Mayr, 1866 [*Macromischa*]; see under ***TETRAMORIUM.***

*aculeatinodis. Leptothorax* (*Goniothorax*) *aculeatinodis* Emery, 1896g: 60 (w.) COSTA RICA. Subspecies/race of *echinatinodis:* Forel, 1899c: 55; Wheeler, W.M. 1921f: 158. Junior synonym of *echinatinodis:* Kempf, 1959c: 425.

*acutinodis. Leptothorax tuberum* subsp. *acutinodis* Arnol'di, 1977a: 204 (w.) ARMENIA. [*Leptothorax tuberum* subsp. *acutinodis* Arnol'di, 1948: 211. *Nomen nudum.*] Junior synonym of *tuberum:* Arakelian, 1994: 58.

#**aeolius.** *Leptothorax bulgaricus* subsp. *aeolius* Forel, 1911d: 334 (w.) TURKEY.

**affinis.** *Leptothorax affinis* Mayr, 1855: 442 (w.) AUSTRIA. Forel, 1874: 88 (q.); Stitz, 1914: 62 (m.). Subspecies/race of *tuberum:* Forel, 1874: 85; Forel, 1892i: 314; Bondroit, 1910: 496; Emery, 1916b: 174; Emery, 1922f: 256. Status as species: Bondroit, 1918: 128; Müller, 1923: 93; Finzi, 1933: 164; Stitz, 1939: 173; Bernard, 1956a: 163; Baroni Urbani, 1971c: 102; Kutter, 1977c: 130; Atanassov & Dlussky, 1992: 133. Senior synonym of *tirolensis:* Emery, 1914d: 158 (footnote). Current subspecies: nominal plus *tuberoaffinis.*

*affinis. Macromischa* (*Macromischa*) *affinis* Mann, 1920: 417, fig. 5 (w.) CUBA. [Junior secondary homonym of *affinis* Mayr, above.] Replacement name: *Leptothorax similis* Baroni Urbani, 1978b: 501. [Replacement name unnecessary: see under *laetus.*]

*affinonigriceps. Leptothorax tuberum* var. *affinonigriceps* Forel, 1874: 86. *Nomen nudum.*

*affinotuberum. Leptothorax tuberum* var. *affinotuberum* Forel, 1874: 88. *Nomen nudum.*

*africana* Mayr, 1866 [*Macromischa*]; see under ***TETRAMORIUM.***

**aguayoi.** *Croesomyrmex aguayoi* Wheeler, W.M. 1931b: 25 (w.) CUBA. Baroni Urbani, 1978b: 410 (m.). Combination in *Macromischa* (*Croesomyrmex*): Aguayo, 1931: 179; in *Leptothorax,* and senior synonym of *archeri* (and its junior synonym *bierigi*) and *natenzoni:* Baroni Urbani, 1978b: 409.

#**alaicus.** *Leptothorax tuberum* subsp. *alaicus* Tarbinsky, 1976: 97 (w.) KIRGIZIA.

**alayoi.** *Leptothorax alayoi* Baroni Urbani, 1978b: 411, figs. 26, 83 (w.) CUBA.

**alinae.** *Leptothorax alinae* Radchenko, 1993c (not yet seen). See also: Radchenko, 1994d: 153.

*albipennis. Stenamma albipennis* Curtis, 1854: 218 (w.m.) GREAT BRITAIN. Junior synonym of *unifasciatus:* Nylander, 1856b: 92; of *nylanderi:* Smith, F. 1858b: 120; of *tuberum:* Emery, 1922f: 256.

**albispinus.** *Macromischa albispina* Wheeler, W.M. 1908a: 139, pl. 11, figs. 3, 4 (w.q.) PUERTO RICO. Combination in *M.* (*Antillaemyrmex*): Mann, 1920: 408; in *Antillaemyrmex:* Wheeler, W.M. 1931b: 27; in *Leptothorax:* Baroni Urbani, 1978b: 413. Senior synonym of *pallipes:* Baroni Urbani, 1978b: 413.

#**algiricus.** *Leptothorax angustulus* var. *algiricus* Forel, 1894d: 37 (w.) ALGERIA. Stirps of *angustulus:* Santschi, 1911a: 12.

**allardycei.** *Macromischa* (*Macromischa*) *allardycei* Mann, 1920: 414, fig. 3 (w.q.) BAHAMAS. Combination in *Leptothorax,* and senior synonym of *floridanus:* Baroni Urbani, 1978b: 415.

**alpinus.** *Leptothorax alpinus* Ruzsky, 1902d: 22 (w.) RUSSIA. Subspecies of *tuberum:* Ruzsky, 1905b: 598. Revived status as species: Radchenko, 1994d: 150 (in key).

**ambiguus.** *Leptothorax curvispinosus* subsp. *ambiguus* Emery, 1895c: 320 (w.) U.S.A. Wesson, L.G. & Wesson, R.G. 1940: 97 (q.m.); Wheeler, G.C. & Wheeler, J. 1955b: 22 (l.). Combination in *L.* (*Myrafant*): Smith, D.R. 1979: 1392. Raised to species: Wesson, L.G. & Wesson, R.G. 1940: 97. Current subspecies: nominal plus *foveatus, pinetorum.*

#**amilcaris.** *Leptothorax flavispinus* st. *amilcaris* Santschi, 1939d: 76 (w.) TUNISIA. [First available use of *Leptothorax flavispinus* st. *santschii* var. *amilcaris* Santschi, 1911a: 12; unavailable name.]

**anacanthus.** *Leptothorax hesperius* st. *anacanthus* Santschi, 1912c: 149. Replacement name for *inermis* Santschi, 1909b: 468. [Junior primary homonym of *inermis* Forel, 1902c: 229.] Raised to species: Santschi, 1931a: 6. Current subspecies: nominal plus *maximus.*

**andrei.** *Leptothorax andrei* Emery, 1895c: 322, pl. 8, fig. 15 (w.) U.S.A. Cole, 1958c: 537 (q.m.). Combination in *L.* (*Myrafant*): Smith, D.R. 1979: 1392. See also: Wheeler, W.M. 1903c: 256.

**androsanus.** *Macromischa androsana* Wheeler, W.M. 1905b: 97, fig. K (w.) BAHAMAS. Baroni Urbani, 1978b: 418 (q.m.). Combination in *Leptothorax:* Baroni Urbani, 1978b: 418.

**anduzei.** *Leptothorax* (*Goniothorax*) *anduzei* Weber, 1943b: 72 (w.) VENEZUELA. Kempf, 1975c: 363 (q.). Combination in *L.* (*Nesomyrmex*): Kempf, 1959c: 398.

**anemicus.** *Leptothorax anemicus* Baroni Urbani, 1978b: 420. Replacement name for (and elevation to species of) *Macromischa* (*Croesomyrmex*) *bermudezi* var. *gracilis* Aguayo, 1932: 223. [Junior secondary homonym of **Leptothorax gracilis* Mayr, 1868c: 89.]

**angulatus.** *Leptothorax angulatus* Mayr, 1862: 739 (w.) EGYPT. Santschi, 1914b: 107 (q.). Combination in *L.* (*Goniothorax*): Emery, 1896g: 58 (footnote). Senior synonym of *concolor, ilgii* and *latinodis:* Bolton, 1982: 324. Current subspecies: nominal plus *lybica.*

#**angustinodis.** *Leptothorax angustulus* var. *angustinode* Stitz, 1917: 336, fig. 3 (w.) LIBYA.

*angustinodus. Leptothorax* (*Leptothorax*) *angustinodus* Cole, 1956b: 28 (w.q.) U.S.A. [Junior primary homonym of *angustinodis* Stitz, above.] Replacement name: *stenotyle* Cole, 1956e: 214.

**angustulus.** *Myrmica angustula* Nylander, 1856b: 88, pl. 3, fig. 36 (w.) FRANCE. Emery, 1916b: 183, 187 (q.m., respectively). Combination in *Leptothorax:* Mayr, 1861: 59. Race of *tuberum:* Forel, 1890a: lxxv; Emery, 1891b: 5. Status as species: Dalla Torre, 1893: 123; Forel, 1894d: 37; Forel, 1907b: 206; Emery, 1916b: 175; Bondroit, 1918: 136. Current subspecies: nominal plus *algiricus, angustinodis, atlantis, bordagei, brunea, silvanus, trabutii.*

*annectens. Leptothorax curvispinosus* subsp. *annectens* Wheeler, W.M. 1903c: 242, pl. 12, fig. 13 (w.) U.S.A. Subspecies of *rugatulus:* Wheeler, W.M. 1917a: 510. Junior synonym of *rugatulus:* Creighton, 1950a: 269.

*annectens. Macromischa annectens* Wheeler, W.M. 1931b: 11 (w.) MEXICO. [Junior secondary homonym of *annectens* Wheeler, W.M., above.] Replacement name: *annexus* Baroni Urbani, 1978b: 421.

***annexus.*** *Leptothorax annexus* Baroni Urbani, 1978b: 421. Replacement name for *annectens* Wheeler, W.M. 1931b: 11. [Junior secondary homonym of *annectens* Wheeler, W.M. 1903c: 242.]

***annibalis.*** *Leptothorax rottenbergi* st. *annibalis* Santschi, 1909b: 476, fig. 9 (w.q.) TUNISIA. Raised to species: Santschi, 1929e: 151. Current subspecies: nominal plus *numidicus.*

***anodonta.*** *Leptothorax anodonta* Arnol'di, 1977a: 199 (w.m.) ARMENIA.

***anodontoides.*** *Leptothorax anodontoides* Dlussky & Zabelin, 1985: 227, fig. 5 (w.) TURKMENIA.

*anoplogynus. Leptothorax anoplogynus* Emery, 1869b: 15 (w.q.) ITALY. Variety of *tuberum:* André, 1883a: 300; of *unifasciatus:* Ruzsky, 1902d: 22. Junior synonym of *unifasciatus:* Emery, 1916b: 175.

***antigoni.*** *Leptothorax (Temnothorax) antigoni* Forel, 1911d: 333 (w.) TURKEY.

*antoniensis. Leptothorax asper* var. *antoniensis* Forel, 1912g: 18 (w.) COLOMBIA. Junior synonym of *tristani:* Kempf, 1959c: 415.

***arcanus.*** *Leptothorax arcanus* Kutter, 1973b: 327, figs. 1–7 (m.) SWITZERLAND.

***archangelskiji.*** *Leptothorax bulgaricus* subsp. *archangelskiji* Kuznetsov-Ugamsky, 1926a: 74, figs. 19–26 (w.q.m.) TURKESTAN. Raised to species: Dlussky, 1962: 179.

*archeri. Croesomyrmex aguayoi* var. *archeri* Wheeler, W.M. 1931b: 26 (w.) CUBA. Combination in *Macromischa,* raised to species, and senior synonym of *bierigi:* Brown, 1959e: 74. Junior synonym of *aguayoi:* Baroni Urbani, 1978b: 409.

***arenarius.*** *Leptothorax arenarius* Santschi, 1908: 527 (w.q.l.) TUNISIA. Santschi, 1910b: 62 (m.). Combination in *L. (Temnothorax)*: Santschi, 1918a: 32. Current subspecies: nominal plus *fusciventris, kruegeri.*

***argentinus.*** *Leptothorax (Goniothorax) argentinus* Santschi, 1922d: 251 (w.) ARGENTINA. Combination in *L. (Nesomyrmex)*: Kempf, 1959c: 425.

***argentipes.*** *Leptothorax argentipes* Wheeler, W.M. 1928c: 25 (w.q.) CHINA.

***arimensis.*** *Leptothorax arimensis* Azuma, 1977: 114 (w.) JAPAN. [*Leptothorax arimensis* Azuma, 1953: 3. *Nomen nudum.*] See also: Onoyama, 1980: 194, 212.

*armouri. Leptothorax risii* subsp. *armouri* Wheeler, W.M. 1927g: 111 (w.) CANARY IS. Junior synonym of *risii:* Hohmann, La Roche, *et al.*, 1993: 151.

***arpini.*** *Leptothorax arpini* Tarbinsky, 1976: 91, figs. 148–150, 189 (w.) KIRGIZIA. Senior synonym of *narinicus:* Radchenko, 1994d: 150.

***asper.*** *Leptothorax asper* Mayr, 1887: 618 (w.q.m.) BRAZIL. Combination in *L. (Goniothorax)*: Emery, 1896g: 59 (in key); in *L. (Nesomyrmex)*, and senior synonym of *rufa* and *sulfurea:* Kempf, 1959c: 414.

#***atlantis.*** *Leptothorax atlantis* Santschi, 1911a: 13, fig. 9 (w.) ALGERIA. Stirps of *normandi:* Santschi, 1921b: 426. Revived status as species: Emery, 1922f: 253; Cagniant, 1964: 105. Subspecies of *angustulus:* Cagniant, 1970a: 420.

*atomus. Leptothorax tuberum* r. *tebessae* var. *atomus* Forel, 1905b: 173 (w.) TUNISIA. Unavailable name.

*atrinodis. Macromischa (Macromischa) squamifera* var. *atrinodis* Mann, 1920: 412, fig. 2 (w.q.) CUBA. Junior synonym of *squamifer:* Wheeler, W.M. 1931b: 21; Baroni Urbani, 1978b: 508.

***augusti.*** *Leptothorax augusti* Baroni Urbani, 1978b: 423. Replacement name for *Macromischa foreli* Aguayo, 1931: 182. [Junior secondary homonym of *Leptothorax foreli* Santschi, 1907: 330.]

***auresianus.*** *Leptothorax auresianus* Santschi, 1929e: 158, figs. 11, 12 (w.) ALGERIA.

*australis* Wheeler, W.M. 1934; see under ***VOMBISIDRIS.***

***aveli.*** *Leptothorax aveli* Bondroit, 1918: 127, fig. 62 (w.) FRANCE. Current subspecies: nominal plus *pici.*

***aztecus.*** *Macromischa azteca* Wheeler, W.M. 1931b: 8 (w.q.m.) MEXICO. Combination in *Leptothorax,* and senior synonym of *maya:* Baroni Urbani, 1978b: 425.

*baborensis. Leptothorax spinosus* var. *baborensis* Cagniant, 1970b: 207 (w.) ALGERIA. Unavailable name.

#***babuganicus.*** *Leptothorax pithyusae* subsp. *babuganicus* Arnol'di, 1977a: 201 (w.) UKRAINE. Subspecies of *tauricus:* current status.

***baeticus.*** *Leptothorax baeticus* Emery, 1924c: 165, fig. A (w.) SPAIN. Santschi, 1931a: 5 (q., name misspelled as *boeticus*); Tinaut, 1989: 51 (m.).

*balcanica. Leptothorax rottenbergi* var. *balcanica* Santschi, 1909b: 474, fig. 7 (w.) BULGARIA. Junior synonym of *semiruber:* Agosti & Collingwood, 1987a: 56.

***balchashensis.*** *Leptothorax balchashensis* Arnol'di, 1971: 1822, fig. 2 (w.q.m.) KAZAKHSTAN.

***balli.*** *Leptothorax balli* Santschi, 1939c: 3, fig. 1 (w.) MOROCCO.

#***barbarus.*** *Leptothorax (Temnothorax) recedens* var. *barbarus* Santschi, 1939d: 76 (w.q.) TUNISIA.

***barbouri.*** *Macromischa barbouri* Aguayo, 1931: 176 (w.) CUBA. Combination in *Leptothorax:* Baroni Urbani, 1978b: 427.

***barroi.*** *Macromischa (Croesomyrmex) barroi* Aguayo, 1931: 179 (w.) CUBA. Combination in *Leptothorax:* Baroni Urbani, 1978b: 428.

***barryi.*** *Leptothorax barryi* Cagniant, 1967b: 272, 2 figs. (w.) ALGERIA.

***berlandi.*** *Leptothorax berlandi* Bondroit, 1918: 130 (w.) FRANCE. Junior synonym of *rabaudi:* Bernard, 1956a: 163; Bernard, 1967: 210. Revived from synonymy: Collingwood & Yarrow, 1969: 72.

***bermudezi.*** *Croesomyrmex bermudezi* Wheeler, W.M. 1931b: 26 (w.) CUBA. Wheeler, G.C. & Wheeler, J. 1955b: 18 (l.). Combination in *Leptothorax,* and senior synonym of *mutabilis:* Baroni Urbani, 1978b: 429.

*betulae. Leptothorax muscorum* var. *betulae* Ruzsky, 1916: 6 (w.) RUSSIA. Subspecies of *muscorum:* Ruzsky, 1936: 94.

**beyrichi* Mayr, 1868; see under ***VOLLENHOVIA.***

*bierigi. Macromischa (Croesomyrmex) bierigi* Santschi, 1931c: 273, fig. 7 (w.) CUBA. Junior synonym of *archeri:* Brown, 1959e: 74.

*bilongrudi* Taylor, 1989; see under ***VOMBISIDRIS.***

*boccaris. Leptothorax exilis* var. *boccaris* Santschi, 1923f: 331 (w.) TUNISIA. Junior synonym of *exilis:* Baroni Urbani, 1971d: 1041.

#***boiteli.*** *Leptothorax flavispinus* st. *boiteli* Santschi, 1929e: 152 (w.) ALGERIA.

#***bordagei.*** *Leptothorax angustulus* var. *bordagei* Bondroit, 1918: 137 (w.) FRANCE.

*borgmeieri. Leptothorax sculptiventris* var. *borgmeieri* Wheeler, W.M. 1931b: 31 (in text) (w.) BRAZIL. Junior synonym of *sculptiventris:* Kempf, 1959c: 400.

***bradleyi.*** *Leptothorax bradleyi* Wheeler, W.M. 1913c: 113 (w.) U.S.A. Combination in *L. (Myrafant)*: Smith, D.R. 1979: 1392.

*brasiliensis* Borgmeier, 1955 [*Macromischa*]; see under ***ROGERIA.***

***brasiliensis.*** *Leptothorax brasiliensis* Kempf, 1958c: 92, figs. 10, 18, 21 (w.q.) BRAZIL. Combination in *L. (Nesomyrmex)*: Kempf, 1959c: 411.

***brauneri.*** *Leptothorax serviculus* var. *brauneri* Ruzsky, 1905b: 603 (w.) CAUCASUS. Raised to species: Arnol'di, 1977a: 203.

#***brauneri.*** *Leptothorax (Leptothorax) tuberum* var. *brauneri* Karavaiev, 1937: 172, fig. 2 (w.) RUSSIA. [Unresolved junior primary homonym of *brauneri* Ruzsky, above.]

***braunsi.*** *Dilobocondyla (Tetramyrma) braunsi* Forel, 1912m: 767 (w.) SOUTH AFRICA. Forel, 1913a: 122 (pseudogyne); Arnold, 1917: 359 (q.). Combination in *Tetramyrma:* Forel, 1913a: 122; in *Leptothorax:* Bolton, 1982: 325.

***brimodus.*** *Leptothorax brimodus* **new name**. Replacement name for *timidus* Kempf, 1959c: 420. [Junior primary homonym of *timidus* Santschi, 1912c: 148.]

#***brunea.*** *Leptothorax angustulus* var. *brunea* Santschi, 1918a: 35 (w.) ALGERIA. [First available use of *Leptothorax angustulus* st. *algiricus* var. *brunea* Santschi, 1911a: 13; unavailable name.]

***bruneri.*** *Macromischa (Macromischa) bruneri* Mann, 1924: 19 (w.) CUBA. Combination in *Leptothorax,* and senior synonym of *imitatrix, ornatipes* and *proxima:* Baroni Urbani, 1978b: 431.

*brunneipes. Macromischa scabripes* var. *brunneipes* Wheeler, W.M. 1937b: 449 (w.) CUBA. Junior synonym of *scabripes:* Baroni Urbani, 1978b: 496.

#***brunnescens.*** *Leptothorax rugatulus* subsp. *brunnescens* Wheeler, W.M. 1917a: 510 (w.) U.S.A. Wheeler, G.C. & Wheeler, J. 1955b: 25 (l.). Combination in *L. (Myrafant)*: Smith, D.R. 1979: 1394. Senior synonym of *dakotensis:* Creighton, 1950a: 269.

***bucheti.*** *Leptothorax bucheti* Santschi, 1909b: 480, fig. 10 (w.q.) MOROCCO. Current subspecies: nominal plus *festai, mixtus, tuneticus.*

***bugnioni.*** *Leptothorax bugnioni* Forel, 1894d: 33 (w.q.) ALGERIA. See also: Santschi, 1909b: 463. Current subspecies: nominal plus *pupseli.*

***bulgaricus.*** *Leptothorax bulgaricus* Forel, 1892i: 314, pl. 5, fig. 5 (w.) BULGARIA. Current subspecies: nominal plus *aeolius, cypridis, smyrnensis.* See also: Atanassov & Dlussky, 1992: 135.

**buschingeri.** *Leptothorax buschingeri* Kutter, 1967b: 88, fig. 15 (m.) SWITZERLAND. See also: Kutter, 1977c: 128.

**cabrerae.** *Leptothorax cabrerae* Forel, 1893d: 464 (w.) CANARY IS. [Also described as new by Forel, 1894a: 160.] Santschi, 1909b: 465 (q.). Senior synonym of *stitzi:* Hohmann, La Roche, *et al.*, 1993: 149. Current subspecies: nominal plus *productus.*

**cagnianti.** *Leptothorax cagnianti* Tinaut, 1983: 319, figs. 1–5 (w.m.) SPAIN.

*calderoni. Leptothorax* (*Mychothorax*) *acervorum* r. *canadensis* var. *calderoni* Forel, 1914c: 617 (w.q.) U.S.A. Unavailable name; material referred to *septentrionalis* by Creighton, 1950a: 276.

*canadensis. Leptothorax canadensis* Provancher, 1887: 245 (w.q.m.) CANADA. Wheeler, G.C. & Wheeler, J. 1955b: 21 (l.). Combination in *L.* (*Mychothorax*): Emery, 1922f: 261. Variety/subspecies of *acervorum:* André, 1887: 295; Wheeler, W.M. 1903c: 225. Status as species: Emery, 1895c: 318; Creighton, 1950a: 274. Material of the unavailable name *convivialis* referred here: Creighton, 1950a: 276. Junior synonym of *muscorum:* Brown, 1955a: 47.

**canescens.** *Leptothorax canescens* Santschi, 1908: 526 (q.) CANARY IS. Santschi, 1919g: 407 (w.).

**carinatus.** *Leptothorax* (*Leptothorax*) *carinatus* Cole, 1957d: 213, fig. 1 (w.q.) U.S.A. Wheeler, G.C. & Wheeler, J. 1973b: 70 (l.). Combination in *L.* (*Myrafant*): Smith, D.R. 1979: 1392.

*carinthiacus. Leptothorax carinthiacus* Bernard, 1957a: 46, figs. 1, 2 (w.q.m.) AUSTRIA. Junior synonym of *sordidulus:* Schulz, 1991: 121.

**cataulacoides.** *Leptothorax catalaucoides* Snelling, 1992a: 99, figs. 6–8 (w.) CAMEROUN. Spelling here emended to *cataulacoides* (as Snelling misspells the genus-group name *Cataulacus* as *Catalaucus* throughout the paper).

**caucasicus.** *Leptothorax caucasicus* Arnol'di, 1977a: 198 (w.) GEORGIA. Senior synonym of *hasardaghi:* Radchenko, 1994d: 154 (in key).

**cenatus.** *Leptothorax cenatus* Bolton, 1982: 327, figs. 11, 15 (w.) KENYA.

**cervantesi.** *Leptothorax cervantesi* Santschi, 1919e: 242, fig. 1 (w.) SPAIN.

*chloana. Macromischa chloana* Wheeler, W.M. 1937b: 454 (w.) CUBA. Junior synonym of *dissimilis:* Brown, 1959e: 74; Baroni Urbani, 1978b: 439.

**#chobauti.** *Leptothorax rusticus* var. *chobauti* Santschi, 1923f: 330 (w.) ALGERIA. Subspecies of *oraniensis:* current status.

**ciferrii.** *Macromischa* (*Antillaemyrmex*) *ciferrii* Menozzi & Russo, 1930: 160, fig. 5 (w.) DOMINICA. Combination in *Antillaemyrmex:* Wheeler, W.M. 1931b: 32; in *Leptothorax:* Baroni Urbani, 1978b: 433.

*cingulata. Myrmica cingulata* Schenck, 1852: 104 (w.q.m.) GERMANY. [Also described as new by Schenck, 1853: 188.] Junior synonym of *nylanderi:* Mayr, 1855: 447.

**#ciscaucasicus.** *Leptothorax tuberum* subsp. *ciscaucasicus* Arnol'di, 1977a: 204 (w.q.) RUSSIA.

**clavipilis.** *Nesomyrmex clavipilis* Wheeler, W.M. 1910a: 259, fig. 1 (q.) WEST INDIES. Combination in *Leptothorax* (*Goniothorax*): Emery, 1922f: 250; in *Leptothorax* (*Nesomyrmex*): Kempf, 1959c: 429.

**#clivulus.** *Leptothorax flavispinus* var. *clivulus* Santschi, 1939d: 76 (w.) TUNISIA.

**clypeatus.** *Myrmica clypeata* Mayr, 1853c: 282 (w.) AUSTRIA. Stitz, 1914: 62 (q.m.); Emery, 1916b: 183 (q.). Combination in *Leptothorax:* Mayr, 1855: 436. See also: Stitz, 1939: 177; Baroni Urbani, 1971c: 103; Atanassov & Dlussky, 1992; 135.

*cockerelli. Leptothorax rugatulus* var. *cockerelli* Wheeler, W.M. 1917a: 509 (w.q.) U.S.A. [First available use of *Leptothorax curvispinosus* subsp. *rugatulus* var. *cockerelli* Wheeler, W.M. 1903c: 241; unavailable name.] Junior synonym of *rugatulus:* Creighton, 1950a: 269.

**#colettae.** *Leptothorax laurae* var. *colettae* Santschi, 1915a: 60, fig. 6 (w.) TUNISIA.

*colorata. Leptothorax rottenbergi* st. *annibalis* var. *colorata* Santschi, 1909b: 477 (w.) TUNISIA. Unavailable name.

*concolor. Leptothorax angulatus* var. *concolor* Santschi, 1914b: 107, fig. 15 (w.) KENYA. Emery, 1915g: 16 (q.m.). Junior synonym of *angulatus:* Bolton, 1982: 324.

**confucii.** *Tetramorium confucii* Forel, 1912a: 53 (w.q.) TAIWAN. Combination in *Leptothorax:* Bolton, 1977: 70.

**congruus.** *Leptothorax congruus* Smith, F. 1874: 406 (w.) JAPAN. Wheeler, W.M. 1906c: 317 (q.); Wheeler, G.C. & Wheeler, J. 1955b: 24 (l.); Imai & Kubota, 1972: 197 (k.). Current subspecies: nominal plus *wui.*

*convexotrabutii. Leptothorax angustulus* r. *trabutii* var. *convexotrabutii* Forel, 1894d: 38 (w.) ALGERIA. Unavailable name.

**convexus.** *Leptothorax convexus* Forel, 1894d: 36 (w.) ALGERIA. Current subspecies: nominal plus *timiaus.*

*convivialis. Leptothorax acervorum* subsp. *canadensis* var. *convivialis* Wheeler, W.M. 1903c: 228 (w.) U.S.A. Unavailable name; material referred to *canadensis* by Creighton, 1950a: 276.

*cordieri. Leptothorax cordieri* Bondroit, 1918: 126, fig. 62 (w.) CORSICA. Junior synonym of *unifasciatus:* Casevitz-Weulersse, 1990b: 417.

*cordincola. Leptothorax* (*Goniothorax*) *echinatinodis* subsp. *cordincola* Wheeler, W.M. 1942: 205 (w.) PANAMA. Junior synonym of *echinatinodis:* Kempf, 1959c: 425.

**cornibrevis.** *Leptothorax cornibrevis* Collingwood, 1961a: 63, fig. 8 (w.) AFGHANISTAN.

**corticalis.** *Myrmica corticalis* Schenck, 1852: 100 (w.q.) GERMANY. [Also described as new in Schenck, 1853: 187.] Stitz, 1914: 63 (m.). Combination in *Leptothorax:* Mayr, 1855: 440. Variety/subspecies of *tuberum:* Forel, 1874: 85; André, 1883a: 16; Ruzsky, 1902d: 21; Bondroit, 1910: 497; Stitz, 1914: 63. Status as species: Dalla Torre, 1893: 123; Bondroit, 1912: 351; Donisthorpe, 1915d: 160; Bondroit, 1918: 124; Emery, 1922f: 253; Karavaiev, 1934: 136; Stitz, 1939: 174; Novak & Sadil, 1941: 91; Kutter, 1977c: 131; Collingwood, 1979: 74; Atanassov & Dlussky, 1992: 137. Current subspecies: nominal plus *nylanderocorticalis.*

**costatus.** *Leptothorax* (*Goniothorax*) *costatus* Emery, 1896g: 59 (w.) BRAZIL. Kempf, 1959c: 404 (q.). Combination in *L.* (*Nesomyrmex*): Kempf, 1959c: 402.

**crassipilis.** *Leptothorax* (*Mychothorax*) *acervorum* subsp. *crassipilis* Wheeler, W.M. 1917a: 513 (w.q.m.) U.S.A. Raised to species: Creighton, 1950a: 278. See also: Loiselle, Francoeur, Fischer & Buschinger, 1990: 329.

*crassispina. Leptothorax flavispinus* st. *santschii* var. *crassispina* Santschi, 1911a: 11, fig. 2 (w.) TUNISIA. Unavailable name.

*crassispina. Leptothorax* (*Leptothorax*) *nylanderi* var. *crassispina* Karavaiev, 1926f: 69 (w.q.) UKRAINE. Junior synonym of *nylanderi:* Karavaiev, 1934: 137.

**creightoni.** *Macromischa creightoni* Mann, 1929: 163, fig. 2 (w.q.) CUBA. Combination in *Leptothorax:* Baroni Urbani, 1978b: 434.

**creolus.** *Leptothorax creolus* Baroni Urbani, 1978b: 437, figs. 44, 94 (w.) SANTO DOMINGO.

*cressoni* André, 1887 [*Macromischa*]; see under ***PHEIDOLE.***

*creticus. Leptothorax exilis* subsp. *creticus* Forel, 1910a: 23 (w.q.) CRETE. Menozzi, 1936d: 289 (m.). Junior synonym of *exilis:* Baroni Urbani, 1971d: 1041.

**#csikii.** *Leptothorax melleus* subsp. *csikii* Pisarski, 1969b: 298, figs. 1–4 (w.) MONGOLIA.

**curtulus.** *Leptothorax tebessae* st. *curtulus* Santschi, 1929e: 158, fig. 7 (w.) MOROCCO. Cagniant, 1970a: 426 (q.m.). Raised to species: Cagniant, 1970a: 425. Current subspecies: nominal plus *gentilis.* See also: Cagniant, 1987c: 249.

**curvispinosus.** *Leptothorax curvispinosus* Mayr, 1866a: 508, pl., fig. 13 (w.) U.S.A. Mayr, 1886d: 451 (q.). Combination in *L.* (*Myrafant*): Smith, M.R. 1950: 30. Senior synonym of *gallarum:* Mayr, 1886d: 453. See also: Wheeler, W.M. 1903c: 239.

*curvispinosus. Leptothorax nigrita* var. *curvispinosus* André, 1881b: 72 (w.) ISRAEL. [Junior primary homonym of *curvispinosus* Mayr, above.] Replacement name: *flavispinus* André, 1883a: 302.

*curvithorax. Leptothorax curvithorax* Bondroit, 1918: 130, fig. 63 (w.) FRANCE. Junior synonym of *rabaudi:* Bernard, 1967: 210.

**#cypridis.** *Leptothorax bulgaricus* st. *cypridis* Santschi, 1930d: 266 (w.q.) CYPRUS.

*dakotensis. Leptothorax rugatulus* subsp. *dakotensis* Wheeler, G.C. & Wheeler, E.W. 1944: 247 (w.) U.S.A. Junior synonym of *brunnescens:* Creighton, 1950a: 269.

*dalmasi. Leptothorax echinatinodis* r. *dalmasi* Forel, 1899c: 55 (w.q.) COLOMBIA. Combination in *L.* (*Goniothorax*): Emery, 1922f: 250. Junior synonym of *echinatinodis:* Kempf, 1959c: 425.

*darii. Leptothorax exilis* var. *darii* Forel, 1911d: 334 (w.) TURKEY. Menozzi, 1936d: 287 (q.l.). Subspecies of *exilis:* Cori & Finzi, 1931: 238. Junior synonym of *exilis:* Baroni Urbani, 1971d: 1041.

**darlingtoni.** *Macromischa darlingtoni* Wheeler, W.M. 1937b: 450 (w.) CUBA. Combination in *Leptothorax:* Baroni Urbani, 1978b: 438.

**#davisi.** *Leptothorax texanus* subsp. *davisi* Wheeler, W.M. 1905f: 385 (w.q.) U.S.A. Combination in *L.* (*Myrafant*): Smith, D.R. 1979: 1395. See also: Smith, M.R. 1952b: 104.

**delaparti.** *Leptothorax delaparti* Forel, 1890a: lxxii (w.q.) ALGERIA. Combination in *L.* (*Temnothorax*): Emery, 1891b: 5; in *Temnothorax:* Cagniant, 1970a: 418.

***denticulatus.*** *Leptothorax denticulatus* Mayr, 1901b: 5 (w.q.) SOUTH AFRICA. Combination in *L. (Goniothorax)*: Wheeler, W.M. 1922a: 891. See also: Bolton, 1982: 328.

*denticulatus. Leptothorax denticulatus* Stitz, 1917: 336, fig. 4 (w.) CANARY IS. [Junior primary homonym of *denticulatus* Mayr, above.] Replacement name: *stitzi* Santschi, 1918a: 34.

***desertorum.*** *Temnothorax desertorum* Dlussky & Soyunov, 1988: 35, figs. 1, 2 (w.q.) RUSSIA. Combination in *Leptothorax:* **new combination**.

***desioi.*** *Leptothorax (Leptothorax) desioi* Menozzi, 1939a: 303, figs. 7, 8 (w.q.m.) INDIA. Current subspecies: nominal plus *melanicus*.

***dessyi.*** *Leptothorax (Leptothorax) dessyi* Menozzi, 1936d: 289, fig. 13 (w.) GREECE.

*dichroa. Leptothorax exilis* var. *dichroa* Emery, 1916b: 174 (w.) ITALY. [First available use of *Leptothorax tuberum* subsp. *exilis* var. *dichroa* Emery, 1905a: 452; unavailable name.] Junior synonym of *leviceps:* Emery, 1916b: 174.

***discoloratus.*** *Leptothorax discoloratus* Arnol'di, 1977a: 200 (w.) RUSSIA.

***dissimilis.*** *Macromischa dissimilis* Aguayo, 1932: 220 (w.) CUBA. Combination in *Leptothorax:* Baroni Urbani, 1978b: 439. Senior synonym of *chloana:* Brown, 1959e: 74.

*distincta. Leptothorax sculptiventris* var. *distincta* Crawley, 1922a: 85 (w.) SOUTH AMERICA (no locality given). Junior synonym of *sculptiventris:* Kempf, 1959c: 400.

*diversipilosus* Smith, M.R. 1939; see under ***FORMICOXENUS.***

*dolani. Leptothorax dolani* Menozzi, 1941: 30 (w.) CHINA. *Nomen nudum.*

***duloticus.*** *Leptothorax (Mychothorax) duloticus* Wesson, L.G. 1937: 125, fig. 1 (w.q.) U.S.A. Wesson, L.G. 1940: 81 (m.). See also: Wilson, 1975: 108.

***eburneipes.*** *Leptothorax congruus* var. *eburneipes* Wheeler, W.M. 1927e: 1 (w.) CHINA. Raised to species: Wheeler, W.M. 1929f: 8.

***echinatinodis.*** *Leptothorax echinatinodis* Forel, 1886b: xlviii (w.) BRAZIL. Wheeler, G.C. & Wheeler, J. 1955b: 25 (l.). Combination in *L. (Goniothorax)*: Emery, 1896g: 59 (in key); in *L. (Caulomyrma)*: Forel, 1914d: 233; in *L. (Nesomyrmex)*: Kempf, 1959c: 425. Senior synonym of *aculeatinodis, cordincola, dalmasi, pungentinodis* and *schmidti:* Kempf, 1959c: 425.

*elderensis. Leptothorax elderensis* Dlussky & Zabelin, 1985: 225, fig. 3 (w.) TURKMENIA. Junior synonym of *tesquorum:* Radchenko, 1994d: 150 (in key).

#***eldoradensis.*** *Leptothorax eldoradensis* Wheeler, W.M. 1915b: 414 (w.) U.S.A. Subspecies of *nevadensis:* Creighton, 1950a: 266.

*elongatus. Leptothorax elongatus* Santschi, 1919e: 247 (w.) CANARY IS. Junior synonym of *risii:* Hohmann, La Roche, *et al.*, 1993: 151.

*emersoni* Wheeler, W.M. 1901; see under ***FORMICOXENUS.***

***emeryi.*** *Leptothorax (Temnothorax) emeryi* Santschi, 1923f: 333, fig. 4 (w.) ALGERIA. Combination in *Temnothorax:* Cagniant, 1970a: 419.

*ergatogyna. Leptothorax (Temnothorax) ergatogyna* Bernard, 1950a: 17, fig. 3 (w.q.) FRANCE. Combination in *Temnothorax:* Bernard, 1967: 186. Junior synonym of *recedens:* Dlussky & Soyunov, 1988: 31.

#***ernesti.*** *Leptothorax rottenbergi* st. *ernesti* Santschi, 1909b: 478 (w.) ALGERIA.

*esmirensis. Leptothorax exilis* subsp. *obscurior* var. *esmirensis* Santschi, 1936c: 204 (w.) MOROCCO. Unavailable name; material referred to *exilis* by Baroni Urbani, 1971d: 1041.

*euxanthus. Leptothorax nylanderi* form *euxanthus* Plateaux, 1981: 64. Unavailable name.

***evelynae.*** *Leptothorax (Goniothorax) evelynae* Forel, 1916: 423 (w.q.) ZAIRE. See also: Bolton, 1982: 328.

***exilis.*** *Leptothorax exilis* Emery, 1869b: 15, pl. 1, fig. 9 (w.) ITALY. Emery, 1916b: 183, 186 (q.m., respectively). Variety/subspecies of *tuberum:* Emery & Forel, 1879: 459; Emery, 1891b: 6; Ruzsky, 1902d: 21; Emery, 1915h: 2. Status as species: Dalla Torre, 1893: 124; Forel, 1913d: 433; Emery, 1916b: 174; Santschi, 1923f: 331; Müller, 1923: 89; Finzi, 1924a: 13; Santschi, 1929e: 153; Baroni Urbani, 1971d: 1041. Senior synonym of *ruficornis:* Emery, 1916b: 174; of *boccaris, creticus, darii, nitidulus, obscurior* and *retifer,* and material of the unavailable name *esmirensis* referred here: Baroni Urbani, 1971d: 1041.

***faberi.*** *Leptothorax faberi* Buschinger, 1983: 198, figs. 1–5 (w.q.m.) CANADA. See also: Loiselle, Francoeur, Fischer & Buschinger, 1990: 329.

#***facetus.*** *Leptothorax flavispinus* var. *facetus* Santschi, 1939c: 2 (w.) MOROCCO.

#***fagi.*** *Leptothorax (Mychothorax) muscorum* var. *fagi* Ruzsky, 1905b: 619, fig. 151 (w.) CAUCASUS.

#***festai.*** *Leptothorax bucheti* var. *festai* Emery, 1924a: 9 (w.) LIBYA.

***finzii.*** *Leptothorax (Mymammophilus) finzii* Menozzi, 1925d: 29, fig. 2 (w.q.m.) ITALY.

*firssovi. Leptothorax nassonovi* subsp. *firssovi* Kuznetsov-Ugamsky, 1928b: 28, fig. 5 (w.) RUSSIA. [Misspelled as *frissovi* by Pisarski, 1969b: 302.] Junior synonym of *nassanowi:* Kupyanskaya, 1990: 143.

#***flavescens.*** *Leptothorax muscorum* var. *flavescens* Ruzsky, 1896: 72 (w.) RUSSIA. [First available use of *Leptothorax acervorum* subsp. *muscorum* var. *flavescens* Ruzsky, 1895: 26; unavailable name.] Subspecies of *muscorum:* Ruzsky, 1905b: 621.

***flavicornis.*** *Leptothorax flavicornis* Emery, 1870: 197 (w.q.) ITALY. Emery, 1895b: 69 (m.). Combination in *L. (Mychothorax)*: Forel, 1915d: 26. Race of *acervorum:* Forel, 1874: 84. Revived status as species: Emery & Forel, 1879: 458. See also: Emery, 1916b: 182 (in key); Bondroit, 1918: 122; Baroni Urbani, 1971c: 109; Kutter, 1977c: 131.

***flavidulus.*** *Macromischa flavidula* Wheeler, W.M. & Mann, 1914: 37, fig. 15 (w.q.) HAITI. Combination in *M. (Antillaemyrmex)*: Mann, 1920: 408; in *Antillaemyrmex:* Wheeler, W.M. 1931b: 32; in *Leptothorax:* Baroni Urbani, 1978b: 441.

***flavispinus.*** *Leptothorax nigrita* var. *flavispinus* André, 1883a: 302. Replacement name for *curvispinosus* André, 1881b: 72. [Junior primary homonym of *curvispinosus* Mayr, 1866a: 508.] Menozzi, 1933b: 67 (m.); Kugler, J. & Soussan, 1992: 210 (q.). Raised to species: Emery, 1891b: 5. Current subspecies: nominal plus *amilcaris, boiteli, clivulus, facetus, impurus, rufescens, santschii.*

*flavitarsis. Macromischa (Macromischa) flavitarsis* Mann, 1920: 420 (w.q.) GUATEMALA. Wheeler, W.M. 1931b: 14 (m.). Junior synonym of *subditivus:* Baroni Urbani, 1978b: 512.

*flavus. Leptothorax (Dichothorax) pergandei* subsp. *flavus* Smith, M.R. 1929: 549 (w.q.) U.S.A. Junior synonym of *floridanus:* Creighton, 1950a: 260.

*floridanus. Leptothorax (Dichothorax) floridanus* Emery, 1895c: 324 (w.) U.S.A. Status as species: Wheeler, W.M. 1903c: 259; Wheeler, W.M. 1932a: 11. Subspecies of *pergandei:* Smith, M.R. 1929: 551; Creighton, 1950a: 260. Senior synonym of *flavus,* and material of the unavailable name *spinosus* referred here: Creighton, 1950a: 260. Junior synonym of *pergandei:* MacKay, 1993a: 289.

*floridanus. Antillaemyrmex floridanus* Wheeler, W.M. 1931b: 27 (w.) U.S.A. [Unresolved junior secondary homonym of *Leptothorax floridanus* Emery, above.] Combination in *Macromischa:* Wheeler, W.M. 1937b: 465. Junior synonym of *allardycei:* Baroni Urbani, 1978b: 415.

***foreli.*** *Leptothorax foreli* Santschi, 1907: 330 (w.) TUNISIA. Combination in *L. (Temnothorax)*: Santschi, 1918a: 32.

*foreli. Macromischa foreli* Aguayo, 1931: 182. Replacement name for *Leptothorax petiolatus* Forel, 1901c: 129. [Junior secondary homonym of **Macromischa petiolata* Mayr, 1868c: 85.] [Junior secondary homonym of *Leptothorax foreli* Santschi, above.] Replacement name: *augusti* Baroni Urbani, 1978b: 423.

*formidolosus* Wheeler, W.M. 1915; see under ***FORMICOXENUS.***

***formosus.*** *Leptothorax formosus* Santschi, 1909b: 469 (w.) MOROCCO.

*formosus. Leptothorax (Goniothorax) formosus* Emery, 1915g: 25 (w.) BOLIVIA. [Junior primary homonym of *formosus* Santschi, above.] Replacement name: *pulcher* Emery, 1917a: 96.

*fortinodis. Leptothorax fortinodis* Mayr, 1886d: 452 (w.q.) U.S.A. Combination in *L. (Myrafant)*: Smith, M.R. 1950: 30. Variety of *schaumii:* Wesson L.G. & Wesson, R.G. 1940: 96. See also: Wheeler, W.M. 1903c: 233. Junior synonym of *schaumii:* Creighton, 1950a: 271.

#***foveatus.*** *Leptothorax foveata* Smith, M.R. 1934b: 211 (w.) U.S.A. Combination in *L. (Myrafant)*: Smith, D.R. 1979: 1392. Subspecies of *ambiguus:* Creighton, 1950a: 263.

*frissovi;* see under ***firssovi.***

***fuentei.*** *Leptothorax fuentei* Santschi, 1919e: 243, fig. 2 (w.) SPAIN. Menozzi, 1922b: 328 (m.).

***fultonii.*** *Leptothorax fultonii* Forel, 1902c: 229 (w.) INDIA. See also: Bingham, 1903: 216.

***fumosus.*** *Leptothorax satunini* subsp. *fumosus* Ruzsky, 1923: 4 (w.) TRANSCAUCASUS. Subspecies of *bulgaricus:* Kuznetsov-Ugamsky, 1927d: 40. Raised to species: Dlussky, Soyunov & Zabelin, 1990: 190.

***furunculus.*** *Leptothorax furunculus* Wheeler, W.M. 1909e: 82 (w.) U.S.A. Combination in *L. (Myrafant)*: Smith, D.R. 1979: 1393.

***fuscatus.*** *Macromischa (Macromischa) fuscata* Mann, 1920: 420 (w.) GUATEMALA. Combination in *Leptothorax:* Baroni Urbani, 1978b: 442.

**#fusciventris.** *Leptothorax arenarius* var. *fusciventris* Santschi, 1911f: 85 (w.) TUNISIA. Combination in *L.* (*Temnothorax*): Emery, 1922f: 259.

**gaetulus.** *Leptothorax gaetulus* Santschi, 1923f: 332, fig. 3 (w.) MOROCCO.

*galatica. Leptothorax rottenbergi* st. *semiruber* var. *galatica* Santschi, 1921a: 111 (w.) TURKEY. Unavailable name (Agosti & Collingwood, 1987a: 61).

**galeatus.** *Leptothorax galeatus* Wheeler, W.M. 1927e: 1 (w.) CHINA.

**gallae.** *Leptothorax* (*Leptothorax*) *gallae* Smith, M.R. 1949e: 112 (w.) U.S.A. Combination in *L.* (*Myrafant*): Smith, D.R. 1979: 1393.

*gallarum. Stenamma gallarum* Patton, 1879: 126 (footnote) (w.q.) U.S.A. Junior synonym of *curvispinosus:* Mayr, 1886d: 453.

**#gapsensis.** *Leptothorax tebessae* var. *gapsensis* Santschi, 1929e: 156, fig. 9 (w.) TUNISIA.

**gazella.** *Leptothorax gazella* Santschi, 1923f: 329 (w.) TUNISIA. Current subspecies: nominal plus *monticola.*

**#gentilis.** *Leptothorax tebessae* var. *gentilis* Santschi, 1923f: 330 (w.) TUNISIA. [First available use of *Leptothorax tuberum* st. *tebessae* var. *gentilis* Santschi, 1918a: 36; unavailable name.] Santschi, 1923f: 330 (q.); Cagniant, 1970a: 426 (m.). Subspecies of *curtulus:* Cagniant, 1970a: 426.

*genualia. Leptothorax* (*Goniothorax*) *genualia* Santschi, 1922d: 252 (w.) PARAGUAY. Junior synonym of *spininodis:* Kempf, 1959c: 427.

**gibbifer.** *Leptothorax gibbifer* Baroni Urbani, 1978b: 444, figs. 56, 132 (w.) CUBA.

*gilvus. Leptothorax fortinodis* var. *gilvus* Wheeler, W.M. 1903c: 235 (w.q.) U.S.A. Junior synonym of *schaumii:* Creighton, 1950a: 271.

*glaber. Leptothorax glaber* Santschi, 1925g: 345 (w.) SPAIN. Junior synonym of *niger:* Espadaler, 1981: 80.

*glacialis* Wheeler, W.M. 1907; see under ***FORMICOXENUS.***

***glaesarius.** **Leptothorax glaesarius* Wheeler, W.M. 1915h: 64 (w.) BALTIC AMBER (Oligocene).

*globulinodis. Leptothorax risii* st. *globulinodis* Santschi, 1909b: 471, fig. 4 (w.) CANARY IS. Junior synonym of *risii:* Barquin, 1981: 256.

*goesswaldi* Kutter, 1967; see under ***DORONOMYRMEX.***

**goniops.** *Leptothorax goniops* Baroni Urbani, 1978b: 445, figs. 65, 110 (w.) MEXICO.

**gracilicornis.** *Leptothorax gracilicornis* Emery, 1882: 450, fig. (w.q.) CANARY IS. Current subspecies: nominal plus *nivarianus.*

***gracilis.** **Leptothorax gracilis* Mayr, 1868c: 89, pl. 5, figs. 89–92 (w.m.) BALTIC AMBER (Oligocene).

*gracilis. Macromischa* (*Croesomyrmex*) *bermudezi* var. *gracilis* Aguayo, 1932: 223 (w.) CUBA. [Junior secondary homonym of *Leptothorax gracilis* Mayr, above.] Replacement name: *anemicus* Baroni Urbani, 1978b: 420.

**graecus.** *Leptothorax bulgaricus* subsp. *graecus* Forel, 1911d: 336 (w.q.) GREECE. Raised to species: Agosti & Collingwood, 1987a: 55.

**gredleri.** *Leptothorax gredleri* Mayr, 1855: 438 (w.q.) AUSTRIA. Schenck, 1861: 196 (m.). Combination in *L.* (*Mychothorax*): Forel, 1915d: 26. Junior synonym of *muscorum:* Roger, 1862c: 296. Revived from synonymy as variety of *muscorum:* Forel, 1874: 87. Variety/subspecies of *muscorum:* Dalla Torre, 1893: 125; Emery, 1916b: 176; Emery, 1922f: 262; Stitz, 1939: 163; Novak & Sadil, 1941: 90. Status as species: Bondroit, 1918: 122; Buschinger, 1966b: 165; Bernard, 1967: 389; Baroni Urbani, 1971c: 98; Pisarski, 1975: 24; Kutter, 1977c: 129.

**gredosi.** *Leptothorax gredosi* Espadaler & Collingwood, 1982: 42, figs. 1–14 (w.q.m.) SPAIN.

**grisoni.** *Leptothorax* (*Goniothorax*) *grisoni* Forel, 1916: 425 (w.q.) ZAIRE. See also: Bolton, 1982: 329.

**grouvellei.** *Leptothorax grouvellei* Bondroit, 1918: 135 (w.) FRANCE. Espadaler, DuMerle & Plateaux, 1983: 275 (q.m.). Variety of *niger:* Emery, 1922f: 255. Revived status as species: Espadaler & Collingwood, 1982: 42.

*guancha. Leptothorax nivarianus* var. *guancha* Santschi, 1925g: 344 (w.q.) CANARY IS. Junior synonym of *nivarianus:* Hohmann, La Roche, *et al.*, 1993: 151.

**gundlachi.** *Macromischa gundlachi* Wheeler, W.M. 1913b: 488 (w.) CUBA. Combination in *M.* (*Croesomyrmex*): Mann, 1920: 408; in *Croesomyrmex:* Wheeler, W.M. 1931b: 32; in *Leptothorax:* Baroni Urbani, 1978b: 446.

**gurnetensis* Cockerell, 1915; see under ****LEUCOTAPHUS.***

**hadrumetensis.** *Leptothorax* (*Temnothorax*) *hadrumetensis* Santschi, 1918a: 31, fig. A (w.q.) TUNISIA.

*hasardaghi. Leptothorax hasardaghi* Dlussky & Zabelin, 1985: 226, fig. 4 (w.q.) TURKMENIA. Junior synonym of *caucasicus:* Radchenko, 1994d: 154 (in key).

*haytiana. Macromischa sallei* subsp. *haytiana* Wheeler, W.M. & Mann, 1914: 33, figs. 12–14 (w.q.m.) HAITI. Junior synonym of *sallei:* Baroni Urbani, 1978b: 490.

*heathii. Leptothorax nitens* var. *heathii* Wheeler, W.M. 1903c: 245 (w.) U.S.A. Subspecies of *nitens:* Creighton, 1950a: 265. Junior synonym of *nitens:* Cole, 1958c: 536.

*herculea. Leptothorax rottenbergi* st. *annibalis* var. *herculea* Santschi, 1922a: 66 (w.) ALGERIA. Unavailable name.

**hesperius.** *Leptothorax hesperius* Santschi, 1909b: 467, fig. 1 (w.q.) CANARY IS.

*hirticornis* Emery, 1895; see under ***FORMICOXENUS.***

*hirtipilis* Wheeler, W.M. 1917; see under ***FORMICOXENUS.***

**hispaniolae.** *Leptothorax hispaniolae* Baroni Urbani, 1978b: 448,figs. 50, 97 (w.q.m.) DOMINICA.

**hispidus.** *Leptothorax* (*Leptothorax*) *hispidus* Cole, 1957b: 42, fig. 1 (w.q.) U.S.A. Wheeler, G.C. & Wheeler, J. 1973b: 71 (l.). Combination in *L.* (*Myrafant*): Smith, D.R. 1979: 1393.

**huehuetenangoi.** *Leptothorax huehuetenangoi* Baroni Urbani, 1978b: 451, figs. 64, 106 (w.) GUATEMALA.

**humerosus.** *Leptothorax* (*Goniothorax*) *humerosus* Emery, 1896g: 62 (w.) EAST AFRICA (no locality given). See also: Bolton, 1982: 329.

*hyperisabellae. Leptothorax hyperisabellae* Baroni Urbani, 1978b: 453, figs. 38, 88 (w.) PUERTO RICO. Junior synonym of *isabellae:* Snelling, 1986: 155.

***hystriculus.** **Leptothorax hystriculus* Wheeler, W.M. 1915h: 65 (w.) BALTIC AMBER (Oligocene).

**ibericus.** *Leptothorax ibericus* Menozzi, 1922b: 329, fig. 2 (w.) SPAIN.

*ilgii. Leptothorax angulatus* r. *ilgii* Forel, 1894b: 82 (w.) ETHIOPIA. Santschi, 1912c: 148 (q.). Combination in *L.* (*Goniothorax*): Wheeler, W.M. 1922a: 891. Junior synonym of *angulatus:* Bolton, 1982: 324.

*imitatrix. Macromischa bruneri* subsp. *imitatrix* Wheeler, W.M. 1937b: 457 (w.) CUBA. Junior synonym of *bruneri:* Baroni Urbani, 1978b: 431.

**#impurus.** *Leptothorax flavispinus* var. *impurus* Santschi, 1939d: 76 (w.) TUNISIA.

**inermis.** *Leptothorax inermis* Forel, 1902c: 229 (w.) INDIA. See also: Bingham, 1903: 215.

*inermis. Leptothorax hesperius* st. *inermis* Santschi, 1909b: 468, fig. 2 (w.q.) MOROCCO. [Junior primary homonym of *inermis* Forel, above.] Replacement name: *anacanthus* Santschi, 1912c: 149.

**innocens.** *Tetramorium* (*Leptothorax?*) *innocens* Forel, 1913b: 317 (w.) ZAIRE. Combination in *Leptothorax:* Forel, 1916: 425. See also: Bolton, 1982: 330.

**interruptus.** *Myrmica interrupta* Schenck, 1852: 106 (w.q.) GERMANY. [Also described as new by Schenck, 1853: 188.] Mayr, 1855: 446 (m.). Combination in *Leptothorax:* Mayr, 1855: 446. Race/subspecies of *tuberum:* Forel, 1874: 85; Emery & Forel, 1879: 458; André, 1883a: 299; Forel, 1890a: lxxiv; Emery, 1891b: 6; Bondroit, 1910: 497; Emery, 1916b: 175; Emery, 1922f: 256; Finzi, 1930d: 315; Stitz, 1939: 171. Status as species: Dalla Torre, 1893: 124; Bondroit, 1911: 12; Donisthorpe, 1915d: 166; Bondroit, 1918: 132; Müller, 1923: 94; Lomnicki, 1928: 3; Novak & Sadil, 1941: 93; Kratochvil, 1944: 42; Bernard, 1967: 212; Kutter, 1977c: 131; Collingwood, 1979: 75; Atanassov & Dlussky, 1992: 138. Senior synonym of *simpliciuscula:* Roger, 1859: 258; Mayr, 1861: 59. Current subspecies: nominal plus *nitidiceps, tuberointerruptus.*

*ionia. Leptothorax bulgaricus* subsp. *smyrnensis* var. *ionia* Forel, 1911d: 336 (w.) GREECE. Unavailable name.

*iowensis. Leptothorax longispinosus* subsp. *iowensis* Buren, 1945: 288. Replacement name for *laeviceps* Buren, 1944a: 287. [Junior primary homonym of *leviceps* Emery, 1898c: 134.] Junior synonym of *longispinosus:* Creighton, 1950a: 264.

**iranicus.** *Leptothorax iranicus* Radchenko, 1993c (not yet seen). See also: Radchenko, 1994d: 150.

**iris.** *Macromischa iris* Roger, 1863a: 188 (w.) CUBA. Combination in *M.* (*Croesomyrmex*): Mann, 1920: 408; in *Croesomyrmex:* Wheeler, W.M. 1931b: 23; in *Leptothorax:* Baroni Urbani, 1978b: 454. Senior synonym of *nigripes:* Wheeler, W.M. 1931b: 24; of *rufithorax* and *tristis:* Baroni Urbani, 1978b: 454.

*irregularis. Leptothorax rottenbergi* st. *scabriosus* var. *irregularis* Santschi, 1909b: 480 (w.) ALGERIA. Unavailable name.

**isabellae.** *Macromischa isabellae* Wheeler, W.M. 1908a: 138, pl. 11, figs. 1, 2 (w.q.m.) PUERTO RICO. Combination in *Leptothorax:* Baroni Urbani, 1978b: 456. Senior synonym of *hyperisabellae* and *mutica:* Snelling, 1986: 155.

**italicus.** *Leptothorax affinis* subsp. *italicus* Consani, in Consani & Zangheri, 1952: 41 (w.) ITALY. Raised to species: Baroni Urbani, 1971c: 110.

**itinerans.** *Leptothorax* (*Nesomyrmex*) *itinerans* Kempf, 1959c: 407, figs. 2, 20, 26, 29, 30, 31 (w.q.m.) VENEZUELA.

***ixili***. *Leptothorax ixili* Baroni Urbani, 1978b: 459, figs. 67, 111 (w.) GUATEMALA.

***jacobsoni***. *Atopula jacobsoni* Forel, 1915a: 25 (w.) SUMATRA. Combination in *Leptothorax:* Forel, 1916: 458 (in text); in *L.* (*Goniothorax*): Emery, 1922f: 250. [Note. This may be a *Vombisidris* species; holotype is currently missing from Forel collection.]

***jailensis***. *Leptothorax jailensis* Arnol'di, 1977a: 200 (w.) UKRAINE.

***janushevi***. *Leptothorax janushevi* Radchenko, 1993c (not yet seen). See also: Radchenko, 1994d: 151.

*jaumei*. *Macromischa* (*Macromischa*) *jaumei* Santschi, 1930c: 78 (w.) CUBA. Variety of *porphyritis:* Wheeler, W.M. 1931b: 18. Junior synonym of *porphyritis:* Baroni Urbani, 1978b: 480.

#***jesus***. *Leptothorax rottenbergi* var. *jesus* Forel, 1913d: 432 (w.) ISRAEL.

#***juglandeti***. *Leptothorax oxianus* subsp. *juglandeti* Arnol'di, in Tarbinsky, 1976: 103, figs. 196–171, 190 (w.) KIRGIZIA.

*junipereti*. *Leptothorax junipereti* Arnol'di, 1977a: 203 (w.) UKRAINE. Junior synonym of *knipovitshi:* Radchenko, 1994d: 156 (in key).

*kabyla*. *Leptothorax rottenbergi* st. *scabrosus* var. *kabyla* Santschi, 1918a: 33 (w.q.) TUNISIA. Unavailable name.

*kamtschaticum*. *Leptothorax acervorum* subsp. *kamtschaticum* Ruzsky, 1920: 78 (w.) RUSSIA. Junior synonym of *acervorum:* Kupyanskaya, 1986b: 96.

***kaszabi***. *Leptothorax kaszabi* Pisarski, 1969b: 301, figs. 11–17 (w.q.) MONGOLIA.

#***kefensis***. *Leptothorax naeviventris* st. *kefensis* Santschi, 1929e: 153 (w.q.) TUNISIA.

***kemali***. *Leptothorax kemali* Santschi, 1934d: 277 (w.) TURKEY.

*kincaidi*. *Leptothorax yankee* var. *kincaidi* Pergande, 1900: 520 (w.q.) U.S.A. Junior synonym of *muscorum:* Brown, 1955a: 47.

***kirghizicus***. *Leptothorax kirghizicus* Tarbinsky, 1976: 100, figs. 134–137 (w.) KIRGIZIA.

#***kirillovi***. *Leptothorax tuberum* subsp. *kirillovi* Ruzsky, 1905b: 591, fig. 140 (w.) TURKESTAN.

***knipovitshi***. *Leptothorax tuberum* subsp. *knipovitshi* Karavaiev, 1916: 499 (w.q.) RUSSIA. Raised to species, and senior synonym of *junipereti* and *nikitae:* Radchenko, 1994d: 156 (in key).

***korbi***. *Leptothorax corticalis* var. *korbi* Emery, 1922f: 254 (w.q.) TRANSCAUCASIA. [First available use of *Leptothorax tuberum* subsp. *corticalis* var. *korbi* Emery, 1898c: 133; unavailable name.] Raised to species: Arnol'di, 1948: 7.

***koreanus***. *Leptothorax* (*Nesomyrmex*) *koreanus* Teranishi, 1940: 16 (w.) KOREA.

***kraussei***. *Leptothorax angustulus* var. *kraussei* Emery, 1916b: 175 (w.) SARDINIA. Bondroit, 1918: 135 (q.m.). [First available use of *Leptothorax angustulus* var. *kraussei* Krausse, 1912b: 162; *nomen nudum*. *Leptothorax tuberum* subsp. *angustulus* var. *kraussei* Emery, 1915a: 260; unavailable name.] Raised to species: Bondroit, 1918: 135. Variety of *angustulus:* Emery, 1922f: 252. Junior synonym of *angustulus:* Bernard, 1967: 202. Revived from synonymy, with status as species: Baroni Urbani, 1971c: 111; Espadaler & Collingwood, 1982: 41; Casevitz-Weulersse, 1990b: 418.

#***kruegeri***. *Leptothorax* (*Myrmammophilus*) *arenarius* var. *kruegeri* Kutter, 1928: 66 (q.) LIBYA.

***kurilensis***. *Leptothorax kurilensis* Radchenko, 1993c (not yet seen). See also: Radchenko, 1994d: 148.

*kutteri* Buschinger, 1966; see under ***DORONOMYRMEX***.

***laciniatus***. *Leptothorax laciniatus* Stitz, 1917: 338, fig. 5 (w.) ALGERIA.

*lacteipennis*. *Myrmica lacteipennis* Zetterstedt, 1838: 452 (m.) SWEDEN. Junior synonym of *acervorum:* Nylander, 1846a: 936.

***laestrygon***. *Leptothorax niger* st. *laestrygon* Santschi, 1931a: 7 (w.q.m.) SICILY. Raised to species: Baroni Urbani, 1971c: 112. Current subspecies: nominal plus *splendidiceps*.

***laetus***. *Macromischa affinis* subsp. *laeta* Wheeler, W.M. 1937b: 456 (w.) CUBA. Combination in *Leptothorax*, and junior synonym of *similis:* Baroni Urbani, 1978b: 501. Revived from synonymy, raised to species, and senior synonym of *maerens* and *similis:* Snelling, 1986: 154.

*laeviceps*. *Leptothorax* (*Leptothorax*) *longispinosus* subsp. *laeviceps* Buren, 1944a: 287 (w.) U.S.A. [Junior primary homonym of *leviceps* Emery, 1898c: 134.] Replacement name: *iowensis* Buren, 1945: 288.

*laevissima*. *Macromischa laevissima* Wheeler, W.M. 1911h: 205 (w.) MEXICO. Junior synonym of *subditivus:* Baroni Urbani, 1978b: 512.

***lagrecai***. *Leptothorax* (*Leptothorax*) *lagrecai* Baroni Urbani, 1964b: 48, figs. 12, 13 (w.q.) ITALY.

*latinodis*. *Leptothorax latinodis* Mayr, 1895: 130 (w.) MOZAMBIQUE. Combination in *L.* (*Goniothorax*): Emery, 1896g: 58 (footnote). Junior synonym of *angulatus:* Bolton, 1982: 324.

*latispina*. *Macromischa porphyritis* var. *latispina* Wheeler, W.M. 1931b: 18 (w.) CUBA. Junior synonym of *porphyritis:* Baroni Urbani, 1978b: 480.

***laurae***. *Leptothorax laurae* Emery, 1884a: 380 (w.q.) TUNISIA. Santschi, 1907: 332 (m.). Current subspecies: nominal plus *colettae*, *rosae*.

***leoni***. *Leptothorax leoni* Arnol'di, 1971: 1820, fig. 1 (w.q.) KAZAKHSTAN.

***lereddei***. *Leptothorax lereddei* Bernard, 1953a: 174, fig. 6 (w.) ALGERIA.

***leucacanthus***. *Leptothorax leucacanthus* Baroni Urbani, 1978b: 460, figs. 61, 103 (w.q.) MEXICO.

***leviceps***. *Leptothorax tuberum* var. *leviceps* Emery, 1898c: 134 (w.) ITALY. Variety of *exilis:* Emery, 1916b: 174; Menozzi, 1925d: 27. Junior synonym of *exilis:* Baroni Urbani, 1971d: 1041. Revived from synonymy, and raised to species: Collingwood, 1978: 70. Senior synonym of *dichroa:* Emery, 1916b: 174.

***lichtensteini***. *Leptothorax lichtensteini* Bondroit, 1918: 123 (w.q.m.) FRANCE. Variety/subspecies of *nylanderi:* Menozzi, 1921: 28; Menozzi, 1925d: 27. Revived status as species: Bernard, 1956a: 162; Bernard, 1967: 205; Collingwood & Yarrow, 1969: 70.

***lindbergi***. *Leptothorax lindbergi* Santschi, 1931a: 8, fig. 1 (w.) MOROCCO.

*littoralis*. *Leptothorax rottenbergi* st. *annibalis* var. *littoralis* Santschi, 1918a: 33 (w.) TUNISIA. Unavailable name.

****longaevus***. **Leptothorax longaevus* Wheeler, W.M. 1915h: 65 (w.) BALTIC AMBER (Oligocene).

***longipilosus***. *Leptothorax longipilosus* Santschi, 1912f: 174, fig. 4 (w.) TUNISIA. Santschi, 1923f: 334 (q.). Combination in *L.* (*Temnothorax*): Santschi, 1918a: 32.

*longipilosus* Tarbinsky, 1976; see under ***CHALEPOXENUS***.

***longispinosus***. *Leptothorax longispinosus* Roger, 1863a: 180 (w.) U.S.A. Emery, 1895c: 321 (q.); Wheeler, W.M. 1903c: 238 (m.); Crozier, 1970: 117 (k.); Wheeler, G.C. & Wheeler, J. 1989a: 323 (l.). Combination in *L.* (*Myrafant*): Smith, M.R. 1950: 30. Senior synonym of *iowensis:* Creighton, 1950a: 264.

*lucayensis*. *Macromischa lucayensis* Forel, 1901e: 340 (w.) BAHAMAS. Junior synonym of *pastinifer:* Baroni Urbani, 1978b: 471.

*luciliae*. *Macromischa luciliae* Mann, 1935, 35, fig. 1 (w.) GUATEMALA. Junior synonym of *subditivus:* Baroni Urbani, 1978b: 512.

*lugens*. *Macromischa lugens* Roger, 1863a: 188 (w.) CUBA. Aguayo, 1932: 221 (q.). Combination in *Croesomyrmex:* Wheeler, W.M. 1931b: 21. Senior synonym of *nigra:* Borgmeier, 1955a: 109. Junior synonym of *versicolor:* Baroni Urbani, 1978b: 523. See also: Mann, 1920: 419.

***luteus***. *Leptothorax tuberum* r. *luteus* Forel, 1874: 85 (w.) SWITZERLAND. André, 1883a: 304 (q.); Bernard, 1956a: 162 (m.). Variety of *unifasciatus:* Dalla Torre, 1893: 129; of *tuberum:* Forel, 1913d: 432. Raised to species: Emery, 1916b: 173. See also: Menozzi, 1918: 85; Bondroit, 1918: 133; Stitz, 1939: 176; Bernard, 1967: 208; Kutter, 1977c: 131; Arnol'di & Dlussky, 1978: 542.

#***lybica***. *Leptothorax* (*Goniothorax*) *angulatus* var. *lybica* Menozzi, 1934: 155 (w.) LIBYA.

***madecassus***. *Leptothorax madecassus* Forel, 1892l: 265 (w.q.) MADAGASCAR. Combination in *L.* (*Goniothorax*): Emery, 1896g: 58 (footnote).

*maerens*. *Macromischa affinis* subsp. *maerens* Wheeler, W.M. 1937b: 456 (w.) CUBA. Junior synonym of *similis:* Baroni Urbani, 1978b: 501; of *laetus:* Snelling, 1986: 154.

*major*. *Leptothorax sculptiventris* var. *major* Forel, 1899c: 54 (footnote) (w.) BRAZIL. Junior synonym of *sculptiventris:* Kempf, 1959c: 400.

***manni***. *Leptothorax manni* Wheeler, W.M. 1914b: 53 (w.q.m.) MEXICO. Combination in *L.* (*Myrafant*): Kempf, 1972a: 132.

*manni*. *Macromischa manni* Wheeler, W.M. 1931b: 19 (w.) CUBA. Wheeler, G.C. & Wheeler, J. 1960b: 24 (l.). [Junior secondary homonym of *Leptothorax manni* Wheeler, above.] Replacement name: *villarensis* Aguayo, 1931: 177 (junior synonym and hence first available replacement name): Snelling, 1986: 155.

*manni*. *Leptothorax* (*Dichothorax*) *manni* Wesson, L.G. 1935: 208 (w.q.m.) U.S.A. [Unresolved junior primary homonym of *manni* Wheeler, W.M. 1914b: 53.] Junior synonym of *pergandei:* Wesson, L.G. 1939: 180.

***mariposa***. *Leptothorax nitens* var. *mariposa* Wheeler, W.M. 1917a: 507 (w.) U.S.A. Combination in *L.* (*Myrafant*): Smith, D.R. 1979: 1393. Junior synonym of *nitens:* Creighton, 1950a: 265. Revived from synonymy, and raised to species: Cole, 1958c: 536.

***marocana.*** *Leptothorax bugnioni* var. *marocana* Santschi, 1909b: 463 (q.m.) MOROCCO. Raised to species: Santschi, 1936c: 203.

*martellii. Leptothorax bugnioni* subsp. *pupseli* var. *martellii* Menozzi, 1940: 269 (w.q.) LIBYA. Unavailable name.

***massiliensis.*** *Leptothorax massiliensis* Bondroit, 1918: 133 (w.) FRANCE. Bernard, 1967: 209 (q.).

***mauritanicus.*** *Leptothorax cabrerae* st. *mauritanicus* Santschi, 1909b: 466 (w.q.m.) MOROCCO. Raised to species: Santschi, 1939c: 2.

***maurus.*** *Leptothorax maurus* Santschi, 1921b: 427 (w.) MOROCCO.

#***maximus.*** *Leptothorax anacanthus* st. *maximus* Santschi, 1931a: 6 (w.) MOROCCO.

*maya. Macromischa azteca* var. *maya* Wheeler, W.M. 1937b: 458 (w.q.) GUATEMALA. Junior synonym of *aztecus:* Baroni Urbani, 1978b: 425.

*mediorufus. Leptothorax rugatulus* var. *mediorufus* Wheeler, W.M. 1917a: 510 (w.q.) U.S.A. Junior synonym of *rugatulus:* Creighton, 1950a: 269.

***megalops.*** *Leptothorax* (*Icothorax*) *megalops* Hamann & Klemm, 1967: 417, fig. 1 (w.q.) SUDAN. See also: Bolton, 1982: 331.

#***melanderi.*** *Leptothorax melanderi* Wheeler, W.M. 1909e: 81 (w.) U.S.A. Subspecies of *nevadensis:* Creighton, 1950a: 266.

#***melanicus.*** *Leptothorax desioi* var. *melanica* Menozzi, 1939a: 307 (w.) INDIA.

*melanocephala. Myrmica melanocephala* Mayr, 1855: 411. *Nomen nudum;* attributed to Koch. Material referred to *acervorum* by Mayr, 1855: 411.

*melanocephalus. Leptothorax melanocephalus* Emery, 1870: 197 (w.) CORSICA. Variety/race of *tuberum:* André, 1883a: 298; Forel, 1890a: lxxiv; Emery, 1891b: 5; Santschi, 1919e: 242. Status as species: Bondroit, 1920a: 154; Baroni Urbani, 1971c: 115; Collingwood, 1978: 70; Atanassov & Dlussky, 1992: 138. Junior synonym of *tuberum:* Casevitz-Weulersse, 1990b: 420.

*melanocephala. Macromischa melanocephala* Wheeler, W.M. 1931b: 15 (w.q.) CUBA. [Junior secondary homonym of *Leptothorax melanocephalus* Emery, above.] Replacement name: *nigricans* Baroni Urbani, 1978b: 467.

*melanoticus. Leptothorax fortinodis* var. *melanoticus* Wheeler, W.M. 1903c: 235 (w.q.) U.S.A. Wesson, L.G. & Wesson, R.G. 1940: 94 (m.). Subspecies of *fortinodis:* Buren, 1944a: 287. Junior synonym of *schaumii:* Creighton, 1950a: 271.

***melas.*** *Leptothorax melas* Espadaler, Plateaux & Casevitz-Weulersse, 1984: 123, figs. 1–15 (w.q.m.) CORSICA.

***melleus.*** *Leptothorax bulgaricus* var. *melleus* Forel, 1904b: 375 (w.q.) UZBEKISTAN. Subspecies of *bulgaricus:* Kuznetsov-Ugamsky, 1926a: 75. Raised to species: Pisarski, 1969b: 298. Current subspecies: nominal plus *csikii.*

***melnikovi.*** *Leptothorax melnikovi* Ruzsky, 1905b: 599, figs. 143, 144 (w.) KAZAN.

***minozzii.*** *Leptothorax minozzii* Santschi, 1922a: 68. Replacement name for *Leptothorax silvestrii* Menozzi, 1921: 27. [Junior secondary homonym of *Tetramorium silvestrii* Santschi, 1911d: 6.] See also: Baroni Urbani, 1971c: 116.

***minutissimus.*** *Leptothorax minutissimus* Smith, M.R. 1942a: 59, pl. 6 (q.) U.S.A. Combination in *L.* (*Myrafant*): Smith, D.R. 1979: 1393.

***mirassolis.*** *Leptothorax* (*Nesomyrmex*) *mirassolis* Diniz, 1975: 79, figs. 1–3 (w.q.) BRAZIL.

***miserabilis.*** *Leptothorax* (*Temnothorax*) *miserabilis* Santschi, 1918a: 32, fig. B (w.) TUNISIA.

#***mixtus.*** *Leptothorax bucheti* var. *mixtus* Santschi, 1929e: 151 (w.) MOROCCO.

***mongolicus.*** *Leptothorax serviculus* subsp. *mongolicus* Pisarski, 1969b: 299, figs. 5–10 (w.) MONGOLIA. Raised to species: Radchenko, 1994d: 156 (in key).

***monjauzei.*** *Leptothorax monjauzei* Cagniant, 1968b: 83, figs. (w.q.m.) ALGERIA.

#***monticola.*** *Leptothorax gazella* var. *monticola* Santschi, 1923f: 330 (w.) TUNISIA.

*mordax. Leptothorax* (*Temnothorax*) *mordax* Santschi, 1919e: 242 (q.) SPAIN. Junior synonym of *recedens:* Dlussky & Soyunov, 1988: 31.

***mortoni.*** *Macromischa mortoni* Aguayo, 1937: 235, fig. 1 (w.) CUBA. Combination in *Leptothorax:* Baroni Urbani, 1978b: 462.

*mulleri. Leptothorax mulleri* Santschi, 1926f: 293. Unnecessary replacement name for *sordidulus* Müller, 1923: 96. [The name is misrepresented as *sordidus* by Santschi.] Junior synonym of *sordidulus:* **new synonymy**.

*muellerianus* Finzi, 1922; see under ***CHALEPOXENUS***.

***muscorum.*** *Myrmica muscorum* Nylander, 1846b: 1054 (w.q.m.) FINLAND. Combination in *Leptothorax:* Mayr, 1855: 439; in *L.* (*Mychothorax*): Forel, 1915d: 26. Race of *acervorum:* Forel, 1874: 84. Status as species: André, 1883a: 294; Ruzsky, 1905b: 616. Senior synonym of *canadensis, kincaidi, septentrionalis* and *yankee* (and its junior synonyms *obscurus* and *sordidus*): Brown, 1955a: 47. Current subspecies: nominal plus *betulae, fagi, flavescens, uvicensis.* See also: Buschinger, 1966b: 165; Baroni Urbani, 1971c: 99; Kutter, 1977c: 130; Collingwood, 1979: 72; Smith, D.R. 1979: 1397; Loiselle, Francoeur, *et al.* 1990: 323; Kupyanskaya, 1990: 139; Atanassov & Dlussky, 1992: 132.

*mutabilis. Croesomyrmex bermudezi* var. *mutabilis* Wheeler, W.M. 1931b: 27 (w.) CUBA. Junior synonym of *bermudezi:* Baroni Urbani, 1978b: 429.

*mutica. Macromischa isabellae* subsp. *mutica* Smith, M.R. 1937: 847 (w.) PUERTO RICO. Combination in *Leptothorax,* and raised to species: Baroni Urbani, 1978b: 464. Junior synonym of *isabellae:* Snelling, 1986: 155.

***myersi.*** *Macromischa myersi* Wheeler, W.M. 1931b: 20 (w.) CUBA. Combination in *Leptothorax:* Baroni Urbani, 1978b: 465.

***nadigi.*** *Leptothorax nadigi* Kutter, 1925: 409, figs. (w.q.) SWITZERLAND. Espadaler, 1984: 136 (m.). See also: Stitz, 1939: 178; Kutter, 1977c: 132.

***naeviventris.*** *Leptothorax naeviventris* Santschi, 1910b: 62 (w.q.m.) TUNISIA. Current subspecies: nominal plus *kefensis.*

*narinicus. Leptothorax narinicus* Tarbinsky, 1976: 93, figs. 151–153, 187 (w.) KIRGIZIA. Junior synonym of *arpini:* Radchenko, 1994d: 150 (in key).

***nassanowi.*** *Leptothorax nassanowi* Ruzsky, 1895: 26 (w.) RUSSIA. [Also described as new by Ruzsky, 1896: 72.] Kupyanskaya, 1990: 143 (q.m.). Senior synonym of *firssovi:* Kupyanskaya, 1990: 143. See also: Tarbinsky, 1976: 94. [This name is frequently misspelled as *nassonovi,* for example Ruzsky, 1905b: 582; Emery, 1922f: 255; Radchenko, 1994d: 155.]

*natenzoni. Macromischa* (*Croesomyrmex*) *aguayoi* subsp. *natenzoni* Aguayo, 1931: 179 (w.) CUBA. Junior synonym of *aguayoi:* Baroni Urbani, 1978b: 409.

#***neomexicanus.*** *Leptothorax neomexicanus* Wheeler, W.M. 1903c: 248, pl. 12, fig. 18 (w.) U.S.A. Subspecies of *tricarinatus:* Creighton, 1950a: 273; Smith, M.R. 1952b: 100.

***nevadensis.*** *Leptothorax nevadensis* Wheeler, W.M. 1903c: 252, pl. 12, fig. 20 (w.q.m.) U.S.A. Wheeler, G.C. & Wheeler, J. 1973b: 71 (l.). Combination in *L.* (*Myrafant*): Smith, D.R. 1979: 1393. Current subspecies: nominal plus *eldoradensis, melanderi, rudis* (unresolved junior homonym).

*nevodovskii* Karavaiev, 1926; see under ***MYRMICA***.

***niger.*** *Leptothorax tuberum* r. *niger* Forel, 1894d: 42 (w.) FRANCE. Forel, 1913h: 359 (q.); Kutter, 1927: 100 (m.). Raised to species: Forel, 1913h: 359. Senior synonym of *glaber:* Espadaler, 1981: 80. See also: Bondroit, 1918: 134; Santschi, 1927a: 58.

*nigra. Macromischa* (*Croesomyrmex*) *lugens* var. *nigra* Santschi, 1931c: 273, fig. 8 (w.) CUBA. [Unresolved junior secondary homonym of *niger* Forel, above.] Junior synonym of *lugens:* Borgmeier, 1955a: 109.

#***nigrescens.*** *Leptothorax acervorum* var. *nigrescens* Ruzsky, 1905b: 614 (w.) RUSSIA. Subspecies of *acervorum:* Ruzsky, 1936: 94; Pisarski, 1969b: 297.

***nigricans.*** *Leptothorax nigricans* Baroni Urbani, 1978b: 467. Replacement name for *Macromischa melanocephala* Wheeler, W.M. 1931b: 15. [Junior secondary homonym of *Leptothorax melanocephalus* Emery, 1870: 197.]

#***nigricephalus.*** *Leptothorax* (*Leptothorax*) *tuberum* var. *nigricephala* Karavaiev, 1930b: 146 (w.q.) SWEDEN. [Also described as new by Karavaiev, 1931e: 212.]

***nigriceps.*** *Leptothorax nigriceps* Mayr, 1855: 441 (w.) AUSTRIA. Forel, 1874: 87 (q.); Bondroit, 1920b: 304 (q.m.); Finzi, 1939: 90 (m.); Hauschteck, 1962: 219 (k.). Variety/race/subspecies of *tuberum:* Forel, 1874: 85; Ruzsky, 1902d: 21; Bondroit, 1910: 497; Santschi, 1910e: 648; Stitz, 1914: 63; Ruzsky, 1925b: 45. Status as species: Bondroit, 1912: 351; Bondroit, 1918: 125; Müller, 1923: 93; Finzi, 1924b: 122; Finzi, 1939: 88; Boven, 1947: 179; Bernard, 1956a: 162; Baroni Urbani, 1971c: 118; Kutter, 1977c: 132; Boven, 1977: 105; Arnol'di & Dlussky, 1978: 542. Current subspecies: nominal plus *pyrenaeus.*

*nigricipoides. Leptothorax tuberum* r. *unifasciatus* var. *nigricipoides* Forel, 1874: 86 (w.) SWITZERLAND. Unavailable name.

*nigripes. Macromischa* (*Croesomyrmex*) *iris* var. *nigripes* Santschi, 1930c: 78 (w.) CUBA. Junior synonym of *iris:* Wheeler, W.M. 1931b: 24.

***nigritus.*** *Leptothorax nigrita* Emery, 1878b: 51 (footnote) fig. (w.) ALGERIA. Forel, 1890a: lxxv (q.). Current subspecies: nominal plus *rupestris, saharensis, salambo.*

***LEPTOTHORAX** (continued)*

*nigroscutellata. Leptothorax tuberum* var. *nigroscutellata* Costa, 1884: 34. *Nomen nudum;* material referred to *tuberum* by Emery, 1914e: 2.

*nikitae. Leptothorax nikitae* Arnol'di, 1977a: 202 (w.) UKRAINE. Junior synonym of *knipovitshi:* Radchenko, 1994d: 156 (in key).

***nitens.** Leptothorax nitens* Emery, 1895c: 322, pl. 8, fig. 16 (w.) U.S.A. Cole, 1958c: 536 (q.m.); Wheeler, G.C. & Wheeler, J. 1973b: 71 (l.). Combination in *L.* (*Myrafant*): Smith, D.R. 1979: 1394. Senior synonym of *occidentalis:* Creighton, 1950a: 265; of *heathii:* Cole, 1958c: 536. See also: Wheeler, W.M. 1903c: 244.

**#*nitidiceps.*** *Leptothorax interruptus* var. *nitidiceps* Dalla Torre, 1893: 124 (w.) ALGERIA. [First available use of *Leptothorax tuberum* r. *interruptus* var. *nitidiceps* Forel, 1890a: lxxiv; unavailable name (Agosti & Collingwood, 1987a: 61).] Variety of *spinosus:* Cagniant, 1970b: 207; of *interruptus:* Baroni Urbani, 1971c: 110.

*nitidulus. Leptothorax exilis* var. *nitidulus* Santschi, 1923f: 331 (w.q.) ALGERIA. [First available use of *Leptothorax tuberum* r. *obscurior* var. *nitidulus* Forel, 1894d: 39; unavailable name.] Junior synonym of *exilis:* Baroni Urbani, 1971d: 1041.

*nivalis. Leptothorax tuberum* subsp. *spinosus* var. *nivalis* Forel, 1909e: 377 (w.q.) ALGERIA. Unavailable name. [*Leptothorax spinosus* var. *nivalis* Cagniant, 1970b: 207; unavailable name.]

**#*nivarianus.*** *Leptothorax nivarianus* Santschi, 1925g: 343 (w.q.) CANARY IS. Subspecies of *gracilicornis,* and senior synonym of *guancha:* Barquin, 1981: 283.

***normandi.** Leptothorax normandi* Santschi, 1912f: 173, fig. 3 (w.) TUNISIA. Santschi, 1929e: 155 (q.).

*nudus* Mayr, 1866; see under ***CARDIOCONDYLA.***

**#*numidicus.*** *Leptothorax annibalis* st. *numidicus* Santschi, 1929e: 151 (w.q.) TUNISIA.

***nylanderi.** Myrmica nylanderi* Foerster, 1850a: 53 (m.) GERMANY. Mayr, 1855: 447 (w.q.). Combination in *Leptothorax:* Mayr, 1855: 447. Subspecies/race of *tuberum:* Forel, 1874: 84; Emery & Forel, 1879: 459; André, 1883a: 300; Forel, 1890a: lxxv; Emery, 1891b: 6; Ruzsky, 1905b: 597; Bondroit, 1910: 496. Status as species: Saunders, E. 1880: 219; Nasonov, 1889: 32; Dalla Torre, 1893: 125; Ruzsky, 1902d: 23; Bondroit, 1911: 12; Stitz, 1914: 64; Forel, 1915d: 22; Donisthorpe, 1915d: 155; Bondroit, 1918: 122; Menozzi, 1922b: 329; Karavaiev, 1926f: 69; Finzi, 1930d: 315; Karavaiev, 1934: 137; Stitz, 1939: 179; Boven, 1947: 178; Kutter, 1977c: 133; Collingwood, 1979: 74; Atanassov & Dlussky, 1992: 139. Senior synonym of *cingulata:* Mayr, 1855: 447; of *crassispina:* Karavaiev, 1934: 137. Current subspecies: nominal plus *nylanderonigriceps, nylanderotuberum.*

**#*nylanderocorticalis.*** *Leptothorax tuberum* var. *nylanderocorticalis* Forel, 1874: 86 (w.) SWITZERLAND. Emery, 1916b: 183, 186 (in keys) (w.q., respectively). Raised to species: Bondroit, 1918: 124. Variety of *corticalis:* Forel, 1915d: 22 (in key); Emery, 1922f: 254; Stitz, 1939: 176; Novak & Sadil, 1941: 92.

**#*nylanderonigriceps.*** *Leptothorax nylanderi* var. *nylanderonigriceps* Stitz, 1939: 180 (w.q.) GERMANY. [First available use of *Leptothorax tuberum* subsp. *nylanderi* var. *nylanderonigriceps* Gösswald, 1932: 149; unavailable name.] Variety of *nylanderi:* Novak & Sadil, 1941: 91.

*nylanderotebessae. Leptothorax tuberum* r. *tebessae* var. *nylanderotebessae* Forel, 1894d: 38 (w.) ALGERIA. Unavailable name.

**#*nylanderotuberum.*** *Leptothorax nylanderi* var. *nylanderotuberum* Ruzsky, 1902d: 23 (w.) RUSSIA.

**#*nylanderounifasciatus.*** *Leptothorax tuberum* var. *nylanderounifasciatus* Forel, 1874: 228 (w.) SWITZERLAND.

**#*obenbergeri.*** *Leptothorax unifasciatus* var. *obenbergeri* Sadil, 1939: 30 (w.q.) CZECHOSLOVAKIA.

***obliquicanthus.** Leptothorax* (*Myrafant*) *obliquicanthus* Cole, 1953c: 28, figs. 1, 2 (w.) U.S.A.

*obscurior. Leptothorax tuberum* var. *obscurior* Dalla Torre, 1893: 128 (w.) ALGERIA. [First available use of *Leptothorax tuberum* r. *melanocephalus* var. *obscurior* Forel, 1890a: lxxiv; unavailable name.] Cagniant, 1970a: 421 (q.m.). Variety/race of *tuberum:* Forel, 1894d: 38; of *exilis:* Santschi, 1911a: 12; Santschi, 1929e: 153. Raised to species: Cagniant, 1970a: 421. Junior synonym of *exilis:* Baroni Urbani, 1971d: 1041.

**#*obscurior.*** *Macromischa salvini* var. *obscurior* Forel, 1899c: 57 (w.) PANAMA. [Unresolved junior secondary homonym of *obscurior* Dalla Torre, above.] Combination in *Leptothorax:* Baroni Urbani, 1978b: 495.

*obscurus. Leptothorax canadensis* subsp. *obscurus* Viereck, 1903: 72 (w.) U.S.A. Junior synonym of *yankee:* Creighton, 1950a: 277.

***obturator.** Leptothorax obturator* Wheeler, W.M. 1903c: 249, pl. 12, fig. 19 (w.q.m.) U.S.A. Wheeler, W.M. 1903f: 662 (gynandromorph); Wheeler, G.C. & Wheeler, J. 1955b: 24 (l.). Combination in *L.* (*Myrafant*): Smith, D.R. 1979: 1394.

***ocarinae.** Leptothorax ocarinae* Baroni Urbani, 1978b: 469, figs. 58, 131 (w.) MEXICO.

*occidentalis. Leptothorax nitens* subsp. *occidentalis* Wheeler, W.M. 1903c: 245 (w.) U.S.A. Junior synonym of *nitens:* Creighton, 1950a: 265.

***oceanicus.** Mychothorax muscorum* subsp. *oceanicum* Kuznetsov-Ugamsky, 1928b: 29 (w.) RUSSIA. Kupyanskaya, 1990: 140 (q.). Combination in *Leptothorax:* Kupyanskaya, 1990: 140. Raised to species: Heinze, Schulz & Radchenko, 1993: 178; Radchenko, 1994d: 147 (in key).

*opacinoda. Macromischa sallei* subsp. *opacinoda* Wheeler, W.M. 1931b: 7 (w.) HAITI. Junior synonym of *sallei:* Baroni Urbani, 1978b: 490.

*opacipes. Macromischa pastinifera* var. *opacipes* Wheeler, W.M. 1905b: 96, fig. I (w.) BAHAMAS. Junior synonym of *pastinifer:* Smith, M.R. 1954c: 6.

***opalinus.** Macromischa opalina* Wheeler, W.M. 1937b: 451 (w.) CUBA. Combination in *Leptothorax:* Baroni Urbani, 1978b: 470. Junior synonym of *darlingtoni:* Brown, 1959e: 73. Revived from synonymy: Baroni Urbani, 1978b: 470.

***oraniensis.** Leptothorax oraniensis* Forel, 1894d: 34 (w.q.) ALGERIA. Senior synonym of *rusticus:* Cagniant, 1970a: 419. Current subspecies: nominal plus *chobauti, sergensis.*

*orientalis. Mychothorax acervorum* subsp. *orientalis* Kuznetsov-Ugamsky, 1928b: 31 (w.) RUSSIA. Junior synonym of *acervorum:* Kupyanskaya, 1990: 137.

*ornatipes. Macromischa bruneri* subsp. *ornatipes* Wheeler, W.M. 1937b: 457 (w.) CUBA. Junior synonym of *bruneri:* Baroni Urbani, 1978b: 431.

***oxianus.** Leptothorax tuberum* subsp. *oxianus* Ruzsky, 1905b: 592 (w.) UZBEKISTAN. Subspecies of *bulgaricus:* Kuznetsov-Ugamsky, 1927d: 40. Raised to species: Tarbinsky, 1976: 97. Senior synonym of *talassicus:* Radchenko, 1994d: 151 (in key). Current subspecies: nominal plus *juglandeti.*

*pallens. Leptothorax nylanderi* morph *pallens* Plateaux, 1981: 64 (w.q.m.) no locality. Unavailable name. [Also referred to as *nylanderi* form *pallens* Plateaux, 1981: 64; unavailable name.]

***pallidipes.** Leptothorax pallidipes* Santschi, 1910b: 61 (w.q.) TUNISIA. Combination in *L.* (*Temnothorax*): Santschi, 1918a: 32.

***pallidus.** Leptothorax pallidus* Collingwood, 1961a: 63, fig. 7 (w.) AFGHANISTAN.

*pallipes. Antillaemyrmex albispina* subsp. *pallipes* Wheeler, W.M. 1931b: 27 (w.q.) PUERTO RICO (attributed to Mann). [*Macromischa albispina* var. *pallipes* Mann, 1920: 424. *Nomen nudum.*] Subspecies of *albispinus:* Smith, M.R. 1937: 851. Junior synonym of *albispinus:* Baroni Urbani, 1978b: 413.

***pamiricus.** Leptothorax pamiricus* Ruzsky, 1902b: 478 (w.) RUSSIA. Menozzi, 1939a: 303 (q.m.). Subspecies of *bulgaricus:* Forel, 1904b: 375; Kuznetsov-Ugamsky, 1926a: 75; Menozzi, 1939a: 303. Revived status as species: Radchenko, 1994d: 152 (in key).

***pan.** Leptothorax pan* Santschi, 1936c: 203, fig. 2 (w.) MOROCCO.

*paolii. Leptothorax tuberum* st. *unifasciatus* var. *paolii* Santschi, 1923f: 331 (w.q.m.) CORSICA. Unavailable name; material referred to *unifasciatus* by Casevitz-Weulersse, 1990b: 417.

***paraxenus.** Leptothorax paraxenus* Heinze & Alloway, 1992: 195, figs. 1–4 (q.m.) CANADA.

***pardoi.** Leptothorax pardoi* Tinaut, 1987: 315, figs. 1, 2 (w.q.m.) SPAIN.

*parkeri. Leptothorax fuentei* var. *parkeri* Espadaler & DuMerle, 1989: 121 (w.) SPAIN. Unavailable name.

***parvulus.** Myrmica parvula* Schenck, 1852: 103 (w.) GERMANY. [Also described as new by Schenck, 1853: 187.] Mayr, 1855: 448 (q.); Novak, in Kratochvil, Novak & Snoflak, 1944: 116 (m.). Combination in *Leptothorax:* Mayr, 1855: 448. Junior synonym of *cingulata:* Nylander, 1856b: 93; of *nylanderi:* Smith, F. 1858b: 8. Revived from synonymy as variety of *nylanderi:* Forel, 1874: 86. Variety/subspecies of *nylanderi:* Ruzsky, 1902d: 23; Forel, 1915d: 22; Menozzi, 1918: 85; Emery, 1922f: 255; Müller, 1923: 95. Revived status as species: Bernard, 1956a: 161; Bernard, 1967: 203; Baroni Urbani, 1971c: 120; Boven, 1977: 105; Agosti & Collingwood, 1987a: 56; Atanassov & Dlussky, 1992: 140.

***pastinifer.** Macromischa pastinifera* Emery, 1894c: 164, pl. 1, fig. 18 (w.) BAHAMAS. Mann, 1929: 161 (q.m.). Combination in *Leptothorax:* Baroni Urbani, 1978b: 470. Senior synonym of *opacipes:* Smith,

M.R. 1954c: 6; of *lucayensis:* Baroni Urbani, 1978b: 471.

**pastoris.** *Leptothorax pastoris* Baroni Urbani, 1978b: 475 (w.) CUBA.

**pelagosanus.** *Leptothorax pelagosanus* Müller, 1923: 88 (w.) ITALY. Junior synonym of *massiliensis:* Baroni Urbani, 1962: 131. Revived from synonymy: Agosti & Collingwood, 1987a: 56.

**peninsularis.** *Leptothorax (Goniothorax) peninsularis* Wheeler, W.M. 1934f: 134 (q.) MEXICO.

**pergandei.** *Leptothorax (Dichothorax) pergandei* Emery, 1895c: 323, pl. 8, fig. 13 (w.) U.S.A. Wheeler, W.M. 1903c: 257 (q.m.); Wheeler, G.C. & Wheeler, J. 1960b: 23 (l.). Senior synonym of *manni:* Wesson, L.G. 1939: 180; of *floridanus* (and its junior synonym *flavus*): MacKay, 1993a: 289.

**personatus.** *Leptothorax personatus* Cagniant, 1987c: 243, figs. 1–3 (w.q.m.) MOROCCO.

**petiolata* Mayr, 1868; see under ***NOTHOMYRMICA.***

*petiolatus. Leptothorax petiolatus* Forel, 1901c: 129 (w.) MEXICO. Wheeler, W.M. 1931b: 13 (q.). Combination in *Macromischa:* Emery, 1922f: 247. [Junior primary homonym of **petiolata* Mayr, 1868c: 85, above.] Replacement name: *foreli* Aguayo, 1931: 182.

*petri. Macromischa wheeleri* subsp. *petri* Aguayo, 1931: 181 (w.) CUBA. Junior synonym of *wheeleri:* Brown, 1959e: 73.

**peyerimhoffi.** *Leptothorax peyerimhoffi* Santschi, 1929c: 101, fig. 3 (w.) ALGERIA. [Also described as new by Santschi, 1934f: 169.]

*picea* Wasmann, 1906; see under ***FORMICOXENUS.***

#***pici.*** *Leptothorax aveli* var. *pici* Bondroit, 1918: 128 (w.) FRANCE.

*pictus. Leptothorax (Temnothorax) pictus* Emery, 1924c: 166 (w.) SPAIN. Junior synonym of *recedens:* Collingwood, 1978: 69.

*pilifer* Roger, 1863; see under ***PHEIDOLE.***

#***pinetorum.*** *Leptothorax ambiguus* var. *pinetorum* Wesson L.G. & Wesson, R.G. 1940: 97 (w.q.m.) U.S.A. Combination in *L. (Myrafant):* Smith, D.R. 1979: 1392. Subspecies of *ambiguus:* Creighton, 1950a: 263.

*pithyusae. Leptothorax pithyusae* Arnol'di, 1977a: 201 (w.q.) RUSSIA. Junior synonym of *tauricus:* Radchenko, 1994d: 157 (in key).

**pittieri.** *Leptothorax pittieri* Forel, 1899c: 56 (w.) COSTA RICA. Kempf, 1959c: 411 (q.). Combination in *L. (Goniothorax):* Emery, 1922f: 250; in *L. (Nesomyrmex):* Kempf, 1959c: 411.

***placivus.** **Leptothorax placivus* Wheeler, W.M. 1915h: 66 (w.) BALTIC AMBER (Oligocene).

**platycnemis.** *Macromischa platycnemis* Wheeler, W.M. 1937b: 452 (w.) CUBA. Combination in *Leptothorax:* Baroni Urbani, 1978b: 476.

**pleuriticus.** *Leptothorax (Nesomyrmex) pleuriticus* Kempf, 1959c: 417 figs. 2, 22, 27 (w.q.m.) GUIANA. [First available use of *Leptothorax (Goniothorax) echinatinodis* subsp. *aculeatinodis* var. *pleuriticus* Wheeler, W.M. 1921f: 158; unavailable name.]

**poeyi.** *Macromischa poeyi* Wheeler, W.M. 1913b: 489 (w.) CUBA. Combination in *M. (Croesomyrmex):* Mann, 1920: 408; in *Croesomyrmex:* Wheeler, W.M. 1931b: 26; in *Leptothorax:* Baroni Urbani, 1978b: 477.

**polita.** *Macromischa polita* Smith, M.R. 1939e: 506, pl. 1, fig. B (w.) U.S.A. Combination in *Leptothorax:* Baroni Urbani, 1978b: 478.

**porphyritis.** *Macromischa porphyritis* Roger, 1863a: 185 (w.) CUBA. Mann, 1929: 162 (q.). Combination in *Leptothorax,* and senior synonym of *jaumei* and *latispina:* Baroni Urbani, 1978b: 480. See also: Mann, 1920: 413; Wheeler, W.M. 1931b: 17.

***praecreolus.** **Leptothorax (Macromischa) praecreolus* De Andrade, 1992: 344, figs. 1, 2 (w.) DOMINICAN AMBER (Miocene).

*pratostepposus. Leptothorax pratostepposus* Arnol'di, 1977a: 202 (w.q.) RUSSIA. Junior synonym of *tuberum:* Radchenko, 1994d: 157 (in key).

**prisca* André, 1895 [*Macromischa*); see under ***VOLLENHOVIA.***

#***productus.*** *Leptothorax cabrerae* st. *productus* Santschi, 1918a: 35, fig. C (w.q.) MOROCCO.

*provancheri* Emery, 1895; see under ***FORMICOXENUS.***

*proxima. Macromischa bruneri* subsp. *proxima* Wheeler, W.M. 1937b: 457 (w.) CUBA. Junior synonym of *bruneri:* Baroni Urbani, 1978b: 431.

*pseudoxanthus. Leptothorax nylanderi form pseudoxanthus* Plateaux, 1981: 64. Unavailable name.

**pulchellus.** *Macromischa pulchella* Emery, 1894c: 162 (w.q.) ANTILLES IS. Combination in *M. (Antillaemyrmex):* Mann, 1920: 408; in *Antillaemyrmex:* Wheeler, W.M. 1931b: 32; in *Leptothorax:* Baroni Urbani, 1978b: 482.

**pulcher.** *Leptothorax (Goniothorax) pulcher* Emery, 1917a: 96. Replacement name for *formosus* Emery, 1915g: 25. [Junior primary homonym of *formosus* Santschi, 1909b: 469.] See also: Kempf, 1959c: 395.

*pungentinodis. Leptothorax pungentinodis* Emery, 1896b: 2 (q.) PANAMA. Combination in *L. (Goniothorax):* Emery, 1896g: 59 (in key). Subspecies of *echinatinodis:* Emery, 1922f: 250. Junior synonym of *echinatinodis:* Kempf, 1959c: 425.

**punicans.** *Macromischa punicans* Roger, 1863a: 189 (w.) CUBA. Combination in *Aphaenogaster:* Emery, 1896a: 103; in *Pheidole:* Emery, 1915d: 67; in *Macromischa:* Mann, 1920: 418; in *Croesomyrmex:* Wheeler, W.M. 1931b: 22; in *Leptothorax:* Baroni Urbani, 1978b: 484.

#***pupseli.*** *Leptothorax bugnioni* st. *pupseli* Santschi, 1909b: 464 (w.) TUNISIA.

**purpuratus.** *Macromischa purpurata* Roger, 1863a: 184 (w.) CUBA. Mann, 1920: 411 (q.). Combination in *Leptothorax:* Baroni Urbani, 1978b: 486.

#***pyrenaeus.*** *Leptothorax nigriceps* var. *pyrenaeus* Bondroit, 1918: 126 (w.q.) FRANCE.

**rabaudi.** *Leptothorax rabaudi* Bondroit, 1918: 129, fig. 64 (w.q.) FRANCE. Bernard, 1967: 210 (m.). Subspecies of *affinis:* Menozzi, 1925d: 28; Finzi, 1933: 164. Status as species: Bernard, 1956a: 163; Bernard, 1967: 210; Collingwood, 1976: 304. Senior synonym of *curvithorax:* Bernard, 1967: 210.

**racovitzai.** *Leptothorax racovitzai* Bondroit, 1918: 133 (w.q.) FRANCE. Kutter, 1977c: 133 (m.). Junior synonym of *parvulus:* Bernard, 1967: 203. Revived from synonymy: Collingwood & Yarrow, 1969: 72; Baroni Urbani, 1971c: 121; Kutter, 1977c: 133.

**recedens.** *Myrmica recedens* Nylander, 1856b: 94, pl. 3, fig. 41 (w.) FRANCE. Roger, 1859: 258 (q.); Forel, 1874: 83 (m.). Combination in *Temnothorax:* Mayr, 1861: 68; in *Leptothorax (Temnothorax):* Forel, 1890a: lxxii. Senior synonym of *pictus* and *suberis:* Collingwood, 1978: 69; of *ergatogyna, mordax* and *rogeri:* Dlussky & Soyunov, 1988: 31. Current subspecies: nominal plus *barbarus.* See also: Forel, 1915d: 27; Emery, 1916b: 186; Bernard, 1967: 186; Kutter, 1977c: 135; Atanassov & Dlussky, 1992: 126.

**reduncus.** *Tetramorium reduncum* Wang & Wu, in Wang, Xiao & Wu, 1988: 268, figs. 21, 22 (w.) CHINA. Combination in *Leptothorax:* Wang (pers. comm.), **new combination**.

*renateae* Taylor, 1989; see under ***VOMBISIDRIS.***

*retifer. Leptothorax exilis* var. *retifer* Santschi, 1929e: 153 (w.) MOROCCO. Junior synonym of *exilis:* Baroni Urbani, 1971d: 1041.

**retractus.** *Leptothorax retractus* Francoeur, 1986a: 1152, figs. 1–17 (w.q.m.) CANADA. See also: Loiselle, Francoeur, *et al.* 1990: 329.

**retusispinosus.** *Leptothorax retusispinosus* Forel, 1892l: 267 (w.) MADAGASCAR. Combination in *L. (Goniothorax):* Emery, 1896g: 58 (footnote).

**risii.** *Leptothorax risii* Forel, 1892g: 456 (w.) CANARY IS. Santschi, 1909b: 470 (q.); Espadaler & Rodriguez, 1989: 141 (m.). Senior synonym of *globulinodis:* Barquin, 1981: 256; of *armouri* and *elongatus:* Hohmann, La Roche, *et al.,* 1993: 151.

*rogeri. Temnothorax rogeri* Emery, 1869b: 18 (footnote) (w.) GREECE. André, 1883a: 309 (q.); Forel, 1886e: clxvi (q.). Combination in *Leptothorax (Temnothorax):* Forel, 1890a: lxxii. Status as species: Dalla Torre, 1893: 126; Ruzsky, 1902d: 25; Emery, 1922f: 260. Subspecies/race of *recedens:* Emery & Forel, 1879: 459; Forel, 1892i: 315; Ruzsky, 1905b: 609; Karavaiev, 1927a: 293; Finzi, 1930d: 314. Junior synonym of *recedens:* Dlussky & Soyunov, 1988: 31.

#***rosae.*** *Leptothorax laurae* st. *rosae* Santschi, 1915a: 60, fig. 7 (w.) TUNISIA.

**rothneyi.** *Leptothorax rothneyi* Forel, 1902c: 230 (w.) INDIA. See also: Bingham, 1903: 217. Current subspecies: nominal plus *simlensis.*

**rottenbergii.** *Macromischa rottenbergii* Emery, 1870: 199 (w.q.) SICILY. Emery, 1895b: 69 (m.). Combination in *Leptothorax:* Emery, 1878b: 51. Current subspecies: nominal plus *ernesti, jesus, scabriosus.* See also: Emery, 1916b: 187.

#***rougeti.*** *Leptothorax cordieri* var. *rougeti* Bondroit, 1918: 127 (w.) FRANCE. Subspecies of *unifasciatus:* current status.

**rudis* Mayr, 1868; see under ***NOTHOMYRMICA.***

#***rudis.*** *Leptothorax nevadensis* subsp. *rudis* Wheeler, W.M. 1917a: 508 (w.q.) U.S.A. [Unresolved junior primary homonym of **rudis* Mayr, 1868c: 85, above.]

*rufa. Leptothorax (Goniothorax) asper* var. *rufa* Emery, 1896g: 61 (w.q.) BRAZIL. Junior synonym of *asper:* Kempf, 1959c: 414.

#***rufescens.*** *Leptothorax flavispinus* var. *rufescens* Forel, 1907b: 206 (w.) TUNISIA.

*ruficornis. Leptothorax tuberum* subsp. *ruficornis* Emery, 1914d: 157 (w.) FRANCE ? [First available use of *Leptothorax tuberum* subsp. *exilis* var. *ruficornis* Emery, 1898c: 135; unavailable name.] Junior synonym of *exilis:* Emery, 1916b: 174.

**LEPTOTHORAX** *(continued)*

*rufithorax. Croesomyrmex iris* var. *rufithorax* Wheeler, W.M. 1931b: 24 (w.) CUBA. Junior synonym of *iris:* Baroni Urbani, 1978b: 454.

***rugatulus.*** *Leptothorax rugatulus* Emery, 1895c: 321 (w.) U.S.A. Taber & Cokendolpher, 1988: 95 (k.). Combination in *L.* (*Myrafant*): Smith, M.R. 1950: 30. Subspecies of *curvispinosus:* Wheeler, W.M. 1903c: 241. Revived status as species: Wheeler, W.M. 1917a: 509. Senior synonym of *annectens, cockerelli* and *mediorufus:* Creighton, 1950a: 269. Current subspecies: nominal plus *brunnescens.*

***rugiceps.*** *Macromischa* (*Croesomyrmex*) *poeyi* subsp. *rugiceps* Aguayo, 1932: 222 (w.) CUBA. Combination in *Leptothorax,* and raised to species: Baroni Urbani, 1978b: 489.

**rugosostriata* Mayr, 1868; see under ***NOTHOMYRMICA.***

#***rupestris.*** *Leptothorax nigrita* var. *rupestris* Forel, 1909e: 377 (w.q.) ALGERIA.

*russula. Myrmica russula* Nylander, 1856b: 82 (footnote) (w.q.) "Santo Domingo". Junior synonym of *sallei:* Roger, 1862c: 293.

*rusticus. Leptothorax rusticus* Santschi, 1918a: 36, fig. D (w.) TUNISIA. Junior synonym of *oraniensis:* Cagniant, 1970a: 419.

***rutilans.*** *Leptothorax rutilans* Kempf, 1958c: 100, figs. 11, 16, 20 (w.) GUIANA. Kempf, 1959c: 424 (q.). Combination in *L.* (*Nesomyrmex*): Kempf, 1959c: 423.

#***sachalinensis.*** *Leptothorax tuberum* subsp. *sachalinensis* Kupyanskaya, 1990: 142 (w.q.) RUSSIA.

#***saharensis.*** *Leptothorax nigrita* var. *saharensis* Santschi, 1929c: 102, fig. 4 (q.) ALGERIA. [Also described as new by Santschi, 1934f: 170.]

#***salambo.*** *Leptothorax nigrita* var. *salambo* Santschi, 1912f: 174 (w.) TUNISIA.

#***salinus.*** *Leptothorax tuberum* var. *salina* Karavaiev, 1937: 174, fig. 2 (w.) UKRAINE.

***sallei.*** *Myrmica sallei* Guérin-Méneville, 1852: 73, pl. 3, figs. 1–4 (w.q.m.) HAITI. Combination in *Pogonomyrmex:* Dalla Torre, 1893: 120; in *Macromischa:* Forel, 1899c: 57 (footnote); in *Leptothorax:* Baroni Urbani, 1978b: 489. Senior synonym of *russula:* Roger, 1862c: 293; of *haytiana* and *opacinoda:* Baroni Urbani, 1978b: 490.

***salvini.*** *Macromischa salvini* Forel, 1899c: 57, pl. 3, fig. 18 (w.) PANAMA. Combination in *Leptothorax:* Baroni Urbani, 1978b: 494. Current subspecies: nominal plus *obscurior* (unresolved junior homonym).

#***santschii.*** *Leptothorax flavispinus* r. *santschii* Forel, 1905b: 173 (w.q.m.) TUNISIA.

***sardous.*** *Leptothorax rottenbergi* var. *sardoa* Santschi, 1909b: 474, fig. 6 (w.) SARDINIA. Bondroit, 1918: 134 (q.m.). Raised to species: Bondroit, 1918: 134.

***satunini.*** *Leptothorax satunini* Ruzsky, 1902b: 477 (w.) CAUCASUS. Subspecies/race of *bulgaricus:* Forel, 1904b: 375; Emery, 1922f: 253; Kuznetsov-Ugamsky, 1926a: 75. Status as species: Ruzsky, 1905b: 575; Tarbinsky, 1976: 102; Arnol'di, 1977a: 203.

***scabripes.*** *Macromischa* (*Macromischa*) *scabripes* Mann, 1920: 416, fig. 4 (w.) CUBA. Combination in *Leptothorax,* and senior synonym of *brunneipes:* Baroni Urbani, 1978b: 496.

#***scabriosus.*** *Leptothorax rottenbergi* st. *scabriosus* Santschi, 1909b: 479 (w.) ALGERIA. [Variant spelling in the same publication: *scabrosus.*] Raised to species: Cagniant, 1964: 105. Subspecies of *rottenbergii:* Barquin, 1981: 273.

***scamni.*** *Leptothorax* (*Mychothorax*) *muscorum* var. *scamni* Ruzsky, 1905b: 620, fig. 151 (w.) CAUCASUS. Heinze, Schulz & Radchenko, 1993: 180 (q.). Raised to species: Arnol'di, 1977a: 204.

*scandens* Mann, 1922; see under ***ROGERIA.***

***schaufussi.*** *Aphaenogaster schaufussi* Forel, 1879b: 465 (w.) SPAIN. Combination in *Leptothorax:* Forel, 1890a: lxxii.

***schaumii.*** *Leptothorax schaumii* Roger, 1863a: 180 (w.) U.S.A. Mayr, 1886d: 451 (m.); Wesson L.G. & Wesson, R.G. 1940: 95 (q.); Wheeler, G.C. & Wheeler, J. 1960b: 23 (l.). Combination in *L.* (*Myrafant*): Smith, D.R. 1979: 1394. Senior synonym of *fortinodis, gilvus* and *melanoticus:* Creighton, 1950a: 271. See also: Wheeler, W.M. 1903c: 232.

*schmidti. Leptothorax* (*Goniothorax*) *echinatinodis* subsp. *schmidti* Menozzi, 1927c: 275 (w.q.m.) COSTA RICA. Junior synonym of *echinatinodis:* Kempf, 1959c: 425.

***schmittii.*** *Leptothorax schmittii* Wheeler, W.M. 1903c: 242, pl. 12, fig. 14 (w.) U.S.A. Combination in *L.* (*Myrafant*): Smith, D.R. 1979: 1395.

***schurri.*** *Leptothorax rothneyi* r. *schurri* Forel, 1902c: 231 (w.) INDIA. Subspecies of *schurri:* Emery, 1922f: 257. Raised to species: Bingham, 1903: 218; current status.

***schwarzi.*** *Macromischa* (*Macromischa*) *schwarzi* Mann, 1920: 415 (q.) CUBA. Baroni Urbani, 1978b: 497 (w.). Combination in *Leptothorax:* Baroni Urbani, 1978b: 497.

***schwebeli.*** *Leptothorax schwebeli* Forel, 1913l: 216 (w.) BRAZIL. Combination in *L.* (*Goniothorax*): Emery, 1922f: 250; in *L.* (*Nesomyrmex*): Kempf, 1959c: 407.

***sculptiventris.*** *Leptothorax sculptiventris* Mayr, 1887: 620 (w.) BRAZIL. Combination in *L.* (*Nesomyrmex*), and senior synonym of *borgmeieri, distincta* and *major:* Kempf, 1959c: 400.

***semenovi.*** *Leptothorax semenovi* Ruzsky, 1903b: 311 (w.) TRANSCASPIA. Dlussky & Soyunov, 1988: 33 (q.m.).

***semiruber.*** *Leptothorax rottenbergi* var. *semiruber* André, 1881b: 72 (w.) ISRAEL. Stirps/subspecies of *rottenbergi:* Santschi, 1909b: 475; Santschi, 1921a: 111. Raised to species, and senior synonym of *balcanica:* Agosti & Collingwood, 1987a: 56. See also: Atanassov & Dlussky, 1992: 141.

***senectutis.*** *Leptothorax senectutis* Baroni Urbani, 1978b: 499, figs. 24, 121 (w.) CUBA.

*septentrionalis. Leptothorax* (*Mychothorax*) *muscorum* var. *septentrionalis* Wheeler, W.M. 1917a: 511 (w.q.m.) U.S.A. Material of the unavailable name *calderoni* referred here by Creighton, 1950a: 276. Junior synonym of *muscorum:* Brown, 1955a: 47.

#***sergensis.*** *Leptothorax rusticus* var. *sergensis* Santschi, 1929e: 154 (w.q.) TUNISIA. Subspecies of *oraniensis:* current status.

*sericeiventris. Leptothorax cabrerae* st. *productus* var. *sericeiventris* Santschi, 1918a: 35 (w.q.) MOROCCO. Unavailable name.

***serviculus.*** *Leptothorax serviculus* Ruzsky, 1902b: 476 (w.) RUSSIA. Ruzsky, 1905b: 768 (q.). See also: Kupyanskaya, 1990: 140.

***sevanensis.*** *Leptothorax sevanensis* Arnol'di, 1977a: 210 (w.q.) ARMENIA.

***shelkovnikovi.*** *Leptothorax* (*Leptothorax*) *shelkovnikovi* Karavaiev, 1926c: 166, fig. 4 (w.) CAUCASUS.

*sibiricus* Forel, 1899; see under ***FORMICOXENUS.***

***sikorai.*** *Leptothorax* (*Goniothorax*) *sikorai* Emery, 1896g: 63 (w.) MADAGASCAR.

#***silvanus.*** *Leptothorax angustulus* subsp. *silvanus* Forel, 1907b: 206 (w.) ALGERIA.

***silvestrii.*** *Tetramorium silvestrii* Santschi, 1911d: 6 (w.) U.S.A. Creighton, 1953f: 5 (q.m.). Combination in *Leptothorax:* Santschi, 1922a: 68; in *L.* (*Myrafant*): Smith, D.R. 1979: 1395.

*silvestrii. Leptothorax silvestrii* Menozzi, 1921: 27, fig. 1 (w.q.) ITALY. [Junior secondary homonym of *silvestrii* Santschi, above.] Replacement name: *minozzii* Santschi, 1922a: 68.

*similis. Leptothorax similis* Baroni Urbani, 1978b: 501. [Unnecessary replacement name for *affinis* Mann, 1920: 417.] Junior synonym of *laetus:* Snelling, 1986: 154.

#***simlensis.*** *Leptothorax rothneyi* var. *simlensis* Forel, 1904c: 22 (w.) INDIA.

***simoni.*** *Tetramorium simoni* Emery, 1895h: 35, pl. 2, fig. 22 (w.) SOUTH AFRICA. Combination in *Tetramyrma:* Emery, 1922f: 291; in *Leptothorax:* Bolton, 1982: 331.

*simpliciuscula. Myrmica simpliciuscula* Nylander, 1856b: 92, pl. 3, fig. 40 (w.) FRANCE. Junior synonym of *interruptus:* Roger, 1859: 258; Mayr, 1861: 59.

***singularis.*** *Leptothorax singularis* Radchenko, 1993c (not yet seen). See also: Radchenko, 1994d: 150.

***skwarrae.*** *Macromischa skwarrae* Wheeler, W.M. 1931b: 10 (w.q.) MEXICO. Combination in *Leptothorax:* Baroni Urbani, 1978b: 503.

***smithi.*** *Leptothorax smithi* Baroni Urbani, 1978b: 557. Replacement name for *Leptothorax wheeleri* Smith, M.R. 1929: 547. [Junior secondary homonym of *Macromischa wheeleri* Mann, 1920: 422.

#***smyrnensis.*** *Leptothorax bulgaricus* subsp. *smyrnensis* Forel, 1911d: 335 (w.q.) TURKEY.

***solerii.*** *Leptothorax* (*Temnothorax*) *solerii* Menozzi, 1936d: 291, fig. 13 (w.) ITALY.

***sordidulus.*** *Leptothorax sordidulus* Müller, 1923: 96 (w.) ITALY. Novak, in Kratochvil, Novak & Snoflak, 1944: 118 (m.). Senior synonym of *carinthiacus:* Schulz, 1991: 121; of *mulleri:* **new synonymy**. Current subspecies: nominal plus *tergestinus.*

*sordidus. Leptothorax muscorum* var. *sordidus* Wheeler, W.M. 1903c: 224, pl. 12, fig. 2 (w.) U.S.A. Junior synonym of *yankee:* Creighton, 1950a: 277.

***specularis.*** *Leptothorax exilis* var. *specularis* Emery, 1916b: 174 (w.) ITALY. [First available use of *Leptothorax tuberum* subsp. *exilis* var. *specularis* Emery, 1898c: 135; unavailable name.] Santschi, 1923a: 135 (m.). Junior synonym of *exilis:* Baroni Urbani, 1971d: 1041. Revived from synonymy, and raised to species: Collingwood, 1978: 70.

***sphagnicola.*** *Leptothorax sphagnicolus* Francoeur, 1986a: 1158, fig. 18–30 (w.q.m.) CANADA. See also: Loiselle, Francoeur, *et al.* 1990: 329.

***spininodis.*** *Leptothorax spininodis* Mayr, 1887: 617 (w.m.) BRAZIL. Forel, 1912g: 18 (q.). Combination in *L.* (*Goniothorax*): Emery, 1896g: 59 (in key); in *L.* (*Nesomyrmex*): Kempf, 1959c: 427. Race/subspecies of *echinatinodis:* Forel, 1899c: 55; Emery, 1922f: 250. Revived status as species, and senior synonym of *genualia* and *umbratilis:* Kempf, 1959c: 427.

***spinosior.*** *Leptothorax congruus* var. *spinosior* Forel, 1901e: 371 (w.) JAPAN. Wheeler, W.M. 1906c: 317 (q.); Imai, 1966: 119 (k.). Raised to species: Terayama & Satoh, 1990b: 532.

***spinosus.*** *Leptothorax tuberum* r. *spinosus* Forel, 1909e: 377 (w.) ALGERIA. [First available use of *Leptothorax tuberum* r. *unifasciatus* var. *spinosus* Forel, 1894d: 38; unavailable name.] Cagniant, 1970b: 203 (q.m.). Raised to species: Cagniant, 1970b: 201; Cagniant, 1987c: 249.

*spinosus. Leptothorax* (*Dichothorax*) *pergandei* subsp. *floridanus* var. *spinosus* Smith, M.R. 1929: 551 (w.) U.S.A. Unavailable name; material referred to *floridanus* by Creighton, 1950a: 260.

***splendens.*** *Macromischa splendens* Wheeler, W.M. 1905b: 98, pl. 7, fig. 9 (w.q.m.) BAHAMAS. Combination in *Leptothorax:* Baroni Urbani, 1978b: 505.

#***splendidiceps.*** *Leptothorax* (*Myrafant*) *niger* subsp. *splendidiceps* Baroni Urbani, 1968b: 458, figs. 33–36 (w.q.l.) MALTA. Schembri & Collingwood, 1981: 433 (m.). Subspecies of *laestrygon:* Baroni Urbani, 1971c: 112.

***squamifer.*** *Macromischa squamifera* Roger, 1863a: 186 (w.) CUBA. Baroni Urbani, 1978b: 509 (q.m.). Combination in *Leptothorax:* Baroni Urbani, 1978b: 508. Senior synonym of *atrinodis:* Wheeler, W.M. 1931b: 21.

#***staegeri.*** *Leptothorax unifasciatus* var. *staegeri* Bondroit, 1918: 131 (w.q.m.) SWITZERLAND. [First available use of *Leptothorax tuberum* r. *unifasciatus* var. *staegeri* Forel, 1917: 251; unavailable name.]

***steinbergi.*** *Leptothorax steinbergi* Arnol'di, 1971: 1823, fig. 1 (w.) KAZAKHSTAN.

***stenotyle.*** *Leptothorax stenotyle* Cole, 1956e: 214. Replacement name for *angustinodus* Cole, 1956b: 28. [Junior primary homonym of *angustinode:* Stitz, 1917: 336.] Combination in *L.* (*Myrafant*): Smith, D.R. 1979: 1395.

***stipaceus.*** *Leptothorax tuberum* subsp. *stipaceus* Ruzsky, 1905b: 586 (w.q.m.) RUSSIA. Raised to species: Ruzsky, 1925b: 45. See also: Arnol'di, 1971: 1825.

*stitzi. Leptothorax cabrerae* var. *stitzi* Santschi, 1918a: 34. Replacement name for *denticulatus* Stitz, 1917: 336. [Junior primary homonym of *denticulatus* Mayr, 1901b: 5.] Junior synonym of *cabrerae:* Hohmann, La Roche, *et al.*, 1993: 149.

***stollii.*** *Leptothorax stollii* Forel, 1885a: 352 (w.m.) GUATEMALA. Forel, 1899c: 54 (q.). Combination in *L.* (*Myrafant*): Kempf, 1972a: 132.

***stramineus.*** *Limnomyrmex stramineus* Arnold, 1948: 223, fig. 10 (w.) SOUTH AFRICA. Combination in *Leptothorax:* Brown, 1971a: 4. See also: Bolton, 1982: 332.

***striatulus.*** *Leptothorax striatulus* Stitz, 1937: 133 (w.) MEXICO. Combination in *L.* (*Myrafant*): Kempf, 1972a: 132.

***subcingulatus.*** *Leptothorax luteus* var. *subcingulata* Emery, 1924c: 166 (w.) SPAIN. Menozzi, 1927b: 90 (q.). Raised to species: Collingwood, 1978: 70.

***subditivus.*** *Macromischa subditiva* Wheeler, W.M. 1903b: 99, fig. 5 (w.) U.S.A. Wheeler, W.M. 1931b: 15 (q.); Smith, M.R. 1939e: 504 (m.). Combination in *Leptothorax,* and senior synonym of *flavitarsis, laevissima* and *luciliae:* Baroni Urbani, 1978b: 512. See also: Creighton, 1966a: 282.

*suberis. Leptothorax* (*Temnothorax*) *recedens* var. *suberis* Forel, 1894d: 33 (w.) ALGERIA. Junior synonym of *recedens:* Collingwood, 1978: 69.

***submuticus.*** *Leptothorax submuticus* Emery, 1915a: 261 (w.q.) MOROCCO.

*sulfurea. Leptothorax asper* var. *sufurea* Forel, 1912g: 18 (w.) BRAZIL. [Justified emendation of spelling to *sulfurea* by Emery, 1922f: 250.] Junior synonym of *asper:* Kempf, 1959c: 414.

#***superus.*** *Leptothorax* (*Mychothorax*) *acervorum* var. *superus* Ruzsky, 1905b: 614 (w.m.) RUSSIA. [Later misspelled as *superbus* by Ruzsky, 1936: 94.]

***susamyri.*** *Leptothorax* (*Leptothorax*) *susamyri* Dlussky, 1965b: 28 (w.) KIRGIZIA. See also: Tarbinsky, 1976: 89.

*suturalis. Leptothorax normandi* st. *atlantis* var. *suturalis* Santschi, 1921b: 426 (w.) MOROCCO. Unavailable name.

***svartshevskii.*** *Leptothorax svartshevskii* Karavaiev, 1916: 500, fig. 2 (w.q.) RUSSIA.

***taivanensis.*** *Leptothorax taivanensis* Wheeler, W.M. 1929g: 54 (w.) TAIWAN.

*talassicus. Leptothorax talassicus* Tarbinsky, 1976: 101, figs. 142–144 (w.) KIRGIZIA. Junior synonym of *oxianus:* Radchenko, 1994d: 151.

***tamarae.*** *Leptothorax tamarae* Radchenko, 1993c (not yet seen). See also: Radchenko, 1994d: 150.

***tauricus.*** *Leptothorax unifasciatus* var. *tauricus* Ruzsky, 1902d: 22 (w.) UKRAINE. Karavaiev, 1927a: 292 (q.). Subspecies of *unifasciatus:* Arnol'di, 1977a: 202. Raised to species, and senior synonym of *pithyusae.* Radchenko, 1994d: 157 (in key). Current subspecies: nominal plus *babuganicus.*

*taylori* Forel, 1912; see under ***PARATOPULA.***

***tebessae.*** *Leptothorax tuberum* r. *tebessae* Forel, 1890a: lxxiv (w.q.) ALGERIA. Cagniant, 1970a: 424 (m.). Raised to species: Dalla Torre, 1893: 127. Subspecies of *nylanderi:* Emery, 1922f: 255. Status as species: Santschi, 1923f: 330; Santschi, 1929e: 156; Cagniant, 1964: 105; Cagniant, 1987c: 249. Current subspecies: nominal plus *gapsensis.*

*tellensis. Leptothorax bucheti* st. *tuneticus* var. *tellensis* Santschi, 1939d: 74 (w.) ALGERIA. Unavailable name.

*tennesseensis* Cole, 1938; see under ***PHEIDOLE.***

***tenuisculptus.*** *Leptothorax tenuisculptus* Baroni Urbani, 1978b: 517, figs. 59, 99 (w.) MEXICO.

***tenuispinus.*** *Leptothorax tenuispinus* Santschi, 1911i: 285 (q.) TUNISIA. Combination in *L.* (*Temnothorax*): Santschi, 1918a: 32.

*tenuispinus. Leptothorax bucheti* st. *tuneticus* var. *tenuispinus* Santschi, 1939d: 74 (w.) ALGERIA. Unavailable name.

#***tergestinus.*** *Leptothorax sordidulus* var. *tergestina* Finzi, 1928b: 129 (w.) ITALY.

***terricola.*** *Macromischa* (*Antillaemyrmex*) *terricola* Mann, 1920: 423, fig. 7 (w.q.) CUBA. Combination in *Antillaemyrmex:* Wheeler, W.M. 1931b: 32; in *Leptothorax:* Baroni Urbani, 1978b: 518.

***terrigena.*** *Leptothorax terrigena* Wheeler, W.M. 1903c: 254, pl. 12, fig. 21 (w.q.) U.S.A. Combination in *L.* (*Myrafant*): Smith, D.R. 1979: 1395.

***tesquorum.*** *Leptothorax tesquorum* Arnol'di, 1977a: 201 (w.) GEORGIA. Senior synonym of *elderensis:* Radchenko, 1994d: 150 (in key).

*testaceus. Leptothorax vicinus* var. *testaceus* Emery, 1894c: 164 (w.q.) BRAZIL. Junior synonym of *vicinus:* Kempf, 1959c: 415.

***texanus.*** *Leptothorax texanus* Wheeler, W.M. 1903c: 245, pl. 12, fig. 16 (w.q.m.) U.S.A. Wheeler, G.C. & Wheeler, J. 1960b: 24 (l.). Current subspecies: nominal plus *davisi.* See also: Smith, M.R. 1952b: 102.

***theryi.*** *Leptothorax theryi* Santschi, 1921b: 425, fig. 1 (w.) MOROCCO.

***tianschanicus.*** *Leptothorax tianschanicus* Tarbinsky, 1976: 100, figs. 138–141 (w.) KIRGIZIA.

#***timidus.*** *Leptothorax convexus* var. *timida* Santschi, 1912c: 148 (w.m.) MOROCCO.

*timidus. Leptothorax* (*Nesomyrmex*) *timidus* Kempf, 1959c: 420, figs. 10, 14 (w.) VENEZUELA. [Junior primary homonym of *timidus* Santschi, above.] Replacement name: *brimodus* **new name**.

*tirolensis. Leptothorax tirolensis* Gredler, 1858: 26 (w.) AUSTRIA. Race of *tuberum:* Emery & Forel, 1879: 459. Status as species: André, 1883a: 297; Dalla Torre, 1893: 127. Junior synonym of *affinis:* Emery, 1914d: 158 (footnote).

***tonsuratus.*** *Leptothorax* (*Nesomyrmex*) *tonsuratus* Kempf, 1959c: 412, figs. 18, 19, 21 (q.) MEXICO. Kempf, 1975c: 364 (w.).

***torrei.*** *Macromischa* (*Antillaemyrmex*) *torrei* Aguayo, 1931: 178 (w.) CUBA. Combination in *Leptothorax:* Baroni Urbani, 1978b: 520.

***totonicapani.*** *Leptothorax totonicapani* Baroni Urbani, 1978b: 521, figs. 64, 102 (w.) GUATEMALA.

*tozae. Leptothorax tuberum* var. *tozae* Schmitz, 1950: 16 PORTUGAL. *Nomen nudum.*

#***trabutii.*** *Leptothorax angustulus* r. *trabutii* Forel, 1894d: 37 (w.) ALGERIA. Stirps of *angustulus:* Santschi, 1911a: 12.

*transitoria. Leptothorax rottenbergi* st. *annibalis* var. *transitoria* Santschi, 1909b: 478 (w.) TUNISIA. Unavailable name.

***tricarinatus.*** *Leptothorax tricarinatus* Emery, 1895c: 321, pl. 8, fig. 14 (w.) U.S.A. Smith, M.R. 1952b: 99 (m.). Combination in *L.* (*Myrafant*): Smith, M.R. 1950: 30. Current subspecies: nominal plus *neomexicanus.* See also: Wheeler, W.M. 1903c: 247.

***tricolor.*** *Leptothorax tricolor* Santschi, 1925g: 346 (w.) MOROCCO. Combination in *Temnothorax:* Cagniant, 1964: 104.

***tristani.*** *Leptothorax* (*Goniothorax*) *tristani* Emery, 1896g: 61 (w.q.) COSTA RICA. Combination in *L.* (*Nesomyrmex*), and senior synonym of *antoniensis:* Kempf, 1959c: 415.

***tristis.*** *Leptothorax tristis* Bondroit, 1918: 129, fig. 62 (w.) FRANCE.

***LEPTOTHORAX*** *(continued)*

*tristis. Croesomyrmex iris* var. *tristis* Wheeler, W.M. 1931b: 25 (w.) CUBA. [Unresolved junior secondary homonym of *tristis* Bondroit, above.] Junior synonym of *iris:* Baroni Urbani, 1978b: 454.

#***tuberoaffinis.*** *Leptothorax tuberoaffinis* Bondroit, 1918: 131 (w.q.) SWITZERLAND. [First available use of *Leptothorax tuberum* var. *tuberoaffinis* Forel, 1874: 86; *nomen nudum. Leptothorax tuberum* r. *affinis* var. *tuberoaffinis* Forel, 1915d: 23 (in key); unavailable name.] Variety/subspecies of *affinis:* Stitz, 1939: 174; Novak & Sadil, 1941: 92.

#***tuberointerruptus.*** *Leptothorax tuberointerruptus* Bondroit, 1918: 126 (w.q.) SWITZERLAND. [First available use of *Leptothorax tuberum* var. *tuberointerruptus* Forel, 1874: 86; *nomen nudum. Leptothorax tuberum* r. *interruptus* var. *tuberointerruptus* Forel, 1915d: 24 (in key); unavailable name.] Status as species: Betrem, 1926: 217. Variety of *interruptus:* Novak & Sadil, 1941: 93.

#***tuberonigriceps.*** *Leptothorax tuberum* var. *tuberonigriceps* Forel, 1874: 88 (w.q.m.) SWITZERLAND.

*tuberosa. Formica tuberosa* Latreille, 1802c: 259 (w.q.) FRANCE. Erroneous rendering and hence junior synonym of *tuberum:* Stephens, 1829: 356; Dalla Torre, 1893: 127.

***tuberum.*** *Formica tuberum* Fabricius, 1775: 393 (w.) SWEDEN. Lepeletier, 1835: 183 (q.m.); Nylander, 1846a: 939 (q.m.); Adlerz, 1886: 82 (gynandromorph); Hauschteck, 1962: 219 (k.). Senior synonym of *tuberosa:* Stephens, 1829: 356; Dalla Torre, 1893: 127; of *albipennis:* Emery, 1922f: 256; of *melanocephalus:* Casevitz-Weulersse, 1990b: 420; of *acutinodis:* Arakelian, 1994: 58; of *pratostepposus:* Radchenko, 1994d: 157 (in key). Material of the *nomen nudum nigroscutellata* referred here: Emery, 1914e: 2. Current subspecies: nominal plus *alaicus, brauneri* (unresolved junior homonym), *ciscaucasicus, kirillovi, nigricephalus, nylanderounifasciatus, sachalinensis, salinus, tuberonigriceps.* See also: Baroni Urbani, 1971c: 124; Tarbinsky, 1976: 96; Kutter, 1977c: 134; Collingwood, 1979: 75; Atanassov & Dlussky, 1992: 142.

#***tuneticus.*** *Leptothorax bucheti* var. *tunetica* Santschi, 1918a: 34 (w.q.m.) TUNISIA. Stirps of *bucheti:* Santschi, 1939d: 74.

***turcicus.*** *Leptothorax turcicus* Santschi, 1934d: 278 (w.) TURKEY.

***tuscaloosae.*** *Leptothorax tuscaloosae* Wilson, 1951: 128 (w.q.) U.S.A.

***tyndalei.*** *Leptothorax tuberum* subsp. *tyndalei* Forel, 1909e: 373 (w.) ALGERIA. Cagniant, 1966a: 18 (q.m.). Raised to species: Cagniant, 1966a: 17.

#***ucrainicus.*** *Leptothorax unifasciatus* subsp. *ucrainicus* Arnol'di, 1977a: 204 (w.) UKRAINE.

*umbratilis. Leptothorax* (*Goniothorax*) *umbratilis* Wheeler, W.M. 1921f: 160 (w.q.) GUIANA. Junior synonym of *spininodis:* Kempf, 1959c: 427.

***umbratipes.*** *Macromischa affinis* subsp. *umbratipes* Wheeler, W.M. 1937b: 455 (w.) CUBA. Combination in *Leptothorax,* and raised to species: Baroni Urbani, 1978b: 522.

#***unifasciatointerruptus.*** *Leptothorax tuberum* var. *unifasciatointerruptus* Forel, 1874: 86 (m.) SWITZERLAND. Variety of *unifasciatus:* Bondroit, 1918: 131.

#***unifasciatonigriceps.*** *Leptothorax unifasciatus* var. *unifasciatonigriceps* Novak & Sadil, 1941: 93 (w.q.) GERMANY. [First available use of *Leptothorax tuberum* subsp. *nigriceps* var. *unifasciatonigriceps* Gösswald, 1932: 149; unavailable name.]

*unifasciatonigricipoides. Leptothorax tuberum* var. *unifasciatonigricipoides* Dalla Torre, 1893: 128. *Nomen nudum;* attributed to Forel.

***unifasciatus.*** *Formica unifasciata* Latreille, 1798: 47 (w.q.m.) FRANCE. Wheeler, G.C. & Wheeler, J. 1955b: 25 (l.). Race/subspecies of *tuberum:* Forel, 1874: 85; Emery & Forel, 1879: 459; André, 1883a: 299; Forel, 1890a: lxxv; Emery, 1891b: 5; Bondroit, 1910: 497; Emery, 1916b: 175; Finzi, 1924a: 13; Menozzi, 1925d: 28; Karavaiev, 1926c: 163. Status as species: Saunders, E. 1880: 220; Nasonov, 1889: 32; Dalla Torre, 1893: 129; Ruzsky, 1902d: 21; Bondroit, 1912: 351; Bondroit, 1918: 131; Müller, 1923: 91; Finzi, 1928b: 130; Finzi, 1930d: 314; Finzi, 1933: 164; Boven, 1947: 179; Bernard, 1967: 211; Kutter, 1977c: 134; Collingwood, 1979: 76; Atanassov & Dlussky, 1992: 144. Senior synonym of *anoplogynus:* Emery, 1916b: 175; of *cordieri,* and material of the unavailable name *paolii* referred here: Casevitz-Weulersse, 1990b: 417. Current subspecies: nominal plus *obenbergeri, rougeti, staegeri, ucrainicus, unifasciatointerruptus, unifasciatonigriceps.*

***usunkul.*** *Leptothorax usunkul* Ruzsky, 1924: 152 (w.) RUSSIA.

#***uvicensis.*** *Leptothorax muscorum* subsp. *uvicensis* Blacker, 1992: 7, fig. 2 (w.) CANADA.

#***vandeli.*** *Mychothorax acervorum* var. *vandeli* Bondroit, 1920a: 153 (w.q.) FRANCE. [Also described as new by Bondroit, 1920b: 303.]

*veneris. Leptothorax angustus* st. *trabuti* var. *veneris* Santschi, 1918a: 35 (w.q.) TUNISIA. Unavailable name.

***versicolor.*** *Macromischa versicolor* Roger, 1863a: 187 (w.) CUBA. Mann, 1929: 166 (q.); Baroni Urbani, 1978b: 524 (m.). Combination in *M.* (*Croesomyrmex*): Mann, 1920: 421; in *Croesomyrmex:* Wheeler, W.M. 1931b: 32; in *Leptothorax:* Baroni Urbani, 1978b: 523. Senior synonym of *lugens* (and its junior synonym *nigra*): Baroni Urbani, 1978b: 523.

***vicinus.*** *Leptothorax vicinus* Mayr, 1887: 620 (w.) BRAZIL. Forel, 1912g: 18 (q.). Combination in *L.* (*Goniothorax*): Emery, 1896g: 58 (in key); in *L.* (*Nesomyrmex*), and senior synonym of *testaceus:* Kempf, 1959c: 415.

***villarensis.*** *Macromischa manni* var. *villarensis* Aguayo, 1931: 177 (w.) CUBA. Raised to species, first available replacement name for *manni* Wheeler, W.M. 1931b: 19 [Junior secondary homonym of *manni* Wheeler, W.M. 1914b: 53.], and senior synonym of *williami:* Snelling, 1986: 155.

***violaceus.*** *Macromischa* (*Macromischa*) *violacea* Mann, 1924: 20 (w.) CUBA. Combination in *Leptothorax:* Baroni Urbani, 1978b: 526.

***volgensis.*** *Leptothorax nassonovi* subsp. *volgensis* Ruzsky, 1905b: 582 (w.) RUSSIA. Kupyanskaya, 1990: 142 (q.). Raised to species: Arnol'di, 1977a: 203.

***volubilis.*** *Leptothorax volubilis* Santschi, 1929e: 155, fig. 5 (w.q.) MOROCCO.

*wasmanni* Forel, 1901 [*Macromischa*]; see under ***TETRAMORIUM.***

***werneri.*** *Leptothorax werneri* Radchenko, 1993c (not yet seen). See also: Radchenko, 1994d: 149.

***wheeleri.*** *Macromischa* (*Croesomyrmex*) *wheeleri* Mann, 1920: 422, fig. 6 (w.q.) CUBA. Wheeler, G.C. & Wheeler, J. 1955b: 18 (l.); Baroni Urbani, 1978b: 529 (m.). Combination in *Croesomyrmex:* Wheeler, W.M. 1931b: 32; in *Leptothorax:* Baroni Urbani, 1978b: 527. Senior synonym of *petri:* Brown, 1959e: 73.

*wheeleri. Leptothorax wheeleri* Smith, M.R. 1929: 547, fig. 1 (w.q.) U.S.A. Wheeler, G.C. & Wheeler, J. 1989a: 323 (l.). Combination in *L.* (*Myrafant*): Smith, D.R. 1979: 1395. [Junior secondary homonym of *wheeleri* Mann, above.] Replacement name: *smithi* Baroni Urbani, 1978b: 557.

***wilda.*** *Leptothorax* (*Goniothorax*) *wilda* Smith, M.R. 1943d: 155 (w.q.) U.S.A. Francoeur & Loiselle, 1988b: 44 (m.). Combination in *L.* (*Nesomyrmex*): Smith, M.R. 1951a: 817. See also: Kempf, 1959c: 421.

*williami. Leptothorax williami* Baroni Urbani, 1978b: 530. [Unnecessary replacement name for *manni* Wheeler, 1931b: 19.] Junior synonym of *villarensis:* Snelling, 1986: 155.

***wilsoni.*** *Leptothorax wilsoni* Heinze, 1989: 50, figs. 1–4 (q.m.) U.S.A.

***wollastoni.*** *Leptothorax wollastoni* Donisthorpe, 1940a: 45 (w.) MADEIRA.

***wroughtonii.*** *Leptothorax wroughtonii* Forel, 1904c: 22 (w.) INDIA.

#***wui.*** *Leptothorax congruus* var. *wui* Wheeler, W.M. 1929f: 7 (w.q.m.) CHINA.

*yankee. Leptothorax canadensis* var. *yankee* Emery, 1895c: 319 (w.q.) U.S.A. Wheeler, W.M. 1917a: 513 (m.); Wheeler, G.C. & Wheeler, J. 1955b: 21 (l.). Subspecies of *canadensis,* and senior synonym of *obscurus* and *sordidus:* Creighton, 1950a: 277. Junior synonym of *muscorum:* Brown, 1955a: 47.

*LEPTOXENUS: nomen nudum.*

****LEUCOTAPHUS*** (Great Britain)

****cockerelli.*** **Leucotaphus cockerelli* Donisthorpe, 1920: 90, pl. 5, fig. 10 (m.) GREAT BRITAIN (Oligocene).

****gurnetensis.*** **Leptothorax gurnetensis* Cockerell, 1915: 485, pl 65, figs. 4, 5 (wing) GREAT BRITAIN (Oligocene). Combination in **Leucotaphus:* Donisthorpe, 1920: 89.

****permancus.*** **Leucotaphus permancus* Cockerell, 1927: 165 (footnote) GREAT BRITAIN (Oligocene).

*LILIDRIS:* see under ***SOLENOPSIS.***

*LIMNOMYRMEX:* see under ***LEPTOTHORAX.***

***LINEPITHEMA*** (Neotropical)

#***angulatum.*** *Iridomyrmex humilis* subsp. *angulatus* Emery, 1894c: 165, pl. 1, fig. 3 (w.q.) BOLIVIA. Combination in *Linepithema:* Shattuck, 1992a: 16.

#***arrogans.*** *Iridomyrmex humilis* var. *arrogans* Chopard, 1921: 237 (footnote), :241, figs. 1–31 (w.q.m.l.) FRANCE. [Also described as new by Santschi, 1929d: 306 (in key).] Combination in *Linepithema:*

Shattuck, 1992a: 16. Junior synonym of *humile:* Bernard, 1967: 251. Revived from synonymy as subspecies of *humile:* Shattuck, 1992a: 16.

**aspidocoptum.** *Iridomyrmex aspidocoptus* Kempf, 1969: 295, figs. 10, 11 (w.) BRAZIL. Combination in *Linepithema:* Shattuck, 1992a: 16.

#**bicolor.** *Iridomyrmex iniquus* var. *bicolor* Forel, 1912h: 47 (w.) MARTINIQUE I. Combination in *Linepithema:* Shattuck, 1992a: 16.

#**breviscapum.** *Iridomyrmex humilis* st. *breviscapa* Santschi, 1929d: 306 (w.) ARGENTINA. [First available use of *Iridomyrmex humilis* r. *platensis* var. *breviscapa* Forel, 1914d: 286; unavailable name.] Santschi, 1916e: 390 (m.). Combination in *Linepithema:* Shattuck, 1992a: 16.

**dispertitum.** *Iridomyrmex dispertitus* Forel, 1885a: 351 (w.m.) GUATEMALA. Combination in *Linepithema:* Shattuck, 1992a: 16. Current subspecies: nominal plus *micans.*

#**dominicense.** *Iridomyrmex melleus* var. *dominicensis* Wheeler, W.M. 1913d: 242 (w.) DOMINICA. Combination in *Linepithema:* Shattuck, 1992a: 16.

#**flavescens.** *Iridomyrmex keiteli* var. *flavescens* Wheeler, W.M. & Mann, 1914: 43 (w.) HAITI. Combination in *Linepithema:* Shattuck, 1992a: 16.

#**fuscescens.** *Iridomyrmex melleus* var. *fuscescens* Wheeler, W.M. 1908a: 153 (w.) PUERTO RICO. Combination in *Linepithema:* Shattuck, 1992a: 16.

**fuscum.** *Linepithema fuscum* Mayr, 1866a: 497, pl. fig. 7 (m.) PERU.

#**gallardoi.** *Dorymyrmex gallardoi* Brèthes, 1914a: 95 (w.) ARGENTINA. Santschi, 1929d: 306 (q.). Combination in *Iridomyrmex:* Gallardo, 1916a: 105; in *Linepithema:* Shattuck, 1992a: 16. Subspecies of *humile:* Gallardo, 1916a: 105; Santschi, 1929d: 306.

*gallardoi. Linepithema gallardoi* Kusnezov, 1969: 37 (m.) ARGENTINA. [Junior secondary homonym of *gallardoi* Brèthes, above.] Replacement name: *inacatum* **new name**.

***hispaniolae.** **Iridomyrmex hispaniolae* Wilson, 1985c: 32, fig. 10 (w.) DOMINICAN AMBER. Combination in *Linepithema:* Shattuck, 1992a: 16.

**humile.** *Hypoclinea humilis* Mayr, 1868b: 164 (w.) ARGENTINA. Forel, 1908c: 395 (m.); Newall, 1908: 28 (q.); Wheeler, G.C. & Wheeler, J. 1951: 186 (l.); Crozier, 1969: 250 (k.). Combination in *H.* (*Iridomyrmex*): Mayr, 1870b: 959; in *Iridomyrmex:* Emery, 1888d: 386; in *Linepithema:* Shattuck, 1992a: 16. Current subspecies: nominal plus *angulatum, arrogans, breviscapum, gallardoi, platense, scotti.* See also: Gallardo, 1916a: 97; Bernard, 1967: 251; Collingwood, 1979: 33; Smith, D.R. 1979: 1418; Ward, 1987: 1; Wheeler, G.C. & Wheeler, J. 1990a: 465; Shattuck, 1994: 123.

***humiloides.** **Iridomyrmex humiloides* Wilson, 1985c: 33, fig. 11 (w.) DOMINICAN AMBER. Combination in *Linepithema:* Shattuck, 1992a: 16.

**impotens.** *Iridomyrmex impotens* Santschi, 1923d: 67 (w.) BRAZIL. Combination in *Linepithema:* Shattuck, 1992a: 16.

**inacatum.** *Linepithema inacatum* **new name**. Replacement name for *gallardoi* Kusnezov, 1969: 37. [Junior secondary homonym of *gallardoi* Brèthes, 1914a: 95.]

**iniquum.** *Hypoclinea iniqua* Mayr, 1870a: 392, fig. 10 (w.) COLOMBIA. Forel, 1908c: 395 (m.). Combination in *Iridomyrmex:* Emery, 1890b: 56; in *Linepithema:* Shattuck, 1992a: 16. Current subspecies: nominal plus *bicolor, nigellum, succineum.*

**keiteli.** *Iridomyrmex keiteli* Forel, 1907e: 8 (w.q.) HAITI. Combination in *Linepithema:* Shattuck, 1992a: 16. Current subspecies: nominal plus *flavescens, subfasciatum.*

**leucomelas.** *Iridomyrmex leucomelas* Emery, 1894g: 378 (footnote) (w.q.) BRAZIL. Combination in *Linepithema:* Shattuck, 1992a: 16.

**melleum.** *Iridomyrmex melleus* Wheeler, W.M. 1908a: 151, pl. 12, figs. 15, 16 (w.q.m.) PUERTO RICO. Wheeler, G.C. & Wheeler, J. 1974b: 397 (l.). Combination in *Linepithema:* Shattuck, 1992a: 16. Current subspecies: nominal plus *dominicense, fuscescens.*

#**micans.** *Iridomyrmex dispertitus* subsp. *micans* Forel, 1908c: 394 (w.m.) BRAZIL. Forel, 1911c: 306 (q.). Combination in *Linepithema:* Shattuck, 1992a: 16.

#**nigellum.** *Iridomyrmex iniquus* var. *nigellus* Emery, 1890b: 56 (w.) COSTA RICA. [Also described as new by Emery, 1894k: 61.] Combination in *Linepithema:* Shattuck, 1992a: 16. Subspecies/race of *dispertitum:* Forel, 1901c: 134; Forel, 1912h: 47; of *iniquum:* Creighton, 1950a: 342.

**oblongum.** *Iridomyrmex humilis* var. *oblongus* Santschi, 1929d: 306 (w.) BRAZIL. Combination in *Linepithema:* Shattuck, 1992a: 16. Raised to species: Snelling & Hunt, 1976: 90.

*paranaensis. Iridomyrmex melleus* st. *succineus* var. *paranaensis* Santschi, 1929d: 305 (w.) BRAZIL. Unavailable name (Shattuck, 1994: 127).

*pertaesta. Iridomyrmex humilis* st. *angulata* var. *pertaesta* Santschi, 1916e: 390 (w.m.) ARGENTINA. Unavailable name (Shattuck, 1994: 127).

**piliferum.** *Hypoclinea pilifera* Mayr, 1870a: 393 (w.q.) COLOMBIA. Crozier, 1970: 113 (k.). Combination in *Iridomyrmex:* Dalla Torre, 1893: 169; in *Linepithema:* Shattuck, 1992a: 16.

#**platense.** *Iridomyrmex humilis* r. *platensis* Forel, 1912h: 46 (w.q.m.) ARGENTINA. Combination in *Linepithema:* Shattuck, 1992a: 16.

**pordescens.** *Iridomyrmex pordescens* Wheeler, W.M. 1942: 214 (w.) COSTA RICA. Combination in *Linepithema:* Shattuck, 1992a: 16.

*praenotandum. Iridomyrmex humilis* var. *praenotandum* Schmitz, 1950: 14, PORTUGAL. *Nomen nudum.*

*pulex. Iridomyrmex humilis* st. *angulata* var. *pulex* Santschi, 1923d: 68 (w.) BRAZIL. Unavailable name (Shattuck, 1994: 127).

**riograndense.** *Iridomyrmex riograndensis* Borgmeier, 1928b: 64 (w.) BRAZIL. Combination in *Linepithema:* Shattuck, 1992a: 16.

#**scotti.** *Iridomyrmex humilis* st. *scotti* Santschi, 1919f: 52 (w.) ARGENTINA. Combination in *Linepithema:* Shattuck, 1992a: 16.

#**subfasciatum.** *Iridomyrmex keiteli* var. *subfasciatus* Wheeler, W.M. & Mann, 1914: 43 (w.) HAITI. Combination in *Linepithema:* Shattuck, 1992a: 16.

#**succineum.** *Iridomyrmex melleus* subsp. *succineus* Forel, 1908c: 396 (w.) BRAZIL. Combination in *Linepithema:* Shattuck, 1992a: 16. Subspecies of *melleum:* Santschi, 1929d: 305; of *iniquum:* Forel, 1912h: 47; Kempf, 1972a: 125.

*transiens. Iridomyrmex humilis* r. *platensis* var. *transiens* Forel, 1913l: 242 (w.q.m.) ARGENTINA. Unavailable name (Shattuck, 1994: 127).

***LIOMETOPUM*** (Holarctic, Oriental)

*aeneum* Mayr, 1862; see under ***IRIDOMYRMEX.***

***antiquum.** **Liometopum antiquum* Mayr, 1867b: 60, pl. fig. 10 (q.) YUGOSLAVIA (Miocene).

**apiculatum.** *Liometopum apiculatum* Mayr, 1870b: 961 (w.) MEXICO. Emery, 1895c: 331 (q.); Wheeler, W.M. 1905e: 324 (m.); Wheeler, G.C. & Wheeler, J. 1951: 181 (l.). See also: Shattuck, 1994: 128.

*austriaca. Formica austriaca* Mayr, 1853a: 144 (w.) AUSTRIA. Junior synonym of *microcephalum:* Mayr, 1861: 39.

*brevicorne* Mayr, 1878; see under ***AZTECA.***

***goepperti.** **Hypoclinea goepperti* Mayr, 1868c: 56, pl. 1, figs. 3–7; pl. 2, figs. 42–46 (w.q.m.) BALTIC AMBER (Oligocene). Combination in *Bothriomyrmex:* Dalla Torre, 1893: 170; in *Iridomyrmex:* Wheeler, W.M. 1915h: 90; in *Liometopum:* Shattuck, 1992a: 15. Senior synonym of *succinalis:* Brown, 1977: 214.

***imhoffii.** **Formica imhoffii* Heer, 1850: 138, pl. 10, fig. 10 (m.) YUGOSLAVIA (Miocene). Combination in **Poneropsis:* Heer, 1867: 24; in *Liometopum:* Mayr, 1867b: 55. Senior synonym of *schmidtii:* Mayr, 1867b: 56. Included by Taylor, 1964: 138 as FORMICIDAE, *incertae generis.* Combination in *Liometopum:* Shattuck, 1994: 129.

**lindgreeni.** *Liometopum lindgreeni* Forel, 1902d: 293 (w.) INDIA.

**luctuosum.** *Liometopum apiculatum* subsp. *luctuosum* Wheeler, W.M. 1905e: 325 (w.) U.S.A. Forel, 1914c: 619 (m.). Subspecies of *occidentale:* Creighton, 1950a: 339. Raised to species: Wheeler, G.C. & Wheeler, J. 1986g: 55; MacKay, Lowrie, *et al.* 1988: 102.

**masonium.** *Formica masonia* Buckley, 1866: 165 (w.) U.S.A. Combination possibly in *Liometopum:* Wheeler, W.M. 1902f: 23. *Incertae sedis* in Dolichoderinae: Shattuck, 1994: 171. *Incertae sedis* in *Liometopum:* current combination.

**microcephalum.** *Formica microcephala* Panzer, 1798: no. 1652 (m.) AUSTRIA. Mayr, 1861: 39 (w.q.). Combination in *Liometopum,* and senior synonym of *austriaca:* Mayr, 1861: 39. See also: Baroni Urbani, 1971c: 158; Atanassov & Dlussky, 1992: 189; Shattuck, 1994: 129.

***miocenicum.** **Liometopum miocenicum* Carpenter, 1930: 46, pl. 3, fig. 6; pl. 8, fig. 3; pl. 9, fig. 8; pl. 10, fig. 8; pl. 11, fig. 7 (w.q.m.) U.S.A. (Miocene).

**occidentale.** *Liometopum microcephalum* var. *occidentale* Emery, 1895c: 330 (w.m.) U.S.A. Wheeler, W.M. 1917a: 522 (q.). Variety of *apiculatum:* Wheeler, W.M. 1905e: 324. Raised to species: Wheeler, W.M. 1917a: 521.

***oligocenicum.** **Liometopum oligocenicum* Wheeler, W.M. 1915h: 95, fig. 46 (w.) BALTIC AMBER (Oligocene).

**orientale.** *Liometopum microcephalum* var. *orientalis* Karavaiev, 1927d: 339 (w.) RUSSIA. Kupyanskaya, 1988: 31 (q.m.). Raised to species: Kupyanskaya, 1988: 29. See also: Kupyanskaya, 1990: 157.

**pingue* Scudder, 1877; see under ****EOFORMICA.***

**schmidtii.* **Formica schmidtii* Heer, 1850: 138, pl. 11, fig. 5 (m.) YUGOSLAVIA (Miocene). Combination in

***LIOMETOPUM** (continued)*

*Poneropsis:* Heer, 1867: 24. Junior synonym of *imhoffii:* Mayr, 1867b: 56.
***scudderi.** *Liometopum scudderi* Carpenter, 1930: 47, pl. 4, fig. 4; pl. 9, fig. 4 (w.q.) U.S.A. (Miocene).
**#*sericatum.*** *Liometopum sinense* var. *sericatum* Wheeler, W.M. 1921c: 541 (w.) CHINA.
***sinense.*** *Liometopum sinense* Wheeler, W.M. 1921c: 540 (w.) CHINA. Current subspecies: nominal plus *sericatum.*
**succinalis.* **Ctenobethylus succinalis* Brues, 1939: 262, fig. 7 (w.) BALTIC AMBER (Oligocene). Junior synonym of **goepperti:* Brown, 1977: 214.
*xanthochroum* Roger, 1863; see under ***AZTECA.***

***LIOMYRMEX*** (Oriental, Indo-Australian)
***aurianus.*** *Liomyrmex aurianus* Emery, 1889b: 504 (w.) BURMA. Wheeler, G.C. & Wheeler, J. 1955c: 128 (l.). Senior synonym of *butteli:* Forel, 1913k: 26. See also: Wheeler, W.M. 1914c: 75.
*butteli. Promyrma butteli* Forel, 1912m: 764 (q.) MALAYSIA. Junior synonym of *aurianus:* Forel, 1913k: 26.
*caeca. Myrmica caeca* Smith, F. 1860b: 108 (w.) NEW GUINEA. Combination in *Liomyrmex:* Mayr, 1865: 23 (footnote). [Junior primary homonym of *Myrmica caeca* Jerdon, 1851: 116.] Replacement name: *gestroi* Emery, 1887b: 461; earliest junior synonym of *caeca* (**new synonymy**) and hence first available replacement name.
***carinatus.*** *Liomyrmex carinata* Stitz, 1911a: 366, figs. 13, 14 (q.) NEW GUINEA.
*dispar* Forel, 1895; see under ***MACHOMYRMA.***
***froggatti.*** *Liomyrmex froggatti* Donisthorpe, 1940a: 40, figs. 1–3 (q.m.) NEW GUINEA. Current subspecies: nominal plus *major.*
***gestroi.*** *Laparomyrmex gestroi* Emery, 1887b: 461, pl. 2, fig. 16 (q.) NEW GUINEA. Combination in *Liomyrmex:* Dalla Torre, 1893: 63. Junior synonym of *caeca* (**new synonymy**, type and other material compared), first available replacement name for *Myrmica caeca* Smith, F. 1860b: 108. [Junior primary homonym of *Myrmica caeca* Jerdon, 1851: 116.]
**#*major.*** *Liomyrmex froggatti* subsp. *major* Donisthorpe, 1941j: 204 (q.) PHILIPPINES.
***reneae.*** *Liomyrmex reneae* Donisthorpe, 1948a: 293, fig. 1 (w.) NEW GUINEA.
***tagalanus.*** *Liomyrmex tagalanus* Menozzi, 1925c: 445, fig. 5 (q.) PHILIPPINES.
***taylori.*** *Liomyrmex taylori* Tiwari & Jonathan, 1986a: 88, fig. 1 (w.) ANDAMAN IS.

*LIOPONERA:* see under ***CERAPACHYS.***

*LITHOMYRMEX:* see under ***AMBLYOPONE.***

**LITHOMYRMEX:* see under ***EULITHOMYRMEX.***

*LOBOGNATHUS:* see under ***MESSOR.***

*LOBOMYRMEX:* see under ***TETRAMORIUM.***

*LOBOPELTA:* see under ***LEPTOGENYS.***

***LONCHOMYRMEX*** (Europe)
***freyeri.*** **Formica freyeri* Heer, 1867: 10, pl. 1, fig. 9 (q.) YUGOSLAVIA (Miocene). Combination in *Lonchomyrmex:* Mayr, 1867b: 61.
***nigritus.*** **Lonchomyrmex nigritus* Assmann, 1870: 39, pl. 1, fig. 2 (wing) GERMANY (Miocene).

*LONCYDA:* see under ***CARDIOCONDYLA.***

***LOPHOMYRMEX*** (Oriental, Indo-Australian)
***ambiguus.*** *Lophomyrmex ambiguus* Rigato, 1994a: 52, figs. 14, 15 (w.) INDIA.
***bedoti.*** *Lophomyrmex bedoti* Emery, 1893e: 192, pl. 8, fig. 17 (w.q.) SUMATRA. Rigato, 1994a: 54 (m.); Imai, Baroni Urbani, *et al.* 1984: 7 (k.). See also: Bingham, 1903: 196; Moffett, 1986a: 207; Rigato, 1994a: 54.
***birmanus.*** *Lophomyrmex birmanus* Emery, 1893e: 192 (diagnosis in key) (w.) BURMA. See also: Emery, 1895k: 466; Bingham, 1903: 196; Rigato, 1994a: 58.
*javana. Lophomyrmex quadrispinosus* var. *javana* Karavaiev, 1933c: 270 (w.) JAVA. Junior synonym of *opaciceps:* Rigato, 1994a: 58.
***kali.*** *Lophomyrmex kali* Rigato, 1994a: 58, figs. 26–28 (w.) INDIA.
***longicornis.*** *Lophomyrmex longicornis* Rigato, 1994a: 55, figs. 11, 12 (w.) BORNEO.
***lucidus.*** *Lophomyrmex bedoti* var. *lucida* Menozzi, 1930d: 328 (w.) WEST MALAYSIA. Rigato, 1994a: 56 (q.m.). Raised to species: Rigato, 1994a: 55.
***opaciceps.*** *Lophomyrmex quadrispinosus* var. *opaciceps* Viehmeyer, 1922: 211 (w.) JAVA. Raised to species, and senior synonym of *javana:* Rigato, 1994a: 58.
***quadrispinosus.*** *Oecodoma quadrispinosa* Jerdon, 1851: 111 (w.) INDIA. Rigato, 1994a: 59 (q.m.); Wheeler, G.C. & Wheeler, J. 1954a: 131 (l.). Combination in *Pheidole:* Smith, F. 1858b: 174; in *Lophomyrmex:* Emery, 1892a: 114. Senior synonym of *taprobanae:* Rigato, 1994a: 59. See also: Bingham, 1903: 195.
***striatulus.*** *Lophomyrmex striatulus* Rigato, 1994a: 56, figs. 16, 17 (w.) THAILAND.
***taivanae.*** *Lophomyrmex quadrispinosus* r. *taivanae* Forel, 1912a: 55 (w.) TAIWAN. Raised to species: Ettershank, 1966: 134; Rigato, 1994a: 60.
*taprobanae. Lophomyrmex quadrispinosus* var. *taprobanae* Forel, 1911i: 223 (w.q.m.) SRI LANKA. Raised to species: Ettershank, 1966: 134. Junior synonym of *quadrispinosus:* Rigato, 1994a: 59.

***LORDOMYRMA*** (Oriental, Indo-Australian, Australasian)
**#*accuminata.*** *Lordomyrma cryptocera* var. *accuminata* Stitz, 1912: 504 (w.) NEW GUINEA. See also: Wheeler, W.M. 1919f: 102.
***azumai.*** *Rogeria* (*Rogeria*) *azumai* Santschi, 1941: 275, fig. 3 (w.) JAPAN. Combination in *Lordomyrma,* and senior synonym of *nobilis:* Brown, 1952c: 124.
***bensoni.*** *Lordomyrma bensoni* Donisthorpe, 1949d: 94, figs. 1, 2 (w.) NEW GUINEA.
***caledonica.*** *Podomyrma caledonica* André, 1889: 225 (w.) NEW CALEDONIA. Combination in *Lordomyrma:* Emery, 1897c: 591. See also: Wheeler, W.M. 1919f: 99.
***crawleyi.*** *Lordomyrma crawleyi* Menozzi, 1923b: 209, fig. 1 (w.) NEW GUINEA.
***cryptocera.*** *Lordomyrma cryptocera* Emery, 1897c: 592, pl. 15, fig. 34 (w.) NEW GUINEA. Current subspecies: nominal plus *accuminata.* See also: Wheeler, W.M. 1919f: 101.
***epinotalis.*** *Rogeria epinotalis* Mann, 1919: 343 (w.q.) SOLOMON IS. Combination in *Lordomyrma:* Kugler, C. **new combination**.
***furcifera.*** *Lordomyrma furcifera* Emery, 1897c: 591, pl. 15, figs. 32, 33 (w.) NEW GUINEA. See also: Wheeler, W.M. 1919f: 100.
***infundibuli.*** *Lordomyrma infundibuli* Donisthorpe, 1940b: 45, figs. (w.q.) NEW GUINEA.
***leae.*** *Lordomyrma leae* Wheeler, W.M. 1919f: 102, fig. 3 (w.m.) AUSTRALIA.
**#*levifrons.*** *Rogeria* (*Irogera*) *tortuosa* subsp. *levifrons* Mann, 1921: 453 (w.) FIJI IS. Combination in *Lordomyrma:* Kugler, C. **new combination**.
***nigra.*** *Lordomyrma niger* Donisthorpe, 1941a: 36 (w.) NEW GUINEA.
*nobilis. Lordomyrma nobilis* Yasumatsu, 1950: 75, fig. 2 (w.m.) JAPAN. Junior synonym of *azumai:* Brown, 1952c: 124.
**#*polita.*** *Rogeria* (*Irogera*) *tortuosa* subsp. *polita* Mann, 1921: 453 (w.) FIJI IS. Combination in *Lordomyrma:* Kugler, C. **new combination**.
***punctiventris.*** *Lordomyrma punctiventris* Wheeler, W.M. 1919f: 105, fig. 4 (w.) AUSTRALIA.
***rouxi.*** *Promeranoplus rouxi* Emery, 1914f: 413, pl. 13, fig. 8 (w.) NEW CALEDONIA. Combination in *Lordomyrma:* **new combination**.
***rugosa.*** *Rogeria* (*Irogera*) *rugosa* Mann, 1921: 455, fig. 20 (w.) FIJI IS. Combination in *Lordomyrma:* Kugler, C. **new combination**.
*rugosa* Clark, 1934: see under ***PODOMYRMA.***
***sarasini.*** *Prodicroaspis sarasini* Emery, 1914f: 414, pl. 13, fig. 9 (w.) NEW CALEDONIA. Combination in *Lordomyrma:* **new combination**.
**#*stoneri.*** *Rogeria* (*Irogera*) *tortuosa* subsp. *stoneri* Mann, 1925b: 5 (w.) FIJI IS. Combination in *Lordomyrma:* Kugler, C. **new combination**.
***striatella.*** *Rogeria* (*Irogera*) *striatella* Mann, 1921: 454, fig. 19 (w.m.) FIJI IS. Combination in *Lordomyrma:* Kugler, C. **new combination**.
***tortuosa.*** *Rogeria* (*Irogera*) *tortuosa* Mann, 1921: 452, fig. 18 (w.) FIJI IS. Combination in *Lordomyrma:* Kugler, C. **new combination**. Current subspecies: nominal plus *levifrons, polita, stoneri.*

***LOWERIELLA*** (Borneo)
***boltoni.*** *Loweriella boltoni* Shattuck, 1992c: 129, figs. 118–120 (w.) BORNEO.

*LUNDELLA:* see under ***HYLOMYRMA.***

*MACHAEROGENYS:* see under ***LEPTOGENYS.***

*MACHAEROMYRMA:* see under ***CATAGLYPHIS.***

***MACHOMYRMA*** (Australia)
***dispar.*** *Liomyrmex* (*Machomyrma*) *dispar* Forel, 1895f: 425 (s.w.q.) AUSTRALIA. Forel, 1913k: 27 (footnote) (q.). Combination in *Machomyrma:* Forel, 1913k: 27.
*silvestrii* Emery, 1914; see under ***ADLERZIA.***

*MACROMISCHA:* see under ***LEPTOTHORAX.***

*MACROMISCHOIDES* : see under ***TETRAMORIUM.***

*MACROPHEIDOLE:* see under ***PHEIDOLE.***

**#*MALACOMYRMA:*** see under ***ACROPYGA.***

***MANICA*** (Holarctic)

*aldrichi*. *Myrmica* (*Oreomyrma*) *aldrichi* Wheeler, W.M. 1914d: 120, fig. 1 (w.) U.S.A. Combination in *Myrmica* (*Neomyrma*): Wheeler, W.M. 1917a: 507; in *M.* (*Manica*): Emery, 1921f: 42; in *Manica:* Weber, 1947: 440; Creighton, 1950a: 108. Junior synonym of *hunteri:* Cole, 1956h: 262.

***bradleyi.*** *Myrmica bradleyi* Wheeler, W.M. 1909e: 77 (w.) U.S.A. Cole, 1957c: 210 (q.m.); Wheeler, G.C. & Wheeler, J. 1960b: 6 (l.). Combination in *Myrmica* (*Oreomyrma*): Wheeler, W.M. 1914d: 120; in *M.* (*Neomyrma*): Emery, 1915d: 69 (footnote); in *M.* (*Manica*): Emery, 1921f: 43; in *Manica:* Weber, 1947: 440; Creighton, 1950a: 108. Senior synonym of *calderoni:* Wheeler, W.M. 1915a: 50.

*calderoni*. *Aphaenogaster* (*Neomyrma*) *calderoni* Forel, 1914a: 275 (w.) U.S.A. Combination in *Myrmica* (*Neomyrma*): Emery, 1915d: 69 (footnote). Junior synonym of *bradleyi:* Wheeler, W.M. 1915a: 50.

***hunteri.*** *Myrmica* (*Oreomyrma*) *hunteri* Wheeler, W.M. 1914d: 121, fig. 1 (w.) U.S.A. Cole, 1957c: 212 (m.); Wheeler, G.C. & Wheeler, J. 1972b: 235 (l.). Combination in *Myrmica* (*Neomyrma*): Wheeler, W.M. 1917a: 507; in *M.* (*Manica*): Emery, 1921f: 43; in *Manica:* Weber, 1947: 440; Creighton, 1950a: 109. Senior synonym of *aldrichi:* Cole, 1956h: 262.

***invidia.*** *Manica invidia* **new name**. Replacement name for *Myrmica mutica* Emery, 1895c: 311. [Junior primary homonym of *Myrmica mutica:* Nylander, 1849: 39 (now in *Messor*).] For previous history of name see under *mutica,* below.

*leonina*. *Myrmica leonina* Losana, 1834: 332, pl. 36, fig. 7 (w.) ITALY. Junior synonym of *rubida:* Roger, 1859: 252.

*montana*. *Myrmica montana* Labram & Imhoff, 1838, pl. 36, figs. 1–3 (w.q.m.) SWITZERLAND. Mayr, 1853b: 110 (w.q.m.). Junior synonym of *rubida:* Mayr, 1855: 400.

*mutica*. *Myrmica mutica* Emery, 1895c: 311 (w.) U.S.A. [Junior primary homonym of *Myrmica mutica* Nylander, 1849: 39 (now in *Messor*).] Replacement name: *Manica invidia:* **new name**. Wheeler, W.M. 1914d: 120 (q.m.); Wheeler, G.C. & Wheeler, J. 1960b: 5 (l.). Combination in *Myrmica* (*Oreomyrma*): Wheeler, W.M. 1914d: 119; in *M.* (*Neomyrma*): Emery, 1915d: 69 (footnote); in *M.* (*Manica*): Emery, 1921f: 43; in *Manica:* Wheeler, G.C. & Wheeler, E.W. 1944: 244. See also: Creighton, 1950a: 109; Wheeler, G.C. & Wheeler, J. 1972b: 235.

***parasitica.*** *Myrmica* (*Manica*) *parasitica* Creighton, 1934: 185 (w.) U.S.A. Combination in *Manica:* Weber, 1947: 440.

*rhynchophora*. *Myrmica rhynchophora* Foerster, 1850b: 488 (m.) ALGERIA. Junior synonym of *rubida:* Mayr, 1861: 63.

***rubida.*** *Formica rubida* Latreille, 1802c: 267, pl. 10, fig. 65 (q.) FRANCE. Mayr, 1855: 400 (w.m.); Hauschteck, 1965: 325 (k.); Wheeler, G.C. & Wheeler, J. 1972b: 236 (l.). Combination in *Manica:* Jurine, 1807: 279; in *Myrmica* (*Oreomyrma*): Wheeler, W.M. 1914d: 118; in *Myrmica* (*Neomyrma*): Emery, 1915d: 69 (footnote); in *Neomyrma:* Bondroit, 1918: 97; in *Myrmica* (*Manica*): Emery, 1921f: 43; in *Manica:* Weber, 1947: 440. Senior synonym of *montana:* Mayr, 1855: 400; of *leonina:* Roger, 1859: 252; of *rhynchophora:* Mayr, 1861: 63. See also: Ruzsky, 1902d: 29; Stitz, 1914: 70; Emery, 1916b: 121; Finzi, 1926: 81; Stitz, 1939: 68; Bernard, 1967: 125; Kutter, 1977c: 40; Arnol'di & Dlussky, 1978: 535 (in key); Atanassov & Dlussky, 1992: 77.

***yessensis.*** *Manica yessensis* Azuma, 1955: 2 (footnote) (w.) JAPAN.

***#MANNIELLA:*** see under ***CAMPONOTUS.***

*MARTIA:* see under ***OXYEPOECUS.***

***#MAYRIA:*** see under ***CAMPONOTUS.***

***MAYRIELLA.*** (Oriental, Indo-Australian, Australasian)

***abstinens.*** *Mayriella abstinens* Forel, 1902h: 452 (w.) AUSTRALIA. Imai, Crozier & Taylor, 1977: 349 (k.). Senior synonym of *overbecki:* Wheeler, W.M. 1935e: 151 (footnote). See also: Wheeler, W.M. 1935e: 154.

***hackeri.*** *Mayriella abstinens* subsp. *hackeri* Wheeler, W.M. 1935e: 157, fig. 1 (w.q.) AUSTRALIA. Wheeler, G.C. & Wheeler, J. 1973a: 36 (l.). Raised to species: Taylor, 1991b: 602.

*overbecki*. *Mayriella overbecki* Viehmeyer, 1925a: 26 (w.q.) AUSTRALIA. Junior synonym of *abstinens:* Wheeler, W.M. 1935e: 151 (footnote).

***spinosior.*** *Mayriella spinosior* Wheeler, W.M. 1935e: 159, fig. 1 (w.) AUSTRALIA.

***transfuga.*** *Mayriella transfuga* Baroni Urbani, 1977a: 411, figs. 1, 2 (w.) NEPAL.

***venustula.*** *Mayriella abstinens* subsp. *venustula* Wheeler, W.M. 1935e: 158 (w.) AUSTRALIA. Raised to species: Taylor, 1991b: 602.

*MAYROMYRMEX:* see under ***ECITON.***

***MEGALOMYRMEX*** (Neotropical)

***acauna.*** *Megalomyrmex acauna* Brandão, 1990: 429, figs. 39, 56 (w.) BRAZIL.

***ayri.*** *Megalomyrmex ayri* Brandão, 1990: 420, figs. 1, 12, 26, 36 (w.q.) BRAZIL.

***balzani.*** *Megalomyrmex balzani* Emery, 1894c: 153 (w.) BOLIVIA. Viehmeyer, 1922: 209 (m.). Senior synonym of *duckei:* Brandão, 1990: 431. See also: Wheeler, W.M. 1925a: 29.

*bicolor* Ettershank, 1965; see under ***NOTHIDRIS.***

***bituberculatus.*** *Formica bituberculata* Fabricius, 1798: 280 (w.) SURINAM. Combination in *Megalomyrmex:* Emery, 1890b: 47. See also: Brandão, 1990: 413 (*species inquirenda*).

*brasiliensis*. *Megalomyrmex* (*Wheelerimyrmex*) *brasiliensis* Borgmeier, 1930: 34, pl. 5, figs. 33, 35 (w.) BRAZIL. Junior synonym of *silvestrii:* Kempf & Brown, 1968: 97.

***caete.*** *Megalomyrmex caete* Brandão, 1990: 422, figs. 2, 9, 27 (w.q.) ECUADOR.

***cuatiara.*** *Megalomyrmex cuatiara* Brandão, 1990: 440, figs. 93, 94, 103 (w.q.) COLOMBIA.

***cupecuara.*** *Megalomyrmex cupecuara* Brandão, 1990: 423, figs. 3, 11, 21–23 (w.m.) COLOMBIA.

***cyendyra.*** *Megalomyrmex cyendyra* Brandão, 1990: 433, figs. 44, 54 (w.) COLOMBIA.

***drifti.*** *Megalomyrmex drifti* Kempf, 1961b: 504, figs. 9–11 (w.q.) SURINAM. Brandão, 1990: 448 (m.). See also: Kempf, 1970b: 362.

*duckei*. *Megalomyrmex duckei* Forel, 1912g: 13 (w.) BRAZIL. Junior synonym of *balzani:* Brandão, 1990: 431.

***emeryi.*** *Megalomyrmex emeryi* Forel, 1904d: 174 (w.) SURINAM. Brandão, 1990: 434 (m.).

***foreli.*** *Megalomyrmex foreli* Emery, 1890b: 46, pl. 5, fig. 3 (w.) COSTA RICA. [Also described as new by Emery, 1894k: 53.] Forel, 1899c: 58 (m.). See also: Brandão, 1990: 434.

***glaesarius.*** *Megalomyrmex glaesarius* Kempf, 1970b: 355, figs. 1, 2 (w.) PERU. See also: Brandão, 1990: 435.

***gnomus.*** *Megalomyrmex gnomus* Kempf, 1970b: 359, figs. 5–8 (w.) BRAZIL. See also: Brandão, 1990: 451.

***goeldii.*** *Megalomyrmex goeldii* Forel, 1912g: 14 (w.) BRAZIL. Wheeler, W.M. 1925a: 29 (m.); Borgmeier, 1930: 34 (q.). See also: Brandão, 1990: 423.

*humilis* Borgmeier, 1930; see under ***PHEIDOLE.***

***iheringi.*** *Megalomyrmex iheringi* Forel, 1911c: 304 (w.) BRAZIL. Borgmeier, 1930: 33 (q.). See also: Brandão, 1990: 424.

***incisus.*** *Megalomyrmex* (*Wheelerimyrmex*) *incisus* Smith, M.R. 1947e: 102 (w.) PANAMA. Brandão, 1990: 451 (m.).

*langi*. *Megalomyrmex* (*Wheelerimyrmex*) *sjostedti* var. *langi* Wheeler, W.M. 1925a: 31 (w.m.) GUIANA. Junior synonym of *silvestrii:* Kempf & Brown, 1968: 97.

***latreillei.*** *Megalomyrmex latreillei* Emery, 1890b: 47 (footnote), pl. 5, fig. 4 (w.) (diagnosis in key) PERU. Brandão, 1990: 436 (m.).

***leoninus.*** *Megalomyrmex leoninus* Forel, 1885a: 372 (w.) COLOMBIA. See also: Brandão, 1990: 436.

***miri.*** *Megalomyrmex miri* Brandão, 1990: 452, figs. 117, 118 (w.) COSTA RICA.

*misionensis*. *Megalomyrmex* (*Wheelerimyrmex*) *brasiliensis* var. *misionensis* Santschi, 1936d: 405 (w.) ARGENTINA. Junior synonym of *silvestrii:* Kempf & Brown, 1968: 97.

***modestus.*** *Megalomyrmex modestus* Emery, 1896g: 94 (w.) COSTA RICA. Brandão, 1990: 426 (q.m.).

***mondabora.*** *Megalomyrmex mondabora* Brandão, 1990: 442, fig. 95, 96 (w.) COSTA RICA.

***myops.*** *Megalomyrmex* (*Wheelerimyrmex*) *myops* Santschi, 1925d: 237 (w.) BRAZIL. See also: Brandão, 1990: 452.

*nasutus*. *Megalomyrmex leoninus* r. *nasutus* Forel, 1912g: 13 (w.) BRAZIL. Junior synonym of *staudingeri:* Brandão, 1990: 438.

***pacova.*** *Megalomyrmex pacova* Brandão, 1990: 437, figs. 46, 55 (w.) U.S.A. [Intercepted in quarantine, from Neotropics.]

***piriana.*** *Megalomyrmex piriana* Brandão, 1990: 443, figs. 97, 98 (w.) ECUADOR.

***poatan.*** *Megalomyrmex poatan* Brandão, 1990: 443, figs. 99, 100 (w.) COLOMBIA.

***pusillus.*** *Megalomyrmex pusillus* Forel, 1912g: 15 (w.) BRAZIL. Brandão, 1990: 453 (q.).

***silvestrii.*** *Megalomyrmex silvestrii* Wheeler, W.M. 1909b: 235 (w.) MEXICO. Combination in *M.* (*Wheelerimyrmex*): Mann, 1922: 29. Senior synonym of *brasiliensis, langi, misionensis, sjostedti* and *wheeleri:* Kempf & Brown, 1968: 97. See also: Brandão, 1990: 443.

***MEGALOMYRMEX** (continued)*

*sjostedti. Megalomyrmex* (*Wheelerimyrmex*) *sjostedti* Wheeler, W.M. 1925a: 30 (w.q.) PERU. Junior synonym of *silvestrii:* Kempf & Brown, 1968: 97.

***staudingeri.*** *Megalomyrmex staudingeri* Emery, 1890b: 47 (footnote; diagnosis in key) (w.) PERU. Senior synonym of *nasutus:* Brandão, 1990: 438.

***symmetochus.*** *Megalomyrmex* (*Cepobroticus*) *symmetochus* Wheeler, W.M. 1925d: 168, fig. 5 (w.q.m.) PANAMA. Wheeler, G.C. & Wheeler, J. 1955c: 126 (l.). See also: Brandão, 1990: 445.

***tasyba.*** *Megalomyrmex tasyba* Brandão, 1990: 446, figs. 91, 92 (w.) ECUADOR.

***timbira.*** *Megalomyrmex timbira* Brandão, 1990: 439, figs. 45, 49 (w.) VENEZUELA.

***wallacei.*** *Megalomyrmex wallacei* Mann, 1916: 445, pl. 3, fig. 28 (w.q.) BRAZIL. See also: Brandão, 1990: 426.

***weyrauchi.*** *Megalomyrmex weyrauchi* Kempf, 1970b: 357, figs. 3, 4 (w.) PERU. See also: Brandão, 1990: 427.

*wheeleri. Megalomyrmex* (*Cepobroticus*) *wheeleri* Weber, 1940a: 425, figs. 15, 16 (w.q.) PANAMA. Junior synonym of *silvestrii:* Kempf & Brown, 1968: 97.

*MEGALOPONERA:* see under ***PACHYCONDYLA.***

*MEGAPONERA:* see under ***PACHYCONDYLA.***

**MEGAPTERITES:* see under ****FORMICIUM.***

*MEIA:* see under ***LEPTOTHORAX.***

***MELISSOTARSUS*** (Afrotropical, Malagasy)

***beccarii.*** *Melissotarsus beccarii* Emery, 1877b: 379, fig. (w.) ETHIOPIA. Forel, 1914d: 243 (q.m.). See also: Arnold, 1916: 189; Bolton, 1982: 336.

*compressus. Melissotarsus compressus* Weber, 1952: 1, figs. 28, 29 (q.) CENTRAL AFRICAN REPUBLIC. Junior synonym of *emeryi:* Bolton, 1982: 337.

***emeryi.*** *Melissotarsus emeryi* Forel, 1907c: 133 (w.) ETHIOPIA. Bolton, 1982: 335 (m.). Senior synonym of *compressus* and *pilipes:* Bolton, 1982: 337.

***insularis.*** *Melissotarsus insularis* Santschi, 1911e: 122, fig. 2 (w.m.) MADAGASCAR.

*major. Melissotarsus major* Santschi, 1919h: 85 (w.) ZAIRE. Santschi, 1923e: 277 (q.m.). Junior synonym of *weissi:* Bolton, 1982: 337.

*pilipes. Melissotarsus emeryi* var. *pilipes* Santschi, 1914b: 71 (w.) KENYA. Junior synonym of *emeryi:* Bolton, 1982: 337.

*titubans. Melissotarsus titubans* Delage-Darchen, 1972a: 216, figs. 1–10 (w.q.m.) IVORY COAST. Wheeler, G.C. & Wheeler, J. 1980: 531 (l.). Junior synonym of *beccarii:* Bolton, 1982: 336; of *weissi:* **new synonymy** (syntypes examined).

***weissi.*** *Melissotarsus weissi* Santschi, 1910c: 356, fig. 3 (q.) CONGO. Bolton, 1982: 335 (m.). Senior synonym of *major:* Bolton, 1982: 337; of *titubans:* **new synonymy**.

***MELOPHORUS*** (Australia)

***aeneovirens.*** *Formica aeneovirens* Lowne, 1865a: 276 (s.w.) AUSTRALIA. Combination in *Cataglyphis:* Mayr, 1876: 78; in *Melophorus:* Forel, 1902h: 488.

#***aesopus.*** *Melophorus turneri* subsp. *aesopus* Forel, 1910b: 64 (s.w.q.m.) AUSTRALIA.

***bagoti.*** *Melophorus bagoti* Lubbock, 1883: 52, pl. 2, figs. 1–10 (w.) AUSTRALIA. Forel, 1886f: 213 (s.); Forel, 1910b: 60 (q.); Wheeler, G.C. & Wheeler, J. 1953c: 129 (l.).

*bidens* Emery, 1895; see under ***LASIOPHANES.***

***biroi.*** *Melophorus biroi* Forel, 1907a: 29 (s.) AUSTRALIA. Combination in *M.* (*Erimelophorus*): Wheeler, W.M. 1935c: 71.

*breviscapa. Melophorus ludius* r. *sulla* var. *breviscapa* Forel, 1915b: 89 (w.) AUSTRALIA. Unavailable name (Taylor, 1986: 34).

***bruneus.*** *Melophorus* (*Melophorus*) *brunea* McAreavey, 1949: 20, figs. 57–65 (s.w.q.m.) AUSTRALIA.

#***candidus.*** *Melophorus turneri* st. *candida* Santschi, 1919a: 328, fig. 2 (w.) AUSTRALIA.

***constans.*** *Melophorus constans* Santschi, 1928e: 475 (w.q.) AUSTRALIA.

***curtus.*** *Melophorus curtus* Forel, 1902h: 485 (s.w.q.) AUSTRALIA.

*depressiceps* Emery, 1914; see under ***PROLASIUS.***

***fieldi.*** *Melophorus fieldi* Forel, 1910b: 62 (s.) AUSTRALIA. Combination in *M.* (*Erimelophorus*): Wheeler, W.M. 1935c: 71. Current subspecies: nominal plus *major, propinquus.*

*formicoides* Forel, 1902; see under ***PROLASIUS.***

#***fraudatrix.*** *Melophorus iridescens* var. *fraudatrix* Forel, 1915b: 87 (footnote) (w.) AUSTRALIA. Subspecies of *iridescens:* Taylor & Brown, D.R. 1985: 123.

#***froggatti.*** *Melophorus iridescens* var. *froggatti* Forel, 1902h: 487 (s.w.q.) AUSTRALIA. Subspecies of *iridescens:* Forel, 1907a: 29.

***fulvohirtus.*** *Melophorus fulvohirtus* Clark, 1941: 88, pl. 13, figs. 19–21 (s.w.) AUSTRALIA.

***hirsutus.*** *Melophorus hirsutus* Forel, 1902h: 488 (w.) AUSTRALIA. Combination in *M.* (*Trichomelophorus*): Wheeler, W.M. 1935c: 71.

*hoffmanni* Forel, 1903; see under ***LASIOPHANES.***

***insularis.*** *Melophorus insularis* Wheeler, W.M. 1934d: 151 (s.w.) AUSTRALIA.

***iridescens.*** *Myrmecocystus iridescens* Emery, 1887a: 247 (w.) AUSTRALIA. Combination in *Melophorus:* Forel, 1902h: 487. Current subspecies: nominal plus *fraudatrix, froggatti.*

***laticeps.*** *Melophorus laticeps* Wheeler, W.M. 1915g: 813, pl. 66, fig. 2 (q.) AUSTRALIA. Combination in *M.* (*Erimelophorus*): Wheeler, W.M. 1935c: 71.

***ludius.*** *Melophorus ludius* Forel, 1902h: 484 (s.w.) AUSTRALIA. Current subspecies: nominal plus *sulla.*

#***major.*** *Melophorus fieldi* var. *major* Forel, 1915b: 87 (s.w.) AUSTRALIA. Subspecies of *fieldi:* Taylor & Brown, D.R. 1985: 123.

***marius.*** *Melophorus marius* Forel, 1910b: 66 (s.) AUSTRALIA. Combination in *M.* (*Erimelophorus*): Wheeler, W.M. 1935c: 71.

***mjobergi.*** *Melophorus mjobergi* Forel, 1915b: 88 (w.) AUSTRALIA. Combination in *M.* (*Erimelophorus*): Wheeler, W.M. 1935c: 71.

***omniparens.*** *Melophorus omniparens* Forel, 1915b: 85 (w.) AUSTRALIA. Combination in *M.* (*Erimelophorus*): Wheeler, W.M. 1935c: 71.

*perplexus* Santschi, 1920; see under ***LASIOPHANES.***

#***perthensis.*** *Melophorus turneri* subsp. *perthensis* Wheeler, W.M. 1934d: 152 (s.w.) AUSTRALIA.

***pillipes.*** *Melophorus pillipes* Santschi, 1919a: 329, fig. 2 (w.) AUSTRALIA. Combination in *M.* (*Erimelophorus*): Wheeler, W.M. 1935c: 71.

*pilosulus* Emery, 1922; see under ***LASIOPHANES.***

***potteri.*** *Melophorus potteri* McAreavey, 1947: 25, fig. 1 (s.w.q.) AUSTRALIA.

#***propinquus.*** *Melophorus fieldi* var. *propinqua* Viehmeyer, 1925a: 36 (w.) AUSTRALIA. Subspecies of *fieldi:* Taylor & Brown, D.R. 1985: 123.

*sauberi* Forel, 1903; see under ***LASIOPHANES.***

***scipio.*** *Melophorus scipio* Forel, 1915b: 86 (w.) AUSTRALIA. Combination in *Prolasius:* Wheeler, W.M. 1935c: 71; in *Melophorus:* Taylor & Brown, D.R. 1985: 124.

*spinisquamis* André, 1896; see under ***NOTONCUS.***

#***sulla.*** *Melophorus ludius* subsp. *sulla* Forel, 1910b: 66 (s.q.m.) AUSTRALIA.

***turneri.*** *Melophorus turneri* Forel, 1910b: 63 (s.w.) AUSTRALIA. Viehmeyer, 1914c: 43 (q.); Wheeler, G.C. & Wheeler, J. 1968: 206 (l.). Combination in *M.* (*Erimelophorus*): Wheeler, W.M. 1935c: 71. Current subspecies: nominal plus *aesopus, candidus, perthensis.*

*uxorius* Emery, 1922; see under ***LASIOPHANES.***

*valdiviensis* Forel, 1904; see under ***LASIOPHANES.***

***wheeleri.*** *Melophorus wheeleri* Forel, 1910b: 60 (s.w.m.) AUSTRALIA. Viehmeyer, 1914c: 43 (q.). Combination in *M.* (*Erimelophorus*): Wheeler, W.M. 1935c: 71. See also: Forel, 1915b: 88.

***MERANOPLUS*** (Old World tropics).

*affinis. Meranoplus nanus* subsp. *affinis* Baroni Urbani, 1971b: 361. Replacement name for *similis* Karavaiev, 1931d: 44. [Junior primary homonym of *similis* Viehmeyer, 1922: 208.] Junior synonym of *inermis:* Bolton, 1981a: 52.

***ajax.*** *Meranoplus mars* r. *ajax* Forel, 1915b: 44 (w.) AUSTRALIA. Raised to species: Taylor, 1990c: 34.

***armatus.*** *Meranoplus armatus* Smith, F. 1862d: 413, pl. 12, fig. 7 (w.) SUMATRA. Senior synonym of *rugifrons:* Taylor, 1990c: 35.

***astericus.*** *Meranoplus astericus* Donisthorpe, 1947d: 190 (w.) NEW GUINEA.

*attenuatus* Smith, F. 1876; see under ***PROCRYPTOCERUS.***

***aureolus.*** *Meranoplus aureolus* Crawley, 1921: 91 (w.) AUSTRALIA.

***barretti.*** *Meranoplus barretti* Santschi, 1928e: 468 (w.) AUSTRALIA.

***bellii.*** *Meranoplus bellii* Forel, 1902c: 240 (w.) INDIA. Current subspecies: nominal plus *javanus.*

***bicolor.*** *Cryptocerus bicolor* Guérin-Méneville, 1844a: 425 (w.) INDIA. Smith, F. 1875: 34 (q.m.); Imai, Baroni Urbani, *et al.* 1984: 6 (k.). Combination in *Meranoplus:* Smith, F. 1853: 224. Senior synonym of *dimicans:* Bingham, 1903: 168; Donisthorpe, 1932b: 576; of *tarda:* Emery, 1892b: 166; of *villosus:* Roger, 1863b: 39. Current subspecies: nominal plus *fuscescens, lucidus.*

*bimaculatus. Meranoplus hirsutus* st. *minor* var. *bimaculatus* Santschi, 1928e: 470 (w.) AUSTRALIA. Unavailable name (Taylor, 1986: 34).

*bondroiti. Meranoplus bondroiti* Santschi, 1915c: 254, fig. 4 (w.) ZAIRE. Junior synonym of *magrettii:* Bolton, 1981a: 55.

***carinatus.*** *Meranoplus carinatus* Donisthorpe, 1942d: 456 (q.) INDIA.

***castaneus.*** *Meranoplus castaneus* Smith, F. 1857a: 81, pl. 2, fig. 7 (q.) BORNEO. Forel, 1910a: 29 (w.). Senior synonym of *cordatus:* Forel, 1912n: 61. Current subspecies: nominal plus *hammaceros.*

*christmasensis. Meranoplus pubescens* r. *fenestratus* var. *christmasensis* Forel, 1915b: 48 (w.) AUSTRALIA. Unavailable name (Taylor, 1986: 34).

***clypeatus.*** *Meranoplus clypeatus* Bernard, 1953b: 244, fig. 13 (w.q.) GUINEA. See also: Bolton, 1981a: 52.

*cordatus. Meranoplus cordatus* Smith, F. 1857a: 82, pl. 2, fig. 5 (w.) BORNEO. Junior synonym of *castaneus:* Forel, 1912n: 61.

*crawleyi. Meranoplus crawleyi* Viehmeyer, 1925a: 27. Second (unnecessary) replacement name for *minor* Crawley, 1918: 89. [Junior primary homonym of *minor* Forel, 1902h: 457.] Junior synonym of *minimus:* **new synonymy**.

*curvispina. Meranoplus puryi* var. *curvispina* Forel, 1910b: 47 (w.) AUSTRALIA. Junior synonym of *puryi:* Taylor, 1990c: 39.

***dichrous.*** *Meranoplus dichrous* Forel, 1907h: 274 (q.) AUSTRALIA. *Species inquirenda:* Taylor, 1990c: 33.

*difformis* Smith, F. 1858; see under ***CYPHOMYRMEX.***

*dimicans. Meranoplus dimicans* Walker, 1859: 375 (w.) SRI LANKA. Junior synonym of *bicolor:* Bingham, 1903: 168; Donisthorpe, 1932b: 576.

***dimidiatus.*** *Meranoplus dimidiatus* Smith, F. 1867: 527, pl. 26, fig. 8 (w.) AUSTRALIA. Wheeler, G.C. & Wheeler, J. 1973a: 37 (l.).

*diversipilosus. Meranoplus simoni* var. *diversipilosus* Santschi, 1932a: 388 (w.) ZIMBABWE. Junior synonym of *magrettii:* Bolton, 1981a: 55.

***diversus.*** *Meranoplus diversus* Smith, F. 1867: 527, pl. 26, fig. 2 (w.) AUSTRALIA. Forel, 1915b: 44 (m.).

***doddi.*** *Meranoplus aureolus* st. *doddi* Santschi, 1928e: 469 (q.) AUSTRALIA (name misspelled as *aerolus* in this publication). Raised to species, but *species inquirenda:* Taylor, 1990c: 38.

***duyfkeni.*** *Meranoplus diversus* r. *duyfkeni* Forel, 1915b: 45 (w.q.) AUSTRALIA. Raised to species: Taylor, 1990c: 34.

***excavatus.*** *Meranoplus excavatus* Clark, 1938: 367, fig. 6 (w.) AUSTRALIA.

*excisus. Meranoplus excisus* Arnold, 1914: 29, fig. 9 (w.) SOUTH AFRICA. Junior synonym of *peringueyi:* Bolton, 1981a: 56.

***fenestratus.*** *Meranoplus fenestratus* Smith, F. 1867: 526, pl. 26, fig. 6 (w.) AUSTRALIA. Junior synonym of *pubescens:* Dalla Torre, 1893: 137. Revived from synonymy as variety/race of *pubescens:* Forel, 1907a: 12; Forel, 1910b: 47; Forel, 1915b: 48; Emery, 1922f: 229. Revived status as species: Taylor & Brown, D.R. 1985: 67.

***ferrugineus.*** *Meranoplus ferrugineus* Crawley, 1922b: 444, fig. 8 (w.) AUSTRALIA.

***flaviventris.*** *Meranoplus flaviventris* Donisthorpe, 1943b: 202 (w.) INDIA.

***froggatti.*** *Meranoplus froggatti* Forel, 1913g: 183 (w.) AUSTRALIA.

#***fuscescens.*** *Meranoplus bicolor* var. *fuscescens* Wheeler, W.M. 1930a: 101 (w.) TAIWAN.

***glaber.*** *Meranoplus spininodis* var. *glaber* Arnold, 1926: 285, fig. 82 (w.) ZIMBABWE. Raised to species: Bolton, 1981a: 49.

*gracilis* Smith, F. 1858; see under ***PROCRYPTOCERUS.***

#***hammaceros.*** *Meranoplus castaneus* subsp. *hammaceros* Forel, 1912n: 60 (w.) SUMATRA.

***hilli.*** *Meranoplus hilli* Crawley, 1922b: 445, fig. 9 (w.) AUSTRALIA.

***hirsutus.*** *Meranoplus hirsutus* Mayr, 1876: 112 (w.) AUSTRALIA.

***hospes.*** *Meranoplus hospes* Forel, 1910b: 48 (q.) AUSTRALIA. *Species inquirenda:* Taylor, 1990c: 33.

***inermis.*** *Meranoplus inermis* Emery, 1895h: 41, pl. 2, fig. 24 (w.) SOUTH AFRICA. Subspecies of *nanus:* Wheeler, W.M. 1922a: 888. Revived status as species, and senior synonym of *affinis, kiboshana, nanior* and *soriculus:* Bolton, 1981a: 52.

*intrudens* Smith, F. 1876; see under ***CATAULACUS.***

#***javanus.*** *Meranoplus bellii* subsp. *javanus* Karavaiev, 1935a: 98, fig. 22 (w.) JAVA.

*kiboshana. Meranoplus nanus* var. *kiboshana* Forel, 1907a: 12 (w.) KENYA. Junior synonym of *inermis:* Bolton, 1981a: 52.

***laeviventris.*** *Meranoplus laeviventris* Emery, 1889b: 506, pl. 10, fig. 16 (w.) BURMA. Current subspecies: nominal plus *punctulatus.*

***leveillei.*** *Meranoplus leveillei* Emery, 1883: 151, fig. (w.) NEW CALEDONIA. André, 1889: 231 (m.).

***levis.*** *Meranoplus levis* Donisthorpe, 1942d: 455 (w.) INDIA.

***linae.*** *Meranoplus aureolus* st. *linae* Santschi, 1928e: 469 (w.) AUSTRALIA (name misspelled as *aerolus* in publication). Raised to species: Taylor, 1990c: 38.

#***lucidus.*** *Meranoplus bicolor* var. *lucida* Forel, 1903a: 706 (w.) BURMA. Forel, 1909d: 224 (m.).

***magrettii.*** *Meranoplus magrettii* André, 1884a: 245 (w.) SUDAN. [Also described as new by André, 1884b: 543.] Menozzi & Consani, 1952: 66 (q.). Senior synonym of *bondroiti, diversipilosus, nitidiventris, simoni, springvalensis* and *suturalis:* Bolton, 1981a: 55.

***mars.*** *Meranoplus mars* Forel, 1902h: 454 (w.) AUSTRALIA.

***mayri.*** *Meranoplus mayri* Forel, 1910a: 19 (w.) MADAGASCAR.

***minimus.*** *Meranoplus minimus* Crawley, 1922b: 445. Replacement name for *minor* Crawley, 1918: 89. [Junior primary homonym of *minor* Forel, 1902h: 457.] Senior synonym of *crawleyi* (unnecessary second replacement name): **new synonymy**.

***minor.*** *Meranoplus hirsutus* r. *minor* Forel, 1902h: 457 (w.q.) AUSTRALIA. Imai, Crozier & Taylor, 1977: 349 (k.). Raised to species: Taylor, 1990c: 38.

*minor. Meranoplus minor* Crawley, 1918: 89 (w.) AUSTRALIA. [Junior primary homonym of *minor* Forel, above.] Replacement name: *minimus:* Crawley, 1922b: 445.

***mjobergi.*** *Meranoplus mjobergi* Forel, 1915b: 46 (w.) AUSTRALIA.

***mucronatus.*** *Meranoplus mucronatus* Smith, F. 1857a: 82, pl. 2, fig. 6 (w.) WEST MALAYSIA. Viehmeyer, 1916a: 129 (m.).

*nanior. Meranoplus nanus* subsp. *nanior* Forel, 1907a: 12 (w.) KENYA. Viehmeyer, 1922: 209 (q.). Junior synonym of *inermis:* Bolton, 1981a: 52.

***nanus.*** *Meranoplus nanus* André, 1892a: 55 (w.) GABON. Santschi, 1914d: 351 (q.). See also: Bolton, 1981a: 53.

***niger.*** *Meranoplus niger* Donisthorpe, 1949b: 494 (q.m.) NEW GUINEA.

*nitidiventris. Meranoplus simoni* subsp. *nitidiventris* Mayr, 1901b: 26 (w.) SOUTH AFRICA. Arnold, 1917: 371 (q.m.). Junior synonym of *magrettii:* Bolton, 1981a: 55.

***oceanicus.*** *Meranoplus oceanicus* Smith, F. 1862d: 414, pl. 12, fig. 5 (w.) AUSTRALIA. Wheeler, G.C. & Wheeler, J. 1954d: 443 (l.).

***oxleyi.*** *Meranoplus diversus* r. *oxleyi* Forel, 1915b: 45 (w.) AUSTRALIA. Raised to species: Taylor, 1990c: 34.

***peringueyi.*** *Meranoplus peringueyi* Emery, 1886: 365, pl. 17, fig. 12 (w.) SOUTH AFRICA. Senior synonym of *excisus:* Bolton, 1981a: 56.

*petiolatus* Smith, F. 1853; see under ***PROCRYPTOCERUS.***

***pubescens.*** *Cryptocerus pubescens* Smith, F. 1853: 223, pl. 20, fig. 4 (q.) AUSTRALIA. Combination in *Meranoplus:* Smith, F. 1862d: 413.

*puncticeps* Smith, F. 1876; see under ***PROCRYPTOCERUS.***

#***punctulatus.*** *Meranoplus laeviventris* var. *punctulata* Emery, 1895k: 472 (w.) BURMA.

***puryi.*** *Meranoplus puryi* Forel, 1902h: 456 (w.) AUSTRALIA. Senior synonym of *curvispina:* Taylor, 1990c: 39.

***radamae.*** *Meranoplus radamae* Forel, 1891b: 148, pl. 4, fig. 10 (w.) MADAGASCAR.

***raripilis.*** *Meranoplus raripilis* Donisthorpe, 1938e: 501 (w.) NEW GUINEA.

***rothneyi.*** *Meranoplus rothneyi* Forel, 1902c: 241 (w.) INDIA.

*rugifrons. Meranoplus spinosus* subsp. *rugifrons* Emery, 1897d: 569, pl. 1, figs. 13–15 (w.) NEW GUINEA. Viehmeyer, 1914b: 525 (q.). Raised to species: Donisthorpe, 1941c: 136. Junior synonym of *armatus:* Taylor, 1990c: 35.

***rugosus.*** *Meranoplus hirsutus* r. *rugosa* Crawley, 1922b: 443 (w.) AUSTRALIA. Raised to species: Taylor, 1990c: 39.

***sabronensis.*** *Meranoplus sabronensis* Donisthorpe, 1941c: 136 (w.) NEW GUINEA.

***similis.*** *Meranoplus similis* Viehmeyer, 1922: 208 (w.) AUSTRALIA.

*similis. Meranoplus nanus* subsp. *similis* Karavaiev, 1931d: 44, fig. 3 (w.) KENYA. [Junior primary homonym of *similis* Viehmeyer, above.] Replacement name: *affinis* Baroni Urbani, 1971b: 361.

*simoni. Meranopus simoni* Emery, 1895h: 41, pl. 2, fig. 23 (w.) SOUTH AFRICA. Junior synonym of *magrettii:* Bolton, 1981a: 55.

*singularis* Smith, F. 1858; see under ***BASICEROS.***

*soriculus. Meranoplus nanus* subsp. *soriculus* Wheeler, W.M. 1922a: 184, fig. 45 (w.q.m.) ZAIRE. Junior synonym of *inermis:* Bolton, 1981a: 52.

***spininodis.*** *Meranoplus spininodis* Arnold, 1917: 366 (w.) ZIMBABWE. Arnold, 1926: 284 (q.m.). See also: Bolton, 1981a: 50.

*spinosus* Smith, F. 1859; see under ***PROCRYPTOCERUS.***

***MERANOPLUS*** *(continued)*

*springvalensis. Meranoplus simoni* var. *springvalensis* Arnold, 1917: 372 (w.) ZIMBABWE. Junior synonym of *magrettii:* Bolton, 1981a: 55.
***sthenus.*** *Meranoplus sthenus* Bolton, 1981a: 50, figs. 1, 9 (w.q.) ZIMBABWE.
*striatus* Smith, F. 1860; see under ***PROCRYPTOCERUS.***
*subpilosus* Smith, F. 1860; see under ***PROCRYPTOCERUS.***
*suturalis. Meranoplus simoni* subsp. *suturalis* Forel, 1910e: 424 (w.) SOUTH AFRICA. Junior synonym of *magrettii:* Bolton, 1981a: 55.
*tarda. Myrmica tarda* Jerdon, 1851: 115 (w.) INDIA. Junior synonym of *bicolor:* Emery, 1892b: 166.
***testudineus.*** *Meranoplus testudineus* McAreavey, 1956: 148, fig. 1 (w.) AUSTRALIA.
***unicolor.*** *Meranoplus diversus* r. *unicolor* Forel, 1902h: 455 (w.) AUSTRALIA. Raised to species: Taylor, 1990c: 34.
***vestigator.*** *Meranoplus vestigator* Smith, F. 1876d: 610, pl. 11, fig. 8 (w.) INDONESIA (Martabello I.).
*villosus. Meranoplus villosus* Motschoulsky, 1859: 115 (q.) SRI LANKA. Junior synonym of *bicolor:* Roger, 1863b: 39.

#***MESANOPLOLEPIS:*** see under ***ANOPLOLEPIS.***

#***MESOCREMA:*** see under ***CREMATOGASTER.***

*MESOMYRMA:* see under ***DILOBOCONDYLA.***

*MESOPONERA:* see under ***PACHYCONDYLA.***

***MESOSTRUMA*** (Australia)
***browni.*** *Mesostruma browni* Taylor, 1962: 1, figs. 1–5 (w.q.m.) AUSTRALIA. Wheeler, G.C. & Wheeler, J. 1973c: 212 (l.).
***eccentrica.*** *Mesostruma eccentrica* Taylor, 1973: 31, pl. 2 (w.q.) AUSTRALIA.
***exolympica.*** *Mesostruma exolympica* Taylor, 1973: 35, pl. 6 (w.q.) AUSTRALIA.
***laevigata.*** *Mesostruma laevigata* Brown, 1952g: 12, fig. 1 (w.q.) AUSTRALIA. Wheeler, G.C. & Wheeler, J. 1955a: 130 (l.). See also: Taylor, 1973: 28.
***loweryi.*** *Mesostruma loweryi* Taylor, 1973: 35, pl. 4 (w.) AUSTRALIA.
***turneri.*** *Strumigenys* (*Epopostruma*) *turneri* Forel, 1895f: 424 (w.) AUSTRALIA. Combination in *Epopostruma:* Emery, 1922f: 330; in *Mesostruma:* Brown, 1948e: 119. See also: Taylor, 1973: 32.

*MESOXENA:* see under ***ECHINOPLA.***

***MESSOR*** (Holarctic, Afrotropical, Oriental)
***abdelazizi.*** *Messor barbarus* st. *abdelazizi* Santschi, 1921d: 70 (s.) MOROCCO. Santschi, 1931a: 4 (w.q.). Stirps of *structor:* Santschi, 1923f: 320; Santschi, 1929e: 142; Santschi, 1936c: 200. Status as species: Santschi, 1931a: 3; Cagniant, 1964: 87.
***aciculatus.*** *Aphaenogaster aciculata* Smith, F. 1874: 405 (w.q.) JAPAN. Imai, 1966: 119 (k.); Onoyama, 1982: 325 (l.); Kupyanskaya, 1990: 122 (m.). Combination in *Stenamma* (*Messor*): Forel, 1901h: 60; in *Aphaenogaster* (*Attomyrma*): Emery, 1921f: 57; in *Messor:* Forel, 1922: 93. Senior synonym of *brunneicorne:* Yano, 1910: 420; Wheeler, W.M. 1928d: 107 (in text); of *lobulifera:* Wheeler, W.M. 1928c: 9 (in text); Wheeler, W.M. 1928d: 107 (in text). Current subspecies: nominal plus *risianus.*
*aedificator. Formica aedificator* Schilling, 1839: 56 (w.) POLAND. Junior synonym of *structor:* Mayr, 1855: 464.
#***aegaeus.*** *Messor structor* var. *aegaea* Menozzi, 1928a: 126 (s.w.q.m.) TURKEY. [First available use of *Messor barbarus* subsp. *structor* var. *aegaea* Emery, 1921b: 213; unavailable name.]
***aegyptiacus.*** *Aphaenogaster barbara* var. *aegyptiaca* Emery, 1878b: 57 (w.q.) EGYPT. Emery, 1908e: 452 (m.). Combination in *A.* (*Messor*): Forel, 1890a: lxix; in *Stenamma* (*Messor*): Forel, 1902a: 148; in *Messor:* Ruzsky, 1905b: 751. Race of *barbarus:* Forel, 1890a: lxix. Raised to species: Santschi, 1923f: 320. Current subspecies: nominal plus *brevispinosus, canaliculatus, felah, fossulatus, hoggarensis, phasianicus, striatulus, surcoufi, tunetinus.* See also: Santschi, 1929e: 142; Finzi, 1936: 157.
*airensis. Messor galla* var. *airensis* Bernard, 1950b: 286 (w.q.) NIGER. Junior synonym of *galla:* Bolton, 1982: 350.
#***alaschanicus.*** *Messor excursionis* subsp. *alaschanicus* Ruzsky, 1915a: 442, figs. 25–27 (w.) MONGOLIA. See also: Stärcke, 1935: 268.
***alexandri.*** *Messor alexandri* Tohmé, G. & Tohmé, H. 1981: 148 (w.) SYRIA. [First available use of *Messor barbarus* st. *structor* var. *alexandri* Santschi, 1917e: 91; unavailable name.]
#***alfierii.*** *Messor arenarius* st. *alfierii* Santschi, 1938a: 34, fig. 4 (w.) EGYPT.
#***amasiensis.*** *Messor oertzeni* var. *amasiensis* Emery, 1921b: 216 (q.) TURKEY.
*amaurocyclia. Myrmica amaurocyclia* Foerster, 1850b: 486 (w.) ALGERIA. Junior synonym of *arenarius:* Roger, 1862c: 292.
*ambiguus. Messor barbarus* var. *ambiguus* Santschi, 1925g: 342 (w.) TUNISIA. Junior synonym of *barbarus:* Collingwood, 1978: 68.
*amphigea. Messor oertzeni* var. *amphigea* Forel, 1911d: 347 (s.w.m.) GREECE. Junior synonym of *oertzeni:* Agosti & Collingwood, 1987a: 54.
***andrei.*** *Aphaenogaster andrei* Mayr, 1886d: 448 (w.) U.S.A. Emery, 1895c: 306 (q.); Wheeler, G.C. & Wheeler, J. 1972b: 240 (l.); Taber & Cokendolpher, 1988: 95 (k.). Combination in *Stenamma* (*Messor*): Emery, 1895c: 306; in *Novomessor:* Emery, 1915d: 73; in *N.* (*Veromessor*): Forel, 1917: 235; in *Veromessor:* Wheeler, W.M. & Creighton, 1934: 362; in *Messor:* Bolton, 1982: 341. Senior synonym of *castaneus* (and its junior synonym *flavus*): Creighton, 1953a: 3. See also: Smith, D.R. 1979: 1364.
***angularis.*** *Messor galla* var. *angularis* Santschi, 1928f: 202 (w.) KENYA. [First available use of *Messor barbarus* st. *semirufus* var. *angularis* Santschi, 1914b: 75; unavailable name.] Raised to species: Bolton, 1982: 344.
***antennatus.*** *Messor antennatus* Emery, 1908e: 441, fig. 3 (w.) MOROCCO. Menozzi, 1922a: 354 (q.). Current subspecies: nominal plus *fodorii, personatus.*
***aphaenogasteroides.*** *Messor aphaenogasteroides* Pisarski, 1967: 381, figs. 10–15 (w.q.) AFGHANISTAN.
***aralocaspius.*** *Aphaenogaster barbara* var. *aralocaspia* Ruzsky, 1902c: 20 (w.) RUSSIA. Karavaiev, 1910b: 63 (q.m.); Stitz, 1930: 240 (q.). Combination in *Messor:* Ruzsky, 1905b: 745. Stirps of *minor:* Santschi, 1927c: 239. Status as species: Pisarski, 1967: 384; Pisarski, 1969b: 306; Tarbinsky, 1976: 65; Arnol'di, 1977b: 1642. Material of the unavailable name *tamerlani* referred here: Dlussky, Soyunov & Zabelin, 1990: 218. Current subspecies: nominal plus *infumatus.*
*arcistriatus. Messor arcistriatus* Santschi, 1928f: 202 (w.) SOUTH AFRICA. Junior synonym of *decipiens:* Bolton, 1982: 348.
***arenarius.*** *Formica arenaria* Fabricius, 1787: 310 (w.) NORTH AFRICA. André, 1883b: 366 (q.); Emery, 1891b: 12 (q.m.); Forel, 1902a: 148 (m.). Combination in *Aphaenogaster:* Roger, 1863b: 29; in *A.* (*Messor*): Forel, 1890a: lxx; in *Messor:* Forel, 1894d: 45; in *Stenamma* (*Messor*): Emery, 1898c: 125; in *Messor:* Emery, 1908e: 439. Senior synonym of *amaurocyclia* and *scalpturata:* Roger, 1862c: 292; of *bugnioni:* Forel, 1904f: 424. Current subspecies: nominal plus *alfierii, diabolus, ratus, thebaicus.* See also: Emery, 1908e: 439; Santschi, 1938a: 32.
*armata. Messor barbarus* subsp. *galla* var. *armata* Emery, 1922c: 98 (w.) GHANA. Unavailable name; material referred to *galla* by Bolton, 1982: 350.
***atanassovii.*** *Messor atanassovii* Atanassov, 1982: 209, figs. 1–3 (w.q.) BULGARIA. [Incorrectly regarded as unavailable name: Baroni Urbani, Aktac & Camiltepe, 1989: 300.] See also: Atanassov & Dlussky, 1992: 119.
*barbarastructor. Aphaenogaster barbara* var. *barbarastructor* Dalla Torre, 1893: 100 (w.) RUSSIA. *Nomen nudum;* attributed to Mayr, 1880: 33, where the name is not mentioned. [The name persists in Ruzsky, 1902d: 27; Forel, 1904b: 373; Forel, 1904d: 175; Forel, 1906c: 189; Emery, 1921f: 73; all *nomina nuda.*]
***barbarus.*** *Formica barbara* Linnaeus, 1767: 962 (w.) MOROCCO. Mayr, 1861: 67 (q.m.); Wheeler, G.C. & Wheeler, J. 1953b: 65 (l.). Combination in *Aphaenogaster:* Roger, 1863b: 29; in *A.* (*Messor*): Forel, 1890a: lxviii; in *Messor:* Forel, 1894d: 8; in *Stenamma* (*Messor*): Emery, 1898c: 125; in *Messor:* Ruzsky, 1903b: 315; Bingham, 1903: 278. Senior synonym of *binodis* and *juvenalis:* Smith, F. 1858b: 162; of *megacephala:* Mayr, 1863: 395; of *rufitarsis* Foerster: Roger, 1863b: 29; of *ambiguus* and *nigriceps:* Collingwood, 1978: 68. Current subspecies: nominal plus *gallaoides, mediosanguineus, nigricans, politus, sahlbergi, santschii, varrialei.*
#***batnensis.*** *Messor lobicornis* var. *batnensis* Forel, 1909e: 371 (w.q.) ALGERIA.
***beduinus.*** *Messor barbarus* st. *beduinus* Emery, 1922c: 93 (w.) TUNISIA. [First available use of *Messor barbarus* subsp. *striativentris* var. *beduina* Emery, 1908e: 453; unavailable name.] Santschi, 1910a: 45 (q.); Santschi, 1926e: 234 (m.). Raised to species: Santschi, 1923a: 134; Santschi, 1923f: 320. See also: Finzi, 1940: 157.
***berbericus.*** *Messor berbericus* Bernard, 1955a: 357, fig. B (w.) ALGERIA.
***bernardi.*** *Messor bernardi* Cagniant, 1967a: 131, figs. 1–11 (w.q.m.) ALGERIA.
*binodis. Formica binodis* Fabricius, 1775: 393 (w.) EGYPT. [Unresolved junior primary homonym of *Formica binodis* Linnaeus, 1763: 413 (now in

*Tetramorium*).] Junior synonym of *barbarus:* Smith, F. 1858b: 162.

**#*bodenheimeri.*** *Messor rugosus* subsp. *bodenheimeri* Menozzi, 1933b: 54 (w.) ISRAEL.

***bouvieri.*** *Messor bouvieri* Bondroit, 1918: 154, fig. 75 (s.w.) FRANCE. Baroni Urbani, 1964b: 31 (q.). Variety of *instabilis:* Menozzi, 1926b: 181. Junior synonym of *sanctus:* Bernard, 1967: 148. Subspecies of *sanctus:* Baroni Urbani, 1964b: 30; Baroni Urbani, 1968b: 430; Baroni Urbani, 1971c: 65. Revived status as species: Collingwood & Yarrow, 1969: 63; Collingwood, 1978: 68; Schembri & Collingwood, 1981: 426. Material of the unavailable name *grandiceps* referred here: Collingwood, 1978: 68.

*braunsi. Messor braunsi* Forel, 1913a: 138 (w.) SOUTH AFRICA. Stitz, 1923: 149 (q.). Junior synonym of *capensis:* Bolton, 1982: 345.

**#*brevispinosus.*** *Messor aegyptiacus* var. *brevispinosus* Santschi, 1923f: 322 (w.) MOROCCO. [First available use of *Messor barbarus* r. *striaticeps* var. *brevispinosus* Stitz, 1917: 344; unavailable name.] Stirps of *aegyptiacus:* Santschi, 1934f: 166.

*brunneicorne. Stenamma* (*Messor*) *aciculatum* var. *brunneicorne* Forel, 1901h: 60 (w.) JAPAN. Combination in *Messor*, and raised to species: Emery, 1908e: 459. Junior synonym of *aciculatus:* Yano, 1910: 420.

*brunni. Messor denticornis* var. *brunni* Forel, 1910e: 444 (w.) NAMIBIA. Junior synonym of *denticornis:* Bolton, 1982: 349.

***buettikeri.*** *Messor buettikeri* Collingwood, 1985: 249, fig. 22 (w.q.m.) SAUDI ARABIA.

*bugnioni. Stenamma* (*Messor*) *bugnioni* Forel, 1904c: 13 (w.) ALGERIA. Junior synonym of *arenarius:* Forel, 1904f: 424.

***caducus.*** *Formica caduca* Victor, 1839: 47, pl. 2, fig. d (w.) RUSSIA. Combination in *Aphaenogaster:* Roger, 1863b: 29; in *Stenamma* (*Messor*): Emery, 1898c: 125; in *Messor:* Ruzsky, 1905b: 754. Junior synonym of *barbarus:* Roger, 1863b: 29; Emery & Forel, 1879: 461; Nasonov, 1889: 37. Revived from synonymy as race of *barbarus:* Emery, 1891b: 12. Revived status as species: Ruzsky, 1905b: 754. Material of the unavailable name *semirubra* referred here: Arnol'di, 1977b: 1638. Senior synonym of *meridionalis:* Atanassov & Dlussky, 1992: 121; of *caucasicola:* Arakelian, 1994: 34. Current subspecies: nominal plus *glabrescens.* See also: Agosti & Collingwood, 1987a: 54; Baroni Urbani, Aktac & Camiltepe, 1989: 291.

**#*calabricus.*** *Messor minor* var. *calabricus* Santschi, 1927c: 234 (w.) ITALY. Subspecies of *minor:* Baroni Urbani, 1964c: 151; Baroni Urbani, 1971c: 62.

**#*canaliculatus.*** *Messor aegyptiacus* subsp. *canaliculatus* Donisthorpe, 1944a: 1, fig. 1 (s.w.) ALGERIA.

***capensis.*** *Atta capensis* Mayr, 1862: 743 (w.) SOUTH AFRICA. Mayr, 1866b: 896 (q.m.). Combination in *Aphaenogaster:* Roger, 1863b: 30; in *Stenamma* (*Messor*): Emery, 1895h: 35; in *Messor:* Forel, 1910e: 444. Subspecies/race of *barbarus:* Emery, 1891b: 12; Forel, 1910e: 444; Arnold, 1920a: 405; Emery, 1922c: 98; Stitz, 1923: 150; Prins, 1963: 100. Status as species: Santschi, 1917e: 94; Wheeler, W.M. 1922a: 804; Bolton, 1982: 345. Senior synonym of *braunsi, donisthorpei, pseudoaegyptiaca* and *schencki:* Bolton, 1982: 345.

***capitatus.*** *Formica capitata* Latreille, 1798: 46 (w.) FRANCE. Latreille, 1802c: 234 (q.); Mayr, 1855: 463 (m.). Combination in *Atta:* Mayr, 1855: 462; in *Aphaenogaster:* Roger, 1863b: 29; in *Stenamma* (*Messor*): Forel, 1902a: 148; in *Messor:* Ruzsky, 1905b: 744. Junior synonym of *barbarus:* Smith, F 1858b: 162; Emery & Forel, 1879: 461; Nasonov, 1889: 37. Race/subspecies of *barbarus:* Emery, 1891b: 12; Emery, 1908e: 449; Emery, 1921f: 69; Santschi, 1923f: 325; Santschi, 1925g: 342; Finzi, 1929: 80; Santschi, 1931a: 4. Status as species: Dalla Torre, 1893: 101; Bondroit, 1918: 153; Baroni Urbani, 1964b: 40; Bernard, 1967: 147; Baroni Urbani, 1968b: 427; Baroni Urbani, 1971c: 57; Schembri & Collingwood, 1981: 425. Senior synonym of *nigra:* Menozzi, 1921: 26; Emery, 1922c: 97; of *galbula* and *huberiana:* Emery, 1921f: 70.

**#*capreensis.*** *Messor minor* var. *capreensis* Santschi, 1927c: 234, fig. 17 (w.) ITALY.

**#*carpathous.*** *Messor oertzeni* var. *carpathous* Menozzi, 1936d: 277, fig. 6 (w.) GREECE.

***carthaginensis.*** *Messor carthaginensis* Bernard, 1980: 268, fig. (w.) TUNISIA.

*castaneus. Veromessor andrei* subsp. *castaneus* Wheeler, W.M. & Creighton, 1934: 365 (w.) U.S.A. Senior synonym of *flavus:* Creighton, 1950a: 159. Junior synonym of *andrei:* Creighton, 1953a: 3.

*caucasicola. Messor caducus* subsp. *caucasicola* Arnol'di, 1977b: 1639 (w.q.m.) CAUCASUS. Junior synonym of *caducus:* Arakelian, 1994: 34.

***caviceps.*** *Stenamma* (*Messor*) *caviceps* Forel, 1902a: 149 (w.) ALGERIA. Combination in *Messor:* Emery, 1908e: 457.

***celiae.*** *Messor celiae* Reyes, 1985: 255, figs. 1–3 (w.m.) SPAIN.

***cephalotes.*** *Stenamma* (*Messor*) *barbarum* subsp. *cephalotes* Emery, 1895a: 179 (w.) ETHIOPIA. Menozzi, 1939c: 102 (q.m.). Combination in *Messor:* Emery, 1908e: 443; in *Sphaeromessor:* Bernard, 1985: 48; in *Messor:* **revived combination**. Raised to species: Emery, 1908e: 443. Senior synonym of *plinii:* Bolton, 1982: 346.

***ceresis.*** *Messor ceresis* Santschi, 1934d: 274 (w.) LEBANON.

***chamberlini.*** *Messor chamberlini* Wheeler, W.M. 1915b: 410 (w.) U.S.A. Wheeler, G.C. & Wheeler, J. 1972b: 240 (l.). Combination in *Veromessor:* Wheeler, W.M. & Creighton, 1934: 366; in *Messor:* Bolton, 1982: 341. Subspecies of *andrei:* Enzmann, J. 1947b: 152. Revived status as species: Creighton, 1950a: 159.

**#*chicoensis.*** *Veromessor stoddardi* subsp. *chicoensis* Smith, M.R. 1956a: 37 (w.) U.S.A. Combination in *Messor:* **new combination**.

**#*clarior.*** *Veromessor julianus* subsp. *clarior* Wheeler, W.M. & Creighton, 1934: 370 (w.) U.S.A. Combination in *Messor:* **new combination**.

*clivorum. Messor structor* var. *clivorum* Ruzsky, 1905b: 735 (s.w.) TURKESTAN. Karavaiev, 1927a: 286 (q.m.). Subspecies of *barbarus:* Emery, 1908e: 454; Forel, 1910a: 24. Raised to species: Emery, 1921b: 210 (footnote); Dlussky, 1962: 179; Tarbinsky, 1976: 56; Arnol'di, 1977b: 1639. Senior synonym of *rubella:* Emery, 1921b: 210; of *subpolitus:* Tarbinsky, 1976: 56. Junior synonym of *structor:* Dlussky, Soyunov & Zabelin, 1990: 224; Atanassov & Dlussky, 1992: 114.

***clypeatus.*** *Messor similis* var. *clypeatus* Kuznetsov-Ugamsky, 1927a: 93 (in key) (w.) TURKESTAN. Subspecies of *rufus:* Kuznetsov-Ugamsky, 1929d: 9. Raised to species: Arnol'di, 1977b: 1646.

***collingwoodi.*** *Messor collingwoodi* Bolton, 1982: 346, fig. 30 (w.) NIGER.

***concolor.*** *Messor semirufus* var. *concolor* Santschi, 1927c: 229 (s.w.m.) SYRIA. [First available use of *Messor barbarus* subsp. *semirufus* var. *concolor* Emery, 1908e: 448; unavailable name.] Subspecies of *semirufus:* Baroni Urbani, 1964a: 1. Junior synonym of *semirufus:* Baroni Urbani, 1974: 227. Revived from synonymy, and raised to species: Tohmé, G. & Tohmé, H. 1981: 143.

**#*crawleyi.*** *Messor rugosus* st. *crawleyi* Santschi, 1928f: 203 (w.) EGYPT.

*curvispina. Messor barbarus* subsp. *striaticeps* var. *curvispina* Karavaiev, 1912a: 10 (s.w.) ALGERIA. Unavailable name; material referred to *striatulus* by Santschi, 1923f: 321 (in text).

*darianus. Messor rufitarsis* subsp. *darianus* Pisarski, 1967: 383 (w.) TURKESTAN. [First available use of *Messor structor* st. *rufitarsis* var. *darianus* Santschi, 1926f: 291; unavailable name.] Junior synonym of *turanicus:* Pisarski, 1967: 383; Pisarski, 1970: 306; Tarbinsky, 1976: 54; Arnol'di, 1977b: 1644 (in all these *darianus* is given seniority over *turanicus*, but the latter is the first available name).

***decipiens.*** *Messor capensis* st. *decipiens* Santschi, 1917e: 94 (w.q.) SOUTH AFRICA. [First available use of *Stenamma* (*Messor*) *barbarum* r. *capense* var. *decipiens* Forel, 1905b: 177; unavailable name.] Raised to species, and senior synonym of *arcistriatus* and *probus:* Bolton, 1982: 348.

***dentatus.*** *Messor semirufus* var. *dentatus* Santschi, 1927c: 228 (w.q.m.) ISRAEL. [First available use of *Messor barbarus* subsp. *semirufus* var. *dentata* Forel, 1910a: 10; unavailable name.] Junior synonym of *semirufus:* Baroni Urbani, 1974: 227. Revived from synonymy, and raised to species: Tohmé, G. & Tohmé, H. 1981: 143.

***denticornis.*** *Messor denticornis* Forel, 1910f: 14 (w.q.m.) NAMIBIA. Senior synonym of *brunni* and *parvidens:* Bolton, 1982: 349.

***denticulatus.*** *Messor minor* st. *denticulatus* Santschi, 1927d: 247 (w.) TURKESTAN. [First available use of *Messor barbarus* subsp. *meridionalis* var. *denticulatus* Kuznetsov-Ugamsky, 1927a: 90 (in key); unavailable name.] Raised to species: Tarbinsky, 1976: 61; Arnol'di, 1977b: 1641; Arnol'di & Dlussky, 1978: 537 (in key). Material of the unavailable name *glabriusculus* referred here by Dlussky, Soyunov & Zabelin, 1990: 220.

*dentiscapus. Messor barbarus* var. *dentiscapus* Finzi, 1930b: 15 (w.) ALGERIA. [First available use of *Messor barbarus* subsp. *capitatus* var. *dentiscapa* Forel, 1909e: 371; unavailable name.] Material referred to *politus* by Emery, 1922c: 97, hence junior synonym of *politus* as the latter has priority: **new synonymy**.

***diabarensis.*** *Messor diabarensis* Arnol'di, 1969: 75, fig. 2 (s.q.m.) TURKESTAN. [*Messor structor* subsp. *diabarensis* Arnol'di, 1948: 211. *Nomen nudum.*]

***MESSOR** (continued)*

**#diabolus.** *Messor arenarius* var. *diabolus* Santschi, 1938a: 35, fig. 5 (w.) EGYPT.

*donisthorpei. Messor donisthorpei* Santschi, 1937a: 51 (w.q.) NAMIBIA. Junior synonym of *capensis:* Bolton, 1982: 345.

**ebeninus.** *Messor semirufus* var. *ebeninus* Santschi, 1927c: 229 (w.) LEBANON. [First available use of *Messor barbarus* subsp. *semirufus* var. *ebenina* Forel, 1910a: 10; unavailable name.] Finzi, 1936: 160 (q.). Raised to species: Tohmé, G. 1971: 569. Junior synonym of *semirufus:* Baroni Urbani, 1974: 227. Revived from synonymy: Tohmé, G. & Tohmé, H. 1981: 142; Collingwood, 1985: 249. See also: Tohmé, G. 1975: 171.

**#emeryi.** *Messor semirufus* var. *emeryi* Santschi, 1927c: 247 (w.) ISRAEL.

**excursionis.** *Messor lobuliferus* var. *excursionis* Ruzsky, 1905b: 739 (w.) TURKESTAN. Arnol'di, 1969: 86 (q.m.). Raised to species: Karavaiev, 1910b: 68. Current subspecies: nominal plus *alaschanicus, sillemi.* See also: Stärcke, 1935: 262; Arnol'di, 1977b: 1646.

*falcistriae. Messor aegyptiacus* st. *striativentris* var. *falcistriae* Santschi, 1939d: 70 (w.m.) ALGERIA. Unavailable name.

**#felah.** *Messor aegyptiacus* var. *felah* Santschi, 1923f: 321, fig. 1 (s.w.) EGYPT.

**ferreri.** *Messor ferreri* Collingwood, 1993: 229, figs. 13–15 (w.) KENYA.

*flavus. Veromessor andrei* subsp. *flavus* Wheeler, W.M. & Creighton, 1934: 366 (w.) U.S.A. Junior synonym of *castaneus:* Creighton, 1950a: 159.

**#fodorii.** *Messor antennatus* var. *fodorii* Röszler, 1942a: 55 (w.) GREECE. [Also described as new by Röszler, 1951: 92.]

**foreli.** *Messor aegyptiacus* var. *foreli* Santschi, 1923f: 322, fig. 1 (w.) TUNISIA. Raised to species: Bernard, 1981b: 143.

**#fossulatus.** *Messor aegyptiacus* var. *fossulatus* Santschi, 1923f: 323, fig. 1 (w.q.m.) TUNISIA. Subspecies of *aegyptiacus:* Menozzi, 1940: 267.

**fraternus.** *Messor lobuliferus* var. *fraternus* Ruzsky, 1905b: 740 (w.) RUSSIA. Raised to species: Ruzsky, 1915a: 444.

*galbula. Myrmica galbula* Losana, 1834: 326, pl. 36, fig. 5 (w.) ITALY. Junior synonym of *barbarus:* Roger, 1863b: 29; of *capitatus:* Emery, 1921f: 70.

**galla.** *Stenamma* (*Messor*) *barbarum* var. *galla* Mayr, 1904b: 5 (w.) ETHIOPIA. [First available use of *Stenamma* (*Messor*) *barbarum* subsp. *caduca* var. *galla* Emery, 1895a: 179; unavailable name.] Finzi, 1939a: 155 (q.m.). Combination in *Messor:* Santschi, 1914d: 335. Stirps of *barbarus:* Santschi, 1910f: 234; Santschi, 1914d: 335. Raised to species: Santschi, 1923f: 326; Santschi, 1928f: 201. Senior synonym of *latinoda:* Viehmeyer, 1923: 87; Bolton, 1982: 350; of *airensis, nobilis, rufula,* and material of the unavailable names *armata, rufa* and *triempressa* referred here: Bolton, 1982: 350. Current subspecies: nominal plus *obscurus.*

**#gallaoides.** *Messor barbarus* var. *gallaoides* Santschi, 1929e: 141 (s.) MOROCCO.

*gallica. Messor barbarus* subsp. *rufitarsis* var. *gallica* Emery, 1921b: 212 (w.) TURKEY. Unavailable name.

**#glabrescens.** *Messor caducus* subsp. *glabrescens* Arnol'di, 1977b: 1639 (w.) TURKESTAN.

*glabriusculus. Messor barbarus* subsp. *meridionalis* var. *glabriusculus* Kuznetsov-Ugamsky, 1927a: 90 (w.) TURKESTAN. Unavailable name; material referred to *denticulatus* by Dlussky, Soyunov & Zabelin, 1990: 220.

*gracilinodis. Aphaenogaster gracilinodis* Emery, 1878b: 55 (w.) SYRIA. Variety of *rufotestaceus:* Emery, 1880: 395. Junior synonym of *rufotestaceus:* Dalla Torre, 1893: 104.

*grandiceps. Messor barbarus* r. *capitatus* var. *grandiceps* Stitz, 1917: 343 (w.) SPAIN. Unavailable name; material referred to *bouvieri* by Collingwood, 1978: 68.

**#grandinidus.** *Messor barbarus* subsp. *grandinida* Emery, 1912f: 93 (w.q.m.) TUNISIA. [First available use of *Messor barbarus* st. *meridionalis* var. *grandinida* Santschi, 1910a: 43; unavailable name.] Subspecies of *semirufus:* Menozii, 1927: 379; Santschi, 1927c: 231.

*hamulifera. Messor barbarus* subsp. *hamulifera* Karavaiev, 1926b: 99 (s.w.) RUSSIA. Raised to species: Arnol'di, 1948: 211. Junior synonym of *incorruptus:* Arnol'di, 1977b: 1640.

**hebraeus.** *Messor semirufus* var. *hebraeus* Santschi, 1927c: 229, fig. 7 (w.) ISRAEL. Junior synonym of *semirufus:* Baroni Urbani, 1974: 227. Revived from synonymy, and raised to species: Tohmé, G. & Tohmé, H. 1981: 143.

**hellenius.** *Messor hellenius* Agosti & Collingwood, 1987a: 54 (w.q.) GREECE. [First available use of *Messor barbarus* st. *varrialei* var. *hellenis* Santschi, 1926f: 290; unavailable name.]

**#hesperius.** *Messor minor* var. *hesperius* Santschi, 1927c: 235 (w.q.m.) CANARY IS.

**himalayanus.** *Stenamma* (*Messor*) *barbarum* r. *himalayanum* Forel, 1902c: 221 (s.w.q.m.) INDIA. Combination in *Messor,* and raised to species: Bingham, 1903: 279. Subspecies/race of *barbarus:* Forel, 1904d: 175; Viehmeyer, 1914c: 40; Emery, 1922c: 95; Kuznetsov-Ugamsky, 1929d: 15. Status as species: Stärcke, 1935: 261; Menozzi, 1939a: 297; Arnol'di, 1977b: 1643.

**hispanicus.** *Messor barbarus* st. *hispanicus* Santschi, 1919e: 244 (w.) SPAIN. Menozzi, 1922b: 326 (q.); Gonzalez, Lopez & Espadaler, 1983: 38 (m.). Stirps of *structor:* Santschi, 1929e: 144. Raised to species: Santschi, 1927c: 225; Collingwood & Yarrow, 1969: 63.

**#hoggarensis.** *Messor sublaeviceps* st. *hoggarensis* Santschi, 1929c: 98, fig. 1 (w.q.) ALGERIA. [Also described as new by Santschi, 1934f: 166.] Subspecies of *meridionalis:* Bernard, 1948: 140; of *instabilis:* Bernard, 1950b: 285; of *aegyptiacus:* Bernard, 1953a: 142 (in key).

*huberiana. Formica huberiana* Leach, 1825: 290 (w.q.m.) FRANCE. Junior synonym of *barbarus:* Mayr, 1863: 395; of *capitatus:* Emery, 1921f: 70.

**ibericus.** *Messor structor* var. *ibericus* Santschi, 1931a: 4 (w.m.) SPAIN. [First available use of *Messor barbarus* subsp. *structor* var. *iberica* Emery, 1922c: 92; unavailable name.] Raised to species: Collingwood, 1978: 68.

**incisus.** *Messor incisus* Stitz, 1923: 149 (q.) NAMIBIA. See also: Bolton, 1982: 351.

**incorruptus.** *Messor barbarus* subsp. *incorruptus* Kuznetsov-Ugamsky, 1929d: 16 (w.) TURKESTAN. [First available use of *Messor barbarus* subsp. *aegyptiacus* var. *incorruptus* Ruzsky, 1905b: 751; unavailable name.] Junior synonym of *caducus:* Kolossov, 1932: 118; Donisthorpe, 1941k: 238. Revived from synonymy as subspecies of *semirufus,* and senior synonym of *hamulifera:* Arnol'di, 1977b: 1640. Raised to species: Bernard, 1981b: 145; Tohmé G. & Tohmé, H. 1981: 144.

**inermis.** *Messor barbarus* subsp. *inermis* Kuznetsov-Ugamsky, 1929d: 21 (w.) TURKESTAN. [First available use of *Messor barbarus* subsp. *meridionalis* var. *inermis* Kuznetsov-Ugamsky, 1927a: 90 (in key); unavailable name.] Raised to species: Tarbinsky, 1976: 65; Arnol'di, 1977b: 1641.

**#infumatus.** *Messor barbarus* subsp. *infumatus* Kuznetsov-Ugamsky, 1929d: 19 (w.) KAZAKHSTAN. [First available use of *Messor barbarus* subsp. *meridionalis* var. *infumatus* Kuznetsov-Ugamsky, 1927a: 91 (in key); unavailable name.] Subspecies of *meridionalis:* Pisarski, 1967: 384. Raised to species: Tarbinsky, 1976: 57. Subspecies of *aralocaspius:* Arnol'di, 1977b: 1642.

**instabilis.** *Atta instabilis* Smith, F. 1858b: 163 (w.m.) INDIA. Bingham, 1903: 279 (s.q.). Combination in *Messor:* Bingham, 1903: 278. Variety/subspecies of *capitatus:* Dalla Torre, 1893: 101; of *barbarus:* Bingham, 1903: 278; Forel, 1906b: 86; Emery, 1922c: 94; Emery, 1924a: 8; Karavaiev, 1926b: 100. Status as species: Santschi, 1923f: 325; Santschi, 1927c: 247 (in key); Wheeler, W.M. 1927g: 103; Santschi, 1931a: 4; Santschi, 1936c: 200; Menozzi, 1939a: 297. Senior synonym of *punctata:* Bingham, 1903: 279; Pisarski, 1970: 306.

**intermedius.** *Messor semirufus* var. *intermedius* Santschi, 1927c: 229 (w.) SYRIA. [First available use of *Messor barbarus* subsp. *semirufus* var. *intermedia* Forel, 1910a: 10; unavailable name.] Variety of *semirufus:* Menozzi, 1927e: 117. Junior synonym of *semirufus:* Baroni Urbani, 1974: 227. Revived from synonymy, and raised to species: Arnol'di, 1977b: 1640 (in key). See also: Tohmé, G. & Tohmé, H. 1981: 142; Dlussky & Zabelin, 1985: 211.

*jakowlevi. Messor rufitarsis* subsp. *jakowlevi* Arnol'di, 1977b: 1643 (w.) UKRAINE. [First available use of *Messor barbarus* subsp. *capitatus* var. *jakowlevi* Ruzsky, 1905b: 750; unavailable name.] Material of the unavailable name *salina* referred here by Arnol'di, 1977b: 1643. Junior synonym of *structor:* Dlussky, Soyunov & Zabelin, 1990: 224; Atanassov & Dlussky, 1992: 114.

**julianus.** *Aphaenogaster juliana* Pergande, 1894b: 164 (w.) MEXICO. Combination in *Stenamma* (*Messor*): Emery, 1895c: 308; in *Aphaenogaster* (*Messor*): Forel, 1899c: 59; in *Novomessor:* Emery, 1915d: 73; in *Veromessor:* Forel, 1917: 235; in *Messor:* Bolton, 1982: 341. Current subspecies: nominal plus *clarior, manni.* See also: Wheeler, W.M. & Creighton, 1934: 368.

*juvenilis. Formica juvenilis* Fabricius, 1804: 405 (w.) FRANCE. Junior synonym of *barbarus:* Smith, F. 1858b: 162.

*karawajevi. Messor karawajevi* Tarbinsky, 1976: 59 (w.) TURKESTAN. [First available use of *Messor barbarus* st. *clivorum* var. *karawayewi* Santschi, 1917e: 92; unavailable name.] Junior synonym of *rufus:* Arnol'di, 1977b: 1642.

***kasakorum.*** *Messor kasakorum* Arnol'di, 1969: 82, fig. 6 (w.m.) KAZAKHSTAN. Current subspecies: nominal plus *nadezhdae.*

***kisilkumensis.*** *Messor kisilkumensis* Arnol'di, 1969: 84, figs. 5, 6 (w.q.m.) UZBEKISTAN.

*korborum. Messor barbarus* subsp. *structor* var. *korborum* Emery, 1921b: 214 (s.w.q.) TURKEY. Unavailable name.

*kutteri. Messor semirufus* var. *kutteri* Santschi, 1934d: 275 (w.) ITALY. Junior synonym of *semirufus:* Baroni Urbani, 1974: 227.

#***laboriosus.*** *Messor minor* st. *laboriosus* Santschi, 1927c: 240, fig. 3 (w.q.) TURKESTAN.

*laeviceps. Messor barbarus* r. *meridionalis* var. *laeviceps* Stitz, 1917: 344 (w.) MOROCCO. Unavailable name.

*laevifrons. Messor denticornis* var. *laevifrons* Stitz, 1923: 148 (w.) NAMIBIA. Junior synonym of *tropicorum:* Bolton, 1982: 354.

***lamellicornis.*** *Messor lamellicornis* Arnol'di, 1968: 1807, figs. 7, 8 (s.w.m.) TURKESTAN.

*lapidum. Formica lapidum* Fabricius, 1804: 407 (w.) AUSTRIA. Junior synonym of *structor:* Nylander, 1856b: 85.

***lariversi.*** *Veromessor lariversi* Smith, M.R. 1951c: 94 (w.) U.S.A. Cole, 1963: 680 (q.m.). Combination in *Messor:* Bolton, 1982: 341.

*latinoda. Messor barbarus* st. *latinoda* Santschi, 1917e: 93, fig. 2 (s.w.) EAST AFRICA. Junior synonym of *galla:* Viehmeyer, 1923: 87; Bolton, 1982: 350.

#***laurenti.*** *Messor lobicornis* var. *laurenti* Santschi, 1939d: 72 (w.) ALGERIA.

***lobicornis.*** *Messor lobicornis* Forel, 1894d: 31 (w.) ALGERIA. Current subspecies: nominal plus *batnensis, laurenti, normandi, rugulosus, submuticus.* See also: Collingwood & Yarrow, 1969: 64.

***lobognathus.*** *Messor lobognathus* Andrews, 1916: 82, fig. 1 (w.) U.S.A. Wheeler, G.C. & Wheeler, J. 1957c: 142 (l.); Cole, 1963: 680 (q.m.). Combination in *Veromessor:* Wheeler, W.M. & Creighton, 1934: 371; in *Messor:* Bolton, 1982: 341. See also: Gregg, 1955b: 45.

*lobulifera. Messor barbarus* var. *lobulifera* Emery, 1901i: 159 (w.) MONGOLIA. Santschi, 1925f: 93 (q.m.). Raised to species: Ruzsky, 1905b: 739. [Misspelled as *lobifer* by Santschi, 1925f: 93.] Junior synonym of *aciculatus:* Wheeler, W.M. 1928c: 9 (in text); Wheeler, W.M. 1928d: 107. See also: Santschi, 1928b: 31.

***luebberti.*** *Messor barbarus* subsp. *luebberti* Forel, 1910f: 13 (w.) NAMIBIA. Raised to species: Emery, 1922c: 99; Bolton, 1982: 351.

***luridus.*** *Messor luridus* Santschi, 1927c: 245 (w.) TURKEY. [First available use of *Stenamma* (*Messor*) *barbarum* subsp. *caducum* var. *lurida* Emery, 1898c: 139; unavailable name.] Current subspecies: nominal plus *virginalis.*

***lusitanicus.*** *Messor lusitanicus* Tinaut, 1985: 85 (w.q.m.) PORTUGAL. [First available use of *Messor structor* st. *hispanicus* var. *lusitanicus* Santschi, 1929e: 144; unavailable name.]

*macrops. Messor excursionis* var. *macrops* Santschi, 1927d: 248 (w.) TURKESTAN. Junior synonym of *variabilis:* Arnol'di, 1977b: 1646.

***maculifrons.*** *Messor semirufus* var. *maculifrons* Santschi, 1927c: 228 (w.) SYRIA. [First available use of *Messor barbarus* st. *semirufus* var. *maculifrons* Santschi, 1917e: 91; unavailable name.] Junior synonym of *semirufus:* Baroni Urbani, 1974: 227. Revived from synonymy, and raised to species: Tohmé, G. & Tohmé, H. 1981: 143.

#***manni.*** *Veromessor julianus* subsp. *manni* Wheeler, W.M. & Creighton, 1934: 371 (w.) U.S.A. Combination in *Messor:* **new combination**.

***marikovskii.*** *Messor marikovskii* Arnol'di, 1969: 79, figs. 3, 4 (w.q.m.) KAZAKHSTAN. See also: Tarbinsky, 1976: 52.

***marocanus.*** *Messor minor* st. *marocanus* Santschi, 1927c: 248 (in key) (w.) MOROCCO. [First available use of *Messor barbarus* subsp. *meridionalis* var. *maroccana* Emery, 1908e: 451; unavailable name.] Stirps of *instabilis:* Santschi, 1929e: 142. Raised to species: Collingwood & Yarrow, 1969: 64.

#***maurus.*** *Messor minor* subsp. *maurus* Barquin, 1981: 81 (w.q.) CANARY IS. [First available use of *Messor instabilis* st. *minor* var. *maura* Santschi, 1923f: 325 (w.q.); unavailable name.] Unresolved junior primary homonym of *maurus* Santschi, below.

#***maurus.*** *Messor sublaeviceps* var. *maurus* Santschi, 1927c: 244 (w.q.) TUNISIA. See also: Santschi, 1936c: 200. Subspecies of *medioruber:* current status.

***medioruber.*** *Messor barbarus* st. *mediorubra* Santschi, 1910a: 44 (w.q.m.) TUNISIA. [First available use of *Stenamma* (*Messor*) *barbarum* r. *capitatum* var. *mediorubra* Forel, 1905b: 176; unavailable name.] Stirps of *instabilis:* Santschi, 1923f: 325; of *minor:* Santschi, 1927c: 237. Raised to species: Cagniant, 1968a: 143; Cagniant, 1970a: 415. Current subspecies: nominal plus *maurus, montanus, postquadratus, sublaeviceps.*

#***mediosanguineus.*** *Messor barbarus* subsp. *mediosanguineus* Donisthorpe, 1946a: 53 (s.w.) TURKEY.

*megacephala. Formica megacephala* Leach, 1825: 292 (w.q.m.) FRANCE. [Unresolved junior primary homonym of *Formica megacephala* Fabicius, 1793: 361 (now in *Pheidole*).] Junior synonym of *barbarus:* Mayr, 1863: 395.

***melancholicus.*** *Messor melancholicus* Arnol'di, 1977b: 1639 (s.w.) CAUCASUS. [First available use of *Messor structor* subsp. *striaticeps* var. *melancholica* Karavaiev, 1926b: 104; unavailable name.] Senior synonym of *schelkovnikovi:* Arakelian, 1994: 36.

*meridionalis. Aphaenogaster barbara* var. *meridionalis* André, 1883b: 355 (w.) GREECE. Ruzsky, 1905b: 747 (m.); Emery, 1908e: 450 (q.). Combination in *Messor:* Emery, 1897f: 239; in *Stenamma* (*Messor*): Emery, 1898c: 125; in *Messor:* Ruzsky, 1903b: 315; Ruzsky, 1905b: 746. Subspecies of *barbarus:* Emery, 1908e: 450; Forel, 1910a: 24; Santschi, 1921a: 110; of *semirufus:* Santschi, 1927c: 230; Finzi, 1928c: 788; of *minor:* Arnol'di, 1948: 211. Status as species: Bondroit, 1918: 155; Müller, 1923: 64; Collingwood, 1961a: 58; Pisarski, 1970: 306. Junior synonym of *semirufus:* Baroni Urbani, 1974: 227. Revived from synonymy, status as species: Collingwood, 1985: 251. Junior synonym of *caducus:* Atanassov & Dlussky, 1992: 121.

***minor.*** *Aphaenogaster barbara* var. *minor* André, 1883b: 355 (w.q.) ITALY. Emery, 1908e: 451 (m.). Combination in *Messor:* Emery, 1908e: 451. Variety/subspecies of *capitatus:* Dalla Torre, 1893: 101; of *barbarus:* Emery, 1908e: 451; Forel, 1910a: 9; Santschi, 1921d: 69; Finzi, 1924a: 13; of *instabilis:* Santschi, 1923f: 325. Status as species: Bondroit, 1918: 155; Santschi, 1927c: 232; Kutter, 1927: 99; Santschi, 1934d: 273; Bernard, 1967: 149; Arnol'di, 1977b: 1640. Senior synonym of *pompeianus:* Baroni Urbani, 1971c: 61. Current subspecies: nominal plus *calabricus, capreensis, hesperius, laboriosus.*

*mixtus. Messor semirufus* st. *grandinidus* var. *mixtus* Santschi, 1927c: 232 (w.) TUNISIA. Unavailable name.

#***montanus.*** *Messor sublaeviceps* var. *montanus* Santschi, 1927c: 242 (w.) ALGERIA. Subspecies of *medioruber:* Cagniant, 1968a: 143.

*mutica. Myrmica mutica* Nylander, 1849: 39 (w.q.m.) RUSSIA. Combination in *Aphaenogaster:* Roger, 1863b: 29; in *Stenamma* (*Messor*): Emery, 1898c: 126; in *Messor:* Ruzsky, 1903b: 315. Junior synonym of *structor:* Mayr, 1855: 464; Nylander, 1856b: 85; Smith, F. 1858b: 162. Revived from synonymy as variety/subspecies of *structor:* Emery, 1898c: 126; Ruzsky, 1904b: 4; Bondroit, 1918: 151; Santschi, 1934d: 275. Junior synonym of *rufitarsis:* Bernard, 1967: 144. Revived from synonymy as subspecies of *rufitarsis:* Arnol'di, 1977b: 1643. Revived status as species: Agosti & Collingwood, 1987a: 54. Junior synonym of *structor:* Atanassov & Dlussky, 1992: 114.

#***nadezhdae.*** *Messor kasakorum* subsp. *nadezhdae* Arnol'di, 1969: 83 (s.m.) TURKESTAN.

***nahali.*** *Messor nahali* Tohmé, G. & Tohmé, H. 1981: 151, fig. 22 (w.) SYRIA.

*nigra. Aphaenogaster barbara* var. *nigra* André, 1883b: 355 (w.q.) FRANCE. Emery, 1908e: 447 (m.). Combination in *Stenamma* (*Messor*): Ruzsky, 1902d: 28; in *Messor:* Emery, 1908e: 446. Junior synonym of *capitatus:* Menozzi, 1921: 26; Emery, 1922c: 97.

#***nigricans.*** *Messor barbarus* var. *nigricans* Santschi, 1929e: 142 (q.) MOROCCO.

*nigricans. Messor semirufus* var. *nigricans* Santschi, 1932f: 517. Replacement name for *nigriceps* Santschi, 1929e: 144. [Junior primary homonym of *nigriceps* Santschi, 1925g: 341.] [Unresolved junior primary homonym of *nigricans* Santschi, above.] Junior synonym of *semirufus:* Baroni Urbani, 1974: 227.

*nigriceps. Messor barbarus* var. *nigriceps* Santschi, 1925g: 341 (w.q.) SPAIN. Junior synonym of *barbarus:* Collingwood, 1978: 68.

*nigriceps. Messor semirufus* var. *nigriceps* Santschi, 1929e: 144 (w.) ISRAEL. [Junior primary homonym of *nigriceps* Santschi, above.] Replacement name: *nigricans* Santschi, 1932f: 517.

*nigriventris. Messor braunsi* var. *nigriventris* Stitz, 1923: 150 (w.) NAMIBIA. Junior synonym of *tropicorum:* Bolton, 1982: 354.

***niloticus.*** *Messor niloticus* Santschi, 1938a: 36, fig. 6 (s.w.) EGYPT.

*nobilis. Messor galla* st. *nobilis* Santschi, 1928f: 201 (w.q.) ETHIOPIA. Junior synonym of *galla:* Bolton, 1982: 350.

**#*normandi.*** *Messor lobicornis* st. *normandi* Santschi, 1918a: 37 (w.q.m.) TUNISIA.

*novaki. Messor structor* subsp. *novaki* Finzi, 1929: 92 (w.) YUGOSLAVIA. Junior synonym of *structor:* Atanassov & Dlussky, 1992: 114.

*oblica. Messor sublaeviceps* st. *hoggarensis* var. *oblica* Santschi, 1939d: 72 (w.q.) ALGERIA. Unavailable name.

*obscurior. Messor barbarus* r. *semirufus* var. *obscurior* Crawley, 1920a: 164 (s.w.) IRAQ. Unavailable name; material referred to *semirufus* by Baroni Urbani, 1974: 227.

*obscuriventris. Messor barbarus* subsp. *capitatus* var. *obscuriventris* Karavaiev, 1912a: 9 (s.w.) ALGERIA. Unavailable name.

**#*obscurus.*** *Messor galla* var. *obscurus* Menozzi & Consani, 1952: 61 (w.) ETHIOPIA. [*Messor galla* var. *obscurus* Menozzi, 1939c: 101. *Nomen nudum.*]

***oertzeni.*** *Messor oertzeni* Forel, 1910a: 25 (w.) TURKEY. Senior synonym of *amphigea:* Agosti & Collingwood, 1987a: 54. Current subspecies: nominal plus *amasiensis, carpathous.* See also: Atanassov & Dlussky, 1992: 117.

***olegianus.*** *Messor olegianus* Arnol'di, 1969: 77, figs. 3, 4 (w.m.) TADZHIKISTAN. See also: Tarbinsky, 1976: 62.

***orientalis.*** *Stenamma (Messor) structor* var. *orientalis* Emery, 1898c: 143 (w.q.m.) TURKEY. Combination in *Messor:* Ruzsky, 1903b: 315. Junior synonym of *structor:* Kuznetsov-Ugamsky, 1929d: 12. Revived from synonymy, and raised to species: Collingwood, 1961a: 61. Junior synonym of *darianus:* Pisarski, 1967: 383; Pisarski, 1969b: 306. Revived from synonymy as subspecies of *structor:* Aktac, 1977: 120. Raised to species: Tohmé, G. & Tohmé, H. 1981: 150; Collingwood, 1985: 251.

*parvidens. Messor denticornis* var. *parvidens* Forel, 1910f: 15 (w.) NAMIBIA. Junior synonym of *denticornis:* Bolton, 1982: 349.

***perantennatus.*** *Messor perantennatus* Arnol'di, 1969: 85, figs. 5, 6 (s.m.) TURKESTAN.

***pergandei.*** *Aphaenogaster pergandei* Mayr, 1886d: 448 (w.) U.S.A. Wheeler, W.M. & Creighton, 1934: 376 (s.q.m.); Wheeler, G.C. & Wheeler, J. 1972b: 240 (l.). Combination in *Stenamma (Messor):* Emery, 1895c: 307; in *Novomessor:* Emery, 1915d: 73; in *N. (Veromessor):* Forel, 1917: 234; in *Veromessor:* Wheeler, W.M. & Creighton, 1934: 374; in *Messor:* Bolton, 1982: 341. See also: Smith, D.R. 1979: 1365.

*persicus. Stenamma (Messor) barbarus* r. *himalayanus* var. *persicus* Forel, 1904d: 175 (w.) IRAN. Unavailable name.

**#*personatus.*** *Messor antennatus* st. *personatus* Santschi, 1931a: 3 (w.) MOROCCO.

**#*phasianicus.*** *Messor aegyptiacus* var. *phasianica* Bernard, 1953a: 143 (in key) (w.) ALGERIA.

***piceus.*** *Messor piceus* Stitz, 1923: 150 (w.q.) NAMIBIA. See also: Bolton, 1982: 352.

***picturatus.*** *Messor sublaeviceps* var. *picturatus* Santschi, 1927c: 243 (w.q.) ALGERIA. [First available use of *Messor barbarus* st. *minor* var. *picturata* Santschi, 1921d: 69; unavailable name.] Santschi, 1939d: 71 (q.m.). Subspecies of *minor:* Cagniant, 1968a: 143; of *medioruber:* Cagniant, 1970a: 415. Raised to species: Collingwood, 1985: 251.

***planiceps.*** *Messor planiceps* Stitz, 1917: 344, fig. 11 (w.) ALGERIA.

**#*platyceras.*** *Messor platyceras* Crawley, 1920a: 163, pl. 5, fig. 3 (s.w.) IRAN. Variety of *clivorum:* Emery, 1921b: 210. Subspecies of *structor:* current status.

*plinii. Messor plinii* Santschi, 1912b: 165 (s.w.) KENYA. Combination in *Sphaeromessor:* Bernard, 1985: 48; in *Messor:* **revived combination**. Stirps of *cephalotes:* Santschi, 1917e: 94. Junior synonym of *cephalotes:* Bolton, 1982: 346.

**#*politus.*** *Messor barbarus* var. *politus* Karavaiev, 1912a: 6 (s.w.) ALGERIA. Senior synonym of *dentiscapus:* **new synonymy**.

*pompeianus. Messor minor* var. *pompeianus* Santschi, 1927c: 233 (w.) ITALY. Finzi, 1929: 86 (q.). Junior synonym of *minor:* Baroni Urbani, 1971c: 61.

***postpetiolatus.*** *Messor barbarus* st. *postpetiolatus* Santschi, 1917e: 90, fig. 1 (s.w.) TUNISIA. Raised to species: Bernard, 1981b: 143.

**#*postquadratus.*** *Mesor sublaeviceps* st. *postquadratus* Santschi, 1932f: 516 (w.) ALGERIA. Subspecies of *medioruber:* current status.

*probus. Messor capensis* var. *probus* Wheeler, W.M. 1922a: 805 (w.) SOUTH AFRICA. [First available use of *Messor barbarus* subsp. *capensis* var. *proba* Forel, 1911e: 266; unavailable name.] Junior synonym of *decipiens:* Bolton, 1982: 348.

*pseudoaegyptiaca. Aphaenogaster pseudoaegyptiaca* Emery, 1884a: 384 (w.) SOUTH AFRICA. Arnold, 1920a: 409 (q.m.). Combination in *Stenamma (Messor):* Emery, 1895h: 35; in *Messor:* Santschi, 1917e: 94. Stirps of *capensis:* Santschi, 1917e: 94. Junior synonym of *capensis:* Bolton, 1982: 345.

*punctata. Aphaenogaster barbara* var. *punctata* Forel, 1886d: 248 (w.) INDIA. Junior synonym of *instabilis:* Bingham, 1903: 279; Pisarski, 1970: 306.

*punctaticeps. Messor barbarus* st. *mediorubra* var. *punctaticeps* Santschi, 1910a: 45 (w.) TUNISIA. Unavailable name.

*punctatohimalayanum. Stenamma (Messor) barbarum* r. *himalayanum* var. *punctatohimalayanum* Forel, 1903a: 695 (w.) INDIA. Unavailable name.

**#*ratus.*** *Messor arenarius* var. *ratus* Menozzi, 1933b: 51 (w.) ISRAEL.

***regalis.*** *Cratomyrmex regalis* Emery, 1892d: 572, pl. 15, fig. 16 (q.) NIGERIA. Santschi, 1913c: 307 (w.). Combination in *Messor (Cratomyrmex):* Santschi, 1920d: 377. Senior synonym of *rubea* and *sculpturatus:* Bolton, 1982: 352.

***reticuliventris.*** *Messor barbarus* subsp. *reticuliventris* Karavaiev, 1910b: 67, fig. 33 (w.) TURKESTAN. Raised to species: Collingwood, 1961a: 60; Pisarski, 1967: 384; Pisarski, 1970: 307; Arnol'di, 1977b: 1644.

**#*risianus.*** *Messor aciculatus* var. *risiana* Forel, 1922: 93 (w.) CHINA.

*romana. Messor barbarus* st. *structor* var. *romana* Santschi, 1917e: 91 (w.q.) ROMANIA. Unavailable name (Agosti & Collingwood, 1987a: 61). Material referred to *structor:* Atanassov & Dlussky, 1992: 114.

*rubea. Cratomyrmex regalis* var. *rubea* Santschi, 1913c: 308 (w.) BENIN. [Misspelled as *rubra* by Emery, 1921f: 43.] Junior synonym of *regalis:* Bolton, 1982: 352.

*rubella. Messor platyceras* var. *rubella* Crawley, 1920a: 164 (s.w.m.) IRAN. Junior synonym of *clivorum:* Emery, 1921b: 210.

*rufa. Messor barbarus* subsp. *semirufus* var. *rufa* Forel, 1910c: 250 (w.) ETHIOPIA. Unavailable name; material referred to *galla* by Bolton, 1982: 350.

*rufitarsis. Formica rufitarsis* Fabricius, 1804: 406 (w.) AUSTRIA. Combination in *Aphaenogaster:* Roger, 1863b: 29; in *Messor:* Emery, 1921b: 211. Junior synonym of *structor:* Nylander, 1856b: 85; Smith, F. 1858b: 162; Mayr, 1863: 397; Nasonov, 1889: 37; Emery, 1921f: 73. Revived from synonymy as subspecies of *barbarus:* Emery, 1921b: 211. Subspecies of *structor:* Santschi, 1926f: 290; Menozzi, 1933b: 54. Status as species: Röszler, 1942a: 51; Bernard, 1955a: 361 (in key); Bernard, 1967: 144; Baroni Urbani, 1971c: 63; Kutter, 1977c: 82; Arnol'di, 1977b: 1643; Arnol'di & Dlussky, 1978: 537 (in key). Junior synonym of *structor:* Agosti & Collingwood, 1987a: 54; Casevitz-Weulersse, 1990a: 154.

*rufitarsis. Myrmica rufitarsis* Foerster, 1850b: 485 (w.) ALGERIA. [Unresolved junior primary homonym of *rufitarsis* Fabricius, above.] Junior synonym of *barbarus:* Roger, 1863b: 29.

***rufotestaceus.*** *Myrmica rufotestacea* Foerster, 1850b: 489 (w.) ALGERIA. Emery, 1908e: 439 (s.q.); Forel, 1910a: 9 (m.). Combination in *Aphaenogaster:* Roger, 1863b: 30; in *Messor:* Emery, 1908e: 439. Senior synonym of *thoracica:* Roger, 1863b: 30; of *gracilinodis:* Dalla Torre, 1893: 104; Emery, 1921f: 74; Menozzi, 1933b: 56.

*rufula. Messor galla* var. *rufula* Finzi, 1939a: 162 (w.) ETHIOPIA. [First avialable use of *Messor barbarus* r. *semirufus* var. *rufula* Forel, 1918b: 156; unavailable name.] Junior synonym of *galla:* Bolton, 1982: 350.

***rufus.*** *Messor rufus* Santschi, 1923f: 326 (w.) RUSSIA. [First available use of *Messor barbarus* subsp. *meridionalis* var. *rufa* Karavaiev, 1910a: 272; unavailable name.] Kuznetsov-Ugamsky, 1929d: 7 (q.m.). Stirps of *minor:* Santschi, 1927c: 240; Santschi, 1927d: 247. Revived status as species: Kuznetsov-Ugamsky, 1929b: 45; Arnol'di, 1977b: 1642. Senior synonym of *similis:* Santschi, 1927d: 248; of *karawajevi:* Arnol'di, 1977b: 1642. Current subspecies: nominal plus *striolatus.*

***ruginodis.*** *Messor barbarus* r. *ruginodis* Stitz, 1916: 374, fig. 1 (w.) CONGO. Raised to species (*nomen dubium*): Bolton, 1982: 353.

***rugosus.*** *Aphaenogaster barbara* var. *rugosa* André, 1881b: 75 (w.m.) ISRAEL. Combination in *Messor,* and raised to species: Emery, 1908e: 458. Current subspecies: nominal plus *bodenheimeri, crawleyi.*

*rugosa. Messor lobicornis* var. *rugosa* Karavaiev, 1912a: 10 (w.) ALGERIA. [Junior secondary homonym of *rugosa* André, above.] Replacement name: *rugulosus* Santschi, below.

**#*rugulosus.*** *Messor lobicornis* var. *rugulosus* Santschi, 1926e: 234. Replacement name for *rugosa* Karavaiev, 1912a: 10. [Junior secondary homonym of *rugosa* André, 1881b: 75.]

**#sahlbergi.** *Messor barbarus* r. *sahlbergi* Forel, 1913d: 428 (w.) EGYPT.

*salina. Messor structor* subsp. *striaticeps* var. *salina* Karavaiev, 1937: 172, fig. 1 (s.w.) TURKESTAN. Unavailable name; material referred to *jakowlevi* by Arnol'di, 1977b: 1643.

**sanctus.** *Messor barbarus* subsp. *sanctus* Emery, 1921f: 71 (w.q.) TUNISIA. [First available use of *Stenamma (Messor) barbarum* r. *capitatum* var. *sancta* Forel, 1905b: 176; unavailable name.] Forel, 1907b: 204 (m.); Karavaiev, 1912a: 7 (q.m.). Subspecies/stirps of *minor:* Santschi, 1927c: 235; Finzi, 1929: 88; of *instabilis:* Santschi, 1932c: 70. Raised to species: Bernard, 1956b: 258; Cagniant, 1964: 87; Bernard, 1967: 148.

**#santschii.** *Messor barbarus* var. *santschii* Emery, 1908e: 447 (w.) TUNISIA. [First available use of *Stenamma (Messor) barbarum* r. *capitatum* var. *santschii* Forel, 1905b: 176; unavailable name.] Forel, 1907b: 204 (q.m.); Karavaiev, 1912a: 6 (q.m.).

*scalpturata. Myrmica scalpturata* Nylander, 1856b: 83 (w.) ALGERIA. Junior synonym of *arenarius:* Roger, 1862c: 292.

*schelkovnikovi. Messor melancholicus* subsp. *schelkovnikovi* Arnol'di, 1977b: 1639 (w.q.) TURKESTAN. Junior synonym of *melancholicus:* Arakelian, 1994: 36.

*schencki. Messor capensis* var. *schencki* Wheeler, W.M. 1922a: 805 (w.) NAMIBIA. [First available use of *Messor barbarus* subsp. *capensis* var. *schencki* Forel, 1910f: 15; unavailable name.] Junior synonym of *capensis:* Bolton, 1982: 345.

*sculpturatus. Cratomyrmex sculpturatus* Stitz, 1916: 377, fig. 2 (w.) CONGO. Variety of *regalis:* Santschi, 1920d: 378 (in text); Wheeler, W.M. 1922a: 802. Junior synonym of *regalis:* Bolton, 1982: 352.

***sculpturatus.** **Messor sculpturatus* Carpenter, 1930: 32, pl. 4, fig. 5; pl. 11, fig. 5 (q.) U.S.A. [Unresolved junior secondary homonym of *sculpturatus* Stitz, above.] Generic combination equivocal: Wheeler, W.M. & Creighton, 1934: 360; Bolton, 1982: 341; *incertae sedis* in *Messor.*

*semirubra. Messor barbarus* subsp. *instabilis* var. *semirubra* Karavaiev, 1926b: 100 (s.w.q.) RUSSIA. Unavailable name; material referred to *caducus* by Arnol'di, 1977b: 1638.

**semirufus.** *Aphaenogaster barbara* var. *semirufa* André, 1883b: 355 (w.q.) vicinity of CASPIAN SEA. Combination in *Messor:* Ruzsky, 1905b: 750. Junior synonym of *caducus:* Emery, 1891b: 11. Revived from synonymy as subspecies of *barbarus:* Emery, 1908e: 447; Forel, 1910a: 9; Santschi, 1921a: 110. Raised to species: Santschi, 1927c: 227; Menozzi, 1933b: 52; Collingwood, 1985: 252. Senior synonym of *kutteri* and *nigricans,* and material of the unavailable name *obscurior* referred here: Baroni Urbani, 1974: 227. Current subspecies: nominal plus *emeryi, grandinidus.*

**semoni.** *Stenamma (Messor) barbarum* var. *semoni* Forel, 1906c: 190 (footnote) (w.) ALGERIA. Karavaiev, 1912a: 10 (q.). Combination in *Messor:* Emery, 1908e: 452. Raised to species: Cagniant, 1964: 101.

*sevani. Messor clivorum* subsp. *sevani* Arnol'di, 1977b: 1644 (s.w.) CAUCASUS. [First available use of *Messor structor* subsp. *striaticeps* var. *sevany* Karavaiev, 1926b: 103; unavailable name.] Junior synonym of *structor:* Arakelian, 1994: 39.

**#sillemi.** *Messor excursionis* var. *sillemi* Stärcke, 1935: 262 (w.) TURKESTAN.

*similis. Messor similis* Kuznetsov-Ugamsky, 1927a: 93 (in key) (w.) KIRGIZIA. Junior synonym of *rufus:* Santschi, 1927d: 248; of *karawajevi:* Tarbinsky, 1976: 59; of *rufus:* Arnol'di, 1977b: 1642.

**smithi.** *Veromessor smithi* Cole, 1963: 678, figs. 1, 2, 3 (w.q.m.) U.S.A. Wheeler, G.C. & Wheeler, J. 1972b: 240 (l.). Combination in *Messor:* Bolton, 1982: 341.

**sordidus.** *Aphaenogaster (Messor) barbara* r. *sordida* Forel, 1892g: 452 (w.) SPAIN. Santschi, 1917e: 89 (s.). Combination in *Messor:* Emery, 1897f: 239. Subspecies/stirps of *barbarus:* Santschi, 1925g: 342 (in key); Wheeler, W.M. 1927g: 104. Raised to species: Bondroit, 1918: 152; Cagniant, 1964: 88; Bernard, 1967: 145.

*splendens. Messor barbarus* subsp. *capitatus* var. *splendens* Karavaiev, 1912a: 8 (s.w.) ALGERIA. Unavailable name.

**stoddardi.** *Stenamma (Messor) stoddardi* Emery, 1895c 307 (w.) U.S.A. Combination in *Novomessor:* Emery, 1915d: 73; in *N. (Veromessor)*: Forel, 1917: 235; in *Veromessor:* Wheeler, W.M. & Creighton, 1934: 385; in *Messor:* Bolton, 1982: 341. Current subspecies: nominal plus *chicoensis.*

**striatellus.** *Messor striatellus* Arnol'di, 1969: 81, figs. 4, 5 (w.) TADZHIKISTAN.

**striaticeps.** *Aphaenogaster barbara* var. *striaticeps* André, 1883b: 356 (w.) CAUCASUS. Forel, 1890a: lxix (q.); Karavaiev, 1926b: 102 (m.); Wheeler, G.C. & Wheeler, J. 1953b: 68 (l.). Combination in *A. (Messor)*: Forel, 1890a: lxix; in *Stenamma (Messor)*: Forel, 1902a: 149; Ruzsky, 1902d: 28; in *Messor:* Ruzsky, 1905b: 753. Race of *barbarus:* Forel, 1890a: lxix; Emery, 1891b: 10; Dalla Torre, 1893: 105; of *structor:* Santschi, 1923f: 318; Karavaiev, 1926b: 101; Karavaiev, 1927c: 261. Raised to species: Bernard, 1955a: 355; Cagniant, 1968a: 142; Cagniant, 1970a: 415.

**striatifrons.** *Messor denticornis* var. *striatifrons* Stitz, 1923: 149 (w.) NAMIBIA. Raised to species: Bolton, 1982: 353.

**striativentris.** *Messor barbarus* subsp. *striativentris* Emery, 1908e: 453 (w.) ALGERIA. [First available use of *Messor barbarus* r. *striaticeps* var. *striativentris* Forel, 1894d: 31; unavailable name.] Stirps of *aegyptiacus:* Santschi, 1929e: 142; Santschi, 1939d: 69. Raised to species: Santschi, 1923f: 320; Finzi, 1930b: 15; Cagniant, 1966b: 279.

**#striatulus.** *Aphaenogaster capitatus* var. *striatula* Emery, 1891b: 11 (in text) (w.) TUNISIA. Combination in *Messor:* Emery, 1908e: 453. Variety of *aegyptiacus:* Santschi, 1936c: 200 (misspelled here as *striatus*). Material of the unavailable name *curvispina* referred here: Santschi, 1923f: 321 (in text).

**#striolatus.** *Messor similis* var. *striolatus* Kuznetsov-Ugamsky, 1927a: 92 (in key) (w.) TURKESTAN. Variety of *rufus:* Kuznetsov-Ugamsky, 1929d: 9.

**structor.** *Formica structor* Latreille, 1798: 46 (w.m.) FRANCE. Latreille, 1802c: 237 (q.). Combination in *Atta:* Mayr, 1855: 464; in *Myrmica:* Nylander, 1856b: 85; in *Aphaenogaster:* Roger, 1863b: 29; in *A. (Messor)*: Forel, 1892i: 316; in *Stenamma (Messor)*: Emery, 1898c: 126; in *Messor:* Emery, 1897f: 238; Ruzsky, 1903b: 315. Subspecies of *barbarus:* Emery, 1908e: 455; Forel, 1910a: 9; Emery, 1914d: 156; Santschi, 1917e: 91; Menozzi, 1918: 83; Santschi, 1921a: 110; Emery, 1924b: 12. Status as species: Nasonov, 1889: 37; Ruzsky, 1905b: 728; Forel, 1915d: 35; Bondroit, 1918: 150; Finzi, 1924a: 13; Menozzi, 1926b: 181; Kuznetsov-Ugamsky, 1927a: 92; Finzi, 1930d: 312; Karavaiev, 1934: 108; Bernard, 1967: 144; Baroni Urbani, 1971c: 66. Senior synonym of *aedificator:* Mayr, 1855: 464; of *lapidum:* Nylander, 1856b: 85; of *tyrrhena:* Emery, 1921b: 210; of *rufitarsis:* Nylander, 1856b: 85; Smith, F. 1858b: 162; Mayr, 1863: 397; Nasonov, 1889: 37; Emery, 1921f: 73; Agosti & Collingwood, 1987a: 54; Casevitz-Weulersse, 1990a: 154; Atanassov & Dlussky, 1992: 115; of *clivorum* (and its junior synonyms *rubella,* and *subpolitus*), *jakowlevi, tataricus, turanicus* (and its junior synonym *darianus*), and material of the unavailable name *romana* referred here: Dlussky, Soyunov & Zabelin, 1990: 224; of *mutica, novaki* and *varrialei:* Atanassov & Dlussky, 1992: 114; of *sevani:* Arakelian, 1994: 39. Current subspecies: nominal plus *aegaeus, platyceras, tadzhikorum.*

**subgracilinodis.** *Messor subgracilinoclis* Arnol'di, 1969: 73, fig. 1 (w.q.) RUSSIA. Spelling emended to *subgracilinodis* in Arnol'di, 1969: 87 (caption of fig. 1), and in Arnol'di, 1977b: 1642.

**#sublaeviceps.** *Messor sublaeviceps* Santschi, 1927c: 241 (w.q.) TUNISIA. [First available use of *Messor barbarus* st. *meridionalis* var. *sublaeviceps* Santschi, 1910a: 45; unavailable name.] Subspecies of *medioruber:* Cagniant, 1970a: 416.

**#submuticus.** *Messor lobicornis* var. *submutica* Emery, 1908e: 459 (w.) TUNISIA.

*subpolitus. Messor structor* var. *subpolitus* Kuznetsov-Ugamsky, 1927a: 92 (in key) (w.) TURKESTAN. Junior synonym of *clivorum:* Tarbinsky, 1976: 56.

**sultanus.** *Messor barbarus* var. *sultana* Santschi, 1917e: 89 (w.) ISRAEL. Tohmé, G. & Tohmé, H. 1981: 149 (q.m.). Subspecies/stirps of *barbarus:* Emery, 1922c: 95; Santschi, 1925g: 342 (in key); Santschi, 1927c: 246. Raised to species: Tohmé, G. & Tohmé, H. 1981: 148; Agosti & Collingwood, 1987a: 54.

**#surcoufi.** *Messor aegyptiacus* var. *surcoufi* Santschi, 1923f: 324 (w.) ALGERIA.

**syriacus.** *Messor syriacus* Tohmé, G. 1969: 9 (w.m.) SYRIA. [First available use of *Messor minor* st. *laboriosus* var. *syriacus* Santschi, 1927c: 241; unavailable name.] Tohmé, G. & Tohmé, H. 1981: 146. Subspecies of *minor:* Arnol'di, 1977b: 1640; Arakelian, 1994: 37. Revived status as species: Tohmé, G. & Tohmé, H. 1981: 145; Collingwood, 1985: 252.

**#tadzhikorum.** *Messor rufitarsis* subsp. *tadzhikorum* Arnol'di, 1969: 79 (s.w.q.) TADZHIKISTAN. Subspecies of *structor:* current status.

*tamerlani. Messor minor* st. *aralocaspius* var. *tamerlani* Santschi, 1927c: 240 (w.) TURKESTAN. Unavailable name; material referred to *aralocaspius* by Dlussky, Soyunov & Zabelin, 1990: 218.

*tataricus. Messor tataricus* Ruzsky, 1905b: 738 (w.) TURKESTAN. Subspecies of *clivorum:* Arnol'di, 1977b: 1644. Junior synonym of *structor:* Dlussky, Soyunov & Zabelin, 1990: 224; Atanassov & Dlussky, 1992: 114.

*MESSOR (continued)*

***testaceus.*** *Messor testaceus* Donisthorpe, 1950c: 68 (s.w.) TURKEY.

#***thebaicus.*** *Messor arenarius* st. *thebaicum* Santschi, 1912b: 164 (s.w.) EGYPT.

*thoracica. Atta thoracica* Mayr, 1862: 742, pl. 19, fig. 12 (w.) SYRIA. Junior synonym of *rufotestaceus:* Roger, 1863b: 30.

*tingitanus. Messor barbarus* st. *sordidus* var. *tingitanus* Emery, in Santschi, 1925g: 342 (in key) (w.) MOROCCO. Unavailable name.

*triempressa. Messor barbarus* st. *galla* var. *triempressa* Santschi, 1917e: 92 (w.) CHAD. Unavailable name; material referred to *galla* by Bolton, 1982: 350.

***tropicorum.*** *Messor capensis* var. *tropicorum* Wheeler, W.M. 1922a: 805 (w.) ANGOLA. [First available use of *Messor barbarus* subsp. *capensis* var. *tropicorum* Forel, 1910e: 444; unavailable name.] Raised to species, and senior synonym of *laevifrons* and *nigriventris:* Bolton, 1982: 354.

#***tunetinus.*** *Messor aegyptiacus* var. *tunetinus* Santschi, 1923f: 321 (w.) TUNISIA. [First available use of *Messor barbarus* subsp. *aegyptiaca* var. *tunetina* Emery, 1922c: 96; unavailable name.] Finzi, 1930b: 14 (q.).

*turanicus. Messor structor* subsp. *turanicus* Kuznetsov-Ugamsky, 1927a: 91 (in key) (w.) UZBEKISTAN. Senior synonym of *darianus:* Pisarski, 1967: 383; Pisarski, 1970: 306; Tarbinsky, 1976: 54; Arnol'di, 1977b: 1644 (all these authors give *darianus* as the senior name, but *turanicus* has priority). Junior synonym of *structor:* Dlussky, Soyunov & Zabelin, 1990: 224.

***turcmenochorassanicus.*** *Messor clivorum* subsp. *turcmenochorassanicus* Arnol'di, 1977b: 1639 (w.q.) TURKESTAN. Raised to species: Dlussky, Soyunov & Zabelin, 1990: 227.

*tyrrhena. Stenamma (Messor) structor* var. *tyrrhena* Emery, 1898c: 141 (w.q.m.) ITALY. Combination in *Messor:* Emery, 1908e: 456. Junior synonym of *structor:* Emery, 1921b: 210.

***valentinae.*** *Messor valentinae* Arnol'di, 1969: 75, fig. 1 (s.w.) TADZHIKISTAN. See also: Tarbinsky, 1976: 64.

***variabilis.*** *Messor variabilis* Kuznetsov-Ugamsky, 1927a: 94 (w.) TURKESTAN. Arnol'di, 1969: 87 (m.). Senior synonym of *macrops:* Arnol'di, 1977b: 1646.

*varrialei. Messor barbarus* subsp. *varrialei* Emery, 1921b: 215, fig. 6 (s.) TURKEY. Junior synonym of *structor:* Atanassov & Dlussky, 1992: 114.

***vaucheri.*** *Messor vaucheri* Emery, 1908e: 441, fig. 2 (w.) MOROCCO.

***vicinus.*** *Messor vicinus* Kuznetsov-Ugamsky, 1927a: 94 (in key) (w.) TURKESTAN.

#***virginalis.*** *Messor luridus* var. *virginalis* Santschi, 1927c: 245 (w.) SYRIA.

*vizir. Messor barbarus* st. *sultana* var. *vizir* Santschi, 1927c: 246 (w.) GREECE. Unavailable name.

***wasmanni.*** *Messor barbarus* subsp. *wasmanni* Krausse, 1910: 524 (w.) SARDINIA. [Also described as new by Krausse, 1911: 15.] Finzi, 1929: 83 (q.); Finzi, 1933: 163 (m.). Variety/subspecies of *semirufus:* Santschi, 1927c: 230; Finzi, 1928c: 788; Finzi, 1930d: 312; of *meridionalis:* Baroni Urbani, 1964b: 30; Baroni Urbani, 1968b: 434. Junior synonym of *semirufus:* Baroni Urbani, 1974: 227; of *meridionalis:* Collingwood, 1978: 68. Revived from synonymy, and raised to species: Casevitz-Weulersse, 1990a: 154.

*werneri. Messor barbarus* subsp. *varrialei* var. *werneri* Finzi, 1928c: 789 (w.) GREECE. Unavailable name.

*METACYLINDROMYRMEX:* see under ***CYLINDROMYRMEX.***

***METAPONE*** (Malagasy, Oriental, Indo-Australian, Australasian)

***bakeri.*** *Metapone bakeri* Wheeler, W.M. 1916b: 10, fig. 1 (q.) PHILIPPINES. See also: Wheeler, W.M. 1919h: 186.

***emersoni.*** *Metapone emersoni* Gregg, 1958: 115, fig. 2 (w.) MADAGASCAR.

***gracilis.*** *Metapone gracilis* Wheeler, W.M. 1935b: 38 (q.) PHILIPPINES.

***greeni.*** *Metapone greeni* Forel, 1911h: 449, pl. 14, figs. 1–10 (w.q.m.l.) SRI LANKA. See also: Emery, 1912b: 95; Wheeler, W.M. 1919h: 179; Wheeler, G.C. & Wheeler, J. 1953d: 186.

***hewitti.*** *Metapone hewitti* Wheeler, W.M. 1919e: 62 (m.) BORNEO.

***jacobsoni.*** *Metapone jacobsoni* Crawley, 1924: 389 (q.) SUMATRA.

***johni.*** *Metapone johni* Karavaiev, 1933b: 115, fig. 10 (w.q.m.) SRI LANKA.

***krombeini.*** *Metapone krombeini* Smith, M.R. 1947d: 76 (q.) NEW GUINEA.

***leae.*** *Metapone leae* Wheeler, W.M. 1919h: 183, figs. 3, 4 (q.) AUSTRALIA.

***madagascarica.*** *Metapone madagascarica* Gregg, 1958: 111, fig. 1 (w.q.) MADAGASCAR.

***mjobergi.*** *Metapone mjobergi* Forel, 1915b: 36 (w.q.) AUSTRALIA. See also: Wheeler, W.M. 1919h: 181.

***nicobarensis.*** *Metapone nicobarensis* Tiwari & Jonathan, 1986b: 150, fig. 1 (w.q.) NICOBAR IS.

***sauteri.*** *Metapone sauteri* Forel, 1912m: 763 (q.) TAIWAN. [Also described as new by Forel, 1913f: 189.] See also: Wheeler, W.M. 1919h: 182.

***tillyardi.*** *Metapone tillyardi* Wheeler, W.M. 1919h: 187, fig. 6 (w.) AUSTRALIA.

***tricolor.*** *Metapone tricolor* McAreavey, 1949: 4, figs. 17–19 (q.) AUSTRALIA.

***truki.*** *Metapone truki* Smith, M.R. 1953d: 135 (w.) TRUK I.

****MIANEURETUS*** (U.S.A.)

****mirabilis.*** **Mianeuretus mirabilis* Carpenter, 1930: 38, pl. 3, fig. 5; pl. 10, fig. 1 (q.) U.S.A. (Miocene).

*MICCOSTRUMA:* see under ***SMITHISTRUMA.***

*MICROBOLBOS:* see under ***LEPTOGENYS.***

***MICRODACETON*** (Afrotropical)

***exornatum.*** *Microdaceton exornatum* Santschi, 1913g: 478 (w.) SOUTH AFRICA. [Also described as new by Santschi, 1914e: 33.] Arnold, 1948: 225 (q.). Senior synonym of *laevior* and *leakeyi:* Bolton, 1983: 402.

*laevior. Microdaceton exornatum* var. *laevior* Arnold, 1948: 225 (w.q.) SOUTH AFRICA. Junior synonym of *exornatum:* Bolton, 1983: 402.

*leakeyi. Microdaceton leakeyi* Patrizi, 1947: 219, figs. 1, 2 (q.) KENYA. Bernard, 1953b: 253 (w.). Junior synonym of *exornatum:* Bolton, 1983: 402.

***tibialis.*** *Microdaceton tibialis* Weber, 1952: 30, fig. 25 (w.) ZAIRE. See also: Bolton, 1983: 403.

*MICROMYRMA:* see under ***TAPINOMA.***

*MICTOPONERA:* see under ***GNAMPTOGENYS.***

****MIOMYRMEX*** (U.S.A.)

****impactus.*** **Formica impacta* Cockerell, 1927: 165 (q.) U.S.A. (Miocene). Combination in **Miomyrmex:* Carpenter, 1930: 52.

****striatus.*** **Miomyrmex striatus* Carpenter, 1930: 54, pl. 8, fig. 6; pl. 10, fig. 9 (w.m.) U.S.A. (Miocene).

*MITARA:* see under ***MONOMORIUM.***

#***MOELLERIUS:*** see under ***ACROMYRMEX.***

*MONACIS:* see under ***DOLICHODERUS.***

*MONOCERATOCLINEA:* see under ***DOLICHODERUS.***

*MONOCOMBUS:* see under ***CATAGLYPHIS.***

***MONOMORIUM*** (World wide)

***abeillei.*** *Monomorium abeillei* André, in Emery, 1881b: 531 (footnote) (w.) ISRAEL. [Also described as new by André, 1881b: 67.] Combination in *M. (Xeromyrmex):* Emery, 1922e: 177. Subspecies of *salomonis:* Forel, 1910a: 23; Hamann & Klemm, 1967: 413. Revived status as species: Collingwood, 1985: 269.

***aberrans.*** *Monomorium aberrans* Forel, 1902c: 209 (w.) INDIA. Combination in *M. (Xeromyrmex):* Emery, 1922e: 176. See also: Bolton, 1987: 297.

***abyssinicum.*** *Holcomyrmex abyssinicus* Forel, 1894b: 83 (w.) ETHIOPIA. Viehmeyer, 1923: 91 (q.). Combination in *Monomorium:* Forel, 1910f: 250. See also: Bolton, 1987: 321.

#***adoneum.*** *Monomorium (Xeromyrmex) subopacum* var. *adoneum* Santschi, 1936a: 41, fig. 22 (w.) LEBANON (caption to fig. 22 is spelled *adonis*). [*Monomorium (Xeromyrmex) subopacum* var. *adoneum* Santschi, 1934d: 277. *Nomen nudum.*]

*adulatrix. Wheeleriella adulatrix* Santschi, 1913f: 229 (q.) TUNISIA. Combination in *Monomorium:* Ettershank, 1966: 87. Stirps of *santschii:* Santschi, 1926e: 234. Junior synonym of *santschii:* Bolton, 1987: 291.

***advena.*** *Monomorium advena* Brown & Wilson, 1957b: 244. Replacement name for *Epixenus andrei* Emery, 1908f: 557. [Junior secondary homonym of *Monomorium andrei* Saunders, E. 1890: 204.] Tohmé, H. & Tohmé, G. 1980a: 1097 (w.).

*aequatoriale. Monomorium (Xeromyrmex) bicolor* var. *aequatoriale* Santschi, 1926b: 240 (w.) CAMEROUN. Junior synonym of *bicolor:* Bolton, 1987: 338.

*aequum. Monomorium aequum* Santschi, 1928f: 195, fig. 3 (w.) ZAIRE. Junior synonym of *leopoldinum:* Bolton, 1987: 397.

***affabile.*** *Monomorium affabile* Santschi, 1926b: 235, fig. 3 (w.) ZAIRE. See also: Bolton, 1987: 375.

*africanum. Monomorium orientale* var. *africanum* Forel, 1907g: 78. *Nomen nudum;* attributed to Mayr. See also: Bolton, 1987: 430.

***afrum.*** *Monomorium afrum* André, 1884a: 244 (w.) SUDAN. [Also described as new by André, 1884b: 540.] Arnold, 1926: 226 (q.); Wheeler, G.C. & Wheeler, J. 1955c: 124 (l.). Combination in *M.*

(*Xeromyrmex*): Wheeler, W.M. 1922a: 868. Senior synonym of *asmarensis* and *fultor:* Bolton, 1987: 334.

#***aharonii.*** *Monomorium atomus* var. *aharonii* Menozzi, 1933b: 64 (w.) ISRAEL. [First available use of *Monomorium* (*Martia*) *atomus* r. *mictilis* var. *aharonii* Forel, 1913d: 438; unavailable name.]

#***ajjer.*** *Monomorium* (*Equestrimessor*) *chobauti* subsp. *ajjer* Bernard, 1953a: 165 (w.) ALGERIA.

*ajjer. Monomorium* (*Monomorium*) *andrei* subsp. *ajjer* Bernard, 1953a: 169, fig. 4 (w.) ALGERIA. [Junior primary homonym of *ajjer* Bernard, above.] Replacement name: *bernardi* Ettershank, 1966: 87.

***alamarum.*** *Monomorium alamarum* Bolton, 1987: 367, fig. 58 (w.) NAMIBIA.

***albopilosum.*** *Monomorium albopilosum* Emery, 1895h: 24 (w.) SOUTH AFRICA. Combination in *M.* (*Xeromyrmex*): Santschi, 1919b: 235. Senior synonym of *clarithorax, fingo, paucipilosa* and *thales:* Bolton, 1987: 335.

***algiricum.*** *Epixenus algiricus* Bernard, 1955b: 274, figs. 1, 2 (w.q.l.) ALGERIA. Tinaut & Ortiz, 1988: 167 (m.). Combination in *Monomorium:* Brown & Wilson, 1957b: 245.

***altinode.*** *Monomorium rhopalocerum* var. *altinodis* Santschi, 1910c: 359, fig. 4 (w.) CONGO. Raised to species: Santschi, 1914e: 18. Senior synonym of *bondroiti:* Bolton, 1987: 375.

*altinode. Tetramorium altinode* Santschi, 1935a: 266, fig. 10 (w.) ZAIRE. Combination in *Monomorium:* Bolton, 1980: 199. [Junior secondary homonym of *altinode* Santschi, above.] Replacement name: *malatu* Bolton, 1987: 399.

*amblyops* Emery, 1894; see under ***TRANOPELTA.***

***anceps.*** *Monomorium subopacum* var. *anceps* Emery, 1895h: 24 (w.) SOUTH AFRICA. Combination in *M.* (*Xeromyrmex*): Wheeler, W.M. 1922a: 872. Raised to species: Bolton, 1987: 336.

***andrei.*** *Monomorium andrei* Saunders, E. 1890: 204 (w.) GIBRALTAR. Junior synonym of *minutum:* Collingwood, 1978: 69. Revived from synonymy: **revived status**. Current subspecies: nominal plus *bernardi, fur, furunculum.*

*andrei. Epixenus andrei* Emery, 1908f: 557, fig. 5 (q.m.) ISRAEL. [Junior secondary homonym of *andrei* Saunders, above.] Replacement name: *advena* Brown & Wilson, 1957b: 244.

*angusticlava. Monomorium* (*Monomorium*) *angusticlava* Donisthorpe, 1947d: 189 (w.) NEW GUINEA. Junior synonym of *floricola:* Bolton, 1987: 390.

***angustinode.*** *Monomorium angustinode* Forel, 1913b: 334 (w.) ZAIRE. See also: Bolton, 1987: 376.

***annamense.*** *Monomorium* (*Monomorium*) *annamense* Donisthorpe, 1941k: 238. Replacement name for *longiceps* Karavaiev, 1935a: 94. [Junior primary homonym of *longiceps* Wheeler, W.M. 1934d: 146.]

***antarcticum.*** *Atta antarctica* Smith, F. 1858b: 167 [not White, A. 1846; see Bolton, 1987: 288] (q.) NEW ZEALAND. Emery, 1887b: 458 (w.m.); Wheeler, G.C. & Wheeler, J. 1955c: 123 (l.). Combination in *Monomorium:* Mayr, 1886c: 360; in *M.* (*Notomyrmex*): Emery, 1915i: 190; in *Notomyrmex:* Kusnezov, 1960b: 343; in *Chelaner:* Ettershank, 1966: 96; in *Monomorium:* Bolton, 1987: 286. Senior synonym of *fulvum:* Emery, 1887b: 458; of *integrum, nitidum, succineum* and *suteri:* Brown, 1958h: 29; of *rapaense:* Wilson & Taylor, 1967: 62. See also: Wheeler, G.C. & Wheeler, J. 1960b: 16.

*antiguensis. Formica antiguensis* Fabricius, 1793: 357 (w.) WEST INDIES. Junior synonym of *pharaonis:* Roger, 1862c: 294.

***antipodum.*** *Monomorium antipodum* Forel, 1901e: 377 (w.) NEW ZEALAND. See also: Brown, 1958h: 35.

***aper.*** *Monomorium aper* Emery, 1914f: 409, pl. 13, fig. 6 (w.m.) NEW CALEDONIA. Combination in *M.* (*Notomyrmex*): Emery, 1922e: 170; in *Chelaner:* Ettershank, 1966: 96; in *Monomorium:* Taylor, 1987b: 4. Current subspecies: nominal plus *dubium.*

#***apuleii.*** *Monomorium* (*Xeromyrmex*) *subopacum* var. *apuleii* Santschi, 1927d: 243 (w.q.m.) TUNISIA.

***arboreum.*** *Monomorium* (*Monomorium*) *minutum* subsp. *arboreum* Weber, 1943c: 360 (w.q.) SUDAN. Raised to species: Bolton, 1987: 377.

*arenarium. Monomorium* (*Xeromyrmex*) *salomonis* st. *arenarium* Santschi, 1938a: 39 (w.) EGYPT. *Nomen nudum.* [Also *arenarium* Santschi, 1939c: 2; *nomen nudum.* The name may be an error for *areniphilum.*]

***areniphilum.*** *Monomorium salomonis* var. *areniphila* Santschi, 1911f: 84 (w.) TUNISIA. Emery, 1915e: 378 (q.). Combination in *M.* (*Xeromyrmex*): Emery, 1922e: 177. Stirps of *salomonis:* Santschi, 1936a: 50. Raised to species: Collingwood, 1985: 269. Senior synonym of *lepineyi* and *pullula:* Bolton, 1987: 336.

***armstrongi.*** *Monomorium* (*Holcomyrmex*) *armstrongi* McAreavey, 1949: 10, figs. 30–35 (s.w.q.m.) AUSTRALIA. Combination in *Chelaner:* Ettershank, 1966: 96; in *Monomorium:* Taylor, 1987b: 2.

***arnoldi.*** *Monomorium arnoldi* Forel, 1913a: 137 (w.) ZIMBABWE. See also: Bolton, 1987: 378.

*arnoldi. Syllophopsis arnoldi* Santschi, 1921c: 120, fig. 2 (w.) SOUTH AFRICA (name misspelled as *Syllopsis* in publication). [Junior secondary homonym of *arnoldi* Forel, above.] Replacement name: *jonesi* Arnold, 1952a: 465.

*asmarensis. Monomorium afrum* var. *asmarensis* Forel, 1910c: 250 (w.m.) ETHIOPIA. Junior synonym of *afrum:* Bolton, 1987: 334.

*atomaria. Myrmica atomaria* Gerstäcker, 1859: 263 (w.) MOZAMBIQUE. [Also described as new by Gerstäcker, 1862: 518.] Junior synonym of *ominosa:* Roger, 1863b: 31.

***atomum.*** *Monomorium atomus* Forel, 1902c: 210 (w.) INDIA. Bingham, 1903: 208 (q.). Combination in *M.* (*Martia*): Forel, 1910c: 252. Current subspecies: nominal plus *aharonii, cooperi, integrium, procax.*

*atra. Myrmica* (*Monomorium*) *atra* Buckley, 1867: 342 (q.) U.S.A. Junior synonym of *minimum:* Emery, 1895c: 274.

*atrata. Monomorium salomonis* var. *atrata* Santschi, 1912c: 147 (w.) TUNISIA. Junior synonym of *sommieri:* Emery, 1916b: 161; Santschi, 1921e: 170.

***australe.*** *Monomorium subopacum* r. *australe* Emery, 1886: 363 (w.q.) SOUTH AFRICA. Forel, 1913j: 216 (q.). Combination in *M.* (*Parholcomyrmex*): Santschi, 1917b: 282; Wheeler, W.M. 1922a: 873. Subspecies/race of *salomonis:* Forel, 1913j: 216; Arnold, 1916: 225. Raised to species: Dalla Torre, 1893: 65; Santschi, 1917b: 282; Bolton, 1987: 337. Material of the unavailable name *laeviceps* referred here: Bolton, 1987: 337.

***australicum.*** *Monomorium subcoecum* subsp. *australicum* Forel, 1907a: 20 (w.) AUSTRALIA. Raised to species: Ettershank, 1966: 87.

***baal.*** *Monomorium* (*Holcomyrmex*) *dentigerum* var. *baal* Wheeler, W.M. & Mann, 1916: 171 (w.) SYRIA. Raised to species: Agosti & Collingwood, 1987a: 55.

***balathir.*** *Monomorium balathir* Bolton, 1987: 378 (w.q.) BURKINA FASO.

***banksi.*** *Monomorium* (*Martia*) *banksi* Forel, 1910d: 123 (w.q.) PHILIPPINES. Combination in *M.* (*Lampromyrmex*): Emery, 1922e: 183.

***barbatulum.*** *Monomorium barbatulum* Mayr, 1877: 17 (w.) TURKESTAN. [Redescribed in Latin: Mayr, 1880: 36.] Emery, 1889a: 441 (q.m.). Combination in *M.* (*Xeromyrmex*): Emery, 1922e: 176. See also: Tarbinsky, 1976: 79.

#***barretti.*** *Monomorium* (*Lampromyrmex*) *fraterculus* var. *barretti* Santschi, 1928e: 467 (w.) AUSTRALIA. Subspecies of *fraterculum:* Taylor & Brown, D.R. 1985: 70.

*basalis. Myrmica basalis* Smith, F. 1858b: 125 (w.) SRI LANKA. Combination in *Monomorium:* Mayr, 1865: 92. Junior synonym of *destructor:* Forel, 1894b: 86.

#***bedui.*** *Monomorium* (*Xeromyrmex*) *salomonis* var. *bedui* Santschi, 1936c: 201 (w.q.) MOROCCO. [First available use of *Monomorium* (*Xeromyrmex*) *salomonis* st. *areniphilum* var. *bedui* Santschi, 1936a: 50; unavailable name.]

*belli. Monomorium salomonis* r. *herrero* var. *belli* Forel, 1914d: 245 (w.) SOUTH AFRICA. Unavailable name; material referred to *willowmorense* by Bolton, 1987: 364.

***bequaerti.*** *Monomorium* (*Martia*) *bequaerti* Forel, 1913b: 334 (w.) ZAIRE. Combination in *M.* (*Lampromyrmex*): Wheeler, W.M. 1922a: 875. See also: Bolton, 1987: 379.

#***bernardi.*** *Monomorium andrei* subsp. *bernardi* Ettershank, 1966: 87. Replacement name for *ajjer* Bernard, 1953a: 169. [Junior primary homonym of *ajjer* Bernard, 1953a: 165.

***bevisi.*** *Monomorium bevisi* Arnold, 1944: 10, fig. 16 (w.) LESOTHO. See also: Bolton, 1987: 379.

***bicolor.*** *Monomorium bicolor* Emery, 1877b: 368 (w.) ETHIOPIA. Mayr, 1901b: 7 (q.). Combination in *M.* (*Xeromyrmex*): Wheeler, W.M. 1922a: 868. Stirps of *subopacum:* Santschi, 1927d: 244; Santschi, 1935b: 268. Revived status as species: Santschi, 1937d: 222; Bolton, 1987: 338. Senior synonym of *coerulescens:* Santschi, 1914d: 353 (in key); Bolton, 1987: 338; of *aequatoriale, rufibasis, uelense* and *tropicale:* Bolton, 1987: 338. Current subspecies: nominal plus *judaicum.*

***bicorne.*** *Monomorium bicorne* Forel, 1907h: 276 (q.) AUSTRALIA. Combination in *M.* (*Notomyrmex*): Emery, 1922e: 170; in *Chelaner:* Ettershank, 1966: 96; in *Monomorium:* Taylor, 1987b: 2.

*bidentatum* Mayr, 1887; see under ***ANTICHTHONIDRIS.***

*bimaculatoides. Monomorium bicolor* subsp. *personatum* var. *bimaculatoides* Ettershank, 1966: 87. [Unnecessary replacement name for *bimaculatum* Santschi]. Unavailable name.

***bimaculatum.*** *Monomorium bimaculatum* Wheeler, W.M. 1928c: 19 (w.) HONG KONG.

*bimaculatum. Monomorium* (*Xeromyrmex*) *bicolor* st. *personatum* var. *bimaculatum* Santschi, 1937d: 223 (w.)

ANGOLA. Unavailable name; material referred to *personatum* by Bolton, 1987: 356.

***binatu.*** *Monomorium binatu* Bolton, 1987: 380, fig. 78 (w.) ZIMBABWE.

***biroi.*** *Monomorium biroi* Forel, 1907a: 19 (w.) INDIA.

*biroi. Epixenus biroi* Forel, 1910a: 21 (q.) CRETE. Combination in *Monomorium:* Brown & Wilson, 1957b: 244. [Junior secondary homonym of *biroi* Forel, 1907a: 19, above.] Replacement name: *biroianum* **new name**.

***biroianum.*** *Monomorium biroianum* **new name**. Replacement name for *Epixenus biroi* Forel, 1910a: 21. [Junior secondary homonym of *Monomrium biroi* Forel, 1907a: 19.]

***bodenheimeri.*** *Monomorium* (*Equestrimessor*) *bodenheimeri* Menozzi, 1929e: 125, figs. H, I (w.) ISRAEL.

***boerorum.*** *Monomorium minutum* var. *boerorum* Forel, 1910e: 442 (w.q.) SOUTH AFRICA. Raised to species: Bolton, 1987: 381.

*boerorum. Monomorium* (*Syllophopsis*) *modestum* var. *boerorum* Santschi, 1915c: 260, fig. 9 (w.) SOUTH AFRICA. [Junior primary homonym of *boerorum* Forel, above.] Replacement name: *transwaalensis:* Emery, 1922e: 175.

#***bogischi.*** *Monomorium subapterum* var. *bogischi* Wheeler, W.M. 1917b: 112 (w.q.) AUSTRALIA. Combination in *M.* (*Notomyrmex*): Emery, 1922e: 170; in *Chelaner:* Ettershank, 1966: 96; in *Monomorium:* Taylor, 1987b: 4.

***boltoni.*** *Monomorium boltoni* Espadaler & Agosti, 1987: 295, fig. 1 (w.q.) CAPE VERDE IS.

*bondroiti. Monomorium altinode* var. *bondroiti* Santschi, 1920b: 10, fig. 1 (w.) ZAIRE. Junior synonym of *altinode:* Bolton, 1987: 375.

***borlei.*** *Monomorium* (*Monomorium*) *springvalense* var. *borlei* Santschi, 1937d: 227, figs. 22–24 (w.) ANGOLA. Raised to species: Bolton, 1987: 381.

***brasiliense.*** *Monomorium minutum* subsp. *brasiliense* Forel, 1908c: 361 (w.) BRAZIL. Subspecies of *monomorium:* Brandão, 1991: 357. Raised to species: **new status**.

***braunsi.*** *Monomorium braunsi* Mayr, 1901b: 7 (w.) SOUTH AFRICA. See also: Bolton, 1987: 382.

*brevicorne* Emery, 1893; see under ***VOLLENHOVIA.***

***broomense.*** *Monomorium* (*Mitara*) *laeve* r. *broomense* Forel, 1915b: 74 (w.) AUSTRALIA. Raised to species: Taylor & Brown, D.R. 1985: 70.

*bruneum. Monomorium latinode* var. *bruneum* Emery, 1893f: 243 (w.) SRI LANKA. Junior synonym of *latinode:* Bolton, 1987: 429.

*bulawayensis. Monomorium* (*Mitara*) *exiguum* var. *bulawayensis* Forel, 1913j: 217 (w.) ZIMBABWE. Combination in *M.* (*Lampromyrmex*): Wheeler, W.M. 1922a: 876. Junior synonym of *exiguum:* Bolton, 1987: 388.

*bulawayense. Monomorium amblyops* r. *bulawayense* Forel, 1914d: 247 (w.) ZIMBABWE. Arnold, 1916: 237 (q.m.). [Junior primary homonym of *bulawayensis* Forel, above.] Replacement name: *prossae* Forel, 1916: 418.

***butteli.*** *Monomorium butteli* Forel, 1913k: 54 (w.q.) SUMATRA. Current subspecies: nominal plus *demochrum.*

***buxtoni.*** *Monomorium buxtoni* Crawley, 1920a: 165, pl. 5, fig. 1 (w.) IRAQ.

*cabrerae. Paraphacota cabrerae* Santschi, 1919g: 405, fig. 1 (m.) CANARY IS. Santschi, 1927d: 242 (w.q.). Combination in *Monomorium,* and variety of *subopacum:* Santschi, 1927d: 241. Junior synonym of *subopacum:* Bolton, 1987: 360.

***captator.*** *Monomorium* (*Monomorium*) *captator* Santschi, 1936a: 43 (w.) ZAIRE. [*Monomorium captator* Santschi, 1932a: 385; *nomen nudum.*] See also: Bolton, 1987: 382.

***carbo.*** *Monomorium salomonis* var. *carbo* Forel, 1910c: 251 (w.) ETHIOPIA. Combination in *M.* (*Xeromyrmex*): Wheeler, W.M. 1922a: 870. Raised to species: Bolton, 1987: 339.

***carbonarium.*** *Myrmica carbonaria* Smith, F. 1858b: 127 (w.q.m.) MADEIRA I. Combination in *Monomorium:* Roger, 1863b: 31. Junior synonym of *minutum:* Mayr, 1862: 753 (in text). Variety/race of *minutum:* Emery, 1893c: 82; Dalla Torre, 1893: 67; Emery, 1908h: 681. Status as species: Emery, 1890b: 48; Forel, 1899c: 78; Santschi, 1933a: 21; Ettershank, 1966: 88; Collingwood, 1985: 270.

***centrale.*** *Monomorium centrale* Forel, 1910b: 28 (w.) AUSTRALIA. Combination in *M.* (*Notomyrmex*): Emery, 1922e: 169; in *Chelaner:* Ettershank, 1966: 96; in *Monomorium:* Taylor, 1987b: 2.

***chinense.*** *Monomorium minutum* var. *chinensis* Santschi, 1925f: 86 (w.) CHINA. Wheeler, W.M. 1929f: 6 (q.m.). Raised to species: Bolton, 1987: 288 (in text).

***chobauti.*** *Holcomyrmex chobauti* Emery, 1896j: 418, fig. (w.) ALGERIA. Combination in *Monomorium:* Emery, 1908h: 672; in *M.* (*Equestrimessor*): Santschi, 1919d: 92. Senior synonym of *faf:* Emery, 1922e: 176. See also: Bolton, 1987: 297. Current subspecies: nominal plus *ajjer.*

#***cinctum.*** *Monomorium rubriceps* var. *cinctum* Wheeler, W.M. 1917b: 113, fig. 3 (w.q.) AUSTRALIA. Combination in *Chelaner:* Ettershank, 1966: 96; in *Monomorium:* Taylor, 1987b: 3. Subspecies of *rubriceps:* Taylor & Brown, D.R. 1985: 58.

*cinnabari. Monomorium cinnabari* Roger, 1863a: 199 (w.) CUBA. Junior synonym of *floricola:* Wheeler, W.M. 1913b: 486.

*clarithorax. Monomorium* (*Xeromyrmex*) *albopilosum* var. *clarithorax* Santschi, 1919b: 235 (w.) SOUTH AFRICA. Junior synonym of *albopilosum:* Bolton, 1987: 335.

#***clarum.*** *Holcomyrmex glaber* var. *clarus* Forel, 1902c: 219 (w.) INDIA. Combination in *Monomorium:* Emery, 1922e: 181.

*claveaui. Monomorium salomonis* subsp. *subopacum* var. *claveaui* Emery, 1922e: 178. [Unnecessary replacement name for *senegalensis* Santschi.] Unavailable name.

***clavicorne.*** *Monomorium clavicorne* André, 1881b: 68, pl. 3, fig. 9 (w.) ISRAEL. Combination in *M.* (*Lampromyrmex*): Wheeler, W.M. 1922a: 876. Variety of *orientale:* Mayr, 1904b: 4; Emery, 1908h: 685. Revived status as species: Santschi, 1915a: 58; Emery, 1922e: 183. Current subspecies: nominal plus *punicum.*

*coecum* Forel, 1911; see under ***BONDROITIA.***

*coerulescens. Monomorium bicolor* var. *coerulescens* Santschi, 1912c: 148 (w.) DJIBOUTI. Junior synonym of *bicolor:* Santschi, 1914d: 353 (in key); Bolton, 1987: 338.

***compressum.*** *Monomorium minimum* subsp. *compressum* Wheeler, W.M. 1914b: 43 (w.q.) MEXICO. Raised to species: DuBois, 1986: 93.

***consternens.*** *Myrmica consternens* Walker, 1859: 374 (w.m.) SRI LANKA. Combination in *Monomorium:* Donisthorpe, 1932b: 576.

*contigua. Myrmica contigua* Smith, F. 1858b: 125 (q.) SRI LANKA. Junior synonym of *pharaonis:* Mayr, 1886c: 359.

#***cooperi.*** *Monomorium* (*Lampromyrmex*) *atomus* subsp. *cooperi* Donisthorpe, 1942a: 29 (w.q.m.) EGYPT.

*cotterelli. Diplomorium cotterelli* Donisthorpe, 1942f: 217 (w.) GHANA. Junior synonym of *rosae:* Bolton, 1987: 408.

***crawleyi.*** *Monomorium* (*Monomorium*) *crawleyi* Santschi, 1930a: 66 (w.) ETHIOPIA. See also: Bolton, 1987: 383.

***creticum.*** *Monomorium abeillei* var. *creticum* Emery, 1895b: 66 (w.) CRETE. Karavaiev, 1927b: 107 (q.). Combination in *M.* (*Xeromyrmex*): Emery, 1922e: 178. Subspecies of *salomonis:* Forel, 1910a: 23 (in text). Raised to species: Karavaiev, 1927b: 107.

***creticum.*** *Epixenus creticus* Emery, 1908f: 558, fig. 5 (m.) CRETE. [Unresolved junior secondary homonym of *creticum* Emery, above.] Combination in *Monomorium:* Brown & Wilson, 1957b: 245.

***criniceps.*** *Holcomyrmex criniceps* Mayr, 1879: 672 (w.) INDIA. Bingham, 1903: 282 (s.). Combination in *Monomorium:* Emery, 1922e: 181. Current subspecies: nominal plus *nigrum, rubrum.*

#***crinicipitoscabriceps.*** *Holcomyrmex scabriceps* var. *crinicipitoscabriceps* Forel, 1902c: 220 (w.) INDIA. Combination in *Monomorium:* Emery, 1922e: 182.

*cristatum* Santschi, 1912; see under ***CARDIOCONDYLA.***

***croceiventre.*** *Monomorium croceiventre* Emery, 1914f: 407 (w.) NEW CALEDONIA. Combination in *M.* (*Notomyrmex*): Emery, 1922e: 169; in *Chelaner:* Ettershank, 1966: 96; in *Monomorium:* Taylor, 1987b: 4.

***cryptobium.*** *Syllophopsis cryptobia* Santschi, 1921c: 119, fig. 2 (w.) ZAIRE. Combination in *Monomorium:* Bolton, 1987: 421.

*cupidum. Monomorium* (*Xeromyrmex*) *salomonis* st. *sommieri* var. *cupidum* Santschi, 1936a: 59, fig. 1 (w.q.) TUNISIA. Unavailable name.

***cyaneum.*** *Monomorium minimum* subsp. *cyaneum* Wheeler, W.M. 1914b: 43 (w.q.) MEXICO. DuBois, 1986: 84 (m.). Raised to species: DuBois, 1981: 35. Senior synonym of *emersoni:* DuBois, 1986: 82.

***dakarense.*** *Monomorium bicolor* st. *dakarense* Santschi, 1914d: 353 (diagnosis in key) (w.) SENEGAL. Stirps of *subopacum:* Santschi, 1927d: 245. Raised to species: Bolton, 1987: 339.

***damarense.*** *Monomorium salomonis* subsp. *damarense* Forel, 1910f: 17 (w.) NAMIBIA. Raised to species, and senior synonym of *unicolor:* Bolton, 1987: 340. See also: Santschi, 1936a: 39.

*debilis*. *Aphaenogaster debilis* Walker, 1871: 11 (w.q.) EGYPT. Junior synonym of *salomonis:* Emery & Forel, 1879: 456.

*decamerum* Emery, 1901; see under ***ANILLOMYRMA***.

***delagoense.*** *Monomorium salomonis* r. *delagoense* Forel, 1894b: 87 (w.) MOZAMBIQUE. Forel, 1910e: 441 (q.m.). Combination in *M.* (*Xeromyrmex*): Wheeler, W.M. 1922a: 870. Raised to species: Santschi, 1928f: 192. Senior synonym of *grahamstownense* and *lacrymans:* Bolton, 1987: 341.

***demisum.*** *Monomorium* (*Xeromyrmex*) *demisum* Santschi, 1936a: 36, fig. 26 (w.) TRANSCASPIA.

#***demochrum.*** *Monomorium butteli* subsp. *demochrum* Viehmeyer, 1916a: 131 (w.) SINGAPORE.

*denticulatum* Mayr, 1887; see under ***ANTI-CHTHONIDRIS***.

***dentigerum.*** *Atta dentigera* Roger, 1862b: 259 (w.) SYRIA. Emery, 1908h: 668 (q.m.). Combination in *Aphaenogaster:* Roger, 1863b: 30; in *Holcomyrmex:* André, 1883b: 346; in *Monomorium:* Emery, 1908h: 667.

*despecta*. *Monomorium destructor* var. *despecta* Menozzi, 1931a: 154 (w.) ETHIOPIA. [First available use of *Monomorium destructor* subsp. *kalahariense* var. *despecta* Forel, 1910c: 252; unavailable name.] Forel, 1913b: 331 (q.). Junior synonym of *oscaris:* Bolton, 1987: 326.

***destructor.*** *Atta destructor* Jerdon, 1851: 105 (w.) INDIA. Bingham, 1903: 209 (q.m.). Combination in *Monomorium:* Dalla Torre, 1893: 66; in *M.* (*Parholcomyrmex*): Wheeler, W.M. 1922a: 874. Senior synonym of *ominosa* (and its junior synonym *atomaria*): Dalla Torre, 1893: 66; of *basalis:* Forel, 1894b: 86; of *vexator:* Donisthorpe, 1932c: 468; of *gracillima:* Bolton, 1987: 324. See also: Smith, D.R. 1979: 1382; Bolton, 1987: 324.

***dichroum.*** *Monomorium dichroum* Forel, 1902c: 212 (w.q.) INDIA. Imai, Baroni Urbani, *et al.* 1984: 7 (k.).

***dictator.*** *Monomorium* (*Xeromyrmex*) *bicolor* st. *dictator* Santschi, 1937d: 224, figs. 20, 21 (w.q.) ANGOLA. Raised to species, and material of the unavailable name *impuriceps* referred here: Bolton, 1987: 341.

#***didonis.*** *Monomorium* (*Xeromyrmex*) *salomonis* var. *didonis* Santschi, 1921b: 425 (w.) TUNISIA. Santschi, 1921e: 168 (q.m.). Stirps of *salomonis:* Santschi, 1936a: 54.

***dilatatum.*** *Monomorium* (*Parholcomyrmex*) *dilatatum* Bernard, 1977: 30, fig. 2 (w.) ALGERIA.

***disertum.*** *Monomorium salomonis* var. *diserta* Forel, 1913j: 216 (w.) ZIMBABWE. Combination in *M.* (*Xeromyrmex*): Wheeler, W.M. 1922a: 872. Stirps of *termitarium:* Santschi, 1928f: 194. Raised to species, and material of the unavailable name *petulans* referred here: Bolton, 1987: 342.

***disoriente.*** *Monomorium disoriente* Bolton, 1987: 383, fig. 83 (w.) TANZANIA.

*dispar*. *Monomorium dispar* Emery, 1895h: 24 (w.) SOUTH AFRICA. Combination in *M.* (*Parholcomyrmex*): Wheeler, W.M. 1922a: 874. Junior synonym of *oscaris:* Bolton, 1987: 326.

*distinctum*. *Monomorium* (*Xeromyrmex*) *distinctum* Arnold, 1944: 11, fig. 18 (w.m.) SOUTH AFRICA. Junior synonym of *havilandi:* Bolton, 1987: 368.

#***doddi.*** *Monomorium* (*Parholcomyrmex*) *rothsteini* st. *doddi* Santschi, 1919a: 328 (w.) AUSTRALIA. Combination in *Chelaner:* Ettershank, 1966: 96; in *Monomorium:* Taylor, 1987b: 3.

***dolatu.*** *Monomorium dolatu* Bolton, 1987: 384 (w.) CAMEROUN.

*dolichops*. *Monomorium* (*Xeromyrmex*) *setuliferum* var. *dolichops* Santschi, 1928f: 194 (w.) ZIMBABWE. Junior synonym of *notulum:* Bolton, 1987: 370.

*domestica*. *Myrmica domestica* Shuckard, 1838: 627 (w.q.) GREAT BRITAIN. Smith, F. 1851: 119 (m.). Junior synonym of *pharaonis:* Roger, 1862c: 294. See also: Curtis, 1854: 217; Bolton, 1987: 356.

***donisthorpei.*** *Monomorium* (*Mitara*) *donisthorpei* Crawley, 1915a: 134 (w.) AUSTRALIA. Combination in *M.* (*Lampromyrmex*): Emery, 1922e: 183.

***drapenum.*** *Monomorium drapenum* Bolton, 1987: 342, fig. 48 (w.) NAMIBIA.

***draxocum.*** *Monomorium draxocum* Bolton, 1987: 385 (w.) CAMEROUN.

#***dubium.*** *Monomorium aper* var. *dubia* Emery, 1914f: 410 (w.) NEW CALEDONIA. Combination in *M.* (*Notomyrmex*): Emery, 1922e: 170; in *Chelaner:* Ettershank, 1966: 96; in *Monomorium:* Taylor, 1987b: 4.

***ebangaense.*** *Monomorium* (*Xeromyrmex*) *bicolor* st. *ebangaense* Santschi, 1937d: 225, figs. 17–19 (w.) ANGOLA. Raised to species, and senior synonym of *nyasae:* Bolton, 1987: 367.

***ebeninum.*** *Monomorium minutum* r. *ebeninum* Forel, 1891b: 165 (in text) (w.) ANTILLES. Forel, 1893g: 388 (q.); Forel, 1897b: 300 (m.); Wheeler, W.M. 1905b: 88 (q.m.). Subspecies/race of *carbonarium:* Wheeler, W.M. 1905b: 88; Forel, 1912g: 3; Mann, 1922: 29; Wheeler, W.M. 1942: 198. Raised to species: Forel, 1913l: 218; DuBois, 1986: 103.

#***ebraicum.*** *Monomorium* (*Xeromyrmex*) *subopacum* var. *ebraicum* Menozzi, 1933b: 62 (w.m.) ISRAEL.

***edentatum.*** *Monomorium edentatum* Emery, 1897d: 562 (w.) NEW GUINEA. Combination in *M.* (*Notomyrmex*): Emery, 1922e: 169; in *Chelaner:* Ettershank, 1966: 96; in *Monomorium:* **revived combination**.

***effractor.*** *Monomorium effractor* Bolton, 1987: 292. Replacement name for *Wheeleriella wroughtoni* Forel, 1910a: 7. [Junior secondary homonym of *Monomorium wroughtoni* Forel, 1902c: 209.]

***egens.*** *Monomorium egens* Forel, 1910e: 443 (w.) CAMEROUN. Senior synonym of *jucundum* and *longiusculum:* Bolton, 1987: 385.

***elgonense.*** *Syllophopsis elgonensis* Santschi, 1935b: 267, fig. 4 (w.) KENYA. Combination in *Monomorium:* Bolton, 1987: 422.

***elongatum.*** *Monomorium elongatum* Smith, F. 1876a: 448 (w.q.) RODRIGUEZ I.

***emarginatum.*** *Monomorium emarginatum* DuBois, 1986: 101, figs. 111–118 (w.q.) U.S.A.

*emersoni*. *Monomorium minimum* subsp. *emersoni* Gregg, 1945b: 66 (w.q.) U.S.A. Junior synonym of *minimum:* Creighton, 1950a: 219; Smith, D.R. 1979: 1382; of *cyaneum:* DuBois, 1986: 82.

***emeryi.*** *Monomorium emeryi* Mayr, 1895: 132 (w.) MOZAMBIQUE. Arnold, 1916: 213 (q.m.). Current subspecies: nominal plus *laevior.* See also: Bolton, 1987: 325.

***epinotale.*** *Monomorium* (*Parholcomyrmex*) *epinotale* Santschi, 1923e: 281 (w.) ZAIRE. See also: Bolton, 1987: 326.

***ergatogyna.*** *Monomorium minutum* subsp. *ergatogyna* Wheeler, W.M. 1904d: 269 (w.q.) U.S.A. DuBois, 1986: 88 (m.). Subspecies of *minimum:* Wheeler, W.M. 1914b: 42. Raised to species: Essig, 1926: 857. Junior synonym of *minimum:* Creighton, 1950a: 219. Revived from synonymy, with status as species: DuBois, 1986: 86.

***esharre.*** *Monomorium esharre* Bolton, 1987: 343 (w.) NAMIBIA.

*estherae*. *Monomorium* (*Monomorium*) *estherae* Weber, 1943c: 361, pl. 15, fig. 18 (w.) SUDAN. Junior synonym of *leopoldinum:* Bolton, 1987: 397.

***evansi.*** *Monomorium* (*Holcomyrmex*) *evansi* Donisthorpe, 1918b: 199 (w.q.m.) IRAQ. See also: Bolton, 1987: 321.

***excelsior.*** *Monomorium tchelichofi* var. *excelsior* Arnold, 1926: 227 (w.m.) SOUTH AFRICA. Raised to species, and senior synonym of *speculiceps:* Bolton, 1987: 344.

***excensurae.*** *Monomorium oscaris* var. *excensurae* Forel, 1915c: 342 (w.) SOUTH AFRICA. Raised to species: Bolton, 1987: 386.

***exchao.*** *Monomorium exchao* Santschi, 1926b: 235 (w.) SOUTH AFRICA. See also: Bolton, 1987: 387.

***exiguum.*** *Monomorium exiguum* Forel, 1894b: 85 (w.) ETHIOPIA. Combination in *M.* (*Martia*): Forel, 1913h: 351; in *M.* (*Lampromyrmex*): Wheeler, W.M. 1922a: 876. Senior synonym of *bulawayensis, faurei* and *flavescens:* Bolton, 1987: 388.

*explorator*. *Monomorium explorator* Santschi, 1920b: 12, fig. 1 (w.) GABON. Junior synonym of *leopoldinum:* Bolton, 1987: 397.

#***extreminigrum.*** *Monomorium rubriceps* r. *extreminigrum* Forel, 1915b: 73 (w.) AUSTRALIA. Combination in *Chelaner:* Ettershank, 1966: 96; in *Monomorium:* Taylor, 1987b: 3.

*faf*. *Holcomyrmex faf* Forel, 1902a: 151 (q.m.) ALGERIA. Combination in *Monomorium:* Emery, 1908h: 673. Junior synonym of *chobauti:* Emery, 1922e: 176.

***falcatum.*** *Schizopelta falcata* McAreavey, 1949: 15, figs. 42–47 (w.q.m.) AUSTRALIA. Combination in *Chelaner:* Ettershank, 1966: 96; in *Monomorium:* Taylor, 1987b: 2.

***fasciatum.*** *Monomorium fasciatum* Santschi, 1920b: 10, fig. 1 (w.) KENYA. See also: Bolton, 1987: 388.

***fastidium.*** *Monomorium fastidium* Bolton, 1987: 389 (w.q.) SOUTH AFRICA.

*faurei*. *Monomorium* (*Mitara*) *faurei* Santschi, 1915c: 260, fig. 10 (w.) GABON. Combination in *M.* (*Lampromyrmex*): Wheeler, W.M. 1922a: 876. Junior synonym of *exiguum:* Bolton, 1987: 388.

*fezzanense*. *Monomorium* (*Xeromyrmex*) *salomonis* st. *areniphilum* var. *fezzanense* Santschi, 1936a: 50, figs. 27, 31 (w.m.) LIBYA. Unavailable name.

***fieldi.*** *Monomorium* (*Martia*) *fieldi* Forel, 1910b: 30 (w.m.) AUSTRALIA. Combination in *M.* (*Lampromyrmex*): Emery, 1922e: 184.

*fingo*. *Monomorium albopilosum* subsp. *fingo* Arnold, 1946: 61, fig. 13 (w.) SOUTH AFRICA. Junior synonym of *albopilosum:* Bolton, 1987: 335.

***firmum.*** *Monomorium firmum* Santschi, 1926b: 231 (w.) ZIMBABWE. See also: Bolton, 1987: 389.

*flavescens. Monomorium (Mitara) exiguum* r. *flavescens* Forel, 1916: 418 (w.) ZAIRE. Junior synonym of *exiguum:* Bolton, 1987: 388.

***flavigaster.*** *Xiphomyrmex flavigaster* Clark, 1938: 366, fig. 5 (w.) AUSTRALIA. Combination in *Chelaner:* Bolton, 1976: 294; in *Monomorium:* Taylor, 1987b: 2.

***flavipes.*** *Monomorium (Notomyrmex) flavipes* Clark, 1938: 369, fig. 8 (w.q.) AUSTRALIA. Combination in *Chelaner:* Ettershank, 1966: 96; in *Monomorium:* Taylor, 1987b: 2.

***flavum.*** *Monomorium (Monomorium) flavum* Collingwood, 1961a: 62, fig. 6 (w.) AFGHANISTAN. Pisarski, 1967: 399 (q.).

*floreanum. Monomorium floreanum* Stitz, 1932a: 368, fig. 1 (w.q.) GALAPAGOS IS. Junior synonym of *floricola:* Brown, in Linsley & Usinger, 1966: 175.

***floricola.*** *Atta floricola* Jerdon, 1851: 107 (w.) INDIA. Forel, 1893g: 388 (q.m.); Wheeler, W.M. 1905b: 88 (q.m.); Donisthorpe, 1914: 136 (gynandromorph); Crawley, 1920d: 217 (gynandromorph); Wheeler, G.C. & Wheeler, J. 1955c: 121 (l.). Combination in *Monomorium:* Mayr, 1879: 671. Senior synonym of *poecilum:* Emery, 1894c: 151; of *specularis:* Mayr, 1879: 671; of *cinnabari:* Wheeler, W.M. 1913b: 486; of *floreanum:* Brown, in Linsley & Usinger, 1966: 175; of *angusticlava* and *impressum:* Bolton, 1987: 390. Current subspecies: nominal plus *furinum, philippinense.*

***forcipatum.*** *Monomorium (Chelaner) forcipatum* Emery, 1914f: 410, pl. 13, fig. 7 (w.q.m.) NEW CALEDONIA. Combination in *Chelaner:* Ettershank, 1966: 96; in *Monomorium:* Taylor, 1987b: 4.

***foreli.*** *Monomorium (Holcomyrmex) foreli* Viehmeyer, 1914c: 32 (w.) AUSTRALIA. Combination in *Chelaner:* Ettershank, 1966: 97; in *Monomorium:* Taylor, 1987b: 2.

*fossulatum. Monomorium fossulatum* Emery, 1895k: 465 (w.q.) BURMA. Synonym of *sechellense:* Wilson & Taylor, 1967: 64 [*fossulatum* is given as senior name, but *sechellense* has priority and hence is the valid name for this taxon].

*fragilis. Myrmica fragilis* Smith, F. 1858b: 124 (w.) SINGAPORE. Junior synonym of *pharaonis:* Mayr, 1886c: 359.

***fraterculum.*** *Monomorium (Mitara) laeve* st. *fraterculus* Santschi, 1919a: 328, fig. 1 (w.) AUSTRALIA. Combination in *M. (Lampromyrmex):* Wheeler, W.M. 1927i: 140. Raised to species: Santschi, 1928e: 467. Current subspecies: nominal plus *barretti.*

***fridae.*** *Monomorium medinae* r. *fridae* Forel, 1905b: 183 (w.) SOUTH AFRICA. Combination in *M. (Xeromyrmex):* Wheeler, W.M. 1922a: 870. Raised to species: Bolton, 1987: 344.

*froggatti* Forel, 1902; see under ***ADLERZIA.***

***fugelanum.*** *Monomorium fugelanum* Bolton, 1987: 391, fig. 88 (w.) BOTSWANA.

*fultor. Monomorium afrum* var. *fultor* Forel, 1913b: 332 (w.) ZAIRE. Junior synonym of *afrum:* Bolton, 1987: 334.

*fulvum. Monomorium fulvum* Mayr, 1862: 754 (w.) NEW ZEALAND. Junior synonym of *antarcticum:* Emery, 1887b: 458.

***#fur.*** *Monomorium andrei* r. *fur* Forel, 1894d: 19 (w.q.) ALGERIA.

***#furinum.*** *Monomorium floricola* var. *furina* Forel, 1911i: 221 (w.) SRI LANKA.

***#furunculum.*** *Monomorium (Monomorium) andrei* var. *furunculus* Emery, 1924c: 165 (w.) SPAIN.

***gabrielense.*** *Monomorium rhopalocerum* var. *gabrielensis* Forel, 1916: 418 (w.q.) ZAIRE. Raised to species: Bolton, 1987: 392.

***gilberti.*** *Monomorium gilberti* Forel, 1902h: 440 (w.) AUSTRALIA. Combination in *M. (Notomyrmex):* Emery, 1922e: 169; in *Chelaner:* Ettershank, 1966: 97; in *Monomorium:* Taylor, 1987b: 2. Current subspecies: nominal plus *mediorubrum.*

***#glabrocriniceps.*** *Holcomyrmex glaber* var. *glabrocriniceps* Forel, 1902c: 220 (w.) INDIA.

***glabrum.*** *Holcomyrmex glaber* André, 1883b: 345 (footnote) (w.) INDIA. Imai, Baroni Urbani, *et al.* 1984: 8 (k.). Combination in *Monomorium:* Emery, 1922e: 181. Current subspecies: nominal plus *clarum, glabrocriniceps.*

*glyciphila. Myrmica glyciphila* Smith, F. 1858b: 125 (w.) SRI LANKA. Combination in *Monomorium:* Roger, 1863b: 32; in *M. (Xeromyrmex):* Emery, 1922e: 179. Junior synonym of *subopacum:* Bolton, 1987: 360.

***#gracilicorne.*** *Monomorium niloticum* r. *gracilicorne* Stitz, 1917: 346 (w.) ALGERIA.

*gracillima. Myrmica gracillima* Smith, F. 1861a: 34 (w.) ISRAEL. Emery, 1877b: 369 (footnote) (q.m.). Combination in *Monomorium:* Mayr, 1862: 753; in *M. (Parholcomyrmex):* Emery, 1915i: 190. Race of *destructor:* Forel, 1913d: 437. Junior synonym of *destructor:* Bolton, 1987: 324.

**gracillimus. *Lampromyrmex gracillimus* Mayr, 1868c: 95, pl. 5, figs. 97, 98 (w.) BALTIC AMBER (Oligocene). [Junior secondary homonym of *gracillima* Smith, above.] Replacement name: **mayrianum* Wheeler, W.M. 1915h: 45.

*gracilum. Monomorium (Parholcomyrmex) nipponense* var. *gracilum* Teranishi, 1940: 29 (q.) JAPAN. Junior synonym of *intrudens:* Ogata & Bolton, 1989: 460.

*grahamstownense. Monomorium delagoense* var. *grahamstownense* Santschi, 1936a: 39 (w.) SOUTH AFRICA. [First available use of *Monomorium salomonis* r. *delagoense* var. *grahamstownensis* Forel, 1914d: 245; unavailable name.] Junior synonym of *delagoense:* Bolton, 1987: 341.

*grandis. Monomorium venustum* subsp. *niloticum* var. *grandis* Forel, 1910a: 6 (w.) OMAN. Unavailable name.

***grassei.*** *Epixenus grassei* Tohmé, H. & Tohmé, G. 1980a: 1088, figs. 1–4 (w.q.m.) LEBANON. Combination in *Monomorium:* Bolton, 1987: 292.

***guillarmodi.*** *Monomorium (Lampromyrmex) guillarmodi* Arnold, 1946: 63, fig. 15 (w.q.) LESOTHO. See also: Bolton, 1987: 392.

***guineense.*** *Epixenus guineensis* Bernard, 1953b: 238, fig. 10 (w.) GUINEA. Combination in *Monomorium:* Brown & Wilson, 1957b: 245. See also: Bolton, 1987: 426.

***hanneli.*** *Monomorium hanneli* Forel, 1907a: 18 (w.) KENYA. Senior synonym of *moestum:* Bolton, 1987: 426.

***hannonis.*** *Monomorium hannonis* Santschi, 1910c: 358 (w.) CONGO. [Misspelled as *hammonis* by Emery, 1922e: 176.] See also: Bolton, 1987: 368.

***havilandi.*** *Monomorium havilandi* Forel, 1910e: 443 (w.q.) SOUTH AFRICA. Combination in *M. (Parholcomyrmex),* and stirps of *australe:* Santschi, 1917b: 282. Revived status as species, and senior synonym of *distinctum* and *leviceps:* Bolton, 1987: 368.

*hemiphaeum. Monomorium (Notomyrmex) hemiphaeum* Clark, 1934c: 61, pl. 4, figs. 19, 20 (w.q.) AUSTRALIA. Junior synonym of *leae:* Ettershank, 1966: 97.

***hercules.*** *Monomorium (Xeromyrmex) salomonis* subsp. *hercules* Viehmeyer, 1923: 91, fig. 3 (w.) SUDAN. Raised to species: Santschi, 1936a: 39.

***herero.*** *Monomorium salomonis* subsp. *herero* Forel, 1910f: 16 (w.q.) NAMIBIA. Combination in *M. (Xeromyrmex):* Wheeler, W.M. 1922a: 870. Stirps of *salomonis:* Santschi, 1917b: 284; of *delagoense:* Santschi, 1936a: 39. Raised to species: Bolton, 1987: 345.

***hesperium.*** *Monomorium hesperium* Emery, 1895b: 66, fig. 3 (w.) CANARY IS. Espadaler & Agosti, 1985: 295 (q.). See also: Bolton, 1987: 294. Current subspecies: nominal plus *lanzarotense.*

*heyeri* Forel, 1901; see under ***TRANOPELTA.***

***hildebrandti.*** *Monomorium minutum* r. *hildebrandti* Forel, 1892l: 256 (q.) MADAGASCAR. Santschi, 1926b: 237 (w.). Subspecies of *minutum:* Wheeler, W.M. 1922a: 1027. Raised to species: Dalla Torre, 1893: 67; and current status.

***hirsutum.*** *Monomorium bicolor* subsp. *hirsutum* Forel, 1910c: 251 (w.) ETHIOPIA. Subspecies of *subopacum:* Santschi, 1927d: 245. Raised to species: Bolton, 1987: 345.

***holothir.*** *Monomorium holothir* Bolton, 1987: 393 (w.) KENYA.

***hospitum.*** *Monomorium (Corynomyrmex) hospitum* Viehmeyer, 1916a: 133 (q.m.) SINGAPORE. See also: Bolton, 1987: 299.

*hottentota. Monomorium minutum* subsp. *hottentota* Emery, 1895h: 26 (q.) SOUTH AFRICA. Junior synonym of *rhopalocerum:* Bolton, 1987: 407.

***howense.*** *Monomorium (Notomyrmex) howense* Wheeler, W.M. 1927i: 138, fig. 5 (w.q.) AUSTRALIA. Combination in *Chelaner:* Ettershank, 1966: 97; in *Monomorium:* Taylor, 1987b: 2.

***#humilior.*** *Monomorium rothsteini* var. *humilior* Forel, 1910b: 27 (w.) AUSTRALIA. Wheeler, W.M. 1917b: 112 (q.). Combination in *Chelaner:* Ettershank, 1966: 97; in *Monomorium:* Taylor, 1987b: 3. Subspecies of *rothsteini:* Taylor & Brown, D.R. 1985: 58.

***ilgii.*** *Monomorium ilgii* Forel, 1894b: 84 (w.) ETHIOPIA. Combination in *M. (Xeromyrmex):* Emery, 1922e: 177. See also: Bolton, 1987: 346.

***ilium.*** *Monomorium (Martia) ilia* Forel, 1907h: 277 (w.q.) AUSTRALIA. Forel, 1910b: 30 (q.); Wheeler, G.C. & Wheeler, J. 1973a: 30 (l.). Combination in *M. (Lampromyrmex):* Emery, 1922e: 184. Current subspecies: nominal plus *lamingtonense.*

***imerinense.*** *Monomorium minutum* r. *imerinense* Forel, 1892l: 257 (q.m.) MADAGASCAR. Subspecies of *minutum:* Wheeler, W.M. 1922a: 1027. Raised to species: Dalla Torre, 1893: 67; and current status.

***impexum.*** *Monomorium impexum* Wheeler, W.M. 1928c: 18 (w.) CHINA.

*impressum. Monomorium impressum* Smith, F. 1876a: 447 (q.m.) RODRIGUEZ I. Junior synonym of *floricola:* Bolton, 1987: 390.

*impuriceps. Monomorium (Xeromyrmex) bicolor* st. *personatum* var. *impuriceps* Santschi, 1937d: 224 (w.) ANGOLA. Unavailable name; material referred to *dictator:* Bolton, 1987: 341.

*indicus. Holcomyrmex indicus* Smith, F. 1873: ix, INDIA. *Nomen nudum;* attributed to Mayr. [*Holcomyrmex indicus* Rothney, 1889: 367; *nomen nudum.*]

***indicum.*** *Monomorium salomonis* r. *indicum* Forel, 1902c: 213 (w.q.m.) INDIA. Imai, Baroni Urbani, *et al.* 1984: 7 (k.). Combination in *M. (Xeromyrmex)*: Emery, 1922e: 178. Stirps/subspecies of *salomonis:* Forel, 1907a: 18; Santschi, 1936a: 49. Raised to species: Bingham, 1903: 205; Collingwood, 1961a: 61; Bolton, 1987: 292. Current subspecies: nominal plus *kusnezowi.*

*inerme* Borgmeier, 1948; see under ***ANTICHTHONIDRIS.***

***inquietum.*** *Monomorium inquietum* Santschi, 1926b: 233, fig. 3 (w.) ZAIRE. See also: Bolton, 1987: 394.

***inquilinum.*** *Monomorium inquilinum* DuBois, 1981: 34, fig. 2 (q.) MEXICO. See also: DuBois, 1986: 116.

*insidiosa. Wheeleriella santschii* st. *insidiosa* Santschi, 1926e: 233 (q.m.) TUNISIA. Combination in *Monomorium:* Ettershank, 1966: 89. Junior synonym of *santschii:* Bolton, 1987: 291.

***insolescens.*** *Monomorium (Notomyrmex) insolescens* Wheeler, W.M. 1934d: 145 (w.m.) AUSTRALIA. Combination in *Chelaner:* Ettershank, 1966: 97; in *Monomorium:* Taylor, 1987b: 3.

***insulare.*** *Monomorium (Notomyrmex) insularis* Clark, 1938: 368, fig. 7 (w.q.) AUSTRALIA. Combination in *Chelaner:* Ettershank, 1966: 97; in *Monomorium:* Taylor, 1987b: 3.

#***integrium.*** *Monomorium atomus* var. *integrius* Forel, 1902c: 210 (w.q.) INDIA. Combination in *M. (Lampromyrmex)*: Emery, 1922e: 183.

*integrum. Monomorium integrum* Forel, 1894e: 228 (w.q.m.) NEW ZEALAND. Junior synonym of *antarcticum:* Brown, 1958h: 29.

*intermedium. Monomorium (Xeromyrmex) subopacum* var. *intermedium* Santschi, 1927d: 242 (w.) CANARY IS. [First available use of *Monomorium (Xeromyrmex) salomonis* subsp. *subopacum* var. *intermedium* Wheeler, W.M. 1927g: 108; unavailable name.] Junior synonym of *subopacum:* Hohmann, La Roche, *et al.*, 1993: 155.

***intrudens.*** *Monomorium intrudens* Smith, F. 1874: 406 (w.) JAPAN. Junior synonym of *floricola:* Wheeler, W.M. 1906c: 310. Revived from synonymy, and senior synonym of *nipponense:* Bolton, 1987: 390 (in text). Senior synonym of *gracilum, robustum* and *satoi:* Ogata & Bolton, 1989: 460. Current subspecies: nominal plus *pieli.*

***invidium.*** *Monomorium invidium* Bolton, 1987: 427, figs. 97, 98 (w.q.) NIGERIA.

*italica. Monomorium (Xeromyrmex) subopacum* subsp. *italica* Baroni Urbani, 1964c: 154, figs. 2, 3 (w.) ITALY. Junior synonym of *subopacum:* Baroni Urbani, 1968b: 450.

***iyenasu.*** *Monomorium iyenasu* Bolton, 1987: 394 (w.) TANZANIA.

***jacksoni.*** *Monomorium jacksoni* Bolton, 1987: 427 (w.q.) CAMEROUN.

*javana. Monomorium minutum* subsp. *liliuokalanii* var. *javana* Forel, 1909d: 225 (w.) JAVA. Unavailable name.

***jonesi.*** *Monomorium (Syllophopsis) jonesi* Arnold, 1952a: 465. Replacement name for *Syllophopsis arnoldi* Santschi, 1921c: 120. [Junior secondary homonym of *Monomorium arnoldi* Forel, 1913a: 137.] See also: Bolton, 1987: 422.

*jucundum. Monomorium jucundum* Santschi, 1926b: 232 (w.) ZAIRE. Junior synonym of *egens:* Bolton, 1987: 385.

#***judaicum.*** *Monomorium (Xeromyrmex) bicolor* subsp. *judaicum* Menozzi, 1933b: 63, fig. 2 (w.) ISRAEL.

***junodi.*** *Monomorium salomonis* subsp. *junodi* Forel, 1910e: 441 (w.) SOUTH AFRICA. Combination in *M. (Xeromyrmex)*: Santschi, 1919b: 235. Subspecies/stirps of *delagoense:* Santschi, 1928f: 192; Arnold, 1944: 14. Raised to species, and senior synonym of *pretoriensis:* Bolton, 1987: 346.

*kalahariense. Monomorium destructor* subsp. *kalahariense* Forel, 1910f: 18 (w.) BOTSWANA. Junior synonym of *oscaris:* Bolton, 1987: 326.

*karawajewi. Monomorium (Parholcomyrmex) gracillimum* var. *karawajewi* Wheeler, W.M. 1922a: 875 (w.) SUDAN. [First available use of *Monomorium destructor* r. *gracillimum* var. *karawajewi* Forel, 1913d: 437: unavailable name.] Raised to species: Collingwood, 1985: 270. Junior synonym of *mayri:* Bolton, 1987: 326.

***katir.*** *Monomorium katir* Bolton, 1987: 395, figs. 72, 74 (w.) NAMIBIA.

***kelapre.*** *Monomorium kelapre* Bolton, 1987: 395 (w.) SOUTH AFRICA.

***kempi.*** *Monomorium (Parholcomyrmex) kempi* Mukerjee, 1930: 154, fig. 2 (w.) INDIA.

***kiliani.*** *Monomorium kiliani* Forel, 1902h: 441 (w.) AUSTRALIA. Combination in *M. (Notomyrmex)*: Emery, 1922e: 169; in *Chelaner:* Ettershank, 1966: 97; in *Monomorium:* Taylor, 1987b: 3. Current subspecies: nominal plus *obscurellum, tambourinense.*

***kineti.*** *Monomorium (Monomorium) minutum* subsp. *kineti* Weber, 1943c: 359, pl. 15, figs. 10, 19 (w.q.m.) SUDAN. Raised to species: Bolton, 1987: 396.

***kitectum.*** *Monomorium kitectum* Bolton, 1987: 347, fig. 49 (w.) NAMIBIA.

***kugitangi.*** *Monomorium kugitangi* Dlussky, in Dlussky, Soyunov & Zabelin, 1990: 236, fig. 31 (w.) TURKMENISTAN.

#***kusnezowi.*** *Monomorium (Xeromyrmex) kusnezowi* Santschi, 1928b: 42 (w.) TURKESTAN. Subspecies of *indicum,* and senior synonym of *turkmenicum:* Dlussky & Zabelin, 1985: 215.

*lacrymans. Monomorium delagoense* var. *lacrymans* Arnold, 1944: 15, fig. 19 (w.) SOUTH AFRICA. Junior synonym of *delagoense:* Bolton, 1987: 341.

***laeve.*** *Monomorium laeve* Mayr, 1876: 101 (w.) AUSTRALIA. Combination in *M. (Mitara)*: Emery, 1913: 261; in *M. (Lampromyrmex)*: Emery, 1922e: 184. Current subspecies: nominal plus *nigrium.*

*laeviceps. Monomorium subopacum* r. *australe* var. *laeviceps* Emery, 1886: 364 (w.) SOUTH AFRICA. Unavailable name; material referred to *australe* by Santschi, 1917c: 282.

#***laevior.*** *Monomorium emeryi* subsp. *laevior* Mayr, 1897: 427 (w.) SRI LANKA.

***lameerei.*** *Holcomyrmex lameerei* Forel, 1902a: 150 (w.) ALGERIA. Santschi, 1907: 327 (q.m.). Combination in *Monomorium:* Emery, 1908h: 673; in *M. (Equestrimessor)*: Santschi, 1919d: 92; in *M. (Xeromyrmex)*: Emery, 1922e: 177. See also: Bolton, 1987: 297. Current subspecies: nominal plus *stauderi.*

#***lamingtonense.*** *Monomorium (Mitara) ilia* var. *lamingtonensis* Forel, 1915b: 73 (w.q.) AUSTRALIA. Combination in *M. (Lampromyrmex)*: Emery, 1922e: 184. Subspecies of *ilium:* Taylor & Brown, D.R. 1985: 71.

#***lanzarotense.*** *Monomorium (Xeromyrmex) hesperium* var. *lanzarotense* Wheeler, W.M. 1927g: 107 (w.) CANARY IS.

*latastei* Emery, 1895; see under ***NOTHIDRIS.***

***latinode.*** *Monomorium latinode* Mayr, 1872: 152 (w.) BORNEO. Imai, Baroni Urbani, *et al.* 1984: 8 (k.). Senior synonym of *bruneum* and *voeltzkowi:* Bolton, 1987: 429.

***latinodoides.*** *Monomorium latinodoides* Wheeler, W.M. 1928c: 17 (w.q.) CHINA.

*latior. Monomorium (Xeromyrmex) setuliferum* var. *latior* Santschi, 1928f: 195 (w.) ANGOLA. Junior synonym of *notulum:* Bolton, 1987: 370.

***leae.*** *Monomorium leae* Forel, 1913g: 185 (w.q.) AUSTRALIA. Combination in *M. (Notomyrmex)*: Emery, 1922e: 169; in *Chelaner:* Ettershank, 1966: 97; in *Monomorium:* Taylor, 1987b: 3. Senior synonym of *hemiphaeum:* Ettershank, 1966: 97.

#***leda.*** *Monomorium rothsteini* var. *leda* Forel, 1915b: 71 (w.m.) AUSTRALIA. Combination in *Chelaner:* Ettershank, 1966: 97; in *Monomorium:* Taylor, 1987b: 3. Subspecies of *rothsteini:* Taylor & Brown, D.R. 1985: 58.

*leimbachi. Monomorium leimbachi* Forel, 1914d: 246 (w.) SOUTH AFRICA. Junior synonym of *rhopalocerum:* Bolton, 1987: 407.

***lene.*** *Monomorium lene* Santschi, 1920b: 11, fig. 2 (w.) ZIMBABWE. See also: Bolton, 1987: 397.

***leopoldinum.*** *Monomorium minutum* var. *leopoldinum* Forel, 1905b: 179 (w.) ZAIRE. Raised to species, and senior synonym of *aequum, estherae* and *explorator:* Bolton, 1987: 397.

*lepineyi. Monomorium (Xeromyrmex) salomonis* var. *lepineyi* Santschi, 1934b: 34, figs. 5, 6 (w.) SUDAN. Junior synonym of *areniphilum:* Bolton, 1987: 336.

*leviceps. Monomorium distinctum* var. *leviceps* Arnold, 1958: 119 (w.) SOUTH AFRICA. Junior synonym of *havilandi:* Bolton, 1987: 368.

***libanicum.*** *Epixenus libanicus* Tohmé, H. & Tohmé, G. 1980a: 1103, figs. 8, 9 (w.q.) LEBANON. Combination in *Monomorium:* Bolton, 1987: 293.

*liberta. Monomorium (Xeromyrmex) subopacum* var. *liberta* Santschi, 1927d: 243 (w.) SENEGAL. [First available use of *Monomorium (Xeromyrmex) salomonis* var. *liberta* Santschi, 1921e: 170; unavailable name.] Junior synonym of *subopacum:* Bolton, 1987: 360.

***liliuokalanii.*** *Monomorium minutum* var. *liliuokalanii* Forel, 1899a: 119 (w.) HAWAII. Subspecies of *minutum:* Forel, 1909d: 225. Junior synonym of *minutum:* Wilson & Taylor, 1967: 67. Revived from synonymy, raised to species, and senior synonym of *samoanum:* Bolton, 1987: 288.

***lindbergi.*** *Monomorium lindbergi* Pisarski, 1967: 397, figs. 50, 51 (w.) AFGHANISTAN.

*lippulum* Wheeler, W.M. 1927; see under ***ANISOPHEIDOLE.***

***longi.*** *Monomorium longi* Forel, 1902c: 211 (w.) INDIA. Combination in *M.* (*Xeromyrmex*): Emery, 1922e: 177.

***longiceps.*** *Monomorium* (*Notomyrmex*) *longiceps* Wheeler, W.M. 1934d: 146 (w.) AUSTRALIA. Combination in *Chelaner:* Ettershank, 1966: 97; in *Monomorium:* Taylor, 1987b: 3.

*longiceps. Monomorium* (*Monomorium*) *longiceps* Karavaiev, 1935a: 94, fig. 20 (w.) VIETNAM. [Junior primary homonym of *longiceps* Wheeler, W.M., above.] Replacement name: *annamense* Donisthorpe, 1941k: 238.

***longipes.*** *Monomorium* (*Chelaner*) *longipes* Emery, 1914f: 411 (w.) NEW CALEDONIA. Combination in *Chelaner:* Ettershank, 1966: 97; in *Monomorium:* Taylor, 1987b: 4.

*longiusculum. Monomorium longiusculum* Santschi, 1926b: 237, fig. 3 (w.) ZAIRE. Junior synonym of *egens:* Bolton, 1987: 385.

***lubricum.*** *Monomorium lubricum* Arnold, 1948: 218, fig. 7 (w.) SOUTH AFRICA. See also: Bolton, 1987: 398.

*luederwaldti* Forel, 1913; see under ***OXYEPOECUS.***

***luisae.*** *Monomorium luisae* Forel, 1904c: 25 (w.q.) INDIA.

***luteum.*** *Monomorium luteum* Emery, 1881b: 533 (w.) YEMEN. Combination in *M.* (*Xeromyrmex*): Emery, 1922e: 177. See also: Santschi, 1936a: 36.

***macareaveyi.*** *Chelaner macareaveyi* Ettershank, 1966: 97. Replacement name for *Monomorium niger* McAreavey, 1949: 12. [Junior secondary homonym of *Holcomyrmex criniceps* var. *niger* Forel, 1902c: 220.] Combination in *Monomorium:* Taylor, 1987b: 3.

***macrops.*** *Monomorium mediocre* subsp. *macrops* Arnold, 1944: 11, fig. 17 (w.) SOUTH AFRICA. Raised to species: Bolton, 1987: 369.

***madecassum.*** *Monomorium minutum* r. *madecassum* Forel, 1892l: 255 (w.q.m.) MADAGASCAR. Race of *minutum:* Forel, 1907g: 77. Raised to species: Dalla Torre, 1893: 67; and current status.

***major.*** *Monomorium* (*Isolcomyrmex*) *santschii* var. *major* Bernard, 1953a: 168 (w.) ALGERIA. Raised to species: **new status**. First available replacement name for *Holcomyrmex santschii* Forel, 1907b: 203. [Junior secondary homonym of *Wheeleria santschii* Forel, 1905b: 171.] Senior synonym of *santschianum:* **new synonymy**.

***malamixtum.*** *Monomorium malamixtum* Bolton, 1987: 423 (w.q.) IVORY COAST.

***malatu.*** *Monomorium malatu* Bolton, 1987: 399. Replacement name for *Tetramorium altinode* Santschi, 1935a: 266. [Junior secondary homonym of *Monomorium rhopalocerum* var. *altinodis* Santschi, 1910c: 359.

*mandibulare* Emery, 1913; see under ***OXYEPOECUS.***

***manir.*** *Monomorium manir* Bolton, 1987: 400 (w.) KENYA.

***mantazenum.*** *Monomorium mantazenum* Bolton, 1987: 348 (w.) NAMIBIA.

***marjoriae.*** *Monomorium marjoriae* DuBois, 1986: 111, figs. 149–154 (w.q.) MEXICO.

***marshi.*** *Monomorium marshi* Bolton, 1987: 349, figs. 50, 53 (w.) NAMIBIA.

***mavide.*** *Monomorium mavide* Bolton, 1987: 400 (w.) SOUTH AFRICA.

***mayri.*** *Monomorium gracillimum* var. *mayri* Forel, 1902c: 209 (w.) INDIA. Wheeler, W.M. 1923b: 3 (m.). Combination in *M.* (*Parholcomyrmex*): Emery, 1922e: 180. Subspecies of *destructor:* Forel, 1911a: 24; Viehmeyer, 1916a: 132. Raised to species, and senior synonym of *karawajewi:* Bolton, 1987: 326.

****mayrianum.*** **Monomorium mayrianum* Wheeler, W.M. 1915h: 45. Replacement name for **Lampromyrmex gracillimus* Mayr, 1868c: 95. [Junior secondary homonym of *Myrmica gracillima* Smith, F. 1861a: 34.]

***medinae.*** *Monomorium medinae* Forel, 1892g: 454 (w.) CANARY IS. Combination in *M.* (*Xeromyrmex*): Wheeler, W.M. 1922a: 869. Senior synonym of *mystes:* Espadaler, 1982b: 112. See also: Bolton, 1987: 294.

***mediocre.*** *Monomorium mediocre* Santschi, 1920d: 376, fig. 13 (w.) SOUTH AFRICA. See also: Bolton, 1987: 349.

#***mediorubrum.*** *Monomorium gilberti* var. *mediorubra* Forel, 1915b: 72 (w.) AUSTRALIA. Combination in *M.* (*Notomyrmex*): Emery, 1922e: 169; in *Chelaner:* Ettershank, 1966: 97; in *Monomorium:* Taylor, 1987b: 2. Subspecies of *gilberti:* Taylor & Brown, D.R. 1985: 56.

*mediterraneum. Monomorium mediterraneum* Mayr, 1861: 72 (diagnosis in key) (w.q.) SPAIN. Santschi, 1936a: 40 (m.). Variety of *subopacum:* Santschi, 1936a: 40. Junior synonym of *subopacum:* Mayr, 1862: 753; Roger, 1862c: 294; Bolton, 1987: 360.

***melleum.*** *Monomorium melleum* Emery, 1914f: 408, pl. 13, fig. 5 (w.q.) NEW CALEDONIA. Combination in *M.* (*Notomyrmex*): Emery, 1922e: 169; in *Chelaner:* Ettershank, 1966: 97; in *Monomorium:* Taylor, 1987b: 4.

*metoecus. Monomorium metoecus* Brown & Wilson, 1957b: 239, fig. 1 (q.) U.S.A. Wilson & Brown, 1958: 33 (w.). Junior synonym of *minimum:* Brown, in Ettershank, 1966: 90.

***micron.*** *Monomorium micron* Crawley, 1925b: 593 (w.q.) AUSTRALIA.

***micropacum.*** *Monomorium micropacum* Bolton, 1987: 350 (w.) SOUTH AFRICA.

***mictile.*** *Monomorium* (*Martia*) *atomus* subsp. *mictilis* Forel, 1910c: 252 (w.q.) ETHIOPIA. Combination in *M.* (*Lampromyrmex*), and subspecies of *exiguum:* Wheeler, W.M. 1922a: 876. Raised to species, senior synonym of *minutissimum* and material of the unavailable name *sudanicum* referred here: Bolton, 1987: 401.

***minimum.*** *Myrmica* (*Monomorium*) *minima* Buckley, 1867: 338 (w.q.) U.S.A. Emery, 1895c: 274 (m.); Wheeler, G.C. & Wheeler, J. 1955c: 122 (l.); Crozier, 1970: 116 (k.). Combination in *Monomorium:* Emery, 1895c: 274. Variety of *minutum:* Emery, 1895c: 274; Wheeler, W.M. 1904e: 301; Wheeler, W.M. 1906b: 2. Revived status as species: Wheeler, W.M. 1910g: 562; Wheeler, W.M. 1913c: 115; Wheeler, W.M. 1914b: 42; Dennis, 1938: 279. Senior synonym of *atra:* Emery, 1895c: 274; of *metoecus:* Brown, in Ettershank, 1966: 90. See also: Wheeler, G.C. & Wheeler, J. 1973a: 30; DuBois, 1986: 76.

***minor.*** *Monomorium salomonis* var. *minor* Stitz, 1923: 156 (w.) NAMIBIA. Raised to species: Bolton, 1987: 350.

*minuta. Atta minuta* Jerdon, 1851: 105 (w.q.) INDIA. Junior synonym of *pharaonis:* Emery, 1892b: 165; Bolton, 1987: 288, 356.

*minutissimum. Monomorium* (*Lampromyrmex*) *minutissimum* Santschi, 1937d: 227, figs. 27, 28 (w.) ANGOLA. Junior synonym of *mictile:* Bolton, 1987: 401.

*minutum. Monomorium minutum* Mayr, 1855: 453 (w.) ITALY. Mayr, 1865: 91 (q.m.). [Junior secondary homonym of *Atta minuta* Jerdon, 1851: 105, above.] Replacement name: *monomorium* Bolton, 1987: 287.

***mirandum.*** *Monomorium* (*Monomorium*) *mirandum* Arnold, 1955: 734, fig. 2 (w.) KENYA. See also: Bolton, 1987: 401.

***modestum.*** *Monomorium modestum* Santschi, 1914e: 17 (w.) SOUTH AFRICA. Combination in *Syllophopsis:* Santschi, 1915c: 259; in *Monomorium:* Bolton, 1987: 423. Senior synonym of *smutsi* and *transwaalensis:* Bolton, 1987: 423.

*moestum. Monomorium moestum* Santschi, 1914b: 74, fig. 7 (w.) KENYA. Combination in *M.* (*Notomyrmex*): Emery, 1922e: 170. Junior synonym of *hanneli:* Bolton, 1987: 426.

#***molestum.*** *Monomorium* (*Xeromyrmex*) *salomonis* var. *molestum* Santschi, 1936c: 201 (w.q.) ALGERIA. First available use of *Monomorium* (*Xeromyrmex*) *salomonis* st. *didonis* var. *molestum* Santschi, 1936a: 55; unavailable name.]

*monardi. Monomorium* (*Xeromyrmex*) *monardi* Santschi, 1937d: 226, figs. 15, 16 (w.) ANGOLA. Junior synonym of *rufulum:* Bolton, 1987: 359.

***monomorium.*** *Monomorium monomorium* Bolton, 1987: 287. Replacement name for *Monomorium minutum* Mayr, 1855: 453. [Junior secondary homonym of *Atta minuta* Jerdon, 1851: 105.]

***musicum.*** *Monomorium oscaris* subsp. *musicum* Forel, 1910e: 442 (w.q.) SOUTH AFRICA. Raised to species: Emery, 1922e: 173; Bolton, 1987: 402.

***muticum.*** *Holcomyrmex muticus* Emery, 1887b: 457 (w.) BURMA. Combination in *Monomorium:* Emery, 1922e: 182.

*myops. Ireneidris myops* Donisthorpe, 1943c: 81 (w.) NEW GUINEA. Junior synonym of *talpa:* Ettershank, 1966: 92.

*mystes. Xenhyboma mystes* Santschi, 1919g: 405, fig. 2 (q.) CANARY IS. Combination in *Monomorium:* Ettershank, 1966: 91. Junior synonym of *medinae:* Espadaler, 1982b: 112. See also: Kutter, 1973a: 321; Bolton, 1987: 293.

*mzabicum. Monomorium* (*Xeromyrmex*) *salomonis* st. *targui* var. *mzabicum* Santschi, 1936a: 54 (w.) ALGERIA. Unavailable name.

*navarinensis* Forel, 1904; see under ***ANTI-CHTHONIDRIS.***

***nengraharicum.*** *Monomorium (Parholcomyrmex) nengraharicum* Pisarski, 1970: 309, figs. 1–9 (w.q.) AFGHANISTAN.

***#nigellum.*** *Monomorium (Mitara) sydneyense* subsp. *nigella* Emery, 1914b: 184, fig. 3 (w.) AUSTRALIA. Viehmeyer, 1925a: 28 (q.). Combination in *M. (Lampromyrmex)*: Emery, 1922e: 184.

*niger. Monomorium (Holcomyrmex) niger* McAreavey, 1949: 12, figs. 36–41 (w.q.m.) AUSTRALIA. [Junior secondary homonym of *Holcomyrmex criniceps* var. *niger* Forel, 1902c: 220, below.] Replacement name: *Monomorium macareaveyi* Ettershank, 1966: 97.

***#nigrium.*** *Monomorium (Mitara) laeve* r. *nigrius* Forel, 1915b: 74 (w.q.) AUSTRALIA.

***#nigriventre.*** *Monomorium sordidum* var. *nigriventris* Forel, 1910b: 29 (w.q.) AUSTRALIA. Combination in *Chelaner:* Ettershank, 1966: 97; in *Monomorium:* Taylor, 1987b: 4. Subspecies of *sordidum:* Taylor & Brown, D.R. 1985: 59.

***#nigrum.*** *Holcomyrmex criniceps* var. *niger* Forel, 1902c: 220 (w.) INDIA. Combination in *Monomorium:* Emery, 1922e: 181.

***#niloticoides.*** *Monomorium venustum* var. *niloticoides* Forel, 1910a: 6 (w.) ISRAEL. Stirps of *niloticum:* Santschi, 1936a: 37.

***niloticum.*** *Monomorium niloticum* Emery, 1881b: 533 (w.) EGYPT. Combination in *M. (Xeromyrmex)*: Emery, 1922e: 179. Subspecies of *venustum:* Forel, 1910a: 6; Wheeler, W.M. & Mann, 1916: 170; Stitz, 1917: 346; Finzi, 1936: 175. Revived status as species: Santschi, 1936a: 37. Current subspecies: nominal plus *gracilicorne, niloticoides.*

*nipponense. Monomorium nipponense* Wheeler, W.M. 1906c: 310 (w.) JAPAN. Santschi, 1928b: 34 (q.m.). Junior synonym of *intrudens:* Bolton, 1987: 390 (in text); Ogata & Bolton, 1989: 460.

***nirvanum.*** *Monomorium nirvanum* Bolton, 1987: 351 (w.) NAMIBIA.

***nitidiventre.*** *Monomorium bicolor* subsp. *nitidiventris* Emery, 1893f: 256 (w.) EGYPT. Mayr, 1901b: 7 (q.); Karavaiev, 1911: 5 (m.). Combination in *M. (Xeromyrmex)*: Wheeler, W.M. 1922a: 869. Variety/subspecies of *bicolor:* Santschi, 1938a: 39; of *subopacum:* Santschi, 1927d: 245; Menozzi, 1932c: 94; Bernard, 1953a: 159; Hamann & Klemm, 1967: 413. Raised to species: Collingwood, 1985: 272. See also: Wheeler, W.M. & Mann, 1916: 171.

*nitidum. Tetramorium nitidum* Smith, F. 1876b: 480 (w.q.m.) NEW ZEALAND. Combination in *Monomorium:* Mayr, 1886c: 363; in *M. (Notomyrmex)*: Emery, 1922e: 169. Junior synonym of *antarcticum:* Brown, 1958h: 29.

***notulum.*** *Monomorium setuliferum* var. *notula* Forel, 1910e: 441 (w.m.) SOUTH AFRICA. Raised to species, and senior synonym of *dolichops* and *latior:* Bolton, 1987: 370.

***noualhieri.*** *Phacota noualhieri* Emery, 1895b: 67, fig. 1 (w.) ALGERIA. Combination in *Paraphacota:* Santschi, 1919d: 92; Santschi, 1927d: 245; in *Monomorium:* Ettershank, 1966: 91. See also: Emery, 1908f: 555; Bolton, 1987: 282.

***noxitum.*** *Monomorium noxitum* Bolton, 1987: 403 (w.) CAMEROUN.

***nuptualis.*** *Monomorium oscaris* var. *nuptualis* Forel, 1913j: 216 (w.) ZIMBABWE. Raised to species: Bolton, 1987: 403.

*nyasae. Monomorium (Xeromyrmex) nyasae* Arnold, 1946: 63, fig. 14 (w.) MALAWI. Junior synonym of *ebangaense:* Bolton, 1987: 367.

*obscuratum. Monomorium salomonis* var. *obscuratum* Stitz, 1917: 346, fig. 12 (w.) ALGERIA. Santschi, 1921e: 169 (m.); Emery, 1924a: 9 (q.). [Misspelled as *obscurium* by Santschi, 1929c: 100.] Junior synonym of *salomonis:* Collingwood, 1978: 69.

***#obscurellum.*** *Monomorium kiliani* var. *obscurella* Viehmeyer, 1925a: 27 (w.) AUSTRALIA. Combination in *Chelaner:* Ettershank, 1966: 97; in *Monomorium:* Taylor, 1987b: 3. Subspecies of *kiliani:* Taylor & Brown, D.R. 1985: 57.

***#obscuriceps.*** *Monomorium (Xeromyrmex) salomonis* var. *obscuriceps* Santschi, 1921e: 169 (w.q.) MOROCCO. Senior synonym of *pestiferum:* Ettershank, 1966: 91.

*obscuripes. Paraphacota cabrerae* st. *obscuripes* Santschi, 1921b: 424 (m.) CANARY IS. Santschi, 1927d: 241 (w.q.). Combination in *Monomorium:* Santschi, 1927d: 241. Variety of *subopacum:* Santschi, 1936a: 40. Junior synonym of *subopacum:* Santschi, 1927d: 241; Bolton, 1987: 360.

***occidaneum.*** *Monomorium occidaneus* Crawley, 1922b: 447, fig. 10 (w.q.) AUSTRALIA. Combination in *Chelaner:* Ettershank, 1966: 97; in *Monomorium:* Taylor, 1987b: 3.

***occidentale.*** *Monomorium (Monomorium) occidentale* Bernard, 1953b: 236, fig. 10 (w.q.) GUINEA. See also: Bolton, 1987: 404.

***ocellatum.*** *Monomorium (Xeromyrmex) salomonis* st. *ocelletum* Arnold, in Santschi, 1920d: 377 (w.) SOUTH AFRICA. Arnold, 1926: 229 (q.). Raised to species: Santschi, 1936a: 39; Arnold, 1944: 14; Bolton, 1987: 352.

*ominosa. Myrmica ominosa* Gerstäcker, 1859: 263 (w.) MOZAMBIQUE. [Also described as new by Gerstäcker, 1862: 517.] Combination in *Monomorium,* and senior synonym of *atomaria:* Roger, 1863b: 31. Junior synonym of *destructor:* Dalla Torre, 1893: 66.

***opacior.*** *Monomorium salomonis* var. *opacior* Forel, 1913j: 216 (in text) (w.) ZIMBABWE. [First available use of *Monomorium salomonis* r. *junodi* var. *opacior* Forel, 1913a: 136; unavailable name.] Arnold, 1916: 220 (q.). Variety of *salomonis:* Arnold, 1916: 220; Stitz, 1923: 156. Raised to species: **new status**. Senior synonym of *opacior* Bolton: **new synonymy**.

*opacior. Monomorium opacior* Bolton, 1987: 352 (w.) ZIMBABWE. Material of the unavailable name *serenum* referred here: Bolton, 1987: 352. Junior synonym of *opacior* Forel: **new synonymy**.

***opacum.*** *Monomorium opacum* Forel, 1913b: 333 (w.) ZAIRE. See also: Bolton, 1987: 353.

***ophthalmicum.*** *Monomorium ophthalmicum* Forel, 1894b: 87 (w.) ETHIOPIA. See also: Bolton, 1987: 353.

***orangiae.*** *Monomorium (Xeromyrmex) orangiae* Arnold, 1956: 67, fig. 16 (w.) SOUTH AFRICA. See also: Bolton, 1987: 354.

***orientale.*** *Monomorium orientale* Mayr, 1879: 670 (w.) INDIA. Emery, 1908h: 685 (q.); Imai, Baroni Urbani, *et al.* 1984: 7 (k.). Combination in *M. (Mitara)*: Santschi, 1915a: 58; in *M. (Lampromyrmex)*: Wheeler, W.M. 1922a: 876. See also: Brown, 1958h: 36.

***oscaris.*** *Monomorium oscaris* Forel, 1894b: 86 (w.) ETHIOPIA. Senior synonym of *despecta, dispar, kalahariense, prossae* and *solleri:* Bolton, 1987: 326.

***osiridis.*** *Monomorium osiridis* Santschi, 1915c: 258, fig. 7 (w.) KENYA. See also: Bolton, 1987: 354.

***pacis.*** *Monomorium pacis* Forel, 1915c: 343 (w.) SOUTH AFRICA. See also: Bolton, 1987: 404.

***pallidipes.*** *Monomorium minutum* var. *pallidipes* Forel, 1910c: 252 (w.) ETHIOPIA. Forel, 1913d: 430 (q.). Raised to species: Bolton, 1987: 405.

***pallidum.*** *Monomorium (Parholcomyrmex) destructor* var. *pallidus* Donisthorpe, 1918b: 166 (w.) IRAQ. Crawley, 1920b: 178 (q.). Raised to species: Crawley, 1920b: 178.

*papuasiae. Monomorium fossulatum* subsp. *sechellense* var. *papuasiae* Emery, 1900c: 326 (w.) NEW GUINEA. Unavailable name.

***parvinode.*** *Monomorium salomonis* var. *parvinode* Forel, 1894b: 88 (w.) ETHIOPIA. Raised to species: Bolton, 1987: 355.

***paternum.*** *Monomorium paternum* Bolton, 1987: 405, fig. 89 (w.) SOUTH AFRICA.

*paterna. Monomorium oscaris* r. *springvalense* var. *paterna* Forel, 1914d: 248 (w.) SOUTH AFRICA. Unavailable name.

*paucipilosa. Monomorium (Xeromyrmex) albopilosum* st. *paucipilosa* Santschi, 1919b: 235 (w.) SOUTH AFRICA. Junior synonym of *albopilosum:* Bolton, 1987: 335.

*peninsulatum. Monomorium peninsulatum* Gregg, 1945b: 62, pl. 9 (w.q.) U.S.A. Subspecies of *viride:* Creighton, 1950a: 223. Junior synonym of *viride:* DuBois, 1986: 96.

***pergandei.*** *Epoecus pergandei* Emery, 1893a: cclxxvi (q.m.) U.S.A. [Also described as new by Emery, 1895c: 273.] Combination in *Monomorium:* Ettershank, 1966: 91. See also: DuBois, 1986: 113; Bolton, 1987: 299.

***personatum.*** *Monomorium (Xeromyrmex) bicolor* st. *personatum* Santschi, 1937d: 222, fig. 29 (w.) ANGOLA. Raised to species, and material of the unavailable names *bimaculatum* and *bimaculatoides* referred here: Bolton, 1987: 356.

*pestiferum. Monomorium (Xeromyrmex) salomonis* st. *pestiferum* Santschi, 1936a: 57, figs. 20, 28, 30 (w.q.m.) ALGERIA. Junior synonym of *obscuriceps:* Ettershank, 1966: 91.

*petulans. Monomorium (Xeromyrmex) termitarium* st. *disertum* var. *petulans* Santschi, 1928f: 194 (w.) ZIMBABWE. Unavailable name; material referred to *disertum* by Bolton, 1987: 342.

***pharaonis.*** *Formica pharaonis* Linnaeus, 1758: 580 (w.) EGYPT. Mayr, 1865: 90 (m.); André, 1883a: 338 (q.); Forel, 1891b: 164 (w.q.m.); Wheeler, G.C. & Wheeler, J. 1955c: 121 (l.); Petralia & Vinson, 1908: 383 (l.); Imai, 1966: 119 (k.). Combination in *Monomorium:* Mayr, 1862: 752. Senior synonym of *antiguensis* and *domestica:* Roger, 1862c: 294; of *contigua* and *fragilis:* Mayr, 1886c: 359; of *minuta:* Emery, 1892b: 165; Bolton, 1987: 288; of *vastator:* Donisthorpe, 1932c: 449. See also: Smith, D.R. 1979: 1383; Bolton, 1987: 356; Atanassov & Dlussky, 1992: 164.

***MONOMORIUM** (continued)*

**#philippinense.** *Monomorium floricola* var. *philippinensis* Forel, 1910d: 123 (w.q.) PHILIPPINES.

**phoenicum.** *Monomorium* (*Xeromyrmex*) *subopacum* var. *phoenicum* Santschi, 1927d: 242 (w.q.) LEBANON. [First available use of *Monomorium salomonis* subsp. *subopacum* var. *phoenicia* Emery, 1908h: 677; unavailable name.] Wheeler, W.M. & Mann, 1916: 170 (m.). Raised to species: Collingwood, 1985: 272; Agosti & Collingwood, 1987a: 55.

*picea* Emery, 1906; see under ***ANTICHTHONIDRIS.***

*piceonigrum* Borgmeier, 1948; see under ***ANTICHTHONIDRIS.***

*pictum. Monomorium* (*Xeromyrmex*) *salomonis* st. *didonis* var. *pictum* Santschi, 1936a: 57 (w.q.m.) TUNISIA. Unavailable name.

**#pieli.** *Monomorium pieli* Santschi, 1925f: 87 (w.) CHINA. Subspecies of *nipponense:* Wheeler, W.M. 1930h: 67; of *intrudens:* current status.

*pietschmanni. Monomorium pietschmanni* Ettershank, 1966: 89. *Nomen nudum;* attributed to Viehmeyer.

***pilipes.** **Monomorium pilipes* Mayr, 1868c: 91, pl. 5, figs. 93, 94 (w.) BALTIC AMBER (Oligocene).

**#planidorsum.** *Monomorium salomonis* subsp. *planidorsa* Emery, 1915e: 378 (w.) TUNISIA. Stirps of *subopacum:* Santschi, 1927d: 244.

*poecilum. Monomorium poecilum* Roger, 1863a: 199 (w.q.) CUBA. Junior synonym of *floricola:* Emery, 1894c: 151.

*pretoriensis. Monomorium delagoense* var. *pretoriensis* Arnold, 1944: 15 (q.) SOUTH AFRICA. Junior synonym of *junodi:* Bolton, 1987: 346.

**#procax.** *Monomorium* (*Martia*) *atomus* var. *procax* Forel, 1911a: 23 (w.q.) SINGAPORE. Combination in *M.* (*Lampromyrmex*): Emery, 1922e: 183.

*prossae. Monomorium amblyops* r. *prossae* Forel, 1916: 418. Replacement name for *bulawayense* Forel, 1914d: 247. [Junior primary homonym of *bulawayensis* Forel, 1913j: 217.] Combination in *M.*(*Parholcomyrmex*): Wheeler, W.M. 1922a: 873. Raised to species: Emery, 1922e: 173. Junior synonym of *oscaris:* Bolton, 1987: 326.

**pulchrum.** *Monomorium* (*Lampromyrmex*) *pulchrum* Santschi, 1926b: 238, fig. 3 (w.) ZIMBABWE. See also: Bolton, 1987: 406.

*pullula. Monomorium* (*Xeromyrmex*) *salomonis* var. *pullula* Santschi, 1919: 235 (w.) SENEGAL. Stirps of *salomonis:* Santschi, 1934b: 33. Junior synonym of *areniphilum:* Bolton, 1987: 336.

*punctifrons* Borgmeier, 1927; see under ***OXYEPOECUS.***

**#punicum.** *Monomorium* (*Mitara*) *clavicorne* var. *punica* Santschi, 1915a: 58, fig. 4 (w.) TUNISIA.

**rabirium.** *Monomorium rabirium* Bolton, 1987: 358 (w.) BOTSWANA.

*rapaense. Monomorium* (*Notomyrmex*) *rapaense* Wheeler, W.M. 1936f: 10 (w.) RAPA I. Combination in *Chelaner:* Ettershank, 1966: 97. Junior synonym of *antarcticum:* Wilson & Taylor, 1967: 62.

*rastratum* Mayr, 1887; see under ***OXYEPOECUS.***

**rastractum.** *Monomorium rastractum* Bolton, 1987: 406 (w.) KENYA.

**rhopalocerum.** *Monomorium rhopalocerum* Emery, 1895h: 25, pl. 2, fig 29 (w.) SOUTH AFRICA. Senior synonym of *hottentota* and *leimbachi:* Bolton, 1987: 407.

**robustior.** *Monomorium gracillimum* r. *robustior* Forel, 1892e: 352 (w.) SOMALIA. Combination in *M.* (*Parholcomyrmex*): Wheeler, W.M. 1922a: 875. Raised to species: Bolton, 1987: 328. [Name misspelled as *robustius* by Emery, 1893f: 256; Forel, 1894e: 228.]

*robustum. Monomorium* (*Parholcomyrmex*) *nipponense* var. *robustum* Teranishi, 1940: 29 (q.) JAPAN. Junior synonym of *intrudens:* Ogata & Bolton, 1989: 460.

**rogeri.** *Trichomyrmex rogeri* Mayr, 1865: 19 (q.) SRI LANKA. Combination in *Monomorium:* Ettershank, 1966: 91.

**rosae.** *Monomorium* (*Lampromyrmex*) *rosae* Santschi, 1920b: 13, fig. 2 (w.) ZAIRE. Senior synonym of *cotterelli:* Bolton, 1987: 408.

**rothsteini.** *Monomorium rothsteini* Forel, 1902h: 444 (w.) AUSTRALIA. Imai, Crozier & Taylor, 1977: 348 (k.). Combination in *M.* (*Parholcomyrmex*): Santschi, 1919a: 328; in *Chelaner:* Ettershank, 1966: 97; in *Monomorium:* Taylor, 1987b: 3. Current subspecies: nominal plus *doddi, humilior, leda, squamigena, tostum.*

**rotundatum.** *Monomorium rotundatum* Santschi, 1920b: 14, fig. 2 (w.) SOUTH AFRICA. See also: Bolton, 1987: 409.

**rubriceps.** *Monomorium rubriceps* Mayr, 1876: 101 (w.m.) AUSTRALIA. Emery, 1914b: 184 (q.). Combination in *M.* (*Notomyrmex*): Emery, 1922e: 170; in *Chelaner:* Ettershank, 1966: 97; in *Monomorium:* Bolton, 1987: 284. Current subspecies: nominal plus *cinctum, extreminigrum, rubrum* (junior homonym).

**#rubrum.** *Holcomyrmex criniceps* var. *ruber* Forel, 1903a: 692 (w.) SRI LANKA. Combination in *Monomorium:* Emery, 1922e: 181.

**#rubrum.** *Monomorium rubriceps* var. *rubra* Forel, 1913g: 184 (w.q.m.) AUSTRALIA. [Also described as new by Forel, 1915b: 72 (footnote).] [Unresolved junior secondary homonym of *rubrum* Forel, above.] Subspecies of *rubriceps:* Taylor & Brown, D.R. 1985: 58.

*rufescens. Wheeleriella santschii* var. *rufescens* Santschi, 1926e: 233 (q.m.) TUNISIA. Combination in *Monomorium:* Ettershank, 1966: 92. Junior synonym of *santschii:* Bolton, 1987: 291.

*rufibasis. Monomorium bicolor* var. *rufibasis* Santschi, 1914d: 353 (diagnosis in key) (w.) EGYPT. Santschi, 1927d: 244 (q.m.). [Misspelled as *rufobasalis* by Santschi, 1926b: 240.] Junior synonym of *bicolor:* Bolton, 1987: 338.

**rufulum.** *Monomorium salomonis* var. *rufula* Stitz, 1923: 156 (w.) NAMIBIA. Raised to species, and senior synonym of *monardi:* Bolton, 1987: 359. See also: Bolton, 1986b: 268.

**ruzskyi.** *Monomorium* (*Xeromyrmex*) *ruzskyi* Dlussky & Zabelin, 1985: 218, fig. 1 (w.q.m.) TURKESTAN.

**sagei.** *Monomorium sagei* Forel, 1902c: 211 (w.) INDIA. Combination in *M.* (*Xeromyrmex*): Emery, 1922e: 177.

**sahlbergi.** *Monomorium sahlbergi* Emery, 1898c: 131, fig. (w., ergatoid q.) ISRAEL.

**sakalavum.** *Monomorium sakalavum* Santschi, 1928f: 196 (w.) MADAGASCAR.

*salaco. Monomorium* (*Xeromyrmex*) *salomonis* st. *areniphilum* var. *salaco* Santschi, 1936a: 52 (w.q.m.) TUNISIA. Unavailable name.

**salomonis.** *Formica salomonis* Linnaeus, 1758: 580 (w.) EGYPT. André, 1883a: 342 (m.); Emery, 1908h: 674 (w.q.m.). Combination in *Monomorium:* Roger, 1862c: 294; in *M.* (*Xeromyrmex*): Emery, 1915i: 190. Senior synonym of *debilis* and *thorense:* Emery & Forel, 1879: 456; of *obscuratum:* Collingwood, 1978: 69. Current subspecies: nominal plus *bedui, didonis, molestum, obscuriceps, scipionis, sommieri, subnitidum, targui, transversale, volubile, zanoni.*

*samoanum. Monomorium minutum* var. *samoanum* Santschi, 1928a: 46, fig. 2 (w.) SAMOA. Junior synonym of *minutum:* Wilson & Taylor, 1967: 67; of *liliuokalanii:* Bolton, 1987: 288.

**sanguinolentum.** *Monomorium* (*Notomyrmex*) *sanguinolentum* Wheeler, W.M. 1927i: 135, fig. 4 (w.m.) AUSTRALIA. Combination in *Chelaner:* Ettershank, 1966: 97; in *Monomorium:* Bolton, 1987: 284.

*santschianum. Monomorium santschianum* Ettershank, 1966: 92. Replacement name for *Holcomyrmex santschii* Forel, 1907b: 203. [Junior secondary homonym of *Wheeleria santschii* Forel, 1905b: 171.] Junior synonym of *major:* **new synonymy**. See also: Bolton, 1987: 296.

*santschiellum. Monomorium* (*Xeromyrmex*) *salomonis* subsp. *subopacum* var. *santschiellum* Wheeler, W.M. 1922a: 872. [Unnecessary replacement name for *senegalensis* Santschi.] Unavailable name.

**santschii.** *Wheeleria santschii* Forel, 1905b: 171 (w.) TUNISIA. Forel, 1906a: 52 (m.). Combination in *Wheeleriella:* Emery, 1908f: 556; in *Monomorium:* Ettershank, 1966: 92. Senior synonym of *adulatrix, insidiosa* and *rufescens:* Bolton, 1987: 291.

*santschii. Holcomyrmex santschii* Forel, 1907b: 203 (w.) TUNISIA. Santschi, 1911f: 84 (m.); Karavaiev, 1912a: 3 (q.). [Junior secondary homonym of *Wheeleria santschii* Forel, above.] Combination in *Monomorium:* Emery, 1908h: 669; in *M.* (*Isolcomyrmex*): Santschi, 1917c: 296; in *M.* (*Parholcomyrmex*): Emery, 1922e: 180. Replacement name: *santschianum* Ettershank, 1966: 92.

*satoi. Monomorium* (*Parholcomyrmex*) *nipponense* var. *satoi* Teranishi, 1940: 30 (w.q.) JAPAN. Junior synonym of *intrudens:* Ogata & Bolton, 1989: 460.

**scabriceps.** *Holcomyrmex scabriceps* Mayr, 1879: 672 (w.) INDIA. Bingham, 1903: 283 (s.q.m.); Imai, Baroni Urbani, *et al.* 1984: 7 (k.). Current subspecies: nominal plus *crinicipitoscabriceps.*

**schultzei.** *Monomorium schultzei* Forel, 1910f: 18 (w.q.) SOUTH AFRICA. See also: Bolton, 1987: 410.

**schurri.** *Monomorium schurri* Forel, 1902c: 212 (w.q.) INDIA. Combination in *M.* (*Xeromyrmex*): Emery, 1922e: 178.

**#scipionis.** *Monomorium* (*Xeromyrmex*) *salomonis* var. *scipionis* Santschi, 1921e: 169 (w.q.) TUNISIA. Santschi, 1936a: 60 (m.).

**sculpturatum.** *Monomorium* (*Notomyrmex*) *sculpturatum* Clark, 1934c: 59, pl. 4, figs. 17, 18 (w.q.) AUSTRALIA. Combination in *Chelaner:* Ettershank, 1966: 97; in *Monomorium:* Taylor, 1987b: 4.

**sechellense.** *Monomorium fossulatum* subsp. *sechellense* Emery, 1894a: 69, fig. (w.q.) SEYCHELLE IS. Synonym of *fossulatum:* Wilson & Taylor, 1967: 64 [*fossulatum* is given as senior name, but *sechellense* has priority].

***senegalense.*** *Monomorium senegalense* Roger, 1862c: 294 (footnote) (w.) SENEGAL. Combination in *M.* (*Xeromyrmex*): Wheeler, W.M. 1922a: 872. See also: Bolton, 1987: 359.

*senegalensis. Monomorium salomonis* st. *subopacum* var. *senegalensis* Santschi, 1913c: 306 (w.) SENEGAL. Unavailable name; material referred to *subopacum* by Bolton, 1987: 360 [along with both unnecessary (unavailable) proposed replacement names, *claveaui* and *santschiellum*].

*serenum. Monomorium* (*Xeromyrmex*) *delagoense* st. *junodi* var. *serenum* Santschi, 1928f: 192 (w.q.m.) ZIMBABWE. Unavailable name; material referred to *opacior* by Bolton, 1987: 352.

***sersalatum.*** *Monomorium sersalatum* Bolton, 1987: 424 (w.q.) RWANDA.

***setuliferum.*** *Monomorium setuliferum* Forel, 1910f: 16 (w.) BOTSWANA. See also: Bolton, 1987: 370.

***shilohense.*** *Monomorium braunsi* var. *shilohensis* Forel, 1913j: 217 (w.) ZIMBABWE. Raised to species: Bolton, 1987: 410.

***shuckardi.*** *Monomorium shuckardi* Forel, 1895c: 251 (w.) MADAGASCAR. Combination in *M.* (*Notomyrmex*): Emery, 1922e: 170.

***silvestrii.*** *Monomorium silvestrii* Wheeler, W.M. 1927h: 87, fig. 2 (w.) VIETNAM.

***smithii.*** *Monomorium smithii* Forel, 1892c: 342 (w.q.m.) NEW ZEALAND. Combination in *M.* (*Notomyrmex*): Emery, 1922e: 170; in *Chelaner:* Ettershank, 1966: 97; in *Monomorium:* Taylor, 1987b: 5. See also: Brown, 1958h: 34.

*smutsi. Monomorium* (*Syllophopsis*) *modestum* var. *smutsi* Wheeler, W.M. 1922a: 867. Unnecessary (second) replacement name for *boerorum* Santschi, 1915c: 260. Junior synonym of *modestum:* Bolton, 1987: 423.

*solleri. Rhoptromyrmex solleri* Forel, 1910e: 430 (q.) SENEGAL. Combination in *Monomorium:* Ettershank & Brown, 1964: 18. Junior synonym of *oscaris:* Bolton, 1987: 326.

#***sommieri.*** *Monomorium salomonis* var. *sommieri* Emery, 1908h: 676 (w.) TUNISIA. Stirps of *salomonis:* Santschi, 1936a: 58. Senior synonym of *atrata:* Emery, 1916b: 161; Santschi, 1921e: 170.

***sordidum.*** *Monomorium sordidum* Forel, 1902h: 443 (w.) AUSTRALIA. Combination in *Chelaner:* Ettershank, 1966: 97; in *Monomorium:* Taylor, 1987b: 4. Current subspecies: nominal plus *nigriventre.*

***spatulicorne.*** *Monomorium spatulicorne* Kuznetsov-Ugamsky, 1926a: 73, figs. 16–18 (w.) TURKESTAN.

***spectrum.*** *Monomorium spectrum* Bolton, 1987: 411 (w.) GABON.

*specularis. Monomorium specularis* Mayr, 1866a: 509 (w.) SAMOA. Junior synonym of *floricola:* Mayr, 1879: 671.

*speculiceps. Monomorium* (*Xeromyrmex*) *speculiceps* Santschi, 1928f: 191, fig. 3 (w.) SOUTH AFRICA. Junior synonym of *excelsior:* Bolton, 1987: 344.

***speluncarum.*** *Monomorium rhopalocerum* st. *speluncarum* Santschi, 1914b: 72, fig. 6 (w.) KENYA. Raised to species: Bolton, 1987: 411.

***springvalense.*** *Monomorium oscaris* r. *springvalense* Forel, 1913a: 136 (w.) ZIMBABWE. Subspecies of *musicum:* Emery, 1922e: 173. Raised to species: Santschi, 1926b: 236; Bolton, 1987: 412.

#***squamigena.*** *Monomorium rothsteini* var. *squamigena* Viehmeyer, 1925a: 28 (w.) AUSTRALIA. Combination in *Chelaner:* Ettershank, 1966: 97; in *Monomorium:* Taylor, 1987b: 3. Subspecies of *rothsteini:* Taylor & Brown, D.R. 1985: 58.

***sryetum.*** *Monomorium sryetum* Bolton, 1987: 413 (w.) BOTSWANA.

#***stauderi.*** *Monomorium lameerei* var. *stauderi* Wolf, 1914: 42, fig. 2 (w.) ALGERIA.

***strangulatum.*** *Monomorium strangulatum* Santschi, 1921c: 121, fig. 3 (w.) TANZANIA. See also: Bolton, 1987: 413.

***subapterum.*** *Monomorium subapterum* Wheeler, W.M. 1917b: 112, fig. 2 (w.q.m.) AUSTRALIA. Combination in *M.* (*Notomyrmex*): Emery, 1922e: 170; in *Chelaner:* Ettershank, 1966: 97; in *Monomorium:* Taylor, 1987b: 4. Current subspecies: nominal plus *bogischi.*

***subcoecum.*** *Monomorium subcoecum* Emery, 1894c: 150 (w.q.) ANTILLES.

***subdentatum.*** *Monomorium subdentatum* Forel, 1913b: 332 (w.) ZAIRE. See also: Bolton, 1987: 359.

#***subnitidum.*** *Monomorium salomonis* var. *subnitidum* Emery, 1895b: 66 (w.q.) ALGERIA. Forel, 1908a: 13 (m.). Stirps of *salomonis:* Santschi, 1936a: 59.

***subopacum.*** *Myrmica subopaca* Smith, F. 1858b: 127 (w.q.) MADEIRA. Combination in *Monomorium:* Mayr, 1862: 753; in *M.* (*Xeromyrmex*): Wheeler, W.M. 1922a: 871. Variety/subspecies of *salomonis:* Forel, 1890a: lxxv; Forel, 1907a: 18; Emery, 1908h: 676; Forel, 1910a: 23; Arnold, 1916: 224; Santschi, 1921e: 170; Santschi, 1923e: 281; Menozzi, 1926b: 182; Wheeler, W.M. 1927g: 107. Status as species: Emery, 1881b: 531; Emery, 1893c: 82; Mayr, 1895: 133; Bondroit, 1918: 143; Santschi, 1927d: 240; Santschi, 1931a: 4; Bolton, 1987: 360. Senior synonym of *mediterraneum:* Mayr, 1862: 753; Bolton, 1987: 360; of *obscuripes:* Santschi, 1927d: 241; of *italica:* Baroni Urbani, 1968b: 450; of *cabrerae, glyciphila, liberta* and *surcoufi,* and material of the unavailable name *senegalensis* referred here: Bolton, 1987: 360; of *intermedium* Hohmann, La Roche, *et al.,* 1993: 155. Current subspecies: nominal plus *adoneum, apuleii, ebraicum, planidorsum.*

*subterraneum* Mann, 1916; see under ***TRANOPELTA.***

*succineum. Monomorium succineum* Stitz, 1911a: 367 (q.) NEW ZEALAND. Combination in *M.* (*Notomyrmex*): Emery, 1922e: 170. Junior synonym of *antarcticum:* Brown, 1958h: 29.

*sudanicum. Monomorium* (*Lampromyrmex*) *exiguum* st. *mictile* var. *sudanicum* Santschi, 1930a: 67, figs. 22–24 (w.) SUDAN. Unavailable name; material referred to *mictile* by Bolton, 1987: 401.

*surcoufi. Paraphacota surcoufi* Santschi, 1919d: 90, fig. 1 (m.) ALGERIA. Santschi, 1927d: 243 (w.q.). Combination in *Monomorium,* and variety of *subopacum:* Santschi, 1927d: 243. Junior synonym of *subopacum:* Bolton, 1987: 360.

*suteri. Monomorium suteri* Forel, 1892c: 340 (w.q.m.) NEW ZEALAND. Combination in *M.* (*Notomyrmex*): Emery, 1922e: 170. Junior synonym of *antarcticum:* Brown, 1958h: 29.

***sutu.*** *Monomorium sutu* Bolton, 1987: 361 (w.) KENYA.

***sydneyense.*** *Monomorium sydneyense* Forel, 1902h: 442 (w.) AUSTRALIA. Combination in *M.* (*Mitara*): Emery, 1914b: 184; in *M.* (*Lampromyrmex*): Emery, 1922e: 184. Current subspecies: nominal plus *nigellum.*

***symmotu.*** *Monomorium symmotu* Bolton, 1987: 414, fig. 79 (w.) ZIMBABWE.

***syriacum.*** *Epixenus syriaca* Tohmé, H. & Tohmé, G. 1980a: 1100, fig. 7 (w.q.) SYRIA. Combination in *Monomorium:* Bolton, 1987: 292.

***tablense.*** *Monomorium altinode* st. *tablensis* Santschi, 1932a: 384, figs. 6, 7 (w.q.) SOUTH AFRICA. Raised to species: Bolton, 1987: 414.

***taedium.*** *Monomorium taedium* Bolton, 1987: 415 (w.) SOUTH AFRICA.

***talbotae.*** *Monomorium talbotae* DuBois, 1981: 31, fig. 1 (q.m.) U.S.A. See also: DuBois, 1986: 112.

***talpa.*** *Monomorium talpa* Emery, 1911c: 252 (w.) NEW GUINEA. Mann, 1919: 329 (q.). Senior synonym of *myops:* Ettershank, 1966: 92.

#***tambourinense.*** *Monomorium kiliani* var. *tambourinensis* Forel, 1915b: 71 (w.q.) AUSTRALIA. Wheeler, G.C. & Wheeler, J. 1960b: 17 (l.). Combination in *M.* (*Notomyrmex*): Emery, 1922e: 169; in *Chelaner:* Ettershank, 1966: 97; in *Monomorium:* Taylor, 1987b: 3. Raised to species: Wheeler, G.C. & Wheeler, J. 1960b: 17. Revived status as subspecies of *kiliani:* Taylor & Brown, D.R. 1985: 57.

*tanit. Monomorium* (*Xeromyrmex*) *salomonis* st. *sommieri* var. *tanit* Santschi, 1936a: 58, figs. 4, 18 (w.q.) TUNISIA. Unavailable name.

***tanysum.*** *Monomorium tanysum* Bolton, 1987: 416 (w.) GHANA.

***taprobanae.*** *Monomorium minutum* var. *taprobanae* Forel, 1913k: 53 (w.q.) SRI LANKA. Raised to species: **new status**.

#***targui.*** *Monomorium* (*Xeromyrmex*) *salomonis* st. *targui* Santschi, 1936a: 53 (w.) ALGERIA.

***tchelichofi.*** *Monomorium tchelichofi* Forel, 1914d: 244 (w.) SOUTH AFRICA. See also: Bolton, 1987: 362.

***termitarium.*** *Monomorium salomonis* subsp. *termitarium* Forel, 1910f: 17 (w.q.) BOTSWANA. Raised to species: Santschi, 1928f: 194; Bolton, 1987: 362.

***termitobium.*** *Monomorium termitobium* Forel, 1892k: 522 (w.) MADAGASCAR.

*thales. Monomorium albopilosum* var. *thales* Forel, 1913a: 136 (w.) ZIMBABWE. Arnold, 1916: 208 (q.m.). Junior synonym of *albopilosum:* Bolton, 1987: 335.

*thorense. Monomorium thorense* Mayr, 1862: 753 (q.) EGYPT. Junior synonym of *salomonis:* Emery & Forel, 1879: 456.

***thrascoleptum.*** *Monomorium thrascoleptum* Bolton, 1987: 424, figs. 93, 94 (w.) IVORY COAST.

***torvicte.*** *Monomorium torvicte* Bolton, 1987: 416 (w.) SOUTH AFRICA.

#***tostum.*** *Monomorium rothsteini* var. *tostum* Wheeler, W.M. 1915g: 806 (w.) AUSTRALIA. Combination in *Chelaner:* Ettershank, 1966: 97; in *Monomorium:* Taylor, 1987b: 3. Subspecies of *rothsteini:* Taylor & Brown, D.R. 1985: 58.

***trageri.*** *Monomorium trageri* DuBois, 1986: 109, figs. 7, 20, 133–148 (w.q.m.) U.S.A.

**trake.** *Monomorium trake* Bolton, 1987: 417 (w.) GHANA.

**#transversale.** *Monomorium* (*Xeromyrmex*) *salomonis* var. *transversalis* Santschi, 1921e: 169 (w.q.) MOROCCO.

*transwaalensis. Monomorium* (*Syllophopsis*) *modestum* var. *transwaalensis* Emery, 1922e: 175. Replacement name for *boerorum* Santschi, 1915c: 260. [Junior primary homonym of *boerorum* Forel, 1910e: 442.] Junior synonym of *modestum:* Bolton, 1987: 423.

**tricolor.** *Monomorium tricolor* Emery, 1914f: 407 (w.) NEW CALEDONIA. Combination in *M.* (*Notomyrmex*): Emery, 1922e: 170; in *Chelaner:* Ettershank, 1966: 97; in *Monomorium:* Taylor, 1987b: 4.

**triviale.** *Monomorium triviale* Wheeler, W.M. 1906c: 311 (w.) JAPAN.

*tropicale. Monomorium* (*Xeromyrmex*) *bicolor* var. *tropicale* Santschi, 1926b: 240 (w.q.m.) ZAIRE. Junior synonym of *bicolor:* Bolton, 1987: 338.

*tuneticum. Monomorium* (*Xeromyrmex*) *salomonis* st. *areniphilum* var. *tuneticum* Santschi, 1936a: 52 (w.q.m.) TUNISIA. Unavailable name.

*turkmenicum. Monomorium* (*Xeromyrmex*) *venustum* st. *turkmenicum* Santschi, 1928b: 43 (w.) TURKESTAN. [Misspelled as *turkestanicum* by Santschi, 1936a: 49.] Variety of *indicum:* Pisarski, 1967: 398. Junior synonym of *kusnezowi:* Dlussky & Zabelin, 1985: 215.

**turneri.** *Vollenhovia turneri* Forel, 1910b: 26 (w.) AUSTRALIA. Combination in *Monomorium:* Emery, 1914f: 407 (footnote); in *M.* (*Notomyrmex*): Emery, 1922e: 169; in *Chelaner:* Ettershank, 1966: 97; in *Monomorium:* Bolton, 1987: 284.

**tynsorum.** *Monomorium tynsorum* Bolton, 1987: 417, figs. 62, 87 (w.) ANGOLA.

*uelense. Monomorium* (*Xeromyrmex*) *bicolor* var. *uelense* Santschi, 1926b: 239 (w.) ZAIRE. [Misspelled as *uluense* by Santschi, 1926b: 240.] Junior synonym of *bicolor:* Bolton, 1987: 338.

*unicolor. Monomorium salomonis* var. *unicolor* Stitz, 1923: 156 (w.) NAMIBIA. Junior synonym of *damarense:* Bolton, 1987: 340.

**vaguum.** *Monomorium* (*Lampromyrmex*) *vaguum* Santschi, 1930a: 68, figs. 26–29 (w.q.) ZAIRE. See also: Bolton, 1987: 41.

**valtinum.** *Monomorium valtinum* Bolton, 1987: 428 (w.) KENYA.

*vastator. Myrmica vastator* Smith, F. 1857a: 71 (w.) SINGAPORE. Combination in *Monomorium:* Mayr, 1886c: 359. Junior synonym of *destructor:* Dalla Torre, 1893: 66; of *pharaonis:* Donisthorpe, 1932c: 449.

**vatranum.** *Monomorium vatranum* Bolton, 1987: 362, fig. 55 (w.) NAMIBIA.

**vecte.** *Monomorium vecte* Bolton, 1987: 419, fig. 68 (w.) ZIMBABWE.

**venustum.** *Myrmica venusta* Smith, F. 1858b: 126 (w.) SYRIA. André, 1881b: 66 (q.m.). Combination in *Monomorium:* Roger, 1863b: 32; in *M.* (*Xeromyrmex*): Wheeler, W.M. 1922a: 873.

*vexator. Myrmica vexator* Smith, F. 1861b: 47 (w.) INDONESIA (Ternate I.). Junior synonym of *destructor:* Donisthorpe, 1932c: 468.

*vezenyii* Forel, 1907; see under ***OXYEPOECUS.***

**viator.** *Monomorium* (*Xeromyrmex*) *viator* Santschi, 1923e: 280, fig. 3 (w.m.) NAMIBIA. See also: Bolton, 1987: 363.

**viride.** *Monomorium viridis* Brown, 1943: 243 (w.q.) U.S.A. Crozier, 1970: 116 (k.); DuBois, 1986: 98 (m.). Senior synonym of *peninsulatum:* DuBois, 1986: 96.

**vitiense.** *Monomorium* (*Monomorium*) *vitiensis* Mann, 1921: 444 (w.) FIJI IS.

*voeltzkowi. Monomorium voeltzkowi* Forel, 1907g: 78 (w.) TANZANIA. Junior synonym of *latinode:* Bolton, 1987: 429.

**#volubile.** *Monomorium* (*Xeromyrmex*) *volubilis* Santschi, 1936: 201c (w.q.) MOROCCO. [First available use of *Monomorium* (*Xeromyrmex*) *salomonis* st. *didonis* var. *volubile* Santschi, 1936a: 56; unavailable name.]

**vonatu.** *Monomorium vonatu* Bolton, 1987: 420 (w.) GHANA.

**westi.** *Monomorium westi* Bolton, 1987: 363, figs. 37, 43 (w.) KENYA.

**wheelerorum.** *Monomorium wheelerorum* DuBois, 1986: 91, figs. 11, 17, 80, 81–86 (w.q.) U.S.A.

**whitei.** *Monomorium* (*Holcomyrmex*) *whitei* Wheeler, W.M. 1915g: 807, pl. 64, fig. 1 (w.) AUSTRALIA. Imai, Crozier & Taylor, 1977: 349 (k.). Combination in *Chelaner:* Ettershank, 1966: 97; in *Monomorium:* Taylor, 1987b: 4.

**willowmorense.** *Monomorium willowmorense* Bolton, 1987: 364, fig. 54 (w.) SOUTH AFRICA. Material of the unavailable names *willowmorensis* and *belli* referred here: Bolton, 1987: 364.

*willowmorensis. Monomorium salomonis* r. *herrero* var. *willowmorensis* Forel, 1914d: 245 (w.) SOUTH AFRICA. Unavailable name; material referred to *willowmorense* by Bolton, 1987: 364.

**wroughtoni.** *Monomorium wroughtoni* Forel, 1902c: 209 (w.) INDIA.

*wroughtoni. Wheeleriella wroughtoni* Forel, 1910a: 7 (q.m.) INDIA. [Junior secondary homonym of *Monomorium wroughtoni* Forel, above.] Replacement name: *Monomorium effractor* Bolton, 1987: 292. Combination in *Monomorium:* Ettershank, 1966: 93.

*wroughtoni. Trichomyrmex wroughtoni* Forel, 1911h: 453 (q.) INDIA. [Junior secondary homonym of *Monomorium wroughtoni* Forel, 1902c: 209.] Replacement name: *Monomorium wroughtonianum* Ettershank, 1966: 93.

**wroughtonianum.** *Monomorium wroughtonianum* Ettershank, 1966: 93. Replacement name for *Trichomyrmex wroughtoni* Forel, 1911h: 453. [Junior secondary homonym of *Monomorium wroughtoni* Forel, 1902c: 209.]

**xanthognathum.** *Monomorium xanthognathum* Arnold, 1944: 9, fig. 15 (w.) SOUTH AFRICA. See also: Bolton, 1987: 371.

**#zanoni.** *Monomorium* (*Xeromyrmex*) *subopacum* var. *zanoni* Santschi, 1927d: 242 (w.q.) LIBYA. [First available use of *Monomorium salomonis* subsp. *subopacum* var. *zanoni* Emery, 1924a: 9; unavailable name.] Santschi, 1927d: 242 (m.). Variety of *salomonis:* Menozzi, 1940: 268.

**zulu.** *Monomorium zulu* Santschi, 1914e: 18 (w.) SOUTH AFRICA. See also: Bolton, 1987: 365.

*MORLEYIDRIS:* see under ***POLYRHACHIS.***

***#MYAGROTERAS:*** see under ***MYRMOTERAS.***

***MYCETAROTES*** (Neotropical)

*luederwaldti. Atta* (*Mycocepurus*) *luederwaldti* Forel, 1911c: 293 (w.q.) BRAZIL. Combination in *Cyphomyrmex* (*Mycetarotes*): Emery, 1913: 251; in *Mycocepurus:* Luederwaldt, 1918: 39. Junior synonym of *parallelus:* Kempf, 1960d: 279.

**parallelus.** *Cyphomyrmex parallelus* Emery, 1906c: 161, fig. 23 (w.q.) BRAZIL. Kempf, 1960d: 280 (m.). Combination in *C.* (*Mycetarotes*): Emery, 1913: 251; in *Mycetarotes,* and senior synonym of *luederwaldti:* Kempf, 1960d: 279.

**senticosus.** *Mycetarotes senticosus* Kempf, 1960d: 282, figs. 4–6 (w.) BRAZIL.

***MYCETOPHYLAX*** (Neotropical)

**#arenicola.** *Myrmicocrypta emeryi* var. *arenicola* Forel, 1912e: 189 (w.q.) ARGENTINA. Combination in *Cyphomyrmex* (*Mycetophylax*): Emery, 1922f: 343; in *Myrmicocrypta* (*Mycetophylax*): Santschi, 1922b: 355; in *Mycetophylax:* Kempf, 1972a: 146.

**#argentina.** *Myrmicocrypta* (*Mycetophylax*) *emeryi* var. *argentina* Santschi, 1916e: 383 (w.) ARGENTINA. Combination in *Cyphomyrmex* (*Mycetophylax*): Emery, 1922f: 343; in *Mycetophylax:* Kempf, 1972a: 146.

**#bolivari.** *Mycetophylax bolivari* Weber, 1948b: 84 (w.) VENEZUELA. Subspecies of *emeryi:* Weber, 1958d: 263.

*brittoni. Myrmicocrypta brittoni* Wheeler, W.M. 1907c: 728, pl. 50, figs. 18, 19 (w.) PUERTO RICO. Wheeler, W.M. 1911b: 170 (m.). Combination in *Cyphomyrmex* (*Mycetophylax*): Emery, 1913: 251; in *Mycetophylax:* Weber, 1937: 401. Junior synonym of *conformis:* Kempf, 1962b: 34.

**bruchi.** *Sericomyrmex bruchi* Santschi, 1916e: 383, fig. (w.) ARGENTINA. Combination in *Myrmicocrypta* (*Mycetophylax*): Santschi, 1922b: 355; in *Paramycetophylax:* Kusnezov, 1956: 24; in *Mycetophylax:* Santschi, 1923c: 268; Weber, 1958d: 262. Current subspecies: nominal plus *pauper.*

**conformis.** *Cyphomyrmex conformis* Mayr, 1884: 38 (w.) FRENCH GUIANA. Combination in *Mycetophylax,* and senior synonym of *brittoni* and *littoralis:* Kempf, 1962b: 34.

**cristulatus.** *Myrmicocrypta* (*Mycetophylax*) *cristulata* Santschi, 1922b: 356 (w.q.m.) ARGENTINA. Combination in *Mycetophylax:* Santschi, 1929d: 304. Current subspecies: nominal plus *emmae.*

**emeryi.** *Myrmicocrypta emeryi* Forel, 1907f: 144 (w.) COLOMBIA. Forel, 1912e: 189 (q.m.). Combination in *Cyphomyrmex* (*Mycetophylax*): Emery, 1913: 251; in *Myrmicocrypta* (*Mycetophylax*): Santschi, 1916e: 383; in *Mycetophylax:* Santschi, 1925e: 163. Senior synonym of *hummelincki:* Weber, 1958d: 263. Current subspecies: nominal plus *arenicola, argentina, bolivari, fortis, gallardoi, hubrichi, weiseri.*

**#emmae.** *Mycetophylax cristulatus* var. *emmae* Santschi, 1929d: 304 (w.) BRAZIL.

**#fortis.** *Myrmicocrypta emeryi* var. *fortis* Forel, 1912e: 189 (w.) ARGENTINA. Combination in *M.* (*Mycetophylax*): Santschi, 1922b: 355; in *Cyphomyrmex* (*Mycetophylax*): Emery, 1922f: 343; in *Mycetophylax:* Kempf, 1972a: 146.

**#gallardoi.** *Myrmicocrypta* (*Mycetophylax*) *emeryi* st. *gallardoi* Santschi, 1922b: 354 (w.) ARGENTINA. Combination in *Mycetophylax:* Kempf, 1972a: 146.

**glaber.** *Mycetophylax glaber* Weber, 1948b: 85 (w.) BOLIVIA.

**#hubrichi.** *Mycetophylax emeryi* st. *hubrichi* Santschi, 1925e: 163, fig. 6 (w.) ARGENTINA.

*hummelincki. Mycetophylax hummelincki* Weber, 1948b: 84 (w.) VENEZUELA. Junior synonym of *emeryi:* Weber, 1958d: 263.

*littoralis. Mycetophylax brittoni* var. *littoralis* Weber, 1937: 401 (w.m.) TRINIDAD. Junior synonym of *conformis:* Kempf, 1962b: 34.

**#pauper.** *Mycetophylax bruchi* var. *pauper* Santschi, 1923c: 268. Replacement name for *simplex* Santschi, 1922b: 355. [Junior secondary homonym of *simplex* Emery, 1888c: 361.]

**simplex.** *Cyphomyrmex simplex* Emery, 1888c: 361 (w.). BRAZIL. Combination in *C.* (*Mycetophylax*): Emery, 1913: 251; in *Mycetophylax:* Kempf, 1972a: 146.

*simplex. Myrmicocrypta* (*Mycetophylax*) *bruchi* var. *simplex* Santschi, 1922b: 355 (w.) ARGENTINA. [Junior secondary homonym of *simplex* Emery, above.] Replacement name: *pauper* Santschi, 1923c: 268.

**#weiseri.** *Mycetophylax emeryi* st. *weiseri* Santschi, 1929d: 303 (w.) BRAZIL.

***MYCETOSORITIS*** (Neotropical, s. Nearctic)

**aspera.** *Cyphomyrmex asper* Mayr, 1887: 561 (q.) BRAZIL. Emery, 1906c: 162 (w.). Combination in *C.* (*Mycetosoritis*): Emery, 1922f: 344; in *Mycetosoritis:* Kempf, 1972a: 146.

**clorindae.** *Cyphomyrmex* (*Mycetosoritis*) *clorindae* Kusnezov, 1949d: 444, figs. 4, 5 (w.) ARGENTINA. Combination in *Mycetosoritis:* Kempf, 1968b: 401.

**explicata.** *Mycetosoritis explicata* Kempf, 1968b: 401, figs. 14–16 (w.) BRAZIL.

**hartmanni.** *Atta* (*Mycetosoritis*) *hartmanni* Wheeler, W.M. 1907c: 714, pl. 49, figs. 6, 7 (w.q.m.) U.S.A. Wheeler, G.C. & Wheeler, J. 1974d: 77 (l.). Combination in *Cyphomyrmex* (*Mycetosoritis*): Emery, 1913: 251; in *Mycetosoritis:* Creighton, 1950a: 318.

*MYCHOTHORAX:* see under ***LEPTOTHORAX.***

***MYCOCEPURUS*** (Neotropical)

*attaxenus. Trachymyrmex attaxenus* Menozzi, 1936b: 85 (w.q.) BRAZIL. Junior synonym of *smithii:* Kempf, 1963b: 425.

*bolivianus. Mycocepurus bolivianus* Weber, 1938b: 155, fig. 8 (w.) BOLIVIA. Junior synonym of *smithii:* Kempf, 1963b: 425.

*borinquenensis. Atta* (*Mycocepurus*) *smithi* var. *borinquenensis* Wheeler, W.M. 1907c: 718 (w.) PUERTO RICO. Combination in *Mycocepurus:* Emery, 1922f: 335. Junior synonym of *smithii:* Kempf, 1963b: 425.

*eucarnitae. Mycocepurus smithii* var. *eucarnitae* Forel, 1913l: 235 (w.) CUBA. Junior synonym of *smithii:* Kempf, 1963b: 425.

*gentilis. Mycocepurus goeldii* st. *gentilis* Santschi, 1925b: 17 (w.) BRAZIL. Junior synonym of *goeldii:* Kempf, 1963b: 420.

**goeldii.** *Atta* (*Mycocepurus*) *goeldii* Forel, 1893g: 370 (footnote) (w.) BRAZIL. Forel, 1908c: 353 (q.m.). Combination in *Mycocepurus:* Luederwaldt, 1918: 39. Senior synonym of *gentilis, ogloblini* Santschi, *ogloblini* Kusnezov, *schuppi:* Kempf, 1963b: 420.

*manni. Mycocepurus manni* Weber, 1938b: 156, figs. 1, 2 (q.) BOLIVIA. Junior synonym of *smithii:* Kempf, 1963b: 425.

**obsoletus.** *Mycocepurus obsoletus* Emery, 1913: 252, fig. 1 (w.) BRAZIL.

*ogloblini. Mycocepurus ogloblini* Santschi, 1933e: 119, figs. 12, 13 (w.q.) ARGENTINA. Junior synonym of *goeldii:* Kempf, 1963b: 420.

*ogloblini. Descolemyrma ogloblini* Kusnezov, 1951d: 460, pl. 1, fig. 1 (m.) ARGENTINA. [Unresolved junior secondary homonym of *ogloblini* Santschi, above.] Junior synonym of *goeldii:* Kempf, 1963b: 420.

*reconditus. Mycocepurus reconditus* Borgmeier, 1937b: 246, figs. 34–36 (w.q.) BRAZIL. Junior synonym of *smithii:* Kempf, 1963b: 425.

*schuppi. Atta* (*Mycocepurus*) *goeldii* var. *schuppi* Forel, 1901d: 301 (w.) BRAZIL. Combination in *Mycocepurus:* Emery, 1922f: 335. Junior synonym of *goeldii:* Kempf, 1963b: 420.

**smithii.** *Atta* (*Mycocepurus*) *smithii* Forel, 1893g: 370 (w.) ANTILLES. Kempf, 1963b: 425 (q.). Combination in *Mycocepurus:* Wheeler, W.M. & Mann, 1914: 42. Senior synonym of *attaxenus, bolivianus, borinquenensis, eucarnitae, manni, reconditus, tolteca, trinidadensis:* Kempf, 1963b: 425.

**tardus.** *Mycocepurus tardus* Weber, 1940a: 416, fig. 13 (w.) PANAMA. See also: Kempf, 1963b: 430.

*tolteca. Atta* (*Mycocepurus*) *smithi* var. *tolteca* Wheeler, W.M. 1907c: 718 (w.) MEXICO. Junior synonym of *smithii:* Kempf, 1963b: 425.

*trinidadensis. Mycocepurus smithi* var. *trinidadensis* Weber, 1937: 378, fig. 1 (w.q.) TRINIDAD. Junior synonym of *smithii:* Kempf, 1963b: 425.

***MYOPIAS*** (Oriental, Indo-Australian, Australasian)

**amblyops.** *Myopias amblyops* Roger, 1861a: 39 (w.q.) SRI LANKA.

**bidens.** *Trapeziopelta bidens* Emery, 1900c: 313 (diagnosis in key) (w.) SUMATRA. [Also described as new by Emery, 1900d: 664.] Combination in *Myopias:* **new combination**. Current subspecies: nominal plus *polita*.

**breviloba.** *Trapeziopelta breviloba* Wheeler, W.M. 1919e: 143 (q.) BORNEO. Combination in *Myopias:* **new combination**.

**castaneicola.** *Trapeziopelta castaneicola* Donisthorpe, 1938d: 596 (w.) NEW GUINEA. Combination in *Myopias:* **new combination**.

**chapmani.** *Myopias chapmani* Willey & Brown, 1983: 264, figs. 10, 26 (w.q.) AUSTRALIA.

**concava.** *Myopias concava* Willey & Brown, 1983: 258, figs. 4, 18 (w.q.) NEW GUINEA.

**crawleyi.** *Trapeziopelta crawleyi* Donisthorpe, 1941k: 237. Replacement name for *Trapeziopelta nitida* Crawley, 1924: 384. [Junior secondary homonym of *Ponera nitida* Smith, F. 1861b: 45.] Combination in *Myopias:* **new combination**.

**cribriceps.** *Myopias cribriceps* Emery, 1901h: 156 (w.q.) NEW GUINEA. Wheeler, G.C. & Wheeler, J. 1976a: 59 (l.).

**delta.** *Myopias delta* Willey & Brown, 1983: 281, figs. 8, 21 (w.q.) NEW GUINEA.

**densesticta.** *Myopias densesticta* Willey & Brown, 1983: 268, figs. 11, 29 (w.q.) AUSTRALIA.

*diadela. Trapeziopelta diadela* Clark, 1934c: 54, pl. 4, figs. 7, 8 (w.q.) AUSTRALIA. Junior synonym of *tasmaniensis:* Brown, 1953i: 51.

**emeryi.** *Trapeziopelta emeryi* Forel, 1913k: 16, fig. E (w.) JAVA. Combination in *Myopias:* **new combination**.

*fulvescens. Trapeziopelta tenuis* var. *fulvescens* Emery, 1901h: 155 (w.q.) NEW GUINEA. Junior synonym of *tenuis:* Willey & Brown, 1983: 270.

**gigas.** *Myopias gigas* Willey & Brown, 1983: 251, figs. 1, 12 (w.) NEW GUINEA.

**hollandi.** *Trapeziopelta hollandi* Forel, 1901f: 325 (w.q.) SUMATRA. Combination in *Myopias:* **new combination**.

**julivora.** *Myopias julivora* Willey & Brown, 1983: 254, figs. 5, 22 (w.q.) NEW GUINEA.

**kuehni.** *Trapeziopelta kuehni* Forel, 1902b: 179 (q.) INDONESIA (Key I.). Combination in *Myopias:* **new combination**.

**latinoda.** *Trapeziopelta latinoda* Emery, 1897d: 555, pl. 1, figs. 3, 4 (q.) NEW GUINEA. Combination in *Myopias:* **new combination**.

**levigata.** *Trapeziopelta levigata* Emery, 1901h: 155 (w.) NEW GUINEA. Combination in *Myopias:* **new combination**.

**lobosa.** *Myopias lobosa* Willey & Brown, 1983: 277, figs. 2, 3, 25 (w.q.) PHILIPPINES.

*longitarsis. Bradyponera longitarsis* Smith, F. 1873: viii. *Nomen nudum;* attributed to Mayr.

**loriai.** *Trapeziopelta loriai* Emery, 1897d: 554, pl. 1, figs. 1, 2 (w.) NEW GUINEA. Combination in *Myopias:* **new combination**.

**maligna.** *Ponera maligna* Smith, F. 1861b: 44 (w.q.) SULAWESI. Combination in *Trapeziopelta:* Mayr, 1862: 715; in *Myopias:* **new combination**. Current subspecies: nominal plus *punctigera*.

**mandibularis.** *Trapeziopelta mandibularis* Crawley, 1924: 386, fig. 3 (w.) SUMATRA. Combination in *Myopias:* **new combinatoin**.

**mayri.** *Trapeziopelta mayri* Donisthorpe, 1932c: 468. Replacement name for *Ponera nitida* Smith, F. 1861b: 45. [Junior primary homonym of **Ponera nitida* Heer, 1850: 149.] Combination in *Myopias:* **new combination**.

**media.** *Myopias media* Willey & Brown, 1983: 257, figs. 6, 23 (w.) NEW GUINEA.

**modiglianii.** *Trapeziopelta modiglianii* Emery, 1900c: 313 (diagnosis in key) (w.) SUMATRA. Combination in *Myopias:* **new combination**. See also: Emery, 1900d: 663.

*nitida. Ponera nitida* Smith, F. 1861b: 45 (w.) SULAWESI. [Junior primary homonym of **Ponera nitida* Heer, 1850: 149.] Replacement name: *Trapeziopelta mayri* Donisthorpe, 1932c: 468. Combination in *Bradyponera:* Mayr, 1886c: 362.

*nitida. Trapeziopelta nitida* Crawley, 1924: 384, fig. 2 (w.) SUMATRA. [Junior secondary homonym of *nitida* Smith, F., above.] Replacement name: *crawleyi* Donisthorpe, 1941k: 237.

**nops.** *Myopias nops* Willey & Brown, 1983: 279, figs. 9, 28 (w.) TAIWAN.

**philippinensis.** *Trapeziopelta philippinensis* Menozzi, 1925c: 441, pl. 1, fig. 3 (w.) PHILIPPINES. Combination in *Myopias:* **new combination**.

*MYOPIAS (continued)*

**#polita.** *Trapeziopelta bidens* var. *polita* Stitz, 1925: 112 (w.q.) PHILIPPINES. Combination in *Myopias:* **new combination**.

**#punctigera.** *Trapeziopelta maligna* var. *punctigera* Emery, 1900d: 663 (w.) INDONESIA (Mentawei I.). Combination in *Myopias:* **new combination**.

**punctigera.** *Trapeziopelta punctigera* Emery, 1901h: 155 (w.) NEW GUINEA. [Unresolved junior primary homonym of *punctigera* Emery, above.] Combination in *Myopias:* **new combination**.

**ruthae.** *Myopias ruthae* Willey & Brown, 1983: 274, figs. 7, 20 (w.) NEW GUINEA.

**santschii.** *Trapeziopelta santschii* Viehmeyer, 1914b: 515 (w.) NEW GUINEA. Combination in *Myopias:* **new combination**.

*silvestrii* Santschi, 1914; see under ***CENTROMYRMEX.***

**tasmaniensis.** *Myopias tasmaniensis* Wheeler, W.M. 1923e: 177, fig. 1 (w.) TASMANIA. Wheeler, G.C. & Wheeler, J. 1964b: 454 (l.). Senior synonym of *diadela:* Brown, 1953i: 51.

**tenuis.** *Trapeziopelta tenuis* Emery, 1900c: 313, pl. 8, fig. 3 (q.) NEW GUINEA. Emery, 1901h: 155 (w.). Combination in *Myopias,* and senior synonym of *fulvescens:* Willey & Brown, 1983: 270.

**trumani.** *Pseudoponera trumani* Donisthorpe, 1949g: 406, fig. 1 (w.) NEW GUINEA. Combination in *Myopias:* **new combination**.

**xiphias.** *Trapeziopelta xiphias* Emery, 1900c: 313, pl. 8, fig. 2 (q.) NEW GUINEA. Combination in *Myopias:* **new combination**.

***MYOPOPONE*** (Oriental, Indo-Australian)

*bakeri. Myopopone castanea* var. *bakeri* Viehmeyer, 1916b: 283 (w.) PHILIPPINES. Variety of *moelleri,* and senior synonym of *proxima:* Donisthorpe, 1942c: 31. Junior synonym of *castanea:* Brown, 1960a: 215.

*beccarii. Myopopone beccarii* Emery, 1887b: 447 (w.) INDONESIA (Ternate I.). Variety of *castanea:* Dalla Torre, 1893: 15; Emery, 1911d: 26. Junior synonym of *castanea:* Brown, 1960a: 215.

*bugnioni. Myopopone castanea* var. *bugnioni* Forel, 1913k: 5 (footnote) (q.m.) SRI LANKA. Variety of *moelleri:* Donisthorpe, 1942c: 31. Junior synonym of *castanea:* Brown, 1960a: 215.

**castanea.** *Amblyopone castaneus* Smith, F. 1860b: 105, pl. 1, fig. 6 (w.) INDONESIA (Batjan I.). Mayr, 1867a: 90 (q.); Wheeler, G.C. & Wheeler, J. 1964b: 446 (l.). Combination in *Myopopone:* Roger, 1862c: 292. Senior synonym of *rufula:* Roger, 1862c: 292; of *picea, rossi, similis, smithi* and *wollastoni:* Wilson, 1958a: 144; of *bakeri* (and its junior synonym *proxima*), *beccarii, bugnioni, maculata, moelleri* and *striatifrons:* Brown, 1960a: 215.

*maculata. Myopopone maculata* Roger, 1861a: 50 (w.q.) SRI LANKA. Donisthorpe, 1942c: 29 (m.). Variety/subspecies of *castanea:* Dalla Torre, 1893: 15; Forel, 1900c: 54; Forel, 1907a: 17; Wheeler, W.M. 1919e: 50. Junior synonym of *castanea:* Bingham, 1903: 33; Brown, 1960a: 215.

*moelleri. Myopopone moelleri* Bingham, 1903: 34 (q.) INDIA. Donisthorpe, 1942c: 30 (m.). Junior synonym of *castanea:* Brown, 1960a: 215.

*picea. Myopopone picea* Donisthorpe, 1938e: 498 (w.) BORNEO. Junior synonym of *castanea:* Wilson, 1958a: 144.

*proxima. Myopopone castanea* var. *proxima* Stitz, 1925: 110 (w.) PHILIPPINES. Junior synonym of *bakeri:* Donisthorpe, 1942c: 31. See also: Wilson, 1958a: 145.

*rossi. Myopopone rossi* Donisthorpe, 1948b: 297 (w.) NEW GUINEA. Junior synonym of *castanea:* Wilson, 1958a: 144.

*rufula. Myopopone rufula* Roger, 1861a: 52 (w.) INDONESIA (Batjan I.). Junior synonym of *castanea:* Roger, 1862c: 292; Brown, 1960a: 215.

*similis. Myopopone similis* Donisthorpe, 1949b: 488 (q.) NEW GUINEA. Junior synonym of *castanea:* Wilson, 1958a: 144.

*smithi. Myopopone smithi* Donisthorpe, 1947c: 577 (q.) NEW GUINEA. Junior synonym of *castanea:* Wilson, 1958a: 144.

*striatifrons. Myopopone moelleri* var. *striatifrons* Stitz, 1925: 110 (q.) INDONESIA (Lombok I.). Junior synonym of *castanea:* Brown, 1960a: 215.

*wollastoni. Myopopone wollastoni* Donisthorpe, 1942c: 29 (q.) NEW GUINEA. Junior synonym of *castanea:* Wilson, 1958a: 144.

*MYRAFANT:* see under ***LEPTOTHORAX.***

***MYRCIDRIS*** (Brazil)

**epicharis.** *Myrcidris epicharis* Ward, 1990: 467, figs. 1–6, 21–25, 29–32 (w.q.m.) BRAZIL.

***#MYRMA:*** see under ***POLYRHACHIS.***

***#MYRMACANTHA:*** see under ***PHASMOMYRMEX.***

***#MYRMACRHAPHE:*** see under ***CAMPONOTUS.***

***#MYRMAMBLYS:*** see under ***CAMPONOTUS.***

*MYRMAMMOPHILUS:* see under ***LEPTOTHORAX.***

*MYRMAPATETES:* see under ***ANOCHETUS.***

***#MYRMAPHAENUS:*** see under ***CAMPONOTUS.***

***#MYRMATOPA:*** see under ***POLYRHACHIS.***

***MYRMECIA*** (Australasian)

**aberrans.** *Myrmecia aberrans* Forel, 1900b: 54 (w.) AUSTRALIA. Clark, 1934a: 10 (q.m.). Combination in *M.* (*Promyrmecia*): Emery, 1911d: 19; in *Promyrmecia:* Clark, 1943: 93; in *Myrmecia:* Taylor & Brown, D.R. 1985: 6. See also: Wheeler, W.M. 1933i: 50; Wheeler, W.M. 1937c: 42; Clark, 1951: 122; Ogata, 1991a: 356; Ogata & Taylor, 1992: 1643.

**acuta.** *Myrmecia acuta* Ogata & Taylor, 1992: 1664, figs. 44, 45 (w.) AUSTRALIA.

*affinis. Myrmecia affinis* Mayr, 1862: 725 (diagnosis in key) (w.) AUSTRALIA. Junior synonym of *simillima:* Forel, 1907h: 265; Clark, 1951: 89.

**analis.** *Myrmecia analis* Mayr, 1862: 728 (w.) AUSTRALIA. Clark, 1927: 34 (q.). Senior synonym of *atriscapa:* Clark, 1927: 34; Clark, 1951: 54.

**apicalis.** *Myrmecia apicalis* Emery, 1883: 150 (w.) NEW CALEDONIA. See also: Clark, 1951: 111.

**arnoldi.** *Myrmecia arnoldi* Clark, 1951: 36, fig. 14 (w.) AUSTRALIA. Wheeler, G.C. & Wheeler, J. 1971d: 246 (l.).

**athertonensis.** *Myrmecia auriventris* var. *athertonensis* Forel, 1915b: 8 (w.m.) AUSTRALIA. Variety of *auriventris:* Clark, 1951: 42. Junior synonym of *auriventris:* Brown, 1953j: 10. Revived from synonymy, and raised to species: Ogata & Taylor, 1992: 1645.

*atrata. Myrmecia atrata* Clark, 1951: 77, fig. 55 (w.) AUSTRALIA. Junior synonym of *pavida:* Ogata & Taylor, 1992: 1659.

*atriscapa. Myrmecia atriscapa* Crawley, 1925b: 580 (w.) AUSTRALIA. Junior synonym of *analis:* Clark, 1927: 34; Clark, 1951: 54.

*aureorufa. Myrmecia mandibularis* subsp. *aureorufa* Forel, 1910b: 6 (w.) AUSTRALIA. Junior synonym of *mandibularis:* Clark, 1943: 137.

**auriventris.** *Myrmecia auriventris* Mayr, 1870b: 968 (w.) AUSTRALIA. Clark, 1951: 41 (m.). See also: Ogata & Taylor, 1992: 1645.

*barbata. Myrmecia* (*Promyrmecia*) *fulvipes* subsp. *barbata* Wheeler, W.M. 1933i: 71 (w.q.) AUSTRALIA. Junior synonym of *femorata:* Clark, 1951: 158.

*basirufa. Myrmecia vindex* var. *basirufa* Forel, 1907h: 264 (w.) AUSTRALIA. Clark, 1951: 71 (q.m.). Junior synonym of *vindex:* Ogata & Taylor, 1992: 1661.

**borealis.** *Myrmecia borealis* Ogata & Taylor, 1992: 1647 (w.) AUSTRALIA.

**brevinoda.** *Myrmecia forficata* var. *brevinoda* Forel, 1910b: 2 (w.q.) AUSTRALIA. Wheeler, G.C. & Wheeler, J. 1971d: 247 (l.); Imai, Crozier & Taylor, 1977: 345 (k.). Raised to species: Clark, 1951: 96. Senior synonym of *gigas* (and its junior synonym *eudoxia*): Brown, 1953j: 22; of *decipians* and *longinodis:* Ogata & Taylor, 1992: 1653. See also: Brown, 1958h: 10.

**browningi.** *Myrmecia browningi* Ogata & Taylor, 1992: 1654, figs. 39, 40 (w.q.m.) AUSTRALIA.

**callima.** *Promyrmecia callima* Clark, 1943: 125, pl. 14, fig. 36 (w.) AUSTRALIA. Combination in *Myrmecia:* Taylor & Brown, D.R. 1985: 7.

*cardigaster. Myrmecia cardigaster* Brown, 1953j: 28. Replacement name for *cordata* Clark, 1951: 116. [Junior primary homonym of *cordata* Fabricius, 1804: 425]. Junior synonym of *rowlandi:* Ogata & Taylor, 1992: 1660.

*celaena. Promyrmecia celaena* Clark, 1943: 120, pl. 14, fig. 42 (w.) AUSTRALIA. Combination in *Myrmecia:* Brown, 1953j: 17. Junior synonym of *harderi:* Ogata & Taylor, 1992: 1661.

**cephalotes.** *Promyrmecia cephalotes* Clark, 1943: 123, pl. 14, figs. 32–34 (w.q.m.) AUSTRALIA. Imai, Crozier & Taylor, 1977: 345 (k.). Combination in *Myrmecia:* Taylor & Brown, D.R. 1985: 7. See also: Ogata, 1991a: 357.

**chasei.** *Myrmecia chasei* Forel, 1894e: 235 (w.) AUSTRALIA. Clark, 1943: 116 (q.m.); Wheeler, G.C. & Wheeler, J. 1971d: 247 (l.). Combination in *Promyrmecia:* Clark, 1943: 115; in *Myrmecia:* Taylor & Brown, D.R. 1985: 7. Senior synonym of *mediorubra:* Clark, 1943: 115. See also: Clark, 1951: 212.

**chrysogaster.** *Promyrmecia chrysogaster* Clark, 1943: 114, pl. 13, fig. 23 (w.) AUSTRALIA. Combination in *Myrmecia:* Taylor & Brown, D.R. 1985: 7.

**clarki.** *Myrmecia clarki* Crawley, 1922b: 432 (w.) AUSTRALIA. Wheeler, W.M. 1933i: 61 (q.); Clark, 1943: 132 (m.); Wheeler, G.C. & Wheeler, J. 1971d: 247

(l.). Combination in *M.* (*Promyrmecia*): Wheeler, W.M. 1933i: 61; in *Promyrmecia:* Clark, 1943: 132; in *Myrmecia:* Taylor & Brown, D.R. 1985: 7. See also: Clark, 1951: 174.

*coelatinoda. Myrmecia* (*Promyrmecia*) *fulvipes* subsp. *coelatinoda* Wheeler, W.M. 1933i: 72 (w.) AUSTRALIA. Combination in *Promyrmecia:* Clark, 1943: 147. Junior synonym of *mandibularis:* Brown, 1953j: 4.

***comata.*** *Myrmecia comata* Clark, 1951: 43, fig. 19 (w.) AUSTRALIA. Wheeler, G.C. & Wheeler, J. 1971d: 247 (l.).

*cordata* Fabricius, 1804; see under ***DACETON.***

*cordata. Myrmecia cordata* Clark, 1951: 116, fig. 90 (w.) AUSTRALIA. [Junior primary homonym of *cordata* Fabricius, 1804: 425, above.] Replacement name: *cardigaster* Brown, 1953j: 28.

*crassinoda. Myrmecia crassinoda* Clark, 1934c: 50, pl. 4, fig. 2 (w.q.) AUSTRALIA. Junior synonym of *pulchra:* Brown, 1953j: 27.

***croslandi.*** *Myrmecia croslandi* Taylor, 1991c: 288 (w.k.) AUSTRALIA.

*crudelis. Myrmecia crudelis* Smith, F. 1858b: 147 (w.q.) AUSTRALIA. Junior synonym of *simillima:* Brown, 1953j: 12. See also: Crawley, 1926: 374; Clark, 1951: 35.

***cydista.*** *Promyrmecia cydista* Clark, 1943: 115, pl. 13, fig. 24 (w.) AUSTRALIA. Combination in *Myrmecia:* Taylor & Brown, D.R. 1985: 8.

*decipians. Myrmecia decipians* Clark, 1951: 86, fig. 66 (w.) AUSTRALIA. Junior synonym of *brevinoda:* Ogata & Taylor, 1992: 1653.

***desertorum.*** *Myrmecia vindex* var. *desertorum* Wheeler, W.M. 1915g: 805 (w.) AUSTRALIA. Clark, 1925b: 144 (q.m.). Raised to species: Clark, 1930c: 22. Senior synonym of *lutea* and *princeps:* Brown, 1953j: 25. See also: Clark, 1951: 59.

***dichospila.*** *Myrmecia* (*Promyrmecia*) *dichospila* Clark, 1938: 359, fig. 2 (w.q.m.) AUSTRALIA. Combination in *Promyrmecia:* Clark, 1943: 108; in *Myrmecia:* Taylor & Brown, D.R. 1985: 8.

***dimidiata.*** *Myrmecia dimidiata* Clark, 1951: 71, fig. 49 (w.) AUSTRALIA.

***dispar.*** *Promyrmecia dispar* Clark, 1951: 226, fig. 193 (w.) AUSTRALIA. Combination in *Myrmecia:* Taylor & Brown, D.R. 1985: 8.

*dixoni. Promyrmecia dixoni* Clark, 1943: 135, pl. 16, figs. 77, 78 (w.q.) AUSTRALIA. Wheeler, G.C. & Wheeler, J. 1971d: 247 (l.). Combination in *Myrmecia:* Taylor & Brown, D.R. 1985: 8. Junior synonym of *testaceipes:* Ogata & Taylor, 1992: 1666.

***elegans.*** *Promyrmecia elegans* Clark, 1943: 122, pl. 14, figs. 44, 45 (w.q.) AUSTRALIA. Wheeler, G.C. & Wheeler, J. 1971d: 248 (l.). Combination in *Myrmecia:* Brown, 1953j: 15.

*emarginata* Fabricius, 1804; see under ***ANOCHETUS.***

***erecta.*** *Myrmecia erecta* Ogata & Taylor, 1992: 1655, figs. 41–43 (w.q.) AUSTRALIA.

***esuriens.*** *Myrmecia esuriens* Fabricius, 1804: 424 (w.) AUSTRALIA. Clark, 1934a: 16 (q.m.). Senior synonym of *tasmaniensis:* Roger, 1861a: 35; of *walkeri:* Emery, 1911d: 20; Clark, 1934a: 16. See also: Clark, 1951: 84.

*eudoxia. Myrmecia forficata* var. *eudoxia* Forel, 1915b: 8 (w.) AUSTRALIA. Raised to species: Wheeler, W.M. 1933i: 35. Junior synonym of *gigas:* Clark, 1951: 104.

***eungellensis.*** *Myrmecia eungellensis* Ogata & Taylor, 1992: 1648, figs. 33–35 (w.) AUSTRALIA.

*eupoecila. Promyrmecia eupoecila* Clark, 1943: 98, pl. 12, fig. 8 (q.) AUSTRALIA. Clark, 1951: 134 (w.). Combination in *Myrmecia:* Brown, 1953j: 18. Junior synonym of *froggatti:* Ogata & Taylor, 1992: 1643.

*excavata. Promyrmecia excavata* Clark, 1951: 137, figs. 105, 106 (w.q.) AUSTRALIA. Combination in *Myrmecia:* Brown, 1953j: 18. Junior synonym of *froggatti:* Ogata & Taylor, 1992: 1643.

***exigua.*** *Promyrmecia exigua* Clark, 1943: 107, pl. 15, fig. 64 (w.) AUSTRALIA. Combination in *Myrmecia:* Taylor & Brown, D.R. 1985: 9.

***fabricii.*** *Myrmecia fabricii* Ogata & Taylor, 1992: 1650, fig. 35 (w.) AUSTRALIA.

*fallax. Myrmecia fallax* Clark, 1951: 79, fig. 57 (w.) AUSTRALIA. Junior synonym of *pulchra:* Brown, 1953j: 27.

*fasciata. Myrmecia fasciata* Clark, 1951: 63, fig. 40 (w.) AUSTRALIA. Junior synonym of *nigriceps:* Ogata & Taylor, 1992: 1659.

*femorata. Myrmecia* (*Pristomyrmecia*) *piliventris* var. *femorata* Santschi, 1928e: 466 (w.) AUSTRALIA. Wheeler, G.C. & Wheeler, J. 1971d: 252 (l.). Combination in *M.* (*Promyrmecia*): Wheeler, W.M. 1933i: 70; in *Promyrmecia:* Clark, 1943: 142. Senior synonym of *barbata:* Clark, 1951: 158. Junior synonym of *fulvipes:* Brown, 1953j: 21.

***ferruginea.*** *Myrmecia nigriceps* var. *ferruginea* Mayr, 1876: 95 (w.) AUSTRALIA. Junior synonym of *vindex:* Forel, 1907h: 264. Revived from synonymy, and raised to species: Brown, 1953j: 22.

***flammicollis.*** *Myrmecia flammicollis* Brown, 1953j: 23 (w.) AUSTRALIA.

***flavicoma.*** *Myrmecia flavicoma* Roger, 1861b: 171 (w.) AUSTRALIA. Viehmeyer, 1924a: 221 (q.). See also: Clark, 1951: 74.

***forceps.*** *Myrmecia forceps* Roger, 1861a: 34 (w.) AUSTRALIA. Clark, 1951: 25 (q.); Wheeler, G.C. & Wheeler, J. 1971d: 248 (l.). Senior synonym of *obscuriceps:* Clark, 1951: 24; of *singularis:* Brown, 1953j: 7.

***forficata.*** *Formica forficata* Fabricius, 1787: 310 (w.) TASMANIA. Mayr, 1876: 95 (q.); Clark, 1927: 35 (m.); Wheeler, G.C. & Wheeler, J. 1971d: 249 (l.); Imai, Crozier & Taylor, 1977: 345 (k.). Combination in *Myrmecia:* Fabricius, 1804: 424. Senior synonym of *lucida:* Clark, 1951: 93; Ogata & Taylor, 1992: 1629; of *rubra* (and its junior synonym *violacea*): Brown, 1953j: 28. See also: Forel, 1913g: 173; Wheeler, W.M. 1933i: 31.

***formosa.*** *Myrmecia* (*Promyrmecia*) *aberrans* subsp. *formosa* Wheeler, W.M. 1933i: 52 (w.) AUSTRALIA. Combination in *Promyrmecia:* Clark, 1943: 94; in *Myrmecia,* and subspecies of *maura:* Brown, 1953j: 19. Raised to species: Clark, 1951: 131; Ogata & Taylor, 1992: 1632 (in key). Senior synonym of *haematosticta:* Brown, 1953j: 19.

***froggatti.*** *Myrmecia froggatti* Forel, 1910b: 9 (w.) AUSTRALIA. Clark, 1951: 129 (q.). Combination in *M.* (*Promyrmecia*): Emery, 1911d: 19; in *Promyrmecia:* Clark, 1943: 96; in *Myrmecia:* Brown, 1953j: 17. Subspecies of *aberrans:* Wheeler, W.M. 1933i: 52. Revived status as species: Clark, 1943: 96. Senior synonym of *sericata* and *taylori* Brown, 1953j: 17; of *eupoecila, excavata* and *greavesi:* Ogata & Taylor, 1992: 1643.

***fucosa.*** *Myrmecia* (*Promyrmecia*) *fucosa* Clark, 1934a: 15, pl. 1, figs. 5, 6 (w.q.) AUSTRALIA. Clark, 1943: 103 (m.); Wheeler, G.C. & Wheeler, J. 1971d: 249 (l.). Combination in *Promyrmecia:* Clark, 1943: 103; in *Myrmecia:* Taylor & Brown, D.R. 1985: 10. See also: Clark, 1951: 142.

***fulgida.*** *Myrmecia fulgida* Clark, 1951: 73, fig. 52 (w.) AUSTRALIA. Senior synonym of *suttoni:* Ogata & Taylor, 1992: 1657.

***fulviculis.*** *Myrmecia* (*Pristomyrmecia*) *fulvipes* r. *fulviculis* Forel, 1913g: 174 (w.) AUSTRALIA. Clark, 1951: 163 (q.m.). Combination in *M.* (*Promyrmecia*): Wheeler, W.M. 1933i: 70; in *Promyrmecia:* Clark, 1943: 144; in *Myrmecia:* Taylor & Brown, D.R. 1985: 10. Raised to species: Clark, 1943: 144.

***fulvipes.*** *Myrmecia fulvipes* Roger, 1861a: 36 (w.) AUSTRALIA. Clark, 1943: 145 (q.m.); Wheeler, G.C. & Wheeler, J. 1971d: 249 (l.); Imai, Crozier & Taylor, 1977: 345 (k.). Combination in *M.* (*Pristomyrmecia*): Emery, 1911d: 21; in *M.* (*Promyrmecia*): Wheeler, W.M. 1933i: 69; in *Promyrmecia:* Clark, 1943: 145; in *Myrmecia:* Brown, 1953j: 21. Senior synonym of *femorata* (and its junior synonym *barbata*): Brown, 1953j: 21. See also: Clark, 1951: 165.

***fuscipes.*** *Myrmecia fuscipes* Clark, 1951: 62, fig. 39 (w.) AUSTRALIA. Junior synonym of *desertorum:* Brown, 1953j: 25. Revived from synonymy: Ogata & Taylor, 1992: 1657.

*gigas. Myrmecia pyriformis* r. *gigas* Forel, 1913b: 310 (q.) AUSTRALIA. Forel, 1915: 7 (w.); Clark, 1951: 107 (m.). Raised to species, and senior synonym of *eudoxia:* Clark, 1951: 104. Junior synonym of *brevinoda:* Brown, 1953j: 22.

***gilberti.*** *Myrmecia fulvipes* var. *gilberti* Forel, 1910b: 6 (w.) AUSTRALIA. Wheeler, W.M. 1933i: 73 (m.). Combination in *M.* (*Pristomyrmecia*): Emery, 1911d: 21; in *M.* (*Promyrmecia*): Wheeler, W.M. 1933i: 72; in *Promyrmecia:* Clark, 1943: 136; in *Myrmecia:* Taylor & Brown, D.R. 1985: 10. Race of *fulvipes:* Forel, 1913g: 173. Raised to species: Wheeler, W.M. 1933i: 72; Clark, 1951: 169. Senior synonym of *regina:* Clark, 1943: 136.

*goudiei. Promyrmecia goudiei* Clark, 1943: 129, pl. 15, figs. 48, 49 (w.q.) AUSTRALIA. Junior synonym of *varians:* Brown, 1953j: 15.

*gracilis. Myrmecia gracilis* Emery, 1898b: 232, fig. 2 (w.) AUSTRALIA. Clark, 1938: 358 (q.m.); Wheeler, G.C. & Wheeler, J. 1971d: 250 (l.). Variety of *crudelis:* Emery, 1911d: 19. Revived status as species: Clark, 1938: 357; Clark, 1951: 39. Junior synonym of *rufinodis:* Brown, 1953j: 8.

***gratiosa.*** *Myrmecia gratiosa* Clark, 1951: 66, fig. 44 (w.) AUSTRALIA. Wheeler, G.C. & Wheeler, J. 1971d: 250 (l.).

*greavesi. Promyrmecia greavesi* Clark, 1943: 99, pl. 12, fig. 9 (q.) AUSTRALIA. Combination in *Myrmecia:* Brown, 1953j: 18. Junior synonym of *froggatti:* Ogata & Taylor, 1992: 1643.

***gulosa.*** *Formica gulosa* Fabricius, 1775: 395 (w.) AUSTRALIA. Smith, 1858b: 143 (m.); Mayr, 1876: 95 (q.); Wheeler, G.C. & Wheeler, J. 1971d: 246 (l.);

Imai, Crozier & Taylor, 1977: 345 (k.). Combination in *Myrmecia:* Fabricius, 1804: 424. Senior synonym of *obscurior:* Clark, 1951: 49. See also: Ogata, 1991a: 358.

*haematosticta. Myrmecia* (*Promyrmecia*) *aberrans* subsp. *haematosticta* Wheeler, W.M. 1933i: 51 (w.) AUSTRALIA. Raised to species: Clark, 1951: 130. Junior synonym of *formosa:* Brown, 1953j: 19.

***harderi.*** *Myrmecia harderi* Forel, 1910b: 8 (w.) AUSTRALIA. Clark, 1943: 118 (q.); Wheeler, G.C. & Wheeler, J. 1971d: 250 (l.). Combination in *Promyrmecia:* Clark, 1943: 118; in *Myrmecia:* Brown, 1953j: 16. Senior synonym of *maloni* and *scabra:* Brown, 1953j: 16; of *celaena:* Ogata & Taylor, 1992: 1661. See also: Clark, 1951: 215.

*hastata* Fabricius, 1804; see under ***ODONTOMACHUS.***

***hilli.*** *Promyrmecia hilli* Clark, 1943: 125, pl. 14, fig. 35 (w.) AUSTRALIA. Combination in *Myrmecia:* Taylor & Brown, D.R. 1985: 11.

***hirsuta.*** *Myrmecia hirsuta* Clark, 1951: 109, fig. 84 (w.) AUSTRALIA.

***infima.*** *Myrmecia picta* var. *infima* Forel, 1900b: 54 (w.) AUSTRALIA. Clark, 1943: 105 (q.m.). Combination in *M.* (*Promyrmecia*): Wheeler, W.M. 1933i: 62; in *Promyrmecia:* Clark, 1943: 105; in *Myrmecia:* Taylor & Brown, D.R. 1985: 11. Raised to species: Wheeler, W.M. 1933i: 62; Clark, 1951: 192.

***inquilina.*** *Myrmecia inquilina* Douglas & Brown, 1959: 13, fig. 1 (q.) AUSTRALIA.

*laevinodis. Promyrmecia laevinodis* Clark, 1943: 139, pl. 17, figs. 94, 95 (w.q.) AUSTRALIA. Junior synonym of *mandibularis:* Brown, 1953j: 4.

*longinodis. Myrmecia longinodis* Clark, 1951: 87, fig. 67 (w.) AUSTRALIA. Junior synonym of *brevinoda:* Ogata & Taylor, 1992: 1653.

***loweryi.*** *Myrmecia loweryi* Ogata & Taylor, 1992: 1669, figs. 46, 47 (w.q.m.) AUSTRALIA.

*lucida. Myrmecia lucida* Forel, 1893d: 457 (w.) AUSTRALIA. Wheeler, G.C. & Wheeler, J. 1971d: 250 (l.). Junior synonym of *forficata:* Clark, 1951: 93; Ogata & Taylor, 1992: 1629.

***ludlowi.*** *Myrmecia chasei* var. *ludlowi* Crawley, 1922b: 431 (w.) AUSTRALIA. Crawley, 1925b: 578 (m.). Combination in *Promyrmecia:* Clark, 1943: 117; in *Myrmecia:* Taylor & Brown, D.R. 1985: 7. Subspecies of *chasei:* Clark, 1951: 215. Raised to species: Ogata & Taylor, 1992: 1662.

*lutea. Myrmecia lutea* Crawley, 1922b: 429 (w.) AUSTRALIA. Junior synonym of *desertorum:* Brown, 1953j: 25.

***luteiforceps.*** *Myrmecia* (*Promyrmecia*) *gilberti* var. *luteiforceps* Wheeler, W.M. 1933i: 74 (w.) AUSTRALIA. [First available use of *Myrmecia* (*Pristomyrmecia*) *fulvipes* r. *gilberti* var. *luteiforceps* Forel, 1915b: 9; unavailable name.] Combination in *Promyrmecia:* Clark, 1943: 143; in *Myrmecia:* Taylor & Brown, D.R. 1985: 12. Raised to species: Clark, 1943: 143; Clark, 1951: 161.

*malandensis. Myrmecia tarsata* r. *malandensis* Forel, 1915b: 9 (w.m.) AUSTRALIA. Subspecies of *tarsata:* Clark, 1951: 32. Junior synonym of *rowlandi:* Brown, 1953j: 10.

*maloni. Promyrmecia maloni* Clark, 1943: 121, pl. 14, fig. 43 (w.) AUSTRALIA. Junior synonym of *harderi:* Brown, 1953j: 16.

***mandibularis.*** *Myrmecia mandibularis* Smith, F. 1858b: 145 (w.) AUSTRALIA. Clark, 1943: 138 (q.m.). Combination in *M.* (*Pristomyrmecia*): Emery, 1911d: 21; in *M.* (*Promyrmecia*): Wheeler, W.M. 1933i: 64; in *Promyrmecia:* Clark, 1943: 137; in *Myrmecia:* Brown, 1953j: 4. Senior synonym of *aureorufa* and *postpetiolaris:* Clark, 1943: 137; of *coelatinoda* and *laevinodis:* Brown, 1953j: 4. See also: Crawley, 1926: 385; Clark, 1951: 151; Ogata, 1991a: 359.

*marmorata. Promyrmecia marmorata* Clark, 1951: 188, fig. 157 (w.) AUSTRALIA. Junior synonym of *varians:* Brown, 1953j: 15.

***maura.*** *Myrmecia* (*Promyrmecia*) *aberrans* subsp. *maura* Wheeler, W.M. 1933i: 51 (w.) AUSTRALIA. Clark, 1943: 96 (q.). Combination in *Promyrmecia:* Clark, 1943: 95; in *Myrmecia:* Brown, 1953j: 18. Raised to species: Clark, 1943: 95; Clark, 1951: 132; Brown, 1953j: 18.

*maxima. Formica maxima* Moore, 1842: 52 (w.) AUSTRALIA. Combination in *Myrmecia,* and application for suppression of name: Ride & Taylor, 1973: 58.

*mediorubra. Myrmecia pilosula* subsp. *mediorubra* Forel, 1910b: 7 (w.) AUSTRALIA. Wheeler, W.M. 1933i: 59 (w.). Combination in *M.* (*Promyrmecia*), and raised to species: Wheeler, W.M. 1933i: 58. Junior synonym of *chasei:* Clark, 1943: 115.

***michaelseni.*** *Myrmecia michaelseni* Forel, 1907h: 267 (w.) AUSTRALIA. Clark, 1943: 112 (q.); Wheeler, G.C. & Wheeler, J. 1971d: 250 (l.). Combination in *M.* (*Promyrmecia*): Wheeler, W.M. 1933i: 59; in *Promyrmecia:* Clark, 1943: 111; in *Myrmecia:* Taylor & Brown, D.R. 1985: 12. Senior synonym of *perthensis:* Clark, 1943: 111. See also: Clark, 1951: 204.

***midas.*** *Myrmecia midas* Clark, 1951: 55, figs. 34, 35 (w.q.) AUSTRALIA.

***minuscula.*** *Myrmecia flavicoma* var. *minuscula* Forel, 1915b: 8 (w.) AUSTRALIA. Raised to species: Ogata & Taylor, 1992: 1658.

***mjobergi.*** *Myrmecia mjobergi* Forel, 1915b: 5, fig. 2 (w.q.m.) AUSTRALIA. See also: Clark, 1951: 101; Brown, 1953j: 11.

*murina. Myrmecia murina* Clark, 1951: 80, figs. 58–60 (w.q.m.) AUSTRALIA. Wheeler, G.C. & Wheeler, J. 1971d: 251 (l.). Junior synonym of *pulchra:* Brown, 1953j: 27; Ogata & Taylor, 1992: 1629.

***nigra.*** *Myrmecia picta* var. *nigra* Forel, 1907h: 267 (w.) AUSTRALIA. Clark, 1943: 106 (q.). Combination in *Promyrmecia:* Clark, 1943: 106; in *Myrmecia:* Taylor & Brown, D.R. 1985: 13. Variety of *infima:* Wheeler, W.M. 1933i: 63. Junior synonym of *infima:* Clark, 1934a: 14. Revived from synonymy, and raised to species: Clark, 1943: 106; Clark, 1951: 195. Subspecies of *picta:* Taylor & Brown, D.R. 1985: 13. Revived status as species: Ogata & Taylor, 1992: 1631.

***nigriceps.*** *Myrmecia nigriceps* Mayr, 1862: 728 (w.) AUSTRALIA. Forel, 1910b: 3 (q.); Viehmeyer, 1914c: 28 (m.). Variety of *vindex:* Forel, 1907h: 264. Revived status as species: Wheeler, W.M. 1933i: 38. Senior synonym of *fasciata:* Ogata & Taylor, 1992: 1659. See also: Clark, 1951: 64.

***nigriscapa.*** *Myrmecia nigriscapa* Roger, 1861a: 33 (w.) AUSTRALIA. Clark, 1925b: 142 (q.m.). See also: Wheeler, W.M. 1933i: 30; Clark, 1951: 51.

*nigriventris. Myrmecia nigriventris* Mayr, 1862: 727 (w.) AUSTRALIA. Variety/subspecies of *tricolor:* Forel, 1910b: 7; Emery, 1911d: 20; Clark, 1951: 45. Senior synonym of *spadicea:* Forel, 1910b: 7; Clark, 1951: 45. Junior synonym of *simillima:* Brown, 1953j: 12.

***nigrocincta.*** *Myrmecia nigrocincta* Smith, F. 1858b: 147 (w.) AUSTRALIA. Roger, 1861a: 36 (q.); Mayr, 1876: 95 (m.); Wheeler, G.C. & Wheeler, J. 1971d: 251 (l.); Imai, Crozier & Taylor, 1977: 345 (k.). See also: Crawley, 1926: 381; Clark, 1951: 113; Ogata, 1991a: 359.

***nobilis.*** *Promyrmecia nobilis* Clark, 1943: 97, pl. 12, figs. 2–4 (w.q.m.) AUSTRALIA. Combination in *Myrmecia:* Brown, 1953j: 19.

*obscuriceps. Myrmecia forceps* var. *obscuriceps* Viehmeyer, 1924a: 222 (w.) AUSTRALIA. Junior synonym of *forceps:* Clark, 1951: 24.

*obscurior. Myrmecia gulosa* var. *obscurior* Forel, 1922: 87 (w.) AUSTRALIA. Junior synonym of *gulosa:* Clark, 1951: 49.

***occidentalis.*** *Promyrmecia occidentalis* Clark, 1943: 119, pl. 13, figs. 30, 31 (w.q.) AUSTRALIA. Combination in *Myrmecia:* Taylor & Brown, D.R. 1985: 13. Senior synonym of *opaca:* Ogata & Taylor, 1992: 1663.

*opaca. Promyrmecia opaca* Clark, 1943: 123, pl. 14, figs. 46, 47 (w.q.) AUSTRALIA. Combination in *Myrmecia:* Taylor & Brown, D.R. 1985: 13. Junior synonym of *occidentalis:* Ogata & Taylor, 1992: 1663.

*overbecki. Myrmecia michaelseni* subsp. *overbecki* Viehmeyer, 1924a: 222 (w.q.) AUSTRALIA. Junior synonym of *queenslandica:* Clark, 1943: 112.

*paucidens. Myrmecia paucidens* Forel, 1910b: 5 (w.) TASMANIA. Junior synonym of *simillima:* Brown, 1953j: 12.

***pavida.*** *Myrmecia pavida* Clark, 1951: 76, fig. 54 (w.) AUSTRALIA. Senior synonym of *atrata:* Ogata & Taylor, 1992: 1659.

*perthensis. Myrmecia michaelseni* var. *perthensis* Crawley, 1922b: 431 (w.) AUSTRALIA. Junior synonym of *michaelseni:* Clark, 1943: 111.

***petiolata.*** *Myrmecia petiolata* Emery, 1895g: 345 (w.) AUSTRALIA. See also: Clark, 1951: 112.

***picta.*** *Myrmecia picta* Smith, F. 1858b: 146 (w.q.) AUSTRALIA. Clark, 1927: 39 (m.); Wheeler, G.C. & Wheeler, J. 1971d: 251 (l). Combination in *M.* (*Promyrmecia*): Clark, 1927: 39; in *Promyrmecia:* Clark, 1943: 100; in *Myrmecia:* Taylor & Brown, D.R. 1985: 14. See also: Crawley, 1926: 382; Clark, 1951: 139; Ogata, 1991a: 360.

***picticeps.*** *Myrmecia picticeps* Clark, 1951: 47, fig. 25 (w.) AUSTRALIA.

***piliventris.*** *Myrmecia piliventris* Smith, F. 1858b: 146 (w.) AUSTRALIA. Mayr, 1876: 93 (q.); Clark, 1943: 141 (m.); Wheeler, G.C. & Wheeler, J. 1971d: 251 (l.); Imai & Taylor, 1986: 59 (k.). Combination in *M.* (*Pristomyrmecia*): Emery, 1911d: 21; in *M.* (*Promyrmecia*): Wheeler, W.M. 1933i: 67; in *Promyrmecia:* Clark, 1943: 140; in *Myrmecia:* Brown, 1953j: 20. Senior synonym of *rectidens:* Brown, 1953j: 20. See also: Crawley, 1926: 385; Clark, 1951: 156.

***pilosula.*** *Myrmecia pilosula* Smith, F. 1858b: 146 (w.q.m.) AUSTRALIA. Wheeler, G.C. & Wheeler, J.

1971d: 252 (l.); Imai, Crozier & Taylor, 1977: 345 (k.). Combination in *M.* (*Halmamyrmecia*): Wheeler, W.M. 1922b: 195; in *M.* (*Promyrmecia*): Wheeler, W.M. 1933i: 56; in *Promyrmecia:* Clark, 1943: 109; in *Myrmecia:* Brown, 1953j: 6. Senior synonym of *ruginoda:* Brown, 1953j: 6. See also: Crawley, 1926: 383; Clark, 1951: 202; Crosland, Crozier & Imai, 1988: 13; Ogata, 1991a: 361.

*postpetiolaris. Myrmecia mandibularis* subsp. *postpetiolaris* Wheeler, W.M. 1933i: 65 (w.m.) AUSTRALIA. Junior synonym of *mandibularis:* Clark, 1943: 137.

***potteri.*** *Promyrmecia potteri* Clark, 1951: 168, fig. 136 (w.) AUSTRALIA. Combination in *Myrmecia:* Taylor & Brown, D.R. 1985: 14.

*princeps. Myrmecia princeps* Clark, 1951: 46, fig. 24 (w.) AUSTRALIA. Junior synonym of *desertorum:* Brown, 1953j: 25.

***pulchra.*** *Myrmecia pulchra* Clark, 1929: 119, pl. 1, fig. 3 (w.q.) AUSTRALIA. Senior synonym of *crassinoda, fallax* and *murina:* Brown, 1953j: 27; Ogata & Taylor, 1992: 1629.

*pumilio. Myrmecia pumilio* Mayr, 1866b: 896 (w.) AUSTRALIA. Junior synonym of *picta:* Mayr, 1876: 94; of *urens:* Clark, 1943: 103.

***pyriformis.*** *Myrmecia pyriformis* Smith, F. 1858b: 144, pl. 10, figs. 1–6 (w.q.m.) AUSTRALIA. Wheeler, G.C. & Wheeler, J. 1971d: 252 (l.); Imai, Crozier & Taylor, 1977: 345 (k.). Race of *forficata:* Forel, 1910b: 2. Revived status as species: Clark, 1927: 36. Senior synonym of *sanguinea:* Brown, 1953j: 9. See also: Crawley, 1926: 377; Clark, 1951: 99.

***queenslandica.*** *Myrmecia michaelseni* r. *queenslandica* Forel, 1915b: 4 (w.) AUSTRALIA. Combination in *Promyrmecia:* Clark, 1943: 112; in *Myrmecia:* Taylor & Brown, D.R. 1985: 12. Senior synonym of *overbecki:* Clark, 1943: 112. Raised to species: Ogata & Taylor, 1992: 1663.

*rectidens. Myrmecia piliventris* var. *rectidens* Forel, 1910b: 5 (w.) AUSTRALIA. Clark, 1951: 160 (q.). Combination in *M.* (*Pristomyrmecia*): Emery, 1911d: 21; in *Promyrmecia:* Clark, 1943: 141. Subspecies of *piliventris:* Clark, 1943: 141. Raised to species: Clark, 1951: 159. Junior synonym of *piliventris:* Brown, 1953j: 20.

*regina. Myrmecia* (*Pristomyrmecia*) *regina* Santschi, 1928e: 465 (w.) AUSTRALIA. Junior synonym of *gilberti:* Clark, 1943: 136.

***regularis.*** *Myrmecia regularis* Crawley, 1925b: 579 (w.) AUSTRALIA. Clark, 1951: 92 (q.m.). Junior synonym of *lucida:* Clark, 1927: 34. Revived from synonymy: Wheeler, W.M. 1933i: 25. See also: Clark, 1951: 93.

*rogeri. Myrmecia tricolor* var. *rogeri* Emery, 1914b: 181 (w.) AUSTRALIA. Viehmeyer, 1924a: 222 (q.). Raised to species: Clark, 1951: 32. Junior synonym of *simillima:* Brown, 1953j: 12.

***rowlandi.*** *Myrmecia tarsata* subsp. *rowlandi* Forel, 1910b: 4 (w.) AUSTRALIA. Raised to species: Clark, 1951: 78. Senior synonym of *malandensis:* Brown, 1953j: 10; of *cardigaster:* Ogata & Taylor, 1992: 1660.

***rubicunda.*** *Promyrmecia rubicunda* Clark, 1943: 107, pl. 15, fig. 65 (w.) AUSTRALIA. Combination in *Myrmecia:* Taylor & Brown, D.R. 1985: 15.

*rubra. Myrmecia forficata* var. *rubra* Forel, 1910b: 3 (w.) AUSTRALIA. Clark, 1927: 36 (q.). Raised to species: Clark, 1927: 36. Senior synonym of *violacea:* Clark, 1951: 98. Junior synonym of *forficata:* Brown, 1953j: 28.

***rubripes.*** *Myrmecia rubripes* Clark, 1951: 34, fig. 12 (w.) AUSTRALIA.

***rufinodis.*** *Myrmecia rufinodis* Smith, F. 1858b: 145 (w.) AUSTRALIA. Clark, 1951: 38 (q.m.). Senior synonym of *gracilis:* Brown, 1953j: 8.

*rufonigra. Myrmecia rufonigra* Crawley, 1921: 87 (w.) AUSTRALIA. Junior synonym of *varians:* Clark, 1943: 126.

*ruginoda. Ponera ruginoda* Smith, F. 1858b: 93 (m.) AUSTRALIA. Crozier, 1970: 123 (k.). Combination in *Ectatomma:* Roger, 1861b: 168; in *Rhytidoponera:* Emery, 1911d: 38; in *Myrmecia,* and junior synonym of *pilosula:* Brown, 1953j: 6.

*ruginodis. Promyrmecia ruginodis* Clark, 1943: 113, pl. 13, figs. 20–22 (w.q.m.) AUSTRALIA. Junior synonym of *rugosa:* Brown, 1953j: 5.

***rugosa.*** *Myrmecia* (*Promyrmecia*) *michaelseni* subsp. *rugosa* Wheeler, W.M. 1933i: 60 (w.) AUSTRALIA. Raised to species, and senior synonym of *ruginodis:* Brown, 1953j: 5.

*sanguinea. Myrmecia sanguinea* Smith, F. 1858b: 148 (w.) TASMANIA. Mayr, 1876: 94 (q.); Forel, 1910b: 3 (m.); Wheeler, G.C. & Wheeler, J. 1952a: 112 (l.). Junior synonym of *forficata:* Clark, 1951: 93; of *pyriformis:* Brown, 1953j: 9. See also: Crawley, 1926: 378.

*scabra. Promyrmecia scabra* Clark, 1943: 118, pl. 14, figs. 40, 41 (w.q.) AUSTRALIA. Junior synonym of *harderi:* Brown, 1953j: 16.

*sericata. Myrmecia* (*Promyrmecia*) *aberrans* subsp. *sericata* Wheeler, W.M. 1933i: 53 (w.) AUSTRALIA. Combination in *Promyrmecia:* Clark, 1943: 94. Junior synonym of *froggatti:* Brown, 1953j: 17.

*shepherdi. Promyrmecia shepherdi* Clark, 1943: 128, pl. 15, figs. 50–52 (w.q.m.) AUSTRALIA. Junior synonym of *varians:* Brown, 1953j: 15.

***simillima.*** *Myrmecia simillima* Smith, F. 1858b: 144 (w.) AUSTRALIA. Forel, 1907h: 265 (q.); Wheeler, G.C. & Wheeler, J. 1971d: 253 (l.). Junior synonym of *forficata:* Mayr, 1876: 95. Revived from synonymy as variety/subspecies of *forficata:* Dalla Torre, 1893: 20; Forel, 1907h: 265; Forel, 1910b: 3; Forel, 1915b: 5. Revived status as species: Clark, 1927: 37; Wheeler, W.M. 1933i: 34; Clark, 1951: 89. Senior synonym of *affinis:* Forel, 1907h: 265; Clark, 1951: 89; of *crudelis, nigriventris* (and its junior synonym *spadicea*), *paucidens, rogeri* and *tricolor:* Brown, 1953j: 12. See also: Crawley, 1926: 376.

*singularis. Myrmecia singularis* Clark, 1951: 26, fig. 5 (w.) AUSTRALIA. Junior synonym of *forceps:* Brown, 1953j: 7.

*spadicea. Myrmecia spadicea* Mayr, 1862: 724 (diagnosis in key) (w.) AUSTRALIA. Junior synonym of *tricolor:* Emery, 1911d: 20; of *nigriventris:* Forel, 1910b: 7; Clark, 1951: 45.

***subfasciata.*** *Myrmecia subfasciata* Viehmeyer, 1924a: 221 (w.) AUSTRALIA.

*suttoni. Myrmecia suttoni* Clark, 1951: 72, figs. 50, 51 (w.q.) AUSTRALIA. Junior synonym of *fulgida:* Ogata & Taylor, 1992: 1657.

***swalei.*** *Myrmecia harderi* r. *swalei* Crawley, 1922b: 429 (w.) AUSTRALIA. Clark, 1943: 134 (q.); Wheeler, G.C. & Wheeler, J. 1971d: 253 (l.). Combination in *Promyrmecia:* Clark, 1943: 133; in *Myrmecia:* Taylor & Brown, D.R. 1985: 16. Raised to species: Clark, 1943: 133; Clark, 1951: 176. See also: Ogata & Taylor, 1992: 1665.

***tarsata.*** *Myrmecia tarsata* Smith, F. 1858b: 145 (w.) AUSTRALIA. Roger, 1861a: 33 (q.); Clark, 1951: 30 (m.). See also: Crawley, 1926: 379; Clark, 1927: 34; Clark, 1951: 28.

*tasmaniensis. Myrmecia tasmaniensis* Smith, F. 1858b: 147 (w.) TASMANIA. Junior synonym of *esuriens:* Roger, 1861a: 35.

*taylori. Myrmecia* (*Promyrmecia*) *aberrans* subsp. *taylori* Wheeler, W.M. 1933i: 53 (w.) AUSTRALIA. Combination in *Promyrmecia:* Clark, 1943: 95. Junior synonym of *froggatti:* Brown, 1953j: 17.

***tepperi.*** *Myrmecia tepperi* Emery, 1898b: 231, fig. 1 (q.m.) AUSTRALIA. Clark, 1943: 130 (w.); Wheeler, G.C. & Wheeler, J. 1971d: 253 (l.). Combination in *Promyrmecia:* Clark, 1943: 130; in *Myrmecia:* Taylor & Brown, D.R. 1985: 17. See also: Clark, 1951: 172; Ogata, 1991a: 361.

***testaceipes.*** *Promyrmecia testaceipes* Clark, 1943: 134, pl. 15, fig. 69 (w.) AUSTRALIA. Combination in *Myrmecia:* Taylor & Brown, D.R. 1985: 17. Senior synonym of *dixoni:* Ogata & Taylor, 1992: 1666.

*tricolor. Myrmecia tricolor* Mayr, 1862: 724 (diagnosis in key) (w.) AUSTRALIA. Clark, 1951: 44 (q.). Junior synonym of *simillima:* Brown, 1953j: 12.

***tridentata.*** *Myrmecia tridentata* Ogata & Taylor, 1992: 1651, figs. 36–38 (w.) AUSTRALIA.

***urens.*** *Myrmecia urens* Lowne, 1865b: 336 (w.) AUSTRALIA. Clark, 1943: 104 (q.); Wheeler, G.C. & Wheeler, J. 1971d: 254 (l.). Combination in *Promyrmecia:* Clark, 1943: 103; in *Myrmecia:* Taylor & Brown, D.R. 1985: 17. Junior synonym of *picta:* Mayr, 1876: 94. Revived from synonymy: Clark, 1927: 39. Senior synonym of *pumilio:* Clark, 1943: 103. See also: Ogata, 1991a: 362; Ogata & Taylor, 1992: 1667.

***varians.*** *Myrmecia varians* Mayr, 1876: 94 (w.) AUSTRALIA. Clark, 1943: 127 (q.); Wheeler, G.C. & Wheeler, J. 1971d: 254 (l.). Combination in *Promyrmecia:* Clark, 1943: 126; in *Myrmecia:* Brown, 1953j: 15. Senior synonym of *rufonigra:* Clark, 1943: 126; of *goudiei, marmorata, shepherdi* and *wilsoni:* Brown, 1953j: 15. See also: Clark, 1951: 181.

***vindex.*** *Myrmecia vindex* Smith, F. 1858b: 144 (w.) AUSTRALIA. Roger, 1861a: 34 (q.); Clark, 1925b: 143 (q.m.); Wheeler, G.C. & Wheeler, J. 1971d: 254 (l.). Senior synonym of *basirufa:* Ogata & Taylor, 1992: 1661. See also: Forel, 1907h: 264; Crawley, 1926: 376; Clark, 1951: 67; Douglas & Brown, 1959: 14; Haskins & Haskins, 1974: 258.

*violacea. Myrmecia simillima* var. *violacea* Clark, 1927: 37 (w.) AUSTRALIA. [First available use of *Myrmecia forficata* r. *simillima* var. *violacea* Forel, 1915b: 5; unavailable name (Taylor, 1986: 34).] Junior synonym of *rubra:* Clark, 1951: 98.

*walkeri. Myrmecia walkeri* Forel, 1893d: 456 (w.) TASMANIA. Junior synonym of *esuriens:* Emery, 1911d: 20; Clark, 1934a: 16.

*wilsoni. Promyrmecia wilsoni* Clark, 1943: 127, pl. 15, fig. 66 (w.) AUSTRALIA. Junior synonym of *varians:* Brown, 1953j: 15.

***MYRMECINA*** (Nearctic, Old World except Afrotropical and Malagasy)

**americana.** *Myrmecina latreillei* subsp. *americana* Emery, 1895c: 271 (w.) U.S.A. Wheeler, G.C. & Wheeler, J. 1954c: 130 (l.). Subspecies of *graminicola:* Wheeler, W.M. 1904e: 300; Wheeler, W.M. 1906b: 1. Raised to species: Brown, 1949a: 44; Creighton, 1950a: 248. Senior synonym of *brevispinosa:* Smith, D.R. 1979: 1399; of *quadrispina:* Brown, 1949a: 44; of *texana:* Brown, 1951: 103; Brown, 1967c: 236; of *californica:* Brown, 1967c: 236. See also: Snelling, 1965b: 101.

#**atlantis.** *Myrmecina graminicola* var. *atlantis* Santschi, 1939d: 77 (w.) ALGERIA.

**australis.** *Myrmecina australis* Wheeler, G.C. & Wheeler, J. 1973a: 34 (l.) AUSTRALIA (attributed to Forel).

**bandarensis.** *Myrmecina bandarensis* Forel, 1913k: 72 (w.q.) SUMATRA.

*bidens. Myrmica bidens* Foerster, 1850a: 50 (w.q.) GERMANY. Schenck, 1852: 95 (m.). Junior synonym of *graminicola:* Mayr, 1855: 421.

**brevicornis.** *Myrmecina brevicornis* Emery, 1897c: 582 (w.) INDONESIA (Halmahera I.)

*brevispinosa. Myrmecina americana* subsp. *brevispinosa* Creighton, 1950a: 249 (w.q.m.) U.S.A. [First available use of *Myrmecina latreillei* subsp. *americana* var. *brevispinosa* Emery, 1895c: 271; unavailable name.] Subspecies of *americana:* Snelling, 1965b: 105. Junior synonym of *americana:* Smith, D.R. 1979: 1399.

**butteli.** *Myrmecina butteli* Forel, 1913k: 71, fig. U (w.) SUMATRA.

**cacabau.** *Archaeomyrmex cacabau* Mann, 1921: 449, fig. 17 (w.) FIJI IS. Combination in *Myrmecina:* Brown, 1971a: 1. See also: Taylor, 1980d: 122.

*californica. Myrmecina californica* Smith, M.R. 1948: 239 (w.) U.S.A. Subspecies of *americana:* Snelling, 1965b: 104. Junior synonym of *americana:* Brown, 1967c: 236.

**curtisi.** *Myrmecina curtisi* Donisthorpe, 1949b: 497 (w.) NEW GUINEA.

#**dentata.** *Myrmecina graminicola* var. *dentata* Santschi, 1939d: 77 (w.) TUNISIA.

**flava.** *Myrmecina flava* Terayama, 1985a: 35, figs. 1–5 (w.q.) JAPAN.

#**gotlandica.** *Myrmecina graminicola* subsp. *gotlandica* Karavaiev, 1930b: 146 (w.) SWEDEN.

**graminicola.** *Formica graminicola* Latreille, 1802c: 255 (w.q.m.) DENMARK. Senior synonym of *bidens* and *latreillei:* Mayr, 1855: 421; of *striatula:* Mayr, 1855: 421; Nylander, 1856b: 96; Forel, 1915d: 20; Donisthorpe, 1915d: 76; Brown, 1951: 106; of *grouvellei:* Müller, 1923: 81; Brown, 1951: 106; Agosti & Collingwood, 1987a: 54; of *kutteri:* Brown, 1951: 106; Baroni Urbani, 1971c: 95. Current subspecies: nominal plus *atlantis, dentata, gotlandica, nipponica, oelandica, sinensis.* See also: Atanassov & Dlussky, 1992: 177.

*gratiosa* Smith, F. 1858; see under ***PODOMYRMA.***

*grouvellei. Myrmecina graminicola* var. *grouvellei* Bondroit, 1918: 116, fig. 58 (w.) FRANCE. Junior synonym of *graminicola:* Müller, 1923: 81.

**harrisoni.** *Myrmecina harrisoni* Brown, 1967c: 237, fig. 3 (w.) MEXICO.

*kutteri. Myrmecina kutteri* Forel, 1914b: 1 (w.) ITALY. [Also described as new by Forel, 1915d: 21 (w.q.m.).] Variety of *graminicola:* Emery, 1916b: 171. Junior synonym of *graminicola:* Brown, 1951: 106.

*latreillei. Myrmecina latreillei* Curtis, 1829: 265 (m.) GREAT BRITAIN. Curtis, 1854: 219 (q.). Junior synonym of *graminicola:* Mayr, 1855: 421.

**mandibularis.** *Myrmecina mandibularis* Viehmeyer, 1914d: 610 (w.) NEW GUINEA.

**modesta.** *Myrmecina modesta* Mann, 1919: 335 (w.m.) SOLOMON IS. Current subspecies: nominal plus *subarmata.*

**nesaea.** *Myrmecina nesaea* Wheeler, W.M. 1924b: 247 (q.) INDONESIA (Sebesi I.).

#**nipponica.** *Myrmecina graminicola* subsp. *nipponica* Wheeler, W.M. 1906c: 307 (w.) JAPAN.

#**oelandica.** *Myrmecina graminicola* subsp. *oelandica* Karavaiev, 1930b: 146 (w.q.) SWEDEN.

**opaciventris.** *Myrmecina opaciventris* Emery, 1897c: 581 (w.) NEW GUINEA.

**pilicornis.** *Myrmecina pilicornis* Smith, F. 1858b: 133 (m.) INDIA.

**polita.** *Myrmecina polita* Emery, 1897d: 557 (w.) NEW GUINEA.

**punctata.** *Myrmecina punctata* Emery, 1897c: 583 (w.m.) NEW GUINEA.

*quadrispina. Myrmecina graminicola* subsp. *quadrispina* Enzmann, J. 1946a: 13, figs. 1, 2 (w.) U.S.A. Junior synonym of *americana:* Brown, 1949a: 44.

**rugosa.** *Myrmecina rugosa* Forel, 1902h: 438 (w.m.) AUSTRALIA.

**sauteri.** *Myrmecina sauteri* Forel, 1912a: 54 (w.q.) TAIWAN.

**semipolita.** *Myrmecina semipolita* Forel, 1905c: 15 (w.) JAVA.

**sicula.** *Myrmecina latreillei* var. *sicula* André, 1882c: 275 (w.) SICILY. Raised to species: Emery, 1916a: 59. See also: Baroni Urbani, 1971c: 96.

#**sinensis.** *Myrmecina graminicola* subsp. *sinensis* Wheeler, W.M. 1921c: 539 (w.) CHINA.

**striata.** *Myrmecina striata* Emery, 1889b: 500 (w.) BURMA.

*striatula. Myrmica striatula* Nylander, 1849: 40 (w.) RUSSIA. Subspecies of *graminicola:* Emery, 1898c: 131. Junior synonym of *graminicola:* Mayr, 1855: 421; Nylander, 1856b: 96; Forel, 1915d: 20; Donisthorpe, 1915d: 76; Brown, 1951: 106.

#**subarmata.** *Myrmecina modesta* subsp. *subarmata* Mann, 1919: 337 (w.) SOLOMON IS.

**sulcata.** *Myrmecina sulcata* Emery, 1887b: 449, pl. 1, fig. 11 (w.m.) SULAWESI.

**taiwana.** *Myrmecina taiwana* Terayama, 1985a: 38, figs. 6–10 (w.q.) TAIWAN.

*texana. Myrmecina graminicola* subsp. *texana* Wheeler, W.M. 1908e: 422 (w.) U.S.A. Subspecies of *americana:* Creighton, 1950a: 250. Junior synonym of *americana:* Brown, 1967c: 236.

**transversa.** *Myrmecina transversa* Emery, 1897c: 582 (w.) NEW GUINEA.

**undulata.** *Myrmecina undulata* Emery, 1900d: 678 (w.q.) SUMATRA.

*MYRMECINELLA:* see under ***XENOMYRMEX.***

***MYRMECOCYSTUS*** (U.S.A., Mexico)

*abyssinicus* Forel, 1904; see under ***CATAGLYPHIS.***

*adenensis* Forel, 1904; see under ***CATAGLYPHIS.***

*altisquamis* André, 1881; see under ***CATAGLYPHIS.***

*arabica* Emery, 1906; see under ***CATAGLYPHIS.***

*arenaria* Finzi, 1940; see under ***CATAGLYPHIS.***

*arenarius. Myrmecocystus (Eremnocystus) arenarius* Snelling, 1982b: 71, figs. 1–6, 13, 15, 19, 21, 23 (w.q.m.) U.S.A. [Junior primary homonym of *arenaria* Finzi, 1940: 164, above.] Replacement name: *snellingi* **new name**.

*aterrima* Karavaiev, 1916; see under ***CATAGLYPHIS.***

*aurata* Karavaiev, 1911; see under ***CATAGLYPHIS.***

*bellicosus* Karavaiev, 1924; see under ***CATAGLYPHIS.***

*bruneipes* Santschi, 1911; see under ***CATAGLYPHIS.***

*bucharica* Forel, 1904; see under ***CATAGLYPHIS.***

*bugnioni* Forel, 1908; see under ***CATAGLYPHIS.***

*californicus* Cole, 1936; see under ***FORMICA.***

*cana* Emery, 1906; see under ***CATAGLYPHIS.***

*caspius* Ruzsky, 1902; see under ***CATAGLYPHIS.***

**christineae.** *Myrmecocystus (Myrmecocystus) christineae* Snelling, 1982b: 80, figs. 7–12, 14, 16, 20, 22, 24 (w.q.m.) U.S.A.

*cinnamomea* Karavaiev, 1910; see under ***CATAGLYPHIS.***

**colei.** *Myrmecocystus (Eremnocystus) colei* Snelling, 1976: 94, figs. 203–211, 255–258 (w.q.m.) U.S.A.

*comatus. Myrmecocystus comatus* Creighton, 1950a: 442 (w.q.m.) U.S.A. [First available use of *Myrmecocystus melliger* subsp. *mendax* var. *comatus* Wheeler, W.M. 1908d: 352; unavailable name.] Junior synonym of *melliger:* Snelling, 1969a: 4.

*congolensis* Stitz, 1916; see under ***CATAGLYPHIS.***

**creightoni.** *Myrmecocystus creightoni* Snelling, 1971a: 6, fig. 2 (w.q.m.) U.S.A. See also: Snelling, 1976: 98.

*cretica* Emery, 1906; see under ***CATAGLYPHIS.***

*cubicus* Forel, 1903; see under ***CATAGLYPHIS.***

*cuneinodis* Karavaiev, 1924; see under ***CATAGLYPHIS.***

**depilis.** *Myrmecocystus melliger* var. *depilis* Forel, 1901c: 135 (w.) MEXICO. Snelling, 1976: 52 (q.m.). Raised to species: Snelling, in Hunt & Snelling, 1975: 22.

*desertorum* Forel, 1894; see under ***CATAGLYPHIS.***

*diehlii* Forel, 1902; see under ***CATAGLYPHIS.***

*emeryi* Karavaiev, 1910; see under ***CATAGLYPHIS.***

**ewarti.** *Myrmecocystus ewarti* Snelling, 1971a: 2, fig. 1 (w.q.m.) U.S.A. See also: Snelling, 1976: 129.

**flaviceps.** *Myrmecocystus yuma* var. *flaviceps* Wheeler, W.M. 1912d: 177 (w.) U.S.A. Snelling, 1976: 84 (q.m.). Raised to species: Creighton, 1950a: 443.

*flavigastra* Karavaiev, 1924; see under ***CATAGLYPHIS.***

*foreli* Ruzsky, 1903; see under ***CATAGLYPHIS.***

*fortis* Forel, 1902; see under ***CATAGLYPHIS.***

*frigidus* André, 1881; see under ***CATAGLYPHIS.***

*gracilens* Ruzsky, 1905; see under ***CATAGLYPHIS.***

**hammettensis.** *Myrmecocystus hammettensis* Cole, 1938b: 678 (w.q.m.) U.S.A. See also: Snelling, 1976: 101.

*hellenicus* Forel, 1886; see under ***CATAGLYPHIS.***

*hispanica* Forel, 1903; see under ***CATAGLYPHIS.***

*hortusdeorum. Myrmecocystus melliger* var. *hortusdeorum* McCook, 1881b: 69, pl. 7, figs. 37–39, 47, 48, 50; pl. 10, figs. 72, 73, 75–81 (w.q.m.) U.S.A. Forel, 1886f: 202 (q.m.). Raised to species: Forel, 1886f: 202. Variety/subspecies of *mexicanus:* Emery, 1893i: 666;

Creighton, 1950a: 446. Junior synonym of *mexicanus:* Snelling, 1976: 120.
*ibericus* Emery, 1906; see under ***CATAGLYPHIS.***
*idahoensis. Myrmecocystus mexicanus* subsp. *idahoensis* Cole, 1936b: 118 (w.q.m.) U.S.A. Junior synonym of *mojave:* Creighton, 1950a: 448. Revived from synonymy, and raised to species: Smith, M.R. 1951a: 854. Junior synonym of *testaceus:* Snelling, 1976: 135.
***intonsus.*** *Myrmecocystus (Endiodioctes) intonsus* Snelling, 1976: 30, figs. 70–72 (w.) MEXICO.
*iridescens* Emery, 1887; see under ***MELOPHORUS.***
*isis* Forel, 1909; see under ***CATAGLYPHIS.***
*italicus* Emery, 1906; see under ***CATAGLYPHIS.***
*jakobsoni* Ruzsky, 1905; see under ***CATAGLYPHIS.***
*jesuita. Myrmecocystus melliger* subsp. *mimicus* var. *jesuita* Wheeler, W.M. 1908d: 354 (w.) U.S.A. Unavailable name; material referred to *mimicus* by Creighton, 1950a: 446, and Snelling, 1976: 54.
***kathjuli.*** *Myrmecocystus (Endiodioctes) kathjuli* Snelling, 1976: 59, figs. 92–99, 160, 172, 185, 193 (w.q.m.) U.S.A.
***kennedyi.*** *Myrmecocystus kennedyi* Snelling, 1969a: 6 (w.q.m.) U.S.A. [First available use of *Myrmecocystus melliger* subsp. *semirufus* var. *kennedyi* Cole, 1936b: 119; unavailable name.] See also: Snelling, 1976: 62.
***koso.*** *Myrmecocystus (Endiodioctes) koso* Snelling, 1976: 74, figs. 127–135, 164, 176, 199, 200 (w.q.m.) U.S.A.
*laevior* Stitz, 1917; see under ***CATAGLYPHIS.***
*lameerei* Forel, 1902; see under ***CATAGLYPHIS.***
*lividus* André, 1881; see under ***CATAGLYPHIS.***
*lomaensis. Myrmecocystus melliger* subsp. *lomaensis* Wheeler, W.M. 1912d: 175 (w.q.) U.S.A. Junior synonym of *mimicus:* Creighton, 1950a: 446; Snelling, 1976: 54.
*lucasi* Emery, 1898; see under ***CATAGLYPHIS.***
***lugubris.*** *Myrmecocystus lugubris* Wheeler, W.M. 1909f: 98 (w.) U.S.A. Wheeler, G.C. & Wheeler, J. 1968: 213 (l.). See also: Creighton, 1956: 1; Snelling, 1976: 104.
*lutea* Emery, 1906; see under ***CATAGLYPHIS.***
*mauritanicus* Emery, 1906; see under ***CATAGLYPHIS.***
***melanoticus.*** *Myrmecocystus mexicanus* var. *melanoticus* Wheeler, W.M. 1914b: 57 (w.m.) MEXICO. Raised to species: Snelling, 1976: 116.
***melliger.*** *Myrmecocystus melliger* Forel, 1886f: 201 (w.) MEXICO. Wheeler, W.M. 1912d: 174 (q.m.); Wheeler, G.C. & Wheeler, J. 1968: 211 (l.). Senior synonym of *comatus:* Snelling, 1969a: 4. See also: Snelling, 1976: 31.
*melligera. Formica melligera* Llave, 1832: 463 (w.) MEXICO. Unidentifiable, possibly cognate with *melliger; incertae sedis* in *Myrmecocystus.* See also: Snelling, 1976: 31.
***mendax.*** *Myrmecocystus melliger* subsp. *mendax* Wheeler, W.M. 1908d: 351, fig. 4 (w.q.m.) U.S.A. Raised to species: Creighton, 1950a: 445. Senior synonym of *orbiceps:* Snelling, 1976: 35.
***mexicanus.*** *Myrmecocystus mexicanus* Wesmael, 1838: 770 (w.) MEXICO. Wheeler, G.C. & Wheeler, J. 1968: 213 (l.). Junior synonym of *melliger:* Forel, 1886f: 201; Mayr, 1886d: 424. Revived from synonymy: Wheeler, W.M. 1908d: 356. Senior synonym of *hortusdeorum:* Snelling, 1976: 120.
***mimicus.*** *Myrmecocystus melliger* subsp. *mimicus* Wheeler, W.M. 1908d: 353 (w.q.m.) U.S.A. Wheeler, G.C. & Wheeler, J. 1968: 213 (l.). Raised to species: Creighton, 1950a: 446. Senior synonym of *lomaensis,* and material of the unavailable name *jesuita* referred here: Creighton, 1950a: 446; Snelling, 1976: 54.
*mixtus* Forel, 1895; see under ***CATAGLYPHIS.***
*mojave. Myrmecocystus mexicanus* subsp. *mojave* Wheeler, W.M. 1908d: 360 (w.) U.S.A. Wheeler, W.M. 1912d: 180 (q.m.); Wheeler, G.C. & Wheeler, J. 1968: 213 (l.). Raised to species: Creighton, 1950a: 448; Smith, M.R. 1951a: 856. Junior synonym of *testaceus:* Snelling, 1969a: 6.
***navajo.*** *Myrmecocystus mexicanus* subsp. *navajo* Wheeler, W.M. 1908d: 360, fig. 11 (w.q.) U.S.A. Snelling, 1976: 126 (m.). Raised to species: Creighton, 1950a: 449.
***nequazcatl.*** *Myrmecocystus nequazcatl* Snelling, 1976: 67, figs. 109–116, 162, 174, 186, 194 (w.q.m.) MEXICO.
*niger* André, 1881; see under ***CATAGLYPHIS.***
*orbiceps. Myrmecocystus melliger* subsp. *orbiceps* Wheeler, W.M. 1908d: 349, fig. 3 (w.q.m.) U.S.A. Junior synonym of *placodops:* Snelling, 1969a: 6; of *mendax:* Snelling, 1976: 35.
*orientalis* Forel, 1895; see under ***CATAGLYPHIS.***
***perimeces.*** *Myrmecocystus perimeces* Snelling, 1976: 106, figs. 234–240 (w.q.) MEXICO.
*persica* Emery, 1906; see under ***CATAGLYPHIS.***
*piliscapa* Forel, 1901; see under ***CATAGLYPHIS.***
***placodops.*** *Myrmecocystus melliger* var. *placodops* Forel, 1908b: 70 (w.) MEXICO. Snelling, 1976: 42 (w.m.). Raised to species: Snelling, 1969a: 6.
***pyramicus.*** *Myrmecocystus pyramicus* Smith, M.R. 1951c: 91, figs. 1–3 (w.) U.S.A. Cole, 1957b: 129 (q.m.). See also: Snelling, 1976: 132.
*rockingeri* Forel, 1911; see under ***CATAGLYPHIS.***
***romainei.*** *Myrmecocystus romainei* Snelling, in Hunt & Snelling, 1975: 22 (w.) U.S.A. [First available use of *Myrmecocystus melliger* subsp. *semirufus* var. *romainei* Cole, 1936b: 120; unavailable name.] Snelling, 1976: 78 (q.m.).
*rosenhaueri* Emery, 1906; see under ***CATAGLYPHIS.***
*rotundinodis* Karavaiev, 1912; see under ***CATAGLYPHIS.***
*ruber* Forel, 1903; see under ***CATAGLYPHIS.***
*rufiventris* Forel, 1911; see under ***CATAGLYPHIS.***
***semirufus.*** *Myrmecocystus melliger* var. *semirufus* Emery, 1893i: 667 (w.m.) U.S.A. Wheeler, G.C. & Wheeler, J. 1968: 213 (l.); Snelling, 1976: 46 (q.). Subspecies of *melliger:* Wheeler, W.M. 1908d: 355. Raised to species: Creighton, 1950a: 449.
*seticornis* Emery, 1898; see under ***CATAGLYPHIS.***
*setipedesertorum* Ruzsky, 1905; see under ***CATAGLYPHIS.***
*setipes* Forel, 1894; see under ***CATAGLYPHIS.***
***snellingi.*** *Myrmecocystus snellingi* **new name**. Replacement name for *arenarius* Snelling, 1982b: 71. [Junior primary homonym of *arenaria* Finzi, 1940: 164.]
*sudanica* Karavaiev, 1912; see under ***CATAGLYPHIS.***
*tancrei* Forel, 1901; see under ***CATAGLYPHIS.***
***tenuinodis.*** *Myrmecocystus (Eremnocystus) tenuinodis* Snelling, 1976: 107, figs. 241–247 (w.q.) U.S.A.
***testaceus.*** *Myrmecocystus melliger* var. *testaceus* Emery, 1893i: 667 (w.) U.S.A. Snelling, 1976: 136 (q.m.). Junior synonym of *semirufus:* Creighton, 1950a: 449. Revived from synonymy, raised to species, and senior synonym of *mojave:* Snelling, 1969a: 6. Senior synonym of *idahoensis:* Snelling, 1976: 135.
*turcomanica* Emery, 1898; see under ***CATAGLYPHIS.***
*vaucheri* Emery, 1906; see under ***CATAGLYPHIS.***
*viaticoides* André, 1881; see under ***CATAGLYPHIS.***
***wheeleri.*** *Myrmecocystus wheeleri* Snelling, 1971a: 11, fig. 3 (w.q.m.) U.S.A. See also: Snelling, 1976: 70.
***yuma.*** *Myrmecocystus yuma* Wheeler, W.M. 1912d: 176 (w.) U.S.A. Snelling, 1976: 110 (q.). Junior synonym of *lugubris:* Creighton, 1956: 1. Revived from synonymy: Snelling, in Hunt & Snelling, 1975: 23.

*MYRMECOPSIS:* see under ***OPISTHOPSIS.***

***MYRMECORHYNCHUS*** (Australia)
***carteri.*** *Myrmecorhynchus carteri* Clark, 1934b: 43, pl. 3, figs. 11, 12 (s.w.) AUSTRALIA. Wheeler, G.C. & Wheeler, J. 1970: 648 (l.).
***emeryi.*** *Myrmecorhynchus emeryi* André, 1896b: 254 (w.) AUSTRALIA. Emery, 1898b: 238 (s.); Wheeler, W.M. 1917f: 17 (l.); Wheeler, G.C. & Wheeler, J. 1968: 208 (l.); Clark, 1934b: 41 (q.m.).
***musgravei.*** *Myrmecorhynchus musgravei* Clark, 1934b: 43, pl. 3, fig. 10 (m.) AUSTRALIA.
***nitidus.*** *Myrmecorhynchus nitidus* Clark, 1934b: 44, pl. 3, figs. 13–16 (s.w.q.m.) AUSTRALIA.
***rufithorax.*** *Myrmecorhynchus rufithorax* Clark, 1934b: 46,pl. 3, figs. 17, 18 (s.w.) AUSTRALIA.

***MYRMELACHISTA*** (Neotropical)
***ambigua.*** *Myrmelachista ambigua* Forel, 1893g: 350 (w.) ANTILLES. Combination in *M. (Decamera)*: Bruch, 1914: 227; in *M. (Myrmelachista)*: Kempf, 1972a: 147.
***amicta.*** *Myrmelachista amicta* Wheeler, W.M. 1934g: 195 (w.) MEXICO.
***arborea.*** *Myrmelachista bambusarum* var. *arborea* Forel, 1909a: 263 (q.) PARAGUAY. Forel, 1912i: 60 (w.). Combination in *M. (Decamera)*: Forel, 1912i: 60; in *M. (Hincksidris)*: Kempf, 1972a: 148. Raised to species: Forel, 1912i: 60. Current subspecies: nominal plus *nasuta.*
***arthuri.*** *Myrmelachista arthuri* Forel, 1903c: 263 (diagnosis in key) (w.) BRAZIL. Combination in *M. (Decamera)*: Emery, 1925b: 33; in *M. (Hincksidris)*: Kempf, 1972a: 148. Senior synonym of *ute:* Kempf, 1978: 36. Current subspecies: nominal plus *brunneiceps.*
***bambusarum.*** *Myrmelachista bambusarum* Forel, 1903c: 264 (diagnosis in key) (w.) BRAZIL. Combination in *M. (Decamera)*: Forel, 1912i: 59; in *M. (Hincksidris)*: Kempf, 1972a: 148.
***bettinae.*** *Myrmelachista bettinae* Forel, 1903c: 261 (diagnosis in key) (w.); 265 (in text) (m.) BRAZIL. Combination in *M. (Decamera)*: Forel, 1912i: 59; in *M. (Hincksidris)*: Kempf, 1972a: 148.
***brevicornis.*** *Myrmelachista brevicornis* Wheeler, W.M. 1934g: 199 (w.) BRAZIL.
***bruchi.*** *Myrmelachista (Decamera) bruchi* Santschi, 1922d: 258 (w.q.m.) ARGENTINA. Combination in *M. (Hincksidris)*: Kempf, 1972a: 149.

***MYRMELACHISTA** (continued)*

**#*brunneiceps.*** *Myrmelachista arthuri* var. *brunneiceps* Forel, 1908c: 397 (w.) BRAZIL.

***catharinae.*** *Myrmelachista catharinae* Mayr, 1887: 527 (w.q.m.) BRAZIL. Combination in *M.* (*Decamera*): Emery, 1925b: 33; in *M.* (*Hincksidris*): Kempf, 1972a: 149. Current subspecies: nominal plus *intermedia, maior, petropolitana.*

***chilensis.*** *Myrmelachista chilensis* Forel, 1904e: 704 (footnote) (w.) CHILE. Forel, 1908c: 399 (m.). Combination in *M.* (*Decamera*): Emery, 1925b: 33; in *M.* (*Hincksidris*): Kempf, 1972a: 149.

***cooperi.*** *Camponotus* (*Myrmostenus*) *cooperi* Gregg, 1951: 80, figs. 2, 3 (q.) COSTA RICA. Combination in *Aphomomyrmex:* Gregg, 1954a: 104; in *Neaphomus:* Kempf, 1972a: 152; in *Myrmelachista:* Snelling & Hunt, 1976: 110.

**#*cordincola.*** *Myrmelachista* (*Decamera*) *schumanni* var. *cordincola* Wheeler, W.M. 1934g: 202 (w.) BOLIVIA. [Also described as new by Wheeler, W.M. 1942: 255.]

***costaricensis.*** *Myrmelachista costaricensis* Wheeler, W.M. 1934g: 196, fig. 3 (w.m.) COSTA RICA.

***dalmasi.*** *Myrmelachista* (*Decamera*) *dalmasi* Forel, 1912i: 60 (w.) COLOMBIA. Combination in *M.* (*Hincksidris*): Kempf, 1972a: 149.

***donisthorpei.*** *Myrmelachista* (*Decamera*) *donisthorpei* Wheeler, W.M. 1934g: 204 (w.) GUIANA. Combination in *M.* (*Hincksidris*): Kempf, 1972a: 149.

**#*dubia.*** *Myrmelachista* (*Decamera*) *ulei* r. *dubia* Forel, 1912i: 59 (w.) BRAZIL. Combination in *M.* (*Hincksidris*): Kempf, 1972a: 150.

***elata.*** *Myrmelachista* (*Decamera*) *elata* Santschi, 1922b: 376 (w.) BRAZIL. Combination in *M.* (*Hincksidris*): Kempf, 1972a: 149.

***elongata.*** *Myrmelachista gallicola* st. *elongata* Santschi, 1925e: 168 (w.) ARGENTINA. Combination in *M.* (*Hincksidris*): Kempf, 1972a: 149. Raised to species: Santschi, 1936d: 419.

**#*flavicornis.*** *Myrmelachista nodigera* var. *flavicornis* Emery, 1896h: 638, fig. F (w.q.) PARAGUAY. Combination in *M.* (*Hincksidris*): Kempf, 1972a: 150.

***flavida.*** *Myrmelachista* (*Decamera*) *flavida* Wheeler, W.M. 1934g: 205 (w.) GUIANA. Combination in *M.* (*Hincksidris*): Kempf, 1972a: 149.

**#*fortior.*** *Myrmelachista ramulorum* subsp. *fortior* Wheeler, W.M. 1934g: 189 (w.) PUERTO RICO.

***gagates.*** *Myrmelachista gagates* Wheeler, W.M. 1936b: 204 (w.) HAITI.

***gagatina.*** *Myrmelachista gagatina* Emery, 1894g: 377 (footnote) (w.) BRAZIL. Combination in *M.* (*Decamera*): Santschi, 1922b: 376; in *M.* (*Hincksidris*): Kempf, 1972a: 149. Senior synonym of *ina:* Kempf, 1978: 37.

***gallicola.*** *Myrmelachista gallicola* Mayr, 1887: 528 (w.) URUGUAY. Gallardo, 1915: 27 (q.m.); Santschi, 1916e: 395 (q.m.). Combination in *M.* (*Decamera*): Forel, 1913l: 245; in *M.* (*Hincksidris*): Kempf, 1972a: 149. Senior synonym of *roveretoi:* Santschi, 1916e: 395; of *rubiginosa:* Santschi, 1936d: 417.

***goeldii.*** *Myrmelachista goeldii* Forel, 1903c: 262 (diagnosis in key) (w.) BRAZIL. Combination in *M.* (*Decamera*): Emery, 1925b: 33; in *M.* (*Hincksidris*): Kempf, 1972a: 149.

***goetschi.*** *Aphomomyrmex* (*Neaphomus*) *goetschi* Menozzi, 1935c: 324, figs. 1–3 (q.m.) CHILE. Combination in *Neaphomus:* Kempf, 1972a: 152; in *Myrmelachista:* Snelling & Hunt, 1976: 110.

***guyanensis.*** *Myrmelachista guyanensis* Wheeler, W.M. 1934g: 198 (w.q.) GUIANA.

***hoffmanni.*** *Myrmelachista hoffmanni* Forel, 1903c: 260 (diagnosis in key) (w.); 265 (in text) (q.m.) CHILE. Forel, 1908c: 399 (m.). Combination in *M.* (*Decamera*): Emery, 1925b: 33; in *M.* (*Hincksidris*): Kempf, 1972a: 149. Senior synonym of *rectinota:* Snelling & Hunt, 1976: 113.

*ina. Myrmelachista ina* Kusnezov, 1969: 33 (w.) ARGENTINA. Junior synonym of *gagatina:* Kempf, 1978: 37.

**#*intermedia.*** *Myrmelachista* (*Decamera*) *catharinae* st. *intermedia* Santschi, 1929d: 308 (w.) BRAZIL. Combination in *M.* (*Hincksidris*): Kempf, 1972a: 149.

***kloetersi.*** *Myrmelachista kloetersi* Forel, 1903c: 262 (diagnosis in key) (w.) BRAZIL. Combination in *M.* (*Decamera*): Emery, 1925b: 33; in *M.* (*Hincksidris*): Kempf, 1972a: 149.

***kraatzii.*** *Myrmelachista kraatzii* Roger, 1863a: 163 (w.) CUBA.

**#*laeta.*** *Myrmelachista skwarrae* subsp. *laeta* Wheeler, W.M. 1934g: 195 (w.) MEXICO.

***longinoda.*** *Myrmelachista longinoda* Forel, 1899c: 124, pl. 4, fig. 18 (q.) PANAMA. Combination in *M.* (*Decamera*): Emery, 1925b: 33; in *M.* (*Hincksidris*): Kempf, 1972a: 149.

**#*maior.*** *Myrmelachista* (*Decamera*) *catharinae* var. *maior* Santschi, 1936b: 202 (w.) BRAZIL. Combination in *M.* (*Hincksidris*): Kempf, 1972a: 149.

**#*manni.*** *Myrmelachista rogeri* var. *manni* Wheeler, W.M. 1934g: 190 (w.q.) CUBA.

***mayri.*** *Myrmelachista* (*Decamera*) *mayri* Forel, 1886f: 214 (q.) CHILE. Combination in *M.* (*Hincksidris*): Kempf, 1972a: 149. Senior synonym of *monticola:* Snelling & Hunt, 1976: 113.

***mexicana.*** *Myrmelachista* (*Decamera*) *mexicana* Wheeler, W.M. 1934g: 200, fig. 4 (w.q.m.) MEXICO. Combination in *M.* (*Hincksidris*): Kempf, 1972a: 149.

*monticola. Myrmelachista mayri* var. *monticola* Mayr, 1887: 526 (w.) CHILE. Junior synonym of *mayri:* Snelling & Hunt, 1976: 113.

***muelleri.*** *Myrmelachista muelleri* Forel, 1903c: 259 (q.) BRAZIL. Forel, 1912i: 59 (m.). Combination in *M.* (*Decamera*): Forel, 1912i: 59; in *M.* (*Hincksidris*): Kempf, 1972a: 149.

**#*nasuta.*** *Myrmelachista* (*Decamera*) *arborea* var. *nasuta* Forel, 1912i: 60 (q.) BRAZIL. Combination in *M.* (*Hincksidris*): Kempf, 1972a: 148.

***nigella.*** *Decamera nigella* Roger, 1863a: 167 (w.q.) VENEZUELA. Combination in *Myrmelachista:* Dalla Torre, 1893: 175; in *M.* (*Decamera*): Emery, 1925b: 33; in *M.* (*Hincksidris*): Kempf, 1972a: 149. Current subspecies: nominal plus *ruficornis.*

***nodigera.*** *Myrmelachista nodigera* Mayr, 1887: 529 (w.) BRAZIL. Combination in *M.* (*Decamera*): Emery, 1925b: 33; in *M.* (*Hincksidris*): Kempf, 1972a: 150. Current subspecies: nominal plus *flavicornis, pallida.*

***paderewskii.*** *Myrmelachista paderewskii* Forel, 1908c: 397 (w.q.m.) BRAZIL. Combination in *M.* (*Decamera*): Emery, 1925b: 33; in *M.* (*Hincksidris*): Kempf, 1972a: 150.

**#*pallida.*** *Myrmelachista nodigera* subsp. *pallida* Forel, 1909a: 263 (w.) PARAGUAY. Combination in *M.* (*Decamera*): Emery, 1925b: 33; in *M.* (*Hincksidris*): Kempf, 1972a: 150.

**#*petropolitana.*** *Myrmelachista* (*Decamera*) *catharinae* var. *petropolitana* Santschi, 1936b: 204 (w.) BRAZIL. Combination in *M.* (*Hincksidris*): Kempf, 1972a: 149.

**#*picea.*** *Myrmelachista skwarrae* subsp. *picea* Wheeler, W.M. 1934g: 194 (w.) MEXICO.

***plebecula.*** *Myrmelachista plebecula* Menozzi, 1927d: 338 (w.) COSTA RICA.

***ramulorum.*** *Myrmelachista ambigua* subsp. *ramulorum* Wheeler, W.M. 1908a: 155, pl. 11, figs. 9, 10 (w.q.m.) PUERTO RICO. Wheeler, G.C. & Wheeler, J. 1953c: 138 (l.). Raised to species: Wheeler, W.M. 1934g: 189. Current subspecies: nominal plus *fortior.*

***reclusi.*** *Myrmelachista reclusi* Forel, 1903c: 263 (diagnosis in key) (w.) COLOMBIA. Combination in *M.* (*Decamera*): Forel, 1912i: 59; in *M.* (*Hincksidris*): Kempf, 1972a: 150.

*rectinota. Myrmelachista rectinota* Forel, 1904e: 705 (footnote) (w.) CHILE. Junior synonym of *hoffmanni:* Snelling & Hunt, 1976: 113.

***reichenspergeri.*** *Myrmelachista reichenspergeri* Santschi, 1922b: 376 (w.) BRAZIL. Combination in *M.* (*Hincksidris*): Kempf, 1972a: 150.

***reticulata.*** *Myrmelachista reticulata* Borgmeier, 1928a: 37, figs. 6, 7 (w.m.) BRAZIL. Combination in *M.* (*Hincksidris*): Kempf, 1972a: 150.

***rogeri.*** *Myrmelachista rogeri* André, 1887: 288 (w.) CUBA. Wheeler, W.M. 1934g: 191 (q.). Current subspecies: nominal plus *manni, rubriceps.*

*roveretoi. Myrmelachista* (*Decamera*) *roveretoi* Forel, 1913l: 245 (m.) BRAZIL. Junior synonym of *gallicola:* Santschi, 1916e: 395.

*rubiginosa. Myrmelachista rubiginosa* Emery, 1906c: 185 (w.) ARGENTINA. Junior synonym of *gallicola:* Santschi, 1936d: 417.

**#*rubriceps.*** *Myrmelachista rogeri* var. *rubriceps* Mann, 1920: 434 (w.) CUBA.

***rudolphi.*** *Myrmelachista rudolphi* Forel, 1903c: 261 (diagnosis in key) (w.) BRAZIL. Combination in *M.* (*Decamera*): Emery, 1925b: 33; in *M.* (*Hincksidris*): Kempf, 1972a: 150.

**#*ruficornis.*** *Myrmelachista* (*Decamera*) *nigella* var. *ruficornis* Santschi, 1936d: 420 (w.) ARGENTINA.

***ruszkii.*** *Myrmelachista ruszkii* Forel, 1903c: 265 (diagnosis in key) (w.) BRAZIL. Combination in *M.* (*Decamera*): Emery, 1925b: 33; in *M.* (*Hincksidris*): Kempf, 1972a: 150. [Sometimes misspelled as *ruzskyi,* for example Forel, 1904c: 1.]

***schachovskoi.*** *Myrmelachista* (*Decamera*) *schachovskoi* Kusnezov, 1951b: 359, figs. 1, 3, 10 (w.l.) ARGENTINA. Kusnezov, 1960b: 384 (q.). Combination in *M.* (*Hincksidris*): Kempf, 1972a: 150.

***schumanni.*** *Myrmelachista* (*Decamera*) *schumanni* Emery, 1890a: 70 (footnote) (w.) COLOMBIA. Combination in *M.* (*Hincksidris*): Kempf, 1972a: 150. Current subspecies: nominal plus *cordincola.*

***skwarrae.*** *Myrmelachista skwarrae* Wheeler, W.M. 1934g: 192, fig. 2 (w.q.m.) MEXICO. Current subspecies: nominal plus *laeta, picea.*

**#*thiemei.*** *Myrmelachista zeledoni* var. *thiemei* Emery, 1906c: 185 (footnote) (w.) VENEZUELA.

**ulei.** *Myrmelachista ulei* Forel, 1904e: 704 (w.) PERU. Combination in *M.* (*Decamera*): Forel, 1912i: 59; in *M.* (*Hincksidris*): Kempf, 1972a: 150. Current subspecies: nominal plus *dubia*.

*ute. Myrmelachista ute* Kusnezov, 1969: 34 (w.) ARGENTINA. Junior synonym of *arthuri:* Kempf, 1978: 36.

**vicina.** *Myrmelachista* (*Decamera*) *vicina* Kusnezov, 1951b: 360, figs. 2, 4, 5–9 (w.q.m.) ARGENTINA. Combination in *M.* (*Hincksidris*): Kempf, 1972a: 150.

**zeledoni.** *Myrmelachista zeledoni* Emery, 1896g: 102 (w.) COLOMBIA. Wheeler, W.M. 1934g: 203 (m.); Wheeler, G.C. & Wheeler, J. 1953c: 138 (l.). Combination in *M.* (*Decamera*): Emery, 1925b: 33; in *M.* (*Hincksidris*): Kempf, 1972a: 150. Current subspecies: nominal plus *thiemei*.

#*MYRMENTOMA:* see under ***CAMPONOTUS***

#*MYRMEPINOTUS:* see under ***CAMPONOTUS.***

#*MYRMEPOMIS:* see under ***CAMPONOTUS.***

#*MYRMESPERA:* see under ***CAMPONOTUS.***

*MYRMETAERUS:* see under ***EPIMYRMA.***

#*MYRMEURYNOTA:* see under ***CAMPONOTUS.***

*MYRMEX:* see under ***PSEUDOMYRMEX.***

#*MYRMHOPLA:* see under ***POLYRHACHIS.***

**MYRMICA** (Holarctic, n. Oriental)

**aborigenica.** *Myrmica aborigenica* Zhigul'skaya, 1991: 58, figs. (w.q.m.) RUSSIA. See also: Zhigul'skaya, Kipyatkov & Kipyatkova, 1992: 72.

*adelaidae* Smith, F. 1858; see under ***PODOMYRMA.***

**ademonia.** *Myrmica ademonia* **new name**. Replacement name for *aspersa* Kupyanskaya, 1990: 105. [Junior primary homonym of *aspersa* Smith, F. 1865: 72.]

**aemula.* **Myrmica aemula* Heer, 1850: 161, pl. 13, fig. 4 (q.) SWITZERLAND (Miocene).

*agilis* Smith, F. 1857; see under ***PHEIDOLE.***

*ahngeri. Myrmica scabrinodis* var. *ahngeri* Karavaiev, 1926f: 66, fig. 3 (w.) RUSSIA. Subspecies of *scabrinodis:* Arnol'di, 1970: 1841; Arnol'di & Dlussky, 1978: 534. Junior synonym of *specioides:* Seifert, 1988b: 16.

**aimonissabaudiae.** *Myrmica aimonissabaudiae* Menozzi, 1939a: 286, figs. 1, 3 (w.q.) INDIA. Collingwood, 1961a: 56 (m.). [Misspelled as *amedeussabaudiae* by Menozzi, 1939a: 286.]

#**alajensis.** *Myrmica tianshanica* subsp. *alajensis* Arnol'di, 1976a: 550 (w.) TURKESTAN.

**alaskensis.** *Myrmica brevinodis* var. *alaskensis* Wheeler, W.M. 1917a: 503 (w.) U.S.A. Junior synonym of *kuschei:* Creighton, 1950a: 97; of *brevinodis:* Smith, M.R. 1951a: 789. Revived from synonymy, raised to species, and senior synonym of *kuschei:* Francoeur, 1986b: 57.

*albuferensis. Myrmica albuferensis* Lomnicki, 1925b: 2 (w.) SPAIN (Balearic Is.). Junior synonym of *rolandi:* Finzi, 1926: 89; Bernard, 1956b: 257; of *aloba:* Santschi, 1931b: 340; Seifert, 1988b: 29.

*aldrichi* Wheeler, W.M. 1914; see under ***MANICA.***

*algirica* Lucas, 1849; see under ***CREMATOGASTER.***

**aloba.** *Myrmica scabrinodis* var. *aloba* Forel, 1909c: 103 (w.) SPAIN. Santschi, 1925g: 340 (q.m.); Finzi, 1926: 95 (m.); Tinaut & Espadaler, 1987: 61 (m.). Variety of *rugulosoides:* Finzi, 1926: 95. Raised to species: Santschi, 1931b: 340; Santschi, 1936c: 199. Senior synonym of *albuferensis:* Santschi, 1931b: 340; of *rolandi:* Seifert, 1988b: 29. See also: Collingwood & Yarrow, 1969: 56.

*alpestris. Myrmica lobicornis* subsp. *alpestris* Arnol'di, 1934: 168, figs. 23–25 (w.m.) RUSSIA. Junior synonym of *lobicornis:* Seifert, 1988b: 38.

*alpina. Myrmica lobicornis* subsp. *alpina* Stärcke, 1927: 80, figs. 1, 3 (w.q.) ITALY. Junior synonym of *lobicornis:* Bernard, 1967: 122; Seifert, 1988b: 38.

*amaurocyclia* Foerster, 1850; see under ***MESSOR.***

**americana.** *Myrmica sabuleti* subsp. *americana* Weber 1939b: 144 (w.q.m.) U.S.A. Wheeler, G.C. & Wheeler, J. 1953a: 116 (l.). Raised to species: Creighton, 1950a: 94.

**angulinodis.** *Myrmica scabrinodis* subsp. *angulinodis* Ruzsky, 1905b: 689 (w.q.) RUSSIA. Raised to species: Collingwood, 1962: 217; Collingwood, 1976: 302.

*****angusticollis.*** **Myrmica angusticollis* Heer, 1850: 162, pl. 13, fig. 7 (q.) SWITZERLAND (Miocene).

*angustifrons. Myrmica lobicornis* subsp. *angustifrons* Stärcke, 1927: 81, figs. 1, 3 (w.q.) GREAT BRITAIN. Junior synonym of *lobicornis:* Seifert, 1988b: 38.

*angustulus* Nylander, 1856; see under ***LEPTOTHORAX.***

*apennina. Myrmica lobicornis* subsp. *apennina* Stärcke, 1927: 82, fig. 1 (w.) ITALY. Junior synonym of *lobicornis:* Seifert, 1988b: 38.

*aquia* Buckley, 1867; see under ***APHAENOGASTER.***

*****archaica.*** **Myrmica archaica* Meunier, 1915: 212, pl. 22, fig. 2 (m.) GERMANY (Oligocene).

*arduennae. Myrmica lobicornis* var. *arduennae* Bondroit, 1911: 12 (w.q.m.) BELGIUM. Raised to species: Bondroit, 1918: 105. Subspecies of *lobicornis:* Emery, 1921f: 38; Menozzi, 1925d: 24; Finzi, 1926: 108; Sadil, 1952: 263. Junior synonym of *lobicornis:* Donisthorpe, 1915d: 134; Bernard, 1967: 122; Boven, 1977: 119; Seifert, 1988b: 38.

#**arisana.** *Myrmica rugosa* var. *arisana* Wheeler, W.M. 1930a: 95 (w.) TAIWAN. Subspecies of *rugosa:* Weber, 1947: 462.

**arnoldii.** *Myrmica arnoldii* Dlussky, 1963: 191, figs. 1, 4, 5 (w.) RUSSIA. Arnol'di, 1968: 1803 (q.m.). Combination in *M.* (*Dodecamyrmica*): Arnol'di, 1968: 1803. See also: Francoeur, 1981: 759.

*aspersa* Smith, F. 1865; see under ***TETRAMORIUM.***

*aspersa. Myrmica aspersa* Kupyanskaya, 1990: 105, figs. 16, 17, 18 (w.q.m.) RUSSIA. [Junior primary homonym of *Myrmica aspersa* Smith, F. 1865: 72, above.] Replacement name: *ademonia* **new name**.

*assimilis. Myrmica assimilis* Spinola, 1851b: 50 (w.) BRAZIL. [Also described as new by Spinola, 1853: 66.] Unidentifiable to genus; *incertae sedis* in *Myrmica.*

*atlantica. Myrmica scabrinodis* var. *atlantica* Stärcke, 1942c: xxvii. *Nomen nudum;* see under *puerilis.*

*atomaria* Gerstäcker, 1859; see under ***MONOMORIUM.***

*atra* Buckley, 1867; see under ***MONOMORIUM.***

*atratula* Schenck, 1852; see under ***ANERGATES.***

#**bactriana.** *Myrmica smythiesi* var. *bactriana* Ruzsky, 1915a: 438, figs. 16–18 (w.m.) TIBET.

#**baikalensis.** *Myrmica saposhnikovi* subsp. *baikalensis* Pisarski, 1969a: 228 (w.q.) RUSSIA. [First available use of *Myrmica scabrinodis* subsp. *saposhnikovi* var. *baikalensis* Karavaiev, 1931b: 28; unavailable name.]

**bakurianica.** *Myrmica pilosiscapus* subsp. *bakurianica* Arnol'di, 1970: 1842 (w.q.m.) TRANSCAUCASUS. Raised to species: Seifert, 1988b: 22.

*balcanica. Myrmica* (*Myrmica*) *balcanica* Sadil, 1952: 253, figs. V, 1–18; VIII, 24; IX, 21; X, 8 (w.q.m.) CZECHOSLOVAKIA. Junior synonym of *sancta:* Arnol'di, 1970: 1840; of *specioides:* Pisarski, 1975: 12; Kutter, 1977c: 70; Seifert, 1988b: 16; of *bessarabica:* Dlussky, Soyunov & Zabelin, 1990: 182; Atanassov & Dlussky, 1992: 93.

*barbata* Smith, F. 1858; see under ***POGONOMYRMEX.***

*barchanica. Myrmica bergi* var. *barchanica* Ruzsky, 1905b: 678 (w.) KIRGHIZIA. Junior synonym of *bergi:* Arnol'di, 1970: 1839.

*basalis* Smith, F. 1858; see under ***MONOMORIUM.***

**beesoni.** *Myrmica beesoni* Mukerjee, 1934: 7, fig. 1 (w.) INDIA.

**bergi.** *Myrmica bergi* Ruzsky, 1902b: 473 (w.) TURKESTAN. [Also described as new by Ruzsky, 1902c: 12.] Ruzsky, 1905b: 676 (q.m.); Emery, 1908a: 172 (q.m.). Senior synonym of *barchanica:* Arnol'di, 1970: 1839; of *kamyschiensis* and *kirgisica:* Seifert, 1988b: 11. Current subspecies: *divergens, persiana.* See also: Tarbinsky, 1976: 45.

#**bergiana.** *Myrmica saposhnikovi* subsp. *bergiana* Tarbinsky, 1976: 38, figs. 37–39 (w.) KIRGIZIA.

**bessarabica.** *Myrmica scabrinodis* var. *bessarabicus* Nasonov, 1889: 36 (w.) UKRAINE. Junior synonym of *sabuleti:* Weber, 1948a: 288; Sadil, 1952: 246. Revived from synonymy, and senior synonym of *balcanica, caucasica, rugulososcabrinodis, sancta* and *tschuliensis:* Dlussky, Soyunov & Zabelin, 1990: 182; of *scabrinodoides* and *specioides* (and its junior synonyms *ahngeri, puerilis* and *striata*): Atanassov & Dlussky, 1992: 93; of *nevodovskii:* Arakelian, 1994: 23. Current subspecies: nominal plus *dolens.*

**bibikoffi.** *Myrmica bibikoffi* Kutter, 1963: 130, figs. 1–11 (w.q.m.) SWITZERLAND. See also: Kutter, 1973c: 255; Kutter, 1977c: 65.

*bicarinata* Nylander, 1846; see under ***TETRAMORIUM.***

**bicolor. *Myrmica bicolor* Heer, 1867: 31, pl. 2, fig. 11 (q.?) YUGOSLAVIA. Junior synonym of **tertiaria:* Dalla Torre, 1893: 117. See also: Mayr, 1867b: 60.

*bicolor. Myrmica bicolor* Kupyanskaya, 1986b: 94, figs. 1–5 (w.q.m.) RUSSIA. [Junior primary homonym of **bicolor* Heer, above.] Replacement name: *displicentia* **new name**. See also: Kupyanskaya, 1990: 106.

*bidens* Foerster, 1850; see under ***MYRMECINA.***

*bidentata. Myrmica bidentata* Smith, F. 1858b: 124 (w.) INDIA. Unidentifiable to genus; *incertae sedis* in *Myrmica.* [Bingham, 1903: 212, suggests that this may belong in *Solenopsis.*]

*blanda* Smith, F. 1858; see under ***ROGERIA.***

*bradleyi* Wheeler, W.M. 1909; see under ***MANICA.***

*****bremii.*** **Myrmica bremii* Heer, 1850: 161, pl. 13, fig. 5 (q.) YUGOSLAVIA (Miocene).

**breviceps.** *Myrmica breviceps* Smith, F. 1878b: 12 (q.) INDIA.

*brevinodis. Myrmica rubra* subsp. *brevinodis* Emery, 1895c: 312 (w.m.) U.S.A. Wheeler, W.M. 1907b: 74 (q.); Wheeler, G.C. & Wheeler, J. 1953a: 116 (l.). Raised to species: Wheeler, W.M. 1914b: 53 (in text); Wheeler, W.M. 1917a: 502. Senior synonym of *canadensis* and *subalpina:* Weber, 1950b: 189.

Junior synonym of *incompleta:* Francoeur & Béique, 1966: 141.

*brevipennis* Smith, F. 1858; see under ***POGONOMYRMEX.***

*brevispina. Myrmica rugosa* subsp. *dshungarica* var. *brevispina* Kuznetsov-Ugamsky, 1927e: 193 (w.) TURKESTAN. Unavailable name; material referred to *dshungarica* by Tarbinsky, 1976: 29.

***brevispinosa.*** *Myrmica brevinodis* var. *brevispinosa* Wheeler, W.M. 1917a: 502 (w.q.m.) U.S.A. [First available use of *Myrmica rubra* subsp. *brevinodis* var. *brevispinosa* Wheeler, W.M. 1907b: 74; unavailable name.] Wheeler, G.C. & Wheeler, J. 1953a: 117 (l.). Subspecies of *brevinodis:* Wheeler, G.C. & Wheeler, E.W. 1944: 241. Raised to species: Creighton, 1950a: 97. Senior synonym of *decedens:* Weber, 1950b: 198.

*bruesi. Myrmica laevinodis* var. *bruesi* Weber, 1947: 453 (w.q.m.) U.S.A. [First available use of *Myrmica rubra* subsp. *laevinodis* var. *bruesi* Wheeler, W.M. 1906a: 38; unavailable name.] Junior synonym of *laevinodis:* Creighton, 1950a: 104.

#***brunescens.*** *Myrmica (Myrmica) schencki* var. *brunescens* Karavaiev, 1929b: 208 (w.) CAUCASUS. Variety/subspecies of *lobicornis:* Santschi, 1931b: 350; Arnol'di, 1934: 167.

*burtshakabramovitshi. Myrmica (Myrmica) schencki* var. *burtshakabramovitshi* Karavaiev, 1929b: 209, fig. 4 (w.) TURKESTAN. Variety/subspecies of *lobicornis:* Santschi, 1931b: 351; Arnol'di, 1934: 167; Karavaiev, 1934: 92. Junior synonym of *lobicornis:* Arnol'di, 1970: 1842.

*busoni. Myrmica busoni* Chapman & Capco, 1951: 125. *Nomen nudum;* attributed to Ruzsky.

#***cachmiriensis.*** *Myrmica smythiesii* r. *cachmiriensis* Forel, 1904c: 23 (w.) INDIA. Menozzi, 1939a: 292 (m.).

*caeca. Myrmica caeca* Jerdon, 1851: 116 (w.) INDIA. Unidentifiable to genus; *incertae sedis* in *Myrmica.*

*caeca* Smith, F. 1860; see under ***LIOMYRMEX.***

*californica* Buckley, 1867; see under ***POGONOMYRMEX.***

*canadensis. Myrmica brevinodis* var. *canadensis* Wheeler, W.M. 1916m: 587 (w.q.m.) CANADA. [First available use of *Myrmica rubra* subsp. *brevinodis* var. *canadensis* Wheeler, W.M. 1907b: 76; unavailable name.] Junior synonym of *brevinodis:* Weber, 1950b: 189.

*carbonaria* Smith, F. 1858; see under ***MONOMORIUM.***

#***carbonaria.*** *Myrmica smythiesii* subsp. *carbonaria* Forel, 1902c: 227 (w.) INDIA. [Unresolved junior primary homonym of *carbonaria* Smith, F. 1858b: 127, above.]

*carinata* Smith, F. 1859; see under ***TETRAMORIUM.***

*carinata. Myrmica carinata* Kupyanskaya, 1990: 114, figs. 16, 19, 20 (w.q.m.) RUSSIA. [Junior primary homonym of *carinata* Smith, F. 1859a: 148, above.] Replacement name: *taediosa* **new name**.

*cariniceps* Guérin-Méneville, 1852; see under ***TETRAMORIUM.***

*castanea* Smith, F. 1858; see under ***PHEIDOLE.***

*caucasica. Myrmica rugulosa* subsp. *caucasica* Arnol'di, 1934: 165, figs. 17–20 (w.m.) CAUCASUS. Seifert, 1988b: 19 (q.). Raised to species: Seifert, 1988b: 19. Junior synonym of *bessarabica:* Dlussky, Soyunov & Zabelin, 1990: 182; Atanassov & Dlussky, 1992: 93.

***caucasicola.*** *Myrmica schencki* nat. *caucasicola* Arnol'di, 1934: 172, fig. 38–40 (w.m.) CAUCASUS. Raised to species: Arakelian, 1994: 24; Radchenko, 1994a: 44; Radchenko, 1994c: 140 (in key).

*cerasi* Fitch, 1854; see under ***CREMATOGASTER.***

*champlaini. Myrmica rubra* r. *champlaini* Forel, 1901h: 80 (w.) CANADA. Subspecies of *laevinodis:* Weber, 1947: 454; of *rubra:* Creighton, 1950a: 103. Junior synonym of *laevinodis:* Smith, M.R. 1951a: 789.

*chersonensis. Myrmica rugulosa* subsp. *limanica* nat. *chersonensis* Arnol'di, 1934: 164, figs. 14, 15 (w.m.) RUSSIA. Unavailable name; material referred to *limanica* by Arnol'di, 1970: 1840. [Also described as new by Arnol'di, in Karavaiev, 1934: 76; unavailable name.]

***chinensis.*** *Myrmica chinensis* Viehmeyer, 1922: 204 (q.m.) CHINA. Subspecies of *tibetana:* Weber, 1947: 466. Revived status as species: Radchenko, 1994a: 41.

*cingulata* Schenck, 1852; see under ***LEPTOTHORAX.***

*clandestina. Myrmica clandestina* Foerster, 1850a: 63 (w.) GERMANY. Schenck, 1852: 85 (q.m.). Junior synonym of *rugulosa:* Mayr, 1855: 405.

*clypeata* Mayr, 1853; see under ***LEPTOTHORAX.***

*coeca* Buckley, 1867; see under ***NEIVAMYRMEX.***

***colax.*** *Paramyrmica colax* Cole, 1957a: 37, fig. 1 (w.q.) U.S.A. Wheeler, G.C. & Wheeler, J. 1959: 219 (l.). Combination in *Myrmica:* Bolton, 1988a: 4.

*columbiana* Buckley, 1867; see under ***CREMATOGASTER.***

***commarginata.*** *Myrmica commarginata* Ruzsky, 1905b: 708 (w.) RUSSIA. See also: Radchenko, 1994c: 138.

****concinna.*** **Myrmica concinna* Heer, 1867: 32, pl. 2, fig. 13 (caste?) YUGOSLAVIA (Miocene).

*consternens* Walker, 1859; see under ***MONOMORIUM.***

*constricta. Myrmica rugulosa* var. *constricta* Karavaiev, 1934: 74. Replacement name for *minuta* Karavaiev, 1929b: 204. [Junior primary homonym of *minuta* Ruzsky, 1905b: 670.] Junior synonym of *slobodensis:* Karavaiev, 1936: 274.

*contigua* Smith, F. 1858; see under ***MONOMORIUM.***

*corrugata. Myrmica corrugata* Say, 1836: 291 (q.m.) U.S.A. Unidentifiable to genus; *incertae sedis* in *Myrmica.*

*corticalis* Schenck, 1852; see under ***LEPTOTHORAX.***

***cursor.*** *Myrmica cursor* Smith, F. 1878b: 11 (q.) CHINA.

***curvithorax.*** *Myrmica curvithorax* Bondroit, 1920a: 148, fig. 1 (w.) SPAIN.

*daurica. Myrmica daurica* Motschoulsky, 1860: 500. *Nomen nudum.*

#***debilior.*** *Myrmica rugosa* var. *debilior* Forel, 1902c: 228 (w.) INDIA. Senior synonym of *rugososmythiesi:* Forel, 1902c: 228.

*debilis* Foerster, 1850; see under ***STENAMMA.***

*decedens. Myrmica brevinodis* var. *decedens* Wheeler, W.M. 1917a: 503 (w.m.) U.S.A. [First available use of *Myrmica rubra* subsp. *brevinodis* var. *decedens* Wheeler, W.M. 1907b: 75; unavailable name.] Junior synonym of *brevispinosa:* Weber, 1950b: 198.

*dentata. Myrmica bergi* subsp. *dshungarica* var. *dentata* Kuznetsov-Ugamsky, 1927e: 192 (m.) TURKESTAN. Unavailable name.

*denticornis. Myrmica denticornis* Curtis, 1854: 215, figs. 18, 19 (w.m.) GREAT BRITAIN. Smith, F. 1855b: 121 (q.). Junior synonym of *lobicornis:* Mayr, 1861: 63.

***deplanata.*** *Myrmica lobicornis* var. *deplanata* Emery, 1921f: 38 (w.) RUSSIA. [First available use of *Myrmica scabrinodis* subsp. *lobicornis* var. *deplanata* Ruzsky, 1905b: 700; unavailable name.] Variety of *lobicornis:* Karavaiev, 1927a: 283; of *schencki:* Finzi, 1926: 111. Raised to species: Arnol'di, 1934: 169; Karavaiev, 1934: 83; Stitz, 1939: 105; Tarbinsky, 1976: 31. Senior synonym of *moravica* and *plana:* Karavaiev, 1934: 83; Arnol'di, 1970: 1843; Seifert, 1988b: 36.

*derthavini. Myrmica sulcinodis* subsp. *derthavini* Ruzsky, 1920: 77 (w.) RUSSIA. Junior synonym of *sulcinodis:* Kupyanskaya, 1986b: 94.

*detritinodis. Myrmica scabrinodis* var. *detritinodis* Emery, 1921f: 40 (w.) U.S.A. [First available use of *Myrmica rubra* subsp. *scabrinodis* var. *detritinodis* Emery, 1895c: 316; unavailable name.] Junior synonym of *fracticornis:* Weber, 1948a: 276; Smith, M.R. 1951a: 790 (misspelled as *detrinodis*).

***dicaporiaccoi.*** *Myrmica dicaporiaccoi* Menozzi, 1939a: 289, figs. 5, 6 (w.) INDIA.

*diffusa* Jerdon, 1851; see under ***CREMATOGASTER.***

*diligens* Smith, F. 1865; see under ***TETRAMORIUM.***

*diluta. Myrmica diluta* Nylander, 1849: 41 (w.) RUSSIA. Junior synonym of *ruginodis:* Mayr, 1861: 63.

*dimidiata. Myrmica dimidiata* Say, 1836: 293 (q.) U.S.A. Junior synonym of *ruginodis:* Weber, 1947: 448.

***discontinua.*** *Myrmica brevinodis* subsp. *discontinua* Weber, 1939b: 150 (w.) U.S.A. Subspecies of *brevispinosa:* Creighton, 1950a: 98. Raised to species: Talbot, 1976: 245; Francoeur, 1977b: 206.

***displicentia.*** *Myrmica displicentia* **new name**. Replacement name for *bicolor* Kupyanskaya, 1986b: 94. [Junior primary homonym of **bicolor* Heer, 1867: 31.]

#***divergens.*** *Myrmica bergi* var. *divergens* Karavaiev, 1931c: 105, fig. 1 (w.) RUSSIA. Subspecies of *bergi:* Pisarski, 1969a: 227; Dlussky & Pisarski, 1970: 85.

*diversa. Myrmica (Monomorium) diversa* Buckley, 1867: 337 (w.q.) U.S.A. Unidentifiable to genus; *incertae sedis* in *Myrmica.* See also: Emery, 1895c: 338; Wheeler, W.M. 1902f: 26.

#***dolens.*** *Myrmica puerilis* ab. *dolens* Stärcke, 1942c: xxvii (w.q.m.) NETHERLANDS. Subspecies of *bessarabica:* current status.

*domestica* Shuckard, 1838; see under ***MONOMORIUM.***

***dshungarica.*** *Myrmica rugosa* subsp. *dshungarica* Ruzsky, 1905b: 661 (w.) RUSSIA. Weber, 1947: 459 (q.m.); Wheeler, G.C. & Wheeler, J. 1953a: 123 (l.). Subspecies of *smythiesii:* Emery, 1908a: 169; Weber, 1947: 458. Raised to species: Arnol'di, 1976a: 551; Tarbinsky, 1976: 29; Radchenko, 1994c: 137 (in key). Senior synonym of *ruginodiformis:* Arnol'di, 1976a: 551. Material of the unavailable names *brevispina* and *infuscata* referred here by Tarbinsky, 1976: 29.

**duisburgi* Mayr, 1868; see under ****AGROECOMYRMEX.***

#***eidmanni.*** *Myrmica eidmanni* Menozzi, 1930d: 331, fig. 3 (w.) RUSSIA. Subspecies of *scabrinodis:* Weber, 1950b: 204.

*emeryana. Myrmica schencki* subsp. *emeryana* Cole, 1942: 368 (w.q.m.) U.S.A. [First available use of *Myrmica scabrinodis* r. *schencki* var. *emeryana* Forel, 1914c:

617; unavailable name.] Wheeler, G.C. & Wheeler, J. 1953a: 114 (l.). Subspecies of *schencki:* Buren, 1944a: 283. Raised to species: Creighton, 1950a: 98; Cole, 1953d: 243. Synonym of *latifrons:* Weber, 1948a: 298; Smith, M.R. 1954d: 138; Wheeler, G.C. & Wheeler, J. 1986g: 12. [All these authors give *emeryana* as the senior synonym, but *latifrons* has priority and is the first available name for this taxon.]

**ereptrix.** *Myrmica ereptrix* Bolton, 1988a: 2, figs. 1, 5 (q.) INDIA.

*erythrothorax. Myrmica erythrothorax* Lund, 1831a: 116 (footnote) (w.) BRAZIL. *Nomen nudum.*

**#europaea.** *Myrmica laevinodis* var. *europaea* Finzi, 1926: 84 (w.) NORWAY. [First available use of *Myrmica rubra* subsp. *champlaini* var. *europaea* Forel, 1911h: 457; unavailable name.] Santschi, 1931b: 339 (m.). Variety/subspecies of *laevinodis:* Stitz, 1939: 83; of *rubra:* current status.

**everesti.** *Myrmica everesti* Donisthorpe, 1929a: 445 (w.) TIBET.

**excelsa.** *Myrmica excelsa* Kupyanskaya, 1990: 115, figs. 16, 19, 20 (w.q.m.) RUSSIA.

*exigua* Buckley, 1867; see under ***SOLENOPSIS.***

**#exigua.** *Myrmica smythiesi* subsp. *exigua* Ruzsky, 1915a: 440, figs. 19, 20 (w.m.) TIBET. [Unresolved junior primary homonym of *exigua* Buckley, 1867: 342, above.]

**#eximia.** *Myrmica sulcinodis* subsp. *eximia* Kupyanskaya, 1990: 104, figs. 17, 18 (w.q.m.) RUSSIA.

**faniensis.** *Myrmica faniensis* Boven, 1970: 127, figs. 1, 2 (q.) BELGIUM. See also: Kutter, 1973c: 256; Boven, 1977: 114; Bolton, 1988a: 4.

**ferganensis.** *Myrmica laevinodis* var. *ferganensis* Finzi, 1926: 84 (w.) TURKESTAN. [First available use of *Myrmica rubra* subsp. *laevinodis* var. *ferganensis* Karavaiev, 1916: 503; unavailable name.] Junior synonym of *ruginodolaevinodis:* Sadil, 1952: 241. Revived from synonymy, and raised to species: Arnol'di, 1976a: 550. Junior synonym of *rubra:* Tarbinsky, 1976: 40. Revived from synonymy and raised to species: Radchenko, 1994a: 41; Radchenko, 1994c: 137 (in key).

*flavidula* Nylander, 1849; see under ***SOLENOPSIS.***

*fodicus* Jerdon, 1851; see under ***MYRMICARIA.***

**forcipata.** *Myrmica* (*Myrmica*) *forcipata* Karavaiev, 1931c: 105, fig. 2 (w.) RUSSIA. Kupyanskaya, 1990: 108 (q.m.).

*foreli. Myrmica lobicornis* st. *foreli* Santschi, 1931b: 348, fig. 6 (w.q.) AUSTRIA. Junior synonym of *lobicornis:* Bernard, 1967: 122; Seifert, 1988b: 38.

**#foreliella.** *Myrmica lobicornis* subsp. *foreliella* Arnol'di, 1976a: 553 (w.) RUSSIA.

*formosa* Smith, F. 1858; see under ***PODOMYRMA.***

**formosae.** *Myrmica margaritae* var. *formosae* Wheeler, W.M. 1929g: 37 (w.) TAIWAN. [Unresolved junior primary homonym of *formosa* Smith, F. 1858b: 128, above.] [*Myrmica margaritae* var. *formosae* Wheeler, W.M. 1928c: 9. *Nomen nudum.*] Subspecies of *ritae:* Weber, 1950b: 220. Raised to species: Radchenko, 1994a: 44.

**#fortior.** *Myrmica smythiesii* var. *fortior* Forel, 1904c: 22 (w.) INDIA.

*fortior. Myrmica bergi* var. *fortior* Crawley, 1920a: 163 (w.) IRAN. [Junior primary homonym of *fortior* Forel, above.] Replacement name: *persiana* Weber, 1947: 474.

**fracticornis.** *Myrmica lobicornis* var. *fracticornis* Forel, 1901e: 371 (in text) (w.) U.S.A. [First available use of *Myrmica rubra* subsp. *scabrinodis* var. *fracticornis* Emery, 1895c: 313; unavailable name.] Weber, 1948a: 280 (q.m.); Wheeler, G.C. & Wheeler, J. 1953a: 117 (l.). Variety/subspecies of *scabrinodis:* Wheeler, W.M. 1916m: 587; of *lobicornis:* Cole, 1942: 368; Wheeler, G.C. & Wheeler, E.W. 1944: 241; Creighton, 1950a: 100. Raised to species: Talbot, 1976: 245; Francoeur, 1977b: 206; Wheeler, G.C. & Wheeler, J. 1986g: 12. Senior synonym of *detritinodis:* Weber, 1948a: 276; Smith, M.R. 1951a: 790.

*fragilis* Smith, F. 1858; see under ***MONOMORIUM.***

*frigida. Myrmica brevinodis* var. *frigida* Wheeler, W.M. 1917a: 502 (w.) CANADA. [First available use of *Myrmica rubra* subsp. *brevinodis* var. *frigida* Forel, 1902i: 699; unavailable name.] Junior synonym of *sulcinodoides:* Wheeler, W.M. 1917e: 16; Weber, 1950b: 195.

*furva. Myrmica tibetana* var. *furva* Ruzsky, 1915a: 441 (w.) TURKESTAN. Junior synonym of *tibetana:* Radchenko, 1994a: 41.

*fuscipennis* Smith, F. 1861; see under ***PRISTOMYRMEX.***

*fuscula* Nylander, 1846; see under ***TETRAMORIUM.***

*galbula* Losana, 1834; see under ***MESSOR.***

**gallienii.** *Myrmica gallienii* Bondroit, 1920a: 150, fig. 1 (w.) FRANCE. [Also described as new by Bondroit, 1920b: 302.] Collingwood, 1979: 48 (q.m.). Junior synonym of *sulcinodis:* Bernard, 1967: 121. Revived from synonymy: Collingwood, 1978: 67. Senior synonym of *jacobsoni* and *limanica:* Collingwood, 1979: 48; Seifert, 1988b: 9. Current subspecies: nominal plus *obensis.* See also: Atanassov & Dlussky, 1992: 91.

*gayi* Spinola, 1851; see under ***SOLENOPSIS.***

*georgica. Myrmica georgica* Seifert, 1987: 183, figs. 1–10 (w.m.) GEORGIA. Junior synonym of *turcica:* Seifert, 1988b: 18.

*glaber* Smith, F. 1862; see under ***SOLENOPSIS.***

*glacialis. Myrmica lobicornis* var. *glacialis* Emery, 1921f: 38 (w.) CANADA. [First available use of *Myrmica rubra* r. *scabrinodis* var. *glacialis* Forel, 1904a: 154; unavailable name.] Junior synonym of *fracticornis:* Weber, 1948a: 276; of *lobifrons:* Creighton, 1950a: 100.

*globularia* Smith, F. 1858; see under ***SOLENOPSIS.***

*glyciphila* Smith, F. 1858; see under ***MONOMORIUM.***

*gracilescens* Smith, F. 1860; see under ***PHEIDOLE.***

*gracillima* Smith, F. 1861; see under ***MONOMORIUM.***

*grandis. Myrmica scabrinodis* subsp. *lobicornis* var. *grandis* Kuznetsov-Ugamsky, 1927e: 196 (w.) TURKESTAN. Unavailable name.

*granulinodis. Myrmica granulinodis* Nylander, 1846b: 1060 (q.m.) RUSSIA. Variety/subspecies of *scabrinodis:* André, 1883a: 319; Weber, 1950b: 205. Junior synonym of *sabuleti:* Emery, 1908a: 177 (footnote); Sadil, 1952: 246 [if synonymy correct then *granulinodis* has priority].

**hamulata.** *Myrmica sabuleti* subsp. *hamulata* Weber, 1939b: 146 (w.q.m.) U.S.A. Raised to species: Creighton, 1950a: 99. Current subspecies: nominal plus *trullicornis.*

**#hecate.** *Myrmica smythiesi* subsp. *hecate* Weber, 1947: 461 (w.) INDIA.

**hellenica.** *Myrmica rugulosa* var. *hellenica* Finzi, 1926: 93 (w.q.m.) GREECE. [First available use of *Myrmica scabrinodis* r. *rugulosa* var. *hellenica* Forel, 1913d: 431; unavailable name.] Finzi, 1926: 93 (m.). Raised to species: Agosti & Collingwood, 1987b: 267; Seifert, 1988b: 13.

**helleri.** *Myrmica helleri* Viehmeyer, 1922: 204 (w.) CHINA. Subspecies of *kurokii:* Weber, 1947: 469. Revived status as species: Radchenko, 1994a: 42.

**#himalayana.** *Myrmica smythiesi* subsp. *himalayana* Weber, 1947: 460 (w.) INDIA.

**hirsuta.** *Myrmica hirsuta* Elmes, 1978: 131, fig. 2 (q.m.) GREAT BRITAIN. See also: Collingwood, 1979: 51; Bolton, 1988a: 4.

*hirtula* Nylander, 1849; see under ***HARPAGOXENUS.***

*humilis. Myrmica humilis* Smith, F. 1858b: 123 (q.) INDIA. Unidentifiable to genus; *incertae sedis* in *Myrmica.*

*hunteri* Wheeler, W.M. 1914; see under ***MANICA.***

*impura* Foerster, 1850; see under ***TETRAMORIUM.***

*incerta* Smith, F. 1863; see under ***PHEIDOLE.***

**incompleta.** *Myrmica incompleta* Provancher, 1881b: 359 (w.q.m.) CANADA. [Also described as new by Provancher, 1883: 602.] Junior synonym of *laevinodis:* Provancher, 1887: 247; Dalla Torre, 1893: 111; Donisthorpe, 1915d: 110; Smith, M.R. 1951a: 789. Revived from synonymy, and senior synonym of *brevinodis* (and its junior synonyms *canadensis* and *subalpina*): Francoeur & Béique, 1966: 141.

**incurvata.** *Myrmica incurvata* Collingwood, 1976: 301, figs. 1–3 (w.) KOREA.

**indica.** *Myrmica ritae* subsp. *indica* Weber, 1950b: 221 (w.) INDIA. Raised to species: Radchenko, 1994a: 44.

**inezae.** *Myrmica inezae* Forel, 1902c: 226 (w.) INDIA. See also: Bingham, 1903: 268.

*inflecta. Myrmica inflecta* Say, 1836: 292 (m.) U.S.A. Unidentifiable to genus; *incertae sedis* in *Myrmica.*

*infuscata. Myrmica rugosa* subsp. *dshungarica* var. *infuscata* Kuznetsov-Ugamsky, 1927e: 192 (w.q.) TURKESTAN. Unavailable name; material referred to *dshungarica* by Tarbinsky, 1976: 29.

*inornata. Myrmica margaritae* subsp. *inornata* Menozzi, 1941 :15 (w.) CHINA. *Nomen nudum.*

*insolens* Smith, F. 1861; see under ***TETRAMORIUM.***

**#intermedia.** *Myrmica scabrinodis* subsp. *intermedia* Kuznetsov-Ugamsky, 1927e: 194 (w.) TURKESTAN.

*interrupta* Schenck, 1852; see under ***LEPTOTHORAX.***

*iranica. Myrmica rugulosa* subsp. *iranica* Arnol'di, 1948: 211. *Nomen nudum.*

*irritans* Smith, F. 1858; see under ***PHEIDOLE.***

*isfahani. Myrmica punctiventris* var. *isfahani* Forel, 1922: 92 (w.q.) U.S.A. Junior synonym of *punctiventris:* Weber, 1950b: 512.

**#iskanderi.** *Myrmica minuta* subsp. *iskanderi* Arnol'di, 1976a: 548, fig. 2 (w.q.m.) RUSSIA.

*jacobsoni. Myrmica jacobsoni* Kutter, 1963: 133, figs. 12–20 (w.q.m.) LATVIA. Subspecies of *limanica:* Arnol'di, 1970: 1840. Revived status as species: Kutter, 1977c: 65. Junior synonym of *gallienii:* Collingwood, 1979: 48; Seifert, 1988b: 9.

**jessensis.** *Myrmica lobicornis* var. *jessensis* Forel, 1901e: 371 (w.) JAPAN. Subspecies of *lobicornis:* Weber, 1948a: 286. Raised to species: Collingwood, 1976:

302; Kupyanskaya, 1986a: 85; Kupyanskaya, 1990: 110. Senior synonym of *littoralis:* Kupyanskaya, 1986a: 85.

*jucunda. Myrmica jucunda* Smith, F. 1861a: 34 (w.) LEBANON. Unidentifiable to genus; *incertae sedis* in *Myrmica.*

***juglandeti.*** *Myrmica juglandeti* Arnol'di, 1976a: 549, figs. 3, 4 (w.m.) KIRGHIZIA. See also: Tarbinsky, 1976: 22.

*****jurinei.*** **Myrmica jurinei* Heer, 1850: 163, pl. 13, fig. 8 (q.) YUGOSLAVIA (Miocene). See also: Mayr, 1867b: 59.

***kabylica.*** *Sifolinia kabylica* Cagniant, 1970d: 40, figs. 1–6 (q.m.) ALGERIA. Combination in *Myrmica:* Bolton, 1988a: 4. See also: Kutter, 1973c: 256 (misspelled as *cabylica*).

***kamtschatica.*** *Myrmica kamtschatica* Kupyanskaya, 1986a: 88, figs. 3, 6, 9, 10, 14 (w.q.m.) RUSSIA. See also: Kupyanskaya, 1990: 112.

*kamyschiensis. Myrmica kamyschiensis* Arnol'di, 1934: 159, figs. 2, 3 (w.) RUSSIA. Karavaiev, 1936: 273 (q.). Subspecies of *bergi:* Karavaiev, 1936: 272. Junior synonym of *bergi:* Seifert, 1988b: 11.

***karavajevi.*** *Symbiomyrma karavajevi* Arnol'di, 1930c: 269, figs. 1–4 (q.m.) RUSSIA. [Also described as new by Arnol'di, 1933a: 41.] Combination in *Sifolinia:* Samsinak, 1964: 156; in *Myrmica:* Bolton, 1988a: 4. Senior synonym of *pechi:* Samsinak, 1964: 156. See also: Kutter, 1973c: 256 (misspelled as *karawajewi*); Collingwood, 1979: 58.

***kasczenkoi.*** *Myrmica scabrinodis* subsp. *kasczenkoi* Ruzsky, 1905b: 702 (w.m.) RUSSIA. Kupyanskaya, 1990: 113 (q.). Raised to species: Pisarski, 1969b: 296; Dlussky & Pisarski, 1970: 85; Collingwood, 1976: 302.

#***khamensis.*** *Myrmica ruginodis* var. *khamensis* Ruzsky, 1915a: 441, figs. 22–24 (w.) TIBET. Variety of *rubra:* Weber, 1947: 449.

*kieviensis. Myrmica (Myrmica) lobicornis* var. *kieviensis* Karavaiev, 1934: 91, fig. 24 (w.) UKRAINE. Junior syonym of *lobicornis:* Arnol'di, 1970: 1842.

*kirbii. Myrmica kirbii* Stephens, 1829: 356. *Nomen nudum.*

*kirbii* Sykes, 1835; see under ***CREMATOGASTER.***

***kirghisorum.*** *Myrmica lobicornis* subsp. *kirghisorum* Arnol'di, 1976a: 553 (w.m.) KIRGHIZIA. Raised to species: Radchenko, 1994a: 43; Radchenko, 1994c: 142 (in key). See also: Tarbinsky, 1976: 33.

*kirgisica. Myrmica rugosa* var. *kirgisica* Ruzsky, 1903b: 314 (w.) KIRGHIZIA. Subspecies of *bergi:* Arnol'di, 1970: 1839. Junior synonym of *bergi:* Seifert, 1988b: 11.

*kollari* Mayr, 1853; see under ***TETRAMORIUM.***

***kotokui.*** *Myrmica rubra* subsp. *kotokui* Forel, 1911e: 267 (w.q.m.) JAPAN. Raised to species: Collingwood, 1976: 300.

***kozlovi.*** *Myrmica kozlovi* Ruzsky, 1915a: 435, figs. 10, 11 (w.) TIBET. Current subspecies: nominal plus *mekongi, ruzskyi, subalpina, subbrevispinosa.*

***kryzhanovskii.*** *Myrmica kryzhanovskii* Arnol'di, 1976a: 551, fig. 6 (w.q.m.) RUSSIA.

***kurokii.*** *Myrmica rubra* subsp. *kurokii* Forel, 1907e: 18 (w.) JAPAN. Raised to species: Emery, 1908a: 171; Weber, 1947: 469. Current subspecies: nominal plus *sontica, tipuna.* See also: Kupyanskaya, 1990: 102.

*kuschei. Myrmica brevinodis* var. *kuschei* Wheeler, W.M. 1917e: 17 (w.q.) U.S.A. Subspecies of *brevinodis:* Weber, 1950b: 197. Raised to species: Francoeur, 1977b: 206. Junior synonym of *alaskensis:* Francoeur, 1986b: 57.

*kutteri. Myrmica schencki* var. *kutteri* Finzi, 1926: 111 (w.m.) SWITZERLAND. Junior synonym of *schencki:* Bernard, 1967: 118.

*lacteipennis* Zetterstedt, 1838; see under ***LEPTOTHORAX.***

***lacustris.*** *Myrmica scabrinodis* var. *lacustris* Ruzsky, 1905b: 686 (w.) RUSSIA. Raised to species: Radchenko, 1994a: 44; Radchenko, 1994c: 140 (in key).

*laevigata* Smith, F. 1855; see under ***PHEIDOLE.***

*laevinodis. Myrmica laevinodis* Nylander, 1846a: 927, pl. 18, figs. 5, 31 (w.q.m.) FINLAND. Subspecies/race of *rubra:* Forel, 1874: 76; Emery & Forel, 1879: 460; Wheeler, W.M. 1906d: 315; Emery, 1914d: 156; Forel, 1915d: 28; Menozzi, 1918: 82; Karavaiev, 1927c: 259; Creighton, 1950a: 104. Status as species: Saunders, E. 1880: 215; André, 1883a: 316; Dalla Torre, 1893: 110; Ruzsky, 1905b: 662; Bondroit, 1912: 351; Donisthorpe, 1915d: 110; Bondroit, 1918: 104; Müller, 1923: 40; Finzi, 1926: 83; Stitz, 1939: 78; Holgersen, 1940: 184; Novak & Sadil, 1941: 76; Weber, 1947: 441; Bernard, 1967: 119; Baroni Urbani, 1971c: 22; Kutter, 1977c: 65. Senior synonym of *longiscapus:* Mayr, 1863: 433; of *champlaini:* Smith, M.R. 1951a: 789; of *bruesi:* Creighton, 1950a: 104. Junior synonym of *rubra:* Yarrow, 1955b: 114; Arnol'di, 1970: 1839; Arnol'di & Dlussky, 1978: 530; Collingwood, 1979: 52; Seifert, 1988b: 5.

*laevissima* Smith, F. 1860; see under ***SOLENOPSIS.***

*laeviuscula* Foerster, 1850; see under ***FORMICOXENUS.***

***lampra.*** *Myrmica lampra* Francoeur, 1968: 729, figs. 1–5 (q.m.) CANADA. See also: Kutter, 1973c: 256; Francoeur, 1981: 756; Bolton, 1988a: 4.

***latifrons.*** *Myrmica schencki* subsp. *latifrons* Stärcke, 1927: 84 (w.) U.S.A. Raised to species: **new status**. Senior synonym of *emeryana:* Weber, 1948a: 298; Smith, M.R. 1954d: 138; Wheeler, G.C. & Wheeler, J. 1986g: 12. [All give *latifrons* as the junior synonym, but it has priority over *emeryana* and is the first available name for this taxon.]

***laurae.*** *Sifolinia laurae* Emery, 1907: 49, figs. (q.) ITALY. Combination in *Myrmica:* Bolton, 1988a: 4. See also: Emery, 1908f: 550; Kutter, 1973c: 256.

***lemasnei.*** *Myrmica lemasnei* Bernard, 1967: 123, fig. 127 (q.) FRANCE. Combination in *Sifolinia:* Kutter, 1973c: 253; in *Myrmica:* Bolton, 1988a: 4.

*leonina* Losana, 1834; see under ***MANICA.***

*limanica. Myrmica (Myrmica) rugulosa* subsp. *limanica* Karavaiev, 1934: 75, fig. 20 (w.m.) RUSSIA. [Also described as new by Arnol'di, 1934: 162.] Raised to species, and material of the unavailable names *chersonensis* and *strandi* referred here: Arnol'di, 1970: 1840. Junior synonym of *gallienii:* Collingwood, 1979: 48; Seifert, 1988b: 9.

*lineolata* Say, 1836; see under ***CREMATOGASTER.***

*lineolata. Myrmica (Monomorium) lineolata* Buckley, 1867: 340 (w.q.) U.S.A. [Unresolved junior primary homonym of *lineolta* Say, 1836: 290.] Unidentifiable to genus; *incertae sedis* in *Myrmica.*

*lippula* Nylander, 1849; see under ***STENAMMA.***

*lissahorensis. Myrmica lobicornis* subsp. *lissahorensis* Stitz, 1939: 100 (w.) CZECHOSLOVAKIA. [First available use of *Myrmica lobicornis* subsp. *lobicornis* var. *lissahorensis* Stärcke, 1927: 79; unavailable name.] Subspecies of *lobicornis:* Weber, 1948a: 287. Junior synonym of *lobicornis:* Bernard, 1967: 122; Seifert, 1988b: 38.

*littoralis. Myrmica lobicornis* subsp. *littoralis* Weber, 1948a: 287 (w.q.m.) RUSSIA. [First available use of *Myrmica scabrinodis* subsp. *lobicornis* var. *littoralis* Kuznetsov-Ugamsky, 1928b: 33; unavailable name.] Junior synonym of *jessensis:* Kupyanskaya, 1986a: 85.

*lobata. Myrmica lobicornis* var. *lobata* Kutter, 1977c: 66. *Nomen nudum;* attributed to Finzi.

***lobicornis.*** *Myrmica lobicornis* Nylander, 1846a: 932, pl. 18, figs. 32, 33 (w.q.) FINLAND. Nylander, 1849: 31 (m.); Hauschteck, 1965: 325 (k.). Variety/subspecies of *rubra:* Forel, 1874: 76; Emery & Forel, 1879: 463; Wheeler, W.M. 1906c: 316; of *scabrinodis:* Mayr, 1886d: 451; Ruzsky, 1905b: 693; Emery, 1908a: 179. Status as species: Saunders, E. 1880: 216; André, 1883a: 318; Nasonov, 1889: 35; Forel, 1892i: 315; Donisthorpe, 1915d: 134; Forel, 1915d: 28; Emery, 1916b: 120; Bondroit, 1918: 105; Finzi, 1926: 106; Stärcke, 1927: 76; Santschi, 1931b: 347; Collingwood, 1958: 73; Bernard, 1967: 122; Kutter, 1977c: 66; Arnol'di & Dlussky, 1978: 534; Collingwood, 1979: 51; Seifert, 1988b: 38; Atanassov & Dlussky, 1992: 101. Senior synonym of *alpestris:* Seifert, 1988b: 38; of *alpina:* Bernard, 1967: 122; Seifert, 1988b: 38; of *angustifrons* and *apennina:* Seifert, 1988b: 38; of *arduennae:* Donisthorpe, 1915d: 134; Bernard, 1967: 122; Boven, 1977: 119; Seifert, 1988b: 38; of *burtshakabramovitshi:* Arnol'di, 1970: 1842; of *denticornis:* Mayr, 1861: 63; Seifert, 1988b: 38; of *foreli:* Bernard, 1967: 122; Seifert, 1988b: 38; of *kieviensis:* Arnol'di, 1970: 1842; of *lissahorensis* and *lobulicornis:* Bernard, 1967: 122; Seifert, 1988b: 38; of *obscura* and *pyrenaea:* Seifert, 1988b: 38; of *starki:* Arnol'di, 1970: 1842; Seifert, 1988b: 38. Material of the *nomen nudum nodicornis* referred here by Donisthorpe, 1915d: 134. Current subspecies: nominal plus *brunescens, foreliella.*

*lobicornosulcinodis. Myrmica rubra* var. *lobicornosulcinodis* Forel, 1874: 77 (w.) SWITZERLAND. *Nomen nudum.*

***lobifrons.*** *Myrmica sabuleti* var. *lobifrons* Pergande, 1900: 521 (w.) U.S.A. Variety/subspecies of *scabrinodis:* Emery, 1921f: 40; of *lobicornis:* Creighton, 1950a: 100. Junior synonym of *fracticornis:* Smith, M.R. 1951a: 790; Cole, 1953d: 242. Revived from synonymy as subspecies of *lobicornis:* Gregg, 1961: 212. Raised to species: Yensen, Clark & Francoeur, 1977: 182. Senior synonym of *glacialis:* Wheeler, W.M. 1917e: 21; Creighton, 1950a: 100.

*lobulicornis. Myrmica lobicornis* var. *lobulicornis* Nylander, 1857: lxxix (w.) FRANCE. Raised to species: Bondroit, 1920a: 151. Variety/subspecies of *lobicornis:* Emery, 1921f: 38; Weber, 1948a: 287.

Junior synonym of *lobicornis:* Bernard, 1967: 122; Seifert, 1988b: 38.

*lonae. Myrmica scabrinodis* subsp. *lonae* Finzi, 1926: 103, fig. 11 (w.q.m.) FINLAND. Stirps/subspecies of *sabuleti:* Santschi, 1931b: 346; Stitz, 1939: 96. Junior synonym of *scabrinodolobicornis:* Sadil, 1952: 249 (Sadil gives *lonae* as senior name, but *scabrinodolobicornis* has priority). Junior synonym of *sabuleti:* Seifert, 1988b: 31.

*longiceps* Smith, F. 1858; see under ***APHAENOGASTER.***

*longipes* Smith, F. 1857; see under ***PHEIDOLE.***

*longiscapus. Myrmica longiscapus* Curtis, 1854: 213, pl. 23, figs. 11–14 (w.q.m.) GREAT BRITAIN. Junior synonym of *laevinodis:* Mayr, 1863: 433.

***longispinosa.** **Myrmica longispinosa* Mayr, 1868c: 87, pl. 4, fig. 86 (w.) BALTIC AMBER (Oligocene).

*lucidula* Smith, F. 1858; see under ***FORMICOXENUS.***

**luctuosa.** *Myrmica luctuosa* Smith, F. 1878b: 12 (m.) INDIA.

*lulakeranica. Myrmica deplanata* susbp. *lulakeranica* Arnol'di, 1948: 211. *Nomen nudum.*

*lundii* Guérin-Méneville, 1838; see under ***ACROMYRMEX.***

**luteola.** *Myrmica luteola* Kupyanskaya, 1990: 103, FIGS. 16, 17 (w.q.) RUSSIA.

**#lutescens.** *Myrmica smythiesi* var. *lutescens* Menozzi, 1939a: 292 (w.q.) INDIA. [First available use of *Myrmica smythiesii.* r. *cachmiriensis* var. *lutescens* Forel, 1904c: 23; unavailable name.]

*lyncea* Spinola, 1851; see under ***PSEUDOMYRMEX.***

***macrocephala.** **Myrmica macrocephala* Heer, 1850: 158, pl. 12, fig. 14 (q.) SWITZERLAND (Miocene).

*macrogyna. Myrmica rubra* var. *macrogyna* Brian, M.V. & Brian, A.D. 1949: 397 (q.m.) GREAT BRITAIN. Junior synonym of of *ruginodis:* **new synonymy**.

*maligna* Smith, F. 1865; see under ***PODOMYRMA.***

**margaritae.** *Myrmica margaritae* Emery, 1889b: 502 (w.) BURMA. Current subspecies: nominal plus *pulchella.* See also: Bingham, 1903: 267.

*marylandica* Buckley, 1867; see under ***CREMATOGASTER.***

*medvedevi. Myrmica stangeana* subsp. *medvedevi* Arnol'di, 1970: 1841 (w.) UKRAINE. Junior synonym of *stangeana:* Seifert, 1988b: 23.

**#mekongi.** *Myrmica kozlovi* subsp. *mekongi* Ruzsky, 1915a: 437, fig. 12 (w.) TIBET.

*melanocephala* Mayr, 1855; see under ***LEPTOTHORAX.***

*mellea* Smith, F. 1859; see under ***SOLENOPSIS.***

**mexicana.** *Myrmica mexicana* Wheeler, W.M. 1914b: 52 (w.q.m.) MEXICO. Subspecies of *scabrinodis:* Cole, 1942: 368; Weber, 1950b: 206. Revived status as species: Gregg, 1961: 211.

*microgyna. Myrmica rubra* var. *microgyna* Brian, M.V. & Brian, A.D. 1949: 397 (q.m.) GREAT BRITAIN. Junior synonym of *ruginodis:* **new synonymy**.

*minima* Buckley, 1867; see under ***MONOMORIUM.***

*minkii* Foerster, 1850; see under ***STENAMMA.***

*minuta* Say, 1836; see under ***SOLENOPSIS.***

**minuta.** *Myrmica laevinodis* var. *minuta* Ruzsky, 1905b: 670 (w.) RUSSIA. [Unresolved junior primary homonym of *minuta* Say, 1836: 294, above.] Raised to species: Arnol'di, 1976a: 548. Current subspecies: nominal plus *iskanderi, tarbinskii.*

*minuta. Myrmica rugulosa* var. *minuta* Karavaiev, 1929b: 204 (w.m.) RUSSIA. [Junior primary homonym of *minuta* Say, 1836: 294, above.] Replacement name: *constricta* Karavaiev, 1934: 74.

*modesta* Foerster, 1850; see under ***TETRAMORIUM.***

*modesta* Smith, F. 1860; see under ***TETRAMORIUM.***

*moesta* Smith, F. 1863; see under ***VOLLENHOVIA.***

***molassica.** **Myrmica molassica* Heer, 1850: 162, pl. 13, fig. 6 (q.) SWITZERLAND (Miocene).

*molefaciens* Buckley, 1860; see under ***POGONOMYRMEX.***

*molesta* Say, 1836; see under ***SOLENOPSIS.***

*montana* Labram & Imhoff, 1838; see under ***MANICA.***

*montana. Myrmica (Monomorium) montana* Buckley, 1867: 339 (w.) U.S.A. [Unresolved junior primary homonym of *montana* Labram & Imhoff, 1838, pl. 36, figs. 1–3, above.] Unidentifiable to genus; *incertae sedis* in *Myrmica.* See also: Emery, 1895c: 338; Wheeler, W.M. 1902f: 27.

**monticola.** *Myrmica monticola* Creighton, 1950a: 101 (w.m.) U.S.A. [First available use of *Myrmica scabrinodis* subsp. *schencki* var. *monticola* Wheeler, W.M. 1917a: 505; unavailable name.] Wheeler, G.C. & Wheeler, J. 1953a: 117 (l.).

*moravica. Myrmica moravica* Soudek, 1922: 45, fig. 21 (w.q.) CZECHOSLOVAKIA. Soudek, 1925a: 35 (m.). [Also described as new by Soudek, 1923: 107.] Status as species: Finzi, 1926: 104; Weber, 1950b: 213. Subspecies of *deplanata:* Novak & Sadil, 1941: 80; Kratochvil, in Kratochvil, Novak & Snoflak, 1944: 40; Sadil, 1952: 263. Junior synonym of *deplanata:* Karavaiev, 1934: 83; Arnol'di, 1970: 1843; Seifert, 1988b: 36.

*muscorum* Nylander, 1846; see under ***LEPTOTHORAX.***

*mutata. Myrmica rubra* var. *mutata* Sadil, 1952: 242, fig. I (w.) CZECHOSLOVAKIA. Junior synonym of *ruginodis:* Seifert, 1988b: 6.

*mutica* Nylander, 1849; see under ***MESSOR.***

*mutica* Emery, 1895; see under ***MANICA.***

*myrmecophila. Myrmica myrmecophila* Wasmann, 1910: 516 (ergatoid q.) AUSTRIA. Junior synonym of *sulcinodis:* Boven, 1970: 130; Seifert, 1988b: 8.

**myrmicoxena.** *Myrmica myrmicoxena* Forel, 1895i: 143 (q.) SWITZERLAND. Emery, 1908a: 181 (m.). See also: Forel, 1910a: 29; Finzi, 1926: 114; Kutter, 1973c: 255; Kutter, 1977c: 67.

*natalensis* Santschi, 1925; see under ***MYRMICARIA.***

**nearctica.** *Myrmica sabuleti* subsp. *nearctica* Weber, 1939b: 148 (w.q.m.) U.S.A. Junior synonym of *monticola:* Creighton, 1950a: 101. Revived from synonymy and raised to species: Francoeur, 1977b: 206.

***nebulosa.** **Myrmica nebulosa* Novak, O. 1878: 91, pl. 3, fig. 1 (m.) GERMANY (Oligocene).

*neglecta. Myrmica neglecta* Stärcke, 1942c: xxvii. *Nomen nudum;* material referred to *puerilis* by Stärcke, 1942c: xxvii.

**#neolaevinodis.** *Myrmica rubra* r. *neolaevinodis* Forel, 1901h: 80 (w.) U.S.A. Variety/subspecies of *laevinodis:* Weber, 1947: 455; of *rubra:* Smith, D.R. 1979: 1351.

*nevodovskii. Leptothorax (Leptothorax) nevodovskii* Karavaiev, 1926c: 164, fig. 3 (w.) CAUCASUS. Junior synonym of *bessarabica:* Arakelian, 1994: 23.

*nigripes. Myrmica sulcinodis* var. *nigripes* Ruzsky, 1895: 32 (w.q.m.) RUSSIA. [Also described as new by Ruzsky, 1896: 73.] Subspecies of *sulcinodis:* Ruzsky, 1936: 95; Weber, 1948a: 268; Sadil, 1952: 244. Junior synonym of *sulcinodis:* Seifert, 1988b: 7.

*nitidula* Nylander, 1846; see under ***FORMICOXENUS.***

*nodicornis. Myrmica nodicornis* Chappell, 1886: 58 (w.m.) GREAT BRITAIN. *Nomen nudum;* material referred to *lobicornis* by Donisthorpe, 1915d: 134.

*novaeboracensis* Buckley, 1867; see under ***CREMATOGASTER.***

*nylanderi* Foerster, 1850; see under ***LEPTOTHORAX.***

**#obensis.** *Myrmica limanica* subsp. *obensis* Arnol'di, 1970: 1840 (w.m.) RUSSIA. Subspecies of *gallienii:* current status.

*oblonga* Smith, F. 1860; see under ***VOLLENHOVIA.***

*obscura. Myrmica schencki* var. *obscura* Finzi, 1926: 111 (w.q.m.) ITALY. Junior synonym of *schencki:* Bernard, 1967: 118; of *lobicornis:* Seifert, 1988b: 38.

*obscurata. Myrmica obscurata* Motschoulsky, 1863: 16 (w.) SRI LANKA. Unidentifiable to genus; *incertae sedis* in *Myrmica.*

***obsoleta.** **Myrmica obsoleta* Heer, 1850: 160, pl. 13, fig. 3 (q.) SWITZERLAND (Miocene).

*occidentalis* Cresson, 1865; see under ***POGONOMYRMEX.***

*ominosa* Gerstäcker, 1859; see under ***MONOMORIUM.***

*opaca* Smith, F. 1861; see under ***ROMBLONELLA.***

*opposita* Say, 1836; see under ***APHAENOGASTER.***

**orientalis.** *Myrmica ruginodis* subsp. *orientalis* Karavaiev, 1926f: 65, fig. 1 (w.) RUSSIA. Kuznetsov-Ugamsky, 1928b: 37 (q.m.). Variety of *rubra:* Weber, 1947: 450. Raised to species: Kupyanskaya, 1990: 101.

**orthostyla.** *Myrmica orthostyla* Arnol'di, 1976a: 552, fig. 8 (w.q.m.) RUSSIA. See also: Tarbinsky, 1976: 27.

**pachei.** *Myrmica pachei* Forel, 1906b: 79 (w.q.m.) NEPAL.

*paleata* Lund, 1831; see under ***SOLENOPSIS.***

*pallida* Nylander, 1849; see under ***APHAENOGASTER.***

*pallidinodis:* see under *pilinodis.*

*pallinodis:* see under *pilinodis.*

*pallidula* Nylander, 1849; see under ***PHEIDOLE.***

*parallela* Smith, F. 1859; see under ***TETRAMORIUM.***

*parasitica* Creighton, 1934; see under ***MANICA.***

*parvula* Schenck, 1852; see under ***LEPTOTHORAX.***

*pechei. Sifolinia pechei* Samsinak, 1957: 167, 2 figs. (q.) CZECHOSLOVAKIA. Pisarski, 1962: 367 (m.). Junior synonym of *karavajevi:* Samsinak, 1964: 156.

*pedestris* Smith, F. 1861; see under ***VOLLENHOVIA.***

*pellucida* Smith, F. 1857; see under ***TAPINOMA.***

*perelegans. Myrmica perelegans* Curtis, 1854: 214, pl. 23, figs. 15–17 (w.q.m.) GREAT BRITAIN. Junior synonym of *sulcinodis:* Mayr, 1855: 408; Smith, F. 1855b: 120.

**#persiana.** *Myrmica bergi* subsp. *persiana* Weber, 1947: 474. Replacement name for *fortior* Crawley, 1920a: 163. [Junior primary homonym of *fortior* Forel, 1904c: 22.]

*pertinax* Smith, F. 1861; see under ***VOLLENHOVIA.***

*pilinodis. Myrmica pilinodis* Motschoulsky, 1863: 16 (w.) SRI LANKA. Unidentifiable to genus; *incertae sedis* in *Myrmica.* [Misspelled as *pallidinodis* or *pallinodis:* Dalla Torre, 1893: 113; Emery, 1922f: 356.]

*pilosiscapus. Myrmica pilosiscapus* Bondroit, 1920a: 147, fig. 1 (w.q.m.) BELGIUM. [Also described as new by Bondroit, 1920b: 301.] Variety/subspecies of *scabrinodis:* Finzi, 1926: 102; Santschi, 1931b: 343; Stitz, 1939: 94. Status as species: Sadil, 1952: 256; Arnol'di, 1970: 1842; Baroni Urbani, 1971c: 26. Junior synonym of *sabuleti:* Bernard, 1967: 117; Boven, 1977: 120; of *scabrinodis:* Kutter, 1977c: 69; Collingwood, 1979: 55; Seifert, 1988b: 27; Casevitz-Weulersse, 1990a: 137.

***pinetorum.*** *Myrmica punctiventris* subsp. *pinetorum* Wheeler, W.M. 1905f: 384 (w.q.) U.S.A. Weber, 1950b: 217 (m.). Raised to species: Creighton, 1950a: 102.

***pisarskii.*** *Myrmica pisarskii* Radchenko, 1994e: 208, figs. 1–12 (w.q.m.) RUSSIA.

*plana. Myrmica lobicornis* var. *plana* Karavaiev, 1927a: 283, fig. 1 (w.) UKRAINE. Arnol'di, 1934: 170 (q.m.). Subspecies of *schencki:* Karavaiev, 1929b: 208; of *deplanata:* Arnol'di, 1934: 170. Junior synonym of *deplanata:* Karavaiev, 1934: 83; Arnol'di, 1970: 1843; Seifert, 1988b: 36.

*polita* Smith, F. 1860; see under ***PHEIDOLOGETON.***

*polita* Smith, F. 1862; see under ***SOLENOPSIS.***

*poneroides* Smith, F. 1860; see under ***VOLLENHOVIA.***

*puerilis. Myrmica puerilis* Stärcke, 1942c: xxvii, figs. 1–3, 5 (w.q.m.) NETHERLANDS. Material of the *nomina nuda atlantica* and *neglecta* referred here: Stärcke, 1942c: xxvii. Junior synonym of *specioides:* Collingwood & Yarrow, 1969: 57; Seifert, 1988b: 16.

#***pulchella.*** *Myrmica margaritae* var. *pulchella* Santschi, 1937h: 368 (w.) TAIWAN.

*punctata* Smith, F. 1860; see under ***PRISTOMYRMEX.***

***punctiventris.*** *Myrmica punctiventris* Roger, 1863a: 190 (w.) U.S.A. Mayr, 1886d: 450 (q.); Emery, 1895c: 312 (m.). Senior synonym of *isfahani:* Weber, 1950b: 512.

*punica* Smith, 1861; see under ***TETRAMORIUM.***

**pusilla* Heer, 1850; see under ***CREMATOGASTER.***

*pyrenaea. Myrmica arduennae* var. *pyrenaea* Bondroit, 1918: 106, fig. 53 (w.) FRANCE. Raised to species: Bondroit, 1920a: 151. Variety/stirps of *lobicornis:* Finzi, 1926: 108; Santschi, 1931b: 350. Junior synonym of *lobicornis:* Seifert, 1988b: 38.

*quadrispinosa* Smith, F. 1865; see under ***PHEIDOLE.***

***quebecensis.*** *Myrmica quebecensis* Francoeur, 1981: 756, figs. 5, 6 (q.) CANADA. Francoeur & Loiselle, 1984: 4 (m.).

***ravasinii.*** *Myrmica ravasinii* Finzi, 1923: 2, fig. (w.) ALBANIA. Arakelian, 1989: 734 (q.m.). See also: Finzi, 1926: 112.

*recedens* Nylander, 1856; see under ***LEPTOTHORAX.***

*rediana* Gené, 1841; see under ***CREMATOGASTER.***

*reticulata* Smith, F. 1862; see under ***TETRAMORIUM.***

*reticulata. Myrmica scabrinodis* st. *rolandi* var. *reticulata* Santschi, 1931b: 344 (w.q.m.) SPAIN. Unavailable name; material referred to *scabrinodis* by Seifert, 1988b: 27.

*rhynchophora* Foerster, 1850; see under ***MANICA.***

***ritae.*** *Myrmica ritae* Emery, 1889b: 501, pl. 11, fig. 27 (w.) BURMA. See also: Bingham, 1903: 267.

*rolandi. Myrmica rolandi* Bondroit, 1918: 101 (w.q.m.) FRANCE. Variety/subspecies of *sulcinodis:* Santschi, 1919e: 244; Menozzi, 1922b: 325; of *scabrinodis:* Emery, 1921f: 40; Santschi, 1925g: 340; Santschi, 1931b: 344. Revived status as species: Finzi, 1926: 89; Bernard, 1967: 115. Junior synonym of *aloba:* Seifert, 1988b: 29.

***rubra.*** *Formica rubra* Linnaeus, 1758: 580 (w.) EUROPE. Latreille, 1802c: 248 (q.m.); Wheeler, G.C. & Wheeler, J. 1953a: 118 (l.); Hauschteck, 1965: 325 (k.). Combination in *Myrmica:* Latreille, 1804: 179. Senior synonym of *laevinodis* (and its junior synonyms *bruesi, champlaini* and *longiscapus*): Yarrow, 1955b: 114; Arnol'di, 1970: 1839; Boven, 1977: 115; Arnol'di & Dlussky, 1978: 530; Collingwood, 1979: 52; Seifert, 1988b: 5. Current subspecies: nominal plus *europaea, khamensis, neolaevinodis.* See also: Emery, 1908a: 169; Brian, M.V. & Brian, A.D. 1949: 393; Smith, D.R. 1979: 1350; Pearson, 1981: 75; Seifert, 1988b: 5; Onoyama, 1989a: 131; Atanassov & Dlussky, 1992: 83.

*rubra* Buckley, 1867; see under ***LABIDUS.***

*rubriceps* Nylander, 1849; see under ***CREMATOGASTER.***

*rufa* Jerdon, 1851; see under ***CREMATOGASTER.***

*ruficeps* Smith, F. 1861; see under ***PHEIDOLE.***

*rufitarsis* Foerster, 1850; see under ***MESSOR.***

*rufotestacea* Foerster, 1850; see under ***MESSOR.***

**rugiceps* Heer, 1850; see under ***PHEIDOLOGETON.***

***rugifrons.*** *Myrmica rugifrons* Smith, F. 1858b: 124 (q.) INDIA.

*ruginodiformis. Myrmica rugulosa* var. *ruginodiformis* Karavaiev, 1929b: 204 (w.q.m.) CAUCASUS. Junior synonym of *dshungarica:* Arnol'di, 1976a: 551.

***ruginodis.*** *Myrmica ruginodis* Nylander, 1846a: 929, pl. 18, figs. 5, 30 (w.q.m.) FINLAND. Hauschteck, 1965: 325 (k.). Variety/subspecies of *rubra:* Forel, 1874: 76; Emery & Forel, 1879: 460; Ruzsky, 1904a: 288; Bondroit, 1910: 498; Forel, 1915d: 28; Menozzi, 1918: 82; Karavaiev, 1927c: 258; of *laevinodis:* Mayr, 1886d: 450; Ruzsky, 1902d: 29. Junior synonym of *rubra:* Santschi, 1931b: 339. Status as species: Saunders, E. 1880: 214; André, 1883a: 317; Nasonov, 1889: 33; Forel, 1892i: 315; Bondroit, 1912: 351; Donisthorpe, 1915d: 115; Bondroit, 1918: 103; Santschi, 1919e: 244; Müller, 1923: 41; Finzi, 1926: 85; Stitz, 1939: 83; Novak & Sadil, 1941: 76; Holgersen, 1942: 8; Collingwood, 1958: 68; Bernard, 1967: 120; Collingwood & Yarrow, 1969: 56; Kutter, 1977c: 67; Arnol'di & Dlussky, 1978: 530; Collingwood; 1979: 53; Seifert, 1988b: 6; Atanassov & Dlussky, 1992: 86. Senior synonym of *diluta:* Mayr, 1861: 63; of *dimidiata:* Weber, 1947: 448; of *ruginodolaevinodis:* Bernard, 1967: 120; Boven, 1977: 115; of *mutata:* Seifert, 1988b: 6; of *macrogyna* and *microgyna:* **new synonymy.**

*ruginodolaevinodis. Myrmica rubra* var. *ruginodolaevinodis* Forel, 1874: 78 (q.m.) SWITZERLAND. Raised to species: Stitz, 1917: 347. Variety of *rubra:* Finzi, 1926: 86; Stitz, 1939: 84; Sadil, 1952: 241. Junior synonym of *ruginodis:* Bernard, 1967: 120; Boven, 1977: 115.

***rugiventris.*** *Tetramorium rugiventris* Smith, M.R. 1943b: 4 (w.) U.S.A. Gregg, 1961: 218 (q.m.). Combination in *Myrmica:* Brown, 1957d: 6; in *Paramyrmica:* Gregg, 1961: 215; in *Myrmica:* Bolton, 1988a: 6.

*rugosa. Myrmica rugosa* Mayr, 1855: 411. *Nomen nudum,* attributed to Koch; material referred to *scabrinodis* by Mayr, 1855: 411.

***rugosa.*** *Myrmica rugosa* Mayr, 1865: 19 (footnote) (w.) INDIA. Current subspecies: nominal plus *arisana, debilior.* See also: Bingham, 1903: 268.

*rugososmythiesi. Myrmica rugosa* var. *rugososmythiesi* Forel, 1902c: 228 (w.) INDIA. Junior synonym of *debilior:* Forel, 1902c: 228.

***rugulosa.*** *Myrmica rugulosa* Nylander, 1849: 32 (w.q.) FINLAND. Mayr, 1855: 405 (m.); Lomnicki, 1924: 817 (gynandromorph). Race/subspecies of *rubra:* Forel, 1874: 77; Emery & Forel, 1879: 40; of *scabrinodis:* Forel, 1892i: 315; Ruzsky, 1905b: 690; Emery, 1908a: 177; Emery, 1916b: 120. Status as species: André, 1883a: 317; Ruzsky, 1902d: 30; Bondroit, 1918: 100; Finzi, 1926: 91; Karavaiev, 1929b: 203; Holgersen, 1940: 185; Boven, 1947: 175; Bernard, 1967: 115; Kutter, 1977c: 68; Arnol'di & Dlussky, 1978: 532; Collingwood, 1979: 54; Seifert, 1988b: 11; Atanassov & Dlussky, 1992: 89. Senior synonym of *clandestina:* Mayr, 1855: 405; of *slobodensis* (and its junior synonym *constricta*): Arnol'di, 1970: 1840. Current subspecies: nominal plus *sulcinodorugulosa.*

*rugulosoides. Myrmica scabrinodis* var. *rugulosoides* Forel, 1915d: 29 (w.) SWITZERLAND. Kutter, 1924: 8 (q.m.). Variety/subspecies of *scabrinodis:* Kutter, 1924: 8; Santschi, 1931b: 342; Stitz, 1939: 93; Novak & Sadil, 1941: 79. Status as species: Bondroit, 1918: 102; Müller, 1923: 42; Finzi, 1926: 94; Baroni Urbani, 1971c: 30; Kutter, 1977c: 68. Junior synonym of *scabrinodis:* Bernard, 1967: 116; Banert & Pisarski, 1972: 350; Collingwood, 1979: 55; Seifert, 1984b: 1; Seifert, 1988b: 27.

*rugulososcabrinodis. Myrmica* (*Myrmica*) *rugulosa* var. *rugulososcabrinodis* Karavaiev, 1929b: 205, fig. 1 (w.m.) CAUCASUS. [*Myrmica rubra* var. *rugulososcabrinodis* Forel, 1874: 226. *Nomen nudum.*] Junior synonym of *bessarabica:* Dlussky, Soyunov & Zabelin, 1990: 182; Atanassov & Dlussky, 1992: 93.

#***rupestris.*** *Myrmica smythiesii* var. *rupestris* Forel, 1902c: 227 (w.) INDIA.

*russula* Nylander, 1856; see under ***LEPTOTHORAX.***

#***ruzskyi.*** *Myrmica kozlovi* subsp. *ruzskyi* Kiseleva, 1925: 75, fig. (w.) RUSSIA.

*ruzskyi. Myrmica kozlovi* subsp. *ruzskyi* Weber, 1947: 471. [Unresolved junior primary homonym of *ruzskyi* Kiseleva, above.] Unnecessary replacement name for, and hence junior synonym of, *subalpina* Ruzsky, 1915a: 438: **new synonymy.**

*sabeana* Buckley, 1867; see under ***SOLENOPSIS.***

***sabuleti.*** *Myrmica sabuleti* Meinert, 1861: 327 (w.m.) DENMARK. Emery, 1908a: 176 (q.); Hauschteck, 1965: 325 (k.); Donisthorpe, 1915b: 258 (gynandromorph). Junior synonym of *lobicornis:* Dalla Torre, 1893: 111. Variety/subspecies of *scabrinodis:* Emery, 1908a: 176; Bondroit, 1910: 495; Emery, 1916b: 120; Menozzi, 1922b: 325; Finzi, 1924a: 12; Finzi, 1926: 101; Menozzi, 1936d: 270. Status as species: Bondroit, 1918: 102; Santschi, 1931b: 345; Karavaiev, 1934: 80; Stitz, 1939: 94; Weber, 1939b: 144; Holgersen, 1940: 184; Novak & Sadil, 1941: 79; Bernard, 1967: 117; Baroni Urbani, 1971c: 31; Kutter, 1977c: 68; Arnol'di & Dlussky, 1978: 534; Collingwood, 1979: 54; Seifert, 1988b: 31; Atanassov & Dlussky, 1992: 97. Senior synonym of *granulinodis:* Emery, 1908a: 177 (footnote); Sadil, 1952: 246 [if synonymy correct then *granulinodis* has priority]; of *lonae:* Bernard, 1967: 117; Seifert, 1988: 31; of *spinosior:* Seifert, 1988b: 31; Casevitz-Weulersse, 1990a: 137; of *scabrinodolobicornis:* **new synonymy**.

*saevissima* Smith, F. 1855; see under ***SOLENOPSIS.***

**salina.** *Myrmica scabrinodis* var. *salina* Ruzsky, 1905b: 687 (w.q.m.) RUSSIA. Variety of *schencki:* Weber, 1948a: 302. Junior synonym of *lonae:* Sadil, 1952: 249. Revived from synonymy, and raised to species: Arnol'di, 1970: 1842; Arnol'di & Dlussky, 1978: 534. Senior synonym of *slovaca:* Seifert, 1988b: 25.

*sallei* Guérin-Méneville, 1852; see under ***LEPTOTHORAX.***

**samnitica.** *Myrmica samnitica* Mei, 1987: 457, figs. 1–15 (q.m.) ITALY.

*sancta. Myrmica scabrinodis* var. *sancta* Karavaiev, 1926f: 67, fig. 4 (w.) UKRAINE. Seifert, 1988b: 21 (q.m.). Junior synonym of *scabrinodis:* Stitz, 1939: 90; Sadil, 1952: 249; of *specioides:* Kutter, 1977c: 70. Revived from synonymy, and raised to species: Arnol'di, 1970: 1840; Arnol'di & Dlussky, 1978: 532; Seifert, 1988b: 21. Junior synonym of *bessarabica:* Dlussky, Soyunov & Zabelin, 1990: 182; Atanassov & Dlussky, 1992: 93.

**saposhnikovi.** *Myrmica lobicornis* r. *saposhnikovi* Ruzsky, 1904b: 3 (w.) RUSSIA. Teranishi, 1940: 15 (q.). Subspecies of *scabrinodis:* Ruzsky, 1905b: 701; Emery, 1908a: 180. Raised to species: Pisarski, 1969b: 296; Arnol'di, 1970: 553; Tarbinsky, 1976: 35. Current subspecies: nominal plus *baikalensis, bergiana.*

*saxicola* Buckley, 1867; see under ***SOLENOPSIS.***

*scabrata. Myrmica scabrata* Buckley, 1867: 343 (w.) U.S.A. Unidentifiable to genus; *incertae sedis* in *Myrmica.*

**scabrinodis.** *Myrmica scabrinodis* Nylander, 1846a: 930 (w.q.m.) FINLAND. Wheeler, G.C. & Wheeler, J. 1953a: 119 (l.); Hauschteck, 1965: 325 (k.); Wasmann, 1891: 298 (gynandromorph). Race/subspecies of *rubra:* Forel, 1874: 76; Emery & Forel, 1879: 460; Emery, 1895c: 313; Forel, 1904b: 374; Wheeler, W.M. 1908g: 406. Status as species: Saunders, E. 1880: 215; Nasonov, 1889: 36; Emery, 1898c: 126; Emery, 1908a: 174; Bondroit, 1912: 351; Stitz, 1914: 71; Donisthorpe, 1915d: 125; Forel, 1915d: 29; Karavaiev, 1916: 504; Emery, 1916b: 120; Wheeler, W.M. 1917a: 504; Bondroit, 1918: 101; Santschi, 1921a: 110; Menozzi, 1922b: 325; Müller, 1923: 43; Finzi, 1926: 98; Karavaiev, 1926b: 95; Santschi, 1931b: 341; Bernard, 1967: 116; Tarbinsky, 1976: 41; Kutter, 1977c: 69; Arnol'di & Dlussky, 1978: 534; Collingwood, 1979: 55; Seifert, 1988b: 27; Atanassov & Dlussky, 1992: 95. Senior synonym of *pilosiscapus:* Kutter, 1977c: 69; Collingwood, 1979: 55; Seifert, 1988b: 27; Casevitz-Weulersse, 1990a: 137; of *rugulosoides:* Bernard, 1967: 116; Banert & Pisarski, 1972: 350; Collingwood, 1979: 55; Seifert, 1984b: 1. Material of the *nomen nudum rugosa* referred here: Mayr, 1855: 411; and of the unavailable name *reticulata:* Seifert, 1988b: 27. Current subspecies: nominal plus *eidmanni, intermedia, scabrinodosabuleti, ussuriensis.*

*scabrinodoides. Myrmica (Myrmica) balcanica* var. *scabrinodoides* Sadil, 1952: 255, figs. V, 9–13; VII, 37, IX, 34; XI, 10, 11 (w.q.m.) CZECHOSLOVAKIA. Junior synonym of *specioides:* Pisarski, 1975: 12; Kutter, 1977c: 70; Seifert, 1988b: 16; of *bessarabica:* Atanassov & Dlussky, 1992: 93.

*scabrinodolobicornis. Myrmica rubra* var. *scabrinodolobicornis* Forel, 1874: 77 (w.q.m.) SWITZERLAND. Variety/subspecies of *scabrinodis:* Emery, 1921f: 40; of *sabuleti:* Weber, 1948a: 296. Junior synonym of *lonae:* Sadil, 1952: 249 (note that *scabrinodolobicornis* has priority over *lonae*). Junior synonym of *sabuleti:* **new synonymy**.

**#scabrinodosabuleti.** *Myrmica (Myrmica) scabrinodis* var. *scabrinodosabuleti* Sadil, 1952: 253, figs. IV, 13–15; VII, 17–19 (w.m.) CZECHOSLOVAKIA.

*scabrosa* Smith, F. 1859; see under ***TETRAMORIUM.***

*scalpturata* Nylander, 1856; see under ***MESSOR.***

**schencki.** *Myrmica rubra* var. *schencki* Viereck, 1903: 72 (w.q.m.) EUROPE (see Baroni Urbani, 1971c: 35 (note)). [First available use of *Myrmica rubra* subsp. *scabrinodis* var. *schencki* Emery, 1895c: 315; unavailable name.] Wheeler, G.C. & Wheeler, J. 1953a: 120 (l.). Subspecies of *scabrinodis:* Emery, 1908a: 178; Forel, 1915d: 29. Status as species: Bondroit, 1911: 11; Bondroit, 1912: 351; Donisthorpe, 1915c: 265; Bondroit, 1918: 103; Müller, 1923: 44; Finzi, 1926: 109; Karavaiev, 1926b: 95; Stärcke, 1927: 84; Santschi, 1931b: 351; Karavaiev, 1934: 93; Bernard, 1967: 118; Arnol'di, 1970: 1843; Tarbinsky, 1976: 42; Kutter, 1977c: 70; Arnol'di & Dlussky, 1978: 535; Collingwood, 1979: 56; Seifert, 1988b: 35; Kupyanskaya, 1990: 109; Atanassov & Dlussky, 1992: 99. Senior synonym of *kutteri:* Bernard, 1967: 118; of *subopaca:* Arnol'di, 1970: 1843.

*seminigra* Cresson, 1865; see under ***POGONOMYRMEX.***

*semipolita* Nylander, 1856; see under ***APHAENOGASTER.***

**serica.** *Myrmica margaritae* var. *serica* Wheeler, W.M. 1928c: 8 (w.) CHINA. Weber, 1950b: 223 (q.). Subspecies of *ritae:* Weber, 1950b: 222. Revived status as species: Radchenko, 1994a: 44.

**silvestrii.** *Myrmica ruginodis* var. *silvestrii* Wheeler, W.M. 1928d: 100 (w.) JAPAN. Variety of *rubra:* Weber, 1947: 451. Raised to species: Collingwood, 1976: 301.

*simillima* Smith, F. 1851; see under ***TETRAMORIUM.***

*simpliciuscula* Nylander, 1856; see under ***LEPTOTHORAX.***

*slobodensis. Myrmica rugulosa* var. *slobodensis* Arnol'di, 1934: 162 (w.m.) RUSSIA. Senior synonym of *constricta:* Karavaiev, 1936: 274. Junior synonym of *rugulosa:* Arnol'di, 1970: 1840.

*slovaca. Myrmica (Myrmica) slovaca* Sadil, 1952: 259, figs. VI, 8–14; VIII, 35, 36; IX, 35; X, 10, XI, 13 (w.q.m.) CZECHOSLOVAKIA. Junior synonym of *salina:* Seifert, 1988b: 25. See also: Kutter, 1977c: 71; Arnol'di & Dlussky, 1978: 534.

**smythiesii.** *Myrmica smythiesii* Forel, 1902c: 226 (w.) INDIA. Current subspecies: nominal plus *bactriana, cachmiriensis, carbonaria* (junior homonym), *exigua* (junior homonym), *fortior, hecate, himalayana, lutescens, rupestris.* See also: Bingham, 1903: 269.

**#sontica.** *Myrmica kurokii* var. *sontica* Santschi, 1937h: 367 (w.) JAPAN. Subspecies of *kurokii:* Weber, 1947: 470.

*sordidula* Nylander, 1849; see under ***CREMATOGASTER.***

**spatulata.** *Myrmica schencki* var. *spatulata* Smith, M.R. 1930b: 566, figs. 1–4 (w.q.) U.S.A. Raised to species: Creighton, 1950a: 105.

*specioides. Myrmica specioides* Bondroit, 1918: 100 (w.q.m.) BELGIUM. Junior synonym of *scabrinodis:* Emery, 1921f: 40; Sadil, 1952: 249; Bernard, 1967: 116; of *rugulosoides:* Santschi, 1931b: 342. Revived from synonymy: Collingwood & Yarrow, 1969: 57; Collingwood, 1979: 56; Seifert, 1988b: 16. Senior synonym of *puerilis:* Collingwood & Yarrow, 1969: 57; of *balcanica:* Pisarski, 1975: 12; Kutter, 1977c: 70; Seifert, 1988b: 16; of *scabrinodoides* and *striata:* Collingwood, 1979: 56; of *ahngeri:* Seifert, 1988b: 16. Junior synonym of *bessarabica:* Atanassov & Dlussky, 1992: 93.

*speciolosa. Myrmica speciolosa* Seifert, 1988b: 14. Unavailable name.

**specularis.** *Myrmica specularis* Donisthorpe, 1929a: 446 (w.) TIBET.

*spinosior. Myrmica sabuleti* var. *spinosior* Santschi, 1931a: 3 (w.q.m.) FRANCE. [Also described as new by Santschi, 1931b: 346.] Junior synonym of *sabuleti:* Seifert, 1988b: 31.

**stangeana.** *Myrmica stangeana* Ruzsky, 1902a: 234 (w.) RUSSIA. Seifert, 1988b: 23 (q.m.). [Also described as new by Ruzsky, 1902b: 474.] Subspecies of *bergi:* Ruzsky, 1903b: 314; Ruzsky, 1905b: 678; of *scabrinodis:* Emery, 1908a: 178. Revived status as species: Arnol'di, 1970: 1841; Arnol'di & Dlussky, 1978: 533. Senior synonym of *medvedevi:* Seifert, 1988b: 23.

*starki. Myrmica (Myrmica) schencki* var. *starki* Karavaiev, 1929b: 208 (w.) RUSSIA. Junior synonym of *lobicornis:* Arnol'di, 1970: 1842. See also: Karavaiev, 1931b: 29.

*strandi. Myrmica rugulosa* subsp. *limanica* var. *strandi* Arnol'di, 1934: 164, fig. 16 (w.) RUSSIA. Unavailable name; material referred to *limanica* by Arnol'di, 1970: 1840.

*striata. Myrmica rugulosoides* var. *striata* Finzi, 1926: 96, fig. 7 (w.q.m.) ITALY. Stirps of *scabrinodis:* Santschi, 1931b: 344. Junior synonym of *specioides:* Collingwood, 1979: 56.

*striatula* Nylander, 1849; see under ***MYRMECINA.***

**striolagaster.** *Myrmica striolagaster* Cole, 1953a: 81 (w.) U.S.A. Cole, 1957c: 208 (q.m.); Wheeler, G.C. & Wheeler, J. 1972b: 234 (l.).

*subacuta. Myrmica rugosa* subsp. *dshungarica* var. *subacuta* Kuznetsov-Ugamsky, 1927e: 192 (m.) TURKESTAN. Unavailable name.

**#subalpina.** *Myrmica kozlovi* subsp. *subalpina* Ruzsky, 1915a: 438, fig. 14 (w.) TIBET. Senior synonym of *ruzskyi* Weber (unnecessary replacement name): **new synonymy**. See also: Menozzi, 1939a: 294.

*subalpina. Myrmica brevinodis* var. *subalpina* Wheeler, W.M. 1917a: 503 (w.q.m.) U.S.A. [First available use of *Myrmica rubra* subsp. *brevinodis* var. *subalpina* Wheeler, W.M. 1907b: 77; unavailable name.] [Unresolved junior primary homonym of *subalpina*

Ruzsky, above.] Junior synonym of *brevinodis:* Weber, 1950b: 189.

*subaltaica. Myrmica scabrinodis* subsp. *kasczenkoi* var. *subaltaica* Ruzsky, 1925b: 45 (w.) RUSSIA. Unavailable name.

#***subbrevispinosa.*** *Myrmica kozlovi* subsp. *subbrevispinosa* Ruzsky, 1915a: 437, fig. 13 (w.) TIBET.

*sublaevis* Nylander, 1849; see under ***HARPAGOXENUS.***

*sublanuginosa. Myrmica (Atta) sublanuginosa* Buckley, 1867: 343 (w.) U.S.A. Unidentifiable to genus; *incertae sedis* in *Myrmica.*

*subopaca* Smith, F. 1858; see under ***MONOMORIUM.***

*subopaca. Myrmica schencki* nat. *subopaca* Arnol'di, 1934: 172, figs. 36, 37 (w.) RUSSIA. [Also described as new by Arnol'di, in Karavaiev, 1934: 96.] [Unresolved junior primary homonym of *subopaca* Smith, F. 1858b: 127, above.] Junior synonym of *schencki:* Arnol'di, 1970: 1843.

*subrubra* Buckley, 1867; see under ***APHAENOGASTER.***

***sulcinodis.*** *Myrmica sulcinodis* Nylander, 1846a: 934 (w.q.) FINLAND. Mayr, 1855: 409 (m.); Hauschteck, 1965: 323 (k.); Donisthorpe, 1919: 1 (ergatandromorph); Santschi, 1931b: 351 (ergatandromorph). Race/subspecies of *rubra:* Forel, 1874: 76; Emery & Forel, 1879: 460; Wheeler, W.M. 1908g: 406. Status as species: Saunders, E. 1880: 215; André, 1883a: 317; Bondroit, 1912: 351; Emery, 1908a: 173; Forel, 1915d: 29; Donisthorpe, 1915d: 122; Emery, 1916b: 119; Finzi, 1926: 86; Bernard, 1967: 121; Kutter, 1977c: 70; Arnol'di & Dlussky, 1978: 531; Collingwood, 1979: 57; Seifert, 1988b: 7; Atanassov & Dlussky, 1992: 87. Senior synonym of *derthavini:* Kupyanskaya, 1986b: 94; of *myrmecophila:* Boven, 1970: 130; Seifert, 1988b: 8; of *nigripes:* Seifert, 1988b: 7; of *perelegans:* Mayr, 1855: 408; Smith, F. 1855b: 120; of *sulcinodoruginodis* and *sulcinodoscabrinodis:* Bernard, 1967: 121. Current subspecies: nominal plus *eximia, vicaria.*

*sulcinodoides. Myrmica brevinodis* var. *sulcinodoides* Wheeler, W.M. 1917a: 502 (w.) U.S.A. [First available use of *Myrmica rubra* subsp. *brevinodis* var. *sulcinodoides* Emery, 1895c: 313; unavailable name.] Wheeler, W.M. 1907b: 76 (q.m.). Junior synonym of *brevinodis:* Creighton, 1950a: 96. Revived from synonymy, and raised to species: Hunt & Snelling, 1975: 21; MacKay, Lowrie, *et al.* 1988: 92 (in key). Senior synonym of *frigida:* Wheeler, W.M. 1917e: 16; Weber, 1950b: 195; of *whymperi:* Wheeler, W.M. 1917a: 502. [Wheeler gives *sulcinodoides* as senior synonym, but *whymperi* has priority over *sulcinodoides* and *frigida,* and hence is first available name for this taxon.]

*sulcinodoruginodis. Myrmica ruginodis* var. *sulcinodoruginodis* Donisthorpe, 1915d: 121 (w.) GREAT BRITAIN. [*Myrmica rubra* var. *sulcinodoruginodis* Emery & Forel, 1879: 461; *nomen nudum,* attributed to Forel.] Junior synonym of *sulcinodis:* Bernard, 1967: 121.

#***sulcinodorugulosa.*** *Myrmica rugulosa* var. *sulcinodorugulosa* Nasonov, 1889: 35 (w.) RUSSIA.

*sulcinodoscabrinodis. Myrmica sulcinodis* var. *sulcinodoscabrinodis* Forel, 1915d: 29 (diagnosis in key) (w.) SWITZERLAND. [*Myrmica rubra* var. *sulcinodoscabrinodis* Forel, 1874: 77; *nomen nudum.*] Junior synonym of *sulcinodis:* Bernard, 1967: 121.

*suspiciosa* Smith, F. 1859; see under ***PHEIDOLE.***

***symbiotica.*** *Sommimyrma symbiotica* Menozzi, 1925d: 25, fig. 1 (w.) ITALY. Combination in *Myrmica:* Bolton, 1988a: 4. See also: Kutter, 1973c: 256.

***taediosa.*** *Myrmica taediosa* **new name**. Replacement name for *carinata* Kupyanskaya, 1990: 114. [Junior primary homonym of *carinata* Smith, F. 1859a: 148.]

***tahoensis.*** *Myrmica schencki* var. *tahoensis* Weber, 1948a: 304 (w.q.m.) U.S.A. [First available use of *Myrmica scabrinodis* subsp. *schencki* var. *tahoensis* Wheeler, W.M. 1917a: 504; unavailable name.] Subspecies of *emeryana:* Creighton, 1950a: 99. Raised to species: Yensen, Clark & Francoeur, 1977: 182.

#***tarbinskii.*** *Myrmica minuta* subsp. *tarbinskii* Arnol'di, 1976a: 549 (w.) RUSSIA.

*tarda* Jerdon, 1851; see under ***MERANOPLUS.***

***tenuispina.*** *Myrmica laevinodis* var. *tenuispina* Ruzsky, 1905b: 670 (w.) UZBEKISTAN. [First available use of *Myrmica rubra* r. *laevinodis* var. *tenuispina* Forel, 1904b: 374; unavailable name.] Arnol'di, 1976a: 552 (q.). Raised to species: Arnol'di, 1976a: 551; Tarbinsky, 1976: 25.

*terminalis* Shuckard, 1838; see under ***CREMATOGASTER.***

****tertiaria.*** **Myrmica tertiaria* Heer, 1850: 159, pl. 13, fig. 1 (q.) YUGOSLAVIA (Miocene). Senior synonym of **bicolor:* Dalla Torre, 1893: 117. See also: Mayr, 1867b: 58.

*testaceopilosa* Lucas, 1849; see under ***APHAENOGASTER.***

*testaceus. Myrmica balcanica* var. *testaceus* Sadil, 1953c: 207. *Nomen nudum.*

*texana* Buckley, 1860; see under ***ATTA.***

*thoracica* Smith, 1859; see under ***CARDIOCONDYLA.***

***tianshanica.*** *Myrmica tianshanica* Arnol'di, 1976a: 550, fig. 5 (w.m.) TURKESTAN. Current subspecies: nominal plus *alajensis.*

***tibetana.*** *Myrmica tibetana* Mayr, 1889: 279 (w.) TIBET. Ruzsky, 1915a: 440 (m.). Senior synonym of *furva:* Radchenko, 1994a: 41.

*tigreensis* Guérin-Méneville, 1849; see under ***MYRMICARIA.***

#***tipuna.*** *Myrmica kurokii* st. *tipuna* Santschi, 1937h: 367 (w.) TAIWAN.

*trachylissa* Smith, F. 1858; see under ***PRISTOMYRMEX.***

***transsibirica.*** *Myrmica transsibirica* Radchenko, 1994e: 212, figs. 23–32 (w.q.m.) RUSSIA.

*transversa* Smith, F. 1858; see under ***POGONOMYRMEX.***

*transversinodis* Enzmann, 1946; see under ***TETRAMORIUM.***

*trinodis* Losana, 1834; see under ***PHEIDOLE.***

*triplarina* Weddell, 1850; see under ***PSEUDOMYRMEX.***

*tristis* Smith, F. 1858; see under ***PHEIDOLE.***

#***trullicornis.*** *Myrmica sabuleti* subsp. *trullicornis* Buren, 1944a: 281 (w.q.) U.S.A. Junior synonym of *americana:* Weber, 1948a: 288. Revived from synonymy as subspecies of *hamulata:* Creighton, 1950a: 99.

***tschekanovskii.*** *Myrmica tschekanovskii* Radchenko, 1994e: 215, figs. 33–39 (w.m.) RUSSIA.

*tshuliensis. Myrmica sancta* subsp. *tshuliensis* Arnol'di, 1976a: 553 (w.q.m.) RUSSIA. Junior synonym of *bessarabica:* Dlussky, Soyunov & Zabelin, 1990: 182.

***turcica.*** *Myrmica scabrinodis* var. *turcica* Santschi, 1931b: 343 (w.q.) TURKEY. Raised to species, and senior synonym of *georgica:* Seifert, 1988b: 18.

*typhlops. Myrmica typhlops* Lund, 1831a: 128 (w.) BRAZIL. *Nomen nudum.*

*umbripennis* Smith, F. 1863; see under ***PRISTOMYRMEX.***

*unifasciata* Bostock, 1838; see Bolton, 1987: 356.

#***ussuriensis.*** *Myrmica scabrinodis* subsp. *ussuriensis* Kuznetsov-Ugamsky, 1928b: 36 (w.) RUSSIA.

*vagans* Curtis, 1854; see under ***FORMICA.***

***vandeli.*** *Myrmica vandeli* Bondroit, 1920a: 148, figs. 1, 2 (q.m.) FRANCE. [Also described as new by Bondroit, 1920b: 301.] Kutter, 1977c: 71 (w.). Variety of *scabrinodis:* Weber, 1950b: 211. Revived status as species: Bernard, 1967: 122; Kutter, 1977c: 71; Seifert, 1988b: 33. See also: Elmes & Thomas, 1985: 97.

*vastator* Smith, F. 1857; see under ***MONOMORIUM.***

*venusta* Smith, F. 1858; see under ***MONOMORIUM.***

****venusta.*** **Myrmica venusta* Heer, 1867: 31, pl. 2, fig. 12 (caste?) YUGOSLAVIA (Miocene). [Unresolved junior primary homonym of *venusta* Smith, F. 1858b: 126, above.] Possible combination in *Solenopsis:* Mayr, 1867b: 60; Handlirsch, 1907: 873; unconfirmed.

*vexator* Smith, F. 1861; see under ***MONOMORIUM.***

#***vicaria.*** *Myrmica sulcinodis* subsp. *vicaria* Kuznetsov-Ugamsky, 1928b: 33 (w.) RUSSIA.

*virulens* Smith, F. 1858; see under ***SOLENOPSIS.***

***wesmaeli.*** *Myrmica wesmaeli* Bondroit, 1918: 106, fig. 54 (w.) FRANCE. Finzi, 1926: 97 (q.m.). Subspecies of *sulcinodis:* Emery, 1921f: 42; Santschi, 1931b: 340; of *scabrinodis:* Weber, 1950b: 211. Status as species: Finzi, 1926: 97; Novak & Sadil, 1941: 78. Junior synonym of *lobicornis:* Bernard, 1967: 122. Revived from synonymy, and revived status as species: Seifert, 1988b: 39.

***wheeleri.*** *Myrmica wheeleri* Weber, 1939b: 150 (w.q.m.) U.S.A.

***whymperi.*** *Myrmica brevinodis* var. *whymperi* Forel, 1913l: 215 (w.) U.S.A. [First available use of *Myrmica rubra* r. *brevinodis* var. *whymperi* Forel, 1904a: 154; unavailable name.] Junior synonym of *sulcinodoides:* Wheeler, W.M. 1917a: 502. [Wheeler gives *sulcinodoides* as the senior synonym, but *whymperi* has priority and hence is senior synonym of *sulcinodoides* and its junior synonym *frigida.*]

***winterae.*** *Sifolinia winterae* Kutter, 1973c: 263, figs. 1, 2, 8, 9, 13–15 (q.m.) SWITZERLAND. Combination in *Myrmica:* Bolton, 1988a: 4.

***yoshiokai.*** *Myrmica rubra* subsp. *yoshiokai* Weber, 1947: 451 (w.) JAPAN. Raised to species: Collingwood, 1981: 26.

***zojae.*** *Myrmica zojae* Radchenko, 1994e: 210, figs. 13–22 (w.q.m.) RUSSIA.

**MYRMICARIA** (Afrotropical, Oriental, Indo-Australian)

**#abyssinica.** *Myrmicaria distincta* st. *abyssinica* Santschi, 1925c: 163, fig. 13 (w.) ETHIOPIA.

**#adpressipilosa.** *Myrmicaria (Heptacondylus) arachnoidea* var. *adpressipilosa* Santschi, 1928h: 130 (w.) JAVA.

**#angustior.** *Myrmicaria rustica* st. *angustior* Santschi, 1925c: 158 (w.) TANZANIA.

*angustithorax. Myrmicaria striata* r. *pilosa* var. *angustithorax* Arnold, 1926: 240 (w.) ZIMBABWE. Unavailable name.

**anomala.** *Myrmicaria anomala* Arnold, 1960a: 81, fig. 2 (w.) SOUTH AFRICA.

**arachnoides.** *Heptacondylus arachnoides* Smith, F. 1857a: 72 (w.q.) BORNEO. Mayr, 1867a: 114 (m.). Combination in *Myrmicaria:* Mayr, 1867a: 113. Senior synonym of *longipes:* Mayr, 1863: 423. Current subspecies: nominal plus *adpressipilosa, latinoda, lutea, luteiventris.* See also: Karavaiev, 1935a: 77.

**arnoldi.** *Myrmicaria arnoldi* Santschi, 1925c: 159 (w.q.m.) ZIMBABWE.

**basutorum.** *Myrmicaria basutorum* Arnold, 1960b: 452, fig. 1 (w.) LESOTHO.

**baumi.** *Myrmicaria baumi* Forel, 1901d: 307 (w.) ANGOLA. [Also described as new by Forel, 1903e: 562.] Forel, 1910f: 5 (m.). Current subspecies: nominal plus *occidentalis.*

**birmana.** *Myrmicaria arachnoides* r. *birmana* Forel, 1902c: 243 (w.) BURMA. Raised to species: Bingham, 1903: 119.

**brunnea.** *Myrmicaria brunnea* Saunders, W.W. 1842: 57, pl. 5, fig. 2 (m.) INDIA. Bingham, 1903: 118 (w.q.); Imai, Baroni Urbani, *et al.* 1984: 6 (k.). See also: Forel, 1903a: 708; Karavaiev, 1935a: 80. Current subspecies: nominal plus *flava, subcarinata.*

*brunnea. Myrmicaria nitida* var. *brunnea* Santschi, 1915c: 254 (w.) IVORY COAST. [Junior primary homonym of *brunnea* Saunders, above.] Replacement name: *fumata* Santschi, 1916d: 242.

**#buttgenbachi.** *Myrmicaria striata* r. *buttgenbachi* Forel, 1913b: 337 (w.) ZAIRE.

**carinata.** *Heptacondylus carinatus* Smith, F. 1857a: 73 (w.) BORNEO. Mayr, 1867a: 112 (q.); Wheeler, W.M. 1919e: 69 (m.). Combination in *Myrmicaria:* Mayr, 1867a: 111. Race of *fodica:* Emery, 1893e: 219. Revived status as species: Dalla Torre, 1893: 155; Wheeler, W.M. 1919e: 69. Stirps of *brunnea:* Santschi, 1925c: 176 (in key). Revived status as species: Stärcke: 1938: xxxv. Senior synonym of *dromedarius:* Emery, 1893e: 219; Wheeler, W.M. 1919e: 69. Current subspecies: nominal plus *gagatina, jacobsoni.*

**castanea.** *Myrmicaria castanea* Crawley, 1924: 394 (w.) SUMATRA.

**#congolensis.** *Myrmicaria eumenoides* var. *congolensis* Forel, 1909b: 59 (w.) ZAIRE. Forel, 1913h: 354 (m.); Santschi, 1925c: 150 (q.). Stirps/subspecies of *eumenoides:* Santschi, 1914b: 115; Santschi, 1914d: 342; of *natalensis:* Emery, 1922e: 123; of *opaciventris:* Santschi, 1925c: 149. Senior synonym of *nitida:* Forel, 1910e: 421.

**#consanguinea.** *Myrmicaria fusca* var. *consanguinea* Santschi, 1925c: 166 (w.q.) KENYA. [First available use of *Myrmicaria eumenoides* st. *congolensis* var. *consanguinea* Santschi, 1914b: 115; unavailable name.]

*crinita. Physatta crinita* Smith, F. 1858b: 171 (q.) INDIA. Junior synonym of *fodica:* Emery, 1893e: 219.

**#crucheti.** *Myrmicaria opaciventris* var. *crucheti* Santschi, 1925c: 151 (w.q.) ANGOLA. [First available use of *Myrmicaria eumenoides* st. *congolensis* var. *crucheti* Santschi, 1913c: 311; unavailable name.] [Also described as new by Santschi, 1914b: 116; unavailable name.]

**distincta.** *Myrmicaria distincta* Santschi, 1925c: 161 (w.) KENYA. Current subspecies: nominal plus *abyssinica, vorax.*

*dromedarius. Physatta dromedarius* Smith, F. 1857a: 78, pl. 1, figs. 15–19 (q.) BORNEO. Combination in *Heptacondylus:* Mayr, 1862: 757; in *Myrmicaria:* Mayr, 1867a: 112. Junior synonym of *subcarinata:* Mayr, 1867a: 112; of *carinata:* Emery, 1893e: 219. Revived from synonymy as subspecies of *subcarinata:* Forel, 1913k: 74. Junior synonym of *carinata:* Wheeler, W.M. 1919e: 69.

**#eumenoides.** *Heptacondylus eumenoides* Gerstäcker, 1859: 263 (q.) MOZAMBIQUE. [Also described as new by Gerstäcker, 1862: 514.] Forel, 1910e: 446 (w.); Santschi, 1914e: 34 (m.). Subspecies of *fodica:* Emery, 1893e: 219. Revived from synonymy: Dalla Torre, 1893: 155; Forel, 1894b: 83; Arnold, 1916: 263; Wheeler, W.M. 1922a: 812. Junior synonym of *natalensis:* Emery, 1922e: 123. Revived from synonymy as variety of *natalensis:* Santschi, 1925c: 145.

**exigua.** *Myrmicaria exigua* André, 1890: 320 (w.m.) SIERRA LEONE. Current subspecies: nominal plus *gracilis, kisangani, obscura, pulla, rufiventris, simplex.*

**faurei.** *Myrmicaria faurei* Arnold, 1947: 134, fig. 3 (w.) SOUTH AFRICA.

**#flava.** *Myrmicaria brunnea* var. *flava* Forel, 1913k: 74 (m.) WEST MALAYSIA.

**fodica.** *Myrmica fodicus* Jerdon, 1851: 115 (w.q.) INDIA. Combination in *Myrmicaria:* Emery, 1892b: 166, and spelling emended to *fodiens* (maintained as such by subsequent authors): unjustified emendation. Junior synonym of *carinata:* Emery, 1892b: 166. Revived from synonymy, and senior synonym of *crinita:* Emery, 1893e: 219. Junior synonym of *brunnea:* Forel, 1903a: 708; Bingham, 1903: 118. Revived from synonymy as variety of *brunnea:* Santschi, 1925c: 176. Revived status as species: Stitz, 1925: 118.

**foreli.** *Myrmicaria foreli* Santschi, 1925c: 163 (w.) SOUTH AFRICA. Current subspecies: nominal plus *pallida.*

**fumata.** *Myrmicaria nitida* var. *fumata* Santschi, 1916d: 242. Replacement name for *brunnea* Santschi, 1915c: 254. [Junior primary homonym of *brunnea* Saunders, W.W. 1841: 57.] Raised to species: Santschi, 1925c: 160. Current subspecies: nominal plus *linearis.*

**fusca.** *Myrmicaria eumenoides* var. *fusca* Stitz, 1911b: 385 (w.) ZAIRE. Variety of *natalensis:* Emery, 1922e: 123. Raised to species: Santschi, 1925c: 166. Current subspecies:nominal plus *consanguinea, nigerrima.*

**#gagatina.** *Myrmicaria carinata* subsp. *gagatina* Wheeler, W.M. 1919e: 69 (w.) BORNEO.

*gibbosa. Physatta gibbosa* Smith, F. 1858b: 172 (q.) JAVA. Junior synonym of *subcarinata:* Emery, 1893e: 219.

**#gracilis.** *Myrmicaria gracilis* Stitz, 1910: 133 (w.) CAMEROUN. Variety of *exigua:* Santschi, 1920c: 120; Santschi, 1925c: 169.

**#insularis.** *Myrmicaria striata* st. *insularis* Santschi, 1920i: 2 (w.m.) SAO THOME I.

**irregularis.** *Myrmicaria irregularis* Santschi, 1925c: 147, figs. 2, 15, 20 (w.q.m.) ZAIRE.

**#jacobsoni.** *Myrmicaria carinata* var. *jacobsoni* Stärcke, 1938: xxxv (w.q.) SUMATRA.

**#kisangani.** *Myrmicaria exigua* subsp. *kisangani* Wheeler, W.M. 1922a: 148 (w.) ZAIRE. See also: Santschi, 1925c: 169.

*laevifrons. Myrmicaria fusca* st. *nigerrima* var. *laevifrons* Santschi, 1925c: 167 (w.) MOZAMBIQUE. Unavailable name.

**laevior.** *Myrmicaria laevior* Forel, 1910e: 445 (w.) SOUTH AFRICA.

**#latinoda.** *Myrmicaria arachnoidea* var. *latinoda* Stitz, 1938: 101, fig. 2 (w.q.m.) BORNEO.

**#linearis.** *Myrmicaria fumata* st. *linearis* Santschi, 1925c: 161, fig. 21 (w.) ZAIRE.

*longipes. Heptacondylus longipes* Smith, F. 1858b: 142 (w.) BORNEO. Junior synonym of *arachnoides:* Mayr, 1863: 423.

**#lutea.** *Myrmicaria arachnoides* var. *lutea* Emery, 1900d: 692 (w.) JAVA. Viehmeyer, 1922: 207 (q.m.). See also: Santschi, 1928h: 130; Karavaiev, 1935a: 80.

**#luteiventris.** *Myrmicaria arachnoides* var. *luteiventris* Emery, 1900d: 692 (w.) SUMATRA. Viehmeyer, 1922: 207 (q.).

**melanogaster.** *Myrmicaria melanogaster* Emery, 1900d: 692 (footnote) (w.) BORNEO. Forel, 1911a: 23 (q.m.). Subspecies of *arachnoides:* Forel, 1911a: 23; Wheeler, W.M. 1919e: 71. Revived status as species: Emery, 1922e: 122.

**#mesonotalis.** *Myrmicaria opaciventris* var. *mesonotalis* Santschi, 1925c: 151 (w.q.m.) ZAIRE.

**natalensis.** *Physatta natalensis* Smith, F. 1858b: 172 (q.) SOUTH AFRICA. Santschi, 1925c: 143 (q.m.). Combination in *Myrmicaria:* Mayr, 1866b: 905. Junior synonym of *eumenoides:* Mayr, 1866b: 905. Revived from synonymy, and senior synonym of *sulcatus:* Emery, 1922e: 123. Current subspecies: nominal plus *eumenoides, navicula, nigriventris, obscuriceps, taeniata, verticalis.* [*M. natalensis* referred by error to *Myrmica* by Santschi, 1925c: 143.]

**#navicula.** *Myrmicaria natalensis* var. *navicula* Santschi, 1925c: 146 (w.) TANZANIA.

**#nigerrima.** *Myrmicaria nigerrima* Arnold, 1916: 270, pl. 7, fig. 92 (w.) ZIMBABWE. Stirps of *fusca:* Santschi, 1925c: 167.

**nigra.** *Heptacondylus niger* Mayr, 1862: 755 (w.) SOUTH AFRICA. Arnold, 1916: 270 (m.). Combination in *Myrmicaria:* Dalla Torre, 1893: 155. See also: Mayr, 1865: 110.

**#nigriventris.** *Myrmicaria natalensis* var. *nigriventris* Santschi, 1930b: 61 (w.) ANGOLA.

*nitida. Myrmicaria nitida* Stitz, 1910: 133 (w.) CAMEROUN. Junior synonym of *congolensis:* Forel, 1910e: 421.

**#obscura.** *Myrmicaria exigua* var. *obscura* Santschi, 1920c: 120 (w.) ZAIRE.

**#obscuriceps.** *Myrmicaria natalensis* var. *obscuriceps* Santschi, 1937d: 228 (w.) SOUTH AFRICA. Stirps of *natalensis:* Santschi, 1939b: 238.

**#obscuripes.** *Myrmicaria opaciventris* var. *obscuripes* Santschi, 1937a: 57 (w.) ZAIRE.
**#occidentalis.** *Myrmicaria eumenoides* st. *occidentalis* Santschi, 1920i: 3 (w.) IVORY COAST. Raised to species: Santschi, 1925c: 156. Stirps of *baumi:* Santschi, 1933b: 105.
**opaciventris.** *Myrmicaria opaciventris* Emery, 1893e: 221, pl. 8, figs. 15, 16 (w.q.m.) ANGOLA. Wheeler, G.C. & Wheeler, J. 1953d: 187 (l.). Variety/subspecies of *eumenoides:* Forel, 1911f: 276; Stitz, 1916: 393; Wheeler, W.M. 1922a: 148; of *natalensis:* Emery, 1922e: 123. Revived status as species: Santschi, 1925c: 148. Current subspecies: nominal plus *congolensis, crucheti, mesonotalis, obscuripes.*
**#pallida.** *Myrmicaria foreli* var. *pallida* Arnold, in Santschi, 1925c: 165 (w.m.) SOUTH AFRICA.
**#pilosa.** *Myrmicaria striata* r. *pilosa* Arnold, 1926: 239, fig. 67 (w.) MOZAMBIQUE.
**#pulla.** *Myrmicaria exigua* var. *pulla* Santschi, 1920c: 119 (q.) ZAIRE.
**reichenspergeri.** *Myrmicaria reichenspergeri* Santschi, 1925c: 159, fig. 22 (w.) TANZANIA.
**rhodesiae.** *Myrmicaria faurei* var. *rhodesiae* Arnold, 1958: 119 (w.) ZIMBABWE. Raised to species: Arnold, 1960b: 453.
**#rufiventris.** *Myrmicaria exigua* var. *rufiventris* Forel, 1915c: 345 (w.) ZAIRE.
**rugosa.** *Heptacondylus rugosus* Smith, F, 1860b: 110 (w.) INDONESIA (Batjan I.). Combination in *Myrmicaria:* Smith, F. 1865: 73.
**rustica.** *Myrmicaria rustica* Santschi, 1925c: 157, fig. 5 (w.q.) TANZANIA. Current subspecies: nominal plus *angustior.*
**salambo.** *Myrmicaria salambo* Wheeler, W.M. 1922a: 147, fig. 37 (w.) ZAIRE.
**#simplex.** *Myrmicaria exigua* var. *simplex* Stitz, in Santschi, 1925c: 170 (footnote) (w.) CAMEROUN.
**striata.** *Myrmicaria striata* Stitz, 1911b: 382, fig. 5 (w.) TANZANIA. Arnold, 1916: 268 (q.m.). Current subspecies: nominal plus *buttgenbachi, insularis, pilosa.*
**striatula.** *Myrmicaria striatula* Santschi, 1925c: 152 (w., not m. as stated) TANZANIA. [*Myrmicaria striatula* Santschi, 1914b: 116. *Nomen nudum,* attributed to Stitz.]
**#subcarinata.** *Heptacondylus subcarinatus* Smith, F. 1857a: 73 (w.) BORNEO. Mayr, 1867a: 112 (q.); Forel, 1913k: 73 (m.). Combination in *Myrmicaria:* Mayr, 1867a: 112. Status as species: Forel, 1909d: 226; Forel, 1913k: 73; Wheeler, W.M. 1919e: 70. Subspecies of *fodica:* Emery, 1893e: 219; Stitz, 1925: 118; of *brunnea:* Bingham, 1903: 119; Emery, 1922e: 122; Santschi, 1925c: 176 (in key); Menozzi, 1932d: 6; Karavaiev, 1935a: 82. Senior synonym of *gibbosa:* Emery, 1893e: 219.
*sulcatus. Heptacondylus sulcatus* Mayr, 1862: 756 (w.) SOUTH AFRICA (not Cuba as stated; see Mayr, 1893: 195). Junior synonym of *eumenoides:* Dalla Torre, 1893: 155; of *natalensis:* Emery, 1922e: 123.
**#taeniata.** *Myrmicaria natalensis* var. *taeniata* Santschi, 1930d: 269 (w.) ZAIRE.
**tigreensis.** *Myrmica tigreensis* Guérin-Méneville, 1849: 351, pl. 7, fig. 7 (m.) ETHIOPIA. Combination in *Myrmicaria:* Smith, F. 1858b: 141.
**#verticalis.** *Myrmicaria striata* var. *verticalis* Santschi, 1920i: 3 (w.) ZIMBABWE. Stirps of *natalensis:* Santschi, 1925c: 146.
**vidua.** *Myrmicaria vidua* Smith, F. 1858b: 141 (m.) JAVA.
**#vorax.** *Myrmicaria distincta* var. *vorax* Santschi, 1933b: 105 (w.) UGANDA.

*MYRMICITES:* unavailable name.

***MYRMICIUM** (Great Britain)
**boreale* Heer, 1870; see under ***PROMYRMICIUM.**
***heerii.** **Myrmicium heerii* Westwood, 1854: 396, pl. 18, fig. 21 (wing) GREAT BRITAIN.

**MYRMICIUM* Heer, 1870: see under ***PROMYRMICIUM.**

**MYRMICOCRYPTA** (Neotropical)
*arenicola* Forel, 1912; see under ***MYCETOPHYLAX.***
*argentina* Santschi, 1916; see under ***MYCETOPHYLAX.***
**boliviana.** *Myrmicocrypta boliviana* Weber, 1938b: 161 (w.) BOLIVIA.
*brittoni* Wheeler, W.M. 1907; see under ***MYCETOPHYLAX.***
**bruchi.** *Myrmicocrypta bruchi* Santschi, 1936d: 409, figs. 13, 14 (w.q.) ARGENTINA.
**buenzlii.** *Myrmicocrypta buenzlii* Borgmeier, 1934: 104 (w.) SURINAM. Junior synonym of *squamosa:* Weber, 1958d: 262. Revived from synonymy: Kempf, 1961b: 517.
**collaris.** *Myrmicocrypta collaris* Emery, 1913: 252, fig. 2, 3 (q.) PERU. Senior synonym of *corniculata:* Weber, 1958d: 261.
*corniculata. Myrmicocrypta corniculata* Emery, 1913: 253, fig. 2 (q.) PERU. Junior synonym of *collaris:* Weber, 1958d: 261.
**#cornuta.** *Myrmicocrypta dilacerata* r. *cornuta* Forel, 1899c: 38 (m.) MEXICO.
*cristulata* Santschi, 1922; see under ***MYCETOPHYLAX.***
*cucumis* Mann, 1922; see under ***TRACHYMYRMEX.***
**dilacerata.** *Glyptomyrmex dilaceratum* Forel, 1885a: 365 (m.) MEXICO. Combination in *Myrmicocrypta:* Forel, 1899c: 38. Current subspecies: nominal plus *cornuta.*
**ednaella.** *Myrmicocrypta ednaella* Mann, 1922: 46, fig. 20 (w.q.) HONDURAS.
**elisabethae.** *Myrmicocrypta elisabethae* Weber, 1937: 383 (q.) GUIANA.
*emeryi* Forel, 1907; see under ***MYCETOPHYLAX.***
**foreli.** *Myrmicocrypta foreli* Mann, 1916: 457, pl. 4, figs. 32–34 (w.q.) BRAZIL. Current subspecies: nominal plus *surianensis.*
*fortis* Forel, 1912; see under ***MYCETOPHYLAX.***
*gallardoi* Santschi, 1922; see under ***MYCETOPHYLAX.***
**godmani.** *Myrmicocrypta godmani* Forel, 1899c: 39 (m.) NICARAGUA.
**guianensis.** *Myrmicocrypta guianensis* Weber, 1937: 385 (w.q.) GUIANA.
**infuscata.** *Myrmicocrypta infuscata* Weber, 1946b: 131 (w.) GUIANA.
**longinoda.** *Myrmicocrypta longinoda* Weber, 1938b: 158 (w.) BRAZIL.
**microphthalma.** *Myrmicocrypta microphthalma* Borgmeier, 1948a: 202, figs. 27, 28 (w.q.) PERU.
**occipitalis.** *Myrmicocrypta occipitalis* Weber, 1938b: 159 (w.m.) BOLIVIA.
**ogloblini.** *Myrmicocrypta ogloblini* Santschi, 1936d: 411, figs. 17–20 (w.) ARGENTINA.
**#peruviana.** *Myrmicocrypta triangulata* var. *peruviana* Emery, 1913: 254, fig. 5 (m.) PERU.
**rudiscapa.** *Myrmicocrypta rudiscapus* Emery, 1913: 253, figs. 2, 3, 4 (q.) BOLIVIA.
*simplex* Santschi, 1922; see under ***MYCETOPHYLAX.***
**spinosa.** *Myrmicocrypta spinosa* Weber, 1937: 382 (w.) GUIANA. Wheeler, G.C. 1949: 667 (l.).
**squamosa.** *Myrmicocrypta squamosa* Smith, F. 1860c: 74, pl. 4, figs. 14–17 (q.) BRAZIL. Forel, 1911c: 295 (w.). Current subspecies: nominal plus *uncinata.*
**subnitida.** *Myrmicocrypta subnitida* Forel, 1899c: 39 (m.) PANAMA.
**#surianensis.** *Myrmicocrypta foreli* subsp. *surianensis* Wheeler, W.M. 1925a: 45 (w.) BOLIVIA.
**triangulata.** *Myrmicocrypta triangulata* Forel, 1912e: 190 (m.) GUATEMALA. Current subspecies: nominal plus *peruviana.*
**tuberculata.** *Myrmicocrypta tuberculata* Weber, 1938b: 157, fig. 9 (w.) BOLIVIA.
**#uncinata.** *Apterostigma uncinatum* Mayr, 1887: 554 (w.) BRAZIL. Emery, 1890b: 70 (q.m.). Combination in *Glyptomyrmex:* Emery, 1890b: 70; in *Myrmicocrypta:* Forel, 1911c: 296. Variety of *squamosa:* Forel, 1911c: 296.
**unidentata.** *Myrmicocrypta unidentata* Weber, 1937: 384 (w.) GUIANA.
**urichi.** *Myrmicocrypta urichi* Weber, 1937: 379, figs. 2, 3 (w.q.m.) TRINIDAD. Wheeler, G.C. 1949: 666 (l.).
**weyrauchi.** *Myrmicocrypta weyrauchi* Borgmeier, 1948a: 204, figs. 29, 30 (w.q.) PERU.

**#MYRMISOLEPIS:** see under **CAMPONOTUS.**

**#MYRMOBRACHYS:** see under **CAMPONOTUS.**

*MYRMOCAMELUS:* see under **CAMPONOTUS.**

**#MYRMOCLADOECUS:** see under **CAMPONOTUS.**

**#MYRMODIRACHIS:** see under **CAMPONOTUS.**

*MYRMOGIGAS:* see under **CAMPONOTUS.**

**#MYRMOGONIA:** see under **CAMPONOTUS.**

*MYRMOLOPHUS:* see under **CAMPONOTUS.**

**#MYRMOMALIS:** see under **CAMPONOTUS.**

**#MYRMONESITES:** see under **CAMPONOTUS.**

**#MYRMOPALPELLA:** see under **CAMPONOTUS.**

**#MYRMOPELTA:** see under **CAMPONOTUS.**

**#MYRMOPHYMA:** see under **CAMPONOTUS.**

**#MYRMOPIROMIS:** see under **CAMPONOTUS.**

**#MYRMOPLATYPUS:** see under **CAMPONOTUS.**

**#MYRMOPLATYS:** see under **CAMPONOTUS.**

**#MYRMOPSAMMA:** see under **CAMPONOTUS.**

**#MYRMOPYTIA:** see under **CAMPONOTUS.**

**#MYRMORHACHIS:** see under **PHASMOMYRMEX.**

**#MYRMOSAGA:** see under **CAMPONOTUS.**

**#MYRMOSAULUS:** see under **CAMPONOTUS.**

**#MYRMOSERICUS:** see under **CAMPONOTUS.**

**#MYRMOSPHINCTA:** see under **CAMPONOTUS.**

**#MYRMOSTENUS:** see under **CAMPONOTUS.**

**#MYRMOTARSUS:** see under **CAMPONOTUS.**

**#MYRMOTEMNUS:** see under **CAMPONOTUS.**

**MYRMOTERAS** (Oriental, Indo-Australian)

***arcoelinae.*** *Myrmoteras* (*Myagroteras*) *arcoelinae* Agosti, 1992: 413, figs. 9, 15, 16 (w.) BORNEO.

***bakeri.*** *Myrmoteras bakeri* Wheeler, W.M. 1919e: 145 (q.m.) BORNEO. Moffett, 1985b: 32 (w.). Combination in *M.* (*Myagroteras*): Moffett, 1985b: 32. See also: Agosti, 1992: 415.

***barbouri.*** *Myrmoteras barbouri* Creighton, 1930a: 185, fig. 2; pl. 11, fig. 6 (w.) JAVA. Moffett, 1985b: 21 (q.). Combination in *M.* (*Myrmoteras*), and senior synonym of *kemneri:* Moffett, 1985b: 20. See also: Moffett, 1986b: 85.

***baslerorum.*** *Myrmoteras* (*Myrmoteras*) *baslerorum* Agosti, 1992: 412, figs. 3, 34, 41, 42 (w.) SUMATRA.

***binghamii.*** *Myrmoteras binghamii* Forel, 1893f: 608 (w.) BURMA. [Also described as new by Forel, 1894c: 419.] See also: Moffett, 1985b: 24.

***brachygnathum.*** *Myrmoteras* (*Myrmoteras*) *brachygnathum* Moffett, 1985b: 26, figs. 19, 22 (w.q.) INDIA.

***brigitteae.*** *Myrmoteras* (*Myagroteras*) *brigitteae* Agosti, 1992: 415, figs. 21, 27, 28 (w.q.) INDONESIA (Lombok I.)

***ceylonica.*** *Myrmoteras ceylonica* Gregg, 1957: 41, fig. 1 (w.) SRI LANKA. See also: Moffett, 1985b: 27.

***chondrogastrum.*** *Myrmoteras* (*Myagroteras*) *chondrogastrum* Moffett, 1985b: 39, figs. 31, 34 (w.q.) BORNEO.

***danieli.*** *Myrmoteras* (*Myagroteras*) *danieli* Agosti, 1992: 422, figs. 31, 35, 36 (w.q.m.) WEST MALAYSIA.

***diastematum.*** *Myrmoteras* (*Myagroteras*) *diastematum* Moffett, 1985b: 36, figs. 26, 29 (w.q.) BORNEO.

***donisthorpei.*** *Myrmoteras donisthorpei* Wheeler, W.M. 1916b: 14, fig. 3 (q.) BORNEO. Moffett, 1985b: 42 (w.m.). Combination in *M.* (*Myagroteras*): Moffett, 1985b: 42.

***elfeorum.*** *Myrmoteras* (*Myagroteras*) *elfeorum* Agosti, 1992: 423, figs. 7, 11, 12 (w.) SULAWESI.

***estrudae.*** *Myrmoteras* (*Myagroteras*) *estrudae* Agosti, 1992: 424, figs. 20, 25, 26 (w.q.) SUMATRA.

***indicum.*** *Myrmoteras* (*Myagroteras*) *indicum* Moffett, 1985b: 37, figs. 4, 7–10, 27, 30 (w.q.m.) INDIA.

***insulcatum.*** *Myrmoteras* (*Myagroteras*) *insulcatum* Moffett, 1985b: 45, figs. 33, 36, 37 (q.) PHILIPPINES.

***iriodum.*** *Myrmoteras* (*Myrmoteras*) *iriodum* Moffett, 1985b: 24, figs. 13, 17 (w.q.) BORNEO.

***ivani.*** *Myrmoteras* (*Myagroteras*) *ivani* Agosti, 1992: 424, figs. 10, 17, 18 (w.q.) SULAWESI.

***jacquelinae.*** *Myrmoteras* (*Myagroteras*) *jacquelinae* Agosti, 1992: 425, figs. 1, 2, 4–6 (w.) SULAWESI.

***karnyi.*** *Myrmoteras karnyi* Gregg, 1954b: 23, fig. 1 (w.) INDONESIA (Mentawei I.) Combination in *M.* (*Myagroteras*): Moffett, 1985b: 43.

*kemneri. Myrmoteras kemneri* Wheeler, W.M. 1933e: 73, fig. 1 (w.) JAVA. Junior synonym of *barbouri:* Moffett, 1985b: 20.

***marianneae.*** *Myrmoteras* (*Myagroteras*) *marianneae* Agosti, 1992: 425, figs. 19, 23, 24 (w.) SULAWESI.

***maudeae.*** *Myrmoteras* (*Myagroteras*) *maudeae* Agosti, 1992: 426, figs. 32, 37, 38 (w.) SULAWESI.

***mjoebergi.*** *Myrmoteras mjoebergi* Wheeler, W.M., in Creighton, 1930a: 188, fig. 2; pl. 11, fig. 1 (w.) BORNEO. Combination in *M.* (*Myrmoteras*): Moffett, 1985b: 25.

***morowali.*** *Myrmoteras* (*Myagroteras*) *morowali* Moffett, 1985b: 48, figs. 38, 42 (w.q.) SULAWESI.

***nicoletteae.*** *Myrmoteras* (*Myagroteras*) *nicoletteae* Agosti, 1992: 426, figs. 22, 29, 30 (w.q.) SULAWESI.

***scabrum.*** *Myrmoteras* (*Myrmoteras*) *scabrum* Moffett, 1985b: 30, figs. 21, 24 (w.) INDIA.

***susanneae.*** *Myrmoteras* (*Myagroteras*) *susanneae* Agosti, 1992: 427, figs. 33, 39, 40 (w.q.) SULAWESI.

***tonboli.*** *Myrmoteras* (*Myagroteras*) *tonboli* Agosti, 1992: 427, figs. 8, 13, 14 (w.) BORNEO.

***toro.*** *Myrmoteras* (*Myagroteras*) *toro* Moffett, 1985b: 49, figs. 1, 39, 43, 46, 47 (w.q.) SULAWESI. See also: Moffett, 1986b: 85.

***williamsi.*** *Myrmoteras williamsi* Wheeler, W.M. 1919e: 146 (q.m.) PHILIPPINES. Creighton, 1930a: 189 (w.). Combination in *M.* (*Myagroteras*): Moffett, 1985b: 50.

***wolasi.*** *Myrmoteras* (*Myagroteras*) *wolasi* Moffett, 1985b: 52, figs. 41, 45 (w.) SULAWESI.

**#MYRMOTHRINAX:** see under **POLYRHACHIS.**

**#MYRMOTHRIX:** see under **CAMPONOTUS.**

**#MYRMOTREMA:** see under **CAMPONOTUS.**

*MYRMOTURBA:* see under **CAMPONOTUS.**

*MYRMOXENUS:* see under **EPIMYRMA.**

**#MYRMOXYGENYS:** see under **CAMPONOTUS.**

*MYRMUS:* see under **STRONGYLOGNATHUS.**

*MYRTOTERAS:* see under **ODONTOMACHUS.**

**MYSTRIUM** (Malagasy, West Africa, Oriental, Indo-Australian)

***camillae.*** *Mystrium camillae* Emery, 1889b: 491, pl. 10, figs. 1–3 (w.q.) BURMA. Wheeler, G.C. & Wheeler, J. 1952a: 113 (l.); Tjan, Imai, *et al.* 1986: 57 (k.). Senior synonym of *javana:* Brown, 1960a: 170. See also: Menozzi, 1929d: 535.

***fallax.*** *Mystrium fallax* Forel, 1897c: 192 (w.m.) MADAGASCAR. Variety of *voeltzkowi:* Emery, 1911d: 23. Revived status as species: Menozzi, 1929d: 532; Brown, 1960a: 170.

*javana. Mystrium camillae* subsp. *javana* Karavaiev, 1925a: 73, fig. 1 (w.) JAVA. Junior synonym of *camillae:* Brown, 1960a: 170.

***mysticum.*** *Mystrium mysticum* Roger, 1862a: 247, pl. 1, fig. 15 (q.) MADAGASCAR. Forel, 1891b: 117 (w.m.); Wheeler, G.C. & Wheeler, J. 1980: 528 (l.). See also: Forel, 1899b: 304; Emery, 1899f: 265; Menozzi, 1929d: 522; Wheeler, G.C. & Wheeler, J. 1986c: 85.

***oberthueri.*** *Mystrium oberthueri* Forel, 1897c: 192 (w.) MADAGASCAR. [*Mystrium oberthueri* Wasmann, 1897: 250. *Nomen nudum,* attributed to Forel.] Emery, 1899f: 267 (m.). See also: Menozzi, 1929d: 530.

***rogeri.*** *Mystrium rogeri* Forel, 1899b: 304 (w.) MADAGASCAR. Menozzi, 1929d: 527 (q.m.).

***silvestrii.*** *Mystrium silvestrii* Santschi, 1914d: 310, fig. 1 (w.q.) CAMEROUN. See also: Menozzi, 1929d: 534.

***stadelmanni.*** *Mystrium stadelmanni* Forel, 1895c: 251 (w.) MADAGASCAR. See also: Menozzi, 1929d: 524.

***voeltzkowi.*** *Mystrium voeltzkowi* Forel, 1897c: 189, fig. 1 (w.m.) MADAGASCAR. See also: Menozzi, 1929d: 528.

*NEAPHOMUS:* see under **MYRMELACHISTA.**

**NEIVAMYRMEX** (New World Tropics and Subtropics)

*abstinens. Eciton* (*Acamatus*) *abstinens* Ihering, 1912: 232. Unnecessary replacement name for, and hence junior synonym of, *raptans:* Forel, 1913l: 213.

***adnepos.*** *Eciton* (*Acamatus*) *adnepos* Wheeler, W.M. 1922c: 2 (w.) TRINIDAD. Combination in *Neivamyrmex:* Borgmeier, 1953: 18. Senior synonym of *marginatum:* Borgmeier, 1953: 5. See also: Borgmeier, 1955b: 562.

***agilis.*** *Neivamyrmex agilis* Borgmeier, 1953: 45 (w.) MEXICO. Material of the *nomen nudum cocula* referred here: Borgmeier, 1955b: 406.

***alfaroi.*** *Eciton alfaroi* Emery, 1890b: 39 (w.) COSTA RICA. [Also described as new by Emery, 1894k: 45.] Borgmeier, 1949: 202 (q.). Combination in *E.* (*Acamatus*): Emery, 1894c: 182 (in text); in *E.* (*Neivamyrmex*): Borgmeier, 1949: 201; in *Neivamyrmex:* Borgmeier, 1953: 6. See also: Borgmeier, 1955b: 533.

*amplipennis. Labidus amplipennis* Smith, F. 1859b: 6, pl. 2, fig. 14 (m.) COLOMBIA. Combination in *Eciton:* Mayr, 1886a: 33; in *E.* (*Acamatus*): Emery, 1900a: 525; in *Neivamyrmex:* Borgmeier, 1953: 16. Junior synonym of *halidaii:* Borgmeier, 1955b: 437.

***andrei.*** *Eciton andrei* Emery, 1901d: 53, fig. 6 (m.) MEXICO. Combination in *Neivamyrmex,* and senior synonym of *oslari:* Borgmeier, 1953: 19. See also: Borgmeier, 1955b: 451.

***angulimandibulatus.*** *Neivamyrmex angulimandibulatus* Watkins, 1974: 309, figs. 1–8 (m.) MEXICO.

***angustinodis.*** *Eciton angustinode* Emery, 1888e: 691 (w.) ARGENTINA. Borgmeier, 1930: 24 (q.). Combination in *E.* (*Acamatus*): Emery, 1894c: 183 (in key); in *Neivamyrmex:* Borgmeier, 1953: 6. See also: Bruch, 1934b: 132; Borgmeier, 1955b: 549.

*angustius. Eciton pilosum* var. *angustius* Forel, 1909a: 256 (w.) PARAGUAY. Combination in *Neivamyrmex:* Borgmeier, 1953: 8. Junior synonym of *pilosus:* Borgmeier, 1955b: 361.

***antillanus.*** *Eciton antillanum* Forel, 1897b: 299 (w.) ANTILLES IS. Combination in *E.* (*Acamatus*): Emery, 1910b: 24; in *E.* (*Neivamyrmex*): Borgmeier, 1948a: 197; in *Neivamyrmex:* Borgmeier, 1953: 8. See also: Borgmeier, 1955b: 418.

*arizonense. Eciton* (*Acamatus*) *arizonense* Wheeler, W.M. 1908e: 414, pl. 26, fig. 5 (m.) U.S.A. Combination in *E.* (*Neivamyrmex*): Smith, M.R. 1942c: 581; in *Neivamyrmex:* Borgmeier, 1953: 18. Subspecies of *swainsonii:* Borgmeier, 1953: 19. Junior synonym of *swainsonii:* Borgmeier, 1955b: 454.

***asper.*** *Neivamyrmex asper* Borgmeier, 1955b: 547 (w.) COSTA RICA.

*aztecum. Eciton aztecum* Forel, 1901h: 49 (m.) GUATEMALA. Variety of *mexicanus:* Emery, 1910b: 26; of

***NIEVAMYRMEX** (continued)*

*pilosus:* Wheeler, W.M. 1921d: 314. Junior synonym of *mexicanus:* Borgmeier, 1936: 60.

***balzani.*** *Eciton (Acamatus) balzani* Emery, 1894c: 182 (w.) BOLIVIA. Combination in *Neivamyrmex:* Borgmeier, 1953: 6. See also: Borgmeier, 1955b: 543.

***baylori.*** *Neivamyrmex baylori* Watkins, 1973: 430, figs. 1–8 (m.) U.S.A.

#***beebei.*** *Eciton (Acamatus) pilosum* var. *beebei* Wheeler, W.M. 1921d: 312, fig. 7 (w.m.) GUIANA. Combination in *Neivamyrmex:* Borgmeier, 1953: 19. Subspecies of *pilosus:* Borgmeier, 1953: 19; Borgmeier, 1955b: 373.

***bohlsi.*** *Eciton (Acamatus) bohlsi* Emery, 1896h: 626 (w.) PARAGUAY. Combination in *E. (Neivamyrmex):* Borgmeier, 1948b: 462; in *Neivamyrmex:* Borgmeier, 1953: 6. See also: Borgmeier, 1955b: 545.

***bruchi.*** *Eciton (Acamatus) bruchi* Forel, 1912g: 30 (w.) ARGENTINA. Combination in *Neivamyrmex:* Borgmeier, 1953: 8. See also: Borgmeier, 1955b: 428.

*bureni. Woitkowskia bureni* Enzmann, E.V. 1952: 447 (m.) PERU. Combination in *Neivamyrmex:* Borgmeier, 1955b: 460 (*species inquirendae*).

***californicus.*** *Eciton californicum* Mayr, 1870b: 969 (w.) U.S.A. Watkins, 1972: 364 (q.). Combination in *E. (Acamatus):* Emery, 1894c: 182 (in text); in *E. (Neivamyrmex):* Smith, M.R. 1942c: 560; in *Neivamyrmex:* Borgmeier, 1953: 8. Senior synonym of *obscura:* Borgmeier, 1955b: 517.

***carettei.*** *Eciton (Acamatus) carettei* Forel, 1913l: 208 (w.) ARGENTINA. Combination in *Neivamyrmex:* Borgmeier, 1953: 8. See also: Borgmeier, 1955b: 568.

***carinifrons.*** *Neivamyrmex carinifrons* Borgmeier, 1953: 24, figs. 3, 5, 8 (m.) BRAZIL. See also: Borgmeier, 1955b: 576.

***carolinensis.*** *Eciton (Acamatus) carolinensis* Emery, 1894c: 184 (diagnosis in key) (w.) U.S.A. Wheeler, W.M. 1921d: 314 (q.m.). Combination in *E. (Neivamyrmex):* Smith, M.R. 1942c: 564; in *Neivamyrmex:* Borgmeier, 1953: 6. See also: Borgmeier, 1955b: 507; Smith, D.R. 1979: 1330.

*castaneum. Eciton (Acamatus) opacithorax* var. *castaneum* Borgmeier, 1939: 416 (w.) COSTA RICA. Borgmeier, 1948a: 192 (q.m.). Combination in *E. (Neivamyrmex):* Borgmeier, 1948a: 191; in *Neivamyrmex:* Borgmeier, 1953: 5. Subspecies of *opacithorax:* Borgmeier, 1948a: 191. Junior synonym of *opacithorax:* Borgmeier, 1955b: 502.

***chamelensis.*** *Neivamyrmex chamelensis* Watkins, 1986: 361, figs. 1–11 (w.q.) MEXICO.

*clavicornis. Eciton clavicornis* Norton, 1868b: 46 (w.) MEXICO. Junior synonym of *pilosus:* Dalla Torre, 1893: 5; Emery, 1910b: 25; of *mexicanus:* Mayr, 1886b: 120; Borgmeier, 1955b: 361.

***clavifemur.*** *Neivamyrmex clavifemur* Borgmeier, 1953: 30, figs. 20, 27 (m.) BRAZIL. See also: Borgmeier, 1955b: 596.

***cloosae.*** *Eciton (Acamatus) cloosae* Forel, 1912c: 48 (m.) MEXICO. Combination in *Neivamyrmex:* Borgmeier, 1953: 8. See also: Borgmeier, 1955b: 637.

*cocula. Eciton (Acamatus) cocula* Mann, 1925a: 77 (w.) MEXICO. *Nomen nudum;* material referred to *agilis* by Borgmeier, 1955b: 406.

*coeca. Myrmica (Monomorium) coeca* Buckley, 1867: 339 (w.) U.S.A. [Unresolved junior primary homonym of *Myrmica caeca* Jerdon, 1851: 116.] Combination in *Eciton:* Emery, 1895c: 260; possibly synonymous with *opacithorax* or *schmitti:* Wheeler, W.M. 1902f: 27; unplaced taxon of Ecitoninae: Smith, D.R. 1979: 1332. Unidentifiable taxon; *incertae sedis* in *Neivamyrmex.*

*colombi. Eciton (Acamatus) colombi* Santschi, 1921g: 94 (m.) COLOMBIA. Combination in *Neivamyrmex:* Borgmeier, 1953: 13. Junior synonym of *halidaii:* Borgmeier, 1955b: 437.

*columbicum. Eciton (Acamatus) fiebrigi* r. *columbicum* Forel, 1912c: 46 (w.) COLOMBIA. Junior synonym of *commutatum:* Borgmeier, 1953: 9.

*commutatum. Eciton commutatum* Emery, 1900a: 184 (w.) COLOMBIA. Combination in *E. (Acamatus):* Emery, 1900a: 187; in *E. (Neivamyrmex):* Smith, M.R. 1942c: 568; in *Neivamyrmex:* Borgmeier, 1953: 6. Senior synonym of *columbicum:* Borgmeier, 1953: 9. Junior synonym of *spinolae:* Borgmeier, 1955b: 409.

***compressinodis.*** *Neivamyrmex compressinodis* Borgmeier, 1953: 50 (w.) COSTA RICA. See also: Borgmeier, 1955b: 558.

*connectens. Woitkowskia connectens* Enzmann, E.V. 1952: 443, figs. 4, 7, 9 (m.) BOLIVIA. Junior synonym of *walkerii:* Borgmeier, 1955b: 445.

***cornutus.*** *Neivamyrmex cornutus* Watkins, 1975a: 92, figs. 1–4 (w.) MEXICO.

***crassiscapus.*** *Neivamyrmex crassiscapus* Watkins, 1990: 348, figs. 1–8 (m.) MEXICO.

***cratensis.*** *Neivamyrmex cratensis* Borgmeier, 1953: 38, figs. 11, 19 (m.) BRAZIL. See also: Borgmeier, 1955b: 404.

*crenulatum. Eciton (Acamatus) legionis* subsp. *crenulatum* Mann, 1916: 422, pl. 1, fig. 1 (w.) BRAZIL. Junior synonym of *cristatus:* Borgmeier, 1953: 10.

***cristatus.*** *Eciton cristatum* André, 1889: 223 (w.) SOUTH AMERICA. Combination in *E. (Acamatus):* Emery, 1894c: 183 (in key); in *Neivamyrmex,* and senior synonym of *crenulatum:* Borgmeier, 1953: 10. See also: Borgmeier, 1955b: 342.

***curvinotus.*** *Neivamyrmex curvinotus* Watkins, 1994: 411, figs. 1–3 (w.) PERU.

***densepunctatus.*** *Eciton (Acamatus) densepunctatum* Borgmeier, 1933d: 162 (w.) BRAZIL. Combination in *Neivamyrmex:* Borgmeier, 1953: 5. See also: Borgmeier, 1955b: 486.

*desartsi. Eciton (Acamatus) melshaemeri* subsp. *desartsi* Forel, 1907e: 2 (m.) PARAGUAY. Junior synonym of *romandii:* Borgmeier, 1953: 9.

***detectus.*** *Neivamyrmex detectus* Borgmeier, 1953: 25, figs. 7, 15 (m.) BRAZIL. See also: Borgmeier, 1955b: 449.

***diabolus.*** *Eciton (Acamatus) diabolus* Forel, 1912c: 47 (m.) MEXICO. Combination in *Neivamyrmex:* Borgmeier, 1953: 8. See also: Borgmeier, 1955b: 526.

***diana.*** *Eciton (Acamatus) diana* Forel, 1912g: 31 (w.) BRAZIL. Borgmeier, 1933d: 161 (m.); Borgmeier, 1939: 410 (q.). Combination in *Neivamyrmex:* Borgmeier, 1953: 8. See also: Borgmeier, 1955b: 390.

***digitistipus.*** *Neivamyrmex digitistipus* Watkins, 1975b: 203, figs. 1–6 (m.) COSTA RICA.

#***distans.*** *Neivamyrmex klugii* subsp. *distans* Borgmeier, 1953: 32 (m.) COSTA RICA. See also: Borgmeier, 1955b: 599.

***diversinodis.*** *Eciton (Acamatus) diversinode* Borgmeier, 1933d: 163 (w.) BRAZIL. Combination in *E. (Neivamyrmex):* Borgmeier, 1948b: 462; in *Neivamyrmex:* Borgmeier, 1953: 5. Junior synonym of *bohlsi:* Borgmeier, 1948b: 462. Revived from synonymy: Borgmeier, 1953: 5; Borgmeier, 1955b: 467.

***dorbignii.*** *Labidus dorbignii* Shuckard, 1840b: 259 (m.) no locality given (loc. Bolivia, see Emery, 1900a: 187). Combination in *Eciton:* Dalla Torre, 1893: 2 (with spelling emended to *dorbignyi;* unjustified emendation); in *Eciton:* Forel, 1895b: 119; in *E. (Acamatus):* Emery, 1900a: 187; in *Neivamyrmex:* Borgmeier, 1953: 14. Senior synonym of *spegazzinii* (and its junior synonym *obscurum*): Santschi, 1921g: 92; Borgmeier, 1955b: 461.

*dubia. Eciton (Acamatus) wheeleri* subsp. *dubia* Creighton, 1932: 75, pl. 3, figs. 1–3 (w.q.) U.S.A. Junior synonym of *wheeleri:* Smith, M.R. 1942c: 561.

****ectopus.*** **Neivamyrmex ectopus* Wilson, 1985b: 11, figs. 1, 2 (w.) DOMINICAN AMBER.

***emersoni.*** *Eciton (Acamatus) angustinode* subsp. *emersoni* Wheeler, W.M. 1921d: 311 (w.) GUIANA. Combination in *Neivamyrmex,* and raised to species: Borgmeier, 1953: 18. Senior synonym of *haskinsi:* Borgmeier, 1955b: 555.

***emeryi.*** *Eciton (Acamatus) emeryi* Santschi, 1921g: 92 (footnote) (m.) BOLIVIA. Combination in *Neivamyrmex:* Borgmeier, 1953: 13. Senior synonym of *exagona:* Borgmeier, 1955b: 317.

***erichsonii.*** *Labidus erichsonii* Westwood, 1842: 77 (m.) BRAZIL. Combination in *Eciton:* Dalla Torre, 1893: 2; in *E. (Labidus):* Emery, 1900a: 187; in *Neivamyrmex:* Borgmeier, 1953: 17. See also: Borgmeier, 1955b: 618.

*exagona. Eciton (Acamatus) emeryi* var. *exagona* Santschi, 1923g: 1261 (m.) PERU. Junior synonym of *emeryi:* Borgmeier, 1955b: 317.

***falcifer.*** *Eciton falciferum* Emery, 1900a: 182, fig. 17 (m.) BOLIVIA. Combination in *E. (Acamatus):* Emery, 1900a: 187; in *Neivamyrmex:* Borgmeier, 1953: 357. Senior synonym of *mesognathum:* Borgmeier, 1955b: 357 (misspelled as *falciferus*).

***fallax.*** *Neivamyrmex fallax* Borgmeier, 1953: 48, figs. 31, 33 (w.) U.S.A. See also: Borgmeier, 1955b: 425.

*fiebrigi. Eciton (Acamatus) fiebrigi* Forel, 1909a: 255 (w.) PARAGUAY. Junior synonym of *spinolae:* Borgmeier, 1953: 9.

*filicorne. Eciton (Acamatus) filicorne* Santschi, 1916e: 369 (m.) ARGENTINA. Subspecies of *halidaii:* Santschi, 1921g: 93. Junior synonym of *jerrmanni:* Borgmeier, 1953: 14.

*fonscolombii. Labidus fonscolombii* Westwood, 1842: 76 (m.) BRAZIL. Combination in *Eciton:* Dalla Torre, 1893: 2; in *E. (Acamatus):* Emery, 1900a: 187; in *Neivamyrmex:* Borgmeier, 1953: 11. Junior synonym of *pertii:* Forel, 1911e: 250. Revived from synonymy as subspecies of *pertii,* and senior synonym of *strobeli* (and its junior synonym *nitens*): Borgmeier, 1953: 11. Junior synonym of *pertii:* Borgmeier, 1955b: 303.

***foveolatus.*** *Neivamyrmex foveolatus* Borgmeier, 1953: 41, fig. 22 (m.) PANAMA. See also: Borgmeier, 1955b: 629.

*frontalis. Eciton (Labidus) frontalis* Menozzi, 1924a: 30 (m.) BRAZIL. Junior synonym of *halidaii:* Borgmeier, 1953: 12.

***fumosus.*** *Eciton (Acamatus) minus* var. *fumosa* Forel, 1913l: 211 (m.) GUATEMALA. Combination in *Neivamyrmex:* Borgmeier, 1953: 8. Raised to species: Borgmeier, 1955b: 625.

***fuscipennis.*** *Eciton (Neivamyrmex) fuscipennis* Smith, M.R. 1942c: 578 (m.) U.S.A. [*Acamatus fuscipennis* Wheeler, W.M. 1908e: 417. *Nomen nudum;* attributed to Cresson.] Combination in *Neivamyrmex:* Borgmeier, 1953: 18. See also: Borgmeier, 1955b: 642; Watkins, 1975d: 85; Smith, D.R. 1979: 1330.

*garbei. Eciton (Acamatus) pseudops* r. *garbei* Forel, 1912c: 47 (w.) BRAZIL. Borgmeier, 1923: 48 (q.). Junior synonym of *pseudops:* Borgmeier, 1953: 9.

***genalis.*** *Neivamyrmex genalis* Borgmeier, 1953: 34, fig. 9 (m.) BOLIVIA. See also: Borgmeier, 1955b: 613.

***gibbatus.*** *Neivamyrmex gibbatus* Borgmeier, 1953: 45, fig. 30 (w.q.) COSTA RICA. See also: Borgmeier, 1955b: 347.

***goeldii.*** *Eciton (Acamatus) goeldii* Forel, 1901f: 352 (w.) BRAZIL. Combination in *E. (Neivamyrmex):* Borgmeier, 1948b: 462; in *Neivamyrmex:* Borgmeier, 1955b: 396.

***graciellae.*** *Eciton gracellae* Mann, 1926: 97 (w.) MEXICO. Watkins & Coody, 1986: 257 (q.). Combination in *Neivamyrmex,* and spelling emended to *graciellae:* Borgmeier, 1953: 10 (see also note in Borgmeier, 1955b: 389).

***gracilis.*** *Neivamyrmex gracilis* Borgmeier, 1955b: 590, pl. 51, figs. 3, 9, 21; pl. 75, fig. 2 (m.) BRAZIL.

***gradualis.*** *Neivamyrmex gradualis* Borgmeier, 1953: 46, fig. 32 (w.) BRAZIL. See also: Borgmeier, 1955b: 345.

*grandipseudops. Eciton (Acamatus) pseudops* r. *grandipseudops* Forel, 1915c: 353 (w.) ARGENTINA. Bruch, 1928: 352 (q.). Junior synonym of *pseudops:* Borgmeier, 1936: 64.

*gravenhorstii. Labidus gravenhorstii* Westwood, 1842: 76 (m.) BRAZIL. Combination in *Eciton:* Dalla Torre, 1893: 3; in *E. (Acamatus):* Emery, 1900a: 187. Junior synonym of *halidaii:* Borgmeier, 1953: 18.

***guerinii.*** *Labidus guerinii* Shuckard, 1840d: 397 (m.) BRAZIL. Combination in *Eciton:* Dalla Torre, 1893: 3; in *E. (Acamatus):* Emery, 1900a: 186; in *Neivamyrmex:* Borgmeier, 1953: 14. See also: Borgmeier, 1955b: 587.

***guyanensis.*** *Eciton (Acamatus) latiscapum* var. *guyanensis* Santschi, 1916e: 369 (m.) FRENCH GUIANA. Combination in *Neivamyrmex:* Borgmeier, 1953: 13. Raised to species: Borgmeier, 1955b: 580. Senior synonym of *nocua:* Borgmeier, 1953: 14.

***halidaii.*** *Labidus halidaii* Shuckard, 1840a: 200 (m.) BRAZIL. Combination in *Eciton:* Forel, 1895b: 119; in *E. (Acamatus):* Emery, 1900a: 187; in *Neivamyrmex:* Borgmeier, 1953: 13. Senior synonym of *frontalis* and *gravenhorstii:* Borgmeier, 1953: 12, 18, respectively; of *amplipennis, colombi* and *lemoulti:* Borgmeier, 1955b: 437. [Name misspelled as *halidayi* by Dalla Torre, 1893: 3, and subsequent authors.]

***harrisii.*** *Labidus harrissii* Haldeman, 1852: 367, pl. 9, figs. 4–6 (m.) U.S.A. Wheeler, G.C. & Wheeler, J. 1984: 273 (l.). Combination in *Eciton (Labidus):* Mayr, 1886d: 441; in *E. (Acamatus):* Emery, 1900a: 187; in *E. (Neivamyrmex):* Smith, M.R. 1942c: 572; in *Neivamyrmex:* Borgmeier, 1955b: 522. Senior synonym of *wheeleri* (and its junior synonym *dubia*) Watkins, 1968: 275. See also: Forbes & Do-Van-Quy, 1965: 95.

*haskinsi. Eciton (Neivamyrmex) angustinode* subsp. *haskinsi* Weber, 1943b: 68, fig. 1 (w.) TRINIDAD. Junior synonym of *emersoni:* Borgmeier, 1955b: 555.

***hetschkoi.*** *Eciton hetschkoi* Mayr, 1886a: 33 (w.m.) BRAZIL. Bruch, 1934b: 128 (q.). Combination in *E. (Labidus):* Mayr, 1886d: 441; in *E. (Acamatus):* Emery, 1894c: 184 (in key); in *Neivamyrmex:* Borgmeier, 1953: 11. Senior synonym of *taunayi:* Borgmeier, 1953: 472; of *ogloblini* and *silvestrii:* Borgmeier, 1955b: 472.

***hopei.*** *Labidus hopei* Shuckard, 1840b: 258 (m.) BRAZIL. Combination in *Eciton (Labidus):* Mayr, 1886d: 441; in *E. (Acamatus):* Emery, 1900a: 187; in *Neivamyrmex:* Borgmeier, 1953: 14. See also: Borgmeier, 1955b: 573.

*hospes. Eciton latiscapum* subsp. *hospes* Emery, 1906c: 111 (m.) ARGENTINA. Junior synonym of *spegazzinii:* Gallardo, 1920: 357; of *spinolae:* Borgmeier, 1953: 7.

***humilis.*** *Eciton (Acamatus) humile* Borgmeier, 1939: 415 (w.) COSTA RICA. Borgmeier, 1948a: 194 (q.m.). Combination in *E. (Neivamyrmex):* Borgmeier, 1948a: 194; in *Neivamyrmex:* Borgmeier, 1953: 5. See also: Borgmeier, 1955b: 401.

*ierense. Eciton (Acamatus) ierense* Weber, 1938a: 207 (w.) TRINIDAD. Junior synonym of *spinolae:* Borgmeier, 1955b: 409.

***iheringi.*** *Eciton iheringi* Forel, 1908c: 347 (m.) BRAZIL. Combination in *Neivamyrmex:* Borgmeier, 1953: 8. See also: Borgmeier, 1955b: 623. [Name misspelled as *jheringi* by Borgmeier, 1955b: 623, and most previous authors.]

*illigeri. Labidus illigeri* Shuckard, 1840d: 397 (m.) BRAZIL. Combination in *Eciton:* Dalla Torre, 1893: 4; in *E. (Acamatus):* Emery, 1900a: 187; in *E. (Neivamyrmex):* Borgmeier, 1948b: 462. Junior synonym of *pertii:* Borgmeier, 1953: 12.

***imbellis.*** *Eciton klugi* var. *imbellis* Emery, 1900a: 181 (m.) PERU. Combination in *E. (Acamatus):* Emery, 1900a: 187; in *Neivamyrmex,* and raised to species: Borgmeier, 1953: 7. See also: Borgmeier, 1955b: 601.

***impudens.*** *Eciton (Acamatus) impudens* Mann, 1922: 19, fig. 9 (w.) HONDURAS. Combination in *Neivamyrmex:* Borgmeier, 1953: 10. See also: Borgmeier, 1955b: 381.

***inca.*** *Eciton (Acamatus) inca* Santschi, 1921g: 93 (m.) PERU. Combination in *Neivamyrmex:* Borgmeier, 1953: 13. See also: Borgmeier, 1955b: 325.

***inflatus.*** *Neivamyrmex inflatus* Borgmeier, 1958: 205, figs. 6–8 (m.) MEXICO.

***iridescens.*** *Neivamyrmex iridescens* Borgmeier, 1950c: 623, figs. 1–5 (w.) GUIANA. See also: Borgmeier, 1955b: 540.

***jerrmanni.*** *Eciton jerrmanni* Forel, 1901h: 48 (m.) PARAGUAY. Combination in *E. (Acamatus):* Emery, 1910b: 26; in *Neivamyrmex:* Borgmeier, 1953: 8. Senior synonym of *luederwaldti:* Borgmeier, 1933d: 167; of *filicorne:* Borgmeier, 1953: 14. See also: Borgmeier, 1955b: 320.

***klugii.*** *Labidus klugii* Shuckard, 1840b: 260 (m.) ANTILLES IS. Combination in *Eciton:* Dalla Torre, 1893: 4; in *E. (Labidus):* Forel, 1893g: 369; in *Neivamyrmex:* Borgmeier, 1953: 6. Current subspecies: nominal plus *distans.* See also: Borgmeier, 1955b: 598.

*kuertii. Woitkowskia kuertii* Enzmann, E.V. 1952: 446, figs. 1, 2, 5 (m.) PERU. Combination in *Neivamyrmex:* Borgmeier, 1955b: 459 (*species inquirendae*).

***laevigatus.*** *Eciton (Neivamyrmex) laevigatum* Borgmeier, 1948b: 460, figs. 1–4 (w.) ARGENTINA. Combination in *Neivamyrmex:* Borgmeier, 1953: 5. See also: Borgmeier, 1955b: 399.

***latiscapus.*** *Eciton latiscapum* Emery, 1901d: 54, fig. 7 (m.) BRAZIL. Combination in *Neivamyrmex:* Borgmeier, 1953: 6. See also: Borgmeier, 1955b: 578.

*latreillii. Labidus latreillii* Perty, 1833: 138, pl.27, fig. 11 (m.) BRAZIL. [Junior primary homonym of *Labidus latreillii* Jurine, 1807: 283.] Replacement name: *pertii* Shuckard, 1840b: 262.

***legionis.*** *Eciton legionis* Smith, F. 1855c: 164 (w.) BRAZIL. Emery, 1900a: 521 (m.); Reichensperger, 1930: 321 (q.); Borgmeier, 1930: 22 (q.). Combination in *E. (Acamatus):* Emery, 1894c: 183 (in key); in *Neivamyrmex:* Borgmeier, 1953: 10. See also: Borgmeier, 1955b: 329.

*lemoulti. Eciton (Acamatus) lemoulti* Santschi, 1912e: 524 (m.) FRENCH GUIANA. Combination in *Neivamyrmex,* and race of *halidaii:* Borgmeier, 1953: 14. Junior synonym of *halidaii:* Borgmeier, 1955b: 437.

***leonardi.*** *Eciton (Acamatus) leonardi* Wheeler, W.M. 1915b: 392 (w.) U.S.A. Combination in *E. (Neivamyrmex):* Smith, M.R. 1942c: 570; in *Neivamyrmex:* Borgmeier, 1953: 10. Senior synonym of *peninsulare:* Borgmeier, 1955b: 431.

***leptognathus.*** *Eciton leptognathum* Emery, 1900a: 182, fig. 15 (m.) BOLIVIA. Combination in *E. (Acamatus):* Emery, 1900a: 187; in *Neivamyrmex:* Borgmeier, 1953: 6. See also: Borgmeier, 1955b: 354.

***lieselae.*** *Eciton (Acamatus) lieselae* Forel, 1913l: 209 (m.) ARGENTINA. Combination in *Neivamyrmex:* Borgmeier, 1953: 8. See also: Borgmeier, 1955b: 584.

***longiscapus.*** *Neivamyrmex longiscapus* Borgmeier, 1953: 27, figs. 4, 25, 26 (m.) COSTA RICA. See also: Borgmeier, 1955b: 383.

*luederwaldti. Eciton (Acamatus) luederwaldti* Emery, 1911a: 220 (m.) BRAZIL. Junior synonym of *jerrmanni:* Borgmeier, 1933d: 167.

***macrodentatus.*** *Eciton (Acamatus) macrodentatus* Menozzi, 1931b: 260, fig. 1 (w.) COSTA RICA. Combination in *Neivamyrmex:* Borgmeier, 1953: 11. Material of the *nomen nudum tristani* referred here by Borgmeier, 1939: 414.

***macropterus.*** *Neivamyrmex macropterus* Borgmeier, 1953: 40, figs. 14, 16 (m.) MEXICO. See also: Borgmeier, 1955b: 645.

#***mandibularis.*** *Eciton (Neivamyrmex) pilosum* subsp. *mandibulare* Smith, M.R. 1942c: 548, pl. 3, fig. 14 (m.) U.S.A. Combination in *Neivamyrmex:* Borgmeier, 1955b: 375.

***manni.*** *Eciton (Acamatus) manni* Wheeler, W.M. 1914b: 41 (w.) MEXICO. Combination in *Neivamyrmex:* Borgmeier, 1953: 18. See also: Borgmeier, 1955b: 519.

*marginatum. Eciton (Acamatus) marginatum* Borgmeier, 1939: 414 (w.) COSTA RICA. Combination in *E. (Neivamyrmex)*: Borgmeier, 1948a: 198. Junior synonym of *adnepos:* Borgmeier, 1953: 5.

***maxillosus.*** *Eciton maxillosum* Emery, 1900a: 181, figs. 13, 14 (m.) BRAZIL. Combination in *E. (Acamatus)*: Emery, 1900a: 187; in *Neivamyrmex:* Borgmeier, 1953: 6. See also: Borgmeier, 1955b: 350.

***megathrix.*** *Neivamyrmex megathrix* Kempf, 1961b: 486, figs. 1–3 (w.) SURINAM.

***melanocephalus.*** *Eciton melanocephalum* Emery, 1895c: 260 (footnote) (w.) MEXICO. Combination in *E. (Neivamyrmex)*: Smith, M.R. 1942c: 549; in *Neivamyrmex:* Borgmeier, 1953: 6. Senior synonym of *xipe:* Smith, M.R. 1942c: 549; Borgmeier, 1953: 19. See also: Borgmeier, 1955b: 385.

***melshaemeri.*** *Labidus melshaemeri* Haldeman, 1852: 368, pl. 9, figs. 7–9 (m.) U.S.A. Combination in *Eciton (Labidus)*: Mayr, 1886d: 442; in *E. (Acamatus)*: Emery, 1900a: 187; in *E. (Neivamyrmex)*: Smith, M.R. 1942c: 576; in *Neivamyrmex:* Borgmeier, 1953: 8. See also: Borgmeier, 1955b: 647.

*mesognathum. Eciton (Acamatus) leptognathum* var. *mesognathum* Santschi, 1920d: 368, fig. 5 (m.) FRENCH GUIANA. Combination in *Neivamyrmex:* Borgmeier, 1953: 13. Junior synonym of *falcifer:* Borgmeier, 1955b: 357.

#***mexicanus.*** *Labidus mexicanus* Smith, F. 1859b: 7 (m.) MEXICO. Forel, 1899c: 27 (w.); Reichensperger, 1939: 297 (q.). Combination in *Eciton:* Dalla Torre, 1893: 4; in *E. (Labidus)*: Emery, 1895c: 260; in *E. (Acamatus)*: Emery, 1900a: 187; in *E. (Neivamyrmex)*: Smith, M.R. 1942c: 544; in *Neivamyrmex:* Borgmeier, 1953: 8. Variety of *pilosus:* Borgmeier, 1936: 59. Junior synonym of *pilosus:* Smith, M.R. 1942c: 544. Revived from synonymy as subspecies of *pilosus:* Borgmeier, 1953: 14; Borgmeier, 1955b: 361. Senior synonym of *clavicornis:* Mayr, 1886b: 120; of *subsulcatum:* Dalla Torre, 1893: 4; of *aztecum:* Borgmeier, 1936: 60; of *militarium:* Borgmeier, 1953: 14. See also: Borgmeier, 1955b: 361.

*mexicana. Woitkowskia mexicana* Enzmann, E.V. 1952: 445, figs. 3, 8 (m.) MEXICO. [Unresolved junior secondary homonym of *mexicanus* Smith, above.] Combination in *Neivamyrmex:* Borgmeier, 1955b: 459 (*species inquirendae*).

***micans.*** *Neivamyrmex micans* Borgmeier, 1953: 32, figs. 10, 12, 23 (m.) BRAZIL. See also: Borgmeier, 1955b: 470.

***microps.*** *Neivamyrmex microps* Borgmeier, 1955b: 635, pl. 56, figs. 6, 10; pl. 80, fig. 4 (m.) U.S.A.

*militarium. Eciton (Labidus) spininode* st. *militarium* Santschi, 1929f: 84, fig. 1 (w.) PANAMA. Junior synonym of *mexicanus:* Borgmeier, 1953: 14.

***minensis.*** *Eciton (Acamatus) balzanii* subsp. *minenses* Borgmeier, 1928b: 58 (w.) BRAZIL. Borgmeier, 1939: 410 (q.). Combination in *Neivamyrmex*, and spelling emended to *minensis:* Borgmeier, 1953: 5. Subspecies of *balzani:* Borgmeier, 1953: 5. Raised to species: Borgmeier, 1939: 410; Borgmeier, 1955b: 537.

***minor.*** *Labidus minor* Cresson, 1872: 195 (m.) U.S.A. Combination in *Eciton (Labidus)*: Mayr, 1886d: 441; in *E. (Acamatus)*: Emery, 1900a: 187; in *E. (Neivamyrmex)*: Smith, M.R. 1942c: 574; in *Neivamyrmex:* Borgmeier, 1953: 8. [Misspelled as *minus* by Dalla Torre, 1893: 4.] See also: Borgmeier, 1955b: 630.

***modestus.*** *Eciton (Acamatus) modestum* Borgmeier, 1933d: 165 (w.) BRAZIL. Combination in *Neivamyrmex:* Borgmeier, 1953: 5. See also: Borgmeier, 1955b: 552.

*modificata. Eciton (Acamatus) richteri* var. *modificata* Forel, 1913l: 212 (m.) ARGENTINA. Variety of *sulcatus:* Santschi, 1916e: 368. Junior synonym of *sulcatus:* Borgmeier, 1953: 9

***mojave.*** *Eciton (Acamatus) mojave* Smith, M.R. 1943e: 196, fig. 1 (m.) U.S.A. Combination in *Neivamyrmex:* Borgmeier, 1955b: 633.

***moseri.*** *Neivamyrmex moseri* Watkins, 1969: 529, figs. 1–4 (w.) U.S.A. Watkins, 1971: 96 (q.).

***nigrescens.*** *Labidus nigrescens* Cresson, 1872: 194 (m.) U.S.A. Wheeler, W.M. 1900a: 563 (q.); Smith, M.R. 1942c: 551 (m.); Wheeler, G.C. & Wheeler, J. 1984: 273 (l.). Combination in *Eciton:* Dalla Torre, 1893: 5; in *E. (Labidus)*: Emery, 1895c: 261; in *E. (Acamatus)*: Emery, 1900a: 187; in *E. (Neivamyrmex)*: Smith, M.R. 1942c: 550; in *Neivamyrmex:* Borgmeier, 1955b: 494. Senior synonym of *schmitti:* Smith, M.R. 1938b: 160; Borgmeier, 1955b: 494. See also: Watkins, 1972: 358; Smith, D.R. 1979: 1331; Petralia & Vinson, 1980: 378.

*nitens. Eciton nitens* Mayr, 1868b: 168 (w.) ARGENTINA. Combination in *E. (Acamatus)*: Emery, 1894c: 181 (in text). Junior synonym of *spinolae:* Emery, 1900a: 522; of *strobeli:* Bruch, 1916: 294.

*nocua. Eciton (Acamatus) latiscapum* var. *nocua* Santschi, 1921g: 94 (m.) FRENCH GUIANA. Junior synonym of *guyanensis:* Borgmeier, 1953: 14.

***nordenskioldii.*** *Leptanilla nordenskioldii* Holmgren, 1908: 347 (w.) PERU. Combination in *Eciton (Acamatus)*: Emery, 1910b: 25; in *Neivamyrmex:* Borgmeier, 1953: 10. See also: Borgmeier, 1955b: 314.

***nyensis.*** *Neivamyrmex nyensis* Watkins, 1977b: 421, figs. 1–4 (w.) U.S.A.

*obscurum. Eciton spinolae* subsp. *obscurum* Forel, 1911e: 250 (m.) ARGENTINA. Junior synonym of *spegazzinii:* Santschi, 1916e: 368.

*obscura. Eciton (Acamatus) californicum* var. *obscura* Forel, 1914d: 265 (w.) U.S.A. [Unresolved junior primary homonym of *obscurum* Forel, above.] Junior synonym of *californicus:* Borgmeier, 1955b: 517.

*ogloblini. Eciton (Acamatus) hetschkoi* var. *ogloblini* Santschi, 1934c: 25, pl. 3, figs. 2, 11, 15 (w.) ARGENTINA. Bruch, 1934b: 129 (q.). Combination in *Neivamyrmex:* Borgmeier, 1953: 13. Junior synonym of *hetschkoi:* Borgmeier, 1955b: 472.

***opacithorax.*** *Eciton (Acamatus) californicum* subsp. *opacithorax* Emery, 1894c: 184 (diagnosis in key) (w.) U.S.A. Wheeler, W.M. & Long, 1901: 163, 173 (m.q., respectively); Wheeler, G.C. & Wheeler, J. 1984: 273 (l.). Combination in *E. (Neivamyrmex)*: Smith, M.R. 1942c: 555; in *Neivamyrmex:* Borgmeier, 1953: 5. Raised to species: Emery, 1900a: 186. Senior synonym of *castaneum:* Borgmeier, 1955b: 502. See also: Smith, D.R. 1979: 1332; Wheeler, G.C. & Wheeler, J. 1990a: 461.

***orthonotus.*** *Eciton (Acamatus) orthonotum* Borgmeier, 1933d: 165 (w.) BRAZIL. Combination in *Neivamyrmex:* Borgmeier, 1953: 5. See also: Borgmeier, 1955b: 565.

*oslari. Eciton (Acamatus) oslari* Wheeler, W.M. 1908e: 415, pl. 26, fig. 8 (m.) U.S.A. Combination in *E. (Neivamyrmex)*: Smith, M.R. 1942c: 579. Junior synonym of *andrei:* Borgmeier, 1953: 19.

***pacificus.*** *Neivamyrmex pacificus* Borgmeier, 1955b: 536 (w.) PERU.

*paraense. Eciton (Acamatus) paraense* Forel, 1912c: 45 (w.) BRAZIL. Junior synonym of *planidorsus:* Borgmeier, 1953: 9.

***pauxillus.*** *Eciton (Acamatus) pauxillum* Wheeler, W.M. 1903b: 93, fig. 1 (w.) U.S.A. Watkins, 1971: 99 (q.). Combination in *E. (Neivamyrmex)*: Smith, M.R. 1942c: 569; in *Neivamyrmex:* Borgmeier, 1953: 19. See also: Borgmeier, 1955b: 570.

*peninsulare. Eciton (Acamatus) peninsulare* Mann, 1926: 98 (w.) MEXICO. Combination in *Neivamyrmex:* Borgmeier, 1953: 10. Junior synonym of *leonardi:* Borgmeier, 1955b: 431.

***perplexus.*** *Neivamyrmex perplexus* Borgmeier, 1953: 36 (m.) BRAZIL. See also: Borgmeier, 1955b: 616.

***pertii.*** *Labidus pertii* Shuckard, 1840b: 262. Replacement name for *latreillii* Perty, 1833: 138. [Junior primary homonym of *latreillii* Jurine, 1807: 283.] Combination in *Eciton:* Forel, 1895b: 120; in *E. (Acamatus)*: Emery, 1900a: 187; in *Neivamyrmex:* Borgmeier, 1953: 15. Senior synonym of *illigeri* and *fonscolombii* (and its junior synonyms *strobeli* and *nitens*): Borgmeier, 1955b: 303.

***physognathus.*** *Eciton leptognathus* var. *physognatha* Emery, 1900a: 182, fig. 16 (m.) BOLIVIA. Combination in *E. (Acamatus)*: Emery, 1900a: 187; in *Neivamyrmex:* Borgmeier, 1953: 6. Raised to species: Borgmeier, 1955b: 356.

***pilosus.*** *Eciton pilosa* Smith, F. 1858b: 151 (w.) BRAZIL. Reichensperger, 1939: 297 (q.); Smith, M.R. 1942c: 544 (m.); Wheeler, G.C. 1943: 331 (l.); Wheeler, G.C. & Wheeler, J. 1984: 273 (l.). Combination in *E. (Acamatus)*: Emery, 1894c: 183 (in key); in *E. (Neivamyrmex)*: Smith, M.R. 1942c: 544; in *Neivamyrmex:* Borgmeier, 1953: 8. Senior synonym of *angustius* and *porrectognathum:* Borgmeier, 1955b: 361. Current subspecies: nominal plus *beebei, mandibularis, mexicanus.*

***piraticus.*** *Neivamyrmex piraticus* Borgmeier, 1953: 37, figs. 13, 18 (m.) BRAZIL. See also: Borgmeier, 1955b: 621.

**planidens.** *Neivamyrmex planidens* Borgmeier, 1953: 31, figs. 21, 28, 29 (m.) ECUADOR. See also: Borgmeier, 1955b: 603.

**planidorsus.** *Eciton planidorsum* Emery, 1906c: 110 (w.) PARAGUAY. Combination in *E.* (*Acamatus*): Emery, 1910b: 25; in *Neivamyrmex:* Borgmeier, 1953: 6. Senior synonym of *paranense:* Borgmeier, 1953: 9. See also: Borgmeier, 1955b: 421.

*porrectognathum. Eciton* (*Acamatus*) *porrectognathum* Borgmeier, 1933d: 167 (m.) BRAZIL. Combination in *Neivamyrmex,* and subspecies of *pilosus:* Borgmeier, 1953: 5. Junior synonym of *pilosus:* Borgmeier, 1955b: 361.

**postangustatus.** *Eciton* (*Acamatus*) *postangustatum* Borgmeier, 1934: 94, fig. 1 (w.) SURINAM. Combination in *Neivamyrmex:* Borgmeier, 1953: 5. See also: Borgmeier, 1955b: 560.

**postcarinatus.** *Neivamyrmex postcarinatus* Borgmeier, 1953: 43, figs. 34, 35 (w.) PANAMA. Wheeler, G.C. & Wheeler, J. 1984: 272 (l.). See also: Borgmeier, 1955b: 395.

**pseudops.** *Eciton pseudops* Forel, 1909a: 254 (w.) PARAGUAY. Borgmeier, 1936: 64 (q.); Borgmeier, 1939: 408 (m.). Combination in *E.* (*Acamatus*): Emery, 1910b: 25; in *E.* (*Neivamyrmex*): Borgmeier, 1948b: 462; in *Neivamyrmex:* Borgmeier, 1953: 8. Senior synonym of *grandipseudops:* Borgmeier, 1936: 64; of *garbei:* Borgmeier, 1953: 9. See also: Borgmeier, 1955b: 336.

**puerulus.** *Neivamyrmex puerulus* Borgmeier, 1955b: 591, pl. 56, fig. 9; pl. 75, fig. 4 (m.) PANAMA.

**pulchellus.** *Neivamyrmex pulchellus* Borgmeier, 1955b: 595, pl. 56, fig. 7 (m.) PANAMA.

**pullus.** *Neivamyrmex pullus* Borgmeier, 1953: 42, fig. 17 (m.) PANAMA. See also: Borgmeier, 1955b: 627.

**punctaticeps.** *Eciton* (*Acamatus*) *punctaticeps* Emery, 1894c: 181 (w.) BRAZIL. Oldest junior synonym of, and hence first available replacement name for *Labidus spinolae* Westwood, 1842: 77. [Junior primary homonym of *Labidus spinolae* Shuckard, 1840c: 327.] Combination in *Neivamyrmex:* Borgmeier, 1955b: 409. See under *spinolae* below for current synonymy and other references.

**quadratooccipitus.** *Neivamyrmex quadratooccipitus* Watkins, 1975c: 207, figs. 1–6 (m.) EL SALVADOR.

**radoszkowskii.** *Eciton radoszkowskii* Emery, 1900a: 181, figs. 9, 10 (m.) PERU. Combination in *E.* (*Acamatus*): Emery, 1900a: 187; in *Neivamyrmex:* Borgmeier, 1953: 6. See also: Borgmeier, 1955b: 593.

*raptans. Eciton raptans* Forel, 1911d: 395 (footnote). [Unnecessary replacement name for *raptor* Forel, see below.] Bruch, 1924a: 232 (q., see Bruch, 1925b: 125). Combination in *Neivamyrmex:* Borgmeier, 1953: 8. Variety of *hetschkoi:* Bruch, 1928: 350. Revived status as species: Santschi, 1934c: 25. Senior synonym of *abstinens* (unnecessary replacement name): Forel, 1913l: 213; of *rebellatum:* Borgmeier, 1955b: 481. Junior synonym of *raptor:* **new synonymy**.

**raptor.** *Eciton* (*Acamatus*) *raptor* Forel, 1911c: 289 (w.) BRAZIL. Combination in *Neivamyrmex:* Borgmeier, 1955b: 481. Senior synonym of *raptans* (and its junior synonyms *abstinens* and *rebellatum*): **new synonymy**. [The name *raptans* Forel was proposed as a replacement for *raptor* Forel because of the name *Eciton raptor* Smith, F. 1855c: 169. The last name is however an incorrect subsequent spelling, in the figure legends, of *Eciton rapax* Smith, F. 1855c: 163. As an incorrect subsequent spelling *raptor* Smith is unavailable, and therefore does not enter into homonymy.]

*rebellatum. Eciton* (*Acamatus*) *raptans* var. *rebellatum* Santschi, 1934c: 25, pl. 3, figs. 4, 6, 10, 14, 19 (w.q.m.) ARGENTINA. Combination in *Neivamyrmex:* Borgmeier, 1953: 13. Junior synonym of *raptans:* Borgmeier, 1955b: 481.

*richteri. Eciton* (*Acamatus*) *richteri* Forel, 1913l: 211 (m.) ARGENTINA. Junior synonym of *sulcatus:* Santschi, 1916e: 368.

**romandii.** *Labidus romandii* Shuckard, 1840b: 261 (m.) BRAZIL. Combination in *Eciton:* Forel, 1895b: 121; in *E.* (*Acamatus*): Emery, 1900a: 187; in *E.* (*Labidus*): Bruch, 1914: 215; in *E.* (*Neivamyrmex*): Borgmeier, 1948b: 463; in *Neivamyrmex:* Borgmeier, 1953: 15. Senior synonym of *desartsi:* Borgmeier, 1953: 9. See also: Borgmeier, 1955b: 605.

**rosenbergi.** *Eciton* (*Acamatus*) *mexicanum* subsp. *rosebergi* Forel, 1911e: 256 (m.) ECUADOR. Combination in *Neivamyrmex:* Borgmeier, 1953: 8. Raised to species: Wheeler, W.M. 1921d: 314; Borgmeier, 1953: 9; Borgmeier, 1955b: 524.

**rugulosus.** *Neivamyrmex rugulosus* Borgmeier, 1953: 49 (w.) MEXICO. See also: Borgmeier, 1955b: 515.

*schmitti. Eciton* (*Acamatus*) *schmitti* Emery, 1894c: 183 (diagnosis in key) (w.) U.S.A. Emery, 1895c: 258 (w.); Wheeler, W.M. & Long, 1901: 161 (m.); Wheeler, G.C. 1943: 331 (l.). Junior synonym of *nigrescens:* Smith, M.R. 1938b: 160; Borgmeier, 1955: 494.

**scutellaris.** *Neivamyrmex scutellaris* Borgmeier, 1953: 28, fig. 6 (m.) PANAMA. See also: Borgmeier, 1955b: 352.

**shuckardi.** *Eciton shuckardi* Emery, 1900a: 183 (m.) PARAGUAY. Combination in *E.* (*Acamatus*): Emery, 1900a: 187; in *Neivamyrmex:* Borgmeier, 1953: 6. See also: Borgmeier, 1955b: 582.

*silvestrii. Eciton silvestrii* Emery, 1906c: 109 (w.) ARGENTINA. Combination in *E.* (*Acamatus*): Bruch, 1914: 215; in *Neivamyrmex:* Borgmeier, 1953: 6. Junior synonym of *hetschkoi:* Borgmeier, 1955b: 472.

**spatulatus.** *Eciton* (*Acamatus*) *spatulatum* Borgmeier, 1939: 417, fig. 2 (m.) COSTA RICA. Combination in *Neivamyrmex:* Borgmeier, 1953: 5. See also: Borgmeier, 1955b: 529.

*spegazzinii. Eciton spegazzinii* Emery, 1888e: 691 (w.) ARGENTINA. Gallardo, 1915: 5 (m.). Combination in *E.* (*Acamatus*): Emery, 1894c: 184 (in key). Stirps of *spinolae,* and senior synonym of *obscurum:* Santschi, 1916e: 368. Junior synonym of *dorbignii:* Santschi, 1921g: 92; Borgmeier, 1955b: 461.

*spinolae. Labidus spinolae* Westwood, 1842: 77 (m.) BRAZIL. [Junior primary homonym of *Labidus spinolae* Shuckard, 1840c: 327.] Replacement name: *punctaticeps* Emery, 1894c: 181. [Oldest junior synonym of *spinolae* Westwood and hence first available replacement name.] Combination in *Eciton:* Forel, 1895b: 121; in *E.* (*Acamatus*): Emery, 1900a: 187; in *E.* (*Labidus*): Bruch, 1914: 215; in *Neivamyrmex:* Borgmeier, 1953: 17. Senior synonym of *fiebrigi, hospes* and *punctaticeps:* Borgmeier, 1953: 9, 7, 7, respectively; of *commutatum* (and its junior synonym *columbicum*), and *ierense:* Borgmeier, 1955b: 409.

**spoliator.** *Eciton spoliator* Forel, 1899c: 29 (m.) COSTA RICA. Combination in *E.* (*Acamatus*): Emery, 1900a: 187; in *Neivamyrmex:* Borgmeier, 1953: 8. See also: Borgmeier, 1955b: 639.

*strobeli. Labidus strobeli* Mayr, 1868b: 166 (m.) ARGENTINA. Bruch, 1916: 294 (w.); Bruch, 1921: 177 (q.l.). Combination in *E.* (*Labidus*): Mayr, 1886d: 441; in *E.* (*Acamatus*): Emery, 1900a: 187; in *E.* (*Neivamyrmex*): Borgmeier, 1948b: 462. Senior synonym of *nitens:* Bruch, 1916: 294. Junior synonym of *fonscolombii:* Borgmeier, 1953: 11.

*subsulcatum. Eciton* (*Labidus*) *subsulcatum* Mayr, 1886d: 440 (m.) U.S.A. Junior synonym of *mexicanus:* Dalla Torre, 1893: 4; Borgmeier, 1955b: 361.

**sulcatus.** *Labidus sulcatus* Mayr, 1868b: 168 (m.) ARGENTINA. Combination in *Eciton* (*Labidus*): Mayr, 1886d: 441; in *E.* (*Acamatus*): Emery, 1900a: 187; in *Neivamyrmex:* Borgmeier, 1953: 11. Senior synonym of *richteri:* Santschi, 1916e: 368; of *modificata:* Borgmeier, 1953: 9. See also: Borgmeier, 1955b: 610.

**sumichrasti.** *Eciton sumichrasti* Norton, 1868a: 62, pl. 2, fig. 7 (w.) MEXICO. Mann, 1926: 99 (q.). Combination in *E.* (*Acamatus*): Emery, 1894c: 182 (in text); in *Neivamyrmex:* Borgmeier, 1955b: 491.

**swaisonii.** *Labidus swainsonii* Shuckard, 1840a: 201 (m.) BRAZIL. Combination in *Eciton:* Forel, 1895b: 121; in *E.* (*Acamatus*): Emery, 1900a: 187; in *E.* (*Neivamyrmex*): Borgmeier, 1948b: 462; in *Neivamyrmex:* Borgmeier, 1953: 15. Senior synonym of *arizonense:* Borgmeier, 1955b: 454.

*taunayi. Eciton* (*Acamatus*) *taunayi* Borgmeier, 1928b: 55, fig. 1 (m.) BRAZIL. Junior synonym of *hetschkoi:* Borgmeier, 1953: 5.

**tenuis.** *Neivamyrmex tenuis* Borgmeier, 1953: 36 (m.) BRAZIL. See also: Borgmeier, 1955b: 615.

**texanus.** *Neivamyrmex texanus* Watkins, 1972: 353, figs. 2–4, 11, 14, 17–19, 33, 34, 49, 61 (w.q.m.) U.S.A. Wheeler, G.C. & Wheeler, J. 1984: 273 (l.); Taber & Cokendolpher, 1988: 95 (k.).

*tristani. Eciton* (*Acamatus*) *tristani* Borgmeier, 1939: 414 (w.) COSTA RICA. *Nomen nudum;* attributed to Emery. Material referred to *macrodentatus* by Borgmeier, 1939: 414.

**tristis.** *Eciton spoliator* r. *triste* Forel, 1901h: 48 (m.) MEXICO. Combination in *Neivamyrmex,* and raised to species: Borgmeier, 1953: 9. See also: Borgmeier, 1955b: 650.

**vicinus.** *Neivamyrmex vicinus* Borgmeier, 1953: 35, fig. 24 (m.) BRAZIL. See also: Borgmeier, 1955b: 608.

**walkerii.** *Labidus walkerii* Westwood, 1842: 77 (m.) BRAZIL. Combination in *Eciton:* Forel, 1895b: 121; in *E.* (*Acamatus*): Emery, 1900a: 187; in *Neivamyrmex:* Borgmeier, 1953: 17. Senior synonym of *connectens:* Borgmeier, 1955b: 445.

*wheeleri. Eciton wheeleri* Emery, 1901d: 55, fig. 8 (w.) U.S.A. Smith, M.R. 1942c: 561 (q.). Combination in *E.* (*Acamatus*): Wheeler, W.M. 1908e: 412; in *E.* (*Neivamyrmex*): Smith, M.R. 1942c: 561; in *Neivamyrmex:* Borgmeier, 1953: 6. Senior synonym of *dubia:* Smith, M.R. 1942c: 561. Junior synonym

***NIEVAMYRMEX** (continued)*

of *harrisii:* Watkins, 1968: 275. See also: Borgmeier, 1955b: 511.

*xipe. Eciton* (*Acamatus*) *melanocephalum* subsp. *xipe* Wheeler, W.M. 1914b: 41 (w.) MEXICO. Junior synonym of *melanocephalus:* Smith, M.R. 1942c: 549; Borgmeier, 1953: 19.

#***NEMATOCREMA:*** see under ***CREMATOGASTER.***

*NEOAMBLYOPONE:* see under ***AMBLYOPONE.***

*NEOATTA:* see under ***ATTA.***

*NEOCLYSTOPSENELLA:* see under ***TAPINOMA.***

*NEOCOLOBOPSIS:* see under ***CAMPONOTUS.***

#***NEOCREMA:*** see under ***CREMATOGASTER.***

*NEOFORELIUS:* see under ***FORELIUS.***

*NEOFORMICA:* see under ***FORMICA.***

*NEOMYRMA:* see under ***MANICA.***

*NEOMYRMAMBLYS:* see under ***CAMPONOTUS.***

*NEOPHYRACACES:* see under ***CERAPACHYS.***

*NEOPONERA:* see under ***PACHYCONDYLA.***

***NEOSTRUMA*** (Neotropical)

***brevicornis.*** *Strumigenys brevicornis* Mann, 1922: 38 (w.q.) HONDURAS. Combination in *Neostruma:* Brown, 1948e: 111. See also: Brown, 1959b: 10.

***crassicornis.*** *Strumigenys crassicornis* Mayr, 1887: 577 (w.) BRAZIL. Brown, 1959b: 12 (q.). Combination in *Neostruma:* Brown, 1948e: 111.

***metopia.*** *Neostruma metopia* Brown, 1959b: 11, fig. 2 (q.) PANAMA. Kempf, 1975a: 417 (w.).

***mustelina.*** *Strumigenys* (*Strumigenys*) *mustelina* Weber, 1934a: 34, fig. 11 (q.) MEXICO. Brown, 1959b: 7 (w.m.). Combination in *Neostruma:* Brown, 1959b: 7

***myllorhapha.*** *Neostruma myllorhapha* Brown, 1959b: 12, fig. 4 (w.) COSTA RICA.

***zeteki.*** *Neostruma zeteki* Brown, 1959b: 10, figs. 1, 5 (w.q.) PANAMA.

*NESOLASIUS:* see under ***PSEUDOLASIUS.***

*NESOMYRMEX:* see under ***LEPTOTHORAX.***

*NIMBAMYRMA:* see under ***OLIGOMYRMEX.***

***NOMAMYRMEX*** (Neotropical, s. Nearctic)

*afflictus. Acanthostichus afflictus* Gallardo, 1919a: 239, fig. 3 (m.) ARGENTINA. Combination in *Eciton:* Emery, 1920a: 60. Junior synonym of *esenbeckii:* Borgmeier, 1955b: 139.

*apicifer. Eciton* (*Labidus*) *apicifer* Santschi, 1916e: 366, fig. (m.) PARAGUAY. Junior synonym of *hartigii:* Borgmeier, 1953: 14.

#***crassicornis.*** *Eciton crassicornis* Smith, F. 1855c: 163, pl. 13, figs. 1, 2 (w.) BRAZIL. Borgmeier, 1958: 202 (q.). Combination in *E.* (*Labidus*): Wheeler, W.M. 1916d: 324; in *E.* (*Holopone*): Santschi, 1925b: 12; in *E.* (*Nomamyrmex*): Borgmeier, 1936: 55; in *Nomamyrmex:* Borgmeier, 1953: 16. Subspecies of *esenbeckii,* senior synonym of *hirsutipodex* and *intrepidum,* and material of the unavailable name *furiosum* referred here: Borgmeier, 1955b: 139. See also: Watkins, 1977a: 205.

*crudele. Eciton* (*Holopone*) *crassicorne* subsp. *crudele* Santschi, 1929a: 414, figs. c, g (w.) BOLIVIA. Junior synonym of *esenbeckii:* Borgmeier, 1955b: 139.

***esenbeckii.*** *Labidus esenbeckii* Westwood, 1842: 75, pl. 20, fig. 4 (m.) BRAZIL. Combination in *Eciton:* Emery, 1890b: 39; in *E.* (*Labidus*): Santschi, 1920d: 364; in *E.* (*Holopone*): Santschi, 1925b: 11; in *E.* (*Nomamyrmex*): Borgmeier, 1936: 55; in *Nomamyrmex:* Borgmeier, 1953: 14. Senior synonym of *fimbriatum, interrupta* and *terrificum:* Borgmeier, 1953: 14; of *afflictus, crudele* and *indomabile:* Borgmeier, 1955b: 139. Current subspecies: nominal plus *crassicornis, mordax, wilsoni.* See also: Borgmeier, 1958: 201; Watkins, 1977a: 204.

*fimbriatum. Eciton* (*Labidus*) *fimbriatum* Santschi, 1920d: 363, figs. 1, 2 (m.) FRENCH GUIANA. Junior synonym of *esenbeckii:* Borgmeier, 1953: 14.

*furiosum. Eciton* (*Holopone*) *crassicornis* st. *crudele* var. *furiosum* Santschi, 1929f: 85 (w.) PANAMA. Unavailable name; material referred to *crassicornis* by Borgmeier, 1955b: 139.

*hansi. Eciton* (*Labidus*) *hartigi* r. *hansi* Forel, 1912c: 43 (m.) BRAZIL. Junior synonym of *hartigii:* Borgmeier, 1953: 9.

***hartigii.*** *Labidus hartigii* Westwood, 1842: 75, pl. 20, fig. 3 (m.) BRAZIL. Combination in *Eciton:* Dalla Torre, 1893: 3; in *E.* (*Labidus*): Emery, 1910b: 23; in *E.* (*Holopone*): Santschi, 1925b: 11; in *E.* (*Nomamyrmex*): Borgmeier, 1936: 55; in *Nomamyrmex:* Borgmeier, 1953: 8. Senior synonym of *apicifer* and *hansi:* Borgmeier, 1953: 14, 9, respectively; of *schlechtendali:* Borgmeier, 1955b: 155. See also: Watkins, 1977a: 204.

*hirsutipodex. Eciton* (*Labidus*) *esenbecki* st. *hirsutipodex* Santschi, 1920d: 364, fig. 2 (m.) MEXICO. Junior synonym of *crassicornis:* Borgmeier, 1955b: 139.

*indomabile. Eciton* (*Holopone*) *crassicorne* var. *indomabile* Santschi, 1930c: 82 (w.) BRAZIL. Junior synonym of *esenbeckii:* Borgmeier, 1955b: 139.

*interrupta. Eciton fimbriatum* var. *interrupta* Santschi, 1923c: 249 (m.) BRAZIL. Junior synonym of *esenbeckii:* Borgmeier, 1953: 14.

*intrepidum. Eciton* (*Holopone*) *crassicorne* var. *intrepidum* Santschi, 1929a: 414, figs. b, f (w.) VENEZUELA. Junior synonym of *crassicornis:* Borgmeier, 1955b: 139.

#***mordax.*** *Eciton* (*Holopone*) *crassicorne* subsp. *mordax* Santschi, 1929a: 415, figs. d, h (w.) MEXICO. Watkins, 1977a: 210 (m.). Combination in *Nomamyrmex:* Borgmeier, 1953: 13. Junior synonym of *wilsoni:* Borgmeier, 1955b: 139. Revived from synonymy as subspecies of *esenbeckii:* Watkins, 1977a: 205.

*schlechtendali. Eciton schlechtendali* Mayr, 1887: 552 (w.) SOUTH AMERICA. Combination in *E.* (*Labidus*): Emery, 1910b: 23; in *E.* (*Holopone*): Santschi, 1925b: 11; in *E.* (*Nomamyrmex*): Borgmeier, 1936: 55; in *Nomamyrmex:* Borgmeier, 1953: 11. Junior synonym of *hartigii:* Borgmeier, 1955b: 155.

*terrificum. Eciton* (*Holopone*) *fimbriatum* var. *terrificum* Santschi, 1930c: 82 (m.) BRAZIL. Junior synonym of *esenbeckii:* Borgmeier, 1953: 14.

#***wilsoni.*** *Eciton* (*Labidus*) *esenbecki* st. *wilsoni* Santschi, 1920d: 366 (m.) U.S.A. Combination in *E.* (*Nomamyrmex*): Smith, M.R. 1951a: 779; in *Nomamyrmex:* Borgmeier, 1953: 13. See also: Borgmeier, 1955b: 139; Watkins, 1977a: 205.

***NOONILLA*** (Philippines)

***copiosa.*** *Noonilla copiosa* Petersen, 1968: 585, figs. 6–10 (m.) PHILIPPINES.

***NOTHIDRIS*** (Neotropical)

***bicolor.*** *Megalomyrmex bicolor* Ettershank, 1965: 55, figs. 1–5 (w.l.) CHILE. Combination in *Nothidris:* Snelling, 1975: 3. See also: Kempf, 1970b: 359.

***cekalovici.*** *Nothidris cekalovici* Snelling, 1975: 3, figs. 3–4 (w.) CHILE.

***latastei.*** *Monomorium latastei* Emery, 1895i: 10 (w.) CHILE. Wheeler, G.C. & Wheeler, J. 1980: 533 (l.). Combination in *M.* (*Notomyrmex*): Emery, 1922e: 169; in *Nothidris:* Ettershank, 1966: 106.

***NOTHOMYRMECIA*** (Australia)

***macrops.*** *Nothomyrmecia macrops* Clark, 1934a: 19, fig. 1 (w.) AUSTRALIA. Taylor, 1978a: 979 (q.m.); Wheeler, G.C., Wheeler, J. & Taylor, 1980: 131 (l.); Imai, Taylor, *et al.* 1991: 133 (k.). See also: Jaisson, Fresnau, *et al.* 1992: 425; Hölldobler & Taylor, 1983: 382; Ward & Taylor, 1981: 177.

****NOTHOMYRMICA*** (Baltic Amber)

****intermedia.*** **Nothomyrmica intermedia* Wheeler, W.M. 1915h: 61, fig. 24 (w.) BALTIC AMBER (Oligocene).

****petiolata.*** **Macromischa petiolata* Mayr, 1868c: 85, pl. 4, figs. 83, 84 (w.) BALTIC AMBER (Oligocene). Combination in **Nothomyrmica:* Wheeler, W.M. 1915h: 62.

****rudis.*** **Macromischa rudis* Mayr, 1868c: 85, pl. 4, fig. 85 (w.) BALTIC AMBER (Oligocene). Combination in **Nothomyrmica:* Wheeler, W.M. 1915h: 60.

****rugosostriata.*** **Macromischa rugosostriata* Mayr, 1868c: 84, pl. 4, fig. 82 (w.) BALTIC AMBER (Oligocene). Combination in **Nothomyrmica:* Wheeler, W.M. 1915h: 61.

*NOTHOSPHINCTUS:* see under ***SPHINCTOMYRMEX.***

*NOTOMYRMEX:* see under ***MONOMORIUM.***

***NOTONCUS*** (Australia)

*acuminata. Notoncus foreli* var. *acuminata* Viehmeyer, 1925a: 37 (w.) AUSTRALIA. Junior synonym of *ectatommoides:* Brown, 1955d: 486.

*annectens. Notoncus gilberti* subsp. *annectens* Wheeler, W.M. 1934d: 154 (w.) AUSTRALIA. Junior synonym of *gilberti:* Brown, 1955d: 490.

***capitatus.*** *Notoncus capitatus* Forel, 1915b: 90, pl. 1, fig. 8 (w.) AUSTRALIA. Junior synonym of *enormis:* Brown, 1955d: 489. Revived from synonymy, and senior synonym of *minor* and *mjobergi:* Taylor, 1992a: 61.

*dentata. Notoncus foreli* var. *dentata* Forel, 1910b: 68 (w.) AUSTRALIA. Junior synonym of *ectatommoides:* Brown, 1955d: 486.

***ectatommoides.*** *Camponotus ectatommoides* Forel, 1892c: 333 (q.) AUSTRALIA. Emery, 1895g: 353 (w.); Wheeler, G.C. & Wheeler, J. 1968: 206 (l.); Imai, Crozier & Taylor, 1977: 346 (k.). Combination in *Notoncus:* Emery, 1895g: 353. Senior synonym of *acuminata, dentata, foreli, rodwayi* and *subdentata:* Brown, 1955d: 485.

***enormis.*** *Notoncus enormis* Szabó, 1910a: 368, fig. 6 (w.) AUSTRALIA. Wheeler, G.C. & Wheeler, J. 1968: 206 (l.). See also: Brown, 1955d: 489.

*foreli. Notoncus foreli* André, 1896b: 256 (w.) AUSTRALIA. Forel, 1915b: 90 (q.); Wheeler, G.C. & Wheeler, J. 1968: 208 (l.). Junior synonym of *ectatommoides:* Brown, 1955d: 485.

***gilberti.*** *Notoncus gilberti* Forel, 1895f: 418 (w.q.) AUSTRALIA. Senior synonym of *annectens, gracilior* and *politus,* and material of the unavailable name *manni* referred here: Brown, 1955d: 490.

*gracilior. Notoncus gilberti* var. *gracilior* Forel, 1907h: 299 (w.) AUSTRALIA. Junior synonym of *gilberti:* Brown, 1955d: 490. See also: Wheeler, W.M. 1934d: 153.

***hickmani.*** *Notoncus hickmani* Clark, 1930a: 126, fig. 1, no. 14 (w.q.) AUSTRALIA. Combination in *Prolasius:* Wheeler, W.M. 1935c: 71; in *Notoncus,* and senior synonym of *rotundiceps:* Brown, 1955d: 492.

*manni. Notoncus gilberti* subsp. *annectens* var. *manni* Wheeler, W.M. 1934d: 155 (w.) AUSTRALIA. Unavailable name (Taylor, 1986: 34). Material referred to *gilberti* by Brown, 1955d: 490.

*minor. Notoncus capitatus* var. *minor* Viehmeyer, 1925b: 139 (w.q.) AUSTRALIA. Junior synonym of *enormis:* Brown, 1955d: 489; of *capitatus:* Taylor, 1992a: 61.

*mjobergi. Notoncus mjobergi* Forel, 1915b: 91 (w.) AUSTRALIA. Junior synonym of *enormis:* Brown, 1955d: 489; of *capitatus:* Taylor, 1992a: 61.

*politus. Notoncus politus* Viehmeyer, 1925a: 38 (w.q.) AUSTRALIA. Junior synonym of *gilberti:* Brown, 1955d: 490.

*rodwayi. Notoncus rodwayi* Donisthorpe, 1941j: 206 (q.) AUSTRALIA. Junior synonym of *ectatommoides:* Brown, 1955d: 486.

*rotundiceps. Notoncus rotundiceps* Clark, 1930a: 127, fig. 1, no. 15 (w.) AUSTRALIA. Combination in *Prolasius:* Wheeler, W.M. 1935c: 71. Junior synonym of *hickmani:* Brown, 1955d: 492.

***spinisquamis.*** *Melophorus spinisquamis* André, 1896b: 254 (w.q.m.) AUSTRALIA. Wheeler, G.C. & Wheeler, J. 1968: 205 (l.). Combination in *Diodontolepis:* Wheeler, W.M. 1920: 53; in *Notoncus:* Brown, 1955d: 493.

*subdentata. Notoncus foreli* var. *subdentata* Forel, 1910b: 68 (w.) AUSTRALIA. Viehmeyer, 1925a: 37 (q.m.). Junior synonym of *ectatommoides:* Brown, 1955d: 486.

***NOTOSTIGMA*** (Australia)

***carazzii.*** *Camponotus carazzii* Emery, 1895g: 354 (w.) AUSTRALIA. Forel, 1915b: 104 (s.m.); Wheeler, G.C. & Wheeler, J. 1988: 355 (l.). Combination in *C. (Myrmosphincta*): Forel, 1912i: 92; in *Notostigma:* Emery, 1920b: 253. Senior synonym of *podenzanai:* Taylor, 1992a: 62.

***foreli.*** *Notostigma foreli* Emery, 1920b: 253 (w.q.m.) AUSTRALIA.

*podenzanai. Camponotus podenzanai* Emery, 1895g: 355 (w.m.) AUSTRALIA. Combination in *C. (Myrmosphincta*): Forel, 1914a: 273; in *Notostigma:* Emery, 1920b: 254. Junior synonym of *carazzii:* Taylor, 1992a: 62.

*sanguinea* Clark, 1930; see under ***CAMPONOTUS.***

*NOVOMESSOR:* see under ***APHAENOGASTER.***

*NYCTERESIA:* see under ***LABIDUS.***

*NYLANDERIA:* see under ***PARATRECHINA.***

*NYSTALOMYRMA:* see under ***APHAENOGASTER.***

***OCHETELLUS*** (Palaearctic, Oriental, Indo-Australian, Australasian)

*abbotti. Iridomyrmex itoi* subsp. *abbotti* Wheeler, W.M. 1906c: 318, pl. 41, fig. 3 (w.) JAPAN. Junior synonym of *itoi:* Yano, 1910: 420.

#***clarithorax.*** *Iridomyrmex glaber* var. *clarithorax* Forel, 1902h: 473 (w.q.) AUSTRALIA. Combination in *Ochetellus:* Shattuck, 1992a: 17.

#***consimilis.*** *Iridomyrmex glaber* var. *consimilis* Viehmeyer, 1914b: 529 (w.) NEW GUINEA. Combination in *Ochetellus:* Shattuck, 1992a: 17.

***epinotalis.*** *Iridomyrmex epinotalis* Viehmeyer, 1914b: 530 (w.) NEW GUINEA. Combination in *Ochetellus:* Shattuck, 1992a: 17.

***flavipes.*** *Hypoclinea flavipes* Kirby, W.F. 1896: 206 (w.) AUSTRALIA. Combination in *Iridomyrmex:* Clark, 1930b: 268; Clark, 1930c: 21; in *Ochetellus:* Shattuck, 1992a: 17. Senior synonym of *rostrinotus:* Clark, 1930b: 268; Clark, 1930c: 21.

***glaber.*** *Hypoclinea glabra* Mayr, 1862: 705 (w.m.) AUSTRALIA. Mayr, 1876: 81 (q.); Crozier, 1969: 251 (k.); Wheeler, G.C. & Wheeler, J. 1974b: 398 (l.). Combination in *Iridomyrmex:* Mayr, 1865: 61; in *Ochetellus:* Shattuck, 1992a: 17. See also: Brown, 1958h: 40. Current subspecies: nominal plus *clarithorax, consimilis, sommeri.*

***itoi.*** *Iridomyrmex itoi* Forel, 1900e: 269 (w.) JAPAN. Wheeler, W.M. 1928d: 116 (q.); Wheeler, G.C. & Wheeler, J. 1951: 185 (l.); Imai & Yosida, 1964: 64 (k.). Combination in *Ochetellus:* Shattuck, 1992a: 17. Senior synonym of *abbotti:* Yano, 1910: 420.

*janthina. Iridomyrmex glaber* subsp. *sommeri* var. *janthina* Emery, 1914f: 420 (w.q.m.) NEW CALEDONIA. Unavailable name (Taylor, 1986: 34).

***punctatissimus.*** *Iridomyrmex punctatissimus* Emery, 1887a: 251, pl. 4, fig. 27 (w.) AUSTRALIA. Wheeler, G.C. & Wheeler, J. 1966: 729 (l.). Combination in *Ochetellus:* Shattuck, 1992a: 17.

*rostrinotus. Iridomyrmex rostrinotus* Forel, 1910b: 53 (w.q.m.) AUSTRALIA. Junior synonym of *flavipes:* Clark, 1930b: 268; Clark, 1930c: 21.

#***sommeri.*** *Iridomyrmex glaber* var. *sommeri* Forel, 1902h: 473 (w.) NEW CALEDONIA. Combination in *Ochetellus:* Shattuck, 1992a: 17. See also: Emery, 1914f: 420.

***sororis.*** *Iridomyrmex sororis* Mann, 1921: 469, fig. 26 (w.q.m.) FIJI IS. Combination in *Ochetellus:* Shattuck, 1992a: 17.

***vinsoni.*** *Iridomyrmex vinsoni* Donisthorpe, 1946c: 781 (w.) MAURITIUS. Combination in *Ochetellus:* Shattuck, 1992a: 17.

***OCHETOMYRMEX*** (Neotropical)

***argentinus.*** *Brownidris argentinus* Kusnezov, 1957a: 276, pl. 4, figs. 1–13 (w.q.m.) ARGENTINA. Combination in *Ochetomyrmex:* Kempf, 1975c: 358.

***bolivianus.*** *Brownidris bolivianus* Kusnezov, 1962b: 155, figs. 16, 26 (q.) BOLIVIA. Combination in *Ochetomyrmex:* Kempf, 1975c: 358.

***mayri.*** *Ochetomyrmex mayri* Forel, 1908c: 360 (w.) BRAZIL.

*rugosus* Forel, 1886; see under ***WASMANNIA.***

***semipolitus.*** *Ochetomyrmex semipolitus* Mayr, 1878: 872 (w.) BRAZIL.

***subpolitus.*** *Wasmannia subpolita* Wheeler, W.M. 1916c: 8 (w.) GUIANA. Kempf, 1975c: 361 (q.m.); Wheeler, G.C. & Wheeler, J. 1977: 595 (l.). Combination in *Ochetomyrmex:* Kempf, 1961b: 511.

*tucumanus. Brownidris tucumanus* Kusnezov, 1958b: 190. *Nomen nudum.*

*OCTELLA:* see under ***SOLENOPSIS.***

***OCTOSTRUMA*** (Neotropical)

***balzani.*** *Rhopalothrix balzani* Emery, 1894c: 217, pl. 1, fig. 10 (w.q.) BOLIVIA. Combination in *Octostruma:* Brown, 1949f: 92. Senior synonym of *barberi, equilatera* and *lutzi:* Brown & Kempf, 1960: 194.

*barberi. Rhopalothrix (Octostruma) barberi* Mann, 1922: 42 (w.) GUATEMALA. Combination in *Octostruma:* Brown, 1949f: 92. Junior synonym of *balzani:* Brown & Kempf, 1960: 194.

***batesi.*** *Rhopalothrix batesi* Emery, 1894c: 218, pl. 1, fig. 11 (w.) BRAZIL. Combination in *Octostruma:* Brown, 1949f: 92. See also: Brown & Kempf, 1960: 201; Hölldobler & Wilson, 1986: 17.

***betschi.*** *Octostruma betschi* Perrault, 1988b: 303, 2 figs (w.q.) FRENCH GUIANA.

*equilatera. Rhopalothrix (Octostruma) equilatera* Weber, 1934a: 52, fig. 10 (w.) NICARAGUA. Combination in *Octostruma:* Brown, 1949f: 92. Junior synonym of *balzani:* Brown & Kempf, 1960: 194.

*godmani. Rhopalothrix godmani* Forel, 1899c: 41, pl. 3, fig. 4 (q.) PANAMA. Combination in *Octostruma:* Brown, 1949f: 92. Junior synonym of *iheringi:* Brown & Kempf, 1960: 187.

***iheringi.*** *Rhopalothrix iheringi* Emery, 1888c: 361 (q.) BRAZIL. Combination in *Octostruma:* Brown, 1949f: 92. Senior synonym of *godmani, simoni, spei* and *wighti,* and material of the unavailable name *sulcata* referred here: Brown & Kempf, 1960: 187.

***inca.*** *Octostruma inca* Brown & Kempf, 1960: 185, figs. 16, 29 (w.q.) PERU. Wheeler, G.C. & Wheeler, J. 1977: 600 (l.). See also: Kempf, 1968b: 400.

*lutzi. Rhopalothrix (Octostruma) lutzi* Wheeler, W.M. 1913d: 241 (w.q.) DOMINICA. Combination in *Octostruma:* Brown, 1949f: 92. Junior synonym of *balzani:* Brown & Kempf, 1960: 194.

***petiolata.*** *Rhopalothrix petiolata* Mayr, 1887: 580 (q.) BRAZIL. Emery, 1894c: 217 (w.). Combination in *Octostruma:* Brown, 1949f: 92. See also: Brown & Kempf, 1960: 182.

***rugifera.*** *Rhopalothrix rugifer* Mayr, 1887: 579 (w.) BRAZIL. Combination in *Octostruma:* Brown, 1949f: 92. Senior synonym of *truncata:* Brown & Kempf, 1960: 199.

***rugiferoides.*** *Octostruma rugiferoides* Brown & Kempf, 1960: 200, fig. 20 (w.) MEXICO.

*simoni. Rhopalothrix simoni* Emery, 1890a: 67 (w.) VENEZUELA. Combination in *R. (Octostruma*): Forel, 1912e: 196; in *Octostruma:* Brown, 1949f: 92. Junior synonym of *iheringi:* Brown & Kempf, 1960: 187.

*spei. Rhopalothrix (Octostruma) simoni* r. *spei* Forel, 1912e: 196 (w.) COLOMBIA. Combination in *Octostruma:* Brown, 1949f: 92. Junior synonym of *iheringi:* Brown & Kempf, 1960: 187.

*OCTOSTRUMA (continued)*

***stenognatha.*** *Octostruma stenognatha* Brown & Kempf, 1960: 196, figs. 25, 28 (w.q.) BRAZIL.

*sulcata. Rhopalothrix* (*Octostruma*) *simoni* st. *spei* var. *sulcata* Santschi, 1936b: 201 (w.) PANAMA. Unavailable name; material referred to *iheringi* by Brown & Kempf, 1960: 187.

*truncata. Rhopalothrix* (*Octostruma*) *truncata* Forel, 1912e: 196 (q.) BRAZIL. Combination in *Octostruma:* Brown, 1949f: 92. Junior synonym of *rugifera:* Brown & Kempf, 1960: 199.

***wheeleri.*** *Rhopalothrix* (*Octostruma*) *wheeleri* Mann, 1922: 43 (w.) GUATEMALA. Combination in *Octostruma:* Brown, 1949f: 92. See also: Brown & Kempf, 1960: 191.

*wighti. Rhopalothrix simoni* var. *wighti* Wheeler, W.M. 1908b: 161 (w.) JAMAICA. Combination in *Octostruma:* Brown, 1949f: 92. Junior synonym of *iheringi:* Brown & Kempf, 1960: 187.

***OCYMYRMEX*** (Afrotropical)

*abdominalis. Ocymyrmex weitzeckeri* st. *abdominalis* Santschi, 1914e: 16 (w.) SOUTH AFRICA. Junior synonym of *arnoldi:* Arnold, 1916: 197.

***afradu.*** *Ocymyrmex afradu* Bolton & Marsh, 1989: 1288 (w.) NAMIBIA.

***alacer.*** *Ocymyrmex alacer* Bolton & Marsh, 1989: 1290, fig. 9 (w.q.) SOUTH AFRICA.

***ankhu.*** *Ocymyrmex ankhu* Bolton, 1981b: 265 (w.q.) ANGOLA.

*arnoldi. Ocymyrmex arnoldi* Forel, 1913a: 138 (w.m.) ZIMBABWE. Wheeler, G.C. & Wheeler, J. 1973b: 76 (l.). Variety/race of *weitzeckeri:* Forel, 1913j: 213; Arnold, 1916: 197. Senior synonym of *abdominalis:* Arnold, 1916: 197. Junior synonym of *fortior:* Bolton, 1981b: 269.

*barbatus* Emery; see under ***barbiger.***

***barbiger.*** *Ocymyrmex barbiger* Emery, 1886: 364, pl. 17, figs. 9–11 (w.m.) SOUTH AFRICA. [Misspelled as *barbatus* by Emery, 1892a: 114.] Bolton, 1981b: 261 (q.). See also: Bolton & Marsh, 1989: 1291.

*carpenteri. Ocymyrmex carpenteri* Donisthorpe, 1933b: 195 (q., not w.) BOTSWANA. Junior synonym of *picardi:* Bolton, 1981b: 275.

***cavatodorsatus.*** *Ocymyrmex cavatodorsatus* Prins, 1965c: 1021, figs. 1, 2 (w.) SOUTH AFRICA. Bolton & Marsh, 1989: 1278 (q.). See also: Bolton, 1981b: 266.

***celer.*** *Ocymyrmex weitzeckeri* subsp. *celer* Weber, 1943c: 368 (w.) SUDAN. Bolton & Marsh, 1989: 1278 (q.). Raised to species: Bolton, 1981b: 267.

***cilliei.*** *Ocymyrmex cilliei* Prins & Roux, in Bolton & Marsh, 1989: 1292 (m.) SOUTH AFRICA.

***cursor.*** *Ocymyrmex cursor* Bolton, 1981b: 267 (w.) ANGOLA.

***dekerus.*** *Ocymyrmex dekerus* Bolton & Marsh, 1989: 1292 (w.) NAMIBIA.

***engytachys.*** *Ocymyrmex engytachys* Bolton & Marsh, 1989: 1294 (w.) NAMIBIA.

***flavescens.*** *Ocymyrmex barbiger* var. *flavescens* Stitz, 1923: 147 (w.) NAMIBIA. Junior synonym of *barbiger:* Bolton, 1981b: 265. Revived from synonymy, and raised to species: Bolton & Marsh, 1989: 1294.

***flaviventris.*** *Ocymyrmex hirsutus* var. *flaviventris* Santschi, 1914a: 431 (w.) TANZANIA. Bolton, 1981b: 261 (q.). Raised to species: Bolton, 1981b: 268. See also: Bolton & Marsh, 1989: 1295.

***foreli.*** *Ocymyrmex weitzeckeri* var. *foreli* Arnold, 1916: 197 (w.) ZIMBABWE. Bolton, 1981b: 261 (q.); Bolton & Marsh, 1989: 1279 (m.). Raised to species: Bolton, 1981b: 269. See also: Forder & Marsh, 1989: 106.

***fortior.*** *Ocymyrmex weitzeckeri* st. *fortior* Santschi, 1911g: 209 (w.) ANGOLA. Bolton, 1981b: 261 (q.m.). Raised to species, and senior synonym of *arnoldi* (and its junior synonym *abdominalis*), *transversus* and *usakosensis:* Bolton, 1981b: 269. See also: Bolton & Marsh, 1989: 1296.

***gariepensis.*** *Ocymyrmex gariepensis* Prins & Roux, in Bolton & Marsh, 1989: 1282 (diagnosis in key) (w.) SOUTH AFRICA.

***gordoni.*** *Ocymyrmex gordoni* Prins & Roux, in Bolton & Marsh, 1989: 1282 (diagnosis in key) (w.) SOUTH AFRICA.

***hirsutus.*** *Ocymyrmex weitzeckeri* subsp. *hirsutus* Forel, 1910f: 13 (w.) BOTSWANA. Bolton & Marsh, 1989: 1278 (q.). Raised to species: Santschi, 1914a: 431. See also: Bolton, 1981b: 271.

***ignotus.*** *Ocymyrmex ignotus* Bolton & Marsh, 1989: 1297 (w.q.) SOUTH AFRICA.

***kahas.*** *Ocymyrmex kahas* Bolton & Marsh, 1989: 1298, figs. 11, 16 (w.) NAMIBIA.

***laticeps.*** *Ocymyrmex laticeps* Forel, 1901d: 306 (w.) ANGOLA. [Also described as new by Forel, 1903e: 561.] See also: Bolton, 1981b: 271.

***micans.*** *Ocymyrmex weitzeckeri* var. *micans* Forel, 1910f: 12 (w.) NAMIBIA. Raised to species: Bolton, 1981b: 272.

***monardi.*** *Ocymyrmex weitzeckeri* st. *monardi* Santschi, 1930b: 68 (w.) ANGOLA. Raised to species: Bolton, 1981b: 273.

***nitidulus.*** *Ocymyrmex robecchii* r. *nitidulus* Emery, 1892a: 116 (q., not w.) SOMALIA. Emery, 1896i: 157 (m.); Bolton, 1981b: 274 (w.). Raised to species: Bolton, 1981b: 274.

***okys.*** *Ocymyrmex okys* Bolton & Marsh, 1989: 1299 (w.q.) NAMIBIA.

***phraxus.*** *Ocymyrmex phraxus* Bolton, 1981b: 274, fig. 27 (w.) TANZANIA. See also: Bolton & Marsh, 1989: 1301.

***picardi.*** *Ocymyrmex picardi* Forel, 1901d: 306 (w.) ANGOLA. [Also described as new by Forel, 1903e: 561.] Senior synonym of *carpenteri:* Bolton, 1981b: 275.

***resekhes.*** *Ocymyrmex resekhes* Bolton & Marsh, 1989: 1301, fig. 17 (w.q.) SOUTH AFRICA.

***robecchii.*** *Ocymyrmex robecchii* Emery, 1892a: 114, fig. (w.) SOMALIA. Menozzi, 1922a: 353 (m.). See also: Bolton, 1981b: 276.

***robustior.*** *Ocymyrmex barbiger* var. *robustior* Stitz, 1923: 146 (w.) NAMIBIA. Bolton & Marsh, 1989: 1279 (q.m.). Junior synonym of *barbiger:* Bolton, 1981b: 265. Revived from synonymy, and raised to species: Marsh, 1986: 339. See also: Marsh, 1985: 286; Bolton & Marsh, 1989: 1303.

***shushan.*** *Ocymyrmex shushan* Bolton, 1981b: 277, fig. 31 (w.) NAMIBIA.

***sobek.*** *Ocymyrmex sobek* Bolton, 1981b: 277, fig. 20 (w.q.) BOTSWANA.

***sphinx.*** *Ocymyrmex sphinx* Bolton, 1981b: 278, fig. 21 (w.) BOTSWANA. Bolton & Marsh, 1989: 1278 (q.).

***tachys.*** *Ocymyrmex tachys* Bolton & Marsh, 1989: 1304 (w.) NAMIBIA.

*transversus. Ocymyrmex weitzeckeri* st. *transversus* Santschi, 1911g: 209 (q., not w.) ANGOLA. Junior synonym of *fortior:* Bolton, 1981b: 269.

***turneri.*** *Ocymyrmex turneri* Donisthorpe, 1931b: 499 (w.) NAMIBIA. Bolton & Marsh, 1989: 1278 (q.). See also: Bolton, 1981b: 279; Bolton & Marsh, 1989: 1305.

*usakosensis. Ocymyrmex weitzeckeri* var. *usakosensis* Stitz, 1923: 146 (w.) NAMIBIA. Junior synonym of *fortior:* Bolton, 1981b: 269.

***velox.*** *Ocymyrmex velox* Santschi, 1932a: 387 (w.) NAMIBIA. Bolton, 1981b: 261 (q.). See also: Bolton, 1981b: 280; Bolton & Marsh, 1989: 1306.

***weitzeckeri.*** *Ocymyrmex weitzeckeri* Emery, 1892a: 116 (footnote) (w.) LESOTHO. Arnold, 1916: 197 (m.); Bolton, 1981b: 280 (q.). Senior synonym of *wroughtoni:* Bolton, 1981b: 280. See also: Bolton & Marsh, 1989: 1306.

*wroughtoni. Ocymyrmex weitzeckeri* subsp. *wroughtoni* Forel, 1910f: 13 (w.m.) SOUTH AFRICA. Junior synonym of *weitzeckeri:* Bolton, 1981b: 280.

***zekhem.*** *Ocymyrmex zekhem* Bolton, 1981b: 281 (w.) NAMIBIA. Bolton & Marsh, 1989: 1278 (q.). See also: Bolton & Marsh, 1989: 1306.

***ODONTOMACHUS*** (World Tropics and Subtropics)

***aciculatus.*** *Odontomachus aciculatus* Smith, F. 1863: 19 (w.) INDONESIA (Misool I.). Variety of *cephalotes:* Emery, 1887b: 428. Revived status as species: Wilson, 1959a: 489. Senior synonym of *aeneus:* Brown, 1976a: 102. Material of the unavailable name *verticillatus* referred here: Wilson, 1959a: 489. See also: Brown, 1976a: 123.

*acutidens. Odontomachus ruficeps* r. *acutidens* Forel, 1900b: 56 (w.) AUSTRALIA. Crawley, 1921: 90 (q.m.). Junior synonym of *ruficeps:* Brown, 1976a: 105.

*aeneus. Odontomachus aeneus* Emery, 1911b: 534 (w.) NEW GUINEA. Junior synonym of *aciculatus:* Brown, 1976a: 102.

***affinis.*** *Odontomachus affinis* Guérin-Méneville, 1844a: 423 (w.) BRAZIL. Roger, 1861a: 27 (q.). See also: Brown, 1976a: 124.

*ajax. Odontomachus ajax* Clark, 1928b: 169 (w.) AUSTRALIA. [First available use of *Odontomachus ruficeps* subsp. *turneri* var. *ajax* Forel, 1910b: 10; unavailable name (Taylor, 1986: 34).] Junior synonym of *cephalotes:* Brown, 1976a: 103.

***allolabis.*** *Odontomachus allolabis* Kempf, 1974c: 551, figs. 1–4 (q.) BRAZIL.

***angulatus.*** *Odontomachus angulatus* Mayr, 1866a: 500, pl., fig. 10 (w.) FIJI IS. Mann, 1921: 427 (q.). Senior synonym of *politus:* Brown, 1976a: 102. See also: Brown, 1976a: 124.

***animosus.*** *Odontomachus animosus* Smith, F. 1860b: 102 (w.) NEW GUINEA. See also: Wilson, 1959a: 490; Brown, 1976a: 125.

*aruanus. Odontomachus ruficeps* subsp. *aruanus* Karavaiev, 1925c: 295, fig. 14 (w.m.) INDONESIA

(Aru I.). Junior synonym of *cephalotes:* Wilson, 1959a: 491.

***assiniensis.*** *Odontomachus assiniensis* Emery, 1892d: 558 (w.) IVORY COAST. Santschi, 1910c: 350 (q.); Forel, 1913e: 666 (q.m.); Santschi, 1914b: 57 (m.). Senior synonym of *intermedius:* Wheeler, W.M. 1922a: 793; Brown, 1976a: 102; of *aterrimus, caffrorum, fauconneti, flavescens, furvior, fuscus* Stitz, 1916, and *tropicalis:* Brown, 1976a: 102. See also: Brown, 1976a: 128.

*aterrimus. Odontomachus assiniensis* var. *aterrimus* Wheeler, W.M. 1922a: 102 (w.) ZAIRE. Junior synonym of *assiniensis:* Brown, 1976a: 102.

***banksi.*** *Odontomachus banksi* Forel, 1910d: 121 (w.) PHILIPPINES. Viehmeyer, 1916b: 284 (q.). See also: Brown, 1976a: 125.

***bauri.*** *Odontomachus bauri* Emery, 1892d: 561 (diagnosis in key) (w.) GALAPAGOS IS. Emery, 1893d: 90 (w.); Wheeler, W.M. 1919c: 267 (q.m.). Subspecies of *haematodus:* Emery, 1911d: 115. Revived status as species: Taylor & Wilson, 1962: 142. Senior synonym of *microcephalus, notata, paucidens, pubescens, rugisquama* and *sericeus:* Brown, 1976a: 102. See also: Brown, 1976a: 129.

***biolleyi.*** *Odontomachus biolleyi* Forel, 1908b: 35 (w.q.) COSTA RICA. Wheeler, G.C. & Wheeler, J. 1952c: 651 (l.).

*bispinosus* Smith, F. 1858; see under ***ANOCHETUS.***

***biumbonatus.*** *Odontomachus biumbonatus* Brown, 1976a: 131, fig. 22, pl. 1B (w.q.) ECUADOR.

***bradleyi.*** *Odontomachus bradleyi* Brown, 1976a: 133 (w.) PERU.

*breviceps. Odontomachus haematoda* var. *breviceps* Crawley, 1915b: 239 (w.) CHRISTMAS I. Junior synonym of *simillimus:* Brown, 1976a: 106.

*bruneipes. Odontomachus haematodus* r. *pubescens* var. *bruneipes* Emery, 1893d: 91 (footnote) (w.) BRAZIL. Unavailable name; material referred to *haematodus* by Brown, 1976a: 104.

***brunneus.*** *Atta brunnea* Patton, 1894: 618 (w.) U.S.A. Deyrup, Trager & Carlin, 1985: 191 (m.). Combination in *Odontomachus,* and junior synonym of *insularis:* Emery, 1895c: 268. Revived from synonymy, and senior synonym of *fuscus* Stitz, 1925: Brown, 1976a: 103. See also: Brown, 1976a: 139.

***caelatus.*** *Odontomachus caelatus* Brown, 1976a: 140, fig. 23, pl. 1A (w.m.) BRAZIL.

*caffrorum. Odontomachus assiniensis* st. *caffrorum* Santschi, 1914e: 14 (w.) SOUTH AFRICA. Junior synonym of *assiniensis:* Brown, 1976a: 102.

***cephalotes.*** *Odontomachus cephalotes* Smith, F. 1863: 19 (w.) INDONESIA (Seram I.). Forel, 1915b: 34 (q.m.); Wheeler, G.C. & Wheeler, J. 1971b: 1212 (l.). Subspecies of *ruficeps:* Forel, 1911e: 251; Forel, 1915b: 34. Revived status as species: Wilson, 1959a: 490. Senior synonym of *aruanus,* and material of the unavailable names *fusca, cruenta, longitudinalis, tamensis* and *ternatensis* referred here: Wilson, 1959a: 491. Senior synonym of *ajax,* and material of the unavailable names *cooktownensis, obtusa* and *yorkensis* referred here: Brown, 1976a: 103. See also: Crawley, 1922b: 441; Brown, 1976a: 120.

***chelifer.*** *Formica chelifera* Latreille, 1802c: 188, pl. 8, fig. 51 (w.) SOUTH AMERICA (no locality given). Forel, 1909a: 252 (q.m.); Wheeler, G.C. & Wheeler, J. 1952c: 651 (l.). Combination in *Odontomachus:* Latreille, 1809: 128. Senior synonym of *emacerata:* Brown, 1976a: 103; of *leptocephalus:* Emery, 1911d: 114; of *theresiae:* Forel, 1899c: 19; Brown, 1976a: 103.

***circulus.*** *Odontomachus circulus* Wang, M. 1993a: 220, fig. 1 (w.) CHINA.

*clarionensis. Odontomachus haematoda* subsp. *clarionensis* Wheeler, W.M. 1934f: 141 (w.) MEXICO. Junior synonym of *clarus:* Taylor & Wilson, 1962: 142.

***clarus.*** *Odontomachus clarus* Roger, 1861a: 26 (w.) U.S.A. Patton, 1894: 618 (m.); Petralia & Vinson, 1980: 381 (l.). Subspecies of *haematodus:* Emery, 1911d: 115; Wheeler, W.M. 1914b: 40; Smith, M.R. 1939d: 129. Revived status as species: Taylor & Wilson, 1962: 142; Brown, 1976a: 136. Senior synonym of *clarionensis:* Taylor & Wilson, 1962: 142; of *coninodis* and *desertorum:* Brown, 1976a: 103; of *texana:* Wheeler, W.M. 1902f: 26. See also: Deyrup, Trager & Carlin, 1985: 192.

*concentrica. Odontomachus papuanus* var. *concentrica* Emery, 1897d: 557 (w.) NEW GUINEA. Junior synonym of *papuanus:* Wilson, 1959a: 497.

*conifera. Odontomachus rixosus* var. *conifera* Forel, 1913k: 19 (w.q.) JAVA. Junior synonym of *rixosus:* Brown, 1976a: 105.

*coninodis. Odontomachus haematodus* subsp. *coninodis* Wheeler, W.M. 1915b: 391 (w.q.) U.S.A. Junior synonym of *clarus:* Brown, 1976a: 103. See also: Smith, M.R. 1939d: 128.

*cooktownensis. Odontomachus ruficeps* r. *cephalotes* var. *cooktownensis* Forel, 1915b: 35 (w.m.) AUSTRALIA. Unavailable name (Taylor, 1986: 34); material referred to *cephalotes* by Brown, 1976a: 103.

***coquereli.*** *Odontomachus coquereli* Roger, 1861a: 30 (w.) MADAGASCAR. Combination in *Champsomyrmex.* Emery, 1892d: 558. Senior synonym of *minor:* Brown, 1976a: 103. See also: Forel, 1891b: 105; Brown, 1976a: 143.

*coriarius. Odontomachus coriarius* Mayr, 1876: 85 (w.m.) AUSTRALIA. Subspecies of *ruficeps:* Emery, 1911d: 115; of *haematodus:* Wheeler, W.M. 1915g: 806. Junior synonym of *ruficeps:* Brown, 1976a: 105.

***cornutus.*** *Odontomachus cornutus* Stitz, 1933: 67, fig. 1 (w.) ECUADOR. Kugler, C. 1980: 225 (q.m.). See also: Brown, 1976a: 144.

*cruenta. Odontomachus ruficeps* susbp. *cephalotes* var. *cruenta* Emery, 1911c: 251 (w.q.) NEW GUINEA. Unavailable name; material referred to *cephalotes* by Wilson, 1959a: 491.

*desertorum. Odontomachus haematoda* subsp. *desertorum* Wheeler, W.M. 1915b: 391 (w.) U.S.A. Junior synonym of *clarus:* Brown, 1976a: 103. See also: Smith, M.R. 1939d: 128.

*dulcis. Odontomachus haematoda* subsp. *dulcis* Mann, 1922: 18 (w.) HONDURAS. Junior synonym of *minuta:* Brown, 1976a: 104.

*emacerata. Odontomachus chelifer* var. *emacerata* Santschi, 1922d: 243 (w.) BRAZIL. Junior synonym of *chelifer:* Brown, 1976a: 103.

*emeryi. Odontomachus imperator* subsp. *emeryi* Mann, 1919: 303, fig. 12 (w.q.m.) SOLOMON IS. Raised to species: Wilson, 1959a: 492. Junior synonym of *rufithorax:* Brown, 1976a: 106.

***erythrocephalus.*** *Odontomachus haematodes* r. *erythrocephalus* Emery, 1890b: 44 (w.) COSTA RICA. [Also described as new by Emery, 1894k: 50.] Forel, 1914e: 9 (q.); Wheeler, G.C. & Wheeler, J. 1952c: 650 (l.). Raised to species: Kempf, 1972a: 170; Brown, 1976a: 144.

*fauconneti. Odontomachus assiniensis* subsp. *fauconneti* Forel, 1910a: 27 (w.) ZAIRE. Junior synonym of *assiniensis:* Brown, 1976a: 102.

*flavescens. Odontomachus assiniensis* subsp. *flavescens* Bernard, 1953b: 215 (w.q.) GUINEA. Junior synonym of *assiniensis:* Brown, 1976a: 102.

***floresensis.*** *Odontomachus floresensis* Brown, 1976a: 146 (w.) INDONESIA (Flores I.)

*formosae. Odontomachus monticola* var. *formosae* Forel, 1912a: 46 (w.q.m.) TAIWAN. Stirps of *monticola:* Santschi, 1925f: 82. Senior synonym of *kuroiwae:* Teranishi, 1940: 61. Junior synonym of *monticola:* Yasumatsu, 1962: 93.

***fulgidus.*** *Odontomachus fulgidus* Wang, M. 1993a: 222, fig. 3 (w.) CHINA.

*furvior. Odontomachus assiniensis* var. *furvior* Wheeler, W.M. 1922a: 101 (w.q.) ZAIRE. Junior synonym of *assiniensis:* Brown, 1976a: 102.

*fusca. Odontomachus ruficeps* subsp. *cephalotes* var. *fusca* Emery, 1911c: 251 (w.) NEW GUINEA. Unavailable name; material referred to *cephalotes* by Wilson, 1959a: 491.

*fuscipennis. Odontomachus haematodes* var. *fuscipennis* Forel, 1913k: 19 (w.q.m.) SRI LANKA. Junior synonym of *simillimus:* Wilson, 1959a: 499.

*fuscus. Odontomachus assiniensis* var. *fuscus* Stitz, 1916: 372 (w.q.) ZAIRE. Junior synonym of *assiniensis:* Brown, 1976a: 102.

*fuscus. Odontomachus haematodes* var. *fuscus* Stitz, 1925: 115, fig. 2 (w.) CUBA. [Unresolved junior primary homonym of *fuscus* Stitz, 1916, above. Note that the proposed replacement name *repetita* Baroni Urbani, 1971b: 361 is unavailable.] Junior synonym of *brunneus:* Brown, 1976a: 103.

*ghilianii* Spinola, 1851; see under ***ANOCHETUS.***

*gladiator* Donisthorpe, 1932; see under ***ANOCHETUS.***

***granatus.*** *Odontomachus granatus* Wang, M. 1993a: 224, fig. 5 (w.) CHINA.

*gressitti. Odontomachus gressitti* Wilson, 1959a: 492, fig. 4 (w.) NEW GUINEA. Junior synonym of *rufithorax:* Brown, 1976a: 106.

*gulosus. Odontomachus gulosus* Emery, 1901h: 160 (w.) NEW GUINEA. Junior synonym of *testaceus:* Wilson, 1959a: 500.

***haematodus.*** *Formica haematoda* Linnaeus, 1758: 582 (q.) "America meridionali". Emery, 1899c: 5 (l.); Wheeler, G.C. & Wheeler, J. 1952c: 646 (l.). Combination in *Odontomachus:* Latreille, 1804: 179. Senior synonym of *maxillosa:* Retzius, 1783: 75; Olivier, 1792: 502; Latreille, 1802c: 192; of *hirsutiusculus:* Roger, 1863b: 22; Brown, 1976a: 104; of *pallipes,* and material of the unavailable name *bruneipes* referred here: Brown, 1976a: 104.

*hainanensis. Odontomachus monticola* var. *hainanensis* Stitz, 1925: 115, fig. 3 (w.) CHINA. Junior synonym of *monticola:* Brown, 1976a: 105.

***hastatus.*** *Myrmecia hastata* Fabricius, 1804: 426 (w.) CENTRAL AMERICA. Wheeler, G.C. & Wheeler, J. 1971b: 1212 (l.). Combination in *Odontomachus:* Illiger, 1807: 194. Senior synonym of *maxillaris:* Roger, 1863b: 22.

*hirsutiusculus. Odontomachus hirsutiusculus* Smith, F. 1858b: 78 (w.) BRAZIL. Forel, 1909a: 252 (q.). Variety/subspecies of *insularis:* Dalla Torre, 1893: 51; of *haematodus:* Roger, 1861a: 24; Forel, 1907e: 1; Wheeler, W.M. 1911a: 22. Junior synonym of *haematodus:* Roger, 1863b: 22; Brown, 1976a: 104.

***imperator.*** *Odontomachus imperator* Emery, 1887b: 429, pl. 1, fig. 1 (w.) NEW GUINEA. See also: Brown, 1976a: 152.

***infandus.*** *Odontomachus infandus* Smith, F. 1858b: 81 (w.) PHILIPPINES. Senior synonym of *philippinus* and *striaticeps:* Brown, 1976a: 104.

***insularis.*** *Odontomachus insularis* Guérin-Méneville, 1844a: 423 (w.m.) CUBA. Forel, 1897b: 298 (q.). Race/subspecies of *haematodus:* Emery, 1890b: 44 (footnote); Smith, M.R. 1939d: 127. Revived status as species: Taylor & Wilson, 1962: 142. Material of the unavailable names *pallens* and *wheeleri* referred here by Brown, 1976a: 104.

*intermedius. Odontomachus intermedius* Stitz, 1911b: 378 (w.) TANZANIA. Subspecies of *assiniensis:* Santschi, 1914e: 14. Junior synonym of *assiniensis:* Wheeler, W.M. 1922a: 793; Brown, 1976a: 102.

*kuroiwae. Myrtoteras kuroiwae* Matsumura, 1912: 192, pl. 54, fig. 1 (w.) JAPAN. Combination in *Myrmoteras:* Matsumura & Uchida, 1926: 51; in *Odontomachus:* Crieghton, 1939: 40. Junior synonym of *formosae:* Teranishi, 1940: 61.

***laticeps.*** *Odontomachus haematodus* var. *laticeps* Roger, 1861a: 25 (w.) MEXICO. Subspecies/race of *haematodus:* Emery, 1890b: 44; Emery, 1911d: 115. Raised to species: Dalla Torre, 1893: 51; Kempf, 1972a: 172. Senior synonym of *striativentris:* Brown, 1976a: 104.

***latidens.*** *Odontomachus latidens* Mayr, 1867a: 80 (q.) JAVA. Emery, 1892d: 560 (w.); Imai, Brown, *et al.* 1984: 67 (k.). Senior synonym of *procerus* and *sumatranus:* Brown, 1976a: 104. See also: Karavaiev, 1925c: 292.

***latissimus.*** *Odontomachus latissimus* Viehmeyer, 1914c: 31, fig. 5 (w.) NEW GUINEA. See also: Brown, 1976a: 128.

*leptocephalus. Odontomachus chelifer* var. *leptocephalus* Emery, 1890b: 45, pl. 5, fig. 2 (w.) BRAZIL. Junior synonym of *chelifer:* Emery, 1911d: 114.

*linae. Odontomachus saevissimus* var. *linae* Donisthorpe, 1940c: 107 (w.q.) NEW GUINEA. Raised to species: Wilson, 1959a: 494. Junior synonym of *saevissimus:* Brown, 1976a: 106.

*longi. Odontomachus monticola* var. *longi* Forel, 1900c: 58 (w.) INDIA. Junior synonym of *monticola:* Brown, 1976a: 105.

*longitudinalis. Odontomachus ruficeps* subsp. *cephalotes* var. *longitudinalis* Donisthorpe, 1940c: 108, fig. 1 (w.) NEW GUINEA. Unavailable name; material referred to *cephalotes* by Wilson, 1959a: 491.

*macrorhynchus. Pedetes macrorhynchus* Bernstein, 1861: 8 (q.) JAVA. Combination in *Odontomachus:* Dalla Torre, 1893: 51. [Brown, 1976a: 104 gives this name as a *nomen nudum,* which is incorrect as Bernstein provides an indication.] Unidentifiable taxon; *incertae sedis* in *Odontomachus.* See also: Brown, 1976a: 159.

*magnus. Odontomachus coriarius* var. *magnus* Mayr, 1876: 85 (w.) AUSTRALIA. Junior synonym of *ruficeps:* Brown, 1976a: 105.

*major. Odontomachus monticola* var. *major* Forel, 1913f: 183 (w.) TAIWAN. Junior synonym of *monticola:* Brown, 1976a: 105.

***malignus.*** *Odontomachus malignus* Smith, F. 1859a: 144 (w.) NEW GUINEA. Senior synonym of *retrolatior:* Brown, 1976a: 104; of *tuberculatus:* Emery, 1887b: 429; Wilson, 1959a: 495. See also: Brown, 1976a: 159.

*maxillaris. Odontomachus maxillaris* Smith, F. 1858b: 77, pl. 5, figs. 12–14 (w.q.) BRAZIL. Junior synonym of *hastatus:* Roger, 1863b: 22.

*maxillosa. Formica maxillosa* De Geer, 1773: 601, pl. 31, figs. 3–5 (q.) SURINAM. Junior synonym of *haematodus:* Retzius, 1783: 75; Olivier, 1792: 502; Latreille, 1802c: 192.

***mayi.*** *Odontomachus affinis* subsp. *mayi* Mann, 1912: 39 (w.) BRAZIL. Raised to species: Kempf, 1972a: 172; Brown, 1976a: 124.

***meinerti.*** *Odontomachus haematodes* r. *meinerti* Forel, 1905b: 156 (w.) VENEZUELA. Raised to species: **new status**. Synonym of *minuta:* Brown, 1976a: 104. [Brown gives *minuta* as senior name but *meinerti* has priority and is the first available name for this taxon.]

*microcephalus. Odontomachus haematodes* var. *microcephalus* Emery, 1890b: 45, pl. 5, fig. 1 (w.) COSTA RICA. [Also described as new by Emery, 1894k: 50.] Junior synonym of *bauri:* Brown, 1976a: 102.

*minor. Champsomyrmex coquereli* var. *minor* Emery, 1899f: 273 (w.q.) MADAGASCAR. Junior synonym of *coquereli:* Brown, 1976a: 103.

*minuta. Odontomachus haematoda* subsp. *minuta* Luederwaldt, 1918: 36 (w.) BRAZIL. [First available use of *Odontomachus haematodes* subsp. *pubescens* var. *minuta* Emery, 1894c: 146; unavailable name.] Raised to species: Kempf, 1972a: 172. Senior synonym of *dulcis* and *meinerti:* Brown, 1976a: 104. [Brown gives *minuta* as senior name, but *meinerti* has priority.]

***montanus.*** *Odontomachus imperator* var. *montanus* Stitz, 1925: 116 (w.) NEW GUINEA. Raised to species: Wilson, 1959a: 495.

***monticola.*** *Odontomachus monticola* Emery, 1892d: 560 (diagnosis in key) (w.) VIETNAM. Emery, 1895k: 463 (q.); Ogata, 1987: 126 (m.). Senior synonym of *formosae* (and its junior synonym *kuroiwae*) and *pauperculus:* Yasumatsu, 1962: 93; of *hainanensis, longi, major, punctulatus* and *striata:* Brown, 1976a: 105.

***mormo.*** *Odontomachus mormo* Brown, 1976a: 161, figs. 10, 15 (w.) ECUADOR.

*nietneri* Roger, 1861; see under ***ANOCHETUS.***

***nigriceps.*** *Odontomachus nigriceps* Smith, F. 1860b: 103, pl. 1, fig. 9 (w.) NEW GUINEA. Variety of *tyrannicus:* Emery, 1897d: 557; Donisthorpe, 1947c: 579. Status as species: Donisthorpe, 1932c: 461; Wilson, 1959a: 496; Brown, 1976a: 105. Senior synonym of *praefectus:* Wilson, 1959a: 496.

*nigrifrons. Odontomachus nigrifrons* Donisthorpe, 1940c: 106 (w.) NEW GUINEA. Junior synonym of *testaceus:* Wilson, 1959a: 500.

*notata. Odontomachus haematoda* var. *notata* Mann, 1920: 404 (w.) PUERTO RICO. Junior synonym of *bauri:* Brown, 1976a: 102.

*nubila. Odontomachus gulosus* var. *nubila* Emery, 1911c: 250 (w.) NEW GUINEA. Junior synonym of *testaceus:* Wilson, 1959a: 500.

*oblita. Odontomachus assiniensis* var. *oblita* Baroni Urbani, 1971b: 360. Unavailable name (and unnecessary replacement name for *fuscus* Stitz, 1916: 372).

*obscura. Odontomachus coriarius* var. *obscura* Crawley, 1922b: 437, fig. 1 (w.) AUSTRALIA. Junior synonym of *ruficeps:* Brown, 1976a: 105.

*obscurior. Odontomachus rixosus* var. *obscurior* Forel, 1900c: 58 (w.) BURMA. Viehmeyer, 1916a: 116 (q.). Junior synonym of *rixosus:* Brown, 1976a: 105.

*obsolescens. Odontomachus tyrannicus* var. *obsolescens* Donisthorpe, 1940c: 106 (w.m.) NEW GUINEA. Junior synonym of *tyrannicus:* Wilson, 1959a: 501.

*obtusa. Odontomachus ruficeps* subsp. *cephalotes* var. *obtusa* Emery, 1911c: 251 (w.) NEW GUINEA. Unavailable name; material referred to *cephalotes* by Brown, 1976a: 103.

***opaciventris.*** *Odontomachus haematodus* r. *opaciventris* Forel, 1899c: 21, pl. 1, fig. 14 (w.q.) MEXICO. Raised to species: Kempf, 1972a: 172; Brown, 1976a: 105.

***opaculus.*** *Odontomachus imperator* subsp. *opaculus* Viehmeyer, 1912: 6, fig. 6 (w.) NEW GUINEA. Raised to species: Wilson, 1959a: 496.

*pallens. Odontomachus haematodes* subsp. *insularis* var. *pallens* Wheeler, W.M. 1905b: 82 (w.q.m.) BAHAMAS. Unavailable name; material referred to *insularis* by Brown, 1976a: 104.

*pallidicornis. Ponera pallidicornis* Smith, F. 1860a: 73 (m.) SULAWESI. Combination in *Euponera* (*Brachyponera*): Donisthorpe, 1932c: 458. Junior synonym of *simillimus:* Brown, 1976a: 106.

*pallipes. Odontomachus haematoda* var. *pallipes* Crawley, 1916b: 368 (w.) GUIANA. Junior synonym of *haematodus:* Brown, 1976a: 104.

***panamensis.*** *Odontomachus affinis* r. *panamensis* Forel, 1899c: 19, pl. 1, fig. 13 (w.q.m.) PANAMA. Raised to species: Kempf, 1972a: 172; Brown, 1976a: 105.

***papuanus.*** *Odontomachus rixosus* var. *papuanus* Emery, 1887b: 429 (w.) NEW GUINEA. Viehmeyer, 1916b: 284 (footnote) (q.). Raised to species: Emery, 1892d: 559 (in key); Wilson, 1959a: 496. Senior synonym of *concentrica:* Wilson, 1959a: 497.

*paucidens. Odontomachus haematodus* var. *paucidens* Emery, 1893d: 91 (footnote) (w.) HAITI. Menozzi & Russo, 1930: 151 (m.). Subspecies of *haematodus:* Wheeler, W.M. 1936b: 196. Junior synonym of *bauri:* Brown, 1976a: 102.

*pauperculus. Odontomachus monticola* subsp. *pauperculus* Wheeler, W.M. 1921c: 530 (w.) CHINA. Junior synonym of *monticola:* Yasumatsu, 1962: 93.

***peruanus.*** *Odontomachus haematodes* var. *peruana* Stitz, 1933: 68 (q.) PERU. Raised to species, *nomen dubium:* Brown, 1976a: 105. See also: Brown, 1976a: 163.

*philippinus*. *Odontomachus papuanus* r. *philippinus* Emery, 1893e: 203 (w.) PHILIPPINES. Junior synonym of *infandus:* Brown, 1976a: 104.

*politus*. *Odontomachus politus* Stitz, 1925: 116 (w.) FIJI IS. Junior synonym of *angulatus:* Brown, 1976a: 102.

*praefectus*. *Odontomachus angulatus* subsp. *praefectus* Forel, 1911e: 251 (w.) NEW GUINEA. Junior synonym of *nigriceps:* Wilson, 1959a: 496.

*procerus*. *Odontomachus latidens* r. *procerus* Emery, 1893e: 203 (q.) WEST MALAYSIA. Junior synonym of *latidens:* Brown, 1976a: 104.

****pseudobauri***. **Odonotomachus pseudobauri* De Andrade, 1994: 9, fig. 14 (w.) DOMINICAN AMBER.

*pubescens*. *Odontomachus haematodus* var. *pubescens* Roger, 1861a: 25 (w.) VENEZUELA. Forel, 1909a: 252 (q.). Raised to species: Dalla Torre, 1893: 52. Race/subspecies of *haematodus:* Emery, 1890b: 44 (footnote); Emery, 1911d: 115. Junior synonym of *bauri:* Brown, 1976a: 102.

*punctulatus*. *Odontomachus monticola* r. *punctulatus* Forel, 1900c: 58 (w.) INDIA. Raised to species: Bingham, 1903: 49; Wheeler, W.M. 1930h: 61; Wu, C.F. 1941: 149. Junior synonym of *monticola:* Brown, 1976a: 105.

*quadrispinosus* Smith, F. 1858; see under ***ANOCHETUS***.

*repetita*. *Odontomachus haematoda* var. *repetita* Baroni Urbani, 1971b: 361. Unavailable name (see under *fuscus*).

*retrolatior*. *Odontomachus retrolatior* Viehmeyer, 1914a: 113 (w.) NEW GUINEA. Junior synonym of *malignus:* Brown, 1976a: 104.

***rixosus***. *Odontomachus rixosus* Smith, F. 1857a: 64 (w.) SINGAPORE. Crawley, 1924: 388 (q.); Wheeler, G.C. & Wheeler, J. 1952c: 651 (l.); Imai, Brown, *et al* 1984: 67 (k.). Senior synonym of *conifera* and *obscurior:* Brown, 1976a: 105. See also: Brown, 1976a: 163.

*rossi* Donisthorpe, 1947; see under ***ANOCHETUS***.

*rubriceps*. *Odontomachus ruficeps* var. *rubriceps* Forel, 1915b: 33 (w.m.) AUSTRALIA. Junior synonym of *ruficeps:* Brown, 1976a: 105.

*rufescens*. *Odontomachus ruficeps* var. *rufescens* Forel, 1915b: 34 (w.) AUSTRALIA. Junior synonym of *ruficeps:* Brown, 1976a: 105.

***ruficeps***. *Odontomachus ruficeps* Smith, F. 1858b: 81 (w.) AUSTRALIA. Forel, 1910b: 10 (m.). Senior synonym of *acutidens, coriarius, magnus, obscura, rubriceps, rufescens, semicircularis, septentrionalis* and *sharpei:* Brown, 1976a: 105. See also: Crawley, 1922b: 438; Brown, 1976a: 120.

***rufithorax***. *Odontomachus imperator* subsp. *rufithorax* Emery, 1911b: 534 (w.) NEW GUINEA. Raised to species: Wilson, 1959a: 497. Senior synonym of *emeryi* and *gressitti:* Brown, 1976a: 106.

*rufus* Jerdon, 1851; see under ***ANOCHETUS***.

***ruginodis***. *Odontomachus haematodus* var. *ruginodis* Smith, M.R. 1937: 828 (w.q.) BAHAMAS. [First available use of *Odontomachus haematodes* subsp. *insularis* var. *ruginodis* Wheeler, W.M. 1905b: 82; unavailable name.] Deyrup, Trager & Carlin, 1985: 192 (m.). Raised to species: Wilson, 1964b: 4. Junior synonym of *brunneus:* Brown, 1976a: 103. Revived from synonymy: Deyrup, Trager & Carlin, 1985: 192.

*rugisquama*. *Odontomachus haematodes* var. *rugisquama* Forel, 1908b: 35 (w.) COSTA RICA. Forel, 1912c: 28 (q.). Junior synonym of *bauri:* Brown, 1976a: 102.

*rugosus* Smith, F. 1857; see under ***ANOCHETUS***.

***saevissimus***. *Odontomachus saevissimus* Smith, F. 1858b: 80, pl. 5, fig. 10 (w.) INDONESIA (Seram I.). Smith, F. 1860b: 102 (q.); Karavaiev, 1925c: 291 (m.). Senior synonym of *tauerni* and *transversostriatus:* Wilson, 1959a: 498; of *linae:* Brown, 1976a: 106. See also: Brown, 1976a: 152.

***scalptus***. *Odontomachus scalptus* Brown, 1978b: 281 (w.) ECUADOR.

*semicircularis*. *Odontomachus coriarius* var. *semicircularis* Mayr, 1876: 85 (w.) AUSTRALIA. Junior synonym of *ruficeps:* Brown, 1976a: 105.

*septentrionalis*. *Odontomachus septentrionalis* Crawley, 1915a: 130 (w.) AUSTRALIA. Junior synonym of *ruficeps:* Brown, 1976a: 105.

*sericeus*. *Odontomachus sericeus* Provancher, 1895: 97 (q.) TRINIDAD. Junior synonym of *bauri:* Brown, 1976a: 102.

*sharpei*. *Odontomachus sharpei* Forel, 1893d: 458 (q.) AUSTRALIA. Junior synonym of *ruficeps:* Brown, 1976a: 105.

***silvestrii***. *Odontomachus silvestrii* Wheeler, W.M. 1927h: 85, fig. 1 (w.) VIETNAM. Senior synonym of *substriatus:* Brown, 1976a: 106.

***simillimus***. *Odontomachus simillimus* Smith, F. 1858b: 80, pl. 5, figs. 8, 9 (q.) FIJI IS. Mayr, 1867a: 79 (w.); Karavaiev, 1925c: 294 (m.); Wheeler, G.C. & Wheeler, J. 1980: 530 (l.); Imai, Brown, *et al.* 1984: 67 (k.); Tjan, Imai, *et al.* 1986: 57 (k.). Junior synonym of *haematodus:* Roger, 1861a: 24; Mayr, 1865: 64; Emery, 1911d: 114. Revived from synonymy: Wilson, 1959a: 499; Brown, 1976a: 106. Senior synonym of *fuscipennis:* Wilson, 1959a: 499; of *breviceps* and *pallidicornis:* Brown, 1976a: 106. See also: Brown, 1976a: 165.

*smithii* Roger, 1863; see under ***ANOCHETUS***.

****spinifer***. **Odontomachus spinifer* De Andrade, 1994: 7, fig. 13 (w.) DOMINICAN AMBER.

***spissus***. *Odontomachus spissus* Kempf, 1962b: 17, figs. 5–7 (w.) BRAZIL. Kempf, 1975c: 345 (q.).

*stanleyi*. *Odontomachus haematodus* var. *stanleyi* Wheeler, W.M. 1922a: 102 (w.) ZAIRE. Junior synonym of *troglodytes:* Brown, 1976a: 106.

*striata*. *Odontomachus latidens* susbp. *striata* Menozzi, 1930d: 329 (w.) CHINA. Junior synonym of *monticola:* Brown, 1976a: 105.

*striaticeps*. *Odontomachus infandus* r. *striaticeps* Stitz, 1925: 117 (w.) PHILIPPINES. Junior synonym of *infandus:* Brown, 1976a: 104.

*striativentris*. *Odontomachus haematodes* r. *striativentris* Emery, 1890b: 44 (w.) COSTA RICA. [Also described as new by Emery, 1894k: 50.] Subspecies of *haematodus:* Emery, 1911d: 115. Raised to species: Dalla Torre, 1893: 52; Kempf, 1972a: 172. Junior synonym of *laticeps:* Brown, 1976a: 104.

*substriatus*. *Odontomachus silvestrii* var. *substriatus* Wheeler, W.M. 1927h: 86 (w.) VIETNAM. Junior synonym of *silvestrii:* Brown, 1976a: 106.

*sumatranus*. *Odontomachus latidens* subsp. *sumatranus* Emery, 1900d: 671, fig. 4 (w.q.) SUMATRA. Junior synonym of *latidens:* Brown, 1976a: 104.

***sumbensis***. *Odontomachus sumbensis* Brown, 1976a: 166 (w.) INDONESIA (Sumba I.).

*tamensis*. *Odontomachus ruficeps* r. *cephalotes* var. *tamensis* Stitz, 1912: 503, fig. 7 (w.) NEW GUINEA. Unavailable name; material referred to *cephalotes* by Wilson, 1959a: 491.

*tauerni*. *Odontomachus imperator* var. *tauerni* Stitz, 1925: 116 (w.) INDONESIA (Seram I.). Junior synonym of *saevissimus:* Wilson, 1959a: 498.

***tensus***. *Odontomachus tensus* Wang, M. 1993a: 223, fig. 4 (w.) CHINA.

*ternatensis*. *Odontomachus ruficeps* subsp. *cephalotes* var. *ternatensis* Forel, 1911e: 252 (w.) INDONESIA (Ternate I.). Unavailable name; material referred to *cephalotes* by Wilson, 1959a: 491.

***testaceus***. *Odontomachus tyrannicus* var. *testacea* Emery, 1897d: 557 (w.) NEW GUINEA. Raised to species, and senior synonym of *gulosus, nigrifrons* and *nubila:* Wilson, 1959a: 500.

*texana*. *Odontomachus texana* Buckley, 1867: 335 (w.) U.S.A. Junior synonym of *insularis:* Emery, 1895c: 268; of *clarus:* Wheeler, W.M. 1902f: 26.

*theresiae*. *Odontomachus haematodes* r. *theresiae* Forel, 1895a: 41 (w.) MEXICO. Emery, 1901a: 53 (m.). Variety/subspecies of *chelifer:* Emery, 1901a: 53; Forel, 1912c: 28; Kempf, 1972a: 170. Junior synonym of *chelifer:* Forel, 1899c: 19; Brown, 1976a: 103.

*transversostriatus*. *Odontomachus transversostriatus* Donisthorpe, 1941h: 53 (w.) NEW GUINEA. Junior synonym of *saevissimus:* Wilson, 1959a: 498.

***troglodytes***. *Odontomachus haematodes* var. *troglodytes* Santschi, 1914b: 58 (w.) KENYA. André, 1887: 290 (m.); Arnold, 1915: 109 (q.); Menozzi, 1944: 437 (l.). Raised to species, and senior synonym of *stanleyi:* Brown, 1976a: 106. See also: Brown, 1976a: 167.

*tropicalis*. *Odontomachus assiniensis* var. *tropicalis* Menozzi, 1922a: 352 (w.q.m.) PRINCIPE I. Junior synonym of *assiniensis:* Brown, 1976a: 102.

*tuberculatus*. *Odontomachus tuberculatus* Roger, 1861a: 28 (w.) ASIA (no locality given). Subspecies of *malignus:* Mann, 1919: 305. Junior synonym of *malignus:* Emery, 1887b: 429; Emery, 1911d: 113; Wilson, 1959a: 495.

***turneri***. *Odontomachus ruficeps* r. *turneri* Forel, 1900b: 56 (w.) AUSTRALIA. Junior synonym of *cephalotes:* Brown, 1976a: 103. Revived from synonymy and raised to species: Brown, 1978b: 283.

***tyrannicus***. *Odontomachus tyrannicus* Smith, F. 1859a: 144 (w.) INDONESIA (Aru I.). Emery, 1897d: 557 (q.); Wheeler, G.C. & Wheeler, J. 1952c: 652 (l.). Senior synonym of *obsolescens:* Wilson, 1959a: 501.

*tyrannicus* Smith, F. 1861; see under ***ANOCHETUS***.

*unispinosa*. *Formica unispinosa* Fabricius, 1793: 359 (w.) GUADELUPE I. Combination in *Odontomachus:* Lepeletier, 1835: 187. Junior synonym of *haematodus:* Roger, 1861a: 24; Mayr, 1865: 63; possible synonym (senior) of *bauri:* Brown, 1976a: 102. Unidentifiable to species; *incertae sedis* in *Odontomachus:* current status.

*verticillatus*. *Odontomachus ruficeps* r. *cephalotes* var. *verticillatus* Stitz, 1912: 501, figs. 5, 6 (w.) NEW GUINEA. Unavailable name; material referred to *aciculatus* by Wilson, 1959a: 489.

*wheeleri. Odontomachus haematodus* subsp. *insularis* var. *wheeleri* Mann, 1920: 405 (w.) CUBA. Unavailable name; material referred to *insularis* by Brown, 1976a: 104.

***xizangensis.*** *Odontomachus xizangensis* Wang, M. 1993a: 222, fig. 2 (w.) CHINA.

*yorkensis. Odontomachus ruficeps* r. *acutidens* var. *yorkensis* Stitz, 1911a: 356 (w.) AUSTRALIA. Unavailable name (Taylor, 1986: 34); material referred to *cephalotes* by Brown, 1976a: 103.

***yucatecus.*** *Odontomachus yucatecus* Brown, 1976a: 169 (w.) MEXICO.

*ODONTOMYRMEX:* see under ***PRISTOMYRMEX.***

*ODONTOPELTA:* see under ***LEPTOGENYS.***

***ODONTOPONERA*** (Oriental, Indo-Australian)

#***biconcentrica.*** *Odontoponera transversa* var. *biconcentrica* Wheeler, W.M. & Chapman, 1925: 62, pl. 2, fig. 6 (w.) PHILIPPINES.

*denticulata. Ponera denticulata* Smith, F. 1858b: 90, pl. 6, figs. 13, 14 (q.) SINGAPORE (not South Africa as originally stated; see Donisthorpe, 1943f: 677). Roger, 1861a: 12 (w.). Combination in *Odontoponera:* Mayr, 1862: 717. Junior synonym of *transversa:* Dalla Torre, 1893: 30; Forel, 1900d: 314.

#***infuscata.*** *Odontoponera transversa* var. *infuscata* Creighton, 1929: 153 (w.) JAVA.

#***nitens.*** *Odontoponera transversa* subsp. *nitens* Creighton, 1929: 154 (w.) BORNEO.

*reticulata. Ponera reticulata* Smith, F. 1858b: 85 (m.) BURMA. Junior synonym of *transversa:* Donisthorpe, 1932c: 474.

***transversa.*** *Ponera transversa* Smith, F. 1857a: 68 (w.) SINGAPORE. Bingham, 1903: 73 (q.); Wheeler, G.C. & Wheeler, J. 1952c: 606 (l); Imai, Baroni Urbani, *et al.* 1984: 5 (k.). Combination in *Odontoponera:* Emery, 1893g: 262. Senior synonym of *denticulata:* Dalla Torre, 1893: 30; Forel, 1900d: 314; of *reticulata:* Donisthorpe, 1932c: 474. Current subspecies: nominal plus *biconcentrica, infuscata, nitens.*

*OECODOMA:* see under ***ATTA.***

*OECOPHTHORA:* see under ***PHEIDOLE.***

***OECOPHYLLA*** (Old World Tropics)

#***annectens.*** *Oecophylla longinoda* var. *annectens* Wheeler, W.M. 1922a: 230 (w.q.m.) ZAIRE.

****atavina.*** **Oecophylla atavina* Cockerell, 1915: 485, pl. 64, fig. 7 (q.) GREAT BRITAIN.

****bartoniana.*** **Oecophylla bartoniana* Cockerell, 1920: 277, pl. 16, fig. 4 (wing) GREAT BRITAIN (Eocene).

*brevinodis. Oecophylla brevinodis* André, 1890: 313 (w.) SIERRA LEONE. Santschi, 1919a: 345 (q.). Variety/race of *longinoda:* Dalla Torre, 1893: 176; of *smaragdina:* Stitz, 1916: 396. Junior synonym of *longinoda:* Wheeler, W.M. 1922a: 945.

**brevinodis* **Oecophylla brevinodis* Wheeler, W.M. 1915h: 116, fig. 56 (w.) BALTIC AMBER (Oligocene). [Junior primary homonym of **brevinodis* André, above.] Replacement name: **crassinoda* Wheeler, W.M. 1922a: 227.

****brischkei.*** **Oecophylla brischkei* Mayr, 1868c: 31, pl. 1, figs. 12, 13 (w.) BALTIC AMBER (Oligocene). Wheeler, W.M. 1915h: 114 (m.).

#***claridens.*** *Oecophylla longinoda* var. *claridens* Santschi, 1928f: 211 (w.q.) IVORY COAST. Stirps of *longinoda:* Santschi, 1935b: 274.

****crassinoda.*** **Oecophylla crassinoda* Wheeler, W.M. 1922a: 227. Replacement name for **brevinodis* Wheeler, W.M. 1915h: 116. [Junior primary homonym of *brevinodis* André, 1890: 313.]

#***fusca.*** *Oecophylla fusca* Emery, 1899e: 496 (w.q.) CAMEROUN. Wheeler, W.M. 1922a: 229 (m.). Variety/subspecies of *smaragdina:* Emery, 1921c: 103; of *longinoda:* Wheeler, W.M. 1922a: 229.

#***fuscoides.*** *Oecophylla smaragdina* var. *fuscoides* Karavaiev, 1933a: 315 (w.) JAVA. [Also described as new by Karavaiev, 1935a: 112.]

#***gracilior.*** *Oecophylla smaragdina* var. *gracilior* Forel, 1911b: 208 (w.q.) JAVA.

#***gracillima.*** *Oecophylla smaragdina* var. *gracillima* Emery, 1893e: 195 (w.) INDONESIA (Batjan I.).

*krampsi. Oecophylla smaragdina* subsp. *subnitida* var. *krampsi* Donisthorpe, 1938d: 598 (w.q.) NEW GUINEA. Unavailable name.

****leakeyi.*** **Oecophylla leakeyi* Wilson & Taylor, 1964: 93, fig. 2, pl. 2 (w.l.) KENYA.

***longinoda.*** *Formica longinoda* Latreille, 1802c: 184, pl. 11, fig. 72 (w.) SENEGAL. Emery, 1892d: 564 (q.); Forel, 1913b: 339 (m.). Combination in *Oecophylla:* Mayr, 1863: 439. Junior synonym of *virescens:* Smith, F. 1858b: 29. Subspecies/race of *smaragdina:* Emery, 1892d: 564; Forel, 1907e: 15; Santschi, 1914b: 128; Santschi, 1919a: 345; Emery, 1925b: 52; Prins, 1965b: 77. Status as species: Dalla Torre, 1893: 176; Emery, 1921c: 102; Wheeler, W.M. 1922a: 227; Santschi, 1928f: 211; and current status. Senior synonym of *brevinodis:* Wheeler, W.M. 1922a: 945. Current subspecies: nominal plus *annectens, claridens, fusca, rubriceps, rufescens, taeniata, textor.* See also: Gotwald, 1973: 72; Hölldobler & Wilson, 1978: 19.

*macra. Formica macra* Guérin-Méneville, 1831, pl. 8, fig. 1 (w.) "Offack". Junior synonym of *virescens:* Smith, F. 1858b: 29; of *smaragdina:* Roger, 1863b: 10; Dalla Torre, 1893: 176; Arnold, 1922: 609.

****megarche.*** **Oecophylla megarche* Cockerell, 1915: 486, pl. 65, figs. 1–3 (q.) GREAT BRITAIN.

****obesa.*** **Formica obesa* Heer, 1850: 108, pl. 8, fig. 1 (q.) YUGOSLAVIA (Miocene). Combination in *Oecophylla:* Mayr, 1867b: 50.

****perdita.*** **Oecophylla perdita* Cockerell, 1915: 485, pl. 64, figs. 5, 6 (q.m.) GREAT BRITAIN.

****praeclara.*** **Oecophylla praeclara* Förster, 1891: 432, pl. 13, fig. 6 (q.) GERMANY (Oligocene).

#***rubriceps.*** *Oecophylla longinoda* var. *rubriceps* Wheeler, W.M. 1922a: 229 (w.q.m.) ZAIRE. [First available use of *Oecophylla smaragdina* r. *fusca* var. *rubriceps* Forel, 1913h: 354; unavailable name.]

#***rufescens.*** *Oecophylla longinoda* var. *rufescens* Santschi, 1928f: 212 (w.q.) CONGO.

#***selebensis.*** *Oecophylla smaragdina* var. *selebensis* Emery, 1893e: 195 (w.) SULAWESI. Emery, 1921c: 102 (m.).

****sicula.*** **Oecophylla sicula* Emery, 1891a: 156, pl. 3, figs. 36, 37 (w.) SICILIAN AMBER (Miocene).

***smaragdina.*** *Formica smaragdina* Fabricius, 1775: 828 (q.) INDIA. Jerdon, 1851: 121 (w.m.); Wheeler, G.C. & Wheeler, J. 1953e: 176 (l.); Crozier, 1970: 115 (k.). Combination in *Oecophylla:* Smith, F. 1860b: 102. Senior synonym of *viridis:* Smith, F. 1857a: 53; Taylor & Brown, D.R. 1985: 127; of *macra* and *zonata:* Roger, 1863b: 10; Dalla Torre, 1893: 176; of *virescens:* Mayr, 1872: 143; Taylor & Brown, D.R. 1985: 127. Current subspecies: nominal plus *fuscoides, gracilior, gracillima, selebensis, subnitida.*

#***subnitida.*** *Oecophylla smaragdina* var. *subnitida* Emery, 1892d: 565 (w.) NEW GUINEA. Donisthorpe, 1942b: 70 (q.). Subspecies of *smaragdina:* Emery, 1897d: 573; Wheeler, W.M. 1922a: 229 (in key).

****superba.*** **Oecophylla superba* Théobald, 1937b: 212, pl. 4, fig. 17; pl. 15, fig. 1 (q.) FRANCE (Oligocene).

#***taeniata.*** *Oecophylla longinoda* var. *taeniata* Santschi, 1928f: 212 (q.) ZAIRE.

#***textor.*** *Oecophylla longinoda* var. *textor* Wheeler, W.M. 1922a: 229 (w.q.) ZAIRE. [First available use of *Oecophylla smaragdina* st. *longinoda* var. *textor* Santschi, 1914b: 128; unavailable name.]

*virescens. Formica virescens* Fabricius, 1775: 392 (w.) AUSTRALIA. Smith, F. 1858b: 30 (m.). Combination in *Oecophylla:* Smith, F. 1860b: 102. Status as species: Dalla Torre, 1893: 177; Emery, 1921c: 102. Variety/race of *smaragdina:* Emery, 1887a: 242; Forel, 1915b: 95; Wheeler, W.M. 1922a: 228 (in key); Emery, 1925b: 52; Karavaiev, 1933a: 315. Junior synonym of *smaragdina:* Mayr, 1872: 143; Taylor & Brown, D.R. 1985: 127.

*viridis. Formica viridis* Kirby, W. 1819: 478 (w.) AUSTRALIA. Junior synonym of *virescens:* Roger, 1863b: 10; Dalla Torre, 1893: 177; Emery, 1925b: 52; of *smaragdina:* Smith, F. 1857a: 53; Taylor & Brown, D.R. 1985: 127.

*zonata. Formica zonata* Guérin-Méneville, 1838: 205 (q.) "Port Praslin". Junior synonym of *smaragdina:* Roger, 1863b: 10; Dalla Torre, 1893: 176.

*OEDALEOCERUS:* see under ***SOLENOPSIS.***

***OLIGOMYRMEX*** (World Tropics and Subtropics)

***aborensis.*** *Pheidologeton* (*Aneleus*) *aborensis* Wheeler, W.M. 1913e: 234 (s.w.) INDIA. Combination in *Aneleus:* Emery, 1922f: 214; in *Oligomyrmex:* Ettershank, 1966: 123.

***acutus.*** *Oligomyrmex* (*Aeromyrma*) *acutus* Weber, 1952: 12, figs. 26, 30 (s.w.) ZAIRE.

#***aeolius.*** *Oligomyrmex oertzeni* var. *aeolia* Forel, 1911d: 338 (q.m.) TURKEY.

***afghanus.*** *Oligomyrmex afghanus* Pisarski, 1970: 311, figs. 10–23 (w.m.) AFGHANISTAN.

***africanus.*** *Aeromyrma africana* Forel, 1910f: 15 (s.) BOTSWANA. Combination in *Oligomyrmex* (*Aeromyrma*): Emery, 1922f: 216.

***alluaudi.*** *Oligomyrmex alluaudi* Santschi, 1913e: 459 (diagnosis in key) (s.w.) KENYA. [Also described as new by Santschi, 1914b: 81 (s.w., descriptions).] Senior synonym of *kenyensis* (unnecessary replacement name): **new synonymy**. Current subspecies: nominal plus *cataractae.* See also: Weber, 1950a: 3.

***alpha.*** *Oligomyrmex alpha* Forel, 1905c: 21 (w.) JAVA.

***amius.*** *Solenopsis amia* Forel, 1913f: 191 (q.) TAIWAN. Combination in *Aneleus:* Emery, 1923: 60; in *Oligomyrmex:* Ettershank, 1966: 123.

***angolensis.*** *Oligomyrmex angolensis* Santschi, 1914d: 358, fig. 21 (s.w.) ANGOLA. Current subspecies: nominal plus *congolensis.*

*anophthalmus* Emery, 1906; see under ***CAREBARA.***

****antiquus.*** **Pheidologeton antiquus* Mayr, 1868c: 93, pl. 5, figs. 95, 96 (q.) BALTIC AMBER (Oligocene). Wheeler, W.M. 1915h: 46 (w.m.). Combination in *Aeromyrma:* Emery, 1891a: 577; in *Erebomyrma:* Wheeler, W.M. 1915h: 46; in *Oligomyrmex:* Ettershank, 1966: 123.

***armatus.*** *Aneleus armatus* Donisthorpe, 1948d: 596 (w.) NEW GUINEA. Combination in *Oligomyrmex:* Ettershank, 1966: 123.

*arnoldi* Forel, 1913; see under ***CAREBARA.***

*arnoldi. Pheidologeton perpusillum* r. *arnoldi* Forel, 1914d: 242 (w.) ZIMBABWE. [Junior secondary homonym of *Oligomyrmex arnoldi* Forel, 1913a: 123, above.] Replacement name: *arnoldianus* Ettershank, 1966: 123. Arnold, 1916: 255 (s.); Arnold, 1926: 238 (q.). Combination in *Aneleus:* Arnold, 1916: 255; in *Oligomyrmex:* Ettershank, 1966: 123.

#***arnoldianus.*** *Oligomyrmex perpusillus* subsp. *arnoldianus* Ettershank, 1966: 123. Replacement name for *arnoldi* Forel, 1914d: 242. [Junior secondary homonym of *arnoldi* Forel, 1913a: 123.]

***arnoldiellus.*** *Oligomyrmex (Aeromyrma) arnoldiella* Santschi, 1919a: 340, fig. 3 (s.w.) SOUTH AFRICA. Combination in *Aeromyrma:* Wheeler, W.M. 1922a: 882; in *Oligomyrmex:* Ettershank, 1966: 123.

***asinus.*** *Oligomyrmex asinus* Forel, 1902c: 214 (s.w.q.) INDIA. Imai, Baroni Urbani, *et al.* 1984: 7 (k.).

***atomus.*** *Oligomyrmex atomus* Emery, 1900c: 328, pl. 8, fig. 30 (s.w.) NEW GUINEA. Mann, 1919: 331 (q.). See also: Taylor, 1991b: 606.

***bengalensis.*** *Oligomyrmex bengalensis* Forel, 1902c: 217 (q.m.) INDIA.

***beta.*** *Oligomyrmex beta* Forel, 1905c: 22 (s.) JAVA.

****bohemicus.*** **Pheidologeton bohemicus* Novak, O. 1878: 90, pl. 3, fig. 3 (m. ?) CZECHOSLOVAKIA (Miocene). Combination in *Aeromyrma:* Dalla Torre, 1893: 78; in *Oligomyrmex:* **new combination**.

***bouvardi.*** *Oligomyrmex bouvardi* Santschi, 1913d: 457, fig. 1 (s.w.m.) VIETNAM.

***bruchi.*** *Oligomyrmex bruchi* Santschi, 1933e: 116 (q.) ARGENTINA. Combination in *Erebomyrma:* Branão, 1991: 343; in *Oligomyrmex:* **revived combination**.

***bruni.*** *Oligomyrmex bruni* Forel, 1913k: 61, fig. T (s.w.) SRI LANKA.

***butteli.*** *Pheidologeton (Lecanomyrma) butteli* Forel, 1913k: 56, fig. S (s.w.) SRI LANKA. Combination in *Aneleus:* Emery, 1922f: 215; in *Oligomyrmex:* Ettershank, 1966: 123.

***capreolus.*** *Oligomyrmex (Hendecatella) capreolus* Wheeler, W.M. 1927h: 93, fig. 5 (s.w.m.) VIETNAM. Current subspecies: nominal plus *laeviceps.*

#***cataractae.*** *Oligomyrmex alluaudi* var. *cataractae* Santschi, 1919a: 337 (s.w.) ZIMBABWE.

#***concedens.*** *Aneleus perpusillus* st. *concedens* Santschi, 1914b: 77 (s.w.) KENYA. Combination in *Oligomyrmex:* Ettershank, 1966: 123.

***concinnus.*** *Oligomyrmex concinnus* Mayr, 1867a: 111 (q.) INDONESIA.

#***congolensis.*** *Oligomyrmex angolensis* r. *congolensis* Forel, 1916: 417 (s.w.) ZAIRE.

***convexus.*** *Oligomyrmex convexus* Weber, 1950a: 6 (s.) UGANDA.

***corniger.*** *Oligomyrmex corniger* Forel, 1902h: 449 (s.w.q.m.) AUSTRALIA. Wheeler, G.C. & Wheeler, J. 1973a: 33 (l.). Senior synonym of *parvicornis:* Taylor, 1991b: 607.

***crassiusculus.*** *Pheidologeton (Aneleus) similis* var. *crassiuscula* Emery, 1900c: 328 (w.) NEW GUINEA. Combination in *Aneleus:* Emery, 1922f: 214; in *Oligomyrmex,* and raised to species: Ettershank, 1966: 123. See also: Taylor, 1991b: 607.

***cribriceps.*** *Aneleus cribriceps* Wheeler, W.M. 1927h: 90, fig. 4 (s.w.) VIETNAM. Combination in *Oligomyrmex:* Ettershank, 1966: 123.

***debilis.*** *Oligomyrmex debilis* Santschi, 1913e: 459 (w.) GUINEA. Santschi, 1914d: 360 (s.w.m.). Combination in *O. (Aeromyrma)*: Santschi, 1919a: 337.

***deponens.*** *Crematogaster deponens* Walker, 1859: 374 (q.) SRI LANKA. Combination in *Oligomyrmex:* Donisthorpe, 1932b: 576.

***diabolicus.*** *Oligomyrmex diabolicus* Baroni Urbani, 1969b: 74, fig. 1 (s.) CHAD.

***diabolus.*** *Oligomyrmex diabolus* Santschi, 1913e: 459 (diagnosis in key) (s.w.) CAMEROUN. Santschi, 1914d: 355 (s.w., descriptions). Combination in *Aneleus:* Santschi, 1914d: 355; in *Oligomyrmex:* Ettershank, 1966: 123.

***donisthorpei.*** *Oligomyrmex donisthorpei* Weber, 1950a: 7, fig. (s.w.) CENTRAL AFRICAN REPUBLIC.

***eidmanni.*** *Erebomyrma eidmanni* Menozzi, 1936a: 47 (s.w.q.) BRAZIL. Combination in *Oligomyrmex:* Ettershank, 1966: 123; in *Erebomyrma:* Wilson, 1986b: 61; in *Oligomyrmex:* **revived combination**.

*eidmanni. Aneleus eidmanni* Menozzi, 1942: 170, fig. 1 (s.w.q.) FERNANDO PO I. [Junior secondary homonym of *Erebomyrma eidmanni* Menozzi, above.] Replacement name: *menozzii* Ettershank, 1966: 124.

***elmenteitae.*** *Solenopsis (Crateropsis) elmenteitae* Patrizi, 1948: 176, figs. 1, 2 (s.) KENYA. Combination in *Oligomyrmex:* Ettershank, 1966: 123.

***erythraeus.*** *Oligomyrmex erythraeus* Emery, 1915g: 14, fig. 7 (s.w.) ETHIOPIA.

***frontalis.*** *Oligomyrmex frontalis* Weber, 1950a: 10, figs. 5, 6, 8 (s.w.) ZAIRE.

***grandidieri.*** *Oligomyrmex grandidieri* Forel, 1891b: 201 (q.) MADAGASCAR. Forel, 1895c: 250 (m.).

*hewitti. Oligomyrmex (Aeromyrma) hewitti* Santschi, 1919a: 341, fig. 3 (s.w.) SOUTH AFRICA. Combination in *Aeromyrma:* Wheeler, W.M. 1922a: 882; in *Oligomyrmex:* Ettershank, 1966: 123. Junior synonym of *semilaevis:* Arnold, 1926: 236.

***incertus.*** *Oligomyrmex (Aeromyrma) incertus* Santschi, 1919a: 340, fig. 3 (w.) ZIMBABWE. Combination in *Aeromyrma:* Wheeler, W.M. 1922a: 882; in *Oligomyrmex:* Ettershank, 1966: 123.

*infimus* Santschi, 1913; see under ***PAEDALGUS.***

***jacobsoni.*** *Oligomyrmex jacobsoni* Forel, 1911b: 199 (s.w.) JAVA. Wheeler, G.C. & Wheeler, J. 1954a: 140 (l).

***jeanneli.*** *Oligomyrmex jeanneli* Santschi, 1913e: 459 (diagnosis in key) (s.w.) KENYA. Santschi, 1914b: 83 (s.w., descriptions).

*kenyensis. Oligomyrmex kenyensis* Ettershank, 1966: 123. Unnecessary replacement name for *alluaudi* Santschi, 1914b: 81. Junior synonym of *alluaudi:* **new synonymy**.

***khamiensis.*** *Oligomyrmex (Aeromyrma) khamiensis* Arnold, 1952a: 463, fig. 1 (w.) ZIMBABWE.

#***laeviceps.*** *Oligomyrmex (Hendecatella) capreolus* subsp. *laeviceps* Wheeler, W.M. 1928c: 24 (s.) CHINA.

***lamellifrons.*** *Pheidologeton lamellifrons* Forel, 1902c: 219 (q.) INDIA. Combination in *P. (Lecanomyrma)*: Forel, 1913k: 58; in *Aneleus (Lecanomyrma)*: Emery, 1922f: 215; in *Oligomyrmex:* Ettershank, 1966: 123.

***latro.*** *Oligomyrmex (Oligomyrmex) latro* Santschi, 1937g: 81, fig. 6 (s.w.) ZAIRE.

***leei.*** *Oligomyrmex leei* Forel, 1902c: 216 (q.) INDIA.

***longiceps.*** *Oligomyrmex longiceps* Santschi, 1929d: 295, figs. 33, 34 (w.m.) ARGENTINA. Combination in *Erebomyrma:* Brandão, 1991: 343; in *Oligomyrmex:* **revived combination**.

***longii.*** *Erebomyrma longii* Wheeler, W.M. 1903a: 140, figs. 1–5 (w.q.m.) U.S.A. Combination in *Oligomyrmex:* Ettershank, 1966: 123; in *Erebomyrma:* Wilson, 1986b: 61; in *Oligomyrmex:* **revived combination**.

***lucidus.*** *Oligomyrmex (Aeromyrma) lucidus* Santschi, 1917b: 284 (w.) ZIMBABWE. Combination in *Aeromyrma:* Wheeler, W.M. 1922a: 882; in *Oligomyrmex:* Ettershank, 1966: 124.

***lusciosus.*** *Oligomyrmex lusciosus* Wheeler, W.M. 1928c: 22 (s.w.) CHINA.

***manni.*** *Oligomyrmex (Oligomyrmex) manni* Donisthorpe, 1941c: 135 (q.) NEW GUINEA.

***menozzii.*** *Oligomyrmex menozzii* Ettershank, 1966: 124. Replacement name for *Aneleus eidmanni* Menozzi, 1942: 170. [Junior secondary homonym of *Erebomyrma eidmanni* Menozzi, 1936a: 47.]

***minimus.*** *Pheidologeton (Aneleus) minimus* Emery, 1900c: 327 (s.w.) NEW GUINEA. Combination in *Aneleus:* Emery, 1922f: 214; in *Oligomyrmex:* Ettershank, 1966: 124.

***mjobergi.*** *Oligomyrmex mjobergi* Forel, 1915b: 69, pl. 1, fig. 4 (s.w.) AUSTRALIA. Wheeler, G.C. & Wheeler, J. 1954a: 140 (l.); Wheeler, G.C.& Wheeler, J. 1973a: 33 (l.).

***morai.*** *Erebomyrma morai* Menozzi, 1931b: 271, fig. 7 (w.) COSTA RICA. Combination in *Oligomyrmex:* Ettershank, 1966: 124; in *Erebomyrma:* Brandão, 1991: 343; in *Oligomyrmex:* **revived combination**.

***nanus.*** *Oligomyrmex (Aeromyrma) nanus* Santschi, 1919a: 338, fig. 3 (s.w.) MOZAMBIQUE. Combination in *Aeromyrma:* Wheeler, W.M. 1922a: 882; in *Oligomyrmex:* Ettershank, 1966: 124.

***nevermanni.*** *Erebomyrma nevermanni* Mann, 1926: 103 (w.) COSTA RICA. Combination in *Oligomyrmex:* Ettershank, 1966: 124; in *Erebomyrma:* Wilson, 1986b: 61; in *Oligomyrmex:* **revived combination**.

#***nicotianae.*** *Sporocleptes nicotianae* Arnold, 1948: 220, fig. 8 (w.) ZIMBABWE. Combination in *Oligomyrmex:* Ettershank, 1966: 124. Subspecies of *politus:* Arnold, 1952a: 460.

***OLIGOMYRMEX*** *(continued)*

***norfolkensis.*** *Oligomyrmex manni* var. *norfolkensis* Donisthorpe, 1941e: 92 (q.m.) AUSTRALIA. Raised to species: Taylor & Brown, D.R. 1985: 72.

***nosindambo.*** *Aeromyrma nosindambo* Forel, 1891b: 199, pl. 6, fig. 1 (q.m.) MADAGASCAR. Forel, 1892k: 523 (s.w.). Combination in *Oligomyrmex:* Santschi, 1919a: 339; Ettershank, 1966: 124.

***oertzeni.*** *Oligomyrmex oertzeni* Forel, 1886e: clxv, fig. 3 (q.) GREECE. Current subspecies: nominal plus *aeolius.*

***overbecki.*** *Oligomyrmex overbecki* Viehmeyer, 1916a: 136, fig. 5 (s.w.m.) SINGAPORE. See also: Moffett, 1986d: 107.

*pachycerus* Forel, 1915; see under ***SOLENOPSIS.***

***paetus.*** *Aneleus paetus* Santschi, 1937d: 230, figs. 11, 12 (q.) ANGOLA. Combination in *Oligomyrmex:* Ettershank, 1966: 124.

***panamensis.*** *Oligomyrmex panamensis* Wheeler, W.M. 1925d: 175, fig. 7 (s.w.) PANAMA. Combination in *Erebomyrma:* Wilson, 1986b: 61; in *Oligomyrmex:* **revived combination**.

*parvicornis. Oligomyrmex corniger* var. *parvicornis* Forel, 1915b: 70 (s.w.m.) AUSTRALIA. Wheeler, G.C. & Wheeler, J. 1954a: 138 (l.). Raised to species: Wheeler, G.C. & Wheeler, J. 1954a: 138. Subspecies of *corniger:* Taylor & Brown, D.R. 1985: 72. Junior synonym of *corniger:* Taylor, 1991b: 607.

***perpusillus.*** *Pheidologeton perpusillum* Emery, 1895h: 26, pl. 2, figs. 8–11 (s.w.) SOUTH AFRICA. Combination in *P.* (*Aneleus*): Emery, 1900c: 327; in *Aneleus:* Santschi, 1914b: 77; in *Oligomyrmex:* Ettershank, 1966: 124. Current subspecies: nominal plus *arnoldianus, concedens, spinosus.* See also: Arnold, 1916: 254.

***peruvianus.*** *Erebomyrma peruviana* Emery, 1906c: 139 (footnote) (q.) PERU. Combination in *Oligomyrmex:* Ettershank, 1966: 124; in *Erebomyrma:* Brandão, 1991: 343; in *Oligomyrmex:* **revived combination**.

***petulcus.*** *Aeromyrma petulca* Wheeler, W.M. 1922a: 166, fig. 40 (s.) ZAIRE. Combination in *Oligomyrmex:* Ettershank, 1966: 124.

***politus.*** *Aneleus politus* Santschi, 1914b: 79, fig. 8 (w.) KENYA. Consani, 1951: 170 (q.). Combination in *Oligomyrmex:* Ettershank, 1966: 124. Current subspecies: nominal plus *nicotianae.*

***polyphemus.*** *Oligomyrmex polyphemus* Wheeler, W.M. 1928c: 21 (s.) CHINA.

***punctatus.*** *Aneleus* (*Aneleus*) *punctatus* Karavaiev, 1931d: 43, fig. 2 (w.) KENYA. Combination in *Oligomyrmex:* Ettershank, 1966: 124.

***raja.*** *Oligomyrmex raja* Forel, 1902c: 216 (q.) INDIA.

***rothneyi.*** *Oligomyrmex rothneyi* Forel, 1902c: 218 (q.m.) INDIA.

***rugatus.*** *Oligomyrmex rugatus* Forel, 1913k: 60 (s.m.) JAVA.

***santschii.*** *Oligomyrmex santschii* Weber, 1943c: 363, pl. 15, figs. 4, 16 (s.w.) SUDAN.

***sarasinorum.*** *Pheidologeton* (*Aneleus*) *sarasinorum* Emery, 1901g: 578, fig. D (s.w.q.) SULAWESI. [*P.* (*A.*) *sarasinorum* Emery, 1900c: 327. *Nomen nudum.*]

***satanus.*** *Oligomyrmex* (*Oligomyrmex*) *satanus* Karavaiev, 1935a: 96, fig. 21 (s.) SUMATRA.

***sauteri.*** *Oligomyrmex sauteri* Forel, 1912a: 56 (s.) TAIWAN. Hung, Imai & Kubota, 1972: 1024 (k.).

***semilaevis.*** *Solenopsis semilaevis* Mayr, 1901b: 15 (w.) SOUTH AFRICA. Combination in *Oligomyrmex* (*Aeromyrma*): Emery, 1915c: 59 (in text); in *Aeromyrma:* Wheeler, W.M. 1922a: 882; in *Oligomyrmex:* Ettershank, 1966: 124. Senior synonym of *hewitti:* Arnold, 1926: 236.

***silvestrii.*** *Aneleus silvestrii* Santschi, 1914d: 357, fig. 20 (s.w.) GHANA. Bernard, 1953b: 239 (m.). Combination in *Oligomyrmex:* Ettershank, 1966: 124.

*silvestrii. Oligomyrmex silvestrii* Wheeler, W.M. 1928c: 23 (s.w.) CHINA. [Junior secondary homonym of *silvestrii* Santschi, above.] Replacement name: *wheeleri* Ettershank, 1966: 124.

***simalurensis.*** *Pheidologeton* (*Aneleus*) *pygmaeus* var. *simalurensis* Forel, 1915a: 27 (s.w.q.) SUMATRA. Wheeler, W.M. 1924b: 246 (m.). Combination in *Aneleus:* Emery, 1922f: 214; in *Oligomyrmex,* and raised to species: Ettershank, 1966: 124.

***similis.*** *Solenopsis similis* Mayr, 1862: 751 (w.) NICOBAR IS. Combination in *Pheidologeton* (*Aneleus*): Emery, 1900c: 327; in *Aneleus:* Emery, 1922f: 214; in *Oligomyrmex:* Ettershank, 1966: 124.

***sodalis.*** *Oligomyrmex corniger* subsp. *sodalis* Emery, 1914f: 412 (s.w.q.m.) NEW CALEDONIA. Raised to species: Taylor, 1991b: 607.

**sophiae.* **Aeromyrma sophiae* Emery, 1891a: 151, pl. 2, figs. 17, 18 (m.) SICILIAN AMBER (Miocene). Combination in *Oligomyrmex:* **new combination**.

#***spinosus.*** *Pheidologeton* (*Aneleus*) *perpusillus* subsp. *spinosus* Forel, 1907a: 17 (s.) TANZANIA. Combination in *Aneleus:* Emery, 1922f: 214; in *Oligomyrmex:* **new combination**.

***stenopterus.*** *Oligomyrmex stenopterus* Kusnezov, 1952f: 184, fig. 2 (q.) ARGENTINA. Combination in *Erebomyrma:* Brandão, 1991: 343; in *Oligomyrmex:* **revived combination**.

***sublatro.*** *Oligomyrmex sublatro* Forel, 1913k: 59 (s.w.) JAVA.

***subreptor.*** *Oligomyrmex subreptor* Emery, 1900c: 329 (w.) NEW GUINEA.

***sundaicus.*** *Aeromyrma sundaica* Forel, 1913k: 63 (w.) SUMATRA. Wheeler, G.C. & Wheeler, J. 1954a: 140 (l.). Combination in *Oligomyrmex* (*Aeromyrma*): Emery, 1922f: 216.

***tahitiensis.*** *Oligomyrmex tahitiensis* Wheeler, W.M. 1936f: 11 (q.m.) TAHITIAN IS. Perrault, 1976: 304 (s.w.).

***taiponicus.*** *Oligomyrmex silvestrii* subsp. *taiponicus* Wheeler, W.M. 1928c: 24 (s.) CHINA. Raised to species: **new status**.

***taprobanae.*** *Oligomyrmex taprobanae* Forel, 1911i: 219 (s.w.) SRI LANKA.

***thoracicus.*** *Oligomyrmex thoracicus* Weber, 1950a: 13, figs. 11, 14, 15, 19 (s.w.) UGANDA. See also: Weber, 1952: 9.

**thorali.* **Erebomyrma thorali* Théobald, 1937a: 157, fig. 1 (m.) FRANCE (Oligocene). Combination in *Oligomyrmex:* **new combination**.

***traegaordhi.*** *Aeromyrma traegaordhi* Santschi, 1914e: 20, fig. 2 (w.) SOUTH AFRICA. Combination in *Oligomyrmex* (*Aeromyrma*): Santschi, 1919a: 338. Variety of *semilaevis:* Santschi, 1917b: 285. Revived status as species: Santschi, 1919a: 338; Emery, 1922f: 216; Arnold, 1926: 233.

***ugandanus.*** *Aeromyrma ugandana* Santschi, 1923e: 279 (q.) UGANDA. Combination in *Oligomyrmex:* Ettershank, 1966: 124.

***urichi.*** *Spelaeomyrmex urichi* Wheeler, W.M. 1922c: 45, fig. 1 (w.) TRINIDAD. Combination in *Erebomyrma:* Wilson, 1962a: 63; Wilson, 1986b: 61; in *Oligomyrmex:* Ettershank, 1966: 124, and current combination.

***viehmeyeri.*** *Oligomyrmex viehmeyeri* Mann, 1919: 331, fig. 25 (s.w.q.m.) SOLOMON IS.

***villiersi.*** *Nimbamyrma villiersi* Bernard, 1953b: 241, fig. 12 (w.) GUINEA. Combination in *Oligomyrmex:* Ettershank, 1966: 124.

***voeltzkowi.*** *Oligomyrmex voeltzkowi* Forel, 1907g: 77 (q.) MADAGASCAR.

***vorax.*** *Aeromyrma vorax* Santschi, 1914d: 358 (w.) GHANA. Combination in *Oligomyrmex* (*Aeromyrma*): Emery, 1922f: 216.

***wheeleri.*** *Oligomyrmex wheeleri* Ettershank, 1966: 124. Replacement name for *Oligomyrmex silvestrii* Wheeler, 1928c: 23. [Junior secondary homonym of *Aneleus silvestrii* Santschi, 1914d: 357.]

***weyeri.*** *Solenopsis* (*Solenops*) *weyeri* Karavaiev, 1930a: 207, fig. 1 (w.q.) INDONESIA (Ambon I.). Combination in *Oligomyrmex:* Ettershank, 1966: 124.

***wroughtonii.*** *Solenopsis wroughtonii* Forel, 1902c: 214 (w.) INDIA. Combination in *Oligomyrmex* (*Aeromyrma*): Emery, 1915c: 59 (in text); Ettershank, 1966: 124.

***ONYCHOMYRMEX*** (Australia)

***doddi.*** *Onychomyrmex doddi* Wheeler, W.M. 1916a: 53, pl. 2, figs. 3–5 (w.q.) AUSTRALIA.

***hedleyi.*** *Onychomyrmex hedleyi* Emery, 1895g: 350, fig. 2 (w.) AUSTRALIA. Forel, 1915b: 4 (m.); Wheeler, G.C. & Wheeler, J. 1964b: 454 (l.). See also: Wheeler, W.M. 1916a: 50.

***mjobergi.*** *Onychomyrmex mjobergi* Forel, 1915b: 2, fig. 1; pl. 1, fig. 3 (w.) AUSTRALIA. Wheeler, W.M. 1916a: 52 (q.l.); Wheeler, G.C. & Wheeler, J. 1952c: 638 (l.).

*OOCERAEA:* see under ***CERAPACHYS.***

*OPHTHALMOPONE:* see under ***PACHYCONDYLA.***

***OPISTHOPSIS*** (Indo-Australian, Australasian)

#***aruanus.*** *Opisthopsis manni* var. *aruana* Karavaiev, 1933a: 316 (w.) INDONESIA (Aru I.).

#***bimaculatus.*** *Opisthopsis pictus* var. *bimaculatus* Wheeler, W.M. 1918d: 352 (w.) AUSTRALIA.

***diadematus.*** *Opisthopsis diadematus* Wheeler, W.M. 1918d: 357, pl. 3, figs. 20, 21 (w.) AUSTRALIA. Current subspecies: nominal plus *dubius.*

#***dubius.*** *Opisthopsis diadematus* var. *dubius* Wheeler, W.M. 1918d: 358 (w.q.) AUSTRALIA.

***haddoni.*** *Opisthopsis haddoni* Emery, 1893e: 226 (w.) AUSTRALIA. Wheeler, W.M. 1918d: 355 (q.m.); Wheeler, G.C. & Wheeler, J. 1968: 215 (l.). Current subspecies: nominal plus *rufoniger.*

***halmaherae.*** *Opisthopsis halmaherae* Karavaiev, 1930a: 210, fig. 2 (w.q.m.) INDONESIA (Halmahera I.)

***jocosus.*** *Opisthopsis jocosus* Wheeler, W.M. 1918d: 359, pl. 3, figs. 22–24 (w.) AUSTRALIA.

#***lepidus.*** *Opisthopsis pictus* var. *lepidus* Wheeler, W.M. 1918d: 352, pl. 1, figs. 3, 4 (w.) AUSTRALIA.
***lienosus.*** *Opisthopsis lienosus* Wheeler, W.M. 1918d: 356, pl. 2, figs. 9, 10 (w.) AUSTRALIA.
***linnaei.*** *Opisthopsis linnaei* Forel, 1901b: 26 (w.) BISMARCK ARCHIPELAGO.
***major.*** *Opisthopsis major* Forel, 1902h: 492 (w.q.) AUSTRALIA. See also: Wheeler, W.M. 1918d: 358.
***manni.*** *Opisthopsis manni* Wheeler, W.M. 1918d: 361, pl. 3, figs. 25, 26 (w.q.) SOLOMON IS. Current subspecies: nominal plus *aruanus*.
***maurus.*** *Opisthopsis maurus* Wheeler, W.M. 1918d: 350, pl. 3, figs. 18, 19 (w.) AUSTRALIA.
#***moestus.*** *Opisthopsis respiciens* var. *moestus* Wheeler, W.M. 1918d: 348 (w.q.m.) AUSTRALIA.
*oculata. Formica oculata* Smith, F. 1859a: 137 (w.) INDONESIA (Aru I.) [Junior primary homonym of **Formica oculata* Heer, 1850: 143.] Combination in *Opisthopsis:* Donisthorpe, 1932c: 452. Replacement name: *panops:* **new name**.
#***palliatus.*** *Opisthopsis pictus* var. *palliatus* Wheeler, W.M. 1918d: 352, pl. 3, fig. 28 (w.) AUSTRALIA.
***panops.*** *Opisthopsis panops* **new name**. Replacement name for *Formica oculata* Smith, F. 1859a: 137. [Junior primary homonym of **Formica oculata* Heer, 1850: 143.]
***pictus.*** *Opisthopsis pictus* Emery, 1895g: 354, fig. 5 (w.) AUSTRALIA. Current subspecies: nominal plus *bimaculatus, lepidus, palliatus*. See also: Wheeler, W.M. 1918d: 350.
***respiciens.*** *Formica (Myrmecopsis) respiciens* Smith, F. 1865: 68, pl. 4, fig. 3 (w.) NEW GUINEA. Forel, 1893d: 461 (q.). Combination in *Myrmecopsis:* Mayr, 1876: 76; in *Opisthopsis:* Dalla Torre, 1893: 219. Current subspecies: nominal plus *moestus*. See also: Wheeler, W.M. 1918d: 347.
***rufithorax.*** *Opisthopsis rufithorax* Emery, 1895g: 354, fig. 5 (w.) AUSTRALIA. Forel, 1915b: 95 (q.); Wheeler, G.C. & Wheeler, J. 1970: 648 (l.); Imai, Crozier & Taylor, 1977: 347 (k.). See also: Wheeler, W.M. 1918d: 353.
#***rufoniger.*** *Opisthopsis haddoni* subsp. *rufoniger* Forel, 1910b: 70 (w.) AUSTRALIA. Wheeler, W.M. 1918d: 356 (q.m.); Wheeler, G.C. & Wheeler, J. 1968: 216 (l.).

*OPISTHOSCYPHUS:* see under ***GNAMPTOGENYS.***

***ORECTOGNATHUS*** (Australasian, New Guinea)
***alligator.*** *Orectognathus alligator* Taylor, 1980a: 778, figs. 4–6 (w.) AUSTRALIA. [*O. alligator* Taylor, 1978b: 9; unavailable name.]
***antennatus.*** *Orectognathus antennatus* Smith, F. 1853: 228, pl. 21, fig. 9 (w.) NEW ZEALAND. Brown, 1953k: 99 (q.m.); Wheeler, G.C. & Wheeler, J. 1973c: 213 (l.). Senior synonym of *septentrionalis:* Brown, 1953k: 99. See also: Brown, 1958h: 36.
***biroi.*** *Orectognathus biroi* Szabó, 1926: 351, fig. B (w.) NEW GUINEA. Combination in *Arnoldidris:* Brown, 1950c: 145; in *Orectognathus:* Taylor, 1977: 583.
***chyzeri.*** *Orectognathus chyzeri* Emery, 1897c: 571, pl. 14, figs. 1, 2 (w.) NEW GUINEA. Combination in *Arnoldidris:* Brown, 1950c: 145; in *Orectognathus:* Taylor, 1977: 583.
***clarki.*** *Orectognathus clarki* Brown, 1953k: 94, figs. 1, 2 (w.q.m.) AUSTRALIA. Wheeler, G.C. & Wheeler, J. 1955a: 126 (l.); Crozier, 1968b: 87 (k.).
***coccinatus.*** *Orectognathus coccinatus* Taylor, 1980a: 779, figs. 7–9 (w.q.) AUSTRALIA. [*O. coccinatus* Taylor, 1978b: 9; unavailable name.]
***csikii.*** *Orectognathus csikii* Szabó, 1926: 351, fig. A (w.) NEW GUINEA. See also: Brown, 1953k: 100.
***darlingtoni.*** *Orectognathus darlingtoni* Taylor, 1977: 606, figs. 31–36 (w.q.k.) AUSTRALIA. Imai, Crozier & Taylor, 1977: 349 (k.).
***echinus.*** *Orectognathus echinus* Taylor & Lowery, 1972: 310, pl. 2 (w.q.) NEW GUINEA. Wheeler, G.C. & Wheeler, J. 1986f: 495 (l.).
***elegantulus.*** *Orectognathus elegantulus* Taylor, 1977: 589, figs. 11–13 (w.q.) AUSTRALIA.
***horvathi.*** *Orectognathus horvathi* Szabó, 1926: 349, fig. C (w.) NEW GUINEA. Combination in *Arnoldidris:* Brown, 1950c: 145; in *Orectognathus:* Taylor, 1977: 583.
***howensis.*** *Orectognathus antennatus* var. *howensis* Wheeler, W.M. 1927i: 145, fig. 7 (w.) AUSTRALIA. Raised to species: Brown, 1953k: 100.
***hystrix.*** *Orectognathus hystrix* Taylor & Lowery, 1972: 307, pl. 1 (w.m.) NEW GUINEA.
***kanangra.*** *Orectognathus kanangra* Taylor, 1980a: 776, figs. 1–3 (w.) AUSTRALIA. [*O. kanangra* Taylor, 1978b: 9; unavailable name.]
***longispinosus.*** *Orectognathus longispinosus* Donisthorpe, 1941h: 58 (w.) NEW GUINEA. Combination in *Arnoldidris:* Brown, 1950c: 145; in *Orectognathus:* Taylor, 1977: 583.
***mjobergi.*** *Orectognathus mjobergi* Forel, 1915b: 38, pl. 2, figs. 3, 4 (w.q.) AUSTRALIA. Wheeler, G.C. & Wheeler, J. 1955a: 127 (l.); Wheeler, G.C. & Wheeler, J. 1973c: 213 (l.). Senior synonym of *unicolor:* Brown, 1953k: 98.
***nanus.*** *Orectognathus nanus* Taylor, 1977: 605, figs. 20–22 (w.q.) AUSTRALIA.
***nigriventris.*** *Orectognathus nigriventris* Mercovich, 1958: 195, fig. (w.) AUSTRALIA. Wheeler, G.C. & Wheeler, J. 1973c: 213 (l.). See also: Brown, 1958b: 28 (in key).
***parvispinus.*** *Orectognathus parvispinus* Taylor, 1977: 603, figs. 17–19 (w.q.) AUSTRALIA.
*perplexus* Smith, F. 1876; see under ***STRUMIGENYS.***
***phyllobates.*** *Orectognathus phyllobates* Brown, 1958b: 25, fig. 4 (w.) AUSTRALIA.
***robustus.*** *Orectognathus robustus* Taylor, 1977: 599, figs. 14–16 (w.q.) AUSTRALIA.
***roomi.*** *Orectognathus roomi* Taylor, 1977: 606, figs. 24–27 (w.) NEW GUINEA.
***rostratus.*** *Orectognathus rostratus* Lowery, 1967: 137, figs. 1, 2 (w.q.) AUSTRALIA. Wheeler, G.C. & Wheeler, J. 1973c: 214 (l.).
***sarasini.*** *Orectognathus sarasini* Emery, 1914f: 416 (w.m.) NEW CALEDONIA. Variety of *antennatus:* Wheeler, W.M. 1927i: 125. Revived status as species: Brown, 1953k: 100. See also: Brown, 1958b: 23
***satan.*** *Orectognathus satan* Brown, 1953k: 102, fig. 3 (w.q.) AUSTRALIA. Wheeler, G.C. & Wheeler, J. 1955a: 127 (l.).
*septentrionalis. Orectognathus antennatus* var. *septentrionalis* Forel, 1910b: 51 (w.) AUSTRALIA Junior synonym of *antennatus:* Brown, 1953k: 99.
***sexspinosus.*** *Orectognathus sexspinosus* Forel, 1915b: 39, pl. 2, figs. 1, 2 (w.q.m.) AUSTRALIA.
***szentivanyi.*** *Arnoldidris szentivanyi* Brown, 1958b: 18, figs. 1, 2 (w.q.) NEW GUINEA. Combination in *Orectognathus:* Taylor, 1977: 583.
*unicolor. Orectognathus mjobergi* var. *unicolor* Forel, 1915b: 39 (w.) AUSTRALIA. Junior synonym of *mjobergi:* Brown, 1953k: 98.
***velutinus.*** *Orectognathus velutinus* Taylor, 1977: 610, figs. 37–42 (w.) NEW GUINEA.
***versicolor.*** *Orectognathus versicolor* Donisthorpe, 1940a: 46 (w.) AUSTRALIA. Brown, 1953k: 93 (q.); Wheeler, G.C. & Wheeler, J. 1955a: 128 (l.); Imai, Crozier & Brown, 1977: 349 (k.).

*OREOMYRMA:* see under ***MANICA.***

*ORNATINODA:* see under ***PSEUDOMYRMEX.***

#***ORTHOCREMA:*** see under ***CREMATOGASTER.***

#***ORTHONOTOMYRMEX:*** see under ***CAMPONOTUS.***

*ORTHONOTUS:* see under ***CAMPONOTUS.***

*OTOMYRMEX:* see under ***CATAULACUS.***

***OVERBECKIA*** (Singapore)
***subclavata.*** *Overbeckia subclavata* Viehmeyer, 1916a: 152, figs. 11, 12 (w.q.m.) SINGAPORE.

***OXYEPOECUS*** (Neotropical)
***bruchi.*** *Oxyepoecus bruchi* Santschi, 1926d: 6, figs. A-D (q.) ARGENTINA. Santschi, 1929d: 295 (w.). Combination in *Mitara:* Santschi, 1927d: 246; in *Martia:* Santschi, 1929d: 295; in *Oxyepoecus:* Ettershank, 1966: 146. Senior synonym of *minuta:* Kempf, 1974b: 478.
***crassinodus.*** *Oxyepoecus crassinodus* Kempf, 1974b: 482, figs. 2, 8, 15, 21 (w.) BRAZIL.
***daguerrei.*** *Martia daguerrei* Santschi, 1933e: 111 (w.) ARGENTINA. Combination in *Oxyepoecus:* Ettershank, 1966: 146. See also: Kempf, 1974b: 486.
***inquilinus.*** *Martia inquilina* Kusnezov, 1952h: 720 (diagnosis in key) (w.) ARGENTINA. Combination in *Oxyepoecus:* Ettershank, 1966: 146. Senior synonym of *turgidus:* Kempf, 1974b: 489.
*luederwaldti. Monomorium (Martia) rastratum* var. *luederwaldti* Forel, 1913l: 219 (w.) BRAZIL. Combination in *Martia:* Kusnezov, 1952h: 722; in *Oxyepoecus:* Ettershank, 1966: 146. Junior synonym of *rastratus:* Kempf, 1974b: 498.
***mandibularis.*** *Monomorium (Martia) mandibulare* Emery, 1913: 261, fig. 12 (w.) BOLIVIA. Combination in *Martia:* Kusnezov, 1952h: 722; in *Oxyepoecus:* Ettershank, 1966: 146. See also: Kempf, 1974b: 491.
*minuta. Martia minuta* Kusnezov, 1952h: 721 (diagnosis in key) (w.) ARGENTINA. Combination in *Oxyepoecus:* Ettershank, 1966: 146. Junior synonym of *bruchi:* Kempf, 1974b: 478.
***plaumanni.*** *Oxyepoecus plaumanni* Kempf, 1974b: 492, figs. 27–30 (w.q.) BRAZIL.
***punctifrons.*** *Monomorium punctifrons* Borgmeier, 1927b: 63 (w.) BRAZIL. [Also described as new by Borgmeier, 1928a: 39.] Wheeler, G.C. & Wheeler, J. 1977: 586 (l.). Combination in *Martia:* Kusnezov, 1952h: 722; in *Oxyepoecus:* Ettershank, 1966: 146. See also: Kempf, 1974b: 495.

*OXYEPOECUS (continued)*

***rastratus.*** *Monomorium rastratum* Mayr, 1887: 615 (w.) BRAZIL. Kempf, 1974b: 498 (q.m.); Wheeler, G.C. & Wheeler, J. 1980: 534 (l.). Combination in *Monomorium (Martia)*: Forel, 1912g: 3; in *Martia:* Kusnezov, 1952h: 722; in *Oxyepoecus:* Ettershank, 1966: 146. Senior synonym of *luederwaldti:* Kempf, 1974b: 498.

***reticulatus.*** *Oxyepoecus reticulatus* Kempf, 1974b: 502, figs. 23–26 (w.q.) BRAZIL.

*turgidus. Oxyepoecus turgidus* Kempf, 1969: 278, figs. 1–3 (w.) BRAZIL. Junior synonym of *inquilinus:* Kempf, 1974b: 489.

***vezenyii.*** *Monomorium (Martia) vezenyii* Forel, 1907a: 20 (w.) PARAGUAY. Kempf, 1969: 277 (q.). Combination in *Martia:* Kusnezov, 1952h: 722; in *Oxyepoecus:* Ettershank, 1966: 145. See also: Kempf, 1974b: 505.

***vivax.*** *Oxyepoecus vivax* Kempf, 1974b: 508, figs. 31–33, 38, 39 (q.) BRAZIL.

#***OXYGYNE:*** see under ***CREMATOGASTER.***

****OXYIDRIS*** (Dominican Amber)

****antillana.*** **Oxyidris antillana* Wilson, 1985a: 5, fig. 3 (w.) DOMINICAN AMBER.

***OXYOPOMYRMEX*** (West Palaearctic)

#***brunnescens.*** *Oxyopomyrmex emeryi* var. *brunnescens* Santschi, 1929e: 147 (w.) TUNISIA.

*cabrerae. Oxyopomyrmex saulcyi* var. *cabrerae* Forel, 1897a: 133 (q.) SPAIN. [Also described as new by Forel, 1897c: 207 (footnote).] Junior synonym of *saulcyi:* Collingwood, 1978: 68.

***emeryi.*** *Oxyopomyrmex emeryi* Santschi, 1908: 524, fig. 9 (w.) TUNISIA. Delye, 1971: 211 (q.m.). Current subspecies: nominal plus *brunnescens.*

***gaetulus.*** *Oxyopomyrmex gaetulus* Santschi, 1929e: 146 (q.m.) MOROCCO.

***insularis.*** *Oxyopomyrmex insularis* Santschi, 1908: 523, fig. 7 (w.m.) CANARY IS. Senior synonym of *major:* Barquin, 1981: 103.

***krueperi.*** *Oxyopomyrmex krueperi* Forel, 1911d: 344 (w.) GREECE.

***lagoi.*** *Oxyopomyrmex lagoi* Menozzi, 1936d: 278, fig. 7 (w.) GREECE.

*laticeps. Oxyopomyrmex emeryi* st. *sabulonis* var. *laticeps* Santschi, 1915a: 62 (w.) TUNISIA. Unavailable name.

#***latinodis.*** *Oxyopomyrmex saulcyi* var. *latinodis* Santschi, 1939c: 2 (w.q.) MOROCCO.

*major. Oxyopomyrmex insularis* var. *major* Santschi, 1923f: 326 (w.) CANARY IS. Junior synonym of *insularis:* Barquin, 1981: 103.

#***nigripes.*** *Oxyopomyrmex santschii* var. *nigripes* Santschi, 1907: 329 (w.q.m.) TUNISIA.

#***nitidior.*** *Oxyopomyrmex santschii* var. *nitidior* Santschi, 1910a: 46 (w.) TUNISIA.

***oculatus.*** *Oxyopomyrmex oculatus* André, 1881b: 73, pl. 3, figs. 1–3 (w.) ISRAEL.

*punicus* Forel, 1907; see under ***GONIOMMA.***

#***rugocciput.*** *Oxyopomyrmex sabulonis* var. *rugocciput* Santschi, 1923f: 326 (w.) TUNISIA.

***sabulonis.*** *Oxyopomyrmex emeryi* st. *sabulonis* Santschi, 1915a: 62 (w.) TUNISIA. Raised to species: Santschi, 1923f: 326; Collingwood, 1985: 246. Current subspecies: nominal plus *rugocciput.*

***santschii.*** *Oxyopomyrmex santschii* Forel, 1904c: 8 (w.q.m.) TUNISIA. Senior synonym of *siciliana:* Emery, 1916b: 146. Current subspecies: nominal plus *nigripes, nitidior.* See also: Baroni Urbani, 1971c: 68.

***saulcyi.*** *Oxyopomyrmex saulcyi* Emery, 1889a: 440 (w.q.m.) FRANCE. Senior synonym of *cabrerae:* Collingwood, 1978: 68. Current subspecies: nominal plus *latinodis.*

*siciliana. Oxyopomyrmex santschii* var. *siciliana* Karavaiev, 1912a: 11 (w.) SICILY. Junior synonym of *santschii:* Emery, 1916b: 146.

*thoracica* Santschi, 1907; see under ***GONIOMMA.***

*tuneticus* Forel, 1905; see under ***GONIOMMA.***

***PACHYCONDYLA*** (World Tropics and Subtropics)

*abyssinica. Ponera abyssinica* Guérin-Méneville, 1849: 352, pl. 7, fig. 6 (w.) ETHIOPIA. Junior synonym of *foetens:* Roger, 1860: 310.

***aciculata.*** *Pachycondyla (Ectomomyrmex) aciculatus* Emery, 1901h: 157 (w.) NEW GUINEA. Combination in *Ectomomyrmex:* Wilson, 1958d: 363; in *Pachycondyla:* Brown, **revived combination**.

***acuta.*** *Pachycondyla (Ectomomyrmex) acuta* Emery, 1900c: 314 (w.) NEW GUINEA. Combination in *Ectomomyrmex:* Wilson, 1958d: 364; in *Pachycondyla:* Brown, **revived combination**. Senior synonym of *dahlii:* Wilson, 1958d: 364.

#***acuticostata.*** *Ponera (Bothroponera) excavata* var. *acuticostata* Forel, 1900b: 64 (w.) AUSTRALIA. Combination in *Pachycondyla:* Emery, 1901a: 45; in *Bothroponera:* Taylor & Brown, D.R. 1985: 21; in *Pachycondyla:* Brown, **revived combination**.

#***admista.*** *Bothroponera perroti* r. *admista* Forel, 1892l: 251 (w.) MADAGASCAR. Combination in *Pachycondyla (Bothroponera)*: Emery, 1901a: 45. Raised to species: Dalla Torre, 1893: 35. Variety/subspecies of *perroti:* Wasmann, 1897: 250; Emery, 1901a: 45.

***aenescens.*** *Pachycondyla aenescens* Mayr, 1870a: 396 (w.) COLOMBIA. Emery, 1890b: 42 (q.); Forel, 1899c: 10 (m.). Combination in *Euponera (Mesoponera)*: Emery, 1901a: 47; in *Mesoponera:* Kempf, 1972a: 141; in *Pachycondyla:* Brown, **revived combination**.

***aenigmatica.*** *Ponera aenigmatica* Arnold, 1949: 265, fig. 5 (w.) SOUTH AFRICA. Combination in *Pachycondyla:* Brown, **new combination**.

***aequalis.*** *Ectomomyrmex exarata* subsp. *aequalis* Mann, 1919: 289, fig. 7 (w.) SOLOMON IS. Combination in *Pachycondyla:* Brown, **new combination**. Raised to species: Wilson, 1958d: 364.

#***affinis.*** *Euponera (Mesoponera) caffraria* var. *affinis* Santschi, 1935b: 258 (w.) CONGO. Combination in *Pachycondyla:* Brown, **new combination**.

#***africana.*** *Euponera (Pseudoponera) darwini* var. *africana* Forel, 1909b: 51 (q.) ZAIRE. Santschi, 1914b: 51 (m.). Combination in *Euponera (Trachymesopus)*: Emery, 1911d: 86; in *Pachycondyla:* Brown, **new combination**. Stirps of *darwinii:* Santschi, 1937b: 95.

***agilis.*** *Neoponera (Eumecopone) agilis* Forel, 1901e: 336 (w.) U.S.A. (locality in error). Combination in *Pachycondyla:* Brown, **new combination**.

#***amazon.*** *Megaponera foetens* var. *amazon* Santschi, 1935b: 257, fig. 2 (m.) ETHIOPIA. Combination in *Pachycondyla:* Brown, **new combination**. Subspecies of *analis:* current status.

***ambigua.*** *Pachycondyla ambigua* André, 1890: 316 (q.) SIERRA LEONE. Combination in *Euponera (Brachyponera)*: Emery, 1901a: 47; in *Pachycondyla:* Brown, **new combination**.

***ambigua.*** *Ponera ambigua* Weber, 1942a: 46, fig. 6 (w.) SUDAN. [Unresolved junior secondary homonym of *Pachycondyla ambigua* André, above.] Combination in *Pachycondyla:* Brown, **new combination**.

***amblyops.*** *Ponera amblyops* Emery, 1887b: 434 (w.) SUMATRA. Emery, 1889b: 497 (q.); Santschi, 1928h: 122 (m.). Combination in *Trapeziopelta:* Emery, 1889b: 497; in *Belonopelta:* Emery, 1897d: 553; in *Pachycondyla (Pseudoponera)*: Emery, 1900d: 668; in *Pseudoponera:* Bingham, 1903: 92; in *Euponera (Pseudoponera)*: Forel, 1905c: 6; in *Pachycondyla:* Brown, **revived combination**. Current subspecies: nominal plus *oculatior.*

*americana. Ponera americana* Mayr, 1862: 722 (w.) COLOMBIA. Junior synonym of *stigma:* Emery, 1887b: 434.

*amplinoda. Ponera amplinoda* Buckley, 1866: 171 (w.) U.S.A. Junior synonym of *harpax:* Emery, 1895c: 266; of *villosa:* Forel, 1899c: 14; of *montezumia:* Emery, 1911d: 75.

***analis.*** *Formica analis* Latreille, 1802c: 282. Replacement name for *Formica foetens* Fabricius, 1793: 354. [Junior primary homonym of *Formica foetens* Olivier, 1792: 503.] Combination in *Pachycondyla:* Brown, **new combination**. [For earlier references and synonymy see under *foetens* Fabricius.] Current subspecies: nominal plus *amazon, crassicornis, rapax, subpilosa, termitivora.*

#***ancilla.*** *Ponera (Bothroponera) soror* var. *ancilla* Emery, 1899e: 472 (w.) CONGO. Combination in *Pachycondyla (Bothroponera)*: Emery, 1901a: 46; in *Bothroponera:* Wheeler, W.M. 1922a: 74; in *Pachycondyla:* Brown, **revived combination**.

*andicola. Pachycondyla fuscoatra* var. *andicola* Santschi, 1913h: 34 (w.) ECUADOR. Junior synonym of *impressa:* Kempf, 1961c: 195.

*andrieui. Bothroponera crassa* st. *crassior* var. *andrieui* Santschi, 1930a: 54 (w.) SUDAN. Unavailable name.

***annamita.*** *Ponera annamita* André, 1892a: 48 (w.q.m.) VIETNAM. Combination in *Bothroponera:* Dalla Torre, 1893: 35; in *Ectomomyrmex:* Forel, 1900d: 322; Bingham, 1903: 87; in *Pachycondyla (Ectomomyrmex)*: Emery, 1901a: 46. Current subspecies: nominal plus *arcuata.*

***apicalis.*** *Formica apicalis* Latreille, 1802c: 204, pl. 7, fig. 42 (w.) SOUTH AMERICA. Forel, 1899c: 11 (q.); Wheeler, G.C. & Wheeler, J. 1952c: 615 (l.). Combination in *Pachycondyla:* Mayr, 1863: 439; in *Neoponera:* Emery, 1901a: 47; in *Pachycondyla:* Brown, **revived combination**. Senior synonym of *latreillei* and *verenae:* Brown, 1957e: 230.

***apicalis.*** *Ponera apicalis* Smith, F. 1857a: 66 (q.) BORNEO. [Unresolved junior secondary homonym of *Formica apicalis* Latreille, above.] Combination in *Euponera (Mesoponera)*: Donisthorpe, 1932c: 447; in *Pachycondyla:* Brown, **new combination**.

**#arcuata.** *Ectomomyrmex annamitus* var. *arcuatus* Forel, 1900d: 322 (q.) INDIA. Stärcke, 1930: 371 (w.). Combination in *Pachycondyla* (*Ectomomyrmex*): Emery, 1901a: 46.

**arcuata.** *Euponera* (*Brachyponera*) *luteipes* var. *arcuata* Karavaiev, 1925b: 125, figs. 3, 4 (w.q.m.) JAVA. [Unresolved junior secondary homonym of *arcuata* Forel, above.] Combination in *Brachyponera:* Wilson, 1958d: 347; in *Pachycondyla:* Brown, **new combination**. Raised to species: Wilson, 1958d: 347. See also: Santschi, 1928h: 124.

**arhuaca.** *Euponera* (*Mesoponera*) *arhuaca* Forel, 1901f: 339 (w.m.) COLOMBIA. Combination in *Mesoponera:* Kempf, 1972a: 141.

*armata* Santschi, 1919; see under ***PHRYNOPONERA***.

*arnoldi. Ponera* (*Xiphopelta*) *arnoldi* Forel, 1913a: 108 (w.) ZIMBABWE. Junior synonym of *rotundata:* Wheeler, W.M. 1922a: 776.

**#asina.** *Pachycondyla* (*Bothroponera*) *asina* Santschi, 1912b: 153 (w.) KENYA. Combination in *Bothroponera:* Wheeler, W.M. 1922a: 770; in *Pachycondyla:* Brown, **revived combination**. Stirps of *kruegeri:* Santschi, 1914e: 4.

**#aspera.** *Bothroponera laevissima* var. *aspera* Arnold, 1962: 844 (w.) SOUTH AFRICA. Combination in *Pachycondyla:* Brown, **new combination**.

**astuta.** *Pachycondyla astuta* Smith, F. 1858b: 107 (w.) AUSTRALIA. Karavaiev, 1925b: 120 (q.); Imai, Kubota, *et al.* 1985: 46 (k.). Combination in *P.* (*Bothroponera*): Emery, 1901a: 45; in *P.* (*Ectomomyrmex*): Forel, 1905c: 6; in *Ectomomyrmex:* Bingham, 1903: 86; Wheeler, W.M. 1927h: 84; in *Bothroponera:* Taylor & Brown, D.R. 1985: 21; in *Pachycondyla:* Brown, **revived combination**. Current subspecies: nominal plus *obscura*.

**atrata.** *Euponera* (*Brachyponera*) *atrata* Karavaiev, 1925b: 126 (w.) INDONESIA (Ambon I.). Karavaiev, 1926d: 418 (q.). Combination in *Brachyponera:* Wilson, 1958d: 347; in *Pachycondyla:* Brown, **new combination**.

**atrovirens.** *Pachycondyla atrovirens* Mayr, 1866b: 890 (w.q.) COLOMBIA. [Also described as new by Mayr, 1870a: 397 (in key).] Combination in *Euponera* (*Mesoponera*): Emery, 1901a: 47; in *Mesoponera:* Kempf, 1972a: 141; in *Pachycondyla:* Brown, **revived combination**. Current subspecies: nominal plus *splendida*.

**#attenata.** *Pachycondyla* (*Bothroponera*) *pachyderma* var. *attenata* Santschi, 1920b: 7 (w.) ZAIRE.

**#attrita.** *Ponera stigma* var. *attrita* Forel, 1893g: 362 (w.q.m.) ANTILLES. Combination in *Pachycondyla* (*Pseudoponera*): Emery, 1901a: 46; in *Euponera* (*Trachymesopus*): Emery, 1911d: 85; in *Trachymesopus:* Kempf, 1972a: 251; in *Pachycondyla:* Brown, **revived combination**.

**australis.** *Ponera melanaria* r. *australis* Forel, 1900b: 62 (w.) AUSTRALIA. Forel, 1915b: 32 (q.); Wheeler, G.C. & Wheeler, J. 1971b: 1206 (l.). Combination in *Euponera* (*Mesoponera*): Emery, 1901a: 46; in *Mesoponera:* Wheeler, G.C. & Wheeler, J. 1971b: 1206; in *Pachycondyla:* Brown, **new combination**. Raised to species: Taylor & Brown, D.R. 1985: 35.

**#azteca.** *Pachycondyla carinulata* r. *azteca* Forel, 1899c: 11 (w.) PANAMA. Combination in *Neoponera:* Emery, 1901a: 47; in *Pachycondyla:* Brown, **new combination**.

*bakeri. Neoponera* (*Neoponera*) *bakeri* Mann, 1916: 411, pl. 1, fig. 9 (w.q.) BRAZIL. Junior synonym of *striatinodis:* Kempf, 1960e: 391.

**barbata.** *Pachycondyla* (*Bothroponera*) *barbata* Stitz, 1911a: 355, fig. 6 (w.) AUSTRALIA.

*bayoni. Euponera* (*Trachymesopus*) *bayoni* Menozzi, 1933a: 97 (w.) UGANDA. Junior synonym of *brunoi:* Brown, 1963: 8.

**berthoudi.** *Ophthalmopone berthoudi* Forel, 1890b: cxii (w.) SOUTH AFRICA. Forel, 1894b: 76 (m.); Wheeler, G.C. & Wheeler, J. 1971b: 1203 (l.). Combination in *Pachycondyla:* Brown, **new combination**. Current subspecies: nominal plus *pubescens*. See also: Arnold, 1915: 50; Peeters & Crewe, 1985: 29; Peeters & Crewe, 1987: 201.

*berthoudi. Pachycondyla* (*Bothroponera*) *berthoudi* Forel, 1901f: 344 (w.) SOUTH AFRICA. [Unresolved junior secondary homonym of *berthoudi* Forel, above.] Forel, 1913b: 306 (m.). Race of *pumicosa:* Forel, 1913a: 109. Junior synonym of *strigulosa:* Arnold, 1952a: 460.

*bicolor. Ponera bicolor* Guérin-Méneville, 1844a: 424 (w.) MEXICO. Junior synonym of *villosa:* Roger, 1861a: 1; Roger, 1862c: 288.

*bicolor. Ponera bicolor* Donisthorpe, 1949b: 492 (w.) NEW GUINEA. [Unresolved junior primary homonym of *bicolor* Guérin-Méneville, above.] Junior synonym of *croceicornis:* Wilson, 1958d: 347.

**bispinosa.** *Pachycondyla bispinosa* Smith, F. 1858b: 107 (w.) INDIA (not Australia). Combination in *Bothroponera:* Emery, 1889b: 495; Bingham, 1903: 97; in *Pachycondyla* (*Bothroponera*): Emery, 1901a: 45. Junior synonym of *rufipes:* Forel, 1886d: 246. Revived from synonymy: Emery, 1889b: 495; Forel, 1900d: 326; Bingham, 1903: 97.

**#brevior.** *Ponera* (*Bothroponera*) *insularis* var. *brevior* Forel, 1901h: 45 (w.) BORNEO. Combination in *Bothroponera:* Wheeler, W.M. 1919e: 55; in *Pachycondyla:* Brown, **new combination**.

*brunneus. Euponera* (*Trachymesopus*) *brunneus* Donisthorpe, 1948b: 300 (w.q.) NEW GUINEA. Junior synonym of *stigma:* Wilson, 1958d: 355.

**brunoi.** *Pachycondyla* (*Ectomomyrmex*) *brunoi* Forel, 1913j: 205 (w.) ZIMBABWE. Arnold, 1915: 54 (m.); Arnold, 1926: 201 (q.); Wheeler, G.C. & Wheeler, J. 1971b: 1207 (l.). Combination in *Ectomomyrmex:* Wheeler, W.M. 1922a: 774; in *Trachymesopus:* Brown, 1963: 8; in *Pachycondyla:* Brown, **revived combination**. Senior synonym of *bayoni, dentis, katangana, lamottei* and *nigeriensis:* Brown, 1963: 8.

**bucki.** *Leptogenys bucki* Borgmeier, 1927b: 57 (w.) BRAZIL. [Also described as new by Borgmeier, 1928a: 39.] Combination in *Euponera:* Borgmeier, 1959a: 315; in *Mesoponera:* Kempf, 1972a: 141; in *Pachycondyla:* Brown, **new combination**.

**#bugabensis.** *Pachycondyla theresiae* var. *bugabensis* Forel, 1899c: 14 (w.) PANAMA. Combination in *Neoponera:* Emery, 1901a: 47; in *Pachycondyla:* Brown, **revived combination**.

**butteli.** *Pseudoponera butteli* Forel, 1913k: 8, fig. B (w.) JAVA. Combination in *Pachycondyla:* Brown, **new combination**.

**#caffra.** *Euponera* (*Mesoponera*) *caffraria* var. *caffra* Santschi, 1935b: 259 (w.) GUINEA. [*E.* (*M.*) *caffra* Santschi, 1914d: 315. *Nomen nudum;* attributed to Smith.] Combination in *Pachycondyla:* Brown, **new combination**.

**caffraria.** *Ponera caffraria* Smith, F. 1858b: 91 (q.) SOUTH AFRICA. Forel, 1913a: 108 (w.); Arnold, 1915: 65 (w.m.); Emery, 1899c: 5 (l.); Wheeler, G.C. & Wheeler, J. 1971b: 1206 (l.). Combination in *Euponera* (*Mesoponera*): Emery, 1901a: 46; in *Mesoponera:* Wheeler, G.C. & Wheeler, J. 1971b: 1206; in *Pachycondyla:* Brown, **new combination**. Senior synonym of *guineensis:* Mayr, 1895: 125; Emery, 1911d: 81; Wheeler, W.M. 1922a: 775. Current subspecies: nominal plus *affinis, caffra*.

***calcarea.** **Euponera calcarea* Théobald, 1937b: 198, pl. 4, fig. 8; pl. 13, figs. 17, 18 (q.m.) FRANCE (Oligocene). Combination in *Pachycondyla:* **new combination**.

*cambodjana. Pachycondyla* (*Ectomomyrmex*) *astuta* subsp. *cambodjana* Forel, 1911e: 253 (w.) CAMBODIA. Junior synonym of *javana:* Yasumatsu, 1962: 94.

**cambouei.** *Bothroponera cambouei* Forel, 1891b: 133, pl. 4, fig. 7 (w.q.) MADAGASCAR. Combination in *Pachycondyla* (*Bothroponera*): Emery, 1901a: 45.

**carbonaria.** *Ponera carbonaria* Smith, F. 1858b: 97 (w.) ECUADOR. Combination in *Pachycondyla:* Mayr, 1886c: 358; in *Euponera* (*Mesoponera*): Emery, 1901a: 47; in *Mesoponera:* Kempf, 1972a: 141; in *Pachycondyla:* Brown, **revived combination**.

**carinulata.** *Ponera carinulata* Roger, 1861a: 4 (q.) GUIANA. Emery, 1890a: 73 (w.); Mann, 1916: 412 (w.). Combination in *Pachycondyla:* Roger, 1863b: 18; in *Neoponera:* Emery, 1901a: 47; in *Pachycondyla:* Brown, **revived combination**. Current subspecies: nominal plus *azteca, gibbinota*.

**cariosa.** *Bothroponera cariosa* Emery, 1895h: 20 (w.) MOZAMBIQUE. Wheeler, G.C. & Wheeler, J. 1971c: 389 (l.). Combination in *Pachycondyla* (*Bothroponera*): Emery, 1901a: 45; in *Bothroponera:* Wheeler, W.M. 1922a: 72; in *Pachycondyla:* Brown, **revived combination**.

**castanea.** *Ponera castanea* Mayr, 1865: 69 (w.) NEW ZEALAND. Combination in *Euponera* (*Mesoponera*): Emery, 1901a: 46; in *Mesoponera:* Brown, 1958h: 20; in *Pachycondyla:* Brown, **new combination**. Senior synonym of *striata* Stitz: Brown, 1958h: 20.

*castanea. Ponera castanea* Smith, 1876c: 489 (w.q.) NEW ZEALAND. [Junior primary homonym of *castanea* Mayr, above.] Replacement name: *castaneicolor:* Dalla Torre, 1893: 38.

**castaneicolor.** *Ponera castaneicolor* Dalla Torre, 1893: 38. Replacement name for *castanea* Smith, F. 1876c: 489. [Junior primary homonym of *castanea* Mayr, 1865: 69.] Combination in *Euponera* (*Mesoponera*): Emery, 1911d: 81; in *Mesoponera:* Wilson & Taylor, 1967: 103; in *Pachycondyla:* Brown, **new combination**.

**cauta.** *Euponera* (*Trachymesopus*) *cauta* Mann, 1922: 8 (w.) HONDURAS. Combination in *Trachymesopus:* Kempf, 1960f: 423; in *Pachycondyla:* Brown, **new combination**. Senior synonym of *compressinodis:* Brown, 1963: 7.

**cavernosa.** *Ponera cavernosa* Roger, 1860: 288 (w.) SOUTH AFRICA. Combination in *Bothroponera:*

Mayr, 1862: 717; Wheeler, W.M. 1922a: 72; in *Pachycondyla* (*Bothroponera*): Emery, 1901a: 45; Brown, **revived combination**. Current subspecies: nominal plus *montivaga*.

***cavinodis.*** *Neoponera* (*Neoponera*) *cavinodis* Mann, 1916: 414, pl. 2, fig. 14 (w.) BRAZIL. Wheeler, G.C. & Wheeler, J. 1971b: 1205 (l.). Combination in *Pachycondyla:* Brown, **new combination**.

*cearensis. Pachycondyla fuscoatra* var. *cearensis* Forel, 1901e: 336 (w.q.) BRAZIL. Junior synonym of *impressa:* Kempf, 1961c: 195.

#***ceylonensis.*** *Pachycondyla* (*Bothroponera*) *rufipes* subsp. *ceylonensis* Forel, 1911i: 216 (w.) SRI LANKA.

*chelifer* Santschi, 1928; see under ***LEPTOGENYS.***

***chinensis.*** *Ponera nigrita* subsp. *chinensis* Emery, 1895k: 460 (w.) CHINA. Wheeler, W.M. 1921c: 530 (q.); Ogata, 1987: 116 (m.); Wheeler, G.C. & Wheeler, J. 1986c: 88 (l.); Imai & Kubota, 1972: 194 (k., misspelled as *sinensis*). Combination in *Euponera* (*Brachyponera*): Emery, 1909c: 367; in *Brachyponera:* Brown, 1958h: 22; in *Pachycondyla:* Brown, **new combination**. Junior synonym of *solitaria* Smith, F. 1874: 404 [Junior primary homonym of *solitaria* Smith, F. 1860b: 103.], and hence first available replacement name: Brown, 1958h: 22.

*chosonensis. Euponera* (*Trachymesopus*) *chosonensis* Teranishi, 1940: 8 (w.) KOREA. Junior synonym of *pilosior:* Brown, 1963: 8.

***christmasi.*** *Euponera* (*Mesoponera*) *christmasi* Donisthorpe, 1935: 630 (w.) CHRISTMAS I. Combination in *Pachycondyla:* Brown, **new combination**.

***chyzeri.*** *Euponera* (*Mesoponera*) *chyzeri* Forel, 1907a: 5 (w.) COLOMBIA. Combination in *Mesoponera:* Kempf, 1972a: 141; in *Pachycondyla:* Brown, **new combination**.

*clara. Euponera* (*Brachyponera*) *lutea* var. *clara* Crawley, 1915a: 133 (w.) AUSTRALIA. Combination in *Brachyponera:* Taylor & Brown, D.R. 1985: 23. Junior synonym of *lutea:* **new synonymy**.

***clarki.*** *Euponera* (*Trachymesopus*) *clarki* Wheeler, W.M. 1934d: 140 (w.q.) AUSTRALIA. Combination in *Trachymesopus:* Taylor & Brown, D.R. 1985: 52; in *Pachycondyla:* Brown, **new combination**. Junior synonym of *rufonigra:* Brown, 1952f: 138. Revived from synonymy: Taylor & Brown, D.R. 1985: 52.

***claudata.*** *Ectomomyrmex claudatus* Menozzi, 1926e: 93 (q.) PHILIPPINES. Combination in *Pachycondyla:* Brown, **new combination**.

*clavicornis. Psalidomyrmex clavicornis* Bernard, 1953b: 209, fig. 5 (q.) GUINEA. Junior synonym of *talpa:* Bolton, 1975b: 6.

***cognata.*** *Ponera cognata* Emery, 1896g: 56 (w.q.) COSTA RICA. Combination in *Pachycondyla* (*Pseudoponera*): Emery, 1901a: 46; in *Euponera* (*Trachymesopus*): Emery, 1911d: 85; in *Trachymesopus:* Kempf, 1960f: 423; in *Pachycondyla:* Brown, **new combination**.

***commutata.*** *Ponera commutata* Roger, 1860: 311 (w.) SOUTH AMERICA. Wheeler, W.M. 1936d: 170 (q.); Borgmeier, 1959a: 313 (m.). Combination in *Pachycondyla:* Emery, 1890a: 72; in *Neoponera:* Emery, 1901a: 47; in *Termitopone* (*Syntermitopone*): Wheeler, W.M. 1936d: 169; in *Pachycondyla:* Brown, **revived combination**. [Emery, 1892b: 163, suggests that this name may be synonymous with *Formica ovata* Reich, 1793: 132. The latter is here regarded as unidentifiable to genus; see under *Formica.*]

***comorensis.*** *Ponera comorensis* André, 1887: 292 (w.) MADAGASCAR. Forel, 1891b: 124 (w.). Combination in *Bothroponera:* Forel, 1891b: 124; in *Pachycondyla* (*Bothroponera*): Emery, 1901a: 45.

*compressinodis. Euponera* (*Trachymesopus*) *stigma* subsp. *compressinodis* Borgmeier, 1928b: 62 (w.) BRAZIL. Kempf, 1960f: 425 (q.). Combination in *Trachymesopus,* and raised to species: Kempf, 1960f: 424. Junior synonym of *cauta:* Brown, 1963: 7.

*concinna. Pachycondyla harpax* var. *concinna* Wheeler, W.M. 1925a: 5 (w.q.) BRAZIL. Junior synonym of *harpax:* Brown, 1950e: 247.

*confusa. Neoponera crenata* st. *confusa* Santschi, 1921g: 86 (w.) BRAZIL. Junior synonym of *crenata:* Brown, 1957e: 234.

***constricta.*** *Ponera constricta* Mayr, 1884: 31 (w.) FRENCH GUIANA. Forel, 1908b: 37 (q.m.); Wheeler, G.C. & Wheeler, J. 1952c: 624 (l.); Wheeler, G.C. & Wheeler, J. 1976a: 53 (l.). Combination in *Euponera* (*Mesoponera*): Emery, 1901a: 46; in *Mesoponera:* Kempf, 1972a: 141; in *Pachycondyla:* Brown, **new combination**. Senior synonym of *josephi:* Dalla Torre, 1893: 39.

#***continentalis.*** *Euponera* (*Brachyponera*) *luteipes* var. *continentalis* Karavaiev, 1925b: 125, fig. 3 (w.) INDIA. Combination in *Pachycondyla:* Brown, **new combination**.

***crassa.*** *Ponera crassa* Emery, 1877b: 366, fig. (w.) ETHIOPIA. Combination in *Bothroponera:* Emery, 1892a: 111; in *Pachycondyla* (*Bothroponera*): Emery, 1895a: 177; Emery, 1901a: 46; in *Bothroponera:* Wheeler, W.M. 1922a: 73; Santschi, 1930a: 53; in *Pachycondyla:* Brown, **revived combination**. Current subspecies: nominal plus *crassior, ilgii* (junior homonym).

#***crassicornis.*** *Ponera crassicornis* Gerstäcker, 1859: 262 (w.) MOZAMBIQUE. [Also described as new by Gerstäcker, 1862: 506.] Junior synonym of *foetens:* Mayr, 1907b: 9; Wheeler, W.M. 1922a: 767. Revived from synonymy as variety of *foetens:* Santschi, 1930b: 59 (in key). Subspecies of *analis:* current status.

***crassinoda.*** *Formica crassinoda* Latreille, 1802c: 198, pl. 7, fig. 41 (w.q.) SURINAM. Wheeler, G.C. & Wheeler, J. 1952c: 617 (l.). Combination in *Pachycondyla:* Smith, F. 1858b: 105.

#***crassior.*** *Bothroponera crassa* var. *crassior* Santschi, 1930a: 53 (w.) KENYA. Santschi, 1930d: 267 (m.). Combination in *Pachycondyla:* Brown, **new combination**. See also: Santschi, 1932a: 382.

****crawleyi.*** **Euponera* (*Mesoponera*) *crawleyi* Donisthorpe, 1920: 85, pl. 5, fig. 3 (q.) GREAT BRITAIN (Oligocene). Combination in *Pachycondyla:* **new combination**.

***crenata.*** *Ponera crenata* Roger, 1861a: 3. Replacement name for *Ponera pallipes* Smith, F. 1858b: 98. [Junior primary homonym of *Ponera pallipes* Smith, F. 1858b: 87.] Mayr, 1887: 534 (w.); Santschi, 1921g: 85 (m.); Wheeler, G.C. & Wheeler, J. 1974g: 278 (l.). Combination in *Pachycondyla:* Roger, 1863b: 18; in *Neoponera:* Emery, 1911d: 73; in *Pachycondyla:* Brown, **revived combination**. Senior synonym of *moesta:* Mayr, 1887: 534; Brown, 1957e: 234; of *confusa, fiebrigi, stipitum* and *sulcatula,* and material of the unavailable name *lata* referred here: Brown, 1957e: 234.

***cribrata.*** *Pachycondyla* (*Bothroponera*) *cribrata* Santschi, 1910c: 349 (w.) CONGO. Combination in *Bothroponera:* Wheeler, W.M. 1922a: 72; in *Pachycondyla:* Brown, **revived combination**.

***croceicornis.*** *Euponera* (*Brachyponera*) *luteipes* var. *croceicornis* Emery, 1900c: 315 (w.q.) NEW GUINEA. Combination in *Brachyponera:* Wilson, 1958d: 347; in *Pachycondyla:* Brown, **new combination**. Raised to species, and senior synonym of *bicolor* and *inops:* Wilson, 1958d: 347.

#***crudelis.*** *Euponera* (*Pseudoponera*) *wroughtonii* var. *crudelis* Forel, 1901f: 343 (w.m.) SOUTH AFRICA. Combination in *E.* (*Trachymesopus*): Emery, 1911d: 85; in *Pachycondyla:* Brown, **new combination**.

*curvinodis. Pachycondyla villosa* r. *curvinodis* Forel, 1899c: 15, pl. 1, fig. 12 (w.) GUATEMALA. Combination in *Neoponera:* Emery, 1901a: 47. Junior synonym of *inversa:* Emery, 1911d: 73.

*dahlii. Ponera* (*Ectomomyrmex*) *dahlii* Forel, 1901b: 9 (w.) BISMARCK ARCHIPELAGO. Combination in *Pachycondyla* (*Ectomomyrmex*): Emery, 1911d: 79. Junior synonym of *acuta:* Wilson, 1958d: 364.

***darwinii.*** *Belonopelta darwinii* Forel, 1893d: 460 (q.) AUSTRALIA. Terayama, 1990: 897 (w.). Combination in *Pachycondyla* (*Pseudoponera*): Emery, 1900d: 668; Emery, 1901a: 46; in *Pseudoponera:* Bingham, 1903: 93; in *Euponera* (*Trachymesopus*): Emery, 1911d: 86; in *Trachymesopus:* Wilson, 1958d: 353; in *Pachycondyla:* Brown, **new combination**. Senior synonym of *lamarki* and *rufotestaceus:* Brown, 1963: 6. Current subspecies: nominal plus *africana, indica, madecassa.*

#***debilior.*** *Ponera* (*Bothroponera*) *tridentata* r. *debilior* Forel, 1901h: 46 (w.) BORNEO. Combination in *Pachycondyla:* Forel, 1911e: 253; in *Bothroponera:* Wheeler, W.M. 1919e: 55; in *Pachycondyla:* Brown, **revived combination**.

#***decolor.*** *Euponera* (*Brachyponera*) *sennaarensis* var. *decolor* Santschi, 1921c: 114 (w.) CONGO. Combination in *Pachycondyla:* Brown, **new combination**.

#***delagoensis.*** *Paltothyreus tarsatus* var. *delagoensis* Emery, 1899e: 468 (w.q.) MOZAMBIQUE. Subspecies of *tarsata:* Wheeler, W.M. 1922a: 765. Combination in *Pachycondyla:* Brown, **new combination**.

*denticeps. Ectomomyrmex denticeps* Wheeler, W.M. 1929g: 32, fig. 2 (w.) TAIWAN. Junior synonym of *javana:* Yasumatsu, 1962: 94.

***denticulata.*** *Bothroponera denticulata* Kirby, W.F. 1896: 206 (w.) AUSTRALIA. Wheeler, G.C. & Wheeler, J. 1971c: 387 (l.). Combination in *Pachycondyla:* Brown, **new combination**. See also: Clark, 1930c: 21.

*dentis. Euponera* (*Mesoponera*) *dentis* Weber, 1942a: 43, fig. 9 (q.) SUDAN. Junior synonym of *brunoi:* Brown, 1963: 8.

***depilis.*** *Ophthalmopone depilis* Emery, 1902c: 29 (w.) SAO THOME I. Combination in *Pachycondyla:* Brown, **new combination**.

*dibullana*. *Pachycondyla harpax* var. *dibullana* Forel, 1901f: 347 (w.) COLOMBIA. Forel, 1912c: 39 (q.). Junior synonym of *harpax:* Brown, 1950e: 247.

#***divaricata***. *Euponera* (*Xiphopelta*) *elisae* subsp. *divaricata* Emery, 1915g: 6, fig. 3 (w.) ETHIOPIA. Combination in *E.* (*Mesoponera*): Wheeler, W.M. 1922a: 775; in *Pachycondyla:* Brown, **new combination**.

*dohrni*. *Megaponera dohrni* Emery, 1902c: 30 (w.) GHANA. Junior synonym of *foetens:* Forel, 1909b: 64.

****dubia***. **Brachyponera dubia* Théobald, 1937b: 196, pl. 4, fig. 15 (q.) FRANCE (Oligocene). Combination in *Pachycondyla:* **new combination**.

***dubitata***. *Ponera* (*Bothroponera*) *dubitata* Forel, 1900b: 63 (w.) AUSTRALIA. Combination in *Pachycondyla* (*Bothroponera*): Emery, 1901a: 45; in *Bothroponera:* Taylor & Brown, D.R. 1985: 21; in *Pachycondyla:* Brown, **revived combination**.

*eburneipes*. *Neoponera unidentata* var. *eburneipes* Wheeler, W.M. & Bequaert, 1929: 29 (w.q.) BRAZIL. Junior synonym of *unidentata:* Brown, 1957e: 232.

***eleonorae***. *Mesoponera eleonorae* Forel, 1921b: 131 (w.) ECUADOR. Combination in *Pachycondyla:* Brown, **new combination**.

***elisae***. *Ponera elisae* Forel, 1891b: 221, pl. 5, fig. 10 (w.) MADAGASCAR. Combination in *Euponera* (*Mesoponera*): Emery, 1901a: 46; in *E.* (*Xiphopelta*): Forel, 1913j: 206; in *Pachycondyla:* Brown, **new combination**. Current subspecies: nominal plus *divaricata, redbankensis, rotundata*.

***emiliae***. *Neoponera emiliae* Forel, 1901f: 349 (w.) VENEZUELA. Combination in *Pachycondyla:* Brown, **new combination**.

***escherichi***. *Pachycondyla* (*Bothroponera*) *escherichi* Forel, 1910c: 244 (w.) ETHIOPIA. Combination in *Euponera* (*Mesoponera*): Emery, 1911d: 82; in *Pachycondyla:* Brown, **revived combination**.

***exarata***. *Pachycondyla* (*Ectomomyrmex*) *exarata* Emery, 1901h: 156 (w.) NEW GUINEA. Combination in *Ectomomyrmex:* Mann, 1919: 289; in *Pachycondyla:* Brown, **revived combination**.

#***exasperans***. *Pachycondyla* (*Bothroponera*) *tridentata* var. *exasperans* Forel, 1911e: 253 (w.) WEST MALAYSIA.

***excavata***. *Bothroponera excavata* Emery, 1893e: 201 (w.) AUSTRALIA. Combination in *Pachycondyla* (*Bothroponera*): Emery, 1901a: 45; in *Bothroponera:* Taylor & Brown, D.R. 1985: 21; in *Pachycondyla:* Brown, **revived combination**. Current subspecies: nominal plus *acuticostata*.

***fauveli***. *Pachycondyla fauveli* Emery, 1896k: 175 (footnote) (w.) BOLIVIA. Wheeler, W.M. 1925a: 6 (q.m.); Wheeler, G.C. & Wheeler, J. 1971b: 1206 (l.). Combination in *Euponera* (*Mesoponera*): Forel, 1901f: 341; in *Mesoponera:* Wheeler, G.C. & Wheeler, J. 1971b: 1206; in *Pachycondyla:* Brown, **revived combination**.

***ferruginea***. *Ponera ferruginea* Smith, F. 1858b: 100 (q.) MEXICO. Wheeler, G.C. & Wheeler, J. 1976a: 54 (l.). Combination in *Euponera* (*Mesoponera*): Emery, 1901a: 47; in *Trachymesopus:* Brown, 1963: 7; in *Mesoponera:* Wheeler, G.C. & Wheeler, J. 1976a: 54; in *Pachycondyla:* Brown, **new combination**. Current subspecies: nominal plus *panamensis*.

*fiebrigi*. *Neoponera crenata* r. *fiebrigi* Forel, 1912c: 37 (w.) PANAMA. Junior synonym of *crenata:* Brown, 1957e: 234.

***flavopilosa***. *Euponera* (*Mesoponera*) *flavopilosa* Weber, 1942a: 43, fig. 10 (q.) SUDAN. Combination in *Pachycondyla:* Brown, **new combination**.

#***fochi***. *Euponera* (*Hagensia*) *havilandi* var. *fochi* Forel, 1918b: 155 (w.) SOUTH AFRICA. Combination in *Pachycondyla:* Brown, **new combination**.

*foetens*. *Formica foetens* Fabricius, 1793: 354 (w.) GUINEA. [Junior primary homonym of *Formica foetens* Olivier, 1792: 503.] Replacement name: *Formica analis* Latreille, 1802c: 282. Emery, 1897e: 597 (m.); Arnold, 1915: 48 (q.); Wheeler, W.M. 1918c: 298 (l.); Wheeler, G.C. & Wheeler, J. 1952c: 612 (l.). Combination in *Megaponera:* Mayr, 1862: 753. Senior synonym of *abyssinica* and *laeviuscula:* Roger, 1860: 310; of *dohrni:* Forel, 1909b: 64. See also: Wheeler, W.M. 1922a: 64; Longhurst & Howse, 1978: 1213. For subspecies see under *analis*.

***foetida***. *Formica foetida* Linnaeus, 1758: 582 (q) "America meridionali". Combination in *Pachycondyla:* Roger, 1863b: 18; in *Neoponera:* Emery, 1901a: 47. Senior synonym of *lobata:* Retzius, 1783: 75; Olivier, 1792: 502; Latreille, 1802c: 206. See also: Linnaeus, 1767: 965.

***fossigera***. *Euponera fossigera* Mayr, 1901b: 3 (w.) SOUTH AFRICA. Combination in *E.* (*Mesoponera*): Arnold, 1915: 71; in *Pachycondyla:* Brown, **new combination**.

#***fossulata***. *Ponera* (*Bothroponera*) *sulcata* var. *fossulata* Forel, 1900d: 323 (diagnosis in key) (w.) INDIA. Combination in *Pachycondyla* (*Bothroponera*): Emery, 1901a: 46.

***fugax***. *Pachycondyla* (*Bothroponera*) *fugax* Forel, 1907a: 7 (w.) TANZANIA. Combination in *Bothroponera:* Wheeler, W.M. 1922a: 73; in *Pachycondyla:* Brown, **revived combination**.

#***funerea***. *Bothroponera pachyderma* var. *funerea* Wheeler, W.M. 1922a: 73 (q.) ZAIRE. Combination in *Pachycondyla:* Brown, **new combination**.

***fuscoatra***. *Ponera* (*Pachycondyla*) *fuscoatra* Roger, 1861a: 8 (w.) COLOMBIA. Combination in *Pachycondyla:* Mayr, 1863: 439.

*gabonensis* André, 1892 [*Bothroponera*]; see under ***PHRYNOPONERA***.

*gagates*. *Ponera gagates* Guérin-Méneville, 1844a: 423 (w.) SENEGAL. Junior synonym of *tarsata:* Roger, 1860: 310; Roger, 1863b: 17.

*gagatina*. *Pachycondyla gagatina* Emery, 1890a: 75 (w.) COSTA RICA. [Also described as new by Emery, 1894k: 48.] Junior synonym of *laevigata:* Emery, 1892b: 167.

#***gibbinota***. *Neoponera carinulata* subsp. *gibbinota* Forel, 1909a: 246 (w.) GUATEMALA. Combination in *Pachycondyla:* Brown, **new combination**.

***gilberti***. *Trachymesopus gilberti* Kempf, 1960f: 425, fig. 6 (w.q.) BRAZIL. Wheeler, G.C. & Wheeler, J. 1971b: 1206 (l.). Combination in *Mesoponera:* Wheeler, G.C. & Wheeler, J. 1971b: 1206; in *Pachycondyla:* Brown, **new combination**.

#***glabricollis***. *Euponera* (*Brachyponera*) *jerdoni* var. *glabricollis* Stärcke, 1930: 373 (w.) INDONESIA (Berhala I.). Combination in *Pachycondyla:* Brown, **new combination**.

***glabripes***. *Bothroponera glabripes* Emery, 1893g: 262 (footnote) (w.) PHILIPPINES. Combination in *Pachycondyla* (*Ectomomyrmex*): Emery, 1901a: 46; in *Bothroponera:* Wheeler, W.M. & Chapman, 1925: 67; in *Pachycondyla:* Brown, **revived combination**.

****globiventris***. **Euponera globiventris* Théobald, 1937b: 198, pl. 13, fig. 16 (q.) FRANCE (Oligocene). Combination in *Pachycondyla:* **new combination**.

#***godfreyi***. *Hagensia havilandi* r. *godfreyi* Arnold, 1926: 203 (w.m.) SOUTH AFRICA. Combination in *Pachycondyla:* Brown, **new combination**.

***goeldii***. *Neoponera goeldii* Forel, 1912c: 36 (w.) BRAZIL. Combination in *Pachycondyla:* Brown, **new combination**.

*golbachi*. *Neoponera* (*Eumecopone*) *golbachi* Kusnezov, 1969: 36 (w.) ARGENTINA. Junior synonym of *goyana:* Kempf, 1978: 36.

***goyana***. *Neoponera* (*Eumecopone*) *goyana* Borgmeier, 1937b: 230, figs. 11–14 (w.) BRAZIL. Combination in *Pachycondyla:* Brown, **new combination**. Senior synonym of *golbachi:* Kempf, 1978: 36.

****gracilicornis***. **Ponera gracilicornis* Mayr, 1868c: 72 (footnote) (w.) BALTIC AMBER. *Incertae sedis* in *Ponera:* Wheeler, W.M. 1915h: 40; in Ponerinae: Taylor, 1964: 139. Combination in *Pachycondyla:* **new combination**.

***grandis***. *Ponera grandis* Donisthorpe, 1947a: 283 (w.) CHINA. [Unresolved junior primary homonym of *Ponera grandis* Guérin-Méneville, 1838: 206.] Combination in *Pachycondyla:* Brown, **new combination**.

***granosa***. *Ponera granosa* Roger, 1860: 290 (w.) SOUTH AFRICA. Arnold, 1926: 201 (m.). Combination in *Bothroponera:* Mayr, 1862: 717; in *Pachycondyla* (*Bothroponera*): Emery, 1901a: 45; in *Bothroponera:* Wheeler, W.M. 1922a: 72; in *Pachycondyla:* Brown, **revived combination**.

***guianensis***. *Wadeura guianensis* Weber, 1939a: 103, figs. 5, 6 (w.q.) GUIANA. Combination in *Pachycondyla:* Brown, **new combination**.

*guineensis*. *Ponera guineensis* André, 1890: 318 (w.) SIERRA LEONE. Combination in *Euponera* (*Mesoponera*), and variety of *caffraria:* Santschi, 1935b: 258. Junior synonym of *caffraria:* Mayr, 1895: 125; Emery, 1911d: 81; Wheeler, W.M. 1922a: 775; and current status.

*harnedi* Smith, M.R. 1929 [*Euponera*]; see under ***CRYPTOPONE***.

***harpax***. *Formica harpax* Fabricius, 1804: 401 (w.) SOUTH AMERICA. Wheeler, W.M. 1900b: 19 (q.m.); Wheeler, G.C. & Wheeler, J. 1952c: 617 (l.). Combination in *Pachycondyla:* Roger, 1863b: 18. Senior synonym of *concinna, dibullana, irina* and *montezumia* (and its junior synonyms *amplinoda* and *orizabana*): Brown, 1950e: 247.

***haskinsi***. *Wadeura haskinsi* Weber, 1939a: 104, fig. 7 (w.) PANAMA. Combination in *Pachycondyla:* Brown, **new combination**. See also: Kempf, 1958e: 175.

***havilandi***. *Megaloponera* (*Hagensia*) *havilandi* Forel, 1901f: 333 (w.) SOUTH AFRICA. Arnold, 1915: 68 (m.). Combination in *Euponera* (*Mesoponera*): Arnold, 1915: 67; in *Hagensia:* Arnold, 1926: 203; in *Pachycondyla:* Brown, **new combination**. Senior synonym of *sulcigera:* Arnold, 1915: 67. Current subspecies: nominal plus *fochi, godfreyi, marleyi*.

***havilandi.*** *Pachycondyla* (*Bothroponera*) *havilandi* Forel, 1901f: 345 (w.) SINGAPORE. [Unresolved junior secondary homonym of *havilandi* Forel, above.]

***henryi.*** *Pachycondyla* (*Bothroponera*) *henryi* Donisthorpe, 1942d: 449 (w.) INDIA.

***holmgreni.*** *Euponera* (*Trachymesopus*) *holmgreni* Wheeler, W.M. 1925a: 6 (w.) PERU. Combination in *Trachymesopus:* Kempf, 1960f: 424; in *Pachycondyla:* Brown, **new combination**. See also: Kempf, 1961b: 494.

*horni. Pachycondyla* (*Ectomomyrmex*) *horni* Forel, 1913f: 185 (w.q.) TAIWAN. Combination in *Ectomomyrmex:* Santschi, 1925f: 82. Junior synonym of *javana:* Yasumatsu, 1962: 94.

***hottentota.*** *Pachycondyla hottentota* Emery, 1886: 360 (w.) SOUTH AFRICA. Arnold, 1915: 52 (m.). Combination in *Ophthalmopone:* Emery, 1902c: 30; in *Pachycondyla:* Brown, **revived combination**. Senior synonym of *lanceolata:* Emery, 1902c: 30.

***ilgii.*** *Ophthalmopone ilgii* Forel, 1894b: 76 (w.) ETHIOPIA. Combination in *Pachycondyla:* Brown, **new combination**.

#***ilgii.*** *Pachycondyla* (*Bothroponera*) *crassa* var. *ilgii* Forel, 1910c: 244 (w.) ETHIOPIA. [Unresolved junior secondary homonym of *ilgii* Forel, above.]

***impressa.*** *Ponera* (*Pachycondyla*) *impressa* Roger, 1861a: 6 (w.) COLOMBIA. Combination in *Pachycondyla:* Roger, 1863b: 18. Senior synonym of *andicola, cearensis, inca, montana, purpurascens* and *transversa:* Kempf, 1961c: 195.

*inca. Pachycondyla fuscoatra* subsp.*inca* Emery, 1901a: 48 (w.q.) PERU. Junior synonym of *impressa:* Kempf, 1961c: 195.

***incisa.*** *Pachycondyla* (*Bothroponera*) *incisa* Emery, 1911b: 533, fig. 1 (w.) NEW GUINEA. Combination in *Bothroponera:* Wilson, 1958d: 362; in *Pachycondyla:* Brown, **revived combination**.

#***indica.*** *Belonopelta darwini* var. *indica* Emery, 1899f: 268 (q.) BURMA. Combination in *Pachycondyla* (*Pseudoponera*): Emery, 1901a: 46; in *Euponera* (*Pseudoponera*): Forel, 1901f: 343; in *Pseudoponera:* Bingham, 1903: 93; in *Euponera* (*Trachymesopus*): Emery, 1911d: 86; in *Pachycondyla:* Brown, **revived combination**. Senior synonym of *indica* Forel: Emery, 1911d: 86.

*indica. Belonopelta darwinii* var. *indica* Forel, 1900d: 320 (q.) INDIA. Junior synonym of *indica* Emery: Emery, 1911d: 86.

***ingesta.*** *Euponera* (*Mesoponera*) *ingesta* Wheeler, W.M. 1922a: 82, fig. 14 (w.) ZAIRE. Combination in *Pachycondyla:* Brown, **new combination**.

*inops. Euponera* (*Brachyponera*) *luteipes* subsp. *inops* Forel, 1910b: 17 (w.) AUSTRALIA. Junior synonym of *croceicornis:* Wilson, 1958d: 347.

***insulana.*** *Ponera insulana* Mayr, 1876: 87 (w.) SAMOA IS. Combination in *Pachycondyla* (*Ectomomyrmex*): Emery, 1901a: 46; in *Ectomomyrmex:* Wilson, 1958d: 364; in *Pachycondyla:* Brown, **revived combination**.

***insularis.*** *Bothroponera insularis* Emery, 1889b: 495 (footnote) (w.) JAVA. Combination in *Pachycondyla* (*Bothroponera*): Emery, 1900d: 667; in *Bothroponera:* Wheeler, W.M. 1919e: 55; in *Pachycondyla:* Brown, **revived combination**. Current subspecies: nominal plus *brevior.*

#***intermedia.*** *Ponera* (*Bothroponera*) *piliventris* var. *intermedia* Forel, 1900b: 63 (w.m.) AUSTRALIA. Combination in *Pachycondyla* (*Bothroponera*): Emery, 1901a: 45; in *Bothroponera:* Taylor & Brown, D.R. 1985: 22; in *Pachycondyla:* Brown, **revived combination**.

#***inversa.*** *Ponera inversa* Smith, F. 1858b: 96 (w.) SOUTH AMERICA. Wheeler, G.C. & Wheeler, J. 1952c: 615 (l.). Combination in *Pachycondyla:* Mayr, 1886c: 358; in *Neoponera:* Emery, 1904b: 597; in *Pachycondyla:* Brown, **revived combination**. Subspecies of *villosa:* Emery, 1904b: 597; Mann, 1922: 7. Senior synonym of *curvinodis:* Emery, 1911d: 73.

*irina. Pachycondyla harpax* var. *irina* Wheeler, W.M. 1925a: 5 (w.) GUATEMALA. Junior synonym of *harpax:* Brown, 1950e: 247.

#***jacobsoni.*** *Pachycondyla* (*Ectomomyrmex*) *leeuwenhoeki* st. *jacobsoni* Forel, 1915a: 22 (w.) INDONESIA (Simalur I.).

*japonica. Pachycondyla* (*Ectomomyrmex*) *japonica* Emery, 1902c: 31 (w.) JAPAN. Teranishi, 1940: 6 (m.). Combination in *Ectomomyrmex:* Santschi, 1925f: 82. Junior synonym of *javana:* Yasumatsu, 1962: 94.

***javana.*** *Ectomomyrmex javanus* Mayr, 1867a: 84, pl. 2, fig. 9 (w.q.) JAVA. Ogata, 1987: 113 (m.). Combination in *Pachycondyla* (*Ectomomyrmex*): Emery, 1900d: 667; in *Ectomomyrmex:* Bingham, 1903: 86; Wheeler, W.M. 1921c: 530; in *Pachycondyla:* Brown, **revived combination**. Senior synonym of *cambodjana, denticeps, horni, japonica, maternus, sundaicus* and *tonkina:* Yasumatsu, 1962: 94. See also: Bingham, 1903: 87.

#***javana.*** *Euponera* (*Mesoponera*) *rubra* var. *javana* Forel, 1905c: 6 (q.m.) JAVA. [Unresolved junior secondary homonym of *javanus* Mayr, above.] Combination in *Pachycondyla:* Brown, **new combination**.

***jerdonii.*** *Ponera jerdonii* Forel, 1900d: 327 (w.) INDIA. Combination in *Euponera* (*Brachyponera*): Emery, 1901a: 47; in *Brachyponera:* Bingham, 1903: 102; in *Pachycondyla:* Brown, **new combination**. Current subspecies: nominal plus *glabricollis.*

***jonesii.*** *Lobopelta jonesii* Forel, 1891b: 219 (m.) MADAGASCAR. Combination in *Leptogenys:* Emery, 1911d: 102; in *Pachycondyla:* Bolton, 1975a: 298.

*josephi. Ponera josephi* Forel, 1886b: xl (w.) BRAZIL. Junior synonym of *constricta:* Dalla Torre, 1893: 39.

*katangana. Ectomomyrmex nigeriensis* var. *katangana* Santschi, 1933b: 96 (w.) ZAIRE. Junior synonym of *brunoi:* Brown, 1963: 8.

***kenyensis.*** *Bothroponera kenyensis* Santschi, 1937a: 47, figs. 1–3 (w.) KENYA. Combination in *Pachycondyla:* Brown, **new combination**.

***kruegeri.*** *Pachycondyla* (*Bothroponera*) *kruegeri* Forel, 1910a: 28 (w.) SOUTH AFRICA. Combination in *Bothroponera:* Wheeler, W.M. 1922a: 22; in *Pachycondyla:* Brown, **revived combination**. Current subspecies: nominal plus *asina, rhodesiana.* See also: Arnold, 1915: 56.

#***kurandensis.*** *Pachycondyla* (*Bothroponera*) *sublevis* subsp. *kurandensis* Forel, 1910b: 16 (w.) AUSTRALIA. Combination in *Bothroponera:* Taylor & Brown, D.R. 1985: 22; in *Pachycondyla:* Brown, **revived combination**.

***laevigata.*** *Ponera laevigata* Smith, F. 1858b: 98 (w.) BRAZIL. Wheeler, W.M. 1936d: 162 (q.m.). Combination in *Pachycondyla:* Mayr, 1886c: 358; in *Euponera* (*Mesoponera*): Emery, 1901a: 47; in *Termitopone:* Wheeler, W.M. 1936d: 159; in *Pachycondyla:* Brown, **revived combination**. Senior synonym of *gagatina:* Emery, 1892b: 167; of *whelpleyi:* Wheeler, W.M. 1936d: 164.

*laevinodis. Neoponera laevinodis* André, 1902: 14 (w.q.) PERU. Junior synonym of *rugosula:* Emery, 1911d: 72.

***laevissima.*** *Pachycondyla* (*Bothroponera*) *laevissima* Arnold, 1915: 58, pl. 2, fig. 15 (w.) SOUTH AFRICA. Combination in *Bothroponera:* Wheeler, W.M. 1922a: 73; in *Pachycondyla:* Brown, **revived combination**. Current subspecies: nominal plus *aspera.*

*laeviuscula. Ponera laeviuscula* Gerstäcker, 1859: 262 (w.) MOZAMBIQUE. [Also described as new by Gerstäcker, 1862: 505.] Junior synonym of *foetens:* Roger, 1860: 310.

*lamarki. Euponera* (*Trachymesopus*) *lamarki* Santschi, 1913c: 303 (q.) CONGO. Junior synonym of *darwinii:* Brown, 1963: 6.

***lamottei.*** *Bothroponera lamottei* Bernard, 1953b: 188 (w.) GUINEA. Combination in *Pachycondyla:* Brown, **new combination**.

*lamottei. Euponera* (*Trachymesopus*) *lamottei* Bernard, 1953b: 195, fig. 2 (q.) GUINEA. [Unresolved junior secondary homonym of *lamottei,* above.] Junior synonym of *brunoi:* Brown, 1963: 8.

*lanceolata. Ophthalmopone lanceolata* Mayr, 1895: 125 (w.) SOUTH AFRICA. Junior synonym of *hottentota:* Emery, 1902c: 30.

*lata. Neoponera crenata* st. *confusa* var. *lata* Santschi, 1921g: 86 (w.) ARGENTINA. Unavailable name; material referred to *crenata* by Brown, 1957e: 234.

#***latocciput.*** *Neoponera obscuricornis* r. *latocciput* Forel, 1921b: 132 (w.q.) ECUADOR. Combination in *Pachycondyla:* Brown, **new combination**.

*latreillei. Neoponera latreillei* Forel, 1905b: 161 (w.q.) SURINAM. Wheeler, G.C. & Wheeler, J. 1952c: 613 (l.). Variety of *obscuricornis:* Emery, 1911d: 72. Junior synonym of *apicalis:* Brown, 1957e: 230.

***leeuwenhoeki.*** *Ponera leeuwenhoeki* Forel, 1886d: 244 (w.) INDIA. Imai, Brown, *et al.* 1984: 67 (k.). Combination in *Bothroponera:* Forel, 1891b: 124; in *Ectomomyrmex:* Forel, 1900d: 321; in *Pachycondyla* (*Ectomomyrmex*): Emery, 1901a: 46. Current subspecies: nominal plus *jacobsoni, sumatrensis.* See also: Bingham, 1903: 88.

***lenis.*** *Pachycondyla lenis* Kempf, 1961c: 197, figs. 4, 6 (w.q.) BRAZIL.

***lenkoi.*** *Pachycondyla lenkoi* Kempf, 1962b: 4, figs. 1–4 (w.) BRAZIL.

***leveillei.*** *Ponera leveillei* Emery, 1890a: 61 (w.) VENEZUELA. Combination in *Euponera* (*Mesoponera*): Emery, 1901a: 46; in *Mesoponera:* Kempf, 1972a: 141; in *Pachycondyla:* Brown, **new combination**.

***lineaticeps.*** *Pachycondyla lineaticeps* Mayr, 1866a: 502 (w.) MEXICO. Combination in *Neoponera:* Emery, 1901a: 47; in *Pachycondyla:* Brown, **new combination**.

*lobata. Formica lobata* De Geer, 1773: 602, pl. 31, figs. 6–8 (q.) SURINAM. Junior synonym of *foetida:* Retzius, 1783: 75; Olivier, 1792: 502; Latreille, 1802c: 206.

*lubbocki* Donisthorpe, 1938 [*Pseudoponera*]; see under ***LEPTOGENYS.***

***lunaris.*** *Ponera lunaris* Emery, 1896g: 55, pl. 1, fig. 12 (w.) PARAGUAY. Combination in *Pachycondyla:* Emery, 1901a: 45; in *Euponera* (*Trachymesopus*): Emery, 1911d: 85; in *Trachymesopus:* Kempf, 1960f: 424; in *Pachycondyla:* Brown, **revived combination.**

***lutea.*** *Ponera lutea* Mayr, 1862: 721 (w.q.) AUSTRALIA. Mayr, 1865: 66 (m.); Crawley, 1918: 86 (q.); Wheeler, G.C. & Wheeler, J. 1971b: 1207 (l.); Imai, Crozier & Taylor, 1977: 347 (k.). Combination in *Euponera* (*Brachyponera*): Emery, 1901a: 47; in *Brachyponera:* Wheeler, G.C. & Wheeler, J. 1971b: 1207; in *Pachycondyla:* Brown, **new combination.** Senior synonym of *socialis:* Emery, 1911d: 84; of *clara:* **new synonymy**. See also: Wheeler, W.M. 1933i: 93.

*luteipedojerdoni. Euponera* (*Brachyponera*) *luteipes* var. *luteipedojerdoni* Forel, 1913f: 184 (w.) TAIWAN. *Nomen nudum.*

***luteipes.*** *Ponera luteipes* Mayr, 1862: 722 (w.q.) NICOBAR IS. Forel, 1900d: 326 (m.); Imai & Yosida, 1964: 64 (k.); Wheeler, G.C. & Wheeler, J. 1976a: 47 (l.). Combination in *Euponera* (*Brachyponera*): Emery, 1900c: 315; Emery, 1911d: 84; Wheeler, W.M. 1919e: 56; in *Brachyponera:* Bingham, 1903: 101; in *Pachycondyla:* Brown, **new combination.** Current subspecies: nominal plus *continentalis.*

***luteola.*** *Ponera luteola* Roger, 1861b: 166 (w.) BRAZIL. Combination in *Pachycondyla:* Roger, 1863b: 18; in *Neoponera:* Emery, 1901a: 47; in *Pachycondyla:* Brown, **revived combination.**

***lydiae.*** *Neoponera lydiae* Santschi, 1920d: 361 (q.) FRENCH GUIANA. Combination in *Pachycondyla:* Brown, **new combination.**

#***macra.*** *Ponera melanaria* var. *macra* Emery, 1894a: 68 (w.) SEYCHELLE IS. Combination in *Euponera* (*Mesoponera*): Emery, 1901a: 46; in *Pachycondyla:* Brown, **new combination.**

#***madecassa.*** *Belonopelta darwini* var. *madecassa* Emery, 1899f: 268 (q.) MADAGASCAR. Combination in *Pachycondyla* (*Pseudoponera*): Emery, 1901a: 46; in *Euponera* (*Trachymesopus*): Emery, 1911d: 86; in *Pachycondyla:* Brown, **revived combination.**

***magnifica.*** *Pachycondyla magnifica* Borgmeier, 1929: 196 (w.) BRAZIL. See also: Kempf, 1961c: 198.

***malayana.*** *Ectomomyrmex malayanus* Wheeler, W.M. 1929g: 34 (w.) WEST MALAYSIA. Combination in *Trachymesopus:* Brown, 1963: 7; in *Pachycondyla:* Brown, **new combination.**

***manni.*** *Euponera* (*Mesoponera*) *melanaria* subsp. *manni* Viehmeyer, 1924a: 228 (w.) SOLOMON IS. Combination in *Mesoponera:* Wilson, 1958d: 349; in *Pachycondyla:* Brown, **new combination.** Raised to species, and senior synonym of *robiginosa:* Wilson, 1958d: 349.

***marginata.*** *Ponera marginata* Roger, 1861a: 8 (w.q.m.) BRAZIL. Combination in *Pachycondyla:* Roger, 1863b: 18; in *Euponera* (*Mesoponera*): Emery, 1901a: 47; in *Termitopone:* Wheeler, W.M. 1936d: 166; in *Pachycondyla:* Brown, **revived combination.** See also: Gallardo, 1918b: 67.

#***marleyi.*** *Hagensia havilandi* r. *marleyi* Arnold, 1926: 203 (w.) SOUTH AFRICA. Combination in *Pachycondyla:* Brown, **new combination.**

*maternus. Ectomomyrmex javanus* r. *maternus* Forel, 1900d: 321 (diagnosis in key) (w.) INDIA. Forel, 1922: 90 (q.). Combination in *Pachycondyla* (*Ectomomyrmex*): Emery, 1901a: 46. Raised to species: Bingham, 1903: 87. Subspecies of *javana:* Emery, 1911d: 79; Wheeler, W.M. 1913e: 233; Forel, 1922: 90. Junior synonym of *javana:* Yasumatsu, 1962: 94.

*maya. Neoponera unidentata* var. *maya* Wheeler, W.M. & Bequaert, 1929: 30 (w.) GUATEMALA. Junior synonym of *unidentata:* Brown, 1957e: 232.

***mayri.*** *Bothroponera mayri* Emery, 1887b: 442 (footnote) (w.) AUSTRALIA. Forel, 1900b: 64 (m.); Wheeler, G.C. & Wheeler, J. 1971c: 392 (l.). Combination in *Pachycondyla* (*Bothroponera*): Emery, 1901a: 45; in *Bothroponera:* Taylor & Brown, D.R. 1985: 22; in *Pachycondyla:* Brown, **revived combination.**

#***mediana.*** *Paltothyreus tarsatus* var. *mediana* Santschi, 1919h: 80 (w.q.m.) CONGO. Combination in *Pachycondyla:* Brown, **new combination.**

***melanaria.*** *Ponera melanaria* Emery, 1893g: 260 (footnote) (w.) SRI LANKA. Bingham, 1903: 100 (q.); Donisthorpe, 1942d: 450 (m.); Wheeler, G.C. & Wheeler, J. 1971b: 1206 (l.). Combination in *Euponera* (*Mesoponera*): Emery, 1900d: 668; in *Mesoponera:* Bingham, 1903: 100; in *Pachycondyla:* Brown, **new combination.** Current subspecies: nominal plus *macra.*

***melancholica.*** *Pachycondyla melancholica* Smith, F. 1865: 71 (w.) INDONESIA (Morotai I.). Combination in *Platythyrea:* Emery, 1897c: 575; in *Pachycondyla:* Brown, 1975: 47. Junior synonym of *Platythyrea parallela:* Wilson, 1958d: 151. Revived from synonymy: Brown, 1975: 47.

***mesonotalis.*** *Neoponera mesonotalis* Santschi, 1923c: 246 (w.m.) BRAZIL (originally misspelled as *mesorotalis*). Combination in *Pachycondyla:* Brown, **new combination.**

***mesoponeroides.*** *Brachyponera mesoponeroides* Radchenko, 1993a: 81, figs. 6, 7 (w.) VIETNAM. Combination in *Pachycondyla:* **new combination.**

***metanotalis.*** *Pachycondyla metanotalis* Luederwaldt, 1918: 54 (w.) BRAZIL. See also: Kempf, 1961c: 200.

***mlanjiensis.*** *Bothroponera mlanjiensis* Arnold, 1946: 55, figs. 8, 9 (w.q.) MALAWI. Combination in *Pachycondyla:* Brown, **new combination.**

***mocquerysi.*** *Ophthalmopone mocquerysi* Emery, 1902c: 29 (w.) SAO THOME I. Combination in *Pachycondyla:* Brown, **new combination.**

***modiglianii.*** *Pachycondyla* (*Ectomomyrmex*) *modiglianii* Emery, 1900d: 667 (w.) SUMATRA.

*moesta. Pachycondyla moesta* Mayr, 1870a: 395 (w.) COLOMBIA. Forel, 1909a: 252 (q.); Wheeler, G.C. & Wheeler, J. 1971b: 1205 (l.). Combination in *Neoponera:* Emery, 1901a: 47. Variety/ race of *pallipes:* Emery, 1890b: 42; of *crenata:* Forel, 1899c: 13; Forel, 1912c: 37; Santschi, 1921g: 86. Junior synonym of *crenata:* Mayr, 1887: 534; Brown, 1957e: 234.

*montana. Pachycondyla fuscoatra* var. *montana* Forel, 1912c: 39 (w.) COLOMBIA. Junior synonym of *impressa:* Kempf, 1961c: 195.

*montezumia. Pachycondyla montezumia* Smith, F. 1858b: 108 (q.m.) MEXICO. Forel, 1899c: 12 (w.); Wheeler, G.C. & Wheeler, J. 1952c: 617 (l.). Junior synonym of *harpax:* Roger, 1862c: 288; Roger, 1863b: 18; Dalla Torre, 1893: 34. Revived from synonym as subspecies of *harpax:* Forel, 1899c: 12; Emery, 1911d: 75. Senior synonym of *amplinoda* and *orizabana:* Emery, 1911d: 75. Junior synonym of *harpax:* Brown, 1950e: 247.

#***montivaga.*** *Bothroponera cavernosa* var. *montivaga* Arnold, 1947: 132 (w.) SOUTH AFRICA. Combination in *Pachycondyla:* Brown, **new combination.**

#***murina.*** *Pachycondyla* (*Bothroponera*) *sublevis* var. *murina* Forel, 1910b: 17 (w.) AUSTRALIA.

*myopa. Ponera myopa* Roger, 1860: 285 (w.) no locality given. Junior synonym of *quadridentata:* Roger, 1860: 285.

***myropola.*** *Euponera* (*Trachymesopus*) *myropola* Menozzi, 1925c: 444 (q.) PHILIPPINES. Combination in *Pachycondyla:* Brown, **new combination.**

***nasica.*** *Pachycondyla* (*Bothroponera*) *nasica* Santschi, 1920b: 6 (w.) GABON.

*niger* Donisthorpe, 1949 [*Euponera*]; see under ***LEPTOGENYS.***

*nigeriensis. Euponera* (*Trachymesopus*) *nigeriensis* Santschi, 1914d: 316 (w.) NIGERIA. Combination in *Ectomomyrmex:* Santschi, 1933b: 96. Junior synonym of *brunoi:* Brown, 1963: 8.

***nigrita.*** *Ponera nigrita* Emery, 1895k: 459 (w.) BURMA. Combination in *Euponera* (*Brachyponera*): Emery, 1900d: 668; in *Brachyponera:* Bingham, 1903: 102; in *Pachycondyla:* Brown, **new combination.** Current subspecies: nominal plus *nigritella.*

#***nigritella.*** *Euponera* (*Brachyponera*) *nigrita* subsp. *nigritella* Stärcke, 1930: 379 (w.) SUMATRA. Combination in *Pachycondyla:* Brown, **new combination.**

#***nimba.*** *Bothroponera silvestrii* r. *nimba* Bernard, 1953b: 188 (w.) GUINEA. Combination in *Pachycondyla:* Brown, **new combination.**

***nimba.*** *Euponera* (*Mesoponera*) *nimba* Bernard, 1953b: 190, fig. 2 (w.q.) GUINEA. [Unresolved junior secondary homonym of *nimba* Bernard, above.] Combination in *Pachycondyla:* Brown, **new combination.**

*nitidiventris. Pachycondyla striata* var. *nitidiventris* Santschi, 1921g: 87 (w.) URUGUAY. Junior synonym of *striata:* Kempf, 1961c: 201.

*nixoni. Euponera* (*Trachymesopus*) *nixoni* Donisthorpe, 1943d: 441 (q.) NEW GUINEA. Donisthorpe, 1948b: 299 (w.). Junior synonym of *stigma:* Wilson, 1958d: 355.

***novemdentata.*** *Euponera* (*Xiphopelta*) *novemdentata* Bernard, 1953b: 194, fig. 2 (w.) GUINEA. Combination in *Pachycondyla:* Brown, **new combination.**

***oberthueri.*** *Pachycondyla oberthueri* Emery, 1890a: 74 (w.) BRAZIL. Combination in *Neoponera:* Emery, 1901a: 47; in *Pachycondyla:* Brown, **revived combination.**

***obesa.*** *Ponera* (*Bothroponera*) *obesa* Emery, 1897d: 551 (w.) NEW GUINEA. Combination in *Pachycondyla* (*Bothroponera*): Emery, 1901a: 46; in *Bothroponera:*

Wilson, 1958d: 362; in *Pachycondyla:* Brown, **revived combination**.

#***obscura.*** *Pachycondyla* (*Ectomomyrmex*) *astuta* subsp. *obscura* Karavaiev, 1935a: 71 (w.) VIETNAM.

***obscurans.*** *Formica obscurans* Walker, 1859: 372 (q.) SRI LANKA. Combination in *Ponera:* Donisthorpe, 1932b: 575; in *Pachycondyla:* Brown, **new combination**.

***obscuricornis.*** *Pachycondyla flavicornis* var. *obscuricornis* Emery, 1890a: 58 (w.) BRAZIL. Combination in *Neoponera:* Emery, 1901a: 47; in *Pachycondyla:* Brown, **revived combination**. Raised to species: Emery, 1911d: 72. Current subspecies: nominal plus *latocciput.*

***obsoleta.*** *Euponera* (*Trachymesopus*) *obsoleta* Menozzi, 1931d: 196, fig. 5 (w.) COSTA RICA. Combination in *Trachymesopus:* Kempf, 1960f: 424; in *Pachycondyla:* Brown, **new combination**.

***obtusa.*** *Pachycondyla* (*Ectomomyrmex*) *obtusa* Emery, 1900d: 667 (footnote) (w.) BORNEO. Combination in *Ectomomyrmex:* Wheeler, W.M. 1919e: 55; in *Pachycondyla:* Brown, **revived combination**.

***oculata.*** *Ponera oculata* Smith, F. 1858b: 93 (m.) AUSTRALIA. Combination in *Pachycondyla:* Brown, **new combination**.

#***oculatior.*** *Euponera* (*Pseudoponera*) *amblyops* var. *oculatior* Forel, 1909d: 221 (q.) JAVA. Combination in *Pseudoponera:* Emery, 1911d: 87; in *Pachycondyla:* Brown, **new combination**.

*orizabana. Pachycondyla orizabana* Norton, 1868c: 8 (w.) MEXICO. [*P. orizabana* Norton, 1868a: 64. *Nomen nudum.*] Junior synonym of *harpax:* Emery, 1892b: 167; of *montezumia:* Emery, 1911d: 75.

***overbecki.*** *Pachycondyla* (*Ectomomyrmex*) *overbecki* Viehmeyer, 1916a: 113 (w.m.) SINGAPORE.

***pachyderma.*** *Pachycondyla* (*Bothroponera*) *pachyderma* Emery, 1901a: 49 (w.q.) CAMEROUN. Combination in *Bothroponera:* Wheeler, W.M. 1922a: 72; in *Pachycondyla:* Brown, **revived combination**. Current subspecies: nominal plus *attenata, funerea, postsquamosa.*

***pachynoda.*** *Euponera* (*Trachymesopus*) *pachynoda* Clark, 1930c: 7, fig. 1 (w.) AUSTRALIA. Combination in *Trachymesopus:* Brown, 1963: 7; in *Pachycondyla:* Brown, **new combination**.

***pallida.*** *Ponera pallida* Smith, F. 1858b: 85 (m.) BURMA. Combination in *Leptogenys:* Donisthorpe, 1932c: 474; in *Pachycondyla:* Brown, **new combination**.

***pallidipennis.*** *Ponera pallidipennis* Smith, F. 1860a: 73 (m.) SULAWESI. Combination in *Euponera* (*Trachymesopus*): Donisthorpe, 1932c: 458; in *Pachycondyla:* Brown, **new combination**.

*pallipes. Ponera pallipes* Smith, F. 1858b: 98 (q.) BRAZIL. [Junior primary homonym of *Ponera pallipes* Smith, F. 1858b: 87.] Replacement name: *crenata* Roger, 1861a: 3. [Misspelled as *pallidipes* by Dalla Torre, 1893: 34.] See also: Emery, 1901a: 47; Forel, 1901f: 349.

#***panamensis.*** *Ponera ferruginea* var. *panamensis* Forel, 1899c: 15 (q.) PANAMA. Combination in *Euponera* (*Mesoponera*): Emery, 1911d: 82; in *Trachymesopus:* Kempf, 1972a: 251; in *Pachycondyla:* Brown, **new combination**.

***papuana.*** *Euponera* (*Mesoponera*) *papuana* Viehmeyer, 1914d: 608 (w.) NEW GUINEA. Combination in *Mesoponera:* Wilson, 1958d: 350; in *Pachycondyla:* Brown, **new combination**. Subspecies of *melanaria:* Viehmeyer, 1924a: 228. Revived status as species, and senior synonym of *pulchella* and *viehmeyeri:* Wilson, 1958d: 350. See also: Mann, 1919: 290.

*pedunculata. Ponera pedunculata* Smith, F. 1858b: 96, pl. 6, fig. 25 (w.) BRAZIL. Junior synonym of *villosa:* Roger, 1861a: 1; Roger, 1862c: 288.

***pergandei.*** *Euponera* (*Mesoponera*) *pergandei* Forel, 1909a: 245 (w.) GUATEMALA. Wheeler, G.C. & Wheeler, J. 1974g: 279 (l.). Combination in *Mesoponera:* Kempf, 1972a: 141; in *Pachycondyla:* Brown, **new combination**.

***peringueyi.*** *Ponera peringueyi* Emery, 1899e: 474 (w.) SOUTH AFRICA. Arnold, 1951: 53 (m.); Wheeler, G.C. & Wheeler, J. 1971b: 1203 (l.). Combination in *Euponera* (*Brachyponera*): Emery, 1901a: 47; in *E.* (*Mesoponera*): Emery, 1911d: 81; in *Hagensia:* Arnold, 1926: 202; in *Pachycondyla:* Brown, **new combination**. Current subspecies: nominal plus *saldanhae.* See also: Arnold, 1915: 66.

***perroti.*** *Bothroponera perroti* Forel, 1891b: 131, pl. 4, fig. 6 (w.) MADAGASCAR. Combination in *Pachycondyla* (*Bothroponera*): Emery, 1901a: 45. Current subspecies: nominal plus *admista.*

*pestilentia. Ponera pestilentia* Smith, F. 1858b: 92 (w.) SIERRA LEONE. Junior synonym of *tarsata:* Roger, 1860: 310.

***picardi.*** *Pachycondyla* (*Bothroponera*) *picardi* Forel, 1901d: 304 (w.) ANGOLA. [Also described as new by Forel, 1903e: 559.] Combination in *Bothroponera:* Wheeler, W.M. 1922a: 72; in *Pachycondyla:* Brown, **revived combination**.

***picea.*** *Euponera* (*Xiphopelta*) *picea* Bernard, 1953b: 192, fig. 2 (w.) GUINEA. Combination in *Pachycondyla:* Brown, **new combination**.

***piliventris.*** *Pachycondyla piliventris* Smith, F. 1858b: 107 (w.) AUSTRALIA. Wheeler, G.C. & Wheeler, J. 1971c: 389 (l.). Combination in *Bothroponera:* Mayr, 1876: 87; Taylor & Brown, D.R. 1985: 22; in *Pachycondyla:* Brown, **revived combination**. Current subspecies: nominal plus *intermedia, regularis.*

*pilosa. Ponera pilosa* Smith, F. 1858b: 95 (m.) BRAZIL. Junior synonym of *villosa:* Roger, 1861a: 1; Roger, 1862c: 288.

***pilosior.*** *Euponera* (*Trachymesopus*) *sharpi* subsp. *pilosior* Wheeler, W.M. 1928d: 98 (w.) JAPAN. Ogata, 1987: 114 (q.m.). Combination in *Trachymesopus:* Brown, 1963: 8; in *Pachycondyla:* Brown, **new combination**. Raised to species, and senior synonym of *chosonensis:* Brown, 1963: 8.

***porcata.*** *Ponera* (*Bothroponera*) *porcata* Emery, 1897d: 552 (footnote) (w.) AUSTRALIA. Wheeler, G.C. & Wheeler, J. 1971c: 389 (l.). Combination in *Pachycondyla* (*Bothroponera*): Emery, 1901a: 45; in *Bothroponera:* Taylor & Brown, D.R. 1985: 22; in *Pachycondyla:* Brown, **revived combination**.

#***postsquamosa.*** *Pachycondyla* (*Bothroponera*) *pachyderma* var. *postsquamosa* Santschi, 1920b: 7 (w.) CONGO.

***procidua.*** *Pachycondyla procidua* Emery, 1890a: 75 (q.) SURINAM. Santschi, 1921g: 87 (w.). See also: Kempf, 1964e: 50.

#***pubescens.*** *Ophthalmopone berthoudi* subsp. *pubescens* Weber, 1942a: 42 (w.) SUDAN. Combination in *Pachycondyla:* Brown, **new combination**.

*pulchella. Euponera* (*Mesoponera*) *pulchella* Donisthorpe, 1941c: 133 (w.q.) NEW GUINEA. Junior synonym of *papuana:* Wilson, 1958d: 350.

***pumicosa.*** *Ponera pumicosa* Roger, 1860: 290 (w.) SOUTH AFRICA. Forel, 1901f: 344 (m.); Wheeler, G.C. & Wheeler, J. 1971c: 390 (l.). Combination in *Bothroponera:* Mayr, 1862: 717; in *Pachycondyla* (*Bothroponera*): Emery, 1901a: 45; in *Bothroponera:* Wheeler, W.M. 1922a: 72; in *Pachycondyla:* Brown, **revived combination**. [Misspelled as *pumicata* by Raffray, 1887: 21.] Current subspecies: nominal plus *sculpturata.* See also: Arnold, 1915: 62.

*punctata* Smith, F. 1858; see under ***PLATYTHYREA***.

***punctata.*** *Pachycondyla* (*Ectomomyrmex*) *punctata* Karavaiev, 1935a: 71, fig. 5 (w.q.) VIETNAM. [Unresolved junior primary homonym of *Pachycondyla punctata* Smith, F. 1858b: 108.]

*purpurascens. Pachycondyla purpurascens* Forel, 1899c: 12, pl. 1, fig. 10 (w.) COSTA RICA. Subspecies of *fuscoatra:* Emery, 1901a: 45. Junior synonym of *impressa:* Kempf, 1961c: 195.

*quadridentata. Ponera quadridentata* Smith, F. 1859a: 143 (q.) INDONESIA (Aru I.). Roger, 1860: 285 (w.); Emery, 1900c: 314 (m.). Combination in *Pachycondyla* (*Pseudoponera*): Emery, 1900c: 314; in *Euponera* (*Trachymesopus*): Emery, 1911d: 85. Variety/subspecies of *stigma:* Emery, 1897d: 552; Karavaiev, 1926d: 419. Senior synonym of *myopa:* Roger, 1860: 285. Junior synonym of *stigma:* Emery, 1887b: 434; Wilson, 1958d: 355.

#***rapax.*** *Megaponera foetens* var. *rapax* Santschi, 1914b: 49 (w.) TANZANIA. Combination in *Pachycondyla:* Brown, **new combination**. Subspecies of *analis:* current status.

#***redbankensis.*** *Ponera* (*Xiphopelta*) *arnoldi* var. *redbankensis* Forel, 1913a: 109 (w.) ZIMBABWE. Combination in *Euponera* (*Mesoponera*): Arnold, 1915: 71; in *Pachycondyla:* Brown, **new combination**. Variety of *elisae:* Wheeler, W.M. 1922a: 775.

#***regularis.*** *Pachycondyla* (*Bothroponera*) *piliventris* subsp. *regularis* Forel, 1907h: 271 (w.) AUSTRALIA. Combination in *Bothroponera:* Taylor & Brown, D.R. 1985: 22; in *Pachycondyla:* Brown, **revived combination**.

#***reticulata.*** *Ponera* (*Bothroponera*) *sublaevis* r. *reticulata* Forel, 1900b: 62 (w.m.) AUSTRALIA. [Unresolved junior primary homonym of *Ponera reticulata* Smith, F. 1858b: 85.] Combination in *Pachycondyla* (*Bothroponera*): Emery, 1901a: 45; in *Bothroponera:* Taylor & Brown, D.R. 1985: 22; in *Pachycondyla:* Brown, **revived combination**.

#***rhodesiana.*** *Pachycondyla* (*Bothroponera*) *kruegeri* var. *rhodesiana* Forel, 1913a: 109 (w.) ZIMBABWE. Arnold, 1915: 57 (m.). Combination in *Bothroponera:* Wheeler, W.M. 1922a: 770; in *Pachycondyla:* Brown, **revived combination**.

*robiginosa. Euponera* (*Mesoponera*) *robiginosa* Donisthorpe, 1941c: 131 (w.) NEW GUINEA. Junior synonym of *manni:* Wilson, 1958d: 349.

**#robusta.** *Paltothyreus tarsatus* var. *robusta* Santschi, 1919h: 81 (w.q.) SOMALIA. Combination in *Pachycondyla:* Brown, **new combination**.

*robustior* Santschi, 1919; see under ***PHRYNOPONERA.***

**rostrata.** *Pachycondyla rostrata* Emery, 1890a: 59 (q.) VENEZUELA. Combination in *Neoponera:* Emery, 1901a: 47; in *N.* (*Eumecopone*): Emery, 1911d: 71; in *Pachycondyla:* Brown, **revived combination**.

**#rotundata.** *Ponera elisae* var. *rotundata* Emery, 1895h: 20 (w.) SOUTH AFRICA. Arnold, 1915: 71 (q.). Combination in *Euponera* (*Mesoponera*): Emery, 1901a: 46; in *E.* (*Xiphopelta*): Forel, 1913j: 206; in *Pachycondyla:* Brown, **new combination**. Stirps of *elisae:* Santschi, 1937d: 217. Senior synonym of *arnoldi:* Wheeler, W.M. 1922a: 776.

*rotundiceps* Emery, 1914 [*Euponera*]; see under ***CRYPTOPONE.***

**rubescens.** *Bothroponera rubescens* Santschi, 1937a: 48 (q.) ZAIRE. Combination in *Pachycondyla:* Brown, **new combination**.

**#rubicunda.** *Bothroponera sublaevis* var. *rubicunda* Emery, 1893e: 210 (w.) AUSTRALIA. Combination in *Pachycondyla* (*Bothroponera*): Emery, 1901a: 45; in *Bothroponera:* Taylor & Brown, D.R. 1985: 22; in *Pachycondyla:* Brown, **revived combination**.

**rubiginosa.** *Ponera rubiginosa* Emery, 1889b: 498 (w.) BURMA. Wheeler, W.M. 1928c: 5 (q.); Imai, Baroni Urbani, *et al.* 1984: 5 (k.). Combination in *Bothroponera:* Dalla Torre, 1893: 36; in *Pachycondyla* (*Bothroponera*): Emery, 1901a: 46; in *Bothroponera:* Wheeler, W.M. 1928c: 5; in *Pachycondyla:* Brown, **revived combination**. See also: Bingham, 1903: 99.

**rubra.** *Ponera rubra* Smith, F. 1857a: 66 (q.) SINGAPORE. Imai, Brown, *et al.* 1984: 67 (k.); Tjan, Imai, *et al.* 1986: 57 (k.). Combination in *Euponera* (*Mesoponera*): Emery, 1901a: 46; in *Pachycondyla:* Brown, **new combination**. Current subspecies: nominal plus *javana* (junior homonym).

**#rufescens.** *Euponera* (*Trachymesopus*) *stigma* var. *rufescens* Wheeler, W.M. & Mann, 1914: 13 (w.q.) HAITI. Combination in *Trachymesopus:* Kempf, 1972a: 251; in *Pachycondyla:* Brown, **new combination**.

**ruficornis.** *Ectomomyrmex ruficornis* Clark, 1934b: 31, pl. 2, fig. 14 (w.) AUSTRALIA. Combination in *Pachycondyla:* Brown, **new combination**.

**rufipes.** *Ponera rufipes* Jerdon, 1851: 119 (w.) INDIA. Combination in *Pachycondyla:* Smith, F. 1858b: 106; in *P.* (*Bothroponera*): Emery, 1901a: 45; in *Bothroponera:* Mayr, 1867a: 83; Wheeler, W.M. 1919e: 55; in *Pachycondyla:* Brown, **revived combination**. Current subspecies: nominal plus *ceylonensis.* See also: Bingham, 1903: 96.

**rufonigra.** *Euponera* (*Brachyponera*) *rufonigra* Clark, 1934b: 30, pl. 2, figs. 12, 13 (w.q.) AUSTRALIA. Combination in *Trachymesopus:* Brown, 1963: 7; in *Pachycondyla:* Brown, **new combination**.

*rufotestaceus. Cryptopone rufotestaceus* Donisthorpe, 1943b: 197 (q.) INDIA. Combination in *Trachymesopus:* Wilson, 1958d: 361 (footnote). Junior synonym of *darwinii:* Brown, 1963: 6.

**#rugatula.** *Neoponera unidentata* var. *rugatula* Santschi, 1919f: 38 (w.) ARGENTINA (attributed to Emery). Combination in *Pachycondyla:* Brown, **new combination**.

**#ruginota.** *Euponera* (*Brachyponera*) *sennaarensis* var. *ruginota* Stitz, 1916: 372 (w.) CAMEROUN. Combination in *Pachycondyla:* Brown, **new combination**.

**#rugosinodis.** *Neoponera rugosinodis* André, 1902: 15 (w.) PERU. Combination in *Pachycondyla:* Brown, **new combination**. Variety of *striatinodis:* Emery, 1911d: 72.

*rugosula. Neoponera unidentata* var. *rugosula* Emery, 1902c: 30 (w.) PERU. Senior synonym of *laevinodis:* Emery, 1911d: 72. Junior synonym of *unidentata:* Brown, 1957e: 232.

**#saldanhae.** *Hagensia peringueyi* r. *saldanhae* Arnold, 1951: 53, fig. 4 (w.) SOUTH AFRICA. Combination in *Pachycondyla:* Brown, **new combination**.

**sandakana.** *Bothroponera sandakana* Wheeler, W.M. 1919e: 141 (q.) BORNEO. Combination in *Pachycondyla:* Brown, **new combination**.

**sanguinea.** *Pachycondyla* (*Bothroponera*) *sanguinea* Santschi, 1920b: 7 (w.) ZAIRE. Combination in *Bothroponera:* Wheeler, W.M. 1922a: 772; in *Pachycondyla:* Brown, **new combination**.

*sauteri* Wheeler, W.M. 1906; see under ***CRYPTOPONE.***

**sauteri.** *Pachycondyla* (*Ectomomyrmex*) *sauteri* Forel 1912a: 49 (w.q.) TAIWAN. [Unresolved junior primary homonym of *Pachycondyla* (*Pseudoponera*) *sauteri* Wheeler, W.M. 1906c: 304, above.]

*schaeferi. Euponera* (*Trachymesopus*) *schaeferi* Menozzi, 1941: 12 (w.) CHINA. *Nomen nudum.*

**scobina.** *Ectomomyrmex scobinus* Wilson, 1958d: 365. Replacement name for *Bothroponera striata* Donisthorpe, 1949b: 489. [Junior secondary homonym of *Pachycondyla striata* Smith, F. 1858b: 106.] Combination in *Pachycondyla:* Brown, **new combination**.

**scolopax.** *Ponera scolopax* Emery, 1899e: 473 (w.) CAMEROUN. Combination in *Euponera* (*Mesoponera*): Emery, 1901a: 46; in *Pachycondyla.* Brown, **new combination**.

**#sculpturata.** *Pachycondyla* (*Bothroponera*) *sculpturata* Santschi, 1912b: 151 (q.) ZIMBABWE. Variety of *pumicosa:* Wheeler, W.M. 1922a: 772.

**sculpturata.** *Pachycondyla* (*Ectomomyrmex*) *sculpturata* Karavaiev, 1925b: 122 (w.) SUMATRA. [Unresolved junior primary homonym of *sculpturata* Santschi, above.]

**senegalensis.** *Euponera* (*Mesoponera*) *senegalensis* Santschi, 1914d: 315, fig. 5 (w.) SENEGAL. Combination in *E.* (*Xiphopelta*): Bernard, 1953b: 194; in *Pachycondyla:* Brown, **new combination**.

**sennaarensis.** *Ponera sennaarensis* Mayr, 1862: 721 (w.) SUDAN. Santschi, 1910c: 350 (q.); Forel, 1910c: 245 (m.); Wheeler, G.C. & Wheeler, J. 1971b: 1207 (l.). Combination in *Euponera* (*Brachyponera*): Emery, 1901a: 47; in *Brachyponera:* Wheeler, G.C. & Wheeler, J. 1971b: 1207; in *Pachycondyla:* Brown, **new combination**. Senior synonym of *sorghi:* Wheeler, W.M. 1922a: 777; Menozzi, 1930b: 80. Current subspecies: nominal plus *decolor, ruginota.* See also: Arnold, 1915: 73.

*sexdentatus. Euponera* (*Trachymesopus*) *sexdentatus* Donisthorpe, 1949c: 746 (w.) NEW GUINEA. Junior synonym of *stigma:* Wilson, 1958d: 355.

**sharpi.** *Euponera* (*Pseudoponera*) *sharpi* Forel, 1901e: 379 (w.q.m.) SINGAPORE. Combination in *E.* (*Trachymesopus*): Emery, 1911d: 85; in *Trachymesopus:* Brown, 1963: 5; in *Pachycondyla:* Brown, **new combination**.

**sheldoni.** *Euponera* (*Trachymesopus*) *sheldoni* Mann, 1919: 292 (w.) SOLOMON IS. Combination in *Trachymesopus:* Wilson, 1958d: 354; in *Pachycondyla:* Brown, **new combination**.

*sicula* Emery, 1909 [*Euponera*]; see under ***CRYPTOPONE.***

**sikorae.** *Ponera* (*Euponera*) *sikorae* Forel, 1891b: 127, pl. 5, fig. 1 (w.) MADAGASCAR. Combination in *Euponera:* Emery, 1901a: 46; in *Pachycondyla:* Brown, **new combination**.

**silvestrii.** *Pachycondyla* (*Bothroponera*) *silvestrii* Santschi, 1914d: 313 (w.) GHANA. Combination in *Bothroponera:* Wheeler, W.M. 1922a: 72; in *Pachycondyla:* Brown, **revived combination**. Current subspecies: nominal plus *nimba.*

*simillima. Pachycondyla simillima* Smith, F. 1858b: 105, pl. 7, fig. 17 (q.) SOUTH AFRICA. Combination in *Paltothyreus:* Mayr, 1886c: 358. Junior synonym of *tarsata:* Emery, 1892d: 557.

**simillima.** *Bothroponera simillima* Donisthorpe, 1949b: 490 (w.q.) NEW GUINEA. [Unresolved junior secondary homonym of *Pachycondyla simillima* Smith, above.] Combination in *Pachycondyla* (*Bothroponera*): Donisthorpe, 1949c: 746; in *Ectomomyrmex:* Wilson, 1958d: 365; in *Pachycondyla:* Brown, **revived combination**.

*sinensis;* see under ***chinensis.***

**sjostedti.** *Ponera sjostedti* Mayr, 1896: 231 (w.q.m.) CAMEROUN. Wheeler, G.C. & Wheeler, J. 1971c: 392 (l.). Combination in *Pachycondyla* (*Bothroponera*): Emery, 1901a: 45; in *Bothroponera:* Wheeler, W.M. 1922a: 73; in *Pachycondyla:* Brown, **revived combination**.

*socialis. Ectatomma socialis* MacLeay, 1873: 369 (w.) AUSTRALIA. Junior synonym of *lutea:* Emery, 1911d: 84.

**solitaria.** *Ponera solitaria* Smith, F. 1860b: 103 (m.) INDONESIA (Batjan I.). Combination in *Pachycondyla:* Donisthorpe, 1932c: 461; in *Bothroponera:* Wilson, 1958d: 362; in *Pachycondyla:* Brown, **revived combination**.

*solitaria. Ponera solitaria* Smith, F. 1874: 404 (w.) JAPAN. [Junior primary homonym of *solitaria* Smith, above.] Forel, 1900e: 267 (q.). Combination in *Euponera* (*Brachyponera*): Emery, 1901a: 47. Senior synonym of *chinensis* and hence the latter the first available replacement name: Brown, 1958h: 22.

*sorghi. Ponera sorghi* Roger, 1863a: 169 (w.) SUDAN. Combination in *Euponera* (*Brachyponera*): Emery, 1911d: 84. Junior synonym of *sennaarensis:* Wheeler, W.M. 1922a: 777.

**soror.** *Ponera* (*Bothroponera*) *soror* Emery, 1899e: 472 (w.q.) CAMEROUN. Menozzi, 1944: 432 (m.l.); Wheeler, G.C. & Wheeler, J. 1952c: 623 (l.); Wheeler, G.C. & Wheeler, J. 1971c: 390 (l.). Combination in *Pachycondyla* (*Bothroponera*): Emery, 1901a: 46; in *Bothroponera:* Wheeler, W.M. 1922a: 72; in *Pachycondyla:* Brown, **revived combination**. Current subspecies: nominal plus *ancilla, suturalis.* See also: Arnold, 1915: 59; Wheeler, W.M. 1922a: 72.

**#splendida.** *Euponera* (*Mesoponera*) *atrovirens* r. *splendida* Forel, 1901f: 340 (w.) ECUADOR. Combination in

***PACHYCONDYLA** (continued)*

*Mesoponera:* Kempf, 1972a: 141; in *Pachycondyla:* Brown, **new combination**.

*spiniventris. Ponera spiniventris* Smith, F. 1858b: 92 (m.) SIERRA LEONE. Junior synonym of *tarsata:* Roger, 1860: 310; Roger, 1863b: 17.

***stigma.*** *Formica stigma* Fabricius, 1804: 400 (q.) SOUTH AMERICA. Wheeler, G.C. & Wheeler, J. 1952c: 627 (l.); Wheeler, G.C. & Wheeler, J. 1976a: 55 (l.). Combination in *Ponera:* Mayr, 1863: 450; in *Pachycondyla* (*Pseudoponera*): Emery, 1900c: 314; Emery, 1901a: 46; in *Euponera* (*Pseudoponera*): Forel, 1901g: 398; in *E.* (*Trachymesopus*): Emery, 1911d: 85; in *Trachymesopus:* Wilson, 1958d: 355; Kempf, 1960f: 424; in *Mesoponera:* Wheeler, G.C. & Wheeler, J. 1976a: 55; in *Pachycondyla:* Brown, **revived combination**. Senior synonym of *americana* and *quadridentata* (and its junior synonym *myopa*): Emery, 1887b: 434; of *brunneus, nixoni, quadridentata* and *sexdentatus:* Wilson, 1958d: 355. Current subspecies: nominal plus *attrita, rufescens.*

*stipitum. Neoponera stipitum* Forel, 1901f: 348 (w.q.) COLOMBIA. Junior synonym of *crenata:* Brown, 1957e: 234.

***striata.*** *Pachycondyla striata* Smith, F. 1858b: 106 (w.) BRAZIL. Roger, 1861a: 6 (q.); Santschi, 1921g: 86 (m.); Wheeler, G.C. & Wheeler, J. 1952c: 616 (l.). Senior synonym of *nitidiventris:* Kempf, 1961c: 210.

*striata. Euponera* (*Mesoponera*) *castanea* var. *striata* Stitz, 1911a: 356 (w.) NEW ZEALAND. [Unresolved junior secondary homonym of *striata* Smith, above.] Junior synonym of *castanea:* Brown, 1958h: 20.

*striata. Pachycondyla* (*Ectomomyrmex*) *striata* Karavaiev, 1927e: 5 (w.) INDONESIA (Ambon I.) [Junior primary homonym of *striata* Smith, F. 1858b: 106.] Replacement name: *striatula* Karavaiev, 1935a: 70.

#***striata.*** *Paltothyreus tarsatus* var. *striatus* Santschi, 1930a: 53 (w.) BENIN. [Unresolved junior secondary homonym of *striata* Smith, F. 1858b: 106.] Combination in *Pachycondyla:* Brown, **new combination**.

*striata. Bothroponera striata* Donisthorpe, 1949b: 489 (w.) NEW GUINEA. [Junior secondary homonym of *striata* Smith, 1858b: 106.] Replacement name: *Ectomomyrmex scobinus* Wilson, 1958d: 365.] Combination in *Pachycondyla* (*Bothroponera*): Donisthorpe, 1949c: 745.

*striatidens* Santschi, 1914; see under ***PHRYNOPONERA.***

#***striatidens.*** *Paltothyreus tarsatus* var. *striatidens* Santschi, 1919h: 81 (w.q.) KENYA. [Unresolved junior secondary homonym of *striatidens* Santschi, 1914d: 315, above.] Combination in *Pachycondyla:* Brown, **new combination**.

***striatinodis.*** *Pachycondyla striatinodis* Emery, 1890a: 75 (w.) COSTA RICA. [Also described as new by Emery, 1894k: 48.] Combination in *Neoponera:* Emery, 1901a: 47; in *Pachycondyla:* Brown, **new combination**. Senior synonym of *bakeri:* Kempf, 1960e: 391. Current subspecies: nominal plus *rugosinodis.*

***striatula.*** *Pachycondyla* (*Ectomomyrmex*) *striatula* Karavaiev, 1935a: 70. Replacement name for *striata* Karavaiev, 1927e: 5. [Junior primary homonym of *striata* Smith, F. 1858b: 106.] Combination in *Ectomomyrmex:* Wilson, 1958d: 365; in *Pachycondyla:* Brown, **revived combination**.

***strigulosa.*** *Bothroponera strigulosa* Emery, 1895h: 19 (w.) SOUTH AFRICA. Combination in *Pachycondyla* (*Bothroponera*): Emery, 1901a: 45; in *Bothroponera:* Wheeler, W.M. 1922a: 72; in *Pachycondyla:* Brown, **revived combination**. Senior synonym of *berthoudi:* Arnold, 1952a: 460.

***striolata.*** *Pachycondyla* (*Ectomomyrmex*) *striolata* Donisthorpe, 1933b: 194 (w.) INDIA. [Misspelled as *Ectomyrmex* in original description.]

*subcarinata. Neoponera crenata* st. *moesta* var. *subcarinata* Santschi, 1939e: 313 (w.) BRAZIL. Unavailable name.

***subiridescens.*** *Euponera* (*Mesoponera*) *subiridescens* Wheeler, W.M. 1922a: 83, fig. 15 (w.) ZAIRE. Combination in *Pachycondyla:* Brown, **new combination**.

***sublaevis.*** *Bothroponera sublaevis* Emery, 1887b: 442 (w.) AUSTRALIA. Wheeler, W.M. 1918c: 299 (l.); Wheeler, G.C. & Wheeler, J. 1952c: 621 (l.); Wheeler, G.C. & Wheeler, J. 1971c: 387 (l.). Combination in *Pachycondyla* (*Bothroponera*): Emery, 1901a: 45; in *Bothroponera:* Taylor & Brown, D.R. 1985: 22; in *Pachycondyla:* Brown, **revived combination**. Current subspecies: nominal plus *kurandensis, murina, reticulata* (junior homonym), *rubicunda.*

#***subopaca.*** *Paltothyreus tarsatus* var. *subopaca* Santschi, 1919h: 80 (w.q.) GABON. Combination in *Pachycondyla:* Brown, **new combination**.

#***subpilosa.*** *Megaponera foetens* var. *subpilosa* Santschi, 1937d: 217 (w.) ANGOLA. Combination in *Pachycondyla:* Brown, **new combination**. Subspecies of *analis:* current status.

***succedanea.*** *Ponera succedanea* Roger, 1863a: 170 (w.q.m.) CUBA. Combination in *Pachycondyla* (*Pseudoponera*): Emery, 1901a: 46; in *Euponera* (*Trachymesopus*): Emery, 1911d: 85; in *Trachymesopus:* Kempf, 1960f: 424; in *Pachycondyla:* Brown, **revived combination**.

****succinea.*** **Ponera succinea* Mayr, 1868c: 72 (q.) BALTIC AMBER (Oligocene). Combination in *Euponera* (*Trachymesopus*): Wheeler, W.M. 1915h: 37; in *Pachycondyla:* **new combination**.

***sulcata.*** *Ponera sulcata* Mayr, 1867c: 441 (w.) ("Vaterland unbekannt") [Emery, 1911d: 78 gives INDIA, possibly based on the collection data in Forel, 1900d: 326, and Bingham, 1903: 99.]. [Unresolved junior primary homonym of *Ponera sulcata* Smith, F. 1858b: 99 (now in *Gnamptogenys*).] Forel, 1900d: 326 (m.). Combination in *Bothroponera:* Forel, 1891b: 124; in *Pachycondyla* (*Bothroponera*): Emery, 1901a: 46. Current subspecies: nominal plus *fossulata, sulcatotesserinoda.*

#***sulcatotesserinoda.*** *Ponera* (*Bothroponera*) *sulcata* var. *sulcatotesserinoda* Forel, 1900d: 323 (diagnosis in key) (w.) INDIA. [Also described as new by Forel, 1922: 88.] Combination in *Pachycondyla* (*Bothroponera*): Emery, 1901a: 46.

*sulcatula. Neoponera unidentata* st. *sulcatula* Santschi, 1919f: 38, fig. 2 (w.) ARGENTINA. Stirps of *carinulata:* Santschi, 1939e: 313. Junior synonym of *crenata:* Brown, 1957e: 234.

*sulcigera. Euponera* (*Mesoponera*) *sulcigera* Mayr, 1904a: 593 (w.) SOUTH AFRICA. Junior synonym of *havilandi:* Arnold, 1915: 67.

#***sumatrensis.*** *Pachycondyla* (*Ectomomyrmex*) *leeuwenhoeki* var. *sumatrensis* Forel, 1901f: 347 (w.) SUMATRA.

*sundaicus. Ectomomyrmex sundaicus* Mayr, 1867a: 85 (q.) JAVA. Junior synonym of *astuta:* Bingham, 1903: 86; of *javana:* Yasumatsu, 1962: 94.

***suspecta.*** *Euponera* (*Trachymesopus*) *suspecta* Santschi, 1914b: 51, fig. 1 (w.) TANZANIA. Combination in *Pachycondyla:* Brown, **new combination**.

#***suturalis.*** *Pachycondyla* (*Bothroponera*) *soror* subsp. *suturalis* Forel, 1907c: 133 (w.) ETHIOPIA.

*sveni* Forel, 1916; see under ***PHRYNOPONERA.***

*taivanae* Forel, 1913 [*Euponera*]; see under ***CRYPTOPONE.***

*takahashii* Wheeler, W.M. 1930 [*Euponera*]; see under ***CRYPTOPONE.***

***talpa.*** *Bothroponera talpa* André, 1890: 316 (w.) SIERRA LEONE. Emery, 1899e: 473 (q.); Santschi, 1914d: 314 (q.). Combination in *Pachycondyla* (*Bothroponera*): Emery, 1901a: 45; in *Bothroponera:* Wheeler, W.M. 1922a: 72; in *Pachycondyla:* Brown, **revived combination**. Senior synonym of *clavicornis:* Bolton, 1975b: 6. Current subspecies: nominal plus *variolata.*

***tarsata.*** *Formica tarsata* Fabricius, 1798: 280 (w.) SENEGAL. Latreille, 1802c: 736 (q.); Mayr, 1866b: 893 (m.). Combination in *Paltothyreus:* Mayr, 1862: 736; in *Pachycondyla:* Brown, **new combination**. Senior synonym of *gagates, pestilentia* and *spiniventris:* Roger, 1860: 310; Roger, 1863b: 17; of *simillima:* Emery, 1892d: 557. Current subspecies: nominal plus *delagoensis, mediana, robusta, striata* (junior homonym), *striatidens* (junior homonym), *subopaca.* See also: Forel, 1891b: 136; Arnold, 1915: 44; Wheeler, W.M. 1922a: 60; Hölldobler, 1980: 86.

*tasmaniensis* Forel, 1913; see under ***PLATYTHYREA.***

#***termitivora.*** *Megaloponera foetens* var. *termitivora* Santschi, 1930b: 60 (diagnosis in key) (w.) ZAIRE. Combination in *Pachycondyla:* Brown, **new combination**. Subspecies of *analis:* current status.

***tesseronoda.*** *Ponera tesseronoda* Emery, 1877b: 368 (w.) INDIA. Forel, 1900d: 352 (m.); Wheeler, G.C. & Wheeler, J. 1976a: 46 (l.). [Name misspelled as *tesserinoda* by Forel, 1900d: 325; Bingham, 1903: 97, and subsequent authors.] Combination in *Bothroponera:* Forel, 1891b: 124; in *Pachycondyla* (*Bothroponera*): Emery, 1901a: 46. See also: Bingham, 1903: 97; Maschwitz, Hölldobler & Möglich, 1974: 113.

***testacea.*** *Euponera* (*Mesoponera*) *testacea* Bernard, 1953b: 191, fig. 2 (w.) GUINEA. Combination in *Pachycondyla:* Brown, **new combination**.

***theresiae.*** *Pachycondyla theresiae* Forel, 1899c: 13, pl. 1, fig. 11 (w.) PANAMA. Combination in *Neoponera:* Emery, 1901a: 47; in *Pachycondyla:* Brown, **revived combination**. Current subspecies: nominal plus *bugabensis.*

***tonkina.*** *Pachycondyla* (*Ectomomyrmex*) *tonkina* Santschi, 1920h: 159 (w.) VIETNAM. [Also described as new by Santschi, 1924c: 96.] Combination in *Ectomomyrmex:* Wheeler, W.M. 1927d: 1; in *Pachycondyla:* Brown, **revived combination**.

*transversa*. *Pachycondyla fuscoatra* r. *transversa* Emery, 1890a: 58 (w.) COSTA RICA. [Also described as new by Emery, 1890b: 42, and by Emery, 1894k: 48.] Forel, 1908b: 38 (q.). Raised to species: Dalla Torre, 1893: 35. Subspecies of *fuscoatra:* Forel, 1908b: 38. Junior synonym of *impressa:* Kempf, 1961c: 195.

***tridentata***. *Pachycondyla tridentata* Smith, F. 1858b: 106 (w.) BORNEO. Wheeler, W.M. 1919e: 142 (m.); Wheeler, G.C. & Wheeler, J. 1976a: 45; Imai, Brown, *et al.* 1984: 67 (k.). Combination in *Pachycondyla* (*Bothroponera*): Emery, 1900d: 667; in *Bothroponera:* Mayr, 1872: 149; Wheeler, W.M. 1919e: 55; in *Pachycondyla:* Brown, **revived combination**. Current subspecies: nominal plus *debilior, exasperans.*

*trinidadensis*. *Neoponera unidentata* var. *trinidadensis* Wheeler, W.M. & Bequaert, 1929: 30 (w.) TRINIDAD. Junior synonym of *unidentata:* Brown, 1957e: 232.

*trumani* Donisthorpe, 1949 [*Pseudoponera*]; see under ***MYOPIAS.***

***unicolor***. *Ponera unicolor* Smith, F. 1860a: 73 (m.) SULAWESI. Combination in *Pachycondyla* (*Bothroponera*): Donisthorpe, 1932c: 458.

***unidentata***. *Pachycondyla unidentata* Mayr, 1862: 720 (w.q.) BRAZIL. Combination in *Neoponera:* Emery, 1901a: 47; in *Pachycondyla:* Brown, **revived combination**. Senior synonym of *eburneipes, maya, rugosula* (and its junior synonym *laevinodis*) and *trinidadensis:* Brown, 1957e: 232. Current subspecies: nominal plus *rugatula.*

#***variolata***. *Pachycondyla* (*Bothroponera*) *variolata* Santschi, 1912b: 151 (w.) CONGO. Race of *talpa:* Santschi, 1914d: 314.

***variolosa***. *Bothroponera variolosa* Arnold, 1947: 131, fig. 1 (w.) SOUTH AFRICA. Combination in *Pachycondyla:* Brown, **new combination**.

***venusta***. *Neoponera venusta* Forel, 1912c: 38 (w.) BRAZIL. Combination in *Pachycondyla:* Brown, **new combination**.

***verecundae***. *Pseudoneoponera verecundae* Donisthorpe, 1943d: 439 (m.) NEW GUINEA. Combination in *Bothroponera:* Wilson, 1958d: 362; in *Pachycondyla:* Brown, **new combination**.

*verenae*. *Neoponera apicalis* var. *verenae* Forel, 1922: 90 (w.) PANAMA. Junior synonym of *apicalis:* Brown, 1957e: 230.

***vermiculata***. *Ponera* (*Bothroponera*) *vermiculata* Emery, 1897d: 552 (footnote) (q.) INDONESIA (?). Combination in *Pachycondyla* (*Ectomomyrmex*): Emery, 1901a: 46; in *Ectomomyrmex:* Wilson, 1958d: 366; in *Pachycondyla:* Brown, **revived combination**.

***vidua***. *Ponera vidua* Smith, F. 1857a: 68 (m.) BORNEO. Combination in *Pachycondyla:* Donisthorpe, 1932c: 448.

*viehmeyeri*. *Euponera* (*Mesoponera*) *viehmeyeri* Donisthorpe, 1948g: 132 (w.) NEW GUINEA. Junior synonym of *papuana:* Wilson, 1958d: 350.

***villiersi***. *Euponera* (*Xiphopelta*) *villiersi* Bernard, 1953b: 193, fig. 2 (w.) GUINEA. Combination in *Pachycondyla:* Brown, **new combination**.

***villosa***. *Formica villosa* Fabricius, 1804: 409 (w.) CENTRAL AMERICA. Roger, 1861a: 2 (w.q.); Wheeler, G.C. & Wheeler, J. 1952c: 615 (l.); Petralia & Vinson, 1980: 378 (l.). Combination in *Pachycondyla:* Mayr, 1862: 720; in *Neoponera:* Emery, 1901a: 43 in *Pachycondyla:* Brown, **revived combination**. Senior synonym of *bicolor, pedunculata* and *pilosa:* Roger, 1861: 1; Roger, 1862c: 288. Current subspecies: nominal plus *inversa.* See also: Wheeler, W.M. 1908e: 404; Gallardo, 1918b: 56.

***wasmannii***. *Bothroponera wasmannii* Forel, 1887: 383 (w.) MADAGASCAR. Forel, 1897c: 196 (q.m.). Combination in *Pachycondyla* (*Bothroponera*): Emery, 1901a: 45.

***weberi***. *Euponera* (*Xiphopelta*) *weberi* Bernard, 1953b: 194, fig. 2 (w.) GUINEA. Combination in *Pachycondyla:* Brown, **new combination**.

*whelpleyi*. *Euponera* (*Mesoponera*) *laevigata* var. *whelpleyi* Wheeler, W.M. 1922c: 3 (w.) TRINIDAD. Junior synonym of *laevigata:* Wheeler, W.M. 1936d: 164.

***williamsi***. *Bothroponera williamsi* Wheeler, W.M. & Chapman, 1925: 67, pl. 2, figs. 13, 14 (w.) PHILIPPINES. Combination in *Pachycondyla:* Brown, **new combination**.

***wroughtonii***. *Euponera* (*Pseudoponera*) *wroughtonii* Forel, 1901f: 341 (w.m.) SOUTH AFRICA. Arnold, 1926: 204 (q.); Wheeler, G.C. & Wheeler, J. 1971b: 1206 (l.). Combination in *E.* (*Trachymesopus*): Emery, 1911d: 85; in *Mesoponera:* Wheeler, G.C. & Wheeler, J. 1971b: 1206; in *Pachycondyla:* Brown, **new combination**. Current subspecies: nominal plus *crudelis.*

***zumpti***. *Bothroponera zumpti* Santschi, 1937b: 94, fig. 1 (w.) CAMEROUN. Combination in *Pachycondyla:* Brown, **new combination**.

*PACHYSIMA:* see under ***TETRAPONERA.***

**PAEDALGUS** (Afrotropical, Sri Lanka)

***distinctus***. *Paedalgus distinctus* Bolton & Belshaw, 1993: 185, figs. 2, 3 (w.m.) GHANA.

***escherichi***. *Paedalgus escherichi* Forel, 1911i: 218 (w.q.) SRI LANKA. See also: Bolton & Belshaw, 1993: 185.

***infimus***. *Oligomyrmex infimus* Santschi, 1913e: 459 (diagnosis in key) (w.) GUINEA. Combination in *Paedalgus:* Santschi, 1914d: 364. See also: Bolton & Belshaw, 1993: 186.

***octatus***. *Paedalgus octatus* Bolton & Belshaw, 1993: 186 (w.) KENYA.

***pisinnus***. *Paedalgus pisinnus* Bolton & Belshaw, 1993: 187, fig. 5 (w.m.) KENYA.

***rarus***. *Paedalgus rarus* Bolton & Belshaw, 1993: 187, fig. 4 (w.) KENYA.

***robertsoni***. *Paedalgus robertsoni* Bolton & Belshaw, 1993: 188, fig. 1 (w.q.) SOUTH AFRICA.

***saritus***. *Paedalgus saritus* Bolton & Belshaw, 1993: 188, fig. 6 (w.) GHANA.

***sudanensis***. *Paedalgus sudanensis* Weber, 1943c: 365, pl. 15, figs. 12, 15 (q.) SUDAN. See also: Bolton & Belshaw, 1993: 188.

***termitolestes***. *Paedalgus termitolestes* Wheeler, W.M. 1918c: 301, fig. 5 (w.l.) ZAIRE. Bolton, 1969: 14 (m.); Wheeler, G.C. & Wheeler, J. 1954a: 143 (l.). See also: Wheeler, W.M. 1922a: 177; Bolton & Belshaw, 1993: 189.

*PALAEATTA:* see under ***ATTA.***

**PALAEOMYRMEX* Heer, 1865: transferred to **HOMOPTERA**.

**PALAEOMYRMEX* Dlussky, 1975: see under ***DLUSSKYIDRIS.***

****PALEOSMINTHURUS*** (U.S.A.)

****juliae***. **Paleosminthurus juliae* Pierce & Gibron, 1962: 147, fig. 4 (m.?) U.S.A. (Miocene) [in Order Collembola, Family *Paleosminthuridae]. Transferred to Formicidae: Najt, 1987: 152.

*PALTOTHYREUS:* see under ***PACHYCONDYLA.***

***PAPYRIUS*** (Indo-Australian, Australasian)

#***clitellarius***. *Iridomyrmex nitidus* subsp. *clitellarius* Viehmeyer, 1925a: 32 (w.) AUSTRALIA. Combination in *Papyrius:* Shattuck, 1992a: 17.

***flavus***. *Iridomyrmex flavus* Mayr, 1865: 60, pl. 3, fig. 17 (w.) AUSTRALIA. Combination in *Papyrius:* Shattuck, 1992a: 17.

***nitidus***. *Iridomyrmex nitida* Mayr, 1862: 702 (w.) AUSTRALIA. Forel, 1901b: 20 (q.m.); Wheeler, G.C. & Wheeler, J. 1951: 186 (l.); Crozier, 1969: 250 (k.). Combination in *Papyrius:* Shattuck, 1992a: 17. Senior synonym of *tuberculatus:* Mayr, 1870b: 955. Current subspecies: nominal plus *clitellarius, oceanicus, queenslandensis.* See also: Wheeler, W.M. 1909a: 28; Wheeler, G.C. & Wheeler, J. 1974b: 400; Shattuck, 1994: 134.

#***oceanicus***. *Iridomyrmex nitidus* r. *oceanicus* Forel, 1901b: 20 (w.q.m.) BISMARCK ARCHIPELAGO. Combination in *Papyrius:* Shattuck, 1992a: 17.

#***queenslandensis***. *Iridomyrmex nitida* var. *queenslandensis* Forel, 1901b: 21 (w.m.) AUSTRALIA. Forel, 1907h: 288 (q.). Race of *nitidus:* Forel, 1915b: 79. Combination in *Papyrius:* Shattuck, 1992a: 17.

*tuberculatus*. *Acantholepis tuberculatus* Lowne, 1865b: 332 (w.) AUSTRALIA. Junior synonym of *nitidus:* Mayr, 1870b: 955.

*victorianus*. *Iridomyrmex nitidus* r. *oceanicus* var. *victorianus* Stitz, 1911a: 368 (w.q.m.) AUSTRALIA. Unavailable name (Taylor, 1986: 34).

*PARACOLOBOPSIS:* see under ***CAMPONOTUS.***

#***PARACREMA:*** see under ***CREMATOGASTER.***

*PARACRYPTOCERUS:* see under ***ZACRYPTOCERUS.***

*PARAENICTUS:* see under ***AENICTUS.***

*PARAFORMICA:* see under ***CATAGLYPHIS.***

****PARAMERANOPLUS*** (Baltic Amber)

****primaevus***. **Parameranoplus primaevus* Wheeler, W.M. 1915h: 69, fig. 28 (w.) BALTIC AMBER (Oligocene).

*PARAMYCETOPHYLAX:* see under ***MYCETOPHYLAX.***

#***PARAMYRMAMBLYS:*** see under ***CAMPONOTUS.***

*PARAMYRMICA:* see under ***MYRMICA.***

*PARANAMYRMA:* see under ***SOLENOPSIS.***

****PARANEURETUS*** (Baltic Amber)

****longicornis***. **Paraneuretus longicornis* Wheeler, W.M. 1915h: 77 (m.) BALTIC AMBER (Oligocene).

****tornquisti***. **Paraneuretus tornquisti* Wheeler, W.M. 1915h: 75, fig. 32 (w.m.) BALTIC AMBER (Oligocene).

*PARANOMOPONE:* see under ***HETEROPONERA.***

*PARAPARATRECHINA:* see under ***PARATRECHINA.***

*PARAPHACOTA:* see under ***MONOMORIUM.***

***PARAPHAENOGASTER*** (Russia, China)
***microphthalma.** *Paraphaenogaster microphthalmus* Dlussky, 1981b: 69, fig. 51 (q.) RUSSIA (Miocene).
***shanwangensis.** *Paraphaenogaster shanwangense* Hong, 1984: 8, pl. 3, fig. 1 (q.) CHINA (Miocene). [Misspelled as *Paraphaeogaster* in original description.]

*PARAPHEIDOLE:* see under ***PHEIDOLE.***

*PARAPLAGIOLEPIS:* see under ***PLAGIOLEPIS.***

***PARAPONERA*** (Neotropical)
*aculeata. Formica aculeata* Olivier, 1792: 498 (w.) FRENCH GUIANA. Junior synonym of *clavata:* Latreille, 1802c: 207.
***clavata.*** *Formica clavata* Fabricius, 1775: 394 (w.) India (locality in error). Lepeletier, 1835: 189 (w.q.); Smith, F. 1858b: 101 (m.); Mayr, 1862: 731 (q.m.); Wheeler, G.C. & Wheeler, J. 1952a: 117 (l.). Combination in *Paraponera:* Smith, F. 1858b: 100. Senior synonym of *aculeata,* and material of the unavailable name *spininoda* referred here: Latreille, 1802c: 207; of *tarsalis:* Smith, F. 1858b: 100. See also: Brown, 1958g: 205.
***dieteri.** *Paraponera dieteri* Baroni Urbani, 1994: 5, figs. 1, 2 (w.) DOMINICAN AMBER.
*spininoda. Formica spininoda* Latreille, 1802c: 207, pl. 7, fig. 45 (w.q.) SURINAM. Unavailable name (proposed in synonymy); material referred to *clavata:* Latreille, 1802c: 207.
*tarsalis. Ponera tarsalis* Perty, 1833: 135, pl. 27, fig. 2 (w.) BRAZIL. Junior synonym of *clavata:* Smith, F. 1858b: 100.

***PARAPRIONOPELTA*** (Neotropical)
***minima.*** *Paraprionopelta minima* Kusnezov, 1955: 271, figs. 1, 2, 5a (m.) ARGENTINA. Probable combination in *Amblyopone:* Brown, 1973b: 183. [*Paraprionopelta minima* Kusnezov, 1954b: 34. *Nomen nudum.*]

*PARASIMA:* see under ***TETRAPONERA.***

*PARASYSCIA:* see under ***CERAPACHYS.***

***PARATOPULA*** (Oriental, Indo-Australian)
***andamanensis.*** *Sima andamanensis* Forel, 1903d: 403 (m.) ANDAMAN IS. Combination in *Paratopula:* Bolton, 1988b: 134.
***ankistra.*** *Paratopula ankistra* Bolton, 1988b: 134, figs. 1, 6 (w.q.) BORNEO.
***catocha.*** *Paratopula catocha* Bolton, 1988b: 136, fig. 4 (w.) SULAWESI.
***ceylonica.*** *Atopomyrmex ceylonicus* Emery, 1901f: 114, fig. (q.) SRI LANKA. Forel, 1913f: 196 (w.); Bolton, 1988b: 138 (m.). Combination in *Atopula:* Emery, 1912b: 104; in *Leptothorax:* Forel, 1916: 459; in *Paratopula:* Wheeler, W.M. 1919e: 144. Senior synonym of *taylori:* Bolton, 1988b: 138.
***demeta.*** *Paratopula demeta* Bolton, 1988b: 139, fig. 3 (w.) BORNEO.
***longispina.*** *Atopula longispina* Stitz, 1938: 103, fig. 4 (q.) BORNEO. Combination in *Paratopula:* Bolton, 1976: 363.
***macta.*** *Paratopula macta* Bolton, 1988b: 140, fig. 2 (w.q.) BORNEO.
***oculata.*** *Cerapachys oculatus* Smith, F. 1857a: 74, pl. 1, figs. 20–24 (m.) BORNEO. Bolton, 1988b: 141 (q.). Combination in *Sima:* Emery, 1887b: 445 (footnote); in *Rhopalothrix:* Donisthorpe, 1932c: 450; in *Acanthomyrmex:* Brown, 1975: 60; in *Paratopula:* Bolton, 1988b: 141.
*srilankensis. Atopomyrmex srilankensis* Cheng, Ye & Yang, 1992: 244 (w.) CHINA. *Nomen nudum.* [The name appears to be a gross hypercorrective error for *Atopomyrmex ceylonicus* Emery, 1901f: 114; see above.]
***sumatrensis.*** *Atopula ceylonica* var. *sumatrensis* Forel, 1913k: 64 (q.) SUMATRA. Crawley, 1924: 399 (w.). Combination in *Paratopula:* Wheeler, W.M. 1919e: 144. Raised to species: Bolton, 1988b: 141.
*taylori. Leptothorax taylori* Forel, 1902c: 228 (w.q.) INDIA. Combination in *Atopula:* Forel, 1913f: 197; in *Leptothorax:* Forel, 1916: 459. Variety of *ceylonica:* Forel, 1913f: 197. Junior synonym of *ceylonica:* Bolton, 1988b: 138.

***PARATRECHINA*** (World tropics and subtropics)
***albipes.*** *Prenolepis albipes* Emery, 1899e: 497 (w.) CAMEROUN. Combination in *Pr. (Nylanderia)*: Emery, 1914f: 422; in *Paratrechina (Nylanderia)*: Emery, 1925b: 218.
***amblyops.*** *Prenolepis amblyops* Forel, 1892l: 240 (w.) MADAGASCAR. Forel, 1892k: 516 (q.); Emery, 1899f: 289 (m.). Combination in *Pr. (Nylanderia)*: Emery, 1914f: 422; in *Paratrechina (Nylanderia)*: Emery, 1925b: 218. Current subspecies: nominal plus *rubescens.*
#***amia.*** *Prenolepis (Nylanderia) bourbonica* var. *amia* Forel, 1913f: 197 (w.q.) TAIWAN. Combination in *Paratechina (Nylanderia)*: Emery, 1925b: 219.
***anthracina.*** *Prenolepis anthracina* Roger, 1863a: 161 (w.) CUBA. Santschi, 1930c: 81 (q.m.). Combination in *Pr. (Nylanderia)*: Wheeler, W.M. 1913b: 501; in *Paratrechina (Nylanderia)*: Emery, 1925b: 221; in *Nylanderia:* Kempf, 1972a: 166; in *Paratrechina:* Brandão, 1991: 366.
#***antillana.*** *Prenolepis guatemalensis* r. *antillana* Forel, 1893g: 340 (w.q.m.) ANTILLES IS. Combination in *Pr. (Nylanderia)*: Forel, 1912i: 66; in *Paratrechina (Nylanderia)*: Emery, 1925b: 223; in *Nylanderia:* Kempf, 1972a: 168; in *Paratrechina:* Brandão, 1991: 368. Variety of *vividula:* Forel, 1912i: 66.
***arenivaga.*** *Prenolepis arenivaga* Wheeler, W.M. 1905f: 391, fig. 3 (w.m.) U.S.A. Combination in *Pr. (Nylanderia)*: Forel, 1922: 98; in *Paratrechina (Nylanderia)*: Emery, 1925b: 221. Subspecies of *melanderi:* Creighton, 1950a: 408. Revived status as species: Trager, 1984b: 119.
***arlesi.*** *Paratrechina (Nylanderia) arlesi* Bernard, 1953b: 257, fig. 15 (w.) GUINEA.
***aseta.*** *Prenolepis aseta* Forel, 1902d: 292 (w.) INDIA. Combination in *Pr. (Nylanderia)*: Emery, 1914f: 422; in *Paratrechina (Nylanderia)*: Emery, 1925b: 219.
***assimilis.*** *Formica assimilis* Jerdon, 1851: 125 (w.) INDIA. Combination in *Prenolepis:* Dalla Torre, 1893: 177; in *Paratrechina (Nylanderia)*: Emery, 1925b: 219.
*atomus. Prenolepis minutula* r. *atomus* Forel, 1901b: 25 (w.) BISMARCK ARCHIPELAGO. Viehmeyer, 1916a: 148 (q.m.). Combination in *Pr. (Nylanderia)*: Viehmeyer, 1916a: 148; in *Paratrechina (Nylanderia)*: Emery, 1925b: 221. Raised to species: Donisthorpe, 1941a: 143; Donisthorpe, 1941c: 143. Junior synonym of *minutula:* Wilson & Taylor, 1967: 89.
#***australis.*** *Paratrechina (Nylanderia) vividula* var. *australis* Santschi, 1929d: 313 (w.) BRAZIL. Combination in *Nylanderia:* Kempf, 1972a: 168; in *Paratrechina:* Brandão, 1991: 368.
***austroccidua.*** *Paratrechina austroccidua* Trager, 1984b: 113, fig. 17, 20 (w.q.m.) U.S.A.
*bengalensis. Prenolepis bourbonica* r. *bengalensis* Forel, 1894c: 408, fig. 3 (w.q.m.) INDIA. Combination in *Pr. (Nylanderia)*: Forel, 1912a: 73; in *Paratrechina (Nylanderia)*: Emery, 1925b: 219; in *Nylanderia:* Wheeler, W.M. 1936f: 16. Raised to species: Bingham, 1903: 329. Variety of *bourbonica:* Mann, 1921: 476. Junior synonym of *bourbonica:* Wilson & Taylor, 1967: 88.
#***biolleyi.*** *Prenolepis (Nylanderia) fulva* subsp. *biolleyi* Forel, 1908b: 67 (w.m.) COSTA RICA. Forel, 1908c: 401. Combination in *Paratrechina (Nylanderia)*: Emery, 1925b: 222; in *Nylanderia:* Kempf, 1972a: 166; in *Paratrechina:* Brandão, 1991: 366.
***birmana.*** *Prenolepis birmana* Forel, 1902d: 291 (w.) BURMA. Combination in *Paratrechina (Nylanderia)*: Emery, 1925b: 219. Current subspecies: nominal plus *hodgsoni.*
#***bismarckensis.*** *Prenolepis obscura* r. *bismarckensis* Forel, 1901b: 26 (w.q.m.) BISMARCK ARCHIPELAGO. Combination in *Paratrechina (Nylanderia)*: Emery, 1925b: 221.
*borcardi. Prenolepis (Nylanderia) jaegerskioeldi* var. *borcardi* Santschi, 1908: 533, fig, 12 (w.q.m.) NORTH AFRICA. Junior synonym of *jaegerskioeldi:* Emery, 1910a: 130.
***bourbonica.*** *Prenolepis nodifera* r. *bourbonica* Forel, 1886f: 210 (w.q.m.) REUNION I. Combination in *Pr. (Nylanderia)*: Forel, 1912a: 73; in *Paratrechina (Nylanderia)*: Emery, 1925b: 219; in *Nylanderia:* Wheeler, W.M. 1936f: 16; in *Paratrechina:* Trager, 1984b: 147. Raised to species: Forel, 1891b: 82. Senior synonym of *bengalensis, hawaiensis* and *skottsbergi:* Wilson & Taylor, 1967: 88. Current subspecies: nominal plus *amia, farquharensis, ngasiyana.*
***brasiliensis.*** *Prenolepis brasiliensis* Mayr, 1862: 697 (q.) BRAZIL. Combination in *Paratrechina (Nylanderia)*: Emery, 1925b: 222; in *Nylanderia:* Kempf, 1972a: 166; in *Paratrechina:* Brandão, 1991: 366.
***braueri.*** *Prenolepis braueri* Mayr, 1868a: 49 (w.) AUSTRALIA. Combination in *Pr. (Nylanderia)*: Forel, 1908b: 64; in *Paratrechina (Nylanderia)*: Emery, 1925b: 221. Current subspecies: nominal plus *donisthorpei, glabrior.*
***bruesii.*** *Prenolepis bruesii* Wheeler, W.M. 1903b: 106, fig. 9 (w.q.m.) U.S.A. Combination in *Pr. (Nylanderia)*: Emery, 1914f: 422; in *Paratrechina (Nylanderia)*: Emery, 1925b: 222. See also: Trager, 1984b: 129.
#***bryanti.*** *Prenolepis (Nylanderia) butteli* r. *bryanti* Forel 1916: 439 (w.) BORNEO. Combination in *Paratrechina (Nylanderia)*: Emery, 1925b: 220. See also: Wheeler, W.M. 1919e: 104.

**burgesi.** *Paratrechina burgesi* Trager, 1984a: 483, figs. 1–4 (w.) PERU.

**butteli.** *Prenolepis (Nylanderia) butteli* Forel, 1913k: 103 (w.) SUMATRA. Combination in *Paratrechina (Nylanderia)*: Emery, 1925b: 220. Current subspecies: nominal plus *bryanti, malaccana.*

#**buxtoni.** *Paratrechina (Nylanderia) minutula* var. *buxtoni* Santschi, 1928e: 479 (w.) VANUATU.

**caeciliae.** *Prenolepis caeciliae* Forel, 1899c: 126 (w.) COSTA RICA. Combination in *Pr. (Nylanderia)*: Forel, 1908b: 65; in *Paratrechina (Nylanderia)*: Emery, 1925b: 222; in *Nylanderia:* Kempf, 1972a: 166; in *Paratrechina:* Brandão, 1991: 366. Current subspecies: nominal plus *elevata.*

**caledonica.** *Prenolepis caledonica* Forel, 1902h: 491 (w.) NEW CALEDONIA. Emery, 1914f: 422 (m.). Combination in *Pr. (Nylanderia)*: Emery, 1914f: 421; in *Paratrechina (Nylanderia)*: Emery, 1925b: 221.

#**celebensis.** *Paratrechina (Nylanderia) obscura* subsp. *celebensis* Karavaiev, 1933b: 109 (w.) SULAWESI.

**cisipa.** *Paratrechina (Nylanderia) cisipa* Smith, D.R. & Lavigne, 1973: 184, figs. 7–12 (w.q.m.) PUERTO RICO.

**clandestina.** *Prenolepis clandestina* Mayr, 1870b: 948 (diagnosis in key) (w.) JAVA. Viehmeyer, 1916a: 147 (q.m.). Combination in *Pr. (Nylanderia)*: Viehmeyer, 1916a: 147; in *Paratrechina (Nylanderia)*: Emery, 1925b: 220.

#**cocoensis.** *Prenolepis guatemalensis* var. *cocoensis* Forel, 1902b: 178 (w.) COCOS IS. Combination in *Paratrechina (Nylanderia)*: Emery, 1925b: 223; in *Nylanderia:* Kempf, 1972a: 167; in *Paratrechina:* Brandão, 1991: 367.

**colchica.** *Paratrechina colchica* Pisarski, 1960: 352, figs. 6–10, 17–19 (w.q.m.) GEORGIA.

**comorensis.** *Prenolepis comorensis* Forel, 1907g: 87 (w.) COMORO IS. Combination in *Pr. (Nylanderia)*: Emery, 1914f: 422; in *Paratrechina (Nylanderia)*: Emery, 1925b: 219.

**concinna.** *Paratrechina concinna* Trager, 1984b: 88, figs. 11, 14, 38 (w.q.m.) U.S.A.

**consuta.** *Nylanderia consuta* Wheeler, W.M. 1935c: 72. Replacement name for *Nylanderia dichroa* Wheeler, W.M. 1934a: 181. [Junior secondary homonym of *Paratrechina (Nylanderia) dichroa* Karavaiev, 1933b: 106.] Combination in *Paratrechina:* **new combination.**

*crassipilis. Paratrechina (Nylanderia) vaga* var. *crassipilis* Santschi, 1928c: 71 (w.) FIJI IS. Junior synonym of *vaga:* Wilson & Taylor, 1967: 90.

#**cubana.** *Paratrechina (Nylanderia) fulva* st. *cubana* Santschi, 1930c: 81 (w.) CUBA. Combination in *Nylanderia:* Kempf, 1972a: 167; in *Paratrechina:* Brandão, 1991: 366.

*currens. Paratrechina currens* Motschoulsky, 1863: 14 (w.) SRI LANKA. Junior synonym of *longicornis:* Emery, 1892b: 166.

**darlingtoni.** *Nylanderia darlingtoni* Wheeler, W.M. 1936b: 210 (w.) HAITI. Combination in *Paratrechina:* Brandão, 1991: 366.

#**dibullana.** *Prenolepis (Nylanderia) johannae* r. *dibullana* Forel, 1912i: 67 (w.) COLOMBIA. Combination in *Paratrechina (Nylanderia)*: Emery, 1925b: 222; in *Nylanderia:* Kempf, 1972a: 167; in *Paratrechina:* Brandão, 1991: 367.

**dichroa.** *Paratrechina (Nylanderia) dichroa* Karavaiev, 1933b: 106 (w.m.) INDONESIA (Prinsen I.).

*dichroa. Nylanderia dichroa* Wheeler, W.M. 1934a: 181 (w.) SOLOMON IS. [Junior secondary homonym of *dichroa* Karavaiev, above.] Replacement name: *consuta* Wheeler, 1935c: 72.

**dispar.** *Prenolepis (Nylanderia) dispar* Forel, 1909a: 251 (w.q.) GUATEMALA. Combination in *Paratrechina (Nylanderia)*: Emery, 1925b: 222; in *Nylanderia:* Kempf, 1972a: 166; in *Paratrechina:* Brandão, 1991: 366.

#**docilis.** *Prenolepis vividula* subsp. *docilis* Forel, 1908c: 402 (w.) BRAZIL. Combination in *Paratrechina (Nylanderia)*: Emery, 1925b: 223; in *Nylanderia:* Kempf, 1972a: 168; in *Paratrechina:* Brandão, 1991: 368.

#**donisthorpei.** *Prenolepis (Nylanderia) braueri* subsp. *donisthorpei* Forel, 1908b: 64 (w.) GREAT BRITAIN. Combination in *Paratrechina (Nylanderia)*: Emery, 1925b: 221.

**dugasi.** *Prenolepis (Nylanderia) dugasi* Forel, 1911i: 227 (footnote) (w.) VIETNAM. Combination in *Paratrechina (Nylanderia)*: Emery, 1925b: 220.

*edenensis. Prenolepis (Nylanderia) vividula* subsp. *guatemalensis* var. *edenensis* Wheeler, W.M. 1924a: 111 (w.q.) GALAPAGOS IS. Unavailable name.

#**elevata.** *Prenolepis (Nylanderia) caeciliae* subsp. *elevata* Forel, 1908b: 65 (w.m.) COSTA RICA. Combination in *Paratrechina (Nylanderia)*: Emery, 1925b: 222; in *Nylanderia:* Kempf, 1972a: 166; in *Paratrechina:* Brandão, 1991: 366.

#**ellisii.** *Prenolepis ellisii* Forel, 1891b: 87, pl. 2, fig. 10 (w.q.m.) MADAGASCAR. Combination in *Pr. (Nylanderia)*: Emery, 1914f: 422; in *Paratrechina (Nylanderia)*: Emery, 1925b: 219. Subspecies of *madagascarensis:* Wheeler, W.M. 1922a: 1038.

**emarginata.** *Prenolepis (Nylanderia) emarginata* Forel, 1913k: 104 (w.q.m.) SUMATRA. Combination in *Paratrechina (Nylanderia)*: Emery, 1925b: 220.

**faisonensis.** *Prenolepis (Nylanderia) arenivaga* var. *faisonensis* Forel, 1922: 98 (w.) U.S.A. Combination in *Paratrechina (Nylanderia)*: Emery, 1925b: 221. Junior synonym of *arenivaga:* Creighton, 1950a: 408. Revived from synonymy, and raised to species: Trager, 1984b: 93.

#**farquharensis.** *Prenolepis bourbonica* var. *farquharensis* Forel, 1907d: 92 (w.) MADAGASCAR (Farquhar I.). Combination in *Paratrechina (Nylanderia)*: Emery, 1925b: 219.

**flavipes.** *Tapinoma flavipes* Smith, F. 1874: 404 (w.q.) JAPAN. Forel, 1900e: 269 (m.). Combination in *Prenolepis:* Mayr, 1886c: 362; in *Pr. (Nylanderia)*: Emery, 1910a: 132; in *Paratrechina (Nylanderia)*: Emery, 1925b: 220. See also: Wheeler, W.M. 1906c: 320; Trager, 1984b: 97; Kupyanskaya, 1990: 233.

**foreli.** *Prenolepis (Nylanderia) foreli* Emery, 1914f: 422, pl. 13, fig. 15 (w.q.m.) NEW CALEDONIA. Combination in *Paratrechina (Nylanderia)*: Emery, 1925b: 221. Current subspecies: nominal plus *nigriventris.*

**formosae.** *Prenolepis (Nylanderia) formosae* Forel, 1912a: 73 (w.q.) TAIWAN. Combination in *Paratrechina (Nylanderia)*: Emery, 1925b: 220.

**fulva.** *Prenolepis fulva* Mayr, 1862: 698 (w.q.) BRAZIL. Forel, 1891b: 94 (m.); Forel, 1912i: 67 (m.). Combination in *Pr. (Nylanderia)*: Forel, 1908b: 67; in *Paratrechina (Nylanderia)*: Emery, 1925b: 222; in *Nylanderia:* Kempf, 1972a: 166; in *Paratrechina:* Snelling & Hunt, 1976: 122. Current subspecies: nominal plus *biolleyi, cubana, fumata, fumatipennis, incisa, longiscapa, nesiotis.*

#**fumata.** *Prenolepis fulva* var. *fumata* Forel, 1909a: 264 (w.) PARAGUAY. Forel, 1912i: 67 (q.). Combination in *Pr. (Nylanderia)*: Forel, 1913l: 246; in *Paratrechina (Nylanderia)*: Emery, 1925b: 222; in *Nylanderia:* Kempf, 1972a: 166; in *Paratrechina:* Brandão, 1991: 366.

#**fumatipennis.** *Prenolepis (Nylanderia) fulva* var. *fumatipennis* Forel, 1915c: 362 (w.q.m.) ARGENTINA. Combination in *Paratrechina (Nylanderia)*: Emery, 1925b: 222; in *Nylanderia:* Kempf, 1972a: 166; in *Paratrechina:* Brandão, 1991: 366.

#**fuscula.** *Paratrechina (Nylanderia) grisoni* var. *fuscula* Menozzi, 1942: 178 (w.) FERNANDO PO I.

**glabra.** *Prenolepis glabra* Forel, 1891b: 92 (w.) MADAGASCAR. Combination in *Pr. (Nylanderia)*: Emery, 1914f: 422; in *Paratrechina (Nylanderia)*: Emery, 1925b: 219.

#**glabrior.** *Prenolepis braueri* var. *glabrior* Forel, 1902h: 490 (w.q.m.) AUSTRALIA. Combination in *Paratrechina (Nylanderia)*: Emery, 1925b: 221.

**goeldii.** *Prenolepis (Nylanderia) goeldii* Forel, 1912i: 68 (w.) BRAZIL. Combination in *Paratrechina (Nylanderia)*: Emery, 1925b: 222; in *Nylanderia:* Kempf, 1972a: 167; in *Paratrechina:* Brandão, 1991: 367.

*gracilescens. Formica gracilescens* Nylander, 1856a: xxviii (w.) FRANCE. [Also described as new by Nylander, 1856b: 73.] Junior synonym of *longicornis:* Roger, 1863b: 10.

**gracilis.** *Prenolepis gracilis* Forel, 1892k: 517 (w.) MADAGASCAR. Combination in *Pr. (Nylanderia)*: Wheeler, W.M. 1922a: 1038; in *Paratrechina (Nylanderia)*: Emery, 1925b: 219.

**grisoni.** *Prenolepis (Nylanderia) grisoni* Forel, 1916: 440, fig. 5 (w.) ZAIRE. Combination in *Paratrechina (Nylanderia)*: Emery, 1925b: 218. Current subspecies: nominal plus *fuscula.*

**guatemalensis.** *Prenolepis vividula* var. *guatemalensis* Forel, 1885a: 348 (w.q.) GUATEMALA. Wheeler, W.M. 1905f: 392 (m.); Wheeler, G.C. & Wheeler, J. 1986d: 336 (l.). Combination in *Pr. (Nylanderia)*: Forel, 1912i: 66; in *Paratrechina (Nylanderia)*: Emery, 1925b: 223; in *Nylanderia:* Kempf, 1972a: 167; in *Paratrechina:* Trager, 1984b: 139. Subspecies/race of *vividula:* Forel, 1912i: 66; Wheeler, W.M. 1924a: 111. Raised to species: Forel, 1893g: 340; Forel, 1902b: 178; Smith, M.R. 1954c: 14; Trager, 1984b: 139. Current subspecies: nominal plus *cocoensis, itinerans.*

#**hagemanni.** *Prenolepis longicornis* var. *hagemanni* Forel, 1901h: 65 (w.) ZAIRE. Combination in *Paratrechina:* Emery, 1925b: 217. Junior synonym of *longicornis:* Wheeler, W.M. 1922a: 942. Revived from synonymy: Emery, 1925b: 217.

*hawaiensis. Prenolepis bourbonica* r. *hawaiensis* Forel, 1899a: 120 (w.q.m.) HAWAII. Combination in *Paratrechina (Nylanderia)*: Emery, 1925b: 219. Junior synonym of *bourbonica:* Wilson & Taylor, 1967: 88.

**#*hodgsoni.*** *Prenolepis birmana* var. *hodgsoni* Forel, 1902d: 291 (w.) BURMA. Combination in *Paratrechina* (*Nylanderia*): Emery, 1925b: 219.

***hubrechti.*** *Andragnathus hubrechti* Emery, 1922d: 111 (m.) JAVA. Combination in *Paratrechina:* Agosti & Bolton, 1990a: 76.

***humbloti.*** *Prenolepis braueri* r. *humbloti* Forel, 1891b: 90, pl. 2, fig. 11 (w.q.m.) MADAGASCAR. Combination in *Pr.* (*Nylanderia*): Wheeler, W.M. 1922a: 1038; in *Paratrechina* (*Nylanderia*): Emery, 1925b: 219. Raised to species: Emery, 1895f: 337.

***hystrix.*** *Paratrechina hystrix* Trager, 1984b: 134, figs. 28, 30, 40 (w.q.m.) U.S.A.

***incallida.*** *Prenolepis* (*Nylanderia*) *incallida* Santschi, 1915c: 263, fig. 12 (w.) SAO THOME I. Combination in *Paratrechina* (*Nylanderia*): Emery, 1925b: 218.

**#*incisa.*** *Prenolepis* (*Nylanderia*) *fulva* subsp. *incisa* Forel, 1908b: 68 (w.q.m.) COSTA RICA. Combination in *Paratrechina* (*Nylanderia*): Emery, 1925b: 222; in *Nylanderia:* Kempf, 1972a: 167; in *Paratrechina:* Brandão, 1991: 366.

***indica.*** *Prenolepis indica* Forel, 1894c: 409, fig. 4 (w.q.m.) INDIA. Imai, Baroni Urbani, *et al.* 1984: 9 (k.). Combination in *Paratrechina* (*Nylanderia*): Emery, 1925b: 220.

***iridescens.*** *Nylanderia iridescens* Donisthorpe, 1942b: 71 (w.) PHILIPPINES. Combination in *Paratrechina:* **new combination**.

*irritans.* *Paratrechina* (*Nylanderia*) *vaga* var. *irritans* Santschi, 1928a: 55, figs. 7, 8 (w.m.) SAMOA. Junior synonym of *vaga:* Wilson & Taylor, 1967: 90.

**#*itinerans.*** *Prenolepis guatemalensis* var. *itinerans* Forel, 1901h: 81 (w.) BRAZIL (intercepted in quarantine in Germany). Combination in *Paratrechina* (*Nylanderia*): Emery, 1925b: 223; in *Nylanderia:* Kempf, 1972a: 167; in *Paratrechina:* Brandão, 1991: 367.

***jaegerskioeldi.*** *Prenolepis jaegerskioeldi* Mayr, 1904b: 8 (w.) EGYPT. Emery, 1910a: 130 (q.m.). Combination in *Pr.* (*Nylanderia*): Santschi, 1914b: 128; in *Paratrechina* (*Nylanderia*): Emery, 1925b: 218. Senior synonym of *borcardi:* Emery, 1910a: 130.

***johannae.*** *Prenolepis* (*Nylanderia*) *johannae* Forel, 1912i: 66 (w.) COLOMBIA. Combination in *Paratrechina* (*Nylanderia*): Emery, 1925b: 222; in *Nylanderia:* Kempf, 1972a: 167; in *Paratrechina:* Brandão, 1991: 367. Current subspecies: nominal plus *dibullana*.

*kincaidi.* *Prenolepis kincaidi* Wheeler, W.M. 1906e: 350, fig. 1 (w.q.m.) BERMUDA. Combination in *Paratrechina* (*Nylanderia*): Emery, 1925b: 222; in *Nylanderia:* Kempf, 1972a: 167. Junior synonym of *vividula:* Trager, 1984b: 75.

***kohli.*** *Prenolepis kohli* Forel, 1916: 438 (w.q.) ZAIRE. Combination in *Paratrechina* (*Nylanderia*): Emery, 1925b: 218.

***koningsbergeri.*** *Paratrechina* (*Nylanderia*) *koningsbergeri* Karavaiev, 1933b: 108 (w.m.) JAVA.

***kraepelini.*** *Prenolepis kraepelini* Forel, 1905c: 24 (w.q.m.) JAVA. Combination in *Pr.* (*Nylanderia*): Forel, 1912a: 73; in *Paratrechina* (*Nylanderia*): Emery, 1925b: 220.

**#*kuenzleri.*** *Prenolepis vividula* var. *kuenzleri* Forel, 1909a: 268 (w.q.m.) ARGENTINA. Combination in *Pr.* (*Nylanderia*): Forel, 1913l: 246; in *Paratrechina* (*Nylanderia*): Emery, 1925b: 222; in *Nylanderia:* Kempf, 1972a: 168; in *Paratrechina:* Brandão, 1991: 367. Variety/subspecies of *silvestrii:* Forel, 1913l: 246; Emery, 1914f: 422 (footnote).

***lecamopteridis.*** *Paratrechina* (*Nylanderia*) *lecamopteridis* Donisthorpe, 1941c: 142 (w.) NEW GUINEA.

***lepida.*** *Prenolepis* (*Nylanderia*) *lepida* Santschi, 1915c: 262, fig. 11 (w.) CAMEROUN. Combination in *Paratrechina* (*Nylanderia*): Emery, 1925b: 218.

**#*levis.*** *Prenolepis* (*Nylanderia*) *taylori* r. *levis* Forel, 1913k: 104 (w.) SRI LANKA. Combination in *Paratrechina* (*Nylanderia*): Emery, 1925b: 220.

***lietzi.*** *Prenolepis* (*Nylanderia*) *lietzi* Forel, 1908b: 66 (w.q.) COSTA RICA. Combination in *Paratrechina* (*Nylanderia*): Emery, 1925b: 222; in *Nylanderia:* Kempf, 1972a: 167; in *Paratrechina:* Brandão, 1991: 367.

***longicornis.*** *Formica longicornis* Latreille, 1802c: 113 (w.) SENEGAL. Jerdon, 1851: 124 (q.); André, 1881b: 60 (m.); Hung, Imai & Kubota, 1972: 1024 (k.); Wheeler, G.C. & Wheeler, J. 1986d: 336 (l.). Combination in *Prenolepis:* Roger, 1863b: 10; in *Pr.* (*Nylanderia*): Emery, 1910a: 129; in *Paratrechina:* Wheeler, W.M. 1921e: 112. Senior synonym of *currens:* Emery, 1892b: 166; of *gracilescens:* Roger, 1863b: 10; of *vagans:* Dalla Torre, 1893: 179. Current subspecies: nominal plus *hagemanni.* See also: Mayr, 1865: 50; Forel, 1891b: 81; Forel, 1894c: 406; Emery, 1910a: 129; Trager, 1984b: 153.

**#*longiscapa.*** *Prenolepis* (*Nylanderia*) *fulva* subsp. *longiscapa* Forel, 1908b: 69 (w.) COSTA RICA. Combination in *Paratrechina* (*Nylanderia*): Emery, 1925b: 222; in *Nylanderia:* Kempf, 1972a: 167; in *Paratrechina:* Brandão, 1991: 367.

***madagascarensis.*** *Prenolepis vividula* var. *madagascarensis* Forel, 1886f: 212 (w.) MADAGASCAR. Forel, 1892l: 238 (q.m.). Combination in *Pr.* (*Nylanderia*): Emery, 1914f: 422; in *Paratrechina* (*Nylanderia*): Emery, 1925b: 219. Race of *vividula:* Forel, 1891b: 87. Raised to species: Forel, 1892l: 238; Wheeler, W.M. 1922a: 1038. Current subspecies: nominal plus *ellisii, rufescens, sechellensis.*

**#*malaccana.*** *Prenolepis* (*Nylanderia*) *butteli* subsp. *malaccana* Viehmeyer, 1916a: 147 (w.) WEST MALAYSIA. Combination in *Paratrechina* (*Nylanderia*): Emery, 1925b: 220.

***manni.*** *Paratrechina* (*Euprenolepis*) *manni* Donisthorpe, 1941b: 41 (w.) SOLOMON IS.

*melanderi.* *Prenolepis melanderi* Wheeler, W.M. 1903b: 104, fig. 8 (w.q.m.) U.S.A. Wheeler, G.C. & Wheeler, J. 1953c: 144 (l.). Combination in *Pr.* (*Nylanderia*): Emery, 1914f: 422; in *Paratrechina* (*Nylanderia*): Emery, 1925b: 223. Subspecies of *vividula:* Smith, M.R. 1951a: 849. Junior synonym of *terricola:* Trager, 1984b: 81.

***mendica.*** *Paratrechina* (*Nylanderia*) *mendica* Menozzi, 1942: 179, fig. 3 (w.q.) FERNANDO PO I.

**#*metallescens.*** *Prenolepis opaca* subsp. *metallescens* Emery, 1900d: 698 (w.) JAVA. Combination in *Paratrechina* (*Nylanderia*): Emery, 1925b: 220.

***mexicana.*** *Prenolepis mexicana* Forel, 1899c: 126 (w.m.) MEXICO. Wheeler, W.M. 1914b: 55 (q.). Combination in *Pr.* (*Nylanderia*): Forel, 1908b: 64; in *Paratrechina* (*Nylanderia*): Emery, 1925b: 222; in *Nylanderia:* Kempf, 1972a: 167; in *Paratrechina:* Brandão, 1991: 367.

***microps.*** *Prenolepis* (*Nylanderia*) *microps* Smith, M.R. 1937: 868 (w.) PUERTO RICO. Combination in *Nylanderia:* Kempf, 1972a: 167; in *Paratrechina:* Brandão, 1991: 367.

**#*minor.*** *Paratrechina* (*Nylanderia*) *obscura* subsp. *minor* Karavaiev, 1933b: 110, fig. 8 (w.m.) INDONESIA (Wammer I.).

**#*minuta.*** *Prenolepis steinheili* var. *minuta* Forel, 1893g: 343 (w.q.m.) ANTILLES IS. [Unresolved junior primary homonym of *Prenolepis imparis* var. *minuta* Emery, 1893i: 636.] Combination in *Pr.* (*Nylanderia*): Forel, 1912i: 66; in *Paratrechina* (*Nylanderia*): Emery, 1925b: 223; in *Nylanderia:* Kempf, 1972a: 168; in *Paratrechina:* Brandão, 1991: 367.

***minutula.*** *Prenolepis minutula* Forel, 1901b: 25 (footnote) (w.) AUSTRALIA. Wheeler, W.M. 1927i: 150 (q.m.). Combination in *Pr.* (*Nylanderia*): Emery, 1914f: 422; in *Paratrechina* (*Nylanderia*): Emery, 1925b: 221. Senior synonym of *atomus:* Wilson & Taylor, 1967: 89. Current subspecies: nominal plus *buxtoni.*

***mixta.*** *Prenolepis mixta* Forel, 1897c: 203, fig. 3 (w.m.) SEYCHELLE IS. Forel, 1912k: 165 (m.). Combination in *Pr.* (*Nylanderia*): Wheeler, W.M. 1922a: 1039; in *Paratrechina* (*Nylanderia*): Emery, 1925b: 219.

**#*mjobergi.*** *Prenolepis* (*Nylanderia*) *vividula* var. *mjobergi* Forel, 1908b: 64 (w.) SWEDEN. Combination in *Paratrechina* (*Nylanderia*): Emery, 1925b: 223.

***myops.*** *Prenolepis* (*Nylanderia*) *myops* Mann, 1920: 432 (w.q.m.) CUBA. Combination in *Paratrechina* (*Nylanderia*): Emery, 1925b: 222; in *Nylanderia:* Brown, 1955g: 135; in *Paratrechina:* Brandão, 1991: 367. Senior synonym of *troglodytes:* Brown, 1955g: 135.

***nana.*** *Paratrechina* (*Nylanderia*) *nana* Santschi, 1928e: 478 (w.m.) AUSTRALIA.

**#*natalensis.*** *Prenolepis* (*Nylanderia*) *traegaordhi* var. *natalensis* Forel, 1915c: 348 (w.) SOUTH AFRICA. Combination in *Paratrechina* (*Nylanderia*): Emery, 1925b: 218.

**#*nesiotis.*** *Prenolepis* (*Nylanderia*) *fulva* subsp. *nesiotis* Wheeler, W.M. 1919c: 277 (w.q.m.) GALAPAGOS IS. Combination in *Nylanderia:* Kempf, 1972a: 167; in *Paratrechina:* Brandão, 1991: 367.

***nettae.*** *Prenolepis* (*Nylanderia*) *nettae* Forel, 1911d: 381 (w.) "Pays des Moïs". Combination in *Paratrechina* (*Nylanderia*): Emery, 1925b: 220.

**#*ngasiyana.*** *Prenolepis bourbonica* subsp. *ngasiyana* Forel, 1907g: 87 (w.) COMORO IS. Combination in *Paratrechina* (*Nylanderia*): Emery, 1925b: 219.

**#*nigriventris.*** *Prenolepis* (*Nylanderia*) *foreli* var. *nigriventris* Emery, 1914f: 424 (w.) NEW CALEDONIA. Combination in *Paratrechina* (*Nylanderia*): Emery, 1925b: 221.

**#*nimba.*** *Paratrechina weissi* subsp. *nimba* Bernard, 1953b: 258, fig. 15 (w.q.) GUINEA.

***nodifera.*** *Prenolepis nodifera* Mayr, 1870a: 388, fig. 7 (w.) COLOMBIA. Pergande, 1896: 860 (q.m.). Combination in *Paratrechina* (*Nylanderia*): Emery, 1925b: 222; in *Nylanderia:* Kempf, 1972a: 167; in *Paratrechina:* Brandão, 1991: 367. Variety of *anthracina:* Pergande, 1896: 860. Revived status as species: Emery, 1925b: 222.

***nuggeti.*** *Paratrechina (Euprenolepis) nuggeti* Donisthorpe, 1941c: 143 (w.) NEW GUINEA. Donisthorpe, 1941h: 64 (m.).

***obscura.*** *Prenolepis obscura* Mayr, 1862: 698 (w.q.) AUSTRALIA. Forel, 1891b: 93 (m.). Combination in *Pr. (Nylanderia)*: Emery, 1914f: 422; in *Paratrechina (Nylanderia)*: Emery, 1925b: 221. Current subspecies: nominal plus *bismarckensis, celebensis, minor, papuana.*

***oceanica.*** *Prenolepis (Nylanderia) oceanica* Mann, 1921: 476 (w.) FIJI IS. Combination in *Paratrechina (Nylanderia)*: Emery, 1925b: 221.

***opaca.*** *Prenolepis clandestina* var. *opaca* Emery, 1887a: 243 (w.) JAVA. Combination in *Pr. (Nylanderia)*: Emery, 1914f: 422; in *Paratrechina (Nylanderia)*: Emery, 1925b: 220. Raised to species: Emery, 1900d: 698. Current subspecies: nominal plus *metallescens.*

***pallida.*** *Paratrechina (Paraparatrechina) pallida* Donisthorpe, 1947d: 192, fig. 2 (w.) NEW GUINEA.

***#papuana.*** *Prenolepis obscura* var. *papuana* Forel, 1901b: 26 (footnote) (w.q.m.) NEW GUINEA. Combination in *Paratrechina (Nylanderia)*: Emery, 1925b: 221.

***parvula.*** *Prenolepis parvula* Mayr, 1870b: 948 (w.q.m.) U.S.A. Wheeler, G.C. & Wheeler, J. 1968: 211 (l.). Combination in *Pr. (Nylanderia)*: Forel, 1922: 98; in *Paratrechina (Nylanderia)*: Emery, 1925b: 222. Variety/race of *vividula:* Forel, 1885a: 348; Mayr, 1886d: 431. Revived status as species: Emery, 1893i: 636. See also: Trager, 1984b: 104.

***pearsei.*** *Nylanderia pearsei* Wheeler, W.M. 1938: 254 (w.) MEXICO. Combination in *Paratrechina:* Brandão, 1991: 367.

*perminuta. Formica perminuta* Buckley, 1866: 162 (w.) U.S.A. Combination in *Brachymyrmex:* Emery, 1895c: 338; in *Prenolepis:* Mayr, 1876: 78; Wheeler, W.M. 1902f: 21. Unidentifiable taxon; *incertae sedis* in *Paratrechina.* [Trager, 1984b: 119 suggests possible synonymy with *arenivaga,* a junior name, but the original description of *perminuta* does not justify this.]

***phantasma.*** *Paratrechina phantasma* Trager, 1984b: 124, figs. 14, 16, 48 (w.q.m.) U.S.A.

*picea. Formica picea* Buckley, 1866: 163 (w.) U.S.A. Junior synonym of *vividula:* Mayr, 1876: 78; Mayr, 1886d: 431.

***picta.*** *Paratrechina (Nylanderia) picta* Wheeler, W.M. 1927h: 104, fig. 9 (w.m.) VIETNAM.

***pieli.*** *Paratrechina (Nylanderia) pieli* Santschi, 1928b: 37 (w.q.m.) CHINA.

***pubens.*** *Prenolepis fulva* r. *pubens* Forel, 1893g: 338 (w.q.m.) ANTILLES IS. Combination in *Paratrechina (Nylanderia)*: Emery, 1925b: 222; in *Nylanderia:* Kempf, 1972a: 167; in *Paratrechina:* Trager, 1984b: 143. Junior synonym of *fulva:* Creighton, 1950a: 406. Revived from synonymy, and raised to species: Trager, 1984b: 143.

***pusillima.*** *Paratrechina pusillima* Emery, 1922d: 110, fig. 3 (w.q.) NEW GUINEA.

***rosae.*** *Prenolepis rosae* Forel, 1902h: 489 (w.q.m.) AUSTRALIA. Combination in *Paratrechina (Nylanderia)*: Emery, 1925b: 221.

***#rubescens.*** *Prenolepis amblyops* r. *rubescens* Forel, 1892k: 516 (w.) MADAGASCAR. Combination in *Paratrechina (Nylanderia)*: Emery, 1925b: 218.

***#rufescens.*** *Prenolepis (Nylanderia) madagascarensis* var. *rufescens* Wheeler, W.M. 1922a: 1038 (w.) MADAGASCAR. [First available use of *Prenolepis ellisii* subsp. *madagascarensis* var. *rufescens* Forel, 1907g: 88; unavailable name.] Combination in *Paratrechina (Nylanderia)*: Emery, 1925b: 219.

***sakurae.*** *Prenolepis (Nylanderia) sakurae* Ito, 1914: 43, figs. 1–3 (w.) JAPAN. Combination in *Paratrechina (Nylanderia)*: Emery, 1925b: 220.

***sauteri.*** *Prenolepis (Nylanderia) minutula* r. *sauteri* Forel, 1913f: 198 (w.) TAIWAN. Combination in *Paratrechina (Nylanderia)*: Emery, 1925b: 221. Raised to species: Collingwood, 1976: 306.

***#sechellensis.*** *Prenolepis madagascarensis* var. *sechellensis* Emery, 1894a: 71, fig. (w.) SEYCHELLE IS. Combination in *Paratrechina (Nylanderia)*: Emery, 1925b 219.

***sharpii.*** *Prenolepis sharpii* Forel, 1899a: 121, fig. 1 (w.q.m.) HAWAII. Combination in *Paratrechina (Nylanderia)*: Emery, 1925b: 221.

***sikorae.*** *Prenolepis sikorae* Forel, 1892l: 238 (q.m.) MADAGASCAR. Combination in *Pr. (Nylanderia)*: Wheeler, W.M. 1922a: 1039; in *Paratrechina (Nylanderia)*: Emery, 1925b: 219.

***silvestrii.*** *Prenolepis silvestrii* Emery, 1906c: 186, figs. 45, 46 (w.q.m.) ARGENTINA. Combination in *Pr. (Nylanderia)*: Forel, 1913l: 246; in *Paratrechina (Nylanderia)*: Emery, 1925b: 222; in *Nylanderia:* Kempf, 1972a: 167; in *Paratrechina:* Brandão, 1991: 367. Current subspecies: nominal plus *kuenzleri.*

***simpliciuscula.*** *Prenopelis simpliciuscula* Emery, 1896f: 247, fig. (q.m.) JAVA. Combination in *Paratrechina (Nylanderia)*: Emery, 1925b: 220.

***sindbadi.*** *Paratrechina sindbadi* Pisarski, 1960: 349, figs. 1–3, 11–13 (q.m.) IRAQ.

*skottsbergi. Prenolepis (Nylanderia) bourbonica* subsp. *skottsbergi* Wheeler, W.M. 1922f: 318 (w.) EASTER I. Combination in *Paratrechina (Nylanderia)*: Emery, 1925b: 219. Junior synonym of *bourbonica:* Wilson & Taylor, 1967: 88.

***smythiesii.*** *Prenolepis smythiesii* Forel, 1894c: 410, fig. 5 (q.m.) INDIA. Combination in *Paratrechina (Nylanderia)*: Emery, 1925b: 220.

***staudingeri.*** *Prenolepis (Nylanderia) staudingeri* Forel, 1912i: 67 (w.q.) PERU. Combination in *Paratrechina (Nylanderia)*: Emery, 1925b: 222; in *Nylanderia:* Kempf, 1972a: 168; in *Paratrechina:* Brandão, 1991: 367.

***steinheili.*** *Prenolepis steinheili* Forel, 1893g: 342 (w.) ANTILLES IS. Forel, 1908b: 64 (q.); Forel, 1912i: 66 (m.). Combination in *Pr. (Nylanderia)*: Forel, 1912i: 66; in *Paratrechina (Nylanderia)*: Emery, 1925b: 223; in *Nylanderia:* Kempf, 1972a: 168; in *Paratrechina:* Brandão, 1991: 367. Current subspecies: nominal plus *minuta* (junior homonym).

***subtilis.*** *Prenolepis (Nylanderia) subtilis* Santschi, 1920g: 174 (w.) SENEGAL. Combination in *Paratrechina (Nylanderia)*: Emery, 1925b: 218. Current subspecies: nominal plus *termitophila.*

***tapinomoides.*** *Prenolepis tapinomoides* Forel, 1905c: 25 (w.) JAVA. Combination in *Paratrechina (Nylanderia)*: Emery, 1925b: 220.

***tasmaniensis.*** *Prenolepis (Nylanderia) tasmaniensis* Forel, 1913g: 190 (w.m.) TASMANIA. Combination in *Paratrechina (Nylanderia)*: Emery, 1925b: 221.

***taylori.*** *Prenolepis taylori* Forel, 1894c: 410, fig. 2 (w.q.m.) INDIA. Combination in *Pr. (Nylanderia)*: Forel, 1913d: 438; in *Paratrechina (Nylanderia)*: Emery, 1925b: 220. Current subspecies: nominal plus *levis.*

***teranishii.*** *Paratrechina (Nylanderia) teranishii* Santschi, 1937h: 386 (w.q.) JAPAN.

***#termitophila.*** *Prenolepis (Nylanderia) subtilis* st. *termitophila* Santschi, 1921c: 122 (w.) CAMEROUN. Combination in *Paratrechina (Nylanderia)*: Emery, 1925b: 218.

***terricola.*** *Formica terricola* Buckley, 1866: 168 (w.q.m.) U.S.A. Combination in *Prenolepis:* Mayr, 1876: 78; in *Paratrechina:* Emery, 1925b: 223. Junior synonym of *vividula:* Mayr, 1876: 78; Smith, M.R. 1951a: 849. Revived from synonymy, and senior synonym of *melanderi:* Trager, 1984b: 81.

***tjibodana.*** *Paratrechina (Nylanderia) tjibodana* Karavaiev, 1929c: 245, fig. 6 (w.q.) JAVA.

***tococae.*** *Paratrechina (Nylanderia) tococae* Wheeler, W.M. & Bequaert, 1929: 38 (w.m.) BRAZIL. Combination in *Nylanderia:* Kempf, 1972a: 168; in *Paratrechina:* Brandão, 1991: 368.

***traegaordhi.*** *Prenolepis traegaordhi* Forel, 1904h: 14 (w.m.) SUDAN. Combination in *Pr. (Nylanderia)*: Forel, 1915c: 348; in *Paratrechina (Nylanderia)*: Emery, 1925b: 218. Current subspecies: nominal plus *natalensis.*

*troglodytes. Paratrechina (Nylanderia) troglodytes* Weber, 1934a: 58, fig. 7 (w.) CUBA. Junior synonym of *myops:* Brown, 1955g: 135.

***vaga.*** *Prenolepis obscura* r. *vaga* Forel, 1901b: 26 (w.) BISMARCK ARCHIPELAGO. Donisthorpe, 1948g: 142 (m.). Combination in *Pr. (Nylanderia)*: Emery, 1914f: 422; in *Paratrechina (Nylanderia)*: Emery, 1925b: 221; in *Nylanderia:* Wheeler, W.M. 1936f: 17; in *Paratrechina:* **revived combination**. Raised to species: Emery, 1914f: 424. Senior synonym of *crassipilis* and *irritans:* Wilson & Taylor, 1967: 90.

***vagabunda.*** *Paratrechina vagabunda* Motschoulsky, 1863: 13, pl. 2, fig. 2 (w.) SRI LANKA.

*vagans. Formica vagans* Jerdon, 1851: 124 (w.q.) INDIA. [Unresolved junior primary homonym of *Formica vagans* Olivier, 1792: 501.] Junior synonym of *longicornis:* Dalla Torre, 1893: 179; Forel, 1894c: 408.

***vitiensis.*** *Prenolepis (Nylanderia) vitiensis* Mann, 1921: 474, fig. 28 (w.m.) FIJI IS. Combination in *Paratrechina (Nylanderia)*: Emery, 1925b: 221.

***vividula.*** *Formica vividula* Nylander, 1846a: 900, pl. 18, figs. 2, 10–14 (w.q.m.) FINLAND. Combination in *Prenolepis:* Mayr, 1861: 52; in *Pr. (Nylanderia)*: Emery, 1906b: 134; in *Paratrechina (Nylanderia)*: Emery, 1925b: 223; in *Nylanderia:* Kempf, 1972a: 168; in *Paratrechina:* Trager, 1984b: 131. Senior synonym of *picea:* Mayr, 1876: 78; Mayr, 1886d: 431; of *kincaidi* Trager, 1984b: 75. Current subspecies: nominal plus *antillana, australis, docilis, mjobergi.* See also: Emery, 1910a: 131.

***waelbroecki.*** *Prenolepis waelbroecki* Emery, 1899e: 496, fig. (q.m.) ZAIRE. Forel, 1911f: 280 (w.). Combination in *Pr.* (*Nylanderia*): Wheeler, W.M. 1922a: 943; in *Paratrechina* (*Nylanderia*): Emery, 1925b: 218.

***weissi.*** *Prenolepis* (*Nylanderia*) *weissi* Santschi, 1911g: 210 (w.) CONGO. Combination in *Paratrechina* (*Nylanderia*): Emery, 1925b: 218. Current subspecies: nominal plus *nimba*.

***wojciki.*** *Paratrechina wojciki* Trager, 1984b: 108, figs. 18, 21 (w.q.m.) U.S.A. Wheeler, G.C. & Wheeler, J. 1986d: 337 (l.).

***yerburyi.*** *Prenolepis yerburyi* Forel, 1894c: 409, fig. 1 (w.q.m.) SRI LANKA. Combination in *Pr.* (*Nylanderia*): Viehmeyer, 1916a: 147; in *Paratrechina* (*Nylanderia*): Wheeler, W.M. 1921e: 112.

***zelotypa.*** *Prenolepis* (*Nylanderia*) *jaegerskioeldi* var. *zelotypa* Santschi, 1915c: 264 (w.) KENYA. Combination in *Paratrechina* (*Nylanderia*): Emery, 1925b: 218. Raised to species: Santschi, 1919b: 238.

*PARECTATOMMA:* see under ***GNAMPTOGENYS.***

*PARHOLCOMYRMEX:* see under ***MONOMORIUM.***

*PEDETES:* see under ***ODONTOMACHUS.***

***PENTASTRUMA*** (Japan, Taiwan)

***canina.*** *Pentastruma canina* Brown & Boisvert, 1979: 203, figs. 2–4 (w.q.m.) JAPAN.

***sauteri.*** *Pentastruma sauteri* Forel, 1912a: 51 (w.) TAIWAN. See also: Brown & Boisvert, 1979: 203.

*PERGANDEA: nomen nudum.*

***PERISSOMYRMEX*** (Guatemala, Bhutan)

***monticola.*** *Perissomyrmex monticola* de Andrade, in Baroni Urbani & de Andrade, 1993: 90, figs. 1–3 (w.q.) BHUTAN.

***snyderi.*** *Perissomyrmex snyderi* Smith, M.R. 1947i: 282, figs. 1, 2 (w.) GUATEMALA. See also: Bolton, 1981b: 283.

***PERONOMYRMEX*** (Australia)

***overbecki.*** *Peronomyrmex overbecki* Viehmeyer, 1922: 213, fig. 2 (w.) AUSTRALIA. See also: Taylor, 1970b: 209.

***PETALOMYRMEX*** (West Africa)

***phylax.*** *Petalomyrmex phylax* Snelling, 1979b: 5, figs. 1–7, (w.q.m.) CAMEROUN. Wheeler, G.C. & Wheeler, J. 1980: 544 (l.). See also: McKey, 1984: 81.

***PETRAEOMYRMEX** (U.S.A.)

***minimus.** **Petraeomyrmex minimus* Carpenter, 1930: 55, pl. 4, fig. 2 (q.) U.S.A. (Miocene).

***PETROPONE** (Russia)

***petiolata.** **Petropone petiolata* Dlussky, 1975: 119, fig. 131 (w.) RUSSIA (Cretaceous).

***PHACOTA*** (Spain)

***sichelii.*** *Phacota sichelii* Roger, 1862b: 262, pl. 1, fig. 20 (w.; may be ergatoid q.) SPAIN. Combination in *Monomorium:* Ettershank, 1966: 92; in *Phacota:* Bolton, 1987: 281. See also: Emery, 1908f: 554; Bolton, 1987: 281.

***PHALACROMYRMEX*** (Brazil)

***fugax.*** *Phalacromyrmex fugax* Kempf, 1960b: 90, figs. 1–6 (w.) BRAZIL. See also: Bolton, 1984: 377.

*PHARAOPHANES: nomen nudum.*

***PHASMOMYRMEX*** (Afrotropical)

***aberrans.*** *Camponotus aberrans* Mayr, 1895: 152 (q.) GHANA. Santschi, 1915c: 264 (w.). Combination in *C.* (*Myrmorhachis*): Forel, 1914a: 274; Wheeler, W.M. 1922a: 707; in *C.* (*Myrmacantha*): Emery, 1920b: 246; in *Phasmomyrmex* (*Myrmacantha*): Emery, 1925b: 58.

***buchneri.*** *Camponotus buchneri* Forel, 1886f: 183 (w.) ANGOLA. Combination in *C.* (*Phasmomyrmex*): Forel, 1912i: 90; in *Phasmomyrmex:* Emery, 1920b: 252. Senior synonym of *sericeus:* Forel, 1910e: 421. Current subspecies: nominal plus *griseus*.

**#*cupreus.*** *Phasmomyrmex paradoxus* var. *cupreus* Santschi, 1923e: 293 (w.) CONGO.

**#*griseus.*** *Phasmomyrmex buchneri* var. *griseus* Santschi, 1937b: 103 (w.) CAMEROUN.

***paradoxus.*** *Polyrhachis paradoxa* André, 1892a: 46 (w.) GABON. Combination in *Phasmomyrmex* (*Myrmorhachis*): Emery, 1925b: 58. Senior synonym of *polyrhachioides:* Emery, 1900b: 15; Emery, 1925b: 58. Current subspecies: nominal plus *cupreus*.

*polyrhachioides. Camponotus polyrhachioides* Emery, 1898a: 227. [Unnecessary replacement name for *paradoxus*, above.] Emery, 1920c: 30 (q.). Combination in *C.* (*Myrmacantha*): Emery, 1920b: 258; in *C.* (*Myrmorhachis*): Forel, 1912i: 92; Wheeler, W.M. 1922a: 707. Junior synonym of *paradoxus:* Emery, 1900b: 15; Emery, 1925b: 58.

*sericeus. Phasmomyrmex sericeus* Stitz, 1910: 146, fig. 11 (w.) CAMEROUN. Junior synonym of *buchneri:* Forel, 1910e: 421.

***wolfi.*** *Camponotus* (*Myrmacantha*) *wolfi* Emery, 1920c: 30 (q.) CAMEROUN. Combination in *Phasmomyrmex* (*Myrmacantha*): Emery, 1925b: 58.

***PHAULOMYRMA*** (Java)

***javana.*** *Phaulomyrma javana* Wheeler, G.C. & Wheeler, E.W. 1930: 193, figs. 1, 2 (m.) JAVA. See also: Petersen, 1968: 593.

*PHEIDOLACANTHINUS:* see under ***PHEIDOLE.***

***PHEIDOLE*** (World wide)

***aana.*** *Pheidole aana* Wilson & Taylor, 1967: 42, fig. 31 (s.w.) SAMOA.

***aberrans.*** *Pheidole aberrans* Mayr, 1868b: 172 (s.w.) ARGENTINA. Emery, 1906c: 142 (q.); Forel, 1913l: 232 (m.). Combination in *P.* (*Elasmopheidole*): Forel, 1913k: 43. Current subspecies: nominal plus *diversiceps, fartilia, minensis, mutica, obscurifrons, weiseri*. See also: Bruch, 1914: 219; Bruch, 1916: 312.

***absurda.*** *Pheidole absurda* Forel, 1886b: xlvii (s.) GUATEMALA. Emery, 1890b: 48 (w.). See also: Emery, 1894k: 54.

**#*abyssinica.*** *Pheidole caffra* subsp. *abyssinica* Forel, 1910c: 254 (s.w.) ETHIOPIA.

***accinota.*** *Pheidole accinota* Wheeler, W.M. 1925a: 16 (s.w.) PERU.

**#*achantella.*** *Pheidole pulchella* var. *achantella* Santschi, 1939b: 242 (s.) CONGO.

*acolhua. Pheidole* (*Cardiopheidole*) *vasliti* var. *acolhua* Wheeler, W.M. 1914b: 48 (s.w.q.m.) MEXICO. Junior synonym of *hirtula:* Creighton, 1958: 211.

**#*acuta.*** *Pheidole radowszkowskii* var. *acuta* Emery, 1894c: 154 (s.w.) BOLIVIA. Subspecies of *radoszkowskii:* Forel, 1909a: 258; Wheeler, W.M. 1925a: 21.

***acutidens.*** *Bruchomyrma acutidens* Santschi, 1922d: 249, figs. A-D (q.) ARGENTINA. Bruch, 1931: 31 (m.). Combination in *Pheidole:* Wilson, 1984: 327. See also: Wheeler, W.M. 1937c: 52.

**#*adami.*** *Pheidole occipitalis* var. *adami* Santschi, 1939f: 161 (w.) GUINEA.

*adamsoni. Pheidole* (*Pheidolacanthinus*) *sexspinosa* subsp. *adamsoni* Wheeler, W.M. 1932c: 157 (s.w.q.m.) MARQUESAS IS. Junior synonym of *sexspinosa:* Wilson & Taylor, 1967: 52.

***adrianoi.*** *Pheidole adrianoi* Naves, 1985: 56, figs. 25, 26, 33, 53, 54 (s.w.q.) U.S.A.

***aeberlii.*** *Pheidole aeberlii* Forel, 1894b: 91 (s.w.m.) SENEGAL. Current subspecies: nominal plus *erythraea*.

**#*aechmeae.*** *Pheidole floridana* subsp. *aechmeae* Wheeler, W.M. 1934g: 166 (s.w.) MEXICO.

***aequiseta.*** *Pheidole aequiseta* Santschi, 1923d: 57 (w.) BOLIVIA.

***agilis.*** *Myrmica agilis* Smith, F. 1857a: 71 (w.) WEST MALAYSIA. Combination in *Pheidole:* Donisthorpe, 1932c: 449.

***aglae.*** *Pheidole aglae* Forel, 1913k: 32 (s.w.q.) JAVA.

***ajax.*** *Pheidole ajax* Forel, 1899c: 71 (s.w.) COSTA RICA.

***akermani.*** *Pheidole akermani* Arnold, 1920a: 471 (s.w.) SOUTH AFRICA.

**#*alacris.*** *Pheidole flavens* st. *alacris* Santschi, 1923d: 61 (s.w.) ARGENTINA.

***albidula.*** *Pheidole albidula* Santschi, 1928f: 205 (s.w.) ZAIRE.

***alfaroi.*** *Pheidole alfaroi* Emery, 1896g: 70 (s.w.q.m.) COSTA RICA.

*alicensis. Pheidole variabilis* r. *ocior* var. *alicensis* Forel, 1915b: 59 (s.w.) AUSTRALIA. Unavailable name (Taylor, 1986: 34.).

**#*aliena.*** *Pheidole variabilis* subsp. *aliena* Emery, 1914f: 401 (s.) NEW CALEDONIA.

**#*alienata.*** *Pheidole bergi* subsp. *alienata* Borgmeier, 1929: 205 (s.w.) BRAZIL.

***allani.*** *Pheidole allani* Bingham, 1903: 264 (s.w.) BURMA.

**#*alpinensis.*** *Pheidole emeryi* var. *alpinensis* Forel, 1912f: 231 (s.w.q.) BRAZIL. Subspecies of *tristis:* current status.

**#*alsia.*** *Pheidole emeryi* var. *alsia* Santschi, 1923d: 58 (s.w.) BRAZIL. Subspecies of *tristis:* current status.

*amalricae. Pheidole aurivillii* r. *kasaiensis* var. *amalricae* Forel, 1915c: 335 (s.w.) ZAIRE. Unavailable name.

***amata.*** *Pheidole amata* Forel, 1901e: 360 (s.w.) COLOMBIA.

***amber.*** *Pheidole* (*Pheidolacanthinus*) *amber* Donisthorpe, 1941h: 55 (w.) NEW GUINEA.

***ambonensis.*** *Pheidole* (*Elasmopheidole*) *ambonensis* Karavaiev, 1930a: 209 (w.q.) INDONESIA (Ambon I.).

**#*ambulans.*** *Pheidole triconstricta* var. *ambulans* Emery, 1906c: 153 (s.w.) ARGENTINA. Forel, 1913l: 230 (m.); Santschi, 1919f: 40 (q.).

***amia.*** *Pheidole amia* Forel, 1912a: 60 (s.w.) TAIWAN.

**#*amoena.*** *Pheidole caffra* var. *amoena* Forel, 1911d: 365 (s.) ETHIOPIA.

**ampla.** *Pheidole variabilis* r. *ampla* Forel, 1893d: 462 (s.) AUSTRALIA. Forel, 1901e: 377 (w.q.). Raised to species: Forel, 1901e: 377. Current subspecies: nominal plus *mackayensis, parviceps, perthensis.*

**amplificata.** *Pheidole amplificata* Viehmeyer, 1914b: 517, fig. 1 (s.w.) NEW GUINEA.

**anastasii.** *Pheidole anastasii* Emery, 1896g: 76 (s.w.) COSTA RICA. Forel, 1901h: 78 (q.). Current subspecies: nominal plus *cellarum, johnsoni, sospes, venezuelana.*

**andrieui.** *Pheidole andrieui* Santschi, 1930a: 61, fig. 21 (s.w.q.m.) SUDAN.

**androsana.** *Pheidole androsana* Wheeler, W.M. 1905b: 90, fig. F (s.w.) BAHAMAS. Current subspecies: nominal plus *bakeri.*

*anfracta. Pheidole pilifera* subsp. *anfracta* Cole, 1952c: 278 (s.w.) U.S.A. Junior synonym of *senex:* Cole, 1953e: 298; Gregg, 1959: 30.

*angulata. Pheidole megacephala* r. *punctulata* var. *angulata* Stitz, 1911b: 385 (s.w.) TANZANIA. Unavailable name.

**angusta.** *Pheidole angusta* Forel, 1908c: 373 (s.w.) BRAZIL. Current subspecies: nominal plus *idulis.*

*angustella. Pheidole biconstricta* r. *hybrida* var. *angustella* Forel, 1912f: 222 (s.w.m.) BRAZIL. Unavailable name.

**#angustinoda.** *Pheidole veteratrix* var. *angustinoda* Forel, 1892k: 526 (s.w.) MADAGASCAR.

**#angustior.** *Pheidole latinoda* r. *angustior* Forel, 1902c: 170 (s.), 189 (w.) INDIA. [Also described as new by Forel, 1902f: 540. Misspelled as *angusta* by Forel, 1903c: 253.] Raised to species: Bingham, 1903: 236. Subspecies of *latinoda:* Emery, 1921f: 92.

**#annectens.** *Pheidole punctatissima* subsp. *annectens* Wheeler, W.M. 1905b: 93 (s.w.q.) BAHAMAS.

**annemariae.** *Pheidole annemariae* Forel, 1918b: 152 (s.w.q.) MADAGASCAR.

**anthracina.** *Pheidole anthracina* Forel, 1902h: 419 (s.w.q.) AUSTRALIA. Current subspecies: nominal plus *grandii, orba.*

**#antillana.** *Pheidole guilelmimuelleri* r. *antillana* Forel, 1893g: 401 (s.w.q.m.) ANTILLES IS.

**#antillensis.** *Pheidole jelskii* var. *antillensis* Forel, 1901e: 356 (s.) ANTILLES IS. Subspecies of *fallax:* current status.

**#antoniensis.** *Pheidole floridana* var. *antoniensis* Forel, 1901e: 364 (s.w.) COLOMBIA.

**aper.** *Pheidole aper* Forel, 1912f: 227 (s.w.q.) BRAZIL. Combination in *P.* (*Trachypheidole*): Emery, 1915i: 190.

**#apterostigmoides.** *Pheidole opaca* subsp. *apterostigmoides* Weber, 1943b: 71, fig. 3 (s.w.) TRINIDAD. Subspecies of *cephalica:* Kempf, 1972a: 188.

**#arabs.** *Pheidole rugaticeps* var. *arabs* Emery, 1881b: 535 (s.w.q.) YEMEN. Stirps of *rugaticeps:* Santschi, 1914d: 336.

**arcifera.** *Pheidole arcifera* Santschi, 1925d: 226 (s.w.) BRAZIL. See also: Kempf, 1964e: 61.

**arciruga.** *Pheidole arciruga* Forel, 1908c: 381 (w.) BRAZIL. Combination in *P.* (*Elasmopheidole*): Bruch, 1914: 219.

*arenarum. Pheidole pallidula* var. *arenarum* Ruzsky, 1905b: 647 (s.w.m.) TURKESTAN. Karavaiev, 1926b: 106 (q.). Subspecies of *pallidula:* Emery, 1915j: 226; Emery, 1916b: 149. Raised to species: Collingwood, 1961a: 64; Collingwood, 1961b: 289. Junior synonym of *pallidula:* Atanassov & Dlussky, 1992: 123.

**#arenicola.** *Pheidole jelskii* var. *arenicola* Emery, 1894g: 390 (footnote) (s.w.) BRAZIL. Subspecies of *fallax:* Forel, 1908c: 370; Santschi, 1923d: 58.

*arenicola. Pheidole arenicola* Forel, 1910f: 11 (s.w.q.m.) BOTSWANA. [Junior primary homonym of *arenicola* Emery, above.] Replacement name: *areniphila* Forel, 1910a: 30 (footnote).

**areniphila.** *Pheidole areniphila* Forel, 1910a: 30 (footnote). Replacement name for *arenicola* Forel, 1910f: 11. [Junior primary homonym of *arenicola* Emery, 1894g: 390.] Current subspecies: nominal plus *aurora.*

**#areolata.** *Pheidole sculpturata* var. *areolata* Forel, 1911e: 269 (s.w.) SOUTH AFRICA.

**#ares.** *Pheidole floridana* subsp. *ares* Forel, 1908b: 57 (s.w.m.) COSTA RICA.

**argentina.** *Gallardomyrma argentina* Bruch, 1932: 273, figs. 1–8 (q.) ARGENTINA. Combination in *Pheidole:* Wilson, 1984: 327.

**arhuaca.** *Pheidole arhuaca* Forel, 1901e: 364 (s.w.q.) COLOMBIA. Current subspecies: nominal plus *bimons.*

**aristotelis.** *Pheidole aristotelis* Forel, 1911a: 43 (s.w.m.) BORNEO.

*arizonica. Pheidole arizonica* Santschi, 1911d: 3, fig. (s.) U.S.A. Subspecies of *vasliti:* Creighton, 1950a: 192. Junior synonym of *subdentata:* Creighton, 1958: 211; Gregg, 1959: 31.

*armatus. Pheidolacanthinus armatus* Smith, F. 1865: 75, pl. 4, fig. 8 (s.) NEW GUINEA. Junior synonym of *quadrispinosa:* Donisthorpe, 1938a: 144.

**arnoldi.** *Pheidole arnoldi* Forel, 1913a: 130 (s.w.) ZIMBABWE. Arnold, 1920a: 447 (q.m.). Current subspecies: nominal plus *ballaensis, rufescens* (junior homonym).

**#artemisia.** *Pheidole pilifera* subsp. *artemisia* Cole, 1933: 616 (s.w.) U.S.A. Cole, 1938c: 372 (q.).

**#ascara.** *Pheidole speculifera* var. *ascarus* Emery, 1901e: 62 (s.) ETHIOPIA. Santschi, 1930d: 268 (w.m.).

**#aspatha.** *Pheidole spathifera* var. *aspatha* Forel, 1902c: 169 (s.), 188 (w.) INDIA. [Also described as new by Forel, 1902f: 540.]

**aspera.** *Pheidole aspera* Mayr, 1862: 746 (s.w.) SOUTH AFRICA. See also: Mayr, 1865: 97.

**asperata.** *Pheidole capellinii* var. *asperata* Emery, 1895k: 469 (s.w.) BURMA. Raised to species: Emery, 1900d: 682 (in text); Emery, 1921f: 90. See also: Bingham, 1903: 246.

**#asperithorax.** *Pheidole minutula* var. *asperithorax* Emery, 1894c: 160 (s.w.) BOLIVIA. Subspecies of *flavens:* Emery, 1896g: 77.

**athertonensis.** *Pheidole athertonensis* Forel, 1915b: 62 (s.w.q.) AUSTRALIA. Current subspecies: nominal plus *cedarensis, tambourinensis.*

**#atricolor.** *Pheidole susannae* r. *atricolor* Forel, 1901e: 356 (s.w.q.m.) COLOMBIA.

*atrocior. Pheidole megacephala* subsp. *impressifrons* var. *atrocior* Emery, 1915j: 243, fig. 7 (s.w.) SIERRA LEONE (attributed to Santschi). Unavailable name.

*atrophispina. Pheidole variabilis* r. *rugocciput* var. *atrophispina* Forel, 1915b: 58 (s.w.q.) AUSTRALIA. Unavailable name (Taylor, 1986: 34; misspelled as *athrospina*).

**#atrox.** *Pheidole punctulata* r. *atrox* Forel, 1913b: 328 (s.w.q.) ZAIRE. Variety/stirps of *megacephala:* Santschi, 1930b: 67; of *punctulata:* Santschi, 1935a: 256. Senior synonym of *inquilina:* Arnold, 1920a: 425 (in text).

**#attenuata.** *Pheidole aurivillii* var. *attenuata* Santschi, 1910c: 370 (s.w.) CONGO.

**#atticola.** *Pheidole nana* r. *atticola* Forel, 1912f: 236 (s.w.) ARGENTINA.

**attila.** *Pheidole attila* Forel, 1913k: 34, figs. G, H (s.w.) SUMATRA.

**atua.** *Pheidole atua* Wilson & Taylor, 1967: 43, fig. 32 (s.w.) SAMOA.

**aurivillii.** *Pheidole aurivillii* Mayr, 1896: 238 (s.w.) CAMEROUN. Santschi, 1910c: 369 (m.). Current subspecies: nominal plus *attenuata, kasaiensis, rubricalva.*

**auropilosa.** *Pheidole auropilosa* Mayr, 1887: 596 (s.), 606 (w.), 608 (q.m.) BRAZIL.

**#aurora.** *Pheidole areniphila* var. *aurora* Santschi, 1928f: 204 (s.) TANZANIA.

**#australis.** *Pheidole radowszkowskii* r. *australis* Emery, 1890b: 50 (footnote) (s.w.) BRAZIL. Raised to species: Dalla Torre, 1893: 88. Subspecies of *radoszkowskii:* Emery, 1922e: 99.

**#avia.** *Pheidole guilelmimuelleri* subsp. *avia* Forel, 1908c: 380 (w.) BRAZIL. Forel, 1911c: 303 (s.).

**#azumai.** *Pheidole nodus* st. *azumai* Santschi, 1941: 274, fig. 1 (s.w.) JAPAN.

**bahai.** *Tetramorium* (*Cephalomorium*) *bahai* Forel, 1922: 91 (w.) U.S.A. Combination in *Pheidole* (*Hendecapheidole*): Santschi, 1925d: 228.

**#bakeri.** *Pheidole bakeri* Forel, 1912b: 82 (s.w.) CUBA. Subspecies of *androsana:* Wheeler, W.M. 1913b: 491.

**#ballaensis.** *Pheidole arnoldi* var. *ballaensis* Arnold, 1920a: 448 (s.w.) ZIMBABWE.

**balzani.** *Pheidole balzani* Emery, 1894c: 161 (s.w.) BOLIVIA.

**bambusarum.** *Pheidole bambusarum* Forel, 1908c: 376 (s.w.q.) BRAZIL. Senior synonym of *ogloblini:* Kempf, 1972b: 450.

**barbata.** *Pheidole barbata* Wheeler, W.M. 1908e: 448 (s.w.) U.S.A.

*barbouri. Pheidole punctatissima* subsp. *jamaicensis* var. *barbouri* Wheeler, W.M. 1913b: 492 (s.w.) CUBA. Unavailable name.

**barreleti.** *Pheidole barreleti* Forel, 1903c: 252 (s.w.) SRI LANKA.

**barumtaun.** *Pheidole* (*Pheidolacanthinus*) *barumtaun* Donisthorpe, 1938a: 141, figs. 1, 2 (s.w.m.) NEW GUINEA.

*batrachorum. Pheidole batrachorum* Wheeler, W.M. 1922a: 128 (s.w.) ZAIRE. Junior synonym of *dea:* Santschi, 1930a: 59.

**#baucis.** *Pheidole bos* subsp. *baucis* Forel, 1910b: 37 (s.w.q.) AUSTRALIA.

**#bayeri.** *Pheidole caffra* r. *bayeri* Forel, 1916: 413 (s.w.) ZAIRE.

**beauforti.** *Pheidole beauforti* Emery, 1911b: 535 (s.) NEW GUINEA.

**belli.** *Pheidole* (*Pheidolacanthinus*) *belli* Mann, 1919: 306, fig. 14 (w.) SOLOMON IS.

*PHEIDOLE (continued)*

#*bengalensis*. *Pheidole* (*Ceratopheidole*) *smythiesii* var. *bengalensis* Forel, 1902c: 186 (w.) INDIA. Combination in *Ceratopheidole:* Emery, 1922e: 113; in *Pheidole:* **revived combination**.

*bequaerti*. *Pheidole bequaerti* Forel, 1913b: 330 (s.w.) ZAIRE.

*bergi*. *Pheidole bergi* Mayr, 1887: 593 (s.), 605 (w.) BRAZIL. Current subspecies: nominal plus *alienata, goetschi, pulliventris, subparallela.*

#*bernhardae*. *Pheidole picata* var. *bernhardae* Emery, 1915j: 245 (s.w.) MADAGASCAR. [First available use of *Pheidole punctulata* r. *spinosa* var. *bernhardae* Forel, 1905b: 164; unavailable name.]

#*berthoudi*. *Pheidole sculpturata* r. *berthoudi* Forel, 1894b: 89 (s.w.) SOUTH AFRICA. Raised to species: Santschi, 1921c: 116. Subspecies of *sculpturata:* Emery, 1921f: 89; Wheeler, W.M. 1922a: 819.

*bessonii*. *Pheidole oswaldi* r. *bessonii* Forel, 1891b: 176, pl. 5, fig. 3 (w.) MADAGASCAR. Forel, 1905b: 162 (s.). Raised to species: Dalla Torre, 1893: 88; Forel, 1905b: 162.

*bhavanae*. *Pheidole bhavanae* Bingham, 1903: 228 (s.w.) INDIA. Combination in *Ceratopheidole:* Emery, 1922e: 113; in *Pheidole:* **revived combination**.

*bicarinata*. *Pheidole bicarinata* Mayr, 1870b: 989 (s.) U.S.A. Senior synonym of *hayesi:* Creighton, 1950a: 171. Current subspecies: nominal plus *vinelandica.*

#*bicolor*. *Pheidole biconstricta* r. *bicolor* Emery, 1890b: 50 (s.w.) COSTA RICA. [Also described as new by Emery, 1894k: 55.] Raised to species: Dalla Torre, 1893: 88. Subspecies/race of *biconstricta:* Forel, 1899c: 64; Emery, 1922e: 98.

*biconstricta*. *Pheidole biconstricta* Mayr, 1870a: 399 (s.w.) COLOMBIA. Current subspecies: nominal plus *bicolor, burtoni, hybrida, lallemandi, rubicunda, simplex, socrates, surda.*

*bicornis*. *Pheidole bicornis* Forel, 1899c: 74, pl. 3, fig. 24 (s.) PANAMA. Combination in *P.* (*Trachypheidole*): Emery, 1915i: 190.

*bifurca*. *Pheidole* (*Pheidolacanthinus*) *bifurcus* Donisthorpe, 1941b: 36 (s.w.) NEW GUINEA.

*bilimeki*. *Pheidole bilimeki* Mayr, 1870b: 985 (s.q.) MEXICO.

*biloba*. *Conothorax bilobum* Karavaiev, 1935a: 75, fig. 8 (w.) CAMBODIA. Combination in *Conothoracoides:* Strand, 1935: 176; in *Pheidole:* Smith, D.R. 1979: 1366.

#*bimons*. *Pheidole arhuaca* r. *bimons* Forel, 1912f: 234 (s.w.) BRAZIL.

*binghamii*. *Pheidole binghamii* Forel, 1902c: 184 (s.), 198 (w.) INDIA. [Also described as new by Forel, 1902f: 546.] Imai, Kubota, *et al.* 1985: 47 (k.).

*biolleyi*. *Pheidole biolleyi* Forel, 1908b: 48 (s.w.q.m.) COSTA RICA. Current subspecies: nominal plus *tristani.*

#*biroi*. *Pheidole sexspinosa* subsp. *biroi* Emery, 1900c: 323 (s.w.m.) NEW GUINEA. Combination in *P.* (*Pheidolacanthinus*): Emery, 1921f: 82.

#*bispecula*. *Pheidole speculifera* var. *bispecula* Santschi, 1930a: 58 (w.) ZAIRE.

*blumenauensis*. *Pheidole blumenauensis* Kempf, 1964e: 64. Replacement name for *Pheidole diligens* Santschi, 1923d: 50. [Junior secondary homonym of *Atta diligens* Smith, F. 1858b: 168.]

*bluntschlii*. *Pheidole* (*Ceratopheidole*) *bluntschlii* Forel, 1911d: 373 (w.) SUMATRA. Combination in *Ceratopheidole:* Emery, 1922e: 113; in *Pheidole:* **revived combination**.

#*bombalensis*. *Pheidole proxima* subsp. *bombalensis* Forel, 1910b: 43 (s.w.) AUSTRALIA. Raised to species: Clark, 1929: 121. Subspecies of *proxima:* Taylor & Brown, D.R. 1985: 77.

*boraborensis*. *Pheidole oceanica* var. *boraborensis* Wheeler, W.M. 1932d: 15 (s.w.) SOCIETY IS. Junior synonym of *oceanica:* Wilson & Taylor, 1967: 48.

*borgmeieri*. *Pheidole borgmeieri* Kempf, 1972b: 461, figs. 11–14 (s.w.) BRAZIL.

*borinquenensis*. *Pheidole subarmata* var. *borinquenensis* Wheeler, W.M. 1908a: 133, pl. 12, figs. 19–21 (s.w.) PUERTO RICO. Junior synonym of *subarmata:* Brown, 1981: 526.

*bos*. *Pheidole bos* Forel, 1893d: 463 (s.) AUSTRALIA. Forel, 1902h: 432 (w.q.). Current subspecies: *baucis, eubos.*

#*bothae*. *Pheidole foreli* r. *bothae* Forel, 1901h: 62 (s.w.) SOUTH AFRICA. Santschi, 1914e: 22 (q.). Race of *tenuinodis:* Santschi, 1914e: 22. Junior synonym of *tenuinodis:* Emery, 1921f: 87. Revived from synonymy as subspecies of *tenuinodis:* Wheeler, W.M. 1922a: 822.

*braueri*. *Pheidole braueri* Forel, 1897c: 204 (s.w.q.) SEYCHELLE IS.

*brevicona*. *Pheidole breviconus* Mayr, 1887: 585 (s.), 601 (w.) BRAZIL. Current subspecies: nominal plus *sarcina.*

*brevicornis*. *Pheidole brevicornis* Mayr, 1876: 106 (s.w.) AUSTRALIA. Wheeler, G.C. & Wheeler, J. 1972b: 243 (l.).

#*breviscapa*. *Pheidole flavens* var. *breviscapa* Forel, 1899c: 77 (s.) GUATEMALA.

*breviseta*. *Pheidole flavens* st. *breviseta* Santschi, 1919f: 39 (w.) ARGENTINA. Santschi, 1923d: 60 (s.). Raised to species: Santschi, 1923d: 60.

*brevispinosa* Donisthorpe, 1948; see under ***CARDIOCONDYLA***.

#*britoi*. *Pheidole fallax* var. *britoi* Forel, 1912f: 221 (s.) COLOMBIA.

#*bruchella*. *Pheidole lignicola* var. *bruchella* Forel, 1915c: 355 (s.w.) ARGENTINA.

*bruchi*. *Pheidole bruchi* Forel, 1914d: 274 (s.w.) ARGENTINA. Current subspecies: nominal plus *brunnescens.*

#*bruesi*. *Pheidole triconstricta* var. *bruesi* Wheeler, W.M. 1911b: 169 (s.w.) GRENADA I.

#*bruesi*. *Pheidole* (*Ceratopheidole*) *hecate* subsp. *bruesi* Wheeler, W.M. 1917g: 460 (s.w.) JAMAICA. [Unresolved junior primary homonym of *bruesi* Wheeler, above.] Combination in *Ceratopheidole:* Emery, 1922e: 113; in *Pheidole:* Kempf, 1972a: 183.

#*brunnescens*. *Pheidole bruchi* var. *brunnescens* Santschi, 1929d: 285 (s.) BRAZIL.

#*bruta*. *Pheidole* (*Pheidole*) *spininodis* var. *bruta* Santschi, 1934c: 30 (s.) ARGENTINA.

*buccalis*. *Pheidole vinelandica* subsp. *buccalis* Wheeler, W.M. 1908e: 454 (s.w.q.) U.S.A. Subspecies of *bicarinata:* Creighton, 1950a: 171. Material of the unavailable name *castanea* referred here: Creighton, 1950a: 171; of the unavailable name *huachucana* referred here: Smith, M.R. 1958c: 120. Junior synonym of *bicarinata:* Cole, 1956c: 114; of *vinelandica:* Gregg, 1959: 18.

#*bucculenta*. *Pheidole guilelmimuelleri* subsp. *bucculenta* Forel, 1908c: 379 (s.w.) BRAZIL.

*buchholzi*. *Pheidole buchholzi* Mayr, 1901a: 276 (s.w.) CAMEROUN. Forel, 1916: 413 (q.).

*buckleyi*. *Pheidole buckleyi* Smith, M.R. 1951a: 806. Replacement name for *Atta pennsylvanica* Buckley, 1867: 345. [Junior secondary homonym of *Pheidole pennsylvanica* Roger, 1863a: 199.] Unplaced taxon: Smith, D.R. 1979: 1414.

*bucolica*. *Pheidole bucolica* Santschi, 1929d: 289, figs. 19–21 (s.) BRAZIL.

*bugi*. *Pheidole bugi* Wheeler, W.M. 1919e: 66 (s.w.) BORNEO.

#*bulawayensis*. *Pheidole xocensis* var. *bulawayensis* Forel, 1913j: 215 (s.w.) ZIMBABWE.

#*burtoni*. *Pheidole* (*Pheidole*) *biconstricta* subsp. *burtoni* Mann, 1916: 436, pl. 3, fig. 24 (s.) BRAZIL.

*butteli*. *Pheidole butteli* Forel, 1913k: 36 (s.w.q.m.) SUMATRA.

*caffra*. *Pheidole caffra* Emery, 1895h: 33 (s.) SOUTH AFRICA. Santschi, 1939b: 239 (w.). Current subspecies: nominal plus *abyssinica, amoena, bayeri, montivaga, senilifrons.*

*cairnsiana*. *Pheidole javana* var. *cairnsiana* Forel, 1902h: 438 (s.w.) AUSTRALIA. Raised to species: Taylor & Brown, D.R. 1985: 76.

*caldwelli*. *Pheidole caldwelli* Mann, 1921: 434, fig. 14 (s.w.q.) FIJI IS.

*calens*. *Pheidole carbonaria* r. *calens* Forel, 1901c: 130 (s.w.) MEXICO. Subspecies of *tepicana:* Emery, 1922e: 106. Raised to species: Kempf, 1972a: 188.

*californica*. *Pheidole californica* Mayr, 1870b: 987 (s.w.) U.S.A. Wheeler, W.M. 1915b: 406 (q.); Wheeler, G.C. & Wheeler, J. 1972b: 243 (l.). Senior synonym of *incenata* and *satura:* Creighton, 1950a: 173; of *pyramidensis:* Wheeler, G.C. & Wheeler, J. 1986g: 13. Current subspecies: nominal plus *oregonica.*

*cameroni*. *Pheidole cameroni* Mayr, 1887: 595 (s.), 605 (w.q.) ECUADOR. Senior synonym of *monticola:* Cameron, 1892: 67; Dalla Torre, 1893: 89.

*campestris*. *Pheidole sitarches* subsp. *campestris* Creighton, 1950a: 189 (s.w.) U.S.A. [First available use of *Pheidole sitarches* subsp. *rufescens* var. *campestris* Wheeler, W.M. 1908e: 443; unavailable name.] Taber & Cokendolpher, 1988: 95 (k.). Junior synonym of *soritis:* Cole, 1953e: 298. Revived from synonymy as subspecies of *sitarches:* Cole, 1956c: 115 (footnote); Gregg, 1959: 30. Synonym of *rufescens:* Creighton, 1950a: 189; Gregg, 1959: 30. [Both references give *campestris* as senior synonym, but *rufescens* has priority and is the first available name for this taxon.]

*camptostela*. *Pheidole camptostela* Kempf, 1972b: 457, figs. 7–10 (s.w.) BRAZIL.

*capellinii*. *Pheidole capellinii* Emery, 1887b: 463 (footnote), pl. 1, fig. 14 (s.) JAVA. Imai, Kubota, *et al.* 1985: 47 (k.).

*capensis*. *Pheidole capensis* Mayr, 1862: 746 (s.w.) SOUTH AFRICA. Forel, 1913a: 130 (q.). Subspecies of *megacephala:* Emery, 1895h: 34. Revived status as species: Forel, 1913a: 130; Arnold, 1920a: 441. Current subspecies: nominal plus *dregei, modestior, reddenburgensis.*

#***capillata.*** *Pheidole gertrudae* subsp. *capillata* Emery, 1906c: 151 (s.w.) BRAZIL.

***carapuna.*** *Pheidole (Pheidole) carapuna* Mann, 1916: 432, pl. 3, fig. 22 (s.w.q.) BRAZIL. Current subspecies: nominal plus *chaquimayensis.*

***carapunco.*** *Pheidole carapunco* Kusnezov, 1952b: 82 (s.w.q.m.) ARGENTINA.

*carbonaria. Pheidole carbonaria* Pergande, 1896: 881 (s.w.) MEXICO. Junior synonym of *tepicana:* Emery, 1901b: 119.

***caribbaea.*** *Pheidole caribbaea* Wheeler, W.M. 1911a: 23 (s.w.q.m.) JAMAICA. Current subspecies: nominal plus *sloanei.*

***carrolli.*** *Pheidole carrolli* Naves, 1985: 58, figs. 14, 43 (s.w.) U.S.A.

***casta.*** *Pheidole casta* Wheeler, W.M. 1908e: 454, pl. 26, figs, 22, 23 (s.w.) U.S.A.

***castanea.*** *Myrmica castanea* Smith, F. 1858b: 131 (q.) BRAZIL. Combination in *Aphaenogaster:* Forel, 1895b: 129; in *Pheidole:* Emery, 1915d: 69.

*castanea. Pheidole vinelandica* subsp. *longula* var. *castanea* Wheeler, W.M. 1915b: 405 (s.w.) U.S.A. Unavailable name; material referred to *buccalis* by Creighton, 1950a: 171.

***cataractae.*** *Pheidole cataractae* Wheeler, W.M. 1916c: 6 (s.w.) GUIANA.

#***catella.*** *Pheidole minima* var. *catella* Santschi, 1914d: 339 (s.w.) NIGERIA.

*cavannae. Pheidole cavannae* Emery, 1887b: 464 (footnote) (s.) NEW CALEDONIA. Subspecies of *oceanica:* Emery, 1914f: 401. Junior synonym of *fervens:* Wilson & Taylor, 1967: 45.

***cavifrons.*** *Pheidole cavifrons* Emery, 1906c: 144, fig. 21 (s.w.) URUGUAY. Combination in *P. (Elasmopheidole)*: Emery, 1915i: 190. Senior synonym of *fuscipunctis:* Gallardo, 1932c: 182. Current subspecies: nominal plus *longior.*

*cavigenis. Pheidole tepicana* subsp. *cavigenis* Wheeler, W.M. 1915b: 403 (s.) U.S.A. Junior synonym of *sorditis:* Gregg, 1959: 30.

#***cearensis.*** *Pheidole pubiventris* r. *cearensis* Forel, 1901e: 353 (s.w.) BRAZIL.

#***cedarensis.*** *Pheidole athertonensis* r. *cedarensis* Forel, 1915b: 64 (s.w.m.) AUSTRALIA.

#***cellarum.*** *Pheidole anastasii* var. *cellarum* Forel, 1908b: 55 (s.w.) COSTA RICA. Forel, 1915d: 34 (q.).

***centeotl.*** *Pheidole (Allopheidole) centeotl* Wheeler, W.M. 1914b: 46 (s.w.q.) MEXICO.

***cephalica.*** *Pheidole cephalica* Smith, F. 1858b: 177, pl. 9, figs. 21–23 (s.w.) BRAZIL. Senior synonym of *opaca:* Kempf, 1965: 183. Current subspecies: nominal plus *apterostigmoides, incrustata, sarrita.*

*cerdo. Pheidole crassinoda* st. *ruspolii* var. *cerdo* Santschi, 1937a: 54 (s.w.q.m.) SOUTH AFRICA. Unavailable name.

***cerebrosior.*** *Pheidole vinelandica* subsp. *cerebrosior* Wheeler, W.M. 1915b: 405 (s.w.) U.S.A. Raised to species: Creighton, 1950a: 175.

***ceres.*** *Pheidole ceres* Wheeler, W.M. 1904a: 10 (s.w.q.m.) U.S.A. Current subspecies: nominal plus *tepaneca.*

***cervicornis.*** *Pheidole cervicornis* Emery, 1900c: 322, pl. 8, fig. 25 (w.) NEW GUINEA. Combination in *P. (Pheidolacanthinus)*: Emery, 1921f: 82; Donisthorpe, 1947b: 174.

***ceylonica.*** *Oecophthora ceylonica* Motschoulsky, 1863: 18 (s.w.q.m.) SRI LANKA. Combination in *Pheidole:* Mayr, 1863: 438.

***chalca.*** *Pheidole chalca* Wheeler, W.M. 1914b: 44 (s.w.q.) MEXICO.

***championi.*** *Pheidole championi* Forel, 1899c: 72, pl. 3, fig. 22 (s.q.) PANAMA. Current subspecies: nominal plus *sima.*

#***chaquimayensis.*** *Pheidole carapuna* var. *chaquimayensis* Wheeler, W.M. 1925a: 20 (s.m.) PERU.

***cheesmannae.*** *Pheidole (Pheidole) cheesmannae* Donisthorpe, 1941h: 55 (q.) NEW GUINEA.

***chilensis.*** *Pheidole chilensis* Mayr, 1862: 748 (s.w.q.m.) CHILE. See also: Mayr, 1865: 95.

#***chloe.*** *Pheidole goeldii* subsp. *chloe* Forel, 1908b: 56 (s.w.) COSTA RICA.

***christopherseni.*** *Pheidole christopherseni* Forel, 1912f: 229 (s.w.) PANAMA.

*cicatricosa. Pheidole pallidula* var. *cicatricosa* Stitz, 1917: 340, fig. 7 (s.w.) ALGERIA. Santschi, 1932g: 3 (q.). Junior synonym of *pallidula:* Collingwood, 1978: 68.

***cingulata.*** *Atta cingulata* Smith, F. 1857a: 77 (s.w.) BORNEO. Combination in *Aphaenogaster:* Dalla Torre, 1893: 102; in *Pheidole:* Emery, 1915d: 69.

*clauda. Pheidole anastasii* st. *sospes* var. *clauda* Santschi, 1929d: 291 (w.) BRAZIL. Unavailable name.

***clavata.*** *Aphaenogaster clavata* Emery, 1877b: 372 (w.) ETHIOPIA. Combination in *Pheidole:* Emery, 1892a: 114.

***claviscapa.*** *Pheidole claviscapa* Santschi, 1925d: 224 (s.w.) BRAZIL. See also: Kempf, 1964e: 56.

***clementensis.*** *Pheidole clementensis* Gregg, 1969: 93, fig. 1 (s.w.) U.S.A.

***clydei.*** *Pheidole (Ceratopheidole) clydei* Gregg, 1950: 89, fig. 1 (w.) U.S.A. Gregg, 1953: 1 (s.). See also: Creighton, 1965a: 169.

***cocciphaga.*** *Pheidole cocciphaga* Borgmeier, 1934: 99, fig. 3 (s.w.q.) SURINAM.

***cockerelli.*** *Pheidole cockerelli* Wheeler, W.M. 1908e: 464 (s.w.) U.S.A.

***coffeicola.*** *Pheidole coffeicola* Borgmeier, 1934: 98, fig. 2 (s.w.q.) SURINAM.

***colaensis.*** *Pheidole (Electropheidole) colaensis* Mann, 1921: 441 (s.w.) FIJI IS.

***colobopsis.*** *Pheidole (Pheidole) colobopsis* Mann, 1916: 439, pl. 3, figs. 25, 26 (q.) BRAZIL. Combination in *P. (Scrobopheidole)*: Emery, 1922e: 112.

#***coloradensis.*** *Pheidole pilifera* var. *coloradensis* Emery, 1895c: 290 (s.w.) U.S.A. Wheeler, W.M. 1908e: 435 (q.m.). Subspecies of *pilifera:* Wheeler, W.M. 1908e: 435; Wheeler, G.C. & Wheeler, E.W. 1944: 244. Material of the unavailable name *neomexicana* referred here: Creighton, 1950a: 187.

#***columbica.*** *Pheidole fallax* r. *columbica* Forel, 1886b: xliv (s.) COLOMBIA. Raised to species: Dalla Torre, 1893: 89; Forel, 1899c: 66. Subspecies of *fallax:* Wheeler, W.M. 1908a: 133 (in key); Emery, 1922e: 101.

*comanche. Pheidole desertorum* var. *comanche* Wheeler, W.M. 1906d: 339 (s.w.) U.S.A. Junior synonym of *desertorum:* Creighton, 1950a: 178.

***comata.*** *Pheidole comata* Smith, F. 1858b: 176 (w.) BORNEO. Mayr, 1886c: 360 (s.).

#***commista.*** *Pheidole impressiceps* r. *commista* Forel, 1901b: 13 (s.w.q.m.) BISMARCK ARCHIPELAGO.

*commutata. Pheidole commutata* Mayr, 1886d: 459 (s.w.) U.S.A. Variety of *dentata:* Forel, 1901e: 352; Wheeler, W.M. 1908e: 460. Revived status as species: Emery, 1922e: 104. Senior synonym of *tennesseensis:* Cole, 1948: 82. Junior synonym of *dentata:* Creighton, 1950a: 178.

***concentrica.*** *Pheidole concentrica* Forel, 1902h: 416 (s.w.) AUSTRALIA. Current subspecies: nominal plus *recurva.*

***concinna.*** *Pheidole concinna* Santschi, 1910c: 362 (s.w.q.) CONGO.

*concinna. Pheidole concinna* Wheeler, W.M. 1928c: 13, fig. 1 (s.w.) CHINA. [Junior primary homonym of *concinna* Santschi, above.] Replacement name: *tsailuni* Wheeler, W.M. 1929f: 2.

#***concordia.*** *Pheidole exasperata* var. *concordia* Santschi, 1916d: 242. Relacement name for *polita* Viehmeyer, 1914d: 606. [Junior primary homonym of *polita* Emery, 1894c: 155.]

***conficta.*** *Pheidole conficta* Forel, 1902h: 417 (s.w.) AUSTRALIA.

***confoedusta.*** *Pheidole confoedusta* Wheeler, W.M. 1909b: 233 (s.) MEXICO.

#***conicollis.*** *Ischnomyrmex longipes* var. *conicollis* Emery, 1900d: 681, fig. 7 (w.) SUMATRA. Forel, 1913k: 50 (s.q.). Combination in *Pheidole (Ischnomyrmex)*: Forel, 1913k: 50; Viehmeyer, 1916a: 122.

#***conigera.*** *Pheidole spinulosa* subsp. *conigera* Forel, 1910e: 440 (s.) SOUTH AFRICA.

***constanciae.*** *Pheidole constanciae* Forel, 1902c: 177 (s.), 194 (w.q.) INDIA. [Also described as new by Forel, 1902f: 543.] Current subspecies: nominal plus *nigra.*

***constipata.*** *Pheidole constipata* Wheeler, W.M. 1908e: 468 (s.w.q.m.) U.S.A.

#***continentis.*** *Pheidole tasmaniensis* var. *continentis* Forel, 1902h: 437 (s.w.q.) AUSTRALIA.

#***continentis.*** *Aphaenogaster (Ischnomyrmex) longipes* var. *continentis* Forel, 1911a: 24 (w.) BURMA. [Unresolved junior secondary homonym of *continentis* Forel, above.] Viehmeyer, 1914d: 608 (s.). Combination in *Ischnomyrmex:* Emery, 1922e: 114; in *Pheidole:* **new combination**.

#***coonoorensis.*** *Pheidole indica* var. *coonoorensis* Forel, 1902c: 185 (s.), 199 (w.) INDIA. [Also described as new by Forel, 1902f: 546.]

**cordata.* **Formica cordata* Holl, 1829: 140 (s.) BALTIC AMBER (Oligocene). Combination in *Pheidole:* Mayr, 1868c: 17. [This ant is noted and figured in Schweigger, 1819: 119, pl. 18, fig. 70, but is not named.]

***cordiceps.*** *Pheidole cordiceps* Mayr, 1868b: 174 (s.w.) ARGENTINA. Emery, 1906c: 149 (q.). Current subspecies: nominal plus *lilianae.*

*cornutula. Pheidole cornutula* Emery, 1890b: 52 (footnote), pl. 5, figs. 14, 15 (s.) PARAGUAY. Forel, 1901e: 365 (w.); Santschi, 1923d: 62 (q.m.); Goni, Zolessi & Imai, 1983: 365 (k.). Junior synonym of *subarmata:* Brown, 1981: 526.

***corticicola.*** *Pheidole minima* st. *corticicola* Santschi, 1910c: 366 (s.w.q.m.) CONGO. Raised to species: Santschi, 1937d: 219.

**#*costauriensis*.** *Pheidole rotundata* st. *costauriensis* Santschi, 1914a: 433 (s.) GHANA. Santschi, 1915c: 250 (w.). Subspecies of *megacephala:* current status.

***cramptoni*.** *Pheidole cramptoni* Wheeler, W.M. 1916c: 4 (s.w.) GUIANA. Senior synonym of *petiolicola:* Kempf & Brown, 1968: 97.

***crassicornis*.** *Pheidole crassicornis* Emery, 1895c: 296 (s.) U.S.A. Forel, 1901e: 350 (w.m.). Current subspecies: nominal plus *tetra.*

***crassinoda*.** *Pheidole crassinoda* Emery, 1895h: 32 (s.w.) SOUTH AFRICA. Mayr, 1907b: 12 (q.m.). [*P. crassinoda* Forel, 1894b: 94. *Nomen nudum;* attributed to Emery.] Current subspecies: nominal plus *pluto, ruspolii, sordidula.*

***crassipes*.** *Pheidole crassipes* Mayr, 1887: 590 (s.), 600 (w.) BRAZIL. Current subspecies: nominal plus *germaini, grantae.*

***creightoni*.** *Pheidole creightoni* Gregg, 1955a: 19, pl. 2, figs. a–c; pl. 3, figs. a-c (s.w.q.m.) U.S.A.

**#*creola*.** *Pheidole moerens* subsp. *creola* Wheeler, W.M. & Mann, 1914: 25 (s.w.) HAITI. Wheeler, W.M. 1936: 198 (q.).

*cressoni. Macromischa cressoni* André, 1887: 296 (w.) MEXICO. Combination in *Aphaenogaster:* Emery, 1896a: 103; in *Pheidole:* Emery, 1915d: 67. Junior synonym of *ursus:* Brown, 1981: 528.

**#*croceithorax*.** *Pheidole mediofusca* var. *croceithorax* Emery, 1914f: 402 (s.) NEW CALEDONIA. Subspecies of *variabilis:* current status.

***cryptocera*.** *Pheidole cryptocera* Emery, 1900c: 324, pl. 8, figs. 26, 27 (s.w.q.) NEW GUINEA.

***cubaensis*.** *Pheidole cubaensis* Mayr, 1862: 747 (s.w.) CUBA. Current subspecies: nominal plus *grayi* (junior homonym).

**#*cubangensis*.** *Pheidole crassinoda* var. *cubangensis* Forel, 1901d: 306 (s.) ANGOLA. [Also described as new, misspelled as *kubangensis,* by Forel, 1903e: 561.] Variety of *speculifera:* Forel, 1916: 413.

***cuitensis*.** *Pheidole cuitensis* Forel, 1910e: 437 (s.w.) ANGOLA. Combination in *P.* (*Allopheidole*): Forel, 1913a: 135.

**#*daguerrei*.** *Pheidole nitidula* var. *daguerrei* Santschi, 1931e: 275 (s.w.) ARGENTINA.

***dammermani*.** *Pheidole dammermani* Wheeler, W.M. 1924b: 244 (s.w.) INDONESIA (Sebesi I.).

***darlingtoni*.** *Pheidole darlingtoni* Wheeler, W.M. 1936b: 198 (w.) HAITI.

***davisi*.** *Pheidole davisi* Wheeler, W.M. 1905f: 380 (s.w.) U.S.A.

***dea*.** *Pheidole dea* Santschi, 1921c: 115 (w.) ZAIRE. Senior synonym of *batrachorum:* Santschi, 1930a: 59.

**#*decanica*.** *Pheidole parva* var. *decanica* Forel, 1902c: 175 (s.), 192 (w.q.m.) INDIA. [Also described as new by Forel, 1902f: 542.]

***decarinata*.** *Pheidole decarinata* Santschi, 1929f: 86 (s.w.q.) SUDAN.

***decem*.** *Pheidole decem* Forel, 1901e: 366 (s.) TRINIDAD. Combination in *P.* (*Decapheidole*): Forel, 1912f: 237.

***deceptrix*.** *Pheidole deceptrix* Forel, 1899c: 66, pl. 3, fig. 20 (s.) GUATEMALA.

**#*decollata*.** *Pheidole oswaldi* r. *decollata* Forel, 1892k: 527 (s.w.) MADAGASCAR.

***defecta*.** *Pheidole defecta* Santschi, 1923d: 54 (s.) GUATEMALA. See also: Hölldobler & Wilson, 1992: 16.

*deflexa. Pheidole lemur* st. *rochai* var. *deflexa* Santschi, 1923d: 46 (s.w.) PARAGUAY. Unavailable name.

***delecta*.** *Pheidole delecta* Forel, 1899c: 67 (s.) MEXICO.

***dentata*.** *Pheidole morrisi* var. *dentata* Mayr, 1886d: 457 (s.w.m.) U.S.A. Forel, 1901e: 351 (q.); Wheeler, G.C. & Wheeler, J. 1953b: 71 (l.); Crozier, 1970: 117 (k.). Raised to species: Forel, 1901e: 351. Senior synonym of *faisonsica* and *commutata* (and its junior synonym *tennesseensis*): Creighton, 1950a: 178. See also: Wilson, 1976b: 63.

***dentigula*.** *Pheidole dentigula* Smith, M.R. 1927: 310 (s.w.) U.S.A. Smith, M.R. 1928b: 245 (q.); Wheeler, G.C. & Wheeler, J. 1960b: 12 (l.); Crozier, 1970: 117 (k.).

*dentimentum. Pheidole cornutula* var. *dentimentum* Santschi, 1929d: 285 (s.) BRAZIL. Junior synonym of *subarmata:* Brown, 1981: 526.

**#*deplanata*.** *Pheidole floridana* var. *deplanata* Pergande, 1896: 883 (s.w.) MEXICO.

***descolei*.** *Pheidole descolei* Kusnezov, 1952b: 76 (s.w.) ARGENTINA.

***deserticola*.** *Pheidole deserticola* Forel, 1910b: 35 (s.w.q.) AUSTRALIA. Current subspecies: nominal plus *foveifrons.*

***desertorum*.** *Pheidole desertorum* Wheeler, W.M. 1906d: 337 (s.w.q.m.) U.S.A. Taber & Cokendolpher, 1988: 95 (k.). Senior synonym of *comanche* and *maricopa:* Creighton, 1950a: 178.

**#*desucta*.** *Pheidole javana* var. *desucta* Wheeler, W.M. 1929f: 2 (s.w.q.) CHINA. Subspecies of *fervens:* current status.

**#*detrita*.** *Pheidole impressa* var. *detrita* Emery, 1890a: 66 (s.) VENEZUELA.

**#*dharmsalana*.** *Pheidole javana* var. *dharmsalana* Forel, 1902c: 184 (s.), 198 (w.) INDIA. [Also described as new by Forel, 1902f: 546.] Subspecies of *fervens:* current status.

***diana*.** *Pheidole diana* Forel, 1908b: 51 (s.w.q.) COSTA RICA.

***diffidens*.** *Formica diffidens* Walker, 1859: 372 (w.; not q.) SRI LANKA. Combination in *Pheidole:* Donisthorpe, 1932b: 575.

***diffusa*.** *Oecodoma diffusa* Jerdon, 1851: 109 (s.w.) INDIA. Combination in *Pheidole:* Smith, F. 1858b: 173.

**#*dignata*.** *Pheidole sculpturata* st. *dignata* Santschi, 1915c: 250 (s.w.) CONGO.

***diligens*.** *Atta diligens* Smith, F. 1858b: 168 (s.w.) BRAZIL. Combination in *Pheidole:* Mayr, 1886c: 360.

*diligens. Pheidole diligens* Santschi, 1923d: 50 (s.w.) BRAZIL. [Junior secondary homonym of *diligens* Smith, above.] Replacement name: *blumenauensis* Kempf, 1964e: 64.

*diligentissima* Dalla Torre, 1892; see under ***TETRAMORIUM.***

***dimidiata*.** *Pheidole dimidiata* Emery, 1894c: 160 (s.w.) BOLIVIA. Current subspecies: nominal plus *nitidicollis, schmalzi.*

**#*dione*.** *Pheidole rufipilis* var. *dione* Forel, 1913l: 225 (s.w.) ARGENTINA.

**#*discursans*.** *Pheidole radoszkowskii* var. *discursans* Forel, 1912f: 223 (s.q.) PARAGUAY. [First available use of *Pheidole radoszkowskii* subsp. *acuta* var. *discursans* Forel, 1909a: 258; unavailable name.]

***distincta*.** *Pheidole* (*Pheidole*) *distincta* Donisthorpe, 1943d: 447 (w.) NEW GUINEA.

***distorta*.** *Pheidole distorta* Forel, 1899c: 74 (footnote) (s.w.) COLOMBIA. Combination in *P.* (*Trachypheidole*): Emery, 1915i: 190. Current subspecies: nominal plus *reclusi.*

*divergens. Pheidole divergens* Mayr, 1867a: 97 (s.w.q.m.) INDONESIA. Junior synonym of *plagiaria:* Mayr, 1879: 675 (in key).

*diversa. Pheidole diversa* Smith, F. 1860c: 74 (s.w.) BRAZIL. Junior synonym of *fimbriata:* Kempf, 1965: 183.

**#*diversiceps*.** *Pheidole* (*Elasmopheidole*) *aberrans* var. *diversiceps* Santschi, 1916e: 374 (s.w.q.) ARGENTINA.

***diversipilosa*.** *Pheidole crassicornis* var. *diversipilosa* Wheeler, W.M. 1908e: 467 (s.w.q.) U.S.A. Junior synonym of *crassicornis:* Creighton, 1950a: 175; Gregg, 1959: 20. Revived from synonymy, and raised to species: Naves, 1985: 61.

**#*divexa*.** *Pheidole rufipilis* var. *divexa* Forel, 1908c: 372 (s.w.m.) BRAZIL.

*divinans. Pheidole indica* subsp. *rotchana* var. *divinans* Forel, 1911i: 222 (s.w.) SRI LANKA. Unavailable name.

**#*doddi*.** *Pheidole longiceps* subsp. *doddi* Forel, 1910b: 38 (s.w.q.) AUSTRALIA.

**#*dolenda*.** *Pheidole javana* var. *dolenda* Forel, 1912a: 60 (s.w.) TAIWAN. Subspecies of *fervens:* current status.

*dolichocephala. Pheidole dolichocephala* André, 1896b: 262 (s.) AUSTRALIA. Junior synonym of *vigilans:* Brown, 1971b: 13.

*dominicensis. Pheidole moerens* subsp. *dominicensis* Wheeler, W.M. 1913d: 241 (s.) DOMINICA. Junior synonym of *moerens:* Naves, 1985: 65.

**#*dregei*.** *Pheidole megacephala* subsp. *dregei* Emery, 1895h: 34 (s.) SOUTH AFRICA. Variety of *capensis:* Emery, 1915j: 246.

*dromedarius* Emery, 1900 [*Ischnomyrmex*]; see under ***APHAENOGASTER.***

*dryas. Pheidole guilelmimuelleri* st. *heyeri* var. *dryas* Santschi, 1929d: 288 (s.w.) BRAZIL. Unavailable name.

***dugasi*.** *Pheidole dugasi* Forel, 1911i: 222 (footnote) (w.) VIETNAM.

**#*duplex*.** *Pheidole megacephala* var. *duplex* Santschi, 1937d: 220 (s.w.q.) ANGOLA.

**#*durionei*.** *Pheidole wolfringi* st. *durionei* Santschi, 1923d: 51 (s.w.) ARGENTINA.

***dwyeri*.** *Pheidole dwyeri* Gregg, 1969: 97, fig. 2 (s.w.) MEXICO.

***dyctiota*.** *Pheidole dyctiota* Kempf, 1972b: 451, figs. 1–5 (s.w.q.) BRAZIL.

**#*ebangana*.** *Pheidole schultzei* st. *ebangana* Santschi, 1937d: 222, figs. 7, 8 (w.) ANGOLA.

*ecitonodora. Pheidole hyatti* var. *ecitonodora* Wheeler, W.M. 1908e: 463 (s.w.) U.S.A. [Misspelled as *ecitodora* by Emery, 1922e: 101.] Junior synonym of *hyatti:* Creighton, 1950a: 180.

*edax. Formica edax* Forskål, 1775: 84 (w.) EGYPT. Junior synonym of *megacephala:* Emery, 1892b: 160; Dalla Torre, 1892: 90.

***eidmanni.*** *Pheidole eidmanni* Menozzi, 1926c: 69, fig. 1 (s.w.) BRAZIL.

***elecebra.*** *Sympheidole elecebra* Wheeler, W.M. 1904a: 8, pl. 2, figs. 8–11 (q.m.) U.S.A. Combination in *Pheidole:* Smith, D.R. 1979: 1370. See also: Wilson, 1984: 328.

***elegans.*** *Pheidole* (*Pheidolacanthinus*) *elegans* Donisthorpe, 1938a: 144, figs. 3, 4 (w.) NEW GUINEA.

***elisae.*** *Pheidole elisae* Emery, 1900d: 686 (s.) INDONESIA (Engano I.). Current subspecies: nominal plus *nenia.*

*elongatula. Pheidole subarmata* var. *elongatula* Forel, 1893g: 408 (s.w.q.m.) ANTILLES IS. Junior synonym of *subarmata:* Brown, 1981: 526.

***embolopyx.*** *Pheidole embolopyx* Brown, 1968a: 332, figs. 1–7 (s.w.q.) BRAZIL.

***emersoni.*** *Pheidole* (*Hendecapheidole*) *emersoni* Wheeler, W.M. 1922d: 4, fig. 2 (w.q.m.) GUIANA.

*emeryi. Pheidole emeryi* Mayr, 1887: 589 (s.), 599 (w.) BRAZIL. Junior synonym of *tristis:* Kempf, 1965: 183.

*emeryi. Pheidole pallidula* var. *emeryi* Krausse, 1912a: 169 (s.w.) SARDINIA. [Junior primary homonym of *emeryi* Mayr, above.] Replacement name: *obscura* Santschi, 1936c: 200.

**#*emiliae.*** *Pheidole fallax* r. *emiliae* Forel, 1901e: 352 (s.w.q.) BRAZIL.

***emmae.*** *Pheidole* (*Ceratopheidole*) *emmae* Forel, 1905c: 20 (w.) JAVA. Combination in *Ceratopheidole:* Emery, 1922e: 113; in *Pheidole:* **revived combination**.

***ensifera.*** *Pheidole ensifera* Forel, 1897c: 197 (s.w.) MADAGASCAR.

***erato.*** *Pheidole* (*Pheidolacanthinus*) *erato* Mann, 1919: 307 (s.w.q.) SOLOMON IS.

***ernsti.*** *Pheidole ernsti* Forel, 1912a: 67 (s.w.) TAIWAN.

*erubens. Pheidole radoszkowskii* r. *parvinoda* var. *erubens* Forel, 1912f: 224 (s.w.) BRAZIL. Unavailable name.

**#*erythraea.*** *Pheidole aeberlii* var. *erythraea* Emery, 1901e: 61 (s.w.) ETHIOPIA.

***escherichii.*** *Pheidole escherichii* Forel, 1910c: 254 (s.w.q.) ETHIOPIA.

**#*eubos.*** *Pheidole bos* var. *eubos* Forel, 1915b: 62 (s.w.) AUSTRALIA. Subspecies of *bos:* Taylor & Brown, D.R. 1985: 75.

**#*euscrobata.*** *Pheidole templaria* r. *euscrobata* Forel, 1913k: 41 (s.) SRI LANKA.

*evoluta. Pheidole incisa* subsp. *evoluta* Borgmeier, 1929: 204, pl. 6, fig. 3 (s.w.) BRAZIL. Junior synonym of *susannae:* Kempf, 1964e: 63.

***exarata.*** *Pheidole exarata* Emery, 1896g: 68 (s.w.) COSTA RICA.

***exasperata.*** *Ischnomyrmex exasperatus* Mayr, 1866a: 506, fig. 12 (w.) "ohne Vaterlandsangabe". Combination in *Pheidole:* Viehmeyer, 1914d: 606. Current subspecies: nominal plus *concordia, fusiformis.*

***excellens.*** *Pheidole excellens* Mayr, 1862: 744 (s.) GHANA. Santschi, 1914d: 335 (w.). Current subspecies: nominal plus *fulvobasalis, weissi.*

**#*exigua.*** *Pheidole exigua* Mayr, 1884: 36 (s.) FRENCH GUIANA. Wheeler, W.M. 1908a: 135 (w.q.). Subspecies of *flavens:* Emery, 1894c: 156.

**#*extensa.*** *Pheidole knowlesi* subsp. *extensa* Mann, 1921: 438 (s.w.) FIJI IS.

***fabricator.*** *Atta fabricator* Smith, F. 1858b: 167 (s.w.q.m.) BRAZIL. Combination in *Pheidole:* Mayr, 1886c: 360. Senior synonym of *nigriventris:* Brown, 1981: 528. [*P. fabricatrix* Schulz, 1906: 155; unjustified emendation.] Current subspecies: nominal plus *nigella, polita.*

*faisonsica. Pheidole dentata* var. *faisonsica* Forel, 1901e: 352 (s.w.) U.S.A. Junior synonym of *dentata:* Creighton, 1950a: 178.

**#*fallacior.*** *Pheidole jelskii* var. *fallacior* Forel, 1901e: 356 (s.) VENEZUELA. Subspecies of *fallax:* current status.

***fallax.*** *Pheidole fallax* Mayr, 1870b: 984 (s.) CUBA. Forel, 1881: 9 (w.); Goni, Zolessi & Imai, 1983: 365 (k.). Senior synonym of *rubens:* Forel, 1901e: 356. Current subspecies: nominal plus *antillensis, arenicola, britoi, columbica, emiliae, fallacior, janeirensis, jelskii, obscurithorax, ovalis, puttemansi.*

*famelicus* Smith, F. 1874 [*Ischnomyrmex*]; see under ***APHAENOGASTER.***

***fantasia.*** *Pheidole fantasia* Chapman, 1963: 255, fig. 6 (s.w.q.) PHILIPPINES.

**#*farquharensis.*** *Pheidole flavens* var. *farquharensis* Forel, 1907d: 91 (w.) MADAGASCAR.

**#*fartilia.*** *Pheidole* (*Elasmopheidole*) *aberrans* var. *fartilia* Forel, 1913l: 233 (s.) ARGENTINA.

***fatigata.*** *Pheidole fatigata* **new name**. Replacement name for *granulata* Donisthorpe, 1949c: 749. [Junior primary homonym of *granulata* Pergande, 1896: 890.]

**#*faurei.*** *Pheidole minima* st. *faurei* Santschi, 1920b: 15 (w.q.) GABON.

***feae.*** *Pheidole feae* Emery, 1895k: 469 (s.w.q.) BURMA.

**#*fera.*** *Pheidole guilelmimuelleri* var. *fera* Santschi, 1925d: 227 (s.w.) BRAZIL.

***fergusoni.*** *Pheidole fergusoni* Forel, 1902c: 169 (s.), 188 (w.) INDIA. [Also described as new by Forel, 1902f: 540.]

***fervens.*** *Pheidole fervens* Smith, F. 1858b: 176 (s.) SINGAPORE. Ogata, 1982: 197 (m.). Senior synonym of *cavannae, javana, nigriscapa,* and material of the unavailable name *tahitiana* referred here: Wilson & Taylor, 1967: 45. Current subspecies: nominal plus *desucta, dharmsalana, dolenda, jacobsoni, jubilans, pectinata, protea, soror.*

***fervida.*** *Pheidole fervida* Smith, F. 1874: 406 (s.q.) JAPAN. Forel, 1900e: 285 (w.); Wheeler, W.M. 1928d: 107 (q.); Ogata, 1982: 195 (m.); Imai, 1966: 119 (k.). Current subspecies: nominal plus *kwazana.* See also: Kupyanskaya, 1990: 124.

**#*festata.*** *Pheidole holmgreni* subsp. *festata* Wheeler, W.M. 1925a: 20 (s.w.) BOLIVIA.

***fimbriata.*** *Pheidole fimbriata* Roger, 1863a: 196 (s.) "Rio Paraguay". Mayr, 1887: 598 (w.). Combination in *P.* (*Macropheidole*): Emery, 1915i: 190. Senior synonym of *diversa* and *smithii* (unnecessary replacement name): Kempf, 1965: 183. Current subspecies: nominal plus *tucumana.*

***fiorii.*** *Pheidole fiorii* Emery, 1890b: 51 (s.) COSTA RICA. Forel, 1899c: 70 (q.); Forel, 1908b: 52 (w.m.). [Also described as new by Emery, 1894k: 56.]

***flavens.*** *Pheidole flavens* Roger, 1863a: 198 (s.w.q.) CUBA. Wheeler, W.M. 1905b: 92 (m.). Current subspecies: nominal plus *alacris, asperithorax, breviscapa, exigua, farquharensis, gracilior, haytiana, mediorubra, navigans, rudigenis, sculptior, spei, thomensis, tuberculata, vincentensis.*

***flavida.*** *Pheidole flavida* Mayr, 1887: 593 (s.), 603 (w.) BRAZIL.

***flavothoracica.*** *Pheidole* (*Pheidolacanthinus*) *flavothoracica* Viehmeyer, 1914d: 610, fig. C (w.) NEW GUINEA.

**#*flebilis.*** *Pheidole nodus* var. *flebilis* Santschi, 1937h: 370 (s.w.) TAIWAN.

***floridana.*** *Pheidole flavens* subsp. *floridana* Emery, 1895c: 293 (s.w.q.) U.S.A. Raised to species: Pergande, 1896: 883; Emery, 1896g: 77. Senior synonym of *lauta:* Gregg, 1959: 21. Current subspecies: nominal plus *aechmeae, antoniensis, ares, deplanata, stomachosa, tillandsiarum.*

**#*foederalis.*** *Pheidole pubiventris* var. *foederalis* Borgmeier, 1928a: 34 (s.) BRAZIL.

**#*folicola.*** *Pheidole minutula* r. *folicola* Forel, 1904e: 681 (s.w.q.m.) BRAZIL.

***foreli.*** *Pheidole foreli* Mayr, 1901b: 8 (s.w.) SOUTH AFRICA. Forel, 1910f: 12 (q.). Current subspecies: nominal plus *pubens.* See also: Arnold, 1920a: 433.

*formosae. Pheidole javana* r. *jubilans* var. *formosae* Forel, 1912a: 60 (s.w.m.) TAIWAN. Unavailable name.

**#*formosensis.*** *Pheidole rhombinoda* var. *formosensis* Forel, 1913f: 193 (s.w.q.m.) TAIWAN. Stirps of *noda:* Santschi, 1937h: 370.

*fortunata. Pheidole susannae* subsp. *atricolor* var. *fortunata* Forel, 1908b: 58 (s.w.q.m.) COSTA RICA. Unavailable name.

**#*fossulata.*** *Pheidole jucunda* r. *fossulata* Forel, 1902c: 181 (s.), 196 (w.) INDIA. Raised to species: Bingham, 1903: 257. Race of *jucunda:* Forel, 1906b: 89; Emery, 1921f: 92.

**#*foveifrons.*** *Pheidole deserticola* subsp. *foveifrons* Viehmeyer, 1924b: 312 (s.w.) AUSTRALIA.

**#*frivola.*** *Pheidole hetschkoi* var. *frivola* Santschi, 1923d: 57 (s.w.) BRAZIL.

*froggatti* Forel, 1902; see under ***ANISOPHEIDOLE.***

**#*frontalis.*** *Pheidole longiceps* var. *frontalis* Forel, 1902h: 436 (s.w.q.) AUSTRALIA. Subspecies of *longiceps:* Forel, 1910b: 38.

**#*fugax.*** *Pheidole strator* r. *fugax* Arnold, 1920a: 449, fig. 25 (s.w.) MOZAMBIQUE.

**#*fulvobasalis.*** *Pheidole excellens* var. *fulvobasalis* Santschi, 1921c: 116 (w.) BENIN.

*fumipennis. Atta fumipennis* Smith, F. 1858b: 169 (w.q.) BRAZIL. Combination in *Aphaenogaster:* Forel, 1895b: 129; in *Pheidole:* Donisthorpe, 1932c: 474. Junior synonym of *tristis:* Brown, 1981: 527.

***funkikoensis.*** *Pheidole funkikoensis* Wheeler, W.M. 1929g: 40, fig. 4 (s.w.) TAIWAN.

*fusca* Emery, 1900 [*Ischnomyrmex*]; see under ***APHAENOGASTER.***

*fuscata. Pheidole biconstricta* r. *rubicunda* var. *fuscata* Emery, 1890b: 51 (s.w.) BRAZIL. Unavailable name.

**#*fuscescens.*** *Pheidole sexspinosa* var. *fuscescens* Emery, 1900c: 323 (s.w.) NEW GUINEA. Combination in *P.* (*Pheidolacanthinus*): Mann, 1919: 307.

**#*fusciclava.*** *Pheidole impariceps* var. *fusciclava* Santschi, 1923d: 48 (s.w.m.) BRAZIL.

*fuscipunctis. Pheidole* (*Elasmopheidole*) *cavifrons* var. *fuscipunctis* Santschi, 1916e: 374, fig. (s.w.q.) ARGENTINA. Junior synonym of *cavifrons:* Gallardo, 1932c: 182.

***PHEIDOLE*** *(continued)*

***#fusciventris.*** *Pheidole umbonata* subsp. *fusciventris* Emery, 1900c: 323 (s.w.) NEW GUINEA.

***fuscula.*** *Pheidole fuscula* Emery, 1900c: 325 (s.w.) NEW GUINEA.

***#fusiformis.*** *Pheidole exasperata* var. *fusiformis* Viehmeyer, 1914d: 607 (s.w.m.) NEW GUINEA.

***gaigei.*** *Pheidole gaigei* Forel, 1914c: 616 (s.w.) COLOMBIA.

*galapagensis. Pheidole galapagensis* Wheeler, W.M. 1919c: 271 (q.) GALAPAGOS IS. *Nomen nudum;* attributed to Ashmead.

***gambogia.*** *Pheidole* (*Pheidole*) *gambogia* Donisthorpe, 1948b: 303 (s.w.) NEW GUINEA.

***gatesi.*** *Aphaenogaster* (*Attomyrma*) *gatesi* Wheeler, W.M. 1927b: 44 (w.) BURMA. Combination in *Pheidole:* Brown, 1967b: 283.

***gauthieri.*** *Pheidole ursus* r. *gauthieri* Forel, 1901e: 367 (w.) COLOMBIA. Raised to species: Forel, 1912f: 233. Current subspecies: nominal plus *oxymora.*

***gavrilovi.*** *Pheidole gavrilovi* Kusnezov, 1952b: 80 (s.w.) ARGENTINA.

***gellibrandi.*** *Pheidole gellibrandi* Clark, 1934c: 58, pl. 4, figs. 15, 16 (s.w.) AUSTRALIA.

*genalis. Pheidole genalis* Borgmeier, 1929: 199, pl. 7, fig. 4 (s.w.) BRAZIL. Junior synonym of *oxyops:* Kempf, 1964e: 58.

***#geraesensis.*** *Pheidole semilaevis* var. *geraesensis* Santschi, 1929d: 287 (s.w.) BRAZIL.

***#germaini.*** *Pheidole crassipes* subsp. *germaini* Emery, 1896g: 68 (s.w.) BRAZIL.

***gertrudae.*** *Pheidole gertrudae* Forel, 1886b: xlii (w.) BRAZIL. Mayr, 1887: 592 (s.). Senior synonym of *humilis:* Kempf, 1960f: 463. Current subspecies: nominal plus *capillata, leonhardi, loretensis.*

***ghatica.*** *Pheidole ghatica* Forel, 1902c: 180 (s.), 196 (w.) INDIA. [Also described as new by Forel, 1902f: 545.]

***ghigii.*** *Pheidole ghigii* Emery, 1900d: 685 (s.w.) SUMATRA.

***gibba.*** *Pheidole gibba* Mayr, 1887: 590 (s.), 604 (w.) BRAZIL. Senior synonym of *lopesi:* Kempf, 1972b: 455.

***#gibbata.*** *Pheidole transversostriata* subsp. *gibbata* Borgmeier, 1934: 101 (s.w.) SURINAM.

*gibbicollis. Pheidole nana* subsp. *subreticulata* var. *gibbicollis* Emery, 1896g: 78 (s.w.q.) BRAZIL. Unavailable name.

***#gietleni.*** *Pheidole punctulata* r. *gietleni* Forel, 1905b: 164 (s.w.) MADAGASCAR. Variety of *picata:* Emery, 1915j: 245.

***gigliolii.*** *Pheidole gigliolii* Menozzi, 1935b: 193, fig. 2 (s.) GUIANA.

***gilvescens.*** *Pheidole gilvescens* Creighton & Gregg, 1955: 5 (s.w.) U.S.A. [First available use of *Pheidole xerophila* subsp. *tucsonica* var. *gilvescens* Wheeler, W.M. 1908e: 448; unavailable name.]

***godmani.*** *Pheidole godmani* Forel, 1893g: 404 (s.w.q.) ANTILLES IS.

***goeldii.*** *Pheidole goeldii* Forel, 1896: 178 (s.w.) BRAZIL. Current subspecies: nominal plus *chloe.*

***#goetschi.*** *Pheidole goetschi* Santschi, 1939e: 314, figs. 1, 2 (s.w.) ARGENTINA. Subspecies of *bergi:* Kusnezov, 1952b: 67 (in key).

*gonacantha* Emery, 1899 [*Ischnomyrmex*]; see under ***APHAENOGASTER.***

***gouldi.*** *Pheidole gouldi* Forel, 1886b: xlii (s.w.q.m.) GUATEMALA.

***gracilescens.*** *Myrmica gracilescens* Smith, F. 1860a: 74 (w.) SULAWESI. Combination in *Pheidole:* Donisthorpe, 1932c: 458.

*gracilinoda. Pheidole ursus* var. *gracilinoda* Forel, 1904d: 172 (w.) MEXICO. Junior synonym of *ursus:* Brown, 1981: 528.

***#gracilior.*** *Pheidole flavens* r. *gracilior* Forel, 1901h: 78 (s.w.q.) GERMANY (intercepted in quarantine, from West Indies).

***gracilipes.*** *Leptomyrma gracilipes* Motschoulsky, 1863: 17, pl. 2, fig. 3 (w.) SRI LANKA. Combination in *Pheidole:* Emery, 1892b: 166.

***grallatrix.*** *Pheidole grallatrix* Emery, 1899f: 278 (s.w.) MADAGASCAR.

*grallipes. Pheidole grallipes* Wheeler, W.M. 1916e: 40. Replacement name for *longipes* Pergande, 1896: 885. [Junior secondary homonym of *longipes* Latreille, 1802c: 233.] Junior synonym of *vistana:* Gregg, 1959: 22.

***#grandii.*** *Pheidole anthracina* subsp. *grandii* Emery, 1914b: 183 (s.w.) AUSTRALIA.

***#grantae.*** *Pheidole crassipes* r. *grantae* Forel, 1908b: 53 (s.w.) COSTA RICA.

***granulata.*** *Pheidole* (*Ceratopheidole*) *granulata* Pergande, 1896: 890 (s.) MEXICO. Combination in *Ceratopheidole:* Emery, 1922e: 113; in *Pheidole:* Kempf, 1972a: 183.

*granulata. Pheidole* (*Pheidolacanthinus*) *granulatus* Donisthorpe, 1949c: 749 (w.) NEW GUINEA. [Junior primary homonym of *granulata* Pergande, above.] Replacement name: *fatigata* **new name**.

*gratiosa. Pheidole nodus* st. *rhombinoda* var. *gratiosa* Santschi, 1937h: 371 (s.w.) INDIA. Unavailable name.

***grayi.*** *Pheidole grayi* Forel, 1902c: 167 (s.), 187 (w.) INDIA. [Also described as new by Forel, 1902f: 539.]

***#grayi.*** *Pheidole cubaensis* var. *grayi* Mann, 1920: 426 (s.w.) CUBA. [Unresolved junior primary homonym of *grayi* Forel, above.]

***greggi.*** *Pheidole greggi* Naves, 1985: 62, figs. 21, 45, 57 (s.w.) U.S.A.

*grenadensis. Pheidole flavens* r. *sculptior* var. *grenadensis* Forel, 1897b: 300 (s.w.q.m.) GRENADA I. Unavailable name.

***grundmanni.*** *Pheidole* (*Ceratopheidole*) *grundmanni* Smith, M.R. 1953e: 144, fig. 1 (w.) U.S.A.

***guilelmimuelleri.*** *Pheidole guilelmimuelleri* Forel, 1886a: 210 (s.w.) BRAZIL. Mayr, 1887: 607 (q.); Wheeler, G.C. & Wheeler, J. 1972b: 243 (l.). Current subspecies: nominal plus *antillana, avia, bucculenta, fera, gustavi, heyeri, mamore, ultrix.*

***guineensis.*** *Formica guineensis* Fabricius, 1793: 357 (w.) GUINEA. Combination in *Tetramorium:* Mayr, 1862: 740; in *Pheidole:* Bolton, 1977: 94.

***#gustavi.*** *Pheidole guilelmimuelleri* var. *gustavi* Forel, 1911c: 302 (s.w.q.) BRAZIL.

***#gwaaiensis.*** *Pheidole schultzei* var. *gwaaiensis* Forel, 1913a: 129 (s.w.) ZIMBABWE. Arnold, 1920a: 459 (q.m.).

*hagermani. Pheidole californica* var. *hagermani* Cole, 1936a: 35 (s.w.) U.S.A. Junior synonym of *oregonica:* Creighton, 1950a: 173.

***hartmeyeri.*** *Pheidole hartmeyeri* Forel, 1907h: 280 (s.w.) AUSTRALIA.

***havilandi.*** *Pheidole havilandi* Forel, 1911a: 38 (s.w.q.m.) BORNEO. Current subspecies: nominal plus *sapuana, selangorensis.*

*hayesi. Pheidole hayesi* Smith, M.R. 1924: 251 (s.w.) U.S.A. Junior synonym of *bicarinata:* Creighton, 1950a: 171.

***#haytiana.*** *Pheidole flavens* var. *haytiana* Forel, 1907e: 6 (w.) HAITI. Wheeler, W.M. & Mann, 1914: 24 (s.q.m.).

***haywardi.*** *Pheidole haywardi* Kusnezov, 1952b: 88 (w.) ARGENTINA.

***#hebe.*** *Pheidole triconstricta* var. *hebe* Santschi, 1923d: 55 (s.w.) BOLIVIA.

***hecate.*** *Pheidole* (*Ceratopheidole*) *hecate* Wheeler, W.M. 1911a: 26 (s.w.) JAMAICA. Combination in *Ceratopheidole:* Emery, 1922e: 113; in *Pheidole:* Kempf, 1972a: 183. Current subspecies: nominal plus *bruesi* (junior homonym), *malevola.*

***#heinzi.*** *Pheidole lutzi* var. *heinzi* Forel, 1913l: 225 (s.w.) BRAZIL.

***hercules.*** *Pheidole* (*Pheidolacanthinus*) *hercules* Donisthorpe, 1941h: 53 (s.w.) NEW GUINEA.

***#heterothrix.*** *Pheidole obtusopilosa* st. *heterothrix* Santschi, 1923d: 59 (s.w.) ARGENTINA.

***hetschkoi.*** *Pheidole hetschkoi* Emery, 1896g: 66 (s.w.) BRAZIL. Current subspecies: nominal plus *frivola, senilis.*

***hewitti.*** *Pheidole hewitti* Santschi, 1932a: 382, figs. 1–4 (s.w.) SOUTH AFRICA. Senior synonym of *hewitti:* Arnold, 1952a: 462.

*hewitti. Pheidole capensis* var. *hewitti* Arnold, 1944: 6 (s.w.) SOUTH AFRICA. Junior synonym of *hewitti:* Arnold, 1952a: 462.

***#heyeri.*** *Pheidole guilelmimuelleri* var. *heyeri* Forel, 1899c: 71 (footnote) (s.w.) BRAZIL. See also: Forel, 1911c: 302; Forel, 1912f: 321.

***#himalayana.*** *Pheidole indica* r. *himalayana* Forel, 1902c: 185 (s.), 199 (w.) INDIA. [Also described as new by Forel, 1902f: 546.] Raised to species: Bingham, 1903: 265. Subspecies of *indica:* Emery, 1921f: 91; Menozzi, 1939a: 298; Pisarski, 1967: 385.

***hirsuta.*** *Pheidole hirsuta* Emery, 1896g: 65 (s.) COSTA RICA.

***hirtula.*** *Pheidole vasliti* var. *hirtula* Forel, 1899c: 65 (s.w.) MEXICO. Wheeler, W.M. 1909b: 232 (q.). Raised to species, and senior synonym of *acolhua:* Creighton, 1950a: 211.

*hohenlohei. Pheidole hohenlohei* Emery, 1888c: 354 (s.w.) BRAZIL. Junior synonym of *spininodis:* Emery, 1906c: 146 (in text).

***holmgreni.*** *Pheidole holmgreni* Wheeler, W.M. 1925a: 18 (s.w.) BOLIVIA. Current subspecies: nominal plus *festata.*

*hondurensis. Pheidole* (*Elasmopheidole*) *hondurensis* Mann, 1922: 25, fig. 12 (s.w.) HONDURAS. Junior synonym of *subarmata:* Brown, 1981: 526.

***#hongkongensis.*** *Pheidole rinae* subsp. *hongkongensis* Wheeler, W.M. 1928c: 11 (s.w.q.) HONG KONG.

***#hoogwerfi.*** *Pheidole sharpi* r. *hoogwerfi* Forel, 1902c: 170 (s.), 189 (w.q.m.) INDIA. [Also described as new by Forel, 1902f: 540.] Raised to species: Bingham, 1903: 235. Subspecies of *sharpi:* Emery, 1921f: 94.

***horni.*** *Pheidole horni* Emery, 1901f: 118 (s.) SRI LANKA.

***hortensis.*** *Pheidole hortensis* Forel, 1913k: 38, fig. J (s.w.q.m.) JAVA. Imai, Kubota, *et al.* 1985: 47 (k.).

***hospes.*** *Pheidole hospes* Smith, F. 1865: 74 (s.w.) NEW GUINEA.

***hospita.*** *Pheidole hospita* Bingham, 1903: 238 (s.) INDIA.

*huachucana. Pheidole vinelandica* subsp. *longula* var. *huachucana* Smith, M.R. 1951a: 805. Unavailable name; material referred to *buccalis* by Smith, M.R. 1958c: 120.

***huberi.*** *Pheidole huberi* Forel, 1911d: 374 (s.w.) SUMATRA. Current subspecies: nominal plus *perakensis.*

***humeralis.*** *Pheidole humeralis* Wheeler, W.M. 1908e: 456, pl. 27, fig. 39 (s.w.) U.S.A.

*humilis. Megalomyrmex* (*Wheelerimyrmex*) *humilis* Borgmeier, 1930: 35, pl. 5, figs. 34, 36 (w.) BRAZIL. Junior synonym of *gertrudae:* Kempf, 1960f: 463.

***hyatti.*** *Pheidole hyatti* Emery, 1895c: 295 (s.w.) U.S.A. Wheeler, W.M. 1908e: 463 (q.m.); Wheeler, G.C. & Wheeler, J. 1953b: 74 (l.); Wheeler, G.C. & Wheeler, J. 1972b: 244 (l.); Taber & Cokendolpher, 1988: 95 (k.). Senior synonym of *ecitonodora:* Creighton, 1950a: 180. Current subspecies: nominal plus *solitanea.*

#***hybrida.*** *Pheidole biconstricta* subsp. *hybrida* Emery, 1894c: 154 (s.w.) BOLIVIA. Borgmeier, 1934: 97 (m.).

#***idiota.*** *Pheidole idiota* Santschi, 1923d: 53 (s.w.q.) ARGENTINA. Stirps of *vafra:* Santschi, 1929d: 284. Material of the unavailable name *maculifrons* referred here: Brown, 1981: 526.

#***idulis.*** *Pheidole angusta* r. *idulis* Forel, 1912f: 235 (s.w.) BRAZIL.

#***ilgi.*** *Pheidole rotundata* subsp. *ilgi* Forel, 1907g: 82 (s.w.) ETHIOPIA. Forel, 1907c: 139 (q.m.). Subspecies of *megacephala:* Emery, 1915j: 236; Wheeler, W.M. 1922a: 132.

#***illota.*** *Pheidole terresi* var. *illota* Wheeler, W.M. & Mann, 1914: 23 (s.w.) HAITI.

*imbecillis. Pheidole cornutula* var. *imbecillis* Emery, 1906c: 151 (s.w.) ARGENTINA. Junior synonym of *subarmata:* Brown, 1981: 526.

*imbellis* Emery, 1900 [*Ischnomyrmex*]; see under ***APHAENOGASTER.***

***impariceps.*** *Pheidole impariceps* Santschi, 1923d: 47 (s.w.) BRAZIL. Current subspecies: nominal plus *fusciclava.*

#***impexa.*** *Pheidole morrisi* var. *impexa* Wheeler, W.M. 1908e: 461, fig. 31 U.S.A. Subspecies of *morrisii:* Creighton, 1950a: 184.

#***impia.*** *Pheidole pubiventris* st. *impia* Santschi, 1923d: 46 (s.) BRAZIL.

***impressa.*** *Pheidole impressa* Mayr, 1870b: 985 (s.) BRAZIL. Current subspecies: nominal plus *detrita.*

***impressiceps.*** *Pheidole impressiceps* Mayr, 1876: 105 (s.w.) AUSTRALIA. Forel, 1902h: 433 (q.). Current subspecies: nominal plus *commista.*

*impressiceps. Pheidole megacephala* subsp. *impressiceps* Wasmann, 1904a: 72 (footnote) (s.w.q.) SOUTH AFRICA. [Junior primary homonym of *impressiceps* Mayr, above.] Replacement name: *impressifrons* Wasmann, 1905: 110.

#***impressifrons.*** *Pheidole megacephala* subsp. *impressifrons* Wasmann, 1905: 110. Replacement name for *impressiceps* Wasmann, 1904a: 72. [Junior primary homonym of *impressiceps* Mayr, 1876: 105.] Forel, 1913j: 213 (q.). Subspecies/race of *punctulata:* Forel, 1913a: 127; Forel, 1913b: 328; of *rotundata:* Forel, 1913j: 213; Arnold, 1920a: 426; of *megacephala:* Emery, 1915j: 236; Emery, 1921f: 85; Wheeler, W.M. 1922a: 814.

*incenata. Pheidole californica* var. *incenata* Wheeler, W.M. 1915b: 407 (s.w.) U.S.A. Junior synonym of *californica:* Creighton, 1950a: 173.

#***incensa.*** *Pheidole rinae* subsp. *incensa* Wheeler, W.M. 1928c: 13 (s.w.) CHINA.

***incerta.*** *Myrmica incerta* Smith, F. 1863: 21 (q.) INDONESIA (Misool I.). Combination in *Pheidole:* Donisthorpe, 1932c: 471.

***incisa.*** *Pheidole incisa* Mayr, 1870a: 401 (s.w.) COLOMBIA.

#***incrustata.*** *Pheidole opaca* var. *incrustata* Forel, 1908b: 59 (w.) COSTA RICA. Subspecies of *cephalica:* Kempf, 1972a: 188.

***incurvata.*** *Pheidole incurvata* Viehmeyer, 1924b: 313 (s.w.) AUSTRALIA.

***indica.*** *Pheidole indica* Mayr, 1879: 679 (s.w.q.) INDIA. Forel, 1902c: 199 (m.); Imai, Baroni Urbani, *et al.* 1984: 6 (k.). Current subspecies: nominal plus *coonoorensis, himalayana, rotschana.* See also: Bingham, 1903: 263; Menozzi, 1939a: 298; Ogata, 1982: 196.

***indistincta.*** *Pheidole indistincta* Forel, 1899c: 75 (s.w.) COSTA RICA.

#***indosinensis.*** *Pheidole sulcaticeps* subsp. *indosinensis* Wheeler, W.M. 1928c: 10 (s.w.) CHINA. [*P. sulcaticeps* subsp. *indosinensis* Wheeler, W.M. 1927h: 87; *nomen nudum.*]

#***industa.*** *Pheidole rufipilis* st. *industa* Santschi, 1939e: 316, fig. 3 (s.w.) URUGUAY.

***inermis.*** *Pheidole inermis* Mayr, 1870b: 984 (s.w.) MEXICO.

#***inermis.*** *Pheidole pallidula* var. *inermis* Stitz, 1917: 340, fig. 6 (s.w.) ALGERIA. [Unresolved junior primary homonym of *inermis* Mayr, above.]

*infans. Pheidole guilelmimuelleri* st. *heyeri* var. *infans* Santschi, 1923d: 56 (s.w.) BRAZIL. Unavailable name.

#***inflexa.*** *Pheidole riveti* var. *inflexa* Santschi, 1923d: 58 (s.) ECUADOR.

#***ingenita.*** *Pheidole prelli* st. *ingenita* Santschi, 1928f: 204 (s.) IVORY COAST.

*injuncta. Pheidole guilelmimuelleri* subsp. *heyeri* var. *injuncta* Forel, 1911c: 302 (s.w.q.m.) BRAZIL. Unavailable name.

***innotata.*** *Pheidole innotata* Mayr, 1866b: 898 (s.) "ohne Vaterlandsangabe".

***innupta.*** *Pheidole innupta* Menozzi, 1931d: 200, fig. 7 (s.w.q.) COSTA RICA.

*inops. Pheidole variabilis* r. *ocyma* var. *inops* Forel, 1915b: 61 (s.w.m.) AUSTRALIA. Unavailable name (Taylor, 1986: 34.).

***inquilina.*** *Epipheidole inquilina* Wheeler, W.M. 1903f: 664, fig. 5 (gynandromorph) U.S.A. Wheeler, W.M. 1904a: 15 (q.m.); Smith, M.R. 1940a: 106 (w.); Cole, 1965: 174 (s.). [Also described as new by Wheeler, W.M. 1904a: 15.] Combination in *Pheidole:* Cole, 1965: 174.

*inquilina. Pheidole inquilina* Forel, 1914d: 240 (w.) ZIMBABWE. [Unresolved junior secondary homonym of *Epipheidole inquilina* Wheeler, above.] Junior synonym of *atrox:* Arnold, 1920a: 425 (in text).

*inquilina. Xenoaphaenogaster inquilina* Baroni Urbani, 1964b: 50, figs. 15, 16 (w.) SICILY. [Unresolved junior secondary homonym of *Epipheidole inquilina* Wheeler, W.M. 1903f: 664.] Junior synonym of *pallidula:* Bolton, 1987: 291.

***inscrobiculata.*** *Pheidole inscrobiculatus* Viehmeyer, 1916a: 120 (s.w.) WEST MALAYSIA.

***insipida.*** *Pheidole kingi* r. *insipida* Forel, 1899c: 76 (s.w.) MEXICO. Raised to species: Kempf, 1972a: 195.

*instabilis. Pheidole kingi* subsp. *instabilis* Emery, 1901b: 120 (s.w.) U.S.A. Wheeler, W.M. 1908e: 433 (q.m.); Wheeler, G.C. & Wheeler, J. 1953b: 74 (l.). Junior synonym of *tepicana:* Creighton & Gregg, 1955: 24.

#***insulana.*** *Pheidole punctatissima* subsp. *insulana* Wheeler, W.M. 1905b: 93 (s.w.) BAHAMAS.

#***inversa.*** *Pheidole radoszkowskii* var. *inversa* Forel, 1901e: 363 (s.w.) COLOMBIA.

***irritans.*** *Myrmica irritans* Smith, F. 1858b: 127 (w.) SOUTH AFRICA. Combination in *Pheidole:* Dixey & Longstaff, 1907: 331.

***isis.*** *Pheidole isis* Mann, 1919: 311, figs. 16–18 (s.w.q.) SOLOMON IS. Current subspecies: nominal plus *taki.*

#***jacobsoni.*** *Pheidole javana* subsp. *jacobsoni* Forel, 1911b: 203 (s.w.q.m.) JAVA. Subspecies of *fervens:* current status.

#***jamaicensis.*** *Pheidole punctatissima* subsp. *jamaicensis* Wheeler, W.M. 1908b: 161 (s.w.) JAMAICA.

#***janeirensis.*** *Pheidole fallax* r. *janeirensis* Forel, 1912f: 222 (s.w.) BRAZIL.

*janus. Pheidole janus* Smith, F. 1858b: 175, pl. 9, figs. 13–17 (s.w.) SRI LANKA. Junior synonym of *pusilla:* Mayr, 1886c: 360.

*javana. Pheidole javana* Mayr, 1867a: 98 (s.w.) JAVA. Viehmeyer, 1916a: 121 (q.m.). Junior synonym of *fervens:* Wilson & Taylor, 1967: 45.

#***jelskii.*** *Pheidole jelskii* Mayr, 1884: 34 (s.w.) FRENCH GUIANA. Forel, 1893g: 400 (q.m.). Variety of *fallax:* Forel, 1893g: 400. Raised to species: Emery, 1894g: 390. Subspecies of *fallax:* Emery, 1906c: 155; Forel, 1912f: 221.

*jheringi. Pheidole flavens* subsp. *exigua* var. *jheringi* Emery, 1894c: 157 (s.w.) BRAZIL. Unavailable name.

#***johnsoni.*** *Pheidole anastasii* var. *johnsoni* Wheeler, W.M. 1907a: 272 (s.w.m.) HONDURAS.

***jonas.*** *Pheidole jonas* Forel, 1907g: 82 (s.w.) COMORO IS.

***jordanica.*** *Pheidole jordanica* Saulcy, 1874: 17 (s.w.) ISRAEL. Finzi, 1936: 161 (q.). Subspecies of *megacephala:* Emery, 1889a: 442. Revived status as species: Dalla Torre, 1893: 91. Senior synonym of *laticeps* and *schmitzi:* Emery, 1915j: 233; of *obtusa:* Finzi, 1936: 161.

#***jubilans.*** *Pheidole javana* subsp. *jubilans* Forel, 1911b: 202 (s.w.) JAVA. Subspecies of *fervens:* current status.

***jucunda.*** *Pheidole jucunda* Forel, 1885b: 179 (s.) INDIA. Forel, 1902f: 545 (w.). Current subspecies: nominal plus *fossulata.*

***jujuyensis.*** *Pheidole oxyops* r. *jujuyensis* Forel, 1913l: 229 (w.) ARGENTINA. Raised to species: Kempf, 1964e: 62.

**#*kasaiensis.*** *Pheidole aurivillii* subsp. *kasaiensis* Forel, 1911f: 279 (s.w.) ZAIRE. Forel, 1916: 415 (q.).

**#*katonae.*** *Pheidole sculpturata* subsp. *katonae* Forel, 1907a: 21 (w.) TANZANIA.

*kingi. Pheidole kingi* André, 1898: 244 (s.w.) MEXICO. Combination in *P.* (*Allopheidole*): Forel, 1912f: 237. Senior synonym of *townsendi:* Emery, 1922e: 105. Junior synonym of *tepicana:* Creighton & Gregg, 1955: 24.

***kitschneri.*** *Pheidole kitschneri* Forel, 1910e: 436 (s.w.) SOUTH AFRICA. Santschi, 1914b: 76 (q.).

***knowlesi.*** *Pheidole knowlesi* Mann, 1921: 436, fig. 13 (s.w.q.) FIJI IS. Current subspecies: nominal plus *extensa.*

***kochi.*** *Acanthomyrmex kochi* Emery, 1911c: 252, fig. 1 (w.) NEW GUINEA. Combination in *Pheidole:* Moffett, 1986: 66.

***kohli.*** *Pheidole kohli* Mayr, 1901b: 11 (footnote) (s.w.) KENYA.

*kohli. Anergatides kohli* Wasmann, 1915b: 283, pls. 7, 8 (q.m.) ZAIRE. [Junior secondary homonym of *kohli* Mayr, above.] Replacement name: *Pheidole neokohli* Wilson, 1984: 328. Wheeler, G.C. & Wheeler, J. 1955c: 130 (l.).

**#*koshewnikovi.*** *Pheidole pallidula* subsp. *koshewnikovi* Ruzsky, 1905b: 648 (s.w.) RUSSIA.

***kraepelini.*** *Pheidole kraepelini* Forel, 1901h: 79 (s.w.) GERMANY (intercepted in quarantine, from Brazil).

**#*kwazana.*** *Pheidole fervida* var. *kwazana* Teranishi, 1940: 13 (w.) KOREA.

*labrata. Pheidole nitidula* st. *strobeli* var. *labrata* Santschi, 1929d: 282 (s.w.) ARGENTINA. Unavailable name.

***lacerta.*** *Pheidole lacerta* Wheeler, W.M. 1922c: 6 (s.) TRINIDAD.

*laevifrons. Pheidolacanthinus laevifrons* Smith, F. 1873: ix INDIA. *Nomen nudum;* attributed to Mayr.

***laevifrons.*** *Pheidole laevifrons* Mayr, 1887: 598 (s.) BRAZIL.

*laevigata. Myrmica laevigata* Smith, F. 1855b: 130, pl. 9, figs. 7, 8 (w.) GREAT BRITAIN. Junior synonym of *pusilla:* Roger, 1859: 259; of *pallidula:* Smith, F. 1858a: 282; of *megacephala:* Roger, 1863b: 30; of *pusilla:* Emery, 1915j: 235.

*laevigata. Pheidole laevigata* Mayr, 1862: 747 (s.) BRAZIL. Junior synonym of *pusilla:* Mayr, 1870b: 981 (footnote).

**#*laevinota.*** *Pheidole rufipilis* var. *laevinota* Forel, 1908c: 371 (w.) BRAZIL.

*laeviuscula. Pheidole vinelandica* subsp. *laeviuscula* Emery, 1895c: 292 (s.w.) U.S.A. Junior synonym of *vinelandica:* Creighton, 1950a: 172.

***laeviventris.*** *Pheidole laeviventris* Mayr, 1870a: 401 (w.) COLOMBIA.

***laevivertex.*** *Pheidole laevivertex* Forel, 1901c: 131 (s.w.) MEXICO.

**#*laidlowi.*** *Pheidole* (*Pheidole*) *triconstricta* var. *laidlowi* Mann, 1916: 434 (s.w.) BRAZIL.

**#*lallemandi.*** *Pheidole radoszkowskii* r. *lallemandi* Forel, 1901c: 133 (s.w.) COLOMBIA. Race of *biconstricta:* Forel, 1901e: 362; Forel, 1912f: 222.

***lamellinoda.*** *Pheidole lamellinoda* Forel, 1902c: 166 (s.), 186 (w.q.m.) INDIA. [Also described as new by Forel, 1902f: 538.]

***lamia.*** *Pheidole lamia* Wheeler, W.M. 1901b: 534, fig. 11 (s.w.) U.S.A. See also: Feener, 1981: 269.

***laminata.*** *Pheidole laminata* Emery, 1900c: 325, pl. 8, figs. 28, 29 (s.w.) NEW GUINEA.

***lanuginosa.*** *Pheidole lanuginosa* Wilson, 1984: 318, figs. 1, 2 (q.m.) INDIA.

*laticeps. Pheidole sinaitica* subsp. *laticeps* Mayr, 1904b: 6 (s.w.) EGYPT. Junior synonym of *jordanica:* Emery, 1915j: 233 (name misspelled by Emery as *latifrons.*)

***laticrista.*** *Pheidole* (*Elasmopheidole*) *laticrista* Santschi, 1916e: 375, 2 figs. (s.w.q.) ARGENTINA.

**#*latigena.*** *Pheidole variabilis* subsp. *latigena* Forel, 1907h: 279 (s.w.) AUSTRALIA.

***latinoda.*** *Pheidole latinoda* Roger, 1863a: 195 (s.) SRI LANKA. Mayr, 1865: 101 (w.); Forel, 1885b: 179 (q.); Imai, Baroni Urbani, *et al.* 1984: 7 (k.). Current subspecies: nominal plus *angustior, major, peradeniyae.* See also: Bingham, 1903: 236.

*lauta. Pheidole lauta* Wheeler, W.M. 1908e: 470 (s.w.q.m.) U.S.A. Subspecies of *floridana:* Creighton, 1950a: 175. Junior synonym of *floridana:* Gregg, 1959: 21.

**#*legitima.*** *Pheidole njassae* var. *legitima* Santschi, 1916d: 242. Replacement name for *sculptior* Viehmeyer, 1914c: 36. [Junior primary homonym of *sculptior* Forel, 1893g: 414.]

***lemur.*** *Pheidole lemur* Forel, 1912f: 225 (s.w.q.) BRAZIL. Current subspecies: nominal plus *rochai.*

**#*leonhardi.*** *Pheidole gertrudae* var. *leonhardi* Forel, 1901h: 62 (s.w.) BOLIVIA.

**#*levocciput.*** *Pheidole lignicola* var. *levocciput* Forel, 1908c: 370 (s.w.m.) PARAGUAY.

**#*liberiensis.*** *Pheidole termitophila* subsp. *liberiensis* Forel, 1911e: 270 (s.w.) LIBERIA.

***liengmei.*** *Pheidole liengmei* Forel, 1894b: 93 (w.) MOZAMBIQUE. Forel, 1913a: 134 (s.); Arnold, 1920a: 467 (q.). Current subspecies: nominal plus *malindana, micrartifex, shinsendensis.*

***lighti.*** *Pheidole lighti* Wheeler, W.M. 1927d: 2 (s.w.) CHINA.

***lignicola.*** *Pheidole lignicola* Mayr, 1887: 586 (s.), 602 (w.) BRAZIL. Current subspecies: nominal plus *bruchella, levocciput, victima.*

**#*lilianae.*** *Pheidole cordiceps* var. *lilianae* Forel, 1912f: 232 (s.w.q.m.) ARGENTINA. Subspecies of *cordiceps:* Kusnezov, 1952b: 66 (in key).

***lilloi.*** *Elasmopheidole lilloi* Kusnezov, 1952b: 74 (s.w.) ARGENTINA. Combination in *Pheidole* (*Elasmopheidole*): Kempf, 1972a: 184.

***liteae.*** *Pheidole liteae* Forel, 1910: 41 (s.w.) TASMANIA.

**#*litigiosa.*** *Pheidole sikorae* var. *litigiosa* Forel, 1892k: 526 (s.w.q.) MADAGASCAR.

***littoralis.*** *Pheidole sitarches* subsp. *littoralis* Cole, 1952b: 443 (s.w.) U.S.A. Raised to species: Naves, 1985: 64.

***lobulata.*** *Pheidole lobulata* Emery, 1900c: 324 (s.w.) NEW GUINEA.

***lokitae.*** *Pheidole* (*Pheidolacanthinus*) *lokitae* Forel, 1913k: 46, figs. L, M (s.w.q.m.) SUMATRA.

***longiceps.*** *Pheidole longiceps* Mayr, 1876: 106 (s.w.) AUSTRALIA. Current subspecies: nominal plus *doddi, frontalis.*

*longiceps. Pheidole longiceps* Aguayo, 1932: 218 (s.) CUBA. [Junior primary homonym of *longiceps* Mayr, above.] Replacement name: *neolongiceps* Brown, 1950e: 250.

***longicornis.*** *Pheidole longicornis* Emery, 1888a: 532 (w.) INDONESIA (Nias I.). Emery, 1901g: 573 (s.q.m.).

**#*longior.*** *Pheidole* (*Elasmopheidole*) *cavifrons* st. *longior* Santschi, 1933e: 110, figs. 1–3 (s.w.) ARGENTINA.

***longipes.*** *Formica longipes* Latreille, 1802c: 233, pl. 11, fig. 68 (w.) FRENCH GUIANA. Combination in *Pheidole:* Emery, 1921a: 25.

***longipes.*** *Myrmica longipes* Smith, F. 1857a: 70, pl. 1, fig. 6 (w.) SINGAPORE. [Unresolved junior secondary homonym of *Formica longipes* Latreille, above.] Combination in *Ischnomyrmex:* Mayr, 1862: 739; Emery, 1922e: 114; in *Aphaenogaster:* Emery, 1895k: 470; Bingham, 1903: 272; in *Pheidole* (*Isopheidole*): Forel, 1912m: 765; in *P.* (*Ischnomyrmex*): Forel, 1913k: 49; Viehmeyer, 1914d: 608; in *Pheidole:* current combination. Crawley, 1924: 391 (s.q.m.); Wheeler, G.C. & Wheeler, J. 1953b: 80 (l.). Current subspecies: nominal plus *conicollis, continentis* (junior homonym), *pseudola.*

*longipes. Pheidole susannae* r. *longipes* Pergande, 1896: 885 (s.w.) MEXICO. [Junior secondary homonym of *Formica longipes* Latreille, 1802c: 233.] Replacement name: *Pheidole grallipes* Wheeler, W.M. 1916e: 40. Wheeler, W.M. 1915b: 397 (q.). Raised to species: Forel, 1899c: 65.

***longiscapa.*** *Pheidole longiscapa* Forel, 1901e: 358 (s.w.m.) VENEZUELA. Current subspecies: nominal plus *martensis.*

***longispinosa.*** *Pheidole longispinosa* Forel, 1891b: 170, pl. 5, fig. 4 (s.w.q.) MADAGASCAR. Current subspecies: nominal plus *scabrata.*

***longula.*** *Pheidole vinelandica* var. *longula* Emery, 1895c: 292 (s.w.) U.S.A. Subspecies of *bicarinata:* Creighton, 1950a: 171. Raised to species: Cole, 1956c: 114.

*lopesi. Pheidole gibba* var. *lopesi* Borgmeier, 1937b: 233 (s.w.) BRAZIL. Junior synonym of *gibba:* Kempf, 1972b: 455.

*loretana. Pheidole* (*Macropheidole*) *fimbriata* st. *tucumana* var. *loretana* Santschi, 1934c: 30 (s.w.) ARGENTINA. Unavailable name.

**#*loretensis.*** *Pheidole* (*Pheidole*) *gertrudae* var. *loretensis* Santschi, 1933e: 108 (w.) ARGENTINA.

***lucida.*** *Pheidole lucida* Forel, 1895c: 248 (w.) MADAGASCAR.

**#*lucifuga.*** *Pheidole spininodis* var. *lucifuga* Santschi, 1923d: 56 (s.w.q.) ARGENTINA.

***lucretii.*** *Pheidole lucretii* Santschi, 1923d: 45 (s.) BRAZIL.

***luteipes.*** *Pheidole luteipes* Emery, 1914f: 402 (s.w.) NEW CALEDONIA. Current subspecies: nominal plus *obvia.*

**#*luteola.*** *Pheidole radoszkowskii* var. *luteola* Forel, 1893g: 406 (s.w.q.m.) ANTILLES IS.

***lutzi.*** *Pheidole lutzi* Forel, 1905b: 168 (s.w.) BRAZIL. Current subspecies: nominal plus *heinzi.*

***macclendoni.*** *Pheidole macclendoni* Wheeler, W.M. 1908e: 450, pl. 27, fig. 36 (s.w.) U.S.A.

**#mackayensis.** *Pheidole ampla* var. *mackayensis* Forel, 1902h: 436 (s.w.) AUSTRALIA. Forel, 1915b: 57 (m.).

*maculifrons. Pheidole vafra* st. *idiota* var. *maculifrons* Santschi, 1929d: 284, fig. 17 (s.w.) ARGENTINA. Unavailable name; material referred to *idiota* by Brown, 1981: 526.

**maculifrons.** *Pheidole maculifrons* Wheeler, W.M. 1929g: 42, fig. 5 (s.) PHILIPPINES.

**madecassa.** *Pheidole madecassa* Forel, 1892k: 525 (s.w.) MADAGASCAR.

**magrettii.** *Pheidole magrettii* Emery, 1887b: 462 (s.w.) JAVA.

**maja.** *Pheidole maja* Forel, 1886b: xlviii (s.w.q.) GUATEMALA.

**#major.** *Pheidole latinoda* var. *major* Forel, 1885b: 178 (s.w.q.) INDIA.

**makilingi.** *Pheidole makilingi* Viehmeyer, 1916b: 285 (s.w.) PHILIPPINES.

**#mala.** *Pheidole rinae* var. *mala* Forel, 1911b: 205 (s.w.) JAVA.

**malabarica.** *Oecodoma malabarica* Jerdon, 1851: 107 (s.w.) INDIA. Combination in *Pheidole:* Smith, F. 1858b: 173.

**#malelana.** *Pheidole minima* subsp. *malelana* Wheeler, W.M. 1922a: 133 (s.w.) ZAIRE.

**#malevola.** *Pheidole* (*Ceratopheidole*) *hecate* subsp. *malevola* Wheeler, W.M. 1911a: 27 (s.w.) JAMAICA. Combination in *Ceratopheidole:* Emery, 1922e: 113; in *Pheidole:* Kempf, 1972a: 183.

**#malindana.** *Pheidole liengmei* var. *malindana* Forel, 1913a: 135 (w.) ZIMBABWE.

**malinsii.** *Pheidole malinsii* Forel, 1902c: 167 (s.), 187 (w.) SRI LANKA. [Also described as new by Forel, 1902f: 539.]

**#mamore.** *Pheidole* (*Pheidole*) *guilelmimuelleri* subsp. *mamore* Mann, 1916: 429 (s.w.q.) BRAZIL.

**manteroi.** *Pheidole manteroi* Emery, 1897d: 564 (s.) NEW GUINEA.

**marcidula.** *Pheidole marcidula* Wheeler, W.M. 1908e: 457 (s.w.) U.S.A.

*maricopa. Pheidole desertorum* var. *maricopa* Wheeler, W.M. 1906d: 339 (s.w.) U.S.A. Junior synonym of *desertorum:* Creighton, 1950a: 178.

**#martensis.** *Pheidole longiscapa* r. *martensis* Forel, 1914c: 615 (s.w.) COLOMBIA.

**maufei.** *Pheidole maufei* Arnold, 1920a: 438 (s.w.) ZIMBABWE.

**#maxwellensis.** *Pheidole nodgii* r. *maxwellensis* Forel, 1913k: 40 (s.w.) WEST MALAYSIA.

**mayri.** *Pheidole mayri* Forel, 1894b: 91 (s.w.) SENEGAL.

*mbalensis. Pheidole excellens* st. *weissi* var. *mbalensis* Santschi, 1930b: 65, fig. 3 (w.) ANGOLA. Unavailable name.

**medioflava.** *Pheidole* (*Pheidolacanthinus*) *medioflava* Donisthorpe, 1941b: 37 (w.) NEW GUINEA.

**#mediofusca.** *Pheidole variabilis* r. *mediofusca* Forel, 1902h: 425 (s.w.) AUSTRALIA. Raised to species: Emery, 1914f: 402. Subspecies of *variabilis:* Emery, 1922e: 97.

**#mediorubra.** *Pheidole flavens* var. *mediorubra* Santschi, 1933e: 107 (s.w.) ARGENTINA.

**megacephala.** *Formica megacephala* Fabricius, 1793: 361 (s.) no locality given. Latreille, 1802c: 232 (q.); Mayr, 1861: 70 (s.w.q.m.); Wheeler, G.C. & Wheeler, J. 1953b: 75 (l.). Combination in *Pheidole:* Roger, 1863b: 30. Senior synonym of *trinodis:* Roger, 1863b: 30; of *edax:* Dalla Torre, 1892: 90; Emery, 1892b: 160; of *perniciosa:* Emery, 1915j: 235; of *pusilla* (and its junior synonyms *janus, laevigata* Smith, and *laevigata* Mayr): Wheeler, W.M. 1922a: 812; of *suspiciosa:* Donisthorpe, 1932c: 455; of *testacea:* Brown, 1981: 530. [*P. megalocephala* Schulz, 1906: 155; unjustified emendation.] Current subspecies: nominal plus *costauriensis, duplex, ilgi, impressifrons, melancholica, nkomoana, rotundata, scabrior, speculifrons, talpa.*

*megacephala* Smith, F. 1860; see under ***PHEIDOLOGETON.***

*megacephalopallidula. Pheidole pallidula* var. *megacephalopallidula* Dalla Torre, 1893: 94. *Nomen nudum.* [*Pheidole megacephala* r. *pallidula* var. *megacephalopallidula* Emery & Forel, 1879: 463; unavailable name.]

**meinerti.** *Pheidole meinerti* Forel, 1905b: 159 (s.) VENEZUELA.

**#melancholica.** *Pheidole punctulata* st. *melancholica* Santschi, 1912b: 164 (s.w.) IVORY COAST. Bernard, 1953b: 225 (q.). Subspecies of *megacephala:* Emery, 1915j: 236; Wheeler, W.M. 1922a: 132.

**melanogaster.** *Pheidole* (*Pheidole*) *melanogaster* Donisthorpe, 1943d: 448 (s.w.) NEW GUINEA.

**mendanai.** *Pheidole mendanai* Mann, 1919: 309, fig. 15 (s.w.q.) SOLOMON IS.

**mendicula.** *Pheidole* (*Hendecapheidole*) *mendicula* Wheeler, W.M. 1925d: 172, fig. 6 (s.w.m.) PANAMA. See also: Smith, M.R. 1955c: 303.

**mentita.** *Pheidole mentita* Santschi, 1914d: 338 (s.w.q.) GUINEA. Current subspecies: nominal plus *pullata.*

**#messalina.** *Pheidole spinulosa* subsp. *messalina* Forel, 1910e: 440 (s.m.) LESOTHO.

**metallescens.** *Pheidole metallescens* Emery, 1895c: 294 (w.) U.S.A. Wheeler, W.M. 1908e: 476 (s.); Wheeler, G.C. & Wheeler, J. 1960b: 12 (l.). Current subspecies: nominal plus *splendidula.*

*mexicanum* Pergande, 1896; see under ***APHAENOGASTER.***

*micantiventris. Pheidole rhombinoda* var. *micantiventris* Mayr, 1897: 427 (s.) SRI LANKA. Junior synonym of *noda:* Yasumatsu, 1962: 96.

**#micrartifex.** *Pheidole liengmei* st. *micrartifex* Forel, 1913j: 215 (s.w.) ZIMBABWE.

**microgyna.** *Pheidole microgyna* Wheeler, W.M. 1928b: 186 (w.q.m.) GUIANA. See also: Wilson, 1984: 323.

**micula.** *Pheidole californica* subsp. *micula* Wheeler, W.M. 1915b: 408 (s.w.) U.S.A. Wheeler, G.C. & Wheeler, J. 1972b: 244 (l.). Raised to species: Gregg, 1959: 24.

*militaris* Smith, F. 1860; see under ***PHEIDOLOGETON.***

*militaris. Pheidole radowszkowskii* r. *militaris* Emery, 1890b: 49 (s.w.) COSTA RICA. [Junior primary homonym of *militaris* Smith, F. 1860a: 74, above.] Replacement name: *pugnax* Dalla Torre, 1892: 91.

**militicida.** *Pheidole militicida* Wheeler, W.M. 1915b: 398 (s.w.) U.S.A. Creighton & Gregg, 1955: 11 (q.m.).

**mimula.** *Pheidole mimula* Wheeler, W.M. 1916c: 5 (s.) GUIANA.

**#minensis.** *Pheidole* (*Elasmopheidole*) *aberrans* var. *minensis* Santschi, 1923d: 64 (s.q.m.) BRAZIL.

**minima.** *Pheidole minima* Mayr, 1901a: 275 (s.w.) CAMEROUN. Current subspecies: nominal plus *catella, faurei, malelana.*

**minor.** *Ocodoma minor* Jerdon, 1851: 110 (w.q.) INDIA. Combination in *Pheidole:* Smith, F. 1858b: 174.

**minuscula.** *Pheidole minuscula* Bernard, 1953b: 226, fig. 8 (s.w.) GUINEA.

**minutissima.** *Pheidole minutissima* Kusnezov, 1952b: 78 (s.w.) ARGENTINA.

**minutula.** *Pheidole minutula* Mayr, 1878: 872 (s.w.) BRAZIL. Wheeler, W.M. 1942: 191 (q.m.). Current subspecies: nominal plus *folicola, personata.*

*misera. Pheidole strobeli* var. *misera* Santschi, 1916e: 373 (s.w.q.) ARGENTINA. Junior synonym of *strobeli:* Santschi, 1929d: 282.

**miseranda.** *Pheidole miseranda* Wheeler, W.M. 1924b: 243 (s.) INDONESIA (Krakatau I.) Wheeler, W.M. 1937a: 22 (w.).

**mjobergi.** *Pheidole* (*Pheidolacanthinus*) *mjobergi* Forel, 1915b: 66, pl. 1, fig. 5 (s.w.) AUSTRALIA.

**#modestior.** *Pheidole capensis* var. *modestior* Santschi, 1919a: 337 (s.w.) SOUTH AFRICA.

**modiglianii.** *Pheidole modiglianii* Emery, 1900d: 684 (s.w.q.) INDONESIA (Mentawei I.).

**moerens.** *Pheidole moerens* Wheeler, W.M. 1908a: 136, pl. 12, figs. 22, 23 (s.w.) PUERTO RICO. Wheeler, G.C. & Wheeler, J. 1972b: 244 (l.). Senior synonym of *dominicensis:* Naves, 1985: 65. Current subspecies: nominal plus *creola.*

**#moica.** *Pheidole plagiaria* var. *moica* Forel, 1911i: 222 (footnote) (s.) VIETNAM.

**#monardi.** *Pheidole monardi* Santschi, 1930b: 65, fig. 4 (w.) ANGOLA. Stirps of *rohani:* Santschi, 1937d: 221.

*monticola. Pheidole monticola* Cameron, 1892: 93, fig. (s.w.q.) ECUADOR. Junior synonym of *cameroni:* Cameron, 1892: 67.

**#montivaga.** *Pheidole caffra* var. *montivaga* Santschi, 1939b: 240, figs. 3, 5 (s.w.q.m.) ZIMBABWE.

*mordax. Pheidole mordax* Smith, F. 1863: 22 (w.) INDONESIA (Misool I.). Junior synonym of *ruficeps:* Mayr, 1867a: 96.

**morrisii.** *Pheidole morrisii* Forel, 1886b: xlvi (s.w.) U.S.A. Forel, 1901e: 350 (q.m.); Wheeler, G.C. & Wheeler, J. 1960b: 12 (l.). Senior synonym of *vanceae:* Creighton, 1950a: 183. Current subspecies: nominal plus *impexa.* See also: Feener, 1987: 569.

**moseni.** *Pheidole moseni* Wheeler, W.M. 1925a: 21 (s.w.q.) BRAZIL.

**multidens.** *Pheidole multidens* Forel, 1902c: 181 (s.), 197 (w.) INDIA.

**mus.** *Pheidole mus* Forel, 1902c: 174 (s.), 191 (w.m.) INDIA. Imai, Baroni Urbani, *et al.* 1984: 6 (k.). [Also described as new by Forel, 1902f: 542.]

**#mutica.** *Pheidole aberrans* var. *mutica* Emery, 1906c: 142, fig. 19 (s.) ARGENTINA. Combination in *P.* (*Elasmopheidole*): Emery, 1922e: 111.

**mylognatha.** *Pheidole mylognatha* Wheeler, W.M. 1922a: 134, fig. 33 (s.w.) ZAIRE.

*myops* Forel, 1902; see under ***ANISOPHEIDOLE.***

**nana.** *Pheidole nana* Emery, 1894c: 158 (s.w.) BRAZIL. Current subspecies: nominal plus *atticola, subreticulata.*

***naoroji.*** *Pheidole naoroji* Forel, 1902c: 168 (s.), 187 (w.) INDIA. [Also described as new by Forel, 1902f: 539.]

*napaea. Pheidole punctatissima* subsp. *napaea* Wheeler, W.M. 1934g: 165 (s.w.) MEXICO. Junior synonym of *punctatissima:* Brown, 1981: 525.

*nassavensis. Pheidole subarmata* var. *nassavensis* Wheeler, W.M. 1905b: 92 (s.w.) BAHAMAS. Junior synonym of *subarmata:* Brown, 1981: 526.

***nasutoides.*** *Pheidole nasutoides* Hölldobler & Wilson, 1992: 19, figs. 1, 2 (s.w.) COSTA RICA.

#***navigans.*** *Pheidole flavens* r. *navigans* Forel, 1901h: 79 (s.w.) GERMANY (intercepted in quarantine, from Mexico).

*nebrascensis. Pheidole* (*Allopheidole*) *vinelandica* var. *nebrascensis* Forel, 1922: 92 (s.w.q.) U.S.A. Junior synonym of *vinelandica:* Gregg, 1959: 18.

*nefasta. Pheidole subarmata* var. *nefasta* Santschi, 1929d: 285 (s.w.) BRAZIL. Junior synonym of *subarmata:* Brown, 1981: 526.

***nemoralis.*** *Pheidole nemoralis* Forel, 1892k: 526 (s.w.) MADAGASCAR. Current subspecies: nominal plus *petax.*

#***nenia.*** *Pheidole elisae* var. *nenia* Forel, 1913k: 30 (s.w.) SUMATRA.

***neokohli.*** *Pheidole neokohli* Wilson, 1984: 328. Replacement name for *Anergatides kohli* Wasmann, 1915b: 283. [Junior secondary homonym of *Pheidole kohli* Mayr, 1901b: 11 (footnote).]

***neolongiceps.*** *Pheidole neolongiceps* Brown, 1950e: 250. Replacement name for *longiceps* Aguayo, 1932: 218. [Junior primary homonym of *longiceps* Mayr, 1876: 106.]

*neomexicana. Pheidole pilifera* subsp. *coloradensis* var. *neomexicana* Wheeler, W.M. 1908e: 436 (s.w.) U.S.A. Unavailable name; material referred to *coloradensis* by Creighton, 1950a: 187.

#***neutralis.*** *Pheidole occipitalis* st. *neutralis* Santschi, 1914d: 336 (s.w.) GUINEA.

#***nevadensis.*** *Pheidole pubiventris* var. *nevadensis* Forel, 1901e: 353 (s.w.q.) COLOMBIA.

*nevadensis. Pheidole californica* subsp. *nevadensis* Wheeler, W.M. 1915b: 408 (s.w.q.) U.S.A. [Junior primary homonym of *nevadensis* Forel, above.] Replacement name: *pyramidensis* Emery, 1922e: 105.

*nexa. Pheidole spinulosa* r. *messalina* var. *nexa* Forel, 1913a: 129 (s.w.) ZIMBABWE. Unavailable name.

*niapuana. Pheidole niapuana* Wheeler, W.M. 1922a: 136, fig. 34 (s.w.) ZAIRE. Junior synonym of *pulchella:* Santschi, 1930a: 59.

***nietneri.*** *Pheidole nietneri* Emery, 1901f: 118 (s.w.) SRI LANKA.

#***nigella.*** *Pheidole fabricator* var. *nigella* Emery, 1894c: 155 (s.w.) BOLIVIA.

***nigeriensis.*** *Pheidole nigeriensis* Santschi, 1914d: 340 (s.) NIGERIA.

#***nigra.*** *Pheidole constanciae* var. *nigra* Forel, 1902c: 177 (s.), 194 (w.q.) INDIA. [Also described as new by Forel, 1902f: 543.]

*nigrescens. Pheidole guilelmimuelleri* st. *antillana* var. *nigrescens* Forel, 1893g: 404 (s.w.q.) ANTILLES IS. Unavailable name.

#***nigridens.*** *Pheidole transversostriata* var. *nigridens* Forel, 1901e: 362 (s.w.) COLOMBIA.

*nigrior. Pheidole* (*Pheidole*) *nitidula* st. *pampana* var. *nigrior* Santschi, 1933e: 108 (s.w.) ARGENTINA. Unavailable name.

*nigriscapa. Pheidole* (*Pheidole*) *oceanica* var. *nigriscapa* Santschi, 1928a: 48, fig. 3 (s.w.) SAMOA. Junior synonym of *fervens:* Wilson & Taylor, 1967: 45.

***nigritella.*** *Pheidole nigritella* Bernard, 1953b: 227, fig. 8 (s.w.) GUINEA.

*nigriventris. Atta nigriventris* Smith, F. 1858b: 169 (w.) BRAZIL. Combination in *Pheidole:* Mayr, 1886c: 360. Junior synonym of *fabricator:* Brown, 1981: 528.

***nimba.*** *Pheidole nimba* Bernard, 1953b: 224, fig. 8 (s.w.) GUINEA.

***nindi.*** *Pheidole nindi* Mann, 1919: 314, figs. 19, 20 (s.w.q.) SOLOMON IS.

#***nitidicollis.*** *Pheidole dimidiata* var. *nitidicollis* Emery, 1896g: 79 (s.w.) COSTA RICA. Subspecies of *dimidiata:* Emery, 1922e: 107.

***nitidula.*** *Pheidole triconstricta* var. *nitidula* Emery, 1888e: 693 (s.w.) ARGENTINA. Santschi, 1929d: 281 (q.m.). Raised to species: Emery, 1906c: 154. Current subspecies: nominal plus *daguerrei, pampana, perversa, richteri, silvicola, strobeli.*

***njassae.*** *Pheidole njassae* Viehmeyer, 1914c: 35 (s.w.) TANZANIA. Current subspecies: nominal plus *legitima.*

#***nkomoana.*** *Pheidole megacephala* r. *nkomoana* Forel, 1916: 415 (s.w.q.m.) ZAIRE.

***noda.*** *Pheidole nodus* Smith, F. 1874: 407 (s.) JAPAN. Forel, 1900e: 268 (w.); Wheeler, W.M. 1906c: 309 (q.); Ogata, 1982: 196 (m.); Wheeler, G.C. & Wheeler, J. 1953b: 75 (l.); Imai & Kubota, 1972: 197 (k.); Imai & Kubota, 1975: 391 (k.). Senior synonym of *micantiventris, praevexata* and *rhombinoda:* Yasumatsu, 1962: 96. Current subspecies: nominal plus *azumai, flebilis, formosensis, stella, taprobanae* (junior homonym).

***nodgii.*** *Pheidole nodgii* Forel, 1905c: 16 (s.w.q.) JAVA. Viehmeyer, 1916a: 119 (m.). [Misspelled as *noggii* by Wheeler, W.M. 1930h: 63.] Current subspecies: nominal plus *maxwellensis, tjibodana, verlatenensis, zoceana.*

***nodifera.*** *Atta nodifer* Smith, F. 1858b: 165 (w.) CHINA. Combination in *Pheidole:* Mayr, 1886c: 360.

*norfolkensis. Pheidole ampla* subsp. *norfolkensis* Wheeler, W.M. 1927i: 134, fig. 3 (s.w.) AUSTRALIA. Donisthorpe, 1941e: 91 (q.m.). Junior synonym of *vigilans:* Brown, 1971b: 13.

*notabilis* Smith, F. 1860; see under ***ACANTHOMYRMEX.***

#***nubila.*** *Pheidole triconstricta* var. *nubila* Emery, 1906c: 153 (s.w.) ARGENTINA. Santschi, 1919f: 40.

***nuculiceps.*** *Pheidole nuculiceps* Wheeler, W.M. 1908e: 473 (s.w.) U.S.A.

*nugax. Pheidole flavens* subsp. *asperithorax* var. *nugax* Forel, 1908c: 380 (s.w.) BRAZIL. Unavailable name.

***obnixa.*** *Pheidole arciruga* r. *obnixa* Forel, 1912f: 236 (w.) ARGENTINA. Forel, 1913l: 231 (s.q.). Combination in *P.* (*Elasmopheidole*): Bruch, 1914: 219. Raised to species: Santschi, 1925e: 157.

#***obscura.*** *Pheidole pallidula* var. *obscura* Santschi, 1936c: 200. Replacement name for *emeryi* Krausse, 1912a: 169. [Junior primary homonym of *emeryi* Mayr, 1887: 589.]

#***obscurifrons.*** *Pheidole* (*Elasmopheidole*) *aberrans* var. *obscurifrons* Santschi, 1925e: 158 (s.w.) ARGENTINA.

#***obscurior.*** *Pheidole susannae* r. *obscurior* Forel, 1886b: xliv (s.w.) BRAZIL. Forel, 1893g: 410 (q.m.).

#***obscurithorax.*** *Pheidole fallax* subsp. *obscurithorax* Naves, 1985: 61 (s.w.) ARGENTINA. [First available use of *Pheidole fallax* st. *arenicola* var. *obscurithorax* Santschi, 1923d: 58; unavailable name.]

*obtusa. Pheidole megacephala* r. *obtusa* Stitz, 1917: 340, fig. 8 (s.) EGYPT. Junior synonym of *jordanica:* Finzi, 1936: 161.

***obtusopilosa.*** *Pheidole obtusopilosa* Mayr, 1887: 586 (s.), 602 (w.) URUGUAY. Santschi, 1917f: 278 (m.); Santschi, 1926d: 7 (q.). Current subspecies: nominal plus *heterothrix.*

***obtusospinosa.*** *Pheidole obtusospinosa* Pergande, 1896: 889 (s.) MEXICO. Junior synonym of *subdentata* Pergande: Wheeler, W.M. 1914b: 50; hence first available replacement name for *Pheidole subdentata* Pergande, 1896: 888. [Junior secondary homonym of *Oecophthora subdentata* Mayr, 1853a: 145.]

#***obvia.*** *Pheidole luteipes* subsp. *obvia* Emery, 1914f: 403 (s.w.) NEW CALEDONIA.

***occipitalis.*** *Pheidole occipitalis* André, 1890: 321 (s.w.) SIERRA LEONE. Current subspecies: nominal plus *adami, neutralis.*

***oceanica.*** *Pheidole oceanica* Mayr, 1866a: 510 (s.w.q.m.) FIJI IS. Senior synonym of *boraborensis, pattesoni, upoluana:* Wilson & Taylor, 1967: 48.

*ocellifera* Smith, F. 1858; see under ***PHEIDOLOGETON.***

***ocior.*** *Pheidole variabilis* r. *ocior* Forel, 1915b: 58 (s.w.) AUSTRALIA.

***oculata.*** *Aphaenogaster oculata* Emery, 1899f: 276 (q.) MADAGASCAR. Combination in *Parapheidole:* Emery, 1915d: 68; in *Pheidole:* Smith, D.R. 1979: 1366.

#***ocyma.*** *Pheidole variabilis* r. *ocyma* Forel, 1915b: 59 (s.w.q.) AUSTRALIA.

*ogloblini. Pheidole ogloblini* Kusnezov, 1952b: 85 (s.w.q.) ARGENTINA. Junior synonym of *bambusarum:* Kempf, 1972b: 450.

***onifera.*** *Pheidole onifera* Mann, 1921: 427, fig. 12 (s.w.) FIJI IS.

*opaca. Pheidole opaca* Mayr, 1862: 749 (s.w.) BRAZIL. Forel, 1899c: 69 (q.); Borgmeier, 1934: 97 (q.). Junior synonym of *cephalica:* Kempf, 1965: 183.

*opacior. Pheidole proxima* subsp. *transversa* var. *opacior* Forel, 1910b: 44 (s.w.) AUSTRALIA. Unavailable name (Taylor, 1986: 34).

#***opacissima.*** *Pheidole radoszkowskii* var. *opacissima* Forel, 1901e: 364 (s.w.) JAMAICA.

***opaciventris.*** *Pheidole opaciventris* Mayr, 1876: 105 (s.w.) AUSTRALIA.

***optiva.*** *Pheidole optiva* Forel, 1901c: 132 (s.w.) MEXICO.

#***orba.*** *Pheidole anthracina* r. *orba* Forel, 1902h: 421 (s.w.q.) AUSTRALIA.

***orbica.*** *Pheidole orbica* Forel, 1893g: 415 (s.w.q.m.) ANTILLES IS.

#***oregonica.*** *Pheidole oregonica* Emery, 1895c: 291 (s.w.) U.S.A. Wheeler, W.M. 1915b: 407 (q.). Subspecies of *californica:* Wheeler, W.M. 1915b: 407; Creighton, 1950a: 173. Senior synonym of *hagermani:* Creighton, 1950a: 173; of *shoshoni:* Gregg, 1959: 19.

*orientalis. Pheidole pallidula* subsp. *orientalis* Müller, 1923: 69 (s.w.) RUSSIA. [First available use of *Pheidole pallidula* subsp. *arenarum* var. *orientalis* Emery, 1915j: 230; unavailable name.] Karavaiev, 1926b: 107 (q.m.); Menozzi, 1936d: 280 (l.). Junior synonym of *pallidula:* Baroni Urbani, 1964a: 3; Atanassov & Dlussky, 1992: 123.

***oswaldi.*** *Pheidole oswaldi* Forel, 1891b: 173 (s.w.q.m.) MADAGASCAR. Current subspecies: nominal plus *decollata.*

**#*ovalis.*** *Pheidole fallax* var. *ovalis* Forel, 1912f: 221 (s.) JAMAICA.

**#*oxymora.*** *Pheidole gauthieri* var. *oxymora* Forel, 1912f: 233 (s.w.) PANAMA.

***oxyops.*** *Pheidole oxyops* Forel, 1908c: 377 (s.q.) PARAGUAY. Forel, 1911c: 303 (w.m.). Senior synonym of *genalis* and *regia:* Kempf, 1964e: 58.

*pabulator* Smith, F. 1860; see under ***PHEIDOLOGETON.***

*pachyderma;* see under **trachyderma.**

**#*pacifica.*** *Pheidole xerophila* subsp. *pacifica* Wheeler, W.M. 1915b: 404 (s.w.q.m.) U.S.A. Subspecies of *pilifera:* Creighton, 1950a: 187.

***paiute.*** *Pheidole bicarinata* subsp. *paiute* Gregg, 1959: 17 (s.w.) U.S.A. Raised to species: Wheeler, G.C. & Wheeler, J. 1973e: 75.

**#*palawanica.*** *Pheidole plagiaria* r. *palawanica* Stitz, 1925: 118 (s.) PHILIPPINES.

***pallidula.*** *Myrmica pallidula* Nylander, 1849: 42 (w.) SICILY. Mayr, 1855: 457 (s.q.m.); Vandel, 1931: 114 (gynandromorph); Hauschteck, 1961: 221 (k.). Combination in *Oecophthora:* Mayr, 1855: 455; in *Pheidole:* Smith, 1858b: 173. Subspecies of *megacephala:* Emery & Forel, 1879: 463; Forel, 1890a: lxxvi; Emery, 1891b: 13. Revived status as species: Dalla Torre, 1893: 94; Ruzsky, 1902d: 26. Senior synonym of *subdentata:* Mayr, 1855: 456; of *orientalis:* Baroni Urbani, 1964a: 3; Atanassov & Dlussky, 1992: 123; of *cicatricosa:* Collingwood, 1978: 68; of *inquilina:* Bolton, 1987: 291; of *arenarum:* Atanassov & Dlussky, 1992: 123. Current subspecies: nominal plus *inermis* (junior homonym), *koshewnikovi, obscura, recticeps, tristis* (junior homonym). See also: Emery, 1915j: 226; Wheeler, W.M. 1937c: 38; Baroni Urbani, 1971c: 70.

**#*pampana.*** *Pheidole nitidula* st. *pampana* Santschi, 1929d: 283, figs. 6–8 (s.w.m.) ARGENTINA.

*parallela. Pheidole ampla* var. *parallela* Forel, 1902h: 435 (s.w.m.) AUSTRALIA. Junior synonym of *vigilans:* Brown, 1971b: 13.

**#*paranana.*** *Pheidole triconstricta* var. *paranana* Santschi, 1925b: 13 (s.) BRAZIL.

***parasitica.*** *Pheidole parasitica* Wilson, 1984: 321, fig. 3 (q.) INDIA.

**#*particeps.*** *Pheidole berthoudi* var. *particeps* Santschi, 1921c: 116 (w.) TANZANIA. Subspecies of *sculpturata:* current status.

***partita.*** *Pheidole partita* Mayr, 1887: 590 (s.), 604 (w.) BRAZIL.

***parva.*** *Pheidole parva* Mayr, 1865: 98, pl. 4, fig. 28 (s.w.) SRI LANKA. Bingham, 1903: 245 (q.). Current subspecies: nominal plus *decanica.*

**#*parviceps.*** *Pheidole ampla* var. *parviceps* Forel, 1915b: 57 (s.) AUSTRALIA.

**#*parvinoda.*** *Pheidole radoszkowskii* r. *parvinoda* Forel, 1912f: 223 (s.w.) BRAZIL.

**#*parvispina.*** *Pheidole variabilis* r. *parvispina* Forel, 1902h: 424 (s.w.m.) AUSTRALIA.

*pattesoni. Pheidole oceanica* var. *pattesoni* Mann, 1919: 317 (s.w.) SOLOMON IS. Junior synonym of *oceanica:* Wilson & Taylor, 1967: 48.

**#*pectinata.*** *Pheidole javana* var. *pectinata* Stitz, 1912: 504 (s.) INDONESIA (Seram I.). Subspecies of *fervens:* current status.

***peguensis.*** *Pheidole latinoda* subsp. *peguensis* Emery, 1895k: 468 (s.w.) BURMA. Raised to species: Forel, 1902c: 182. Current subspecies: nominal plus *yomensis.*

**#*pellax.*** *Pheidole rohani* st. *pellax* Santschi, 1930b: 63, fig. 2 (s.w.) ANGOLA.

**#*pencosensis.*** *Pheidole spininodis* var. *pencosensis* Forel, 1914d: 271 (s.w.q.m.) ARGENTINA.

***penetralis.*** *Pheidole penetralis* Smith, F. 1863: 23 (w.) INDONESIA (Misool I.).

*pennsylvanica. Pheidole pennsylvanica* Roger, 1863a: 199 (s) U.S.A. Mayr, 1886d: 456 (s.w.q.m.). Junior synonym of *pilifera:* Emery, 1895c: 290.

*pennsylvanica. Atta pennsylvanica* Buckley, 1867: 345 (s.w.) U.S.A. Combination in *Pheidole:* Emery, 1895c: 338. [Junior secondary homonym of *pennsylvanica* Roger, above.] Replacement name: *buckleyi:* Smith, M.R. 1951a: 806.

**#*peradeniyae.*** *Pheidole latinoda* var. *peradeniyae* Forel, 1911i: 222 (s.w.q.) SRI LANKA.

**#*perakensis.*** *Pheidole huberi* var. *perakensis* Forel, 1911d: 376 (s.w.) WEST MALAYSIA. Forel, 1911d: 386 (q.).

***peregrina.*** *Pheidole peregrina* Wheeler, W.M. 1916l: 181 (s.w.q.) BRAZIL.

*pergandei* Donisthorpe, 1950 [*Ceratopheidole*]; see under ***APHAENOGASTER.***

*perniciosa. Oecophthora perniciosa* Gerstäcker, 1859: 263 (w.) MOZAMBIQUE. [Also described as new by Gerstäcker, 1862: 516.] Combination in *Pheidole:* Roger, 1863b: 31. Junior synonym of *megacephala:* Emery, 1915j: 235.

***perpusilla.*** *Pheidole flavens* subsp. *perpusilla* Emery, 1894c: 157 (s.w.q.) BRAZIL. Combination in *P.* (*Decapheidole*): Forel, 1912f: 237. Raised to species: Emery, 1896g: 80.

**#*personata.*** *Pheidole minutula* var. *personata* Santschi, 1929d: 291 (s.w.) GUIANA.

**#*perthensis.*** *Pheidole ampla* r. *perthensis* Crawley, 1922c: 24 (s.w.q.) AUSTRALIA.

**#*perversa.*** *Pheidole perversa* Forel, 1908c: 373 (s.w.) BRAZIL. Subspecies of *strobeli:* Emery, 1922e: 101; of *nitidula:* Kempf, 1972a: 197.

**#*petax.*** *Pheidole nemoralis* var. *petax* Forel, 1895h: 488 (s.w.) MADAGASCAR.

*petiolicola. Pheidole cramptoni* subsp. *petiolicola* Wheeler, W.M. 1921f: 147 (s.w.) GUIANA. Junior synonym of *cramptoni:* Kempf & Brown, 1968: 97.

***philemon.*** *Pheidole philemon* Forel, 1910b: 44 (s.w.m.) SOLOMON IS.

***philippi.*** *Pheidole philippi* Emery, 1915g: 8 (s.w.) ETHIOPIA.

***phipsoni.*** *Pheidole phipsoni* Forel, 1902c: 172 (s.), 190 (w.) INDIA. [Also described as new by Forel, 1902f: 541.]

***picata.*** *Pheidole megacephala* var. *picata* Forel, 1891b: 178 (s.w.) MADAGASCAR. Race/subspecies of *megacephala:* Forel, 1895a: 49; of *punctulata:* Forel, 1897c: 186; Forel, 1905b: 163; Santschi, 1910c: 370. Raised to species: Emery, 1915j: 245; Wheeler, W.M. 1922a: 1019. Current subspecies: nominal plus *bernhardae, gietleni.*

*picea. Atta picea* Buckley, 1867: 344 (s.w.) U.S.A. Combination in *Pheidole:* Emery, 1895c: 338. Unrecognisable taxon; *incertae sedis* in *Pheidole:* Smith, M.R. 1951a: 806.

*picea. Pheidole picea* Mayr, 1870b: 988 (s.w.) MEXICO. [Junior secondary homonym of *picea* Buckley, above.] Replacement name: *piceonigra* Emery, 1922e: 109.

***piceonigra.*** *Pheidole piceonigra* Emery, 1922e: 109. Replacement name for *picea* Mayr, 1870b: 988. [Junior secondary homonym of *Atta picea* Buckley, 1967: 344.]

***pieli.*** *Pheidole pieli* Santschi, 1925f: 83 (s.w.) CHINA. Ogata, 1982: 195 (m.).

***pilifera.*** *Leptothorax pilifer* Roger, 1863a: 180 (w.) U.S.A. Emery, 1895c: 290 (s.); Wheeler, G.C. & Wheeler, J. 1953b: 79 (l.). Combination in *Pheidole:* Emery, 1895c: 290. Senior synonym of *pennsylvanica:* Emery, 1895c: 290; of *septentrionalis,* and material of the unavailable name *simulans* referred here: Creighton, 1950a: 186. Current subspecies: nominal plus *artemisia, coloradensis, pacifica.*

***piliventris.*** *Atta piliventris* Smith, F. 1858b: 169 (q.) BRAZIL. Combination in *Pheidole:* Mayr, 1886c: 360.

***pinealis.*** *Pheidole pinealis* Wheeler, W.M. 1908e: 459, pl. 27, fig. 38 (s.w.) U.S.A.

***plagiaria.*** *Pheidole plagiaria* Smith, F. 1860b: 112 (s.w.) INDONESIA (Batjan I.). Senior synonym of *divergens:* Mayr, 1879: 675 (in key). Current subspecies: nominal plus *moica, palawanica, rectilineata.*

***planifrons.*** *Pheidole planifrons* Santschi, 1920h: 166, fig. 1 (s.w.) VIETNAM. [Also described as new by Santschi, 1924c: 105.]

*platensis. Pheidole flavens* r. *tuberculata* var. *platensis* Forel, 1914d: 273 (s.w.m.) ARGENTINA. Unavailable name.

**#*platycephala.*** *Pheidole platycephala* Stitz, 1916: 380, fig. 3 (s.w.) ZAIRE. Variety/subspecies of *schoutedeni:* Arnold, 1920a: 450 (in text); Wheeler, W.M. 1922a: 819.

***platypus.*** *Pheidole platypus* Crawley, 1915b: 234 (s.w.) AUSTRALIA.

***plebecula.*** *Pheidole plebecula* Forel, 1899c: 68 (s.) COSTA RICA.

***plinii.*** *Pheidole plinii* Forel, 1911a: 40 (s.w.q.) BORNEO.

**#*pluto.*** *Pheidole crassinoda* var. *pluto* Arnold, 1920a: 473 (s.w.) MOZAMBIQUE.

**#*polita.*** *Pheidole fabricator* var. *polita* Emery, 1894c: 155 (s.w.) BOLIVIA.

*polita. Pheidole exasperata* var. *polita* Viehmeyer, 1914d: 606 (s.w.) SINGAPORE. [Junior primary homonym of *polita* Emery, above.] Replacement name: *concordia* Santschi, 1916d: 242.

*politocciput. Pheidole prelli* st. *redbankensis* var. *politocciput* Santschi, 1916b: 503 (s.w.) ZIMBABWE. Unavailable name.

***porcula.*** *Pheidole crassicornis* subsp. *porcula* Wheeler, W.M. 1908e: 466, pl. 27, fig. 35 (s.w.) U.S.A. Taber & Cokendolpher, 1988: 95 (k.). Raised to species: Creighton, 1950a: 187.

*poweri. Pheidole spinulosa* r. *messalina* var. *poweri* Forel, 1914d: 241 (s.w.) SOUTH AFRICA. Unavailable name.

#***praedo.*** *Pheidole variabilis* r. *praedo* Forel, 1902h: 426 (s.w.) AUSTRALIA.

*praetermissa. Pheidole punctatissima* subsp. *jamaicensis* var. *praetermissa* Wheeler, W.M. 1937b: 446 (s.w.) CUBA. Unavailable name.

***praeusta.*** *Pheidole praeusta* Roger, 1863a: 197 (s.) COLOMBIA. Forel, 1901e: 361 (w.).

*praevexata. Pheidole nodus* var. *praevexata* Wheeler, W.M. 1929f: 3 (s.w.q.) JAPAN. Junior synonym of *noda:* Yasumatsu, 1962: 96.

***prelli.*** *Pheidole prelli* Forel, 1911d: 362 (s.w.q.m.) KENYA. Current subspecies: nominal plus *ingenita, redbankensis.*

***pronotalis.*** *Pheidole pronotalis* Forel, 1902c: 173 (s.), 190 (w.) SRI LANKA. [Also described as new by Forel, 1902f: 541.]

*proserpina. Pheidole proserpina* Wheeler, W.M. 1908e: 437 (s.w.) U.S.A. Junior synonym of *sciophila:* Creighton & Gregg, 1955: 19.

#***protea.*** *Pheidole javana* subsp. *proteus* Forel, 1912n: 55 (s.) SUMATRA. Subspecies of *fervens:* current status.

***providens.*** *Atta providens* Sykes, 1835: 103, pl. 13, fig. 5 (s.) INDIA. Combination in *Pheidole:* Westwood, 1839: 219. [Bingham, 1903: 220 (footnote) suggests that this name may be synonymous with *indica.*]

***proxima.*** *Pheidole proxima* Mayr, 1876: 104 (s.w.) AUSTRALIA. Current subspecies: nominal plus *bombalensis, transversa.*

***psammophila.*** *Pheidole psammophila* Creighton & Gregg, 1955: 15, figs. 4, 5 (s.w.) U.S.A.

#***pseudola.*** *Pheidole* (*Ischnomyrmex*) *longipes* r. *pseudola* Forel, 1915a: 28 (s.w.q.m.) SUMATRA.

#***pubens.*** *Pheidole foreli* var. *pubens* Forel, 1910f: 12 (s.w.) SOUTH AFRICA. Forel, 1910e: 436 (q.m.).

***pubiventris.*** *Pheidole pubiventris* Mayr, 1887: 595 (s.), 604 (w.), 607 (q.) BRAZIL. Current subspecies: nominal plus *cearensis, foederalis, impia, nevadensis, timmii, variegata.*

#***pugnax.*** *Pheidole radoszkowskii* subsp. *pugnax* Dalla Torre, 1892: 91. Replacement name for *militaris* Emery, 1890b: 49. [Junior primary homonym of *militaris* Smith, F. 1860a: 74.]

***pulchella.*** *Pheidole pulchella* Santschi, 1910c: 360 (w.) CONGO. [Misspelled as *putchella* by Emery, 1921f: 89.] Senior synonym of *niapuana:* Santschi, 1930a: 59. Current subspecies: nominal plus *achantella.*

#***pullata.*** *Pheidole mentita* var. *pullata* Santschi, 1914d: 339 (s.w.) GUINEA.

#***pulliventris.*** *Pheidole bergi* st. *pulliventris* Santschi, 1929d: 288 (s.) BRAZIL.

#***pullula.*** *Pheidole silvestrii* var. *pullula* Santschi, 1911b: 53 (s.w.q.) ARGENTINA.

***punctatissima.*** *Pheidole punctatissima* Mayr, 1870a: 400 (s.w.) MEXICO. Forel, 1908b: 52 (q.). Senior synonym of *napaea:* Brown, 1981: 525. Current subspecies: nominal plus *annectens, insulana, jamaicensis.*

***punctithorax.*** *Pheidole punctithorax* Borgmeier, 1929: 210, pl. 7, fig. 5 (s.w.) BRAZIL.

***punctulata.*** *Pheidole punctulata* Mayr, 1866b: 899 (s.) SOUTH AFRICA. Arnold, 1920a: 421 (w.q.m.). Subspecies of *megacephala:* Forel, 1894b: 94; Stitz, 1911b: 385; Emery, 1915j: 235; Emery, 1921f: 85; Wheeler, W.M. 1922a: 132; Santschi, 1930b: 67. Status as species: Forel, 1910e: 441; Santschi, 1914b: 76; Forel, 1914d: 240; Arnold, 1920a: 420; Santschi, 1935a: 256. Current subspecies: nominal plus *atrox, spinosa, subatrox.*

#***punensis.*** *Pheidole sulcaticeps* var. *punensis* Forel, 1902c: 179 (s.), 195 (w.) INDIA.

***pungens.*** *Solenopsis pungens* Smith, F. 1861b: 48 (s.w.) SULAWESI. Combination in *Pheidologeton:* Donisthorpe, 1932c: 469; in *Pheidole:* **new combination.**

***purpurascens.*** *Pheidole purpurascens* Emery, 1897d: 565 (w.) NEW GUINEA. Combination in *P.* (*Pheidolacanthinus*): Emery, 1921f: 82.

*pusilla. Oecophthora pusilla* Heer, 1852: 15, pl. 1, figs. 1–4 (s.w.q.m.) MADEIRA. Combination in *Pheidole:* Smith, F. 1858b: 173. Subspecies of *megacephala:* Emery, 1915j: 235. Senior synonym of *janus:* Mayr, 1886c: 360; of *laevigata* Smith: Roger, 1859: 259; Emery, 1915j: 235; of *laevigata* Mayr: Mayr, 1870b: 981 (footnote). Junior synonym of *megacephala:* Wheeler, W.M. 1922a: 812.

*putata. Pheidole flavens* r. *tuberculata* var. *putata* Forel, 1912f: 235 (s.) BRAZIL. Unavailable name.

#***puttemansi.*** *Pheidole fallax* subsp. *puttemansi* Forel, 1911c: 304 (s.) BRAZIL.

*pyramidensis. Pheidole californica* subsp. *pyramidensis* Emery, 1922e: 105. Replacement name for *nevadensis* Wheeler, W.M. 1915b: 408. [Junior primary homonym of *nevadensis* Forel, 1901e: 353.] Junior synonym of *californica:* Wheeler, G.C. & Wheeler, J. 1986g: 13.

***pyriformis.*** *Pheidole pyriformis* Clark, 1938: 371, figs. 10, 11 (s.w.) AUSTRALIA.

***quadrensis.*** *Pheidole quadrensis* Forel, 1900a: 25 (s.) SUMATRA. Emery, 1900d: 682 (w.). Combination in *P.* (*Pheidolacanthinus*): Forel, 1913k: 45.

***quadricuspis.*** *Pheidole quadricuspis* Emery, 1900d: 683 (s.w.) SUMATRA. Combination in *P.* (*Pheidolacanthinus*): Emery, 1921f: 83.

***quadriprojecta.*** *Pheidole quadriprojectus* Smith, M.R. 1947c: 74, figs. 1, 2 (s.) NEW GUINEA.

***quadrispinosa.*** *Myrmica quadrispinosa* Smith, F. 1865: 72, pl. 4, fig. 6 (w.) INDONESIA (Salwatty I.). Donisthorpe, 1941b: 38 (q.m.). Combination in *Pheidole:* Mayr, 1878: 677; in *P.* (*Pheidolacanthinus*): Emery, 1921f: 82. Senior synonym of *smithi* (unnecessary replacement name): Dalla Tore, 1893: 95; of *armatus:* Donisthorpe, 1938a: 144.

***quiaccana.*** *Pheidole quiaccana* Wheeler, W.M. 1925a: 13 (s.w.m.) BOLIVIA.

***rabo.*** *Pheidole rabo* Forel, 1913k: 28 (s.w.q.) SUMATRA.

***radoszkowskii.*** *Pheidole radoszkowskii* Mayr, 1884: 35 (s.w.) FRENCH GUIANA. Current subspecies: nominal plus *acuta, australis, discursans, inversa, luteola, opacissima, parvinoda, pugnax, reflexans, saviozae.*

*ralumensis. Pheidole sexspinosa* r. *biroi* var. *ralumensis* Forel, 1901b: 14 (w.) BISMARCK ARCHIPELAGO. Unavailable name.

***rata.*** *Pheidole rata* Borgmeier, 1929: 206, pl. 6, fig. 2; pl. 7, fig. 6 (s.w.) BRAZIL.

#***reclusi.*** *Pheidole distorta* r. *reclusi* Forel, 1899c: 75 (footnote) (s.w.) COLOMBIA. Forel, 1901e: 356 (q.). Combination in *P.* (*Trachypheidole*): Emery, 1922e: 111.

#***recticeps.*** *Pheidole pallidula* var. *recticeps* Menozzi, 1932e: 452 (s.w.) EGYPT. [First available use of *Pheidole pallidula* subsp. *tristis* var. *recticeps* Forel, 1909e: 391; unavailable name.]

#***rectilineata.*** *Pheidole plagiaria* var. *rectilineata* Viehmeyer, 1916b: 288 (s.) SULAWESI (in copal).

#***recurva.*** *Pheidole concentrica* subsp. *recurva* Forel, 1910b: 39 (s.w.q.m.) TASMANIA.

#***redbankensis.*** *Pheidole prelli* r. *redbankensis* Forel, 1913a: 128 (s.w.) ZIMBABWE.

#***reddenburgensis.*** *Pheidole* (*Allopheidole*) *cuitensis* r. *reddenburgensis* Forel, 1913a: 135 (s.w.q.m.) SOUTH AFRICA. Subspecies of *capensis:* Emery, 1915j: 246; Wheeler, W.M. 1922a: 809. [Spelling emended to *reddersbergensis* by Arnold, 1920a: 450; emendation not accepted by Wheeler, W.M. 1922a: 809.]

#***redunca.*** *Pheidole variabilis* var. *redunca* Crawley, 1915b: 235 (s.w.) AUSTRALIA.

#***reflexans.*** *Pheidole* (*Pheidole*) *radoszkowskii* var. *reflexans* Santschi, 1933e: 109 (s.w.) ARGENTINA.

*regia. Pheidole oxyops* subsp. *regia* Forel, 1908c: 378 (s.w.) BRAZIL. Forel, 1909a: 258 (q.). Junior synonym of *oxyops:* Kempf, 1964e: 58.

*regina. Pheidole biconstricta* subsp. *bicolor* var. *regina* Forel, 1908b: 52 (s.w.) COSTA RICA. Unavailable name.

***rehi.*** *Pheidole rehi* Forel, 1900e: 285 (s.w.q.m.) VENEZUELA.

***reichenspergeri.*** *Pheidole reichenspergeri* Santschi, 1923d: 49 (s.w.) BRAZIL.

*retifera. Pheidole fallax* subsp. *emiliae* var. *retifera* Emery, 1906c: 155 (s.w.) ARGENTINA. Unavailable name.

***retronitens.*** *Pheidole retronitens* Santschi, 1930a: 59 (s.w.) ZAIRE.

*rhaphidiiceps* Mayr, 1877 [*Ischnomyrmex*]; see under ***APHAENOGASTER.***

***rhea.*** *Pheidole rhea* Wheeler, W.M. 1908e: 452 (q.) U.S.A. Smith, M.R. 1943c: 7 (s.w.). Combination in *P.* (*Macropheidole*): Smith, M.R. 1943c: 5. Junior synonym of *fimbriata:* Wheeler, W.M. 1915b: 403; Emery, 1921f: 81. Revived from synonymy: Smith, M.R. 1943c: 7. See also: Gregg, 1949a: 70; Creighton, 1950a: 168.

***rhinoceros.*** *Pheidole rhinoceros* Forel, 1899c: 73, pl. 3, fig. 23 (s.w.) PANAMA.

#***rhodesiana.*** *Pheidole excellens* r. *rhodesiana* Forel, 1913a: 131 (s.w.) ZIMBABWE. Arnold, 1920a: 470 (m.). Stirps of *sculpturata:* Santschi, 1930b: 63 (footnote).

*rhombinoda. Pheidole rhombinoda* Mayr, 1879: 678 (s.) INDIA. Bingham, 1903: 251 (q.). Subspecies of *noda:* Wheeler, W.M. 1929f: 3; Santschi, 1937h: 371. Junior synonym of *noda:* Yasumatsu, 1962: 96. [Misspelled as *rhomboida* by Santschi, 1925f: 83.]

**#richteri.** *Pheidole perversa* subsp. *richteri* Forel, 1909a: 266 (s.w.q.) ARGENTINA. Forel, 1914d: 273 (m.). Variety/subspecies of *strobeli:* Santschi, 1916e: 373; of *nitidula:* Santschi, 1929d: 281.

**ridicula.** *Pheidole ridicula* Wheeler, W.M. 1916f: 29, fig. 1 (s.) U.S.A. See also: Creighton, 1966b: 1.

**rinae.** *Pheidole rinae* Emery, 1900d: 687 (s.) SUMATRA. Current subspecies: nominal plus *hongkongensis, incensa, mala, taipoana, tipuna.*

**risii.** *Pheidole risii* Forel, 1892a: 38 (s.w.q.m.) ARGENTINA. Current subspecies: nominal plus *rugatula, trachyderma.*

**riveti.** *Pheidole riveti* Santschi, 1911i: 278 (s.w.q.) ECUADOR. [Also described as new by Santschi, 1913h: 37.] Current subspecies: nominal plus *inflexa.*

**roberti.** *Pheidole roberti* Forel, 1902c: 183 (s.), 198 (w.) INDIA. [Also described as new by Forel, 1902f: 545.]

**#robusta.** *Pheidole tenuinodis* var. *robusta* Stitz, 1923: 154 (s.w.q.m.) NAMIBIA.

**#rochai.** *Pheidole lemur* r. *rochai* Forel, 1912f: 226 (s.w.) BRAZIL.

**rogeri.** *Pheidole rogeri* Emery, 1896g: 64 (s.) COSTA RICA.

**rogersi.** *Pheidole rogersi* Forel, 1902c: 182 (s.), 197 (w.) INDIA. [Also described as new by Forel, 1902f: 545.] Current subspecies: nominal plus *taylori.*

**rohani.** *Pheidole rohani* Santschi, 1925h: 160, fig. 11 (w.) ANGOLA. Santschi, 1930b: 62 (s.). Current subspecies: nominal plus *monardi, pellax.*

**roosevelti.** *Pheidole (Electropheidole) roosevelti* Mann, 1921: 438, fig. 15 (s.w.q.) FIJI IS.

**rosae.** *Pheidole rosae* Forel, 1901h: 63 (s.) BRAZIL.

**#rosariensis.** *Pheidole triconstricta* var. *rosariensis* Forel, 1913l: 230 (s.w.) ARGENTINA.

**#rotschana.** *Pheidole indica* r. *rotschana* Forel, 1902c: 185 (s.), 199 (w.m.) INDIA. Imai, Baroni Urbani, *et al.* 1984: 6 (k.). [Also described as new by Forel, 1902f: 546.] Raised to species: Bingham, 1903: 264. Subspecies of *indica:* Forel, 1909e: 394; Forel, 1911i: 222.

**#rotundata.** *Pheidole rotundata* Forel, 1894b: 92 (s.w.) MOZAMBIQUE. Arnold, 1920a: 426 (m.); Santschi, 1932a: 384 (q.). Status as species: Arnold, 1920a: 426. Subspecies/race of *megacephala:* Emery, 1915j: 236; Emery, 1921f: 86; Wheeler, W.M. 1922a: 816; Stitz, 1923: 154; Santschi, 1932a: 384.

*rubens. Pheidole fallax* var. *rubens* Forel, 1899c: 66 (s.) COLOMBIA. Junior synonym of *fallax:* Forel, 1901e: 365.

**#rubicunda.** *Pheidole biconstricta* r. *rubicunda* Emery, 1890b: 50 (s.w.) COSTA RICA. [Also described as new by Emery, 1894k: 55.] Raised to species: Dalla Torre, 1893: 96. Race of *biconstricta:* Forel, 1899c: 64.

*rubra. Atta rubra* Smith, F. 1858b: 168 (q.) BRAZIL. Combination in *Pheidole:* Mayr, 1886c: 360. Junior synonym of *tristis:* Kempf, 1965: 183.

*rubra* Smith, F. 1860; see under ***PHEIDOLOGETON.***

**#rubricalva.** *Pheidole aurivillii* var. *rubricalva* Forel, 1915c: 337 (s.w.) ZAIRE.

*rubrobrunnea* Dalla Torre, 1892; see under ***PHEIDOLOGETON.***

**#rudigenis.** *Pheidole flavens* subsp. *rudigenis* Emery, 1906c: 155 (s.) ARGENTINA.

**#rufescens.** *Pheidole sitarches* subsp. *rufescens* Wheeler, W.M. 1908e: 443 (s.w.q.) U.S.A. Junior synonym of *campestris:* Creighton, 1950a: 189; of *soritis:* Cole. 1953e: 298; of *campestris:* Gregg, 1959: 30. [Although both Creighton and Gregg give *campestris* as the senior name, *rufescens* has priority.] Subspecies of *sitarches:* current status.

**#rufescens.** *Pheidole arnoldi* var. *rufescens* Arnold, 1920a: 449 (s.w.) ZIMBABWE. [Unresolved junior primary homonym of *rufescens* Wheeler, above.] [The proposed replacement name *sipapomae* Baroni Urbani, 1971b: 361, is unavailable.]

**ruficeps.** *Myrmica ruficeps* Smith, F. 1861b: 46 (q.) SULAWESI. Combination in *Pheidole,* and senior synonym of *mordax:* Mayr, 1867a: 96.

**rufipilis.** *Pheidole rufipilis* Forel, 1908c: 371 (s.w.) BRAZIL. Current subspecies: nominal plus *dione, diversa, industa, laevinota.*

**rugaticeps.** *Pheidole rugaticeps* Emery, 1877b: 375 (s.w.q.) ETHIOPIA. Current subspecies: nominal plus *arabs.*

**#rugatula.** *Pheidole (Pheidole) risi* var. *rugatula* Santschi, 1933e: 109 (s.w.) ARGENTINA.

*rugifrons. Pheidole rugifrons* Pergande, 1896: 880 (s.) MEXICO. Junior synonym of *tepicana:* Emery, 1901b: 119.

**#rugocciput.** *Pheidole variabilis* var. *rugocciput* Forel, 1902h: 423 (s.w.) AUSTRALIA.

**rugosa.** *Pheidole rugosa* Smith, F. 1858b: 175 (w.) SRI LANKA. See also: Forel, 1902c: 194.

**#rugosula.** *Pheidole variabilis* var. *rugosula* Forel, 1902h: 423 (s.w.) AUSTRALIA.

**rugulosa.** *Pheidole rugulosa* Gregg, 1959: 26 (s.w.q.m.) U.S.A.

*rupicapra* Stitz, 1938; see under ***ANCYRIDRIS.***

**#ruspolii.** *Pheidole crassinoda* subsp. *ruspolii* Emery, 1897e: 597 (s.w.) ETHIOPIA. Forel, 1907c: 138 (m.); Arnold, 1920a: 475 (q.).

**ryukyuensis.** *Pheidole ryukyuensis* Ogata, 1982: 193, figs. 1–5 (s.w.q.) JAPAN.

**sagana.** *Pheidole sagana* Wheeler, W.M. 1934g: 169 (s.w.) MEXICO.

**sagei.** *Pheidole sagei* Forel, 1902c: 174 (s.), 192 (w.) INDIA. [Also described as new by Forel, 1902f: 542.]

**#santschii.** *Pheidole santschii* Forel, 1907b: 204 (s.w.) TUNISIA. Variety of *sinaitica:* Emery, 1915j: 234.

**#sapuana.** *Pheidole havilandi* var. *sapuana* Forel, 1911d: 373 (s.w.) SUMATRA.

**#sarawakana.** *Pheidole sauberi* subsp. *sarawakana* Forel, 1911a: 45 (s.w.) BORNEO.

**#sarcina.** *Pheidole breviconus* var. *sarcina* Forel, 1912f: 230 (s.w.) BRAZIL.

**#sarrita.** *Pheidole opaca* subsp. *sarrita* Forel, 1908b: 59 (w.) COSTA RICA. Subspecies of *cephalica:* Kempf, 1972a: 188.

*satura. Pheidole californica* var. *satura* Wheeler, W.M. 1915b: 407 (s.w.) U.S.A. Junior synonym of *californica:* Creighton, 1950a: 173.

**sauberi.** *Pheidole sauberi* Forel, 1905c: 18 (s.w.) JAVA. Current subspecies: nominal plus *sarawakana.*

**sauteri.** *Pheidole sauteri* Wheeler, W.M. 1909d: 334 (s.w.) TAIWAN.

**#saviozae.** *Pheidole radoszkowskii* var. *saviozae* Forel, 1911c: 304 (s.) BRAZIL.

**saxicola.** *Pheidole saxicola* Wheeler, W.M. 1922a: 138, fig. 35 (s.w.) ZAIRE.

**#scabrata.** *Pheidole longispinosa* r. *scabrata* Forel, 1895c: 249 (s.w.) MADAGASCAR.

**#scabrior.** *Pheidole megacephala* var. *scabrior* Forel, 1891b: 178 (s.w.) MADAGASCAR. See also: Forel, 1897c: 188.

**scabriuscula.** *Pheidole scabriuscula* Gerstäcker, 1871: 356 (w.) KENYA.

*scabrocciput. Pheidole longiscapa* r. *martensis* var. *scabrocciput* Forel, 1914c: 615 (s.w.) COLOMBIA. Unavailable name.

**scapulata.** *Pheidole scapulata* Santschi, 1923d: 63 (w.) ARGENTINA. Combination in *P.* (*Trachypheidole*): Kempf, 1972a: 205.

**#schmalzi.** *Pheidole dimidiata* var. *schmalzi* Emery, 1894c: 161 (s.w.) BRAZIL.

*schmitzi. Pheidole schmitzi* Forel, 1911h: 455 (s.w.q.) ISRAEL. Junior synonym of *jordanica:* Emery, 1915j: 233.

**schoutedeni.** *Pheidole schoutedeni* Forel, 1913b: 329 (s.w.) ZAIRE. Current subspecies: nominal plus *platycephala.*

**schultzei.** *Pheidole schultzei* Forel, 1910f: 9, fig. 1 (s.w.q.m.) BOTSWANA. Current subspecies: nominal plus *ebangana, gwaaiensis, woodvalensis.*

**schwarzmaieri.** *Pheidole schwarzmaieri* Borgmeier, 1939: 420, fig. 6 (s.w.) BRAZIL.

**sciara.** *Pheidole sciara* Cole, 1955a: 47 (s.w.) U.S.A.

**sciophila.** *Pheidole sciophila* Wheeler, W.M. 1908e: 443, pl. 26, figs. 18, 19 (s.w.q.m.) U.S.A. Senior synonym of *proserpina* and *semilaevicephala:* Creighton & Gregg, 1955: 19.

**scrobifera.** *Pheidole scrobifera* Emery, 1896g: 73, pl. 1, fig. 14 (s.w.q.m.) COSTA RICA. Combination in *P.* (*Scrobopheidole*): Emery, 1915i: 190.

**#sculptior.** *Pheidole flavens* r. *sculptior* Forel, 1893g: 414 (s.w.q.) ANTILLES IS.

*sculptior. Pheidole njassae* var. *sculptior* Viehmeyer, 1914c: 36 (s.w.) TANZANIA. [Junior primary homonym of *sculptior* Forel, above.] Replacement name: *legitima* Santschi, 1916d: 242.

**sculpturata.** *Pheidole sculpturata* Mayr, 1866b: 897 (s.) SOUTH AFRICA. Emery, 1895h: 31 (w.). Current subspecies: nominal plus *areolata, berthoudi, dignata, katonae, particeps, rhodesiana, welgelegenensis, zambesiana.* See also: Arnold, 1920a: 462.

**seeldrayersi.** *Pheidole seeldrayersi* Forel, 1910a: 30 (s.w.q.) COLOMBIA.

**#selangorensis.** *Pheidole havilandi* var. *selangorensis* Forel, 1913k: 31 (s.w.q.m.) WEST MALAYSIA.

*semilaevicephala. Pheidole sciophila* var. *semilaevicephala* Smith, M.R. 1934a: 385 (s.) U.S.A. Subspecies of *sciophila:* Creighton, 1950a: 188. Junior synonym of *sciophila:* Creighton & Gregg, 1955: 19.

**semilaevis.** *Pheidole radoszkowskii* var. *semilaevis* Forel, 1901e: 363 (s.) BRAZIL. Raised to species: Santschi, 1929d: 287. Current subspecies: nominal plus *geraesensis.*

*semipolita. Pheidole flavens* subsp. *asperithorax* var. *semipolita* Emery, 1896g: 78 (s.w.) BRAZIL. Unavailable name.

**senex.** *Pheidole senex* Gregg, 1952a: 1 (s.w.) U.S.A. Senior synonym of *anfracta:* Cole, 1953e: 298; Gregg, 1959: 30.

**#senilifrons.** *Pheidole caffra* subsp. *senilifrons* Wheeler, W.M. 1922a: 130, fig. 32 (s.w.) ZAIRE.

***PHEIDOLE** (continued)*

**#senilis.** *Pheidole hetschkoi* var. *senilis* Santschi, 1929d: 289 (w.) BRAZIL.

**sensitiva.** *Pheidole sensitiva* Borgmeier, 1959a: 316, figs. 3–9 (s.w.) BRAZIL.

*septentrionalis. Pheidole pilifera* subsp. *septentrionalis* Wheeler, W.M. 1908e: 437 (s.w.) U.S.A. Junior synonym of *pilifera:* Creighton, 1950a: 186.

**sepulchralis.** *Pheidole sepulchralis* Bingham, 1903: 255 (s.) INDIA.

**sericella.** *Pheidole sericella* Viehmeyer, 1914c: 33 (s.w.) NEW GUINEA.

**severini.** *Pheidole severini* Forel, 1904d: 171 (s.w.) SURINAM.

**sexdentata.** *Pheidole* (*Pheidolacanthinus*) *sexdentatus* Donisthorpe, 1948d: 592 (w.) NEW GUINEA.

**sexspinosa.** *Pheidole sexspinosa* Mayr, 1870b: 977 (s.w.) TUVALU. Combination in *P.* (*Pheidolacanthinus*): Mann, 1919: 307. Senior synonym of *adamsoni:* Wilson & Taylor, 1967: 52. Current subspecies: nominal plus *biroi, fuscescens.*

**#seymourensis.** *Pheidole williamsi* var. *seymourensis* Wheeler, W.M. 1924a: 108 (s.w.) GALAPAGOS IS.

**sharpi.** *Pheidole sharpi* Forel, 1902c: 169 (s.), 188 (w.) INDIA. [Also described as new by Forel, 1902f: 540.] Current subspecies: nominal plus *hoogwerfi.*

**#shinsendensis.** *Pheidole liengmei* var. *shinsendensis* Forel, 1913b: 327 (w.) ZAIRE. Arnold, 1920a: 468 (s.).

*shoshoni. Pheidole californica* var. *shoshoni* Cole, 1933: 618 (s.w.) U.S.A. Subspecies of *californica:* Creighton, 1950a: 174. Junior synonym of *oregonica:* Gregg, 1959: 19.

**sikorae.** *Pheidole sikorae* Forel, 1891b: 223 (s.w.) MADAGASCAR. Forel, 1892l: 258 (q.). Current subspecies: nominal plus *litigiosa.*

*silenus* Smith, F. 1858; see under ***PHEIDOLOGETON.***

**silvestrii.** *Pheidole silvestrii* Emery, 1906c: 146 (s.w.q.) ARGENTINA. Current subspecies: nominal plus *pullula.*

**#silvicola.** *Pheidole strobeli* subsp. *silvicola* Borgmeier, 1927b: 61 (s.w.) BRAZIL. [Also described as new by Borgmeier, 1928a: 39.] Stirps of *nitidula:* Santschi, 1929d: 283.

**#sima.** *Pheidole championii* var. *sima* Forel, 1912f: 234 (s.w.) COSTA RICA.

**similigena.** *Pheidole similigena* Wheeler, W.M. 1937b: 446 (s.w.) CUBA.

**simoni.** *Pheidole simoni* Emery, 1893g: 266 (s.) PHILIPPINES.

**#simplex.** *Pheidole biconstricta* subsp. *simplex* Wheeler, W.M. 1925a: 21 (s.) PERU.

*simulans. Pheidole pilifera* subsp. *coloradensis* var. *simulans* Wheeler, W.M. 1908e: 436 (s.w.) U.S.A. Unavailable name; material referred to *pilifera* by Creighton, 1950a: 186.

**sinaitica.** *Pheidole sinaitica* Mayr, 1862: 745 (s.w.) EGYPT. Magretti, 1884: 542 (q.); Finzi, 1936: 163 (q.). Current subspecies: nominal plus *santschii.*

**singularis.** *Pheidole singularis* Smith, F. 1863: 22 (w.) INDONESIA (Misool I.). Combination in *P.* (*Pheidolacanthinus*): Emery, 1921f: 82.

**sinica.** *Ceratopheidole sinica* Wu & Wang, 1992: 1306, fig. 4580 (s.w.) CHINA. Combination in *Pheidole:* **new combination**.

**#sipapomae.** *Pheidole tenuinodis* var. *sipapomae* Arnold, 1920a: 430 (s.w.) ZIMBABWE.

*sipapomae. Pheidole arnoldi* var. *sipapomae* Baroni Urbani, 1971b: 361. Unavailable name (incorrectly proposed replacement name; see under *rufescens*).

**sitarches.** *Pheidole sitarches* Wheeler, W.M. 1908e: 440 (s.w.q.) U.S.A. Senior synonym of *transvarians:* Creighton, 1950a: 189; Gregg, 1959: 30. Current subspecies: nominal plus *rufescens, soritis.*

**skwarrae.** *Pheidole skwarrae* Wheeler, W.M. 1934g: 163 (s.w.q.) MEXICO.

**#sloanei.** *Pheidole caribbaea* subsp. *sloanei* Wheeler, W.M. 1917g: 459 (s.w.q.) JAMAICA.

*smithi. Pheidole smithi* Emery, 1887b: 464. [Unnecessary replacement name for *quadrispinosa.*] Junior synonym of *quadrispinosa:* Dalla Torre, 1893: 95.

*smithii. Pheidole smithii* Dalla Torre, 1892: 90. [Unnecessary replacement name for *diversa.*] Junior synonym of *fimbriata:* Kempf, 1965: 183.

**smythiesii.** *Pheidole* (*Ceratopheidole*) *smythiesii* Forel, 1902c: 165 (s.), 185 (w.q.m.) INDIA. [Also described as new by Forel, 1902f: 38.] Combination in *Ceratopheidole:* Emery, 1922e: 113; in *Pheidole:* **revived combination**. Current subspecies: nominal plus *bengalensis.*

**#socrates.** *Pheidole biconstricta* r. *socrates* Forel, 1912f: 223 (s.w.) BRAZIL.

*socrus. Pheidole guilelmimuelleri* subsp. *bucculenta* var. *socrus* Forel, 1914e: 12 (w.) COLOMBIA. Unavailable name.

**#solaris.** *Pheidole spininodis* var. *solaris* Santschi, 1929d: 287 (s.w.q.m.) BRAZIL.

**#solitanea.** *Pheidole hyatti* subsp. *solitanea* Wheeler, W.M. 1915b: 409 (s.w.q.) U.S.A.

**#sordidula.** *Pheidole crassinoda* var. *sordidula* Santschi, 1937a: 54 (s.w.m.) TANZANIA.

**#soritis.** *Pheidole soritis* Wheeler, W.M. 1908e: 439 (s.w.) U.S.A. Subspecies of *sitarches:* Creighton, 1950a: 190. Revived status as species: Cole, 1953e: 298. Junior synonym of *sitarches:* Cole, 1956c: 115. Revived from synonymy as subspecies of *sitarches:* Gregg, 1959: 30; Smith, D.R. 1979: 1373. Senior synonym of *cavigenis:* Gregg, 1959: 30.

**#soror.** *Pheidole javana* st. *soror* Santschi, 1937h: 369, fig. 6 (s.w.) TAIWAN. Subspecies of *fervens:* current status.

**#sospes.** *Pheidole anastasii* var. *sospes* Forel, 1908c: 381 (s.w.) BRAZIL. Stirps of *anastasii:* Santschi, 1929d: 291.

**spadonia.** *Pheidole spadonia* Wheeler, W.M. 1915b: 400 (s.w.) U.S.A. Creighton & Gregg, 1955: 23 (q.).

**spathifera.** *Pheidole spathifera* Forel, 1902c: 168 (s.), 187 (w.q.m.) INDIA. [Also describrd as new by Forel, 1902f: 539.] Current subspecies: nominal plus *aspatha, yerburyi.*

**speculifera.** *Pheidole speculifera* Emery, 1877b: 373 (s.w.) ETHIOPIA. Santschi, 1930d: 268 (m.); Bernard, 1953b: 223 (q.). [Misspelled as *speculifrons* by Dalla Torre, 1893: 96.] Current subspecies: nominal plus *ascara, bispecula, cubangensis.*

**#speculifrons.** *Pheidole megacephala* var. *speculifrons* Stitz, 1911b: 386 (s.w.) TANZANIA.

**#spei.** *Pheidole flavens* st. *spei* Santschi, 1930c: 77 (s.w.) CUBA.

**sperata.** *Pheidole sperata* Forel, 1915a: 29 (s.w.) SUMATRA.

**spielbergi.** *Pheidole spielbergi* Emery, 1888c: 354 (s.w.) BRAZIL.

**spininodis.** *Pheidole spininodis* Mayr, 1887: 591 (s.), 605 (w.) ARGENTINA. Goni, Zolessi & Imai, 1983: 365 (k.). Senior synonym of *hohenlohei:* Emery, 1906c: 146 (in text). Current subspecies: nominal plus *bruta, lucifuga, pencosensis, solaris.*

**spinoda.** *Atta spinoda* Smith, F. 1858b: 166 (q.) AUSTRALIA. Combination in *Aphaenogaster:* Dalla Torre, 1893: 104 (misspelled as *spininoda*); in *Podomyrma:* Emery, 1915d: 68; in *Pheidole:* Taylor & Brown, D.R. 1985: 77.

**#spinosa.** *Pheidole megacephala* var. *spinosa* Forel, 1891b: 178 (s.w.) MADAGASCAR. Race of *megacephala:* Forel, 1897c: 186; Forel, 1904b: 373; of *punctulata:* Forel, 1905b: 164.

**spinulosa.** *Pheidole spinulosa* Forel, 1910e: 438 (s.w.q.m.) LESOTHO. Current subspecies: nominal plus *conigera, messalina.*

**#splendidula.** *Pheidole metallescens* subsp. *splendidula* Wheeler, W.M. 1908e: 474, pl. 26, figs. 20, 21 (s.w.q.m.) U.S.A.

**squalida.** *Pheidole squalida* Santschi, 1910c: 364 (s.w.) CONGO.

**steinheili.** *Pheidole steinheili* Forel, 1901e: 353 (s.w.q.m.) COLOMBIA.

**#stella.** *Pheidole rhombinoda* subsp. *stella* Forel, 1911d: 380 (s.) INDIA. Subspecies of *noda:* Wheeler, W.M. 1929f: 3.

**#stomachosa.** *Pheidole floridana* subsp. *stomachosa* Wheeler, W.M. 1917g: 459 (s.w.q.) JAMAICA.

**strator.** *Pheidole strator* Forel, 1910c: 256 (s.w.) ETHIOPIA. Current subspecies: nominal plus *fugax, tabida.*

**striata.** *Pheidole* (*Pheidolacanthinus*) *striatus* Donisthorpe, 1947b: 172, fig. 1 (s.w.m.) NEW GUINEA.

**striaticeps.** *Pheidole striaticeps* Mayr, 1870b: 987 (s.w.) MEXICO.

**striativentris.** *Pheidole striativentris* Mayr, 1879: 678 (s.) INDIA. Forel, 1902c: 195 (w.q.).

**#strobeli.** *Pheidole strobeli* Emery, 1906c: 149 (s.w.) ARGENTINA. Goni, Zolessi & Imai, 1983: 365 (k.). Subspecies/stirps of *cordiceps:* Santschi, 1912e: 528; Bruch, 1915: 531. Status as species: Emery, 1922e: 101; Borgmeier, 1927b: 61. Subspecies/stirps of *nitidula:* Santschi, 1929d: 282; Kempf, 1972a: 197. Senior synonym of *misera:* Santschi, 1929d: 282.

**stulta.** *Pheidole stulta* Forel, 1886b: xlvi (s.) BRAZIL.

**subaberrans.** *Elasmopheidole subaberrans* Kusnezov, 1952b: 71 (s.w.) ARGENTINA. Combination in *Pheidole* (*Elasmopheidole*): Kempf, 1972a: 184.

**subarmata.** *Pheidole subarmata* Mayr, 1884: 37 (s.w.) FRENCH GUIANA. Senior synonym of *borinquenensis, cornutula, dentimentum, elongatula, hondurensis, imbecillis, nassavensis* and *nefasta:* Brown, 1981: 526.

**#subatrox.** *Pheidole punctulata* var. *subatrox* Santschi, 1937g: 76 (s.w.) ZAIRE.

*subdentata. Oecophthora subdentata* Mayr, 1853a: 145 (s.w.) AUSTRIA. Junior synonym of *pallidula:* Mayr, 1855: 456.

*subdentata*. *Pheidole subdentata* Pergande, 1896: 888 (w.) MEXICO. [Junior secondary homonym of *subdentata* Mayr, above.] Subspecies of *vasliti:* Wheeler, W.M. 1914b: 50. Revived status as species: Creighton, 1958: 211. Senior synonym of *obtusospinosa:* Wheeler, W.M. 1914b: 50; of *arizcnica:* Creighton, 1958: 211. Replacement name: *obtusospinosa* Pergande, 1896: 889; first available replacement name for *subdentata* Pergande.

#***subparallela.*** *Pheidole bergi* subsp. *subparallela* Emery, 1906c: 145 (s.w.) URUGUAY.

#***subreticulata.*** *Pheidole nana* var. *subreticulata* Emery, 1894c: 159 (s.w.) BRAZIL. Subspecies of *nana:* Emery, 1896g: 78.

***sulcaticeps.*** *Pheidole sulcaticeps* Roger, 1863a: 193 (s.w.q.m.) SRI LANKA. Current subspecies: nominal plus *indosinensis, punensis, vellicans, yeensis.*

#***surda.*** *Pheidole biconstricta* var. *surda* Forel, 1912f: 222 (s.w.) COSTA RICA.

***susannae.*** *Pheidole susannae* Forel, 1886b: xliii (s.w.) GUATEMALA. Senior synonym of *evoluta:* Kempf, 1964e: 63. Current subspecies: nominal plus *atricolor, obscurior.*

*suspiciosa*. *Myrmica suspiciosa* Smith, F. 1859a: 148 (w.) INDONESIA (Aru I.). Junior synonym of *megacephala:* Donisthorpe, 1932c: 455.

***sykesii.*** *Pheidole sykesii* Forel, 1902c: 173 (s.), 190 (w.q.m.) INDIA. [Also described as new by Forel, 1902f: 541.]

***symbiotica.*** *Pheidole symbiotica* Wasmann, 1909: 693 (footnote), figs. 1, 2 (w.) PORTUGAL. Wasmann, 1910: 515 (q.m.). Junior synonym of *pallidula:* Forel, 1913d: 430. Revived from synonymy as variety of *pallidula:* Emery, 1915j: 226. Revived status as species: Collingwood, 1978: 68.

***symbiotica.*** *Eriopheidole symbiotica* Kusnezov, 1952a: 16, figs. 1–9 (q.m.) ARGENTINA. [Unresolved junior secondary homonym of *Pheidole symbiotica* Wasmann, above.] Combination in *Pheidole:* Wilson, 1984: 328.

#***tabida.*** *Pheidole strator* var. *tabida* Menozzi, 1933a: 102 (s.) UGANDA.

***tachigaliae.*** *Pheidole tachigaliae* Wheeler, W.M. 1921f: 148 (s.w.q.) GUIANA. Combination in *P.* (*Hendecapheidole*): Wheeler, W.M. 1922d: 3. See also: Smith, M.R. 1955c: 304.

*tahitiana*. *Pheidole oceanica* subsp. *nigriscapa* var. *tahitiana* Santschi, in Cheesman & Crawley, 1928: 516 (w.) TAHITI. Unavailable name; material referred to *fervens* by Wilson & Taylor, 1967: 45.

#***taina.*** *Pheidole teneriffana* subsp. *taina* Aguayo, 1932: 219 (s.) CUBA.

*taipingensis*. *Pheidole javana* r. *jacobsoni* var. *taipingensis* Forel, 1913k: 28 (s.w.) WEST MALAYSIA. Unavailable name.

#***taipoana.*** *Pheidole rinae* subsp. *taipoana* Wheeler, W.M. 1928c: 12 (s.w.) CHINA.

***taivanensis.*** *Pheidole taivanensis* Forel, 1912a: 59 (s.) TAIWAN.

#***taki.*** *Pheidole isis* var. *taki* Mann, 1919: 314 (s.) SOLOMON IS.

#***talpa.*** *Pheidole talpa* Gerstäcker, 1871: 356 (s.) KENYA. Santschi, 1930b: 67 (w.q.). Junior synonym of *punctulata:* Dalla Torre, 1893: 95; Emery, 1915j: 235. Revived from synonymy as variety of *megacephala:* Santschi, 1925h: 160.

#***tambourinensis.*** *Pheidole athertonensis* r. *tambourinensis* Forel, 1915b: 65 (s.w.q.m.) AUSTRALIA.

***tandjongensis.*** *Pheidole tandjongensis* Forel, 1913k: 42 (s.w.) SUMATRA.

*taprobanae* Smith, F. 1858; see under ***PHEIDOLOGETON.***

#***taprobanae.*** *Pheidole rhombinoda* var. *taprobanae* Forel, 1902c: 178 (s.), 195 (w.) SRI LANKA. [Unresolved junior primary homonym of *taprobanae* Smith, F. 1858b: 175, above.] [Also described as new by Forel, 1902f: 544.] Race/stirps of *rhombinoda:* Forel, 1913e: 662; of *noda:* Santschi, 1937h: 371.

***tarda.*** *Pheidole* (*Pheidole*) *tardus* Donisthorpe, 1947a: 285 (q.) MAURITIUS.

***tasmaniensis.*** *Pheidole tasmaniensis* Mayr, 1866a: 511 (s.) TASMANIA. Current subspecies: nominal plus *continentis.*

*taunggyiana*. *Pheidole latinoda* r. *angusta* var. *taunggyiana* Forel, 1903c: 253 (s.) BURMA. Unavailable name. [The name *angusta* is a misspelling of *angustior.* Emery, 1921f: 92, gives the misspelling *tanngiana.*]

***taurus.*** *Pheidole taurus* Emery, 1906c: 142, fig. 20 (s.w.q.) ARGENTINA. Combination in *P.* (*Elasmopheidole*): Forel, 1913k: 43.

#***taylori.*** *Pheidole rogersi* var. *taylori* Forel, 1902c: 182 (s.), 198 (w.) INDIA. [Also described as new by Forel, 1902f: 545.]

***templaria.*** *Pheidole templaria* Forel, 1902c: 175 (s.) INDIA. [Also described as new by Forel, 1902f: 542.] Bingham, 1903: 244 (w.). Current subspecies: nominal plus *euscrobata.*

***tenerescens.*** *Pheidole tenerescens* Wheeler, W.M. 1922c: 7 (s.) TRINIDAD.

***teneriffana.*** *Pheidole teneriffana* Forel, 1893d: 465 (s.w.) CANARY IS. [Also described as new by Forel, 1894a: 160.] Santschi, 1908: 521 (q.). Current subspecies: nominal plus *taina.* See also: Baroni Urbani, 1968b: 438; Snelling, 1992b: 121.

*tennesseensis*. *Leptothorax tennesseensis* Cole, 1938a: 238, fig. 1 (w.) U.S.A. Junior synonym of *commutata:* Cole, 1948: 82.

***tenuiclavata.*** *Pheidole* (*Pheidole*) *tenuiclavata* Donisthorpe, 1943d: 449 (s.w.) NEW GUINEA.

***tenuinodis.*** *Pheidole tenuinodis* Mayr, 1901b: 10 (s.w.) SOUTH AFRICA. Arnold, 1920a: 430 (q.). Current subspecies: nominal plus *bothae, robusta, sipapomae.*

#***tepaneca.*** *Pheidole ceres* subsp. *tepaneca* Wheeler, W.M. 1914b: 46 (s.w.q.) MEXICO.

***tepicana.*** *Pheidole tepicana* Pergande, 1896: 878 (s.w.) MEXICO. Taber & Cokendolpher, 1988: 95 (k.). Senior synonym of *carbonaria* and *rugifrons:* Emery, 1901b: 119; of *instabilis, kingi* (and its junior synonym *townsendi*) and *torpescens:* Creighton & Gregg, 1955: 24.

***termitobia.*** *Pheidole termitobia* Forel, 1901g: 390 (s.w.q.) BRAZIL.

***termitophila.*** *Pheidole termitophila* Forel, 1904h: 13 (s.w.q.) SUDAN. Current subspecies: nominal plus *liberiensis.*

***terresi.*** *Pheidole terresi* Wheeler, W.M. & Mann, 1914: 22 (s.w.q.) HAITI. Current subspecies: nominal plus *illota.*

****tertiaria.*** **Pheidole tertiaria* Carpenter, 1930: 33, pl. 5, fig. 2; pl. 11, fig. 2 (q.) U.S.A. (Miocene).

*testacea*. *Atta testacea* Smith, F. 1858b: 168 (s.w.) BRAZIL. Combination in *Pheidole:* Mayr, 1886c: 360. Junior synonym of *megacephala:* Brown, 1981: 530.

****tethepa.*** **Pheidole tethepa* Wilson, 1985a: 5, figs. 4, 5 (w.) DOMINICAN AMBER.

#***tetra.*** *Pheidole crassicornis* subsp. *tetra* Creighton, 1950a: 176 (s.w.) U.S.A. [First available use of *Pheidole crassicornis* subsp. *porcula* var. *tetra* Wheeler, W.M. 1908e: 467; unavailable name.]

***tetracantha.*** *Pheidole tetracantha* Emery, 1897d: 564 (s.) NEW GUINEA. Combination in *P.* (*Pheidolacanthinus*): Emery, 1921f: 82.

***tetrica.*** *Pheidole tetrica* Forel, 1913l: 226 (s.w.q.) BRAZIL.

***texana.*** *Pheidole texana* Wheeler, W.M. 1903b: 97, fig. 4 (s.w.) U.S.A.

#***thomensis.*** *Pheidole flavens* var. *thomensis* Emery, 1894c: 156 (s.w.q.) ANTILLES IS.

*thysvillensis*. *Pheidole caffra* subsp. *bayeri* var. *thysvillensis* Wheeler, W.M. 1922a: 130 (s.w.) ZAIRE. Unavailable name.

***tijucana.*** *Pheidole tijucana* Borgmeier, 1927b: 59 (s.w.) BRAZIL. [Also described as new by Borgmeier, 1928a: 39.]

#***tillandsiarum.*** *Pheidole floridana* subsp. *tillandsiarum* Wheeler, W.M. 1934g: 166 (s.w.) MEXICO.

#***timmii.*** *Pheidole pubiventris* r. *timmii* Forel, 1901h: 62 (s.w.) U.S.A.

#***tipuna.*** *Pheidole rinae* r. *tipuna* Forel, 1912a: 68 (s.w.) TAIWAN.

***tisiphone.*** *Pheidole tisiphone* Wheeler, W.M. 1911h: 203 (s.) MEXICO. Combination in *P.* (*Elasmopheidole*): Emery, 1915i: 190.

***titanis.*** *Pheidole titanis* Wheeler, W.M. 1903b: 95, fig. 3 (s.w.) U.S.A.

#***tjibodana.*** *Pheidole nodgii* var. *tjibodana* Forel, 1905c: 16 (s.w.q.) JAVA.

***tolteca.*** *Pheidole tolteca* Forel, 1901c: 133 (w.) MEXICO.

*torpescens*. *Pheidole kingi* subsp. *torpescens* Wheeler, W.M. 1915b: 404 (s.w.) U.S.A. Junior synonym of *tepicana:* Creighton & Gregg, 1955: 24.

*townsendi*. *Pheidole townsendi* André, 1898: 246 (s.w.) MEXICO. Junior synonym of *kingi:* Emery, 1922e: 105.

#***trachyderma.*** *Pheidole risi* subsp. *trachyderma* Emery, 1906c: 148 (s.w.q.) ARGENTINA. [Misspelled as *pachyderma* by Santschi, 1912e: 529.]

***tragica.*** *Pheidole tragica* Wheeler, W.M. 1934g: 167 (s.w.) MEXICO.

***transfigens.*** *Pheidole transfigens* Forel, 1911e: 268 (w.) NEW GUINEA. Combination in *P.* (*Pheidolacanthinus*): Emery, 1921f: 82.

*transvarians*. *Pheidole sitarches* var. *transvarians* Wheeler, W.M. 1908e: 442 (s.w.) U.S.A. Junior synonym of *sitarches:* Creighton, 1950a: 189; Gregg, 1959: 30.

#***transversa.*** *Pheidole proxima* r. *transversa* Forel, 1902h: 428 (s.w.q.m.) AUSTRALIA.

***transversostriata.*** *Pheidole transversostriata* Mayr, 1887: 584 (s.) GUIANA. Current subspecies: nominal plus *gibbata, nigridens.*

***trapezoidea.*** *Pheidole trapezoidea* Viehmeyer, 1914c: 36 (s.) AUSTRALIA.

***treubi.*** *Pheidole treubi* Forel, 1905c: 19 (s.q.) JAVA.

***PHEIDOLE** (continued)*

***tricarinata.*** *Pheidole tricarinata* Santschi, 1914d: 340 (s.) GUINEA.

***tricolor.*** *Pheidole (Pheidolacanthinus) tricolor* Donisthorpe, 1949g: 408 (w.) NEW GUINEA.

***triconstricta.*** *Pheidole triconstricta* Forel, 1886b: xlv (s.) ARGENTINA. Emery, 1906c: 152 (q.). Current subspecies: nominal plus *ambulans, bruesi, hebe, laidlowi, nubila, paranana, rosariensis.*

*trinodis. Myrmica trinodis* Losana, 1834: 327, pl. 36, fig. 6 (w.) ITALY. Junior synonym of *megacephala:* Roger, 1863b: 30.

#***tristani.*** *Pheidole biolleyi* subsp. *tristani* Forel, 1908b: 50 (s.w.) COSTA RICA.

*tristior. Pheidole pallidula* subsp. *tristior* Schmitz, 1950: 16 PORTUGAL. *Nomen nudum.*

***tristis.*** *Myrmica tristis* Smith, F. 1858b: 132 (w.) BRAZIL. Combination in *Pheidole:* Mayr, 1886c: 360. Senior synonym of *emeryi* and *rubra:* Kempf, 1965: 183; of *fumipennis:* Brown, 1981: 527. Current subspecies: nominal plus *alpinensis, alsia.*

#***tristis.*** *Pheidole pallidula* var. *tristis* Forel, 1907b: 204 (q.) TUNISIA. [Unresolved junior secondary homonym of *Myrmica tristis* Smith, above.] Santschi, 1908: 521 (s.q.m.). See also: Emery, 1915j: 226; Karavaiev, 1926b: 104; Santschi, 1932g: 3.

***tsailuni.*** *Pheidole tsailuni* Wheeler, W.M. 1929f: 2. Replacement name for *concinna* Wheeler, W.M. 1928c: 13. [Junior primary homonym of *concinna* Santschi, 1910c: 362.]

#***tuberculata.*** *Pheidole exigua* var. *tuberculata* Mayr, 1887: 585 (s.) BRAZIL. Subspecies of *flavens:* Emery, 1894c: 157.

***tucsonica.*** *Pheidole xerophila* subsp. *tucsonica* Wheeler, W.M. 1908e: 448 (s.w.) U.S.A. Creighton & Gregg, 1955: 42 (q.). Raised to species: Wheeler, G.C. & Wheeler, J. 1986g: 13.

#***tucumana.*** *Pheidole fimbriata* var. *tucumana* Forel, 1913l: 228 (s.) ARGENTINA. Forel, 1915c: 355 (w.). Combination in *P. (Macropheidole)*: Emery, 1921f: 81.

***turneri.*** *Pheidole turneri* Forel, 1902h: 430 (s.w.) AUSTRALIA.

***tysoni.*** *Pheidole tysoni* Forel, 1901e: 348 (s.w.m.) U.S.A.

#***ultrix.*** *Pheidole guilelmimuelleri* var. *ultrix* Forel, 1912f: 221 (s.) BRAZIL.

***umbonata.*** *Pheidole umbonata* Mayr, 1870b: 978 (s.w.) TONGA. Mann, 1921: 431 (q.). Senior synonym of *zimmermani:* Wilson & Taylor, 1967: 50. Current subspecies: nominal plus *fusciventris.*

***upeneci.*** *Pheidole (Elasmopheidole) upeneci* Forel, 1913k: 43, fig. K (s.w.) JAVA. Combination in *P. (Stegopheidole)*: Emery, 1915i: 190.

*upoluana. Pheidole (Pheidole) oceanica* var. *upoluana* Santschi, 1928a: 47, fig. 3 (s.w.q.) SAMOA. Junior synonym of *oceanica:* Wilson & Taylor, 1967: 48.

***ursus.*** *Pheidole ursus* Mayr, 1870b: 986 (s.w.) MEXICO. Senior synonym of *cressoni* and *gracilinoda:* Brown, 1981: 528.

***vafella.*** *Pheidole vafella* Wheeler, W.M. 1925a: 15 (s.w.) BOLIVIA.

***vafra.*** *Pheidole vafra* Santschi, 1923d: 51 (s.w.) BRAZIL. Current subspecies: nominal plus *idiota.*

***vallicola.*** *Pheidole crassicornis* subsp. *vallicola* Wheeler, W.M. 1915b: 409 (s.w.) U.S.A. Raised to species: Creighton, 1950a: 191.

***vallifica.*** *Pheidole vallifica* Forel, 1901e: 357 (s.w.) COLOMBIA. Combination in *P. (Elasmopheidole)*: Forel, 1913k: 43.

*vanceae. Pheidole morrisii* var. *vanceae* Forel, 1901e: 351 (s.w.q.m.) U.S.A. Junior synonym of *morrisii:* Creighton, 1950a: 183.

***vanderveldi.*** *Pheidole vanderveldi* Forel, 1913b: 326 (s.w.) ZAIRE.

***variabilis.*** *Pheidole variabilis* Mayr, 1876: 103 (s.w.q.m.) AUSTRALIA. Current subspecies: nominal plus *aliena, croceithorax, latigena, mediofusca, ocior, ocyma, parvispina, praedo, redunca, rugocciput, rugosula.*

#***variegata.*** *Pheidole pubiventris* subsp. *variegata* Emery, 1896g: 72 (s.w.q.) COSTA RICA. Forel, 1908b: 54 (m.).

***variolosa.*** *Pheidole variolosa* Emery, 1892a: 113 (w.) SOMALIA. Menozzi, 1930b: 82 (s.).

***vaslitii.*** *Pheidole vaslitii* Pergande, 1896: 883 (s.w.) MEXICO. Combination in *P. (Allopheidole)*: Forel, 1912f: 237; in *P. (Cardiopheidole)*: Wheeler, W.M. 1914b: 48. Junior synonym of *obtusospinosa:* Forel, 1901c: 130. Revived from synonymy: Wheeler, W.M. 1914b: 48. See also: Creighton, 1958: 203.

***vatu.*** *Pheidole vatu* Mann, 1921: 431, fig. 13 (s.w.) FIJI IS.

#***vellicans.*** *Pheidole sulcaticeps* subsp. *vellicans* Forel, 1911i: 222 (s.w.) SRI LANKA.

***velox.*** *Pheidole velox* Emery, 1887b: 462 (w.) INDONESIA (Ternate I.).

#***venezuelana.*** *Pheidole anastasii* var. *venezuelana* Forel, 1905b: 159 (s.m.) VENEZUELA.

#***verlatenensis.*** *Pheidole nodgii* subsp. *verlatenensis* Wheeler, W.M. 1937a: 22 (s.) INDONESIA (Verlaten I.).

***veteratrix.*** *Pheidole veteratrix* Forel, 1891b: 225 (s.w.) MADAGASCAR. Current subspecies: nominal plus *angustinoda.*

*vicina. Pheidole nitidula* st. *strobeli* var. *vicina* Santschi, 1929d: 282 (s.w.) ARGENTINA. Unavailable name.

#***victima.*** *Pheidole lignicola* var. *victima* Santschi, 1929d: 290 (s.) BRAZIL.

***victoris.*** *Pheidole victoris* Forel, 1913a: 127 (s.w.) ZIMBABWE. See also: Arnold, 1920a: 430.

***vigilans.*** *Atta vigilans* Smith, F. 1858b: 166 (w.) AUSTRALIA. Combination in *Aphaenogaster:* Dalla Torre, 1893:108; in *Pheidole:* Emery, 1915d: 69. Senior synonym of *dolichocephala, norfolkensis, parallela* and *yarrensis:* Brown, 1971b: 13.

#***vincentensis.*** *Pheidole flavens* var. *vincentensis* Forel, 1893g: 411 (s.w.q.m.) ANTILLES IS.

#***vinelandica.*** *Pheidole bicarinata* r. *vinelandica* Forel, 1886b: xlv (s.w.q.m.) U.S.A. Combination in *P. (Allopheidole)*: Forel, 1912f: 237. Raised to species: Mayr, 1886d: 458; Dalla Torre, 1893: 97. Subspecies of *bicarinata:* Creighton, 1950a: 172; Gregg, 1959: 18. Senior synonym of *laeviuscula:* Creighton, 1950a: 172; of *buccalis* and *nebrascensis:* Gregg, 1959: 18.

***virago.*** *Pheidole virago* Wheeler, W.M. 1915b: 401 (s.w.) U.S.A.

***vistana.*** *Pheidole longipes* var. *vistana* Forel, 1914d: 272 (w.) U.S.A. Raised to species: Smith, D.R. 1979: 1374. Senior synonym of *grallipes:* Gregg, 1959: 22.

***voeltzkowii.*** *Pheidole voeltzkowii* Forel, 1894e: 227 (s.w.m.) MADAGASCAR. Forel, 1897c: 207 (q.).

***vorax.*** *Formica vorax* Fabricius, 1804: 412 (w.) CENTRAL AMERICA. Combination in *Aphaenogaster:* Roger, 1863b: 30; in *Pheidole:* Emery, 1915d: 69.

***walkeri.*** *Pheidole walkeri* Mann, 1922: 27, fig. 13 (s.w.) HONDURAS.

***wallacei.*** *Pheidole wallacei* Mann, 1916: 437 (s.w.) BRAZIL.

***watsoni.*** *Pheidole watsoni* Forel, 1902c: 171 (s.), 189 (w.) BURMA. [Also described as new by Forel, 1902f: 541.]

#***weiseri.*** *Pheidole (Elasmopheidole) aberrans* st. *weiseri* Santschi, 1923d: 64 (s.w.q.) ARGENTINA.

#***weissi.*** *Pheidole excellens* st. *weissi* Santschi, 1910c: 361 (s.) CONGO. Santschi, 1914d: 335 (q.); Santschi, 1930b: 65 (footnote) (w.).

#***welgelegenensis.*** *Pheidole sculpturata* r. *welgelegenensis* Forel, 1913b: 327 (s.w.) ZAIRE. Forel, 1913h: 354 (q.).

***wheeleri.*** *Pheidole (Pheidole) wheeleri* Mann, 1916: 431, pl. 3, fig. 23 (s.w.) BRAZIL.

***wheelerorum.*** *Pheidole wheelerorum* MacKay, in MacKay, Lowrie, Fisher, *et al.* 1988: 96, figs. 29, 30 (s.w.) U.S.A.

***wiesei.*** *Pheidole wiesei* Forel, 1910b: 40 (s.w.q.m.) AUSTRALIA.

***williamsi.*** *Pheidole williamsi* Wheeler, W.M. 1919c: 267 (s.w.q.m.) GALAPAGOS IS. Current subspecies: nominal plus *seymourensis.*

***wilsoni.*** *Pheidole wilsoni* Mann, 1921: 433 (s.w.) FIJI IS.

***wolfringi.*** *Pheidole wolfringi* Forel, 1908c: 374 (s.w.) BRAZIL. Current subspecies: nominal plus *durionei.*

***woodmasoni.*** *Pheidole woodmasoni* Forel, 1885b: 180 (s.w.) INDIA. Forel, 1902c: 191 (q.m.); Imai, Baroni Urbani, *et al.* 1984: 6 (k.).

#***woodvalensis.*** *Pheidole schultzei* var. *woodvalensis* Arnold, 1960b: 454, fig. 3 (s.w.) ZIMBABWE.

***wroughtonii.*** *Pheidole wroughtonii* Forel, 1902c: 176 (s.), 193 (w.q.m.) INDIA. [Also described as new by Forel, 1902f: 543.]

***xanthocnemis.*** *Pheidole xanthocnemis* Emery, 1914f: 403 (s.w.) NEW CALEDONIA.

***xerophila.*** *Pheidole xerophila* Wheeler, W.M. 1908e: 446, pl. 27, fig. 37 (s.w.m.) U.S.A.

***xocensis.*** *Pheidole xocensis* Forel, 1913j: 213 (s.w.) ZIMBABWE. Current subspecies: nominal plus *bulawayensis.* See also: Arnold, 1920a: 443.

***yaqui.*** *Pheidole yaqui* Creighton & Gregg, 1955: 43, fig. 10 (s.w.) U.S.A.

*yarrensis. Pheidole ampla* var. *yarrensis* Forel, 1902h: 434 (s.w.q.) AUSTRALIA. Junior synonym of *vigilans:* Brown, 1971b: 13.

#***yeensis.*** *Pheidole sulcaticeps* r. *yeensis* Forel, 1902c: 179 (s.) BURMA. [Also described as new by Forel, 1902f: 544.] Bingham, 1903: 253 (w.). Raised to species: Bingham, 1903: 252. Subspecies of *sulcaticeps:* Emery, 1921f: 94.

#***yerburyi.*** *Pheidole spathifera* var. *yerburyi* Forel, 1902c: 168 (s.), 188 (w.q.) SRI LANKA. [Also described as new by Forel, 1902f: 540.] [*P. spathifera* var. *yerburyi* Emery, 1901f: 118. *Nomen nudum.*]

**#yomensis.** *Pheidole peguensis* r. *yomensis* Forel, 1903c: 253 (s.) BURMA.
**#zambesiana.** *Pheidole sculpturata* r. *zambesiana* Forel, 1913a: 133 (s.w.) ZIMBABWE.
**zeteki.** *Pheidole (Decapheidole) zeteki* Smith, M.R. 1947h: 194 (w.) PANAMA.
*zimmermani. Pheidole umbonata* subsp. *zimmermani* Wheeler, W.M. 1936f: 6 (s.w.q.m.) TAHITI. Junior synonym of *umbonata:* Wilson & Taylor, 1967: 50.
**#zoceana.** *Pheidole nodgii* var. *zoceana* Santschi, 1925f: 83 (s.w.) CHINA (name misspelled as *noggii*).

***PHEIDOLOGETON*** (Old World tropics)
**aberrans.** *Pheidologeton aberrans* Santschi, 1937d: 231, figs. 5, 6 (q.) ANGOLA.
*aborensis* Wheeler, W.M. 1913; see under ***OLIGOMYRMEX.***
**affinis.** *Oecodoma affinis* Jerdon, 1851: 110 (s.w.) INDIA. Emery, 1893e: 212 (q.m.); Wheeler, G.C. & Wheeler, J. 1954a: 138 (l.). Combination in *Pheidole:* Smith, F. 1858b: 174; in *Pheidologeton:* Roger, 1863b: 30. Senior synonym of *bellicosa* and *laboriosa* (and its junior synonym *calida*): Emery, 1893e: 206; of *australis* and *mjobergi:* Taylor, 1991b: 608. Current subspecies: nominal plus *javana, minor, spinosior, sumatrensis.* See also: Bingham, 1903: 164.
**#albipes.** *Pheidologeton pygmaeus* var. *albipes* Emery, 1893g: 266 (w.) PHILIPPINES. Combination in *Aneleus:* Emery, 1922f: 214; in *Pheidologeton:* Ettershank, 1966: 118.
*antiquus* Mayr, 1868; see under ***OLIGOMYRMEX.***
*arnoldi* Forel, 1914; see under ***OLIGOMYRMEX.***
*arnoldi. Pheidologeton arnoldi* Santschi, 1928f: 198, fig. 4 (s.w.q.) NAMIBIA. [Junior primary homonym of *arnoldi* Forel, 1914d: 242, above.] Replacement name: *kunensis* Ettershank, 1966: 119.
*australis. Pheidologeton affinis* var. *australis* Forel, 1915b: 68 (w.q.) AUSTRALIA. Forel, 1918a: 722 (s.). Raised to species: Forel, 1918a: 721. Junior synonym of *affinis:* Taylor, 1991b: 608.
*bellicosa. Atta bellicosa* Smith, F. 1858b: 164 (s.w.) BURMA. Combination in *Pheidologeton:* Mayr, 1886c: 360. Junior synonym of *affinis:* Emery, 1893e: 206.
*bohemicus* Novak, O. 1878; see under ***OLIGOMYRMEX.***
**#bugnioni.** *Pheidologeton (Aneleus) pygmaeus* var. *bugnioni* Forel, 1915a: 28 (s.) SRI LANKA. Combination in *Aneleus:* Emery, 1922f: 214; in *Pheidologeton:* Ettershank, 1966: 118.
*butteli* Forel, 1913; see under ***OLIGOMYRMEX.***
*calida. Solenopsis calida* Smith, F. 1863: 22 (q.) INDONESIA (Seram I.). Combination in *Pheidologeton:* Mayr, 1862: 750. Junior synonym of *laboriosa:* Mayr, 1867a: 100.
**ceylonensis.** *Pheidologeton pygmaeus* subsp. *ceylonensis* Forel, 1911i: 220 (w.q.) SRI LANKA. Forel, 1912n: 53 (s.w.). Combination in *Aneleus:* Emery, 1922f: 214; in *Pheidologeton:* Ettershank, 1966: 118. Raised to species: Forel, 1915a: 28.
*crassiuscula* Emery, 1900; see under ***OLIGOMYRMEX.***
**#densistriatus.** *Pheidologeton (Aneleus) pygmaeus* var. *densistriatus* Stitz, 1925: 120 (w.) no locality given.
**dentiviris.** *Pheidologeton dentiviris* Forel, 1913f: 192 (m.) TAIWAN. Santschi, 1937h: 372 (q.). Combination in *Idrisella:* Santschi, 1937h: 372; in *Pheidologeton:* Ettershank, 1966: 118.
**diversus.** *Oecodoma diversa* Jerdon, 1851: 109 (s.w.) INDIA. Emery, 1893e: 212 (q.m.); Wheeler, G.C. & Wheeler, J. 1954a: 135 (l.); Imai, Baroni Urbani, *et al.* 1984: 7 (k.). Combination in *Pheidole:* Smith, F. 1858b: 174; in *Pheidologeton:* Roger, 1863b: 30. Senior synonym of *megacephala, ocellifera* and *pabulator:* Emery, 1893e: 206; of *megacephalotes:* Dalla Torre, 1893: 73; of *militaris* and *polita:* Donisthorpe, 1932c: 458, 463, respectively. Current subspecies nominal plus *draco, fictus, laotinus, macgregori, philippinus, standfussi, taprobanae, tenuirugosus, williamsi.* See also: Bingham, 1903: 163; Wheeler, W.M. 1919e: 85; Moffett, 1984: 7.
**#draco.** *Pheidologeton diversus* st. *draco* Santschi, 1920h: 163 (s.w.q.) VIETNAM. [Also described as new by Santschi, 1924c: 101.]
**#fictus.** *Pheidologeton diversus* var. *ficta* Forel, 1911d: 386 (w.) VIETNAM.
**hammoniae.** *Pheidologeton hammoniae* Stitz, 1923: 157, fig. 2 (s.w.) NAMIBIA.
**hostilis.** *Atta hostilis* Smith, F. 1858b: 165 (w.q.) SOUTH AFRICA. Combination in *Pheidologeton:* Mayr, 1866b: 899.
**#javana.** *Pheidologeton affinis* var. *javana* Emery, 1893e: 212 (w.) JAVA.
**kunensis.** *Pheidologeton kunensis* Ettershank, 1966: 119. Replacement name for *arnoldi* Santschi, 1928f: 198. [Junior primary homonym of *arnoldi* Forel, 1914d: 242.]
*laboriosa. Solenopsis laboriosa* Smith, F. 1861b: 48 (s.w.) SULAWESI. Mayr, 1867a: 101 (q.). Combination in *Pheidologeton:* Mayr, 1862: 750. Senior synonym of *calida:* Mayr, 1867a: 100. Junior synonym of *affinis:* Emery, 1893e: 206.
*lamellifrons* Forel, 1902; see under ***OLIGOMYRMEX.***
**#laotinus.** *Pheidologeton diversus* var. *laotina* Santschi, 1920h: 162 (s.w.q.) LAOS. [Also described as new by Santschi, 1924c: 100.]
**maccus.** *Pheidologeton maccus* Wheeler, W.M. 1929g: 52 (s.w.) PHILIPPINES.
**#macgregori.** *Pheidologeton diversus* subsp. *macgregori* Wheeler, W.M. 1929g: 50, fig. 6 (w.) PHILIPPINES.
**mayri.** *Pheidologeton mayri* Santschi, 1928f: 201 (q.) no locality given (Afrotropical).
*megacephala. Pheidole megacephala* Smith, F. 1860b: 112 (w.) INDONESIA (Batjan I.). Combination in *Pheidologeton:* Roger, 1863b: 30. Junior synonym of *diversus:* Emery, 1893e: 206.
*megacephalotes. Pheidole megacephalotes* Dalla Torre, 1892: 90. [Unnecessary replacement name for *megacephala* Smith, above.] Junior synonym of *megacephala:* Dalla Torre, 1893: 73.
**melanocephalus.** *Pheidologeton melanocephalus* Donisthorpe, 1948g: 134 (s.w.) NEW GUINEA.
*militaris. Pheidole militaris* Smith, F. 1860a: 74 (s.w.) SULAWESI. Junior synonym of *diversus:* Donisthorpe, 1932c: 458.
*minimus* Emery, 1900; see under ***OLIGOMYRMEX.***
**#minor.** *Pheidologeton affinis* var. *minor* Emery, 1900c: 326 (s.q.) NEW GUINEA.
*mjobergi. Pheidologeton australis* var. *mjobergi* Forel, 1918a: 723 (q.) AUSTRALIA. Junior synonym of *affinis:* Taylor, 1991b: 608.
**nanus.** *Pheidologeton nanus* Roger, 1863a: 191 (s.w.q.) SRI LANKA. See also: Mayr, 1897: 429.
**obscurus.** *Pheidologeton obscurus* Viehmeyer, 1914c: 37 (w.) JAVA.
*ocellifera. Pheidole ocellifera* Smith, F. 1858b: 174 (w.) BURMA. Combination in *Pheidologeton:* Mayr, 1862: 750. Junior synonym of *diversus:* Emery, 1893e: 206.
*pabulator. Pheidole pabulator* Smith, F. 1860b: 112 (s.w.) INDONESIA (Batjan I.). Combination in *Pheidologeton:* Mayr, 1886b: 362. Junior synonym of *diversus:* Emery, 1893e: 206.
*perpusillum* Emery, 1895; see under ***OLIGOMYRMEX.***
**petulens.** *Pheidologeton petulens* Santschi, 1920h: 164 (s.w.q.) LAOS. [Also described as new by Santschi, 1924c: 103.]
**#philippinus.** *Pheidologeton diversus* subsp. *philippinus* Wheeler, W.M. 1929g: 46 (s.w.q.m.) PHILIPPINES.
*pilosior. Pheidologeton diversus* subsp. *fictus* var. *pilosior* Wheeler, W.M. 1929g: 45 (s.) PHILIPPINES. Unavailable name.
*polita. Myrmica polita* Smith, F. 1860b: 108 (w.) INDONESIA (Batjan I.). Junior synonym of *diversus:* Donisthorpe, 1932c: 463.
**pullatus.** *Pheidologeton pullatus* Santschi, 1920h: 165 (q.) JAVA. [Also described as new by Santschi, 1924c: 104.]
**pungens.** *Solenopsis pungens* Smith, F. 1861b: 48 (s.w.) SULAWESI. Combination in *Pheidologeton:* Emery, 1922f: 213.
**pygmaeus.** *Pheidologeton pygmaeus* Emery, 1887b: 465 (s.w.) INDONESIA (Ternate I.). Combination in *P. (Aneleus):* Emery, 1900c: 327; in *Aneleus:* Emery, 1922f: 214; in *Pheidologeton:* Ettershank, 1966: 119. Current subspecies: nominal plus *albipes, bugnioni, densistriatus.*
**ruber.** *Pheidole rubra* Smith, F. 1860b: 111 (s.w.) INDONESIA (Batjan I.). Combination in *Pheidologeton:* Donisthorpe, 1932c: 464. Senior synonym of *rubrobrunea:* **new synonymy**.
*rubrobrunea. Pheidole rubrobrunea* Dalla Torre, 1892: 91. [Unnecessary replacement name for *Pheidole rubra* Smith, F. 1860b: 111, above.] Junior synonym of *ruber:* **new synonymy**.
***rugiceps.** *Myrmica rugiceps* Heer, 1850: 160, pl. 13, fig. 2 (q.) SWITZERLAND (Miocene). Combination in *Pheidologeton:* Assmann, 1870: 40.
**rugosus.** *Pheidologeton rugosus* Karavaiev, 1935a: 95 (s.w.) JAVA.
*sarasinorum* Emery, 1901; see under ***OLIGOMYRMEX.***
***schossnicensis.** *Pheidologeton schossnicensis* Assmann, 1870: 40, pl. 1, fig. 3 (wing) GERMANY (Miocene).
**silenus.** *Pheidole silenus* Smith, F. 1858b: 176 (s.w.) SINGAPORE. Combination in *Pheidologeton:* Mayr, 1862: 751. Senior synonym of *speculifrons:* Ettershank, 1966: 119.
**silvestrii.** *Pheidologeton silvestrii* Wheeler, W.M. 1929g: 51 (s.) PHILIPPINES.
*simalurensis* Forel, 1915; see under ***OLIGOMYRMEX.***
**solitarius.** *Pheidologeton solitarius* Stitz, 1910: 141 (q.) TOGO.
*speculifrons. Amauromyrmex speculifrons* Wheeler, W.M. 1929b: 1, fig. 1 (w.) BORNEO. Junior synonym of *silenus:* Ettershank, 1966: 119.
**#spinosior.** *Pheidologeton affinis* var. *spinosior* Forel, 1911d: 372 (w.) SUMATRA.

***PHEIDOLOGETON** (continued)*

*spinosus* Forel, 1907; see under ***OLIGOMYRMEX.***
**#*standfussi.*** *Pheidologeton diversus* subsp. *standfussi* Forel, 1911d: 386 (w.) GUINEA.
**#*sumatrensis.*** *Pheidologeton affinis* var. *sumatrensis* Forel, 1913k: 55 (w.q.) SUMATRA.
*tagalus. Pheidologeton diversus* st. *macgregori* var. *tagalus* Santschi, 1933d: 1 (q.) PHILIPPINES. Unavailable name.
**#*taprobanae.*** *Pheidole taprobanae* Smith, F. 1858b: 175 (q.) SRI LANKA. Emery, 1901f: 119 (s.w.). Combination in *Pheidologeton:* Roger, 1863b: 30. Variety of *diversus:* Emery, 1901f: 119. Junior synonym of *diversus:* Bingham, 1903: 162. Revived from synonymy as variety of *diversus:* Forel, 1911i: 220; Emery, 1922f: 212.
**#*tenuirugosus.*** *Pheidologeton diversus* subsp. *tenuirugosus* Wheeler, W.M. 1929g: 44 (s.w.) PHILIPPINES.
***transversalis.*** *Solenopsis transversalis* Smith, F. 1860a: 74 (s.) SULAWESI. Combination in *Pheidologeton:* Mayr, 1862: 751.
***varius.*** *Pheidologeton varius* Santschi, 1920h: 163 (s.w.m.) VIETNAM. [Also described as new by Santschi, 1924c: 102.]
***vespillo.*** *Pheidologeton vespillo* Wheeler, W.M. 1921c: 533 (s.w.) CHINA.
***volsellata.*** *Pheidologeton volsellata* Santschi, 1937b: 100 (m.) CAMEROUN.
**#*williamsi.*** *Pheidologeton diversus* subsp. *williamsi* Wheeler, W.M. 1929g: 49 (s.w.m.) PHILIPPINES.
***yanoi.*** *Pheidologeton yanoi* Forel, 1912a: 57 (w.q.) TAIWAN. Forel, 1913f: 192 (m.). See also: Wheeler, W.M. 1930a: 100.

***PHILIDRIS*** (Oriental, Indo-Australian)
**#*agnata.*** *Iridomyrmex nagasau* subsp. *agnatus* Mann, 1921: 472 (w.) FIJI IS. Combination in *Philidris:* Shattuck, 1992a: 18.
**#*alticola.*** *Iridomyrmex nagasau* subsp. *alticola* Mann, 1921: 472 (w.) FIJI IS. Combination in *Philidris:* Shattuck, 1992a: 18.
**#*andamanensis.*** *Iridomyrmex myrmecodiae* var. *andamanensis* Forel, 1903d: 407 (w.m.) ANDAMAN IS. Combination in *Philidris:* Shattuck, 1992a: 18.
***brunnea.*** *Pseudolasius brunneus* Donisthorpe, 1949b: 504 (w.) NEW GUINEA. Combination in *Philidris:* Shattuck, 1992a: 18.
*butteli. Iridomyrmex cordatus* r. *protensus* var. *butteli* Forel, 1913k: 90 (w.q.) SUMATRA. Unavailable name (Shattuck, 1994: 138).
***cordata.*** *Formica cordata* Smith, F. 1859a: 137 (w.) INDONESIA (Aru I.). [Unresolved junior primary homonym of **Formica cordata* Holl, 1829: 140.] Emery, 1887a: 250 (q.m.); Imai, Brown, *et al.* 1984: 68 (k.). Combination in *Hypoclinea:* Mayr, 1879: 659; in *Iridomyrmex:* Emery, 1887a: 249; in *Philidris:* Shattuck, 1992a: 18. Current subspecies: nominal plus *fusca, protensa, stewartii.* See also: Karavaiev, 1926d: 435; Tjan, Imai, *et al.* 1986: 58; Shattuck, 1994: 136.
***cruda.*** *Formica cruda* Smith, F. 1860b: 95 (w.) INDONESIA (Batjan I.). Combination in *Iridomyrmex:* Emery, 1887a: 249; in *Philidris:* Shattuck, 1992a: 18. Junior synonym of *cordata:* Emery, 1887a: 249. Revived from synonymy: Emery, 1912g: 24.
*decipiens. Iridomyrmex myrmecodiae* var. *decipiens* Forel, 1901b: 23 (w.q.m.) BISMARCK ARCHIPELAGO. *Nomen nudum;* material referred to *cordata* by Dahl, 1901: 13.
**#*fusca.*** *Iridomyrmex cordatus* r. *fuscus* Forel, 1901b: 23 (w.q.) BISMARCK ARCHIPELAGO. Combination in *Philidris:* Shattuck, 1992a: 18. Junior synonym of *cordata:* Dahl, 1901: 13. Revived from synonymy as subspecies of *cordata:* Emery, 1912g: 24.
*jactans. Iridomyrmex cordatus* subsp. *fuscus* var. *jactans* Forel, 1911a: 48 (w.q.m.) SINGAPORE. Unavailable name (Shattuck, 1994: 138).
***laevigata.*** *Iridomyrmex laevigatus* Emery, 1895k: 475 (w.) BURMA. Viehmeyer, 1916a: 142 (q.m.). Combination in *Philidris:* Shattuck, 1992a: 18.
**#*mandibularis.*** *Iridomyrmex myrmecodiae* var. *mandibularis* Santschi, 1928h: 131 (m.) SUMATRA (name misspelled as *myrmecordia*). Combination in *Philidris:* Shattuck, 1992a: 18.
***myrmecodiae.*** *Iridomyrmex cordatus* var. *myrmecodiae* Emery, 1887a: 249 (w.) WEST MALAYSIA. Combination in *Philidris:* Shattuck, 1992a: 18. Raised to species: Emery, 1900d: 694. Junior synonym of *cordata:* Dahl, 1901: 13. Revived from synonymy, status as species: Forel, 1903d: 407; Emery, 1912g: 25; Mann, 1919: 362. Current subspecies: nominal plus *andamanensis, mandibularis, nigriventris.*
***nagasau.*** *Iridomyrmex nagasau* Mann, 1921: 470, fig. 27 (w.) FIJI IS. Combination in *Philidris:* Shattuck, 1992a: 18. Current subspecies: nominal plus *agnata, alticola.*
**#*nigriventris.*** *Iridomyrmex myrmecodiae* var. *nigriventris* Donisthorpe, 1941h: 60 (w.) NEW GUINEA. Donisthorpe, 1943d: 455 (m.). Combination in *Philidris:* Shattuck, 1992a: 18.
**#*protensa.*** *Iridomyrmex cordatus* subsp. *protensus* Forel, 1911a: 47 (w.q.m.) BORNEO. Combination in *Philidris:* Shattuck, 1992a: 18.
***pubescens.*** *Iridomyrmex pubescens* Donisthorpe, 1949b: 499 (q.) NEW GUINEA. Combination in *Philidris:* Shattuck, 1992a: 18.
*simalurana. Iridomyrmex cordatus* r. *fuscus* var. *simalurana* Forel, 1915a: 34 (w.) SUMATRA. Unavailable name (Shattuck, 1994: 139).
**#*stewartii.*** *Iridomyrmex cordatus* var. *stewartii* Forel, 1893d: 456 (w.) AUSTRALIA. Combination in *Philidris:* Shattuck, 1992a: 18. Variety/subspecies of *myrmecodiae:* Forel, 1902h: 541; of *cordata:* Taylor & Brown, D.R. 1985: 97.
*waldoi. Iridomyrmex cordatus* subsp. *fuscus* var. *waldoi* Forel, 1911a: 48 (w.) THAILAND. [Also described as new by Forel, 1911g: 285; unavailable name.] Unavailable name (Shattuck, 1994: 139).

***PHRYNOPONERA*** (Afrotropical)
***armata.*** *Pachycondyla* (*Bothroponera*) *armata* Santschi, 1919h: 82 (w.) ZAIRE. Combination in *Phrynoponera:* Wheeler, W.M. 1922a: 773.
***bequaerti.*** *Phrynoponera bequaerti* Wheeler, W.M. 1922a: 79, fig. 12 (q.) ZAIRE.
*esta. Phrynoponera gabonensis* var. *esta* Wheeler, W.M. 1922a: 77 (w.q.) ZAIRE. Junior synonym of *gabonensis:* Brown, 1950e: 246.
*fecunda. Phrynoponera gabonensis* var. *fecunda* Wheeler, W.M. 1922a: 78 (w.q.) ZAIRE. Junior synonym of *gabonensis:* Brown, 1950e: 246.
***gabonensis.*** *Bothroponera gabonensis* André, 1892a: 50 (w.) GABON. Stitz, 1910: 130 (q.). Combination in *Pachycondyla* (*Bothroponera*): Emery, 1901a: 45; in *Phrynoponera:* Wheeler, W.M. 1920: 53. Senior synonym of *esta, fecunda, robustior, striatidens* and *umbrosa:* Brown, 1950e: 246.
***heteroda.*** *Phrynoponera heterodus* Wheeler, W.M. 1922a: 78 (q.) ZAIRE.
*robustior. Pachycondyla* (*Bothroponera*) *gabonensis* var. *robustior* Santschi, 1919h: 82 (w.) ZAIRE. Combination in *Phrynoponera:* Wheeler, W.M. 1922a: 774. Junior synonym of *gabonensis:* Brown, 1950e: 246.
*striatidens. Pachycondyla* (*Bothroponera*) *gabonensis* var. *striatidens* Santschi, 1914d: 315, fig. 4 (w.) CAMEROUN. Combination in *Phrynoponera:* Wheeler, W.M. 1922a: 78. Junior synonym of *gabonensis:* Brown, 1950e: 246.
***sveni.*** *Pachycondyla* (*Bothroponera*) *sveni* Forel, 1916: 398 (w.) ZAIRE. Combination in *Phrynoponera:* Wheeler, W.M. 1922a: 80.
*umbrosa. Phrynoponera gabonensis* var. *umbrosa* Wheeler, W.M. 1922a: 78 (w.) ZAIRE. Junior synonym of *gabonensis:* Brown, 1950e: 246.

*PHYRACACES:* see under ***CERAPACHYS.***

*PHYSATTA:* see under ***MYRMICARIA.***

**#*PHYSOCREMA:*** see under ***CREMATOGASTER.***

***PILOTROCHUS*** (Madagascar)
***besmerus.*** *Pilotrochus besmerus* Brown, 1978a: 221, figs. 1, 2 (w.) MADAGASCAR.

****PITYOMYRMEX*** (Baltic Amber)
****tornquisti.*** **Pityomyrmex tornquisti* Wheeler, W.M. 1915h: 99, fig. 48 (w.) BALTIC AMBER (Oligocene).

***PLAGIOLEPIS*** (Old World tropical and temperate)
***abyssinica.*** *Plagiolepis exigua* r. *abyssinica* Forel, 1894b: 73 (w.) ETHIOPIA. Raised to species: Collingwood, 1985: 297.
***adynata.*** *Plagiolepis adynata* **new name**. Replacement name for *flava* Karavaiev, 1935a: 110. [Junior primary homonym of *flava* Mayr, 1862: 699.]
***alluaudi.*** *Plagiolepis alluaudi* Emery, 1894a: 71 (w.) SEYCHELLE IS. Forel, 1907a: 30 (q.). Senior synonym of *augusti, foreli, mactavishi* and *ornata:* Smith, M.R. 1958b: 196. See also: Smith, D.R. 1979: 1423.
***ampeloni.*** *Aporomyrmex ampeloni* Faber, 1969: 54, figs. 3–7, 14–16, 19, 21, 23, 24 (q.m.) AUSTRIA. Combination in *Plagiolepis:* **new combination**.
***ancyrensis.*** *Plagiolepis maura* var. *ancyrensis* Santschi, 1920g: 171, fig. 1 (w.q.) TURKEY. Finzi, 1936: 184 (m.). Variety of *pallescens:* Emery, 1925b: 21; Menozzi, 1933b: 78. Raised to species: Tohmé, G. 1969: 10.
***arnoldii.*** *Plagiolepis arnoldii* Dlussky, in Dlussky, Soyunov & Zabelin, 1990: 161, fig. 16 (w.q.) TURKMENISTAN.
**#*atlantis.*** *Plagiolepis maura* var. *atlantis* Santschi, 1920g: 171 (w.q.m.) TUNISIA. Subspecies of *pallescens:* current status.
*augusti. Plagiolepis augusti* Emery, 1921d: 317. Replacement name for *foreli* Mann, 1921: 473. [Junior pri-

mary homonym of *foreli* Santschi, 1920g: 165.] Junior synonym of *alluaudi:* Smith, M.R. 1958b: 196.

***balestrierii.*** *Plagiolepis balestrierii* Menozzi, 1939a: 310, figs. 11, 12 (w.q.) INDIA.

*barbara. Plagiolepis pygmaea* var. *barbara* Santschi, 1911i: 286 (w.m.) TUNISIA. Santschi, 1920g: 172 (q.). Raised to species: Santschi, 1920g: 172. Junior synonym of *schmitzii:* Collingwood, 1978: 71.

***bicolor.*** *Plagiolepis bicolor* Forel, 1901b: 24 (w.) BISMARCK ARCHIPELAGO. Forel, 1912d: 109 (q.). Combination in *Acropyga:* Dahl, 1901: 12; in *Plagiolepis:* Forel, 1912d: 109.

*bothae* Forel, 1907; see under ***ANOPLOLEPIS.***

*braunsi* Forel, 1913; see under ***ANOPLOLEPIS.***

***brunni.*** *Plagiolepis brunni* Mayr, 1895: 148 (w.) MOZAMBIQUE. Forel, 1914d: 248 (q.m.). Current subspecies: nominal plus *nilotica, pubescens.* See also: Arnold, 1922: 604.

#***bulawayensis.*** *Plagiolepis (Plagiolepis) pygmaea* var. *bulawayensis* Arnold, 1922: 601 (w.q.) ZIMBABWE.

#***canariensis.*** *Plagiolepis barbara* var. *canariensis* Santschi, 1920g: 173 (w.q.m.) CANARY IS. Subspecies of *schmitzii:* current status.

***capensis.*** *Plagiolepis capensis* Mayr, 1865: 55 (q.) SOUTH AFRICA.

*carinata* Emery, 1899; see under ***ANOPLOLEPIS.***

***chirindensis.*** *Plagiolepis (Anacantholepis) chirindensis* Arnold, 1949: 269, fig. 9 (w.) ZIMBABWE.

***clarki.*** *Plagiolepis clarki* Wheeler, W.M. 1934d: 157 (w.q.m.) AUSTRALIA. Current subspecies: nominal plus *impasta.*

*crosi. Plagiolepis schmitzi* st. *crosi* Santschi, 1920g: 168, fig. 1 (w.q.) ALGERIA. Wheeler, G.C. & Wheeler, J. 1953c: 135 (l.). Raised to species: Emery, 1925b: 20; Santschi, 1931a: 10. Junior synonym of *schmitzii:* Collingwood, 1978: 71.

*deceptor* Arnold, 1922; see under ***ANOPLOLEPIS.***

*decolor* Emery, 1895; see under ***ANOPLOLEPIS.***

***decora.*** *Plagiolepis (Anacantholepis) decora* Santschi, 1914e: 36, fig. 9 (w.) SOUTH AFRICA. Arnold, 1922: 582 (q.).

***demangei.*** *Plagiolepis (Anacantholepis) demangei* Santschi, 1920h: 172 (w.) VIETNAM. [Also described as new by Santschi, 1924c: 112.]

*detrita* Emery, 1892; see under ***ANOPLOLEPIS.***

***deweti.*** *Plagiolepis deweti* Forel, 1904f: 423 (w.q.) SOUTH AFRICA. Combination in *P. (Anacantholepis)*: Forel, 1914d: 249.

***dichroa.*** *Plagiolepis dichroa* Forel, 1902c: 235 (footnote) (w.) INDIA.

*dufourii. Micromyrma dufourii* Perris, 1878: 382 (w.) FRANCE. Junior synonym of *Tapinoma erraticum:* Dalla Torre, 1893: 165; Emery, 1912g: 40; Forel, 1915d: 43; of *pygmaea:* Menozzi, 1925a: 19 (footnote).

#***elguerrhae.*** *Plagiolepis (Plagiolepis) pallescens* var. *elguerrhae* Karavaiev, 1931a: 311, fig. 2 (w.q.) ALGERIA.

***exigua.*** *Plagiolepis exigua* Forel, 1894c: 417 (w.q.) INDIA. Viehmeyer, 1916a: 145 (m.). Current subspecies: nominal plus *quadrimaculata.*

*fallax* Mayr, 1865; see under ***ANOPLOLEPIS.***

*flava* Mayr, 1862; see under ***ACROPYGA.***

*flava. Plagiolepis (Anacantholepis) flava* Karavaiev, 1935a: 110 (w.) VIETNAM. [Junior primary homonym of *flava* Mayr, 1862: 699, above.] Replacement name: *adynata* **new name**.

***flavescens.*** *Plagiolepis flavescens* Collingwood, 1976: 307, fig. 5 (w.) KOREA.

*flavidula* Roger, 1863; see under ***BRACHYMYRMEX.***

*foreli. Plagiolepis foreli* Santschi, 1920g: 165, fig. 1 (w.) SWITZERLAND. Variety of *alluaudi:* Emery, 1921d: 318. Junior synonym of *alluaudi:* Smith, M.R. 1958b: 196.

*foreli. Plagiolepis foreli* Mann, 1921: 473 (w.) FIJI IS. [Junior primary homonym of *foreli* Santschi, above.] Replacement name: *augusti* Emery, 1921d: 317.

***funicularis.*** *Plagiolepis funicularis* Santschi, 1919a: 344 (w.) SOUTH AFRICA.

***fuscula.*** *Plagiolepis fuscula* Emery, 1895h: 45 (w.) SOUTH AFRICA. Combination in *P. (Anacantholepis)*: Emery, 1925b: 22.

*gaetula. Plagiolepis schmitzi* st. *crossi* var. *gaetula* Santschi, 1936c: 205, fig. 4 (w.) MOROCCO. Unavailable name.

*gertrudae* Forel, 1900; see under ***ANOPLOLEPIS.***

***grassei.*** *Plagiolepis grassei* Le Masne, 1956: 673 (w.q.m.) FRANCE. See also: Passera, 1969: 327.

*hirsuta* Emery, 1892; see under ***ANOPLOLEPIS.***

***hoggarensis.*** *Plagiolepis hoggarensis* Bernard, 1981a: 172, fig. 1 (w.) ALGERIA.

#***impasta.*** *Plagiolepis clarki* subsp. *impasta* Wheeler, W.M. 1934d: 158 (w.) AUSTRALIA.

***intermedia.*** *Plagiolepis pygmaea* var. *intermedia* Emery, 1895h: 45 (w.q.) SOUTH AFRICA. Raised to species: Emery, 1921d: 316. Current subspecies: nominal plus *minutula.*

#***isis.*** *Plagiolepis pallescens* var. *isis* Santschi, 1938a: 39, fig. 8 (w.q.) EGYPT,

***jerdonii.*** *Plagiolepis jerdonii* Forel, 1894c: 416 (w.) INDIA.

***jouberti.*** *Plagiolepis jouberti* Forel, 1910f: 23 (w.) SOUTH AFRICA. Combination in *P. (Anacantholepis)*: Santschi, 1914b: 122.

#***kabyla.*** *Plagiolepis maura* var. *kabyla* Santschi, 1920g: 170, fig. 1 (w.m.) TUNISIA. Subspecies of *pallescens:* current status.

*karavaevi. Plagiolepis karavaevi* Arnol'di, 1948: 212. *Nomen nudum.*

***karawajewi.*** *Plagiolepis karawajewi* Radchenko, 1989b: 154, figs. 1–6 (w.q.m.) UKRAINE.

*karrooensis* Arnold, 1922; see under ***ANOPLOLEPIS.***

****klinsmanni.*** **Plagiolepis klinsmanni* Mayr, 1868c: 37, pl. 1, figs. 19, 20 (w.) BALTIC AMBER (Oligocene). Wheeler, W.M. 1915h: 101 (m.).

****kuenowi.*** **Plagiolepis kuenowi* Mayr, 1868c: 39, pl. 1, figs. 22, 23 (w.) BALTIC AMBER (Oligocene).

****labilis.*** **Plagiolepis labilis* Emery, 1891a: 154, pl. 3, figs. 29–32 (w.) SICILIAN AMBER (Miocene).

***livingstonei.*** *Plagiolepis livingstonei* Santschi, 1926b: 245 (w.q.m.) ZAMBIA.

***lucidula.*** *Plagiolepis lucidula* Wheeler, W.M. 1934d: 155 (w.) AUSTRALIA.

*macgregori* Arnold, 1922; see under ***ANOPLOLEPIS.***

*mactavishi. Plagiolepis mactavishi* Wheeler, W.M. 1908c: 166 (w.) SOCIETY IS. Wheeler, W.M. 1922f: 318 (q.); Santschi, 1928b: 37 (m.). Junior synonym of *alluaudi:* Smith, M.R. 1958b: 196.

***madecassa.*** *Plagiolepis pygmaea* r. *madecassa* Forel, 1892l: 241 (w.) MADAGASCAR. Forel, 1895c: 245 (q.m.). Raised to species: Forel, 1892k: 519.

#***madeirensis.*** *Plagiolepis barbara* var. *madeirensis* Emery, 1921d: 315, fig. I (w.) MADEIRA. Subspecies of *schmitzii:* current status.

***manczshurica.*** *Plagiolepis pygmaea* var. *manczshurica* Ruzsky, 1905b: 467 (w.) RUSSIA. Wheeler, W.M. 1929f: 8 (q.). Raised to species: Santschi, 1925f: 95.

#***maura.*** *Plagiolepis maura* Santschi, 1920g: 169, fig. 1 (w.q.m.) MOROCCO. Subspecies/stirps of *pallescens:* Emery, 1924a: 8; Santschi, 1936c: 206.

***mediorufa.*** *Plagiolepis pygmaea* var. *mediorufa* Forel, 1916: 437 (w.) ZAIRE. Raised to species: Wheeler, W.M. 1922a: 213.

*mediterranea* Mayr, 1866; see under ***ANOPLOLEPIS.***

*melanaria* Arnold, 1922; see under ***ANOPLOLEPIS.***

#***mima.*** *Plagiolepis (Plagiolepis) pygmaea* var. *mima* Arnold, 1922: 602 (w.) ZIMBABWE.

#***minu.*** *Plagiolepis pygmaea* var. *minu* Forel, 1911d: 350 (w.) GREECE.

#***minutula.*** *Plagiolepis intermedia* var. *minutula* Emery, 1921d: 316, fig. 2 (w.q.) SOUTH AFRICA.

***moelleri.*** *Plagiolepis moelleri* Bingham, 1903: 321 (w.) INDIA. Combination in *P.(Anacantholepis)*: Emery, 1925b: 23.

***montivaga.*** *Plagiolepis montivaga* Arnold, 1958: 124, fig. 5 (w.) LESOTHO.

#***nilotica.*** *Plagiolepis brunni* var. *nilotica* Mayr, 1904b: 7 (w.) SUDAN. Subspecies of *brunni:* Forel, 1907a: 30.

***nitida.*** *Plagiolepis (Plagiolepis) nitida* Karavaiev, 1935a: 109 (w.) CAMBODIA.

*nuptialis* Santschi, 1917; see under ***ANOPLOLEPIS.***

***nynganensis.*** *Plagiolepis nynganensis* McAreavey, 1949: 23, figs. 66, 67 (w.) AUSTRALIA.

***obscuriscapa.*** *Plagiolepis pygmaea* var. *obscuriscapus* Santschi, 1923a: 137 (w.) ITALY. Stärcke, 1936: 278 (q.). Raised to species: Agosti & Collingwood, 1987a: 57.

*opaciventris* Emery, 1899; see under ***ANOPLOLEPIS.***

*ornata. Plagiolepis foreli* var. *ornata* Santschi, 1920g: 166 (w.) NEW CALEDONIA. Variety of *alluaudi:* Emery, 1921d: 318. Junior synonym of *alluaudi:* Smith, M.R. 1958b: 196.

***pallescens.*** *Plagiolepis pygmaea* var. *pallescens* Forel, 1889: 265 (w.) GREECE. Santschi, 1911i: 287 (q.); Finzi, 1928c: 790 (m.). Variety of *pygmaea:* Ruzsky, 1902d: 18; Santschi, 1920g: 168. Raised to species: Emery, 1921d: 315; Finzi, 1928c: 790; Karavaiev, 1931a: 311. Current subspecies: nominal plus *atlantis, elguerrhae, isis, kabyla, maura, polygyna, sordida.* See also: Menozzi, 1936d: 297.

*pernix* Viehmeyer, 1923; see under ***ANOPLOLEPIS.***

***pictipes.*** *Plagiolepis (Anacantholepis) pictipes* Santschi, 1914b: 122, fig. 25 (w.) KENYA.

*pilipes* Emery, 1892; see under ***ANOPLOLEPIS.***

***pissina.*** *Plagiolepis pissina* Roger, 1863a: 162 (w.) SRI LANKA.

#***polita.*** *Plagiolepis (Anacantholepis) vanderkeleni* st. *polita* Santschi, 1914e: 37, fig. 10 (w.) SOUTH AFRICA.

#***polygyna.*** *Plagiolepis maura* var. *palygyna* Santschi, 1922a: 67 (w.q.) TUNISIA. [Justified emendation of

spelling to *polygyna:* Emery, 1925b: 21.] Subspecies of *pallescens:* current status.
**pontii.** *Plagiolepis pontii* Menozzi, 1939a: 309, figs. 9, 10 (w.) INDIA.
#**pubescens.** *Plagiolepis brunni* var. *pubescens* Forel, 1913a: 140 (w.) ZIMBABWE.
**puncta.** *Plagiolepis pygmaea* var. *punctum* Forel, 1910e: 448 (w.m.) SOUTH AFRICA. Arnold, 1922: 602 (q.). Variety of *pygmaea:* Arnold, 1922: 602; Wheeler, W.M. 1922a: 930. Raised to species: Emery, 1921d: 317; Emery, 1925b: 21.
**pygmaea.** *Formica pygmaea* Latreille, 1798: 45 (w.q.) FRANCE. Schenck, 1852: 69 (m.); Wheeler, G.C. & Wheeler, J. 1953c: 135 (l.). Combination in *Plagiolepis:* Mayr, 1861: 43. Senior synonym of *dufourii:* Menozzi, 1925a: 19 (footnote). Current subspecies: nominal plus *bulawayensis, mima, minu.* See also: Stitz, 1939: 231; Kutter, 1977c: 185.
#**pyrenaica.** *Plagiolepis barbara* var. *pyrenaica* Emery, 1921d: 315, fig. I (w.) FRANCE. Subspecies of *schmitzii:* current status.
#**quadrimaculata.** *Plagiolepis exigua* var. *quadrimaculata* Forel, 1902h: 483 (w.m.) AUSTRALIA.
**regis.** *Plagiolepis (Plagiolepis) regis* Karavaiev, 1931a: 312, fig. 3 (w.q.m.) TURKESTAN. Combination in *Aporomyrmex:* Faber, 1969: 63; in *Plagiolepis:* Dlussky & Zabelin, 1985: 234.
**rogeri.** *Plagiolepis rogeri* Forel, 1894c: 417 (w.) INDIA.
*rothneyi* Forel, 1894; see under ***LEPISIOTA.***
*rufescens* Santschi, 1917; see under ***ANOPLOLEPIS.***
**satunini.** *Plagiolepis (Plagiolepis) satunini* Karavaiev, 1931a: 314, fig. 4 (w.) ARMENIA.
**schmitzii.** *Plagiolepis pygmaea* var. *schmitzii* Forel, 1895d: 231 (w.q.) MADEIRA. Santschi, 1911i: 286 (m.). Raised to species: Santschi, 1920g: 168. Senior synonym of *barbara* and *crosi:* Collingwood, 1978: 71. Current subspecies: nominal plus *canariensis, madeirensis, pyrenaica, tingitana.*
**simoni.** *Plagiolepis simoni* Emery, 1921d: 317, fig. II (w.) LESOTHO.
*simulans* Santschi, 1908; see under ***ANOPLOLEPIS.***
***singularis.** **Plagiolepis singularis* Mayr, 1868c: 38 (q.) BALTIC AMBER (Oligocene).
***solitaria.** **Plagiolepis solitaria* Mayr, 1868c: 40 (m.) BALTIC AMBER (Oligocene).
#**sordida.** *Plagiolepis maura* var. *sordida* Santschi, 1920g: 170, fig. 1 (w.q.) TUNISIA. Subspecies of *pallescens:* current status.
*splendida* Viehmeyer, 1916; see under ***LEPISIOTA.***
***squamifera.** **Plagiolepis squamifera* Mayr, 1868c: 40, pl. 1, fig. 24 (w.) BALTIC AMBER (Oligocene).
**squamulosa.** *Plagiolepis squamulosa* Wheeler, W.M. 1934d: 156 (w.) AUSTRALIA.
*steingroeveri* Forel, 1894; see under ***ANOPLOLEPIS.***
***succini.** **Plagiolepis succini* André, 1895b: 81 (w.) BALTIC AMBER (Oligocene).
**sudanica.** *Plagiolepis (Plagiolepis) sudanica* Weber, 1943c: 381, pl. 16, fig. 38 (w.) SUDAN.
*sundaica* Emery, 1921; see under ***LEPISIOTA.***
*taivanae* Forel, 1913; see under ***LEPISIOTA.***
**taurica.** *Plagiolepis maura* var. *taurica* Santschi, 1920g: 171 (w.) UKRAINE. Variety of *pallescens:* Emery, 1921d: 314; Santschi, 1922a: 67; of *pygmaea:* Karavaiev, 1931a: 309. Raised to species: Radchenko, 1989b: 155.
*tenella* Santschi, 1911; see under ***ANOPLOLEPIS.***
#**tingitana.** *Plagiolepis schmitzi* var. *tingitana* Santschi, 1936c: 205, fig. 3 (w.) MOROCCO.
#**tricolor.** *Plagiolepis vanderkelleni* var. *tricolor* Forel, 1910f: 24 (w.) BOTSWANA.
*trimenii* Forel, 1895; see under ***ANOPLOLEPIS.***
*tumidula* Emery, 1915; see under ***ANOPLOLEPIS.***
**vanderkelleni.** *Plagiolepis vanderkelleni* Forel, 1901d: 310 (w.) ANGOLA. [Also described as new by Forel, 1903e: 564.] Combination in *P. (Anacantholepis):* Santschi, 1914e: 37. Current subspecies: nominal plus *polita, tricolor.*
**vindobonensis.** *Plagiolepis vindobenensis* Lomnicki, 1925c: 77, figs. 1–4 (w.) AUSTRIA. Santschi, 1926f: 293 (q.); Boven, 1946: 9 (m.). Junior synonym of *pygmaea:* Bernard, 1967: 277. Revived from synonymy: Kutter, 1977c: 189. See also: Collingwood, 1979: 110; Atanassov & Dlussky, 1992: 203.
*watsonii* Forel, 1894; see under ***LEPISIOTA.***
*wroughtoni* Forel, 1902; see under ***LEPISIOTA.***
**xene.** *Plagiolepis xene* Stärcke, 1936: 279 (q.) ITALY. Kutter, 1952: 61 (m.). Combination in *P. (Paraplagiolepis):* Faber, 1969: 65. See also: Bernard, 1967: 278.

*PLANIMYRMA:* see under ***APHAENOGASTER.***

*PLATYSTRUMA:* see under ***SMITHISTRUMA.***

***PLATYTHYREA*** (World tropics)
**angusta.** *Platythyrea angusta* Forel, 1901f: 336 (w.) TRINIDAD.
*annamita. Platythyrea coxalis* var. *annamita* Forel, 1911i: 215 (footnote) (w.) VIETNAM. Junior synonym of *parallela:* Brown, 1975: 8.
*apicalis. Platythyrea lamellosa* st. *apicalis* Santschi, 1937g: 71, fig. 1 (w.) ZAIRE. Junior synonym of *arnoldi:* Brown, 1975: 8.
**arnoldi.** *Platythyrea arnoldi* Forel, 1913j: 206 (w.) ZIMBABWE. Arnold, 1926: 197 (q.m.). Senior synonym of *apicalis:* Brown, 1975: 8.
**arthuri.** *Platythyrea arthuri* Forel, 1910a: 15 (w.) MADAGASCAR.
*aruana. Platythyrea melancholica* var. *aruana* Karavaiev, 1925a: 75 (w.) INDONESIA (Aru I.). Junior synonym of *parallela:* Wilson, 1958a: 151; Brown, 1975: 8.
*australis. Platythyrea pusilla* var. *australis* Forel, 1915b: 10 (w.) AUSTRALIA. Wheeler, G.C. & Wheeler, J. 1971b: 1198 (l.). Junior synonym of *parallela:* Brown, 1975: 8.
*bequaerti. Platythyrea schultzei* var. *bequaerti* Forel, 1913b: 307 (w.m.) ZAIRE. Subspecies of *schultzei:* Santschi, 1937g: 72. Junior synonym of *schultzei:* Brown, 1975: 9.
*bicolor. Eubothroponera bicolor* Clark, 1930c: 11, fig. 1 (w.) AUSTRALIA. Junior synonym of *turneri:* Brown, 1975: 9.
**bicuspis.** *Platythyrea bicuspis* Emery, 1899f: 268 (w.m.) MADAGASCAR. See also: Brown, 1975: 52.
**bidentata.** *Platythyrea bidentata* Brown, 1975: 51, figs. 26, 28 (w.) PHILIPPINES.
*boliviana. Platythyrea meinerti* st. *boliviana* Santschi, 1921g: 83 (w.) BOLIVIA. Junior synonym of *sinuata:* Brown, 1975: 9.
*brevidentata. Platythyrea cribrinodis* var. *brevidentata* Wheeler, W.M. 1922a: 759. Replacement name for *punctata* Arnold, 1915: 24. [Junior secondary homonym of *Pachycondyla punctata* Smith, F. 1858b: 108.] Junior synonym of *cribrinodis:* Brown, 1975: 8.
*brevinodis. Platythyrea cribrinodis* var. *brevinodis* Santschi, 1914b: 47 (w.) KENYA. Santschi, 1935b: 257 (m.). Junior synonym of *cribrinodis:* Brown, 1975: 8.
**brunnipes.** *Eubothroponera brunnipes* Clark, 1938: 361, fig. 3 (w.) AUSTRALIA. Combination in *Platythyrea:* Brown, 1975: 8.
*cephalotes. Platythyrea cephalotes* Viehmeyer, 1924a: 224 (w.) AUSTRALIA. Junior synonym of *parallela:* Brown, 1975: 8.
*ceylonensis. Platythyrea ceylonensis* Donisthorpe, 1941j: 202 (w.q.) SRI LANKA. Junior synonym of *parallela:* Brown, 1975: 8.
*cineracea. Platythyrea cineracea* Forel, 1886b: xxxix (w.) GUATEMALA. Forel, 1899c: 4 (m.). Variety of *punctata:* Forel, 1893g: 359; Forel, 1912c: 35. Junior synonym of *punctata:* Brown, 1975: 9.
**clypeata.** *Platythyrea clypeata* Forel, 1911d: 378 (q.) "Pays de Mois" (S.E. Asia). Senior synonym of *thwaitesi:* Brown, 1975: 8.
**conradti.** *Platythyrea conradti* Emery, 1899e: 464 (w.m.) CAMEROUN. Senior synonym of *monodi:* Brown, 1975: 8.
**cooperi.** *Platythyrea cooperi* Arnold, 1915: 29 (w.m.) SOUTH AFRICA.
*coxalis. Platythyrea coxalis* Emery, 1893e: 189 (footnote) (w.) WEST MALAYSIA. Viehmeyer, 1916a: 111 (q.). Junior synonym of *parallela:* Wilson, 1958a: 151.
**cribrinodis.** *Ponera cribrinodis* Gerstäcker, 1859: 262 (w.) MOZAMBIQUE. [Also described as new by Gerstäcker, 1862: 504.] Emery, 1896i: 154 (m.); Wheeler, G.C. & Wheeler, J. 1971b: 1198 (l.). Combination in *Platythyrea:* Roger, 1863a: 173. Senior synonym of *brevidentata* and *brevinodis:* Brown, 1975: 8.
**crucheti.** *Platythyrea crucheti* Santschi, 1911g: 205 (w.) ANGOLA.
*cylindrica. Platythyrea coxalis* var. *cylindrica* Forel, 1913k: 5, fig. A (q.) SUMATRA. Junior synonym of *parallela:* Brown, 1975: 8.
*cyriluli. Platythyrea cyriluli* Forel, 1922: 88 (w.) ANGOLA. Junior synonym of *schultzei:* Brown, 1975: 9.
*debilior. Platythyrea mocquerysi* var. *debilior* Forel, 1907g: 76 (w.) MADAGASCAR. Junior synonym of *mocquerysi:* Brown, 1975: 8.
**dentinodis.** *Eubothroponera dentinodis* Clark, 1930c: 9, fig. 1 (w.) AUSTRALIA. Combination in *Platythyrea:* Brown, 1975: 8.
*egena. Platythyrea pusilla* var. *egena* Viehmeyer, 1916a: 111 (w.q.) SINGAPORE. Junior synonym of *parallela:* Brown, 1975: 8.
**exigua.** *Platythyrea exigua* Kempf, 1964a: 142, figs. 1–3 (w.q.) BRAZIL.
**frontalis.** *Platythyrea frontalis* Emery, 1899e: 466 (w.q.) CAMEROUN. See also: Brown, 1975: 46.

***gracillima.*** *Platythyrea gracillima* Wheeler, W.M. 1922a: 59, fig. 5 (w.) ZAIRE. [Misspelled as *gracilliens* by Santschi, 1937g: 73.] See also: Brown, 1975: 52.

*grisea. Sima aethiops* subsp. *grisea* Forel, 1910f: 5 (m.) BOTSWANA. Junior synonym of *lamellosa:* Forel, 1913b: 307.

*incerta. Platythyrea incerta* Emery, 1890a: 56 (w.) VENEZUELA. Wheeler, G.C. & Wheeler, J. 1971b: 1198 (l.). Race of *punctata:* Forel, 1912c: 36. Junior synonym of *pilosula:* Brown, 1975: 9.

*inconspicua. Platythyrea inconspicua* Mayr, 1870b: 961 (q.) SRI LANKA. Emery, 1890a: 56 (w.). Junior synonym of *punctata:* Forel, 1901f: 335; Creighton, 1950a: 34; of *parallela:* Brown, 1975: 8.

***inermis.*** *Platythyrea inermis* Forel, 1910d: 122 (w.) PHILIPPINES. Wheeler, G.C. & Wheeler, J. 1952a: 118 (l.). Senior synonym of *maquilingi:* Brown, 1975: 8.

*javana. Platythyrea coxalis* var. *javana* Forel, 1905c: 5 (w.q.) JAVA. Junior synonym of *parallela:* Brown, 1975: 8.

***lamellosa.*** *Ponera lamellosa* Roger, 1860: 295 (w.) SOUTH AFRICA. Wheeler, G.C. & Wheeler, J. 1989b: 51 (l.). Combination in *Platythyrea:* Roger, 1863a: 173. Senior synonym of *grisea:* Forel, 1913b: 307; of *longinoda, rhodesiana* and *suturalis:* Brown, 1975: 8.

*lata. Platythyrea schultzei* var. *lata* Santschi, 1930a: 53 (w.) ZIMBABWE. Junior synonym of *schultzei:* Brown, 1975: 9.

*longinoda. Platythyrea lamellosa* r. *longinoda* Forel, 1894b: 75 (w.) SOUTH AFRICA. Junior synonym of *lamellosa:* Brown, 1975: 8.

*maquilingi. Platythyrea maquilingi* Wheeler, W.M. & Chapman, 1925: 58, pl. 2, figs. 1, 2 (w.) PHILIPPINES. Junior synonym of *inermis:* Brown, 1975: 8.

***matopoensis.*** *Platythyrea mocquerysi* var. *matopoensis* Arnold, 1915: 32 (q.) ZIMBABWE. Arnold, 1926: 196 (m.). Raised to species: Brown, 1975: 8.

*meinerti. Platythyrea meinerti* Forel, 1905b: 156 (w.) VENEZUELA. Junior synonym of *sinuata:* Brown, 1975: 9.

***micans.*** *Eubothroponera micans* Clark, 1930c: 10, fig. 1 (w.) AUSTRALIA. Combination in *Platythyrea:* Brown, 1975: 8.

***mocquerysi.*** *Platythyrea mocquerysi* Emery, 1899f: 270 (w.q.) MADAGASCAR. Senior synonym of *debilior:* Brown, 1975: 8.

***modesta.*** *Platythyrea modesta* Emery, 1899e: 467 (w.) CAMEROUN. Wheeler, G.C. & Wheeler, J. 1976a: 41 (l.). See also: Brown, 1975: 45.

*monodi. Platythyrea monodi* Bernard, 1953b: 185, fig. 1 (w.) GUINEA. Junior synonym of *conradti:* Brown, 1975: 8.

***nicobarensis.*** *Platythyrea nicobarensis* Forel, 1905b: 165 (w.) NICOBAR IS. See also: Brown, 1975: 49.

***occidentalis.*** *Platythyrea occidentalis* André, 1890: 315 (q.) SIERRA LEONE. André, 1895a: 3 (m.); Arnold, 1954: 291 (w.). See also: Brown, 1975: 46.

*pacifica. Platythyrea pusilla* var. *pacifica* Santschi, 1928a: 43 (w.) SAMOA. Junior synonym of *parallela:* Wilson & Taylor, 1967: 20.

***parallela.*** *Ponera parallela* Smith, F. 1859a: 143 (w.) INDONESIA (Aru I.). Donisthorpe, 1943d: 434 (m.); Wheeler, G.C. & Wheeler, J. 1976a: 42 (l.). Combination in *Platythyrea:* Donisthorpe, 1932c: 454. Senior synonym of *aruana, coxalis* and *pusilla:* Wilson, 1958a: 151; of *pacifica:* Wilson & Taylor, 1967: 20; of *annamita, australis, cephalotes, ceylonensis, cylindrica, egena, inconspicua, javana, parva, philippinensis, pulchella, sechellensis, subtilis, tritschleri, victoriae* and*wroughtonii:* Brown, 1975: 8.

*parva. Platythyrea parva* Crawley, 1915a: 133 (w.) AUSTRALIA. Junior synonym of *parallela:* Brown, 1975: 8.

*penangensis. Platythyrea tricuspidata* subsp. *penangensis* Wheeler, W.M. 1929g: 29 (q.) WEST MALAYSIA. Junior synonym of *tricuspidata:* Brown, 1975: 9.

*philippinensis. Platythyrea coxalis* var. *philippinensis* Viehmeyer, 1916b: 283 (w.) PHILIPPINES. Junior synonym of *parallela:* Brown, 1975: 8.

***pilosula.*** *Ponera pilosula* Smith, F. 1858b: 95 (w.) BRAZIL. Combination in *Platythyrea:* Mayr, 1886c: 358; in *Leptogenys:* Emery, 1911d: 106; in *Platythyrea:* Kempf, 1965: 181. Senior synonym of *incerta:* Brown, 1975: 9.

*****primaeva.*** **Platythyrea primaeva* Wheeler, W.M. 1915h: 36, fig. 7 (w.q.) BALTIC AMBER (Oligocene).

***prizo.*** *Platythyrea prizo* Kugler, C. 1977: 216, figs. 1–3 (w.) COSTA RICA.

*pruinosa. Platythyrea pruinosa* Mayr, 1870b: 962 (w.) MEXICO. Forel, 1893g: 358 (m.). Variety of *punctata:* Forel, 1901f: 335; Wheeler, W.M. 1917g: 457; Wheeler, W.M. 1923c: 3. Junior synonym of *punctata:* Wheeler, W.M. 1908a: 123; Brown, 1975: 9.

*pulchella. Platythyrea pulchella* Santschi, 1928h: 121 (w.) SUMATRA. Junior synonym of *parallela:* Brown, 1975: 8.

***punctata.*** *Pachycondyla punctata* Smith, F. 1858b: 108 (w.m.) CENTRAL AMERICA. Wheeler, W.M. 1905b: 81 (q.l.). Combination in *Platythyrea:* Roger, 1863a: 173. Senior synonym of *cineracea:* Brown, 1975: 9; of *pruinosa:* Wheeler, W.M. 1908a: 123; Brown, 1975: 9.

*punctata. Platythyrea cribrinodis* var. *punctata* Arnold, 1915: 24 (w.) ZIMBABWE. [Junior secondary homonym of *punctata* Smith, above.] Replacement name: *brevidentata* Wheeler, W.M. 1922a: 759.

*pusilla. Platythyrea pusilla* Emery, 1893e: 188 (w.) INDONESIA (Ambon I.). Wheeler, W.M. 1919e: 50 (q.). Junior synonym of *parallela:* Wilson, 1958a: 151.

***quadridenta.*** *Platythyrea quadridenta* Donisthorpe, 1941c: 134 (w.) NEW GUINEA. Imai, Brown, *et al.* 1984: 67 (k.).

*recta. Platythyrea schultzei* st. *bequaerti* var. *recta* Santschi, 1937g: 72 (w.) ZAIRE. Unavailable name; material referred to *schultzei* by Brown, 1975: 9.

*reticulata. Eubothroponera reticulata* Clark, 1934b: 33, pl. 2, fig. 16 (w.) AUSTRALIA. Junior synonym of *turneri:* Brown, 1975: 9.

*rhodesiana. Platythyrea lamellosa* var. *rhodesiana* Forel, 1913a: 110 (w.m.) ZIMBABWE. Forel, 1913h: 347 (q.). Junior synonym of *lamellosa:* Brown, 1975: 8.

*ruficornis. Ponera ruficornis* Spinola, 1851b: 53 (w.) BRAZIL. [Also described as new by Spinola, 1853: 69.] [Specimen consists of head and prothorax of a *Dolichoderus* species, remainder of body is of a *Platythyrea* species: Emery, 1893h: 2; Emery, 1911d: 30 (footnote).] See also: Brown, 1975: 9.

***sagei.*** *Platythyrea sagei* Forel, 1900d: 315 (w.) INDIA. See also: Brown, 1975: 52.

***schultzei.*** *Platythyrea schultzei* Forel, 1910f: 3 (w.) NAMIBIA. Arnold, 1926: 195 (m.); Wheeler, G.C. & Wheeler, J. 1989b: 51 (l.). Senior synonym of *bequaerti, cyriluli* and *lata,* and material of the unavailable name *recta* referred here: Brown, 1975: 9.

*sechellensis. Platythyrea wroughtoni* subsp. *sechellensis* Forel, 1912k: 159 (w.) SEYCHELLE IS. Junior synonym of *parallela:* Brown, 1975: 8.

*septentrionalis. Eubothroponera septentrionalis* Clark, 1934b: 34, pl. 2, fig. 17 (w.) AUSTRALIA. Junior synonym of *turneri:* Brown, 1975: 9.

***sinuata.*** *Ponera sinuata* Roger, 1860: 297 (w.m.) SURINAM. Combination in *Platythyrea:* Roger, 1863a: 173. Senior synonym of *boliviana* and *meinerti:* Brown, 1975: 9.

***strenua.*** *Platythyrea strenua* Wheeler, W.M. & Mann, 1914: 6, fig. 1 (w.) HAITI.

*subtilis. Platythyrea subtilis* Emery, 1900d: 666 (footnote) (w.) BORNEO. Junior synonym of *parallela:* Brown, 1975: 8.

*suturalis. Platythyrea lamellosa* subsp. *suturalis* Forel, 1909b: 64 (w.) ANGOLA. Junior synonym of *lamellosa:* Brown, 1975: 8.

*tasmaniensis. Pachycondyla* (*Bothroponera*) *tasmaniensis* Forel, 1913g: 176 (w.) TASMANIA. Wheeler, G.C. & Wheeler, J. 1971b: 1200 (l.). Combination in *Eubothroponera:* Clark, 1934b: 32. Junior synonym of *turneri:* Brown, 1975: 9.

***tenuis.*** *Platythyrea tenuis* Emery, 1899e: 467 (w.q.) CAMEROUN. See also: Brown, 1975: 45.

*thwaitesi. Platythyrea thwaitesi* Donisthorpe, 1931b: 496 (q.) SRI LANKA. Junior synonym of *clypeata:* Brown, 1975: 8.

***tricuspidata.*** *Platythyrea tricuspidata* Emery, 1900d: 665 (w.) SUMATRA. Imai, Brown, *et al.* 1984: 67 (k.). Senior synonym of *penangensis:* Brown, 1975: 9.

*tritschleri. Platythyrea coxalis* var. *tritschleri* Forel, 1901f: 338 (w.) SUMATRA. Junior synonym of *parallela:* Brown, 1975: 8.

***turneri.*** *Platythyrea turneri* Forel, 1895f: 420 (w.) AUSTRALIA. Senior synonym of *bicolor, reticulata, septentrionalis* and *tasmaniensis:* Brown, 1975: 9.

*victoriae. Platythyrea wroughtoni* r. *victoriae* Forel, 1900d: 316 (w.) INDIA. Raised to species: Bingham, 1903: 75. Junior synonym of *parallela:* Brown, 1975: 8.

***viehmeyeri.*** *Platythyrea viehmeyeri* Santschi, 1914a: 429 (w.) TANZANIA.

*wroughtonii. Platythyrea wroughtonii* Forel, 1900d: 315 (w.) INDIA. Junior synonym of *parallela:* Brown, 1975: 8.

***zodion.*** *Platythyrea zodion* Brown, 1975: 55, figs. 34–36 (w.) ECUADOR.

***PLECTROCTENA*** (Afrotropical)

***anops.*** *Plectroctena anops* Bolton, 1974c: 319, figs. 4, 7 (w.) GHANA.

*caffra. Ponera caffra* Spinola, 1851b: 53; Spinola, 1853: 69. *Nomen nudum;* attributed to Klug. Material referred to *mandibularis* by Roger, 1861a: 41.

***conjugata.*** *Plectroctena minor* st. *conjugata* Santschi, 1914e: 8 (w.q.) SOUTH AFRICA. Santschi, 1924a: 166 (m.); Wheeler, G.C. & Wheeler, J. 1989b: 52 (l.). Raised to species: Santschi, 1924a: 166. Junior

***PLECTROCTENA** (continued)*

synonym of *mandibularis:* Arnold, 1926: 209. Revived from synonymy, status as species: Bolton, 1974c: 326.

***cristata.*** *Plectroctena cristata* Emery, 1899e: 470 (w.) CAMEROUN. Bolton, 1974c: 322 (q.). Senior synonym of *semileavis:* Bolton, 1974c: 321.

***cryptica.*** *Plectroctena cryptica* Bolton, 1974c: 327, fig. 5 (w.q.) GHANA.

***dentata.*** *Plectroctena minor* var. *dentata* Santschi, 1912b: 150 (w.) ANGOLA. Bolton, 1974c: 322 (q.). Raised to species: Santschi, 1924a: 164. Senior synonym of *emeryi:* Bolton, 1974c: 322.

*emeryi. Plectroctena emeryi* Santschi, 1924a: 164 (ergatoid q., not w.) ZAIRE. Junior synonym of *dentata:* Bolton, 1974c: 322.

***gabonensis.*** *Plectroctena gabonensis* Santschi, 1919a: 336 (w.) GABON. [Also described as new by Santschi, 1919h: 90.] Bolton, 1974c: 323 (q.). Combination in *Myopias:* Wheeler, W.M. 1922a: 785; in *Plectroctena:* Santschi, 1924a: 170. Stirps of *subterranea:* Santschi, 1919h: 90. Revived status as species: Santschi, 1924a: 170; Bolton, 1974c: 323.

***gestroi.*** *Plectroctena gestroi* Menozzi, 1922a: 348, fig. 1 (w.q.) PRINCIPE I. See also: Bolton, 1974c: 328.

***hastifera.*** *Cacopone hastifer* Santschi, 1914d: 325, fig. 11 (w.) GHANA. Combination in *Plectroctena:* Bolton, 1974c: 320.

*insularis. Plectroctena minor* var. *insularis* Santschi, 1924a: 169, fig. 3 (w.) FERNANDO PO I. Junior synonym of *minor:* Bolton, 1974c: 324.

*integra. Plectroctena mandibularis* var. *integra* Santschi, 1924a: 161 (w.m.) KENYA. Junior synonym of *mandibularis:* Bolton, 1974c: 330.

***laevior.*** *Plectroctena mandibularis* st. *laevior* Stitz, in Santschi, 1924a: 163, fig. 1 (w.) TANZANIA. Raised to species: Bolton, 1974c: 329.

***latinodis.*** *Plectroctena latinodis* Santschi, 1924a: 165, fig. 2 (w.q.) ZAIRE. See also: Bolton, 1974c: 323.

*liberiana. Plectroctena minor* var. *liberiana* Santschi, 1924a: 169, fig. 2 (w.) LIBERIA. Junior synonym of *minor:* Bolton, 1974c: 324.

***lygaria.*** *Plectroctena lygaria* Bolton, Gotwald & Leroux, 1979: 373, figs. 1, 2 (w.q.m.) IVORY COAST.

*mabirensis* Arnold, 1954; see under ***PSALIDOMYRMEX.***

***macgeei.*** *Plectroctena macgeei* Bolton, 1974c: 330, figs. 1, 8 (w.) NIGERIA.

*major. Plectroctena caffra* r. *major* Forel, 1894b: 74 (ergatoid q., not w.) MOZAMBIQUE. Arnold, 1915: 86 (w.). Variety of *mandibularis:* Emery, 1911d: 95; Santschi, 1914b: 54. Junior synonym of *mandibularis:* Emery, 1899e: 469; Bolton, 1974c: 330.

***mandibularis.*** *Plectroctena mandibularis* Smith, F. 1858b: 101, pl. 7, figs. 1–5 (m., ergatoid q., not w.) SOUTH AFRICA. Forel, 1913a: 108 (w.); Arnold, 1915: 86 (w.). Material of the *nomen nudum caffra* referred here: Roger, 1861a: 41. Senior synonym of *major:* Emery, 1899e: 469; Bolton, 1974c: 330; of *integra,* and material of the unavailable name *strialiventris* referred here: Bolton, 1974c: 330.

***minor.*** *Plectroctena minor* Emery, 1892d: 556, pl. 15, figs. 1, 2 (q.) IVORY COAST. Emery, 1902c: 32 (w.); Wheeler, W.M. 1922a: 88 (m.). Senior synonym of *insularis, liberiana* and *perusta:* Bolton, 1974c: 324.

*perusta. Plectroctena minor* var. *perusta* Santschi, 1924a: 168, fig. 2 (w.) CAMEROUN. Junior synonym of *minor:* Bolton, 1974c: 324.

*punctatus. Plectroctena punctatus* Santschi, 1924a: 170 (m.) KENYA. Junior synonym of *subterranea:* Bolton, 1974c: 333.

*semileavis. Plectroctena cristata* var. *semileavis* Santschi, 1924a: 163 (w.) ZAIRE. Junior synonym of *cristata:* Bolton, 1974c: 321.

*strialiventris. Plectroctena mandibularis* st. *strigosa* var. *strialiventris* Stitz, in Santschi, 1924a: 162 (w.) MALAWI. Unavailable name; material referred to *mandibularis* by Bolton, 1974c: 330. [Stitz, in Santschi, 1924a: 162 also variously spells this unavailable name *strativentris* and *striativentris.*]

***strigosa.*** *Plectroctena mandibularis* var. *strigosa* Emery, 1899e: 469 (w.) SOUTH AFRICA. Stirps of *cristata:* Santschi, 1924a: 161. Raised to species: Bolton, 1974c: 332.

***subterranea.*** *Plectroctena subterranea* Arnold, 1915: 84, pl. 3, fig. 23 (w.q.) ZIMBABWE. Combination in *Myopias:* Wheeler, W.M. 1922a: 785; in *Plectroctena:* Santschi, 1924a: 171. Senior synonym of *punctatus:* Bolton, 1974c: 333.

***ugandensis.*** *Plectroctena ugandensis* Menozzi, 1933a: 99, fig. 2 (q.) UGANDA. See also: Bolton, 1974c: 334.

***PODOMYRMA*** (Indo-Australian, Australasian)

***abdominalis.*** *Podomyrma abdominalis* Emery, 1887b: 459 (w.) INDONESIA (Ternate I.). Subspecies of *ruficeps:* Viehmeyer, 1914b: 523 (in key); Emery, 1922f: 238. Revived status as species: Taylor & Brown, D.R. 1985: 80. Current subspecies: nominal plus *pulchra.*

***adelaidae.*** *Myrmica adelaidae* Smith, F. 1858b: 128 (w.) AUSTRALIA. Wheeler, G.C. & Wheeler, J. 1954c: 127 (l.); Imai, Crozier & Taylor, 1977: 348 (k.). Combination in *Podomyrma:* Dalla Torre, 1893: 60. Senior synonym of *bimaculata:* Forel, 1915b: 50; Emery, 1922f: 237; of *sericeiventris:* Emery, 1922f: 237. Current subspecies: nominal plus *brevidentata, obscurior.*

***alae.*** *Podomyrma alae* Donisthorpe, 1949c: 751 (w.) NEW GUINEA.

***albertisi.*** *Podomyrma albertisi* Emery, 1887b: 460 (w.) NEW GUINEA.

***basalis.*** *Podomyrma basalis* Smith, F. 1859a: 147 (w.) INDONESIA (Aru I.). Smith, F. 1863: 19 (q.). Current subspecies: nominal plus *brunnea, nigrescens, reyi, salomo, woodfordi.*

***#bicolor.*** *Podomyrma silvicola* subsp. *bicolor* Emery, 1897d: 559 (w.) NEW GUINEA.

*bimaculata. Podomyrma bimaculata* Forel, 1901h: 57 (w.q.) AUSTRALIA. Junior synonym of *adelaidae:* Forel, 1915b: 50.

***bispinosa.*** *Podomyrma bispinosa* Forel, 1901h: 56 (w.) AUSTRALIA.

***#brevidentata.*** *Podomyrma bimaculata* var. *brevidentata* Forel, 1915b: 49 (w.) AUSTRALIA. Variety/subspecies of *adelaidae:* Emery, 1922f: 237; Taylor & Brown, D.R. 1985: 80.

***#brunnea.*** *Podomyrma basalis* var. *brunneus* Donisthorpe, 1943d: 453 (w.) NEW GUINEA.

*caledonica* André, 1889; see under ***LORDOMYRMA.***

***carinata.*** *Podomyrma carinata* Donisthorpe, 1947c: 584 (w.m.) NEW GUINEA.

*castanea. Podomyrma castanea* Stitz, 1911a: 358, fig. 7 (w.) AUSTRALIA. Junior synonym of *striata:* Viehmeyer, 1914b: 520.

*cataulacoidea* Stitz, 1911; see under ***DILOBOCONDYLA.***

***chasei.*** *Podomyrma chasei* Forel, 1901h: 58 (w.m.) AUSTRALIA.

***christae.*** *Dacryon christae* Forel, 1907a: 16 (w.) AUSTRALIA. Combination in *Podomyrma:* Taylor & Brown, D.R. 1985: 81.

***clarki.*** *Pseudopodomyrma clarki* Crawley, 1925a: 40 (w.) AUSTRALIA. Clark, 1926: 49 (q.). Combination in *Podomyrma:* Taylor & Brown, D.R. 1985: 81.

***convergens.*** *Podomyrma convergens* Forel 1895f: 427 (q.) AUSTRALIA.

***delbrueckii.*** *Podomyrma delbrueckii* Forel, 1901h: 58 (w.) AUSTRALIA.

***densestrigosa.*** *Podomyrma densestrigosa* Viehmeyer, 1924b: 316 (w.) AUSTRALIA. Current subspecies: nominal plus *teres.*

***#dimidiata.*** *Podomyrma silvicola* var. *dimidiata* Forel, 1904d: 170 (w.) BISMARCK ARCHIPELAGO.

***#dohertyi.*** *Podomyrma dohertyi* Emery, 1897d: 558 (footnote) (w.) INDONESIA (Moluccas). Subspecies of *ruficeps:* Emery, 1922f: 238.

***elongata.*** *Podomyrma elongata* Forel, 1895f: 428 (w.) AUSTRALIA. Senior synonym of *parva:* Brown, 1953b: 3.

***femorata.*** *Podomyrma femorata* Smith, F. 1859a: 145 (w.q.) INDONESIA (Aru I.).

***ferruginea.*** *Dacryon ferruginea* Clark, 1934b: 37, pl. 3, fig. 2 (w.) AUSTRALIA. Combination in *Podomyrma:* Taylor & Brown, D.R. 1985: 82.

***formosa.*** *Myrmica formosa* Smith, F. 1858b: 128 (w.) AUSTRALIA. Combination in *Podomyrma:* Taylor & Brown, D.R. 1985: 82.

***fortirugis.*** *Podomyrma fortirugis* Viehmeyer, 1924b: 315 (w.q.m.) AUSTRALIA.

***#gastralis.*** *Podomyrma ruficeps* var. *gastralis* Emery, 1897d: 558 (w.) NEW GUINEA.

***gibbula.*** *Podomyrma gibbula* Viehmeyer, 1914b: 523 (w.) NEW GUINEA.

***gracilis.*** *Podomyrma gracilis* Emery, 1887b: 460, pl. 2, fig. 15 (w.) NEW GUINEA. Subspecies of *laeviceps:* Viehmeyer, 1914b: 521 (in key); Donisthorpe, 1949b: 498. Revived status as species: Taylor & Brown, D.R. 1985: 82. Current subspecies: nominal plus *nugenti.*

***gratiosa.*** *Myrmecina gratiosa* Smith, F. 1858b: 133 (w.q.) AUSTRALIA. Viehmeyer, 1924b: 315 (m.). Combination in *Podomyrma:* Roger, 1863b: 27. Current subspecies: nominal plus *papuana.*

***grossestriata.*** *Podomyrma elongata* r. *grossestriata* Forel, 1915b: 50 (w.) AUSTRALIA. Raised to species: Brown, 1953b: 3.

***inermis.*** *Podomyrma inermis* Mayr, 1876: 111 (w.) AUSTRALIA.

***keysseri.*** *Podomyrma keysseri* Viehmeyer, 1914b: 524 (w.) NEW GUINEA.

***kitschneri***. *Dacryon kitschneri* Forel, 1915b: 52 (w.q.) AUSTRALIA. Combination in *Podomyrma:* Taylor & Brown, D.R. 1985: 82.

***kraepelini***. *Podomyrma kraepelini* Forel, 1901h: 59 (q.) AUSTRALIA.

***laeviceps***. *Podomyrma laeviceps* Mayr, 1867a: 108 (w.) NEW GUINEA. [Misspelled as *leviceps* by Dalla Torre, 1893: 60.]

***laevifrons***. *Podomyrma laevifrons* Smith, 1859a: 146 (w.) INDONESIA (Aru I.).

*laevifrons*. *Podomyrma gratiosa* r. *papuana* var. *laevifrons* Stitz, 1938: 103 (w.) NEW GUINEA. Unavailable name.

***laevissima***. *Podomyrma laevissima* Smith, F. 1863: 20 (w.) INDONESIA (Misool I.).

***lampros***. *Podomyrma lampros* Viehmeyer, 1924b: 317 (w.) AUSTRALIA.

***libra***. *Dacryon liber* Forel, 1907h: 275 (w.) AUSTRALIA. Combination in *Podomyrma:* Taylor & Brown, D.R. 1985: 83.

*longicephala* Terayama & Yamane, 1991; see under ***VOLLENHOVIA***.

***macrophthalma***. *Podomyrma macrophthalma* Viehmeyer, 1925a: 25 (w.) AUSTRALIA.

#***maculiventris***. *Podomyrma micans* var. *maculiventris* Emery, 1887b: 459 (w.) AUSTRALIA.

***maligna***. *Myrmica maligna* Smith, F. 1865: 72 (q.) INDONESIA (Morotai I.). Combination in *Podomyrma:* Donisthorpe, 1932c: 473.

***marginata***. *Dacryon marginatus* McAreavey, 1949: 8, figs. 26–29 (w.) AUSTRALIA. Combination in *Podomyrma:* Taylor & Brown, D.R. 1985: 83.

****mayri***. **Podomyrma* (*Acrostigma*) *mayri* Emery, 1891a: 150, pl. 2, figs. 12–16 (w.) SICILIAN AMBER.

***micans***. *Podomyrma micans* Mayr, 1876: 111 (w.) AUSTRALIA. Current subspecies: nominal plus *maculiventris*.

***minor***. *Podomyrma minor* Donisthorpe, 1949b: 498 (w.) NEW GUINEA.

***mjobergi***. *Dacryon mjobergi* Forel, 1915b: 51 (w.) AUSTRALIA. Combination in *Podomyrma:* Taylor & Brown, D.R. 1985: 83.

***muckeli***. *Podomyrma muckeli* Forel, 1910b: 25 (w.) AUSTRALIA.

#***nigra***. *Podomyrma striata* var. *nigra* Stitz, 1938: 102 (w.) NEW GUINEA.

#***nigrescens***. *Podomyrma striata* var. *nigrescens* Stitz, 1912: 504 (w.) NEW GUINEA. Subspecies of *basalis:* Viehmeyer, 1914b: 521 (in key).

*nitida* Smith, F. 1860; see under ***VOLLENHOVIA***.

***nitida***. *Dacryon nitida* Clark, 1938: 364, fig. 4 (w.q.m.) AUSTRALIA. [Unresolved junior secondary homonym of *Podomyrma nitida* Smith, F. 1860b: 110, above.] Combination in *Podomyrma:* Taylor & Brown, D.R. 1985: 83.

***novemdentata***. *Podomyrma novemdentata* Forel, 1901h: 55 (w.q.) AUSTRALIA.

***nuda***. *Podomyrma nuda* Crawley, 1922b: 441, fig. 7 (w.) AUSTRALIA. Crawley, 1925b: 591 (q.).

#***nugenti***. *Podomyrma gracilis* var. *nugenti* Forel, 1901h: 54 (w.) AUSTRALIA.

***obscura***. *Podomyrma obscura* Stitz, 1911a: 362, fig. 9 (w.) AUSTRALIA.

#***obscurior***. *Podomyrma bimaculata* var. *obscurior* Forel, 1915b: 50 (w.) AUSTRALIA. Variety/subspecies of *adelaidae:* Emery, 1922f: 237; Taylor & Brown, D.R. 1985: 80.

***octodentata***. *Podomyrma octodentata* Forel, 1901h: 54 (w.) AUSTRALIA.

***odae***. *Podomyrma odae* Forel, 1910b: 23 (w.) AUSTRALIA.

***omniparens***. *Dacryon omniparens* Forel, 1895f: 421 (w.) AUSTRALIA. Combination in *Podomyrma:* Taylor & Brown, D.R. 1985: 84.

*opacula*. *Podomyrma ruficeps* subsp. *abdominalis* var. *opacula* Viehmeyer, 1914b: 525 (w.) NEW GUINEA. Unavailable name.

***overbecki***. *Podomyrma overbecki* Viehmeyer, 1924b: 318 (w.) AUSTRALIA. Current subspecies: nominal plus *varicolor*.

#***papuana***. *Podomyrma gratiosa* r. *papuana* Stitz, 1938: 102 (w.) NEW GUINEA.

*parva*. *Podomyrma parva* Crawley, 1925b: 592 (w.) AUSTRALIA. Junior synonym of *elongata:* Brown, 1953b: 3.

***pulchella***. *Podomyrma pulchellus* Donisthorpe, 1938a: 145, fig. 5 (w.) NEW GUINEA.

#***pulchra***. *Podomyrma abdominalis* r. *pulchra* Forel, 1901h: 54 (w.) AUSTRALIA. Subspecies/variety of *ruficeps:* Emery, 1922f: 238; of *abdominalis:* Forel, 1915b: 50; Taylor & Brown, D.R. 1985: 80.

#***reyi***. *Podomyrma basalis* subsp. *reyi* Karavaiev, 1935a: 99 (w.) INDONESIA (Ambon I.).

***ruficeps***. *Podomyrma ruficeps* Smith, F. 1863: 20 (w.) INDONESIA (Misool I.). Current subspecies: nominal plus *dohertyi, gastralis, thoracica*.

***rugosa***. *Lordomyrma rugosa* Clark, 1934b: 38, pl. 3, figs. 3, 4 (w.q.) AUSTRALIA. [Unresolved junior secondary homonym of *Rogeria rugosa* Mann, 1921: 455 (now in *Lordomyrma*).] Combination in *Dacryon:* Wheeler, G.C. & Wheeler, J. 1954a: 135; in *Podomyrma:* Taylor & Brown, D.R. 1985: 84. Junior synonym of *christae:* Brown, 1957a: 49. Revived from synonymy: Taylor & Brown, D.R. 1985: 84.

#***salomo***. *Podomyrma basalis* subsp. *salomo* Mann, 1919: 333 (w.) SOLOMON IS.

*sericeiventris*. *Podomyrma micans* subsp. *sericeiventris* Emery, 1898b: 235 (w.q.) AUSTRALIA. Junior synonym of *adelaidae:* Emery, 1922f: 237.

***silvicola***. *Podomyrma silvicola* Smith, F. 1860b: 110 (w.) INDONESIA (Batjan I.). [Misspelled as *sylvicola* by Mayr, 1863: 443.] Current subspecies: nominal plus *bicolor, dimidiata*.

***simillima***. *Podomyrma simillima* Smith, F. 1860b: 111 (w.) INDONESIA (Batjan I.).

***striata***. *Podomyrma striata* Smith, F. 1859a: 146 (w.) INDONESIA (Aru I.). Senior synonym of *castanea:* Viehmeyer, 1914b: 520. Current subspecies: nominal plus *nigra*.

#***teres***. *Podomyrma densestrigosa* subsp. *teres* Viehmeyer, 1924b: 317 (w.) AUSTRALIA.

***testacea***. *Podomyrma testacea* Donisthorpe, 1949g: 413 (w.) NEW GUINEA.

#***thoracica***. *Podomyrma ruficeps* var. *thoracica* Stitz, 1911a: 360, fig. 8 (w.) NEW GUINEA.

***tricolor***. *Podomyrma tricolor* Clark, 1934b: 36, pl. 3, fig. 1 (w.) AUSTRALIA.

***tristis***. *Podomyrma tristis* Karavaiev, 1935a: 102 (w.) INDONESIA (Aru I.).

***turneri***. *Dacryon turneri* Forel, 1901h: 60 (w.q.) AUSTRALIA. Combination in *Podomyrma:* Taylor & Brown, D.R. 1985: 85.

#***varicolor***. *Podomyrma overbecki* var. *varicolor* Viehmeyer, 1925a: 25 (w.) AUSTRALIA.

***vidua***. *Podomyrma vidua* Santschi, 1932b: 15, fig. 2 (q.) INDONESIA (Weeim I.).

#***woodfordi***. *Podomyrma basalis* subsp. *woodfordi* Mann, 1919: 334, fig. 26 (w.q.) SOLOMON IS.

***POECILOMYRMA*** (Fiji Is.)

#***myrmecodiae***. *Poecilomyrma senirewae* subsp. *myrmecodiae* Mann, 1921: 448 (w.) FIJI IS.

***senirewae***. *Poecilomyrma senirewae* Mann, 1921: 446, fig. 16 (w.) FIJI IS. Current subspecies: nominal plus *myrmecodiae*.

***POGONOMYRMEX*** (Nearctic, Neotropical)

***abdominalis***. *Pogonomyrmex* (*Ephebomyrmex*) *naegeli* st. *abdominalis* Santschi, 1929d: 278 (w.q.) BRAZIL. Combination in *Ephebomyrmex:* Kempf, 1972a: 106; in *Pogonomyrmex:* **revived combination**. Raised to species: Kusnezov, 1951a: 250.

#***andinus***. *Pogonomyrmex longibarbis* subsp. *andinus* Kusnezov, 1951a: 276, fig. 12 (w.q.m.) ARGENTINA.

***anergismus***. *Pogonomyrmex* (*Pogonomyrmex*) *anergismus* Cole, 1954b: 115 (q.m.) U.S.A. See also: Cole, 1968: 46.

***angustus***. *Pogonomyrmex angustus* Mayr, 1870b: 970 (w.) CHILE. Mayr, 1887: 612 (q.m.). Combination in *P.* (*Ephebomyrmex*): Emery, 1921f: 48; in *Ephebomyrmex:* Kusnezov, 1960b: 353; in *Pogonomyrmex:* Snelling & Hunt, 1976: 72. See also: Gallardo, 1932b: 103.

***anzensis***. *Pogonomyrmex* (*Pogonomyrmex*) *anzensis* Cole, 1968: 87, pl. 3, fig. 13; pl. 4, fig. 11; pl. 6, fig. 12; pl. 7, fig. 15 (w.) U.S.A.

***apache***. *Pogonomyrmex apache* Wheeler, W.M. 1902c: 392 (w.) U.S.A. Cole, 1954e: 269 (q.m.); Taber, Cokendolpher & Francke, 1988: 51 (k.). Senior synonym of *sanctihyacinthi:* Cole, 1954e: 267. See also: Cole, 1968: 49.

*ater*. *Pogonomyrmex* (*Pogonomyrmex*) *ater* Donisthorpe, 1933c: 532 (w.q.m.) ARGENTINA. Junior synonym of *brevibarbis:* Kusnezov, 1951a: 250.

#***aterrimus***. *Pogonomyrmex* (*Ephebomyrmex*) *schmitti* subsp. *aterrimus* Wheeler, W.M. 1936b: 197 (w.) HAITI. Combination in *Ephebomyrmex:* Kempf, 1972a: 106; in *Pogonomyrmex:* **revived combination**.

***atratus***. *Pogonomyrmex vermiculatus* var. *atrata* Santschi, 1922b: 347 (w.) ARGENTINA. Raised to species: Kusnezov, 1951a: 251.

***badius***. *Formica badia* Latreille, 1802c: 238, pl. 11, fig. 71 (w.q.) U.S.A. Lepeletier, 1835: 175 (m.); Wheeler, G.C. & Wheeler, J. 1960b: 2 (l.); Taber, Cokendolpher & Francke, 1988: 51 (k.). Combination in *Aphaenogaster:* Roger, 1863b: 30; in *Pogonomyrmex:* Mayr, 1870b: 971. Senior synonym of *transversa* (and its junior synonyms *brevipennis* and *crudelis*): Emery, 1895c: 310. See also: Cole, 1968: 148; Gordon, 1984: 413; Hölldobler & Wilson, 1971: 385; Smith, D.R. 1979: 1356.

***barbatus***. *Myrmica barbata* Smith, F. 1858b: 130 (q.) MEXICO. Mayr, 1870b: 971 (w.); Mayr, 1887: 610

(m.); Wheeler, W.M. 1900b: 20 (l.); Taber, Cokendolpher & Francke, 1988: 51 (k.). Combination in *Pogonomyrmex:* Mayr, 1868b: 170. Senior synonym of *molefaciens:* Mayr, 1886c: 365; Creighton, 1950a: 119; Cole, 1968: 56; of *nigrescens:* Cole, 1968: 56. See also: Wheeler, G.C. & Wheeler, J. 1953a: 107; Petralia & Vinson, 1980: 381; Smith, D.R. 1979: 1353.

*barnesi. Pogonomyrmex californicus* subsp. *barnesi* Smith, M.R. 1929: 546 (w.) U.S.A. Subspecies of *maricopa:* Creighton, 1950a: 127. Junior synonym of *maricopa:* Cole, 1968: 138.

***bicolor.*** *Pogonomyrmex (Pogonomyrmex) bicolor* Cole, 1968: 59, pl. 3, fig. 15; pl. 4, fig. 17; pl. 5, fig. 4; pl. 7, fig. 21; pl. 8, fig. 14; pl. 10, fig. 14; pl. 11, fig. 15 (w.q.m.) U.S.A.

***bigbendensis.*** *Pogonomyrmex (Pogonomyrmex) bigbendensis* Francke & Merickel, 1982: 371, figs. 1, 2, 9, 12, 15 (w.) U.S.A.

***bispinosus.*** *Atta bispinosa* Spinola, 1851a: 244 (w.q.) CHILE. Combination in *Pogonomyrmex:* Mayr, 1868b: 170. [Misspelled as *brevispinosus* by Santschi, 1925d: 223.] See also: Gallardo, 1932b: 130.

***brevibarbis.*** *Pogonomyrmex brevibarbis* Emery, 1906c: 158 (w.) ARGENTINA. Santschi, 1925d: 223 (m.). Senior synonym of *ater* and *carettei:* Kusnezov, 1951a: 250. Current subspecies: nominal plus *niger, silvestrii.*

*brevipennis. Myrmica brevipennis* Smith, F. 1858b: 130 (m.) U.S.A. Junior synonym of *transversa:* Mayr, 1886c: 359.

***brevispinosus.*** *Pogonomyrmex (Pogonomyrmex) brevispinosus* Cole, 1968: 89, pl.3, fig. 8; pl. 4, fig. 6; pl. 6, fig. 6; pl. 7, fig. 6; pl. 8, fig. 6; pl. 10, fig. 5; pl. 11, fig. 5 (w.q.m.) U.S.A. Taber, Cokendolpher & Francke, 1988: 51 (k.).

*brevispinus. Pogonomyrmex cunicularis* var. *brevispinus* Santschi, 1931e: 275 (w.q.m.) ARGENTINA. Junior synonym of *cunicularius:* Kusnezov, 1951a: 251.

***bruchi.*** *Pogonomyrmex coarctatus* r. *bruchi* Forel, 1913l: 217 (w.) ARGENTINA. Raised to species: Forel, 1914d: 268.

***californicus.*** *Myrmica californica* Buckley, 1867: 336 (w.) U.S.A. Wheeler, W.M. 1914e: 153 (q.m.); Taber, Cokendolpher & Francke, 1988: 51 (k.). Combination in *Pogonomyrmex:* Emery, 1895c: 311. Senior synonym of *estebanius, longinodis* and *nitratus:* Cole, 1968: 120; of *hindleyi:* Snelling, 1982a: 98. See also: Tulloch, 1930: 61.

***carbonarius.*** *Pogonomyrmex carbonarius* Mayr, 1868b: 172 (w.) ARGENTINA. Junior synonym of *rastratus:* Dalla Torre, 1893: 119. Revived from synonymy as variety of *rastratus:* Emery, 1906c: 157; Gallardo, 1932b: 136. Revived status as species: Kusnezov, 1951a: 251. Current subspecies: nominal plus *sanmartini.*

*carettei. Pogonomyrmex carettei* Santschi, 1922b: 351 (w.q.m.) ARGENTINA. Junior synonym of *brevibarbis:* Kusnezov, 1951a: 250.

*carnivora. Pogonomyrmex cunicularis* var. *carnivora* Santschi, 1925e: 155, fig. 2 (w.) ARGENTINA. Junior synonym of *cunicularius:* Kusnezov, 1951a: 251. Revived from synonymy, and raised to species: Wheeler, G.C. & Wheeler, J. 1972b: 236. Resynonymised under *cunicularius:* **revived synonymy**.

***catanlilensis.*** *Pogonomyrmex catanlilensis* Gallardo, 1931a: 187, fig. 2 (w.) ARGENTINA.

#***chubutensis.*** *Pogonomyrmex vermiculatus* var. *chubutensis* Forel, 1913l: 218 (w.) ARGENTINA. Subspecies of *vermiculatus,* and senior synonym of *jorgenseni:* Kusnezov, 1951a: 253.

***coarctatus.*** *Pogonomyrmex coarctatus* Mayr, 1868b: 170 (w.) ARGENTINA. Mayr, 1887: 614 (q.); Bruch, 1917a: 303 (m.). Senior synonym of *striaticeps:* Kusnezov, 1951a: 253. See also: Gallardo, 1932b: 150.

***colei.*** *Pogonomyrmex (Pogonomyrmex) colei* Snelling, 1982a: 99, figs. 1–11 (q.m.) U.S.A. See also: Rissing, 1983: 321.

*columbicus* Forel, 1912; see under ***HYLOMYRMA.***

***comanche.*** *Pogonomyrmex occidentalis* subsp. *comanche* Wheeler, W.M. 1902c: 392 (w.) U.S.A. Cole, 1968: 126 (q.m.); Taber, Cokendolpher & Francke, 1988: 51 (k.). Raised to species: Wheeler, W.M. 1914e: 156; Olsen, 1934: 498. Subspecies of *occidentalis:* Creighton, 1950a: 128. Revived status as species: Smith, M.R. 1951a: 793; Cole, 1968: 126.

*crudelis. Atta crudelis* Smith, F. 1858b: 170 (w.q.) U.S.A. Mayr, 1862: 741 (m.). Combination in *Myrmica:* Mayr, 1862: 740; in *Pogonomyrmex:* Mayr, 1868b: 170. Junior synonym of *transversa:* Mayr, 1886c: 359.

***cunicularius.*** *Pogonomyrmex cunicularius* Mayr, 1887: 613 (w.m.) ARGENTINA. [Misspelled as *cunicularis* by Santschi, 1922b: 349, and some subsequent authors.] Senior synonym of *brevispinus, carnivora, pencosensis* and *serpens,* and material of the unavailable name *dubia* referred here: Kusnezov, 1951a: 251.

*curvispinosus. Pogonomyrmex barbatus* subsp. *curvispinosus* Cole, 1936b: 120 (w.) U.S.A. Junior synonym of *fuscatus:* Creighton, 1950a: 119.

#***darlingtoni.*** *Pogonomyrmex (Ephebomyrmex) schmitti* subsp. *darlingtoni* Wheeler, W.M. 1936b: 197 (w.) HAITI. Combination in *Ephebomyrmex:* Kempf, 1972a: 106; in *Pogonomyrmex:* **revived combination.**

*dentatus. Pogonomyrmex dentatus* Olsen, 1934: 505 (w.) MEXICO. Junior synonym of *tenuispinus:* Cole, 1968: 77.

***desertorum.*** *Pogonomyrmex desertorum* Wheeler, W.M. 1902c: 387 (w.) U.S.A. Cole, 1968: 67 (q.m.); Taber, Cokendolpher & Francke, 1988: 51 (k.). Senior synonym of *ferrugineus:* Cole, 1968: 65.

*dubia. Pogonomyrmex cunicularius* r. *pencosensis* var. *dubia* Forel, 1914d: 267 (w.) ARGENTINA. Unavailable name; material referred to *cunicularius* by Kusnezov, 1951a: 251.

*estebanius. Pogonomyrmex badius* subsp. *estebanius* Pergande, 1894a: 33 (w.q.) MEXICO. Wheeler, W.M. 1914e: 154 (m.). Variety of *californicus:* Emery, 1895c: 311; Wheeler, W.M. 1902c: 391. Junior synonym of *californicus:* Cole, 1968: 120.

*ferrugineus. Pogonomyrmex desertorum* var. *ferrugineus* Olsen, 1934: 506 (w.) U.S.A. Junior synonym of *desertorum:* Cole, 1968: 65.

***fossilis.** **Pogonomyrmex fossilis* Carpenter, 1930: 33, pl. 9, fig. 6 (w.) U.S.A. (Miocene).

*fuscatus. Pogonomyrmex barbatus* var. *fuscatus* Emery, 1895c: 309 (w.q.) U.S.A. Subspecies of *barbatus,* and senior synonym of *curvispinosus* and *marfensis:* Creighton, 1950a: 119. Junior synonym of *barbatus:* Cole, 1954b: 117; of *rugosus:* Cole, 1968: 70.

*goeldii* Forel, 1912; see under ***HYLOMYRMA.***

***guatemaltecus.*** *Pogonomyrmex guatemaltecus* Wheeler, W.M. 1914e: 149 (w.q.) GUATEMALA.

*hindleyi. Pogonomyrmex californius* var. *hindleyi* Forel, 1914d: 270 (w.) U.S.A. Raised to species: Creighton, 1950a: 125. Junior synonym of *californicus:* Snelling, 1982a: 98.

***huachucanus.*** *Pogonomyrmex huachucanus* Wheeler, W.M. 1914e: 151 (w.) U.S.A. Creighton, 1952b: 79 (q.m.); Wheeler, G.C. & Wheeler, J. 1972b: 237 (l.); Taber, Cokendolpher & Francke, 1988: 51 (k.). Combination in *P. (Ephebomyrmex)*: Cole, 1968: 157.

***imberbiculus.*** *Pogonomyrmex imberbiculus* Wheeler, W.M. 1902a: 87, figs. 1, 2 (w.) U.S.A. Creighton, 1957a: 63 (q.); Cole, 1968: 171 (m.); Wheeler, G.C. & Wheeler, J. 1972b: 237 (l.); Taber, Cokendolpher & Francke, 1988: 51 (k.). Combination in *P. (Ephebomyrmex)*: Wheeler, W.M. 1902c: 390. Senior synonym of *townsendi:* Creighton, 1957a: 56.

*incultus. Pogonomyrmex carettei* var. *incultus* Santschi, 1925e: 157, fig. 1 (w.q.) ARGENTINA. Gallardo, 1932b: 118 (m.). Junior synonym of *niger:* Kusnezov, 1951a: 250.

***inermis.*** *Pogonomyrmex inermis* Forel, 1914d: 267 (w.) ARGENTINA. See also: Gallardo, 1932b: 128.

*intermedia. Pogonomyrmex bispinosus* var. *intermedia* Menozzi, 1935c: 320 (w.) CHILE. Junior synonym of *vermiculatus:* Snelling & Hunt, 1976: 75.

*jorgenseni. Pogonomyrmex vermiculatus* var. *jorgenseni* Forel, 1915c: 353 (w.) ARGENTINA. Junior synonym of *chubutensis:* Kusnezov, 1951a: 253.

***laevigatus.*** *Pogonomyrmex (Ephebomyrmex) laevigatus* Santschi, 1921g: 97 (w.) CHILE. Combination in *Ephebomyrmex:* Kusnezov, 1960b: 354; in *Pogonomyrmex:* Snelling & Hunt, 1976: 75.

***laevinodis.*** *Pogonomyrmex (Ephebomyrmex) laevinodis* Snelling, 1982a: 110, fig. 20–22 (w.) MEXICO.

***laticeps.*** *Pogonomyrmex laticeps* Santschi, 1922b: 350 (w.) ARGENTINA.

#***leonis.*** *Pogonomyrmex meridionalis* subsp. *leonis* Kusnezov, 1951a: 278 (w.) ARGENTINA.

***lobatus.*** *Pogonomyrmex lobatus* Santschi, 1921g: 96 (w.) ARGENTINA.

***longibarbis.*** *Pogonomyrmex longibarbis* Gallardo, 1931a: 185, fig. 1 (w.) ARGENTINA. Current subspecies: nominal plus *andinus.*

*longiceps. Pogonomyrmex brevibarbis* st. *longiceps* Santschi, 1917f: 277 (w.) ARGENTINA. Junior synonym of *silvestrii:* Kusnezov, 1951a: 250.

*longinodis. Pogonomyrmex californicus* subsp. *longinodis* Emery, 1895c: 311 (w.) U.S.A. Raised to species: Creighton, 1950a: 126. Junior synonym of *californicus:* Cole, 1968: 120.

***magnacanthus.*** *Pogonomyrmex (Pogonomyrmex) magnacanthus* Cole, 1968: 133, pl. 2, fig. 5; pl. 3, fig. 12; pl. 4, fig. 10; pl. 6, fig. 9; pl. 7, fig. 16; pl. 8, fig. 10; pl. 11, fig. 10 (w.q.m.) U.S.A. Taber, Cokendolpher & Francke, 1988: 51 (k.).

***marcusi.*** *Pogonomyrmex marcusi* Kusnezov, 1951a: 260 (w.) BOLIVIA. Wheeler, G.C. & Wheeler, J. 1960b: 3 (l.).

*marfensis. Pogonomyrmex barbatus* var. *marfensis* Wheeler, W.M. 1902a: 98 (diagnosis in key) (w.) U.S.A. Junior synonym of *fuscatus:* Creighton, 1950a: 119.

***maricopa.*** *Pogonomyrmex californicus* subsp. *maricopa* Wheeler, W.M. 1914e: 155 (w.q.) U.S.A. Cole, 1968: 140 (m.); Taber, Cokendolpher & Francke, 1988: 51 (k.). Raised to species: Creighton, 1950a: 126. Senior synonym of *barnesi:* Cole, 1968: 138; of *sinaloanus:* Brown, in Cole, 1954b: 120.

***mayri.*** *Pogonomyrmex* (*Janetia*) *mayri* Forel, 1899c: 61 (footnote) (w.m.) COLOMBIA. Kugler, C. 1979a: 170 (q.); Wheeler, G.C. & Wheeler, J. 1953a: 112 (l.). Combination in *P.* (*Forelomyrmex*): Wheeler, W.M. 1913a: 80. See also: Kugler, C. & Hincapié, 1983: 190.

***meridionalis.*** *Pogonomyrmex meridionalis* Kusnezov, 1951a: 277 (w.) ARGENTINA. Current subspecies: nominal plus *leonis.*

***micans.*** *Pogonomyrmex bruchi* r. *micans* Forel, 1914d: 268 (w.m.) ARGENTINA. Raised to species: Kusnezov, 1951a: 254.

*molefaciens. Myrmica* (*Atta*) *molefaciens* Buckley, 1860b: 445 (w.q.) U.S.A. Emery, 1895c: 308 (m.). Combination in *Pogonomyrmex:* Mayr, 1886c: 365. Variety/race of *barbatus:* Forel, 1886b: xlii; Emery, 1895c: 308; Forel, 1901c: 128. Junior synonym of *barbatus:* Mayr, 1886c: 365; Creighton, 1950a: 119; Cole, 1968: 56.

***montanus.*** *Pogonomyrmex montanus* MacKay, 1980: 151, figs. 1–7 (w.q.m.) U.S.A. Taber, Cokendolpher & Francke, 1988: 51 (k.). See also: MacKay, 1981: 25; Shattuck, 1987: 172.

***naegelii.*** *Pogonomyrmex naegelii* Forel, in Emery, 1878a: x (w.) BRAZIL. Mayr, 1887: 612 (q.m.); Wheeler, G.C. & Wheeler, J. 1972b: 237 (l.). Combination in *P.* (*Ephebomyrmex*): Wheeler, W.M. 1902c: 390; in *Ephebomyrmex:* Kempf, 1972a: 106; in *Pogonomyrmex:* **revived combination**. Senior synonym of *rupununi* and *venezuelensis:* Kempf, 1960f: 428. See also: Forel, 1886b: xli; Gallardo, 1932b: 109.

*neuquensis. Pogonomyrmex weiseri* var. *neuquensis* Santschi, 1922d: 243 (w.) ARGENTINA. Junior synonym of *rastratus:* Kusnezov, 1951a: 252.

***#niger.*** *Pogonomyrmex silvestrii* var. *niger* Santschi, 1925e: 156, fig. 16 (w.m.) ARGENTINA. Subspecies of *brevibarbis,* and senior synonym of *incultus:* Kusnezov, 1951a: 250.

*nigrescens. Pogonomyrmex barbatus* var. *nigrescens* Wheeler, W.M. 1902c: 389 (w.) U.S.A. Junior synonym of *fuscatus:* Creighton, 1950a: 119; of *barbatus:* Cole, 1968: 56.

*nitratus. Pogonomyrmex californicus* subsp. *nitratus* Cook, 1953: 99, fig. (w.) U.S.A. Junior synonym of *californicus:* Cole, 1968: 120.

***occidentalis.*** *Myrmica occidentalis* Cresson, 1865: 426 (w.q.) U.S.A. Cole, 1968: 94 (m.); Wheeler, G.C. & Wheeler, J. 1953a: 110 (l.); Taber, Cokendolpher & Francke, 1988: 51 (k.); Taber & Francke, 1986: 274 (gynandromorph). Combination in *Pogonomyrmex:* Cresson, 1879: 22. Senior synonym of *opaciceps* and *seminigra:* Mayr, 1886d: 449; Forel, 1886b: xlii; of *ruthveni:* Olsen, 1934: 507; of *utahensis:* Cole, 1968: 95. See also: Smith, D.R. 1979: 1355; Taber & Francke, 1986: 274; Shattuck, 1987: 172.

***odoratus.*** *Pogonomyrmex* (*Ephebomyrmex*) *odoratus* Kusnezov, 1949c: 302, figs. 1, 2 (w.q.m.) ARGENTINA. Combination in *Ephebomyrmex:* Kusnezov, 1960b: 354; in *Pogonomyrmex:* Snelling & Hunt, 1976: 75.

*opaciceps. Pogonomyrmex opaciceps* Mayr, 1870b: 971 (diagnosis in key) (w.) U.S.A. Junior synonym of *occidentalis:* Mayr, 1886d: 449; Forel, 1886b: xlii.

*owyheei. Pogonomyrmex owyheei* Cole, 1938a: 240 (w.q ) U.S.A. Cole, 1968: 102 (m.). Raised to species: Creighton, 1950a: 130. Junior synonym of *salinus:* Shattuck, 1987: 173.

*pencosensis. Pogonomyrmex cunicularius* r. *pencosensis* Forel, 1914d: 265 (w.m.) ARGENTINA. Junior synonym of *cunicularius:* Kusnezov, 1951a: 251.

***pima.*** *Pogonomyrmex* (*Ephebomyrmex*) *pima* Wheeler, W.M. 1909e: 79 (w.) U.S.A. Cole, 1968: 175 (q.m.).

***pronotalis.*** *Pogonomyrmex pronotalis* Santschi, 1922b: 350 (w.) ARGENTINA.

***#pulchellus.*** *Pogonomyrmex rastratus* var. *pulchellus* Santschi, 1925d: 223 (w.) ARGENTINA.

***rastratus.*** *Pogonomyrmex rastratus* Mayr, 1868b: 171 (w.) ARGENTINA. Mayr, 1887: 611 (m.). Senior synonym of *neuquensis* and *weiseri:* Kusnezov, 1951a: 252. Current subspecies: nominal plus *pulchellus.*

***rugosus.*** *Pogonomyrmex barbatus* subsp. *rugosus* Emery, 1895c: 309 (w.m.) U.S.A. Cole, 1968: 72 (q.); Taber, Cokendolpher & Francke, 1988: 51 (k.). Raised to species: Cole, 1968: 70. Senior synonym of *similis:* Creighton, 1950a: 120; of *fuscatus* (and its junior synonyms *curvispinosus* and *marfensis*) and *spadix:* Cole, 1968: 70.

*rupununi. Pogonomyrmex* (*Ephebomyrmex*) *venezuelensis* var. *rupununi* Weber, 1943b: 71 (w.) GUIANA. Junior synonym of *naegelii:* Kempf, 1960f: 428.

*ruthveni. Pogonomyrmex occidentalis* subsp. *ruthveni* Gaige, 1914: 93 (w.q.m.) U.S.A. Junior synonym of *occidentalis:* Olsen, 1934: 507.

***salinus.*** *Pogonomyrmex salinus* Olsen, 1934: 510, pl. 5, fig. 3 (w.) U.S.A. Cole, 1968: 108 (q.m.); Wheeler, G.C. & Wheeler, J. 1972b: 237 (l.). Junior synonym of *occidentalis:* Smith, M.R. 1953g: 132. Revived from synonymy: Cole, 1968: 106. Senior synonym of *owyheei:* Shattuck, 1987: 173.

*sanctihyacinthi. Pogonomyrmex sanctihyacinthi* Wheeler, W.M. 1902c: 388 (w.) U.S.A. Junior synonym of *apache:* Cole, 1954e: 267.

***#sanmartini.*** *Pogonomyrmex carbonarius* subsp. *sanmartini* Kusnezov, 1953a: 110 (diagnosis in key) (w.) ARGENTINA.

***saucius.*** *Pogonomyrmex* (*Ephebomyrmex*) *saucius* Wheeler, W.M. & Mann, 1914: 29, figs. 10, 11 (w.m.) HAITI. Combination in *Ephebomyrmex:* Kempf, 1972a: 106; in *Pogonomyrmex:* **revived combination**.

***schmitti.*** *Pogonomyrmex schmitti* Forel, 1901e: 339 (w.) HAITI. Wheeler, W.M. & Mann, 1914: 28 (q.). Combination in *P.* (*Ephebomyrmex*): Wheeler, W.M. 1902c: 390; in *Ephebomyrmex:* Kempf, 1972a: 106; in *Pogonomyrmex:* **revived combination**. Current subspecies: nominal plus *aterrimus, darlingtoni, sublaevigatus.*

*seminigra. Myrmica seminigra* Cresson, 1865: 427 (m.) U.S.A. Junior synonym of *occidentalis:* Mayr, 1886d: 449; Forel, 1886b: xlii.

*semistriata. Pogonomyrmex bispinosus* var. *semistriata* Emery, 1906c: 158 (w.q.) CHILE. Junior synonym of *vermiculatus:* Snelling & Hunt, 1976: 75.

*serpens. Pogonomyrmex cunicularis* var. *serpens* Santschi, 1922b: 349 (w.) BOLIVIA. Junior synonym of *cunicularius:* Kusnezov, 1951a: 251.

***#silvestrii.*** *Pogonomyrmex silvestrii* Emery, 1906c: 159, fig. 22 (w.) ARGENTINA. Subspecies of *brevibarbis,* and senior synonym of *longiceps:* Kusnezov, 1951a: 250.

*similis. Pogonomyrmex similis* Olsen, 1934: 512, pl. 6, fig. 2 (w.) U.S.A. Junior synonym of *rugosus:* Creighton, 1950a: 120.

*sinaloanus. Pogonomyrmex californicus* subsp. *sinaloanus* Olsen, 1934: 504 (w.) MEXICO. Junior synonym of *maricopa:* Brown, in Cole, 1954b: 120.

*spadix. Pogonomyrmex barbatus* subsp. *spadix* Cook, 1953: 98, figs. (w.) U.S.A. Junior synonym of *rugosus:* Cole, 1968: 70.

*spinolae. Pogonomyrmex bispinosus* var. *spinolae* Emery, 1906c: 158 (q.) CHILE. Santschi, 1925d: 223 (w.). Junior synonym of *vermiculatus:* Snelling & Hunt, 1976: 75.

*striaticeps. Pogonomyrmex coarctatus* var. *striaticeps* Emery, 1906c: 157 (footnote) (w.) ARGENTINA. Junior synonym of *coarctatus:* Kusnezov, 1951a: 253.

***subdentatus.*** *Pogonomyrmex subdentatus* Mayr, 1870b: 971 (diagnosis in key) (w.) U.S.A. Cole, 1968: 111 (q.m.). See also: Shattuck, 1987: 175.

***#sublaevigatus.*** *Pogonomyrmex* (*Ephebomyrmex*) *schmitti* var. *sublaevigatus* Wheeler, W.M. & Mann, 1914: 29 (w.q.) HAITI. Combination in *Ephebomyrmex:* Kempf, 1972a: 106; in *Pogonomyrmex:* **revived combination**. Subspecies of *schmitti:* Wheeler, W.M. 1936b: 197.

***subnitidus.*** *Pogonomyrmex occidentalis* var. *subnitidus* Emery, 1895c: 310 (w.) U.S.A. Cole, 1968: 115 (q.m.); Taber, Cokendolpher & Francke, 1988: 51 (k.). Raised to species: Wheeler, W.M. 1914e: 156. See also: MacKay, 1981: 25.

***sylvestris.*** *Pogonomyrmex sylvestris* Lattke, 1990c: 305, figs. 1, 2 (w.) VENEZUELA.

***tenuipubens.*** *Pogonomyrmex* (*Ephebomyrmex*) *tenuipubens* Santschi, 1936d: 403, fig. 4 (w.) ARGENTINA. Combination in *Ephebomyrmex:* Kempf, 1972a: 106; in *Pogonomyrmex:* **revived combination**.

***tenuispinus.*** *Pogonomyrmex desertorum* var. *tenuispina* Forel, 1914d: 269 (w.) U.S.A. Subspecies of *desertorum:* Creighton, 1950a: 125. Raised to species, and senior synonym of *dentatus:* Cole, 1968: 77.

***texanus.*** *Pogonomyrmex* (*Pogonomyrmex*) *texanus* Francke & Merickel, 1982: 375, figs. 4, 7, 8, 11, 13, 14, 16 (w.m.) U.S.A. Taber, 1988: 244 (q.).

***theresiae.*** *Pogonomyrmex theresiae* Forel, 1899b: 303 (w.) ECUADOR. [Also described as new by Forel, 1899d: 274.]

*townsendi. Pogonomyrmex* (*Ephebomyrmex*) *townsendi* Wheeler, W.M. 1909e: 80 (w.) MEXICO. Subspecies of *imberbiculus:* Creighton, 1950a: 132. Junior synonym of *imberbiculus:* Creighton, 1957a: 56.

*transversa. Myrmica transversa* Smith, F. 1858b: 129 (w.) U.S.A. Combination in *Pogonomyrmex:* Mayr, 1886c: 359. Senior synonym of *brevipennis* and *crudelis:* Mayr, 1886d: 450. Junior synonym of *badius:* Emery, 1895c: 310.

***uruguayensis.*** *Pogonomyrmex uruguayensis* Mayr, 1887: 614 (w.) URUGUAY. See also: Gallardo, 1932b: 148.

*utahensis. Pogonomyrmex occidentalis* var. *utahensis* Olsen, 1934: 509 (w.q.m.) U.S.A. Subspecies of *occidentalis:* Cole, 1942: 365. Junior synonym of *comanche:* Creighton, 1950a: 128; of *occidentalis:* Cole, 1968: 95.

***variabilis.*** *Pogonomyrmex vermiculatus* var. *variabilis* Santschi, 1916b: 511 (w.m.) ARGENTINA. Raised to species: Kusnezov, 1951a: 253.

*venezuelensis. Pogonomyrmex* (*Ephebomyrmex*) *venezuelensis* Weber, 1943b: 69, fig. 2 (w.) VENEZUELA. Junior synonym of *naegelii:* Kempf, 1960f: 428.

***vermiculatus.*** *Pogonomyrmex vermiculatus* Emery, 1906c: 157 (w.) ARGENTINA. Senior synonym of *intermedia, semistriata* and *spinolae:* Snelling & Hunt, 1976: 75. Current subspecies: nominal plus *chubutensis.*

*weiseri. Pogonomyrmex weiseri* Santschi, 1922b: 348 (w.q.m.) ARGENTINA. Junior synonym of *rastratus:* Kusnezov, 1951a: 252.

***wheeleri.*** *Pogonomyrmex wheeleri* Olsen, 1934: 511, pl. 6, fig. 1 (w.) MEXICO. Snelling, 1982a: 106 (q.m.). See also: Cole, 1968: 81.

***POLYERGUS*** (Holarctic)

*bicolor. Polyergus rufescens* subsp. *bicolor* Wasmann, 1901: 369 (w.q.m.) U.S.A. Junior synonym of *breviceps:* Wheeler, J. 1968: 163.

***breviceps.*** *Polyergus rufescens* subsp. *breviceps* Emery, 1893i: 666 (w.) U.S.A. Raised to species: Kannowski, 1956: 185; Wheeler, J. 1968: 163. Senior synonym of *bicolor* and *umbratus:* Wheeler, J. 1968: 163. Material of the unavailable names *fusciventris, montezuma* and *silvestrii* referred here by Creighton, 1950a: 559. See also: Hölldobler, 1985: 225.

*fusciventris. Polyergus rufescens* subsp. *breviceps* var. *fusciventris* Wheeler, W.M. 1917a: 555 (w.) U.S.A. Unavailable name; material referred to *breviceps* by Creighton, 1950a: 559.

#***laeviceps.*** *Polyergus rufescens* subsp. *laeviceps* Wheeler, W.M. 1915b: 420 (w.) U.S.A. See also: Creighton, 1950a: 559.

#***longicornis.*** *Polyergus lucidus* subsp. *longicornis* Smith, M.R. 1947g: 155 (w.) U.S.A.

***lucidus.*** *Polyergus lucidus* Mayr, 1870b: 952 (w.q.m.) U.S.A. Wheeler, G.C. & Wheeler, J. 1968: 214 (l.); Wheeler, W.M. 1903f: 659 (gynandromorph). Race of *rufescens:* Forel, 1886f: 200. Revived status as species: Dalla Torre, 1893: 214; Wheeler, W.M. 1917i: 465; Smith, M.R. 1947g: 152; Creighton, 1950a: 557. Senior synonym of *montivagus:* Creighton, 1950a: 557. Current subspecies: nominal plus *longicornis.* See also: Talbot, 1968: 299.

#***mandarin.*** *Polyergus samurai* subsp. *mandarin* Wheeler, W.M. 1927e: 4 (w.) CHINA.

#***mexicanus.*** *Polyergus rufescens* r. *mexicanus* Forel, 1899c: 129 (w.) MEXICO.

*montezuma. Polyergus rufescens* subsp. *breviceps* var. *montezuma* Wheeler, W.M. 1914b: 56 (w.q.m.) MEXICO. Unavailable name; material referred to *breviceps* by Crieghton, 1950: 559.

*montivagus. Polyergus lucidus* subsp. *montivagus* Wheeler, W.M. 1915b: 419 (w.q.m.) U.S.A. Junior synonym of *lucidus:* Creighton, 1950a: 557.

***nigerrimus.*** *Polyergus nigerrimus* Marikovsky, 1963: 110, figs. 1–7 (w.q.m.) RUSSIA. See also: Kupyanskaya, 1990: 208.

***rufescens.*** *Formica rufescens* Latreille, 1798: 44 (w.q.) FRANCE. Schenck, 1852: 70 (m.); Forel, 1874: 137 (gynandromorph); André, 1882b: 163 (w.q.m.); Wheeler, G.C. & Wheeler, J. 1968: 214 (l). Combination in *Polyergus:* Latreille, 1804: 179. Senior synonym of *testacea:* Smith, F. 1858b: 57. Current subspecies: nominal plus *laeviceps, mexicanus, tianschanicus.* See also: Stitz, 1939: 369; Hölldobler, 1985: 225.

***samurai.*** *Polyergus rufescens* subsp. *samurai* Yano, 1911: 110, fig. 1 (w.q.) JAPAN. Terayama, Yamaguchi & Hasegawa, 1993: 511 (ergatoid q.); Imai, 1966: 119 (k.). Raised to species: Emery, 1925b: 269; Wheeler, W.M. 1927e: 3. Current subspecies: nominal plus *mandarin.* See also: Kupyanskaya, 1990: 209.

*silvestrii. Polyergus rufescens* subsp. *breviceps* var. *silvestrii* Santschi, 1911d: 7 (w.m.) U.S.A. Unavailable name; material referred to *breviceps* by Creighton, 1950a: 559.

*testacea. Formica testacea* Fabricius, 1804: 400 (q.) CZECHOSLOVAKIA. [Unresolved junior primary homonym of *Formica testacea* Gmelin, in Linnaeus, 1790: 2804.] Junior synonym of *rufescens:* Smith, F. 1858b: 57.

*texana. Polyergus texana* Buckley, 1866: 170 (q.) U.S.A. Unidentifiable to genus; *incertae sedis* in *Polyergus.* See also: Wheeler, W.M. 1902f: 24.

#***tianschanicus.*** *Polyergus rufescens* subsp. *tianschanicus* Kuznetsov-Ugamsky, 1927c: 41, figs. 1–6 (w.q.m.) TURKESTAN.

*umbratus. Polyergus rufescens* subsp. *umbratus* Creighton, 1950a: 560 (w.) U.S.A. [First available use of *Polyergus rufescens* subsp. *breviceps* var. *umbratus* Wheeler, W.M. 1915b: 419; unavailable name.] Junior synonym of *breviceps:* Wheeler, J. 1968: 163.

*POLYHOMOA:* see under ***KYIDRIS.***

***POLYRHACHIS*** (Old World Tropics except Malagasy)

*abdominalis. Polyrhachis abdominalis* Smith, F. 1858b: 63 (w.) BURMA. Combination in *P.* (*Myrmhopla*): Wheeler, W.M. 1919e: 130. Senior synonym of *achilles:* Bingham, 1903: 398. Junior synonym of *hector:* Bolton, 1974b: 174.

***abnormis.*** *Polyrhachis* (*Myrmothrinax*) *abnormis* Donisthorpe, 1948g: 141, fig. 1 (w.) NEW GUINEA.

***abrupta.*** *Polyrhachis abrupta* Mayr, 1867a: 62 (w.) INDONESIA (Halmahera I.). Combination in *P.* (*Myrma*), and senior synonym of *halmaheirae:* Emery, 1925b: 204.

*abrupta. Polyrhachis* (*Chariomyrma*) *abrupta* Karavaiev, 1927e: 19, fig. 10 (w.) INDONESIA (Aru I.). [Junior primary homonym of *abrupta* Mayr, above.] Replacement name: *verticalis* Santschi, 1928h: 139.

*acantha. Polyrhachis acantha* Smith, F. 1860b: 98, pl. 1, fig. 16 (w.) INDONESIA (Batjan I.). Combination in *P.* (*Myrmhopla*): Wheeler, W.M. 1919e: 132. Junior synonym of *dives:* Mayr, 1879: 649; Wheeler, W.M. 1919e: 132. Revived from synonymy: Emery, 1925b: 194; Karavaiev, 1927e: 33. Junior synonym of *saevissima:* Bolton, 1974b: 178.

*acasta. Polyrhachis acasta* Smith, F. 1860b: 100, pl. 1, fig. 23 (w.) INDONESIA (Batjan I.). Combination in *P.* (*Myrmhopla*): Emery, 1925b: 194. Variety of *acantha:* Emery, 1900d: 717. Revived status as species: Forel, 1901b: 33. Junior synonym of *tibialis:* Bingham, 1903: 397. Revived from synonymy as variety of *acantha:* Emery, 1925b: 194. Junior synonym of *saevissima:* Bolton, 1974b: 178.

*acheron. Polyrhachis gagates* r. *indefinita* var. *acheron* Arnold, 1924: 746 (w.) ZIMBABWE. Unavailable name; material referred to *gagates* by Bolton, 1973b: 305.

*achilles. Polyrhachis achilles* Forel, 1893c: 24 (diagnosis in key) (w.) BURMA. Junior synonym of *abdominalis:* Bingham, 1903: 398.

*aciculatus* Smith, F. 1858; see under ***ECHINOPLA.***

#***aciculata.*** *Polyrhachis antennata* var. *aciculata* Viehmeyer, 1912: 13, fig. 17 (w.) NEW GUINEA. [Unresolved junior primary homonym of *aciculatus* Smith, F. 1858b: 70, above.] Combination in *P.* (*Chariomyrma*): Emery, 1925b: 187.

***aculeata.*** *Polyrhachis aculeata* Mayr, 1879: 657 (w.) INDIA. Emery, 1887a: 235 (q.); Forel, 1893c: 28 (m.). Combination in *Hemioptica:* Bingham, 1903: 382; in *P.* (*Hemioptica*): Wheeler, W.M. 1919e: 126; in *P.* (*Myrma*): Emery, 1925b: 205. Current subspecies: nominal plus *cybele, gibbosa.*

#***acutinota.*** *Polyrhachis arcuata* var. *acutinota* Forel, 1901b: 32 (w.) BISMARCK ARCHIPELAGO. Combination in *P.* (*Chariomyrma*): Emery, 1925b: 186.

#***addax.*** *Polyrhachis* (*Myrmhopla*) *tibialis* var. *addax* Santschi, 1928h: 136 (w.) JAVA.

***aedipus.*** *Polyrhachis aedipus* Forel, 1893c: 31 (w.) SRI LANKA. Combination in *P.* (*Myrmhopla*): Emery, 1925b: 193. [Misspelled as *oedipus* by Emery, 1893f: 255, and subsequent authors.] Current subspecies: nominal plus *curvispina.*

*aegra. Polyrhachis* (*Chariomyrma*) *gab* var. *aegra* Forel, 1915b: 109 (w.) AUSTRALIA. Raised to species: Kohout, 1988a: 50. Junior synonym of *lata:* Kohout & Taylor, 1990: 515.

***aenescens.*** *Polyrhachis aenescens* Stitz, 1910: 151 (w.) CAMEROUN. Combination in *P.* (*Myrma*): Wheeler, W.M. 1922a: 993. See also: Bolton, 1973b: 332.

#***aequalis.*** *Polyrhachis textor* var. *aequalis* Forel, 1910d: 129 (w.q.) PHILIPPINES. Combination in *P.* (*Myrmothrinax*): Emery, 1925b: 183.

#***aequicuspis.*** *Polyrhachis* (*Myrmothrinax*) *aequicuspis* Wheeler, W.M. 1919e: 127. Replacement name for *constructor* Smith, F. 1858b: 68. [Junior primary homonym of *constructor* Smith, F. 1857a: 60.] Variety of *atossa:* Emery, 1925b: 183; Donisthorpe, 1932c: 445.

*aerea. Polyrhachis hookeri* var. *aerea* Forel, 1902h: 521 (w.q.m.) AUSTRALIA. Combination in *P. (Chariomyrma)*: Emery, 1925b: 186. Junior synonym of *hookeri:* Kohout & Taylor, 1990: 515.

***aerope.*** *Polyrhachis aerope* Wheeler, W.M. 1922a: 265, fig. 72 (w.) ZAIRE. See also: Bolton, 1973b: 296.

*aeschyle. Polyrhachis (Hedomyrma) aeschyle* Forel, 1915b: 111 (w.) AUSTRALIA. Junior synonym of *erato:* Kohout, 1988c: 434.

*affinis. Formica affinis* Le Guillou, 1842: 314 (q.) BORNEO. [Unresolved junior primary homonym of *Formica affinis* Leach, 1825: 290.] Combination in *Polyrhachis:* Roger, 1863b: 7. Junior synonym of *bihamata:* Mayr, 1872: 139; Dalla Torre, 1893: 259.

*affinis. Polyrhachis affinis* Smith, F. 1858b: 63 (w.) BURMA. [Junior secondary homonym of *affinis* Guillou, above.] Replacement name: *vicina* Roger, 1863b: 7.

***agesilas.*** *Polyrhachis agesilas* Forel, 1913k: 137, fig. P 1 (q.) JAVA. Unplaced to subgenus: Emery, 1925b: 209.

*ajax. Polyrhachis mutata* r. *ajax* Forel, 1893c: 32 (w.) BURMA. Junior synonym of *mutatus:* Bingham, 1903: 399.

**#*alata.*** *Polyrhachis gracilis* r. *alata* Forel, 1904d: 177 (w.) SULAWESI. Combination in *P. (Myrmatopa)*, and variety of *schang:* Emery, 1925b: 181.

**#*alatisquamis.*** *Polyrhachis pubescens* var. *alatisquamis* Forel, 1893c: 17 (diagnosis in key) (w.) BURMA. Combination in *P. (Myrma)*: Emery, 1925b: 205.

***albertisi.*** *Polyrhachis albertisi* Emery, 1887a: 240 (w.q.) NEW GUINEA. Combination in *P. (Cyrtomyrma)*: Emery, 1925b: 207.

***alexandri.*** *Polyrhachis alexandri* Karavaiev, 1906: 375, fig. 1 (q.) JAVA. Combination in *P. (Myrmhopla)*: Emery, 1925b: 192.

***alexisi.*** *Polyrhachis (Cyrtomyrma) alexisi* Forel, 1916: 455, fig. 7 (w.q.) ZAIRE. Combination in *P. (Pseudocyrtomyrma)*: Emery, 1925b: 206; in *P. (Myrma)*: Bolton, 1973b: 346.

***alluaudi.*** *Polyrhachis alluaudi* Emery, 1892d: 567, pl. 15, figs. 9, 10 (w.) IVORY COAST. Bolton, 1973b: 297 (q.). Combination in *P. (Myrma)*: Forel, 1916: 448. Senior synonym of *anteplana:* Bolton, 1973b: 297.

*aloseana. Polyrhachis relucens* r. *litigiosa* var. *aloseana* Forel, 1901b: 28 (w.) BISMARCK ARCHIPELAGO. Unavailable name.

***alphea.*** *Polyrhachis alpheus* Smith, F. 1863: 14 (w.) NEW GUINEA. Combination in *P. (Myrmatopa)*: Emery, 1925b: 180. Current subspecies: nominal plus *rufiventris* (junior homonym).

***alphena.*** *Polyrhachis alphenus* Smith, F. 1860b: 100, pl. 1, fig. 24 (q.) INDONESIA (Batjan I.). Combination in *P. (Myrmhopla)*: Emery, 1925b: 197.

***amana.*** *Polyrhachis amanus* Smith, F. 1861b: 41, pl. 1, fig. 13 (w.) SULAWESI. Combination in *P. (Myrmhopla)*: Emery, 1925b: 194.

**#*amboinae.*** *Polyrhachis (Myrmatopa) schang* var. *amboinae* Santschi, 1928h: 139. Replacement name for *gracilior* Karavaiev, 1927e: 11. [Junior primary homonym of *gracilior* Forel, 1893c: 25.]

***ammon.*** *Formica ammon* Fabricius, 1775: 394 (w.) AUSTRALIA. Mayr, 1876: 72 (q.m.); Imai, Crozier & Taylor, 1977: 347 (k.). Combination in *Polyrhachis:* Smith, F. 1858b: 73; in *P. (Hagiomyrma)*: Wheeler, W.M. 1911c: 860. Senior synonym of *angustata:* Kohout, 1988c: 430.

***ammonoeides.*** *Polyrhachis ammonoeides* Roger, 1863a: 157 (w.) AUSTRALIA. Combination in *P. (Hagiomyrma)*: Forel, 1915b: 108. Stirps of *ammon:* Forel, 1879a: 116. Revived status as species: Dalla Torre, 1893: 258. Senior synonym of *chalchas:* Kohout, 1994a: 135.

***andrei.*** *Polyrhachis (Myrma) andrei* Emery, 1921e: 22, fig. 1 (q.) CAMEROUN. See also: Bolton, 1973b: 298.

***andromache.*** *Polyrhachis andromache* Roger, 1863b: 8. Replacement name for *hector* Smith, F. 1859a: 142. [Junior primary homonym of *hector* Smith, F. 1857a: 62.] Donisthorpe, 1949g: 420 (m.); Kohout, 1989: 510 (w.q.). Combination in *P. (Myrma)*: Forel, 1915b: 110. Subspecies/race of *relucens:* Emery, 1897d: 579; Forel, 1901b: 28. Senior synonym of *connectens:* Emery, 1897d: 580. Junior synonym of *ithona:* Bolton, 1974b: 177. Revived from synonymy, revived status as species, and senior synonym of *australiae:* Kohout, 1988c: 430. Current subspecies: nominal plus *semitestacea.*

*andromeda. Polyrhachis (Myrma) relucens* r. *andromache* var. *andromeda* Forel, 1915b: 110 (w.) AUSTRALIA. Unavailable name (Taylor, 1986: 34).

*anguliceps. Polyrhachis (Hedomyrma) anguliceps* Viehmeyer, 1925b: 148 (w.) AUSTRALIA. Junior synonym of *mjobergi:* Kohout, 1988c: 435.

***angusta.*** *Polyrhachis ammon* r. *angusta* Forel, 1902h: 524 (w.q.m.) AUSTRALIA. Combination in *P. (Hagiomyrma)*: Forel, 1915b: 108. Raised to species: Kohout, 1988c: 431.

*angustata. Polyrhachis ammon* var. *angustata* Forel, 1902h: 525 (w.) AUSTRALIA. Combination in *P. (Hagiomyrma)*: Emery, 1925b: 184. Junior synonym of *ammon:* Kohout, 1988c: 430.

**#*angustior.*** *Polyrhachis similis* var. *angustior* Viehmeyer, 1912: 8, fig. 9 (w.) NEW GUINEA. Combination in *P. (Myrma)*: Emery, 1925b: 203.

***annae.*** *Polyrhachis (Hedomyrma) annae* Mann, 1919: 377, fig. 48 (w.) SOLOMON IS. Wheeler, W.M. 1934a: 180 (q.).

***antennata.*** *Polyrhachis antennata* Viehmeyer, 1912: 13 (w.) NEW GUINEA. Combination in *P. (Chariomyrma)*: Emery, 1925b: 187. Current subspecies: nominal plus *aciculata* (junior homonym).

*anteplana. Polyrhachis (Myrma) alluaudi* var. *anteplana* Forel, 1916: 448 (w.) ZAIRE. Junior synonym of *alluaudi:* Bolton, 1973b: 297.

*antinorii. Polyrhachis antinorii* Emery, 1877b: 365, fig. (w.) ETHIOPIA. Junior synonym of *viscosa:* Dalla Torre, 1893: 271.

***antoniae.*** *Polyrhachis antoniae* Stitz, 1911a: 372, figs. 19, 20 (w.) NEW GUINEA. Combination in *P. (Myrmatopa)*: Emery, 1925b: 181.

***appendiculata.*** *Polyrhachis appendiculata* Emery, 1893e: 227, pl. 8, fig. 9 (w.) AUSTRALIA. Combination in *P. (Chariomyrma)*: Emery, 1925b: 186.

***arachne.*** *Polyrhachis arachne* Emery, 1896f: 249 (w.q.) JAVA. Wheeler, G.C. & Wheeler, J. 1990b: 758 (l.). Combination in *P. (Myrmhopla)*: Emery, 1925b: 192. Senior synonym of *uncinata:* Emery, 1898a: 230. See also: Bingham, 1903: 390; Dorow & Maschwitz, 1990: 73.

*arboricola* Norton, 1868; see under ***DOLICHODERUS.***

*architecta. Polyrhachis (Myrma) laboriosa* var. *architecta* Santschi, 1924b: 224 (w.) ZAIRE. Junior synonym of *laboriosa:* Bolton, 1973b: 308.

*arcispina. Polyrhachis variolosa* var. *arcispina* Santschi, 1916d: 243. Replacement name for *curvispina* Stitz, 1911a: 379. [Junior primary homonym of *curvispina* Forel, 1908a: 8.] Junior synonym of *bubastes:* Bolton, 1975c: 6.

***arcuata.*** *Formica arcuata* Le Guillou, 1842: 315 (w.q.) BORNEO. Combination in *Polyrhachis:* Mayr, 1863: 443; in *P. (Chariomyrma)*: Viehmeyer, 1916a: 166. Junior synonym of *latreillei:* Dalla Torre, 1893: 264; Emery, 1896d: 380. Revived from synonymy, and senior synonym of *latifrons* (and its junior synonym *modiglianii*): Emery, 1897d: 583. Current subspecies: nominal plus *acutinota, aruana, continentis, denselineata.*

*arcuata. Polyrhachis hostilis* var. *arcuata* Stitz, 1911a: 378 (w.) NEW GUINEA. [Junior secondary homonym of *arcuata* Guillou, above.] Replacement name: *circumflexa* Santschi, 1916d: 243.

*arcuata. Polyrhachis (Myrmatopa) fruhstorferi* var. *arcuata* Karavaiev, 1927e: 9 (w.) INDONESIA (Ambon I.). [Junior secondary homonym of *arcuata* Guillou, above.] Replacement name: *torta* Santschi, 1928h: 139.

*arcuspinosa. Polyrhachis (Myrmhopla) arcuspinosa* Donisthorpe, 1941c: 140, fig. 1 (w.) NEW GUINEA. Junior synonym of *sexspinosa:* Bolton, 1975c: 12.

*argentata. Formica argentata* Fabricius, 1804: 413 (w.) NEW CALEDONIA. Combination in *Polyrhachis:* Smith, F. 1858b: 73. Junior synonym of *sexspinosa:* Lepeletier, 1835: 219; Roger, 1863b: 6.

*argentatus. Polyrhachis militaris* r. *cupreopubescens* var. *argentatus* Stitz, 1910: 150 (w.) CAMEROUN. Unavailable name; material referred to *militaris* by Bolton, 1973b: 313.

**#*argentea.*** *Polyrhachis argenteus* Mayr, 1862: 682 (w.) PHILIPPINES. Combination in *P. (Myrmhopla)*: Wheeler, W.M. 1919e: 131. Junior synonym of *acasta:* Dalla Torre, 1893: 257; Mayr, 1893: 197. Revived from synonymy as variety of *acantha:* Emery, 1900d: 717. Junior synonym of *tibialis:* Bingham, 1903: 397. Revived from synonymy, and revived status as species: Wheeler, W.M. 1919e: 131. Variety of *acantha:* Emery, 1925b: 194. Subspecies of *saevissima:* current status.

***argenteosignata.*** *Polyrhachis argenteosignata* Emery, 1900c: 335 (w.) NEW GUINEA. Combination in *P. (Chariomyrma)*: Emery, 1925b: 186.

***argentosa.*** *Polyrhachis daemeli* r. *argentosa* Forel, 1902h: 515 (w.) AUSTRALIA. Combination in *P. (Hedomyrma)*: Emery, 1925b: 189. Raised to species: Kohout, 1988c: 431.

***armata.*** *Formica armata* Le Guillou, 1842: 313 (q.) PHILIPPINES. Mayr, 1867a: 46 (w.); Karavaiev, 1927e: 29 (m.); Wheeler, G.C. & Wheeler, J. 1990b: 759 (l.). Combination in *Polyrhachis:* Roger, 1863b: 9; in *P. (Myrmhopla)*: Forel, 1915b: 107. Senior synonym of *pandarus:* Roger, 1863b: 9. Current subspecies: nominal plus *defensa.* See also: Bingham, 1903: 393.

***POLYRHACHIS** (continued)*

***arnoldi***. *Polyrhachis (Myrma) arnoldi* Forel, 1914d: 263 (w.) SOUTH AFRICA. See also: Arnold, 1924: 750; Bolton, 1973b: 324.

*arthurimuelleri*. *Polyrhachis arthurimuelleri* Forel, 1915a: 43 (w.) SUMATRA. *Nomen nudum.*

#***aruana***. *Polyrhachis (Chariomyrma) arcuata* var. *aruana* Karavaiev, 1927e: 16 (w.) INDONESIA (Aru I.).

#***aruensis***. *Polyrhachis laevissima* var. *aruensis* Viehmeyer, 1912: 9 (w.) NEW GUINEA. Combination in *P. (Cyrtomyrma)*: Emery, 1925b: 207.

***asomaningi***. *Polyrhachis asomaningi* Bolton, 1973b: 298, figs. 44, 53 (w.) GHANA.

***aspasia***. *Polyrhachis aspasia* Forel, 1911a: 59 (w.) BORNEO. Combination in *P. (Myrmhopla)*: Wheeler, W.M. 1919e: 132.

#***assamensis***. *Polyrhachis striata* r. *assamensis* Forel, 1902d: 289 (w.) INDIA. Combination in *P. (Myrma)*: Viehemeyer, 1916: 166.

*atalanta*. *Polyrhachis atalanta* Emery, 1898b: 243, figs. 15, 16 (w.) NEW GUINEA. Junior synonym of *nigriceps:* Emery, 1925b: 192.

*atalanta*. *Polyrhachis atalanta* Wheeler, W.M. 1922a: 263, fig. 71 (q.) ZAIRE. [Junior primary homonym of *atalanta* Emery, above.] Replacement name: *latispina* Emery, 1925b: 206.

***atossa***. *Polyrhachis constructor* var. *atossa* Forel, 1913k: 134 (q.) SUMATRA. Combination in *P. (Myrmothrinax)*: Emery, 1925b: 183. Raised to species: Emery, 1925b: 183; Donisthorpe, 1932c: 445. Current subspecies: nominal plus *aequicuspis*.

#***atra***. *Polyrhachis biroi* var. *atra* Viehmeyer, 1914c: 50 (w.) NEW GUINEA. Combination in *P. (Myrma)*: Emery, 1925b: 200.

#***atrocastanea***. *Polyrhachis (Myrmhopla) bicolor* var. *atrocastanea* Karavaiev, 1927e: 34 (w.q.m.) JAVA.

*atrociliata*. *Polyrhachis (Myrma) schistacea* st. *atrociliata* Santschi, 1913c: 314 (w.) ZAIRE. [Also described as new by Santschi, 1914b: 141.] Raised to species: Santschi, 1933b: 108. Junior synonym of *schistacea:* Bolton, 1973b: 318.

***atropos***. *Polyrhachis atropos* Smith, F. 1860b: 100, pl. 1, fig. 22 (w.) NEW GUINEA. Combination in *P. (Hedomyrma)*: Emery, 1925b: 189. Senior synonym of *eucharis:* Kohout & Taylor, 1990: 513. Current subspecies: nominal plus *circumdata, tersa.*

***atrovirens***. *Polyrhachis atrovirens* Emery, 1900d: 718, fig. 16 (w.q.) INDONESIA (Engano I.). Combination in *P. (Myrmhopla)*: Wheeler, W.M. 1919e: 135.

#***aurata***. *Polyrhachis (Myrmhopla) bicolor* var. *aurata* Karavaiev, 1935a: 114 (w.) JAVA.

***aurea***. *Polyrhachis guerini* var. *aurea* Mayr, 1876: 74 (w.q.) AUSTRALIA. Combination in *P. (Chariomyrma)*: Forel, 1915b: 108. Raised to species: Emery, 1897d: 584 (in key). Current subspecies: nominal plus *costulata, fiorii, radicicola.*

***aureovestita***. *Polyrhachis (Myrmhopla) aureovestitus* Donisthorpe, 1937c: 274 (w.) NEW GUINEA. See also: Bolton, 1975c: 4.

*aurichalceus*. *Polyrhachis aurichalceus* Mayr, 1862: 684 (w.) JAVA. Mayr, 1865: 43 (q.). Junior synonym of *zopyra:* Emery, 1925b: 204.

***auriformis***. *Polyrhachis (Chariomyrma) auriformis* Donisthorpe, 1943d: 462 (w.) NEW GUINEA.

#***aurinasis***. *Polyrhachis bicolor* var. *aurinasis* Forel, 1901h: 77 (w.) BORNEO. Forel, 1911e: 296 (q.). Combination in *P. (Myrmhopla)*: Wheeler, W.M. 1919e: 132.

***aurita***. *Polyrhachis aurita* Emery, 1911b: 538 (w.) NEW GUINEA. Combination in *P. (Chariomyrma)*: Emery, 1925b: 187. Current subspecies: nominal plus *longispina, schlaginhaufeni.*

*australiae*. *Polyrhachis connectens* var. *australiae* Emery, 1887a: 231 (w.) AUSTRALIA. Combination in *P. (Myrma)*: Emery, 1925b: 202. Junior synonym of *relucens:* Emery, 1893e: 228; of *andromache:* Emery, 1896d: 379. Revived from synonymy as subspecies of *relucens:* Taylor & Brown, D.R. 1985: 140. Junior synonym of *andromache:* Kohout, 1988c: 431.

***australis***. *Polyrhachis australis* Mayr, 1870b: 945 (w.) AUSTRALIA. Junior synonym of *levior:* Mayr, 1876: 71; Dalla Torre, 1893: 264. Combination in *P. (Cyrtomyrma)*, revived from synonymy, and senior synonym of *nox:* Kohout & Taylor, 1990: 513.

#***baduri***. *Polyrhachis (Cyrtomyrma) rastellata* var. *baduri* Donisthorpe, 1941h: 63 (w.q.) NEW GUINEA.

***bakeri***. *Polyrhachis (Campomyrma) bakeri* Viehmeyer, 1916b: 286, fig. 1 (w.) PHILIPPINES. Combination in *P. (Myrma)*: Emery, 1925b: 204.

*balli*. *Polyrhachis (Pseudocyrtomyrma) revoili* st. *balli* Santschi, 1939c: 10 (w.) ZAIRE. Junior synonym of *weissi:* Bolton, 1973b: 342.

***bamaga***. *Polyrhachis bamaga* Kohout, 1990: 500, fig. 4 (w.) AUSTRALIA.

***banghaasi***. *Polyrhachis (Myrmhopla) banghaasi* Viehmeyer, 1922: 219 (w.) BRUNEI.

#***bankensis***. *Polyrhachis furcata* var. *bankensis* Forel, 1911e: 297 (w.) INDONESIA (Bangka I.).

*barnardi*. *Polyrhachis (Myrmhopla) barnardi* Clark, 1928a: 39, pl. 1, figs. 37, 38 (w.) AUSTRALIA. Clark, 1930c: 14 (q.). Junior synonym of *sexspinosa:* Kohout & Taylor, 1990: 519. See also: Bolton, 1975c: 6.

***barretti***. *Polyrhachis (Hedomyrma) barretti* Clark, 1928b: 170, fig. 1 (w.) AUSTRALIA.

***basirufa***. *Polyrhachis basirufa* Emery, 1900d: 715 (w.) INDONESIA (Metawei I.). Combination in *P. (Myrmhopla)*: Emery, 1925b: 194.

***batesi***. *Polyrhachis batesi* Forel, 1911e: 301 (w.) INDONESIA (Seram I.). Combination in *P. (Myrmhopla)*: Emery, 1925b: 194.

***beauforti***. *Polyrhachis beauforti* Emery, 1911b: 538 (w.) NEW GUINEA. Combination in *P. (Chariomyrma)*: Emery, 1925b: 187. Current subspecies: nominal plus *punctinota.*

***beccarii***. *Polyrhachis beccarii* Mayr, 1872: 141 (w.) BORNEO. Combination in *P. (Myrma)*: Wheeler, W.M. 1919e: 123.

***bedoti***. *Polyrhachis bedoti* Forel, 1902h: 518 (q.) "probablement d'Australie ou de Nouvelle Guinée". Combination in *P. (Chariomyrma)*: Emery, 1925b: 187.

*bellendenensis*. *Polyrhachis (Chariomyrma) hookeri* r. *obscura* var. *bellendenensis* Forel, 1915b: 109 (w.) AUSTRALIA. Unavailable name (Taylor, 1986: 34).

#***belli***. *Polyrhachis dives* subsp. *belli* Forel, 1912n: 74 (w.) INDIA. Combination in *P. (Myrmhopla)*: Emery, 1925b: 195.

***bellicosa***. *Polyrhachis bellicosus* Smith, F. 1859a: 142 (w.) INDONESIA (Aru I.). Hung, 1970: 7 (m.); Kohout, 1988b: 418 (q.); Wheeler, G.C. & Wheeler, J. 1990b: 762 (l). Variety of *bihamata:* Mayr, 1862: 678. Revived status as species: Mayr, 1867a: 50. Senior synonym of *crudelis:* Hung, 1970: 5.

*benguelensis*. *Polyrhachis (Myrma) schistacea* st. *atrociliata* var. *benguelensis* Santschi, 1913c: 314 (w.) ANGOLA. Unavailable name; material referred to *schistacea* by Bolton, 1973b: 318.

*bequaerti*. *Polyrhachis (Myrma) bequaerti* Wheeler, W.M. 1922a: 267, fig. 76 (w.) ZAIRE. Santschi, 1923e: 294 (q.). Junior synonym of *fissa:* Bolton, 1973b: 304.

***bicolor***. *Polyrhachis bicolor* Smith, F. 1858b: 65 (q.) BURMA. Mayr, 1862: 681 (w.). Combination in *P. (Myrmhopla)*: Viehmeyer, 1916a: 167. Current subspecies: nominal plus *atrocastanea, aurata, aurinasis, brachyacantha, comata, concolor, erecta, exflavicornis, fumata, nigripes, weyeri.*

*bicolor*. *Polyrhachis (Hedomyrma) bicolor* Karavaiev, 1927e: 21, fig. 11 (w.) INDONESIA (Ambon I.). [Junior primary homonym of *bicolor* Smith, above.] Replacement name: *indocilis* Santschi, 1928h: 139.

#***bidentata***. *Polyrhachis biroi* var. *bidentata* Stitz, 1912: 512 (w.) NEW GUINEA. Combination in *P. (Myrma)*: Emery, 1925b: 200.

***bihamata***. *Formica bihamata* Drury, 1773: 73, pl. 38, figs. 7, 8 (w.) MADAGASCAR (Johanna I.). [Locality in error, see Bolton, 1973b: 352.] Emery, 1887a: 238 (q.); Donisthorpe, 1942b: 70 (m.); Wheeler, G.C. & Wheeler, J. 1990b: 763 (l.). Combination in *Polyrhachis:* Smith, F. 1857a: 59. Senior synonym of *affinis* Guillou: Mayr, 1872: 139; Dalla Torre, 1893: 259; of *minor* Karavaiev, *perplexa* and *tonsilis:* Hung, 1970: 16.

*biloba*. *Polyrhachis biloba* Forel, 1911a: 58 (w.) BORNEO. Junior synonym of *equina:* Wheeler, W.M. 1919e: 122.

***binghamii***. *Polyrhachis binghamii* Forel, 1893c: 25 (diagnosis in key) (w.) BURMA. Forel, 1893c: 33 (q.). Combination in *P. (Myrmhopla)*: Emery, 1925b: 193.

***biroi***. *Polyrhachis biroi* Forel, 1907a: 40 (w.) INDONESIA (Ambon I.). Combination in *P. (Myrma)*: Emery, 1925b: 200. Current subspecies: nominal plus *atra, bidentata, paprika.*

#***bismarckensis***. *Polyrhachis mucronata* var. *bismarckensis* Forel, 1901b: 33 (w.) BISMARCK ARCHIPELAGO. Combination in *P. (Myrmhopla)*: Emery, 1925b: 196.

*bismarckensis*. *Polyrhachis (Myrma) conops* var. *bismarckensis* Karavaiev, 1927e: 46 (w.) BISMARCK ARCHIPELAGO. [Junior primary homonym of *bismarckensis* Forel, above.] Replacement name: *stitzi* Santschi, 1928h: 139.

#***boettcheri***. *Polyrhachis (Myrmhopla) hippomanes* r. *boettcheri* Stitz, 1925: 131 (w.) PHILIPPINES.

***bouvieri***. *Polyrhachis (Myrmatopa) bouvieri* Santschi, 1928g: 250, fig. D (q.) SUMATRA.

#***brachyacantha***. *Polyrhachis bicolor* var. *brachyacantha* Karavaiev, 1935a: 114 (w.) CAMBODIA.

***braxa***. *Polyrhachis braxa* Bolton, 1973b: 333, figs. 50, 56 (w.) GHANA.

#***breviorspinosa***. *Polyrhachis (Myrma) relucens* var. *breviorspinosa* Donisthorpe, 1947d: 194 (w.) NEW GUINEA.

**#brunneipes.** *Polyrhachis (Myrma) labella* var. *brunneipes* Wheeler, W.M. 1934a: 179 (w.) SOLOMON IS.

**#brunneogaster.** *Polyrhachis (Myrmothrinax) textor* var. *brunneogaster* Donisthorpe, 1937a: 626 (w.) NICOBAR IS.

*bruta. Polyrhachis militaris* st. *bruta* Santschi, 1912b: 166 (q.) ZAIRE. Santschi, 1924b: 223 (w.). Combination in *P. (Myrma)*: Wheeler, W.M. 1922a: 998. Junior synonym of *militaris:* Bolton, 1973b: 313.

*bryanti. Polyrhachis (Myrma) bryanti* Donisthorpe, 1942e: 707 (w.) BORNEO. Junior synonym of *inermis:* Bolton, 1974b: 176.

**bubala.** *Polyrhachis (Myrmhopla) bubalus* Stitz, 1925: 129, fig. 8 (w.) PHILIPPINES.

**bubastes.** *Polyrhachis bubastes* Smith, F. 1863: 15 (w.q.) NEW GUINEA. Combination in *P. (Myrmhopla)*: Emery, 1925b: 191. Senior synonym of *arcispina, kellyi, rugousissima, spinosa, variolosa* and *waigiouensis:* Bolton, 1975c: 6.

**bugnioni.** *Polyrhachis (Hemioptica) bugnioni* Forel, 1908a: 11, pl. 1 (w.q.) SRI LANKA.

**burmanensis.** *Polyrhachis (Cyrtomyrma) burmanensis* Donisthorpe, 1938c: 257, fig. 8 (w.) BURMA.

*busiris. Polyrhachis busiris* Smith, F. 1860b: 98, pl. 1, fig. 15 (w.) NEW GUINEA. Junior synonym of *rastellata:* Mayr, 1862: 688; Roger, 1863b: 6.

**caeciliae.** *Polyrhachis caeciliae* Forel, 1912n: 76 (w.) SUMATRA. Combination in *P. (Myrmhopla)*: Emery, 1925b: 194.

*cafrorum. Polyrhachis militaris* r. *cafrorum* Forel, 1879a: 120 (w.q.m.) SOUTH AFRICA. Raised to species: Dalla Torre, 1893: 260. Junior synonym of *rugulosus:* Mayr, 1893: 197; Forel, 1894b: 72.

*calabarica. Polyrhachis militaris* var. *calabarica* Forel, 1907a: 38 (w.) NIGERIA. Junior synonym of *militaris:* Bolton, 1973b: 313.

**#caligata.** *Polyrhachis caligata* Emery, 1895k: 482 (w.q.) BURMA. Combination in *P. (Myrmhopla)*: Emery, 1925b: 196. Junior synonym of *tibialis:* Forel, 1895e: 457; Bingham, 1903: 397. Revived from synonymy as variety of *tibialis:* Emery, 1900d: 717; Emery, 1925b: 196.

**calliope.** *Polyrhachis calliope* Emery, 1900c: 335 (w.) NEW GUINEA. Combination in *P. (Hedomyrma)*: Emery, 1925b: 189.

**calypso.** *Polyrhachis spinosa* subsp. *calypso* Forel, 1911d: 394 (w.) SUMATRA. Combination in *P. (Myrmhopla)*: Emery, 1925b: 191. Raised to species, and senior synonym of *capra* and *malaccana:* Bolton, 1975c: 7.

**#campbelli.** *Polyrhachis (Hedomyrma) santschii* subsp. *campbelli* Mann, 1919: 376 (w.) SOLOMON IS.

*capra. Polyrhachis (Myrmhopla) capra* Karavaiev, 1927e: 27, fig. 14 (w.) JAVA. Junior synonym of *calypso:* Bolton, 1975c: 7.

**carbonaria.** *Polyrhachis carbonarius* Smith, F. 1857a: 60 (w.) WEST MALAYSIA. Combination in *P. (Aulacomyrma)*: Donisthorpe, 1932c: 445.

**carinata.** *Formica carinata* Fabricius, 1804: 413 (w.) NEW CALEDONIA. Combination in *Polyrhachis:* Smith, F. 1857a: 59. Unplaced to subgenus: Emery, 1925b: 209.

*carinatus. Polyrhachis carinatus* Smith, F. 1858b: 71, pl. 4, figs. 48, 49 (w.) SOUTH AFRICA. [Junior secondary homonym of *carinata* Fabricius, above.] Senior synonym of *rugulosus:* Emery, 1896d: 379; hence the latter is first available replacement name for this taxon.

**#castanea.** *Polyrhachis (Myrmothrinax) thrinax* var. *castanea* Stitz, 1925: 136 (w.) PHILIPPINES. Senior synonym of *castanella:* Donisthorpe, 1943e: 116.

**castaneiventris.** *Polyrhachis castaneiventris* Smith, F. 1858b: 67 (q.) BORNEO. Combination in *P. (Myrmhopla)*: Emery, 1925b: 197.

*castanella. Polyrhachis (Myrmothrinax) thrinax* var. *castanella* Santschi, 1928h: 140. [Unnecessary replacement name for *castanea,* above.] Junior synonym of *castanea:* Donisthorpe, 1943e: 116.

*cataulacoidea. Polyrhachis cataulacoidea* Stitz, 1911a: 377, fig. 23 (w.) AUSTRALIA. Combination in *P. (Chariomyrma)*: Emery, 1925b: 186. Junior synonym of *hookeri:* Kohout, 1994a: 136.

**caulomma.** *Polyrhachis caulomma* Viehmeyer, 1914c: 57, fig. 12 (w.) NEW GUINEA. Combination in *P. (Chariomyrma)*: Emery, 1925b: 187. Current subspecies: nominal plus *parallela.*

**cedarensis.** *Polyrhachis (Campomyrma) leae* r. *cedarensis* Forel, 1915b: 114 (w.q.) AUSTRALIA. Raised to species: Kohout & Taylor, 1990: 511. See also: Taylor, 1989a: 27.

**#celebensis.** *Polyrhachis rastellata* var. *celebensis* Viehmeyer, 1913: 155 (w.) SULAWESI. [Also described as new by Viehmeyer, 1914c: 51.] Combination in *P. (Cyrtomyrma)*: Emery, 1925b: 208.

**cephalotes.** *Polyrhachis cephalotes* Emery, 1893e: 199, pl. 8, fig. 6 (w.) SUMATRA. Forel, 1913e: 665 (q.m.). Combination in *P. (Myrmhopla)*: Wheeler, W.M. 1919e: 132.

**ceramensis.** *Polyrhachis ceramensis* Mayr, 1883: 246 (w.) INDONESIA (Seram I.). Combination in *P. (Myrma)*: Emery, 1925b: 200.

**#ceylonensis.** *Polyrhachis hippomanes* subsp. *ceylonensis* Emery, 1893f: 254 (w.q.) SRI LANKA. Combination in *P. (Myrmhopla)*: Viehmeyer, 1916a: 166. Raised to species: Bingham, 1903: 400. Subspecies of *hippomanes:* Viehmeyer, 1916a: 166.

**#ceylonensis.** *Polyrhachis (Cyrtomyrma) rastellata* var. *ceylonensis* Donisthorpe, 1938c: 256 (w.) SRI LANKA. [Unresolved junior primary homonym of *ceylonensis* Emery, above.]

*chalchas. Polyrhachis chalchas* Forel, 1907h: 307 (w.q.) AUSTRALIA. Combination in *P. (Hagiomyrma)*: Forel, 1915b: 108. Junior synonym of *ammonoeides:* Kohout, 1994a: 135.

**chalybea.** *Polyrhachis chalybeus* Smith, F. 1857a: 61 (w.) SINGAPORE. Combination in *P. (Myrmhopla)*: Wheeler, W.M. 1919e: 131.

*chaonia. Polyrhachis chaonia* Smith, F. 1861b: 42, pl. 1, fig. 18 (q.) INDONESIA (Halmahera I.). Combination in *P. (Myrma)*: Emery, 1925b: 200. Junior synonym of *ithona:* Bolton, 1974b: 177.

**charaxa.** *Polyrhachis charaxus* Smith, F. 1860b: 98, pl. 1, fig. 14 (w.) INDONESIA (Batjan I.). Combination in *P. (Myrmatopa)*: Emery, 1925b: 181.

**#charpillioni.** *Polyrhachis textor* var. *charpillioni* Forel, 1911d: 392 (q.) SUMATRA. Combination in *P. (Myrmothrinax)*: Emery, 1925b: 183.

**chartifex.** *Polyrhachis chartifex* Emery, 1900c: 334 (w.q.) NEW GUINEA. Combination in *P. (Myrmatopa)*: Emery, 1925b: 181.

**cheesmanae.** *Polyrhachis (Evelyna) cheesmanae* Donisthorpe, 1937c: 273, figs. 1–3 (q.) NEW GUINEA. Combination in *P. (Myrmothrinax)*: current combination.

*cheesmanae. Polyrhachis (Campomyrma) cheesmanae* Donisthorpe, 1943d: 459 (q.) NEW GUINEA. [Junior primary homonym of *cheesmanae* Donisthorpe, above.] Replacement name: *horacei* Hung, 1967a: 201.

*chlorizans. Polyrhachis creusa* var. *chlorizans* Forel, 1901b: 30 (w.q.) BISMARCK ARCHIPELAGO. Combination in *P. (Campomyrma)*: Emery, 1925b: 179. Junior synonym of *creusa:* Kohout & Taylor, 1990: 514.

**#chrysophanes.** *Polyrhachis acantha* var. *chrysophanes* Emery, 1900d: 718 (w.) INDONESIA (Tenimber I.). Combination in *P. (Myrmhopla)*: Emery, 1925b: 194. Subspecies of *saevissima:* current status.

*chrysothorax. Polyrhachis (Hedomyrma) chrysothorax* Viehmeyer, 1925b: 148 (w.) AUSTRALIA. Wheeler, G.C. & Wheeler, J. 1974a: 62 (l.). Junior synonym of *ornata:* Kohout, 1988c: 435.

**#cincta.** *Polyrhachis dahli* var. *cincta* Viehmeyer, 1913: 149 (w.) SULAWESI (in copal). Combination in *P. (Myrmothrinax)*: Emery, 1925b: 183.

**cingula.** *Polyrhachis (Chariomyrma) cingula* Donisthorpe, 1947c: 590 (w.q.) NEW GUINEA.

**#circumdata.** *Polyrhachis circumdata* Viehmeyer, 1913: 152, fig. (w.q.) SULAWESI (in copal). Combination in *P. (Hedomyrma)*: Emery, 1925b: 189. Variety of *atropos:* Viehmeyer, 1914c: 52.

**#circumflexa.** *Polyrhachis hostilis* var. *circumflexa* Santschi, 1916d: 243. Replacement name for *arcuata* Stitz, 1911a: 378. [Junior secondary homonym of *arcuata* Le Guillou, 1842: 315.]

*clariseta. Polyrhachis nigriseta* var. *clariseta* Santschi, 1910c: 400 (w.) CONGO. Combination in *P. (Myrma)*: Santschi, 1914b: 140. Junior synonym of *gagates:* Bolton, 1973b: 305.

**clarkei.** *Polyrhachis clarkei* Donisthorpe, 1949b: 502 (q.) NEW GUINEA.

**cleopatra.** *Polyrhachis cleopatra* Forel, 1902h: 513 (w.) AUSTRALIA. Combination in *P. (Hedomyrma)*: Emery, 1925b: 189.

**cleophanes.** *Polyrhachis cleophanes* Smith, F. 1861b: 41, pl. 1, fig. 14 (w.) SULAWESI. Combination in *P. (Myrmhopla)*: Emery, 1925b: 193. Senior synonym of *vibidia:* Forel, 1911e: 299.

**clio.** *Polyrhachis clio* Forel, 1902h: 515 (w.) AUSTRALIA. Combination in *P. (Hedomyrma)*: Emery, 1925b: 189.

**clotho.** *Polyrhachis clotho* Forel, 1902h: 525 (w.) AUSTRALIA. Combination in *P. (Myrmhopla)*: Emery, 1925b: 195; in *P. (Hedomyrma)*: Kohout & Taylor, 1990: 512.

*clypeata. Polyrhachis clypeatus* Mayr, 1862: 683 (q.) SRI LANKA. Combination in *P. (Campomyrma)*: Wheeler, W.M. 1911c: 860. Senior synonym of *indica:* Dalla Torre, 1893: 260; Forel, 1893c: 29. Junior synonym of *exercita:* Donisthorpe, 1932b: 575.

**#cnemidata.** *Polyrhachis gracilis* var. *cnemidata* Emery, 1900d: 710 (w.q.) INDONESIA (Mentawei I.). Combination in *P. (Myrmatopa)*, and variety of *schang:* Emery, 1925b: 181.

*POLYRHACHIS (continued)*

**coerulescens.** *Polyrhachis coerulescens* Emery, 1897d: 588 (w.) NEW GUINEA. Combination in *P.* (*Chariomyrma*): Emery, 1925b: 187. Current subspecies: nominal plus *nigronitens, strigifrons.*

#**comata.** *Polyrhachis bicolor* subsp. *comata* Emery, 1911b: 538 (w.) NEW GUINEA. Combination in *P.* (*Myrmhopla*): Emery, 1925b: 194.

*comata. Polyrhachis comata* Crawley, 1915b: 237 (w.) AUSTRALIA. [Junior primary homonym of *comata* Emery, above.] Replacement name: *crawleyella* Santschi, 1916d: 243.

#**completa.** *Polyrhachis* (*Myrmhopla*) *tibialis* var. *completa* Santschi, 1928h: 135 (w.) SUMATRA.

**compressicornis.** *Polyrhachis compressicornis* Smith, F. 1860a: 69 (w.q.) SULAWESI. Combination in *P.* (*Myrma*): Emery, 1925b: 200.

**concava.** *Polyrhachis concava* André, 1889: 218 (w.) SIERRA LEONE. André, 1892a: 45 (q.). Combination in *P.* (*Myrma*): Forel, 1916: 448. See also: Bolton, 1973b: 299.

#**conclusa.** *Polyrhachis restituta* var. *conclusa* Viehmeyer, 1913: 151, fig. (w.) SULAWESI (in copal). Combination in *P.* (*Myrma*): Emery, 1925b: 202.

#**concolor.** *Polyrhachis bicolor* var. *concolor* Forel, 1910d: 129 (w.q.m.) PHILIPPINES. Combination in *P.* (*Myrmhopla*): Emery, 1925b: 194.

*conduensis. Polyrhachis* (*Myrma*) *revoili* var. *conduensis* Forel, 1915c: 351 (w.) ZAIRE. Combination in *P.* (*Pseudocyrtomyrma*): Emery, 1925b: 207. Junior synonym of *weissi:* Bolton, 1973b: 342.

*confinis. Polyrhachis achilles* var. *confinis* Forel, 1912n: 76 (w.q.) SUMATRA. Combination in *P.* (*Myrmhopla*): Emery, 1925b: 192. Junior synonym of *hector:* Bolton, 1974b: 174.

#**congener.** *Polyrhachis* (*Cyrtomyrma*) *rastellata* var. *congener* Santschi, 1928h: 135, fig. 2 (w.) SUMATRA.

*congolensis. Polyrhachis gagates* var. *congolensis* Santschi, 1910c: 399 (w.) CONGO. Combination in *P.* (*Myrma*): Santschi, 1914b: 139. Race/subspecies of *schistacea:* Forel, 1913h: 357; Wheeler, W.M. 1922a: 1001. Senior synonym of *indefinita:* Santschi, 1924b: 224. Junior synonym of *gagates:* Bolton, 1973b: 305.

*connectens. Polyrhachis connectens* Emery, 1887a: 230, pl. 4, fig. 19 (w.) NEW GUINEA. Variety of *relucens:* Dalla Torre, 1893: 267; Emery, 1896d: 379. Junior synonym of *andromache:* Emery, 1897d: 580.

#**conophthalma.** *Polyrhachis nigropilosa* var. *conophthalma* Emery, 1900d: 713 (w.) SUMATRA. Combination in *P.* (*Myrma*): Wheeler, W.M. 1919e: 124.

**conops.** *Polyrhachis conops* Forel, 1901b: 28 (w.q.m.) BISMARCK ARCHIPELAGO. Combination in *P.* (*Myrma*): Emery, 1925b: 200. Current subspecies: nominal plus *cuspidata* (junior homonym), *simpla, spinifera, stitzi.*

*conradti. Polyrhachis* (*Pseudocyrtomyrma*) *lanuginosa* st. *conradti* Santschi, 1923e: 293. Replacement name for *santschii* Emery, 1921e: 24. [Junior primary homonym of *santschii* Mann, 1919: 375.] Junior synonym of *lanuginosa:* Bolton, 1973b: 335.

**consimilis.** *Polyrhachis consimilis* Smith, F. 1858b: 73, pl. 4, figs. 30, 31 (w.) AUSTRALIA (not Sierra Leone; see Bolton, 1973b: 352; Kohout, 1988c: 432). Combination in *P.* (*Myrma*): Wheeler, W.M. 1922a: 994; in *P.* (*Hagiomyrma*): Emery, 1925b: 185; in *P.* (*Hedomyrma*): Kohout & Taylor, 1990: 512.

**constricta.** *Polyrhachis constricta* Emery, 1897d: 584 (footnote), pl. 1, fig. 21 (w.) AUSTRALIA. Combination in *P.* (*Chariomyrma*): Emery, 1925b: 186.

**constructor.** *Polyrhachis constructor* Smith, F. 1857a: 60 (q.) BORNEO. Combination in *P.* (*Myrmatopa*): Donisthorpe, 1932c: 445. [*P. constructrix:* Schulz, 1906: 156; unjustified emendation.]

*constructor. Polyrhachis constructor* Smith, F. 1858b: 68, pl. 4, fig. 23 (q.) BORNEO. [Junior primary homonym of *constructor* Smith, above.] Replacement name: *aequicuspis* Wheeler, W.M. 1919e: 127.

**contemta.** *Polyrhachis contemta* Mayr, 1876: 74 (w.) AUSTRALIA. Combination in *P.* (*Chariomyrma*): Emery, 1925b: 186.

#**continentis.** *Polyrhachis arcuata* var. *continentis* Forel, 1905b: 179 (w.) WEST MALAYSIA. Combination in *P.* (*Chariomyrma*): Emery, 1925b: 186.

**continua.** *Polyrhachis continua* Emery, 1887a: 235, pl. 4, fig. 21 (w.) INDONESIA (Ternate I.). Combination in *P.* (*Myrma*): Emery, 1925b: 200. Current subspecies: nominal plus *hirsutula, procera, revocata.*

**convexa.** *Polyrhachis convexa* Roger, 1863a: 153 (w.) SRI LANKA. Combination in *P.* (*Myrma*): Emery, 1925b: 204. Current subspecies: nominal plus *isabellae.*

*convexa. Polyrhachis* (*Chariomyrma*) *convexa* Karavaiev, 1927e: 17, fig. 7 (w.) INDONESIA (Ambon I.). [Junior primary homonym of *convexa* Roger, above.] Replacement name: *karawaiewi* Santschi, 1928h: 139.

**cornuta.** *Polyrhachis cornuta* Stitz, 1910: 150 (w.) ZAIRE. Combination in *P.* (*Myrma*): Wheeler, W.M. 1922a: 994. See also: Bolton, 1973b: 300.

**coronata.** *Polyrhachis* (*Cyrtomyrma*) *coronata* Santschi, 1928h: 140. Replacement name for *obsidiana* Karavaiev, 1927e: 50. [Junior primary homonym of *obsidiana* Emery, 1921e: 21.

#**corporaali.** *Polyrhachis* (*Cyrtomyrma*) *rastellata* var. *corporaali* Santschi, 1928h: 134, fig. 2 (w.q.) SUMATRA.

#**costulata.** *Polyrhachis aurea* subsp. *costulata* Emery, 1897d: 590 (w.) NEW GUINEA. Combination in *P.* (*Chariomyrma*): Emery, 1925b: 186. Raised to species: Dahl, 1901: 15. Subspecies/race of *aurea:* Forel, 1901b: 32; Emery, 1925b: 186.

**craddocki.** *Polyrhachis craddocki* Bingham, 1903: 403, fig. 138 (w.) BURMA. See also: Hung, 1970: 31.

*crassa. Polyrhachis* (*Pseudocyrtomyrma*) *revoili* subsp. *crassa* Emery, 1921e: 23, fig. 2 (w.q.) CAMEROUN. Junior synonym of *weissi:* Bolton, 1973b: 342.

**crassispinosa.** *Polyrhachis crassispinosa* Viehmeyer, 1914c: 49, fig. 9 (w.) NEW GUINEA. Combination in *P.* (*Myrma*), and misspelled as *crassispina:* Emery, 1925b: 201.

#**crassisquama.** *Polyrhachis tibialis* var. *crassisquama* Forel, 1913k: 139 (w.) WEST MALAYSIA. Combination in *P.* (*Myrmhopla*): Emery, 1925b: 196.

*crawleyella. Polyrhachis crawleyella* Santschi, 1916d: 243. Replacement name for *comata* Crawley, 1915b: 237. [Junior primary homonym of *comata* Emery, 1911b: 538.] Combination in *P.* (*Chariomyrma*), and misspelled as *crawleyana:* Emery, 1925b: 186. Junior synonym of *gab:* Bolton, 1974b: 173; of *senilis:* Kohout, 1988a: 50.

**crawleyi.** *Polyrhachis* (*Hagiomyrma*) *ammonoeides* var. *crawleyi* Forel, 1916: 447 (w.) AUSTRALIA. Raised to species: Kohout, 1988c: 433.

**creusa.** *Polyrhachis creusa* Emery, 1897d: 577 (w.) NEW GUINEA. Combination in *P.* (*Campomyrma*): Emery, 1925b: 179. Senior synonym of *chlorizans* and *hecuba:* Kohout & Taylor, 1990: 514. Current subspecies: nominal plus *distinguenda.*

**croceiventris.** *Polyrhachis croceiventris* Emery, 1900c: 336 (w.) NEW GUINEA. Viehmeyer, 1914b: 535 (q.). Combination in *P.* (*Myrmhopla*): Emery, 1925b: 192.

*crudelis. Polyrhachis bellicosa* var. *crudelis* Emery, 1887a: 238 (w.) INDONESIA (Morotai I.). Junior synonym of *bellicosa:* Hung, 1970: 5.

**cryptoceroides.** *Polyrhachis cryptoceroides* Emery, 1887a: 228, pl. 3, fig. 14 (w.) SULAWESI. Combination in *P.* (*Myrmhopla*): Emery, 1925b: 190.

**cubaensis.** *Polyrhachis cubaensis* Mayr, 1862: 687 (w.) SOUTH AFRICA (not Cuba; see Mayr, 1893: 195). Combination in *P.* (*Myrma*): Wheeler, W.M. 1922a: 994. Senior synonym of *gerstaeckeri, striolatorugosa* and *wilmsi:* Bolton, 1973b: 325.

**cupreata.** *Polyrhachis hermione* var. *cupreata* Emery, 1895g: 357 (w.) AUSTRALIA. Combination in *P.* (*Hedomyrma*): Emery, 1925b: 190. Raised to species, and senior synonym of *exlex:* Kohout, 1988c: 433.

*cupreopubescens. Polyrhachis militaris* r. *cupreopubescens* Forel, 1879a: 120 (q.) "AFRICA". Forel, 1907a: 38 (w.); Wheeler, G.C. & Wheeler, J. 1953e: 207 (l.). Combination in *P.* (*Myrma*): Santschi, 1914b: 140. Raised to species: Dalla Torre, 1893: 261. Subspecies of *militaris:* Mayr, 1895: 154; Wheeler, W.M. 1922a: 998. Junior synonym of *militaris:* Bolton, 1973b: 313.

**curta.** *Polyrhachis curta* André, 1890: 312 (w.) SIERRA LEONE. Bolton, 1973b: 347 (q.). Combination in *P.* (*Pseudocyrtomyrma*): Emery, 1925b: 206; in *P.* (*Myrma*): Wheeler, W.M. 1922a: 995; Bolton, 1973b: 346. Senior synonym of *lyrifera* and *maynei:* Bolton, 1973b: 346.

#**curvispina.** *Polyrhachis oedipus* var. *curvispina* Forel, 1908a: 8 (w.) SRI LANKA. Combination in *P.* (*Myrmhopla*): Emery, 1925b: 193.

*curvispina. Polyrhachis variolosa* var. *curvispina* Stitz, 1911a: 379 (w.) NEW GUINEA. [Junior primary homonym of *curvispina* Forel, above.] Replacement name: *arcispina* Santschi, 1916d: 243.

*cuspidatus* Smith, F. 1857; see under ***DOLICHODERUS.***

#**cuspidata.** *Polyrhachis conops* var. *cuspidata* Stitz, 1911a: 376, fig. 22 (w.) NEW GUINEA. [Unresolved junior primary homonym of *Polyrhachis cuspidatus* Smith, F. 1857a: 63, above.] Combination in *P.* (*Myrma*): Emery, 1925b: 200.

*cyaneus. Polyrhachis cyaneus* Mayr, 1862: 684, pl. 19, fig. 6 (w.) PHILIPPINES. Junior synonym of *cyaniventris:* Roger, 1863b: 9.

**cyaniventris.** *Polyrhachis cyaniventris* Smith, F. 1858b: 70, pl. 4, fig. 47 (w.) PHILIPPINES. Forel, 1910d: 128 (q.). Combination in *P.* (*Myrma*): Emery, 1925b: 201. Senior synonym of *cyaneus:* Roger, 1863b: 9.

**#cybele.** *Polyrhachis (Hemioptica) aculeata* subsp. *cybele* Wheeler, W.M. 1919e: 126 (w.) BORNEO. Combination in *P. (Myrma)*: Emery, 1925b: 205.

**cyrtomyrmoides.** *Polyrhachis (Myrmhopla) cyrtomyrmoides* Donisthorpe, 1947d: 195 (w.q.) NEW GUINEA.

**cyrus.** *Polyrhachis cyrus* Forel, 1901b: 32 (w.q.m.) BISMARCK ARCHIPELAGO. Combination in *P. (Chariomyrma)*: Emery, 1925b: 186.

**daemeli.** *Polyrhachis daemeli* Mayr, 1876: 72 (w.) AUSTRALIA. Forel, 1915b: 110 (m.). Combination in *P. (Hedomyrma)*: Forel, 1915b: 110. Current subspecies: nominal plus *sulcativentris.*

**dahlii.** *Polyrhachis dahlii* Forel, 1901b: 30 (w.) BISMARCK ARCHIPELAGO. Combination in *P. (Myrmothrinax)*: Mann, 1919: 378. Current subspecies: nominal plus *cincta, unisculpta.*

**daphne.** *Polyrhachis (Myrmhopla) daphne* Wheeler, W.M. 1919e: 133 (w.) BORNEO.

**davydovi.** *Polyrhachis (Myrmhopla) davydovi* Karavaiev, 1927e: 24 (w.) INDONESIA (Aru I.). See also: Kohout, 1990: 501.

**debilis.** *Polyrhachis laevior* var. *debilis* Emery, 1887a: 240 (w.) NEW GUINEA. Donisthorpe, 1938c: 266 (q.). Combination in *P. (Cyrtomyrma)*: Emery, 1925b: 208. Raised to species: Donisthorpe, 1938c: 265. Current subspecies: nominal plus *johnsoni.*

**decellei.** *Polyrhachis decellei* Bolton, 1973b: 301, figs. 46, 54 (w.) GHANA.

**decemdentata.** *Polyrhachis decemdentata* André, 1889: 219 (w.) SIERRA LEONE. Emery, 1892d: 567 (q.); Forel, 1915c: 351 (m.). Combination in *P. (Myrma)*: Forel, 1916: 450. Senior synonym of *fernandensis, flavipes, gustavi* and *tenuistriata:* Bolton, 1973b: 302.

**#decipiens.** *Polyrhachis decipiens* Roger, 1863a: 156 (w.) INDONESIA (Batjan I.). Combination in *P. (Myrma)*: Emery, 1925b: 202. Subspecies of *relucens:* Emery, 1897d: 580.

**#defensa.** *Polyrhachis defensus* Smith, F. 1857a: 59 (w.) SINGAPORE. Combination in *P. (Myrmhopla)*: Emery, 1925b: 192. Junior synonym of *armata:* Roger, 1863b: 9; Dalla Torre, 1893: 258; Wheeler, W.M. 1919e: 129. Revived from synonymy as variety of *armata:* Emery, 1925b: 192; Donisthorpe, 1932c: 445. Senior synonym of *minor:* Bingham, 1903: 393.

*degener. Polyrhachis sokolova* var. *degener* Forel, 1910b: 84 (w.) AUSTRALIA. Combination in *P. (Hagiomyrma)*: Emery, 1925b: 185. Junior synonym of *sokolova:* Kohout, 1988c: 436.

**delicata.** *Polyrhachis delicata* Crawley, 1915b: 238 (w.) AUSTRALIA. Junior synonym of *queenslandica:* Crawley, 1921: 96. Revived from synonymy, and senior synonym of *lysistrata:* Kohout, 1994a: 135.

**demangei.** *Polyrhachis rastellata* subsp. *demangei* Santschi, 1910h: 284, fig. 3 (w.) VIETNAM. Combination in *P. (Cyrtomyrma)*: Emery, 1925b: 208. Raised to species: Donisthorpe, 1938c: 264.

*democles. Polyrhachis democles* Smith, F. 1861b: 40, pl. 1, fig. 9 (q.) SULAWESI. Junior synonym of *dives:* Forel, 1911e: 298.

**#denselineata.** *Polyrhachis arcuata* var. *denselineata* Viehmeyer, 1914b: 533 (w.) NEW GUINEA. Combination in *P. (Chariomyrma)*: Emery, 1925b: 186.

**dentata.** *Polyrhachis (Cyrtomyrma) dentata* Donisthorpe, 1947d: 196, fig. 3 (w.) NEW GUINEA.

**denticulata.** *Polyrhachis (Hagiomyrma) denticulata* Karavaiev, 1927e: 13, fig. 4 (w.) INDONESIA (Ambon I.).

*dentinasis. Polyrhachis (Campomyrma) micans* st. *ops* var. *dentinasis* Santschi, 1920g: 185 (w.q.) AUSTRALIA. Unavailable name (Taylor, 1986: 34).

**#dentulata.** *Polyrhachis (Myrma) orsyllus* var. *dentulata* Stitz, 1925: 135 (w.) PHILIPPINES.

*depilis. Polyrhachis aurea* var. *depilis* Emery, 1897d: 589 (footnote) (w.) AUSTRALIA. Combination in *P. (Chariomyrma)*: Emery, 1925b: 186. Junior synonym of *pallescens:* Kohout & Taylor, 1990: 516.

*derecyna. Polyrhachis derecynus* Smith, F. 1871: 316. Replacement name for *dolomedes* Smith, F. 1863: 16. [Junior primary homonym of *dolomedes* Smith, F. 1863: 14.] Combination in *P. (Myrma)*: Donisthorpe, 1932c: 470; in *P. (Myrmatopa)*: current combination. Senior synonym of *pseuadonyma:* Dalla Torre, 1893: 261; of *taurus:* Bolton, 1974b: 173.

**diana.** *Polyrhachis diana* Wheeler, W.M. 1909d: 343 (w.) PHILIPPINES. Combination in *P. (Myrma)*: Emery, 1925b: 201.

**diaphanta.** *Polyrhachis diaphantus* Smith, F. 1861b: 40, pl. 1, fig. 12 (w.) SULAWESI. Combination in *P. (Myrmhopla)*: Emery, 1925b: 194. Junior synonym of *acantha:* Emery, 1900d: 717. Revived from synonymy as variety of *acantha:* Emery, 1925b: 194. Revived status as species: Donisthorpe, 1932c: 466.

**#dichroa.** *Polyrhachis laevissima* var. *dichrous* Forel, 1893c: 21 (diagnosis in key) (w.) INDIA. Combination in *P. (Cyrtomyrma)*: Emery, 1925b: 207.

*dichroa. Polyrhachis (Myrmhopla) acantha* subsp. *dichroa* Karavaiev, 1927e: 33 (w.) JAVA. [Junior primary homonym of *dichroa* Forel, above.] Replacement name: *romanovi* Santschi, 1928h: 139.

*dido. Polyrhachis (Myrma) militaris* subsp. *cupreopubescens* var. *dido* Wheeler, W.M. 1922a: 261. Unavailable name. [Unnecessary replacement name for *argentatus* Stitz, 1910: 150; unavailable name.] Material referred to *militaris* by Bolton, 1973b: 313.

**diotima.** *Polyrhachis diotima* Forel, 1911a: 60 (w.) BORNEO. Combination in *P. (Myrmhopla)*: Wheeler, W.M. 1919e: 132. See also: Forel, 1911e: 297.

*discrepans. Polyrhachis achilles* subsp. *discrepans* Forel, 1912n: 76 (w.) SUMATRA. Viehmeyer, 1922: 220 (q.). Combination in *P. (Myrmhopla)*: Emery, 1925b: 192. Junior synonym of *hector:* Bolton, 1974b: 174.

**distincta.** *Polyrhachis (Myrmhopla) distincta* Karavaiev, 1927e: 40, fig. 18 (w.) INDONESIA (Ambon I.).

**#distinguenda.** *Polyrhachis (Campomyrma) creusa* subsp. *distinguenda* Karavaiev, 1927e: 7, fig. 1 (w.) INDONESIA (Ambon I.).

**dives.** *Polyrhachis dives* Smith, F. 1857a: 64 (w.) SINGAPORE. Mayr, 1867a: 49 (q.m.); Wheeler, G.C. & Wheeler, J. 1953e: 208 (l.); Hung, Imai & Kubota, 1972: 1025 (k.). Combination in *P. (Myrmhopla)*: Wheeler, W.M. 1919e: 132. Senior synonym of *democles:* Forel, 1911e: 298; of *euclides* and *mutiliae:* Bolton, 1974b: 173; of *exulans:* Kohout, 1988c: 433; of *vicina:* Wang & Wu, 1991: 599. Current subspecies: nominal plus *belli, rectispina, siwiensis.* See also: Bingham, 1903: 396.

*divina. Polyrhachis schistacea* var. *divina* Forel, 1913b: 348 (in text) (w.) TANZANIA. Combination in *P. (Myrma)*: Santschi, 1914b: 141. Junior synonym of *schistacea:* Bolton, 1973b: 318.

*divinoides. Polyrhachis schistacea* var. *divinoides* Arnold, 1924: 744 (w.) ZAIRE. [First available use of *Polyrhachis schistacea* r. *rugulosa* var. *divinoides* Forel, 1913b: 348; unavailable name.] Junior synonym of *schistacea:* Bolton, 1973b: 318.

**doddi.** *Polyrhachis (Cyrtomyrma) doddi* Donisthorpe, 1938c: 263, fig. 13 (w.q.) AUSTRALIA.

**dohrni.** *Polyrhachis dohrni* Forel, 1901b: 34 (w.) BISMARCK ARCHIPELAGO. Combination in *P. (Aulacomyrma)*: Emery, 1921e: 17.

**dolichocephala.** *Polyrhachis dolichocephala* Viehmeyer, 1914b: 532, fig. 5 (w.) NEW GUINEA. Combination in *P. (Dolichorhachis)*: Emery, 1925b: 189.

**dolomedes.** *Polyrhachis dolomedes* Smith, F. 1863: 14 (w.q.) INDONESIA (Seram I.). Combination in *P. (Myrmatopa)*: Donisthorpe, 1932c: 469.

*dolomedes. Polyrhachis dolomedes* Smith, F. 1863: 16 (w.) NEW GUINEA. [Junior primary homonym of *dolomedes* Smith, above.] Replacement name: *derecyna* Smith, F. 1871: 316.

*donisthorpei. Polyrhachis (Myrma) revoili* var. *donisthorpei* Forel, 1916: 453 (w.) ZAMBIA. [Also described as new by Forel, 1917: 252.] Combination in *P. (Pseudocyrtomyrma)*: Emery, 1925b: 207. Junior synonym of *revoili:* Bolton, 1973b: 338.

**dorsorugosa.** *Polyrhachis latona* var. *dorsorugosa* Forel, 1913f: 202 (w.m.) TAIWAN. Combination in *P. (Myrma)*: Emery, 1925b: 201. Raised to species: Wang & Wu, 1991: 599.

*duodentata. Polyrhachis (Myrma) duodentata* Donisthorpe, 1942d: 461, pl. 2, fig. 2 (q.) INDIA. Junior synonym of *illaudata:* Bolton, 1974b: 176.

**durbanensis.** *Polyrhachis (Myrma) cubaensis* r. *durbanensis* Forel 1914d: 262 (w.q.) SOUTH AFRICA. Raised to species: Bolton, 1973b: 327.

**durvillei.** *Polyrhachis (Myrmothrinax) durvillei* Donisthorpe, 1938a: 147 (w.q.) NEW GUINEA.

**#edentula.** *Polyrhachis zopyrus* var. *edentula* Emery, 1900d: 712 (w.) SUMATRA. Forel, 1911b: 212 (q.m.). Combination in *P. (Myrma)*: Emery, 1925b: 204.

**edwardi.** *Polyrhachis (Myrmatopa) edwardi* Donisthorpe, 1948b: 314 (w.) NEW GUINEA.

*elegans. Polyrhachis terpsichore* var. *elegans* Forel, 1910b: 84 (w.) AUSTRALIA. Combination in *P. (Hedomyrma)*: Emery, 1925b: 190. Junior synonym of *rufifemur:* Kohout, 1988c: 436.

**elii.** *Polyrhachis elii* Emery, 1900d: 711 (w.) INDONESIA (Mentawei I.). Combination in *P. (Myrmatopa)*: Emery, 1925b: 180.

**emeryana.** *Polyrhachis (Cyrtomyrma) emeryana* Mann, 1919: 390, fig. 59 (w.) SOLOMON IS.

*emeryi. Camponotus emeryi* Forel, 1879a: 113, pl. 1, fig. 4 (q.) AUSTRALIA. Junior synonym of *femorata:* Forel, 1894b: 72; Taylor, 1989a: 23.

**emmae.** *Polyrhachis (Myrmhopla) emmae* Santschi, 1920h: 175, fig. 3 (w.) LAOS. [Also described as new by Santschi, 1924c: 115.]

*epinotalis. Polyrhachis militaris* r. *cupreopubescens* var. *epinotalis* Forel, 1913h: 357 (w.) ZAIRE. Unavailable name; material referred to *militaris* by Bolton, 1973b: 313.

**equina.** *Polyrhachis equinus* Smith, F. 1857a: 63 (w.) BORNEO. Combination in *P.* (*Campomyrma*), and senior synonym of *biloba:* Wheeler, W.M. 1919e: 122.

**erato.** *Polyrhachis erato* Forel, 1902h: 512 (w.) AUSTRALIA. Combination in *P.* (*Hedomyrma*): Emery, 1925b: 189. Senior synonym of *aeschyle:* Kohout, 1988c: 434.

**#erecta.** *Polyrhachis* (*Myrmhopla*) *bicolor* var. *erecta* Karavaiev, 1935a: 114 (w.) JAVA.

**eremita.** *Polyrhachis eremita* Kohout, 1990: 502 (w.) AUSTRALIA.

**erosispina.** *Polyrhachis bellicosa* var. *erosispina* Emery, 1900d: 713 (footnote) (w.) NEW GUINEA. Karavaiev, 1927e: 12 (q.). Junior synonym of *bellicosa:* Hung, 1970: 5. Revived from synonymy, and raised to species: Kohout, 1988b: 419.

**esarata.** *Polyrhachis esarata* Bolton, 1973b: 303, figs. 45, 55 (w.) GHANA.

**escherichi.** *Polyrhachis escherichi* Viehmeyer, 1914b: 534, fig. 6 (w.q.) NEW GUINEA. Combination in *P.* (*Chariomyrma*): Emery, 1925b: 187.

*esuriens. Polyrhachis sexspinosa* var. *esuriens* Emery, 1897d: 591 (w.) NEW GUINEA. Combination in *P.* (*Myrmhopla*): Emery, 1925b: 191. Junior synonym of *sexspinosa:* Bolton, 1975c: 12.

**etheli.** *Polyrhachis* (*Myrmhopla*) *etheli* Chapman, 1963: 260, fig. 8 (w.q.) PHILIPPINES.

*eucharis. Polyrhachis* (*Hedomyrma*) *eucharis* Karavaiev, 1927e: 22, fig. 12 (w.q.) INDONESIA (Aru I.). Junior synonym of *atropos:* Kohout & Taylor, 1990: 513.

*euclides. Polyrhachis dives* var. *euclides* Forel, 1913f: 202 (w.q.m.) TAIWAN. Junior synonym of *dives:* Bolton, 1974b: 173.

**eudora.** *Polyrhachis eudora* Smith, F. 1860b: 99, pl. 1, fig. 19 (w.q.) INDONESIA (Batjan I.). Combination in *P.* (*Myrmothrinax*): Emery, 1925b: 183.

**euryala.** *Polyrhachis euryalus* Smith, F. 1863: 16 (w.) INDONESIA (Misool I.). Combination in *P.* (*Cyrtomyrma*): Emery, 1925b: 208. Junior synonym of *rastellata:* Mayr, 1872: 138; Bingham, 1903: 414. Revived from synonymy as subspecies of *rastellata:* Emery, 1900d: 720; Emery, 1925b: 208. Revived status as species: Donisthorpe, 1938c: 259. Senior synonym of *torricelliana:* Viehmeyer, 1914c: 50. Current subspecies: nominal plus *goramensis, javana.*

**eurynota.** *Polyrhachis eurynota* Emery, 1897d: 587, pl. 1, figs. 26, 27 (w.) NEW GUINEA. Combination in *P.* (*Chariomyrma*): Emery, 1925b: 187.

*eurytus. Polyrhachis eurytus* Smith, F. 1861b: 43, pl. 1, fig. 24 (q.) SULAWESI. Combination in *P.* (*Myrma*): Donisthorpe, 1932c: 467. Junior synonym of *olena:* Bolton, 1974b: 177.

**euterpe.** *Polyrhachis euterpe* Forel, 1902h: 511 (w.) AUSTRALIA. Combination in *P.* (*Hedomyrma*): Emery, 1925b: 189.

**exarata.** *Polyrhachis exarata* Emery, 1887a: 226 (w.) INDONESIA (Misool I.). Combination in *P.* (*Aulacomyrma*): Emery, 1921e: 17.

*exasperatus. Polyrhachis exasperatus* Smith, F. 1861b: 41, pl. 1, fig. 16 (w.) SULAWESI. Viehmeyer, 1916b: 288 (q.). Combination in *P.* (*Myrmhopla*): Emery, 1925b: 193. Junior synonym of *rufipes:* Bolton, 1974b: 178.

**excellens.** *Polyrhachis excellens* Viehmeyer, 1912: 14, fig. 19 (w.) NEW GUINEA. Combination in *P.* (*Aulacomyrma*): Emery, 1921e: 17.

**excisa.** *Polyrhachis excisa* Mayr, 1867a: 64, pl. 2, fig. 5 (q.) JAVA. Subgenus indeterminate: Emery, 1925b: 209.

**#excitata.** *Polyrhachis excitata* Viehmeyer, 1913: 147, fig. (w.) SULAWESI (in copal). Combination in *P.* (*Myrmatopa*): Emery, 1925b: 181. Variety of *schang:* Viehmeyer, 1914c: 48.

**exercita.** *Formica exercita* Walker, 1859: 370 (q.m.) SRI LANKA. Combination in *Polyrhachis:* Donisthorpe, 1932b: 575; in *P.* (*Campomyrma*): current combination. Senior synonym of *clypeata* (and its junior synonym *indica*): Donisthorpe, 1932b: 575. Current subspecies: nominal plus *lucidiventris, obtusisquama, rastrata.*

**#exflavicornis.** *Polyrhachis* (*Myrmhopla*) *bicolor* subsp. *exflavicornis* Baltazar, 1966: 282. Replacement name for *flavicornis* Stitz, 1925: 132. [Junior primary homonym of *flavicornis* Smith, F. 1857a: 63.]

*exlex. Polyrhachis* (*Hedomyrma*) *daemeli* var. *exlex* Forel, 1915b: 110 (w.) AUSTRALIA. Junior synonym of *cupreata:* Kohout, 1988c: 433.

**#exophthalma.** *Polyrhachis striatorugosa* var. *exophthalma* Forel, 1913k: 136 (w.q.) SUMATRA. Combination in *P.* (*Myrma*): Emery, 1925b: 203. Subspecies of *sumatrensis:* current status.

**exotica.** *Polyrhachis exotica* Kohout, 1987: 170, figs. 9, 11, 15 (w.q.) PHILIPPINES.

*exul. Polyrhachis exul* Emery, 1883: 147 (w.) NEW CALEDONIA. Junior synonym of *guerini:* Emery, 1897d: 584 (in key).

*exulans. Polyrhachis* (*Myrmhopla*) *exulans* Clark, 1941: 91, pl. 13, fig. 24 (w.) AUSTRALIA. Junior synonym of *dives:* Kohout, 1988c: 433.

*felici. Polyrhachis* (*Pseudocyrtomyrma*) *lanuginosa* subsp. *felici* Emery, 1925b: 206. Unnecessary (second) replacement name for *santschii* Emery, 1921e: 24. [Junior primary homonym of *santschii* Mann, 1919: 375.] Junior synonym of *lanuginosa:* Bolton, 1973b: 335.

**femorata.** *Polyrhachis femoratus* Smith, F. 1858b: 73, pl. 4, fig. 46 (w.) AUSTRALIA. Wheeler, G.C. & Wheeler, J. 1953e: 207 (l.). Combination in *P.* (*Campomyrma*): Forel, 1915b: 113. Senior synonym of *emeryi:* Forel, 1894b: 72; Taylor, 1989a: 23.

**#fergusoni.** *Polyrhachis punctillata* r. *fergusoni* Forel, 1902d: 289 (w.) INDIA. Combination in *P.* (*Myrma*): Emery, 1925b: 204. Junior synonym of *punctillata:* Bingham, 1903: 410. Subspecies of *punctillata:* Emery, 1925b: 204.

*fernandensis. Polyrhachis decemdentata* var. *fernandensis* Forel, 1901e: 377 (w.) EQUATORIAL GUINEA. Santschi, 1910c: 396 (q.). Junior synonym of *decemdentata:* Bolton, 1973b: 302.

**fervens.** *Polyrhachis fervens* Smith, F. 1860b: 101, pl. 1, fig. 26 (w.) INDONESIA (Ambon I.). Combination in *P.* (*Dolichorhachis*): Emery, 1925b: 189; in *P.* (*Hedomyrma*): current combination. Senior synonym of *indocilis, kershawi* and *valerus:* Kohout, 1988c: 434.

**#fiorii.** *Polyrhachis aurea* subsp. *fiorii* Emery, 1914f: 429 (w.) VANUATU. Combination in *P.* (*Chariomyrma*): Emery, 1925b: 186.

**fissa.** *Polyrhachis fissus* Mayr, 1902: 301 (w.q.) CAMEROUN. Combination in *P.* (*Myrma*): Wheeler, W.M. 1922a: 995. Senior synonym of *bequaerti* and *ugandensis:* Bolton, 1973b: 304.

**flavibasis.** *Polyrhachis* (*Campomyrma*) *flavibasis* Clark, 1930c: 16, fig. 1 (w.q.) AUSTRALIA.

**flavicornis.** *Polyrhachis flavicornis* Smith, F. 1857a: 63 (w.q.) SINGAPORE. Combination in *P.* (*Myrmatopa*): Donisthorpe, 1932c: 446.

*flavicornis. Polyrhachis* (*Myrmhopla*) *bicolor* var. *flavicornis* Stitz, 1925: 132 (w.) PHILIPPINES. [Junior primary homonym of *flavicornis* Smith, above.] Replacement name: *exflavicornis* Baltazar, 1966: 282. [A second proposed replacement name, *rubricornis* Baroni Urbani, 1971b: 362, is both unnecessary and unavailable.]

*flavipes. Polyrhachis decemdentata* var. *flavipes* Stitz, 1910: 149 (w.q.) CAMEROUN. Junior synonym of *decemdentata:* Bolton, 1973b: 302.

**flavoflagellata.** *Polyrhachis* (*Myrmhopla*) *flavoflagellata* Karavaiev, 1927e: 35, fig. 16 (w.) JAVA.

**follicula.** *Polyrhachis* (*Myrmatopa*) *follicula* Menozzi, 1926e: 101 (w.) PHILIPPINES.

**foreli.** *Polyrhachis foreli* Kohout, 1989: 510, figs. 2, 7, 11 (w.q.) AUSTRALIA.

**fornicata.** *Polyrhachis rastellata* subsp. *fornicata* Emery, 1900d: 720 (w.) SULAWESI. Combination in *P.* (*Cyrtomyrma*): Emery, 1925b: 208. Raised to species: Donisthorpe, 1938c: 261.

**fortis.** *Polyrhachis fortis* Emery, 1893e: 228, pl. 8, fig. 5 (w.) BURMA. Combination in *P.* (*Myrmhopla*): Emery, 1925b: 195.

*foveolatus* Lowne, 1865; see under ***DOLICHODERUS.***

*fracta. Polyrhachis* (*Myrma*) *schistacea* st. *fracta* Santschi, 1914b: 141 (w.) KENYA. Junior synonym of *schistacea:* Bolton, 1973b: 318.

**frauenfeldi.** *Polyrhachis frauenfeldi* Mayr, 1862: 687 (w.) JAVA. Mayr, 1865: 45 (m.); Mayr, 1867a: 52 (q.). Combination in *P.* (*Myrmothrinax*): Viehmeyer, 1916a: 165. Current subspecies: nominal plus *sanguinea.*

*froggatti. Polyrhachis froggatti* Forel, 1910b: 89 (w.) AUSTRALIA. Combination in *P.* (*Campomyrma*): Emery, 1925b: 179. Junior synonym of *hexacantha:* Taylor, 1989a: 24.

**fruhstorferi.** *Polyrhachis fruhstorferi* Emery, 1898b: 238, figs. 11, 12 (w.) SULAWESI. Combination in *P.* (*Campomyrma*): Viehmeyer, 1916a: 163; in *P.* (*Myrmatopa*): Emery, 1925b: 180. Current subspecies: nominal plus *torta, varicolor.*

**#fulakora.** *Polyrhachis* (*Cyrtomyrma*) *rastellata* var. *fulakora* Mann, 1919: 389 (w.) SOLOMON IS.

**fulgens.** *Polyrhachis fulgens* Viehmeyer, 1912: 11, fig. 5 (w.) NEW GUINEA. Combination in *P.* (*Chariomyrma*): Emery, 1925b: 187.

**#fumata.** *Polyrhachis* (*Myrmhopla*) *bicolor* var. *fumata* Stitz, 1925: 131 (w.q.) PHILIPPINES.

**furcata.** *Polyrhachis furcatus* Smith, F. 1858b: 64, pl. 4, fig. 20 (w.) BURMA. Emery, 1889b: 518 (q.). Combination in *P.* (*Myrmhopla*): Wheeler, W.M. 1919e:

131. Current subspecies: nominal plus *bankensis, pahangana, tenella.* See also: Bingham, 1903: 388.

**furcula.** *Polyrhachis furcula* Emery, 1911b: 537 (w.) NEW GUINEA. Combination in *P.* (*Myrmatopa*): Emery, 1925b: 181.

*fusca. Polyrhachis relucens* r. *litigiosa* var. *fusca* Stitz, 1911a: 375 (w.) "Neu-Hannover". Unavailable name.

**fuscipes.** *Polyrhachis fuscipes* Mayr, 1862: 679 (w.) AUSTRALIA. Combination in *P.* (*Campomyrma*): Emery, 1925b: 179. Junior synonym of *hexacantha:* Roger, 1863b: 9; Dalla Torre, 1893: 263. Revived from synonymy as subspecies of *hexacantha:* Emery, 1925b: 179. Revived status as species: Taylor & Brown, D.R. 1985: 134. Senior synonym of *hestia:* Taylor, 1989a: 23.

**gab.** *Polyrhachis guerini* r. *gab.* Forel, 1879a: 116 (w.) AUSTRALIA. Combination in *P.* (*Chariomyrma*): Forel, 1915b: 108. Raised to species: Dalla Torre, 1893: 262; Emery, 1897d: 584 (in key). See also: Kohout, 1988a: 50.

**gagates.** *Polyrhachis gagates* Smith, F. 1858b: 71, pl. 4, fig. 14 (w.) SOUTH AFRICA. Forel, 1886f: 194 (q.); Wheeler, G.C. & Wheeler, J. 1953e: 207 (l.). Combination in *P.* (*Myrma*): Santschi, 1914b: 139. Race of *schistacea:* Arnold, 1924: 745. Revived status as species: Emery, 1925b: 199; Bolton, 1973b: 305. Senior synonym of *clariseta, congolensis* (and its junior synonym *indefinita*), *nigriseta* and *obsidiana,* and material of the unavailable name *acheron* referred here: Bolton, 1973b: 305.

*gagatoides. Polyrhachis* (*Myrma*) *schistacea* var. *gagatoides* Santschi, 1913c: 314 (w.) CONGO. [Also described as new by Santschi, 1914b: 142.] Junior synonym of *schistacea:* Bolton, 1973b: 318.

*gallicola. Polyrhachis cubaensis* r. *gallicola* Forel, 1894b: 71 (w.q.) MOZAMBIQUE. Combination in *P.* (*Myrma*): Wheeler, W.M. 1922a: 994. Junior synonym of *spinicola:* Bolton, 1973b: 329.

**gamaii.** *Polyrhachis* (*Myrma*) *gamaii* Santschi, 1917b: 295 (q.) SOUTH AFRICA. Arnold, 1947: 136 (w.). See also: Bolton, 1973b: 351.

**geminata.** *Polyrhachis* (*Hedomyrma*) *geminata* Mann, 1919: 376, fig. 47 (w.) SOLOMON IS.

**geometrica.** *Polyrhachis geometricus* Smith, F. 1859a: 141 (w.) INDONESIA (Aru I.). Combination in *P.* (*Aulacomyrma*): Donisthorpe, 1932c: 453.

*gerstaeckeri. Polyrhachis gerstaeckeri* Forel, 1886f: 197 (w.) TANZANIA. Combination in *P.* (*Myrma*): Wheeler, W.M. 1922a: 994. Subspecies/variety of *cubaensis:* Mayr, 1893: 196; Forel, 1894b: 71; Emery, 1896d: 379. Junior synonym of *cubaensis:* Bolton, 1973b: 325.

**gestroi.** *Polyrhachis gestroi* Emery, 1900d: 714 (w.q.) INDONESIA (Mentawei I.). Combination in *P.* (*Myrmhopla*): Emery, 1925b: 195. Current subspecies: nominal plus *rufiventris.*

**gibba.** *Polyrhachis gibba* Emery, 1901g: 580, fig. E (w.) SULAWESI. Combination in *P.* (*Cyrtomyrma*): Emery, 1925b: 207.

#**gibbosa.** *Polyrhachis aculeata* var. *gibbosa* Forel, 1908a: 9 (w.) SRI LANKA. Forel, 1913k: 139 (footnote) (q.). Combination in *P.* (*Myrma*): Emery, 1925b: 205.

#**glabra.** *Polyrhachis sericata* subsp. *glabra* Forel, 1911e: 295 (w.) NEW GUINEA. Combination in *P.* (*Myrma*): Emery, 1925b: 203.

**glabrinota.** *Polyrhachis* (*Myrmhopla*) *glabrinotum* Clark, 1930c: 13, fig. 1 (w.) AUSTRALIA. See also: Bolton, 1975c: 8.

*globularia. Polyrhachis globularia* Mayr, 1867a: 41 (w.) JAVA. Junior synonym of *laevissima:* Mayr, 1879: 651.

**glykera.** *Polyrhachis glykera* Forel, 1912n: 72 (q.) SUMATRA. Combination in *P.* (*Myrmhopla*): Emery, 1925b: 195.

#**goramensis.** *Polyrhachis rastellata* var. *goramensis* Emery, 1887a: 239 (w.) INDONESIA (Goram I.). Combination in *P.* (*Cyrtomyrma*): Emery, 1925b: 208. Subspecies of *euryala:* Donisthorpe, 1938c: 260.

**gracilior.** *Polyrhachis furcata* r. *gracilior* Forel, 1893c: 25 (diagnosis in key) (w.) INDIA. Combination in *P.* (*Myrmhopla*): Emery, 1925b: 193. Raised to species: Bingham, 1903: 388. Senior synonym of *weberi:* Bolton, 1974b: 174.

*gracilior. Polyrhachis* (*Myrmatopa*) *schang* var. *gracilior* Karavaiev, 1927e: 11 (w.q.m.) INDONESIA (Ambon I.). [Junior primary homonym of *gracilior* Forel, above.] Replacement name: *amboinae* Santschi, 1928h: 139.

*gracilis. Polyrhachis gracilis* Emery, 1887a: 223, pl. 3, fig. 11 (w.q.) INDONESIA (Goram I.). Junior synonym of *schang:* Forel, 1909d: 232.

**grandis.** *Polyrhachis* (*Cyrtomyrma*) *grandis* Donisthorpe, 1949g: 415 (w.q.) NEW GUINEA.

**gravis.** *Polyrhachis* (*Campomyrma*) *gravis* Clark, 1930c: 15, fig. 1 (w.) AUSTRALIA.

**greensladei.** *Polyrhachis greensladei* Kohout, 1990: 503 (w.q.) SOLOMON IS.

**gribodoi.** *Polyrhachis gribodoi* Emery, 1887a: 221 (w.q.) JAVA. Imai, Kubota, *et al.* 1985: 48 (k.). Combination in *P.* (*Campomyrma*): Wheeler, W.M. 1919e: 122.

*grisea. Formica grisea* Le Guillou, 1842: 314 (q.) "Triton Bay". Combination in *Polyrhachis:* Roger, 1863b: 8. Junior synonym of *sericata:* Dalla Torre, 1893: 269; Emery, 1896d: 379.

#**grisescens.** *Polyrhachis simplex* var. *grisescens* Emery, 1895k: 483 (w.q.) BURMA. Combination in *P.* (*Myrmhopla*): Emery, 1925b: 196. Subspecies of *lacteipennis:* current status.

**guerini.** *Polyrhachis guerini* Roger, 1863a: 157 (w.) AUSTRALIA. Combination in *P.* (*Chariomyrma*): Forel, 1915b: 107. Senior synonym of *exul:* Emery, 1897d: 584 (in key).

*gustavi. Polyrhachis* (*Myrma*) *decemdentata* var. *gustavi* Emery, 1921e: 22 (q.) FERNANDO PO I. Junior synonym of *decemdentata:* Bolton, 1973b: 302.

**halidayi.** *Polyrhachis halidayi* Emery, 1889b: 517 (w.) BURMA. Bingham, 1903: 413 (q.). Combination in *P.* (*Campomyrma*): Emery, 1925b: 180.

*halmaheirae. Polyrhachis orsyllus* r. *halmaheirae* Forel, 1886f: 196 (w.) INDONESIA (Halmahera I.). Raised to species: Dalla Torre, 1893: 263. Junior synonym of *abrupta:* Emery, 1925b: 204.

#**hamulata.** *Polyrhachis sumatrensis* r. *hamulata* Emery, 1887a: 234 (w.q.) SULAWESI. Combination in *P.* (*Myrma*): Emery, 1925b: 203. Raised to species: Dalla Torre, 1893: 263; Bingham, 1903: 406. Subspecies of *sumatrensis:* Emery, 1925b: 203.

*harmsi. Polyrhachis* (*Myrma*) *sericata* subsp. *pruinosula* var. *harmsi* Karavaiev, 1930a: 212 (w.) NEW GUINEA. Unavailable name.

**hastata.** *Formica hastata* Latreille, 1802c: 129, pl. 4, fig. 23 (w.) INDONESIA ("Indes orientales"). Combination in *Polyrhachis:* Smith, F. 1858b: 59; in *P.* (*Myrma*): Emery, 1925b: 201.

**hauxwelli.** *Polyrhachis hauxwelli* Bingham, 1903: 394, fig. 133 (w.) BURMA. Combination in *P.* (*Campomyrma*): Emery, 1925b: 180.

*hebes. Polyrhachis* (*Chariomyrma*) *hostilis* subsp. *hebes* Donisthorpe, 1941h: 62 (w.) NEW GUINEA. Junior synonym of *hostilis:* Bolton, 1974b: 175.

**hector.** *Polyrhachis hector* Smith, F. 1857a: 62 (w.) SINGAPORE. Imai, Brown, *et al.* 1984: 68 (k.). Combination in *P.* (*Myrmhopla*): Emery, 1925b: 193. Senior synonym of *abdominalis* (and its junior synonym *achilles*), *confinis, discrepans, malignus, mutatus* (and its junior synonym *ajax*), *phyllophilus* and *reversa:* Bolton, 1974b: 174.

*hector. Polyrhachis hector* Smith, F. 1859a: 142 (w.) INDONESIA (Aru I.). [Junior primary homonym of *hector* Smith, above.] Replacement name: *andromache* Roger, 1863b: 8.

*hecuba. Polyrhachis hecuba* Forel, 1902h: 527 (w.q.m.) AUSTRALIA. Wheeler, G.C. & Wheeler, J. 1970: 649 (l.). Combination in *P.* (*Campomyrma*): Forel, 1915b: 115. Junior synonym of *creusa:* Kohout & Taylor, 1990: 514.

**heinlethii.** *Polyrhachis heinlethii* Forel, 1895a: 47 (w.) AUSTRALIA. Combination in *P.* (*Chariomyrma*): Forel, 1915b: 108. Senior synonym of *sophiae:* Kohout & Taylor, 1990: 514. Current subspecies: nominal plus *papuana.*

**hemiopticoides.** *Polyrhachis* (*Myrma*) *hemiopticoides* Mukerjee, 1930: 161, fig. 5 (w.) INDIA.

**hera.** *Polyrhachis hera* Forel, 1911e: 302 (w.) INDONESIA (Seram I.). Combination in *P.* (*Hedomyrma*): Emery, 1925b: 190.

**hermione.** *Polyrhachis hermione* Emery, 1895g: 357 (w.) AUSTRALIA. Combination in *P.* (*Hedomyrma*): Emery, 1925b: 190.

#**hero.** *Polyrhachis textor* r. *hero* Forel, 1913k: 135 (q.) SUMATRA. Combination in *P.* (*Myrmothrinax*): Emery, 1925b: 183.

*hestia. Polyrhachis semipolita* subsp. *hestia* Forel, 1911e: 295 (w.) AUSTRALIA. Combination in *P.* (*Campomyrma*): Emery, 1925b: 179. Subspecies of *hexacantha:* Emery, 1925b: 179; of *semipolita:* Taylor & Brown, D.R. 1985: 140. Junior synonym of *fuscipes:* Taylor, 1989a: 23.

**hexacantha.** *Formica hexacantha* Erichson, 1842: 260 (w.) TASMANIA. Combination in *Polyrhachis:* Smith, F. 1858b: 74; in *P.* (*Campomyrma*): Emery, 1925b: 179. Senior synonym of *froggatti:* Taylor, 1989a: 24.

**hippomanes.** *Polyrhachis hippomanes* Smith, F. 1861b: 43, pl. 1, fig. 21 (w.) SULAWESI. Wheeler, G.C. & Wheeler, J. 1970: 649 (l.). Combination in *P.* (*Myrmhopla*): Viehmeyer, 1916a: 166. Current subspecies: nominal plus *boettcheri, ceylonensis, hortensis, lucidula.* See also: Santschi, 1928g: 253.

#**hiram.** *Polyrhachis salomo* subsp. *hiram* Forel, 1912a: 80 (w.q.) SOLOMON IS. Combination in *P.* (*Myrma*): Emery, 1925b: 202.

***hirsuta.*** *Polyrhachis hirsuta* Mayr, 1876: 75 (w.) AUSTRALIA. Combination in *P.* (*Campomyrma*): Emery, 1925b: 179. Senior synonym of *quinquedentata:* Taylor, 1989a: 24.

*hirsuta. Polyrhachis hostilis* var. *hirsuta* Viehmeyer, 1914c: 58 (w.) NEW GUINEA. [Junior primary homonym of *hirsuta* Mayr, above.] Replacement name: *intricata* Forel, 1918a: 727. [*P. hirsuta* Emery, 1911b: 532; *nomen nudum.*]

**#*hirsutula.*** *Polyrhachis continua* var. *hirsutula* Emery, 1911c: 256 (w.) NEW GUINEA. Combination in *P.* (*Myrma*): Emery, 1925b: 200.

***hirta.*** *Polyrhachis hirta* Viehmeyer, 1914c: 59 (w.) NEW GUINEA. Kohout, 1990: 503 (q.). Combination in *P.* (*Myrmhopla*): Emery, 1925b: 191.

*hirta. Polyrhachis* (*Myrmhopla*) *hirta* Donisthorpe, 1949g: 418 (w.) NEW GUINEA. [Junior primary homonym of *hirta* Viehmeyer, above.] Replacement name: *kellyi* Hung, 1967a: 201.

***hodgsoni.*** *Polyrhachis hodgsoni* Forel, 1902d: 289 (w.) BURMA. Wheeler, G.C. & Wheeler, J. 1990b: 760 (l.). Combination in *P.* (*Myrmhopla*): Emery, 1925b: 195. See also: Bingahm, 1903: 390; Dorow & Maschwitz, 1990: 73.

***hookeri.*** *Polyrhachis hookeri* Lowne, 1865b: 334 (w.) AUSTRALIA. Wheeler, G.C. & Wheeler, J. 1953e: 206 (l.); Wheeler, G.C. & Wheeler, J. 1968: 220 (l.). Combination in *P.* (*Chariomyrma*): Forel, 1915b: 109. Senior synonym of *aerea:* Kohout & Taylor, 1990: 515; of *cataulacoidea:* Kohout, 1994a: 136.

***horacei.*** *Polyrhachis horacei* Hung, 1967a: 201. Replacement name for *cheesmanae* Donisthorpe, 1943d: 459. [Junior primary homonym of *cheesmanae* Donisthorpe, 1937c: 273.]

***horni.*** *Polyrhachis horni* Emery, 1901f: 122, fig. (w.q.) SRI LANKA. Forel, 1908a: 8 (m.). Combination in *P.* (*Myrma*): Emery, 1925b: 201.

**#*hortensis.*** *Polyrhachis hippomanes* var. *hortensis* Forel, 1913k: 138 (w.q.m.) JAVA. Combination in *P.* (*Myrmhopla*): Emery, 1925b: 195.

*hortulana. Polyrhachis* (*Myrma*) *hortulana* Arnold, 1955: 735, fig. 3 (w.) UGANDA. Junior synonym of *laboriosa:* Bolton, 1973b: 308.

*hosei. Polyrhachis* (*Myrma*) *hosei* Donisthorpe, 1942e: 708 (w.) BORNEO. Junior synonym of *inermis:* Bolton, 1974: 176.

***hostilis.*** *Polyrhachis hostilis* Smith, F. 1859a: 139 (w.) INDONESIA (Aru I.). Combination in *P.* (*Chariomyrma*): Emery, 1925b: 187. Senior synonym of *hebes* and *intricata:* Bolton, 1974b: 175. Current subspecies: nominal plus *circumflexa.*

*humerosa. Polyrhachis* (*Hedomyrma*) *humerosa* Emery, 1921e: 18 (w.) AUSTRALIA. Junior synonym of *ornata:* Kohout, 1988c: 435.

***hungi.*** *Polyrhachis hungi* Bolton, 1974b: 175. Replacement name for *nitens* Donisthorpe, 1944c: 65. [Junior primary homonym of *nitens* Donisthorpe, 1943d: 464.]

*hystrix. Myrma hystrix* Billberg, 1820: 104. *Nomen nudum.*

***ignota.*** *Polyrhachis ignota* Kohout, 1987: 171, figs. 7, 10, 13 (w.) PHILIPPINES.

***illaudata.*** *Polyrhachis illaudatus* Walker, 1859: 373 (w., not m.) SRI LANKA. Imai, Brown, *et al.* 1984: 68 (k.). Combination in *P.* (*Myrma*): Donisthorpe, 1932b: 576. Senior synonym of *mayri:* Donisthorpe, 1932b: 576; of *duodentata* and *latispinosa:* Bolton, 1974b: 176. Current subspecies: nominal plus *intermedia, obesior, pauperata, proximomayri.*

*imatongica. Polyrhachis* (*Myrma*) *cubaensis* subsp. *imatongica* Weber, 1943c: 388, pl. 16, fig. 22 (w.) SUDAN. Junior synonym of *viscosa:* Bolton, 1973b: 330.

**#*imbellis.*** *Polyrhachis imbellis* Emery, 1887a: 225 (w.) JAVA. Combination in *P.* (*Myrma*): Emery, 1925b: 204. Junior synonym of *zopyra:* Emery, 1898a: 228. Revived from synonymy as subspecies of *zopyra:* Emery, 1925b: 204.

***inclusa.*** *Polyrhachis inclusa* Viehmeyer, 1912: 13, fig. 18 (w.) NEW GUINEA. Combination in *P.* (*Chariomyrma*): Emery, 1925b: 188.

***inconspicua.*** *Polyrhachis inconspicua* Emery, 1887a: 225 (w.) AUSTRALIA. Combination in *P.* (*Campomyrma*): Emery, 1925b: 179. Senior synonym of *thalia:* Taylor, 1989a: 24.

**#*inconstans.*** *Polyrhachis* (*Myrmothrinax*) *thrinax* subsp. *inconstans* Viehmeyer, 1916a: 164 (w.) SINGAPORE.

*indefinita. Polyrhachis gagates* r. *indefinita* Forel, 1913b: 349 (w.) ZAIRE. Combination in *P.* (*Myrma*): Santschi, 1914b: 139. Junior synonym of *congolensis:* Santschi, 1924b: 224.

*indica. Polyrhachis indica* Mayr, 1870b: 945 (w.) INDIA. Junior synonym of *clypeata:* Dalla Torre, 1893: 260; Forel, 1893c: 58.

***indificans.*** *Formica indificans* Jerdon, 1851: 125 (w.q.) INDIA. Combination in *Polyrhachis:* Smith, F. 1858b: 59; in *P.* (*Myrmhopla*): Emery, 1925b: 197. [Spelling emended to *nidificans* by Jerdon, 1854: 108, and subsequent authors; unjustified emendation.]

*indigens. Polyrhachis* (*Myrma*) *schlueteri* var. *indigens* Forel, 1914d: 261 (w.) SOUTH AFRICA. Junior synonym of *schlueteri:* Arnold, 1924: 747.

*indocilis. Polyrhachis* (*Hedomyrma*) *indocilis* Santschi, 1928h: 139. Replacement name for *bicolor* Karavaiev, 1927e: 21. [Junior primary homonym of *bicolor* Smith, F. 1858b: 65.] Junior synonym of *fervens:* Kohout, 1988c: 434.

***inermis.*** *Polyrhachis inermis* Smith, F. 1858b: 68, pl. 4, figs. 25, 26 (w.) SULAWESI. Combination in *P.* (*Myrma*): Emery, 1925b: 204. Senior synonym of *bryanti* and *hosei:* Bolton, 1974b: 176.

***insularis.*** *Polyrhachis inconspicua* var. *insularis* Emery, 1887a: 226 (w.) NEW GUINEA. Combination in *P.* (*Campomyrma*): Emery, 1925b: 179. Raised to species, and senior synonym of *subnitens:* Taylor, 1989a: 25.

**#*intermedia.*** *Polyrhachis mayri* subsp. *intermedia* Forel, 1886d: 242 (w.) INDIA. Combination in *P.* (*Myrma*): Emery, 1925b: 201. Raised to species: Dalla Torre, 1893: 264; Bingham, 1903: 405. Subspecies of *mayri:* Emery, 1925b: 201; of *illaudata:* current status.

*intricata. Polyrhachis* (*Chariomyrma*) *hostilis* var. *intricata* Forel, 1918a: 727. Replacement name for *hirsuta* Viehmeyer, 1914c: 58. [Junior primary homonym of *hirsuta* Mayr, 1876: 75.] Junior synonym of *hostilis:* Bolton, 1974b: 175.

***inusitata.*** *Polyrhachis inusitata* Kohout, 1989: 513, figs. 4, 5, 9 (w.) AUSTRALIA.

***io.*** *Polyrhachis* (*Campomyrma*) *thalia* var. *io* Forel, 1915b: 114 (w.) AUSTRALIA. Raised to species: Taylor, 1989a: 25.

*iperpunctata. Polyrhachis* (*Myrma*) *iperpunctata* Menozzi, 1942: 181, fig. 4 (w.) FERNANDO PO I. [Misspelled as *iperstriata* by Menozzi, 1944: 424.] Junior synonym of *latispina:* Bolton, 1973b: 310.

*irritabilis. Polyrhachis irritabilis* Smith, F. 1859a: 141 (q.) INDONESIA (Aru I.). Junior synonym of *sexspinosa:* Roger, 1863b: 6.

**#*isabellae.*** *Polyrhachis convexa* var. *isabellae* Forel, 1908a: 9 (w.q.) SRI LANKA. Combination in *P.* (*Myrma*): Emery, 1925b: 204.

***isacantha.*** *Polyrhachis isacantha* Emery, 1887a: 232, pl. 4, figs. 22, 23 (w.q.) INDONESIA (Goram I.). Combination in *P.* (*Myrma*): Emery, 1925b: 201.

***ithona.*** *Polyrhachis ithonus* Smith, F. 1860b: 99, pl. 1, fig. 18 (w.q.) INDONESIA (Batjan I.). Combination in *P.* (*Myrma*): Emery, 1925b: 202. Variety/subspecies of *relucens:* Dalla Torre, 1893: 268; Emery, 1893e: 228; Emery, 1896d: 379; Emery, 1897d: 579; Stitz, 1911a: 375; Donisthorpe, 1932c: 460. Revived status as species, and senior synonym of *chaonia:* Bolton, 1974b: 177.

***jacksoniana.*** *Polyrhachis jacksoniana* Roger, 1863a: 158 (w.) AUSTRALIA. Combination in *P.* (*Campomyrma*), and subspecies of *hexacantha:* Emery, 1925b: 179. Revived status as species: Taylor & Brown, D.R. 1985: 137; Taylor, 1989a: 25.

***jacobsoni.*** *Polyrhachis jacobsoni* Forel, 1909d: 230 (w.q.m.) JAVA. Combination in *P.* (*Myrmatopa*): Emery, 1925b: 181.

**#*janthinogaster.*** *Polyrhachis mucronata* var. *janthinogaster* Emery, 1911b: 538 (w.q.) NEW GUINEA. Combination in *P.* (*Myrmhopla*): Emery, 1925b: 196.

**#*japensis.*** *Polyrhachis* (*Myrmhopla*) *mucronata* subsp. *japensis* Donisthorpe, 1941h: 63 (w.m.) NEW GUINEA. Donisthorpe, 1943d: 467 (q.).

*javana;* Emery, 1887 and Forel, 1893; see under **#*javanica.***

**#*javana.*** *Polyrhachis rastellata* var. *javana* Crawley, 1923a: 31 (w.) JAVA. [First available use of *Polyrhachis rastellata* subsp. *euryala* var. *javana* Viehmeyer, 1914c: 51; unavailable name.] Crawley, 1924: 409 (q.). Combination in *P.* (*Cyrtomyrma*): Emery, 1925b: 208. Variety of *euryala:* Donisthorpe, 1938c: 260.

*javana. Polyrhachis sculpturata* r. *javana* Stitz, 1925: 134 (w.) JAVA. [Junior primary homonym of *javana* Crawley, above.] Replacement name: *javaniana* Santschi, 1928h: 140.]

*javana. Polyrhachis* (*Myrma*) *orsyllus* var. *javana* Karavaiev, 1927e: 49 (w.) JAVA. [Junior primary homonym of *javana* Crawley, 1923a: 31.] Replacement name: *javanensis* Santschi, 1928h: 140.

**#*javanensis.*** *Polyrhachis* (*Myrma*) *orsyllus* var. *javanensis* Santschi, 1928h: 140. Replacement name for *javana* Karavaiev, 1927e: 49. [Junior primary homonym of *javana* Crawley, 1923a: 31.]

**#*javaniana.*** *Polyrhachis* (*Myrma*) *sculpturata* var. *javaniana* Santschi, 1928h: 140. Replacement name for *javana* Stitz, 1925: 134. [Junior primary homonym of *javana* Crawley, 1923a: 31.]

**#javanica.** *Polyrhachis thrinax* var. *javanica* Mayr, 1867a: 52 (w.) JAVA. Emery, 1887a: 222 (q.); Karavaiev, 1927e: 13 (m.). Combination in *P.* (*Myrmothrinax*): Wheeler, W.M. 1919e: 127. [Misspelled as *javana* by Emery, 1887a: 222 and Forel, 1893c: 19 (in key).] Subspecies of *thrinax:* Emery, 1896d: 378; Wheeler, W.M. 1909d: 343.

**jerdonii.** *Polyrhachis jerdonii* Forel, 1892m: 17 (footnote) (w.) SRI LANKA. Forel, 1893c: 17 (q.). Combination in *P.* (*Myrmhopla*): Emery, 1925b: 191. [Also described as new by Forel, 1893c: 17.]

**jianghuaensis.** *Polyrhachis jianghuaensis* Wang & Wu, 1991: 597, figs. 1, 4, 5, 9 (w.q.) CHINA.

**#johnsoni.** *Polyrhachis* (*Cyrtomyrma*) *rastellata* var. *johnsoni* Mann, 1919: 390 (w.) SOLOMON IS. Variety of *debilis:* Donisthorpe, 1938c: 266.

**jurii.** *Polyrhachis* (*Cyrtomyrma*) *jurii* Karavaiev, 1935a: 116, fig. 30 (w.) CAMBODIA.

*juxtaspinosa. Polyrhachis* (*Myrmhopla*) *juxtaspinosa* Donisthorpe, 1949g: 417, fig. 2 (w.) NEW GUINEA. Junior synonym of *sexspinosa:* Bolton, 1975c: 12.

**kaipi.** *Polyrhachis* (*Chariomyrma*) *kaipi* Mann, 1919: 382, fig. 52 (not 62 as given in text) (w.) SOLOMON IS.

**karawaiewi.** *Polyrhachis* (*Chariomyrma*) *karawaiewi* Santschi, 1928h: 139. Replacement name for *convexa* Karavaiev, 1927e: 17. [Junior primary homonym of *convexa* Roger, 1863a: 153.]

*kellyi. Polyrhachis kellyi* Hung, 1967a: 201. Replacement name for *hirta* Donisthorpe, 1949g: 418. [Junior primary homonym of *hirta* Viehmeyer, 1914c: 59.] Junior synonym of *bubastes:* Bolton, 1975c: 6.

**keratifera.** *Polyrhachis* (*Myrmhopla*) *keratifera* Karavaiev, 1927e: 39, fig. 17 (w.q.) INDONESIA (Ambon I.).

**#kerri.** *Polyrhachis acantha* var. *kerri* Forel, 1911g: 286 (w.) THAILAND. Combination in *P.* (*Myrmhopla*): Emery, 1925b: 194. Subspecies of *saevissima:* current status.

*kershawi. Polyrhachis* (*Hedomyrma*) *kershawi* Clark, 1930c: 12, fig. 1 (w.) AUSTRALIA. Junior synonym of *fervens:* Kohout, 1988a: 434.

**khepra.** *Polyrhachis khepra* Bolton, 1973b: 334, figs. 51, 57 (w.) GHANA.

*kirkae. Polyrhachis* (*Florencea*) *kirkae* Donisthorpe, 1937a: 624, figs. 5, 6 (w.) NEW GUINEA. Junior synonym of *nigriceps:* Hung, 1971: 44.

*kohli. Polyrhachis* (*Cyrtomyrma*) *kohli* Forel, 1916: 454 (w.q.) ZAIRE. Combination in *P.* (*Pseudocyrtomyrma*): Emery, 1925b: 206. Junior synonym of *volkarti:* Bolton, 1973b: 341.

**labella.** *Polyrhachis labella* Smith, F. 1860b: 101, pl. 1, fig. 25 (w.) INDONESIA (Batjan I.). Combination in *P.* (*Myrma*): Emery, 1925b: 201. Current subspecies: nominal plus *brunneipes, obliqua.*

**laboriosa.** *Polyrhachis laboriosus* Smith, F. 1858b: 72, pl. 4, figs. 21, 22 (w.) SIERRA LEONE. Mayr, 1896: 250 (q.m.); Wheeler, G.C. & Wheeler, J. 1953e: 207 (l.). Combination in *P.* (*Myrma*): Santschi, 1914d: 384. Senior synonym of *architecta* and *hortulana:* Bolton, 1973b: 308.

**lachesis.** *Polyrhachis lachesis* Forel, in Emery, 1897d: 582 (footnote) (w.) AUSTRALIA. Combination in *P.* (*Hagiomyrma*): Emery, 1925b: 185.

**laciniata.** *Polyrhachis laciniata* Emery, 1900c: 335, pl. 8, fig. 45 (w.q.) NEW GUINEA. Combination in *P.* (*Chariomyrma*): Emery, 1925b: 187.

**lacteipennis.** *Polyrhachis lacteipennis* Smith, F. 1858b: 60, pl. 4, fig. 40 (q.) INDIA. Combination in *P.* (*Myrmhopla*): Emery, 1925b: 195. Senior synonym of *simplex* (and its junior synonym *spiniger*): Bolton, 1974b: 177. Current subspecies: nominal plus *grisescens, obsoleta.*

*laeta. Polyrhachis* (*Myrma*) *lauta* var. *laeta* Emery, 1921e: 22 (q.) CAMEROUN. Junior synonym of *lauta:* Bolton, 1973b: 311.

**laevigata.** *Polyrhachis laevigatus* Smith, F. 1857a: 62 (w.) WEST MALAYSIA. Combination in *P.* (*Myrmhopla*): Emery, 1925b: 195.

**laevissima.** *Polyrhachis laevissimus* Smith, F. 1858b: 64, pl. 4, fig. 42 (w.) BURMA. Mayr, 1879: 651 (q.); Bingham, 1903: 402 (m.). Combination in *P.* (*Cyrtomyrma*): Wheeler, W.M. 1919e: 137. Senior synonym of *globularia:* Mayr, 1879: 651. Current subspecies: nominal plus *aruensis, dichroa.*

*laevissimus. Polyrhachis laevissimus* Smith, F. 1859a: 141 (w.) INDONESIA (Aru I.). [Junior primary homonym of *laevissima* Smith, above.] Replacement name: *levior* Roger, 1863b: 8.

**lama.** *Polyrhachis lama* Kohout, 1994b: 137, fig. 1 (w.) TIBET.

**lamellidens.** *Polyrhachis lamellidens* Smith, F. 1874: 403 (w.) JAPAN. Donisthorpe, 1937a: 627 (q.); Koriba, 1963: 200 (l.); Wheeler, G.C. & Wheeler, J. 1970: 649 (l.); Hung, 1970: 29 (m.).

**laminata.** *Polyrhachis laminata* Mayr, 1867a: 65, pl.2, fig. 6 (q.) INDONESIA. Combination in *P.* (*Myrmhopla*): Emery, 1925b: 197.

**#lancearia.** *Polyrhachis thrinax* var. *lancearius* Forel, 1893c: 19 (diagnosis in key) (w.q.) INDIA. Combination in *P.* (*Myrmothrinax*): Emery, 1925b: 184.

**lanuginosa.** *Polyrhachis lanuginosa* Santschi, 1910c: 394, fig. 17 (w.) CONGO. Combination in *P.* (*Pseudocyrtomyrma*): Emery, 1921e: 24; in *P.* (*Myrma*): Wheeler, W.M. 1922a: 997. Senior synonym of *conradti* and *felici:* Bolton, 1973b: 335.

**lata.** *Polyrhachis guerini* subsp. *lata* Emery, 1895g: 357 (w.) AUSTRALIA. Combination in *P.* (*Chariomyrma*) Forel, 1915b: 110. Subspecies/race of *aurea:* Emery, 1897d: 584 (in key); Forel, 1915b: 110; of *guerini:* Taylor & Brown, D.R. 1985: 135. Raised to species, and senior synonym of *aegra:* Kohout & Taylor, 1990: 515.

**latharis.** *Polyrhachis latharis* Bolton, 1973b: 348, figs. 49, 60 (w.q.) GHANA.

*latifrons. Polyrhachis latifrons* Roger, 1863a: 155 (w.) INDONESIA (Bintang I.). Senior synonym of *modiglianii:* Emery, 1895g: 357. Junior synonym of *arcuata:* Emery, 1897d: 583.

**latinota.** *Polyrhachis hostilis* var. *latinota* Viehmeyer, 1912: 11, fig. 14 (w.) NEW GUINEA. Combination in *P.* (*Chariomyrma*): Emery, 1925b: 188. Raised to species: Viehmeyer, 1913: 54.

**latispina.** *Polyrhachis* (*Myrma*) *latispina* Emery, 1925b: 206. Replacement name for *atalanta* Wheeler, W.M. 1922a: 263. [Junior primary homonym of *atalanta* Emery, 1898b: 243.] Senior synonym of *iperpunctata:* Bolton, 1973b: 309.

*latispinosa. Polyrhachis* (*Myrma*) *latispinosa* Donisthorpe, 1942d: 460, pl. 2, fig. 1 (q.) INDIA. Junior synonym of *illaudata:* Bolton, 1974b: 176.

**latona.** *Polyrhachis latona* Wheeler, W.M. 1909d: 337 (w.) TAIWAN. Forel, 1912a: 79 (m.). Combination in *P.* (*Myrma*): Emery, 1925b: 201.

**latreillei.** *Formica latreillei* Guérin-Méneville, 1838: 205 (w.) AUSTRALIA. Combination in *Polyrhachis:* Smith, F. 1858b: 73; in *P.* (*Chariomyrma*): Emery, 1925b: 187.

**#laurae.** *Polyrhachis* (*Myrmatopa*) *schang* var. *laurae* Menozzi, 1926e: 100 (w.m.) SINGAPORE.

**lauta.** *Polyrhachis lauta* Santschi, 1910c: 397, fig. 19 (q.) CONGO. Bolton, 1973b: 311 (w.). Combination in *P.* (*Myrma*): Wheeler, W.M. 1922a: 997. Senior synonym of *laeta* and *localis:* Bolton, 1973b: 311.

**leae.** *Polyrhachis leae* Forel, 1913g: 193 (w.) TASMANIA. Combination in *P.* (*Campomyrma*): Forel, 1915b: 114. See also: Taylor, 1989a: 27.

**leonidas.** *Polyrhachis leonidas* Forel, 1901b: 34 (w.q.) BISMARCK ARCHIPELAGO. Combination in *P.* (*Cyrtomyrma*): Emery, 1925b: 208.

**leopoldi.** *Polyrhachis* (*Chariomyrma*) *leopoldi* Santschi, 1932b: 25, fig. 9 (w.) INDONESIA (Ternate I.).

**lestoni.** *Polyrhachis lestoni* Bolton, 1973b: 349, figs. 48, 61 (w.) GHANA.

**levior.** *Polyrhachis levior* Roger, 1863b: 8. Replacement name for *laevissimus* Smith, F. 1859a: 141. [Junior primary homonym of *laevissima* Smith, 1858b: 64.] Combination in *P.* (*Cyrtomyrma*): Forel, 1915b: 110. Race of *rastellata:* Forel, 1902h: 527. Revived status as species: Donisthorpe, 1938c: 248; Taylor & Brown, D.R. 1985: 137.

**#leviuscula.** *Polyrhachis* (*Myrmatopa*) *schang* var. *leviuscula* Viehmeyer, 1916a: 164 (w.) SINGAPORE.

**lilianae.** *Polyrhachis lilianae* Forel, 1911b: 213 (w.) JAVA. Forel, 1912n: 71 (q.). Combination in *P.* (*Myrmatopa*): Emery, 1925b: 181.

**limbata.** *Polyrhachis limbata* Emery, 1897d: 585, pl. 1, fig. 24 (w.) NEW GUINEA. Combination in *P.* (*Chariomyrma*): Emery, 1925b: 188. Senior synonym of *mentor:* Viehmeyer, 1914c: 54.

**limitis.** *Polyrhachis* (*Pseudocyrtomyrma*) *alexisi* st. *limitis* Santschi, 1939c: 12 (w.) ZAIRE. Raised to species: Bolton, 1973b: 350.

**linae.** *Polyrhachis* (*Cyrtomyrma*) *linae* Donisthorpe, 1938c: 262, fig. 12 (w.q.) NEW GUINEA.

**#litigiosa.** *Polyrhachis relucens* subsp. *litigiosa* Emery, 1897d: 581 (w.) NEW GUINEA. Combination in *P.* (*Myrma*): Mann, 1919: 380.

*localis. Polyrhachis lauta* var. *localis* Forel, 1913h: 359 (q.) ZAIRE. Junior synonym of *lauta:* Bolton, 1973b: 311.

**lombokensis.** *Polyrhachis lombokensis* Emery, 1898b: 239 (w.q.) INDONESIA (Lombok I.). Combination in *P.* (*Myrmatopa*): Forel, 1915b: 115.

**longipes.** *Polyrhachis longipes* Smith, F. 1859a: 140 (w.) INDONESIA (Aru I.). Combination in *P.* (*Myrmhopla*): Emery, 1925b: 195.

*longipes. Polyrhachis* (*Campomyrma*) *longipes* Wheeler, W.M. 1915g: 821, pl. 65, figs. 1, 2; pl. 66, fig. 10 (w.) AUSTRALIA. [Junior primary homonym of *longipes* Smith, above.] Replacement name: *macropa* Wheeler, W.M. 1916e: 40.

**#*longispina*.** *Polyrhachis schlaginhaufeni* var. *longispina* Viehmeyer, 1912: 12 (w.) NEW GUINEA. [Misspelled as *longispinosa* by Emery, 1925b: 187.] Subspecies of *aurita:* current status.

***loriai*.** *Polyrhachis loriai* Emery, 1897d: 585, pl. 1, fig. 25 (w.) NEW GUINEA. Combination in *P.* (*Chariomyrma*): Emery, 1925b: 188.

***loweryi*.** *Polyrhachis loweryi* Kohout, 1990: 505 (w.) AUSTRALIA.

***lownei*.** *Polyrhachis hookeri* r. *lownei* Forel, 1895a: 44 (w.) AUSTRALIA. Combination in *P.* (*Chariomyrma*): Forel, 1915b: 109. Raised to species: Kohout & Taylor, 1990: 515.

***lucens*.** *Polyrhachis* (*Myrmhopla*) *lucens* Donisthorpe, 1947d: 194 (q.) NEW GUINEA.

**#*lucida*.** *Polyrhachis thrinax* subsp. *lucida* Emery, 1894b: 74. Replacement name for *lucidula* Emery, 1893g: 269 (footnote). [Junior primary homonym of *lucidula* Emery, 1893f: 254 (footnote).]

**#*lucidiventris*.** *Polyrhachis clypeata* var. *lucidiventris* Forel, 1907a: 39 (w.) INDIA. Combination in *P.* (*Campomyrma*): Emery, 1925b: 179. Subspecies of *exercita:* current status.

**#*lucidula*.** *Polyrhachis hippomanes* subsp. *lucidula* Emery, 1893f: 254 (footnote) (w.) BURMA. Viehmeyer, 1916a: 167 (q.m.). Combination in *P.* (*Myrmhopla*): Viehmeyer, 1916a: 166.

*lucidula*. *Polyrhachis thrinax* subsp. *lucidula* Emery, 1893g: 269 (footnote) (q.) WEST MALAYSIA. [Junior primary homonym of *lucidula* Emery, above.] Replacement name: *lucida* Emery, 1894b: 74.]

***luctuosa*.** *Polyrhachis* (*Cyrtomyrma*) *luctuosa* Emery, 1921e: 25 (w.) NEW GUINEA.

***lugens*.** *Polyrhachis lugens* Mayr, 1867a: 63, pl. 2, fig. 5 (q.) BORNEO. Combination in *P.* (*Myrmhopla*): Wheeler, W.M. 1919e: 130.

***lycidas*.** *Polyrhachis lycidas* Smith, F. 1861b: 43, pl. 1, fig. 23 (w.) SULAWESI. Mayr, 1867a: 58 (q.). Combination in *P.* (*Myrma*): Wheeler, W.M. 1919e: 123.

***lydiae*.** *Polyrhachis schencki* r. *lydiae* Forel, 1902h: 523 (w.q.) AUSTRALIA. Combination in *P.* (*Hagiomyrma*): Emery, 1925b: 185. Raised to species: Kohout, 1988c: 434.

*lyrifera*. *Polyrhachis* (*Pseudocyrtomyrma*) *lyrifera* Stitz, 1933: 73, fig. 5 (w.) CAMEROUN. Junior synonym of *curta:* Bolton, 1973b: 346.

*lysistrata*. *Polyrhachis* (*Myrmothrinax*) *lysistrata* Santschi, 1920a: 569 (w.) AUSTRALIA. Junior synonym of *delicata:* Kohout, 1994a: 135.

***machaon*.** *Polyrhachis* (*Hedomyrma*) *machaon* Santschi, 1920a: 568 (w.) AUSTRALIA.

***mackayi*.** *Polyrhachis* (*Cyrtomyrma*) *mackayi* Donisthorpe, 1938c: 258, fig. 9 (w.q.) AUSTRALIA.

***macropa*.** *Polyrhachis* (*Campomyrma*) *macropus* Wheeler, W.M. 1916e: 40. Replacement name for *longipes* Wheeler, W.M. 1915g: 821. [Junior primary homonym of *longipes* Smith, F. 1859a: 140.] [*P. macropus* also given as replacement name by Wheeler, W.M. 1916i: 37.]

***maculata*.** *Polyrhachis* (*Campomyrma*) *polymnia* var. *maculata* Forel, 1915b: 115 (w.q.) AUSTRALIA. Raised to species: Taylor, 1989a: 25.

*maeandrifera*. *Polyrhachis lachesis* subsp. *maeandrifera* Emery, 1897d: 582 (w.) NEW GUINEA. Junior synonym of *paxilla:* Kohout, 1988c: 435.

***magnifica*.** *Polyrhachis sexspinosa* subsp. *magnifica* Menozzi, 1926e: 98 (w.) PHILIPPINES. Raised to species: Bolton, 1975c: 9.

*major*. *Polyrhachis rastellata* var. *major* Stitz, 1911a: 381 (w.) BISMARCK ARCHIPELAGO. Junior synonym of *ralumensis:* Viehmeyer, 1914c: 50.

*malaccana*. *Polyrhachis* (*Myrmhopla*) *sexspinosa* var. *malaccana* Viehmeyer, 1916a: 167, fig. 15 (w.) WEST MALAYSIA. Junior synonym of *calypso:* Bolton, 1975c: 7.

***malaensis*.** *Polyrhachis* (*Dolichorhachis*) *malaensis* Mann, 1919: 386, fig. 57 (w.q.) SOLOMON IS.

*malignus*. *Polyrhachis malignus* Smith, F. 1858b: 70, pl. 4, fig. 44 (q.) PHILIPPINES. Combination in *P.* (*Myrmhopla*): Emery, 1925b: 195. Junior synonym of *hector:* Bolton, 1974b: 174.

***marginata*.** *Polyrhachis marginatus* Smith, F. 1859a: 139 (w.) INDONESIA (Aru I.). Combination in *P.* (*Chariomyrma*): Emery, 1925b: 188.

*maynei*. *Polyrhachis maynei* Forel, 1911f: 282 (w.) ZAIRE. Combination in *P.* (*Myrma*): Wheeler, W.M. 1922a: 997. Junior synonym of *curta:* Bolton, 1973b: 346.

*mayri*. *Polyrhachis mayri* Roger, 1863b: 7 (w.) SRI LANKA. Bingham, 1903: 405 (q.); Karavaiev, 1927e: 46 (m.). Combination in *P.* (*Myrma*): Viehmeyer, 1916a: 165. Junior synonym of *illaudata:* Donisthorpe, 1932b: 576.

*mayumbensis*. *Polyrhachis rufipalpis* r. *mayumbensis* Forel, 1913h: 358 (w.) ZAIRE. Junior synonym of *rufipalpis:* Bolton, 1973b: 317.

*mediopilosa*. *Polyrhachis* (*Myrma*) *schistacea* st. *atrociliata* var. *mediopilosa* Santschi, 1923e: 295 (w.) ZAIRE. Unavailable name; material referred to *schistacea* by Bolton, 1973b: 318.

***medusa*.** *Polyrhachis schistacea* r. *medusa* Forel, 1897c: 206 (w.q.m.) TANZANIA. Combination in *P.* (*Myrma*): Santschi, 1914b: 140. Subspecies/race of *schistacea:* Forel, 1901h: 78; Wheeler, W.M. 1922a: 1002. Raised to species: Forel, 1907g: 92; Santschi, 1914b: 140; Emery, 1925b: 199. See also: Bolton, 1973b: 312.

***melpomene*.** *Polyrhachis melpomene* Emery, 1897d: 592 (w.q.) NEW GUINEA. Combination in *P.* (*Myrmhopla*): Emery, 1925b: 195.

***menelas*.** *Polyrhachis menelas* Forel, 1904c: 30 (w.) INDIA. Combination in *P.* (*Myrmhopla*): Emery, 1925b: 195.

***menozzii*.** *Polyrhachis* (*Myrmatopa*) *menozzii* Karavaiev, 1927e: 9, fig. 2 (w.q.) INDONESIA (Aru I.).

*mentor*. *Polyrhachis mentor* Forel, 1901b: 31 (w.) BISMARCK ARCHIPELAGO. Junior synonym of *limbata:* Viehmeyer, 1914c: 54.

*merops*. *Polyrhachis merops* Smith, F. 1860b: 98, pl. 1, fig. 17 (w.) INDONESIA (Batjan I.). Combination in *P.* (*Myrma*): Emery, 1925b: 202. Variety of *rufofemorata:* Emery, 1898a: 228; Donisthorpe, 1932c: 460. Junior synonym of *rufofemorata:* Bolton, 1974b: 178.

***metella*.** *Polyrhachis metella* Smith, F. 1860b: 99, pl. 1, figs. 20, 21 (w.) NEW GUINEA. Combination in *P.* (*Hagiomyrma*): Emery, 1925b: 185. See also: Viehmeyer, 1912: 10.

***micans*.** *Polyrhachis micans* Mayr, 1876: 76 (w.q.) AUSTRALIA. Combination in *P.* (*Campomyrma*): Santschi, 1920g: 185.

***militaris*.** *Formica militaris* Fabricius, 1782: 493 (q.) "Tropical Africa". Smith, F. 1858b: 72 (m.); Mayr, 1866b: 886 (w.). Combination in *Myrma:* Billberg, 1820: 104; in *Polyrhachis:* Smith, F. 1858b: 72; in *P.* (*Myrma*): Santschi, 1914b: 140. Senior synonym of *bruta, calabarica, cupreopubescens, ssibangensis* and *striativentris,* and material of the unavailable names *argentatus, dido, epinotalis, nkomoensis, pleurata, sankisiana* and *transversaria* referred here: Bolton, 1973b: 313.

***mindanaensis*.** *Polyrhachis* (*Polyrhachis*) *ypsilon* var. *mindanaensis* Emery, 1923: 62 (w.) PHILIPPINES. Raised to species: Hung, 1970: 20.

*minor*. *Polyrhachis armata* var. *minor* Forel, 1886d: 241 (w.) INDIA. Junior synonym of *defensa:* Bingham, 1903: 393.

*minor*. *Polyrhachis bihamata* var. *minor* Karavaiev, 1927e: 12 (w.) JAVA. [Unresolved junior primary homonym of *minor* Forel, above.] Junior synonym of *bihamata:* Hung, 1970: 16.

***mitrata*.** *Polyrhachis* (*Myrmhopla*) *mitrata* Menozzi, 1932f: 303 (w.q.) PHILIPPINES.

***mjobergi*.** *Polyrhachis* (*Hedomyrma*) *mjobergi* Forel, 1915b: 112 (w.) AUSTRALIA. Senior synonym of *anguliceps:* Kohout, 1988c: 435.

***modesta*.** *Polyrhachis modestus* Smith, F. 1857a: 62 (q.) SINGAPORE. Combination in *P.* (*Chariomyrma*): Emery, 1925b: 187.

*modiglianii*. *Polyrhachis modiglianii* Emery, 1888a: 529, pl. 9, fig. 1 (w.) SUMATRA. Junior synonym of *latifrons:* Emery, 1895g: 357.

***moeschi*.** *Polyrhachis moeschi* Forel, 1912n: 73 (q.) SUMATRA. Crawley, 1923a: 31 (w.). Combination in *P.* (*Myrmhopla*): Emery, 1925b: 196.

*moeschiella*. *Polyrhachis* (*Myrmhopla*) *gestroi* var. *moeschiella* Forel, 1918a: 726. Unnecessary replacement name for, and hence junior synonym of, *rufiventris* Forel, 1911d: 391; **new synonymy**.

***moesta*.** *Polyrhachis hippomanes* var. *moesta* Emery, 1887a: 237 (w.) SUMATRA. Santschi, 1928b: 39 (q.). Combination in *P.* (*Myrmhopla*): Emery, 1925b: 195. Raised to species: Wang & Wu, 1991: 599.

*monacha*. *Polyrhachis monacha* Karavaiev, 1926a: 144. *Nomen nudum;* material referred to *phyllophilus* by Karavaiev, 1927e: 29.

***mondoi*.** *Polyrhachis* (*Cyrtomyrma*) *mondoi* Donisthorpe, 1938c: 250, fig. 3 (w.) NEW GUINEA.

***monista*.** *Polyrhachis monista* Santschi, 1910c: 398, fig. 20 (q.) CONGO. Santschi, 1914d: 384 (w.). Combination in *P.* (*Myrma*): Santschi, 1914d: 384. See also: Bolton, 1973b: 343.

***montana*.** *Polyrhachis montana* Hung, 1970: 23, figs. 16–22 (w.) BORNEO.

*moorei*. *Polyrhachis* (*Myrma*) *moorei* Donisthorpe, 1941j: 208 (w.) PHILIPPINES. Junior synonym of *zopyra:* Bolton, 1974b: 179.

***mucronata*.** *Polyrhachis mucronatus* Smith, F. 1859a: 140 (w.) INDONESIA (Aru I.). Combination in *P.* (*Myrmhopla*): Emery, 1925b: 196. Current subspecies: nominal plus *bismarckensis, janthinogaster, japensis.*

*mucronis*. *Polyrhachis* (*Myrmothrinax*) *thrinax* var. *mucronis* Donisthorpe, 1942d: 460 (w.) INDIA. Junior synonym of *thrinax:* Brown, 1959d: 164.
***muelleri***. *Polyrhachis muelleri* Forel, 1893c: 32 (w.) SINGAPORE. Forel, 1901d: 302 (q.m.); Wheeler, G.C. & Wheeler, J. 1990b: 754 (l.). Combination in *P.* (*Myrmhopla*): Emery, 1925b: 193.
#***multicella***. *Polyrhachis solmsi* var. *multicella* Forel, 1911b: 214 (w.q.m.) JAVA. Combination in *P.* (*Myrmatopa*): Emery, 1925b: 181.
***murina***. *Polyrhachis murina* Emery, 1893e: 198 (w.) BORNEO. Combination in *P.* (*Myrma*): Wheeler, W.M. 1919e: 123. Current subspecies: nominal plus *selecta*.
#***muscula***. *Polyrhachis orsyllus* r. *musculus* Forel, 1901b: 29 (w.) JAVA. Combination in *P.* (*Myrma*): Emery, 1925b: 204.
*mutatus*. *Polyrhachis mutatus* Smith, F. 1858b: 64, pl. 4, figs. 12, 13 (w.) BORNEO. Combination in *P.* (*Myrmhopla*): Emery, 1925b: 193. Senior synonym of *ajax:* Bingham, 1903: 399. Junior synonym of *hector:* Bolton, 1974b: 174.
*mutiliae*. *Polyrhachis mutiliae* Smith, F. 1861b: 39, pl. 1, figs. 15, 20 (w.) SULAWESI. Combination in *P.* (*Myrmhopla*): Emery, 1925b: 197. Junior synonym of *dives:* Bolton, 1974b: 173.
***mystica***. *Polyrhachis* (*Aulacomyrma*) *mystica* Karavaiev, 1927e: 41, fig. 19 (w.q.m.) JAVA.
*natalensis*. *Polyrhachis natalensis* Santschi, 1914e: 41 (w.) SOUTH AFRICA. Combination in *P.* (*Myrma*): Wheeler, W.M. 1922a: 999; in *P.* (*Pseudocyrtomyrma*): Emery, 1925b: 207. Status as species: Wheeler, W.M. 1922a: 999. Variety of *revoili:* Emery, 1925b: 207; Santschi, 1939c: 12. Junior synonym of *revoili:* Forel, 1916: 453; Arnold, 1924: 754; Bolton, 1973b: 338.
***neptunus***. *Polyrhachis neptunus* Smith, F. 1865: 69, pl. 4, fig. 2 (w.) NEW GUINEA. Combination in *P.* (*Myrmothrinax*): Emery, 1925b: 183.
*nesiotis*. *Polyrhachis* (*Myrma*) *relucens* subsp. *andromache* var. *nesiotis* Mann, 1919: 380, fig. 50 (w.) SOLOMON IS. Unavailable name.
*nidificans* Jerdon; see under ***indificans***.
***nigra***. *Polyrhachis niger* Mayr, 1862: 683 (q.) SRI LANKA. Combination in *P.* (*Myrma*): Emery, 1925b: 201.
***nigrescens***. *Polyrhachis* (*Chariomyrma*) *nigrescens* Karavaiev, 1927e: 16, fig. 6 (w.) INDONESIA (Ambon I.).
***nigriceps***. *Polyrhachis nigriceps* Smith, F. 1863: 17 (w.) NEW GUINEA. Combination in *P.* (*Myrmhopla*): Emery, 1925b: 192. Senior synonym of *atalanta:* Emery, 1925b: 192; of *kirkae:* Hung, 1971: 44.
#***nigricornis***. *Polyrhachis* (*Myrmhopla*) *tibialis* var. *nigricornis* Santschi, 1928h: 136 (w.) SUMATRA.
#***nigripes***. *Polyrhachis bicolor* var. *nigripes* Emery, 1897d: 592 (w.) NEW GUINEA. Combination in *P.* (*Myrmhopla*): Emery, 1925b: 194.
#***nigripes***. *Polyrhachis* (*Myrmothrinax*) *thrinax* var. *nigripes* Viehmeyer, 1916a: 164 (w.) SINGAPORE. [Unresolved junior primary homonym of *nigripes* Emery, above.] Viehmeyer, 1922: 220 (q.).
*nigriseta*. *Polyrhachis nigriseta* Santschi, 1910c: 399 (w.) CONGO. Combination in *P.* (*Myrma*): Santschi, 1914b: 140. Race/subspecies of *schistacea:* Forel, 1913h: 357; Wheeler, W.M. 1922a: 1002. Revived status as species: Emery, 1925b: 199. Junior synonym of *gagates:* Bolton, 1973b: 305.
***nigrita***. *Polyrhachis nigrita* Mayr, 1895: 153 (w.) GHANA. Combination in *P.* (*Myrma*): Wheeler, W.M. 1922a: 267. Senior synonym of *schoutedeni:* Bolton, 1973b: 328.
#***nigronitens***. *Polyrhachis coerulescens* var. *nigronitens* Viehmeyer, 1914c: 58 (w.) NEW GUINEA. Combination in *P.* (*Chariomyrma*): Emery, 1925b: 187.
***nigropilosa***. *Polyrhachis nigropilosa* Mayr, 1872: 141 (w.) BORNEO. Menozzi, 1926e: 99 (q.); Karavaiev, 1927e: 47 (m.). Combination in *P.* (*Myrma*): Wheeler, W.M. 1919e: 124. Current subspecies: nominal plus *conophthalma, polluta*.
***nitens***. *Polyrhachis* (*Chariomyrma*) *nitens* Donisthorpe, 1943d: 464 (w.q.m.) NEW GUINEA.
*nitens*. *Polyrhachis* (*Hedomyrma*) *nitens* Donisthorpe, 1944c: 65 (w.) NEW GUINEA. [Junior primary homonym of *nitens* Donisthorpe, above.] Replacement name: *hungi* Bolton, 1974b: 175.
***nitida***. *Polyrhachis nitidus* Smith, F. 1857a: 61 (q.) BORNEO. Combination in *P.* (*Myrmhopla*): Emery, 1925b: 197.
#***nitidissima***. *Polyrhachis* (*Myrma*) *sericata* var. *nitidissima* Karavaiev, 1927e: 48 (w.) INDONESIA (Aru I.).
#***nitidiventris***. *Polyrhachis sericata* var. *nitidiventris* Stitz, 1911a: 375 (w.) NEW GUINEA. Karavaiev, 1927e: 47 (q.); Donisthorpe, 1943a: 173 (m.). Combination in *P.* (*Myrma*): Emery, 1925b: 203.
*nkomoensis*. *Polyrhachis* (*Myrma*) *militaris* r. *cupreopubescens* var. *nkomoensis* Forel, 1916: 447 (w.q.m.) ZAIRE. Unavailable name; material referred to *militaris* by Bolton, 1973b: 313.
#***noesaensis***. *Polyrhachis villipes* var. *noesaensis* Forel, 1915a: 43 (footnote) (w.) JAVA. Combination in *P.* (*Myrma*): Emery, 1925b: 203.
***nofra***. *Polyrhachis nofra* Bolton, 1975c: 9, fig. 8 (w.) SOLOMON IS.
#***nomo***. *Polyrhachis* (*Cyrtomyrma*) *rastellata* var. *nomo* Donisthorpe, 1941c: 142 (w.) NEW GUINEA.
*nox*. *Polyrhachis* (*Cyrtomyrma*) *nox* Donisthorpe, 1938c: 249, fig. 2 (w.) AUSTRALIA. Junior synonym of *australis:* Kohout & Taylor, 1990: 513.
***nudata***. *Polyrhachis nudatus* Smith, F. 1860a: 71 (w.) SULAWESI. Combination in *P.* (*Myrmhopla*): Emery, 1925b: 197.
***numeria***. *Polyrhachis numeria* Smith, F. 1861b: 42, pl. 1, fig. 25 (w.) SULAWESI. Combination in *P.* (*Myrma*): Emery, 1925b: 201.
#***obesior***. *Polyrhachis* (*Myrma*) *mayri* var. *obesior* Viehmeyer, 1916a: 165 (w.) WEST MALAYSIA. Subspecies of *illaudata:* current status.
#***obliqua***. *Polyrhachis labella* var. *obliqua* Stitz, 1911a: 375, fig. 21 (w.) SOLOMON IS. Combination in *P.* (*Myrma*): Emery, 1925b: 201.
*oblisa*. *Polyrhachis exasperata* var. *oblisa* Forel, 1911d: 395 (w.) WEST MALAYSIA. Combination in *P.* (*Myrmhopla*): Emery, 1925b: 193. Junior synonym of *rufipes:* Bolton, 1974b: 178.
***obscura***. *Polyrhachis hookeri* r. *obscura* Forel, 1895a: 44 (w.) AUSTRALIA. Combination in *P.* (*Chariomyrma*): Emery, 1925b: 186. Raised to species: Kohout & Taylor, 1990: 516.
*obsidiana*. *Polyrhachis* (*Myrma*) *gagates* subsp. *obsidiana* Emery, 1921e: 21 (w.) GABON. Junior synonym of *gagates:* Bolton, 1973b: 305.
*obsidiana*. *Polyrhachis* (*Cyrtomyrma*) *obsidiana* Karavaiev, 1927e: 50, fig. 21 (w.q.m.) INDONESIA (Ambon I.). [Junior primary homonym of *obsidiana* Emery, above.] Replacement name: *coronata* Santschi, 1928h: 140.
#***obsoleta***. *Polyrhachis simplex* var. *obsoleta* Forel, 1893c: 26 (diagnosis in key) (w.) INDIA. Combination in *P.* (*Myrmhopla*): Emery, 1925b: 196. Subspecies of *lacteipennis:* current status.
***obtusa***. *Polyrhachis aurea* var. *obtusa* Emery, 1897d: 589 (w.) NEW GUINEA. Combination in *P.* (*Chariomyrma*): Forel, 1915b: 110. Raised to species: Kohout & Taylor, 1990: 516.
*obtusata* Emery, 1895; see under **CAMPONOTUS**.
#***obtusisquama***. *Polyrhachis clypeata* var. *obtusisquama* Forel, 1902d: 289 (w.) INDIA. Combination in *P.* (*Campomyrma*): Emery, 1925b: 179. Subspecies of *exercita:* current status.
***ochracea***. *Polyrhachis* (*Myrmhopla*) *ochracea* Karavaiev, 1927e: 30 (w.) JAVA. Santschi, 1933d: 2 (q.).
***oedacantha***. *Polyrhachis* (*Myrmhopla*) *oedacantha* Wheeler, W.M. 1919e: 135 (w.) BORNEO.
*oedipus;* see under ***aedipus***.
***olena***. *Polyrhachis olenus* Smith, F. 1861b: 39, pl. 1, fig. 8 (w.) SULAWESI. Combination in *P.* (*Myrma*): Emery, 1925b: 201. Senior synonym of *eurytus:* Bolton, 1974b: 177.
***olybria***. *Polyrhachis olybrius* Forel, 1912n: 73 (q.) SUMATRA. Combination in *P.* (*Myrmhopla*): Emery, 1925b: 197.
***omyrmex***. *Dolichoderus* (*Irenia*) *omyrmex* Donisthorpe, 1938e: 502, fig. (w.) NEW GUINEA. Combination in *Polyrhachis* (*Myrmatopa*): **new combination**.
***opalescens***. *Polyrhachis* (*Chariomyrma*) *opalescens* Clark, 1930c: 11, fig. 1 (w.) VANUATU.
***ops***. *Polyrhachis micans* subsp. *ops* Forel, 1907h: 308 (w.) AUSTRALIA. Combination in *P.* (*Campomyrma*): Emery, 1925b: 179. Raised to species: Taylor, 1989a: 26.
#***orientalis***. *Polyrhachis* (*Myrmhopla*) *tibialis* var. *orientalis* Karavaiev, 1927e: 38 (w.) INDONESIA (Ambon I.).
***ornata***. *Polyrhachis ornata* Mayr, 1876: 73 (w.) AUSTRALIA. Combination in *P.* (*Hedomyrma*): Forel, 1915b: 107. Senior synonym of *chrysothorax* and *humerosa:* Kohout, 1988c: 435.
***orpheus***. *Polyrhachis orpheus* Forel, 1911b: 216 (w.q.) JAVA. Combination in *P.* (*Myrmhopla*): Emery, 1925b: 196.
***orsylla***. *Polyrhachis orsyllus* Smith, F. 1861b: 39, pl. 1, figs. 6, 7 (w.) SULAWESI. Viehmeyer, 1916a: 166 (q.). Combination in *P.* (*Myrma*): Viehmeyer, 1916a: 166. Junior synonym of *vindex:* Wheeler, W.M. 1924b: 254. Revived from synonymy: Karavaiev, 1927e: 48. Current subspecies: nominal plus *dentulata, javanensis, muscula, ritsemai, subcarinata*.
***osae***. *Polyrhachis* (*Myrmatopa*) *osae* Mann, 1919: 384, figs. 54–56 (w.q.) SOLOMON IS.
***osiris***. *Polyrhachis osiris* Bolton, 1975c: 10, figs. 4, 13 (w.) PHILIPPINES.
***otleti***. *Polyrhachis* (*Myrma*) *otleti* Forel, 1916: 449 (w.q.m.) ZAIRE. See also: Bolton, 1973b: 336.

**#pagana.** *Polyrhachis (Cyrtomyrma) rastellata* var. *pagana* Santschi, 1928h: 134, fig. 2 (w.) INDIA.

**#pahangana.** *Polyrhachis furcata* subsp. *pahangana* Forel, 1911d: 395 (w.) WEST MALAYSIA. Combination in *P. (Myrmhopla)*: Emery, 1925b: 193.

**pallescens.** *Polyrhachis guerini* var. *pallescens* Mayr, 1876: 74 (w.) AUSTRALIA. Combination in *P. (Chariomyrma)*: Emery, 1925b: 186. Subspecies/variety of *aurea:* Emery, 1897d: 584 (in key); of *guerini:* Taylor & Brown, D.R. 1985: 135. Raised to species, and senior synonym of *depilis:* Kohout & Taylor, 1990: 516.

**pallipes.** *Polyrhachis (Aulacomyrma) pallipes* Donisthorpe, 1948d: 603 (w.) NEW GUINEA.

*pandarus. Polyrhachis pandarus* Smith, F. 1857a: 62 (w.) BORNEO. Mayr, 1862: 679 (q.). Junior synonym of *armata:* Roger, 1863b: 9.

**#paprika.** *Polyrhachis biroi* var. *paprika* Forel, 1911e: 296 (w.) NEW GUINEA. Combination in *P. (Myrma)*: Emery, 1925b: 200.

*papuana. Polyrhachis relucens* subsp. *decipiens* var. *papuana* Emery, 1897d: 580 (w.) NEW GUINEA. Unavailable name.

**#papuana.** *Polyrhachis heinlethi* var. *papuana* Emery, 1900c: 336 (w.) NEW GUINEA.

**parabiotica.** *Polyrhachis (Anoplomyrma) parabiotica* Chapman, 1963: 258, fig. 7 (w.q.) PHILIPPINES.

**paracamponota.** *Polyrhachis paracamponota* Wang & Wu, 1991: 599, figs. 3, 7 (w.) CHINA.

*paradoxa* André, 1892; see under ***PHASMOMYRMEX.***

**#parallela.** *Polyrhachis caulomma* var. *parallela* Viehmeyer, 1914d: 612 (w.) NEW GUINEA. Combination in *P. (Chariomyrma)*: Emery, 1925b: 187.

**paromala.** *Polyrhachis paromalus* Smith, F. 1863: 15 (w.) INDONESIA (Seram I.). Combination in *P. (Myrmhopla)*: Emery, 1925b: 196. Current subspecies: nominal plus *tobias.*

**#parsis.** *Polyrhachis tibialis* var. *parsis* Emery, 1900d: 717 (footnote) (w.) INDIA. Forel, 1908a: 9 (w.q.). Combination in *P. (Myrmhopla)*: Viehmeyer, 1916a: 166.

**#parvicella.** *Polyrhachis schang* var. *parvicella* Forel, 1911b: 214 (w.q.) JAVA. Karavaiev, 1927e: 11 (m.). Combination in *P. (Myrmatopa)*: Emery, 1925b: 181.

**patiens.** *Polyrhachis (Campomyrma) patiens* Santschi, 1920g: 185 (w.) AUSTRALIA.

**#pauperata.** *Polyrhachis mayri* var. *pauperata* Emery, 1889b: 519 (w.) BURMA. Combination in *P. (Myrma)*: Emery, 1925b: 201. Subspecies of *illaudata:* current status.

**paxilla.** *Polyrhachis paxillus* Smith, F. 1863: 17 (w.) INDONESIA (Martabello I.). Combination in *P. (Chariomyrma)*: Emery, 1925b: 188; in *P. (Hagiomyrma)*: Kohout, 1988c: 435. Senior synonym of *maeandrifera:* Kohout, 1988c: 435.

**#pectita.** *Polyrhachis (Myrmhopla) tibialis* var. *pectita* Santschi, 1928h: 136 (w.) SRI LANKA.

**pellita.** *Polyrhachis pellita* Menozzi, 1922a: 356, figs. 3, 4 (w.q.) PHILIPPINES. Combination in *P. (Myrmhopla)*: Emery, 1925b: 196.

**penelope.** *Polyrhachis penelope* Forel, 1895a: 46 (w.) AUSTRALIA. Forel, 1901b: 31 (q.m.). Combination in *P. (Hagiomyrma)*: Forel, 1915b: 108.

**peregrina.** *Polyrhachis peregrinus* Smith, F. 1860a: 71 (q.) SULAWESI. Combination in *P. (Myrmhopla)*: Emery, 1925b: 197.

*perplexa. Polyrhachis Polyrhachis) bihamata* var. *perplexa* Santschi, 1925f: 92 (w.) LAOS. Junior synonym of *bihamata:* Hung, 1970: 16.

**personata.** *Polyrhachis (Myrmhopla) personata* Wheeler, W.M. 1919e: 134 (w.) BORNEO.

*perthensis. Polyrhachis (Campomyrma) sidnica* var. *perthensis* Crawley, 1922c: 36 (w.) AUSTRALIA. Wheeler, W.M. 1934d: 163 (q.). Raised to species: Taylor, 1989a: 26. Junior synonym of *phryne:* Kohout & Taylor, 1990: 517.

*phaenogaster. Polyrhachis (Pseudocyrtomyrma) revoili* subsp. *crassa* var. *phaenogaster* Emery, 1921e: 24 (q.) CAMEROUN. Unavailable name; material referred to *weissi* by Bolton, 1973b:342.

**phalerata.** *Polyrhachis (Myrmatopa) ph lerata* Menozzi, 1926e: 102 (w.) PHILIPPINES.

**phidias.** *Polyrhachis phidias* Forel, 1910e: 450 (w.) "Equatorial Africa". Combination in *P. (Myrma)*: Wheeler, W.M. 1922a: 999. See also: Bolton, 1973b: 316.

**philippinensis.** *Polyrhachis philippinensis* Smith, F. 1858b: 69, pl. 4, fig. 34 (w.) PHILIPPINES. Combination in *P. (Myrma)*: Emery, 1925b: 201.

*phipsoni. Polyrhachis phipsoni* Forel, 1894c: 399 (w.) BURMA. Combination in *P. (Myrmhopla)*: Emery, 1925b: 193. Variety/subspecies of *rufipes:* Forel, 1911e: 297; Emery, 1925b: 193. Junior synonym of *rufipes:* Bolton, 1974b: 178.

**phryne.** *Polyrhachis phryne* Forel, 1907a: 41 (w.) AUSTRALIA. Combination in *P. (Campomyrma)*: Emery, 1925b: 179. Senior synonym of *sempronia:* Taylor, 1989a: 26; of *perthensis:* Kohout & Taylor, 1990: 517.

*phyllophilus. Polyrhachis phyllophilus* Smith, F. 1860a: 69 (w.) SULAWESI. Mayr, 1872: 139 (q.); Karavaiev, 1927e: 28 (m.). Combination in *P. (Myrmhopla)*: Emery, 1925b: 192. Junior synonym of *abdominalis:* Dalla Torre, 1893: 257; Forel, 1893c: 24. Revived from synonymy as variety/subspecies of *abdominalis:* Emery, 1900d: 714; Forel, 1912n: 71; Karavaiev, 1927e: 28. Material of the *nomen nudum monacha* referred here by Karavaiev, 1927e: 29. Junior synonym of *hector:* Bolton, 1974b: 174.

**piliventris.** *Polyrhachis piliventris* Smith, F. 1858b: 60, pl. 4, fig. 24 (q.) SINGAPORE. Combination in *P. (Myrmatopa)*: current combination.

**pilosa.** *Polyrhachis (Cyrtomyrma) rastellata* var. *pilosa* Donisthorpe, 1938c: 256 (w.) AUSTRALIA. [First available use of *Polyrhachis rastellata* r. *laevior* var. *pilosa* Forel, 1902h: 527; unavailable name (Taylor, 1986: 34).] Raised to species: Kohout & Taylor, 1990: 518.

**plato.** *Polyrhachis plato* Forel, 1911d: 393 (q.) SUMATRA. Combination in *P. (Myrmhopla)*: Emery, 1925b: 196.

**platynota.** *Polyrhachis (Myrmhopla) platynota* Stitz, 1933: 74, fig. 8 (q.) BORNEO.

**platyomma.** *Polyrhachis (Pseudocyrtomyrma) platyomma* Emery, 1921e: 24, fig. 3 (w.) CAMEROUN. See also: Bolton, 1973b: 337.

*plebeia. Polyrhachis (Myrma) schlueteri* var. *plebeia* Santschi, 1914b: 143 (w.) KENYA. Junior synonym of *schlueteri:* Bolton, 1973b: 321.

*pleurata. Polyrhachis militaris* st. *cupreopubescens* var. *pleurata* Santschi, 1924b: 223 (w.) ZAIRE. Unavailable name; material referred to *militaris* by Bolton, 1973b: 313.

**#polluta.** *Polyrhachis (Myrma) nigropilosa* var. *polluta* Menozzi, 1926e: 99 (w.q.) PHILIPPINES.

**polymnia.** *Polyrhachis polymnia* Forel, 1902h: 532 (w.q.) AUSTRALIA. Combination in *P. (Campomyrma)*: Forel, 1915b: 115.

**porcata.** *Polyrhachis (Aulacomyrma) porcata* Emery, 1921e: 20 (w.) NEW GUINEA.

**pressa.** *Polyrhachis pressus* Mayr, 1862: 681 (w.q.) JAVA. Combination in *P. (Myrmhopla)*: Wheeler, W.M. 1919e: 130.

**#procera.** *Polyrhachis continua* var. *procera* Emery, 1897d: 581 (w.q.) NEW GUINEA. Combination in *P. (Myrma)*: Emery, 1925b: 200.

**prometheus.** *Polyrhachis (Campomyrma) prometheus* Santschi, 1920a: 566 (w.) AUSTRALIA.

**proxima.** *Polyrhachis proxima* Roger, 1863a: 155 (w.) INDONESIA (Lingga I.) Bingham, 1903: 405 (q.). Combination in *P. (Myrma)*: Viehmeyer, 1916a: 166. Current subspecies: nominal plus *semirufipes.*

**#proximomayri.** *Polyrhachis mayri* var. *proximomayri* Emery, 1896d: 379 (w.) BURMA. [First available use of *Polyrhachis mayri* r. *intermedia* var. *proximomayri* Forel, 1893c: 20; unavailable name.] Subspecies of *illaudata:* current status.

**pruinosa.** *Polyrhachis pruinosa* Mayr, 1872: 142 (w.) BORNEO. Combination in *P. (Myrma)*: Viehmeyer, 1916a: 166.

*pruinosa. Polyrhachis (Myrma) sericata* subsp. *pruinosa* Karavaiev, 1927e: 48 (w.) INDONESIA (Aru I.). [Junior primary homonym of *pruinosa* Mayr, above.] Replacement name: *pruinosula* Santschi, 1928h: 139.

**#pruinosula.** *Polyrhachis (Myrma) sericata* var. *pruinosula* Santschi, 1928h: 139. Replacement name for *pruinosa* Karavaiev, above. [Junior primary homonym of *pruinosa* Mayr, 1872: 142.

*pseuadonyma. Polyrhachis pseuadonyma* Forel, 1886d: 243. [Unnecessary (second) replacement name for *dolomedes* Smith, F. 1863: 16.] Junior synonym of *derecyna:* Dalla Torre, 1893: 261.

**pseudothrinax.** *Polyrhachis pseudothrinax* Hung, 1967a: 199, figs. 1–6 (w.) AUSTRALIA. Combination in *P. (Campomyrma)*: Kohout & Taylor, 1990: 512.

**pubescens.** *Polyrhachis pubescens* Mayr, 1879: 657 (w.) INDIA. Combination in *Hemioptica:* Bingham, 1903: 381; in *P. (Myrma)*: Emery, 1925b: 205. Current subspecies: nominal plus *alatisquamis.*

**#pubiventris.** *Polyrhachis villosa* var. *pubiventris* Viehmeyer, 1914c: 59 (w.) NEW GUINEA. Combination in *P. (Chariomyrma)*: Emery, 1925b: 188.

**punctata.** *Polyrhachis (Myrmhopla) punctata* Karavaiev, 1927e: 36 (w.) JAVA.

**punctillata.** *Polyrhachis punctillata* Roger, 1863a: 152 (w.q.) SRI LANKA. Combination in *P. (Myrma)*: Emery, 1925b: 204. Current subspecies: nominal plus *fergusoni, smythiesii.* See also: Bingham, 1903: 409.

**#punctinota.** *Polyrhachis beauforti* var. *punctinota* Viehmeyer, 1914c: 58 (w.) NEW GUINEA. Combination in *P. (Chariomyrma)*: Emery, 1925b: 187.

***punctiventris.*** *Polyrhachis punctiventris* Mayr, 1876: 73 (w.q.) AUSTRALIA. Combination in *P. (Chariomyrma)*: Forel, 1915b: 108.

***pyrgops.*** *Polyrhachis pyrgops* Viehmeyer, 1912: 9, fig. 10 (w.) NEW GUINEA. [Forel, 1912a: 81 gives type-locality as China.] Combination in *P. (Myrma)*: Emery, 1925b: 202.

***pyrrhus.*** *Polyrhachis pyrrhus* Forel, 1910b: 90 (w.) AUSTRALIA. Combination in *P. (Campomyrma)*: Emery, 1925b: 179.

*quadricuspis. Polyrhachis quadricuspis* Mayr, 1870b: 946 (w.) AUSTRALIA. Combination in *P. (Campomyrma)*: Emery, 1925b: 180. Junior synonym of *sidnica:* Mayr, 1876: 75; Dalla Torre, 1893: 269. Revived from synonymy as variety of *sidnica:* Emery, 1896d: 378. Junior synonym of *sidnica:* Taylor, 1989a: 27; Kohout & Taylor, 1990: 512.

***queenslandica.*** *Polyrhachis queenslandica* Emery, 1895g: 356 (w.) AUSTRALIA. Crawley, 1921: 96 (q.m.). Combination in *P. (Myrmothrinax)*: Emery, 1925b: 183.

*quinquedentata. Polyrhachis (Campomyrma) hirsuta* subsp. *quinquedentata* Viehmeyer, 1925b: 147 (w.) AUSTRALIA. Junior synonym of *hirsuta:* Taylor, 1989a: 24.

#***radicicola.*** *Polyrhachis costulata* var. *radicicola* Dahl, 1901: 15 (w.q.m.) BISMARCK ARCHIPELAGO. [First available use of *Polyrhachis aurea* r. *costulata* var. *radicicola* Forel, 1901b: 32; unavailable name.] Combination in *P. (Chariomyrma)*: Emery, 1925b: 186. Subspecies of *aurea:* current status.

***ralumensis.*** *Polyrhachis ralumensis* Forel, 1901b: 34 (w.) BISMARCK ARCHIPELAGO. Combination in *P. (Cyrtomyrma)*: Emery, 1925b: 208. Senior synonym of *major:* Viehmeyer, 1914c: 50.

***rastellata.*** *Formica rastellata* Latreille, 1802c: 130 (q.) INDONESIA ("Indes orientales"). Mayr, 1862: 688 (w.); Mayr, 1876: 70 (m.); Crozier, 1970: 115 (k.); Wheeler, G.C. & Wheeler, J. 1990b: 756 (l.). Combination in *Polyrhachis:* Smith, F. 1858b: 59; in *P. (Cyrtomyrma)*: Forel, 1915b: 107. Senior synonym of *busiris:* Mayr, 1862: 688; Roger, 1863b: 6. Current subspecies: nominal plus *baduri, celebensis, ceylonensis* (junior homonym), *congener, corporaali, fulakora, nomo, pagana, semiinermis.*

#***rastrata.*** *Polyrhachis rastrata* Emery, 1889b: 517 (w.) BURMA. Combination in *P. (Campomyrma)*: Emery, 1925b: 179. Status as species: Bingham, 1903: 412. Subspecies of *clypeata:* Forel, 1893c: 19; Emery, 1925b: 179; of *exercita:* current status.

***reclinata.*** *Polyrhachis sexspinosa* var. *reclinata* Emery, 1887a: 236 (w.) NEW GUINEA. Karavaiev, 1927e: 25 (q.m.). Combination in *P. (Myrmhopla)*: Emery, 1925b: 191. Raised to species: Emery, 1911c: 259. Junior synonym of *sexspinosa:* Bolton, 1975c: 12. Revived from synonymy: Kohout & Taylor, 1990: 518.

*rectinota. Polyrhachis sexspinosa* var. *rectinota* Forel, 1911e: 299. *Nomen nudum.* [Probably an error for *reclinata,* above.]

#***rectispina.*** *Polyrhachis (Myrmhopla) dives* var. *rectispina* Karavaiev, 1927e: 35 (w.) INDONESIA (Ambon I.).

***regesa.*** *Polyrhachis regesa* Bolton, 1973b: 337, figs. 52, 58 (w.) GHANA.

***regularis.*** *Polyrhachis regularis* Mayr, 1867a: 63, pl. 2, fig. 4 (q.) JAVA. Combination in *P. (Myrmhopla)*: Emery, 1925b: 197.

***relucens.*** *Formica relucens* Latreille, 1802c: 131, pl. 4, fig. 24 (w.) INDONESIA ("Indes orientales"). Combination in *Polyrhachis:* Smith, F. 1857a: 59; in *P. (Myrma)*: Forel, 1915b: 110. Current subspecies: nominal plus *breviorspinosa, decipiens, litigiosa.*

***rere.*** *Polyrhachis (Chariomyrma) rere* Mann, 1919: 381, fig. 51 (w.) SOLOMON IS.

***restituta.*** *Polyrhachis restituta* Viehmeyer, 1913: 149, fig. (w.) SULAWESI (in copal). Combination in *P. (Myrma)*: Emery, 1925b: 202. Current subspecies: nominal plus *conclusa.*

***retrorsa.*** *Polyrhachis retrorsa* Emery, 1900d: 719 (w.q.) INDONESIA (Mentawei I.). Combination in *P. (Myrmhopla)*: Emery, 1925b: 196.

*reversa. Polyrhachis abdominalis* var. *reversa* André, 1896b: 253 (w.) PHILIPPINES. Combination in *P. (Myrmhopla)*: Emery, 1925b: 192. Junior synonym of *hector:* Bolton, 1974b: 174.

#***revocata.*** *Polyrhachis continua* var. *revocata* Viehmeyer, 1913: 151 (w.) SULAWESI (in copal). Combination in *P. (Myrma)*: Emery, 1925b: 200.

***revoili.*** *Polyrhachis revoili* André, 1887: 285 (q.) SOMALIA. Forel, 1894b: 71 (w.); Forel, 1916: 452 (m.). Combination in *P. (Pseudocyrtomyrma)*: Emery, 1921e: 23; Emery, 1925b: 207; in *P. (Myrma)*: Forel, 1916: 452; Wheeler, W.M. 1922a 999; Bolton, 1973b: 338. Senior synonym of *natalensis:* Forel, 1916: 453; Arnold, 1924: 754; of *donisthorpei:* Bolton, 1973b: 338.

***rhea.*** *Polyrhachis sexspinosa* subsp. *rhea* Forel, 1911e: 299 (w.) INDONESIA (Seram I.). Combination in *P. (Myrmhopla)*: Emery, 1925b: 191. Raised to species: Bolton, 1975c: 11.

***ridleyi.*** *Polyrhachis ridleyi* Forel, 1912n: 71 (w.) SUMATRA. Combination in *P. (Myrmhopla)*: Emery, 1925b: 196.

#***ritsemai.*** *Polyrhachis ritsemai* Mayr, 1883: 245 (w.q.) SUMATRA. Combination in *P. (Myrma)*: Viehmeyer, 1916a: 166. Subspecies of *orsylla:* Emery, 1900d: 712; Emery, 1925b: 204.

***rixosa.*** *Polyrhachis rixosus* Smith, F. 1858b: 68 (q.) SULAWESI. Combination in *P. (Myrma)*: Emery, 1925b: 202.

#***robustior.*** *Polyrhachis (Myrmhopla) tibialis* var. *robustior* Karavaiev, 1927e: 39 (w.) INDONESIA (Aru I.).

#***romanovi.*** *Polyrhachis (Myrmhopla) acantha* st. *romanovi* Santschi, 1928h: 139. Replacement name for *dichroa* Karavaiev, 1927e: 33. [Junior primary homonym of *dichrous* Forel, 1893c: 21.] Subspecies of *saevissima:* current status.

***rossi.*** *Polyrhachis (Myrmatopa) rossi* Donisthorpe, 1948b: 315 (w.q.) NEW GUINEA.

***rotumana.*** *Polyrhachis (Chariomyrma) rotumana* Wilson & Taylor, 1967: 99, fig. 83 (w.) ROTUMA I. (Polynesia).

***rotundiceps.*** *Polyrhachis (Chariomyrma) rotundiceps* Karavaiev, 1927e: 18, fig. 9 (w.) INDONESIA (Aru I.).

#***rotundinota.*** *Polyrhachis subcyanea* var. *rotundinota* Viehmeyer, 1914c: 58 (w.) NEW GUINEA. Combination in *P. (Chariomyrma)*: Emery, 1925b: 188.

***rowlandi.*** *Polyrhachis rowlandi* Forel, 1910b: 85 (w.) AUSTRALIA. Combination in *P. (Chariomyrma)*: Forel, 1915b: 108.

***rubigastrica.*** *Polyrhachis rubigastrica* Wu & Wang, in Wang & Wu, 1991: 598, figs. 2, 6, 8 (w.) CHINA.

***rubiginosa.*** *Formica ruiginosa* Le Guillou, 1842: 316 (w.) BORNEO. [Unresolved junior primary homonym of *Formica rubiginosa* Latreille, 1802c: 170.] Wheeler, G.C. & Wheeler, J. 1990b: 760 (l.). Combination in *Polyrhachis,* and spelling emended to *rubiginosa:* Roger, 1863b: 7; in *P. (Myrmhopla)*: Wheeler, W.M. 1919e: 130.

*rubricornis. Polyrhachis bicolor* var. *rubricornis* Baroni Urbani, 1971b: 362. [Unnecessarily proposed replacement name for *flavicornis* Stitz, 1925: 132.] Unavailable name.

*rufa. Polyrhachis (Campomyrma) micans* r. *ops* var. *rufa* Crawley, 1921: 97 (w.) AUSTRALIA. Unavailable name (Taylor, 1986: 34).

***ruficornis.*** *Polyrhachis ruficornis* Smith, F. 1857a: 60 (q.) BORNEO. Combination in *P. (Myrmatopa)*: Donisthorpe, 1932c: 445.

***rufifemur.*** *Polyrhachis terpsichore* var. *rufifemur* Forel, 1907a: 41 (w.) AUSTRALIA. Combination in *P. (Hedomyrma)*: Emery, 1925b: 190. Raised to species, and senior synonym of *elegans:* Kohout, 1988c: 436.

***rufipalpis.*** *Polyrhachis rufipalpis* Santschi, 1910c: 396 (w.) CONGO. Bolton, 1973b: 317 (q.). Combination in *P. (Myrma)*: Wheeler, W.M. 1922a: 1000. Senior synonym of *mayumbensis:* Bolton, 1973b: 317.

***rufipes.*** *Polyrhachis rufipes* Smith, F. 1858b: 66, pl. 4, fig. 28 (w.) BORNEO. Combination in *P. (Myrmhopla)*: Emery, 1925b: 196. Senior synonym of *exasperatus, oblisa* and *phipsoni:* Bolton, 1974b: 178.

#***rufiventris.*** *Polyrhachis gestroi* var. *rufiventris* Forel, 1911d: 391 (q.) SUMATRA. Senior synonym of *moeschiella* (unnecessary replacement name): **new synonymy**.

#***rufiventris.*** *Polyrhachis alpheus* var. *rufiventris* Emery, 1911c: 256 (w.q.) NEW GUINEA. [Unresolved junior primary homonym of *rufiventris* Forel, above.] Combination in *P. (Myrmatopa)*: Emery, 1925b: 180.

***rufofemorata.*** *Polyrhachis rufofemoratus* Smith, F. 1859a: 142 (w.) INDONESIA (Aru I.). Kohout, 1989: 514 (w.q.). Combination in *P. (Myrma)*: Emery, 1925b: 202. Senior synonym of *merops:* Bolton, 1974b: 178.

***rugifrons.*** *Polyrhachis rugifrons* Smith, F. 1860a: 70 (w.q.) SULAWESI. Combination in *P. (Myrmhopla)*: Wheeler, W.M. 1919e: 130. Subspecies of *sexspinosa:* Emery, 1897d: 591. Revived status as species: Wheeler, W.M. 1919e: 130; Emery, 1925b: 191; Karavaiev, 1927e: 25. See also: Bolton, 1975c: 11.

*rugosus* Smith, F. 1858; see under ***DOLICHODERUS.***

*rugousissima. Polyrhachis (Myrmhopla) rugousissima* Donisthorpe, 1943d: 468 (w.) NEW GUINEA. Junior synonym of *bubastes:* Bolton, 1975c: 6.

*rugulosus. Polyrhachis rugulosus* Mayr, 1862: 685, pl. 19, fig. 7 (w.) SOUTH AFRICA. Arnold, 1924: 744 (q.). Junior synonym of, and hence first available replacement name for *carinatus* Smith, F. 1858b: 71 [Junior secondary homonym of *Formica carinata* Fabricius, 1804: 413.]: Emery, 1896d: 379; Wheeler,

W.M. 1922a: 1002. Combination in *P.* (*Myrma*): Santschi, 1914b: 141. Variety/race of *militaris:* Mayr, 1893: 197; of *schistacea:* Stitz, 1910: 151; Wheeler, W.M. 1922a: 1002; Arnold, 1924: 744. Senior synonym of *cafrorum:* Mayr, 1893: 197; Forel, 1894b: 72. Junior synonym of *schistacea:* Bolton, 1973b: 318.

***rupicapra.*** *Polyrhachis rupicapra* Roger, 1863a: 154 (w.q.) SRI LANKA. Combination in *P.* (*Myrmhopla*): Emery, 1925b: 196.

***rustica.*** *Polyrhachis rustica* Kohout, 1990: 505 (w.q.) AUSTRALIA.

***saevissima.*** *Polyrhachis saevissimus* Smith, F. 1860a: 71 (w.) SULAWESI. Combination in *P.* (*Myrmhopla*): Emery, 1925b: 197. Senior synonym of *acantha* and *acasta:* Bolton, 1974b: 178. Current subspecies: nominal plus *argentea, chrysophanes, kerri, romanovi, timorensis.*

***saigonensis.*** *Polyrhachis thrinax* r. *saigonensis* Forel, 1886f: 199 (w.) VIETNAM. Viehmeyer, 1916a: 165 (q.). Combination in *P.* (*Myrmothrinax*): Emery, 1925b: 183. Subspecies of *thrinax:* Emery, 1896d: 379; Wheeler, W.M. 1909d: 343. Raised to species: Forel, 1893c: 28; Emery, 1925b: 183.

***salomo.*** *Polyrhachis salomo* Forel, 1910b: 87 (w.) SOLOMON IS. Combination in *P.* (*Myrma*): Mann, 1919: 379. Current subspecies: nominal plus *hiram.*

#***sanguinea.*** *Polyrhachis frauenfeldi* subsp. *sanguinea* Forel, 1911d: 393 (w.) SUMATRA. Combination in *P.* (*Myrmothrinax*): Emery, 1925b: 183.

*sankisiana. Polyrhachis militaris* r. *cupreopubescens* var. *sankisiana* Forel, 1913b: 348 (w.) ZAIRE. Unavailable name; material referred to *militaris* by Bolton, 1973b: 313.

***santschii.*** *Polyrhachis* (*Hedomyrma*) *santschi* Mann, 1919: 375 (w.) SOLOMON IS. Current subspecies: nominal plus *campbelli.*

*santschii. Polyrhachis* (*Pseudocyrtomyrma*) *lanuginosa* subsp. *santschii* Emery, 1921e: 24 (q.) CAMEROUN. [Junior primary homonym of *santschii* Mann, above.] Replacement name: *conradti* Santschi, 1923e: 293.

***sappho.*** *Polyrhachis sappho* Forel, 1911e: 299 (w.) SUMATRA. Combination in *P.* (*Myrmhopla*): Emery, 1925b: 194.

***scabra.*** *Polyrhachis scabra* Kohout, 1987: 175, figs. 8, 12, 14 (w.q.) PHILIPPINES.

***scapulata.*** *Polyrhachis* (*Chariomyrma*) *scapulata* Santschi, 1932b: 23, fig. 8 (w.) NEW GUINEA.

***schang.*** *Polyrhachis schang* Forel, 1879a: 123 (q.) CHINA. Combination in *P.* (*Myrmatopa*): Forel, 1915b: 107. Senior synonym of *gracilis:* Forel, 1909d: 232. Current subspecies: nominal plus *alata, amboinae, cnemidata, excitata, laurae, leviuscula, parvicella.*

***schenckii.*** *Polyrhachis schenckii* Forel, 1886f: 198 (w.) NEW GUINEA. Wheeler, G.C. & Wheeler, J. 1968: 221 (l.). Combination in *P.* (*Hedomyrma*): Santschi, 1920a: 569; in *P.* (*Hagiomyrma*): Emery, 1925b: 185.

***schistacea.*** *Hoplomyrmus schistaceus* Gerstäcker, 1859: 262 (w.) MOZAMBIQUE. [Also described as new by Gerstäcker, 1862: 508.] Forel, 1907a: 38 (q.). Combination in *Polyrhachis:* Mayr, 1863: 446; in *P.* (*Myrma*): Santschi, 1914b: 140. Subspecies of *militaris:* André, 1887: 288. Revived status as species: Dalla Torre, 1893: 268; Forel, 1907a: 38; Wheeler, W.M. 1922a: 1000; Arnold, 1924: 742; Bolton, 1973b: 318. Senior synonym of *atrociliata, divina, divinoides, fracta, gagatoides* and *rugulosus* (and its junior synonyms *cafrorum* and *carinatus*), and material of the unavailable names *beguelensis, mediopilosa* and *subplana* referred here: Bolton, 1973b: 318.

***schizospina.*** *Polyrhachis* (*Johnia*) *schizospina* Karavaiev, 1927e: 44, fig. 20 (q.) INDONESIA (Prinsen I.). Combination in *P.* (*Aulacomyrma*): Hung, 1967b: 402.

#***schlaginhaufeni.*** *Polyrhachis schlaginhaufeni* Viehmeyer, 1912: 12, fig. 16 (w.) NEW GUINEA. Combination in *P.* (*Chariomyrma*): Emery, 1925b: 187. Subspecies of *aurita:* Viehmeyer, 1914c: 54.

***schlueteri.*** *Polyrhachis militaris* r. *schlueteri* Forel, 1886f: 195 (w.) EAST AFRICA. Arnold, 1924: 748 (q.). Combination in *P.* (*Myrma*): Santschi, 1914b: 143. Raised to species: Dalla Torre, 1893: 268. Subspecies of *schistacea:* Santschi, 1914e: 42; Wheeler, W.M. 1922a: 1003; Arnold, 1924: 747. Status as species: Forel, 1907g: 92; Forel, 1914d: 261; Santschi, 1914b: 143; Emery, 1925b: 200; Bolton, 1973b: 321. Senior synonym of *indigens:* Arnold, 1924: 747; of *plebeia:* Bolton, 1973b: 321.

***schoopae.*** *Polyrhachis appendiculata* r. *schoopae* Forel, 1902h: 520 (w.) AUSTRALIA. Combination in *P.* (*Chariomyrma*), and raised to species: Kohout & Taylor, 1990: 519.

*schoutedeni. Polyrhachis* (*Myrma*) *schoutedeni* Santschi, 1919c: 249 (w.) ZAIRE. Junior synonym of *nigrita:* Bolton, 1973b: 328.

***schwiedlandi.*** *Polyrhachis schwiedlandi* Forel, 1902h: 529 (w.q.) AUSTRALIA. Wheeler, G.C. & Wheeler, J. 1974a: 62 (l.). Combination in *P.* (*Campomyrma*): Emery, 1925b: 179.

***scissa.*** *Hemioptica scissa* Roger, 1862a: 240, pl. 1, fig. 12 (w.q.) SRI LANKA. Forel, 1893c: 27 (m.); Wheeler, G.C. & Wheeler, J. 1990b: 757 (l.). Combination in *Polyrhachis* (*Hemioptica*): Mayr, 1879: 651; in *Hemioptica:* Bingham, 1903: 380; Emery, 1925b: 210.

***sculpta.*** *Polyrhachis sculpta* Emery, 1887a: 226 (w.) NEW GUINEA. Combination in *P.* (*Campomyrma*): Emery, 1925b: 180.

***sculpturata.*** *Polyrhachis sculpturatus* Smith, F. 1860a: 70 (w.q.) SULAWESI. Combination in *P.* (*Myrma*): Wheeler, W.M. 1919e: 124. Current subspecies: nominal plus *javaniana, siamensis.*

***scutulata.*** *Polyrhachis scutulatus* Smith, F. 1859a: 140 (w.) INDONESIA (Aru I.). Combination in *P.* (*Chariomyrma*): Emery, 1925b: 188.

#***selecta.*** *Polyrhachis murina* subsp. *selecta* Forel, 1911b: 215 (w.m.) JAVA. Combination in *P.* (*Myrma*): Emery, 1925b: 205.

*selene* Emery, 1889; see under ***CAMPONOTUS.***

***semiaurata.*** *Polyrhachis semiaurata* Mayr, 1876: 71 (w.) AUSTRALIA. Combination in *P.* (*Hagiomyrma*): Forel, 1915b: 108.

#***semiinermis.*** *Polyrhachis* (*Cyrtomyrma*) *rastellata* var. *semiinermis* Donisthorpe, 1941j: 209 (w.) PHILIPPINES.

***semiobscura.*** *Polyrhachis* (*Hagiomyrma*) *semiobscura* Donisthorpe, 1944c: 65 (w.) BISMARCK ARCHIPELAGO.

***semipolita.*** *Polyrhachis semipolita* André, 1896b: 251 (w.) AUSTRALIA. Combination in *P.* (*Campomyrma*): Emery, 1925b: 179. Subspecies of *hexacantha:* Emery, 1898a: 228. Revived status as species: Clark, 1934c: 72.

#***semirufipes.*** *Polyrhachis* (*Myrma*) *proxima* var. *semirufipes* Donisthorpe, 1943d: 469 (w.) NEW GUINEA.

#***semitestacea.*** *Polyrhachis andromache* var. *semitestacea* Emery, 1900c: 334 (w.) NEW GUINEA. Combination in *P.* (*Myrma*): Emery, 1925b: 202.

*sempronia. Polyrhachis sempronia* Forel, 1907a: 39 (w.) AUSTRALIA. Combination in *P.* (*Campomyrma*): Emery, 1925b: 180. Junior synonym of *phryne:* Taylor, 1989a: 26.

***senilis.*** *Polyrhachis gab* var. *senilis* Forel, 1902h: 520 (w.) AUSTRALIA. Combination in *P.* (*Chariomyrma*): Forel, 1915b: 108. Raised to species, and senior synonym of *crawleyella* and *tripellis:* Kohout, 1988a: 50.

***sericata.*** *Formica sericata* Guérin-Méneville, 1831, pl. 8, fig. 2 (w.) "Offack". Emery, 1887a: 232 (q.). Combination in *Polyrhachis:* Smith, F. 1858b: 36; in *P.* (*Myrma*): Emery, 1925b: 203. Senior synonym of *grisea:* Dalla Torre, 1893: 269; Emery, 1896d: 379. Current subspecies: nominal plus *glabra, nitidissima, nitidiventris, pruinosula.*

*sericea. Polyrhachis* (*Myrmhopla*) *sexspinosa* var. *sericea* Karavaiev, 1927e: 26, fig. 13 (w.q.m.) INDONESIA (Aru I.). [Also described as new by Karavaiev, 1928: 316.] Junior synonym of *sexspinosa:* Bolton, 1975c: 12.

***sericeopubescens.*** *Polyrhachis* (*Myrma*) *sericeopubescens* Donisthorpe, 1941h: 61 (w.q.) NEW GUINEA.

*serratus* Smith, F. 1859; see under ***ECHINOPLA.***

*setulosus. Polyrhachis setulosus* Radoszkovsky, 1881: 197 (w.) ANGOLA. *Nomen nudum.*

***sexspinosa.*** *Formica sexspinosa* Latreille, 1802c: 126, pl. 4, fig. 21 (w.) INDONESIA ("Indes orientales"). Mayr, 1867a: 42 (q.); Karavaiev, 1927e: 26 (m.). Combination in *Polyrhachis:* Smith, F. 1858b: 59; in *P.* (*Myrmhopla*): Viehmeyer, 1916a: 167. Senior synonym of *argentata:* Lepeletier, 1835: 219; of *irritabilis:* Roger, 1863b: 6; of *arcuspinosa, esuriens, juxtaspinosa, sericea* and *waigeuensis:* Bolton, 1975c: 12; of *barnardi:* Kohout & Taylor, 1990: 519.

#***siamensis.*** *Polyrhachis sculpturata* var. *siamensis* Mayr, 1879: 657 (w.) THAILAND. Combination in *P.* (*Myrma*): Emery, 1925b: 203.

***sidnica.*** *Polyrhachis sidnica* Mayr, 1866b: 886, pl. 20, fig. 1 (w.) AUSTRALIA. Mayr, 1876: 75 (q.). Combination in *P.* (*Campomyrma*): Forel, 1915b: 113. Senior synonym of *quadricuspis:* Taylor, 1989a: 27. See also: Kohout & Taylor, 1990: 512.

*silvatica. Polyrhachis* (*Myrma*) *relucens* st. *ithonus* var. *silvatica* Santschi, 1932b: 21, fig. 5 (w.) INDONESIA (Weeim I.). Unavailable name.

***similis.*** *Polyrhachis similis* Viehmeyer, 1912: 8 (w.) NEW GUINEA. Combination in *P.* (*Myrma*): Mann, 1919: 379. Current subspecies: nominal plus *angustior.*

***simillima.*** *Polyrhachis simillima* Emery, 1900d: 711 (w.) INDONESIA (Mentawei I.). Combination in *P.* (*Myrmatopa*): Emery, 1925b: 181.

**#simpla.** *Polyrhachis* (*Myrma*) *conops* st. *simpla* Santschi, 1928h: 139. Replacement name for *simplex* Karavaiev, 1927e: 45. [Junior primary homonym of *simplex* Mayr, 1862: 682.]

*simplex. Polyrhachis simplex* Mayr, 1862: 682 (q.) INDIA. Wheeler, G.C. & Wheeler, J. 1970: 649 (l.); Imai, Baroni Urbani, *et al.* 1984: 9 (k.). Combination in *P.* (*Myrmhopla*): Emery, 1925b: 196. Senior synonym of *spiniger:* Forel, 1893c: 36. Junior synonym of *lacteipennis:* Bolton, 1974b: 177. See also: Bingham, 1903: 394; Ofer, 1970: 49.

*simplex. Polyrhachis* (*Myrma*) *conops* subsp. *simplex* Karavaiev, 1927e: 45 (w.q.m.) INDONESIA (Aru I.). [Junior primary homonym of *simplex* Mayr, above.] Replacement name: *simpla* Santschi, 1928h: 139.

**#siwiensis.** *Polyrhachis* (*Myrmhopla*) *dives* var. *siwiensis* Santschi, 1932b: 20 (q.) NEW GUINEA.

**smithi.** *Polyrhachis smithi* Emery, 1901g: 579 (w.) SULAWESI. Combination in *P.* (*Myrmhopla*): Emery, 1925b: 196.

**#smythiesii.** *Polyrhachis punctillata* var. *smythiesii* Forel, 1895e: 456 (w.) INDIA. Combination in *P.* (*Myrma*): Emery, 1925b: 204. Junior synonym of *punctillata:* Bingham, 1903: 409. Revived from synonymy as variety of *punctillata:* Emery, 1925b: 204.

**sokolova.** *Polyrhachis sokolova* Forel, 1902h: 522 (w.) AUSTRALIA. Combination in *P.* (*Hagiomyrma*): Emery, 1925b: 185. Senior synonym of *degener:* Kohout, 1988c: 436.

**solivaga.** *Polyrhachis* (*Myrmatopa*) *solivaga* Menozzi, 1926e: 100 (w.) PHILIPPINES.

**solmsi.** *Polyrhachis solmsi* Emery, 1887a: 224 (footnote), pl. 3, fig. 12 (w.q.m.) JAVA. Santschi, 1928g: 247 (l.). Combination in *P.* (*Myrmatopa*): Emery, 1925b: 181. Current subspecies: nominal plus *multicella.*

*sophiae. Polyrhachis heinlethi* var. *sophiae* Forel, 1902h: 521 (w.) AUSTRALIA. Combination in *P.* (*Chariomyrma*): Forel, 1915b: 108. Junior synonym of *heinlethii:* Kohout & Taylor, 1990: 514.

**sophocles.** *Polyrhachis sophocles* Forel, 1908a: 10 (w.) SRI LANKA. Combination in *P.* (*Myrmhopla*): Emery, 1925b: 196.

**sparaxes.** *Polyrhachis sparaxes* Smith, F. 1863: 16 (q.) INDONESIA (Misool I.). Combination in *P.* (*Myrmothrinax*): Emery, 1925b: 183.

**spengeli.** *Polyrhachis spengeli* Forel, 1912n: 69 (w.) SUMATRA. Combination in *P.* (*Campomyrma*): Emery, 1925b: 180.

**spinicola.** *Polyrhachis spinicola* Forel, 1894b: 70 (w.q.) MOZAMBIQUE. Combination in *P.* (*Myrma*): Wheeler, W.M. 1922a: 1004. Senior synonym of *gallicola:* Bolton, 1973b: 329. See also: Arnold, 1924: 751.

**#spinifera.** *Polyrhachis conops* var. *spinifera* Stitz, 1911a: 376, fig. 22 (w.) NEW GUINEA. Combination in *P.* (*Myrma*): Emery, 1922b: 200.

*spiniger. Polyrhachis spiniger* Mayr, 1879: 653 (w.m.) INDIA. Junior synonym of *simpex:* Forel, 1893c: 36. [*P. spiniger* Smith, F. 1873: viii. *Nomen nudum*, attributed to Mayr.]

*spinosa. Polyrhachis spinosa* Mayr, 1867a: 43, pl. 2, fig. 2 (w.q.) INDONESIA (Halmahera I.). Combination in *P.* (*Myrmhopla*): Emery, 1925b: 191. Junior synonym of *bubastes:* Bolton, 1975c: 6.

**spitteleri.** *Polyrhachis* (*Myrma*) *spitelleri* Forel, 1916: 450, fig. 6 (w.) ZAIRE. Combination in *P.* (*Pseudocyrtomyrma*): Emery, 1925b: 207. See also: Bolton, 1973b: 344.

**splendens.** *Polyrhachis* (*Chariomyrma*) *splendens* Santschi, 1932b: 22, fig. 7 (w.) NEW GUINEA.

*spretula. Polyrhachis* (*Myrma*) *viscosa* var. *spretula* Santschi, 1923e: 294 (w.q.) ZAIRE. Junior synonym of *viscosa:* Bolton, 1973b: 330.

*ssibangensis. Polyrhachis militaris* var. *ssibangensis* Forel, 1907a: 38 (w.) GABON. Santschi, 1910c: 400 (q.). Combination in *P.* (*Myrma*): Wheeler, W.M. 1922a: 998. Junior synonym of *militaris:* Bolton, 1973b: 313.

**stigmatifera.** *Polyrhachis stigmatifera* Kohout, 1990: 507 (w.) INDONESIA (Seram I.).

**#stitzi.** *Polyrhachis* (*Myrma*) *conops* var. *stitzi* Santschi, 1928h: 139. Replacement name for *bismarckensis* Karavaiev, 1927e: 46. [Junior primary homonym of *bismarckensis* Forel, 1901b: 33.]

**striata.** *Polyrhachis striatus* Mayr, 1862: 686, pl. 19, fig. 8 (w.) JAVA. Forel, 1886d: 242 (q.). Combination in *P.* (*Myrma*): Viehmeyer, 1916a: 166. Current subspecies: nominal plus *assamensis, tritschleri.*

*striativentris. Polyrhachis militaris* r. *striativentris* Emery, 1892d: 566 (w.) IVORY COAST. Santschi, 1912b: 166 (q.). Combination in *P.* (*Myrma*): Wheeler, W.M. 1922a: 999. Subspecies of *militaris:* Emery, 1896d: 777; Wheeler, W.M. 1922a: 999. Junior synonym of *militaris:* Bolton, 1973b: 313.

**#striatorugosa.** *Polyrhachis striatorugosus* Mayr, 1862: 686, pl. 19, fig. 9 (w.) JAVA. Mayr, 1867a: 60 (q.). Combination in *P.* (*Myrma*): Emery, 1925b: 203. Status as species: Dalla Torre, 1893: 270; Forel, 1913k: 136. Subspecies of *sumatrensis:* Emery, 1887a: 234 (footnote); Emery, 1925b: 203.

**strictifrons.** *Polyrhachis strictifrons* Emery, 1898b: 242, fig. 14 (w.) SULAWESI. Combination in *P.* (*Myrmhopla*): Emery, 1925b: 196.

*strigata* Norton, 1868; see under ***GNAMPTOGENYS.***

**#strigifrons.** *Polyrhachis coerulescens* var. *strigifrons* Viehmeyer, 1914b: 535 (w.) NEW GUINEA. Combination in *P.* (*Chariomyrma*): Emery, 1925b: 187.

*striolatorugosa. Polyrhachis cubaensis* var. *striolatorugosa* Mayr, 1893: 195 (w.) TANZANIA. Combination in *P.* (*Myrma*): Wheeler, W.M. 1922a: 994. Junior synonym of *cubaensis:* Bolton, 1973b: 325.

**stylifera.** *Polyrhachis* (*Cephalomyrma*) *stylifera* Karavaiev, 1935a: 115, fig. 29 (w.) CAMBODIA.

**subaenescens.** *Polyrhachis subaenescens* Viehmeyer, 1912: 10, fig. 13 (w.) NEW GUINEA. Combination in *P.* (*Chariomyrma*): Emery, 1925b: 188.

**#subcarinata.** *Polyrhachis orsyllus* subsp. *subcarinata* Emery, 1900d: 712 (w.) INDONESIA (Mentawei I.). Combination in *P.* (*Myrma*): Emery, 1925b: 204.

**subcyanea.** *Polyrhachis subcyanea* Emery, 1897d: 586 (w.) NEW GUINEA. Combination in *P.* (*Chariomyrma*): Emery, 1925b: 188. Current subspecies: nominal plus *rotundinota.*

**subfossa.** *Polyrhachis subfossa* Viehmeyer, 1913: 154, fig. (w.) SULAWESI (in copal). Combination in *P.* (*Myrmhopla*): Emery, 1925b: 196.

**subfossoides.** *Polyrhachis* (*Myrmhopla*) *subfossoides* Karavaiev, 1927e: 37 (w.) INDONESIA (Sumbawa I.).

*subnitens. Polyrhachis inconspicua* var. *subnitens* Emery, 1895g: 357 (w.) AUSTRALIA. Combination in *P.* (*Campomyrma*): Emery, 1925b: 179. Junior synonym of *insularis:* Taylor, 1989a: 25.

**subpilosa.** *Polyrhachis subpilosa* Emery, 1895k: 480 (w.) BURMA. Combination in *P.* (*Myrma*): Emery, 1925b: 204. Junior synonym of *punctillata:* Bingham, 1903: 409. Revived from synonymy: Emery, 1925b: 204.

*subplana. Polyrhachis* (*Myrma*) *schistacea* st. *fracta* var. *subplana* Santschi, 1914b: 142 (w.) KENYA. Unavailable name; material referred to *schistacea* by Bolton, 1973b: 318.

**subtridens.** *Polyrhachis subtridens* Emery, 1900d: 711 (w.) INDONESIA (Mentawei I.). Combination in *P.* (*Myrmatopa*): Emery, 1925b: 181.

**sulcata.** *Polyrhachis sulcata* André, 1895a: 1 (q.) CONGO. Bolton, 1973b: 322 (w.). Combination in *P.* (*Myrma*): Wheeler, W.M. 1922a: 1004.

**#sulcativentris.** *Polyrhachis* (*Hedomyrma*) *daemeli* var. *sulcativentris* Emery, in Forel, 1915b: 111 (w.) AUSTRALIA.

**sumatrensis.** *Polyrhachis sumatrensis* Smith, F. 1858b: 65, pl. 4, fig. 43 (q.) SUMATRA. Emery, 1887a: 234 (w.). Combination in *P.* (*Myrma*): Wheeler, W.M. 1919e: 125. Current subspecies: nominal plus *exophthalma, hamulata, striatorugosa.*

**sylvicola.** *Formica siloicola* Jerdon, 1851: 126 (w.q.) INDIA. [Spelling emended to *sylvicola* by Jerdon, 1854: 108.] Combination in *Polyrhachis:* Smith, F. 1858b: 59; in *P.* (*Myrmhopla*): Emery, 1925b: 197.

*synacantha. Polyrhachis* (*Polyrhachis*) *ypsilon* ab. *synacantha* Santschi, 1933d: 2 (w.) BORNEO. Junior synonym of *ypsilon:* Hung, 1970: 19.

**tambourinensis.** *Polyrhachis* (*Campomyrma*) *sidnica* var. *tambourinensis* Forel, 1915b: 113 (w.) AUSTRALIA. Raised to species: Kohout & Taylor, 1990: 512. See also: Taylor, 1989a: 27.

*taurus. Polyrhachis* (*Myrmatopa*) *taurus* Donisthorpe, 1937c: 274, figs. 4, 5 (q.) NEW GUINEA. Donisthorpe, 1943a: 172 (w.); Donisthorpe, 1943d: 470 (m.). Junior synonym of *derecyna:* Bolton, 1974b: 173.

**taylori.** *Polyrhachis taylori* Kohout, 1988b: 422, fig. 4 (w.q.) NEW GUINEA.

**templi.** *Polyrhachis templi* Forel, 1902h: 531 (w.) AUSTRALIA. Combination in *P.* (*Campomyrma*): Emery, 1925b: 180.

**#tenella.** *Polyrhachis furcata* var. *tenella* Forel, 1902d: 289 (w.) BURMA. Combination in *P.* (*Myrmhopla*): Emery, 1925b: 193.

*tenuistriata. Polyrhachis* (*Myrma*) *decemdentata* subsp. *tenuistriata* Menozzi, 1933a: 114 (w.) UGANDA. Santschi, 1937g: 85 (q.). Junior synonym of *decemdentata:* Bolton, 1973b: 302.

**ternatae.** *Polyrhachis* (*Myrmothrinax*) *ternatae* Karavaiev, 1933b: 105, fig. 7 (w.) INDONESIA (Ternate I.).

**terpsichore.** *Polyrhachis terpsichore* Forel, 1893d: 455 (q.) AUSTRALIA. Combination in *P.* (*Hedomyrma*): Emery, 1925b: 190.

**#tersa.** *Polyrhachis atropos* var. *tersa* Viehmeyer, 1914c: 52 (w.) NEW GUINEA. Viehmeyer, 1914b: 532 (q.). Combination in *P.* (*Hedomyrma*): Emery, 1925b: 189.

**textor.** *Polyrhachis textor* Smith, F. 1857a: 60 (w.) WEST MALAYSIA. Combination in *P.* (*Myrmothrinax*):

Viehmeyer, 1916a: 165. [*P. textrix:* Schulz, 1906: 156; unjustified emendation.] Current subspecies: nominal plus *aequalis, brunneogaster, charpillioni, hero.*

***thais.*** *Polyrhachis thais* Forel, 1910b: 86 (w.) AUSTRALIA. Combination in *P.* (*Hedomyrma*): Emery, 1925b: 190.

*thalia. Polyrhachis thalia* Forel, 1902h: 530 (q.) AUSTRALIA. Combination in *P.* (*Campomyrma*): Forel, 1915b: 114. Junior synonym of *inconspicua:* Taylor, 1989a: 24.

***thompsoni.*** *Polyrhachis thompsoni* Bingham, 1903: 391, fig. 130 (w.) BURMA. Combination in *P.* (*Myrmhopla*): Emery, 1925b: 196.

***thrinax.*** *Polyrhachis thrinax* Roger, 1863a: 152 (w.) SRI LANKA. Forel, 1893c: 29 (q.m.); Wheeler, G.C. & Wheeler, J. 1990b: 761 (l.). Combination in *P.* (*Myrmothrinax*): Forel, 1915b: 107. [Misspelled as *schrinax* by Rothney, 1889: 352.] Senior synonym of *mucronis:* Brown, 1959d: 164. Current subspecies: nominal plus *castanea, inconstans, javanica, lancearia, lucida, nigripes* (junior homonym).

***thusnelda.*** *Polyrhachis thusnelda* Forel, 1902h: 509 (w.q.m.) AUSTRALIA. Combination in *P.* (*Hagiomyrma*): Emery, 1925b: 185.

***tibialis.*** *Polyrhachis tibialis* Smith, F. 1858b: 63 (w.) BURMA. Forel, 1893c: 34 (q.m.). Combination in *P.* (*Myrmhopla*): Viehmeyer, 1916a: 166. Current subspecies: nominal plus *addax, caligata, completa, crassisquama, nigricornis, orientalis, parsis, pectita, robustior.* See also: Bingham, 1903: 396.

#***timorensis.*** *Polyrhachis acantha* r. *timorensis* Forel, 1913e: 664 (w.) INDONESIA (Timor I.). Combination in *P.* (*Myrmhopla*): Emery, 1925b: 194. Subspecies of *saevissima:* current status.

#***tobias.*** *Polyrhachis paromalus* subsp. *tobias* Forel, 1911d: 391 (w.) SUMATRA. Combination in *P.* (*Myrmhopla*): Emery, 1925b: 196.

*tonsilis. Polyrhachis* (*Polyrhachis*) *bihamata* var. *tonsilis* Santschi, 1928h: 133 (w.) SUMATRA. Junior synonym of *bihamata:* Hung, 1970: 16.

*torricelliana. Polyrhachis rastellata* var. *torricelliana* Viehmeyer, 1912: 9, fig. 11 (w.) NEW GUINEA. Junior synonym of *euryala:* Viehmeyer, 1914c: 50.

#***torta.*** *Polyrhachis* (*Myrmatopa*) *fruhstorferi* var. *torta* Santschi, 1928h: 139. Replacement name for *arcuata* Karavaiev, 1927e: 9. [Junior secondary homonym of *arcuata* Le Guillou, 1842: 315.

***townsvillei.*** *Polyrhachis* (*Cyrtomyrma*) *townsvillei* Donisthorpe, 1938c: 251, fig. 4 (w.q.) AUSTRALIA.

***tragos.*** *Polyrhachis* (*Myrmhopla*) *tragos* Stitz, 1925: 133, fig. 9 (w.) PHILIPPINES.

***transiens.*** *Polyrhachis transiens* Bolton, 1973b: 340, figs. 47, 59 (w.q.) UGANDA.

*transversaria. Polyrhachis militaris* r. *cupreopubescens* var. *transversaria* Forel, 1901h: 77 (w.) LIBERIA. Unavailable name; material referred to *militaris* by Bolton,1973: 313.

***trapezoidea.*** *Polyrhachis trapezoidea* Mayr, 1876: 72 (w.q.m.) AUSTRALIA. Combination in *P.* (*Hagiomyrma*): Forel, 1915b: 108. Race of *ammon:* Forel, 1915b: 108. Revived status as species: Emery, 1925b: 185; Taylor & Brown, D.R. 1985: 142.

***triaena.*** *Polyrhachis* (*Myrmothrinax*) *triaena* Wheeler, W.M. 1919e: 127 (w.) BORNEO.

***tricuspis.*** *Polyrhachis tricuspis* André, 1887: 284 (q.) INDONESIA (Ambon I.). Combination in *P.* (*Myrmothrinax*): Emery, 1925b: 184.

***trina.*** *Polyrhachis* (*Morleyidris*) *trina* Donisthorpe, 1944c: 64, fig. 1 (w.) NEW GUINEA. Combination in *P.* (*Hedomyrma*): current combination.

*tripellis. Polyrhachis* (*Chariomyrma*) *gab* var. *tripellis* Forel, 1915b: 108 (w.q.) AUSTRALIA. Junior synonym of *gab:* Bolton, 1974b: 173; of *senilis:* Kohout, 1988a: 50.

***trispinosa.*** *Polyrhachis trispinosus* Smith, F. 1861b: 40, pl. 1, fig. 11 (q.) SULAWESI. Combination in *P.* (*Myrmothrinax*): Emery, 1925b: 184.

***tristis.*** *Polyrhachis tristis* Mayr, 1867a: 46 (w.) JAVA. Combination in *P.* (*Myrmhopla*): Emery, 1925b: 196.

#***tritschleri.*** *Polyrhachis striata* r. *tritschleri* Forel, 1912d: 111 (footnote) (w.) SUMATRA. Combination in *P.* (*Myrma*): Viehmeyer, 1916a: 166.

***trophima.*** *Polyrhachis trophimus* Smith, F. 1863: 14 (w.) INDONESIA (Seram I.). Combination in *P.* (*Hagiomyrma*): Emery, 1925b: 185; in *P.* (*Chariomyrma*): Donisthorpe, 1932c: 469.

***tschu.*** *Polyrhachis tschu* Forel, 1879a: 122 (q.) CHINA. Combination in *P.* (*Myrmhopla*): Emery, 1925b: 191. Subspecies of *sexspinosa:* Forel, 1909d: 232. Revived status as species: Bolton, 1975c: 13.

***tubericeps.*** *Polyrhachis tubericeps* Forel, 1893c: 33 (w.) INDIA. Combination in *P.* (*Myrmhopla*): Emery, 1925b: 196. See also: Bingham, 1903: 391.

***tubifera.*** *Polyrhachis tubifera* Forel, 1902h: 517 (w.m.) AUSTRALIA. Combination in *P.* (*Hagiomyrma*): Emery, 1925b: 185.

***tubifex.*** *Polyrhachis* (*Myrmhopla*) *tubifex* Karavaiev, 1926a: 144 (w.q.m.) JAVA. [Also described as new by Karavaiev, 1927e: 31.]

***turneri.*** *Polyrhachis turneri* Forel, 1895a: 45 (w.) AUSTRALIA. Wheeler, G.C. & Wheeler, J. 1974a: 62 (l.). Combination in *P.* (*Hedomyrma*): Emery, 1925b: 190.

***tyrannica.*** *Polyrhachis tyrannicus* Smith, F. 1858b: 69, pl. 4, fig. 29 (w.) CHINA. Combination in *P.* (*Myrma*): Emery, 1925b: 204.

*ugandensis. Polyrhachis* (*Myrma*) *fissa* subsp. *ugandensis* Arnold, 1954: 294, fig. 6 (w.) UGANDA. Junior synonym of *fissa:* Bolton, 1973b: 304.

***ugiensis.*** *Polyrhachis* (*Cyrtomyrma*) *rastellata* subsp. *ugiensis* Mann, 1919: 389 (w.) SOLOMON IS. Raised to species: Donisthorpe, 1938c: 260.

***ulysses.*** *Polyrhachis ulysses* Forel, 1910b: 91 (w.q.m.) SOLOMON IS. Mann, 1919: 383 (q.). Combination in *P.* (*Myrmatopa*): Mann, 1919: 383.

*uncinata. Polyrhachis uncinata* André, 1896b: 252 (w.) THAILAND. Junior synonym of *arachne:* Emery, 1898a: 230.

#***unicolor.*** *Polyrhachis vestita* var. *unicolor* Emery, 1898b: 242 (w.) SULAWESI. Combination in *P.* (*Myrma*): Emery, 1925b: 203.

***unicuspis.*** *Polyrhachis unicuspis* Emery, 1898b: 240, fig. 13 (w.) SULAWESI. Combination in *P.* (*Myrmothrinax*): Emery, 1925b: 184.

#***unisculpta.*** *Polyrhachis dahli* var. *unisculpta* Viehmeyer, 1914c: 48 (w.q.) NEW GUINEA. Combination in *P.* (*Myrmothrinax*): Emery, 1925b: 183.

***urania.*** *Polyrhachis urania* Forel, 1902h: 516 (w.) AUSTRALIA. Combination in *P.* (*Chariomyrma*): Emery, 1925b: 187.

*vaga. Polyrhachis* (*Myrma*) *relucens* st. *andromache* var. *vaga* Santschi, 1932b: 21, fig. 6 (w.) INDONESIA (Halmahera I.). Unavailable name.

*valerus. Polyrhachis valerus* Smith, F. 1861b: 40, pl. 1, fig. 10 (w.) SULAWESI. Combination in *P.* (*Hedomyrma*): Emery, 1925b: 190. Junior synonym of *fervens:* Kohout, 1988c: 434.

#***varicolor.*** *Polyrhachis* (*Campomyrma*) *fruhstorferi* subsp. *varicolor* Viehmeyer, 1916a: 163 (w.q.) SINGAPORE. Combination in *P.* (*Myrmatopa*): Emery, 1925b: 180.

*variolosa. Polyrhachis variolosa* Emery, 1887a: 236, pl. 4, fig. 24 (w.) NEW GUINEA. Combination in *P.* (*Myrmhopla*): Emery, 1925b: 192. Junior synonym of *bubastes:* Bolton, 1975c: 6.

***venus.*** *Polyrhachis venus* Forel, 1893c: 31 (w.) BURMA. Combination in *P.* (*Myrmhopla*): Emery, 1925b: 194. See also: Bingham, 1903: 392.

***vermiculosa.*** *Polyrhachis guerini* var. *vermiculosa* Mayr, 1876: 74 (w.q.m.) AUSTRALIA. Combination in *P.* (*Chariomyrma*): Emery, 1925b: 186. Variety/subspecies of *aurea:* Emery, 1897d: 584 (in key); of *guerini:* Taylor & Brown, D.R. 1985: 135. Raised to species: Kohout & Taylor, 1990: 520.

***verticalis.*** *Polyrhachis* (*Chariomyrma*) *verticalis* Santschi, 1928h: 139. Replacement name for *abrupta* Karavaiev, 1927e: 19. [Junior primary homonym of *abrupta* Mayr, 1867a: 62.]

***vestita.*** *Polyrhachis vestitus* Smith, F. 1860a: 71 (w.) SULAWESI. Combination in *P.* (*Myrma*): Emery, 1925b: 203. Current subspecies: nominal plus *unicolor.*

*vibidia. Polyrhachis vibidia* Smith, F. 1861b: 42, pl. 1, fig. 17 (q.) SULAWESI. Junior synonym of *cleophanes:* Forel, 1911e: 299.

*vicina. Polyrhachis vicina* Roger, 1863b: 7. Replacement name for *affinis* Smith, F. 1858b: 63. [Junior secondary homonym of *affinis* Le Guillou, 1842: 314.] Combination in *P.* (*Myrmhopla*): Emery, 1925b: 197. Junior synonym of *dives:* Wang & Wu, 1991: 599.

*victoris. Polyrhachis ypsilon* var. *victoris* Santschi, 1925f: 93 (w.) SUMATRA. Junior synonym of *ypsilon:* Hung, 1970: 19.

***viehmeyeri.*** *Polyrhachis* (*Myrmhopla*) *viehmeyeri* Emery, 1921e: 19 (w.) NEW GUINEA. Kohout, 1990: 508 (q.).

***vigilans.*** *Polyrhachis vigilans* Smith, F. 1858b: 69, pl. 4, fig. 39 (q.) CHINA. Donisthorpe, 1937a: 626 (w.). Combination in *P.* (*Myrma*): Emery, 1925b: 204.

***villipes.*** *Polyrhachis villipes* Smith, F. 1857a: 61 (w.) BORNEO. Mayr, 1872: 140 (q.); Forel, 1915a: 43 (m.). Combination in *P.* (*Myrma*): Viehmeyer, 1916a: 166. Current subspecies: nominal plus *noesaensis.*

***villosa.*** *Polyrhachis hostilis* var. *villosa* Emery, 1897d: 584, pl. 1, figs. 22, 23 (w.q.) NEW GUINEA. Combination in *P.* (*Chariomyrma*): Emery, 1925b: 188. Raised to species: Viehmeyer, 1914c: 59. Current subspecies: nominal plus *pubiventris.*

***vindex.*** *Polyrhachis vindex* Smith, F. 1857a: 64 (w.) BORNEO. Combination in *P.* (*Myrma*): Wheeler, W.M. 1919e: 125.

***violaceonigra.*** *Polyrhachis violaceonigra* Viehmeyer, 1914c: 53 (w.) NEW GUINEA. Combination in *P.* (*Hedomyrma*): Emery, 1925b: 190.

***viscosa.*** *Polyrhachis viscosus* Smith, F. 1858b: 71, pl. 4, fig. 41 (w.) SOUTH AFRICA. Combination in *P.* (*Myrma*): Santschi, 1914b: 143. Senior synonym of *antinorii:* Dalla Torre, 1893: 271; of *imatongica* and *spretula:* Bolton, 1973b: 330. See also: Arnold, 1924: 749.

***vitalisi.*** *Polyrhachis* (*Cyrtomyrma*) *vitalisi* Santschi, 1920a: 567 (w.q.m.) LAOS.

***volkarti.*** *Polyrhachis* (*Myrma*) *revoili* r. *volkarti* Forel, 1916: 453 (q.) ZAIRE. Combination in *P.* (*Pseudocyrtomyrma*): Emery, 1925b: 207. Raised to species, and senior synonym of *kohli:* Bolton, 1973b: 341.

***wagneri.*** *Polyrhachis wagneri* Viehmeyer, 1914c: 51, fig. 10 (w.) NEW GUINEA. Combination in *P.* (*Cyrtomyrma*): Emery, 1925b: 208.

*waigeuensis. Polyrhachis* (*Myrmhopla*) *arcuspinosa* subsp. *waigeuensis* Donisthorpe, 1943d: 467 (w.) NEW GUINEA. Junior synonym of *sexspinosa:* Bolton, 1975c: 12.

*waigiouensis. Polyrhachis variolosa* var. *waigiouensis* Forel, 1911e: 299 (w.) NEW GUINEA. Junior synonym of *bubastes:* Bolton, 1975c: 6.

***wallacei.*** *Polyrhachis wallacei* Emery, 1887a: 223, pl. 3, fig. 10 (w.q.) SULAWESI. Combination in *P.* (*Myrmatopa*): Forel, 1915b: 107. Current subspecies: nominal plus *warburgi.*

#***warburgi.*** *Polyrhachis wallacei* r. *wartburgi* Forel, 1901h: 76 (w.) JAVA. Combination in *P.* (*Myrmatopa*), and emendation of spelling to *warburgi:* Emery, 1925b: 180.

*weberi. Polyrhachis* (*Myrmhopla*) *weberi* Donisthorpe, 1943b: 206 (w.) INDIA. Junior synonym of *gracilior:* Bolton, 1974b: 174.

***weissi.*** *Polyrhachis weissi* Santschi, 1910c: 395, fig. 18 (w.) CONGO. Combination in *P.* (*Pseudocyrtomyrma*): Emery, 1925b: 207; in *P.* (*Myrma*): Wheeler, W.M. 1922a: 1004; Bolton, 1973b: 342. Variety/stirps of *revoili:* Forel, 1916: 453; Emery, 1925b: 207; Santschi, 1939c: 12. Status as species: Wheeler, W.M. 1922a: 1004; Bolton, 1973b: 342. Senior synonym of *balli, conduensis* and *crassa,* and material of the unavailable name *phaenogaster* referred here: Bolton, 1973b: 342.

***wellmani.*** *Polyrhachis wellmani* Forel, 1909b: 68 (w.) ANGOLA. Combination in *P.* (*Myrma*): Wheeler, W.M. 1922a: 1004. See also: Bolton, 1973b: 323.

#***weyeri.*** *Polyrhachis* (*Myrmhopla*) *bicolor* var. *weyeri* Karavaiev, 1930a: 212 (w.m.) INDONESIA (Ambon I.).

***wheeleri.*** *Polyrhachis* (*Myrmhopla*) *wheeleri* Mann, 1919: 387, fig. 58 (w.) SOLOMON IS. Donisthorpe, 1941h: 63 (q.); Wheeler, G.C. & Wheeler, J. 1970: 649 (l.).

*wilmsi. Polyrhachis cubaensis* subsp. *wilmsi* Forel, 1910f: 30 (w.) MOZAMBIQUE. Combination in *P.* (*Myrma*): Wheeler, W.M. 1922a: 995. Junior synonym of *cubaensis:* Bolton, 1973b: 325.

***wolfi.*** *Polyrhachis wolfi* Forel, 1912a: 79 (w.) TAIWAN. Combination in *P.* (*Myrma*): Emery, 1925b: 204.

***wroughtonii.*** *Polyrhachis wroughtonii* Forel, 1894c: 398 (w.) INDIA. Forel, 1895e: 457 (q.m.). Combination in *P.* (*Myrmhopla*): Emery, 1925b: 191.

***xanthippe.*** *Polyrhachis xanthippe* Forel, 1911a: 61 (w.) SRI LANKA. Combination in *P.* (*Myrmhopla*): Emery, 1925b: 197.

***xiphias.*** *Polyrhachis xiphias* Smith, F. 1863: 16 (q.) NEW GUINEA. Combination in *P.* (*Hagiomyrma*): Emery, 1925b: 185.

***yarrabahensis.*** *Polyrhachis* (*Myrmatopa*) *lombokensis* var. *yarrabahensis* Forel, 1915b: 115 (w.q.) AUSTRALIA. Raised to species: Kohout & Taylor, 1990: 520.

***yerburyi.*** *Polyrhachis yerburyi* Forel, 1893c: 29 (w.q.) SRI LANKA. Combination in *P.* (*Myrma*): Emery, 1925b: 203. [*P. yerburyi* Emery, 1893f: 250. *Nomen nudum;* attributed to Forel.]

***yorkana.*** *Polyrhachis* (*Cyrtomyrma*) *rastellata* var. *yorkana* Forel, 1915b: 110 (w.) AUSTRALIA. Raised to species: Kohout & Taylor, 1990: 521.

***ypsilon.*** *Polyrhachis ypsilon* Emery, 1887a: 239 (w.) SINGAPORE. Wheeler, G.C. & Wheeler, J. 1990b: 764 (l.). Senior synonym of *synacantha* and *victoris:* Hung, 1970: 19.

***zimmerae.*** *Polyrhachis* (*Campomyrma*) *zimmerae* Clark, 1941: 92, pl. 13, fig. 25 (w.) AUSTRALIA.

***zopyra.*** *Polyrhachis zopyrus* Smith, F. 1861b: 43, pl. 1, fig. 22 (w.) SULAWESI. Combination in *P.* (*Myrma*), and senior synonym of *aurichalceus:* Emery, 1925b: 204. Senior synonym of *moorei:* Bolton, 1974b: 179. Current subspecies: nominal plus *edentula, imbellis.*

---

**EXCURSUS**

**Species-rank taxa of *POLYRHACHIS,* grouped by subgenus.**

The subgenera of *Polyrhachis* are arranged alphabetically below in ***BOLD ITALIC CAPITALS.*** The species currently assigned to each subgenus follow the subgeneric name in alphabetical order.

#***ANOPLOMYRMA:*** *parabiotica, paracamponota.*

#***AULACOMYRMA:*** *carbonaria, dohrni, exarata, excellens, geometrica, mystica, pallipes, porcata, schizospina.*

#***CAMPOMYRMA:*** *cedarensis, creusa, equina, exercita, femorata, flavibasis, fuscipes, gravis, gribodoi, halidayi, hauxwelli, hexacantha, hirsuta, horacei, inconspicua, insularis, io, jacksoniana, leae, macropa, maculata, micans, ops, patiens, phryne, polymnia, prometheus, pseudothrinax, pyrrhus, schwiedlandi, sculpta, semipolita, sidnica, spengeli, tambourinensis, templi, zimmerae.*

#***CHARIOMYRMA:*** *antennata, appendiculata, arcuata, argenteosignata, aurea, auriformis, aurita, beauforti, bedoti, caulomma, cingula, coerulescens, constricta, contemta, cyrus, escherichi, eurynota, fulgens, gab, guerini, heinlethii, hookeri, hostilis, inclusa, kaipi, karawaiewi, laciniata, lata, latinota, latreillei, leopoldi, limbata, loriae, lownei, marginata, modesta, nigrescens, nitens, obscura, obtusa, opalescens, pallescens, punctiventris, rere, rotumana, rotundiceps, rowlandi, scapulata, schoopae, scutulata, senilis, splendens, subaenescens, subcyanea, trophima, urania, vermiculosa, verticalis, villosa.*

#***CYRTOMYRMA:*** *albertisi, australis, burmanensis, coronata, debilis, demangei, dentata, doddi, emeryana, euryala, fornicata, gibba, grandis, jurii, laevissima, leonidas, levior, linae, luctuosa, mackayi, mondoi, pilosa, ralumensis, rastellata, townsvillei, ugiensis, vitalisi, wagneri, yorkana.*

#***DOLICHORHACHIS:*** *dolichocephala, malaensis.*

#***HAGIOMYRMA:*** *ammon, ammonoeides, angusta, crawleyi, denticulata, lachesis, lydiae, metella, paxilla, penelope, schenckii, semiaurata, semiobscura, sokolova, thusnelda, trapezoidea, tubifera, xiphias.*

#***HEDOMYRMA:*** *annae, argentosa, atropos, barretti, calliope, cleopatra, clio, clotho, consimilis, cupreata, daemeli, erato, euterpe, fervens, geminata, hera, hermione, hungi, machaon, mjobergi, ornata, rufifemur, santschii, terpsichore, thais, trina, turneri, violaceonigra.*

#***HEMIOPTICA:*** *bugnioni, scissa.*

#***MYRMA:*** *abrupta, aculeata, aenescens, aerope, alexisi, alluaudi, andrei, andromache, arnoldi, asomaningi, bakeri, beccarii, biroi, braxa, ceramensis, compressicornis, concava, conops, continua, convexa, cornuta, crassispinosa, cubaensis, curta, cyaniventris, decellei, decemdentata, diana, dorsorugosa, durbanensis, esarata, fissa, foreli, gagates, gamaii, hastata, hemiopticoides, horni, illaudata, inermis, inusitata, isacantha, ithona, khepra, labella, laboriosa, lanuginosa, latharis, latispina, latona, lauta, lestoni, limitis, lycidas, medusa, militaris, monista, murina, nigra, nigrita, nigropilosa, numeria, olena, orsylla, otleti, phidias, philippinensis, platyomma, proxima, pruinosa, pubescens, punctillata, pyrgops, regesa, relucens, restituta, revoili, rixosa, rufipalpis, rufofemorata, salomo, schistacea, schlueteri, sculpturata, sericata, sericeopubescens, similis, spinicola, spitteleri, striata, subpilosa, sulcata, sumatrensis, transiens, tyrannica, vestita, vigilans, villipes, vindex, viscosa, volkarti, weissi, wellmani, wolfi, yerburyi, zopyra.*

#***MYRMATOPA:*** *alphea, antoniae, bouvieri, charaxa, chartifex, constructor, derecyna, dolomedes, edwardi, elii, flavicornis, follicula, fruhstorferi, furcula, jacobsoni, lilianae, lombokensis, menozzii, omyrmex, osae, phalerata, piliventris, rossi, ruficornis, schang, simillima, solivaga, solmsi, subtridens, ulysses, wallacei, yarrabahensis.*

#***MYRMHOPLA:*** *aedipus, alexandri, alphena, amana, arachne, armata, aspasia, atrovirens, aureovestita, bamaga, banghaasi, basirufa, batesi, bicolor, binghamii, bubala, bubastes, caeciliae, calypso, castaneiventris, cephalotes, chalybea, cleophanes, croceiventris, cryptoceroides, cyrtomyrmoides, daphne, davydovi, diaphanta, diotima, distincta, dives, emmae, eremita, etheli, exotica, flavoflagellata, fortis, furcata, gestroi, glabrinota, glykera, gracilior, greensladei, hector, hippomanes, hirta, hodgsoni, ignota, indificans, jerdonii, jianghuaensis, keratifera, lacteipennis, laevigata, lama, laminata, longipes, loweryi, lucens, lugens, magnifica, melpomene, menelas, mitrata, moeschi, moesta, mucronata, muelleri, nigriceps, nitida, nofra, nudata, ochracea, oedacantha, olybria, orpheus, osiris, paromala, pellita, peregrina, personata, plato, platynota, pressa, punctata, reclinata, regularis, retrorsa, rhea, ridleyi, rubigastrica, rubiginosa, rufipes, rugifrons, rupicapra, rustica, saevissima, sappho, scabra, sexspinosa, smithi, sophocles, stigmatifera, strictifrons, stylifera, subfossa, sub-*

***POLYRHACHIS*** *(continued)*

*fossoides, sylvicola, thompsoni, tibialis, tragos, tristis, tschu, tubericeps, tubifex, venus, viehmeyeri, wheeleri, wroughtonii, xanthippe.*

***#MYRMOTHRINAX:*** *abnormis, atossa, cheesmanae, dahlii, delicata, durvillei, eudora, frauenfeldi, neptunus, queenslandica, saigonensis, sparaxes, ternatae, textor, thrinax, triaena, tricuspis, trispinosa, unicuspis.*

***#POLYRHACHIS:*** *bellicosa, bihamata, craddocki, erosispina, lamellidens, mindanaensis, montana, taylori, ypsilon.*

**SUBGENUS UNKNOWN OR UNPLACED TO SUBGENUS**: *agesilas, carinata, clarkei, excisa.*

---

***PONERA*** (Holarctic, Oriental, Indo-Australian, Australasian)

*abeillei* André, 1881; see under ***HYPOPONERA.***

*abyssinica* Guérin-Méneville, 1849; see under ***PACHYCONDYLA.***

*abyssinica* Santschi, 1938; see under ***HYPOPONERA.***

*acuticostata* Forel, 1900; see under ***PACHYCONDYLA.***

*aemula* Santschi, 1911; see under ***HYPOPONERA.***

*aenigmatica* Arnold, 1949; see under ***PACHYCONDYLA.***

*aethiopica* Smith, F. 1858; see under ***STREBLOGNATHUS.***

*aethiopica* Forel, 1907; see under ***HYPOPONERA.***

**affinis* Heer, 1850; see under ****PONEROPSIS.***

*affinis. Ponera affinis* Jerdon, 1851: 118 (w.) INDIA. [Unresolved junior primary homonym of **affinis* Heer, 1850: 147, above.] Unidentifiable to genus; *incertae sedis* in *Ponera.*

*agilis* Borgmeier, 1934; see under ***HYPOPONERA.***

*aitkenii* Forel, 1900; see under ***HYPOPONERA.***

*albopubescens* Menozzi, 1939; see under ***HYPOPONERA.***

*aliena* Smith, F. 1858; see under ***HYPOPONERA.***

***alisana.*** *Ponera alisana* Terayama, 1986: 591, figs. 1–5 (w.q.) TAIWAN.

***alpha.*** *Ponera alpha* Taylor, 1967a: 38, figs. 23, 24 (w.q.l.) NEW GUINEA.

*ambigua* Weber, 1942; see under ***PACHYCONDYLA.***

*amblyops* Emery, 1887; see under ***PACHYCONDYLA.***

*americana* Mayr, 1862; see under ***PACHYCONDYLA.***

*amplinoda* Buckley, 1866; see under ***PACHYCONDYLA.***

*ancilla* Emery, 1899; see under ***PACHYCONDYLA.***

*andrei* Emery, 1900; see under ***HYPOPONERA.***

*androgyna* Roger, 1859; see under ***HYPOPONERA.***

*annamita* André, 1892; see under ***PACHYCONDYLA.***

*anommata* Donisthorpe, 1948; see under ***CRYPTOPONE.***

*antipodum* Forel, 1895; see under ***HYPOPONERA.***

*antoniensis* Forel, 1912; see under ***HYPOPONERA.***

*apicalis* Smith, F. 1857; see under ***PACHYCONDYLA.***

*araneoides* Le Guillou, 1842; see under ***RHYTIDOPONERA.***

*argentina* Santschi, 1922; see under ***HYPOPONERA.***

*argonautorum* Arnol'di, 1932; see under ***HYPOPONERA.***

*arnoldi* Forel, 1913; see under ***PACHYCONDYLA.***

*aspera* Roger, 1860; see under ***RHYTIDOPONERA.***

*assmuthi* Forel, 1905; see under ***HYPOPONERA.***

****atavia.*** **Ponera atavia* Mayr, 1868c: 72, pl. 4, figs. 66–69 (q.m.) BALTIC AMBER (Oligocene). Wheeler, W.M. 1915h: 38 (w.). See also: Taylor, 1967a: 29.

*atlantis. Ponera coarctata* var. *atlantis* Santschi, 1921e: 166, fig. 2 (w.) TUNISIA. Junior synonym of *coarctata:* Taylor, 1967a: 21.

*attenuata* Smith, F. 1858; see under ***LEPTOGENYS.***

*attrita* Forel, 1893; see under ***PACHYCONDYLA.***

***augusta.*** *Ponera augusta* Taylor, 1967a: 68, figs. 35, 59, 60 (w.) NEW GUINEA.

*aurata* Roger, 1861; see under ***RHYTIDOPONERA.***

*australis* Forel, 1900; see under ***PACHYCONDYLA.***

***bableti.*** *Ponera bableti* Perrault, 1993: 334 (w.) MURUROA (Fangataufa Atoll).

*beebei* Wheeler, W.M. 1924; see under ***HYPOPONERA.***

*bellicosa* Forel, 1895; see under ***HYPOPONERA.***

*bicolor* Guérin-Méneville, 1844; see under ***PACHYCONDYLA.***

*bicolor* Donisthorpe, 1949; see under ***PACHYCONDYLA.***

*biroi* Emery, 1900; see under ***HYPOPONERA.***

*bispinosa* Le Guillou, 1842; see under ***DIACAMMA.***

*boerorum* Forel, 1901; see under ***HYPOPONERA.***

*bondroiti* Forel, 1911; see under ***HYPOPONERA.***

***borneensis.*** *Ponera borneensis* Taylor, 1967a: 69, figs. 61, 62 (w.q.) BORNEO.

*brachycola* Roger, 1861; see under ***CENTROMYRMEX.***

*breviceps* Bernard, 1953; see under ***HYPOPONERA.***

*brevior* Forel, 1901; see under ***PACHYCONDYLA.***

*brevis* Santschi, 1921; see under ***HYPOPONERA.***

*bugnioni* Forel, 1912; see under ***HYPOPONERA.***

*bulawayensis* Forel, 1913; see under ***HYPOPONERA.***

*butteli* Forel, 1913; see under ***HYPOPONERA.***

*caeca* Donisthorpe, 1949; see under ***PROCERATIUM.***

*caffra* Spinola, 1851; see under ***PLECTROCTENA.***

*caffraria* Smith, F. 1858; see under ***PACHYCONDYLA.***

*caledonica. Ponera caledonica* Wilson, 1957b: 361, fig. 2 (w.q.) NEW CALEDONIA. Junior synonym of *leae:* Taylor, 1967a: 89.

*cammerunensis* Santschi, 1914; see under ***HYPOPONERA.***

*carbonaria* Smith, F. 1858; see under ***PACHYCONDYLA.***

*carinulata* Roger, 1861; see under ***PACHYCONDYLA.***

*castanea* Mayr, 1865; see under ***PACHYCONDYLA.***

*castanea* Smith, F. 1876; see under ***PACHYCONDYLA.***

*castaneicolor* Dalla Torre, 1893; see under ***PACHYCONDYLA.***

*cauta* Forel, 1912; see under ***HYPOPONERA.***

*cavernosa* Roger, 1860; see under ***PACHYCONDYLA.***

*ceylonensis* Mayr, 1897; see under ***HYPOPONERA.***

***chapmani.*** *Ponera chapmani* Taylor, 1967a: 47, figs. 6, 29, 30 (w.q.) PHILIPPINES.

*chilensis* Forel, 1914; see under ***HYPOPONERA.***

*chinensis* Emery, 1895; see under ***PACHYCONDYLA.***

***chiponensis.*** *Ponera chiponensis* Terayama, 1986: 593, figs. 6–10 (w.q.) TAIWAN.

*clavatula* Emery, 1906; see under ***HYPOPONERA.***

***clavicornis.*** *Ponera clavicornis* Emery, 1900c: 317, pl. 8, figs. 7, 8 (w.) NEW GUINEA. Taylor, 1967a: 74 (q.). Combination in *Selenopone:* Wheeler, W.M. 1933g: 22; in *Ponera (Hypoponera):* Santschi, 1938b: 79; in *Ponera:* Wilson, 1957b: 377; Taylor, 1967a: 73.

***coarctata.*** *Formica coarctata* Latreille, 1802b: 65, pl. 3, fig. 1 (w.q.) LUXEMBOURG. Foerster, 1850a: 46 (m.). Combination in *Ponera:* Latreille, 1804: 178. Senior synonym of *contracta:* Roger, 1863b: 16. Senior synonym of *atlantis, crassisquama, lucida* and *testacea,* and material of the unavailable names *colchica* and *taurica* referred here: Taylor, 1967a: 21. See also: Emery, 1909c: 368; Donisthorpe, 1915d: 67; Stitz, 1939: 58.

*coeca* Santschi, 1914; see under ***HYPOPONERA.***

*cognata* Emery, 1896; see under ***PACHYCONDYLA.***

*cognata* Santschi, 1912; see under ***HYPOPONERA.***

***colaensis.*** *Ponera colaensis* Mann, 1921: 417 (w.) FIJI IS. Santschi, 1928c: 68 (q.). See also: Taylor, 1967a: 58.

*colchica. Ponera coarctata* subsp. *coarctata* nat. *colchica* Arnol'di, 1932b: 63 (w.) CAUCASUS. Unavailable name; material referred to *coarctata* by Taylor, 1967a: 21.

*collegiana* Santschi, 1925; see under ***HYPOPONERA.***

*commutata* Roger, 1860; see under ***PACHYCONDYLA.***

*comorensis* André, 1887; see under ***PACHYCONDYLA.***

*confinis* Roger, 1860; see under ***HYPOPONERA.***

*congrua* Wheeler, W.M. 1934; see under ***HYPOPONERA.***

*constricta* Mayr, 1884; see under ***PACHYCONDYLA.***

*contracta. Formica contracta* Latreille, 1802c: 195, pl. 7, fig. 40 (w.q.) LUXEMBOURG. [Unnecessary replacement name for *coarctata* Latreille, 1802b: 65.] Junior synonym of *coarctata:* Roger, 1863b: 16.

*convexiuscula* Forel, 1900; see under ***HYPOPONERA.***

*coxalis* Roger, 1860; see under ***GNAMPTOGENYS.***

*crassa* Emery, 1877; see under ***PACHYCONDYLA.***

*crassicornis* Gerstäcker, 1859; see under ***PACHYCONDYLA.***

**crassinervis* Heer, 1850; see under ****PONEROPSIS.***

*crassisquama. Ponera coarctata* var. *crassisquama* Emery, 1916a: 54, fig. 1 (w.q.) ITALY. [Also described as new by Emery, 1916b: 108.] Junior synonym of *coarctata:* Taylor, 1967a: 21.

*crenata* Roger, 1861; see under ***PACHYCONDYLA.***

*creola* Menozzi, 1931; see under ***HYPOPONERA.***

*cribrinodis* Gerstäcker, 1859; see under ***PLATYTHYREA.***

**croatica* Heer, 1850; see under ****PONEROPSIS.***

*crocea* Roger, 1860; see under ***PROCERATIUM.***

*crocea. Ponera japonica* var. *crocea* Santschi, 1941: 273 (w.) JAPAN. [Unresolved junior primary homonym of *crocea* Roger, 1860: 288, above.] Junior synonym of *japonica:* Taylor, 1967a: 76.

*crudelis* Smith, F. 1858; see under ***LEPTOGENYS.***

*cubana* Santschi, 1930; see under ***HYPOPONERA.***

*cuprea* Smith, F. 1860; see under ***DIACAMMA.***

*dahlii* Forel, 1901; see under ***PACHYCONDYLA.***

*debilior* Forel, 1901; see under ***PACHYCONDYLA.***

*decipiens* Forel, 1899; see under ***HYPOPONERA.***

*decora* Clark, 1934; see under ***HYPOPONERA.***

*denticulata* Smith, 1858; see under ***ODONTOPONERA.***

*devota* Santschi, 1914; see under ***HYPOPONERA.***

*dideroti* Forel, 1913; see under ***HYPOPONERA.***

*diminuta* Smith, F. 1857; see under ***LEPTOGENYS.***

*dispar* Santschi, 1924; see under ***HYPOPONERA.***

*distinguenda* Emery, 1890; see under ***HYPOPONERA.***

*dolo* Roger, 1860; see under ***HETEROPONERA.***

*dubitata* Forel, 1900; see under ***PACHYCONDYLA.***

*dulcis* Forel, 1907; see under ***HYPOPONERA.***

*durbanensis* Forel, 1914; see under ***HYPOPONERA.***
*eduardi* Forel, 1894; see under ***HYPOPONERA.***
***elegantissima.*** **Ponera elegantissima* Meunier, 1923: 609 (caste ?) GERMANY (Oligocene).
***elegantula.*** *Ponera elegantula* Wilson, 1957b: 379 (w.q.) NEW GUINEA. Taylor, 1967a: 72 (m.l.).
*elisae* Forel, 1891; see under ***PACHYCONDYLA.***
*elliptica* Forel, 1900; see under ***HYPOPONERA.***
*elongata* Buckley, 1866; see under ***LEPTOGENYS.***
**elongatula* Heer, 1850; see under ****PONEROPSIS.***
*emeryi* Donisthorpe, 1943; see under ***HYPOPONERA.***
*epinotalis* Viehmeyer, 1916; see under ***HYPOPONERA.***
*ergatandria* Forel, 1893; see under ***HYPOPONERA.***
*eutrepta* Wilson, 1958; see under ***HYPOPONERA.***
*exacta* Santschi, 1923; see under ***HYPOPONERA.***
*excedra. Ponera excedra* Wilson, 1957b: 364 (w.q.) AUSTRALIA. Junior synonym of *leae:* Taylor, 1967a: 89
*excelsior* Forel, 1892; see under ***HYPOPONERA.***
*exoecata* Wheeler, W.M. 1928; see under ***HYPOPONERA.***
***exotica.*** *Ponera exotica* Smith, M.R. 1962b: 378, fig. 1 (w.q.) U.S.A. See also: Taylor, 1967a: 96; Johnson, 1987: 358.
*faceta* Menozzi, 1931; see under ***HYPOPONERA.***
*fallax* Forel, 1909; see under ***HYPOPONERA.***
*famini* Forel, 1912; see under ***HYPOPONERA.***
*fenestralis* Gallardo, 1918; see under ***HYPOPONERA.***
*ferox* Smith, F. 1865; see under ***LEPTOGENYS.***
*ferruginea* Smith, F. 1858; see under ***PACHYCONDYLA.***
*fidelis* Santschi, 1926; see under ***HYPOPONERA.***
*fiebrigi* Forel, 1908; see under ***HYPOPONERA.***
*foeda* Forel, 1893; see under ***HYPOPONERA.***
*foreli* Mayr, 1887; see under ***HYPOPONERA.***
*formosae* Forel, 1913; see under ***HYPOPONERA.***
*fossulata* Forel, 1900; see under ***PACHYCONDYLA.***
*fugitans* Forel, 1912; see under ***HYPOPONERA.***
**fuliginosa* Herr, 1850; see under ****PONEROPSIS.***
*fuscoatra* Roger, 1861; see under ***PACHYCONDYLA.***
*fuscula* Emery, 1902; see under ***HYPOPONERA.***
*gagates* Guérin-Méneville, 1844; see under ***PACHYCONDYLA.***
*gaigei* Forel, 1914; see under ***HYPOPONERA.***
*geometrica* Smith, F. 1857; see under ***DIACAMMA.***
*gibbinota* Forel, 1912; see under ***HYPOPONERA.***
*gigantea* Perty, 1833; see under ***DINOPONERA.***
*gilva* Roger, 1863; see under ***CRYPTOPONE.***
*glabrata* Forel, 1895; see under ***HYPOPONERA.***
*gleadowi* Forel, 1895; see under ***HYPOPONERA.***
**globosa* Heer, 1850; see under ****PONEROPSIS.***
**gracilicornis* Mayr, 1868; see under ***PACHYCONDYLA.***
*gracilicornis* Menozzi, 1931; see under ***HYPOPONERA.***
*grandidieri* Santschi, 1921; see under ***HYPOPONERA.***
*grandis* Guérin-Méneville, 1838; see under ***DINOPONERA.***
*grandis* Donisthorpe, 1947; see under ***PACHYCONDYLA.***
*granosa* Roger, 1860; see under ***PACHYCONDYLA.***
*guatemalensis* Forel, 1899; see under ***CRYPTOPONE.***
*guineensis* André, 1890; see under ***PACHYCONDYLA.***
*gyptis* Santschi, 1921; see under ***HYPOPONERA.***
**hendersoni* Cockerell, 1906; see under ****PROTAZTECA.***
*herbertonensis* Forel, 1915; see under ***HYPOPONERA.***
*histrio* Forel, 1912; see under ***HYPOPONERA.***
*holoserica* Roger, 1860; see under ***DIACAMMA.***
*huonica. Ponera huonica* Wilson, 1957b: 365, fig. 2 (w.q.m.) NEW GUINEA. Junior synonym of *tenuis:* Taylor, 1967a: 98.
**hypolitha* Cockerell, 1915; see under ****PONEROPSIS.***
*idelettae* Santschi, 1923; see under ***HYPOPONERA.***
*ignigera* Menozzi, 1927; see under ***HYPOPONERA.***
*iheringi* Forel, 1908; see under ***HYPOPONERA.***
*imatongica* Weber, 1942; see under ***HYPOPONERA.***
*impressa* Roger, 1861; see under ***PACHYCONDYLA.***
*inaudax* Santschi, 1919; see under ***HYPOPONERA.***
***incerta.*** *Pseudocryptopone incerta* Wheeler, W.M. 1933g: 672, fig. 7 (w.q.) JAVA. Taylor, 1967a: 82 (m.l.). Combination in *Ponera:* Wilson, 1957b: 367. Senior synonym of *ratardorum:* Taylor, 1967a: 80.
*incisa* Santschi, 1914; see under ***HYPOPONERA.***
*indifferens* Forel, 1895; see under ***HYPOPONERA.***
*indigens* Forel, 1895; see under ***HYPOPONERA.***
*inexorata* Wheeler, W.M. 1903; see under ***HYPOPONERA.***
*inexpedita* Forel, 1911; see under ***HYPOPONERA.***
*insulana* Mayr, 1876; see under ***PACHYCONDYLA.***
*intermedia* Forel, 1900; see under ***PACHYCONDYLA.***
*intermedia* Bernard, 1953; see under ***HYPOPONERA.***
*intricata* Smith, F. 1857; see under ***DIACAMMA.***
*inversa* Smith, F. 1858; see under ***PACHYCONDYLA.***
*iridescens* Smith, F. 1857; see under ***LEPTOGENYS.***
*iridipennis* Smith, F. 1858; see under ***LEPTOGENYS.***
*jamaicensis* Aguayo, 1932; see under ***HYPOPONERA.***
***japonica.*** *Ponera japonica* Wheeler, W.M. 1906c: 306 (w.q.) JAPAN. Imai, Brown, *et al.* 1984: 67 (k.). Senior synonym of *crocea:* Taylor, 1967a: 76. See also: Kupyanskaya, 1990: 86.
*javana* Forel, 1905; see under ***HYPOPONERA.***
*jeanneli* Santschi, 1935; see under ***HYPOPONERA.***
*jerdonii* Forel, 1900; see under ***PACHYCONDYLA.***
*johannae* Forel, 1891; see under ***HYPOPONERA.***
*josephi* Forel, 1886; see under ***PACHYCONDYLA.***
*jugata* Forel, 1892; see under ***HYPOPONERA.***
*kalakauae* Forel, 1899; see under ***HYPOPONERA.***
*laeviceps* Smith, F. 1857; see under ***LEPTOGENYS.***
*laevigata* Smith, F. 1858; see under ***PACHYCONDYLA.***
*laeviuscula* Gerstäcker, 1859; see under ***PACHYCONDYLA.***
*lamellosa* Roger, 1860; see under ***PLATYTHYREA.***
*lamellosa* Forel, 1907; see under ***HYPOPONERA.***
*lamottei* Bernard, 1953; see under ***HYPOPONERA.***
*lea* Santschi, 1937; see under ***HYPOPONERA.***
***leae.*** *Ponera leae* Forel, 1913g: 175 (w.) TASMANIA. Wheeler, G.C. & Wheeler, J. 1971b: 1210 (l.). Senior synonym of *caledonica, excedra* and *norfolkensis:* Taylor, 1967a: 89.
*leeuwenhoeki* Forel, 1886; see under ***PACHYCONDYLA.***
*leninei* Santschi, 1925; see under ***HYPOPONERA.***
****leptocephala.*** **Ponera leptocephala* Emery, 1891a: 146, pl. 1, figs. 3, 4 (q.) SICILIAN AMBER (Miocene).
*lesnei* Bondroit, 1916; see under ***HYPOPONERA.***
*leveillei* Emery, 1890; see under ***PACHYCONDYLA.***
*libani. Ponera libani* Tohmé, G. 1969: 7, figs. 4–6 (q.) LEBANON. Transferred to, and *incertae sedis* in, **Family MUTILLIDAE**.
*lincecumii* Buckley, 1866; see under ***PSEUDOMYRMEX.***
*linearis* Smith, F. 1858; see under ***LEPTOGENYS.***
***loi.*** *Ponera loi* Taylor, 1967a: 60, figs. 52–55 (w.q.) SAMOA.
**longaeva* Heer, 1850; see under ****PONEROPSIS.***
*longiceps* Forel, 1913; see under ***HYPOPONERA.***
*longiceps* Menozzi, 1931; see under ***HYPOPONERA.***
*lotti* Weber, 1942; see under ***HYPOPONERA.***
*lucida. Ponera coarctata* var. *lucida* Emery, 1898c: 130 (w.) TURKESTAN. [Misspelled as *lucidula* by Emery, 1909c: 370.] Junior synonym of *coarctata:* Taylor, 1967a: 21.
*ludovicae* Forel, 1892; see under ***HYPOPONERA.***
*lumpurensis* Forel, 1907; see under ***HYPOPONERA.***
*lunaris* Emery, 1896; see under ***PACHYCONDYLA.***
*lutea* Mayr, 1862; see under ***PACHYCONDYLA.***
*luteipes* Mayr, 1862; see under ***PACHYCONDYLA.***
*luteola* Roger, 1861; see under ***PACHYCONDYLA.***
*mackayensis* Forel, 1900; see under ***HYPOPONERA.***
*macra* Emery, 1894; see under ***PACHYCONDYLA.***
*macradelphe* Wilson, 1958; see under ***HYPOPONERA.***
*madecassa* Santschi, 1938; see under ***HYPOPONERA.***
*malayana* Wheeler, W.M. 1929; see under ***HYPOPONERA.***
*maligna* Smith, F. 1861; see under ***MYOPIAS.***
*mandibularis* Bernard, 1953; see under ***HYPOPONERA.***
***manni.*** *Ponera manni* Taylor, 1967a: 86, figs. 76, 77 (w.) FIJI IS.
*marginata* Roger, 1861; see under ***PACHYCONDYLA.***
*massiliensis* Bondroit, 1920; see under ***HYPOPONERA.***
*maxillosa* Smith, F. 1858; see under ***LEPTOGENYS.***
*melanaria* Emery, 1893; see under ***PACHYCONDYLA.***
*melina* Roger, 1860; see under ***PROCERATIUM.***
*menozzii* Santschi, 1932; see under ***HYPOPONERA.***
*mesoepinotalis* Weber, 1942; see under ***HYPOPONERA.***
*metallica* Smith, F. 1858; see under ***RHYTIDOPONERA.***
*mina* Wheeler, W.M. 1927; see under ***HYPOPONERA.***
****minuta.*** **Ponera minuta* Donisthorpe, 1920: 85, pl. 5, fig. 4 (m.?) GREAT BRITAIN (Oligocene). Formicidae *incertae generis:* Taylor, 1964: 139.
*mjobergi* Forel, 1915; see under ***CRYPTOPONE.***
*mocsaryi* Emery, 1900; see under ***HYPOPONERA.***
*mocsaryi. Cryptopone mocsaryi* Szabó, 1910b: 186, 2 figs. (w.) NEW GUINEA. Combination in *Pseudocryptopone:* Wheeler, W.M. 1933g: 14. [Junior secondary homonym of *Ponera mocsaryi* Emery, 1900c: 320, above.] Replacement name: *Ponera szaboi* Wilson, 1957b: 371.
*monticola* Mann, 1921; see under ***HYPOPONERA.***
*mordax* Smith, F. 1858; see under ***GNAMPTOGENYS.***
*mucronata* Roger, 1860; see under ***ACANTHOPONERA.***
*mumfordi* Wheeler, W.M. 1933; see under ***HYPOPONERA.***
*muscicola* Weber, 1942; see under ***HYPOPONERA.***
*mutabilis* Smith, F. 1861; see under ***LEPTOGENYS.***
*myopa* Roger, 1860; see under ***PACHYCONDYLA.***
*myrmicariae* Wasmann, 1918; see under ***HYPOPONERA.***
*natalensis* Santschi, 1914; see under ***HYPOPONERA.***
*nautarum* Santschi, 1920; see under ***HYPOPONERA.***
*neglecta* Santschi, 1923; see under ***HYPOPONERA.***
*nigrita* Emery, 1895; see under ***PACHYCONDYLA.***
*nippona* Santschi, 1937; see under ***HYPOPONERA.***
**nitida* Heer, 1850; see under ***DOLICHODERUS.***

***PONERA** (continued)*

*nitida* Smith, F. 1858; see under ***LEPTOGENYS.***
*nitida* Smith, F. 1861; see under ***MYOPIAS.***
*nitidula* Emery, 1890; see under ***HYPOPONERA.***
*nivariana* Santschi, 1908; see under ***HYPOPONERA.***
*norfolkensis. Ponera leae* subsp. *norfolkensis* Wheeler, W.M. 1935c: 72. [Also described as new by Wheeler, W.M. 1935g: 13.] Replacement name for *oculata* Wheeler, W.M. 1927i: 130. [Junior primary homonym of *oculata* Smith, F. 1858b: 93.] Junior synonym of *leae:* Taylor, 1967a: 89. See also: Taylor, 1960: 179.
*obesa* Emery, 1897; see under ***PACHYCONDYLA.***
*oblongiceps* Smith, M.R. 1939; see under ***HYPOPONERA.***
*occidentalis* Bernard, 1953; see under ***HYPOPONERA.***
*ocellifera* Roger, 1861; see under ***LEPTOGENYS.***
*ochracea* Mayr, 1855; see under ***CRYPTOPONE.***
*oculata* Smith, F. 1858; see under ***PACHYCONDYLA.***
*oculata. Ponera leae* subsp. *oculata* Wheeler, W.M. 1927i: 130, fig. 1 (w.q.) AUSTRALIA. [Junior primary homonym of *oculata* Smith, F. 1858b: 93, above.] Replacement name: *norfolkensis* Wheeler, W.M. 1935c: 72.
*opaciceps* Mayr, 1887; see under ***HYPOPONERA.***
*opacior* Forel, 1893; see under ***HYPOPONERA.***
*opaciventris* Roger, 1861; see under ***ECTATOMMA.***
*orba* Emery, 1915; see under ***HYPOPONERA.***
***oreas.*** *Selenopone oreas* Wheeler, W.M. 1933g: 20, fig. 8 (w.) PHILIPPINES. Taylor, 1967a: 52 (q.m.). Combination in *Ponera:* Wilson, 1957b: 381.
*pallida* Smith, F. 1858; see under ***PACHYCONDYLA.***
*pallidicornis* Smith, F. 1860; see under ***ODONTOMACHUS.***
*pallidipennis* Smith, F. 1860; see under ***PACHYCONDYLA.***
*pallidula* Emery, 1900; see under ***HYPOPONERA.***
*pallipes* Smith, 1858b: 87; see under ***HARPEGNATHOS.***
*pallipes* Smith, 1858b: 98; see under ***PACHYCONDYLA.***
*pampana* Santschi, 1925; see under ***HYPOPONERA.***
*panamensis* Forel, 1899; see under ***PACHYCONDYLA.***
*papuana* Emery, 1900; see under ***HYPOPONERA.***
*parallela* Smith, F. 1859; see under ***PLATYTHYREA.***
*paranensis* Santschi, 1925; see under ***HYPOPONERA.***
*parva* Forel, 1909; see under ***HYPOPONERA.***
*parva* Bondroit, 1918; see under ***HYPOPONERA.***
*paulina* Forel, 1913; see under ***HYPOPONERA.***
*pedunculata* Smith, F. 1858; see under ***PACHYCONDYLA.***
***pennsylvanica.*** *Ponera pennsylvanica* Buckley, 1866: 171 (w.) U.S.A. Emery, 1895c: 267 (q.m.); Wheeler, G.C. & Wheeler, J. 1952c: 631 (l.). Subspecies of *coarctata:* Emery, 1895c: 267; Dennis, 1938: 277; Creighton, 1950a: 48. Revived status as species: Taylor, 1967a: 29.
*peringueyi* Emery, 1899; see under ***PACHYCONDYLA.***
*perkinsi* Forel, 1899; see under ***HYPOPONERA.***
*perplexa* Mann, 1922; see under ***HYPOPONERA.***
*pestilentia* Smith, F. 1858; see under ***PACHYCONDYLA.***
***petila.*** *Ponera petila* Wilson, 1957b: 368, fig. 2 (w.) NEW GUINEA. See also: Taylor, 1967a: 102.
*petiolata* Bernard, 1953; see under ***HYPOPONERA.***
*petri* Forel, 1916; see under ***HYPOPONERA.***
*pia* Forel, 1901; see under ***HYPOPONERA.***
*pilosa* Smith, F. 1858; see under ***PACHYCONDYLA.***
*pilosula* Smith, F. 1858; see under ***PLATYTHYREA.***
*pompiloides* Smith, F. 1857; see under ***LEPTOGENYS.***
*porcata* Emery, 1897; see under ***PACHYCONDYLA.***
*postangustata* Forel, 1908; see under ***HYPOPONERA.***
*processionalis* Jerdon, 1851; see under ***LEPTOGENYS.***
*pruinosa* Emery, 1900; see under ***HYPOPONERA.***
*pumicosa* Roger, 1860; see under ***PACHYCONDYLA.***
*pumila. Ponera pumila* Jerdon, 1851: 119 (w.) INDIA. Unidentifiable to genus; *incertae sedis* in *Ponera.*
*punctatissima* Roger, 1859; see under ***HYPOPONERA.***
*punctiventris* Emery, 1902; see under ***HYPOPONERA.***
*purpurea* Smith, F. 1863; see under ***DIACAMMA.***
*pygmaea* Forel, 1907; see under ***HYPOPONERA.***
*quadridentata* Smith, F. 1859; see under ***PACHYCONDYLA.***
*queenslandensis* Forel, 1900; see under ***HYPOPONERA.***
*ragusae* Emery, 1894; see under ***HYPOPONERA.***
*ratardorum. Ponera ratardorum* Wilson, 1957b: 369, figs. 1, 2 (w.) NEW BRITAIN. Junior synonym of *incerta:* Taylor, 1967a: 80.
*rectidens* Clark, 1934; see under ***HYPOPONERA.***
*redbankensis* Forel, 1913; see under ***PACHYCONDYLA.***
*reichenspergeri* Santschi, 1923; see under ***HYPOPONERA.***
*reticulata* Smith, F. 1858; see under ***ODONTOPONERA.***
*reticulata* Forel, 1900; see under ***PACHYCONDYLA.***
**rhenana* Meunier, 1917; see under ****PONEROPSIS.***
*rimulosa* Roger, 1861; see under ***GNAMPTOGENYS.***
*rothkirchi* Wasmann, 1918; see under ***HYPOPONERA.***
*rotundata* Emery, 1895; see under ***PACHYCONDYLA.***
*rubiginosa* Emery, 1889; see under ***PACHYCONDYLA.***
*rubra* Smith, F. 1857; see under ***PACHYCONDYLA.***
*ruficornis* Spinola, 1851; see under ***DOLICHODERUS*** and under ***PLATYTHYREA.***
*rufipes* Jerdon, 1851; see under ***PACHYCONDYLA.***
*ruginoda* Smith, F. 1858; see under ***MYRMECIA.***
*rugosa* Le Guillou, 1842; see under ***DIACAMMA.***
*rugosa* Smith, F. 1857; see under ***GNAMPTOGENYS.***
*rugosa* Mann, 1921; see under ***HYPOPONERA.***
*ruida* Roger, 1860; see under ***ECTATOMMA.***
*sabronae* Donisthorpe, 1941; see under ***HYPOPONERA.***
*sakalava* Forel, 1891; see under ***HYPOPONERA.***
*sancta* Forel, 1914; see under ***HYPOPONERA.***
*santschii* Emery, 1909; see under ***HYPOPONERA.***
*saroltae* Forel, 1912; see under ***HYPOPONERA.***
*sauteri* Forel, 1912; see under ***HYPOPONERA.***
***scabra.*** *Ponera scabra* Wheeler, W.M. 1928d: 99 (w.q.) JAPAN. Santschi, 1937h: 364 (gynecoid w.); Santschi, 1941: 273 (q.); Ogata, 1987: 121 (m.); Imai & Kubota, 1972: 194 (k.). See also: Wilson, 1957b: 381; Taylor, 1967a: 49.
*scalprata* Smith, F. 1858; see under ***DIACAMMA.***
*schauinslandi* Emery, 1899; see under ***HYPOPONERA.***
*schmalzi* Emery, 1896; see under ***HYPOPONERA.***
*schwebeli* Forel, 1913; see under ***HYPOPONERA.***
*scitula* Clark, 1934; see under ***HYPOPONERA.***
*scolopax* Emery, 1899; see under ***PACHYCONDYLA.***
*sculpta* Jerdon, 1851; see under ***DIACAMMA.***
*sculpturata* Smith, F. 1859; see under ***DIACAMMA.***
***selenophora.*** *Ponera selenophora* Emery, 1900c: 317, pl. 8, figs. 4–6 (w.) NEW GUINEA. Combination in *Selenopone:* Wheeler, W.M. 1933g: 21; in *Ponera:* Wilson, 1957b: 382. See also: Taylor, 1967a: 41.
*sennaarensis* Mayr, 1862; see under ***PACHYCONDYLA.***
*sikorae* Forel, 1891; see under ***PACHYCONDYLA.***
*silvestrii* Donisthorpe, 1947; see under ***HYPOPONERA.***
*simillima* Smith, F. 1860b: 104; see under ***LEPTOGENYS.***
*simillima* Smith, F. 1860b: 105; see under ***PRIONOPELTA.***
***sinensis.*** *Ponera sinensis* Wheeler, W.M. 1928c: 6 (w.) HONG KONG. See also: Taylor, 1967a: 53.
*singaporensis* Viehmeyer, 1916; see under ***HYPOPONERA.***
*sinuata* Roger, 1860; see under ***PLATYTHYREA.***
*sinuosa* Bernard, 1953; see under ***HYPOPONERA.***
*siremps* Forel, 1901; see under ***HYPOPONERA.***
*sjostedti* Mayr, 1896; see under ***PACHYCONDYLA.***
*slamatana* Forel, 1913; see under ***HYPOPONERA.***
*solitaria* Smith, F. 1860; see under ***PACHYCONDYLA.***
*solitaria* Smith, F. 1874; see under ***PACHYCONDYLA.***
*sordida* Santschi, 1914; see under ***HYPOPONERA.***
*sorghi* Roger, 1863; see under ***PACHYCONDYLA.***
*soror* Emery, 1899; see under ***PACHYCONDYLA.***
*sororcula* Wilson, 1958; see under ***HYPOPONERA.***
*spei* Forel, 1910; see under ***HYPOPONERA.***
*spiniventris* Smith, F. 1858; see under ***PACHYCONDYLA.***
*stenocheilos* Jerdon, 1851; see under ***LEPTOGENYS.***
*stoica* Santschi, 1912; see under ***HYPOPONERA.***
*striata* Smith, F. 1860; see under ***DIACAMMA.***
*strigata. Ponera strigata* Norton, 1868a: 61, MEXICO. *Nomen nudum.*
*succedanea* Roger, 1863; see under ***PACHYCONDYLA.***
**succinea* Mayr, 1868; see under ***PACHYCONDYLA.***
*sulcata* Smith, F. 1858; see under ***GNAMPTOGENYS.***
*sulcata* Mayr, in Frauenfeld, 1867; see under ***PACHYCONDYLA.***
*sulcatinasis* Santschi, 1914; see under ***HYPOPONERA.***
*sulcatotesserinoda* Forel, 1900; see under ***PACHYCONDYLA.***
*sulciceps* Clark, 1928; see under ***HYPOPONERA.***
***swezeyi.*** *Pseudocryptopone swezeyi* Wheeler, W.M. 1933g: 16, fig. 6 (w.q.) HAWAII. Combination in *Ponera:* Wilson, 1957b: 370. See also: Taylor, 1967a: 85.
***syscena.*** *Ponera syscena* Wilson, 1957b: 384 (w.) NEW GUINEA. See also: Taylor, 1967a: 57.
***sysphinctoides.*** *Ponera sysphinctoides* Bernard, 1950a: 4, fig. 1 (m.) FRANCE.
***szaboi.*** *Ponera szaboi* Wilson, 1957b: 371. Replacement name for *Cryptopone mocsaryi* Szabó, 1910b: 186. [Junior secondary homonym of *Ponera mocsaryi* Emery, 1900c: 320.] See also: Taylor, 1967a: 103.
***szentivanyi.*** *Ponera szentivanyi* Wilson, 1957b: 372, figs. 1, 2 (w.) NEW GUINEA. See also: Taylor, 1967a: 103.
***taipingensis.*** *Ponera taipingensis* Forel, 1913k: 12 (w.) WEST MALAYSIA. See also: Taylor, 1967a: 56.
*taprobanae* Forel, 1913; see under ***HYPOPONERA.***
*tarda* Charsley, 1877; see under ***HYPOPONERA.***
*tarsalis* Perty, 1833; see under ***PARAPONERA.***
*taurica. Ponera coarctata* subsp. *testacea* natio *taurica* Arnol'di, 1932b: 65 (w.) RUSSIA. Unavailable name; material referred to *coarctata* by Taylor, 1967a: 21.

*tenella* Emery, 1902; see under ***HYPOPONERA.***
***tenuis.*** *Cryptopone tenuis* Emery, 1900c: 321, pl. 8, figs. 21, 22 (w.q.) NEW GUINEA. Combination in *Pseudocryptopone:* Wheeler, W.M. 1933g: 13; in *Ponera:* Wilson, 1957b: 373. Senior synonym of *huonica:* Taylor, 1967a: 98.
*tesseronoda* Emery, 1877; see under ***PACHYCONDYLA.***
*testacea. Ponera coarctata* var. *testacea* Emery, 1895b: 62 (w.q.) FRANCE. Junior synonym of *coarctata:* Taylor, 1967a: 21.
*texana* Buckley, 1866; see under ***LEPTOGENYS.***
*tornata* Roger, 1861; see under ***GNAMPTOGENYS.***
*tortuolosa* Smith, F. 1858; see under ***GNAMPTOGENYS.***
*tortuolosa* Smith, F. 1863; see under ***DIACAMMA.***
*traegaordhi* Santschi, 1914; see under ***HYPOPONERA.***
*transiens* Santschi, 1925; see under ***HYPOPONERA.***
*transvaalensis* Arnold, 1947; see under ***HYPOPONERA.***
*transversa* Smith, F. 1857; see under ***ODONTOPONERA.***
*trigona* Mayr, 1887; see under ***HYPOPONERA.***
*truncata* Smith, F. 1860; see under ***HYPOPONERA.***
*turaga* Mann, 1921; see under ***HYPOPONERA.***
*typhlos* Karavaiev, 1935; see under ***CRYPTOPONE.***
**umbra* Popov, 1933; see under ****PONERITES.***
*uncta* Santschi, 1914; see under ***HYPOPONERA.***
*unicolor* Smith, F. 1860; see under ***PACHYCONDYLA.***
*ursa* Santschi, 1924; see under ***HYPOPONERA.***
*ursoidea* Bernard, 1953; see under ***HYPOPONERA.***
*vagans* Smith, F. 1860; see under ***DIACAMMA.***
*vana* Forel, 1909; see under ***HYPOPONERA.***
*vanreesi* Forel, 1912; see under ***HYPOPONERA.***
**veneraria* Heer, 1865; see under ****PONEROPSIS.***
**ventrosa* Heer, 1850; see under ****PONEROPSIS.***
*vermiculata* Emery, 1897; see under ***PACHYCONDYLA.***
*vernacula* Kempf, 1962; see under ***HYPOPONERA.***
*versicolor* Smith, F. 1857; see under ***DIACAMMA.***
*vidua* Smith, F. 1857; see under ***PACHYCONDYLA.***
*villiersi* Bernard, 1953; see under ***HYPOPONERA.***
*viri* Santschi, 1923; see under ***HYPOPONERA.***
*vitiensis* Mann, 1921; see under ***HYPOPONERA.***
*wilsoni* Santschi, 1925; see under ***HYPOPONERA.***
***woodwardi.*** *Ponera woodwardi* Taylor, 1967a: 62, figs. 7, 55–58 (w.q.m.l.) SAMOA.
*wroughtonii* Forel, 1900; see under ***HYPOPONERA.***
***xenagos.*** *Ponera xenagos* Wilson, 1957b: 385 (w.) NEW GUINEA. See also: Taylor, 1967a: 43.
***yakushimensis.*** *Ponera yakushimensis* Tanaka, 1974b: 32, fig. 1 (w.q.) JAPAN.
*zwaluwenburgi* Wheeler, W.M. 1933 [*Pseudocryptopone*]; see under ***HYPOPONERA.***

*PONERACANTHA:* see under ***GNAMPTOGENYS.***

****PONERITES*** (Palaearctic)
****karaganensis.*** **Ponerites karaganensis* Dlussky, 1981b: 68, fig. 50 (wing) RUSSIA (Miocene).
****stauropolitanus.*** **Ponerites stauropolitanus* Dlussky, 1981b: 67, fig. 50 (wing) RUSSIA (Miocene).
****umbra.*** **Ponera umbra* Popov, 1933: 17, fig. 1 (q.) CAUCASUS. Combination in **Ponerites:* Dlussky, 1981b: 67.

****PONEROPSIS*** (Palaearctic)
****affinis.*** **Ponera affinis* Heer, 1850: 147, pl. 12, fig. 2 (q.) YUGOSLAVIA (Miocene). Combination in **Poneropsis:* Heer, 1867: 19. Formicidae *incertae sedis:* Taylor, 1964: 138.
**anthracina* Heer, 1867; see under ***DOLICHODERUS.***
****brunascens.*** **Poneropsis brunascens* Heer, 1867: 27, pl. 2, fig. 7 (m.) YUGOSLAVIA (Miocene).
****crassinervis.*** **Ponera crassinervis* Heer, 1850: 150, pl. 12, fig. 6 (q.) SWITZERLAND (Miocene). Combination in **Poneropsis:* Taylor, 1964: 138.
****croatica.*** **Ponera croatica* Heer, 1850: 148, pl. 12, fig 3 (q.) YUGOSLAVIA (Miocene). Combination in **Poneropsis:* Taylor, 1964: 138.
****elongata.*** **Poneropsis elongata* Heer, 1867: 19, pl. 1, fig. 21 (q.) YUGOSLAVIA (Miocene). Formicidae *incertae generis:* Taylor, 1964: 138.
****elongatula.*** **Ponera elongatula* Heer, 1850: 150, pl. 12, fig. 7 (q.) YUGOSLAVIA (Miocene). Combination in **Poneropsis:* Heer, 1867: 22. Formicidae *incertae generis:* Taylor, 1964: 138.
****escheri.*** **Poneropsis escheri* Heer, 1867: 20, pl. 2, fig. 22 (m.) SWITZERLAND (Miocene).
****fuliginosa.*** **Ponera fuliginosa* Heer, 1850: 145, pl. 12, fig. 1 (q.) SWITZERLAND (Miocene). Combination in *Aphaenogaster:* Mayr, 1867b: 57; Scudder, 1891: 688; Handlirsch, 1907: 874; in **Poneropsis:* Heer, 1867: 19; Taylor, 1964: 138.
****globosa.*** **Ponera globosa* Heer, 1850: 151, pl. 12, fig. 9 (q.) SWITZERLAND (Miocene). Combination in **Poneropsis:* Taylor, 1964: 138.
****hypolitha.*** **Ponera hypolitha* Cockerell, 1915: 483, pl. 64, figs. 3, 4 (wing) GREAT BRITAIN. Combination in **Poneropsis:* Taylor, 1964: 139.
**livida* Heer, 1867; see under ***APHAENOGASTER.***
****longaeva.*** **Ponera longaeva* Heer, 1850: 148, pl. 12, fig. 5 (q.) SWITZERLAND (Miocene). Combination in **Poneropsis:* Taylor, 1964: 138.
****lugubris.*** **Poneropsis lugubris* Heer, 1867: 21, pl. 1, fig. 23 (m.) YUGOSLAVIA (Miocene). Current subspecies: nominal plus **minor.*
*#***minor.*** **Poneropsis lugubris* var. *minor* Heer, 1867: 21 (m.) YUGOSLAVIA.
****morio.*** **Poneropsis morio* Heer, 1867: 26, pl. 2, fig. 6 (caste?) YUGOSLAVIA (Miocene). Current subspecies: nominal plus **pallens.*
*#***pallens.*** **Poneropsis morio* subsp. *pallens* Heer, 1867: 267, pl. 2, fig. 6 (caste?) YUGOSLAVIA (Miocene).
****pallida.*** **Poneropsis pallida* Heer, 1867: 23, pl. 2, fig. 2 (caste?) YUGOSLAVIA (Miocene).
****rhenana.*** **Ponera rhenana* Meunier, 1917: 8, fig. 6 (wing) GERMANY (Tertiary). Combination in **Poneropsis:* Taylor, 1964: 139.
****stygia.*** **Poneropsis stygia* Heer, 1867: 27, pl. 1, fig. 20 (q.?) SWITZERLAND. Formicidae *incertae generis:* Taylor, 1964: 138.
****tenuis.*** **Poneropsis tenuis* Heer, 1867: 22, pl. 2, fig. 1 (m.) YUGOSLAVIA (Miocene).
****veneraria.*** **Ponera veneraria* Heer, 1865: 386, fig. 288 (q.m.) SWITZERLAND. Combination in **Poneropsis:* Heer, 1879. Formicidae *incertae generis:* Taylor, 1964: 138.
****ventrosa.*** **Ponera ventrosa* Heer, 1850: 151, pl. 12, fig. 8 (q.) SWITZERLAND (Miocene). Combination in **Poneropsis:* Taylor, 1964: 138.

****PONEROPTERUS*** (Russia)
****sphecoides.*** **Poneropterus sphecoides* Dlussky, 1983: 73, fig. 7 (m.) RUSSIA (Cretaceous).

***PRENOLEPIS*** (Neotropical, Holarctic, Oriental, Indo-Australian)
***acuminata.*** *Prenolepis acuminata* Forel, 1911c: 309 (w.) MEXICO.
*adlerzii* Forel, 1886; see under ***CAMPONOTUS.***
#***albimaculata.*** *Prenolepis gibberosa* st. *albimaculata* Santschi, 1930c: 81 (w.) CUBA.
*albipes* Emery, 1899; see under ***PARATRECHINA.***
*amblyops* Forel, 1892; see under ***PARATRECHINA.***
*americana. Prenolepis nitens* var. *americana* Forel, 1891b: 94, pl. 3, fig. 4 (m.) U.S.A. Junior synonym of *imparis:* Emery, 1893i: 635; Dalla Torre, 1893: 178.
*amia* Forel, 1913; see under ***PARATRECHINA.***
*anthracina* Roger, 1863; see under ***PARATRECHINA.***
*antillana* Forel, 1893; see under ***PARATRECHINA.***
*arenivaga* Wheeler, W.M. 1905; see under ***PARATRECHINA.***
#***arizonica.*** *Prenolepis imparis* var. *arizonica* Wheeler, W.M. 1930b: 22 (w.q.m.) U.S.A. Subspecies of *imparis:* Creighton, 1950a: 414.
*aseta* Forel, 1902; see under ***PARATRECHINA.***
*atomus* Forel, 1901; see under ***PARATRECHINA.***
*bengalensis* Forel, 1894; see under ***PARATRECHINA.***
*biolleyi* Forel, 1908; see under ***PARATRECHINA.***
*birmana* Forel, 1902; see under ***PARATRECHINA.***
*bismarckensis* Forel, 1901; see under ***PARATRECHINA.***
*bolivari* Santschi, 1916; see under ***LASIOPHANES.***
*borcardi* Santschi, 1908; see under ***PARATRECHINA.***
*bourbonica* Forel, 1886; see under ***PARATRECHINA.***
*brasiliensis* Mayr, 1862; see under ***PARATRECHINA.***
*braueri* Mayr, 1868; see under ***PARATRECHINA.***
*bruchi* Forel, 1915; see under ***LASIOPHANES.***
*bruesii* Wheeler, W.M. 1903; see under ***PARATRECHINA.***
*bryanti* Forel, 1916; see under ***PARATRECHINA.***
*butteli* Forel, 1913; see under ***PARATRECHINA.***
*caeciliae* Forel, 1899; see under ***PARATRECHINA.***
*caledonica* Forel, 1902; see under ***PARATRECHINA.***
*californica. Prenolepis imparis* var. *californica* Wheeler, W.M. 1930b: 23, fig. 4 (w.q.m.) U.S.A. Subspecies of *imparis:* Creighton, 1950a: 414. Junior synonym of *imparis:* Wheeler, G.C. & Wheeler, J. 1986g: 14.
#***carinifrons.*** *Prenolepis melanogaster* st. *carinifrons* Santschi, 1920g: 173 (w.) VIETNAM.
*clandestina* Mayr, 1870; see under ***PARATRECHINA.***
*cocoensis* Forel, 1902; see under ***PARATRECHINA.***
#***colimana.*** *Prenolepis imparis* var. *colimana* Wheeler, W.M. 1930b: 24 (w.) MEXICO.
#***coloradensis.*** *Prenolepis imparis* var. *coloradensis* Wheeler, W.M. 1930b: 22, fig. 4 (w.) U.S.A. Subspecies of *imparis:* Creighton, 1950a: 415.
*comorensis* Forel, 1907; see under ***PARATRECHINA.***
*crepusculascens. Formica crepusculascens* Roger, 1859: 238 (q.) TURKEY. Junior synonym of *nitens:* Roger, 1862b: 256.
*dibullana* Forel, 1912; see under ***PARATRECHINA.***
*discoidalis* Donisthorpe, 1947; see under ***IRIDOMYRMEX.***
*dispar* Forel, 1909; see under ***PARATRECHINA.***
*docilis* Forel, 1908; see under ***PARATRECHINA.***
*donisthorpei* Forel, 1908; see under ***PARATRECHINA.***
*dugasi* Forel, 1911; see under ***PARATRECHINA.***
*edenensis* Wheeler, W.M. 1924; see under ***PARATRECHINA.***
*elevata* Forel, 1908; see under ***PARATRECHINA.***

***PRENOLEPIS** (continued)*

*ellisii* Forel, 1891; see under ***PARATRECHINA.***
*emarginata* Forel, 1913; see under ***PARATRECHINA.***
**emmae.** *Prenolepis emmae* Forel, 1894c: 406 (footnote) (w.) HONG KONG. Combination in *P.* (*Nylanderia*): Emery, 1906b: 133; in *Paratrechina* (*Euprenolepis*): Emery, 1925b: 224; in *Prenolepis:* Brown, 1953h: 7 (where the name was wrongly rendered as *P. geei;* see Wheeler, W.M. 1930h: 80 for correspondence of data). See also: Wheeler, W.M. 1921c: 542.
*faisonensis* Forel, 1922; see under ***PARATRECHINA.***
*farquharensis* Forel, 1907; see under ***PARATRECHINA.***
*foreli* Emery, 1914; see under ***PARATRECHINA.***
*formosae* Forel, 1912; see under ***PARATRECHINA.***
*fullawayi* Wheeler, W.M. 1912; see under ***TAPINOMA.***
*fulva* Mayr, 1862; see under ***PARATRECHINA.***
*fumata* Forel, 1909; see under ***PARATRECHINA.***
*fumatipennis* Forel, 1915; see under ***PARATRECHINA.***
*fuscula* Emery, 1869; see under ***LASIUS.***
*geei* Brown, 1953h: 7; see under ***emmae,*** above.
**gibberosa.** *Prenolepis gibberosa* Roger, 1863a: 161 (w.) CUBA. Current subspecies: nominal plus *albimaculata, rogeri.* See also: Mann, 1920: 433.
*glabra* Forel, 1891; see under ***PARATRECHINA.***
*glabrior* Forel, 1902; see under ***PARATRECHINA.***
*goeldii* Forel, 1912; see under ***PARATRECHINA.***
*gracilis* Forel, 1892; see under ***PARATRECHINA.***
*grandula* Forel, 1922; see under ***DORYMYRMEX.***
*grisoni* Forel, 1916; see under ***PARATRECHINA.***
*guatemalensis* Forel, 1885; see under ***PARATRECHINA.***
*hagemanni* Forel, 1901; see under ***PARATRECHINA.***
*hawaiensis* Forel, 1899; see under ***PARATRECHINA.***
*helleri* Viehmeyer, 1913; see under ***EUPRENOLEPIS.***
***henschei.** **Prenolepis henschei* Mayr, 1868c: 34, pl. 1, figs. 14–17 (w.m.) BALTIC AMBER (Oligocene). Wheeler, W.M. 1915h: 118 (q.).
*hodgsoni* Forel, 1902; see under ***PARATRECHINA.***
*humbloti* Forel, 1891; see under ***PARATRECHINA.***
**imparis.** *Formica imparis* Say, 1836: 287 (q.m.) U.S.A. Wheeler, G.C. & Wheeler, J. 1953c: 142 (l.); Hauschteck, 1962: 219 (k.). Combination in *Prenolepis:* Mayr, 1886d: 431. Senior synonym of *americana:* Emery, 1893i: 635; Dalla Torre, 1893: 178; of *wichita:* Dalla Torre, 1893: 178; Wheeler, W.M. 1930b: 15; of *minuta, pumila* and *testacea:* Creighton, 1950a: 414; of *californica:* Wheeler, G.C. & Wheeler, J. 1986g: 14. Current subspecies: nominal plus *arizonica, colimana, coloradensis, veracruzensis.* See also: Wheeler, W.M. 1930b: 16.
*incallida* Santschi, 1915; see under ***PARATRECHINA.***
*incisa* Forel, 1908; see under ***PARATRECHINA.***
*indica* Forel, 1894; see under ***PARATRECHINA.***
*itinerans* Forel, 1901; see under ***PARATRECHINA.***
**jacobsoni.** *Prenolepis jacobsoni* Crawley, 1923a: 30 (w.) SUMATRA.
*jaegerskioeldi* Mayr, 1904; see under ***PARATRECHINA.***
**jerdoni.** *Prenolepis jerdoni* Emery, 1893e: 222, pl. 8, fig. 20 (w.) WEST MALAYSIA. Imai, Brown, *et al.* 1984: 68 (k.). Current subspecies: nominal plus *subopaca.*
*johannae* Forel, 1912; see under ***PARATRECHINA.***
*kincaidi* Wheeler, W.M. 1906; see under ***PARATRECHINA.***
*kohli* Forel, 1916; see under ***PARATRECHINA.***
*kraepelini* Forel, 1905; see under ***PARATRECHINA.***
*kuenzleri* Forel, 1909; see under ***PARATRECHINA.***
*lasioides* Emery, 1869; see under ***LASIUS.***
*lepida* Santschi, 1915; see under ***PARATRECHINA.***
*levis* Forel, 1913; see under ***PARATRECHINA.***
*liburnica. Prenolepis imparis* subsp. *nitens* var. *liburnica* Röszler, 1942a: 58 (w.q.m.) ITALY. Unavailable name.
*lietzi* Forel, 1908; see under ***PARATRECHINA.***
*longiscapa* Forel, 1908; see under ***PARATRECHINA.***
*madagascarensis* Forel, 1886; see under ***PARATRECHINA.***
*malaccana* Viehmeyer, 1916; see under ***PARATRECHINA.***
*melanderi* Wheeler, W.M. 1903; see under ***PARATRECHINA.***
**melanogaster.** *Prenolepis melanogaster* Emery, 1893e: 223 (w.) BURMA. [Also described as new by Emery, 1895k: 476.] Current subspecies: nominal plus *carinifrons.*
*metallescens* Emery, 1900; see under ***PARATRECHINA.***
*mexicana* Forel, 1899; see under ***PARATRECHINA.***
*microps* Smith, M.R. 1937; see under ***PARATRECHINA.***
*minuta. Prenolepis imparis* var. *minuta* Emery, 1893i: 636 (w.m.) U.S.A. Wheeler, W.M. 1930b: 21 (q.). Junior synonym of *imparis:* Creighton, 1950a: 414.
*minuta* Forel, 1893; see under ***PARATRECHINA.***
*minutula* Forel, 1901; see under ***PARATRECHINA.***
*mixta* Forel, 1897; see under ***PARATRECHINA.***
*mjobergi* Forel, 1908; see under ***PARATRECHINA.***
*mjobergi* Forel, 1915; see under ***PROLASIUS.***
*mjoebergella* Forel, 1916; see under ***PROLASIUS.***
*myops* Mann, 1920; see under ***PARATRECHINA.***
**naoroji.** *Prenolepis naoroji* Forel, 1902d: 290 (w.) INDIA.
*natalensis* Forel, 1915; see under ***PARATRECHINA.***
*nesiotis* Wheeler, W.M. 1919; see under ***PARATRECHINA.***
*nettae* Forel, 1911; see under ***PARATRECHINA.***
*ngasiyana* Forel, 1907; see under ***PARATRECHINA.***
*nigriventris* Emery, 1914; see under ***PARATRECHINA.***
**nitens.** *Tapinoma nitens* Mayr, 1853a: 144 (w.) AUSTRIA. André, 1882b: 205 (q.m.). Combination in *Prenolepis:* Mayr, 1861: 52. Variety/subspecies of *imparis:* Dalla Torre, 1893: 178; Forel, 1895d: 229; Ruzsky, 1905b: 262; Emery, 1910a: 128; Emery, 1914d: 159; Wheeler, W.M. 1930b: 19; Novak & Sadil, 1941: 99; Röszler, 1942a: 56. Status as species: Finzi, 1922: 120; Müller, 1923: 118; Emery, 1925b: 225; Baroni Urbani, 1962: 136; Arnol'di & Dlussky, 1978: 556. Senior synonym of *crepusculascens:* Roger, 1862b: 256; of *polita:* Roger, 1863b: 10.
*nodifera* Mayr, 1870; see under ***PARATRECHINA.***
*obscura* Mayr, 1862; see under ***PARATRECHINA.***
*oceanica* Mann, 1921; see under ***PARATRECHINA.***
*opaca* Emery, 1887; see under ***PARATRECHINA.***
*papuana* Forel, 1901; see under ***PARATRECHINA.***
*parvula* Mayr, 1870; see under ***PARATRECHINA.***
*polita. Tapinoma polita* Smith, F. 1855a: 112 (w.) GREAT BRITAIN. Junior synonym of *nitens:* Roger, 1863b: 10.
*procera* Emery, 1900; see under ***EUPRENOLEPIS.***
*pubens* Forel, 1893; see under ***PARATRECHINA.***
*pumila. Prenolepis imparis* var. *pumila* Wheeler, W.M. 1930b: 21 (w.m.) U.S.A. Junior synonym of *imparis:* Creighton, 1950a: 414.
***pygmaea.** **Prenolepis pygmaea* Mayr, 1868c: 36, pl. 1, fig. 18 (q.m.) BALTIC AMBER (Oligocene).
*pyramica* Roger, 1863; see under ***DORYMYRMEX.***
#**rogeri.** *Prenolepis* (*Prenolepis*) *gibberosa* var. *rogeri* Mann, 1920: 434 (w.) CUBA. Aguayo, 1932: 225 (q.).
*rosae* Forel, 1912; see under ***PARATRECHINA.***
*rubescens* Forel, 1892; see under ***PARATRECHINA.***
*rufescens* Forel, 1907; see under ***PARATRECHINA.***
*sauteri* Forel, 1913; see under ***PARATRECHINA.***
*sechellensis* Emery, 1894; see under ***PARATRECHINA.***
*sharpii* Forel, 1899; see under ***PARATRECHINA.***
*sikorae* Forel, 1892; see under ***PARATRECHINA.***
*silvestrii* Emery, 1906; see under ***PARATRECHINA.***
*simpliciuscula* Emery, 1896; see under ***PARATRECHINA.***
*skottsbergi* Wheeler, W.M. 1922; see under ***PARATRECHINA.***
*smythiesii* Forel, 1894; see under ***PARATRECHINA.***
*staudingeri* Forel, 1912; see under ***PARATRECHINA.***
*steinheili* Forel, 1893; see under ***PARATRECHINA.***
*steeli* Forel, 1910; see under ***EUPRENOLEPIS.***
*stigmaticus* Mann, 1919; see under ***EUPRENOLEPIS.***
#**subopaca.** *Prenolepis jerdoni* var. *subopaca* Emery, 1900d: 699 (w.) SUMATRA.
*subtilis* Santschi, 1920; see under ***PARATRECHINA.***
*sumatrensis* Mayr, 1883; see under ***PSEUDOLASIUS.***
*tapinomoides* Forel, 1905; see under ***PARATRECHINA.***
*tasmaniensis* Forel, 1913; see under ***PARATRECHINA.***
*taylori* Forel, 1894; see under ***PARATRECHINA.***
*termitophila* Santschi, 1921; see under ***PARATRECHINA.***
*testacea. Prenolepis imparis* var. *testacea* Emery, 1893i: 636 (q.m.) U.S.A. Wheeler, W.M. 1905f: 390 (w.). Junior synonym of *imparis:* Creighton, 1950a: 414. See also: Wheeler, W.M. 1930b: 20.
*traegaordhi* Forel, 1904; see under ***PARATRECHINA.***
*vaga* Forel, 1901; see under ***PARATRECHINA.***
#**veracruzensis.** *Prenolepis imparis* var. *veracruzensis* Wheeler, W.M. 1930b: 24 (w.) MEXICO.
*vitiensis* Mann, 1921; see under ***PARATRECHINA.***
*waelbroecki* Emery, 1899; see under ***PARATRECHINA.***
*weissi* Santschi, 1911; see under ***PARATRECHINA.***
*wichita. Formica* (*Tapinoma*) *wichita* Buckley, 1866: 169 (w.) U.S.A. Combination in *Prenolepis:* Mayr, 1886d: 431. Junior synonym of *nitens:* Mayr, 1886d: 431; of *imparis:* Dalla Torre, 1893: 178; Wheeler, W.M. 1930b: 15.
*yerburyi* Forel, 1894; see under ***PARATRECHINA.***
*zelotypa* Santschi, 1915; see under ***PARATRECHINA.***

*PRIONOGENYS:* see under ***LEPTOGENYS.***

****PRIONOMYRMEX*** (Baltic Amber)
***longiceps.** **Prionomyrmex longiceps* Mayr, 1868c: 78, pl. 4, figs. 74, 75 (w.) BALTIC AMBER (Oligocene). Wheeler, W.M. 1915h: 27 (m.).

***PRIONOPELTA*** (World tropics)
**aethiopica.** *Prionopelta descarpentriesi* subsp. *aethiopica* Arnold, 1949: 263, fig. 4 (w.) SOUTH AFRICA. Raised to species: Brown, 1960a: 177.

***amabilis***. *Prionopelta amabilis* Borgmeier, 1949: 203, figs. 3–5 (w.) COSTA RICA. See also: Brown, 1960a: 217; Hölldobler & Wilson, 1986b: 45.

***amieti***. *Prionopelta amieti* Terron, 1974: 106, figs. 1–4, 7 (w.q.m.) CAMEROUN.

***antillana***. *Prionopelta punctulata* subsp. *antillana* Forel, 1909a: 239 (w.) ANTILLES IS. Raised to species: Brown, 1960a: 177.

***brocha***. *Prionopelta brocha* Wilson, 1958a: 147, fig. 7 (w.) NEW CALEDONIA.

*bruchi*. *Prionopelta bruchi* Santschi, 1923c: 245 (w.) ARGENTINA. Junior synonym of *punctulata:* Brown, 1960a: 178.

*churchilli*. *Examblyopone churchilli* Donisthorpe, 1949g: 401 (q.) NEW GUINEA. Junior synonym of *majuscula:* Brown, 1951: 102.

***descarpentriesi***. *Prionopelta descarpentriesi* Santschi, 1924b: 195 (w.) MADAGASCAR.

***humicola***. *Prionopelta humicola* Terron, 1974: 115, figs. 1–3 (w.q.) CAMEROUN.

***kraepelini***. *Prionopelta kraepelini* Forel, 1905c: 3 (w.q.) JAVA.

***majuscula***. *Prionopelta majuscula* Emery, 1897c: 595 (w.q.) NEW GUINEA. Senior synonym of *churchilli:* Brown, 1951: 102; of *poultoni* and *testacea:* Brown, 1953c: 12.

***marthae***. *Prionopelta marthae* Forel, 1909a: 240 (w.) VENEZUELA. Combination in *Typhlomyrmex:* Brown, 1953a: 104; in *Prionopelta:* Brown, 1965c: 77.

*mayri*. *Prionopelta mayri* Forel, 1909a: 239 (w.) BRAZIL. Junior synonym of *punctulata:* Brown, 1960a: 178.

*mocsaryi*. *Prionopelta mocsaryi* Forel, 1907a: 1 (w.) PARAGUAY. Junior synonym of *opaca:* Brown, in Wilson, 1958a: 149.

***modesta***. *Prionopelta modesta* Forel, 1909a: 241 (w.) GUATEMALA. Wheeler, G.C. & Wheeler, J. 1964b: 447 (l.). See also: Brown, 1960a: 218.

***opaca***. *Prionopelta opaca* Emery, 1897c: 596, pl. 15, figs. 44, 45 (w.q.m.) NEW GUINEA. Wheeler, G.C. & Wheeler, J. 1964b: 447 (l.). Senior synonym of *mocsaryi:* Brown, in Wilson, 1958a: 149.

*poultoni*. *Prionopelta poultoni* Donisthorpe, 1932c: 462. Replacement name for *Ponera simillima* Smith, F. 1860b: 105. [Junior primary homonym of *Ponera simillima* Smith, F. 1860b: 104.] Junior synonym of *majuscula:* Brown, 1953c: 12.

***punctulata***. *Prionopelta punctulata* Mayr, 1866a: 505 (q.) BRAZIL. Mayr, 1887: 549 (w.); Forel, 1893g: 367 (m.); Wheeler, G.C. & Wheeler, J. 1952a: 120 (l.). Senior synonym of *bruchi* and *mayri:* Brown, 1960a: 178.

*simillima*. *Ponera simillima* Smith, F. 1860b: 105 (q.) NEW GUINEA. Combination in *Rhopalopone:* Emery, 1900c: 311 (footnote). [Junior primary homonym of *Ponera simillima* Smith, F. 1860b: 104.] Replacement name: *Prionopelta poultoni* Donisthorpe, 1932c: 462.

*testacea*. *Renea testacea* Donisthorpe, 1947d: 183, fig. 1 (w.m.) NEW GUINEA. Donisthorpe, 1948d: 590 (q.). Junior synonym of *majuscula:* Brown, 1953c: 12.

*PRISTOMYRMECIA:* see under ***MYRMECIA***.

***PRISTOMYRMEX*** (Old World tropics and subtropics)

***africanus***. *Pristomyrmex africanus* Karavaiev, 1931d: 47, fig. 5 (w.) KENYA. Combination in *Hylidris:* Weber, 1952: 17; in *Pristomyrmex:* Bolton, 1981b: 284. Senior synonym of *beni, mbomu, myersi* and *primus:* Bolton, 1981b: 284.

#***aruensis***. *Pristomyrmex quadridens* var. *aruensis* Karavaiev, 1933c: 270 (w.) INDONESIA (Aru I.). See also: Karavaiev, 1935a: 99.

*beni*. *Hylidris myersi* subsp. *beni* Weber, 1952: 20 (w.) ZAIRE. Junior synonym of *africanus:* Bolton, 1981b: 284.

#***bicolor***. *Pristomyrmex trachylissa* var. *bicolor* Emery, 1900d: 678 (w.) SUMATRA.

***bispinosus***. *Dodous bispinosus* Donisthorpe, 1949e: 272 (w.) MAURITIUS. Combination in *Pristomyrmex:* Brown, 1971a: 3.

***brevispinosus***. *Pristomyrmex brevispinosus* Emery, 1887b: 451 (w.m.) SUMATRA. Current subspecies: nominal plus *sulcatus*.

***castaneicolor***. *Pristomyrmex castaneicolor* Donisthorpe, 1949g: 412 (w.) NEW GUINEA.

***castor***. *Pristomyrmex castor* Donisthorpe, 1944d: 81 (q.m.) NEW GUINEA.

***coggii***. *Pristomyrmex coggii* Emery, 1897c: 584 (w.) NEW GUINEA.

***cribrarius***. *Pristomyrmex cribrarius* Arnold, 1926: 281, fig. 81 (q.) MOZAMBIQUE. Arnold, 1948: 222 (w.). Combination in *Hylidris:* Weber, 1952: 17; in *Pristomyrmex:* Bolton, 1981b: 285.

***eduardi***. *Pristomyrmex eduardi* Forel, 1914d: 232 (footnote) (w.) SUMATRA.

***erythropus***. *Pristomyrmex erythropus* Taylor, 1968c: 65 (w.q.) AUSTRALIA.

*formosae*. *Pristomyrmex brevispinosus* r. *sulcatus* var. *formosae* Forel, 1912a: 54 (w.) TAIWAN. Unavailable name.

***fossulatus***. *Tetramorium* (*Xiphomyrmex*) *fossulatum* Forel, 1910e: 428 (w.) SOUTH AFRICA. Menozzi, 1942: 172 (q.). Combination in *Pristomyrmex:* Santschi, 1916a: 51; in *Hylidris:* Weber, 1952: 17; in *Pristomyrmex:* Bolton, 1981b: 285.

***foveolatus***. *Pristomyrmex foveolatus* Taylor, 1965b: 38, figs. 1, 2 (w.) AUSTRALIA.

***fuscipennis***. *Myrmica fuscipennis* Smith, F. 1861b: 46 (q.) SULAWESI. Combination in *Pristomyrmex:* Emery, 1901g: 567; Donisthorpe, 1932c: 468.

*japonicus*. *Pristomyrmex japonicus* Forel, 1900e: 268 (w.) JAPAN. Junior synonym of *pungens:* Viehmeyer, 1922: 207.

*laevigatus*. *Hylidris laevigatus* Weber, 1952: 20, fig. 12 (w.) ZAIRE. Junior synonym of *orbiceps:* Bolton, 1981b: 286.

***laevis***. *Solenopsis laevis* Smith, F. 1865: 75 (w.) INDONESIA (Morotai I.). Combination in *Pheidologeton:* Emery, 1922f: 213; in *Pristomyrmex:* Donisthorpe, 1932c: 473.

***levigatus***. *Pristomyrmex levigatus* Emery, 1897c: 583, pl. 15, fig. 24 (w.) NEW GUINEA.

***lucidus***. *Pristomyrmex lucidus* Emery, 1897c: 584 (w.) NEW GUINEA. Donisthorpe, 1948b: 305 (q.).

***mandibularis***. *Pristomyrmex mandibularis* Mann, 1921: 444 (w.) FIJI IS.

*mbomu*. *Hylidris myersi* subsp. *mbomu* Weber, 1952: 19 (w.) CENTRAL AFRICAN REPUBLIC. Junior synonym of *africanus:* Bolton, 1981b: 284.

#***melanoticus***. *Pristomyrmex obesus* subsp. *melanoticus* Mann, 1919: 340 (w.) SOLOMON IS.

***mendanai***. *Pristomyrmex mendanai* Mann, 1919: 341 (w.) SOLOMON IS.

*myersi*. *Hylidris myersi* Weber, 1941a: 190, figs. 8, 9 (w.) SUDAN. Weber, 1952: 18 (q.). Junior synonym of *africanus:* Bolton, 1981b: 284.

***nitidissimus***. *Pristomyrmex nitidissimus* Donisthorpe, 1949g: 411 (w.) NEW GUINEA.

***obesus***. *Pristomyrmex obesus* Mann, 1919: 339 (w.q.m.) SOLOMON IS. Current subspecies: nominal plus *melanoticus*.

***orbiceps***. *Xiphomyrmex orbiceps* Santschi, 1914d: 367, fig. 30 (w.) CAMEROUN. Combination in *Pristomyrmex:* Santschi, 1916a: 51; in *Hylidris:* Weber, 1952: 18; in *Pristomyrmex:* Bolton, 1981b: 286. Senior synonym of *laevigatus:* Bolton, 1981b: 286.

***orbiculatus***. *Pristomyrmex orbiculatus* Donisthorpe, 1948b: 306 (w.q.m.) NEW GUINEA.

***parumpunctatus***. *Pristomyrmex parumpunctatus* Emery, 1887b: 452 (w.q.) NEW GUINEA.

*parvispina*. *Pristomyrmex parvispina* Emery, 1900d: 678 (w.) SUMATRA. *Nomen nudum*.

***pegasus***. *Pristomyrmex pegasus* Mann, 1919: 338, fig. 27 (w.) SOLOMON IS.

***picteti***. *Pristomyrmex picteti* Emery, 1893e: 190 (w.) SUMATRA. Forel, 1912d: 107 (q.m.). Current subspecies: nominal plus *tingiana*.

***pollux***. *Pristomyrmex pollux* Donisthorpe, 1944d: 83 (m.) WEST MALAYSIA.

*primus*. *Hylidris myersi* subsp. *primus* Weber, 1952: 19 (w.) ZAIRE. Junior synonym of *africanus:* Bolton, 1981b: 284.

***punctatus***. *Myrmica punctata* Smith, F. 1860b: 108 (w.) INDONESIA (Batjan I.). Combination in *Pristomyrmex:* Mayr, 1886c: 361.

***pungens***. *Pristomyrmex pungens* Mayr, 1866b: 904, pl. 20, fig. 13 (w.) WEST MALAYSIA. Wheeler, W.M. 1928d: 114 (m.); Wheeler, G.C. & Wheeler, J. 1954c: 131 (l.); Imai, 1966: 119 (k.). Senior syonym of *japonicus:* Viehmeyer, 1922: 207. See also: Itow, Kobayashi, *et al.* 1984: 87.

***quadridens***. *Pristomyrmex quadridens* Emery, 1897c: 584, pl. 15, fig. 25 (w.q.) NEW GUINEA. Current subspecies: nominal plus *aruensis*.

***quadridentatus***. *Odontomyrmex quadridentatus* André, 1905: 208 (w.) AUSTRALIA. Taylor, 1965b: 47 (q.); Wheeler, G.C. & Wheeler, J. 1973a: 34 (l.). Combination in *Pristomyrmex:* Forel, 1915b: 53. Senior synonym of *queenslandensis:* Taylor, 1965b: 43.

*queenslandensis*. *Pristomyrmex* (*Odontomyrmex*) *quadridentatus* var. *queenslandensis* Forel, 1915b: 53 (w.) AUSTRALIA. Junior synonym of *quadridentatus:* Taylor, 1965b: 43.

***reticulatus***. *Pristomyrmex reticulatus* Donisthorpe, 1949c: 750 (m.) NEW GUINEA.

#***sulcatus***. *Pristomyrmex brevispinosus* subsp. *sulcatus* Emery, 1895k: 464 (w.q.) BURMA.

***taurus***. *Pristomyrmex taurus* Stitz, 1925: 120, fig. 5 (w.) PHILIPPINES.

***PRISTOMYRMEX*** *(continued)*

***thoracicus.*** *Pristomyrmex thoracicus* Taylor, 1965b: 41, figs. 3–6 (w.q.) AUSTRALIA.

#***tingiana.*** *Pristomyrmex picteti* var. *tingiana* Stitz, 1925: 120, fig. 4 (w.) PHILIPPINES.

***trachylissus.*** *Myrmica trachylissa* Smith, F. 1858b: 126 (q.) BORNEO. Mayr, 1886c: 359 (w.). Combination in *Pristomyrmex:* Mayr, 1886c: 359. Current subspecies: nominal plus *bicolor.* See also: Emery, 1887b: 450.

***trispinosus.*** *Dodous trispinosus* Donisthorpe, 1946f: 145, figs. 1–4 (w.m.) MAURITIUS. Combination in *Pristomyrmex:* Brown, 1971a: 3.

***trogor.*** *Pristomyrmex trogor* Bolton, 1981b: 287, figs. 33, 36 (w.m.) ZAIRE.

***umbripennis.*** *Myrmica umbripennis* Smith, F. 1863: 21 (q.) INDONESIA (Misool I.). Combination in *Pristomyrmex:* Donisthorpe, 1932c: 471.

***wheeleri.*** *Pristomyrmex wheeleri* Taylor, 1965b: 48, figs. 11–13; pl. 3 (w.q.) AUSTRALIA.

***wilsoni.*** *Pristomyrmex wilsoni* Taylor, 1968c: 63, figs. 1, 2 (w.) AUSTRALIA.

***PROATTA*** (Indo-Australian)

***butteli.*** *Proatta butteli* Forel, 1912m: 769 (w.m.) SUMATRA. Crawley, 1924: 401 (q.); Wheeler, G.C. & Wheeler, J. 1986b: 447 (l.). [Also described as new by Forel, 1913k: 86.] See also: Moffett, 1987: 444.

***PROBOLOMYRMEX*** (World tropics, Japan)

*angusticeps. Probolomyrmex angusticeps* Smith, M.R. 1949b: 39 (w.) PANAMA. Taylor, 1965d: 360 (q.l.). Junior synonym of *boliviensis:* Brown, 1975: 11.

***bidens.*** *Probolomyrmex bidens* Brown, 1975: 56, figs. 7, 9 (w.) INDIA.

***boliviensis.*** *Probolomyrmex boliviensis* Mann, 1923: 16, fig. 2 (q.) BOLIVIA. Senior synonym of *angusticeps:* Brown, 1975: 11. See also: Taylor, 1965d: 361.

***brevirostris.*** *Escherichia brevirostris* Forel, 1910c: 246 (w.) ETHIOPIA. Combination in *Probolomyrmex:* Taylor, 1965d: 355. Senior synonym of *parvus:* Taylor, 1965d: 355; Brown, 1975: 11.

***dammermani.*** *Probolomyrmex dammermani* Wheeler, W.M. 1928e: 7, fig. 1 (w.) JAVA. See also: Taylor, 1965d: 356.

***filiformis.*** *Probolomyrmex filiformis* Mayr, 1901b: 3 (w.) SOUTH AFRICA. See also: Taylor, 1965d: 353.

***greavesi.*** *Probolomyrmex greavesi* Taylor, 1965d: 358, figs. 2, 22–27 (w.q.m.) AUSTRALIA.

***guineensis.*** *Probolomyrmex guineensis* Taylor, 1965d: 353, figs. 11, 12 (w.) GUINEA.

***longinodus.*** *Probolomyrmex longinodus* Terayama & Ogata, 1988: 592, figs. 6–8 (w.) JAPAN.

***okinawensis.*** *Probolomyrmex okinawensis* Terayama & Ogata, 1988: 591, figs. 1–5 (w.q.) JAPAN.

*palauensis* Smith, M.R. 1953; see under ***LEPTANILLA.***

*parvus. Probolomyrmex parvus* Weber, 1949b: 3, fig. 2 (q.) UGANDA. Junior synonym of *brevirostris:* Taylor, 1965d: 355; Brown, 1975: 11.

***petiolatus.*** *Probolomyrmex petiolatus* Weber, 1940b: 76, fig. 1 (w.) PANAMA. See also: Taylor, 1965d: 363; Brown, 1975: 57.

***procne.*** *Probolomyrmex procne* Brown, 1975: 56, figs. 8, 10 (w.) INDIA.

***salomonis.*** *Probolomyrmex salomonis* Taylor, 1965d: 358, figs. 20, 21 (w.) SOLOMON IS.

***watanabei.*** *Probolomyrmex watanabei* Tanaka, 1974a: 35, pl. 3 (w.) WEST MALAYSIA.

**PROCERAPACHYS:* see under ***CERAPACHYS.***

***PROCERATIUM*** (World wide)

***algiricum.*** *Proceratium (Sysphingta) algiricum* Forel, 1899b: 305 (w.) ALGERIA. Santschi, 1929e: 139 (q.). Combination in *Sysphincta:* Emery, 1909c: 362; in *Proceratium:* Brown, 1958g: 247. Senior synonym of *mayri:* Baroni Urbani, 1977d: 92.

***arnoldi.*** *Proceratium (Sysphincta) arnoldi* Forel, 1913j: 210 (w.) ZIMBABWE. Combination in *Sysphincta:* Arnold, 1915: 35; in *Proceratium:* Brown, 1958g: 247.

***avium.*** *Proceratium avium* Brown, 1974a: 72, figs. 1, 2 (w.q.m.) MAURITIUS.

***boltoni.*** *Proceratium boltoni* Leston, 1971: 118, figs. 1, 2 (w.q.) GHANA.

***brasiliense.*** *Proceratium brasiliense* Borgmeier, 1959a: 309, figs. 1, 2 (w.) BRAZIL.

*caeca. Ponera caeca* Donisthorpe, 1949b: 491 (w.) NEW GUINEA. [Unresolved junior primary homonym of *Ponera coeca* Santschi, 1914d: 322.] Junior synonym of *papuanum:* Brown, 1958g: 248.

***californicum.*** *Proceratium californicum* Cook, 1953: 45, fig. (m.) U.S.A. Snelling, 1967: 1 (q.); Ward, 1988: 109 (w.). See also: Brown, 1958g: 332.

***carinifrons.*** *Proceratium carinifrons* Menozzi, 1939b: 175, fig. 1 (w.q.) SUMATRA.

*cavernicola. Sysphincta cavernicola* Borgmeier, 1937b: 221, figs. 1–4 (q.) PANAMA. Weber, 1940b: 79 (w.). Junior synonym of *micrommatum:* Borgmeier, 1957: 118.

*coecum. Proceratium coecum* Terron, 1981: 96, fig. 2 (w.) CAMEROUN. [Junior secondary homonym of *Ponera caeca* Donisthorpe, 1949b: 491, above.] Replacement name: *terroni* **new name**.

***compitale.*** *Proceratium compitale* Ward, 1988: 113, fig. 6 (w.q.) U.S.A.

*convexiceps. Proceratium convexiceps* Borgmeier, 1957: 120, figs. 34, 35 (w.) COSTA RICA. Junior synonym of *micrommatum:* Brown, 1980b: 342.

*crassicorne. Proceratium crassicorne* Emery, 1895c: 265, pl. 8, fig. 9 (w.) U.S.A. Smith, M.R. 1928b: 244 (q.); Wheeler, G.C. & Wheeler, J. 1952a: 137 (l.). Junior synonym of *silaceum:* Creighton, 1950a: 40; Brown, 1958g: 248.

***croceum.*** *Ponera crocea* Roger, 1860: 288 (q.) U.S.A. Emery, 1895c: 265 (w.); Smith, M.R. 1930a: 390 (m.); Wheeler, G.C. & Wheeler, J. 1952a: 135 (l.). Combination in *Sysphingta:* Mayr, 1866a: 501; in *Proceratium:* Mayr, 1886d: 437.

***deelemani.*** *Proceratium deelemani* Perrault, 1981: 189, figs. 1–6 (w.) BORNEO.

****denticulatum.*** **Proceratium denticulatum* Lattke, 1991a: 101, figs. 1, 2 (q.) DOMINICAN AMBER.

***diplopyx.*** *Proceratium diplopyx* Brown, 1980b: 337, figs. 1, 2 (w.) MADAGASCAR.

*europaea. Sysphingta europaea* Forel, 1886e: clxiii, fig. 2 (w.) GREECE. Forel, 1889: 259 (q.); Forel, 1905b: 177 (m.). Combination in *Proceratium:* Forel, 1889: 258; in *Sysphincta:* Emery, 1909c: 359. See also: Emery, 1916b: 103; Lomnicki, 1922: 1; Menozzi, 1925d: 24. Junior synonym of *melinum:* Brown, 1958g: 248.

*fialai. Sysphincta fialai* Kratochvil, 1944: 54, figs. 1–4 (w.q.m.) CZECHOSLOVAKIA. Junior synonym of *melinum:* Brown, 1958g: 248.

*formosicola. Proceratium formosicola* Terayama, 1985c: 406, figs. 1–4 (w.q.) TAIWAN. Junior synonym of *japonicum:* Onoyama, 1991: 695.

***goliath.*** *Proceratium goliath* Kempf & Brown, 1968: 94, figs. 1, 2 (w.) COSTA RICA.

***itoi.*** *Sysphincta itoi* Forel, 1918a: 717 (w.) JAPAN. Ogata, 1987: 107 (m.). Combination in *Proceratium:* Brown, 1958g: 247.

***japonicum.*** *Proceratium japonicum* Santschi, 1937h: 362, fig. 1 (w.) JAPAN. Senior synonym of *formosicola:* Onoyama, 1991: 695.

***lombokense.*** *Proceratium lombokense* Emery, 1897c: 593 (footnote) (q.) INDONESIA (Lombok I.).

***longigaster.*** *Proceratium longigaster* Karavaiev, 1935a: 59, fig. 2 (w.) VIETNAM.

***lunatum.*** *Proceratium lunatum* Terron, 1981: 96, fig. 1 (w.) CAMEROUN.

*mancum. Proceratium mancum* Mann, 1922: 6 (w.q.) HONDURAS. Junior synonym of *silaceum:* Brown, 1974a: 82.

*mayri. Proceratium (Sysphincta) mayri* Forel, 1899b: 306 (w.) GREECE. Emery, 1909c: 361 (q.). Combination in *Sysphincta:* Emery, 1909c: 361; in *Proceratium:* Brown, 1958g: 248. Junior synonym of *algiricum:* Baroni Urbani, 1977d: 92.

***melinum.*** *Ponera melina* Roger, 1860: 291 (w.q.m.) EUROPE (not U.S.A.; see Baroni Urbani, 1977d: 91). Combination in *Proceratium:* Mayr, 1886d: 438; in *Sysphincta:* Emery, 1895c: 263; in *Proceratium:* Brown, 1958g: 248. Senior synonym of *europaea, fialai* and *rossica:* Brown, 1958g: 248. See also: Atanassov & Dlussky, 1992: 64.

***micrommatum.*** *Sysphingta micrommata* Roger, 1863a: 176 (w.) SOUTH AMERICA. Combination in *Proceratium:* Dalla Torre, 1893: 18; Forel, 1895b: 111. Senior synonym of *cavernicola:* Borgmeier, 1957: 118; of *convexiceps:* Brown, 1980b: 342. See also: Brown, 1958g: 333.

*normandi. Proceratium normandi* Santschi, 1929e: 138 (w.q.) TUNISIA. Junior synonym of *numidicum:* Brown, 1974a: 82.

***numidicum.*** *Proceratium numidicum* Santschi, 1912f: 172, figs. 1, 2 (w.q.) TUNISIA. Senior synonym of *normandi:* Brown, 1974a: 82. See also: Brown, 1958g: 334; Brown, 1980b: 343.

***papuanum.*** *Proceratium papuanum* Emery, 1897c: 592, pl. 15, fig. 38 (q.) NEW GUINEA. Senior synonym of *caeca:* Brown, 1958g: 248.

***pergandei.*** *Sysphincta pergandei* Emery, 1895c: 264, pl. 8, fig. 4 (w.) U.S.A. Smith, M.R. 1928b: 242 (m.). Combination in *Proceratium:* Brown, 1958g: 248.

***relictum.*** *Proceratium relictum* Mann, 1921: 413, fig. 3 (w.q.) FIJI IS.

*rossica. Sysphincta europaea* subsp. *rossica* Arnol'di, 1930b: 144, figs. 1–3 (w.q.) RUSSIA. [Also described as new by Arnol'di, 1930d: 157.] Junior synonym of *melinum:* Brown, 1958g: 248.

*rugulosum. Proceratium silaceum* subsp. *rugulosum* Wheeler, W.M. 1915b: 390 (w.q.) U.S.A. Junior synonym of *silaceum:* Creighton, 1950a: 40.

**silaceum.** *Proceratium silaceum* Roger, 1863a: 172 (w.) U.S.A. Emery, 1895c: 265 (q.); Kennedy & Talbot, 1939: 202 (m.); Crozier, 1970: 116 (k.). Senior synonym of *crassicorne, rugulosum* and *vestitum:* Creighton, 1950a: 40; of *mancum:* Brown, 1974a: 82. See also: Brown, 1958g: 330.

**stictum.** *Proceratium stictum* Brown, 1958g: 336, figs. 45, 46 (w.) AUSTRALIA.

*tio. Proceratium tio* Snelling & Cover, 1992: 50, fig. 1 (w.) MEXICO.

**terroni.** *Proceratium terroni* **new name.** Replacement name for *Proceratium coecum* Terron, 1981: 91. [Junior secondary homonym of *Ponera caeca* Donisthorpe, 1949b: 491.]

**toschii.** *Sysphincta toschii* Consani, 1951: 167, fig. 1 (w.) KENYA. Combination in *Proceratium:* Brown, 1958g: 248.

*vestitum. Proceratium crassicorne* var. *vestitum* Emery, 1895c: 266 (w.) U.S.A. Junior synonym of *silaceum:* Creighton, 1950a: 40.

**watasei.** *Sysphincta watasei* Wheeler, W.M. 1906c: 303, pl. 41, fig. 5 (w.q.) JAPAN. Ogata, 1987: 107 (m.). Combination in *Proceratium:* Brown, 1958g: 248.

***PROCRYPTOCERUS*** (Neotropical)

**adlerzi.** *Cataulacus adlerzi* Mayr, 1887: 562 (w.q.m.) BRAZIL. Wheeler, G.C. & Wheeler, J. 1973b: 79 (l.). Combination in *Procryptocerus:* Dalla Torre, 1893: 139. Subspecies of *striatus:* Emery, 1894c: 199. Revived status as species, senior synonym of *latitans*, and material of the unavailable name *muelleri* referred here: Kempf, 1951: 79.

**attenuatus.** *Meranoplus attenuatus* Smith, F. 1876d: 610, pl. 11, fig. 9 (q.) BRAZIL. Combination in *Cataulacus:* Mayr, 1886c: 364; in *Procryptocerus:* Emery, 1887b: 470. Subspecies of *subpilosus:* Forel, 1911e: 263; Kempf, 1951: 61. Revived status as species: Kempf, 1963c: 436. Senior synonym of *puncticeps:* Forel, 1911e: 263; of *guianensis:* Kempf, 1963c: 436.

**balzani.** *Procryptocerus balzani* Emery, 1894c: 199 (w.q.) BOLIVIA. See also: Kempf, 1951: 84.

**batesi.** *Procryptocerus batesi* Forel, 1899c: 43 (footnote) (w.) COLOMBIA. See also: Kempf, 1951: 22.

**belti.** *Procryptocerus belti* Forel, 1899c: 46, pl. 3, fig. 6 (w.q.m.) COSTA RICA. See also: Kempf, 1951: 31.

**carbonarius.** *Cataulacus carbonarius* Mayr, 1870a: 414 (w.) COLOMBIA. Combination in *Procryptocerus:* Emery, 1887b: 470. Senior synonym of *laeviventris:* Kempf, 1951: 24.

**clathratus.** *Procryptocerus clathratus* Emery, 1896g: 94 (q.w.) BRAZIL. See also: Kempf, 1951: 27; Kempf, 1957: 395.

*concentricus. Procryptocerus striatus* subsp. *convergens* var. *concentricus* Emery, 1894c: 197 (w.) BRAZIL. Unavailable name; material referred to *regularis* by Kempf, 1951: 74.

**convergens.** *Cataulacus convergens* Mayr, 1887: 564 (w.m.; not q., see Kempf, 1951: 97) BRAZIL. Kempf, 1957: 399 (q.). Combination in *Procryptocerus:* Emery, 1888c: 362. Subspecies of *striatus:* Emery, 1894c: 197; Forel, 1912e: 207; Santschi, 1933e: 117. Revived status as species, and senior synonym of *odiosa:* Kempf, 1951: 95.

**#convexus.** *Procryptocerus hirsutus* r. *convexus* Forel, 1904c: 34 (w.) BRAZIL. See also: Kempf, 1951: 35.

**coriarius.** *Cataulacus coriarius* Mayr, 1870a: 414 (w.) COLOMBIA. Kempf, 1951: 38 (q.). Combination in *Procryptocerus:* Emery, 1887b: 470.

**curvistriatus.** *Procryptocerus sulcatus* subsp. *curvistriatus* Kempf, 1949: 425, fig. 5 (w.) BRAZIL. Raised to species: Kempf, 1964b: 243. See also: Kempf, 1951: 72.

**elegans.** *Procryptocerus elegans* Santschi, 1921g: 98 (w.) BRAZIL. See also: Kempf, 1951: 102; Kempf, 1964b: 245.

**ferreri.** *Procryptocerus ferreri* Forel, 1912e: 208 (w.) COLOMBIA. See also: Kempf, 1951: 66.

**gibbosus.** *Procryptocerus gibbosus* Kempf, 1949: 423, figs. 1–4 (w.) BRAZIL. See also: Kempf, 1951: 105.

**goeldii.** *Procryptocerus goeldii* Forel, 1899c: 45 (footnote) (w.) BRAZIL. Forel, 1907a: 12 (q.); Kempf, 1951: 48 (m.).

**gracilis.** *Meranoplus gracilis* Smith, F. 1858b: 194 (w.) BRAZIL. Combination in *Cataulacus:* Mayr, 1886c: 361; in *Procryptocerus:* Emery, 1887b: 470. See also: Kempf, 1951: 52.

*guianensis. Procryptocerus goeldii* subsp. *guianensis* Weber, 1938a: 208 (q.) GUIANA. Junior synonym of *attenuatus:* Kempf, 1963c: 436.

**hirsutus.** *Procryptocerus hirsutus* Emery, 1896g: 96 (w.) BRAZIL. Current subspecies: nominal plus *convexus.* See also: Kempf, 1951: 34.

**hylaeus.** *Procryptocerus goeldii* subsp. *hylaeus* Kempf, 1951: 50 (w.) COLOMBIA. Raised to species: Kempf, 1964b: 247.

*impressus. Procryptocerus puncticeps* r. *impressus* Forel, 1899c: 47 (w.q.) PANAMA. Subspecies of *subpilosus:* Forel, 1911e: 263. Junior synonym of *subpilosus:* Kempf, 1963c: 435.

*laeviventris. Procryptocerus carbonarius* var. *laeviventris* Forel, 1899c: 47 (w.) PANAMA. Junior synonym of *carbonarius:* Kempf, 1951: 24.

*latitans. Procryptocerus striatus* r. *latitans* Forel, 1912e: 206 (w.) BRAZIL. Junior synonym of *adlerzi:* Kempf, 1951: 79.

**lenkoi.** *Procryptocerus lenkoi* Kempf, 1969: 283, figs. 4–6 (w.q.) BRAZIL.

**lepidus.** *Procryptocerus subpilosus* st. *lepidus* Forel, 1908c: 355 (w.) BRAZIL. Raised to species: Kempf, 1964b: 247.

*loretensis. Procryptocerus striatus* st. *convergens* var. *loretensis* Santschi, 1933e: 117 (w.) ARGENTINA. Unavailable name; material referred to *regularis* by Kempf, 1951: 74.

**marginatus.** *Procryptocerus marginatus* Borgmeier, 1948a: 201, fig. 26 (w.q.m.) BRAZIL. See also: Kempf, 1951: 58.

**mayri.** *Procryptocerus mayri* Forel, 1899c: 43 (footnote) (w.) COLOMBIA. See also: Kempf, 1951: 103. Current subspecies: nominal plus *reichenspergeri.*

**montanus.** *Procryptocerus montanus* Kempf, 1957: 400, figs. 6–10 (w.q.m.) BRAZIL.

*muelleri. Procryptocerus striatus* r. *latitans* var. *muelleri* Forel, 1912e: 207 (w.) BRAZIL. Unavailable name; material referred to *adlerzi* by Kempf, 1951: 79.

*odiosa. Procryptocerus striatus* var. *odiosa* Forel, 1912e: 206 (w.) BRAZIL. Junior synonym of *convergens:* Kempf, 1951: 95.

**paleatus.** *Procryptocerus paleatus* Emery, 1896g: 97 (w.) COSTA RICA. See also: Kempf, 1951: 53.

*parva. Procryptocerus striatus* subsp. *scabriuscula* var. *parva* Menozzi, 1935b: 196 (w.) GUIANA. Unavailable name; material referred to *pictipes* by Kempf, 1951: 42.

**petiolatus.** *Meranoplus petiolatus* Smith, F. 1853: 224, pl. 20, fig. 7 (q.) BRAZIL. Combination in *Procryptocerus:* Emery, 1887b: 470.

**pictipes.** *Procryptocerus pictipes* Emery, 1896g: 98 (w.) COSTA RICA. Kempf, 1951: 44 (q.); Wheeler, G.C. & Wheeler, J. 1954b: 151 (l.). Material of the unavailable name *parva* referred here by Kempf, 1951: 42.

*puncticeps. Meranoplus puncticeps* Smith, F. 1876d: 610, pl. 11, fig. 10 (w.) BRAZIL. Combination in *Cataulacus:* Mayr, 1886c: 364; in *Procryptocerus:* Emery, 1887b: 470. Junior synonym of *attenuatus:* Forel, 1911e: 263.

**regularis.** *Procryptocerus convergens* r. *regularis* Emery, 1888c: 362 (w.) BRAZIL. Kempf, 1951: 77 (q.); Wheeler, G.C. & Wheeler, J. 1973b: 81 (l.). Subspecies/race of *striatus:* Forel, 1901h: 50; Emery, 1922f: 302. Raised to species: Dalla Torre, 1893: 140; Kempf, 1951: 74. Material of the unavailable names *concentricus, loretensis* and *rotundiceps* referred here: Kempf, 1951: 74.

**#reichenspergeri.** *Paracryptocerus mayri* st. *reichenspergeri* Santschi, 1921g: 98 (w.) BRAZIL. Combination in *Procryptocerus:* Kempf, 1951: 105.

*rotundiceps. Procryptocerus striatus* subsp. *regularis* var. *rotundiceps* Forel, 1908c: 356 (w.) BRAZIL. Unavailable name; material referred to *regularis* by Kempf, 1951: 74.

**rudis.** *Cataulacus rudis* Mayr, 1870a: 414 (w.) COLOMBIA. Combination in *Procryptocerus:* Emery, 1887b: 470. See also: Kempf, 1951: 28.

**sampaioi.** *Procryptocerus sampaioi* Forel, 1912e: 207 (w.) BRAZIL. See also: Kempf, 1951: 68.

**scabriusculus.** *Procryptocerus striatus* r. *scabriusculus* Forel, 1899c: 45 (w.) COSTA RICA. [First available use of *Procryptocerus striatus* subsp. *schmalzi* var. *scabriusculus* Emery, 1894c: 198; unavailable name.] Kempf, 1951: 91 (q.); Snelling, 1968a: 2 (m.); Wheeler, G.C. & Wheeler, J. 1973b: 82 (l.). Raised to species: Kempf, 1951: 89.

**schmalzi.** *Procryptocerus striatus* subsp. *schmalzi* Emery, 1894c: 198 (w.q.) BRAZIL. Wheeler, G.C. & Wheeler, J. 1954b: 152 (l.). Raised to species: Kempf, 1951: 98.

**schmitti.** *Procryptocerus schmitti* Forel, 1901e: 338 (w.) BRAZIL. Kempf, 1951: 40 (q.m.).

**seabrai.** *Procryptocerus seabrai* Kempf, 1964b: 248, figs. 1–8 (w.) BRAZIL.

**spiniperdus.** *Procryptocerus spiniperdus* Forel, 1899c: 43 (footnote) (w.m.) TRINIDAD. Wheeler, W.M. 1922c: 11 (q.). See also: Kempf, 1951: 55.

**spinosus.** *Meranoplus spinosus* Smith, F. 1859a: 150 (w.) INDONESIA (Aru I.). Combination in *Procryptocerus:* Donisthorpe, 1932c: 456. [Specimen supposed to be holotype is a queen of a *Procryptocerus* species. Therefore either locality in error or specimen has been changed, or both.]

**striatus.** *Meranoplus striatus* Smith, F. 1860c: 77, pl. 4, fig. 1 (w.) BRAZIL. Combination in *Cataulacus:*

*PROCRYPTOCERUS* (*continued*)

Mayr, 1886c: 361; in *Procryptocerus:* Emery, 1887b: 470. See also: Kempf, 1951: 93.

***subpilosus.*** *Meranoplus subpilosus* Smith, F. 1860c: 78, pl. 4, fig. 2 (w.) BRAZIL. Combination in *Cataulacus:* Mayr, 1886c: 361; in *Procryptocerus:* Emery, 1887b: 470. Senior synonym of *impressus:* Kempf, 1963c: 435. See also: Kempf, 1951: 60.

***sulcatus.*** *Procryptocerus sulcatus* Emery, 1894c: 200 (w.) BRAZIL. Kempf, 1951: 72 (q.).

***victoris.*** *Procryptocerus victoris* Kempf, 1960f: 437, figs. 14–18 (w.) BRAZIL.

***virgatus.*** *Procryptocerus virgatus* Kempf, 1964b: 250, figs. 9–12 (w.) ECUADOR.

*PRODICROASPIS:* see under ***LORDOMYRMA.***

*****PRODIMORPHOMYRMEX*** (Baltic Amber)

*****primigenius.*** **Prodimorphomyrmex primigenius* Wheeler, W.M. 1915h: 112, fig. 54 (w.) BALTIC AMBER (Oligocene).

*PRODISCOTHYREA:* see under ***DISCOTHYREA.***

***PROFORMICA*** (Palaearctic)

*aerea. Formica aerea* Roger, 1859: 237 (w.) GREECE. Junior synonym of *nasuta:* Mayr, 1863: 402.

***alaica.*** *Proformica alaica* Kuznetsov-Ugamsky, 1926a: 72, figs. 6–11 (w.) TURKESTAN.

*alpina. Proformica nitida* var. *alpina* Kuznetsov-Ugamsky, 1927b: 26 (w.) RUSSIA. Kuznetsov-Ugamsky, 1928a: 22 (m.). Senior synonym of *orites* (unnecessary replacement name): Dlussky, 1969a: 222. Junior synonym of *nitida:* Dlussky, 1969a: 222; Tarbinsky, 1976: 166.

*barbatula. Proformica nasuta* n. *barbatula* Arnol'di, 1948: 212. *Nomen nudum;* see Dlussky, 1969a: 228.

***buddhaensis.*** *Proformica mongolica* subsp. *buddhaensis* Ruzsky, 1915a: 432 (w.) CHINA. Raised to species: Dlussky, 1969a: 226; Pisarski, 1969b: 307. Senior synonym of *lefevrei:* Dlussky, 1969a: 226.

***caucasea.*** *Cataglyphis* (*Paraformica*) *caucasea* Santschi, 1925g: 353 (w.) CAUCASUS. Combination in *Paraformica:* Karavaiev, 1926e: 199; in *Formica* (*Proformica*): Santschi, 1926f: 293; in *Proformica:* Arnol'di & Dlussky, 1978: 554.

***coriacea.*** *Proformica coriacea* Kuznetsov-Ugamsky, 1927b: 26, fig. 1 (s.w.) TURKESTAN. Senior synonym of *frigida:* Dlussky, 1969a: 229.

*depilis. Formica* (*Proformica*) *nasuta* var. *depilis* Santschi, 1925g: 353 (w.) FRANCE. Raised to species: Collingwood & Yarrow, 1969: 86. Junior synonym of *nasuta:* Espadaler & Cagniant, 1987: 136.

*deserta* Kuznetsov-Ugamsky, 1926; see under ***CATAGLYPHIS.***

***dolichocephala.*** *Proformica dolichocephala* Kuznetsov-Ugamsky, 1927b: 27, figs. 6–8 (s.) TURKESTAN.

***epinotalis.*** *Proformica epinotalis* Kuznetsov-Ugamsky, 1927b: 27, figs. 2–4 (s.w.) TURKESTAN. Kupyanskaya, 1990: 206 (q.m.). See also: Dlussky, 1969a: 226.

***ferreri.*** *Proformica ferreri* Bondroit, 1918: 41 (w.m.) SPAIN. Wheeler, G.C. & Wheeler, J. 1982: 176 (l.). Variety of *nasuta:* Santschi, 1919e: 246; Menozzi, 1922b: 332; Santschi, 1931a: 13. Revived status as species: Bernard, 1967: 285.

***flavosetosa.*** *Formica* (*Proformica*) *flavosetosa* Viehmeyer, 1922: 215 (w.) CHINA. Combination in *Proformica:* **new combination**.

*frigida. Proformica frigida* Dlussky, 1965b: 28 (w.) RUSSIA. Junior synonym of *coriacea:* Dlussky, 1969a: 229.

***jacoti.*** *Formica* (*Proformica*) *jacoti* Wheeler, W.M. 1923b: 4 (w.) CHINA. Wheeler, W.M. 1929f: 11 (q.). Combination in *Proformica:* Dlussky, 1969a: 230.

***kaszabi.*** *Proformica kaszabi* Dlussky, 1969a: 228 (w.) KIRGHIZIA.

***kobachidzei.*** *Proformica kobachidzei* Arnol'di, 1968: 1818, fig. 18 (w.) GEORGIA. See also: Atanassov & Dlussky, 1992: 288.

***korbi.*** *Formica* (*Proformica*) *korbi* Emery, 1909b: 202 (w.q.) TURKEY. Combination in *Proformica:* Dlussky, 1969a: 230.

***kosswigi.*** *Acanthomyops* (*Donisthorpea*) *kosswigi* Donisthorpe, 1950d: 640 (w.) TURKEY. Combination in *Proformica:* **new combination**.

***kusnezowi.*** *Formica* (*Proformica*) *kusnezowi* Santschi, 1928b: 43 (w.) TURKESTAN. Combination in *Proformica:* **new combination**.

*lefevrei. Formica* (*Proformica*) *lefevrei* Wheeler, W.M. 1929f: 11 (w.) CHINA. Junior synonym of *buddhaensis:* Dlussky, 1969a: 226.

***longiseta.*** *Proformica longiseta* Collingwood, 1978: 92 (diagnosis in key) (w.) SPAIN.

*menozzii* Santschi, 1923; see under ***BAJCARIDRIS.***

**#*metalica.*** *Proformica nasuta* var. *metalica* Kuznetsov-Ugamsky, 1923: 246 (w.) TURKESTAN.

***mongolica.*** *Formica nasuta* subsp. *mongolica* Emery, 1901i: 159 (w.) MONGOLIA. Combination in *Formica* (*Proformica*): Emery, 1909b: 202; in *Proformica:* Ruzsky, 1915a: 432. Raised to species: Emery, 1909b: 202; Ruzsky, 1915a: 432. See also: Dlussky, 1969a: 225; Pisarski, 1969b: 307.

***nasuta.*** *Formica nasuta* Nylander, 1856b: 66 (w.) FRANCE. Forel, 1886f: 205 (q.); Ruzsky, 1903b: 304 (m.). Combination in *Myrmecocystus:* Emery & Forel, 1879: 449; in *Formica* (*Proformica*): Ruzsky, 1902d: 13; in *Proformica:* Emery, 1912f: 100; Ruzsky, 1915a: 433 (footnote); Bondroit, 1918: 40. Senior synonym of *aerea:* Mayr, 1863: 402; of *depilis:* Espadaler & Cagniant, 1987: 136. Current subspecies: nominal plus *metalica, syrdariana.* See also: Emery, 1909b: 201; Wheeler, W.M. 1913f: 543; Kuznetsov-Ugamsky, 1928a: 18; Stumper, 1957: 87; Dlussky, 1969a: 224.

***nitida.*** *Proformica nitida* Kuznetsov-Ugamsky, 1923: 246 (w.) TURKESTAN. [Also described as new by Kuznetsov-Ugamsky, 1927b: 26.] Senior synonym of *alpina* (and its junior synonym *orites*): Dlussky, 1969a: 222; Tarbinsky, 1976: 166.

***oculatissima.*** *Formica oculatissima* Forel, 1886e: clxii (m.) GREECE. Combination in *F.* (*Proformica*): Emery, 1909b: 203; in *Proformica:* Agosti & Collingwood, 1987a: 61.

*orites. Formica* (*Proformica*) *nitida* var. *orites* Santschi, 1928b: 46. Unnecessary replacement name for, and hence junior synonym of, *alpina* Kuznetsov-Ugamsky: Dlussky, 1969a: 222.

***pilosiscapa.*** *Proformica pilosiscapa* Dlussky, 1969a: 228 (w.q.) ARMENIA. Junior synonym of *caucasea:* Arnol'di & Dlussky, 1978: 554. Revived from synonymy: Atanassov & Dlussky, 1992: 290.

***seraphimi.*** *Proformica* (*Proformica*) *seraphimi* Tarbinsky, 1970: 309, figs. 1–4 (w.q.) TURKESTAN.

***similis.*** *Proformica similis* Dlussky, 1969a: 228 (w.q.) TURKESTAN.

***splendida.*** *Proformica splendida* Dlussky, 1965b: 28 (w.) KIRGHIZIA. See also: Dlussky, 1969a: 220; Tarbinsky, 1976: 167.

***striaticeps.*** *Formica* (*Proformica*) *nasuta* var. *striaticeps* Forel, 1911d: 352 (w.) GREECE. Combination in *Proformica,* and raised to species: Dlussky, 1969a: 224. See also: Atanassov & Dlussky, 1992: 287.

**#*syrdariana.*** *Formica* (*Proformica*) *nasuta* var. *syrdariana* Santschi, 1928b: 45 (w.) TURKESTAN. Combination in *Proformica:* **new combination**.

***PROLASIUS*** (Australasian)

***abruptus.*** *Prolasius abruptus* Clark, 1934c: 66, pl. 4, fig. 25 (w.) AUSTRALIA.

***advenus.*** *Formica advena* Smith, F. 1862c: 53 (w.q.) NEW ZEALAND. Forel, 1892c: 332 (m.). Combination in *Prenolepis:* Mayr, 1886c: 362; in *Lasius* (*Prolasius*): Forel, 1892c: 332; in *Prolasius:* Wheeler, W.M. 1935c: 71. Senior synonym of *zealandica:* Brown, 1958h: 43. See also: McAreavey, 1947: 14.

***antennatus.*** *Prolasius antennata* McAreavey, 1947: 13, pl. 1, figs. 4, 5 (w.) AUSTRALIA.

***bruneus.*** *Prolasius brunea* McAreavey, 1947: 16, pl. 1, figs. 22, 23 (w.) AUSTRALIA.

***clarki.*** *Prolasius clarki* McAreavey, 1947: 15, pl. 1, figs. 26, 27 (w.q.) AUSTRALIA.

***convexus.*** *Prolasius convexa* McAreavey, 1947: 15, pl. 1, figs. 6, 7 (w.) AUSTRALIA. Senior synonym of *nigriventris:* Taylor, 1992a: 65.

***depressiceps.*** *Melophorus depressiceps* Emery, 1914b: 186, fig. 5 (w.) AUSTRALIA. Combination in *Prolasius:* Wheeler, W.M. 1935c: 71. See also: McAreavey, 1947: 23.

***flavicornis.*** *Prolasius flavicornis* Clark, 1934c: 69, pl. 4, figs. 31, 32 (w.q.) AUSTRALIA. Senior synonym of *flavidiscus* and *minor:* Taylor, 1992a: 65.

*flavidiscus. Prolasius flavidiscus* McAreavey, 1947: 21, pl. 1, figs. 20, 21 (w.q.) AUSTRALIA. Junior synonym of *flavicornis:* Taylor, 1992a: 65.

***formicoides.*** *Melophorus formicoides* Forel, 1902h: 483 (w.q.) AUSTRALIA. Combination in *Prolasius:* Wheeler, W.M. 1935c: 71. Variety of *nitidissimus:* McAreavey, 1947: 24. Revived status as species: Taylor, 1992a: 65.

***hellenae.*** *Prolasius hellenae* McAreavey, 1947: 13, pl. 1, figs. 8, 9 (w.) AUSTRALIA.

***hemiflavus.*** *Prolasius hemiflavus* Clark, 1934c: 68, pl. 4, figs. 28, 29 (w.q.) AUSTRALIA.

*minor. Prolasius flavicornis* var. *minor* McAreavey, 1947: 21 (w.) AUSTRALIA. Junior synonym of *flavicornis:* Taylor, 1992a: 65.

*mjobergi. Prenolepis mjobergi* Forel, 1915b: 93, pl. 2, fig. 6 (w.) AUSTRALIA. [Junior primary homonym of *Prenolepis vividula* var. *mjobergi* Forel, 1908b: 64.] Replacement name: *Prenolepis mjoebergella* Forel, in Santschi, 1916d: 242.]

***mjoebergella.*** *Prenolepis mjoebergella* Forel, in Santschi, 1916d: 242. Replacement name for *mjobergi* Forel, 1915b: 93, above. [Junior primary homonym of *mjobergi* Forel, 1908b: 64.] Combination in *Melophorus:* Emery, 1925b: 12; in *Prolasius:* Wheeler, W.M. 1935c: 71. See also: McAreavey, 1947: 16.

*niger. Prolasius niger* Clark, 1934c: 68, pl. 4, fig. 30 (w.) AUSTRALIA. Junior synonym of *nitidissimus:* Taylor, 1992a: 65.

*nigriventris. Prolasius nigriventris* McAreavey, 1947: 17, pl. 1, figs. 12, 13 (w.m.) AUSTRALIA. Junior synonym of *convexus:* Taylor, 1992a: 65.

***nitidissimus.*** *Formica nitidissima* André, 1896b: 255 (w.) AUSTRALIA. Combination in *Melophorus:* Forel, 1902h: 483; in *Prolasius:* Wheeler, W.M. 1935c: 71. Senior synonym of *niger* and *similis:* Taylor, 1992a: 65.

***pallidus.*** *Prolasius pallidus* Clark, 1934c: 67, pl. 4, figs. 26, 27 (w.q.) AUSTRALIA.

***quadratus.*** *Prolasius quadrata* McAreavey, 1947: 19, pl. 1, figs. 10, 11 (w.) AUSTRALIA.

***reticulatus.*** *Prolasius reticulata* McAreavey, 1947: 22, pl. 1, figs. 14, 15 (w.) AUSTRALIA.

***robustus.*** *Prolasius robustus* McAreavey, 1947: 20, pl. 1, figs. 18, 19 (w.) AUSTRALIA.

*similis. Prolasius depressiceps* var. *similis* McAreavey, 1947: 23 (w.) AUSTRALIA. Junior synonym of *nitidissimus:* Taylor, 1992a: 65.

***wheeleri.*** *Prolasius wheeleri* McAreavey, 1947: 22 (w.) AUSTRALIA.

***wilsoni.*** *Prolasius hemiflavus* var. *wilsoni* McAreavey, 1947: 18, pl. 1, figs. 16, 17 (w.) AUSTRALIA. Raised to species: Taylor, 1992a: 66.

*zealandica. Formica zealandica* Smith, F. 1878a: 6 (q.) NEW ZEALAND. McAreavey, 1947: 14 (w.). Combination in *Prolasius:* McAreavey, 1947: 14. Junior synonym of *advenus:* Brown, 1958h: 43.

*PROMERANOPLUS:* see under ***LORDOMYRMA.***

*PROMYOPIAS:* see under ***CENTROMYRMEX.***

*PROMYRMA:* see under ***LIOMYRMEX.***

*PROMYRMECIA:* see under ***MYRMECIA.***

**PROMYRMICIUM*** (Palaearctic)

**boreale.*** **Myrmicium boreale* Heer, 1870: 78, pl. 16, fig. 46 (wing) NORWAY (Miocene). Combination in **Promyrmicium:* Baroni Urbani, 1971b: 362.

**PROPODOMYRMA:* see under ***VOLLENHOVIA.***

*PROSCOPOMYRMEX:* see under ***STRUMIGENYS.***

*PROSOPIDRIS:* see under ***CARDIOCONDYLA.***

***PROTALARIDRIS*** (Neotropical)

***armata.*** *Protalaridris armata* Brown, 1980a: 37, figs. 1–8 (w.) ECUADOR. See also: Hölldobler & Wilson, 1986a: 16.

*PROTAMBLYOPONE:* see under ***AMBLYOPONE.***

**PROTAMBLYOPONE:* see under ****CASALEIA.***

****PROTANEURETUS*** (Baltic Amber)

****succineus.*** **Protaneuretus succineus* Wheeler, W.M. 1915h: 72, fig. 31 (w.q.) BALTIC AMBER (Oligocene).

***PROTANILLA*** (Indo-Australian)

***rafflesi.*** *Protanilla rafflesi* Taylor, in Bolton, 1990b: 279 (w.) SINGAPORE.

****PROTAZTECA*** (Nearctic)

****capitata.*** **Protazteca capitata* Carpenter, 1930: 45, pl. 3, fig. 3; pl. 9, fig. 10 (w.q.) U.S.A. (Miocene).

****elongata.*** **Protazteca elongata* Carpenter, 1930: 43, pl. 2, fig. 3; pl. 8, fig. 5; pl. 9, fig. 3; pl. 10, fig. 4 (w.q.m.) U.S.A. (Miocene).

****hendersoni.*** **Ponera hendersoni* Cockerell, 1906: 28 (q.) U.S.A. (Miocene). Combination in *Euponera:* Cockerell, 1927: 164; in **Protazteca:* Carpenter, 1930: 42.

****quadrata.*** **Protazteca quadrata* Carpenter, 1930: 44, pl. 3, fig. 1; pl. 6, fig. 7; pl. 10, fig. 5 (w.q.) U.S.A. (Miocene).

*PROTHOLCOMYRMEX:* see under ***MONOMORIUM.***

****PROTOFORMICA*** (Baltic Amber)

****proformicoides.*** **Formica (Protoformica) proformicoides* Dlussky, 1967b: 85, fig. 2 (w.) BALTIC AMBER (Oligocene). Combination in **Protoformica:* Dlussky & Fedoseeva, 1987: 77.

***PROTOMOGNATHUS*** (Nearctic)

***americanus.*** *Tomognathus americanus* Emery, 1895c: 272 (w.) U.S.A. Creighton, 1927: 27 (q.m.); Wheeler, G.C. & Wheeler, J. 1955b: 26 (l.). Combination in *T. (Protomognathus)*: Wheeler, W.M. 1905a: 3; in *Harpagoxenus:* Wheeler, W.M. 1910g: 494; in *Protomognathus:* Cover, in Hölldobler & Wilson, 1990: 65. See also: Smith, M.R. 1939b: 166.

****PROTOPONE*** (Russia)

****primigena.*** **Protopone primigena* Dlussky, 1988: 53, fig. 2 (w.) RUSSIA (Palaeocene).

****PROTRECHINA*** (Nearctic)

****carpenteri.*** **Protrechina carpenteri* Wilson, 1985d: 213, fig. 6 (w.) U.S.A. (Eocene).

***PSALIDOMYRMEX*** (Afrotropical)

*clavicornis* Bernard, 1953; see under ***PACHYCONDYLA.***

*collarti. Psalidomyrmex procerus* st. *collarti* Santschi, 1937g: 74 (w.) ZAIRE. Junior synonym of *procerus:* Bolton, 1975b: 12.

***feae.*** *Psalidomyrmex feae* Menozzi, 1922a: 349 (w.q.m.) PRINCIPE I. Senior synonym of *impressa:* Bolton, 1975b: 11.

***foveolatus.*** *Psalidomyrmex foveolatus* André, 1890: 314 (w.) SIERRA LEONE. Emery, 1899e: 471 (q.m.); Stitz, 1910: 129 (m.). See also: Bolton, 1975b: 8.

*impressa. Psalidomyrmex feae* var. *impressa* Menozzi, 1922a: 352 (w.q.m.) PRINCIPE I. Junior synonym of *feae:* Bolton, 1975b: 11.

*longiscapus. Psalidomyrmex longiscapus* Santschi, 1920b: 8 (q.) GABON. Junior synonym of *procerus:* Bolton, 1975b: 12.

*mabirensis. Plectroctena mandibularis* subsp. *mabirensis* Arnold, 1954: 293, fig. 3 (w.) UGANDA. Combination in *Psalidomyrmex,* and raised to species: Bolton, 1974c: 334. Junior synonym of *reichenspergeri:* Bolton, 1975b: 8.

*obesus. Psalidomyrmex obesus* Wheeler, W.M. 1922a: 92, fig. 19 (w.) ZAIRE. Variety of *procerus:* Santschi, 1937g: 74. Junior synonym of *procerus:* Bolton, 1975b: 12.

***procerus.*** *Psalidomyrmex procerus* Emery, 1901a: 50 (w.q.m.) CAMEROUN. Wheeler, G.C. & Wheeler, J. 1964b: 454 (l.). Senior synonym of *collarti, longiscapus* and *obesus:* Bolton, 1975b: 12.

***reichenspergeri.*** *Psalidomyrmex reichenspergeri* Santschi, 1913c: 302 (w.) CAMEROUN. Santschi, 1914c: 288 (m.). Senior synonym of *mabirensis:* Bolton, 1975b: 8.

***sallyae.*** *Psalidomyrmex sallyae* Bolton, 1975b: 9, fig. 5 (w.) GHANA.

***wheeleri.*** *Psalidomyrmex wheeleri* Santschi, 1923e: 263 (w.) ZAIRE. See also: Bolton, 1975b: 13.

*PSAMMOMYRMA:* see under ***DORYMYRMEX.***

*PSEUDACANTHOLEPIS: nomen nudum.*

***PSEUDAPHOMOMYRMEX*** (Philippines)

***emeryi.*** *Aphomyrmex emeryi* Ashmead, 1905a: 111 (q.) PHILIPPINES. [Name of genus a misspelling of *Aphomomyrmex.*] Combination in *Pseudaphomomyrmex:* Wheeler, W.M. 1920: 53.

****PSEUDARMANIA*** (Russia)

****aberrans.*** **Pseudarmania aberrans* Dlussky, 1983: 70, fig. 4 (q.) RUSSIA (Cretaceous).

****rasnitsyni.*** **Pseudarmania rasnitsyni* Dlussky, 1983: 69, fig. 3 (q.) RUSSIA (Cretaceous).

***PSEUDOATTA*** (Neotropical)

***argentina.*** *Pseudoatta argentina* Gallardo, 1916d: 321, figs. 1, 2 (q.m.) ARGENTINA. See also: Bruch, 1928: 341; Gallardo, 1929: 197; Wheeler, W.M. 1937c: 52. Current subspecies: nominal plus *platensis.*

***#platensis.*** *Pseudoatta argentina* var. *platensis* Santschi, 1926d: 7 (q.m.) ARGENTINA.

****PSEUDOCAMPONOTUS*** (U.S.A.)

****elkoanus.*** **Pseudocamponotus elkoanus* Carpenter, 1930: 22, pl. 2, fig. 2 (q.) U.S.A. (Miocene).

*PSEUDOCOLOBOPSIS:* see under ***CAMPONOTUS.***

*PSEUDOCRYPTOPONE:* see under ***PONERA.***

*PSEUDOCYRTOMYRMA:* see under ***POLYRHACHIS.***

*PSEUDODICHTHADIA:* see under ***LABIDUS.***

***PSEUDOLASIUS*** (Old World tropics except Malagasy)

***#abhorrens.*** *Pseudolasius butteli* st. *abhorrens* Santschi, 1932b: 28, fig. 12 (m.) NEW GUINEA.

***#adolphi.*** *Pseudolasius mayri* r. *adolphi* Forel, 1913k: 108 (footnote) (q.m.) JAVA.

*allax;* see ***fallax,*** below.

***amaurops.*** *Pseudolasius amaurops* Emery, 1922d: 109, fig. 2 (w.) NEW GUINEA. Donisthorpe, 1949b: 503 (m.).

***amblyops.*** *Pseudolasius amblyops* Forel, 1901b: 25 (w.) BISMARCK ARCHIPELAGO. Current subspecies: nominal plus *sundaicus.*

***australis.*** *Pseudolasius australis* Forel, 1915b: 94 (w.) AUSTRALIA.

***badius.*** *Pseudolasius badius* Viehmeyer, 1916a: 150 (q.) SINGAPORE.

***#bandarensis.*** *Pseudolasius minutus* var. *bandarensis* Forel, 1913k: 120 (q.m.) SUMATRA.

***bayonii.*** *Pseudolasius bayonii* Menozzi, 1924b: 224, figs. 5, 6 (s.) UGANDA. Senior synonym of *myersi* and *occipitalis:* Brown, 1957h: 146.

***#bedoti.*** *Pseudolasius bedoti* Emery, 1911a: 217 (q.m.) INDONESIA. Variety of *mayri:* Forel, 1913k: 106.

***PSEUDOLASIUS*** *(continued)*

***binghami.*** *Pseudolasius binghami* Emery, 1911a: 216 (w.) BURMA. Current subspecies: nominal plus *taivanae.*

***boreus.** **Pseudolasius boreus* Wheeler, W.M. 1915h: 133, fig. 64 (s.w.) BALTIC AMBER (Oligocene).

***breviceps.*** *Pseudolasius breviceps* Emery, 1887a: 246 (s.q.m.) INDONESIA (Ambon I.). Karavaiev, 1929a: 41 (w.). Senior synonym of *signatus:* Viehmeyer, 1914c: 43.

*brunneus* Donisthorpe, 1949; see under ***PHILIDRIS.***

***bucculentus.*** *Pseudolasius bucculentus* Wheeler, W.M. 1922a: 222, fig. 56 (s.w.m.) ZAIRE.

***bufonus.*** *Pseudolasius bufonum* Wheeler, W.M. 1922a: 220, fig. 55 (s.w.q.m.) ZAIRE.

***butteli.*** *Pseudolasius butteli* Forel, 1913k: 115, fig. J1 (s.m.) WEST MALAYSIA. Current subspecies: nominal plus *abhorrens.*

***caecus.*** *Pseudolasius caecus* Donisthorpe, 1949c: 758 (w.) NEW GUINEA.

***carinatus.*** *Pseudolasius carinatus* Karavaiev, 1929a: 44, fig. 2 (w.) INDONESIA (Aru I.).

***cibdelus.*** *Pseudolasius cibdelus* Wu & Wang, 1992: 1312, fig. 4607 (s.w.q.) CHINA.

***circularis.*** *Pseudolasius circularis* Viehmeyer, 1916a: 149 (q.) SINGAPORE.

**#*crassicornis.*** *Pseudolasius minutus* var. *crassicornis* Forel, 1913k: 120 (q.m.) SUMATRA.

*dodo* Donisthorpe, 1946; see under ***ACROPYGA.***

**#*duplicatus.*** *Pseudolasius mayri* var. *duplicata* Viehmeyer, 1916a: 150 (q.) SINGAPORE. [Misspelled as *deplanata* by Emery, 1925b: 215.]

**#*elisae.*** *Pseudolasius mayri* var. *elisae* Forel, 1913k: 107 (q.m.) SUMATRA.

***emeryi.*** *Pseudolasius emeryi* Forel, 1911g: 286 (w.) BURMA. Wheeler, W.M. 1928c: 33 (s.w.q.m.).

***fallax.*** *Pseudolasius fallax* Emery, 1911a: 216 (w.) JAVA. [Bold print in original description has *allax.* Printer's error as name in preceding key (p. 214) and following check-list (p. 218) of same publication show the spelling *fallax.*] Forel, 1912d: 109 (q.m.). Current subspecies: nominal plus *slamatensis.*

***familiaris.*** *Formica familiaris* Smith, F. 1860a: 68 (q.) SULAWESI. Mayr, 1867a: 75 (m.); Emery, 1887a: 244 (footnote) (w.); Forel, 1894c: 405 (w.). Combination in *Lasius:* Mayr, 1867a: 74; in *Pseudolasius:* Emery, 1887a: 244 (footnote).

***gowdeyi.*** *Pseudolasius gowdeyi* Wheeler, W.M. 1922a: 223, fig. 57 (s.w.) UGANDA.

***hummeli.*** *Pseudolasius hummeli* Stitz, 1934: 5, figs. 3, 4 (q.m.) CHINA.

***isabellae.*** *Pseudolasius isabellae* Forel, 1908a: 4 (q.m.) SRI LANKA. Current subspecies: nominal plus *simaluranus.*

***jacobsoni.*** *Pseudolasius jacobsoni* Crawley, 1924: 403, fig. 4 (s.w.) SUMATRA.

***karawajewi.*** *Pseudolasius karawajewi* Donisthorpe, 1942h: 166 (w.m.) NEW GUINEA.

***lasioides.*** *Pseudolasius lasioides* Wheeler, W.M. 1927h: 104, fig. 8 (w.) VIETNAM.

***leopoldi.*** *Pseudolasius leopoldi* Santschi, 1932b: 26, fig. 10 (m.) BORNEO.

***liliputi.*** *Pseudolasius liliputi* Forel, 1913k: 121 (q.) SUMATRA.

***ludovici.*** *Pseudolasius ludovici* Forel, 1913k: 110, figs. G1, H1 (s.w.q.m.) SUMATRA. Current subspecies: nominal plus *malaccanus, papuanus.*

**#*malaccanus.*** *Pseudolasius ludovici* var. *malaccana* Forel, 1913k: 114 (s.w.q.) WEST MALAYSIA.

***martini.*** *Pseudolasius martini* Forel, 1911d: 389 (q.m.) WEST MALAYSIA.

***mayri.*** *Pseudolasius mayri* Emery, 1911a: 215 (w.q.m.) JAVA. Current subspecies: nominal plus *adolphi, bedoti, duplicatus, elisae, sarawakanus, sericeus.*

***minor.*** *Pseudolasius minor* Donisthorpe, 1947c: 593 (q.) NEW GUINEA.

***minutissimus.*** *Pseudolasius minutissimus* Forel, 1913k: 122 (m.) SUMATRA.

***minutus.*** *Pseudolasius minutus* Emery, 1896f: 248 (q.m.) JAVA. Current subspecies: nominal plus *bandarensis, crassicornis.*

*myersi. Pseudolasius myersi* Weber, 1943c: 389, pl. 16, fig. 35 (w.) SUDAN. Junior synonym of *bayonii:* Brown, 1957h: 146.

*occipitalis. Pseudolasius myersi* subsp. *occipitalis* Weber & Anderson, 1950: 2, figs. 1–18 (s.w.l.) UGANDA. Junior synonym of *bayonii:* Brown, 1957h: 146.

***overbecki.*** *Pseudolasius overbecki* Viehmeyer, 1914c: 44 (q.m.) JAVA.

***pallidus.*** *Pseudolasius pallidus* Donisthorpe, 1949b: 505 (q.) NEW GUINEA.

**#*papuanus.*** *Pseudolasius ludovici* st. *papuanus* Santschi, 1932b: 27, fig. 11 (m.) NEW GUINEA.

***pheidolinus.*** *Pseudolasius pheidolinus* Emery, 1887a: 245 (s.w.q.) JAVA. Forel, 1913k: 110 (m.).

***pygmaeus.*** *Pseudolasius pygmaeus* Forel, 1913k: 117, fig. K1 (s.w.q.m.) WEST MALAYSIA.

***risii.*** *Pseudolasius risii* Forel, 1894c: 405 (footnote) (w.q.) HONG KONG. Wheeler, W.M. 1927d: 12 (m.); Wheeler, W.M. 1928c: 37 (s.).

***salvazai.*** *Pseudolasius salvazai* Santschi, 1920a: 565, fig. 1 (s.w.) VIETNAM.

**#*sarawakanus.*** *Pseudolasius mayri* var. *sarawakana* Forel, 1913k: 107 (footnote) (q.m.) BORNEO.

***sauteri.*** *Pseudolasius sauteri* Forel, 1913f: 199 (q.) TAIWAN.

**#*sericeus.*** *Pseudolasius mayri* r. *sericeus* Forel, 1912d: 110 (q.m.) JAVA. [Misspelled as *sericca* by Emery, 1925b: 215.]

***sexdentatus.*** *Pseudolasius sexdentatus* Donisthorpe, 1949c: 757 (q.) NEW GUINEA. Donisthorpe, 1949g: 421 (m.).

*signatus. Pseudolasius signatus* Stitz, 1911a: 370, fig. 17 (q.m.) NEW GUINEA. Junior synonym of *breviceps:* Viehmeyer, 1914c: 43.

***silvestrii.*** *Pseudolasius silvestrii* Wheeler, W.M. 1927h: 102, fig. 8 (s.) VIETNAM.

**#*simaluranus.*** *Pseudolasius isabellae* var. *simalurana* Forel, 1915a: 35 (w.q.) SUMATRA.

**#*slamatensis.*** *Pseudolasius fallax* var. *slamatensis* Forel, 1913k: 108 (q.m.) SUMATRA.

**#*sordidus.*** *Pseudolasius weissi* var. *sordida* Santschi, 1914d: 378 (w.) GHANA. Wheeler, W.M. 1922a: 219 (q.).

***streesemanni.*** *Pseudolasius streesemanni* Viehmeyer, 1914a: 114 (q.) WEST MALAYSIA.

***sumatrensis.*** *Prenolepis sumatrensis* Mayr, 1883: 247 (q.m.) SUMATRA. Combination in *Pseudolasius:* Forel, 1913k: 114.

***sunda.*** *Pseudolasius sunda* Karavaiev, 1929a: 51, fig. 6 (q.) INDONESIA (Prince I.).

**#*sundaicus.*** *Pseudolasius amblyops* var. *sundaica* Forel, 1913k: 122 (w.) SUMATRA.

**#*taivanae.*** *Pseudolasius binghami* r. *taivanae* Forel, 1912a: 74 (w.q.) TAIWAN.

***tenuicornis.*** *Pseudolasius tenuicornis* Emery, 1897d: 573 (q.) NEW GUINEA.

***trimorphus.*** *Pseudolasius trimorphus* Karavaiev, 1929a: 52, fig. 7 (s.w.q.m.) JAVA.

***typhlops.*** *Pseudolasius* (*Nesolasius*) *typhlops* Wheeler, W.M. 1935b: 50, fig. 3 (s.) PHILIPPINES.

*undecemus* Donisthorpe, 1949; see under ***ACROPYGA.***

***waigeuensis.*** *Pseudolasius waigeuensis* Donisthorpe, 1943d: 473 (w.) NEW GUINEA.

***weissi.*** *Pseudolasius weissi* Santschi, 1910c: 391, fig. 16 (w.q.) CONGO. Current subspecies: nominal plus *sordidus.*

*PSEUDOMYRMA:* see under ***PSEUDOMYRMEX.***

**PSEUDOMYRMEX** (Neotropical, Nearctic)

*acaciarum. Pseudomyrma sericea* var. *acaciarum* Wheeler, W.M. 1942: 176 (w.q.m.) PANAMA. Combination in *Pseudomyrmex:* Kempf, 1972a: 223. Senior synonym of *acaciorum:* Brown, 1949a: 43. Junior synonym of *ita:* Ward, 1993: 158.

*acaciorum. Pseudomyrma sericea* var. *acaciorum* Enzmann, E.V. 1944: 90, pl. 3, fig. 9 (w.) PANAMA. Junior synonym of *acaciarum:* Brown, 1949a: 43.

***acanthobius.*** *Pseudomyrma acanthobia* Emery, 1896h: 628, fig. A (w.q.m.) PARAGUAY. Combination in *Pseudomyrmex:* Kusnezov, 1953e: 214. Current subspecies: nominal plus *cocae, fuscatus, virgo.* See also: Gallardo, 1932a: 63.

***adustus.*** *Pseudomyrma pallens* var. *adusta* Borgmeier, 1929: 198 (w.) BRAZIL. Kempf, 1961a: 398 (q.); Wheeler, G.C. & Wheeler, J. 1973d: 207 (l.). Combination in *Pseudomyrmex:* Kempf, 1960: 7. Raised to species: Kempf, 1961a: 396.

*advena. Pseudomyrma advena* Smith, F. 1855c: 157 (q.) BRAZIL. Junior synonym of *oculatus:* Forel, 1911e: 277.

*affinis. Pseudomyrma haytiana* var. *affinis* Wheeler, W.M. 1936b: 196 (w.q.) HAITI. [First available use of *Pseudomyrma championi* subsp. *haytiana* var. *affinis* Wheeler, W.M. & Mann, 1914: 18; unavailable name (Ward, 1990: 487).] Junior synonym of *haytiana:* Kempf, 1961a: 393.

*agilis. Pseudomyrma agilis* Smith, F. 1860c: 70 (w.) BRAZIL. Kempf, 1958f: 448 (q.). Combination in *Pseudomyrmex:* Kempf, 1958f: 446. Subspecies of *gracilis:* Emery, 1890b: 61. Revived status as species: Dalla Torre, 1893: 55; Wheeler, W.M. 1923a: 3. Senior synonym of *carapuna* and *peruviana* Enzmann: Kempf, 1958f: 446. Junior synonym of *venustus:* Kempf, 1967c: 10.

*alfari. Pseudomyrma alfari* Forel, 1906d: 228 (w.) COSTA RICA. Combination in *Pseudomyrmex:* Kempf, 1972a: 215. Junior synonym of *nigrocinctus:* Ward, 1993: 145.

*allaborans* Walker, 1859 [*Pseudomyrma*]; see under ***TETRAPONERA.***

*allidora. Pseudomyrma allidora* Enzmann, E.V. 1944: 77 (w.) PANAMA. Junior synonym of *alliodorae:* Brown, 1949a: 42.

*alliodorae*. *Pseudomyrma alliodorae* Wheeler, W.M. 1942: 157, pl. 47, fig. a (w.q.) PANAMA. Wheeler, G.C. & Wheeler, J. 1956: 379 (l.). Combination in *Pseudomyrmex:* Wheeler, G.C. & Wheeler, J. 1956: 379. Senior synonym of *allidora:* Brown, 1949a: 42. Junior synonym of *elongatus:* Ward, 1989: 419.

***alternans***. *Pseudomyrma alternans* Santschi, 1936b: 197, figs. 1, 2 (w.) FRENCH GUIANA. Combination in *Pseudomyrmex:* Kempf, 1958f: 434.

*altinoda*. *Pseudomyrma sericea* var. *altinoda* Mann, 1916: 427 (w.) BRAZIL. Combination in *Pseudomyrmex:* Kempf, 1972a: 223. Junior synonym of *oculatus:* Ward, 1989: 422.

***alustratus***. *Pseudomyrmex alustratus* Ward, 1989: 409, fig. 13 (w.q.) PERU.

***alvarengai***. *Pseudomyrmex alvarengai* Kempf, 1961a: 372, figs. 1–3 (w.) BRAZIL.

*anceps*. *Pseudomyrma unicolor* var. *anceps* Santschi, 1925d: 222 (w.) BRAZIL. Junior synonym of *unicolor:* Kempf, 1958f: 439.

*andina*. *Pseudomyrma tenuis* var. *andina* Enzmann, E.V. 1944: 92 (w.) PERU. Junior synonym of *tenuis:* Kempf, 1960a: 26.

***antiguanus***. *Pseudomyrma antiguana* Enzmann, E.V. 1944: 78 (w.) GUATEMALA. Combination in *Pseudomyrmex:* Kempf, 1972a: 215. Current subspecies: nominal plus *brunnipes*.

****antiquus***. **Pseudomyrmex antiquus* Ward, 1992: 57, figs. 1, 12 (w.) DOMINICAN AMBER (Miocene).

***apache***. *Pseudomyrmex apache* Creighton, 1953b: 134, pl. 12, figs. 1–7 (w.q.m.) U.S.A. Wheeler, G.C. & Wheeler, J. 1956: 380 (l.). See also: Creighton, 1954: 9; Ward, 1985b: 229.

*arborissanctae*. *Pseudomyrma arborissanctae* Emery, 1894c: 147 (w.q.) BOLIVIA. Junior synonym of *triplarinus:* Wheeler, W.M. 1942: 186.

#***argentinus***. *Pseudomyrma gracilis* var. *argentina* Santschi, 1934c: 27 (w.) ARGENTINA. Combination in *Pseudomyrmex:* Kempf, 1958f: 434.

*atra*. *Pseudomyrma atra* Stitz, 1933: 69 (w.) BRAZIL. Junior synonym of *unicolor:* Kempf, 1969: 276.

#***atrinodus***. *Pseudomyrma gracilis* var. *atrinoda* Santschi, 1934c: 26 (w.q.) BRAZIL. Combination in *Pseudomyrmex:* Kempf, 1958f: 434.

***atripes***. *Pseudomyrma atripes* Smith, F. 1860c: 70 (w.) BRAZIL. Combination in *Pseudomyrmex*, and senior synonym of *kitschelti:* Kempf, 1967c: 2.

*atrox*. *Pseudomyrma spinicola* r. *atrox* Forel, 1912g: 24 (w.) PANAMA. Combination in *Pseudomyrmex:* Kempf, 1972a: 224. Junior synonym of *spinicola:* Ward, 1993: 149.

*auripes*. *Pseudomyrma auripes* Wheeler, W.M. 1922c: 5 (q.) TRINIDAD. Combination in *Pseudomyrmex:* Kempf, 1961a: 400. Junior synonym of *malignus:* Ward, 1989: 440.

****avitus***. **Pseudomyrmex avitus* Ward, 1992: 59, fig. 11 (w.m.) DOMINICAN AMBER (Miocene).

*baileyi*. *Pseudomyrma triplaridis* subsp. *baileyi* Wheeler, W.M. 1942: 185, pl. 50 (w.q.) GUIANA. Combination in *Pseudomyrmex:* Kempf, 1961a: 400. Senior synonym of *biolleyi:* Brown, 1949a: 43. Junior synonym of *viduus:* Ward, 1989: 445.

****baros***. **Pseudomyrmex baros* Ward, 1992: 63, figs. 2, 13 (w.) DOMINICAN AMBER (Miocene).

***beccarii***. *Pseudomyrma beccarii* Menozzi, 1935b: 192, fig. 1 (w.) GUIANA. Combination in *Pseudomyrmex:* Kempf, 1972a: 215.

#***belgranoi***. *Pseudomyrma solisi* var. *belgranoi* Santschi, 1916e: 372, figs. 2, 3 (w.) ARGENTINA. Combination in *Pseudomyrmex:* Kempf, 1972a: 223.

*belti*. *Pseudomyrma belti* Emery, 1890b: 63, pl. 7, fig. 1 (w.q.) COSTA RICA. [Also described as new by Emery, 1894k: 52.] Wheeler, G.C. & Wheeler, J. 1956: 380 (l.). Combination in *Pseudomyrmex:* Wheeler, G.C. & Wheeler, J. 1956: 380. Junior synonym of *flavicornis:* Ward, 1989: 438. See also: Wheeler, W.M. 1942: 159.

*bequaerti*. *Pseudomyrma belti* subsp. *bequaerti* Wheeler, W.M. 1942: 164 (w.) HONDURAS. Combination in *Pseudomyrmex:* Kempf, 1972a: 216. Senior synonym of *bequaerti* Enzmann: Brown, 1949a: 42. Junior synonym of *ferrugineus:* Ward, 1989: 437.

*bequaerti*. *Pseudomyrma belti* subsp. *bequaerti* Enzmann, E.V. 1944: 80, pl. 2, fig. 17 (w.) HONDURAS. Junior synonym of *bequaerti* Wheeler: Brown, 1949a: 42.

*bicincta*. *Pseudomyrma nigrocincta* var. *bicincta* Santschi, 1922b: 347 (w.) COSTA RICA. Combination in *Pseudomyrmex:* Kempf, 1972a: 221. Junior synonym of *nigrocinctus:* Ward, 1993: 145.

*bicolor*. *Pseudomyrma bicolor* Guérin-Méneville, 1844a: 427 (w.) COLOMBIA. Combination in *Pseudomyrmex:* Kempf, 1958f: 434. Junior synonym of *gracilis:* Roger, 1862c: 289; Mayr, 1863: 452. Revived from synonymy as variety/subspecies of *gracilis:* Emery, 1890b: 59; Santschi, 1925d: 223; Wheeler, W.M. 1942: 166. Junior synonym of *gracilis:* Ward, 1993: 155.

*biconvexa*. *Pseudomyrma biconvexa* Forel, 1899c: 95, pl. 4, fig. 10 (w.) GUATEMALA. Junior synonym of *filiformis:* Wheeler, W.M. 1919i: 125.

*bierigi*. *Pseudomyrma kuenckeli* var. *bierigi* Santschi, 1932e: 412 (w.) PANAMA. Junior synonym of *kuenckeli:* Kempf, 1961a: 402.

*biolleyi*. *Pseudomyrma triplaridis* subsp. *biolleyi* Enzmann, E.V. 1944: 93 (w.) GUIANA. Junior synonym of *baileyi:* Brown, 1949a: 43.

***boopis***. *Pseudomyrma boopis* Roger, 1863b: 25. Replacement name for *modesta* Smith, F. 1862b: 32. [Junior primary homonym of *modesta* Smith, F. 1860b: 106.] Combination in *Pseudomyrmex:* Kempf, 1967c: 2. Senior synonym of *excavata* (and its junior synonyms *flaviventris, fusciceps* and *guatemalensis*): Kempf, 1967c: 2; of *thoracica:* Ward, 1993: 154.

*boxi*. *Pseudomyrma triplaridis* subsp. *boxi* Wheeler, W.M. 1942: 184 (w.q.m.) GUIANA. Combination in *Pseudomyrmex:* Kempf, 1961a: 400. Senior synonym of *boxi* Enzmann: Brown, 1949a: 43. Junior synonym of *triplaridis:* Ward, 1989: 444.

*boxi*. *Pseudomyrma triplaridis* subsp. *boxi* Enzmann, E.V. 1944: 94 (w.) GUIANA. Junior synonym of *boxi* Wheeler: Brown, 1949a: 43.

*bradleyi*. *Pseudomyrma latinoda* subsp. *bradleyi* Wheeler, W.M. 1942: 169 (w.) PERU. Combination in *Pseudomyrmex:* Kempf, 1961a: 400. Senior synonym of *bradleyi* Enzmann: Brown, 1949a: 42. Junior synonym of *tachigaliae:* Ward, 1989: 444.

*bradleyi*. *Pseudomyrma bradleyi* Enzmann, E.V. 1944: 82, pl. 2, fig. 20 (w.) PERU. Junior synonym of *bradleyi* Wheeler: Brown, 1949a: 42.

*breviceps*. *Pseudomyrma elegans* r. *breviceps* Forel, 1912g: 20 (w.) TRINIDAD. Junior synonym of *termitarius:* Kempf, 1960a: 8.

***browni***. *Pseudomyrmex browni* Kempf, 1967c: 3 (w.) PANAMA.

***brunneus***. *Pseudomyrma brunnea* Smith, 1877b: 63 (w.) MEXICO. Emery, 1895c: 270 (q.); Wheeler, G.C. & Wheeler, J. 1956: 382 (l.). Combination in *Leptalea:* Smith, M.R. 1951a: 788; in *Pseudomyrmex:* Wheeler, G.C. & Wheeler, J. 1956: 382. Senior synonym of *nigrita:* Ward, 1985b: 231.

#***brunnipes***. *Pseudomyrma antiguana* var. *brunnipes* Enzmann, E.V. 1944: 79 (w.) GUATEMALA. Combination in *Pseudomyrmex* Kempf, 1972a: 215.

***caeciliae***. *Pseudomyrma eduardi* r. *caeciliae* Forel, 1913l: 214 (w.) GUATEMALA. Combination in *Pseudomyrmex:* Kempf, 1972a: 217. Raised to species: Ward, 1989: 410.

*canescens*. *Pseudomyrma canescens* Smith, F. 1877b: 66 (q.) BRAZIL. Combination in *Pseudomyrmex:* Kempf, 1958f: 434. Junior synonym of *gracilis:* Ward, 1993: 155.

*canescens*. *Pseudomyrma canescens* Wasmann, 1915a: 321, pl. 14, figs. 1–4 (w.q.m.) MEXICO. [Junior primary homonym of *canescens* Smith, above.] Replacement name: *wasmanni* Wheeler, W.M. 1921b: 92 (footnote).

*capensis* Smith, F. 1858; see under ***TETRAPONERA***.

*capperi*. *Pseudomyrma delicatula* var. *capperi* Forel, 1899c: 93 (w.) JAMAICA. Combination in *Pseudomyrmex:* Kempf, 1972a: 215. Variety of *flavidula:* Wheeler, W.M. 1913d: 240. Junior synonym of *simplex:* Ward, 1985b: 238.

*carapuna*. *Pseudomyrma gracilis* subsp. *carapuna* Mann, 1916: 428 (w.) BRAZIL. Junior synonym of *agilis:* Kempf, 1958f: 446.

*carbonaria* Smith, F. 1863; see under ***TETRAPONERA***.

*caroli*. *Pseudomyrma caroli* Forel, 1899c: 89 (w.q.) COSTA RICA. Wheeler, G.C. & Wheeler, J. 1956: 382 (l.). Combination in *Pseudomyrmex:* Wheeler, G.C. & Wheeler, J. 1956: 382. Junior synonym of *viduus:* Ward, 1989: 445.

#***castus***. *Pseudomyrma picta* subsp. *casta* Wheeler, W.M. 1942: 173 (w.) PERU. Combination in *Pseudomyrmex:* Kempf, 1972a: 222.

*cephalica*. *Pseudomyrma cephalica* Smith, F. 1855c: 168 (w.q.m.) BRAZIL. Junior synonym of *filiformis:* Roger, 1862c: 289.

***championi***. *Pseudomyrma championi* Forel, 1899c: 96 (w.) MEXICO. Wheeler, G.C. & Wheeler, J. 1956: 382 (l.); Kempf, 1961a: 392 (q.). Combination in *Pseudomyrmex:* Wheeler, G.C. & Wheeler, J. 1956: 382. Senior synonym of *leonhardi:* Kempf, 1961a: 391.

*chodati*. *Pseudomyrma chodati* Forel, 1921a: 201 (w.) PARAGUAY. Combination in *Pseudomyrmex:* Kempf, 1972a: 217. Junior synonym of *urbanus:* Ward, 1989: 427.

*cholerica*. *Pseudomyrma maligna* var. *cholerica* Wheeler, W.M. 1921f: 146 (w.) GUIANA. Combination in *Pseudomyrmex:* Kempf, 1961a: 400. Junior synonym of *malignus:* Ward, 1989: 440.

***cladoicus***. *Pseudomyrma cladoica* Smith, F. 1858b: 157, pl. 13, fig. 12 (w.) BRAZIL. Combination in

*Pseudomyrmex:* Kempf, 1958f: 434. See also: Kempf, 1967c: 4.

#***cocae.*** *Pseudomyrma acanthobia* var. *cocae* Santschi, 1922d: 243 (w.q.m.) ARGENTINA. Combination in *Pseudomyrmex:* Kempf, 1972a: 215.

***colei.*** *Pseudomyrma colei* Enzmann, E.V. 1944: 83, pl. 3, fig. 17 (w.) PERU. Combination in *Pseudomyrmex:* Kempf, 1972a: 217. Current subspecies: nominal plus *vistanus*.

***concolor.*** *Pseudomyrma concolor* Smith, F. 1860c: 70 (q.) BRAZIL. Combination in *Pseudomyrmex:* Kempf, 1961a: 400. Senior synonym of *penetrator:* Kempf, 1967c: 5; of *latinoda* (and its junior synonym *damnosa*): Ward, 1989: 435.

*confusior. Pseudomyrma schuppi* var. *confusior* Forel, 1901d: 299 (w.) BRAZIL. Junior synonym of *schuppi:* Kempf, 1961a: 380.

*convarians. Pseudomyrma spinicola* r. *convarians* Forel, 1913l: 214 (w.) GUATEMALA. Combination in *Pseudomyrmex:* Kempf, 1972a: 224. Junior synonym of *peperi:* Ward, 1989: 452.

#***cordiae.*** *Pseudomyrma sericea* var. *cordiae* Forel, 1904e: 690 (w.q.) PERU. Forel, 1921a: 202 (m.). Combination in *Pseudomyrmex:* Kempf, 1972a: 223.

#***cordobensis.*** *Pseudomyrma arborissanctae* var. *cordobensis* Forel, 1914d: 265 (w.) ARGENTINA. Combination in *Pseudomyrmex:* Kempf, 1961a: 400. Variety of *triplarinus:* Wheeler, W.M. 1942: 188.

***coronatus.*** *Pseudomyrma latinoda* var. *coronata* Wheeler, W.M. 1942: 167 (w.m.) GUIANA. Combination in *Pseudomyrmex:* Kempf, 1961a: 400. Senior synonym of *coronata* Enzmann: Brown, 1949a: 42. Subspecies of *concolor:* Ward, 1989: 436. Raised to species: Ward, 1990: 484.

*coronata. Pseudomyrma latinoda* var. *coronata* Enzmann, E.V. 1944: 88 (w.) GUIANA. Junior synonym of *coronatus* Wheeler: Brown, 1949a: 42.

****coruscus.*** **Pseudomyrmex coruscus* Ward, 1992: 65, fig. 3 (w.) DOMINICAN AMBER (Miocene).

#***costaricensis.*** *Pseudomyrma voytowskii* var. *costaricensis* Enzmann, E.V. 1944: 96 (w.) COSTA RICA. Combination in *Pseudomyrmex:* Kempf, 1972a: 226.

*crenulata. Pseudomyrma crenulata* Enzmann, E.V. 1944: 84, pl. 2, fig. 22 (w.) MEXICO. Junior synonym of *kuenckeli:* Kempf, 1961a: 402.

***cretus.*** *Pseudomyrmex cretus* Ward, 1989: 412, figs. 9, 26, 27 (w.q.) COSTA RICA.

*crucians. Pseudomyrma maligna* var. *crucians* Wheeler, W.M. 1921f: 147 (w.) GUIANA. Combination in *Pseudomyrmex:* Kempf, 1961a: 400. Junior synonym of *malignus:* Ward, 1989: 440.

***cubaensis.*** *Pseudomyrma elongata* var. *cubaensis* Forel, 1901e: 342 (w.) CUBA. Forel, 1913l: 215 (q.); Wheeler, G.C. & Wheeler, J. 1956: 384 (l., misidentified as *elongatus*). Combination in *Pseudomyrmex,* and junior synonym of *elongatus:* Creighton, 1957b: 18. Revived from synonymy, and raised to species: Ward, 1985b: 226. See also: Ward, 1989: 413.

*culmicola. Pseudomyrma culmicola* Forel, 1912g: 27 (w.) TRINIDAD. Combination in *Pseudomyrmex:* Kempf, 1972a: 217. Junior synonym of *tenuissimus:* Ward, 1989: 432.

***curacaensis.*** *Pseudomyrma dolichopsis* var. *curacaensis* Forel, 1912g: 23 (w.) SURINAM. Combination in *Pseudomyrmex,* and raised to species: Kempf, 1961a: 379 (misspelled as *curacaoensis*).

*damnosa. Pseudomyrma damnosa* Wheeler, W.M. 1921f: 139, fig. 13 (w.q.m.) GUIANA. Junior synonym of *latinoda:* Kempf, 1961a: 406.

*decipiens. Pseudomyrma decipiens* Forel, 1899c: 95, pl. 4, fig. 9 (w.q.) MEXICO. Wheeler, G.C. & Wheeler, J. 1956: 382 (l.). Combination in *Pseudomyrmex:* Creighton, 1953b: 132. Junior synonym of *elongatulus:* Kempf, 1967c: 6.

*delicatula. Pseudomyrma delicatula* Forel, 1899c: 93, pl. 4, fig. 6 (w.q.) MEXICO. Wheeler, W.M. 1913b: 484 (m.). Combination in *Pseudomyrmex:* Kempf, 1972a: 215. Variety/race of *acanthobius:* Forel, 1907e: 7; of *flavidulus:* Wheeler, W.M. 1913b: 484; Wheeler, W.M. 1913d: 240; Wheeler, W.M. & Mann, 1914: 17. Junior synonym of *simplex:* Ward, 1985b: 238.

***dendroicus.*** *Pseudomyrma dendroica* Forel, 1904c: 40 (w.m.) BRAZIL. Combination in *Pseudomyrmex,* and junior synonym of *triplarinus:* Kempf, 1961a: 408. Revived from synonymy, and senior synonym of *emarginata:* Ward, 1989: 436.

***denticollis.*** *Pseudomyrma denticollis* Emery, 1890b: 66, pl. 5, fig. 26 (w.) PARAGUAY. Kempf, 1960a: 15 (q.). Combination in *Pseudomyrmex:* Kusnezov, 1953e: 214. See also: Gallardo, 1932a: 56.

***depressus.*** *Pseudomyrma depressa* Forel, 1906d: 225 (w.) COLOMBIA. Combination in *Pseudomyrmex:* Kempf, 1972a: 217.

*dichroa. Pseudomyrma kuenckeli* var. *dichroa* Forel, 1904c: 41 (w.) COLOMBIA. Junior synonym of *kuenckeli:* Kempf, 1961a: 402.

*dimidiata. Pseudomyrma dimidiata* Roger, 1863a: 177 (w.) COLOMBIA. Combination in *Pseudomyrmex:* Kempf, 1958f: 434. Variety of *gracilis:* Mayr, 1870a: 406; Forel, 1912g: 19. Junior synonym of *gracilis:* Ward, 1993: 155.

***distinctus.*** *Pseudomyrma distincta* Smith, F. 1877b: 63 (w.) MEXICO. Combination in *Pseudomyrmex,* and senior synonym of *pulchella:* Kempf, 1967c: 5.

*dolichopsis. Pseudomyrma dolichopsis* Forel, 1899c: 87 (w.) COSTA RICA. Junior synonym of *oculatus:* Kempf, 1961a: 375.

***duckei.*** *Pseudomyrma duckei* Forel, 1906d: 227 (w.) BRAZIL. Combination in *Pseudomyrmex:* Kusnezov, 1964: 83.

*ecuadoriana. Pseudomyrma arborissanctae* var. *ecuadoriana* Enzmann, E.V. 1944: 79 (w.) ECUADOR. Junior synonym of *triplarinus:* Brown, 1949a: 44.

***eduardi.*** *Pseudomyrma eduardi* Forel, 1912g: 29 (w.) JAMAICA. Combination in *Pseudomyrmex:* Kempf, 1972a: 217. See also: Ward, 1989: 418.

***ejectus.*** *Pseudomyrma ejecta* Smith, F. 1858b: 157 (w.) "probably Brazil". Emery, 1894c: 147 (q.). Combination in *Pseudomyrmex:* Kempf, 1972a: 217. Senior synonym of *lincecumii:* Ward, 1989: 436. See also: Ward, 1985b: 232.

*elegans. Pseudomyrma elegans* Smith, F. 1855c: 159 (w.) BRAZIL. Forel, 1912g: 19 (q.); Wheeler, G.C. & Wheeler, J. 1956: 382 (l.). Combination in *Pseudomyrmex:* Kusnezov, 1953e: 214. Junior synonym of *termitarius:* Kempf, 1960a: 8.

*elongata. Pseudomyrma elongata* Smith, F. 1877b: 67 (w.) MEXICO. [Junior primary homonym of *elongata* Mayr, 1870a: 413, below.] Replacement name: *elongatula* Dalla Torre, 1892: 89.

***elongatulus.*** *Pseudomyrma elongatula* Dalla Torre, 1892: 89. Replacement name for *elongata* Smith, above. [Junior primary homonym of *elongata* Mayr, 1870a: 413, below.] Combination in *Pseudomyrmex,* and senior synonym of *decipiens:* Kempf, 1967c: 6. Current subspecies: nominal plus *longus*.

***elongatus.*** *Pseudomyrma elongata* Mayr, 1870a: 413 (w.) COLOMBIA. Wheeler, W.M. 1905b: 86 (q.m.); Wheeler, G.C. & Wheeler, J. 1973d: 207 (l.). Combination in *Leptalea:* Smith, M.R. 1951a: 788; in *Pseudomyrmex:* Wheeler, G.C. & Wheeler, J. 1956: 384. Senior synonym of *tandem:* Creighton, 1957b: 18; of *alliodorae* (and its junior synonym *allidora*): Ward, 1989: 419. See also: Ward, 1985b: 227.

*emarginata. Pseudomyrma dendroica* var. *emarginata* Forel, 1904e: 684 (w.q.) BRAZIL. Junior synonym of *triplarinus:* Kempf, 1961a: 408; of *dendroicus:* Ward, 1989: 436.

***endophytus.*** *Pseudomyrma latinoda* var. *endophyta* Forel, 1912g: 22 (w.) BRAZIL. Combination in *Pseudomyrmex:* Kempf, 1961a: 400. Raised to species: Ward, 1990: 484.

***ethicus.*** *Pseudomyrma ethica* Forel, 1911e: 280 (w.) BRAZIL. Combination in *Pseudomyrmex:* Kusnezov, 1964: 83.

***euryblemma.*** *Pseudomyrma euryblemma* Forel, 1899c: 90 (w.) COSTA RICA. Combination in *Pseudomyrmex:* Kempf, 1972a: 218. Senior synonym of *kurokii:* Ward, 1989: 437.

*excavata. Pseudomyrma excavata* Mayr, 1870a: 410 (w.) COLOMBIA. Kempf, 1960a: 23 (q.). Combination in *Pseudomyrmex:* Kempf, 1960a: 7. Senior synonym of *flaviventris, fusciceps* and *guatemalensis:* Kempf, 1960a: 22. Junior synonym of *boopis:* Kempf, 1967c: 2.

***excisus.*** *Pseudomyrma excisa* Mayr, 1870a: 410 (w.) COLOMBIA. Combination in *Pseudomyrmex:* Kempf, 1958f: 434.

****extinctus.*** **Pseudomyrma extincta* Carpenter, 1930: 29, pl. 3, fig. 4 (q.) U.S.A. (Miocene). Combination in *Pseudomyrmex:* Ward, 1990: 487.

***faber.*** *Pseudomyrma faber* Smith, F. 1858b: 157, pl. 13, fig. 11 (w.) BRAZIL. Combination in *Pseudomyrmex:* Kempf, 1958f: 449. Senior synonym of *perforator:* Perrault, 1988a: 384.

*fellosa. Pseudomyrma belti* subsp. *fellosa* Wheeler, W.M. 1942: 160 (w.) NICARAGUA. Combination in *Pseudomyrmex:* Kempf, 1972a: 216. Junior synonym of *flavicornis:* Ward, 1989: 438.

***ferrugineus.*** *Pseudomyrma ferruginea* Smith, F. 1877b: 64 (w.) MEXICO. Combination in *Pseudomyrmex:* Kempf, 1972a: 218. Senior synonym of *bequaerti* Wheeler (and its junior synony *bequaerti* Enzmann), *fulvescens, hondurana, saffordi* Wheeler, *vesana* and *wasmanni:* Ward, 1989: 437. See also: Ward, 1993: 136.

***fervidus.*** *Pseudomyrma fervida* Smith, F. 1877b: 65 (w.) MEXICO. Combination in *Pseudomyrmex:* Kempf, 1972a: 218.

**fiebrigi.** *Pseudomyrma fiebrigi* Forel, 1908c: 383 (w.q.) PARAGUAY. Forel, 1909a: 260 (m.). Combination in *Pseudomyrmex:* Kempf, 1972a: 218.

**filiformis.** *Formica filiformis* Fabricius, 1804: 405 (q.) CENTRAL AMERICA. Wheeler, W.M. 1919i: 125 (w.m.); Wheeler, G.C. & Wheeler, J. 1956: 384 (l.). Combination in *Leptalea:* Erichson, 1839: 309; in *Pseudomyrma:* Roger, 1862c: 289; in *Pseudomyrmex:* Wheeler, G.C. & Wheeler, J. 1956: 384. Senior synonym of *cephalica:* Roger, 1862c: 289; of *biconvexa:* Wheeler, W.M. 1919i: 125; of *longiceps* Forel, and *longiceps* Stitz: Ward, 1989: 438.

**flavicornis.** *Pseudomyrma flavicornis* Smith, F. 1877b: 67 (w.) NICARAGUA. Combination in *Pseudomyrmex:* Kempf, 1972a: 218. Senior synonym of *belti, fellosa* and *obnubila:* Ward, 1989: 438. See also: Ward, 1993: 141.

**flavidulus.** *Pseudomyrma flavidula* Smith, F. 1858b: 157 (w.) BRAZIL. Wheeler, W.M. 1905b: 84 (q.m.); Emery, 1899c: 6 (l.). Combination in *Leptalea:* Smith, M.R. 1951a: 788; in *Pseudomyrmex:* Kusnezov, 1953e: 214. Current subspecies: nominal plus *lizeri.* See also: Smith, M.R. 1958c: 112.

*flaviventris. Pseudomyrma excavata* var. *flaviventris* Emery, 1896b: 2 (w.) PANAMA. Junior synonym of *excavata:* Kempf, 1960a: 22.

#**fortis.** *Pseudomyrma sericea* var. *fortis* Forel, 1899c: 89 (w.) MEXICO. Combination in *Pseudomyrmex:* Kempf, 1972a: 223.

*fulvescens. Pseudomyrma belti* r. *fulvescens* Emery, 1890b: 64 (w.) GUATEMALA. Forel, 1907e: 7 (m.); Wheeler, W.M. 1942: 160 (q.). Combination in *Pseudomyrmex:* Kempf, 1972a: 216. Raised to species: Dalla Torre, 1893: 57. Race of *belti:* Forel, 1899c: 87. Junior synonym of *ferrugineus:* Ward, 1989: 437.

#**fuscatus.** *Pseudomyrma acanthobia* var. *fuscata* Emery, 1896h: 629 (w.q.m.) PARAGUAY. Combination in *Pseudomyrmex:* Kempf, 1972a: 215.

*fusciceps. Pseudomyrma excavata* var. *fusciceps* Santschi, 1931c: 271 (w.) PANAMA. Junior synonym of *excavata:* Kempf, 1960a: 22.

*gaigei. Pseudomyrma spinicola* r. *gaigei* Forel, 1914c: 615 (w.) COLOMBIA. Combination in *Pseudomyrmex:* Kempf, 1972a: 224. Junior synonym of *spinicola:* Ward, 1993: 150.

**gebellii.** *Pseudomyrma gebellii* Forel, 1899b: 303 (q.) COLOMBIA. [Also described as new by Forel, 1899d: 275.] Forel, 1912g: 25 (w.). Combination in *Pseudomyrmex:* Kempf, 1972a: 219.

*geraensis. Pseudomyrma schuppi* var. *geraensis* Forel, 1912g: 23 (w.) BRAZIL. Junior synonym of *schuppi:* Kempf, 1961a: 380.

**gibbinotus.** *Pseudomyrma pallens* var. *gibbinota* Forel, 1908c: 384 (w.) BRAZIL. Kempf, 1961a: 385 (q.). Combination in *Pseudomyrmex:* Kempf, 1960a: 7. Raised to species: Kempf, 1961a: 384.

*glabriventris. Pseudomyrma gracilis* var. *glabriventris* Santschi, 1922b: 345 (w.) BOLIVIA. Combination in *Pseudomyrmex:* Kempf, 1958f: 434. Junior synonym of *gracilis:* Ward, 1993: 156.

**godmani.** *Pseudomyrma godmani* Forel, 1899c: 95, pl. 4, fig. 11 (w.) PANAMA. Wheeler, G.C. & Wheeler, J. 1956: 385 (l.). Combination in *Pseudomyrmex:* Wheeler, G.C. & Wheeler, J. 1956: 385. Senior synonym of *semotus:* Kempf, 1967c: 6. See also: Kempf, 1958f: 436.

**goeldii.** *Pseudomyrma goeldii* Forel, 1912g: 28 (w.) BRAZIL. Combination in *Pseudomyrmex:* Kempf, 1972a: 219.

**gracilis.** *Formica gracilis* Fabricius, 1804: 405 (w.) CENTRAL AMERICA. Wheeler, W.M. & Bailey, 1920: 256 (l.); Wheeler, G.C. & Wheeler, J. 1956: 385 (l.). Combination in *Leptalea:* Erichson, 1839: 309; in *Pseudomyrma:* Roger, 1862c: 289; in *P.* (*Clavanoda*): Enzmann, E.V. 1944: 61; in *Pseudomyrmex:* Kusnezov, 1953e: 214. Senior synonym of *longinoda:* Brown, 1949a: 43; Kempf, 1961a: 370, of *variabilis:* Ward, 1989: 439; of *bicolor, canescens, dimidiata, glabriventris, mexicana, pilosula, sericata, velifera* and *volatilis,* and material of the unavailable name *guayaquilensis* referred here: Ward, 1993: 155. Current subspecies: nominal plus *argentinus, atrinodus.*

*guatemalensis. Pseudomyrma tenuis* var. *guatemalensis* Enzmann, E.V. 1944: 92 (w.) GUATEMALA. Junior synonym of *excavata:* Kempf, 1960a: 22.

*guayaquilensis. Pseudomyrma gracilis* subsp. *mexicana* var. *guayaquilensis* Forel, 1907e: 7 (w.) ECUADOR Unavailable name (Ward, 1990: 487); material referred to *gracilis* by Ward, 1993: 156.

**haytianus.** *Pseudomyrma championi* r. *haytiana* Forel, 1901e: 342 (w.) HAITI. Combination in *Pseudomyrmex:* Kempf, 1961a: 393. Raised to species: Wheeler, W.M. 1936b: 196. Senior synonym of *affinis* and *torquata:* Kempf, 1961a: 393.

**hesperius.** *Pseudomyrmex hesperius* Ward, 1993: 157, fig. 4 (w.) MEXICO.

#**heterogyna.** *Pseudomyrma picta* var. *heterogyna* Wheeler, W.M. & Mann, in Wheeler, W.M. 1942: 172 (w.q.) BOLIVIA. Combination in *Pseudomyrmex:* Kusnezov, 1953e: 214.

**holmgreni.** *Pseudomyrma holmgreni* Wheeler, W.M. 1925a: 11 (w.) PERU. Combination in *Pseudomyrmex:* Kempf, 1972a: 220. See also: Ward, 1989: 440.

*hondurana. Pseudomyrma kuenckeli* var. *hondurana* Enzmann, E.V. 1944: 87 (w.) HONDURAS. Junior synonym of *kuenckeli:* Kempf, 1961a: 402; of *ferrugineus:* Ward, 1989: 437.

#**huberi.** *Pseudomyrma sericea* var. *huberi* Santschi, 1922b: 346 (w.) BRAZIL. Combination in *Pseudomyrmex:* Kempf, 1972a: 223.

#**humboldi.** *Pseudomyrma picta* var. *humboldi* Enzmann, E.V. 1944: 75 (diagnosis in key), pl. 3, fig. 21 (w.) BRAZIL. Combination in *Pseudomyrmex:* Kempf, 1972a: 222.

*icterica. Pseudomyrma icterica* Wheeler, W.M. 1922c: 4 (w.) TRINIDAD. Combination in *Pseudomyrmex:* Kempf, 1961a: 399. Junior synonym of *perboscii:* Ward, 1989: 442.

*implicata. Pseudomyrma dolichopsis* var. *implicata* Forel, 1911e: 277 (w.) BRAZIL. Junior synonym of *oculatus:* Kempf, 1961a: 375.

**incurrens.** *Pseudomyrma championi* r. *incurrens* Forel, 1912g: 21 (w.) BRAZIL. Combination in *Pseudomyrmex,* and raised to species: Kempf, 1961a: 396.

*infernalis. Pseudomyrma spinicola* subsp. *infernalis* Wheeler, W.M. 1942: 180 (w.q.m.) PANAMA. Combination in *Pseudomyrmex:* Kempf, 1972a: 224. Senior synonym of *infernalis* Enzmann: Brown, 1949a: 43. Junior synonym of *spinicola:* Ward, 1993: 150.

*infernalis. Pseudomyrma spinolae* var. *infernalis* Enzmann, E.V. 1944: 91 (w.q.m.) PANAMA. [The name *spinolae* is a misspelling of *spinicola.*] Junior synonym of *infernalis* Wheeler: Brown, 1949a: 43.

*infusca. Pseudomyrma denticollis* var. *infusca* Forel, 1908c: 382 (w.) BRAZIL. Junior synonym of *termitarius:* Kempf, 1960a: 8.

*insularis. Pseudomyrma laevigata* subsp. *insularis* Enzmann, E.V. 1944: 88 (w.) PANAMA. Junior synonym of *faber:* Kempf, 1958f: 449; of *osurus:* Ward, 1989: 441.

**ita.** *Pseudomyrma sericea* var. *ita* Forel, 1906d: 230 (w.) COSTA RICA. Wheeler, W.M. 1942: 177 (q.). Combination in *Pseudomyrmex:* Kempf, 1972a: 223. Raised to species, and senior synonym of *acaciarum* (and its junior synonym *acaciorum*): Ward, 1993: 158.

**janzeni.** *Pseudomyrmex janzeni* Ward, 1993: 143, figs. 19, 29, 64, 70 (w.q.m.) MEXICO.

*jaumei. Pseudomyrma flavidula* var. *jaumei* Aguayo, 1932: 217 (w.) CUBA. Combination in *Pseudomyrmex:* Kempf, 1972a: 218. Junior synonym of *pazosi:* Ward, 1989: 441.

*kitschelti. Pseudomyrma levigata* subsp. *kitschelti* Forel, 1911e: 279 (w.) COLOMBIA. Combination in *Pseudomyrmex,* and raised to species: Kempf, 1958f: 444. Junior synonym of *atripes:* Kempf, 1967c: 2.

**kuenckeli.** *Pseudomyrma kuenckeli* Emery, 1890b: 62, pl. 6, fig. 5 (w.q.) COSTA RICA. [Also described as new by Emery, 1894k: 51.] Combination in *Pseudomyrmex:* Kusnezov, 1953e: 214. Senior synonym of *bierigi, crenulata* and *dichroa:* Kempf, 1961a: 402. See also: Ward, 1993: 158.

*kurokii. Pseudomyrma kurokii* Forel, 1906d: 226 (w.) COLOMBIA. Combination in *Pseudomyrmex:* Kempf, 1972a: 220. Junior synonym of *euryblemma:* Ward, 1989: 437.

*laeviceps* Smith, F. 1859; see under ***TETRAPONERA.***

*laeviceps. Pseudomyrma laeviceps* Smith, F. 1877b: 63 (w.) BRAZIL. [Misspelled as *leviceps* by Dalla Torre, 1893: 57.] Combination in *Pseudomyrmex:* Kusnezov, 1964: 83. [Junior primary homonym of *laeviceps* Smith, F. 1859a: 145, above.]. Replacement name: *laevifrons* Ward, 1989: 440.

**laevifrons.** *Pseudomyrmex laevifrons* Ward, 1989: 440. Replacement name for *laeviceps* Smith, F. 1877b: 63, above. [Junior primary homonym of *laeviceps* Smith, F. 1859a: 145.]

**laevigatus.** *Pseudomyrma laevigata* Smith, F. 1877b: 62 (w.) BRAZIL. Combination in *Pseudomyrmex:* Kempf, 1958f: 443. Race of *gracilis:* Emery, 1890b: 61. Revived status as species: Forel, 1899c: 91.

**laevivertex.** *Pseudomyrma acanthobia* r. *laevivertex* Forel, 1906d: 227 (w.) COLOMBIA. Combination in *Pseudomyrmex:* Kempf, 1972a: 219. Race of *flavidulus:* Forel, 1912g: 26. Raised to species: Ward, 1992: 80.

#**landolti.** *Pseudomyrma pallens* var. *landolti* Forel, 1912g: 26 (w.) COLOMBIA. Combination in *Pseudomyrmex:* Kempf, 1960a: 7.

*laticeps. Pseudomyrma nigropilosa* r. *laticeps* Forel, 1906d: 229 (w.) TRINIDAD. Junior synonym of *maculatus:* Kempf, 1958f: 454.

*latinoda. Pseudomyrma latinoda* Mayr, 1878: 877 (w.) BRAZIL. Combination in *P. (Latinoda)*: Enzmann, 1944: 62; in *Pseudomyrmex*, and senior synonym of *damnosa:* Kempf, 1961a: 406. Junior synonym of *concolor:* Ward, 1989: 435.

*leonhardi. Pseudomyrma leonhardi* Stitz, 1937: 132, figs. (w.) MEXICO. Junior synonym of *championi:* Kempf, 1961a: 391.

***leptosus.*** *Pseudomyrmex leptosus* Ward, 1985b: 233, figs. 19, 26, 34, 35, 38 (q.m.) U.S.A.

*ligniseca. Pseudomyrma ligniseca* Smith, F. 1858b: 158 (q.) BRAZIL. Junior synonym of *tenuis:* Roger, 1862c: 289.

*limai. Pseudomyrma acanthobia* subsp. *deliculata* var. *limai* Enzmann, E.V. 1944: 77, pl. 4, fig. 24 (w.) PERU. Unavailable name (Ward, 1990: 487).

*lincecumii. Ponera (Ectatomma) lincecumii* Buckley, 1866: 172 (w.) U.S.A. Combination in *Pseudomyrma:* Mayr, 1886d: 464; in *Leptalea:* Smith, M.R. 1951a: 788. See also: Emery, 1895c: 270. Junior synonym of *ejectus:* Ward, 1989: 436.

**#lisa.** *Pseudomyrma sericea* var. *lisa* Enzmann, E.V. 1944: 90 (w.) PERU. Combination in *Pseudomyrmex:* Kempf, 1972a: 223.

**#lizeri.** *Pseudomyrma flavidula* st. *lizeri* Santschi, 1922b: 346 (w.) BOLIVIA. Combination in *Pseudomyrmex:* Kusnezov, 1953e: 214.

*loewensohni. Pseudomyrma arborissanctae* subsp. *symbiotica* var. *loewensohni* Forel, 1918a: 719. Unavailable name (Ward, 1990: 487).

*longiceps. Pseudomyrma biconvexa* var. *longiceps* Forel, 1906d: 229 (w.) COLOMBIA. Combination in *Pseudomyrmex:* Kempf, 1972a: 218. Variety of *filiformis:* Wheeler, W.M. 1919i: 131. Junior synonym of *filiformis:* Ward, 1989: 438.

*longiceps. Pseudomyrma longiceps* Stitz, 1933: 68 (q.) VENEZUELA. [Unresolved junior primary homonym of *longiceps* Wheeler, above.] Combination in *Pseudomyrmex:* Kempf, 1972a: 221. Junior synonym of *filiformis:* Ward, 1989: 438.

*longinoda. Pseudomyrma gracilis* var. *longinoda* Enzmann, E.V. 1944: 87 (w.) PERU. Junior synonym of *gracilis:* Brown, 1949a: 43; Kempf, 1961a: 370.

**#longior.** *Pseudomyrma sericea* var. *longior* Forel, 1904e: 690 (w.) PERU. Combination in *Pseudomyrmex:* Kempf, 1972a: 223.

**#longus.** *Pseudomyrma decipiens* r. *longa* Forel, 1912g: 24 (w.) BRAZIL. Combination in *Pseudomyrmex*, and subspecies of *elongatulus:* Kempf, 1967c: 6.

***lynceus.*** *Myrmica lyncea* Spinola, 1851a: 241 (w.) CHILE. Combination in *Pseudomyrma:* Mayr, 1870b: 972; in *Pseudomyrmex:* Kempf, 1961a: 390.

****macrops.*** **Pseudomyrmex macrops* Ward, 1992: 66, figs. 4, 14 (w.) DOMINICAN AMBER (Miocene).

***maculatus.*** *Pseudomyrma maculata* Smith, F. 1855c: 158 (w.) BRAZIL. Combination in *Pseudomyrmex:* Kempf, 1958f: 454. Race of *gracilis:* Emery, 1890b: 60. Revived status as species: Dalla Torre, 1893: 57; Mann, 1916: 426. Senior synonym of *laticeps* and *wagneri:* Kempf, 1958f: 454.

***major.*** *Pseudomyrma gracilis* var. *major* Forel, 1899c: 91 (w.) MEXICO. Combination in *Pseudomyrmex:* Kempf, 1958f: 434. Raised to species: Ward, 1993: 156 (in text).

***malignus.*** *Pseudomyrma maligna* Wheeler, W.M. 1921f: 143, fig. 14 (w.q.m.) GUIANA. Wheeler, G.C. & Wheeler, J. 1990a: 464 (l.). Combination in *Pseudomyrmex:* Kempf, 1961a: 400. Senior synonym of *auripes, cholerica* and *crucians:* Ward, 1989: 440.

***mandibularis.*** *Leptalaea mandibularis* Spinola, 1851b: 52 (w.) BRAZIL. [Also described as new by Spinola, 1853: 68.] Combination in *Pseudomyrma:* Smith, F. 1858b: 156; in *Pseudomyrmex:* Kusnezov, 1964: 83.

*mexicana. Pseudomyrma mexicana* Roger, 1863a: 178 (w.) MEXICO. Wheeler, G.C. & Wheeler, J. 1956: 385 (l.); Petralia & Vinson, 1980: 381 (l.). Combination in *Leptalea:* Smith, M.R. 1951a: 788; in *Pseudomyrmex:* Kusnezov, 1953e: 214. Variety/subspecies of *gracilis:* Mayr, 1870a: 409; Emery, 1890b: 45; Forel, 1907e: 7; Creighton, 1950a: 80. Revived status as species: Whitcomb, Denmark, *et al.* 1972: 31. Junior synonym of *gracilis:* Ward, 1993: 155. See also: Ward, 1985b: 225.

***mixtecus.*** *Pseudomyrmex mixtecus* Ward, 1993: 144, figs. 15, 26, 55, 61, 69 (w.q.m.) MEXICO.

*modesta* Smith, F. 1860; see under ***TETRAPONERA.***

*modesta. Pseudomyrma modesta* Smith, F. 1862b: 32 (w.) PANAMA. [Junior primary homonym of *modesta* Smith, F. 1860b: 106, above.] Replacement name: *boopis* Roger, 1863b: 25.

***monochrous.*** *Pseudomyrma monochrous* Dalla Torre, 1892: 89. Replacement name for *unicolor* Smith, F. 1877b: 68. [Junior primary homonym of *unicolor* Smith, F. 1855c: 158.] Combination in *Pseudomyrmex:* Kempf, 1972a: 221.

*mutica. Pseudomyrma mutica* Mayr, 1887: 627 (w.q.m.) BRAZIL. Combination in *Pseudomyrmex:* Kempf, 1960a: 7. Senior synonym of *paulina:* Kempf, 1961a: 386. Junior synonym of *phyllophilus:* Kempf, 1967c: 8. See also: Gallardo, 1932a: 58.

*mutilloides. Pseudomyrma mutilloides* Emery, 1890b: 61, pl. 5, fig. 23 (w.) BRAZIL. Junior synonym of *unicolor:* Kempf, 1958f: 439.

*natalensis* Smith, F. 1858; see under ***TETRAPONERA.***

****nexilis.*** **Pseudomyrmex nexilis* Ward, 1992: 68, figs. 5, 17 (w.) DOMINICAN AMBER (Miocene).

***niger.*** *Pseudomyrma nigra* Donisthorpe, 1940a: 39 (w.) GUIANA. Combination in *Pseudomyrmex:* Kempf, 1958f: 438.

***nigrescens.*** *Pseudomyrma latinoda* var. *nigrescens* Forel, 1904c: 38 (w.) BRAZIL. Combination in *Pseudomyrmex:* Kempf, 1961a: 400. Raised to species: Ward, 1990: 485.

*nigriceps. Pseudomyrma nigriceps* Smith, F. 1855c: 159 (w.) BRAZIL. Combination in *Pseudomyrmex:* Kusnezov, 1953e: 214. Stirps of *tenuis:* Santschi, 1922b: 345. Junior synonym of *termitarius:* Perrault, 1988a: 384. See also: Kempf, 1960a: 7.

*nigrita. Pseudomyrma brunnea* var. *nigrita* Enzmann, E.V. 1944: 82, pl. 2, fig. 12 (w.) MEXICO. Combination in *Pseudomyrmex:* Kempf, 1972a: 216. Junior synonym of *brunneus:* Ward, 1985b: 231.

***nigrocinctus.*** *Pseudomyrma nigrocincta* Emery, 1890b: 64, pl. 6, fig. 3 (w.q.) COSTA RICA. [Also described as new by Emery, 1894k: 52.] Combination in *Pseudomyrmex:* Kempf, 1972a: 221. Senior synonym of *alfari, bicincta* and *peltata:* Ward, 1993: 144.

***nigropilosus.*** *Pseudomyrma nigropilosa* Emery, 1890b: 62, pl. 5, fig. 24 (w.) COSTA RICA. [Also described as new by Emery, 1894k: 51.] Combination in *Pseudomyrmex:* Kempf, 1958f: 453. See also: Ward, 1993: 159.

*nitida* Smith, F. 1860; see under ***TETRAPONERA.***

*obnubila. Pseudomyrma belti* var. *obnubila* Menozzi, 1927c: 273 (w.q.) COSTA RICA. Combination in *Pseudomyrmex:* Kempf, 1972a: 216. Junior synonym of *flavicornis:* Ward, 1989: 438.

***oculatus.*** *Pseudomyrma oculata* Smith, F. 1855c: 159, pl. 13 (w.q.) BRAZIL. Combination in *Pseudomyrmex:* Kempf, 1961a: 375. Senior synonym of *advena:* Forel, 1911e: 277; of *dolichopsis* and *implicata:* Kempf, 1961a: 375; of *altinoda, tuberculata* and *wessoni:* Ward, 1989: 422.

*ogloblini. Pseudomyrma ogloblini* Santschi, 1936d: 402, figs. 1–3 (w.) ARGENTINA. Combination in *Pseudomyrmex:* Kempf, 1972a: 222. Junior synonym of *urbanus:* Ward, 1989: 427.

***oki.*** *Pseudomyrma oki* Forel, 1906d: 225 (w.) COLOMBIA. Forel, 1911e: 275 (q.). Combination in *Pseudomyrmex:* Kempf, 1961a: 399.

***opaciceps.*** *Pseudomyrmex opaciceps* Ward, 1993: 159, fig. 5 (w.q.m.) GUATEMALA.

***opacior.*** *Pseudomyrma latinoda* var. *opacior* Forel, 1904d: 170 (w.) CUBA. Combination in *Pseudomyrmex:* Kempf, 1961a: 400. Subspecies of *latinoda:* Wheeler, W.M. 1942: 169. Raised to species: Ward, 1990: 485.

****oryctus.*** **Pseudomyrmex oryctus* Ward, 1992: 69, figs. 7, 15 (w.) DOMINICAN AMBER (Miocene).

***osurus.*** *Pseudomyrma levigata* subsp. *osura* Forel, 1911e: 279 (w.) COSTA RICA. Junior synonym of *faber:* Kempf, 1958f: 449. Combination in *Pseudomyrmex*, revived from synonymy, raised to species, and senior synonym of *insularis:* Ward, 1989: 441.

***pallens.*** *Pseudomyrma pallens* Mayr, 1870a: 411 (w.q.) COLOMBIA. Combination in *Pseudomyrmex:* Kempf, 1960a: 7. Current subspecies: nominal plus *landolti.*

***pallidus.*** *Pseudomyrma pallidus* Smith, F. 1855c: 160 (w.) U.S.A. Wheeler, G.C. & Wheeler, J. 1956: 386 (l.); Petralia & Vinson, 1980: 381 (l.). Combination in *Leptalea:* Smith, M.R. 1951a: 788; in *Pseudomyrmex:* Wheeler, G.C. & Wheeler, J. 1956: 386. Variety of *flavidulus:* Enzmann, E.V. 1944: 66. Revived status as species: Smith, M.R. 1947f: 544; Creighton, 1950a: 80. See also: Blum & Callahan, 1963: 69; Ward, 1985b: 235.

*panamensis. Pseudomyrma delicatula* var. *panamensis* Forel, 1899c: 93 (w.) GUATEMALA. Combination in *Pseudomyrmex:* Kempf, 1972a: 215. Junior synonym of *simplex:* Ward, 1985b: 238.

*panamensis. Pseudomyrma arborissanctae* r. *symbiotica* var. *panamensis* Forel, 1912g: 22 (w.q.m.) PANAMA. Unavailable name (Ward, 1990: 487).

*paraensis. Pseudomyrma tenuis* var. *paraensis* Forel, 1912g: 19 (w.) BRAZIL. Junior synonym of *tenuis:* Kempf, 1960a: 26.

***particeps.*** *Pseudomyrmex particeps* Ward, 1993: 146, figs. 14, 23, 54, 57, 72 (w.q.m.) COSTA RICA.

*paulina. Pseudomyrma championi* var. *paulina* Wheeler, W.M. 1916d: 324 (w.) BRAZIL. [First available use of *Pseudomyrma championii* subsp. *haytiana* var. *paulina* Forel, 1911c: 305; unavailable name (Ward, 1990: 487).] Junior synonym of *muticus:* Kempf, 1961a: 386.

***pazosi.*** *Pseudomyrma pazosi* Santschi, 1909a: 309 (w.q.) CUBA. Wheeler, G.C. & Wheeler, J. 1956: 386 (l.). Combination in *Pseudomyrmex:* Wheeler, G.C. & Wheeler, J. 1956: 386. Variety of *flavidulus:* Wheeler, W.M. 1913b: 484; Mann, 1920: 406. Revived status as species, and senior synonym of *jaumei:* Ward, 1989: 441.

*peltata. Pseudomyrma peltata* Menozzi, 1927c: 273, fig. 2 (w.m.) COSTA RICA. Combination in *Pseudomyrmex:* Kempf, 1972a: 222. Junior synonym of *nigrocinctus:* Ward, 1993: 145.

*penetrator. Pseudomyrma penetrator* Smith, F. 1877b: 66 (q.) BRAZIL. Junior synonym of *concolor:* Kempf, 1967c: 5.

***peperi.*** *Pseudomyrma peperi* Forel, 1913l: 213 (w.) GUATEMALA. Combination in *Pseudomyrmex:* Kempf, 1972a: 222. Senior synonym of *convarians* and *saffordi* Enzmann: Ward, 1989: 452. See also: Ward, 1993: 146.

***perboscii.*** *Myrmex perboscii* Guérin-Méneville, 1844a: 428 (q.) MEXICO. Forel, 1899c: 86 (w.). Combination in *Pseudomyrma:* Smith, F. 1858b: 156; in *Pseudomyrmex:* Kempf, 1972a: 222. Senior synonym of *icterica, simoides* and *testacea:* Ward, 1989: 442.

*perforator. Pseudomyrma perforator* Smith, F. 1860c: 69 (w.) BRAZIL. [Misspelled as *perforatrix* by Forel, 1911e: 276.] Combination in *Pseudomyrmex:* Kempf, 1958f: 434. Junior synonym of *faber:* Perrault, 1988a: 384.

***peruvianus.*** *Pseudomyrma ejecta* var. *peruviana* Wheeler, W.M. 1925a: 11 (w.q.) PERU. Combination in *Pseudomyrmex:* Kempf, 1972a: 217. Raised to species: Ward, 1985b: 232. See also: Ward, 1990: 486.

*peruviana. Pseudomyrma gracilis* var. *peruviana* Enzmann, E.V. 1944: 86, pl. 4, fig. 28 (w.) PERU. [Unresolved junior primary homonym of *peruviana* Wheeler, above.] Junior synonym of *agilis:* Kempf, 1958f: 446.

***phyllophilus.*** *Pseudomyrma phyllophila* Smith, F. 1858b: 156 (w.) BRAZIL. Combination in *Pseudomyrmex,* and senior synonym of *mutica* (and its junior synonym *paulina*): Kempf, 1967c: 8.

***pictus.*** *Pseudomyrma picta* Stitz, 1913: 209, fig. 2 (w.) BRAZIL. Combination in *Pseudomyrmex:* Kempf, 1961a: 399. Current subspecies: nominal plus *castus, heterogyna, humboldi.* See also: Wheeler, W.M. 1942: 171.

*pilosa* Smith, F. 1858; see under ***TETRAPONERA.***

*pilosula. Pseudomyrma pilosula* Smith, F. 1877b: 62 (w.) BARBADOS. Combination in *Pseudomyrmex:* Kempf, 1958f: 435. Junior synonym of *gracilis:* Ward, 1993: 155.

***pisinnus.*** *Pseudomyrmex pisinnus* Ward, 1989: 425, fig. 11 (w.) BRAZIL.

*pittieri. Pseudomyrma tenuis* var. *pittieri* Forel, 1906d: 230 (w.) COSTA RICA. Junior synonym of *tenuis:* Kempf, 1960a: 26.

****prioris.*** **Pseudomyrmex prioris* Ward, 1992: 70, figs. 8, 16 (w.) DOMINICAN AMBER (Miocene).

*pruinosa. Lepalaea pruinosa* Spinola, 1851b: 53 (w.) BRAZIL. [Repeated in Spinola, 1853: 69.] *Nomen nudum;* attributed to Klug.

*pulchella. Pseudomyrma distincta* var. *pulchella* Forel, 1899c: 94, pl. 4, fig. 7 (w.q.) MEXICO. Junior synonym of *distinctus:* Kempf, 1967c: 5.

***pupa.*** *Pseudomyrma mutilloides* subsp. *pupa* Forel, 1911e: 274 (w.) BRAZIL. Combination in *Pseudomyrmex,* and raised to species: Kempf, 1958f 441.

***reconditus.*** *Pseudomyrmex reconditus* Ward, 1993: 160, fig. 8 (w.q.) NICARAGUA.

***rochai.*** *Pseudomyrma rochai* Forel, 1912g: 27 (w.) BRAZIL. Combination in *Pseudomyrmex:* Kempf, 1972a: 222.

#***rubiginosus.*** *Pseudomyrma sericea* var. *rubiginosa* Stitz, 1913: 211 (w.) BRAZIL. Combination in *Pseudomyrmex:* Kempf, 1972a: 223.

*rufa. Pseudomyrma rufa* Smith, F. 1877b: 64 (w.) BRAZIL. Mann, 1916: 425 (q.). Junior synonym of *tenuis:* Kempf, 1960a: 26.

***rufiventris.*** *Pseudomyrma kurokii* var. *rufiventris* Forel, 1911e: 275 (q.) PARAGUAY. Combination in *Pseudomyrmex:* Kempf, 1972a: 220. Raised to species: Ward, 1989: 442.

***rufomedius.*** *Pseudomyrma rufomedia* Smith, F. 1877b: 66 (q.) GUATEMALA. Combination in *Pseudomyrmex:* Wheeler, G.C. & Wheeler, J. 1956: 386. Senior synonym of *stolli:* Ward, 1989: 443. See also: Kempf, 1960a: 7.

#***rurrenabaquensis.*** *Pseudomyrma triplarina* var. *rurrenabaquensis* Wheeler, W.M. & Mann, in Wheeler, W.M. 1942: 188 (w.) BOLIVIA. Combination in *Pseudomyrmex:* Kempf, 1961a: 401.

*saffordi. Pseudomyrma belti* subsp. *saffordi* Wheeler, W.M. 1942: 162 (w.) MEXICO. Combination in *Pseudomyrmex:* Kempf, 1972a: 216. Junior synonym of *ferrugineus:* Ward, 1989: 437.

*saffordi. Pseudomyrma sabanica* var. *saffordi* Enzmann, E.V. 1944: 89 (w.) MEXICO. [The name *sabanica* is a misspelling of *satanica.*] [Unresolved junior primary homonym of *saffordi* Wheeler, above.] Junior synonym of *saffordi* Wheeler: Brown, 1949a: 42; of *peperi:* Ward, 1989: 452.

***salvini.*** *Pseudomyrma salvini* Forel, 1899c: 94, pl. 4, fig. 8 (w.q.m.) MEXICO. Combination in *Pseudomyrmex:* Kempf, 1960a: 30.

***santschii.*** *Pseudomyrma santschii* Enzmann, E.V. 1944: 89 (w.) PERU. Combination in *Pseudomyrmex:* Kempf, 1958f: 435.

*sapii. Pseudomyrma caroli* var. *sapii* Forel, 1904e: 688 (w.q.) BRAZIL. Combination in *Pseudomyrmex:* Kempf, 1961a: 400. Junior synonym of *viduus:* Ward, 1989: 445.

***satanicus.*** *Pseudomyrma satanica* Wheeler, W.M. 1942: 174, pl. 47, fig. b (w.q.m.) PANAMA. Combination in *Pseudomyrmex:* Kempf, 1972a: 223. See also: Ward, 1993: 149.

*scelerosa. Pseudomyrma spinicola* subsp. *scelerosa* Wheeler, W.M. 1942: 181 (w.) NICARAGUA. Combination in *Pseudomyrmex:* Kempf, 1972a: 224. Senior synonym of *scelerosa* Enzmann: Brown, 1949a: 43. Junior synonym of *spinicola:* Ward, 1993: 150.

*scelerosa. Pseudomyrma spinolae* var. *scelerosa* Enzmann, E.V. 1944: 91 (w.) NICARAGUA. [The name *spinolae* is a misspelling of *spinicola.*] Junior synonym of *scelerosa* Wheeler: Brown, 1949a: 43.

***schuppi.*** *Pseudomyrma schuppi* Forel, 1901d: 298 (w.) BRAZIL. Kempf, 1961a: 381 (q.); Wheeler, G.C. & Wheeler, J. 1973d: 207 (l.). Combination in *Pseudomyrmex,* and senior synonym of *confusior* and *geraensis:* Kempf, 1961a: 380. See also: Ward, 1989: 426.

*sedula. Pseudomyrma sedula* Smith, F. 1877b: 67 (m.) BRAZIL. Combination in *Pseudomyrmex:* Kempf, 1958f: 435. Junior synonym of *venustus:* Kempf, 1967c: 10.

***seminole.*** *Pseudomyrmex seminole* Ward, 1985b: 237, figs. 18, 21, 28, 44, 45 (w.q.m.) U.S.A. Wheeler, G.C. & Wheeler, J. 1991b: 90 (l.).

*semotus. Pseudomyrmex semotus* Kempf, 1960a: 392, figs. 4–6 (w.) BRAZIL. Junior synonym of *godmani:* Kempf, 1967c: 6.

*sericata. Pseudomyrma sericata* Smith, F. 1855c: 159 (w.) BRAZIL. Gallardo, 1932a: 50 (m.). Combination in *Pseudomyrmex:* Kempf, 1958f: 434. Status as species: Forel, 1912g: 19. Subspecies/variety of *gracilis:* Emery, 1890b: 60; Santschi, 1916e: 370; Gallardo, 1932a: 48. Junior synonym of *gracilis:* Ward, 1993: 155.

***sericeus.*** *Pseudomyrma sericea* Mayr, 1870a: 412 (w.) COLOMBIA. Forel, 1904e: 691 (q.); Wheeler, G.C. & Wheeler, J. 1956: 386 (l.). Combination in *P.* (*Apedunculata*): Enzmann, E.V. 1944: 62; in *Pseudomyrmex:* Wheeler, G.C. & Wheeler, J. 1956: 385. Current subspecies: nominal plus *acaciarum, cordiae, fortis, huberi, itus, lisa, longior, rubiginosus, vinneni.*

*simoides. Pseudomyrma simoides* Forel, 1911e: 281 (w.) BRAZIL. Combination in *Pseudomyrmex:* Kempf, 1961a: 399. Junior synonym of *perboscii:* Ward, 1989: 442.

***simplex.*** *Pseudomyrma simplex* Smith, F. 1877b: 64 (w.) BRAZIL. Combination in *Pseudomyrmex:* Kempf, 1972a: 223. Senior synonym of *capperi, delicatula* and *panamensis,* and material of the unavailable name *vittata* referred here: Ward, 1985b: 238. See also: Ward, 1992: 75 (specimen in amber).

***simulans.*** *Pseudomyrmex simulans* Kempf, 1958f: 459, figs. 31–33 (w.q.) PANAMA. See also: Ward, 1993: 161.

***solisi.*** *Pseudomyrma solisi* Santschi, 1916e: 371, fig. 1 (w.) ARGENTINA. Combination in *Pseudomyrmex:* Kusnezov, 1964: 102. Current subspecies: nominal plus *belgranoi.* See also: Gallardo, 1932a: 68.

***spiculus.*** *Pseudomyrmex spiculus* Ward, 1989: 430, figs. 31–33 (w.q.) COSTA RICA.

***spinicola.*** *Pseudomyrma spinicola* Emery, 1890b: 64, pl. 6, fig. 2 (w.q.m.) COSTA RICA. [Also described as new by Emery, 1894k: 52.] Combination in *P.* (*Triangulinoda*): Enzmann, E.V. 1944: 61; in *Pseudomyrmex:* Wheeler, G.C. & Wheeler, J. 1956: 386. Senior synonym of *atrox, gaigei, infernalis* Wheeler (and its junior synonym *infernalis* Enzmann) and *scelerosa* Wheeler (and its junior synonym *scelerosa* Enzmann): Ward, 1993: 149.

***squamifer.*** *Pseudomyrma gracilis* subsp. *squamifera* Emery, 1890b: 60, pl. 5, fig. 20 (w.) BRAZIL. Combination in *Pseudomyrmex:* Kempf, 1958f: 457. Raised to species: Dalla Torre, 1893: 59; Forel, 1895b: 123; Kempf, 1958f: 457.

*stolli. Pseudomyrma stolli* Forel, 1912g: 20 (w.q.m.) GUATEMALA. Combination in *Pseudomyrmex:* Kempf, 1972a: 224. Junior synonym of *rufomedius:* Ward, 1989: 443.

***subater.*** *Pseudomyrma elongata* subsp. *subatra* Wheeler, W.M. & Mann, 1914: 19 (w.) HAITI. Combination in *Pseudomyrmex,* and raised to species: Kempf, 1972a: 224. See also: Ward, 1985b: 228; Ward, 1990: 486.

***subtilissimus.*** *Pseudomyrma subtilissima* Emery, 1890b: 65, pl. 6, fig. 7 (w.q.) COSTA RICA. [Also described as new by Emery, 1894k: 53.] Combination in *Pseudomyrmex:* Kempf, 1972a: 224. See also: Ward, 1989: 432; Ward, 1993: 161.

*succinea. Leptalaea succinea* Spinola, 1851b: 53 (w.) BRAZIL. [Repeated in Spinola, 1853: 69.] *Nomen nudum;* attributed to Klug.

*****succinus.*** **Pseudomyrmex succinus* Ward, 1992: 71, figs. 9, 18 (w.q.) DOMINICAN AMBER (Miocene).

#***symbioticus.*** *Pseudomyrma arborissanctae* r. *symbiotica* Forel, 1904c: 38 (w.m.) COLOMBIA. Combination in *Pseudomyrmex:* Kempf, 1961a: 401. Subspecies of *triplarinus:* Wheeler, W.M. & Darlington, 1930: 109; Wheeler, W.M. 1942: 189.

***tachigaliae.*** *Pseudomyrma latinoda* r. *tachigaliae* Forel, 1904e: 686 (w.q.m.) BRAZIL. Combination in *Pseudomyrmex:* Kempf, 1961a: 400. Raised to species, and senior synonym of *bradleyi* Wheeler (and its junior synonym *bradleyi* Enzmann): Ward, 1989: 444. See also: Wheeler, W.M. 1942: 170.

*tandem. Pseudomyrma elongata* var. *tandem* Forel, 1906d: 228 (w.) COSTA RICA. Junior synonym of *elongatus:* Creighton, 1957b: 18.

***tenuis.*** *Formica tenuis* Fabricius, 1804: 405 (w.) BRAZIL. Kempf, 1960a: 28 (q.). Combination in *Pseudomyrma:* Roger, 1862c: 289; in *P.* (*Ornatinoda*): Enzmann, 1944: 61; in *Pseudomyrmex:* Kempf, 1960a: 7. Senior synonym of *ligniseca:* Roger, 1862c: 289; of *andina, paraensis, pittieri* and *rufa:* Kempf, 1960a: 26. See also: Perrault, 1988a: 384.

***tenuissimus.*** *Pseudomyrma subtilissima* subsp. *tenuissima* Emery, 1906c: 119 (w.) BRAZIL. Wheeler, G.C. & Wheeler, J. 1973d: 207 (l.). Combination in *Pseudomyrmex:* Kempf, 1972a: 224. Raised to species, and senior synonym of *culmicola:* Ward, 1989: 432.

***terminalis.*** *Pseudomyrma terminalis* Smith, F. 1877b: 64 (q.) BRAZIL. Combination in *Pseudomyrmex:* Kusnezov, 1964: 83.

***termitarius.*** *Pseudomyrma termitaria* Smith, F. 1855c: 158 (w.q.) BRAZIL. Wheeler, G.C. & Wheeler, J. 1973d: 207 (l.). Combination in *Pseudomyrmex:* Kempf, 1960a: 8. Junior synonym of *gracilis:* Forel, 1911e: 276. Revived from synonymy as variety of *gracilis:* Forel, 1912g: 19. Variety of *denticollis:* Gallardo, 1932a: 56. Revived status as species: Kempf, 1960a: 8. Senior synonym of *breviceps, elegans* and *infusca:* Kempf, 1960a: 8; of *nigriceps:* Perrault, 1988a: 384.

*testacea. Tetraponera testacea* Smith, F. 1852: 45 (footnote) (w.) SOUTH AMERICA. Smith, F. 1855c: 168 (w.). Combination in *Pseudomyrma:* Smith, F. 1855c: 168; in *Pseudomyrmex:* Kempf, 1961a: 400. Junior synonym of *perboscii:* Ward, 1989: 442.

*****thecolor.*** **Pseudomyrmex thecolor* Ward, 1992: 73, figs. 6, 19 (w.) DOMINICAN AMBER (Miocene).

*thoracica. Pseudomyrma thoracica* Norton, 1868c: 8 (w.) MEXICO. Combination in *Pseudomyrmex:* Kempf, 1972a: 225. [*P. thoracica* Norton, 1868a: 66; *nomen nudum.*] Junior synonym of *boopis:* Ward, 1993: 154.

*tigrina. Pseudomyrma triplaridis* subsp. *tigrina* Wheeler, W.M. 1942: 186 (w.m.) GUIANA. Combination in *Pseudomyrmex:* Kempf, 1961a: 400. Senior synonym of *trigona:* Brown, 1949a: 43. Junior synonym of *viduus:* Ward, 1989: 445.

*torquata. Pseudomyrma haytiana* var. *torquata* Wheeler, W.M. 1936b: 196 (w.) HAITI. [First available use of *Pseudomyrma championi* subsp. *haytiana* var. *torquata* Wheeler, W.M. & Mann, 1914: 18; unavailable name (Ward, 1990: 487).] Junior synonym of *haytiana:* Kempf, 1961a: 393.

*trigona. Pseudomyrma triplaridis* subsp. *trigona* Enzmann, E.V. 1944: 94 (w.) GUIANA. Junior synonym of *tigrina:* Brown, 1949a: 43.

***triplaridis.*** *Pseudomyrma triplaridis* Forel, 1904e: 684 (w.q.m.) BRAZIL. Combination in *Pseudomyrmex:* Kusnezov, 1953e: 214. Senior synonym of *boxi* Wheeler (and its junior synonym *boxi* Enzmann): Ward, 1989: 444. See also: Wheeler, W.M. 1942: 182; Kempf, 1961a: 400.

***triplarinus.*** *Myrmica triplarina* Weddell, 1850: 263 (footnote) (w.) BRAZIL. Wheeler, G.C. & Wheeler, J. 1990a: 464 (l.). Combination in *Pseudomyrma:* Wheeler, W.M. 1942: 186; in *Pseudomyrmex:* Wheeler, G.C. & Wheeler, J. 1956: 386. Senior synonym of *arborissanctae:* Wheeler, W.M. 1942: 186; of *ecuadoriana:* Brown, 1949a: 44. Current subspecies: nominal plus *cordobensis, rurrenabaquensis, symbioticus.* See also: Kempf, 1961a: 408.

*tuberculata. Pseudomyrma wessoni* var. *tuberculata* Enzmann, E.V. 1944: 98, pl. 4, fig. 25 (w.) PERU. Combination in *Pseudomyrmex:* Kempf, 1972a: 226. Junior synonym of *oculatus:* Ward, 1989: 422.

*ulei. Pseudomyrma ulei* Forel, 1904e: 689 (w.q.m.) BRAZIL. Combination in *Pseudomyrmex:* Kempf, 1961a: 401. Junior synonym of *viduus:* Ward, 1989: 445.

***unicolor.*** *Pseudomyrma unicolor* Smith, F. 1855c: 158 (w.) BRAZIL. Combination in *Pseudomyrmex:* Kempf, 1958f: 439. Senior synonym of *anceps* and *mutilloides:* Kempf, 1958f: 439; of *atra:* Kempf, 1969: 276.

*unicolor. Pseudomyrma unicolor* Smith, F. 1877b: 68 (w.) BRAZIL. [Junior primary homonym of *unicolor* Smith, above.] Replacement name: *monochroa* Dalla Torre, 1892: 89.

***urbanus.*** *Pseudomyrma urbana* Smith, F. 1877b: 65 (w.) BRAZIL. Combination in *Pseudomyrmex:* Kusnezov, 1964: 83. Senior synonym of *chodati* and *ogloblini:* Ward, 1989: 427.

*variabilis. Pseudomyrma variabilis* Smith, F. 1877b: 62 (w.) BARBADOS. Combination in *Pseudomyrmex:* Kempf, 1958f: 435. Junior synonym of *gracilis:* Ward, 1989: 439.

*velifera. Pseudomyrma gracilis* var. *velifera* Stitz, 1933: 68 (q.) GUATEMALA. Combination in *Pseudomyrmex:* Kempf, 1958f: 434. Junior synonym of *gracilis:* Ward, 1993: 156.

***veneficus.*** *Pseudomyrma belti* subsp. *venefica* Wheeler, W.M. 1942: 162 (w.q.m.) MEXICO. Combination in *Pseudomyrmex:* Kempf, 1972a: 216. Senior synonym of *venifica* Enzmann: Brown, 1949a: 42. Raised to species: Ward, 1989: 439. See also: Ward, 1993: 152.

*venifica. Pseudomyrma belti* subsp. *venifica* Enzmann, E.V. 1944: 81 (w.q.) MEXICO. Junior synonym of *venefica:* Brown, 1949a: 42.

***venustus.*** *Pseudomyrma venusta* Smith, F. 1858b: 158 (q.) BRAZIL. Combination in *Pseudomyrmex:* Kusnezov, 1964: 83. Senior synonym of *agilis* (and its junior synonyms *carapuna* and *peruviana* Enzmann) and *sedula:* Kempf, 1967c: 10.

*vesana. Pseudomyrma belti* subsp. *vesana* Wheeler, W.M. 1942: 163 (w.) MEXICO. Combination in *Pseudomyrmex:* Kempf, 1972a: 216. Junior synonym of *ferrugineus:* Ward, 1989: 437.

*****vicinus.*** **Pseudomyrmex vicinus* Ward, 1992: 74, fig. 20 (w.) DOMINICAN AMBER (Miocene).

***viduus.*** *Pseudomyrma vidua* Smith, F. 1858b: 158 (m.) BRAZIL. Combination in *Pseudomyrmex:* Kusnezov, 1964: 83. Senior synonym of *baileyi* (and its junior synonym *biolleyi*), *caroli, sapii, tigrina* (and its junior synonym *trigona*) and *ulei:* Ward, 1989: 445.

***villosus.*** *Pseudomyrmex villosus* Ward, 1989: 434, figs. 39, 39 (w.) BRAZIL.

#***vinneni.*** *Pseudomyrma sericea* var. *vinneni* Forel, 1906d: 230 (w.) BRAZIL. Combination in *Pseudomyrmex:* Kempf, 1972a: 223.

#***virgo.*** *Pseudomyrma acanthobia* var. *virgo* Santschi, 1922b: 345 (w.) BRAZIL. Combination in *Pseudomyrmex:* Kempf, 1972a: 215.

#***vistanus.*** *Pseudomyrma colei* var. *vistana* Enzmann, E.V. 1944: 84 (w.) PERU. Combination in *Pseudomyrmex:* Kempf, 1972a: 217.

*vittata. Pseudomyrma acanthobia* r. *delicatula* var. *vittata* Forel, 1912g: 26 (w.) BRAZIL. Unavailable name (Ward, 1990: 487); material referred to *simplex* by Ward, 1985b: 238.

*volatilis. Pseudomyrma volatilis* Smith, F. 1877b: 65 (m.) MEXICO. Combination in *Pseudomyrmex:* Kempf, 1958f: 435. Junior synonym of *gracilis:* Ward, 1993: 155.

***voytowskii.*** *Pseudomyrma voytowskii* Enzmann, E.V. 1944: 95 (w.) PERU. Combination in *Pseudomyrmex:* Kempf, 1972a: 95. Current subspecies: nominal plus *costaricensis.*

*wagneri. Pseudomyrma nigropilosa* st. *wagneri* Santschi, 1922b: 347 (w.) ARGENTINA. Gallardo, 1932a: 53 (q.). Junior synonym of *maculatus:* Kempf, 1958f: 454.

*wasmanni. Pseudomyrma wasmanni* Wheeler, W.M. 1921b: 92 (footnote). Replacement name for *canescens* Wasmann, 1915a: 321. [Junior primary homonym of *canescens* Smith, F. 1877b: 66.] Variety/subspecies of *belti:* Stitz, 1937: 133; Wheeler, W.M. 1942: 164. Junior synonym of *ferrugineus:* Ward, 1989: 437.

***weberi.*** *Pseudomyrma weberi* Enzmann, E.V. 1944: 96, pl. 3, fig. 16 (w.) GUATEMALA. Combination in *Pseudomyrmex:* Kempf, 1972a: 226.

*wessoni. Pseudomyrma wessoni* Enzmann, E.V. 1944: 97, pl. 3, fig. 10 (w.) PERU. Combination in

*Pseudomyrmex:* Kempf, 1972a: 226. Junior synonym of *oculatus:* Ward, 1989: 422.
**wheeleri.** *Pseudomyrma wheeleri* Enzmann, E.V. 1944: 98 (q.m.) PERU. Combination in *Pseudomyrmex:* Kempf, 1972a: 226.

*PSEUDONEOPONERA:* see under ***PACHYCONDYLA.***

***PSEUDONOTONCUS*** (Australia)
**hirsutus.** *Pseudonotoncus hirsutus* Clark, 1934c: 65, pl. 4, figs. 23, 24 (w.q.) AUSTRALIA.
**turneri.** *Pseudonotoncus turneri* Donisthorpe, 1937a: 619, figs. 1, 2 (w.) AUSTRALIA.

*PSEUDOPODOMYRMA:* see under ***PODOMYRMA.***

*PSEUDOPONERA:* see under ***PACHYCONDYLA.***

*PSEUDOSPHINCTA:* see under ***DISCOTHYREA.***

*PSEUDOSTIGMACROS:* see under ***STIGMACROS.***

*PSEUDOSYSPHINCTA:* see under ***DISCOTHYREA.***

*PTEROPONERA:* see under ***PONERA.***

*PYRAMICA:* see under ***STRUMIGENYS.***

***QUADRISTRUMA*** (Tropics)
*clypeatus. Epitritus clypeatus* Szabó, 1909: 27, fig. 1 (w.) NEW GUINEA. Junior synonym of *emmae:* Brown, 1949b: 48.
**emmae.** *Epitritus emmae* Emery, 1890b: 70, pl. 8, fig. 6 (w.) ANTILLES IS. Wheeler, W.M. 1908a: 149 (q.). Combination in *Quadristruma,* and senior synonym of *clypeatus, malesiana* and *wheeleri:* Brown, 1949b: 48. See also: Bolton, 1983: 400.
**eurycera.** *Epitritus eurycera* Emery, 1897c: 581, pl. 14, fig. 17 (w.) NEW GUINEA. Combination in *Quadristruma:* Brown, 1949b: 50.
*malesiana. Epitritus clypeatus* var. *malesiana* Forel, 1913k: 83 (w.q.) SUMATRA. Junior synonym of *emmae:* Brown, 1949b: 48.
*wheeleri. Epitritus wheeleri* Donisthorpe, 1916a: 121 (w.) HAWAII. Junior synonym of *emmae:* Brown, 1949b: 48.

*RAPTIFORMICA:* see under ***FORMICA.***

***RECURVIDRIS*** (Palaearctic, Oriental, Indo-Australian)
**browni.** *Recurvidris browni* Bolton, 1992: 43, figs. 1, 3 (w.) BORNEO.
**hebe.** *Recurvidris hebe* Bolton, 1992: 44, figs. 7, 10 (w.) SULAWESI.
**kemneri.** *Trigonogaster recurvispinosa* subsp. *kemneri* Wheeler, G.C. & Wheeler, J. 1954a: 133 (l.) JAVA. Bolton, 1992: 41 (w.). [*Trigonogaster recurvispinosa* subsp. *kemneri* Wheeler, W.M. 1936d: 220; *nomen nudum.*] Combination in *Recurvidris,* and raised to species: Bolton, 1992: 40.
**pickburni.** *Recurvidris pickburni* Bolton, 1992: 45, fig. 5 (w.) SRI LANKA.
**proles.** *Recurvidris proles* Bolton, 1992: 41, figs. 2, 8 (w.) SULAWESI.
**recurvispinosa.** *Trigonogaster recurvispinosus* Forel, 1890b: cix, fig. (w.) INDIA. Wheeler, W.M. 1927d: 5 (m.). Combination in *Recurvidris:* Bolton, 1992: 46.
**williami.** *Recurvidris williami* Bolton, 1992: 47, fig. 9 (w.) INDONESIA (Flores I.).

*RENEA:* see under ***PRIONOPELTA.***

*#RHACHIOCREMA:* see under ***CREMATOGASTER.***

*#RHINOMYRMEX:* see under ***CAMPONOTUS.***

*#RHIZOMYRMA:* see under ***ACROPYGA.***

*#RHOGMUS:* see under ***DORYLUS.***

***RHOPALOMASTIX*** (Oriental, Indo-Australian)
**escherichi.** *Rhopalomastix escherichi* Forel, 1911i: 217 (q.) SRI LANKA.
**janeti.** *Rhopalomastix janeti* Donisthorpe, 1936a: 55 (w.q.m.) THAILAND.
**#javana.** *Rhopalomastix rothneyi* subsp. *javana* Wheeler, W.M. 1929d: 96, fig. 1 (w.q.m.) JAVA.
**#johorensis.** *Rhopalomastix rothneyi* subsp. *johorensis* Wheeler, W.M. 1929d: 96 (w.) SINGAPORE.
**rothneyi.** *Rhopalomastix rothneyi* Forel, 1900a: 24 (q.) INDIA. Forel, 1911i: 217 (w.m.); Wheeler, G.C. & Wheeler, J. 1953d: 185 (l.). Current subspecies: nominal plus *javana, johorensis.*

****RHOPALOMYRMEX*** (Baltic Amber)
***pygmaeus.** **Rhopalomyrmex pygmaeus* Mayr, 1868c: 42, pl. 2, figs. 25, 26 (w.) BALTIC AMBER (Oligocene).

*RHOPALOPONE:* see under ***GNAMPTOGENYS.***

***RHOPALOTHRIX*** (Neotropical, Indo-Australian, Australasian)
**acutipilis.** *Rhopalothrix acutipilis* Kempf, 1962b: 28, fig. 28 (q.) BRAZIL.
*amoena* Mann, 1922; see under ***EURHOPALOTHRIX.***
*angulinodis* Stitz, 1925; see under ***EURHOPALOTHRIX.***
*ballionii* Forel, 1904; see under ***EURHOPALOTHRIX.***
*balzani* Emery, 1894; see under ***OCTOSTRUMA.***
*barberi* Mann, 1922; see under ***OCTOSTRUMA.***
*batesi* Emery, 1894; see under ***OCTOSTRUMA.***
*biroi* Szabó, 1910; see under ***EURHOPALOTHRIX.***
*bolaui* Mayr, 1870; see under ***EURHOPALOTHRIX.***
*borneensis* Wheeler, W.M. 1919; see under ***EURHOPALOTHRIX.***
*brevicornis* Emery, 1897; see under ***EURHOPALOTHRIX.***
*bruchi* Santschi, 1922; see under ***EURHOPALOTHRIX.***
**ciliata.** *Rhopalothrix ciliata* Mayr, 1870a: 415 (w.) COLOMBIA. See also: Brown & Kempf, 1960: 232.
**diadema.** *Rhopalothrix diadema* Brown & Kempf, 1960: 239, fig. 59 (w.) NEW GUINEA.
*elegans* Mann, 1921; see under ***EURHOPALOTHRIX.***
*emeryi* Forel, 1912; see under ***EURHOPALOTHRIX.***
*equilatera* Weber, 1934; see under ***OCTOSTRUMA.***
*godmani* Forel, 1899; see under ***OCTOSTRUMA.***
*gravis* Mann, 1922; see under ***EURHOPALOTHRIX.***
*iheringi* Emery, 1888; see under ***OCTOSTRUMA.***
**inopinata.** *Rhopalothrix inopinata* De Andrade, in Baroni Urbani & De Andrade, 1994: 54, figs. 31, 32 (w.) SRI LANKA.
*isabellae* Mann, 1919; see under ***EURHOPALOTHRIX.***
**isthmica.** *Acanthidris isthmicus* Weber, 1941a: 188, figs. 4–7 (w.) PANAMA. Combination in *Rhopalothrix:* Brown & Kempf, 1960: 235.
*kokodensis* Donisthorpe, 1936; see under ***EURHOPALOTHRIX.***
**kusnezovi.** *Rhopalothrix kusnezovi* Brown & Kempf, 1960: 238, fig. 60 (q.) ARGENTINA.
*lutzi* Wheeler, W.M. 1913; see under ***OCTOSTRUMA.***
*malua* Mann, 1919; see under ***EURHOPALOTHRIX.***
*manni* Menozzi, 1923; see under ***EURHOPALOTHRIX.***
*melanotica* Mann, 1919; see under ***EURHOPALOTHRIX.***
*mixta* Szabó, 1910; see under ***EURHOPALOTHRIX.***
**orbis.** *Rhopalothrix orbis* Taylor, 1968b: 336, figs. 1–3 (w.) AUSTRALIA. Taylor, 1970a: 49 (q.).
*petiolata* Mayr, 1887; see under ***OCTOSTRUMA.***
**plaumanni.** *Rhopalothrix plaumanni* Brown & Kempf, 1960: 235, fig. 58 (w.) BRAZIL.
*procera* Emery, 1897; see under ***EURHOPALOTHRIX.***
*punctata* Szabó, 1910; see under ***EURHOPALOTHRIX.***
*redux* Donisthorpe, 1939; see under ***BASICEROS.***
*reichenspergeri* Santschi, 1923; see under ***EURHOPALOTHRIX.***
*rugifer* Mayr, 1887; see under ***OCTOSTRUMA.***
*samoana* Santschi, 1928; see under ***EURHOPALOTHRIX.***
*schmidti* Menozzi, 1936; see under ***EURHOPALOTHRIX.***
*simoni* Emery, 1890; see under ***OCTOSTRUMA.***
*spei* Forel, 1912; see under ***OCTOSTRUMA.***
**stannardi.** *Rhopalothrix stannardi* Brown & Kempf, 1960: 236, fig. 57 (w.) MEXICO.
*subdentatus* Donisthorpe, 1942; see under ***EURHOPALOTHRIX.***
*sulcata* Santschi, 1936; see under ***OCTOSTRUMA.***
*truncata* Forel, 1912; see under ***OCTOSTRUMA.***
**weberi.** *Rhopalothrix weberi* Brown & Kempf, 1960: 234. Replacement name for *Heptastruma wheeleri* Weber, 1934a: 55. [Junior secondary homonym of *Rhopalothrix wheeleri* Mann, 1922: 43.]
*wheeleri* Mann, 1922; see under ***OCTOSTRUMA.***
*wheeleri. Heptastruma wheeleri* Weber, 1934a: 55, fig. 13 (w.) CUBA. [Junior secondary homonym of *Rhopalothrix wheeleri* Mann, 1922: 43, above.] Replacement name: *Rhopalothrix weberi* Brown & Kempf, 1960: 234.
*wighti* Wheeler, W.M. 1908; see under ***OCTOSTRUMA.***

***RHOPTROMYRMEX*** (Old World tropics)
*alberti. Rhoptromyrmex globulinodis* r. *alberti* Forel, 1916: 419 (w.) ZAIRE. Junior synonym of *globulinodis:* Brown, 1964c: 17.
*arnoldi* Santschi, 1916; see under ***TETRAMORIUM.***
**caritus.** *Rhoptromyrmex caritus* Bolton, 1986a: 12, figs. 15, 17 (q.) UGANDA.
**critchleyi.** *Rhoptromyrmex critchleyi* Bolton, 1976: 303, figs. 3, 12 (w.) NIGERIA. See also: Bolton, 1986a: 13.
*esta. Rhoptromyrmex opacus* var. *esta* Forel, 1909b: 59 (w.q.m.) ZAIRE. Junior synonym of *opacus:* Brown, 1964c: 15.
**globulinodis.** *Rhoptromyrmex globulinodis* Mayr, 1901b: 20 (w.q.m.) SOUTH AFRICA. Senior synonym of *alberti* and *obscurus:* Brown, 1964c: 17. See also: Bolton, 1986a: 13.
*intermedia. Rhoptromyrmex rothneyi* var. *intermedia* Forel, 1913k: 80 (w.) SUMATRA. Junior synonym of *wroughtonii:* Brown, 1964c: 14.
*laeviceps. Rhoptromyrmex opacus* var. *laeviceps* Santschi, 1916b: 504 (w.) ZAIRE. Junior synonym of *opacus:* Brown, 1964c: 15.

***RHOPTROMYRMEX*** *(continued)*

*leno. Rhoptromyrmex rothneyi* subsp. *leno* Viehmeyer, 1914a: 113 (w.) WEST MALAYSIA. Junior synonym of *wroughtonii:* Brown, 1964c: 14.

*longi. Rhoptromyrmex wroughtonii* r. *rothneyi* var. *longi* Forel, 1902c: 232 (w.) INDIA. Unavailable name; material referred to *wroughtonii* by Brown, 1964c: 14.

***mayri.*** *Rhoptromyrmex mayri* Forel, 1912n: 57 (q.) INDIA. Combination in *Hagioxenus:* Brown, 1964c: 19; in *Rhoptromyrmex:* Bolton, 1979: 173. See also: Bolton, 1986a: 14.

***melleus.*** *Tetramorium melleum* Emery, 1897c: 586, pl. 15, figs. 29, 30 (w.) NEW GUINEA. Brown, 1964c: 13. Combination in *Rhoptromyrmex:* Emery, 1922f: 290. Senior synonym of *papuensis:* Bolton, 1976: 302. See also: Bolton, 1986a: 14.

*monodi. Rhoptromyrmex opacus* r. *monodi* Bernard, 1953b: 251, fig. 14 (w.) GUINEA. Junior synonym of *opacus:* Brown, 1964c: 15.

*obscurus. Rhoptromyrmex globulinodis* var. *obscurus* Santschi, 1932a: 389 (w.m.) ZIMBABWE. Junior synonym of *globulinodis:* Brown, 1964c: 17.

***opacus.*** *Rhoptromyrmex opacus* Emery, in Forel, 1909b: 59 (footnote) (w.) CAMEROUN. Bolton, 1986a: 15 (q.m.). Senior synonym of *esta, laeviceps* and *monodi:* Brown, 1964c: 15.

*papuensis. Ireneella papuensis* Donisthorpe, 1941i: 175 (m.) NEW GUINEA. Junior synonym of *melleus:* Bolton, 1976: 302.

*pretoriae. Rhoptromyrmex transversinodis* var. *pretoriae* Arnold, 1926: 282 (w.q.m.) SOUTH AFRICA. Junior synonym of *transversinodis:* Brown, 1964c: 16.

***rawlinsoni.*** *Rhoptromyrmex rawlinsoni* Taylor, 1992b: 125, figs. 1–4 (q.) INDONESIA (Anak Krakatau I.).

*rothneyi. Rhoptromyrmex wroughtonii* r. *rothneyi* Forel, 1902c: 232 (w.) INDIA. Raised to species: Bingham, 1903: 177; Wheeler, W.M. 1930a: 103. Subspecies of *wroughtonii:* Wheeler, W.M. 1928c: 26. Junior synonym of *wroughtonii:* Brown, 1964c: 14.

***schmitzi.*** *Hagioxenus schmitzi* Forel, 1910a: 8 (q.) ISRAEL. Combination in *Rhoptromyrmex:* Bolton, 1986a: 15.

*solleri* Forel, 1910; see under ***MONOMORIUM.***

*steini. Rhoptromyrmex steini* Forel, 1913a: 122 (w.) SOUTH AFRICA. Junior synonym of *transversinodis:* Brown, 1964c: 16.

*sumatrensis. Rhoptromyrmex rothneyi* r. *sumatrensis* Forel, 1913k: 80, fig. W (w.) SUMATRA. Junior synonym of *wroughtonii:* Brown, 1964c: 14.

*taivanensis. Rhoptromyrmex* (*Acidomyrmex*) *rothneyi* var. *taivanensis* Wheeler, W.M. 1930a: 103 (w.) TAIWAN. Junior synonym of *wroughtonii:* Brown, 1964c: 14.

*tessmanni* Forel, 1910; see under ***TETRAMORIUM.***

***transversinodis.*** *Rhoptromyrmex transversinodis* Mayr, 1901b: 22 (w.) SOUTH AFRICA. Arnold, 1917: 356 (q.); Bolton, 1986a: 16 (m.). Senior synonym of *pretoriae* and *steini:* Brown, 1964c: 16.

***wroughtonii.*** *Rhoptromyrmex wroughtonii* Forel, 1902c: 231 (w.m.) INDIA. Bolton, 1986a: 16 (q.). Combination in *R.* (*Acidomyrmex*): Emery, 1915i: 191. Senior synonym of *intermedia, leno, rothneyi, sumatrensis* and *taivanensis,* and material of the unavailable name *longi* referred here: Brown, 1964c: 14.

***RHYTIDOPONERA*** (Indo-Australian, Australasian)

***abdominalis.*** *Rhytidoponera subcyanea* subsp. *abdominalis* Viehmeyer, 1912: 4, fig. 1 (w.) NEW GUINEA. Raised to species: Wilson, 1958d: 306; Brown, 1958g: 202.

***acanthoponeroides.*** *Rhytidoponera acanthoponeroides* Viehmeyer, 1924a: 227 (w.) NEW CALEDONIA. Brown, 1958g: 290 (m.). See also: Ward, 1984: 138.

***aciculata.*** *Ectatomma aciculata* Smith, F. 1858b: 104 (w.) AUSTRALIA. Clark, 1936: 56 (m.); Imai, Crozier & Taylor, 1977: 347 (k.). Combination in *E.* (*Rhytidoponera*): Dalla Torre, 1893: 23; in *Rhytidoponera:* Emery, 1911d: 38. Subspecies of *punctata:* Forel, 1907h: 268; Forel, 1913g: 174. Revived status as species: Clark, 1936: 55. Senior synonym of *caro,* and material of the unavailable name *laevior* referred here: Brown, 1958g: 202.

*acupuncta. Rhytidoponera* (*Chalcoponera*) *acupuncta* Emery, 1914f: 396 (w.m.) NEW CALEDONIA. Junior synonym of *numeensis:* Brown, 1958g: 204.

***aenescens.*** *Rhytidoponera aenescens* Emery, 1900c: 312 (w.) NEW GUINEA. Combination in *R.* (*Chalcoponera*): Emery, 1911d: 39. See also: Ward, 1980: 492.

***anceps.*** *Rhytidoponera anceps* Emery, 1898b: 233, fig. 3 (w.) AUSTRALIA. Combination in *R.* (*Chalcoponera*): Emery, 1911d: 39. See also: Brown, 1958g: 264.

*andrei. Chalcoponera victoriae* var. *andrei* Wheeler, W.M. & Chapman, 1925: 59, pl. 2, figs. 3, 4 (w.) AUSTRALIA. Junior synonym of *croesus:* Brown, 1954g: 10.

***aquila.*** *Rhytidoponera aquila* Ward, 1984: 139, figs. 15, 16, 21, 22 (w.m.) NEW CALEDONIA.

***araneoides.*** *Ponera araneoides* Le Guillou, 1842: 317 (w.) SOLOMON IS. Combination in *Ectatomma:* Mayr, 1876: 92; in *Rhytidoponera:* Emery, 1897d: 548. Senior synonym of *rugosa:* Mayr, 1876: 92; Donisthorpe, 1932c: 454; of *ceranensis, froggatti* and *impressinodis:* Wilson, in Brown, 1958g: 202; Wilson, 1958d: 307; of *arcuata:* Brown, 1958g: 202.

***arborea.*** *Rhytidoponera arborea* Ward, 1984: 153, figs. 41, 42, 47, 48 (w.) NEW CALEDONIA.

*arcuata. Rhytidoponera araneoides* var. *arcuata* Stitz, 1911a: 352 (w.) AUSTRALIA. Junior synonym of *araneoides:* Brown, 1958g: 202.

*arnoldi. Rhytidoponera* (*Chalcoponera*) *arnoldi* Forel, 1915b: 14 (footnote) (w.) AUSTRALIA. Junior synonym of *aspera:* Brown, 1954g: 9.

*aruana. Rhytidoponera* (*Rhytidoponera*) *subcyanea* var. *aruana* Karavaiev, 1925a: 78 (w.) INDONESIA (Aru I.). Junior synonym of *subcyanea:* Wilson, in Brown, 1958g: 205; Wilson, 1958d: 319.

***aspera.*** *Ponera metallica* var. *aspera* Roger, 1860: 308 (w.) AUSTRALIA. Crawley, 1922b: 435 (q.m.); Wheeler, G.C. & Wheeler, J. 1964b: 448 (l.). Combination in *Ectatomma:* Mayr, 1863: 410; in *Rhytidoponera* (*Chalcoponera*): Emery, 1911d: 39. Raised to species: Emery, 1911d: 39. Senior synonym of *arnoldi:* Brown, 1954g: 9. See also: Emery, 1912a: 77.

***atropurpurea.*** *Rhytidoponera* (*Chalcoponera*) *atropurpurea* Emery, 1914f: 396 (w.m.) NEW CALEDONIA. See also: Ward, 1984: 143.

***aurata.*** *Ponera* (*Ectatomma*) *aurata* Roger, 1861b: 169 (w.) AUSTRALIA (attributed to Guillou). Clark, 1936: 28 (m.). Combination in *Ectatomma:* Mayr, 1876: 91; in *E.* (*Rhytidoponera*): Dalla Torre, 1893: 23; in *Rhytidoponera:* Emery, 1911d: 37. Senior synonym of *flava:* Clark, 1936: 28.

***barnardi.*** *Rhytidoponera barnardi* Clark, 1936: 54, pl. 4, fig. 31 (w.) AUSTRALIA.

***barretti.*** *Rhytidoponera barretti* Clark, 1941: 81, pl. 13, fig. 10 (w.) AUSTRALIA.

***borealis.*** *Rhytidoponera* (*Chalcoponera*) *numeensis* subsp. *borealis* Crawley, 1918: 88 (w.) AUSTRALIA. Raised to species, and senior synonym of *brunnea:* Brown, 1958g: 202.

*brevior. Rhytidoponera cerastes* var. *brevior* Crawley, 1925b: 586 (w.m.) AUSTRALIA. Junior synonym of *taurus:* Clark, 1936: 25.

*brunnea. Chalcoponera brunnea* Clark, 1941: 86, pl. 13, fig. 17 (w.) AUSTRALIA. Junior synonym of *borealis:* Brown, 1958g: 202.

*caeciliae. Rhytidoponera* (*Chalcoponera*) *caeciliae* Viehmeyer, 1924a: 227 (w.q.) AUSTRALIA. Junior synonym of *metallica:* Brown, 1958g: 204.

*carbonaria. Chalcoponera metallica* var. *carbonaria* Wheeler, W.M. 1934d: 139 (w.) AUSTRALIA. Junior synonym of *inornata:* Brown, 1958g: 203.

***carinata.*** *Rhytidoponera carinata* Clark, 1936: 54, pl. 4, fig. 32 (w.) AUSTRALIA.

*caro. Ectatomma* (*Rhytidoponera*) *cristatum* subsp. *caro* Forel, 1910b: 11 (w.) AUSTRALIA. Combination in *Rhytidoponera:* Emery, 1911d: 38. Junior synonym of *aciculata:* Clark, 1936: 55.

*castanea. Rhytidoponera castanea* Crawley, 1925b: 589 (w.) AUSTRALIA. Junior synonym of *rufiventris:* Clark, 1936: 83.

*cedarensis. Rhytidoponera* (*Chalcoponera*) *victoriae* var. *cedarensis* Forel, 1915b: 15 (w.m.) AUSTRALIA. Junior synonym of *victoriae:* Brown, 1958g: 205.

***celtinodis.*** *Rhytidoponera celtinodis* Wilson, 1958d: 309 (w.) NEW GUINEA.

*ceramensis. Rhytidoponera araneoides* var. *ceramensis* Viehmeyer, 1914a: 112 (w.) INDONESIA (Seram I.). Junior synonym of *araneoides:* Wilson, in Brown, 1958g: 202; Wilson, 1958d: 307.

***cerastes.*** *Rhytidoponera cerastes* Crawley, 1925b: 584, fig. 1 (w.) AUSTRALIA. Wheeler, G.C. & Wheeler, J. 1964b: 448 (l.).

***chalybaea.*** *Rhytidoponera impressa* var. *chalybaea* Emery, 1901a: 51 (w.) AUSTRALIA. Wheeler, G.C. & Wheeler, J. 1971b: 1201 (l.); Imai, Crozier & Taylor, 1977: 347 (k.). Combination in *R.* (*Chalcoponera*): Emery, 1911d: 39. Raised to species, and senior synonym of *cyrus:* Brown, 1954g: 4. See also: Ward, 1980: 478.

***chnoopyx.*** *Rhytidoponera chnoopyx* Brown, 1958g: 269 (w.) AUSTRALIA.

***clarki.*** *Rhytidoponera* (*Chalcoponera*) *clarki* Donisthorpe, 1943e: 115. Replacement name for *Chalcoponera hilli* Clark, 1941: 85. [Junior secondary homonym of *Rhytidoponera hilli* Crawley, 1915a: 131.] Senior synonym of *obscurum:* Brown, 1958g: 203.

***confusa.*** *Rhytidoponera confusa* Ward, 1980: 482, figs. 1–3, 16, 17, 26, 28 (w.) AUSTRALIA.

***convexa.*** *Ectatomma convexum* Mayr, 1876: 92 (w.m.) AUSTRALIA. Wheeler, G.C. & Wheeler, J. 1952a: 127 (l.). Combination in *E.* (*Rhytidoponera*): Dalla

Torre, 1893: 24; in *Rhytidoponera:* Emery, 1897d: 548. Material of the unavailable name *subumbrata* referred here: Clark, 1936: 84; Senior synonym of *nigra:* Brown, 1958g: 272. See also: Brown, 1958g: 266.

**cornuta.** *Ectatomma* (*Rhytidoponera*) *cornutum* Emery, 1895g: 347 (w.) AUSTRALIA. Forel, 1915b: 11 (m.). Combination in *Rhytidoponera:* Emery, 1911d: 38.

**crassinoda.** *Ectatomma* (*Rhytidoponera*) *crassinode* Forel, 1907h: 270 (w.) AUSTRALIA. Forel, 1915b: 11 (m.). Combination in *Rhytidoponera:* Emery, 1911d: 38.

**cristata.** *Ectatomma cristatum* Mayr, 1876: 91 (w.) AUSTRALIA. Clark, 1936: 50 (m.); Wheeler, G.C. & Wheeler, J. 1952a: 126 (l.). Combination in *E.* (*Rhytidoponera*): Dalla Torre, 1893: 24; in *Rhytidoponera:* Emery, 1897d: 548.

*cristulatum. Ectatomma* (*Rhytidoponera*) *metallicum* var. *cristulatum* Forel, 1900b: 59 (w.) AUSTRALIA. Forel, 1907a: 2 (m.); Crawley, 1918: 88 (m.). Combination in *Rhytidoponera* (*Chalcoponera*): Emery, 1911d: 39. Subspecies of *metallica:* Forel, 1907a: 1. Junior synonym of *tasmaniensis:* Brown, 1958g: 205.

**croesus.** *Rhytidoponera croesus* Emery, 1901a: 50 (w.) AUSTRALIA. Wheeler, G.C. & Wheeler, J. 1964b: 448 (l.). Combination in *R.* (*Chalcoponera*): Emery, 1911d: 39. Senior synonym of *andrei* and *fastuosa:* Brown, 1954g: 10.

*curvata. Rhytidoponera strigosa* var. *curvata* Stitz, 1912: 499, fig. 3 (w.) NEW GUINEA. Karavaiev, 1925a: 77 (m.). Junior synonym of *strigosa:* Wilson, in Brown, 1958g: 204; Wilson, 1958d: 317.

*cyrus. Ectatomma* (*Rhytidoponera*) *cyrus* Forel, 1910b: 13 (w.q.) AUSTRALIA. Combination in *Rhytidoponera* (*Chalcoponera*): Emery, 1911d: 39. Junior synonym of *chalybaea:* Brown, 1954g: 4.

**depilis.** *Rhytidoponera depilis* Ward, 1984: 157, figs. 39, 40, 45, 46 (w.) NEW CALEDONIA.

*dixoni. Rhytidoponera dixoni* Clark, 1936: 46, pl. 4, figs. 22, 23 (w.m.) AUSTRALIA. Junior synonym of *mayri:* Brown, 1958g: 203.

*douglasi. Rhytidoponera douglasi* Brown, 1952f: 137. Unnecessary replacement name for *levior* Crawley, 1925b: 581. Junior synonym of *levior:* Taylor, 1986: 35.

**dubia.** *Rhytidoponera* (*Chalcoponera*) *dubia* Crawley, 1915a: 132 (w.) AUSTRALIA.

**enigmatica.** *Rhytidoponera enigmatica* Ward, 1980: 484, figs. 7–9, 20, 21, 27, 30 (w.) AUSTRALIA.

**eremita.** *Rhytidoponera eremita* Clark, 1936: 78, pl. 6, fig. 56 (w.) AUSTRALIA.

*fastuosa. Rhytidoponera* (*Chalcoponera*) *fastuosa* Santschi, 1916c: 174 (w.q.m.) AUSTRALIA. Junior synonym of *croesus:* Brown, 1954g: 10.

**ferruginea.** *Rhytidoponera ferruginea* Clark, 1936: 48, pl. 4, fig. 25 (w.) AUSTRALIA.

*flava. Rhytidoponera* (*Rhytidoponera*) *flava* Crawley, 1915b: 232 (m.) AUSTRALIA. Junior synonym of *aurata:* Clark, 1936: 28.

**flavicornis.** *Rhytidoponera flavicornis* Clark, 1936: 64, fig. 4 (w.) AUSTRALIA.

**flavipes.** *Chalcoponera flavipes* Clark, 1941: 84, pl. 13, fig. 14 (w.) AUSTRALIA. Combination in *Rhytidoponera:* Brown, 1958g: 203.

**flindersi.** *Rhytidoponera flindersi* Clark, 1936: 60, pl. 5, fig. 37 (w.) AUSTRALIA.

**foreli.** *Rhytidoponera* (*Rhytidoponera*) *foreli* Crawley, 1918: 87 (w.) AUSTRALIA.

**foveolata.** *Rhytidoponera foveolata* Crawley, 1925b: 581 (w.) AUSTRALIA. Clark, 1936: 70 (m.).

*froggatti. Ectatomma* (*Rhytidoponera*) *froggatti* Forel, 1910b: 10 (w.) SOLOMON IS. Mann, 1919: 286 (m.); Wheeler, G.C. & Wheeler, J. 1952a: 127 (l.). Combination in *Rhytidoponera:* Emery, 1911d: 38. Junior synonym of *araneoides:* Wilson, in Brown, 1958g: 202; Wilson, 1958d: 307.

**fulgens.** *Ectatomma fulgens* Emery, 1883: 148 (w.) NEW CALEDONIA. Emery, 1914f: 395 (m.). Combination in *E.* (*Rhytidoponera*): Dalla Torre, 1893: 24; in *Rhytidoponera* (*Chalcoponera*): Emery, 1911d: 39. Senior synonym of *socrula:* Brown, 1958g: 203. See also: Ward, 1984: 144.

**fuliginosa.** *Rhytidoponera fuliginosa* Clark, 1936: 79, pl. 6, fig. 57 (w.) AUSTRALIA.

*fusciventris. Rhytidoponera cornuta* var. *fusciventris* Stitz, 1911a: 352 (w.) AUSTRALIA. Junior synonym of *taurus:* Clark, 1936: 25.

*gagates. Rhytidoponera* (*Rhytidoponera*) *gagates* Donisthorpe, 1941h: 51 (w.) NEW GUINEA. Junior synonym of *nexa:* Wilson, in Brown, 1958g: 204; Wilson, 1958d: 312.

*gemma. Ectatomma* (*Rhytidoponera*) *convexum* var. *gemma* Forel, 1907h: 269 (w.m.) AUSTRALIA. Combination in *Rhytidoponera:* Emery, 1911d: 37. Junior synonym of *violacea:* Clark, 1936: 87.

*glabrius. Ectatomma* (*Rhytidoponera*) *mayri* subsp. *glabrius* Forel, 1907h: 268 (w.) AUSTRALIA. Forel, 1913g: 174 (m.). Combination in *Rhytidoponera:* Emery, 1911d: 38 (name misspelled as *glabrior*). Raised to species: Clark, 1936: 34. Junior synonym of *mayri:* Brown, 1958g: 203.

**greavesi.** *Rhytidoponera greavesi* Clark, 1941: 81, pl. 13, fig. 11 (w.) AUSTRALIA.

**gregoryi.** *Rhytidoponera gregoryi* Clark, 1936: 47, pl. 4, fig. 24 (w.) AUSTRALIA.

**haeckeli.** *Ectatomma* (*Rhytidoponera*) *haeckeli* Forel, 1910b: 15 (w.) AUSTRALIA. Combination in *Rhytidoponera* (*Chalcoponera*): Emery, 1911d: 39. See also: Brown, 1958g: 284.

**hanieli.** *Rhytidoponera hanieli* Forel, 1913e: 660 (w.) INDONESIA (Timor I.).

**hilli.** *Rhytidoponera* (*Rhytidoponera*) *hilli* Crawley, 1915a: 131 (w.) AUSTRALIA.

*hilli. Chalcoponera hilli* Clark, 1941: 85, pl. 13, fig. 15 (w.) AUSTRALIA. [Junior secondary homonym of *hilli* Crawley, above.] Replacement name: *clarki:* Donisthorpe, 1943e: 115.

**impressa.** *Ectatomma impressum* Mayr, 1876: 92 (w.q.) AUSTRALIA. Wheeler, G.C. & Wheeler, J. 1952a: 124 (l.); Imai, Crozier & Taylor, 1977: 347 (k.). Combination in *E.* (*Rhytidoponera*): Dalla Torre, 1893: 24; in *Rhytidoponera* (*Chalcoponera*): Emery, 1897d: 548. See also: Brown, 1954g: 5; Ward, 1980: 487.

*impressinodis. Rhytidoponera araneoides* var. *impressinodis* Stitz, 1912: 498, fig. 2 (w.) INDONESIA (Seram I.). Junior synonym of *araneoides:* Wilson, in Brown, 1958g: 202; Wilson, 1958d: 307.

**incisa.** *Rhytidoponera* (*Rhytidoponera*) *incisa* Crawley, 1915a: 132 (w.) AUSTRALIA.

**inops.** *Rhytidoponera inops* Emery, 1900c: 312 (w.) NEW GUINEA. Combination in *R.* (*Chalcoponera*): Emery, 1911d: 39. Senior syonym of *striata:* Wilson, in Brown, 1958g: 203; Wilson, 1958d: 310.

**inornata.** *Rhytidoponera* (*Chalcoponera*) *metallica* var. *inornata* Crawley, 1922b: 436 (w.) AUSTRALIA. Crawley, 1925b: 591 (m.); Wheeler, G.C. & Wheeler, J. 1971b: 1201 (l.). Raised to species, and senior synonym of *carbonaria:* Brown, 1958g: 203.

**insularis.** *Rhytidoponera insularis* Ward, 1984: 157, figs. 53, 54, 59, 60 (w.) NEW CALEDONIA.

*intricata. Rhytidoponera subcyanea* subsp. *intricata* Emery, 1911b: 533 (w.) NEW GUINEA. Junior synonym of *strigosa:* Wilson, in Brown, 1958g: 204; Wilson, 1958d: 317.

***kirghizorum.** **Rhytidoponera kirghizorum* Dlussky, 1981b: 66, fig. 50 (w.) RUSSIA (Miocene).

**koumensis.** *Rhytidoponera koumensis* Ward, 1984: 146, figs. 31, 32, 37, 38 (w.m.) NEW CALEDONIA.

**kurandensis.** *Rhytidoponera kurandensis* Brown, 1958g: 267 (w.) AUSTRALIA.

**laciniosa.** *Rhytidoponera laciniosa* Viehmeyer, 1912: 5, fig. 3 (w.) NEW GUINEA. Senior synonym of *petiolata:* Wilson, in Brown, 1958g: 203; Wilson, 1958d: 311.

*laevior. Rhytidoponera mayri* r. *glabrius* var. *laevior* Stitz, 1911a: 352, figs. 1, 2 (w.) AUSTRALIA. Unavailable name (Taylor, 1986: 34); material referred to *aciculata:* Brown, 1958g: 202.

**lamellinodis.** *Rhytidoponera* (*Chalcoponera*) *lamellinodis* Santschi, 1919a: 327 (w.) AUSTRALIA.

**laticeps.** *Rhytidoponera laticeps* Forel, 1915b: 12, fig. 3 (w.m.) AUSTRALIA.

**levior.** *Rhytidoponera punctata* var. *levior* Crawley, 1925b: 581 (w.) AUSTRALIA. Clark, 1936: 61 (m.). Raised to species: Clark, 1936: 61. Senior synonym of *douglasi:* Taylor, 1986: 35.

**litoralis.** *Rhytidoponera litoralis* Ward, 1984: 159, figs. 65, 66, 71, 72 (w.) NEW CALEDONIA.

**luteipes.** *Rhytidoponera luteipes* Ward, 1984: 160, figs. 51, 52, 57, 58 (w.m.) NEW CALEDONIA.

*major. Rhytidoponera strigosa* var. *major* Stitz, 1912: 501 (w.) NEW GUINEA. Junior synonym of *nexa:* Wilson, in Brown, 1958g: 204; Wilson, 1958d: 312.

*malandensis. Rhytidoponera laciniosa* r. *malandensis* Forel, 1915b: 10 (w.m.) AUSTRALIA. Raised to species: Clark, 1936: 73. Junior synonym of *scaberrima:* Brown, 1958g: 204.

**maledicta.** *Rhytidoponera* (*Chalcoponera*) *victoriae* r. *maledicta* Forel, 1915b: 15 (w.q.m.) AUSTRALIA. Raised to species: Brown, 1958g: 203.

**maniae.** *Ectatomma* (*Rhytidoponera*) *maniae* Forel, 1900b: 57 (w.) AUSTRALIA. Imai, Crozier & Taylor, 1977: 347 (k.). Combination in *Rhytidoponera:* Emery, 1911d: 38. Material of the unavailable name *rotundata* referred here: Clark, 1936: 65; Senior synonym of *spatiatum:* Brown, 1958g: 203.

**mayri.** *Ectatomma mayri* Emery, 1883: 150 (footnote) (w.) AUSTRALIA. Clark, 1936: 36 (m.); Wheeler, G.C. & Wheeler, J. 1964b: 448 (l.); Imai, Crozier & Taylor, 1977: 347 (k.). Combination in *Rhytidoponera:* Emery, 1911d: 38. Senior synonym of *dixoni, glabrius, occidentalis, petiolata, quadriceps* and *stridulator:* Brown, 1958g: 203.

*mediana. Rhytidoponera rothneyi* subsp. *mediana* Viehmeyer, 1924a: 224 (w.m.) AUSTRALIA. Junior synonym of *nodifera:* Clark, 1936: 68.

***RHYTIDOPONERA*** *(continued)*

***metallica.*** *Ponera metallica* Smith, F. 1858b: 94 (w.q.) AUSTRALIA. Mayr, 1866b: 891 (m.); Wheeler, G.C. & Wheeler, J. 1952a: 125 (l.); Wheeler, G.C. & Wheeler, J. 1964b: 449 (l.); Crozier, 1969: 241 (k.); Imai, Crozier & Taylor, 1977: 347 (k.). Combination in *Ecataomma* (*Rhytidoponera*): Mayr, 1862: 732; in *Rhytidoponera* (*Chalcoponera*): Emery, 1897d: 548. Senior synonym of *caeciliae, pulchra, purpurascens* and *varians:* Brown, 1958g: 204.

***micans.*** *Rhytidoponera micans* Clark, 1936: 62, pl. 5, figs. 40, 41 (w.m.) AUSTRALIA.

***mimica.*** *Rhytidoponera mimica* Ward, 1984: 163, figs. 67, 68, 73, 74 (w.) NEW CALEDONIA.

***mirabilis.*** *Rhytidoponera mirabilis* Clark, 1936: 29, pl. 3, fig. 7 (w.) AUSTRALIA.

*modestum. Ectatomma* (*Rhytidoponera*) *metallicum* var. *modestum* Emery, 1895g: 348 (w.) AUSTRALIA. Combination in *Rhytidoponera* (*Chalcoponera*): Emery, 1911d: 39. Variety of *victoriae:* Emery, 1911d: 39; Emery, 1912a: 80. Junior synonym of *victoriae:* Brown, 1958g: 205.

***nexa.*** *Rhytidoponera strigosa* var. *nexa* Stitz, 1912: 500, fig. 4 (w.) NEW GUINEA. Raised to species: Wilson, in Brown, 1958g: 204. Senior synonym of *gagates, major* and *waigeuensis:* Wilson, in Brown, 1958g: 204; Wilson, 1958d: 312.

*nigra. Rhytidoponera nigra* Clark, 1936: 81, pl. 6, fig. 59 (w.) AUSTRALIA. Junior synonym of *convexa:* Brown, 1958g: 272.

*nitens. Rhytidoponera* (*Rhytidoponera*) *nitens* Donisthorpe, 1949g: 403 (w.m.) NEW GUINEA. Junior synonym of *strigosa:* Wilson, in Brown, 1958g: 204; Wilson, 1958d: 317.

***nitida.*** *Rhytidoponera nitida* Clark, 1936: 45, pl. 4, fig. 21 (w.) AUSTRALIA.

***nitidiventris.*** *Rhytidoponera nitidiventris* Ward, 1984: 164, figs. 63, 64, 69, 70 (w.) NEW CALEDONIA.

***nodifera.*** *Ectatomma* (*Rhytidoponera*) *convexum* var. *nodiferum* Emery, 1895g: 348 (w.) AUSTRALIA. Clark, 1936: 69 (m.). Combination in *Rhytidoponera:* Emery, 1911d: 38. Raised to species, and senior synonym of *mediana, pronotalis* and *rothneyi:* Clark, 1936: 68.

***nudata.*** *Ectatomma nudatum* Mayr, 1876: 91 (w.) AUSTRALIA. Combination in *E.* (*Rhytidoponera*): Dalla Torre, 1893: 25; in *Rhytidoponera:* Emery, 1911d: 38.

***numeensis.*** *Ectatomma* (*Rhytidoponera*) *numeensis* André, 1889: 221 (w.) NEW CALEDONIA. Emery, 1914f: 395 (m.). Combination in *Rhytidoponera* (*Chalcoponera*): Emery, 1911d: 39. Senior synonym of *acupuncta:* Brown, 1958g: 204. See also: Brown, 1958g: 291; Ward, 1984: 149.

*obscurum. Ectatomma* (*Rhytidoponera*) *metallicum* var. *obscurum* Forel, 1900b: 60 (w.q.m.) AUSTRALIA. [Unresolved junior primary homonym of *Ectatomma obscurum* Emery, 1896g: 48.] Combination in *Rhytidoponera* (*Chalcoponera*): Emery, 1911d: 39. Raised to species: Emery, 1911d: 39; Emery, 1912a: 79. Junior synonym of *clarki:* Brown, 1958g: 203.

*occidentalis. Rhytidoponera occidentalis* Clark, 1936: 39, fig. 2 (w.) AUSTRALIA. Junior synonym of *mayri:* Brown, 1958g: 203.

*opacior. Rhytidoponera convexa* subsp. *opacior* Clark, 1936: 86 (w.) AUSTRALIA. [First available use of *Rhytidoponera convexa* subsp. *violacea* var. *opacior* Crawley, 1925b: 583; unavailable name (Taylor, 1986: 34).] Junior synonym of *violacea:* Brown, 1958g: 205.

***opaciventris.*** *Rhytidoponera opaciventris* Ward, 1984: 145, figs. 17, 18, 23, 24 (w.m.) NEW CALEDONIA.

***peninsularis.*** *Rhytidoponera peninsularis* Brown, 1958g: 280 (w.) AUSTRALIA.

*petiolata. Rhytidoponera laciniosa* subsp. *petiolata* Viehmeyer, 1912: 5, fig. 4 (w.) NEW GUINEA. Junior synonym of *laciniosa:* Wilson, in Brown, 1958g: 203; Wilson, 1958d: 311.

*petiolata. Rhytidoponera petiolata* Clark, 1936: 41, pl. 4, fig. 18 (w.) AUSTRALIA. [Unresolved junior primary homonym of *petiolata* Viehmeyer, above.] Junior synonym of *mayri:* Brown, 1958g: 203.

***pilosula.*** *Rhytidoponera pilosula* Clark, 1936: 80, pl. 6, fig. 58 (w.) AUSTRALIA.

*pronotalis. Rhytidoponera pronotalis* Crawley, 1925b: 588 (w.) AUSTRALIA. Junior synonym of *nodifera:* Clark, 1936: 68.

***pulchella.*** *Ectatomma pulchellum* Emery, 1883: 149 (w.) NEW CALEDONIA. Emery, 1914f: 396 (m.). Combination in *E.* (*Rhytidoponera*): Dalla Torre, 1893: 25; in *Rhytidoponera* (*Chalcoponera*): Emery, 1897d: 548. See also: Brown, 1958g: 291; Ward, 1984: 167.

*pulchra. Chalcoponera pulchra* Clark, 1941: 86, pl. 13, fig. 16 (w.) AUSTRALIA. Junior synonym of *metallica:* Brown, 1958g: 204.

***punctata.*** *Ectatomma punctata* Smith, F. 1858b: 104 (w.) AUSTRALIA. Clark, 1938: 363 (m.). Combination in *E.* (*Rhytidoponera*): Dalla Torre, 1893: 25; in *Rhytidoponera:* Emery, 1911d: 38. See also: Brown, 1958g: 270.

***punctigera.*** *Rhytidoponera punctigera* Crawley, 1925b: 582 (w.) AUSTRALIA.

***punctiventris.*** *Ectatomma* (*Rhytidoponera*) *cristatum* var. *punctiventris* Forel, 1900b: 56 (w.) AUSTRALIA. Clark, 1936: 52 (m.). Combination in *Rhytidoponera:* Emery, 1911d: 38. Raised to species: Brown, 1958g: 204.

*purpurascens. Rhytidoponera* (*Chalcoponera*) *metallica* var. *purpurascens* Wheeler, W.M. 1915g: 805 (w.) AUSTRALIA. Junior synonym of *metallica:* Brown, 1958g: 204.

***purpurea.*** *Ectatomma impressum* var. *purpureum* Emery, 1887b: 444 (w.q.) NEW GUINEA. Imai, Crozier & Taylor, 1977: 347 (k.). Combination in *Rhytidoponera* (*Chalcoponera*): Emery, 1897d: 549. Raised to species, and senior synonym of *splendidum:* Brown, 1954g: 7.

*quadriceps. Rhytidoponera quadriceps* Clark, 1936: 30, pl. 3, fig. 8 (w.) AUSTRALIA. Junior synonym of *mayri:* Brown, 1958g: 203.

***reflexa.*** *Rhytidoponera reflexa* Clark, 1936: 76, pl. 6, fig. 54 (w.) AUSTRALIA. See also: Brown, 1958g: 277.

***reticulata.*** *Ectatomma* (*Rhytidoponera*) *reticulatum* Forel, 1893d: 459 (w.) AUSTRALIA. Crawley, 1921: 88 (m.). Combination in *Rhytidoponera* (*Chalcoponera*): Emery, 1911d: 39.

*rothneyi. Ectatomma* (*Rhytidoponera*) *rothneyi* Forel, 1900b: 56 (w.) AUSTRALIA. Combination in *Rhytidoponera:* Emery, 1911d: 38. Junior synonym of *nodifera:* Clark, 1936: 68.

*rotundata. Rhytidoponera mayri* r. *glabrius* var. *rotundata* Stitz, 1911a: 353, fig. 3 (w.) AUSTRALIA. Unavailable name (Taylor, 1986: 34). Material referred to *maniae* by Clark, 1936: 65.

***rotundiceps.*** *Rhytidoponera* (*Rhytidoponera*) *rotundiceps* Viehmeyer, 1914c: 28, fig. 2 (w.) NEW GUINEA.

***rufescens.*** *Ectatomma* (*Rhytidoponera*) *convexum* var. *rufescens* Forel, 1900b: 58 (w.) AUSTRALIA. Clark, 1936: 88 (m.). Combination in *Rhytidoponera:* Emery, 1911d: 38. Subspecies of *convexa:* Clark, 1936: 85. Raised to species: Brown, 1958g: 204.

***rufithorax.*** *Rhytidoponera rufithorax* Clark, 1941: 82, pl. 13, fig. 12 (w.) AUSTRALIA.

***rufiventris.*** *Rhytidoponera convexa* var. *rufiventris* Forel, 1915b: 11 (w.m.) AUSTRALIA. Raised to species, and senior synonym of *castanea:* Clark, 1936: 83. See also: Brown, 1958g: 278.

***rufonigra.*** *Rhytidoponera rufonigra* Clark, 1936: 58, pl. 5, figs. 35, 36 (w.m.) AUSTRALIA.

*rugosa. Ectatomma rugosa* Smith, F. 1859a: 143 (w.m.) INDONESIA (Aru I.). Junior synonym of *araneoides:* Mayr, 1876: 92; Donisthorpe, 1932c: 454.

***scaberrima.*** *Ectatomma* (*Rhytidoponera*) *scaberrimum* Emery, 1895g: 347 (w.) AUSTRALIA. Combination in *Rhytidoponera* (*Chalcoponera*): Emery, 1897d: 548. Senior synonym of *malandensis:* Brown, 1958g: 204.

***scabra.*** *Ectatomma scabrum* Mayr, 1876: 90 (w.m.) AUSTRALIA. Combination in *E.* (*Rhytidoponera*): Dalla Torre, 1893: 26; in *Rhytidoponera:* Emery, 1897d: 548.

***scabrior.*** *Rhytidoponera* (*Chalcoponera*) *aspera* var. *scabrior* Crawley, 1925b: 590 (w.) AUSTRALIA. Junior synonym of *chalybaea:* Brown, 1954g: 4. Revived from synonymy, and raised to species: Ward, 1980: 493.

*schlaginhaufeni. Rhytidoponera schlaginhaufeni* Viehmeyer, 1912: 4, fig. 2 (w.) NEW GUINEA. Junior synonym of *strigosa:* Wilson, in Brown, 1958g: 204; Wilson, 1958d: 317.

*scrobiculatum. Ectatomma* (*Rhytidoponera*) *metallicum* var. *scrobiculatum* Forel, 1900b: 59 (w.q.m.) AUSTRALIA. Combination in *Rhytidoponera* (*Chalcoponera*): Emery, 1911d: 39. Variety of *victoriae:* Emery, 1911d: 39; Emery, 1912a: 80. Junior synonym of *victoriae:* Brown, 1958g: 205.

*socrula. Rhytidoponera* (*Chalcoponera*) *fulgens* var. *socrula* Emery, 1914f: 395 (w.) NEW CALEDONIA. Junior synonym of *fulgens:* Brown, 1958g: 203.

***socra.*** *Ectatomma* (*Rhytidoponera*) *socrus* Forel, 1894e: 236 (w.) AUSTRALIA. Wheeler, W.M. 1909a: 25 (m.). Combination in *Rhytidoponera:* Emery, 1911d: 38.

*spatiatum. Ectatomma* (*Rhytidoponera*) *convexum* var. *spatiatum* Forel, 1900b: 58 (w.) AUSTRALIA. Clark, 1936: 67 (m.). Combination in *Rhytidoponera:* Emery, 1911d: 38. Raised to species: Clark, 1936: 66. Junior synonym of *maniae:* Brown, 1958g: 203.

*splendidum. Ectatomma* (*Rhytidoponera*) *impressum* var. *splendidum* Forel, 1910b: 12 (w.) AUSTRALIA. [Unresolved junior primary homonym of *Ectatomma rimulosum* var. *splendidum* Pergande, 1896: 871.] Combination in *Rhytidoponera* (*Chalcoponera*):

Emery, 1911d: 39. Junior synonym of *purpurea:* Brown, 1954g: 7.

***spoliata.*** *Ectatomma (Rhytidoponera) spoliatum* Emery, 1895g: 348 (w.) AUSTRALIA. Combination in *Rhytidoponera:* Emery, 1911d: 38. See also: Brown, 1958g: 279.

*striata. Rhytidoponera (Chalcoponera) striata* Donisthorpe, 1949c: 744 (w.) NEW GUINEA. Junior synonym of *inops:* Wilson, in Brown, 1958g: 203; Wilson, 1958d: 310.

*stridulator. Rhytidoponera stridulator* Clark, 1936: 37, fig. 1 (w.m.) AUSTRALIA. Junior synonym of *mayri:* Brown, 1958g: 203.

***strigosa.*** *Ectatomma araneoides* var. *strigosum* Emery, 1887b: 444 (w.) NEW GUINEA. Donisthorpe, 1943d: 437 (m.). Combination in *Rhytidoponera:* Emery, 1911d: 38. Subspecies of *araneoides:* Forel, 1907a: 2. Raised to species: Emery, 1911d: 38. Senior synonym of *curvata, intricata, nitens* and *schlaginhaufeni:* Wilson, in Brown, 1958g: 204; Wilson, 1958d: 317.

***subcyanea.*** *Rhytidoponera subcyanea* Emery, 1897d: 548 (w.) NEW GUINEA. Senior synonym of *aruana, transversiruga* and *wallacei:* Wilson, in Brown, 1958g: 205; Wilson, 1958d: 319.

*subumbrata. Rhytidoponera convexa* r. *violacea* var. *subumbrata* Crawley, 1922b: 434 (w.) AUSTRALIA. Unavailable name (Taylor, 1986: 34); material referred to *convexa* by Clark, 1936: 84.

***tasmaniensis.*** *Rhytidoponera metallica* var. *tasmaniensis* Emery, 1898b: 232 (w.) TASMANIA. Wheeler, G.C. & Wheeler, J. 1964b: 449 (l.); Crozier, 1969: 241 (k.). Raised to species, and senior synonym of *cristulatum:* Brown, 1958g: 205.

***taurus.*** *Ectatomma (Rhytidoponera) cornutum* subsp. *taurus* Forel, 1910b: 12 (w.) AUSTRALIA. Forel, 1915b: 12 (m.). Combination in *Rhytidoponera:* Emery, 1911d: 38. Raised to species, and senior synonym of *brevior* and *fusciventris:* Clark, 1936: 25.

***tenuis.*** *Ectatomma (Rhytidoponera) tenue* Forel, 1900b: 58 (w.) AUSTRALIA. Combination in *Rhytidoponera (Chalcoponera):* Emery, 1911d: 39.

***terrestris.*** *Rhytidoponera terrestris* Ward, 1984: 168, figs. 43, 44, 49, 50 (w.m.) NEW CALEDONIA.

***trachypyx.*** *Rhytidoponera trachypyx* Brown, 1958g: 281 (w.) AUSTRALIA.

*transversiruga. Rhytidoponera subcyanea* subsp. *transversiruga* Emery, 1911b: 532 (w.) NEW GUINEA. Junior synonym of *subcyanea:* Wilson, in Brown, 1958g: 205; Wilson, 1958d: 319.

***turneri.*** *Ectatomma (Rhytidoponera) turneri* Forel, 1910b: 14 (w.) AUSTRALIA. Combination in *Rhytidoponera (Chalcoponera):* Emery, 1911d: 39.

***tyloxys.*** *Rhytidoponera tyloxys* Brown & Douglas, in Brown, 1958g: 282, figs. 36, 37 (w.) AUSTRALIA.

*varians. Rhytidoponera (Chalcoponera) metallica* var. *varians* Crawley, 1922b: 436 (w.) AUSTRALIA. Crawley, 1925b: 591 (m.). Junior synonym of *metallica:* Brown, 1958g: 204.

***versicolor.*** *Rhytidoponera versicolor* Brown, 1958g: 292, fig. 40 (w.) NEW CALEDONIA. See also: Ward, 1984: 169.

***victoriae.*** *Ectatomma (Rhytidoponera) victoriae* André, 1896b: 261 (w.) AUSTRALIA. Crawley, in Poulton & Crawley, 1922: 122 (m.); Wheeler, G.C. & Wheeler, J. 1964b: 449 (l.); Crozier, 1969: 241 (k.). Combination in *Rhytidoponera (Chalcoponera):* Emery, 1911d: 39. Senior synonym of *cedarensis, modestum* and *scrobiculatum:* Brown, 1958g: 205.

***violacea.*** *Ectatomma (Rhytidoponera) convexum* subsp. *violaceum* Forel, 1907h: 269 (w.m.) AUSTRALIA. Combination in *Rhytidoponera:* Emery, 1911d: 38. Raised to species: Brown, 1958g: 205. Senior synonym of *gemma:* Clark, 1936: 87; of *opacior:* Brown, 1958g: 205.

***viridis.*** *Chalcoponera viridis* Clark, 1941: 83, pl. 13, fig. 13 (w.) AUSTRALIA. Combination in *Rhytidoponera:* Brown, 1958g: 205.

*waigeuensis. Rhytidoponera (Rhytidoponera) gagates* subsp. *waigeuensis* Donisthorpe, 1942e: 703 (w.) NEW GUINEA. Donisthorpe, 1943d: 435 (m.). Junior synonym of *nexa:* Wilson, in Brown, 1958g: 204; Wilson, 1958d: 312.

*wallacei. Rhytidoponera (Rhytidoponera) wallacei* Donisthorpe, 1932c: 474 (w.) INDONESIA (Aru I.). Junior synonym of *subcyanea:* Wilson, in Brown, 1958g: 205; Wilson, 1958d: 319.

***wilsoni.*** *Rhytidoponera wilsoni* Brown, 1958g: 294, fig. 38 (w.) NEW CALEDONIA. See also: Ward, 1984: 150.

***yorkensis.*** *Rhytidoponera cristata* r. *yorkensis* Forel, 1915b: 12 (w.) AUSTRALIA. Raised to species: Clark, 1936: 53.

***ROGERIA*** (New World, Indo-Australian)

*azumai* Santschi, 1941; see under ***LORDOMYRMA.***

***belti.*** *Rogeria belti* Mann, 1922: 31 (w.) HONDURAS. Wheeler, G.C. & Wheeler, J. 1989a: 324 (l.).

***blanda.*** *Myrmica blanda* Smith, F. 1858b: 131 (w.) BRAZIL. Wheeler, G.C. & Wheeler, J. 1989a: 325 (l.). Combination in *Tetramorium:* Mayr, 1886c: 359; in *Rogeria:* Emery, 1894c: 189. Senior synonym of *foveata:* Kempf, 1965: 185.

*brasiliensis. Macromischa brasiliensis* Borgmeier, 1955a: 107, figs. 1–4 (w.) BRAZIL. Junior synonym of *procera:* Kempf, 1961d: 437.

***bruchi.*** *Rogeria bruchi* Santschi, 1922b: 352 (w.) ARGENTINA.

***brunnea.*** *Rogeria (Rogeria) curvipubens* st. *brunnea* Santschi, 1930c: 79 (w.) CUBA. Raised to species: Santschi, 1936b: 201.

***caraiba.*** *Rogeria (Irogera) caraiba* Santschi, 1936b: 198, fig. 5 (w.) CUBA.

***creightoni.*** *Rogeria creightoni* Snelling, 1973a: 2, fig. 1 (w.) U.S.A.

***cubensis.*** *Rogeria (Irogera) cubensis* Santschi, 1936b: 199, figs. 3, 4 (w.) CUBA. Current subspecies: nominal plus *habanica.*

***curvipubens.*** *Rogeria curvipubens* Emery, 1894c: 190 (w.q.) ANTILLES IS.

*epinotalis* Mann, 1919; see under ***LORDOMYRMA.***

***exsulans.*** *Rogeria exsulans* Wilson & Taylor, 1967: 74, fig. 60 (w.) SAMOA.

***foreli.*** *Rogeria foreli* Emery, 1894c: 191 (w.) ANTILLES IS. [Previously referred to as *Tetramorium foreli* Emery by Forel, 1893g: 383; *nomen nudum.*] Current subspecies: nominal plus *gaigei.*

*foveata. Irogera foveata* Kempf, 1964e: 64, figs. 19, 20 (w.) BRAZIL. Junior synonym of *blanda:* Kempf, 1965: 185.

***#gaigei.*** *Rogeria foreli* r. *gaigei* Forel, 1914c: 617 (w.) COLOMBIA.

***germaini.*** *Rogeria germaini* Emery, 1894c: 189 (w.) BRAZIL. See also: Kempf, 1962b: 20.

***#habanica.*** *Rogeria (Irogera) cubensis* var. *habanica* Santschi, 1936b: 200 (w.) CUBA.

***huachucana.*** *Rogeria huachucana* Snelling, 1973a: 4, fig. 1 (w.) U.S.A.

***inermis.*** *Rogeria inermis* Mann, 1922: 32, fig. 15 (w.) HONDURAS.

***lacertosa.*** *Rogeria lacertosa* Kempf, 1963a: 194, fig. 5, 6 (w.) BRAZIL.

*levifrons* Mann, 1921; see under ***LORDOMYRMA.***

***manni.*** *Rogeria manni* Santschi, 1922b: 353 (footnote) (w.) SOLOMON IS.

***micromma.*** *Rogeria micromma* Kempf, 1961b: 509, figs. 12, 13 (w.) SURINAM.

***minensis.*** *Rogeria germaini* st. *minensis* Santschi, 1923g: 1262 (w.) BRAZIL. Raised to species: Kempf, 1963a: 189.

***minima.*** *Rogeria minima* Kusnezov, 1958a: 44, figs. 1–3 (q.) ARGENTINA. [*R. minima* Kusnezov, 1954b: 35; *nomen nudum.*]

***pellecta.*** *Rogeria pellecta* Kempf, 1963a: 191, figs. 3, 4 (w.) BRAZIL.

*polita* Mann, 1921; see under ***LORDOMYRMA.***

***procera.*** *Rogeria procera* Emery, 1896g: 92, pl. 1, fig. 19 (w.) BRAZIL. Wheeler, G.C. & Wheeler, J. 1973b: 74 (l.). Combination in *R. (Irogera):* Emery, 1915i: 191; in *Irogera:* Brown, 1953h: 4 (in text); in *Rogeria:* Kempf, 1965: 185. Senior synonym of *brasiliensis:* Kempf, 1961d: 437.

*rugosa* Mann, 1921; see under ***LORDOMYRMA.***

***scabra.*** *Rogeria (Rogeria) scabra* Weber, 1934a: 27, fig. 2 (w.q.) CUBA.

***scandens.*** *Macromischa scandens* Mann, 1922: 30, fig. 14 (w.) HONDURAS. Wheeler, G.C. & Wheeler, J. 1986f: 493 (l.). Combination in *Irogera:* Kempf, 1961d: 438; in *Rogeria:* Kempf, 1965: 185.

***sicaria.*** *Rogeria sicaria* Kempf, 1962b: 22, figs. 20, 21 (w.) BRAZIL.

***stigmatica.*** *Rogeria stigmatica* Emery, 1897c: 589 (w.) NEW GUINEA. Mann, 1919: 342 (q.m.).

*stoneri* Mann, 1925; see under ***LORDOMYRMA.***

*striatella* Mann, 1921; see under ***LORDOMYRMA.***

***subarmata.*** *Irogera subarmata* Kempf, 1961d: 438, figs. 1–4 (w.) BRAZIL. Combination in *Rogeria:* Kempf, 1965: 185.

***sublevinodis.*** *Rogeria stigmatica* subsp. *sublevinodis* Emery, 1914f: 415 (w.) LOYAUTE IS. Wheeler, G.C. & Wheeler, J. 1955b: 28 (l.). Raised to species: Wilson & Taylor, 1967: 76.

***tonduzi.*** *Rogeria tonduzi* Forel, 1899c: 53 (w.) COSTA RICA. Combination in *R. (Irogera):* Emery, 1915i: 191; in *Irogera:* Brown, 1953h: 4 (in text); in *Rogeria:* Kempf, 1965: 185.

*tortuosa* Mann, 1921; see under ***LORDOMYRMA.***

***ROMBLONELLA*** (Indo-Australian, Australasian)

***elysii.*** *Tetramorium scrobiferum* var. *elysii* Mann, 1919: 346, fig. 29 (w.) SOLOMON IS. Combination in *Romblonella,* and raised to species: Smith, M.R. 1956b: 18.

*grandinodis. Romblonella grandinodis* Wheeler, W.M. 1935a: 7, fig. 2 (w.) PHILIPPINES. Junior synonym of *opaca:* Bolton, 1976: 294.

***heatwolei.*** *Romblonella heatwolei* Taylor, 1991a: 282, figs. 1–3 (w.m.) AUSTRALIA.

***ROMBLONELLA*** *(continued)*

**#*liogaster*.** *Tetramorium scrobiferum* var. *liogaster* Santschi, 1928c: 69 (w.) FIJI IS. Combination in *Romblonella:* **new combination**.

***opaca*.** *Myrmica opaca* Smith, F. 1861b: 47 (w.) SULAWESI. Combination in *Tetramorium:* Emery, 1901g: 567; Donisthorpe, 1932c: 469; in *Romblonella*, and senior synonym of *grandinodis:* Bolton, 1976: 294.

***palauensis*.** *Romblonella palauensis* Smith, M.R. 1953f: 164 (w.m.) PALAU IS (Micronesia).

***scrobifera*.** *Tetramorium scrobiferum* Emery, 1897c: 587, pl. 15, fig. 31 (w.) NEW GUINEA. Combination in *Romblonella:* Smith, M.R. 1956b: 18. Current subspecies: nominal plus *liogaster.*

***townesi*.** *Romblonella townesi* Smith, M.R. 1953a: 80 (w.) MARIANAS IS.

***vitiensis*.** *Romblonella vitiensis* Smith, M.R. 1953a: 79 (w.) FIJI IS.

***yapensis*.** *Romblonella yapensis* Smith, M.R. 1953a: 78 (w.) CAROLINE IS.

***ROSSOMYRMEX*** (Palaearctic)

***minuchae*.** *Rossomyrmex minuchae* Tinaut, 1981: 195, figs. 1–3 (w.q.) SPAIN.

***proformicarum*.** *Rossomyrmex proformicarum* Arnol'di, 1928c: 300, figs. 1–6 (w.) RUSSIA. See also: Arnol'di, 1932a: 319.

***ROSTROMYRMEX*** (Malaysia)

***pasohensis*.** *Rostromyrmex pasohensis* Rosciszewski, 1994: 163, figs. 1–19 (w.q.m.) WEST MALAYSIA.

***ROTASTRUMA*** (Oriental, Indo-Australian)

***recava*.** *Rotastruma recava* Bolton, 1991: 9, figs. 12, 14 (w.q.) BORNEO.

***stenoceps*.** *Rotastruma stenoceps* Bolton, 1991: 9, figs. 13, 15 (w.) CHINA.

***SANTSCHIELLA*** (Afrotropical)

***kohli*.** *Santschiella kohli* Forel, 1916: 435, figs. 2, 3 (w.) ZAIRE.

*SCHIZOPELTA:* see under ***MONOMORIUM***.

*SCROBOPHEIDOLE:* see under ***PHEIDOLE***.

***SCYPHODON*** (Indo-Australian)

***anomalum*.** *Scyphodon anomalum* Brues, 1925: 95 (m.) SUMATRA. See also: Petersen, 1968: 591.

***SECOSTRUMA*** (Indo-Australian)

***lethifera*.** *Secostruma lethifera* Bolton, 1988c: 265, figs. 1–4 (w.) BORNEO.

*SELENOPONE:* see under ***PONERA***.

*SEMONIUS:* see under ***TAPINOMA***.

***SERICOMYRMEX*** (Neotropical)

***amabilis*.** *Sericomyrmex amabilis* Wheeler, W.M. 1925d: 166, fig. 8 (not fig. 4) (w.) PANAMA. Wheeler, G.C. 1949: 671 (l.). Senior synonym of *bierigi:* Weber, 1958d: 263.

**#*arawakensis*.** *Sericomyrmex harekulli* subsp. *arawakensis* Weber, 1937: 399, fig. 11 (w.q.) GUIANA.

***aztecus*.** *Sericomyrmex aztecus* Forel, 1885a: 363 (w.) MEXICO.

***beniensis*.** *Sericomyrmex beniensis* Weber, 1938b: 182 (w.) BOLIVIA.

*bierigi. Sericomyrmex bierigi* Santschi, 1931c: 279 (w.) PANAMA. Junior synonym of *amabilis:* Weber, 1958d: 263.

***bondari*.** *Sericomyrmex bondari* Borgmeier, 1937b: 248, figs. 37, 38 (w.) BRAZIL.

*bruchi* Santschi, 1916; see under ***MYCETOPHYLAX***.

***burchelli*.** *Sericomyrmex burchelli* Forel, 1905b: 183 (q.m.) BRAZIL.

***diego*.** *Sericomyrmex diego* Forel, 1912e: 192 (w.m.) COLOMBIA.

*gallardoi* Santschi, 1920; see under ***ACROMYRMEX***.

***harekulli*.** *Sericomyrmex harekulli* Weber, 1937: 398, fig. 10 (w.) GUIANA. Current subspecies: nominal plus *arawakensis.*

***impexus*.** *Sericomyrmex impexus* Wheeler, W.M. 1925a: 54 (w.) GUIANA.

***luederwaldti*.** *Sericomyrmex luederwaldti* Santschi, 1925b: 15 (w.) BRAZIL.

***lutzi*.** *Sericomyrmex lutzi* Wheeler, W.M. 1916c: 9 (w.q.m.) GUIANA.

**#*maracas*.** *Sericomyrmex urichi* subsp. *maracas* Weber, 1937: 395, fig. 7 (w.q.m.) TRINIDAD.

***mayri*.** *Sericomyrmex mayri* Forel, 1912e: 194 (w.) BRAZIL.

***moreirai*.** *Sericomyrmex moreirai* Santschi, 1925b: 16 (w.q.m.) BRAZIL.

**#*muelleri*.** *Sericomyrmex opacus* var. *muelleri* Forel, 1912e: 195 (q.) BRAZIL.

***myersi*.** *Sericomyrmex myersi* Weber, 1937: 400 (w.) SURINAM.

***opacus*.** *Sericomyrmex opacus* Mayr, 1865: 84, pl. 3, fig. 22 (q.) BRAZIL. Mayr, 1866a: 505 (m.); Wheeler, W.M. 1916c: 11 (w.). Current subspecies: nominal plus *muelleri.*

*opulenta* Mann, 1922; see under ***TRACHYMYRMEX***.

*pakeelai* Weber, 1937; see under ***TRACHYMYRMEX***.

***parvulus*.** *Sericomyrmex parvulus* Forel, 1912e: 193 (w.) BRAZIL.

***saussurei*.** *Sericomyrmex saussurei* Emery, 1894c: 223 (w.) BRAZIL.

***scrobifer*.** *Sericomyrmex scrobifer* Forel, 1911c: 296 (w.) BRAZIL.

***urichi*.** *Sericomyrmex urichi* Forel, 1912e: 193 (w.) TRINIDAD. Current subspecies: nominal plus *maracas.*

*wheeleri* Weber, 1937; see under ***TRACHYMYRMEX***.

***zacapanus*.** *Sericomyrmex zacapanus* Wheeler, W.M. 1925a: 54 (w.) GUATEMALA.

***SERRASTRUMA*** (Afrotropical, Malagasy)

*aequalis. Strumigenys* (*Cephaloxys*) *aequalis* Menozzi, 1942: 177 (w.q.) FERNANDO PO I. Junior synonym of *lujae:* Brown, 1952e: 78. See also: Menozzi, 1944: 457.

*alluaudi. Strumigenys alluaudi* Santschi, 1911c: 360 (w.q.) TANZANIA. Combination in *S.* (*Trichoscapa*): Santschi, 1913b: 258 (in key); in *S.* (*Cephaloxys*): Wheeler, W.M. 1922a: 918; in *Serrastruma*, and senior synonym of *rothkirchi:* Brown, 1952e: 75. Junior synonym of *ludovici:* Bolton, 1983: 343.

***bequaerti*.** *Strumigenys* (*Cephaloxys*) *bequaerti* Santschi, 1923e: 286 (w.) ZAIRE. Consani, 1951: 171 (m.). Combination in *Serrastruma:* Brown, 1952e: 80. Junior synonym of *lujae:* Bolton, 1983: 345. Revived from synonymy: **revived status**.

*biconvexa. Strumigenys* (*Trichoscapa*) *biconvexa* Santschi, 1913b: 258 (in key) (w.q.) KENYA. Santschi, 1914b: 111 (w.q.). Combination in *S.* (*Cephaloxys*): Wheeler, W.M. 1922a: 919. Junior synonym of *simoni:* Brown, 1952e: 82.

*boerorum. Strumigenys* (*Trichoscapa*) *cognata* st. *boerorum* Santschi, 1913b: 259 (in key) (w.) SOUTH AFRICA. Santschi, 1914e: 32 (q.). Combination in *S.* (*Cephaloxys*): Wheeler, W.M. 1922a: 919. Stirps of *escherichi:* Santschi, 1914e: 32. Junior synonym of *simoni:* Brown, 1952e: 82.

*calypso. Strumigenys* (*Cephaloxys*) *calypso* Santschi, 1923e: 288, fig. 4 (w.) TANZANIA. Combination in *Serrastruma:* Brown, 1952e: 85. Junior synonym of *lujae:* Bolton, 1983: 345.

*cliens. Strumigenys escherichi* var. *cliens* Forel, 1913b: 317 (w.) ZAIRE. Combination in *S.* (*Cephaloxys*): Wheeler, W.M. 1922a: 919. Junior synonym of *simoni:* Brown, 1952e: 82.

*cognata. Strumigenys cognata* Santschi, 1911c: 362 (w.q.) ANGOLA. Santschi, 1913b: 258 (q.). Combination in *S.* (*Trichoscapa*): Santschi, 1913b: 258 (in key); in *S.* (*Cephaloxys*): Wheeler, W.M. 1922a: 919. Race of *escherichi:* Forel, 1913b: 317 (in text); Forel, 1916: 427. Junior synonym of *simoni:* Brown, 1952e: 82.

***concolor*.** *Strumigenys* (*Trichoscapa*) *concolor* Santschi, 1914d: 375 (w.q.) GHANA. Combination in *S.* (*Cephaloxys*): Wheeler, W.M. 1922a: 919; in *Serrastruma:* Brown, 1952e: 81. Junior synonym of *serrula:* Brown, 1952e: 81. Revived from synonymy: Bolton, 1983: 338.

***dotaja*.** *Serrastruma dotaja* Bolton, 1983: 339, figs. 35, 38 (w.) CAMEROUN.

*escherichi. Strumigenys escherichi* Forel, 1910c: 261 (w.) ETHIOPIA. Combination in *S.* (*Trichoscapa*): Santschi, 1913b: 258 (in key); in *S.* (*Cephaloxys*): Wheeler, W.M. 1922a: 919. Junior synonym of *simoni:* Brown, 1952e: 82.

*fusciventris. Strumigenys* (*Cephaloxys*) *escherichi* var. *fusciventris* Emery, 1922f: 324, (w.) NIGERIA. [First available use of *Strumigenys* (*Trichoscapa*) *escherichi* st. *cognata* var. *fusciventris* Santschi, 1915c: 261; unavailable name.] Junior synonym of *simoni:* Brown, 1952e: 82.

***geoterra*.** *Serrastruma geoterra* Bolton, 1983: 341, fig. 36 (w.) CAMEROUN.

*gerardi. Strumigenys* (*Cephaloxys*) *gerardi* Santschi, 1923e: 287 (w.) ZAIRE. Junior synonym of *lujae:* Brown, 1952e: 78.

*glanduscula. Strumigenys* (*Cephaloxys*) *glanduscula* Santschi, 1919h: 88 (w.) ZAIRE. Junior synonym of *lujae:* Brown, 1952e: 78.

***inquilina*.** *Serrastruma inquilina* Bolton, 1983: 342 (q.) ZAIRE.

*latiuscula. Strumigenys* (*Trichoscapa*) *maynei* var. *latiuscula* Forel, 1916: 428 (w.) ZAIRE. Combination in *S.* (*Cephaloxys*): Wheeler, W.M. 1922a: 920. Junior synonym of *maynei:* Brown, 1952e: 77.

*limbata. Strumigenys escherichi* r. *limbata* Forel, 1913j: 222 (w.) ZIMBABWE. Arnold, 1917: 382 (q.). Combination in *S.* (*Cephaloxys*): Wheeler, W.M. 1922a: 920. Junior synonym of *simoni:* Brown, 1952e: 82.

*lotti*. *Strumigenys (Cephaloxys) escherichi* subsp. *lotti* Weber, 1943c: 378, pl. 15, fig. 13 (w.q.) SUDAN. Combination in *Serrastruma*, and raised to species: Brown, 1952e: 76. Junior synonym of *ludovici*: Bolton, 1983: 343.

***ludovici***. *Strumigenys ludovici* Forel, 1904b: 369 (w.) MADAGASCAR. Combination in *Serrastruma*: Brown, 1952e: 85. Senior synonym of *alluaudi* (and its junior synonym *rothkirchi*), *lotti* and *nigeriensis*: Bolton, 1983: 343.

***lujae***. *Strumigenys lujae* Forel, 1902e: 294 (footnote), pl. 1, fig. 1 (w.) MOZAMBIQUE. Combination in *S. (Trichoscapa)*: Santschi, 1913b: 258 (in key); in *S. (Cephaloxys)*: Wheeler, W.M. 1922a: 920; in *Serrastruma*: Brown, 1952e: 78. Senior synonym of *aequalis, gerardi, glanduscula* and *reticulata*: Brown, 1952e: 78; of *calypso*: Bolton, 1983: 345.

***maynei***. *Strumigenys (Trichoscapa) maynei* Forel, 1916: 427 (w.q.m.) ZAIRE. Combination in *S. (Cephaloxys)*: Wheeler, W.M. 1922a: 920; in *Serrastruma*: Brown, 1952e: 77. Senior synonym of *latiuscula*: Brown, 1952e: 77. See also: Bolton, 1983: 347.

***miccata***. *Serrastruma miccata* Bolton, 1983: 348, fig. 39 (w.) GHANA.

*nigeriensis*. *Strumigenys (Trichoscapa) alluaudi* st. *nigeriensis* Santschi, 1914d: 376 (w.) NIGERIA. Combination in *S. (Cephaloxys)*: Wheeler, W.M. 1922a: 919. Junior synonym of *simoni*: Brown, 1952e: 83; of *ludovici*: Bolton, 1983: 343.

*obscuriventris*. *Strumigenys (Trichoscapa) escherichi* st. *cognata* var. *obscuriventris* Santschi, 1914d: 376 (w.) NIGERIA. Unavailable name; material referred to *simoni* by Brown, 1952e: 82.

*raymondi*. *Strumigenys (Cephaloxys) raymondi* Donisthorpe, 1946c: 778 (w.q.) MAURITIUS. Junior synonym of *alluaudi*: Brown, 1952e: 75; of *simoni*: Bolton, 1983: 350.

*reticulata*. *Strumigenys reticulata* Stitz, 1910: 141 (w.) CAMEROUN. Junior synonym of *lujae*: Brown, 1952e: 78.

*rothkirchi*. *Strumigenys rothkirchi* Wasmann, 1918b: 142, pl. 2, figs. 9, 10 (w.) CAMEROUN. Combination in *S. (Cephaloxys)*: Wheeler, W.M. 1922a: 920. Junior synonym of *alluaudi*: Brown, 1952e: 75.

***serrula***. *Strumigenys lujae* var. *serrula* Santschi, 1910c: 390 (w.) CONGO. Combination in *S. (Trichoscapa)*: Santschi, 1913b: 258 (in key); in *S. (Cephaloxys)*: Wheeler, W.M. 1922a: 920; in *Serrastruma*: Brown, 1952e: 81. Raised to species: Santschi, 1911c: 361. Senior synonym of *uelensis*: Brown, 1952e: 81. See also: Bolton, 1983: 349.

***simoni***. *Strumigenys simoni* Emery, 1895h: 42, pl. 2, fig. 21 (w.) SOUTH AFRICA. Combination in *S. (Trichoscapa)*: Santschi, 1913b: 258 (in key); in *S. (Cephaloxys)*: Wheeler, W.M. 1922a: 920; in *Smithistruma (Serrastruma)*: Brown, 1948e: 107; in *Serrastruma*: Brown, 1952e: 82. Senior synonym of *biconvexa, boerorum, cliens, cognata, escherichi, fusciventris* and *limbata*, and material of the unavailable name *obscuriventris* referred here: Brown, 1952e: 82; of *raymondi*: Bolton, 1983: 350.

***sulumana***. *Serrastruma sulumana* Bolton, 1983: 352 (w.) CAMEROUN.

*uelensis*. *Strumigenys (Cephaloxys) uelensis* Santschi, 1923e: 289, fig. 4 (w.) ZAIRE. Junior synonym of *serrula*: Brown, 1952e: 81.

*SERVIFORMICA*: see under ***FORMICA***.

*SHUCKARDIA*: see under ***DORYLUS***.

*****SICILOMYRMEX*** (Sicilian Amber)

****corniger***. **Gesomyrmex corniger* Emery, 1891a: 155, pl. 3, figs. 33–35 (w.) SICILIAN AMBER (Miocene). Combination in **Sicelomyrmex*: Wheeler, W.M. 1915h: 111. Spelling emended to **Sicilomyrmex*: Brown & Carpenter, 1979: 423.

*SIFOLINIA*: see under ***MYRMICA***.

*SIMA*: see under ***TETRAPONERA***.

***SIMOPELTA*** (Neotropical)

***bicolor***. *Simopelta bicolor* Borgmeier, 1950a: 377, figs. 13–20 (w.) BRAZIL. Combination in *Belonopelta*: Baroni Urbani, 1975b: 299; in *Simopelta*: **revived combination**. See also: Gotwald & Brown, 1967: 273.

***curvata***. *Belonopelta curvata* Mayr, 1887: 532 (w.) BRAZIL. Combination in *B. (Simopelta)*: Mann, 1922: 10; in *Simopelta*: Wheeler, W.M. 1935d: 13; in *Belonopelta*: Baroni Urbani, 1975b: 300; in *Simopelta*: **revived combination**. See also: Gotwald & Brown, 1967: 265.

***jeckylli***. *Belonopelta jeckylli* Mann, 1916: 415, pl. 2, figs. 12, 13 (w.) BRAZIL. Combination in *B. (Simopelta)*: Mann, 1922: 10; in *Simopelta*: Wheeler, W.M. 1935d: 14; in *Belonopelta*: Baroni Urbani, 1975b: 300; in *Simopelta*: **revived combination**. See also: Gotwald & Brown, 1967; 267.

***laticeps***. *Simopelta laticeps* Gotwald & Brown, 1967: 273, figs. 15, 16 (w.) PERU. Combination in *Belonopelta*: Baroni Urbani, 1975b: 299; in *Simopelta*: **revived combination**.

***manni***. *Simopelta manni* Wheeler, W.M. 1935d: 17, fig. 4 (w.) ECUADOR. Combination in *Belonopelta*: Baroni Urbani, 1975b: 299; in *Simopelta*: **revived combination**. See also: Gotwald & Brown, 1967: 273.

***minima***. *Belonopelta minima* Brandão, 1989: 136, figs. 1–4 (w.) BRAZIL. Combination in *Simopelta*: **new combination**.

***oculata***. *Simopelta oculata* Gotwald & Brown, 1967: 267, figs. 1–14 (w.q.) COSTA RICA. Combination in *Belonopelta*: Baroni Urbani, 1975b: 300; in *Simopelta*: **revived combination**.

***paeminosa***. *Simopelta paeminosa* Snelling, 1971b: 17, fig. 1 (w.) COSTA RICA. Combination in *Belonopelta*: Baroni Urbani, 1975b: 300; in *Simopelta*: **revived combination**.

***pergandei***. *Belonopelta pergandei* Forel, 1909a: 242 (w.) GUATEMALA. Borgmeier, 1950a: 374 (q.l.); Wheeler, G.C. & Wheeler, J. 1957b: 192 (l.). Combination in *B. (Simopelta)*: Mann, 1922: 10; in *Simopelta*: Wheeler, W.M. 1935d: 11; in *Belonopelta*: Baroni Urbani, 1975b: 300; in *Simopelta*: **revived combination**. See also: Gotwald & Brown, 1967: 265.

***williamsi***. *Simopelta williamsi* Wheeler, W.M. 1935d: 14, fig. 3 (w.) ECUADOR. Combination in *Belonopelta*: Baroni Urbani, 1975b: 300; in *Simopelta*: **revived combination**. See also: Gotwald & Brown, 1967: 267.

***SIMOPONE*** (Old World tropics)

***annettae***. *Simopone annettae* Kutter, 1976b: 273, figs. 1–8 (q.) CAMEROUN.

***bakeri***. *Simopone bakeri* Menozzi, 1926e: 92 (w.) SINGAPORE. See also: Taylor, 1966b: 288.

***chapmani***. *Simopone chapmani* Taylor, 1966b: 287 (w.) PHILIPPINES.

***conciliatrix***. *Simopone conciliatrix* Brown, 1975: 79, figs. 11–13 (w.q.) GHANA.

***conradti***. *Simopone conradti* Emery, 1899e: 475 (w.) CAMEROUN.

***emeryi***. *Simopone (Cerapachys) emeryi* Forel, 1892l: 247 (w.) MADAGASCAR.

***fulvinodis***. *Simopone fulvinodis* Santschi, 1923e: 262, fig. 1 (w.) ZAIRE.

***grandidieri***. *Simopone grandidieri* Forel, 1891b: 141, pl. 4, fig. 8 (w.) MADAGASCAR.

***grandis***. *Simopone grandis* Santschi, 1923d: 259 (w.) ZAIRE. See also: Brown, 1975: 79.

***gressitti***. *Simopone gressitti* Taylor, 1965a: 3, figs. 1, 2 (w.) NEW GUINEA.

***laevissima***. *Simopone laevissima* Arnold, 1954: 291, fig. 1 (w.) UGANDA.

***marleyi***. *Simopone marleyi* Arnold, 1915: 20 (w.) SOUTH AFRICA.

***matthiasi***. *Simopone matthiasi* Kutter, 1977b: 173, figs. 1–6 (q.) CAMEROUN.

*mayri*. *Cerapachys mayri* Emery, 1899f: 264 (m.) MADAGASCAR. Combination in *Simopone*: Emery, 1911d: 16. [Junior primary homonym of *Cerapachys mayri* Forel, 1892l: 244.] Replacement name: *satagia* **new name**.

***oculata***. *Simopone oculata* Radchenko, 1993b: 45, figs. 4–6 (w.) VIETNAM.

***satagia***. *Simopone satagia* **new name**. Replacement name for *mayri* Emery, 1899f: 264. [Junior primary homonym of *mayri* Forel, 1892l: 244.]

***schoutedeni***. *Simopone schoutedeni* Santschi, 1923e: 260, fig. 1 (w.) ZAIRE. Senior synonym of *wilburi*: Brown, 1975: 36.

*wilburi*. *Simopone wilburi* Weber, 1949a: 7, figs. 6, 7 (w.) ZAIRE. Junior synonym of *schoutedeni*: Brown, 1975: 36.

***SMITHISTRUMA*** (World tropics, subtropics and temperates)

***abdita***. *Strumigenys (Cephaloxys) abdita* Wesson L.G. & Wesson, R.G. 1939: 106, pl. 3, fig. 6 (w.) U.S.A. Brown, 1953g: 89 (q.); Brown, 1964a: 189 (m.). Combination in *Smithistruma*: Brown, 1953g: 89.

***alberti***. *Strumigenys alberti* Forel, 1893g: 380 (w.q.) ANTILLES IS. Brown, 1953g: 95 (q.); Wheeler, G.C. & Wheeler, J. 1960b: 27 (l.). Combination in *S. (Cephaloxys)*: Emery, 1922f: 325; in *Smithistruma*: Brown, 1953g: 93. Senior synonym of *guianensis* and *intermedia*: Brown, 1953g: 94.

*alegrensis*. *Smithistruma (Smithistruma) alegrensis* Brown, 1953g: 103, pl. 2, fig. 22 (w.) BRAZIL. Junior synonym of *emiliae*: Brown, 1964a: 193.

***anarta***. *Smithistruma anarta* Bolton, 1983: 314, fig. 18 (w.) SOUTH AFRICA.

***angulata***. *Strumigenys (Cephaloxys) angulata* Smith, M.R. 1931c: 697, pl. 1, fig. 3 (w.) U.S.A. Brown,

1953g: 54 (q.). Combination in *Smithistruma* (*Wessonistruma*): Brown, 1948e: 106.

***arahana.*** *Smithistruma arahana* Bolton, 1983: 300 (w.) CAMEROUN.

***arizonica.*** *Smithistruma arizonica* Ward, 1988: 121 (w.m.) U.S.A.

***assamensis.*** *Strumigenys assamensis* De Andrade, in Baroni Urbani & De Andrade, 1994: 61, figs. 35, 36 (w.) INDIA. Combination in *Smithistruma:* **new combination**.

***baudueri.*** *Epitritus baudueri* Emery, 1875a: 76 (w.) FRANCE. [Also described as new by Emery, 1875b: 474.] André, 1883b: 403 (q.); Medina, 1892: 70 (m.); Emery, 1916b: 204 (w.m.). Combination in *Strumigenys:* Emery, 1887b: 468 (footnote); in *S.* (*Trichoscapa*): Santschi, 1913b: 258 (in key); in *Smithistruma:* Brown, 1948e: 105. Senior synonym of *normandi:* Emery, 1922f: 323. See also: Menozzi, 1928a: 128; Brown, 1953g: 131.

***behasyla.*** *Smithistruma behasyla* Bolton, 1983: 286, fig. 4 (w.) CAMEROUN.

***bimarginata.*** *Strumigenys* (*Cephaloxys*) *bimarginata* Wesson L.G. & Wesson, R.G. 1939: 95, pl. 3, fig. 2 (w.q.) U.S.A. Combination in *Smithistruma:* Brown, 1953g: 62.

*brevisetosa. Strumigenys* (*Cephaloxys*) *clypeata* var. *brevisetosa* Smith, M.R. 1935: 215 (w.) U.S.A. Brown, 1953g: 60 (q.). Combination in *Smithistruma:* Brown, 1953g: 59. Raised to species: Wesson L.G. & Wesson, R.G. 1939: 108. Junior synonym of *pilinasis:* Brown, 1964a: 197. See also: Wilson, 1954: 485.

***bunki.*** *Smithistruma* (*Smithistruma*) *bunki* Brown, 1950b: 41, pl. 3, figs. 2, 3 (w.) U.S.A. Brown, 1953: 83 (q.).

***californica.*** *Smithistruma* (*Smithistruma*) *californica* Brown, 1950b: 40 (q.) U.S.A. Ward, 1988: 120 (w.). See also: Brown, 1953g: 85.

***capitata.*** *Cephaloxys capitata* Smith, F. 1865: 77, pl. 4, fig. 5 (q.) NEW GUINEA. Emery, 1887b: 468 (w.; not m., see Brown, 1953g: 118.) Combination in *Strumigenys* (*Cephaloxys*): Emery, 1922f: 324; in *Smithistruma:* Brown, 1948e: 105. See also: Donisthorpe, 1948h: 80; Brown, 1953g: 117.

***carolinensis.*** *Smithistruma carolinensis* Brown, 1964a: 185, pl. 16, fig. 2 (w.) U.S.A.

***castanea.*** *Smithistruma* (*Smithistruma*) *castanea* Brown, 1953g: 107 (w.) COLOMBIA.

***cavinasis.*** *Smithistruma* (*Smithistruma*) *cavinasis* Brown, 1950b: 42 (w.) ZAIRE. See also: Brown, 1953g: 129; Bolton, 1983: 287.

***chiricahua.*** *Smithistruma chiricahua* Ward, 1988: 119, fig. 8 (w.) U.S.A.

***chyatha.*** *Smithistruma chyatha* Bolton, 1983: 288, fig. 5 (w.) GHANA.

***cincinnata.*** *Smithistruma cincinnata* Kempf, 1975a: 419, figs. 8–10 (w.) BRAZIL.

***cloydi.*** *Smithistruma cloydi* Pfitzer, 1951: 198, figs. (w.) U.S.A.

***clypeata.*** *Strumigenys clypeata* Roger, 1863a: 213 (w.) U.S.A. [Misspelled as *clipeata* in original description.] Emery, 1895c: 328 (q.m.). Combination in *S.* (*Cephaloxys*): Emery, 1922f: 325; in *Smithistruma:* Brown, 1953g: 58. See also: Wilson, 1954: 486; Smith, D.R. 1979: 1406.

***conspersa.*** *Strumigenys conspersa* Emery, 1906c: 169, fig. 29 (w.) ARGENTINA. Santschi, 1929d: 302 (q.). Combination in *S.* (*Cephaloxys*): Emery, 1922f: 325; in *Smithistruma:* Brown, 1953g: 100.

***creightoni.*** *Strumigenys* (*Cephaloxys*) *creightoni* Smith, M.R. 1931c: 705, pl. 4, fig. 16 (w.) U.S.A. Brown, 1953g: 80 (q.); Brown, 1964a: 191 (m.). Combination in *Smithistruma:* Brown, 1953g: 80.

***datissa.*** *Smithistruma datissa* Bolton, 1983: 289, fig. 9 (w.) RWANDA.

***dendexa.*** *Smithistruma dendexa* Bolton, 1983: 290 (w.) CAMEROUN.

***depressiceps.*** *Strumigenys* (*Cephaloxys*) *depressiceps* Weber, 1934a: 47, figs. 6, 14 (w.m.) GUIANA. Combination in *Smithistruma* (*Platystruma*): Brown, 1953g: 112.

***dietrichi.*** *Strumigenys* (*Cephaloxys*) *dietrichi* Smith, M.R. 1931c: 696, pl. 2, fig. 6 (w.) U.S.A. Brown, 1953g: 67 (q.). Combination in *Smithistruma:* Brown, 1953g: 67. See also: Kennedy & Schramm, 1933: 100; Wilson, 1954: 487.

***dohertyi.*** *Strumigenys dohertyi* Emery, 1897c: 576 (w.) BURMA. Combination in *S.* (*Cephaloxys*): Emery, 1922f: 325; in *Smithistruma:* Brown, 1948e: 105. See also: Brown, 1953g: 121.

***dubia.*** *Smithistruma* (*Smithistruma*) *dubia* Brown, 1953g: 122, pl. 1, fig. 13 (w.q.) PALAU IS.

*****electrina.*** **Strumigenys electrina* De Andrade, in Baroni Urbani & De Andrade, 1994: 38, figs. 17, 18, 24, 25 (w.) DOMINICAN AMBER. Combination in *Smithistruma:* **new combination**.

***elegantula.*** *Smithistruma elegantula* Terayama & Kubota, 1989: 788, figs. 23–27 (w.q.) TAIWAN.

***emarginata.*** *Strumigenys emarginata* Mayr, 1901b: 26 (w.) SOUTH AFRICA. Combination in *S.* (*Trichoscapa*): Santschi, 1913b: 257; in *S.* (*Cephaloxys*): Wheeler, W.M. 1922a: 919; in *Smithistruma:* Brown, 1948e: 105. See also: Brown, 1953g: 126; Bolton, 1983: 291.

*emeryi. Strumigenys* (*Cephaloxys*) *emeryi* Karavaiev, 1935a: 106, fig. 25 (q.) SUMATRA. [Junior primary homonym of *Strumigenys emeryi* Mann, 1922: 37.] Replacement name: *karawajewi* Brown, 1948b: 44.

***emiliae.*** *Strumigenys emiliae* Forel, 1907a: 11 (w.) PARAGUAY. Combination in *S.* (*Cephaloxys*): Emery, 1922f: 325; in *Smithistruma:* Brown, 1953g: 104. Senior synonym of *alegrensis:* Brown, 1964a: 193.

***enkara.*** *Smithistruma enkara* Bolton, 1983: 301, fig. 15 (w.) IVORY COAST.

***epinotalis.*** *Strumigenys* (*Cephaloxys*) *studiosi* subsp. *epinotalis* Weber, 1934a: 46 (w.q.) COSTA RICA. Combination in *Smithistruma,* raised to species, and senior synonym of *skwarrae:* Brown, 1953g: 101.

***fenkara.*** *Smithistruma fenkara* Bolton, 1983: 302 (w.) ANGOLA.

***filirrhina.*** *Smithistruma* (*Smithistruma*) *filirrhina* Brown, 1950b: 37 (w.) U.S.A. See also: Brown, 1953g: 63.

***filitalpa.*** *Smithistruma* (*Smithistruma*) *filitalpa* Brown, 1950b: 39, pl. 3, fig. 4 (w.) U.S.A. See also: Brown, 1953g: 79.

***fridericimuelleri.*** *Strumigenys fridericimuelleri* Forel, 1886a: 213 (w.) BRAZIL. Combination in *S.* (*Cephaloxys*): Emery, 1922f: 325; in *Smithistruma:* Brown, 1953g: 98.

***fulda.*** *Smithistruma fulda* Bolton, 1983: 282, fig. 1 (w.) IVORY COAST.

***gatuda.*** *Smithistruma gatuda* Bolton, 1983: 292, fig. 11 (w.) RWANDA.

*guianensis. Strumigenys* (*Cephaloxys*) *alberti* subsp. *guianensis* Weber, 1934a: 50 (w.) GUIANA. Combination in *Smithistruma:* Brown, 1948e: 106. Junior synonym of *alberti:* Brown, 1953g: 94.

*habei. Smithistruma* (*Smithistruma*) *habei* Azuma, 1951: 89, figs. A, B (w.) JAPAN. Junior synonym of *incerta:* Onoyama, 1980: 194.

***hensekta.*** *Smithistruma hensekta* Bolton, 1983: 293, fig. 12 (w.) GHANA.

***hyphata.*** *Smithistruma* (*Smithistruma*) *hyphata* Brown, 1953g: 110 (w.) TRINIDAD.

***impidora.*** *Smithistruma impidora* Bolton, 1983: 294, fig. 6 (w.) IVORY COAST.

***incerta.*** *Smithistruma* (*Smithistruma*) *incerta* Brown, 1949d: 10 (w.q.) JAPAN. Senior synonym of *habei:* Onoyama, 1980: 194.

***inezae.*** *Strumigenys inezae* Forel, 1905c: 12 (w.) JAVA. Combination in *S.* (*Cephaloxys*): Emery, 1922f: 325; in *Smithistruma:* Brown, 1948e: 105. See also: Brown, 1953g: 121.

*intermedia. Strumigenys alberti* var. *intermedia* Wheeler, W.M. 1913d: 242 (w.) DOMINICA. Combination in *S.* (*Cephaloxys*): Emery, 1922f: 325. Junior synonym of *alberti:* Brown, 1953g: 94.

***inusitata.*** *Trichoscapa inusitata* Lattke, 1992b: 142, fig. 1 (w.) VENEZUELA. Combination in *Smithistruma:* **new combination**.

***jacobsoni.*** *Strumigenys* (*Cephaloxys*) *jacobsoni* Menozzi, 1939b: 180, fig. 4 (w.) SUMATRA. Brown, 1949d: 8 (q.). Combination in *Smithistruma:* Brown, 1948e: 105; in *Weberistruma:* Brown, 1949d: 8.

***japonica.*** *Strumigenys japonica* Ito, 1914: 40 (w.) JAPAN. Wheeler, W.M. 1928d: 116 (q.). Combination in *S.* (*Cephaloxys*): Emery, 1922f: 325; in *Smithistruma:* Brown, 1948e: 105; in *Weberistruma:* Brown, 1953g: 27. See also: Brown, 1949d: 10.

***karawajewi.*** *Strumigenys* (*Trichoscapa*) *karawajewi* Brown, 1948b: 44. Replacement name for *emeryi* Karavaiev, 1935a: 106. [Junior primary homonym of *emeryi* Mann, 1922: 37.] Combination in *Smithistruma:* Brown, 1948e: 105. See also: Brown, 1953g: 124.

***kempfi.*** *Smithistruma kempfi* Taylor & Brown, 1978: 35, figs. 1–6 (w.q.) BORNEO.

***kerasma.*** *Smithistruma kerasma* Bolton, 1983: 303, fig. 16 (w.) CAMEROUN.

***kyidriformis.*** *Smithistruma kyidriformis* Brown, 1964a: 188,pl. 16, figs. 4, 5 (w.) ARGENTINA.

***laevinasis.*** *Strumigenys* (*Cephaloxys*) *clypeata* var. *laevinasis* Smith, M.R. 1931c: 701, pl. 3, fig. 11 (w.) U.S.A. Brown, 1953g: 62 (q.). Combination in *Smithistruma:* Brown, 1953g: 61. Subspecies of *pilinasis:* Creighton, 1950a: 308. Raised to species: Brown, 1953g: 61.

***lamellignatha.*** *Smithistruma* (*Smithistruma*) *lamellignatha* Brown, 1953g: 119, pl. 3, fig. 33 (w.q.m.) BORNEO.

***leptothrix.*** *Strumigenys* (*Cephaloxys*) *leptothrix* Wheeler, W.M. 1929g: 55, fig. 7 (w.) TAIWAN. Terayama &

Kubota, 1989: 787 (q.). Combination in *Smithistruma (Weberistruma)*: Brown, 1948e: 107; in *Weberistruma:* Brown, 1949d: 8.

*longipilis. Strumigenys (Cephaloxys) alberti* subsp. *longipilis* Weber, 1934a: 50 (w.) CUBA. Combination in *Smithistruma:* Brown, 1948e: 106. Junior synonym of *nigrescens:* Brown, 1953g: 97.

***malaplax.*** *Smithistruma malaplax* Bolton, 1983: 304 (w.) ANGOLA.

***mandibularis.*** *Epitritus mandibularis* Szabó, 1909: 27, fig. 2 (w.) TANZANIA. Santschi, 1914e: 34 (q.). Combination in *Miccostruma:* Brown, 1948e: 123; in *Smithistruma:* Bolton, 1983: 283.

*manni. Strumigenys (Cephaloxys) manni* Wesson L.G. & Wesson, R.G. 1939: 97, pl. 3, fig. 3 (w.) U.S.A. Junior synonym of *ohioensis:* Smith, M.R. 1951a: 828; Brown, 1953g: 87.

***margaritae.*** *Strumigenys margaritae* Forel, 1893g: 378 (w.q.m.) ANTILLES IS. Combination in *S. (Cephaloxys)*: Emery, 1922f: 325; in *Smithistruma:* Brown, 1953g: 108.

***marginata.*** *Epitritus marginatus* Santschi, 1914b: 114, fig. 21 (w.q.) KENYA. Combination in *Miccostruma:* Brown, 1948e: 123; in *Smithistruma:* Bolton, 1983: 312.

*medialis. Strumigenys (Cephaloxys) medialis* Wesson L.G. & Wesson, R.G. 1939: 94, pl. 3, fig. 1 (w.q.) U.S.A. Junior synonym of *pilinasis:* Smith, M.R. 1951a: 828; Brown, 1953g: 60.

***mekaha.*** *Smithistruma mekaha* Bolton, 1983: 305 (w.) CAMEROUN.

***microthrix.*** *Smithistruma microthrix* Kempf, 1975a: 422, figs. 11, 12 (w.) COSTA RICA.

***minkara.*** *Smithistruma minkara* Bolton, 1983: 306, fig. 14 (w.q.) IVORY COAST.

***missouriensis.*** *Strumigenys (Cephaloxys) missouriensis* Smith, M.R. 1931c: 701, pl. 4, fig. 14 (w.) U.S.A. Brown, 1953g: 72 (q.). Combination in *Smithistruma:* Brown, 1953g: 72. Senior synonym of *sculpturata:* Smith, M.R. 1951a: 828; Brown, 1953g: 72. See also: Wilson, 1954: 485.

***mumfordi.*** *Strumigenys (Cephaloxys) inezi* subsp. *mumfordi* Wheeler, W.M. 1932c: 160 (w.) MARQUESAS IS. Combination in *Smithistruma:* Brown, 1948e: 105. Raised to species: Brown, 1953g: 124. See also: Brown, 1964a: 195.

*nana. Strumigenys (Cephaloxys) alberti* var. *nana* Santschi, 1930c: 80 (w.) CUBA. Junior synonym of *nigrescens:* Brown, 1953g: 97.

***nepalensis.*** *Strumigenys nepalensis* De Andrade, in Baroni Urbani & De Andrade, 1994: 57, figs. 33, 34 (w.q.) NEPAL. Combination in *Smithistruma:* **new combination**.

***nigrescens.*** *Strumigenys alberti* var. *nigrescens* Wheeler, W.M. 1911a: 28 (w.) JAMAICA. Brown, 1953g: 98 (q.); Wheeler, G.C. & Wheeler, J. 1955a: 143 (l.). Combination in *S. (Cephaloxys)*: Emery, 1922f: 325; in *Smithistruma,* raised to species, and senior synonym of *longipilis* and *nana:* Brown, 1953g: 96.

***ninda.*** *Smithistruma ninda* Bolton, 1983: 284, fig. 2 (w.) CAMEROUN.

*normandi. Strumigenys normandi* Santschi, 1910d: 71, fig. (w.) TUNISIA. Variety of *baudueri:* Santschi, 1913b: 258 (in key). Junior synonym of *baudueri:* Emery, 1922f: 323.

***nykara.*** *Smithistruma nykara* Bolton, 1983: 307 (w.) ZIMBABWE.

***ohioensis.*** *Strumigenys ohioensis* Kennedy & Schramm, 1933: 98, figs. 1, 2 (w.) U.S.A. Brown, 1953g: 87 (q.m.). Combination in *Smithistruma:* Brown, 1953g: 87. Senior synonym of *manni:* Smith, M.R. 1951a: 828; Brown, 1953g: 87.

***orchibia.*** *Smithistruma (Smithistruma) orchibia* Brown, 1953g: 105, pl. 2, fig. 24 (w.m.) VENEZUELA.

***ornata.*** *Strumigenys ornata* Mayr, 1887: 571 (footnote) (w.) U.S.A. Brown, 1953g: 66 (q.). Combination in *S. (Cephaloxys)*: Emery, 1922f: 325; in *Smithistruma:* Brown, 1953g: 65.

***oxysma.*** *Smithistruma oxysma* Bolton, 1983: 315, fig. 19 (w.q.) SOUTH AFRICA.

***pedunculata.*** *Smithistruma (Smithistruma) pedunculata* Brown, 1953g: 118, pl. 1, fig. 14 (w.q.) PHILIPPINES.

***pergandei.*** *Strumigenys pergandei* Emery, 1895c: 326 pl. 8, figs. 17, 18 (w.q.m.) U.S.A. Wheeler, G.C. & Wheeler, J. 1955a: 144 (l.). Combination in *S. (Cephaloxys)*: Emery, 1922f: 325; in *Smithistruma (Wessonistruma)*: Brown, 1948e: 106. See also: Brown, 1953g: 51.

***pilinasis.*** *Strumigenys clypeata* var. *pilinasis* Forel, 1901e: 339 (w.) U.S.A. Combination in *S. (Cephaloxys)*: Emery, 1922f: 325; in *Smithistruma:* Brown, 1953g: 60. Raised to species: Wesson L.G. & Wesson, R.G. 1939: 109. Senior synonym of *medialis:* Smith, M.R. 1951a: 828; Brown, 1953g: 60; of *brevisetosa:* Brown, 1964a: 197.

***placora.*** *Smithistruma placora* Bolton, 1983: 308 (w.) CAMEROUN.

***probatrix.*** *Smithistruma probatrix* Brown, 1964a: 186 pl. 16, fig. 3 (w.) MEXICO.

***pulchella.*** *Strumigenys pulchella* Emery, 1895c: 327, pl. 8, fig. 19 (w.) U.S.A. Brown, 1953g: 71 (q.m.). Combination in *S. (Cephaloxys)*: Emery, 1922f: 325; in *Smithistruma:* Brown, 1953g: 70.

***reflexa.*** *Strumigenys (Cephaloxys) reflexa* Wesson L.G. & Wesson, R.G. 1939: 102, pl. 3, fig. 4 (w.) U.S.A. Brown, 1953g: 74 (q.m.). Combination in *Smithistruma:* Brown, 1953g: 73.

***reliquia.*** *Smithistruma reliquia* Ward, 1988: 117, fig. 7 (w.) U.S.A.

***rohweri.*** *Strumigenys (Cephaloxys) rohweri* Smith, M.R. 1935: 214 (w.) U.S.A. Combination in *Smithistruma:* Brown, 1953g: 57.

***rostrata.*** *Strumigenys rostrata* Emery, 1895c: 329, pl. 8. figs. 23, 24 (w.q.m.) U.S.A. Wheeler, G.C. & Wheeler, J. 1955a: 144 (l.). Combination in *S. (Cephaloxys)*: Emery, 1922f: 325; in *Smithistruma:* Brown, 1953g: 84. See also: Wilson, 1954: 485.

***rostrataeformis.*** *Smithistruma rostrataeformis* Brown, 1949d: 12 (w.) JAPAN.

***rudinodis.*** *Strumigenys (Cephaloxys) inezae* var. *rudinodis* Stärcke, 1941: ii (w.) JAVA. Combination in *Smithistruma:* Brown, 1948e: 105; in *Weberistruma,* and raised to species: Brown, 1953g: 26.

***rusta.*** *Smithistruma rusta* Bolton, 1983: 313, fig. 17 (w.) ZIMBABWE.

***schulzi.*** *Strumigenys schulzi* Emery, 1894c: 213, pl. 1, fig. 7 (w.) BRAZIL. Wheeler, G.C. & Wheeler, J. 1955a: 144 (l.). Combination in *S. (Cephaloxys)*: Emery, 1922f: 325; in *Smithistruma:* Brown, 1953g: 108.

*sculpturata. Strumigenys (Cephaloxys) sculpturata* Smith, M.R. 1931c: 706, pl. 4, fig. 15 (w.) U.S.A. Junior synonym of *missouriensis:* Smith, M.R. 1951a: 828; Brown, 1953g: 72.

***sharra.*** *Smithistruma sharra* Bolton, 1983: 295, fig. 3 (w.q.) IVORY COAST.

****schleeorum.*** **Strumigenys schleeorum* Baroni Urbani, in Baroni Urbani & De Andrade, 1994: 35, figs. 19, 22, 23 (q.) DOMINICAN AMBER. Combination in *Smithistruma:* **new combination**.

*skwarrae. Strumigenys (Cephaloxys) skwarrae* Wheeler, W.M. 1934g: 177 (w.) MEXICO. Junior synonym of *epinotalis:* Brown, 1953g: 101.

***studiosi.*** *Strumigenys (Cephaloxys) studiosi* Weber, 1934a: 44, fig. 5 (w.) COSTA RICA. Wheeler, G.C. & Wheeler, J. 1955a: 144 (l.). Combination in *Smithistruma:* Brown, 1953g: 104.

***sublucida.*** *Smithistruma (Smithistruma) sublucida* Brown, 1953g: 99 (w.) BRAZIL.

***synkara.*** *Smithistruma synkara* Bolton, 1983: 309 (w.) GABON.

***tacta.*** *Smithistruma tacta* Bolton, 1983: 317 (w.) IVORY COAST.

***taipingensis.*** *Strumigenys inezae* var. *taipingensis* Forel, 1913k: 83 (w.) WEST MALAYSIA. Imai, Brown, *et al.* 1984: 68 (k.). Combination in *S. (Cephaloxys)*: Emery, 1922f: 325; in *Smithistruma:* Brown, 1948e: 105. Raised to species: Brown, 1953g: 122.

***talpa.*** *Strumigenys (Cephaloxys) talpa* Weber, 1934b: 63, fig. 1 (w.) U.S.A. Brown, 1953g: 77 (q.m.); Wheeler, G.C. & Wheeler, J. 1955a: 141 (l.). Combination in *Smithistruma:* Brown, 1953g: 76. Senior synonym of *venatrix:* Wesson, 1949: 21. See also: Wilson, 1954: 486.

***tanymastax.*** *Smithistruma tanymastax* Brown, 1964a: 183, pl. 16, fig. 1 (w.) BRAZIL.

***tenuipilis.*** *Strumigenys baudueri* var. *tenuipilis* Emery, 1915a: 264 (w.) ITALY. Combination in *S. (Cephaloxys)*: Emery, 1922f: 323; in *Smithistruma:* Brown, 1948e: 105. Raised to species: Brown, 1953g: 132.

***tenuissima.*** *Smithistruma (Smithistruma) tenuissima* Brown, 1953g: 133 (w.) GREECE.

***terroni.*** *Smithistruma terroni* Bolton, 1983: 299, fig. 7 (w.) CAMEROUN.

***tigrilla.*** *Miccostruma tigrilla* Brown, 1973a: 32, figs. 1, 2 (w.) IVORY COAST. Combination in *Smithistruma:* Bolton, 1983: 284.

***tolomyla.*** *Smithistruma tolomyla* Bolton, 1983: 310 (w.) CAMEROUN.

***transversa.*** *Strumigenys (Trichoscapa) transversa* Santschi, 1913b: 258 (diagnosis in key) (w.) SOUTH AFRICA. Santschi, 1914e: 31 (w.). Combination in *S. (Cephaloxys)*: Wheeler, W.M. 1922a: 920; in *Smithistruma:* Brown, 1948e: 105. See also: Brown, 1953g: 127; Bolton, 1983: 297.

***truncatidens.*** *Smithistruma (Smithistruma) truncatidens* Brown, 1950b: 43, pl. 3, fig. 1 (w.) TANZANIA. See also: Brown, 1953g: 127; Bolton, 1983: 296.

*venatrix. Strumigenys (Cephaloxys) venatrix* Wesson L.G. & Wesson, R.G. 1939: 103, pl. 3, fig. 5 (w.) U.S.A. Junior synonym of *talpa:* Wesson, 1949: 21.

***SMITHISTRUMA** (continued)*

***vodensa.*** *Smithistruma vodensa* Bolton, 1983: 317, fig. 20 (w.) CAMEROUN.

***weberi.*** *Smithistruma weberi* Brown, 1959c: 7, fig. 4 (w.) ZAIRE. See also: Bolton, 1983: 311.

***wrayi.*** *Smithistruma (Smithistruma ) wrayi* Brown, 1950b: 38, pl. 3, fig. 5 (w.) U.S.A. See also: Brown, 1953g: 78.

*SOLENOMYRMA:* see under ***VOLLENHOVIA.***

*SOLENOPS:* see under ***OLIGOMYRMEX.***

**SOLENOPSIS** (World wide)

***abdita.*** *Solenopsis abdita* Thompson, 1989: 275, figs. 1–4 (w.q.m.) U.S.A. See also: Thompson & Johnson, 1989: 698.

#***abjecta.*** *Solenopsis decipiens* subsp. *abjecta* Emery, 1906c: 128, fig. 10 (w.) ARGENTINA. Forel, 1912g: 8 (q.).

*abjectior. Solenopsis decipiens* subsp. *abjecta* var. *abjectior* Forel, 1909a: 266 (w.) ARGENTINA. Unavailable name.

*acuminata. Labauchena acuminata* Borgmeier, 1949: 208, figs. 10–14 (w.) ARGENTINA. Junior synonym of *daguerrei:* Borgmeier, 1959a: 318.

**acuta.* **Attopsis acuta* Förster, 1891: 447, pl. 13, fig. 17 (w.) GERMANY (Oligocene). Junior synonym of **blanda:* Théobald, 1937b: 204.

***africana.*** *Solenopsis fugax* st. *africana* Santschi, 1914b: 81 (w.m.) KENYA. Raised to species: Emery, 1915c: 59 (in text); Emery, 1922e: 201.

***albidula.*** *Solenopsis albidula* Emery, 1906c: 129, fig. 12 (w.q.m.) ARGENTINA. Current subspecies: nominal plus *flava, postbrunnea.*

***alecto.*** *Solenopsis (Diplorhoptrum) alecto* Santschi, 1934e: 575, figs. 5, 19, 32, 49 (w.q.) TUNISIA.

***altinodis.*** *Solenopsis altinodis* Forel, 1912g: 10 (w.) VENEZUELA. See also: Wheeler, W.M. 1921f: 154.

#***amazonensis.*** *Solenopsis corticalis* r. *amazonensis* Forel, 1904e: 680 (w.q.) PERU.

***amblychila.*** *Solenopsis aurea* subsp. *amblychila* Wheeler, W.M. 1915b: 394 (w.q.m.) U.S.A. Subspecies of *xyloni:* Creighton, 1930b: 104; of *aurea:* Creighton, 1950a: 230. Junior synonym of *aurea:* Snelling, 1963: 7; Wheeler, G.C. & Wheeler, J. 1986g: 13. Revived from synonymy, and raised to species: Trager, 1991: 168.

*amia* Forel, 1913; see under ***OLIGOMYRMEX.***

*amica. Solenopsis amica* Menozzi, 1936a: 46 (w.q.) BRAZIL. *Nomen nudum.*

***andina.*** *Solenopsis andina* Santschi, 1923c: 262, fig. 3 (w.) ARGENTINA. Combination in *S. (Euophthalma)*: Creighton, 1930b: 107.

***angulata.*** *Solenopsis angulata* Emery, 1894g: 393 (footnote) (w.q.) BRAZIL. Combination in *S. (Diplorhoptrum)*: Santschi, 1925e: 161; in *S. (Oedaleocerus)*: Creighton, 1930b: 146. [Misspelled as *angustata* by Santschi, 1912e: 526.] Current subspecies: nominal plus *carettei, dolichops, huasanensis, mendozensis, nigelloides.*

*angustata* Santschi; see ***angulata,*** above.

#***antoniensis.*** *Solenopsis laeviceps* var. *antoniensis* Forel, 1912g: 5 (w.) COLOMBIA.

*ardua. Solenopsis angulata* st.*carettei* var. *ardua* Santschi, 1929d: 298 (w.q.) BRAZIL. Unavailable name.

#***arga.*** *Solenopsis metanotalis* var. *argus* Santschi, 1923c: 260 (w.m.) ARGENTINA.

***atlantis.*** *Solenopsis (Diplorhoptrum) atlantis* Santschi, 1934e: 573, figs. 10, 16, 17, 42, 45 (w.q.) TUNISIA.

#***atticola.*** *Solenopsis westwoodi* var. *atticola* Forel, 1912g: 7 (w.) ARGENTINA. Santschi, 1929d: 298 (q.).

*aurata. Solenopsis latro* subsp. *aurata* Karavaiev, 1912a: 11, fig. 4 (w.) ALGERIA. Junior synonym of *lou:* Karavaiev, 1927b: 108.

***aurea.*** *Solenopsis geminata* var. *aurea* Wheeler, W.M. 1906d: 336 (w.q.m.) U.S.A. Crozier, 1970: 116 (k.). Raised to species: Forel, 1909a: 269. Subspecies of *xyloni:* Creighton, 1930b: 103. Revived status as species: Creighton, 1950a: 230; Snelling, 1963: 6. Senior synonym of *huachucana:* Trager, 1991: 170.

***avia.*** *Diplorhoptrum avium* Bernard, 1978a: 568, fig. 15 (w.q.m.) FRANCE. Combination in *Solenopsis:* **new combination**.

***azteca.*** *Solenopsis azteca* Forel, 1893g: 390 (w.) ANTILLES IS. Current subspecies: nominal plus *pallida.*

*bahiaensis. Solenopsis edouardi* var. *bahiaensis* Santschi, 1925d: 237 (w.) BRAZIL. Junior synonym of *medusa:* Creighton, 1930b: 68.

***balachowskyi.*** *Solenopsis balachowskyi* Bernard, 1959: 346, fig. 3 (w.q.m.) FRANCE.

***banyulensis.*** *Solenopsis emeryi* subsp. *banyulensis* Bernard, 1950a: 10 (w.q.m.) FRANCE. Raised to species: Bernard, 1967: 177; Bernard, 1978a: 556.

#***barbara.*** *Solenopsis (Diplorhoptrum) longiceps* var. *barbara* Santschi, 1934e: 572, figs. 12, 27, 30 (q.) TUNISIA.

***basalis.*** *Solenopsis basalis* Forel, 1896: 178 (w.) BRAZIL. Forel, 1913l: 219 (q.m.). Current subspecies: nominal plus *major, oculatior, raptor, urichi, vittata.*

***belisarius.*** *Solenopsis belisarius* Forel, 1907h: 278 (w.m.) AUSTRALIA.

#***binotata.*** *Solenopsis corticalis* var. *binotata* Mann, 1920: 428 (w.) CUBA.

****blanda.*** **Attopsis blanda* Förster, 1891: 446, pl. 13, fig. 16 (q.) GERMANY (Oligocene). Combination in *Solenopsis,* and senior synonym of *acuta:* Théobald, 1937b: 204.

*blumi. Solenopsis blumi* Buren, 1972: 20, fig. 5 (w.) URUGUAY. Junior synonym of *quinquecuspis:* Trager, 1991: 185.

*bogatshevi. Solenopsis latroides* subsp. *bogatshevi* Arnol'di, 1948: 211. *Nomen nudum.*

*bondari. Solenopsis bondari* Santschi, 1925d: 236 (w.) BRAZIL. Junior synonym of *virulens:* Trager, 1991: 155.

*borinquenensis. Solenopsis globularia* var. *borinquenensis* Wheeler, W.M. 1908a: 131, pl. 11, fig. 7 (w.) PUERTO RICO. Junior synonym of *steinheili:* Creighton, 1930b: 116.

***brasiliana.*** *Solenopsis brasiliana* Santschi, 1925d: 235 (w.) BRAZIL.

*brazoensis. Atta brazoensis* Buckley, 1867: 345 (s.w.) U.S.A. Junior synonym of *geminata:* Emery, 1895c: 276. Unidentifiable taxon; *incertae sedis* in *Solenopsis.*

***brevicornis.*** *Solenopsis brevicornis* Emery, 1888c: 356 (w.q.) BRAZIL. Current subspecies: nominal plus *medioclara, petropolitana.*

***brevipes.*** *Solenopsis brevipes* Emery, 1906c: 135, fig. 16 (w.) ARGENTINA.

#***bruchi.*** *Solenopsis clytemnestra* r. *bruchi* Forel, 1912g: 6 (w.m.) ARGENTINA. Santschi, 1933e: 115 (q.).

*bruchi. Solenopsis (Synsolenopsis) bruchi* Forel, 1918b: 156 (q.) ARGENTINA. Combination in *Synsolenopsis:* Kusnezov, 1953c: 347. [Junior primary homonym of *bruchi* Forel, above.] Replacement name: *bruchiella* Emery, 1922e: 199.

***bruchiella.*** *Solenopsis bruchiella* Emery, 1922e: 199. Replacement name for *bruchi* Forel, 1918b: 156. [Junior primary homonym of *bruchi* Forel, 1912g: 6.]

***bruesi.*** *Solenopsis (Solenopsis) gayi* subsp. *bruesi* Creighton, 1930b: 52, pl. 4, figs. 4, 5 (w.) PERU. Raised to species: Trager, 1991: 172.

***bucki.*** *Solenopsis bucki* Kempf, 1973a: 30, figs. 1–3 (w.) BRAZIL.

#***caffra.*** *Solenopsis punctaticeps* var. *caffra* Forel, 1894b: 99 (w.q.) SOUTH AFRICA. Senior synonym of *cyclops* and *diversipilosa:* Emery, 1915c: 65.

*calida* Smith, F. 1863; see under ***PHEIDOLOGETON.***

***canariensis.*** *Solenopsis orbuloides* r. *canariensis* Forel, 1893d: 466 (w.) CANARY IS. [Also described as new by Forel, 1894a: 161.] Raised to species: Emery, 1909a: 35.

***capensis.*** *Solenopsis capensis* Mayr, 1866b: 905 (q.) SOUTH AFRICA.

#***carettei.*** *Solenopsis angulata* r. *carettei* Forel, 1913l: 223 (w.) ARGENTINA.

#***carhuensis.*** *Solenopsis hammari* var. *carhuensis* Forel, 1912g: 8 (w.) ARGENTINA.

***carolinensis.*** *Solenopsis texana* r. *carolinensis* Forel, 1901e: 345 (w.q.m.) U.S.A. Combination in *S. (Diplorhoptrum)*: Creighton, 1950a: 236. Raised to species: Creighton, 1950a: 236. See also: Thompson & Johnson, 1989: 698.

*castanea. Solenopsis molesta* var. *castanea* Wheeler, W.M. 1908e: 430 (w.) U.S.A. Junior synonym of *truncorum:* Creighton, 1950a: 239.

***castor.*** *Solenopsis castor* Forel, 1893g: 391 (w.q.) ANTILLES IS.

#***catalinae.*** *Solenopsis texana* subsp. *catalinae* Wheeler, W.M. 1904d: 269 (w.q) U.S.A.

***celata.*** *Diplorhoptrum celatum* Dlussky & Zabelin, 1985: 222, fig. 2 (q.m.) TURKESTAN. Dlussky & Radchenko, 1994: 103 (w.). Combination in *Solenopsis:* **new combination**.

*cephalotes. Solenopsis cephalotes* Smith, F. 1859a: 149 (s.w.) INDONESIA (Aru I.). Junior synonym of *geminata:* Roger, 1863b: 32; of *rufa:* Creighton, 1950a: 231; of *geminata:* Trager, 1991: 163.

***clarki.*** *Solenopsis clarki* Crawley, 1922c: 16 (s.w.) AUSTRALIA.

*cleptis. Solenopsis cleptis* Mann, 1919: 330 (w.) SOLOMON IS. Junior synonym of *papuana:* Wilson & Taylor, 1967: 60.

#***cleptomana.*** *Solenopsis cleptomana* Santschi, 1914b: 80 (w.) KENYA. Variety of *punctaticeps:* Emery, 1915c: 65.

*clypeata. Atta clypeata* Smith, F. 1858b: 169 (q.m.) MEXICO. Junior synonym of *geminata:* Roger, 1862c: 289.

***clytemnestra.*** *Solenopsis clytemnestra* Emery, 1896g: 87 (w.) BRAZIL. Emery, 1906c: 124 (m.). Current subspecies: nominal plus *bruchi, leda, orestes, strangulata.*

*coloradensis*. *Atta coloradensis* Buckley, 1867: 346 (s.w.) U.S.A. Junior synonym of *geminata:* Mayr, 1886c: 365.

#***congolensis***. *Solenopsis ugandensis* st. *congolensis* Santschi, 1935a: 263 (w.) ZAIRE.

***conjurata***. *Solenopsis conjurata* Wheeler, W.M. 1925d: 178, fig. 4 (not fig. 8 as stated) (w.) PANAMA.

***cooperi***. *Solenopsis cooperi* Donisthorpe, 1947e: 110 (q.m.) LIBYA.

***corticalis***. *Solenopsis corticalis* Forel, 1881: 13 (w.q.) ANTILLES IS. Current subspecies: nominal plus *amazonensis, binotata, margotae, virgula*. See also: Thompson, 1989: 281.

***crivellarii***. *Solenopsis crivellarii* Menozzi, 1936d: 284, fig. 10 (w.) GREECE.

*cubaensis*. *Solenopsis globularia* var. *cubaensis* Wheeler, W.M. 1913b: 493 (w.) CUBA. Junior synonym of *steinheili:* Creighton, 1930b: 116.

*curta*. *Solenopsis globularia* r. *lucayensis* var. *curta* Forel, 1912g: 4 (w.) COLOMBIA. Unavailable name; material referred to *globularia* by Creighton, 1930b: 110.

#***cuspisior***. *Solenopsis joergenseni* var. *cuspisior* Santschi, 1923c: 254 (w.) ARGENTINA.

*cyclops*. *Solenopsis punctaticeps* st. *cyclops* Santschi, 1914e: 19, fig. 1 (s.w.) SOUTH AFRICA. Junior synonym of *caffra:* Emery, 1915c: 65.

#***cypridis***. *Solenopsis* (*Diplorhoptrum*) *fugax* var. *cypridis* Santschi, 1934e: 584, fig. 47 (w.q.m.) CYPRUS.

***daguerrei***. *Labauchena daguerrei* Santschi, 1930e: 81, fig. 12 (q.m.) ARGENTINA. Combination in *Solenopsis:* Ettershank, 1966: 140. Senior synonym of *acuminata:* Borgmeier, 1959a: 318.

*dahlii*. *Solenopsis dahlii* Forel, 1901b: 14 (w.q.m.) BISMARCK ARCHIPELAGO. Junior synonym of *papuana:* Wilson & Taylor, 1967: 60.

***dalli***. *Bisolenopsis dalli* Kusnezov, 1969: 36 (diagnosis in key) (q.) ARGENTINA. Combination in *Solenopsis:* Brandão, 1991: 378.

*debilior*. *Solenopsis* (*Diplorhoptrum*) *fugax* var. *debilior* Santschi, 1934e: 583 (w.q.m.) CORSICA. Junior synonym of *fugax:* Casevitz-Weulersse, 1990a: 138.

*debilis*. *Solenopsis debilis* Mayr, 1886d: 461 (w.q.m.) U.S.A. Junior synonym of *molesta:* Emery, 1895c: 277; Emery, 1896g: 85.

***decipiens***. *Solenopsis decipiens* Emery, 1906c: 126, fig. 8 (w.q.m.) ARGENTINA. Current subspecies: nominal plus *abjecta, scelesta*.

#***delfinoi***. *Solenopsis tenuis* r. *delfinoi* Forel, 1913l: 222 (w.) ARGENTINA. Forel, 1914d: 278 (q.).

***delta***. *Diplorhoptrum delta* Bernard, 1978a: 565, figs. 8, 13 (w.m.) FRANCE. Combination in *Solenopsis:* **new combination**.

#***desecheoensis***. *Solenopsis globularia* var. *desecheoensis* Mann, 1920: 428 (w.) PUERTO RICO.

***deserticola***. *Solenopsis deserticola* Ruzsky, 1905b: 515 (q.m.) TURKESTAN. Senior synonym of *oblongior:* Dlussky & Zabelin, 1985: 222.

*diabola*. *Solenopsis geminata* var. *diabola* Wheeler, W.M. 1908e: 424 (w.q.m.) U.S.A. Junior synonym of *rufa:* Creighton, 1930b: 66.

*diversipilosa*. *Solenopsis punctaticeps* var. *diversipilosa* Mayr, 1901b: 15 (w.) SOUTH AFRICA. Junior synonym of *caffra:* Emery, 1915c: 65.

#***dolichops***. *Solenopsis angulata* var. *dolichops* Emery, 1906c: 123 (w.q.) ARGENTINA.

*drewseni*. *Diplorhoptrum drewseni* Mayr, 1861: 73 (diagnosis in key) (w.) ITALY. Junior synonym of *geminata:* Mayr, 1870b: 996; Mayr, 1886d: 460.

***duboscqui***. *Solenopsis duboscqui* Bernard, 1950a: 10, figs. 2, 9 (w.m.) FRANCE.

***dysderces***. *Solenopsis dysderces* Snelling, 1975: 2, figs. 1, 2 (w.) CHILE.

#***edentula***. *Solenopsis* (*Diplorhoptrum*) *joergenseni* var. *edentula* Santschi, 1933e: 115 (w.) ARGENTINA.

*eduardi*. *Solenopsis eduardi* Forel, 1912g: 12 (w.) COLOMBIA. Subspecies of *geminata*, and senior synonym of *perversa:* Creighton, 1930b: 67. Junior synonym of *geminata:* Trager, 1991: 163.

***egregia***. *Synsolenopsis egregia* Kusnezov, 1953c: 342, figs. 1–3 (s.w.q.) ARGENTINA. Combination in *Solenopsis:* Ettershank, 1966: 140.

***electra***. *Solenopsis pylades* r. *electra* Forel, 1914d: 275 (w.) ARGENTINA. Creighton, 1930b: 93 (q.). Subspecies/stirps of *geminata:* Wheeler, W.M. 1915b: 397; of *saevissima:* Santschi, 1923c: 266; Santschi, 1929d: 299; Creighton, 1930b: 92. Junior synonym of *saevissima:* Wilson, 1952: 55. Revived from synonymy, and raised to species: Trager, 1991: 192.

*elmenteitae* Patrizi, 1948; see under ***OLIGOMYRMEX***.

***emeryi***. *Solenopsis* (*Diplorhoptrum*) *emeryi* Santschi, 1934e: 579, figs. 9, 21, 25, 44 (q.m.) TUNISIA.

#***emesa***. *Solenopsis fugax* subsp. *emesus* Thomé, H. & Thomé, G. 1980b: 135, fig. 5 (w.) SYRIA.

#***emiliae***. *Solenopsis metanotalis* var. *emiliae* Santschi, 1912e: 526 (w.) URUGUAY.

#***erythraea***. *Solenopsis punctaticeps* subsp. *erythraea* Emery, 1915c: 64, fig. 5 (w.) ETHIOPIA. [Also described as new by Emery, 1915g: 12.]

*exigua*. *Myrmica* (*Tetramorium*) *exigua* Buckley, 1867: 342 (w.q.) U.S.A. Junior synonym of *molesta:* Emery, 1895c: 277. See also: Wheeler, W.M. 1902f: 28.

*exigua*. *Solenopsis exigua* Forel, 1893g: 395 (w.) ANTILLES IS. Junior secondary homonym of *exigua* Buckley, above.] Replacement name: *pygmaea:* Forel, 1901e: 345.

***eximia***. *Synsolenopsis eximia* Kusnezov, 1953c: 346, fig. 4 (w.) ARGENTINA. Combination in *Solenopsis:* Ettershank, 1966: 140.

**extensa*. **Attopsis extensa* Förster, 1891: 438, pl. 13, fig. 9 (m.) GERMANY (Oligocene). Junior synonym of **superba:* Théobald, 1937b: 203.

***fairchildi***. *Solenopsis latro* subsp. *fairchildi* Wheeler, W.M. 1926: 3 (w.q.m.) SPAIN. Raised to species: Collingwood & Yarrow, 1969: 67.

*fazi*. *Solenopsis gayi* var. *fazi* Santschi, 1923c: 261 (w.) CHILE. Creighton, 1930b: 51 (q.). Junior synonym of *gayi:* Brown, 1950e: 248.

#***flava***. *Solenopsis albidula* var. *flava* Santschi, 1929d: 298 (w.q.) ARGENTINA.

#***flaveolens***. *Solenopsis schmalzi* st. *flaveolens* Forel, 1901d: 298 (w.m.) BRAZIL. Santschi, 1923c: 255 (q.).

*flavidula*. *Myrmica flavidula* Nylander, 1849: 43 (w.) RUSSIA. Junior synonym of *fugax:* Mayr, 1855: 450; Smith, F. 1855b: 121.

#***foederata***. *Solenopsis stricta* st. *foederata* Santschi, 1923c: 255 (w.q.) BRAZIL.

****foersteri***. **Solenopsis foersteri* Théobald, 1937b: 203, pl. 14, fig. 5 (m.) FRANCE (Oligocene).

***franki***. *Solenopsis franki* Forel, 1908c: 364 (w.) BRAZIL. Forel, 1909a: 259 (q.). Current subspecies: nominal plus *idae*.

***froggatti***. *Solenopsis froggatti* Forel, 1913g: 187 (w.) TASMANIA. Viehmeyer, 1925a: 28 (q.).

***fugax***. *Formica fugax* Latreille, 1798: 46 (w.q.m.) FRANCE. Hauschteck, 1961: 221 (k.); Santschi, 1910e: 649 (gynandromorph). Combination in *Diplorhoptrum:* Mayr, 1855: 450; Baroni Urbani, 1968a: 63; Kutter, 1977c: 102; Bernard, 1978a: 574; Arnol'di & Dlussky, 1978: 539; Collingwood, 1979: 64; Atanassov & Dlussky, 1992: 168; in *Solenopsis:* Mayr, 1862: 751; Emery, 1909a: 30; Emery, 1922e: 201; Bernard, 1967: 182; Bolton, 1987: 285; current combination. Senior synonym of *flavidula:* Mayr, 1855: 450; Smith, F. 1855b: 121; of *orientalis:* Karavaiev, 1910b: 49; Karavaiev, 1934: 122; Dlussky & Zabelin, 1985: 221; of *kasalinensis* and *pontica:* Dlussky & Zabelin, 1985: 221; of *debilior:* Casevitz-Weulersse, 1990a: 138; of *latroides* and *scythica:* Dlussky & Radchenko, 1994: 109; Arakelian, 1994: 50. Current subspecies: nominal plus *cypridis, emesa, furtiva, karaki, tisiphone*. See also: Donisthorpe, 1915d: 102; Boven, 1977: 85; Bolton, 1987: 285.

#***fur***. *Solenopsis punctaticeps* st. *fur* Santschi, 1926b: 230 (w.q.m.) ZIMBABWE.

#***furtiva***. *Solenopsis* (*Diplorhoptrum*) *fugax* var. *furtiva* Santschi, 1934e: 584 (w.q.m.) FRANCE.

***fusciventris***. *Solenopsis fusciventris* Clark, 1934c: 62, pl. 4, fig. 21 (w.) AUSTRALIA.

#***gaetula***. *Solenopsis* (*Diplorhoptrum*) *lusitanica* st. *gaetula* Santschi, 1936c: 201, fig. 1 (w.) MOROCCO.

*galapageia*. *Solenopsis geminata* var. *galapageia* Wheeler, W.M. 1919c: 272 (w.q.) GALAPAGOS IS. Subspecies of *geminata:* Creighton, 1930b: 65. Junior synonym of *geminata:* Trager, 1991: 163.

***gallardoi***. *Solenopsis* (*Diplorhoptrum*) *gallardoi* Santschi, 1925e: 160, fig. 5 (w.) ARGENTINA.

***gallica***. *Solenopsis* (*Diplorhoptrum*) *gallica* Santschi, 1934e: 581 (q.) FRANCE.

***gayi***. *Myrmica gayi* Spinola, 1851a: 242 (w.) CHILE. Combination in *Solenopsis:* Mayr, 1886c: 365. Junior synonym of *geminata:* Dalla Torre, 1893: 76. Revived from synonymy as subspecies of *geminata:* Emery, 1895i: 11; Emery, 1906c: 121. Revived status as species: Forel, 1909a: 269; Santschi, 1923c: 261; Creighton, 1930b: 48. Senior synonym of *fazi:* Brown, 1950e: 248. See also: Trager, 1991: 170.

***geminata***. *Atta geminata* Fabricius, 1804: 423 (q.) CENTRAL AMERICA. Roger, 1862c: 289 (w.m.); Mayr, 1867a: 110 (w.q.m.); Wheeler, W.M. 1900b: 21 (l.); Wheeler, G.C. & Wheeler, J. 1955c: 132 (l.); Crozier, 1970: 116 (k.). Combination in *Solenopsis:* Mayr, 1863: 453. Senior synonym of *clypeata* and *mandibularis:* Roger, 1862c: 289; of *paleata* and *cephalotes:* Roger, 1863b: 32; Trager, 1991: 163; of *glaber, laboriosus* and *polita:* Mayr, 1863: 453; Mayr, 1886c: 362; Trager, 1991: 163; of *drewseni:* Mayr, 1870b: 996 (footnote); Mayr, 1886d: 460; of *coloradensis:* Mayr, 1886c: 365; of *lincecumii* and *saxicola:* Emery, 1895c: 276; of *innota:* Wheeler, W.M. 1922a: 877; of *nigra:* Creighton, 1930b: 59; of *mellea* and *laevissima:* Donisthorpe, 1932c: 455, 463 respectively; of *rufa* (and its junior synonym *diabola*): Ettershank, 1966: 141; of *eduardi* (and its junior synonym *perversa*), *medusa* (and its junior synonym *bahiaensis*) and *galapageia:* Trager, 1991: 163. See also: Smith, D.R. 1979: 1386. Current subspecies: nominal plus *micans*.

#***gensterblumi.*** *Solenopsis picta* r. *gensterblumi* Forel, 1901d: 298 (w.) BRAZIL. [Misspelled as *gerstenblumi* by Emery, 1922e: 200.] Subspecies of *nigella,* and senior synonym of *prevalens:* Creighton, 1930b: 132.

***georgica.*** *Solenopsis georgica* Menozzi, 1942: 169 (w.) FERNANDO PO I. See also: Menozzi, 1944: 452.

***germaini.*** *Solenopsis germaini* Emery, 1895i: 12, figs. (w.) CHILE. Senior synonym of *schedingi:* Snelling & Hunt, 1976: 84.

*glaber. Myrmica glaber* Smith, F. 1862b: 34 (w.) PANAMA. Junior synonym of *geminata:* Mayr, 1886c: 362; Trager, 1991: 163.

***globularia.*** *Myrmica globularia* Smith, F. 1858b: 131 (w.q.m.) BRAZIL. Combination in *Solenopsis:* Roger, 1863b: 32; in *S.* (*Euophthalma*): Creighton, 1930b: 110. Material of the unavailable name *curta* referred here by Creighton, 1930b: 110. Current subspecies: nominal plus *desecheoensis, littoralis, lucayensis, pacifica, steinheili.*

***gnomula.*** *Solenopsis gnomula* Emery, 1915g: 13, fig. 6 (w.) ETHIOPIA.

***goeldii.*** *Solenopsis goeldii* Forel, 1912g: 9 (w.m.) BRAZIL.

*gracilior. Solenopsis moelleri* var. *gracilior* Forel, 1904d: 174 (w.) BRAZIL. Junior synonym of *saevissima:* Creighton, 1930b: 80.

***granivora.*** *Solenopsis* (*Granisolenopsis*) *granivora* Kusnezov, 1957a: 277, pl. 6, figs. 1–10 (w.q.) ARGENTINA.

***hammari.*** *Solenopsis hammari* Mayr, 1903: 400 (w.q.) BRAZIL. Current subspecies: nominal plus *carhuensis.*

***hayemi.*** *Solenopsis hayemi* Forel, 1908b: 45 (w.) COSTA RICA.

***helena.*** *Solenopsis helena* Emery, 1895i: 14, figs. (w.q.) CHILE. Current subspecies: nominal plus *hermione, ultrix.*

#***hermione.*** *Solenopsis helena* subsp. *hermione* Wheeler, W.M. 1921f: 157 (w.q.) GUIANA.

*hoffmanni. Solenopsis latastei* var. *hoffmanni* Forel, 1912g: 7 (w.q.) CHILE. Junior synonym of *latastei:* Snelling & Hunt, 1976: 87.

***hostilis.*** *Labauchena hostilis* Borgmeier, 1959a: 318, figs. 10, 11 (q.) BRAZIL. Combination in *Solenopsis:* Ettershank, 1966: 141.

*huachucana. Solenopsis huachucana* Wheeler, W.M. 1915b: 393 (w.q.) U.S.A. Junior synonym of *aurea:* Trager, 1991: 170.

#***huasanensis.*** *Solenopsis angulata* r. *huasanensis* Forel, 1912g: 8 (w.) ARGENTINA.

#***idae.*** *Solenopsis franki* subsp. *idae* Forel, 1908c: 365 (w.) BRAZIL.

*ignobilis. Solenopsis decipiens* r. *abjecta* var. *ignobilis* Forel, 1913l: 220 (w.) BRAZIL. Unavailable name.

***iheringi.*** *Solenopsis iheringi* Forel, 1908c: 362 (w.m.) BRAZIL. [Sometimes misspelled as *jheringi,* for example Kempf, 1972a: 237.]

***ilinei.*** *Solenopsis* (*Diplorhoptrum*) *orbula* var. *ilinei* Santschi, 1936c: 202. Replacement name for *oculata* Karavaiev, 1926c: 162. [Junior primary homonym of *oculatus* Santschi, 1925e: 161.] Dlussky & Radchenko, 1994: 105 (q.m.). Raised to species: Dlussky & Radchenko, 1994: 105.

*incrassata. Solenopsis geminata* var. *incrassata* Forel, 1908c: 362 (w.) BRAZIL. Variety of *pylades:* Forel, 1909a: 268; of *saevissima:* Gallardo, 1919b: 247. Junior synonym of *moelleri:* Creighton, 1930b: 83.

***indagatrix.*** *Solenopsis indagatrix* Wheeler, W.M. 1928c: 20 (w.) CHINA.

#***indocilis.*** *Solenopsis punctaticeps* var. *indocilis* Santschi, 1914d: 355 (w.q.) GUINEA. Subspecies of *punctaticeps:* Emery, 1915c: 65.

*inermiceps. Solenopsis inermiceps* Wheeler, W.M. & Mann, 1914: 20, fig. 7 (w.) HAITI. Junior synonym of *succinea:* Creighton, 1930b: 139.

*innocens. Solenopsis wasmanni* r. *transformis* var. *innocens* Forel, 1915c: 354 (w.) ARGENTINA. Unavailable name.

*innota. Solenopsis geminata* var. *innota* Santschi, 1915c: 257, fig. 6 (s.w.q.m.) GABON. Junior synonym of *geminata:* Wheeler, W.M. 1922a: 877.

***insculpta.*** *Solenopsis insculptus* Clark, 1938: 370, fig. 9 (w.) AUSTRALIA.

***insinuans.*** *Solenopsis insinuans* Santschi, 1933b: 103 (w.) UGANDA.

***insularis.*** *Diplorhoptrum insulare* Bernard, 1978a: 569 (w.q.m.) FRANCE. Combination in *Solenopsis:* **new combination**.

***interrupta.*** *Solenopsis saevissima* var. *interrupta* Santschi, 1916e: 379 (w.) ARGENTINA. Subspecies of *saevissima:* Creighton, 1930b: 89. Raised to species: Wilson, 1952: 51. See also: Buren, 1972: 22; Trager, 1991: 178.

*invicta. Solenopsis invicta* Buren, 1972: 9, fig. 2 (w.q.m.) BRAZIL. Wheeler, G.C. & Wheeler, J. 1977: 588 (l.). Junior synonym of *wagneri:* **new synonymy**. [Trager, 1991: 173 incorrectly gives *wagneri* as an unavailable name; the name is available and has priority over *invicta,* see note under *wagneri.*] See also: Rhoades, 1977: 1; Smith, D.R. 1979: 1386.

#***itinerans.*** *Solenopsis pylades* var. *itinerans* Forel, 1911f: 276 (w.) KENYA. Variety of *saevissima:* Wheeler, W.M. 1922a: 879.

***jacoti.*** *Solenopsis jacoti* Wheeler, W.M. 1923b: 2 (w.) CHINA. Current subspecies: nominal plus *pekingensis.*

***jalalabadica.*** *Solenopsis jalalabadica* Pisarski, 1970: 314, figs. 24–40 (w.m.) AFGHANISTAN. Junior synonym of *deserticola:* Dlussky & Zabelin, 1985: 222. Revived from synonymy: Dlussky & Radchenko, 1994: 105.

***japonica.*** *Solenopsis fugax* var. *japonica* Wheeler, W.M. 1928d: 113 (w.) JAPAN. Santschi, 1934e: 585 (q.m.). Raised to species: Collingwood, 1976: 303; Kupyanskaya, 1990: 133; Ogata, 1991b: 110.

***joergenseni.*** *Solenopsis joergenseni* Santschi, 1919f: 42 (w.) ARGENTINA. Current subspecies: nominal plus *cuspisior, edentula.*

#***juba.*** *Solenopsis punctaticeps* subsp. *juba* Weber, 1943c: 362, pl. 15, figs. 5, 8, 14 (w.q.) SUDAN.

***juliae.*** *Diplorhoptrum juliae* Arakelian, 1991: 93, figs. (w.q.m.) ARMENIA. Combination in *Solenopsis:* **new combination**.

***kabylica.*** *Solenopsis* (*Diplorhoptrum*) *kabylica* Santschi, 1934e: 572, fig. 8, 43 (q.) ALGERIA.

#***karaki.*** *Solenopsis fugax* subsp. *karaki* Thomé, H. & Thomé, G. 1980b: 134, fig. 4 (w.m.) SYRIA.

*kasalinensis. Solenopsis fugax* var. *kasalinensis* Emery, 1909a: 31 (w.) TURKESTAN. [First available use of *Solenopsis fugax* subsp. *orientalis* var. *kasalinensis* Ruzsky, 1905b: 513; unavailable name.] Junior synonym of *fugax:* Dlussky & Zabelin, 1985: 221.

#***kibaliensis.*** *Solenopsis punctaticeps* subsp. *kibaliensis* Wheeler, W.M. 1922a: 164 (w.m.) ZAIRE.

***knuti.*** *Solenopsis orbula* subsp. *knuti* Pisarski, 1967: 400, figs. 56–61 (w.q.) AFGHANISTAN. Raised to species: Dlussky & Radchenko, 1994: 105.

#***kochi.*** *Solenopsis orbula* subsp. *kochi* Finzi, 1936: 178, fig. 8 (w.q.m.) EGYPT.

***krockowi.*** *Solenopsis krockowi* Wheeler, W.M. 1908e: 428, pl. 26, figs. 28, 29 (w.q.) U.S.A.

*laboriosus. Crematogaster laboriosus* Smith, F. 1860b: 109 (q.) INDONESIA (Batjan I.). Combination in *Solenopsis:* Mayr, 1863: 453. Junior synonym of *rufa:* Creighton, 1950a: 231; of *geminata:* Trager, 1991: 163.

*laboriosa* Smith, F. 1861; see under ***PHEIDOLOGETON.***

***laeviceps.*** *Solenopsis laeviceps* Mayr, 1870a: 406 (w.) COLOMBIA. Current subspecies: nominal plus *antoniensis.*

*laevis* Smith, F. 1865; see under ***PRISTOMYRMEX.***

*laevissima. Myrmica laevissima* Smith, F. 1860b: 108 (w.) INDONESIA (Batjan I.). Junior synonym of *geminata:* Donisthorpe, 1932c: 463.

***laevithorax.*** *Solenopsis laevithorax* Bernard, 1950a: 12, figs. (w.m.) FRANCE.

***latastei.*** *Solenopsis latastei* Emery, 1895i: 13, figs. (w.q.) CHILE. Menozzi, 1935c: 320 (m.). Senior synonym of *hoffmanni:* Snelling & Hunt, 1976: 87. Current subspecies: nominal plus *masora.*

***latro.*** *Solenopsis latro* Forel, 1894d: 21 (w.) ALGERIA. Current subspecies: nominal plus *msilana, sicula.*

*latroides. Solenopsis orbula* var. *latroides* Ruzsky, 1905b: 514 (w.) CAUCASUS. Stirps of *latro:* Santschi, 1934e: 568. Raised to species: Arnol'di, 1948: 211. Junior synonym of *fugax:* Arakelian, 1994: 50.

#***leda.*** *Solenopsis clytemnestra* var. *leda* Forel, 1913l: 221 (w.) BRAZIL.

#***lehmannnitschei.*** *Solenopsis tridens* var. *lehmannnitschei* Santschi, 1916e: 378 (w.) ARGENTINA.

***leptanilloides.*** *Solenopsis* (*Diplorhoptrum*) *leptanilloides* Santschi, 1925e: 159, fig. 4 (w.) ARGENTINA.

*lincecumii. Atta lincecumii* Buckley, 1867: 344 (s.w.) U.S.A. Junior synonym of *geminata:* Emery, 1895c: 276.

#***littoralis.*** *Solenopsis* (*Euophthalma*) *globularia* subsp. *littoralis* Creighton, 1930b: 113, pl. 6, fig. 3 (w.q.m.) U.S.A. Wheeler, G.C. & Wheeler, J. 1960b: 22 (l.). Senior synonym of *mobilensis:* Smith, M.R. 1951a: 813.

***longiceps.*** *Solenopsis longiceps* Forel, 1907b: 202 (q.) TUNISIA. Emery, 1909a: 36 (w.). Current subspecies: nominal plus *barbara.*

*longiceps. Solenopsis* (*Diplorhoptrum*) *longiceps* Smith, M.R. 1943a: 210 (w.) U.S.A. [Junior primary homonym of *longiceps* Forel, above.] Replacement name: *tennesseensis:* Smith, M.R. 1951a: 814.

***loretana.*** *Solenopsis* (*Oedaleocerus*) *loretana* Santschi, 1936d: 406, figs. 6–8 (w.) ARGENTINA.

***lotophaga.*** *Solenopsis lotophaga* Santschi, 1911f: 81, fig. 15 (w.m.) TUNISIA.

***lou.*** *Solenopsis lou* Forel, 1902a: 152 (q.m.) ALGERIA. [Also described as new by Forel, 1903b: 459.] Santschi, 1911f: 81 (w.). Senior synonym of *aurata:* Karavaiev, 1927b: 108. See also: Karavaiev, 1912a: 12.

#***lucayensis.*** *Solenopsis globularia* subsp. *lucayensis* Wheeler, W.M. 1908a: 131, pl. 11, fig. 8 (w.) BAHAMAS.

***lusitanica.*** *Solenopsis latro* subsp. *lusitanica* Emery, 1915a: 259, pl. 4, fig. 13 (w.q.) PORTUGAL. Santschi, 1934e: 578 (m.). Raised to species: Bondroit, 1918: 145. Current subspecies: nominal plus *gaetula.*

***macdonaghi.*** *Solenopsis saevissima* var. *macdonaghi* Santschi, 1916e: 379 (w.q.) ARGENTINA. Junior synonym of *interrupta:* Creighton, 1930b: 89. Revived from synonymy, and raised to species: Trager, 1991: 179.

***macrops.*** *Solenopsis macrops* Santschi, 1917f: 280 (w.) ARGENTINA. Combination in *S.* (*Euophthalma*): Creighton, 1930b: 120.

*madara. Solenopsis madara* Roger, 1863a: 200 (w.) U.S.A. Unrecognisable taxon: Smith, M.R. 1951a: 814.

#***major.*** *Solenopsis basalis* var. *major* Forel, 1913l: 220 (w.q.) ARGENTINA.

*****major.*** **Solenopsis major* Théobald, 1937b: 201, pl. 4, fig. 16; pl. 14, fig. 4 (m.) FRANCE (Oligocene). [Unresolved junior primary homonym of *major* Forel, above.]

***maligna.*** *Solenopsis maligna* Santschi, 1910c: 359, fig. 5 (w.) CONGO.

***mameti.*** *Solenopsis mameti* Donisthorpe, 1946c: 777 (w.q.) MAURITIUS.

*mandibularis. Solenopsis mandibularis* Westwood, 1840b: 87, pl. 2, fig. 5 (w.) "America aequinoctiali". Junior synonym of *geminata:* Roger, 1862c: 289.

*maniosa. Solenopsis geminata* subsp. *maniosa* Wheeler, W.M. 1915b: 396 (w.q.m.) U.S.A. Variety of *xyloni:* Creighton, 1930b: 102. Junior synonym of *xyloni:* Creighton, 1950a: 232. Revived from synonymy, and raised to species: Wheeler, G.C. & Wheeler, J. 1986g: 13. Junior synonym of *xyloni:* Trager, 1991: 166.

#***margotae.*** *Solenopsis corticalis* subsp. *margotae* Forel, 1908c: 364 (w.q.) BRAZIL.

*maroccana. Diplorhoptrum maroccana* Bernard, 1950a: 10. *Nomen nudum.*

***marxi.*** *Solenopsis marxi* Forel, 1915c: 354. Replacement name for *pygmaea* Forel, 1905c: 21. [Junior primary homonym of *pygmaea* Forel, 1901e: 345.]

#***masora.*** *Solenopsis latastei* var. *masora* Forel, 1912g: 7 (w.) "Pemnes" (Chile ?).

***maxillosa.*** *Solenopsis maxillosa* Emery, 1900c: 329, pl. 8, figs. 31, 32 (w.q.) NEW GUINEA.

*****maxima.*** **Attopsis maxima* Förster, 1891: 443, pl. 13, fig. 13 (q.) GERMANY (Oligocene). Combination in *Solenopsis:* Théobald, 1937b: 200.

#***medeis.*** *Solenopsis patagonica* r. *medeis* Forel, 1912g: 10 (w.m.) BRAZIL.

#***medioclara.*** *Solenopsis brevicornis* var. *medioclara* Santschi, 1923c: 254 (w.) BRAZIL.

*medusa. Solenopsis geminata* subsp. *medusa* Mann, 1916: 447, pl. 4, fig. 31 (w.) BRAZIL. Creighton, 1930b: 69 (q.). Senior synonym of *bahiaensis:* Creighton, 1930b: 68. Junior synonym of *geminata:* Trager, 1991: 163.

***megera.*** *Solenopsis* (*Diplorhoptrum*) *megera* Santschi, 1934e: 576, figs. 6, 20, 32, 53 (w.q.) TUNISIA.

***megergates.*** *Solenopsis megergates* Trager, 1991: 181, figs. 68, 69, 72, 73 (w.q.) BRAZIL.

*mellea. Myrmica mellea* Smith, F. 1859a: 148 (w.) INDONESIA (Aru I.). Junior synonym of *geminata:* Donisthorpe, 1932c: 455.

#***mendozensis.*** *Solenopsis angulata* r. *mendozensis* Forel, 1914d: 277 (w.) ARGENTINA. [First available use of *Solenopsis angulata* r. *carettei* var. *mendozensis* Forel, 1913l: 223; unavailable name.]

***metanotalis.*** *Solenopsis metanotalis* Emery, 1896g: 86 (w.) ARGENTINA. Combination in *S.* (*Euophthalma*): Creighton, 1930b: 124. Current subspecies: nominal plus *arga, emiliae, pelotana, picturata, shiptoni, steigeri.*

***metatarsalis.*** *Lilidris metatarsalis* Kusnezov, 1957a: 274, pl. 3, figs. 1–6 (q.) ARGENTINA. Combination in *Solenopsis:* Ettershank, 1966: 142.

#***micans.*** *Solenopsis geminata* var. *micans* Stitz, 1912: 506,fig. 8 (w.) INDONESIA (Seram I.).

#***minuiscens.*** *Solenopsis tenuis* var. *minuiscens* Forel, 1912g: 8 (w.) BRAZIL.

*minuta. Myrmica minuta* Say, 1836: 294 (w.) U.S.A. Junior synonym of *molesta:* Emery, 1895c: 277.

***minutissima.*** *Solenopsis minutissima* Emery, 1906c: 133, fig. 15 (w.m.) ARGENTINA. Wheeler, W.M. 1914b: 44 (q.).

*mobilensis. Solenopsis globularia* subsp. *mobilensis* Smith, M.R. 1931a: 20 (w.) U.S.A. (attributed to Creighton). Junior synonym of *littoralis:* Smith, M.R. 1951a: 813.

*moelleri. Solenopsis moelleri* Forel, 1904d: 173 (w.q.m.) BRAZIL. Variety of *saevissima,* and senior synonym of *incrassata* and *morosa:* Creighton, 1930b: 83. Junior synonym of *saevissima:* Wilson, 1952: 55.

*moerens. Solenopsis picta* var. *moerens* Wheeler, W.M. 1915b: 393 (w.) U.S.A. Junior synonym of *picta:* Creighton, 1950a: 237.

*****moesta.*** **Attopsis moesta* Förster, 1891: 439, pl. 13, fig. 10 (m.) GERMANY (Oligocene). Combination in *Solenopsis:* Théobald, 1937b: 203.

***molesta.*** *Myrmica molesta* Say, 1836: 293 (q.) U.S.A. Wheeler, G.C. & Wheeler, J. 1955c: 134 (l.); Petralia & Vinson, 1980: 383 (l.); Crozier, 1970: 116 (k.). Combination in *Solenopsis:* Mayr, 1863: 407; in *S.* (*Diplorhoptrum*): Creighton, 1950a: 237. Senior synonym of *debilis, exigua* and *minuta:* Emery, 1895c: 277; Emery, 1896g: 85. Current subspecies: nominal plus *validiuscula.*

***monticola.*** *Solenopsis monticola* Bernard, 1950a: 13, figs. 2, 4, 11 (w.q.m.) FRANCE.

*morosa. Solenopsis saevissima* var. *morosa* Santschi, 1916e: 380 (w.) BRAZIL. Santschi, 1923c: 265 (q.). Junior synonym of *moelleri:* Creighton, 1930b: 83.

***mozabensis.*** *Diplorhoptrum mozabense* Bernard, 1977: 31, fig. 2 (w.) ALGERIA.

#***msilana.*** *Solenopsis oraniensis* var. *msilana* Forel, 1894d: 22 (w.) ALGERIA. Subspecies of *latro:* Emery, 1909a: 34.

***nicaeensis.*** *Solenopsis nicaeensis* Bernard, 1950a: 11, figs. 2, 5 (w.q.m.) FRANCE.

#***nicai.*** *Solenopsis succinea* r. *nicai* Forel, 1913l: 222 (w.) BRAZIL. Santschi, 1923c: 267 (q.m.). Combination in *S.* (*Diagyne*): Santschi, 1923c: 267.

***nickersoni.*** *Solenopsis* (*Diplorhoptrum*) *nickersoni* Thompson, 1982: 486, fig. 1 (w.) U.S.A. Thompson, 1989: 281 (q.).

***nigella.*** *Solenopsis nigella* Emery, 1888c: 355 (w.) BRAZIL. Forel, 1912g: 5 (q.). Combination in *S.* (*Euophthalma*): Creighton, 1930b: 130. Current subspecies: nominal plus *gensterblumi.*

#***nigelloides.*** *Solenopsis angulata* r. *nigelloides* Forel, 1913l: 223 (w.q.m.) ARGENTINA.

*nigra. Solenopsis geminata* var. *nigra* Forel, 1908b: 45 (w.) COSTA RICA. Junior synonym of *geminata:* Creighton, 1930b: 59.

***nitens.*** *Solenopsis nitens* Bingham, 1903: 160 (w.) SRI LANKA.

***nitidum.*** *Diplorhoptrum nitidum* Dlussky & Radchenko, 1994: 105, fig. 4 (w.q.) ARMENIA.

***normandi.*** *Solenopsis* (*Diplorhoptrum*) *normandi* Santschi, 1934e: 568, figs. 14, 48 (w.m.) TUNISIA.

***novemmaculata.*** *Solenopsis novemmaculata* Wheeler, W.M. 1925a: 35 (w.) BOLIVIA.

*oblongiceps. Solenopsis saevissima* st. *oblongiceps* Santschi, 1936d: 405, fig. 5 (w.) ARGENTINA. Junior synonym of *saevissima:* Wilson, 1952: 55; of *richteri:* Trager, 1991: 187.

*oblongior. Solenopsis orbula* subsp. *oblongior* Karavaiev, 1926c: 161, fig. 1 (w.) CAUCASUS. Junior synonym of *deserticola:* Dlussky & Zabelin, 1985: 222.

***occipitalis.*** *Solenopsis occipitalis* Santschi, 1911f: 83 (w.) TUNISIA. Combination in *S.* (*Diplorhoptrum*): Santschi, 1934e: 575.

***oculata.*** *Solenopsis* (*Diplorhoptrum*) *angulata* st. *oculatus* Santschi, 1925e: 161 (w.) ARGENTINA. Combination in *S.* (*Euophthalma*), and raised to species: Creighton, 1930b: 133.

*oculata. Solenopsis orbula* var. *oculata* Karavaiev, 1926c: 162, fig. 2 (w.) CAUCASUS. [Junior primary homonym of *oculatus* Santschi, above.] Replacement name: *ilinei:* Santschi, 1936c: 202.

#***oculatior.*** *Solenopsis basalis* r. *oculatior* Forel, 1913l: 220 (w.) ARGENTINA.

***oraniensis.*** *Solenopsis oraniensis* Forel, 1894d: 22 (w.q.) ALGERIA. Combination in *S.* (*Diplorhoptrum*): Santschi, 1934e: 568. Subspecies of *latro:* Emery, 1909a: 34. Revived status as species: Santschi, 1934e: 568.

***orbula.*** *Solenopsis orbula* Emery, 1875a: 74 (w.) CORSICA. [Also described as new by Emery, 1875b: 472.] André, 1883b: 389 (q.); Tohmé, H. & Tohmé, G. 1980b: 133 (m.). Combination in *S.* (*Diplorhoptrum*): Santschi, 1934e: 574. Current subspecies: nominal plus *kochi, terniensis.* See also: Poldi, 1992: 91.

***orbuloides.*** *Solenopsis orbuloides* André, 1890: 321 (w.) SIERRA LEONE. Santschi, 1914d: 354 (q.).

#***orestes.*** *Solenopsis clytemnestra* r. *orestes* Forel, 1903c: 256 (w.m.) BRAZIL. Forel, 1912g: 6 (q.).

*orientalis. Solenopsis fugax* subsp. *orientalis* Ruzsky, 1905b: 509, figs. 111, 112 (w.q.m.) RUSSIA. Junior synonym of *fugax:* Karavaiev, 1910b: 49; Karavaiev, 1934: 122. Revived from synonymy: Tarbinsky, 1976: 80. Junior synonym of *fugax:* Dlussky & Zabelin, 1985: 221.

**overbecki.** *Solenopsis overbecki* Viehmeyer, 1916a: 135 (w.q.) SINGAPORE.

**pachycera.** *Oligomyrmex (Octella) pachycerus* Forel, 1915b: 69 (w.) AUSTRALIA. Combination in *Solenopsis:* Taylor, 1991b: 611.

**#pacifica.** *Solenopsis globularia* subsp. *pacifica* Wheeler, W.M. 1919c: 273 (w.m.) GALAPAGOS IS. Wheeler, W.M. 1924a: 108 (q.).

*paleata. Myrmica paleata* Lund, 1831a: 116 (footnote) (w.) BRAZIL. Junior synonym of *geminata:* Roger, 1863b: 32.

**#pallida.** *Solenopsis azteca* var. *pallida* Wheeler, W.M. 1908a: 131 (w.) PUERTO RICO.

**papuana.** *Solenopsis papuana* Emery, 1900c: 330 (w.q.) NEW GUINEA. Senior synonym of *cleptis, dahlii* and *vitiensis:* Wilson & Taylor, 1967: 60.

**parabiotica.** *Solenopsis parabiotica* Weber, 1943a: 90, fig. 2 (w.q.) GUIANA.

**parva.** *Solenopsis parva* Mayr, 1868b: 175 (w.) ARGENTINA.

**patagonica.** *Solenopsis patagonica* Emery, 1906c: 132, fig. 14 (w.) ARGENTINA. Gallardo, 1919b: 246 (q.m.). Senior synonym of *thoracica:* Kusnezov, 1960b: 338. Current subspecies: nominal plus *medeis.* See also: Kusnezov, 1949b: 281.

**pawaensis.** *Solenopsis pawaensis* Mann, 1919: 329 (w.) SOLOMON IS.

**#pekingensis.** *Solenopsis jacoti* subsp. *pekingensis* Wheeler, W.M. 1923b: 2 (w.) CHINA.

**#pelotana.** *Solenopsis metanotalis* var. *pelotana* Forel, 1912g: 5 (w.q.) BRAZIL. Subspecies of *metanotalis:* Creighton, 1930b: 127.

*perfida. Solenopsis saevissima* var. *perfida* Santschi, 1923c: 266 (w.) BRAZIL. Junior synonym of *saevissima:* Wilson, 1952: 55.

**pergandei.** *Solenopsis pergandei* Forel, 1901e: 343 (w.q.m.) U.S.A. Wheeler, G.C. & Wheeler, J. 1960b: 21 (l.). Combination in *S. (Diplorhoptrum)*: Smith, M.R. 1947f: 568. See also: Thompson & Johnson, 1989: 698.

*perversa. Solenopsis edouardi* var. *perversa* Santschi, 1925b: 13 (s.w.q.) BRAZIL. Junior synonym of *eduardi:* Creighton, 1930b: 67.

**#petropolitana.** *Solenopsis brevicornis* var. *petropolitana* Borgmeier, 1928a: 35, figs. 1–3 (w.) BRAZIL.

**photophila.** *Solenopsis (Synsolenopsis) photophila* Santschi, 1923c: 250, fig. 1 (s.w.) ARGENTINA. Combination in *Synsolenopsis:* Kusnezov, 1953c: 347; in *Solenopsis:* Ettershank, 1966: 142.

**picea.** *Solenopsis picea* Emery, 1896g: 89 (w.q.) COSTA RICA. Wheeler, G.C. & Wheeler, J. 1955c: 135 (l.). Race of *tenuis:* Forel, 1912g: 8. Revived status as species: Mann, 1922: 30. Current subspecies: nominal plus *reducta, subadpressa.*

*picea. Solenopsis geminata* subsp. *saevissima* var. *picea* Wasmann, 1918a: 70 (w.) BRAZIL. Unavailable name; material referred to *saevissima* by Trager, 1991: 189.

**picquarti.** *Solenopsis picquarti* Forel, 1899c: 80 (w.) COSTA RICA. Combination in *S. (Euophthalma)*: Creighton, 1930b: 135.

**picta.** *Solenopsis picta* Emery, 1895c: 278 (w.) U.S.A. Wheeler, G.C. & Wheeler, J. 1960b: 22 (l.). Senior synonym of *moerens:* Creighton, 1950a: 237. See also: Thompson & Johnson, 1989: 697.

**#picturata.** *Solenopsis (Euophthalma) metanotalis* var. *picturata* Santschi, 1931e: 276 (w.) ARGENTINA. Santschi, 1933e: 114 (q.).

**#pieli.** *Solenopsis soochowensis* st. *pieli* Santschi, 1925f: 86 (q.m.) CHINA. Santschi, 1928b: 35 (w.).

**pilosa.** *Diplorhoptrum pilosum* Bernard, 1978a: 562, fig. 1, 15 (w.q.m.) FRANCE. Combination in *Solenopsis:* **new combination**.

**pilosula.** *Solenopsis pilosula* Wheeler, W.M. 1908e: 426, pl. 26, figs. 26, 27 (w.m.) U.S.A.

**#platensis.** *Solenopsis westwoodi* var. *platensis* Emery, 1906c: 125, fig. 7 (w.q.) ARGENTINA.

*polita. Myrmica polita* Smith, F. 1862b: 34 (w.) PANAMA. [Unresolved junior primary homonym of *Myrmica polita* Smith, F. 1860b: 108 (now in *Pheidologeton*).] Junior synonym of *geminata:* Mayr, 1863: 453; Mayr, 1886c: 362.

**pollux.** *Solenopsis pollux* Forel, 1893g: 393 (w.q.m.) ANTILLES IS.

*pontica. Solenopsis (Diplorhoptrum) fugax* var. *pontica* Santschi, 1934e: 584, fig. 37 (w.q.m.) RUSSIA. Junior synonym of *fugax:* Dlussky & Zabelin, 1985: 221.

**#postbrunnea.** *Solenopsis albidula* var. *postbrunnea* Forel, 1913l: 220 (w.) BRAZIL.

*prevalens. Solenopsis nigella* st. *prevalens* Santschi, 1923c: 257 (s.w.) ARGENTINA. Santschi, 1929d: 297 (m.). Junior synonym of *gensterblumi:* Creighton, 1930b: 132.

***privata.** **Attopsis privata* Förster, 1891: 444, pl. 13, fig. 14 (q.) GERMANY (Oligocene). Combination in *Solenopsis:* Théobald, 1937b: 201.

**provincialis.** *Solenopsis provincialis* Bernard, 1950a: 12 (w.) FRANCE.

**punctaticeps.** *Solenopsis punctaticeps* Mayr, 1865: 109 (w.) SOUTH AFRICA. Forel, 1910e: 431 (q.m.). Current subspecies: nominal plus *caffra, cleptomana, erythraea, fur, indocilis, juba, kibaliensis.* See also: Arnold, 1916: 243.

**puncticeps.** *Solenopsis (Diplorhoptrum) puncticeps* MacKay & Vinson, 1989a: 177, figs. 5–8 (w.) U.S.A.

*pungens* Smith, F. 1861; see under ***PHEIDOLE.***

**pusillignis.** *Solenopsis pusillignis* Trager, 1991: 194, figs. 40, 41, 44, 45, 84 (w.) BRAZIL.

**pygmaea.** *Solenopsis pygmaea* Forel, 1901e: 345. [Originally misspelled as *pygmaca;* justified emendation to *pygmaea* by Forel, 1915c: 354.] Replacement name for *exigua* Forel, 1893g: 395. [Junior secondary homonym of *exigua* Buckley, 1867: 342.]

*pygmaea. Solenopsis pygmaea* Forel, 1905c: 21 (w.) JAVA. [Junior primary homonym of *pygmaea* Forel, above.] Replacement name: *marxi* Forel, 1915c: 354.

*pygmaea. Solenopsis pygmaea* Bernard, 1950a: 15 (w.) FRANCE. Bernard, 1978a: 571 (q.m.). [Junior primary homonym of *pygmaea* Forel, 1901e: 345.] Replacement name: *tertialis* Ettershank, 1966: 144.

*pylades. Solenopsis geminata* r. *pylades* Forel, 1904d: 172 (q.) MEXICO. Forel, 1909a: 267 (w.); Forel, 1911c: 297 (m.); Bruch, 1916: 315 (m.). Raised to species: Forel, 1909a: 267; Bruch, 1914: 223; Forel, 1916: 459. Variety of *saevissima:* Santschi, 1923c: 267. Junior synonym of *geminata:* Emery, 1922e: 198; of *saevissima:* Wheeler, W.M. 1915b: 395; Wheeler, W.M. 1916g: 143; Wheeler, W.M. 1919c: 272; Creighton, 1930b: 70; of *xyloni:* Trager, 1991: 166.

**pythia.** *Solenopsis (Solenopsis) pythia* Santschi, 1934c: 30 (q.) ARGENTINA. Trager, 1991: 182 (w.). See also: Wilson, 1952: 50.

**quinquecuspis.** *Solenopsis pylades* var. *quinquecuspis* Forel, 1913l: 224 (w.) ARGENTINA. Variety of *saevissima:* Santschi, 1916e: 379; Creighton, 1930b: 86. Junior synonym of *richteri:* Wilson, 1952: 57. Revived from synonymy, and raised to species: Buren, 1972: 17. Senior synonym of *blumi:* Trager, 1991: 185.

**#raptor.** *Solenopsis basalis* var. *raptor* Santschi, 1919f: 42 (w.) ARGENTINA.

**#reducta.** *Solenopsis picea* subsp. *reducta* Menozzi, 1927c: 276 (w.) COSTA RICA.

**reichenspergeri.** *Solenopsis reichenspergeri* Santschi, 1923c: 256 (w.) BRAZIL. Borgmeier, 1929: 211 (q.).

**richardi.** *Solenopsis richardi* Bernard, 1950a: 15 (w.) FRANCE.

**richteri.** *Solenopsis pylades* var. *richteri* Forel, 1909a: 267 (w.q.) ARGENTINA. Creighton, 1930b: 87 (m.); Wheeler, G.C. & Wheeler, J. 1977: 589 (l.). Variety/subspecies of *saevissima:* Santschi, 1916e: 379; Creighton, 1950a: 232; Wilson, 1952: 58. Raised to species: Buren, 1972: 4. Senior synonym of *tricuspis:* Creighton, 1930b: 87; of *oblongiceps:* Trager, 1991: 187. See also: Rhoades, 1977: 1.

**robusta.** *Solenopsis robusta* Bernard, 1950a: 12 (w.) FRANCE.

*robustior. Solenopsis wasmanni* st. *transformis* var. *robustior* Santschi, 1923c: 265 (w.) ARGENTINA. Unavailable name.

*rosella. Solenopsis rosella* Kennedy, 1938: 232, figs. 1–3, 5, 6, 8, 10 (w.q.m.) U.S.A. Junior synonym of *texana:* Creighton, 1950a: 238.

*rubida. Solenopsis globularia* subsp. *pacifica* var. *rubida* Wheeler, W.M. 1919c: 273 (w.) GALAPAGOS IS. Unavailable name.

*rufa. Atta rufa* Jerdon, 1851: 106 (w.q.) INDIA. Bingham, 1903: 158 (m.). Combination in *Solenopsis:* Emery, 1892b: 166. Junior synonym of *geminata:* Dalla Torre, 1893: 76. Revived from synonymy as subspecies of *geminata:* Forel, 1899a: 119; Wheeler, W.M. 1915b: 397; Creighton, 1930b: 66. Senior synonym of *diabola:* Creighton, 1930b: 66. Junior synonym of *geminata:* Ettershank, 1966: 141.

**rugiceps.** *Solenopsis rugiceps* Mayr, 1870a: 406 (w.) COLOMBIA.

**rugosa.** *Solenopsis rugosa* Bernard, 1950a: 14 (w.m.) FRANCE.

*sabeana. Myrmica (Atta) sabeana* Buckley, 1867: 343 (w.) U.S.A. Junior synonym of *geminata:* Emery, 1895c: 276. Synonymy with *xyloni* postulated but cannot be proved: Wheeler, W.M. 1915b: 395; Creighton, 1930b: 98; Smith, D.R. 1979: 1389; Trager, 1991: 167. Unidentifiable taxon; *incertae sedis* in *Solenopsis.*

**saevissima.** *Myrmica saevissima* Smith, F. 1855c: 166, pl. 13, fig. 18 (w.) BRAZIL. Combination in *Solenopsis:* Mayr, 1862: 751. Goni, Zolessi & Imai, 1983: 365 (k.). Junior synonym of *geminata:* Mayr, 1865: 109; Dalla Torre, 1893: 76. Revived from synonymy as subspecies of *geminata:* Wheeler, W.M. 1915b: 395. Revived status as species: Mann, 1916: 447;

Santschi, 1916e: 378; Santschi, 1923c: 265; Creighton, 1930b: 80. Senior synonym of *gracilior:* Creighton, 1930b: 80; of *perfida* and *moelleri* (and its junior synonyms *incrassata* and *morosa*): Wilson, 1952: 55. Material of the unavailable name *picea* referred here by Trager, 1991: 189. Current subspecies: nominal plus *itinerans.* See also: Buren, 1972: 15; Trager, 1991: 188.

***salina.*** *Solenopsis salina* Wheeler, W.M. 1908e: 427, pl. 26, figs. 24, 25 (w.) U.S.A.

***santschii.*** *Solenopsis santschii* Forel, 1905b: 174 (w.q.) TUNISIA. Baroni Urbani, 1968b: 451 (q.l.).

*saxicola. Myrmica (Monomorium) saxicola* Buckley, 1867: 341 (w.) U.S.A. Junior synonym of *geminata:* Emery, 1895c: 276.

#***scelesta.*** *Solenopsis decipiens* var. *scelesta* Forel, 1908c: 364 (w.) BRAZIL.

*schedingi. Solenopsis germaini* subsp. *schedingi* Forel, 1907e: 4 (w.q.m.) CHILE. Junior synonym of *germaini:* Snelling & Hunt, 1976: 84.

***schilleri.*** *Solenopsis schilleri* Santschi, 1923c: 263, fig. 3 (w.) ARGENTINA. Combination in *S.* (*Euophthalma*): Creighton, 1930b: 136.

***schmalzi.*** *Solenopsis schmalzi* Forel, 1901d: 297 (w.) BRAZIL. Current subspecies: nominal plus *flaveolens.*

***scipio.*** *Solenopsis scipio* Santschi, 1911f: 80, fig. 11 (w.) TUNISIA. See also: Santschi, 1934e: 575.

*scythica. Solenopsis (Diplorhoptrum) fugax* var. *scythica* Santschi, 1934e: 585 (w.) RUSSIA. Junior synonym of *fugax:* Arakelian, 1994: 50.

***sea.*** *Bisolenopsis sea* Kusnezov, 1953d: 9, pl. 1, figs. 1–3; pl. 2, fig. 8, pl. 3, fig. 13 (q.m.) ARGENTINA. Combination in *Solenopsis:* Ettershank, 1966: 143.

*semilaevis* Mayr, 1901; see under ***OLIGOMYRMEX.***

***seychellensis.*** *Solenopsis seychellensis* Forel, 1909b: 55 (w.) SEYCHELLES IS. Forel, 1912k: 164 (q.).

#***shiptoni.*** *Solenopsis shiptoni* Forel, 1914d: 276 (w.) ARGENTINA. Subspecies of *metanotalis:* Creighton, 1930b: 127.

#***sicula.*** *Solenopsis latro* subsp. *sicula* Emery, 1915a: 259, pl. 4, figs. 11, 12 (w.) SICILY.

***silvestrii.*** *Solenopsis silvestrii* Emery, 1906c: 120 (w.q.) URUGUAY. Combination in *S.* (*Euophthalma*): Creighton, 1930b: 137.

*similis* Mayr, 1862; see under ***OLIGOMYRMEX.***

***solenopsidis.*** *Paranamyrma solenopsidis* Kusnezov, 1953d: 17, pl. 3, fig. 16; pl. 4, fig. 23 (q.m.) ARGENTINA. [Also described as new by Kusnezov, 1954a: 12.] Combination in *Solenopsis:* Ettershank, 1966: 143.

***soochowensis.*** *Solenopsis soochowensis* Wheeler, W.M. 1921c: 531 (q.m.) CHINA. Current subspecies: nominal plus *pieli.*

*specularis. Solenopsis stricta* st. *foederata* var. *specularis* Santschi, 1923c: 256 (w.) BRAZIL. Unavailable name.

***spei.*** *Solenopsis spei* Forel, 1912g: 11 (w.) COLOMBIA.

#***steigeri.*** *Solenopsis shiptoni* var. *steigeri* Santschi, 1916e: 378 (w.) ARGENTINA. Subspecies of *metanotalis:* current status.

#***steinheili.*** *Solenopsis steinheili* Forel, 1881: 11 (w.) ANTILLES IS. Junior synonym of *globularia:* Wheeler, W.M. 1908a: 130; Emery, 1922e: 199. Revived from synonymy as subspecies of *globularia,* and senior synonym of *borinquenensis* and *cubaensis:* Creighton, 1930b: 116.

#***strangulata.*** *Solenopsis clytemnestra* r. *strangulata* Forel, 1913l: 221 (w.) BRAZIL.

***stricta.*** *Solenopsis stricta* Emery, 1896g: 90, pl. 1, fig. 15 (w.) BOLIVIA. Current subspecies: nominal plus *foederata.*

#***subadpressa.*** *Solenopsis picea* r. *subadpressa* Forel, 1903c: 257 (w.) BRAZIL.

***substituta.*** *Solenopsis tridens* var. *substituta* Santschi, 1925d: 236 (w.q.) BRAZIL. Raised to species: Trager, 1991: 161.

***subterranea.*** *Solenopsis (Diplorhoptrum) subterranea* MacKay & Vinson, 1989a: 175, figs. 1–4 (w.) U.S.A.

***subtilis.*** *Solenopsis subtilis* Emery, 1896g: 90, pl. 1, fig. 16 (w.) PARAGUAY.

***succinea.*** *Solenopsis succinea* Emery, 1890b: 52 (w.q.m.) COSTA RICA. [Also described as new by Emery, 1894k: 56.] Combination in *S.* (*Diagyne*): Santschi, 1923c: 267. Senior synonym of *inermiceps:* Creighton, 1930b: 139. Current subspecies: nominal plus *nicai.*

***sulfurea.*** *Diplorhoptrum sulfureum* Roger, 1862c: 296 (w.) SOUTH AMERICA. Combination in *Solenopsis:* Mayr, 1863: 407.

*****superba.*** **Attopsis superba* Förster, 1891: 445, pl. 13, fig. 15 (m.) GERMANY (Oligocene). Combination in *Solenopsis,* and senior synonym of **extensa:* Théobald, 1937b: 203.

***targuia.*** *Solenopsis targuia* Bernard, 1953a: 171, fig. 5 (q.) ALGERIA.

***tennesseensis.*** *Solenopsis (Diplorhoptrum) tennesseensis* Smith, M.R. 1951a: 814. Replacement name for *longiceps* Smith, M.R. 1943a: 210. [Junior primary homonym of *longiceps* Forel, 1907b: 202.] See also: Thompson & Johnson, 1989: 698.

***tenuis.*** *Solenopsis tenuis* Mayr, 1878: 874 (w.q.) BRAZIL. Forel, 1913l: 222 (m.); Wheeler, G.C. & Wheeler, J. 1955c: 135 (l.). Current subspecies: nominal plus *delfinoi, minuiscens.*

#***terniensis.*** *Solenopsis orbula* r. *terniensis* Forel, 1905b: 175 (w.q.) ALGERIA. Emery, 1909a: 33 (m.).

***terricola.*** *Solenopsis terricola* Menozzi, 1931b: 267, fig. 5 (w.q.) COSTA RICA.

***tertialis.*** *Solenopsis tertialis* Ettershank, 1966: 144. Replacement name for *pygmaea* Bernard, 1950a: 15. [Junior primary homony of *pygmaea* Forel, 1901e 345.]

***tetracantha.*** *Solenopsis tetracantha* Emery, 1906c: 131, fig. 13 (w.q.) ARGENTINA. Santschi, 1917f: 281 (m.). Current subspecies: nominal plus *videns.*

***texana.*** *Solenopsis pollux* var. *texana* Emery, 1895c: 278 (w.) U.S.A. Forel, 1901e: 345 (q.m.); Wheeler, G.C. & Wheeler, J. 1955c: 136 (l.). Combination in *S.* (*Diplorhoptrum*): Creighton, 1950a: 238. Raised to species: Forel, 1901e: 345. Senior synonym of *rosella:* Creighton, 1950a: 238. Current subspecies: nominal plus *catalinae.*

*thoracica. Solenopsis thoracica* Santschi, 1923c: 261 (w.m.) CHILE. Junior synonym of *patagonica:* Kusnezov, 1960b: 338.

***tipuna.*** *Solenopsis tipuna* Forel, 1912a: 56 (w.) TAIWAN.

#***tisiphone.*** *Solenopsis (Diplorhoptrum) fugax* st. *tisiphone* Santschi, 1934e: 586, fig. 1, 34 (q.) TUNISIA.

***tonsa.*** *Solenopsis tonsa* Thompson, 1989: 272, figs. 29–32 (w.q.m.) U.S.A.

#***transformis.*** *Solenopsis wasmanni* subsp. *transformis* Forel, 1911c: 298 (w.q.) PARAGUAY.

*transversalis* Smith, F. 1860; see under ***PHEIDOLOGETON.***

*tricuspis. Solenopsis pylades* var. *tricuspis* Forel, 1912g: 4 (w.) ARGENTINA. Variety of *saevissima:* Santschi, 1916e: 379. Junior synonym of *richteri:* Creighton, 1930b: 87.

***tridens.*** *Solenopsis tridens* Forel, 1911c: 298 (w.) BRAZIL. Current subspecies: nominal plus *lehmannnitschei.* See also: Trager, 1991: 162.

***trihasta.*** *Solenopsis trihasta* Santschi, 1923c: 252, fig. 2 (w.) ARGENTINA.

***truncorum.*** *Solenopsis texana* r. *truncorum* Forel, 1901e: 346 (w.q.) U.S.A. Raised to species, and senior synonym of *castanea:* Creighton, 1950a: 239. See also: Thompson & Johnson, 1989: 697.

*tucumana. Solenopsis clytemnestra* r. *orestes* var. *tucumana* Forel, 1914d: 278 (w.) ARGENTINA. Unavailable name.

***ugandensis.*** *Solenopsis ugandensis* Santschi, 1933b: 102 (w.) UGANDA. Current subspecies: nominal plus *congolensis.*

#***ultrix.*** *Solenopsis helena* subsp. *ultrix* Wheeler, W.M. 1921f: 157 (w.q.) GUIANA.

#***urichi.*** *Solenopsis basalis* var. *urichi* Forel, 1912g: 6 (w.q.) TRINIDAD.

*****valida.*** **Attopsis valida* Förster, 1891: 447, pl. 13, fig. 18 (w.) GERMANY (Oligocene). Théobald, 1937b: 201 (m.). Combination in *Solenopsis:* Théobald, 1937b: 201.

#***validiuscula.*** *Solenopsis molesta* var. *validiuscula* Emery, 1895c: 278 (w.) U.S.A. Wheeler, G.C. & Wheeler, J. 1955c: 135 (l.). Subspecies of *molesta:* Creighton, 1950a: 237.

#***videns.*** *Solenopsis tetracantha* var. *videns* Forel, 1912g: 4 (w.) ARGENTINA.

#***virgula.*** *Solenopsis corticalis* var. *virgula* Forel, 1904d: 172 (w.) CUBA. Menozzi, 1929a: 2 (q.).

***virulens.*** *Myrmica virulens* Smith, F. 1858b: 132 (w.) BRAZIL. Combination in *Solenopsis:* Dalla Torre, 1893: 76. Junior synonym of *geminata:* Dalla Torre, 1893: 76; Creighton, 1930b: 59; Creighton, 1950a: 231; of *bondari:* Kempf & Brown, 1968: 99. Revived from synonymy, and senior synonym of *bondari:* Trager, 1991: 155.

*vitiensis. Solenopsis cleptis* var. *vitiensis* Mann, 1921: 444 (w.) FIJI IS. Junior synonym of *papuana:* Wilson & Taylor, 1967: 60.

#***vittata.*** *Solenopsis basalis* var. *vittata* Forel, 1912g: 6 (w.) BRAZIL.

***vorax.*** *Solenopsis (Diplorhoptrum) vorax* Santschi, 1934e: 580, figs. 23, 29, 40 (q.m.) RUSSIA.

***wagneri.*** *Solenopsis saevissima* var. *wagneri* Santschi, 1916e: 380 (w.) ARGENTINA. Junior synonym of *saevissima:* Wilson, 1952: 55. Revived from synonymy, raised to species, and senior synonym of *invicta:* **new status, new synonymy**. [Trager, 1991: 173 wrongly states that *wagneri* is an unavailable name, incorrectly citing its original notation as *S. saevissima* subsp. *electra* var. *wagneri,* and referring its material to *invicta.* But *wagneri* is available, with the original notation cited above, and has priority over *invicta.*]

***SOLENOPSIS*** *(continued)*

***wasmannii.*** *Solenopsis wasmannii* Emery, 1894c: 151 (w.) PARAGUAY. Emery, 1906c: 122 (q.m.); Forel, 1911c: 297 (q.m.). Current subspecies: nominal plus *transformis*.
***weiseri.*** *Solenopsis tenuis* r. *weiseri* Forel, 1914d: 278 (w.q.m.) ARGENTINA. Raised to species: Santschi, 1923c: 258.
***westwoodi.*** *Solenopsis westwoodi* Forel, 1894b: 100 (w.) BRAZIL. Current subspecies: nominal plus *atticola, platensis*.
*weyeri* Karavaiev, 1930; see under ***OLIGOMYRMEX.***
***weyrauchi.*** *Solenopsis weyrauchi* Trager, 1991: 190, figs. 76, 77, 80, 81 (w.) PERU.
***wolfi.*** *Solenopsis wolfi* Emery, 1915a: 259, pl. 4, fig. 10 (w.) YUGOSLAVIA. Stirps of *fugax:* Müller, 1923: 79; Santschi, 1934e: 567. Revived status as species: Agosti & Collingwood, 1987a: 55.
*wroughtonii* Forel, 1902; see under ***OLIGOMYRMEX.***
***xyloni.*** *Solenopsis xyloni* McCook, 1879: 188, figs. 37, 38 (s.w.q.) U.S.A. Wheeler, W.M. 1915b: 396 (m.); Wheeler, G.C. & Wheeler, J. 1955c: 133 (l.); Taber & Cokendolpher, 1988: 95 (k.). Junior synonym of *geminata:* Dalla Torre, 1893: 76. Revived from synonymy as subspecies of *geminata:* Wheeler, W.M. 1910g: 563; Wheeler, W.M. 1915b: 395. Revived status as species: Creighton, 1930b: 98; Creighton, 1950a: 232; Snelling, 1963: 9. Senior synonym of *maniosa* and *pylades:* Trager, 1991: 166.
***zambesiae.*** *Solenopsis zambesiae* Arnold, 1926: 232, fig. 66 (w.) ZIMBABWE.
***zeteki.*** *Solenopsis zeteki* Wheeler, W.M. 1942: 204 (w.) PANAMA.

*SOMMIMYRMA:* see under ***MYRMICA.***

*SPALACOMYRMEX:* see under ***CENTROMYRMEX.***

*SPANIOPONE:* see under ***GNAMPTOGENYS.***

*SPELAEOMYRMEX:* see under ***OLIGOMYRMEX.***

**#*SPHAEROCREMA:*** see under ***CREMATOGASTER.***

*SPHAEROMESSOR:* see under ***MESSOR.***

****SPHECOMYRMA*** (Nearctic)
****canadensis.*** **Sphecomyrma canadensis* Wilson, 1985d: 206, figs. 1, 2 (w.) CANADA (Cretaceous).
****freyi.*** **Sphecomyrma freyi* Wilson & Brown, in Wilson, Carpenter & Brown, 1967: 8, pls. 1–4, figs. 1, 2 (w.) U.S.A. (Cretaceous).

*SPHEGOMYRMEX:* see under ***DORYLUS.***

***SPHINCTOMYRMEX*** (World tropics)
***asper.*** *Sphinctomyrmex asper* Brown, 1975: 78 (w.q.m.) AUSTRALIA.
*brunnicornis. Eusphinctus (Nothosphinctus) brunnicornis* Clark, 1930c: 2, fig. 1 (w.) AUSTRALIA. Junior synonym of *imbecilis:* Brown, 1975: 33.
***caledonicus.*** *Sphinctomyrmex caledonicus* Wilson, 1957a: 8 (w.q.) NEW CALEDONIA. See also: Wilson, 1958c: 136.
***cedaris.*** *Sphinctomyrmex (Eusphinctus) fallax* var. *cedaris* Forel, 1915b: 16 (w.) AUSTRALIA. Wheeler, W.M. 1918a: 229 (q.). Combination in *Eusphinctus (Eusphinctus)*: Wheeler, W.M. 1918a: 229; in *Sphinctomyrmex*, and raised to species: Brown, 1975: 33.
***chariensis.*** *Sphinctomyrmex rufiventris* st. *chariensis* Santschi, 1915c: 245 (m.) CENTRAL AFRICAN REPUBLIC. Combination in *Aethiopopone:* Santschi, 1930a: 50; in *Sphinctomyrmex*, and raised to species: Brown, 1975: 33.
***clarus.*** *Cerapachys emeryi* r. *clarus* Forel, 1893d: 462 (w.) AUSTRALIA. Combination in *Sphinctomyrmex:* Forel, 1902h: 537; in *Eusphinctus (Nothosphinctus)*: Wheeler, W.M. 1918a: 235; in *Sphinctomyrmex:* Brown, 1975: 33. Raised to species: Forel, 1900b: 72.
***cribratus.*** *Sphinctomyrmex (Eusphinctus) cribratus* Emery, 1897c: 594 (w.) NEW GUINEA.
***duchaussoyi.*** *Eusphinctus duchaussoyi* André, 1905: 205 (w.q.) AUSTRALIA. Combination in *Sphinctomyrmex (Eusphinctus)*: Emery, 1911d: 7; in *Eusphinctus (Eusphinctus)*: Wheeler, W.M. 1918a: 229; in *Sphinctomyrmex:* Brown, 1975: 33. Variety of *steinheili:* Wheeler, W.M. 1918a: 229. Revived status as species, and senior synonym of *hackeri:* Brown, 1975: 33.
***emeryi.*** *Cerapachys emeryi* Forel, 1893d: 461 (w.) AUSTRALIA. Combination in *Sphinctomyrmex:* Emery, 1895k: 456; in *Eusphinctus (Nothosphinctus)*: Wheeler, W.M. 1918a: 235. Senior synonym of *perstictus* (unnecessary replacement name): **new synonymy.**
*fallax. Sphinctomyrmex (Eusphinctus) fallax* Forel, 1900b: 73 (w.) AUSTRALIA. Junior synonym of *steinheili:* Wheeler, W.M. 1918a: 225.
***froggatti.*** *Sphinctomyrmex froggatti* Forel, 1900b: 71 (w.) AUSTRALIA. Combination in *Eusphinctus (Nothosphinctus)*: Wheeler, W.M. 1918a: 233; in *Sphinctomyrmex:* Brown, 1975: 33.
*fulvidus. Eusphinctus (Nothosphinctus) fulvidus* Clark, 1923: 75, fig. 2 (w.q.l.) AUSTRALIA. Junior synonym of *imbecilis:* Brown, 1975: 33.
*fulvipes. Eusphinctus fulvipes* Clark, 1934c: 49, pl. 4, fig. 1 (w.q.) AUSTRALIA. Junior synonym of *steinheili:* Brown, 1975: 33.
***furcatus.*** *Eusphinctus furcatus* Emery, 1893a: cclxxv (w.) BURMA. [Also described as new by Emery, 1895k: 457.] Combination in *Sphinctomyrmex (Eusphinctus)*: Emery, 1895k: 457. See also: Bingham, 1903: 25; Brown, 1975: 75.
*hackeri. Eusphinctus (Eusphinctus) hackeri* Wheeler, W.M. 1918a: 229, fig. 3 (w.q.) AUSTRALIA. Junior synonym of *duchaussoyi:* Brown, 1975: 33.
*hedwigae. Sphinctomyrmex (Eusphinctus) fallax* subsp. *hedwigae* Forel, 1910b: 21 (w.q.) AUSTRALIA. Variety of *steinheili:* Wheeler, W.M. 1918a: 228. Junior synonym of *steinheili:* Brown, 1975: 33.
*hirsutus. Eusphinctus hirsutus* Clark, 1929: 118, pl. 1, figs. 1, 2 (w.q.) AUSTRALIA. Junior synonym of *steinheili:* Brown, 1975: 33.
***imbecilis.*** *Sphinctomyrmex froggatti* subsp. *imbecilis* Forel, 1907h: 272 (w.q.) AUSTRALIA. Combination in *Eusphinctus (Nothosphinctus)*: Wheeler, W.M. 1918a: 235; in *Sphinctomyrmex:* Brown, 1975: 33. Raised to species: Wheeler, W.M. 1918a: 235. Senior synonym of *brunnicornis, fulvidus, manni* and *silaceus:* Brown, 1975: 33.
*manni. Eusphinctus (Nothosphinctus) manni* Wheeler, W.M. 1918a: 236, fig. 6 (w.q.) AUSTRALIA. Junior synonym of *imbecilis:* Brown, 1975: 33.
***mjobergi.*** *Sphinctomyrmex clarus* r. *mjobergi* Forel, 1915b: 16 (w.) AUSTRALIA. Combination in *Eusphinctus (Nothosphinctus)*: Wheeler, W.M. 1918a: 239; in *Sphinctomyrmex:* Brown, 1975: 33. Raised to species: Wheeler, W.M. 1918a: 239.
***myops.*** *Sphinctomyrmex emeryi* var. *myops* Forel, 1895f: 421 (q.) AUSTRALIA. Combination in *Eusphinctus (Nothosphinctus)*: Wheeler, W.M. 1918a: 235; in *Sphinctomyrmex:* Brown, 1975: 33. Raised to species: Wheeler, W.M. 1918a: 235.
***nigricans.*** *Eusphinctus (Nothosphinctus) nigricans* Clark, 1926: 44, pl. 6, fig. 1 (w.) AUSTRALIA. Combination in *Sphinctomyrmex:* Brown, 1975: 33.
***occidentalis.*** *Eusphinctus (Eusphinctus) occidentalis* Clark, 1923: 74, fig. 1 (w.q.l.) AUSTRALIA. Combination in *Sphinctomyrmex:* Brown, 1975: 33.
*perstictus. Sphinctomyrmex perstictus* Brown, 1975: 33. Unnecessary replacement name for, and hence junior synonym of *emeryi:* **new synonymy.**
***rufiventris.*** *Sphinctomyrmex rufiventris* Santschi, 1915c: 244, fig. 1 (m.) BENIN. Combination in *Aethiopopone:* Santschi, 1930a: 49; in *Sphinctomyrmex:* Brown, 1975: 33.
***septentrionalis.*** *Eusphinctus (Nothosphinctus) septentrionalis* Crawley, 1925b: 577 (w.) AUSTRALIA. Combination in *Sphinctomyrmex:* Brown, 1975: 33.
*silaceus. Eusphinctus (Nothosphinctus) silaceus* Clark, 1923: 77, fig. 3 (w.) AUSTRALIA. Junior synonym of *imbecilis:* Brown, 1975: 33.
***stali.*** *Sphinctomyrmex stali* Mayr, 1866b: 895, pl. 20, fig. 8 (q.) BRAZIL. Borgmeier, 1957: 105 (w.).
***steinheili.*** *Sphinctomyrmex (Eusphinctus) steinheili* Forel, 1900b: 72 (q., not w. as stated; see Clark, 1925a: 62) AUSTRALIA. Wheeler, W.M. 1918a: 225 (w.l.); Wheeler, G.C. 1950: 104 (l.); Imai, Crozier & Taylor, 1977: 348 (k.). Combination in *Eusphinctus (Eusphinctus)*: Wheeler, W.M. 1918a: 225; in *Sphinctomyrmex:* Brown, 1975: 33. Senior synonym of *fallax:* Wheeler, W.M. 1918a: 225; of *fulvipes, hedwigae* and *hirsutus:* Brown, 1975: 33. See also: Wilson, 1958c: 136; Buschinger, Peeters & Crozier, 1990: 287.
***taylori.*** *Sphinctomyrmex (Eusphinctus) taylori* Forel, 1900d: 328 (w.) INDIA.
***trux.*** *Sphinctomyrmex trux* Brown, 1975: 77, figs. 2, 101, 102, 105–113 (w.) AUSTRALIA.
***turneri.*** *Sphinctomyrmex turneri* Forel, 1900b: 70 (w.) AUSTRALIA. Combination in *Eusphinctus (Zasphinctus)*: Wheeler, W.M. 1918a: 231; in *Sphinctomyrmex:* Brown, 1975: 33.

*SPINOMYRMA:* see under ***DORYMYRMEX.***

*SPOROCLEPTES:* see under ***OLIGOMYRMEX.***

***STEGOMYRMEX*** (Neotropical)
***connectens.*** *Stegomyrmex connectens* Emery, 1912b: 100, fig. 5 (q.m.) PERU. See also: Hölldobler & Wilson, 1986a: 16; Diniz, 1990: 284.
***manni.*** *Stegomyrmex manni* Smith, M.R. 1946: 288 (w.) PANAMA. See also: Diniz, 1990: 288.

***vizottoi.*** *Stegomyrmex vizottoi* Diniz, 1990: 290, figs. 7–9, 11, 12, 18–21, 25–28, 32–35 (w.q.) BRAZIL. See also: Diniz & Brandão, 1993: 301.

*STEGOPHEIDOLE:* see under ***PHEIDOLE.***

***STENAMMA*** (Neotropical, Holarctic, Oriental, Indo-Australian)

***africanum.*** *Stenamma africanum* Santschi, 1939d: 66, fig. 2 (w.q.) TUNISIA. Cagniant, 1971: 98 (m.). Current subspecies: nominal plus *submuticum.*

*albipennis* Curtis, 1854; see under ***LEPTOTHORAX.***

*ashmeadi* Emery, 1895; see under ***APHAENOGASTER.***

#***asiaticum.*** *Stenamma westwoodi* subsp. *asiaticum* Ruzsky, 1905b: 711 (w.) RUSSIA.

*aztecum* Emery, 1895; see under ***APHAENOGASTER.***

*bakeri* Wheeler, W.M. 1904; see under ***APHAENOGASTER.***

*balcanica* Emery, 1898; see under ***APHAENOGASTER.***

*barbarostructor* Forel, 1904; see under ***MESSOR.***

****berendti.*** **Aphaenogaster berendti* Mayr, 1868c: 82, pl. 4, fig. 78, 79 (m.) BALTIC AMBER (Oligocene). Combination in *Stenamma:* Wheeler, W.M. 1915h: 53.

***bhutanense.*** *Stenamma bhutanense* Baroni Urbani, 1977b: 420, figs. 2, 4 (w.) BHUTAN.

***brevicorne.*** *Aphaenogaster brevicornis* Mayr, 1886d: 447 (w.q.) U.S.A. Emery, 1895c: 299 (m.); Smith, M.R. 1957b: 167 (m.); Hauschteck, 1962: 219 (k.). Combination in *Stenamma:* Emery, 1895c: 299. Senior synonym of *neoarcticum:* Smith, M.R. 1957b: 165.

*brunneicorne* Forel, 1901; see under ***MESSOR.***

*bugnioni* Forel, 1904; see under ***MESSOR.***

***californicum.*** *Stenamma californicum* Snelling, 1973c: 16, figs. 16, 17, 24, 25, 27 (w.q.) U.S.A.

*canescens* Emery, 1895; see under ***APHAENOGASTER.***

***carolinense.*** *Stenamma carolinense* Smith, M.R. 1951b: 156, figs. 1, 2 (w.) U.S.A.

*caucasicum. Stenamma caucasicum* Arnol'di, 1975: 1823 (page number misprinted as 1283 in text), fig. 3 (w.q.) CAUCASUS. Junior synonym of *lippulum:* DuBois, 1993: 321.

*caviceps* Forel, 1902; see under ***MESSOR.***

*cephalotes* Emery, 1895; see under ***MESSOR.***

***chiricahua.*** *Stenamma chiricahua* Snelling, 1973c: 7, figs. 6–11 (w.q.m.) U.S.A.

*cristatus* Forel, 1902; see under ***APHAENOGASTER.***

***debile.*** *Myrmica debilis* Foerster, 1850a: 52 (m.) GERMANY. DuBois, 1993: 314 (w.q.). Combination in *Stenamma,* and junior synonym of *westwoodii:* Mayr, 1863: 454. Revived from synonymy, and senior synonym of *minkii* and *polonicum:* DuBois, 1993: 314.

*decipiens* Forel, 1905; see under ***MESSOR.***

***diecki.*** *Stenamma westwoodi* subsp. *diecki* Emery, 1895c: 300 (w.q.) U.S.A. Smith, M.R. 1957b: 161 (m.); Wheeler, G.C. & Wheeler, J. 1953b: 50 (l.). Subspecies of *brevicorne:* Forel, 1901e: 347. Raised to species: Creighton, 1950a: 136. Senior synonym of *impressum:* Smith, M.R. 1957b: 158. See also: Snelling, 1973c: 18.

***diversum.*** *Stenamma diversum* Mann, 1922: 20 (w.) HONDURAS.

***dyscheres.*** *Stenamma dyscheres* Snelling, 1973c: 21, figs. 28, 29, 34, 35 (w.q.m.) U.S.A.

*ecalcaratum* Emery, 1895; see under ***APHAENOGASTER.***

*epirotes* Emery, 1895; see under ***APHAENOGASTER.***

***exasperatum.*** *Stenamma exasperatum* Snelling, 1973c: 28, figs. 40, 41, 46, 47, 63 (w.) U.S.A.

***expolitum.*** *Stenamma expolitum* Smith, M.R. 1962a: 36 (w.q.) COSTA RICA.

***felixi.*** *Stenamma felixi* Mann, 1922: 21, fig. 10 (w.) HONDURAS.

***fovolocephalum.*** *Stenamma fovolocephala* Smith, M.R. 1930b: 564 (w.) U.S.A.

*galla* Emery, 1895; see under ***MESSOR.***

*gallarum* Patton, 1879; see under ***LEPTOTHORAX.***

***georgii.*** *Stenamma georgii* Arnol'di, 1975: 1823 (page number misprinted as 1283 in text), fig. 3 (w.q.) RUSSIA.

*golosejevi. Stenamma golosejevi* Karavaiev, 1926f: 68, fig. 5 (w.) UKRAINE. Subspecies of *westwoodii:* Karavaiev, 1934: 102. Revived status as species: Arnol'di & Dlussky, 1978: 536. Junior synonym of *westwoodii:* Atanassov & Dlussky, 1992: 103.

***heathi.*** *Stenamma brevicorne* subsp. *heathi* Wheeler, W.M. 1915b: 410 (w.) U.S.A. Snelling, 1973c: 31 (q.). Raised to species: Creighton, 1950a: 137. See also: Smith, M.R. 1957b: 154.

*himalayanum* Forel, 1902; see under ***MESSOR.***

*hirtula. Stenamma westwoodi* var. *hirtula* Emery, 1898c: 135 (w.) RUSSIA. Arnol'di, 1975: 1826 (m.). Raised to species: Emery, 1908c: 308. Junior synonym of *lippulum:* DuBois, 1993: 321. See also: Arnol'di, 1928b: 208.

***hissarianum.*** *Stenamma hissarianum* Arnol'di, 1975: 1824, fig. 5 (w.q.m.) RUSSIA.

*holtzi* Emery, 1898; see under ***APHAENOGASTER.***

***huachucanum.*** *Stenamma huachucanum* Smith, M.R. 1957b: 153, pl. 2, fig. 8 (w.) U.S.A.

***impar.*** *Stenamma brevicorne* r. *impar* Forel, 1901e: 347 (w.q.) U.S.A. Smith, M.R. 1957b: 152. Raised to species: Creighton, 1950a: 137.

*impressum. Stenamma brevicorne* subsp. *impressum* Buren, 1944a: 284 (w.) U.S.A. [First available use of *Stenamma westwoodi* subsp. *diecki* var. *impressum* Emery, 1895c: 301; unavailable name.] Raised to species: Creighton, 1950a: 137. Junior synonym of *diecki:* Smith, M.R. 1957b: 158.

***kashmirense.*** *Stenamma kashmirense* Baroni Urbani, 1977b: 415, figs. 1, 3 (w.) INDIA.

*knowltoni. Stenamma knowltoni* Gregg, 1972: 35, fig. 1 (w.q.) U.S.A. Junior synonym of *smithi:* Snelling, 1973c: 12.

***kurilense.*** *Stenamma kurilense* Arnol'di, 1975: 1825, fig. 2 (w.) RUSSIA.

***lippulum.*** *Myrmica lippula* Nylander, 1849: 41 (w.) RUSSIA. Smith, F. 1858a: 280 (q.); DuBois, 1993: 322 (w.q.m.). Combination in *Tetramorium:* Roger, 1859: 258; in *Asemorhoptrum:* Mayr, 1861: 76; in *Stenamma:* Forel, 1889: 258. Junior synonym of *westwoodii:* Forel, 1889: 258. Revived from synonymy, and senior synonym of *caucasicum* and *hirtula:* DuBois, 1993: 321.

*loriai* Emery, 1897; see under ***APHAENOGASTER.***

*lurida* Emery, 1898; see under ***MESSOR.***

***manni.*** *Stenamma manni* Wheeler, W.M. 1914b: 51 (w.q.) MEXICO. Wheeler, G.C. & Wheeler, J. 1972b: 237 (l.). See also: Smith, M.R. 1962a: 35.

*marocana* Forel, 1903; see under ***APHAENOGASTER.***

*mediorubra* Forel, 1905; see under ***MESSOR.***

***meridionale.*** *Stenamma meridionale* Smith, M.R. 1957b: 169, pl. 4, fig. 15 (w.q.) U.S.A.

*minkii. Myrmica minkii* Foerster, 1850a: 63 (w.) GERMANY. Junior synonym of *lippulum:* Mayr, 1863: 395; of *debile:* DuBois, 1993: 314.

#***msilanum.*** *Stenamma* (*Stenamma*) *westwoodi* var. *msilanum* Forel, 1901e: 347 (q.) ALGERIA.

*myops. Theryella myops* Santschi, 1921d: 68 (w.) MOROCCO. Junior synonym of *punctiventre:* Santschi, 1923a: 136.

*neoarcticum. Stenamma neoarcticum* Mayr, 1886d: 454 (q.m., not w.; see Creighton, 1950a: 138) U.S.A. Emery, 1895c: 299 (w.). Junior synonym of *brevicorne:* Mayr, 1887: 628. Revived from synonymy: Dalla Torre, 1893: 121. Subspecies of *westwoodii:* Emery, 1895c: 299. Revived status as species: Forel, 1901e: 347; Creighton, 1950a: 138. Junior synonym of *brevicorne:* Smith, M.R. 1957b: 165.

*nicobarensis* Forel, 1903; see under ***APHAENOGASTER.***

***nipponense.*** *Stenamma nipponense* Yasumatsu & Murakami, 1960: 28, fig. 1; pl. 7, fig. 1 (w.q.) JAPAN.

*nitidum* Emery, 1895; see under ***APHAENOGASTER.***

*occidentale* Emery, 1895; see under ***APHAENOGASTER.***

*occidentale. Stenamma occidentale* Smith, M.R. 1957b: 146, pl. 2, fig. 7 (w.q.m.) U.S.A. [Junior primary homonym of *occidentale* Emery, 1895c: 301, above.] Replacement name: *snellingi* **new name.**

*orientalis* Emery, 1898; see under ***MESSOR.***

***orousseti.*** *Stenamma orousseti* Casevitz-Weulersse, 1990a: 141, figs. 1–6 (w.q.m.) CORSICA.

*ovaticeps* Emery, 1898; see under ***APHAENOGASTER.***

***owstoni.*** *Stenamma owstoni* Wheeler, W.M. 1906c: 314 (w.) JAPAN. Yasumatsu & Murakami, 1960: 30 (q.).

*pachei* Forel, 1906; see under ***APHAENOGASTER.***

*persicus* Forel, 1904; see under ***MESSOR.***

***petiolatum.*** *Stenamma petiolatum* Emery, 1897a: 12, fig. (q.) ITALY. Emery, 1915a: 256 (w.); Kutter, 1971: 264 (m.). See also: Emery, 1916b: 128.

***picetojuglandeti.*** *Stenamma picetojuglandeti* Arnol'di, 1975: 1821, figs. 1, 6 (w.q.m.) TURKESTAN. See also: Tarbinsky, 1976: 46.

*piceum* Emery, 1895; see under ***APHAENOGASTER.***

*polonicum. Stenamma westwoodi* subsp. *polonicum* Begdon, 1932: 118, fig. 1 (w.) POLAND. Junior synonym of *westwoodii:* Pisarski, 1975: 15; of *debile:* DuBois, 1993: 314.

*prudens* Forel, 1902; see under ***APHAENOGASTER.***

*punctatohimalayanum* Forel, 1902; see under ***MESSOR.***

***punctatoventre.*** *Stenamma punctatoventre* Snelling, 1973c: 10, figs. 12, 13, 20, 21, 55 (w.q.m.) U.S.A.

***punctiventre.*** *Stenamma punctiventre* Emery, 1908c: 309 (q.) MOROCCO. Senior synonym of *myops:* Santschi, 1923a: 136.

*pusillum* Emery, 1895; see under ***APHAENOGASTER.***

*romanum* Emery, 1985; see under ***APHAENOGASTER.***

*rothneyi* Forel, 1902; see under ***APHAENOGASTER.***

*rude* Emery, 1895; see under ***APHAENOGASTER.***

*ruginota* Forel, 1902; see under ***APHAENOGASTER.***

*sagei* Forel, 1902; see under ***APHAENOGASTER.***

*sancta* Forel, 1905; see under ***MESSOR.***

*sangiorgii* Emery, 1901; see under ***APHAENOGASTER.***

***STENAMMA** (continued)*

*santschii* Forel, 1905; see under ***MESSOR.***
***sardoum.*** *Stenamma sardoum* Emery, 1915a: 255, pl. 4, figs. 5, 6 (w.q.) SARDINIA.
***schmidti.*** *Stenamma schmidti* Menozzi, 1931d: 198, fig. 6 (w.q.) COSTA RICA.
***schmitti.*** *Stenamma brevicorne* subsp. *schmitti* Wheeler, W.M. 1903d: 167 (diagnosis in key) (w.) U.S.A. Smith, M.R. 1957b: 144 (q.m.). Raised to species: Creighton, 1950a: 138.
*schurri* Forel, 1902; see under ***APHAENOGASTER.***
*semoni* Forel, 1906; see under ***MESSOR.***
***sequoiarum.*** *Stenamma brevicorne* subsp. *sequoiarum* Wheeler, W.M. 1917a: 520 (w.q.) U.S.A. Subspecies of *diecki:* Creighton, 1950a: 137. Raised to species: Smith, M.R. 1957b: 156.
***smithi.*** *Stenamma smithi* Cole, 1966: 7, fig. (w.) U.S.A. Senior synonym of *knowltoni:* Snelling, 1973c: 12.
*smythiesii* Forel, 1902; see under ***APHAENOGASTER.***
***snellingi.*** *Stenamma snellingi* **new name**. Replacement name for *occidentale* Smith, M.R. 1957b: 146. [Junior primary homonym of *occidentale* Emery, 1895c: 301.]
***sogdianum.*** *Stenamma sogdianum* Arnol'di, 1975: 1824, figs. 2, 5 (w.) TURKESTAN.
*stoddardi* Emery, 1895; see under ***MESSOR.***
***striatulum.*** *Stenamma westwoodi* var. *striatulum* Emery, 1895c: 300 (footnote) (w.q.) ITALY. Kutter, 1971: 259 (m.). Raised to species: Müller, 1923: 46. Senior synonym of *tscherkessicum:* Arnol'di, 1975: 1822. See also: Kutter, 1971: 263; Kutter, 1977c: 74.
#***submuticum.*** *Stenamma africanum* var. *submuticum* Santschi, 1939d: 67, fig. 3 (w.) ALGERIA.
*texanum* Emery, 1895; see under ***APHAENOGASTER.***
*tscherkessicum. Stenamma westwoodi* var. *tscherkessicum* Arnol'di, 1928b: 214 (q.) RUSSIA. Junior synonym of *striatulum:* Arnol'di, 1975: 1822.
*tyrrhena* Emery, 1898; see under ***MESSOR.***
*ucrainicum. Stenamma ucrainicum* Arnol'di, 1928b: 209 (w.q.m.) UKRAINE. Subspecies of *westwoodii:* Karavaiev, 1934: 100. Revived status as species: Arnol'di, 1975: 1826; Arnol'di & Dlussky, 1978: 536. Junior synonym of *westwoodii:* Atanassov & Dlussky, 1992: 103.
***ussuriense.*** *Stenamma ussuriense* Arnol'di, 1975: 1825, figs. 2, 6 (w.q.m.) RUSSIA. See also: Kupyanskaya, 1990: 117.
***westwoodii.*** *Stenamma westwoodii* Westwood, 1839: 219, fig. 86 (m.) GREAT BRITAIN. Mayr, 1861: 56 (w.q.); Perkins, 1891: 123 (gynandromorph). Senior synonym of *golosejevi* and *ucrainicum:* Atanassov & Dlussky, 1992: 103. Current subspecies: nominal plus *asiaticum, msilanum.* See also: Donisthorpe, 1915d: 139; Arnol'di, 1928b: 206; DuBois, 1993: 307.
***wheelerorum.*** *Stenamma wheelerorum* Snelling, 1973c: 34, figs. 50–54 (w.q.) U.S.A.

*STENOMYRMEX:* see under ***ANOCHETUS.***

*STENOTHORAX:* see under ***ADLERZIA.***

***STEREOMYRMEX*** (Sri Lanka)
***horni.*** *Stereomyrmex horni* Emery, 1901f: 116, fig. (w.m.) SRI LANKA. Forel, 1908a: 3 (q.).

*STICTOPONERA:* see under ***GNAMPTOGENYS.***

***STIGMACROS*** (Australia)
***aciculata.*** *Stigmacros (Cyrtostigmacros) aciculata* McAreavey, 1957: 50, figs. 171–173 (w.) AUSTRALIA.
***acuta.*** *Stigmacros (Stigmacros) acuta* McAreavey, 1957: 12, figs. 28–30 (w.) AUSTRALIA. Wheeler, G.C. & Wheeler, J. 1968: 209 (l.).
***aemula.*** *Acantholepis (Stigmacros) aemula* Forel, 1907h: 298 (w.) AUSTRALIA. Wheeler, W.M. 1934d: 159 (q.). Combination in *Stigmacros:* Emery, 1925b: 34; in *S. (Campostigmacros)*: McAreavey, 1957: 32.
***anthracina.*** *Stigmacros (Campostigmacros) anthracina* McAreavey, 1957: 29, figs. 95–97 (w.) AUSTRALIA. Wheeler, G.C. & Wheeler, J. 1974a: 60 (l.).
***armstrongi.*** *Stigmacros (Cyrtostigmacros) armstrongi* McAreavey, 1957: 52, figs. 180–183 (w.q.) AUSTRALIA.
***australis.*** *Acantholepis (Acrostigma) australis* Forel, 1902h: 479 (w.) AUSTRALIA. Combination in *A. (Stigmacros)*: Forel, 1905b: 179; in *Stigmacros:* Emery, 1925b: 34; in *S. (Cyrtostigmacros)*: McAreavey, 1957: 36.
***barretti.*** *Stigmacros barretti* Santschi, 1928e: 477, fig. 2 (w.) AUSTRALIA. McAreavey, 1957: 21 (q.m.); Wheeler, G.C. & Wheeler, J. 1968: 210 (l.). Combination in *S. (Hagiostigmacros)*: McAreavey, 1957: 19.
***bosii.*** *Acantholepis (Acrostigma) bosii* Forel, 1902h: 481 (w.q.) AUSTRALIA. Combination in *A. (Stigmacros)*: Forel, 1905b: 179; in *Stigmacros:* Emery, 1925b: 34.
***brachytera.*** *Stigmacros (Campostigmacros) brachytera* McAreavey, 1957: 27, figs. 85–88 (w.q.) AUSTRALIA.
***brevispina.*** *Stigmacros (Stigmacros) brevispina* McAreavey, 1957: 14, figs. 34–36 (w.) AUSTRALIA.
***brooksi.*** *Stigmacros (Cyrtostigmacros) brooksi* McAreavey, 1957: 42, figs. 140–144 (w.q.m.) AUSTRALIA.
***castanea.*** *Stigmacros (Cyrtostigmacros) castanea* McAreavey, 1957: 49, figs. 165–170 (w.q.m.) AUSTRALIA.
***clarki.*** *Stigmacros (Cyrtostigmacros) clarki* McAreavey, 1957: 41, figs. 137–139 (w.) AUSTRALIA.
***clivispina.*** *Acantholepis (Acrostigma) clivispina* Forel, 1902h: 482 (w.) AUSTRALIA. McAreavey, 1957: 47 (q.m.). Combination in *A. (Stigmacros)*: Forel, 1905b: 179; in *Stigmacros:* Emery, 1925b: 34; in *S. (Cyrtostigmacros)*: McAreavey, 1957: 46.
***debilis.*** *Stigmacros debilis* **new name**. Replacement name for *foreli* Viehmeyer, 1925a: 34. [Junior primary homonym of *foreli* Arnold, 1920a: 564.]
***elegans.*** *Stigmacros elegans* McAreavey, 1949: 3, figs. 68–70 (w.q.) AUSTRALIA. Combination in *S. (Campostigmacros)*: McAreavey, 1957: 31.
***epinotalis.*** *Stigmacros (Campostigmacros) epinotalis* McAreavey, 1957: 28, figs. 89–91 (w.) AUSTRALIA.
***extreminigra.*** *Stigmacros (Cyrtostigmacros) extreminigra* McAreavey, 1957: 48, figs. 162–164 (w.) AUSTRALIA.
***ferruginea.*** *Stigmacros (Cyrtostigmacros) ferruginea* McAreavey, 1957: 46, figs. 152–154 (w.) AUSTRALIA.
***flava.*** *Stigmacros (Cyrtostigmacros) flava* McAreavey, 1957: 40, figs. 131–133 (w.) AUSTRALIA.
***flavinodis.*** *Stigmacros flavinodis* Clark, 1938: 375, fig. 13 (w.) AUSTRALIA. Combination in *S. (Campostigmacros)*: McAreavey, 1957: 26.
*foreli. Acantholepis (Stigmacros) foreli* Viehmeyer, 1925a: 34 (w.q.m.) AUSTRALIA. [Junior primary homonym of *Acantholepis foreli* Arnold, 1920a: 564 (now in *Lepisiota*).] Junior synonym of *froggatti:* McAreavey, 1957: 10. Revived from synonymy: Taylor, 1992a: 67. Replacement name: *debilis* **new name**.
***fossulata.*** *Acantholepis (Stigmacros) fossulata* Viehmeyer, 1925a: 34 (w.) AUSTRALIA. Junior synonym of *froggatti:* McAreavey, 1957: 10. Revived from synonymy: Taylor, 1992a: 67.
***froggatti.*** *Acantholepis (Acrostigma) froggatti* Forel, 1902h: 478 (w.q.m.) AUSTRALIA. Combination in *Stigmacros:* Emery, 1925b: 34.
***glauerti.*** *Stigmacros (Cyrtostigmacros) glauerti* McAreavey, 1957: 41, figs. 134–136 (w.) AUSTRALIA.
***hirsuta.*** *Stigmacros (Chariostigmacros) hirsuta* McAreavey, 1957: 23, figs. 72–75 (w.) AUSTRALIA.
***impressa.*** *Stigmacros (Stigmacros) impressa* McAreavey, 1957: 14, figs. 37–39 (w.) AUSTRALIA.
***inermis.*** *Stigmacros (Pseudostigmacros) inermis* McAreavey, 1957: 24, figs. 76–78 (w.) AUSTRALIA.
***intacta.*** *Acantholepis (Stigmacros) aemula* var. *intacta* Viehmeyer, 1925a: 34 (w.) AUSTRALIA. Combination in *Stigmacros (Campostigmacros)*, and raised to species: McAreavey, 1957: 33.
***lanaris.*** *Stigmacros (Cyrtostigmacros) lanaris* McAreavey, 1957: 43, figs. 145–148 (w.q.) AUSTRALIA.
***major.*** *Stigmacros (Cyrtostigmacros) major* McAreavey, 1957: 39, figs. 128–130 (w.) AUSTRALIA.
***marginata.*** *Stigmacros (Campostigmacros) marginata* McAreavey, 1957: 27, figs. 82–84 (w.) AUSTRALIA.
***medioreticulata.*** *Acantholepis (Stigmacros) medioreticulata:* Viehmeyer, 1925a: 32 (w.) AUSTRALIA. Combination in *Stigmacros:* McAreavey, 1957: 53.
***minor.*** *Stigmacros (Stigmacros) minor* McAreavey, 1957: 17, figs. 48–50 (w.) AUSTRALIA.
***nitida.*** *Stigmacros (Campostigmacros) nitida* McAreavey, 1957: 30, figs. 98–100 (w.) AUSTRALIA.
***occidentalis.*** *Acantholepis (Stigmacros) occidentalis* Crawley, 1922c: 30 (w.) AUSTRALIA. McAreavey, 1957: 44 (q.). Combination in *Stigmacros (Cyrtostigmacros)*: McAreavey, 1957: 44.
***pilosella.*** *Acantholepis (Stigmacros) pilosella* Viehmeyer, 1925a: 33 (w.) AUSTRALIA. Combination in *Stigmacros (Campostigmacros)*: McAreavey, 1957: 30.
***proxima.*** *Stigmacros (Cyrtostigmacros) proxima* McAreavey, 1957: 51, figs. 174–176 (w.) AUSTRALIA.
***punctatissima.*** *Stigmacros (Hagiostigmacros) punctatissima* McAreavey, 1957: 22, figs. 68–71 (w.) AUSTRALIA.
***pusilla.*** *Stigmacros (Stigmacros) pusilla* McAreavey, 1957: 16, figs. 45–47 (w.) AUSTRALIA.
***rectangularis.*** *Stigmacros (Stigmacros) rectangularis* McAreavey, 1957: 15, figs. 40–44 (w.m.) AUSTRALIA.
***reticulata.*** *Stigmacros reticulata* Clark, 1930a: 127, fig. 1 (w.) AUSTRALIA. McAreavey, 1957: 29 (q.). Combination in *S. (Campostigmacros)*: McAreavey, 1957: 29.

**rufa.** *Stigmacros (Stigmacros) rufa* McAreavey, 1957: 13, figs. 31–33 (w.) AUSTRALIA.
**sordida.** *Stigmacros (Cyrtostigmacros) sordida* McAreavey, 1957: 52, figs. 177–179 (w.) AUSTRALIA.
**spinosa.** *Stigmacros (Hagiostigmacros) spinosa* McAreavey, 1957: 19, figs. 55–59 (w.q.) AUSTRALIA.
**stanleyi.** *Stigmacros (Campostigmacros) stanleyi* McAreavey, 1957: 34, figs. 115–117 (w.) AUSTRALIA.
**striata.** *Stigmacros (Cyrtostigmacros) striata* McAreavey, 1957: 38, figs. 125–127 (w.q.m.) AUSTRALIA.
**termitoxena.** *Stigmacros termitoxenus* Wheeler, W.M. 1936d: 215, fig. 9 (w.q.) AUSTRALIA. Combination in *S. (Cyrtostigmacros)*: McAreavey, 1957: 37.
**wilsoni.** *Stigmacros (Stigmacros) wilsoni* McAreavey, 1957: 11, figs. 25–27 (w.) AUSTRALIA.

*STIGMATOMMA:* see under ***AMBLYOPONE.***

***STIGMOMYRMEX** (Baltic Amber)
**robustus* Mayr, 1868; see under ***STIPHROMYRMEX.***
***venustus.** **Stigmomyrmex venustus* Mayr, 1868c: 97, pl. 5, figs. 99, 100 (w.) BALTIC AMBER (Oligocene).

***STIPHROMYRMEX** (Baltic Amber)
***robustus.** **Stigmomyrmex robustus* Mayr, 1868c: 97, pl. 5, fig. 101 (w.) BALTIC AMBER (Oligocene). Combination in **Stiphromyrmex:* Wheeler, W.M. 1915h: 67.

**STREBLOGNATHUS** (Afrotropical)
**aethiopicus.** *Ponera aethiopica* Smith, F. 1858b: 91, pl. 6, figs. 19, 20 (w.) SOUTH AFRICA. Emery, 1899e: 468 (m.); Wheeler, G.C. & Wheeler, J. 1989b: 53 (l.). Combination in *Streblognathus:* Mayr, 1862: 716. See also: Arnold, 1915: 41.

**STRONGYLOGNATHUS** (Palaearctic)
**afer.** *Strongylognathus afer* Emery, 1884a: 380 (footnote) (q.) ALGERIA. Forel, 1900e: 279 (w.); Santschi, 1910d: 71 (m.). Variety/race of *huberi:* Medina, 1889: 31; Dalla Torre, 1893: 130; Forel, 1900e: 279. Revived status as species: Santschi, 1910d: 71; Emery, 1922f: 285; Bolton, 1976: 305.
**alboini.** *Strongylognathus huberi* subsp. *alboini* Finzi, 1924b: 121 (w.) ITALY. Baroni Urbani, 1969c: 141 (q.m.). Raised to species: Baroni Urbani, 1969c: 141; Kutter, 1977c: 161.
**alpinus.** *Strongylognathus huberi* subsp. *alpinus* Wheeler, W.M. 1909g: 178, fig. 1 (w.q.m.) SWITZERLAND. Hauschteck, 1962: 219 (k.). Race of *huberi:* Kutter, 1920: 528; Kutter, 1923: 387. Raised to species: Bondroit, 1918: 111; Baroni Urbani, 1969c: 140. See also: Kutter, 1977c: 165.
**arnoldii.** *Strongylognathus arnoldii* Radchenko, 1985a: 1514, fig. 1 (w.q.m.) RUSSIA.
*barnassarensis. Strongylognathus testaceus* subsp. *barnassarensis* Arnol'di, 1948: 211. *Nomen nudum.*
*bulgaricus. Strongylognathus rehbinderi* subsp. *bulgaricus* Pisarski, 1966: 515 (w.q.m.) BULGARIA. [First available use of *Strongylognathus huberi* subsp. *rehbinderi* var. *bulgarica* Viehmeyer, 1922: 211; unavailable name.] Raised to species: Bolton, 1976: 306; Atanassov & Dlussky, 1992: 158. Synonym of *kratochvili:* Pisarski, 1966: 515. [Pisarski gives *bulgaricus* as senior synonym, but *kratochvili* has priority.]
**caeciliae.** *Strongylognathus caeciliae* Forel, 1897a: 132 (q.) SPAIN. Forel, 1900e: 287 (m.); Baroni Urbani, 1969c: 138 (w.). Subspecies of *afer:* Emery, 1909d: 712; Wheeler, W.M. 1909g: 186. Revived status as species: Baroni Urbani, 1969c: 138.
**cecconii.** *Strongylognathus huberi* subsp. *cecconii* Emery, 1916b: 199 (q.) ITALY. [First available use of *Strongylognathus huberi* subsp. *rehbinderi* var. *cecconii* Emery, 1908d: 24; unavailable name.] Raised to species: Baroni Urbani, 1969c: 133.
**chelifer.** *Strongylognathus cheliferus* Radchenko, 1985a: 1516, fig. 2 (q.m.) RUSSIA.
**christophi.** *Strongylognathus christophi* Emery, 1889c: 439 (q.) RUSSIA. Forel, 1900e: 280 (w.); Radchenko, 1985a: 1519 (m.). Subspecies/variety of *huberi:* Emery, 1898c: 145; Emery, 1909d: 709; Wheeler, W.M. 1909g: 186; Emery, 1922f: 285. Status as species: Forel, 1904c: 1; Ruzsky, 1905b: 545; Pisarski, 1966: 510; Bolton, 1976: 305; Tarbinsky, 1976: 116; Radchenko, 1985a: 1519.
**dalmaticus.** *Strongylognathus dalmaticus* Baroni Urbani, 1969c: 154, figs. 2–4 (w.) YUGOSLAVIA.
**destefanii.** *Strongylognathus destefanii* Emery, 1915a: 263 (q.) SICILY. Donisthorpe, 1927a: 6 (w.). Material of the unavailable name *kutteri* referred here by Baroni Urbani, 1964b: 54.
*diabaransis. Strongylognathus rehbinderi* subsp. *diabaransis* Arnol'di, 1948: 211. *Nomen nudum.*
*diveri. Strongylognathus diveri* Donisthorpe, 1936c: 113, figs. 1, 2, 7 (w.) GREAT BRITAIN. Junior synonym of *testaceus:* Brown, 1955c: 113.
*emarginatus. Myrmus emarginatus* Schenck, 1853: 188 (w.) GERMANY. Junior synonym of *testaceus:* Mayr, 1853d: 390.
**emeryi.** *Strongylognathus emeryi* Menozzi, 1921: 30, fig. 3 (q.) ITALY.
**foreli.** *Strongylognathus huberi* subsp. *foreli* Emery, 1922f: 286 (w.) ALGERIA. [First available use of *Strongylognathus huberi* subsp. *huberi* var. *foreli* Emery, 1909d: 709; unavailable name.] Subspecies of *huberi:* Kutter, 1923: 388. Raised to species: Bolton, 1976: 308.
*gallica. Strongylognathus gallica* Bondroit, 1918: 111 (w.) FRANCE. [First available use of *Strongylognathus huberi* subsp. *huberi* var. *gallica* Emery, 1909d: 709; unavailable name.] Variety of *huberi:* Kutter, 1923: 388. Junior synonym of *huberi:* Baroni Urbani, 1962: 132.
**huberi.** *Strongylognathus huberi* Forel, 1874: 71 (see also: 94) (w.) SWITZERLAND. André, 1885: 851 (q.); Forel, 1900e: 277 (m.); Hauschteck, 1962: 213 (k.). Senior synonym of *gallica:* Baroni Urbani, 1962: 132. See also: Emery, 1916b: 201; Baroni Urbani, 1969c: 137.
*hyrcanus. Strongylognathus rehbinderi* subsp. *hyrcanus* Arnol'di, 1948: 211. *Nomen nudum.*
**insularis.** *Strongylognathus insularis* Baroni Urbani, 1968b: 470, figs. 38, 39 (w.) MALTA.
**italicus.** *Strongylognathus huberi* subsp. *italica* Finzi, 1924a: 14 (q.) ITALY. Raised to species: Bolton, 1976: 308.
**karawajewi.** *Strongylognathus karawajewi* Pisarski, 1966: 521, figs. 40–43 (w.) UKRAINE. Radchenko, 1991: 88 (m.). See also: Radchenko, 1985a: 1519.
**kervillei.** *Strongylognathus kervillei* Santschi, 1921a: 113 (w.) TURKEY.
**koreanus.** *Strongylognathus koreanus* Pisarski, 1966: 519, figs. 36–39 (w.) KOREA.
**kratochvili.** *Strongylognathus kratochvili* Silhavy, 1937: 5, 6 figs. (w.q.) CZECHOSLOVAKIA. Synonym of *bulgaricus:* Pisarski, 1966: 515. [Pisarski gives *bulgaricus* as senior name, but *kratochvili* has priority.]
*kutteri. Strongylognathus huberi* st. *cecconii* var. *kutteri* Santschi, 1927a: 58 (q.) SICILY. Unavailable name; material referred to *destefanii* by Baroni Urbani, 1964b: 54.
**minutus.** *Strongylognathus minutus* Radchenko, 1991: 84, fig. 1 (w.q.m.) TURKMENIA.
**palaestinensis.** *Strongylognathus palaestinensis* Menozzi, 1933b: 75, fig. 5 (q.) ISRAEL.
**pisarskii.** *Strongylognathus pisarskii* Poldi, 1994: 187, fig. 1 (w.) ITALY.
**rehbinderi.** *Strongylognathus christophi* var. *rehbinderi* Forel, 1904c: 2 (w.) CAUCASUS. Emery, 1909d: 710 (q.); Pisarski, 1966: 512 (m.). Subspecies of *huberi:* Emery, 1909d: 710; Wheeler, W.M. 1909g: 186. Raised to species: Arnol'di, 1948: 211; Pisarski, 1966: 512; Arnol'di & Dlussky, 1978: 546.
**ruzskyi.** *Strongylognathus huberi* subsp. *ruzskyi* Emery, 1909d: 711 (w.) RUSSIA. Radchenko, 1985a: 1517 (q.m.). Raised to species: Bolton, 1976: 309; Arnol'di & Dlussky, 1978: 546; Dlussky & Zabelin, 1985: 233.
*scabrinodoides. Strongylognathus scabrinodoides* Sadil, 1953c: 207. *Nomen nudum.*
**silvestrii.** *Strongylognathus silvestrii* Menozzi, 1936d: 292, fig. 14 (w.q.) GREECE.
**testaceus.** *Eciton testaceum* Schenck, 1852: 117 (w.q.m.) GERMANY. Combination in *Strongylognathus:* Mayr, 1853d: 390. Senior synonym of *emarginatus:* Mayr, 1853d: 390; of *diveri:* Brown, 1955c: 113. See also: Baroni Urbani, 1971c: 151; Kutter, 1977c: 166; Atanassov & Dlussky, 1992: 156.

**STRUMIGENYS** (World tropics and subtropics)
*abdita* Wesson, L.G. & Wesson, R.G. 1939; see under ***SMITHISTRUMA.***
**adrasora.** *Strumigenys adrasora* Bolton, 1983: 364, fig. 68 (w.) RWANDA.
*aequalis* Menozzi, 1942; see under ***SERRASTRUMA.***
*alberti* Forel, 1893; see under ***SMITHISTRUMA.***
*alluaudi* Santschi, 1910; see under ***SERRASTRUMA.***
**anetes.** *Strumigenys anetes* Brown, 1988f: 42, 2 figs. (w.) AUSTRALIA.
*angulata* Smith, M.R. 1931; see under ***SMITHISTRUMA.***
*angusticeps. Strumigenys saliens* var. *angusticeps* Forel, 1912e: 198 (w.) BRAZIL. Junior synonym of *saliens:* Brown, 1954l: 55.
**antaeus.** *Strumigenys antaeus* Mann, 1919: 355, fig. 34 (w.) SOLOMON IS. Current subspecies: nominal plus *fuscior.*
*antarctica. Strumigenys antarctica* Forel, 1892c: 338 (w.) NEW ZEALAND. Junior synonym of *perplexa:* Emery, 1922f: 321.
**arnoldi.** *Strumigenys arnoldi* Forel, 1913a: 114 (w.) ZIMBABWE. See also: Brown, 1954k: 26; Bolton, 1983: 365.

***STRUMIGENYS*** *(continued)*

*assamensis* De Andrade, in Baroni Urbani & De Andrade, 1994; see under ***SMITHISTRUMA.***

*australis. Strumigenys szalayi* subsp. *australis* Forel, 1910b: 50 (w.m.) AUSTRALIA. Wheeler, G.C. & Wheeler, J. 1955a: 137 (l.). Junior synonym of *szalayi:* Brown, 1971c: 75.

*banillensis. Strumigenys eggersi* var. *banillensis* Santschi, 1930c: 80 (w.) CUBA. Junior synonym of *gundlachi:* Brown, 1960b: 40.

*batesi. Strumigenys batesi* Forel, 1911e: 264 (w.) BRAZIL. Junior synonym of *mandibularis:* Brown, 1953d: 53.

*bequaerti* Santschi, 1923; see under ***SERRASTRUMA.***

*berlesei. Strumigenys eggersi* var. *berlesei* Weber, 1934a: 36 (q.) CUBA. Junior synonym of *gundlachi:* Brown, 1960b: 40.

***bernardi.*** *Strumigenys bernardi* Brown, 1960c: 206 (w.) ZAIRE. See also: Bolton, 1983: 356.

*biconvexa* Santschi, 1913; see under ***SERRASTRUMA.***

*bierigi. Strumigenys bierigi* Santschi, 1930c: 80 (w.) CUBA. Junior synonym of *gundlachi:* Brown, 1960b: 40.

*bimarginata* Wesson, L.G. & Wesson, R.G. 1939; see under ***SMITHISTRUMA.***

***biolleyi.*** *Strumigenys biolleyi* Forel, 1908b: 43 (w.) COSTA RICA. Brown, 1953f: 103 (q.); Wheeler, G.C. & Wheeler, J. 1960b: 26 (l.). Senior synonym of *luctuosa* and *tridens:* Brown, 1953f: 101.

***biroi.*** *Strumigenys biroi* Emery, 1897c: 580, pl. 14, fig. 16 (w.) NEW GUINEA.

*bismarckensis. Strumigenys mayri* var. *bismarckensis* Forel, 1901b: 10 (w.) BISMARCK ARCHIPELAGO. Junior synonym of *mayri:* Brown, 1973c: 264.

***bitheria.*** *Strumigenys bitheria* Bolton, 1983: 367 (w.) CAMEROUN.

*boerorum* Santschi, 1913; see under ***SERRASTRUMA.***

***boneti.*** *Strumigenys boneti* Brown, 1959a: 103 (w.) MEXICO.

***borgmeieri.*** *Strumigenys borgmeieri* Brown, 1954l: 57 (w.) BRAZIL.

*brevicornis* Mann, 1922; see under ***NEOSTRUMA.***

*brevisetosa* Smith, M.R. 1935; see under ***SMITHISTRUMA.***

*bruchi. Strumigenys bruchi* Forel, 1912e: 197 (w.m.) ARGENTINA. Junior synonym of *louisianae:* Brown, 1953g: 28.

***bryanti.*** *Strumigenys bryanti* Wheeler, W.M. 1919e: 95 (q.) BORNEO. See also: Brown, 1954f: 164.

***buleru.*** *Strumigenys buleru* Brown, 1988c: 39, fig. (w.q.) AUSTRALIA.

*butteli. Strumigenys godeffroyi* var. *butteli* Forel, 1913k: 83 (w.q.) SRI LANKA. Junior synonym of *godeffroyi:* Brown, 1949d: 17.

***cacaoensis.*** *Strumigenys cacaoensis* Bolton, 1971: 59, fig. 1 (w.q.) NIGERIA. See also: Bolton, 1983: 367.

*calypso* Santschi, 1923; see under ***SERRASTRUMA.***

*caribbea. Strumigenys (Strumigenys) caribbea* Weber, 1934a: 43, fig. 12 (w.q.) CUBA. Junior synonym of *silvestrii:* Brown, 1959f: 25.

***carinithorax.*** *Strumigenys carinithorax* Borgmeier, 1934: 103, fig. 4 (not 6 as stated); pl. 1, fig. 3 (w.q.m.) SURINAM. See also: Brown, 1959f: 29.

***chapmani.*** *Strumigenys chapmani* Brown, 1954j: 68, fig. 1 (w.m.) PHILIPPINES.

***chernovi.*** *Strumigenys chernovi* Dlussky, 1993: 57, figs. 2, 3 (w.q.) FIJI IS.

***chyzeri.*** *Strumigenys chyzeri* Emery, 1897c: 576, pl. 14, figs. 4–6 (w.q.m.) NEW GUINEA.

*clasmospongia. Strumigenys clasmospongia* Brown, 1953e: 2 (w.) BRAZIL. Junior synonym of *louisianae:* Brown, 1961a: 64.

*clavata. Strumigenys (Strumigenys) clavata* Weber, 1934a: 32, fig. 8 (w.) GUATEMALA. Junior synonym of *subedentata:* Brown, 1960b: 48.

*cliens* Forel, 1913; see under ***SERRASTRUMA.***

*clypeata* Roger, 1863; see under ***SMITHISTRUMA.***

***cochlearis.*** *Strumigenys cochlearis* Brown, 1988d: 40, 2 figs. (w.) AUSTRALIA.

*cognata* Santschi, 1911; see under ***SERRASTRUMA.***

*concolor* Santschi, 1914; see under ***SERRASTRUMA.***

***connectens.*** *Strumigenys connectens* Kempf, 1958b: 59, figs. 1–3 (w.) COLOMBIA.

***consanii.*** *Strumigenys consanii* Brown, 1954c: 192 (w.) COSTA RICA.

*conspersa* Emery, 1906; see under ***SMITHISTRUMA.***

*convexiceps* Santschi, 1931; see under ***GLAMYROMYRMEX.***

***cordovensis.*** *Strumigenys cordovensis* Mayr, 1887: 573 (w.) MEXICO. See also: Brown, 1958c: 218.

***cosmostela.*** *Strumigenys cosmostela* Kempf, 1975a: 413, figs. 1, 2 (w.) BRAZIL.

*costaricensis. Strumigenys (Strumigenys) louisianae* subsp. *costaricensis* Weber, 1934a: 39 (w.) COSTA RICA. Junior synonym of *louisianae:* Brown, 1953g: 28.

*crassicornis* Mayr, 1887; see under ***NEOSTRUMA.***

*creightoni* Smith, M.R. 1931; see under ***SMITHISTRUMA.***

*cubaensis. Strumigenys eggersi* var. *cubaensis* Mann, 1920: 430 (w.) CUBA. Junior synonym of *louisianae:* Brown, 1953g: 28.

***cultrigera.*** *Strumigenys cultriger* Mayr, 1887: 571 (w.) BRAZIL. See also: Brown, 1957c: 97.

***decollata.*** *Strumigenys decollata* Mann, 1919: 353, figs. 32, 33 (w.m.) SOLOMON IS. See also: Brown, 1958i: 109.

***deltisquama.*** *Strumigenys deltisquama* Brown, 1957c: 99, fig. 1 (w.) PANAMA.

***denticulata.*** *Strumigenys denticulata* Mayr, 1887: 576 (w.) BRAZIL. Brown, 1960b: 47 (q.).

*depressiceps* Weber, 1934; see under ***SMITHISTRUMA.***

***dextra.*** *Strumigenys dextra* Brown, 1954k: 27 (w.) UGANDA. See also: Bolton, 1983: 368.

*dietrichi* Smith, M.R. 1931; see under ***SMITHISTRUMA.***

***disarmata.*** *Strumigenys disarmata* Brown, 1971c: 80, fig. 6 (w.q.) NEW GUINEA.

*dohertyi* Emery, 1897; see under ***SMITHISTRUMA.***

***dolichognatha.*** *Strumigenys (Strumigenys) cordovensis* subsp. *dolichognatha* Weber, 1934a: 40 (w.m.) GUIANA. Raised to species: Brown, 1958c: 223.

***doriae.*** *Strumigenys doriae* Emery, 1887b: 469, pl. 2, fig. 22 (w.) INDONESIA (Ambon I.). Imai, Kubota, *et al.* 1985: 47 (k.). See also: Brown, 1954f: 166.

***dromoshaula.*** *Strumigenys dromoshaula* Bolton, 1983: 369 (w.) BURUNDI.

***dyak.*** *Strumigenys dyak* Brown, 1959g: 81 (w.) BORNEO.

***dyshaula.*** *Strumigenys dyshaula* Bolton, 1983: 370, fig. 55 (w.q.) ZIMBABWE.

***ebbae.*** *Strumigenys ebbae* Forel, 1905c: 11 (w.) JAVA.

***ecliptacoca.*** *Strumigenys ecliptacoca* Brown, 1958i: 111 (w.q.) NEW GUINEA.

***eggersi.*** *Strumigenys eggersi* Emery, 1890b: 69, pl. 7, fig. 9 (w.q.) ANTILLES IS. See also: Brown, 1960b: 46.

**electrina* De Andrade, in Baroni Urbani & De Andrade, 1994; see under ***SMITHISTRUMA.***

***elongata.*** *Strumigenys elongata* Roger, 1863a: 212 (w.) PANAMA. Brown, 1954c: 191 (q.m.); Wheeler, G.C. & Wheeler, J. 1955a: 138 (l.). Senior synonym of *imitator* and *nicaraguensis:* Brown, 1954c: 189.

*emarginata* Mayr, 1901; see under ***SMITHISTRUMA.***

***emdeni.*** *Strumigenys emdeni* Forel, 1915b: 41 (w.) AUSTRALIA. Combination in *Labidogenys:* Brown, 1948e: 103; in *Strumigenys:* Taylor & Brown, D.R. 1985: 87.

***emeryi.*** *Strumigenys (Strumigenys) emeryi* Mann, 1922: 37, fig. 18 (w.q.) HONDURAS. See also: Brown, 1959a: 97.

*emeryi* Karavaiev, 1935; see under ***SMITHISTRUMA.***

*emiliae* Forel, 1907; see under ***SMITHISTRUMA.***

*epinotalis* Weber, 1934; see under ***SMITHISTRUMA.***

*escherichi* Forel, 1910; see under ***SERRASTRUMA.***

***esrossi.*** *Strumigenys esrossi* Brown, 1957g: 114 (w.q.) PHILIPPINES.

***ettillax.*** *Strumigenys ettillax* Bolton, 1983: 371 (w.) CAMEROUN.

*excisa. Eneria excisa* Donisthorpe, 1948d: 598, fig. 1 (w.q.) NEW GUINEA. Junior synonym of *loriae:* Brown, 1949d: 15.

***fairchildi.*** *Strumigenys fairchildi* Brown, 1961a: 60 (w.) PANAMA.

***faurei.*** *Strumigenys faurei* Arnold, 1948: 226, fig. 12 (w.) SOUTH AFRICA. Junior synonym of *rufobrunea:* Brown, 1954k: 17. Revived from synonymy: Bolton, 1983: 371.

***feae.*** *Strumigenys feae* Emery, 1895k: 473 (w.q.) BURMA.

***ferocior.*** *Strumigenys ferocior* Brown, 1973c: 266, fig. 3 (w.q.) AUSTRALIA.

*ferox. Strumigenys (Strumigenys) ferox* Weber, 1934a: 41, fig. 4 (w.) GUIANA. Junior synonym of *godmani:* Brown, 1953d: 55.

*foochowensis* Wheeler, W.M. 1928; see under ***TRICHOSCAPA.***

***forficata.*** *Strumigenys forficata* Brown, 1959g: 86 (w.) BORNEO.

***formosensis.*** *Strumigenys feae* var. *formosensis* Forel, 1912a: 52 (w.) TAIWAN. Brown, 1949d: 24 (q.). Raised to species: Brown, 1949d: 19. See also: Terayama & Kubota, 1989: 781.

*fridericimuelleri* Forel, 1886; see under ***SMITHISTRUMA.***

***friedae.*** *Strumigenys friedae* Forel, 1915b: 42 (w.q.) AUSTRALIA. Imai, Crozier & Taylor, 1977: 349 (k.).

***frivaldszkyi.*** *Strumigenys frivaldszkyi* Emery, 1897c: 580, pl. 14, fig. 9 (w.) NEW GUINEA.

*fusca. Strumigenys fusca* Emery, 1894c: 215, pl. 1, fig. 8 (w.) BRAZIL. Junior synonym of *louisianae:* Brown, 1961a: 64.

#***fuscior.*** *Strumigenys antaeus* var. *fuscior* Mann, 1919: 357 (w.) SOLOMON IS.

*fusciventris* Santschi, 1915; see under ***SERRASTRUMA.***

***gemella.*** *Strumigenys gemella* Kempf, 1975a: 416, fig. 5 (w.) COLOMBIA.

*gerardi* Santschi, 1923; see under ***SERRASTRUMA.***

*glanduscula* Santschi, 1919; see under ***SERRASTRUMA.***

***godeffroyi.*** *Strumigenys godeffroyi* Mayr, 1866a: 516 (w.) SAMOA. Mayr, 1887: 569 (q.); Imai, Brown, *et al.* 1984: 68 (k.). Senior synonym of *butteli* and *indica:* Brown, 1949d: 17. See also: Bingham, 1903: 149.

***godmani.*** *Strumigenys godmani* Forel, 1899c: 42, pl. 3, fig. 5 (q.) PANAMA. Wheeler, G.C. & Wheeler, J. 1960b: 26 (l.). Senior synonym of *ferox:* Brown, 1953d: 55.

***grandidieri.*** *Strumigenys grandidieri* Forel, 1892k: 517 (w.) MADAGASCAR. See also: Brown, 1954k: 12.

*guatemalensis. Strumigenys (Strumigenys) louisianae* subsp. *guatemalensis* Weber, 1934a: 39 (w.) GUATEMALA. Junior synonym of *louisianae:* Brown, 1953g: 28.

*guianensis* Weber, 1934; see under ***SMITHISTRUMA.***

***gundlachi.*** *Pyramica gundlachi* Roger, 1862a: 253, pl. 1, fig. 18 (w.) CUBA. Brown, 1960b: 44 (q.). Combination in *Strumigenys:* Roger, 1863b: 40. Senior synonym of *banillensis, berlesei, bierigi, infuscata, isthmica* and *vincentensis:* Brown, 1960b: 40.

***guttulata.*** *Strumigenys guttulata* Forel, 1902h: 458 (w.) AUSTRALIA. See also: Brown, 1973c: 264.

***hastyla.*** *Strumigenys hastyla* Bolton, 1983: 372 (w.) BURUNDI.

***havilandi.*** *Strumigenys havilandi* Forel, 1905c: 13 (footnote) (w.q.) SOUTH AFRICA. See also: Brown, 1954k: 25; Bolton, 1983: 373.

***helytruga.*** *Strumigenys helytruga* Bolton, 1983: 374 (w.) ANGOLA.

***hemidisca.*** *Strumigenys hemidisca* Brown, 1953f: 107 (w.) VENEZUELA.

***hemichlaena.*** *Strumigenys hemichlaena* Brown, 1971c: 78 (w.) NEW GUINEA.

***hindenburgi.*** *Strumigenys hindenburgi* Forel, 1915c: 357 (w.) ARGENTINA. Brown, 1961a: 63 (q.).

***hoplites.*** *Strumigenys hoplites* Brown, 1973c: 266, fig. 2 (w.) NEW GUINEA.

***horvathi.*** *Strumigenys horvathi* Emery, 1897c: 577, pl. 14, fig. 8 (w.) NEW GUINEA. Brown, 1973c: 261 (q.).

*imitator. Strumigenys imitator* Mayr, 1887: 572 (w.m.) BRAZIL. Junior synonym of *elongata:* Brown, 1954c: 189.

*inaequalis. Strumigenys smithi* var. *inaequalis* Emery, 1890b: 67, pl. 7, fig. 3 (w.) BRAZIL. Junior synonym of *smithii:* Brown, 1953f: 104.

*incisa. Strumigenys incisa* Godfrey, 1907: 102 (w.) GREAT BRITAIN. Junior synonym of *rogeri:* Donisthorpe, 1915d: 341.

*indica. Strumigemys godefroyi* var. *indica* Forel, 1902c: 243 (q.) INDIA. Junior synonym of *godeffroyi:* Brown, 1949d: 17.

***indigatrix.*** *Strumigenys indigatrix* Wheeler, W.M. 1919e: 94 (w.) BORNEO. See also: Brown, 1959g: 86.

*inezae* Forel, 1905; see under ***SMITHISTRUMA.***

*infidelis. Strumigenys infidelis* Santschi, 1919f: 48 (w.) ARGENTINA. Junior synonym of *louisianae:* Brown, 1953g: 28.

*infuscata. Strumigenys (Strumigenys) eggersi* subsp. *infuscata* Weber, 1934a: 35 (w.q.) CUBA. Junior synonym of *gundlachi:* Brown, 1960b: 40.

*intermedia* Wheeler, W.M. 1913; see under ***SMITHISTRUMA.***

***irrorata.*** *Strumigenys (Strumigenys) irrorata* Santschi, 1913b: 257 (diagnosis in key) (w.) SOUTH AFRICA. Santschi, 1914e: 29 (w.). See also: Brown, 1954k: 33; Bolton, 1983: 375.

*isthmica. Strumigenys (Strumigenys) eggersi* var. *isthmica* Santschi, 1931c: 276 (w.) PANAMA. Junior synonym of *gundlachi:* Brown, 1960b: 40.

*jacobsoni* Menozzi, 1939; see under ***SMITHISTRUMA.***

***jamaicensis.*** *Strumigenys jamaicensis* Brown, 1959c: 6 (w.) JAMAICA. See also: Brown, 1960b: 45.

*japonica* Ito, 1914; see under ***SMITHISTRUMA.***

***jepsoni.*** *Strumigenys jepsoni* Mann, 1921: 462, fig. 22 (w.) FIJI IS.

***juliae.*** *Strumigenys juliae* Forel, 1905c: 12 (w.) JAVA.

*karawajewi* Brown, 1948; see under ***SMITHISTRUMA.***

***katapelta.*** *Strumigenys katapelta* Bolton, 1983: 375, figs. 61, 76 (w.) BURUNDI.

***koningsbergeri.*** *Strumigenys koningsbergeri* Forel, 1905c: 9 (w.q.) JAVA.

***korahyla.*** *Strumigenys korahyla* Bolton, 1983: 376, fig. 64 (w.) CAMEROUN.

***kraepelini.*** *Strumigenys kraepelini* Forel, 1905c: 8 (w.) JAVA.

***lacacoca.*** *Strumigenys lacacoca* Brown, 1959a: 101 (w.) PANAMA. See also: Brown, 1961a: 58.

*laevinasis* Smith, M.R. 1931; see under ***SMITHISTRUMA.***

***laevipleura.*** *Strumigenys laevipleura* Kempf, 1958b: 64, figs. 5–7 (w.) COLOMBIA.

***lanuginosa.*** *Strumigenys lanuginosa* Wheeler, W.M. 1905b: 104, fig. M (w.q.) BAHAMAS. See also: Brown, 1961a: 62.

*laticephala. Strumigenys louisianae* subsp. *laticephala* Smith, M.R. 1931c: 690, pl. 1, fig. 2 (w.) U.S.A. Haug, 1932: 170 (m.). Junior synonym of *louisianae:* Smith, M.R. 1951a: 826; Brown, 1953g: 28.

*latiuscula* Forel, 1916; see under ***SERRASTRUMA.***

*leae. Strumigenys leae* Forel, 1913g: 182 (w.q.) TASMANIA. Junior synonym of *perplexa:* Brown, 1958h: 38.

*leptothrix* Wheeler, W.M. 1929; see under ***SMITHISTRUMA.***

***lewisi.*** *Strumigenys lewisi* Cameron, 1886: 229 (w.q.) JAPAN. Wheeler, G.C. & Wheeler, J. 1955a: 140 (l.). Subspecies/variety of *godeffroyi:* Mayr, 1887: 569 (footnote); Wheeler, W.M. 1906c: 318; Emery, 1922f: 321; Wheeler, W.M. 1923b: 4; Wheeler, W.M. 1928d: 115. Revived status as species: Bingham, 1903: 149; Brown, 1949d: 16.

*limbata* Forel, 1913; see under ***SERRASTRUMA.***

***liukueiensis.*** *Strumigenys liukueiensis* Terayama & Kubota, 1989: 785, figs. 18–22 (w.q.) TAIWAN.

***londianensis.*** *Proscopomyrmex londianensis* Patrizi, 1946: 295, figs. 1, 2 (w.) KENYA. Combination in *Strumigenys:* Arnold, 1948: 227; Brown, 1949d: 15. See also: Brown, 1954k: 14; Bolton, 1983: 377.

*longicornis. Strumigenys unispinulosa* var. *longicornis* Emery, 1894c: 214 (w.) BOLIVIA. Junior synonym of *louisianae:* Brown, 1953g: 28.

*longipilis* Weber, 1934; see under ***SMITHISTRUMA.***

***longispinosa.*** *Strumigenys longispinosa* Brown, 1958e: 123, figs. 1, 2 (w.) PANAMA.

***lopotyle.*** *Strumigenys lopotyle* Brown, 1969a: 27, figs. 1–3 (w.q.) NEW GUINEA.

***loriae.*** *Strumigenys loriae* Emery, 1897c: 576, pl. 14, fig. 3 (w.) NEW GUINEA. Szabó, 1910a: 368 (q.). Senior synonym of *excisa:* Brown, 1949d: 15.

*lotti* Weber, 1943; see under ***SERRASTRUMA.***

***louisianae.*** *Strumigenys louisianae* Roger, 1863a: 211 (w.) U.S.A. Emery, 1895c: 326 (q.); Haug, 1932: 170 (m.); Wheeler, G.C. & Wheeler, J. 1955a: 136 (l.). Senior synonym of *unispinulosa:* Emery, 1895c: 326; Brown, 1953g: 28; of *laticephala:* Smith, M.R. 1951a: 826; Brown, 1953g: 28; of *bruchi, costaricensis, cubaensis, guatemalensis, infidelis, longicornis, obscuriventris* and *soledadensis:* Brown, 1953g: 28; of *clasmospongia, fusca* and *unidentata:* Brown, 1961a: 64. See also: Wilson, 1954: 481.

*luctuosa. Strumigenys (Strumigenys) luctuosa* Menozzi, 1936c: 81, fig. 1 (w.) COSTA RICA. Junior synonym of *biolleyi:* Brown, 1953f: 101.

***ludia.*** *Strumigenys (Strumigenys) ludia* Mann, 1922: 35, fig. 17 (w.) HONDURAS. Brown, 1954c: 195 (q.). Senior synonym of *tenuis:* Brown, 1954c: 194.

*ludovici* Forel, 1904; see under ***SERRASTRUMA.***

*lujae* Forel, 1902; see under ***SERRASTRUMA.***

***lyroessa.*** *Labidogenys lyroessa* Roger, 1862a: 251, pl. 1, fig. 17 (w.) SRI LANKA. Combination in *Strumigenys:* Roger, 1863b: 40; in *Labidogenys:* Brown, 1948e: 103; in *Strumigenys:* current combination.

***mailei.*** *Strumigenys mailei* Wilson & Taylor, 1967: 38, fig. 28 (w.) SAMOA. Dlussky, 1993: 59 (q.).

***mandibularis.*** *Strumigenys mandibularis* Smith, F. 1860c: 72, pl. 4, figs. 6, 8, 10 (q., not w.; see Mayr, 1887: 574) BRAZIL. Senior synonym of *batesi:* Brown, 1953d: 53. See also: Donisthorpe, 1948h: 78.

*manni* Wesson, L.G. & Wesson, R.G. 1939; see under ***SMITHISTRUMA.***

*margaritae* Forel, 1893; see under ***SMITHISTRUMA.***

***marginiventris.*** *Strumigenys (Strumigenys) fusca* st. *marginiventris* Santschi, 1931c: 275 (w.) PANAMA. Brown, 1958e: 127 (q.). Raised to species: Brown, 1958e: 126.

*marioni* Wheeler, W.M. 1933; see under ***TRICHOSCAPA.***

***marleyi.*** *Strumigenys havilandi* r. *marleyi* Arnold, 1914: 31, fig. 10 (w.) SOUTH AFRICA. Raised to species: Arnold, 1926: 286. See also: Brown, 1954k: 25; Bolton, 1983: 378.

*maynei* Forel, 1916; see under ***SERRASTRUMA.***

***mayri.*** *Strumigenys mayri* Emery, 1897c: 579, pl. 14, fig. 12 (w.q.) NEW GUINEA. Brown, 1973c: 265. Senior synonym of *bismarckensis:* Brown, 1973c: 264.

*medialis* Wesson, L.G. & Wesson, R.G. 1939; see under ***SMITHISTRUMA.***

*membranifera* Emery, 1869; see under ***TRICHOSCAPA.***

***mesahyla.*** *Strumigenys mesahyla* Bolton, 1983: 379, fig. 69 (w.) ZIMBABWE.

***micretes.*** *Strumigenys micretes* Brown, 1959a: 100 (w.) COSTA RICA.

***minutula.*** *Strumigenys minutula* Terayama & Kubota, 1989: 782, figs. 13–17 (w.q.) TAIWAN.

*mirabilis* Mann, 1926; see under ***TINGIMYRMEX.***

*missouriensis* Smith, M.R. 1931; see under ***SMITHISTRUMA.***

***mixta.*** *Strumigenys mixta* Brown, 1953e: 4 (w.) GUATEMALA.

***mjoebergi.*** *Strumigenys mjoebergi* Brown, 1959g: 83 (w.q.) BORNEO.

***moczaryi.*** *Strumigenys moczaryi* Emery, 1897c: 580, pl. 14, fig. 15 (w.) NEW GUINEA.

***mokensis.*** *Strumigenys cordovensis* var. *mokensis* Forel, 1905b: 158 (w.) VENEZUELA. Raised to species: Brown, 1958c: 221.

*mumfordi* Wheeler, W.M. 1932; see under ***SMITHISTRUMA.***

***murshila.*** *Strumigenys murshila* Bolton, 1983: 380, fig. 77 (w.) RWANDA.

*mustelina* Weber, 1934; see under ***NEOSTRUMA.***

*nana* Santschi, 1930; see under ***SMITHISTRUMA.***

*nepalensis* De Andrade, in Baroni Urbani & De Andrade, 1994; see under ***SMITHISTRUMA.***

***nevermanni.*** *Strumigenys nevermanni* Brown, 1959a: 99 (w.q.) COSTA RICA.

*nicaraguensis. Strumigenys (Strumigenys) elongata* subsp. *nicaraguensis* Weber, 1934a: 31 (w.) NICARAGUA. Junior synonym of *elongata:* Brown, 1954c: 189.

***nidifex.*** *Strumigenys nidifex* Mann, 1921: 464, fig. 23 (w.q.m.) FIJI IS. Wheeler, G.C. & Wheeler, J. 1955a: 140 (l.). See also: Brown, 1971c: 81.

*nigeriensis* Santschi, 1914; see under ***SERRASTRUMA.***

***nigra.*** *Strumigenys nigra* Brown, 1971c: 79 (w.) NEW GUINEA.

*nigrescens* Wheeler, W.M. 1911; see under ***SMITHISTRUMA.***

***nimbrata.*** *Strumigenys nimbrata* Bolton, 1983: 381 (w.q.) IVORY COAST.

*nitens* Santschi, 1932; see under ***DORISIDRIS.***

*normandi* Santschi, 1910; see under ***SMITHISTRUMA.***

*obscuriventris. Strumigenys louisianae* var. *obscuriventris* Wheeler, W.M. 1908a: 145, pl. 12, fig. 14 (w.) PUERTO RICA. Junior synonym of *louisianae:* Brown, 1953g: 28.

*obscuriventris* Santschi, 1914; see under ***SERRASTRUMA.***

***ogloblini.*** *Strumigenys (Strumigenys) ogloblini* Santschi, 1936d: 409, figs. 15, 16 (w.) ARGENTINA. Brown, 1958f: 136 (q.).

*ohioensis* Kennedy & Schramm, 1933; see under ***SMITHISTRUMA.***

***omalyx.*** *Strumigenys omalyx* Bolton, 1983: 382, fig. 63 (w.q.) KENYA.

***opaca.*** *Strumigenys opaca* Brown, 1954a: 86 (w.) AUSTRALIA.

*ornata* Mayr, 1887; see under ***SMITHISTRUMA.***

***pallestes.*** *Strumigenys pallestes* Bolton, 1971: 62, figs. 2, 3 (w.q.) GHANA. See also: Bolton, 1983: 383.

***paranax.*** *Strumigenys paranax* Bolton, 1983: 383 (w.) CAMEROUN.

***paranetes.*** *Strumigenys paranetes* Brown, 1988g: 43, 2 figs. (w.) AUSTRALIA.

*pergandei* Emery, 1895; see under ***SMITHISTRUMA.***

***perparva.*** *Strumigenys perparva* Brown, 1958f: 133, fig. 1 (w.q.) TRINIDAD.

***perplexa.*** *Orectognathus perplexus* Smith, F. 1876c: 491 (w.q.) NEW ZEALAND. Wheeler, G.C. & Wheeler, J. 1955a: 140 (l.); Wheeler, G.C. & Wheeler, J. 1973c: 214 (l.). Combination in *Strumigenys:* Emery, 1897c: 575. Senior synonym of *antarctica:* Emery, 1922f: 321; of *leae:* Brown, 1958h: 38.

***petiolata.*** *Strumigenys petiolata* Bernard, 1953b: 254, fig. 14 (w.) GUINEA. Junior synonym of *rufobrunea:* Brown, 1954k: 17. Revived from synonymy: Bolton, 1983: 384.

***philiporum.*** *Strumigenys philiporum* Brown, 1988b: 38, fig. 1 (w.) AUSTRALIA.

***phytibia.*** *Strumigenys phytibia* Brown, 1957g: 117 (w.q.) PHILIPPINES.

*pilinasis* Forel, 1901; see under ***SMITHISTRUMA.***

***planeti.*** *Strumigenys planeti* Brown, 1953d: 57 (w.q.) BOLIVIA.

***precava.*** *Strumigenys precava* Brown, 1954c: 196 (w.q.) GUIANA.

***pretoriae.*** *Strumigenys pretoriae* Arnold, 1949: 267, fig. 8 (w.) SOUTH AFRICA. See also: Brown, 1954k: 15; Bolton, 1983: 385.

***princeps.*** *Strumigenys princeps* Kempf & Brown, 1969: 21, figs. 4, 5 (w.q.) BRAZIL.

*procera. Strumigenys saliens* var. *procera* Emery, 1894c: 215, pl. 1, fig. 9 (q.) BRAZIL. Junior synonym of *saliens:* Brown, 1954l: 55.

***producta.*** *Strumigenys producta* Brown, 1953e: 3 (w.) BOLIVIA.

***prospiciens.*** *Strumigenys smithi* subsp. *prospiciens* Emery, 1906c: 167, fig. 26 (w.) ARGENTINA. Brown, 1953f: 109 (q.). Junior synonym of *smithii:* Forel, 1911e: 263; of *mandibularis:* Emery, 1922f: 322. Revived from synonymy, and raised to species: Brown, 1953f: 108.

*pulchella* Emery, 1895; see under ***SMITHISTRUMA.***

*quadrispinosa* Forel, 1895; see under ***EPOPOSTRUMA.***

***quinquedentata.*** *Strumigenys quinquedentata* Crawley, 1923b: 177 (w.) AUSTRALIA.

*raymondi* Donisthorpe, 1946; see under ***SERRASTRUMA.***

***rectidens.*** *Strumigenys rectidens* Brown, 1966: 23, fig. 1 (w.) INDONESIA (Soemba I.).

*reflexa* Wesson, L.G. & Wesson, R.G. 1939; see under ***SMITHISTRUMA.***

***rehi.*** *Strumigenys rehi* Forel, 1907e: 3 (w.) BRAZIL. See also: Brown, 1958c: 222.

***relahyla.*** *Strumigenys relahyla* Bolton, 1983: 386, figs. 57, 74 (w.) ANGOLA.

*reticulata* Stitz, 1910; see under ***SERRASTRUMA.***

***rogeri.*** *Strumigenys rogeri* Emery, 1890b: 68, pl. 7, fig. 6 (w.) ANTILLES IS. Forel, 1893g: 378 (q.). Senior synonym of *incisa:* Donisthorpe, 1915d: 341; of *sulfurea:* Brown, 1954k: 20. See also: Bolton, 1983: 387.

*rohweri* Smith, M.R. 1935; see under ***SMITHISTRUMA.***

*rostrata* Emery, 1895; see under ***SMITHISTRUMA.***

*rothkirchi* Wasmann, 1918; see under ***SERRASTRUMA.***

*rudinodis* Stärcke, 1941; see under ***SMITHISTRUMA.***

***rufobrunea.*** *Strumigenys rufobrunea* Santschi, 1914d: 373 (w.q.) GUINEA. See also: Brown, 1954k: 17; Bolton, 1983: 389.

***rukha.*** *Strumigenys rukha* Bolton, 1983: 389 (w.q.) KENYA.

***saliens.*** *Strumigenys saliens* Mayr, 1887: 574 (w.q.) BRAZIL. Wheeler, G.C. & Wheeler, J. 1955a: 140 (l.). Senior synonym of *angusticeps* and *procera:* Brown, 1954l: 55.

***sanctipauli.*** *Strumigenys sanctipauli* Kempf, 1958g: 556, figs. 1–4 (w.) BRAZIL.

*santschii* Forel, 1904; see under ***TRICHOSCAPA.***

***sarissa.*** *Strumigenys sarissa* Bolton, 1983: 390, fig. 50 (w.q.) BURUNDI.

***scelesta.*** *Strumigenys scelestus* Mann, 1921: 463, fig. 22 (w.) FIJI IS.

**schleeorum* Baroni Urbani, in Baroni Urbani & De Andrade, 1994; see under ***SMITHISTRUMA.***

***schmalzi.*** *Strumigenys schmalzi* Emery, 1906c: 169 (footnote), fig. 28 (w.) BRAZIL. See also: Brown, 1959f: 28.

*schulzi* Emery, 1894; see under ***SMITHISTRUMA.***

***scotti.*** *Strumigenys scotti* Forel, 1912k: 159 (w.) SEYCHELLES IS. See also: Brown, 1954k: 23; Bolton, 1983: 391.

*sculpturata* Smith, M.R. 1931; see under ***SMITHISTRUMA.***

*serrula* Santschi, 1910; see under ***SERRASTRUMA.***

***shaula.*** *Strumigenys shaula* Bolton, 1983: 392 (w.) ZIMBABWE.

***signeae.*** *Strumigenys signeae* Forel, 1905c: 10 (w.) JAVA.

*silvestriana* Wheeler, W.M. 1928; see under ***TRICHOSCAPA.***

***silvestrii.*** *Strumigenys silvestrii* Emery, 1906c: 168, fig. 27 (w.q.) ARGENTINA. Senior synonym of *caribbea:* Brown, 1959f: 25.

*simillima* Emery, 1890; see under ***TRICHOSCAPA.***

*simoni* Emery, 1895; see under ***SERRASTRUMA.***

***sisyrata.*** *Strumigenys sisyrata* Brown, 1968b: 24, figs. 1, 2 (w.) NEW GUINEA.

*skwarrae* Wheeler, W.M. 1934; see under ***SMITHISTRUMA.***

***smithii.*** *Strumigenys smithii* Forel, 1886a: 215 (w.) BRAZIL. Forel, 1893g: 375 (q.m.). Senior synonym of *inaequalis:* Brown, 1953f: 104.

***smythiesii.*** *Strumigenys smythiesii* Forel, 1902c: 242 (w.q.) INDIA.

*soledadensis. Strumigenys (Strumigenys) louisianae* subsp. *soledadensis* Weber, 1934a: 38 (w.q.) CUBA. Junior synonym of *louisianae:* Brown, 1953g: 28.

***solifontis.*** *Strumigenys (Strumigenys) solifontis* Brown, 1949d: 18 (w.q.) JAPAN.

***spathoda.*** *Strumigenys spathoda* Bolton, 1983: 393, fig. 62 (w.) TOGO.

***stemonixys.*** *Strumigenys stemonixys* Brown, 1971c: 81, figs. 2, 4 (w.) PALAU IS.

*studiosi* Weber, 1934; see under ***SMITHISTRUMA.***

***stygia.*** *Strumigenys (Strumigenys) stygia* Santschi, 1913b: 257 (diagnosis in key) (w.) KENYA. Santschi, 1914b: 113 (w.). See also: Brown, 1954k: 29; Bolton, 1983: 394.

***subedentata.*** *Strumigenys subedentata* Mayr, 1887: 575 (w.) BRAZIL. Brown, 1960b: 49 (q.m.). Senior synonym of *clavata* and *tristani:* Brown, 1960b: 48.

***sublaminata.*** *Strumigenys sublaminata* Brown, 1959g: 84 (w.q.) BORNEO.

***sublonga.*** *Strumigenys sublonga* Brown, 1958c: 221, fig. 1 (w.q.) BOLIVIA.

*sulfurea. Strumigenys sulfurea* Santschi, 1915c: 261 (w.) GABON. Junior synonym of *rogeri:* Brown, 1954k: 20.

**szalayi.** *Strumigenys szalayi* Emery, 1897c: 578, pl. 14, figs. 10, 11 (w.q.) NEW GUINEA. Senior synonym of *australis:* Brown, 1971c: 75.
*taipingensis* Forel, 1913; see under ***SMITHISTRUMA.***
*talpa* Weber, 1934; see under ***SMITHISTRUMA.***
*tenuipilis* Emery, 1915; see under ***SMITHISTRUMA.***
*tenuis. Strumigenys (Strumigenys) ludia* subsp. *tenuis* Weber, 1934a: 31 (w.) NICARAGUA. Junior synonym of *ludia:* Brown, 1954c: 194.
**tetraphanes.** *Strumigenys tetraphanes* Brown, 1954k: 30 (w.) UGANDA. See also: Bolton, 1983: 395.
**thomae.** *Strumigenys thomae* Kempf, 1976b: 39, figs. 1–3 (w.) BRAZIL.
**tigris.** *Strumigenys tigris* Brown, 1971c: 82, fig. 7 (w.q.) NEW GUINEA.
**tococae.** *Strumigenys tococae* Wheeler, W.M. & Bequaert, 1929: 31 (w.) BRAZIL. See also: Brown, 1957c: 101.
**totyla.** *Strumigenys totyla* Bolton, 1983: 395, fig. 56 (w.) CAMEROUN.
**traegaordhi.** *Strumigenys (Strumigenys) traegaordhi* Santschi, 1913b: 257 (diagnosis in key) (w.) SOUTH AFRICA. Santschi, 1914e: 28 (w.); Emery, 1915g: 19 (q.). See also: Brown, 1954k: 27; Bolton, 1983: 396.
*transversa* Santschi, 1913; see under ***SMITHISTRUMA.***
*tridens. Strumigenys (Strumigenys) tridens* Weber, 1934a: 29, fig. 3 (w.) PANAMA. Junior synonym of *biolleyi:* Brown, 1953f: 101.
**trieces.** *Strumigenys trieces* Brown, 1960b: 50, fig. 8 (w.) COSTA RICA.
**trinidadensis.** *Strumigenys trinidadensis* Wheeler, W.M. 1922c: 12 (w.) TRINIDAD. Brown, 1954l: 61 (m.).
*tristani. Strumigenys tristani* Menozzi, 1931b: 273, fig. 8 (w.) COSTA RICA. Junior synonym of *subedentata:* Brown, 1960b: 48.
**trudifera.** *Strumigenys trudifera* Kempf & Brown, 1969: 18, figs. 1–3 (w.q.) BRAZIL.
*turneri* Forel, 1895; see under ***MESOSTRUMA.***
*uelensis* Santschi, 1923; see under ***SERRASTRUMA.***
**uichancoi.** *Strumigenys uichancoi* Brown, 1957g: 116 (w.) PHILIPPINES.
**ulcerosa.** *Strumigenys ulcerosa* Brown, 1954f: 162 (w.) INDONESIA (Sumba I.).
*unidentata. Strumigenys unidentata* Mayr, 1887: 575 (w.) BRAZIL. Junior synonym of *louisianae:* Brown, 1961a: 64.
*unispinulosa. Strumigenys unispinulosa* Emery, 1890b: 67, pl. 7, fig. 5 (w.q.) COSTA RICA. [Also described as new by Emery, 1894k: 58.] Junior synonym of *louisianae:* Emery, 1895c: 326; Brown, 1953g: 28.
**vazerka.** *Strumigenys vazerka* Bolton, 1983: 397, fig. 52 (w.q.) IVORY COAST.
*venatrix* Wesson, L.G. & Wesson, R.G. 1939; see under ***SMITHISTRUMA.***
*vincentensis. Strumigenys eggersi* var. *vincentensis* Forel, 1893g: 378 (w.) ANTILLES IS. Junior synonym of *gundlachi:* Brown, 1960b: 40.
*vitiensis* Mann, 1921; see under ***TRICHOSCAPA.***
**wallacei.** *Strumigenys wallacei* Emery, 1897c: 578, pl. 14, fig. 7 (w.) NEW GUINEA. See also: Brown, 1973c: 267.
**wheeleri.** *Strumigenys wheeleri* Mann, 1921: 466, fig. 24 (w.q.) FIJI IS.
*williamsi* Wheeler, W.M. 1933; see under ***TRICHOSCAPA.***
**wilsoni.** *Strumigenys wilsoni* Brown, 1969b: 28, figs. (w.) NEW GUINEA.
**xenognatha.** *Strumigenys xenognatha* Kempf, 1958b: 64, fig. 4 (q.) COLOMBIA.
**xenohyla.** *Strumigenys xenohyla* Bolton, 1983: 398, figs. 54, 73 (w.) CAMEROUN.
**xenos.** *Strumigenys xenos* Brown, 1955h: 182 (q.) AUSTRALIA.
**yaleopleura.** *Strumigenys yaleopleura* Brown, 1988e: 41 (w.q.) AUSTRALIA.
**yasumatsui.** *Strumigenys yasumatsui* Brown, 1971c: 84, figs. 8, 9 (w.q.) NEW GUINEA.
**zakharovi.** *Strumigenys zakharovi* Dlussky, 1993: 60, fig. 2 (w.) TONGA.
**zandala.** *Strumigenys zandala* Bolton, 1983: 399,fig. 66 (w.) EQUATORIAL GUINEA.

*SULCOMYRMEX:* see under ***TETRAMORIUM.***

*SYLLOPHOPSIS:* see under ***MONOMORIUM.***

*SYMBIOMYRMA:* see under ***MYRMICA.***

*SYMMYRMICA:* see under ***FORMICOXENUS.***

*SYMPHEIDOLE:* see under ***PHEIDOLE.***

*SYNSOLENOPSIS:* see under ***SOLENOPSIS.***

****SYNTAPHUS*** (Great Britain)
****wheeleri.*** **Syntaphus wheeleri* Donisthorpe, 1920: 84, pl. 5, figs. 1, 2 (q.) GREAT BRITAIN (Oligocene).

*SYNTERMITOPONE:* see under ***PACHYCONDYLA.***

*SYSCIA:* see under ***CERAPACHYS.***

*SYSPHINGTA:* see under ***PROCERATIUM.***

***TALARIDRIS*** (Neotropical)
**mandibularis.** *Talaridris mandibularis* Weber, 1941a: 185, figs. 1–3 (w.q.) TRINIDAD. See also: Brown & Kempf, 1960: 241.

*TAMMOTECA:* see under ***GNAMPTOGENYS.***

***#TANAEMYRMEX:*** see under ***CAMPONOTUS.***

***#TAPINOLEPIS:*** see under ***ANOPLOLEPIS.***

***TAPINOMA*** (World wide)
**acuminatum.** *Tapinoma acuminatum* Forel, 1907g: 85 (w.q.) KENYA.
*albitarse* Motschoulsky, 1863; see under ***TECHNOMYRMEX.***
**amazone.** *Tapinoma amazonae* Wheeler, W.M. 1934g: 185 (w.) BRAZIL.
**ambiguum.** *Tapinoma erraticum* subsp. *ambiguum* Emery, 1925d: 57, fig. 9 (w.q.m.) FRANCE. Junior synonym of *erraticum:* Baroni Urbani, 1964b: 55. Revived from synonymy, and raised to species: Kutter, 1977c: 181; Arnol'di & Dlussky, 1978: 547; Seifert, 1984a: 151.
*anale* André, 1893; see under ***FORELIUS.***
**andamanense.** *Tapinoma andamanensis* Forel, 1903d: 407 (q.) ANDAMAN IS. Current subspecies: nominal plus *capsincola.*
**annandalei.** *Zatapinoma annandalei* Wheeler, W.M. 1928a: 20, fig. 1 (s.w.) INDIA. Combination in *Tapinoma:* Shattuck, 1992c: 152.
**#annellatum.** *Tapinoma ramulorum* subsp. *annellatum* Wheeler, W.M. 1934g: 180 (w.q.) COSTA RICA.
**antarcticum.** *Tapinoma antarcticum* Forel, 1904c: 17 (w.q.) CHILE. Senior synonym of *eidmanni:* Menozzi, 1935c: 321; Goetsch & Menozzi, 1935: 98; of *fazi:* Menozzi, 1935c: 321.
**arnoldi.** *Tapinoma arnoldi* Forel, 1913a: 139 (w.) ZIMBABWE. Arnold, 1915: 154 (q.m.). Current subspecies: nominal plus *tectum.*
**#atomum.** *Formica atomus* Latreille, 1798: 44 (w.) FRANCE. Combination in *Tapinoma:* Roger, 1863b: 13. Junior synonym of *erraticum:* Roger, 1863b: 13. Revived from synonymy as variety/subspecies of *erraticum:* Dalla Torre, 1893: 165; Emery, 1912g: 40.
**atriceps.** *Tapinoma (Micromyrma) atriceps* Emery, 1888c: 363 (w.q.m.) BRAZIL. Current subspecies: nominal plus *breviscapum.*
*australis. Tapinoma (Micromyrma) melanocephalum* var. *australis* Santschi, 1928a: 53 (w.) SAMOA. Junior synonym of *melanocephalum:* Wilson & Taylor, 1967: 80.
*australe. Tapinoma (Micromyrma) melanocephalum* var. *australe* Santschi, 1928e: 475 (w.) VANUATU. Junior synonym of *melanocephalum:* **new synonymy**. [Note. Shattuck, 1994: 148 treats *australis* and *australe* as variant spellings, but the type-localities of the two are different and the names must be treated as separate.]
*azerbeidzhanica. Tapinoma simrothi* subsp. *azerbeidzhanica* Karavaiev, 1932: 248, fig. 1 (w.m.) AZERBAIDJAN. Junior synonym of *karavaievi:* Dlussky, Soyunov & Zabelin, 1990: 170.
**#bevisi.** *Tapinoma danitschi* var. *bevisi* Forel, 1915c: 346 (w.q.m.) SOUTH AFRICA.
*bononiensis. Tapinoma erraticum* var. *bononiensis* Emery, 1925d: 55 (w.q.) ITALY. Junior synonym of *erraticum:* Atanassov & Dlussky, 1992: 192.
*boreale. Tapinoma boreale* Roger, 1863a: 165 (w.q.) NORTH AMERICA. Subspecies of *sessile:* Dalla Torre, 1893: 164; Ruzsky, 1925b: 45; Ruzsky, 1936: 95. Senior synonym of *boreale* Provancher: Dalla Torre, 1893: 164. Junior synonym of *sessile:* Mayr, 1886d: 434; Creighton, 1950a: 353. [Note. Many publications refer to *T. boreale* Mayr, 1886d: 434, for example Creighton, 1950a: 342. This reference is to *T. boreale* Roger as redescribed by Mayr; there is no *T. boreale* Mayr.]
*boreale. Tapinoma boreale* Provancher, 1887: 238 (w.q.) CANADA. Junior synonym of *boreale* Roger: Dalla Torre, 1893: 164.
*breve. Tapinoma breve* Emery, 1925d: 60, fig. 12 (w.) TURKESTAN. Subspecies of *erraticum:* Kuznetsov-Ugamsky, 1927d: 34. Junior synonym of *erraticum:* Dlussky, Soyunov & Zabelin, 1990: 169.
**#breviscapum.** *Tapinoma atriceps* var. *breviscapa* Forel, 1908c: 384 (w.q.) BRAZIL.
**#broomense.** *Tapinoma minutum* var. *broomensis* Forel, 1915b: 83 (w.q.) AUSTRALIA.
*caerulescens. Formica caerulescens* Losana, 1834: 314, fig. 3 (w.) ITALY. Junior synonym of *erraticum:* Mayr, 1865: 62.
*canalis. Tapinoma canalis* Wheeler, W.M. 1942: 251 (w.q.m.) PANAMA. Junior synonym of *panamense:* Shattuck, 1992c: 153.
**#capsincola.** *Tapinoma andamanense* subsp. *capsincola* Forel, 1911b: 205 (w.q.) JAVA.

***TAPINOMA** (continued)*

***carininotum.*** *Tapinoma carininotum* Weber, 1943c: 379, pl. 16, fig. 32 (w.) SUDAN.

*cearense. Tapinoma ramulorum* r. *inrectum* var. *cearense* Forel, 1912h: 57 (w.m.) BRAZIL. Unavailable name (Shattuck, 1994: 156).

#***cephalicum.*** *Tapinoma (Micromyrma) minutum* var. *cephalicum* Santschi, 1928e: 472, fig. 1 (w.q.m.) AUSTRALIA.

***chiaromontei.*** *Tapinoma chiaromontei* Menozzi, 1930b: 113, fig. 4 (w.) SOMALIA.

***christophi.*** *Tapinoma christophi* Emery, 1925d: 60, fig. 13 (m.) LEBANON.

*collina. Tapinoma collina* Foerster, 1850a: 43 (w.q.) GERMANY. Junior synonym of *erraticum:* Schenck, 1852: 67.

*confusum. Tapinoma confusum* Smith, M.R. 1951a: 837. Replacement name for *dufouri* Donisthorpe, 1943f: 662. [Unnecessary replacement name for *pygmaea* Dufour, 1857: 61.] Junior synonym of *pygmaeum:* Smith, M.R. 1951a: 837.

#***connexum.*** *Tapinoma luridum* st. *connexum* Santschi, 1914b: 118, fig. 22 (w.m.) KENYA.

#***coronatum.*** *Tapinoma melanocephalum* var. *coronatum* Forel, 1908b: 62 (w.) COSTA RICA.

#***cubaense.*** *Tapinoma litorale* var. *cubaensis* Wheeler, W.M. 1913b: 498 (w.) CUBA.

#***dagestanicum.*** *Tapinoma emeryanum* subsp. *dagestanicum* Kuznetsov-Ugamsky, 1927d: 36 (w.) TURKESTAN.

***danitschi.*** *Tapinoma danitschi* Forel, 1915c: 345 (w.) SOUTH AFRICA. Current subspecies: nominal plus *bevisi.*

*darwinianus* Forel, 1907; see under ***DOLEROMYRMA.***

***demissum.*** *Tapinoma demissum* **new name**. Replacement name for *Tapinoma gracile* Forel, 1913a: 140. [Junior secondary homonym of *Formica gracilis* Buckley, 1866: 158.]

*dimmocki. Bothriomyrmex dimmocki* Wheeler, W.M. 1915b: 417 (w.q.) U.S.A. Combination in *Tapinoma:* Emery, 1925e: 19. Junior synonym of *sessile:* Shattuck, 1992c: 153.

*dufourii* Perris, 1878 [*Micromyrma*]; see under ***PLAGIOLEPIS.***

*dufouri. Micromyrma dufouri* Donisthorpe, 1943f: 662. Unnecessary replacement name for *Micromyrma pygmaea* Dufour, 1857: 61. [Junior primary homonym of *dufourii* Perris, above.] Replacement name: *Tapinoma confusum* Smith, M.R. 1951a: 837.

*eidmanni. Forelius eidmanni* Goetsch, 1933: 28, fig. 2 (w.q.) CHILE (attributed to Menozzi). Junior synonym of *antarcticum:* Menozzi, 1935c: 321; Goetsch & Menozzi, 1935: 98.

***emeryanum.*** *Tapinoma emeryanum* Kuznetsov-Ugamsky, 1927d: 35 (w.q.m.) TURKESTAN. Subspecies of *sinense:* Dlussky, 1962: 180. Revived status as species: Tarbinsky, 1976: 119. Current subspecies: nominal plus *dagestanicum.*

#***emeryi.*** *Technomyrmex luteus* subsp. *emeryi* Forel, 1910e: 447 (w.m.) SOUTH AFRICA. Arnold, 1915: 153 (q.). Combination in *Tapinoma:* Emery, 1912g: 42.

***epinotale.*** *Tapinoma epinotalis* Karavaiev, 1935a: 109, fig. 27 (w.) VIETNAM.

***erraticum.*** *Formica erratica* Latreille, 1798: 44 (w.q.m.) FRANCE. Wheeler, G.C. & Wheeler, J. 1951: 197 (l.). Combination in *Tapinoma:* Smith, F. 1855a: 111. Senior synonym of *collina:* Schenck, 1852: 67; of *glabrella:* Smith, F. 1855a: 111; Mayr, 1855: 373; of *caerulescens:* Mayr, 1865: 62; of *breve, nigerrima* (and its junior synonyms *ibericum* and *magnum*), *tauridis* and *transcaucasica:* Dlussky, Soyunov & Zabelin, 1990: 169; of *bononiensis:* Atanassov & Dlussky, 1992: 192. Current subspecies: nominal plus *atomum, madeirense, platyops.* See also: Emery, 1925d: 53; Kutter, 1977c: 181; Seifert, 1984a: 151; Shattuck, 1994: 143.

*erraticonigerrima. Tapinoma nigerrima* var. *erraticonigerrima* Forel, 1894d: 13 (w.q.m.) ALGERIA. *Nomen nudum.* [*Nomen nudum* repeated in Forel, 1902a: 154; Forel, 1904b: 376; Forel, 1904c: 12.] See also: Emery, 1925d: 49.

*familiaris. Formica familiaris* Smith, F. 1860b: 96 (w.) INDONESIA (Batjan I.). [Unresolved junior primary homonym of *Formica familiaris* Smith, F. 1860a: 68 (now in *Pseudolasius*).] Junior synonym of *melanocephalum:* Forel, 1899c: 101.

*fazi. Tapinoma fazi* Santschi, 1923c: 270 (w.m.) CHILE. Junior synonym of *antarcticum:* Menozzi, 1935c: 321.

#***festae.*** *Tapinoma simrothi* subsp. *festae* Emery, 1925d: 51, fig. 4 (q.m.) GREECE.

*fida* Forel, 1907; see under ***DOLEROMYRMA.***

***flavidum.*** *Tapinoma flavidum* André, 1892b: 51 (w.) BORNEO.

*flavipes* Smith, F. 1874; see under ***PARATRECHINA.***

***fragile.*** *Tapinoma fragile* Smith, F. 1876a: 447 (w.) RODRIGUEZ I.

*fullawayi. Prenolepis minutula* subsp. *atomus* var. *fullawayi* Wheeler, W.M. 1912a: 46 (w.) GUAM I. Unavailable name; material referred to *minutum* by Wilson & Taylor, 1967: 82.

*fulvum* Wheeler, W.M. 1934; see under ***TECHNOMYRMEX.***

***funiculare.*** *Tapinoma (Micromyrma) funiculare* Santschi, 1928b: 35 (w.) CHINA.

***geei.*** *Tapinoma geei* Wheeler, W.M. 1927d: 8 (w.) CHINA. Current subspecies: nominal plus *tinctum.*

***gibbosum.*** *Tapinoma gibbosum* Stitz, 1933: 70, fig. 2 (w.) BRAZIL.

*glabrata* Smith, F. 1857; see under ***ANONYCHOMYRMA.***

*glabrella. Formica glabrella* Nylander, 1849: 38 (footnote) (w.) RUSSIA. Junior synonym of *erraticum:* Smith, F. 1855a: 111; Mayr, 1855: 373.

***glaucum.*** *Semonius glaucus* Viehmeyer, 1916a: 144 (w.) SINGAPORE. Combination in *Tapinoma:* Shattuck, 1992c: 152.

*gracilis. Formica gracilis* Buckley, 1866: 158 (w.q.) U.S.A. [Unresolved junior primary homonym of *Formica gracilis* Fabricius, 1804: 405 (now in *Pseudomyrmex*).] Combination in *Tapinoma:* Emery, 1893i: 337. Junior synonym of *sessile:* Emery, 1895c: 337; Wheeler, W.M. 1902f: 20.

*gracile. Tapinoma gracile* Forel, 1913a: 140 (w.) ZIMBABWE. [Junior secondary homonym of *gracilis* Buckley, above.] Replacement name: *demissum* **new name**.

***heyeri.*** *Tapinoma heyeri* Forel, 1902e: 296 (footnote), pl. 1, fig. 11 (w.) BRAZIL. Current subspecies: nominal plus *risii.*

*ibericum. Tapinoma nigerrima* var. *ibericum* Santschi, 1925g: 349 (m.) SPAIN. Junior synonym of *nigerrima:* Collingwood, 1978: 71.

***indicum.*** *Tapinoma indicum* Forel, 1895e: 472 (w.) INDIA. Viehmeyer, 1916a: 142 (q.m.); Imai, Baroni Urbani, *et al.* 1984: 8 (k.). Combination in *T. (Micromyrma)*: Santschi, 1928e: 473. Subspecies of *melanocephalum:* Forel, 1909d: 228. Revived status as species: Emery, 1912g: 41; Wheeler, W.M. 1919e: 100; Karavaiev, 1926d: 441; Santschi, 1928e: 473. Current subspecies: nominal plus *timidum.*

#***inrectum.*** *Tapinoma inrectum* Forel, 1908b: 63 (w.) COSTA RICA. Race of *ramulorum:* Forel, 1912h: 57.

*instabilis* Smith, F. 1862; see under ***AZTECA.***

#***integrum.*** *Tapinoma minutum* var. *integrum* Forel, 1902h: 476 (w.) AUSTRALIA. Forel, 1915b: 83 (q.).

***israele.*** *Tapinoma erraticum* r. *israelis* Forel, 1904c: 16 (w.) ISRAEL. Emery, 1925d: 58 (q.). Raised to species: Emery, 1925d: 58.

*javanum* Forel, 1905; see under ***TECHNOMYRMEX.***

***karavaievi.*** *Tapinoma simrothi* subsp. *karavaievi* Emery, 1925d: 52, fig. 5 (w.q.m.) TURKESTAN. Raised to species: Arnol'di & Dlussky, 1978: 547. Senior synonym of *azerbeidzhanica:* Dlussky, Soyunov & Zabelin, 1990: 170.

***kinburni.*** *Tapinoma kinburni* Karavaiev, 1937: 174, figs. 3, 4 (w.) UKRAINE. Radchenko, 1983: 1905 (q.m.).

*laurenti* Emery, 1899; see under ***TECHNOMYRMEX.***

***litorale.*** *Tapinoma litorale* Wheeler, W.M. 1905b: 109 (w.q.m.) BAHAMAS. Current subspecies: nominal plus *cubaense.*

#***longiceps.*** *Tapinoma luridum* subsp. *longiceps* Wheeler, W.M. 1922a: 209 (w.) ZAIRE.

***luffae.*** *Neoclystopsenella luffae* Kurian, 1955: 133, figs. 249–255 (m.) INDIA. Combination in *Tapinoma:* Brown, 1988a: 337.

***lugubre.*** *Tapinoma gracile* var. *lugubre* Santschi, 1917b: 288 (w.m.) ZIMBABWE. Raised to species: Arnold, 1926: 222.

***luridum.*** *Tapinoma luridum* Emery, 1908b: 188 (w.) ZAIRE. Current subspecies: nominal plus *connexum, longiceps, sokolovi.*

***luteum.*** *Technomyrmex luteus* Emery, 1895h: 43 (w.) SOUTH AFRICA. Mayr, 1907b: 18 (q.); Wheeler, G.C. & Wheeler, J. 1966: 730 (l.). Combination in *Tapinoma:* Mayr, 1907b: 18. Current subspecies: nominal plus *emeryi.*

#***madeirense.*** *Tapinoma erraticum* var. *madeirense* Forel, 1895d: 231 (w.) MADEIRA I. Emery, 1925d: 56 (q.).

*magnum. Tapinoma magnum* Mayr, 1861: 41 (q.m.) ITALY. Junior synonym of *nigerrima:* Forel, 1876: 61.

#***malesianum.*** *Tapinoma melanocephalum* var. *malesianum* Forel, 1913k: 93 (w.q.) SUMATRA.

***melanocephalum.*** *Formica melanocephala* Fabricius, 1793: 353 (w.) FRENCH GUIANA. Emery, 1887a: 249 (m.); Forel, 1891b: 102 (q.); Wheeler, G.C. & Wheeler, J. 1951: 197 (l.); Crozier, 1970: 119 (k.). Combination in *Micromyrma:* Roger, 1862b: 258; in *Tapinoma:* Mayr, 1862: 651; in *Tapinoma (Micromyrma)*: Santschi, 1928e: 475. Senior synonym of *pellucida:* Mayr, 1886c: 359; of *nana:*

Emery, 1892b: 166; of *familiaris:* Forel, 1899c: 101; of *australis:* Wilson & Taylor, 1967: 80; of *australe:* **new synonymy**. Current subspecies: nominal plus *coronatum, malesianum.* See also: Smith, D.R. 1979: 1421; Shattuck, 1994: 148.

***menozzii.*** *Tapinoma menozzii* Donisthorpe, 1936e: 526 (w.) ZIMBABWE.

*meridionale* Roger, 1863; see under ***BOTHRIOMYRMEX.***

***minimum.*** *Tapinoma minimum* Mayr, 1895: 147 (w.q.) TANZANIA.

***minor.*** *Tapinoma minor* Bernard, 1945: 135 (w.) MOROCCO.

****minutissimum.*** **Tapinoma minutissimum* Emery, 1891a: 153, pl. 3, figs. 24, 25 (m.) SICILIAN AMBER (Miocene).

***minutum.*** *Tapinoma minutum* Mayr, 1862: 703 (w.) AUSTRALIA. Wheeler, W.M. 1909a: 27 (q.). Combination in *T.* (*Micromyrma*): Santschi, 1928e: 472. Material of the unavailable name *fullawayi* referred here by Wilson & Taylor, 1967: 82. Current subspecies: nominal plus *broomense, cephalicum, integrum.*

***modestum.*** *Tapinoma* (*Micromyrma*) *modestum* Santschi, 1932a: 390 (w.) ZIMBABWE.

***muelleri.*** *Tapinoma muelleri* Karavaiev, 1926d: 441 (w.) INDONESIA (Aru I.).

*nana. Formica nana* Jerdon, 1851: 125 (w.) INDIA. [Unresolved junior primary homonym of *Formica nana* Latreille, 1802c: 263.] Junior synonym of *melanocephalum:* Emery, 1892b: 166.

*nigerrima. Formica nigerrima* Nylander, 1856b: 71 (w.) FRANCE. [Unresolved junior primary homonym of *Formica nigerrima* Christ, 1791: 513 (now in *Lasius*).] Emery, 1869b: 10 (q.m.); Wheeler, G.C. & Wheeler, J. 1951: 199 (l.). Combination in *Tapinoma:* Mayr, 1861: 41. Junior synonym of *erraticum:* Mayr, 1865: 62. Revived from synonymy as race of *erraticum:* Emery & Forel, 1879: 454. Subspecies/race of *erraticum:* André, 1882b: 224; Forel, 1902a: 154; Forel, 1904b: 376; Ruzsky, 1905b: 478; Emery, 1916b: 215; Finzi, 1924a: 14; Kutter, 1928: 65. Status as species: Ruzsky, 1902d: 20; Bondroit, 1918: 89; Emery, 1925d: 46; Finzi, 1930d: 316; Santschi, 1931a: 10; Bernard, 1967: 258; Kutter, 1977c: 181. Senior synonym of *magnum:* Forel, 1876: 61; of *ibericum:* Collingwood, 1978: 71. Junior synonym of *erraticum:* Dlussky, Soyunov & Zabelin, 1990: 169; Atanassov & Dlussky, 1992: 192. See also: Shattuck, 1994: 150. [Note. Shattuck retains *nigerrima* as a valid species, but was unaware of the works of Dlussky, Soyunov & Zabelin, and Dlussky & Atanassov, which are omitted from his bibliography. Their treatment of *nigerrima* is accepted here.]

*nigrum* Mayr, 1862; see under ***TECHNOMYRMEX.***

*nitens* Mayr, 1853; see under ***PRENOLEPIS.***

*nitida* Smith, F. 1860; see under ***CAMPONOTUS.***

***opacum.*** *Tapinoma opacum* Wheeler, W.M. & Mann, 1914: 43 (w.m.) HAITI.

***orthocephalum.*** *Tapinoma orthocephalum* Stitz, 1934: 3, fig. 1 (w.) MONGOLIA.

***pallipes.*** *Tapinoma pallipes* Smith, F. 1876a: 447 (w.) RODRIGUEZ I. [Misspelled as *pallidipes* by Dalla Torre, 1893: 166.]

***panamense.*** *Tapinoma panamense* Wheeler, W.M. 1934g: 182 (w.q.m.) PANAMA. Senior synonym of *canalis:* Shattuck, 1992c: 153.

*parva. Formica parva* Buckley, 1866: 159 (w.) U.S.A. Junior synonym of *sessile:* Mayr, 1886d: 434.

*pellucida. Myrmica pellucida* Smith, F. 1857a: 71 (w.) SINGAPORE. Junior synonym of *melanocephalum:* Mayr, 1886c: 359.

***philippinense.*** *Tapinoma philippinensis* Donisthorpe, 1942b: 69 (q.) PHILIPPINES.

#***phoeniceum.*** *Tapinoma simrothi* var. *phoenicea* Emery, 1925d: 50, fig. 3 (w.q.m.) CYPRUS. See also: Shattuck, 1994: 154.

*piceata* Norton, 1868; see under ***CAMPONOTUS.***

#***platyops.*** *Tapinoma erraticum* var. *platyops* Emery, 1925d: 56, fig. 7 (w.q.m.) SWITZERLAND.

*polita* Smith, F. 1855; see under ***PRENOLEPIS.***

***pomone.*** *Tapinoma pomonae* Donisthorpe, 1947a: 285 (w.) MAURITIUS.

*pratensis* Smith, F. 1860; see under ***TECHNOMYRMEX.***

*pruinosum* Roger, 1863; see under ***FORELIUS.***

*pusillum* Mayr, 1876; see under ***BOTHRIOMYRMEX.***

***pygmaeum.*** *Micromyrma pygmaea* Dufour, 1857: 61 (w.) FRANCE. Espadaler, 1978: 122 (q.m.). Combination in *Tapinoma:* Roger, 1859: 240; in *T.* (*Micromyrma*): Menozzi, 1925a: 17. Junior synonym of *erraticum:* Dalla Torre, 1893: 165; Emery, 1912g: 40. Revived from synonymy: Menozzi, 1925a: 17; Emery, 1925d: 62. Senior synonym of *confusum:* Smith, M.R. 1951a: 837. See also: Smith, M.R. 1958c: 141; Kutter, 1977c: 182.

***ramulorum.*** *Tapinoma ramulorum* Emery, 1896g: 101 (w.q.m.) COSTA RICA. Current subspecies: nominal plus *annellatum, inrectum, saga, satullum, toltecum.* See also: Wheeler, W.M. 1934g: 178.

***rasenum.*** *Tapinoma rasenum* Smith, D.R. & Lavigne, 1973: 182, figs. 1–6 (w.m.) PUERTO RICO.

***rectinotum.*** *Tapinoma rectinotum* Wheeler, W.M. 1927d: 8 (w.) CHINA.

*revolutionis. Tapinoma erraticum* subsp. *ambiguum* var. *revolutionis* Karavaiev, 1927c: 273 (w.q.m.) RUSSIA. Unavailable name (Shattuck, 1994: 156).

*rhodesiae* Forel, 1913; see under ***TECHNOMYRMEX.***

#***risii.*** *Tapinoma heyeri* var. *risii* Forel, 1912h: 58 (w.) URUGUAY.

***rottnestense.*** *Tapinoma* (*Micromyrma*) *rottnestense* Wheeler, W.M. 1934d: 150 (w.) AUSTRALIA.

#***saga.*** *Tapinoma ramulorum* var. *saga* Forel, 1912h: 58 (w.q.) TRINIDAD.

#***satullum.*** *Tapinoma ramulorum* var. *satullum* Wheeler, W.M. 1934g: 179 (w.) MEXICO.

***schultzei.*** *Semonius schultzei* Forel, 1910f: 21 (w.) BOTSWANA. Santschi, 1914b: 120 (m.); Arnold, 1915: 158 (q.). Combination in *Tapinoma:* Shattuck, 1992c: 152.

***sessile.*** *Formica sessilis* Say, 1836: 287 (w.q.) U.S.A. Emery, 1895c: 333 (q.m.); Wheeler, G.C. & Wheeler, J. 1951: 196 (l.); Crozier, 1970: 119 (k.); Taber & Cokendolpher, 1988: 95 (k.). Combination in *Tapinoma:* Smith, F. 1858b: 57. Senior synonym of *parva* and *boreale* Roger (and its junior synonym *boreale* Provancher): Mayr, 1886d: 434; Creighton, 1950a: 353; of *gracilis:* Emery, 1895c: 337; Wheeler W.M. 1902f: 20; Creighton, 1950a: 353; of *dimmocki:* Shattuck, 1992c: 153. See also: Smith, M.R. 1928a: 307; Smith, D.R. 1979: 1422; Kupyanskaya, 1990: 156; Shattuck, 1994: 153.

*setiferum* Emery, 1900; see under ***TECHNOMYRMEX.***

***silvestrii.*** *Tapinoma silvestrii* Wheeler, W.M. 1928c: 30 (w.) CHINA.

***simrothi.*** *Tapinoma erraticum* var. *simrothi* Krausse, 1911: 18 (w.) SARDINIA. Emery, 1925d: 49 (q.m.). Raised to species: Emery, 1925d: 48; Menozzi, 1933b: 77; Santschi, 1936c: 204. Junior synonym of *erraticum:* Baroni Urbani, 1964b: 55. Revived from synonymy, status as species: Bernard, 1967: 259; Kutter, 1977c: 182. Current subspecies: nominal plus *festae, phoeniceum.* See also: Shattuck, 1994: 154.

***sinense.*** *Tapinoma sinense* Emery, 1925d: 61, fig. 14 (w.) CHINA. Santschi, 1925f: 94 (q.m.). See also: Collingwood, 1976: 299.

#***sokolovi.*** *Tapinoma luridum* subsp. *sokolovi* Karavaiev, 1931d: 50, fig. 7 (w.) KENYA.

*sublucidum* Wheeler, W.M. 1934; see under ***TECHNOMYRMEX.***

*subnigrum. Tapinoma ramulorum* subsp. *inrectum* var. *subnigrum* Wheeler, W.M. 1934g: 181 (w.q.) MEXICO. Unavailable name (Shattuck, 1994: 156).

***subtile.*** *Tapinoma subtile* Santschi, 1911e: 125 (w.m.) MADAGASCAR.

*sundaicum* Emery, 1900; see under ***TECHNOMYRMEX.***

*tauridis. Tapinoma tauridis* Emery, 1925d: 59, fig. 11 (w.q.m.) UKRAINE. Junior synonym of *erraticum:* Dlussky, Soyunov & Zabelin, 1990: 169. See also: Karavaiev, 1927d: 336.

#***tectum.*** *Tapinoma tenue* var. *tectum* Santschi, 1917b: 288 (w.) SOUTH AFRICA. Variety of *arnoldi:* Wheeler, W.M. 1922a: 923.

*tenue. Tapinoma tenue* Santschi, 1917c: 288 (attributed to Forel). *Nomen nudum;* error for *arnoldi,* see Wheeler, W.M. 1922a: 923.

*thoracica* Smith, F. 1860; see under ***DOLICHODERUS.***

#***timidum.*** *Tapinoma* (*Micromyrma*) *indicum* var. *timidum* Santschi, 1928e: 473, fig. 1 (w.m.) VANUATU.

#***tinctum.*** *Tapinoma geei* var. *tinctum* Wheeler, W.M. 1927e: 2 (w.) CHINA.

#***toltecum.*** *Tapinoma ramulorum* subsp. *toltecum* Wheeler, W.M. 1934g: 180 (w.) MEXICO.

*tomentosa* Norton, 1868; see under ***CAMPONOTUS.***

*transcaucasica. Tapinoma tauridis* subsp. *transcaucasica* Karavaiev, 1927d: 337 (w.m.) TURKESTAN. [First available use of *Tapinoma simrothi* subsp. *karavaievi* var. *transcaucasica* Karavaiev, 1926e: 187; unavailable name.] Junior synonym of *erraticum:* Dlussky, Soyunov, & Zabelin, 1990: 169.

****troche.*** **Tapinoma trochis* Wilson, 1985c: 34, fig. 12 (w.) DOMINICAN AMBER (Miocene).

***vexatum.*** *Tapinoma vexatum* Santschi, 1919i: 220 (footnote) (m.) MOROCCO. Combination in *T.* (*Tapinoptera*): Santschi, 1925g: 348.

*voeltzkowi* Forel, 1907; see under ***TECHNOMYRMEX.***

***wheeleri.*** *Zatapinoma wheeleri* Mann, 1935: 36, fig. 2 (s.) SAMOA. Combination in *Tapinoma:* Shattuck, 1992c: 152.

***williamsi.*** *Zatapinoma williamsi* Wheeler, W.M. 1935b: 44 (s.w.) PHILIPPINES. Combination in *Tapinoma:* Shattuck, 1992c: 152.

***wroughtonii.*** *Tapinoma wroughtonii* Forel, 1904c: 26 (w.) INDIA.

*TAPINOPTERA:* see under ***TAPINOMA.***

***TATUIDRIS*** (Neotropical)
***tatusia.*** *Tatuidris tatusia* Brown & Kempf, 1968: 187, figs. 1–4 (w.) EL SALVADOR.

***TECHNOMYRMEX*** (World tropics)
***aberrans.*** *Technomyrmex aberrans* Santschi, 1911e: 127 (q.) MADAGASCAR.
#***affinis.*** *Technomyrmex foreli* var. *affinis* Santschi, 1930a: 71 (w.) KENYA. [First available use of *Technomyrmex albipes* st. *foreli* var. *affinis* Santschi, 1914b: 119; unavailable name.]
#***albicoxis.*** *Technomyrmex cheesmanae* var. *albicoxis* Donisthorpe, 1945b: 58 (w.) NEW GUINEA.
#***albinasis.*** *Technomyrmex nigriventris* subsp. *albinasis* Forel, 1910f: 22 (w.) SOUTH AFRICA.
***albipes.*** *Formica (Tapinoma) albipes* Smith, F. 1861b: 38 (w.) SULAWESI. Forel, 1891b: 98 (q.); Forel, 1908a: 21 (ergatoid m.); Karavaiev, 1926d: 441 (m.); Wheeler, G.C. & Wheeler, J. 1951: 205 (l.); Crozier, 1969: 245 (k.). Combination in *Tapinoma:* Mayr, 1863: 455; in *Technomyrmex:* Emery, 1888d: 392. Senior synonym of *nigrum:* Mayr, 1872: 147; Mayr, 1876: 83; of *albitarse:* Emery, 1893f: 249; of *rufescens* and *vitiensis:* Wilson & Taylor, 1967: 82. Junior synonym of *detorquens:* Donisthorpe, 1932b: 575. Revived from synonymy: Wilson & Taylor, 1967: 82. Current subspecies: nominal plus *bruneipes, brunneus, cedarensis, cordiformis, rotundiceps, truncicola, wedda.* See also: Tsuji, Furukawa, *et al.* 1991: 413; Shattuck, 1994: 157.
*albitarse. Tapinoma albitarse* Motschoulsky, 1863: 14 (w.q.) SRI LANKA. Junior synonym of *albipes:* Emery, 1893f: 249.
***albomaculatus.*** *Technomyrmex albomaculatus* Karavaiev, 1926d: 444, fig. 12 (q.) SULAWESI.
***allectus.*** *Engramma allecta* Stitz, 1916: 394, fig. 9 (w.) ZAIRE. Combination in *Technomyrmex:* Shattuck, 1992c: 160.
***andrei.*** *Technomyrmex andrei* Emery, 1899e: 488, fig. (w.) GABON. Bernard, 1953b: 255 (q.). Current subspecies: nominal plus *camerunensis, schereri.*
#***angustior.*** *Technomyrmex modiglianii* var. *angustior* Forel, 1912a: 71 (w.) TAIWAN.
#***antonii.*** *Technomyrmex bicolor* var. *antonii* Forel, 1902h: 475 (w.m.) AUSTRALIA.
***arnoldinus.*** *Technomyrmex arnoldinus* Forel, 1913j: 222 (w.) ZIMBABWE.
***atrichosus.*** *Technomyrmex atrichosus* Viehmeyer, 1922: 214 (w.) KENYA.
#***bandarensis.*** *Technomyrmex grandis* var. *bandarensis* Forel, 1913k: 96 (w.) SUMATRA.
***bicolor.*** *Technomyrmex bicolor* Emery, 1893f: 249 (w.) SRI LANKA. Current subspecies: nominal plus *antonii, textor.*
***brevicornis.*** *Technomyrmex brevicornis* Santschi, 1930a: 71, figs. 34–36 (w.) ANGOLA.
#***bruneipes.*** *Technomyrmex albipes* var. *bruneipes* Forel, 1895e: 466 (w.) INDIA.
#***brunneus.*** *Technomyrmex albipes* r. *brunneus* Forel, 1895e: 467 (w.) INDIA. Raised to species: Bingham, 1903: 302. Subspecies of *albipes:* Emery, 1912g: 43.
***butteli.*** *Technomyrmex butteli* Forel, 1913k: 97, fig. C (w.q.m.) WEST MALAYSIA.
#***camerunensis.*** *Technomyrmex andrei* var. *camerunensis* Emery, 1899e: 489 (w.) CAMEROUN.
#***cedarensis.*** *Technomyrmex albipes* var. *cedarensis* Forel, 1915b: 85 (w.q.) AUSTRALIA.
***cheesmanae.*** *Technomyrmex cheesmanae* Donisthorpe, 1945b: 57 (w.) NEW GUINEA. Current subspecies: nominal plus *albicoxis.*
#***congolensis.*** *Engramma laurenti* var. *congolensis* Forel, 1916: 431 (w.q.) ZAIRE. Combination in *Technomyrmex:* Shattuck, 1992c: 161.
*congolensis. Technomyrmex albipes* subsp. *congolensis* Karavaiev, 1926d: 443 (w.) ZAIRE. [Unresolved junior secondary homonym of *congolensis* Forel, above.] Junior synonym of *moerens:* Santschi, 1930a: 69.
***convexifrons.*** *Technomyrmex convexifrons* Karavaiev, 1926d: 443 (w.) SUMATRA.
#***cordiformis.*** *Technomyrmex albipes* var. *cordiformis* Viehmeyer, 1916a: 143 (w.) SINGAPORE.
****deletus.*** **Technomyrmex deletus* Emery, 1891a: 153, pl. 3, figs. 26–28 (w.) SICILIAN AMBER (Miocene).
*denticulatum* Wheeler, W.M. 1922 [*Engramma*]; see under ***AXINIDRIS.***
***detorquens.*** *Formica detorquens* Walker, 1859: 372 (q.) SRI LANKA. Combination in *Technomyrmex,* and senior synonym of *forticulus:* Donisthorpe, 1932b: 575. See also: Wilson & Taylor, 1967: 84.
#***difficilis.*** *Technomyrmex mayri* r. *difficilis* Forel, 1892l: 242 (w.q.) MADAGASCAR.
***elatior.*** *Technomyrmex modiglianii* r. *elatior* Forel, 1902d: 293 (w.) INDIA. Forel, 1908a: 3 (m.). Race/subspecies of *modiglianii:* Forel, 1905c: 23; Wheeler, W.M. 1921c: 541. Raised to species: Bingham, 1903: 302; Mukerjee, 1930: 155.
*emeryi* Forel, 1910; see under ***TAPINOMA.***
***foreli.*** *Technomyrmex albipes* subsp. *foreli* Emery, 1893f: 249 (w.) MADAGASCAR. Forel, 1910e: 446 (q.m.). Raised to species: Santschi, 1930a: 70. Current subspecies: nominal plus *affinis.*
*forticulus. Crematogaster forticulus* Walker, 1859: 372 (q., not w.) SRI LANKA. Junior synonym of *detorquens:* Donisthorpe, 1932b: 576.
***fulvus.*** *Tapinoma fulvum* Wheeler, W.M. 1934g: 184 (w.) PANAMA. Combination in *Technomyrmex:* Shattuck, 1992c: 161. Current subspecies: nominal plus *sublucidus.*
#***fusciventris.*** *Technomyrmex madecassus* var. *fusciventris* Forel, 1907g: 86 (w.) COMORO IS.
***gibbosus.*** *Technomyrmex gibbosus* Wheeler, W.M. 1906c: 319, pl. 41, fig. 4 (w.) JAPAN. Wheeler, G.C. & Wheeler, J. 1951: 205 (l.).
***gilvus.*** *Technomyrmex gilvus* Donisthorpe, 1941j: 205 (w.) NEW GUINEA.
***gowdeyi.*** *Engramma gowdeyi* Wheeler, W.M. 1922a: 207, fig. 53 (w.) ZAIRE. Combination in *Technomyrmex:* Shattuck, 1992c: 160.
***grandis.*** *Technomyrmex grandis* Emery, 1887a: 248 (w.) SUMATRA. Current subspecies: nominal plus *bandarensis.*
***griseopubens.*** *Engramma griseopubens* Wheeler, W.M. 1922a: 206, fig. 52 (w.) ZAIRE. Combination in *Technomyrmex:* Shattuck, 1992c: 160.
***horni.*** *Technomyrmex horni* Forel, 1912a: 71 (w.q.) TAIWAN.
***hypoclinoides.*** *Technomyrmex hypoclinoides* Santschi, 1919h: 89 (w.) ZAIRE.
***ilgi.*** *Engramma ilgi* Forel, 1910c: 264 (w.m.) ETHIOPIA. Combination in *Technomyrmex:* Shattuck, 1992c: 160. Current subspecies: nominal plus *stygius.*
***incisus.*** *Engramma incisum* Mukerjee, 1930: 155, fig. 3 (w.) INDIA. Combination in *Technomyrmex:* Shattuck, 1992c: 161.
*incisus. Technomyrmex incisus* Weber, 1943c: 380, pl. 16, fig. 37 (w.) SUDAN. [Junior secondary homonym of *incisum* Mukerjee, above.] Replacement name: *nequitus* **new name**.
#***javanus.*** *Tapinoma setiferum* var. *javanum* Forel, 1905c: 22 (w.q.m.) JAVA. Combination in *Technomyrmex:* Shattuck, 1992c: 161.
#***javanus.*** *Technomyrmex modiglianii* r. *javanus* Forel, 1905c: 23 (w.) JAVA. [Unresolved junior secondary homonym of *javanum* Forel, above.]
***jocosus.*** *Technomyrmex jocosus* Forel, 1910b: 56 (w.) AUSTRALIA.
***kohli.*** *Engramma kohli* Forel, 1916: 429 (w.q.) ZAIRE. Combination in *Technomyrmex:* Shattuck, 1992c: 161.
***kraepelini.*** *Technomyrmex kraepelini* Forel, 1905c: 23 (w.) JAVA.
***laurenti.*** *Tapinoma laurenti* Emery, 1899e: 487 (w.q.) ZAIRE. Combination in *Engramma:* Emery, 1912g: 38; in *Technomyrmex:* Shattuck, 1992c: 161. Current subspecies: nominal plus *congolensis.*
***lisae.*** *Technomyrmex lisae* Forel, 1913k: 94, fig. B (w.q.) WEST MALAYSIA.
***longiscapus.*** *Technomyrmex longiscapus* Weber, 1943c: 381, pl. 16, fig. 41 (w.) SUDAN.
***lujae.*** *Engramma lujae* Forel, 1905b: 181 (w.m.) ZAIRE. Wheeler, G.C. & Wheeler, J. 1951: 194 (l.). Combination in *Technomyrmex:* Shattuck, 1992c: 161. Current subspecies: nominal plus *pulliceps, wasmanni.*
*luteus* Emery, 1895; see under ***TAPINOMA.***
***madecassus.*** *Technomyrmex madecassus* Forel, 1897c: 199 (w.) MADAGASCAR. Current subspecies: nominal plus *fusciventris.*
***mayri.*** *Technomyrmex mayri* Forel, 1891b: 99, pl. 3, fig. 6 (w.m.) MADAGASCAR. Current subspecies: nominal plus *difficilis, nitidulans.*
***modiglianii.*** *Technomyrmex modiglianii* Emery, 1900d: 696, fig. 12 (w.) SUMATRA. Current subspecies: nominal plus *angustior, javanus* (junior homonym).
***moerens.*** *Technomyrmex moerens* Santschi, 1913c: 312, fig. 4 (w.) CONGO. Senior synonym of *congolensis:* Santschi, 1930a: 69. Current subspecies: nominal plus *nigricans.*
***nequitus.*** *Technomyrmex nequitus* **new name**. Replacement name for *Technomyrmex incisus* Weber, 1943c: 380. [Junior secondary homonym of *Engramma incisum* Mukerjee, 1930: 155.]
***niasensis.*** *Technomyrmex niasensis* Menozzi, 1932d: 8, fig. 3 (w.) INDONESIA (Nias I.).
#***nigricans.*** *Technomyrmex moerens* var. *nigricans* Santschi, 1930a: 69 (w.) ZAIRE.
***nigriventris.*** *Technomyrmex nigriventris* Santschi, in Forel, 1910f: 22 (footnote) (w.) CONGO. Current subspecies: nominal plus *albinasis.*
*nigrum. Tapinoma nigrum* Mayr, 1862: 703 (w.) SRI LANKA. Junior synonym of *albipes:* Mayr, 1876: 83.
#***nitidulans.*** *Technomyrmex mayri* st. *nitidulans* Santschi, 1930a: 72, figs. 31–33 (w.) MADAGASCAR.

**#obscurior.** *Technomyrmex schimmeri* var. *obscurior* Wheeler, W.M. 1928c: 31 (w.) CHINA.
**#okiavoensis.** *Engramma zimmeri* r. *okiavoensis* Forel, 1916: 431 (w.) ZAIRE. Combination in *Technomyrmex:* Shattuck, 1992c: 161.
**pilipes.** *Technomyrmex pilipes* Emery, 1899e: 490 (q.) CAMEROUN.
**pratensis.** *Tapinoma pratensis* Smith, F. 1860b: 97 (w.) INDONESIA (Batjan I.). Combination in *Iridomyrmex:* Donisthorpe, 1932c: 460; in *Technomyrmex:* Shattuck, 1992c: 161.
**primroseae.** *Technomyrmex primroseae* Donisthorpe, 1949e: 273 (w.) MAURITIUS.
**#pulliceps.** *Engramma lujae* var. *pulliceps* Santschi, 1926b: 244 (w.) ZAIRE. Combination in *Technomyrmex:* Shattuck, 1992c: 161.
**quadricolor.** *Aphantolepis quadricolor* Wheeler, W.M. 1930d: 44, fig. 2 (w.) AUSTRALIA. Combination in *Technomyrmex:* Brown, 1953h: 5.
**#rhodesiae.** *Tapinoma voeltzkowi* var. *rhodesiae* Forel, 1913a: 139 (w.) ZIMBABWE. Combination in *Technomyrmex:* Shattuck, 1992c: 161.
**#rotundiceps.** *Technomyrmex albipes* subsp. *rotundiceps* Karavaiev, 1926d: 443 (w.) JAVA.
*rufescens. Technomyrmex albipes* st. *rufescens* Santschi, 1928c: 70, fig. 1 (w.) FIJI IS. Junior synonym of *albipes:* Wilson & Taylor, 1967: 82.
**rusticus.** *Technomyrmex rusticus* Santschi, 1930a: 72, figs. 43–45 (w.) ZAIRE.
**#schereri.** *Technomyrmex andrei* var. *schereri* Forel, 1911e: 283 (w.) LIBERIA.
**schimmeri.** *Technomyrmex schimmeri* Viehmeyer, 1916a: 143, fig. 8 (w.q.) SINGAPORE. Current subspecies: nominal plus *obscurior.*
**schoutedeni.** *Technomyrmex schoutedeni* Forel, 1910e: 447 (w.q.m.) ZAIRE.
**semiruber.** *Technomyrmex semiruber* Emery, 1899e: 489 (w.) CAMEROUN.
**setifer.** *Tapinoma setiferum* Emery, 1900d: 695, fig. 11 (w.) SUMATRA. Combination in *Technomyrmex:* Wheeler, W.M. 1927h: 99 (misspelled as *setiferus*). Current subspecies: nominal plus *javanus.*
**setosus.** *Technomyrmex setosus* Collingwood, 1985: 243, fig. 12 (w.) SAUDI ARABIA.
*smithi. Iridomyrmex smithi* Wheeler, W.M. 1909d: 341 (w.q.) PHILIPPINES. [Junior secondary homonym of *Formica smithii* Lowne, 1865a: 276 (now in *Iridomyrmex*).] Replacement name: *wheeleri* Emery, 1912g: 26.
**sophiae.** *Technomyrmex sophiae* Forel, 1902h: 474 (w.q.) AUSTRALIA.
**strenuus.** *Technomyrmex strenua* Mayr, 1872: 147 (w.q.) BORNEO. Forel, 1911a: 46 (m.).
**#stygius.** *Engramma stygium* Santschi, 1911c: 363 (w.m.) KENYA. Santschi, 1914b: 117 (q.). Combination in *Technomyrmex:* Shattuck, 1992c: 161. Variety of *ilgi:* Santschi, 1914b: 117.
**#sublucidus.** *Tapinoma fulvum* subsp. *sublucidum* Wheeler, W.M. 1934g: 185 (w.) PANAMA. Combination in *Technomyrmex:* Shattuck, 1992c: 161.
**sundaicus.** *Tapinoma sundaicum* Emery, 1900d: 695, fig. 10 (w.) SUMATRA. Combination in *Technomyrmex:* Shattuck, 1992c: 161.
**tailori.** *Engramma tailori* Santschi, 1930d: 269 (w.) MOZAMBIQUE. [Also described as new (spelled *taylori*) by Santschi, in Taylor, J.S. 1931: 42.] Combination in *Technomyrmex:* Shattuck, 1992c: 161.
**#textor.** *Technomyrmex bicolor* subsp. *textor* Forel, 1909d: 228 (w.q.) JAVA.
**transiens.** *Technomyrmex transiens* Forel, 1913k: 96 (w.) SUMATRA.
*tridens* Arnold, 1946; see under ***AXINIDRIS.***
**#truncicola.** *Technomyrmex albipes* subsp. *truncicolus* Weber, 1943c: 380 (w.) SUDAN.
*vitiensis. Technomyrmex albipes* var. *vitiensis* Mann, 1921: 473 (w.) FIJI IS. Santschi, 1928a: 52 (m.). Junior synonym of *albipes:* Wilson & Taylor, 1967: 82.
**voeltzkowi.** *Tapinoma voeltzkowi* Forel, 1907g: 84 (w.m.) KENYA. Combination in *Technomyrmex:* Shattuck, 1992c: 161. Current subspecies: nominal plus *rhodesiae.*
**#wasmanni.** *Engramma lujae* r. *wasmanni* Forel, 1916: 432 (w.q.m.) ZAIRE. Combination in *Technomyrmex:* Shattuck, 1992c: 161.
**#wedda.** *Technomyrmex albipes* r. *wedda* Forel, 1913e: 663 (w.q.) SRI LANKA.
**wheeleri.** *Iridomyrmex wheeleri* Emery, 1912g: 26. Replacement name for *Iridomyrmex smithi* Wheeler, W.M. 1909d: 341. [Junior secondary homonym of *Formica smithii* Lowne, 1865a: 276 (now in *Iridomyrmex*).] Combination in *Technomyrmex:* Shattuck, 1992a: 15.
**wolfi.** *Engramma wolfi* Forel, 1916: 432, fig. 1 (w.) ZAIRE. Wheeler, W.M. 1922a: 204 (q.m.). Combination in *Technomyrmex:* Shattuck, 1992c: 161.
**zimmeri.** *Engramma zimmeri* Forel, 1911d: 370 (w.) TANZANIA. Combination in *Technomyrmex:* Shattuck, 1992c: 161. Current subspecies: nominal plus *okiavoensis.*
**zumpti.** *Technomyrmex zumpti* Santschi, 1937b: 102 (m.) CAMEROUN.

***TELEUTOMYRMEX*** (Palaearctic)
**kutteri.** *Teleutomyrmex kutteri* Tinaut, 1990b: 202, figs. 1–3, photos. 1–2 (q.m.) SPAIN.
**schneideri.** *Teleutomyrmex schneideri* Kutter, 1950a: 82, figs. 1–23 (q.m.) SWITZERLAND. See also: Stumper, 1951: 129; Brun, 1952: 73; Gösswald, 1953: 81; Kutter, 1977c: 167.

*TEMNOTHORAX:* see under ***LEPTOTHORAX.***

***TERATANER*** (Afrotropical, Malagasy)
**alluaudi.** *Atopomyrmex alluaudi* Emery, 1895f: 341, fig. 2 (w.) MADAGASCAR. Wheeler, G.C. & Wheeler, J. 1980: 536 (l.); Alpert, 1992: 123 (q.m.). Combination in *Teratoner:* Emery, 1912b: 103.
**bottegoi.** *Atopomyrmex bottegoi* Emery, 1896i: 155, fig. (w.) SOMALIA. Combination in *Teratoner:* Emery, 1912b: 103. See also: Bolton, 1981b: 291.
**elegans.** *Teratoner elegans* Bernard, 1953b: 243, fig. 13 (q.) GUINEA. Bolton, 1981b: 292 (w.).
**foreli.** *Atopomyrmex foreli* Emery, 1899f: 274, figs. (w.m.) MADAGASCAR. Combination in *Teratoner:* Emery, 1912b: 103.
**luteus.** *Atopomyrmex luteus* Emery, 1899e: 477 (w.) CAMEROUN. Combination in *Teratoner:* Emery, 1912b: 103. See also: Bolton, 1981b: 293.
**piceus.** *Teratoner piceus* Menozzi, 1942: 173 (w.) EQUATORIAL GUINEA. See also: Bolton, 1981b: 293.
**rufipes.** *Teratoner rufipes* Emery, 1912b: 104 (w.) MADAGASCAR.
**scotti.** *Atopomyrmex scotti* Forel, 1912k: 160 (w.m.) SEYCHELLES IS. Combination in *Teratoner:* Emery, 1912b: 103.
**steinheili.** *Atopomyrmex steinheili* Forel, 1895h: 485 (w.) MADAGASCAR. Combination in *Teratoner:* Emery, 1912b: 103.
**transvaalensis.** *Teratoner (Tranetera) transvaalensis* Arnold, 1952b: 130, fig. (w.) SOUTH AFRICA. See also: Bolton, 1981b: 294.
**velatus.** *Teratoner velatus* Bolton, 1981b: 295 (w.) GHANA.
**xaltus.** *Teratoner xaltus* Bolton, 1981b: 296, fig. 52 (w.) MADAGASCAR.

***TERATOMYRMEX*** (Australia)
**greavesi.** *Teratomyrmex greavesi* McAreavey, 1957: 55, figs. 1–3 (w.) AUSTRALIA. Wheeler, G.C. & Wheeler, J. 1974e: 39 (l.).

*TERMITOPONE:* see under ***PACHYCONDYLA.***

***TETHEAMYRMA*** (Indo-Australian)
**subspongia.** *Tetheamyrma subspongia* Bolton, 1991: 10, figs. 16, 17 (w.) BORNEO.

***TETRAMORIUM*** (World wide)
*abdominalis. Macromichoides aculeatus* st. *wasmanni* var. *abdominalis* Santschi, 1924b: 209, fig. 8 (w.) ZAIRE. Unavailable name; material referred to *aculeatum* by Bolton, 1980: 353.
*aciculatum. Tetramorium solidum* r. *glabratum* var. *aciculatum* Stitz, 1923: 162 (w.) NAMIBIA. Unavailable name; material referred to *glabratum* by Bolton, 1980: 246.
**aculeatum.** *Macromischa aculeata* Mayr, 1866a: 507 (w.) GHANA. André, 1889: 224 (q.); Mayr, 1902: 292 (m.); Wheeler, G.C. & Wheeler, J. 1955b: 18 (l.). Combination in *Tetramorium:* Emery, 1896a: 103; in *Macromischoides:* Wheeler, W.M. 1920: 53; in *Tetramorium:* Bolton, 1976: 363 (in text); Bolton, 1980: 353. Senior synonym of *andricum, inermis, major, melanogyne, militaris, pulchellus, rubroflava, viridis, wasmanni* and *zumpti,* and material of the unavailable names *abdominalis* and *gladiator* referred here: Bolton, 1980: 353.
**#acutiseta.** *Tetramorium exasperatum* var. *acutiseta* Santschi, 1921d: 72 (w.q.) MOROCCO. [Also described as new by Santschi, 1923f: 334.]
**adelphon.** *Tetramorium adelphon* Bolton, 1979: 177 (w.) BORNEO. Imai, Kubota, *et al.* 1985: 47 (k.).
**adpressum.** *Triglyphothrix adpressa* Bolton, 1976: 344, fig. 66 (w.) SULAWESI. Combination in *Tetramorium:* Bolton, 1985: 247.
**aegeum.** *Tetramorium aegea* Radchenko, 1992b: 55 (w.q.m.) CAUCASUS. First available replacement name for *laevigatum* Arnol'di, 1948: 211. [Junior primary homonym of *laevigatum* Mayr, 1901b: 25.]
*aequa. Tetramorium ferox* var. *aequa* Baroni Urbani, 1971b: 361. Unavailable name. [Incorrectly proposed replacement name for *levigata* Karavaiev, 1926c: 167; unavailable name.]
**africanum.** *Macromischa africana* Mayr, 1866a: 507 (w.) GHANA. Mayr, 1895: 131 (q.). Combination in *Tetramorium:* Emery, 1896a: 103; in *Macromischoides:* Wheeler, W.M. 1922a: 188; in *Tetramorium:* Bolton,

1976: 363 (in text); Bolton, 1980: 355. Senior synonym of *tessmanni:* Brown, 1964c: 12; of *lamottei:* Bolton, 1980: 355.

***agile.*** *Tetramorium agile* Arnold, 1960b: 455, fig. 5 (w.) ZIMBABWE. See also: Bolton, 1980: 275.

***agnum.*** *Triglyphothrix agna* Santschi, 1935a: 264, figs. 7–9 (w.) ZAIRE. Combination in *Tetramorium:* Bolton, 1985: 247. See also: Bolton, 1976: 317.

*akengensis. Xiphomyrmex occidentalis* subsp. *akengensis* Wheeler, W.M. 1922a: 194 (w.) ZAIRE. Junior synonym of *occidentale:* Bolton, 1980: 229.

***akermani.*** *Tetramorium akermani* Arnold, 1926: 265, fig. 73 (w.) SOUTH AFRICA. Senior synonym of *myersi:* Bolton, 1980: 253.

*akermani. Tetramorium* (*Triglyphothrix*) *marleyi* var. *akermani* Arnold, 1926: 276 (w.) SOUTH AFRICA. [Unresolved junior primary homonym of *akermani* Arnold, above.] Junior synonym of *eminii:* Bolton, 1976: 324.

*algoa. Tetramorium joffrei* var. *algoa* Arnold, 1917: 304 (w.q.) SOUTH AFRICA. Junior synonym of *grassii:* Bolton, 1980: 262.

#***alternans.*** *Tetramorium caespitum* var. *alternans* Santschi, 1929e: 150 (w.) MOROCCO.

*altinode* Santschi, 1935; see under ***MONOMORIUM.***

***altivagans.*** *Tetramorium caespitum* st. *altivagans* Santschi, 1914b: 103 (w.) KENYA. Subspecies of *pusillum:* Emery, 1922f: 281. Raised to species, and senior synonym of *isis:* Bolton, 1980: 305.

***amatongae.*** *Tetramorium amatongae* Bolton, 1980: 358, figs. 133, 137 (w.) MOZAMBIQUE.

*amatongae. Tetramorium setigerum* r. *quaerens* var. *amatongae* Arnold, 1926: 264 (w.) MOZAMBIQUE. Unavailable name.

***amaurum.*** *Tetramorium amaurum* Bolton, 1980: 286, fig. 69 (w.) ZIMBABWE.

*amen. Xiphomyrmex minusculus* subsp. *amen* Weber, 1943c: 376 (w.) SUDAN. Junior synonym of *minusculum:* Bolton, 1980: 240.

***amentete.*** *Tetramorium amentete* Bolton, 1980: 266, fig. 50 (w.) GHANA.

***amium.*** *Tetramorium mixtum* r. *amia* Forel, 1912a: 53 (w.) TAIWAN. Raised to species: Bolton, 1977: 113.

***amissum.*** *Tetramorium amissum* Bolton, 1980: 336, fig. 113 (w.) ZAIRE.

***andrei.*** *Tetramorium* (*Xiphomyrmex*) *andrei* Forel, 1892l: 263 (w.) MADAGASCAR. Combination in *Xiphomyrmex:* Wheeler, W.M. 1922a: 1031; in *Tetramorium:* Bolton, 1979: 143.

*andricum. Tetramorium aculeatum* subsp. *andricum* Emery, 1908b: 187 (w.q.m.) ZAIRE. Combination in *Macromischoides:* Wheeler, W.M. 1922a: 890. Junior synonym of *aculeatum:* Bolton, 1980: 353.

***andrynicum.*** *Tetramorium andrynicum* Bolton, 1977: 143, figs. 66, 67 (w.q.) AUSTRALIA.

*angolense. Tetramorium quadrispinosum* st. *angolense* Santschi, 1930b: 71 (w.q.) ANGOLA. Junior synonym of *quadrispinosum:* Bolton, 1980: 330.

***angulinode.*** *Tetramorium* (*Xiphomyrmex*) *angulinode* Santschi, 1910c: 385, fig. 12 (w.q.m.) CONGO. Combination in *Xiphomyrmex:* Santschi, 1920b: 16; in *Tetramorium:* Bolton, 1980: 237. Senior synonym of *daphnis, humerosum* and *papyri:* Bolton, 1980: 237; of *nullispinum:* **new synonymy**.

*annauensis. Tetramorium caespitum* subsp. *annauensis* Karavaiev, 1931a: 315 (w.q.m.) TRANSCASPIA. Junior synonym of *armatum:* Radchenko, 1992b: 53.

*annectens. Tetramorium annectens* Pisarski, 1969a: 230 (w.) CHINA. [First available use of *Tetramorium caespitum* subsp. *jacoti* var. *annectens* Wheeler, W.M. 1927d: 7; unavailable name.] Junior synonym of *jacoti:* Radchenko. 1992b: 51. See also: Yasumatsu, 1962: 96; Dlussky & Pisarski, 1970: 86.

***anodontion.*** *Tetramorium anodontion* Bolton, 1979: 156, figs. 22, 23 (w.) MADAGASCAR.

***antennatum.*** *Triglyphothrix fulviceps* subsp. *antennatus* Mann, 1919: 350, fig. 30 (w.) SOLOMON IS. Combination in *Tetramorium:* Bolton, 1985: 247. Raised to species: Bolton, 1976: 344.

*anteversa. Tetramorium setigerum* var. *anteversa* Santschi, 1921c: 121 (w.) TANZANIA. Junior synonym of *setigerum:* Bolton, 1980: 283.

***antipodum.*** *Tetramorium antipodum* Wheeler, W.M. 1927i: 143 (w.q.m.) AUSTRALIA. Junior synonym of *caldarium:* Bolton, 1979: 169. Revived from synonymy: **revived status**.

***antrema.*** *Triglyphothrix antrema* Bolton, 1976: 318 (w.) CAMEROUN. Combination in *Tetramorium:* Bolton, 1985: 247.

***anxium.*** *Tetramorium pusillum* var. *anxia* Santschi, 1914d: 365, fig. 28 (w.q.) GUINEA. Raised to species: Bolton, 1980: 305.

***aptum.*** *Tetramorium aptum* Bolton, 1977: 115 (w.) THAILAND.

*arenarium. Tetramorium sericeiventre* var. *arenarium* Santschi, 1918b: 126 (w.) TUNISIA. Santschi, 1937g: 82 (m.). Junior synonym of *sericeiventre:* Bolton, 1980: 332.

*areolatus. Triglyphothrix areolatus* Stitz, 1910: 142, fig. 9 (w.) CAMEROUN. Junior synonym of *gabonense:* Bolton, 1976: 326.

***argenteopilosum.*** *Tetramorium argenteopilosum* Arnold, 1926: 261, fig. 71 (w.q.) ZIMBABWE. See also: Bolton, 1980: 306.

***armatum.*** *Tetramorium inerme* st. *armatum* Santschi, 1927a: 57 (w.q.) TRANSCASPIA. [First available use of *Tetramorium caespitum* subsp. *inerme* var. *armata* Emery, 1925c: 191; unavailable name.] Subspecies of *inerme:* Dlussky, 1962: 179. Raised to species: Pisarski & Krzysztofiak, 1981: 158; Dlussky & Zabelin, 1985: 232. Senior synonym of *annauensis* and *brevinodis,* and material of the unavailable name *ligatum* referred here: Radchenko, 1992b: 53.

***arnoldi.*** *Triglyphothrix arnoldi* Forel, 1913j: 220 (w.) ZIMBABWE. Arnold, 1917: 339 (q.). Combination in *Tetramorium:* Arnold, 1917: 338; Bolton, 1985: 247. See also: Bolton, 1976: 318.

*arnoldi. Rhoptromyrmex arnoldi* Santschi, 1916b: 503 (w.) ZIMBABWE. Combination in *Tetramorium:* Santschi, 1917b: 286 (in text). [Junior secondary homonym of *arnoldi* Forel, above.] Replacement name: *Tetramorium incruentatum* Arnold, 1926: 271. See also: Bolton, 1980: 306; Bolton, 1985: 247.

*aruensis. Xiphomyrmex aruensis* Karavaiev, 1935a: 105, fig. 24 (w.) INDONESIA (Aru I.). Junior synonym of *carinatum:* Bolton, 1977: 86.

***arzi.*** *Tetramorium arzi* Thomé, G. 1969: 4, fig. 2 (w.q.) LEBANON.

***asetyum.*** *Tetramorium asetyum* Bolton, 1980: 324, fig. 101 (w.q.) NIGERIA.

***aspersum.*** *Myrmica aspersa* Smith, F. 1865: 72 (w.) INDONESIA (Morotai I.). Combination in *Xiphomyrmex:* Donisthorpe, 1932c: 473; in *Tetramorium:* Bolton, 1977: 86. Senior synonym of *bismarckii* and *willowsi:* Bolton, 1977: 86.

*aspinatum. Tetramorium aspinatum* Prins, 1973: 12, figs. 10–13, 29 (w.q.) SOUTH AFRICA. Junior synonym of *jordani:* Bolton, 1980: 248.

***ataxium.*** *Tetramorium ataxium* Bolton, 1980: 295, figs. 78, 86 (w.) NIGERIA.

#***atlante.*** *Tetramorium semilaeve* subsp. *atlantis* Cagniant, 1970a: 430 (w.q.) TUNISIA. [First available use of *Tetramorium caespitum* st. *punicum* var. *atlantis* Santschi, 1918c: 155; unavailable name.] [*T. punicum* var. *atlantis* Cagniant, 1964: 106; unavailable name.]

*atomum* Santschi, 1914 [*Xiphomyrmex*]; see under ***WASMANNIA.***

*atrinodis. Xiphomyrmex kivuense* st. *atrinodis* Santschi, 1928f: 208 (w.) KENYA. Junior synonym of *tersum:* Bolton, 1980: 233.

*auropunctatum* Roger, 1863; see under ***WASMANNIA.***

*auropunctatus. Triglyphothrix auropunctatus* Forel, 1910f: 20 (w.q.) SOUTH AFRICA. [Unresolved junior primary homonym of *auropunctatum* Roger, 1863a: 182, above.] Junior synonym of *microps:* Bolton, 1976: 329.

*australis. Triglyphothrix striatidens* var. *australis* Forel, 1902h: 449 (w.q.) AUSTRALIA. Junior synonym of *lanuginosum:* Bolton, 1976: 350.

*australe. Tetramorium australe* Bolton, 1977: 146, fig. 68 (w.) AUSTRALIA. [Junior secondary homonym of *australis* Forel, above.] Replacement name: *taylori* Bolton, 1985: 247.

***avium.*** *Tetramorium avium* Bolton, 1980: 276, figs. 58, 63, 64 (w.q.) SOUTH AFRICA.

*bacchus. Tetramorium bacchus* Forel, 1910e: 426 (w.) SOUTH AFRICA. Junior synonym of *erectum:* Bolton, 1980: 269.

*bahai* Forel, 1922; see under ***PHEIDOLE.***

*balzani* Emery, 1894; see under ***HYLOMYRMA.***

*bantouana. Tetramorium pusillum* var. *bantouana* Santschi, 1910c: 382, fig. 10 (w.q.m.) CONGO. Junior synonym of *simillimum:* Bolton, 1980: 320.

***banyulense.*** *Tetramorium banyulense* Bernard, 1983: 98 (diagnosis in key) (w.m.) FRANCE.

#***barabense.*** *Tetramorium caespitum* var. *barabensis* Ruzsky, 1925a: 288 (w.) RUSSIA.

***barbigerum.*** *Tetramorium barbigerum* Bolton, 1980: 243, figs. 19, 20, 25 (w.) NAMIBIA.

*bariensis. Tetramorium caespitum* var. *bariensis* Forel, 1911d: 331 (w.) ITALY. Junior synonym of *diomedea:* Emery, 1916b: 194; of *ferox:* Bernard, 1967: 235. Revived from synonymy, and raised to species: Baroni Urbani, 1971c: 134. Junior synonym of *ferox:* Radchenko, 1992b: 55.

***barryi.*** *Tetramorium barryi* Mathew, 1981: 425, figs. 1–4 (w.q.) INDIA.

***basum.*** *Tetramorium basum* Bolton, 1977: 104, fig. 30 (w.) NEW GUINEA.

***baufra.*** *Triglyphothrix baufra* Bolton, 1976: 319, figs. 49, 57 (w.) ZIMBABWE. Combination in *Tetramorium:* Bolton, 1985: 247.

*beirae. Tetramorium quadrispinosum* r. *beirae* Arnold, 1926: 252 (w.) MOZAMBIQUE. Junior synonym of *quadrispinosum:* Bolton, 1980: 330.

***belgaense.*** *Tetramorium (Xiphomyrmex) belgaense* Forel, 1902c: 238 (q.) INDIA. Bolton, 1979: 174 (w.). See also: Bolton, 1977: 79.

***bellicosum.*** *Tetramorium bellicosum* Bolton, 1980: 296 (w.) NIGERIA.

*bellii. Tetramorium (Xiphomyrmex) tortuosum* var. *bellii* Forel, 1902c: 239 (w.) INDIA. Junior synonym of *tortuosum:* Bolton, 1977: 83.

*benguelense. Tetramorium quadrispinosum* st. *elegans* var. *benguelense* Santschi, 1937d: 234 (w.) ANGOLA. Unavailable name; material referred to *quadrispinosum* by Bolton, 1980: 330.

***bequaerti.*** *Tetramorium bequaerti* Forel, 1913b: 318 (w.) ZAIRE. Senior synonym of *humile:* Bolton, 1980: 325.

***berbiculum.*** *Tetramorium berbiculum* Bolton, 1980: 307, fig. 92 (w.) ZIMBABWE.

***bessonii.*** *Tetramorium (Xiphomyrmex) bessonii* Forel, 1891b: 156, pl. 4, fig. 13 (w.) MADAGASCAR. Combination in *Xiphomyrmex:* Wheeler, W.M. 1922a: 1031; in *Tetramorium,* and senior synonym of *orientale:* Bolton, 1979: 141.

***bevisi.*** *Tetramorium bevisi* Arnold, 1958: 120, fig. 1 (w.q.m.) LESOTHO. See also: Bolton, 1980: 308.

***bicarinatum.*** *Myrmica bicarinata* Nylander, 1846b: 1061 (w.q.) U.S.A. Forel, 1891b: 151 (m., misidentified as *guineense*); Wheeler, G.C. & Wheeler, J. 1954d: 449 (l., misidentified as *guineense*); Wheeler, W.M. 1924c: 136 (gynandromorph, misidentified as *guineense*). Combination in *Tetramorium:* Mayr, 1862: 740. Junior synonym of *Formica guineensis* Fabricius, 1793: 357 (now in *Pheidole*): Mayr, 1862: 740. Revived from synonymy, combination in *Tetramorium,* and senior synonym of *cariniceps* (and its junior synonym *kollari*), *modesta* Smith and *reticulata:* Bolton, 1977: 94. [Note. The names *cariniceps, kollari* and *reticulata* had previously been incorrectly synonymised with *guineensis* Fabricius by Roger, 1862c: 293; *modesta* Smith was wrongly synonymised with *guineensis* by Donisthorpe, 1932c: 463.] See also: Bolton, 1979: 164.

***bicolor.*** *Tetramorium bicolor* Viehmeyer, 1914c: 39 (w.) NEW GUINEA. See also: Bolton, 1977: 97.

*binodis. Formica binodis* Linnaeus, 1763: 413 (q.) JAMAICA. Junior synonym of *caespitum:* Linnaeus, 1767: 963; Olivier, 1792: 496 (synonymy probably incorrect). Unidentifiable taxon; *incertae sedis* in *Tetramorium.*

*bipartita. Tetramorium sericeiventre* var. *bipartita* Santschi, 1918b: 126 (w.) KENYA. Junior synonym of *sericeiventre:* Bolton, 1980: 332.

*biroi. Tetramorium caespitum* subsp. *pyrenaeica* var. *biroi* Röszler, 1937: 199 (w.q.m.) HUNGARY. Unavailable name.

***biskrense.*** *Tetramorium caespitum* var. *biskrensis* Forel, 1904c: 13 (w.q.m.) ALGERIA. Stirps of *caespitum:* Santschi, 1918c: 155; Santschi, 1921a: 112. Raised to species: Menozzi, 1933b: 75 (in key); Finzi, 1940: 161; Cagniant, 1964: 89; Baroni Urbani, 1971c: 134. Current subspecies: nominal plus *jarbas, kahenae.*

*bismarckii. Tetramorium (Xiphomyrmex) bismarckii* Forel, 1901b: 11 (w.) BISMARCK ARCHIPELAGO. Junior synonym of *aspersum:* Bolton, 1977: 86.

*blochmannii. Tetramorium blochmannii* Forel, 1887: 384 (w.) MADAGASCAR. Forel, 1910f: 19 (q.m.). Stirps of *sericeiventre:* Santschi, 1918b: 124 (in key). Junior synonym of *sericeiventre:* Bolton, 1979: 155.

***bothae.*** *Tetramorium simillimum* subsp. *bothae* Forel, 1910e: 425 (w.q.m.) SOUTH AFRICA. Raised to species, and senior synonym of *guillarmodi:* Bolton, 1980: 309.

*boulognei. Triglyphothrix gabonensis* var. *boulognei* Forel, 1916: 423 (w.) ZAIRE. Junior synonym of *gabonense:* Bolton, 1976: 326.

*braunsi. Tetramorium braunsi* Forel, 1913a: 119 (w.) SOUTH AFRICA. Variety of *capense:* Forel, 1913j: 220. Junior synonym of *capense:* Santschi, 1914a: 435; Bolton, 1980: 359.

*breve. Tetramorium pusillum* st. *bantuala* var. *breve* Santschi, 1924b: 213 (w.) ZAIRE. Unavailable name; material referred to *simillimum* by Bolton, 1980: 320.

***brevicorne.*** *Tetramorium caespitum* var. *brevicorne* Bondroit, 1918: 108 (w.q.) CORSICA. [First available use of *Tetramorium caespitum* subsp. *caespitum* var. *brevicornis* Emery, 1916b: 196; unavailable name.] Emery, 1925c: 185 (m.). Raised to species: Baroni Urbani, 1964b: 53. See also: Casevitz-Weulersse, 1990b: 421.

***brevidentatum.*** *Triglyphothrix brevidentata* Kutter, 1932: 208, fig. (w.) JAVA. Imai, Kubota, *et al.* 1985: 47 (k.). Combination in *Tetramorium:* Bolton, 1985: 247. See also: Bolton, 1976: 345.

*brevinodis. Tetramorium caespitum* var. *brevinodis* Karavaiev, 1927b: 109 (w.q.m.) TURKESTAN. Junior synonym of *armatum:* Radchenko, 1992b: 53.

*brevis. Tetramorium brevis* Weber, 1943c: 370 (w.) SUDAN. Junior synonym of *nigrum:* Bolton, 1980: 316.

***brevispinosum.*** *Triglyphothrix gabonensis* var. *brevispinosus* Stitz, 1910: 144 (w.) CAMEROUN. Combination in *Tetramorium:* Bolton, 1985: 247. Raised to species, and senior synonym of *nion:* Bolton, 1976: 320.

*brevispinosa. Wasmannia auropunctata* subsp. *brevispinosa* Borgmeier, 1928a: 36, figs. 4, 5 (w.) BRAZIL. [Unresolved junior secondary homonym of *brevispinosus* Stitz, above.] Junior synonym of *simillimum:* Borgmeier, 1937b: 241.

***browni.*** *Tetramorium browni* Bolton, 1980: 337 (w.) GHANA.

*bruni. Tetramorium bequaerti* st. *bruni* Santschi, 1917b: 285 (w.) ZIMBABWE. Junior synonym of *bulawayense:* Bolton, 1980: 326.

***bulawayense.*** *Tetramorium bequaerti* r. *bulawayensis* Forel, 1913a: 119 (w.) ZIMBABWE. Arnold, 1917: 282 (q.). Raised to species, senior synonym of *bruni* and material of the unavailable name *mashona* referred here: Bolton, 1980: 326.

*bulawayensis. Tetramorium (Triglyphothrix) auropunctatus* var. *bulawayensis* Arnold, 1917: 335 (w.) ZIMBABWE. Combination in *Triglyphothrix:* Wheeler, W.M. 1922a: 909. [Unresolved junior primary homonym of *bulawayense* Forel, above.] Junior synonym of *microps:* Bolton, 1976: 329.

*burgeoni. Triglyphothrix areolata* var. *burgeoni* Santschi, 1935a: 265 (w.) ZAIRE. Junior synonym of *gabonense:* Bolton, 1976: 326.

***bursakovi.*** *Tetramorium bursakovi* Radchenko, 1992a: 41, fig. 2 (w.q.m.) KAZAKHSTAN.

***buthrum.*** *Tetramorium buthrum* Bolton, 1980: 309 (w.) CENTRAL AFRICAN REPUBLIC.

#***caespitomoravicum.*** *Tetramorium moravicum* var. *caespitomoravicum* Kratochvil, in Novak & Sadil, 1941: 86 (diagnosis in key) (w.) CZECHOSLOVAKIA. Variety of *caespitum:* Kratochvil, in Kratochvil, Novak & Snoflak, 1944: 44.

***caespitum.*** *Formica caespitum* Linnaeus, 1758: 581 (w.) EUROPE. Latreille, 1798: 50 (q.m.); Mayr, 1861: 62 (q.m.); Wheeler, G.C. & Wheeler, J. 1954d: 445 (l.); Hauschteck, 1961: 221 (k.); Imai, 1966: 119 (k.). Combination in *Tetramorium:* Mayr, 1855: 426. Senior synonym of *fuscula:* Smith, F. 1851: 118; of *modesta* Foerster: Curtis, 1854: 215; Mayr, 1855: 426; of *fusca:* Dalla Torre, 1893: 132; of *transversinodis:* Brown, 1949a: 47; of *immigans:* Bolton, 1979: 171; of *himalayanum, indocile* and *transbaicalense:* Radchenko, 1992b: 50; of *hammi* **new synonymy**. Current subspecies: nominal plus *alternans, barabense, caespitomoravicum, flavidulum, japonicum, oxyomma, pallidum, penninum, rhodium, typicum.* See also: Emery, 1909d: 697; Bondroit, 1918: 107; Emery, 1925c: 177; Baroni Urbani, 1971c: 135; Kutter, 1977c: 157; Arnol'di & Dlussky, 1978: 544; Smith, D.R. 1979: 1400; Collingwood, 1979: 84; Kupyanskaya, 1990: 151; López, 1991a: 31; López, 1991b: 73.

***caldarium.*** *Tetrogmus caldarius* Roger, 1857: 12 (w.q.) GERMANY. Meinert, 1861: 331 (m.). Junior synonym of *simillimum:* Roger, 1862c: 297. Revived from synonymy, combination in *Tetramorium,* and senior synonym of *hemisi* and *minutum:* Bolton, 1979: 169. Senior synonym of *transformans:* Bolton, 1980: 310.

***calidum.*** *Tetramorium caespitum* var. *calidum* Forel, 1907a: 15 (w.) MUSCAT. Raised to species: Collingwood, 1985: 263.

***calinum.*** *Tetramorium calinum* Bolton, 1980: 239, figs. 15, 16 (w.) GHANA.

*calvum. Tetramorium blochmanni* var. *calvum* Stitz, 1923: 162 (w.) NAMIBIA. Junior synonym of *quadrispinosum:* Bolton, 1980: 330.

***camerunense.*** *Tetramorium camerunense* Mayr, 1895: 129 (w.) CAMEROUN. See also: Bolton, 1980: 338.

***candidum.*** *Tetramorium candidum* Bolton, 1980: 345, fig. 124 (w.m.) ZAIRE.

***capense.*** *Tetramorium capense* Mayr, 1865: 89 (w.) SOUTH AFRICA. Emery, 1895h: 36 (q.m.). Senior synonym of *braunsi:* Santschi, 1914a: 435; Bolton, 1980: 359; of *popovici:* Bolton, 1980: 359. See also: Prins, 1973: 8.

***capillosum.*** *Tetramorium capillosum* Bolton, 1980: 236, fig. 13 (w.) GABON.

***capitale.*** *Xiphomyrmex capitalis* McAreavey, 1949: 6, figs. 20–25 (w.q.) AUSTRALIA. Combination in *Tetramorium:* Bolton, 1977: 136.

***carinatum.*** *Myrmica carinata* Smith, F. 1859a: 148 (w.) INDONESIA (Aru I.). Combination in *Xiphomyrmex:*

Donisthorpe, 1932c: 455; in *Tetramorium:* Bolton, 1977: 86. Senior synonym of *aruensis, costatus, deficiens* and *flavescens:* Bolton, 1977: 86.
*cariniceps. Myrmica cariniceps* Guérin-Méneville, 1852: 79 (w.) DOMINICA. Combination in *Tetramorium,* and senior synonym of *kollari:* Roger, 1861b: 171. Junior synonym of *guineense:* Roger, 1862c: 293; of *bicarinatum:* Bolton, 1977: 94.
*caspium. Tetramorium punicum* subsp. *caspium* Arnol'di, 1948: 211. *Nomen nudum.*
*cataractae. Tetramorium intextum* var. *cataractae* Santschi, 1916b: 506, fig. B (w.) ZIMBABWE. Junior synonym of *delagoense:* Bolton, 1980: 311; of *intextum:* **new synonymy**.
***centum.*** *Tetramorium centum* Bolton, 1977: 105 (w.q.) NEW GUINEA.
*ceramensis. Triglyphothrix ceramensis* Stitz, 1912: 506 (w.) INDONESIA (Seram I.). Junior synonym of *lanuginosum:* Bolton, 1976: 350.
***chapmani.*** *Tetramorium chapmani* Bolton, 1977: 120, fig. 46, 47 (w.) PHILIPPINES.
***chefteki.*** *Tetramorium caespitum* var. *chefteki* Forel, 1911d: 332 (w.) TURKEY. Raised to species: Agosti & Collingwood, 1987a: 56.
***chepocha.*** *Triglyphothrix chepocha* Bolton, 1976: 345, fig. 72 (w.) BORNEO. Combination in *Tetramorium:* Bolton, 1985: 247.
*chinensis. Tetramorium caespitum* st. *jacoti* var. *chinensis* Santschi, 1928b: 35 (w.) CHINA. Unavailable name; material referred to *caespitum* by Yasumatsu, 1962: 96; to *jacoti* by Radchenko, 1992b: 51.
***chloe.*** *Xiphomyrmex angulinode* var. *chloe* Santschi, 1920b: 17, fig. 3 (w.) ZIMBABWE. Combination in *Tetramorium,* and raised to species: Bolton, 1980: 239.
***christiei.*** *Tetramorium christiei* Forel, 1902c: 232 (w.) INDIA. See also: Bolton, 1977: 124.
***ciliatum.*** *Tetramorium ciliatum* Bolton, 1977: 121, fig. 49 (w.q.) THAILAND.
*cinereus. Triglyphothrix cinereus* Weber, 1943c: 377, pl. 16, fig. 25 (w.) SUDAN. Junior synonym of *eminii:* Bolton, 1976: 324.
*cinnamomeum. Tetramorium sericeiventre* st. *cinnamomeum* Santschi, 1918b: 124 (diagnosis in key) (w.) MOZAMBIQUE (attributed to Arnold). [Also described as new by Arnold, 1926: 249.] Junior synonym of *sericeiventre:* Bolton, 1980: 332.
***clunum.*** *Tetramorium setuliferum* st. *cluna* Forel, 1913j: 218 (w.) SOUTH AFRICA. Raised to species: Bolton, 1980: 244.
*coeleste. Tetramorium caespitum* st. *tsushimae* var. *coeleste* Santschi, 1927a: 54 (w.) CHINA. Unavailable name.
***cognatum.*** *Tetramorium cognatum* Bolton, 1979: 135, figs. 1, 2 (w.q.) MADAGASCAR.
***coillum.*** *Tetramorium coillum* Bolton, 1979: 147, fig. 21 (w.q.) MADAGASCAR.
*colluta. Tetramorium sericeiventre* st. *femoratum* var. *colluta* Santschi, 1918b: 129 (w.) SOUTH AFRICA. Unavailable name; material referred to *sericeiventre* by Bolton, 1980: 332.
***coloreum.*** *Tetramorium coloreum* Mayr, 1901a: 273 (w.) CAMEROUN. Senior synonym of *muscicola:* Bolton, 1980: 297.
*commodum. Tetramorium commodum* Santschi, 1924b: 215 (w.q.) ZAIRE. Junior synonym of *quadridentatum:* Bolton, 1980: 364.
***concaviceps.*** *Tetramorium concaviceps* Bursakov, 1984: 402, fig. 2 (w.q.m.) KAZAKHSTAN.
***confine.*** *Tetramorium confinis* Radchenko & Arakelian, 1990: 374, fig. 2 (w.q.m.) ARMENIA.
*confucii* Forel, 1912; see under ***LEPTOTHORAX.***
***confusum.*** *Tetramorium confusum* Bolton, 1977: 143, fig. 63 (w.) AUSTRALIA.
***constanciae.*** *Tetramorium (Triglyphothrix) constanciae* Arnold, 1917: 339, pl. 7, fig. 94 (w.) ZIMBABWE. Combination in *Triglyphothrix:* Wheeler, W.M. 1922a: 910; in *Tetramorium:* Bolton, 1985: 247. Senior synonym of *longispinosa:* Bolton, 1976: 321.
*continentis. Tetramorium blochmanni* subsp. *continentis* Forel, 1910e: 426 (w.) SOUTH AFRICA. Forel, 1913b: 319 (q.); Arnold, 1917: 279 (m.). Stirps/race of *sericeiventre:* Santschi, 1918b: 130; Arnold, 1926: 250. Junior synonym of *sericeiventre:* Bolton, 1980: 332.
***convexum.*** *Tetramorium convexum* Bolton, 1980: 322 (w.) GHANA.
***coonoorense.*** *Tetramorium coonoorense* Forel, 1902c: 237 (w.q.) INDIA. Combination in *Triglyphothrix:* Bolton, 1976: 346; in *Tetramorium:* Bolton, 1985: 247.
*costatus. Xiphomyrmex costatus* Emery, 1897c: 587, pl. 15, fig. 26 (w.) NEW GUINEA. Junior synonym of *carinatum:* Bolton, 1977: 86.
***crepum.*** *Tetramorium crepum* Wang & Wu, in Wang, M., Xiao & Wu, 1988: 269, fig. 28 (w.) CHINA.
***cristatum.*** *Tetramorium guineense* var. *cristatum* Stitz, 1910: 144 (w.) TOGO. Raised to species, senior synonym of *medje,* and material of the unavailable name *ebangense* referred here: Bolton, 1980: 268.
*cristulatum. Tetramorium emeryi* st. *cristulatum* Forel, 1913j: 218 (w.m.) SOUTH AFRICA. Arnold, 1917: 302 (q.). Junior synonym of *emeryi:* Bolton, 1980: 269.
***crypticum.*** *Triglyphothrix cryptica* Bolton, 1976: 321 (w.) GHANA. Combination in *Tetramorium:* Bolton, 1985: 247.
*cucalense. Tetramorium setuliferum* var. *cucalense* Santschi, 1911c: 356 (w.) ANGOLA. Santschi, 1930b: 71 (q.). Senior synonym of *triptolemus:* Santschi, 1930b: 71. Junior synonym of *setuliferum:* Bolton, 1980: 250.
***cuneinode.*** *Tetramorium cuneinode* Bolton, 1977: 126, fig. 56 (w.) THAILAND.
***curtulum.*** *Tetramorium curtulum* Emery, 1895k: 471 (w.) BURMA. See also: Bolton, 1977: 116.
***curvispinosum.*** *Tetramorium curvispinosum* Mayr, 1897: 430 (w.) SRI LANKA. See also: Bolton, 1977: 121.
***cynicum.*** *Tetramorium cynicum* Bolton, 1977: 97, figs. 19, 20 (w.) PHILIPPINES.
*cyprium. Tetramorium semilaeve* st. *syriacum* var. *cyprium* Santschi, 1934d: 279 (w.q.) CYPRUS. Unavailable name.
**#*cyrenaicum.*** *Tetramorium punicum* var. *cyrenaica* Menozzi, 1927g: 379 (w.q.m.) LIBYA. [First available use of *Tetramorium caespitum* subsp. *punicum* var. *cyrenaica* Emery, 1924a: 10; unavailable name.]
*daphnis. Xiphomyrmex angulinode* var. *daphnis* Santschi, 1920b: 16, fig. 3 (w.q.) ZIMBABWE. Junior synonym of *angulinode:* Bolton, 1980: 237.
***davidi.*** *Tetramorium caespitum* subsp. *davidi* Forel, 1911g: 285 (w.) ISRAEL. Forel, 1913d: 431 (q.). Subspecies of *ferox:* Menozzi, 1933b: 75 (in key). Raised to species: Agosti & Collingwood, 1987a: 56.
*debile. Tetramorium sericeiventre* var. *debile* Forel, 1894b: 80 (w.) ETHIOPIA. Junior synonym of *sericeiventre:* Bolton, 1980: 332.
*debilis. Tetramorium caespitum* var. *debilis* Emery, 1922f: 276 (w.q.m.) EGYPT. [First available use of *Tetramorium caespitum* subsp. *caespitum* var. *debilis* Emery, 1909d: 701; unavailable name.] [Junior primary homonym of *debile* Forel, above.] Replacement name: *taueret* **new name**. Variety of *biskrense:* Emery, 1925c: 182; Finzi, 1936: 181. Raised to species: Bernard, 1967: 235.
***decamerum.*** *Triglyphothrix decamera* Forel, 1902c: 240 (q.) INDIA. Combination in *Tetramorium:* Bolton, 1985: 247. See also: Bolton, 1976: 347.
*decem* Forel, 1913; see under ***DECAMORIUM.***
***deceptum.*** *Tetramorium deceptum* Bolton, 1977: 146 (w.) AUSTRALIA.
***dedefra.*** *Triglyphothrix dedefra* Bolton, 1976: 322, figs. 48, 56 (w.) ANGOLA. Combination in *Tetramorium:* Bolton, 1985: 247.
*deficiens. Xiphomyrmex costatus* subsp. *deficiens* Emery, 1897c: 588 (w.) NEW GUINEA. Junior synonym of *carinatum:* Bolton, 1977: 86.
*defricta. Tetramorium sericeiventre* st. *inversa* var. *defricta* Santschi, 1918b: 129 (w.) ZIMBABWE. Unavailable name; material referred to *sericeiventre* by Bolton, 1980: 332.
***degener.*** *Tetramorium (Xiphomyrmex) degener* Santschi, 1911e: 124 (w.) MADAGASCAR. See also: Bolton, 1979: 148.
***delagoense.*** *Tetramorium simillimum* st. *delagoense* Forel, 1894b: 80 (w.q.m.) MOZAMBIQUE. Raised to species, and senior synonym of *madecassum:* Bolton, 1979: 156.
***densopilosum.*** *Tetramorium densopilosus* Radchenko & Arakelian, 1990: 372, fig. 1 (w.q.m.) ARMENIA.
*denticulatum. Tetramorium simillimum* r. *denticulatum* Forel, 1902c: 235 (w.) INDIA. Junior synonym of *simillimum:* Bolton, 1977: 131.
***depressiceps.*** *Tetramorium semilaeve* subsp. *depressiceps* Menozzi, 1933b: 71, fig. 3 (w.q.m.) ISRAEL. Raised to species: Collingwood, 1985: 264.
**#*depressum.*** *Tetramorium caespitum* r. *depressum* Forel, 1892g: 455 (w.) CANARY IS. Emery, 1893c: 84 (q.m.). Stirps of *semilaeve:* Santschi, 1936c: 202. Senior synonym of *guancha,* and material of the unavailable name *palmense* referred here: Hohmann, La Roche, *et al.,* 1993: 158. See also: Santschi, 1921b: 430.
***desertorum.*** *Triglyphothrix desertorum* Forel, 1910f: 20 (q.m.) BOTSWANA. Combination in *Tetramorium:* Arnold, 1917: 344. See also: Bolton, 1976: 323; Bolton, 1985: 247.
***dichroum.*** *Tetramorium solidum* st. *dichroum* Santschi, 1932a: 388 (w.) SOUTH AFRICA. Subspecies of *peringueyi:* Prins, 1973: 11. Raised to species: Bolton, 1980: 245.

***difficile***. *Tetramorium difficile* Bolton, 1977: 126 (w.) NEPAL.

***diligens***. *Myrmica diligens* Smith, F. 1865: 73 (w.) NEW GUINEA. Combination in *Xiphomyrmex:* Donisthorpe, 1932c: 473; in *Tetramorium:* Bolton, 1977: 87. Senior synonym of *diligentissima:* **new synonymy**.

*diligentissima*. *Pheidole diligentissima* Dalla Torre, 1892: 90. [Unnecessary replacement name for *diligens* Smith, above.] Junior synonym of *diligens:* **new synonymy**.

***diomandei***. *Tetramorium diomandei* Bolton, 1980: 268, figs. 67, 71, 73 (w.) IVORY COAST.

*diomedea*. *Tetramorium caespitum* var. *diomedea* Emery, 1908d: 24 (w.q.) ITALY. Subspecies/variety of *ferox:* Grandi, 1935: 102; Baroni Urbani, 1971c: 138. Raised to species: Schembri & Collingwood, 1981: 435; Agosti & Collingwood, 1987a: 56. Junior synonym of *ferox:* Bernard, 1967: 235; Dlussky, Soyunov & Zabelin, 1990: 201; Radchenko, 1992b: 55.

***distinctum***. *Triglyphothrix distincta* Bolton, 1976: 323 (w.q.) GHANA. Combination in *Tetramorium:* Bolton, 1985: 247.

***do***. *Tetramorium squaminode* r. *do* Forel, 1914d: 224 (w.q.m.) ZIMBABWE. Raised to species: Arnold, 1960a: 82. See also: Bolton, 1980: 254.

***dogieli***. *Tetramorium dogieli* Karavaiev, 1931d: 48, fig. 6 (w.) KENYA. See also: Bolton, 1980: 254.

***dolichosum***. *Tetramorium dolichosum* Bolton, 1980: 277 (w.) ZAIRE.

***dominum***. *Tetramorium dominum* Bolton, 1980: 360, fig. 136, 139 (w.) SOUTH AFRICA.

***doriae***. *Tetramorium doriae* Emery, 1881b: 530 (w.) ETHIOPIA. See also: Bolton, 1980: 277.

*drakensbergensis*. *Tetramorium akermani* var. *drakensbergensis* Arnold, 1926: 267 (w.q.) SOUTH AFRICA. Junior synonym of *frigidum:* Bolton, 1980: 256.

***dumezi***. *Tetramorium simillimum* var. *dumezi* Menozzi, 1942: 177 (w.) ZAIRE. [First available use of *Tetramorium simillimum* st. *isipingense* var. *dumezi* Forel, 1916: 422; unavailable name.] Raised to species, and senior synonym of *dumezi* Bolton: **new status, new synonymy**.

*dumezi*. *Tetramorium dumezi* Bolton, 1980: 346 (w.) GHANA. Junior synonym of *dumezi* Menozzi: **new synonymy**.

***dysalum***. *Tetramorium dysalum* Bolton, 1979: 141, fig. 24 (w.q.) MADAGASCAR.

***dysderke***. *Tetramorium dysderke* Bolton, 1980: 287 (w.) NIGERIA.

*ebangense*. *Tetramorium guineense* st. *cristatum* var. *ebangense* Santschi, 1937d: 235, figs. 3, 4 (w.) ANGOLA. Unavailable name; material referred to *cristatum* by Bolton, 1980: 268.

*ebeninum*. *Tetramorium* (*Xiphomyrmex*) *ebeninum* Arnold, 1926: 277, fig. 80 (w.) SOUTH AFRICA. Junior synonym of *weitzeckeri:* Bolton, 1980: 233.

*edithae*. *Xiphomyrmex weitzeckeri* subsp. *edithae* Weber, 1943c: 375 (w.) SUDAN. Junior synonym of *weitzeckeri:* Bolton, 233.

***edouardi***. *Tetramorium* (*Xiphomyrmex*) *edouardi* Forel, 1894b: 82 (w.) ETHIOPIA. Santschi, 1919a: 344 (q.). Combination in *Xiphomyrmex:* Santschi, 1919a: 344; in *Tetramorium:* Bolton, 1980: 225.

*eidmanni*. *Tetramorium eidmanni* Menozzi, 1941: 21 (w.) CHINA. *Nomen nudum;* material referred to *kraepelini* by Bolton, 1977: 117.

***eleates***. *Tetramorium* (*Xiphomyrmex*) *tortuosum* var. *eleates* Forel, 1913k: 82 (w.q.) SUMATRA. Imai, Brown, *et al.* 1984: 68 (k.). Raised to species: Bolton, 1977: 79.

***electrum***. *Tetramorium electrum* Bolton, 1979: 144, fig. 20 (w.) MADAGASCAR.

*elegans*. *Tetramorium quadrispinosum* st. *elegans* Santschi, 1918b: 125 (w.) SOUTH AFRICA. Junior synonym of *eudoxia:* Arnold, 1926: 252.

***elidisum***. *Tetramorium elidisum* Bolton, 1980: 346, fig. 125 (w.) GHANA.

***elisabethae***. *Tetramorium elisabethae* Forel, 1904c: 20 (w.) INDIA. See also: Bolton, 1977: 111.

*elysii* Mann, 1919; see under ***ROMBLONELLA***.

***emeryi***. *Tetramorium emeryi* Mayr, 1901b: 23 (w.q.m.) SOUTH AFRICA. Senior synonym of *cristulatum:* Bolton, 1980: 269.

***eminii***. *Triglyphothrix eminii* Forel, 1894: 79 (w.) ETHIOPIA. Combination in *Tetramorium:* Bolton, 1985: 247. Senior synonym of *akermani, cinereus, guillodi, marleyi, marthae* and *mus:* Bolton, 1976: 324.

***erectum***. *Tetramorium guineense* var. *erectum* Emery, 1895h: 37 (w.) SOUTH AFRICA. Raised to species, and senior synonym of *bacchus:* Bolton, 1980: 269.

***ericae***. *Tetramorium ericae* Arnold, 1917: 332 (w.) ZIMBABWE. Junior synonymy of *pauper* Santschi: Bolton, 1976: 333. Hence first available replacement name for *Triglyphothrix pauper* Santschi, 1917b: 286. [Junior secondary homonym of *Tetramorium pauper* Forel, 1907a: 14.] See also: Bolton, 1985: 247.

*ernesti*. *Tetramorium caespitum* st. *semilaeve* var. *ernesti* Santschi, 1921b: 431 (w.q.) FRANCE. Unavailable name.

*escherichi*. *Tetramorium* (*Xiphomyrmex*) *escherichi* Forel, 1910c: 259 (w.q.) ETHIOPIA. Combination in *Xiphomyrmex:* Emery, 1915g: 4. Junior synonym of *weitzeckeri:* Bolton, 1980: 233.

*ethica*. *Tetramorium* (*Xiphomyrmex*) *tortuosum* var. *ethica* Forel, 1911i: 225 (w.) SRI LANKA. Junior synonym of *tortuosum:* Bolton, 1977: 83.

***etiolatum***. *Tetramorium etiolatum* Bolton, 1977: 105, fig. 29 (w.q.m.) NEW GUINEA.

*eudoxia*. *Tetramorium quadrispinosum* st. *eudoxia* Santschi, 1918b: 122 (w.) SOUTH AFRICA. [First available use of *Tetramorium blochmanni* r. *continentis* var. *eudoxia* Forel, 1914d: 231; unavailable name.] Senior synonym of *elegans:* Arnold, 1926: 252. Junior synonym of *quadrispinosum:* Bolton, 1980: 330.

*evidens*. *Tetramorium sericeiventre* st. *inversum* var. *evidens* Santschi, 1928f: 206 (w.) ZAIRE. Unavailable name; material referred to *sericeiventre* by Bolton, 1980: 332.

***exasperatum***. *Tetramorium caespitum* r. *exasperatum* Emery, 1891b: 3 (w.q.) TUNISIA. Raised to species: Dalla Torre, 1893: 133. Current subspecies: nominal plus *acutiseta*. See also: Emery, 1909d: 705.

*exoleta*. *Tetramorium pusillum* var. *exoleta* Santschi, 1914d: 366 (w.) NIGERIA. Junior synonym of *simillimum:* Bolton, 1980: 320.

*felix*. *Triglyphothrix striatidens* var. *felix* Forel, 1912k: 160 (w.) SEYCHELLES IS. Junior synonym of *lanuginosum:* Bolton, 1976: 350.

*femoratum*. *Tetramorium sericeiventre* subsp. *femoratum* Emery, 1895h: 37 (w.) SOUTH AFRICA. Junior synonym of *sericeiventre:* Bolton, 1980: 332.

***fergusoni***. *Tetramorium fergusoni* Forel, 1902c: 234 (w.) INDIA. See also: Bolton, 1977: 91.

*fernandensis*. *Tetramorium fernandensis* Menozzi, 1942: 174, fig. 2 FERNANDO PO I. [Also described as new by Menozzi, 1944: 454.] Junior synonym of *pullulum:* Bolton, 1980: 273.

***ferox***. *Tetramorium caespitum* var. *ferox* Ruzsky, 1903b: 309 (w.q.m.) RUSSIA. Subspecies of *caespitum:* Emery, 1909d: 703; Emery, 1916b: 194. Raised to species: Cori & Finzi, 1931: 239; Soudek, 1931: 10; Menozzi, 1933b: 75 (in key); Kratochvil, 1944: 72; Tarbinsky, 1976: 112; Atanassov & Dlussky, 1992: 155. Senior synonym of *diomedea* and *silhavyi:* Bernard, 1967: 235; Dlussky, Soyunov & Zabelin, 1990: 201; Radchenko, 1992b: 55; of *bariensis*, and material of the unavailable name *striaticeps* referred here: Radchenko, 1992b: 55. Current subspecies: nominal plus *laevior, sarkissiani*.

***feroxoide***. *Tetramorium feroxoides* Dlussky & Zabelin, 1985: 230, fig. 7 (w.q.) TURKESTAN. Synonym of *perspicax:* Radchenko, 1992b: 56. [Radchenko gives *perspicax* as the senior name, but *feroxoide* has priority.]

***fezzanense***. *Tetramorium fezzanense* Bernard, 1948: 149, figs. 7, 8 (w.q.m.) ALGERIA.

***flabellum***. *Tetramorium flabellum* Bolton, 1980: 298 (w.) GHANA.

***flagellatum***. *Tetramorium flagellatum* Bolton, 1977: 122, fig. 48 (w.) BORNEO.

*flavescens*. *Xiphomyrmex costatus* subsp. *flavescens* Emery, 1897c: 588 (w.) NEW GUINEA. Mann, 1919: 348 (m.). Junior synonym of *carinatum:* Bolton, 1977: 86.

*flavescens*. *Triglyphothrix striatidens* var. *flavescens* Wheeler, W.M. 1929g: 55 (w.) SINGAPORE. [Unresolved junior secondary homonym of *flavescens* Emery, above.] Junior synonym of *lanuginosum:* Bolton, 1976: 350.

***flaviceps***. *Tetramorium do* var. *flaviceps* Arnold, 1960a: 82 (w.) ZIMBABWE. [First available use of *Tetramorium squaminode* r. *do* var. *flaviceps* Arnold, 1917: 316; unavailable name.] Raised to species, and senior synonym of *mus:* Bolton, 1980: 255.

#***flavidulum***. *Tetramorium caespitum* var. *flavidula* Emery, 1922f: 276 (w.) TURKEY. [First available use of *Tetramorium caespitum* subsp. *caespitum* var. *flavidula* Emery, 1909d: 702; unavailable name.]

*flavigaster* Clark, 1938 [*Xiphomyrmex*]; see under ***MONOMORIUM***.

***flavipes***. *Tetramorium* (*Xiphomyrmex*) *flavipes* Emery, 1893f: 247 (footnote) (w.) THAILAND. See also: Bolton, 1977: 81.

***flavithorax***. *Xiphomyrmex muralti* st. *flavithorax* Santschi, 1914d: 369, fig. 31 (w.) GHANA. Combination in *Tetramorium*, and raised to species: Bolton, 1980: 226.

*flavus*. *Xiphomyrmex gambogecus* var. *flavus* Donisthorpe, 1941h: 58 (w.) NEW GUINEA. Junior synonym of *gambogecum:* Bolton, 1977: 88.

*foreli*. *Tetramorium foreli* Forel, 1893g: 383 (w.) ANTILLES IS. *Nomen nudum;* attributed to Emery.

***TETRAMORIUM** (continued)*

*formosae. Triglyphothrix parvispina* var. *formosae* Forel, 1912a: 52 (w.) TAIWAN. Junior synonym of *parvispinum:* Bolton, 1976: 355.

***forte.*** *Tetramorium caespitum* var. *forte* Forel, 1904b: 371 (w.q.m.) FRANCE. Race/stirps of *caespitum:* Forel, 1905: 173; of *ferox:* Santschi, 1936c: 203. Raised to species: Collingwood, 1978: 71; Casevitz-Weulersse, 1990b: 420. Senior synonym of *silvestrianum:* Collingwood, 1978: 71; of *moravicum:* Bernard, 1967: 233; of *moravicum* and *taurocaucasicum:* Dlussky, Soyunov & Zabelin, 1990: 202; Radchenko, 1992b: 51; Atanassov & Dlussky, 1992: 152. [Name misspelled as *jorte* by Ruzsky, 1905b: 534.]

*fortunatarum. Tetramorium caespitum* subsp. *semilaeve* var. *fortunatarum* Emery, 1925c: 190, fig. D (w.q.m.) CANARY IS. Unavailable name.

*fossulatum* Forel, 1910; see under ***PRISTOMYRMEX.***

***frenchi.*** *Tetramorium frenchi* Forel, 1914d: 229 (w.) SOUTH AFRICA. See also: Bolton, 1980: 278.

***frigidum.*** *Tetramorium akermani* var. *frigidum* Arnold, 1926: 266 (w.) SOUTH AFRICA. Raised to species, and senior synonym of *drakensbergensis:* Bolton, 1980: 256.

***fulviceps.*** *Triglyphothrix fulviceps* Emery, 1897c: 588 (w.) NEW GUINEA. Viehmeyer, 1914b: 529 (q.). Combination in *Tetramorium:* Bolton, 1985: 247. See also: Bolton, 1976: 347.

***furtivum.*** *Triglyphothrix furtiva* Arnold, 1956: 69, fig. 18 (w.q.) ZIMBABWE. Combination in *Tetramorium:* Bolton, 1985: 247. See also: Bolton, 1976: 325.

*fusca. Formica fusca* Leach, 1825: 290 (q.m.) FRANCE. [Unresolved junior primary homonym of *Formica fusca* Linnaeus, 1758: 580.] Junior synonym of *caespitum:* Dalla Torre, 1893: 132.

*fusciclava. Tetramorium caespitum* subsp. *caespitum* var. *fusciclava* Emery, 1925c: 187 (w.) ITALY. Unavailable name.

***fuscipes.*** *Xiphomyrmex turneri* subsp. *fuscipes* Viehmeyer, 1925a: 29 (w.) AUSTRALIA. Combination in *Tetramorium,* and raised to species: Bolton, 1977: 138.

*fusciventris. Triglyphothrix auropunctatus* var. *fusciventris* Forel, 1913a: 116 (w.) ZIMBABWE. Combination in *Tetramorium:* Arnold, 1917: 336. Junior synonym of *microps:* Bolton, 1976: 329.

*fuscula. Myrmica fuscula* Nylander, 1846a: 935, pl. 18, figs. 34, 36 (w.q.m.) FINLAND. Junior synonym of *caespitum:* Smith, F. 1851: 118.

***gabonense.*** *Triglyphothrix gabonensis* André, 1892a: 53 (w.q.m.) GABON. Combination in *Tetramorium:* Bolton, 1985: 247. Senior synonym of *areolatus, boulognei, burgeoni, kamerunensis, mucidus* and *soyauxi:* Bolton, 1976: 326.

**#*gaetulum.*** *Tetramorium semilaeve* var. *gaetulum* Santschi, 1936c: 203 (w.) MOROCCO. [First available use of *Tetramorium semilaeve* st. *guancha* var. *gaetulum* Santschi, 1929e: 150; unavailable name.]

**#*galaticum.*** *Tetramorium semilaeve* var. *galatica* Menozzi, 1936d: 292 (w.q.) TURKEY. [First available use of *Tetramorium caespitum* st. *biskrensis* var. *galatica* Santschi, 1921a: 112; unavailable name.]

***galoasanum.*** *Tetramorium setuliferum* var. *galoasana* Santschi, 1910c: 381 (w.q.m.) CONGO. Raised to species: Bolton, 1980: 245.

*gamaii. Tetramorium sericeiventre* var. *gamaii* Santschi, 1918b: 128 (w.) ZIMBABWE. Junior synonym of *sericeiventre:* Bolton, 1980: 332.

***gambogecum.*** *Xiphomyrmex gambogecus* Donisthorpe, 1941h: 57 (w.) NEW GUINEA. Combination in *Tetramorium,* and senior synonym of *flavus:* Bolton, 1977: 88.

***gazense.*** *Tetramorium guineense* subsp. *gazensis* Arnold, 1958: 122, fig. 3 (w.) ZIMBABWE. Raised to species: Bolton, 1980: 270.

*geei. Tetramorium caespitum* subsp. *jacoti* var. *geei* Wheeler, W.M. 1927d: 7 (w.) CHINA. Unavailable name; material referred to *caespitum* by Yasumatsu, 1962: 96; to *jacoti* by Radchenko, 1992b: 51.

***gegaimi.*** *Tetramorium camerunense* var. *gegaimi* Forel, 1916: 421 (w.q.) ZAIRE. Raised to species: Bolton, 1980: 338.

***geminatum.*** *Tetramorium geminatum* Bolton, 1980: 298 (w.) GABON.

*georgei. Tetramorium sericeiventre* st. *continentis* var. *georgei* Santschi, 1918b: 131 (w.q.m.) ZIMBABWE. Unavailable name; material referred to *sericeiventre* by Bolton, 1980: 332.

***gestroi.*** *Triglyphothrix gestroi* Menozzi, 1933a: 105, fig. 3 (w.) UGANDA. Combination in *Tetramorium:* Bolton, 1985: 247. See also: Bolton, 1976: 326.

***ghindanum.*** *Tetramorium caespitum* subsp. *ghindanum* Forel, 1910c: 260 (w.) ETHIOPIA. Subspecies of *pusillum:* Forel, 1914d: 223. Raised to species: Bolton, 1980: 312.

***glabratum.*** *Tetramorium solidum* st. *glabratum* Stitz, 1923: 162 (w.) NAMIBIA. Raised to species, senior synonym of *rutilum,* and material of the unavailable name *aciculatum* referred here: Bolton, 1980: 246.

*gladiator. Tetramorium aculeatum* st. *andricum* var. *gladiator* Santschi, 1919c: 248 (w.) ZAIRE. Unavailable name; material referred to *aculeatum* by Bolton, 1980: 353.

*gladiator. Tetramorium sericeiventre* st. *continentis* var. *gladiator* Santschi, 1928f: 206 (w.) ZIMBABWE. Unavailable name; material referred to *sericeiventre* by Bolton, 1980: 332.

***gladstonei.*** *Tetramorium gladstonei* Forel, 1913j: 219 (w.) ZIMBABWE. Stirps of *sericeiventre:* Santschi, 1918b: 125. Revived status as species: Santschi, 1918b: 131; Wheeler, W.M. 1922a: 895. See also: Arnold, 1917: 284; Bolton, 1980: 326.

***goniommoide.*** *Tetramorium goniommoide* Poldi, 1979: 499, fig. 1 (w.) TURKEY.

***gracile.*** *Tetramorium gracile* Forel, 1894b: 81 (w.) ETHIOPIA. Forel, 1907c: 136 (m.). See also: Bolton, 1980: 279.

***grandinode.*** *Tetramorium grandinode* Santschi, 1913c: 308 (w.) SOUTH AFRICA. Senior synonym of *hopensis:* Bolton, 1980: 247.

***granulatum.*** *Tetramorium granulatum* Bolton, 1980: 299, fig. 83 (w.) NIGERIA.

***grassii.*** *Tetramorium grassii* Emery, 1895h: 37 (w.) SOUTH AFRICA. Mayr, 1901b: 25 (q.). Senior synonym of *algoa, joffrei, laevigatum, mayri* and *simulans:* Bolton, 1980: 262.

**#*gregori.*** *Tetramorium staerckei* var. *gregori* Kratochvil, in Kratochvil, Novak & Snoflak, 1944: 66, fig. 7 (w.q.m.) CZECHOSLOVAKIA. Subspecies of *impurum:* current status.

*grootensis. Tetramorium solidum* var. *grootensis* Forel, 1913a: 118 (q.m.) SOUTH AFRICA. Junior synonym of *signatum:* Bolton, 1980: 251.

*guancha. Tetramorium semilaeve* st. *guancha* Santschi, 1929e: 150 (w.) CANARY IS. [First available use of *Tetramorium caespitum* st. *semilaeve* var. *guancha* Santschi, 1921b: 431; unavailable name.] Junior synonym of *depressum:* Hohmann, La Roche, *et al.,* 1993: 158.

*guillarmodi. Tetramorium guillarmodi* Arnold, 1960b: 454, fig. 4 (w.q.) LESOTHO. Junior synonym of *bothae:* Bolton, 1980: 309.

*guillodi. Triglyphothrix guillodi* Santschi, 1937d: 233 (w.) ANGOLA. Junior synonym of *eminii:* Bolton, 1976: 324.

*guineensis* Fabricius, 1793 [*Formica*]; see under ***PHEIDOLE.***

***guineense.*** *Xiphomyrmex weitzeckeri* r. *guineensis* Bernard, 1953b: 251, fig. 14 (w.) GUINEA. Combination in *Tetramorium,* and raised to species: Bolton, 1980: 227.

*haltrichi. Tetramorium caespitum* subsp. *hungarica* var. *haltrichi* Röszler, 1936b: 61 (w.q.m.) HUNGARY. Unavailable name.

*hammi. Tetramorium caespitum* var. *hammi* Donisthorpe, 1915d: 178 (w.) GREAT BRITAIN. Junior synonym of *caespitum:* **new synonymy**.

***hapale.*** *Tetramorium hapale* Bolton, 1980: 339 (w.) IVORY COAST.

*hemisi. Tetramorium pusillum* var. *hemisi* Wheeler, W.M. 1922a: 193 (w.) ZAIRE. Junior synonym of *caldarium:* Bolton, 1979: 169.

*hepburni. Tetramorium* (*Triglyphothrix*) *hepburni* Arnold, 1917: 340 (w.q.m.) ZIMBABWE. Combination in *Triglyphothrix:* Wheeler, W.M. 1922a: 910. Junior synonym of *inezulae:* Bolton, 1976: 328.

*hertigi. Tetramorium guineense* st. *hertigi* Santschi, 1937d: 236 (w.) ANGOLA. Junior synonym of *phasias:* Bolton, 1980: 273.

*himalayanum. Tetramorium caespitum* subsp. *himalayanum* Viehmeyer, 1914c: 38 (w.q.m.) INDIA. Junior synonym of *caespitum:* Radchenko, 1992b: 50.

***hippocrate.*** *Tetramorium hippocratis* Agosti & Collingwood, 1987a: 56 (w.) TURKEY. [First available use of *Tetramorium caespitum* subsp. *semilaeve* var. *hippocratis* Emery, 1921b: 217; unavailable name.]

*hispanicum. Tetramorium caespitum* var. *hispanicum* Bondroit, 1918: 108 (q.) SPAIN. [First available use of *Tetramorium caespitum* subsp. *caespitum* var. *hispanica* Emery, 1909d: 701; unavailable name.] Bondroit, 1920a: 153 (w.). Stirps of *caespitum:* Santschi, 1931a: 9. Raised to species: Bondroit, 1920a: 153; Collingwood & Yarrow, 1969: 73. Synonym of *ruginode:* Collingwood, 1978: 71. [Collingwood gives *hispanicum* as senior synonym, but *ruginode* has priority and is first available name for this taxon.] See also: López, 1991a: 153.

***hispidum.*** *Xiphomyrmex spinosus* subsp. *hispidus* Wheeler, W.M. 1915b: 415 (w.) U.S.A. Combination in *Tetramorium,* and raised to species: Bolton, 1979: 161.

**#hoggarense.** *Tetramorium semilaeve* var. *hoggarense* Santschi, 1929c: 103, fig. 2 (w.) ALGERIA. Santschi, 1929g: 165 (q.). [Also described as new by Santschi, 1934f: 170.]

*hopensis. Tetramorium grandinode* var. *hopensis* Forel, 1914d: 223 (w.q.) SOUTH AFRICA. Junior synonym of *grandinode:* Bolton, 1980: 247.

*hori. Tetramorium sericeiventre* var. *hori* Santschi, 1918b: 125 (w.) SUDAN. Junior synonym of *sericeiventre:* Bolton, 1980: 332.

*hortensis. Atopula hortensis* Bernard, 1948: 173, fig. 9 (w.q.m.) LIBYA. Junior synonym of *sericeiventre:* Bolton, 1976: 363; Bolton, 1980: 333.

**hortorum.** *Tetramorium hortorum* Arnold, 1958: 121, fig. 2 (w.) ZIMBABWE. See also: Bolton, 1980: 327.

**humbloti.** *Tetramorium (Xiphomyrmex) humbloti* Forel, 1891b: 154, pl. 4, fig. 12 (w.) COMORO IS. Emery, 1899f: 285 (q.). Combination in *Xiphomyrmex:* Emery, 1899f: 285; in *Tetramorium:* Bolton, 1979: 142. Senior synonym of *pembensis* and *victoriensis:* Bolton, 1980: 228.

*humerosum. Tetramorium humerosum* Bernard, 1953b: 246, fig. 13 (w.q.) CAMEROUN. Junior synonym of *angulinode:* Bolton, 1980: 237.

*humile. Tetramorium humile* Santschi, 1914a: 434 (w.q.) TANZANIA. Junior synonym of *bequaerti:* Bolton, 1980: 325.

**hungaricum.** *Tetramorium caespitum* subsp. *hungarica* Röszler, 1935: 78, figs. (w.q.) HUNGARY. Variety of *semilaeve:* Novak & Sadil, 1941: 86. Raised to species: Röszler, 1951: 88.

**ibycterum.** *Tetramorium ibycterum* Bolton, 1979: 148, fig. 26 (w.) MADAGASCAR.

**ictidum.** *Tetramorium ictidum* Bolton, 1980: 339 (w.) CAMEROUN.

**imbelle.** *Triglyphothrix imbellis* Emery, 1915g: 18, fig. 9 (w.) ETHIOPIA. Combination in *Tetramorium:* Bolton, 1985: 247.

*immigrans. Tetramorium caespitum* var. *immigrans* Santschi, 1927a: 54 (w.) CHILE. Junior synonym of *caespitum:* Bolton, 1979: 171.

**impressum.** *Xiphomyrmex impressus* Viehmeyer, 1925a: 30 (w.) AUSTRALIA. Combination in *Tetramorium:* Bolton, 1977: 138.

**impurum.** *Myrmica impura* Foerster, 1850a: 48 (w.) GERMANY. Emery, 1925c: 184 (q.); Kutter, 1977c: 151 (m.). Combination in *Tetramorium:* Mayr, 1855: 426. Junior synonym of *caespitum:* Curtis, 1854: 215; Mayr, 1855: 426. Revived from synonymy: Kutter, 1977c: 159. Senior synonym of *staerckei:* Kutter, 1977c: 159. Current subspecies: nominal plus *gregori.* See also: Pasteels, Verhaeghe, *et al.* 1981: 675.

**incruentatum.** *Tetramorium incruentatum* Arnold, 1926: 271. Replacement name for *Rhoptromyrmex arnoldi* Santschi, 1916b: 503. [Junior secondary homonym of *Triglyphothrix arnoldi* Forel, 1913j: 220.] See also: Bolton, 1980: 306; Bolton, 1985: 247.

**indicum.** *Tetramorium guineense* var. *indica* Forel, 1913k: 81 (w.q.) SUMATRA. Raised to species: Bolton, 1977: 98.

*indocile. Tetramorium caespitum* var. *indocile* Santschi, 1927a: 53 (w.q.m.) TURKESTAN. Subspecies of *caespitum:* Pisarski, 1967: 402. Raised to species: Pisarski, 1969b: 304. Junior synonym of *caespitum:* Radchenko, 1992b: 50.

**indosinense.** *Tetramorium indosinense* Wheeler, W.M. 1927h: 97, fig. 6 (w.) VIETNAM. Combination in *Triglyphothrix:* Bolton, 1976: 348; in *Tetramorium:* Bolton, 1985: 247.

**inerme.** *Tetramorium caespitum* var. *inerme* Mayr, 1877: 17 (w.) TURKESTAN. [Redescribed in Latin: Mayr, 1880: 19.] Ruzsky, 1905b: 540 (q.). Subspecies/race of *caespitum:* Ruzsky, 1903b: 309; Forel, 1904b: 371; Emery, 1909d: 705. Raised to species: Dalla Torre, 1893: 133; Santschi, 1927a: 57. Senior synonym of *picta, plana* and *reticuliventre:* Radchenko, 1992b: 53.

*inermis. Macromischoides aculeatus* r. *inermis* Bernard, 1953b: 249 (w.) GUINEA. [Unresolved junior secondary homonym of *inerme* Mayr, above.] Junior synonym of *aculeatum:* Bolton, 1980: 353.

**inezulae.** *Triglyphothrix inezulae* Forel, 1914d: 221 (w.m.) SOUTH AFRICA. Combination in *Tetramorium:* Arnold, 1917: 343. Senior synonym of *hepburni* and *mashonana:* Bolton, 1976: 328. See also: Bolton, 1985: 247.

**infraspinosum.** *Tetramorium infraspinosum* Karavaiev, 1935a: 104 (w.) VIETNAM. See also: Bolton, 1977: 132.

**infraspinum.** *Tetramorium infraspinum* Forel, 1905c: 14 (w.) JAVA. See also: Bolton, 1977: 127.

**inglebyi.** *Tetramorium inglebyi* Forel, 1902c: 233 (w.) INDIA. See also: Bolton, 1977: 111.

*innocens* Forel, 1913; see under ***LEPTOTHORAX.***

*inscia. Tetramorium subcoecum* var. *inscia* Forel, 1913j: 218 (w.) ZIMBABWE. Junior synonym of *subcoecum:* Bolton, 1980: 291.

**insolens.** *Myrmica insolens* Smith, F. 1861b: 47 (q., not w.) SULAWESI. Bolton, 1977: 99 (w.); Imai, Kubota, *et al.* 1985: 47 (k.). Combination in *Tetramorium:* Emery, 1901g: 567; Donisthorpe, 1932c: 468. Senior synonym of *macra, pallidiventre* and *wilsoni:* Bolton, 1977: 99.

*insons. Xiphomyrmex spinosus* subsp. *insons* Wheeler, W.M. 1915b: 416 (w.q.m.) U.S.A. Junior synonym of *spinosum:* Bolton, 1979: 163.

*insularis. Xiphomyrmex insularis* Menozzi, 1924b: 223, fig. 4 (w.) PRINCIPE I. Junior synonym of *occidentale:* Bolton, 1980: 229.

*insulare. Tetramorium simillimum* var. *insulare* Santschi, 1928c: 69 (w.) FIJI IS. [Unresolved junior secondary homonym of *insularis* Menozzi, above.] Junior synonym of *simillimum:* Yarrow, 1967: 28; Bolton, 1977: 131.

**intextum.** *Tetramorium intextum* Santschi, 1914b: 104, fig. 14 (w.) KENYA. Junior synonym of *delagoense:* Bolton, 1980: 311. Revived from synonymy, and senior synonym of *cataractae:* **revived status, new synonymy**.

**intonsum.** *Tetramorium intonsum* Bolton, 1980: 288, fig. 68 (w.) GHANA.

*inversa. Tetramorium sericeiventre* var. *inversa* Santschi, 1910c: 384 (w.) CONGO. Junior synonym of *sericeiventre:* Bolton, 1980: 332.

**invictum.** *Tetramorium invictum* Bolton, 1980: 300, fig. 84 (w.) IVORY COAST.

**isectum.** *Tetramorium isectum* Bolton, 1979: 145, figs. 17, 18 (w.) MADAGASCAR.

**isipingense.** *Tetramorium simillimum* r. *isipingense* Forel, 1914d: 225 (w.) SOUTH AFRICA. Raised to species: Bolton, 1980: 347.

*isis. Tetramorium simillimum* subsp. *isis* Weber, 1943c: 373 (w.) SUDAN. Junior synonym of *altivagans:* Bolton, 1980: 305.

**#italicum.** *Tetramorium semilaevis* susbp. *italica* Menozzi, 1932a: 11 (w.) ITALY.

*jacobsoni* Forel, 1915 [*Atopula*]; see under ***LEPTOTHORAX.***

*jacoti. Tetramorium caespitum* subsp. *jacoti* Wheeler, W.M. 1927d: 7 (w.) CHINA. [First available use of *Tetramorium caespitum* subsp. *semilaeve* var. *jacoti* Wheeler, W.M. 1923b: 3; unavailable name.] Synonym of *tsushimae:* Wheeler, W.M. 1927d: 7. [Wheeler gives *jacoti* as senior synonym, but *tsushimae* has priority and is first available name for this taxon.] Raised to species: Pisarski, 1969b: 303; Radchenko, 1992b: 51. Senior synonym of *annectens* and *tsushimae,* and material of the unavailable names *chinensis* and *geei* referred here: Radchenko, 1992b: 51.

**#japonicum.** *Tetramorium caespitum* var. *japonica* Röszler, 1936a: 121 (w.q.) JAPAN.

**#jarbas.** *Tetramorium biskrense* subsp. *jarbas* Cagniant, 1970a: 430 (w.q.m.) TUNISIA. [First available use of *Tetramorium caespitum* st. *punicum* var. *jarbas* Santschi, 1921b: 433; unavailable name.]

*jasonis. Tetramorium sericeiventre* var. *jasonis* Santschi, 1918b: 127 (w.q.) IVORY COAST. Junior synonym of *sericeiventre:* Bolton, 1980: 332.

**jauresi.** *Tetramorium jauresi* Forel, 1914d: 226 (w.) SOUTH AFRICA. Senior synonym of *latens:* Bolton, 1980: 348. See also: Prins, 1973: 17.

*jeanae. Tetramorium jeanae* Weber, 1943c: 371, pl. 16, fig. 29 (w.) SUDAN. Junior synonym of *laevithorax:* Bolton, 1980: 279.

**jejunum.** *Tetramorium jejunum* Arnold, 1926: 267, fig. 74 (w.) ZIMBABWE. See also: Bolton, 1980: 256.

**jiangxiense.** *Tetramorium jiangxiense* Wang & Xiao, in Wang, M., Xiao & Wu, 1988: 269, figs. 24, 25 (w.) CHINA.

**jizane.** *Tetramorium jizani* Collingwood, 1985: 264, fig. 48 (w.q.) SAUDI ARABIA.

*joffrei. Tetramorium joffrei* Forel, 1914d: 228 (w.q.) SOUTH AFRICA. Junior synonym of *grassii:* Bolton, 1980: 262.

**jordani.** *Tetramorium jordani* Santschi, 1937a: 62 (w.) NAMIBIA. Senior synonym of *aspinatum:* Bolton, 1980: 248.

*jorte* Ruzsky, 1905; see under ***forte.***

**juba.** *Tetramorium juba* Collingwood, 1985: 264 (w.) ALGERIA. [First available use of *Tetramorium caespitum* st. *judas* var. *juba* Santschi, 1921b: 433; unavailable name.]

**#judas.** *Tetramorium caespitum* subsp. *judas* Wheeler, W.M. & Mann, 1916: 172 (w.) ISRAEL. Menozzi, 1933b: 70 (q.m.). Subspecies of *semilaeve:* Menozzi, 1933b: 69.

**jugatum.** *Tetramorium jugatum* Bolton, 1980: 289, fig. 75 (w.) IVORY COAST.

**#jugurtha.** *Tetramorium semilaeve* var. *jugurtha* Menozzi, 1934: 162 (w.) TUNISIA. [First available use of *Tetramorium caespitum* st. *semilaeve* var. *jugurtha*

***TETRAMORIUM** (continued)*

Santschi, 1921b: 430; unavailable name.] Subspecies of *semilaeve:* Finzi, 1940: 162.

***kabulistanicum.*** *Tetramorium striativentre* subsp. *kabulistanicum* Pisarski, 1967: 403, figs. 62–72 (w.q.m.) AFGHANISTAN. Raised to species: Radchenko, 1992b: 53.

#***kahenae.*** *Tetramorium biskrensis* var. *kahenae* Menozzi, 1934: 162 (w.q.) TUNISIA. [First available use of *Tetramorium caespitum* st. *biskrensis* var. *kahenae* Santschi, 1918c: 155; unavailable name.] Variety of *biskrense:* Menozzi, 1940: 270.

*kamerunensis. Triglyphothrix gabonensis* var. *kamerunensis* Santschi, 1937b: 102 (w.) CAMEROUN. Junior synonym of *gabonense:* Bolton, 1976: 326.

*kanariense. Tetramorium (Xiphomyrmex) smithi* var. *kanariense* Forel, 1902c: 703 (w.) INDIA. Viehmeyer, 1916a: 139 (q.). Junior synonym of *smithi:* Bolton, 1977: 90.

***karakalense.*** *Tetramorium karakalense* Dlussky & Zabelin, 1985: 229, fig. 6 (q.) TURKMENIA.

***katypum.*** *Triglyphothrix katypa* Bolton, 1976: 349 (w.) PHILIPPINES. Combination in *Tetramorium:* Bolton, 1985: 247.

***kelleri.*** *Tetramorium (Xiphomyrmex) kelleri* Forel, 1887: 385 (w.) MADAGASCAR. See also: Bolton, 1979: 145.

*kenyense. Tetramorium sericeiventre* st. *femoratum* var. *kenyense* Santschi, 1933b: 106 (w.q.) KENYA. Unavailable name; material referred to *sericeiventre* by Bolton, 1980: 332.

***kestrum.*** *Tetramorium kestrum* Bolton, 1980: 300 (w.) UGANDA.

***kheperra.*** *Triglyphothrix kheperra* Bolton, 1976: 349, fig. 71 (w.) JAVA. Imai, Kubota, *et al.* 1985: 46 (k.). Combination in *Tetramorium:* Bolton, 1985: 247.

***khnum.*** *Tetramorium khnum* Bolton, 1977: 122, fig. 52 (w.) PHILIPPINES.

***khyarum.*** *Tetramorium khyarum* Bolton, 1980: 327, fig. 111 (w.) NIGERIA.

***kisilkumense.*** *Tetramorium kisilkumense* Dlussky, in Dlussky, Soyunov & Zabelin, 1990: 205, fig. 53 (w.) TURKMENISTAN.

*kivuense. Tetramorium (Xiphomyrmex) kivuense* Stitz, 1911b: 386, fig. 6 (w.) ZAIRE. Santschi, 1924b: 216 (q.). Combination in *Xiphomyrmex:* Wheeler, W.M. 1922a: 907. Junior synonym of *tersum:* Bolton, 1980: 233.

*kollari. Myrmica kollari* Mayr, 1853c: 283 (w.q.m.) AUSTRIA. Junior synonym of *cariniceps:* Roger, 1861b: 171.

*kosmalianicum. Tetramorium semilaeve* subsp. *kosmalianicum* Arnol'di, 1948: 211. *Nomen nudum.*

***kraepelini.*** *Tetramorium kraepelini* Forel, 1905c: 15 (w.) JAVA. Senior synonym of *yanoi,* and material of the *nomen nudum eidmanni* referred here: Bolton, 1977: 117.

***krynitum.*** *Tetramorium krynitum* Bolton, 1980: 313, fig. 94 (w.) NAMIBIA.

#***kutteri.*** *Tetramorium semilaeve* var. *kutteri* Santschi, 1927a: 57 (w.) SWITZERLAND.

***kydelphon.*** *Tetramorium kydelphon* Bolton, 1979: 177 (w.) NEW GUINEA.

*ladismithensis. Tetramorium caespitum* r. *ladismithensis* Forel, 1913a: 117 (w.q.) SOUTH AFRICA. Subspecies of *pusillum:* Forel, 1914d: 223. Junior synonym of *pusillum:* Bolton, 1980: 318.

*laevidens. Triglyphothrix striatidens* var. *laevidens* Forel, 1900e: 284 (w.) MEXICO. Junior synonym of *lanuginosum:* Bolton, 1976: 350.

*laevigatum. Tetramorium grassii* var. *laevigatum* Mayr, 1901b: 25 (w.) SOUTH AFRICA. Junior synonym of *grassii:* Bolton, 1980: 262.

*laevigatum. Tetramorium inerme* subsp. *laevigatum* Arnol'di, 1948: 211 (w.q.m.) CAUCASUS. [First available use of *Tetramorium caespitum* subsp. *ferox* var. *levigata* Karavaiev, 1926c: 167; unavailable name.] [Junior primary homonym of *laevigatum* Mayr, above.] First available replacement name: *aegeum* Radchenko, 1992b: 55. [Note that a previously proposed replacement name for the unavailable *levigata* Karavaiev, *aequa* Baroni Urbani, 1971b: 361, is unavailable.] See also: Radchenko & Arakelian, 1990: 376.

*laevinode. Tetramorium simillimum* r. *laevinode* Forel, 1902c: 235 (w.) INDIA. Junior synonym of *smithi:* Bolton, 1977: 90.

#***laevior.*** *Tetramorium ferox* var. *laevior* Menozzi, 1936d: 292 (w.) TURKEY. [First available use of *Tetramorium caespitum* subsp. *ferox* var. *laevior* Forel, 1911d: 333; unavailable name.] Finzi, 1939c: 157 (m.).

***laevithorax.*** *Tetramorium laevithorax* Emery, 1895h: 39 (w.) SOUTH AFRICA. Senior synonym of *jeanae:* Bolton, 1980: 279.

*lamottei. Tetramorium lamottei* Bernard, 1953b: 247, fig. 13 (q.) GUINEA. Junior synonym of *africanum:* Bolton, 1980: 355.

***lanuginosum.*** *Tetramorium lanuginosum* Mayr, 1870b: 976 (w.) JAVA. Viehmeyer, 1916a: 140 (q.); Imai, Baroni Urbani, *et al.* 1984: 8 (k.). Combination in *Triglyphothrix:* Emery, 1891b: 4 (footnote); in *Tetramorium:* Bolton, 1985: 247. Senior synonym of *australis, ceramensis, felix, flavescens, laevidens, mauricei, orissana, striatidens* and *tricolor:* Bolton, 1976: 350.

***laparum.*** *Tetramorium laparum* Bolton, 1977: 127 (w.) PHILIPPINES.

*latens. Tetramorium latens* Arnold, 1948: 224, fig. 11 (w.) ZIMBABWE. Race of *jauresi:* Arnold, 1952a: 463. Junior synonym of *jauresi:* Bolton, 1980: 348.

***laticephalum.*** *Tetramorium laticephalum* Bolton, 1977: 139 (w.) AUSTRALIA.

*latior. Xiphomyrmex latior* Santschi, 1926b: 243 (w.) MADAGASCAR. Junior synonym of *sikorae:* Bolton, 1979: 138.

***latreillei.*** *Tetramorium (Xiphomyrmex) latreillei* Forel, 1895c: 247 (w.) MADAGASCAR. See also: Bolton, 1979: 146.

***legone.*** *Tetramorium legone* Bolton, 1980: 240 (w.) GHANA.

*levigata* Karavaiev, 1926; see *laevigatum,* above.

*ligatum. Tetramorium inerme* st. *armatum* var. *ligatum* Santschi, 1927a: 57 (w.q.) TURKESTAN. Unavailable name; material referred to *armatum* by Radchenko, 1992b: 53.

*liogaster* Santschi, 1928; see under ***ROMBLONELLA.***

#***liparaeum.*** *Tetramorium semilaeve* var. *liparaeum* Santschi, 1927a: 55 (w.) SICILY.

***lobulicorne.*** *Tetramorium lobulicorne* Santschi, 1916b: 504 (w.) ZIMBABWE. Arnold, 1917: 299 (q.m.); Wheeler, G.C. & Wheeler, J. 1989a: 325 (l.). See also: Bolton, 1980: 361.

*longicarinum. Tetramorium longicarinum* Donisthorpe, 1941h: 57 (w.) NEW GUINEA. Junior synonym of *validiusculum:* Bolton, 1977: 104.

***longicorne.*** *Tetramorium longicorne* Forel, 1907a: 13 (w.) KENYA. See also: Bolton, 1980: 328.

*longispina. Tetramorium striativentre* subsp. *schneideri* var. *longispina* Karavaiev, 1912b: 585 (w.) TRANSCASPIA. Unavailable name; material referred to *schneideri* by Dlussky, Soyunov & Zabelin, 1990: 207.

*longispina* Stitz, 1938 [*Atopula*]; see under ***PARATOPULA.***

*longispinosa. Triglyphothrix longispinosa* Arnold, 1956: 68, fig. 17 (w.) ZIMBABWE. Junior synonym of *constanciae:* Bolton, 1976: 321.

***longoi.*** *Tetramorium longoi* Forel, 1915c: 344 (w.) SOUTH AFRICA. See also: Bolton, 1980: 362.

***lucayanum.*** *Tetramorium lucayanum* Wheeler, W.M. 1905b: 100, fig. L (w.) BAHAMAS. Mann, 1920: 406 (q.). Senior synonym of *rectinodis:* Brown, 1964b: 131; Bolton, 1979: 172; of *sexdens* and *waelbroeki:* Brown, 1964b: 131. See also: Bolton, 1980: 340.

***lucidulum.*** *Tetramorium punicum* var. *lucidulum* Menozzi, 1933b: 69 (w.) SYRIA. [First available use of *Tetramorium caespitum* subsp. *punicum* var. *lucidula* Emery, 1909d: 704.] Raised to species: Agosti & Collingwood, 1987a: 56.

*lugubre. Tetramorium solidum* subsp. *lugubre* Forel, 1910e: 425 (w.) ANGOLA. Junior synonym of *signatum:* Bolton, 1980: 251.

***luteipes.*** *Tetramorium grassii* st. *luteipes* Santschi, 1910c: 383, fig. 11 (w.q.m.) CONGO. Raised to species: Santschi, 1914e: 24. See also: Bolton, 1980: 341.

***luteolum.*** *Tetramorium incruentatum* var. *luteolum* Arnold, 1926: 272, fig. 27 (w.) ZIMBABWE. Raised to species: Bolton, 1980: 313.

*macra. Tetramorium guineense* var. *macra* Emery, 1914f: 415 (w.) NEW CALEDONIA. Junior synonym of *insolens:* Bolton, 1977: 99.

*madecassum. Tetramorium simillimum* var. *madecassum* Forel, 1895c: 248 (w.) MADAGASCAR. Junior synonym of *delagoense:* Bolton, 1979: 156.

*magitae. Tetramorium magitae* Forel, 1911i: 224 (w.) SRI LANKA. [Misspelled as *mayitae* in original.] Junior synonym of *tonganum:* Bolton, 1977: 129.

***magnificum.*** *Tetramorium magnificum* Bolton, 1980: 363, fig. 144 (w.) IVORY COAST.

***mai.*** *Tetramorium mai* Wang, M. 1993b: 433, fig. 1 (w.) CHINA.

*major. Tetramorium aculeatum* var. *major* Forel, 1915c: 344 (w.) ZAIRE. Combination in *Macromischoides:* Wheeler, W.M. 1922a: 889. Junior synonym of *aculeatum:* Bolton, 1980: 353.

***manni.*** *Tetramorium manni* Bolton, 1985: 247. Replacement name for *Triglyphothrix pacifica* Mann, 1921: 460. [Junior secondary homonym of *Tetramorium pacificum* Mayr, 1870b: 976.]

***marginatum.*** *Tetramorium (Xiphomyrmex) marginatum* Forel, 1895h: 485 (w.) MADAGASCAR. See also: Bolton, 1979: 142.

*marleyi*. *Triglyphothrix marleyi* Forel, 1914d: 221 (w.) SOUTH AFRICA. Combination in *Tetramorium:* Arnold, 1917: 342. Junior synonym of *eminii:* Bolton, 1976: 324.

*marocana*. *Tetramorium caespitum* st. *ferox* var. *marocana* Santschi, 1921e: 170 (w.) MOROCCO. Unavailable name. [*T. ferox* var. *marocanum* Cagniant, 1964: 106; unavailable name.]

*marthae*. *Triglyphothrix marthae* Forel, 1911e: 265 (w.) TANZANIA. Junior synonym of *eminii:* Bolton, 1976: 324.

*mashona*. *Tetramorium bequaerti* r. *bruni* var. *mashona* Arnold, 1926: 254 (w.) ZIMBABWE. Unavailable name; material referred to *bulawayense* by Bolton, 1980: 326.

*mashonana*. *Triglyphothrix hepburni* subsp. *mashonana* Arnold, 1949: 266, fig. 7 (w.) ZIMBABWE. Junior synonym of *inezulae:* Bolton, 1976: 328.

***matopoense***. *Tetramorium matopoensis* Arnold, 1926: 254, fig. 68 (w.) ZIMBABWE. See also: Bolton, 1980: 257.

*mauricei*. *Triglyphothrix mauricei* Donisthorpe, 1946c: 778 (w.) MAURITIUS. Junior synonym of *lanuginosum:* Bolton, 1976: 350.

***maurum***. *Tetramorium caespitum* st. *maura* Santschi, 1918c: 154, fig. 1 (w.q.) TUNISIA. Raised to species: Santschi, 1929e: 150; Santschi, 1931a: 10 (in text); Cagniant, 1964: 89. Current subspecies: nominal plus *tingitanum*.

***mayri***. *Triglyphothrix mayri* Mann, 1919: 351 (w.) SOLOMON IS. Combination in *Tetramorium:* Bolton, 1985: 247. See also: Bolton, 1976: 352.

*mayri*. *Tetramorium grazsii* var. *mayri* Emery, 1922f: 281 (q.) SOUTH AFRICA. [Unresolved junior secondary homonym of *mayri* Mann, above.] Junior synonym of *grassii:* Bolton, 1980: 262.

*medje*. *Tetramorium guineense* subsp. *medje* Wheeler, W.M. 1922a: 192 (w.) ZAIRE. Junior synonym of *cristatum:* Bolton, 1980: 268.

***megalops***. *Tetramorium megalops* Bolton, 1977: 139, fig. 61 (w.) AUSTRALIA.

***melanogyna***. *Tetramorium melanogyna* Mann, 1919: 345, fig. 28 (w.q.) SOLOMON IS. Junior synonym of *insolens:* Bolton, 1977: 99. Revived from synonymy: Bolton, 1979: 173.

*melanogyne*. *Macromischoides aculeatus* var. *melanogyne* Santschi, 1923e: 285 (w.q.) CONGO. [Unresolved junior secondary homonym of *melanogyna* Mann, above.] Junior synonym of *aculeatum:* Bolton, 1980: 353.

*melleum* Emery, 1897; see under ***RHOPTROMYRMEX***.

***menkaura***. *Triglyphothrix menkaura* Bolton, 1976: 328, figs. 41, 46, 47, 55 (w.q.) GHANA. Combination in *Tetramorium:* Bolton, 1985: 247.

***meressei***. *Tetramorium meressei* Forel, 1916: 422 (w.) ZAIRE. See also: Bolton, 1980: 349.

***meridionale***. *Tetramorium meridionale* Emery, 1870: 198 (w.q.) ITALY. Variety of *caespitum:* Emery & Forel, 1879: 458; Ruzsky, 1902d: 32. Revived status as species: Emery, 1909d: 705; Collingwood & Yarrow, 1969: 74.

***meshena***. *Triglyphothrix meshena* Bolton, 1976: 353, figs. 51, 67 (w.) WEST MALAYSIA. Combination in *Tetramorium:* Bolton, 1985: 247.

***metactum***. *Tetramorium metactum* Bolton, 1980: 280, figs. 54, 57 (w.) KENYA.

***mexicanum***. *Tetramorium mexicanum* Bolton, 1979: 161, figs. 52–54 (w.) MEXICO.

***microgyna***. *Tetramorium microgyna* Santschi, 1918b: 132 (q., not w.) SOUTH AFRICA. See also: Arnold, 1926: 253; Santschi, 1928f: 207; Bolton, 1980: 329.

***microps***. *Triglyphothrix microps* Mayr, 1901b: 25 (w.q.) SOUTH AFRICA. Combination in *Tetramorium:* Arnold, 1917: 337; Bolton, 1985: 247. Senior synonym of *auropunctatus, bulawayensis, fusciventris, pallens* and *rhodesiana:* Bolton, 1976: 329.

*militaris*. *Macromischoides aculeatus* st. *militaris* Santschi, 1924b: 209, fig. 8 (w.) ZAIRE. Junior synonym of *aculeatum:* Bolton, 1980: 353.

***minimum***. *Triglyphothrix minima* Bolton, 1976: 330 (w.) GHANA. Combination in *Tetramorium:* Bolton, 1985: 247.

***minusculum***. *Xiphomyrmex minusculus* Santschi, 1914d: 369, fig. 32 (w.) CAMEROUN. Combination in *Tetramorium*, and senior synonym of *amen:* Bolton, 1980: 240.

*minutum*. *Tetramorium minutum* Donisthorpe, 1942a: 30 (q.) EGYPT. Junior synonym of *caldarium:* Bolton, 1979: 169.

***miserabile***. *Tetramorium miserabilis* Santschi, 1918c: 153 (w.) KENYA. See also: Bolton, 1980: 342.

***mixtum***. *Tetramorium mixtum* Forel, 1902c: 236 (w.) INDIA. See also: Bolton, 1977: 113.

*modesta*. *Myrmica modesta* Foerster, 1850a: 49 (w.) GERMANY. Junior synonym of *caespitum:* Curtis, 1854: 215; Mayr, 1855: 426.

*modesta*. *Myrmica modesta* Smith, F. 1860b: 108 (w.) INDONESIA (Batjan I.). [Unresolved junior primary homonym of *modesta* Foerster, above.] Combination in *Tetramorium:* Donisthorpe, 1932c: 463. Junior synonym of *guineense:* Donisthorpe, 1932c: 463; of *bicarinatum:* Bolton, 1977: 94.

***monardi***. *Triglyphothrix monardi* Santschi, 1937d: 232, figs. 1, 2 (w.) ANGOLA. Combination in *Tetramorium:* Bolton, 1985: 247. See also: Bolton, 1976: 331.

*montanum*. *Tetramorium blochmannii* var. *montanum* Forel, 1891b: 152, pl. 5, fig. 2 (w.q.) MADAGASCAR. Stirps of *quadrispinosum:* Santschi, 1918b: 122 (in key). Junior synonym of *quadrispinosum:* Bolton, 1979: 155.

*moravicum*. *Tetramorium moravicum* Kratochvil, in Novak & Sadil, 1941: 86 (diagnosis in key) (w.) CZECHOSLOVAKIA. Kratochvil, in Kratochvil, Novak & Snoflak, 1944: 71 (q.m.). Junior synonym of *forte:* Bernard, 1967: 233; Radchenko, 1992b: 51; Atanassov & Dlussky, 1992: 152.

***mossamedense***. *Tetramorium caespitum* var. *mossamedensis* Forel, 1901d: 306 (w.) ANGOLA. [Also described as new by Forel, 1903e: 560.] Stirps of *pusillum:* Santschi, 1917c: 285. Raised to species, senior synonym of *schultzei*, and material of the unavailable name *tristis* referred here: Bolton, 1980: 314.

*mucidus*. *Triglyphothrix mucidus* Forel, 1909b: 71 (w.) ZAIRE. Junior synonym of *gabonense:* Bolton, 1976: 326.

*munda*. *Tetramorium sericeiventre* var. *munda* Santschi, 1918b: 127 (w.) GUINEA. Junior synonym of *sericeiventre:* Bolton, 1980: 332.

***muralti***. *Tetramorium* (*Xiphomyrmex*) *muralti* Forel, 1910e: 429 (w.) CAMEROUN. Santschi, 1914d: 368 (q.). Combination in *Xiphomyrmex:* Santschi, 1919h: 88; in *Tetramorium*, and senior synonym of *trilineata:* Bolton, 1980: 229.

*mus*. *Triglyphothrix guillodi* var. *mus* Santschi, 1937d: 234 (w.) ANGOLA. Junior synonym of *eminii:* Bolton, 1976: 324.

*mus*. *Tetramorium do* var. *mus* Arnold, 1960a: 82 (w.) ZIMBABWE. [First available use of *Tetramorium squaminode* r. *do* var. *mus* Arnold, 1917: 316; unavailable name.] [Unresolved junior secondary homonym of *mus* Santschi, above.] Junior synonym of *flaviceps:* Bolton, 1980: 255.

*muscicola*. *Tetramorium humerosum* r. *muscicola* Bernard, 1953b: 246 (w.) GUINEA. Junior synonym of *coloreum:* Bolton, 1980: 297.

***muscorum***. *Tetramorium* (*Triglyphothrix*) *muscorum* Arnold, 1926: 274, fig. 78 (w.) MOZAMBIQUE. Combination in *Triglyphothrix:* Bolton, 1976: 331; in *Tetramorium:* Bolton, 1985: 247.

*musculus*. *Triglyphothrix musculus* Forel, 1902c: 239 (w.) INDIA. Junior synonym of *walshi:* Bolton, 1976: 358.

***mutatum***. *Tetramorium mutatum* Bolton, 1985: 247. Replacement name for *Triglyphothrix pulchella* Mann, 1919: 352. [Junior secondary homonym of *Tetramorium pulchellum* Emery, 1897c: 586.]

*myersi*. *Tetramorium akermani* var. *myersi* Arnold, 1958: 123 (w.) SOUTH AFRICA. Junior synonym of *akermani:* Bolton, 1980: 253.

***myops***. *Tetramorium myops* Bolton, 1977: 111, figs. 36, 37 (w.) INDIA.

***nacta***. *Triglyphothrix nacta* Bolton, 1976: 353 (w.) THAILAND. Combination in *Tetramorium:* Bolton, 1985: 247.

***naganum***. *Tetramorium naganum* Bolton, 1979: 136, fig. 7 (w.q.) MADAGASCAR.

*nassonowii*. *Tetramorium* (*Xiphomyrmex*) *nassonowii* Forel, 1892k: 521 (w.) MADAGASCAR. Combination in *Xiphomyrmex:* Wheeler, W.M. 1922a: 1031. Junior synonym of *schaufussii:* Bolton, 1979: 137.

***nautarum***. *Tetramorium caespitum* st. *nautarum* Santschi, 1918c: 156 (w.) ANNOBON I. Raised to species: Bolton, 1980: 321.

***navum***. *Tetramorium navum* Bolton, 1977: 106, fig. 27 (w.) NEW GUINEA.

***nefassitense***. *Tetramorium caespitum* var. *nefassitensis* Forel, 1910c: 260 (w.) ETHIOPIA. Raised to species: Bolton, 1980: 315.

*neuvillei*. *Tetramorium neuvillei* Forel, 1907c: 135 (w.) ETHIOPIA. Junior synonym of *sericeiventre:* Bolton, 1980: 332.

*nigellus*. *Xiphomyrmex weitzeckeri* var. *nigellus* Santschi, 1932a: 389 (w.q.) ZIMBABWE. Junior synonym of *weitzeckeri:* Bolton, 1980: 233.

*nigriventre*. *Tetramorium blochmanni* var. *nigriventre* Stitz, 1910: 144 (w.) CAMEROUN. Variety of *sericeiventre:* Santschi, 1918b: 126. Junior synonym of *sericeiventre:* Bolton, 1980: 332.

***nigrum***. *Tetramorium pauper* subsp. *nigrum* Forel, 1907a: 15 (w.) KENYA. Raised to species, and senior synonym of *brevis:* Bolton, 1980: 316.

*nion. Triglyphothrix nion* Bernard, 1953b: 248, fig. 14 (w.) GUINEA. Junior synonym of *brevispinosum:* Bolton, 1976: 320.

***nipponense.*** *Tetramorium guineense* subsp. *nipponense* Wheeler, W.M. 1928d: 115 (w.) JAPAN. Raised to species: Bolton, 1977: 100.

***nitidissimum.*** *Tetramorium punicum* subsp. *nitidissimum* Pisarski, 1967: 402 (w.) TRANSCASPIA. [First available use of *Tetramorium caespitum* subsp. *punicum* var. *nitidissima* Emery, 1925c: 190; unavailable name.] Subspecies of *punicum:* Dlussky & Zabelin, 1985: 232. Raised to species: Dlussky, Soyunov & Zabelin, 1990: 206; Radchenko, 1992b: 54.

*nitidiusculum. Tetramorium caespitum* subsp. *nitidiusculum* Arnol'di, 1948: 211. *Nomen nudum.*

*nitidum* Smith, F. 1876; see under ***MONOMORIUM.***

***nodiferum.*** *Atopomyrmex nodifer* Emery, 1901f: 115 (footnote), fig. (w.q.) CAMEROUN. Combination in *Atopula:* Emery, 1912b: 104; in *Tetramorium:* Bolton, 1976: 362. See also: Bolton, 1980: 349.

***noratum.*** *Tetramorium noratum* Bolton, 1977: 81, fig. 9 (w.) WEST MALAYASIA.

***notiale.*** *Tetramorium notiale* Bolton, 1980: 271. Replacement name for *striatum* Arnold, 1917: 308. [Junior primary homonym of *striatum* Smith, F. 1876b: 481.]

***nube.*** *Tetramorium squaminode* subsp. *nubis* Weber, 1943c: 369, pl. 16, fig. 24 (w.) SUDAN. Raised to species: Bolton, 1980: 257.

*nullispinum. Tetramorium nullispinum* Bolton, 1980: 241 (w.) NIGERIA. Junior synonym of *angulinode:* **new synonymy**.

***nursei.*** *Tetramorium nursei* Bingham, 1903: 181, fig. 67 (w.) PAKISTAN. See also: Bolton, 1977: 93.

***obesum.*** *Tetramorium obesum* André, 1887: 294 (w.) INDIA. Mann, 1919: 349 (m.). Combination in *Triglyphothrix:* Emery, 1891b: 4; in *Tetramorium:* Bolton, 1985: 247. See also: Bolton, 1976: 354.

*obscurius. Tetramorium ornatum* var. *obscurius* Forel, 1901b: 11 (w.) BISMARCK ARCHIPELAGO. Junior synonym of *ornatum:* Bolton, 1977: 107.

***obtusidens.*** *Tetramorium obtusidens* Viehmeyer, 1916a: 138, fig. 6 (w.) SINGAPORE. See also: Bolton, 1977: 101; Bolton, 1979: 178.

***occidentale.*** *Xiphomyrmex occidentalis* Santschi, 1916a: 50, fig. 1 (w.) CAMEROUN. Combination in *Tetramorium,* and senior synonym of *akengensis* and *insularis:* Bolton, 1980: 229.

***ocothrum.*** *Tetramorium ocothrum* Bolton, 1979: 176, fig. 57 (w.) BORNEO.

***oculatum.*** *Tetramorium oculatum* Forel, 1913a: 116 (w.) ZIMBABWE. See also: Bolton, 1980: 316.

*opacior. Tetramorium simillimum* var. *opacior* Forel, 1913k: 81 (w.) SRI LANKA. Junior synonym of *simillimum:* Bolton, 1977: 131.

*orbiceps* Santschi, 1914 [*Xiphomyrmex*]; see under ***PRISTOMYRMEX.***

*orientale. Tetramorium* (*Xiphomyrmex*) *bessonii* var. *orientale* Forel, 1895c: 247 (w.) MADAGASCAR. Junior synonym of *bessonii:* Bolton, 1979: 141.

*orissana. Triglyphothrix striatidens* r. *orissana* Forel, 1902c: 239 (w.) INDIA. Raised to species: Bingham, 1903: 174. Subspecies of *striatidens:* Emery, 1922f: 274. Junior synonym of *lanuginosum:* Bolton, 1976: 350.

***ornatum.*** *Tetramorium ornatum* Emery, 1897c: 585, pl. 15, fig. 27 (w.) NEW GUINEA. Senior synonym of *obscurius:* Bolton, 1977: 107.

***osiris.*** *Triglyphothrix osiris* Bolton, 1976: 332, fig. 61 (w.) GHANA. Combination in *Tetramorium:* Bolton, 1985: 247.

*otaviensis. Tetramorium quadrispinosum* r. *otaviensis* Arnold, 1926: 253 (w.) NAMIBIA. Junior synonym of *quadrispinosum:* Bolton, 1980: 330.

#***oxyomma.*** *Tetramorium caespitum* var. *oxyomma* Karavaiev, 1912a: 13 (w.q.m.) ALGERIA.

***pacificum.*** *Tetramorium pacificum* Mayr, 1870b: 976 (w.q.) TONGA. Imai, Kubota, *et al.* 1985: 47 (k.). Senior synonym of *scabrum* and *subscabrum:* Bolton, 1977: 102.

*pacifica. Triglyphothrix pacifica* Mann, 1921: 460 (w.q.) FIJI IS. [Junior secondary homonym of *pacificum* Mayr, above.] Replacement name: *Tetramorium manni* Bolton, 1985: 247. See also: Bolton, 1976: 355.

***palaense.*** *Tetramorium palaense* Bolton, 1979: 175, fig. 56 (w.) BORNEO.

*pallens. Triglyphothrix auropunctatus* var. *pallens* Forel, 1910e: 424 (w.) SOUTH AFRICA. Junior synonym of *microps:* Bolton, 1976: 329.

#***pallidum.*** *Tetramorium caespitum* var. *pallida* Stitz, 1934: 2 (w.) CHINA.

*pallidiventre. Tetramorium melanogyna* var. *pallidiventre* Wheeler, W.M. 1934a: 177 (w.) SOLOMON IS. Junior synonym of *insolens:* Bolton, 1977: 99.

*palmense. Tetramorium caespitum* subsp. *semilaeve* var. *palmense* Wheeler, W.M. 1927g: 113 (w.q.) CANARY IS. Unavailable name; material referred to *depressum* by Hohmann, La Roche, *et al.*, 1993: 158.

*papuanum. Tetramorium papuanum* Emery, 1887b: 452 (w.) NEW GUINEA. Junior synonym of *scabrosum:* Bolton, 1977: 119.

*papyri. Xiphomyrmex papyri* Weber, 1943c: 374, pl. 16, fig. 36 (q.) SUDAN. Junior synonym of *angulinode:* Bolton, 1980: 237.

*parallela. Myrmica parallela* Smith, F. 1859a: 147 (w.) INDONESIA (Aru I.). Combination in *Tetramorium:* Donisthorpe, 1932c: 455. Junior synonym of *simillimum:* Bolton, 1977: 131.

***parasiticum.*** *Tetramorium parasiticum* Bolton, 1980: 281, figs. 65, 66 (q.) SOUTH AFRICA.

***parvispinum.*** *Triglyphothrix parvispina* Emery, 1893e: 214 (diagnosis in key) (w.) BORNEO. Combination in *Tetramorium:* Bolton, 1985: 247. Senior synonym of *formosae:* Bolton, 1976: 355.

***parvum.*** *Tetramorium parvum* Bolton, 1977: 117 (w.) JAVA.

***pauper.*** *Tetramorium pauper* Forel, 1907a: 14 (w.) KENYA. See also: Bolton, 1980: 317.

*pauper. Triglyphothrix pauper* Santschi, 1917b: 286. [Junior secondary homonym of *pauper* Forel, above.] Replacement name: *ericae:* Arnold, 1917: 332. [Junior synonym of *pauper* Santschi, hence first available replacement name.]

*pembensis. Tetramorium* (*Xiphomyrmex*) *humbloti* var. *pembensis* Forel, 1907g: 83 (q.m.) TANZANIA. Forel, 1913a: 120 (w.). Combination in *Xiphomyrmex:* Wheeler, W.M. 1922a: 907. Junior synonym of *humbloti:* Bolton, 1980: 228.

#***penninum.*** *Tetramorium caespitum* var. *penninum* Santschi, 1927a: 54 (w.) SWITZERLAND.

***peringueyi.*** *Tetramorium peringueyi* Arnold, 1926: 260 (w.) SOUTH AFRICA. See also: Prins, 1973: 9; Bolton, 1980: 248.

**peritulum* Cockerell, 1927; see under ***LASIUS.***

***perlongum.*** *Tetramorium perlongum* Santschi, 1923e: 284 (w.) ANGOLA. [Also described as new by Santschi, 1925h: 165.] See also: Bolton, 1980: 281.

***persignatum.*** *Tetramorium persignatum* **new name**. Replacement name for *signatum* Menozzi, 1933b: 72. [Junior primary homonym of *signatum* Emery, 1895h: 35.]

*perspicax. Tetramorium perspicax* Radchenko, 1992b: 56 (w.q.) TURKEY. [First available use of *Tetramorium caespitum* st. *ferox* var. *perspicax* Santschi, 1921a: 111; unavailable name.] Synonym of *feroxoide:* Radchenko, 1992b: 56. [Radchenko gives *perspicax* as senior name, but *feroxoide* has priority.]

***petersi.*** *Tetramorium blochmanni* subsp. *petersi* Forel, 1910f: 19 (w.) NAMIBIA. Subspecies of *sericeiventre:* Wheeler, W.M. 1922a: 902. Raised to species: Bolton, 1980: 329.

***peutli.*** *Tetramorium guineense* r. *peutli* Forel, 1916: 419 (w.q.) ZAIRE. Raised to species: Bolton, 1980: 272.

***phasias.*** *Tetramorium guineense* var. *phasias* Forel, 1914d: 226 (w.) SOUTH AFRICA. Raised to species, and senior synonym of *hertigi:* Bolton, 1980: 273.

***pialtum.*** *Tetramorium pialtum* Bolton, 1980: 350 (w.) GHANA.

*picta. Tetramorium caespitum* var. *picta* Karavaiev, 1912b: 584 (w.q.) TRANSCASPIA. Junior synonym of *inerme:* Radchenko, 1992b: 53.

***pilosum.*** *Tetramorium* (*Xiphomyrmex*) *pilosum* Emery, 1893f: 247 (w.) SRI LANKA. See also: Bolton, 1977: 82.

***pinnipilum.*** *Tetramorium pinnipilum* Bolton, 1980: 230, figs. 3, 6 (w.) ANGOLA.

***placidum.*** *Tetramorium placidum* Bolton, 1979: 162, fig. 46 (w.) MEXICO.

*plana. Tetramorium caespitum* var. *plana* Karavaiev, 1927b: 109, fig. 5 (w.q.m.) TURKESTAN. Junior synonym of *inerme:* Radchenko, 1992b: 53.

*platonis. Tetramorium sericeiventre* st. *continentis* var. *platonis* Santschi, 1918b: 130 (w.) BOTSWANA. Unavailable name; material referred to *sericeiventre* by Bolton, 1980: 332.

***platynode.*** *Tetramorium platynode* Bolton, 1980: 258, figs. 35, 38 (w.) SOUTH AFRICA.

***pleganon.*** *Tetramorium pleganon* Bolton, 1979: 146 (w.) MADAGASCAR.

***plesiarum.*** *Tetramorium plesiarum* Bolton, 1979: 150, fig. 27 (w.) MADAGASCAR.

***plumosum.*** *Tetramorium plumosum* Bolton, 1980: 263, fig. 43 (w.) SWAZILAND.

***pnyxis.*** *Triglyphothrix pnyxis* Bolton, 1976: 356, fig. 70 (w.) WEST MALAYSIA. Imai, Brown, *et al.* 1984: 68 (k.). Combination in *Tetramorium:* Bolton, 1985: 247.

***pogonion.*** *Tetramorium pogonion* Bolton, 1980: 249, fig. 21 (w.q.m.) NAMIBIA.

***politum.*** *Tetramorium politum* Emery, 1897d: 568 (w.) NEW GUINEA. See also: Bolton, 1977: 107.

*politum*. *Tetramorium semireticulatum* var. *politum* Arnold, 1948: 225 (w.) ZIMBABWE. [Unresolved junior primary homonym of *politum* Emery, above.] Junior synonym of *semireticulatum:* Bolton, 1980: 361.

*popovici*. *Tetramorium popovici* Forel, 1914d: 230 (w.) SOUTH AFRICA. Arnold, 1917: 312 (q.). Junior synonym of *capense:* Bolton, 1980: 359.

***postpetiolatum***. *Tetramorium coloreum* var. *postpetiolata* Santschi, 1919h: 88 (w.) ZAIRE. Raised to species: Brown, 1957b: 75. See also: Bolton, 1980: 301.

***poweri***. *Tetramorium simillimum* var. *poweri* Forel, 1914d: 225 (w.) SOUTH AFRICA. Raised to species: Bolton, 1980: 317.

***praetextum***. *Tetramorium praetextum* Bolton, 1980: 282, fig. 52 (w.) NAMIBIA.

***proximum***. *Tetramorium proximum* Bolton, 1979: 137, figs. 5, 6 (w.q.) MADAGASCAR.

***psymanum***. *Tetramorium psymanum* Bolton, 1980: 351 (w.) GHANA.

***pulchellum***. *Tetramorium pulchellum* Emery, 1897c: 586, pl. 15, fig. 28 (w.) NEW GUINEA. See also: Bolton, 1977: 118.

*pulchella*. *Triglyphothrix pulchella* Mann, 1919: 352, fig. 31 (w.q.) SOLOMON IS. [Junior secondary homonym of *pulchellum* Emery, above.] Replacement name: *mutatum* Bolton, 1985: 247. See also: Bolton, 1976: 356.

*pulchellus*. *Macromischoides aculeatus* var. *pulchellus* Santschi, 1924b: 208, fig. 8 (w.q.m.) ZAIRE. [Unresolved junior secondary homonym of *pulchellum* Emery, above.] Junior synonym of *aculeatum:* Bolton, 1980: 353.

***pulcherrimum***. *Triglyphothrix pulcherrima* Donisthorpe, 1945c: 76 (w.) UGANDA. Combination in *Tetramorium:* Bolton, 1985: 247. See also: Bolton, 1976: 333.

***pullulum***. *Tetramorium guineense* st. *pullulum* Santschi, 1924b: 211, fig. 9 (w.) ZAIRE. Raised to species, and senior synonym of *fernandensis* and *uelensis:* Bolton, 1980: 273.

*pullum*. *Tetramorium caespitum* st. *tsushimaeum* var. *pullum* Santschi, 1941: 277 (w.) JAPAN. Unavailable name.

#***punctatum***. *Tetramorium semilaeve* var. *punctatum* Santschi, 1927a: 55 (w.q.) SICILY.

***punctiventre***. *Tetramorium punctiventre* Emery, 1887b: 453 (q.) NEW GUINEA. See also: Bolton, 1977: 118.

***punicum***. *Myrmica punica* Smith, F. 1861a: 34 (w.) ISRAEL. Finzi, 1936: 182 (m.); Wheeler, G.C. & Wheeler, J. 1954d: 449 (l.). Combination in *Tetramorium:* Mayr, 1870b: 975 (in text). Variety/subspecies of *caespitum:* André, 1883a: 287; Forel, 1904c: 5; Forel, 1905b: 173; Emery, 1909d: 704; Forel, 1910a: 23; Santschi, 1921a: 111; Emery, 1925c: 190; Finzi, 1930d: 315. Status as species: Dalla Torre, 1893: 134; Menozzi, 1933b: 69; Baroni Urbani, 1971c: 141; Collingwood, 1978: 71; Casevitz-Weulersse, 1990b: 422. Current subspecies: nominal plus *cyrenaicum*.

***pusillum***. *Tetramorium pusillum* Emery, 1895h: 38 (w.q.) SOUTH AFRICA. Senior synonym of *ladismithensis* and *tablensis:* Bolton, 1980: 318.

*pygmaeum*. *Tetramorium pygmaeum* Emery, 1877b: 371 (q.) ETHIOPIA. Emery, 1901e: 62 (m.); Emery, 1915g: 17 (w.). Junior synonym of *simillimum:* Forel, 1916: 421.

***pylacum***. *Tetramorium pylacum* Bolton, 1980: 301, figs. 82, 85 (w.) IVORY COAST.

***pyrenaeicum***. *Tetramorium caespitum* subsp. *pyrenaeica* Röszler, 1937: 198, fig. 6 (w.q.m.) ANDORRA. Raised to species: Röszler, 1951: 88.

***quadridentatum***. *Tetramorium quadridentatum* Stitz, 1910: 144 (w.) CAMEROUN. Senior synonym of *commodum:* Bolton, 1980: 364.

***quadrispinosum***. *Tetramorium quadrispinosum* Emery, 1886: 362, pl. 17, fig. 8 (w.) SOUTH AFRICA. Senior synonym of *montanum:* Bolton, 1979: 155; of *angolense, beirae, calvum, eudoxia* (and its junior synonym *elegans*), *otaviensis* and *repertum*, and material of the unavailable name *benguelense* referred here: Bolton, 1980: 330.

*quaerens*. *Tetramorium setigerum* r. *quaerens* Forel, 1914d: 226 (w.) ZIMBABWE. Junior synonym of *setigerum:* Bolton, 1980: 283.

***qualarum***. *Tetramorium qualarum* Bolton, 1980: 351, figs. 122, 127 (w.) GHANA.

***quasirum***. *Tetramorium quasirum* Bolton, 1979: 150 (w.q.) MADAGASCAR.

***ranarum***. *Tetramorium* (*Xiphomyrmex*) *ranarum* Forel, 1895h: 486 (w.) MADAGASCAR. Combination in *Xiphomyrmex:* Wheeler, W.M. 1922a: 1032; in *Tetramorium:* Bolton, 1979: 151.

*rectinodis*. *Tetramorium rectinodis* Menozzi, 1942: 176, fig. 2 (w.) FERNANDO PO I. Junior synonym of *lucayanum:* Brown, 1964b: 131; Bolton, 1979: 172.

*reduncum* Wang & Wu, in Wang, M., Xiao & Wu, 1988; see under ***LEPTOTHORAX.***

***regulare***. *Tetramorium regulare* Bolton, 1980: 263, fig 45 (w.m.) SOUTH AFRICA.

*reitteri* Mayr, 1887; see under ***HYLOMYRMA.***

***rekhefe***. *Tetramorium rekhefe* Bolton, 1979: 175, fig. 58 (w.) INDONESIA (Timor I.).

***repentinum***. *Tetramorium repentinum* Arnold, 1926: 257, fig. 70 (w.) ZIMBABWE. See also: Bolton, 1980: 258.

*repertum*. *Tetramorium sericeiventre* var. *repertum* Santschi, 1926b: 242 (w.) MOZAMBIQUE. Junior synonym of *quadrispinosum:* Bolton, 1980: 330.

***repletum***. *Tetramorium repletum* Wang & Xiao, in Wang, M., Xiao & Wu, 1988: 166, figs. 8, 9 (w.) CHINA.

***reptana***. *Triglyphothrix reptana* Bolton, 1976: 334 (w.) UGANDA. Combination in *Tetramorium:* Bolton, 1985: 247.

*reticulata*. *Myrmica reticulata* Smith, F. 1862b: 33 (w.) PANAMA. Combination in *Tetramorium:* Mayr, 1862: 740. Junior synonym of *guineense:* Roger, 1862c: 293; of *bicarinatum:* Bolton, 1977: 94.

***reticuligerum***. *Tetramorium reticuligerum* Bursakov, 1984: 400, fig. 1 (w.q.m.) KAZAKHSTAN. See also: Radchenko, 1992a: 45.

*reticuliventre*. *Tetramorium caespitum* var. *reticuliventre* Ruzsky, 1902b: 481 (w.) TURKESTAN. [Also described as new by Ruzsky, 1902c: 14.] Ruzsky, 1905b: 767 (q.). Junior synonym of *inerme:* Radchenko, 1992b: 53.

***rhetidum***. *Tetramorium rhetidum* Bolton, 1980: 318, fig. 99 (w.q.) GHANA.

*rhodesiana*. *Triglyphothrix auropunctata* var. *rhodesiana* Forel, 1913j: 221 (q.) ZIMBABWE. Combination in *Tetramorium:* Arnold, 1917: 336. Junior synonym of *microps:* Bolton, 1976: 329.

#***rhodium***. *Tetramorium caespitum* var. *rhodia* Emery, 1922f: 277 (w.) GREECE. [First available use of *Tetramorium caespitum* subsp. *caespitum* var. *rhodia* Emery, 1915h: 3; unavailable name.] Subspecies of *caespitum:* Emery, 1925c: 179.

***rigidum***. *Tetramorium rigidum* Bolton, 1977: 108, fig. 28 (w.) NEW GUINEA.

***rimytyum***. *Tetramorium rimytyum* Bolton, 1980: 356, fig. 129 (w.) GHANA.

***rinatum***. *Tetramorium rinatum* Bolton, 1977: 82, fig. 11 (w.) PHILIPPINES.

***robustior***. *Tetramorium* (*Xiphomyrmex*) *andrei* r. *robustior* Forel, 1892k: 521 (w.) MADAGASCAR. Raised to species: Bolton, 1979: 147.

***rogatum***. *Tetramorium rogatum* Bolton, 1980: 231, fig. 10 (w.) ANGOLA.

*romana*. *Tetramorium caespitum* st. *semilaeve* var. *romana* Santschi, 1921b: 431 (w.) no locality given. Unavailable name.

***rossi***. *Triglyphothrix rossi* Bolton, 1976: 357 (w.q.) INDIA. Combination in *Tetramorium:* Bolton, 1985: 247.

***rothschildi***. *Triglyphothrix rothschildi* Forel, 1907c: 134 (w.) ETHIOPIA. Menozzi & Consani, 1952: 67 (q.m.). Combination in *Tetramorium:* Bolton, 1985: 247. See also: Bolton, 1976: 334.

***rotundatum***. *Macromischoides africanus* var. *rotundatus* Santschi, 1924b: 209 (q.) ZAIRE. Combination in *Tetramorium*, and raised to species: Bolton, 1980: 357.

*rubroflava*. *Tetramorium aculeatum* var. *rubroflava* Forel, 1916: 420 (w.) ZAIRE. Junior synonym of *aculeatum:* Bolton, 1980: 353.

***rufescens***. *Tetramorium solidum* st. *rufescens* Stitz, 1923: 163 (w.) NAMIBIA. Raised to species: Bolton, 1980: 249.

*rufitarsis*. *Tetramorium caespitum* subsp. *hungarica* var. *rufitarsis* Röszler, 1936b: 58, fig. 1 (w.q.) ANDORRA. Unavailable name.

***rugigaster***. *Tetramorium rugigaster* Bolton, 1977: 114, fig. 39 (w.q.m.) INDIA.

***ruginode***. *Tetramorium caespitum* var. *ruginode* Stitz, 1917: 339 (w.) SPAIN. Santschi, 1931a: 9 (q.). Synonym of *hispanicum:* Collingwood, 1978: 71. [Collingwood gives *hispanicum* as senior name, but *ruginode* has priority.] Raised to species: **new status**.

*rugiventris* Smith, M.R. 1943; see under ***MYRMICA.***

*rutilum*. *Tetramorium rutilum* Prins, 1973: 14, figs. 14–18, 30 (w.q.) SOUTH AFRICA. Junior synonym of *glabratum:* Bolton, 1980: 246.

***saginatum***. *Tetramorium saginatum* Bolton, 1980: 302 (w.) ANGOLA.

***sahlbergi***. *Tetramorium punicum* var. *sahlbergi* Finzi, 1936: 182 (w.q.) EGYPT. [First available use of *Tetramorium caespitum* r. *punicum* var. *sahlbergi* Forel, 1913d: 431; unavailable name.] Raised to species: Agosti & Collingwood, 1987a: 56.

***salomo***. *Tetramorium salomo* Mann, 1919: 344 (w.) SOLOMON IS. See also: Bolton, 1977: 109.

***salvatum***. *Tetramorium salvatum* Forel, 1902c: 235 (w.) INDIA. See also: Bolton, 1977: 128.

#***sarkissiani***. *Tetramorium caespitum* var. *sarkissiani* Forel, 1911d: 332 (w.) TURKEY. Variety of *ferox:* Menozzi, 1934: 163.

***scabrosum.*** *Myrmica scabrosa* Smith, F. 1859a: 147 (w.) INDONESIA (Aru I.). Combination in *Tetramorium:* Donisthorpe, 1932c: 455. Senior synonym of *papuanum:* Bolton, 1977: 119.

*scabrum. Tetramorium scabrum* Mayr, 1879: 672 (w.) BORNEO. Forel, 1913k: 81 (q.m.). Subspecies of *pacificum:* Emery, 1922f: 284. Junior synonym of *pacificum:* Bolton, 1977: 102.

***schaufussii.*** *Tetramorium (Xiphomyrmex) schaufussii* Forel, 1891b: 158 (w.) MADAGASCAR. Forel, 1892l: 263 (q.). Senior synonym of *nassonowii:* Bolton, 1979: 137.

***schmidti.*** *Tetramorium caespitum* var. *schmidti* Forel, 1904c: 15 (w.) ISRAEL. Karavaiev, 1912a: 14 (q.); Wheeler, G.C. & Wheeler, J. 1954d: 449. Subspecies of *biskrense:* Menozzi, 1933b: 75 (in key). Raised to species: Aktaç, 1977: 123.

***schneideri.*** *Tetramorium schneideri* Emery, 1898c: 145 (w.) TURKESTAN. Subspecies of *striativentre:* Ruzsky, 1905a: 518; Ruzsky, 1905b: 767; Emery, 1909d: 706. Revived status as species: Dlussky & Zabelin, 1985: 232. Material of the unavailable name *longispina* referred here by Dlussky, Soyunov & Zabelin, 1990: 207.

***schoutedeni.*** *Tetramorium schoutedeni* Santschi, 1924b: 213, fig. 9 (w.) ZAIRE. See also: Bolton, 1980: 231.

*schultzei. Tetramorium caespitum* subsp. *schultzei* Forel, 1910f: 19 (w.) BOTSWANA. Junior synonym of *mossamedense:* Bolton, 1980: 314.

*scrobiferum* Emery, 1897; see under ***ROMBLONELLA.***

***sculptatum.*** *Tetramorium sculptatum* Bolton, 1977: 109, fig. 26 (w.) NEW GUINEA.

*sculptum. Tetramorium sculptum* Aktaç, 1977: 122. *Nomen nudum;* attributed to Poldi.

***scytalum.*** *Tetramorium scytalum* Bolton, 1979: 157, fig. 29 (w.) MADAGASCAR.

***semilaeve.*** *Tetramorium caespitum* var. *semilaeve* André, 1883a: 286 (w.) Mediterranean region. Emery, 1891b: 2 (q.); Forel, 1902a: 148 (m.). Race/stirps/subspecies of *caespitum:* Forel, 1902a: 148; Forel, 1904b: 371; Emery, 1909d: 703; Emery, 1916b: 194; Santschi, 1921a: 111; Santschi, 1923a: 135; Menozzi, 1926b: 182; Emery, 1924a: 8. Status as species: Dalla Torre, 1893: 134; Bondroit, 1918: 109; Müller, 1923: 104; Santschi, 1927a: 54; Santschi, 1929e: 150; Finzi, 1930d: 315; Santschi, 1931a: 10; Bernard, 1967: 234; Arnol'di & Dlussky, 1978: 545. Senior synonym of *tenuicornis:* Radchenko, 1992b: 51. Current subspecies: *atlante, depressum, gaetulum, galaticum, guancha, hoggarense, italicum, judas, jugurtha, kutteri, liparaeum, punctatum, siciliense.* See also: Emery, 1925c: 179; López, 1991a: 36; López, 1991b: 76; Atanassov & Dlussky, 1992: 153; Radchenko, 1992b: 51.

***semireticulatum.*** *Tetramorium semireticulatum* Arnold, 1917: 319 (w.m.) ZIMBABWE. Arnold, 1926: 265 (q.). Senior synonym of *politum:* Bolton, 1980: 361.

***seneb.*** *Tetramorium seneb* Bolton, 1977: 128 (w.) WEST MALAYSIA. Imai, Brown, *et al.* 1984: 68 (k.).

***sepositum.*** *Tetramorium gladstonei* var. *seposita* Santschi, 1918b: 131 (w.) ZIMBABWE. Raised to species: Bolton, 1980: 331.

***sepultum.*** *Tetramorium sepultum* Bolton, 1980: 232 (w.) SWAZILAND.

***sericeiventre.*** *Tetramorium sericeiventre* Emery, 1877b: 370 (w.) ETHIOPIA. Senior synonym of *hortensis:* Bolton, 1976: 363; of *blochmannii:* Bolton, 1979: 155; of *arenarium, bipartita, cinnamomeum, continentis, debile, femoratum, gamaii, hori, inversa, jasonis, munda, neuvillei, nigriventre, vascoi* and *vividum,* and material of the unavailable names *colluta, defricta, evidens, georgei, gladiator, kenyense, platonis* and *transversa* referred here: Bolton, 1980: 332.

***sericeum.*** *Tetramorium (Triglyphothrix) hepburni* var. *sericeus* Arnold, 1926: 277 (w.) ZIMBABWE. Combination in *Triglyphothrix:* Bolton, 1976: 335; in *Tetramorium:* Bolton, 1985: 247. Raised to species: Bolton, 1976: 335.

***setigerum.*** *Tetramorium setigerum* Mayr, 1901b: 22 (w.) SOUTH AFRICA. Forel, 1910e: 424 (q.m.). Senior synonym of *anteversa* and *quaerens:* Bolton, 1980: 283.

***setuliferum.*** *Tetramorium setuliferum* Emery, 1895h: 36 (w.) SOUTH AFRICA. Forel, 1910f: 19 (q.); Arnold, 1917: 290 (m.). Material of the *nomen nudum squamiferum* referred here by Wheeler, W.M. 1922a: 903. Senior synonym of *cucalense* (and its junior syonym *triptolemus*): Bolton, 1980: 250.

***severini.*** *Xiphomyrmex severini* Emery, 1895f: 343 (w.) MADAGASCAR. Combination in *Tetramorium:* Bolton, 1979: 138.

*sexdens. Tetramorium lucayanum* var. *sexdens* Forel, 1915c: 357 (w.) IRELAND. Junior synonym of *lucayanum:* Brown, 1964b: 131; Bolton, 1979: 172.

***shensiense.*** *Tetramorium shensiense* Bolton, 1977: 83, fig. 10 (w.) CHINA.

***shilohense.*** *Tetramorium simillimum* var. *shilohensis* Forel, 1913j: 218 (w.) ZIMBABWE. Raised to species: Bernard, 1953b: 248. See also: Bolton, 1980: 290.

#***siciliense.*** *Tetramorium semilaeve* var. *siciliense* Santschi, 1927a: 56 (w.q.) SICILY.

***sigillum.*** *Tetramorium sigillum* Bolton, 1980: 303, fig. 79 (w.) IVORY COAST.

*sigmoideum* Mayr, 1884; see under ***WASMANNIA.***

***signatum.*** *Tetramorium solidum* var. *signatum* Emery, 1895h: 35 (w.) SOUTH AFRICA. Raised to species, and senior synonym of *grootensis, lugubre* and *tuckeri:* Bolton, 1980: 251.

*signatum. Tetramorium signatum* Menozzi, 1933b: 72, fig. 4 ISRAEL. [Junior primary homonym of *signatum* Emery, above.] Replacement name: *persignatum* **new name**.

***sikorae.*** *Tetramorium (Xiphomyrmex) sikorae* Forel, 1892k: 522 (w.) MADAGASCAR. Combination in *Xiphomyrmex:* Emery, 1895f: 343; in *Tetramorium:* Bolton, 1979: 138. Senior synonym of *latior:* Bolton, 1979: 138.

*silhavyi. Tetramorium (Lobomyrmex) ferox* subsp. *silhavyi* Kratochvil, in Novak & Sadil, 1941: 84, fig. 1 (w.q.) CZECHOSLOVAKIA. [Also described as new by Kratochvil, in Kratochvil, Novak & Snoflak, 1944: 72.] Junior synonym of *ferox:* Bernard, 1967: 235; Radchenko, 1992b: 55.

*silvestrianum. Tetramorium silvestrianum* Emery, 1924c: 167, fig. A1 (w.) SPAIN. Junior synonym of *forte:* Collingwood, 1978: 71.

*silvestrii* Santschi, 1911; see under ***LEPTOTHORAX.***

*silvestrii. Triglyphothrix silvestrii* Emery, 1915g: 17, fig. 8 (w.) ETHIOPIA. [Junior secondary homonym of *Tetramorium silvestrii* Santschi, 1911d: 6.] Replacement name: *Tetramorium surrogatum* Bolton, 1985: 247. See also: Bolton, 1976: 336.

***simillimum.*** *Myrmica simillima* Smith, F. 1851: 118 (w.) GREAT BRITAIN. Meinert, 1861: 331 (gynandromorph); André, 1883a: 289 (q.m.); Imai, Baroni Urbani, *et al.* 1984: 8 (k.). Combination in *Tetramorium:* Mayr, 1861: 61. Senior synonym of *pygmaeum:* Forel, 1916: 421; of *brevispinosa:* Borgmeier, 1937b: 241; of *insulare:* Yarrow, 1967: 28; of *denticulatum, opacior* and *parallela:* Bolton, 1977: 131; of *bantouana* and *exoleta,* and material of the unavailable name *breve* referred here: Bolton, 1980: 320. See also: Emery, 1909d: 696.

*simoni* Emery, 1895; see under ***LEPTOTHORAX.***

*simulans. Tetramorium grassii* var. *simulans* Santschi, 1914e: 24 (w.) SOUTH AFRICA. Junior synonym of *grassii:* Bolton, 1980: 262.

***simulator.*** *Tetramorium simulator* Arnold, 1917: 297, pl. 7, fig. 102 (w.) ZIMBABWE. See also: Bolton, 1980: 365.

***sitefrum.*** *Tetramorium sitefrum* Bolton, 1980: 259 (w.) GHANA.

***sjostedti.*** *Tetramorium (Xiphomyrmex) sjostedti* Forel, 1915b: 48 (w.) AUSTRALIA. Combination in *Xiphomyrmex:* Viehmeyer, 1925a: 28; in *Tetramorium:* Bolton, 1977: 140.

***smithi.*** *Tetramorium smithi* Mayr, 1879: 673 (w.) INDIA. Imai, Baroni Urbani, *et al.* 1984: 8 (k.). Senior synonym of *kanariense* and *laevinode:* Bolton, 1977: 90.

***solidum.*** *Tetramorium solidum* Emery, 1886: 362, pl. 17, fig. 7 (w.q.) SOUTH AFRICA. See also: Prins, 1973: 4; Bolton, 1980: 252.

***somniculosum.*** *Tetramorium somniculosum* Arnold, 1926: 262, fig. 72 (w.q.m.) MOZAMBIQUE. See also: Bolton, 1980: 290.

*soyauxi. Triglyphothrix gabonensis* st. *soyauxi* Forel, 1901h: 53 (w.) GABON. [Misspelled as *soyansi* by Emery, 1922f: 273.] Junior synonym of *gabonense:* Bolton, 1976: 326.

***spininode.*** *Tetramorium spininode* Bolton, 1977: 140, fig. 62 (w.q.) AUSTRALIA.

***spinosum.*** *Xiphomyrmex spinosus* Pergande, 1896: 894 (w.) MEXICO. Cole, 1957c: 209 (m.); Taber & Cokendolpher, 1988: 95 (k.). Combination in *Tetramorium,* and senior synonym of *insons* and *wheeleri:* Bolton, 1979: 163. See also: Smith, M.R. 1938a: 127.

***splendens.*** *Tetramorium caespitum* var. *splendens* Ruzsky, 1902d: 33 (w.) CAUCASUS. Variety/subspecies of *semilaeve:* Finzi, 1928c: 789; of *caespitum:* Karavaiev, 1937: 174. Raised to species: Arnol'di & Dlussky, 1978: 545.

***splendidior.*** *Xiphomyrmex striolatus* subsp. *splendidior* Viehmeyer, 1925a: 29 (w.) AUSTRALIA. Combination in *Tetramorium,* and raised to species: Bolton, 1977: 144.

*spuria. Triglyphothrix walshi* var. *spuria* Forel, 1912n: 58 (w.) SINGAPORE. Junior synonym of *walshi:* Bolton, 1976: 358.

*squamiferum. Tetramorium squamiferum* Forel, 1894b: 80. *Nomen nudum;* attributed to Emery. Material referred to *setuliferum* by Wheeler, W.M. 1922a: 903.

***squaminode.*** *Tetramorium squaminode* Santschi, 1911c: 356, fig. (w.) TANZANIA. Santschi, 1914b: 102 (q.m.). See also: Bolton, 1980: 260.

*staerckei Tetramorium staerckei* Kratochvil, in Kratochvil, Novak & Snoflak, 1944: 65 (w.q.m.) HUNGARY. [First available use of *Tetramorium caespitum* subsp. *hungarica* var. *staerckei* Röszler, 1936b: 60; unavailable name.] Junior synonym of *impurum:* Kutter, 1977c: 159.

***steinheili.*** *Tetramorium (Xiphomyrmex) steinheili* Forel, 1892k: 520 (w.q.) MADAGASCAR. See also: Bolton, 1979: 143.

*striaticeps. Tetramorium caespitum* subsp. *ferox* var. *striaticeps* Emery, 1925c: 188, fig. C (q.) TURKESTAN. Unavailable name; material referred to *ferox* by Radchenko, 1992b: 55.

*striatidens. Tetramorium obesum* r. *striatidens* Emery, 1889b: 501 (w.) BURMA. Wheeler, G.C. & Wheeler, J. 1973b: 78 (l.). Combination in *Triglyphothrix:* Emery, 1891b: 4. Raised to species: Forel, 1903a: 704. Junior synonym of *lanuginosum:* Bolton, 1976: 350.

*striatis. Tetramorium caespitum* subsp. *hungarica* var. *striatis* Röszler, 1936b: 58, fig. 2 (w.q.m.) HUNGARY. Unavailable name.

***striativentre.*** *Tetramorium caespitum* var. *striativentre* Mayr, 1877: 17 (q.) TURKESTAN. [Redescribed in Latin: Mayr, 1880: 20.] André, 1883a: 286 (w.); Wheeler, G.C. & Wheeler, J. 1954d: 450 (l.). Race of *caespitum:* Emery, 1891b: 3. Raised to species: Dalla Torre, 1893: 135; Ruzsky, 1905a: 518; Ruzsky, 1905b: 766; Emery, 1909d: 706; Collingwood, 1961b: 289; Radchenko, 1992a: 45 (in key).

*striatum* Smith, F. 1876; see under ***HUBERIA.***

*striatum. Tetramorium guineense* r. *striatum* Arnold, 1917: 308 (w.q.m.) ZIMBABWE (attributed to Stitz). [Junior primary homonym of *striatum* Smith, F. 1876b: 481, above.] Replacement name: *notiale* Bolton, 1980: 271.

***strictum.*** *Tetramorium strictum* Bolton, 1977: 144, fig. 64 (w.) AUSTRALIA.

***striolatum.*** *Tetramorium (Xiphomyrmex) viehmeyeri* var. *striolatus* Viehmeyer, 1914c: 39 (w.) AUSTRALIA. Combination in *Xiphomyrmex:* Viehmeyer, 1925a: 28; in *Tetramorium:* Bolton, 1977: 141. Raised to species: Viehmeyer, 1925a: 28.

***subcoecum.*** *Tetramorium subcoecum* Forel, 1907c: 137 (w.) KENYA. Senior synonym of *inscia:* Bolton, 1980: 291.

*subscabrum. Tetramorium pacificum* var. *subscabrum* Emery, 1893f: 246 (w.) SRI LANKA. Subspecies of *pacificum:* Wheeler, W.M. 1909d: 340. Junior synonym of *pacificum:* Bolton, 1977: 102.

***sudanense.*** *Xiphomyrmex sudanensis* Weber, 1943c: 373, pl. 16, fig. 40 (w.) SUDAN. Combination in *Tetramorium:* Bolton, 1980: 241.

*sulcinode. Tetramorium caespitum* var. *sulcinode* Santschi, 1927a: 53 (w.) TURKESTAN. Subspecies of *caespitum:* Pisarski, 1970: 318. Junior synonym of *turcomanicum:* Radchenko, 1992b: 52.

*sumatrensis* Forel, 1913 [*Atopula*]; see under ***PARATOPULA.***

***surrogatum.*** *Tetramorium surrogatum* Bolton, 1985: 247. Replacement name for *Triglyphothrix silvestrii* Emery, 1915g: 17. [Junior secondary homonym of *Tetramorium silvestrii* Santschi, 1911d: 6.]

***syriacum.*** *Tetramorium caespitum* var. *syriaca* Emery, 1922f: 277 (w.) SYRIA. [First available use of *Tetramorium caespitum* subsp. *caespitum* var. *syriaca* Emery, 1909d: 702; unavailable name.] Subspecies of *caespitum:* Emery, 1925c: 179; of *semilaeve:* Santschi, 1934d: 279. Raised to species: Collingwood, 1985: 265.

*szaboi. Tetramorium caespitum* subsp. *hungarica* var. *szaboi* Röszler, 1936b: 59, fig. 3 (w.q.m.) HUNGARY. Unavailable name.

***tabarum.*** *Tetramorium tabarum* Bolton, 1980: 236, fig. 14 (w.) ZAIRE.

*tablensis. Tetramorium pusillum* var. *tablensis* Forel, 1914d: 223 (w.q.m.) SOUTH AFRICA. Junior synonym of *pusillum:* Bolton, 1980: 318.

***talpa.*** *Triglyphothrix talpa* Bolton, 1976: 336 (w.) ZAIRE. Combination in *Tetramorium:* Bolton, 1985: 247.

***tanakai.*** *Tetramorium tanakai* Bolton, 1977: 119 (w.q.) JAPAN.

***tantillum.*** *Tetramorium tantillum* Bolton, 1979: 152, figs. 30, 31 (w.) MADAGASCAR.

***taueret.*** *Tetramorium taueret* **new name**. Replacement name for *debilis* Emery, 1922f: 276. [Junior primary homonym of *debile* Forel, 1894b: 80.]

*taurocaucasicum. Tetramorium taurocaucasicum* Arnol'di, 1968: 1813, fig. 12 (w.q.m.) UKRAINE. Junior synonym of *forte:* Radchenko, 1992b: 51; Atanassov & Dlussky, 1992: 152.

***taylori.*** *Tetramorium taylori* Bolton, 1985: 247. Replacement name for *Tetramorium australe* Bolton, 1977: 146. [Junior secondary homonym of *Triglyphothrix striatidens* var. *australis* Forel, 1902h: 449.]

***tenebrosum.*** *Tetramorium (Triglyphothrix) constanciae* var. *tenebrosus* Arnold, 1926: 275, fig. 79 (w.q.m.) MOZAMBIQUE. Combination in *Triglyphothrix:* Bolton, 1976: 337; in *Tetramorium:* Bolton, 1985: 247. Raised to species: Bolton, 1976: 337.

*tenuicornis. Tetramorium caespitum* subsp. *tenuicornis* Emery, 1925c: 188, figs. B, D (w.m.) TRANSCASPIA. Junior synonym of *semilaeve:* Radchenko, 1992b: 51.

***tenuicrine.*** *Xiphomyrmex tenuicrinis* Emery, 1914f: 416 (w.) NEW CALEDONIA. Combination in *Tetramorium:* Bolton, 1977: 88.

***termitobium.*** *Tetramorium termitobium* Emery, 1908b: 186 (w.) ZAIRE. See also: Bolton, 1980: 292.

***tersum.*** *Tetramorium tersum* Santschi, 1911c: 357, fig. 1 (w.) KENYA. Senior synonym of *atrinodis* and *kivuense:* Bolton, 1980: 233.

*tessmanni. Rhoptromyrmex tessmanni* Forel, 1910e: 421 (w.) EQUATORIAL GUINEA. Combination in *Macromischoides,* and junior synonym of *africanum:* Brown, 1964c: 12.

***thalidum.*** *Tetramorium thalidum* Bolton, 1977: 141, fig. 70 (w.) AUSTRALIA.

***thoth.*** *Triglyphothrix thoth* Bolton, 1976: 337 (w.) SOUTH AFRICA. Combination in *Tetramorium:* Bolton, 1985: 247.

***#tingitanum.*** *Tetramorium maurum* st. *tingitanum* Santschi, 1929e: 150 (w.q.) MOROCCO. [First available use of *Tetramorium caespitum* st. *maura* var. *tingitana* Santschi, 1921b: 434; unavailable name.]

***titus.*** *Tetramorium titus* Forel, 1910e: 427 (w.) SOUTH AFRICA. See also: Bolton, 1980: 264.

***tonganum.*** *Tetramorium tonganum* Mayr, 1870b: 976 (w.) TONGA. Santschi, 1928a: 50 (q.). Senior synonym of *magitae:* Bolton, 1977: 129.

***tortuosum.*** *Tetramorium tortuosum* Roger, 1863a: 181 (w.q.) SRI LANKA. Senior synonym of *bellii* and *ethica:* Bolton, 1977: 83.

***tosii.*** *Tetramorium tosii* Emery, 1899f: 284, fig. (w.) MADAGASCAR. See also: Bolton, 1979: 153.

***traegaordhi.*** *Tetramorium traegaordhi* Santschi, 1914e: 23 (w.q.) SOUTH AFRICA. See also: Bolton, 1980: 292.

*transbaicalense. Tetramorium semilaeve* subsp. *transbaicalense* Ruzsky, 1936: 93 (w.) TURKESTAN. Junior synonym of *caespitum:* Radchenko, 1992b: 50.

*transformans. Tetramorium pauper* st. *transformans* Santschi, 1914b: 104 (w.) KENYA. Junior synonym of *caldarium:* Bolton, 1980: 310.

*transversa. Tetramorium sericeiventre* st. *femoratum* var. *transversa* Santschi, 1918b: 128 (w.) SOUTH AFRICA. Unavailable name; material referred to *sericeiventre* by Bolton, 1980: 332.

***transversarium.*** *Tetramorium transversarium* Roger, 1863a: 181 (w.) SRI LANKA. See also: Bolton, 1977: 115.

*transversinodis. Myrmica (Myrmica) brevinodis* var. *transversinodis* Enzmann, J. 1946b: 47, figs. 1,2 (w.) U.S.A. Junior synonym of *caespitum:* Brown, 1949a: 47.

***tricarinatum.*** *Tetramorium bicolor* subsp. *tricarinatum* Viehmeyer, 1914b: 529, fig. 4 (w.) NEW GUINEA. Raised to species, and senior synonym of *tricolor:* Bolton, 1977: 103.

*tricolor. Triglyphothrix tricolor* Donisthorpe, 1948g: 136 (w.q.) NEW GUINEA. Junior synonym of *lanuginosum:* Bolton, 1976: 350.

*tricolor. Xiphomyrmex tricolor* Donisthorpe, 1949c: 753 (w.) NEW GUINEA. [Unresolved junior secondary homonym of *Triglyphothrix tricolor* Donisthorpe, above.] Junior synonym of *tricarinatum:* Bolton, 1977: 103.

*trilineata. Xiphomyrmex muralti* var. *trilineata* Santschi, 1919h: 88 (w.q.) GHANA. Junior synonym of *muralti:* Bolton, 1980: 229.

***trimeni.*** *Triglyphothrix trimeni* Emery, 1895h: 40 (w.q.) SOUTH AFRICA. Combination in *Tetramorium:* Arnold, 1917: 337. See also: Bolton, 1976: 338; Bolton, 1985: 247.

*triptolemus. Tetramorium setuliferum* var. *triptolemus* Arnold, 1917: 292 (w.) ZAMBIA. Junior synonym of *cucalense:* Santschi, 1930b: 71.

*tristis. Tetramorium pusillum* st. *mossamedense* var. *tristis* Santschi, 1917b: 285 (w.) ZIMBABWE. Unavailable name; material referred to *mossamedense* by Bolton, 1980: 314.

***tsushimae.*** *Tetramorium caespitum* subsp. *tsushimae* Emery, 1925c: 187, fig. C (w.q.) JAPAN. Synonym of *jacoti:* Wheeler, W.M. 1927d: 7. [Wheeler gives *jacoti* as senior synonym, but *tsushimae* has priority and is the first available name for this taxon.] Raised to species: **new status**. [For current synonymy of *tsushimae* see under its junior synonym *jacoti.*]

*tuckeri. Tetramorium solidum* var. *tuckeri* Arnold, 1926: 259 (w.q.) SOUTH AFRICA. Subspecies of *solidum:* Prins, 1973: 5. Junior synonym of *signatum:* Bolton, 1980: 251.

***turcomanicum.*** *Tetramorium caespitum* st. *turcomanicum* Santschi, 1921a: 111 (w.q.) TURKESTAN. [First available use of *Tetramorium caespitum* subsp. *caespitum* var. *turcomanica* Emery, 1909d: 702; unavailable name.] Raised to species: Tarbinsky, 1976: 109. Senior synonym of *sulcinode:* Radchenko, 1992b: 52.

***turneri.*** *Tetramorium* (*Xiphomyrmex*) *turneri* Forel, 1902h: 447 (w.q.) AUSTRALIA. Wheeler, G.C. & Wheeler, J. 1954d: 451 (l.). Combination in *Xiphomyrmex:* Viehmeyer, 1925a: 29; in *Tetramorium:* Bolton, 1977: 145.

***tychadion.*** *Tetramorium tychadion* Bolton, 1980: 342 (w.) TANZANIA.

***tylinum.*** *Tetramorium tylinum* Bolton, 1977: 123, fig. 53 (w.) BORNEO.

***typhlops.*** *Tetramorium typhlops* Bolton, 1980: 293, fig. 70 (w.) IVORY COAST.

#***typicum.*** *Tetramorium caespitum* subsp. *typicum* Ruzsky, 1902c: 14 (w.q.m.) TURKESTAN.

***ubangense.*** *Tetramorium camerunense* var. *ubangense* Santschi, 1937g: 82 (w.) ZAIRE. Raised to species: Bolton, 1980: 343.

*uelense* Santschi, 1923; see under ***DECAMORIUM.***

*uelensis. Xiphomyrmex uelensis* Santschi, 1935a: 267 (w.) ZAIRE. [Unresovled junior primary homonym of *uelense* Santschi, 1923e: 285, above.] Junior synonym of *pullulum:* Bolton, 1980: 273.

*ultor* Forel, 1913; see under ***DECAMORIUM.***

***umtaliense.*** *Tetramorium umtaliensis* Arnold, 1926: 256, fig. 69 (w.) ZIMBABWE. See also: Bolton, 1980: 261.

***unicum.*** *Tetramorium unicum* Bolton, 1980: 366 (w.) ZAIRE.

***urbanii.*** *Tetramorium urbanii* Bolton, 1977: 84, fig. 12 (w.) BHUTAN.

***validiusculum.*** *Tetramorium pacificum* subsp. *validiusculum* Emery, 1897c: 585 (w.) NEW GUINEA. Raised to species: Wilson & Taylor, 1967: 73. Senior synonym of *longicarinum:* Bolton, 1977: 104.

***vandalum.*** *Tetramorium vandalum* Bolton, 1977: 130, fig. 57 (w.q.) NEW GUINEA.

*vascoi. Tetramorium sericeiventre* var. *vascoi* Santschi, 1918b: 128 (w.q.) ZIMBABWE. Junior synonym of *sericeiventre:* Bolton, 1980: 332.

*venustus. Xiphomyrmex viehmeyeri* var. *venustus* Wheeler, W.M. 1934d: 147 (q.) AUSTRALIA. Junior synonym of *viehmeyeri:* Bolton, 1977: 142.

***vernicosum.*** *Tetramorium vernicosus* Radchenko, 1992a: 40, fig. 1 (w.q.) TADZHIKISTAN.

***versiculum.*** *Tetramorium versiculum* Bolton, 1980: 343 (w.) GHANA.

***vertigum.*** *Tetramorium vertigum* Bolton, 1977: 84, fig. 8 (w.) SULAWESI.

***vexator.*** *Tetramorium vexator* Arnold, 1926: 269, fig. 76 (w.) SOUTH AFRICA. See also: Bolton, 1980: 265.

*victoriensis. Tetramorium* (*Xiphomyrmex*) *humblotii* var. *victoriensis* Forel, 1913a: 120 (w.) ZIMBABWE. Arnold, 1917: 347 (q.m.). Junior synonym of *humbloti:* Bolton, 1980: 228.

***viehmeyeri.*** *Tetramorium* (*Xiphomyrmex*) *viehmeyeri* Forel, 1907h: 275 (q.) AUSTRALIA. Bolton, 1977: 142 (w.). Combination in *Xiphomyrmex:* Viehmeyer, 1925a: 29; in *Tetramorium*, and senior synonym of *venustus:* Bolton, 1977: 142.

*viridis. Macromischoides viridis* Weber, 1943c: 367, pl. 15, fig. 9 (w.q.m.) SUDAN. Junior synonym of *aculeatum:* Bolton, 1980: 353.

***viticola.*** *Tetramorium viticola* Weber, 1943c: 372, pl. 16, fig. 31 (w.) SUDAN. See also: Bolton, 1980: 366.

*vividum. Tetramorium sericeiventre* var. *vividum* Santschi, 1926b: 242 (w.) MOZAMBIQUE. Junior synonym of *sericeiventre:* Bolton, 1980: 332.

***vombis.*** *Triglyphothrix vombis* Bolton, 1976: 358, fig. 73 (w.q.) SOLOMON IS. Mann, 1919: 349 (m.). Combination in *Tetramorium:* Bolton, 1985: 247.

***wadje.*** *Tetramorium wadje* Bolton, 1980: 322 (w.q.) GHANA.

*waelbroeki. Tetramorium camerunense* var. *waelbroeki* Forel, 1909b: 53 (w.) ZAIRE. Junior synonym of *lucayanum:* Brown, 1964b: 131.

***wagneri.*** *Tetramorium wagneri* Viehmeyer, 1914b: 528, fig. 3 (w.) NEW GUINEA. See also: Bolton, 1977: 110.

***walshi.*** *Triglyphothrix walshi* Forel, 1890b: cvii (w.q.) INDIA. Imai, Baroni Urbani, *et al.* 1984: 8 (k.). Combination in *Tetramorium:* Bolton, 1985: 247. Senior synonym of *musculus* and *spuria:* Bolton, 1976: 358.

***warreni.*** *Tetramorium warreni* Arnold, 1926: 268, fig. 75 (w.) SOUTH AFRICA. See also: Bolton, 1980: 294.

*wasmanni. Macromischa wasmanni* Forel, 1901d: 300 (w.) ZAIRE. Mayr, 1902: 292 (q.m.). Combination in *Tetramorium:* Emery, 1908b: 187 (footnote); in *Macromischoides:* Wheeler, W.M. 1922a: 190. Subspecies/variety of *aculeatum:* Emery, 1908b: 187 (footnote); Forel, 1909b: 71; Santschi, 1910c: 385; Santschi, 1924b: 208. Junior synonym of *aculeatum:* Bolton, 1980: 353.

***weitzeckeri.*** *Tetramorium* (*Xiphomyrmex*) *weitzeckeri* Emery, 1895h: 39 (w.) SOUTH AFRICA. Arnold, 1917: 346 (q.). Combination in *Xiphomyrmex:* Wheeler, W.M. 1922a: 908; in *Tetramorium:* Bolton, 1980: 233. Senior synonym of *ebeninum, edithae, escherichi* and *nigellus:* Bolton, 1980: 233.

*wheeleri. Tetramorium* (*Xiphomyrmex*) *wheeleri* Forel, 1901c: 128 (w.) MEXICO. Subspecies of *spinosum:* Wheeler, W.M. 1915b: 416; Smith, M.R. 1938a: 130. Junior synonym of *spinosum:* Bolton, 1979: 163.

*willowsi. Xiphomyrmex costatus* subsp. *willowsi* Wheeler, W.M. 1934a: 177 (w.) SOLOMON IS. Junior synonym of *aspersum:* Bolton, 1977: 86.

*wilsoni. Tetramorium pacificum* var. *wilsoni* Mann, 1921: 460 (w.) FIJI IS. Raised to species: Taylor, 1967b: 1093. Junior synonym of *insolens:* Bolton, 1977: 99.

***xanthogaster.*** *Tetramorium* (*Xiphomyrmex*) *sikorae* var. *xanthogaster* Santschi, 1911e: 124 (w.) MADAGASCAR. [Original spelling of *xantogaster* justifiably emended to *xanthogaster* by Wheeler, W.M. 1922a: 1032.] Raised to species: Bolton, 1979: 139.

***xuthum.*** *Tetramorium xuthum* Bolton, 1980: 335 (w.) GHANA.

*yanoi. Tetramorium yanoi* Santschi, 1937h: 376, figs. 8–10 (w.) JAPAN. Junior synonym of *kraepelini:* Bolton, 1977: 117.

***yarthiellum.*** *Triglyphothrix yarthiella* Bolton, 1976: 338 (w.) ZAMBIA. Combination in *Tetramorium:* Bolton, 1985: 247.

***yerburyi.*** *Tetramorium* (*Xiphomyrmex*) *pilosum* r. *yerburyi* Forel, 1902c: 238 (w.) SRI LANKA. Raised to species: Bingham, 1903: 187. See also: Bolton, 1977: 85.

***youngi.*** *Tetramorium youngi* Bolton, 1980: 284 (w.) ANGOLA.

***zahrae.*** *Tetramorium zahrae* Santschi, 1923a: 135 (q.) MOROCCO.

***zambezium.*** *Tetramorium zambezium* Santschi, 1939b: 244 (w.) ZIMBABWE. Junior synonym of *delagoense:* Bolton, 1980: 311. Revived from synonymy: **revived status.**

***zapyrum.*** *Tetramorium zapyrum* Bolton, 1980: 242 (w.) GHANA.

***zenatum.*** *Tetramorium zenatum* Bolton, 1979: 151, fig. 19 (w.q.) MADAGASCAR.

***zonacaciae.*** *Xiphomyrmex zonacaciae* Weber, 1943c: 376, pl. 16, fig. 34 (w.) SUDAN. Combination in *Tetramorium:* Bolton, 1980: 234.

*zumpti. Macromischoides zumpti* Santschi, 1937b: 101, fig. 4 (w.) CAMEROUN. Junior synonym of *aculeatum:* Bolton, 1980: 353.

***zypidum.*** *Tetramorium zypidum* Bolton, 1977: 123, fig. 51 (w.) PHILIPPINES.

---

**EXCURSUS. Species-rank taxa of the *Tetramorium caespitum* group.**

Components of all species-groups in the genus *TETRAMORIUM*, except for the *caespitum* group, are noted in the various revisionary studies of Bolton (1976, 1977, 1979, 1980). Currently valid species-rank taxa in the *caespitum* group are as follows.

*aegeum, armatum, arzi, banyulense, biskrense, brevicorne, bursakovi, caespitum, calidum, chefteki, concaviceps, davidi, densopilosum, depressiceps, exasperatum, ferox, feroxoide, fezzanense, forte, goniommoide, hippocrate, hungaricum, impurum, inerme, jizane, juba, kabulistanicum, karakalense, kisilkumense, lucidulum, mai, maurum, meridionale, nautarum, nitidissimum, nursei, persignatum, punicum, pyrenaeicum, reticuligerum, ruginode, sahlbergi, schmidti, schneideri, semilaeve, splendens, striativentre, syriacum, taueret, tsushimae, turcomanicum, vernicosum, zahrae.*

---

*TETRAMYRMA:* see under ***LEPTOTHORAX.***

***TETRAPONERA*** (Old World tropics)

***aethiops.*** *Tetraponera aethiops* Smith, F. 1877b: 71 (w.) SOUTH AFRICA. Santschi, 1911g: 207 (m.); Forel, 1913c: 352 (q.); Wheeler, G.C. & Wheeler, J. 1956: 392 (l.). Combination in *Sima:* Dalla Torre, 1893: 53; in *Sima* (*Pachysima*): Emery, 1912b: 97; in *Pachysima:* Wheeler, W.M. 1918c: 305; in *Tetraponera:* Ward, 1990: 487. Senior synonym of *spininoda:* Emery, 1912b: 97; Wheeler, W.M. 1922a: 801. See also: Arnold, 1916: 175; Wheeler, W.M. 1922a: 114.

***aitkenii.*** *Sima aitkenii* Forel, 1902c: 245 (w.) INDIA. Wheeler, G.C. & Wheeler, J. 1956: 388 (l.). Combination in *S.* (*Tetraponera*): Emery, 1921f: 25; in *Tetraponera:* Chapman & Capco, 1951: 78.

***allaborans.*** *Pseudomyrma allaborans* Walker, 1859: 375 (w.q.m.) SRI LANKA. Wheeler, G.C. & Wheeler, J. 1956: 390 (l.); Hung, Imai & Kubota, 1972: 1024 (k.). Combination in *Tetraponera:* Smith, F. 1877b: 69; in *Sima:* Dalla Torre, 1893: 53; in *Sima* (*Tetraponera*): Viehmeyer, 1916a: 117; Karavaiev, 1933c: 260; in *Tetraponera:* Wheeler, W.M. 1921e: 110; Wheeler, W.M. 1930h: 61; Donisthorpe, 1948d: 592; Ward, 1990: 487. Senior synonym of *compressa:* Smith, F. 1877b: 69; of *compressa* (and its junior synonyms *ceylonica* and *femoralis*): Dalla Torre, 1893: 53; Bingham, 1903: 113; of *subtilis:* Dalla Torre, 1893: 53; Forel, 1903a: 709; Bingham, 1903: 113. Current subspecies: nominal plus *crassiuscula, longinoda, sumatrensis.* See also: Wheeler, W.M. & Bailey, 1920: 257.

***ambigua.*** *Sima ambigua* Emery, 1895h: 23 (w.q.) SOUTH AFRICA. Forel, 1910f: 5 (m.). Combination in *S.* (*Tetraponera*): Arnold, 1916: 185; in *Tetraponera:* Wheeler, W.M. 1922a: 796. [*Sima ambigua* Forel, 1894b: 89; *nomen nudum*, attributed to Emery.] Current subspecies: nominal plus *occidentalis, rhodesiana.*

*andamanensis* Forel, 1903 [*Sima*]; see under ***PARATOPULA.***

***andrei.*** *Sima andrei* Mayr, 1895: 144 (w.) MOZAMBIQUE. Combination in *Tetraponera:* Wheeler, W.M. 1922a: 796.

***#angolensis.*** *Tetraponera ophthalmica* st. *angolensis* Santschi, 1930b: 61, fig. 1 (w.) ANGOLA.

***angusta.*** *Sima angusta* Arnold, 1949: 266, fig. 6 (w.) ZIMBABWE. Combination in *Tetraponera:* Ward, 1990: 487.

****angustata.*** **Sima angustata* Mayr, 1868c: 102, pl. 5, fig. 106 (w.) BALTIC AMBER (Oligocene). Combination in *Tetrapomera:* Ward, 1990: 489.

***#angusticeps.*** *Sima* (*Tetraponera*) *bidentata* var. *angusticeps* Karavaiev, 1933c: 266, fig. 14 (w.) JAVA. Combination in *Tetraponera:* Chapman & Capco, 1951: 79.

***anthracina.*** *Sima anthracina* Santschi, 1910c: 355 (w.) CONGO. Terron, 1968: 339 (q.m.). Combination in *Sima:* Santschi, 1928d: 58; in *Tetraponera:* Wheeler, W.M. 1922a: 106; Menozzi, 1942: 166; Terron, 1968: 339. See also: Terron, 1977: 115.

***arrogans.*** *Sima arrogans* Santschi, 1911e: 117 (q.) MADAGASCAR. Combination in *S.* (*Tetraponera*): Emery, 1921f: 27; in *Tetraponera:* Wheeler, W.M. 1922a: 1014.

***atra.*** *Tetraponera atra* Donisthorpe, 1949b: 493 (q.) NEW GUINEA.

*atrata. Tetraponera atrata* Smith, F. 1852: 44, 3 figs. (w.) INDIA. Junior synonym of *nigra:* Dalla Torre, 1893: 54; Emery, 1895k: 464.

***attenuata.*** *Tetraponera attenuata* Smith, F. 1877b: 71 (w.) BORNEO. Wheeler, W.M. 1919e: 144 (q.m.). Combination in *Sima:* André, 1892b: 53; in *Tetraponera:* Wheeler, W.M. 1919e: 65. Senior synonym of *tenuissima:* Forel, 1912n: 55.

***bidentata.*** *Sima* (*Tetraponera*) *bidentata* Karavaiev, 1933c: 264, fig. 13 (w.) JAVA. Combination in *Tetraponera:* Chapman & Capco, 1951: 78. Current subspecies: nominal plus *angusticeps.* See also: Ward, 1990: 488.

***bifoveolata.*** *Sima bifoveolata* Mayr, 1895: 146 (w.) MOZAMBIQUE. Menozzi, 1926a: 37 (q.). Combination in *S.* (*Tetraponera*): Arnold, 1916: 184; in *Tetraponera:* Wheeler, W.M. 1922a: 796. Current subspecies: nominal plus *maculifrons, syriaca.*

***binghami.*** *Sima binghami* Forel, 1902c: 243 (w.q.m.) BURMA. Combination in *S.* (*Tetraponera*): Emery, 1921f: 25; in *Tetraponera:* Wheeler, W.M. 1921c: 531. Current subspecies: nominal plus *lindgreeni.*

***#biozellata.*** *Sima* (*Sima*) *mocquerysi* var. *biozellata* Karavaiev, 1931d: 42, fig. 1 (w.q.) KENYA. Combination in *Tetraponera:* Ward, 1990: 488.

***birmana.*** *Sima birmana* Forel, 1902c: 245 (q.) BURMA. Bingham, 1903: 112 (w.). Combination in *S.* (*Tetraponera*): Emery, 1921f: 25; in *Tetraponera:* Chapman & Capco, 1951: 79.

***braunsi.*** *Sima braunsi* Forel, 1913a: 112 (w.q.) SOUTH AFRICA. Combination in *S.* (*Tetraponera*): Arnold, 1916: 184; in *Tetraponera:* Wheeler, W.M. 1922a: 797. Subspecies/race of *clypeata:* Arnold, 1916: 184; Wheeler, W.M. 1922a: 797. Revived status as species: Ward, 1990: 488. Current subspecies: nominal plus *durbanensis.*

***brevicornis.*** *Sima brevicornis* Emery, 1900d: 675 (footnote) (w.) PHILIPPINES. Combination in *S.* (*Tetraponera*): Emery, 1921f: 25; in *Tetraponera:* Chapman & Capco, 1951: 79.

*bulawayana. Sima natalensis* r. *cuitensis* var. *bulawayana* Forel, 1913a: 112 (w.) ZIMBABWE. Unavailable name (Ward, 1990: 489.).

***#caffra.*** *Sima natalensis* st. *caffra* Santschi, 1914e: 15 (w.q.m.) SOUTH AFRICA. Combination in *Tetraponera:* Wheeler, W.M. 1922a: 799.

***capensis.*** *Pseudomyrma capensis* Smith, F. 1858b: 160 (w.) SOUTH AFRICA. Combination in *Sima:* Mayr, 1866b: 906; in *Tetraponera:* Wheeler, W.M. 1922a: 797.

***carbonaria.*** *Pseudomyrma carbonaria* Smith, F. 1863: 20 (w.q.) INDONESIA (Bouru I.). Combination in *Tetraponera:* Smith, F. 1877b: 69; in *Sima:* Dalla Torre, 1893: 54; in *Sima* (*Tetraponera*): Emery, 1921f: 25; in *Tetraponera:* Donisthorpe, 1932c: 471. Junior synonym of *nigra:* Dalla Torre, 1893: 54; Forel, 1903a: 709; Bingham, 1903: 110. Revived from synonymy as variety of *laeviceps:* Emery, 1900d: 674. Revived status as species: Emery, 1921f: 25. See also: Ward, 1990: 488.

*castanea. Viticicola tessmanni* var. *castanea* Wheeler, W.M. 1922a: 112 (w.q.) ZAIRE. Junior synonym of *tessmanni:* Brown, 1950e: 248.

***#ceylonensis.*** *Sima rufonigra* var. *ceylonensis* Forel, 1909e: 394 (w.q.) SRI LANKA. Combination in *Tetraponera:* Chapman & Capco, 1951: 81.

*ceylonica. Cerapachys ceylonica* Motschoulsky, 1863: 22 (w.) SRI LANKA. Junior synonym of *compressa:* Emery, 1892b: 167.

***claveaui.*** *Sima claveaui* Santschi, 1913c: 304 (w.) SENEGAL. Combination in *S.* (*Tetraponera*): Emery, 1921f: 27; in *Tetraponera:* Wheeler, W.M. 1922a: 797.

***clypeata.*** *Sima clypeata* Emery, 1886: 361, pl. 17, figs. 4, 5 (q.) SOUTH AFRICA. Emery, 1895h: 24 (w.); Emery, 1899c: 6 (l.); Wheeler, G.C. & Wheeler, J. 1956: 390 (l.). Combination in *S.* (*Tetraponera*): Arnold, 1916: 182; in *Tetraponera:* Wheeler, W.M. 1922a: 797.

*compressa. Sima compressa* Roger, 1863a: 179 (w.) SRI LANKA. Junior synonym of *allaborans:* Smith, F. 1877b: 69. Revived from synonymy, and senior synonym of *ceylonica* and *femoralis:* Emery, 1892b: 167; of *subtilis:* Emery, 1895k: 464. Junior synonym of *allaborans:* Dalla Torre, 1893: 53; Forel, 1903a: 709; Bingham, 1903: 113.

***#continua.*** *Sima penzigi* subsp. *continua* Forel, 1907c: 138 (w.) ETHIOPIA. Combination in *S.* (*Tetraponera*): Emery, 1921f: 28; in *Tetraponera:* Wheeler, W.M. 1922a: 800.

***#crassiuscula.*** *Sima allaborans* subsp. *crassiuscula* Emery, 1900d: 677, fig. 6 (w.) SUMATRA. Combination in *S.* (*Tetraponera*): Viehmeyer, 1916a: 117; Emery, 1921f: 25; in *Tetraponera:* Crawley, 1924: 390; Chapman & Capco, 1951: 78.

***#cuitensis.*** *Sima natalensis* subsp. *cuitensis* Forel, 1911d: 368 (w.) ANGOLA. Combination in *Tetraponera:* Wheeler, W.M. 1922a: 799.

***demens.*** *Sima demens* Santschi, 1911i: 282 (w.) MADAGASCAR. Combination in *S.* (*Tetraponera*): Emery, 1921f: 27; in *Tetraponera:* Wheeler, W.M. 1922a: 1014.

***dentifera.*** *Sima* (*Tetraponera*) *dentifera* Karavaiev, 1933c: 266 (q.) INDONESIA (Aru I.). Combination in *Tetraponera:* Chapman & Capco, 1951: 79.

***#deplanata.*** *Sima sahlbergi* r. *deplanata* Forel, 1904b: 375 (w.) MADAGASCAR. Combination in *S.* (*Tetraponera*): Emery, 1921f: 28; in *Tetraponera:* Wheeler, W.M. 1922a: 1016.

***diana.*** *Sima diana* Santschi, 1911e: 119, fig. 1 (q.) MADAGASCAR. Combination in *S.* (*Tetraponera*): Emery, 1921f: 27; in *Tetraponera:* Wheeler, W.M. 1922a: 1014.

***difficilis.*** *Sima difficilis* Emery, 1900d: 677 (w.) SUMATRA. Wheeler, W.M. 1919e: 144 (q.). Combination in *Tetraponera:* Wheeler, W.M. 1919e: 65. Current subspecies: nominal plus *longiceps.*

***dilatata.*** *Sima* (*Tetraponera*) *dilatata* Karavaiev, 1933c: 267, fig. 15 (w.) INDONESIA (Prinsen I.). Combination in *Tetraponera:* Chapman & Capco, 1951: 80.

***#dimidiata.*** *Sima hysterica* r. *dimidiata* Forel, 1895h: 487 (w.) MADAGASCAR. Combination in *S.* (*Tetraponera*): Emery, 1921f: 27; in *Tetraponera:* Wheeler, W.M. 1922a: 1015.

***#durbanensis.*** *Sima braunsi* var. *durbanensis* Forel, 1914d: 218 (w.) SOUTH AFRICA. Combination in *Tetraponera:* Wheeler, W.M. 1922a: 797.

***#elongata.*** *Sima mocquerysi* var. *elongata* Stitz, 1911b: 378, fig. 2 (w.) TANZANIA. Combination in *Tetraponera:* Wheeler, W.M. 1922a: 798.

***emacerata.*** *Sima mocquerysi* st. *emacerata* Santschi, 1911c: 352, fig. 1 (w.) KENYA. Santschi, 1914b: 70 (q.). Combination in *Tetraponera:* Wheeler, W.M. 1922a: 107. Subspecies of *mocquerysi:* Wheeler, W.M. 1922a: 798; Santschi, 1933b: 100. Raised to species: Santschi, 1920b: 9; Santschi, 1928d: 57; Bernard, 1953b: 221. Current subspecies: nominal plus *oberbecki, odiosa.*

***emeryi.*** *Sima emeryi* Forel, 1911d: 367 (w.) SOUTH AFRICA. Combination in *S.* (*Tetraponera*): Arnold,

***TETRAPONERA** (continued)*

1916: 187; in *Tetraponera:* Wheeler, W.M. 1922a: 797.

***encephala.*** *Sima encephala* Santschi, 1919h: 84 (q.) SENEGAL. Combination in *Tetraponera:* Wheeler, W.M. 1922a: 797.

*equidentata. Sima (Tetraponera) clypeata* r. *braunsi* var. *equidentata* Arnold, 1916: 184, pl. 5, fig. 48 (w.) SOUTH AFRICA. Unavailable name (Ward, 1990: 489).

***erythraea.*** *Sima ambigua* subsp. *erythraea* Emery, 1895h: 23 (w.) YEMEN. [*Sima ambigua* susbp. *erythraea* Emery, 1893f: 256; *nomen nudum.*] Combination in *S.* (*Tetraponera*): Emery, 1921f: 27; in *Tetraponera:* Wheeler, W.M. 1922a: 796. Raised to species: Collingwood, 1985: 242.

***exasciata.*** *Sima exasciata* Forel, 1892l: 261 (w.q.) MADAGASCAR. Combination in *S.* (*Tetraponera*): Emery, 1921f: 27; in *Tetraponera:* Wheeler, W.M. 1922a: 1014.

*femoralis. Cerapachys femoralis* Motschoulsky, 1863: 21 (w.) SRI LANKA. Junior synonym of *compressa:* Emery, 1892b: 167.

#***fergusoni.*** *Sima nigra* r. *fergusoni* Forel, 1902c: 248 (w.) INDIA. Combination in *S.* (*Tetraponera*): Emery, 1921f: 26; in *Tetraponera:* Chapman & Capco, 1951: 81. Raised to species: Bingham, 1903: 114; Collingwood, 1962: 225. Subspecies of *nigra:* Emery, 1921f: 26; Ward, 1990: 488.

***fictrix.*** *Sima fictrix* Forel, 1897c: 198 (q.) MADAGASCAR. Combination in *S.* (*Tetraponera*): Emery, 1921f: 27; in *Tetraponera:* Wheeler, W.M. 1922a: 1014.

***flexuosa.*** *Sima flexuosa* Santschi, 1911e: 120 (w.) MADAGASCAR. Combination in *S.* (*Tetraponera*): Emery, 1921f: 27; in *Tetraponera:* Wheeler, W.M. 1922a: 1014.

***fulva.*** *Sima* (*Tetraponera*) *fulva* Viehmeyer, 1916a: 117, fig. 2 (w.q.) SINGAPORE. Combination in *Tetraponera:* Chapman & Capco, 1951: 80.

***gerdae.*** *Sima gerdae* Stitz, 1911b: 381, fig. 4 (q.) TANZANIA. Combination in *S.* (*Tetraponera*): Emery, 1921f: 27; in *Tetraponera:* Wheeler, W.M. 1922a: 797.

***grandidieri.*** *Sima grandidieri* Forel, 1891b: 203, pl. 5, fig. 3 (w.) MADAGASCAR. Forel, 1891b: 229 (q.m.). Combination in *Tetraponera:* Wheeler, W.M. 1922a: 1014. Current subspecies: nominal plus *hildebrandti, variegata.*

*grisea* Forel, 1910 [*Sima*]; see under ***PLATYTHYREA.***

#***hildebrandti.*** *Sima grandidieri* var. *hildebrandti* Forel, 1891b: 204 (w.) MADAGASCAR. Combination in *Tetraponera:* Wheeler, W.M. 1922a: 1014. Raised to species: Dalla Torre, 1893: 54. Variety of *grandidieri:* Wheeler, W.M. 1922a: 1014.

***humerosa.*** *Sima humerosa* Emery, 1900d: 674 (footnote) (w.) NEW GUINEA. Combination in *S.* (*Tetraponera*): Mann, 1919: 306; in *Tetraponera:* Chapman & Capco, 1951: 80.

***hysterica.*** *Sima hysterica* Forel, 1892l: 258 (w.) MADAGASCAR. Combination in *S.* (*Tetraponera*): Emery, 1921f: 27; in *Tetraponera:* Wheeler, W.M. 1922a: 1015. Current subspecies: nominal plus *dimidiata, inflata.*

#***illota.*** *Sima triangularis* st. *illota* Santschi, 1914d: 334 (w.q.) NIGERIA. Combination in *Tetraponera:* Wheeler, W.M. 1922a: 801.

#***inflata.*** *Sima sahlbergi* var. *inflata* Emery, 1899f: 273 (w.) MADAGASCAR. Combination in *S.* (*Tetraponera*): Emery, 1921f: 27; in *Tetraponera:* Wheeler, W.M. 1922a: 1015. Variety of *hysterica:* Forel, 1904b: 376.

#***insularis.*** *Sima nigra* var. *insularis* Emery, 1901f: 113 (w.q.) SRI LANKA. Combination in *S.* (*Tetraponera*): Emery, 1921f: 26; in *Tetraponera:* Chapman & Capco, 1951: 80.

#***kimberleyensis.*** *Sima punctulata* var. *kimberleyensis* Forel, 1915b: 37 (w.) AUSTRALIA. Combination in *S.* (*Tetraponera*): Emery, 1921f: 26; in *Tetraponera:* Taylor & Brown, D.R. 1985: 18.

****klebsi.*** **Sima klebsi* Wheeler, W.M. 1915h: 41, fig. 10 (w.) BALTIC AMBER (Oligocene). Combination in *Tetraponera:* Ward, 1990: 489.

#***krama.*** *Sima nigra* var. *krama* Forel, 1912d: 105 (q.) JAVA. Combination in *S.* (*Tetraponera*): Emery, 1921f: 26; in *Tetraponera:* Chapman & Capco, 1951: 80.

****lacrimarum.*** **Sima lacrimarum* Wheeler, W.M. 1915h: 44, fig. 13 (w.) BALTIC AMBER (Oligocene). Combination in *Tetraponera:* Ward, 1990: 489.

***laeviceps.*** *Pseudomyrma laeviceps* Smith, F. 1859a: 145 (w.) INDONESIA (Aru I.). Combination in *Sima:* Mayr, 1876: 111; in *S.* (*Tetraponera*): Emery, 1921f: 26; in *Tetraponera:* Smith, F. 1877b: 69; Donisthorpe, 1932c: 455. [Misspelled as *leviceps* by Dalla Torre, 1893: 54.]

***latifrons.*** *Sima* (*Pachysima*) *latifrons* Emery, 1912b: 98, fig. 4 (q.) GABON. Santschi, 1914c: 288 (w.m.); Wheeler, G.C. & Wheeler, J. 1956: 396 (l.). Combination in *Pachysima:* Wheeler, W.M. 1918c: 305; in *Tetraponera:* Ward, 1990: 488. See also: Wheeler, W.M. 1922a: 308.

***ledouxi.*** *Tetraponera ledouxi* Terron, 1969: 629, figs. 1–5 (w.q.m.) CAMEROUN.

***lemoulti.*** *Sima lemoulti* Santschi, 1920d: 375 (q.) CONGO. Combination in *Tetraponera:* Wheeler, W.M. 1922a: 798.

#***lepida.*** *Tetraponera mocquerysi* var. *lepida* Wheeler, W.M. 1922a: 106 (w.q.) ZAIRE.

***liengmei.*** *Sima liengmei* Forel, 1894b: 88 (w.m.) MOZAMBIQUE. Combination in *S.* (*Tetraponera*): Arnold, 1916: 181; in *Tetraponera:* Wheeler, W.M. 1922a: 798.

#***lindgreeni.*** *Sima binghami* var. *lindgreeni* Forel, 1902c: 245 (w.) INDIA. Combination in *S.* (*Tetraponera*): Emery, 1921f: 25; in *Tetraponera:* Chapman & Capco, 1951: 79.

#***longiceps.*** *Sima difficilis* r. *longiceps* Forel, 1902c: 247 (w.) INDIA. Viehmeyer, 1916a: 118 (m.). Combination in *S.* (*Tetraponera*): Viehmeyer, 1916a: 118; Emery, 1921f: 26; in *Tetraponera:* Chapman & Capco, 1951: 79. Raised to species: Bingham, 1903: 115. Subspecies of *difficilis:* Emery, 1921f: 26; Ward, 1990: 488.

#***longinoda.*** *Sima allaborans* var. *longinoda* Forel, 1909e: 395 (w.) SRI LANKA. Combination in *S.* (*Tetraponera*): Emery, 1921f: 25; in *Tetraponera:* Chapman & Capco, 1951: 78.

#***longula.*** *Sima sahlbergi* var. *longula* Emery, 1895f: 340 (w.) MADAGASCAR. Combination in *S.* (*Tetraponera*): Emery, 1921f: 28; in *Tetraponera:* Wheeler, W.M. 1922a: 1016.

#***lutea.*** *Sima mocquerysi* var. *lutea* Stitz, 1911b: 381, fig. 3 (w.q.) ZAIRE. Combination in *Tetraponera:* Wheeler, W.M. 1922a: 798.

#***maculifrons.*** *Sima bifoveolata* st. *maculifrons* Santschi, 1912b: 162 (w.) SOMALIA. [Name misspelled as *foveolata* in original description.] Combination in *Sima* (*Tetraponera*): Emery, 1921f: 27; in *Tetraponera:* Wheeler, W.M. 1922a: 796.

***maffini.*** *Tetraponera* (*Tetraponera*) *maffini* Donisthorpe, 1948d: 591 (q.) NEW GUINEA.

***mandibularis.*** *Sima mandibularis* Emery, 1895f: 340, fig. 1 (q.) MADAGASCAR. Combination in *S.* (*Tetraponera*): Emery, 1921f: 27; in *Tetraponera:* Wheeler, W.M. 1922a: 1015.

***mayri.*** *Sima mayri* Forel, 1901h: 53 (q.) CAMEROUN. Combination in *S.* (*Tetraponera*): Emery, 1921f: 27; in *Tetraponera:* Wheeler, W.M. 1922a: 798.

***microcarpa.*** *Tetraponera microcarpa* Wu & Wang, 1990: 515, fig. 1 (w.) CHINA.

***minuta.*** *Eciton minutum* Jerdon, 1851: 112 (w.) INDIA. Combination in *Pseudomyrma:* Smith, F. 1858b: 159; in *Sima:* Dalla Torre, 1893: 54; in *Sima* (*Tetraponera*): Emery, 1921f: 26; in *Tetraponera:* Chapman & Capco, 1951: 80. See also: Bingham, 1903: 116.

***mocquerysi.*** *Sima mocquerysi* André, 1890: 319 (w.) SIERRA LEONE. Combination in *Tetraponera:* Wheeler, W.M. 1922a: 106. Current subspecies: nominal plus *biozellata, elongata, lepida, lutea.*

***modesta.*** *Pseudomyrma modesta* Smith, F. 1860b: 106 (w.) INDONESIA (Batjan I.). Combination in *Sima:* Dalla Torre, 1893: 54; in *S.* (*Tetraponera*): Emery, 1921f: 26; in *Tetraponera:* Donisthorpe, 1932c: 462.

***monardi.*** *Sima monardi* Santschi, 1937d: 218, figs. 13, 14 (w.) ANGOLA. Combination in *Tetraponera:* Ward, 1990: 488.

***morondaviensis.*** *Sima sahlbergii* r. *morondaviensis* Forel, 1891b: 206 (w.) MADAGASCAR. Combination in *S.* (*Tetraponera*): Emery, 1921f: 28; in *Tetraponera:* Wheeler, W.M. 1922a: 1016. Subspecies of *sahlbergii:* Emery, 1921f: 28; Wheeler, W.M. 1922a: 1016. Raised to species: Dalla Torre, 1893: 54; Ward, 1991: 338, 346.

***nasuta.*** *Tetraponera nasuta* Bernard, 1953b: 222, fig. 7 (q.) GUINEA. Terron, 1971: 74 (s.w.m.).

***natalensis.*** *Pseudomyrma natalensis* Smith, F. 1858b: 160 (q.) SOUTH AFRICA. Emery, 1895h: 22 (w.); Emery, 1899c: 6 (l.); Wheeler, G.C. & Wheeler, J. 1973d: 208 (l.). Combination in *Sima:* Emery, 1892e: 237; in *Tetraponera:* Smith, F. 1877b: 69; Wheeler, W.M. 1922a: 798. Junior synonym of *capensis:* Gerstäcker, 1871: 355. Revived from synonymy: Smith, F. 1877b: 69. Current subspecies: nominal plus *caffra, cuitensis, obscurata, usambarensis.* See also: Arnold, 1916: 176.

#***nebulosa.*** *Sima siggii* var. *nebulosa* Forel, 1903d: 404 (w.) NICOBAR IS. Combination in *S.* (*Tetraponera*): Emery, 1921f: 27; in *Tetraponera:* Chapman & Capco, 1951: 82.

***nicobarensis.*** *Sima nicobarensis* Forel, 1903d: 402 (w.) NICOBAR IS. Combination in *Tetraponera:* Chapman & Capco, 1951: 80.

***nigra.*** *Eciton nigrum* Jerdon, 1851: 112 (w.q.) INDIA. Combination in *Pseudomyrma:* Smith, F. 1858b: 159;

in *Sima:* Roger, 1863b: 25; Stitz, 1925: 117; in *Tetraponera:* Smith, F. 1877b: 68; Wheeler, W.M. 1919e: 64; Donisthorpe, 1942d: 454. Senior synonym of *atrata:* Dalla Torre, 1893: 54; Emery, 1895k: 464. Current subspecies: nominal plus *fergusoni, insularis, krama.*

**nitens.** *Sima nitens* Stitz, 1925: 117 (w.q.) PHILIPPINES. Combination in *Tetraponera:* Chapman & Capco, 1951: 81.

**nitida.** *Pseudomyrma nitida* Smith, F. 1860b: 106 (w.) INDONESIA (Batjan I.). Combination in *Sima:* Dalla Torre, 1893: 54; in *S. (Tetraponera)*: Emery, 1921f: 26; in *Tetraponera:* Donisthorpe, 1932c: 462.

#**oberbecki.** *Sima oberbecki* Forel, 1911f: 275 (w.) ZAIRE. Forel, 1916: 403 (q.). Combination in *S. (Tetraponera)*: Emery, 1921f: 28; in *Tetraponera:* Wheeler, W.M. 1922a: 800. Variety of *emacerata:* Santschi, 1928d: 58.

#**obscurata.** *Sima natalensis* subsp. *obscurata* Emery, 1895h: 22 (w.) SOUTH AFRICA. Arnold, 1916: 177 (q.m.). Combination in *Tetraponera:* Wheeler, W.M. 1922a: 799.

***ocellata.** **Sima ocellata* Mayr, 1868c: 101, pl. 5, figs. 104, 105 (w.) BALTIC AMBER (Oligocene). Combination in *Tetraponera:* Ward, 1990: 489.

#**occidentalis.** *Tetraponera ambigua* var. *occidentalis* Menozzi, 1934: 154 (w.) ALGERIA. [First available use of *Sima ambigua* r. *erythraea* var. *occidentalis* Stitz, 1917: 336. Unavailable name (Ward, 1990: 489).]

#**odiosa.** *Sima prelli* var. *odiosa* Forel, 1916: 403 (w.) ZAIRE. Combination in *S. (Tetraponera)*: Emery, 1921f: 28; in *Tetraponera:* Wheeler, W.M. 1922a: 800. Variety of *emacerata:* Santschi, 1928d: 57.

***oligocenica.** **Sima oligocenica* Théobald, 1937b: 127, pl. 10, fig. 16 (caste?) FRANCE (Oligocene). Combination in *Tetraponera:* Ward, 1990: 489.

**ophthalmica.** *Sima ophthalmica* Emery, 1912b: 98 (w.) CAMEROUN. Combination in *S. (Tetraponera)*: Emery, 1921f: 28; in *Tetraponera:* Wheeler, W.M. 1922a: 107. Current subspecies: nominal plus *angolensis, tenebrosa, unidens.*

**penzigi.** *Sima penzigi* Mayr, 1907b: 10 (w.q.m.) TANZANIA. Combination in *S. (Tetraponera)*: Arnold, 1916: 180; in *Tetraponera:* Wheeler, W.M. 1922a: 800. Current subspecies: nominal plus *continua, praestigiatrix.*

**perlonga.** *Tetraponera perlonga* Santschi, 1928d: 59, fig. 2 (q.) MADAGASCAR.

**petiolata.** *Tetraponera petiolata* Smith, 1877b: 70 (w.q.m.) SRI LANKA. Combination in *Sima:* Emery, 1887b: 445 (footnote); in *Sima (Tetraponera)*: Emery, 1921f: 26; in *Tetraponera:* Chapman & Capco, 1951: 81. Junior synonym of *nigra:* Forel, 1903a: 709. Revived from synonymy: Bingham, 1903: 113.

**pilosa.** *Pseudomyrma pilosa* Smith, F. 1858b: 160 (w.) BORNEO. Forel, 1915a: 23 (q.); Viehmeyer, 1916a: 117 (m.). Combination in *Sima:* Emery, 1887b: 449; in *Tetraponera:* Wheeler, W.M. 1919e: 65.

**platynota.** *Sima (Tetraponera) platynota* Karavaiev, 1933c: 269, fig. 16 (w.) INDONESIA (Wammer I.). Combination in *Tetraponera:* Chapman & Capco, 1951: 81.

**plicatidens.** *Sima (Tetraponera) plicatidens* Santschi, 1926a: 26 (q.m.) MADAGASCAR. Combination in *Tetraponera:* Ward, 1990: 489.

**poultoni.** *Tetraponera poultoni* Donisthorpe, 1931b: 497 (w.) UGANDA.

#**praestigiatrix.** *Tetraponera penzigi* var. *praestigiatrix* Santschi, 1937a: 50 (q.) TANZANIA.

**prelli.** *Sima prelli* Forel, 1911d: 365 (w.) TANZANIA Combination in *S. (Tetraponera)*: Emery, 1921f: 28; in *Tetraponera:* Wheeler, W.M. 1922a: 800.

**punctulata.** *Tetraponera punctulata* Smith, F. 1877b: 72 (q.) AUSTRALIA. Combination in *Sima:* Dalla Torre, 1893: 55; in *S. (Tetraponera)*: Emery, 1921f: 26; in *Tetraponera:* Chapman & Capco, 1951: 81. Current subspecies: nominal plus *kimberleyensis.*

*quaniama. Sima natalensis* st. *cuitensis* var. *quaniama* Santschi, 1937d: 218 (w.) ANGOLA. Unavailable name (Ward, 1990: 489).

**rakotonis.** *Sima rakotonis* Forel, 1891b: 206, pl. 5, fig. 7 (w.) MADAGASCAR. Combination in *S. (Tetraponera)*: Emery, 1921f: 28; in *Tetraponera:* Wheeler, W.M. 1922a: 1015.

#**rhodesiana.** *Sima ambigua* r. *rhodesiana* Forel, 1913a: 112 (w.) ZIMBABWE. Arnold, 1914: 31 (s.w.); Arnold, 1916: 187 (q.). Combination in *S. (Tetraponera)*: Arnold, 1916: 186; in *Tetraponera:* Wheeler, W.M. 1922a: 796.

**rufipes.** *Eciton rufipes* Jerdon, 1851: 112 (w.) INDIA. Combination in *Pseudomyrma:* Smith, F. 1858b: 159; in *Sima:* Dalla Torre, 1893: 55; in *S. (Tetraponera)*: Emery, 1921f: 26; in *Tetraponera:* Chapman & Capco, 1951: 81. Junior synonym of *allaborans:* Forel, 1903a: 709. Revived from synonymy: Emery, 1921f: 26.

**rufonigra.** *Eciton rufonigrum* Jerdon, 1851: 111 (w.) INDIA. Smith, F. 1875: 35 (q.). Combination in *Pseudomyrma:* Smith, F. 1858b: 159; in *Sima:* Roger, 1863b: 25; Emery, 1921f: 24; in *Tetraponera:* Smith, F. 1877b: 68; Wheeler, W.M. 1922a: 1015; Ward, 1990: 489. Current subspecies: nominal plus *ceylonensis, testaceonigra, yeensis.* See also: Bingham, 1903: 108.

**sahlbergii.** *Sima sahlbergii* Forel, 1887: 386 (w.) MADAGASCAR. Forel, 1892l: 260 (q.). Combination in *S. (Tetraponera)*: Emery, 1921f: 28; in *Tetraponera:* Wheeler, W.M. 1922a: 1015. Current subspecies: nominal plus *deplanata, longula, spuria.*

**schulthessi.** *Sima schulthessi* Santschi, 1915c: 249 (q.) MOZAMBIQUE. Santschi, 1928d: 58 (w.). Combination in *Tetraponera:* Wheeler, W.M. 1922a: 800.

**scotti.** *Tetraponera scotti* Donisthorpe, 1931b: 498 (w.q.) ETHIOPIA.

#**setifera.** *Sima (Tetraponera) siggi* var. *setifera* Viehmeyer, 1916a: 119 (w.) SINGAPORE. Combination in *Tetraponera:* Chapman & Capco, 1951: 82.

**siggi.** *Sima siggi* Forel, 1902c: 246 (w.) THAILAND. Forel, 1909d: 226 (q.). Combination in *S. (Tetraponera)*: Viehmeyer, 1916a: 119; in *Tetraponera:* Chapman & Capco, 1951: 82. Current subspecies: nominal plus *nebulosa, setifera.*

***simplex.** **Sima simplex* Mayr, 1868c: 102 (w.) BALTIC AMBER (Oligocene). Combination in *Tetraponera:* Ward, 1990: 489.

*spininoda. Sima spininoda* André, 1892a: 51 (w.q.) GABON. Junior synonym of *aethiops:* Emery, 1912b: 97.

#**spuria.** *Sima sahlbergi* r. *spuria* Forel, 1897c: 199 (w.m.) MADAGASCAR. Combination in *S. (Tetraponera)*: Emery, 1921f: 28; in *Tetraponera:* Wheeler, W.M. 1922a: 1016.

**stipitum.** *Sima stipitum* Forel, 1912n: 54 (w.) SINGAPORE. Combination in *S. (Tetraponera)*: Emery, 1921f: 27; in *Tetraponera:* Chapman & Capco, 1951: 82.

*subtilis. Sima subtilis* Emery, 1899b: 500, pl. 11, figs. 24, 25 (w.) BURMA. Junior synonym of *compressa:* Emery, 1895k: 464; of *allaborans:* Dalla Torre, 1893: 53; Forel, 1903a: 709; Bingham, 1903: 113.

#**sumatrensis.** *Sima allaborans* var. *sumatrensis* Emery, 1900d: 676, fig. 6 (w.q.) SUMATRA. Viehmeyer, 1916a: 117 (q.). Combination in *S. (Tetraponera)*: Viehmeyer, 1916a: 117; in *Tetraponera:* Chapman & Capco, 1951: 78.

#**syriaca.** *Sima bifoveolata* var. *syriaca* Wheeler, W.M. & Mann, 1916: 167 (w.) EGYPT. Combination in *S. (Tetraponera)*: Emery, 1921f: 27; in *Tetraponera:* Ward, 1990: 489.

#**tenebrosa.** *Tetraponera ophthalmica* var. *tenebrosa* Santschi, 1928d: 61 (w.) ZAIRE.

*tenuissima. Sima attenuata* var. *tenuissima* Emery, 1900d: 675, fig. 5 (w.q.) SUMATRA. Junior synonym of *attenuata:* Forel, 1912n: 55.

**tessmanni.** *Sima tessmanni* Stitz, 1910: 131, fig. 2 (w.) EQUATORIAL GUINEA. Santschi, 1919h: 84 (q.); Wheeler, W.M. 1922a: 110 (m.); Wheeler, W.M. 1918c: 304 (l.); Wheeler, G.C. & Wheeler, J. 1956: 391 (l.). Combination in *Tetraponera:* Wheeler, W.M. 1918c: 303; in *Viticicola:* Wheeler, W.M. 1919i: 130; Wheeler, W.M. 1922a: 109; in *Sima (Tetraponera)*: Emery, 1921f: 28; in *Tetraponera:* Ward, 1990: 489. Senior synonym of *castanea:* Brown, 1950e: 248.

*testacea* Smith, F. 1852; see under ***PSEUDOMYRMEX.***

#**testaceonigra.** *Sima rufonigra* var. *testaceonigra* Forel, 1903d: 402 (w.) NICOBAR IS. Combination in *Tetraponera:* Chapman & Capco, 1951: 82.

**thagatensis.** *Sima attenuata* var. *thagatensis* Forel, 1902c: 249 (w.) INDIA. Combination in *S. (Tetraponera)*: Emery, 1921f: 27; in *Tetraponera:* Wheeler, W.M. 1924b: 5; Chapman & Capco, 1951: 82. Junior synonym of *attenuata:* Bingham, 1903: 112. Revived from synonymy as subspecies of *nigra:* Forel, 1909d: 226. Raised to species: Forel, 1912n: 55.

**triangularis.** *Sima triangularis* Stitz, 1910: 131 (w.) EQUATORIAL GUINEA. Combination in *Tetraponera:* Wheeler, W.M. 1922a: 800. Current subspecies: nominal plus *illota.*

*umbrata. Sima (Tetraponera) bifoveolata* st. *maculifrons* var. *umbrata* Santschi, 1929c: 98 (w.) ALGERIA. Unavailable name (Ward, 1990: 489.).

#**unidens.** *Tetraponera ophthalmica* st. *unidens* Santschi, 1928d: 60, fig. 2 (q.) ZAIRE.

#**usambarensis.** *Sima natalensis* var. *usambarensis* Forel, 1911d: 367 (w.) TANZANIA. Santschi, 1914b: 70 (q.). Combination in *Tetraponera:* Wheeler, W.M. 1922a: 799.

#**variegata.** *Sima grandidieri* var. *variegata* Forel, 1895h: 487 (w.) MADAGASCAR. Santschi, 1926a: 27 (q.). Combination in *Tetraponera:* Wheeler, W.M. 1922a: 1014.

#**yeensis.** *Sima rufonigra* var. *yeensis* Forel, 1902c: 248 (q.) BURMA. Combination in *Tetraponera:* Chapman & Capco, 1951: 82.

***zavattarii.*** *Sima* (*Tetraponera*) *zavattarii* Menozzi, 1939c: 99, fig. 1 (w.) ETHIOPIA. Combination in *Tetraponera:* Ward, 1990: 489.

*TETROGMUS:* see under ***TETRAMORIUM.***

***THAUMATOMYRMEX*** (Neotropical)

***atrox.*** *Thaumatomyrmex atrox* Weber, 1939a: 98, fig. 3 (w.) GUIANA. Senior synonym of *manni* and *zeteki:* Longino, 1988: 38. See also: Kempf, 1975b: 122; Brandão, Diniz & Tomotake, 1991: 339.

***cochlearis.*** *Thaumatomyrmex cochlearis* Creighton, 1928: 163, fig. 1 (w.) CUBA. See also: Kempf, 1975b: 110.

***contumax.*** *Thaumatomyrmex contumax* Kempf, 1975b: 108, figs. 13, 22, 33 (w.) BRAZIL. See also: Brandão, Diniz & Tomotake, 1991: 336.

***ferox.*** *Thaumatomyrmex ferox* Mann, 1922: 3, fig. 1 (w.) HONDURAS. Senior synonym of *paludis:* Longino, 1988: 38. See also: Kempf, 1975b: 119.

*manni. Thaumatomyrmex manni* Weber, 1939a: 99 (w.) BOLIVIA. Junior synonym of *atrox:* Longino, 1988: 38. See also: Kempf, 1975b: 123.

***mutilatus.*** *Thaumatomyrmex mutilatus* Mayr, 1887: 531 (w.) BRAZIL. Kempf, 1954: 48 (m.); Wheeler, G.C. & Wheeler, J. 1964b: 450 (l.). See also: Kempf, 1975b: 103.

*paludis. Thaumatomyrmex paludis* Weber, 1942b: 68, figs. 1, 2 (w.) VENEZUELA. Junior synonym of *ferox:* Longino, 1988: 38. See also: Kempf, 1975b: 114.

*zeteki. Thaumatomyrmex zeteki* Smith, M.R. 1944b: 98 (w.) PANAMA. Junior synonym of *atrox:* Longino, 1988: 38. See also: Kempf, 1975b: 116.

*THERYELLA:* see under ***STENAMMA.***

#***THLIPSEPINOTUS:*** see under ***CAMPONOTUS.***

***TINGIMYRMEX*** (Neotropical)

***mirabilis.*** *Strumigenys* (*Tingimyrmex*) *mirabilis* Mann, 1926: 105, fig. 1 (w.) BOLIVIA. Combination in *Tingimyrmex:* Brown, 1948e: 111.

*TITUSIA: nomen nudum.*

*TOMOGNATHUS:* see under ***HARPAGOXENUS.***

*TRACHYMESOPUS:* see under ***PACHYCONDYLA.***

***TRACHYMYRMEX*** (Neotropical, Nearctic)

*abortivus. Trachymyrmex abortivus* Wheeler, W.M. 1925a: 40 (w.) GUIANA. Junior synonym of *bugnioni:* Weber, 1967: 273.

***agudensis.*** *Trachymyrmex agudensis* Kempf, 1967a: 124, figs. 4–6 (w.) BRAZIL.

*annulatus. Trachymyrmex annulatus* Santschi, 1936b: 201, figs. 6–9 (w.) PANAMA. Junior synonym of *cornetzi:* Weber, 1958b: 49.

#***antiguensis.*** *Trachymyrmex jamaicensis* subsp. *antiguensis* Weber, 1938b: 196 (w.) ANTILLES IS.

***arizonensis.*** *Atta* (*Trachymyrmex*) *arizonensis* Wheeler, W.M. 1907c: 710, pl. 49, figs. 9, 10 (q.m.) U.S.A. Wheeler, W.M. 1911e: 93 (w.). Combination in *Trachymyrmex:* Gallardo, 1916b: 242.

*attaxenus* Menozzi, 1936; see under ***MYCOCEPURUS.***

*balboai. Trachymyrmex balboai* Weber, 1940a: 424 (w.) PANAMA. Junior synonym of *zeteki:* Weber, 1958b: 55.

*bivittatus. Trachymyrmex cornetzi* var. *bivittatus* Wheeler, W.M. 1922c: 13 (w.) TRINIDAD. Subspecies of *cornetzi:* Weber, 1945: 54. Junior synonym of *cornetzi:* Weber, 1958b: 49.

*brevispinosus. Trachymyrmex cornetzi* subsp. *brevispinosus* Weber, 1945: 55 (footnote) (w.) COLOMBIA. Junior synonym of *cornetzi:* Weber, 1958b: 49.

***bugnioni.*** *Atta* (*Trachymyrmex*) *bugnioni* Forel, 1912e: 186 (w.) COLOMBIA. Combination in *Trachymyrmex:* Gallardo, 1916b: 242. Senior synonym of *abortivus, cucumis, humilis, morgani, schomburgki* and *tucuche:* Weber, 1967: 273.

***carib.*** *Trachymyrmex carib* Weber, 1945: 61 (w.) TRINIDAD.

#***caroli.*** *Atta* (*Trachymyrmex*) *turrifex* subsp. *caroli* Wheeler, W.M. 1911g: 248 (w.) U.S.A. Combination in *Trachymyrmex:* Creighton, 1950a: 324.

#***cordovanus.*** *Acromyrmex* (*Trachymyrmex*) *iheringi* var. *cordovana* Bruch, 1921: 202 (w.) ARGENTINA. Combination in *Trachymyrmex:* Kempf, 1972a: 254. Stirps of *tucumanus:* Santschi, 1922b: 358.

***cornetzi.*** *Atta* (*Trachymyrmex*) *cornetzi* Forel, 1912e: 183 (w.q.) COLOMBIA. Combination in *Trachymyrmex:* Gallardo, 1916b: 242. Senior synonym of *annulatus, bivittatus, brevispinosus, gatun, naranjo* and *uncifer:* Weber, 1958b: 49.

*crystallina. Atta* (*Trachymyrmex*) *septentrionalis* subsp. *obscurior* var. *crystallina* Wheeler, W.M. 1911g: 247 (w.q.m.) U.S.A. Unavailable name; material referred to *obscurior* by Creighton, 1950a: 324.

#***cubaensis.*** *Trachymyrmex jamaicensis* subsp. *cubaensis* Wheeler, W.M. 1937b: 459 (w.) CUBA.

*cucumis. Myrmicocrypta cucumis* Mann, 1922: 45 (w.q.) HONDURAS. Combination in *Trachymyrmex:* Weber, 1958b: 50. Junior synonym of *bugnioni:* Weber, 1967: 273.

***desertorum.*** *Atta* (*Trachymyrmex*) *desertorum* Wheeler, W.M. 1911e: 98, fig. 2 (w.) U.S.A. Combination in *Trachymyrmex:* Creighton, 1950a: 321.

***dichrous.*** *Trachymyrmex dichrous* Kempf, 1967a: 126, figs. 7–9 (w.) BRAZIL.

***diversus.*** *Trachymyrmex diversus* Mann, 1916: 454 (w.) BRAZIL. Wheeler, G.C. & Wheeler, J. 1974d: 79 (l.). Senior synonym of *phippsi:* Weber, 1967: 274.

***echinus.*** *Trachymyrmex echinus* Weber, 1938b: 191, fig. 3 (w.) BOLIVIA.

***farinosus.*** *Atta* (*Trachymyrmex*) *farinosa* Emery, 1894c: 221 (w.) BRAZIL. Combination in *Trachymyrmex:* Gallardo, 1916b: 242. Senior synonym of *trifurcatus:* Weber, 1958b: 51.

***fiebrigi.*** *Cyphomyrmex* (*Trachymyrmex*) *fiebrigi* Santschi, 1916e: 384, fig. (w.) PARAGUAY. Combination in *Trachymyrmex:* Kempf, 1972a: 252.

#***fitzgeraldi.*** *Trachymyrmex relictus* subsp. *fitzgeraldi* Weber, 1937: 401 (w.) TRINIDAD.

#***fracticornis.*** *Trachymyrmex tucumana* var. *fracticornis* Santschi, 1925e: 164 (w.) ARGENTINA.

#***frontalis.*** *Trachymyrmex jamaicensis* var. *frontalis* Santschi, 1925d: 238 (w.) HAITI.

*fusca. Atta* (*Trachymyrmex*) *urichi* subsp. *fusca* Emery, 1894c: 222 (w.) BRAZIL. Junior synonym of *urichii:* Weber, 1958b: 53.

***gaigei.*** *Acromyrmex* (*Trachymyrmex*) *gaigei* Forel, 1914c: 617 (w.) COLOMBIA. Combination in *Trachymyrmex:* Kempf, 1972a: 253.

*gatun. Trachymyrmex cornetzi* subsp. *gatun* Weber, 1940a: 420 (w.q.) PANAMA. Junior synonym of *cornetzi:* Weber, 1958b: 49.

***guianensis.*** *Trachymyrmex guianensis* Weber, 1937: 404 (w.) GUIANA.

#***haytianus.*** *Atta* (*Trachymyrmex*) *jamaicensis* subsp. *haytiana* Wheeler, W.M. & Mann, 1914: 41, fig. 18 (w.) HAITI. Combination in *Trachymyrmex:* Kempf, 1972a: 253.

***holmgreni.*** *Trachymyrmex holmgreni* Wheeler, W.M. 1925a: 37 (w.) BOLIVIA.

*humilis. Trachymyrmex humilis* Wheeler, W.M. 1916d: 325 (w.) TRINIDAD. Junior synonym of *bugnioni:* Weber, 1967: 273.

***iheringi.*** *Atta* (*Acromyrmex*) *iheringi* Emery, 1888c: 359 (w.q.m.) BRAZIL. Combination in *Atta* (*Trachymyrmex*): Forel, 1893e: 601; in *Acromyrmex* (*Trachymyrmex*): Forel, 1914d: 282; in *Trachymyrmex:* Gallardo, 1916b: 242. [Name sometimes misspelled as *jheringi*, for example by Kempf, 1972a: 253.]

***intermedius.*** *Atta* (*Trachymyrmex*) *intermedia* Forel, 1909a: 249 (w.) GUATEMALA. Combination in *Trachymyrmex:* Kempf, 1972a: 253.

***irmgardae.*** *Atta* (*Trachymyrmex*) *irmgardae* Forel, 1912e: 185 (w.) COLOMBIA. Combination in *Trachymyrmex:* Gallardo, 1916b: 242.

*irrorata. Atta* (*Trachymyrmex*) *septentrionalis* subsp. *obscurior* var. *irrorata* Wheeler, W.M. 1911g: 247 (w.) U.S.A. Unavailable name; material referred to *obscurior* by Creighton, 1950a: 324.

***isthmicus.*** *Trachymyrmex isthmicus* Santschi, 1931c: 280, figs. 13–15 (w.) PANAMA.

***jamaicensis.*** *Atta* (*Acromyrmex*) *jamaicensis* André, 1893b: 149 (w.) JAMAICA. Wheeler, W.M. 1907c: 713 (q.m.); Wheeler, G.C. 1949: 674 (l.). Combination in *Atta* (*Trachymyrmex*): Wheeler, W.M. 1907c: 712; in *Trachymyrmex:* Gallardo, 1916b: 242. Current subspecies: nominal plus *antiguensis, cubaensis, frontalis, haytianus, maritimus.* See also: Wheeler, W.M. & Mann, 1914: 41.

***kempfi.*** *Trachymyrmex kempfi* Fowler, 1982: 70, figs. 1–3 (w.) PARAGUAY.

***levis.*** *Trachymyrmex levis* Weber, 1938b: 192 (w.) BOLIVIA.

***mandibularis.*** *Trachymyrmex mandibularis* Weber, 1938b: 194 (w.q.) BOLIVIA.

#***maritimus.*** *Atta* (*Trachymyrmex*) *maritima* Wheeler, W.M. 1905b: 107, pl. 7, figs. 7, 8 (w.) BAHAMAS. Combination in *Trachymyrmex:* Kempf, 1972a: 253. Junior synonym of *jamaicensis:* Wheeler, W.M. 1907c: 712. Revived from synonymy as variety of *jamaicensis:* Mann, 1920: 428.

*marthae. Atta* (*Trachymyrmex*) *urichii* r. *marthae* Forel, 1912e: 183 (w.q.) COLOMBIA. Junior synonym of *urichii:* Weber, 1958b: 53.

*missionensis. Trachymyrmex tucumana* st. *cordovana* var. *missionensis* Santschi, 1934c: 31 (w.) ARGENTINA. Unavailable name.

*morgani. Trachymyrmex morgani* Weber, 1940a: 421 (w.) PANAMA. Junior synonym of *bugnioni:* Weber, 1967: 273.

*naranjo. Atta (Trachymyrmex) cornetzi* var. *naranjo* Forel, 1912e: 185 (w.) COLOMBIA. Junior synonym of *cornetzi:* Weber, 1958b: 49.

#***neomexicanus.*** *Trachymyrmex smithi* subsp. *neomexicanus* Cole, 1952a: 159, fig. 1 (w.) U.S.A. Cole, 1953f: 300 (q.m.).

***nogalensis.*** *Trachymyrmex nogalensis* Byars, 1951: 109, figs. 1, 2 (w.) U.S.A.

*obscurior. Atta (Trachymyrmex) septentrionalis* var. *obscurior* Wheeler, W.M. 1907c: 709 (w.) U.S.A. Wheeler, W.M. 1911g: 246 (q.). Combination in *Trachymyrmex:* Creighton, 1950a: 324. Subspecies of *septentrionalis:* Wheeler, W.M. 1911g: 247. Material of the unavailable names *crystallina* and *irrorata* referred here by Creighton, 1950a: 324. Junior synonym of *septentrionalis:* Weber, 1958b: 53.

***oetkeri.*** *Atta (Trachymyrmex) oetkeri* Forel, 1908c: 352 (w.) BRAZIL. Combination in *Trachymyrmex:* Gallardo, 1916b: 242.

***opulentus.*** *Sericomyrmex opulenta* Mann, 1922: 48, fig. 21 (w.) HONDURAS. Combination in *Trachymyrmex:* Weber, 1958b: 51.

#***pakeelai.*** *Sericomyrmex wheeleri* subsp. *pakeelai* Weber, 1937: 398, fig. 9 (w.) GUIANA. Combination in *Trachymyrmex:* Kempf, 1972a: 254.

*panamensis. Trachymyrmex urichi* subsp. *panamensis* Wheeler, W.M. 1925a: 38 (w.) PANAMA. Junior synonym of *urichii:* Weber, 1958b: 53.

***papulatus.*** *Trachymyrmex papulatus* Santschi, 1922b: 359 (w.) ARGENTINA.

***phaleratus.*** *Trachymyrmex phaleratus* Wheeler, W.M. 1925a: 39 (w.q.) GUIANA.

*phippsi. Trachymyrmex phippsi* Weber, 1937: 405 (w.q.m.) GUIANA. Junior synonym of *diversus:* Weber, 1967: 274.

****primaevus.*** **Trachymyrmex primaevus* Baroni Urbani, 1980b: 4, figs. 1–14 (w.) DOMINICAN AMBER.

***pruinosus.*** *Atta (Trachymyrmex) pruinosa* Emery, 1906c: 163, fig. 25 (w.) ARGENTINA. Gallardo, 1916b: 249 (q.m.). Combination in *Trachymyrmex:* Gallardo, 1916b: 248. Current subspecies: nominal plus *spinosior.*

*radicis. Trachymyrmex urichi* subsp. *radicis* Weber, 1938b: 197 (w.) TRINIDAD. Junior synonym of *urichii:* Weber, 1958b: 53.

***relictus.*** *Trachymyrmex relictus* Borgmeier, 1934: 107, figs. 6, 7 (w.q.) SURINAM. Current subspecies: nominal plus *fitzgeraldi.*

***ruthae.*** *Trachymyrmex ruthae* Weber, 1937: 402 (w.q.m.) TRINIDAD.

***saussurei.*** *Atta (Acromyrmex) tardigrada* st. *saussurei* Forel, 1885a: 361 (w.) MEXICO. Combination in *Atta (Trachymyrmex)*: Forel, 1893e: 601; in *Trachymyrmex:* Gallardo, 1916b: 242. Raised to species: Forel, 1893e: 601.

*schomburgki. Trachymyrmex schomburgki* Weber, 1938b: 195 (w.) GUIANA. Junior synonym of *bugnioni:* Weber, 1967: 273.

*seminole. Trachymyrmex septentrionalis* subsp. *seminole* Creighton, 1950a: 324 (w.q.m.) U.S.A. [First available use of *Atta (Trachymyrmex) septentrionalis* subsp. *obscurior* var. *seminole* Wheeler, W.M. 1911g: 247; unavailable name.] Junior synonym of *septentrionalis:* Weber, 1958b: 53.

***septentrionalis.*** *Atta septentrionalis* McCook, 1881a: 362 (w.) U.S.A. Wheeler, W.M. 1907c: 707 (q.m.); Wheeler, G.C. 1949: 673 (l.). Combination in *Atta (Trachymyrmex)*: Forel, 1912e: 182; in *Trachymyrmex:* Gallardo, 1916b: 242. Senior synonym of *virginiana:* Wheeler, W.M. 1902f: 29; of *vertebrata:* Creighton, 1950a: 323; of *obscurior* and *seminole:* Weber, 1958b: 53. See also: Wheeler, W.M. 1911g: 246; Smith, D.R. 1979: 1411; Petralia & Vinson, 1980: 383.

***sharpii.*** *Atta (Trachymyrmex) sharpii* Forel, 1893g: 372 (w.) ANTILLES IS. Combination in *Trachymyrmex:* Kempf, 1972a: 254. Junior synonym of *jamaicensis:* Wheeler, W.M. 1907c: 712. Revived from synonymy: Kempf, 1972a: 254.

***smithi.*** *Trachymyrmex smithi* Buren, 1944b: 5 (w.) MEXICO. Current subspecies: nominal plus *neomexicanus.*

#***spinosior.*** *Trachymyrmex pruinosus* var. *spinosior* Santschi, 1922b: 359 (w.) ARGENTINA.

***squamulifer.*** *Atta (Trachymyrmex) squamulifera* Emery, 1896g: 100 (w.) COSTA RICA. Combination in *Trachymyrmex:* Gallardo, 1916b: 242.

*trifurcatus. Trachymyrmex trifurcatus* Weber, 1938b: 199 (w.) GUIANA. Junior synonym of *farinosus:* Weber, 1958b: 51.

*tucuche. Trachymyrmex tucuche* Weber, 1938b: 197 (w.) TRINIDAD. Junior synonym of *bugnioni:* Weber, 1967: 273.

***tucumanus.*** *Acromyrmex (Trachymyrmex) iheringi* var. *tucumana* Forel, 1914d: 282 (w.) ARGENTINA. Bruch, 1921: 202 (q.). Combination in *Trachymyrmex:* Gallardo, 1916d: 325. Raised to species: Santschi, 1925e: 164. Current subspecies: nominal plus *cordovanus, fracticornis, weiseri.*

***turrifex.*** *Atta (Trachymyrmex) turrifex* Wheeler, W.M. 1903b: 100, fig. 6 (w.q.) U.S.A. Combination in *Trachymyrmex:* Gallardo, 1916b: 242. Current subspecies: nominal plus *caroli.*

*uncifer. Trachymyrmex uncifer* Santschi, 1931c: 281, figs. 16, 17 (w.) PANAMA. Junior synonym of *cornetzi:* Weber, 1958b: 49.

***urichii.*** *Atta (Trachymyrmex) urichii* Forel, 1893e: 601 (w.m.) TRINIDAD. Emery, 1894c: 222 (q.). Combination in *Trachymyrmex:* Gallardo, 1916b: 242; Wheeler, W.M. 1916d: 325. Senior synonym of *fusca, marthae, panamensis* and *radicis:* Weber, 1958b: 53.

***verrucosus.*** *Trachymyrmex verrucosus* Borgmeier, 1948a: 205, figs. 31, 32 (w.q.) SURINAM.

*vertebrata. Atta (Trachymyrmex) septentrionalis* var. *vertebrata* Wheeler, W.M. 1911g: 246 (w.q.) U.S.A. Junior synonym of *septentrionalis:* Creighton, 1950a: 323.

*virginiana. Oecodoma virginiana* Buckley, 1867: 346 (w.) U.S.A. Junior synonym of *septentrionalis:* Wheeler, W.M. 1902f: 29.

#***weiseri.*** *Trachymyrmex tucumana* var. *weiseri* Santschi, 1925e: 164 (w.) ARGENTINA.

***wheeleri.*** *Sericomyrmex wheeleri* Weber, 1937: 396, fig. 8 (w.q.) GUIANA. Wheeler, G.C. 1949: 670 (l.). Combination in *Trachymyrmex:* Weber, 1958b: 54. Current subspecies: nominal plus *pakeelai.*

***zeteki.*** *Trachymyrmex zeteki* Weber, 1940a: 422 (w.) PANAMA. Senior synonym of *balboai:* Weber, 1958b: 55.

*TRACHYPHEIDOLE:* see under ***PHEIDOLE.***

*TRANETERA:* see under ***TERATANER.***

***TRANOPELTA*** (Neotropical)

*albida. Tranopelta gilva* var. *albida* Mann, 1916: 446 (w.) BRAZIL. Junior synonym of *gilva:* Emery, 1919b: 61; of *columbica:* Emery, 1922e: 193.

***amblyops.*** *Monomorium amblyops* Emery, 1894c: 148 (w.) BRAZIL. Santschi, 1922d: 251 (q.m.). Combination in *Tranopelta:* Emery, 1919b: 61. Variety of *gilva:* Wheeler, 1922e: 8. Revived status as species: Santschi, 1922d: 251; Emery, 1922e: 193. Senior synonym of *brunnea:* Emery, 1919b: 61; Emery, 1922e: 193.

*brunnea. Tranopelta gilva* var. *brunnea* Forel, 1909a: 259 (w.q.m.) PARAGUAY. Junior synonym of *amblyops:* Emery, 1919b: 61.

#***columbica.*** *Tranopelta heyeri* var. *columbica* Forel, 1912g: 3 (w.) COLOMBIA. Junior synonym of *gilva:* Emery, 1919b: 61. Revived from synonymy, and raised to species: Mann, 1922: 29. Variety of *heyeri,* and senior synonym of *albida:* Emery, 1922e: 193.

***gilva.*** *Tranopelta gilva* Mayr, 1866a: 514 (q.m.) VENEZUELA. Wheeler, W.M. 1922e: 6 (w.); Wheeler, G.C. & Wheeler, J. 1955c: 131 (l.).

***heyeri.*** *Monomorium heyeri* Forel, 1901g: 389 (w.) BRAZIL. Combination in *M. (Martia)*: Forel, 1908c: 361; in *Tranopelta:* Forel, 1911d: 396. Current subspecies: nominal plus *columbica.*

*huberi* Forel, 1907; see under ***CREMATOGASTER.***

*mayri* Forel, 1901; see under ***CAREBARA.***

***subterranea.*** *Monomorium (Mitara) subterraneum* Mann, 1916: 444, pl. 4, figs. 29, 30 (w.) BRAZIL. Combination in *Tranopelta,* and junior synonym of *huberi:* Emery, 1919b: 61. Revived from synonymy: Emery, 1922e: 193.

*TRANOPELTOIDES:* see under ***CREMATOGASTER.***

*TRAPEZIOPELTA:* see under ***MYOPIAS.***

*TRIANGULINODA:* see under ***PSEUDOMYRMEX.***

*TRICHOMELOPHORUS:* see under ***MELOPHORUS.***

*TRICHOMYRMEX:* see under ***MONOMORIUM.***

***TRICHOSCAPA*** (Tropical and temperate tramp)

*foochowensis. Strumigenys (Cephaloxys) foochowensis* Wheeler, W.M. 1928c: 28 (q.) CHINA. Junior synonym of *membranifera:* Brown, 1948e: 114.

*inusitata* Lattke, 1992; see under ***SMITHISTRUMA.***

*marioni. Strumigenys (Cephaloxys) membranifera* var. *marioni* Wheeler, W.M. 1933h: 276 (w.) U.S.A. Junior synonym of *membranifera:* Brown, 1948e: 114.

***membranifera.*** *Strumigenys (Trichoscapa) membranifera* Emery, 1869b: 24, fig. 11 (w.) ITALY. Emery, 1916b: 205 (q.); Wheeler, G.C. & Wheeler, J. 1991b: 93 (l.). Combination in *S. (Cephaloxys)*: Emery, 1922f: 323; in *Trichoscapa:* Brown, 1948e: 113. Senior synonym of *foochowensis, marioni, santschii, silvestriana, simillima, vitiensis* and *williamsi:* Brown, 1948e: 114. See also: Brown, 1949d: 6; Wilson, 1954: 483; Bolton, 1983: 319.

*santschii. Strumigenys membranifera* var. *santschii* Forel, 1904c: 6 (w.) TUNISIA. Junior synonym of *membranifera:* Brown, 1948e: 114.

***TRICHOSCAPA* (continued)**

*silvestriana. Strumigenys (Cephaloxys) silvestriana* Wheeler, W.M. 1928c: 27 (w.) CHINA. Junior synonym of *membranifera:* Brown, 1948e: 114.
*simillima. Strumigenys membranifera* r. *simillima* Emery, 1890b: 69, pl. 8, fig. 5 (w.) ANTILLES IS. Junior synonym of *membranifera:* Brown, 1948e: 114.
*vitiensis. Strumigenys (Cephaloxys) vitiensis* Mann, 1921: 461, fig. 22 (w.) FIJI IS. Junior synonym of *membranifera:* Brown, 1948e: 114.
*williamsi. Strumigenys (Cephaloxys) membranifera* var. *williamsi* Wheeler, W.M. 1933h: 276 (w.q.) HAWAII. Junior synonym of *membranifera:* Brown, 1948e: 114.

***TRICYTARUS*** (New Guinea)
***parviumgulatus.*** *Tricytarus parviumgulatus* Donisthorpe, 1947d: 188 (m.) NEW GUINEA.

*TRIGLYPHOTHRIX:* see under ***TETRAMORIUM.***

*TRIGONOGASTER:* see under ***RECURVIDRIS.***

***TURNERIA*** (Indo-Australian, Australasian)
***arbusta.*** *Turneria arbusta* Shattuck, 1990: 106, figs. 4, 5 (w.q.m.) NEW GUINEA.
***bidentata.*** *Turneria bidentata* Forel, 1895f: 419 (w.) AUSTRALIA. Shattuck, 1990: 108 (q.m.).
*butteli* Forel, 1913; see under ***IRIDOMYRMEX.***
***collina.*** *Turneria collina* Shattuck, 1990: 109, figs. 8, 9 (w.) NEW GUINEA.
***dahlii.*** *Turneria dahlii* Forel, 1901b: 17 (w.) BISMARCK ARCHIPELAGO. Wilson, 1962c: 17 (q.). See also: Shattuck, 1990: 109.
***frenchi.*** *Turneria frenchi* Forel, 1911b: 207 (footnote) (w.) AUSTRALIA. Combination in *Stigmacros:* Shattuck, 1990: 105; in *Turneria:* Taylor, 1992a: 67.
***pacifica.*** *Turneria pacifica* Mann, 1919: 361, figs. 39, 40 (w.) SOLOMON IS. Wilson, 1962c: 17 (q.); Shattuck, 1990: 112 (m.).
***postomma.*** *Turneria postomma* Shattuck, 1990: 112, figs. 15, 16 (w.) NEW GUINEA.

*TYPHLATTA:* see under ***AENICTUS.***

***TYPHLOMYRMEX*** (Neotropical)
*bruchi. Typhlomyrmex bruchi* Santschi, 1922d: 242. *Nomen nudum.*
***clavicornis.*** *Typhlomyrmex clavicornis* Emery, 1906c: 112 (q.) BOLIVIA. Brown, 1965c: 71 (w.). Senior synonym of *divergens* and *richardsi:* Brown, 1965c: 71.
*divergens. Typhlomyrmex clavicornis* var. *divergens* Forel, 1906d: 248 (q.m.) PARAGUAY. Junior synonym of *clavicornis:* Brown, 1965c: 71.
***foreli.*** *Typhlomyrmex foreli* Santschi, 1925b: 6 (q.) BRAZIL. See also: Brown, 1965c: 76.
***major.*** *Typhlomyrmex pusillus* st. *major* Santschi, 1923c: 246 (w.) BRAZIL. Brown, 1965c: 72 (q.). Raised to species: Brown, 1965c: 71.
*manco. Typhlomyrmex robustus* subsp. *manco* Wheeler, W.M. 1925a: 2 (w.m.) PERU. Junior synonym of *rogenhoferi:* Brown, 1957e: 229.
***prolatus.*** *Typhlomyrmex prolatus* Brown, 1965c: 72, fig. 6 (q.) COSTA RICA.
***pusillus.*** *Typhlomyrmex pusillus* Emery, 1894c: 141, pl. 1, fig. 2 (w.) BOLIVIA. Emery, 1906c: 111 (q.); Wheeler, G.C. & Wheeler, J. 1964b: 447 (l.). Senior synonym of *schmidti:* Brown, 1965c: 73. See also: Gallardo, 1918b: 16.
*richardsi. Typhlomyrmex richardsi* Donisthorpe, 1939b: 161 (m.) GUIANA. Junior synonym of *clavicornis:* Brown, 1965c: 71.
*robustus. Typhlomyrmex rogenhoferi* r. *robustus* Emery, 1890b: 40 (w.) COSTA RICA. [Also described as new by Emery, 1894k: 46.] Forel, 1899c: 3 (q.); Wheeler, G.C. & Wheeler, J. 1952a: 121 (l.). Raised to species: Dalla Torre, 1893: 16; Emery, 1911d: 24. Junior synonym of *rogenhoferi:* Brown, 1957e: 229.
***rogenhoferi.*** *Typhlomyrmex rogenhoferi* Mayr, 1862: 737 (w.) BRAZIL. Mayr, 1887: 538 (q.m.); Wheeler, G.C. & Wheeler, J. 1976a: 42 (l.). Senior synonym of *manco* and *robustus:* Brown, 1957e: 229; of *sculpturatus:* Brown, 1965c: 74.
*schmidti. Typhlomyrmex schmidti* Menozzi, 1927c: 268 (q.m.) COSTA RICA. Junior synonym of *pusillis:* Brown, 1965c: 73.
*sculpturatus. Centromyrmex sculpturatus* Santschi, 1931c: 266 (q.) PANAMA. Junior synonym of *rogenhoferi:* Brown, 1965c: 74.

*#TYPHLOPONE:* see under ***DORYLUS***

*TYPHLOTERAS:* see under ***CENTROMYRMEX.***

*VEROMESSOR:* see under ***MESSOR.***

*VITICICOLA:* see under ***TETRAPONERA.***

***VOLLENHOVIA*** (Old World tropics and subtropics except Africa)
***acanthina.*** *Solenomyrma acanthina* Karavaiev, 1935a: 103, fig. 23 (w.) VIETNAM. Combination in *Gauromyrmex:* Brown, 1953c: 10; in *Vollenhovia:* **new combination**. Senior synonym of *donisthorpei:* Brown, 1953c: 10.
**#*affinis.*** *Vollenhovia subtilis* var. *affinis* Emery, 1887b: 454 (w.q.) NEW GUINEA.
*agilis. Vollenhovia agilis* Santschi, 1928a: 44, fig. 1 (q.) SAMOA. Junior synonym of *samoensis:* Wilson & Taylor, 1967: 57.
**#*alluaudi.*** *Vollenhovia laevithorax* subsp. *alluaudi* Emery, 1894a: 68 (w.) SEYCHELLES IS. Forel, 1912k: 162 (q.). Subspecies of *oblonga:* Emery, 1914f: 406; of *laevithorax:* Wheeler, W.M. 1922a: 1026; of *oblonga:* current status.
*ambigua. Vollenhovia ambigua* Santschi, 1916d: 242. [Unnecessary (second) replacement name for *brevicornis* Emery, 1897d: 560.] Junior synonym of *brachycera:* Emery, 1917a: 96.
***ambitiosa.*** *Vollenhovia ambitiosa* Menozzi, 1925c: 446 (q.) PHILIPPINES.
**#*bandarensis.*** *Vollenhovia pedestris* var. *bandarensis* Forel, 1913k: 67 (w.q.m.) SUMATRA. Subspecies of *oblonga:* current status.
***banksi.*** *Vollenhovia banksi* Forel, 1910d: 126 (w.) PHILIPPINES. Current subspecies: nominal plus *kuchingensis.*
***bengakalisi.*** *Gauromyrmex bengakalisi* Menozzi, 1933c: 146, figs. 1, 2 (w.) SUMATRA. Combination in *Vollenhovia:* **new combination**.
****beyrichi.*** **Macromischa beyrichi* Mayr, 1868c: 84, pl. 4, figs. 80, 81 (w.) BALTIC AMBER (Oligocene). Combination in *Vollenhovia,* and senior synonym of **samlandica:* Wheeler, W.M. 1915h: 51.
***brachycera.*** *Vollenhovia brachycera* Emery, 1914f: 407 (footnote). Replacement name for *brevicornis* Emery, 1897d: 560. [Junior secondary homonym of *brevicorne* Emery, below.] Senior synonym of *ambigua:* Emery, 1917a: 96.
***brevicornis.*** *Monomorium brevicorne* Emery, 1893e: 203 (w.) SUMATRA. Combination in *Vollenhovia:* Emery, 1914f: 406 (footnote). Current subspecies: nominal plus *minuta.*
*brevicornis. Vollenhovia brevicornis* Emery, 1897d: 560 (w.) NEW GUINEA. [Junior secondary homonym of *brevicorne* Emery, above.] Replacement name: *brachycera* Emery, 1914f: 407 (footnote).
***brunnea.*** *Vollenhovia brunnea* Donisthorpe, 1947c: 582 (q.) NEW GUINEA.
***butteli.*** *Vollenhovia butteli* Forel, 1913k: 68 (w.q.) WEST MALAYSIA. Current subspecies: nominal plus *jacobsoni, seeliebi.*
**#*chosenica.*** *Vollenhovia emeryi* subsp. *chosenica* Wheeler, W.M. 1928d: 113 (w.) KOREA.
***cristata.*** *Aratromyrmex cristatus* Stitz, 1938: 108 (q.) PALAWAN I. Combination in *Vollenhovia:* Baroni Urbani, 1980a: 95.
***dentata.*** *Vollenhovia dentata* Mann, 1919: 325 (w.m.) SOLOMON IS. Current subspecies: nominal plus *marginata.*
***denticulata.*** *Vollenhovia denticulata* Emery, 1914f: 405, pl. 13, fig. 4 (w.) NEW CALEDONIA.
**#*dispar.*** *Vollenhovia oblonga* subsp. *dispar* Forel, 1910d: 125 (w.q.m.) PHILIPPINES.
*donisthorpei. Acalama donisthorpei* Smith, M.R. 1949a: 207, figs. 1, 2 (w.) INDIA. Junior synonym of *acanthina:* Brown, 1953c: 10.
***duodecimalis.*** *Vollenhovia duodecimalis* Donisthorpe, 1948d: 595 (q.) NEW GUINEA.
***elysii.*** *Vollenhovia elysii* Mann, 1919: 327 (w.) SOLOMON IS.
***emeryi.*** *Vollenhovia emeryi* Wheeler, W.M. 1906c: 312, pl. 41, figs. 10, 11 (w.q.) JAPAN. Wheeler, W.M. 1928d: 112 (m.); Imai, 1966: 119 (k.). Current subspecies: nominal plus *chosenica.*
***escherichi.*** *Vollenhovia escherichi* Forel, 1911b: 198 (footnote) (w.) SRI LANKA.
***foveaceps.*** *Vollenhovia foveaceps* Mann, 1919: 328 (w.q.) SOLOMON IS.
***fridae.*** *Vollenhovia fridae* Forel, 1913k: 65 (w.) SUMATRA.
***hewitti.*** *Vollenhovia hewitti* Wheeler, W.M. 1919e: 82 (w.) BORNEO.
***irenea.*** *Dyomorium ireneum* Donisthorpe, 1947d: 191 (w.) NEW GUINEA. Combination in *Vollenhovia:* Ettershank, 1966: 149.
**#*jacobsoni.*** *Vollenhovia butteli* var. *jacobsoni* Forel, 1915a: 25 (w.) SUMATRA.
*kappa. Vollenhovia oblonga* r. *rufescens* var. *kappa* Forel, 1915a: 25 (w.) SUMATRA. Unavailable name.
**#*kuchingensis.*** *Vollenhovia banksi* subsp. *kuchingensis* Wheeler, W.M. 1919e: 83 (w.q.) BORNEO.
**#*laevithorax.*** *Vollenhovia laevithorax* Emery, 1889b: 501 (w.) BURMA. Subspecies/race of *oblonga:* Emery, 1900d: 680; Forel, 1903d: 404; Emery, 1914f: 406.
***loboii.*** *Vollenhovia loboii* Mann, 1919: 324 (w.) SOLOMON IS.

***longicephala.*** *Podomyrma longicephala* Terayama & Yamane, 1991: 69, figs. 1–5 (w.) SUMATRA. Combination in *Vollenhovia:* **new combination**.

***longiceps.*** *Vollenhovia longiceps* Emery, 1893e: 205 (w.) SUMATRA.

***luctuosa.*** *Aratromyrmex luctuosus* Stitz, 1938: 106, fig. 5 (q.) NEW GUINEA. Combination in *Vollenhovia:* Baroni Urbani, 1980a: 96. Current subspecies: nominal plus *major.*

***#magna.*** *Vollenhovia subtilis* subsp. *magna* Viehmeyer, 1914b: 525 (w.) NEW GUINEA.

***#major.*** *Aratromyrmex luctuosus* var. *major* Stitz, 1938: 108 (q.) PALAWAN I. Combination in *Vollenhovia:* Baroni Urbani, 1980a: 95.

***#marginata.*** *Vollenhovia dentata* subsp. *marginata* Mann, 1919: 327 (w.) SOLOMON IS.

***#minuta.*** *Vollenhovia brevicornis* var. *minuta* Viehmeyer, 1916a: 129 (w.) WEST MALAYSIA.

***modiglianii.*** *Vollenhovia modiglianii* Emery, 1900d: 679 (w.q.) SUMATRA.

***moesta.*** *Myrmica moesta* Smith, F. 1863: 21 (w.) INDONESIA (Martabello I.). Combination in *Vollenhovia:* Donisthorpe, 1932c: 471.

***#ngoko.*** *Vollenhovia piroskae* var. *ngoko* Forel, 1912d: 108 (q.) JAVA.

***#nigrescens.*** *Vollenhovia subtilis* r. *nigrescens* Forel, 1913k: 69 (w.q.m.) JAVA.

***nipponica.*** *Vollenhovia nipponica* Kinomura & Yamauchi, 1992: 203, figs. 1, 2 (q.m.) JAPAN.

***nitida.*** *Podomyrma nitida* Smith, F. 1860b: 110 (w.) NEW GUINEA. Combination in *Vollenhovia:* **new combination**.

***novobritainae.*** *Dorothea novobritainae* Donisthorpe, 1948c: 65 (w.) NEW BRITAIN. Combination in *Vollenhovia:* Ettershank, 1966: 150.

***oblonga.*** *Myrmica oblonga* Smith, F. 1860b: 107 (w.) INDONESIA (Batjan I.). Wheeler, W.M. 1919e: 144 (q.). Combination in *Vollenhovia:* Emery, 1897d: 560. Senior synonym of *poneroides:* Emery, 1897d: 560; Donisthorpe, 1932c: 463. Current subspecies: nominal plus *alluaudi, bandarensis, dispar, laevithorax, pedestris, rufescens.*

***opacinoda.*** *Vollenhovia opacinoda* Forel, 1913k: 66 (w.) SUMATRA.

***overbecki.*** *Vollenhovia overbecki* Viehmeyer, 1916a: 129, fig. 4 (w.m.) SINGAPORE.

***pacifica.*** *Vollenhovia pacifica* Wilson & Taylor, 1967: 56, fig. 41 (w.q.m.) SAMOA.

***papuana.*** *Vollenhovia papuana* Viehmeyer, 1914b: 527 (w.) NEW GUINEA.

***#pedestris.*** *Myrmica pedestris* Smith, F. 1861b: 46 (q.) SULAWESI. Wheeler, G.C. & Wheeler, J. 1955c: 120 (l.). Combination in *Vollenhovia:* Mayr, 1867a: 94. Status as species: Mann, 1919: 324. Subspecies of *oblonga:* Emery, 1914f: 405; Donisthorpe, 1932c: 468; Donisthorpe, 1948g: 133.

***penetrans.*** *Atta penetrans* Smith, F. 1857a: 77 (q.) BORNEO. Combination in *Aphaenogaster:* Dalla Torre, 1893: 104; in *Vollenhovia:* Donisthorpe, 1932c: 450.

***pertinax.*** *Myrmica pertinax* Smith, F. 1861b: 46 (w.) SULAWESI. Combination in *Vollenhovia:* Donisthorpe, 1932c: 468.

***piroskae.*** *Vollenhovia piroskae* Forel, 1912k: 162 (w.m.) SEYCHELLES IS. Current subspecies: nominal plus *ngoko, taipingensis.*

*poneroides. Myrmica poneroides* Smith, F. 1860b: 107 (q.) INDONESIA (Batjan I.). Combination in *Vollenhovia:* Mayr, 1886c: 361. Junior synonym of *oblonga:* Emery, 1897d: 560; Donisthorpe, 1932c: 463.

***prisca.** **Macromischa prisca* André, 1895b: 83 (w.) BALTIC AMBER (Oligocene). Combination in *Vollenhovia:* Wheeler, W.M. 1915h: 52.

***punctata.*** *Vollenhovia punctata* Viehmeyer, 1914b: 526 (w.) NEW GUINEA.

***punctatostriata.*** *Vollenhovia punctatostriata* Mayr, 1865: 21 (footnote) (q.) JAVA.

***pyrrhoria.*** *Vollenhovia pyrrhoria* Wu & Xiao, 1989: 239, fig. 1 (w.q.) CHINA.

*reepeni. Vollenhovia oblonga* r. *rufescens* var. *reepeni* Forel, 1913k: 65 (w.m.) WEST MALAYSIA. Unavailable name.

***#rufescens.*** *Vollenhovia laevithorax* var. *rufescens* Emery, 1894a: 69 (w.) BORNEO. Race of *oblonga:* Forel, 1913k: 64; Forel, 1915a: 25.

***rufipes.*** *Vollenhovia rufipes* Donisthorpe, 1949g: 409 (w.q.) NEW GUINEA.

***rufiventris.*** *Vollenhovia rufiventris* Forel, 1901e: 374 (w.) BORNEO. Viehmeyer, 1916a: 130 (q.m.). Combination in *Monomorium:* Emery, 1914f: 407 (footnote); in *Heteromyrmex:* Wheeler, W.M. 1920: 53; in *Vollenhovia:* Viehmeyer, 1916a: 130; Ettershank, 1966: 150. [*Vollenhovia rufiventris* Emery, 1900d: 680; *nomen nudum,* attributed to Forel.] Current subspecies: nominal plus *simalurana.* See also: Wheeler, W.M. 1919e: 79.

**samlandica. *Propodomyrma samlandica* Wheeler, W.M. 1910g: 163, fig. 93 (w.) BALTIC AMBER (Oligocene). Junior synonym of **beyrichi:* Wheeler, W.M. 1915h: 51.

***samoensis.*** *Vollenhovia samoensis* Mayr, 1876: 99 (w.m.) SAMOA. Senior synonym of *agilis:* Wilson & Taylor, 1967: 57.

***satoi.*** *Vollenhovia satoi* Santschi, 1937h: 373, fig. 7 (w.q.) TAIWAN.

***#seeliebi.*** *Vollenhovia butteli* r. *seeliebi* Forel, 1913k: 69 (w.q.) WEST MALAYSIA.

***#simalurana.*** *Vollenhovia rufiventris* var. *simalurana* Forel, 1915a: 23 (w.q.m.) SUMATRA.

***simoides.*** *Vollenhovia simoides* Emery, 1897d: 561, pl. 1, figs. 7, 8 (w.q.) NEW GUINEA.

***soleaferrea.*** *Vollenhovia soleaferrea* Donisthorpe, 1942b: 66 (q.) PHILIPPINES.

***subtilis.*** *Vollenhovia subtilis* Emery, 1887b: 454 (w.) INDONESIA (Key I.). Current subspecies: nominal plus *affinis, magna, nigrescens.*

***#taipingensis.*** *Vollenhovia piroskae* var. *taipingensis* Forel, 1913k: 64 (w.) WEST MALAYSIA.

*turneri* Forel, 1910; see under ***MONOMORIUM.***

***umbilicata.*** *Vollenhovia umbilicata* Donisthorpe, 1941h: 56 (q.) NEW GUINEA.

***undecimalis.*** *Vollenhovia undecimalis* Donisthorpe, 1948d: 594 (q.) NEW GUINEA. Donisthorpe, 1949b: 494 (w.).

***VOMBISIDRIS*** (Oriental, Indo-Australian)

***acherdos.*** *Vombisidris acherdos* Bolton, 1991: 5, fig. 7 (w.) NEW GUINEA.

***australis.*** *Leptothorax* (*Goniothorax*) *australis* Wheeler, W.M. 1934b: 60 (w.) AUSTRALIA. Taylor, 1989b: 608 (q.). Combination in *Vombisidris:* Bolton, 1991: 7.

***bilongrudi.*** *Leptothorax bilongrudi* Taylor, 1989b: 605, figs. 1–3 (w.) NEW GUINEA. Combination in *Vombisidris:* Bolton, 1991: 5.

***dryas.*** *Vombisidris dryas* Bolton, 1991: 7, fig. 8 (w.) BORNEO.

***harpeza.*** *Vombisidris harpeza* Bolton, 1991: 7, fig. 5 (w.) BORNEO.

***lochme.*** *Vombisidris lochme* Bolton, 1991: 7, fig. 9 (w.) SULAWESI.

***nahet.*** *Vombisidris nahet* Bolton, 1991: 5, fig. 3 (w.) SULAWESI.

***occidua.*** *Vombisidris occidua* Bolton, 1991: 8, fig. 6 (w.) INDIA.

***philax.*** *Vombisidris philax* Bolton, 1991: 6, fig. 4 (w.) SULAWESI.

***regina.*** *Vombisidris regina* Bolton, 1991: 6, figs. 1, 2 (w.) BORNEO.

***renateae.*** *Leptothorax renateae* Taylor, 1989b: 609, figs. 4–7 (w.q.) AUSTRALIA. Combination in *Vombisidris:* Bolton, 1991: 6.

***xylochos.*** *Vombisidris xylochos* Bolton, 1991: 7, figs. 10, 11 (w.) BORNEO.

*WADEURA:* see under ***PACHYCONDYLA.***

***WASMANNIA*** (Neotropical)

***affinis.*** *Wasmannia affinis* Santschi, 1929d: 300, figs. 25, 26 (w.) BRAZIL.

*atomum. Xiphomyrmex atomum* Santschi, 1914d: 370 (w.) GABON. Combination in *Wasmannia,* and variety of *auropunctata:* Santschi, 1916b: 504. Junior synonym of *auropunctata:* Wheeler, W.M. 1922a: 912.

***auropunctata.*** *Tetramorium auropunctatum* Roger, 1863a: 182 (w.q.m.) CUBA. Wheeler, G.C. & Wheeler, J. 1954d: 444 (l.). Combination in *Ochetomyrmex:* Forel, 1886b: xlix; in *Wasmannia:* Forel, 1893g: 383. Senior synonym of *atomum:* Wheeler, W.M. 1922a: 912; of *glabra:* Kempf, 1964e: 66; of *panamana:* Brown, 1948d: 102. Current subspecies: nominal plus *australis, laevifrons, nigricans, obscura, pulla, rugosa.*

***#australis.*** *Wasmannia auropunctata* var. *australis* Emery, 1894c: 193 (w.) BRAZIL.

*brevispinosa* Borgmeier, 1928; see under ***TETRAMORIUM.***

*bruchi. Wasmannia bruchi* Santschi, 1919f: 43 (w.) ARGENTINA. Junior synonym of *sulcaticeps:* Kusnezov, 1952e: 174 (in text).

*glabra. Wasmannia glabra* Santschi, 1931c: 272 (q.) FRENCH GUIANA. Junior synonym of *auropunctata:* Kempf, 1964e: 66.

***iheringi.*** *Wasmannia iheringi* Forel, 1908c: 359 (w.q.) BRAZIL.

***#laevifrons.*** *Wasmannia auropunctata* var. *laevifrons* Emery, 1894c: 193 (w.) BOLIVIA. Wheeler, W.M. 1925a: 36 (q.).

***lutzi.*** *Wasmannia lutzi* Forel, 1908c: 357 (w.q.m.) BRAZIL.

***#nigricans.*** *Wasmannia auropunctata* var. *nigricans* Emery, 1906c: 160 (w.) PARAGUAY.

***#obscura.*** *Wasmannia auropunctata* var. *obscura* Forel, 1912g: 1 (w.) COLOMBIA.

*panamana. Hercynia panamana* Enzmann, J. 1947a: 44, pl. 5 (w.q.) PANAMA. Junior synonym of *auropunctata:* Brown, 1948d: 102.

***WASMANNIA*** *(continued)*

**#*pulla*.** *Wasmannia auropunctata* st. *pulla* Santschi, 1931c: 272 (w.) PANAMA.

***rochai*.** *Wasmannia rochai* Forel, 1912g: 1 (w.) BRAZIL.

**#*rugosa*.** *Ochetomyrmex auropunctatus* var. *rugosus* Forel, 1886b: xlix (w.) GUATEMALA. Santschi, 1929d: 299 (q.m.). Combination in *Wasmannia:* Forel, 1901c: 128. Race of *auropunctata:* Emery, 1888c: 353.

***scrobifera*.** *Wasmannia scrobifera* Kempf, 1961b: 512, figs. 14, 15 (w.) SURINAM.

***sigmoidea*.** *Tetramorium sigmoideum* Mayr, 1884: 13 (w.) FRENCH GUIANA. Forel, 1893g: 386 (q.m.). Combination in *Wasmannia:* Forel, 1893g: 386.

*subpolita* Wheeler, W.M. 1916; see under ***OCHETOMYRMEX***.

***sulcaticeps*.** *Wasmannia sulcaticeps* Emery, 1894c: 195 (w.) ARGENTINA. Forel, 1914d: 280 (q.m.). Senior synonym of *bruchi:* Kusnezov, 1952e: 174 (in text). Current subspecies: nominal plus *weiseri.*

***villosa*.** *Wasmannia villosa* Emery, 1894c: 194 (q.) BRAZIL.

***williamsoni*.** *Wasmannia williamsoni* Kusnezov, 1952e: 181, figs. 1, 4–7 (w.q.) ARGENTINA.

**#*weiseri*.** *Wasmannia sulcaticeps* var. *weiseri* Forel, 1914d: 281 (q.m.) ARGENTINA.

*WEBERIDRIS:* see under ***CALYPTOMYRMEX***.

*WEBERISTRUMA:* see under ***SMITHISTRUMA***.

*WESSONISTRUMA:* see under ***SMITHISTRUMA***.

*WHEELERIA:* see under ***MONOMORIUM***.

*WHEELERIELLA:* see under ***MONOMORIUM***.

*WHEELERIMYRMEX:* see under ***MEGALOMYRMEX***.

*WHEELERIPONE:* see under ***GNAMPTOGENYS***.

***WILLOWSIELLA*** (Indo-Australian, Australasia)

***anderseni*.** *Willowsiella anderseni* Taylor, 1991a: 285, figs. 6–9 (w.) AUSTRALIA.

***dispar*.** *Willowsiella dispar* Wheeler, W.M. 1934a: 175, fig. 1 (w.) SOLOMON IS.

*WOITKOWSKIA:* see under ***NEIVAMYRMEX***.

*XENHYBOMA:* see under ***MONOMORIUM***.

*XENOAPHAENOGASTER:* see under ***PHEIDOLE***.

*XENOMETRA:* see under ***CARDIOCONDYLA***.

***XENOMYRMEX*** (Neotropical, Nearctic)

*castus. Xenomyrmex stolli* subsp. *castus* Wheeler, W.M. 1931a: 138 (w.q.m.) MEXICO. Junior synonym of *skwarrae:* Creighton, 1957c: 8.

*cubanus. Xenomyrmex stolli* subsp. *cubanus* Wheeler, W.M. 1931a: 134 (w.) CUBA. Junior synonym of *floridanus:* Creighton, 1957c: 6.

***floridanus*.** *Xenomyrmex stolli* subsp. *floridanus* Emery, 1895c: 275 (w.m.) U.S.A. Wheeler, W.M. 1931a: 136 (q.). Raised to species: Creighton, 1957c: 6. Senior synonym of *rufescens:* Creighton, 1950a: 226; of *cubanus* and *lucayanus:* Creighton, 1957c: 6. Current subspecies: nominal plus *skwarrae.*

*lucayanus. Xenomyrmex stolli* subsp. *lucayanus* Wheeler, W.M. 1931a: 137 (w.) BAHAMAS. [First available use of *Xenomyrmex stolli* subsp. *floridanus* var. *lucayanus* Wheeler, W.M. 1905b: 87; unavailable name.] Junior synonym of *floridanus:* Creighton, 1957c: 6.

*mexicanus. Xenomyrmex stolli* subsp. *mexicanus* Wheeler, W.M. 1931a: 134 (w.) MEXICO. Wheeler, G.C. & Wheeler, J. 1960b: 20 (l.). Junior synonym of *stollii:* Creighton, 1957c: 10.

*nodosus. Xenomyrmex stolli* subsp. *nodosus* Creighton, 1953e: 2, fig. 1 (w.q.) MEXICO. Junior synonym of *stollii:* Creighton, 1957c: 10.

***panamanus*.** *Myrmecinella panamana* Wheeler, W.M. 1922d: 1, fig. 1 (w.) PANAMA. Combination in *Xenomyrmex:* Wheeler, W.M. 1931a: 133.

*rufescens. Xenomyrmex stolli* subsp. *rufescens* Wheeler, W.M. 1931a: 137 (q.) U.S.A. Junior synonym of *floridanus:* Creighton, 1950a: 226.

**#*skwarrae*.** *Xenomyrmex stolli* subsp. *skwarrae* Wheeler, W.M. 1931a: 137, fig. 2 (w.q.m.) MEXICO. Subspecies of *floridanus,* and senior synonym of *castus:* Creighton, 1957c: 8.

***stollii*.** *Xenomyrmex stollii* Forel, 1885a: 370 (w.) GUATEMALA. Senior synonym of *mexicanus* and *nodosus:* Creighton, 1957c: 10.

*XEROMYRMEX:* see under ***MONOMORIUM***.

**#*XIPHOCREMA:*** see under ***CREMATOGASTER***.

*XIPHOMYRMEX:* see under ***TETRAMORIUM***.

*XIPHOPELTA:* see under ***PACHYCONDYLA***.

*XYMMER:* see under ***AMBLYOPONE***.

***YAVNELLA*** (Palaearctic, Oriental)

***argamani*.** *Yavnella argamani* Kugler, J. 1987: 53, figs. 14–18 (m.) ISRAEL.

***indica*.** *Yavnella indica* Kugler, J. 1987: 55, figs. 19–22 (m.) INDIA.

***ZACRYPTOCERUS*** (Neotropical, Nearctic)

***adolphi*.** *Cryptocerus adolphi* Emery, 1906c: 172, fig. 33 (w.) BRAZIL. Combination in *Paracryptocerus* (*Harnedia*): Kempf, 1958a: 86; in *Zacryptocerus:* Brandão, 1991: 384. See also: Kempf, 1960f: 439.

**#*affinis*.** *Hypocryptocerus haemorrhoidalis* subsp. *affinis* Wheeler, W.M. 1936b: 202 (w.) HAITI. Combination in *Zacryptocerus:* Brandão, 1991: 384. See also: Kempf, 1951: 147.

*amazonensis. Cryptocerus multispinus* var. *amazonensis* Forel, 1911e: 261 (w.) BRAZIL. Combination in *C.* (*Paracryptocerus*): Emery, 1922f: 307. Junior synonym of *complanatus:* Kempf, 1951: 193.

*angulatus. Cryptocerus angulatus* Smith, F. 1858b: 189 (w.) BRAZIL. Junior synonym of *complanatus:* Emery, 1922f: 307.

*angulosus. Cryptocerus angulosus* Smith, F. 1867: 525, pl. 26, fig. 7 (w.) MEXICO. Wheeler, W.M. 1907a: 273 (s.m.). Senior synonym of *jucundus:* Emery, 1890b: 76. Junior synonym of *scutulatus:* Emery, 1922f: 311; Kempf, 1958d: 110.

***angustus*.** *Cryptocerus angustus* Mayr, 1862: 759 (s.q.) BRAZIL. Mayr, 1887: 566 (w.m.). Combination in *Paracryptocerus* (*Harnedia*): Kempf, 1958a: 20; in *Zacryptocerus:* Brandão, 1991: 384. Senior synonym of *clarior:* Kempf, 1958a: 74.

*araneolus. Cryptocerus araneolus* Smith, F. 1853: 223, pl. 19, fig. 4 (w.) ANTILLES IS. Variety of *discocephalus:* Forel, 1893g: 382. Revived status as species: Forel, 1897b: 300. Variety of *pallens:* Forel, 1899c: 51. Junior synonym of *pallens:* Emery, 1894c: 212; Kempf, 1958a: 151.

***argentatus*.** *Cryptocerus argentatus* Smith, F. 1853: 218, pl. 19, fig. 7 (q.) COLOMBIA. Combination in *Paracryptocerus:* Kempf, 1951: 232; in *Zacryptocerus:* Brandão, 1991: 384.

**#*auricomus*.** *Hypocryptocerus haemorroidalis* susbp. *auricomus* Wheeler, W.M. 1936b: 202 (w.) DOMINICA. Combination in *Zacryptocerus:* Brandão, 1991: 384. See also: Kempf, 1951: 148.

***aztecus*.** *Cryptocerus aztecus* Forel, 1901c: 125 (s.w.) MEXICO. Combination in *Paracryptocerus* (*Harnedia*): Kempf, 1958a: 109; in *Zacryptocerus:* Brandão, 1991: 384.

***basalis*.** *Cryptocerus basalis* Smith, F. 1876d: 608 (q.) BRAZIL. Combination in *Paracryptocerus:* Kempf, 1951: 232; in *Zacryptocerus:* Brandão, 1991: 384.

**#*biguttatus*.** *Cryptocerus gibbosus* r. *biguttatus* Emery, 1890b: 73, pl. 9, fig. 3 (s.w.q.) COSTA RICA. [Also described as new by Emery, 1894k: 59.] Combination in *Paracryptocerus:* Kempf, 1951: 226; in *Zacryptocerus:* Brandão, 1991: 387. Race/subspecies of *multispinosus:* Forel, 1899c: 50; Kempf, 1972a: 179.

***bimaculatus*.** *Cryptocerus bimaculatus* Smith, F. 1860c: 77, pl. 4, fig. 4 (q.) MEXICO. Combination in *Paracryptocerus* (*Harnedia*): Kempf, 1958a: 136; in *Zacryptocerus:* Brandão, 1991: 384.

***bivestitus*.** *Cryptocerus bivestitus* Santschi, 1922d: 254 (s.w.q.) ARGENTINA. Combination in *Paracryptocerus* (*Harnedia*): Kempf, 1958a: 38; in *Zacryptocerus:* Brandão, 1991: 384. Material of the unavaiable name *tucumana* referred here by Kempf, 1958a: 38.

***bohlsi*.** *Cryptocerus bohlsi* Emery, 1896h: 631, fig. C (s.w.) PARAGUAY. Combination in *Paracryptocerus* (*Harnedia*): Kempf, 1958a: 101; in *Zacryptocerus:* Brandão, 1991: 385. Senior synonym of *medusa:* Kempf, 1958a: 101.

*boliviensis. Cryptocerus cordatus* var. *boliviensis* Santschi, 1921h: 126, fig. 2 (w.q.) BOLIVIA. Junior synonym of *cordatus:* Kempf, 1951: 213.

***borgmeieri*.** *Paracryptocerus* (*Paracryptocerus*) *borgmeieri* Kempf, 1951: 211, fig. 147 (w.q.m.) ARGENTINA. Combination in *Zacryptocerus:* Brandão, 1991: 385. See also: Kempf, 1969: 285.

*brevispinosa. Cryptocerus pusillus* var. *brevispinosa* Santschi, 1921h: 127, fig. 2 (w.) ARGENTINA. Junior synonym of *pusillus:* Kempf, 1951: 176.

***bruchi*.** *Cryptocerus bruchi* Forel, 1912e: 202 (w.) ARGENTINA. Combination in *Paracryptocerus:* Kusnezov, 1953b: 338; in *Zacryptocerus:* Brandão, 1991: 385. Senior synonym of *pampaensis* and *ridiculus:* Kempf, 1958a: 44.

*caustica. Formica caustica* Kollar, in Pohl & Kollar, 1832: 17, fig. 12 (w.) BRAZIL. Combination in *Cryptocerus:* Guérin-Méneville, 1844a: 426. Junior synonym of *minutus:* Emery, 1890b: 74; Forel, 1895b: 134; of *pusillus:* Emery, 1922f: 307; Kempf, 1951: 176.

*cearensis*. *Cryptocerus maculatus* var. *cearensis* Forel, 1908e: 69. *Nomen nudum;* material referred to *maculatus* by Kempf, 1952: 19.

***christopherseni***. *Cryptocerus laminatus* r. *christopherseni* Forel, 1912e: 204 (s.w.) COLOMBIA. Wheeler, G.C. & Wheeler, J. 1986f: 494 (l.). Combination in *C.* (*Paracryptocerus*): Emery, 1922f: 307; in *Paracryptocerus:* Kempf, 1951: 166; in *Zacryptocerus:* Brandão, 1991: 385. Raised to species: Kempf, 1972a: 176.

*clarior*. *Cryptocerus angustus* var. *clarior* Forel, 1912e: 201 (w.) BRAZIL. Junior synonym of *angustus:* Kempf, 1958a: 74.

***clypeatus***. *Cryptocerus clypeatus* Fabricius, 1804: 420 (w.) CENTRAL AMERICA. Smith, F. 1853: 217 (q.m.); Forel, 1906d: 235 (s.); Wheeler, G.C. & Wheeler, J. 1954b: 152 (l.). Combination in *Cephalotes:* Emery, 1914c: 39; in *Cryptocerus* (*Zacryptocerus*): Wheeler, W.M. 1916d: 326; in *Zacryptocerus:* Wheeler, W.M. 1911f: 175; Emery, 1915i: 192; Santschi, 1916e: 283. See also: Kempf, 1951: 136; Kempf, 1973c: 456.

***coffeae***. *Paracryptocerus* (*Harnedia*) *coffeae* Kempf, 1953: 81, figs. 1–13 (s.w.q.m.) COLOMBIA. Combination in *Zacryptocerus:* Brandão, 1991: 385. See also: Kempf, 1958a: 104.

***cognatus***. *Cryptocerus cognatus* Smith, F. 1862d: 411, pl. 13, fig. 4 (q.) BRAZIL. Combination in *C.* (*Paracryptocerus*): Emery, 1922f: 307; in *Paracryptocerus* (*Harnedia*): Kempf, 1958a: 137; in *Zacryptocerus:* Brandão, 1991: 385. Variety of *minutus:* Emery, 1922f: 307. Revived status as species: Kempf, 1958a: 136.

#***columbicus***. *Cryptocerus pusillus* var. *columbica* Forel, 1912e: 201 (s.w.) COLOMBIA. Combination in *Paracryptocerus:* Kempf, 1951: 183; in *Zacryptocerus:* Brandão, 1991: 388.

***complanatus***. *Cryptocerus complanatus* Guérin-Méneville, 1844a: 424 (w.) FRENCH GUIANA. Mann, 1916: 451 (s.). Combination in *C.* (*Paracryptocerus*): Emery, 1915i: 192; in *Paracryptocerus:* Kempf, 1951: 193; in *Zacryptocerus:* Brandão, 1991: 385. Senior synonym of *angulatus:* Emery, 1922f: 307; of *amazonensis:* Kempf, 1951: 193.

***conspersus***. *Cryptocerus conspersus* Smith, F. 1867: 523, pl. 26, fig. 1 (w.) BRAZIL. Combination in *Paracryptocerus:* Kempf, 1951: 232; in *Zacryptocerus:* Brandão, 1991: 385. Senior synonym of *targionii* (and its junior synonyms *denticulatus* and *variegata*): Kempf, 1958d: 108.

*convexus*. *Cryptocerus convexus* Santschi, 1916e: 382, fig. (w.) ARGENTINA. Junior synonym of *quadratus:* Kempf, 1958a: 33.

***cordatus***. *Cryptocerus cordatus* Smith, F. 1853: 220, pl. 21, fig. 3 (w.) BRAZIL. Emery, 1894c: 202 (s.). Combination in *Paracryptocerus:* Kempf, 1951: 213; in *Zacryptocerus:* Brandão, 1991: 385. Senior synonym of *boliviensis:* Kempf, 1951: 213.

***cordiae***. *Cryptocerus cordiae* Stitz, 1913: 207, fig. 1 (s.w.) BRAZIL. Kempf, 1951: 201 (q.). Combination in *C.* (*Paracryptocerus*): Emery, 1922f: 307; in *Paracryptocerus:* Kempf, 1951: 199; in *Zacryptocerus:* Brandão, 1991: 385.

*cordiventris*. *Cryptocerus* (*Paracryptocerus*) *multispinus* var. *cordiventris* Santschi, 1931c: 274 (s.w.) PANAMA. Junior synonym of *multispinus:* Kempf, 1951: 204.

***crenaticeps***. *Cryptocerus crenaticeps* Mayr, 1866a: 515, pl., fig. 15 (q.) COLOMBIA. Combination in *Paracryptocerus* (*Harnedia*): Kempf, 1958a: 138; in *Zacryptocerus:* Brandão, 1991: 385.

***cristatus***. *Cryptocerus cristatus* Emery, 1890b: 72, pl. 9, fig. 2 (s.w.q.) COSTA RICA. [Also described as new by Emery, 1894k: 60.] Combination in *Paracryptocerus:* Kempf, 1951: 216; in *Zacryptocerus:* Brandão, 1991: 385. Senior synonym of *ruficeps:* Kempf, 1951: 216.

***curvistriatus***. *Cryptocerus curvistriatus* Forel, 1899c: 52 (w.) COSTA RICA. Kempf, 1958a: 115 (s.q.m.). Combination in *Paracryptocerus* (*Harnedia*): Kempf, 1958a: 114; in *Zacryptocerus:* Brandão, 1991: 385.

*denticulatus*. *Cryptocerus denticulatus* Emery, 1894c: 206, pl. 3, fig. 17 (w.) BOLIVIA. Junior synonym of *targionii:* Kempf, 1958a: 94.

***depressus***. *Cryptocerus depressus* Klug, 1824: 204 (w.) BRAZIL. Emery, 1906c: 170 (s.q.m.); Forel, 1906d: 235 (q.m.). Combination in *Paracryptocerus:* Kempf, 1951: 218; in *Zacryptocerus:* Brandão, 1991: 385. Junior synonym of *pavonii:* Roger, 1861b: 173; Forel, 1895b: 134. Revived from synonymy: Forel, 1912e: 200. Senior synonym of *sorocabensis:* Kempf, 1951: 218.

*discocephalus*. *Cryptocerus discocephalus* Smith, F. 1853: 222, pl. 20, fig. 2 (w.) BRAZIL. Smith, F. 1858b: 191 (s.m.); Smith, F. 1876d: 608 (q.). Combination in *C.* (*Cyathocephalus*): Emery, 1922f: 311. Variety of *pallens:* Forel, 1911e: 257. Junior synonym of *pallens:* Emery, 1894c: 212; Kempf, 1958a: 151.

***dorbignyanus***. *Cryptocerus dorbignyanus* Smith, F. 1853: 218, pl. 19, fig. 5 (q.) SOUTH AMERICA. Combination in *Paracryptocerus:* Kempf, 1951: 232; in *Zacryptocerus:* Brandão, 1991: 385. Junior synonym of *pavonii:* Roger, 1861b: 173; Forel, 1895b: 134. Revived from synonymy: Emery, 1922f: 309.

***duckei***. *Cryptocerus duckei* Forel, 1906d: 233, fig. 1 (w.) BRAZIL. Combination in *C.* (*Paracryptocerus*): Emery, 1922f: 307; in *Paracryptocerus:* Kempf, 1951: 231; in *Zacryptocerus:* Kempf, 1974a: 69. See also: Kempf, 1967e: 363.

***eduarduli***. *Cryptocerus eduarduli* Forel, 1921a: 204 (s.w.q.) PARAGUAY. Combination in *Paracryptocerus:* Kempf, 1951: 221; in *Zacryptocerus:* Brandão, 1991: 386.

*elegans*. *Cryptocerus elegans* Smith, F. 1853: 222, pl. 19, fig. 3 (w.) COLOMBIA. Junior synonym of *umbraculatus:* Emery, 1890b: 72.

*ellenriederi*. *Cryptocerus peltatus* subsp. *ellenriederi* Forel, 1911e: 258 (s.w.q.) ARGENTINA. Combination in *Paracryptocerus:* Kusnezov, 1953b: 338. Junior synonym if *jheringi:* Kempf, 1958a: 50.

*elongatus*. *Cryptocerus elongatus* Klug, 1824: 214 (q.) BRAZIL. Smith, F. 1860c: 75 (w.); Smith, F. 1862d: 409 (s.). Junior synonym of *pusillus:* Roger, 1862c: 297; Kempf, 1951: 176.

***emeryi***. *Cryptocerus emeryi* Forel, 1912e: 203 (w.) CURACAO I. Combination in *Paracryptocerus:* Kempf, 1951: 233; in *Zacryptocerus:* Brandão, 1991: 386.

*exiguus*. *Cryptocerus exiguus* Smith, F. 1867: 524, pl. 26, fig. 4 (w.) MEXICO. Combination in *C.* (*Paracryptocerus*): Emery, 1922f: 307; in *Paracryptocerus:* Kempf, 1951: 232. Junior synonym of *minutus:* Kempf, 1958a: 110.

***femoralis***. *Cryptocerus femoralis* Smith, F. 1853: 219, pl. 20, fig. 3 (w.) COLOMBIA. Combination in *Paracryptocerus:* Kempf, 1951: 202; in *Zacryptocerus:* Brandão, 1991: 386. Subspecies of *complanatus:* Emery, 1890b: 75. Junior synonym of *complanatus:* Roger, 1861b: 173; Dalla Torre, 1893: 142; Emery, 1922f: 307. Revived from synonymy, revived status as species, and senior synonym of *silvae:* Kempf, 1951: 202.

*fervidus*. *Cryptocerus fervidus* Smith, F. 1876d: 605, pl. 11, fig. 1 (q.) BRAZIL. Junior synonym of *membranaceus:* Kempf, 1958a: 135.

*fides*. *Cryptocerus lobigaster* var. *fides* Santschi, 1925e: 162 (s.w.q.) ARGENTINA. Junior synonym of *liogaster:* Kempf, 1958a: 25.

***fiebrigi***. *Cryptocerus pilosus* r. *fiebrigi* Forel, 1906d: 235 (s.w.) PARAGUAY. Combination in *C.* (*Paracryptocerus*): Santschi, 1919f: 45; in *Paracryptocerus* (*Harnedia*): Kempf, 1958a: 28; in *Zacryptocerus:* Brandão, 1991: 386. Raised to species, and senior synonym of *guttifer* and *trivialis:* Kempf, 1958a: 28.

*flavomaculatus*. *Cryptocerus flavomaculatus* Mayr, 1862: 757 (q.) BRAZIL. Junior synonym of *umbraculatus:* Emery, 1890b: 72.

*fleddermanni*. *Paracryptocerus* (*Harnedia*) *fleddermanni* Kempf, 1958a: 87, pl. 2, fig. 3; pl. 3, fig. 4; pl. 5, fig. 1 (s.w.q.) BRAZIL. Junior synonym of *notatus:* Kempf, 1967e: 367.

***foliaceus***. *Cryptocerus foliaceus* Emery, 1906c: 172 (footnote), fig. 32 (w.) PERU. Kempf, 1952: 8 (s.). Combination in *Paracryptocerus* (*Harnedia*): Kempf, 1952: 7; in *Zacryptocerus:* Brandão, 1991: 386.

***fossithorax***. *Cryptocerus fossithorax* Santschi, 1921h: 125, fig. 1 (s.w.) ARGENTINA. Combination in *Paracryptocerus* (*Harnedia*): Kempf, 1958a: 58; in *Zacryptocerus:* Brandão, 1991: 386.

***frigidus***. *Paracryptocerus* (*Harnedia*) *frigidus* Kempf, 1960f: 440, fig. 19 (w.) BRAZIL. Combination in *Zacryptocerus:* Brandão, 1991: 386.

*gaudens*. *Cryptocerus peltatus* st. *ellenriederi* var. *gaudens* Santschi, 1922d: 255 (s.w.q) ARGENTINA. Unavailable name; material referred to *jheringi* by Kempf, 1958a: 50.

*gibbosus*. *Cryptocerus gibbosus* Smith, F. 1876d: 605, pl. 2, fig. 3 (not 2) (w.) MEXICO. Emery, 1890b: 73 (s.). Junior synonym of *multispinosus:* Emery, 1892b: 167.

***goeldii***. *Cryptocerus goeldii* Forel, 1912e: 205 (w.) BRAZIL. Combination in *Paracryptocerus* (*Harnedia*): Kempf, 1958a: 93; in *Zacryptocerus:* Brandão, 1991: 386.

***grandinosus***. *Cryptocerus grandinosus* Smith, F. 1860c: 76, pl. 4, fig. 5 (w.) BRAZIL. Emery, 1894c: 219 (s.q.). Combination in *C.* (*Cyathocephalus*): Santschi, 1919f: 47; in *Paracryptocerus* (*Harnedia*): Kempf, 1952: 11; in *Zacryptocerus:* Brandão, 1991: 386. Junior synonym of *pinelii:* Emery, 1890b: 75; Forel, 1895b: 134. Revived from synonymy: Luederwaldt, 1918: 41. Senior synonym of *magdalenensis* and *nevadensis:* Kempf, 1952: 11.

*guttifer*. *Cryptocerus* (*Paracryptocerus*) *guttifer* Santschi, 1919f: 45 (s.q.) ARGENTINA. Santschi, 1922d: 253 (w.m.). [Misspelled as *guttatus* by Santschi, 1929d: 301.] Junior synonym of *fiebrigi:* Kempf, 1958a: 28.

***haemorrhoidalis.*** *Formica haemorrhoidalis* Latreille, 1802c: 276 (w.) HAITI. Combination in *Cryptocerus:* Lepeletier, 1835: 172; in *C.* (*Paracryptocerus*): Emery, 1915i: 192; in *C.* (*Hypocryptocerus*): Wheeler, W.M. 1920: 53; in *Hypocryptocerus:* Wheeler, W.M. 1936b: 200; in *Zacryptocerus:* Brandão, 1991: 384. Current subspecies: nominal plus *affinis, auricomus, hamulus, signatus, unimaculatus, vinosus.* See also: Kempf, 1951: 146.

#***hamulus.*** *Cryptocerus hamulus* Roger, 1863a: 209 (w.) DOMINICA. Combination in *C.* (*Paracryptocerus*): Emery, 1922f: 307; in *Hypocryptocerus:* Wheeler, W.M. 1936b: 201; in *Zacryptocerus:* Brandão, 1991: 384. Junior synonym of *haemorrhoidalis:* Emery, 1922f: 307; Menozzi & Russo, 1930: 162. Revived from synonymy, status as subspecies of *haemorrhoidalis:* Wheeler, W.M. 1936b: 201. Senior synonym of *haytianus:* Kempf, 1951: 149.

*haytianus. Cryptocerus hamulus* var. *haytianus* Forel, 1901e: 337 (w.) HAITI. Junior synonym of *haemorrhoidalis:* Wheeler, W.M. & Mann, 1914: 38; of *hamulus:* Kempf, 1951: 149.

***inaequalis.*** *Cryptocerus* (*Cryptocerus*) *inaequalis* Mann, 1916: 449 (w.) BRAZIL. Combination in *C.* (*Paracryptocerus*): Emery, 1922f: 307; in *Paracryptocerus:* Kempf, 1951: 163; in *Zacryptocerus:* Brandão, 1991: 386.

***inca.*** *Cryptocerus multispinus* st. *inca* Santschi, 1911i: 278 (w.) PERU. [Also described as new by Santschi, 1913h: 40.] Combination in *C.* (*Paracryptocerus*): Emery, 1922f: 307; in *Paracryptocerus:* Kempf, 1951: 208; in *Zacryptocerus:* Brandão, 1991: 386. Raised to species: Kempf, 1967e: 362.

***incertus.*** *Cryptocerus incertus* Emery, 1906c: 171, fig. 31 (w.) ARGENTINA. Kempf, 1952: 16 (q.); Kempf, 1974a: 71 (s.). Combination in *C.* (*Cyathocephalus*): Santschi, 1921h: 125; in *Paracryptocerus* (*Harnedia*): Kempf, 1952: 15; in *Zacryptocerus:* Kempf, 1974a: 71.

***insularis.*** *Cryptocerus pilosus* subsp. *insularis* Wheeler, W.M. 1934f: 140 (w.) MEXICO. Combination in *Paracryptocerus* (*Harnedia*): Kempf, 1958a: 128; in *Zacryptocerus:* Brandão, 1991: 386. Raised to species: Kempf, 1958a: 128.

*jamaicensis. Cryptocerus* (*Cyathocephalus*) *varians* var. *jamaicensis* Forel, 1922: 97 (w.) JAMAICA. Junior synonym of *varians:* Kempf, 1958a: 155.

***jheringi.*** *Cryptocerus jheringi* Emery, 1894c: 205, pl. 3, figs. 13, 14 (s.w.) BRAZIL. [Also described as new by Emery, 1894k: 384 (footnote).] Combination in *Paracryptocerus:* Kusnezov, 1953b: 338; in *Zacryptocerus:* Brandão, 1991: 386. Senior synonym of *ellenriederi* and *peltatus,* and material of the unavailable names *gaudens* and *jocans* referred here: Kempf, 1958a: 50.

*jocans. Cryptocerus peltatus* r. *ellenriederi* var. *jocans* Forel, 1913l: 235 (s.) ARGENTINA. Unavailable name; material referred to *jheringi* by Kempf, 1958a: 50.

*jucundus. Cryptocerus jucundus* Smith, F. 1876d: 606, pl. 11, fig. 2 (w.) MEXICO. Junior synonym of *angulosus:* Emery, 1890b: 76.

***klugi.*** *Cryptocerus klugi* Emery, 1894c: 210, pl. 4, figs. 27, 28 (q.) BRAZIL. Combination in *Paracryptocerus* (*Harnedia*): Kempf, 1958a: 139; in *Zacryptocerus:* Brandão, 1991: 386.

***laminatus.*** *Cryptocerus laminatus* Smith, F. 1860c: 76, pl. 4, fig. 3 (w.) BRAZIL. Emery, 1894c: 203 (s.). Combination in *C.* (*Paracryptocerus*): Emery, 1915i: 192; in *Paracryptocerus:* Kempf, 1951: 164; in *Zacryptocerus:* Brandão, 1991: 386.

*lanuginosus. Cryptocerus* (*Paracryptocerus*) *lanuginosus* Santschi, 1919f: 46 (w.) ARGENTINA. Junior synonym of *liogaster:* Kempf, 1958a: 25.

***liogaster.*** *Cryptocerus liogaster* Santschi, 1916e: 381, fig. (w.) ARGENTINA. Combination in *Paracryptocerus* (*Harnedia*): Kempf, 1958a: 25; in *Zacryptocerus:* Brandão, 1991: 386. Senior synonym of *fides, lanuginosus* and *lobigaster:* Kempf, 1958a: 25.

*lobigaster. Cryptocerus* (*Paracryptocerus*) *lobigaster* Santschi, 1919f: 47, fig. 4 (w.) ARGENTINA. Junior synonym of *liogaster:* Kempf, 1958a: 25.

***maculatus.*** *Cryptocerus maculatus* Smith, F. 1876d: 607, pl. 11, fig. 6 (q.) BRAZIL. Emery, 1894c: 208 (s.w.); Forel, 1905b: 158 (m.); Wheeler, G.C. & Wheeler, J. 1954b: 156 (l.). Combination in *Paracryptocerus* (*Harnedia*): Kempf, 1952: 18; in *Zacryptocerus:* Brandão, 1991: 387. Senior synonym of *nanus,* and material of the unavailable name *cearensis* referred here: Kempf, 1952: 19. See also: Forel, 1911e: 258; Menozzi, 1927d: 336.

*magdalenensis. Cryptocerus grandinosus* r. *magdalenensis* Forel, 1899b: 303 (w.) COLOMBIA. [Also described as new by Forel, 1899d: 275.] Junior synonym of *grandinosus:* Kempf, 1952: 11.

***manni.*** *Paracryptocerus* (*Paracryptocerus*) *manni* Kempf, 1951: 228, fig. 127, 128 (s.w.) BRAZIL. Combination in *Zacryptocerus:* Brandão, 1991: 387.

*marginata. Cryptocerus varians* subsp. *marginata* Wheeler, W.M. & Mann, 1914: 39, fig. 17 (s.w.q.) HAITI. [Unresolved junior primary homonym of *Cryptocerus marginatus* Fabricius, 1804: 419 (now in *Cephalotes*).] Junior synonym of *pallens:* Kempf, 1958a: 151.

*medusa. Cryptocerus* (*Paracryptocerus*) *bohlsi* var. *medusa* Santschi, 1919f: 45 (s.) BRAZIL. Junior synonym of *bohlsi:* Kempf, 1958a: 101.

***membranaceus.*** *Cryptocerus membranaceus* Klug, 1824: 208 (w.) BRAZIL. Kempf, 1973c: 453 (s.). Combination in *Cephalotes:* Emery, 1914c: 39; in *Zacryptocerus:* Emery, 1915i: 192. Senior synonym of *fervidus:* Kempf, 1958a: 135. See also: Kempf, 1951: 141.

***minutus.*** *Cryptocerus minutus* Fabricius, 1804: 420 (w.) CENTRAL AMERICA. Smith, F. 1862d: 409 (s.); Emery, 1890b: 74 (q.); Wheeler, G.C. & Wheeler, J. 1954b: 155 (l.). Combination in *C.* (*Paracryptocerus*): Emery, 1915i: 192; in *Paracryptocerus:* Kempf, 1951: 169; in *Zacryptocerus:* Brandão, 1991: 387. Senior synonym of *quadrimaculatus* and *volxemi:* Emery, 1890a: 68; Emery, 1890b: 74; of *exiguus:* Kempf, 1958a: 110.

***multispinosus.*** *Cryptocerus multispinosus* Norton, 1868a: 72, pl. 2, fig. 11 (s.) MEXICO. Norton, 1868c: 9 (s.). Combination in *Paracryptocerus:* Kempf, 1951: 224; in *Zacryptocerus:* Brandão, 1991: 387. Senior synonym of *gibbosus:* Emery, 1892b: 167. Current subspecies: nominal plus *biguttatus.*

***multispinus.*** *Cryptocerus cordatus* r. *multispinus* Emery, 1890b: 75, pl. 9, fig. 5 (w.) COSTA RICA. [Also described as new by Emery, 1894k: 59.] Emery, 1894c: 202 (s.); Forel, 1899c: 49 (q.); Wheeler, G.C. & Wheeler, J. 1954b: 155 (l.). Combination in *C.* (*Paracryptocerus*): Emery, 1915i: 192; in *Paracryptocerus:* Kempf, 1951: 204; in *Zacryptocerus:* Brandão, 1991: 387. Raised to species: Dalla Torre, 1893: 143; Emery, 1894c: 201. Senior synonym of *cordiventris:* Kempf, 1951: 204.

*multistrigus. Zacryptocerus multistrigus* Ashmead, 1905b: 384. *Nomen nudum.*

*nanus. Cryptocerus maculatus* r. *nanus* Forel, 1912e: 202 (w.) TRINIDAD. Junior synonym of *maculatus:* Kempf, 1952: 19.

*nevadensis. Cryptocerus grandinosus* var. *nevadensis* Forel, 1912e: 202 (w.) COLOMBIA. Junior synonym of *grandinosus:* Kempf, 1952: 11.

***notatus.*** *Cryptocerus notatus* Mayr, 1866b: 907, pl. 20, fig. 16 (q.) BRAZIL. Combination in *Paracryptocerus* (*Harnedia*): Kempf, 1958a: 140; in *Zacryptocerus:* Brandão, 1991: 387. Senior synonym of *fleddermanni:* Kempf, 1967e: 367.

*obtusus. Cryptocerus obtusus* Smith, F. 1858b: 191 (w.) BRAZIL. Junior synonym of *pusillus:* Forel, 1911e: 262.

*orbis. Cryptocerus setulifer* r. *orbis* Forel, 1906d: 231 (s.w.) COSTA RICA. Combination in *C.* (*Cyathocephalus*): Emery, 1922f: 311. Junior synonym of *setulifer:* Kempf, 1958a: 146.

***pallens.*** *Cryptocerus pallens* Klug, 1824: 206 (w.) BRAZIL. Emery, 1894c: 212 (s.q.); Emery, 1896h: 635 (m.); Wheeler, G.C. & Wheeler, J. 1973b: 79 (l.). Combination in *C.* (*Cyathocephalus*): Emery, 1915i: 192; in *C.* (*Cyathomyrmex*): Creighton, 1933: 100; in *Zacryptocerus:* Brandão, 1991: 387. Senior synonym of *araneolus* and *discocephalus:* Emery, 1894c: 212; Kempf, 1958a: 151; of *marginata, patellaris* and *porrasi:* Kempf, 1958a: 151.

*pallidicephalus. Cryptocerus pallidicephalus* Smith, F. 1876d: 606, pl. 11, fig. 5 (q.) MEXICO. Junior synonym of *scutulatus:* Kempf, 1963c: 438.

*pampaensis. Cryptocerus* (*Cryptocerus*) *iheringi* st. *pampaensis* Santschi, 1931e: 277 (s.w.) ARGENTINA. Junior synonym of *bruchi:* Kempf, 1958a: 44.

***patei.*** *Paracryptocerus* (*Harnedia*) *patei* Kempf, 1951: 235, figs. 151, 153–155 (s.w.q.m.) COLOMBIA. Combination in *Zacryptocerus:* Brandão, 1991: 387.

*patellaris. Cryptocerus patellaris* Mayr, 1866b: 907, pl. 20, fig. 15 (q.) BRAZIL. Combination in *C.* (*Cyathocephalus*): Emery, 1922f: 311. Variety of *pallens:* Emery, 1894c: 213. Junior synonym of *pallens:* Kempf, 1958a: 151.

***pavonii.*** *Cryptocerus pavonii* Latreille, 1809: 132 (w.q.) SOUTH AMERICA. Emery, 1890b: 73 (s.w.q.); Kempf, 1951: 224 (s.). Combination in *Paracryptocerus:* Kempf, 1951: 222; in *Zacryptocerus:* Brandão, 1991: 387.

*peltatus. Cryptocerus peltatus* Emery, 1896h: 633, fig. D (s.q.) PARAGUAY. Combination in *Paracryptocerus:* Kusnezov, 1953b: 338. Junior synonym of *jheringi:* Kempf, 1958a: 50.

*peruvianus. Cryptocerus laminatus* subsp. *peruvianus* Forel, 1911c: 297 (s.) PERU. Combination in *C.* (*Paracryptocerus*): Emery, 1922f: 308; in *Paracryptocerus:* Kempf, 1951: 191. Subspecies of *spinosus:* Forel, 1911e: 260. Junior synonym of *spinosus:* Kempf, 1967e: 361.

***pilosus.*** *Cryptocerus pilosus* Emery, 1896h: 630, fig. B (s.w.) PARAGUAY. Santschi, 1921h: 127 (q.). Combination in *Paracryptocerus* (*Harnedia*): Kempf, 1958a: 20; in *Zacryptocerus:* Brandão, 1991: 387.

***pinelii.*** *Cryptocerus pinelii* Guérin-Méneville, 1844a: 425 (w.) BRAZIL. Emery, 1890b: 75 (s.q.); Emery, 1894c: 208 (s.); Forel, 1908c: 354 (m.). Combination in *C.* (*Cyathocephalus*): Santschi, 1916e: 283; in *Paracryptocerus* (*Harnedia*): Kempf, 1952: 22; in *Zacryptocerus:* Brandão, 1991: 388.

*porrasi. Cryptocerus* (*Cyathocephalus*) *pallens* var. *porrasi* Wheeler, W.M. 1942: 210, pl. 52 (s.w.q.m.) PANAMA. Junior synonym of *pallens:* Kempf, 1958a: 151.

***prodigiosus.*** *Cryptocerus* (*Cyathocephalus*) *prodigiosus* Santschi, 1921h: 124 (s.q.) ARGENTINA. Combination in *Paracryptocerus* (*Harnedia*): Kempf, 1958a: 49; in *Zacryptocerus:* Brandão, 1991: 388. See also: Kempf, 1964b: 252.

*punctatus. Cryptocerus punctatus* Mayr, 1862: 762 (w.) BRAZIL. Junior synonym of *spinosus:* Dalla Torre, 1893: 144; Emery, 1922f: 308.

***pusillus.*** *Cryptocerus pusillus* Klug, 1824: 202 (w.) BRAZIL. Emery, 1890b: 75 (s.q.); Emery, 1896h: 636 (m.); Wheeler, G.C. & Wheeler, J. 1954b: 156 (l.). Combination in *C.* (*Paracryptocerus*): Emery, 1915i: 192; in *Paracryptocerus:* Kempf, 1951: 176; in *Zacryptocerus:* Brandão, 1991: 388. Senior synonym of *elongatus:* Roger, 1862c: 297; of *obtusus:* Forel, 1911e: 262; of *caustica:* Emery, 1922f: 307; of *brevispinosa,* and material of the unavailable name *rufescens* referred here: Kempf, 1951: 176. Current subspecies: nominal plus *columbicus.*

***quadratus.*** *Cryptocerus quadratus* Mayr, 1868b: 175 (s.) ARGENTINA. Emery, 1896h: 634 (w.q.). Combination in *Paracryptocerus* (*Harnedia*): Kempf, 1958a: 33; in *Zacryptocerus:* Brandão, 1991: 388. Senior synonym of *convexus:* Kempf, 1958a: 33.

*quadriguttatus. Cryptocerus quadriguttatus* Guérin-Méneville, 1844a: 425 (w.) COLOMBIA. Junior synonym of *umbraculatus:* Emery, 1890b: 72.

*quadrimaculatus. Cryptocerus quadrimaculatus* Klug, 1824: 215 (q.) BRAZIL. Smith, F. 1853: 219 (q.); Smith, F. 1860c: 74 (w.). Junior synonym of *minutus:* Emery, 1890a: 68; Emery, 1890b: 74.

***ramiphilus.*** *Cryptocerus complanatus* r. *ramiphilus* Forel, 1904e: 678 (s.w.) BRAZIL. Combination in *C.* (*Paracryptocerus*): Emery, 1922f: 307; in *Paracryptocerus:* Kempf, 1951: 196; in *Zacryptocerus:* Brandão, 1991: 388. Raised to species: Kempf, 1967e: 362.

*reichenspergeri* Santschi, 1921 [*Paracryptocerus*]; see under ***PROCRYPTOCERUS.***

*ridiculus. Cryptocerus ridiculus* Santschi, 1915b: 207, fig. 1 (s.) ARGENTINA. Bruch, 1917b: 269 (q.). Junior synonym of *bruchi:* Kempf, 1958a: 44.

***rohweri.*** *Cryptocerus* (*Cyathocephalus*) *rohweri* Wheeler, W.M. 1916f: 32, fig. 2 (s.w.) U.S.A. Smith, M.R. 1947a: 35 (q.). Combination in *Paracryptocerus* (*Harnedia*): Smith, M.R. 1951a: 825; in *Zacryptocerus:* Smith, D.R. 1979: 1403. See also: Kempf, 1958a: 129; Creighton & Nutting, 1965: 59.

*rufescens. Cryptocerus pusillus* var. *brevispinosus* ab. *rufescens* Santachi, 1921: 128 (w.) ARGENTINA. Unavailable name; material referred to *pusillus* by Kempf, 1951: 176.

*ruficeps. Cryptocerus* (*Cryptocerus*) *cristatus* var. *ruficeps* Santschi, 1929d: 301 (s.) BRAZIL. Junior synonym of *cristatus:* Kempf, 1951: 216.

***scutulatus.*** *Cryptocerus scutulatus* Smith, F. 1867: 524, pl. 26, fig. 3 (s.) MEXICO. Wheeler, W.M. 1907a: 273 (s.w.m.). Combination in *Paracryptocerus* (*Harnedia*): Kempf, 1952: 26; in *Zacryptocerus:* Brandão, 1991: 388. Senior synonym of *angulosus* (and its junior synonym *jucundus*): Emery, 1922f: 311; Kempf, 1958a: 110; of *pallidicephalus:* Kempf, 1963c: 438.

***setulifer.*** *Cryptocerus setulifer* Emery, 1894c: 211, pl. 4, figs. 34–37 (w.q.) COSTA RICA. Wheeler, W.M. 1942: 211 (s.m.). Combination in *C.* (*Cyathocephalus*): Emery, 1915i: 192; in *Paracryptocerus* (*Cyathomyrmex*): Kempf, 1958a: 146; in *Zacryptocerus:* Brandão, 1991: 388. Senior synonym of *orbis:* Kempf, 1958a: 146.

#***signatus.*** *Hypocryptocerus haemorrhoidalis* subsp. *signatus* Wheeler, W.M. 1936b: 203 (w.) HAITI. Combination in *Zacryptocerus:* Brandão, 1991: 384. See also: Kempf, 1951: 150.

*silvae. Cryptocerus silvae* Forel, 1906d: 233 (w.) COLOMBIA. Combination in *C.* (*Paracryptocerus*): Emery, 1922f: 307. Junior synonym of *femoralis:* Kempf, 1951: 202.

***simillimus.*** *Paracryptocerus* (*Paracryptocerus*) *simillimus* Kempf, 1951: 184, figs. 125, 126 (w.) GUIANA. Combination in *Zacryptocerus:* Brandão, 1991: 388.

***sobrius.*** *Paracryptocerus* (*Harnedia*) *sobrius* Kempf, 1958a: 119, pl. 4, fig. 6; pl. 5, fig. 6; pl. 6, fig. 1 (s.q.) GUATEMALA. Combination in *Zacryptocerus:* Brandão, 1991: 388.

***solidus.*** *Zacryptocerus solidus* Kempf, 1974a: 73, figs. 3, 4 (w.) BRAZIL.

*sorocabensis. Cryptocerus depressus* var. *sorocabensis* Forel 1912e: 200 (s.) BRAZIL. Junior synonym of *depressus:* Kempf, 1951: 218.

***spinosus.*** *Cryptocerus spinosus* Mayr, 1862: 761 (w.) BRAZIL. Forel, 1911e: 260 (s.); Kempf, 1951: 187 (q.). Combination in *C.* (*Paracryptocerus*): Emery, 1915i: 192; in *Paracryptocerus:* Kempf, 1951: 187; in *Zacryptocerus:* Brandão, 1991: 388. Senior synonym of *punctatus:* Dalla Torre, 1893: 144; Emery, 1922f: 308; of *peruvianus:* Kempf, 1967e: 361.

*steinheili. Cryptocerus hamulus* subsp. *steinheili* Forel, 1908c: 355 (w.) ANTILLES IS. Combination in *C.* (*Paracryptocerus*), and subspecies of *haemorrhoidalis:* Emery, 1922f: 307. Junior synonym of *unimaculatus:* Kempf, 1951: 151.

***striativentris.*** *Cryptocerus striativentris* Emery, 1894c: 203, pl. 3, figs. 10–12 (s.w.) BRAZIL. [Also described as new by Emery, 1894g: 384 (footnote).] Kempf, 1958a: 83 (q.). Combination in *Paracryptocerus* (*Harnedia*): Kempf, 1958a: 80; in *Zacryptocerus:* Brandão, 1991: 388.

*targionii. Cryptocerus targionii* Emery, 1894c: 205, pl. 3, figs. 15, 16 (s.) BRAZIL. Kempf, 1958a: 98 (q.). Combination in *Paracryptocerus* (*Harnedia*), and senior synonym of *denticulatus* and *variegata:* Kempf, 1958a: 94. Junior synonym of *conspersus:* Kempf, 1958d: 108.

***texanus.*** *Cryptocerus texanus* Santschi, 1915b: 208, fig. 2 (s.w.) U.S.A. Smith, M.R. 1947a: 38 (m.); Creighton & Gregg, 1954: 51 (q.). Combination in *Paracryptocerus* (*Harnedia*): Smith, M.R. 1951a: 825; in *Zacryptocerus:* Smith, D.R. 1979: 1403. See also: Kempf, 1958a: 123.

*trivialis. Cryptocerus* (*Cryptocerus*) *guttatus* var. *trivialis* Santschi, 1929d: 301 (w.) ARGENTINA. Junior synonym of *fiebrigi:* Kempf, 1958a: 28. [The name *guttatus* is a misspelling of *guttifer.*]

*tucumana. Cryptocerus peltatus* r. *ellenriederi* var. *tucumana* Forel, 1914d: 281 (w.) ARGENTINA. Unavailable name; material referred to *bivestitus* by Kempf, 1958a: 38.

***umbraculatus.*** *Cryptocerus umbraculatus* Fabricius, 1804: 420 (w.) CENTRAL AMERICA. Klug, 1824: 211 (q.); Emery, 1890b: 72 (s.); Forel, 1899c: 51 (m.); Wheeler, G.C. & Wheeler, J. 1954b: 157 (l.). Combination in *Paracryptocerus* (*Harnedia*): Smith, M.R. 1949c: 20; in *Zacryptocerus:* Brandão, 1991: 388. Senior synonym of *elegans, flavomaculatus* and *quadriguttatus:* Emery, 1890b: 72; Kempf, 1958a: 142.

#***unimaculatus.*** *Cryptocerus unimaculatus* Smith, F. 1853: 221, pl. 19, fig. 9 (w.) BRAZIL. Combination in *C.* (*Paracryptocerus*): Emery, 1922f: 308; in *Hypocryptocerus:* Kempf, 1951: 151; in *Zacryptocerus:* Brandão, 1991: 384. Subspecies of *haemorrhoidalis,* and senior synonym of *steinheili:* Kempf, 1951: 151.

***ustus.*** *Zacryptocerus ustus* Kempf, 1973c: 450, figs. 1–4 (w.) BRAZIL.

***varians.*** *Cryptocerus varians* Smith, F. 1876d: 606, pl. 11, fig. 6 (w.) CUBA. Wheeler, W.M. 1905b: 102 (s.q.m.); Wheeler, G.C. & Wheeler, J. 1954b: 156 (l.). Combination in *C.* (*Cyathocephalus*): Emery, 1915i: 192; in *C.* (*Cyathomyrmex*): Wheeler, W.M. 1937b: 459; in *Paracryptocerus* (*Cyathomyrmex*): Smith, M.R. 1951a: 825; in *Zacryptocerus:* Smith, D.R. 1979: 1403. Senior synonym of *jamaicensis:* Kempf, 1958a: 155. See also: Wilson, 1976a: 354.

*variegata. Cryptocerus denticulatus* var. *variegata* Forel, 1911e: 262 (w.) BRAZIL. Junior synonym of *targionii:* Kempf, 1958a: 94.

#***vinosus.*** *Hypocryptocerus haemorrhoidalis* subsp. *vinosus* Wheeler, W.M. 1936b: 202 (w.) HAITI. Combination in *Zacryptocerus:* Brandão, 1991: 384. See also: Kempf, 1951: 152.

*volxemi. Cryptocerus volxemi* Emery, 1878a: ix (w.) BRAZIL. Junior synonym of *minutus:* Emery, 1890a: 68; Emery, 1890b: 74.

***wheeleri.*** *Cryptocerus wheeleri* Forel, 1901c: 126 (s.) MEXICO. Wheeler, G.C. & Wheeler, J. 1954b: 157 (l.). Combination in *Paracryptocerus* (*Harnedia*): Kempf, 1958a: 132; in *Zacryptocerus:* Brandão, 1991: 389.

*ZASPHINCTUS:* see under ***SPHINCTOMYRMEX.***

*ZATAPINOMA:* see under ***TAPINOMA.***

#***ZEALLEYELLA:*** see under ***ANOPLOLEPIS.***

****ZHERICHINIUS*** (Russia)

****horribilis.*** **Zherichinius horribilis* Dlussky, 1988: 57, figs. 4, 5 (w.) RUSSIA (Paleocene).

****rapax.*** **Zherichinius rapax* Dlussky, 1988: 57, figs. 4, 5 (w.) RUSSIA (Paleocene).

# References

References are given in alphabetical order by author, and by date order under each author's name. The full title of each reference is given and the names of serial publications are fully written out. The date of publication of each entry which contains critical taxonomic data, as accurately as I have been able to determine it, is recorded in square brackets at the end of each entry. Standard Arabic numerals are used to indicate day and year, the number of the month is given in Roman numerals; hence [21.x.1946.] = twenty-first of October, 1946. In publications where only month and year of issue are given, the date of publication is taken as the last day of that month. In such cases the final day of the month is given in parentheses: [(30).ix.1980.] = (thirtieth) of September, 1980. This is to distinguish the entry from [30.ix.1980.], where the thirtieth of September is the actual date of publication. Similarly, in publications where only the year of issue is recorded, the date of publication is taken as the last day of that year, and both day and month are in parentheses [(31.xii).1923.], to distinguish it from an actual publication date of thirty-first of December, [31.xii.1923.]. For references which do not contain critical taxonomic data but which do contain important secondary information, only the year of publication is stated. In cases where I have not been able to see a particular reference, merely the year of publication is given followed by the statement "not seen" and if possible, an indication of where the reference was acquired, all enclosed within square brackets. When the real date of publication is later than that indicated on the title page, the year in which the publication actually appeared is given immediately after the author's name, and the apparent year is recorded in parentheses after the number of the volume or part in question.


Abe, M. & Smith, D.R. 1991. The genus-group names of Symphyta and their type species. *Esakia* **31**: 1–115. [31.vii.1991.]

Acosta Salmerón, F.J. 1982. Sobre los caracteres morfólogicos de *Goniomma*, con algunas sugerencias sobre su taxonomia. *EOS. Revista Española de Entomologia* **57** (1981): 7–14. [28.iv.1982.]

Adlerz, G. 1896. Myrmekologiska Studier. 3. *Tomognathus sublaevis* Mayr. *Bihang till Kongl. Svenska Vetenskaps-Akademiens Handlingar* **21** (4): 1–76. [1896.]

Adlerz, G. 1902. Myrmekologiska Studier. 4. *Formica suecica* n. sp., eine neue schwedische Ameise. *Öfversigt af Kongl. Vetenskaps-Akademiens Förhandlingar* **59**: 263–265. [12.ix.1902.]

Adlerz, G. 1908. Zwei Gynandromorphen von *Anergates atratulus* Schenck. *Arkiv för Zoologi* **5** (2): 1–6. [12.xi.1908.]

Agassiz, J.L.R., 1848. *Nomenclatoris Zoologici Index Universalis*, continens nomina systematica classium, ordinum, familiarum, et genera animalium omnium, tam viventium quam fossilium: 1135 pp. Soloduri. [(31.xii).1848.]

Agassiz, J.L.R., 1850. In Dixon, F. *The geology and fossils of the Tertiary and Cretaceous formation of Sussex:* 422 pp. London. [(31.xii).1850.]

Agosti, D. 1990a. Review and reclassification of *Cataglyphis*. *Journal of Natural History* **24**: 1457–1505. [19.x.1990.]

Agosti, D. 1990b. What makes the Formicini the Formicini? *Actes des Colloques Insectes Sociaux* **6**: 295–303. [1990.]

Agosti, D. 1991. Revision of the Oriental ant genus *Cladomyrma*, with an outline of the higher classification of the Formicinae. *Systematic Entomology* **16**: 293–310. [(31).vii.1991.]

Agosti, D. 1992. Revision of the ant genus *Myrmoteras* in the Malay Archipelago. *Revue Suisse de Zoologie* **99**: 405–429. [(30).vi.1992.]

Agosti, D. 1994a. The phylogeny of the ant tribe Formicini, with the description of a new genus. *Systematic Entomology* **19**: 93–117. [1.viii.1994.]

Agosti, D. 1994b. A new inquiline ant in *Cataglyphis* and its phylogenetic relationship. *Journal of Natural History* **28**: 913–919. [29.vii.1994.]

Agosti, D. & Bolton, B. 1990a. The identity of *Andragnathus*, a forgotten formicine ant genus. *Entomologist's Monthly Magazine* **126**: 75–77. [12.iv.1990.]

Agosti, D. & Bolton, B. 1990b. New characters to differentiate the ant genera *Lasius* F. and *Formica* L. *Entomologist's Gazette* **41**: 149–156. [30.ix.1990.]

Agosti, D. & Collingwood, C.A. 1987a. A provisional list of the Balkan ants and a key to the worker caste. 1. Synonymic list. *Mitteilungen der Schweizerischen Entomologischen Gesellschaft* **60**: 51–62. [31.viii.1987.]

Agosti, D. & Collingwood, C.A. 1987b. A provisional list of the Balkan ants with a key to the worker caste. 2. Key to the worker caste, including the European species without the Iberian. *Mitteilungen der Schweizerischen Entomologischen Gesellschaft* **60**: 261–293. [15.xii.1987.]

Agosti, D. & Hauschteck-Jungen, E. 1988. Polymorphism of males in *Formica exsecta* Nyl. *Insectes Sociaux* **34** (1987): 280–290. [(30).vi.1988.]

Aguayo, C.G. 1931. New ants of the genus *Macromischa*. *Psyche* **38**: 175–185. [(31).xii.1931.]

Aguayo, C.G. 1932. Notes on West Indian ants. *Bulletin of the Brooklyn Entomological Society* **27**: 215–227. [29.xi.1932.]

Aguayo, C.G. 1937. Nueva especie de hormiga cubana. *Memorias de la Sociedad Cubana de Historia Natural* **11**: 235–236. [10.vii.1937.]

Aktaç, N. 1977. Studies on the myrmecofauna of Turkey. 1. Ants of Siirt, Bodrum, and Trabzon. *Istanbul Universitesi Fen Fakültesi Mecmuasi.* Seri B. **41** (1976): 115–135. [1977.]

Alayo, P.D. 1974. Introduccion al estudio de los Himenopteros de Cuba. Superfamilia Formicoidea. *Academia de Ciencias de Cuba. Instituto de Zoologia. Serie Biologica* no.**53**: 58 pp. La Habana. [(31.xii).1974.]

Alayo, P.D. & Zayas Montero, L. 1977. Estudios sobre los himenópteros de Cuba. 7. Dos nuevas especies para la fauna mirmecológica Cubana. *Poeyana* **174**: 1–5. [4.x.1977.]

Almeida, F.A.J. 1986. Descriçao de quatro machos do gênero *Ectatomma* Smith, 1858. *Quid* **6**: 24–38. [Not seen. Reference from Brandão, 1991: 394.]

Almeida, F.A.J. 1987. Descriçao de seis fêmeas do gênero *Ectatomma* Smith, 1858. *Anais Soc. Nord. Zool.* **1**: 175–183. [Not seen. Reference from Brandão, 1991: 394.]

Alpert, G.D. 1992. Observations on the genus *Terataner* in Madagascar. *Psyche* **99**: 117–127. [1992.]

André, E. 1874. Description des fourmis d'Europe pour servir à l'étude des insectes myrmécophiles. *Revue et Magasin de Zoologie* (3) **2**: 152–335. [(31.xii).1874.]

André, E. 1881a. Untitled contribution in "séance du 13 avril 1881," introduced by, "M. Ernest André, de Gray, adresse les descriptions de trois nouvelles espèces de fourmis." *Bulletin Bimensuel de la Société Entomologique de France* **1881, no. 7**: 60–62. [23.iv.1881.] Reissued with same title in *Annales de la Société Entomologique de France* (*Bulletin des Séances*) (6) **1**: xlviii-l. [Page xlviii published 22.vii.1881; pp. xlix-l published 12.x.1881.]

André, E. 1881b. Catalogue raisonné des formicides provenant du Voyage en Orient de M. Abeille de Perrin et description des espèces nouvelles. *Annales de la Société Entomologique de France* (6) **1**: 53–78. [22.vii.1881.]

André, E. 1882a. *Species des Hyménoptères d'Europe et d'Algérie* **2**: 81–152. Beaune. [(30).iv.1882.]

André, E. 1882b. *Species des Hyménoptères d'Europe et d'Algérie* **2**: 153–232. Beaune. [(31).vii.1882.]

André, E. 1882c. *Species des Hyménoptères d'Europe et d'Algérie* **2**: 233–280. Beaune. [(31).x.1882.]

André, E. 1883a. *Species des Hyménoptères d'Europe et d'Algérie* **2**: 281–344. Beaune. [(31).i.1883.]

André, E. 1883b. *Species des Hyménoptères d'Europe et d'Algérie* **2**: 345–404. Beaune. [(30).iv.1883.]

André, E. 1884a. In Magretti, P. Raccolte imenotterologiche nell'Africa orientale. *Bullettino della Società Entomologica Italiana* **15** (1883): 241–253. [15.iv.1884.]

André, E. 1884b. In Magretti, P. Risultati di raccolte imenotterologiche nell'Africa orientale. *Annali del Museo Civico di Storia Naturale di Genova* (2) **1** [21]: 523–636. [15.xi.1884.]

André, E. 1885. *Species des Hyménoptères d'Europe et d'Algérie* **2** (supplément aux fourmis): 833–854. Beaune. [(30).xi.1885.]

André, E. 1886. *Species des Hyménoptères d'Europe et d'Algérie* **2** (appendice aux fourmis): 854–859. [(31).i.1886.]

André, E. 1887. Description de quelques fourmis nouvelles ou imparfaitement connues. *Revue d'Entomologie* **6**: 280–298. [(30).xi.1887.]

André, E. 1889. Hyménoptères nouveaux appartenant au groupe des formicides. *Revue d'Entomologie* **8**: 217–231. [(31).viii.1889.]

André, E. 1890. Matériaux pour servir à la faune myrmécologique de Sierra Leone (Afrique occidentale). *Revue d'Entomologie* **9**: 311–327. [(30).xi.1890.]

André, E. 1892a. Matériaux myrmécologiques. *Revue d'Entomologie* **11**: 45–56. [(28).ii.1892.]

André, E. 1892b. Voyage de M. Chaper à Borneo. Catalogue des fourmis et description des espèces nouvelles. *Mémoires de la Société Zoologique de France* **5**: 46–55. [(31.xii).1892.]

André, E. 1893a. Untitled contribution introduced by, "M. E. André (de Gray) envoie la description d'une nouvelle espèce de Fourmi de Tunisie." *Bulletin Bimensuel de la Société Entomologique de France* **1893, no. 8**: cxci-cxcii. [6.v.1893.] (Note. Article repeated in *Annales de la Société Entomologique de France (Bulletin)* **62**: cxci-cxcii. [30.xii.1893.])

André, E. 1893b. Description de quatre espèces nouvelles de fourmis d'Amerique. *Revue d'Entomologie* **12**: 148–152. [(31).vii.1893.]

André, E. 1895a. Formicides de l'Ogooué (Congo Français). *Revue d'Entomologie* **14**: 1–5. [(31).i.1895.]

André, E. 1895b. Notice sur les fourmis fossiles de l'ambre de la Baltique et description de deux espèces nouvelles. *Bulletin de la Société Zoologique de France* **20**: 80–84. [(31).iii.1895.]

André, E. 1896a. Description d'une nouvelle fourmi de France. *Bulletin de la Société Entomologique de France* **1896**: 367–368. [28.x.1896.]

André, E. 1896b. Fourmis nouvelles d'Asie et d'Australie. *Revue d'Entomologie* **15**: 251–265. [(30).xi.1896.]

André, E. 1898. Description de deux nouvelles fourmis de Mexique. *Bulletin de la Société Entomologique de France* **1898**: 244–247. [22.vi.1898.]

André, E. 1902. Description de deux nouvelles fourmis de Pérou. *Bulletin de la Société Entomologique de France* **1902**: 14–17. [1.ii.1902.]

André, E. 1903. Description d'une nouvelle espèce de *Dorymyrmex* et tableau dichotomique des ouvrières de ce genre. *Zeitschrift für Systematische Hymenopterologie und Dipterologie* **3**: 364–365. [1.xi.1903.]

André, E. 1905. Description d'un genre nouveau et deux espèces nouvelles de fourmis d'Australie. *Revue d'Entomologie* **24**: 205–208. [(31).viii.1905.]

Andrews, H. 1916. A new ant of the genus *Messor* from Colorado. *Psyche* **23**: 81–83. [(30).vi.1916.]

Arakelian, G.R. 1989. Neizvestnye nauke polovye osobi murav'ya *Myrmica ravasinii* Finzi iz Armenii. *Biologicheskii Zhurnal Armenii* **42**: 733–736. [1989.]

Arakelian, G.R. 1991. Novyi vid murav'ev roda *Diplorhoptrum* Mayr iz Armenii. *Doklady Akademii Nauk Armenii* **92** (2): 93–96. [25.x.1991.]

Arakelian, G.R. 1994. *Fauna Respubliki Armeniya Nasekomye Pereponchatokrylye* Murav'i (Formicidae): 153 pp. Erevan. [20.vi.1994.]

Arnold, G. 1914. Nest-changing migrations of two species of ants. *Proceedings of the Rhodesia Scientific Association* **13**: 250–32. [(31.xii).1914.]

Arnold, G. 1915. A monograph of the Formicidae of South Africa. Part 1. (Ponerinae; Dorylinae.) *Annals of the South African Museum* **14**: 1–159. [11.ii.1915.]

Arnold, G. 1916. A monograph of the Formicidae of South Africa. Part 2. (Ponerinae; Dorylinae.) *Annals of the South African Museum* **14**: 159–270. [20.v.1916.]

Arnold, G. 1917. A monograph of the Formicidae of South Africa. Part 3. (Myrmicinae.) *Annals of the South African Museum* **14**: 271–402. [9.viii.1917.]

Arnold, G. 1920a. A monograph of the Formicidae of South Africa. Part 4. (Myrmicinae.) *Annals of the South African Museum* **14**: 403–578. [16.iv.1920.]

Arnold, G. 1920b. In Santschi, F. Formicides africains et américains nouveaux. *Annales de la Société Entomologique de France* **88** (1919): 361–390. [30.vi.1920.]

Arnold, G. 1922. A monograph of the Formicidae of South Africa. Part 5. (Myrmicinae.) *Annals of the South African Museum* **14**: 579–674. [(31).x.1922.]

Arnold, G. 1924. A monograph of the Formicidae of South Africa. Part 6. (Camponotinae.) *Annals of the South African Museum* **14**: 675–766. [(30).iv.1924.]

Arnold, G. 1926. A monograph of the Formicidae of South Africa. Appendix. *Annals of the South African Museum* **23**: 191–295. [(28).ii.1926.]

Arnold, G. 1944. New species of African Hymenoptera. No. 5. *Occasional Papers of the National Museum of Southern Rhodesia* **2** (No. 11): 1–38. [17.vii.1944.]

Arnold, G. 1946. New species of African Hymenoptera. No. 6. *Occasional Papers of the National Museum of Southern Rhodesia* **2** (No. 12): 49–97. [17.v.1946.]
Arnold, G. 1947. New species of African Hymenoptera. No. 7. *Occasional Papers of the National Museum of Southern Rhodesia* **2** (No. 13): 131–167. [5.xi.1947.]
Arnold, G. 1948. New species of African Hymenoptera. No. 8. *Occasional Papers of the National Museum of Southern Rhodesia* **2** (No. 14): 213–250. [15.vi.1948.]
Arnold, G. 1949. New species of African Hymenoptera. No. 9. *Occasional Papers of the National Museum of Southern Rhodesia* **2** (No. 15): 261–275. [29.xii.1949.]
Arnold, G. 1951. The genus *Hagensia* Forel. *Journal of the Entomological Society of Southern Africa* **14**: 53–56. [31.v.1951.]
Arnold, G. 1952a. New species of African Hymenoptera. No. 10. *Occasional Papers of the National Museum of Southern Rhodesia* **2** (No. 17): 460–493. [2.vii.1952.]
Arnold, G. 1952b. The genus *Terataner* Emery. *Journal of the Entomological Society of Southern Africa* **15**: 129–131. [30.xi.1952.]
Arnold, G. 1954. New Formicidae from Kenya and Uganda. *Annales du Musée Royal du Congo Belge* (N.S. in quarto). Zoologie **1**: 291–295. [(30).vi.1954.]
Arnold, G. 1955. New species of African Hymenoptera. No. 11. *Occasional Papers of the National Museum of Southern Rhodesia* **2** (No. 20): 733–762. [2.xi.1955.]
Arnold, G. 1956. New species of African Hymenoptera. No. 12. *Occasional Papers of the National Museum of Southern Rhodesia* **No. 21B**: 52–77. [(31).xii.1956.]
Arnold, G. 1958. New species of African Hymenoptera. No. 13. *Occasional Papers of the National Museum of Southern Rhodesia* **No. 22B**: 119–143. [16.vii.1958.]
Arnold, G. 1959. New species of African Hymenoptera. No. 14. *Occasional Papers of the National Museum of Southern Rhodesia* **No. 23B**: 316–339. [27.vii.1959.]
Arnold, G. 1960a.Aculeate Hymenoptera from the Drakensberg Mountains, Natal. *Annals of the Natal Museum* **15**: 79–87. [8.xii.1960.]
Arnold, G. 1960b. New species of African Hymenoptera. No. 15. *Occasional Papers of the National Museum of Southern Rhodesia* **No. 24B**: 452–488. [30.xii.1960.]
Arnold, G. 1962. New species of African Hymenoptera. No. 16. *Occasional Papers of the National Museum of Southern Rhodesia* **No. 26B**: 844–855. [31.xii.1962.]
Arnol'di, K.V. Note. In papers transliterated from Russian his name is rendered Arnol'di. In papers published in German his name is given as Arnoldi. For consistency the Russian version is used throughout this catalogue.
Arnol'di, K.V. 1928a. Studien über die Systematik der Ameisen. 1. Allgemeiner Teil. *Zoologischer Anzeiger* **75**: 123–137. [1.ii.1928.]
Arnol'di, K.V. 1928b. Studien über die Systematik der Ameisen. 2. *Stenamma* Westw. *Zoologischer Anzeiger* **75**: 199–215. [15.ii.1928.]
Arnol'di, K.V. 1928c. Studien über die Systematik der Ameisen. 3. *Rossomyrmex.* Neue Gattung der Ameisen und ihre Beziehungen zu den anderen Gattungen der Formicidae. *Zoologischer Anzeiger* **75**: 299–310. [1.iii.1928.]
Arnol'di, K.V. 1930a. Studien über die Systematik der Ameisen. 4. *Aulacopone*, eine neue Ponerinengattung in Russland. *Zoologischer Anzeiger* **89**: 139–144. [10.vi.1930.]
Arnol'di, K.V. 1930b. Studien über die Systematik der Ameisen. 5. Der erste Vertreter der Tribe Proceratiini in USSR. *Zoologischer Anzeiger* **91**: 143–146. [5.x.1930.]
Arnol'di, K.V. 1930c. Studien über die Systematik der Ameisen. 6. Eine neue parasitische Ameise, mit Bezugnahme auf die Frage nach der Entstehung der Gattungsmerkmale bei den parasitären Ameisen. *Zoologischer Anzeiger* **91**: 267–283. [20.x.1930.]
Arnol'di, K.V. 1930d. O predstavitelyakh dvukh novykh dlya Soyuza SSR trib murav'ev-ponerin. *Russkoe Entomologicheskoe Obozrenie* **24**: 156–161. [(31.xii).1930.]
Arnol'di, K.V. 1932a. Biologische Beobachtungen an der neuen paläarktischen Sklavenhalterameise *Rossomyrmex proformicarum* K. Arn. *Zeitschrift für Morphologie und Ökologie der Tiere* **24**: 319–326. [11.i.1932.]
Arnol'di, K.V. 1932b. Studien über die Systematik der Ameisen. 7. Die russischen Poneriden meiner Sammlung, teilweise biometrisch bearbeitet. *Zoologischer Anzeiger* **98**: 49–68. [15.iii.1932.]
Arnol'di, K.V. 1933a. Ob odnom novom rode murav'ev v svyazi s proiskhozhdeniem rodovykh priznakov murav'ev-parasitov. *Entomologicheskoe Obozrenie* **25**: 40–51. [(31.xii).1933.]
Arnol'di, K.V. 1933b. Formicidae-Murav'i (pp. 594–605). In Filip'eva, I.N. & Ogloblina, D.A. (eds.) *Opredelitel' Nasekomykh:* 820 pp. Moscow-Leningrad. [(31.xii).1933.]
Arnol'di, K.V. 1934. Studien über die Systematik der Ameisen. 8. Vorläufige Ergebnisse einer biometrischen Untersuchung einiger *Myrmica*-arten aus dem europäischen Teile der USSR. *Folia Zoologica et Hydrobiologica* **6**: 151–174. [10.x.1934.]
Arnol'di, K.V. 1948. Murav'i Talysha i Diabarskoi kotloviny. Ikh znachenie dlya kharakteristiki tsenozov nazemnykh bespozvonochnykh i istoricheskogo analiza fauny. *Trudy Zoologicheskogo Instituta Izdatel'stvo Akademii Nauk SSR* **7**: 206–262. [(31.xii).1948.]
Arnol'di, K.V. 1964. Vysshie i spetsializirovannye predstaviteli murav'ev-begunkov i faetonchikov roda *Cataglyphis* v faune SSSR. *Zoologichesky Zhurnal* **43**: 1800–1815. [26.xi.1964.]
Arnol'di, K.V. 1967. Novye dallye o murav'yakh roda *Camponotus* fauny SSSR. *Zoologichesky Zhurnal* **46**: 1815–1830. [6.xii.1967.]
Arnol'di, K.V. 1968. Vazhnye dopolneniya k mirmekofaune SSSR i opisanie novykh form. *Zoologichesky Zhurnal* **47**: 1800–1822. [9.xii.1968.]
Arnol'di, K.V. 1969. Novye vidy i rasy murav'ev roda *Messor. Zoologichesky Zhurnal* **49** (1970): 72–88. [22.xii.1969.]
Arnol'di, K.V. 1970. Obzor murav'ev roda *Myrmica* evropeiskoi chasti SSSR. *Zoologichesky Zhurnal* **49**: 1829–1844. [27.xi.1970.]
Arnol'di, K.V. 1971. Novye vidy i obzor murav'ev roda *Leptothorax* ravninnogo Kazakhstani. *Zoologichesky Zhurnal* **50**: 1818–1826. [2.xii.1971.]
Arnol'di, K.V. 1975. Obzor vidov roda *Stenamma* soyuza SSR i opisanie novykh vidov. *Zoologichesky Zhurnal* **54**: 1819–1829. [8.xii.1975.]
Arnol'di, K.V. 1976a. Murav'i roda *Myrmica* Latr. srednei azii i yuzhnogo Kazakhstana. *Zoologichesky Zhurnal* **55**: 547–558. [18.iii.1976.]
Arnol'di, K.V. 1976b. Obzor roda *Aphaenogaster* SSSR. *Zoologichesky Zhurnal* **55**: 1019–1026. [25.vi.1976.]
Arnol'di, K.V. 1977a. Novye i maloizvestnye vidy murav'ev roda *Leptothorax* Mayr evropeiskoi chasti SSSR i Kavkaza. *Entomologicheskoe Obozrenie* **56**: 198–204. [28.ii.1977.]
Arnol'di, K.V. 1977b. Obzor murav'ev-zhnetsov roda *Messor* fauny SSSR. *Zoologichesky Zhurnal* **56**: 1637–1648. [7.xi.1977.]
Arnol'di, K.V. & Dlussky, G.M. 1978. Formicoidea. In Medvedev, G.S. (ed.). *Opredelitel' Nasekomykh Evropeiskoi Chasti SSSR.* **3**. Pereponchatokrylye, pervaya chast': 519–556. Leningrad. [(31.xii).1978.]
Ashmead, W.H. 1905a. New Hymenoptera from the Philippines. *Proceedings of the United States National Museum* **29**: 107–119. [30.ix.1905.]
Ashmead, W.H. 1905b. A skeleton of a new arrangement of the families, subfamilies, tribes and genera of the ants, or the superfamily Formicoidea. *Canadian Entomologist* **37**: 381–384. [1.xi.1905.]
Ashmead, W.H. 1906. Classification of the foraging and driver ants, or family Dorylidae, with a description of the genus *Ctenopyga* Ashm. *Proceedings of the Entomological Society of Washington* **8**: 21–31. [13.vii.1906.]
Assmann, A. 1870. Die fossilen Insekten des tertiären (miocenen) Thonlagers von Schossnitz bei Kanth. 1 Beitrag (pp. 33–58). In, Beiträge zur Insekten-fauna der Vorwelt. *Zeitschrift für Entomologie* (N.F.) **1**: 1–62. [(31).iii.1870.]

Atanassov, N. 1982. Neue Ameisen aus den Gattungen *Messor* und *Cataglyphis* für die Fauna Bulgariens. *Waldhygiene* **14**: 209–214. [(31.xii).1982.]

Atanassov, N. & Dlussky, G.M. 1992. *Fauna na Bulgariya* **22**. Hymenoptera, Formicidae: 310 pp. Sofia. [(30).xi.1992.]

Azuma, M. 1950. On the myrmecological fauna of Tomogashima, Kii Prov., with the description of a new genus and a new species. *Hyogo Biology* **No. 4**: 34–37. [10.i.1950.]

Azuma, M. 1951. On the myrmecological fauna of Osaka Prefecture, Japan, with description of new species. *Hyogo Biology* **1** (No. 5): 86–90. [20.i.1951.]

Azuma, M. 1953. On the myrmecological fauna of Mt Rokko, Hyogo Prefecture. *Warera* **2**: 1–7 (in Japanese). [Not seen. Reference from Onoyama, 1980. *Kontyu* **48**: 205.]

Azuma, M. 1955. A list of ants from Hokkaido Is. *Hyogo Biology* **3**: 1–2. [(31.xii).1955.]

Azuma, M. 1977. On the myrmecological fauna of Mt Rokko, Hyogo, with description of a new species. *Hyogo Biology* **7**: 114–116. [Not seen. Reference from Onoyama, 1980. *Kontyu* **48**: 212.]

Bachmayer, F. 1960. Insektenreste aus den Congerienschichten (Pannon) von Brunn-Vösendorf (südl. von Wien) Niederösterreich. *Sitzungsberichte der Österreichische Akademie der Wissenschaften, Mathematisch-Naturwissenschaftliche Klasse* **169**: 11–16. [(31.xii).1960.]

Baker, G.L. 1976. The seasonal life cycle of *Anoplolepis longipes* (Jerdon) in a cacao plantation and under brushed rain forest in the northern district of Papua New Guinea. *Insectes Sociaux* **23**: 253–261. [(30).ix.1976.]

Baltazar, C.R. 1966. A catalogue of the Philippine Hymenoptera. (With a bibliography, 1758–1963.) *Pacific Insects Monograph* **8**: 1–488. [30.iv.1966.]

Banert, P. & Pisarski, B. 1972. Mrówki Sudetów. *Fragmenta Faunistica* **18**: 345–359. [25.xi.1972.]

Baroni Urbani, C. 1962. Studi sulla mirmecofauna d'Italia. 1. *Redia* **47**: 129–138. [(31.xii).1962.]

Baroni Urbani, C. 1964a. Su alcune formiche raccolte in Turchia. *Annuario dell'Istituto e Museo di Zoologia della Università di Napoli* **16**: 1–12. [(28).ii.1964.]

Baroni Urbani, C. 1964b. Studi sulla mirmecofauna d'Italia. 2. Formiche di Sicilia. *Atti della Accademia Gioenia di Scienze Naturali in Catania* (6) **16**: 25–66. [30.xii.1964.]

Baroni Urbani, C. 1964c. Studi sulla mirmecofauna d'Italia. 3. Formiche dell'Italia Appenninica. *Memorie del Museo Civico di Storia Naturale di Verona* **12**: 149–172. [31.xii.1964.]

Baroni Urbani, C. 1966. Osservazioni diverse intorno al nomadismo dell'*Aphaenogaster picena* Baroni con particolare riguardo all'orientamento. *Insectes Sociaux* **13**: 69–86. [(30).ix.1966.]

Baroni Urbani, C. 1968a. Über die eigenartige Morphologie der männlichen Genitalien des Genus *Diplorhoptrum* Mayr und die taxonomischen Schlussfolgerungen. *Zeitschrift für Morphologie der Tiere* **63**: 63–74. [10.vi.1968.]

Baroni Urbani, C. 1968b. Studi sulla mirmecofauna d'Italia. 4. La fauna mirmecologica delle isole Maltesi ed il suo significato ecologico e biogeografica. *Annali del Museo Civico di Storia Naturale di Genova* **77**: 408–559. [1.xii.1968.]

Baroni Urbani, C. 1969a. Studi sulla mirmecofauna d'Italia. 8. L'Isola di Giannutri ed alcuni scogli minori dell'arcipelago Toscano. *Atti della Società Toscana di Scienze Naturali. Memorie (B)* **75** (1968): 325–338. [(31.xii).1969.]

Baroni Urbani, C. 1969b. Una nuova specie di *Oligomyrmex* del Sahara meridionale. *Bollettino della Società Entomologica Italiana* **99–101**: 74–77. [20.iv.1969.]

Baroni Urbani, C. 1969c. Gli *Strongylognathus* del gruppo *huberi* nell'Europa occidentale: saggio di una revisione basata sulla casta operaia. *Bollettino della Società Entomologica Italiana* **99–101**: 132–168. [20.x.1969.]

Baroni Urbani, C. 1969d. Una nuova *Cataglyphis* dei monti dell'Anatolia. *Frammenta Entomologica* **6**: 213–222. [10.xi.1969.]

Baroni Urbani, C. 1971a. Studi sulla mirmecofauna d'Italia. 9. Una nuova specie di *Aphaenogaster. Bollettino della Società Entomologica Italiana* **103**: 32–41. [21.ii.1971.]

Baroni Urbani, C. 1971b. Einige Homonymien in der Familie Formicidae. *Mitteilungen der Schweizerischen Entomologischen Gesellschaft* **44**: 360–362. [20.xii.1971.]

Baroni Urbani, C. 1971c. Catalogo delle specie di Formicidae d'Italia. (Studi sulla mirmecofauna d'Italia, 10.) *Memorie della Società Entomologica Italiana* **50**: 5–287. [30 xii.1971.]

Baroni Urbani, C. 1971d. Studien zur Ameisenfauna Italiens. 11. Die Ameisen des Toskanischen Archipels. Betrachtungen zur Herkunft der Inselfaunen. *Revue Suisse de Zoologie* **78**: 1037–1067. [(31).xii.1971.]

Baroni Urbani, C. 1972. Studi sui *Camponotus. Verhandlungen der Naturforschenden Gesellschaft in Basel* **82**: 122–135. [(31).iv.1972.]

Baroni Urbani, C. 1973. Die Gattung *Xenometra,* ein objektives synonym. *Mitteilungen der Schweizerischen Entomologischen Gesellschaft* **46**: 199–201. [28.xii.1973.]

Baroni Urbani, C. 1974. Studi sulla mirmecofauna d'Italia. 12. Le Isole Pontine. *Fragmenta Entomologica* **9**: 225–252. [15.ii.1974.]

Baroni Urbani, C. 1975a. Primi reperti del genere *Calyptomyrmex* Emery nel subcontinente Indiano. *Entomologica Basiliensia* **1**: 395–411. [1.x.1975.]

Baroni Urbani, C. 1975b. Contributo alla conoscenza dei generi *Belonopelta* Mayr e *Leiopelta* gen.n. *Mitteilungen der Schweizerischen Entomologischen Gesellschaft* **48**: 295–310. [28.xii.1975.]

Baroni Urbani, C. 1976. Le formiche dell'Arcipelago della Galita (Tunisia). *Redia* **59**: 207–223. [10.xi.1976.]

Baroni Urbani, C. 1977a. Ergebnisse der Bhutan-Expedition 1972 des Naturhistorischen Museums in Basel. Hymenoptera: Fam. Formicidae. Genus *Mayriella. Entomologica Basiliensia* **2**: 411–414. [7.i.1977.]

Baroni Urbani, C. 1977b. Ergebnisse der Bhutan-Expedition 1972 des Naturhistorischen Museums in Basel. Hymenoptera: Fam. Formicidae. Genus *Stenamma,* con una nuova species del Kashmir. *Entomologica Basiliensia* **2**: 415–422. [7.i.1977.]

Baroni Urbani, C. 1977c. Materiali per una revisione della sottofamiglia Leptanillinae Emery. *Entomologica Basiliensia* **2**: 427–488. [7.i.1977.]

Baroni Urbani, C. 1977d. Les espèces européennes du genre *Proceratium* Roger. *Mitteilungen der Schweizerischen Entomologischen Gesellschaft* **50**: 91–93. [31.iii.1977.]

Baroni Urbani, C. 1977e. Katalog der Typen von Formicidae der Sammlung des Naturhistorischen Museums Basel (2.Theil). *Mitteilungen der Entomologischen Gesellschaft Basel* **27**: 61–102. [(30).ix.1977.]

Baroni Urbani, C. 1977f. *Discothyrea stumperi* n.sp. de Bhutan, premier représentant du genre dans le subcontinent Indien. *Archives. Institut Grand-Ducal de Luxembourg* (N.S.) **37** (1974–1976): 97–100. [(31.xii).1977.]

Baroni Urbani, C. 1978a. Contributo alla conoscenza del genere *Amblyopone* Erichson. *Mitteilungen der Schweizerischen Entomologischen Gesellschaft* **51**: 39–51. [15.ii.1978.]

Baroni Urbani, C. 1978b. Materiali per una revisione dei *Leptothorax* neotropicali al sottogenere *Macromischa* Roger. *Entomologica Basiliensia* **3**: 395–618. [18.x.1978.]

Baroni Urbani, C. 1980a. Die Stellung der Gattung *Aratromyrmex* Stitz. *Mitteilungen aus dem Zoologischen Museum in Berlin* **56**: 95–96. [28.iv.1980.]

Baroni Urbani, C. 1980b. First description of fossil gardening ants. (Amber collection Stuttgart and Natural History Museum Basel; Hymenoptera: Formicidae. 1. Attini.) *Stuttgarter Beiträge zur Naturkunde.* Serie B (Geologie und Paläontologie) **54**: 1–13. [31.vii.1980.]

Baroni Urbani, C. 1980c. *Anochetus corayi* n.sp., the first fossil Odontomachiti ant. (Amber collection Stuttgart: Hym., Formicidae. 2. Odontomachiti.) *Stuttgarter Beiträge zur Naturkunde.* Serie B (Geologie und Paläontologie) **55**: 1–6. [31.vii.1980.]

Baroni Urbani, C. 1980d. The first fossil species of the Australian ant genus *Leptomyrmex* in amber from the Dominican Republic. (Amber collection Stuttgart: Hym., Formicidae. 3. Leptomyrmecini.) *Stuttgarter Beiträge zur Naturkunde.* Serie B (Geologie und Paläontologie) **62**: 1–10. [1.xii.1980.]

Baroni Urbani, C. 1980e. The ant genus *Gnamptogenys* in Dominican Amber. (Amber collection Stuttgart: Hym., Formicidae. 4. Ectatommini.) *Stuttgarter Beiträge zur Naturkunde.* Serie B (Geologie und Paläontologie) **67**: 1–10. [15.xii.1980.]

Baroni Urbani, C. 1984. Clave para la determinación de los géneros de hormigas neotropicales. *Graellsia. Revista de Entomólogos Ibericos* **39** (1983): 73–82. [10.ii.1984.]

Baroni Urbani, C. 1989. Phylogeny and behavioural evolution in ants, with a discussion of the role of behaviour in evolutionary processes. *Ethology Ecology and Evolution* **1**: 137–168. [(31).x.1989.]

Baroni Urbani, C. 1994. The identity of the Dominican *Paraponera.* (Amber Collection Stuttgart: Hymenoptera, Formicidae. V: Ponerinae, partim.) *Stuttgarter Beiträge zur Naturkunde* Serie B (Geologie und Paläontologie) **197**: 1–9. [15.iii.1994.]

Baroni Urbani, C. & de Andrade, M.L. 1993. *Perissomyrmex monticola* n. sp., from Bhutan: the first natural record for a presumed Neotropical genus with a discussion on its taxonomic status. *Tropical Zoology* **6**: 89–95. [(31).v.1993.]

Baroni Urbani, C. & de Andrade, M.L. 1994. First description of fossil Dactini ants with a critical analysis of the current classification of the tribe. (Amber Collection Stuttgart: Hymenoptera, Formicidae. VI: Dacetini.) *Stuttgarter Beiträge zur Naturkunde* Serie B (Geologie und Paläontologie) **198**: 1–65. [20.iv.1994.]

Baroni Urbani, C., Aktaç, N., & Camlitepe, Y. 1989. Disclosing the mystery of *Messor caducus* Motschulsky. *Mitteilungen der Schweizerischen Entomologischen Gesellschaft* **62**: 291–301. [15.xii.1989.]

Baroni Urbani, C., Bolton, B. & Ward, P.S. 1992. The internal phylogeny of ants. *Systematic Entomology* **17**: 301–329. [(30).x.1992.]

Baroni Urbani, C. & Graeser, S. 1987. REM-Analysen an einer pyritisierten Ameise aus Baltischem Bernstein. *Stuttgarter Beiträge zur Naturkunde.* Serie B (Geologie und Paläontologie) **133**: 1–16. [15.xii.1987.]

Baroni Urbani, C. & Wilson, E.O. 1987. The fossil members of the ant tribe Leptomyrmecini. *Psyche* **94**: 1–8. [6.x.1987.]

Barquin, J. 1981. Las hormigas de Canarias. Taxonomiá, ecologiá y distribución de los Formicidae. *Secret. Publ. Univ. La Laguna, Col. Monograf.* **3**: 1–584. [Not seen; reference from Hohmann, La Roche, *et al.*, 1993: 690.]

Beattie, A.J. 1985. *The evolutionary ecology of ant-plant mutualisms:* 182 pp. Cambridge. [1985.]

Begdon, J. 1932. Wymiary i wskazniki niektórych znamion mrówki *Stenamma* Westw. *westwoodi* Arn. (Westw?) *polonicum* nov. subsp., znalezionej na Pomorzu. *Sprawozdanie Komisji Fizjograficznej* **65** (1931): 113–119. [(31.xii).1932.]

Belshaw, R. & Bolton, B. 1994. A new myrmicine ant genus from cocoa leaf litter in Ghana. *Journal of Natural History* **28**: 631–634. [31.v.1994.]

Bequaert, J. 1926. The date of publication of the Hymenoptera and Diptera described by Guérin in Duperrey's "Voyage de La Coquille." *Entomologische Mitteilungen* **15**: 186–195. [20.iii.1926.]

Berendt, G.C. 1830. *Die Insekten im Bernstein. Ein Beitrag zur Thiergeschichte der Vorwelt:* 38 pp. Danzig & Berlin. [(31.xii).1830.]

Bernard, F. 1945. Notes sur l'écologie des fourmis en forêt de Mamora (Maroc). *Bulletin de la Société d'Histoire Naturelle de l'Afrique du Nord* **35** (1944): 125–140. [14.viii.1945.]

Bernard, F. 1948. Les Insectes Sociaux du Fezzân. In *Mission scientifique du Fezzân* (1944–1945). (Institut de Recherches Sahariennes de l'Université d'Alger.) 5 Zoologie: 87–200. [30.iii.1948.]

Bernard, F. 1950a. Notes sur les fourmis de France. 2. Peuplement des montagnes méridionales. *Annales de la Société Entomologique de France* **115** (1946): 1–36. [(31).iii.1950.] (Note. Final page of this volume gives, "Date de publication de ce volume: décembre, 1949." However, end-wrapper states, "Dépot légal 1er trimestre 1950.)

Bernard, F. 1950b. Contribution à l'étude de l'Air. *Mémoires de l'Institut Français d'Afrique Noire* **10**: 284–294. [(31.xii).1950.]

Bernard, F. 1951. Super-famille des Formicoidea (pp. 997–1104). In Grassé, P.-P. *Traité de Zoologie. Anatomie, Systematique, Biologie* **10**. Fascicule 2, Insectes supérieurs et hémipteroides: 976–1948. Paris. [(31.xii).1951.]

Bernard, F. 1953a. Mission scientifique au Tassili des Ajjer (1949). 1. Recherches zoologiques et medicales. Les fourmis du Tassili des Ajjer (Sahara central). *Institut de Recherches Sahariennes de l'Université d'Alger:* 121–250. [31.i.1953.]

Bernard, F. 1953b. La réserve naturelle intégrale du Mt Nimba. 11. Hyménoptères Formicidae. *Mémoires de l'Institut Français d'Afrique Noire* **19** (1952): 165–270. [(31).iii.1953.]

Bernard, F. 1954. Une fourmi nouvelle: *Cataglyphis halophila* nichant au milieu du Chott Djerid. *Bulletin de la Société des Sciences Naturelles de Tunisie* **6** (1953): 47–56. [15.iii.1954.]

Bernard, F. 1955a. Fourmis moissonneuses nouvelles ou peu connues des montagnes d'Algérie et revision des *Messor* du groupe *structor* (Latr.). *Bulletin de la Société d'Histoire Naturelle de l'Afrique du Nord* **45** (1954): 345–365. [29.iii.1955.]

Bernard, F. 1955b. Morphologie et comportement des fourmis lestobiotiques du genre *Epixenus* Emery. *Insectes Sociaux* **2**: 273–283. [(31).xii.1955.]

Bernard, F. 1956a. Révision des *Leptothorax* d'Europe occidentale, bassée sur la biométrie et les genitales males. *Bulletin de la Société Zoologique de France* **81**: 151–165. [15.x.1956.]

Bernard, F. 1956b. Remarques sur le peuplement des Baléares en fourmis. *Bulletin de la Société d'Histoire Naturelle de l'Afrique du Nord* **47**: 254–266. [(31.xii).1956]

Bernard, F. 1956c. Révision des fourmis paléarctiques du genre *Cardiocondyla* Emery. *Bulletin de la Société d'Histoire Naturelle de l'Afrique du Nord* **47**: 299–306. [(31.xii).1956.]

Bernard, F. 1957a. Note sur quelques *Leptothorax* d'Europe centrale, avec description de *L. carinthiacus* n.sp. *Bulletin de la Société Entomologique de France* **62**: 46–53. [28.v.1957.]

Bernard, F. 1957b. *Xenometra* Emery, genre de fourmis parasite nouveau pour l'Ancien Monde. *Bulletin de la Société Entomologique de France* **62**: 100–103. [26.vii.1957.]

Bernard, F. 1959. Les fourmis de l'Ile de Port-Cros. Contribution à l'écologie des anciennes forêts mediterranéennes. *Vie et Milieu* **9** (1958): 340–360. [(31).i.1959.]

Bernard, F. 1960. Fourmis récoltées en Corse par J. Bonfils (1957). *Compte Rendu Sommaire des Séances de la Société de Biogéographie* **36** (1959): 108–114. [(30).iv.1960.]

Bernard, F. 1967. *Faune de l'Europe et du Bassin Méditerranéen* **3.** Les fourmis d'Europe occidentale et septentrionale (1968): 411 pp. Paris. [(30).xi.1967.]

Bernard, F. 1971. Les fourmis de l'île de Djerba (Tunisie). *Bulletin de la Société d'Histoire Naturelle de l'Afrique du Nord* **62**: 3–14. [(31.xii).1971.]

Bernard, F. 1977. Trois fourmis nouvelles du Sahara. *Bulletin de la Société Entomologique de France* **82**: 29–32. [24.ix.1977.]

Bernard, F. 1978a. Révision des *Diplorhoptrum* de France, fourmis plus différenciées par l'écologie que par leurs formes. *Annales de la Société Entomologique de France* (N.S.) **13** (1977): 543–577. [28.ii.1978.]

Bernard, F. 1978b. *Orthocrema estrelana,* espèce nouvelle commune dans l'Estérel. *Bulletin de la Société Entomologique de France* **83**: 43–46. [18.vii.1978.]

Bernard, F. 1980. *Messor carthaginensis* n.sp. de Tunis, et révision des *Messor* du groupe *barbara. Bulletin de la Société Entomologique de France* **84** (1979): 265–269. [25.iv.1980.]

Bernard, F. 1981a. Variabilité des proportions biométriques chez les *Plagiolepis*, avec description de *Pl. hoggarensis* n.sp. *Bulletin de la Société Entomologique de France* **86**: 169–172. [6.xi.1981.]
Bernard, F. 1981b. Revision of the genus *Messor* (Harvesting Ants) on a biometrical basis (pp. 141–145). In Howse, P.E. & Clement, J.-L. *Biosystematics of Social Insects*. Systematics Association Special Volume, **No. 19**: 364 pp. London. [(31.xii).1981.]
Bernard, F. 1983. Les fourmis et leur milieu en France méditerranéenne. *Encyclopedie Entomologique* **45**: 149 pp. Paris. [(31.xii).1983.]
Bernard, F. 1985. Recherches sur l'évolution des fourmis moissonneuses. *Actes des Colloques Insectes Sociaux* **2** (1984): 45–55. [(31).vi.1985.]
Bernard, F. & Cagniant, H. 1963. Capture au Hoggar de trois *Acantholepis* nouveaux pour ce massif, avec observations sur leurs modes de vie. *Bulletin de la Société Entomologique de France* **67** (1962): 161–164. [28.ii.1963.]
Bernstein, A. 1861. Untitled contribution introduced by, "Von Herrn Dr. Agath. Bernstein in Gadok auf Java ist folgendes Schreiben eingegangen." *Verhandlungen der k.-k. Zoologisch-Botanischen Gesellschaft in Wien* **11**: 7–8. [(31.xii).1861.]
Bertkau, P. 1879. Bericht über die wissenschaftilchen Leistungen im Gebiete der Arthropoden während der Jahre 1877–78. *Archiv für Naturgeschichte* **45** (Zweites Band): 1–222. [(31.xii).1879.]
Betrem, J.G. 1926. De Mierenfauna van Meijendel. *Levende Natuur* **31**: 211–220. [1.xi.1926.]
Betrem, J.G. 1953. Enkele opmerkingen omtrent de soorten van de *Formica rufa*-groep. *Entomologische Berichten* **14**: 322–326. [1.viii.1953.]
Betrem, J.G. 1954. De Satermier (*Formica exsecta* Nyl., 1846) en enkele van haar problemen. *Entomologische Berichten* **15**: 224–230. [1.x.1954.]
Betrem, J.G. 1955. De systematische plaats van *Formica congerens* ab. *thijssei* Stärcke. *Entomologische Berichten* **15**: 391–393. [1.v.1955.]
Betrem, J.G. 1960a. *Formica truncorum* F. niet inheems. *Entomologische Berichten* **20**: 130–134. [1.vii.1960.]
Betrem, J.G. 1960b. Ueber die Systematik der *Formica rufa*-gruppe. *Tijdschrift voor Entomologie* **103**: 51–81. [5.viii.1960.]
Betrem, J.G. 1962. Quelques remarques sur l'identité de la *Formica nigricans*. *Entomologische Berichten* **22**: 38. [1.ii.1962.]
Billberg, G.J. 1820. *Enumeratio Insectorum in Museo Gust. Joh. Billberg:* 138 pp. Stockholm. [(31.xii).1820.]
Billen, J. 1989. Morphology of the cloacal gland in the ant *Cataglyphis savignyi*. *Actes des Colloques Insectes Sociaux* **5** (1988): 301–306. [(30).vi.1989.]
Bingham, C.T. 1903. *The fauna of British India, including Ceylon and Burma*. Hymenoptera **2**. Ants and Cuckoo-Wasps: 506 pp. London. [(31.xii).1903.]
Blacker, N.C. 1992. Some ants from southern Vancouver Island, British Columbia. *Journal of the Entomological Society of British Columbia* **89**: 3–12. [(31).xii.1992.]
Blanchard, C.E. 1846. In, *Le Règne Animal* distribué d'après son organisation, pour servir de base à l'histoire naturelle des animaux, et d'introduction à l'anatomie comparée par Georges Cuvier. Les Insectes. **2** Atlas, livraison 234 (Ins. 44): pl. 118. [(31).xi.1846.] (Date of publication, see Cowan, 1976: 32.)
Blum, M.S. 1984. Poisonous ants and their venoms (pp. 225–242). In Tu, A.T. *Handbook of Natural Toxins* **2**. Insect poisons, allergens, and other invertebrate venoms: 732 pp. New York. [1984.]
Blum, M.S. & Callahan, P.S. 1963. The venom and poison glands of *Pseudomyrmex pallidus* (F.Smith). *Psyche* **70**: 69–74. [9.vii.1963.]
Bolton, B. 1969. Male of *Paedalgus termitolestes* W.M. Wheeler. *Nigerian Entomologists' Magazine* **2**: 14–16. [(31).xii.1969.]
Bolton, B. 1971. Two new subarboreal species of the ant genus *Strumigenys* from West Africa. *Entomologist's Monthly Magazine* **107**: 59–64. [27.viii.1971.]
Bolton, B. 1972. Two new species of the ant genus *Epitritus* from Ghana, with a key to the world species. *Entomologist's Monthly Magazine* **107** (1971): 205–208. [9.vi.1972.]
Bolton, B. 1973a. The ant genera of West Africa: a synonymic synopsis with keys. *Bulletin of the British Museum (Natural History)* (Entomology) **27**: 317–368. [18.i.1973.]
Bolton, B. 1973b. The ant genus *Polyrhachis* F. Smith in the Ethiopian region. *Bulletin of the British Museum (Natural History)* (Entomology) **28**: 283–369. [25.vii.1973.]
Bolton, B. 1973c. A remarkable new arboreal ant genus from West Africa. *Entomologist's Monthly Magazine* **108** (1972): 234–237. [25.ix.1973.]
Bolton, B. 1974a. A revision of the Palaeotropical arboreal ant genus *Cataulacus* F. Smith. *Bulletin of the British Museum (Natural History)* (Entomology) **30**: 1–105. [15.i.1974.]
Bolton, B. 1974b. New synonymy and a new name in the ant genus *Polyrhachis* F. Smith. *Entomologist's Monthly Magazine* **109**: 172–180. [3.v.1974.]
Bolton, B. 1974c. A revision of the ponerine ant genus *Plectroctena* F. Smith. *Bulletin of the British Museum (Natural History)* (Entomology) **30**: 309–338. [20.vi.1974.]
Bolton, B. 1975a. A revision of the ant genus *Leptogenys* Roger in the Ethiopian region, with a review of the Malagasy species. *Bulletin of the British Museum (Natural History)* (Entomology) **31**: 235–305. [5.ii.1975.]
Bolton, B. 1975b. A revision of the African ponerine ant genus *Psalidomyrmex* André. *Bulletin of the British Museum (Natural History)* (Entomology) **32**: 1–16. [10.iv.1975.]
Bolton, B. 1975c. The *sexspinosa*-group of the ant genus *Polyrhachis* F. Smith. *Journal of Entomology* (Series B) **44**: 1–14. [27.v.1975.]
Bolton, B. 1976. The ant tribe Tetramoriini. Constituent genera, review of smaller genera and revision of *Triglyphothrix* Forel. *Bulletin of the British Museum (Natural History)* (Entomology) **34**: 281–379. [28.x.1976.]
Bolton, B. 1977. The ant tribe Tetramoriini. The genus *Tetramorium* Mayr in the Oriental and Indo-Australian regions, and in Australia. *Bulletin of the British Museum (Natural History)* (Entomology) **36**: 67–151. [29.ix.1977.]
Bolton, B. 1979. The ant tribe Tetramoriini. The genus *Tetramorium* Mayr in the Malagasy region and in the New World. *Bulletin of the British Museum (Natural History)* (Entomology) **38**: 129–181. [29.iii.1979.]
Bolton, B. 1980. The ant tribe Tetramoriini. The genus *Tetramorium* Mayr in the Ethiopian zoogeographical region. *Bulletin of the British Museum (Natural History)* (Entomology) **40**: 193–384. [31.vii.1980.]
Bolton, B. 1981a. A revision of the ant genera *Meranoplus* F. Smith, *Dicroaspis* Emery and *Calyptomyrmex* Emery in the Ethiopian zoogeographical region. *Bulletin of the British Museum (Natural History)* (Entomology) **42**: 43–81. [26.ii.1981.]
Bolton, B. 1981b. A revision of six minor genera of Myrmicinae in the Ethiopian zoogeographical region. *Bulletin of the British Museum (Natural History)* (Entomology) **43**: 245–307. [26.xi.1981.]
Bolton, B, 1982. Afrotropical species of the myrmicine ant genera *Cardiocondyla, Leptothorax, Melissotarsus, Messor and Cataulacus. Bulletin of the British Museum (Natural History)* (Entomology) **45**: 307–370. [30.ix.1982.]
Bolton, B. 1983. The Afrotropical dacetine ants. *Bulletin of the British Museum (Natural History)* (Entomology) **46**: 267–416. [25.viii.1983.]
Bolton, B. 1984. Diagnosis and relationships of the myrmicine ant genus *Ishakidris* gen.n. *Systematic Entomology* **9**: 373–382. [(31).x.1984.]
Bolton, B. 1985. The ant genus *Triglyphothrix* Forel a synonym of *Tetramorium* Mayr. *Journal of Natural History* **19**: 243–248. [23.v.1985.]
Bolton, B. 1986a. A taxonomic and biological review of the tetramoriine ant genus *Rhoptromyrmex. Systematic Entomology* **11**: 1–17. [(31).i.1986.]
Bolton, B. 1986b. Apterous females and shift of dispersal strategy in the *Monomorium salomonis*-group. *Journal of Natural History* **20**: 267–272. [3.iii.1986.]

Bolton, B. 1987. A review of the *Solenopsis* genus-group and revision of Afrotropical *Monomorium* Mayr. *Bulletin of the British Museum (Natural History)* (Entomology) **54**: 263–452. [25.vi.1987.]
Bolton, B. 1988a. A new socially parasitic *Myrmica,* with a reassessment of the genus. *Systematic Entomology* **13**: 1–11. [(31).i.1988.]
Bolton, B. 1988b. A review of *Paratopula* Wheeler, a forgotten genus of myrmicine ants. *Entomologist's Monthly Magazine* **124**: 125–143. [27.vii.1988.]
Bolton, B. 1988c. *Secostruma,* a new subterranean tetramoriine ant genus. *Systematic Entomology* **13**: 263–270. [(31).vii.1988.]
Bolton, B. 1990a. Abdominal characters and status of the cerapachyine ants. *Journal of Natural History* **24**: 53–68. [22.ii.1990.]
Bolton, B. 1990b. The higher classification of the ant subfamily Leptanillinae. *Systematic Entomology* **15**: 267–282. [(31).vii.1990.]
Bolton, B. 1990c. Army ants reassessed: the phylogeny and classification of the doryline section. *Journal of Natural History* **24**: 1339–1364. [19.x.1990.]
Bolton, B. 1991. New myrmicine ant genera from the Oriental region. *Systematic Entomology* **16**: 1–13. [(31).i.1991.]
Bolton, B. 1992. A review of the ant genus *Recurvidris,* a new name for *Trigonogaster* Forel. *Psyche* **99**: 35–48. [4.xii.1992.]
Bolton, B. 1994. *Identification Guide to the Ant Genera of the World:* 222 pp. Cambridge, Mass. [(31).vi.1994.]
Bolton, B. & Belshaw, R. 1993. Taxonomy and biology of the supposedly lestobiotic ant genus *Paedalgus. Systematic Entomology* **18**: 181–189. [(31).vii.1993.]
Bolton, B. & Collingwood, C.A. 1975. Hymenoptera: Formicidae. *Handbooks for the Identification of British Insects* **6**, part 3 (c): 34 pp. London. [(30).vi.1975.]
Bolton, B., Gotwald, W.H. & Leroux, J.-M. 1979. A new West African ant of the genus *Plectroctena* with ecological notes. *Annales de l'Université d'Abidjan.* Série E (Ecologie) **9** (1976): 371–381. [(31.xii).1979.]
Bolton, B., & Marsh, A.C. 1989. The Afrotropical thermophilic ant genus *Ocymyrmex. Journal of Natural History* **23**: 1267–1308. [9.xi.1989.]
Bondroit, J. 1910. Les fourmis de Belgique. *Annales de la Société Entomologique de Belgique* **53** (1909): 479–500. [5.i.1910.]
Bondroit, J. 1911. Contribution à la faune de Belgique. Notes diverses. *Annales de la Société Entomologique de Belgique* **55**: 8–13. [3.ii.1911.]
Bondroit, J. 1912. Fourmis des Hautes Fagnes. *Annales de la Société Entomologique de Belgique* **56**: 351–352. [31.x.1912.]
Bondroit, J. 1916. Un nouveau *Ponera* de France. *Bulletin de la Société Entomologique de France* **1916**: 211–212. [10.viii.1916.]
Bondroit, J. 1917a. Notes sur quelques Formicidae de France. *Bulletin de la Société Entomologique de France* **1917**: 174–177. [27.vi.1917.]
Bondroit, J. 1917b. Diagnoses de trois nouveaux *Formica* d'Europe. *Bulletin de la Société Entomologique de France* **1917**: 186–187. [17.vii.1917.]
Bondroit, J. 1918. Les fourmis de France et de Belgique. *Annales de la Société Entomologique de France* **87**: 1–174. [24.xii.1918.]
Bondroit, J. 1920a. Notes diverses sur les fourmis d'Europe. *Annales de la Société Entomologique de Belgique* **59** (1919): 143–158. [25.ii.1920.]
Bondroit, J. 1920b. Supplément aux fourmis de France et de Belgique. *Annales de la Société Entomologique de France* **88** (1919): 299–305. [30.vi.1920.]
Borgmeier, T. 1923. Catalogo systematico e synonymico das formigas do Brasil. 1 parte. Subfam. Dorylinae, Cerapachyinae, Ponerinae, Dolichoderinae. *Archivos do Museu Nacional* **24**: 34–103. [(31.xii).1923.]
Borgmeier, T. 1927a. Um caso de trophobiose entre uma formiga e um parasita do caféeiro. *Boletim do Museu Nacional* **3**: 285–289. [(31).xii.1927.]
Borgmeier, T. 1927b. Algumas novas formigas brasileiras. *Archivos do Museu Nacional* **29**: 57–65. [(31.xii).1927.]
Borgmeier, T. 1927c. Catalogo systematico e synonymico das formigas do Brasil. 2a parte. Subf. Pseudomyrminae, Myrmicinae, Formicinae. *Archivos do Museu Nacional* **29**: 69–164. [(31.xii).1927.]
Borgmeier, T. 1928a. Einige neue Ameisen aus Brasilien. *Zoologischer Anzeiger* **75**: 32–39. [2.i.1928.]
Borgmeier, T. 1928b. Algumas formigas do Museo Paulista. *Boletim Biologico* **12**: 55–70. [10.vii.1928.]
Borgmeier, T. 1929. Zur Kenntnis der brasilianischen Ameisen. *Eos. Revista Española de Entomología* **5**: 195–214. [15.ix.1929.]
Borgmeier, T. 1930. Duas rainhas de *Eciton* e algumas outras formigas brasileiras. *Archivos do Instituto Biologico* **3**: 21–40. [(31).viii.1930.]
Borgmeier, T. 1932a. A proposito de *Acropyga pickeli* Borgm. (1927). *Revista de Entomologia* **2**: 238–239. [31.iii.1932.]
Borgmeier, T. 1932b. *Leptogenys crudelis* Fr. Smith, 1858. *Revista de Entomologia* **2**: 485. [28.xii.1932.]
Borgmeier, T. 1933a. A rainha de *Eciton rogeri* Dalla Torre. *Revista de Entomologia* **3**: 92–96. [22.iii.1933.]
Borgmeier, T. 1933b. Uma nove especie do genero *Leptogenys. Revista de Entomologia* **3**: 226–227. [20.vii.1933.]
Borgmeier, T. 1933c. Nota prévia sobre *Acropyga paramaribensis. Revista de Entomologia* **3**: 263. [20.vii.1933.]
Borgmeier, T. 1933d. Sobre algumas especies de formigas do genero *Eciton* Latreille. *Archivos da Escola Superior da Agricultura e Medicina Veterinaria* **10**: 161–168. [(31).xii.1933.]
Borgmeier, T. 1934. Contribuiçao para o conhecimento da fauna mirmecológica dos cafezais de Paramaribo, Guiana Holandesa. *Archivos do Instituto de Biologia Vegetal* **1**: 93–111. [(31).xi.1934.]
Borgmeier, T. 1936. Sobre algumas formigas dos generos *Eciton* e *Cheliomyrmex. Archivos do Instituto de Biologia Vegetal* **3**: 51–68. [(31).xii.1936.]
Borgmeier, T. 1937a. *Cardiocondyla emeryi* Forel no Brasil, e a discoberta do macho ergatoide desta especie. *Revista de Entomologia* **7**: 129–134. [27.vii.1937.]
Borgmeier, T. 1937b. Formigas novas ou pouco conhecidas da America do Sul e Central, principalmente do Brasil. *Archivos do Instituto de Biologia Vegetal* **3**: 217–255. [(31).viii.1937.]
Borgmeier, T. 1939. Nova contribuiçao para o conhecimento das formigas neotropicas. *Revista de Entomologia* **10**: 403–428. [4.ix.1939.]
Borgmeier, T. 1940. Duas notas myrmecologicas. *Revista de Entomologia* **11**: 606. [28.vi.1940.]
Borgmeier, T. 1948a. Die Geschlechtstiere zweier *Eciton*-Arten und einige andere Ameisen aus Mittel-und Suedamerika. *Revista de Entomologia* **19**: 191–206. [1.vi.1948.]
Borgmeier, T, 1948b. Einige Ameisen aus Argentinien. *Revista de Entomologia* **19**: 459–471. [31.xii.1948.]
Borgmeier, T. 1949. Formigas novas ou pouco conhecidas de Costa Rica. *Revista Brasileira de Biologia* **9**: 201–210. [(30).vi.1949.]
Borgmeier, T. 1950a. A fêmea dichthadiiforme e os estadios evolutivos de *Simopelta pergandei* (Forel), e a descriçao de *S. bicolor* n.sp. *Revista de Entomologia* **21**: 369–380. [10.viii.1950.]
Borgmeier, T. 1950b. Bemerkungen zu Dr. Creighton's Werk "The Ants of North America". *Revista de Entomologia* **21**: 381–386. [10.viii.1950.]
Borgmeier, T. 1950c. Uma nova espécie do genero *Neivamyrmex* Borgmeier. *Revista de Entomologia* **21**: 623–624. [30.xii.1950.]
Borgmeier, T. 1950d. Estudos sobre *Atta. Memórias do Instituto Oswaldo Cruz* **48**: 239–263. [(31.xii).1950.]
Borgmeier, T. 1952. Algumas formigas do genero *Macromischa* Roger. *Arquivos do Museu Nacional* **42**: 107–111. [7.iv.1952.]
Borgmeier, T. 1953. Vorarbeiten zu einer Revision der Neotropischen Wanderameisen. *Studia Entomologica* **2**: 1–51. [10.vii.1953.]
Borgmeier, T. 1954a. Uma nova *Discothyrea* com seis articulos antenais. *Revista Brasileira de Entomologia* **1**: 191–194. [(31).iii.1954.]
Borgmeier, T. 1954b. Two interesting dacetine ants from Brazil. *Revista Brasileira de Biologia* **14**: 279–284. [(31).viii.1954.]
Borgmeier, T. 1954c. Aenictini n. trib. und die Tribus-Einteilung der Dorylinen. *Zoologischer Anzeiger* **153**: 211–214. [(30).xi.1954.]
Borgmeier, T. 1955. Die Wanderameisen der Neotropischen Region. *Studia Entomologica* **3**: 1–720. [(31.xii).1955.]

Borgmeier, T. 1957. Myrmecologische Studien. 1. *Anais de Academia Brasileira de Ciencias* **29**: 103–128. [31.iii.1957.]
Borgmeier, T. 1958. Nachtraege zu meiner Monographie der Neotropischen Wanderameisen. *Studia Entomologica* (N.S.) **1**: 197–208. [31.i.1958.]
Borgmeier, T. 1959a. Myrmecologische Studien. 2. *Anais de Academia Brasileira de Ciencias* **31**: 309–319. [30.vi.1959.]
Borgmeier, T. 1959b. Revision der Gattung *Atta* Fabricius. *Studia Entomologica* (N.S.) **2**: 321–390. [29.ix.1959.]
Bourne, R.A. 1973. A taxonomic study of the ant genus *Lasius* Fabricius in the British Isles. *Journal of Entomology* (B) **42**: 17–27. [29.vi.1973.]
Boven, J.K.A. van. 1946. Le mâle de *Plagiolepis vindobonensis* Lomn. *Natuurhistorisch Maandblad* **35**: 9–10. [28.ii.1946.]
Boven, J.K.A. van. 1947. Liste de détermination des principales espèces de fourmis belges. *Bulletin et Annales de la Société Entomologique de Belgique* **83**: 163–190. [3.v.1947.]
Boven, J.K.A. van. 1967. La femelle de *Dorylus fimbriatus* et *termitarius. Natuurhistorisch Maandblad* **56**: 55–60. [26.iv.1967.]
Boven, J.K.A. van. 1968. La reine de *Dorylus (Anomma) kohli* Wasmann. *Naturaliste Canadien* **95**: 731–739. [(30).vi.1968.]
Boven, J.K.A. van. 1970. *Myrmica faniensis,* une nouvelle espèce parasite. *Bulletin et Annales de la Société Royale Entomologique de Belgique* **106**: 127–132. [15.vi.1970.]
Boven, J.K.A. van. 1972. Description de deux reines d'*Anomma. Bulletin et Annales de la Société Royale Entomologique de Belgique* **108**: 133–146. [30.ix.1972.]
Boven, J.K.A. van. 1975. Deux nouvelles reines du genre *Dorylus* Fabricius. *Annales Zoologici* **33**: 189–199. [15.xii.1975.]
Boven, J.K.A. van. 1977. De Mierenfauna van Belgie. *Acta Zoologica et Pathologica Antverpiensia* **67**: 1–191. [(30).vi.1977.]
Boven, J.K.A. van. & Lévieux, J. 1970. Les Dorylinae de la savane de Lamto. *Annales de l'Université d'Abidjan.* Serie E (Ecologie) **1** (1968): 351–358. [(31).v.1970.]
Bradley, J.C. 1975. Dates of publication of Westwood's Arcana Entomologica. *Proceedings of the Biological Society of Washington* **88** (10): 91–94. [23.iv.1975.]
Brandão, C.R.F. 1989. *Belonopelta minima,* a new species from eastern Brazil. *Revista Brasileira de Entomologia* **33**: 135–138. [31.iii.1989.]
Brandão, C.R.F. 1990. Systematic revision of the Neotropical ant genus *Megalomyrmex* Forel, with the description of thirteen new species. *Arquivos de Zoologia* **31**: 411–481. [21.ix.1990.]
Brandão, C.R.F. 1991. Adendos ao catálogo abreviado das formigas da regiao Neotropical. *Revista Brasileira de Entomologia* **35**: 319–412. [31.viii.1991.]
Brandão, C.R.F., Diniz, J.L.M. & Tomotake, E.M. 1991. *Thaumatomyrmex* strips millipedes for prey: a novel predatory behaviour in ants, and the first case of sympatry in the genus. *Insectes Sociaux* **38**: 335–344. [(31.xii).1991.]
Brandão, C.R.F. & Lattke, J.E. 1990. Description of a new Ecuadorian *Gnamptogenys* species, with a discussion on the status of the *Alfaria* group. *Journal of the New York Entomological Society* **98**: 489–494. [14.xi.1990.]
Brandão, C.R.F., Martins-Neto, R.G. & Vulcano, M.A. 1990. The earliest known fossil ant (first southern hemisphere Mesozoic record). *Psyche* **96** (1989): 195–208. [20.iv.1990]
Brauns, J. 1903. Ueber das Weibchen von *Dorylus (Rhogmus) fimbriatus* Shuck. *Zeitschrift für Systematische Hymenopterologie und Dipterologie* **3**: 294–298. [1.ix.1903.]
Brèthes, J. 1914a. Note sur quelques dolichoderines argentines. *Anales del Museo Nacional de Historia Natural de Buenos Aires* **26**: 93–96. [12.v.1914.]
Brèthes, J. 1914b. Sur les formes sexuelles de deux dolichoderines. *Anales del Museo Nacional de Historia Natural de Buenos Aires* **26**: 231–233. [16.xi.1914.]
Brian, M.V. & Brian, A.D. 1949. Observations on the taxonomy of the ants *Myrmica rubra* L. and *M. laevinodis* Nylander. *Transactions of the Royal Entomological Society of London* **100**: 393–409. [30.xii.1949.]
Brown, W.L.,Jr. 1943. A new metallic ant from the Pine Barrens of New Jersey. *Entomological News* **54**: 243–248. [29.xii.1943.]
Brown, W.L.,Jr. 1947. A note upon two neglected species of *Formica* Linn. *Entomological News* **58**: 6–9. [17.iv.1947.]
Brown, W.L.,Jr.1948a. A new *Stictoponera,* with notes on the genus. *Psyche* **54** (1947): 263–264. [17.ii.1948.]
Brown, W.L.,Jr. 1948b. *Strumigenys karawajewi,* new name for a Sumatran ant. *Entomological News* **59**: 44. [22.iv.1948.]
Brown, W.L.,Jr. 1948c. A new *Discothyrea* from New Caledonia. *Psyche* **55**: 38–40. [30.vi.1948.]
Brown, W.L.,Jr. 1948d. The status of the genus *Hercynia* J. Enzmann. *Entomological News* **59**: 102. [2.vii.1948.]
Brown, W.L.,Jr. 1948e. A preliminary generic revision of the higher Dacetini. *Transactions of the American Entomological Society* **74**: 101–129. [27.vii.1948.]
Brown, W.L.,Jr. 1949a. Synonymic and other notes on Formicidae. *Psyche* **56**: 41–49. [17.v.1949.]
Brown, W.L.,Jr. 1949b. Revision of the ant tribe Dacetini. 3. *Epitritus* Emery and *Quadristruma* new genus. *Transactions of the American Entomological Society* **75**: 43–51. [6.vii.1949.]
Brown, W.L.,Jr. 1949c. A new American *Amblyopone,* with notes on the genus. *Psyche* **56**: 81–88. [9.viii.1949.]
Brown, W.L.,Jr. 1949d. Revision of the ant tribe Dacetini. 1. Fauna of Japan, China and Taiwan. *Mushi* **20**: 1–25. [20.ix.1949.]
Brown, W.L.,Jr. 1949e. Notes on Chinese ants. 1. *Crematogaster* Lund. *Mushi* **20**: 37–38. [20.ix.1949.]
Brown, W.L.,Jr. 1949f. Revision of the ant tribe Dacetini. 4. Some genera properly excluded from the Dacetini, with the establishment of the Basicerotini, new tribe. *Transactions of the American Entomological Society* **75**: 83–96. [7.xii.1949.]
Brown, W.L.,Jr. 1950a. Revision of the ant tribe Dacetini. 2. *Glamyromyrmex* Wheeler and closely related small genera. *Transactions of the American Entomological Society* **76**: 27–36. [9.viii.1950.]
Brown, W.L.,Jr. 1950b. Preliminary description of seven new species of the dacetine ant genus *Smithistruma* Brown. *Transactions of the American Entomological Society* **76**: 37–45. [25.viii.1950.]
Brown, W.L.,Jr. 1950c. Revision of the ant tribe Dacetini. 5. The delimitation of *Arnoldidris,* new genus. *Transactions of the American Entomological Society* **76**: 143–145. [8.ix.1950.]
Brown, W.L.,Jr. 1950d. The status of two common North American carpenter ants. *Entomological News* **61**: 157–161. [31.x.1950.]
Brown, W.L.,Jr. 1950e. Morphological, taxonomic and other notes on ants. *Wasmann Journal of Biology* **8**: 241–250. [8.xi.1950.]
Brown, W.L.,Jr. 1951. New synonymy of a few genera and species of ants. *Bulletin of the Brooklyn Entomological Society* **46**: 101–106. [1.xi.1951.]
Brown, W.L.,Jr. 1952a. *Adlerzia froggatti* Forel and some new synonymy. *Psyche* **58** (1951): 110. [7.iv.1952.]
Brown, W.L.,Jr. 1952b. New synonymy in the army ant genus *Aenictus* Shuckard. *Psyche* **58** (1951): 123. [7.iv.1952.]
Brown, W.L.,Jr. 1952c. Synonymous ant names. *Psyche* **58** (1951): 124. [7.iv.1952.]
Brown, W.L.,Jr. 1952d. Correction to the synonymy of the ant *Camponotus formosensis* Wheeler. *Psyche* **59**: 19. [19.viii.1952.]
Brown, W.L.,Jr. 1952e. Revision of the ant genus *Serrastruma. Bulletin of the Museum of Comparative Zoology at Harvard College* **107**: 67–86. [(31).viii.1952.]
Brown, W.L.,Jr. 1952f. Notes on two well-known Australian ant species. *Western Australian Naturalist* **3**: 137–138. [15.ix.1952.]
Brown, W.L.,Jr. 1952g. The dacetine ant genus *Mesostruma* Brown. *Transactions of the Royal Society of South Australia* **75**: 9–13. [(30).ix.1952.]

Brown, W.L.,Jr. 1952h. *Heteroponera* Mayr reinstated. *Psyche* **59**: 70. [16.x.1952.]
Brown, W.L.,Jr. 1952i. The status of some Australian *Amblyopone* species. *Entomological News* **63**: 265–267. [15.xii.1952.]
Brown, W.L.,Jr. 1952j. On the identity of *Adlerzia* Forel. *Pan-Pacific Entomologist* **28**: 173–177. [23.xii.1952.]
Brown, W.L.,Jr. 1953a. Composition of the ant tribe Typhlomyrmicini. *Psyche* **59** (1952): 104. [30.i.1953.]
Brown, W.L.,Jr. 1953b. Notes on Australian *Podomyrma. North Queensland Naturalist* **21**: 3. [1.iii.1953.]
Brown, W.L.,Jr. 1953c. Characters and synonymies among the genera of ants. Part 1. *Breviora* **11**: 1–13. [20.iii.1953.]
Brown, W.L.,Jr. 1953d. The Neotropical species of the ant genus *Strumigenys* Fr. Smith: group of *mandibularis* Fr. Smith. *Journal of the New York Entomological Society* **61**: 53–59. [26.vi.1953.]
Brown, W.L.,Jr. 1953e. Three new ants related to *Strumigenys louisianae* Roger. *Psyche* **60**: 1–5. [26.vi.1953.]
Brown, W.L.,Jr. 1953f. The Neotropical species of the ant genus *Strumigenys* Fr. Smith: group of *smithii* Forel. *Journal of the New York Entomological Society* **61**: 101–110. [30.vii.1953.]
Brown, W.L.,Jr. 1953g. Revisionary studies in the ant tribe Dacetini. *American Midland Naturalist* **50**: 1–137. [10.ix.1953.]
Brown, W.L.,Jr. 1953h. Characters and synonymies among the genera of ants. Part 2. *Breviora* **18**: 1–8. [23.ix.1953.]
Brown, W.L.,Jr. 1953i. An Australian *Trapeziopelta. Psyche* **60**: 51. [28.ix.1953.]
Brown, W.L.,Jr. 1953j. Revisionary notes on the ant genus *Myrmecia* of Australia. *Bulletin of the Museum of Comparative Zoology at Harvard College* **111**: 1–35. [(30).xi.1953.]
Brown, W.L.,Jr. 1953k. A revision of the dacetine ant genus *Orectognathus. Memoirs of the Queensland Museum* **13**: 84–104. [14.xii.1953.]
Brown, W.L.,Jr. 1954a. The Indo-Australian species of the ant genus *Strumigenys* Fr. Smith: *S. wallacei* Emery and relatives. *Psyche* **60** (1953): 85–89. [8.i.1954.]
Brown, W.L.,Jr. 1954b. A preliminary report on dacetine ant studies in Australia. *Annals of the Entomological Society of America* **46** (1953): 465–471. [8.ii.1954.]
Brown, W.L.,Jr. 1954c. The Neotropical species of the ant genus *Strumigenys* Fr. Smith: group of *elongata* Roger. *Journal of the New York Entomological Society* **61** (1953): 189–200. [20.ii.1954.]
Brown, W.L.,Jr. 1954d. The synonymy of the ant *Aphaenogaster lepida* Wheeler. *Pan-Pacific Entomologist* **30**: 10. [30.iii.1954.]
Brown, W.L.,Jr. 1954e. Remarks on the internal phylogeny and subfamily classification of the family Formicidae. *Insectes Sociaux* **1**: 21–31. [(31).iii.1954.]
Brown, W.L.,Jr. 1954f. The Indo-Australian species of the ant genus *Strumigenys* Fr. Smith: group of *doriae* Emery. *Psyche* **60** (1953): 160–166. [27.iv.1954.]
Brown, W.L.,Jr. 1954g. Systematic and other notes on some of the smaller species of the ant genus *Rhytidoponera* Mayr. *Breviora* **33**: 1–11. [14.v.1954.]
Brown, W.L.,Jr. 1954h. A review of the *coxalis* group of the ant genus *Stictoponera. Breviora* **34**: 1–10. [20.vii.1954.]
Brown, W.L.,Jr. 1954i. New synonymy of an Australian *Iridomyrmex. Psyche* **61**: 67. [26.viii.1954.]
Brown, W.L.,Jr. 1954j. The Indo-Australian species of the ant genus *Strumigenys* Fr. Smith: *S. chapmani* new species. *Psyche* **61**: 68–73. [26.viii.1954.]
Brown, W.L.,Jr. 1954k. The ant genus *Strumigenys* Fred. Smith in the Ethiopian and Malagasy regions. *Bulletin of the Museum of Comparative Zoology at Harvard College* **112**: 3–34. [(31).viii.1954.]
Brown, W.L.,Jr. 1954l. The Neotropical species of the ant genus *Strumigenys* Fr. Smith: group of *saliens* Mayr. *Journal of the New York Entomological Society* **62**: 55–62. [29.x.1954.]
Brown, W.L.,Jr. 1955a. The ant *Leptothorax muscorum* (Nylander) in North America. *Entomological News* **66**: 43–47. [28.i.1955.]
Brown, W.L.,Jr. 1955b. *Forelifidis* M.R. Smith a synonym. *Entomological News* **66**: 68. [5.iii.1955.]
Brown, W.L.,Jr. 1955c. The identity of the British *Strongylognathus. Journal of the Society for British Entomology* **5**: 113–114. [22.iii.1955.]
Brown, W.L.,Jr. 1955d. A revision of the Australian ant genus *Notoncus* Emery, with notes on the other genera of Melophorini. *Bulletin of the Museum of Comparative Zoology at Harvard College* **113**: 471–494. [(30).vi.1955.]
Brown, W.L.,Jr. 1955e. The ant *Cerapachys rufithorax* and its synonyms. *Psyche* **62**: 52. [19.vii.1955.]
Brown, W.L.,Jr. 1955f. The ant *Centromyrmex donisthorpei* Menozzi, a synonym. *Psyche* **62**: 103. [8.xi.1955.]
Brown, W.L.,Jr. 1955g. *Nylanderia myops* (Mann), new combination. *Psyche* **62**: 135–136. [8.xi.1955.]
Brown, W.L.,Jr. 1955h. The first social parasite in the ant tribe Dacetini. *Insectes Sociaux* **2**: 181–186. [(31).xii.1955.]
Brown, W.L.,Jr. 1956a. Some synonymies in the ant genus *Camponotus. Psyche* **63**: 38–40. [11.x.1956.]
Brown, W.L.,Jr. 1956b. Notes on the ant genus *Holcoponera* Mayr, with descriptions of two new species. *Insectes Sociaux* **3**: 489–497. [(31).xii.1956.]
Brown, W.L.,Jr. 1957a. The identity of *Lordomyrma rugosa* Clark. *Psyche* **63** (1956): 49. [23.i.1957.]
Brown, W.L.,Jr. 1957b. The relationship of two African *Tetramorium* species. *Psyche* **63** (1956): 75. [23.i.1957.]
Brown, W.L.,Jr. 1957c. The Neotropical species of the ant genus *Strumigenys* Fr. Smith: group of *cultriger* Mayr and *S. tococae* Wheeler. *Journal of the New York Entomological Society* **63** (1955): 97–102. [8.iii.1957.]
Brown, W.L.,Jr. 1957d. Is the ant genus *Tetramorium* native in North America? *Breviora* **72**: 1–8. [29.iii.1957.]
Brown, W.L.,Jr. 1957e. In Paynter, R.A. (ed.). Biological investigations in the Selva Lacandona, Chiapas, Mexico. 4. Ants from Laguna Ocotal. *Bulletin of the Museum of Comparative Zoology at Harvard College* **116**: 228–237. [(30).iv.1957.]
Brown, W.L.,Jr. 1957f. Distribution and variation of the ant *Formica dakotensis* Emery. *Entomological News* **68**: 165–167. [10.vi.1957.]
Brown, W.L.,Jr. 1957g. The Indo-Australian species of the ant genus *Strumigenys* Fr. Smith: three new Philippine species. *Psyche* **63** (1956): 113–118. [18.xi.1957.]
Brown, W.L.,Jr. 1957h. The synonymy and relationships of the ant *Pseudolasius bayoni* Menozzi. *Psyche* **63** (1956): 146. [18.xi.1957.]
Brown, W.L.,Jr. 1957i. The ending for subtribal names in zoology. *Systematic Zoology* **6**: 193–194. [(31).xii.1957.]
Brown, W.L.,Jr. 1958a. The army ant *Aenictus exiguus* Clark a synonym. *Psyche* **64** (1957): 5. [10.i.1958.]
Brown, W.L.,Jr. 1958b. A supplement to the revisions of the dacetine ant genera *Orectognathus* and *Arnoldidris*, with keys to the species. *Psyche* **64** (1957): 17–29. [10.i.1958.]
Brown, W.L.,Jr. 1958c. The Neotropical species of the ant genus *Strumigenys* Fr. Smith: group of *cordovensis* Mayr. *Studia Entomologica* (N.S.) **1**: 217–224. [31.i.1958.]
Brown, W.L.,Jr. 1958d. A new Japanese species of the dacetine ant genus *Epitritus. Mushi* **31**: 69–72. [15.iii.1958.]
Brown, W.L.,Jr. 1958e. The Neotropical species of the ant genus *Strumigenys* Fr. Smith: group of *marginiventris* Santschi. *Journal of the New York Entomological Society* **65** (1957): 123–128. [27.vi.1958.]
Brown, W.L.,Jr. 1958f. The Neotropical species of the ant genus *Strumigenys* Fr. Smith: group of *ogloblini* Santschi. *Journal of the New York Entomological Society* **65** (1957): 133–137. [27.vi.1958.]

Brown, W.L.,Jr. 1958g. Contributions toward a reclassification of the Formicidae. 2. Tribe Ectatommini. *Bulletin of the Museum of Comparative Zoology at Harvard College* **118**: 175–362. [(30).vi.1958.]

Brown, W.L.,Jr. 1958h. A review of the ants of New Zealand. *Acta Hymenopterologica* **1**: 1–50. [30.vii.1958.]

Brown, W.L.,Jr. 1958i. The Indo-Australian species of the ant genus *Strumigenys* Fr. Smith: *S. decollata* Mann and *S. ecliptacoca* new species. *Psyche* **64** (1957): 109–114. [14.viii.1958.]

Brown, W.L.,Jr. 1958j. Predation of arthropod eggs by the ant genera *Proceratium* and *Discothyrea*. *Psyche* **64** (1957): 115. [14.viii.1958.]

Brown, W.L.,Jr. 1959a. The Neotropical species of the ant genus *Strumigenys* Fr. Smith: group of *emeryi* Mann. *Entomological News* **70**: 97–104. [3.iv.1959.]

Brown, W.L.,Jr. 1959b. A revision of the dacetine ant genus *Neostruma*. *Breviora* **107**: 1–13. [6.v.1959.]

Brown, W.L.,Jr. 1959c. Some new species of dacetine ants. *Breviora* **108**: 1–11. [7.v.1959.]

Brown, W.L.,Jr. 1959d. Variation in the ant *Polyrhachis thrinax*. *Entomological News* **70**: 164. [15.vi.1959.]

Brown, W.L.,Jr. 1959e. Synonymy in the ant genus *Macromischa* Roger. *Florida Entomologist* **42**: 73–74. [(30).vi.1959.]

Brown, W.L.,Jr. 1959f. The Neotropical species of the ant genus *Strumigenys* Fr. Smith: group of *silvestrii* Emery. *Studia Entomologica* (N.S.) **2**: 25–30. [29.ix.1959.]

Brown, W.L.,Jr. 1959g. The Indo-Australian species of the ant genus *Strumigenys* Fr. Smith: group of *S. godeffroyi* in Borneo. *Psyche* **65** (1958): 81–89. [3.xii.1959.]

Brown, W.L.,Jr. 1960a. Contributions toward a reclassification of the Formicidae. 3. Tribe Amblyoponini. *Bulletin of the Museum of Comparative Zoology at Harvard College* **122**: 145–230. [(31).iii.1960.]

Brown, W.L.,Jr. 1960b. The Neotropical species of the ant genus *Strumigenys* Fr. Smith: group of *gundlachi* (Roger). *Psyche* **66** (1959): 37–52. [8.ix.1960.]

Brown, W.L.,Jr. 1960c. A new African ant of the genus *Strumigenys*, tribe Dacetini. *Entomological News* **71**: 206. [26.ix.1960.]

Brown, W.L.,Jr. 1961a. The Neotropical species of the ant genus *Strumigenys* Fr. Smith: miscellaneous concluding studies. *Psyche* **68**: 58–69. [29.xii.1961.]

Brown, W.L.,Jr. 1961b. A note on the ant *Gnamptogenys hartmani* Wheeler. *Psyche* **68**: 69. [29.xii.1961.]

Brown, W.L.,Jr. 1962a. A new ant of the genus *Amblyopone* from Panama. *Psyche* **69**: 73–76. [1.vii.1962.]

Brown, W.L.,Jr. 1962b. A new ant of the genus *Epitritus* from south of the Sahara. *Psyche* **69**: 77–80. [1.vii.1962.]

Brown, W.L.,Jr. 1962c. The Neotropical species of the ant genus *Strumigenys* Fr. Smith: synopsis and key to species. *Psyche* **69**: 238–267. [29.xii.1962.]

Brown, W.L.,Jr. 1963. Characters and synonymies among the genera of ants. Part 3. Some members of the tribe Ponerini. *Breviora* **190**: 1–10. [30.ix.1963.]

Brown, W.L.,Jr. 1964a. The ant genus *Smithistruma:* a first supplement to the world revision. *Transactions of the American Entomological Society* **89**: 183–200. [31.i.1964.]

Brown, W.L.,Jr. 1964b. Solution to the problem of *Tetramorium lucayanum*. *Entomological News* **75**: 130–132. [12.v.1964.]

Brown, W.L.,Jr. 1964c. Genus *Rhoptromyrmex*, revision of, and key to species. *Pilot Register of Zoology* **cards 11–19**. [20.v.1964.]

Brown, W.L.,Jr. 1964d. Synonymy and variation of some species of the ant genus *Anochetus*. *Journal of the Kansas Entomological Society* **37**: 212–215. [6.viii.1964.]

Brown, W.L.,Jr. 1965a. *Colobostruma papulata* species nov. *Pilot Register of Zoology* **card 21**. [5.iv.1965.]

Brown, W.L.,Jr. 1965b. *Colobostruma nancyae* species nov. *Pilot Register of Zoology* **card 22**. [5.iv.1965.]

Brown, W.L.,Jr. 1965c. Contributions toward a reclassification of the Formicidae. 4. Tribe Typhlomyrmecini. *Psyche* **72**: 65–78. [25.vi.1965.]

Brown, W.L.,Jr. 1965d. Studies on North American ants. 1. The *Formica integra* subgroup. *Entomological News* **76**: 181–186. [25.vi.1965.]

Brown, W.L.,Jr. 1966. *Strumigenys rectidens* species nov. *Pilot Register of Zoology* **card 23**. [10.iv.1966.]

Brown, W.L.,Jr. 1967a. The ant *Cataglyphis birmana* a synonym. *Psyche* **73** (1966): 277. [24.v.1967.]

Brown, W.L.,Jr. 1967b. The ant *Aphaenogaster gatesi* transferred to *Pheidole*. *Psyche* **73** (1966): 283. [24.v.1967.]

Brown, W.L.,Jr. 1967c. Studies on North American ants. 2. *Myrmecina*. *Entomological News* **78**: 233–240. [9.xi.1967.]

Brown, W.L.,Jr. 1968a. A new *Pheidole* with reversed phragmosis. *Psyche* **74** (1967): 331–339. [12.vi.1968.]

Brown, W.L.,Jr. 1968b. *Strumigenys sisyrata* species nov. *Pilot Register of Zoology* **card 24**. [15.x.1968.]

Brown, W.L.,Jr. 1969a. *Strumigenys lopotyle* species nov. *Pilot Register of Zoology* **card 27**. [12.x.1969.]

Brown, W.L.,Jr. 1969b. *Strumigenys wilsoni* species nov. *Pilot Register of Zoology* **card 28**. [3.xi.1969.]

Brown, W.L.,Jr. 1971a. Characters and synonymies among the genera of ants. Part 4. Some genera of subfamily Myrmicinae. *Breviora* **365**: 1–5. [15.i.1971.]

Brown, W.L.,Jr. 1971b. The identity and synonymy of *Pheidole vigilans*, a common ant of southeastern Australia. *Journal of the Australian Entomological Society* **10**: 13–14. [31.iii.1971.]

Brown, W.L.,Jr. 1971c. The Indo-Australian species of the ant genus *Strumigenys:* group of *szalayi*. *Entomological Essays to Commemorate the Retirement of Professor K. Yasumatsu:* 73–86. [(31.xii).1971.]

Brown, W.L.,Jr. 1972. *Asketogenys acubecca*, a new genus and species of dacetine ants from Malaya. *Psyche* **79**: 23–26. [15.xi.1972.]

Brown, W.L.,Jr. 1973a. A new species of *Miccostruma* from West Africa, with notes on the genus. *Journal of the Kansas Entomological Society* **46**: 32–35. [30.i.1973.]

Brown, W.L.,Jr. 1973b. A comparison of the Hylean and Congo-West African rain forest ant faunas. In Meggers, B.J., Ayensu, E.S. & Duckworth, W.D. (eds.). *Tropical Forest Ecosystems in Africa and South America: a Comparative Review:* 161–185. [(31).i.1973.]

Brown, W.L.,Jr. 1973c. The Indo-Australian species of the ant genus *Strumigenys:* groups of *horvathi, mayri* and *wallacei*. *Pacific Insects* **15**: 259–269. [20.vii.1973.]

Brown, W.L.,Jr. 1974a. A remarkable new island isolate in the ant genus *Proceratium*. *Psyche* **81**: 70–83. [21.v.1974.]

Brown, W.L.,Jr. 1974b. *Novomessor manni* a synonym of *Aphaenogaster ensifera*. *Entomological News* **85**: 45–47. [9.viii.1974.]

Brown, W.L.,Jr. 1974c. A supplement to the revision of the ant genus *Basiceros*. *Journal of the New York Entomological Society* **82**: 131–140. [22.viii.1974.]

Brown, W.L.,Jr. 1974d. *Concoctio* genus nov.; *C. concenta* species nov. *Pilot Register of Zoology* **card 29–30**. [31.viii.1974.]

Brown, W.L.,Jr. 1974e. *Dolioponera* genus. nov.; *D. fustigera* species nov. *Pilot Register of Zoology* **card 31–32**. [20.xi.1974.]

Brown, W.L.,Jr. 1975. Contributions toward a reclassification of the Formicidae. 5. Ponerinae, tribes Platythyreini, Cerapachyini, Cylindromyrmecini, Acanthostichini, and Aenictogitini. *Search Agriculture* **5**. Entomology (Ithaca) 15: 1–115. [(30).vi.1975.]

Brown, W.L.,Jr. 1976a. Contributions toward a reclassification of the Formicidae. Part 6. Ponerinae, tribe Ponerini, subtribe Odontomachiti. Section A. Introduction, subtribal characters, genus *Odontomachus*. *Studia Entomologica* (N.S.) **19**: 67–171. [30.xii.1976.]

Brown, W.L.,Jr. 1976b. *Cladarogenys* genus nov,; *C. lasia* species nov. *Pilot Register of Zoology* **card 33–34**. [30.xii.1976.]

Brown, W.L.,Jr. 1977. *Ctenobethylus* (Bethylidae) a new synonym of *Iridomyrmex*. *Psyche* **83** (1976): 214–216. [24.ii.1977.]

Brown, W.L.,Jr. 1978a. An aberrant new genus of myrmicine ant from Madagascar. *Psyche* **84** (1977): 218–224. [19.vii.1978.]

Brown, W.L.,Jr. 1978b. A supplement to the world revision of *Odontomachus*. *Psyche* **84** (1977): 281–285. [19.vii.1978.]

Brown, W.L.,Jr. 1978c. Contributions toward a reclassification of the Formicidae. Part 6. Ponerinae, tribe Ponerini, subtribe Odontomachiti. Section B. Genus *Anochetus* and bibliography. *Studia Entomologica* (N.S.) **20**: 549–652. [30.viii.1978.]

Brown, W.L.,Jr. 1980a. *Protalaridris* genus nov.;*P. armata* species nov. *Pilot Register of Zoology* **card 36–37**. [31.iii.1980.]

Brown, W.L.,Jr. 1980b. A remarkable new species of *Proceratium*, with dietary and other notes on the genus. *Psyche* **86** (1979): 337–346. [19.xii.1980.]

Brown, W.L.,Jr. 1981. Preliminary contributions toward a revision of the ant genus *Pheidole*. Part 1. *Journal of the Kansas Entomological Society* **54**: 523–530. [17.vii.1981.]

Brown, W.L.,Jr. 1986. *Indomyrma dasypyx*, new genus and species, a myrmicine ant from peninsular India. *Israel Journal of Entomology* **19** (1985): 37–49. [(28).ii.1986.]

Brown, W.L.,Jr. 1988a. *Neoclystopsenella* (Bethylidae) a synonym of *Tapinoma* (Formicidae). *Psyche* **94** (1987): 337. [25.vii.1988.]

Brown, W.L.,Jr. 1988b. *Strumigenys philiporum* species nov. *Pilot Register of Zoology* **card 38**. [15.xi.1988.]

Brown, W.L.,Jr. 1988c. *Strumigenys buleru* species nov. *Pilot Register of Zoology* **card 39**. [15.xi.1988.]

Brown, W.L.,Jr. 1988d. *Strumigenys cochlearis* species nov. *Pilot Register of Zoology* **card 40**. [15.xi.1988.]

Brown, W.L.,Jr. 1988e. *Strumigenys yaleopleura* species nov. *Pilot Register of Zoology* **card 41**. [10.xii.1988.]

Brown, W.L.,Jr. 1988f. *Strumigenys anetes* species nov. *Pilot Register of Zoology* **card 42**. [10.xii.1988.]

Brown, W.L.,Jr. 1988g. *Strumigenys paranetes* species nov. *Pilot Register of Zoology* **card 43**. [10.xii.1988.]

Brown, W.L.,Jr. 1988h. Data on malpighian tubules in ants (pp. 17–27). In Trager, J.C. (ed.). *Advances in Myrmecology:* 551 pp. Leiden. [(31.xii).1988.]

Brown, W.L.,Jr. 1993. Two new species of *Gnamptogenys*, and an account of millipede predation by one of them. *Psyche* **99** (1992): 275–289. [2.vi.1993.]

Brown, W.L.,Jr. & Boisvert, R.G. 1979. The dacetine ant genus *Pentastruma*. *Psyche* **85** (1978): 201–207. [24.iv.1979.]

Brown, W.L.,Jr. & Carpenter, F.M. 1979. A restudy of two ants from the Sicilian amber. *Psyche* **85** (1978): 417–423. [18.ix.1979.]

Brown, W.L.,Jr., Gotwald, W.H. & Lévieux, J. 1971. A new genus of ponerine ants from West Africa, with ecological notes. *Psyche* **77** (1970): 259–275. [31.v.1971.]

Brown, W.L.,Jr. & Kempf, W.W. 1960. A world revision of the ant tribe Basicerotini. *Studia Entomologica* (N.S.) **3**: 161–250. [20.xii.1960.]

Brown, W.L.,Jr. & Kempf, W.W. 1961. The type species of the ant genus *Eurhopalothrix*. *Psyche* **67** (1960): 44. [16.ii.1961.]

Brown, W.L.,Jr. & Kempf, W.W. 1968. *Tatuidris*, a remarkable new genus of Formicidae. *Psyche* **74** (1967): 183–190. [20.ii.1968.]

Brown, W.L.,Jr. & Kempf, W.W. 1969. A revision of the Neotropical dacetine ant genus *Acanthognathus*. *Psyche* **76**: 87–109. [12.xii.1969.]

Brown, W.L.,Jr. & Nutting, W.L. 1950. Wing venation and the phylogeny of the Formicidae. *Transactions of the American Entomological Society* **75**: 113–132. [4.i.1950.]

Brown, W.L.,Jr. & Wilson, E.O. 1957a. *Dacetinops*, a new ant genus from New Guinea. *Breviora* **77**: 1–7. [21.vi.1957.]

Brown, W.L.,Jr. & Wilson, E.O. 1957b. A new parasitic ant of the genus *Monomorium* from Alabama, with a consideration of the status of genus *Epixenus* Emery. *Entomological News* **68**: 239–246. [22.x.1957.]

Brown, W.L.,Jr. & Wilson, E.O. 1959a. The search for *Nothomyrmecia*. *Western Australian Naturalist* **7**: 25–30. [25.viii.1959.]

Brown, W.L.,Jr. & Wilson, E.O. 1959b. The evolution of the dacetine ants. *Quarterly Review of Biology* **34**: 278–294. [(31).xii.1959.]

Brown, W.L.,Jr. & Yasumatsu, K. 1951. On the publication date of *Polyhomoa itoi* Azuma. *Mushi* **22**: 93–95. [15.iv.1951.]

Bruch, C. 1914. Catalogo sistematico de los formicidos argentinos. *Revista del Museo de La Plata* **19**: 211–234. [20.iii.1914.]

Bruch, C. 1915. Suplemento al catalogo de los formicidos argentinos. 1. (Addenda et corrigenda.) *Revista del Museo de La Plata* **19**: 527–537. [24.xii.1915.]

Bruch, C. 1916. Contribución al estudio de las hormigas de la Provincia de San Luis. *Revista del Museo de La Plata* **23**: 291–357. [22.ix.1916.]

Bruch, C. 1917a. Costumbes y nidos de hormigas. *Anales de la Sociedad Científica Argentina* **83**: 302–316. [(30).vi.1917.]

Bruch, C. 1917b. La forma sexuada femenina de *Cryptocerus ridiculus* Santschi. *Physis. Revista de la Sociedad Argentina de Ciencias Naturales* **3**: 269–270. [30.vii.1917.]

Bruch, C. 1917c. Costumbres y nidos de hormigas. 2. *Anales de la Sociedad Científica Argentina* **84**: 154–168. [(31).x.1917.]

Bruch, C. 1917d. Hormigas de Catamarca. *Physis. Revista de la Sociedad Argentina de Ciencias Naturales* **3**: 430–433. [31.xii.1917.]

Bruch, C. 1919. Descripción de una curiosa ponerina de Córdoba. *Discothyrea neotropica* n. sp. *Physis. Revista de la Sociedad Argentina de Ciencias Naturales* **4**: 400–403. [31.xii.1919.]

Bruch, C. 1921. Estudios mirmecológicos. *Revista del Museo de La Plata* **26**: 175–211. [31.xii.1921.]

Bruch, C. 1923. Estudios mirmecológicos. *Revista del Museo de La Plata* **27**: 172–220. [(31.xii).1923.]

Bruch, C. 1924a. Descripción de la reina *Eciton (Acamatus) hetschkoi* Mayr. *Physis. Revista de la Sociedad Argentina de Ciencias Naturales* **7**: 232–235. [25.iii.1924.]

Bruch, C. 1924b. Una nueva ponerina, *Acanthostichus ramosmexiae* n. spec. *Physis. Revista de la Sociedad Argentina de Ciencias Naturales* **7**: 260–261. [25.iii.1924.]

Bruch, C. 1925a. Macho, larva y ninfa de *Acanthostichus ramosmexiae* Bruch. *Physis. Revista de la Sociedad Argentina de Ciencias Naturales* **8**: 110–114. [23.v.1925.]

Bruch, C. 1925b. Rectificación. *Physis. Revista de la Sociedad Argentina de Ciencias Naturales* **8**: 125. [23.v.1925.]

Bruch, C. 1928. Estudios mirmecológicos. *Anales del Museo Nacional de Historia Natural "Bernardino Rivadavia"* **34**: 341–360. [17.ii.1928.]

Bruch, C. 1931. Notas biológicas y sistematicas acerca de *Bruchomyrma acutidens* Santschi. *Revista del Museo de La Plata* **33**: 31–55. [17.iv.1931.]

Bruch, C. 1932. Descripción de un género y especie nueva de una hormiga parasita. *Revista del Museo de La Plata* **33**: 271–275. [29.ii.1932.]

Bruch, C. 1934a. La hembra dictadiforme de *Acanthostichus laticornis* For. v. *obscuridens* Sants. *Revista de la Sociedad Entomologica Argentina* **6**: 3–8. [30.v.1934.]

Bruch, C. 1934b. Las formas femeninas de *Eciton*. Descripcion y redescripcion de algunas especies de la Argentina. *Anales de la Sociedad Científica Argentina* **118**: 113–135. [(30).ix.1934.]

Bruch, C. 1934c. Breves observaciones sobre costumbres de *Eciton (Holopone) dulcius* For. var. *jujuyensis* For. y descripcion del macho. *Revista de la Sociedad Entomologica Argentina* **6**: 261–265. [30.xii.1934.]

Brues, C.T. 1925. *Scyphodon*, an anomalous genus of Hymenoptera of doubtful affinities. *Treubia* **6**: 93–96. [(28).ii.1925.]

Brues, C.T. 1939. New Oligocene Braconidae and Bethylidae from Baltic Amber. *Annals of the Entomological Society of America* **32**: 251–263. [30.vi.1939.]

Brullé, M. 1832. *Expédition Scientifique de Morée.* Section des Sciences Physiques **3**, part 1. Zoologie. Deuxième section -Des animaux articulés: 400 pp. Paris. [(31.xii).1832.]

Brullé, M. 1840. Insectes (pp. 53–95). In Webb, P. Barker & Berthelot, S. *Histoire Naturelles des Iles Canaries* **2** (deuxième partie -Entomologie): 119 pp. Paris. [(31.xii).1840.]

Buckley, S.B. 1860a. The cutting ant of Texas. *Proceedings of the Academy of Natural Sciences of Philadelphia* **12** (1859): 233–236. [before x.1860.]

Buckley, S.B. 1860b. *Myrmica (Atta) molefaciens,* "Stinging Ant", or "Mound-Making Ant", of Texas. *Proceedings of the Academy of Natural Sciences of Philadelphia* **1860**: 445–447. [(31).x.1860.]

Buckley, S.B. 1866. Descriptions of new species of North American Formicidae. *Proceedings of the Entomological Society of Philadelphia* **6**: 152–172. [(31).vii.1866.]

Buckley, S.B. 1867. Descriptions of new species of North American Formicidae. *Proceedings of the Entomological Society of Philadelphia* **6**: 335–350. [(28).ii.1867.]

Bünzli, G.H. 1935. Untersuchungen über coccidophile Ameisen aus den Kaffeefelden von Surinam. *Mitteilungen der Schweizerischen Entomologischen Gesellschaft* **16**: 455–593. [14.viii.1935.]

Buren, W.F. 1941. *Lasius (Acanthomyops) plumopilosus,* a new ant with plumose hairs from Iowa. *Iowa State College Journal of Science* **15**: 231–235. [(30).iv.1941.]

Buren, W.F. 1942. New ants from Minnesota, Iowa, and Wisconsin. *Iowa State College Journal of Science* **16**: 399–408. [(30).iv.1942.]

Buren, W.F. 1944a. A list of Iowa ants. *Iowa State College Journal of Science* **18**: 277–312. [(30).iv.1944.]

Buren, W.F. 1944b. A new fungus growing ant from Mexico. *Psyche* **51**: 5–6. [(30).vi.1944.]

Buren, W.F. 1945. *Leptothorax longispinosus* subsp. *iowensis,* nom. n. *Proceedings of the Entomological Society of Washington* **47**: 288. [28.xii.1945.]

Buren, W.F. 1950. A new *Lasius (Acanthomyops)* with a key to North American females. *Proceedings of the Entomological Society of Washington* **52**: 184–190. [18.viii.1950.]

Buren, W.F. 1959. A review of the species of *Crematogaster, sensu stricto,* in North America. Part 1. *Journal of the New York Entomological Society* **66** (1958): 119–134. [24.iii.1959.]

Buren, W.F. 1968a. Some fundamental taxonomic problems in *Formica. Journal of the Georgia Entomological Society* **3**: 25–40. [(30).iv.1968.]

Buren, W.F. 1968b. A review of the species of *Crematogaster, sensu stricto,* in North America. 2. Descriptions of new species. *Journal of the Georgia Entomological Society* **3**: 91–121. [(31).vii.1968.]

Buren, W.F. 1972. Revisionary studies on the taxonomy of the imported fire ants. *Journal of the Georgia Entomological Society* **7**: 1–27. [(31).i.1972.]

Buren, W.F., Nickerson, J.C. & Thompson, C.R. 1976. Mixed nests of *Conomyrma insana* and *C. flavopecta*-evidence of parasitism. *Psyche* **82** (1975): 306–313. [20.iv.1976.]

Burmeister, C.H.C. 1831. "Derselbe über Bernsteininsecten." *Isis von Oken* **1831**: 1100. [(31.xii).1831.]

Burnham, L. 1979. Survey of social insects in the fossil record. *Psyche* **85** (1978): 85–134. [26.i.1979.]

Bursakov, S.S. 1984. Dva novykh vida murav'ev roda *Tetramorium* iz yugovostochnogo Kazakhstana. *Zoologichesky Zhurnal* **63**: 399–405. [21.ii.1984.]

Buschinger, A. 1966a. *Leptothorax (Mychothorax) kutteri* n.sp., eine sozialparasitische Ameise. *Insectes Sociaux* **12** (1965): 327–334. [(31).iii.1966.]

Buschinger, A. 1966b. *Leptothorax (Mychothorax) muscorum* Nylander und *Leptothorax (M.) gredleri* Mayr, zwei gute Arten. *Insectes Sociaux* **13**: 165–172. [(31).xii.1966.]

Buschinger, A. 1972. Kreuzung zweier sozialparasitischer Ameisenarten, *Doronomyrmex pacis* Kutter und *Leptothorax kutteri* Buschinger. *Zoologischer Anzeiger* **189**: 169–179. [(31).x.1972.]

Buschinger, A. 1979a. Functional monogyny in the American guest ant *Formicoxenus hirticornis* (Emery) (=*Leptothorax hirticornis*). *Insectes Sociaux* **26**: 61–68. [(30).vi.1979.]

Buschinger, A. 1979b. *Doronomyrmex pocahontas,* n. sp., a parasitic ant from Alberta, Canada. *Insectes Sociaux* **26**: 216–222. [(30).ix.1979.]

Buschinger, A. 1981. Biological and systematic relationships of social parasitic Leptothoracini from Europe and North America (pp. 211–222). In Howse, P.E. & Clement, J.-L. (eds.). *Biosystematics of Social Insects.* Systematics Association Special Volume **19**: 346 pp. London. [(31.xii).1981.]

Buschinger, A. 1982. *Epimyrma goesswaldi* Menozzi, 1931 = *Epimyrma ravouxi* (André, 1896)—Morphologischer und biologischer Nachweis der Synonymie. *Zoologischer Anzeiger* **208**: 352–358. [(31.xii).1982.]

Buschinger, A. 1983. *Leptothorax faberi* n. sp., an apparently parasitic ant from Jasper National Park, Canada. *Psyche* **89** (1982): 197–209. [27.iv.1983.]

Buschinger, A. 1987. Synonymy of *Leonomyrma* Arnoldi 1968 with *Chalepoxenus* Menozzi 1922. *Psyche* **94**: 117–126. [6.x.1987.]

Buschinger, A. 1989. Evolution, speciation and inbreeding in the parasitic ant genus *Epimyrma. Journal of Evolutionary Biology* **2**: 265–283. [1989.]

Buschinger, A. 1990. Sympatric speciation and radiative evolution of socially parasitic ants. Heretic hypotheses and their factual background. *Zeitschrift für Zoologische Systematik und Evolutionsforschung* **28**: 241–260. [1990.]

Buschinger, A., Cagniant, H., Erhardt, W., & Heinze, J. 1989. *Chalepoxenus brunneus,* a workerless "degenerate slave-maker" ant. *Psyche* **95** (1988): 253–264. [13.vi.1989.]

Buschinger, A., Erhardt, W., Fischer, K., & Ofer, J. 1988. The slave-making ant genus *Chalepoxenus.* 1. Review of literature, range, slave species. *Zoologische Jahrbücher. Abteilung für Systematik und Geographie der Tiere.* **115**: 383–401. [1988.]

Buschinger, A., Fischer, K., Guthy, H.-P., Jessen, K., & Winter, U. 1987. Biosystematic revision of *Epimyrma kraussei, E. vandeli,* and *E. foreli. Psyche* **93** (1986): 253–276. [17.iii.1987.]

Buschinger, A. & Francoeur, A. 1983. The guest ant *Symmyrmica chamberlini,* rediscovered near Salt Lake City, Utah. *Psyche* **90**: 297–306. [9.xii.1983.]

Buschinger, A., Francoeur, A., & Fischer, K. 1981. Functional monogyny, sexual behavior and karyotype of the guest ant *Leptothorax provancheri* Emery. *Psyche* **87** (1980): 1–12. [27.iv.1981.]

Buschinger, A. & Heinze, J. 1993. *Doronomyrmex pocahontas:* not a workerless parasite but still an enigmatic taxon. *Insectes Sociaux* **40**: 423–432. [(31.xii).1993.]

Buschinger, A., Jessen, K. & Cagniant, H. 1990. The life history of *Epimyrma algeriana,* a slave-making ant with facultative polygyny. *Zoologische Beiträge* (N.F.) **33**: 23–49. [1990.]

Buschinger, A., Peeters, C. & Crozier, R.H. 1990. Life-pattern studies on an Australian *Sphinctomyrmex:* functional polygyny, brood periodicity and raiding behavior. *Psyche* **96** (1989): 287–300. [1990.]

Buschinger, A. & Winter, U. 1985. Life history and male morphology of the workerless parasitic ant *Epimyrma corsica. Entomologia Generalis* **10**: 65–75. [(31).v.1985.]

Buschinger, A., Winter, U. & Faber, W. 1984. The biology of *Myrmoxenus gordiagini* Ruzsky, a slave-making ant. *Psyche* **90** (1983): 335–342. [6.iv.1984.]

Byars, L.F. 1951. A new fungus growing ant from Arizona. *Proceedings of the Entomological Society of Washington* **53**: 109–111. [12.iv.1951.]

Cagniant, H. 1964. Etude de quelques fourmis marocaines. Statistique provisoire des Formicidae du Maroc. *Bulletin de la Société d'Histoire Naturelle de l'Afrique du Nord* **53** (1962): 83–118. [(31.xii).1964.]

Cagniant, H. 1966a. Description des trois castes de *Leptothorax tyndalei* (Forel). *Bulletin de la Société Entomologique de France* **71**: 17–21. [18.vi.1966.]

Cagniant, H. 1966b. Note sur le peuplement en fourmis d'une montagne de la region d'Alger, l'Atlas de Blida. *Bulletin de la Société d'Histoire Naturelle de Toulouse* **102**: 278–284. [30.vi.1966.]

Cagniant, H. 1966c. Nouvelle description d'*Aphaenogaster (Attomyrma) crocea* (André). Representation des trois castes. Notes biologiques. *Bulletin de la Société Zoologique de France* **91**: 61–69. [27.x.1966.]

Cagniant, H. 1967a. Description de *Messor bernardi* n. sp. Representation des trois castes. *Bulletin de la Société Entomologique de France* **72**: 131–139 [27.vi.1967.]

Cagniant, H. 1967b. *Leptothorax barryi* n. sp. Hyménoptère Formicidae Myrmicinae d'Algérie. *Bulletin de la Société Entomologique de France* **72**: 272–275. [29.xii.1967.]

Cagniant, H. 1968a. Liste préliminaire de fourmis forestières d'Algérie. Resultats obtenus de 1963 a 1964. *Bulletin de la Société d'Histoire Naturelle de Toulouse* **104**: 138–147. [30.v.1968.]

Cagniant, H. 1968b. Description de *Leptothorax monjauzei* n. sp. d'Algérie. Representation des trois castes et notes biologiques. *Bulletin de la Société Entomologique de France* **73**: 83–90. [11.vii.1968.]

Cagniant, H. 1968c. Description d'*Epimyrma algeriana* (nov. sp.), fourmi parasite. Representation des trois castes. Quelques observations biologiques, écologiques et éthologiques. *Insectes Sociaux* **15**: 157–170. [(31.xii).1968.]

Cagniant, H. 1969a. Description d'*Aphaenogaster crocea faureli* n. subsp. d'Algérie. *Bulletin de la Société Entomologique de France* **73** (1968): 232–235. [26.ii.1969.]

Cagniant, H. 1969b. Sur deux *Aphaenogaster* rares d'Algérie. *Insectes Sociaux* **16**: 103–114. [(31).xii.1969.]

Cagniant, H. 1970a. Deuxième liste de fourmis d'Algérie récoltées principalement en forêt. *Bulletin de la Société d'Histoire Naturelle de Toulouse* **105** (1969): 405–430. [5.ii.1970.]

Cagniant, H. 1970b. Nouvelle description de *Leptothorax spinosus* Forel d'Algérie. *Bulletin de la Société Entomologique de France* **74** (1969): 201–208. [26.ii.1970.]

Cagniant, H. 1970c. Deuxième liste de fourmis d'Algérie récoltées principalement en forêt. (Deuxième partie.) *Bulletin de la Société d'Histoire Naturelle de Toulouse* **106**: 28–40. [30.vi.1970.]

Cagniant, H. 1970d. Une nouvelle fourmi parasite d'Algérie: *Sifolinia kabylica* (nov. sp.). *Insectes Sociaux* **17**: 39–48. [(30).vi.1970.]

Cagniant, H. 1971. Description du mâle de *Stenamma africanum* Santschi. *Bulletin de la Société Entomologique de France* **76**: 98–101. [28.vii.1971.]

Cagniant, H. 1973. Description et representation des trois castes de *Chalepoxenus kutteri* (nov. sp.). *Insectes Sociaux* **20**: 145–156. [(31).xii.1973.]

Cagniant, H. 1982. Contribution a la connaissance des fourmis marocaines. *Aphaenogaster dejeani,* nov. sp. *Nouvelle Revue d'Entomologie* **12**: 281–286. [30.x.1982.]

Cagniant, H. 1983. Contribution a la connaissance des fourmis marocaines. *Chalepoxenus tramieri,* nov.sp. *Nouvelle Revue d'Entomologie* **13**: 319–322. [30.x.1983.]

Cagniant, H. 1984. Contribution a la connaissance des fourmis marocaines. *Aphaenogaster espadaleri,* n. sp. *Nouvelle Revue d'Entomologie* (N.S.) **1**: 387–395. [30.xii.1984.]

Cagniant, H. 1985a. Contribution a la connaissance des fourmis marocaines. *Chalepoxenus brunneus,* n. sp. *Nouvelle Revue d'Entomologie* (N.S.) **2**: 141–146. [25.vi.1985.]

Cagniant, H. 1985b. Contribution a la connaissance des fourmis marocaines. Etude des *Camponotus* du groupe *cruentatus* au Maroc. *Bulletin de la Société d'Histoire Naturelle de Toulouse* **121**: 77–84. [(30).vi. 1985.]

Cagniant, H. 1986. Contribution a la connaissance des fourmis marocaines. Description des sexués et complément a la définition de l'espèce *Aphaenogaster theryi* Santschi, 1923. *Bulletin de la Société d'Histoire Naturelle de Toulouse* **122**: 139–143. [(31).x.1986.]

Cagniant, 1987a. *Aphaenogaster nadigi* Santschi, bona species (n. status). *Eos* **62** (1986): 31–43. [27.ii.1987.]

Cagniant, H. 1987b. Contribution a la connaissance des fourmis marocaines. *Camponotus vagus ifranensis* n. ssp. *Bulletin de la Société Entomologique de France* **91** (1986): 117–123. [30.iii.1987.]

Cagniant, H. 1987c. Contribution a la connaissance des fourmis marocaines. *Leptothorax personatus,* n. sp. *Bulletin de la Société Entomologique de France* **91** (1986): 243–250. [22.vi.1987.]

Cagniant, H. 1987d. Contribution a la connaissance des fourmis marocaines. Nouvelle description et compléments a la définition de l'espèce *Aphaenogaster praedo* Emery. Problèmes posés par le groupe *praedo. Bulletin de la Société d'Histoire Naturelle de Toulouse* **123**: 159–165. [(31).x.1987.]

Cagniant, H. 1988a. Contribution a la connaissance des fourmis marocaines. Description des trois castes d'*Aphaenogaster torossiani* n. sp. et notes biologiques. *Bulletin de la Société Entomologique de France* **92**: 241–250. [7.iii.1988.]

Cagniant, H. 1988b. Contribution a la connaissance des fourmis marocaines. Description des trois castes d'*Aphaenogaster wilsoni* n. sp. *Nouvelle Revue d'Entomologie* (N.S.) **5**: 49–55. [20.iv.1988.]

Cagniant, H. 1988c. Contribution a la connaissance des fourmis marocaines. *Aphaenogaster baronii* n. sp. *Bulletin de la Société d'Histoire Naturelle de Toulouse* **124**: 43–50. [(31).vii.1988.]

Cagniant, H. 1989. Contribution a la connaissance des fourmis marocaines. Description des trois castes d'*Aphaenogaster weulersseae* n. sp.; notes biologiques et écologiques; étude comparée de trois populations. *Bulletin de la Société Entomologique de France* **94**: 113–125. [30.viii.1989.]

Cagniant, H. 1990a. Contribution a la connaissance des fourmis marocaines. *Aphaenogaster gemella* au Maroc. Nouvelle description d'*Aphaenogaster gemella* ssp. *marocana* Forel (n. status) de la région de Tanger. Problèmes biogéographiques soulevés par l'espèce *Aphaenogaster gemella. Bulletin de la Société d'Histoire Naturelle de Toulouse* **125** (1989): 47–54. [(31).i.1990.]

Cagniant, H. 1990b. Contribution a la connaissance des fourmis marocaines. *Aphaenogaster atlantis* Santschi, 1929 bona species (n. status). *Bulletin du Muséum National d'Histoire Naturelle* (4) **12** Section A (Zoologie, Biologie et Ecologie animales): 123–133. [2.viii.1990.]

Cagniant, H. 1990c. Contribution a la connaissance des fourmis marocaines. *Aphaenogaster sicardi. Bulletin du Muséum National d'Histoire Naturelle* (4) **12** Section A (Zoologie, Biologie et Ecologie animales): 443–453. [24.xi.1990.]

Cagniant, H. 1990d. Contribution a la connaissance des fourmis marocaines. *Aphaenogaster miniata* n. sp. Description des trois castes, populations et justification de son statut de bonne espèce. *Nouvelle Revue d'Entomologie* (N.S.) **7**: 143–154. [30.xi.1990.]

Cagniant, H. 1990e. Contribution a la connaissance des fourmis marocaines. Révision de l'espèce *Aphaenogaster curiosa* Santschi (n. status). *Journal of African Zoology* **104**: 457–475. [30.x.1990.]

Cagniant, H. 1991a. Contribution a la connaissance des fourmis marocaines. *Camponotus hoelldobleri* n. sp. *Bulletin de la Société Zoologique de France* **116**: 37–48. [(31.xii).1991.]

Cagniant, H. 1991b. Etude des populations marocaines de la superespece *Camponotus cruentatus* Latreille. Mise en evidence de *Camponotus obscuriventris. Eos* **67**: 35–54. [15.xi.1991.]

Cagniant, H. 1992a. Contribution a la connaissance des fourmis marocaines. Description des trois castes d'*Aphaenogaster dejeani* Cagniant, 1982. *Bulletin de la Société Zoologique de France* **117**: 65–73. [(31.xii).1992.]

Cagniant, H. 1992b. Contribution a la connaissance des fourmis marocaines. *Aphaenogaster tinauti* n. sp. *Bulletin de la Société Zoologique de France* **117**: 365–374. [(31.xii).1992.]

Cagniant, H. 1992c. Contribution a la connaissance des fourmis marocaines. *Aphaenogaster fallax* sp. n. *Journal of African Zoology* **106**: 197–210. [10.vii.1992.]

Cagniant, H. 1994. Contribution a la connaissance des fourmis marocaines. Description d'*Aphaenogaster rifensis* n. sp. Revision de la superespèce *Aphaenogaster* (supersp.) *gemella* Rog. (n. taxon). *Bulletin de la Société Zoologique de France* **119**: 15–29. [(31.xii).1994.]

Cagniant, H. & Espadaler, X. 1993. *Camponotus guanchus* Santschi, 1908, stat. nov., et étude des populations de *Camponotus sicheli* Mayr, 1866. *Journal of African Zoology* **107**: 419–438. [29.x.1993.]

Cameron, P. 1886. On a new species of *Strumigenys (S. lewisi)* from Japan. *Proceedings of the Manchester Literary and Philosophical Society* **25**: 229–232. [(31.xii).1886.]

Cameron, P. 1891. Appendix. Hymenoptera, Formicidae (pp. 89–95). In Whymper, E. *Travels Amongst the Great Andes of the Equator:* 147 pp. London. [(31.xii).1891.]

Cameron, P. 1892. Synonymical notes on Cynipidae and Formicidae. *Entomologist's Monthly Magazine* (2) **3** [28]: 67. [(31).iii.1892.]

Campione, B.M., Novak, J.A. & Gotwald, W.H.,Jr. 1983. Taxonomy and morphology of the West African army ant *Aenictus asantei* n. sp. *Annals of the Entomological Society of America* **76**: 873–883. [(30).ix.1983.]

Carpenter, F.M. 1929. A fossil ant from the Lower Eocene (Wilcox) of Tennessee. *Journal of the Washington Academy of Sciences* **19**: 300–301. [19.viii.1929.]

Carpenter, F.M. 1930. The fossil ants of North America. *Bulletin of the Museum of Comparative Zoology at Harvard College* **70**: 1–66. [(31).i.1930.]

Carpenter, F.M. 1935. A new name for *Lithomyrmex* Carp. *Psyche* **42**: 91. [(30).vi.1935.]

Carpenter, F.M. 1992. In Kaesler, R.L. (ed.) *Treatise on Invertebrate Paleontology.* Part R. Arthropoda **4**. Volume 4: Superclass Hexapoda: 279–655. Boulder. [Formicidae pp. 490–495.] [(31.xii).1992.]

Casevitz-Weulersse, J. 1990a. Etude systématique de la myrmécofaune corse. (Première partie.) *Bulletin du Muséum National d'Histoire Naturelle* (4) **12** Section A (Zoologie, Biologie et Ecologie animales): 135–163. [2.viii.1990.]

Casevitz-Weulersse, J. 1990b. Etude systématique de la myrmécofaune corse. (Deuxième partie.) *Bulletin du Muséum National d'Histoire Naturelle* (4) **12** Section A (Zoologie, Biologie et Ecologie animales): 415–442. [24.xi.1990.]

Ceballos, G. 1956. Catalogo de los himenopteros de España. *Trabajos del Instituto Español de Entomologia, Madrid:* 554 pp. Madrid. [(31.xii).1956.]

Cerda, X., Retana, J., Carpintero, S. & Cros, S. 1992. Petals as the main resource collected by the ant *Cataglyphis floricola. Sociobiology* **20**: 315–319. [(31.xii).1992.]

Chapman, J.W. 1963. Some new and interesting Philippine ants. *Philippine Journal of Science* **92**: 247–263. [(30).vi.1963.]

Chapman, J.W. & Capco, S.R. 1951. Check list of the ants of Asia. *Monographs of the Institute of Science and Technology, Manila* **1**: 327 pp. [(31.xii).1951.]

Chappell, J. 1886. the insects of Llangollen and its vicinity. *Young Naturalist* **7**: 57–61. [(30).iv.1886.]

Charsley, R.S. 1877. A new species of ant found in Britain. *Entomologist's Monthly Magazine* **14**: 162–163. [(31).xii.1877.]

Cheesman, L.E. & Crawley, W.C. 1928. A contribution towards the insect fauna of French Oceania. Part 3. Formicidae. *Annals and Magazine of Natural History* (10) **2**: 514–525. [(30).xi.1928.]

Cheng, L., Ye, Q. & Yang, Y. 1992. *Atopomyrmex srilankensis*—a new record from China. *Entomotaxonomia* **14**: 244. [(31).xii.1992.]

Cherrett, J.M. & Cherrett, F.J. 1989. A bibliography of the leaf-cutting ants *Atta* spp. and *Acromyrmex* spp., up to 1975. *Overseas Development Natural Resources Institute* Bulletin No. **14**: 58 pp. [1989.]

Chopard, L. 1921. La fourmi d'Argentine, *Iridomyrmex humilis* var. *arrogans* Santschi, dans le Midi de la France. *Annales des Epiphyties* **7**: 237–265. [(31.xii).1921.]

Christ, J.C. 1791. *Naturgeschichte, Klassification und Nomenclatur der Insekten* vom Bienen, Wespen und Ameisengeschlecht: 535 pp. Frankfurt am Main. [(31.xii).1791.]

Clark, J. 1923. Australian Formicidae. *Journal of the Royal Society of Western Australia* **9**: 72–89. [8.v.1923.]

Clark, J. 1924. Australian Formicidae. *Journal of the Royal Society of Western Australia* **10**: 75–89. [30.iv.1924.]

Clark, J. 1925a. The ants of Victoria. (Part 1.) *Victorian Naturalist* **42**: 58–64. [10.vii.1925.]

Clark, J. 1925b. The ants of Victoria. (Part 2.) *Victorian Naturalist* **42**: 135–144. [9.x.1925.]

Clark, J. 1926. Australian Formicidae. *Journal of the Royal Society of Western Australia* **12**: 43–52. [25.i.1926.]

Clark, J. 1927. The ants of Victoria. (Part 3.) *Victorian Naturalist* **44**: 33–39. [7.vi.1927.]

Clark, J. 1928a. Australian Formicidae. *Journal of the Royal Society of Western Australia* **14**: 29–41. [17.iv.1928.]

Clark, J. 1928b. Ants from north Queensland. *Victorian Naturalist* **45**: 169–171. [4.x.1928.]

Clark, J. 1928c. Entomological reports. Formicidae (pp.39–44). In, Report of the Victorian Field Naturalists' Expedition through the Western District of Victoria in October, 1927. *Victorian Naturalist* **45** (supplement): 1–52. [(31).x.1928.]

Clark, J. 1929. Results of a collecting trip to the Cann River, East Gippsland. *Victorian Naturalist* **46**: 115–123. [10.x.1929.]

Clark, J. 1930a. Some new Australian Formicidae. *Proceedings of the Royal Society of Victoria* **42** 116–128. [10.iii.1930.]

Clark, J. 1930b. The Australian ants of the genus *Dolichoderus*, subgenus *Hypoclinea* Mayr. *Australian Zoologist* **6**: 252–268. [20.viii.1930.]

Clark, J. 1930c. New Formicidae, with notes on some little-known species. *Proceedings of the Royal Society of Victoria* **43**: 2–25. [30.viii.1930.]

Clark, J. 1934a. Notes on Australian ants, with descriptions of new species and a new genus. *Memoirs of the National Museum, Victoria* **8**: 5–20. [(30).ix.1934.]

Clark, J. 1934b. New Australian ants. *Memoirs of the National Museum, Victoria* **8**: 21–47, [(30).ix.1934.]

Clark, J. 1934c. Ants from the Otway Ranges. *Memoirs of the National Museum, Victoria* **8**: 48–73. [(30).ix.1934.]

Clark, J. 1936. A revision of Australian species of *Rhytidoponera* Mayr. *Memoirs of the National Museum, Victoria* **9**: 14–88. [(30).xi.1936.]

Clark, J. 1938. Reports of the McCoy Society for field investigation and research. No. 2. Sir Joseph Banks Islands. 10. Formicidae. *Proceedings of the Royal Society of Victoria* **50**: 356–382. [23.v.1938.]

Clark, J. 1941. Australian Formicidae. Notes and new species. *Memoirs of the National Museum, Victoria* **12**: 71–93. [(31).iii.1941.]

Clark, J. 1943. A revision of the genus *Promyrmecia* Emery. *Memoirs of the National Museum, Victoria* **13**: 83–149. [(30).ix.1943.]

Clark, J. 1951. *The Formicidae of Australia.* **1**. Subfamily Myrmeciinae: 230 pp. CSIRO, Melbourne. [(31.xii).1951.]

Cockerell, T.D.A. 1906. A new fossil ant. *Entomological News* **17**: 27–28. [(31).i.1906.]

Cockerell, T.D.A. 1915. British fossil insects. *Proceedings of the United States National Museum* **49**: 469–499. [11.xii.1915.]

Cockerell, T.D.A. 1920. Fossil arthropods in the British Museum. 1. *Annals and Magazine of Natural History* (9) **5**: 273–279. [(31).iii.1920.]

Cockerell, T.D.A. 1921. Some Eocene insects from Colorado and Wyoming. *Proceedings of the United States National Museum* **59** (1922): 29–39. [27.vi.1921.]

Cockerell, T.D.A. 1923a. The earliest known ponerine ant. *Entomologist* **56**: 51–52. [(31).iii.1923.]

Cockerell, T.D.A. 1923b. Fossil insects from the Eocene of Texas. *American Journal of Science* **5**: 397–410. [(31).v.1923.]

Cockerell, T.D.A. 1927. Fossil insects from the Miocene of Colorado. *Annals and Magazine of Natural History* (9) **19**: 161–166. [(28).ii.1927.]

Cokendolpher, J.C. & Franke, O.F. 1985. Karyotype of *Conomyrma flava. Journal of the New York Entomological Society* **92** (1984): 349–351. [5.ii.1985.]

Cole, A.C.,Jr. 1933. Descriptions of two new ants of the genus *Pheidole*. *Annals of the Entomological Society of America* **26**: 616–618. [28.xii.1933.]
Cole, A.C.,Jr. 1936a. An annotated list of the ants of Idaho. *Canadian Entomologist* **68**: 34–39. [26.ii.1936.]
Cole, A.C.,Jr. 1936b. Descriptions of seven new western ants. *Entomological News* **47**: 118–121. [8.v.1936.]
Cole, A.C.,Jr. 1938a. Suggestions concerning taxonomic nomenclature of the hymenopterous family Formicidae, and description of three new ants. *American Midland Naturalist* **19**: 236–241. [15.i.1938.]
Cole, A.C.,Jr. 1938b. A new ant from Idaho's semidesert. *American Midland Naturalist* **19**: 678–681. [25.v.1938.]
Cole, A.C.,Jr. 1938c. Descriptions of new ants from the western United States. *American Midland Naturalist* **20**: 368–373. [29.ix.1938.]
Cole, A.C.,Jr. 1939. A new ant from the Great Smoky Mountains, Tennessee. *American Midland Naturalist* **22**: 413–417. [27.ix.1939.]
Cole, A.C.,Jr. 1940. A new ant from Indiana. *American Midland Naturalist* **23**: 224–226. [6.ii.1940.]
Cole, A.C.,Jr. 1942. The ants of Utah. *American Midland Naturalist* **28**: 358–388. [13.x.1942.]
Cole, A.C.,Jr. 1943a. Synonyms of *Formica difficilis* Emery. *Annals of the Entomological Society of America* **35** (1942): 389. [7.i.1943.]
Cole, A.C.,Jr. 1943b. A new subspecies of *Formica moki* Wheeler. *American Midland Naturalist* **29**: 183–184. [27.ii.1943.]
Cole, A.C.,Jr. 1947. A description of *Formica parcipappa*, a new ant from Idaho. *Annals of the Entomological Society of America* **39** (1946): 616–618. [23.i.1947.]
Cole, A.C.,Jr. 1948. A synonym of *Pheidole dentata* var.*commutata*. *Proceedings of the Entomological Society of Washington* **50**: 82. [30.iv.1948.]
Cole, A.C.,Jr. 1949. A study of the genus *Gesomyrmex* Mayr, and a description of a species new to the genus. *Annals of the Entomological Society of America* **42**: 71–76. [25.v.1949.]
Cole, A.C.,Jr. 1952a A new subspecies of *Trachymyrmex smithi* from New Mexico. *Journal of the Tennessee Academy of Science* **27**: 159–162. [(30).iv.1952.]
Cole, A.C.,Jr. 1952b. A new *Pheidole* from Florida. *Annals of the Entomological Society of America* **45**: 443–444. [25.x.1952.]
Cole, A.C.,Jr. 1952c. Notes on the *Pheidole pilifera* complex and a description of a new subspecies. *Journal of the Tennessee Academy of Science* **27**: 278–280. [(31).x.1952.]
Cole, A.C.,Jr. 1953a. Studies of New Mexico ants. 1. A new species of *Myrmica*. *Journal of the Tennessee Academy of Science* **28**: 81–82. [(31).i.1953.]
Cole, A.C.,Jr. 1953b. Studies of New Mexico ants. 2. A description of a new subspecies of *Aphaenogaster huachucana*. *Journal of the Tennessee Academy of Science* **28**: 82–84. [(31).i.1953.]
Cole, A.C.,Jr. 1953c. Notes on the genus *Leptothorax* in New Mexico and a description of a new species. *Proceedings of the Entomological Society of Washington* **55**: 27–30. [16.ii.1953.]
Cole, A.C.,Jr. 1953d. Studies of New Mexico ants. 4. The genera *Myrmica, Manica, Aphaenogaster* and *Novomessor. Journal of the Tennessee Academy of Science* **28**: 242–244. [(31).vii.1953.]
Cole, A.C.,Jr. 1953e. Studies of New Mexico ants. 5. The genus *Pheidole*, with synonymy. *Journal of the Tennessee Academy of Science* **28**: 297–299. [(31).x.1953.]
Cole, A.C.,Jr. 1953f. Studies of New Mexico ants. 6. The genera *Monomorium, Solenopsis, Myrmecina* and *Trachymyrmex*. *Journal of the Tennessee Academy of Science* **28**: 299–300. [(31).x.1953.]
Cole, A.C.,Jr. 1953g. *Brachymyrmex depilis* subsp. *flavescens* Grundmann a synonym of *Brachymyrmex depilis*. *Entomological News* **64**: 266. [8.xii.1953.]
Cole, A.C.,Jr. 1954a. Studies of New Mexico ants. 8. A solution to the *Formica densiventris* Viereck problem. *Journal of the Tennessee Academy of Science* **29**: 89–90. [(31).i.1954.]
Cole, A.C.,Jr. 1954b. Studies of New Mexico ants. 7. The genus *Pogonomyrmex* with synonymy and a description of a new species. *Journal of the Tennessee Academy of Science* **29**: 115–121. [(30).iv.1954.]
Cole, A.C.,Jr. 1954c. Studies of New Mexico ants. 11. The genus *Formica* with a description of a new species. *Journal of the Tennessee Academy of Science* **29**: 163–167. [(30).iv.1954.]
Cole, A.C.,Jr. 1954d. Studies of New Mexico ants. 10. The genus *Leptothorax*. *Journal of the Tennessee Academy of Science* **29**: 240–241. [(31).vii.1954.]
Cole, A.C.,Jr. 1954e. Studies of New Mexico ants. 9. *Pogonomyrmex apache* Wheeler a synonym of *Pogonomyrmex sanctihyacinthi*. *Journal of the Tennessee Academy of Science* **29**: 266–271. [(31).x.1954.]
Cole, A.C.,Jr. 1955a. Studies of New Mexico ants. 14. A description of a new species of *Pheidole* Westwood. *Journal of the Tennessee Academy of Science* **30**: 47–49. [(31).i.1955.]
Cole, A.C.,Jr. 1955b. Studies of New Mexico ants. 15. Additions, corrections and new synonymy. *Journal of the Tennessee Academy of Science* **30**: 49–50. [(31).i.1955.]
Cole, A.C.,Jr. 1955c. Studies of Nevada ants. 1. Notes on *Veromessor lariversi* M.R. Smith and a description of the queen. *Journal of the Tennessee Academy of Science* **30**: 51–52. [(31).i.1955.]
Cole, A.C.,Jr. 1956a. Studies of Nevada ants. 2. A new species of *Lasius (Chthonolasius)*. *Journal of the Tennessee Academy of Science* **31**: 26–27. [(31).i.1956.]
Cole, A.C.,Jr. 1956b. A new species of *Leptothorax* from Arizona. *Journal of the Tennessee Academy of Science* **31**: 28–31. [(31).i.1956.]
Cole, A.C.,Jr. 1956c. Observations of some members of the genus *Pheidole* in the southwestern United States with synonymy. *Journal of the Tennessee Academy of Science* **31**: 112–118. [(30).iv.1956.]
Cole, A.C.,Jr. 1956d. In the defense of the integrity of an ant. *Journal of the Tennessee Academy of Science* **31**: 212–214. [(31).vii.1956.]
Cole, A.C.,Jr. 1956d. *Leptothorax stenotyle* n. nov. for *Leptothorax angustinodus* Cole. *Journal of the Tennessee Academy of Science* **31**: 214. [(31).vii.1956.]
Cole, A.C.,Jr. 1956e. Studies of Nevada ants. 3. The status of *Formica nevadensis* Wheeler. *Journal of the Tennessee Academy of Science* **31**: 256–257. [(31).x.1956.]
Cole, A.C.,Jr. 1956f. Studies of Nevada ants. 4. Descriptions of sexual castes of three members of the *rufa* group of the genus *Formica*. *Journal of the Tennessee Academy of Science* **31**: 257–260. [(31).x.1956.]
Cole, A.C.,Jr. 1956g. New synonymy in the genus *Manica* Jurine. *Journal of the Tennessee Academy of Science* **31**: 260–262. [(31).x.1956.]
Cole, A.C.,Jr. 1957a. *Paramyrmica*, a new North American genus of ants allied to *Myrmica* Latreille. *Journal of the Tennessee Academy of Science* **32**: 37–42. [(31).i.1957.]
Cole, A.C.,Jr. 1957b. A new *Leptothorax* from Texas. *Journal of the Tennessee Academy of Science* **32**: 42–45. [(31).i.1957.]
Cole, A.C.,Jr. 1957c. Descriptions of sexual castes of some ants in the genera *Myrmica, Manica* and *Xiphomyrmex*. *Journal of the Tennessee Academy of Science* **32**: 208–213. [(31).vii.1957.]
Cole, A.C.,Jr. 1957d. Another new *Leptothorax* from Texas. *Journal of the Tennessee Academy of Science* **32**:213–215. [(31).vii.1957.]
Cole, A.C.,Jr. 1958a. A remarkable new species of *Lasius (Chthonolasius)* from California. *Journal of the Tennessee Academy of Science* **33**: 75–77. [(31).i.1958.]
Cole, A.C.,Jr. 1958b. Notes on western ants. *Journal of the New York Entomological Society* **65** (1957): 129–131. [27.vi.1958.]
Cole, A.C.,Jr. 1958c. North American *Leptothorax* of the *nitens-carinatus* complex. *Annals of the Entomological Society of America* **51**: 535–538. [29.xi.1958.]
Cole, A.C.,Jr. 1963. A new species of *Veromessor* from the Nevada test site and notes on related species. *Annals of the Entomological Society of America* **56**: 678–682. [15.ix.1963.]

Cole, A.C.,Jr. 1965. Discovery of the worker caste of *Pheidole (P.) inquilina*, new combination. *Annals of the Entomological Society of America* **58**: 173–175. [15.iii.1965.]

Cole, A.C.,Jr. 1966. Ants of the Nevada test site. *Brigham Young University Science Bulletin* **7** (No. 3): 1–27. [(30).vi.1966.]

Cole, A.C.,Jr. 1968. *Pogonomyrmex* Harvester Ants. A study of the genus in North America: 222 pp. Knoxville. [(31.xii).1968.]

Collingwood, C.A. 1958. The ants of the genus *Myrmica* in Britain. *Proceedings of the Royal Entomological Society of London* (A) **33**: 65–75. [30.vi.1958.]

Collingwood, C.A. 1961a. The 3rd. Danish expedition to Central Asia. Zoological Results 27. Formicidae from Afghanistan. *Videnskabelige Meddelelser fra Dansk Naturhistorisk Forening* **123** (1960): 51–79. [15.i.1961.]

Collingwood, C.A. 1961b. Ergebnisse der Deutschen Afghanistan-Expedition 1956 der Landessammlungen für Naturkunde Karlsruhe. *Beiträge zur Naturkundlichen Forschung in Südwestdeutschland* **19**: 289–290. [1.iv.1961.]

Collingwood, C.A. 1962. Some ants from North-East Asia. *Entomologisk Tidskrift* **83**: 215–230. [31.xii.1962.]

Collingwood, C.A. 1963. The *Lasius (Chthonolasius) umbratus* species complex in north Europe. *Entomologist* **96**: 145–158. [26.vii.1963.]

Collingwood, C.A. 1976. Ants from North Korea. *Annales Historico-Naturales Musei Nationalis Hungarici* **68**: 295–309. [(31.xii).1976.]

Collingwood, C.A. 1978. A provisional list of Iberian Formicidae with a key to the worker caste. *Eos. Revista Española de Entomologia* **52** (1976): 65–95. [10.v.1978.]

Collingwood, C.A. 1979. The Formicidae of Fennoscandia and Denmark. *Fauna Entomologica Scandinavica* **8**: 1–174. [(31.xii).1979.]

Collingwood, C.A. 1981. Ants from Korea. 2. *Folia Entomologica Hungarica (Rovartani Közlemények)* **42** [34]: 25–30. [30.vi.1981.]

Collingwood, C.A. 1982. Himalayan ants of the genus *Lasius*. *Systmatic Entomology* **7**: 283–296. [(31).vii.1982.]

Collingwood, C.A. 1985. Hymenoptera: Fam. Formicidae of Saudi Arabia. *Fauna of Saudi Arabia* **7**: 230–302. [(31.xii).1985.]

Collingwood, C.A. 1993. Description of the host ant of *Cossyphodinus bremeri* Ferrer, sp. nov. *Entomofauna. Zeitschrift für Entomologie* **14**: 229–231. [10.v.1993.]

Collingwood, C.A. & Yarrow, I.H.H. 1969. A survey of Iberian Formicidae. *Eos. Revista Española de Entomologia* **44**: 53–101. [1.x.1969.]

Consani, M. 1951. Formiche dell'Africa orientale. *Bollettino dell'Istituto di Entomologia della Università degli Studi di Bologna* **18**: 167–172. [15.v.1951.]

Consani, M. 1953. Un caso di omonimia nel genere *Crematogaster* Lund. *Bollettino della Società Entomologica Italiana* **82** (1952): 100. [31.i.1953.]

Consani, M. & Zangheri, P. 1952. Fauna di Romagna. Imenotteri-Formicidi. *Memorie della Società Entomologica Italiana* **31**: 38–48. [5.xi.1952.]

Cook, C.W. 1953. *The ants of California:* 462 pp. Palo Alto. [(31.xii).1953.]

Cori, K. & Finzi, B. 1931. Aufzählung der von Karl Cori 1914 auf süddalmatinischen Inseln gesammelten Ameisen. *Anzeiger der Akademie der Wissenschaften in Wien. Mathematisch-Naturwissenschaftlichen Klasse* **68**: 237–240. [(31.xii).1931.]

Costa, A. 1883. Notizie ed osservazioni sulla geo-fauna Sarda. Memoria seconde. Risultamento di ricerche fatte in Sardegna nella primavera del 1882. *Atti della Reale Accademia delle Scienze Fisiche e Matematiche Napoli* (2) **1** (No. 2): 1–109. [30.vi.1883.]

Costa, A. 1884. Notizie ed osservazioni sulla geo-fauna Sarda. Memoria terza. Risultamento delle ricerche fatte in Sardegna nella estate del 1883. *Atti della Reale Accademia delle Scienze Fisiche e Matematiche Napoli* (2) **1** (No. 9): 1–64. [4.vii.1884.]

Costa Lima, A.de. 1931. A proposito da *Acropya pickeli* Borgm., 1927. *Boletim Biologico, Rio de Janeiro* **17**: 2–9. [10.viii.1931.]

Cotti, G. 1963. Bibliografia ragionata 1930–1961 del gruppo *Formica rufa* in Italiano, Deutsch, English. *Collana Verde* **8**: 413 pp. [7.vi.1963.]

Cowan, C.F. 1970. The insects of the "Coquille" voyage. *Journal of the Society for the Bibliography of Natural History* **5**: 358–360. [(31).x.1970.]

Cowan, C.F. 1971. On Guérin's Iconographie: particularly the insects. *Journal of the Society for the Bibliography of Natural History* **6**: 18–29. [(31).x.1971.]

Cowan, C.F. 1976. On the Disciples' Edition of Cuvier's Règne Animal. *Journal of the Society for the Bibliography of Natural History* **8**: 32–64. [(30).xi.1976.]

Crawley, W.C. 1915a. Ants from north and central Australia, collected by G.F. Hill. Part 1. *Annals and Magazine of Natural History* (8) **15**: 130–136. [(31).i.1915.]

Crawley, W.C. 1915b. Ants from north and south-west Australia (G.F. Hill, Rowland Turner) and Christmas Island, Straights Settlements. Part 2. *Annals and Magazine of Natural History* (8) **15**: 232–239. [(28).ii.1915.]

Crawley, W.C. 1916a. A new species of ponerine ant captured by an *Asilus*. *Entomologist* **49**: 30–31. [(29).ii.1916.]

Crawley, W.C. 1916b. Ants from British Guiana. *Annals and Magazine of Natural History* (8) **17**: 366–378. [(31).v.1916.]

Crawley, W.C. 1918. Some new Australian ants. *Entomologist's Record and Journal of Variation* **30**: 86–92. [15.v.1918.]

Crawley, W.C. 1920a. Ants from Mesopotamia and north-west Persia. *Entomologist's Record and Journal of Variation* **32**: 162–166. [15.ix.1920.]

Crawley, W.C. 1920b. Ants from Mesopotamia and north-west Persia (concluded). *Entomologist's Record and Journal of Variation* **32**: 177–179. [15.x.1920.]

Crawley, W.C. 1920c. A new species of ant imported into England. *Entomologist's Record and Journal of Variation* **32**: 180–181. [15.x.1920.]

Crawley, W.C. 1920d. A gynandromorph of *Monomorium floricola*. *Entomologist's Record and Journal of Variation* **32**: 217–218. [15.xii.1920.]

Crawley, W.C. 1921. New and little-known species of ants from various localities. *Annals and Magazine of Natural History* (9) **7**: 87–97. [(31).i.1921.]

Crawley, W.C. 1922a. Formicidae. A new species and variety. *Entomologist's Record and Journal of Variation* **34**: 85–86. [15.ii.1922.]

Crawley, W.C. 1922b. New ants from Australia. *Annals and Magazine of Natural History* (9) **9**: 427–448. [(30).iv.1922.]

Crawley, W.C. 1922c. New ants from Australia. (Concluded from vol. 9., p. 449.) *Annals and Magazine of Natural History* (9) **10**: 16–36. [(31).vii.1922.]

Crawley, W.C. 1923a. Myrmecological notes. *Entomologist's Record and Journal of Variation* **35**: 29–32. [15.i.1923.]

Crawley, W.C. 1923b. Myrmecological notes. New Australian Formicidae. *Entomologist's Record and Journal of Variation* **35**: 177–179. [15.xii.1923.]

Crawley, W.C. 1924. Ants from Sumatra. With biological notes by Edward Jacobson. *Annals and Magazine of Natural History* (9) **13**: 380–409. [(30).iv.1924.]

Crawley, W.C. 1925a. Formicidae. A new genus. *Entomologist's Record and Journal of Variation* **37**: 40–41. [15.iii.1925.]

Crawley, W.C. 1925b. New ants from Australia. 2. *Annals and Magazine of Natural History* (9) **16**: 577–598. [(30).xi.1925.]

Crawley, W.C. 1926. A revision of some old types of Formicidae. *Transactions of the Entomological Society of London* **1925**: 373–393. [5.ii.1926.]

Creighton, W.S. 1927. The slave-raids of *Harpagoxenus americanus*. *Psyche* **34**: 11–29. [(28).ii.1927.]

Creighton, W.S. 1928. A new species of *Thaumatomyrmex* from Cuba. *Psyche* **35**: 162–166. [(30).ix.1928.]

Creighton, W.S. 1929. New forms of *Odontoponera transversa*. *Psyche* **36**: 150–154. [(30).vi.1929.]

Creighton, W.S. 1930a. A review of the genus *Myrmoteras*. *Journal of the New York Entomological Society* **38**: 177–192. [12.vii.1930.]

Creighton, W.S. 1930b. The New World species of the genus *Solenopsis*. *Proceedings of the American Academy of Arts and Sciences* **66**: 39–151. [(31).xii.1930.]

Creighton, W.S. 1932. A new female of *Acamatus* from Texas. *Psyche* **39**: 73–78. [(30).ix.1932.]

Creighton, W.S. 1933. *Cyathomyrmex,* a new name for the subgenus *Cyathocephalus. Psyche* **40**: 98–100. [(30).ix.1933.]
Creighton, W.S. 1934. Descriptions of three new North American ants with certain ecological observations on previously described forms. *Psyche* **41**: 185–200. [(31).xii.1934.]
Creighton, W.S. 1935. Two new species of *Formica* from western United States. *American Museum Novitates* **773**: 1–8. [28.i.1935.]
Creighton, W.S. 1939a. A generic reallocation for *Myrmoteras kuroiwae. Journal of the New York Entomological Society* **47**: 39–40. [7.iv.1939.]
Creighton, W.S. 1939b. A new subspecies of *Crematogaster minutissima* with revisionary notes concerning that species. *Psyche* **46**: 137–140. [(31).xii.1939.]
Creighton, W.S. 1940a. A revision of the North American variants of the ant *Formica rufa. American Museum Novitates* **1055**: 1–10. [15.iv.1940.]
Creighton, W.S. 1940b. A revision of the forms of *Stigmatomma pallipes. American Museum Novitates* **1079**: 1–8. [17.vii.1940.]
Creighton, W.S. 1945. Observations on the subgenus *Rhachiocrema* with the description of a new species from Borneo. *Psyche* **52**: 109–118. [26.x.1945.]
Creighton, W.S. 1950a. The ants of North America. *Bulletin of the Museum of Comparative Zoology at Harvard College* **104**: 1–585. [(30).iv.1950.]
Creighton, W.S. 1950b. *Polyhomoa* Azuma, a synonym of *Kyidris* Brown. *Psyche* **57**: 93. [29.xii.1950.]
Creighton, W.S. 1951. Studies on Arizona ants. 1. The habits of *Camponotus ulcerosus* Wheeler and its identity with *C. bruesi* Wheeler. *Psyche* **58**: 47–64. [19.xi.1951.]
Creighton, W.S. 1952a. Studies on Arizona ants. 2. New data on the ecology of *Aphaenogaster huachucana* and a description of the sexual forms. *Psyche* **58** (1951): 89–99. [7.iv.1952.]
Creighton, W.S. 1952b. Studies on Arizona ants. 3. The habits of *Pogonomyrmex huachucanus* Wheeler and a description of the sexual castes. *Psyche* **59**: 71–81. [16.x.1952.]
Creighton, W.S. 1953a. New data on the habits of the ants of the genus *Veromessor. American Museum Novitates* **1612**: 1–18. [20.iii.1953.]
Creighton, W.S. 1953b. *Pseudomyrmex apache,* a new species from the southwestern United States. *Psyche* **59** (1952): 131–142. [27.iv.1953.]
Creighton, W.S. 1953c. Studies on Arizona ants. 4. *Camponotus (Colobopsis) papago,* a new species from southern Arizona. *Psyche* **59** (1952): 148–162. [27.iv.1953.]
Creighton, W.S. 1953d. New data on the habits of *Camponotus (Myrmaphaenus) ulcerosus* Wheeler. *Psyche* **60**: 82–84. [28.ix.1953.]
Creighton, W.S. 1953e. A new subspecies of *Xenomyrmex stolli* from northeastern Mexico. *American Museum Novitates* **1634**: 1–5. [14.xii.1953.]
Creighton, W.S. 1953f. The rediscovery of *Leptothorax silvestrii* Santschi. *American Museum Novitates* **1635**: 1–7. [15.xii.1953.]
Creighton, W.S. 1954. Additional studies on *Pseudomyrmex apache. Psyche* **61**: 9–15. [30.vi.1954.]
Creighton, W.S. 1956. *Myrmecocystus lugubris* Wheeler and its synonym *Myrmecocystus yuma* Wheeler. *American Museum Novitates* **1807**: 1–7. [7.xii.1956.]
Creighton, W.S. 1957a. Studies on the North American representatives of *Ephebomyrmex. Psyche* **63** (1956): 54–66. [23.i.1957.]
Creighton, W.S. 1957b. Observations on *Pseudomyrmex elongata* Mayr. *Journal of the New York Entomological Society* **63** (1955): 17–20. [8.iii.1957.]
Creighton, W.S. 1957c. A study of the genus *Xenomyrmex. American Museum Novitates* **1843**: 1–14. [12.ix.1957.]
Creighton, W.S. 1958. A revisionary study of *Pheidole vasliti* Pergande. *Journal of the New York Entomological Society* **65** (1957): 203–212. [27.vi.1958.]
Creighton, W.S. 1965a. The habits of *Pheidole (Ceratopheidole) clydei* Gregg. *Psyche* **71** (1964): 169–173. [6.iii.1965.]
Creighton, W.S. 1965b. Studies on southwestern ants belonging to *Camponotus* subgenus *Myrmobrachys. American Museum Novitates* **2239**: 1–9. [17.xii.1965.]
Creighton, W.S. 1966a. The habits and distribution of *Macromischa subditiva* Wheeler. *Psyche* **72** (1965): 282–286. [10.v.1966.]
Creighton, W.S. 1966b. The habits of *Pheidole ridicula* Wheeler, with remarks on habit patterns in the genus *Pheidole. Psyche* **73**: 1–7. [31.viii.1966.]
Creighton, W.S. & Gregg, R.E. 1954. Studies on the habits and distribution of *Cryptocerus texanus* Santschi. *Psyche* **61**: 41–57. [26.viii.1954.]
Creighton, W.S. & Gregg, R.E. 1955. New and little-known species of *Pheidole* from the southwestern United States and northern Mexico. *University of Colorado Studies. Series in Biology* **3**: 1–46. [(30).ix.1955.]
Creighton, W.S. & Nutting, W.L. 1965. The habits and distribution of *Cryptocerus rohweri* Wheeler. *Psyche* **72**: 59–64. [25.vi.1965.]
Creighton, W.L. & Snelling, R.R. 1967. The rediscovery of *Camponotus (Myrmaphaenus) yogi* Wheeler. *Psyche* **73** (1966): 187–195. [25.i.1967.]
Creighton, W.S. & Tulloch, G.S. 1930. Notes on *Euponera gilva* (Roger). *Psyche* **37**: 71–79. [(31).iii.1930.]
Cresson, E.T. 1865. Catalogue of Hymenoptera in the collection of the Entomological Society of Philadelphia, from Colorado Territory. *Proceedings of the Entomological Society of Philadelphia* **4**: 426–488. [(30).vi.1865.]
Cresson, E.T. 1872. Hymenoptera Texana. *Transactions of the American Entomological Society* **4**: 153–292. [(30).xi.1872.]
Crosland, M.W.J., Crozier, R.H. & Imai, H.T. 1988. Evidence for several sibling biological species centred on *Myrmecia pilosula* (F. Smith). *Journal of the Australian Entomological Society* **27**: 13–14. [29.ii.1988.]
Crozier, R.H. 1968a. Interpopulation karyotype differences in Australian *Iridomyrmex* of the *detectus* group. *Journal of the Australian Entomological Society* **7**: 25–27. [30.vi.1968.]
Crozier, R.H. 1968b. The chromosomes of three Australian dacetine ant species. *Psyche* **75**: 87–90. [6.viii.1968.]
Crozier, R.H. 1969. Cytotaxonomic studies on some Australian dolichoderine ants. *Caryologia* **21** (1968): 241–259. [1969.]
Crozier, R.H. 1970. Karyotypes of twenty-one ant species, with reviews of the known ant karyotypes. *Canadian Journal of Genetics and Cytology* **12**: 109–128. [(31).iii.1970.]
Crozier, R.H., Pamilo, P., Taylor, R.W. & Crozier, Y.C. 1986. Evolutionary patterns in some putative Australian species in the ant genus *Rhytidoponera. Australian Journal of Zoology* **34**: 535–560. [1986.]
Curtis, J. 1829. *British Entomology; being illustrations and descriptions of the genera of insects found in Great Britain and Ireland* **6**: 242–288. London. [(31.xii).1829.]
Curtis, J. 1854. On the genus *Myrmica,* and other indigenous ants. *Transactions of the Linnean Society of London* **21**: 211–220. [21.xi.1854.]
Dahl, F. 1901. Das Leben der Ameisen im Bismarck-Archipel, nach eigenen Beobachtungen vergleichend dargestellt. *Mitteilungen aus dem Zoologischen Museum in Berlin* **2**: 1–70. [3.iv.1901.]
Dalla Torre, C.G.de. 1892. Hymenopterologische Notizen. *Wiener Entomologische Zeitung* **11**: 89–93. [18.iii.1892.]
Dalla Torre, C.G.de. 1893. *Catalogus Hymenopterorum,* hucusque descriptorum systematicus et synonymicus **7**: 289 pp. Lipsiae. [(31.xii).1893.]
Datta, S.K. & Raychaudhuri, D. 1985. A new species of ant from Nagaland, north-east India. *Science & Culture* **51**: 271–273. [(31).viii.1985.]
De Andrade, M.L. 1992. First fossil "true *Macromischa*" in amber from the Dominican Republic. *Mitteilungen der Schweizerischen Entomologischen Gesellschaft* **65**: 341–351. [(31).xii.1992.]
De Andrade, M.L. 1994. Fossil Odontomachiti ants from the Dominican Republic. (Amber Collection Stuttgart: Hymenoptera, Formicidae. VII: Odontomachiti.) *Stuttgarter Beiträge zur Naturkunde. Serie B (Geologie und Paläontologie)* **199**: 1–28. [29.iv.1994.]
De Geer, C. 1773. *Memoirs pour Servir à l'Histoire des Insectes* **3**: 696 pp. Stockholm. [(31.xii).1773.]

De Geer, C. 1778. *Memoirs pour Servir à l'Histoire des Insectes* **7**: 950 pp. Stockholm. [(31.xii).1778.]

Deichmüller, J.V. 1881. Fossile Insecten aus dem Diatomeenschiefer von Kutschlin bei Bilin, Böhmen. *Nova Acta Academiae Caesareae Leopoldino-Carolinae Germanicae Naturae Curiosorum* **42**: 293–330. [(31.xii).1881.]

Delage-Darchen, B. 1972a. Une fourmi de Côte d'Ivoire: *Melissotarsus titubans* Del., n. sp. *Insectes Sociaux* **19**: 213–226. [(31).xii.1972.]

Delage-Darchen, B. 1972b. Le polymorphisme larvaire chez les fourmis *Nematocrema* d'Afrique. *Insectes Sociaux* **19**: 259–278. [(31).xii.1972.]

Delage-Darchen, B. 1973. Evolution de l'aile chez les fourmis *Crematogaster* d'Afrique. *Insectes Sociaux* **20**: 221–242. [(31).xii.1973.]

Delage-Darchen, B. 1978. Les stades larvaires de *Crematogaster (Sphaerocrema) striatula,* fourmi forestière d'Afrique. *Annales de la Société Entomologique de France* (N.S.) **14**: 293–299. [5.xii.1978.]

Delye, G. 1965. *Cataglyphis (Paraformica) emmae* Forel, sexués et soldats. *Bulletin de la Société Entomologique de France* **70**: 52–56. [3.vii.1965.]

Delye, G. 1971. *Oxyopomyrmex emeryi* Santschi dans le grand erg occidental. Description des sexués. *Nouvelle Revue d'Entomologie* **1**: 211–214. [30.vi.1971.]

Dennis, C.A. 1938. The distribution of ant species in Tennessee with reference to ecological factors. *Annals of the Entomological Society of America* **31**: 267–308. [29.vi.1938.]

Deyrup, M. & Trager, J. 1986. Ants of the Archbold Biological Station, Highlands County, Florida. *Florida Entomologist* **69**: 206–228. [23.v.1986.]

Deyrup, M., Trager, J. & Carlin, N. 1985. The genus *Odontomachus* in the southeastern United States. *Entomological News* **96**: 188–195. [1.xii.1985.]

Deyrup, M., Johnson, C., Wheeler, G.C. & Wheeler, J. 1989. A preliminary list of the ants of Florida. *Florida Entomologist* **72**: 91–101. [31.iii.1989.]

Dietz, B.H. & Brandão, C.R.F. 1993. Comportamento de caça e dieta de *Acanthognathus rudis* Brown & Kempf, com comentários sobre a evolução da predação em Dacetini. *Revista Brasileira de Entomologia* **37**: 683–692. [31.xii.1993.]

Diniz, J.L.M. 1975. *Leptothorax (Nesomyrmex) mirassolis,* nova espécie de formiga da regiao norte-ocidental do estado de Sao Paulo, Brasil. *Revista Brasileira de Entomologia* **19**: 79–83. [30.vi.1975.]

Diniz, J.L.M. 1990. Revisão sistematica da tribo Stegomyrmicini, com a descrição de uma nova espécie. *Revista Brasileira de Entomologia* **34**: 277–295. [30.ix.1990.]

Diniz, J.L.M. & Brandão, C.R.F. 1993. Biology and myriopod egg predation by the Neotropical myrmicine ant *Stegomyrmex vizottoi*. *Insectes Sociaux* **40**: 301–311. [(31.xii).1993.]

Dixey, F.A. & Longstaff, G.B. 1907. Entomological observations and captures during the visit of the British Association to South Africa in 1905. *Transactions of the Entomological Society of London* **1907**: 309–381. [26.ix.1907.]

Dlussky, G.M. 1962. Murav'i severnykh sklonov Talasskogo Alatau. *Trudy Instituta Zoologii, Akademiya Nauk Kazakhskoi SSR* **18**: 177–188. [14.iv.1962.]

Dlussky, G.M. 1963. Dva novykh vida murav'ev iz Vostochnogo Zabaikal'ya. *Entomologicheskoe Obozrenie* **42**: 190–194. [15.iii.1963.]

Dlussky, G.M. 1964. Murav'i podroda *Coptoformica* roda *Formica* SSSR. *Zoologichesky Zhurnal* **43**: 1026–1040. [26.vi.1964.]

Dlussky, G.M. 1965a. Ants of the genus *Formica* L. of Mongolia and northeast Tibet. *Annales Zoologici* **23**: 15–43. [20.v.1965.]

Dlussky, G.M. 1965b. Tri novykh vida murav'ev iz Kirgizii. *Entomologicheskie Issledovaniya v Kirgizii* **4**: 27–33. [30.vi.1965.]

Dlussky, G.M. 1967a. Murav'i roda *Formica. Institut Morfologii Zhivotnykh, Akademiya Nauk SSSR:* 236 pp. Moskva. [14.iii.1967.]

Dlussky, G.M. 1967b. Murav'i roda *Formica* iz Baltiiskogo Yantarya. *Paleontologicheskii Zhurnal* **1967** (No. 2): 80–89. [19.v.1967.]

Dlussky, G.M. 1969a. Murav'i roda *Proformica* Ruzs. SSSR i sopredel'nykh stran. *Zoologichesky Zhurnal* **48**: 218–232. [31.i.1969.]

Dlussky, G.M. 1969b. Pervoe nakhozhdenie murav'ya iz podsemeistva Leptanillinae v SSSR. *Zoologichesky Zhurnal* **48**: 1666–1671. [24.x.1969.]

Dlussky, G.M. 1975. Semeistvo Formicidae. In Rasnitsyn, A.P. Vysshie Pereponchatokrylye Mezozoya. *Trudy Paleontologicheskogo Instituta* **147**: 114–122. [7.i.1975.]

Dlussky, G.M. 1981a. *Murav'i Pustyn'.:* 230 pp. Moskva. [21.x.1981.]

Dlussky, G.M. 1981b. Miotsenovye murav'i (pp. 64–83). In Vishnyakova, V.N., Dlussky, G.M. & Pritykina, L.N. Novye iskopaemye nasekomye s territorii SSSR. *Trudy Paleontologicheskogo Instituta* **183**: 1–85. [3.ix.1981.]

Dlussky, G.M. 1983. Novoe semeistvo pozdnemelovykh pereponchatokrylykh—"promezhutochnoe zveno" mezhdu murav'yami i skolioidami. *Paleontologicheskii Zhurnal* **1983** (No.3): 65–78. [29.viii.1983.]

Dlussky, G.M. 1987. Novye Formicoidea pozdnego mela. *Paleontologicheskii Zhurnal* **1987** (No. 1): 131–135. [4.ii.1987.]

Dlussky, G.M. 1988. Murav'i Sakhalinskogo Yantarya (paleotsen?). *Paleontologicheskii Zhurnal* **1988** (No. 1): 50–61. [9.ii.1988.]

Dlussky, G.M. 1993. Murav'i Fidzhi, Tonga i Samoa i problema formirovaniya ostrovnykh faun. 2. Triba Dacetini. *Zoologichesky Zhurnal* **72** (6): 52–65. [19.v.1993.]

Dlussky, G.M. & Fedoseeva, E.B. 1987. Proiskhozhdenie i rannie etapy evolyutsii murav'ev (pp.70–144). In Ponomarenko, A.G. *Melovoi Biotsenoticheskii Krizis i Evolyutsiya Nasekomykh* (1988): 228 pp. Moskva. [22.xii.1987.]

Dlussky, G.M. & Pisarski, B. 1970. Formicidae aus der Mongolei. Ergebnisse der Mongolisch-Deutschen Biologischen Expeditionen Seit 1962. *Mitteilungen aus dem Zoologischen Museum in Berlin* **46**: 85–90. [24.iv.1970.]

Dlussky, G.M. & Pisarski, B. 1971. Rewizja polskich gatunków mrówek z rodzaju *Formica* L. *Fragmenta Faunistica* **16**: 145–224. [25.iii.1971.]

Dlussky, G.M. & Radchenko, A.G. 1994. Murav'i roda *Diplorhoptrum* Mayr tsentral'noi palearktiki. *Zoologichesky Zhurnal* **73** (2): 102–111. [11.i.1994.]

Dlussky, G.M. & Soyunov, O.S. 1988. Murav'i roda *Temnothorax* Mayr SSSR. *Izvestiya Akademii Nauk Turkmenskoi SSR.* Seriya Biologicheskikh Nauk **1988** (No. 4): 29–37. [26.viii.1988.]

Dlussky, G.M., Soyunov, O.S. & Zabelin, S.I. 1990. *Murav'i Turkmenistana* (1989): 273 pp. Ashkhabad. [7.vi.1990.]

Dlussky, G.M. & Zabelin, S.I. 1985. Fauna murav'ev Basseina R. Sumbar (Yugo-zapadnyi Kopetdag) (pp. 208–246). In Nechaevaya, N.T. *Rastitel'nost i Zhivotnyi Mir Zaladnogo Kopetdaga:* 277 pp. Ashkhabad, Ylym. [22.x.1985.]

Donisthorpe, H. Note. In the majority of his publications Donisthorpe uses only the initial H. However, in various papers and books he sometimes uses his full set of initials, H.StJ.K. Throughout the current bibliography only the initial H. is utilised.

Donisthorpe, H. 1914. Myrmecophilous notes for 1913. *Entomologist's Record and Journal of Variation* **26**: 37–45. [15.ii.1914.]

Donisthorpe, H. 1915a. The type of *Camponotus (Myrmoturba) maculatus* F. *Entomologist's Record and Journal of Variation* **27**: 221–222. [15.x.1915.]

Donisthorpe, H. 1915b. Descriptions of a pterergate and two gynandromorphs of *Myrmica sabuleti* Nyl., with a list of all the known cases of the latter. *Entomologist's Record and Journal of Variation* **27**: 258–260. [15.xi.1915.]

Donisthorpe, H. 1915c. *Myrmica schencki* Emery, an ant new to Britain. *Entomologist's Record and Journal of Variation* **27**: 265–266. [15.xii.1915.]

Donisthorpe, H. 1915d. *British Ants, their life-history and classification:* 379 pp. Plymouth. [(31.xii).1915.]

Donisthorpe, H. 1916a. *Epitritus wheeleri*, n. sp., an ant new to science; with notes on the genus *Epitritus* Emery. *Entomologist's Record and Journal of Variation* **28**: 121–122. [15.vi.1915.]

Donisthorpe, H. 1916b. Synonymy of some genera of ants. *Entomologist's Record and Journal of Variation* **28**: 241–244. [15.xi.1916.]

Donisthorpe, H. 1916c. Synonymy of some genera of ants. *Entomologist's Record and Journal of Variation* **28**: 275–277. [15.xii.1916.]

Donisthorpe, H. 1917. *Dolichoderus (Hypoclinea) crawleyi* n. sp., a species of ant new to science; with a few notes on the genus. *Entomologist's Record and Journal of Variation* **29**: 201–202. [15.x.1917.]

Donisthorpe, H. 1918a. Some notes on a paper by Dr. Leach on ants and gnats in 1825. *Entomologist's Record and Journal of Variation* **30**: 8–9. [15.i.1918.]

Donisthorpe, H. 1918b. A list of ants from Mesopotamia; with a description of a new species and a new variety. *Entomologist's Record and Journal of Variation* **30**: 165–168. [15.x.1918.]

Donisthorpe, H. 1919. Myrmecophilous notes for 1918. *Entomologist's Record and Journal of Variation* **31**: 1–5. [17.i.1919.]

Donisthorpe, H. 1920. British Oligocene ants. *Annals and Magazine of Natural History* (9) **6**: 81–94. [(31).vii.1920.]

Donisthorpe, H. 1922. The subfamilies of Formicidae. *Proceedings of the Entomological Society of London* **1921**: xl-xlvi. [13.iv.1922.]

Donisthorpe, H. 1926a. Ants and myrmecophiles at Bordighera. *Entomologist's Record and Journal of Variation* **38**: 5–8. [15.i.1926.]

Donisthorpe, H. 1926b. Ants and myrmecophiles at Bordighera. *Entomologist's Record and Journal of Variation* **38**: 17–18. [15.ii.1926.]

Donisthorpe, H. 1927a. The ants, and some myrmecophiles, of Sicily. *Entomologist's Record and Journal of Variation* **39**: 6–9. [15.i.1927.]

Donisthorpe, H. 1927b. *British Ants, their life-history and classification* (2nd. edition): 436 pp. London. [(31.xii).1927.]

Donisthorpe, H. 1929a. The Formicidae taken by Major R.G.W. Hingston, M.C., I.M.S. (ret.), on the Mount Everest Expedition, 1924. *Annals and Magazine of Natural History* (10) **4**: 444–449. [(30).xi.1929.]

Donisthorpe, H. 1929b. Gynandromorphism in ants. *Zoologischer Anzeiger* (Wasmann-Festband) **82**: 92–96. [1929.]

Donisthorpe, H. 1930a. A new subspecies of *Acanthomyops* from Kashmir. *Annals and Magazine of Natural History* (10) **5**: 225–226. [(31).viii.1930.]

Donisthorpe, H. 1930b. *Cardiocondyla bicolor,* sp. n., a species of myrmicine ant new to science. *Annals and Magazine of Natural History* (10) **5**: 366. [(30).ix.1930.]

Donisthorpe, H. 1931a. *Camponotus (Tanaemyrmex) britteni* sp. n., a formicine ant new to science. *Annals and Magazine of Natural History* (10) **8**: 129–131. [(31).vii.1931.]

Donisthorpe, H. 1931b. Descriptions of some new species of ants. *Annals and Magazine of Natural History* (10) **8**: 494–501. [(30).xi.1931.]

Donisthorpe, H. 1932a. A new species of *Camponotus* from Colombia. *Stylops* **1**: 88–89. [15.iv.1932.]

Donisthorpe, H. 1932b. On the identity of some ants from Ceylon described by F. Walker. *Annals and Magazine of Natural History* (10) **9**: 574–576. [(30).vi.1932.]

Donisthorpe, H. 1932c. On the identity of Smith's types of Formicidae collected by Alfred Russell Wallace in the Malay Archipelago, with descriptions of two new species. *Annals and Magazine of Natural History* (10) **10**: 441–476. [(30).xi.1932.]

Donisthorpe, H. 1933a. A new species of *Aphaenogaster* from India. *Stylops* **2**: 24. [31.i.1933.]

Donisthorpe, H. 1933b. Descriptions of three new species of Formicidae, and a synonymical note. *Annals and Magazine of Natural History* (10) **11**: 194–198. [(28).ii.1933.]

Donisthorpe, H. 1933c. On a small collection of ants made by Dr. F.W. Edwards in Argentina. *Annals and Magazine of Natural History* (10) **12**: 532–538. [(30).xi.1933.]

Donisthorpe, H. 1935. The ants of Christmas Island. *Annals and Magazine of Natural History* (10) **15**: 629–635. [(30).vi.1935.]

Donisthorpe, H. 1936a. *Rhopalomastix janeti,* a species of ant new to science. *Entomologist's Record and Journal of Variation* **58**: 55–56. [15.v.1936.]

Donisthorpe, H. 1936b. *Acropyga (Rhizomyrma) robae* sp. nov., a new S. American ant, with remarks on the genus, etc. *Entomologist* **69**: 108–111. [(31).v.1936.]

Donisthorpe, H. 1936c. *Strongylognathus diveri* sp. n., a genus and species new to the British list, with notes on the genus. *Entomologist's Monthly Magazine* **72**: 111–116. [(31).v.1936.]

Donisthorpe, H. 1936d. A new study of *Mesoxena mistura* Smith. *Proceedings of the Royal Entomological Society of London* (B) **5**: 119–120. [15.vi.1936.]

Donisthorpe, H. 1936e. Five new species of ants from various localities. *Annals and Magazine of Natural History* (10) **18**: 524–530. [(30).xi.1936.]

Donisthorpe, H. 1937a. Some new forms of Formicidae and a correction. *Annals and Magazine of Natural History* (10) **19**: 619–628. [(30).vi.1937.]

Donisthorpe, H. 1937b. A new species of *Harpegnathos* Jerd., with some remarks on the genus, and the other known species. *Entomologist's Monthly Magazine* **73**: 196–201. [(30).ix.1937.]

Donisthorpe, H. 1937c. A new subgenus and three new species of *Polyrhachis* Smith. *Entomologist* **70**: 273–276. [(31).xii.1937.]

Donisthorpe, H. 1938a. Five new species of ant, chiefly from New Guinea. *Annals and Magazine of Natural History* (11) **1**: 140–148. [(31).i.1938.]

Donisthorpe, H. 1938b. Some ants of the subgenus *Planimyrma* Viehmeyer of the genus *Aphaenogaster* Mayr. *Entomologist's Monthly Magazine* **74**: 30–32. [(28).ii.1938.]

Donisthorpe, H. 1938c. The subgenus *Cyrtomyrma* Forel of *Polyrhachis* Smith, with descriptions of new species, etc. *Annals and Magazine of Natural History* (11) **1**: 246–267. [(31).iii.1938.]

Donisthorpe, H. 1938d. New species and varieties of ants from New Guinea. *Annals and Magazine of Natural History* (11) **1**: 593–599. [(30).vi.1938.]

Donisthorpe, H. 1938e. New species of ants and a new subgenus of *Dolichoderus* from various localities. *Annals and Magazine of Natural History* (11) **2**: 498–504. [(30).xi.1938.]

Donisthorpe, H. 1939a. The genus *Lioponera* Mayr, with descriptions of two new species and an ergatandromorph. *Annals and Magazine of Natural History* (11) **3**: 252–257. [(31).iii.1939.]

Donisthorpe, H. 1939b. *Typhlomyrmex richardsi,* a new species of ponerine ant from British Guiana. *Entomologist's Monthly Magazine* **75**: 161–162. [(31).vii.1939.]

Donisthorpe, H. 1939c. Descriptions of several species of ants taken by Dr. O.W. Richards in British Guiana. *Proceedings of the Royal Entomological Society of London* (B) **8**: 152–154. [15.viii.1939.]

Donisthorpe, H. 1940a. Descriptions of new species of ants from various localities. *Annals and Magazine of Natural History* (11) **5**: 39–48. [(31).i.1940.]

Donisthorpe, H. 1940b. *Lordomyrma infundibuli,* a new species of ant from Dutch New Guinea. *Entomologist's Monthly Magazine* **76**: 45–47. [1.ii.1940.]

Donisthorpe, H. 1940c. Some new forms of *Odontomachus. Entomologist* **73**: 106–109. [(31).v.1940.]

Donisthorpe, H. 1941a. *Lordomyrma niger* sp. n., with key and notes on the genus. *Entomologist's Monthly Magazine* **77**: 36–38. [3.ii.1941.]

Donisthorpe, H. 1941b. New ants from Waigeu Island, New Guinea, and the Solomons. *Entomologist* **74**: 36–42. [17.ii.1941.]

Donisthorpe, H. 1941c. Descriptions of new species of ants from New Guinea. *Annals and Magazine of Natural History* (11) **7**: 129–144. [(28).ii.1941.]

Donisthorpe, H. 1941d. Glanures myrmecologiques. *Entomologist's Record and Journal of Variation* **53**: 27–28. [15.iii.1941.]

Donisthorpe, H. 1941e. The ants of Norfolk Island. *Entomologist's Monthly Magazine* **77**: 90–93. [2.iv.1941.]

Donisthorpe, H. 1941f. Glanures myrmecologiques. *Entomologist's Record and Journal of Variation* **53**: 36–38. [15.iv.1941.]

Donisthorpe, H. 1941g. A new species of *Echinopla,* with some notes on the genus. *Entomologist* **74**: 115–116. [9.v.1941.]

Donisthorpe, H. 1941h. The ants of Japen Island, Dutch New Guinea. *Transactions of the Royal Entomological Society of London* **91**: 51–64. [15.vii.1941.]

Donisthorpe, H. 1941i. A new genus and species of Formicidae from Papua. *Entomologist's Monthly Magazine* **77**: 175. [31.viii.1941.]

Donisthorpe, H 1941j. Descriptions of new ants from various localities. *Annals and Magazine of Natural History* (11) **8**: 199–210. [(30).ix.1941.]
Donisthorpe, H. 1941k. Synonymical notes, etc., on Formicidae. *Entomologist's Monthly Magazine* **77**: 237–240. [1.x.1941.]
Donisthorpe, H. 1941l. Description of a new species of *Crematogaster* Lund, subgenus *Physocrema* Forel, with a list of, and a key to, the known species of the subgenus. *Entomologist* **74**: 225–227. [17.x.1941.]
Donisthorpe, H. 1942a. The Formicidae of the Armstrong College expedition to the Siwa Oasis. *Annals and Magazine of Natural History* (11) **9**: 26–33. [(31).i.1942.]
Donisthorpe, H. 1942b. Descriptions of a few ants from the Philippine Islands, and a male of *Polyrhachis bihamata* Drury from India. *Annals and Magazine of Natural History* (11) **9**: 64–72. [(31).i.1942.]
Donisthorpe, H. 1942c. *Myopopone wollastoni* sp. n., with notes on other forms in the genus and descriptions of the males of two species. *Entomologist's Monthly Magazine* **78**: 29–31. [5.ii.1942.]
Donisthorpe, H. 1942d. Ants from the Colombo Museum expedition to southern India, September-October 1938. *Annals and Magazine of Natural History* (11) **9**: 449–461. [(30).vi.1942.]
Donisthorpe, H. 1942e. New species of ants from the Gold Coast, Borneo, Celebes, New Guinea and New Hebrides. *Annals and Magazine of Natural History* (11) **9**: 701–709. [(30).ix.1942.]
Donisthorpe, H. 1942f. A new species of *Diplomorium* with some notes on the genus. *Entomologist* **75**: 217–218. [9.x.1942.]
Donisthorpe, H. 1942g. Notes on the subgenus *Orthonotomyrmex* Ashmead of *Camponotus* Mayr, and description of a new species. *Entomologist's Monthly Magazine* **78**: 248–251. [2.xi.1942.]
Donisthorpe, H. 1942h. Notes on the genus *Pseudolasius* Emery with the description of *Pseudolasius karawajewi* sp. n. *Proceedings of the Royal Entomological Society of London* (B) **11**: 166–169. [14.xi.1942.]
Donisthorpe, H. 1943a. Descriptions of new ants, chiefly from Waigeu Island, N. Dutch New Guinea. *Annals and Magazine of Natural History* (11) **10**: 167–176. [(31).iii.1943.]
Donisthorpe, H. 1943b. Ants from the Colombo Museum expedition to southern India, September-October 1938. *Annals and Magazine of Natural History* (11) **10**: 196–208. [(31).iii.1943.]
Donisthorpe, H. 1943c. *Ireneidris myops* gen. et sp. n. from Waigeu Island, Dutch New Guinea. *Entomologist's Monthly Magazine* **79**: 81–82. [6.iv.1943.]
Donisthorpe, H. 1943d. The ants of Waigeu Island, North Dutch New Guinea. *Annals and Magazine of Natural History* (11) **10**: 433–475. [(31).vii.1943.]
Donisthorpe, H. 1943e. Myrmecological gleanings. *Proceedings of the Royal Entomological Society of London* (B) **12**: 115–116. [24.viii.1943.]
Donisthorpe, H. 1943f. A list of the type-species of the genera and subgenera of Formicidae. *Annals and Magazine of Natural History* (11) **10**: 617–688. [(30).ix.1943.]
Donisthorpe, H. 1943g. A list of the type-species of the genera and subgenera of Formicidae. (Conclusion.) *Annals and Magazine of Natural History* (11) **10**: 721–737. [(30).xi.1943.]
Donisthorpe, H. 1944a. *Messor aegyptiacus* Emery subspecies *canaliculatus* subsp. n. (*striaticeps* Santschi? (1923) nec Er. André (1883)). *Proceedings of the royal Entomological Society of London* (B) **13**: 1–3. [25.ii.1944.]
Donisthorpe, H. 1944b. Notes arising out of Mr. W.D. Hincks's review of Mr. Donisthorpe's "A list of the type-species of the genera and subgenera of Formicidae." *Entomologist's Monthly Magazine* **80**: 59. [29.iii.1944.]
Donisthorpe, H. 1944c. A new subgenus and three new species of *Polyrhachis* F.Smith. *Entomologist's Monthly Magazine* **80**: 64–66. [29.iii.1944.]
Donisthorpe, H. 1944d. Two new species of *Pristomyrmex*, with some notes on the genus. *Proceedings of the Royal Entomological Society of London* (B) **13**: 81–84. [15.viii.1944.]
Donisthorpe, H. 1944e. A new species of *Bothriomyrmex* Emery and some notes on the genus. *Proceedings of the Royal Entomological Society of London* (B) **13**: 100–103. [25.x.1944.]
Donisthorpe, H. 1945a. New and rare species of *Crematogaster* Lund from the Gold Coast, etc. *Entomologist* **78**: 10–11. [16.i.1945.]
Donisthorpe, H. 1945b. A new species and a new variety of *Technomyrmex* Emery from New Guinea. *Entomologist's Monthly Magazine* **81**: 57–58. [16.iii.1945.]
Donisthorpe, H. 1945c. A new species of *Triglyphothrix* from Uganda, with some notes on the genus. *Entomologist's Monthly Magazine* **81**: 76–77. [14.iv.1945.]
Donisthorpe, H. 1945d. On a small collection of ants from West Africa, associated with Coccidae. *Annals and Magazine of Natural History* (11) **12**: 265–273. [18.ix.1945.]
Donisthorpe, H. 1946a. A new subspecies of *Messor* Forel, and a new variety of *Aphaenogaster* Mayr from Turkey. *Proceedings of the Royal Entomological Society of London* (B) **15**: 53–54. [15.vi.1946.]
Donisthorpe, H. 1946b. Undescribed forms of *Camponotus (Colobopsis) vitiensis* from the Fiji Islands. *Proceedings of the Royal Entomological Society of London* (B) **15**: 69–70. [15.vi.1946.]
Donisthorpe, H. 1946c. New species of ants from the Island of Mauritius. *Annals and Magazine of Natural History* (11) **12** (1945): 776–782. [19.ix.1946.]
Donisthorpe, H. 1946d. *Ireneopone gibber*, a new genus and species of myrmicine ant from Mauritius. *Entomologist's Monthly Magazine* **82**: 242–243. [1.x.1946.]
Donisthorpe, H. 1946e. The ants of Mauritius. *Annals and Magazine of Natural History* (11) **13**: 25–35. [31.x.1946.]
Donisthorpe, H. 1946f. A new genus and species of Formicidae from Mauritius. *Proceedings of the Royal Entomological Society of London* (B) **15**: 145–147. [23.xii.1946.]
Donisthorpe, H. 1947a. New species of ants from China and Mauritius. *Annals and Magazine of Natural History* (11) **13** (1946): 283–286. [22.i.1947.]
Donisthorpe, H. 1947b. *Pheidole (Pheidolacanthinus) striatus* sp. n. from New Guinea, with notes on the subgenus *Pheidolacanthinus* F. Smith. *Entomologist's Monthly Magazine* **83**: 172–174. [31.vii.1947.]
Donisthorpe, H. 1947c. Ants from New Guinea, including new species and a new genus. *Annals and Magazine of Natural History* (11) **13** (1946): 577–595. [15.viii.1947.]
Donisthorpe, H. 1947d. Some new ants from New Guinea. *Annals and Magazine of Natural History* (11) **14**: 183–197. [23.xii.1947.]
Donisthorpe, H. 1947e. Results of the Armstrong College expedition to Siwa Oasis (Libyan Desert), 1935. A second list of the Formicidae. *Bulletin de la Société Fouad 1er d'Entomologie* **31**: 109–111. [31.xii.1947.]
Donisthorpe, H. 1948a. *Liomyrmex reneae* sp. n. with a list of the species and some notes on the genus *Liomyrmex* Mayr. *Entomologist's Monthly Magazine* **83** (1947): 293–294. [9.i.1948.]
Donisthorpe, H. 1948b. A second instalment of the Ross Collection of ants from New Guinea. *Annals and Magazine of Natural History* (11) **14** (1947): 297–317. [4.ii.1948.]
Donisthorpe, H. 1948c. A new genus and species of ant from New Britain. *Entomologist's Record and Journal of Variation* **60**: 65–66. [15.v.1948.]
Donisthorpe, H. 1948d. A third instalment of the Ross Collection of ants from New Guinea. *Annals and Magazine of Natural History* (11) **14** (1947): 589–604. [7.vi.1948.]
Donisthorpe, H. 1948e. A new species of *Colobopsis* from New Guinea, with a few notes on the subgenus. *Entomologist's Monthly Magazine* **84**: 121–122. [10.vi.1948.]
Donisthorpe, H. 1948f. *Microbolbos testaceus*, a new genus and species of ponerine ant. *Entomologist* **81**: 170–171. [13.vii.1948.]
Donisthorpe, H. 1948g. A fourth instalment of the Ross Collection of ants from New Guinea. *Annals and Magazine of Natural History* (12) **1**: 131–143. [17.ix.1948.]

Donisthorpe, H. 1948h. A redescription of the types of *Strumigenys mandibularis* F. Smith, and *Cephaloxys capitata* F. Smith. *Psyche* **55**: 78–82. [23.x.1948.]
Donisthorpe, H. 1949a. A new genus and species of dacetine ant from New Guinea. *Entomologist's Monthly Magazine* **84** (1948): 281. [11.i.1949.]
Donisthorpe, H. 1949b. A fifth instalment of the Ross Collection of ants from New Guinea. *Annals and Magazine of Natural History* (12) **1** (1948): 487–506. [14.i.1949.]
Donisthorpe, H. 1949c. A sixth instalment of the Ross Collection of ants from New Guinea. *Annals and Magazine of Natural History* (12) **1** (1948): 744–759. [4.ii.1949.]
Donisthorpe, H. 1949d. *Lordomyrma bensoni*, a species of ant new to science from New Guinea. *Entomologist's Monthly Magazine* **85**: 94. [25.iv.1949.]
Donisthorpe, H. 1949e. A new *Camponotus* from Madagascar and a small collection of ants from Mauritius. *Annals and Magazine of Natural History* (12) **2**: 271–275. [15.vii.1949.]
Donisthorpe, H. 1949f. A species of *Calyptomyrmex* Emery from New Guinea. *Entomologist's Monthly Magazine* **85**: 186. [22.vii.1949.]
Donisthorpe, H. 1949g. A seventh instalment of the Ross Collection of ants from New Guinea. *Annals and Magazine of Natural History* (12) **2**: 401–422. [29.viii.1949.]
Donisthorpe, H. 1950a. An eighth instalment of the Ross Collection of ants from New Guinea. *Annals and Magazine of Natural History* (12) **3**: 338–341. [(30).iv.1950.]
Donisthorpe, H. 1950b. Two new species of ants from Turkey. *Entomologist's Record and Journal of Variation* **62**: 60–61. [15.vi.1950.]
Donisthorpe, H. 1950c. Two more new ants from Turkey. *Entomologist's Record and Journal of Variation* **62**: 68–69. [15.vii.1950.]
Donisthorpe, H. 1950d. Two new species of ants, and a few others from Turkey. *Annals and Magazine of Natural History* (12) **3**: 638–640. [(31).vii.1950.]
Donisthorpe, H. 1950e. A first instalment of the ants of Turkey. *Annals and Magazine of Natural History* (12) **3**: 1057–1067. [(31).xii.1950.]
Dorow, W.H.O. & Maschwitz, U. 1990. The *arachne*-group of *Polyrhachis:* weaver ants cultivating Homoptera in Borneo. *Insectes Sociaux* **37**: 73–89. [(31).iii.1990.]
Douglas, A. & Brown, W.L.,Jr. 1959. *Myrmecia inquilina* new species: the first parasite among the lower ants. *Insectes Sociaux* **6**: 13–19. [(30).ix.1959.]
Douwes, P. 1979. *Formica rufa*-gruppens systematik. *Entomologisk Tidskrift* **100**: 187–191. [10.xii.1979.]
Douwes, P. 1981. Intraspecific and interspecific variation in workers of the *Formica rufa* group in Sweden. *Entomologica Scandinavica, Supplement* **15**: 213–223. [1.vii.1981.]
Douwes, P., Jessen, K. & Buschinger, A. 1988. *Epimyrma adlerzi* sp. n. from Greece: morphology and life history. *Entomologica Scandinavica* **19**: 239–249. [28.ix.1988.]
Drury, D. 1773. *Illustrations of Natural History.* Wherein are exhibited upwards of two hundred and twenty figures of exotic insects. **2**: 90 pp. London. [(31.xii).1773.]
Drury, D. 1782. *Illustrations of Natural History.* Wherein are exhibited upwards of two hundred figures of exotic insects. **3**: 76 pp. London. [(31.xii).1782.]
DuBois, M.B. 1981. Two new species of inquilinous *Monomorium* from North America. *University of Kansas Science Bulletin* **52**: 31–37. [12.vi.1981.]
DuBois, M.B. 1986. A revision of the native New World species of the ant genus *Monomorium* (*minimum* group). *University of Kansas Science Bulletin* **53**: 65–119. [24.iii.1986.]
DuBois, M.B. 1993. What's in a name? A clarification of *Stenamma westwoodi, S. debile,* and *S. lippulum. Sociobiology* **21**: 299–334. [(31.xii).1993.]
Dufour, L. 1857. Mélanges entomologiques. *Annales de la Société Entomologique de France* (3) **5**: 39–70. [24.vi.1857.]
Dufour, L. 1862. Notices entomologiques. 3. Notice sur la *Formica savignyi* Duf. *Annales de la Société Entomologique de France* (4) **2**: 131–148. [5.vi.1862.]
Dufour, L. 1864. Note sur une nouvelle espèce de fourmi. (*Formica vinsonnella.*) *Annales de la Société Entomologique de France* (4) **4**: 210. [12.x.1864.]
Dumeril, C. 1860. Entomologie analytique. *Mémoires de l'Académie des Sciences de l'Institut de France* **31** (2e partie): 665–1340. [(31.xii).1860.]
Dumpert, K. 1978. *Das Sozialleben der Ameisen:* 298 pp. Berlin & Hamburg. [(31.xii).1978.] (English translation: 1981. *The Social Biology of Ants:* 298 pp. Bath.)
Dumpert, K. 1986. *Camponotus (Karavaievia) texens* sp. n. and *C. (K.) gombaki* sp. n. from Malaysia in comparison with the other *Karavaievia* species. *Psyche* **92** (1985): 557–574. [27.iv.1986.]
Dumpert, K., Maschwitz, U., Nässig, W. & Dorow, W. 1989. *Camponotus (Karavaievia) asli* sp. n. and *C. (K.) montanus* sp. n., two weaver ant species from Malaysia. *Zoologische Beiträge* (N.F.) **32**: 217–231. [12.ix.1989.]
Dupuis, C. 1986. Dates de publication de l'Histoire Naturelle Générale et Particulière des Crustacés et des Insectes (1802–1805) par Latreille dans le "Buffon de Sonnini." *Annales de la Société Entomologique de France* (N.S.) **22**: 205–210. [30.vi.1986.]
Eder, J. & Rembold, H. (Eds.). 1987. Chemistry and Biology of Social Insects. *Proceedings of the 10th. International Congress of the IUSSI:* 756 pp. München. [(31.xii).1987.]
Eichwald, E. 1841. Fauna Caspio-Caucasia nonnullis observationibus novis. *Nouveaux Mémoires de la Société Impériale des Naturalistes de Moscou* **7**: 1–290. [(31.xii).1841.]
Eidmann, H. 1936. Ökologisch-faunistische Studien an südbrasilianischen Ameisen. *Arbeiten über Physiologische und angewandte Entomologie aus Berlin-Dahlem* **3**: 26–48. [18.iii.1936.]
Eidmann, H. 1942. Zur Kenntnis der Ameisenfauna des Nanga Parbat. *Zoologische Jahrbücher. Abteilung für Systematik, Ökologie und Geographie der Tiere* **75**: 239–266. [1.vii.1942.]
Eisner, T. 1957. A comparative morphological study of the proventriculus of ants. *Bulletin of the Museum of Comparative Zoology at Harvard College* **116**: 439–490. [(31).vii.1957.]
Eisner, T. & Wilson, E.O. 1952. The morphology of the proventriculus of a formicine ant. *Psyche* **59**: 47–60. [16.x.1952.]
Elmes, G.W. 1978. A morphometric comparison of three closely related species of *Myrmica,* including a new species from England. *Systematic Entomology* **3**: 131–145. [(30).iv.1978.]
Elmes, G.W. & Thomas, J.A. 1985. Morphometrics as a tool in identification: a case study on a *Myrmica* from France. *Actes des Colloques Insectes Sociaux* **2** (1984): 97–108. [(30).vi.1985.]
Emery, C. 1869a. Formicidarum italicorum species duae novae. *Bullettino della Società Entomologica Italiana* **1**: 135–137. [(30).iv.1869.]
Emery, C. 1869b. Enumerazione dei Formicidi che rinvengonsi nei contorni di Napoli. *Annali dell'Accademia degli Aspiranti Naturalisti* (2) **2**: 1–26. [after 5.viii.1869.]
Emery, C. 1869c. Descrizione di una nuova formica italiana. *Annuario del Museo Zoologico della R. Università di Napoli* **5**: 117–118. [(31.xii).1869.]
Emery, C. 1870. Studi mirmecologici. *Bullettino della Società Entomologica Italiana* **2**: 193–201. [(30).vi.1870.]
Emery, C. 1875a. Ueber hypogaeische Ameisen. *Entomologische Zeitung herausgegeben von dem Entomologischen Vereine zu Stettin* **37** (1876): 71–76. [(31).xii.1875.] (The final page of this volume, p. 112, states "Ausgegeben Mitte December 1875".)
Emery, C. 1875b. Le formiche Ipogee con descrizioni di specie nuove o poco note. *Annali del Museo Civico di Storia Naturale di Genova* **7**: 465–474. [(31.xii).1875.]
Emery, C. 1877a. Saggio di un ordinamento naturale dei Mirmicidei e considerazioni sulla filogenesi delle formiche. *Bullettino della Società Entomologica Italiana* **9**: 67–83. [(31).iii.1877.]

Emery, C. 1877b. Catalogo delle formiche esistenti nelle collezioni del Museo Civico di Genova. Parte prima. Formiche provenienti dal Viaggio dei signori Antinori, Beccari e Issel nel Mar Rosso e nel paese dei Bogos. *Annali del Museo Civico di Storia Naturale di Genova* **9**: 363–381. [Pp. 363–368 dated 27.iii.1877; pp. 369–381 dated 3.iv.1877.]
Emery, C. 1878a. Liste des fourmis de la collection de feu Camille van Volxem, avec la description d'une espèce nouvelle. *Annales de la Société Entomologique de Belgique, Comptes-rendus des Séances* **21**: viii-x. [5.i.1878.]
Emery, C. 1878b. Catalogo delle formiche esistenti nelle collezioni del Museo Civico di Genova. Parte seconda. Formiche dell'Europa e delle regioni limitrofe in Africa e in Asia. *Annali del Museo Civico di Storia Naturale di Genova* **12**: 43–59. [Pp. 43–48 dated 12.ii.1878; pp. 49–59 dated 19.iii.1878.]
Emery, C. 1880. Crociera del Violante comandato dal Capitano Armatore Enrico d'Albertis durante l'anno 1877. Formiche. *Annali del Museo Civico di Storia Naturale di Genova* **15**: 389–398. [9.iv.1880.]
Emery, C. 1881a. Spedizione Italiana nell'Africa Equatoriale. Risultati Zoologici. Formiche. *Annali del Museo Civico di Storia Naturale di Genova* **16**: 270–276. [13.i.1881.]
Emery, C. 1881b. Viaggio ad Assab nel Mar Rosso, dei signori G. Doria ed O. Beccari con il R. Avviso "Esploratore" dal 16 Novembre 1879 al 26 Febbaio 1880. 1. Formiche. *Annali del Museo Civico di Storia Naturale di Genova* **16**: 525–535. [9.iii.1881.]
Emery, C. 1882. Le crociere dell'yacht "Corsaro" del capitano armatore Enrico d'Albertis. 2. Formiche. *Annali del Museo Civico di Storia Naturale di Genova* **18**: 448–452. [28.xii.1882.]
Emery, C. 1883. Alcune formiche della Nuova Caledonia. *Bullettino della Società Entomologica Italiana* **15**: 145–151. [25.ix.1883.]
Emery, C. 1884a. Materiali per lo studio della fauna Tunisia raccolti da G. e L. Doria. 3. Rassegna delle formiche della Tunisia. *Annali del Museo Civico di Storia Naturale di Genova* (2) **1** [21]: 373–386. [Pp. 373–384 dated 6.x.1884; pp. 385–386 dated 9.x.1884.]
Emery, C. 1884b. In Beccari, O. *Malesia. Raccolta di osservazioni botaniche intorno alla Piante dell'Archipelago Indo-malese e Papuano* **2**: 340 pp. Genova. [(31).xii.1884.]
Emery, C. 1886. Alcune formiche africane. *Bullettino della Società Entomologica Italiana* **18**: 355–366. [29.vi.1886.]
Emery, C. 1887a. Catalogo delle formiche esistenti nelle collezioni del Museo Civico di Genova. Parte terza. Formiche della regione Indo-Malese e dell'Australia. *Annali del Museo Civico di Storia Naturale di Genova* (2) **4** [24] (1886): 209–258. [Pp. 209–240 dated 20.i.1887; pp. 241–256 dated 24.i.1887; pp. 257–258 dated 26.i.1887.]
Emery, C. 1887b. Catalogo delle formiche esistenti nelle collezioni del Museo Civico di Genova. Parte terza. Formiche della regione Indo-Malese e dell'Australia (continuazione e fine). *Annali del Museo Civico di Storia Naturale di Genova* (2) **5** [25]: 427–473. [Pp. 427–432 dated 25.x.1887; pp. 433–448 dated 14.xi.1887; pp. 449–464 dated 25.xi.1887; pp. 465–473 dated 29.xi.1887.]
Emery, C. 1888a. Catalogo delle formiche esistenti nelle collezioni del Museo Civico di Genova. Parte terza (supplemento). Formiche raccolte dal sig. Elio Modigliani in Sumatra e nell'isola Nias. *Annali del Museo Civico di Storia Naturale di Genova* (2) **5** [25]: 528–534. [P. 528 dated 12.xii.1887; pp. 529–534 dated 2.i.1888.]
Emery, C. 1888b. Le tre forme sessuali del *Dorylus helvolus* L. e degli altri Dorilidi. *Bullettino della Società Entomologica Italiana* **19** (1887): 344–351. [20.i.1888.]
Emery, C. 1888c. Formiche della provincia di Rio Grande do Sul nel Brasile, raccolte dal dott. Hermann von Ihering. *Bullettino della Società Entomologica Italiana* **19** (1887): 352–366. [20.i.1888.]
Emery, C. 1888d. Über den sogenannten Kaumagen einiger Ameisen. *Zeitschrift für Wissenschaftliche Zoologie* **46**: 378–412. [23.v.1888.]
Emery, C. 1888e. Alcune formiche della Repubblica Argentina raccolte dal Dott. C. Spegazzini. *Annali del Museo Civico di Storia Naturale di Genova* (2) **6** [26]: 690–694. [20.xii.1888.]
Emery, C. 1889a. Intorno ad alcune formiche della fauna paleartica. *Annali del Museo Civico di Storia Naturale di Genova* (2) **7** [27]: 439–443. [14.viii.1889.]
Emery, C. 1889b. Viaggio di Leonardo Fea in Birmania e regioni vicine. 20. Formiche di Birmania e del Tenasserim raccolte da Leonardo Fea (1885–87). *Annali del Museo Civico di Storia Naturale di Genova* (2) **7** [27]: 485–520. [Pp. 481–512 dated 4.ix.1889; pp. 513–520 dated 9.ix.1889.]
Emery, C. 1890a. Voyage de M.E. Simon au Venezuela (Décembre 1887-Avril 1888). *Annales de la Société Entomologique de France* (6) **10**: 55–76. [8.viii.1890.]
Emery, C. 1890b. Studi sulle formiche della fauna neotropica. *Bullettino della Società Entomologica Italiana* **22**: 38–80. [15.ix.1890.]
Emery, C. 1891a. Le formiche dell'ambra Siciliana nel Museo Mineralogico dell'Università di Bologna. *Memorie della R. Accademia delle Scienze dell'Istituto di Bologna* (5) **1**: 141–165. (In separate pp. 567–591.) [(31.xii).1891; "sessione" date: 12.iv.1891.]
Emery, C. 1891b. Revision critique des fourmis de la Tunisie. *Exploration Scientifique de la Tunisie.* Zoologie- Hyménoptères: 21 pp. Paris. [(31.xii).1891.]
Emery, C. 1892a. Sopra alcune formiche raccolte dall'Ingegnere L. Bricchetti Robecchi nel paese dei Somali. *Annali del Museo Civico di Storia Naturale di Genova* (2) **12** [32]: 110–122. [Pp. 110–112 dated 27.i.1892; pp. 113–122 dated 28.i.1892.]
Emery, C. 1892b. Note sinonimiche sulle formiche. *Bullettino della Società Entomologica Italiana* **23** (1891): 159–167. [27.ii.1892.]
Emery, C. 1892c. Untitled contribution introduced by, "M.C. Emery, de Bologne, écrit." *Bulletin Bimensuel de la Société Entomologique de France* **1892** No. 4: liii-lv. [5.iii.1892.] (Note. This article repeated in *Annales de la Société Entomologique de France (Bulletin)* **61**: liii-lv. [12.x.1892.])
Emery, C. 1892d. Voyage de M. Ch. Alluaud dans le territoire d'Assinie (Afrique occidentale) en juillet et août 1886. Formicides. *Annales de la Société Entomologique de France* **60** (1891): 553–574. [13.iv.1892.]
Emery, C. 1892e. Älteres über Ameisen in Dornen afrikanischer Akazien. *Zoologischer Anzeiger* **15**: 237. [27.vi.1892.]
Emery, C. 1893a. Untitled contribution introduced by, "M.C. Emery, de Bologne, envoie les diagnoses de cinq nouveaux genres de Formicides." *Bulletin Bimensuel de la Société Entomologique de France* **1892** No. 20: cclxxv-cclxxvii. [7.i.1893.] (Note. This article repeated in *Annales de la Société Entomologique de France (Bulletin)* **61**: cclxxv–cclxxvii. [28.iv.1893.])
Emery, C. 1893b. Studio monografico sul genere *Azteca* Forel. *Memorie della R. Accademia delle Scienze dell'Istituto di Bologna.* Sezione delle Scienze Naturali (5) **3**: 119–152. (In separate pp. 319–352.) [(31.xii).1893; "sessione" date: 27.iii.1893.]
Emery, C. 1893c. Voyage de M. Ch. Alluaud aux îles Canaries. Formicides. *Annales de la Société Entomologique de France* **62**: 81–88. [31.vii.1893.]
Emery, C. 1893d. Notice sur quelques fourmis des îles Galapagos. *Annales de la Société Entomologique de France* **62**: 89–92. [31.vii.1893.]
Emery, C. 1893e. Voyage de MM. Bedot et Pictet dans l'Archipel Malais. Formicides de l'Archipel Malais. *Revue Suisse de Zoologie* **1**: 187–229. [28.ix.1893.]
Emery, C. 1893f. Voyage de M.E. Simon à l'île de Ceylan (janvier- février 1892). Formicides. *Annales de la Société Entomologique de France* **62**: 239–258. [25.x.1893.]
Emery, C. 1893g. Voyage de M.E. Simon aux îles Philippines (mars et avril 1890). Formicides. *Annales de la Société Entomologique de France* **62**: 259–270. [25.x.1893.]
Emery, C. 1893h. Intorno ad alcune formiche della collezione Spinola. *Bollettino dei Musei di Zoologia ed Anatomia Comparata della R. Università di Torino* **8** (No. 163): 1–3. [1.xii.1893.]

Emery, C. 1893i. Beiträge zur Kenntnis der nordamerikanischen Ameisenfauna. *Zoologische Jahrbücher. Abtheilung für Systematik, Geographie und Biologie der Thiere* **7**: 633–682. [23.xii.1893.]

Emery, C. 1894a. Mission scientifique de M. Ch. Alluaud aux îles Séchelles (mars, avril, mai 1892). Formicides. *Annales de la Société Entomologique de France* **63**: 67–72. [10.iv.1894.]

Emery, C. 1894b. Descriptions de deux fourmis nouvelles. *Annales de la Société Entomologique de France* **63**: 72–74. [10.iv.1894.]

Emery, C. 1894c. Studi sulle formiche della fauna Neotropica. *Bullettino della Società Entomologica Italiana* **26**: 137–241. [27.vi.1894.]

Emery, C. 1894d. Viaggio del Dott. E. Festa in Palestina, nel Libano e regioni vicine. 11. Descrizione di un nuovo *Camponotus. Bollettino dei Musei di Zoologia ed Anatomia Comparata della R. Università di Torino* **9** (No. 185): 1–2. [25.x.1894.]

Emery, C. 1894e. Viaggio del dottor Alfredo Borelli nella Repubblica Argentina e nel Paraguay. *Bollettino dei Musei di Zoologia ed Anatomia Comparata della R. Università di Torino* **9** (No. 186): 1–4. [25.x.1894.]

Emery, C. 1894f. *Camponotus sexguttatus* Fab. e *C. sexguttatus* Sm. et auct. *Bollettino dei Musei di Zoologia ed Anatomia Comparata della R. Università di Torino* **9** (No. 187): 1–4. [25.x.1894.]

Emery, C. 1894g. In Jhering, H. von. Die Ameisen von Rio Grande do Sul (pp. 372–401). *Berliner Entomologische Zeitschrift* **39**: 321–446. [(31).x.1894.]

Emery, C. 1894h. Descrizione di una nuova formica di Sicilia. *Naturalista Siciliano. Giornale di Scienze Naturali* **14**: 28. [(31).xii.1894.]

Emery, C. 1894i. Rassegna degl'Imenotteri raccolti nel Mozambico dal Cav. Fornasini, esistenti nel Museo Zoologico della R. Università di Bologna. Memoria di C. Emery, G. Gribodo, e G. Kriechbaumer. *Memorie della R. Accademia delle Scienze dell'Istituto di Bologna* (5) **4**: 45–90. (In separate pp. 111–156.) [(31.xii).1894; "sessione" date: 10.xii.1893.]

Emery, C. 1894j. Alcune formiche dell'isola di Creta. *Bullettino della Società Entomologica Italiana. Resoconti di Adunanze* **26**: 7–11. [(31.xii).1894.]

Emery, C. 1894k. Estudios sobre las hormigas de Costa Rica. *Anales del Museo Nacional, Costa Rica* **1889**: 45–67. [(31.xii).1894.]

Emery, C. 1895a. Esplorazione del Giuba e dei suoi affluenti compiuta dal Cap. V. Bottego durante gli anni 1892–93 sotto gli auspicii della Società Geografica Italiana. Risultati Zoologici. 10. Formiche. *Annali del Museo Civico di Storia Naturale di Genova* (2) **15** [35]: 177–184. [9.iv.1895.]

Emery, C. 1895b. Sopra alcune formiche della fauna Mediterranea. *Memorie della R. Accademia delle Scienze dell'Istituto di Bologna* (5) **5**: 59–75. (Pp. 291–307 in separate.) [(31.xii).1895; "sessione" date: 21.iv.1895.]

Emery, C. 1895c. Beiträge zur Kenntniss der nordamerikanischen Ameisenfauna. (Schluss.) *Zoologische Jahrbücher. Abtheilung für Systematik, Geographie und Biologie der Tiere* **8**: 257–360. [11.v.1895.]

Emery, C. 1895d. Note sur les fourmis du Chili, avec descriptions de deux espèces nouvelles. *Actes de la Société Scientifique du Chili* **4** (1894): 213–216. [29.v.1895.]

Emery, C. 1895e. On a new species of ant from New Zealand. *Transactions of the New Zealand Institute* **27** (1894): 635–636. [(31).v.1895.]

Emery, C. 1895f. Mission scientifique de M. Ch. Alluaud dans le territoire de Diego-Suarez (Madagascar-Nord). (Avril-août 1893.) *Annales de la Société Entomologique de Belgique* **39**: 336–345. [6.vii.1895.]

Emery, C. 1895g. Descriptions de quelques fourmis nouvelles d'Australie. *Annales de la Société Entomologique de Belgique* **39**: 345–358. [6.vii.1895.]

Emery, C. 1895h. Voyage de M.E. Simon dans l'Afrique australe (janvier-avril 1893). Formicides. *Annales de la Société Entomologique de France* **64**: 15–56. [16.viii.1895.]

Emery, C. 1895i. Deuxième note sur les fourmis du Chili. *Actes de la Société Scientifique du Chili* **5**: 10–18. [31.viii.1895.]

Emery, C. 1895j. Die Gattung *Dorylus* Fab. und die systematische Eintheilung der Formiciden. *Zoologische Jahrbücher. Abtheilung für Systematik, Geographie und Biologie der Thiere* **8**: 685–778. [18.x.1895.]

Emery, C. 1895k. Viaggio di Leonardo Fea in Birmania e regioni vicine. 63. Formiche di Birmania, del Tenasserim e dei Monti Carin, raccolte da L. Fea. *Annali del Museo Civico di Storia Naturale di Genova* (2) **14** [34] (1894): 450–483. [(31.xii).1895.] (At base of page 481 of this paper is the date 7.ix.1894, but a publication date of 1895 is given on both the front and back wrappers of the volume. The foreword to the volume, by Doria and Gestro, is dated 20.iv.1895; hence the date of publication given above is perhaps the best that can be achieved.)

Emery, C. 1896a. Sur les fourmis du genre *Macromischa* Roger. *Bulletin de la Société Entomologique de France* **1896**: 102–103. [26.ii.1896.]

Emery, C. 1896b. Formiche raccolte dal dott. E. Festa nei pressi del golfo di Darien. *Bollettino dei Musei di Zoologia ed Anatomia Comparata della R. Università di Torino* **11** (No. 229): 1–4. [1.iii.1896.]

Emery, C. 1896c. Alcune formiche del genere *Azteca* Forel e note biologiche. *Bollettino dei Musei di Zoologia ed Anatomia Comparata della R. Università di Torino* **11** (No. 230): 1–7. [1.iii.1896.]

Emery, C. 1896d. Saggio di un catalogo dei generi *Camponotus, Polyrhachis* e affini. *Memorie della R. Accademia delle Scienze dell'Istituto di Bologna* (5) **5**: 363–382. (Pp. 761–780 in separate.) [(31.xii).1896; "sessione" date: 8.iii.1896.]

Emery, C. 1896e. Clef analytique des genres de la famille des formicides. *Annales de la Société Entomologique de Belgique* **40**: 172–189. [2.v.1896.]

Emery, C, 1896f. Formicides récoltés a Buitenzorg (Java), par M. Massart. *Annales de la Société Entomologique de Belgique* **40**: 245–249. [6.vi.1896.]

Emery, C. 1896g. Studi sulle formiche della fauna Neotropica. *Bullettino della Società Entomologica Italiana* **28**: 33–107. [31.x.1896.]

Emery, C. 1896h. Formiciden, gesammelt in Paraguay von Dr. J. Bohls. *Zoologische Jahrbücher, Abtheilung für Systematik, Geographie und Biologie der Thiere* **9**: 625–638. [15.xi.1896.]

Emery, C. 1896i. Formiche raccolte dal Cap. V. Bottego nella regione dei Somali. *Annali del Museo Civico di Storia Naturale di Genova* (2) **17** [37] (1897): 153–160. [15.xii.1896.]

Emery, C. 1896j. Description d'une fourmi nouvelle d'Algérie. *Bulletin de la Société Entomologique de France* **1896**: 418–419. [23.xii.1896.]

Emery, C. 1896k. In Wasmann, E. Die Ameisen- und Termitengäste von Brasilien. *Verhandlungen der k.-k. Zoologisch- Botanischen Gesellschaft in Wien* **45** (1895): 137–179. [(31.xii).1896.]

Emery, C. 1897a. Descriptions de deux fourmis. *Bulletin de la Société Entomologique de France* **1897**: 12–14. [13.i.1897.]

Emery, C. 1897b. Revisione del genere *Diacamma* Mayr. *Rendiconto delle Sessioni della R. Accademia delle Scienze dell'Istituto di Bologna* (N.S.) **1**: 147–167. [25.iv.1897.]

Emery, C. 1897c. Formicidarum species novae vel minus cognitae in collectione Musaei Nationalis Hungarici, quas in Nova Guinea, colonia germanica, collegit L. Biró. *Természetrajzi Füzetek* **20**: 571–599. [1.xi.1897.]

Emery, C. 1897d. Viaggio di Lamberto Loria nella Papuasia orientale. 18. Formiche raccolte nella Nuova Guinea dal Dott. Lamberto Loria. *Annali del Museo Civico di Storia Naturale di Genova* (2) **18** [38]: 546–594. [Pp. 546–576 dated 16.xi.1897; pp. 577–594 dated 22.xi.1897.]

Emery, C. 1897e. Formiche raccolte da Don Eugenio dei Principi Ruspoli, durante l'ultimo suo viaggio nelle regioni dei Somali e dei Galla. *Annali del Museo Civico di Storia Naturale di Genova* (2) **18** [38]: 595–605. [22.xi.1897.]

Emery, C. 1897f. In Escherich, K. Zur Kenntniss der Myrmekophilen Kleinasiens. 1. Coleoptera. Mit einem Verzeichniss der in Kleinasiens gesammelten Ameisen und einer Neubeschreibung. *Wiener Entomologische Zeitung* **16**: 229–239. [30.xi.1897.]

Emery, C. 1898a. Aggiunte e correzioni al Saggio di un catalogo sistematico dei generi *Camponotus, Polyrhachis* e affini. *Rendiconto delle Sessioni della R. Accademia delle Scienze dell'Istituto di Bologna* (N.S.) **2**: 225–231. [22.v.1898.]

Emery, C. 1898b. Descrizioni di formiche nuove malesi e australiane; note sinonimiche. *Rendiconto delle Sessioni della R. Accademia delle Scienze dell'Istituto di Bologna* (N.S.) **2**: 231–245. [22.v.1898.]

Emery, C. 1898c. Beiträge zur Kenntniss der palaearktischen Ameisen. *Öfversigt af Finska Vetenskaps-Societetens Förhandlingar* **40**: 124–151. [(31.xii).1898.]
Emery, C. 1899a. Formiche dell'ultima spedizione Bottego. *Annali del Museo Civico di Storia Naturale di Genova* (2) **19** [39]: 499–501. [10.i.1899.]
Emery, C. 1899b. Glanures myrmécologiques. *Bulletin de la Société Entomologique de France* **1899**: 17–20. [25.i.1899.]
Emery, C. 1899c. Intorno alle larve di alcune formiche. *Memorie della R. Accademia delle Scienze dell'Istituto di Bologna* (5) **8**: 3–10. [(31.xii).1899; "sessione" date: 7.v.1899.]
Emery, C. 1899d. Ergebnisse einer Reise nach dem Pacific (Schauinsland, 1896–1897). *Zoologische Jahrbücher. Abtheilung für Systematik, Geographie und Biologie der Thiere* **12**: 438–440. [28.ix.1899.]
Emery, C. 1899e. Fourmis d'Afrique. *Annales de la Société Entomologique de Belgique* **43**: 459–504. [6.xi.1899.]
Emery, C. 1899f. Formiche di Madagascar raccolte dal Sig. A. Mocquerys nei pressi della Baia di Antongil (1897–1898). *Bullettino della Società Entomologica Italiana* **31**: 263–290. [31.xii.1899.]
Emery, C. 1900a. Nuovi studi sul genere *Eciton*. *Memorie della R. Accademia delle Scienze dell'Istituto di Bologna* (5) **8**: 173–188. (Pp. 511–526 in separate.) [(31.xii).1900; "sessione" date: 25.iii.1900.]
Emery, C. 1900b. Intorno al torace delle formiche e particolarmente nei neutri. *Bullettino della Società Entomologica Italiana* **32**: 103–119. [20.v.1900.]
Emery, C. 1900c. Formicidarum species novae vel minus cognitae in collectione Musaei Nationalis Hungarici, quas in Nova- Guinea, colonia germanica, collegit L. Biró. Publicatio secunda. *Természetrajzi Füzetek* **23**: 310–338. [Separates issued 1.viii.1900; whole volume dated 15.ix.1900.]
Emery, C. 1900d. Formiche raccolte da Elio Modigliani in Sumatra, Engano e Mentawei. *Annali del Museo Civico di Storia Naturali di Genova* (2) **20** [40]: 661–722. [Pp. 657–688 dated 20.xii.1900; pp. 689–720 dated 21.xii.1900; pp. 721–722 dated 4.ii.1901.]
Emery, C. 1901a. Notes sur les sous-familles des dorylines et ponérines (famille des formicides). *Annales de la Société Entomologique de Belgique* **45**: 32–54. [7.iii.1901.]
Emery, C. 1901b. Remarques sur un petit groupe de *Pheidole* de la région sonorienne. *Bulletin de la Société Entomologique de France* **1901**: 119–121. [13.iii.1901.]
Emery, C. 1901c. Studi sul polimorfismo e la metamorfosi nel genere *Dorylus*. *Memorie della R. Accademia delle Scienze dell'Istituto di Bologna* (5) **9**: 183–201. (Pp. 415–433 in separate.) [(31.xii).1901: "sessione" date: 12.v.1901]
Emery, C. 1901d. Note sulle doriline. *Bullettino della Società Entomologica Italiana* **33**: 43–56. [31.v.1901.]
Emery, C. 1901e. Spicilegio mirmecologico. *Bullettino della Società Entomologica Italiana* **33**: 57–63. [31.v.1901.]
Emery, C. 1901f. Ameisen gesammelt in Ceylon von Dr. W. Horn, 1899. *Deutsche Entomologische Zeitschrift* **1901**: 113–122. [(31).vii.1901.]
Emery, C. 1901g. Formiciden von Celebes. *Zoologische Jahrbücher. Abteilung für Systematik, Geographie und Biologie der Tiere* **14**: 565–580. [6.viii.1901.]
Emery, C. 1901h. Formicidarum species novae vel minus cognitae in collectione Musaei Nationalis Hungarici, quas in Nova- Guinea, colonia germanica, collegit L. Biró. Publicatio tertia. *Természetrajzi Füzetek* **25** (1902): 152–160. [Separates issued 15.xii.1901; whole volume dated 20.iv.1902.]
Emery, C. 1901i. In Mocsáry, S. & Szépligeti, G. Hymenopterák (Hymenopteren) (Pp. 121–169). In Horváth, G. *Zichy Jenő Gróf harmadik ázsiai utazásának állattani eredményei* **2**: 470 pp. Budapest/Leipzig. [(31.xii).1901.]
Emery, C. 1902a. Description d'une nouvelle espèce de fourmi du Brésil. *Bulletin de la Société Entomologique de France* **1902**: 181. [6.vi.1902.]
Emery, C. 1902b. In Wasmann, E. Neues über die zusammengesetzten Nester und gemischten Kolonien der Ameisen. (Fortsetzung aus No. 14/15.) *Allgemeine Zeitschrift für Entomologie* **7**: 293–298. [15.viii.1902.]
Emery, C. 1902c. Note mirmecologiche. *Rendiconto delle Sessioni della R. Accademia delle Scienze dell'Istituto di Bologna* (N.S.) **6** (1901): 22–34. [(31.xii).1902.]
Emery, C. 1903. Intorno ad alcune specie di *Camponotus* dell'America meridionale. *Rendiconto delle Sessioni della R. Accademia della Scienze dell'Istituto di Bologna* (N.S.) **7**: 62–81. [(31.xii).1903.]
Emery, C. 1904a. Le affinità del genere *Leptanilla* e i limiti delle Dorylinae. *Archivio Zoologico* **2**: 107–116. [24.viii.1904.]
Emery, C. 1904b. Zur Kenntniss des Polymorphismus der Ameisen. *Zoologischen Jahrbüchern. Supplement* 7: 587–610. [(31.xii).1904.]
Emery, C. 1905a. In Mantero, G. Materiali per una fauna dell'Archipelago Toscano. Isola del Giglio. *Annali del Museo Civico di Storia Naturale di Genova* (3) **1** [41]: 449–454. [11.i.1905.]
Emery, C. 1905b. Deux fourmis de l'ambre de la Baltique. *Bulletin de la Société Entomologique de France* **1905**: 187–189. [25.ix.1905.]
Emery, C. 1905c. Revisione delle specie del genere *Atta* appartenenti ai sottogeneri *Moellerius* e *Acromyrmex*; con figure intercalate nel testo. *Memorie della R. Accademia delle Scienze dell'Istituto di Bologna* (6) **2**: 39–54. (Pp. 107–122 in separate.) [(30).xi.1905.]
Emery, C. 1905d. Le forme paleartiche del *Camponotus maculatus* F. *Rendiconto delle Sessioni della R. Accademia delle Scienze dell'Istituto di Bologna* (N.S.) **9**: 27–44. [(31.xii).1905.]
Emery, C. 1906a. Über W.H. Ashmeads neues System der Ameisen. *Zoologischer Anzeiger* **29**: 717–718. [23.ii.1906.]
Emery, C. 1906b. Note sur *Prenolepis vividula* Nyl. et sur la classification des espèces du genre *Prenolepis*. *Annales de la Société Entomologique de Belgique* **50**: 130–134. [25.iv.1906.]
Emery, C. 1906c. Studi sulle formiche della fauna Neotropica. *Bullettino della Società Entomologica Italiana* **37** (1905): 107–194. [16.vi.1906.]
Emery, C. 1906d. Rassegna critica delle specie paleartiche del genere *Myrmecocystus*. *Memorie della R. Accademia delle Scienze dell'Istituto di Bologna* (6) **3**: 47–61. (Pp. 173–187 in separate.) [(30).xi.1906.]
Emery, C. 1907. Una formica nuova italiana spetante ad un nuovo genere. *Rendiconto delle Sessioni della R. Accademia delle Scienze dell'Istituto di Bologna* (N.S.) **11**: 49–51. [12.v.1907, "sessione" date.]
Emery, C. 1908a. Beiträge zur Monographie der Formiciden des paläarktischen Faunengebietes. *Deutsche Entomologische Zeitschrift* **1908**: 165–205. [1.iii.1908.]
Emery, C. 1908b. Descriptions d'un genre nouveau et de plusiers formes nouvelles de fourmis du Congo. *Annales de la Société Entomologique de Belgique* **52**: 184–189. [30.iv.1908.]
Emery, C. 1908c. Beiträge zur Monographie der Formiciden des paläarktischen Faunengebietes. 3. Die mit *Aphaenogaster* verwandte Gattungengruppe. *Deutsche Entomologische Zeitschrift* **1908**: 305–338. [1.v.1908.]
Emery, C. 1908d. In Cecconi, G. Contributo alla fauna delle Isole Tremiti. *Bollettino dei Musei di Zoologia ed Anatomia Comparata della R. Università di Torino* **23** (No. 583): 1–53. [20.v.1908.]
Emery, C. 1908e. Beiträge zur Monographie der Formiciden des paläarktischen Faunengebietes. 3. *Deutsche Entomologische Zeitschrift* **1908**: 437–465. [1.vii.1908.]
Emery, C. 1908f. Beiträge zur Monographie der Formiciden des paläarktischen Faunengebietes. 4. Parasitische und Gast- Myrmicinen mit Ausnahme von *Strongylognathus*. *Deutsche Entomologische Zeitschrift* **1908**: 549–558. [1.ix.1908.]
Emery, C. 1908g. *Myrmecocystus viaticus* et formes voisines. *Bulletin de la Société Vaudoise des Sciences Naturelles* **44**: 213–217. [(30).ix.1908.]
Emery, C. 1908h. Beiträge zur Monographie der Formiciden des paläarktischen Faunengebietes. 5. *Deutsche Entomologische Zeitschrift* **1908**: 663–686. [1.xi.1908.]
Emery, C. 1909a. Beiträge zur Monographie der Formiciden des paläarktischen Faunengebietes. 6. *Deutsche Entomologische Zeitschrift* **1909**: 19–37. [3.i.1909.]

Emery, C. 1909b. Beiträge zur Monographie der Formiciden des paläarktischen Faunengebietes. 7. *Deutsche Entomologische Zeitschrift* **1909**: 179–204. [1.iii.1909.]

Emery, C. 1909c. Beiträge zur Monographie der Formiciden des paläarktischen Faunengebietes. 8. *Deutsche Entomologische Zeitschrift* **1909**: 355–376. [1.v.1909.]

Emery, C. 1909d. Beiträge zur Monographie der Formiciden des paläarktischen Faunengebietes. 9. *Deutsche Entomologische Zeitschrift* **1909**: 695–712. [1.xi.1909.]

Emery, C. 1910a. Beiträge zur Monographie der Formiciden des paläarktischen Faunengebietes. 10. *Deutsche Entomologische Zeitschrift* **1910**: 127–132. [5.iii.1910.]

Emery, C. 1910b. In Wytsman, P. *Genera Insectorum*. Hymenoptera, Fam. Formicidae, subfam. Dorylinae. **Fasc. 102**: 34 pp. Bruxelles. [(31.xii).1910.]

Emery, C. 1911a. Fragments myrmécologiques. *Annales de la Société Entomologique de Belgique* **55**: 213–225. [4.viii.1911.]

Emery, C. 1911b. *Nova Guinea*. Résultats de l'expédition scientifique néerlandaise à la Nouvelle-Guinée en 1903 sous les auspices de Arthur Wichmann. **5** Zoologie: 531–539. [3.xi.1911.]

Emery, C. 1911c. *Nova Guinea*. Résultats de l'expédition scientifique néerlandaise à la Nouvelle-Guinée en 1907 et 1909 sous les auspices de Dr. H.A. Lorentz. **9** Zoologie: 249–259. [(31.xii).1911.]

Emery, C. 1911d. In Wytsman, P. *Genera Insectorum*. Hymenoptera, Fam. Formicidae, subfam. Ponerinae. **Fasc. 118**: 124 pp. Bruxelles. [(31.xii).1911.]

Emery, C. 1912a. Revision der *Rhytidoponera* (subg. *Chalcoponera*) der *metallica*-Gruppe. *Deutsche Entomologische Zeitschrift* **1912**: 77–81. [29.i.1912.]

Emery, C. 1912b. Etudes sur les Myrmicinae. *Annales de la Société Entomologique de Belgique* **56**: 94–105. [9.v.1912.]

Emery, C. 1912c. Formiche raccolte duranti i viaggi di S.A.R. la Duchessa Elena d'Aosta nella regione dei grandi laghi dell'Africa equatoriale. *Annuario del Museo Zoologico della R. Università di Napoli* (N.S.) **3** (No. 26): 2 pp. [22.v.1912.]

Emery, C. 1912d. Les espèces-type des genres et sous-genres de la famille des formicides. *Annales de la Société Entomologique de Belgique* **56**: 271–273. [2.viii.1912.]

Emery, C. 1912e. Beiträge zur Monographie der Formiciden des paläarktischen Faunengebietes. 11. *Deutsche Entomologische Zeitschrift* **1912**: 651–672. [30.xi.1912.]

Emery, C. 1912f. Der Wanderzug der Steppen- und Wüstenameisen von Zentral-Asien nach Süd-Europa und Nord-Afrika. *Zoologische Jahrbücher. Supplement* **15**: 95–104. [(31.xii).1912.]

Emery, C. 1912g. In Wytsman, P. *Genera Insectorum*. Hymenoptera, Fam. Formicidae, subfam. Dolichoderinae. **Fasc. 137**: 50 pp. Bruxelles. [(31.xii).1912.]

Emery, C. 1913. Etudes sur les Myrmicinae. *Annales de la Société Entomologique de Belgique* **57**: 250–262. [5.ix.1913.]

Emery, C. 1914a. Intorno alla classificazione dei Myrmicinae. *Rendiconto delle Sessioni della R. Accademia delle Scienze dell'Istituto di Bologna* (N.S.) **18**: 29–42. [11.i.1914, "sessione" date.]

Emery, C. 1914b. Formiche d'Australia e di Samoa raccolte dal Prof. F. Silvestri nel 1913. *Bollettino del Laboratorio di Zoologia generale e agraria della R. Scuola superiore d'Agricoltura in Portici* **8**: 179–186. [30.i.1914.]

Emery, C. 1914c. *Cephalotes* et *Cryptocerus*. Le type du genre *Crematogaster*. *Annales de la Société Entomologique de Belgique* **58**: 37–39. [6.ii.1914.]

Emery, C. 1914d. Wissenschaftliche Ergebnisse der Bearbeitung von O. Leonhards Sammlungen. 5. Südeuropäische Ameisen. *Entomologische Mitteilungen* **3**: 156–159. [1.v.1914.]

Emery, C. 1914e. Note sulle formiche della collezione sarda e della collezione dell'Italia meridionale, radunate da Achille Costa, e conservate nel Museo Zoologico della R. Università di Napoli. *Annuario del Museo Zoologico della R. Università di Napoli* (N.S.) **4**: 1–3. [7.x.1914.]

Emery, C. 1914f. Les fourmis de la Nouvelle-Calédonie et des îles Loyalty. In Sarasin, F. & Roux, J. *Nova Caledonia* Zoologie **1**: 393–437. Wiesbaden. [(31.xii).1914.]

Emery, C. 1915a. Contributo alla conoscenza delle formiche delle isole italiani. Descrizioni di forme mediterranee nuove o critiche. *Annali del Museo Civico di Storia Naturale di Genova* (3) **6** [46]: 244–270. [17.i.1915.]

Emery, C. 1915b. *Sima* oder *Tetraponera*? *Zoologischer Anzeiger* **45**: 265–266. [23.ii.1915.]

Emery, C. 1915c. Le formiche del genere *Solenopsis* abitanti l'Africa. *Rendiconto delle Sessioni della R. Accademia delle Scienze dell'Istituto di Bologna* (N.S.) **19**: 57–66 [7.iii.1915, "sessione" date.]

Emery, C. 1915d. Definizione del genere *Aphaenogaster* e partizione di esso in sottogeneri. *Parapheidole* e *Novomessor* nn. gg. *Rendiconto delle Sessioni della R. Accademia delle Scienze dell'Istituto di Bologna* (N.S.) **19**: 67–75. [21.iii.1915, "sessione" date.]

Emery, C. 1915e. Su due formiche della Tripolitania. *Bollettino del Laboratorio di Zoologia generale e agraria della R. Scuola superiore d'Agricoltura in Portici* **9**: 378. [7.iv.1915.]

Emery, C. 1915f. Sur le type de *Camponotus maculatus* (*Formica maculata* F.). *Bulletin de la Société Entomologique de France* **1915**: 79–80. [13.iv.1915.]

Emery, C. 1915g. Formiche raccolte nell'Eritrea dal Prof. F. Silvestri. *Bollettino del Laboratorio di Zoologia generale e agraria della R. Scuola superiore d'Agricoltura in Portici* **10**: 3–26. [10.v.1915.]

Emery, C. 1915h. Escursioni zoologiche del Dr. Enrico Festa nell'Isola di Rodi. 12. Formiche. *Bollettino dei Musei di Zoologia ed Anatomia Comparata della R. Università di Torino* **30**: 1–7. [30.v.1915.]

Emery, C. 1915i. Noms de sous-genres et de genres proposés pour la sous-famille des Myrmicinae. Modifications à la classification de ce groupe. *Bulletin de la Société Entomologique de France* **1915**: 189–192. [31.vii.1915.]

Emery, C. 1915j. Les *Pheidole* du groupe *megacephala*. *Revue Zoologique Africaine* **4**: 223–250. [15.xii.1915.]

Emery, C. 1916a. Formiche d'Italia nuove o critiche. *Rendiconto delle Sessioni della R. Accademia delle Scienze dell'Istituto di Bologna* (N.S.) **20**: 53–66. [12.iii.1916, "sessione" date.]

Emery, C. 1916b. Fauna Entomologica Italiana. 1. Hymenoptera, Formicidae. *Bullettino della Società Entomologica Italiana* **47** (1915): 79–275. [30.xi.1916.]

Emery, C. 1917a. Questions de nomenclature et synonymies relative à quelques genres et espèces de formicides. *Bulletin de la Société Entomologique de France* **1917**: 94–97. [28.iii.1917.]

Emery, C. 1917b. Formiche Ibride. *Rendiconto delle Sessioni della R. Accademia delle Scienze dell'Istituto di Bologna* (N.S.) **21**: 23–29. [(31.xii).1917.]

Emery, C. 1919a. Errata au mémoire les *Pheidole* du groupe *megacephala*. *Revue Zoologique Africaine* **6**: 169–170. [10.iii.1919.]

Emery, C. 1919b. Sur le genre *Tranopelta* et sur le type du genre *Crematogaster*. *Bulletin de la Société Entomologique de France* **1919**: 60–62. [7.iv.1919.]

Emery, C. 1919c. Notes critiques de myrmécologie. *Annales de la Société Entomologique de Belgique* **59**: 100–107. [1.viii.1919.]

Emery, C. 1920a. Notes critiques de myrmécologie. 6. Réponse aux critiques de M. Bondroit. *Annales de la Société Entomologique de Belgique* **60**: 59–61. [20.v.1920.]

Emery, C. 1920b. Le genre *Camponotus* Mayr. Nouvel essai de sa subdivision en sous-genres. *Revue Zoologique Africaine* **8**: 229–260. [1.x.1920.]

Emery, C. 1920c. Studi sui *Camponotus*. *Bullettino della Società Entomologica Italiana* **52**: 3–49. [6.xii.1920.]

Emery, C. 1921a. Specific names repeated in the Linnean genus *Formica*. *Psyche* **28**: 24–26. [(28).ii.1921.]

Emery, C. 1921b. Formiche raccolte a Budrum (Anatolia) da Raffaele Varriale, Cap. Medico nella R. Marina. *Annali del Museo Civico di Storia Naturale di Genova* (3) **9** [49]: 208–218. [31.iii.1921.]

Emery, C. 1921c. Formiche tessitrici del genere *Oecophylla* fossile e viventi. *Rendiconto delle Sessioni della R. Accademia delle Scienze dell'Istituto di Bologna* (N.S.) **25**: 99–105. [22.v.1921, "sessione" date.]

Emery, C. 1921d. Notes critiques de myrmécologie. 9. Etude sur le genre *Plagiolepis. Annales de la Société Entomologique de Belgique* **61**: 313–319. [2.ix.1921.]

Emery, C. 1921e. Le genre *Polyrhachis.* Classification; espèces nouvelles ou critiques. *Bulletin de la Société Vaudoise des Sciences Naturelles* **54**: 17–25. [16.ix.1921.]

Emery, C. 1921f. In Wytsman, P. *Genera Insectorum.* Hymenoptera, Fam. Formicidae, subfam. Myrmicinae. **Fasc. 174A**: 1–94. Bruxelles. [(31.xii).1921.]

Emery, C. 1922a. Le specie americane del genere *Melophorus (Lasiophanes). Rendiconto delle Sessioni della R. Accademia delle Scienze dell'Istituto di Bologna* (N.S.) **26**: 90–95. [12.ii.1922, "sessione" date.]

Emery, C. 1922b. Il genere *Lasius* (F.) Mayr, e particolarmente le forme mediterranee del gruppo *umbratus* Nyl. *Bollettino della Società Entomologica Italiana* **54**: 9–15. [20.vi.1922.]

Emery, C. 1922c. *Messor barbarus* (L.). Appunti di sinonimia e di systematica. *Bollettino della Società Entomologica Italiana* **54**: 92–99. [22.x.1922.]

Emery, C. 1922d. Quelques fourmis nouvelles minuscules. *Annales Historico-Naturales Musei Nationalis Hungarici* **19**: 107–112. [29.xii.1922.]

Emery, C. 1922e. In Wytsman, P. *Genera Insectorum.* Hymenoptera, Fam. Formicidae, subfam. Myrmicinae. **Fasc. 174B**: 95–206. Bruxelles. [(31.xii).1922.]

Emery, C. 1922f. In Wytsman, P. *Genera Insectorum.* Hymenoptera, Fam. Formicidae, subfam. Myrmicinae. **Fasc. 174C**: 207–397. Bruxelles. [(31.xii).1922.]

Emery, C. 1923. Einige exotische Ameisen des Deutschen Entomologischen Institutes. *Entomologische Mitteilungen* **12**: 60–62. [5.i.1923.]

Emery, C. 1924a. Formiche della Cirenaica raccolte dal Dott. Enrico Festa e dal Prof. Filippo Silvestri. *Bollettino della Società Entomologica Italiana* **56**: 6–11. [24.i.1924.]

Emery, C. 1924b. Alcune formiche di Malta. *Bollettino della Società Entomologica Italiana* **56**: 11–12. [24.i.1924.]

Emery, C. 1924c. Formiche di Spagna, raccolte dal Prof. Filippo Silvestri. *Bollettino del Laboratorio di Zoologia generale e agraria della R. Scuola superiore d'Agricoltura in Portici* **17** (1923): 164–171. [29.iii.1924.]

Emery, C. 1925a. I *Camponotus (Myrmentoma)* paleartici del gruppo *lateralis. Rendiconto delle Sessioni della R. Accademia delle Scienze dell'Istituto di Bologna* (N.S.) **29**: 62–72. [11.i.1925, "sessione" date.]

Emery, C. 1925b. In Wytsman, P. *Genera Insectorum.* Hymenoptera, Fam. Formicidae, subfam. Formicinae. **Fasc. 183**: 302 pp. Bruxelles. [15.i.1925.]

Emery, C. 1925c. Notes critiques de myrmécologie. *Annales de la Société Entomologique de Belgique* **64** (1924): 177–191. [(30).iv.1925.]

Emery, C. 1925d. Revision des espèces paléarctiques du genre *Tapinoma. Revue Suisse de Zoologie* **32**: 45–64. [(31).v.1925.]

Emery, C. 1925e. Les espèces européennes et orientales du genre *Bothriomyrmex. Bulletin de la Société Vaudoise des Sciences Naturelles* **56**: 5–22. [30.xi.1925.]

Emery, C. 1925f. In Santschi, F. Fourmis d'Espagne et autres espèces paléarctiques. *Eos. Revista Española de Entomologia* **1**: 339–360. [31.xii.1925.]

Emery, C. 1926. Ultime note mirmecologiche. *Bollettino della Società Entomologica Italiana* **58**: 1–9. [29.i.1926.]

Emery, C. & Forel, A. 1879. Catalogue des fourmis d'Europe. *Mittheilungen der Schweizerischen Entomologische Gesellschaft* **5**: 441–484. [(28).ii.1879.]

Enzmann, E.V. 1944. Systematic notes on the genus *Pseudomyrma. Psyche* **51**: 59–103. [(31.xii).1944.]

Enzmann, E.V. 1952. *Woitkowskia,* a new genus of army ants. *Proceedings of the Iowa Academy of Science* **59**: 442–447. [(31).xii.1952.]

Enzmann, J. 1946a. A new form of *Myrmecina. Journal of the New York Entomological Society* **54**: 13–15. [17.iv.1946.]

Enzmann, J. 1946b. A new house-invading ant from Massachusetts. *Journal of the New York Entomological Society* **54**: 47–49. [17.iv.1946.]

Enzmann, J. 1946c. *Crematogaster lineolata cerasi,* the Cherry Ant of Asa Fitch (with a survey of the American forms of *Crematogaster,* subgenus *Acrocoelia*). *Journal of the New York Entomological Society* **54**: 89–97. [18.vi.1946.]

Enzmann, J. 1947a. *Hercynia,* a new genus of myrmicine ants. *Journal of the New York Entomological Society* **55**: 43–47. [27.iii.1947.]

Enzmann, J. 1947b. New forms of *Aphaenogaster* and *Novomessor. Journal of the New York Entomological Society* **55**: 147–153. [26.v.1947.]

Erichson, W.F. 1839. Insecten. *Archiv für Naturgeschichte* **5** (2): 281–375. [(31.xii).1839.]

Erichson, W.F. 1842. Beitrag zur Insecten-Fauna von Vandiemensland, mit besonderer Berücksichtigung der geographischen Verbreitung der Insecten. *Archiv für Naturgeschichte* **8**: 83–287. [(31.xii).1842.]

Espadaler, X. Note. In most publications this is the form of his name that is most commonly used. However, in some papers his name appears in the form Espadaler Gelabert, X. The former is used universally in this catalogue.

Espadaler, X. 1978. Descripcion de los sexuados de *Tapinoma pygmaeum* (Dufour, 1857). *Vie et Milieu* **27** (1977): 119–128. [(31).xii.1978.]

Espadaler, X. 1981. Una nueva hormiga de la Peninsula Iberica. *Miscelánea Zoologica* **5** (1979): 77–81. [(31.xii).1981.]

Espadaler, X. 1982a. *Epimyrma bernardi* n. sp., a new parasitic ant. *Spixiana* **5**: 1–6. [1.iii.1982.]

Espadaler, X. 1982b. *Xenhyboma mystes* Santschi, 1919 = *Monomorium medinae* Forel, 1892. Evidence biologique de la synonymie. *Nouvelle Revue d'Entomologie* **12**: 111–113. [30.iii.1982.]

Espadaler, X. 1984. *Leptothorax nadigi* Kutter, 1925 y *Goniomma blanci* (André, 1881): descripción de los machos. *Boletín de la Asociation Española de Entomologia* **8**: 135–141. [(30).vi.1984.]

Espadaler, X. 1986a. *Goniomma kugleri,* a new granivorous ant from the Iberian Peninsula. *Israel Journal of Entomology* **19** (1985): 61–66. [(28).ii.1986.]

Espadaler, X. 1986b. *Formica decipiens* Bondroit, 1918: descripción del macho y dos adiciones a la fauna ibérica. *Boletín de la Asociation Española de Entomologia* **10**: 45–50. [(31).v.1986.]

Espadaler, X. & Agosti, D. 1985. *Monomorium hesperium* Emery: description de la femelle. *Mitteilungen der Schweizerischen Entomologischen Gesellschaft* **58**: 295–297. [15.xii.1985.]

Espadaler, X. & Agosti, D. 1987. *Monomorium boltoni* n. sp. from Sao Nicolau (Cape Verde Islands). *Mitteilungen der Schweizerischen Entomologischen Gesellschaft* **60**: 295–299. [15.xii.1987.]

Espadaler, X. & Batet, J.M. 1979. *Goniomma blanci* (André, 1881): descripción de la hembra. *Boletín de la Asociacion Española de Entomologia* **3**: 11–15. [(30).xi.1979.]

Espadaler, X. & Cagniant, H. 1987. Contribution à la connaissance des fourmis marocaines. Description du mâle de *Proformica theryi* Santschi, 1936. *Nouvelle Revue d'Entomologie* (N.S.) **4**: 133–138. [30.vi.1987.]

Espadaler, X. & Collingwood, C.A. 1982. Notas sobre *Leptothorax* Mayr, 1855, con descripción de *L. gredosi* n. sp. *Boletín de la Asociation Española de Entomologia* **6**: 41–48. [(31).xii.1982.]

Espadaler, X. & Du Merle, P. 1989. *Leptothorax fuentei* Santschi, 1919, en France. *Vie et Milieu* **39**: 121–123. [(31).v.1989.]

Espadaler, X., Du Merle, P. & Plateaux, L. 1983. Redescription de *Leptothorax grouvellei* Bondroit, 1918. Notes biologiques et écologiques. *Insectes Sociaux* **30**: 274–286. [(31).xii.1983.]

Espadaler, X., Plateaux, L. & Casevitz-Weulersse, J. 1984. *Leptothorax melas* n. sp., de Corse. Notes écologiques et biologiques. *Revue Francaise d'Entomologie* (N.S.) **6**: 123–132. [15.ix.1984.]

Espadaler, X. & Riasol, J.M. 1983. Distribución, variabilidad y sinonimias en *Aphaenogaster iberica* Emery, 1908 y dos adiciones a la fauna ibérica. *Actas I Congreso Ibérico de Entomología* **1**: 219–228. [(31.xii).1983.]

Espadaler, X. & Rodriguez, R. 1989. The male of *Leptothorax risi* Forel, 1892. *Orsis* **4**: 141–144. [1989.]

Essig, E.O. 1926. *Insects of Western North America:* 1035 pp. New York. [(30).ix.1926.]

Ettershank, G. 1965. A new species of *Megalomyrmex* from the Chilean Andes. *Psyche* **72**: 55–58. [25.vi.1965.]

Ettershank, G. 1966. A generic revision of the world Myrmicinae related to *Solenopsis* and *Pheidologeton*. *Australian Journal of Zoology* **14**: 73–171. [(28).ii.1966.]

Ettershank, G. & Brown, W.L.,Jr. 1964. *Monomorium solleri* comb. nov. *Pilot Register of Zoology* **card 18**. [20.v.1964.]

Faber, W. 1967. Beiträge zur Kenntnis sozialparasitischer Ameisen. 1. *Lasius (Austrolasius* n. sg.) *reginae* n. sp., eine neue temporär sozialparasitische Erdameise aus Österreich. *Pflanzenschutz Berichte* **36**: 73–107. [(30).xi.1967.]

Faber, W. 1969. Beiträge zur Kenntnis sozialparasitischer Ameisen. 2. *Aporomyrmex ampeloni* nov. gen., nov. spec., ein neuer permanenter Sozialparasit bei *Plagiolepis vindobonensis* Lomnicki aus Österreich. *Pflanzenschutz Berichte* **39**: 39–100. [(28).ii.1969.]

Fabricius, J.C. 1775. *Systema Entomologiae*, sistens insectorum classes, ordines, genera, species, adiectis synonymis, locis, descriptionibus, observationibus: 832 pp. Flensburgi et Lipsiae. [17.iv.1775.]

Fabricius, J.C. 1782. *Species Insectorum* exhibentes eorum differentias specificas, synonyma, autorum, loca natalia, metamorphosin adiectis observationibus, descriptionibus. **1** (1781): 552 pp. Hamburgi et Kilonii. [(31.xii).1782.]

Fabricius, J.C. 1787. *Mantissa Insectorum* **1**: 348 pp. Hafniae. [(31.xii).1787.]

Fabricius, J.C. 1793. *Entomologia Systematica* emendata et aucta. Secundum classes, ordines, genera, species adjectis synonimis, locis, observationibus, descriptionibus **2**: 519 pp. Hafniae. [(31.xii).1793.]

Fabricius, J.C. 1798. *Supplementum Entomologiae Systematicae:* 572 pp. Hafniae. [20.v.1798.]

Fabricius, J.C. 1804. *Systema Piezatorum:* 439 pp. Brunsvigae. [(31.xii).1804.]

Feener, D.H.,Jr. 1981. Notes on the biology of *Pheidole lamia* at its type locality (Austin, Texas). *Journal of the Kansas Entomological Society* **54**: 269–277. [29.iv.1981.]

Feener, D.H.,Jr. 1987. Response of *Pheidole morrisi* to two species of enemy ants, and a general model of defense behavior in *Pheidole*. *Journal of the Kansas Entomological Society* **60**: 569–575. [29.xii.1987.]

Finzi, B. 1922. Primo contributo alla conoscenza della fauna mirmecologica della Venezia Giulia. *Bullettino della Società Entomologica Italiana* **53** (1921): 118–120. [15.ii.1922.]

Finzi, B. 1923. Risultati scientifici della spedizione Ravasini- Lona in Albania. 3. Formiche. *Bollettino della Società Entomologica Italiana* **55**: 1–4. [22.i.1923.]

Finzi, B. 1924a. Formiche dell'Isola d'Elba e Monte Argentario. *Bollettino della Società Entomologica Italiana* **56**: 12–15. [24.i.1924.]

Finzi, B. 1924b. Secondo contributo alla conoscenza della fauna mirmecologica della Venezia Giulia. *Bollettino della Società Entomologica Italiana* **56**: 120–123. [29.x.1924.]

Finzi, B. 1926. Le forme europee del genere *Myrmica* Latr. Primo contributo. *Bollettino della Società Adriatica di Scienze Naturali* **29**: 71–119. [(31.xii).1926.]

Finzi, B. 1927a. Terzo contributo alla conoscenza della fauna mirmecologica della Venezia Giulia. *Bollettino della Società Entomologica Italiana* **59**: 7–10. [31.i.1927.]

Finzi, B. 1927b. Nota sui *Camponotus (Myrmentoma) lateralis, piceus, dalmaticus. Folia Myrmecologica et Termitologica* **1**: 51–52. [(28).ii.1927.]

Finzi, B. 1928a. *Formica cinerea* Mayr e varietà paleartiche. *Bollettino della Società Entomologica Italiana* **60**: 65–75. [31.v.1928.]

Finzi, B. 1928b. Quarto contributo alla conoscenza della fauna mirmecologica della Venezia Giulia. *Bollettino della Società Entomologica Italiana* **60**: 128–130. [24.x.1928.]

Finzi, B. 1928c. In Finzi, B., Adensamer, W., Käufel, F., Strouhal, H. & Priesner, H. Weitere Beiträge zur Kenntnis der Fauna Griechenlands und der Inseln des Aegäischen Meeres. 1. Ameisen aus Griechenland und von den Aegäischen Inseln. *Sitzungsberichte der Akademie der Wissenschaften. Mathematisch-Naturwissenschaftliche Classe. Wien* **137**: 787–791. [(31.xii).1928.]

Finzi, B. 1929. Le forme italiane del genere *Messor. Bollettino della Società Entomologica Italiana* **61**: 75–94. [3.vi.1929.]

Finzi, B. 1930a. Contributo allo studio degli *Aphaenogaster* paleartici. *Bollettino della Società Entomologica Italiana* **62**: 151–156. [15.x.1930.]

Finzi, B. 1930b. Wissenschaftliche Ergebnisse einer zoologischen Forschungsreise nach Westalgerien und Marokko. 3. Ameisen aus Marokko und Westalgerien. *Sitzungsberichte der Akademie der Wissenschaften. Mathematisch-Naturwissenschaftliche Klasse. Wien* **139**: 14–16. [(31.xii).1930.]

Finzi, B. 1930c. Formicidae. In Ebner, R. Hymenopteren aus Palästina und Syrien. (Zoologische Studienreise von R. Ebner 1928 mit Unterstützung der Akademie der Wissenschaften in Wien.) *Sitzungsberichte der Akademie der Wissenschaften. Mathematisch-Naturwissenschaftliche Klasse. Wien* **139**: 22–24. [(31.xii).1930.]

Finzi, B. 1930d. Die Ameisen der Jonischen Inseln. In Beier, M. Zoologische Forschungsreise nach den Jonischen Inseln und dem Peloponnes. *Sitzungsberichte der Akademie der Wissenschaften. Mathematisch-Naturwissenschaftliche Klasse. Wien* **139**: 309–319. [(31.xii).1930.]

Finzi, B. 1932. Sopra alcune formiche dell'isola di Rodi. *Bollettino della Società Entomologica Italiana* **64**: 24–27. [15.ii.1932.]

Finzi, B. 1933. Raccolte entomologiche nell'Isola di Capraia fatte da C. Mancini e F. Capra (1927–1931). 2. Formicidae. *Memorie della Società Entomologica Italiana* **11**: 162–165. [25.i.1933.]

Finzi, B. 1936. Risultati scientifici della spedizione di S.A.S. il Principe Alessandro della Torre e Tasso nell'Egitto e Penisola del Sinai. *Bulletin de la Société Royale Entomologique d'Egypte* **20**: 155–210. [28.xii.1936.]

Finzi, B. 1939a. Materiali zoologici dell'Eritrea raccolti da G. Müller durante la spedizione dell'Istituto Sieroterapico Milanese e conservati al Museo di Trieste. Parte 3. Hymenoptera: Formicidae. *Atti del Museo Civico di Storia Naturale Trieste* **14**: 153–168. [31.i.1939.]

Finzi, B. 1939b. Quinto contributo alla conoscenza della fauna mirmecologica della Venezia Giulia. *Bollettino della Società Entomologica Italiana* **71**: 86–90. [5.vi.1939.]

Finzi, B. 1939c. Ergebnisse der von Franz Werner und Otto v. Wettstein auf den Ägäischen Inseln unternommenen Sammelreisen. *Sitzungsberichte der Akademie der Wissenschaften. Mathematisch-Naturwissenschaftliche Klasse. Wien* **148**: 153–161. [(31.xii).1939.]

Finzi, B. 1940. Formiche della Libia. *Memorie della Società Entomologica Italiana* **18** (1939): 155–166. [5.ii.1940.]

Fitch, A. 1854. First report on the noxious, beneficial and other insects of the State of New York made to the State Agricultural Society, pursuant to an approbation for this purpose from the Legislature of the State. *Transactions of the New York State Agricultural Society* **1854**: 705–880. [(31.xii).1854.]

Foerster, A. 1850a. *Hymenopterologische Studien* **1**. Formicariae: 74 pp. Aachen. [(31.xii).1850.]

Foerster, A. 1850b. Eine Centurie neuer Hymenopteren. Zweite Dekade. *Verhandlungen des Naturhistorischen Vereines der Preussischen Rheinlande und Westphalens* **7**: 485–500. [(31.xii).1850.]

Fonscolombe, B. de. 1846. Notes sur huit espèces nouvelles d'Hyménoptères et de Néuroptères, trouvées aux environs d'Aix. *Annales de la Société Entomologique de France* (2) **4**: 39–51. [8.vii.1846.]

Forbes, J. & Do-Van-Quy, D. 1965. The anatomy and histology of the male reproductive system of the legionary ant *Neivamyrmex harrisi* (Haldeman). *Journal of the New York Entomological Society* **73**: 95–111. [16.vi.1965.]

Forder, J.C. & Marsh, A.C. 1989. Social organisation and reproduction in *Ocymyrmex foreli. Insectes Sociaux* **36**: 106–115. [1989.]

Forel, A. 1870. Notices myrmécologiques. *Mitteilungen der Schweizerischen Entomologischen Gesellschaft* **3**: 306–312. [(31).x.1870.]

Forel, A. 1874. Les fourmis de la Suisse. Systématique. Notices anatomiques et physiologiques. Architecture. Distribution géographique. Nouvelles expériences et observations de moeurs. *Neue Denkschriften der allgemeinen Schweizerischen Gesellschaft für die gesammten Naturwissenschaften* **26**: 1–447. [(31.xii).1874.]

Forel, A. 1876. Etudes myrmécologiques en 1875. *Bulletin de la Société Vaudoise des Sciences Naturelles* **14** (1875): 33–62. [(28).ii.1876.]

Forel, A. 1878. Etudes myrmécologiques en 1878 (première partie) avec l'anatomie du gésier des fourmis. *Bulletin de la Société Vaudoise des Sciences Naturelles* **15**: 337–392. [(31).x.1878.]

Forel, A. 1879a. Etudes myrmécologiques en 1879 (deuxième partie [1re partie en 1878]). *Bulletin de la Société Vaudoise des Sciences Naturelles* **16**: 53–128. [(31).iii.1879.]

Forel, A. 1879b. *Aphaenogaster* (?) *schaufussi* n. sp. *Nunquam Otiosus:* 465–466. [(31.xii).1879.]

Forel, A. 1881. Die Ameisen der Antille St. Thomas. *Mittheilungen des Münchener Entomologischen Vereins* **5**: 1–16. [(30).iv.1881.]

Forel, A. 1885a. Etudes myrmécologiques en 1884; avec une description des organes sensoriels des antennes. *Bulletin de la Société Vaudoise des Sciences Naturelles* **20** (1884): 316–380. [(28).ii.1885.]

Forel, A. 1885b. Indian ants of the Indian Museum, Calcutta. *Journal of the Asiatic Society of Bengal. Part 2. Natural Science* **54**: 176–182. [29.xii.1885.]

Forel, A. 1886a. Einige Ameisen aus Itajahy (Brasilien). *Mittheilungen der Schweizerischen Entomologischen Gesellschaft* **7**: 210–217. [(31).i.1886.]

Forel, A. 1886b. Espèces nouvelles de fourmis Américaines. *Annales de la Société Entomologique de Belgique. Comptes- rendus* **30**: xxxvii-xlix. [6.ii.1886.]

Forel, A. 1886c. Diagnoses provisoires de quelques espèces nouvelles de fourmis de Madagascar, récoltées par M. Grandidier. *Annales de la Société Entomologique de Belgique. Comptes-rendus* **30**: ci-cvii. [1.v.1886.]

Forel, A. 1886d. Indian ants of the Indian Museum, Calcutta. No. 2. *Journal of the Asiatic Society of Bengal. Part 2. Natural Science* **55**: 239–249. [1.x.1886.]

Forel, A. 1886e. Nouvelles fourmis de Grèce récoltées par M.E. von Oertzen. *Annales de la Société Entomologique de Belgique. Comptes-rendus* **30**: clix-clxviii. [2.x.1886.]

Forel, A. 1886f. Etudes myrmécologiques en 1886. *Annales de la Société Entomologique de Belgique* **30**: 131–215. [6.xi.1886.]

Forel, A. 1887. Fourmis récoltées à Madagascar, par le Dr. Conrad Keller. *Mittheilungen der Schweizerischen Entomologischen Gesellschaft* **7**: 381–389. [(31).x.1887.]

Forel, A. 1889. Ameisen aus den Sporaden, den Cykladen und Griechenland, gesammelt 1887 von Herrn v. Oertzen. *Berliner Entomologische Zeitschrift* **32** (1888): 255–265. [(28).ii.1889.]

Forel, A. 1890a. Fourmis de Tunisie et de l'Algérie orientale récoltées et décrites par Auguste Forel. *Annales de la Société Entomologique de Belgique. Comptes-rendus* **34**: lxi–lxxvi. [5.iv.1890.]

Forel, A. 1890b. *Aenictus-Typhlatta* découverte de M. Wroughton. Nouveaux genres de formicides. *Annales de la Société Entomologique de Belgique. Comptes-rendus* **34**: cii-cxiv. [7.vi.1890.]

Forel, A. 1890c. Une nouvelle fourmi. *Le Naturaliste.* (*Revue Illustrée des Sciences Naturelles*) **12**: 217–220. [15.ix.1890.]

Forel, A. 1891a. Un nouveau genre de myrmicides. *Annales de la Société Entomologique de Belgique. Comptes-rendus* **35**: cccvii. [6.vi.1891.]

Forel, A. 1891b. In Grandidier, A. *Histoire Physique, Naturelle et Politique de Madagascar* **20**. Histoire naturelle des Hyménoptères. 2 (fascicule 28). Les Formicides: 1–231. Paris. [(31.xii).1891.]

Forel, A. 1892a. Notes myrmécologiques. *Annales de la Société Entomologique de Belgique* **36**: 38–43. [6.ii.1892.]

Forel, A. 1892b. Die Akazien-*Crematogaster* von Prof. Keller aus dem Somaliland. *Zoologischer Anzeiger* **15**: 140–143. [11.iv.1892.]

Forel, A. 1892c. Die Ameisen Neu-Seelands. *Mitteilungen der Schweizerischen Entomologischen Gesellschaft* **8**: 331–343. [(31).v.1892.]

Forel, A. 1892d. Attini und Cryptocerini. Zwei neue *Apterostigma*- Arten. *Mitteilungen der Schweizerischen Entomologischen Gesellschaft* **8**: 344–349. [(31).v.1892.]

Forel, A. 1892e. Liste der aus dem Somaliland von Hrn. Prof. Dr. Conr. Keller aus der Expedition des Prinzen Ruspoli im August und September 1891 zurückgebrachten Ameisen. *Mitteilungen der Schweizerischen Entomologischen Gesellschaft* **8**: 349–354. [(31).v.1892.]

Forel, A. 1892f. Critique de: Peter Cameron. Hymenoptera, Formicidae; extracted from supplementary appendix to travels amongst the Great Andes of the Equator by Edw. Whymper. London, 1891. *Annales de la Société Entomologique de Belgique* **36**: 255–256. [2.vii.1892.]

Forel, A. 1892g. Quelques fourmis de la faune méditerranéenne. *Annales de la Société Entomologique de Belgique* **36**: 452–457. [3.ix.1892.]

Forel, A. 1892h. Le mâle des *Cardiocondyla* et la reproduction consanguine perpétuée. *Annales de la Société Entomologique de Belgique* **36**: 458–461. [3.ix.1892.]

Forel, A. 1892i. Die Ameisenfauna Bulgariens. (Nebst biologischen Beobachtungen.) *Verhandlungen der k.k. Zoologisch- Botanischen Gesellschaft in Wien* **42**: 305–318. [(30).ix.1892.]

Forel, A. 1892j. Les formicides de l'Empire des Indes et de Ceylan. Part 1. *Journal of the Bombay Natural History Society* **7**: 219–245. [1.x.1892.]

Forel, A. 1892k. Nouvelles espèces de formicides de Madagascar. (Récoltées par M. Sikora.) *Annales de la Société Entomologique de Belgique* **36**: 516–535. [3.xii.1892.]

Forel, A. 1892l. In Grandidier, A. *Histoire Physique, Naturelle et Politique de Madagascar* **20**. Histoire naturelle des Hymémoptères. 2 (supplément au 28 fascicule). Les Formicides: 229–280. Paris. [(31.xii).1892.]

Forel, A. 1892m. Die Nester der Ameisen. *Neujahrsblatt herausgegeben von der Naturforschenden Gesellschaft auf das Jahr 1893, Zürich* **95**: 1–36. [(31.xii).1892.]

Forel, A. 1893a. Sur la classification de la famille des formicides, avec remarques synonymiques. *Annales de la Société Entomologique de Belgique* **37**: 161–167. [1.iv.1893.]

Forel, A. 1893b. Les formicides de l'Empire des Indes et de Ceylan. Part 2. Suite au genre *Camponotus. Journal of the Bombay Natural History Society* **7**: 430–439. [23.iv.1893.]

Forel, A. 1893c. Les formicides de l'Empire des Indes et de Ceylan. Part 3. *Journal of the Bombay Natural History Society* **8**: 17–36. [1.vii.1893.]

Forel, A. 1893d. Nouvelles fourmis d'Australie et des Canaries. *Annales de la Société Entomologique de Belgique* **37**: 454–466. [2.ix.1893.]

Forel, A. 1893e. Note sur les Attini. *Annales de la Société Entomologique de Belgique* **37**: 586–607. [2.xii.1893.]

Forel, A. 1893f. Note préventive sur un nouveau genre et une nouvelle espèce de formicide (camponotide). *Annales de la Société Entomologique de Belgique* **37**: 607–608. [2.xii.1893.]

Forel, A. 1893g. Formicides de l'Antille St. Vincent. Récoltées par Mons. H.H.Smith. *Transactions of the Entomological Society of London* **1893**: 333–418. [29.xii.1893. Date of publication from Wheeler, G. 1912: 759.]

Forel, A. 1894a. Algunas hormigas de Canarias recogidas por el Sr. Cabrera y Diaz. *Actas* [in *Anales*] *de la Sociedad Española de Historia Natural* (2) **2** [22]: 159–162. [31.i.1894.]

Forel, A. 1894b. Abessinische und andere afrikanische Ameisen, gesammelt von Herrn Ingenieur Alfred Ilg, von Herrn Dr. Liengme, von Herrn Pfarrer Missionar P. Berthoud, Herrn Dr. Arth. Müller, etc. *Mitteilungen der Schweizerischen Entomologischen Gesellschaft* **9**: 64–100. [(31).i.1894.]

Forel, A. 1894c. Les formicides de l'Empire des Indes et de Ceylan. Part 4. Adjonction aux genres *Camponotus* Mayr, et *Polyrhachis* Shuck. *Journal of the Bombay Natural History Society* **8**: 396–420. [1.ii.1894.]

Forel, A. 1894d. Les formicides de la province d'Oran (Algérie). *Bulletin de la Société Vaudoise des Sciences Naturelles* **30**: 1–45. [(31).iii.1894.]
Forel, A. 1894e. Quelques fourmis de Madagascar (récoltées par M. le Dr Völtzkow); de Nouvelle Zélande (récoltées par M. W.W. Smith); de Nouvelle Calédonie (récoltées par M. Sommer); de Queensland (Australie) (récoltées par M. Wiederkehr); et de Perth (Australie occidentale) (récoltées par M. Chase). *Annales de la Société Entomologique de Belgique* **38**: 226–237. [5.v.1894.]
Forel, A. 1895a. Nouvelles fourmis de diverses provenances, surtout d'Australie. *Annales de la Société Entomologique de Belgique* **39**: 41–49. [4.i.1895.]
Forel, A. 1895b. A fauna das formigas do Brazil. *Boletim do Museu Paraense de Historia Natural e Ethnographia* **1**: 89–143. [(30).iv. 1895.]
Forel, A. 1895c. Nouvelles fourmis de l'Imerina oriental (Moramanga etc.). *Annales de la Société Entomologique de Belgique* **39**: 243–251. [4.v.1895.]
Forel, A. 1895d. Südpalaearctische Ameisen. *Mittheilungen der Schweizerischen Entomologischen Gesellschaft* **9**: 227–234. [(31).v.1895.]
Forel, A. 1895e. Les formicides de l'Empire des Indes et de Ceylan. Part 5. Adjonction aux Camponotinae. *Journal of the Bombay Natural History Society* **9**: 453–472. [20.vi.1895.]
Forel, A. 1895f. Nouvelles fourmis d'Australie, récoltées a The Ridge, Mackay, Queensland par M. Gilbert Turner. *Annales de la Société Entomologique de Belgique* **39**: 417–428. [7.ix.1895.]
Forel, A. 1895g. Une nouvelle fourmis melligère. *Annales de la Société Entomologique de Belgique* **39**: 429–430. [7.ix.1895.]
Forel, A. 1895h. Quelques fourmis du centre de Madagascar. *Annales de la Société Entomologique de Belgique* **39**: 485–488. [7.xii.1895.]
Forel, A. 1895i. Ueber den Polymorphismus und Ergatomorphismus der Ameisen. Abtheilung für Entomologie (pp. 141–149). *Verhandlungen der Gesellschaft Deutscher Naturforscher und Arzte* (Versammlung zu Wien, 24–28 Sept., 1894) **66**: 1–324. [(31.xii).1895.]
Forel, A. 1896. In Wasmann, E. Die Ameisen- und Termitengäste von Brasilien. 1 Theil. Mit einem Anhange von Dr. August Forel. Anhang (pp. 178–179). Beschreibung einiger neuer brasilianischer Ameisenarten. *Verhandlungen der k.k. Zoologisch-Botanischen Gesellschaft in Wien* **45** (1895): 137–179. [(31.xii).1896.]
Forel, A. 1897a. Deux fourmis d'Espagne. *Annales de la Société Entomologique de Belgique* **41**: 132–133. [1.v.1897.]
Forel, A. 1897b. Quelques formicides de l'Antille de Grenada récoltés par M. H.H. Smith. *Transactions of the Entomological Society of London* **1897**: 297–300. [1.ix.1897.]
Forel, A. 1897c. Ameisen aus Nossi-Bé, Majunga, Juan de Nova (Madagaskar), den Aldabra-Inseln und Sansibar. Gesammelt von Herrn Dr. A. Voeltzkow aus Berlin. *Abhandlungen herausgegeben von der Senckenbergischen Naturforschenden Gesellschaft* **21**: 185–208. [(31.xii).1897.]
Forel, A. 1899a. *Fauna Hawaiiensis, or the Zoology of the Sandwich (Hawaiian) Isles* **1**. Hym. Aculeata: 115–122. Cambridge. [20.iii.1899.]
Forel, A. 1899b. Trois notices myrmécologiques. *Annales de la Société Entomologique de Belgique* **43**: 303–310. [28.vi.1899.]
Forel, A. 1899c. *Biologia Centrali-Americana;* or, contributions to the knowledge of the fauna and flora of Mexico and Central America. Insecta. Hymenoptera. **3** (Formicidae): 169 pp. London. [Pp. 1–24 dated (30).iv.1899; pp. 25–56 dated (30).vi.1899; pp. 57–80 dated (31).viii.1899; pp. 81–104 dated (30).ix.1899; pp. 105–136 dated (30).xi.1899; pp. 137–160 dated (31).xii.1899; pp. 161–169 (index) dated (30).iv.1900.]
Forel, A. 1899d. Von Ihrer Königl. Hoheit der Prinzessin Therese von Bayern auf einer Reise in Südamerika gesammelte Insecten. 1. Hymenoptera. a. Fourmis. *Berliner Entomologische Zeitschrift* **44**: 273–276. [(31).xii.1899.]
Forel, A. 1900a. Un nouveau genre et une nouvelle espèce de myrmicide. *Annales de la Société Entomologique de Belgique* **44**: 24–26. [30.i.1900.]
Forel, A. 1900b. Ponerinae et Dorylinae d'Australie. Récoltés par MM. Turner, Froggatt, Nugent, Chase, Rothney, J.-J. Walker, etc. *Annales de la Société Entomologique de Belgique* **44**: 54–77. [9.iii.1900.]
Forel, A. 1900c. Les formicides de l'Empire des Indes et de Ceylan. Part 6. 3me sous famille Ponerinae. *Journal of the Bombay Natural History Society* **13**: 52–65. [15.iv.1900.]
Forel, A. 1900d. Les formicides de l'Empire des Indes et de Ceylan. Part 7. *Journal of the Bombay Natural History Society* **13**: 303–332. [29.vii.1900.]
Forel, A. 1900e. Fourmis du Japon. Nids en toile. *Strongylognathus huberi* et voisins. Fourmilière triple. *Cyphomyrmex wheeleri.* Fourmis importées. *Mittheilungen der Schweizerischen Entomologischen Gesellschaft* **10**: 267–287. [(31).x.1900.]
Forel, A. 1901a. Les formicides de l'Empire des Indes et de Ceylan. Part 8. Sous famille Dorylinae. *Journal of the Bombay Natural History Society* **13**: 462–477. [20.i.1901.]
Forel, A. 1901b. Formiciden aus dem Bismarck-Archipel, auf Grundlage des von Prof. Dr. F. Dahl gesammelten Materials bearbeitet. *Mitteilungen aus dem Zoologischen Museum in Berlin* **2**: 1–38. [3.iv.1901.]
Forel, A. 1901c. 1, Fourmis mexicaines. Récoltées par M. le Professeur W.M. Wheeler. 2, A propos de la classification des fourmis. *Annales de la Société Entomologique de Belgique* **45**: 123–141. [30.iv.1901.]
Forel, A. 1901d. Einige neue Ameisen aus Südbrasilien, Java, Natal und Mossamedes. *Mittheilungen der Schweizerischen Entomologischen Gesellschaft* **10**: 297–311. [(30).vi.1901.]
Forel, A. 1901e. Variétés myrmécologiques. *Annales de la Société Entomologique de Belgique* **45**: 334–382. [5.xii.1901.]
Forel, A. 1901f. Nouvelles espèces de Ponerinae. (Avec un nouveau sous-genre et une espèce nouvelle d'*Eciton*.) *Revue Suisse de Zoologie* **9**: 325–353. [18.xii.1901.]
Forel, A. 1901g. Fourmis termitophages, lestobiose, *Atta tardigrada,* sous-genres d'*Euponera*. *Annales de la Société Entomologiques de Belgique* **45**: 389–398. [30.xii.1901.]
Forel, A. 1901h. Formiciden des Naturhistorischen Museums zu Hamburg. Neue *Calyptomyrmex, Dacryon, Podomyrma,* und *Echinopla*-Arten. *Jahrbuch der Hamburgischen Wissenschaftlichen Anstalten [Mitteilungen aus dem Naturhistorischen Museum]* **18**: 45–82. [(31.xii).1901.]
Forel, A. 1902a. Les fourmis du Sahara Algérien. Récoltées par M. le Professeur A. Lameere et le Dr. A. Diehl. *Annales de la Société Entomologique de Belgique* **46**: 147–158. [1.v.1902.]
Forel, A. 1902b. Quatre notices myrmécologiques. *Annales de la Société Entomologique de Belgique* **46**: 170–182. [1.v.1902.]
Forel, A. 1902c. Myrmicinae nouveaux de l'Inde et de Ceylan. *Revue Suisse de Zoologie* **10**: 165–249. [30.vi.1902.]
Forel, A. 1902d. Variétés myrmécologique. *Annales de la Société Entomologique de Belgique* **46**: 284–296. [3.vii.1902.]
Forel, A. 1902e. In Wasmann, E. Neues über die zusammengesetzten Nester und gemischten Kolonien der Ameisen. (Fortsetzung aus No. 14/15.) *Allgemeine Zeitschrift für Entomologie* **7**: 293–298. [15.viii.1902.]
Forel, A. 1902f. Les formicides de l'Empire des Indes et de Ceylan. Part 9. Sous famille Myrmicinae. *Journal of the Bombay Natural History Society* **14**: 520–546. [18.x.1902.]
Forel, A. 1902g. Fourmis d'Algérie. Récoltées par M. le Dr K. Escherich. *Annales de la Société Entomologique de Belgique* **46**: 462–463. [4.xii.1902.]
Forel, A. 1902h. Fourmis nouvelles d'Australie. *Revue Suisse de Zoologie* **10**: 405–548. [30.xii.1902.]
Forel, A. 1902i. Descriptions of some ants from the Rocky Mountains of Canada (Alberta and British Columbia). Collected by Edward Whymper. *Transactions of the Entomological Society of London* **1902**: 699–700. [30.xii.1902.]
Forel, A. 1903a. Les formicides de l'Empire des Indes et de Ceylan. Part 10. *Journal of the Bombay Natural History Society* **14**: 679–715. [10.ii.1903.]

Forel, A. 1903b. Die Sitten und Nester einiger Ameisen der Sahara bei Turgut und Biskra. *Mittheilungen der Schweizerischen Entomologischen Gesellschaft* **10**: 453–459. [(28).ii.1903.]

Forel, A. 1903c. Mélanges entomologiques, biologiques et autres. *Annales de la Société Entomologique de Belgique* **47**: 249–268. [2.vii.1903.]

Forel, A. 1903d. Les fourmis des îles Andamans et Nicobars. *Revue Suisse de Zoologie* **11**: 399–411. [5.ix.1903.]

Forel, A. 1903e. Einige neue Ameisen aus Süd-Angola (pp. 559–564). In Baum, H. *Kunene-Sambesi-Expedition, 1903:* 593 pp. Berlin. [(31.xii).1903.]

Forel, A. 1904a. Fourmis de British Columbia. *Annales de la Société Entomologique de Belgique* **48**: 152–155. [29.iii.1904.]

Forel, A. 1904b. Note sur les fourmis du Musée Zoologique de l'Academie des Sciences à St. Pétersbourg. *Ezhegodnik Zoologicheskago Muzeya Imperatorskoi Akademii Nauk* **8** (1903): 368–389. [(31).iii.1904.]

Forel, A. 1904c. Miscellanea myrmécologiques. *Revue Suisse de Zoologie* **12**: 1–52. [18.iv.1904.]

Forel, A. 1904d. Fourmis du Musée de Bruxelles. *Annales de la Société Entomologique de Belgique* **48**: 168–177. [4.v.1904.]

Forel, A. 1904e. In und mit Pflanzen lebende Ameisen aus dem Amazonas-Gebiet und aus Peru, gesammelt von Herrn E. Ule. *Zoologische Jahrbücher. Abtheilung für Systematik, Geographie und Biologie der Thiere* **20**: 677–707. [31.viii.1904.]

Forel, A. 1904f. Dimorphisme du mâle chez les fourmis et quelques autres notices myrmécologiques. *Annales de la Société Entomologique de Belgique* **48**: 421–425. [23.xii.1904.]

Forel, A. 1904g. Formicidae. *Ergebnisse der Hamburger Magalhaensischen Sammelreise* **7** (No. 8): 1–7. [(31.xii).1904.]

Forel, A. 1904h. In Wasmann, E. Termitophilen aus dem Sudan. In Jägerskiöld, L.A. *Results of the Swedish Zoological Expedition to Egypt and the White Nile, 1901* **1** (No. 13): 21 pp. Uppsala. [(31.xii).1904.]

Forel, A. 1905a. A revision of the species of Formicidae (ants) of New Zealand. *Transactions and Proceedings of the New Zealand Institute* **37** (1904): 353–355. [(30).vi.1905.]

Forel, A. 1905b. Miscellanea myrmécologiques, 2. (1905.) *Annales de la Société Entomologique de Belgique* **49**: 155–185. [31.viii.1905.]

Forel, A. 1905c. Ameisen aus Java. Gesammelt von Prof. Karl Kraepelin, 1904. *Jahrbuch der Hamburgischen Wissenschaftlichen Anstalten [Mitteilungen aus dem Naturhistorischen Museum]* **22**: 1–26. [(31.xii).1905.]

Forel, A. 1906a. Moeurs des fourmis parasites des genres *Wheeleria* et *Bothriomyrmex*. *Revue Suisse de Zoologie* **14**: 51–69. [28.iii.1906.]

Forel, A. 1906b. Les fourmis de l'Himalaya. *Bulletin de la Société Vaudoise des Sciences Naturelles* **42**: 79–94. [(31).iii.1906.]

Forel, A. 1906c. Fourmis d'Asie Mineure et de la Dobrudscha récoltées par M. le Dr Oscar Vogt et Mme Cécile Vogt, Dr méd. *Annales de la Société Entomologique de Belgique* **50**: 187–190. [26.vi.1906.]

Forel, A. 1906d. Fourmis Néotropiques nouvelles ou peu connues. 1. Genres *Pseudomyrma, Cryptocerus* et *Azteca*. 2. Fourmis d'autres genres récoltées par M. Fiebrig, à San Bernardino (Paraguay). *Annales de la Société Entomologique de Belgique* **50**: 225–245 (sect. 1.); 246–249 (sect 2.). [30.viii.1906.]

Forel, A. 1907a. Formicides du Musée National Hongrois. *Annales Historico-Naturales Musei Nationalis Hungarici* **5**: 1–42. [30.vi.1907.]

Forel, A. 1907b. Fourmis nouvelles de Kairouan et d'Orient. *Annales de la Société Entomologique de Belgique* **51**: 201–208. [2.viii.1907.]

Forel, A. 1907c. Fourmis d'Ethiopie. Récoltées par M. le baron Maurice de Rothschild en 1905. *Revue d'Entomologie* **26**: 129–144. [(30).ix.1907.]

Forel, A. 1907d. The Percy Sladen Trust Expedition to the Indian Ocean in 1905, under the leadership of Mr. J. Stanley Gardiner, M.A. Fourmis des Seychelles, Amirantes, Farquhar et Chagos. *Transactions of the Linnean Society of London* (2) **12**: 91–94. [(30).ix. 1907.]

Forel, A. 1907e. Formiciden aus dem Naturhistorischen Museum in Hamburg. *Jahrbuch der Hamburgischen Wissenschaftlichen Anstalten [Mitteilungen aus dem Naturhistorischen Museum]* **24**: 1–20. [15.x.1907.]

Forel, A. 1907f. Nova speco kaj nova gentonomo de formikoj. *Internacia Scienca Revuo* **4**: 144–145. [(31.xii).1907.]

Forel, A. 1907g. Ameisen von Madagaskar, den Comoren und Ostafrika. In Voeltzkow, A. *Reise in Ostafrika in den Jahren 1903–1905 mit Mitteln der Hermann und Elise geb. Heckmann Wentzel-Stiftung ausgeführt von Professor Dr. Alfred Voeltzkow. Wissenschaftliche Ergebnisse* **2**. Systematische Arbeiten: 75–92. [(31.xii).1907.]

Forel, A. 1907h. Formicidae. In Michaelsen, W. & Hartmeyer, R. *Die Fauna Südwest-Australiens.* Ergebnisse der Hamburger südwest-australischen Forschungsreise 1905. **1**: 263–310. Jena. [(31.xii).1907.]

Forel, A. 1908a. Fourmis de Ceylan et d'Egypte, récoltées par le Prof. E. Bugnion. *Bulletin de la Société Vaudoise des Sciences Naturelles* **44**: 1–22. [(31).iii.1908.]

Forel, A. 1908b. Fourmis de Costa-Rica, récoltées par M. Paul Biolley. *Bulletin de la Société Vaudoise des Sciences Naturelles* **44**: 35–72. [(31).iii.1908.]

Forel, A. 1908c. Ameisen aus Sao Paulo (Brasilien), Paraguay, etc. Gesammelt von Prof. Herm. v. Ihering, Dr. Lutz, Dr. Fiebrig, etc. *Verhandlungen der k.k. Zoologisch-Botanischen Gesellschaft in Wien* **58**: 340–418. [18.xii.1908.]

Forel, A. 1908d. Formiciden (p. 105). In *Wissenschaftlichen Ergebnisse der Expedition Filchner nach China und Tibet, 1903–05* **10**. Zoologische Sammlungen 1: 244 pp. Berlin. [(31.xii).1908.]

Forel, A. 1908e. Catalogo sytematico da colleçao de formigas do Ceará. *Boletim do Museu Rocha, Fortaleza* **1**: 61–69. [1908. Not seen. Reference from Kempf, 1972a: 278.]

Forel, A. 1909a. Ameisen aus Guatemala usw., Paraguay und Argentinien. *Deutsche Entomologische Zeitschrift* **1909**: 239–269. [1.iii.1909.]

Forel, A. 1909b. Fourmis du Musée de Bruxelles. Fourmis de Benguela récoltées par M. Creighton Wellman, et fourmis du Congo récoltées par MM. Luja, Kohl et Laurent. *Annales de la Société Entomologique de Belgique* **53**: 51–73. [5.iii.1909.]

Forel, A. 1909c. Fourmis d'Espagne. Récoltées par M. O. Vogt et Mme Cécile Vogt, docteurs en médecine. *Annales de la Société Entomologique de Belgique* **53**: 103–106. [2.iv.1909.]

Forel, A. 1909d. Ameisen aus Java und Krakatau beobachtet und gesammelt von Herrn Edward Jacobson. 1. Systematischen Thiel. *Notes from the Leyden Museum* **31**: 221–232. [20.xii.1909.]

Forel, A. 1909e. Etudes myrmécologiques en 1909. Fourmis de Barbarie et de Ceylan. Nidification des *Polyrhachis. Bulletin de la Société Vaudoise des Sciences Naturelles* **45**: 369–407. [(31).xii.1909.]

Forel, A. 1910a. Glanures myrmécologiques. *Annales de la Société Entomologique de Belgique* **54**: 6–32. [4.ii.1910.]

Forel, A. 1910b. Formicides australiens reçus de MM. Froggatt et Rowland Turner. *Revue Suisse de Zoologie* **18**: 1–94. [22.iii.1910.]

Forel, A. 1910c. Ameisen aus der Kolonie Erythräa. Gesammelt von Prof. Dr. K. Escherich (nebst einigen in West-Abessinien von Herrn A. Ilg gesammelten Ameisen). *Zoologische Jahrbücher. Abteilung für Systematik, Geographie und Biologie der Tiere* **29**: 243–274. [22.vii.1910.]

Forel, A. 1910d. Fourmis des Philippines. *Philippine Journal of Science* **5** (sect. D): 121–130. [(31).vii.1910.]

Forel, A. 1910e. Note sur quelques fourmis d'Afrique. *Annales de la Société Entomologique de Belgique* **54**: 421–458. [2.xii.1910.]

Forel, A. 1910f. In Schultze, L. *Zoologische und anthropologische Ergebnisse einer Forschungsreise im westlichen und zentralen Südafrika,* ausgeführt in den Jahren 1903–1905 mit Unterstützung der Kgl. Preussischen Akademie der Wissenschaften zu Berlin **4**. Formicidae. (Fortsetzung von Bd. 2. X. Insecta (3te serie)): 1–30. Jena. [(31.xii).1910. Note. Internal evidence provided by the ant names indicates that this paper was published before 1910e.]

Forel, A. 1911a. Fourmis de Bornéo, Singapore, Ceylan, etc. *Revue Suisse de Zoologie* **19**: 23–62. [(31).i.1911.]

Forel, A. 1911b. Ameisen aus Java beobachtet und gesammelt von Herrn Edward Jacobson. 2 Theil. *Notes from the Leyden Museum* **33**: 193–218. [29.iv.1911.]

Forel, A. 1911c. Ameisen des Herrn Prof. v. Ihering aus Brasilien (Sao Paulo usw.) nebst einigen anderen aus Südamerika und Afrika. *Deutsche Entomologische Zeitschrift* **1911**: 285–312. [4.v.1911.]

Forel, A. 1911d. Fourmis nouvelles ou intéressantes. *Bulletin de la Société Vaudoise des Sciences Naturelles* **47**: 331–400. [(30).vi.1911.]

Forel, A. 1911e. Die Ameisen des K. Zoologischen Museums in München. *Sitzungsberichte der Königlich Bayerischen Akademie der Wissenschaften Mathematisch-Physikalische Klasse* **1911**: 249–303. [(31).vii.1911.]

Forel, A. 1911f. Fourmis d'Afrique et d'Asie. 1. Fourmis d'Afrique surtout du Musée du Congo Belge. *Revue Zoologique Africaine* **1**: 274–283. [31.viii.1911.]

Forel, A. 1911g. Fourmis d'Afrique et d'Asie. Annexe. 2. Quelques fourmis d'Asie. *Revue Zoologique Africaine* **1**: 284–286. [31.viii.1911.]

Forel, A. 1911h. Sur le genre *Metapone* n. g. nouveau groupe des formicides et sur quelques autres formes nouvelles. *Revue Suisse de Zoologie* **19**: 445–459. [(31).xii.1911.]

Forel, A. 1911i. Ameisen aus Ceylon, gesammelt von Prof. K. Escherich (einige von Prof. E. Bugnion). In Escherich, K. *Termitenleben auf Ceylon:* 213–228. Jena. [(31.xii).1911.]

Forel, A. 1912a. H. Sauter's Formosa-Ausbeute: Formicidae. *Entomologische Mitteilungen* **1**: 45–81. [1.iii.1912.]

Forel, A. 1912b. Einige interessante Ameisen des Deutschen Entomologischen Museums zu Berlin-Dahlem. *Entomologische Mitteilungen* **1**: 81–83. [1.iii.1912.]

Forel, A. 1912c. Formicides Néotropiques. Part 1. *Annales de la Société Entomologique de Belgique* **56**: 28–49. [2.iii.1912.]

Forel, A. 1912d. Ameisen aus Java beobachtet und gesammelt von Edward Jacobson. 3 Theil. *Notes from the Leyden Museum* **34**: 97–112. [1.iv.1912.]

Forel, A. 1912e. Formicides Néotropiques. Part 2. 3me sous- famille Myrmicinae Lep. (Attini, Dacetii, Cryptocerini). *Mémoires de la Société Entomologique de Belgique* **19**: 179–209. [15.iv.1912.]

Forel, A. 1912f. Formicides Néotropiques. Part 3. 3me sous- famille Myrmicinae (suite). Genres *Cremastogaster* et *Pheidole*. *Mémoires de la Société Entomologique de Belgique* **19**: 211–237. [15.iv.1912.]

Forel, A. 1912g. Formicides Néotropiques. Part 4. 3me sous- famille Myrmicinae Lep. (suite). *Mémoires de la Société Entomologique de Belgique* **20**: 1–32. [5.vi.1912.]

Forel, A. 1912h. Formicides Néotropiques. Part 5. 4me sous- famille Dolichoderinae Forel. *Mémoires de la Société Entomologique de Belgique* **20**: 33–58. [10.vi.1912.]

Forel, A. 1912i. Formicides Néotropiques. Part 6. 5me sous- famille Camponotinae Forel. *Mémoires de la Société Entomologique de Belgique* **20**: 59–92. [20.vi.1912.]

Forel, A. 1912j. Die Weibchen der "Treiberameisen" *Anomma nigricans* Illiger und *Anomma wilverthi* Emery, nebst einigen anderen Ameisen aus Uganda. *Jahrbuch der Hamburgischen Wissenschaftlichen Anstalten [Mitteilungen aus dem Naturhistorischen Museum]* **29**: 173–180. [15.viii.1912.]

Forel, A. 1912k. The Percy Sladen Trust Expedition to the Indian Ocean in 1905, under the leadership of Mr. J. Stanley Gardiner, M.A. Volume 4, No. 11. Fourmis des Seychelles et des Aldabras, reçues de M. Hugh Scott. *Transactions of the Linnean Society of London Zoology* (2) **15**: 159–167. [(30).ix.1912.]

Forel, A. 1912l. Quelques fourmis de Tokio. *Annales de la Société Entomologique de Belgique* **56**: 339–342. [3.x.1912.]

Forel, A. 1912m. Descriptions provisoires de genres, sous-genres et espèces de formicides des Indes orientales. *Revue Suisse de Zoologie* **20**: 761–774. [(31).xii.1912.]

Forel, A. 1912n. Einige neue und interessante Ameisenformen aus Sumatra etc. *Zoologische Jahrbücher Supplement* **15**: 51–78. [(31.xii).1912.]

Forel, A. 1913a. Fourmis de Rhodesia, etc. récoltées par M. G. Arnold, le Dr H. Brauns et K. Fikendey. *Annales de la Société Entomologique de Belgique* **57**: 108–147. [2.v.1913.]

Forel, A. 1913b. Formicides du Congo Belge. Récoltés par MM. Bequaert, Luja, etc. *Revue Zoologique Africaine* **2**: 306–351. [30.v.1913.]

Forel, A. 1913c. Fourmis de Nigérie. *Revue Zoologique Africaine* **2**: 352–353. [30.v.1913.]

Forel, A. 1913d. Fourmis de la faune méditerranéenne récoltées par MM. U. et J. Sahlberg. *Revue Suisse de Zoologie* **21**: 427–438. [(30).vi.1913.]

Forel, A. 1913e. Quelques fourmis des Indes, du Japon et d'Afrique. *Revue Suisse de Zoologie* **21**: 659–673. [(30).ix.1913.]

Forel, H. 1913f. H. Sauter's Formosa-Ausbeute: Formicidae 2. *Archiv für Naturgeschichte* **79** (A): 183–202. [(30).ix.1913.]

Forel, A. 1913g. Fourmis de Tasmanie et d'Australie récoltées par MM. Lea, Froggatt, etc. *Bulletin de la Société Vaudoise des Sciences Naturelles* **49**: 173–195. [(30).ix.1913.]

Forel, A. 1913h. Quelques fourmis du Musée du Congo Belge. *Annales de la Société Entomologique de Belgique* **57**: 347–359. [5.xii.1913.]

Forel, A. 1913i. Notes sur quelques *Formica*. *Annales de la Société Entomologique de Belgique* **57**: 360–361. [5.xii.1913.]

Forel, A. 1913j. Ameisen aus Rhodesia, Kapland usw. Gesammelt von Herrn G. Arnold, Dr. H. Brauns und anderen. *Deutsche Entomologische Zeitschrift* **1913** (Beiheft): 203–225. [15.xii.1913.]

Forel, A. 1913k. Wissenschaftliche Ergebnisse einer Forschungsreise nach Ostindien, ausgeführt im Auftrage der Kgl. Preuss. Akademie der Wissenschaften zu Berlin von H. v. Buttel-Reepen. 2. Ameisen aus Sumatra, Java, Malacca und Ceylon. Gesammelt von Herrn Prof. Dr. v. Buttel-Reepen in den Jahren 1911–1912. *Zoologische Jahrbücher. Abteilung für Systematik, Geographie und Biologie der Tiere* **36**: 1–148. [19.xii.1913.]

Forel, A. 1913l. Fourmis d'Argentine, du Brésil, du Guatémala & de Cuba. Reçues de MM. Bruch, Prof. v. Ihering, Mlle Baez, M. Peper et M. Rovereto. *Bulletin de la Société Vaudoise des Sciences Naturelles* **49**: 203–250. [(31).xii.1913.]

Forel, A. 1914a. Le genre *Camponotus* Mayr et les genres voisins. *Revue Suisse de Zoologie* **22**: 257–276. [(31).v.1914.]

Forel, A. 1914b. *Deux nouveautés myrmécologiques:* p. 1. (Edition de l'auteur.) [1.ix.1914.]

Forel, A. 1914c. Einige amerikanische Ameisen. *Deutsche Entomologische Zeitschrift* **1914**: 615–620. [10.xii.1914.]

Forel, A. 1914d. Formicides d'Afrique et d'Amerique nouveaux ou peu connus. *Bulletin de la Société Vaudoise des Sciences Naturelles* **50**: 211–288. [(31).xii.1914.]

Forel, A. 1914e. Quelques fourmis de Colombie (pp. 9–14). In Fuhrmann, O. & Mayor, E. Voyage d'Exploration scientifique en Colombie. *Mémoires de la Société Neuchâteloise des Sciences Naturelles* **5**: 1090 pp. [(31.xii).1914.]

Forel, A. 1915a. Fauna Simalurensis. Hymenoptera Aculeata, Fam. Formicidae. *Tijdschrift voor Entomologie* **58**: 22–43. [15.iii.1915.]

Forel, A. 1915b. Results of Dr. E. Mjöberg's Swedish scientific expeditions to Australia, 1910–1913. 2. Ameisen. *Arkiv för Zoologi* **9** (16): 1–119. [16.vi.1915.]

Forel, A. 1915c. Formicides d'Afrique et d'Amerique nouveaux ou peu connus. 2e partie. *Bulletin de la Société Vaudoise des Sciences Naturelles* **50**: 335–364. [(30).vi.1915.]

Forel, A. 1915d. Fauna insectorum helvetiae. Hymenoptera. Formicidae. Die Ameisen ser Schweiz. *Mitteilungen der Schweizerischen Entomologischen Gesellschaft. Beilage zu Heft 7/8 des 12 Bandes:* 77 pp. Dübendorf. [(31.xii).1915.]

Forel, A. 1916. Fourmis du Congo et d'autres provenances récoltées par MM. Hermann Kohl, Luja, Mayné, etc. *Revue Suisse de Zoologie* **24**: 397–460. [(30).iv.1916.]

Forel, A. 1917. Cadre synoptique actuel de la faune universelle des fourmis. *Bulletin de la Société Vaudoise des Sciences Naturelles* **51**: 229–253. [8.iii.1917.]

Forel, A. 1918a. Etudes myrmécologiques en 1917. *Bulletin de la Société Vaudoise des Sciences Naturelles* **51** (1917): 717–727. [5.iv.1918.]

Forel, A. 1918b. Quelques fourmis de Madagascar récoltées par le Dr Friederichs et quelques remarques sur d'autres fourmis. *Bulletin de la Société Vaudoise des Sciences Naturelles* **52**: 151–156. [2.x.1918.]

Forel, A. 1920. Deux fourmis nouvelles du Congo. *Bulletin de la Société Vaudoise des Sciences Naturelles* **52**: 479–481. [16.ii.1920.]

Forel, A. 1921a. Fourmis trouvées dans les galles de *Cordia* et d'*Agonandra*. *Bulletin de la Société Botanique de Genève* (2) **12** (1920): 201–208. [31.viii.1921.]

Forel, A. 1921b. Quelques fourmis des environs de Quito (Ecuador), récoltées par Mme Eléonore Naumann. *Bulletin de la Société Vaudoise des Sciences Naturelles* **54**: 131–135. [15.xi.1921.]

Forel, A. 1922. Glanures myrmécologiques en 1922. *Revue Suisse de Zoologie* **30**: 87–102. [(30).ix.1922.]

Forskål, P. 1775. *Descriptiones animalium, avium, amphibiorum, piscium, insectorum, vermium; quae in itinere orientali observavit Petrus Forskål.* Post mortem auctoris edidit Carsten Niebuhr: xxxiv + 164 pp. Hauniae. [(31.xii).1775.]

Förster, A. See under Foerster, A.

Förster, B. 1891. Die Insekten des "Plattigen Steinmergels" von Brunnstatt. *Abhandlungen zur Geologischen Specialkarte von Elsass-Lothringen* **3**: 333–594. [(31.xii).1891.]

Fowler, H.G. 1982. A new species of *Trachymyrmex* fungus-growing ant from Paraguay. *Journal of the New York Entomological Society* **90**: 70–73. [7.vi.1982.]

Fowler, H.G. 1988. Taxa of the Neotropical grass-cutting ants, *Acromyrmex (Moellerius)*. *Científica, Sao Paulo* **16**: 281–295. [(31.xii).1988.]

Francke, O.F. & Merickel, F.W. 1982. Two new species of *Pogonomyrmex* harvester ants from Texas. *Pan-Pacific Entomologist* **57** (1981): 371–379. [15.iii.1982.]

Francoeur, A. 1968. Une nouvelle espèce du genre *Myrmica* au Québec. *Naturaliste Canadien* **95**: 727–730. [(30).vi.1968.]

Francoeur, A. 1973. Révision taxonomique des espèces Néarctiques du groupe *fusca*, genre *Formica*. *Mémoires de la Société Entomologique du Québec* **3**: 1–316. [(30).ix.1973.]

Francoeur, A. 1977a. The taxonomic status and biogeographic significance of the Sumatran *Formica*. *Psyche* **84**: 11–12. [30.xi.1977.]

Francoeur, A. 1977b. Synopsis taxonomique et économique des fourmis du Québec. *Annales de la Société Entomologique de Québec* **22**: 205–212. [(30).xi.1977.]

Francoeur, A. 1979. Les fourmis du Québec. 1. Introduction. 2. La famille des Formicidae. 3. La sous-famille des Ponerinae. *Annales de la Société Entomologique de Québec* **24**: 48–64 (alternative pagination: 12–47). [(31).i.1979.]

Francoeur, A. 1981. Le groupe néarctique *Myrmica lampra*. *Canadian Entomologist* **113**: 755–759. [23.x.1981.]

Francoeur, A. 1986a. Deux nouvelles fourmis néarctiques: *Leptothorax retractus* et *L. sphagnicolus*. *Canadian Entomologist* **118**: 1151–1164. [(30).xi.1986.]

Francoeur, A. 1986b. Supplément aux fourmis de la presqu'Ile de Forillon, Gaspé-est, Québec. *Fabreries* **12**: 56–58. [(31.xii).1986.]

Francoeur, A. & Beique, R. 1966. Les formicides de Provancher. *Canadian Entomologist* **98**: 140–145. [9.ii.1966.]

Francoeur, A. & Elias, S.A. 1985. *Dolichoderus taschenbergi* Mayr from an early holocene fossil insect assemblage in the Colorado front range. *Psyche* **92**: 303–308. [28.xi.1985.]

Francoeur, A. & Loiselle, R. 1984. Description du mâle et notice sur la biologie de *Myrmica quebecensis* Francoeur. *Revue d'Entomologie du Québec* **29**: 3–11. [(31).i.1984.]

Francoeur, A. & Loiselle, R. 1988a. Evolution du strigile chez les formicides. *Naturaliste Canadien* **115**: 333–353. [(31).xii.1988.]

Francoeur, A. & Loiselle, R. 1988b. The male of *Leptothorax wilda* with notes on the subgenus *Nesomyrmex* (pp. 43–54). In Trager, J.C. (Ed.). *Advances in Myrmecology*: 551 pp. New York. [(31.xii).1988.]

Francoeur, A., Loiselle, R. & Buschinger, A. 1985. Biosystématique de la tribu Leptothoracini. 1. Le genre *Formicoxenus* dans la région holarctique. *Naturaliste Canadien* **112**: 343–403. [(30).ix.1985.]

Francoeur, A. & Snelling, R.R. 1979. Notes for a revision of the ant genus *Formica*. 2. Reidentifications for some specimens from the T.W. Cook collection and new distribution data. *Contributions in Science. Natural History Museum of Los Angeles County* **309**: 1–7. [16.iii.1979.]

Froggatt, W.W. 1896. Honey Ants. In Spencer, B. (Ed.). *Report on the Work of the Horn Scientific Expedition to Central Australia* **2** Zoology: 385–392. Melbourne. [(28).ii.1896.]

Fromantin, J. & Soulié, J. 1961. Notes systématiques sur le genre *Crematogaster* avec description de trois espèces nouvelles du Cambodge. *Bulletin de la Société d'Histoire Naturelle de Toulouse* **96**: 87–112. [30.v.1961.]

Fujiyama, I. 1970. Fossil insects from the Chôjabaru Formation, Iki Island, Japan. *Memoirs of the National Science Museum* **3**: 65–74. [20.x.1970.]

Gaige, S.M. 1914. Description of a new subspecies of *Pogonomyrmex occidentalis* Cresson from Nevada. *Proceedings of the Biological Society of Washington* **27**: 93–96. [11.v.1914.]

Gallardo, A. 1915. Observaciones sobre algunas hormigas de la República Argentina. *Anales del Museo Nacional de Historia Natural de Buenos Aires* **27**: 1–35. [27.v.1915.]

Gallardo, A. 1916a. Las hormigas de la República Argentina. Subfamilia dolicoderinas. *Anales del Museo Nacional de Historia Natural de Buenos Aires* **28**: 1–130. [Pp. 1–128 dated 18.v.1916; pp. 129–130 dated 13.vi.1916.]

Gallardo, A. 1916b. Notas acerca de la hormiga *Trachymyrmex pruinosus* Emery. *Anales del Museo Nacional de Historia Natural de Buenos Aires* **28**: 241–252. [31.viii.1916.]

Gallardo, A. 1916c. Notas complementarias sobre las dolicoderinas argentinas. *Anales del Museo Nacional de Historia Natural de Buenos Aires* **28**: 257–261. [4.ix.1916.]

Gallardo, A. 1916d. Notes systématiques et éthologiques sur les fourmis attines de la République Argentine. *Anales del Museo Nacional de Historia Natural de Buenos Aires* **28**: 317–344. [25.ix.1916.]

Gallardo, A. 1917. Notes critiques sur les "Formicides sud- américains nouveaux ou peu connus du Docteur Santschi." *Physis. Revista de la Sociedad Argentina de Ciencias Naturales* **3**: 48–51. [17.iii.1917.]

Gallardo, A. 1918a. Hormigas dolicoderinas de los Andes de Mendoza. *Physis. Revista de la Sociedad Argentina de Ciencias Naturales* **4**: 28–31. [15.v.1918.]

Gallardo, A. 1918b. Las hormigas de la República Argentina. Subfamilia Ponerinas. *Anales del Museo Nacional de Historia Natural de Buenos Aires* **30**: 1–112. [15.xii.1918.]

Gallardo, A. 1919a. Una nueva prodorilina *Acanthostichus afflictus*. *Anales del Museo Nacional de Historia Natural de Buenos Aires* **30**: 237–242. [22.vii.1919.]

Gallardo, A. 1919b. Hormigas del Neuquén y Rio Negro. *Anales del Museo Nacional de Historia Natural de Buenos Aires* **30**: 243–254. [2.viii.1919.]

Gallardo, A. 1920. Las hormigas de la República Argentina. Subfamilia dorilinas. *Anales del Museo Nacional de Historia Natural de Buenos Aires* **30**: 281–410. [Pp. 281–408 dated 28.ii.1920; pp. 409–410 dated 13.iv.1920.]

Gallardo, A. 1929. Note sur les moeurs de la fourmi *Pseudoatta argentina*. *Revista de la Sociedad Entomológica Argentina* **10**: 197–202. [31.x.1929.]

Gallardo, A. 1930. Sobre el género *Dorymyrmex* Mayr en la Argentina. *Revista Chilena de Historia Natural* **34**: 143–148. [24.xii.1930.]

Gallardo, A. 1931a. Deux nouvelles espèces de *Pogonomyrmex* de la Republique Argentine. *Revista del Museo de la Plata* **33**: 185–188. [18.xi.1931.]

Gallardo, A. 1931b. Algunas formas sexuales aun no descriptas de las hormigas del género *Crematogaster* de la Republica Argentina. *Revista de la Sociedad Entomologica Argentina* **17**: 297–304. [25.xi.1931.]

Gallardo, A. 1932a. Las hormigas de la República Argentina. Subfamilia mirmicinas, sección Promyrmicinae. *Anales del Museo Nacional de Historia Natural "Bernadino Rivadavia"*, **37**: 37–88. [21.v.1932.]

Gallardo, A. 1932b. Las hormigas de la República Argentina. Subfamilia mirmicinas. Segunda sección Eumyrmicinae, género *Pogonomyrmex* Mayr. *Anales del Museo Nacional de Historia Natural "Bernardino Rivadavia"* **37**: 89–169. [30.x.1932.]

Gallardo, A. 1932c. El subgenero *Elasmopheidole* en la Argentina. *Revista Chilena de Historia Natural* **36**: 178–182. [(31.xii).1932.]

Gallardo, A. 1934. Las hormigas de la República Argentina. Subfamilia mirmicinas. Segunda sección Eumyrmicinae. Tribu Crematogastrini. Género *Crematogaster* Lund. *Anales del Museo Nacional de Ciencias Naturales "Bernardino Rivadavia"* **38**: 1–84. [31.viii.1934.]

Geer, C. de. 1773. *Memoirs pour Servir à l'Histoire des Insectes* **3**: 696 pp. Stockholm. [(31.xii).1773.]

Geer, C. de. 1778. *Memoirs pour Servir à l'Histoire des Insectes* **7**: 950 pp. Stockholm. [(31.xii).1778.]

Gené, G. 1841. Adunanza del di 29 Settembre 1841 (pp. 391–413). *Atti della Terza Riunione degli Scienziati Italiani, Firenze:* 791 pp. [(31.xii).1841.]

Geoffroy, E.S.-H. 1785. In Fourcroy, A.F.de, *Entomologia Parisiensis;* sive catalogus insectorum quae in Agro Parisiensi reperiuntur **2**: 233–544. Parisiis. [(31.xii).1785.]

Germar, E.F. 1837. *Fauna Insectorum Europae.* Fasciculus 19. Insectorum protogaeae specimen sistens insecta carbonum fossilium: pl.19, figs.1–25. Halae. [(31.xii).1837.]

Gerstäcker, A. Note. Alternative spellings with and without the umlaut (Gerstäcker/Gerstaecker) occur in his titles. His publications are listed here with his name standardised as the former.

Gerstäcker, A. 1859. Hr. Peters berichtete über sein Reisewerk, von dem die Insecten bis zum 64., die Botanik bis zum 34. Bogen gedruckt sind und theilte den Schluss der Diagnosen der von Hrn. Dr. Gerstäcker bearbeiteten Hymenopteren mit. *Monatsberichte der Königlichen Preuss. Akademie der Wissenschaften zu Berlin, April 1858:* 261–264. [(31.xii).1859.]

Gerstäcker, A. 1862. In Peters, W.C.H. *Naturwissenschaftliche Reise nach Mossambique* auf befehl seiner Majestät des Königs Friedrich Wilhelm IV in den Jahren 1842 bis 1848 ausgeführt. Zoologie **5**. Insecten und Myriapoden. Hymenoptera Hautflügler: 439–526. Berlin. [(31.xii).1862.]

Gerstäcker, A. 1863. Ueber ein merkwürdiges neues Hymenopteren. *Entomologische Zeitung herausgegeben von dem Entomologischen Vereine zu Stettin* **24**: 76–93. [(31).iii.1863.]

Gerstäcker, A. 1871. Beitrag zur Insektenfauna von Zanzibar. *Archiv für Naturgeschichte* **37**: 345–363. [(31.xii).1871.]

Gerstäcker, A. 1872. Hymenopterologische Beiträge. *Entomologische Zeitung herausgegeben von dem Entomologischen Vereine zu Stettin* **33**: 250–308. [(31).vii.1872.]

Giebel, C.G. 1856. *Fauna der Vorwelt* mit steter Berucksichtigung der lebenden Thiere **2**. Die Insecten und Spinnen der Vorwelt: 511 pp. Leipzig. [(31.xii).1856.]

Girard, M. 1879. *Traité élémentaire d'entomology* **2**: 1028 pp. Paris [(31.xii).1879.]

Gistel, J. 1856. *Mysterien der europäischen Insectenwelt:* 530 pp. Kempten. [(31.xii).1856.]

Gmelin, J.F. 1790. In Linnaeus, C. *Systema Naturae.* Editio 13. **1** (part 5) :2225–3020. Lipsiae. [6.xii.1790.]

Godfrey, R. 1907. Notes on the animal life of the hothouses of the Royal Botanic Garden, Edinburgh. *Notes from the Royal Botanic Garden, Edinburgh* **17**: 99–103. [(30).iv.1907.]

Goetsch, W. 1933. Chilenische Wüsten-, Steppen- und Wald-Ameisen. *Sitzungsberichte der Gesellschaft für Morphologie und Physiologie in München* **42**: 27–35. [13.vi.1933.]

Goetsch, W. & Menozzi, C. 1935. Die Ameisen Chiles. *Konowia* **14**: 94–102. [15.iii.1935.]

Gonçalves, C.R. 1942. Contribuiçao para o conhecimento do gênero *Atta* Fabr., das formigas saúvas. *Boletim da Sociedade Brasileira de Agronomia* **5**: 333–358. [(30).ix.1942.]

Gonçalves, C.R. 1944. Descriçao de uma nova sauva Brasileira. *Revista Brasileira de Biologia* **4**: 233–238. [(30).vi.1944.]

Gonçalves, C.R. 1945. Saúvas do sul e centro do Brasil. *Boletim Fitossanitário* **2**: 183–218. [1945.]

Gonçalves, C.R. 1946. Palestras de técnicos da DDSV la reuniao. Formigas cortadeiras. *Boletim Fitossanitário* **3**: 52–54. [1946.]

Gonçalves, C.R. 1961. O gênero *Acromyrmex* no Brasil. *Studia Entomologica* (N.S.) **4**: 113–180. [10.x.1961.]

Gonçalves, C.R. 1967. *Acromyrmex muticinodus* (Forel, 1901), sinônimo de *Acromyrmex niger* (Smith, 1858). *Revista Brasileira de Entomologia* **12**: 17–20. [(28).ii.1967.]

Gonçalves, C.R. 1983a. Descriçao de *Acromyrmex diasi,* uma nova espécie de formiga cortadeira de folhas. *Revista Brasileira de Biologia* **42** (1982): 485–487. [30.i.1983.]

Gonçalves, C.R. 1983b. *Atta silvae,* nova especie de formiga sauva. *Arquivos Univ. fed. rur. Rio de J.* **5**: 173–178. [Not seen. Reference from Zoological Record.]

Goñi, B., Zolessi, L.C. de, & Imai, H.T. 1983. Karyotypes of thirteen ant species from Uruguay. *Caryologia* **36**: 363–371. [1983.]

González, A.R., López, J.R., & Espadaler, X. 1983. Descripción del macho de *Messor hispanicus* Santschi, 1919. *Boletín de la Asociacion Española de Entomologia* **7**: 37–42. [(30).xi.1983.]

Gordon, D.M. 1984. Daily rhythms in social activities of the harvester ant *Pogonomyrmex badius. Psyche* **90** (1983): 413–424. [6.iv.1984.]

Gösswald, K. 1932. Ökologische Studien über die Ameisenfauna des mittleren Maingebietes. *Zeitschrift für Wissenschaftliche Zoologie* **142** (Abt.A): 1–156. [(31).viii.1932.]

Gösswald, K. 1942. Rassenstudien an der roten Waldameise *Formica rufa* L. auf systematischer, ökologischer, physiologischer und biologischer Grundlage. *Zeitschrift für Angewandte Entomologie* **28** (1941): 62–124. [(31.xii).1942.]

Gösswald, K. 1951. Zur Biologie, Ökologie und Morphologie einer neuen Varietät der Kleinen Roten Waldameise: *Formica minor pratensoides. Zeitschrift für Angewandte Entomologie* **32**: 433–457. [(31).v.1951.]

Gösswald, K. 1985. *Organisation und Leben der Ameisen.* 355 pp. Stuttgart. [(31.xii).1985.]

Gösswald, K. 1989. *Die Waldameise.* Band 1. Biologische Grundlagen, Ökologie und Verhalten: 660 pp. Wiesbaden. [(31.xii).1989.]

Gösswald, K. 1990. *Die Waldameise.* Band 2. Die Waldameise im Ökosystem Wald, ihr Nutzen und ihre Hege: 510 pp. Wiesbaden. [(31.xii).1990.]

Gotwald, W.H.,Jr. 1969. Comparative morphological studies of the ants, with particular reference to the mouthparts. *Cornell University Agricultural Experiment Station Memoir* **408**: 1–50. [(31).vii.1969.]

Gotwald, W.H.,Jr. 1970. Mouthpart morphology of the ant *Aneuretus simoni. Annals of the Entomological Society of America* **63**: 950–952. [15.vii.1970.]

Gotwald, W.H.,Jr. 1971. Phylogenetic affinities of the ant genus *Cheliomyrmex. Journal of the New York Entomological Society* **79**: 161–173. [14.x.1971.]

Gotwald, W.H.,Jr. 1973. Mouthpart morphology of the African ant *Oecophylla longinoda* Latreille. *Journal of the New York Entomological Society* **81**: 72–78. [20.viii.1973.]

Gotwald, W.H.,Jr. 1982. Army Ants (pp. 157–254). In Hermann, H.R. (ed.) *Social Insects* **4**: 385pp. New York. [(31.xii.)1982.]

Gotwald, W.H.,Jr. & Barr, D. 1980. Quantitative studies on major workers of the ant genus *Dorylus. Annals of the Entomological Society of America* **73**: 231–238. [15.iii.1980.]

Gotwald, W.H.,Jr. & Brown, W.L.,Jr. 1967. The ant genus *Simopelta. Psyche* **73** (1966): 261–276. [24.v.1967.]

Gotwald, W.H.,Jr. & Cunningham-Van Someren, G.R. 1976. Taxonomic and behavioral notes on the African ant *Aenictus eugenii* Emery, with a descrip-

tion of the queen. *Journal of the New York Entomological Society* **84**: 182–188. [28.x.1976.]
Gotwald, W.H.,Jr. & Kupiec, B.M. 1975. Taxonomic implications of doryline worker ant morphology: *Cheliomyrmex morosus. Annals of the Entomological Society of America* **68**: 961–971. [17.xi.1975.]
Gotwald, W.H.,Jr. & Leroux, J.M. 1980. Taxonomy of the African army ant, *Aenictus decolor* (Mayr), with a description of the queen. *Proceedings of the Entomological Society of Washington* **82**: 599–608. [24.x.1980.]
Gotwald, W.H.,Jr. & Lévieux, J. 1972. Taxonomy and biology of a new West African ant belonging to the genus *Amblyopone. Annals of the Entomological Society of America* **65**: 383–396. [15.iii.1972.]
Gowdey, C.C. 1917. A list of Ugandan Coccidae, their food-plants and natural enemies. *Bulletin of Entomological Research* **8**: 187–189. [(31).xii.1917.]
Grandi, G. 1935. Contributi alla conoscenza degli imenotteri aculeati. 15. (Elenco degli imenotteri aculeati entrati a far parte delle mie collezioni nel 1934, pp. 95–117.) *Bollettino dell'Istituto di Entomologia della R. Universita degli Studi di Bologna* **8**: 27–121. [10.xi.1935.]
Gravenhorst, J.L.C. 1807. *Vergleichende Uebersicht des Linneischen einiger neuern Zoologischen Systeme:* 476 pp. Gottingen. [(31.xii).1807]
Gray, G.R. 1832. In Griffith, E. *The Animal Kingdom,* arranged in conformity with its organization, by the Baron Cuvier **15**, Class Insecta **2**: 796 pp. London. [(31.xii).1832.]
Gray, J.E. 1841. Contributions towards the geographical distribution of the Mammalia of Australia, with notes on some recently discovered species. (Appendix C., pp. 397–414.) In Grey, G. *Journals of Two Expeditions of Discovery in North West and Western Australia during the Years 1837, 38, and 39* **2**: 482 pp. London. [(30).xi.1841.]
Greaves, T. 1971. The distribution of the three forms of the Meat Ant, *Iridomyrmex purpureus* in Australia. *Journal of the Australian Entomological Society* **10**: 15–21. [31.iii.1971.]
Gredler, V.M. 1858. Die Ameisen von Tirol. *8 Programm des Gymnasiums in Bozen:* 1–31. [(31.xii).1858.]
Greenslade, P.J.M. 1974. The identity of *Iridomyrmex purpureus* form *viridiaeneus* Viehmeyer. *Journal of the Australian Entomological Society* **13**: 247–248. [13.ix.1974.]
Greenslade, P.J.M. 1979. A guide to ants of South Australia. *Special Educational Bulletin Series, South Australian Museum, Adelaide:* 44 pp. [(31.xii).1979.]
Gregg, E.V. 1945. A statistical study of taxonomic categories in ants: *Lasius neoniger* and *Lasius americanus. Annals of the Entomological Society of America* **38**: 529–548. [31.xii.1945.]
Gregg, R.E. 1945a. The worker caste of *Harpagoxenus canadensis* Smith. *Canadian Entomologist* **77**: 74–76. [9.viii.1945.]
Gregg, R.E. 1945b. Two new forms of *Monomorium. Psyche* **52**: 62–69. [26.x.1945.]
Gregg, R.E. 1949a. A note on *Pheidole (Macropheidole) rhea* Wheeler. *Psyche* **56**: 70–73. [9.viii.1949.]
Gregg, R.E. 1949b. A new ant from southwestern United States. *Proceedings of the Entomological Society of Washington* **51**: 171–174. [17.viii.1949.]
Gregg, R.E. 1950. A new species of *Pheidole* from the southwest. *Journal of the New York Entomological Society* **58**: 89–93. [23.vi.1950.]
Gregg, R.E. 1951. Two new species of exotic ants. *Psyche* **58**: 77–84. [19.xi.1951.]
Gregg, R.E. 1952a. A new ant of the genus *Pheidole* from Colorado. *American Museum Novitates* **1557**: 1–4. [5.v.1952.]
Gregg, R.E. 1952b. The female of *Formica opaciventris* Emery. *Psyche* **59**: 13–19. [19.viii.1952.]
Gregg, R.E. 1953. The soldier caste of *Pheidole (Ceratopheidole) clydei* Gregg. *American Museum Novitates* **1637**: 1–7 [16.xii.1953.]
Gregg, R.E. 1954a. Taxonomic notes on the ant *Camponotus cooperi* Gregg. *Psyche* **60** (1953): 102–104. [8.i.1954.]
Gregg, R.E. 1954b. Geographical distribution of the genus *Myrmoteras,* including the description of a new species. *Psyche* **61**: 20–30. [30.vi.1954.]
Gregg, R.E. 1955a. A new species of ant belonging to the *Pheidole pilifera* complex. *Psyche* **62**: 19–28. [23.v.1955.]
Gregg, R.E. 1955b. The rediscovery of *Veromessor lobognathus* (Andrews). *Psyche* **62**: 45–51. [19.vii.1955.]
Gregg, R.E. 1957. A new species of *Myrmoteras* from Ceylon. *Psyche* **63** (1956): 41–44. [23.i.1957.]
Gregg, R.E. 1958. Two new species of *Metapone* from Madagascar. *Proceedings of the Entomological Society of Washington* **60**: 111–121. [18.iv.1958.]
Gregg, R.E. 1959. Key to the species of *Pheidole* in the United States. *Journal of the New York Entomological Society* **66** (1958): 7–48. [20.i.1959.]
Gregg, R.E. 1961. The status of certain myrmicine ants in western North America with a consideration of the genus *Paramyrmica* Cole. *Journal of the New York Entomological Society* **69**: 209–220. [(31).xii.1961.]
Gregg, R.E. 1963. *The Ants of Colorado:* 792 pp. Boulder. [(31.xii).1963.]
Gregg, R.E. 1969. New species of *Pheidole* from Pacific Coast Islands. *Entomological News* **80**: 93–101. [2.v.1969.]
Gregg, R.E. 1972. A new species of *Stenamma* from Utah. *Great Basin Naturalist* **32**: 35–39. [31.iii.1972.]
Gregg, R.E. 1973. A new species of *Camponotus* from Nevada. *Southwestern Naturalist* **18**: 39–43. [30.iii.1973.]
Griffin, F.J. 1932. On the dates of publication and contents of the parts of Westwood (J.O.) "Introduction to the Modern Classification of Insects, 1838–1840." *Proceedings of the Entomological Society of London* **6**: 83–84. [22.iii.1932.]
Grundmann, A.W. 1952. A new *Brachymyrmex* from northern Utah. *Journal of the Kansas Entomological Society* **25**: 117. [15.vii.1952.]
Guérin-Méneville, F.E. 1831/1838. Histoire naturelle des Crustacés, Arachnides et Insectes In Duperrey, L.I. *Voyage autour du Monde, exécuté par ordre du Roi, sur la corvette de Sa Majesté, La Coquille, pendent les années 1822, 1823, 1824 et 1825.* Zoologie **2**: 319 pp. Paris. [Publication date of Plate 8 is 15.xi.1831; publication date of text is 31.xii.1838. For dates of publication see Sherborn & Woodward, 1906a: 335; Bequaert, 1926: 186; Cowan, 1970: 358.]
Guérin-Méneville, F.E. 1844a. *Iconographie du régne animal de G. Cuvier,* ou représentation d'après nature de l'une des espèces les plus remarquables et souvent non figurées de chaque genre d'animaux. Insectes (1829–1838): 576 pp. Paris. [7.ix.1844. For date of publication see Cowan, 1971: 18–29.]
Guérin-Méneville, F.E. 1844b. Observations sur un insecte qui attaque les olives, dans nos départements méridionaux, et cause une diminution très-considérable dans la récolte de l'huile. *Comptes Rendus Hebdomadaires des Séances de l'Académie des Sciences* **19**: 1147–1150. [(31.xii).1844.]
Guérin-Méneville (here spelled Menneville), F.E. 1849. Famille des Hétérogynes (pp.351–353). In Lefebvre, T. *Voyage en Abyssinie* exécuté pendant les années 1839–1843. **6** (part 4). Histoire Naturelle-Zoologie: 398 pp. Paris. [Date of publication of pp. 241–398 is (31.xii).1849. See Sherborn & Woodward, 1901: 162.]
Guérin-Méneville, F.E. 1852. Notice sur une nouvelle espèce de fourmi découverte à Saint-Domingue par M. Auguste Sallé, et qui fait son nid dans des plaines marécageuses, sur les buissons. *Revue et Magasin de Zoologie Pure et Appliquée* (2) **4**: 73–79. [(31.xii).1852.]
Hahn, C.W. 1832. Die Wanzenartigen Insecten. *Getreu nach der Natur abgebildet und beschrieben* **1** (3): 80–117. Nürnberg. [(30).xi.1832.]
Haines, I.H. & Haines, J.B. 1978. Colony structure, seasonality and food requirements of the Crazy Ant, *Anoplolepis longipes* (Jerd.) in the Seychelles. *Ecological Entomology* **3**: 109–118. [(31).v.1978.]
Haldeman, S.S. 1844. Descriptions of insects, presumed to be undescribed. *Proceedings of the Academy of Natural Sciences of Philadelphia* **2**: 53–55. [(30).iv.1844.]

Haldeman, S.S. 1849a. On the identity of *Annoma* with *Dorylus*, suggested by specimens which Dr Savage found together, and transmitted to illustrate his paper on the Driver Ants. *Proceedings of the Academy of Natural Sciences of Philadelphia* **4**: 200–202. [(31).viii.1849.]

Haldeman, S.S. 1849b. On several new Hymenoptera of the genera *Ampulex, Sigalphus, Chelonus and Dorylus. Proceedings of the Academy of Natural Sciences of Philadelphia* **4**: 203–204. [(31).viii.1849.]

Haldeman, S.S. 1852. Appendix C.- Insects (pp. 366–378). In Stansbury, H. *An expedition to the Valley of the Great Salt Lake of Utah: including a description of its geography, natural history, and minerals, and an analysis of its waters:* 487 pp. London. [(31.xii).1852.]

Haliday, A.H. 1836. Descriptions of the Hymenoptera. In Curtis, J., Haliday, A.H. & Walker, F. Descriptions etc. of the insects collected by Captain P.P. King R.N., F.R.S., in the survey of the Straits of Magellan. *Transactions of the Linnean Society of London* **17**: 315–331. [24.vi.1836.]

Hamann, H.H.F. 1957. On a new record of *Gesomyrmex* Mayr. *Idea, Bogor* **10**: 1–6. [15.viii.1957.]

Hamann, H.H.F. & Klemm, W. 1967. Ergebnisse der zoologischen Nubien-Expedition 1962. *Annalen des Naturhistorisches Museums in Wien* **70** (1966): 411–421. [(31).x.1967.]

Hamann, H.H.F. & Klemm, W. 1976. Ergebnisse der von Dr O. Paget and Dr E. Kritscher auf Rhodos durchgeführten zoologischen Exkursionen. *Annalen des Naturhistorisches Museums in Wien* **80**: 669–679. [(30).xi.1976.]

Handlirsch, A. 1906. *Die Fossilen Insekten und die Phylogenie der Rezenten Formen. Ein Handbuch für Paläontologen und Zoologen.* 4 Lieferung: 481–640. Leipzig. [(31).x.1906.]

Handlirsch, A. 1907. *Die Fossilen Insekten und die Phylogenie der Rezenten Formen. Ein Handbuch für Paläontologen und Zoologen.* 6 Lieferung: 801–960. Leipzig. [(30).vi.1907.]

Harada, A.Y. 1987. Uma nova espécie do gênero *Monacis* Roger, da Amazônia. *Acta Amazonica* **16/17**: 599–606. [(31).x.1987.]

Harris, R.A. 1979. A glossary of surface sculpturing. *Occasional Papers in Entomology, State of California, Department of Food and Agriculture* **28**: 31 pp. [(30).vi.1979.]

Hashmi, A.A. 1973. A revision of the Neotropical ant subgenus *Myrmothrix* of genus *Camponotus. Studia Entomologica* (N.S.) **16**: 1–140. [31.x.1973.]

Haskins, C.P. & Haskins, E.F. 1974. Notes on necrophoric behavior in the archaic ant *Myrmecia vindex. Psyche* **81**: 258–267. [26.ix.1974.]

Haskins, C.P. & Zahl, P.A. 1971. The reproductive pattern of *Dinoponera grandis* Roger with notes on the ethology of the species. *Psyche* **78**: 1–11. [30.xii.1971.]

Haug, G.W. 1932. Description of the male of *Strumigenys louisianae* subsp. *laticephala* M.R. Smith. *Annals of the Entomological Society of America* **25**: 170–172. [31.iii.1932.]

Hauschteck, E. 1961. Die Chromosomen von fünf Ameisenarten. *Revue Suisse de Zoologie* **68**: 218–223. [(31).vii.1961.]

Hauschteck, E. 1962. Die Chromosomen einiger in der Schweiz vorkommender Ameisenarten. *Vierteljahrsschrift der Naturforschenden Gesellschaft in Zürich* **107**: 213–220. [31.xii.1962.]

Hauschteck, E. 1965. Halbe haploide Chromosomenzahl im Hoden von *Myrmica sulcinodis* Nyl. *Experientia* **21**: 323–327. [15.vi.1965.]

Heer, O. 1850. Die Insektenfauna der Tertiärgebilde von Oeningen und von Radoboj in Croatien. Zweite Abtheilung: Heuschrecken, Florfliegen, Aderflügler, Schmetterlinge und Fliegen. *Neue Denkschriften der allgemeinen Schweizerischen Gesellschaft für die gesammten Naturwissenschaften* (1) **11** (1849): 1–264. [(31.xii).1850.]

Heer, O. 1852. Ueber die Haus-Ameise Madeiras. *An die Zürcherische Jugend auf das Jahr 1852. Von der Naturforschenden Gesellschaft* **54**: 1–24. [(31.xii).1852.]

Heer, O. 1856. On the House Ant of Madeira. *Annals and Magazine of Natural History* (2) **17**: 209–224. [(31).iii.1856.] And pp. 322–334 (conclusion) [(30).iv.1856.]

Heer, O. 1865. *Die Urwelt der Schweiz:* 622 pp. Zürich. [(31.xii).1865.]

Heer, O. 1867. Fossile Hymenopteren aus Oeningen und Radoboj. *Neue Denkschriften der allgemeinen Schweizerischen Gesellschaft für die gesammten Naturwissenschaften* (4) **22**: 1–42. [(31.xii).1867.]

Heer, O. 1870. Die Miocene flora und fauna Spitzbergens. *Kongliga Svenska Vetenskaps-Akademiens Handlingar* **8** (No. 7): 1–98. [(31.xii).1870.]

Heer, O. 1879. *Die Urwelt der Schweiz* (Ed. 2). Zürich. [Not seen.]

Heinze, J. 1989. *Leptothorax wilsoni* n. sp., a new parasitic ant from eastern North America. *Psyche* **96**: 49–61. [18.xii.1989.]

Heinze, J. & Alloway, T.M. 1992. *Leptothorax paraxenus* n. sp., a workerless social parasite from North America. *Psyche* **98** (1991): 195–206. [7.ii.1992.]

Heinze, J., Schulz, A. & Radchenko, A.G. 1993. Redescription of the ant *Leptothorax* (s. str.) *scamni* Ruzsky, 1905. *Psyche* **100**: 177–183. [20.xii.1993.]

Hermann, H.R. (ed.) 1979–1982. *Social Insects.* 4 volumes. Vol. **1** (1979): 437 pp. Vol. **2** (1981): 491 pp. Vol. **3** (1982): 459 pp. Vol. **4** (1982): 385 pp. New York.

Hermann, H.R. (ed.) 1984. *Defensive Mechanisms in Social Insects:* 259 pp. New York.

Hermann, H.R. & Blum, M.S. 1981. Defensive mechanisms in the social Hymenoptera (pp. 77–197). In Hermann, H.R. (ed.). *Social Insects* **2**: 491 pp. New York. [1981.]

Higgins, L.G. 1963. Dates of publication of the Novara Reise. *Journal of the Society for the Bibliography of Natural History* **4**: 153–159. [(30).xi.1963.]

Hohmann, H., La Roche, F., Ortega, G. & Barquin, J. 1993. Bienen, Wespen und Ameisen der Kanarischen Inseln. *Veröffentlichnungen aus dem Übersee-Museum Bremen Naturwissenschaften* **12** (1): 1–465 (Formicidae pp. 145–166). [(31.xii).1993.]

Holgersen, H. 1940. Myrmekologiske notiser. 1. *Norsk Entomologisk Tidsskrift* **5**: 183–187. [(31).i.1940.]

Holgersen, H. 1942. Ants of northern Norway. *Tromso Museums Årshefter (Naturhistorisk Avd. Nr. 24)* **63** (1940): 3–33. [9.x.1942.]

Holgersen, H. 1943. *Formica gagatoides* Ruzs. in Norway. *Tromso Museums Årshefter (Naturhistorisk Avd. Nr. 26)* **64** (1941): 3–17. [(31).v.1943.]

Holl, F. 1829. *Handbuch der Petrefactenkunde* **2**: 117–232. Dresden. [(31.xii).1829.]

Hölldobler, B. 1976. Recruitment behaviour, home range orientation and territoriality in harvester ants, *Pogonomyrmex. Behavioral Ecology and Sociobiology* **1**: 3–44. [1976.]

Hölldobler, B. 1977. Communication in social Hymenoptera. In Sebeok, T.A. (Ed.). *How Animals Communicate:* 418–471. Indiana University Press. [1977.]

Hölldobler, B. 1978. Ethological aspects of chemical communication in ants. *Advances in the Study of Behavior* **8**: 75–115. [1978.]

Hölldobler, B. 1980. Canopy orientation: a new kind of orientation in ants. *Science* **210**: 86–88. [3.x.1980.]

Hölldobler, B. 1982. Communication, raiding behavior and prey storage in *Cerapachys. Psyche* **89**: 3–24. [17.xii.1982.]

Hölldobler, B. 1984. Evolution of insect communication. In *Insect Communication:* 349–377. London. [1984.]

Hölldobler, B. 1985. A new exocrine gland in the slave-raiding ant genus *Polyergus. Psyche* **91** (1984): 225–236. [22.iv.1985.]

Hölldobler, B, & Engel, H. 1979. Tergal and sternal glands in ants. *Psyche* **85** (1978): 285–330. [18.ix.1979.]

Hölldobler, B. & Engel-Siegel, H. 1985. On the metapleural glands of ants. *Psyche* **91** (1984): 201–224. [22.iv.1985.]

Hölldobler, B., Stanton, R. & Engel, H. 1976. A new exocrine gland in *Novomessor* and its possible significance as a taxonomic character. *Psyche* **83**: 32–41. [4.xi.1976.]

Hölldobler, B., Stanton, R.C. & Markl, H. 1978. Recruitment and food-retrieving behavior in *Novomessor.* 1. Chemical signals. *Behavioral Ecology and Sociobiology* **4**: 163–181. [1978.]

Hölldobler, B. & Taylor, R.W. 1984. A behavioral study of the primitive ant *Nothomyrmecia macrops* Clark. *Insectes Sociaux* **30** (1983): 384–401. [1984.]

Hölldobler, B. & Traniello, J.F.A. 1980. The pygidial gland and chemical recruitment in *Pachycondyla* (=*Termitopone*) *laevigata. Journal of Chemical Ecology* **6**: 883–893. [1980.]

Hölldobler, B. & Wilson, E.O. 1971. Recruitment trails in the harvester ant *Pogonomyrmex badius. Psyche* **77** (1970): 385–399. [21.x.1971.]

Hölldobler, B. & Wilson, E.O. 1977. Weaver ants. *Scientific American* (offprints) **237**: 146–154. [xii.1977.]

Hölldobler, B. & Wilson, E.O. 1978. The multiple recruitment systems of the African weaver ant *Oecophylla longinoda* (Latreille). *Behavioral Ecology and Sociobiology* **3**: 19–60. [1978.]

Hölldobler, B. & Wilson, E.O. 1986a. Soil-binding pilosity and camouflage in ants of the tribes Basicerotini and Stegomyrmecini. *Zoomorphology* **106**: 12–20. [1986.]

Hölldobler, B. & Wilson, E.O. 1986b. Ecology and behavior of the primitive cryptobiotic ant *Prionopelta amabilis. Insectes Sociaux* **33**: 45–58. [vi.1986.]

Hölldobler, B. & Wilson, E.O. 1990. *The Ants:* 732 pp. Harvard University Press. [1990.]

Hölldobler, B. & Wilson, E.O. 1992. *Pheidole nasutoides,* a new species of Costa Rican ant that apparently mimics termites. *Psyche* **99**: 15–22. [4.xii.1992.]

Holmgren, N. 1908. Über einige myrmecophile Insekten aus Bolivia und Peru. *Zoologischer Anzeiger* **33**: 337–349. [18.viii.1908.]

Hong, Y.-C., Yang, T.-C., Wang, S.-T., Wang, S.-E., Li, Y.-K., Sun, M.-R., Sun, H.-C. & Tu, N.-C. 1974. Stratigraphy and palaeontology of Fushun coalfield, Liaoning Province. *Acta Geologica Sinica* **1974** (No. 2): 113–149. [(31.xii).1974.]

Hong, Y.-C. 1984. Fossil insects in the diatoms of Shanwang. *Bulletin of the Tianjin Institute of Geology and Mineral Resources* **8** (1983): 1–12. [(31.xii).1984.]

Hung, A.C.F. 1967a. A new species and two new names of the *Polyrhachis* ants. *Mushi* **40**: 199–202. [24.iii.1967.]

Hung, A.C.F. 1967b. A revision of the ant genus *Polyrhachis* at the subgeneric level. *Transactions of the American Entomological Society* **93**: 395–422. [20.xii.1967.]

Hung, A.C.F. 1969. The chromosome numbers of six species of formicine ants. *Annals of the Entomological Society of America* **62**: 455–456. [(31).iii.1969.]

Hung, A.C.F. 1970. A revision of ants of the subgenus *Polyrhachis* Fr. Smith. *Oriental Insects* **4**: 1–36. [(30).iv.1970.]

Hung, A.C.F. 1971. On the taxonomic status of *Polyrhachis kirkae* Donisthorpe and its presumed mimicry. *Entomological News* **82**: 43–47. [28.v.1971.]

Hung, A.C.F. & Brown, W.L.,Jr. 1966. Structure of gastric apex as a subfamily character of the Formicinae. *Journal of the New York Entomological Society* **74**: 198–200. [29.xii.1966.]

Hung, A.C.F., Imai, H.T. & Kubota, M. 1972. The chromosomes of nine ant species from Taiwan, Republic of China. *Annals of the Entomological Society of America* **65**: 1023–1025. [15.ix.1972.]

Hunt, J.H. & Snelling, R.R. 1975. A checklist of the ants of Arizona. *Journal of the Arizona Academy of Science* **10**: 20–23. [(28).ii.1975.]

Ihering, H. von. 1912. Biologie und Verbreitung der brasilianischen Arten von *Eciton. Entomologische Mitteilungen* **1**: 226–235. [1.viii.1912.]

Illiger, K. 1802. Neue Insekten. *Magazin für Insektenkunde* **1**: 163–208. [(31.xii).1802.]

Imai, H.T. 1966. The chromosome observation techniques of ants and the chromosomes of Formicinae and Myrmicinae. *Acta Hymenopterologica* **2**: 119–131. [30.ix.1966.]

Imai, H.T. 1971. Karyological studies of Japanese ants, 2. Species differentiation in *Aphaenogaster;* with special regard to their morphology, distribution and chromosomes. *Mushi* **44**: 137–151. [5.v.1971.]

Imai, H.T., Baroni Urbani, C., Kubota, M., Sharma, G.P., Narasimhanna, M.N., Das, B.C., Sharma, A.K., Sharma, A., Deodikar, G.B., Vaidya, V.G. & Rajasekarasetty, M.R. 1984. Karylogical survey of Indian ants. *Japanese Journal of Genetics* **59**: 1–32. [1984.]

Imai, H.T., Brown, W.L.,Jr., Kubota, M., Yong, H.-S. & Tho, Y.P. 1984. Chromosome observations on tropical ants from western Malaysia. 2. *Annual Report of the National Institute of Genetics, Japan* **34**: 66–69. [1984.]

Imai, H.T., Crozier, R.H. & Taylor, R.W. 1977. Karyotype evolution in Australian ants. *Chromosoma* **59**: 341–393. [1977.]

Imai, H.T. & Kubota, M. 1972. Karyological studies of Japanese ants. 3. Karyotypes of nine species in Ponerinae, Formicinae, and Myrmicinae. *Chromosoma* **37**: 193–200. [1972.]

Imai, H.T. & Kubota, M. 1975. Chromosome polymorphism in the ant *Pheidole nodus. Chromosoma* **51**: 391–399. [1975.]

Imai, H.T., Kubota, M., Brown, W.L.,Jr., Ihara, M. Tohari, M. & Pranata, R.I. 1985. Chromosome observations on tropical ants from Indonesia. *Annual Report of the National Institute of Genetics, Japan* **35** (1984): 46–48. [1985.]

Imai, H.T. & Taylor, R.W. 1986. The exceptionally low chromosome number n=2 in an Australian bulldog ant, *Myrmecia piliventris. Annual Report of the National Institute of Genetics, Japan* **36** (1985): 59–61. [1986.]

Imai, H.T., Taylor, R W., Kubota, M., Ogata, K. & Wada, M.Y. 1991. Notes on the remarkable karyology of the primitive ant *Nothomyrmecia macrops,* and of the related genus *Myrmecia. Psyche* **97** (1990): 133–139. [1991.]

Imai, H.T. & Yosida, T.H. 1964. Chromosome observations in Japanese ants. *Annal Report of the National Institute of Genetics, Japan* **15**: 64–66. [1964.]

Imhoff, L. 1852. Ueber eine Art afrikanischer Ameisen. *Bericht ueber die Verhandlungen der Naturforschenden Gesellschaft in Basel* **10**: 175–177. [(30).vi.1852.]

Ito, T. 1914. Formicidarum Japonicum species novae vel minus cognitae. *Annales de la Société Entomologique de Belgique* **58**: 40–45. [6.ii.1914.]

Itow, T., Kobayashi, K., Kubota, M., Ogata, K., Imai, H.T. & Crozier, R.H. 1984. The reproductive cycle of the queenless ant *Pristomyrmex pungens. Insectes Sociaux* **31**: 87–102. [1984.]

Jaffe, K. 1984. Negentropy and the evolution of chemical recruitment in ants. *Journal of Theoretical Biology* **106**: 587–604. [1984.]

Jaisson, P., Fresneau, D., Taylor, R.W. & Lenoir, A. 1992. Social organisation in some primitive Australian ants. 1. *Nothomyrmecia macrops. Insectes Sociaux* **39**: 425–438. [(31.xii).1992.]

Jayasuriya, A.K. & Traniello, J.F.A. 1986. The biology of the primitive ant *Aneuretus simoni.* 1. Distribution, abundance, colony structure, and foraging ecology. *Insectes Sociaux* **32** (1985): 363–374. [(30).vi.1986.]

Jekel, H. 1854. *Fabricia Entomologica.* Première partie. 1 Livraison: 96 pp. [(31.xii).1854.]

Jell, P.A. & Duncan, P.M. 1986. Invertebrates, mainly insects, from the freshwater, Lower Cretaceous, Koonwarra Fossil Bed (Korumburra Group), South Gippsland, Victoria (pp. 111–205). In Jell, P.A. & Roberts J. (eds.). Plants and invertebrates from the Lower Cretaceous Koonwarra Fossil Bed, South Gippsland, Victoria. *Association of Australasian Palaeontologists Memoir* **3**: 205 pp. Sydney. [(31.xii).1986.]

Jerdon, T.C. 1851. A catalogue of the species of ants found in southern India. *Madras Journal of Literature and Science* **17**: 103–127. [(31.xii).1851.]

Jerdon, T.C. 1854. A catalogue of the species of ants found in southern India. *Annals and Magazine of Natural History* (2) **13**: 45–56. [(28).ii.1854.]

Johnson, C. 1987. Biogeography and habitats of *Ponera exotica. Journal of Entomological Science* **22**: 358–361. [(31).x.1987.]

Johnson, C. 1988. Species identification in the eastern *Crematogaster. Journal of Entomological Science* **23**: 314–332. [3.xi.1988.]

Johnson, C. 1989a. Identification and nesting sites of North American species of *Dolichoderus* Lund. *Insecta Mundi* **3**: 1–9. [(31).iii.1989.]

Johnson, C. 1989b. Taxonomy and diagnosis of *Conomyrma insana* (Buckley) and *C. flava* (McCook). *Insecta Mundi* **3**: 179–194. [(30).ix.1989.]

Jolivet, P. 1986. *Les fourmis et les plantes. Un example de coévolution:* 254 pp. Paris. [1986.]

Jolivet, P. 1992. *Insects and Plants. Parallel Evolution and Adaptations:* 190 pp. Gainesville. [(31.xii).1992.]

Joseph, G. 1882. Systematisches Verzeichniss der in den Tropfstein-Grotten von Krain einheimischen Arthropoden nebst Diagnosen der vom Verfasser entdeckten und bisher noch nicht beschreiben Arten. *Berliner Entomologische Zeitschrift* **26**: 1–50. [(31).xii.1882.]

Jurine, L. 1801. In Panzer, G.W.F. Nachricht von einem neuen entomolischen [sic] Werke, des Hrn. Prof. Jurine in Geneve. *Intelligenzblatt der Litteratur-Zeitung, Erlangen* **1**: 161–165. [30.v.1801.]

Jurine, L. 1807. *Nouvelle Méthode de Classer les Hyménoptères et les Diptères.* Hyménoptères **1**: 319 pp. Geneve. [(31.xii).1807.]

Karavaiev, V. Note. This name appears in many different spellings in the literature (e.g. Karawaiew, Karawajew, Karavaiev, Karavajev, etc.). The name is stabilised here upon the spelling Karavaiev, a modern transliteration from the Ukrainian Cyrillic original version of his name, KAPABAEB.

Karavaiev, V. 1906. Systematisch-Biologisches über drei Ameisen aus Buitenzorg. *Zeitschrift für Wissenschaftliche Insektenbiologie* **2**: 369–376. [30.xii.1906.]

Karavaiev, V. 1910a. Nachtrag zu meinen "Ameisen aus Transcaspien und Turkestan." *Russkoe Entomologicheskoe Obozrenie* **9** (1909): 268–272. [10.iii.1910.]

Karavaiev, V. 1910b. Ameisen aus Transkaspien und Turkestan. *Trudy Russkago Entomologicheskago Obshchestva* **39** (1909): 1–72. [31.xii.1910.]

Karavaiev, V. 1911. Murav'i, sobrannye v Egipte i Sudane. *Russkoe Entomologicheskoe Obozrenie* **11**: 1–12. [29.v.1911.]

Karavaiev, V. 1912a. Ameisen aus Tunesien und Algerien, nebst einigen unterwegs in Italien gesammelten Arten. *Russkoe Entomologicheskoe Obozrenie* **12**: 1–22. [20.iv.1912.]

Karavaiev, V. 1912b. Ameisen aus dem paläarktischen Faunengebiete. *Russkoe Entomologicheskoe Obozrenie* **12**: 581–596. [28.xii.1912.]

Karavaiev, V. 1916. Murav'i iz Gadyachskago uezda Poltavskoi gubernii i iz Ferganskoi oblasti. *Russkoe Entomologicheskoe Obozrenie* **15**: 496–507. [28.iii.1916.]

Karavaiev, V. 1924. Zur Systematik der paläarktischen *Myrmecocystus,* nebst einigen biologischen Notizen. *Konowia* **3**: 301–308. [31.x.1924.]

Karavaiev, V. 1925a. Ponerinen aus dem Indo-Australischen Gebiet. *Konowia* **4**: 69–81. [15.iii.1925.]

Karavaiev, V. 1925b. Ponerinen aus dem Indo-Australischen Gebiet (Fortsetzung). *Konowia* **4**: 115–131. [1.vii.1925.]

Karavaiev, V. 1925c. Ponerinen aus dem Indo-Australischen Gebiet (Schluss). *Konowia* **4**: 276–296. [25.x.1925.]

Karavaiev, V. 1926a. Über den Nestbau von *Polyrhachis* (subg. *Myrmhopla) tubifex* sp. n. *Biologischen Zentralblatt* **46**: 143–145. [(31).iii.1926.]

Karavaiev, V. 1926b. Beiträge zur Ameisenfauna des Kaukasus, nebst einigen Bemerkungen über andere palaearktische Formen. *Konowia* **5**: 93–109. [10.vi.1926.]

Karavaiev, V. 1926c. Beiträge zur Ameisenfauna des Kaukasus, nebst einigen Bemerkungen über andere palaearktische Formen. (Fortsetzung.) *Konowia* **5**: 161–169. [28.vii.1926.]

Karavaiev, V. 1926d. Ameisen aus dem Indo-Australischen Gebiet. *Treubia* **8**: 413–445. [(31).vii.1926.]

Karavaiev, V. 1926e. Beiträge zur Ameisenfauna des Kaukasus, nebst einigen Bemerkungen über andere palaearktische Formen. (Schluss.) *Konowia* **5**: 187–199. [20.xi.1926.]

Karavaiev, V. 1926f. Mirmekolohichni frahmenty. *Trudy Ukrains'ka Akademiya Nauk, Fizichno-Matematichnoho Viddilu [Zbirnyk Prats' Zoolohichnoho Muzeyu* **1**: 47–51.] **4**: 65–69. [(31.xii).1926.]

Karavaiev, V. 1927a. Übersicht der Ameisenfauna der Krim nebst einigen Neubeschreibungen. *Konowia* **5** (1926): 281–303. [28.i.1927.]

Karavaiev, V. 1927b. Myrmekologische miszellen. *Ezhegodnik Zoologicheskogo Muzeya Akademiya Nauk SSSR* **27** (1926): 104–112. [(31).vii.1927.]

Karavaiev, V. 1927c. Favna murashok Ukrainy. *Trudy Ukrains'ka Akademiya Nauk, Fizichno-Matematichnoho Viddilu [Zbirnyk Prats' Zoolohichnoho Muzeyu* **2**: 35–53.] **4** (1926): 247–296. [(31.xii).1927.]

Karavaiev, V. 1927d. Murashky z Palearktychnoho krayu. 2. *Trudy Ukrains'ka Akademiya Nauk, Fizichno-Matematichnoho Viddilu [Zbirnyk Prats' Zoolohichnoho Muzeyu* **2**: 98–104.] **4**: 333–348. [(31.xii).1927.]

Karavaiev, V. 1927e. Murashky z Indo-Avstraliys'koho krayu. 3. *Trudy Ukrains'ka Akademiya Nauk, Fizichno-Matematichnoho Viddilu [Zbirnyk Prats' Zoolohichnoho Muzeyu* **3**: 3–52.] **7**: 3–52. [(31.xii).1927.]

Karavaiev, V. 1928. Murashky z Indo-Avstraliys'koho krayu. 4. Hnizda murashok, perevazhno vydiv *Polyrhachis. Trudy Ukrains'ka Akademiya Nauk, Fizichno-Matematichnoho Viddilu [Zbirnyk Prats' Zoolohichnoho Muzeyu* **5**: 129–150.] **6**: 307–328. [(31.xii).1928.]

Karavaiev, V. 1929a. Murashky z Indo-Avstraliys'koho krayu. 6 [sic]. *Trudy Vseukrains'ka Akademiya Nauk, Fizichno- Matematichnoho Viddilu [Zbirnyk Prats' Zoolohichnoho Muzeyu* **7**: 43–58.] **13**: 41–46. [(31.xii).1929.]

Karavaiev, V. 1929b. Mirmekolohichni frahmenty. 2. *Trudy Vseukrains'ka Akademiya Nauk, Fizichno-Matematichnoho Viddilu [Zbirnyk Prats' Zoolohichnoho Muzeyu* **7**: 205–220.] **13**: 203–218. [(31.xii).1929.]

Karavaiev, V. 1929c. Murashky z Indo-Avstraliys'koho krayu. 6. *Trudy Vseukrains'ka Akademiya Nauk, Fizichno-Matematichnoho Viddilu [Zbirnyk Prats' Zoolohichnoho Muzeyu* **7**: 235–248.] **13**: 233–246. [(31.xii).1929.]

Karavaiev, V. 1930a. Ameisen von den Molukken und Neuguinea. (Ergebnisse der Sunda-Expedition der Notgemeinschaft der deutschen Wissenschaft, 1929/30.) *Zoologischer Anzeiger* **92**: 206–214. [18.x.1930.]

Karavaiev, V. 1930b. Materiyaly do favny murashok Shveds'kych ostroviv Hotlyanda ta Elyada. *Trudy Vseukrains'ka Akademiya Nauk, Fizichno-Matematichnoho Viddilu [Zbirnyk Prats' Zoolohichnoho Muzeyu* **8**: 5–46.] **15**: 109–150. [(31.xii).1930.]

Karavaiev, V. 1931a. Myrmekologische Fragmente. 3. *Zoologischer Anzeiger* **92**: 309–317. [5.i.1931.]

Karavaiev, V. 1931b. Beitrag zur Ameisenfauna der Umgebung des Baikalsees. *Zoologischer Anzeiger* **93**: 28–32. [15.i.1931.]

Karavaiev, V. 1931c. Beitrag zur Ameisenfauna Jakutiens. *Zoologischer Anzeiger* **94**: 104–117. [20.iv.1931.]

Karavaiev, V. 1931d. Ameisen aus Englisch-Ostafrika. *Zoologischer Anzeiger* **95**: 42–51. [15.vi.1931.]

Karavaiev, V. 1931e. Ohlyad favny murashok Shvetsii. *Trudy Vseukrains'ka Akademiya Nauk, Pryrodnycho-Technichnoho Viddilu [Zbirnyk Prats' Zoolohichnoho Muzeyu* **10**: 207–220.] **5**: 207–220. [(31.xii).1931.]

Karavaiev, V. 1932. Zwei neue Ameisen aus Aserbeidschan (Transkaukasien). *Zoologischer Anzeiger* **98**: 248–250. [5.v.1932.]

Karavaiev, V. 1933a. Ameisen aus dem Indo-Australischen Gebiet. 7. *Konowia* **11** (1932): 305–320. [15.iii.1933.]

Karavaiev, V. 1933b. Ameisen aus dem Indo-Australischen Gebiet. 7. (Fortsetzung). *Konowia* **12**: 103–120. [25.viii.1933.]

Karavaiev, V. 1933c. Ameisen aus dem Indo-Australischen Gebiet. 7. (Schluss.) *Konowia* **12**: 260–271. [30.xii.1933.]

Karavaiev, V. 1934. Fauna rodyny Formicidae (murashky) Ukrainy. *Trudy Instytutu Zoolohii ta Biolohii Vseukrains'ka Akademii Nauk.* Seriya **1-a**. *Pratsi z systematyky ta faunistyky:* 1–162. [27.ix.1934.]

Karavaiev, V. 1935a. Neue Ameisen aus dem Indo-Australischen Gebiet, nebst Revision einiger Formen. *Treubia* **15**: 57–117. [(30).vi.1935.]

Karavaiev, V. 1935b. Materialy do fauny murashok Mariupol's'koi okruhy. *Zbirnyk Prats' Zoolohichnoho Muzeyu Ukrains'ka Akademiya Nauk* **16**: 107–111. [(31.xii).1935.]

Karavaiev, V. 1936. Fauna rodyny Formicidae (murashky) Ukrainy. *Trudy Instytutu Zoolohii ta Biolohii Ukrains'ka Akademiya Nauk.* Seriya **1-a**. *Pratsi z systematyky ta faunistyky:* 161–316. [26.v.1936.]

Karavaiev, V. 1937. Murashky, zibrani v zapovidnykach Kinburns'koho pivostrova i Burkutiv. *Zbirnyk Prats' Zoolohichnoho Muzeyu, Instytut Zoolohii ta Biolohii an URSR* **19**: 171–181. [(31.xii).1937.]

Keilbach, R. 1982: Bibliographie und Liste der Arten tierischer Einschlüsse in fossilen Harzen sowie ihrer Aufbewahrungsorte. *Deutsche Entomologische Zeitschrift* (N.F.) **29**: 129–286. [15.iii.1982.]

Kempf, W.W. 1949. A new species and subspecies of *Procryptocerus* from Espírito Santo, Brazil. *Revista de Entomologia* **20**: 423–426. [31.viii.1949.]

Kempf, W.W. 1951. A taxonomic study on the ant tribe Cephalotini. *Revista de Entomologia* **22**: 1–244. [31.xii.1951.]

Kempf, W.W. 1952. A synopsis of the *pinelii*-complex in the genus *Paracryptocerus. Studia Entomologica* **1**: 1–30. [15.v.1952.]

Kempf, W.W. 1953. Uma nova espécie de *Paracryptocerus* da Colômbia, praga do cafeeiro. *Papéis Avulsos de Departamento de Zoologia* **11**: 79–88. [10.ii.1953.]

Kempf, W.W. 1954. A descoberta do primeiro macho do gênero *Thaumatomyrmex* Mayr. *Revista Brasileira de Entomologia* **1**: 47–52. [(31).i.1954.]

Kempf, W.W. 1957. Sôbre algumas espécies de *Procryptocerus* com a descriçao duma espécie nova. *Revista Brasileira de Biologia* **17**: 395–404. [(30).ix.1957.]

Kempf, W.W. 1958a. New studies of the ant tribe Cephalotini. *Studia Entomologica* (N.S.) **1**: 1–176. [31.i.1958.]

Kempf, W.W. 1958b. Three new ants of the genus *Strumigenys* from Colombia. *Revista Brasileira de Entomologia* **8**: 59–68. [15.iii.1958.]

Kempf, W.W. 1958c. Sôbre algumas formigas neotrópicas do gênero *Leptothorax* Mayr. *Anais da Academia Brasileira de Ciencias* **30**: 91–102. [31.iii.1958.]

Kempf, W.W. 1958d. Synonymic note on ants of the genus *Paracryptocerus. Entomological News* **69**: 108–110. [7.v.1958.]

Kempf, W.W. 1958e. Discovery of the ant genus *Wadeura* in Brazil. *Revista Brasileira de Entomologia* **8**: 175–180. [15.v.1958.]

Kempf, W.W. 1958f. Estudos sôbre *Pseudomyrmex*. 2. *Studia Entomologica* (N.S.) **1**: 433–462. [29.xi.1958.]

Kempf, W.W. 1958g. The ants of the tribe Dacetini in the state of Sao Paulo, Brazil, with the description of a new species of *Strumigenys. Studia Entomologica* (N.S.) **1**: 553–560. [29.xi.1958.]

Kempf, W.W. 1959a. Sôbre algumas formigas Cephalotini do Museu de Oxford. *Revista Brasileira de Biologia* **19**: 91–98. [(30).iv.1959.]

Kempf, W.W. 1959b. A revision of the Neotropical ant genus *Monacis* Roger. *Studia Entomologica* (N.S.) **2**: 225–270. [29.ix.1959.]

Kempf, W.W. 1959c. A synopsis of the New World species belonging to the *Nesomyrmex*-group of the ant genus *Leptothorax* Mayr. *Studia Entomologica* (N.S.) **2**: 391–432. [29.ix.1959.]

Kempf, W.W. 1959d. Two new species of *Gymnomyrmex* Borgmeier, 1954, from southern Brazil, with remarks on the genus. *Revista Brasileira de Biologia* **19**: 337–344. [(31).x.1959.]

Kempf, W.W. 1960a. Estudo sôbre *Pseudomyrmex*. 1. *Revista Brasileira de Entomologia* **9**: 5–32. [21.v.1960.]

Kempf, W.W. 1960b. *Phalacromyrmex*, a new ant genus from southern Brazil. *Revista Brasileira de Biologia* **20**: 89–92. [(31).v.1960.]

Kempf, W.W. 1960c. *Tranopeltoides* Wheeler, a synonym of *Crematogaster* Lund. *Entomological News* **71**: 173–175. [14.vii.1960.]

Kempf, W.W. 1960d. A review of the ant genus *Mycetarotes* Emery. *Revista Brasileira de Biologia* **20**: 277–283. [(31).x.1960.]

Kempf, W.W. 1960e. Insecta Amapaensia—Hymenoptera: Formicidae. (Segunda contribuiçao.) *Studia Entomologica* (N.S.) **3**: 385–400. [20.xii.1960.]

Kempf, W.W. 1960f. Miscellaneous studies on Neotropical ants. *Studia Entomologica* (N.S.) **3**: 417–466. [20.xii.1960.]

Kempf, W.W. 1961a. Estudos sôbre *Pseudomyrmex*. 3. *Studia Entomologica* (N.S.) **4**: 369–408. [10.x.1961.]

Kempf, W.W. 1961b. A survey of the ants of the soil fauna in Surinam. *Studia Entomologica* (N.S.) **4**: 481–524. [10.x.1961.]

Kempf, W.W. 1961c. As formigas do gênero *Pachycondyla* Fr. Smith no Brasil. *Revista Brasileira de Entomologia* **10**: 189–204. [26.xii.1961.]

Kempf, W.W. 1961d. Remarks on the ant genus *Irogera* Emery, with the description of a new species. *Revista Brasileira de Biologia* **21**: 435–441. [(31).xii.1961.]

Kempf, W.W. 1962a. Retoques à classificaçao das formigas neotropicas do gênero *Heteroponera. Papeís Avulsos do Departamento de Zoologia* **15**: 29–47. [2.vii.1962.]

Kempf, W.W. 1962b. Miscellaneous studies on Neotropical ants. 2. *Studia Entomologica* (N.S.) **5**: 1–38. [20.x.1962.]

Kempf, W.W. 1963a. Additions to the Neotropical ant genus *Rogeria* Emery, with a key to the hitherto recorded South American species. *Revista Brasileira de Biologia* **23**: 189–196. [(31).viii.1963.]

Kempf, W.W. 1963b. A review of the ant genus *Mycocepurus* Forel, 1893. *Studia Entomologica* (N.S.) **6**: 417–432. [10.xii.1963.]

Kempf, W.W. 1963c. Nota sinonímia acêrca de formigas da tribo Cephalotini. *Revista Brasileira de Biologia* **23**: 435–438. [(31).xii.1963.]

Kempf, W.W. 1964a. Uma nova *Platythyrea* do Brasil. *Revista Brasileira do Entomologia* **11**: 141–144. [15.v.1964.]

Kempf, W.W. 1964b. Additions to the knowledge of the Cephalotini ants. *Papéis Avulsos do Departamento de Zoologia* **16**: 243–255. [22.vii.1964.]

Kempf, W.W. 1964c. A propósito de um estudo sôbre as formigas do gênero *Acanthostichus* Mayr. *Papéis Avulsos do Departamento de Zoologia* **16**: 263–266. [22.vii.1964.]

Kempf, W.W. 1964d. A revision of the Neotropical fungus-growing ants of the genus *Cyphomyrmex* Mayr. Part 1. Group of *strigatus* Mayr. *Studia Entomologica* (N.S.) **7**: 1–44. [10.xii.1964.]

Kempf, W.W. 1964e. Miscellaneous studies on Neotropical ants. 3. *Studia Entomologica* (N.S.) **7**: 45–71. [10.xii.1964.]

Kempf, W.W. 1964f. The ants of the genus *Anochetus (Stenomyrmex)* in Brazil. *Studia Entomologica* (N.S.) **7**: 237–246. [10.xii.1964.]

Kempf, W.W. 1965. Nota preliminar sôbre algumas formigas neotropicas, descritas por Frederick Smith. *Revista Brasileira de Biologia* **25**: 181–186. [(31).vii.1965.]

Kempf, W.W. 1966. A revision of the Neotropical fungus-growing ants of the genus *Cyphomyrmex* Mayr. Part 2. Group of *rimosus* (Spinola). *Studia Entomologica* (N.S.) **8** (1965): 161–200. [28.ii.1966.]

Kempf, W.W. 1967a. New ants from southeastern and central Brazil. *Studia Entomologica* (N.S.) **9** (1966): 121–128. [30.i.1967.]

Kempf, W.W. 1967b. A synopsis of the Neotropical ants of the genus *Centromyrmex* Mayr. *Studia Entomologica* (N.S.) **9** (1966): 401–410. [30.i.1967.]

Kempf, W.W. 1967c. Estudos sôbre *Pseudomyrmex*. 4. *Revista Brasileira de Entomologia* **12**: 1–12. [(28).ii.1967.]

Kempf, W.W. 1967d. Three new South American ants. *Studia Entomologica* (N.S.) **10**: 353–360. [15.xii.1967.]

Kempf, W.W. 1967e. A new revisionary note on the genus *Paracryptocerus* Emery. *Studia Entomologica* (N.S.) **10**: 361–368. [15.xii.1967.]

Kempf, W.W. 1968a. A new species of *Cyphomyrmex* from Colombia, with further remarks on the genus. *Revista Brasileira de Biologia* **28**: 35–41. [(30).iv.1968.]

Kempf, W.W. 1968b. Miscellaneous studies on Neotropical ants. 4. *Studia Entomologica* (N.S.) **11**: 369–415. [30.ix.1968.]

Kempf, W.W. 1969. Miscellaneous studies on Neotropical ants. 5. *Studia Entomologica* (N.S.) **12**: 273–296. [28.xi.1969.]

Kempf, W.W. 1970a. Levantamento das formigas da Mata Amazônica, nos Arredores de Belém do Pará, Brasil. *Studia Entomologica* (N.S.) **13**: 321–344. [30.xi.1970.]

Kempf, W.W. 1970b. Taxonomic notes on ants of the genus *Megalomyrmex* Forel, with the description of new species. *Studia Entomologica* (N.S.) **13**: 353–364. [30.xi.1970.]

Kempf, W.W. 1971. A preliminary review of the ponerine ant genus *Dinoponera* Roger. *Studia Entomologica* (N.S.) **14**: 369–394. [29.xi.1971.]

Kempf, W.W. 1972a. Catálogo abreviado das formigas da Regiao Neotropical. *Studia Entomologica* (N S.) **15**: 3–344. [25.viii.1972.]

Kempf, W.W. 1972b. A study of some Neotropical ants of genus *Pheidole* Westwood. *Studia Entomologica* (N.S.) **15**: 449–464. [25.viii.1972.]

Kempf, W.W. 1972c. A new species of the dolichoderine ant genus *Monacis* Roger, from the Amazon, with further remarks on the genus. *Revista Brasileira de Biologia* **32**: 251–254. [(31).viii.1972.]

Kempf, W.W. 1973a. Uma nova *Solenopsis* do Rio Grande do Sul, Brasil. *Revista Brasileira de Entomologia* **17**: 29–32. [14.v.1973.]

Kempf, W.W. 1973b. A revision of the Neotropical myrmicine ant genus *Hylomyrma* Forel. *Studia Entomologica* (N.S.) **16**: 225–260. [31.x.1973.]

Kempf, W.W. 1973c. A new *Zacryptocerus* from Brazil, with remarks on the generic classification of the tribe Cephalotini. *Studia Entomologica* (N.S.) **16**: 449–462. [31.x.1973.]

Kempf, W.W. 1974a. Taxonomic and faunistic notes on some Neotropical Cephalotini ants. *Revista Brasileira de Entomologia* **18**: 67–76. [15.viii.1974.]

Kempf, W.W. 1974b. A review of the Neotropical ant genus *Oxyepoecus* Santschi. *Studia Entomologica* (N.S.) **17**: 471–512. [31.x.1974.]

Kempf, W.W. 1974c. A remarkable new Neotropical species in the ant genus *Odontomachus* Latreille. *Studia Entomologica* (N.S.) **17**: 551–553. [31.x.1974.]

Kempf, W.W. 1975a. Report on Neotropical dacetine ant studies. *Revista Brasileira de Biologia* **34** (1974): 411–424. [26.iii.1975.]

Kempf, W.W. 1975b. A revision of the Neotropical ponerine ant genus *Thaumatomyrmex* Mayr. *Studia Entomologica* (N.S.) **18**: 95–126. [30.xi.1975.]

Kempf, W.W. 1975c. Miscellaneous studies on Neotropical ants. 6. *Studia Entomologica* (N.S.) **18**: 341–380. [30.xi.1975.]

Kempf, W.W. 1976a. Discovery of a major worker in *Camponotus branneri* (Mann), a new combination. *Psyche* **83**: 106–111. [4.xi.1976.]

Kempf, W.W. 1976b. A new species of *Strumigenys* from the lower Amazon, Brazil. *Studia Entomologica* (N.S.) **19**: 39–44. [30.xii.1976.]

Kempf, W.W. 1978. Five new synonyms for the Argentine ant fauna. *Studia Entomologica* (N.S.) **20**: 35–38. [30.viii.1978.]

Kempf, W.W. & Brown, W.L.,Jr. 1968. Report on some Neotropical ant studies. *Papéis Avulsos de Zoologia* **22**: 89–102. [4.xii.1968.]

Kempf, W.W. & Brown, W.L.,Jr. 1969. Two new *Strumigenys* ants from the Amazon Valley in Brasil. *Revista Brasileira de Biologia* **29**: 17–24. [(30).iv.1969.]

Kempf, W.W. & Brown, W.L.,Jr. 1970. Two new ants of tribe Ectatommini from Colombia. *Studia Entomologica* (N.S.) **13**: 311–320. [30.xi.1970.]

Kempf, W.W. & Lenko, K. 1976. Levantamento da formicifauna no litoral norte e ilhas adjacentes do Estado de Sao Paulo, Brasil. 1. Subfamilias Dorylinae, Ponerinae, e Pseudomyrmecinae. *Studia Entomologica* (N.S.) **19**: 45–66. [30.xii.1976.]

Kennedy, C.H. 1938. *Solenopsis rosella* Kennedy, a new ant from southern Ontario. *Canadian Entomologist* **70**: 232–236. [26.xi.1938.]

Kennedy, C.H. & Dennis, C.A. 1937. New ants from Ohio and Indiana, *Formica prociliata, F. querquetulana, F. postoculata* and *F. lecontei*. *Annals of the Entomological Society of America* **30**: 531–544. [14.x.1937.]

Kennedy, C.H. & Schramm, M.M. 1933. A new *Strumigenys* with notes on Ohio species. *Annals of the Entomological Society of America* **26**: 95–104. [22.iii.1933.]

Kennedy, C.H. & Talbot, M. 1939. Notes on the hypogaeic ant, *Proceratium silaceum*. *Proceedings of the Indiana Academy of Science* **48**: 202–210 [(31.xii).1939.]

Kessler, K. 1868. Materialy dlya poznaniya Onezhskago ozera i Obonezhskago kraya, preimushchestvenno v zoologicheskom' otnoshenii. *Trudy Russkikh Estestvoispytatelei* **1**: 1–144. [(31.xii).1868.]

Kieffer, J.-J. 1896. Diagnose de deux espèces nouvelles de Cécidomyies. *Bulletin de la Société Entomologique de France* **1896**: 236–237. [27.v.1896.]

Kim, B.J. 1988. A systematic study of the subfamily Formicinae from Korea. *Korean Journal of Entomology* **18**: 119. [28.iv.1988.]

Kim, B.J. & Kim, C.-W. 1986. On the one new species, *Camponotus jejuensis* (n. sp.) from Korea. *Korean Journal of Entomology* **16**: 139–141. [25.x.1986.]

Kinomura, K. & Yamauchi, K. 1992. A new workerless socially parasitic species of the genus *Vollenhovia* from Japan. *Japanese Journal of Entomology* **60**: 203–206. [25.iii.1992.]

Kirby, W. 1819. A description of several new species of insects collected in New Holland by Robert Brown, Esq. F.R.S. Lib. Linn. Soc. *Transactions of the Linnean Society of London* **12** (1818): 454–478. [2.vii.1819. For date of publication see Raphael, 1970: 64.]

Kirby, W. 1837. The insects. In Richardson, J. *Fauna Boreali- Americana; or the Zoology of the Northern Parts of British America* **4**: 325 pp. Norwich. [(31.xii).1837.]

Kirby, W.F. 1888. On the insects (exclusive of Coleoptera and Lepidoptera) of Christmas Island. *Proceedings of the Zoological Society of London* **1888**: 546–555. [4.xii.1888.]

Kirby, W.F. 1896. Hymenoptera. In Spencer, B. *Report on the Work of the Horn Scientific Expedition to Central Australia* **1**: 203–209. Melbourne. [(30).ix.1896.]

Kiseleva, E.F. 1925. K faune murav'ev Ussuriiskago kraya. *Izvestiya Tomskogo Gosudarstvennogo Universiteta* **75**: 73–75. [(31.xii).1925.]

Klug, F. 1824. *Entomologische Monographieen:* 242 pp. Berlin. [(31.xii).1824.]

Kohout, R.J. 1987. Three new *Polyrhachis sexspinosa*-group species from the Philippines. *Memoirs of the Queensland Museum* **25**: 169–176. [31.x.1987.]

Kohout, R.J. 1988a. New nomenclature of the Australian ants of the *Polyrhachis gab* Forel species complex. *Australian Entomological Magazine* **15**: 49–52. [15.vi.1988.]

Kohout, R.J. 1988b. A new species of *Polyrhachis (Polyrhachis)* from Papua New Guinea with a review of the New Guinean and Australian species. *Memoirs of the Queensland Museum* **25**: 417–427. [7.xi.1988.]

Kohout, R.J. 1988c. Nomenclatural changes and new Australian records in the ant genus *Polyrhachis* Fr. Smith. *Memoirs of the Queensland Museum* **25**: 429–438. [7.xi.1988.]

Kohout, R.J. 1989. The Australian ants of the *Polyrhachis relucens* species-group. *Memoirs of the Queensland Museum* **27**: 509–516. [13.xi.1989.]

Kohout, R.J. 1990. A review of the *Polyrhachis viehmeyeri* species-group. *Memoirs of the Queensland Museum* **28**: 499–508. [31.viii.1990.]

Kohout, R.J. 1994a. New synonymy of three Australian ants. *Memoirs of the Queensland Museum* **35**: 135–136. [1.vi.1994.]

Kohout, R.J. 1994b. *Polyrhachis lama*, a new ant from the Tibetan plateau. *Memoirs of the Queensland Museum* **35**: 137–138. [1.vi.1994.]

Kohout, R.J. & Taylor, R.W. 1990. Notes on Australian ants of the genus *Polyrhachis* Fr. Smith, with a synonymic list of the species. *Memoirs of the Queensland Museum* **28**: 509–522. [31.viii.1990.]

Kolossov, J. 1932. Synonymische u. systematische Bemerkungen über palaearctische Insekten. *Entomologisches Nachrichtenblatt, Troppau* **6**: 115–118. [(31).xii.1932.]

Kono, H. & Sugihara, Y. 1939. The ants of fir and spruce woodland in Japan. (Translated from Japanese.) *Transactions of the Kansai Entomological Society* **8**: 8–14. [Not seen. Reference from Yasumatsu, 1941: 182.]

Kohriba, O. 1963. A parasitic life of *Polyrhachis lamellidens* F.Smith. First report. *Kontyû* **31**: 200–209. [30.ix.1963.]

Kratochvíl, J. 1941. In Novák, V. & Sadil, J. Klíc k urcování mravencu stredni Evropy se zvlástním zretelem k mravencí zvírene Cech a Moravy. *Entomologické Príručky (Entomologickych Listu v Brne)* **4**: 65–115. [(31.xii).1941.]

Kratochvíl, J., Novák, V. & Snoflák, J. 1944. Mohelno. Soubor praci venovanych studiu vyznamné památky prírodni. 5. Hymenoptera—Aculeata. Formicidae—Apidae—Vespidae. *Archiv Svazu na Ochranu Prírody a Domoviny na Morave* **6**: 1–155. [(31.xii).1944.]

Krausse, A.H. 1910. Über Stridulationstöne bei Ameisen. *Zoologischer Anzeiger* **35**: 523–526. [15.iii.1910.]

Krausse, A.H. 1911. Über *Messor structor* Ltr. und einige andere Ameisen auf Sardinien. *Bullettino della Società Entomologica Italiana* **41** (1909): 14–18. [28.ii.1911.]

Krausse, A.H. 1912a. Eine neue Ameisenform von Sardinien (*Pheidole pallidula* v. n. *emeryi* m.) *Internationale Entomologische Zeitschrift* **6**: 169. [14.ix.1912.]

Krausse, A.H. 1912b. Ueber sardische Ameisen. *Archiv für Naturgeschichte* **78** (A7): 162–166. [(31).x.1912.]

Krausse, A.H. 1922. *Formica rufa pratensis incisa* m. v. n. *Entomologisches Jahrbuch* **31**: 155. [(31.xii).1922.]

Krausse, A.H. 1926a. *Formica rufa* an Weidenkätzchen. *Internationale Entomologische Zeitschrift* **20**: 107–108. [1.vii.1926.]

Krausse, A.H. 1926b. Waldameisen-Varietäten. *Internationale Entomologische Zeitschrift* **20**: 114–115. [8.vii.1926.]

Krausse, A.H. 1926c. *Formica rufa aruncicola menozii* m. f. n. *Entomologische Zeitschrift* **40**: 336. [8.x.1926.]

Krausse, A.H. 1926d. Über einige Formen der *Formica rufa* und *exsecta*. *Internationale Entomologische Zeitschrift* **20**: 264. [1.xi.1926.]

Kroyer, H.N. 1846. *Danmarks Fiske* **3**: 1279 pp. Kjobenhavn [(31.xii).1846.]

Kugler, C. 1977. A new species of *Platythyrea* from Costa Rica. *Psyche* **83** (1976): 217–221. [24.ii.1977.]

Kugler, C. 1978a. A comparative study of the myrmicine sting apparatus. *Studia Entomologica* (N.S.) **20**: 413–548. [30.viii.1978.]

Kugler, C. 1978b. Pygidial glands in myrmicine ants. *Insectes Sociaux* **25**: 267–274. [(30).ix.1978.]

Kugler, C. 1979a. Description of the ergatoid queen of *Pogonomyrmex mayri*, with notes on the worker and male. *Psyche* **85** (1978): 169–183. [24.iv.1979.]

Kugler, C. 1979b. Further studies of the myrmicine sting apparatus: *Eutetramorium, Oxyopomyrmex,* and *Terataner. Psyche* **85** (1978): 255–264. [24.iv.1979.]

Kugler, C. 1980a. *Odontomachus cornutus* rediscovered. *Journal of the Kansas Entomological Society* **53**: 225–229. [2.i.1980.]

Kugler, C. 1980b. The sting apparatus in the primitive ants *Nothomyrmecia* and *Myrmecia. Journal of the Australian Entomological Society* **19**: 263–267. [15.xii.1980.]

Kugler, C. 1986. Stings of ants of the tribe Pheidologetini. *Insecta Mundi* **1**: 221–230. [(31).xii.1986.]

Kugler, C. 1991. Stings of ants of the tribe Ectatommini. *Insecta Mundi* **5**: 153–166. [(31).xii.1991.]

Kugler, C. 1992. Stings of ants of the Leptanillinae. *Psyche* **99**: 103–115. [4.xii.1992.]

Kugler, C. & Brown, W.L.,Jr. 1982. Revisionary and other studies on the ant genus *Ectatomma,* including the descriptions of two new species. *Search Agriculture* **24**: 1–7. [(30).xi.1982.]

Kugler, C. & Hincapié, M. del C. 1983. Ecology of the ant *Pogonomyrmex mayri:* distribution, abundance, nest structure and diet. *Biotropica* **15**: 190–198. [1983.]

Kugler, J. 1981. A new species of *Cataglyphis* Förster from Israel and Sinai. *Israel Journal of Entomology* **15**: 83–88. [(31).xii.1981.]

Kugler, J. 1984. The males of *Cardiocondyla* Emery, with a description of the winged male of *Cardiocondyla wroughtonii* (Forel). *Israel Journal of Entomology* **17** (1983): 1–21. [15.iv.1984.]

Kugler, J. 1987. The Leptanillinae of Israel and a description of a new species from India. *Israel Journal of Entomology* **20** (1986): 45–57. [(31).iii.1987.]

Kugler, J. & Soussan, O. 1992. Redescription and biological observations on *Leptothorax flavispinus* André. *Israel Journal of Entomology* **25–26**: 209–213. [15.iii.1992.]

Kulmatycki, W. 1922. Przyczynek do fauny mrówek Wielkpolski i Pomorza. *Sprawozdanie Komisji Fizjograficznej* **55–56**: 71–86. [(31.xii).1922.]

Kumbkarni, C.G. 1965. Cytological studies in Hymenoptera. Part 3. Cytology of parthenogenesis in the formicid ant, *Camponotus compressus. Caryologia* **18**: 305–312. [10.ix.1965.]

Kupyanskaya, A.N. 1980. Murav'i roda *Formica* Linnaeus Sovetskogo dal'nego Vostoka. In Ler, P.A. *Taksonomiya Nasekomykh dal'nego Vostoka:* 95–108. Vladivostok. [6.ii.1980.]

Kupyanskaya, A.N. 1984. Muravei *Formica yessensis* Forel, 1901 na yuge Primorskogo kraya. In Ler, P.A. *Fauna i Ekologiya Nasekomykh Yuga dal'nego Vostoka:* 98–119 Vladivostok. [6.i.1984.]

Kupyanskaya, A.N. 1986a. Murav'i gruppy *Myrmica lobicornis* na dal'nem Vostoke. In Ler, P.A. & Kupyanskaya, A.N. *Sistematika i Ekologiya Nasekomykh dal'nego Vostoka:* 83–90. Vladivostok. [16.i.1986.]

Kupyanskaya, A.N. 1986b. Murav'i severnoi chasti dal'nego Vostoka. In Ler, P.A. & Kupyanskaya, A.N. *Sistematika i Ekologiya Nasekomykh dal'nego Vostoka:* 91–102. Vladivostok. [16.i.1986.]

Kupyanskaya, A.N. 1988. Dal'nevostochnyi predstavitel' roda *Liometopum. Vestnyk Zoologii* **1988** (No. 1): 29–34. [26.i.1988.]

Kupyanskaya, A.N. 1989. Murav'i podroda *Dendrolasius* Ruzsky,1912 dal'nego Vostoka SSR. *Entomologicheskoe Obozrenie* **68**: 779–789. [8.xii.1989.]

Kupyanskaya, A.N. 1990. *Murav'i Dal'nego Vostoka SSR* (1989): 258 pp. Vladivostok. [13.ix.1990.]

Kurian, C. 1955. Bethyloidea from India. *Agra University Journal of Research* **4**: 67–155. [(31).i.1955.]

Kusnezov, N. Note. This was the name adopted post-1945 by Kuznetsov-Ugamsky, N.N., for which see below. Listings in this catalogue are given separately, according to the name in use at the time of publication of each paper.

Kusnezov, N. 1949a. El genero *Monomorium* en la Argentina. *Acta Zoologica Lilloana* **7**: 423–448. [31.xii.1949.]

Kusnezov, N. 1949b. Sobre la reproduccion de las formas sexuales en *Solenopsis patagonica* Emery. *Acta Zoologica Lilloana* **8**: 281–290. [31.xii.1949.]

Kusnezov, N. 1949c. *Pogonomyrmex* del grupo *Ephebomyrmex* en la fauna de la Patagonia. *Acta Zoologica Lilloana* **8**: 291–307. [31.xii.1949.]

Kusnezov, N. 1949d. El genero *Cyphomyrmex* en la Argentina. *Acta Zoologica Lilloana* **8**: 427–456. [31.xii.1949.]

Kusnezov, N. 1949e. *Crematogaster (Neocrema) descolei* n. sp. *Acta Zoologica Lilloana* **8**: 587–590. [31.xii.1949.]

Kusnezov, N. 1951a. El genero *Pogonomyrmex* Mayr. *Acta Zoologica Lilloana* **11**: 227–333. [13.xi.1951.]

Kusnezov, N. 1951b. *Myrmelachista* en la Patagonia. *Acta Zoologica Lilloana* **11**: 353–365. [13.xi.1951.]

Kusnezov, N. 1951c. *Dorymyrmex emmaericaellus* n. sp. *Folia Universitaria Cochabamba* **5**: 59–61. [(31).x.1951.]

Kusnezov, N. 1951d. *Descolemyrma ogloblini* genero y especie nuevos de la tribu Attini. *Acta Zoologica Lilloana* **11**: 459–465. [13.xi.1951.]

Kusnezov, N. 1952a. Un caso de evolucion eruptiva, *Eriopheidole symbiotica* nov. gen. nov. sp. *Memorias del Museo de Entre Rios* **29** (1951): 6–30. [30.i.1952.]

Kusnezov, N. 1952b. El género *Pheidole* en la Argentina. *Acta Zoologica Lilloana* **12** (1951): 2–88. [15.x.1952.]

Kusnezov, N. 1952c. *Lasiophanes* Emery en la Patagonia. *Acta Zoologica Lilloana* **12** (1951): 89–100. [15.x.1952.]

Kusnezov, N. 1952d. El género *Camponotus* en la Argentina. *Acta Zoologica Lilloana* **12** (1951): 183–252. [15.x.1952.]

Kusnezov, N. 1952e. El género *Wasmannia* en la Argentina. *Acta Zoologica Lilloana* **10**: 173–182. [(31.xii).1952.]

Kusnezov, N. 1952f. El género *Oligomyrmex* Mayr en la Argentina. *Acta Zoologica Lilloana* **10**: 183–187. [(31.xii).1952.]

Kusnezov, N. 1952g. El estado real del grupo *Dorymyrmex* Mayr. *Acta Zoologica Lilloana* **10**: 427–448. [(31.xii).1952.]

Kusnezov, N. 1952h. Acerca de las hormigas simbióticas del género *Martia* Forel. *Acta Zoologica Lilloana* **10**: 717–722. [(31.xii).1952.]
Kusnezov, N. 1953a. Las hormigas en los parques nacionales de la Patagonia y los problemas relacionados. *Anales del Museo Nahuel Huapí* **3**: 105–124. [(30).xi.1953.]
Kusnezov, N. 1953b. Lista de las hormigas de Tucuman con descripcion de dos nuevos generos. *Acta Zoologica Lilloana* **13**: 327–339. [(31).xii.1953.]
Kusnezov, N. 1953c. El genero *Synsolenopsis* Forel. *Acta Zoologica Lilloana* **13**: 341–348. [(31).xii.1953.]
Kusnezov, N. 1953d. *Bisolenopsis sea.* Un nuevo género y especie de hormigas y sus relaciones con los géneros vecinos. *Memorias del Museo de Entre Rios* **31**: 1–44. [(31.xii).1953.]
Kusnezov, N. 1953e. La fauna mirmecológica de Bolivia. *Folia Universitaria Cochabamba* **6**: 211–239. [(31.xii).1953. Not seen. Reference from Kempf, 1972a: 284.]
Kusnezov, N. 1954a. Un genero nuevo de hormigas. (*Paranamyrma solenopsidis* nov. gen. nov. sp.) y los problemas relacionados. *Memorias del Museo de Entre Rios* **30**: 7–21. [(30).iv.1954.]
Kusnezov, N. 1954b. Phyletische Bedeutung der Maxillar- und Labialtaster der Ameisen. *Zoologischer Anzeiger* **153**: 28–38. [(31).vii.1954.]
Kusnezov, N. 1955. Zwei neue Ameisengattungen aus Tucuman (Argentinien). *Zoologischer Anzeiger* **154**: 268–277. [(30).vi.1955.]
Kusnezov, N. 1956. Claves para la identificación de las hormigas de la fauna Argentina. *Idia, agosto-septiembre, Min. Agric. Ganad. Argent.* n. **104–105**: 1–56. [Not seen; reference from Kempf, 1972a: 284.]
Kusnezov, N. 1957a. Die Solenopsidinen-Gattungen von Südamerika. *Zoologischer Anzeiger* **158**: 266–280. [(30).vi.1957.]
Kusnezov, N. 1957b. Nuevas especies de hormigas. *Revista de la Sociedad Uruguaya de Entomologia* **2**: 7–18. [1.xii.1957.]
Kusnezov, N. 1958a. La posición sistemática del género *Rogeria,* con descripción de una nueva especie. *Acta Zoologica Lilloana* **15**: 41–45. [29.viii.1958.]
Kusnezov, N. 1958b. *Lilidris metatarsalis* gen. et spec. nov. *Acta Zoologica Lilloana* **15**: 189–193. [29.viii.1958.]
Kusnezov, N. 1958c. Nota sobre la sinonimia de *Linepithema* Mayr. *Acta Zoologica Lilloana* **16**: 273. [25.ix.1958.]
Kusnezov, N. 1959. Die Dolichoderinen-Gattungen von Süd-Amerika. *Zoologischer Anzeiger* **162**: 38–51. [(31).i.1959.]
Kusnezov, N. 1960a. La posicion sistematica del genero *Metapone* Forel. *Acta Zoologica Lilloana* **17** (1959): 119–126. [10.vi.1960.]
Kusnezov, N. 1960b. La fauna de hormigas en el oeste de la Patagonia. *Acta Zoologica Lilloana* **17** (1959): 321–401. [10.vi.1960.]
Kusnezov, N. 1960c. *Brachymyrmex physogaster* n. sp. aus Argentinien und das Problem der Physogastrie bei den Ameisen. *Zoologischer Anzeiger* **165**: 382–388. [(30).xi.1960.]
Kusnezov, N. 1962a. El género *Acanthostichus* Mayr. *Acta Zoologica Lilloana* **18**: 121–138. [23.v.1962.]
Kusnezov, N. 1962b. Una nueva especie del género *Brownidris* Kusnezov. *Acta Zoologica Lilloana* **18**: 155–161. [23.v.1962.]
Kusnezov, N. 1969. Nuevas especies de hormigas. *Acta Zoologica Lilloana* **24**: 33–38. [(31).vii.1969.]
Kutter, H. 1920. *Strongylognathus huberi* For. r. *alpinus* Wh. eine Sklaven raubende Ameise. *Biologisches Zentralblatt* **40**: 528–538. [15.xii.1920.]
Kutter, H. 1923. Der Sklavenräuber *Strongylognathus huberi* ssp. *alpinus* Wheeler. *Revue Suisse de Zoologie* **30**: 387–424. [(31).xii.1923.]
Kutter, H. 1924. *Myrmica scabrinodis* Nyl. var. *rugulosoides* Forel. *Schweizer Entomologischer Anzeiger* **3**: 8. [1.i.1924.]
Kutter, H. 1925. Eine neue Ameise der Schweiz. *Mitteilungen der Schweizerischen Entomologischen Gesellschaft* **13**: 409–412. [15.vii.1925.]
Kutter, H. 1927. Ein myrmekologischer Streifzug durch Sizilien. *Folia Myrmecologica et Termitologica* **1**: 94–104. [(30).iv.1927.]
Kutter, H. 1928. Beiträge zur Kenntnis nordafrikanischer Hymenopteren von Dr. A. v. Schulthess, unter Mitarbeit von Specialisten. A. Formicidae. *Eos. Revista Española de Entomologia* **4**: 65–67. [28.iv.1928.]
Kutter, H. 1931a. Mit Bananen eingeschleppte Ameisen. *Mitteilungen der Schweizerischen Entomologischen Gesellschaft* **15**: 61–63. [15.iii.1931.]
Kutter, H. 1931b. *Forelophilus,* eine neue Ameisengattung. *Mitteilungen der Schweizerischen Entomologischen Gesellschaft* **15**: 193–194. [15.xii.1931.]
Kutter, H. 1932. Ameisen aus dem Museum zu Dresden. *Mitteilungen der Schweizerischen Entomologischen Gesellschaft* **15**: 207–210. [15.iii.1932.]
Kutter, H. 1945. Eine neue Ameisengattung. *Mitteilungen der Schweizerischen Entomologischen Gesellschaft* **19**: 485–487. [20.x.1945.]
Kutter, H. 1948. Beitrag zur Kenntnis der Leptanillinae. Eine neue Ameisengattung aus süd-Indien. *Mitteilungen der Schweizerischen Entomologischen Gesellschaft* **21**: 286–295. [25.viii.1948.]
Kutter, H. 1950a. Über eine neue, extrem parasitische Ameise. *Mitteilungen der Schweizerischen Entomologischen Gesellschaft* **23**: 81–94. [10.viii.1950.]
Kutter, H. 1950b. Über zwei neue Ameisen. *Mitteilungen der Schweizerischen Entomologischen Gesellschaft* **23**: 337–346. [15.x.1950.]
Kutter, H. 1950c. Über *Doronomyrmex* und verwandte Ameisen. *Mitteilungen der Schweizerischen Entomologischen Gesellschaft* **23**: 347–353. [15.x.1950.]
Kutter, H. 1951. *Epimyrma stumperi* Kutter. 2 Mitteilung. *Mitteilungen der Schweizerischen Entomologischen Gesellschaft* **24**: 155–174. [20.vii.1951.]
Kutter, H. 1952. Über *Plagiolepis xene* Stärcke. *Mitteilungen der Schweizerischen Entomologischen Gesellschaft* **25**: 57–72. [15.v.1952.]
Kutter, H. 1956. Beiträge zur Biologie palaearktischer *Coptoformica. Mitteilungen der Schweizerischen Entomologischen Gesellschaft* **29**: 1–18. [5.iii.1956.]
Kutter, H. 1957. Zur Kenntnis schweizerischer *Coptoformica* Arten. *Mitteilungen der Schweizerischen Entomologischen Gesellschaft* **30**: 1–24. [30.v.1957.]
Kutter, H. 1963. Miscellanea myrmecologica. 1. *Mitteilungen der Schweizerischen Entomologischen Gesellschaft* **36**: 129–137. [30.vi.1963.]
Kutter, H. 1965. *Formica nigricans* Em. (=*cordieri* Bondr.). *Mitteilungen der Schweizerischen Entomologischen Gesellschaft* **37** (1964): 138–150. [15.ii.1965.]
Kutter, H. 1967a. Einige Ergebnisse weiterer *Coptoformica*- studien. *Insectes Sociaux* **13** (1966): 227–240. [(31).iii.1967.]
Kutter, H. 1967b. Beschreibung neuer Sozialparasiten von *Leptothorax acervorum* F. *Mitteilungen der Schweizerischen Entomologischen Gesellschaft* **40**: 78–91. [20.vii.1967.]
Kutter, H. 1968. Die sozialparasitischen Ameisen der Schweiz. *Neujahrsblatt Naturforschenden Gesellschaft in Zürich* **1969**: 62 pp. Zürich. [31.xii.1968.]
Kutter, H. 1971. Taxonomische Studien an Schweizer Ameisen. *Mitteilungen der Schweizerischen Entomologischen Gesellschaft* **43**: 258–271. [15.iii.1971.]
Kutter, H. 1973a. Über *Xenhyboma mystes* Santschi. *Mitteilungen der Schweizerischen Entomologischen Gesellschaft* **45** (1972): 321–324. [28.ii.1973.]
Kutter, H. 1973b. *Leptothorax arcanus* n. sp. (?)—eine neue und Rätselhafte Schmalbrustameise. *Mitteilungen der Schweizerischen Entomologischen Gesellschaft* **45** (1972): 327–328. [28.ii.1973.]
Kutter, H. 1973c. Über die morphologischen Beziehungen der Gattung *Myrmica* zu ihren Satellitengenera *Sifolinia* Em., *Symbiomyrma* Arnoldi und *Sommimyrma* Menozzi. *Mitteilungen der Schweizerischen Entomologischen Gesellschaft* **46**: 253–268. [28.xii.1973.]
Kutter, H. 1973d. Zur Taxonomie der Gattung *Chalepoxenus. Mitteilungen der Schweizerischen Entomologischen Gesellschaft* **46**: 269–280. [28.xii.1973.]
Kutter, H. 1973e. Beitrag zur Lösung Taxonomischer Probleme in der Gattung *Epimyrma. Mitteilungen der Schweizerischen Entomologischen Gesellschaft* **46**: 281–289. [28.xii.1973.]
Kutter, H. 1976a. Beitrag zur Kenntnis der Gattung *Calyptomyrmex* Emery. *Mitteilungen der Schwiezerischen Entomologischen Gesellschaft* **49**: 269–272. [31.xii.1976.]
Kutter, H. 1976b. Beitrag zur Kenntnis der Gattung *Simopone* Forel. *Mitteilungen der Schweizerischen Entomologischen Gesellschaft* **49**: 273–276. [31.xii.1976.]

Kutter, H. 1977a. Zur Kenntnis der Gattung *Hylomyrma* Forel. *Mitteilungen der Schweizerischen Entomologischen Gesellschaft* **50**: 85–89. [31.iii.1977.]

Kutter, H. 1977b. Zweiter Beitrag zur Kenntnis der Gattung *Simopone* Forel. *Mitteilungen der Schweizerischen Entomologischen Gesellschaft* **50**: 173–176. [30.vi.1977.]

Kutter, H. 1977c. *Insecta Helvetica Fauna* **6**. Hymenoptera, Formicidae: 298 pp. Zürich. [(31.xii).1977.]

Kuznetsov-Ugamsky, N.N. Note. This name appears in many different spellings in the literature, e.g. Kuznetzov-Ugamskij, Kusnezow-Ugamsky, Kusnezov-Ugamsky, etc. The name is stabilised here at the above spelling, which is a modern transliteration from the Cyrillic alphabet version of the name. Post-1945 this author used the name Kusnezov, N., for which see above.

Kuznetsov-Ugamsky, N.N. 1923. Mirmekologicheskaya fauna Tashkentskogo uezda. *Trudy Turkestanskogo Nauchnogo Obshchestva* **1**: 239–558. [(31.xii).1923.]

Kuznetsov-Ugamsky, N.N. 1926a. Novye Turkestanskie murav'i. *Russkoe Entomologicheskoe Obozrenie* **20**: 71–77. [(31.xii).1926.]

Kuznetsov-Ugamsky, N.N. 1926b. Materialy po mirmekologii Turkestana. *Russkoe Entomologicheskoe Obozrenie* **20**: 93–100. [(31.xii).1926.]

Kuznetsov-Ugamsky, N.N. 1927a. Vorläufige Uebersicht über die mittelasiatischen Formen der Gatting *Messor. Folia Myrmecologica et Termitologica* **1**: 89–94. [(30).iv.1927.]

Kuznetsov-Ugamsky, N.N. 1927b. Zur Kenntnis der Gattung *Proformica* Ruzsky. *Societas Entomologica* **42**: 26–28. [1.vi.1927.]

Kuznetsov-Ugamsky, N.N. 1927c. *Polyergus rufescens tianschanicus* subsp. nov. aus Turkestan. *Societas Entomologica* **42**: 41–42. [1.xi.1927.]

Kuznetsov-Ugamsky, N.N. 1927d. Materialy po mirmekologii Turkestana. 2. *Russkoe Entomologicheskoe Obozrenie* **21**: 33–42. [(31.xii).1927.]

Kuznetsov-Ugamsky, N.N. 1927e. Materialy po mirmekologii Turkestana. 3. *Russkoe Entomologicheskoe Obozrenie* **21**: 186–196. [(31.xii).1927.]

Kuznetsov-Ugamsky, N.N. 1928a. Beiträge zur Ameisenfauna Mittelasiens. 1. Die Gattung *Proformica* Ruzsky, 1903. *Zoologischer Anzeiger* **75**: 7–23. [2.i.1928.]

Kuznetsov-Ugamsky, N.N. 1928b. Murav'i Yuzhno-Ussuriiskogo kraya. *Zapiski Vladivostokskogo Otdela Gosudarstvennogo, Russkogo Geograficheskogo Obshchestva (Obshchestva Izucheniya Amurskog kraya.)* **1** (18): 1–47. [(31.xii).1928.]

Kuznetsov-Ugamsky, N.N. 1929a. Die Ameisen des Süd-Ussuri- Gebietes. *Zoologischer Anzeiger* **83**: 16–34. [15.vi.1929.]

Kuznetsov-Ugamsky, N.N. 1929b. Die Ameisenfauna Daghestans. *Zoologischer Anzeiger* **83**: 34–45. [15.vi.1929.]

Kuznetsov-Ugamsky, N.N. 1929c. Die Gattung *Acantholepis* in Turkestan. *Zoologischer Anzeiger* **82**: 477–492. [(31.xii).1929.]

Kuznetsov-Ugamsky, N.N. 1929d. Formy roda *Messor* v faune Srednei Azii. *Trudy Sredne-Aziatskogo Gosudarstvennogo Universiteta.* Seriya 8-a, Zoologiya **6**: 1–25. [(31.xii).1929.]

Labram, J.D. & Imhoff, L. 1838. *Insekten der Schweiz* **2**: 84 pls. Basel. [(31.xii).1838.]

Lacordaire, T. 1848. Monographie des coléoptères subpentamères de la famille des Phytophages. *Mémoires de la Société Royale des Sciences de Liège* **5** (tome second): 1–890. [(31).v.1848.]

Lamarck, J.B.P.A. de Monet de. 1817. *Histoire Naturelle des Animaux sans Vertèbres* **4**: 603 pp. Paris. [(31).iii.1817.]

Latreille, P.A. 1798. *Essai sur l'Histoire des Fourmis de la France:* 50 pp. Brive. [(31.xii).1798.]

Latreille, P.A. 1802a. *Histoire Naturelle, Générale et Particulière, des Crustacés et des Insectes* **3**: 467 pp. Paris. [6.xi.1802. Date of publication, see Dupuis, 1986: 208.]

Latreille, P.A. 1802b. Description d'une nouvelle espèce de fourmi. *Bulletin des Sciences par la Société Philomathique* **3**: 65–66. [(31.xii).1802.]

Latreille, P.A. 1802c. *Histoire Naturelle des Fourmis*, et recueil de mémoires et d'observations sur les abeilles, les araignées, les faucheurs, et autres insectes: 445 pp. Paris. [(31.xii).1802.]

Latreille, P.A. 1803. *Histoire Naturelle, Générale et Particulière, des Crustacés et des Insectes* **5**: 406 pp. Paris. [14.iv.1803. Date of publication, see Dupuis, 1986: 208.]

Latreille, P.A. 1804. Tableau méthodique des insectes. Classe huitième. Insectes, Insecta. *Nouveau Dictionnaire d'Histoire Naturelle* **24**: 129–200. [21.iii.1804.]

Latreille, P.A. 1809. *Genera Crusaceorum et Insectorum* secundum ordinem naturalem in familias disposita, iconibus, exemplisque plurimis explicata. **4**: 399 pp. Parisiis et Argentorati. [20.iii.1809.]

Latreille, P.A. 1810. *Considérations générales sur l'ordre naturel des animaux composant les classes des Crustacés, des Arachnides, et des Insectes; avec un tableau méthodique de leurs genres, disposés en famille:* 444 pp. Paris. [23.v.1810.]

Latreille, P.A. 1817. Insectes de l'Amérique equinoxiale, recueilles pendant le voyage de MM. Humboldt et Bonpland. Seconde partie (pp. 9–138). In Humboldt, F.H.A. von & Bonpland, A. *Voyage aux Régions Equinoxiales du Nouveau Continent, fait en 1799–1804. Recueil d'Observations de Zoologie et d'Anatomie Comparée* **2**: 352 pp. Paris. [17.xii.1817.]

Latreille, P.A. 1818. P. 50 in: *Nouveau Dictionnaire d'Histoire Naturelle,* appliquée aux Artes. à l'Agriculture, à l'Economie rurale et domestique, à la Médecine, etc. Nouvelle edition. **23**: 612 pp. Paris. [5.ix.1818.]

Lattke, J.E. 1987a. Two new species of Neotropical *Anochetus* Mayr. *Insectes Sociaux* **33** (1986): 352–358. [(31).iii.1987.]

Lattke, J.E. 1987b. Notes on the ant genus *Hypoclinea* Mayr, with descriptions of three new species. *Revista de Biologia Tropical* **34** (1986): 259–265. [(31).viii.1987.]

Lattke, J.E. 1990a. A new genus of myrmicine ants from Venezuela. *Entomologica Scandinavica* **21**: 173–178. [(30).vi.1990.]

Lattke, J.E. 1990b. Revisión del género *Gnamptogenys* Roger en Venezuela. *Acta Terramaris* **2**: 1–47. [(30).vi.1990.]

Lattke, J.E. 1990c. Una nueva especie de *Pogonomyrmex* Mayr de selva humeda tropical. *Revista de Biologia Tropical* **38** (2A): 305–309. [(30).xi.1990.]

Lattke, J.E. 1991a. The genus *Proceratium* Roger in Dominican Amber. *Caribbean Journal of Science* **26** (1990): 101–103. [(31).iii.1991.]

Lattke, J.E. 1991b. Studies of Neotropical *Amblyopone* Erichson. *Contributions in Science* **428**: 1–7. [4.ix.1991.]

Lattke, J.E. 1992a. Revision of the *minuta*-group of the genus *Gnamptogenys. Deutsche Entomologische Zeitschrift* **39**: 123–129. [6.iii.1992.]

Lattke, J.E. 1992b. *Trichoscapa inusitata,* a remarkable dacetine ant from the Orinoco Watershed. *Tropical Zoology* **5**: 141–144. [(31).xii.1992.]

Lattke, J.E. 1994. Phylogenetic relationships and classification of ectatommine ants. *Entomologica Scandinavica* **25**: 105–119. [(31.xii).1994.]

Laza, J.H. 1982. Signos de actividad atribuibles a *Atta* en el Mioceno de la Provincia de La Pampa, Republica Argentina. Significacion paleozoogeografica. *Ameghiniana* **19**: 109–124. [15.x.1982.]

Leach, W.E. 1815. Entomology. In *Brewster's Edinburgh Encyclopaedia* **9**: 57–172. [(30).iv.1815.]

Leach, W.E. 1825. Descriptions of thirteen species of *Formica* and three species of *Culex,* found in the environs of Nice. *Zoological Journal* **2**: 289–293. [(31).x.1825.]

Le Guillou, E.J.F. 1842. Catalogue raisonné des insectes hyménoptères recueillis dans le voyage de circumnavigation des corvettes l'Astrolabe et La Zélée. *Annales de la Société Entomologique de France* **10** (1841): 311–324. [(31).i.1842]

Le Masne, G. 1956. Recherches sur les fourmis parasites. *Plagiolepis grassei* et l'évolution des *Plagiolepis* parasites. *Comptes Rendus Hebdomadaires des Séances de l'Acadaemie des Sciences* **243** (No. 7): 673–675. [(31.xii).1956.]

Lenko, K. 1964. Uma nova espécie de *Gnamptogenys* de Mato Grosso. *Papéis Avulsos do Departamento de Zoologia* **16**: 257–261. [22.vii.1964.]

Leo, P. & Fancello, L. 1990. Osservazioni sul genere *Leptanilla* Emery in Sardegna e riabilitazione di *L. doderoi* Emery. *Bollettino della Società Entomologica Italiana* **122**: 128–132. [30.ix.1990.]

Lepeletier de Saint-Fargeau, A.L.M. 1835. *Histoire Naturelle des Insectes.* Hyménoptères **1** (1836): 547 pp. Paris. [Handwritten note by Sherborn in BMNH copy states, "This volume was published in the week ending 26 Dec. 1835 (Bibl. France)".]

Leston D. 1971. The Ectatommini of Ghana. *Journal of Entomology* (B) **40**: 117–120. [26.xi.1971.]

Letendre, M. & Huot, L. 1972. Considérations préliminaires en vue de la revision taxonomique des fourmis du groupe *microgyna*, genre *Formica*. *Annales de la Société Entomologique de Québec* **17**: 117–132. [(30).ix.1972.]

Lever, R.J.A.W. 1930. The female of *Ectatomma tuberculatum* var *punctigerum* Emery. *Entomologist's Monthly Magazine* **66**: 214–215. [(30).ix.1930.]

Linnaeus, C. 1758. *Systema Naturae* per regna tria naturae, secundum classes, ordines, genera, species, cum characteribus, differentiis, synonymis, locis. Editio 10. **1**: 823 pp. Holmiae. [1.i.1758.]

Linnaeus, C. 1761. *Fauna Svecica*, sistens animalia Sveciae regni: mammalia, aves, amphibia, pisces, insecta, vermes: 578 pp. Stockholmiae. [(31.xii).1761.]

Linnaeus, C. 1763. Centuria insectorum, quam, praeside D.D. Car. von Linné, proposuit Boas Johansson, Calmariensis (Upsaliae 1763. Junii. 23) (pp. 384–415). *Amoenitates Academicae;* seu dissertationes variae physicae, medicae, botanicae, antehac seorsim editae, nunc collectae et auctae cum tabulis aeneis. **6**: 486 pp. Holmiae. [(31.xii).1763.]

Linnaeus, C. 1764. *Museum S:ae R:ae M:tis Ludovicae Ulricae Reginae Svecorum, Gothorum, Vandalorumque, &c.* In quo animalia rariora, exotica, imprimis. Insecta & Conchilia describuntur & determinantur. Prodromi instar: 719 pp. Holmiae. [(31.xii).1764.]

Linnaeus, C. 1767. *Systema Naturae.* Editio 12. **1** (part 2): 533–1327. Holmiae. [5.xi.1767.]

Linnaeus, C. 1771. *Mantissa Plantarum.* Regni Animalis Appendix: 588 pp. Holmiae. [(31).x.1771.]

Linsley, E.G. & Usinger, R.L. 1966. Insects of the Galapagos Islands. *Proceedings of the California Academy of Sciences* (4) **33**: 113–196. [31.i.1966.]

Llave, P. de la. 1832. Sobre las busileras u hormigas de miel. *Registro Trimestre; o Colección de Memorias de Historia, Literatura, Ciencias y Artes* **1** (4): 455–463. [(31).x.1832.]

Lofgren, C.S. & Vander Meer, R.K. (Eds.). 1986. *Fire Ants and Leaf-Cutting Ants. Biology and Management:* 435 pp. Westview studies in insect biology, Boulder & London.

Lohmander, H. 1949. Eine neue schwedische Ameise. Myrmekologische Fragmente. *Opuscula Entomologica* **14**: 163–167. [31.xii.1949.]

Loiselle, R., Francoeur, A., Fischer, K. & Buschinger, A. 1990. Variations and taxonomic significance of the chromosome numbers in the Nearctic species of the genus *Leptothorax* (*s. s.*). *Caryologia* **43**: 321–334. [1990.]

Lomnicki, J. 1922. Przyczynek do opisu królowej mrówki powolnicy europejskiej (*Sysphincta europaea* For.). *Polskie Pismo Entomologiczne* **1**: 3–4. [(31.xii).1922.]

Lomnicki, J. 1924. O trzech gynandromorfach mrówki wscieklicy marszczystej *Myrmica rugulosa* Nyl. *Kosmos. Czasopismo Polskiego Towarzystwa Przyrodników im Kopernika* **49**: 817–830. [1924.]

Lomnicki, J. 1925a. Przeglad polskich gatunków rodzaju mrówki (*Formica* Linné). *Polskie Pismo Entomologiczne* **3** (1924): 151–182. [1925. Date indeterminate, possibly iii.]

Lomnicki, J. 1925b. Une contribution a la connaissance de la faune des fourmis des îles Baléares. *Polskie Pismo Entomologiczne* **4**: 1–3. [15.iii.1925.]

Lomnicki, J. 1925c. *Plagiolepis vindobonensis* sp. n. *Polskie Pismo Entomologiczne* **4**: 77–79. [15.vi.1925.]

Lomnicki, J. 1928. Spis mrówek Lwowa i Okolicy. *Ksiegi Pamiatkowej (Lecia gimnazjum iv im. Jana Dlugosza we Lwowie)* **50**: 1–10. [(31.xii).1928.]

Longhurst, C. & Howse, P.E. 1978. The use of kairomones by *Megaponera foetens* (F.) in the detection of its termite prey. *Animal Behavior* **26**: 1213–1218. [1978.]

Longhurst, C., Johnson, R.A. & Wood, T.G. 1979. Foraging, recruitment and predation by *Decamorium uelense* (Santschi) on termites in southern Guinea Savannah, Nigeria. *Oecologia* **38**: 83–91. [1979.]

Longino, J.T. 1988. Notes on the taxonomy of the Neotropical ant genus *Thaumatomyrmex* Mayr (pp. 35–42). In Trager, J.C. (Ed.). *Advances in Myrmecology:* 551 pp. New York. [(31.xii).1988.]

Longino, J.T. 1989. Taxonomy of the *Cecropia*-inhabiting ants in the *Azteca alfari* species group: evidence for two broadly sympatric species. *Contributions in Science* **412**: 1–16. [19.xii.1989.]

Longino, J.T. 1991. Taxonomy of the *Cecropia*-inhabiting *Azteca* ants. *Journal of Natural History* **25**: 1571–1602. [31.xii.1991.]

López, F. 1991a. Estudio morfológico y taxonómico de los grupos de especies ibéricas del género *Tetramorium* Mayr, 1855. *Boletin de la Asociacion Española de Entomologia* **15**: 29–52. [30.xii.1991.]

López, F. 1991b. Variabilidad morfológica y problemas taxonómicos en *Tetramorium caespitum* (Linné, 1758) y *Tetramorium semilaeve* André, 1881. *Boletin de la Asociacion Española de Entomologia* **15**: 65–78. [30.xii.1991.]

López, F., Martínez, M.D. & Barandica, J.M. 1994. Four new species of the genus *Leptanilla* from Spain—relationships to other species and ecological issues. *Sociobiology* **24**: 179–212. [(31.xii).1994.]

Losana, M. 1834. Saggio sopra le formiche indigene del Piemonte. *Memorie della Reale Accademia delle Scienze di Torino* **37**: 307–333. [(31.xii).1834.]

Lowery, B.B. 1967. A new ant of the dacetine genus *Orectognathus*. *Journal of the Australian Entomological Society* **6**: 137–140. [31.xii.1967.]

Lowne, B.T. 1865a. Contributions to the natural history of Australian ants. *Entomologist* **2**: 273–280. [(30).ix.1865.]

Lowne, B.T. 1865b. Contributions to the natural history of Australian ants. *Entomologist* **2**: 331–336. [(31).xii.1865.]

Lubbock, J. 1880. Observations on ants, bees, and wasps; with a description of a new species of honey-ant. Part 7. Ants. *Journal of the Linnean Society of London. Zoology* **15**: 167–187. [3.ix.1880.]

Lubbock, J. 1883. Observations on ants, bees and wasps. Part 10. With a description of a new genus of honey-ant. *Journal of the Linnean Society of London. Zoology* **17**: 41–52. [17.iv.1883.]

Lucas, H. 1849. *Exploration Scientifique de l'Algérie pendant les années 1840, 1841, 1842. Sciences Physiques. Zoologie* **3**. Histoire Naturelle des Animaux Articulés. Troisiéme partie—Insectes: 527 pp. Paris. [(31.xii).1849.]

Luederwaldt, H. 1918. Notas myrmecologicas. *Revista do Museu Paulista* **10**: 29–64. [(31.xii).1918.]

Luederwaldt, H. 1920. Neue brasilianische Ameisen. *Weiszflog Irmaos, Sao Paulo:* 14 pp. [1920. Privately printed; not seen. Reference from Kempf, 1972a: 285.]

Lund, P.W. 1831a. Lettre sur les habitudes de quelques fourmis du Brésil, adressée à M. Audouin. *Annales des Sciences Naturelles* **23**: 113–138. [(30).vi.1831.]

Lund, P.W. 1831b. Ueber die Lebenweise einiger Brasilianischen Ameisen. *Notizen aus dem Gebiete der Natur- und Heilkunde* **32**: 97–111. [(30).xi.1831.]

Lutz, H. 1986. Eine neue Unterfamilie der Formicidae aus dem mitteleozänen Ölschiefer der "Grube Messel" bei Darmstadt (Deutschland, S-Hessen). *Senckenbergiana Lethaea* **67**: 177–218. [31.x.1986.]

Lutz, H. 1990. Systematische und palökologische Untersuchungen an Insekten aus dem Mittel-Eozän der Grube Messel bei Darmstadt. *Courier Forschungsinstitut Senckenberg* **124**: 1–164. [28.ix.1990.]

Mabelis, A.A. 1979. Wood ant wars. The relationship between aggression and predation in the red wood ant (*Formica polyctena* Foerst.). *Netherlands Journal of Zoology* **29**: 451–620. [1979.]

MacKay, W.P. 1980. A new harvester ant from the mountains of southern California. *Southwestern Naturalist* (25) **2**: 151–156. [15.v.1980.]

MacKay, W.P. 1981. A comparison of the nest phenologies of three species of *Pogonomyrmex* harvester ants. *Psyche* **88**: 25–74. [28.xii.1981.]
MacKay, W.P. 1985. *Acanthostichus sanchezorum,* una nueva especie de Colombia. *Sociobiology* **11**: 127–131. [(31. xii).1985.]
MacKay, W.P. 1989. A new *Aphaenogaster* from southern New Mexico. *Journal of the New York Entomological Society* **97**: 47–49. [29.iii.1989.]
MacKay, W.P. 1991. *Anochetus brevidentatus,* new species, a second fossil Odontomachiti ant. *Journal of the New York Entomological Society* **99**: 138–140. [14.ii.1991.]
MacKay, W.P. 1993a. The status of the ant *Leptothorax pergandei* Emery. *Sociobiology* **21**: 287–297. [(31.xii).1993.]
MacKay, W.P. 1993b. A review of the New World ants of the genus *Dolichoderus. Sociobiology* **22**: 1–148. [(31.xii).1993.]
MacKay, W.P. & Baena, M.L. 1993. A new "horned" fungus growing ant, *Cyphomyrmex castagnei,* from Colombia. *Sociobiology* **23**: 31–37. [(31.xii).1993.]
MacKay, W.P., Lowrie, D., Fisher, A., MacKay, E., Barnes, F. & Lowrie, D. 1988. The ants of Los Alamos County, New Mexico (pp. 79–131). In Trager, J.C. (Ed.). *Advances in Myrmecology:* 551 pp. New York. [(31.xii).1988.]
MacKay, W.P. & MacKay, E.E. 1994. *Lasius xerophilus,* a new ant species from White Sands National Monument, New Mexico. *Psyche* **101**: 37–43. [1994.]
MacKay, W.P., MacKay, E.E., Dominguez, J.F.P., Sanchez, L.I.V. & Orozco, P.V. 1985. Las hormigas del Estado de Chihuahua, Mexico: el genero *Pogonomyrmex. Sociobiology* **1**: 39–54. [(31.xii).1985.]
MacKay, W.P. & Vinson, S.B. 1989a. Two new ants of the genus *Solenopsis (Diplorhoptrum)* from eastern Texas. *Proceedings of the Entomological Society of Washington* **91**: 175–178. [13.iv.1989.]
MacKay, W.P. & Vinson, S.B. 1989b. A guide to the species identification of the New World ants. *Sociobiology* **16**: 3–47. [(31.xii).1989.]
MacLeay, W. 1873. Miscellanea entomologica. *Transactions of the Entomological Society of New South Wales* **2**: 319–370. [(31.xii).1873.]
Mann, W.M. 1912. The Stanford expedition to Brazil, 1911. J.C. Branner, chief. Parabiosis in Brazilian ants. *Psyche* **19**: 36–41. [(30).iv.1912.]
Mann, W.M. 1915. A new form of a southern ant from Naushon Island, Massachusetts. *Psyche* **22**: 51. [(30).iv.1915.]
Mann, W.M. 1916. The Stanford expedition to Brazil, 1911. John C. Branner, director. The ants of Brazil. *Bulletin of the Museum of Comparative Zoology at Harvard College* **60**: 399–490. [(30).ix.1916.]
Mann, W.M. 1919. The ants of the British Solomon Islands. *Bulletin of the Museum of Comparative Zoology at Harvard College* **63**: 273–391. [(31).xii.1919.]
Mann, W.M. 1920. Additions to the ant fauna of the West Indies and Central America. *Bulletin of the American Museum of Natural History* **42**: 403–439. [20.xii.1920.]
Mann, W.M. 1921. The ants of the Fiji Islands. *Bulletin of the Museum of Comparative Zoology at Harvard College* **64**: 401–499. [(28).ii.1921.]
Mann, W.M. 1922. Ants from Honduras and Guatemala. *Proceedings of the United States National Museum* **61**: 1–54. [14.vi.1922.]
Mann, W.M. 1923. Two new ants from Bolivia. (Results of the Mulford Biological Exploration. Entomology.). *Psyche* **30**: 13–18. [(28).ii.1923.]
Mann, W.M. 1924. Notes on Cuban ants. *Psyche* **31**: 19–23. [(28).ii.1924.]
Mann, W.M. 1925a. New beetle guests of army ants. *Journal of the Washington Academy of Sciences* **15**: 73–77. [19.ii.1925.]
Mann, W.M. 1925b. Ants collected by the University of Iowa Fiji- New Zealand Expedition. *University of Iowa Studies* **11** (No. 4): 5–6. [15.iii.1925.]
Mann, W.M. 1926. Some new Neotropical ants. *Psyche* **33**: 97–107. [(31).x.1926.]
Mann, W.M. 1929. Notes on Cuban ants of the genus *Macromischa. Proceedings of the Entomological Society of Washington* **31**: 161–166. [26.xii.1929.]
Mann, W.M. 1931. A new ant from Porto Rico. *Journal of the Washington Academy of Sciences* **21**: 440–441. [19.x.1931.]
Mann, W.M. 1935. Two new ants collected in quarantine. *Psyche* **42**: 35–37. [(31).iii.1935.]
Marikovsky, P.I. 1963. Novye vid murav'ya *Polyergus nigerrimus* Marik., sp. n., i nekotorye cherty ego biologii. *Entomologicheskoe Obozrenie* **42**: 110–114. [15.iii.1963.]
Marsh, A.C. 1985. Microclimatic factors influencing foraging patterns and success of the thermophilic desert ant, *Ocymyrmex barbiger. Insectes Sociaux* **32**: 286–296. [(31).xii.1985.]
Marsh, A.C. 1986. Checklist, biological notes and distribution of ants in the central Namib Desert. *Madoqua* **14**: 333–344. [(31.xii).1986.]
Maschwitz, U. & Hänel, H. 1985. The migrating herdsman *Dolichoderus (Diabolus) cuspidatus:* an ant with a novel mode of life. *Behavioral Ecology and Sociobiology* **17**: 171–184. [1985.]
Maschwitz, U., Hölldobler, B. & Möglich, M. 1974. Tandemlaufen als Rekrutierungsverhalten bei *Bothroponera tesserinoda* Forel. *Zeitschrift für Tierpsychologie* **35**: 113–123. [1974.]
Maschwitz, U., Steghaus-Kovac, S., Gaube, R. & Hänel, H. 1989. A South East Asian ponerine ant of the genus *Leptogenys* with army ant life habits. *Behavioral Ecology and Sociobiology* **24**: 305–316. [1989.]
Masuko, K. 1987. *Leptanilla japonica:* the first bionomic information on the enigmatic ant subfamily Leptanillae. *Proceedings of the 10th. International Congress of IUSSI, Munich* (1986):597–598. [(31.xii).1987.]
Masuko, K. 1990a. Behavior and ecology of the enigmatic ant *Leptanilla japonica* Baroni Urbani. *Insectes Sociaux* **37**: 31–57. [(31).iii.1990.]
Masuko, K. 1990b. The instars of the ant *Amblyopone silvestrii. Sociobiology* **17**: 221–244. [1990.]
Masuko, K. 1993. Predation of centipedes by the primitive ant *Amblyopone silvestrii. Bulletin of the Association of Natural Science, Senshu* **24**: 35–44. [(31.xii).1993.]
Mathew, R. 1981. Description of a new species of ant from the Khasi Hills, Meghalaya. *Oriental Insects* **14** (1980): 425–427. [(31).i.1981.]
Matsumura, S. 1912. *Thousand Insects of Japan.* Supplement **4**: 247 pp. [10.vii.1912.] (Second edition dated 1918.)
Matsumura, S. & Uchida, T. 1926. Die Hymenopteren-Fauna von den Riukiu-Inseln. *Insecta Matsumurana* **1**: 32–52. [(31).vii.1926.]
Mayr, G. Note. In most of his publications Mayr gives only the initial G. However, in a few papers his full initials of G.L. are noted. Initial G. alone is used universally in this catalogue.
Mayr, G. 1853a. Einige neue Ameisen. *Verhandlungen des Zoologisch-Botanischen Vereins in Wien* **2** (1852): 143–150. [(31.xii).1853.]
Mayr, G. 1853b. Beiträge zur Kenntniss der Ameisen. *Verhandlungen des Zoologisch-Botanischen Vereins in Wien* **3**: 101–114. [(31.xii).1853.]
Mayr, G. 1853c. Beschreibungen einiger neuer Ameisen. *Verhandlungen des Zoologisch-Botanischen Vereins in Wien* **3**: 277–286. [(31.xii).1853.]
Mayr, G. 1853d. Ueber die Abtheilung der Myrmiciden, und eine neue Gattung derselben. *Verhandlungen des Zoologisch- Botanischen Vereins in Wien* **3**: 387–394. [(31.xii).1853.]
Mayr, G. 1855. Formicina austriaca. Beschreibung der bisher im österreichischen Kaiserstaate aufgefundenen Ameisen nebst Hinzufügung jener in Deutschland, in der Schweiz und in Italien vorkommenden Ameisen. *Verhandlungen des Zoologisch- Botanischen Vereins in Wien* **5**: 273–478. [(31.xii).1855.]
Mayr, G. 1856. Untitled contribution introduced by, "M. H. Lucas fait connaître la note suivante. . . .Je communiquerai à la Société un travail de M. Gustav Mayr." *Annales de la Société Entomologique de France. Bulletins Trimestriels* (3) **4**: xix–xxi. [Séance date: 13.ii.1856.]
Mayr, G. 1861. *Die Europäischen Formiciden.* (Ameisen.): 80 pp. Wien. [(31.xii).1861.]
Mayr, G. 1862. Myrmecologische Studien. *Verhandlungen der k.k. Zoologisch-Botanischen Gesellschaft in Wien* **12**: 649–776. [(31).xii.1862.]
Mayr, G. 1863. Formicidarum index synonymicus. *Verhandlungen der k.k. Zoologisch-Botanischen Gesellschaft in Wien* **13**: 385–460. [(31.xii).1863.]

Mayr, G. 1865. *Reise der Österreichischen Fregatte Novara um die Erde in den Jahren 1857, 1858, 1859, unter den befehlen des Commodore B. von Wüllerstorf-Urbair.* Zoologischer Theil. Formicidae: 119 pp. Wien. [14.xii.1865.]

Mayr, G. 1866a. Myrmecologische Beiträge. *Sitzungsberichte der k. Akademie der Wissenschaften. Mathematisch- Naturwissenschaftliche Classe* **53**: 484–517. [14.iv.1866.]

Mayr, G. 1866b. Diagnosen neuer und wenig gekannter Formiciden. *Verhandlungen der k.k. Zoologisch-Botanischen Gesellschaft in Wien* **16**: 885–908. [(31.xii).1866.]

Mayr, G. 1867a. Adnotationes in monographiam formicidarum Indo- Neerlandicarum. *Tijdschrift voor Entomologie* (2) **2** [**10**]: 33–117. [(31.xii).1867.]

Mayr, G. 1867b. Vorläufige Studien über die Radoboj-Formiciden, in der Sammlung der k.k. geologischen Reichsanstalt. *Jahrbuch der k.k. Geologischen Reichsanstalt* **17**: 47–62. [(31.xii).1867.]

Mayr, G. 1867c. In Frauenfeld, G.R.von. Zoologische Miscellen. XI. *Verhandlungen der k.k. Zoologisch-Botanischen Gesellschaft in Wien* **17**: 425–502. [(31.xii).1867.]

Mayr. G. 1868a. In Brauer, F. *Reise der Österreichischen Fregatte Novara um die Erde in den Jahren 1857, 1858, 1859, unter den befehlen des Commodore B. von Wüllerstorf-Urbair.* Zoologischer Theil. 2 Abt. Neuropteren: 107 pp. Wien. [8.x.1868; see Higgins, 1963: 153 for publication date.]

Mayr, G. 1868b. Formicidae novae americanae collectae a Prof. P. de Strobel. *Annuario della Società dei Naturalisti Modena* **3**: 161–178. [(31.xii).1868.]

Mayr, G. 1868c. Die Ameisen des baltischen Bernsteins. *Beiträge zur Naturkunde Preussens. Königlichen Physikalisch- Ökonomischen Gesellschaft zu Königsberg* **1**: 1–102. [(31.xii).1868.]

Mayr, G. 1868d. *Crematogaster ransonneti* n. sp. *Verhandlungen der k.k. Zoologisch-Botanischen Gesellschaft in Wien* **18**: 287–288. [(31.xii).1868.]

Mayr, G. 1870a. Formicidae novogranadenses. *Sitzungsberichte der k. Akademie der Wissenschaften. Mathematisch- Naturwissenschaftliche Classe* **61**: 370–417. [7.iv.1870.]

Mayr, G. 1870b. Neue Formiciden. *Verhandlungen der k.k. Zoologisch-Botanischen Gesellschaft in Wien* **20**: 939–996. [31.xii.1870.]

Mayr, G. 1872. Formicidae Borneenses collectae a J. Doria et O. Beccari in territorio Sarawak annis 1865–1867. *Annali del Museo Civico di Storia Naturale di Genova* **2**: 133–155. [(30).iv.1872.]

Mayr, G. 1876. Die australischen Formiciden. *Journal des Museum Goddefroy* (4) **12**: 56–115. [(31.xii).1876.]

Mayr, G. 1877. In Fedchenko, A.P. Puteshestvie v Turkestan, 2 (5). Formicidae. *Izvestiya Imperatorskago Obshchestva Lyubitelei Estestvoznaniya, Antropologii i Etnografii* **26**: 1–20. [(31.xii).1877.]

Mayr, G. 1878. Formiciden. Gesammelt in Brasilien von Professor Trail. *Verhandlungen der k.k. Zoologisch-Botanischen Gesellschaft in Wien* **27** (1877): 867–878. [(31.xii).1878.]

Mayr, G. 1879. Beiträge zur Ameisen-Fauna Asiens. *Verhandlungen der k.k. Zoologisch-Botanischen Gesellschaft in Wien* **28** (1878): 645–686. [(31.xii).1879.]

Mayr, G. 1880. Die Ameisen Turkestan's gesammelt von A. Fedtschenko. *Tijdschrift voor Entomologie* **23**: 17–40. [(31.xii).1880.]

Mayr, G. 1883. Drei neue Ost-Indische Formiciden-Arten. *Notes from the Leyden Museum* **5**: 245–247. [(31).x.1883.]

Mayr, G. 1884. In Radoszkowsky, O. Fourmis de Cayenne Française. *Trudy Russkago Entomologicheskago Obshchestva* **18**: 30–39. [1.v.1884.]

Mayr, G. 1886a. Ueber *Eciton-Labidus.* Ein myrmecologischer Beitrag. *Wiener Entomologische Zeitung* **5**: 33–36. [20.ii.1886.]

Mayr, G. 1886b. Ueber *Eciton-Labidus* (schluss). *Wiener Entomologische Zeitung* **5**: 115–122. [10.v.1886.]

Mayr, G. 1886c. Notizen über die Formiciden-Sammlung des British Museum in London. *Verhandlungen der k.k. Zoologisch- Botanischen Gesellschaft in Wien* **36**: 353–368. [(30).ix.1886.]

Mayr, G. 1886d. Die Formiciden der Vereinigten Staaten von Nordamerika. *Verhandlungen der k.k. Zoologisch-Botanischen Gesellschaft in Wien* **36**: 419–464. [(31).xii.1886.]

Mayr, G. 1887. Südamerikanische Formiciden. *Verhandlungen der k.k. Zoologisch-Botanischen Gesellschaft in Wien* **37**: 511–632. [1.vi.1887.]

Mayr, G. 1889. Insecta in itinare Cl. Przewalski in Asia centrali novissime lecta. 17. Formiciden aus Tibet. *Trudy Russkago Entomologicheskago Obshchestva* **24**: 278–280. [12.viii.1889.]

Mayr, G. 1893. Formiciden von Herrn Dr. Fr. Stuhlmann in Ost- Afrika gesammelt. *Jahrbuch der Hamburgischen Wissenschaftlichen Anstalten* **10** (1892): 193–201. [(31.xii).1893.]

Mayr, G. 1895. Afrikanische Formiciden. *Annalen des k.k. Naturhistorischen Hofmuseums Wien* **10**: 124–154. [(31.xii).1895.]

Mayr, G. 1896. Beiträge zur Kenntniss der Insektenfauna von Kamerun. 5. Formiciden gesammelt von Herrn Yngve Sjöstedt. *Entomologisk Tidskrift* **17**: 225–252. [26.viii.1896.]

Mayr, G. 1897. Formiciden aus Ceylon und Singapur. *Természetrajzi Füzetek* **20**: 420–436. [6.vi.1897.]

Mayr, G. 1901a. Drei neue Formiciden aus Kamerun gesammelt von Herrn Prof. Dr. Reinhold Buchholz. *Entomologisk Tidskrift* **21** (1900): 273–279. [31.i.1901.]

Mayr, G. 1901b. Südafrikanische Formiciden, gesammelt von Dr. Hans Brauns. *Annalen des k.k. Naturhistorischen Hofmuseums Wien* **16**: 1–30. [(31.xii).1901.]

Mayr, G. 1902. Hymenopterologische Miscellen. *Verhandlungen der k.k. Zoologisch-Botanischen Gesellschaft in Wien* **52**: 287–303. [5.vii.1902.]

Mayr, G. 1903. Hymenopterologische Miszellen. 2. *Verhandlungen der k.k. Zoologisch-Botanischen Gesellschaft in Wien* **53**: 387–403. [10.x.1903.]

Mayr, G. 1904a. Hymenopterologische Miszellen. 3. *Verhandlungen der k.k. Zoologisch-Botanischen Gesellschaft in Wien* **54**: 559–598. [15.x.1904.]

Mayr, G. 1904b. Formiciden aus Ägypten und dem Sudan. In Jägerskiöld, L.A. *Results of the Swedish Zoological Expedition to Egypt and the White Nile, 1901* **1** (no. 9): 11 pp. Uppsala. [(31.xii).1904.]

Mayr, G. 1907a. Ergebnisse der mit Subvention aus der Erbschaft Treitl unternommenen zoologischen Forschungsreise Dr. F. Werner's nach dem äegyptischen Sudan und nach Nord-Uganda. 11. Liste der von Dr. Franz Werner am oberen Nil gesammelten Ameisen nebst Beschreibung einer neuen Art. *Sitzungsberichte der k. Akademie der Wissenschaften. Mathematisch-Naturwissenschaftliche Klasse* **116**: 387–392. [(31).iii.1907.]

Mayr, G. 1907b. In Sjöstedt, Y. *Wissenschaftliche Ergebnisse der Schwedischen Zoologischen Expedition nach dem Kilimandjaro, dem Meru und den umgebenden Massaisteppen Deutsch-Ostafrikas 1905–1906* **2**. 8, Hymenoptera. 2. Formicidae: 7–23. Uppsala & Stockholm. [(30).ix.1907.]

McAreavey, J.J. 1947. New species of the genera *Prolasius* Forel and *Melophorus* Lubbock. *Memoirs of the National Museum of Victoria* **15**: 7–27. [(31).x.1947.]

McAreavey, J.J. 1949. Australian Formicidae. New genera and species. *Proceedings of the Linnean Society of New South Wales* **74**: 1–25. [15.vi.1949.]

McAreavey, J.J. 1956. A new species of the genus *Meranoplus. Memoirs of the Queensland Museum* **13**: 148–150. [26.iv.1956.]

McAreavey, J.J. 1957. Revision of the genus *Stigmacros* Forel. *Memoirs of the National Museum of Victoria* **21**: 7–64. [6.viii.1957.]

McCluskey, E.S. & Brown, W.L.,Jr. 1973. Rhythms and other biology of the giant tropical ant *Paraponera. Psyche* **79** (1972): 335–347. [25.iv.1973.]

McCook, H.C. 1879. Family Formicidae (pp. 182–189). In Comstock, J.H. *Report upon Cotton Insects:* 511 pp. Washington. [(31.xii).1879.]

McCook, H.C. 1881a. Note on a new northern cutting ant, *Atta septentrionalis. Proceedings of the Academy of Natural Sciences of Philadelphia* **32** (1880): 359–363. [22.ii.1881.]

McCook, H.C. 1881b. The honey ants of the Garden of the Gods. *Proceedings of the Academy of Natural Sciences of Philadelphia* **1881** (Part 1): 17–77.

[Pp. 17–24 dated 10.v.1881; pp. 25–56 dated 31.v.1881; pp. 57–88 dated 14.vi.1881.]

McKey, D. 1984. Interaction of the ant-plant *Leonardoxa africana* (Caesalpiniaceae) with its obligate inhabitants in a rainforest in Cameroun. *Biotropica* **16**: 81–99. [1984.]

Medina, M. 1889. Untitled contribution introduced by, "El Sr. Medina leyó la nota siguiente." *Actas de la Sociedad Española de Historia Natural* **18**: 31. [30.iv.1889.]

Medina, M. 1892. Untitled contribution introduced by, "El Sr. Medina leyó la nota siguiente." *Actas de la Sociedad Española de Historia Natural* (2) **1** [21]: 69–70. [1.ix.1892.]

Medina, M. 1893. Formicidos de Pozuelo de Calatrava (Ciudad- Real). *Actas de la Sociedad Española de Historia Natural* (2) **1** [21]: 104–105. [31.i.1893.]

Meer, R.K.Vander, Jaffe, K. & Cedeno, A. (Eds.). 1990. *Applied myrmecology. A world perspective:* 741 pp. Westview Studies in Insect Biology; Boulder & Oxford.

Mei, M. 1987. *Myrmica samnitica* n. sp.: una nuova formica parassita dell'Appennino Abruzzese. *Fragmenta Entomologica, Roma* **19**: 457–469. [15.x.1987.]

Meinert, F. 1861. Bidrag de danske Myrers Naturhistorie. *Kongelige Danske Videnskabernes Selskabs Skrifter* **5**: 275–340. [(31.xii).1861.]

Menozzi, C. Note. In earlier publications his name was frequently spelled Minozzi. In later papers, and universally in this catalogue, the spelling Menozzi is used.

Menozzi, C. 1918. Primo contributo alla conoscenza della fauna mirmecologica del Modenese. *Atti della Società dei Naturalisti e Matematici di Modena* (5) **4**: 81–88. [(31.xii).1918.]

Menozzi, C. 1921. Formichi dei dintorni di Sambiase di Calabria. *Bollettino del Laboratorio di Zoologia generale e agraria della R. Scuola superiore d'Agricoltura in Portici* **15**: 24–32. [25.v.1921.]

Menozzi, C. 1922a. Miscellanea mirmecologica. *Annali del Museo Civico di Storia Naturale di Genova* (3) **9** [49]: 347–358. [Pp. 347–352 dated 28.iv.1922; pp. 353–358 dated 19.vi.1922.]

Menozzi, C. 1922b. Contribution à la faune myrmécologique de l'Espagne. *Boletín de la Real Sociedad Española de Historia Natural* **22**: 324–332. [(31).vii.1922.]

Menozzi, C. 1922c. Nota complementare per la distinzione specifica dei *Camponotus herculeanus* L. et *ligniperda* Latr. *Bollettino della Società Entomologica Italiana* **54**: 141–145. [20.xii.1922.]

Menozzi, C. 1923a. Nota su un nuovo genere e nuova specie di formica parassita. *Atti della Società Italiana di Scienze Naturali* **61** (1922): 256–260. [(28).ii.1923.]

Menozzi, C. 1923b. Trois fourmis nouvelles. *Bulletin de la Société Entomologique de France* **1923**: 209–212. [5.xii.1923.]

Menozzi, C. 1924a. Una specie inedita di *Eciton* Latr. del sottogenere *Labidus* Jur. *Bollettino della Società Entomologica Italiana* **56**: 29–31. [26.ii.1924.]

Menozzi, C. 1924b. Alcune nuove formiche africane. *Annali del Museo Civico di Storia Naturale di Genova* **51**: 220–227. [Pp. 220–224 dated 7.v.1924; pp. 225–227 dated 26.vi.1924.]

Menozzi, C. 1924c. Nuova specie di *Dolichoderus* Lund del Peru. *Bollettino della Società Entomologica Italiana* **56**: 119–120. [29.x.1924.]

Menozzi, C. 1925a. Note staccate di mirmecologia. *Bollettino della Società Entomologica Italiana* **57**: 17–22. [25.ii.1925.]

Menozzi, C. 1925b. Qualche formica nuova od interessante del Deutsch. Ent. Institut di Dahlem. *Entomologische Mitteilungen* **14**: 368–371. [10.x.1925.]

Menozzi, C. 1925c. Nouvelles fourmis des Philippines. *Philippine Journal of Science* **28**: 439–449. [6.xi.1925.]

Menozzi, C. 1925d. Res Mutinenses. *Atti della Società dei Naturalisti e Matematici di Modena* (6) **3** (1924): 22–47. [(31.xii).1925.]

Menozzi, C. 1926a. Formiche dell'Africa centrale. *Bollettino della Società Entomologica Italiana* **58**: 36–41. [31.iii.1926.]

Menozzi, C. 1926b. Zur Kenntnis der Ameisenfauna der Balearen. *Zoologischer Anzeiger* **66**: 180–182. [5.iv.1926.]

Menozzi, C. 1926c. Neue Ameisen aus Brasilien. *Zoologischer Anzeiger* **69**: 68–72. [20.xi.1926.]

Menozzi, C. 1926d. Due nuove specie di *Eciton* Latr. *Folia Myrmecologica et Termitologica* **1**: 29–32. [(31).xii.1926.]

Menozzi, C. 1926e. Nuove formiche delle isole Filippine e di Singapore. *Atti della Società dei Naturalisti e Matematici di Modena* (6) **4** [56] (1925): 92–103. [(31.xii).1926.]

Menozzi, C. 1927a. Zur Kenntnis des Weibchens von *Dorylus (Anomma) nigricans* var. *molesta* Gerst. *Zoologischer Anzeiger* **70**: 263–266. [20.iii.1927.]

Menozzi, C. 1927b. Beitrag zur Ameisenfauna des nördlichen und östlichen Spaniens. *Senckenbergiana* **9**: 89–92. [18.vi.1927.]

Menozzi, C. 1927c. Formiche raccolte dal Sig. H. Schmidt nei dintorni di San José di Costa Rica. *Entomologische Mitteilungen* **16**: 266–277. [1.vii.1927.]

Menozzi, C. 1927d. Formiche raccolte dal Sig. H. Schmidt nei dintorni di San José di Costa Rica. (Schluss.) *Entomologische Mitteilungen* **16**: 336–345. [15.viii.1927.]

Menozzi, C. 1927e. Zur Erforschung des Persischen Golfes. (Beitrag Nr. 12.) Formicidae. *Supplementa Entomologica* **16**: 117–119. [10.xi.1927.]

Menozzi, C. 1927f. Raccolte mirmecologiche dell'Africa orientale conservate nel Museo Civico di Storia Naturale "Giacomo Doria" di Genova. Parte prima. Formiche raccolte dal Marchese Saverio Patrizi nella Somalia italiana ed in alcune località dell'Africa orientale inglese. *Annali del Museo Civico di Storia Naturale di Genova* **52**: 356–362. [10.xi.1927.]

Menozzi, C. 1927g. Risultati zoologici della Missione inviata dalla R. Società Geografica Italiana per l'esplorazione dell'Oasi di Giarabub (1926–1927). Formicidae. *Annali del Museo Civico di Storia Naturale di Genova* **52**: 379–382. [20.xii.1927.]

Menozzi, C. 1928a. Note sulla mirmecofauna Paleartica. *Bollettino del Laboratorio di Zoologia generale ed agraria del R. Istituto superiore agrario di Portici* **21**: 126–129. [22.v.1928.]

Menozzi, C. 1928b. Tre nuove formiche della Sumatra orientale. *Miscellanea Zoologica Sumatrana* **30**: 1–5. [(31).vii.1928.]

Menozzi, C. 1929a. Formiche di Cuba e delle Isole Canarie. *Bollettino del Laboratorio di Zoologia generale ed agraria del R. Istituto superiore agrario di Portici* **23** (1930): 1–5. [31.v.1929.]

Menozzi, C. 1929b. Una nuova specie di formica del genere *Aphaenogaster* Mayr del Nord America. *Bollettino del Laboratorio di Zoologia generale ed agraria del R. Istituto superiore agrario di Portici* **22**: 282–284. [6.viii.1929.]

Menozzi, C. 1929c. A new species of *Camponotus* belonging to the subgenus *Orthonotomyrmex* Ashm. *Annals and Magazine of Natural History* (10) **4**: 430–433. [(30).xi.1929.]

Menozzi, C. 1929d. Revisione delle formiche del genere *Mystrium* Roger. *Zoologischer Anzeiger* **82**: 518–536. [(31.xii).1929.]

Menozzi, C. 1929e. Formiche del Sinai, raccolte dal Dr. F.S. Bodenheimer, con descrizione di una nuova specie di *Monomorium* del sottogenere *Equestrimessor* (pp. 125–128). In Bodenheimer, F.S. & Theodor, O. *Ergebnisse der Sinai- Expedition 1927 der Hebräischen Universität, Jerusalem:* 142 pp. Leipzig. [(31.xii).1929.]

Menozzi, C. 1930a. Formiche di Sumatra raccolte dal Prof. J.C. van der Meer Mohr. *Miscellanea Zoologica Sumatrana* **47**: 1–5. [(31).iii.1930.]

Menozzi, C. 1930b. Formiche della Somalia Italiana meridionale. *Memorie della Società Entomologica Italiana* **9**: 76–129. [2.v.1930.]

Menozzi, C. 1930c. In Gösswald, K. Die Biologie einer neuen *Epimyrma*-Art aus dem mittleren Maingebiet. *Zeitschrift für Wissenschaftliche Zoologie* **136**: 464–484. [(31).vii.1930.]

Menozzi, C. 1930d. Formicidae (pp. 327–332). In Eidmann, H. Entomologische Ergebnisse einer Reise nach Ostasien. *Verhandlungen der Zoologisch-Botanischen Gesellschaft in Wien* **79** (1929): 308–335. [31.x.1930.]

Menozzi, C. 1931a. Spedizione del Barone Raimondo Franchetti in Dancalia. Hymenoptera—Formicidae. *Annali del Museo Civico di Storia Naturale di Genova* **55**: 154–156. [24.iii.1931.]

Menozzi, C. 1931b. Contribuzione alla conoscenza del "microgenton" di Costa Rica. *Bollettino del Laboratorio di Zoologia generale ed agrario del R. Istituto superiore agrario di Portici* **25**: 259–274. [24.viii.1931.]

Menozzi, C. 1931c. Revisione del genere *Epimyrma* Em. e descrizione di una specie inedita di questo genere. *Memorie della Società Entomologica Italiana* **10**: 36–53. [10.ix.1931.]

Menozzi, C. 1931d. Qualche nuova formica di Costa Rica. *Stettiner Entomologische Zeitung* **92**: 188–202. [(31).x.1931.]

Menozzi, C. 1932a. Contributo alla conoscenza della fauna mirmecologica d'Italia. *Bollettino del Laboratorio di Entomologica del R. Istituto superiore agrario di Bologna* **5**: 8–12. [10.iv.1932.]

Menozzi, C. 1932b. Formiche del Nord America raccolte dal Prof. F. Silvestri. *Bollettino del Laboratorio di Zoologia generale ed agraria del R. Istituto superiore agrario di Portici* **26**: 310–312. [25.vi.1932.]

Menozzi, C. 1932c. Missione scientifica del Prof. E. Zavattari nel Fezzan (1931). *Bollettino della Società Entomologica Italiana* **64**: 93–95. [30.vi.1932.]

Menozzi, C. 1932d. Formiche dell'Isola di Nias. *Miscellanea Zoologica Sumatrana* **65**: 1–13. [(30).vi.1932.]

Menozzi, C. 1932e. Spedizione scientifica all'Oasi di Cufra. (Marzo-Luglio 1931.) Formiche. *Annali del Museo Civico di Storia Naturale di Genova* **55**: 451–456. [16.viii.1932.]

Menozzi, C. 1932f. Una nuova specie di *Polyrhachis* e descrizione di tre nidi di formiche appartenenti a questo genere. *Wiener Entomologische Zeitung* **49**: 303–308. [8.xii.1932.]

Menozzi, C. 1933a. Raccolte mirmecologiche dell'Africa orientale conservate nel Museo Civico di Storia Naturale "Giacomo Doria" di Genova. Parte seconda. Formiche dell'Uganda e delle Isole Sesse raccolte dal Dr. E. Bayon. *Annali del Museo Civico di Storia Naturale Giacomo Doria* **56**: 93–114. [Pp. 93–112 dated 10.xii.1932; pp. 113–114 dated 23.i.1933.]

Menozzi, C. 1933b. Le formiche della Palestina. *Memorie della Società Entomologica Italiana* **12**: 49–113. [15.vi.1933.]

Menozzi, C. 1933c. Description préliminaire d'une espèce nouvelle de fourmi constituant un genre nouveau. *Natuurhistorisch Maandblad* **22**: 146–147. [29.xii.1933.]

Menozzi, C. 1933d. In Goetsch, W. Chilenische Wüsten-, Steppen- und Wald-Ameisen. *Sitzungsberichte der Gesellschaft für Morphologie und Physiologie in München* **42**: 27–35. [(31.xii).1933.]

Menozzi, C. 1934. Reperti mirmecofaunistici raccolte dal Prof. L. di Caporiacco nelle Oasi di Cufra e in altre località del deserto Libico. *Atti della Società dei Naturalisti e Matematici di Modena* (6) **13** [65]: 153–166. [(31.xii).1934.]

Menozzi, C. 1935a. Formiche indo-australiane del genere *Crematogaster* Lund raccolte da W. Karawaiew. *Konowia* **14**: 103–116. [15.iii.1935.]

Menozzi, C. 1935b. Spedizione del Prof. Nello Beccari nella Guina Britannica. Hymenoptera, Formicidae. *Redia* **21**: 189–203. [23.iii.1935.]

Menozzi, C. 1935c. Fauna Chilensis. 2 pars. Le formiche del Cile. *Zoologische Jahrbücher. Abteilung für Systematik, Ökologie und Geographie der Tiere* **67**: 319–336. [6.xi.1935.]

Menozzi, C. 1936a. In Eidmann, H. Ökologisch-faunistische Studien an südbrasilianischen Ameisen. *Arbeiten über Physiologie und Angewandte Entomologie aus Berlin-Dahlem* **3**: 26–48. [18.iii.1936.]

Menozzi, C. 1936b. In Eidmann, H. Ökologisch-faunistische Studien an südbrasilianischen Ameisen. *Arbeiten über Physiologie und Angewandte Entomologie aus Berlin-Dahlem* **3**: 81–114. [4.v.1936.]

Menozzi, C. 1936c. Due nuovi Dacetini di Costa Rico e descrizione della larva di uno di essi. *Arbeiten über Morphologische und Taxonomische Entomologie aus Berlin-Dahlem* **3**: 81–85. [8.vi.1936.]

Menozzi, C. 1936d. Nuovi contributi alla conoscenza della fauna delle isole italiane dell'Egeo. 6. Formicidae. *Bollettino del Laboratorio di Zoologia generale ed agraria del R. Istituto superiore agrario di Portici* **29**: 262–311. [24.vi.1936.]

Menozzi, C. 1939a. Formiche dell'Himalaya e del Karakorum raccolte dalla Spedizione Italiana comandata da S.A.R. il Duca di Spoleto (1929). *Atti della Società Italiana di Scienze Naturali* **78**: 285–345. [(31).x.1939.]

Menozzi, C. 1939b. Qualche nuova formica di Sumatra. *Tijdschrift voor Entomologie* **82**: 175–181. [(30).xi.1939.]

Menozzi, C. 1939c. Hymenoptera. Formicidae (pp. 97–110). In *Missione Biologica nel Paese dei Borana* **3**. Raccolte zoologiche, parte seconda: 463 pp. Roma. [(31.xii).1939.]

Menozzi, C. 1940. Contributo alla fauna della Tripolitania. *Bollettino del Laboratorio di Zoologia generale e agraria della Facolta Agraria in Portici* **31**: 238–273. [22.ii.1940.]

Menozzi, C. 1941. In Eidmann, H. Zur Ökologie und Zoogeographie der Ameisenfauna von Westchina und Tibet. Wissenschaftliche Ergebnisse der 2. Brooke Dolan-Expedition, 1934–1935. *Zeitschrift für Morphologie und Ökologie die Tiere* **38**: 1–43. [18.xi.1941.]

Menozzi, C. 1942. Formiche dell'isola Fernando Poo e del territorio del Rio Muni (Guinea Spagnola). 24. Beitrag zu den Wissenschaftlichen Ergebnissen der Forschungsreise H. Eidmann nach Spanisch-Guinea 1939 bis 1940. *Zoologischer Anzeiger* **140**: 164–182. [15.xii.1942.]

Menozzi, C. 1944. In Eidmann, H. Die Ameisenfauna von Fernando Poo. 27. Beitrag zu den Ergebnissen der Westafrika- Expedition Eidmann 1939/40. *Zoologische Jahrbücher. Abteilung für Systematik, Ökologie und Geographie der Tiere* **76** (1943): 413–490. [18.i.1944.]

Menozzi, C. & Consani, M. 1952. Missione Biologica Sagan-Omo, diretta dal Prof. E. Zavattari. Hymenoptera. Formicidae. *Rivista di Biologia Coloniale* **11** (1951): 57–71. [(31).xii.1952.]

Menozzi, C. & Russo, G. 1930. Contributo alla conoscenza della mirmecofauna della Repubblica Dominicana (Antille). *Bollettino del Laboratorio di Zoologia generale ed agraria del R. Istituto superiore agrario di Portici* **24**: 148–173. [1.ix.1930.]

Mercovich, T.C. 1958. A new species of the genus *Orectognathus*. *Memoirs of the Queensland Museum* **13**: 195–198. [28.vii.1958.]

Meunier, F. 1915. Über einige fossile Insekten aus den Braunkohlenschichten (Aquitanien) von Rott (Siebengebirge). *Zeitschrift der Deutschen Geologischen Gesellschaft* A. Abhandlungen **67**: 205–217. [(31.xii).1915.]

Meunier, F. 1917. Sur quelques insectes l'Aquitanien de Rott, Sept Montagnes (Prusse rhénane). *Verhandlingen der Koninklijke Akademie van Wetenschappen te Amsterdam* **20** (Sectie 2): 1–17. [(30).xi.1917.]

Meunier, F. 1923. Sur quelques nouveaux insectes des lignites oligocènes de Rott, Siebengebirge (Rhénanie). *Proceedings. Koninklijke Akademie van Wetenschappen te Amsterdam* **26**: 605–612. [(31.xii).1923.]

Miller, C.D.F. 1957. Taxonomic status of *Formica subnitens* Creighton and *F. integroides* Emery, with a description of the sexuals of *F. subnitens*. *Insectes Sociaux* **4**: 253–257. [(30).vi.1957.]

Milne-Edwards, A. 1879. Description de quelques Crustaces nouveaux. *Bulletin de la Société Philomathique de Paris* (7) **3**: 103–110. [1879.]

Minozzi, C. See under Menozzi, C.

Miré, P. Bruneau de. 1966. Une affection du caféier Robusta dans l'est du Cameroun: la défoliation en mannequin d'osier. *Café, Cacao, Thé* (Paris) **10** (No. 3): 237–242. [(30).ix.1966.]

Moffett, M.W. 1984. Swarm raiding in a myrmicine ant. *Naturwissenschaften* **71**: 588–589. [1984.]

Moffett, M.W. 1985a. Behavioral notes on the asiatic harvesting ants *Acanthomyrmex notabilis* and *A. ferox*. *Psyche* **92**: 165–180. [28.xi.1985.]

Moffett, M.W. 1985b. Revision of the genus *Myrmoteras*. *Bulletin of the Museum of Comparative Zoology* **151**: 1–53. [26.xii.1985.]

Moffett, M.W. 1986a. Observations on *Lophomyrmex* ants from Kalimantan, Java and Malaysia. *Malayan Nature Journal* **39**: 207–211. [(28).ii.1986.]

Moffett, M.W. 1986b. Trap-jaw predation and other observations on two species of *Myrmoteras*. *Insectes Sociaux* **33**: 85–99. [(30).vi.1986.]

Moffett, M.W. 1986c. Revision of the myrmicine genus *Acanthomyrmex*. *Bulletin of the Museum of Comparative Zoology* **151**: 55–89. [15.viii.1986.]

Moffett, M.W. 1986d. Notes on the behavior of the dimorphic ant *Oligomyrmex overbecki*. *Psyche* **93**: 107–116. [10.x.1986.]

Moffett, M.W. 1987. Behavior of the group-predatory ant *Proatta butteli:* an Old World relative of the attine ants. *Insectes Sociaux* **33** (1986): 444–457. [(30).vi.1987.]

Moore, G.F. 1842. *A descriptive vocabulary of the language in common use amongst the aborigines of Western Australia:* 54. London. [Not seen. Reference from Ride & Taylor, 1973: 58.]

Morice, F.D. & Durrant, J.H. 1915. The authorship and first publication of the "Jurinean" genera of Hymenoptera: being a reprint of a long-lost work by Panzer, with a translation into English, an introduction, and bibliographical and critical notes. *Transactions of the Entomological Society of London* **1914**: 339–436. [27.ii.1915.]

Morisita, M., Kubota, M., Onoyama, K., Ogata, K., Terayama, M., Kondoh, M. & Imai, H.T. 1989. A guide for the identification of Japanese ants. 1. Ponerinae, Cerapachyinae, Pseudomyrmecinae, Dorylinae and Leptanillinae. *Myrmecological Society of Japan:* 42 pp. (In Japanese.) [30.xi.1989.]

Morisita, M., Kubota, M., Onoyama, K., Ogata, K., Terayama, M., Yamauchi, K., Sonobe, R., Kondoh, M. & Imai, H.T. 1991. A guide for the identification of Japanese ants. 2. Dolichoderinae and Formicinae. *Myrmecological Society of Japan:* 56 pp. (In Japanese.) [31.iii.1991.]

Morisita, M., Kubota, M., Onoyama, K., Ogata, K., Terayama, M., Yamauchi, K., Sonobe, R., Yamane, S., Kondoh, M. & Imai, H.T. 1992. A guide for the identification of Japanese ants. 3. Myrmicinae and supplement to Leptanillinae. *Myrmecological Society of Japan:* 94 pp. (In Japanese.) [31.iii.1992.]

Motschoulsky, V.de. Note. Alternative spellings of this name include Motschulsky and Motschoulsky. His name is standardised as the latter spelling here. See also under Victor, A., a pen name of this author.

Motschoulsky, V.de. 1859. Entomologie spéciale. Insectes des Indes orientales, et de contrées analogues. *Etudes Entomologiques* **8**: 25–118. [(31.xii).1859.]

Motschoulsky, V.de. 1860. Catalogue des insectes rapportés des environs du fl. Amour, depuis la Schilka jusqu'à Nikolaëvski, examinés et énumérés. *Bulletin de la Société Impériale des Naturalistes de Moscou* **32** (II) (1859): 487–507. [22.ii.1860.]

Motschoulsky, V.de. 1863. Essai d'un catalogue des insectes de l'ile Ceylan. *Bulletin de la Société Impériale des Naturalistes de Moscou* **36**: 1–153. [12.xii.1863.]

Motschoulsky, V.de. 1866. Catalogue des insectes reçus du Japon. *Bulletin de la Société Impériale des Naturalistes de Moscou* **39**: 163–200. [15.iii.1866.]

Mukerjee, D. Note. This name is variously spelled Mukerjee, Mukerji, and Mukherji; it is standardised here as Mukerjee.

Mukerjee, D. 1930. Report on a collection of ants in the Indian Museum, Calcutta. *Journal of the Bombay Natural History Society* **34**: 149–163. [1.iii.1930.]

Mukerjee, D. 1934. Entomological investigations on the spike disease of sandal. Formicidae. *Indian Forest Records* **20** (Part 5): 1–15. [12.v.1934.]

Mukerjee, D. & Ribeiro, S. 1925. On a collection of ants from the Andaman Islands. *Records of the Indian Museum* **27**: 205–209. [22.v.1925.]

Müller, G. 1921. Due nuove formiche della regione adriatica. *Bollettino della Società Adriatica di Scienze Naturali* **27**: 46–49. [(31.xii).1921.]

Müller, G. 1923. Le formiche della Venezia Giulia e della Dalmazia. *Bollettino della Società Adriatica di Scienze Naturali* **28**: 11–180. [(31.xii).1923.]

Mulsant, E. 1842. *Histoire naturelle des coléoptères de France.* Lamellicornes: 623 pp. Paris. [(31).viii.1842.]

Nadig, N. 1918. Alcune note sulla fauna dell'alta Valsesia. *Atti della Società Italiana di Scienze Naturali* **56**: 331–341. [(31).i.1918.]

Najt, J. 1987. Le collembole fossile *Paleosminthurus juliae* est un hyménoptère. *Revue Francaise d'Entomologie* (N.S.) **9**: 152–154. [15.xii.1987.]

Naora, N. 1933. Notes on some fossil insects from East-Asiatic Continent, with descriptions of three new species. *Entomological World* **1**: 208–219. [(30).vi.1933.]

Nasonov, N.V. 1889. Mater'yaly po estestvennoi istorie murav'ev preimushchestvenno Rossii. 1. Mater'yaly po faune murav'ev Rossii. *Izvestiya Imperatorskago Obshchestva Lyubitelei Estestvoznaniya, Antropologii i Etnografii (Trudy Laboratorii Zoologicheskago Muzeya Moskovskago Universiteta 4)* **58**: 1–78. [(31.xii).1889.]

Naumann, I.D. 1993. The supposed Cretaceous ant *Cretacoformica explicata* Jell and Duncan. *Journal of the Australian Entomological Society* **32**: 355–356. [30.xi.1993.]

Naves, M.A. 1985. A monograph of the genus *Pheidole* in Florida, USA. *Insecta Mundi* **1**: 53–90. [(31).iii.1985.]

Newman, E. 1850. Description of an apparently new lepidopterous insect, of the family Glaucopidae, from the Upper Amazons. *Zoologist* **8** (appendix): cxxii-cxxiii. [(31).xii.1850.]

Nishizono, Y. & Yamane, S. 1990. The genus *Aphaenogaster* in Kagoshima-ken, southern Japan. *Report of the Faculty of Science, Kagoshima University* **23**: 23–40. [(31).xii.1990.]

Norton, E. 1868a. Notes on Mexican ants. *American Naturalist* **2**: 57–72. [(30).iv.1868.]

Norton, E. 1868b. Remarks by Edward Norton. *Transactions of the American Entomological Society* **2**: 44–46. [(30).vi.1868.]

Norton, E. 1868c. Description of Mexican ants noticed in the American Naturalist, April, 1868 *Proceedings of the Essex Institute (Communications)* **6**: 1–10. [(31).vii.1868.]

Novak, O. 1878. Fauna der Cyprisschiefer des Egerer Tertiärbeckens. *Sitzungsberichte der k. Akademie der Wissenschaften. Mathematisch-Naturwissenschaftliche Classe* **76**: 71–96. [(31.xii).1878.]

Novak, V. 1941. *Dolichoderus* (subgen. *Hypoclinea*) *quadripunctatus* var. *kratochvili* var. nova. *Casopis Ceské Spolecnosti Entomologické* **38**: 45–48. [15.v.1941.]

Novak, V. & Sadil, J. 1941. Klíc k urcování mravencu strední Evropy se zvlástním zretelem k mravencí zvírene Cech a Moravy. *Entomologické Prírucky (Entomologickych Listu v Brne)* **4**: 65–116. [(31.xii).1941.]

Nylander, W. 1846a. Adnotationes in monographiam formicarum borealium Europae. *Acta Societatis Scientiarum Fennicae* **2**: 875–944. [(31.xii).1846.]

Nylander, W. 1846b. Additamentum adnotationum in monographiam formicarum borealium Europae. *Acta Societatis Scientiarum Fennicae* **2**: 1041–1062. [(31.xii).1846.]

Nylander, W. 1849. Additamentum alterum adnotationum in monographiam formicarum borealium. *Acta Societatis Scientiarum Fennicae* **3**: 25–48. [(31.xii).1849.]

Nylander, W. 1856a. Untitled contribution introduced by, "M. L. Fairmaire communique la note suivante de M. Nylander et la Société en décide l'impression dans le Bulletin." *Annales de la Société Entomologique de France. Bulletins Trimestriels* (3) **4**: xxviii. [23.vii.1856.]

Nylander, W. 1856b. Synopsis des formicides de France et d'Algérie. *Annales des Sciences Naturelles (Zoologie)* (4) **5**: 51–109. [(31.xii).1856.]

Nylander, W. 1857. Untitled note introduced by, "M. L. Fairmaire communique la note suivante de M. Nylander sur les Formicides du Mont-Dore, et la Société en décide l'impression dans le Bulletin." *Annals de la Société Entomologique de France. Bulletins Trimestriels* (3) **4** (1856): lxxviii-lxxix. [28.i.1857.]

Ofer, J. 1970. *Polyrhachis simplex.* The weaver ant of Israel. *Insectes Sociaux* **17**: 49–82. [(30).vi.1970.]

Ogata, K. 1982. Taxonomic study of the ant genus *Pheidole* Westwood of Japan, with a description of a new species. *Kontyû* **50**: 189–197. [25.vi.1982.]

Ogata, K. 1983. The ant genus *Cerapachys* F. Smith of Japan, with description of a new species. *Esakia* **20**: 131–137. [15.xii.1983.]

Ogata, K. 1987. A generic synopsis of the poneroid complex of the family Formicidae in Japan. Part 1. Subfamilies Ponerinae and Cerapachyinae. *Esakia* **25**: 97–132. [31.i.1987.]

Ogata, K. 1990. A new species of the ant genus *Epitritus* Emery from Japan. *Esakia, Special Issue* **1**: 197–199. [20.iv.1990.]

Ogata, K. 1991a. Ants of the genus *Myrmecia* Fabricius: a review of the species groups and their phylogenetic relationships. *Systematic Entomology* **16**: 353–381. [(31).vii.1991.]

Ogata, K. 1991b. A generic synopsis of the poneroid complex of the family Formicidae in Japan. Part 2. Subfamily Myrmicinae. *Bulletin of the Institute of Tropical Agriculture, Kyushu University* **14**: 61–149. [(31).x.1991.]

Ogata, K. & Bolton, B. 1989. A taxonomic note on the ant *Monomorium intrudens* F. Smith. *Japanese Journal of Entomology* **57**: 459–460. [25.vi.1989.]

Ogata, K. & Taylor, R.W. 1991. Ants of the genus *Myrmecia* Fabricius: a preliminary review and key to the named species. *Journal of Natural History* **25**: 1623–1673. [31.xii.1991.]

Olivier, G.A. 1792. *Encyclopédie Méthodique. Histoire Naturelle. Insectes* **6** (part 2): 369–704. Paris. [(31.xii).1792. For date of publication see Sherborn & Woodward, 1906b: 577.]

Olsen, O.W. 1934. Notes on the North American harvesting ants of the genus *Pogonomyrmex. Bulletin of the Museum of Comparative Zoology at Harvard College* **77**: 493–514. [(31).xii.1934.]

Onoyama, K. 1980. An introduction to the ant fauna of Japan, with a check list. *Kontyû* **48**: 193–212. [25.vi.1980.]

Onoyama, K. 1982. Immature stages of the harvester ant *Messor aciculatus. Kontyû* **50**: 324–329. [25.vi.1982.]

Onoyama, K. 1989a. Confirmation of the occurrence of *Myrmica rubra* in Japan, with taxonomic and ecological notes. *Japanese Journal of Entomology* **57**: 131–135. [25.iii.1989.]

Onoyama, K. 1989b. Notes on the ants of the genus *Hypoponera* in Japan. *Edaphologia* **41**: 1–10. [25.ix.1989.]

Onoyama, K. 1991. A new synonym of the ant *Proceratium japonicum. Japanese Journal of Entomology* **59**: 695–696. [25.ix.1991.]

Onoyama, K. & Terayama, M. 1994. *A list of references on Japanese ants:* 50 pp. The Myrmecological Society of Japan. (In Japanese.) [25.iii.1994.]

Ortiz, F.J. & Tinaut, J.A. 1987. Citas nuevas o interesantes de formicidos para Andalucía. *Boletin de la Asociacion Española de Entomologia* **11**: 31–34. [(30).xi.1987.]

Pagliano, G. & Scaramozzino, P. 1990. Elenco dei generi di Hymenoptera del mundo. *Memorie della Società Entomologica Italiana* **68** (1989): 3–210. [10.v.1990.]

Panzer, G.W.F. 1798. *Faunae Insectorum Germanicae initia;* oder Deutschlands Insecten gesammelt und herausgegeben von D.G.W.F. Panzer **54** Heft: Nos. 1651–1652. Nürnberg. [1798. For date of publication see Sherborn, 1923: 566.]

Panzer, G.W.F. 1801. Nachricht von einem neuen entomolischen Werke, des Hrn. Prof. Jurine in Geneve. *Intelligenzblatt der Litteratur-Zeitung Erlangen* **1**: 161–165. [30.v.1801.]

Parapura, E. 1972. An ergatandromorph *Formica exsecta* Nyl. from Poland. *Bulletin de l'Academie Polonaise des Sciences* (2) **20**: 763–767. [(30).xi.1972.]

Paraschivescu, D. 1972. Pozitia sistematica a speciilor *Formica pratensis* Retz. si *F. nigricans* Em. *Studii si Cercetari de Biologie* (seria zoologie) **24**: 527–535. [(31.xii).1972.]

Passera, L. 1969. Observations biologiques sur la fourmi *Plagiolepis grassei* Le-Masne, parasite social de *Plagiolepis pygmaea* Latr. *Insectes Sociaux* **15** (1968): 327–336. [(31).iii.1969.]

Passera, L. 1984. *L'organisation sociale des fourmis:* 360 pp. Toulouse. [(30).ix.1984.]

Patrizi, S. 1946. Contribuzioni alla conoscenza delle formiche e dei mirmecofili dell'Africa orientale. *Bollettino dell'Istituto di Entomologia della Università di Bologna* **15**: 292–296. [24.xii.1946.]

Patrizi, S. 1947. Contribuzioni alla conoscenza delle formiche e dei mirmecofili dell'Africa orientale. 2. *Microdaceton leakeyi* n. sp. *Bollettino dell'Istituto di Entomologia della Università di Bologna* **16**: 219–221. [15.xi.1947.]

Patrizi, S. 1948. Contribuzioni alla conoscenza delle formiche e dei mirmecofili dell'Africa orientale. 6. *Crateropsis elmenteitae,* nuovo sottogenere aberrante di *Solenopsis* Westw. *Bollettino dell'Istituto di Entomologia dell'Università di Bologna* **17**: 174–176. [27.xi.1948.]

Patton, W.H. 1879. A gall-inhabiting ant. *American Naturalist* **13**: 126–127. [4.ii.1879.]

Patton, W.H. 1894. Habits of the leaping-ant of southern Georgia. *American Naturalist* **28**: 618–619. [(31).vii.1894.]

Pavan, M. 1981. Utilità delle formiche del gruppo *Formica rufa. Ministero dell'Agricoltura e delle Foreste. Collana Verde* **59**: 99 pp. [1981.]

Pearson, B. 1981. The electrophoretic determination of *Myrmica rubra* microgynes as a social parasite: possible significance in the evolution of ant social parasites. In Howse, P.E. & Clement, J.-L. (Eds.). Biosystematics of Social Insects. *Systematics Association Special Volume* **19**: 346 pp. London. [(31.xii).1981.]

Peeters, C. & Crewe, R. 1985. Worker reproduction in the ponerine ant *Ophthalmopone berthoudi:* an alternative form of eusocial organization. *Behavioral Ecology and Sociobiology* **18**: 29–37. [1985.]

Peeters, C. & Crewe, R. 1987. Foraging and recruitment in ponerine ants: solitary hunting in the queenless *Ophthalmopone berthoudi. Psyche* **94**: 201–213. [6.x.1987.]

Pergande, T. 1894a. On a collection of Formicidae from Lower California and Sonora, Mexico. *Proceedings of the California Academy of Sciences* (2) **4**: 26–36. [28.ix.1894.]

Pergande, T. 1894b. Formicidae of Lower California, Mexico. *Proceedings of the California Academy of Sciences* (2) **4**: 161–165. [28.ix.1894.]

Pergande, T. 1896. Mexican Formicidae. *Proceedings of the California Academy of Sciences* (2) **5**: 858–896. [18.i.1896.]

Pergande, T. 1900. Papers from the Harriman Alaska Expedition. 17. Entomological results (11): Formicidae. *Proceedings of the Washington Academy of Sciences* **2**: 519–521. [20.xii.1900.]

Perrault, G.H. 1976. Description des ouvrières et des soldats de *Oligomyrmex tahitiensis* Wheeler. Mise au point concernant les sexuées. *Nouvelle Revue d'Entomologie* **6**: 303–307. [30.xi.1976.]

Perrault, G.H. 1981. *Proceratium deelemani,* nouvelle espèce de Kalimantan. *Nouvelle Revue d'Entomologie* **11**: 189–193. [15.vi.1981.]

Perrault, G.H. 1986a. *Gymnomyrmex villiersi,* nouvelle espèce de Guyane. *Revue Francaise d'Entomologie* (N.S.) **8**: 1–4. [10.iv.1986.]

Perrault, G.H. 1986b. *Gnamptogenys falcifera* Kempf, 1967. Description de l'ouvrière et levée d'un doute. *Revue Francaise d'Entomologie* (N.S.) **8**: 157–159. [16.xii.1986.]

Perrault, G.H. 1988a. Notes sur des types de *Pseudomyrmex* décrits par F. Smith. 1. *Nouvelle Revue d'Entomologie* (N.S.) **4** (1987): 381–385. [5.ii.1988.]

Perrault, G.H. 1988b. *Octostruma betschi* n. sp. de Guyane Francaise. *Revue Francaise d'Entomologie* (N.S.) **10**: 303–307. [15.xii.1988.]

Perrault, G.H. 1993. Peuplement en fourmis de l'atoll de Fangataufa. *Bulletin de la Société Entomologique de France* **98**: 323–338. [(30).xi.1993.]

Perris, E. 1878. Rectifications et additions à mes promenades entomologiques. *Annales de la Société Entomologique de France* (5) **7** (1877): 379–386. [10.iv.1878.]

Perty, M. 1833. *Delectus animalium articulatorum,* quae in itinere per Brasiliam annis MDCCCXVII-MDCCCXX jussu et auspiciis Maximiliani Joseph I Bavariae regis augustissimi peracto, collegerunt Dr. J.B. de Spix et Dr. C.F.Ph. Martius. **Fasc. 3**: 125–224. Monachii. [(31.xii).1833.]

Petal, J.M. 1963. *Formica forsslundi* Lohm. ssp. *strawinskii* n. ssp. *Annales Universitatis Mariae Curie-Sklodowska* **17** (Sectio C, Biologia) (1962): 195–202. [16.iii.1963.]

Petersen, B. 1968. Some novelties in presumed males of Leptanillinae. *Entomologiske Meddelelser* **36**: 577–598. [30.xii.1968.]

Petralia, R.S. & Vinson, S.B. 1980. Comparative anatomy of the ventral region of ant larvae and its relation to feeding behavior. *Psyche* **86** (1979): 375–394. [19.xii.1980.]

Petrov, I.Z. & Collingwood, C.A. 1993. *Formica balcanica* sp. n., a new species related to the *Formica cinerea*-group. *European Journal of Entomology* **90**: 349–354. [25.x.1993.]

Pfitzer, D.W. 1951. A new species of *Smithistruma* from Tennessee. *Journal of the Tennessee Academy of Science* **26**: 198–200. [(31).vii.1951.]

Pierce, W.D. & Gibron, J.,Sr. 1962. Fossil arthropods of California. 24. Some unusual fossil arthropods from the Calico Mountains nodules. *Bulletin of the Southern California Academy of Sciences* **61**: 143–151. [24.x.1962.]

Pisarski, B. 1960. Dwa nowe gatunki z rodzaju *Paratrechina* Motschulsky. *Annales Zoologici* **18**: 349–356. [20.iv.1960.]

Pisarski, B. 1961. Nouvelle espèce de *Cataglyphis* Först. de l'Inde. *Bulletin de l'Académie Polonaise des Sciences. Cl. 2* **9**: 515–516. [22.xii.1961.]

Pisarski, B. 1962a. Notes synonymiques sur les espèces balcaniques du genre *Cardiocondyla* Emery. *Bulletin de l'Académie Polonaise des Sciences. Cl. 2* **10**: 331–333. [24.viii.1962.]

Pisarski, B. 1962b. Sur *Sifolinia pechi* Sams. trouvée en Pologne. *Bulletin de l'Académie Polonaise des Sciences. Cl. 2* **10**: 367–369. [28.xi.1962.]

Pisarski, B. 1963. Nouvelle espèce du genre *Harpagoxenus* For. de la Mongolie. *Bulletin de l'Académie Polonaise des Sciences. Cl. 2* (série des sciences biologiques) **11**: 39–41. [1.ii.1963.]

Pisarski, B. 1965. Les fourmis du genre *Cataglyphis* Foerst. en Irak. *Bulletin de l'Académie Polonaise des Sciences. Cl. 2* **13**: 417–422. [3.ix.1965.]

Pisarski, B. 1966. Etudes sur les fourmis du genre *Strongylognathus* Mayr. *Annales Zoologici* **23**: 509–523. [10.xii.1966.]

Pisarski, B. 1967. Fourmis d'Afghanistan récoltées par M. Dr K. Lindberg. *Annales Zoologici* **24**: 375–425. [20.iv.1967.]

Pisarski, B. 1969a. Fourmis de la Mongolie. *Fragmenta Faunistica* **15**: 221–236. [10.x.1969.]

Pisarski, B. 1969b. Myrmicidae und Formicidae. Ergebnisse der zoologischen Forschungen von Dr. Z. Kaszab in der Mongolei. *Faunistische Abhandlungen, Staatliches Museum für Tierkunde in Dresden* **2**: 295–316. [(31.xii).1969.]

Pisarski, B. 1970. Beiträge zur Kenntnis der Fauna Afghanistans. *Casopis Moravského Musea* **54** (supplementum) (1969): 305–325. [15.v.1970.]

Pisarski, B. 1971a. Nouvelles espèces de fourmis du sous-genre *Tanaemyrmex* Ashm. d'Iraq. *Bulletin de l'Académie Polonaise des Sciences. Cl. 2* **19**: 671–675. [8.xi.1971.]

Pisarski, B. 1971b. Les fourmis du genre *Camponotus* Mayr d'Iraq. *Bulletin de l'Académie Polonaise des Sciences. Cl. 6* **19**: 727–731. [22.xi.1971.]

Pisarski, B. 1975. Mrówki, Formicoidea. *Katalog Fauny Polski* **26**: 84 pp. Warszawa. [14.xi.1975.]

Pisarski, B. & Krzysztofiak, L. 1981. Myrmicidae und Formicidae aus der Mongolei. 2. *Folia Entomologica Hungarica (Rovartani Közlemények)* **42** [34]: 155–166. [14.v.1981.]

Piton, L. & Théobald, N. 1935. La faune entomologique des gisements Mio-Pliocènes du Massif Central. *Revue des Sciences Naturelles d'Auvergne* (N.S.) **1**: 65–104. [22.ix.1935.]

Plateaux, L. 1981. The *pallens* morph of the ant *Leptothorax nylanderi:* description, formal genetics, and study of populations (pp. 63–74). In Howse, P.E. & Clement, J.-L. (Eds.). Biosystematics of Social Insects. *Systematics Association Special Volume* **19**: 346 pp. London. [(31.xii).1981.]

Pohl, I.E. & Kollar, V. 1832. *Brasiliens vorzüglich lästige Insecten.* Besonderer Abdruck aus der Reise im Innern von Brasilien von Dr. Pohl: 20 pp. Wien. [(31.xii).1832.]

Poldi, B. 1979. Un nuovo *Tetramorium* dell'Anatolia. *Entomologica Basiliensia* **4**: 499–503. [10.x.1979.]

Poldi, B. 1992. A poorly known taxon: *Diplorhoptrum orbulum* (Emery 1875). *Ethology, Ecology & Evolution. Special Issue No.* **2**: 91–94. [1992.]

Poldi, B. 1994. *Strongylognathus pisarskii* species nova. Ants, fourmis, mrówki. In memory of Professor Dr Bohdan Pisarski. *Memorabilia Zoologica* **48**: 187–191. [(31.xii).1994.]

Popov, V. 1933. Two new fossil ants from Caucasus. *Trudy Paleozoologicheskogo Instituta, Akademiya Nauk SSSR* **2**: 17–21. [17.vi.1933.]

Poulton, E.B. 1933. The gentle Driver ant "Baongo", discovered by Miss Vinall at Bongandanga, Belgian Congo. *Proceedings of the Entomological Society of London* **7**: 58. [21.iv.1933.]

Poulton, E.B. & Crawley, W.C. 1922. Notes on some Australian ants. Biological notes by E.B. Poulton, D.Sc., M.A., F.R.S., and notes and descriptions of new forms by W.C. Crawley, B.A., F.E.S., F.R.M.S. *Entomologist's Monthly Magazine* (3) **8**: 118–126. [Pp. 118–120 dated (31).v.1922; pp. 121–126 dated (30).vi.1922.]

Presl, J.W. 1822. Additamenta ad faunam protogaeam, sistens descriptiones aliquot animalium in succino inclusorum (pp. 191–210). In *Deliciae Pragenses, historiam naturalem spectantes* **1**: 244 pp. Pragae. [(31.xii).1822.]

Prins, A.J. 1963. A list of the ants collected in the Kruger National Park with notes on their distribution. *Koedoe* **6**: 91–108. [(31.xii).1963.]

Prins, A.J. 1965a. Notes on African Formicidae. 1. *Journal of the Entomological Society of Southern Africa* **27**: 153–159. [28.ii.1965.]

Prins, A.J. 1965b. Notes on African Formicidae. 2. *Journal of the Entomological Society of Southern Africa* **28**: 77–79. [(30).xi.1965.]

Prins, A.J. 1965c. African Formicidae. Description of a new species. *South African Journal of Agricultural Science* **8**: 1021–1024. [(31).xii.1965.]

Prins, A.J. 1965d. Ants of the Kruger National Park. *Koedoe* **8**: 104–108. [(31).xii.1965.]

Prins, A.J. 1973. African Formicidae in the South African Museum. Description of four new species and notes on *Tetramorium* Mayr. *Annals of the South African Museum* **63**: 1–40. [(31).viii.1973.]

Prins, A.J. 1982. Review of *Anoplolepis* with reference to male genitalia, and notes on *Acropyga. Annals of the South African Museum* **89**: 215–247. [(30).vi.1982.]

Prins, A.J. 1983. A new ant genus from southern Africa. *Annals of the South African Museum* **94**: 1–11. [(30).xi.1983.]

Prins, A.J. & Roux, A. 1989. In Bolton, B. & Marsh, A.C. The Afrotropical thermophilic ant genus *Ocymyrmex. Journal of Natural History* **23**: 1267–1308. [9.xi.1989.]

Provancher, L. 1881a. Faune canadienne. Les insectes—Hyménoptères. *Naturaliste Canadien* **12**: 193–207. [(28).ii.1881.]

Provancher, L. 1881b. Faune canadienne. *Naturaliste Canadien* **12**: 353–362. [(31).xii.1881.]

Provancher, L. 1883. *Petite Faune Entomologique du Canada.* **2**. Comprenant les orthoptères, les névroptères et les hyménoptères: 879 pp. Québec. [(31.xii).1883.]

Provancher, L. 1887. Fam. 10—Les Formicides (pp. 224–249). In *Additions et Corrections au Volume 2 de la Faune Entomologique du Canada, traitant des Hyménoptères:* 477 pp. Québec. [Pp. 224–228 dated 27.viii.1887; pp. 229–244 dated 28.ix.1887; pp. 245–249 dated 29.x.1887.]

Provancher, L. 1888. Supplément aux additions et corrections aux hyménoptères (Fam. X—Formicidae pp. 408–409). In *Additions et Corrections au volume 2 de la Faune Entomologique du Canada:* 346–438. [P. 408 dated (31).x.1888; p. 409 dated (30).xi.1888.]

Provancher, L. 1895. Les dernieres descriptions de l'Abbé Provancher. *Naturaliste Canadien* **22**: 95–97. [(30).vi.1895.]

Radchenko, A.G. Note. This name is sometimes spelled Radtchenko or Radtschenko in transliteration; the form Radchenko is used universally here.

Radchenko, A.G. 1983. *Tapinoma kinburni,* endemichnyi vid fauny SSSR. *Zoologichesky Zhurnal* **62**: 1904–1907. [(31).xii.1983.]

Radchenko, A.G. 1985a. Murav'i roda *Strongylognathus* evropeiskoi chasti SSR. *Zoologichesky Zhurnal* **64**: 1514–1523. [26.ix.1985.]

Radchenko, A.G. 1985b. Novyi vid murav'ya s Ukrainy. *Vestnik Zoologii* **1985** (No. 6): 64–66. [29.xi.1985.]

Radchenko, A.G. 1989a. Murav'i roda *Chalepoxenus* fauny SSSR. *Vestnik Zoologii* **1989** (No. 2): 37–41. [24.iii.1989.]

Radchenko, A.G. 1989b. Murav'i roda *Plagiolepis* evropeiskoi chasti SSR. *Zoologichesky Zhurnal* **68**: 153–156. [21.viii.1989.]

Radchenko, A.G. 1991. Murav'i roda *Strongylognathus* fauny SSSR. *Zoologichesky Zhurnal* **70** (10): 84–90. [17.ix.1991.]

Radchenko, A.G. 1992a. Murav'i roda *Tetramorium* fauny SSSR. Soobshchenie 1. *Zoologichesky Zhurnal* **71** (8): 39–49. [13.vii.1992.]

Radchenko, A.G. 1992b. Murav'i roda *Tetramorium* fauny SSSR. Soobshchenie 2. *Zoologichesky Zhurnal* **71** (8): 50–58. [13.vii.1992.]

Radchenko, A.G. 1993a. Ants from Vietnam in the collection of the Institute of Zoology, PAS, Warsaw. 1. Pseudomyrmicinae, Dorylinae, Ponerinae. *Annales Zoologici* **44**: 75–82. [15.ii.1993.]

Radchenko, A.G. 1993b. New ants of the subfamily Cerapachyinae from Vietnam. *Zhurnal Ukrains'koho Entomolohichnoho Tovarystva* **1**: 43–47. [(31.xii).1993.]

Radchenko, A.G. 1993c. Novye vidy murav'ev roda *Leptothorax* iz yuzhnoi i vostochnoi palearktiki. *Zhurnal Ukrains'koho Entomolohichnoho Tovarystva* **2**: 23–43. [Not seen.]

Radchenko, A.G. 1994a. Taksonomicheskaya struktura roda *Myrmica* evrazii. Soobshchenie 1. *Zoologichesky Zhurnal* **73** (6): 39–51. [17.v.1994.]

Radchenko, A.G. 1994b. Opredelitel'nye tablitsy murav'ev Yuzhnoi Sibiri (pp. 95–118). In Pereponchatokrylye Sibiri i Dal'nego Vostoka. *Trudy Zapovednika "Daurskii"* **3**: 1–147. [10.vi.1994.]

Radchenko, A.G. 1994c. Opredelitel'naya tablitsa murav'ev roda *Myrmica* tsentral'noi i vostochnoi palearktiki. *Zoologichesky Zhurnal* **73** (7–8): 130–145. [21.vi.1994.]

Radchenko, A.G. 1994d. Opredelitel'naya tablitsa murav'ev roda *Leptothorax* tsentral'noi i vostochnoi palearktiki. *Zoologichesky Zhurnal* **73** (7–8): 146–158. [21.vi.1994.]

Radchenko, A.G. 1994e. New Palaearctic species of the genus *Myrmica* Latr. Ants, fourmis, mrówki. In memory of Professor Dr Bohdan Pisarski. *Memorabilia Zoologica* **48**: 207–217. [(31.xii).1994.]

Radchenko, A.G. & Arakelian, G.R. 1990. Murav'i gruppy *Tetramorium ferox* Ruzky iz Kryma i Kavkaza. *Biologicheskii Zhurnal Armenii* **43**: 371–378. [28.viii.1990.]

Radchenko, A.G. & Arakelian, G.R. 1991. Novye vidy murav'ev iz Armenii. *Vestnik Zoologii* **1991** (No. 5): 72–75.

Radoszkowsky, O. 1876. Compte-rendu des hyménoptères recueilles en Egypte et Abyssinie en 1873. *Trudy Russkago Entomologicheskago Obshchestva* **12**: 111–150. [15.ix.1876.]

Radoszkovsky, O. 1881. Hyménoptères d'Angola. *Jornal de Sciencias Mathematicas Physicas e Naturaes* **8** (No. 31): 197–221. [(31).xii.1881.]

Raffray, A. 1887. Pselaphides nouveaux ou peu connus. *Revue d'Entomologie* **6** (part): 18–37. [(31).i.1887. Note that pp. 37–56 were published ii.1887.]

Rafin, G. 1885. Ignivorous ant. *American Naturalist* **19**: 403. [21.iii.1885.]

Rafinesque, C.S. 1815. *Analyse de la Nature, ou Tableau de l'Univers et des Corps Organisés:* 224 pp. Palerme.

Ragonot, E.L. 1887. *Diagnoses of North American Phycitidae and Galleriidae:* 20 pp. Paris. [(31.xii).1887. Privately published.]

Raignier, A. & Boven, J.K.A. van. 1955. Etude taxonomique, biologique et biométrique des *Dorylus* du sous-genre *Anomma. Annales du Musée Royal du Congo Belge, Tervuren* (N.S. in quarto). Sciences Zoologiques **2**: 1–359. [(31).viii.1955.]

Rao, N.S. & Veeresh, G.K. 1991. Some observations on the biology and behaviour of Crazy Ant, *Anoplolepis longipes* (Jerdon). *Entomon* **16**: 261–267. [(31).xii.1991.]

Raphael, S. 1970. The publication dates of the Transactions of the Linnean Society of London, Series 1, 1791–1875. *Biological Journal of the Linnean Society* **2**: 61–76. [26.iii.1970.]

Razoumowsky, G. de. 1789. *Histoire Naturelle du Jorat et de ses environs; et celle des trois Lacs de Neufchatel, Morat et Bienne* **1**: 322 pp. Lausanne. [(31.xii).1789.]

Regnier, F.E. & Wilson, E.O. 1968. The alarm-defence system of the ant *Acanthomyops claviger. Journal of Insect Physiology* **14**: 955–970. [1968.]

Regnier, F.E. & Wilson, E.O. 1969. The alarm-defence system of the ant *Lasius alienus. Journal of Insect Physiology* **15**: 893–898. [1969.]

Reich, G.C. 1793. Kurze Beschreibung neuen, oder noch wenig bekkanten Thiere, welche Herr Le Blond der naturforschenden Gesellschaft zu Paris aus Cayenne als Geschenk überschikt hat. *Magazin des Thierreichs* **1**: 128–134. [(31.xii).1793.]

Reichensperger, A. 1924. Das Weibchen von *Eciton quadriglume* Hal., einige neue ecitophile Histeriden und allgemeine Bemerkungen. *Zoologischer Anzeiger* **60**: 201–213. [5.viii.1924.]

Reichensperger, A. 1926. Das Weibchen von *Eciton mattogrossensis* Luederw. *Entomologische Mitteilungen* **15**: 401–404. [10.x.1926.]

Reichensperger, A. 1930. Ein Nest und die Königin von *Eciton (Acamatus) legionis* Sm. *Zoologischer Anzeiger* **88**: 321–325. [10.v.1930.]

Reichensperger, A. 1939. Beiträge zur Kenntnis der Myrmecophilenfauna Costa Ricas und Brasiliens 7, nebst Beschreibung der Königin von *Eciton (Acamatus) pilosus. Zoologische Jahrbücher. Abteilung für Systematik, Ökologie und Geographie der tiere* **73**: 261–300. [24.xi.1939.]

Reiskind, J. 1965. A revision of the ant tribe Cardiocondylini. 1. The genus *Prosopidris* Wheeler. *Psyche* **72**: 79–86. [25.vi.1965.]

Rettenmeyer, C.W. 1974. Description of the queen and male with some biological notes on the army ant *Eciton rapax* (pp. 291–302). In Beard, R.L. (ed.) *Connecticut Entomological Society 25th. Anniversary Memoirs:* 322 pp. New Haven.

Retzius, A.J. 1783. *Caroli de Geer. Genera et Species Insectorum e generosissimi auctoris scriptis extraxit, digessit, Latine quoad partem reddidit, et terminologiam insectorum Linneanam addidit:* 220 pp. Lipsiae. [(31.xii).1783.]

Reyes, J.L. 1985. Descripción de *Messor celiae* nov. sp. *Boletin de la Asociacion Española de Entomologia* **9**: 255–261. [(31).v.1985.]

Reyes, J.L., Espadaler, X. & Rodriguez, A. 1987. Descripción de *Goniomma baeticum* nov. sp. *Eos. Revista Española de Entomologia* **63**: 269–276. [15.xii.1987.]

Reza, A. 1925. Recursos alimenticios de Mexico de origen animal poco conocidos. *Memorias y Revista de la Sociedad Cientifica "Antonio Alzate"* **44**: 1–22. [(28).ii.1925.]

Rhoades, R.B. 1977. *Medical Aspects of the Imported Fire Ant:* 75 pp. Gainesville. [1977.]

Ride, W.D.L. & Taylor, R.W. 1973. *Formica maxima* Moore, 1842: proposed suppression under the plenary powers in accordance with article 23 (a-b). ZN (S) 2023. *Bulletin of Zoological Nomenclature* **30**: 58–60. [6.vii.1973.]

Rigato, F. 1994a. Revision of the myrmicine ant genus *Lophomyrmex*, with a review of its taxonomic position. *Systematic Entomology* **19**: 47–60. [12.iii.1994.]

Rigato, F. 1994b. *Dacatria templaris* gen. n., sp. n. A new myrmicine ant from the Republic of Korea. *Deutsche Entomologische Zeitschrift* (N. F.) **1**: 155–162. [4.iii.1994.]

Rissing, S.W. 1983. Natural history of the workerless inquiline ant *Pogonomyrmex colei. Psyche* **90**: 321–332. [9.xii.1983.]

Ritsema, C. 1874. Aanteekeningen betreffende eene kleine Collectie Hymenoptera van Neder-Guinea, en beschrijving van de nieuwe Soorten. *Tijdschrift voor Entomologie* **17**: 175–211. [(31.xii).1874.]

Robertson, H.G. 1990. Unravelling the *Camponotus fulvopilosus* species complex. In Veeresh, G.K., Mallik, B. & Viraktamath, C.A. (Eds.). Social Insects and the Environment. *Proceedings of the 11th. International Congress of IUSSI, 1990:* 327–328. New Delhi. [(31).viii.1990.]

Robertson, H.G. & Villet, M.H. 1989. Colony founding in the ant *Carebara vidua:* the dispelling of a myth. *South African Journal of Science* **85**: 121–122. [(28).ii.1989.]

Roger, J. 1857. Einiges über Ameisen. *Berliner Entomologische Zeitschrift* **1**: 10–20. [(31.xii).1857.]

Roger, J. 1859. Beiträge zur Kenntniss der Ameisenfauna der Mittelmeerländer. Erstes Stück. *Berliner Entomologische Zeitschrift* **3**: 225–259. [(31.xii).1859.]

Roger, J. 1860. Die *Ponera*-artigen Ameisen. *Berliner Entomologische Zeitschrift* **4**: 278–312. [(31.xii).1860.]

Roger, J. 1861a. Die *Ponera*-artigen Ameisen. (Schluss.) *Berliner Entomologische Zeitschrift* **5**: 1–54. [(31.xii).1861.]

Roger, J. 1861b. Myrmicologische Nachlese. *Berliner Entomologische Zeitschrift* **5**: 163–174. [(31.xii).1861.]

Roger, J. 1862a. Einige neue exotische Ameisen-Gattungen und Arten. *Berliner Entomologische Zeitschrift* **6**: 233–254. [(31).v.1862.]

Roger, J. 1862b. Beiträge zur Kenntniss der Ameisen-Fauna der Mittelmeerländer. *Berliner Entomologische Zeitschrift* **6**: 255–262. [(31).v.1862.]

Roger, J. 1862c. Synonymische Bemerkungen. *Berliner Entomologische Zeitschrift* **6**: 283–297. [(31).v.1862.]

Roger, J. 1863a. Die neu aufgeführten Gattungen und Arten meines Formiciden-Verzeichnisses, nebst Ergänzung einiger früher gegeben Beschreibungen. *Berliner Entomologische Zeitschrift* **7**: 131–214. [(30).vi.1863.]

Roger, J. 1863b. Verzeichniss der Formiciden-Gattungen und Arten. *Berliner Entomologische Zeitschrift* **7** (Beilage): 1–65. [(30).vi.1863.]

Romand, M. de. 1846. Notice sur un insecte nouveau. *Bulletin de la Société Entomologique de France* (2) **4**: xxxii. [8.vii.1846.]

Rosciszewski, K. 1994. *Rostromyrmex*, a new genus of myrmicine ants from Peninsular Malaysia. *Entomologica Scandinavica* **25**: 159–168. [(30).vi.1994.]

Röszler, P. 1935. Beiträge zur Kenntnis der Ameisenfauna von Siebenbürgen und Ungarn. *Verhandlungen und Mitteilungen des Siebenbürgischen Vereins für Naturwissenschaften zu Hermannstadt* **83–84** (1933–1934): 72–83. [(31.xii).1935.]

Röszler, P. 1936a. Beitrag zur Kenntnis der Ameisenfauna von Japan. *Entomologischer Anzeiger* **16**: 121–122. [1.vi.1936.]

Röszler, P. 1936b. Beiträge zur Kenntnis der Ameisenfauna von Mitteleuropa. 3. Teil der Arbeit: Ein Versuch der systematischen Einteilung der mitteleuropäischen *Tetramorium*. *Tijdschrift voor Entomologie* **79**: 55–63. [(30).vi.1936.]

Röszler, P. 1937. Beiträge zur Kenntnis der Ameisenfauna von Spanien und anderer mitteleuropäischer Länder. 2. Teil der Arbeit: Ein Versuch zur systematischen Einteilung der mitteleuropäischen *Tetramorium*. *Verhandlungen und Mitteilungen des Siebenbürgischen Vereins für Naturwissenschaften zu Hermannstadt* **85–86** (1935–1936): 195–208. [(31.xii).1937.]

Röszler, P. 1942a. Myrmecologisches 1938. *Tijdschrift voor Entomologie* **85**: 50–71. [(31).xii.1942.]

Röszler, P. 1942b. Myrmekologische Mitteilungen 1939. *Mitteilungen der Arbeitsgemeinschaft für Naturwissenschaften in Fortsetzung der "Verhandlungen und Mitteilungen des Siebenbürgischen Vereins für Naturwissenschaften zu Hermannstadt"* **91–92**: 27–41. [(31.xii).1942.]

Röszler, P. 1943. Ameisen aus Siebenbürgen mit der Beschreibung ihrer Lebenweise. *Zoologischer Anzeiger* **144**: 41–48. [30.x.1943.]

Röszler, P. 1950. Die Ameisenwelt des Nagy Pietrosz, 2305m (Ungarn) und Umgebung. *Zoologischer Anzeiger* **145**: 210–225. [(31).x.1950.]

Röszler, P. 1951. Myrmecologisches aus dem Jahre 1938. *Zoologischer Anzeiger* **146**: 88–96. [(28).ii.1951.]

Rothney, G.A.J. 1889. Notes on Indian ants. *Transactions of the Entomological Society of London* **1889**: 347–374. [26.x.1889.]

Rothney, G.A.J. 1903. The aculeate Hymenoptera of Barrackpore, Bengal. *Transactions of the Entomological Society of London* **1903**: 93–116. [29.iv.1903.]

Ruzsky, M. Note. In some publications Ruzsky's initials are given as M.D. Elsewhere only the initial M. is noted. The latter is used universally in this catalogue.

Ruzsky, M. 1895. Faunisticheskiya izsledovaniya v vostochnoi Rossii. K faune murav'ev vostoka Rossii. *Trudy Obshchestva Estestvoispytatelei pri Imperatorskom Kazanskom Universitete* **28** (5): 1–32. [(31.xii).1895.]

Ruzsky, M. 1896. Verzeichniss der Ameisen des östlichen Russlands und des Uralgebirges. *Berliner Entomologische Zeitschrift* **41**: 67–74. [(31).v.1896.]

Ruzsky, M. 1902a. K faune murav'ev Turgaiskoi oblasti. *Russkoe Entomologicheskoe Obozrenie* **2**: 232–235. [(31).viii.1902.]

Ruzsky, M. 1902b. Neue Ameisen aus Russland. *Zoologische Jahrbücher. Abtheilung für Systematik, Geographie und Biologie der Thiere* **17**: 469–484. [18.xii.1902.]

Ruzsky, M. 1902c. Murav'i okrestnostei Aral'skago Morya. *Izvestiya Turkestanskago Otdela Imperatorskago Russkago Geograficheskago Obshchestva* **3**: 1–24. [(31.xii).1902.]

Ruzsky, M. 1902d. Materialy po mirmekologicheskoi faune Kavkaza i Kryma. *Protokolamy Obshchestva Estestvoispytatelei pri Imperatorskomy Kazanskomy Universitete* **206** (Prilozhenie): 1–33. [(31.xii).1902.]

Ruzsky, M. 1902e. Formicidae (pp. 15–16). In Dopolneniya k spiskam zhivotnykh' Moskovskoi gubernii. *Izvestiya Imperatorskago Obshchestva Lyubitelei Estestvoznaniya, Antropologii i Etnografi* **98**. (*Trudy Zoologicheskago Otdeleniya Obshchestva* 13.) *Dnevnikh' Zoologicheskago Otdeleniya* **3** (No. 4): 6–18. [(31.xii).1902.]

Ruzsky, M. 1903a. Novyi vid murav'ya iz Zakaspiiskoi oblasti. *Russkoe Entomologicheskoe Obozrenie* **3**: 36–37. [(28).ii.1903.]

Ruzsky, M. 1903b. Ocherk mirmekologicheskoi fauny Kirgizskoi steni. *Trudy Russkago Entomologicheskago Obshchestva* **36**: 294–316. [30.vi.1903.]

Ruzsky, M. 1903c. Murav'i iz Zabaikalvskoi Oblasti. *Russkoe Entomologicheskoe Obozrenie* **3**: 205–207. [(31).viii.1903.]

Ruzsky, M. 1904a. O murav'yakh Arkhangel'skoi gubernii. *Zapiski po Obshchei Geografii Imperatorskago Russkago Geograficheskago Obshchestva* **41**: 287–294. [(31.xii).1904.]

Ruzsky, M. 1904b. Murav'i Dzhungarskago Alatau. *Izvestiya Tomskogo Universiteta* **24**: 1–6. [(31.xii).1904.]

Ruzsky, M. 1905a. Über *Tetramorium striativentre* Mayr und *Tetr. schneideri* Em. *Zoologischer Anzeiger* **29**: 517–518. [14.xi.1905.]

Ruzsky, M. 1905b. Murav'i Rossii. (Formicariae Imperii Rossici.) *Trudy Obshchestva Estestvoispytatelei pri Imperatorskom Kazanskom Universitete* **38**: 1–799. [(31.xii).1905.]

Ruzsky. M. 1907. Murav'i Rossii. (Formicariae Imperii Rossici.) *Trudy Obshchestva Estestvoispytatelei pri Imperatorskom Kazanskom Universitete* **40**: 1–122. [(31.xii).1907.]

Ruzsky, M. 1912. Mirmekologicheskiya zametki. *Uchenye Zapiski Kazanskago Veterinarago Instituta* **29**: 629–636. [(31.xii).1912.]

Ruzsky, M. 1914a. Myrmekologische Notizen. *Archiv für Naturgeschichte* **79** (A 9): 58–63. [(31).v.1914.]

Ruzsky, M. 1914b. Murav'i Surgutskago uezda Tobol'skoi gubernii. *Russkoe Entomologicheskoe Obozrenie* **14**: 100–105. [14.vii.1914.]

Ruzsky, M. 1914c. Novaya forma murav'ya iz evropeiskoi Rossii. *Russkoe Entomologicheskoe Obozrenie* **14**: 323–324. [14.x.1914.]

Ruzsky, M. 1915a. O murav'yakh Tibeti i yuzhnoi Gobi. Po materialam sobrannym ekspeditsiei polkovnika P.K. Kozlova. *Ezhegodnik Zoologicheskago Muzeya Imperatorskoi Akademii Nauk* **20**: 418–444. [(31.xii).1915.] (Note. This paper was intended for publication in volume 19 (1914), page 480 onwards, of this journal. but did not appear there.)

Ruzsky, M. 1915b. Materialy po mirmekologii Sibiri. 1. O mirmekologicheskoi faune Tomskoi gubernii i nekotorykh drugikh mestnostei Sibiri (po issledovaniyam 1914–1915). *Izvestiya Imperatorskago Tomskogo Universiteta* **64**: 1–14. [(31.xii).1915. Not seen; reference from Arnol'di, 1967: 1815–1830; Dlussky, 1967a: 214; Tarbinsky, 1976: 214; and elsewhere.]

Ruzsky, M. 1916. O zoologicheskikh issledovaniyakh v Eniseiskoi gubernii, provedennykh letom 1915 g. *Izvestiya Imperatorskago Tomskogo Universiteta* **65**: 1–21. [(31.xii).1916. Not seen; reference from Radchenko, 1994b: 118.]
Ruzsky, M. 1920. Murav'i Kamchatki. *Izvestiya Instituta Issled. Sibiri (Tomsk)* **2**: 76–80. [(31.xii).1920. Not seen. Reference from Dlussky, 1967a: 215.]
Ruzsky, M. 1923. Murav'i ostrova Chelekenya. *Izvestiya Tomskogo Gosudarstvennogo Universiteta* **72**: 1–6. [(31.xii).1923.]
Ruzsky, M. 1924. Novyi vid murav'ya—leptotoraksa v Sibiri. *Izvestiya Tomskogo Gosudarstvennogo Universiteta* **74**: 152. [(31.xii).1924.]
Ruzsky, M. 1925a. Material po faune kurorta "Karachinskoe ozero." *Izvestiya Tomskogo Gosudarstvennogo Universiteta* **75**: 283–290. [(31.xii).1925.]
Ruzsky, M. 1925b. Novye dannye po faune murav'ev Sibiri. *Russkoe Entomologicheskoe Obozrenie* **19**: 41–46. [(31.xii).1925.]
Ruzsky, M. 1926. Sistematicheskii spisok murav'ev, vodyashchikhsya v Sibiri. 1. Obozrenie vidov iz rodov *Camponotus* i *Formica*. *Izvestiya Tomskogo Gosudarstvennogo Universiteta* **77**: 107–111. [(31.xii).1926.]
Ruzsky, M. 1936. Murav'i Zabaikal'ya. (Vostochno-Sibirskii krai, Buryato-Mongol'skaya ASSR.) *Trudy Biologicheskogo Nauchno- Issledovatel'skogo Instituta* **2**: 89–97. [(31.xii).1936.]
Sadil, J.V. 1939. Nová varieta druhu *Leptothorax unifasciatus* Latr. *Casopis Ceské Spolecnosti Entomologické* **36**: 30. [15.iii.1939.]
Sadil, J.V. 1952. A revision of the Czechoslovak forms of the genus *Myrmica* Latr. *Sborník Entomologického Oddelení Národního Musea v Praze* **27** (1951): 233–278. [(31.xii).1952.]
Sadil, J.V. 1953a. *Epimyrma zaleskyi* nov. spec. *Rocenka Ceskoslovenské Spolecnosti Entomologické* **50**: 188–196. [1.x.1953.]
Sadil, J.V. 1953b. Prispevek k Poznáni mravenci zvireny nasich hor. *Rocenka Ceskoslovenské Spolecnosti Entomologické* **50**: 197–202. [1.x.1953.]
Sadil, J.V. 1953c. Mravencí Zvírena nasich vápencovych, skalních stepi. *Rocenka Ceskoslovenské Spolecnosti Entomologické* **50**: 206–209. [1.x.1953.]
Samsinak, K. 1951. *Formica fusca* r. *lemani* Bondr. *Casopis Ceskoslovenské Spolecnosti Entomologické* **48**: 122–126. [30.xi.1951.]
Samsinak, K. 1957. *Sifolinia pechi* n.sp. *Rocenka Ceskoslovenské Spolecnosti Entomologické* **53** (1956): 167–170. [(31.xii).1957.]
Samsinak, K. 1964. Zur Kenntnis der Ameisenfauna der Tschechoslowakei. *Casopis Ceskoslovenské Spolecnosti Entomologické* **61**: 156–158. [20.iv.1964.]
Samsinak, K. 1967. *Camponotus novotnyi* sp.n., eine neue tertiäre Ameise aus Böhmen. *Vestník Ustredního Ustavu Geologického* **42**: 365–366. [(30).ix.1967.]
Santschi, F. 1907. Fourmis de Tunisie. Capturées en 1906. *Revue Suisse de Zoologie* **15**: 305–334. [15.xi.1907.]
Santschi, F. 1908. Nouvelles fourmis de l'Afrique du Nord (Egypte, Canaries, Tunisie). *Annales de la Société Entomologique de France* **77**: 517–534. [23.xii.1908.]
Santschi, F. 1909a. Une nouvelle fourmi de Cuba. *Bulletin de la Société Entomologique de France* **1909**: 309–310. [13.xii.1909.]
Santschi, F. 1909b. *Leptothorax rottenbergi* et espèces voisines. *Revue Suisse de Zoologie* **17**: 459–482. [30.xii.1909.]
Santschi, F. 1910a. Nouvelles fourmis de Tunisie (3e note). *Bulletin de la Société d'Histoire Naturelle de l'Afrique du Nord* **2**: 43–46. [15.i.1910.]
Santschi, F. 1910b. Nouvelles fourmis de Tunisie (suite). *Bulletin de la Société d'Histoire Naturelle de l'Afrique du Nord* **2**: 61–64. [15.ii.1910.]
Santschi, F. 1910c. Formicides nouveaux ou peu connus du Congo Français. *Annales de la Société Entomologique de France* **78** (1909): 349–400. [23.ii.1910.]
Santschi, F. 1910d. Nouvelles fourmis de Tunisie (suite). *Bulletin de la Société d'Histoire Naturelle de l'Afrique du Nord* **2**: 70–72. [15.iii.1910.]
Santschi, F. 1910e. Contributions à la faune entomologique de la Romanie. Formicides capturées par Mr. A.L. Montandon. *Buletinul Societatii de Stiinte din Bucuresti-Romania* **19**: 648–652. [(30).ix.1910.]
Santschi, F. 1910f. Hyménoptères 2e partie. *Actes de la Société Linnéenne de Bordeaux* **64**: 233–234. [8.x.1910.]
Santschi, F. 1910g. Nouveaux dorylines africains. *Revue Suisse de Zoologie* **18**: 737–759. [15.xii.1910.]
Santschi, F. 1910h. Deux nouvelles fourmis du Tonkin. *Naturaliste. (Revue Illustrée des Sciences Naturelles)* (2) **32**: 283–284. [15.xii.1910.]
Santschi, F. 1911a. Formicides nouveaux de l'Afrique Mineure (4e note). *Bulletin de la Société d'Histoire Naturelle de l'Afrique du Nord* **3**: 11–14. [15.i.1911.]
Santschi, F. 1911b. Deux nouvelles fourmis de Buenos-Ayres. *Bulletin de la Société Entomologique de France* **1911**: 52–53. [22.ii.1911.]
Santschi, F. 1911c. Nouvelles fourmis d'Afrique. *Annales de la Société Entomologique de France* **79** (1910): 351–369. [22.ii.1911.]
Santschi, F. 1911d. Formicides recoltés par Mr. le Prof. F. Silvestri aux Etats Unis en 1908. *Bulletino della Società Entomologica Italiana* **41** (1909): 3–7. [28.ii.1911.]
Santschi, F. 1911e. Nouvelles fourmis de Madagascar. *Revue Suisse de Zoologie* **19**: 117–134. [(28).ii.1911.]
Santschi, F. 1911f. Formicides nouveaux de l'Afrique Mineure (4e note suite). *Bulletin de la Société d'Histoire Naturelle de l'Afrique du Nord* **3**: 78–85. [15.v.1911.]
Santschi, F. 1911g. Nouvelles fourmis du Congo et du Benguela. *Revue Zoologique Africaine* **1**: 204–217. [31.viii.1911.]
Santschi, F. 1911h. Une nouvelle espèce d'*Eciton*. *Berliner Entomolgische Zeitschrift* **56**: 113. [(31).viii.1911.]
Santschi, F. 1911i. Formicides de diverses provenances. *Annales de la Société Entomologique de Belgique* **55**: 278–287. [6.x.1911.]
Santschi, F. 1911j. Une nouvelle variété de *Formica rufa* L. *Bulletin de la Société Entomologique de France* **1911**: 349–350. [22.xi.1911.]
Santschi, F. 1912a. Un *Carebara* américain. *Bulletin de la Société Entomologique de France* **1912**: 139–141. [10.iv.1912.]
Santschi, F. 1912b. Fourmis d'Afrique et de Madagascar. *Annales de la Société Entomologique de Belgique* **56**: 150–167. [1.vi.1912.]
Santschi, F. 1912c. Quelques nouvelles variétés de fourmis africaines. *Bulletin de la Société d'Histoire Naturelle de l'Afrique du Nord* **4**: 147–149. [15.vii.1912.]
Santschi, F. 1912d. Deux nouveaux *Carebara* africains. *Bulletin de la Société Entomologique de France* **1912**: 284–286. [29.viii.1912.]
Santschi, F. 1912e. Quelques fourmis de l'Amérique australe. *Revue Suisse de Zoologie* **20**: 519–534. [(30).ix.1912.]
Santschi, F. 1912f. Nouvelles fourmis de Tunisie, récoltées par le Dr Normand. *Bulletin de la Société d'Histoire Naturelle de l'Afrique du Nord* **4**: 172–175. [15.xi.1912.]
Santschi, F. 1912g. Contribution à la faune entomologique de la Roumanie. Description d'une nouvelle espèce de formicide. *Buletinul Societatii de Stiinte din Bucuresti-Romania* **20**: 657–658. [(31).xii.1912.]
Santschi, F. 1913a. *Cremastogaster* du groupe *tricolor—menileki*. *Bulletin de la Société Entomologique de France* **1912**: 411–414. [8.i.1913.]
Santschi, F. 1913b. Clé analytique des fourmis africaines du genre *Strumigenys* Sm. *Bulletin de la Société Entomologique de France* **1913**: 257–259. [11.vi.1913.]
Santschi, F. 1913c. Glanure de fourmis africaines. *Annales de la Société Entomologique de Belgique* **57**: 302–314. [7.xi.1913.]
Santschi, F. 1913d. Un nouvel *Oligomyrmex* de Cochinchine. *Bulletin de la Société Entomologique de France* **1913**: 457–458. [10.xii.1913.]
Santschi, F. 1913e. Clé dichotomique des *Oligomyrmex* africains. *Bulletin de la Société Entomologique de France* **1913**: 459–460. [10.xii.1913.]
Santschi, F. 1913f. Une nouvelle fourmi parasite. *Bulletin de la Société d'Histoire Naturelle de l'Afrique du Nord* **5**: 229–230. [15.xii.1913.]
Santschi, F. 1913g. Genre nouveau et espèce nouvelle de formicides. *Bulletin de la Société Entomologique de France* **1913**: 478. [30.xii.1913.]

Santschi, F. 1913h. *Mission de l'Equateur* **10**. Hyménoptères. Formicides: 33–43. [(31.xii).1913].
Santschi, F. 1914a. Mélanges myrmecologiques. *Annales de la Société Entomologique de Belgique* **57** (1913): 429–437. [2.i.1914.]
Santschi, F. 1914b. *Voyage de Ch. Alluaud et R. Jeannel en Afrique Orientale (1911–1912).* Résultats scientifiques. Insectes Hyménoptères **2**. Formicidae: 41–148. Paris. [25.ii.1914.]
Santschi, F. 1914c. Fam. Formicidae (p. 288). In Schulthess, A.v. Hymenopteren aus Kamerun. *Deutsche Entomologische Zeitschrift* **1914**: 283–297. [1.vi.1914.]
Santschi, F. 1914d. Formicides de l'Afrique occidentale et australe du voyage de Mr. le Professeur F. Silvestri. *Bollettino del Laboratorio di Zoologia generale e agraria della R. Scuola superiore d'Agricoltura in Portici* **8**: 309–385. [29.vii.1914.]
Santschi, F. 1914e. Meddelanden från Göteborgs Musei Zoologiska Afdelning. 3. Fourmis du Natal et du Zoulouland récoltées par le Dr I. Trägårdh. *Göteborgs Kungl. Vetenskaps och Vitterhets-Samhälles Handlingar* **15**: 1–47. [(31.xii).1914.]
Santschi, F. 1915a. Nouvelles fourmis d'Algérie, Tunisie et Syrie. *Bulletin de la Société d'Histoire Naturelle de l'Afrique du Nord* **7**: 54–63. [25.iv.1915.]
Santschi, F. 1915b. Deux *Cryptocerus* nouveaux. *Bulletin de la Société Entomologique de France* **1915**: 207–209. [7.viii.1915.]
Santschi, F. 1915c. Nouvelles fourmis d'Afrique. *Annales de la Société Entomologique de France* **84**: 244–282. [27.x.1915.]
Santschi, F. 1916a. Description d'un nouveau formicide de l'Afrique occidentale. *Bulletin de la Société Entomologique de France* **1916**: 50–51. [23.ii.1916.]
Santschi, F. 1916b. Descriptions de fourmis nouvelles d'Afrique et d'Amérique. *Annales de la Société Entomologique de France* **84** (1915): 497–513. [(30).iv.1916.]
Santschi, F. 1916c. Deux nouvelles fourmis d'Australie. *Bulletin de la Société Entomologique de France* **1916**: 174–175. [20.vi.1916.]
Santschi, F. 1916d. Rectification à la nomenclature de quelques formicides. *Bulletin de la Société Entomologique de France* **1916**: 242–243. [13.xi.1916.]
Santschi, F. 1916e. Formicides Sudaméricains nouveaux ou peu connus. *Physis. Revista de la Sociedad Argentina de Ciencias Naturales* **2**: 365–399. [30.xii.1916.]
Santschi, F. 1917a. Description d'une nouvelle reine de formicide du genre *Aenictus* Shuckard. *Annales de la Société Entomologique de France* **85** (1916): 277–278. [10.i.1917.]
Santschi, F. 1917b. Fourmis nouvelles de la Colonie du Cap, du Natal, et de Rhodesia. *Annales de la Société Entomologique de France* **85** (1916): 279–296. [10.i.1917.]
Santschi, F. 1917c. Note sur *Dorylus affinis* Shuckard mâle et ses variétés. *Bulletin de la Société d'Histoire Naturelle de l'Afrique du Nord* **8**: 18–21. [15.i.1917.]
Santschi, F. 1917d. *Acantholepis frauenfeldi* Mayr et ses variétés. *Bulletin de la Société d'Histoire Naturelle de l'Afrique du Nord* **8**: 42–48. [15.ii.1917.]
Santschi, F. 1917e. Races et variétés nouvelles du *Messor barbarus* L. *Bulletin de la Société d'Histoire Naturelle de l'Afrique du Nord* **8**: 89–94. [15.iv.1917.]
Santschi, F. 1917f. Description de quelques nouvelles fourmis de la République Argentine. *Anales de la Sociedad Científica Argentina* **84**: 277–283. [(31).xii.1917.]
Santschi, F. 1918a. *Leptothorax* nouveaux de l'Afrique mineure. *Bulletin de la Société d'Histoire Naturelle de l'Afrique du Nord* **9**: 31–38. [15.ii.1918.]
Santschi, F. 1918b. Nouveaux *Tetramorium* africains. *Bulletin de la Société d'Histoire Naturelle de l'Afrique du Nord* **9**: 121–132. [15.vi.1918.]
Santschi, F. 1918c. Nouveaux *Tetramorium* africains (suite et fin). *Bulletin de la Société d'Histoire Naturelle de l'Afrique du Nord* **9**: 153–156. [15.vii.1918.]
Santschi, F. 1918d. Sous-genres et synonymies de *Cremastogaster. Bulletin de la Société Entomologique de France* **1918**: 182–185. [27.viii.1918.]
Santschi, F. 1918e. Cinq notes myrmécologiques. *Bulletin de la Société Vaudoise des Sciences Naturelles* **52** (proces- verbaux): 63–64. [(31).xii.1918.]
Santschi, F. 1919a. Cinq notes myrmécologiques. *Bulletin de la Société Vaudoise des Sciences Naturelles* **52**: 325–350. [14.iii.1919.]
Santschi, F. 1919b. Fourmis nouvelles éthiopiennes. *Revue Zoologique Africaine* **6**: 229–240. [1.iv.1919.]
Santschi, F. 1919c. Fourmis nouvelles du Congo. *Revue Zoologique Africaine* **6**: 243–250. [1.iv.1919.]
Santschi, F. 1919d. Nouveaux genre et sous-genre de fourmis barbaresques. *Bulletin de la Société Entomologique de France* **1919**: 90–92. [18.iv.1919.]
Santschi, F. 1919e. Fourmis d'Espagne et des Canaries. *Boletín de la Real Sociedad Española de Historia Natural* **19**: 241–248. [(30).iv.1919.]
Santschi, F. 1919f. Nouveaux formicides de la République Argentine. *Anales de la Sociedad Científica Argentina* **87**: 37–57. [(30).vi.1919.]
Santschi, F. 1919g. Trois nouvelles fourmis des Canaries. *Boletín de la Real Sociedad Española de Historia Natural* **19**: 405–407. [(31).vii.1919.]
Santschi, F. 1919h. Nouvelles fourmis du Congo Belge du Musée du Congo Belge, à Tervuren. *Revue Zoologique Africaine* **7**: 79–91. [15.viii.1919.]
Santschi, F. 1919i. Fourmis du genre *Bothriomyrmex* Emery. (Systématique et moeurs.) *Revue Zoologique Africaine* **7**: 201–224. [15.xii.1919.]
Santschi, F. 1920a. Quelques nouveaux Camponotinae d'Indochine et Australie. *Bulletin de la Société Vaudoise des Sciences Naturelles* **52**: 565–569. [16.ii.1919.]
Santschi, F. 1920b. Formicides nouveaux du Gabon, du Congo, de la Rhodesia et du Natal. *Annales de la Société Entomologique de Belgique* **60**: 6–17. [8.iv.1920.]
Santschi, F. 1920c. Fourmis nouvelles du Congo Belge. *Revue Zoologique Africaine* **8**: 118–120. [15.v.1920.]
Santschi, F. 1920d. Formicides africains et américains nouveaux. *Annales de la Société Entomologique de France* **88** (1919): 361–390. [30.vi.1920.]
Santschi, F. 1920e. Quelques nouvelles fourmis de Bolivie. (Expédition Lizer-Delétang, 1917.) *Anales de la Sociedad Científica Argentina* **89**: 122–126. [(30).vi.1920.]
Santschi, F. 1920f. Nouvelles fourmis du genre *Cephalotes* Latr. *Bulletin de la Société Entomologique de France* **1920**: 147–149. [20.vii.1920.]
Santschi, F. 1920g. Cinq nouvelles notes sur les fourmis. *Bulletin de la Société Vaudoise des Sciences Naturelles* **53**: 163–186. [15.ix.1920.]
Santschi, F. 1920h. Fourmis d'Indo-Chine. *Annales de la Société Entomologique de Belgique* **60**: 158–176. [3.xii.1920.]
Santschi, F. 1920i. Études sur les maladies et les parasites du cacaoyer et d'autres plantes cultivées à S. Thomé. 10. Fourmis de S. Thomé. [Section Technique et de Pathologie Végétale de la "Companhia Agricola Ultramarina."] *Memoires Publiés par la Société Portugaise des Sciences Naturelles:* 1–4. Lisbonne. [(31.xii).1920; reissued in serial, 1922.]
Santschi, F. 1921a. Notes sur les fourmis paléarctiques. 2. Fourmis d'Asie Mineure récoltées par M.H. Gadeau de Kerville. *Boletín de la Real Sociedad Española de Historia Natural* **21**: 110–116. [(28).ii.1921.]
Santschi, F. 1921b. Notes sur les fourmis paléarctiques 1. Quelques fourmis du Nord de l'Afrique et des Canaries. *Real Sociedad Española de Historia Natural.* Tomo del 50° aniv.: 424–436. [15.iii.1921.]
Santschi, F. 1921c. Quelques nouveaux formicides africains. *Annales de la Société Entomologique de Belgique* **61**: 113–122. [2.iv.1921.]
Santschi, F. 1921d. Formicides nouveaux de l'Afrique du Nord. *Bulletin de la Société d'Histoire Naturelle de l'Afrique du Nord* **12**: 68–77. [15.iv.1921.]
Santschi, F. 1921e. Nouvelles fourmis paléarctiques. 3ème note. *Boletín de la Real Sociedad Española de Historia Natural* **21**: 165–170. [(30).iv.1921.]
Santschi, F. 1921f. Retouches aux sous-genres de *Camponotus. Annales de la Société Entomologique de Belgique* **61**: 310–312. [2.ix.1921.]
Santschi, F. 1921g. Ponerinae, Dorylinae et quelques autres formicides néotropiques. *Bulletin de la Société Vaudoise des Sciences Naturelles* **54**: 81–103. [16.ix.1921.]
Santschi, F. 1921h. Quelques nouveaux *Cryptocerus* de l'Argentine et pays voisins. *Anales de la Sociedad Científica Argentina* **92**: 124–128. [(30).ix.1921.]

Santschi, F. 1922a. Quelques nouvelles variétés de fourmis paléarctiques. *Bulletin de la Société d'Histoire Naturelle de l'Afrique du Nord* **13**: 66–68. [15.iii.1922.]

Santschi, F. 1922b. Myrmicines, dolichodérines et autres formicides néotropiques. *Bulletin de la Société Vaudoise des Sciences Naturelles* **54**: 345–378. [15.vii.1922.]

Santschi, F. 1922c. *Camponotus* néotropiques. *Annales de la Société Entomologique de Belgique* **62**: 97–124. [1.viii.1922.]

Santschi, F. 1922d. Description de nouvelles fourmis de l'Argentine et pays limitrophes. *Anales de la Sociedad Científica Argentina* **94**: 241–262. [(31).xii.1922.]

Santschi, F. 1923a. Notes sur les fourmis paléarctiques. 4ème note. *Boletin de la Real Sociedad Española de Historia Natural* **23**: 133–137. [(31).iii.1923.]

Santschi, F. 1923b. Revue des fourmis du genre *Brachymymex* Mayr. *Anales del Museo Nacional de Historia Natural de Buenos Aires* **31**: 650–678. [2.iv.1923.]

Santschi, F. 1923c. *Solenopsis* et autres fourmis néotropicales. *Revue Suisse de Zoologie* **30**: 245–273. [(30).vi.1923.]

Santschi, F. 1923d. *Pheidole* et quelques autres fourmis néotropiques. *Annales de la Société Entomologique de Belgique* **63**: 45–69. [25.vii.1923.]

Santschi, F. 1923e. Descriptions de nouveaux formicides éthiopiens et notes diverses. 1. *Revue Zoologique Africaine* **11**: 259–295. [30.x.1923.]

Santschi, F. 1923f. *Messor* et autres fourmis paléarctiques. *Revue Suisse de Zoologie* **30**: 317–336. [(31).xii.1923.]

Santschi, F. 1923g. Description de quelques nouvelles fourmis du Brésil. *Revista do Museu Paulista* **13** (1922): 1255–1264. [(31.xii).1923.]

Santschi, F. 1924a. Revue du genre *Plectroctena* F.Smith. *Revue Suisse de Zoologie* **31**: 155–173. [(31).v.1924.]

Santschi, F. 1924b. Descriptions de nouveaux formicides africains et notes diverses. 2. *Revue Zoologique Africaine* **12**: 195–224. [1.vi.1924.]

Santschi, F. 1924c. Fourmis d'Indochine. *Faune Entomologique de l'Indochine Française* **8**: 95–117. [(30).ix.1924.]

Santschi, F. 1925a. Revision du genre *Acromyrmex* Mayr. *Revue Suisse de Zoologie* **31**: 355–398. [(31).iii.1925.]

Santschi, F. 1925b. Nouvelles fourmis brésiliennes. *Annales de la Société Entomologique de Belgique* **64** (1924): 5–20. [(30).iv.1925. Date from footnote on p. 179 of this volume.]

Santschi, F. 1925c. Révision des *Myrmicaria* d'Afrique. *Annales de la Société Entomologique de Belgique* **64** (1924): 133–176. [(30).iv.1925. Date from footnote on p. 179 of this volume.]

Santschi, F. 1925d. Nouveaux formicides brésiliens et autres. *Annales et Bulletin de la Société Entomologique de Belgique* **65**: 221–247. [31.vii.1925.]

Santschi, F. 1925e. Fourmis des provinces Argentine de Santa Fe, Catamarca, Santa Cruz, Córdoba et Los Andes. *Comunicaciones del Museo Nacional de Historia Natural "Bernardino Rivadavia"* **2**: 149–168. [24.ix.1925.]

Santschi, F. 1925f. Contribution à la faune myrmécologique de la Chine. *Bulletin de la Société Vaudoise des Sciences Naturelles* **56**: 81–96. [30.xi.1925.]

Santschi, F. 1925g. Fourmis d'Espagne et autres espèces paléarctiques. *Eos. Revista Española de Entomología* **1**: 339–360. [31.xii.1925.]

Santschi, F. 1925h. *Mission Rohan-Chabot en Angola et Rhodesia (1912–1914).* **4**. Formicidae: 159–168. Paris. [(31.xii).1925.]

Santschi, F. 1926a. Trois notes myrmécologiques. *Annales de la Société Entomologique de France* **95**: 13–28. [31.iii.1926.]

Santschi, F. 1926b. Description de nouveaux formicides Ethiopiens (3me. partie). *Revue Zoologique Africaine* **13** (1925): 207–267. [1.iv.1926.]

Santschi, F. 1926c. Nouvelles notes sur les *Camponotus. Revue Suisse de Zoologie* **33**: 597–618. [(31).vii.1926.]

Santschi, F. 1926d. Deux nouvelles fourmis parasites de l'Argentine. *Folia Myrmecologica et Termitologica* **1**: 6–8. [(31).x.1926.]

Santschi, F. 1926e. Quelques fourmis nord-africaines. *Bulletin de la Société d'Histoire Naturelle de l'Afrique du Nord* **17**: 229–236. [15.xi.1926.]

Santschi, F. 1926f. Travaux scientifiques de l'armée d'Orient (1916–1918). Fourmis. *Bulletin du Muséum National d'Histoire Naturelle* **5**: 286–293. [25.xi.1926.]

Santschi, F. 1927a. A propos du *Tetramorium caespitum* L. *Folia Myrmecologica et Termitologica* **1**: 52–58. [(28).ii.1927.]

Santschi, F. 1927b. Notes myrmécologiques. *Bulletin de la Société Entomologique de France* **1927**: 126–128. [20.v.1927.]

Santschi, F. 1927c. Revision des *Messor* du groupe *instabilis* Sm. *Boletín de la Real Sociedad Española de Historia Natural* **27**: 225–250. [(31).v.1927.]

Santschi, F. 1927d. Révision myrmécologique. *Bulletin et Annales de la Société Entomologique de Belgique* **67**: 240–248. [10.xi.1927.]

Santschi, F. 1928a. *Insects of Samoa and Other Samoan Terrestrial Arthropoda* **5**. Hymenoptera, Formicidae: 41–58. London. [25.ii.1928.]

Santschi, F. 1928b. Nouvelles fourmis de Chine et du Turkestan Russe. *Bulletin et Annales de la Société Entomologique de Belgique* **68**: 31–46. [3.iii.1928.]

Santschi, F. 1928c. Fourmis des îles Fidji. *Revue Suisse de Zoologie* **35**: 67–74. [(31).v.1928.]

Santschi, F. 1928d. Descriptions de nouvelles fourmis ethiopiennes. (Quatrième note). *Revue de Zoologie et de Botanique Africaines* **16**: 54–69. [1.vi.1928.]

Santschi, F. 1928e. Nouvelles fourmis d'Australie. *Bulletin de la Société Vaudoise des Sciences Naturelles* **56**: 465–483. [30.viii.1928.]

Santschi, F. 1928f. Descriptions de nouvelles fourmis ethiopiennes. (Suite.) *Revue de Zoologie et de Botanique Africaines* **16**: 191–213. [15.ix.1928.]

Santschi, F. 1928g. Quelques nids de fourmis du Muséum d'Histoire Naturelle de Paris. *Annales des Sciences Naturelles. Annales de Zoologie* (10) **11**: 247–259. [(30).ix.1928.]

Santschi, F. 1928h. Fourmis de Sumatra, récoltées par Mr. J.B. Corporaal. *Tijdschrift voor Entomologie* **71**: 119–140. [31.xii.1928.]

Santschi, F. 1929a. Sur quelques nouvelles fourmis du Brésil. *Deutsche Entomologische Zeitschrift* **1928**: 414–416. [10.i.1929.]

Santschi, F. 1929b. Etude sur les *Cataglyphis. Revue Suisse de Zoologie* **36**: 25–70. [(31).iii.1929.]

Santschi, F. 1929c. Fourmis du Sahara central récoltées par la Mission du Hoggar (Février-Mars 1928). *Bulletin de la Société d'Histoire Naturelle de l'Afrique du Nord* **20**: 97–108. [15.iv.1929.]

Santschi, F. 1929d. Nouvelles fourmis de la République Argentine et du Brésil. *Anales de la Sociedad Científica Argentina* **107**: 273–316. [(30).iv.1929.]

Santschi, F. 1929e. Fourmis du Maroc, d'Algérie et de Tunisie. *Annales et Bulletin de la Société Entomologique de Belgique* **69**: 138–165. [15.vi.1929.]

Santschi, F. 1929f. Mélange myrmecologique. *Wiener Entomologische Zeitung* **46**: 84–93. [15.ix.1929.]

Santschi, F. 1929g. Note additionelle aux fourmis du Sahara central récoltées par la Mission du Hoggar (février-mars 1928). *Bulletin de la Société d'Histoire Naturelle de l'Afrique du Nord* **20**: 164–166. [15.xi.1929.]

Santschi, F. 1929h. Révision du genre *Holcoponera* Mayr. *Zoologischer Anzeiger* **82**: 437–477. [(31.xii).1929.]

Santschi, F. 1930a. Description de formicides éthiopiens nouveaux ou peu connus. *Bulletin et Annales de la Société Entomologique de Belgique* **70**: 49–77. [8.iii.1930.]

Santschi, F. 1930b. Formicides de l'Angola. Résultats de la Mission scientifique suisse en Angola (1928–1929). *Revue Suisse de Zoologie* **37**: 53–81. [(31).iii.1930.]

Santschi, F. 1930c. Quelques fourmis de Cuba et du Brésil. *Bulletin de la Société Royale Entomologique d'Égypte* (N.S.) **14**: 75–83. [(31.xii).1930.]

Santschi, F. 1930d. Trois notes myrmécologiques. *Bulletin et Annales de la Société Entomologique de Belgique* **70**: 263–270. [1.xii.1930.]

Santschi, F. 1930e. Un nouveau genre de fourmi parasite sans ouvrières de l'Argentine. *Revista de la Sociedad Entomológica Argentina* **3**: 81–83. [26.xii.1930.]

Santschi, F. 1931a. Inventa entomologica itineris Hispanici et Maroccani, quod a. 1926 fecerunt Harald et Håkan Lindberg. 9. Fourmis du Bassin Méditerranéen occidental et du Maroc récoltées par MM. Lindberg. *Societas Scientiarum Fennica Commentationes Biologicae* **3**: 1–13. [(28).ii.1931.]
Santschi, F. 1931b. Notes sur le genre *Myrmica* Latreille. *Revue Suisse de Zoologie* **38**: 335–355. [(31).v.1931.]
Santschi, F. 1931c. Fourmis de Cuba et de Panama. *Revista de Entomologia* **1**: 265–282. [5.ix.1931.]
Santschi, F. 1931d. La reine du *Dorylus fulvus* Westw. *Bulletin de la Société d'Histoire Naturelle de l'Afrique du Nord* **22**: 401–408. [15.xi.1931.]
Santschi, F. 1931e. Contribution à l'étude des fourmis de l'Argentine. *Anales de la Sociedad Científica Argentina* **112**: 273–282. [(30).xi.1931.]
Santschi, F. 1932a. Formicides sud-africains. *Société Entomologique de France, Livre du Centenaire* **1932**: 381–392. [30.vi.1932.]
Santschi, F. 1932b. Résultats scientifiques du Voyage aux Indes Orientales Néerlandaises de L.L.A.A.R.R. le Prince et la Princesse Léopold de Belgique. Formicidae. *Mémoires du Musée Royal d'Histoire Naturelle de Belgique* (2) **4**: 11–29. [15.x.1932.]
Santschi, F. 1932c. Liste de fourmis d'Espagne recueillies par Mr. J.M. Dusmet. *Boletín de la Sociedad Entomológica de España* **15**: 69–74. [30.x.1932.]
Santschi, F. 1932d. Etudes sur quelques *Attomyrma* paléarctiques. *Mitteilungen der Schweizerischen Entomologischen Gesellschaft* **15**: 338–346. [15.xii.1932.]
Santschi, F. 1932e. Quelques fourmis inédites de l'Amérique centrale et Cuba. *Revista de Entomologia* **2**: 410–414. [28.xii.1932.]
Santschi, F. 1932f. Notes sur les fourmis du Sahara. *Bulletin du Muséum National d'Histoire Naturelle* (2) **4**: 516–520. [2.viii.1932.]
Santschi, F. 1932g. Fourmis du Portugal. *Memórias e Estudos do Museu Zoológico da Universidade de Coimbra* (1) **59**: 1–3. [(31.xii).1932.]
Santschi, F. 1933a. Insectes des Açores. Fourmis (pp. 21–22). In, Voyage de MM. L. Chopard et A. Méquignon aux Açores. *Annales de la Société Entomologique de France* **102**: 19–24. [30.iii.1933.]
Santschi, F. 1933b. Contribution à l'étude des fourmis de l'Afrique tropicale. *Bulletin et Annales de la Société Entomologique de Belgique* **73**: 95–108. [30.iii.1933.]
Santschi, F. 1933c. Etudes sur le sous-genre *Aphaenogaster* Mayr. *Revue Suisse de Zoologie* **40**: 389–408. [(31).vii.1933.]
Santschi, F. 1933d. Formicides des collections de S.A.R. le Prince Léopold de Belgique. Voyage aux Indes Orientales, 1932. *Bulletin du Musée Royal d'Histoire Naturelle de Belgique* **9** (27): 1–3. [(30).ix.1933.]
Santschi, F. 1933e. Fourmis de la République Argentine, en particulier du territoire de Misiones. *Anales de la Sociedad Científica Argentina* **116**: 105–124. [(30).ix.1933.]
Santschi, F. 1934a. Deux nouveaux *Crematogaster* intéressants. *Bulletin de la Société Vaudoise des Sciences Naturelles* **58**: 187–191. [25.ii.1934.]
Santschi, F. 1934b. Mission J. de Lépiney au Soudan français 1933–34. (Huitième note). Fourmis. *Bulletin de la Société des Sciences Naturelles du Maroc* **14**: 33–34. [31.iii.1934.]
Santschi, F. 1934c. Fourmis de Misiones et du chaco argentin. *Revista de la Sociedad Entomológica Argentina* **6**: 23–34. [30.v.1934.]
Santschi, F. 1934d. Fourmis d'une croisière. *Annales de la Société Entomologique de Belgique* **74**: 273–282. [25.viii.1934.]
Santschi, F. 1934e. Contribution aux *Solenopsis* paléarctiques. *Revue Suisse de Zoologie* **41**: 565–592. [(31).x.1934.]
Santschi, F. 1934f. Fourmis du Sahara central. *Mémoires de la Société d'Histoire Naturelle d'Afrique du Nord* **4**: 165–177. [(31.xii).1934.]
Santschi, F. 1935a. Fourmis du Musée du Congo Belge. *Revue de Zoologie et de Botanique Africaines* **27**: 254–285. [25.ix.1935.]
Santschi, F. 1935b. *Mission Scientifique de l'Omo.* **2**. Zoologie. Fascicule 15, Hymenoptera Formicidae: 255–277. Paris. [(31.xii).1935.]
Santschi, F. 1936a. Étude sur les fourmis du genre *Monomorium* Mayr. *Bulletin de la Société des Sciences Naturelles du Maroc* **16**: 32–64. [31.iii.1936.]
Santschi, F. 1936b. Contribution à l'étude des fourmis de l'Amerique du Sud. *Revista de Entomologia* **6**: 196–218. [15.vii.1936.]
Santschi, F. 1936c Liste et descriptions de fourmis du Maroc. *Bulletin de la Société des Sciences Naturelles du Maroc* **16**: 198–210. [30.ix.1936.]
Santschi, F. 1936d. Fourmis nouvelles ou intéressantes de la République Argentine. *Revista de Entomologia* **6**: 402–421. [30.x.1936.]
Santschi, F. 1936e. Hyménoptères. Fourmis (pp. 81–82). In, Mission au Tibesti (1930–31), dirigée par M. Marius Dalloni. Deuxième mémoire. Zoologie. Étude préliminaire de la faune du Tibesti. *Memoires de l'Academie des Sciences de l'Institut de France* (2) **62**: 41–92. [(31.xii).1936.]
Santschi, F. 1937a. Glanure de fourmis éthiopiennes. *Bulletin et Annales de la Société Entomologique de Belgique* **77**: 47–66. [28.ii.1937.]
Santschi, F. 1937b. Résultats entomologiques d'un voyage au Cameroun. Formicides récoltés par Mr le Dr F. Zumpt. *Mitteilungen der Schweizerischen Entomologischen Gesellschaft* **17**: 93–104. [15.iii.1937.]
Santschi, F. 1937c. Les sexués du genre *Anillidris* Santschi. *Bulletin de la Société Entomologique de France* **42**: 68–70. [21.iv.1937.]
Santschi, F. 1937d. Fourmis angolaises. Résultats de la Mission scientifique suisse en Angola (2me voyage) 1932–1933. *Revue Suisse de Zoologie* **44**: 211–250. [(30).iv.1937.]
Santschi, F. 1937e. Contribution à l'étude des *Crematogaster* paléarctiques. *Mémoires de la Société Vaudoise des Sciences Naturelles* **5**: 295–317. [30.vi.1937.]
Santschi, F. 1937f. Note sur *Acromyrmex subterraneus* Forel. *Revista de Entomologia* **7**: 230–233. [27.vii.1937.]
Santschi, F. 1937g. Fourmis du Congo et régions limitrophes. *Revue de Zoologie et de Botanique Africaines* **30**: 71–85. [30.x.1937.]
Santschi, F. 1937h. Fourmis du Japon et de Formose. *Bulletin et Annales de la Société Entomologique de Belgique* **77**: 361–388. [30.x.1937.]
Santschi, F. 1938a. Quelques nouvelles fourmis d'Égypte. *Bulletin de la Société Royale Entomologique d'Égypte* **21** (1937): 28–44. [31.iii.1938.]
Santschi, F. 1938b. Notes sur quelques *Ponera*. *Bulletin de la Société Entomologique de France* **43**: 78–80. [15.iv.1938.]
Santschi, F. 1939a. Contribution au sous-genre *Alaopone*. *Revue Suisse de Zoologie* **46**: 143–154. [(31).i.1939.]
Santschi, F. 1939b. Fourmis de Rhodesia et du Congo. *Bulletin et Annales de la Société Entomologique de Belgique* **79**: 237–246. [24.iv.1939.]
Santschi, F. 1939c. Trois notes sur quelques fourmis du Musée Royal d'Histoire Naturelle de Belgique. *Bulletin du Musée Royal d'Histoire Naturelle de Belgique* **15** (14): 1–15. [(30).iv.1939.]
Santschi, F. 1939d. Notes sur des *Camponotus* et autres fourmis de l'Afrique mineure. *Bulletin de la Société des Sciences Naturelles du Maroc* **19**: 66–87. [30.vi.1939.]
Santschi, F. 1939e. Etudes et descriptions de fourmis néotropiques. *Revista de Entomologia* **10**: 312–330. [4.ix.1939.]
Santschi, F. 1939f. Résultats scientifiques des croisières du Navire-École Belge "Mercator." 2. Formicidae. *Musée Royal d'Histoire Naturelle de Belgique* (2) **15**: 159–167. [31.xii.1939.]
Santschi, F. 1941. Quelques fourmis japonaises inédites. *Mitteilungen der Schweizerischen Entomologischen Gesellschaft* **18**: 273–279. [15.i.1941.]
Saulcy, F.de. 1874. Species des paussides, clavigéroides, psélaphides & scyménides de l'Europe & des pays circonvoisins. *Bulletin de la Société d'Histoire Naturelle de Metz* **13**: 1–132. [(31.xii).1874.]
Saunders, E. 1880. Synopsis of the British Heterogyna and fossorial Hymenoptera. *Transactions of the Entomological Society of London* **1880**: 201–304. [31.xii.1880.]
Saunders, E. 1890. Aculeate Hymenoptera collected by J.J. Walker, Esq., R.N., F.L.S., at Gibraltar and in North Africa. *Entomologist's Monthly Magazine* **26**: 201–205. [(31).viii.1890.]
Saunders, W.W. 1842. Descriptions of two hymenopterous insects from northern India. *Transactions of the Entomological Society of London* **3** (1841): 57–58. [27.i.1842. Date of publication from Wheeler, G. 1912: 754.]

Savage, T.S. 1849. The driver ants of western Africa. *Proceedings of the Academy of Natural Sciences of Philadelphia* **4**: 195–200. [(31).viii.1849.]

Say, T. 1836. Descriptions of new species of North American Hymenoptera, and observations on some already described. *Boston Journal of Natural History* **1**: 209–305. [(31).v.1836.]

Schembri, S.P. & Collingwood, C.A. 1981. A revision of the myrmecofauna of the Maltese Islands. *Annali del Museo Civico di Storia Naturale di Genova* **83**: 417–442. [1.ix.1981.]

Schenck, C.F. 1852. Beschreibung Nassauischer Ameisenarten. *Jahrbuch des Vereins für Naturkunde im Herzogthum Nassau* **8**: 1–149. [(31.xii).1852.]

Schenck, C.F. 1853. Die Nassauischen Ameisen-Species (Fortsetzung). *Entomologische Zeitung herausgegeben von dem Entomologischen Vereine zu Stettin* **14**: 185–198. [(30).vi.1853.]

Schilling, P.S. 1830. Bericht über die entomologische Section der Schlesischen Gesellschaft für vaterländische Cultur im Jahre 1829. *Uebersicht der Arbeiten und Veränderungen der Schlesischen Gesellschaft für Vaterländische Kultur im Jahre 1829:* 52–56. [(31.xii).1830.]

Schilling, P.S. 1839. Bemerkungen über die in Schlesien und der Grafschaft Glaz vorgefundenen Arten der Ameisen. *Uebersicht der Arbeiten und Veränderungen der Schlesischen Gesellschaft für Vaterländische Kultur im Jahre 1838:* 51–56. [(31.xii).1839.]

Schmitz, H. 1950. Formicidae quaedam a cl. A. Stärcke determinatae, quas in Lusitania collegit. *Bróteria. Ciências Naturais* Série Trimestral **19**: 12–16. [(31.xii).1950.]

Schneirla, T.C. 1971. *Army Ants. A Study in Social Organisation:* 349 pp. San Francisco.

Schoenherr, C.J. 1833. *Genera et Species Curculionidum, cum synonymia hujus familiae* **1** (pars 1): 381 pp. Parisiis. [(31.xii).1833.]

Scholz, E.J.R. 1924. *Formica exsecta* var. *sudetica* nov. var. *Neue Beiträge zur Systematischen Insektenkunde* **3**: 48. [Not seen.]

Schrank, F.von, P.von. 1837. In Gistel, J. von N.F.X. *Faunus* (Neue Folge) **1**: 16. [Not seen.]

Schulz, A. 1991. *Tetramorium semilaeve* und *Bothriomyrmex gibbus* neu für Österreich sowie über die Verbreitung von *Leptothorax sordidulus. Zeitschrift der Arbeitsgemeinschaft Österreichischer Entomologen* **43**: 120–121. [(31).xii.1991.]

Schulz, W.A. 1906. *Spolia Hymenopterologica:* 355 pp. Paderborn. [(31.xii).1906.]

Schumann, K. 1888. Einige neue Ameisenpflanzen. *Jahrbücher für Wissenschaftliche Botanik* **19**: 357–421. [(31.xii).1888.]

Schweigger, A.F. 1819. *Beobachtungen auf naturhistorischen Reisen. Anatomisch-physiologische Untersuchungen über Corallen; nebst einem Anhang, Bemerkungen über den Bernstein enthaltend:* 127 pp. Berlin. [(31.xii).1819.]

Scopoli, J.A. 1763. *Entomologia Carniolica* exhibens insecta carnioliae indigena et distributa in ordines, genera, species, varietates. Methodo Linnaeana: 420 pp. Vindobonae. [(31.xii).1763.]

Scudder, S.H. 1877a. The first discovered traces of fossil insects in the American tertiaries. *Bulletin of the United States Geological and Geographical Survey of the Territories* **3**: 741–762. [15.viii.1877.]

Scudder, S.H. 1877b. Appendix to Mr. George M. Dawson's report. The insects of the Tertiary beds at Quesnel. *Report of Progress. Geological Survey of Canada* **1875–76**: 266–280. [(31.xii).1877.]

Scudder, S.H. 1878. The fossil insects of the Green River shales. *Bulletin of the United States Geological and Geographical Survey of the Territories* **4**: 747–776. [(31.xii).1878.]

Scudder, S.H. 1891. Index to the known fossil insects of the world, including myriapods and arachnids. *Bulletin of the United States Geological Survey* **No. 71**: 744 pp. [(31.xii).1891.]

Seifert, B. 1982. *Lasius (Chthonolasius) jensi* n. sp.—eine neue temporär sozialparasitische Erdameise aus Mitteleuropa. *Reichenbachia* **20**: 85–96. [27.ix.1982.]

Seifert, B. 1983. The taxonomical and ecological status of *Lasius myops* Forel and first description of its males. *Abhandlungen und Berichte des Naturkundemuseums Görlitz* **57**: 1–16. [1.vi.1983.]

Seifert, B. 1984a. A method for differentiation of the female castes of *Tapinoma ambiguum* Emery and *Tapinoma erraticum* (Latr.) and remarks on their distribution in Europe north of the Mediterranean region. *Faunistische Abhandlungen Staatliches Museum für Tierkunde in Dresden* **11**: 151–155. [1.iii.1984.]

Seifert, B. 1984b. Firm evidence for synonymy of *Myrmica rugulosoides* Forel, 1915 and *Myrmica scabrinodis* Nylander, 1846. *Abhandlungen und Berichte des Naturkundemuseums Görlitz* **58**: 1–10. [30.xi.1984.]

Seifert, B. 1987. *Myrmica georgica* n. sp., a new ant from Transcaucasia and North Kazakhstan (USSR). *Reichenbachia* **24**: 183–187. [20.v.1987.]

Seifert, B. 1988a. A revision of the European species of the ant subgenus *Chthonolasius. Entomologische Abhandlungen. Staatliches Museum für Tierkunde Dresden* **51**: 143–180. [20.vii.1988.]

Seifert, B. 1988b. A taxonomic revision of the *Myrmica* species of Europe, Asia Minor, and Caucasus. *Abhandlungen und Berichte des Naturkundemuseums Görlitz* **62**: 1–75. [12.x.1988.]

Seifert, B. 1990. Supplementation to the revision of the European species of the ant subgenus *Chthonolasius. Doriana. Supplemento agli Annali del Museo Civico di Storia Naturale "G. Doria"* **6** (271): 1–13. [15.xi.1990.]

Seifert, B. 1991. *Lasius platythorax* n. sp., a widespread sibling species of *Lasius niger. Entomologia Generalis* **16**: 69–81. [(30).vi.1991.]

Seifert, B. 1992a. *Formica nigricans* Emery, 1909—an ecomorph of *Formica pratensis* Retzius, 1783. *Entomologica Fennica* **2** (1991): 215–226. [8.i.1992.]

Seifert, B. 1992b. A taxonomic revision of the Palaearctic members of the ant subgenus *Lasius* s. str. *Abhandlungen und Berichte des Naturkundemuseums Görlitz* **66**: 1–67. [7.viii.1992.]

Shattuck, S.O. 1987. An analysis of geographic variation in the *Pogonomyrmex occidentalis* complex. *Psyche* **94**: 159–180. [6.x.1987.]

Shattuck, S.O. 1990. Revision of the dolichoderine ant genus *Turneria. Systematic Entomology* **15**: 101–117. [(31).i.1990.]

Shattuck, S.O. 1991. Revision of the dolichoderine ant genus *Axinidris. Systematic Entomology* **16**: 105–120. [(31).i.1991.]

Shattuck, S.O. 1992a. Review of the dolichoderine ant genus *Iridomyrmex* Mayr with descriptions of three new genera. *Journal of the Australian Entomological Society* **31**: 13–18. [28.ii.1992.]

Shattuck, S.O. 1992b. Higher classification of the ant subfamilies Aneuretinae, Dolichoderinae and Formicinae. *Sysytematic Entomology* **17**: 199–206. [(30).iv.1992.]

Shattuck, S.O. 1992c. Generic revision of the ant subfamily Dolichoderinae. *Sociobiogy* **21**: 1–181. [(31.xii).1992.]

Shattuck, S.O. 1993a. Revision of the *Iridomyrmex purpureus* species group. *Invertebrate Taxonomy* **7**: 113–149. [27.iv.1993.]

Shattuck, S.O. 1993b. Revision of the *Iridomyrmex calvus* species- group. *Invertebrate Taxonomy* **7**: 1303–1325. [27.xi.1993.]

Shattuck, S.O. 1994. Taxonomic catalog of the ant subfamilies Aneuretinae and Dolichoderinae. *University of California Publications in Entomology* **112**: 1–241. [(31).i.1994.]

Shattuck, S.O. & McArthur, A.J. (1995). Generic placements of Australian ants described by W.F. Erichson. *Journal of the Australian Entomological Society* (currently in press).

Sherborn, C.D. 1923. On the dates of G.W.F. Panzer's "Fauna Insect. German.", 1792–1844. *Annals and Magazine of Natural History* (9) **11**: 566–568. [(30).iv.1923.]

Sherborn, C.D. & Woodward, B.B. 1901. Dates of publication of the zoological and botanical portions of some French voyages. Part 2. *Annals and Magazine of Natural History* (7) **8**: 161–164. [(31).viii.1901.]

Sherborn, C.D. & Woodward, B.B. 1906a. Notes on the dates of publication of the natural history portions of some French voyages. "Voyage autour du Monde. . .sur. . .La Coquille pendent. . .1822–25. . .par L.J. Duperry."—a correction. *Annals and Magazine of Natural History* (7) **17**: 335–336. [(31).iii.1906.]

Sherborn, C.D. & Woodward, B.B. 1906b. On the dates of publication of the natural history portions of the "Encyclopédie Méthodique." *Annals and Magazine of Natural History* (7) **17**: 577–582. [(30).vi.1906.]

Shuckard, W.E. 1838. Description of a new species of *Myrmica* which has been found in houses both in the metropolis and provinces. *Magazine of Natural History* (N.S.) **2**: 626–627. [(30).xi.1838.]

Shuckard, W.E. 1840a. Monograph of the Dorylidae, a family of the Hymenoptera Heterogyna. *Annals of Natural History; or Magazine of Zoology, Botany and Geology* **5**: 188–201. [(31).v.1840.]

Shuckard, W.E. 1840b. Monograph of the Dorylidae, a family of the Hymenoptera Heterogyna. (Continued from p. 201.) *Annals of Natural History; or Magazine of Zoology, Botany and Geology* **5**: 258–271. [(30).vi.1840.]

Shuckard, W.E. 1840c. Monograph of the Dorylidae, a family of the Hymenoptera Heterogyna. (Concluded from p. 271.) *Annals of Natural History; or Magazine of Zoology, Botany and Geology* **5**: 315–328. [(31).vii.1840.]

Shuckard, W.E. 1840d. Appendix to Mr. Shuckard's monograph of the Dorylidae, containing a description of two new species of *Labidus*. *Annals of Natural History; or Magazine of Zoology, Botany and Geology* **5**: 396–398. [(31).viii.1840.]

Signoret, V. 1847. Description de deux hémiptères-homoptères, tribu des octicelles, groupe des cicadides. *Annales de la Société Entomologique de France* (2) **5**: 293–296. [20.viii.1847.]

Silhavy, V. 1937. *Strongylognathus kratochvili* n. sp., novy praeglaciální mravenec z Moravy. *Sborník Prírodovedeckého Klubu v Trebíci* **1** (1936): 5–12. [(31.xii).1937.]

Sjöstedt, Y. 1908. Akaziengallen und Ameisen auf den Ostafrikanischen Steppen. *Wissenschaftliche Ergebnisse der Schwedischen Zoologischen Expedition nach dem Kilimandjaro, dem Meru und den umgebenden Massaisteppen Deutsch-Ostafrikas, 1905–1906* **2**. 8, Hymenoptera 4: 97–118. Uppsala & Stockholm. [(31).viii.1908.]

Smith, D.R. 1979. Formicoidea (pp. 1323–1467). In Krombein, K.V., Hurd, P.D. Jr., Smith, D.R. & Burks, B.D. *Catalog of Hymenoptera in America North of Mexico* **2** Apocrita (Aculeata): 1199–2209. Washington, D.C. [(31.xii).1979.]

Smith, D.R. & Lavigne, R.J. 1973. Two new species of ants of the genera *Tapinoma* Foerster and *Paratrechina* Motschoulsky from Puerto Rico. *Proceedings of the Entomological Society of Washington* **75**: 181–187. [10.vii.1973.]

Smith, F. 1851. *List of the Specimens of British Animals in the Collection of the British Museum* **6**. Hymenoptera Aculeata: 134 pp. London. [(31.xii).1851.]

Smith, F. 1852. Descriptions of some hymenopterous insects captured in India, with notes on their oeconomy, by Ezra T. Downes, Esq., who presented them to the Honourable the East India Company. *Annals and Magazine of Natural History* (2) **9**: 44–50. [(31).i.1852.]

Smith, F. 1853. Monograph of the genus *Cryptocerus*, belonging to the group Cryptoceridae—family Myrmicidae—division Hymenoptera Heterogyna. *Transactions of the Entomological Society of London* (2) **2** (1854): 213–228. [26.xii.1853. Date of publication from Wheeler, G. 1912: 755.]

Smith, F. 1855a. Essay on the genera and species of British Formicidae. *Transactions of the Entomological Society of London* (2) **3**: 95–112. [(31).i.1855.]

Smith, F. 1855b. Essay on the genera and species of British Formicidae (continued). *Transactions of the Entomological Society of London* (2) **3**: 113–135. [(30).iv.1855.]

Smith, F. 1855c. Descriptions of some species of Brazilian ants belonging to the genera *Pseudomyrma, Eciton* and *Myrmica* (with observations on their economy by Mr. H.W. Bates). *Transactions of the Entomological Society of London* (2) **3**: 156–169. [6.viii.1855.]

Smith, F. 1857a. Catalogue of the hymenopterous insects collected at Sarawak, Borneo; Mount Ophir, Malacca; and at Singapore, by A.R. Wallace. *Journal of the Proceedings of the Linnean Society of London, Zoology* **2**: 42–88. [2.xi.1857.]

Smith, F. 1857b. Notes on aculeate Hymenoptera, with some observations on their economy. *Entomologist's Annual* **1858**: 34–46. [19.xii.1857.]

Smith, F. 1858a. Revision of an essay on the British Formicidae, published in the Transactions of the Society. *Transactions of the Entomological Society of London* (2) **4**: 274–284. [(30).iv.1858.]

Smith, F. 1858b. *Catalogue of Hymenopterous Insects in the Collection of the British Museum* **6** Formicidae: 216 pp. London. [27.iii.1858.]

Smith, F. 1859a. Catalogue of hymenopterous insects collected by Mr. A.R. Wallace at the Islands of Aru and Key. *Journal of the Proceedings of the Linnean Society of London, Zoology* **3**: 132–158. [1.ii.1859.]

Smith, F. 1859b. *Catalogue of Hymenopterous Insects in the Collection of the British Museum* **7** Dorylidae and Thynnidae: 76 pp. London. [23.vii.1859.]

Smith, F. 1860a. Descriptions of new species of hymenopterous insects collected by Mr. A.R. Wallace at Celebes. *Journal of the Proceedings of the Linnean Society, Zoology* **5** (supplement to volume 4): 57–93. [18.vii.1860.]

Smith, F. 1860b. Catalogue of hymenopterous insects collected by Mr. A.R. Wallace in the Islands of Bachian, Kaisaa, Amboyna, Gilolo, and at Dory in New Guinea. *Journal of the Proceedings of the Linnean Society, Zoology* **5** (supplement to volume 4): 93–143. [18.vii.1860.]

Smith, F. 1860c. Descriptions of new genera and species of exotic Hymenoptera. *Journal of Entomology* **1**: 65–84. [(31).x.1860.]

Smith, F. 1861a. Descriptions of some new species of ants from the Holy Land, with a synonymic list of others previously described. *Journal of the Proceedings of the Linnean Society, Zoology* **6**: 31–35. [1.xi.1861.]

Smith, F. 1861b. Catalogue of hymenopterous insects collected by Mr. A.R. Wallace in the Islands of Ceram, Celebes, Ternate, and Gilolo. *Journal of the Proceedings of the Linnean Society, Zoology* **6**: 36–48. [1.xi.1861.]

Smith, F. 1862a. Catalogue of hymenopterous insects collected by Mr. A.R. Wallace in the Islands of Ceram, Celebes, Ternate, and Gilolo (continued). *Journal of the Proceedings of the Linnean Society, Zoology* **6**: 49–66. [1.iii.1862.]

Smith, F. 1862b. Descriptions of new species of aculeate Hymenoptera, collected at Panama by R.W. Stretch, Esq, with a list of described species, and the various localities where they have previously occurred. *Transactions of the Entomological Society of London* (3) **1**: 29–48. [7.iv.1862.]

Smith, F. 1862c. Descriptions of new species of Australian Hymenoptera, and of a species of *Formica* from New Zealand. *Transactions of the Entomological Society of London* (3) **1**: 53–62. [(31).v.1862.]

Smith, F. 1862d. A list of the genera and species belonging to the family Cryptoceridae, with descriptions of new species; also a list of the species of the genus *Echinopla*. *Transactions of the Entomological Society of London* (3) **1**: 407–416. [(30).xi.1862.]

Smith, F. 1863. Catalogue of hymenopterous insects collected by Mr. A.R. Wallace in the Islands of Mysol, Ceram, Waigiou, Bouru and Timor. *Journal of the Proceedings of the Linnean Society, Zoology* **7**: 6–48. [4.iii.1863.]

Smith, F. 1865. Descriptions of new species of hymenopterous insects from the Islands of Sumatra, Sula, Gilolo, Salwatty, and New Guinea, collected by Mr. A.R. Wallace. *Journal of the Proceedings of the Linnean Society, Zoology* **8**: 61–94. [13.i.1865.]

Smith, F. 1867. Descriptions of new species of Cryptoceridae. *Transactions of the Entomological Society of London* (3) **5**: 523–528. [(31).xii.1867.]

Smith, F. 1871. A catalogue of the aculeate Hymenoptera and Ichneumonidae of India and the Eastern Archipelago. *Journal of the Linnean Society (Zoology)* **11**: 285–415. [Pp. 285–348 dated 16.x.1871; pp. 349–415 dated 20.xii.1871.]

Smith, F. 1873. "Exhibitions, &c." *Transactions of the Entomological Society of London* **1873** (Proceedings): viii-ix. [20.v.1873.]

Smith, F. 1874. Descriptions of new species of Tenthredinidae, Ichneumonidae, Chrysididae, Formicidae, &c. of Japan. *Transactions of the Entomological Society of London* (4) **7**: 373–409. [(30).vi.1874.]

Smith, F. 1875. Descriptions of new species of Indian aculeate Hymenoptera, collected by Mr. G.R. James Rothney, member of the Entomological Society. *Transactions of the Entomological Society of London* (4) **8**: 33–51. [(31).v.1875.]
Smith, F. 1876a. Preliminary notice of new species of Hymenoptera, Diptera, and Forficulidae collected in the Island of Rodriguez by the naturalists accompanying the Transit-of-Venus Expedition. *Annals and Magazine of Natural History* (4) **17**: 447–451. [(30).vi.1876.]
Smith, F. 1876b. Descriptions of new species of hymenopterous insects of New Zealand, collected by C.M. Wakefield, Esq., principally in the neighbourhood of Canterbury. *Transactions of the Entomological Society of London* (4) **9**: 473–487. [2.xi.1876.]
Smith, F. 1876c. Descriptions of three new species of Hymenoptera (Formicidae) from New Zealand. *Transactions of the Entomological Society of London* (4) **9**: 489–492. [(31).xii.1876.]
Smith, F. 1876d. Descriptions of new species of Cryptoceridae, belonging to the genera *Cryptocerus, Meranoplus* and *Cataulacus. Transactions of the Entomological Society of London* (4) **9**: 603–612. [(31).xii.1876.]
Smith, F. 1877a. Hymenoptera and Diptera (pp. 82–84). In Günther, A. Account of the zoological collection made during the visit of H.M.S. Peterel to the Galapagos Islands. *Proceedings of the Zoological Society of London* **1877**: 64–93. [6.ii.1877.]
Smith, F. 1877b. Descriptions of new species of the genera *Pseudomyrma* and *Tetraponera*, belonging to the family Myrmicidae. *Transactions of the Entomological Society of London* (4) **10**: 57–72. [2.vii.1877.]
Smith, F. 1878a. Descriptions of new species of hymenopterous insects from New Zealand, collected by Prof. Hutton, at Otago. *Transactions of the Entomological Society of London* (4) **11**: 1–7. [(30).iv.1878.]
Smith, F. 1878b. *Scientific Results of the Second Yarkand Mission;* based upon the collections and notes of the late Ferdinand Stoliczka, Ph.D. Part **9**. Hymenoptera: 22 pp. Calcutta. [(31.xii).1878.]
Smith, F. 1879. Descriptions of new species of aculeate Hymenoptera collected by the Rev. Thos. Blackburn in the Sandwich Islands. *Journal of the Proceedings of the Linnean Society, Zoology* **14**: 674–685. [20.v.1879.]
Smith, M.R. 1923a. Two new Mississippi ants of the subgenus *Colobopsis. Psyche* **30**: 82–88. [(30).iv.1923.]
Smith, M.R. 1923b. Two new varieties of ants. *Entomological News* **34**: 306–308. [15.xii.1923.]
Smith, M.R. 1924. A new species of ant from Kansas. *Entomological News* **35**: 250–253. [3.vii.1924.]
Smith, M.R. 1927. An additional annotated list of ants of Mississippi, with a description of a new species of *Pheidole. Entomological News* **38**: 308–314. [14.xii.1927.]
Smith, M.R. 1928a. The biology of *Tapinoma sessile* Say, an important house-infesting ant. *Annals of the Entomological Society of America* **21**: 307–329. [18.viii.1928.]
Smith, M.R. 1928b. An additional annotated list of the ants of Mississippi. With a description of a new species of *Aphaenogaster. Entomological News* **39**: 242–246. [8.x.1928.]
Smith, M.R. 1928c. An additional annotated list of the ants of Mississippi. With a description of a new species of *Aphaenogaster* (conclusion). *Entomological News* **39**: 275–279. [2.xi.1928.]
Smith, M.R. 1929. Descriptions of five new North American ants, with biological notes. *Annals of the Entomological Society of America* **22**: 543–551. [12.x.1929.]
Smith, M.R. 1930a. A description of the male of *Proceratium croceum* Emery, with remarks. *Annals of the Entomological Society of America* **23**: 390–392. [28.vi.1930.]
Smith, M.R. 1930b. Descriptions of three new North American ants, with biological notes. *Annals of the Entomological Society of America* **23**: 564–568. [18.x.1930.]
Smith, M.R. 1931a. An additional annotated list of the ants of Mississippi. *Entomological News* **42**: 16–24. [13.i.1931.]
Smith, M.R. 1931b. Is *Eciton mexicanum* F. Smith really *Eciton pilosus* F. Smith? *Journal of the New York Entomological Society* **39**: 295–298. [30.ix.1931.]
Smith, M.R. 1931c. A revision of the genus *Strumigenys* of America, north of Mexico, based on a study of the workers. *Annals of the Entomological Society of America* **24**: 686–706. [19.xii.1931.]
Smith, M.R. 1934a. Three new North American ants. *Annals of the Entomological Society of America* **27**: 384–387. [10.x.1934.]
Smith, M.R. 1934b. Two new North American ants. *Psyche* **41**: 211–213. [(31).xii.1934.]
Smith, M.R. 1935. Two new species of North American *Strumigenys. Annals of the Entomological Society of America* **28**: 214–216. [13.vi.1935.]
Smith, M.R. 1936. Ants of the genus *Ponera* in America, north of Mexico. *Annals of the Entomological Society of America* **29**: 420–430. [26.ix.1936.]
Smith, M.R. 1937. The ants of Puerto Rico. *Journal of Agriculture of the University of Puerto Rico* **20** (1936): 819–875. [(31).i.1937.]
Smith, M.R. 1938a. A study of North American ants of the genus *Xiphomyrmex* Forel. *Journal of the Washington Academy of Sciences* **28**: 126–130. [15.iii.1938.]
Smith, M.R. 1938b. Notes on the legionary ants (*Eciton*, subgenus *Acamatus*) with a record of new specific synonymy. *Proceedings of the Entomological Society of Washington* **40**: 157–160. [24.vi.1938.]
Smith, M.R. 1939a. A new species of North American *Ponera*, with an ergatandrous form. *Proceedings of the Entomological Society of Washington* **41**: 76–78. [24.iii.1939.]
Smith, M.R. 1939b. The North American ants of the genus *Harpagoxenus* Forel, with the description of a new species. *Proceedings of the Entomological Society of Washington* **41**: 165–172. [27.v.1939.]
Smith, M.R. 1939c. Notes on *Leptothorax (Mychothorax) hirticornis* Emery, and description of a related new species. *Proceedings of the Entomological Society of Washington* **41**: 176–180. [27.v.1939.]
Smith, M.R. 1939d. A study of the subspecies of *Odontomachus haematoda* (L.) of the United States. *Journal of the New York Entomological Society* **47**: 125–130. [8.vi.1939.]
Smith, M.R. 1939e. Ants of the genus *Macromischa* Roger in the United States. *Annals of the Entomological Society of America* **32**: 502–509. [20.ix.1939.]
Smith, M.R. 1939f. Notes on *Formica (Neoformica) moki* Wheeler, with description of a new subspecies. *Annals of the Entomological Society of America* **32**: 581–584. [20.ix.1939.]
Smith, M.R. 1940a. The discovery of the worker caste of an inquilinous ant, *Epipheidole inquilina* Wheeler. *Proceedings of the Entomological Society of Washington* **42**: 104–109. [28.v.1940.]
Smith, M.R. 1940b. The identity of the ant *Camponotus (Myrmentoma) caryae* (Fitch). *Proceedings of the Entomological Society of Washington* **42**: 137–141. [31.x.1940.]
Smith, M.R. 1941. Two new species of *Aphaenogaster. Great Basin Naturalist* **2**: 118–121. [29.xi.1941.]
Smith, M.R. 1942a. A new, apparently parasitic ant. *Proceedings of the Entomological Society of Washington* **44**: 59–61. [30.iv.1942.]
Smith, M.R. 1942b. The males of two North American cerapachyine ants. *Proceedings of the Entomological Society of Washington* **44**: 62–64. [30.iv.1942.]
Smith, M.R. 1942c. The legionary ants of the United States belonging to *Eciton* subgenus *Neivamyrmex* Borgmeier. *American Midland Naturalist* **27**: 537–590. [30.vi.1942.]
Smith, M.R. 1943a. A new North American *Solenopsis (Diplorhoptrum). Proceedings of the Entomological Society of Washington* **44** (1942): 209–211. [5.i.1943.]
Smith, M.R. 1943b. Ants of the genus *Tetramorium* in the United States, with the description of a new species. *Proceedings of the Entomological Society of Washington* **45**: 1–5. [30.i.1943.]

Smith, M.R. 1943c. *Pheidole (Macropheidole) rhea* Wheeler, a valid species. *Proceedings of the Entomological Society of Washington* **45**: 5–9. [30.i.1943.]
Smith, M.R. 1943d. The first record of *Leptothorax,* subgenus *Goniothorax* Emery, in the United States, with the description of a new species. *Proceedings of the Entomological Society of Washington* **45**: 154–156. [3.v.1943.]
Smith, M.R. 1943e. A new male legionary ant from the Mojave Desert, California. *Lloydia* **6**: 196–197. [(30).ix.1943.]
Smith, M.R. 1943f. A generic and subgeneric synopsis of the male ants of the United States. *American Midland Naturalist* **30**: 273–321. [22.x.1943.]
Smith, M.R. 1944a. Ants of the genus *Cardiocondyla* Emery in the United States. *Proceedings of the Entomological Society of Washington* **46**: 30–41. [13.iii.1944.]
Smith, M.R. 1944b. The ants of the genus *Thaumatomyrmex* Mayr with the description of a new Panamanian species. *Proceedings of the Entomological Society of Washington* **46**: 97–99. [3.v.1944.]
Smith, M.R. 1944c. A key to the genus *Acanthognathus* Mayr, with the description of a new species. *Proceedings of the Entomological Society of Washington* **46**: 150–152. [10.vii.1944.]
Smith, M.R. 1944d. Additional ants recorded from Florida, with descriptions of two new subspecies. *Florida Entomologist* **27**: 14–17. [(31).vii.1944.]
Smith, M.R. 1944e. The genus *Lachnomyrmex,* with the description of a second species. *Proceedings of the Entomological Society of Washington* **46**: 225–228. [21.xi.1944.]
Smith, M.R. 1944f. A second species of *Glamyromyrmex* Wheeler. *Proceedings of the Entomological Society of Washington* **46**: 254–256. [30.xii.1944.]
Smith, M.R. 1946. A second species of *Stegomyrmex,* and the first description of a *Stegomyrmex* worker. *Revista de Entomologia* **17**: 286–289 [26.viii.1946.]
Smith, M.R. 1947a. Ants of the genus *Cryptocerus* F. in the United States. *Proceedings of the Entomological Society of Washington* **49**: 29–40. [15.i.1947.]
Smith, M.R. 1947b. Ants of the genus *Apsychomyrmex. Revista de Entomologia* **17** (1946): 468–473. [20.i.1947.]
Smith, M.R. 1947c. A new and extraordinary *Pheidole* from New Guinea. *Proceedings of the Entomological Society of Washington* **49**: 73–75. [28.ii.1947.]
Smith, M.R. 1947d. A new species of *Metapone* Forel from New Guinea. *Proceedings of the Entomological Society of Washington* **49**: 75–77. [28.ii.1947.]
Smith, M.R. 1947e. A new species of *Megalomyrmex* from Barro Colorado Island, Canal Zone. *Proceedings of the Entomological Society of Washington* **49**: 101–103. [15.iv.1947.]
Smith, M.R. 1947f. A generic and subgeneric synopsis of the United States ants, based on the workers. *American Midland Naturalist* **37**: 521–647. [27.vii.1947.]
Smith, M.R. 1947g. A study of *Polyergus* in the United States, based on the workers. *American Midland Naturalist* **38**: 150–161. [20.viii.1947.]
Smith, M.R. 1947h. Notes on *Pheidole (Decapheidole)* and the description of a new species. *Revista de Entomologia* **18**: 193–196. [26.viii.1947.]
Smith, M.R. 1947i. A new genus and species of ant from Guatemala. *Journal of the New York Entomological Society* **55**: 281–284. [16.xii.1947.]
Smith, M.R. 1948. A new species of *Myrmecina* from California. *Proceedings of the Entomological Society of Washington* **50**: 238–240. [29.xii.1948.]
Smith, M.R. 1949a. A new genus and species of ant from India. *Journal of the New York Entomological Society* **56** (1948): 205–208. [26.i.1949.]
Smith, M.R. 1949b. A new species of *Probolomyrmex* from Barro Colorado Island, Canal Zone. *Proceedings of the Entomological Society of Washington* **51**: 38–40. [21.ii.1949.]
Smith, M.R. 1949c. On the status of *Cryptocerus* Latreille and *Cephalotes* Latreille. *Psyche* **56**: 18–21. [17.v.1949.]
Smith, M.R. 1949d. A new species of *Camponotus,* subg. *Colobopsis,* from Mexico. *Journal of the New York Entomological Society* **57**: 177–181. [12.ix.1949.]
Smith, M.R. 1949e. A new *Leptothorax* commonly inhabiting the Canyon Live Oak of California. *Psyche* **56**: 112–115. [31.x.1949.]
Smith, M.R. 1950. On the status of *Leptothorax* Mayr and some of its subgenera. *Psyche* **57**: 29–30. [16.vi.1950.]
Smith, M.R. 1951a. Formicidae (pp. 778–875). In Muesebeck, C.F.W., Krombein, K.V. & Townes, H.K. Hymenoptera of America North of Mexico. Synoptic Catalog. *United States Department of Agriculture. Agriculture Monograph* **2**: 1420 pp. [(30).iv.1951.]
Smith, M.R. 1951b. A new species of *Stenamma* from North Carolina. *Proceedings of the Entomological Society of Washington* **53**: 156–158. [15.vi.1951.]
Smith, M.R. 1951c. Two new ants from western Nevada. *Great Basin Naturalist* **11**: 91–96. [29.xii.1951.]
Smith, M.R. 1952a. The correct name for the group of ants formerly known as *Pseudomyrma. Proceedings of the Entomological Society of Washington* **54**: 97–98. [25.iv.1952.]
Smith, M.R. 1952b. North American *Leptothorax* of the *tricarinatus—texanus* complex. *Journal of the New York Entomological Society* **60**: 96–106. [23.vi.1952.]
Smith, M.R. 1953a. A revision of the genus *Romblonella* W.M. Wheeler. *Proceedings of the Hawaiian Entomological Society* **15**: 75–80. [27.iii.1953.]
Smith, M.R. 1953b. *Dolichoderus granulatus* Pergande, a synonym. *Proceedings of the Entomological Society of Washington* **55**: 211. [13.xi.1953.]
Smith, M.R. 1953c. A new species of *Probolomyrmex,* and the first description of a *Probolomyrmex* male. *Journal of the New York Entomological Society* **61**: 127–129. [20.xi.1953.]
Smith, M.R. 1953d. A new *Metapone* from the Micronesian Islands. *Journal of the New York Entomological Society* **61**: 135–137. [20.xi.1953.]
Smith, M.R. 1953e. A new *Pheidole* (subg. *Ceratopheidole*) from Utah. *Journal of the New York Entomological Society* **61**: 143–146. [20.xi.1953.]
Smith, M.R. 1953f. A new *Romblonella* from Palau, and the first description of a *Romblonella* male. *Journal of the New York Entomological Society* **61**: 163–167. [20.xi.1953.]
Smith, M.R. 1953g. *Pogonomyrmex salinus* Olsen, a synonym of *Pogonomyrmex occidentalis* (Cress.). *Bulletin of the Brooklyn Entomological Society* **48**: 131–132. [30.xi.1953.]
Smith, M.R. 1954a. A new name for *Martia* Forel. *Bulletin of the Brooklyn Entomological Society* **49**: 17. [17.ii.1954.]
Smith, M.R. 1954b. A new *Camponotus* in California apparently inhabiting Live Oak, *Quercus* sp. *Journal of the New York Entomological Society* **61** (1953): 211–214. [20.ii.1954.]
Smith, M.R. 1954c. Ants of the Bimini Island Group, Bahamas, British West Indies. *American Museum Novitates* **1671**: 1–16. [28.vi.1954.]
Smith, M.R. 1954d. Concerning type locality and type fixation of the North American ant, *Myrmica emeryana* Forel. *Bulletin of the Brooklyn Entomological Society* **49**: 138–140. [3.xii.1954.]
Smith, M.R. 1955a. *Acanthostichus (Ctenopyga) townsendi* (Ashm.), a synonym of *Acanthostichus texanus* Forel. *Bulletin of the Brooklyn Entomological Society* **50**: 48–50. [15.iv.1955.]
Smith, M.R. 1955b. Remarks concerning the types of five species of ants described by Roger or Forel. *Bulletin of the Brooklyn Entomological Society* **50**: 98–99. [4.xi.1955.]
Smith, M.R. 1955c. Ants of the genus *Pheidole,* subgenus *Hendecapheidole. Proceedings of the Entomological Society of Washington* **57**: 301–305. [22.xii.1955.]
Smith, M.R. 1955d. The correct taxonomic status of *Pheidole (Pheidolacanthinus) brevispinosa* Donisthorpe. *Proceedings of the Entomological Society of Washington* **57**: 305. [22.xii.1955.]
Smith, M.R. 1956a. A key to the workers of *Veromessor* Forel of the United States and the description of a new subspecies. *Pan-Pacific Entomologist* **32**: 36–38. [14.iii.1956.]
Smith, M.R. 1956b. A list of the species of *Romblonella* including two generic transfers. *Bulletin of the Brooklyn Entomological Society* **51**: 18. [19.iii.1956.]
Smith, M.R. 1956c. A further contribution to the taxonomy and biology of the inquiline ant, *Leptothorax diversipilosus* Smith. *Proceedings of the Entomological Society of Washington* **58**: 271–575. [23.xi.1956.]

Smith, M.R. 1957a. New synonymy of a New Guinea ant. *Proceedings of the Entomological Society of Washington* **58** (1956): 347. [31.i.1957.]

Smith, M.R. 1957b. Revision of the genus *Stenamma* Westwood in America north of Mexico. *American Midland Naturalist* **57**: 133–174. [(31).i.1957.]

Smith, M.R. 1958a. New synonymy of a North American ant, *Aphaenogaster macrospina* M.R. Smith. *Bulletin of the Brooklyn Entomological Society* **52** (1957): 113. [29.i.1958.]

Smith, M.R. 1958b. A contribution to the taxonomy, distribution and biology of the vagrant ant, *Plagiolepis alluaudi* Emery. *Journal of the New York Entomological Society* **65** (1957): 195–198. [27.vi.1958.]

Smith, M.R. 1958c. Formicidae (pp. 108–162). In Krombein, K.V. Hymenoptera of America North of Mexico. Synoptic Catalog. First Supplement. *United States Department of Agriculture. Agriculture Monograph* **2** (Supplement 1): 305 pp. [(30).ix.1958.]

Smith, M.R. 1961a. Notes on the synonymy of a North American ant. *Proceedings of the Entomological Society of Washington* **62** (1960): 251–252. [4.i.1961.]

Smith, M.R. 1961b. A study of New Guinea ants of the genus *Aphaenogaster* Mayr. *Acta Hymenopterologica* **1**: 213–237. [31.v.1961.]

Smith, M.R. 1962a. A remarkable new *Stenamma* from Costa Rica, with pertinent facts on other Mexican and Central American species. *Journal of the New York Entomological Society* **70**: 33–38. [14.iii.1962.]

Smith, M.R. 1962b. A new species of exotic *Ponera* from North Carolina. *Acta Hymenopterologica* **1**: 377–382. [30.ix.1962.]

Smith, M.R. 1963. A new species of *Aphaenogaster (Attomyrma)* from the western United States. *Journal of the New York Entomological Society* **71**: 244–246. [23.xii.1963.]

Smith, M.R. 1967. Formicidae (pp. 343–374). In Krombein, K.V. & Burks, B.D. Hymenoptera of America North of Mexico. Synoptic Catalog. Second Supplement. *United States Department of Agriculture. Agriculture Monograph* **2** (Supplement 2): 584 pp. [(28).ii.1967.]

Smith, M.R. & Wing, M.W. 1955. Redescription of *Discothyrea testacea* Roger, a little-known North American ant, with notes on the genus. *Journal of the New York Entomological Society* **62** (1954): 105–112. [5.v.1955.]

Snelling, R.R. 1963. The United States species of fire ants of the genus *Solenopsis* Westwood, subgenus *Solenopsis* Westwood, with synonymy of *Solenopsis aurea* Wheeler. *Occasional Papers. Bureau of Entomology. California Department of Agriculture* **3**: 1–11. [14.i.1963.]

Snelling, R.R. 1965a. Studies on California ants. 1. *Leptothorax hirticornis* Emery, a new host and descriptions of the female and ergatoid male. *Bulletin of the Southern California Academy of Sciences* **64**: 16–21. [26.iv.1965.]

Snelling, R.R. 1965b. Studies on California ants. 2. *Myrmecina californica. Bulletin of the Southern California Academy of Sciences* **64**: 101–105. [30.vi.1965.]

Snelling, R.R. 1966. The female of *Eucryptocerus placidus* (F. Smith). *Bulletin of the Southern California Academy of Sciences* **65**: 37–40. [31.iii.1966.]

Snelling, R.R. 1967. Studies on California ants. 3. The taxonomic status of *Proceratium californicum* Cook. *Contributions in Science* **124**: 1–10. [31.v.1967.]

Snelling, R.R. 1968a. Taxonomic notes on some Mexican cephalotine ants. *Contributions in Science* **132**: 1–10. [20.ii.1968.]

Snelling, R.R. 1968b. Studies on California ants. 4. Two species of *Camponotus. Proceedings of the Entomological Society of Washington* **70**: 350–358. [18.xii.1968.]

Snelling, R.R. 1968c. A new species of *Eurhopalothrix* from El Salvador. *Contributions in Science* **154**: 1–4. [31.xii.1968.]

Snelling, R.R. 1969a. Taxonomic notes on the *Myrmecocystus melliger* complex. *Contributions in Science* **170**: 1–9. [30.vi.1969.]

Snelling, R.R. 1969b. Notes on the systematics and dulosis of some western species of *Formica*, subgenus *Raptiformica. Proceedings of the Entomological Society of Washington* **71**: 194–197. [30.vi.1969.]

Snelling, R.R. 1970. Studies on California ants. 5. Revisionary notes on some species of *Camponotus*, subgenus *Tanaemyrmex. Proceedings of the Entomological Society of Washington* **72**: 390–397. [25.ix.1970.]

Snelling, R.R. 1971a. Studies on California ants. 6. Three new species of *Myrmecocystus. Contributions in Science* **214**: 1–16. [28.v.1971.]

Snelling, R.R. 1971b. A new species of *Simopelta* from Costa Rica. *Bulletin of the Southern California Academy of Sciences* **70**: 16–17. [27.viii.1971.]

Snelling, R.R. 1973a. Two ant genera new to the United States. *Contributions in Science* **236**: 1–8. [10.i.1973.]

Snelling, R.R. 1973b. The ant genus *Conomyrma* in the United States. *Contributions in Science* **238**: 1–6. [29.i.1973.]

Snelling, R.R. 1973c. Studies on California ants. 7. The genus *Stenamma. Contributions in Science* **245**: 1–38. [28.vi.1973.]

Snelling, R.R. 1974. Studies on California ants. 8. A new species of *Cardiocondyla. Journal of the New York Entomological Society* **82**: 76–81. [22.viii.1974.]

Snelling, R.R. 1975. Descriptions of new Chilean ant taxa. *Contributions in Science* **274**: 1–19. [23.xii.1975.]

Snelling, R.R. 1976. A revision of the honey ants, genus *Myrmecocystus. Natural History Museum of Los Angeles County Science Bulletin* **24**: 1–163. [5.viii.1976.]

Snelling, R.R. 1979a. Three new species of the palaeotropical arboreal ant genus *Cataulacus. Contributions in Science* **315**: 3–8. [31.vii.1979.]

Snelling, R.R. 1979b. *Aphomomyrmex* and a related new genus of arboreal African ants. *Contributions in Science* **316**: 3–8. [31.vii.1979.]

Snelling, R.R. 1981. Systematics of social Hymenoptera (pp. 369–453). In Hermann, H.R. (Ed.) *Social Insects* **2**: 491 pp. New York. [(31.xii).1981.]

Snelling, R.R. 1982a. The taxonomy and distribution of some North American *Pogonomyrmex* and descriptions of two new species. *Bulletin of the Southern California Academy of Sciences* **80** (1981): 97–112. [16.viii.1982.]

Snelling, R.R. 1982b. A revision of the honey ants, genus *Myrmecocystus*, first supplement. *Bulletin of the Southern California Academy of Sciences* **81**: 69–86. [13.x.1982.]

Snelling, R.R. 1985. In Millar, S.E. The California Channel Islands—past, present and future: an entomological perspective. Pp. 3–28, in Menke, A.S. & Miller, D.R. (Eds.) *Entomology of the Channel Islands. Proceedings of the First Symposium:* 178 pp. Santa Barbara Museum of Natural History. [(31.xii).1985.]

Snelling, R.R. 1986. New synonymy in Caribbean ants of the genus *Leptothorax. Proceedings of the Entomological Society of Washington* **88**: 154–156. [15.i.1986.]

Snelling, R.R. 1988. Taxonomic notes on Nearctic species of *Camponotus*, subgenus *Myrmentoma* (pp. 55–78). In Trager, J.C. (Ed.) *Advances in Myrmecology:* 551 pp. New York. [(31.xii).1988.]

Snelling, R.R. 1992a. Two unusual new myrmicine ants from Cameroon. *Psyche* **99**: 95–101. [1992.]

Snelling, R.R. 1992b. A newly adventive ant of the genus *Pheidole* in southern California. *Bulletin of the Southern California Academy of Sciences* **91**: 121–125. [17.xii.1992.]

Snelling, R.R. & Buren, W.F. 1985. Description of a new species of slave-making ant in the *Formica sanguinea* group. *Great Lakes Entomologist* **18**: 69–78. [3.vi.1985.]

Snelling, R.R. & Cover, S.P. 1992. Description of a new *Proceratium* from Mexico. *Psyche* **99**: 49–53. [4.xii.1992.]

Snelling, R.R. & Hunt, J.H. 1976. The ants of Chile. *Revista Chilena de Entomologia* **9** (1975): 63–129. [(31).v.1976.]

Snelling, R.R. & Longino, J.T. 1992. Revisionary notes on the fungus-growing ants of the genus *Cyphomyrmex, rimosus*-group (pp. 479–494). In Quintero, D. & Aiello, A. (editors). *Insects of Panama and Mesoamerica: selected studies:* 692 pp. Oxford. [(31.xii).1992.]

Sonobe, R. & Dlussky, G.M. 1977. On two ant species of the genus *Formica* from Japan. *Kontyû* **45**: 23–25. [25.iii.1977.]

Soudek, S. 1922. Mravenci. Soustava, zemepisné rozsírení oekologie a urcovací klic mravencu zijících na uzemi Ceskoslovenské republiky. *Ceskoslovenské Spolecnosti Entomologické v Praze:* 100 pp. [(31.xii).1922.]
Soudek, S. 1923. *Myrmica moravica* n. sp., relikt fauny praeglaciální. *Casopis Moravského Zemského Musea* **20–21** (1922–23): 107–134. [(31.xii).1923.]
Soudek, S. 1925a. Four new European ants. *Entomologist's Record and Journal of Variation* **37**: 33–37. [15.iii.1925.]
Soudek, S. 1925b. Dalmatsti mravenci. *Casopisu Ceskoslovenské Spolecnosti Entomologické* **22**: 12–17. [20.v.1925.]
Soudek, S. 1925c. Príspevky k vedeckému poznání Moravského krasu. 3. *Bothriomyrmex meridionalis gibbus* n. ssp., novy mravenec z Moravy. *Casopis Moravského Zemského Musea* **22–23** (1924–25): 216–232. [(31.xii).1925.]
Soudek, S. 1931. Mravenci. "Hádu", jizního vybezku Moravského Krasu. *Zprávy komise na Prírodovedechy vyzkum Moravy a Slezka. Oddelení Zoologické* **19**: 1–30. [(31.xii).1931.]
Soulié, J. 1961. Des considérations écologiques peuvent-elles apporter une contribution à la connaissance du cycle biologique des colonies de *Cremastogaster? Insectes Sociaux* **7**: 283–295. [(30).vi.1961.]
Soulié, J. 1964. Sur la répartition géographique des genres de la tribu des Cremastogastrini. *Bulletin de la Société d'Histoire Naturelle de Toulouse* **99**: 397–409. [30.xii.1964.]
Soulié, J. 1965. Contribution à la systématique des fourmis de la tribu des Cremastogastrini. *Annales de l'Université d'Abidjan* (Sciences) **1**: 69–83. [(31.xii).1965.]
Soulié, J. & Dicko, L.D. 1965. La répartition des genres de fourmis de la tribu des Cremastogastrini dans la faune éthiopienne et malgache. *Annales de l'Université d'Abidjan* (Sciences) **1**: 85–106. [(31.xii).1965.]
Spahr, U. 1987. Ergänzungen und Berichtigungen zu R. Keilbach's Bibliographie und Liste der Bernsteinfossilien—Ordnung Hymenoptera. *Stuttgarter Beiträge zur Naturkunde.* Serie B (Geologie und Paläontologie) **127**: 1–121. [16.iii.1987.]
Spinola, M. 1808. *Insectorum Liguriae species novae aut rariores,* quas in agro ligustico nuper detexit, descripsit, et iconibus illustravit Maximilianus Spinola, adjecto catalogo specierum auctoribus jam enumeratarum, quae in eadem regione passim occurrunt **2**: 262 pp. Genuae. [17.iii.1808.]
Spinola, M. 1851a. Insectos. Orden 7. Himenopteros (pp. 153–569). In Gay, C. *Historia Fisica y Politica de Chile.* Zoologia **6**: 572 pp. Paris. [(31.xii).1851.]
Spinola, M. 1851b. Compte rendu des Hyménoptères inédits provenants du voyage entomologique de M. Ghiliani dans la Para en 1846. *Extrait des Mémoires de l'Académie des Sciences de Turin* (2) **13**: 3–78. [(31.xii).1851. Publication date from Vecht, 1975: 60.]
Spinola, M. 1853. Compte rendu des Hyménoptères inédits provenants du voyage entomologique de M. Ghiliani dans la Para en 1846. *Memorie della Reale Accademia delle Scienze di Torino* (2) **13**: 19–94. [(31.xii).1853.]
Stärcke, A. 1927. Beginnende divergenz bei *Myrmica lobicornis* Nyl. *Tijdschrift voor Entomologie* **70**: 73–84. [(31).viii.1927.]
Stärcke, A. 1930. Verzeichnis der bis jetzt von der Insel Pulau Berhala bekannt gewordenen Ameisen. *Treubia* **12**: 371–381. [(31).xii.1930.]
Stärcke, A. 1933a. Contribution à l'étude de la faune népenthicole. 3. Un nouveau *Camponotus* de Bornéo, habitant les tiges creuses de *Nepenthes,* récolté par J.P. Schuitemaker. *Natuurhistorisch Maandblad* **22**: 29–31. [31.iii.1933.]
Stärcke, A. 1933b. Untitled contribution introduced by,"De heer Stärcke spreekt over de larven der Dolichoderinen." In, Verslag van de zes-en-zestigste Wintervergadering der Nederlandsche Entomologische Vereeniging. *Tijdschrift voor Entomologie* **76**: xxvi-xxxi. [(31).v.1933.]
Stärcke, A. 1934. Un nouveau sous-genre de *Camponotus* de la Malaisie avec description de formes nouvelles, récoltées par M.G.F. Nainggolan. *Zoologische Mededeelingen* **17**: 20–30. [1.viii.1934.]
Stärcke, A. 1935. *Wissenschaftliche Ergebnisse der Niederländischen Expeditionen in den Karakorum und die angrenzenden Gebiete 1922, 1925 und 1929/30.* Zoologie. Formicidae: 260–269. Leipzig. [15.i.1935.]
Stärcke, A. 1936. Retouches sur quelques fourmis d'Europe. 1. *Plagiolepis xene* nov. sp. et *Pl. vindobonensis* Lomnicki. *Entomologische Berichten* **9**: 277–279. [1.xi.1936.]
Stärcke, A. 1937. Retouches sur quelques fourmis d'Europe. *Tijdschrift voor Entomologie* **80**: 38–72. [(31).v.1937.]
Stärcke, A. 1938. Untitled contribution introduced by "De heer Stärcke doet eene mededeeling over gedragen ontwikkeling van enkele Javaansche mieren tijdens hun verblijf hier te lande." *Tijdschrift voor Entomologie* **81**: xxxiii-xli. [(31).v.1938. The paper was later issued as a separate with the title "Fauna Javana."]
Stärcke, A. 1941. Hersenganglion van *Strumigenys;* koloniestichting van *Polyrhachis bicolor. Tijdschrift voor Entomologie* **84**: ii-xv. [1.v.1941.]
Stärcke, A. 1942a. Ants, collected by Dr. C.F. Engelhard at Stockmarknes (Island Hadsekoy, Vesterale Archipelago circa 68 deg. 50 min. Lat. N. Norway) and between 64 deg. and 66 deg. Lat. N. on the western coast of Norway, July 1932, identified by A. Stärcke. *Entomologische Berichten* **11**: 21–23. [1.v.1942.]
Stärcke, A. 1942b. Definities van species (soort), subspecies (ras, stirps), variëteit en aberratie. *Entomologische Berichten* **11**: 40–48. [1.v.1942.]
Stärcke, A. 1942c. Die nog onbeschreven Europeesche miervormen. *Tijdschrift voor Entomologie* **85**: xxiv-xxix. [(31.xii).1942.]
Stärcke, A. 1944a. Retouches sur quelques fourmis d'Europe. 3. Autres *Lasius. Entomologische Berichten* **11**: 153–158. [25.ii.1944.]
Stärcke, A. 1944b. Nederlandsche Adventief-mieren. *Tijdschrift voor Entomologie* **86** verslag (1943): xvii-xxii. [7.iv.1944.]
Stärcke, A. 1947. De boreale vorm van de roode boschmier (*Formica rufa rufa*) op de Hooge Veluwe. *Entomologische Berichten* **12**: 144–146. [7.vi.1947.]
Stefani, R. 1970. Contributo alla conoscenza dei formicidi cavernicoli della Sardegna. *Rendiconti del Seminario della Facoltà di Scienze della Università di Cagliari* **38** (suppl.) (1968): 1–5. [(31).i.1970.]
Stephens, J.F. 1829. *A Systematic Catalogue of British Insects.* Part 1: 416 pp.; Part 2: 388 pp. London. [15.vii.1829.]
Stitz, H. 1909. Eine neue afrikanische *Dichthadia. Zoologischer Anzeiger* **35**: 231–232. [21.xii.1909.]
Stitz, H. 1910. Westafrikanische Ameisen. *Mitteilungen aus dem Zoologischen Museum in Berlin* **5**: 125–151. [(31).viii.1910.]
Stitz, H. 1911a. Australische Ameisen. (Neu-Guinea und Salomons- Inseln, Festland, Neu-Seeland.) *Sitzungsberichte der Gesellschaft Naturforschender Freunde zu Berlin* **1911**: 351–381. [10.x.1911.]
Stitz, H. 1911b. *Wissenschaftliche Ergebnisse der Deutschen Zentral-Afrika-Expedition 1907–1908 unter führung Adolf Friedrichs, Herzogs zu Mecklenburg* **3** Zoologie. 1 Formicidae: 375–392. [(31.xii).1911.]
Stitz, H. 1912. Ameisen aus Ceram und Neu-Guinea. *Sitzungsberichte der Gesellschaft Naturforschender Freunde zu Berlin* **1912**: 498–514. [12.xi.1912.]
Stitz, H. 1913. Ameisen aus Brasilien, gesammelt von Ule. *Deutsche Entomologische Zeitschrift* **1913**: 207–212. [31.iii.1913.]
Stitz, H. 1914. Die Ameisen Mitteleuropas, insbesondere Deutschlands. In Schröder, C. *Die Insekten Mitteleuropas, insbesondere Deutschlands* **2**. Hymenoptera: 1–111. Stuttgart. [(31.xii).1914.]
Stitz, H. 1916. *Ergebnisse der zweiten Deutschen Zentral-Afrika- Expedition 1910–1911.* (Hamburgische Wissenschaftliche Stiftung.) **1** Zoologie. Formicidae: 369–405. Leipzig. [14.x.1916.]
Stitz, H. 1917. Ameisen aus dem westlichen Mittelmeergebiet und von den Kanarischen Inseln. *Mitteilungen aus dem Zoologischen Museums in Berlin* **8**: 333–353. [(30).iv.1917.]
Stitz, H. 1923. *Beiträge zur Kenntnis der Land- und Süsswasserfauna Deutsch-Südwestafrikas.* Ergebnisse der Hamburger deutsch-südwestafrikanischen Studienreise 1911. Hymenoptera **7**. Formicidae: 143–167. Hamburg. [(31.xii).1923.]

Stitz, H. 1925. Ameisen von den Philippinen, den malayischen und ozeanischen Inseln. *Sitzungsberichte der Gesellschaft Naturforschender Freunde zu Berlin* **1923**: 110–136. [19.xii.1925.]

Stitz, H. 1930. Entomologische Ergebnisse der Deutsch-Russischen Alai-Pamir-Expediton 1928. 5. Hymenoptera, Formicidae. *Mitteilungen aus dem Zoologische Museum in Berlin* **16**: 238–240. [30.vi.1930.]

Stitz, H. 1932a. The Norwegian Zoological Expedition to the Galapagos Islands 1925, conducted by Alf Wollebaek. 5. Formicidae. *Meddelelser fra det Zoologiske Museum Oslo* **31** (Nyt Magazin for Naturvidenskaberne B.71): 367–372. [19.x.1932.]

Stitz, H. 1932b. Formicidae der Deutschen Limnologischen Sunda- Expedition. *Archiv für Hydrobiologie* Suppl.-Bd. **9** (Tropische Binnengewässer, 2): 733–737. [(31.xii).1932.]

Stitz, H. 1933. Neue Ameisen des Hamburger Museums. *Mitteilungen der Deutschen Entomologischen Gesellschaft* **4**: 67–75. [(31).v.1933.]

Stitz, H. 1934. Schwedisch-Chinesischen Wissenschaftliche Expedition nach den Nordwestlichen Provinzen Chinas, unter leitung von Dr. Sven Hedin und Prof. Sü Ping-Chang. 25, Hymenoptera. 3, Formicidae. *Arkiv för Zoologi* **27** (A): 1–9. [26.i.1934.]

Stitz, H. 1937. Einige Ameisen aus Mexiko. *Sitzungsberichte der Gesellschaft Naturforschender Freunde zu Berlin* **1937**: 132–136. [3.xi.1937.]

Stitz, H. 1938. Neue Ameisen aus dem indo-malayischen Gebiet. *Sitzungsberichte der Gesellschaft Naturforschender Freunde zu Berlin* **1938**: 99–122. [28.xii.1938.]

Stitz, H. 1939. In Dahl, F. *Die Tierwelt Deutschlands und der angrenzenden Meeresteile, nach ihren Merkmalen und nach ihrer Lebenweise* **37** Hautflügler oder Hymenoptera. 1 Ameisen oder Formicidae: 428 pp. Jena. [(31.xii).1939.]

Strand, E. 1911. Ein bisher unbekanntes Dorylidenweibchen aus Kamerun. *Jahrbücher des Nassauischen Vereins für Naturkunde* **64**: 118–120. [(31.xii).1911.]

Strand, E. 1935. Miscellanea nomenclatorica zoologica et palaeontologica. *Folia Zoologica et Hydrobiologica* **8**: 176. [30.viii.1935.]

Stumper, R. 1951. *Teleutomyrmex schneideri* Kutter. 2 Mitteilung. Über die Lebensweise der neuen Schmarotzerameise. *Mitteilungen der Schweizerischen Entomologischen Gesellschaft* **24**: 129–152. [20.vii.1951.]

Stumper, R. 1957. Sur l'éthologie de la fourmi a miel *Proformica nasuta* Nyl. *Bulletin de la Société des Naturalistes Luxembourgeois* **60** (1955): 87–97. [1957.]

Stumper, R. & Kutter, H. 1951. Sur l'éthologie du nouveau myrmécobionte *Epimyrma stumperi. Comptes Rendus Hebdomadaire des Séances de l'Academie des Sciences* **233**: 983–985. [(31.xii).1951.]

Sturm, J. 1826. *Catalog meiner Insecten-Sammlung* **1**. Käfer: 207 pp. Nürnberg. [(31.xii).1826.]

Sudd, J.H. 1967. *An Introduction to the Behaviour of Ants:* 200 pp. London. [1967.]

Sudd, J.H. & Franks, N.R. 1987. *The Behavioural Ecology of Ants:* 206 pp. Glasgow & London. [1987.]

Swainson, W. & Shuckard, W.E. 1840. *On the History and Natural Arrangement of Insects:* 406 pp. London. [(31.xii).1840.]

Sykes, W.H. 1835. Descriptions of new species of Indian ants. *Transactions of the Entomological Society of London* **1**: 99–107. [2.x.1835. Date of publication from Wheeler, G. 1912: 754.]

Szabó, J. 1909. De duabus speciebus novis formicidarum generis *Epitritus* Emery. *Archivum Zoologicum* **1**: 27–28. [30.xii.1909.]

Szabó, J. 1910a. Formicides nouveaux ou peu connus des collections du Musée National Hongrois. *Annales Historico- Naturales Musei Nationalis Hungarici* **8**: 364–369. [20.vi.1910. Note that p. 369 is first page of next issue and is dated 30.xii.1910; this page contains no new taxonomic data.]

Szabó, J. 1910b. Uj hangya Uj-Guineából. *Rovartani Lapok* **17**: 186. [31.xii.1910.]

Szabó, J. (as Szabó-Patay, J.) 1926. Trois *Orectognathus* nouveaux de la collection du Musée National Hongrois. *Annales Historico-Naturales Musei Nationalis Hungarici* **24**: 348–352. [23.xi.1926.]

Taber, S.W. 1988. The gyne of the harvester ant *Pogonomyrmex texanus. Journal of the Kansas Entomological Society* **61**: 244–246. [2.vi.1988.]

Taber, S.W. 1990. Cladistic phylogeny of the North American species complexes of *Pogonomyrmex. Annals of the Entomological Society of America* **83**: 307–316. [(31).v.1990.]

Taber, S.W. & Cokendolpher, J.C. 1988. Karyotypes of a dozen ant species from the southwestern U.S.A. *Caryologia* **41**: 93–102. [1988.]

Taber, S.W., Cokendolpher, J.C. & Francke, O.F. 1988. Karyological study of North American *Pogonomyrmex. Insectes Sociaux* **35**: 47–60. [(30).ix.1988.]

Taber, S.W. & Francke, O.F. 1986. A bilateral gynandromorph of the western harvester ant, *Pogonomyrmex occidentalis. Southwestern Naturalist* **31**: 274–276. [22.v.1986.]

Tafuri, J.F. 1957. Growth and polymorphism in the larva of the army ant (*Eciton (E.) hamatum* Fabricius). *Journal of the New York Entomological Society* **63** (1955): 21–41. [8.iii.1957.]

Talbot, M. 1968. Slave-raids of the ant *Polyergus lucidus* Mayr. *Psyche* **74** (1967): 299–313. [12.vi.1968.]

Talbot, M. 1976. A list of the ants of the Edwin S. George Reserve, Livingston County, Michigan. *Great Lakes Entomologist* **8** (1975): 245–246. [1.iii.1976.]

Talbot, M. 1977. The natural history of the workerless ant parasite, *Formica talbotae. Psyche* **83** (1976): 282–288. [29.viii.1977.]

Talbot, M. & Kennedy, C.H. 1940. The slave-making ant, *Formica sanguinea subintegra* Emery, its raids, nuptual flights and nest structure. *Annals of the Entomological Society of America* **33**: 560–577. [30.ix.1940.]

Tanaka, M. 1974a. Description of a new species of ant of the genus *Probolomyrmex* from Malaysia. *Entomological Review of Japan* **26**: 35–37. [30.vi.1974.]

Tanaka, M. 1974b. A new species of the ant genus *Ponera* from Yaku Island. *Entomological Review of Japan* **27**: 32–36. [31.xii.1974.]

Tang, C. & Li, S. 1982. Hymenoptera: Formicidae (pp. 371–374). In *Xizang Kunchong (Tibetan Insects)* **2**: 508 pp. Beijing. [(30).iv.1982.]

Tarbinsky, Y.S. 1970. Novyi vid murav'ev roda *Proformica* iz Tyan'-Shanya. *Zoologichesky Zhurnal* **49**: 309–311. [29.i.1970.]

Tarbinsky, Y.S. 1976. *Murav'i Kirgizii:* 217 pp. Frunze. [4.vi.1976.]

Taylor, J.S. 1931. A note on the fauna of mangrove. *Entomologist's Record and Journal of Variation* **43**: 41–42. [15.iii.1931.]

Taylor, R.W. 1960. Taxonomic notes on the ants *Ponera leae* Forel and *Ponera norfolkensis* (Wheeler). *Pacific Science* **14**: 178–180. [(30).iv.1960.]

Taylor, R.W. 1962. New Australian dacetine ants of the genera *Mesostruma* Brown and *Codiomyrmex* Wheeler. *Breviora* **152**: 1–10. [15.i.1962.]

Taylor, R.W. 1964. Taxonomy and parataxonomy of some fossil ants. *Psyche* **71**: 134–141. [30.xii.1964.]

Taylor, R.W. 1965a. New Melanesian ants of the genera *Simopone* and *Amblyopone* of zoogeographical significance. *Breviora* **221**: 1–11. [7.v.1965.]

Taylor, R.W. 1965b. The Australian ants of the genus *Pristomyrmex,* with a case of apparent character displacement. *Psyche* **72**: 35–54. [25.vi.1965.]

Taylor, R.W. 1965c. A new species of the ant genus *Dacetinops* from Sarawak. *Breviora* **237**: 1–4. [15.xii.1965.]

Taylor, R.W. 1965d. A monographic revision of the rare tropicopolitan ant genus *Probolomyrmex* Mayr. *Transactions of the Royal Entomological Society of London* **117**: 345–365. [31.xii.1965.]

Taylor, R.W. 1966a. A second African species of the dacetine ant genus *Codiomyrmex. Psyche* **72** (1965): 225–228. [18.i.1966.]

Taylor, R.W. 1966b. Notes on the Indo-Australian ants of the genus *Simopone* Forel. *Psyche* **72** (1965): 287–290. [10.v.1966.]

Taylor, R.W. 1967a. A monographic revision of the ant genus *Ponera* Latreille. *Pacific Insects Monograph* **13**: 1–112. [30.v.1967.]

Taylor, R.W. 1967b. Entomological survey of the Cook Islands and Niue. 1. Hymenoptera, Formicidae. *New Zealand Journal of Science* **10**: 1092–1095. [(31).xii.1967.]

Taylor, R.W. 1968a. Nomenclature and synonymy of the North American ants of the genera *Ponera* and *Hypoponera. Entomological News* **79**: 63–66. [25.iii.1968.]

Taylor, R.W. 1968b. Notes on the Indo-Australian basicerotine ants. *Australian Journal of Zoology* **16**: 333–348. [(31).v.1968.]

Taylor, R.W. 1968c. A supplement to the revision of Australian *Pristomyrmex* species. *Journal of the Australian Entomological Society* **7**: 63–66. [30.vi.1968.]

Taylor, R.W. 1968d. A new Malayan species of the ant genus *Epitritus*, and a related new genus from Singapore. *Journal of the Australian Entomological Society* **7**: 130–134. [31.xii.1968.]

Taylor, R.W. 1969. The identity of *Dorylozelus mjobergi* Forel. *Journal of the Australian Entomological Society* **8**: 131–133. [31.xii.1969.]

Taylor, R.W. 1970a. Notes on some Australian and Melanesian basicerotine ants. *Journal of the Australian Entomological Society* **9**: 49–52. [30.iv.1970.]

Taylor, R.W. 1970b. Characterization of the Australian endemic ant genus *Peronomyrmex* Viehmeyer. *Journal of the Australian Entomological Society* **9**: 209–211. [31.xii.1970.]

Taylor, R.W. 1973. Ants of the Australian genus *Mesostruma* Brown. *Journal of the Australian Entomological Society* **12**: 24–38. [31.iii.1973.]

Taylor, R.W. 1977. New ants of the Australasian genus *Orectognathus*, with a key to the known species. *Australian Journal of Zoology* **25**: 581–612. [5.viii.1977.]

Taylor, R.W. 1978a. *Nothomyrmecia macrops:* a living-fossil ant rediscovered. *Science* **201**: 979–985. [15.ix.1978.]

Taylor, R.W. 1978b. A taxonomic guide to the ant genus *Orectognathus. CSIRO Division of Entomology* Report No. **3**: 1- 11 (plus microfiche). [(31).x.1978.]

Taylor, R.W. 1979. Melanesian ants of the genus *Amblyopone. Australian Journal of Zoology* **26** (1978): 823–839. [24.i.1979.]

Taylor, R.W. 1980a. New Australian ants of the genus *Orectognathus*, with summary description of the twenty-nine known species. *Australian Journal of Zoology* **27** (1979): 773–788. [15.ii.1980.]

Taylor, R.W. 1980b. Australian and Melanesian ants of the genus *Eurhopalothrix* Brown and Kempf—notes and new species. *Journal of the Australian Entomological Society* **19**: 229–239. [26.ix.1980.]

Taylor, R.W. 1980c. Notes on the Russian endemic ant genus *Aulacopone* Arnoldi. *Psyche* **86** (1979): 353–361. [19.xii.1980.]

Taylor, R.W. 1980d. The rare Fijian ant *Myrmecina* (=*Archaeomyrmex*) *cacabau* (Mann) rediscovered. *New Zealand Entomologist* **7**: 122–123. [(31).xii.1980.]

Taylor, R.W. 1985. The ants of the Papuasian genus *Dacetinops*. In Ball, G.E. (Ed.). *Taxonomy, Phylogeny and Zoogeography of Beetles and Ants:* 41–67. Dordrecht. [(31.xii).1985.]

Taylor, R.W. 1986. The quadrinomial infrasubspecific names of Australian ants. *General and Applied Entomology* **18**: 33–37. [10.x.1986.]

Taylor, R.W. 1987a. A checklist of the ants of Australia, New Caledonia and New Zealand. *CSIRO Division of Entomology* Report No. **41**: 1–92. [(30).vi.1987.]

Taylor, R.W. 1987b. A checklist of the ants of Australia, New Caledonia and New Zealand. (Supplement.) *CSIRO Division of Entomology* Report No. **41** (supplement): 1–5. [(31).vii.1987.]

Taylor, R.W. 1988. The nomenclature and distribution of some Australian and New Caledonian ants of the genus *Leptogenys* Roger (= *Prionogenys* Emery, syn. n.). *General and Applied Entomology* **20**: 33–37. [31.vii.1988.]

Taylor, R.W. 1989a. The nomenclature and distribution of some Australian ants of the genus *Polyrhachis* Fr. Smith. *Journal of the Australian Entomological Society* **28**: 23–27. [28.ii.1989.]

Taylor, R.W. 1989b. Australian ants of the genus *Leptothorax* Mayr. *Memoirs of the Queensland Museum* **27**: 605–610. [13.xi.1989.]

Taylor, R.W. 1990a. In Bolton, B. The higher classification of the ant subfamily Leptanillinae. *Systematic Entomology* **15**: 267–282. [(31).vii.1990.]

Taylor, R.W. 1990b. New Asian ants of the tribe Basicerotini, with an on-line computer interactive key to the twenty-six known Indo-Australian species. *Invertebrate Taxonomy* **4**: 397–425. [12.x.1990.]

Taylor, R.W. 1990c. The nomenclature and distribution of some Australian and New Caledonian ants of the genus *Meranoplus* Fr. Smith. *General and Applied Entomology* **22**: 31–40. [31.x.1990.]

Taylor, R.W. 1991a. Notes on the ant genera *Romblonella* and *Willowsiella*, with comments on their affinities, and the first description of Australian species. *Psyche* **97** (1990): 281–296. [26.ii.1991.]

Taylor, R.W. 1991b. Nomenclature and distribution of some Australasian ants of the Myrmicinae. *Memoirs of the Queensland Museum* **30**: 599–614. [1.viii.1991.]

Taylor, R.W. 1991c. *Myrmecia croslandi* sp. n., a karyologically remarkable new Australian jack-jumper ant. *Journal of the Australian Entomological Society* **30**: 288. [29.xi.1991.]

Taylor, R.W. 1992a. Nomenclature and distribution of some Australian and New Guinean ants of the subfamily Formicinae. *Journal of the Australian Entomological Society* **31**: 57–69. [28.ii.1992.]

Taylor, R.W. 1992b. *Rhoptromyrmex rawlinsoni* sp. nov., a new apparently workerless parasitic ant from Anak Krakatau, Indonesia. *Memoirs of the Museum of Victoria* **53**: 125–128. [30.v.1992.]

Taylor, R.W. & Brown, D.R. 1985. *Zoological Catalogue of Australia* **2**. Hymenoptera: Formicoidea, Vespoidea and Sphecoidea: 381 pp. Canberra. [(31.xii).1985.]

Taylor, R.W. & Brown, W.L.,Jr. 1978. *Smithistruma kempfi* species nov. *Pilot Register of Zoology*, card **35**. [25.ii.1978.]

Taylor, R.W. & Lowery, B.B. 1972. The New Guinean species of the ant genus *Orectognathus* Fr. Smith. *Journal of the Australian Entomological Society* **11**: 306–310. [31.xii.1972.]

Taylor, R.W. & Wilson, E.O. 1962. Ants from three remote oceanic islands. *Psyche* **68** (1961): 137–143. [13.iii.1962.]

Teranishi, C. 1915. A new species of Formicidae from Japan. *Entomological Magazine, Kyoto* 1: 137–138. [(31).xii.1915.]

Teranishi, C. 1936. *Insects of Jehol* (VII). Orders: Coleoptera (II) & Hymenoptera (I). Family Formicidae: Report of first scientific expedition to Manchoukuo (5)1, part 11, article 60; 1–12 (in Japanese). [30.xii.1936.]

Teranishi, C. 1940. *Works of Cho Teranishi. Memorial Volume:* 312 pp. plus posthumous section of 95 pp. Osaka. [(31.xii).1940]

Terayama, M. 1984. A new species of the army ant genus *Aenictus* from Taiwan. *Bulletin of the Biogeographical Society of Japan* **39**: 13–16. [20.xii.1984.]

Terayama, M. 1985a. Two new species of the ant genus *Myrmecina* from Japan and Taiwan. *Edaphologia* **32**: 35–40. [25.vi.1985.]

Terayama, M. 1985b. Two new species of the genus *Acropyga* from Taiwan and Japan. *Kontyû* **53**: 284–289. [25.vi.1985.]

Terayama, M. 1985c. Description of a new species of the genus *Proceratium* Roger from Taiwan. *Kontyû* **53**: 406–408. [25.ix.1985.]

Terayama, M. 1986. Two new ants of the genus *Ponera* from Taiwan. *Kontyû* **54**: 591–595. [25.xii.1986.]

Terayama, M. 1987. A new species of *Amblyopone* from Japan. *Edaphologia* **36**: 31–33. [25.iii.1987.]

Terayama, M. 1989a. The ant tribe Odontomachini from Taiwan, with description of a new species. *Edaphologia* **40**: 25–29. [25.iii.1989.]

Terayama, M. 1989b. The ant tribe Amblyoponini of Taiwan, with description of a new species. *Japanese Journal of Entomology* **57**: 343–346. [25.vi.1989.]

Terayama, M. 1990. Discovery of worker caste in *Trachymesopus darwinii* (Forel, 1893). *Japanese Journal of Entomology* **58**: 897–898. [25.xii.1990.]

Terayama, M. 1991. The subgenus *Paramyrmamblys* of the genus *Camponotus* from Japan, with a description of a new species. *Bulletin of the Biogeographical Society of Japan* **46**: 165–169. [20.xii.1991.]

Terayama, M. & Kubota, S. 1989. The ant tribe Dacetini of Taiwan, with descriptions of three new species. *Japanese Journal of Entomology* **57**: 778–792. [25.xii.1989.]

Terayama, M., Kubota, S., Sakai, H. & Kawazoe, A. 1988. Rediscovery of *Cerapachys sauteri* Forel, 1913, from Taiwan, with notes on the Taiwanese species of the genus *Cerapachys. Bulletin of the Biogeographical Society of Japan* **43**: 35–38. [20.xii.1988.]

Terayama, M. & Ogata, K. 1988. Two new species of the ant genus *Probolomyrmex* from Japan. *Kontyû* **56**: 590–594. [25.ix.1988.]

Terayama, M. & Satoh, T. 1990a. A new species of the genus *Camponotus* from Japan, with notes on two known forms of the subgenus *Myrmamblys. Japanese Journal of Entomology* **58**: 405–414. [25.vi.1990.]

Terayama, M. & Satoh, T. 1990b. Taxonomic notes on two Japanese species of Formicidae. *Japanese Journal of Entomology* **58**: 532. [25.ix.1990.]

Terayama, M. & Satoh, T. 1990c. *Camponotus (Myrmamblys) ogasawarensis* sp. nov. from the Ogasawara Islands, Japan. *Bulletin of the Biogeographical Society of Japan* **45**: 117–121. [20.xii.1990.]

Terayama, M., Yamaguchi, T. & Hasegawa, E. 1993. Ergatoid queens in the slave-making ant *Polyergus samurai* Yano. *Japanese Journal of Entomology* **61**: 511–514. [25.ix.1993.]

Terayama, M. & Yamane, S. 1989. The army ant genus *Aenictus* from Sumatra, with descriptions of three new species. *Japanese Journal of Entomology* **57**: 597–603. [25.ix.1989.]

Terayama, M. & Yamane, S. 1991. A new ant of the genus *Podomyrma* from Sumatra, Indonesia. *Proceedings of the Japanese Society of Systematic Zoology* **44**: 69–72. [25.vi.1991.]

Terron, G. 1968. Description des castes de *Tetraponera anthracina* Santschi. *Insectes Sociaux* **14** (1967): 339–348. [(30).vi.1968.]

Terron, G. 1969. Description de *Tetraponera ledouxi,* espèce nouvelle du Cameroun, parasite temporaire. *Bulletin de l'Institut Fondamental d'Afrique Noire* **31** (Série A): 629–642. [(31).vii.1969.]

Terron, G. 1971. Description des castes de *Tetraponera nasuta* Bernard. *Annales de la Faculté des Sciences, Université Fédérale du Cameroun* **6**: 73–84. [(30).iv.1971.]

Terron, G. 1974. Découverte au Cameroun de deux espèces nouvelles du genre *Prionopelta* Mayr. *Annales de la Faculté des Sciences du Cameroun* **17**: 105–119. [(31).iii.1974.]

Terron, G. 1977. Evolution des colonies de *Tetraponera anthracina* Santschi avec reines. *Bulletin Biologique de la France et de la Belgique* **111**: 115–181. [1977.]

Terron, G. 1981. Deux nouvelles espèces éthiopiennes pour le genre *Proceratium. Annales de la Faculté des Sciences de Yaoundé* **28**: 95–103. [(30).iv.1981.]

Théobald, N. 1937a. Notes complémentaire sur les insectes fossiles oligocènes des gypses d'Aix-en-Provence. *Bulletin Mensuel de la Société des Sciences de Nancy* (N.S. 5) **6**: 157–178. [(30).vi.1937.]

Théobald, N. 1937b. *Les Insectes Fossiles des Terrains Oligocènes de France:* 473 pp. Nancy. [(31.xii).1937.]

Thompson, C.R. 1982. A new *Solenopsis (Diplorhoptrum)* species from Florida. *Journal of the Kansas Entomological Society* **55**: 485–488. [20.vii.1982.]

Thompson, C.R. 1989. The thief ants, *Solenopsis molesta* group, of Florida. *Florida Entomologist* **72**: 268–283. [30.vi.1989.]

Thompson, C.R. & Johnson, C. 1989. Rediscovered species and revised key to the Florida thief ants. *Florida Entomologist* **72**: 697–698. [22.xii.1989.]

Tinaut, A. Note. In most of his publications his name is given as Tinaut, A., but in some it is rendered Tinaut Ranera, A., Tinaut, J.A., or Tinaut Ranera, J.A. The first, most commonly occurring form, is used universally in this catalogue.

Tinaut, A. 1981. *Rossomyrmex minuchae* nov. sp. encontrada en Sierra Nevada, España. *Boletin de la Asociacion Española de Entomologia* **4** (1980): 195–203. [(31).v.1981.]

Tinaut, A. 1983. Descripción de una nueva especie de *Leptothorax* Mayr, 1855, del sur de la Península Ibérica. *Eos. Revista Española de Entomologia* **58** (1982): 319–325. [15.iii.1983.]

Tinaut, A. 1985. Descripción de los sexuados de *Messor lusitanicus* Santschi, 1929. *Nouvelle Revue d'Entomologie* (N.S.) **2**: 85–90. [7.v.1985.]

Tinaut, A. 1987. Descripción de *Leptothorax pardoi* nov. sp. *Eos. Revista Española de Entomologia* **63**: 315–320. [15.xii.1987.]

Tinaut, A. 1989. Descripción del macho de *Leptothorax baeticus* Emery, 1924. *Nouvelle Revue d'Entomologie* (N.S.) **6**: 51–55. [22.vi.1989.]

Tinaut, A. 1990a. Descripción del macho de *Formica subrufa* Roger, 1859, y creación de un nuevo subgénero. *Eos. Revista Española de Entomologia* **65** (1989): 281–291. [(30).vi.1990.]

Tinaut, A. 1990b. *Teleutomyrmex kutteri,* spec. nov. A new species from Sierra Nevada (Granada, Spain). *Spixiana* **13**: 201–208. [31.vii.1990.]

Tinaut, A. 1990c. Taxonomic situation of the genus *Cataglyphis* Förster, 1850 in the Iberian Peninsula. 2. New position for *C. viatica* (Fabricius, 1787) and redescription of *C. velox* Santschi, 1929 stat. n. *Eos. Revista Española de Entomologia* **66**: 49–59. [(31).xii.1990.]

Tinaut, A. 1990d. El género *Amblyopone* Erichson en la Península Ibérica. *Miscellania Zoologica* **12** (1988): 189–193. [(31.xii).1990.]

Tinaut, A. 1991. Situación taxonomíca del genero *Cataglyphis* Förster, 1850 en la Península Ibérica. 3. El grupo de *C. velox* Santschi, 1929 y descripción de *Cataglyphis humeya* sp. n. *Eos. Revista Española de Entomologia* **66** (1990): 215–227. [(31).iii.1991.]

Tinaut, A. 1993. *Cataglyphis floricola* nov. sp. new species for the genus *Cataglyphis* Förster, 1850, in the Iberian Peninsula. *Mitteilungen der Schweizerischen Entomologischen Gesellschaft* **66**: 123–134. [(30).vi.1993.]

Tinaut, A. & Espadaler, X. 1987. Descripción del macho de *Myrmica aloba* Forel, 1909. *Nouvelle Revue d'Entomologie* (N.S.) **4**: 61–69. [29.v.1987.]

Tinaut, A., Espadaler, X. & Jimenez, J.J. 1992. *Camponotus universitatis* Forel, 1891, en la Península Ibérica. Descripción de sus sexuados. *Nouvelle Revue d'Entomologie* **9**: 233–238. [4.xii.1992.]

Tinaut, A. & Jiménez-Rojas, J. 1991. Redescripción de *Aphaenogaster striativentris* Forel, 1895 y consideraciones sobre su polimorfismo. *Eos. Revista Española de Entomologia* **66** (1990): 117–126. [(31).iii.1991.]

Tinaut, A. & Ortiz, F.J. 1988. Descripción del macho de *Monomorium algiricum* (Bernard, 1955) y consideraciones sobre el valor taxonómico del género *Epixenus* Emery, 1908. *Boletín de la Asociación Española de Entomología* **12**: 165–174. [(30).xi.1988.]

Tinaut, A. & Plaza, J.L. 1990. Situación taxonomíca del género *Cataglyphis* Förster, 1850 en la Península Ibérica. 1. Las especies del subgénero *Cataglyphis. Eos. Revista Española de Entomologia* **65** (1989): 189–199. [(31).i.1990.]

Tiwari, R.N. & Jonathan, J.K. 1986a. A new species of *Liomyrmex* Mayr from Andaman Islands. *Record of the Zoological Survey of India* **83**: 87–90. [(31).vii.1986.]

Tiwari, R.N. & Jonathan, J.K. 1986b. A new species of *Metapone* Forel from Nicobar Islands. *Record of the Zoological Survey of India* **83**: 149–153. [(31).vii.1986.]

Tjan, K.N., Imai, H.T., Kubota, M., Brown, W.L.,Jr., Gotwald, W.H.,Jr., Yong, H.-S. & Leh, C. 1986. Chromosome observations of Sarawak ants. *Annual Report of the National Institute of Genetics* **36**: 57–58. [1986.]

Tohmé, G. 1969. Description d'espèces nouvelles de fourmis au Liban. *Publications de l'Université Libanaise. Section des Sciences Naturelles* **7**: 1–15. [30.viii.1969.]

Tohmé, G. 1971. Description de *Messor ebeninus* (Forel). *Bulletin de la Société Entomologique d'Egypte* **54** (1970): 569–577. [(31).xii.1971.]

Tohmé, G. 1975. Ecologie, biologie de la reproducton et éthologie de *Messor ebeninus* Forel. *Bulletin Biologique de la France et de la Belgique* **109**: 171–251. [1975.]

Tohmé, G. & Tohmé, H. 1981. Les fourmis du genre *Messor* en Syrie. Position systématique. Description de quelques ailés et de formes nouvelles.

Répartition géographique. *Ecologia Mediterranea* **7**: 139–153. [(31).xii.1981.]
Tohmé, H. 1981. Ecologie et biologie de la reproduction de la fourmi *Acantholepis frauenfeldi* Mayr. *Publications de l'Université Libanaise. Section des Sciences Naturelles* **12**: 1–195. [1981.]
Tohmé, H. & Tohmé, G. 1980a. Le genre *Epixenus* Emery et ses principaux représentants au Liban et en Syrie. *Bulletin du Muséum National d'Histoire Naturelle, Paris* Section A. Zoologie, Biologie et Ecologie Animales (4) **1** (1979): 1087–1108. [18.iv.1980.]
Tohmé, H. & Tohmé, G. 1980b. Les fourmis du genre *Solenopsis* en Syrie. Description de deux nouvelles sous-espèces et d'ailés inédits. Notes biogéographiques et systématiques. *Revue Française d'Entomologie* (N.S.) **2**: 129–137. [22.ix.1980.]
Tohmé, H. & Tohmé, G. 1980c. Contribution à l'étude systématique et biologique de *Acantholepis syriaca* André. *Bulletin du Muséum National d'Histoire Naturelle, Paris* Section A. Zoologie, Biologie et Ecologie Animales (4) **2**: 517–524. [31.x.1980.]
Tohmé, H. & Tohmé, G. 1981. Contribution à l'étude systématique et biologique de *Bothriomyrmex syrius* (Forel). *Bulletin de la Société Entomologique de France* **86**: 97–103. [11.ix.1981.]
Tohmé, H. & Tohmé, G. 1985. Contribution à l'étude systématique et biologique de *Cataglyphis frigida*. *Revue Française d'Entomologie* (N.S.) **7**: 83–88. [14.vi.1985.]
Trager, J.C. 1984a. A new *Paratrechina* from Machu Picchu, Peru. *Florida Entomologist* **66** (1983): 482–486. [28.ii.1984.]
Trager, J.C. 1984b. A revision of the genus *Paratrechina* of the continental United States. *Sociobiology* **9**: 51–162. [(31.xii).1984.]
Trager, J.C. 1988. A revision of *Conomyrma* from the southeastern United States, especially Florida, with keys to the species. *Florida Entomologist* **71**: 11–29. [18.iii.1988.]
Trager, J.C. 1991. A revision of the fire ants, *Solenopsis geminata* group. *Journal of the New York Entomological Society* **99**: 141–198. [29.v.1991.]
Trager, J.C. & Johnson, C. 1988. The ant genus *Leptogenys* in the United States (pp. 29–34). In Trager, J.C. (Ed.). *Advances in Myrmecology:* 551 pp. New York. [(31.xii).1988.]
Traniello, J.F.A. 1982. Population structure and social organization in the primitive ant *Amblyopone pallipes*. *Psyche* **89**: 65–80. [17.xii.1982.]
Traniello, J.F.A. & Jayasuriya, A.K. 1986. The biology of the primitive ant *Aneuretus simoni* (Emery). 2. The social ethogram and division of labor. *Insectes Sociaux* **32** (1985): 375–388. [(30).vi.1986.]
Tryon, E.H.,Jr. 1986. The striped earwig, and ant predators of sugarcane rootstock borer, in Florida citrus. *Florida Entomologist* **69**: 336–343. [8.viii.1986.]
Tsuji, K., Furukawa, T., Kinomura, K., Takamine, H. & Yamauchi, K. 1991. The caste system of the dolichoderine ant *Technomyrmex albipes:* morphological description of queens, workers and reproductively active intercastes. *Insectes Sociaux* **38**: 413–422. [(31.xii).1991.]
Tulloch, G.S. 1930. An unusual nest of *Pogonomyrmex. Psyche* **37**: 61–70. [(31).iii.1930.]
Tutt, J.W. 1905. Types of the genera of the agdistid, alucitid and orneodid plume moths. *Entomologist's Record and Journal of Variation* **17**: 34–37. [15.ii.1905.]
Valcurone, M.D. & Fanfani, A. 1985. Investigations on Formicidae: Pavan's gland and other glands of the gaster in Dolichoderinae. *Pubblicazioni dell'Istituto di Entomologia dell'Università di Pavia* **31**: 1–20. [1985.]
Van Loon, A.J., Boomsma, J.J. & Andrasfalvy, A. 1990. A new polygynous *Lasius* species from central Europe. 1. Description and general biology. *Insectes Sociaux* **37**: 348–362. [(31).xii.1990.]
Vandel, A. 1926. Fourmis françaises rares ou peu connues. *Bulletin de la Société Entomologique de France* **1926**: 196–198. [30.xii.1926.]
Vashkevich, A.F. 1924a. Novaya forma murav'ya iz Semirechenskoi oblasti. *Izvestiya Tomskogo Gosudarstvennogo Universiteta* **74**: 144–145. [(31.xii).1924.]
Vashkevich, A.F. 1924b. K faune murav'ev severa Tobol'skoi gub. *Izvestiya Tomskogo Gosudarstvennogo Universiteta* **74**: 146–149. [(31.xii).1924.]
Vashkevich, A.F. 1926. Novaya forma murav'ya iz doliny reki Ili. *Izvestiya Tomskogo Gosudarstvennogo Universiteta* **77**: 118. [(3l.xii).1926.]
Vecht, J. van der, 1975. The date of publication of M. Spinola's paper on the Hymenoptera collected by V. Ghiliani in Para, with notes on the Eumenidae described in this work. *Entomologische Berichten* **35**: 60–63. [1.iv.1975.]
Viana, M.J. & Haedo Rossi, J.A. 1957. Primer hallazgo en el hemisferio sur de Formicidae extinguidos y catalogo mundial de los Formicidae fosiles. *Ameghiniana. Revista de la Asociacion Paleontologica Argentina* **1** (1–2): 108–113. [22.i.1957.]
Victor, T. 1839. Insectes du Caucase et des provinces transcaucasiennes. *Bulletin de la Société Impériale des Naturalistes de Moscou* **12**: 44–67. [(3l.xii).1839.]
Viehmeyer, H. 1906. Beiträge zur Ameisenfauna Königreiches Sachsen. *Abhandlungen der Naturwissenschaftlichen Gesellschaft ISIS in Dresden* **1906**: 55–69. [(31.xii).1906.]
Viehmeyer, H. 1912. Ameisen aus Deutsch Neuguinea gesammelt von Dr. O. Schlaginhaufen. Nebst einem Verzeichnisse der papuanischen Ameisen. *Abhandlungen und Berichte des Königl. Zoologischen und Anthropologisch-Ethnographischen Museums zu Dresden* **14**: 1–26. [5.xi.1912.]
Viehmeyer, H. 1913. Ameisen aus dem Kopal von Celebes. *Stettiner Entomologische Zeitung* **74**: 141–155. [1.vii.1913.]
Viehmeyer, H. 1914a. Ameisen aus Perak, Bali und Ceram (Hym.) (Freiburger Molukken-Expedition), gesammelt von E. Streesemann. *Entomologische Mitteilungen* **3**: 112–116. [1.iv.1914.]
Viehmeyer, H. 1914b. Papuanische Ameisen. *Deutsche Entomologische Zeitschrift* **1914**: 515–535. [30.ix.1914.]
Viehmeyer, H. 1914c. Neue und unvollständig bekannte Ameisen der alten Welt. *Archiv für Naturgeschichte* **79** (A.12) (1913): 24–60. [(31).v.1914.]
Viehmeyer, H. 1914d. Mayr's Gattung *Ischnomyrmex* nebst Beschreibung einiger neuer Arten aus anderen Gattungen. *Zoologische Jahrbücher. Abteilung für Systematik, Geographie und Biologie der Tiere* **37**: 601–612. [10.xii.1914.]
Viehmeyer, H. 1916a. Ameisen von Singapore. Beobachtet und gesammelt von H. Overbeck. *Archiv für Naturgeschichte* **81** (A.8) (1915): 108–168. [(30).iv.1916.]
Viehmeyer, H. 1916b. Ameisen von den Philippinen und anderer Herkunft. *Entomologische Mitteilungen* **5**: 283–291. [11.xi.1916.]
Viehmeyer, H. 1922. Neue Ameisen. *Archiv für Naturgeschichte* **88** (A.7): 203–220. [(30).vi.1922.]
Viehmeyer, H. 1923. Wissenschaftliche Ergebnisse der mit unterstützung der Akademie der Wissenschaften in Wien aus der erbschaft treitl von F. Werner unternommenen Zoologischen Expedition nach dem Anglo-Agyptischen Sudan (Kordofan) 1914. 7. Hymenoptera A. Formicidae. *Denkschriften der Akademie der Wissenschaften in Wien (Mathematisch-Naturwissenschaftliche Klasse)* **98**: 83–94. [(31.xii).1923.]
Viehmeyer, H. 1924a. Formiciden der australischen Faunenregion. *Entomologische Mitteilungen* **13**: 219–229. [9.viii.1924.]
Viehmeyer, H. 1924b. Formiciden der australischen Faunenregion. (Forsetzung.) *Entomologische Mitteilungen* **13**: 310–319. [7.xi.1924.]
Viehmeyer, H. 1925a. Formiciden der australischen Faunenregion. (Fortsetzung.) *Entomologische Mitteilungen* **14**: 25–39. [15.i.1925.]
Viehmeyer, H. 1925b. Formiciden der australischen Faunenregion. (Schluss.) *Entomologische Mitteilungen* **14**: 139–149. [10.iv.1925.]
Viereck, H.L. 1903. Hymenoptera of Beulah, New Mexico [part]. *Transactions of the American Entomological Society* **29**: 56–87. [(31).i.1903.]
Walckenaer, C.A. 1802. *Faune Parisienne, Insectes.* Ou histoire abrégée des insectes des environs de Paris, classés d'après le système de Fabricius **2**: 438pp. Paris. [(3l.xii).1802 (= An XI).]

Walker, F. 1859. Characters of some apparently undescribed Ceylon insects. *Annals and Magazine of Natural History* (3) **4**: 370–376. [(30).xi.1859.]

Walker, F. 1860. Characters of some apparently undescribed Ceylon insects. *Annals and Magazine of Natural History* (3) **5**: 304–311. [(30).iv.1860.]

Walker, F. 1871. *A list of Hymenoptera collected by J.K. Lord, esq. in Egypt, in the neighbourhood of the Red Sea, and in Arabia:* 59 pp. London. [(31.xii).1871.]

Walsh, B.D. 1863. On the genera of Aphidae found in the United States. *Proceedings of the Entomological Society of Philadelphia* **1** (1861–63): 294–3ll. [12.i.1863.]

Wang, C. & Wu, J. 1991. Taxonomic studies on the genus *Polyrhachis* of China. *Forest Research* **4**: 596–601. [(31).xii.1991.]

Wang, C. & Wu, J. 1992a. A new species of the ant genus *Acropyga* Roger of China. *Scientia Silvae Sinicae* **28**: 226–229. [(31).v.1992.]

Wang, C. & Wu, J. 1992b. Ants of the Jianfengling forest region in Hainan Province. *Scientia Silvae Sinicae* **28**: 561–564. [(30).xi.1992.]

Wang, C. & Wu, J. 1994. Second revisionary studies on genus *Camponotus* Mayr of China. *Journal of Beijing Forestry University* (English edition) **3**: 23–34. [(31.xii).1994.]

Wang, C., Xiao, G. & Wu, J. 1989a. Taxonomic studies on the genus *Camponotus* in China. *Forest Research* **2**: 221–228. [(30).vi.1989.]

Wang, C., Xiao, G. & Wu, J. 1989b. Taxonomic studies on the genus *Camponotus* in China (conclusion). *Forest Research* **2**: 321–328. [(31).viii.1989.]

Wang, M. 1993a. Taxonomic study of the ant tribe Odontomachini in China. *Scientific Treatise on Systematic and Evolutionary Zoology* **2**: 219–230. [(30).iv.1993.]

Wang, M. 1993b. Two new species and three new records of Myrmicinae from China. *Sinozoologia* **10**: 433–436. [(31).v.1993.]

Wang, M., Xiao, G. & Wu, J. 1988. Taxonomic studies on the genus *Tetramorium* Mayr in China. *Forest Research* **1**: 264–274. [(30).vi.1988.]

Ward, P.S. 1980. A systematic revision of the *Rhytidoponera impressa* group in Australia and New Guinea. *Australian Journal of Zoology* **28**: 475–498. [26.viii.1980.]

Ward, P.S. 1981a. Ecology and life history of the *Rhytidoponera impressa* group. 1. Habitats, nest sites and foraging behaviour. *Psyche* **88**: 89–108. [28.xii.1981.]

Ward, P.S. 1981b. Ecology and life history of the *Rhytidoponera impressa* group. 2. Colony origin, seasonal cycles and reproduction. *Psyche* **88**: 109–126. [28.xii.1981.]

Ward, P.S. 1984. A revision of the ant genus *Rhytidoponera* in New Caledonia. *Australian Journal of Zoology* **32**: 131–175. [9.iii.1984.]

Ward, P.S. 1985a. Taxonomic congruence and disparity in an insular ant fauna: *Rhytidoponera* in New Caledonia. *Systematic Zoology* **34**: 140–151. [18.vii.1985.]

Ward, P.S. 1985b. The Nearctic species of the genus *Pseudomyrmex. Quaestiones Entomologicae* **21**: 209–246. [(31.xii).1985.]

Ward, P.S. 1987. Distribution of the introduced Argentine ant (*Iridomyrmex humilis*) in natural habitats of the Lower Sacramento Valley and its effects on the indigenous ant fauna. *Hilgardia* **55** (2): 1–16. [(30).iv.1987.]

Ward, P.S. 1988. Mesic elements in the western Nearctic ant fauna: taxonomic and biological notes on *Amblyopone, Proceratium* and *Smithistruma. Journal of the Kansas Entomological Society* **61**: 102–124. [4.iii.1988.]

Ward, P.S. 1989. Systematic studies on pseudomyrmecine ants: revision of the *Pseudomyrmex oculatus* and *P. subtilissimus* species groups, with taxonomic comments on other species. *Quaestiones Entomologicae* **25**: 393–468. [(3l).xii.1989.]

Ward, P.S. 1990. The ant subfamily Pseudomyrmecinae: generic revision and relationship to other formicids. *Systematic Entomology* **15**: 449–489. [(30).xi.1990.]

Ward, P.S. 1992. Ants of the genus *Pseudomyrmex* from Dominican Amber, with a synopsis of the extant Antillean species. *Psyche* **99**: 55–85. [4.xii.1992.]

Ward, P.S. 1993. Systematic studies on *Pseudomyrmex* acacia-ants. *Journal of Hymenoptera Research* **2**: 117–168. [17.xi.1993. For date of publication see *Sphecos* **26**: 4 (xii.1993).]

Ward, P.S. 1994. *Adetomyrma,* an enigmatic new ant genus from Madagascar, and its implications for ant phylogeny. *Systematic Entomology* **19**: 159–175. [(30).iv.1994.]

Ward, P.S. & Taylor, R.W. 1981. Allozyme variation, colony structure and genetic relatedness in the primitive ant *Nothomyrmecia macrops* Clark. *Journal of the Australian Entomological Society* **20**: 177–183. [24.viii.1981.]

Wasmann, E. 1891. Einige neue Hermaphroditen von *Myrmica scabrinodis* und *laevinodis. Stettiner Entomologische Zeitung* **51** (1890): 298–299. [(3l).i.1891.]

Wasmann, E. 1894. *Kritisches Verzeichniss der Myrmekophilen und Termitophilen Arthropoden. Mit Angabe der Lebensweise und mit beschreibung neuer Arten* : 23l pp. Berlin. [(31.xii).1894.]

Wasmann, E. 1897. Bemerkungen über einige Ameisen aus Madagascar. *Zoologischer Anzeiger* **20**: 249–250. [19.vii.1897.]

Wasmann, E. 1901. Neues über die zusammengesetzten Nester und gemischten Kolonien der Ameisen. *Allgemeine Zeitschrift für Entomologie* **6**: 369–371. [15.xii.190l.]

Wasmann, E. 1902a. Neues über die zusammengesetzten Nester und gemischten Kolonien der Ameisen. *Allgemeine Zeitschrift für Entomologie* **7**: 1–5. [l.i.1902.]

Wasmann, E. 1902b. Neues über die zusammengesetzten Nester und gemischten Kolonien der Ameisen. *Allgemeine Zeitschrift für Entomologie* **7**: 293–298. [15.viii.1902.]

Wasmann, E. 1904a. Neue beiträge zur Kenntniss der Paussiden, mit biologischen und phylogenetischen bemerkungen. *Notes from the Leyden Museum* **25**: 1–82. [(3l).x.1904.]

Wasmann, E. 1904b. Zur Kenntniss der Gäste der Treiberameisen u. ihrer Wirthe am obern Congo nach den Sammlungen und Beobachtungen von P. Herm. Kohl C.SS.C. bearbeitet. *Zoologische Jahrbücher.* Supplement **7**: 611–682. [(31.xii).1904.]

Wasmann, E. 1905. Berichtigungen zu note 1 dieses Bandes. *Notes from the Leyden Museum* **25**: 110. [15.iv.1905.]

Wasmann, E. 1906. Zur Kenntniss der Ameisen und Ameisengäste von Luxemburg. *Archives Trimestrielles de l'Institut Grand- Ducal, Section des Sciences* (n.s.) **1** (1/2): 104–124. [(30.xii).1906.]

Wasmann, E. 1909. Über den Ursprung des sozialen Parasitismus, der Sklaverei und der Myrmekophilie bei den Ameisen. *Biologischen Centralblatt* **29**: 683–703. [15.xi.1909.]

Wasmann, E. 1910. Nachträge zum sozialen Parasitismus und der Sklaverei bei den Ameisen. *Biologischen Centralblatt* **30**: 515–524. [l.viii.1910.]

Wasmann, E. 1911. Zur Kenntnis der Termiten und Termitengäste vom Belgischen Congo. *Revue Zoologique Africaine* **1**: 91–117. [10.iv.19ll.]

Wasmann, E. 1915a. Eine neue *Pseudomyrma* aus der Ochsenhorndornakazie in Mexico mit Bemerkungen über Ameisen in Akaziendornen und ihre Gäste. Ein krisitscher Beitrag zur Pflanzen-Myrmekophilie. *Tijdschrift voor Entomologie* **58**: 296–325. [15.x.1915.]

Wasmann, E. 1915b. *Anergatides kohli,* eine neue arbeiterlose Schmarotzerameise vom oberen Kongo. *Entomologische Mitteilungen* **4**: 279–288. [27.xii.1915.]

Wasmann, E. 1916. Neue dorylophile Staphyliniden Afrikas. *Entomologische Mitteilungen* **5**: 134–147. [15.vii.1916.]

Wasmann, E. 1917. Neue Anpassungstypen bei Dorylinengästen Afrikas. (218 Beitrag zur Kenntnis der Myrmekophilen.) *Zeitschrift für Wissenschaftliche Zoologie* **117**: 257–360. [(3l.xii).1917.]

Wasmann, E. 1918a. Ueber *Solenopsis geminata saevissima* Sm. und ihre Gäste. *Entomologische Blätter. Zeitschrift für Biologie und Systematik der Käfer* **14**: 69–76. [18.iii.1918.]

Wasmann, E. 1918b. Über die von v. Rothkirch 1912 in Kamerun gesammelten Myrmekophilen. *Entomologische Mitteilungen* **7**: 135–149. [15.viii.1918.]

Wasmann, E. 1931. *Acromyrmex bucki*, n. sp. *Revista de Entomologia* **1**: 106. [25.iv.1931.]

Watkins, J.F. II. 1968. The rearing of the army ant male, *Neivamyrmex harrisi* (Haldeman) from larvae collected from a nest of *N. wheeleri* (Emery). *American Midland Naturalist* **80**: 273–276. [25.vii.1968.]

Watkins, J.F. II. 1969. A new species of *Neivamyrmex* from Louisiana. *Journal of the Kansas Entomological Society* **41** (1968): 528–531. [10.i.1969.]

Watkins, J.F. II. 1971. A taxonomic review of *Neivamyrmex moseri, N. pauxillus*, and *N. leonardi*, including new distribution records and original descriptions of queens of the first two species. *Journal of the Kansas Entomological Society* **44**: 93–103. [26.ii.1971.]

Watkins, J.F. II. 1972. The taxonomy of *Neivamyrmex texanus* n. sp., *N. nigrescens* and *N. californicus*, with distribution map and keys to the species of *Neivamyrmex* of the United States. *Journal of the Kansas Entomological Society* **45**: 347–372. [18.vii.1972.]

Watkins, J.F. II. 1973. *Neivamyrmex baylori* n. sp. from Waco, Texas, U.S.A. *Journal of the Kansas Entomological Society* **46**: 430–433. [3l.vii.1973.]

Watkins, J.F. II. 1974. *Neivamyrmex angulimandibulatus*, new species from Cordoba, Mexico. *Southwestern Naturalist* **19**: 309–312. [23.ix.1974.]

Watkins, J.F. II. 1975a. *Neivamyrmex cornutus* n. sp. from Oaxaca, Mexico. *Journal of the Kansas Entomological Society* **48**: 92–95. [22.i.1975.]

Watkins, J.F. II. 1975b. *Neivamyrmex digitistipus* n. sp. from Costa Rica. *Texas Journal of Science* **26**: 203–206. [(28).ii.1975.]

Watkins, J.F. II. 1975c. *Neivamyrmex quadratooccipitus* n. sp. from El Salvador. *Texas Journal of Science* **26**: 207–211. [(28).ii.1975.]

Watkins, J.F. II. 1975d. The relationship of *Neivamyrmex fuscipennis* to *N. macropterus*. *Southwestern Naturalist* **20**: 85–90. [15.v.1975.]

Watkins, J.F. II. 1976. *The Identification and Distribution of New World Army Ants:* 102 pp. Waco, Texas. [(3l.xii).1976.]

Watkins, J.F. II. 1977a. The species and subspecies of *Nomamyrmex*. *Journal of the Kansas Entomological Society* **50**: 203–214. [29.iv.1977.]

Watkins, J.F. II. 1977b. *Neivamyrmex nyensis*, n. sp. from Nye County, Nevada, U.S.A. *Southwestern Naturalist* **22**: 421–425. [15.x.1977.]

Watkins, J.F. II. 1982. The Army Ants of Mexico. *Journal of the Kansas Entomological Society* **55**: 197–247. [20.iv.1982.]

Watkins, J.F. II. 1986. *Neivamyrmex chamelensis*, n. sp. from Jalisco, Mexico. *Journal of the Kansas Entomological Society* **59**: 361–366. [28.v.1986.]

Watkins, J.F. II. 1990. *Neivamyrmex crassiscapus* n. sp., from Mexico. *Journal of the Kansas Entomological Society* **64**: 348–350. [31.v.1990.]

Watkins, J.F. II. 1994. *Neivamyrmex curvinotus* n. sp. from South America. *Journal of the Kansas Entomological Society* **66** (1993): 411–413. [29.iii.1994.]

Watkins, J.F. II. & Coody, C.J. 1986. The taxonomy of *Neivamyrmex graciellae* (Mann) including an original description of the queen and field observations. *Southwestern Naturalist* **31**: 256–259. [22.v.1986.]

Weber, N.A. 1934a. Notes on Neotropical ants, including the descriptions of new forms. *Revista de Entomologia* **4**: 22–59. [10.iv.1934.]

Weber, N.A. 1934b. A new *Strumigenys* from Illinois. *Psyche* **41**: 63–65. [(30).vi.1934.]

Weber, N.A. 1935. The biology of the thatching ant, *Formica rufa obscuripes* Forel, in North Dakota. *Ecological Monographs* **5**: 165–206. [(30).iv.1935.]

Weber, N.A. 1937. The biology of the fungus-growing ants. Part l. New forms. *Revista de Entomologia* **7**: 378–409. [11.x.1937.]

Weber, N.A. 1938a. New ants from stomachs of *Bufo marinus* L. and *Typhlops reticulatus* (L.). *Annals of the Entomological Society of America* **31**: 207–210. [29.vi.1938.]

Weber, N.A. 1938b. The biology of the fungus-growing ants. Part 4. Additional new forms. Part 5. The Attini of Bolivia. *Revista de Entomologia* **9**: 154–206. [27.ix.1938.]

Weber, N.A. 1939a. New ants of rare genera and a new genus of ponerine ants. *Annals of the Entomological Society of America* **32**: 91–104. [20.iii.1939.]

Weber, N.A. 1939b. Descriptions of new North American species and subspecies of *Myrmica*. *Lloydia* **2**: 144–152. [(30).vi.1939.]

Weber, N.A. 1940a. The biology of the fungus-growing ants. Part 6. Key to *Cyphomyrmex*, new Attini and a new guest ant. *Revista de Entomologia* **11**: 406–427. [28.vi.1940.]

Weber, N.A. 1940b. Rare ponerine genera in Panama and British Guiana. *Psyche* **47**: 75–84. [(30).ix.1940.]

Weber, N.A. 1941a. Four new genera of Ethiopian and Neotropical Formicidae. *Annals of the Entomological Society of America* **34**: 183–194. [7.iv.1941.]

Weber, N.A. 1941b. The biology of the fungus-growing ants. Part 7. The Barro Colorado Island, Canal Zone, species. *Revista de Entomologia* **12**: 93–130. [31.vii.1941.]

Weber, N.A. 1941c. The rediscovery of the queen of *Eciton (Labidus) coecum* Latr. *American Midland Naturalist* **26**: 325–329. [30.ix.1941.]

Weber, N.A. 1942a. New doryline, cerapachyine and ponerine ants from the Imatong Mountains, Anglo-Egyptian Sudan. *Proceedings of the Entomological Society of Washington* **44**: 40–49. [31.iii.1942.]

Weber, N.A. 1942b. The genus *Thaumatomyrmex* Mayr with description of a Venezuelan species. *Boletin de Entomologia Venezolana* **1**: 65–71. [30.viii.1942.]

Weber, N.A. 1943a. The queen of a British Guiana *Eciton* and a new ant garden *Solenopsis*. *Proceedings of the Entomological Society of Washington* **45**: 88–91. [30.iv.1943.]

Weber, N.A. 1943b. New ants from Venezuela and neighboring countries. *Boletin de Entomologia Venezolana* **2**: 67–78. [30.vi.1943.]

Weber, N.A. 1943c. The ants of the Imatong Mountains, Anglo- Egyptian Sudan. *Bulletin of the Museum of Comparative Zoology at Harvard College* **93**: 263–389. [(31).xii.1943.]

Weber, N.A. 1944. The Neotropical coccid-tending ants of the genus *Acropyga* Roger. *Annals of the Entomological Society of America* **37**: 89–122. [10.iv.1944.]

Weber, N.A. 1945. The biology of the fungus-growing ants. Part 8. The Trinidad, B.W.I. species. *Revista de Entomologia* **16**: 1–88. [30.viii.1945.]

Weber, N.A. 1946a. Two common ponerine ants of possible economic significance, *Ectatomma tuberculatum* (Olivier) and *E. ruidum* Roger. *Proceedings of the Entomological Society of Washington* **48**: 1–16. [31.i.1946.]

Weber, N.A. 1946b. The biology of the fungus-growing ants. Part 9. The British Guiana species. *Revista de Entomologia* **17**: 114–172. [26.viii.1946.]

Weber, N.A. 1947. A revision of the North American ants of the genus *Myrmica* Latreille with a synopsis of the Palaearctic species. 1. *Annals of the Entomological Society of America* **40**: 437–474. [21.xi.1947.]

Weber, N.A. 1948a. A revision of the North American ants of the genus *Myrmica* Latreille, with a synopsis of the Palearctic species. 2. *Annals of the Entomological Society of America* **41**: 267–308. [19.x.1948.]

Weber, N.A. 1948b. Studies on the fauna of Curaçao, Aruba, Bonaire and the Venezuelan islands: no. 14. Ants from the Leeward group and some other Caribbean localities. *Natuurwetenschappelijke Studiekring voor Suriname en Curaçao* **5**: 78–86. [(31.xii).1948.]

Weber, N.A. 1949a. New African ants of the genera *Cerapachys, Phyracaces* and *Simopone*. *American Museum Novitates* **1396**: 1–9. [21.i.1949.]

Weber, N.A. 1949b. New ponerine ants from Equatorial Africa. *American Museum Novitates* **1398**: 1–9. [25.i.1949.]

Weber, N.A. 1949c. A new Panama *Eciton*. *American Museum Novitates* **1441**: 1–8. [22.xii.1949.]

Weber, N.A. 1950a. The African species of the genus *Oligomyrmex* Mayr. *American Museum Novitates* **1442**: 1–19. [4.i.1950.]

Weber, N.A. 1950b. A revision of the North American ants of the genus *Myrmica* Latreille with a synopsis of the Palearctic species. 3. *Annals of the Entomological Society of America* **43**: 189–226. [8.viii.1950.]

Weber, N.A. 1950c. New Trinidad Myrmicinae, with a note on *Basiceros* Schulz. *American Museum Novitates* **1465**: 1–6. [16.viii.1950.]
Weber, N.A. 1952. Studies on African Myrmicinae, 1. *American Museum Novitates* **1548**: 1–32. [7.iv.1952.]
Weber, N.A. 1958a. Nomenclatural notes on *Proatta* and *Atta*. *Entomological News* **69**: 7–13. [18.i.1958.]
Weber, N.A. 1958b. Nomenclatural changes in *Trachymyrmex*. *Entomological News* **69**: 49–55. [21.ii.1958.]
Weber, N.A. 1958c. Synonymies and types of *Apterostigma*. *Entomological News* **69**: 243–251. [12.xi.1958.]
Weber, N.A. 1958d. Some attine synonyms and types. *Proceedings of the Entomological Society of Washington* **60**: 259–264. [18.xii.1958.]
Weber, N.A. 1966. The subgenus *Cyphomannia* Weber 1938 of *Cyphomyrmex* Mayr 1862, reinstated, and systematic notes. *Entomological News* **77**: 166–168. [14.vi.1966.]
Weber, N.A. 1967. Synonyms of *Trachymyrmex bugnioni* Forel and *Trachymyrmex diversus* Mann. *Proceedings of the Entomological Society of Washington* **69**: 273–274. [13.x.1967.]
Weber, N.A. 1982. Fungus Ants (pp. 255–363). In Hermann, H.R. (Ed..) *Social Insects* **4**: 385 pp. New York. [(31.xii).1982.]
Weber, N.A. & Anderson, J.L. 1950. Studies on central African ants of the genus *Pseudolasius* Emery. *American Museum Novitates* **1443**: 1–7. [4.i.1950.]
Weddell, H.A. 1850. Additions a la flore de l'Amerique du Sud. (Suite.) *Annales des Sciences Naturelles Botanique* (3) **13** (1849): 249–268. [Pp. 249–256 dated (30).iv.1850; pp. 257–268 dated (31).v.1850.]
Wehner, R. 1983. Taxonomie, Funktionsmorphologie und Zoogeographie der Saharischen Wüstenameise *Cataglyphis fortis* (Forel 1902) stat. nov. *Senckenbergiana Biologica* (N.S.) **64**: 89–132. [30.xi.1983.]
Wehner, R., Harkness, R.D., & Schmid-Hempel, P. 1983. Foraging strategies in individually searching ants. *Cataglyphis bicolor. Information Processing in Animals* **1**: 79 pp. Mainz. [(31.xii).1983.]
Wesmael, C. 1838. Sur une nouvelle espèce de fourmi du Mexique. *Bulletins de l'Académie Royale des Sciences et Belles- Lettres de Bruxelles* **5**: 766–771. [(31.xii).1838.]
Wesson, L.G. 1935. A new species of ant from Tennessee. *Entomological News* **46**: 208–210. [10.x.1935.]
Wesson, L.G. 1937. A slave-making *Leptothorax*. *Entomological News* **48**: 125–129. [13.v.1937.]
Wesson, L.G. 1939. *Leptothorax manni* Wesson synonymous with *L. pergandei* Emery. *Entomological News* **50**: 180. [16.vi.1939.]
Wesson, L.G. 1940. Observations on *Leptothorax duloticus*. *Bulletin of the Brooklyn Entomological Society* **35**: 73–83. [7.vi.1940.]
Wesson, L.G. 1949. *Strumigenys venatrix* Wesson and Wesson synonymous with *S. talpa* Weber. *Psyche* **56**: 21. [17.v.1949.]
Wesson, L.G. & Wesson, R.G. 1939. Notes on *Strumigenys* from southern Ohio, with descriptions of six new species. *Psyche* **46**: 91–112. [(30).ix.1939.]
Wesson, L.G. & Wesson, R.G. 1940. A collection of ants from southcentral Ohio. *American Midland Naturalist* **24**: 89–106. [31.vii.1940.]
Westwood, J.O. 1829. Page 344, in Stephens, J.F. *A Systematic Catalogue of British Insects* Part 2: 388 pp. London. [15.vii.1829.]
Westwood, J.O. 1835. ". . . Hymenopterous Insects, which Mr Westwood regarded as new to science". *Proceedings of the Zoological Society of London* **3**: 68–72. [(31.xii).1835.]
Westwood, J.O. 1839. *An Introduction to the Modern Classification of Insects;* founded on the natural habits and corresponding organisation of the different families **2** (part 11): 193–224. London. [(31).iii.1839. For date of publication see Griffin, 1932: 83.]
Westwood, J.O. 1840a. Synopsis of the Genera of British Insects. In *An Introduction to the Modern Classification of Insects* (above) **2**, synopsis sheet G: 81–96. [(31).i.1840.]
Westwood, J.O. 1840b. Observations on the genus *Typhlopone*, with descriptions of several exotic species of ants. *Annals and Magazine of Natural History* **6**: 81–89. [(31).x.1840.]
Westwood, J.O. 1842. Monograph of the hymenopterous group, Dorylides (No. 5, pp. 73–80). *Arcana Entomologica;* or illustrations of new, rare, and interesting insects **1**: 192 pp. London. [Publication date of No. 5 is 1.i.1842; see Bradley, 1975: 91.]
Westwood, J.O. 1843. Description of a new species of the hymenopterous genus *Aenictus*, belonging to the Dorylidae. *Journal of Proceedings of the Entomological Society of London* **1840–1846**: 85. [(31.xi).1843.]
Westwood, J.O. 1847a. Description of a new dorylideous insect from South Africa, belonging to the genus *Aenictus*. *Transactions of the Entomological Society of London* **4**: 237–238. [3.iii.1847.]
Westwood, J.O. 1847b. Description of the "Driver" ants, described in the preceding article. *Transactions of the Entomological Society of London* **5**: 16–18. [30.vi.1847. Dates of publication of this and the above by Wheeler, G. 1912: 754.]
Westwood, J.O. 1854. Contributions to fossil entomology. *Quarterly Journal of the Geological Society of London* **10**: 378–396. [(31.xii).1854.]
Wheeler, D.E. 1985. Behavior of the ant *Procryptocerus scabriusculus* with comparisons to other cephalotines. *Psyche* **91** (1984): 171–192. [22.iv.1985.]
Wheeler, G. 1912. On the dates of publications of the Entomological Society of London. *Transactions of the Entomological Society of London* **1911**: 750–767. [10.ii.1912.]
Wheeler, G.C. 1928. The larva of *Leptanilla*. *Psyche* **35**: 85–91. [(30).vi.1928.]
Wheeler, G.C. 1935. The larva of *Allomerus*. *Psyche* **42**: 92–98. [(30).vi.1935.]
Wheeler, G.C. 1943. The larvae of the army ants. *Annals of the Entomological Society of America* **36**: 319–332. [1.vii.1943.]
Wheeler, G.C. 1949. The larvae of the fungus-growing ants. *American Midland Naturalist* **40** (1948): 664–689. [24.ii.1949.]
Wheeler, G.C. 1950. Ant larvae of the subfamily Cerpachyinae. *Psyche* **57**: 102–113. [29.xii.1950.]
Wheeler, G.C. & Wheeler, E.W. 1930. Two new ants from Java. *Psyche* **37**: 193–201. [(30).ix.1930.]
Wheeler, G.C. & Wheeler, E.W. 1934. New forms of *Aphaenogaster treatae* Forel from the Southern United States. *Psyche* **41**: 6–12. [(31).iii.1934.]
Wheeler, G.C. & Wheeler, E.W. 1944. The ants of North Dakota. *North Dakota Historical Quarterly* **11**: 231–271. [(31).x.1944.]
Wheeler, G.C. & Wheeler, J. 1951. The ant larvae of the subfamily Dolichoderinae. *Proceedings of the Entomological Society of Washington* **53**: 169–210. [28.viii.1951.]
Wheeler, G.C. & Wheeler, J. 1952a. The ant larvae of the subfamily Ponerinae. Part 1. *American Midland Naturalist* **48**: 111–144. [(31).vii.1952.]
Wheeler, G.C. & Wheeler, J. 1952b. The ant larvae of the myrmicine tribe Crematogastrini. *Journal of the Washington Academy of Sciences* **42**: 248–262. [(31).viii.1952.]
Wheeler, G.C. & Wheeler, J. 1952c. The ant larvae of the subfamily Ponerinae. Part 2. *American Midland Naturalist* **48**: 604–672. [(30).xi.1952.]
Wheeler, G.C. & Wheeler, J. 1953a. The ant larvae of the myrmicine tribe Myrmicini. *Psyche* **59** (1952): 105–125. [30.i.1953.]
Wheeler, G.C. & Wheeler, J. 1953b. The ant larvae of the myrmicine tribe Pheidolini. *Proceedings of the Entomological Society of Washington* **55**: 49–84. [27.iv.1953.]
Wheeler, G.C. & Wheeler, J. 1953c. The ant larvae of the subfamily Formicinae. *Annals of the Entomological Society of America* **46**: 126–171. [14.v.1953.]
Wheeler, G.C. & Wheeler, J. 1953d. The ant larvae of the myrmicine tribes Melissotarsini, Metaponini, Myrmicariini and Cardiocondylini. *Journal of the Washington Academy of Sciences* **43**: 185–189. [(30).vi.1953.]
Wheeler, G.C. & Wheeler, J. 1953e. The ant larvae of the subfamily Formicinae. Part 2. *Annals of the Entomological Society of America* **46**: 175–217. [3.viii.1953.]

Wheeler, G.C. & Wheeler, J. 1954a. The ant larvae of the myrmicine tribe Pheidologetini. *Psyche* **60** (1953): 129–147. [27.iv.1954.]
Wheeler, G.C. & Wheeler, J. 1954b. The ant larvae of the myrmicine tribes Cataulacini and Cephalotini. *Journal of the Washington Academy of Sciences* **44**: 149–157. [(31).v.1954.]
Wheeler, G.C. & Wheeler, J. 1954c. The ant larvae of the myrmicine tribe Myrmecinini. *Proceedings of the Entomological Society of Washington* **56**: 126–138. [15.vi.1954.]
Wheeler, G.C. & Wheeler, J. 1954d. The ant larvae of the myrmicine tribes Meranoplini, Ochetomyrmicini and Tetramoriini. *American Midland Naturalist* **52**: 443–452. [26.xii.1954.]
Wheeler, G.C. & Wheeler, J. 1955a. The ant larvae of the myrmicine tribes Basicerotini and Dacetini. *Psyche* **61** (1954): 111–145. [26.i.1955.]
Wheeler, G.C. & Wheeler, J. 1955b. The ant larvae of the myrmicine tribe Leptothoracini. *Annals of the Entomological Society of America* **48**: 17–29. [18.v.1955.]
Wheeler, G.C. & Wheeler, J. 1955c. The ant larvae of the myrmicine tribe Solenopsidini. *American Midland Naturalist* **54**: 119–141. [27.viii.1955.]
Wheeler, G.C. & Wheeler, J. 1956. The ant larvae of the subfamily Pseudomyrmecinae. *Annals of the Entomological Society of America* **49**: 374–398. [30.viii.1956.]
Wheeler, G.C. & Wheeler, J. 1957a. The larva of the ant genus *Dacetinops. Breviora* **78**: 1–4. [21.vi.1957.]
Wheeler, G.C. & Wheeler, J. 1957b. The larva of *Simopelta. Proceedings of the Entomological Society of Washington* **59**: 191–194. [18.ix.1957.]
Wheeler, G.C. & Wheeler, J. 1957c. *Veromessor lobognathus* in North Dakota. *Psyche* **63** (1956): 140–145. [18.xi.1957.]
Wheeler, G.C. & Wheeler, J. 1959. The larva of *Paramyrmica. Journal of the Tennessee Academy of Science* **34**: 219–220. [(31).x.1959.]
Wheeler, G.C. & Wheeler, J. 1960a. The ant larvae of the subfamily Myrmicinae. *Annals of the Entomological Society of America* **53**: 98–110. [5.ii.1960.]
Wheeler, G.C. & Wheeler, J. 1960b. Supplementary studies on the larvae of the Myrmicinae. *Proceedings of the Entomological Society of Washington* **62**: 1–32. [27.iv.1960.]
Wheeler, G.C. & Wheeler, J. 1963. *The Ants of North Dakota:* 326 pp. Grand Forks. [(31.xii).1963.]
Wheeler, G.C. & Wheeler, J. 1964a. The ant larvae of the subfamily Cerapachyinae: supplement. *Proceedings of the Entomological Society of Washington* **66**: 65–71. [6.vii.1964.]
Wheeler, G.C. & Wheeler, J. 1964b. The ant larvae of the subfamily Ponerinae: supplement. *Annals of the Entomological Society of America* **57**: 443–462. [15.vii.1964.]
Wheeler, G.C. & Wheeler, J. 1964c. The ant larvae of the subfamily Dorylinae: supplement. *Proceedings of the Entomological Society of Washington* **66**: 129–137. [22.x.1964.]
Wheeler, G.C. & Wheeler, J. 1965. The ant larvae of the subfamily Leptanillinae. *Psyche* **72**: 24–34. [25.vi.1965.]
Wheeler, G.C. & Wheeler, J. 1966. The ant larvae of the subfamily Dolichoderinae: supplement. *Annals of the Entomological Society of America* **59**: 726–732. [15.vii.1966.]
Wheeler, G.C. & Wheeler, J. 1968. The ant larvae of the subfamily Formicinae: supplement. *Annals of the Entomological Society of America* **61**: 205–222. [15.i.1968.]
Wheeler, G.C. & Wheeler, J. 1969. The larva of *Acanthognathus. Psyche* **76**: 110–113. [12.xii.1969.]
Wheeler, G.C. & Wheeler, J. 1970. Ant larvae of the subfamily Formicinae: second supplement. *Annals of the Entomological Society of America* **63**: 648–656. [15.v.1970.]
Wheeler, G.C. & Wheeler, J. 1971a. The larva of *Apomyrma. Psyche* **77** (1970): 276–279. [31.v.1971.]
Wheeler, G.C. & Wheeler, J. 1971b. Ant larvae of the subfamily Ponerinae: second supplement. *Annals of the Entomological Society of America* **64**: 1197–1217. [15.xi.1971.]
Wheeler, G.C. & Wheeler, J. 1971c. The larvae of the ant genus *Bothroponera. Proceedings of the Entomological Society of Washington* **73**: 386–394. [17.xii.1971.]
Wheeler, G.C. & Wheeler, J. 1971d. Ant larvae of the subfamily Myrmeciinae. *Pan-Pacific Entomologist* **47**: 245–256. [30.xii.1971.]
Wheeler, G.C. & Wheeler, J. 1972a. The subfamilies of Formicidae. *Proceedings of the Entomological Society of Washington* **74**: 35–45. [28.iii.1972.]
Wheeler, G.C. & Wheeler, J. 1972b. Ant larvae of the subfamily Myrmicinae: second supplement on the tribes Myrmicini and Pheidolini. *Journal of the Georgia Entomological Society* **7**: 233–246. [(31).x.1972.]
Wheeler, G.C. & Wheeler, J. 1973a. The ant larvae of six tribes: second supplement. *Journal of the Georgia Entomological Society* **8**: 27–39. [(31).i.1973.]
Wheeler, G.C. & Wheeler, J. 1973b. Ant larvae of four tribes: second supplement. *Psyche* **80**: 70–82. [7.ix.1973.]
Wheeler, G.C. & Wheeler, J. 1973c. The ant larvae of the tribes Basicerotini and Dacetini: second supplement. *Pan-Pacific Entomologist* **49**: 207–214. [19.x.1973.]
Wheeler, G.C. & Wheeler, J. 1973d. Supplementary studies on ant larvae: Cerpachyinae, Pseudomyrmecinae and Myrmicinae. *Psyche* **80**: 204–211. [6.xii.1973.]
Wheeler, G.C. & Wheeler, J. 1973e. *Ants of Deep Canyon:* 162 pp. Riverside, California. [(31.xii).1973.]
Wheeler, G.C. & Wheeler, J. 1974a. Ant larvae of the subfamily Formicinae: third supplement. *Journal of the Georgia Entomological Society* **9**: 59–64. [(31).i.1974.]
Wheeler, G.C. & Wheeler, J. 1974b. Ant larvae of the subfamily Dolichoderinae: second supplement. *Pan-Pacific Entomologist* **49** (1973): 396–401. [16.iv.1974.]
Wheeler, G.C. & Wheeler, J. 1974c. Ant larvae of the subfamily Dorylinae: second supplement. *Journal of the Kansas Entomological Society* **47**: 166–172. [26.iv.1974.]
Wheeler, G.C. & Wheeler, J. 1974d. Ant larvae of the myrmicine tribe Attini: second supplement. *Proceedings of the Entomological Society of Washington* **76**: 76–81. [30.iv.1974.]
Wheeler, G.C. & Wheeler, J. 1974e. Supplementary studies on ant larvae: *Teratomyrmex. Psyche* **81**: 38–41. [21.v.1974.]
Wheeler, G.C. & Wheeler, J. 1974f. Supplementary studies on ant larvae: *Simopone* and *Turneria. Journal of the New York Entomological Society* **82**: 103–105. [22.viii.1974.]
Wheeler, G.C. & Wheeler, J. 1974g. Ant larvae of the subfamily Ponerinae: third supplement. *Proceedings of the Entomological Society of Washington* **76**: 278–281. [7.xi.1974.]
Wheeler, G.C. & Wheeler, J. 1976a. Supplementary studies on ant larvae: Ponerinae. *Transactions of the American Entomological Society* **102**: 41–64. [27.iv.1976.]
Wheeler, G.C. & Wheeler, J. 1976b. Ant larvae: review and synthesis. *Memoirs of the Entomological Society of Washington* **7**: 1–108. [(31.xii).1976.]
Wheeler, G.C. & Wheeler, J. 1977. Supplementary studies on ant larvae: Myrmicinae. *Transactions of the American Entomological Society* **103**: 581–602. [23.xii.1977.]
Wheeler, G.C. & Wheeler, J. 1980. Supplementary studies on ant larvae: Ponerinae, Myrmicinae and Formicinae. *Transactions of the American Entomological Society* **106**: 527–545. [31.xii.1980.]
Wheeler, G.C. & Wheeler, J. 1982. Supplementary studies on ant larvae: Formicinae. *Psyche* **89**: 175–182. [17.xii.1982.]
Wheeler, G.C. & Wheeler, J. 1983. Supplementary studies on ant larvae: Myrmicinae. *Transactions of the American Entomological Society* **108**: 601–610. [11.i.1983.]

Wheeler, G.C. & Wheeler, J. 1984. The larvae of the army ants: a revision. *Journal of the Kansas Entomological Society* **57**: 263–275. [18.v.1984.]
Wheeler, G.C. & Wheeler, J. 1985. A simplified conspectus of the Formicidae. *Transactions of the American Entomological Society* **111**: 255–264. [30.vii.1985.]
Wheeler, G.C. & Wheeler, J. 1986a. The larva of *Dinoponera. Psyche* **92** (1985): 387–392. [27.iv.1986.]
Wheeler, G.C. & Wheeler, J. 1986b. The larva of *Proatta. Psyche* **92** (1985): 447–450. [27.iv.1986.]
Wheeler, G.C. & Wheeler, J. 1986c. Supplementary studies of ant larvae: Ponerinae. *Transactions of the American Entomological Society* **112**: 85–94. [27.vi.1986.]
Wheeler, G.C. & Wheeler, J. 1986d. Supplementary studies on ant larvae: Formicinae. *Journal of the New York Entomological Society* **94**: 331–341. [2.vii.1986.]
Wheeler, G.C. & Wheeler, J. 1986e. Ten-year supplement to "Ant larvae: review and synthesis". *Proceedings of the Entomological Society of Washington* **88**: 684–702. [14.x.1986.]
Wheeler, G.C. & Wheeler, J. 1986f. Supplementary studies on ant larvae: Myrmicinae. *Journal of the New York Entomological Society* **94**: 489–499. [16.x.1986.]
Wheeler, G.C. & Wheeler, J. 1986g. The ants of Nevada. *Natural History Museum of Los Angeles County:* 138 pp. Los Angeles. [(3l.xii).1986.]
Wheeler, G.C. & Wheeler, J. 1988. The larva of *Notostigma. Journal of the New York Entomological Society* **96**: 355–358. [12.viii.1988.]
Wheeler, G.C. & Wheeler, J. 1989a. Notes on ant larvae: Myrmicinae. *Transactions of the American Entomological Society* **114** (1988): 319–327. [25.i.1989.]
Wheeler, G.C. & Wheeler, J. 1989b. Notes on ant larvae: Ponerinae. *Journal of the New York Entomological Society* **97**: 50–55. [29.iii.1989.]
Wheeler, G.C. & Wheeler, J. 1989c. The larva of *Leptanilla japonica,* with notes on the genus. *Psyche* **95** (1988): 185–189. [13.vi.1989.]
Wheeler, G.C. & Wheeler, J. 1990a. Notes on ant larvae. *Transactions of the American Entomological Society* **115** (1989): 457–473. [15.i.1990.]
Wheeler, G.C. & Wheeler, J. 1990b. Larvae of the formicine ant genus *Polyrhachis. Transactions of the American Entomological Society* **116**: 753–767. [(30).ix.1990.]
Wheeler, G.C. & Wheeler, J. 1991a. The larva of *Blepharidatta. Journal of the New York Entomological Society* **99**: 132–137. [14.ii.1991.]
Wheeler, G.C. & Wheeler, J. 1991b. Instars of three ant species. *Psyche* **98**: 89–99. [16.viii.1991.]
Wheeler, G.C., Wheeler, J. & Taylor, R.W. 1980. The larval and egg stages of the primitive ant *Nothomyrmecia macrops* Clark. *Journal of the Australian Entomological Society* **19**: 131–137. [24.vii.1980.]
Wheeler, J. 1968. Male genitalia and the taxonomy of *Polyergus. Proceedings of the Entomological Society of Washington* **70**: 156–164. [29.vi.1968.]
Wheeler, W.M. 1900a. The female of *Eciton sumichrasti* Norton, with some notes on the habits of Texan ecitons. *American Naturalist* **34**: 563–574. [25.vii.1900.]
Wheeler, W.M. 1900b. A study of some Texan Ponerinae. *Biological Bulletin* **2**: 1–31. [(31).x.1900.]
Wheeler, W.M. 1900c. The habits of *Ponera* and *Stigmatomma. Biological Bulletin* **2**: 43–69. [(30).xi.1900.]
Wheeler, W.M. 1901a. The compound and mixed nests of American ants. *American Naturalist* **35**: 431–448. [28.vi.1901.]
Wheeler, W.M. 1901b. The compound and mixed nests of American ants. Part 2. The known cases of social symbiosis among American ants. *American Naturalist* **35**: 513–539. [31.vii.1901.]
Wheeler, W.M. 1901c. The compound and mixed nests of American ants. Part 2 (continued). *American Naturalist* **35**: 701–724. [12.ix.1901.]
Wheeler, W.M. 1902a. A new agricultural ant from Texas, with remarks on the known North American species. *American Naturalist* **36**: 85–100. [13.ii.1902.]
Wheeler, W.M. 1902b. *Formica fusca* Linn. ssp. *subpolita* Mayr, var. *perpilosa,* n.var. *Memorias y Revista de la Sociedad Científica "Antonio Alzate"* **17**: 141–142. [7.v.1902.]
Wheeler, W.M. 1902c. New agricultural ants from Texas. *Psyche* **9**: 387–393. [(30).ix.1902.]
Wheeler, W.M. 1902d. An American *Cerapachys,* with remarks on the affinities of the Cerapachyinae. *Biological Bulletin* **3**: 181–191. [(31).x.1902.]
Wheeler, W.M. 1902e. The occurrence of *Formica cinerea* Mayr and *Formica rufibarbis* Fabricius in America. *American Naturalist* **36**: 947–952. [24.xii.1902.]
Wheeler, W.M. 1902f. A consideration of S.B. Buckley's "North American Formicidae". *Transactions of the Texas Academy of Science* **4** (1901): 17–31. [(31.xii).1902.]
Wheeler, W.M. 1903a. *Erebomyrma,* a new genus of hypogaeic ants from Texas. *Biological Bulletin* **4**: 137–148. [(28).ii.1903.]
Wheeler, W.M. 1903b. A decad of Texan Formicidae. *Psyche* **10**: 93–111. [(30).vi.1903.]
Wheeler. W.M. 1903c. A revision of the North American ants of the genus *Leptothorax* Mayr. *Proceedings of the Academy of Natural Sciences of Philadelphia* **55**: 215–260. [8.vii.1903.]
Wheeler, W.M. 1903d. The North American ants of the genus *Stenamma* sensu stricto. *Psyche* **10**: 164–168. [(31).viii.1903.]
Wheeler, W.M. 1903e. Extraordinary females in three species of *Formica,* with remarks on mutation in the Formicidae. *Bulletin of the American Museum of Natural History* **19**: 639–651. [21.xi.1903.]
Wheeler, W.M. 1903f. Some new gynandromorphous ants, with a review of the previously recorded cases. *Bulletin of the American Museum of Natural History* **19**: 653–683. [5.xii.1903.]
Wheeler, W.M. 1903g. Some notes on the habits of *Cerapachys augustae. Psyche* **10**: 205–209. [(31).xii.1903.]
Wheeler, W.M. 1904a. Three new genera of inquiline ants from Utah and Colorado. *Bulletin of the American Museum of Natural History* **20**: 1–17. [14.i.1904.]
Wheeler, W.M. 1904b. The American ants of the subgenus *Colobopsis. Bulletin of the American Museum of Natural History* **20**: 139–158. [23.iv.1904.]
Wheeler, W.M. 1904c. A crustacean-eating ant (*Leptogenys elongata* Buckley). *Biological Bulletin* **6**: 251–259. [(31).v.1904.]
Wheeler, W.M. 1904d. Ants from Catalina Island, California. *Bulletin of the American Museum of Natural History* **20**: 269–271. [2.viii.1904.]
Wheeler, W.M. 1904e. The ants of North Carolina. *Bulletin of the American Museum of Natural History* **20**: 299–306. [8.ix.1904.]
Wheeler, W.M. 1904f. A new type of social parasitism among ants. *Bulletin of the American Museum of Natural History* **20**: 347–375. [11.x.1904.]
Wheeler, W.M. 1905a. An interpretation of the slave-making instincts in ants. *Bulletin of the American Museum of Natural History* **21**: 1–16. [14.ii.1905.]
Wheeler, W.M. 1905b. The ants of the Bahamas, with a list of the known West Indian species. *Bulletin of the American Museum of Natural History* **21**: 79–135. [30.vi.1905.]
Wheeler, W.M. 1905c. New species of *Formica. Bulletin of the American Museum of Natural History* **21**: 267–274. [28.ix.1905.]
Wheeler, W.M. 1905d. The North American ants of the genus *Dolichoderus. Bulletin of the American Museum of Natural History* **21**: 305–319. [11.xi.1905.]
Wheeler, W.M. 1905e. The North American ants of the genus *Liometopum. Bulletin of the American Museumof Natural History* **21**: 321–333. [14.xi.1905.]
Wheeler, W.M. 1905f. An annotated list of the ants of New Jersey. *Bulletin of the American Museum of Natural History* **21**: 371–403. [9.xii.1905.]
Wheeler, W.M. 1906a. New ants from New England. *Psyche* **13**: 38–41. [(30).vi.1906.]

Wheeler, W.M. 1906b. Fauna of New England. 7. List of the Formicidae. *Occasional Papers of the Boston Society of Natural History* **7** (No. 7): 1–24. [(31).vii.1906.]
Wheeler, W.M. 1906c. The ants of Japan. *Bulletin of the American Museum of Natural History* **22**: 301–328. [17.ix.1906.]
Wheeler, W.M. 1906d. The ants of the Grand Cañon. *Bulletin of the American Museum of Natural History* **22**: 329–345. [17.ix.1906.]
Wheeler, W.M. 1906e. The ants of the Bermudas. *Bulletin of the American Museum of Natural History* **22**: 347–352. [29.ix.1906.]
Wheeler, W.M. 1907a. A collection of ants from British Honduras. *Bulletin of the American Museum of Natural History* **23**: 271–277. [30.iii.1907.]
Wheeler, W.M. 1907b. Notes on a new guest-ant, *Leptothorax glacialis*, and the varieties of *Myrmica brevinodis* Emery. *Bulletin of the Wisconsin Natural History Society* **5**: 70–83. [(30).iv.1907.]
Wheeler, W.M. 1907c. The fungus-growing ants of North America. *Bulletin of the American Museum of Natural History* **23**: 669–807. [30.ix.1907.]
Wheeler, W.M. 1908a. The ants of Porto Rico and the Virgin Islands. *Bulletin of the American Museum of Natural History* **24**: 117–158. [7.ii.1908.]
Wheeler, W.M. 1908b. The ants of Jamaica. *Bulletin of the American Museum of Natural History* **24**: 159–163. [7.ii.1908.]
Wheeler, W.M. 1908c. Ants from Moorea, Society Islands. *Bulletin of the American Museum of Natural History* **24**: 165–167. [7.ii.1908.]
Wheeler, W.M. 1908d. Honey ants, with a revision of the American Myrmecocysti. *Bulletin of the American Museum of Natural History* **24**: 345–397. [9.v.1908.]
Wheeler, W.M. 1908e. The ants of Texas, New Mexico and Arizona. (Part 1.) *Bulletin of the American Museum of Natural History* **24**: 399–485. [20.v.1908.]
Wheeler, W.M. 1908f. The ants of Casco Bay, Maine, with observations on two races of *Formica sanguinea* Latreille. *Bulletin of the American Museum of Natural History* **24**: 619–645. [25.ix.1908.]
Wheeler, W.M. 1908g. Comparative ethology of the European and North American ants. *Journal für Psychologie und Neurologie* **13**: 404–435. [(31.xii).1908.]
Wheeler, W.M. 1909a. A small collection of ants from Victoria, Australia. *Journal of the New York Entomological Society* **17**: 25–29. [(31).iii.1909.]
Wheeler, W.M. 1909b. Ants collected by Prof. F. Silvestri in Mexico. *Bollettino del Laboratorio di Zoologia generale e agraria della R. Scuola Superiore d'Agricoltura in Portici* **3**: 228–238. [2.iv.1909.]
Wheeler, W.M. 1909c. Ants collected by Prof. F. Silvestri in the Hawaiian Islands. *Bollettino del Laboratorio di Zoologia generale e agraria della R. Scuola Superiore d'Agricoltura in Portici* **3**: 269–272. [8.iv.1909.]
Wheeler, W.M. 1909d. Ants of Formosa and the Philippines. *Bulletin of the American Museum of Natural History* **26**: 333–345. [24.vi.1909.]
Wheeler, W.M. 1909e. A decade of North American Formicidae. *Journal of the New York Entomological Society* **17**: 77–90. [(30).vi.1909.]
Wheeler, W.M. 1909f. A new honey ant from California. *Journal of the New York Entomological Society* **17**: 98–99. [(30).ix.1909.]
Wheeler, W.M. 1909g. Observations on some European ants. *Journal of the New York Entomological Society* **17**: 172–187. [(31).xii.1909.]
Wheeler, W.M. 1910a. Three new genera of myrmicine ants from tropical America. *Bulletin of the American Museum of Natural History* **28**: 259–265. [30.vii.1910.]
Wheeler, W.M. 1910b. An aberrant *Lasius* from Japan. *Biological Bulletin* **19**: 130–137. [(31).vii.1910.]
Wheeler, W.M. 1910c. A new species of *Aphomomyrmex* from Borneo. *Psyche* **17**: 131–135. [(31).viii.1910.]
Wheeler, W.M. 1910d. The North American ants of the genus *Camponotus* Mayr. *Annals of the New York Academy of Sciences* **20**: 295–354. [30.xii.1910.]
Wheeler, W.M. 1910e. The North American forms of *Lasius umbratus* Nylander. *Psyche* **17**: 235–243. [(31).xii.1910.]
Wheeler, W.M. 1910f. The North American forms of *Camponotus fallax* Nylander. *Journal of the New York Entomological Society* **18**: 216–232. [(31).xii.1910.]
Wheeler, W.M. 1910g. Ants, their structure, development and behaviour. *Columbia University Biological Series* **9**: 663 pp. New York. [(31.xii).1910.]
Wheeler, W.M. 1911a. Additions to the ant fauna of Jamaica. *Bulletin of the American Museum of Natural History* **30**: 21–29. [26.v.1911.]
Wheeler, W.M. 1911b. Ants collected in Grenada, W.I. by Mr. C.T. Brues. *Bulletin of the Museum of Comparative Zoology at Harvard college* **54**: 167–172. [(31).v.1911.]
Wheeler, W.M. 1911c. Three formicid names which have been overlooked. *Science. New York* (N.S.) **33**: 858–860. [2.vi.1911.]
Wheeler, W.M. 1911d. A new *Camponotus* from California. *Journal of the New York Entomological Society* **19**: 96–98. [(30).vi.1911.]
Wheeler, W.M. 1911e. Two fungus-growing ants from Arizona. *Psyche* **18**: 93–101. [(30).vi.1911.]
Wheeler, W.M. 1911f. A list of the type species of the genera and subgenera of Formicidae. *Annals of the New York Academy of Sciences* **21**: 157–175. [17.x.1911.]
Wheeler, W.M. 1911g. Descriptions of some new fungus-growing ants from Texas, with Mr. C.G. Hartman's observations on their habits. *Journal of the New York Entomological Society* **19**: 245–255. [(31).xii.1911.]
Wheeler, W.M. 1911h. Three new ants from Mexico and Central America. *Psyche* **18**: 203–208. [(31).xii.1911.]
Wheeler, W.M. 1912a. The ants of Guam. *Journal of the New York Entomological Society* **20**: 44–48. [(31).iii.1912.]
Wheeler, W.M. 1912b. Notes on a mistletoe ant. *Journal of the New York Entomological Society* **20**: 130–133. [(30).vi.1912.]
Wheeler, W.M. 1912c. New names for some ants of the genus *Formica*. *Psyche* **19**: 90. [(30).vi.1912.]
Wheeler, W.M. 1912d. Additions to our knowledge of the ants of the genus *Myrmecocystus* Wesmael. *Psyche* **19**: 172–181. [(31).xii.1912.]
Wheeler, W.M. 1912e. The male of *Eciton vagans* Olivier. *Psyche* **19**: 206–207. [(31).xii.1912.]
Wheeler, W.M. 1913a. Corrections and additions to "List of type species of the genera and subgenera of Formicidae." *Annals of the New York Academy of Sciences* **23**: 77–83. [29.v.1913.]
Wheeler, W.M. 1913b. The ants of Cuba. *Bulletin of the Museum of Comparative Zoology at Harvard College* **54**: 477–505. [(31).v.1913.]
Wheeler, W.M. 1913c. Ants collected in Georgia by Dr. J.C. Bradley and Mr. W.T. Davis. *Psyche* **20**: 112–117. [(30).vi.1913.]
Wheeler, W.M. 1913d. Ants collected in the West Indies. *Bulletin of the American Museum of Natural History* **32**: 239–244. [9.vii.1913.]
Wheeler, W.M. 1913e. Zoological results of the Abor Expedition, 1911–12. Hymenoptera 2. Ants (Formicidae). *Records of the Indian Museum* **8**: 233–237. [(30).ix.1913.]
Wheeler, W.M. 1913f. A revision of the ants of the genus *Formica* (Linné) Mayr. *Bulletin of the Museum of Comparative Zoology at Harvard College* **53**: 379–565. [(31).x.1913.]
Wheeler, W.M. 1914a. *Formica exsecta* in Japan. *Psyche* **21**: 26–27. [(28).ii.1914.]
Wheeler, W.M. 1914b. Ants collected by W.M. Mann in the state of Hidalgo, Mexico. *Journal of the New York Entomological Society* **22**: 37–61. [(31).iii.1914.]
Wheeler, W.M. 1914c. Note on the habits of *Liomyrmex*. *Psyche* **21**: 75–76. [(30).iv.1914.]
Wheeler, W.M. 1914d. The American species of *Myrmica* allied to *M. rubida* Latreille. *Psyche* **21**: 118–122. [(31).viii.1914.]
Wheeler, W.M. 1914e. New and little known harvesting ants of the genus *Pogonomyrmex*. *Psyche* **21**: 149–157. [(31).x.1914.]
Wheeler, W.M. 1915a. *Neomyrma* versus *Oreomyrma*. *Psyche* **22**: 50. [(30).iv.1915.]

Wheeler, W.M. 1915b. Some additions to the North American ant fauna. *Bulletin of the American Museum of Natural History* **34**: 389–421. [4.vi.1915.]
Wheeler, W.M. 1915c. *Paranomopone*, a new genus of ponerine ants from Queensland. *Psyche* **22**: 117–120. [(31).viii.1915.]
Wheeler, W.M. 1915d. The Australian honey-ants of the genus *Leptomyrmex* Mayr. *Proceedings of the American Academy of Arts and Sciences* **51**: 255–286. [(30).xi.1915.]
Wheeler, W.M. 1915e. Two new genera of myrmicine ants from Brazil. *Bulletin of the Museum of Comparative Zoology at Harvard College* **59**: 483–491. [(30).xi.1915.]
Wheeler, W.M. 1915f. A new bog-inhabiting variety of *Formica fusca* L. *Psyche* **22**: 203–206. [(31).xii.1915.]
Wheeler, W.M. 1915g. Hymenoptera. *Transactions and Proceedings of the Royal Society of South Australia* **39**: 805–823. [(31).xii.1915.]
Wheeler, W.M. 1915h. The ants of the Baltic Amber. *Schriften der Physikalisch-ökonomischen Gesellschaft zu Königsberg in Pr.* **55** (1914): 1–142. [(31.xii).1915.]
Wheeler, W.M. 1916a. The Australian ants of the genus *Onychomyrmex. Bulletin of the Museum of Comparative Zoology at Harvard College* **60**: 45–54. [(31).i.1916.]
Wheeler, W.M. 1916b. Four new and interesting ants from the mountains of Borneo and Luzon. *Proceedings of the New England Zoological Club* **6**: 9–18. [10.ii.1916.]
Wheeler, W.M. 1916c. Ants collected in British Guiana by the expedition of the American Museum of Natural History during 1911. *Bulletin of the American Museum of Natural History* **35**: 1–14. [21.ii.1916.]
Wheeler, W.M. 1916d. Ants collected in Trinidad by Professor Roland Thaxter, Mr. F.W. Urich, and others. *Bulletin of the Museum of Comparative Zoology at Harvard College* **60**: 323–330. [(31).iii.1916.]
Wheeler, W.M. 1916e. Some new formicid names. *Psyche* **23**: 40–41. [(30).iv.1916.]
Wheeler, W.M. 1916f. Two new ants from Texas and Arizona. *Proceedings of the New England Zoological Club* **6**: 29–35. [18.x.1916.]
Wheeler, W.M. 1916g. Note on the Brazilian fire-ant, *Solenopsis saevissima* F. Smith. *Psyche* **23**: 142–143. [(31).x.1916.]
Wheeler, W.M. 1916h. An anomalous blind worker ant. *Psyche* **23**: 143–145. [(31).x.1916.]
Wheeler, W.M. 1916i. *Prodiscothyrea*, a new genus of ponerine ants from Queensland. *Transactions of the Royal Society of South Australia* **40**: 33–37. [23.xii.1916.]
Wheeler, W.M. 1916j. The Australian ants of the genus *Aphaenogaster* Mayr. *Transactions of the Royal Society of South Australia* **40**: 213–223. [23.xii.1916.]
Wheeler, W.M. 1916k. Questions of nomenclature connected with the ant genus *Lasius* and its subgenera. *Psyche* **23**: 168–173. [(31).xii.1916.]
Wheeler, W.M. 1916l. Ants carried in a floating log from the Brazilian mainland to San Sebastian Island. *Psyche* **23**: 180–183. [(31).xii.1916.]
Wheeler, W.M. 1916m. Hymenoptera of Connecticut. Formicoidea (pp. 577–601). In Britton, W.E. (director). Guide to the Insects of Connecticut. Part 3. The Hymenoptera, or Wasp-like Insects of Connecticut. *State Geological and Natural History Survey (State of Connecticut Public Document No. 47)* **5**. Bulletin 22: 824 pp. [(31.xii).1916.]
Wheeler, W.M. 1917a. The mountain ants of western North America. *Proceedings of the American Academy of Arts and Sciences* **52**: 457–569. [(31).i.1917.]
Wheeler, W.M. 1917b. The phylogenetic development of subapterous and apterous castes in the Formicidae. *Proceedings of the National Academy of Sciences* **3**: 109–117. [(28).ii.1917.]
Wheeler, W.M. 1917c. The North American ants described by Asa Fitch. *Psyche* **24**: 26–29. [(28).ii.1917.]
Wheeler, W.M. 1917d. A new Malayan ant of the genus *Prodiscothyrea. Psyche* **24**: 29–30. [(28).ii.1917.]
Wheeler, W.M. 1917e. The ants of Alaska. *Bulletin of the Museum of Comparative Zoology at Harvard College* **61**: 13–23. [(31).iii.1917.]
Wheeler, W.M. 1917f. The Australian ant genus *Myrmecorhynchus* Ern. André and its position in the subfamily Camponotinae. *Transactions of the Royal Society of South Australia* **41**: 14–19. [24.xii.1917.]
Wheeler, W.M. 1917g. Jamaican ants collected by Prof. C.T. Brues. *Bulletin of the Museum of Comparative Zoology at Harvard College* **61**: 457–471. [(31).xii.1917.]
Wheeler, W.M. 1917h. The temporary social parasitism of *Lasius subumbratus* Viereck. *Psyche* **24**: 167–176. [(31).xii.1917.]
Wheeler, W.M. 1917i. A list of Indiana ants. *Proceedings of the Indiana Academy of Science* **26**: 460–466. [(31.xii).1917.]
Wheeler, W.M. 1918a. The Australian ants of the ponerine tribe Cerapachyini. *Proceedings of the American Academy of Arts and Sciences* **53**: 215–265. [(31).i.1918.]
Wheeler, W.M. 1918b. Ants collected in British Guiana by Mr. C. William Beebe. *Journal of the New York Entomological Society* **26**: 23–28. [(31).iii.1918.]
Wheeler, W.M. 1918c. A study of some ant larvae, with a consideration of the origin and meaning of the social habit among insects. *Proceedings of the American Philosophical Society* **57**: 293–343. [8.viii.1918.]
Wheeler, W.M. 1918d. The ants of the genus *Opisthopsis* Emery. *Bulletin of the Museum of Comparative Zoology at Harvard College* **62**: 341–362. [(30).xi.1918.]
Wheeler, W.M. 1919a. Two gynandromorphous ants. *Psyche* **26**: 1–8. [(28).ii.1919.]
Wheeler, W.M. 1919b. A new subspecies of *Aphaenogaster treatae* Forel. *Psyche* **26**: 50. [(30).iv.1919.]
Wheeler, W.M. 1919c. Expedition of the California Academy of Sciences to the Galapagos Islands, 1905–1906. Part 14. The ants of the Galapagos Islands. *Proceedings of the California Academy of Sciences* (4) **2** (2): 259–297. [16.vi.1919.]
Wheeler, W.M. 1919d. Expedition of the California Academy of Sciences to the Galapagos Islands, 1905–1906. Part 15. The ants of Cocos Island. *Proceedings of the California Academy of Sciences* (4) **2** (2): 299–308. [16.vi.1919.]
Wheeler, W.M. 1919e. The ants of Borneo. *Bulletin of the Museum of Comparative Zoology at Harvard College* **63**: 43–147. [(31).vii.1919.]
Wheeler, W.M. 1919f. The ant genus *Lordomyrma. Psyche* **26**: 97–106. [(31).viii.1919.]
Wheeler, W.M. 1919g. A new paper-making *Crematogaster* from the southeastern United States. *Psyche* **26**: 107–112. [(31).viii.1919.]
Wheeler, W.M. 1919h. The ants of the genus *Metapone* Forel. *Annals of the Entomological Society of America* **12**: 173–191. [21.x.1919.]
Wheeler, W.M. 1919i. A singular Neotropical ant (*Pseudomyrma filiformis* Fabricius). *Psyche* **26**: 124–131. [(31).x.1919.]
Wheeler, W.M. 1920. The subfamilies of Formicidae, and other taxonomic notes. *Psyche* **27**: 46–55. [(30).vi.1920.]
Wheeler, W.M. 1921a. Professor Emery's subgenera of the genus *Camponotus. Psyche* **28**: 16–19. [(28).ii.1921.]
Wheeler, W.M. 1921b. A new case of parabiosis and the "ant gardens" of British Guiana. *Ecology* **2**: 89–103. [26.iv.1921.]
Wheeler, W.M. 1921c. Chinese ants. *Bulletin of the Museum of Comparative Zoology at Harvard College* **64**: 529–547. [(30).iv.1921.]
Wheeler, W.M. 1921d. Observations on army ants in British Guiana. *Proceedings of the American Academy of Arts and Sciences* **56**: 291–328. [(30).vi.1921.]
Wheeler, W.M. 1921e. Chinese ants collected by Prof. C.W. Howard. *Psyche* **28**: 110–115. [(31).viii.1921.]
Wheeler, W.M. 1921f. The *Tachigalia* ants. *Zoologica. Scientific Contributions of the New York Zoological Society* **3**: 137–168. [24.xii.1921.]
Wheeler, W.M. 1922a. The ants of the Belgian Congo. *Bulletin of the American Museum of Natural History* **45**: 1–1139. [Pp. 13–269 dated 10.ii.1922; pp. 1–11 and pp. 271–1139 dated 25.x.1922.]
Wheeler, W.M. 1922b. Observations on *Gigantiops destructor* Fabricius and other leaping ants. *Biological Bulletin* **42**: 185–201. [(30).iv.1922.]

Wheeler, W.M. 1922c. The ants of Trinidad. *American Museum Novitates* **45**: 1–16. [7.ix.1922.]
Wheeler, W.M. 1922d. A new genus and subgenus of Myrmicinae from tropical America. *American Museum Novitates* **46**: 1–6. [7.ix.1922.]
Wheeler, W.M. 1922e. Neotropical ants of the genera *Carebara, Tranopelta* and *Tranopeltoides* new genus. *American Museum Novitates* **48**: 1–14. [16.x.1922.]
Wheeler, W.M. 1922f. Formicidae from Easter Island and Juan Fernandez. In Skottsberg, C. *The Natural History of Juan Fernandez and Easter Island* **3**: 317–319. Uppsala. [(31.xii).1922.]
Wheeler, W.M. 1923a. Wissenschaftliche Ergebnisse der schwedischen entomologischen Reise des Herrn Dr. A. Roman in Amazonas, 1914–1915. 7. Formicidae. *Arkiv för Zoologi* **15**: 1–6. [31.i.1923.]
Wheeler, W.M. 1923b. Chinese ants collected by Professor S.F. Light and Professor A.P. Jacot. *American Museum Novitates* **69**: 1–6. [20.iv.1923.]
Wheeler, W.M. 1923c. Report on the ants collected by the Barbados-Antigua Expedition from the University of Iowa in 1918. *University of Iowa Studies in Natural History* **10** (No. 3): 3–9. [1.viii.1923.]
Wheeler, W.M. 1923d. The occurrence of winged females in the ant genus *Leptogenys* Roger, with descriptions of new species. *American Museum Novitates* **90**: 1–16. [16.x.1923.]
Wheeler, W.M. 1923e. Ants of the genera *Myopias* and *Acanthoponera. Psyche* **30**: 175–192. [(31).xii.1923.]
Wheeler, W.M. 1923f. *Social Life Among the Insects:* 375 pp. New York. [(31.xii).1923.]
Wheeler, W.M. 1924a. The Formicidae of the Harrison Williams Galapagos Expedition. *Zoologica. New York Zoological Society* **5**: 101–122. [27.ii.1924.]
Wheeler, W.M. 1924b. Ants of Krakatau and other islands in the Sunda Straight. *Treubia* **5**: 239–258. [(28).ii.1924.]
Wheeler, W.M. 1924c. A gynandromorph of *Tetramorium guineense* Fabr. *Psyche* **31**: 136–137. [(31).viii.1924.]
Wheeler, W.M. 1924d. On the ant genus *Chrysapace* Crawley. *Psyche* **31**: 224–225. [(31).x.1924.]
Wheeler, W.M. 1925a. Neotropical ants in the collections of the Royal Museum of Stockholm. *Arkiv för Zoologi* **17** (A 8): 1–55. [7.ii.1925.]
Wheeler, W.M. 1925b. Zoological results of the Swedish expedition to Central Africa. *Arkiv för Zoologi* **17** (A 25): 1–3. [16.iii.1925.]
Wheeler, W.M. 1925c. The finding of the queen of the army ant *Eciton hamatum* Fabricius. *Biological Bulletin* **49**: 139–149. [(30).ix.1925.]
Wheeler, W.M. 1925d. A new guest-ant and other new Formicidae from Barro Colorado Island, Panama. *Biological Bulletin* **49**: 150–181. [(30).ix.1925.]
Wheeler, W.M. 1926. Ants from the Balearic Islands. *Folia Myrmecologica et Termitologica* **1**: 1–6. [(31).x.1926.]
Wheeler, W.M. 1927a. The occurrence of *Formica fusca* L. in Sumatra. *Psyche* **34**: 40–41. [(28).ii.1927.]
Wheeler, W.M. 1927b. Burmese ants collected by Professor G.E. Gates. *Psyche* **34**: 42–46. [(28).ii.1927.]
Wheeler, W.M. 1927c. Ants of the genus *Amblyopone* Erichson. *Proceedings of the American Academy of Arts and Sciences* **62**: 1–29. [(28).ii.1927.]
Wheeler, W.M. 1927d. Chinese ants collected by Professor S.F. Light and Professor N. Gist Gee. *American Museum Novitates* **255**: 1–12. [12.iii.1927.]
Wheeler, W.M. 1927e. A few ants from China and Formosa. *American Museum Novitates* **259**: 1–4. [18.iii.1927.]
Wheeler, W.M. 1927f. The physiognomy of insects. *Quarterly Review of Biology* **2**: 1–36. [(31).iii.1927.]
Wheeler, W.M. 1927g. The ants of the Canary Islands. *Proceedings of the American Academy of Arts and Sciences* **62**: 93–120. [(30).iv.1927.]
Wheeler, W.M. 1927h. Ants collected by Professor F. Silvestri in Indochina. *Bollettino del Laboratorio di Zoologia generale e agraria della R. Scuola Superiore di Agricoltura di Portici* **20**: 83–106. [16.v.1927.]
Wheeler, W.M. 1927i. The ants of Lord Howe Island and Norfolk Island. *Proceedings of the American Academy of Arts and Sciences* **62**: 121–153. [(31).v.1927.]
Wheeler, W.M. 1928a. *Zatapinoma*, a new genus of ants from India. *Proceedings of the New England Zoological Club* **10**: 19–23. [30.i.1928.]
Wheeler, W.M. 1928b. *Mermis* parasitism and intercastes among ants. *Journal of Experimental Zoology* **50**: 165–237. [5.ii.1928.]
Wheeler, W.M. 1928c. Ants collected by Professor F. Silvestri in China. *Bollettino del Laboratorio di Zoologia generale e agraria del R. Istituto Superiore agrario di Portici* **22**: 3–38. [5.iii.1928.]
Wheeler, W.M. 1928d. Ants collected by Professor F. Silvestri in Japan and Korea. *Bollettino del Laboratorio di Zoologia generale e agraria del R. Istituto Superiore agrario di Portici* **22**: 96–125. [16.iii.1928.]
Wheeler, W.M. 1928e. A new species of *Probolomyrmex* from Java. *Psyche* **35**: 7–9. [(31).iii.1928.]
Wheeler, W.M. 1929a. The identity of the ant genera *Gesomyrmex* Mayr and *Dimorphomyrmex* Ernest André. *Psyche* **36**: 1–12. [(31).iii.1929.]
Wheeler, W.M. 1929b. Three new genera of ants from the Dutch East Indies. *American Museum Novitates* **349**: 1–8. [29.iv.1929.]
Wheeler, W.M. 1929c. Note on *Gesomyrmex. Psyche* **36**: 91–92. [(30).vi.1929.]
Wheeler, W.M. 1929d. The ant genus *Rhopalomastix. Psyche* **36**: 95–101. [(30).vi.1929.]
Wheeler, W.M. 1929e. A *Camponotus* mermithergate from Argentina. *Psyche* **36**: 102–106. [(30).vi.1929.]
Wheeler, W.M. 1929f. Some ants from China and Manchuria. *American Museum Novitates* **361**: 1–11. [20.vii.1929.]
Wheeler, W.M. 1929g. Ants collected by Professor F. Silvestri in Formosa, the Malay Peninsula and the Philippines. *Bollettino del Laboratorio di Zoologia generale e agraria del R. Istituto Superiore agrario di Portici* **24**: 27–64. [22.x.1929.]
Wheeler, W.M. 1929h. Two interesting Neotropical myrmecophytes (*Cordia nodosa* and *C. alliodora*) (pp. 342–353). In Jordan, K. & Horn, W. (Eds.). *Fourth International Congress of Entomology* **2**: 1037 pp. Naumburg. [(31).xii.1929.]
Wheeler, W.M. 1930a. Formosan ants collected by Dr. R. Takahashi. *Proceedings of the New England Zoological Club* **11**: 93–106. [7.iii.1930.]
Wheeler, W.M. 1930b. The ant *Prenolepis imparis* Say. *Annals of the Entomological Society of America* **23**: 1–26. [21.iii.1930.]
Wheeler, W.M. 1930c. A second note on *Gesomyrmex. Psyche* **37**: 35–40. [(31).iii.1930.]
Wheeler, W.M. 1930d. Two new genera of ants from Australia and the Philippines. *Psyche* **37**: 41–47. [(31).iii.1930.]
Wheeler, W.M. 1930e. A new parasitic *Crematogaster* from Indiana. *Psyche* **37**: 55–60. [(31).iii.1930.]
Wheeler, W.M. 1930f. A new *Emeryella* from Panama. *Proceedings of the New England Zoological Club* **12**: 9–13. [3.iv.1930.]
Wheeler, W.M. 1930g. Philippine ants of the genus *Aenictus* with descriptions of the females of two species. *Journal of the New York Entomological Society* **38**: 193–212. [7.vii.1930.]
Wheeler, W.M. 1930h. A list of the known Chinese ants. *Peking Natural History Bulletin* **5**: 53–81. [(30).ix.1930.]
Wheeler, W.M. 1931a. Neotropical ants of the genus *Xenomyrmex* Forel. *Revista de Entomologia* **1**: 129–139. [15.vii.1931.]
Wheeler, W.M. 1931b. New and little known ants of the genera *Macromischa, Croesomyrmex* and *Antillaemyrmex. Bulletin of the Museum of Comparative Zoology at Harvard College* **72**: 1 -34. [(31).vii.1931.]
Wheeler, W.M. 1931c. The ant *Camponotus (Myrmepomis) sericeiventris* Guérin. *Psyche* **38**: 86–98. [(30).ix.1931.]
Wheeler, W.M. 1932a. A list of the ants of Florida with descriptions of new forms. *Journal of the New York Entomological Society* **40**: 1–17. [15.iii.1932.]
Wheeler, W.M. 1932b. An Australian *Leptanilla. Psyche* **39**: 53–58. [(30).ix.1932.]

Wheeler, W.M. 1932c. Ants of the Marquesas Islands. *Pacific Entomological Survey Publication* **1** (16): 155–163. [20.xi.1932.] [Also issued in *Bulletin of the Bernice P. Bishop Museum* **98**: 155–163.]

Wheeler, W.M. 1932d. Ants from the Society Islands. *Pacific Entomological Survey Publication* **6** (3): 13–19. [23.xii.1932.] [Later also issued in 1935 *Bulletin of the Bernice P. Bishop Museum* **113**: 13–19. [(31.xii).1935.]]

Wheeler, W.M. 1933a. The Templeton Crocker Expedition of the California Academy of Sciences, 1932. No. 6. Formicidae of the Templeton Crocker Expedition. *Proceedings of the California Academy of Sciences* (4) **21**: 57–64. [22.iii.1933.]

Wheeler, W.M. 1933b. *Mermis* parasitism in some Australian and Mexican ants. *Psyche* **40**: 20–31. [(31).iii.1933.]

Wheeler, W.M. 1933c. New ants from China and Japan. *Psyche* **40**: 65–67. [(30).vi.1933.]

Wheeler, W.M. 1933d. A second parasitic *Crematogaster. Psyche* **40**: 83–86. [(30).vi.1933.]

Wheeler, W.M. 1933e. A new *Myrmoteras* from Java. *Proceedings of the New England Zoological Club* **13**: 73–75. [17.vii.1933.]

Wheeler, W.M. 1933f. A new species of *Ponera* and other records of ants from the Marquesas Islands. *Pacific Entomological Survey Publication* **7** (9): 141–144. [30.viii.1933.] [Later also issued in 1935 *Bulletin of the Bernice P. Bishop Museum* **114**: 141–144. [(31.xii).1935.]]

Wheeler, W.M. 1933g. Three obscure genera of ponerine ants. *American Museum Novitates* **672**: 1–23. [6.xi.1933.]

Wheeler, W.M. 1933h. An ant new to the fauna of the Hawaiian Islands. *Proceedings of the Hawaiian Entomological Society* **8**: 275–278. [(30).xi.1933.]

Wheeler, W.M. 1933i. *Colony-Founding Among Ants,* with an account of some primitive Australian species: 179 pp. Cambridge, Mass. [(31.xii).1933.]

Wheeler, W.M. 1934a. Formicidae of the Templeton Crocker Expedition, 1933. *Proceedings of the California Academy of Sciences* (4) **21**: 173–181. [20.iv.1934.]

Wheeler, W.M. 1934b. An Australian ant of the genus *Leptothorax* Mayr. *Psyche* **41**: 60–62. [(30).vi.1934.]

Wheeler, W.M. 1934c. A second revision of the ants of the genus *Leptomyrmex* Mayr. *Bulletin of the Museum of Comparative Zoology at Harvard College* **77**: 69–118. [(30).vi.1934.]

Wheeler, W.M. 1934d. Contributions to the fauna of Rottnest Island, Western Australia. No. 9. The ants. *Journal of the Royal Society of Western Australia* **20**: 137–163. [5.x.1934.]

Wheeler, W.M. 1934e. Some aberrant species of *Camponotus (Colobopsis)* from the Fiji Islands. *Annals of the Entomological Society of America* **27**: 415–424. [10.x.1934.]

Wheeler, W.M. 1934f. Ants from the islands off the west coast of Lower California and Mexico. *Pan-Pacific Entomologist* **10**: 132–144. [17.x.1934.]

Wheeler, W.M. 1934g. Neotropical ants collected by Dr. Elisabeth Skwarra and others. *Bulletin of the Museum of Comparative Zoology at Harvard College* **77**: 157–240. [(30).xi.1934.]

Wheeler, W.M. 1935a. Two new genera of myrmicine ants from Papua and the Philippines. *Proceedings of the New England Zoological Club* **15**: 1–9. [6.ii.1935.]

Wheeler,W.M. 1935b. New ants from the Philippines. *Psyche* **42**: 38–52. [(31).iii.1935.]

Wheeler, W.M. 1935c. Myrmecological notes. *Psyche* **42**: 68–72. [(31).iii.1935.]

Wheeler, W.M. 1935d. Ants of the genera *Belonopelta* Mayr and *Simopelta* Mann. *Revista de Entomologia* **5**: 8–19. [12.iv.1935.]

Wheeler, W.M. 1935e. The Australian ant genus *Mayriella. Psyche* **42**: 151–160. [(30).ix.1935.]

Wheeler, W.M. 1935f. Ants of the genus *Acropyga* Roger, with description of a new species. *Journal of the New York Entomological Society* **43**: 321–329. [1.x.1935.]

Wheeler, W.M. 1935g. Check list of the ants of Oceania. *Occasional Papers of the Bernice P.Bishop Museum* **11** (No. 11): 56 pp. [(31.xii).1935.]

Wheeler, W.M. 1936a. The Australian ant genus *Froggattella. American Museum Novitates* **842**: 1–11. [13.iv.1936.]

Wheeler, W.M. 1936b. Ants from Hispaniola and Mona Island. *Bulletin of the Museum of Comparative Zoology at Harvard College* **80**: 195–211. [(30).ix.1936.]

Wheeler, W.M. 1936c. A singular *Crematogaster* from Guatemala. *Psyche* **43**: 40–48. [(30).ix.1936.]

Wheeler, W.M. 1936d. Ecological relations of ponerine and other ants to termites. *Proceedings of the American Academy of Arts and Sciences* **71**: 159–243. [(31).x.1936.]

Wheeler, W.M. 1936e. Notes on some aberrant Indonesian ants of the subfamily Formicinae. *Tijdschrift voor Entomologie* **79**: 217–221. [(30).xi.1936.]

Wheeler, W.M. 1936f. Ants from the Society, Austral, Tuamotu and Mangareva Islands. *Occasional Papers of the Bernice P. Bishop Museum* **12** (No. 18): 1–17. [31.xii.1936.]

Wheeler, W.M. 1937a. Additions to the ant fauna of Krakatau and Verlaten Island. *Treubia* **16**: 21–24. [(31).v.1937.]

Wheeler, W.M. 1937b. Ants mostly from the mountains of Cuba. *Bulletin of the Museum of Comparative Zoology at Harvard College* **81**: 439–465. [(31).v.1937.]

Wheeler, W.M. 1937c. *Mosaics and other anomalies among ants:* 95 pp. Cambridge, Mass. [(31.xii).1937.]

Wheeler, W.M. 1938. Ants from the caves of Yucatan. *Carnegie Institution of Washington, Publication* **491**: 251–255. [15.vi.1938.]

Wheeler, W.M. 1942. Studies of Neotropical ant-plants and their ants. *Bulletin of the Museum of Comparative Zoology at Harvard College* **90**: 1–262. [(31).x.1942.]

Wheeler, W.M. & Bailey, I.W. 1920. The feeding habits of the pseudomyrmine and other ants. *Transactions of the American Philosophical Society* (N.S.) **22**: 235–279. [(31.xii).1920.]

Wheeler, W.M. & Bequaert, J.C. 1929. Amazonian myrmecophytes and their ants. *Zoologischer Anzeiger* **82**: 10–39. [(31.xii).1929.]

Wheeler, W.M. & Chapman, J.W. 1925. The ants of the Philippine Islands. Part 1. Dorylinae and Ponerinae. *Philippine Journal of Science* **28**: 49–73. [21.ix.1925.]

Wheeler, W.M. & Creighton, W.S. 1934. A study of the ant genera *Novomessor* and *Veromessor. Proceedings of the American Academy of Arts and Sciences* **69**: 341–387. [(30).ix.1934.]

Wheeler, W.M. & Darlington, P.J. 1930. Ant-tree notes from Rio Frio, Colombia. *Psyche* **37**: 107–117. [(30).vi.1930.]

Wheeler, W.M. & Long, W.H. 1901. The males of some Texan ecitons. *American Naturalist* **35**: 157–173. [15.iii.1901.]

Wheeler, W.M. & Mann, W.M. 1914. The ants of Haiti. *Bulletin of the American Museum of Natural History* **33**: 1–61. [20.ii.1914.]

Wheeler, W.M. & Mann, W.M. 1916. The ants of the Phillips expedition to Palestine during 1914. *Bulletin of the Museum of Comparative Zoology at Harvard College* **60**: 167–174. [(29).ii.1916.]

Wheeler, W.M. & McClendon, J.F. 1903. Dimorphic queens in an American ant. *Biological Bulletin* **4**: 149–163. [(31).iii.1903.]

Whelden, R.M. 1958. Notes on the anatomy of the Formicidae. 1. *Stigmatomma pallipes* (Haldeman). *Journal of the New York Entomological Society* **65** (1957): 1–21. [25.iii.1958.]

Whitcomb, W.H., Denmark, H.A., Buren, W.F. & Carroll, J.F. 1972. Habits and present distribution in Florida of the exotic ant *Pseudomyrmex mexicanus. Florida Entomologist* **55**: 31–33. [1.iii.1972.]

White, A. 1846. *The Zoology of the Voyage of H.M.S. Erebus and Terror under the command of Captain Sir James Clark Ross, R.N., F.R.S., during the years 1839 to 1843.* **2.** Insects: 1–24. [(31.xii).1846.]

White, W.F. 1884. *Ants and their ways.* With illustrations and an appendix giving a complete list of genera and species of the British Isles: 279 pp. London. [(31.xii).1884.]

Willey, R.B. & Brown, W.L., Jr. 1983. New species of the ant genus *Myopias. Psyche* **90**: 249–285. [9.xii.1983.]

Williams, F.X. 1946. *Stigmatomma (Fulakora) zwaluwenburgi,* a new species of ponerine ant from Hawaii. *Proceedings of the Hawaiian Entomological Society* **12** (1945): 639–640. [28.v.1946.]

Wilson, E.O. 1951. A new *Leptothorax* from Alabama. *Psyche* **57** (1950): 128–130. [23.iv.1951.]

Wilson, E.O. 1952. O complexo *Solenopsis saevissima* na America do sul. *Memórias do Instituto Oswaldo Cruz* **50**: 49–68. [(31).iii.1952.]

Wilson, E.O. 1954. The ecology of some North American dacetine ants. *Annals of the Entomological Society of America* **46** (1953): 479–495. [8.ii.1954.]

Wilson, E.O. 1955a. A monographic revision of the ant genus *Lasius. Bulletin of the Museum of Comparative Zoology at Harvard College* **113**: 1–201. [(31).iii.1955.]

Wilson, E.O. 1955b. Ecology and behaviour of the ant *Belonopelta deletrix* Mann. *Psyche* **62**: 82–87. [19.vii.1955.]

Wilson, E.O. 1955c. The status of the ant genus *Microbolbos* Donisthorpe. *Psyche* **62**: 136. [8.xi.1955.]

Wilson, E.O. 1957a. The discovery of cerapachyine ants on New Caledonia, with the description of new species of *Phyracaces* and *Sphinctomyrmex. Breviora* **74**: 1–9. [1.v.1957.]

Wilson, E.O. 1957b. The *tenuis* and *selenophora* groups of the ant genus *Ponera. Bulletin of the Museum of Comparative Zoology at Harvard College* **116**: 355–386. [(31).v.1957.]

Wilson, E.O. 1958a. Studies on the ant fauna of Melanesia. 1. The tribe Leptogenyini. 2. The tribes Amblyoponini and Platythyreini. *Bulletin of the Museum of Comparative Zoology at Harvard College* **118**: 101–153. [(30).iv.1958.]

Wilson, E.O. 1958b. Sympatry of the ants *Conomyrma bicolor* (Wheeler) and *C. pyramica* (Roger). *Psyche* **64** (1957): 76. [1.v.1958.]

Wilson, E.O. 1958c. Observations on the behavior of the cerapachyine ants. *Insectes Sociaux* **5**: 129–140. [(30).vi.1958.]

Wilson, E.O. 1958d. Studies on the ant fauna of Melanesia. 3. *Rhytidoponera* in western Melanesia and the Moluccas. 4. The tribe Ponerini. *Bulletin of the Museum of Comparative Zoology at Harvard College* **119**: 303–371. [(31).viii.1958.]

Wilson, E.O. 1959a. Studies on the ant fauna of Melanesia. 5. The tribe Odontomachini. *Bulletin of the Museum of Comparative Zoology at Harvard College* **120**: 483–510. [(31).v.1959.]

Wilson, E.O. 1959b. Studies on the ant fauna of Melanesia. 6. The tribe Cerapachyini. *Pacific Insects* **1**: 39–57. [15.vii.1959.]

Wilson, E.O. 1962a. The Trinidad cave ant *Erebomyrma* (=*Spelaeomyrmex*) *urichi* (Wheeler), with a comment on cavernicolous ants in general. *Psyche* **69**: 62–72. [1.vii.1962.]

Wilson, E.O. 1962b. Behavior of *Daceton armigerum* (Latreille), with a classification of self-grooming movements in ants. *Bulletin of the Museum of Comparative Zoology at Harvard College* **127**: 403–421. [28.viii.1962.]

Wilson, E.O. 1962c. The ants of Rennell and Bellona Islands. *Natural History of Rennell Island, British Solomon Islands* **4**: 13–23. [(31.xii).1962.]

Wilson, E.O. 1964a. The true army ants of the Indo-Australian area. *Pacific Insects* **6**: 427–483. [10.xi.1964.]

Wilson, E.O. 1964b. The ants of the Florida Keys. *Breviora* **210**: 1–14. [12.xi.1964.]

Wilson, E.O. 1971. *The Insect Societies:* 548 pp. Cambridge, Mass. [(31.xii).1971.]

Wilson, E.O. 1974. The soldier of the ant *Camponotus (Colobopsis) fraxinicola,* as a trophic caste. *Psyche* **81**: 182–188. [21.v.1974.]

Wilson, E.O. 1975. *Leptothorax duloticus* and the beginnings of slavery in ants. *Evolution* **29**: 108–119. [31.iii.1975.]

Wilson, E.O. 1976a. A social ethogram of the Neotropical arboreal ant *Zacryptocerus varians* (Fr. Smith). *Animal Behaviour* **24**: 354–363. [(31).v.1976.]

Wilson, E.O. 1976b. The organization of colony defense in the ant *Pheidole dentata* Mayr. *Behavioral Ecology and Sociobiology* **1**: 63–81. [6.ii.1976.]

Wilson, E.O. 1977. The first workerless parasite in the ant genus *Formica. Psyche* **83** (1976): 277–281. [29.viii.1977.]

Wilson, E.O. 1984. Tropical social parasites in the ant genus *Pheidole,* with an analysis of the anatomical parasitic syndrome. *Insectes Sociaux* **31**: 316–334. [(31).xii.1984.]

Wilson, E.O. 1985a. Ants of the Dominican amber. 1. Two new myrmicine genera and an aberrant *Pheidole. Psyche* **92**: 1–9. [28.vi.1985.]

Wilson, E.O. 1985b. Ants of the Dominican amber. 2. The first fossil army ants. *Psyche* **92**: 11–16. [28.vi.1985.]

Wilson, E.O. 1985c. Ants of the Dominican amber. 3. The subfamily Dolichoderinae. *Psyche* **92**: 17–37. [28.vi.1985.]

Wilson, E.O. 1985d. Ants from the Cretaceous and Eocene amber of North America. *Psyche* **92**: 205–216. [28.xi.1985.]

Wilson, E.O. 1986a. Ants of the Dominican amber. 4. A giant ponerine in the genus *Paraponera. Israel Journal of Entomology* **19** (1985): 197–200. [(28).ii.1986.]

Wilson, E.O. 1986b. Caste and division of labor in *Erebomyrma,* a genus of dimorphic ants. *Insectes Sociaux* **33**: 59–69. [(30).vi.1986.]

Wilson, E.O. 1987. The earliest known ants: an analysis of the Cretaceous species and an inference concerning their social organisation. *Paleobiology* **13**: 44–53. [13.iii.1987.]

Wilson, E.O. 1989. *Chimaeridris,* a new genus of hook-mandibled myrmicine ants from tropical Asia. *Insectes Sociaux* **36**: 62–69. [(30).vi.1989.]

Wilson, E.O. & Brown, W.L.,Jr. 1953. The subspecies concept and its taxonomic application. *Systematic Zoology* **2**: 97–111. [(30).ix.1953.]

Wilson, E.O. & Brown, W.L.,Jr. 1955. Revisionary notes on the *sanguinea* and *neogagates* groups of the ant genus *Formica. Psyche* **62**: 108–129. [8.xi.1955.]

Wilson, E.O. & Brown, W.L.,Jr. 1956. New parasitic ants of the genus *Kyidris,* with notes on ecology and behavior. *Insectes Sociaux* **3**: 439–454. [(30).ix.1956.]

Wilson, E.O. & Brown, W.L.,Jr. 1958. The worker caste of the parasitic ant *Monomorium metoecus* Brown and Wilson, with notes on behavior. *Entomological News* **69**: 33–38. [21.ii.1958.]

Wilson, E.O. & Brown, W.L.,Jr. 1985. Behavior of the cryptobiotic predaceous ant *Eurhopalothrix heliscata* n. sp. *Insectes Sociaux* **31** (1984): 408–428. [(30).iv.1985.]

Wilson, E.O., Carpenter, F.M. & Brown, W.L.,Jr. 1967. The first Mesozoic ants, with the description of a new subfamily. *Psyche* **74**: 1–19. [18.viii.1967.]

Wilson, E.O., Eisner, T., Wheeler, G.C. & Wheeler, J. 1956. *Aneuretus simoni* Emery, a major link in ant evolution. *Bulletin of the Museum of Comparative Zoology at Harvard College* **115**: 81–99. [(31).viii.1956.]

Wilson, E.O. & Hölldobler, B. 1986. Ecology and behavior of the Neotropical cryptobiotic ant *Basiceros manni. Insectes Sociaux* **33**: 70–84. [(30).vi.1986.]

Wilson, E.O. & Taylor, R.W. 1964. A fossil ant colony: new evidence of social antiquity. *Psyche* **71**: 93–103. [24.viii.1964.]

Wilson, E.O. & Taylor, R.W. 1967. The ants of Polynesia. *Pacific Insects Monograph* **14**: 1–109. [15.viii.1967.]

Wing, W.M. 1949. A new *Formica* from northern Maine, with a discussion of its supposed type of social parasitism. *Canadian Entomologist* **81**: 13–17. [28.ii.1949.]

Wing, W.M. 1968. Taxonomic revision of the Nearctic genus *Acanthomyops. Cornell University Agricultural Experiment Station Memoir* **405**: 1–173. [(31).iii.1968.]

Wolf, K. 1915. Studien über palaearktische Formiciden. 1. *Berichte des Naturwissenschaftlich-Medizinischen Vereins in Innsbruck* **35**: 37–52. [(31.xii).1915.]

Wu, C.F. 1941. Superfamily Formicoidea (pp. 141–204). In *Catalogus Insectorum Sinensium* **6**: 333 pp. Peiping. [16.v.1941.]

Wu, J. 1990. Taxonomic studies on the genus *Formica* of China. *Forest Research* **3**: 1–8. [(28).ii.1990.]

Wu, J. & Wang, C. 1990. A taxonomic study on the genus *Tetraponera* Smith in China. *Scientia Silvae Sinicae* **26**: 515–518. [(30).xi.1990.]

Wu, J. & Wang, C. 1992. Formicidae (pp. 1301–1320). In Peng, J. *et al. Iconography of Forest Insects in Hunan, China.* Forest Bureau of Hunan Province: 1473 pp. Hunan Scientific and Technical Publishing House. [(31.xii).1992.]

Wu, J. & Wang, C. 1994. A new genus of ants from Yunnan, China. *Journal of Beijing Forestry University* (English edition) **3**: 35–38. [(31.xii).1994.]

Wu, J. & Xiao, G. 1987. A new species of *Gnamptogenys* from China. *Scientia Silvae Sinicae* **23**: 303–305. [(31).viii.1987.]

Wu, J. & Xiao, G. 1989. A new species of the genus *Vollenhovia* from China. *Entomotaxonomia* **11**: 239–241. [(30).vi.1989.]

Yamauchi, K. 1979. Taxonomical and ecological studies on the ant genus *Lasius* in Japan. 1. Taxonomy. *Science Reports of the Faculty of Education, Gifu University* (Nat. Sci.) **6**: 147–181. [Not seen. Reference from Onoyama, 1980: 211.]

Yamauchi, K. & Hayashida, K. 1968. Taxonomic studies on the genus *Lasius* in Hokkaido, with ethological and ecological notes. 1. The subgenus *Dendrolasius* or jet black ants. *Journal of the Faculty of Science, Hokkaido University* Series 6, Zoology **16**: 396–412. [(30).ix.1968.]

Yamauchi, K. & Hayashida, K. 1970. Taxonomic studies on the genus *Lasius* in Hokkaido, with ethological and ecological notes. 2. The subgenus *Lasius*. *Journal of the Faculty of Science, Hokkaido University* Series 6, Zoology **17**: 501–519. [(30).ix.1970.]

Yano, M. 1910. On the ants of Japan. *Zoological Magazine (Tokyo)* **22**: 416–425. [(31.xii).1910.]

Yano, M. 1911. A new slave-making ant from Japan. *Psyche* **18**: 110–112. [(30).vi.1911.]

Yarrow, I.H.H. 1954a. The British ants allied to *Formica fusca* L. *Transactions of the Society for British Entomology* **11**: 229–244. [14.vii.1954.]

Yarrow, I.H.H. 1954b. Application for the re-examination and re- phrasing of the decision taken by the International Commission regarding the name of the type species of *Formica* Linnaeus, 1758. *Bulletin of Zoological Nomenclature* **9**: 313–318. [30.xii.1954.]

Yarrow, I.H.H. 1955a. The British ants allied to *Formica rufa* L. *Transactions of the Society for British Entomology* **12**: 1–48. [22.iii.1955.]

Yarrow, I.H.H. 1955b. The type species of the ant genus *Myrmica* Latreille. *Proceedings of the Royal Entomological Society of London* (B) **24**: 113–115. [22.vi.1955.]

Yarrow, I.H.H. 1967. On the Formicidae of the Azores. Report No. 51 from the Lund University Expedition in 1957 to the Azores and Madeira. *Boletim do Museu Municipal do Funchal* **21**: 24–32. [(31).xii.1967.]

Yasumatsu, K. 1940a. Materiaux pour servir à la faune myrmecologique des iles de Yaeyama. *Mushi* **13**: 67–70. [(31).iii.1940.]

Yasumatsu, K. 1940b. Beiträge zur Kenntnis der Ameisenfauna Mikronesiens. 1. Die Ameisengattung *Anochetus* Mayr der Karolinen. *Annotationes Zoologicae Japonensis* **19**: 312–315. [10.xii.1940.]

Yasumatsu, K. 1941. On the ants of the genus *Dolichoderus* of Angaran element from the far east. *Kontyû* **14**: 177–183. [25.i.1941.]

Yasumatsu, K. 1950. Discovery of an ant of the genus *Lordomyrma* Emery in eastern Asia. *Insecta Matsumurana* **17**: 73–79. [(30).iv.1950.]

Yasumatsu, K. 1960. The occurrence of the subfamily Leptanillinae in Japan. *Esakia* **1**: 17–20. [20.i.1960.]

Yasumatsu, K. 1962. Notes on synonymies of five ants widely spread in the orient. *Mushi* **36**: 93–97. [15.iii.1962.]

Yasumatsu, K. & Brown, W.L., Jr. 1951. Revisional notes on *Camponotus herculeanus* Linné and close relatives in the Palaearctic regions. *Journal of the Faculty of Agriculture, Kyushu University* **10**: 29–44. [30.x.1951.]

Yasumatsu, K. & Brown, W.L., Jr. 1957. A second look at the ants of the *Camponotus herculeanus* group in eastern Asia. *Journal of the Faculty of Agriculture, Kyushu University* **11**: 45–51. [30.viii.1957.]

Yasumatsu, K. & Murakami, Y. 1960. A revision of the genus *Stenamma* of Japan. *Esakia* **1**: 27–31. [20.i.1960.]

Yensen, N.P., Clark, W.H. & Francoeur, A. 1977. A checklist of Idaho ants. *Pan-Pacific Entomologist* **53**: 181–187. [28.xi.1977.]

Yoshioka, H. 1939. A new ant of the genus *Dolichoderus* from Japan. *Transactions of the Kansai Entomological Society* **8**: 70–71. [Not seen. Reference from Yasumatsu, 1941: 183.]

Zalessky, Y.M. 1949. (Title unknown.) *Sovetskaya Geologiya* **40**: 51. [Not seen.]

Zetterstedt, J.W. 1838. *Insecta Lapponica.* Sectio secunda. Hymenoptera: 317–475. Lipsiae. [(31.xii).1838.]

Zhigul'skaya, Z.A. 1991. Novyi vid murav'ya roda *Myrmica* s verkhnei Kolymy. *Zoologichesky Zhurnal* **70** (5): 58–62. [26.iv.1991.]

Zhigul'skaya, Z.A., Kipyatkov, V.E. & Kipyatkova, T.A. 1992. Sezonnyi tsikl razvitiya murav'ya *Myrmica aborigenica* v verkhov'yakh kolymy. *Zoologichesky Zhurnal* **71** (5): 72–82. [10.iv.1992.]

Zimmermann, S. 1935. Beitrag zur Kenntnis der Ameisenfauna Süddalmatiens. *Verhandlungen der Zoologisch-Botanischen Gesellschaft in Wien* **84** (1934): 1–65. [20.ii.1935.]

Zolessi, L.C. de & Abenante, Y.P. de. 1974. Nidificacion y mesoetologia de *Acromyrmex* en el Uruguay. 3. *Acromyrmex (A.) hispidus* Santschi, 1925. *Revista de Biologia del Uruguay* **1** (1973): 151–165. [(31).v.1974.]

Zolessi, L.C. de, Abenante, Y.P. de & González, L.A. 1978. Descripcion y observaciones bioetologicas sobre una nueva especie de *Brachymyrmex*. *Revista de Biologia del Uruguay* **4** (1976): 21–44. [(30).ix.1978.]

Zolessi, L.C. de, Abenante, Y.P. de & Philippi, M.E. 1989. *Catalogo sistematico de las especies de formicidos del Uruguay:* 40 pp. Montevideo. [(31.xii).1989.]